Optical Trapping and Manipulation of
Neutral Particles
Using Lasers
A Reprint Volume with Commentaries

Optical Trapping and Manipulation of
Neutral Particles
Using Lasers

A Reprint Volume with Commentaries

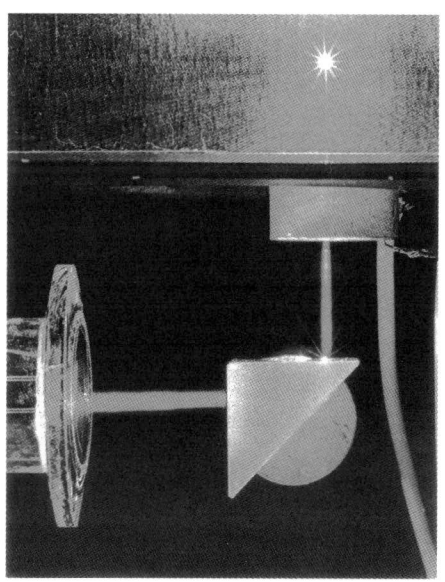

Arthur Ashkin

Retired, former head of Laser Science Research Department
Bell Laboratories, Lucent Technologies, USA

W⬡ World Scientific

NEW JERSEY · LONDON · SINGAPORE · BEIJING · SHANGHAI · HONG KONG · TAIPEI · CHENNAI

Published by

World Scientific Publishing Co. Pte. Ltd.

5 Toh Tuck Link, Singapore 596224

USA office: 27 Warren Street, Suite 401-402, Hackensack, NJ 07601

UK office: 57 Shelton Street, Covent Garden, London WC2H 9HE

British Library Cataloguing-in-Publication Data
A catalogue record for this book is available from the British Library.

ISBN   981-02-4057-0
ISBN   981-02-4058-9 (pbk)

Printed in Singapore by Mainland Press

To my wife, Aline,
and to my children,
Michael, Judith, and Daniel

Preface

This is a book about the *optical* trapping and manipulation of small neutral particles using the forces of radiation pressure from lasers.

This subject traces its origin back to 1969, to the time of the discovery of significant radiation pressure forces on dielectric spheres using lasers and the first experimental observation of optical trapping by Ashkin [1]* (a reference number followed by an asterisk indicates that the paper is reprinted in this volume).

At that time this author also proposed the possibility of stably trapping atoms by radiation pressure [1]*, [2]*. Radiation pressure forces are those arising from the momentum of the light itself. Over the course of the next 35 years or so the techniques of optical trapping and manipulation have made unique and revolutionary contributions to experimental studies in the fields of light scattering, atomic physics, biophysics and other areas of the biological sciences and, more recently, chemistry.

These optical manipulation techniques apply to neutral particles as diverse as: (i) small dielectric spheres in the size range of tens of micrometers down to tens of nanometers, and (ii) atoms and molecules at temperatures ranging from hundreds of degrees Kelvin down to about a nanokelvin, as well as (iii) atomic particles in Bose–Einstein and fermionic condensates in atomic vapors and newly demonstrated atom lasers. Included under (i) above are biological particles such as viruses, bacteria, living cells, organelles within cells and macromolecules.

Among the unique capabilities of optical manipulation is the simple ability to noninvasively hold a single particle fixed in space, free of any mechanical support, using a single beam of light. Submicrometer particles in a compact trap, such as the single-beam gradient or optical tweezers trap [3]*, [4]*, [5]*, can be localized to within a small fraction of a wavelength of light or moved over long distances of many centimeters [6]. Implied in this long distance motion is the capability of separating a single selected particle, such as a bacterium, from a mixed collection, or "*Gemisch*", of particles. Furthermore, trapping, moving, and separation of particles can be done in controlled environments of temperature, pressure, pH, etc.

Using optical manipulation one can continuously apply controlled force to particles by moving the tweezer trapping beam or by applying an external force on a particle held within a fixed optical tweezer trap. One can measure applied forces by directly observing the displacement of particles in a trap due to applied forces or one can use optical feedback techniques, which are extremely sensitive. For physical particles one can measure gravitational, electric, magnetic, or optical forces. The feedback technique was originally conceived in order to apply a form of optical damping for stabilizing optically levitated particles in vacuum [7]*.

Using feedback techniques in biology, one can measure the minute forces of single motor molecules, such as kinesin moving on microtubules or myosin moving on actin filaments. Motor molecules also power chromosome separation in mitosis, cell locomotion through liquids by the action of flagella, and cell motion over surfaces in amoeboid motion or in tissue growth. These single molecular motor forces are in the range of a fraction of a piconewton up to about 100 pN. Large single macromolecules with molecular weights $\geq 10^6$, such as Tobacco Mosaic Virus (TMV) or microtubules, can be manipulated directly. Smaller molecules, for example the molecular motors or mechano-enzymes such as RNA or DNA polymerase, are too small to be directly manipulated by optical tweezers. However, they can be attached to small transparent dielectric spheres that are then optically manipulated. This is called the "handles technique".

Handles and tweezer techniques are useful for studying cell–cell adhesion due to antibody–antigen binding or other lock and key type molecular recognition processes. The handles technique is also useful for applying mechanical forces on single molecules to study the elasticity and compliance of the molecule itself. Recent reviews by the author [8], [9], [10]* give overviews of the above-mentioned applications of optical trapping and manipulation techniques.

There are, of course, other techniques which exist for manipulating small particles and cells, for instance using fine microneedles [11], [12]* controlled by conventional mechanical micromanipulators or using the specialized techniques of Atomic Force Microscopy (AFM) [13], [14]. In Ref. [11] the breakage of the strong monomer–monomer bonds of actin was accomplished with microneedles. In Ref. [15] the motion of myosin on the nanometer scale was observed with needles. In Ref. [13] the breakage of receptor–ligand links was done with AFM. Rief *et al.* [14] used a stiff AFM cantilever to pull on a single titin molecule and observe the successive unfolding of coiled domains. Microneedles and AFM manipulation techniques have their own particular uses, advantages, and ranges of applicability. The measure of a unique technique, however, is that it must have important advantages specific to itself. One of the specific advantages of optical tweezers is the range of optical spring constants available. Tweezers can typically measure forces weaker by far than either needles or AFM. Force constants as low as 0.04 pN/nm have been reported [16]. Furthermore, its optical spring constant is rapidly and continuously adjustable. Overall, it is a better match to the forces of molecular motion than needles or AFM. It is quite simple and can manipulate single particles just by moving a single laser beam. The microneedles technique has the advantage of generating high force with little compliance, but it is much more difficult experimentally and less adaptable to rapid feedback control. Optical tweezers lack the spatial resolution of AFM, although resolution of a fraction of a nm is possible, which is sufficient to resolve the stepping distances of single motor molecules and the stepping of mechano-enzymes over DNA with single-base-pair resolution. See Chapter 8.

The ability to use optical tweezers to manipulate particles and organelles deep within living cells [17]*, [18], even underneath cell structure, as in confocal microscopy, is quite unique [19]*. This latter ability stems from the high numerical aperture of tweezer trapping light. It also involves use of infrared laser light to minimize optical damage to biological cells and particles [20]*, [21]*. Optical tweezers have provided a powerful new way of studying the mechanical properties of the cytoplasm. The elastic and viscoelastic properties of cell membranes and of the polymers making up the cytoskeleton of cells can be readily observed and measured [19]*, [22], [23]. Numerous studies of the elastic properties of long strands of DNA have also been carried out and they have given new information on the packing and unfolding of these complex molecules [22], [24], [25]. As we will see, measurements are even being made of the elastic compliance of parts of various motor molecules in order to understand their detailed functioning.

In physics, it is fair to say that trapping and cooling techniques have revolutionized large areas of experimental atomic physics.

Ashkin's proposal of stable optical trapping in 1970 [1]* and Hänsch and Schawlow's proposal to use molasses for optical cooling of atoms in 1973 [26]* led to the invention, in the late 1970s, of single-beam gradient traps cooled by molasses to the Doppler limit in the regime of hundreds of μK (see Chapters 3 and 4).

The final experimental achievement of cooling and trapping of atoms by Chu, Hollberg, Bjorkholm, Cable, and Ashkin [27]* in 1985 and by Chu, Bjorkholm, Ashkin and Cable [4]* in 1986 marks the beginning of the modern era of optical manipulation of ultracold atoms. Atomic vapors have since been cooled with the help of light to temperatures of about a nanokelvin, which is probably at present the lowest observed kinetic temperature in the universe. Such a low temperature corresponds to an atomic center of mass velocity of a fraction of a mm/s. For comparison, the recoil velocity of a sodium atom emitting just a single photon is about 3 cm/s (see Sec. 5.1 and Chapter 16). With atomic velocities as slow as mm/s one is dealing with atoms that are essentially at rest. Here, gravity becomes a major force. When such atoms are held in optical traps, the trapping forces balance the gravitational force and there are no serious difficulties. Such samples of slow atoms are close to the ultimate for Doppler-free spectroscopy. It is interesting to note that single-photon recoils, called the "Einstein *Rückstoß*", were first observed in 1933 (in pre-laser days) by Frisch [28], using a strongly apertured sodium atomic beam and a sodium lamp light source. It took the unique properties of lasers, however, to exploit these atomic recoils in a usable way.

One does not need such ultralow sub-recoil temperatures for many of the practical applications of optical manipulation of atoms, such as experiments on atomic fountains, atomic clocks, studies of atom–atom collisions, atomic interferometers, atomic gravity measuring devices, atomic lithography, and other so-called atom optics uses. To achieve Bose–Einstein Condensation (BEC), however, does require ultra low atom temperatures; to observe BEC one must also reach a critical phase–space density that depends on temperature and particle density.

The BEC was first successfully achieved in 1995 by Anderson, Ensher, Matthews, Wieman, and Cornell [29]* and by Davis, Mewes, Andrews, van Druten, Durfee, Kurn and Ketterle [30]*, using evaporative cooling of atoms from magnetic traps. This was followed closely thereafter by observation of primitive forms of pulsed atom lasers [31]*, [32], [33], [34]*. A difficulty associated with these early atom lasers was the low intensity of coherent atoms and the pulsed nature of the output. These lasers need more particles for further observations and for many practical applications. Evaporation as currently practiced results in the loss of 90–98.5% of all the original atoms! Proposals have recently been made for achieving true continuous wave (cw) atom laser operation based on optical dipole traps (see Sec. 20.19). In addition to condensing bosons into a BE condensate, it has also been shown, in 1999, that fermions could be cooled to temperatures below their threshold for condensation into a degenerate gas [35]* and into the regime where fermion pairing occurs. See Chapter 25.

At present there are literally thousands of papers and dozens of review articles and reprint volumes on various aspects of the overall subject of optical trapping, cooling, and manipulation of neutral particles. A Nobel Prize in physics was awarded jointly to Chu, Cohen-Tannoudji, and Phillips in 1997 "for development of methods to cool and trap atoms with laser light" [36], [37], [38]. Another Nobel Prize in physics was awarded in 2001 jointly to Cornell, Ketterle, and Wieman "for the achievement of BEC in dilute gases of alkali atoms, and for early fundamental studies of the properties of the condensates" [39]. With a history extending over 30 years, one might reasonably ask, why yet another review and reprint volume such as this one? The answer is that the present book will be substantially different from the previous reviews and books by others, because none of these earlier reviews treats the entire subject of optical manipulation of neutral matter as a single coherent subject. The reviews have become so specialized that readers have barely even been made aware of the existence of other related subfields of the overall subject. To mention just a few, I note the following:

(i) "Trapping of Neutral Atoms", edited by N. R. Newbury and C. Wieman [40] is a reprint volume (Resource Letter TNA-1) published by the American Association of Physics Teachers (AAPT), intended as a guide for university physicists and astronomers, and as a teaching aid. This tutorial volume discusses essentially only the trapping and cooling of atoms. As explained by Newbury and Wieman on p. 18, the review specifically limits its considerations to independent atoms. It mentions two types of neutral atom traps, the dipole or gradient trap, and the spontaneous force trap. As they further point out, principal emphasis is on the so-called Magneto-Optical Trap (MOT) or spontaneous force trap. It should be stressed that the MOT is *not* an all-optical trap, because it is a MOT. There is, therefore, scant information in this resource letter on the origin of all-optical trapping or the developments leading to the use of all-optical trapping in the rest of physics, chemistry, and the biological sciences.

(ii) C. Cohen-Tannoudji and W. D. Phillips, "New Mechanisms of Laser Cooling" [41] is a review article narrowly restricted to the problems of atom cooling, starting with the Hänsch and Schawlow paper in 1975 [26]*.

(iii) "Ultracold atoms and BEC", Volume VII of Trends in Optics and Photonics (TOPS), edited by Keith Burnett, featuring papers from EQEC '96 (the European Quantum Electronics Conference September 8–13, 1996, Hamburg, Germany), published by the Optical Society of America. This volume primarily treats cooling atoms to ultracold temperatures and BEC. Practitioners of BEC essentially consider BEC to be an independent field, separate from laser cooling and trapping of neutral atoms. Even Physical Review Letters for quite a while put papers on BEC in the "General Physics" category, separate from the "Atomic Physics" category. This TOPS review is thus quite restricted in subject matter.

(iv) Koen Visscher and Steven M. Block, "Versatile Optical Traps with Feedback Control" [42], is a review that focuses primarily on the use of optical tweezers and the newly developed feedback control method in biology for measuring the detailed forces and stepping motion of the motor molecule kinesin. The dynamics of other single molecular motors and the so-called mechano-enzymes are also considered, but practically little else.

(v) "Laser Tweezers in Biology", in Methods of Cell Biology, Vol. 55, edited by Michael P. Sheetz, Academic Press, New York, 1998. The subject matter of the articles in this book is restricted to the use of optical tweezers in biology.

As can be seen from the above five examples, the subject of laser trapping and manipulation of neutral particles has become so fragmented that the overall coherence and perspective that existed earlier in the field is almost completely lost. Nowhere in these reviews is made clear the intimate interrelationship between the forces on small macroscopic dielectric particles and biological particles, and the forces on atoms and BE condensates. Indeed, the fact that *the basic optical trapping principles and manipulative techniques closely apply for all particle types* has, over time, become blurred or forgotten. Also lost in this process of specialization and concentration on specific applications is the basic simplicity of the subject.

The thrust of the present book is therefore to present the subject in its simplest form, with emphasis on the origins of the subject and a basic understanding of the physical principles at work. A beautiful aspect of the subject is that simple optical principles (such as ray optics, Fresnel reflection and refraction), an elementary semi-classical view of the atom, and a description of optical forces based on the conservation of momentum are sufficient for an understanding of most of what we now know about optical trapping, cooling, and manipulation of small neutral particles.

Indeed, it was only by resorting to such elementary, simplistic concepts, with the help of simple experiments, physical intuition, and a little luck, that it became possible for Ashkin and collaborators to discover optical traps and optical manipulation in the first place [1]*, [2]*, [3]*, [20]*, [43]*.

One may rightfully ask, for whom is this book written? My answer is that it is written primarily for intelligent college seniors or starting graduate students interested in learning about optical traps and optical manipulation of particles. Because very few practitioners or young students of this subject may have read or have ready access to the early (mostly 25- or 35-year-old) papers, most of them probably know very little about the origins of the field and its relation to the subfields. I also hope that professional scientists and historians of science working in this field will benefit from this book, as well.

Knowing the origin of the field and knowing the interconnection between different parts of the overall field allows one to apply what was learned in one area to others areas. What was considered of only marginal interest in the early work on trapping forces and manipulation of micrometer-sized dielectric spheres [1]*, in time led to the understanding of the optical forces on atoms [3]*, to the invention of optical tweezers [3]*, and to the demonstration of atom cooling and trapping [4]*, [27]*. The discovery of atom trapping and cooling work led quite directly to BE condensation [29]*, [30]*, fermion condensation [35]* and later to the atom laser [31]*, [32], [33]*, [34]* (see also the recent work in Chapter 20). The invention of optical tweezers gave rise to the technique of damage-free trapping of biological particles by Ashkin and Dziedzic [44]*. This technique, in turn, led to the manipulation of single live cells and single molecules, and studies of the mechanical properties of cells, as originally shown by Ashkin and Dziedzic [19]*.

For the most part, the book is organized chronologically. This helps one to follow the development of the subject as it actually happened. It also helps to separate good ideas from bad, since some ideas that were considered impractical or even wrong when originally proposed can later, in new contexts, be modified and made to work. Conversely, ideas and techniques originally touted to be essential have proven to be limiting and in need of replacement. Although there is no serious attempt on my part to be encyclopedic or complete, most of the important papers in the field up to the year 2005 will be alluded to, if only briefly.

General questions about the possibility of trapping neutral particles using radiation pressure forces will be addressed in the context of the much misunderstood Optical Earnshaw Theorem [45]*. The relationship of the Optical Earnshaw Theorem and the conventional Electrostatic Earnshaw Theorem to the trapping of other particles, such as electrons, ions, and neutrons, using other forces, will be clarified. Questions of AC (alternating force) versus DC (cw force) traps will be considered for optical traps, partly optical traps, ion traps, magnetic traps, and other types of traps. The question of the conservative and nonconservative nature of the optical and other trapping forces also will be discussed. This review as outlined above should help to place *optical traps* in overall context with other types of particle traps.

The author feels he writes this review volume from a privileged perspective. In 1970, he discovered optical trapping of neutral particles and also proposed optical trapping of atoms [1]*. In 1978 he discovered how to maximize the forces of radiation pressure on atoms and invented the single-beam gradient trap (now known as "optical tweezers") [3]*; at that time he also developed the concept of cold light, which plays such a big role in BE condensates. His collaboration with Chu, Bjorkholm, Hollberg, and Cable resulted in the first demonstration of optical molasses in 1985 [27]* and the first optical trapping of atoms in 1986 [4]*. In 1987, with Dziedzic and Yamane, he discovered the infrared optical tweezer trap [20]*, and initiated the use of laser trapping in biology. He has followed subsequent developments in optical trapping and manipulation closely and with great interest [8], [9], [10]*, and as recently as 2004 proposed ideas for a viable optical cw atom laser [46]*.

Contents

I
Introduction

CHAPTER

Beginnings

The subject of optical trapping and manipulation of small neutral particles essentially got its start in late 1969 in my attempts to answer the simple question "is it possible to observe significant motion of small particles using the forces of radiation pressure from laser light?" [1]*, [2]*.

1.1. Radiation Pressure Using Microwave Magnetrons

This question was not completely a "bolt out of the blue" for me. Its origins went back a long way, to 1944 or so, during my army wartime experience in the Columbia Radiation Laboratory, where I worked for almost four years on microwave magnetrons. Working for Sid Millman, the inventor of the so-called "rising sun" magnetron, I designed the AX-9, a high-powered megawatt X-band (3 cm) pulsed rising sun magnetron [47].

Although only a college sophomore at the time, I wondered whether one could use a megawatt of X-band peak power radiation at $\lambda \cong 3$ cm for purposes other than radar. I thought of the possibility of using radiation pressure from microwaves to push on metal objects. To check the idea I found an old electromagnetic telephone earpiece consisting of a metallic vibration receiver plate and some magnetic pickup coils. When I shined the one-megawatt peak power radar-type pulses of a microsecond in duration, at a repetition rate of 1000 pulses/s, on the receiver plate, I picked up a clear thousand-cycle signal on an oscilloscope (see Fig. 1.1). I interpreted this as radiation pressure. One could presumably use the technique to measure the peak power, but it was easier to use a matched waveguide water load calorimeter to deduce the power. So, although I was pleased with the result, I did nothing more with it at the time.

1.2. Runners and Bouncers

My next contact with radiation pressure came in the late 1960s at a laser conference, where I heard a talk by Eric Rawson, of the University of Toronto, on what he called "runners and bouncers". These

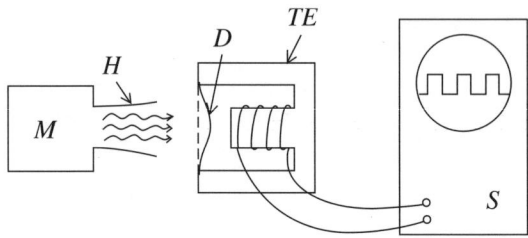

Fig. 1.1. One microsecond pulses of 3 cm wavelength microwave radiation 1000 times a second from magnetron M are coupled out via a waveguide horn H. Light pressure displaces the magnetic metal diaphragm D of a magnetic telephone earpiece TE. The induced electrical signal is detected as a square wave on oscilloscope S.

apparently are small, bright, scattering particles that one sometimes sees floating and moving in the air inside the cavity of a visible laser. He showed movies of their peculiar motion in the light beam. They would shoot to the right or left quite rapidly at various angles, stop, reverse, and even oscillate back and forth across the beam. Hence the name "runners and bouncers" (see Fig. 1.2). Rawson asked the question, "What are these strange particles and what drives them?" He offered a number of suggestions for the driving force, including radiation pressure. This suggestion reminded me of my experiment at Columbia during the war.

I made some simple calculations showing that radiation pressure was too small to account for this strange motion. Considering the internal circulating power in the laser resonator and the typical beam size, the force was inadequate. Besides, the force moving a particle inside a resonator must be a differential force. If the particle is symmetrical, there can be no net force along the beam axis. I decided that much larger forces, such as radiometric forces, were probably the source of the motion. Later, Rawson and his adviser, Professor May published a more complete account of their experiments in a paper in the *Journal of Applied Physics* [48]. They had managed to isolate some of the particles, examined them under a microscope, and found them to be quite irregular and in the

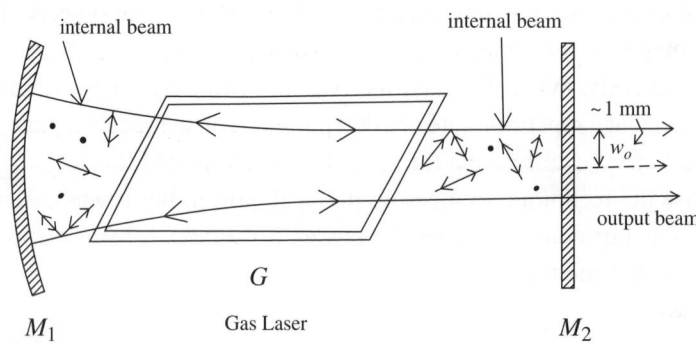

Fig. 1.2. The strange paths followed by runner and bouncer particles moving back and forth within the internal beam of a gas laser cavity exposed to unfiltered room air. M_1 and M_2 are the laser mirrors; G is the gas discharge tube.

micrometer-size regime, thus accounting for the axial asymmetry in axial force and rapid reversals in direction. They concluded that radiation pressure was too small to account for the effect and that it was very likely due to thermal or radiometric forces. They did not pursue the question of radiation pressure any further.

$$E = mc^2 \qquad E = pc = h\nu$$
$$P = h\nu/c$$
$$= mc$$

1.3. Back of the Envelope Calculation of Laser Radiation Pressure

I, in the meantime, following my old interest in radiation pressure, asked an additional question about the possibility of observing particle motion generated by the forces of laser radiation pressure. I did the following "back of the envelope" type calculation for the force, as shown in Fig. 1.3.

In Fig. 1.3, as a preliminary, I show, using a bit of plausible numerology, that the momentum of a single photon is $h\nu/c$. We know the energy of a single photon E_{ph} is $h\nu$. Although we know the photon has no rest mass, we can associate an effective rest mass m_{eff} with the photon as a particle, where $E_{ph} = m_{eff}c^2$, using Einstein's equivalence between mass and energy. Since the momentum of a particle is mass times velocity, we find $m_{eff}c = h\nu/c$. This is the well-known result of quantum mechanics.

Consider first the case of an incoherent light beam hitting a macroscopic, 100% reflecting mirror. If, as in Fig. 1.4, we use a light beam of power P, we have $P/h\nu$ photons in the beam. If the beam hits the 100% reflecting mirror M at normal incidence, it is reflected straight back on itself and each photon transfers a momentum of $2h\nu/c$ to the mirror by conservation of momentum. The total force on the mirror is $2h\nu/c$ times $P/h\nu$, the number of photons per second incident. Thus the force F_{rad} or momentum per second given to the mirror is $2P/c$. The term $h\nu$ cancels since this is basically a classical problem. For a power $P = 1\,\text{W}$, $F_{rad} \cong 10^{-3}$ dyne, using cgs units. This is a very small force when applied to a macroscopic mirror!

Consider now a laser beam with a power of $\sim 1\,\text{W}$. We know that a laser beam can be focused to a spot size $\cong \lambda$ or $\sim 10^{-4}$ cm. Imagine placing a 100% reflecting particle of diameter $\cong \lambda \sim 10^{-4}$ cm at the focus of the beam (see Fig. 1.5). The full force of 10^{-3} dyne is now applied to the particle of mass $\cong 10^{-12}$ g, assuming the particle density is $\sim 1\,\text{g/cm}^3$. This gives an acceleration $A = F_{rad}/m = 10^{-3}\,\text{dyne}/10^{-12}\,\text{g} = 10^9\,\text{cm/s}^2 \cong 10^6\,g$, where g is the acceleration due to gravity. This is a large number! Even if the small particle were to have a reflectivity of $\sim 10\%$ and

Momentum of a single particle

Energy of a photon: $E_{ph} = h\nu$ (Planck's law).
Photon energy in terms of the effective rest mass m_{eff}:

$$E_{ph} = m_{eff}c^2 \quad \text{(Einstein's law)}.$$

Thus the photon momentum m_{eff} times its velocity is:

$$\text{Photon momentum} = m_{eff}c = h\nu/c \quad \text{(as in quantum mechanics)}.$$

Fig. 1.3. Heuristic derivation of the momentum of a single photon.

Force on a 100% reflecting mirror

Momentum of a photon $= h\nu/c$.

Number photons/second $= P/h\nu$.

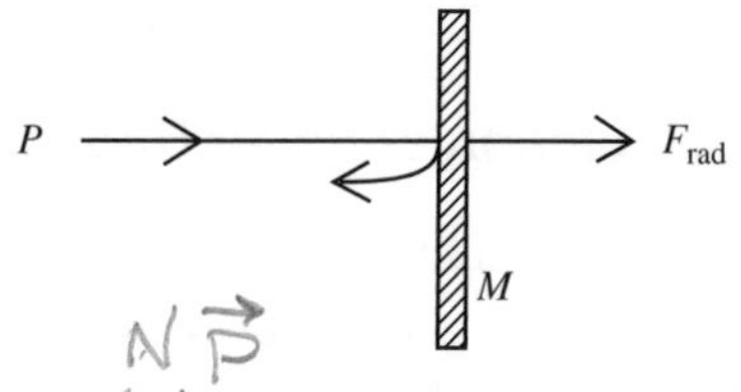

$$F_{rad} = \left(\frac{P}{h\nu}\right)\frac{2h\nu}{c} = \frac{2P}{c}.$$

If $P = 1$ W, then $F_{rad} \cong 10^{-3}$ dyne. This is small!

Fig. 1.4. Radiation pressure force F_{rad} of a light beam of power P on a 100% reflecting mirror, M.

Radiation pressure with a laser

1 W is focused on a 1 μm diameter particle of mass $\cong 10^{-12}$ g.

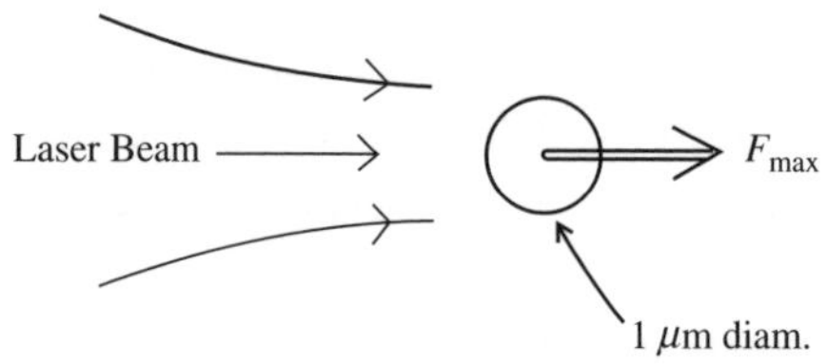

Acceleration: $A = \dfrac{F_{max}}{m} = \dfrac{10^{-3}}{10^{-12}} = 10^9$ cm/s^2,

$A \cong 10^6\, g$, This is big! (g = acceleration of gravity).

Actual $F \sim \dfrac{1}{10}\, F_{max}$ and $A \cong 10^5\, g$, This is still big!

Fig. 1.5. Estimate of the radiation pressure force and acceleration of a focused laser beam on a 1 μm diameter sphere.

the particle were not quite so small as 1 μm, this crude calculation suggests trying some simple experiments.

In the choice of particles, one wants to avoid absorption, which could damage the particle or give rise to unwanted thermal forces. Metal particles, though they are highly reflective, still have absorption of a few percent or more, and so should be avoided. I chose polystyrene latex spheres. These particles are almost ideal for this application: they are quite transparent, are colloidal in nature, come in water, are highly monodisperse, and are available in a variety of sizes.

1.4. First Observation of Laser Radiation Pressure

For the first experiment, shown in Fig. 1.6, I used a greatly diluted solution of 2.5 μm or, at times, 10 μm diameter latex spheres placed in a glass cell made from microscope slides. When a modest laser power of a few hundred milliwatts cw was focused on these spheres, they moved along quite rapidly at velocities given by Stokes' law, using the approximate values of the force calculated above. Particle motion was observed using low power microscopes viewing from the side or along the beam axis as shown in Fig. 1.6.

On closer examination (see Fig. 1.7) it was found that when the laser beam hits a particle near the edge of the beam, the particle was drawn into the beam axis and proceeded to move to the output glass face of the cell, where it stopped. If the power was turned off and the particles wandered into the fringes of the beam, they were quickly pulled back to the beam axis when the power was turned on again. If one wiggled the beam back and forth as a particle was moving along the beam, the particle followed the beam. It was being guided by the light! All these observations suggested that the force had a *transverse* component that was pulling particles into the beam axis! Was this radiation pressure? To answer this question one needs to examine the light force on a sphere more closely.

We first take the case of a sphere, large compared to the wavelength and illuminate it by a plane wave (see Fig. 1.8(a)). Consider a pair of typical rays "a" and "b" located symmetrically about the center of the sphere O. Ray "a" hits the sphere and is refracted by the sphere as shown. Apart from minor surface reflections that we neglect, it emerges from the sphere at the same angle as it entered. By conservation of momentum this gives rise to a force F_a in the direction of the momentum change. Ray "b" gives rise to a symmetrical force F_b. Taking the vector sum of F_a and F_b and other pairs of

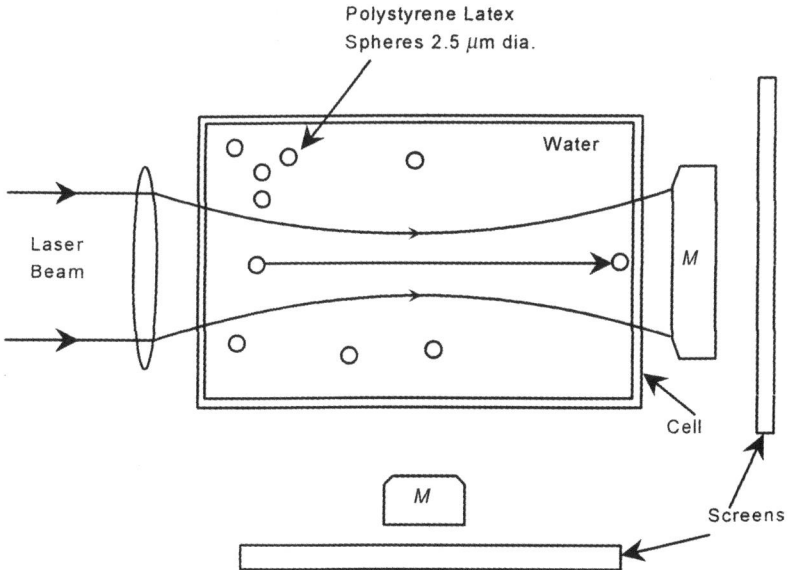

Fig. 1.6. First observation of particle motion using the force of radiation pressure on transparent 2.5 μm diameter polystyrene latex spheres in water.

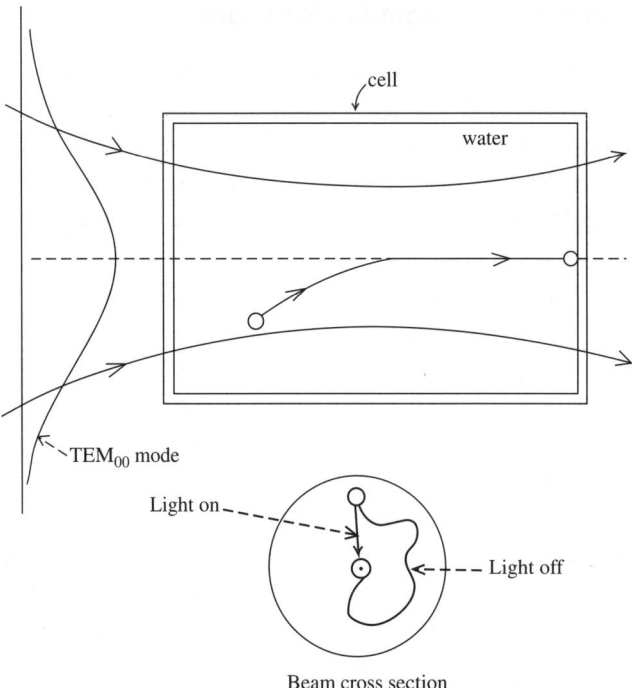

Fig. 1.7. First observation of transverse light forces pulling particles into high intensity region of a TEM$_{00}$ mode Gaussian laser beam.

similar rays, we get a net force on the sphere in the direction of the incident light. We call this F_{scat}, the scattering force, since this arises from the scattering of light momentum.

If we place the sphere off axis in the near-field of a mildly focused Gaussian beam, as shown in Fig. 1.8(b), then we expect that F_a will be larger than F_b, because there is more light incident on the "a" side of the particle than the "b" side. Thus when we take the vector sum of all such pairs of rays we see that in addition to a net forward F_{scat} component in the direction of the light, we should have an additional net inward transverse force component that we call the gradient force component, since it arises from the transverse intensity gradient. This simple ray optic picture for the forces explains all the previously observed phenomena. It applies for high index of refraction particles in a surrounding medium of low index of refraction.

Suppose we had a low index particle in a high index medium? What should happen then? As Fig. 1.9 shows, we expect the refraction at the interfaces to reverse. The transverse force components also reverse and the net gradient force is now away from the high intensity region of the light beam, as shown. At first it seemed difficult, in practice, to find a particle with an index of refraction lower than the index of 1.33 for water. Finally, it occurred to me that I could use micrometer-sized air bubbles, in a viscous medium such as glycerol, as the low-index particles [1]*. I called these the "nothing particles". They have an index of unity, they float upward due to their buoyancy, and they move through fluids obeying Stokes' law. They are easily generated by a common kitchen electric blender. Micrometer-sized bubbles last a long time in a viscous medium. If one shines a focused

Mie Particles $d \gg \lambda$

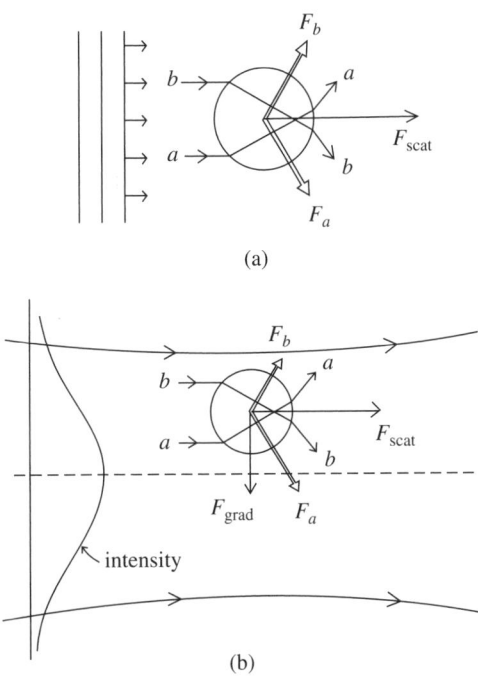

(a)

(b)

Fig. 1.8. Simplified ray optics diagrams of the scattering force and gradient force components of the radiation pressure force on a high index particle with a diameter large compared to the wavelength. (a) Origin of the scattering force F_{scat} in the direction of an incident plane wave beam. (b) Origin of the transverse gradient force component for a particle located off-axis in the near-field of a mildly focused Gaussian beam.

Low index particles in a Gaussian beam

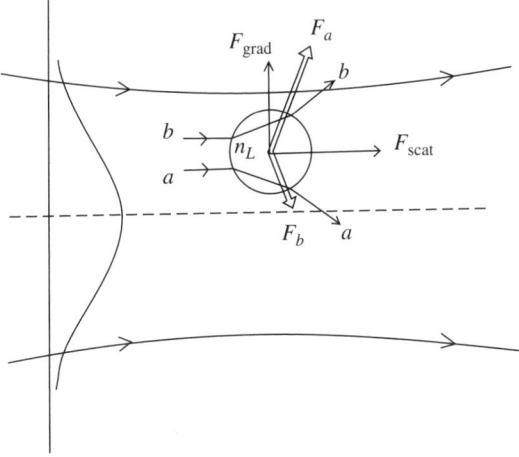

Fig. 1.9. Simplified ray optics diagram shows the directions of the scattering force and gradient force components for a low index particle located off-axis in the near-field of a mildly focused Gaussian beam.

Gaussian laser beam on these bubbles, they are observed to avoid the light and move out of the beam just as our simple theory predicted.

1.5. Observation of the First Three-Dimensional All-Optical Trap

In Fig. 1.7, we essentially observed a case of three-dimensional (3D) particle trapping, where the transverse inward gradient force F_{grad} gave rise to two-dimensional (2D) optical trapping and the axial scattering force F_{scat} was balanced by the mechanical "wall force" at the output face of the chamber. The question now arises, "Can one observe 3D all-optical trapping with just the scattering and gradient forces?" The answer is yes!

Suppose, as in Fig. 1.10, we replace the glass wall of the chamber by a second opposing light beam focused at F_2 and symmetrically located about the point E. If we place a particle at point E, it is clearly in stable equilibrium. The gradient forces of both beams act to push the particle toward the beam axes. If we imagine displacing the particle along the axis to point E', this generates a restoring force back to point E, because the axial scattering force from beam #2 is now larger than that from beam #1. Similarly, a radial displacement to E'' results in a restoring force, since the gradient force components of beams #1 and #2 are radially inward.

To test this supposition experimentally, we simply added a second identical opposing beam in our sample chamber as shown in Fig. 1.10. Looking from the side, we soon see a bright spot appearing near the symmetry point of the two beams. Is the bright spot due to a trapped particle or is it perhaps an artifact, such as a piece of "*schmutz?*" To check, we simply block one of the beams, say #2. The bright spot rapidly moves to the right, driven by beam #1. If we restore beam #2 before the particle has moved too far, it then returns to its former position, only more slowly. If we block #1, the particle shoots to the left and again returns to its original position when beam #1 is restored. One can repeat

First Optical Trap

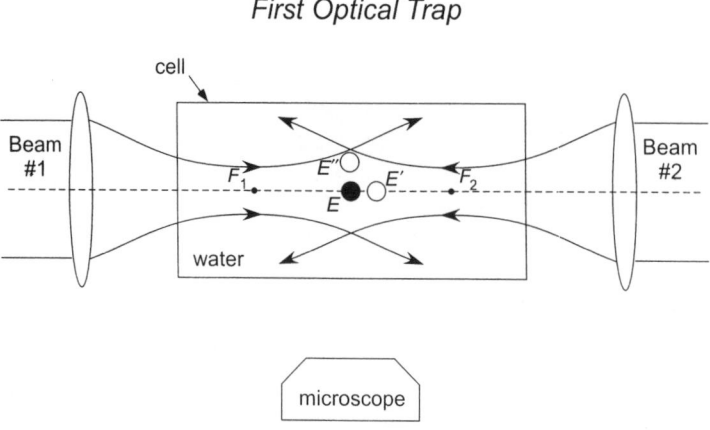

Fig. 1.10. Diagram of the first stable optical trap consisting of two Gaussian beams located symmetrically about the equilibrium point E. Displacement of the particle to E' or E'' results in a restoring force.

the process again and again until one is finally convinced that a true, stable, all-optical trap is being observed [1]*.

In the above description of the total force on a sphere, it was a matter of convenience to resolve the total force into a *scattering force* in the direction of the incident light beam and a *gradient force* in the direction of the intensity gradient. These were the natural force components for a sphere in a mildly focused beam as used in the above experiments. It turns out that the same two components give the most natural description of light forces on atoms and molecules. As will be seen later, it is also possible, with only a slight generalization of the concept of scattering and gradient force components as applied to an entire particle, to consider scattering and gradient force components exerted by each individual ray. Taking a vector sum of the contributions of all incident rays, we are then able to calculate net forces, using much more complicated beam and particle geometry [49]*, [50], [51].

Another important topic that merits discussion is the question of damping. In the above laser experiments, particle motion was heavily damped by the viscosity η of the surrounding liquid. In such a heavily damped medium, in the absence of external forces, the only motion possible is Brownian diffusion. If, however, an external force such as radiation pressure from a laser beam is introduced, a particle diffusing into this beam responds to the light force according to Stokes' law; that is, it moves with a velocity $\mathbf{v} = \mathbf{F}_{rad}/6\pi\eta r$ [52]. In such motion, there are no momentum effects. If the field changes or drops to zero, the particle simply follows $\mathbf{F}$ with no overshoot. Indeed, it was only by virtue of the heavy damping that we were able to capture spheres in the first place and follow their motions in the guiding and trapping experiment described above [1]*.

1.6. Scattering Force on Atoms

After having observed the stable laser trapping of small macroscopic neutral particles by the forces of radiation pressure, it is natural to ask whether similar radiation pressure forces existed on neutral atoms that could give rise to similar types of atom traps. In particular, are there scattering forces and gradient forces on atoms with the same general properties as the forces on macroscopic particles (see Fig. 1.11)?

Considering a neutral atom as a two-level system, we adopt the simplistic rate-equation semiclassical analysis of Einstein for atomic interactions with light [2]*.

Take the case of a plane wave incident on an atom as shown in Fig. 1.12. We use resonant light to get a large interaction. The process of resonant absorption of photons by atoms, followed by an average symmetric spontaneous emission after an average natural lifetime of τ_n, resulting in the scattering of the incident light momentum. This gives rise to an average scattering force in the direction of the incident light. This can be called resonance radiation pressure. At low light intensity the atom spends most of its time in the ground state and has an absorption cross section of about λ^2 [1]*. This is huge! Since a laser beam can be focused to a spot size of $\sim\lambda$, this implies that a single atom is capable of absorbing the entire laser beam! A single atom on resonance is thus the most potent light absorber possible.

Considering a power of 1 W focused on a sodium atom of mass $\cong 23 \times 1.7 \times 10^{-27}$ g, one may naively calculate a resultant acceleration of the atom $A = F_{scat}/m = P/cm \cong 10^{-3}/10^{-26} = 10^{23} \cong 10^{20} g$ (I leave it as an exercise for the reader to figure out the energy achieved by a sodium

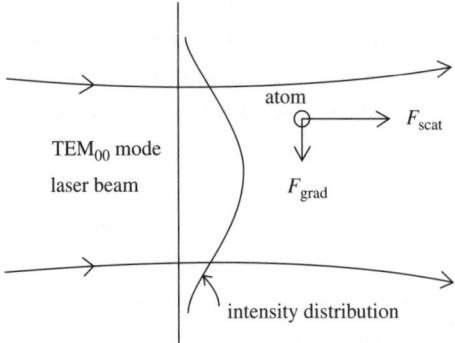

Radiation pressure forces on an atom in
the near field of a Gaussian laser beam

Fig. 1.11. Radiation pressure forces on an atom located off-axis in the near-field of a TEM$_{00}$ mode Gaussian laser beam.

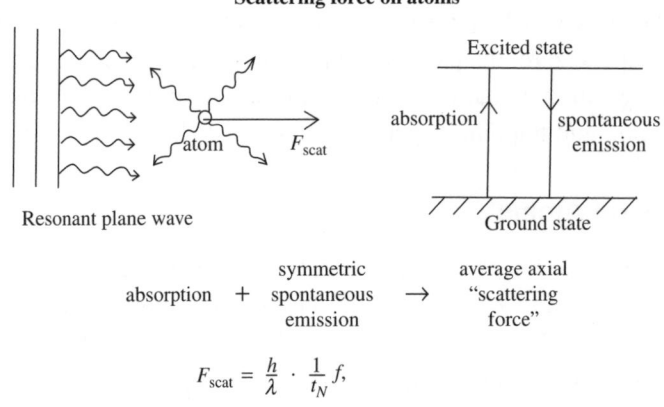

Scattering force on atoms

$$F_{\text{scat}} = \frac{h}{\lambda} \cdot \frac{1}{t_N} f,$$

where h/λ is the momentum of a photon, $1/t_N$ the maximum number of photons scattered/s, f the fraction of time spent in excited state, t_N is the spontaneous emission lifetime, at low power, $f \propto$ intensity — unsaturated, at high power, $f \rightarrow 1/2$ — fully saturated.

Fig. 1.12. Origin of the scattering force on atoms. Resonant light from a plane wave is scattered symmetrically, giving an average axial scattering force in the direction of the incident light.

atom accelerated by $A = 10^{20} g$ for a time equal to $\tau_n = 1.5 \times 10^{-8}$ s). However, the value of acceleration, as calculated above for a sodium atom, is incorrect, because it assumes that the low intensity absorption cross section of λ^2 applies at all light levels. As anyone routinely working with optical lasers knows, atomic transitions are easily saturated. At saturation for a two-level atom, the atom essentially spends half of the time in the excited state and half of the time in the ground state, and the force in that region is constant, independent of power [2]*. The saturation intensity, which is defined as the intensity necessary to reduce the scattering force by a factor of two, is a mere 20 mW/cm^2 for sodium. Thus an atom, when saturated, scatters a number of photons/s equal to

$1/\tau_n$ times $1/2$, as shown in Fig. 1.12. This saturated scattering force F_{scat} is minute compared to the unsaturated value of 10^{20} mg but is still quite substantial, as we shall see shortly.

1.7. Saturation of the Scattering Force on Atoms

Reference [2]* considers the saturation of the scattering force in more detail. It makes use of the well-known Einstein A and B coefficients for the spontaneous emission rate and stimulated emission rate. This early theory, introduced in 1917, treats the equilibrium of a two-level system consisting of a lower level and an excited level at thermal equilibrium in the presence of radiation. It introduces the notion of stimulated emission into the early quantum theory [53] (see also [54]). At thermal equilibrium the number of excited atoms N_2 is related to the number of atoms in the lower level N_1 by the Boltzmann relation $N_2/N_1 = \exp(-h\nu_{21}/kt)$, where $h\nu_{21}$ is the energy difference between the two levels. The dynamic equilibrium can be expressed as

$$W(\nu)B_{12}N_1 = AN_2 + W(\nu)B_{21}N_2 .$$

This states that the stimulated absorption rate for formation of atoms in the excited state from 1 to 2 on the left equals the spontaneous decay plus the stimulated emission rate from atoms in state from 2 to 1 on the right. The stimulated emission terms, of course, contain the factor $W(\nu)$, the energy density per unit frequency interval. From arguments involving "black body" radiation, Einstein concluded that $B_{12} = B_{21} = B$. Calling the total number of atoms $N = N_1 + N_2$, one can then readily solve for the quantity $f = N_2/N$, which represents the fraction of atoms in the excited state as well as the fraction of time a given atom spends in the excited state.

One finds:

$$\frac{N_2}{N} = f = \frac{1}{[2 + A/W(\nu)B]} .$$

At high radiation intensity $W(\nu)B \gg A$ and f saturates to $1/2$. As one sees in Ref. [2]*, using the conservation of momentum, the scattering force on atoms F_{scat} can be written in terms of f as $F_{scat} = (h/\lambda)(1/\tau_n)f$ and the value of F_{scat} saturates to a maximum value of $F_{max} = (h/\lambda)(1/2\tau_n)$. In Ref. [2]* the energy density $W(\nu)$ is rewritten for an incident monochromatic laser beam with a Lorentzian line shape $S(\nu)$. The fraction f can then be expressed in terms of a saturation parameter $p(\nu)$. As we shall see later, these basic saturation formulas play a crucial role in understanding the gradient force on atoms and in achieving stable optical trapping of atoms.

1.8. Gradient (Dipole) Force on Atoms

There also is the equivalent of a gradient force for atoms. This has been known for a long time [8], [9], [55], [56] and arises from the fact that an atom in a gradient of light intensity becomes polarized. As for any polarizable entity, the optically induced dipole has a dipole moment or polarization $\mathbf{d} = \alpha\mathbf{E}$, where α is the polarizability of the particle. In a gradient of an electric field there is a force on the dipole

$$F_{dip} = \mathbf{d} \cdot \nabla\mathbf{E},$$

just as for a particle with a permanent dipole moment (see Sec. 1.9).

Gradient force or dipole force on atoms

Induced dipole moment $d = \alpha E$, where α is the polarizability of the atom

$$\mathbf{F}_{dip} = -\frac{1}{2}\alpha\nabla E^2,$$

$$\alpha = \alpha_o \frac{\nu-\nu_o}{(\nu-\nu_o)^2 + \gamma_N^2/4} \qquad \text{dispersive}$$

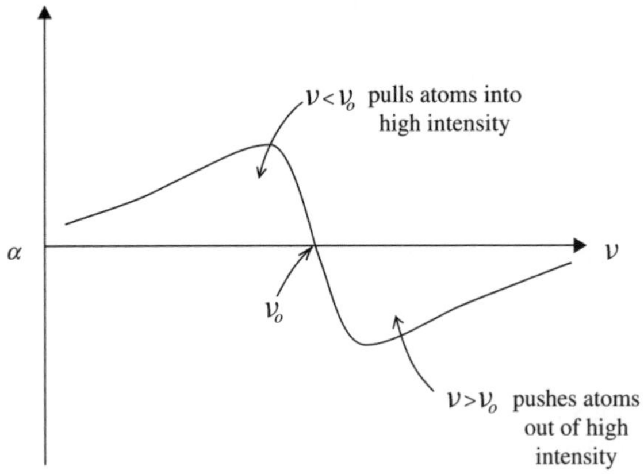

Fig. 1.13. Origin of the gradient or dipole force on atoms. Plot of the polarizability α as a function of frequency ν, showing the dispersive nature of α.

The dipole force on an optically induced dipole does not change sign as the electric field oscillates back and forth because the induced dipole and the electric field gradient both change sign in synchronism, giving a net force whose sign depends only on the sign of the polarizability α. The polarizability α is dispersive in nature and changes sign on either side of the resonance frequency ν_o. For $\nu < \nu_o$ and $\alpha > 0$, the sign of the dipole force pulls the atom into the high intensity region of the light. For $\nu > \nu_o$, atoms are pushed out of the high intensity region of the light (see Fig. 1.13). This is the atomic analog of the behavior of high and low index macroscopic particles. The polarizability of an atom can be calculated traditionally using just a simple harmonic oscillator model for an atom. I will discuss this in some detail below, because it gives considerable insight into atomic processes.

In 1970, I did not yet know how to treat the saturation of the dipole force. Nevertheless, based on the close analogy between the scattering and gradient force components for macroscopic particles and atoms, I made the suggestion in my 1970 paper [1]* that stable all-optical traps should be possible for neutral atoms.

1.9. Dispersive Properties of the Dipole Force on Atoms

To calculate the dispersive properties of an atom using simple classical physics [54], consider the following model. Imagine a point electron of mass m and charge $-e$ bound to a fixed point in space by a spring having a force constant k and a velocity-dependent damping constant γ. If we take the driving electric field as the real part of $Ee^{i\omega t}$ and the amplitude of motion as the real part of $xe^{i\omega t}$, then the equation of motion of the electron is given by

$$m\left(\frac{d^2x}{dt^2}\right) + m\gamma\left(\frac{dx}{dt}\right) + kx = -eEe^{i\omega t}.$$

To solve for x we insert the time derivatives for d^2x/dt^2 and dx/dt and define $\omega_o^2 \equiv k/m$, where ω_o is the resonance frequency of the oscillator.

We find for the frequency-dependent amplitude of oscillation

$$x(\omega) = \frac{-(e/m)E}{\omega_o^2 - \omega^2 + i\omega\gamma}.$$

This gives rise to an oscillating dipole with a dipole moment that is $\mathrm{Re}[-ex(t)]$. One can write the complex polarization of the oscillator as

$$P = -ex = \alpha E,$$

where $\alpha = -ex$ is the complex polarizability. To find the real parts of polarization $\mathrm{Re}[P]$, we need to find the real part of $\mathrm{Re}[\alpha E] = \mathrm{Re}[-ex]$. Since we are considering frequencies close to ω_o, we write $\omega_o^2 - \omega^2$ as $2\omega_o(\omega_o - \omega)$ in the equation for $x(\omega)$ above. We find by rationalizing the denominator and using the notation $\Delta\nu = \gamma/2\pi$ and $\omega = 2\pi\gamma$

$$-ex = \frac{e^2}{m\omega_o\gamma}\left[\frac{(\nu_o - \nu)}{(\nu_o - \nu)^2 + (\Delta\nu/2)^2} - i\frac{\Delta\nu}{(\nu_o - \nu)^2 + (\Delta\nu/2)^2}\right].$$

Thus the real part of $-ex$ gives the polarizability α in $P = \alpha E$. We see that α as obtained from this classical model has the characteristic dispersive shape with varying driving frequency ν, as for real atoms. For real atoms, of course, we use the quantum mechanically calculated value of the polarizability. This is given later in the discussion on saturation of the dipole force. The imaginary part of $-ex$, containing the damping constant γ, gives the correct Lorentzian line shape of atomic absorption [2]*, [54].

If one imagines the electron oscillator to be excited at line center and then the excitation is turned off, one expects the energy to decay exponentially by radiation at the resonance frequency. One can show that this decaying wave has a Lorentzian shape, with a half-width that depends on the damping. So, even here, the simple electron harmonic oscillator behavior agrees qualitatively with the observed spontaneous emission of real atoms [54], [57].

Real atoms, of course, have many atomic levels and many resonances. The mechanical oscillator or classical picture for an atom can be generalized by imagining the atom as a composition of many such oscillators, each with an appropriate oscillator strength f_n. The total absorption of an atom is then distributed over the separate resonances, with the condition $\sum_n f_n = 1$. If the values of f_n are determined experimentally, one gets a credible picture of the absorption and dispersive properties of real atoms.

The quantum treatment of absorption and dispersion is given in the same general manner with the f_n values calculated on the basis of quantum mechanics. The f_n values are closely related to the Einstein B coefficients. The quantum mechanical calculations for polarizability or index of refraction specifically compute quantities such as f_{jk} for transitions from level j up to level k. However, the general case for an atomic gas at high temperature or in a discharge also includes negative induced emission terms such as f_{kj} for transitions from level k to a lower level j. Since $B_{jk} = B_{kj}$, it is not surprising that $f_{jk} = -f_{kj}$ and we see, in the general expression for the index of a gas, that atoms in the excited state contribute negatively to the index. Thus, for the approximate Kramers–Heisenberg dispersion formula relating the index $n - 1$ to the oscillator strengths, one has for the general case:

$$n - 1 = \frac{e^2}{2\pi m} \sum_j N_j \left(\sum_k \frac{f_{jk}}{v_{jk}^2 - v^2} + \sum_i \frac{f_{ji}}{v_{ji}^2 - v^2} \right),$$

where the damping factors have been omitted and all the transitions from lower state to excited states, and vice versa, have been included, see p. 67 of Ref. [57]. Kuhn points out that this dispersion formula was one of the first steps in the evolution of Quantum Mechanics.

As we shall see, the fact that transitions from an excited state to a ground or lower state contribute negatively to the total polarizability or index, compared to a transition from the ground state to an excited state is a key piece of information in understanding how the gradient force on atoms saturates at high laser intensities.

1.10. Applications of the Scattering Force

Applications of the scattering force were considered at the time, in 1970, such as an atomic velocity analyzer and an isotope separation scheme [2]*. In all the applications of the scattering force on atoms, the Doppler shift of the atoms is a prime consideration. With a natural line width of $\sim 10\,\text{MHz}$ in sodium, it does not take much of a velocity shift to put the atoms out of resonance with the light. For beam slowing or stopping, the light is directed opposite to the atoms and the Doppler shift is huge. In the isotope separation scheme and the atom velocity analyzer application, the light was directed at right angles to the atomic motion to avoid these Doppler shifts (see Figs. 1.14 and 1.15).

To give an idea of the magnitude of the saturated scattering force on sodium, for example, one finds that pushing with F_{sat} in a central force field, atoms of thermal velocity $\sim 10^5\,\text{cm/s}$ are turned in a circular orbit of radius of curvature $\rho \cong 20\,\text{cm}$ [2]*. If the saturated force is applied in opposition to the thermal atoms, one can stop the atoms in a distance $\rho/2 = 10\,\text{cm}$, assuming one can somehow compensate for the large Doppler shift of the atomic resonance [38].

Another matter not discussed in the above suggestion on optical trapping of atoms was the important question of how to get the damping needed for filling optical traps. As mentioned earlier, it was the viscous damping of water that dissipated the energy gained by colloidal particles as they were pulled into the two-beam optical trap.

1.11. "It's not Even Wrong!"

The above two earlier papers of mine [1]*, [2]* showed how asking the simple question, "Can one see motion of neutral particles due to radiation pressure?" led not only to the observation of the scattering force component, but also to the discovery of the gradient force component of radiation pressure. These forces are observable because of the ability of laser light to generate high optical intensities and high optical intensity gradients. In 1970, I observed the first guiding of small neutral

Application of the saturated scattering force on atoms
Atom stopping

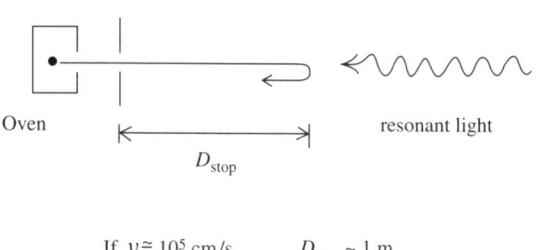

If $v \cong 10^5$ cm/s $\qquad D_{\text{stop}} \sim 1$ m

If the force is continuously applied, there is a huge Doppler shift.

Atom beam deflection and isotope selection

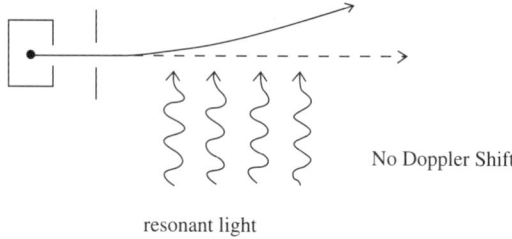

Fig. 1.14. Applications of the saturated scattering force for stopping and deflecting atomic beams.

dielectric particles by light [1]*. This guiding was an example of 2D trapping. More importantly, the first 3D stable all-optical trapping of neutral particles was subsequently demonstrated. Remarkably, it took only simple ray optics, semiclassical physics, and a little luck to make this far from obvious discovery of all-optical trapping. As will be seen, luck (or the prepared mind) played a big role in taking many of the key "forks in the road" in the development of the subject.

It may be interesting and instructive to recall the initial reactions of other scientists to paper [1]*, which described the earliest trapping work. At Bell Labs., before a manuscript could be sent out to a journal it had to undergo an internal review to make sure it would not tarnish the laboratory's excellent reputation in research. Since paper [1]* was intended for *Physical Review Letters*, it was sent to the theoretical physics department for comment. The Bell Labs. internal reviewer made only four points: (i) there was no new physics here, (ii) the reviewer could not actually find anything wrong with the work (this is a reminiscent of the famous Pauli insult, when he commented on some work he thought worthless that "*it is not even wrong!*"), (iii) the work could probably be published somewhere, and (iv) but not in *Phys. Rev. Lett.*

This four-point internal referee report from the theoretical group greatly distressed me, and so I went to my boss, Rudi Kompfner, inventor of the traveling wave tube, whom I greatly admired. Rudi, a man usually slow to anger, simply said, "Hell, just send it in!" As it turned out, I had no problem whatever with the *Physical Review Letters* reviewers. I even received a letter after the paper appeared, saying, "This is one for your grandchildren!"

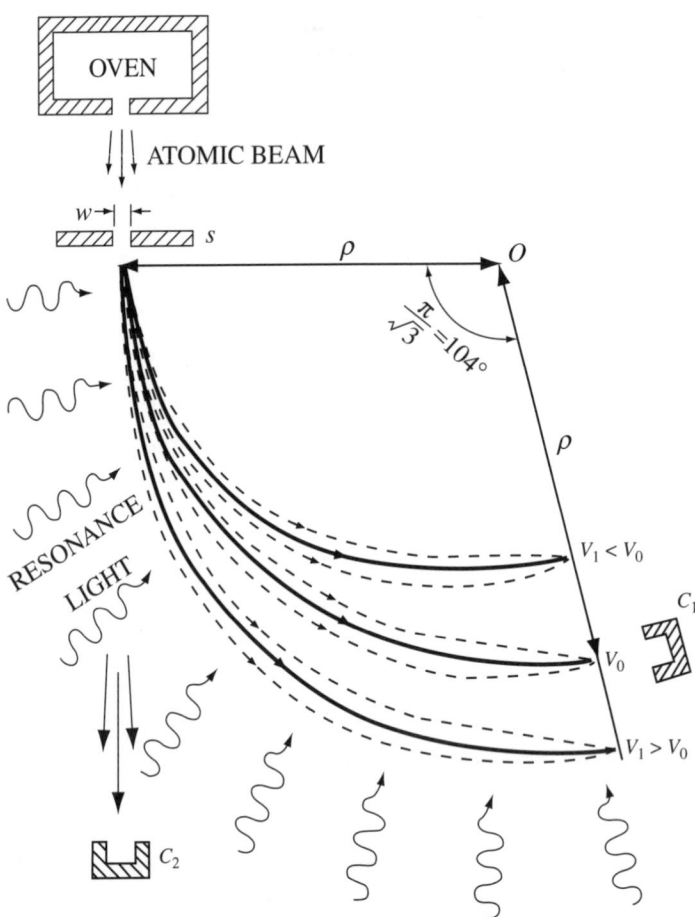

Fig. 1.15. Atomic beam velocity analyzer using radially directed saturated scattering forces in a central force field.

In 1999, paper [1]* had the honor of being selected as one of the 23 seminal papers on atomic physics reprinted in the compilation, *"The Physical Review — The First Hundred Years"*, edited by H. Henry Stroke, American Institute of Physics Press and Springer Verlag (1999) on the occasion of the centennial of the American Physical Society.

To continue with our discussion, at the end of 1970 I decided to pursue some of the new possibilities, consequences, and questions raised by the results of papers [1]* and [2]* on optical manipulation of macroscopic particles and atoms.

Could traps be observed for macroscopic particles in other media such as air or even in a vacuum? Could optical manipulation be used as a practical tool for studying light scattering, for example, and other properties of small macroscopic particles?

Also, there was a great deal more to be learned about the radiation forces on atoms and the properties of atom traps before one could consider actually trapping atoms experimentally. This adventure, started in 1970, has led far beyond those initial goals and questions about optical manipulation, and is still being pursued after 35 years in new areas of optical manipulation of particles with lasers, such as Bose–Einstein condensation, cw atom lasers and single biological molecules, not conceived of in those early times.

The history of this still growing subject falls rather naturally into three time periods: the 1970s–1980s, the 1980s–1990s, and the 1990s to 2006. For convenience, I will therefore divide the present book, somewhat arbitrarily, into three sections, each lasting about a decade or more.

1.12. Optical Traps and the Prepared Mind

In retrospect, it may be useful to consider or speculate whether the concept of optical trapping [1]* was "ripe" for discovery in 1969? Were there many other laser scientists with "prepared minds" ready for an accidental optical trap to manifest itself? Was the concept of optical trapping so trivial that any optical laser researcher could have put one together, had only a need for one arisen? Did any laser scientist suspect, at that time, that laser light forces could significantly modify the dynamics of small macroscopic particles, let alone stably trap them and manipulate them?

In my opinion, stable optical trapping was and still is far from obvious. Just because optical traps can be explained simply, or observed by simple experiments, does not imply they are obvious. If one looks at the explanations of many present day scientists of how optical trapping works, the explanations are often so complex and their focus so narrow that one doubts that these scientists could have conceived of the basic idea of trapping by themselves. The original simple configurations are a far cry from the complex trapping and cooling techniques currently used by the Bose–Einstein condensate atomic physicists, or by the molecular motor community of biologists, or by the single molecule community of chemists.

As we saw, the concept of trapping and manipulation developed in response to some very simple questions and some very simple experiments on the possibility of moving particles with light; and then they just evolved step by step, with the help of some good luck, persistence and imagination. They did not spring, as Athena, "full grown from the head of Zeus".

Knowledge of how the various branches of optical trapping and manipulation evolved out of their simple beginnings and of how the techniques developed in the many separate branches in response to special problems, is an asset whose value is becoming increasingly evident, even as the complexity of the overall subject continues to grow.

II
1969–1979

CHAPTER

Optical Levitation

2.1. Levitation in Air

One of the first questions addressed after my initial papers [1]*, [2]* was about the possibility of trapping a neutral dielectric sphere in a medium less viscous than water, such as air. Applying the trapping principles of paper [1]*, one can conceive of a simple levitation-type trapping scheme as shown in Fig. 2.1 taken from Ref. [58]*. In this scheme a vertically directed, mildly convergent TEM_{00} mode beam is focused from below on a glass sphere sitting on a transparent glass plate.

As the power is raised sufficiently one expects the sphere to rise into the air above the beam focus. Because of the action of the scattering and gradient force components, it should come to rest at an equilibrium point E above the beam focus, where the upward scattering and the downward gravity force balance. The particle should be stable vertically because it sees an increasing scattering force if it drops below point E. Above point E the scattering force decreases, giving a net restoring force. Thus, any vertical displacement gives a restoring force. Transverse stability is the result of the inward gradient force pushing the particle into the high intensity region on the axis of the beam. Overall, the particle is stable.

It is obvious that it is impossible to have vertical stability at any point E anywhere below the beam focus. The base plate is covered by a glass box to avoid disturbing air currents. Glass spheres are sprinkled on the base plate at random. With the overhead viewing microscope one can select a $15\,\mu m$ or $20\,\mu m$ sphere and irradiate it from below using a $5\,cm$ focal length loupé lens. One finds that even if the vertical force on the sphere is many times mg, the particle does not rise. This is due to a strong van der Waals force that causes the small particles to stick to the surface. To break this bond we use a piezoelectric cylinder cemented to the base plate to jiggle the particles loose.

In Ref. [58]* Dziedzic and I tuned an audio-oscillator rapidly through a mechanical resonance of the plate to cause the particles to break free. In later work, we found it better to actuate the piezoelectric ceramic with a short pulse from a pulse generator. This was less of a disturbance to the

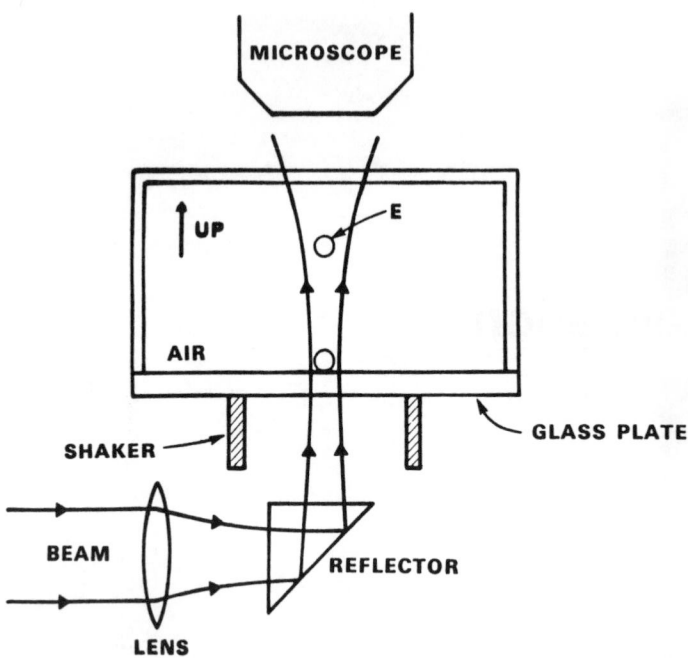

Fig. 2.1. Diagram of the first optical levitation experiment in air.

particles. With the resonance technique the particles tended to bounce off quickly in all directions and become lost.

Once a particle broke free from the base plate, it rose to its equilibrium point E, where it came to rest as sketched in Fig. 2.1. Figure 2.2 shows a photograph of an optically levitated glass sphere trapped in air about a centimeter above the base plate of a glass box. The small, roughly $10\,\mu$m sphere is visible as a star-like object. It scatters the mildly focused 5145 Å argon laser beam into a beautiful Mie ring pattern, part of which is visible on the opaque back wall of the box. The photo appeared in the June 1999 issue of "Optics and Photonics News" as a bit of optical history [59]. See also the photograph of Fig. 2.3, which shows the Mie ring structure in the strong forward-scattered light.

We were able to manipulate the sphere at will by simply moving the lens. We could lift it up into the middle of the cell, where it could be used as an ideal test particle for studying light scattering from an arbitrary second light beam. It could also be attached to the roof of the cell or separated from the untrapped particles as desired. This was the first demonstration of one of the most powerful capabilities of optical trapping, *the ability to sensitively manipulate individual selected particles in space without mechanical contact.* This capability reached a higher level of perfection in the work of Ashkin, Dziedzic, Bjorkholm, and Chu using the so-called single-beam gradient traps [5]* or optical tweezer traps [49]*.

With levitation traps we are dealing with trapping forces of mg [58]*. With optical tweezers the force can be orders of magnitude stronger. As one increases the power of tweezers, the forces increase proportional to the power. In the case of optical levitation, the vertical force on the particle always remains equal to mg as the power increases. The particle simply rises in the beam until the scattering

Fig. 2.2. Photograph of an optically levitated sphere in air and part of the Mie ring pattern of scattered light.

force balances gravity. Moreover, the transverse forces get weaker as the particle moves up in the beam, due to a decrease in the transverse intensity gradient.

This difference in strength of trapping forces between the two traps actually stems from the fact that the levitation trap is not an all-optical trap because it depends on gravity for its stability.

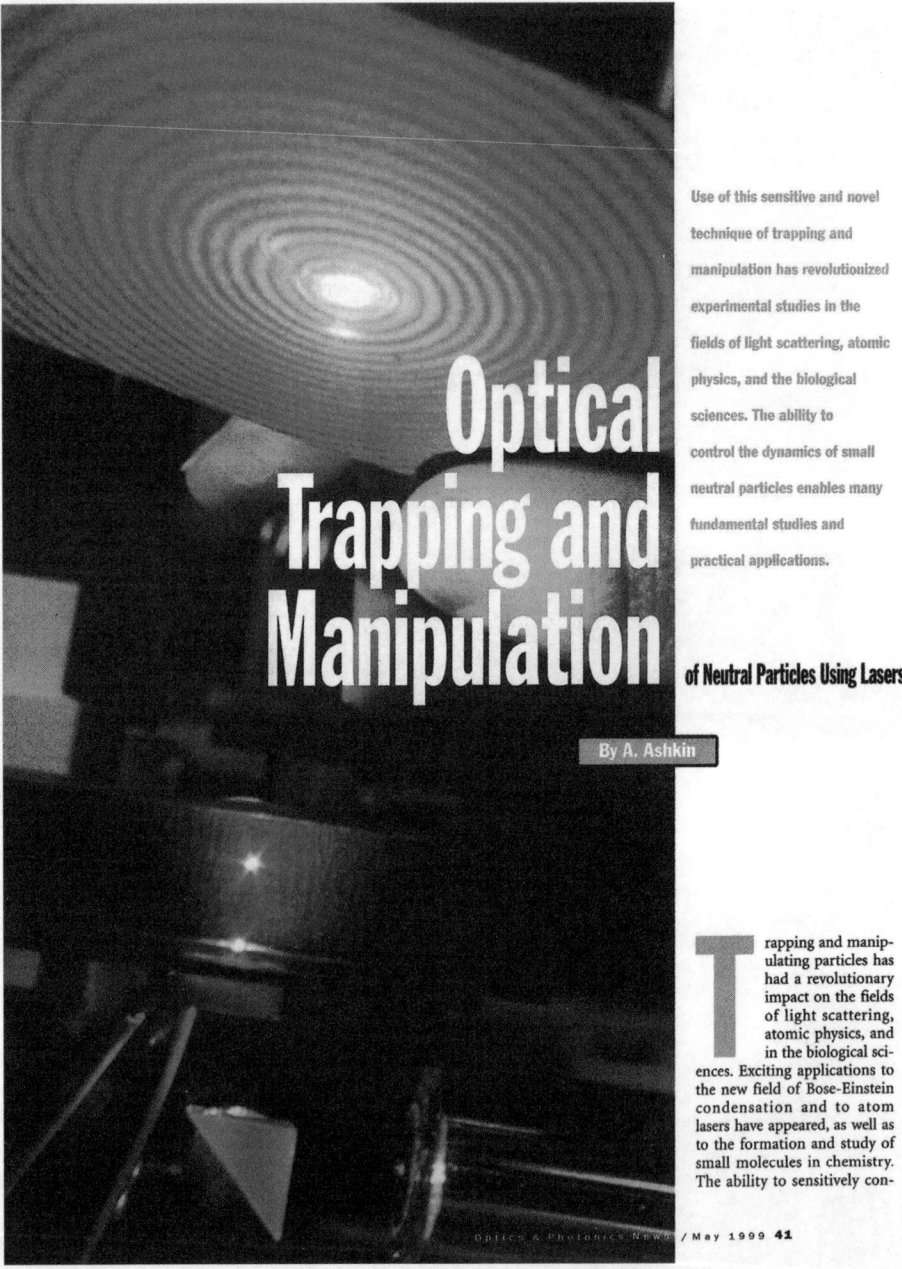

Use of this sensitive and novel technique of trapping and manipulation has revolutionized experimental studies in the fields of light scattering, atomic physics, and the biological sciences. The ability to control the dynamics of small neutral particles enables many fundamental studies and practical applications.

Optical Trapping and Manipulation

of Neutral Particles Using Lasers

By A. Ashkin

Trapping and manipulating particles has had a revolutionary impact on the fields of light scattering, atomic physics, and in the biological sciences. Exciting applications to the new field of Bose-Einstein condensation and to atom lasers have appeared, as well as to the formation and study of small molecules in chemistry. The ability to sensitively con-

Optics & Photonics News / May 1999 **41**

Fig. 2.3. Photograph of an optically levitated sphere in air, showing the Mie ring structure in the forward-scattered light, taken from Ref. [9].

Furthermore, if one turns a levitation trap upside down, the particle is lost. In an all-optical trap for macroscopic particles, such as the single-beam tweezer trap or the two-beam trap of Ref. [1]*, the trapping force simply increases linearly with increasing power. It would be interesting to see if one could avoid the need for a piezoelectric ceramic just by making the tweezer force exceed the

van der Waals force. To lift a sphere off a glass plate and break the van der Waals force requires about 10^4 mg of force, or greater. If possible, this could give a sensitive way to measure and study the van der Waals force. This has not been tried, to my knowledge.

2.2. Scientific American Article of 1973

Shortly after papers [1]*, [2]*, the present author wrote a Scientific American article on "The Pressure of Laser Light" [60] . In this article, I recapitulated the contents of reprinted papers [1]*, [2]*, [58]* in somewhat greater detail. I also speculated on the future of levitation and suggested the use of optical feedback damping to stabilize the position and damp the velocity of a particle in vacuum. Just such a feedback damping and stabilizing system was developed. See Ref. [7]* and Sec. 2.7. Feedback has since become one of the most powerful force measuring techniques for studying motor molecules. See Refs. [61]*, [62], and Sec. 7.13. It has also been suggested that feedback techniques can play a role in some remarkable experiments in which a single atom is trapped by the field of a single photon within an optical resonator. See Refs. [63]*, [64], and Sec. 21.10.

In 1975, Ashkin proposed use of radiation pressure forces to accelerate small particles in the micrometer-size range to high velocities of about 3×10^8 cm/s in vacuum. When two such fast moving particles of $\sim 0.5 \, \mu$m collide (see Fig. 2.4), one generates a power of $\sim 10^{11}$ W for about 100 fs (10^{-13} s), which results in a temperature some 50 times higher than the temperature needed for thermonuclear reaction in deuterium. The length of the laser acceleration apparatus is unfortunately fairly long, but still shorter than linear accelerators such as those at Stanford University. It should probably be a fairly straightforward apparatus to build. This proposal was made rather early in the history of optical manipulation and was rejected by the reviewer when I submitted it for publication (this was my only rejection). He said it was premature to discuss optical acceleration to high velocities, but that I should submit my estimates on temperatures achievable and the thermonuclear reaction in a separate paper. Since this was not my primary interest, I decided to first understand the problems of levitation in vacuum. I never returned to the question of reaching fusion temperatures by collisions of guided radiation pressure-driven particles. When Dziedzic and I finally achieved optical levitation in high vacuum in 1976 [65]*, we only mentioned the specific application of our experiment to the study of high angular velocities due to absorption of optical angular momentum (I leave it to

Optical Linear Accelerator for Small Spheres

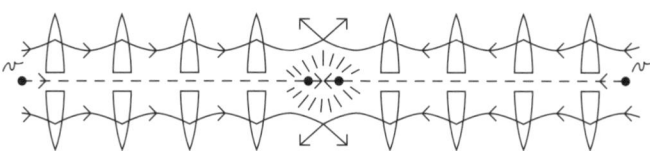

Fig. 2.4. Proposed apparatus for accelerating and colliding two macroscopic micrometer particles to velocities of $\sim 3 \times 10^8$ cm/s to produce a thermonuclear reaction. The opposing pairs of optical linear accelerators are made from a sequence of confocal thin lenses with holes to allow passage of the pair of oppositely moving particles. Mirrors (not shown) can be used to periodically introduce light along the optical path to compensate for optical losses.

the reader to re-evaluate such a thermonuclear power source in the light of twenty-first century technology).

In this early work [2]*, I also cited astronomical observations of radiation pressure in the ultraviolet driving gaseous ions and atoms in clouds around very hot stars to very high velocities. These light forces greatly exceed the stellar gravitational forces of the stars on the ions. Furthermore, due to the continuous nature of the thermal radiation, Doppler shifts do not reduce the driving force as the velocity of the ejected atoms increases to values as high as 3×10^8 cm/s. As we shall see later, just such Doppler shifts figure prominently in laser cooling or acceleration of atoms or ions [26]*.

Another interesting device suggested in my Scientific American article [60]* was the trapping of neutral atoms stably in a storage ring. This was an extension of the atomic beam velocity analyzer, discussed in Ref. [2]*. These rings can range from several centimeters to meters in diameter, depending on the velocity of the circulating atoms.

2.3. Levitation with TEM$_{01}$* Donut Mode Beams

The first demonstration of optical levitation in air, as discussed above [58]*, was made using high index of refraction spheres that were pulled into the high intensity region of a TEM$_{00}$ mode Gaussian beam. The question then arises could one conceivably levitate particles that are pushed out of the high-intensity region of the beam?

One such particle is a metallic sphere. Figure 2.5 shows why a metal sphere is pushed out of a TEM$_{00}$ mode beam. Looking again at the scattering from a typical pair of rays "a" and "b" hitting the sphere symmetrically about the center of the sphere, one sees that F_a is greater than F_b and the metallic sphere acts just like the bubbles in Fig. 1.9. We previously avoided metal spheres in water

Light Pressure on a Metal Sphere

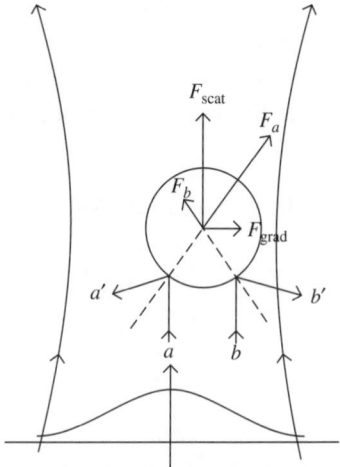

Fig. 2.5. Light pressure forces on a metallic sphere in a TEM$_{00}$ mode Gaussian beam.

for fear of thermal forces arising from optical absorption. It was discovered, however, that there is an interesting type of low loss dielectric particle that moves out of the high-intensity region of a beam. This is a thin-walled hollow glass sphere or microballoon particle, made commercially under the name of Eccospheres [66]*. It was shown in a follow-up paper entitled "Stability of optical levitation by radiation pressure" [66]* that these hollow spheres could be stably levitated in a TEM$_{01}$* donut mode Gaussian beam. Figure 2.6(a) shows a hollow sphere with a typical ray passing through it. One sees for such a thin-walled sphere (with about 45 μm outer diameter and about 1 μm wall) that refraction is negligible. The light scattering is dominated by reflections from the outer and inner surfaces of the thin shell. Since the external reflections are always larger than the internal reflections, this has the same net behavior as that of a metallic sphere and moves away from regions of high light intensity.

It was confirmed experimentally (see Fig. 2.6(b)) that if one places hollow spheres in a TEM$_{01}$* mode (or donut mode beam) that one can in fact achieve stable levitation. The actual donut mode beam used was made by inserting an on-axis "stop" inside the laser resonator. The quality of the resulting donut beam was not optimal, as seen in Fig. 1 of Ref. [66]*. If we were to take advantage of modern techniques for generating the donut mode holographically, we would expect to get a deeper and better trap for hollow spheres. Hollow thin-walled spheres filled with deuterium gas have been used as fusion targets in laser fusion research [67]. The techniques of optical trapping and manipulation of fusion targets may be useful in such fusion experiments.

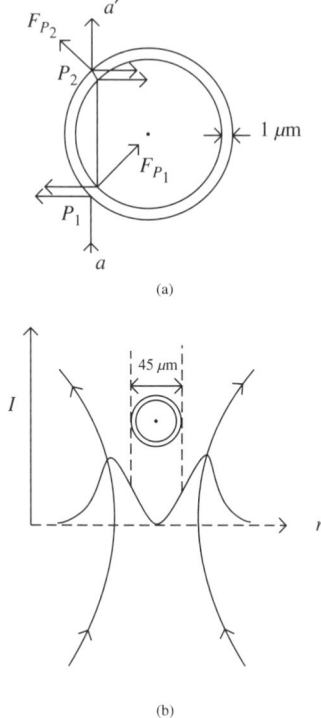

Fig. 2.6. (a) Light pressure forces of a typical ray passing through a transparent thin-walled dielectric sphere. (b) The thin-walled sphere is levitated at an equilibrium position in a TEM$_{01}$* mode beam, where the scattering force and gravity balance.

2.4. Levitation of Liquid Drops

Other types of particles that one can easily levitate in air or in any other gas are liquid drops. A scheme for dropping liquid drops into the waiting arms of a vertically directed levitating beam is shown in Fig. 2.7, taken from Ref. [68]*.

Drops are sprayed into a plastic vessel and allowed to settle under gravity [68]*. Some fall through a small hole located at the bottom into the levitating chamber and enter directly into the levitation beam. Drops in the correct range of sizes for the laser power chosen are readily trapped. Many drops can be trapped simultaneously and arrange themselves in what might be called an optical crystal. The largest particle is at the bottom on axis. This particle generates a ring-shaped beam above it, into which the next largest particle settles at a local intensity maximum. The beam is further distorted and a subsequent particle finds its equilibrium at a somewhat higher point, and so on. These particles are obviously optically coupled to one another. Move one and all those above it move. There also may be electrical coupling between particles if they are charged. If there are electrodes in the chamber, one can turn on a vertical electric field and see them move up or down, depending on their charge.

The spraying process is capable of generating both positively and negatively charged particles. One readily sees coalescence of positively and negatively charged particles as the electric field is increased. These "liquid-drop" crystals or arrays were the first of many types of crystals formed by light. It subsequently was discovered that charged colloidal particles arrange themselves in arrays in the focus of single-beam tweezer traps by virtue of their mutual repulsion [44]*. One can also make arrays of optical traps using the interference patterns of several beams to study artificial crystals of colloidal particles with different symmetries [69]. As we shall see later, atoms can arrange themselves in two- and three-dimensional (2D and 3D) standing wave patterns [3]*, [70], [71]. Ions likewise collect in the arrays in ion traps [72].

One can readily arrange to levitate only a single selected drop by eliminating the other drops. By simply raising or lowering the beam, one can deposit all drops above a given drop on the roof of the

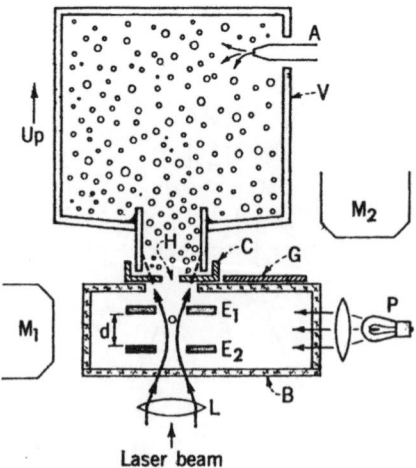

Fig. 2.7. Diagram of apparatus for feeding liquid drops into a levitation chamber in air.

chamber. All drops below a given drop can be deposited on the floor of the chamber. Silicone oil drops are almost ideal particles for light scattering purposes. They are essentially perfect spheres because of strong surface tension and their small radius of curvature. Furthermore, the optical loss of silicone is very low [65]*. They have served as the particle of choice for use in many subsequent experiments. The spraying and levitation technique also works for other liquids, such as salt solutions, volatile oils, and even water, if the surrounding vapor pressure is properly controlled [73].

The optical manipulation techniques for studying liquid drops are superior in many ways to other techniques for studying small micrometer-sized liquid drops. In one technique there is a thin spider web suspension scheme that is quite clever, but it affects the sphericity of the drop. There is the technique of electrodynamic trapping of charged macroscopic particles invented by Wuerker *et al.* [72]. This gives an alternating quadrupole field trap of the type later used by Dehmelt, in his single electron trapping work [74], and by Wineland in his single atomic ion work [75]. There is also an electronic feedback technique used by Arnold and others for trapping small particles and liquid drops [76]. It involves shaped static fields to keep particles transversely centered on a vertical axis, plus an active feedback system that controls a vertical voltage to keep particles at a fixed height. These are excellent techniques, but they lack the manipulative ability of optical traps [8].

2.5. Radiometric or Thermal Forces

Another force that figures very importantly in the early history of radiation pressure and also in our more modern considerations on optical levitation and traps is the radiometric, or thermal, force. This force is due to the interaction of the hot surface of a particle with the surrounding medium, such as water or air. Radiometric forces were studied extensively, both experimentally and theoretically, at the turn of the nineteenth century, and on through to the 1920s [77]. They figured prominently in Crookes's early attempt to observe radiation pressure forces, which resulted in the discovery of what is now called the Crookes' Radiometer. This instrument has four vanes, each with black on one side and silver on the other, mounted on a rotor. When irradiated, the vanes were observed to rotate in the opposite sense from that expected from radiation pressure (see Fig. 2.8).

The culprit here was the large radiometric force pushing normally on the black side of each vane. This force was orders of magnitude larger than the radiation pressure force on the silver side. It was only in experiments in 1901 and 1905 that E. F. Nichols and G. F. Hull succeeded finally in minimizing radiometric forces on the vanes and measured true radiation pressure in agreement with theoretical expectations, to within about a percent. They accomplished this feat by using lower gas pressure than Crookes, a clever vane design and other heat minimizing techniques [78], [79], [80].

There was also an experiment to detect radiation pressure by Lebedev, Russia's most famous scientist of his time. His work preceded that of Nichols and Hull by a few months. Lebedev's work observed the radiation pressure on absorbing gases held in enclosed cells [81], however, it was not as careful as Nichols and Hull.

Poynting, of Poynting vector fame, also carried out a clever experiment in 1904, which completely eliminated the effect of radiometric forces due to heated gas [82]. His method involved measuring the transverse component of radiation pressure of light beams hitting a reflecting surface at an angle of incidence of 45° [83]. Although it may be hard to believe, at the end of the nineteenth century,

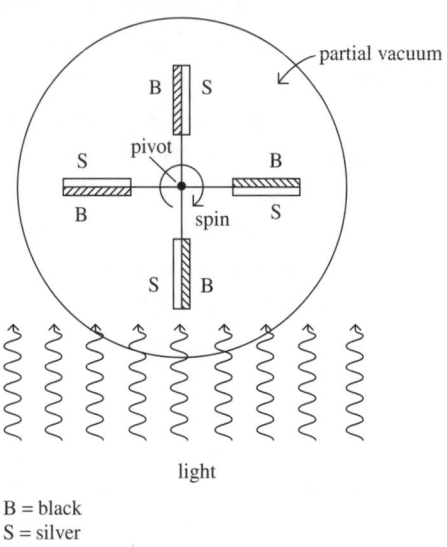

Fig. 2.8. Crookes' radiometer. The rotor spins in response to the large radiometric (thermal) force, which greatly exceeds light pressure.

measuring radiation pressure and radiometric forces was one of the "hot topics" of physics. Radiation pressure featured prominently in Poynting's presidential address to the Physical Society in England in 1905. He said "A very short experience in attempting to measure these light forces is sufficient to make one realise their extreme minuteness — a minuteness which appears to put them beyond consideration in terrestrial affairs, though I have tried to show that they may just come into comparison with radiometric action on very small dust particles".

I quoted Poynting in my Scientific American article in 1972, "The Radiation Pressure of Laser Light" [60]*. His statement stands in strong contrast with the situation in this subject at the turn of the twenty-first century. At the present time, as we observed, the acceleration of radiation pressure from lasers on micrometer-sized particles and atoms can be as high as $10^6\, g$, where g is the acceleration of gravity and does play a considerable role in terrestrial affairs. Poynting's allusion to the size of radiation pressure forces and radiometric forces on dust particles was prophetic and hinted at his later work, in 1920, on the so-called "Poynting–Robertson effect", which explains the absence of interplanetary dust in the solar system [84]. As I will discuss later in connection with atoms, the Poynting–Robertson effect is in reality closely equivalent to "optical molasses" damping.

The theory of radiometric forces was developed in detail in the 1920s [85]. These thermal effects on small particles can be further characterized as thermophoresis and photophoresis.

In thermophoresis, small particles move under the influence of steady-state thermal gradients in a surrounding medium. Tyndall first discovered this in 1870 [86], [87], when he observed what is known as the Tyndall dark space, or a dust-free area, around a hot body in a dusty environment.

Photophoresis forces are of more immediate interest to us in connection with optical levitation. These forces are caused by optically induced temperature gradients generated within an absorbing particle that is located in a surrounding gaseous medium. The term photophoresis was coined by Ehrenhaft, the Austrian physicist, who carefully studied thermal forces on small particles in connection with his work on the charge of the electron [88], [89]. He distinguished positive and negative photophoresis, depending on whether the thermal force was in the direction of the incident light or was in the opposite direction. This depended on the optical absorption coefficient of the particle. For highly absorbing particles the front face of the particle was heated much more than the rear face and the resultant force was positive, see Fig. 2.9(a). For weakly absorbing particles the light penetrated and refracted through the particle and was focused on the back surface, giving a backward directed force, see Fig. 2.9(b). A summary of the many experimental investigations of photophoresis was given by Reiss [89] and Ehrenhaft [90].

As can be seen from Fig. 2.10, the effect of photophoresis, at the equilibrium temperature of a levitated particle, is opposite that of the scattering and gradient forces of radiation pressure. Thus a moderately lossy levitated particle should sit off axis of the TEM_{00} mode beam at a lower height than a lossless particle. For sufficiently lossy particles, levitation by radiation pressure is not possible. One can, however, make a levitation-type trap using the radiometric force from a laser, using a donut mode and a lossy particle [91]. This type of levitation is quite analogous to levitation of reflecting metal particles or hollow glass spheres.

Positive Photophoresis

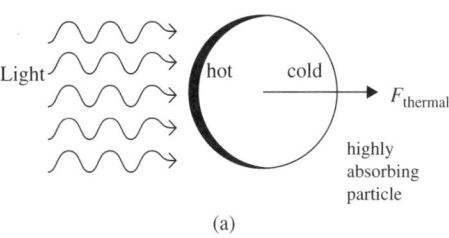

(a)

Negative Photophoresis

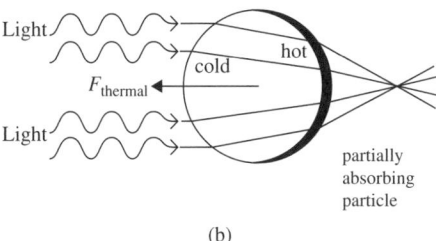

(b)

Fig. 2.9. Origin of photophoresis, or thermal, forces. (a) Positive photophoresis for highly absorbing particles. (b) Negative photophoresis for mildly absorbing particles.

Radiometric (Thermal) Effects in a Gaussian Beam

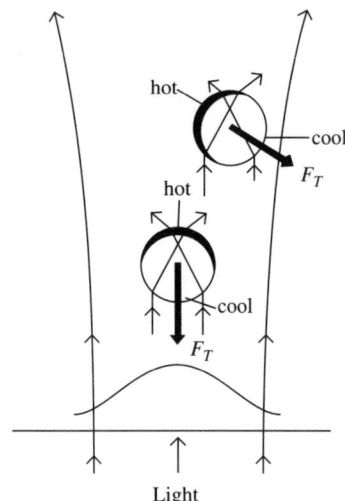

Light

Fig. 2.10. Radiometric effects on particles in a TEM_{00} mode levitating beam give rise to destabilizing thermal forces.

2.6. Levitation at Reduced Air Pressure

The effects of the radiometric force were seen very graphically in Ref. [65]*, where observations were made of levitation at reduced air pressure. These were made with a simple vacuum cell with an adjustable valve.

It is known from the kinetic theory of gases that the radiometric or thermal forces increase as $1/p$, where p is the gas pressure, down to the point where the mean free path of an air molecule is comparable to the particle size. It is also known that the viscosity η of the gas is constant, independent of pressure, down to the same point [65]*. It was found in the first levitation experiments that the particles gradually descend in the levitating beam as the pressure was reduced and became unstable both vertically and horizontally, and finally escaped from the beam at pressures from 1 to 10 Torr, depending on particle size [58]*. This behavior is a sure sign of lossy particles. The high loss probably stems from the fact that these were the so-called "flexo-lite" glass particles. They are doped to give a high index of refraction so that they act as retro-reflecting particles. Such particles find wide use when mixed with paints to make reflective lane markers and traffic signs for roads and highways.

A calculation of the optical absorption loss α in a flexo-lite particle based on the estimated radiometric force gave $\alpha \cong 5 \times 10^{-2}\,\text{cm}^{-1}$, which is quite high for levitation in partial vacuum. Since $\alpha \cong 10^{-4} - 10^{-5}\,\text{cm}^{-1}$ is possible in low loss glass, for example, this thermal limitation at low pressure is not fundamental.

Indeed, in subsequent work in 1976, it was shown that one could levitate down to pressures of 10^{-6} Torr with very low loss silicone oil drops and also low loss silica particles [65]*. Figure 2.11, taken from Ref. [65]*, shows the calculated dependence on pressure of the radiation pressure F_{rad}, the radiometric (thermal) force F_T, the thermal conductivity Λ, the viscosity η and the particle temperature for $20\,\mu\text{m}$ particles with a loss $\alpha \cong 5 \times 10^{-4}\,\text{cm}^{-1}$ and $2 \times 10^{-2}\,\text{cm}^{-1}$. The maximum

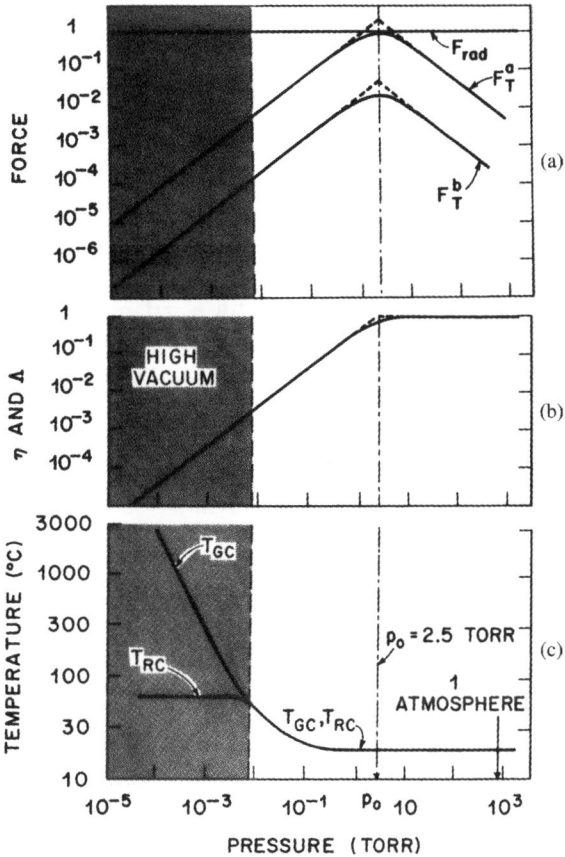

Fig. 2.11. (a) Dependence of the radiation pressure force F_{rad} and the radiometric (thermal) force F_T on gas pressure, for a 20-μ particle. F_T^a is for $\alpha \cong 2 \times 10^{-2}$ cm^{-1} and F_T^b is for $\alpha \cong 5 \times 10^{-4}$ cm^{-1}. (b) Thermal conductivity Λ and viscosity η versus gas pressure. (c) Particle temperature versus gas pressure for a 20-μ particle with $\alpha \cong 5 \times 10^{-4}$ cm^{-1}. T_{GC} is only for gas cooling, T_{RC} is for gas cooling plus radiation cooling.

of the radiometric force at p_o, the pressure at which the mean free path is equal to the particle size, is called "the radiometric barrier". With low loss particles the radiometric barrier is small compared to the radiation pressure force and one can pump down and pass through this radiometric barrier with no difficulty.

From Fig. 2.11 one would expect, at pressures of $\sim 10^{-6}$ Torr, that the radiometric force, the viscosity (or damping), and the thermal conductivity of gas to all be essentially zero, as would be the case for zero pressure, or total vacuum. It was found experimentally that as one pumps out the gas from the chamber very slowly, if the laser power is steady enough, one does not excite any oscillations and particles sit quietly at rest in the trap for long periods of time. This is the case for silicone oil drops or silica spheres with losses $\alpha \cong 5 \times 10^{-4}$ cm^{-1}.

However, silicone oil drops last for only about 10 min, due to what we interpret as a photochemical damage reaction, which increases the loss. Silica spheres are much more stable and stay in the trap for about half an hour, before fluctuations gradually build up and the particle is lost. One can manipulate particles in high vacuum, even when the damping is essentially zero, if the position of the particle is changed very slowly, that is, adiabatically. In our experimental case, this means slow compared with an oscillation period of the particle in the trap, which is about $1/20$ s [65]*.

One conclusion from this experiment, by demonstrating levitation down to pressures where the radiometric forces are less than 10^{-6} of the radiation pressure forces, is that particles can be stably trapped solely by the forces of radiation pressure. This finally puts to rest any questions about a possible role for radiometric forces in the stability of radiation pressure traps. One of my fellow department heads at our Holmdel laboratory, for example, always insisted that radiometric forces were necessary for stable traps.

It is interesting to ask if there is any other additional form of damping besides viscous or radiometric damping in total vacuum. Put another way, will a small particle, oscillating in total vacuum where radiometric damping is zero, continue to do so forever? It was shown theoretically that there is indeed a small intrinsic damping due to the light flux itself that exists at zero pressure. This intrinsic optical damping is due to the Doppler shift of the levitating light. It was theoretically calculated to give a half amplitude decay time of ~ 0.7 year! [65]*. More will be said later about this optical damping on macroscopic particles in connection with "optical molasses" damping of atoms.

2.7. Feedback Damping of Levitated Particles and Automatic Force Measurement

Although we avoided the effects of radiometric forces and succeeded in levitating and adiabatically manipulating particles in high vacuum, the absence of strong damping and residual laser power fluctuations made it difficult to work in this environment. Any sudden motion, vibration, or low power fluctuations can cause particle oscillations. Introducing an optical feedback system solved these problems. Basically what was done was to electronically control the light power not only to lock the position of the levitated sphere to an external reference, but also to oppose any particle velocity directed away from the equilibrium position [7]*.

This is accomplished by having an external split photodiode reference that gives rise to an error signal Δ whenever a fluctuation in the height of the particle occurs (see Fig. 2.12). By amplifying any error signal and feeding back a voltage proportional to Δ to an electro-optic modulator, one can, in essence, lock the particle to the reference height of the photodiode. This is done by automatically adjusting the light power to compensate for all sources of vertical fluctuations. Feedback proportional to Δ acts as a light power stabilizer and also provides a strong additional vertical stability to the intrinsic optical stability designed into the trap. To get damping we take the derivative of the error signal $d\Delta/dt$, which is essentially a velocity term, amplify it and add it to the feedback voltage. This provides a crucial external damping force $F_d = -\gamma v$, where γ is the feedback damping constant and v is the velocity. This counteracts any vertical fluctuations. The above feedback system works beautifully to stabilize particles levitated in essentially total vacuum.

One may wonder why the system works as well as it does, since we are damping only the vertical component of motion. The reason is that all components of motion of a particle in the trap are coupled so that if you damp one component, you damp them all. There is no such thing as purely transverse motion in our trap. For example, as a particle moves off axis transversely, it feels less axial force and drops vertically. It is also true for optical tweezer traps that damping one component damps all

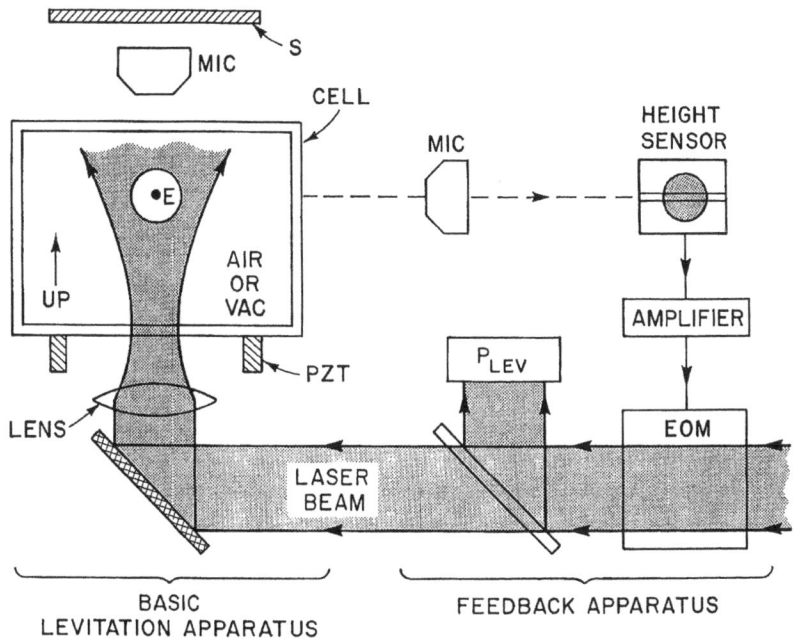

Fig. 2.12. Basic apparatus for levitating and feedback stabilizing dielectric spheres in vacuum or air. PZT is a piezoelectric ceramic shaker. EOM is an electro-optic modulator.

components of motion, unless you happen to be using a specially designed tweezer trap, where the oscillation frequencies in the x, y, and z directions are the same.

Perhaps even more important than the damping and stabilizing capabilities of the feedback system, it was also found that the feedback scheme can be used as an extremely powerful automatic force measuring apparatus [7]*. For example, with a particle held in a feedback-controlled trap, any additional external force you might apply automatically changes the trapping power needed to keep the particle at the reference height. The resulting change in the levitation power is a direct measure of the applied force.

2.8. Feedback Measurement of Axial Scattering Force

One of the first applications of the optical levitation feedback scheme was to measure the axial scattering force of the levitating beam itself [7]*. The axial scattering force had been measured, in part, earlier in a study of the stability of optically levitated particles [66]*.

There were two aspects to these earlier experiments. The first was the sensitivity of the position of levitated particles to changes in the laser power. Not surprisingly, it was found that if the so-called diffraction beam angle $\theta = \lambda/\pi w_o$ was increased, implying smaller beam spot size w_o, then the particle equilibrium position became less sensitive to changes in laser power (and vice versa). Secondly, it was discovered accidentally that as the beam angle was increased to the point where the beam spot size became less than the particle size, an additional regime of stability appeared, in which the particle could be levitated stably below the beam focus.

This was explained in Ref. [66]* by the fact that when the beam spot size is smaller than the particle, the focus is not a maximum of the axial force but a minimum. Indeed, as one moves the particle away from the focus the beam spreads over the full particle surface, giving strong refraction and therefore increased axial force. This results in the additional possibility of levitation below the focus. This was observed in Ref. [66]*. One also sees bi-stable switching of a particle between the two possible equilibrium points, above and below the focus.

Using feedback stabilization [7]*, it was possible to measure the scattering force to high precision (about 1%) for all heights in the stable and also unstable regimes of the levitating beam. The measurements in the unstable region are possible because of the additional stability gained from the feedback scheme. This not only confirmed our earlier picture but also extended the results to larger and smaller beam diffraction angles. At larger diffraction angles, the force minimum near the focus got deeper and deeper and the stability for axial motion for small changes in power increased. These beams with increasing beam angle began to resemble the high convergence beams of optical tweezers, but did not yet display the backward radiation pressure force characteristic of tweezers. For this to occur, the force at the minimum first has to reach zero and then negative with further increase in convergence angle (or numerical aperture).

We also found interesting behaviors at decreased beam angle θ, i.e., beams with large spot sizes [7]*. For example, an $\sim 9.0\,\mu$m particle levitated at the focus of a $70\,\mu$m focal spot diameter moved $\sim 10\,\mu$m for a one part in 10^6 change in levitating force (i.e., the particle's weight). The force of gravity on this particle is $\sim 3 \times 10^{-9}$ dyne or 3×10^{-2} pN. This weakly spreading beam acted like an extremely weak spring. The spring constant, k, in the equation $F = -kx$ is $\sim 3 \times 10^{-3}$ pN/μm for this beam.

There is every indication that we can work with a levitating beam width of 1 mm, for which $k \cong 20\,$fN/μm. This potential for working with feedback control of weak forces on small macroscopic particles has not yet been exploited. I will suggest an interesting possible use shortly.

2.9. Feedback Force Measurement of High-Q Surface Wave Resonances

Another obvious application of the feedback force-measuring scheme is the measurement of the wavelength dependence of the radiation pressure force. At first this did not seem like a very interesting experiment. The scattering force, as described earlier in the ray optics regime, should not at first sight display much wavelength dependence. Figure 2.13 shows a particle being hit by a plane wave. The typical rays "a" and "b" simply undergo Fresnel refraction and reflection, giving the qualitative behavior of the force as explained above. Rays hitting the sphere near the particle axis, such as "c" and "d", are barely reflected or refracted as one expects. At best, one might observe some evidence of weak Fabry–Perot resonances of the axial rays. Rays such as "e" and "f" striking the sphere at grazing incidence are expected to be highly reflected with little change in direction and with only a very small fraction of the energy refracted into the particle. The index of refraction, for most transparent materials, is only weakly wavelength dependent.

The experiment involved capturing a low-loss silicone oil drop [68]* in a feedback levitation trap and recording the laser power as one tunes the wavelength of a dye laser [92]*. Typical results adapted from [92]* are shown in Fig. 2.14, taken from [92]*. When Dziedzic and I saw the results

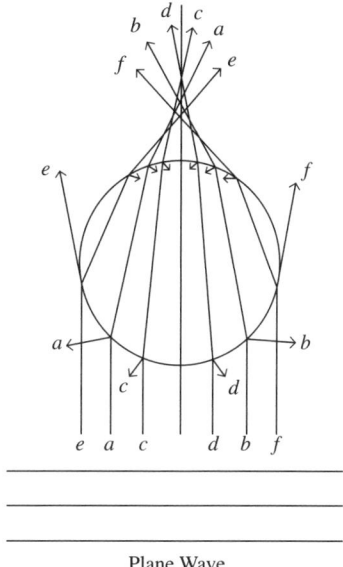

Fig. 2.13. Typical rays passing through a dielectric sphere illuminated by a plane wave.

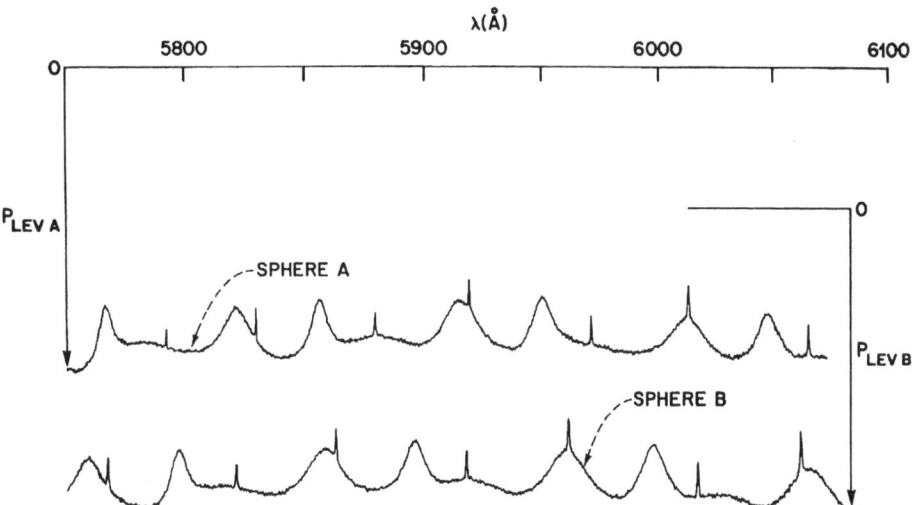

Fig. 2.14. Resonant behavior of light forces on dielectric spheres. The spectra show the variation with wavelength $P_{\text{lev}A}$ and $P_{\text{lev}B}$, the power needed to levitate oil drops A and B, which have index of refraction $n = 1.47$ and slightly different diameters ($\sim 10\,\mu$m). The resonances of sphere A are shifted ~ 50 Å higher in wavelength than the corresponding ones in sphere B.

for sphere A, our first reaction was complete surprise. Where did all those strange wiggles come from? We thought that perhaps the feedback circuit had failed and was oscillating or something else bad had occurred.

Before taking it apart I suggested we compare the data on A with data on a few other spheres. This led to the second surprise: all the spectra were essentially identical except for a shift in wavelength.

For example, the spectrum of particle B, shown in Fig. 2.14, is the same as for particle A, only shifted ~ 50 Å lower on the wavelength scale. This implies that there is nothing wrong with the feedback circuit since the spectral features are real and tied to the particles, which differ from each other only in diameter. These wiggles and spikes are clearly indicative of some sort of resonance phenomenon. But what is it? Somehow it must be contained in the classical Mie–Debye electromagnetic theory of light scattering from spheres. Indeed, Debye mentioned the possibility of resonances in his PhD thesis (see Ref. [1]*).

In 1977, at the time of this experiment [92]*, there were no really high resolution computations of the Mie theory for light scattering from spheres at high "X" number, or "size parameter", where $X = 2\pi a/\lambda$ and a is the radius of the sphere. There was only Irvine's computation [93] using the moderate resolution $\Delta X/X = 10^{-4}$ which he extended up to X values as high as ~ 100. Our own experimental resolution was $\sim 1/20$ Å out of 6000 Å, or $\Delta\lambda/\lambda = \Delta X/X \cong 8 \times 10^{-6}$, which was more than an order of magnitude higher than Irvine's computation. Nevertheless, the general character of the experimentally measured power spectra resembled Irvine's Q_{pr} spectra. So it seemed that the observed sharp resonant features are contained in Mie–Debye theory. The astronomer van de Hulst, in his book "Light Scattering by Small Particles", considering smaller X values, explains the well known ripple structure in the total light scattering from spheres at low X as due to dielectric surface wave resonances of the sphere [94]. It seemed likely that these surface wave resonances extended to higher particle sizes or higher X and explained the experimental findings.

Figure 2.15 shows the essence of van de Hulst's surface wave resonance picture. A ray "a" is incident on a sphere at closely grazing incidence. We might expect most of the light to be reflected into ray a' with only a little energy entering the sphere at roughly the critical angle θ_c, as indicated in Fig. 2.13. However, if the size of the sphere is correct, the light entering the sphere will continue to be reflected around the sphere until it closes on itself in phase. The internal light cuts across the sphere three times and it is called a three-cut resonance. Because of the resonance condition the field inside the sphere builds up to a high intensity. The high intensity circulating internally contributes a field component to the externally reflected wave a', tending to cancel a'. The net result is strong coupling of grazing incidence rays into what we call a surface wave resonance.

Surface Wave Resonance

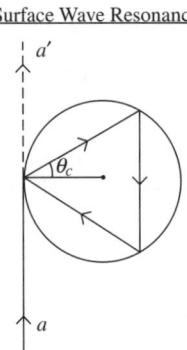

Fig. 2.15. Surface wave resonance. Grazing incidence ray "a" entering the sphere at the critical angle θ_c feeds an internal wave resonance that cancels the externally reflected ray a'.

Off Resonance

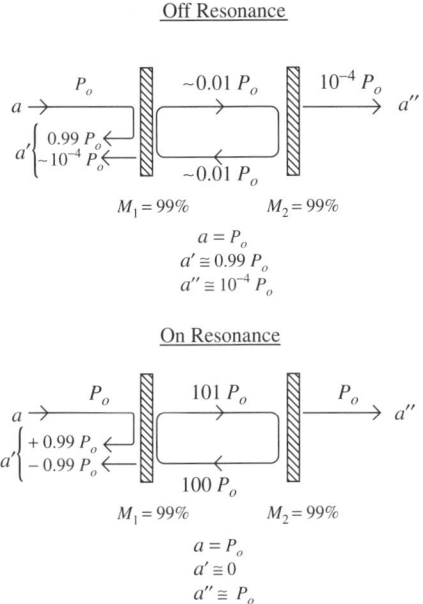

On Resonance

Fig. 2.16. Coupling of an input ray "a" with power P_o to a parallel mirror resonator with two 99% mirrors, when off-resonance and when on-resonance.

This is much like the usual picture of light coupling to a parallel mirror resonator, such as shown in Fig. 2.16. For example, off-resonance for the case of mirror $M_1 = M_2 = 99\%$ reflectivity, only 1% of the incident light gets through M_1 and only 0.01% gets through M_2. The power in the reflected ray a' is $\sim 99\%$ of ray a. If one tunes to resonance with two equal mirrors, then the reflectivity a' is zero and all the incident power is matched into the resonator. The internal power builds up to 101 times the incident power P_o in the forward direction and 99% in the backward direction. The transmitted power of ray a'' equals the input power in ray a. This is all in the absence of any absorption loss.

There are, of course, many resonances of the sphere. As the diameter increases or the wavelength decreases, some of the light can travel a path around the surface on an arc before closing in phase (see Fig. 2.17, where we show a mode with three cuts plus an arc of length α). Any combination of three cuts plus any number of arcs adding up to α is also resonant. We also can have a four-cut mode and four-cut arcs. Propagation on an arc is very much like light propagating along a plane surface in a Goos–Hänchen shift or a surface wave [95] (see Fig. 2.18, showing the Goos–Hänchen geometry, where light striking a surface at the critical angle propagates along the surface varying distances before being reflected back at the critical angle of total internal reflection). The only difference between propagation along a plane surface and along a curved surface like that of a sphere is that the curved surface radiates energy into the low index medium. This is a loss mechanism that reduces the Q of surface wave resonances having large surface paths α.

Let us return to the experiment of Ashkin and Dziedzic [92]*. Comparing the resonances of sphere A with sphere B in Fig. 2.14, one can say that the ratio of the radii of the spheres $r_a/r_b = \lambda_{\text{res}\,a}/\lambda_{\text{res}\,b}$, where $\lambda_{\text{res}\,a}/\lambda_{\text{res}\,b}$ is the ratio of the wavelengths of the same resonances of "a" and "b". Using all

Three-Cut Plus Surface Wave Mode Resonance

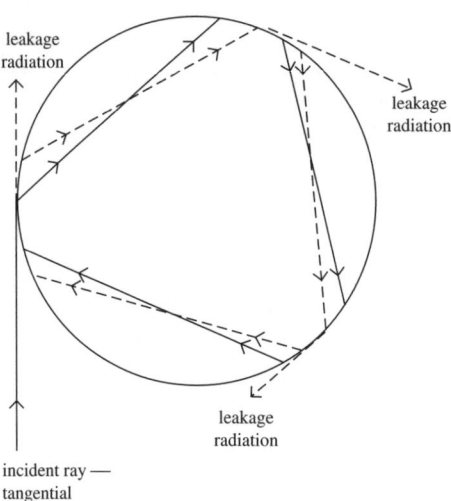

Fig. 2.17. Diagram of a resonance made of three critical-angle cuts plus propagation along an arc of length α on the surface. Any three cuts plus three arcs adding up to α in length have the same total optical path and feed the resonance. Tangential leakage radiating from the surface propagation is indicated schematically.

Goos–Hänchen Shift

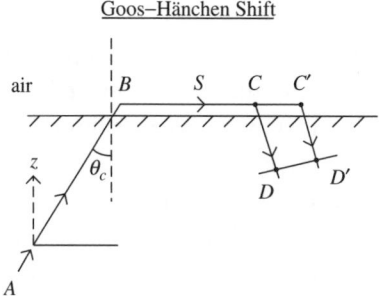

Fig. 2.18. Light incident on a plane surface at the critical angle θ_c propagating along the surface as a surface wave S in the so-called Goos–Hänchen shift in $ABCD$ and $ABC'D'$. Figure adapted from Fig. 2.3 in Ref. [133].

pairs of sharp resonances this measure of relative size is accurate to about one part in 10^5–10^6, a good three orders of magnitude better than conventional Mie ring counting techniques. The Mie ring counting technique is based on the fact that the spacing of the Mie interference rings varies with wavelength. For a given wavelength, the larger the diameter, the higher the ring density. See Ref. [58]* on the first levitation experiments, where this matter is discussed. Subsequently, Chylek and his colleagues made a high resolution calculation of the wavelength dependence of the radiation pressure force resonances based on exact Mie theory [96], showing very impressive agreement with the observed force spectra of Ashkin and Dziedzic [92]*. Up until this time there were questions about whether the partial wave resonances of Mie theory actually corresponded with observable effects in the scattering. Reference [96] made this connection unambiguous.

In 1980 and 1981 my colleague, Dziedzic, and I made a high resolution study of the light scattering of low loss levitated liquid oil drops as a way of studying the resonance behavior of dielectric spheres [97]*. Measurements were made in the near and far-field at 90° and in backscatter.

In the near-field, spectra were taken of light emerging at different locations on the sphere, from the edges and from the axis of the sphere. This type of measurement shows that the resonant light emerges from the edges of the sphere as expected for van de Hulst surface wave resonances. Simultaneous measurements of the radiation pressure spectra and 90° and backscattering spectra were made. The backscattered light is particularly strong and interesting. This is what is known as the "glory" to light scatterers. These measurements [97]* are, to this day, the most detailed light scattering measurements ever made, as a function of λ. Figure 2.19, taken from Fig. 3(f) of Ref. [97]*, shows the comparison between our experimental glory spectrum and Chylek's exact Mie calculations. This probably represents the most precise check on the prediction of Mie theory and shows the power of levitation and feedback for use in light scattering measurements. Note the effect of the acceptance angle. The difference between 4° and 1.5° is dramatic. Perhaps reducing the acceptance angle further will result in an even closer agreement with theory on the high-resonance peaks.

Use of a tunable single frequency laser beam might also result in even closer agreement with theory. A single frequency beam might also make some of the higher Q resonances calculated by Chylek appear. See Fig. 3(e) of Ref. [97]*. Another way to display the variation of scattered light or levitating power with particle size is to hold the particle with a fixed wavelength beam and to change the particle size continuously by evaporation, for example. Results are shown in Fig. 13 of Ref. [97]*.

By changing the wavelength so that one is situated on the side of one of the steepest resonances, one can detect minute changes in size as a shift in levitating power or scattered light. One can calculate that the sensitivity of one part in 10^5–10^6 corresponds to changes of drop size, which is a

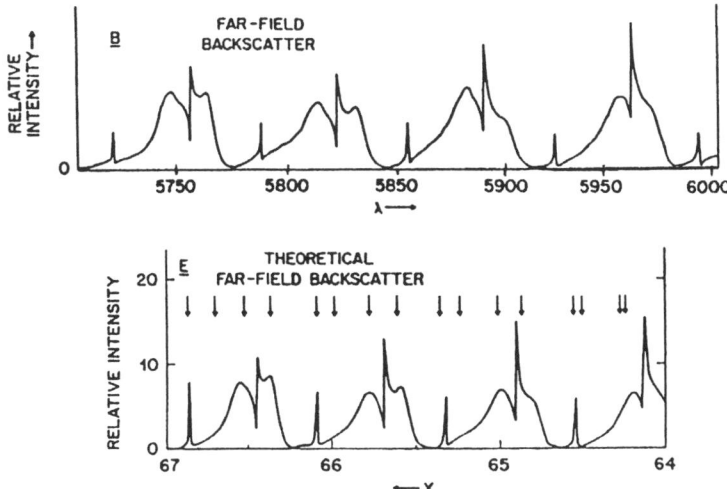

Fig. 2.19. Comparison of the experimental far-field backscatter with the theoretical far-field backscatter versus $x = 2\pi a/\lambda$, where a is the particle radius.

fraction of a monolayer. It was suggested in Ref. [97]* that the experimentally observed spectrum was sufficiently unique in character that it could be used to calculate the absolute size and index of refraction of the particle to very high precision. In 1983 this was done by Chylek, Ramaswamy, Ashkin, and Dziedzic [98] to an accuracy of 5×10^{-5}. This is the most accurate determination of size and refractive index by light scattering. This was based on *a priori* knowledge of the diameter and refractive index with an accuracy of 10^{-1} and 5×10^{-3}, respectively. It is estimated that an order of magnitude improvement in accuracy is possible with improved experimental technique.

As time has gone by, these surface wave resonances, or morphology-dependent resonances (MDR), or nearly total internal reflection, or whispering gallery mode resonances, as they are variously called, are playing increasingly important roles in modern optics. The high Qs of these resonances and the high values of internal intensities have made possible the observation of very low-threshold dye lasers, Raman and Brillouin oscillators, and nonlinear optical effects in spherical cavities [99], [100]. They are being used as the cavities in the field of Cavity Quantum Electrodynamics (CQED) [101], [102]. They are also being used as resonant filters [103] and channel dropping filters [104], [105] in communication research. With cylindrical geometry they have frequently played a role as resonators for semiconductor lasers [106] and quantum cascade lasers [107]. More will be said later about CQED experiments using single atoms.

2.10. Measurement of Electric Forces by Feedback Control of Levitated Particles

Another application of feedback control of levitated particles is to the measurement of electric forces on single electrons in what amounts to a modern version of the Millikan oil drop experiment [108]. In our experiment we placed a feedback controlled silicone oil drop in a vertical electric field and then proceeded to change the electric charge on the sphere by introducing ions into the chamber. These were mostly negative ions generated at a slow rate by shining a UV lamp on the interior surfaces and air surrounding the oil drop. As the number of electrons on the drop increased so too did the upward electric force. To keep the height fixed, the levitating power dropped incrementally. These results are shown in Fig. 2.20, taken from Fig. 5 of my 1980 Science review paper [109]*. The single electron steps were beautifully resolved and showed graphically the uniformity of the electric charge. The charge resolution achieved was about 0.1 electronic charges. No free quarks or one-third electron charges were seen, as in the experiment of La Rue *et al.* [110]. No "sub-electrons" were seen, as reported by the Viennese physicist Ehrenhaft [111].

Ehrenhaft's sub-electrons are not given much credence now. The work of La Rue *et al.*, however, was very carefully done and their results still remain a mystery. For example, they found small niobium spheres, processed in a similar way on a tungsten substrate, with residual charges of $\sim +e/3$, $-e/3$, and ~ 0. The experimental accuracy was high: $(+0.337 \pm 0.009)e$, $(-0.33 \pm 0.070)e$, and $(-0.001 \pm 0.025)e$. These ~ 100–$300\,\mu m$ diameter spheres were measured while being magnetically levitated in the superconducting state at low temperatures. The authors believe that they have understood and eliminated possible errors due to electric dipole and magnetic force effects on the spheres and on the pair of metal condenser plates used in the electric charge measurements. However, if the residual charge could be measured using optical levitation, one could avoid the need for

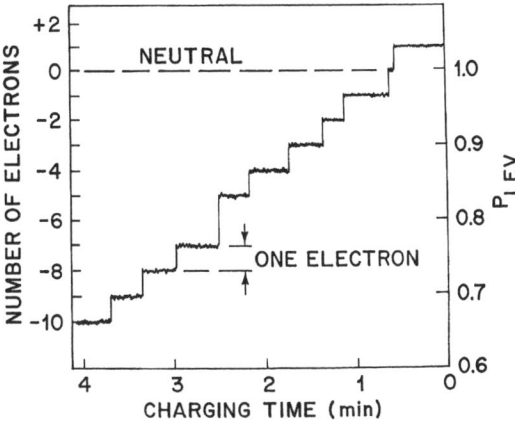

Fig. 2.20. Changes in the optical levitating power caused by the automatic feedback system as the charge on the sphere in an electric field increases by single-electron amounts.

superconducting magnetic levitation and thus avoid any possible errors due to magnetic effects. Also, the electric charge measurement does not require dc electric field gradients. An optical measurement would then be an independent check on this unexplained result. Optical levitation of metal spheres is more difficult than for low loss dielectric spheres, but is still possible. An optical tweezer method based on a focused beam scanning method will be mentioned later in Chapter 13 on microchemistry. This method has successfully levitated aluminum, iron, and even carbon particles [112] and so should work for a shiny metal such as niobium.

It is interesting to note that Millikan's oil drop technique for directly measuring the electronic charge does not figure in the modern determination of e. Millikan's original value of $e = 4.774 \pm 0.005 \times 10^{-10}$ esu was found to be in serious error (see p. 119 of Ref. [108]). Millikan apparently used too low a value of the viscosity of air in his calculations [111]. The currently accepted value of e is known to seven or eight significant figures. The value of e is no longer determined by direct measurement, but is rather derived from other more accurately known physical constants involving the electronic charge. A direct independent measurement of e, accurate to six significant figures, using the optical version of the Millikan oil drop experiment would be very useful in the study of fundamental physical constants. It would affect the accuracy of all other fundamental interrelated constants containing e directly or indirectly, such as the fine structure constant $\alpha = [\mu_o c^2 / 4\pi](e^2 / he)$ or the Josephson frequency–voltage ratio of $2e/h$, etc.

The determination of e using a modern optical version of Millikan's oil drop experiment would not involve the use of the viscosity of air to determine the diameter of the drop. The diameter of the drop can be determined by surface wave resonances of MDRs to accuracies of one part in 10^5–10^6 as discussed earlier. Therefore, in the Millikan balance equation $eE = mg$ we can get m, the mass of the drop, to high accuracy knowing the density of silicone oil and the diameter of the drop. The absolute value of the acceleration due to gravity is currently known to very high accuracy [113]. The number of electrons on the sphere is unequivocally known from measurements, as in Fig. 35 of Ref. [109]*. Finally, one can determine E to very high accuracy using carefully designed condensers with very flat parallel electrodes with guard rings, as used in determining the value of the farad. The farad is currently known to 0.02 ppm [114]. Thus, it seems that a direct measurement of e is now possible to an accuracy of one part in 10^5–10^6.

It is remarkable that it took so long for the existence and usefulness of this complex spectroscopy of spheres to be recognized. The idea of resonance based on what was called the whispering gallery modes is almost as old as the laser itself. Garrett *et al.* [115] observed pulsed lasing on these modes in hand-polished doped crystal spheres several millimeters in diameter. These lasers were difficult to fabricate and not particularly useful. Radar scattering from spheres in the context of Mie scattering from spheres with relatively low values of X was also known. See Ref. [92]*. However, sharp resonances develop only for large X. As discussed above, it finally took the combination of isolated, single, highly precise, levitated liquid drops having size parameters of $X > \sim 20$ with tunable dye lasers and feedback techniques to make the high-Q modes observable [92]*.

CHAPTER

3

Atom Trapping and Manipulation by Radiation Pressure Forces

3.1. Early Concepts and Experiments with Atoms

3.1.1. *Deflection of atoms by the scattering force*

Let us now continue our review of the developments that occurred in the field of atom trapping and manipulation during the approximate time period from 1970 to 1980. Recall that stable atom trapping and manipulation had been proposed by Ashkin in analogy with macroscopic particle trapping and manipulation [1]*, [2]* using the two components of the radiation pressure force, the scattering force and the gradient force. Following this early work on atoms, experiments were performed showing atomic beam deflection by resonance light applied at right angles to the atom beam direction [116], [117]. Ashkin [2]* also suggested Doppler free isotope separation with resonance light in the above geometry and calculated essentially 100% separation of the selected isotope with a yield of ~ 1 mg for a beam power of ~ 100 mW applied for 1 kWh of energy. Bernhardt [118], [119] experimentally demonstrated separation of isotopes using transverse beam deflection. There are other optical techniques for separating isotopes, but this technique might have importance for special cases, such as with short-lived radioactive isotopes.

3.1.2. *Doppler cooling of atoms using the scattering force and fluctuational heating of atoms*

In 1975 Hänsch and Schawlow made the very important suggestion that one could use the strong velocity dependence of the scattering force due to Doppler shift to damp or cool atomic motion [26]*. For example, in one dimension (see Fig. 3.1), with a pair of identical opposing beams tuned below resonance, any atomic motion v along the axis meets a net opposing force due to the strong Doppler shifts of the absorption. For motion to the right as shown, the atom interacts more strongly with the right beam R at frequency v_R and less strongly with left beam L at frequency v_L. With three such

One-Dimensional Optical Damping of Atomic Motion

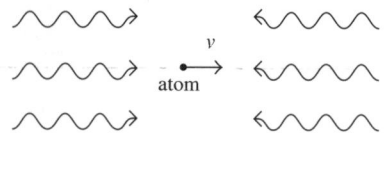

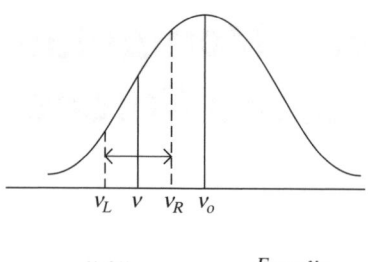

v_L v v_R v_0

$v < v_0$ $F = -\gamma v$

in three dimensions $\longrightarrow$ optical molasses

Fig. 3.1. Hänsch and Schawlow's proposal for Doppler damping of atomic motion. Atoms of velocity v exposed to a pair of opposing beams tuned below resonance feel a net slowing force $F = -\gamma v$ due to Doppler shift.

pairs of orthogonal beams one should be able to damp all three degrees of freedom of the atomic motion. Optimal damping occurs using lasers tuned a half wavelength below resonance, where the change in absorption with frequency is a maximum. This damping is now called optical molasses.

In order to be practical one has to have atoms that are initially slow enough to interact and thermalize within a reasonable cooling volume. This matter was not discussed by Hänsch and Schawlow. For atoms of a given injection velocity there is a "capture length" that brings them to rest within our cooling volume. Since the scattering force near resonance is quite large [2]*, a capture length of ~ 1 cm and a cooling volume of ~ 1 cm^3 seem reasonable. For the case of sodium atoms with a natural lifetime of ~ 16 ns, one can viscously stop most of the atoms entering molasses with a velocity ≤ 104 cm/s. As we shall see later, Phillips and collaborators and Ertmer *et al.* [38], [120], [121], [122] succeeded in slowing atomic beams to zero velocity in one dimension.

The simple viscous damping behavior, of course, requires that the slowing beams are operating at low saturation parameter p. At high p the damping can actually turn into heating. The complete theory for the velocity dependence of an atom moving in a standing-wave field was worked out by Gordon and Ashkin in Ref. [123]*.

I recall meeting Stig Stenholm at about this time at the University of Arizona and he asked me, "Is not the Hänsch and Schawlow paper obvious, based on your work on the scattering force?" I replied, "Yes, it is. They refer to only one paper and that is my 1970 paper, but I wish I had published this idea. It will be important some day". Indeed, Stenholm in later years spent much time working on molasses cooling (see, e.g., his review paper [124]).

Hänsch and Schawlow [26]* estimated that if one tuned a half linewidth below resonance one could cool atoms to an energy $\sim h\gamma_n$, where γ_n is the absorption linewidth. Of course, one does not

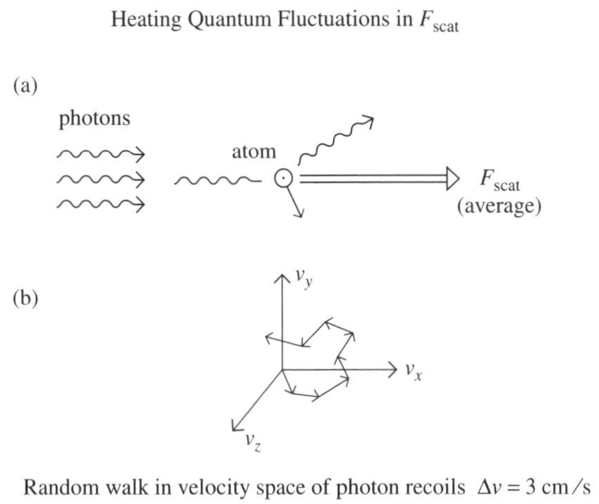

Heating Quantum Fluctuations in F_{scat}

Fig. 3.2. Heating due to quantum fluctuations in F_{scat}. (a) F_{scat} is the average force due to many symmetrical random kicks. (b) Linear heating rate due to a random walk in velocity space.

expect to cool atoms to absolute zero because the cooling process is based on the average behavior of the forces.

As shown in Fig. 3.2(a), the scattering force in reality is an average force in the direction of an incident light beam, which is the sum of all the individual photon recoil kicks. In detail, the force can be viewed as the result of a random walk in velocity space. In sodium each spontaneous photon emitted gives a randomly directed impulse of magnitude 3 cm/s. For the case of molasses (Fig. 3.2(b)) the average value of the Δv impulses is zero. That is, $\langle \Delta v \rangle = 0$. However, the average value of the square of the velocity impulses $\langle \Delta v^2 \rangle$ grows proportionally to the time t. This is just like random walk or diffusion in position space. See Chapter V of Fuchs [52] and Millikan [108]. This implies that $KE_{random} \sim \langle \Delta v^2 \rangle \sim t$ and we have a linear heating rate. Thus, as an atom is being cooled by the scattering forces of molasses, it is also being heated by the same forces. The atom approaches a temperature that is the balance of the cooling and the heating rates, independent of the laser power. In the absence of saturation, lower laser power implies a slower cooling rate, and a slower heating rate. However, the equilibrium temperature after scattering of the requisite number of molasses photons is the same.

The Russian researchers Letokhov, Minogin, and Pavlik [125] were the first ones to calculate this balance based on the scattering force in detail, although they used a somewhat different approach. For a tuning $\gamma_n/2$ below resonance, which gives the optimum cooling rate, they estimated an equilibrium kinetic energy of $\sim h\gamma_n$, much as Hänsch and Schawlow did in their hand-waving estimate. Letokhov and Minogin also then proposed [126] that one could use the same six-beam cooling geometry for stably trapping atoms on the intensity maxima of the standing wave pattern by virtue of the gradient

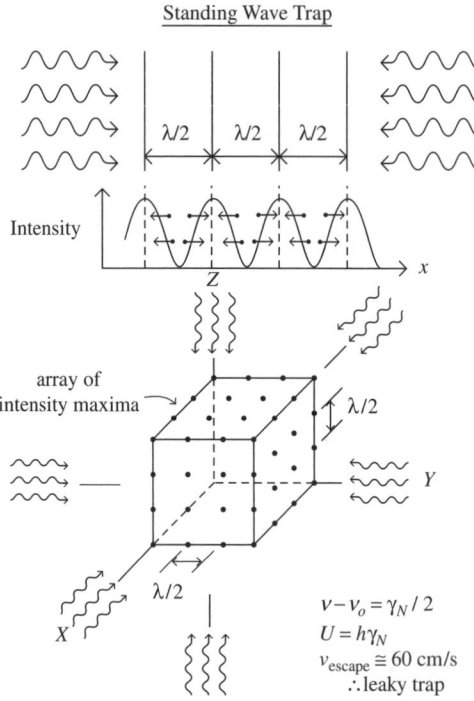

Fig. 3.3. Schematic diagram of the standing wave trap proposed by Letokhov and Minogin.

force (see Fig. 3.3). They estimated a trap depth U that was also $\sim h\gamma_n$, making the trap depth and average kinetic energy of atoms in the trap about the same. This makes a very poor trap, since the atoms boil out of the trap almost instantaneously (see also Sec. 3.1.5).

3.1.3. *Damping of macroscopic particles*

It is interesting to note that the Doppler cooling or "molasses cooling" proposed by Hänsch and Schawlow is highly analogous in its origin to the residual Doppler damping computed by the present author [65]* for a macroscopic particle oscillating in total vacuum about its equilibrium point in a levitation trap, as alluded to earlier in Sec. 2.6. As seen in Ref. [65]*, the radiation force on the oscillating particle increases by a factor $(1 + v/c)$ when the particle motion is opposing the light beam and decreases by the factor $(1 - v/c)$ when it is in the direction of the motion, due to Doppler shift. There are also factors of $(1 \pm v/c)$ due to changes in the light flux hitting the sphere. These two effects give an effective damping force $F_{\text{damp}} = -\gamma v$, which causes an exponential decay of the amplitude of the oscillating particle. The damping is small and the amplitude falls to one half in ~ 0.7 years for macroscopic particles [65]*.

This type of macroscopic single-beam cooling is analogous to six-beam molasses cooling as long as the particle is oscillating in a trap. Ashkin suggested it as a way to cool the single beam gradient (tweezer) trap in 1978 [3]*. One can use only a single beam in place of six beams because the particle motion in the trap couples all three degrees of freedom.

This exponential decay mechanism is purely classical and the kinetic temperature approaches absolute zero with time. What about heating? Is there the equivalent of the quantum or fluctuational heating we saw in atom molasses?

If we treat the sphere as a perfect sphere and the light as a classical wave, the answer is no. However, if we consider the photon or particle nature of the light, we have fluctuations in the number of particles hitting the sphere and being scattered. This gives a fluctuational heating rate which one has to balance against the cooling rate to give a minimum equilibrium temperature (I leave it to the reader as an exercise to calculate this temperature, for the case of a particle in total vacuum and in the presence of no ambient light).

There is yet another classical damping process for small particles, based on the Doppler shift. This is the damping of a dust particle rotating about the sun by the so-called Poynting–Robertson effect. It is well known in astronomy that there are huge clouds of dust and micrometeoroids present around young stars during the process of planet formation. For our sun this dust is largely gone, but remnants remain in the rings around Saturn, Uranus, and Neptune.

According to the Poynting–Robertson effect [84], which was conceived of around 1920, a dust particle orbiting around our sun in a stable orbit must be affected by the gravitational pull of the earth, the radial radiation pressure of sunlight on the particle, and the re-radiation of the thermal energy from the heated particle. The radial radiation pressure tends to cause the particle to spiral outwards, away from the sun, but the re-radiated thermal energy acts as a damping force. For an observer located in the sun's frame of reference the absorbed thermal energy of the particle is re-radiated isotropically outwards in all directions. However, the forward-directed radiation is up-shifted in frequency, whereas the backward-emitted radiation is down-shifted. Since the energy density and momentum of the forward-emitted light is higher than the down-shifted backward light, there is a net backward optical damping force exerted on the particle that tends to cause the particle to spiral inwards. For small particles, the effect of the radially outward radiation pressure dominates and the particle is pushed out into space. For larger particles the damping effect dominates and the net effect is for the particle to spiral inwards and be lost.

This Poynting–Robertson Doppler damping effect figures prominently in planetary science, which is presently having a resurgence by virtue of satellite observations. See, for example, Habing *et al.* [127] on the disappearance of stellar dust debris from main-sequence stars after 400 million years, and "News and Views" from Nature [128] on "Neptune's misbehaving rings", discussing the recent observation of the incomplete faint rings around Neptune by the Voyager satellite. The Habing paper discusses the competition between the Poynting–Robertson effect causing particles to spiral into the sun and radiation pressure pushing particles away from the sun.

These modern observations on planetary dust bring to mind the observation that comets' tails, made up of gaseous material and dust, point away from the sun as they pass through the solar system. Since the days of Kepler, in 1619, this has been attributed to radiation pressure. Nowadays this explanation is considered to be only partially correct. Another force exerted on the comets' tails is from the solar wind and its electrons that emanate from the sun. As we saw, the high intensity and monochromaticity of laser sources have made it possible to observe these light force effects within the confines of the laboratory.

Finally, there is yet another radiation pressure process that is thought to account for the disappearance of micrometer-sized dust grains from interplanetary space, which has nothing to do with the Poynting–Robertson effect. This is the rotational bursting of irregularly shaped dust particles driven by solar radiation pressure as they orbit the sun [129]. There is a strong possibility of generating a net optical torque about the center of mass of an irregularly shaped particle in the micrometer or geometrical optics size regime. In the total vacuum of space such a torqued particle is expected to rotate at increasing angular velocity until it flies apart by centrifugal forces. More will be said about optically produced torques and rotation of optically trapped particles in connection with work in the 1990s on the angular momentum of light [130]*, [131], [132].

3.1.4. *Saturation of the gradient force on atoms*

Returning to atom trapping, Letokhov and Minogin's calculated value of the well depth of $\sim h\gamma_a$ for their trap was very discouraging for the prospects of strong optical atom traps [126]. Therefore, in 1978, I decided to address the problem of saturation of the gradient force on atoms using the same semiclassical rate-equation approach used earlier for understanding saturation of the scattering force [2]*, with the hope of achieving deeper traps [3]*. The new key points were the realization that the classical value of the polarizability α in the gradient force formula

$$F_{\text{grad}} \equiv F_{\text{dip}} = \mathbf{d} \cdot \nabla \mathbf{E} = \frac{1}{2}\alpha \nabla \mathbf{E}^2 \,, \tag{3.1}$$

(where $\mathbf{d}$ is the optically induced dipole moment) applies to an atom in its ground state; and that an atom in its excited state contributes polarizability of the opposite sign in proportion to the fraction of time, f, it spends in the excited state. Thus,

$$\alpha = \alpha_o(1 - f) - \alpha_o f = \alpha_o(1 - 2f) \,, \tag{3.2}$$

where the polarizability in the ground state α_o is given by

$$\alpha_o = -\frac{\lambda_o^3}{16\pi^3} \frac{\gamma_n}{2} \frac{(\nu - \nu_o)}{(\nu - \nu_o)^2 + \gamma_n^2/4} \,. \tag{3.3}$$

This expression for α_o is the standard quantum mechanical formula. See, for example, the Kramers Heisenberg formula in Sec. 1.9, or the closely related formula of laser theory for the susceptibility of an atomic system with populations in ground and excited states (see p. 90 of Ref. [54]). From the rate-equation treatment of the scattering force [2]* we have

$$f = \frac{1}{2} \frac{p(\nu)}{1 + p(\nu)} \,, \tag{3.4}$$

where $p(\nu)$ is the saturation parameter at frequency ν. From Ref. [2]*, $p(\nu)$ is given by

$$2\pi \frac{p(\nu)}{h\nu} = \frac{\lambda_o^2}{4\tau_n} \frac{I(\nu)}{\tau_n} \frac{S(\nu)}{I_{\text{sat}}} \frac{1}{4\tau_n} = I_o S(\nu) \tag{3.5}$$

and

$$\frac{S(\nu)}{4\tau_n} = \frac{\gamma_N^2/4}{(\nu - \nu_o)^2 + \gamma_n^2/4} \,. \tag{3.6}$$

In the above equations $2\pi\tau_n = 1/\gamma_n$ and $S(\nu)$ is the Lorenzian line shape, where γ_n is the full width at half the maximum intensity. I_{sat} is the saturation intensity. For sodium, $I_{\text{sat}} \cong 20\,\text{mW/cm}^2$.

The incident Gaussian beam $I(r) = I_o \exp(-2r^2/w_o^2)$ and has a maximum intensity I_o at the beam focus given by $I_o = 2P_o/\pi w_o^2$, where P_o is the total power of the beam and w_o is the radius of the beam waist (see Ref. [2]*).

This expression for $I(r)$ used here for a Gaussian beam is the one usually used in laser science. The expression for $I(r)$ found on p. 150 of Metcalf and van der Straten's book [135] in connection with the dipole force is not recommended.

The first thing to notice from Eq. (3.4) is that as the incident beam intensity I_o and $p(\nu)$ get very large, f saturates to $1/2$. From Eq. (3.2) this implies that α saturates to zero. This does not mean that $F_{\text{dip}} = \alpha \nabla E^2 /2$ goes to zero at high intensity, since α falls only by virtue of an increase in ∇E^2. F_{dip}, which depends on the product of α and ∇E^2, continues to rise.

Looking at Fig. 3.4, we see a schematic drawing of the polarizability α_o of the ground state and the saturation parameter $p(\nu)$ versus laser tuning ν. The saturation parameter p is proportional to the Einstein B coefficient, which is the stimulated absorption rate. Letokhov and Minogin's choice for laser tuning was a half linewidth below resonance, where the Doppler cooling rate is a maximum. At this frequency $dp/d\nu$ is a maximum. It is also the point where the absorption rate falls to a half. Since the magnitude of the saturation intensity I_{sat} is quite small, one can get very large values of p at fairly modest laser intensities. We see from Eq. (3.5) that this very quickly causes α to saturate to a small fraction of α_o and F_{dip} suffers badly. Thus it is saturation that basically limits the Russian trap to a very low well depth U [126].

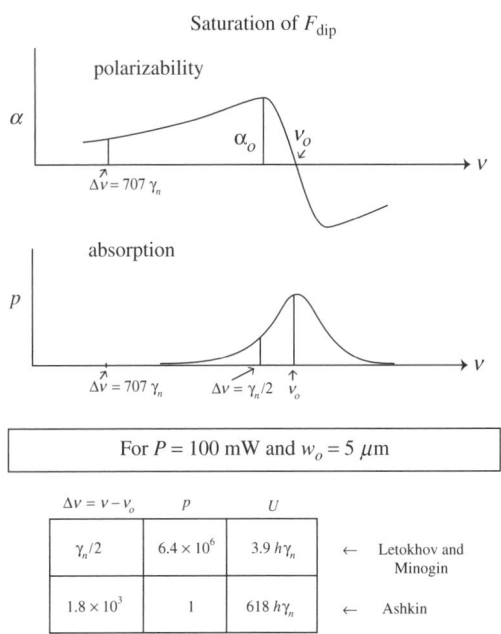

$\Delta \nu = \nu - \nu_o$	p	U	
$\gamma_n/2$	6.4×10^6	$3.9\, h\gamma_n$	$\leftarrow$ Letokhov and Minogin
1.8×10^3	1	$618\, h\gamma_n$	$\leftarrow$ Ashkin

Fig. 3.4. Diagram of the polarizability α of the ground state and the saturation parameter p versus laser tuning ν, showing the tuning choice of Letokhov and Minogin, giving $\Delta \nu = \gamma_n/2$, and ν_A of Ashkin, giving $\Delta \nu = 707\, \gamma_n$. The table shows the values of trapping parameters and potentials U for each case, indicating an improvement of 150 times for the case of detuning from resonance.

Suppose, however, one tunes far off-resonance to a frequency $(v - v_o) \cong 10^3 \gamma_n$, that is well below resonance, as shown in Fig. 3.4. First one sees that the absorption or value of p, which follows a Lorentzian line shape, falls drastically from its value at resonance. Second, the polarizability, which has the usual dispersion shape, falls to a more modest value compared to tuning close to resonance. Overall this implies a much larger value of α and a much larger dipole force.

Having discussed the dipole force qualitatively, we can now derive a general mathematical expression for the potential U of the dipole force on an atom for all frequencies v. From Eqs. (3.4)–(3.6) one can express F_{dip} in terms of $S(v)$ as follows:

$$F_{\mathrm{dip}} = \frac{\alpha_o/2}{1 + p(v)} \cdot \nabla \left[E_{\mathrm{sat}}^2 \frac{(v - v_o)^2 + \gamma_n^2/4}{\gamma_n^2/4} \right] p(v) \equiv -\nabla U, \tag{3.7}$$

where U is a conservative potential. Solving Eq. (3.7) for U, one gets

$$U = \left(\frac{h}{2}\right)(v - v_o) \ln[1 + p(v)] \tag{3.8}$$

(see Refs. [3]*, [123]*).

The potential shown in Eq. (3.8) above is the sought-after function that includes the effect of saturation on the dipole force. Using Eq. (3.8) combined with Eqs. (3.5) and (3.6), one can show explicitly that the potential depth U never decreases with increasing power. We see that for fixed tuning $(v - v_o)$ if we increase the power P_o, then p increases proportionally and U increases logarithmically as $\ln(1 + p)$. If p is very large it is always advantageous, at a fixed power, to increase $(v - v_o)$ and decrease p. Thus the dipole force is different from the scattering force. It does not saturate to a fixed value but always increases with increasing power.

Using the potential formula for U which applies generally to all dipole forces on atoms, we now compare the magnitudes of the dipole forces used by Letokhov and Minogin in their molasses-type trap, which was tuned a half linewidth below resonance with Ashkin's idea of detuning off-resonance to avoid saturation [3]*. Figure 3.4 lists the calculated potential U using modest values of the power, $P_o = 100\,\mathrm{mW}$, and a beam radius at the focus is $5\,\mu m$. For $(v - v_o) = \gamma_n/2$, the saturation parameter p, calculated from Eq. (3.5), is 6.4×10^6. This is enormous over-saturation and gives $U = 3.9\,h\gamma_n$. For the off-resonance case we choose a much more modest value of $p = 1$. From Eq. (3.5) this occurs for $(v - v_o) = 1.8 \times 10^3 \gamma_n$ and gives a potential $U = 618\,h\gamma_n$, which is ~ 160 times larger than the Letokhov and Minogin case.

We thus gain a factor of more than two orders of magnitude in well depth by tuning off resonance. One sees from Eq. (3.4) that some saturation is starting to occur at $p = 1$ and $(v - v_o) = 1.8 \times 10^3 \gamma_n$. The over-saturation corresponding to $p \cong 6 \times 10^6$ at the tuning of $(v - v_o) = \gamma_n/2$ is huge. This makes it clear that one cannot simultaneously cool by molasses and get a reasonable trap depth when operating at $(v - v_o) = \gamma_n/2$. With the far-off-resonance tuning one can readily obtain deep traps of up to $10^3\,h\gamma_n$ and $kT_{\min} \cong h\gamma_n/2 =$ the Doppler limit, using separate trapping and cooling beams as proposed by Ashkin [2]* and Ashkin and Gordon [136]*. However, in practice, one generally does not need values of U/kT_{atom} as high as 10^3. Values of $U/kT_{\mathrm{atom}} = 10$ are often quite adequate. It is desirable, therefore, to tune yet further off-resonance and reduce p values to 10^{-1} or even orders of magnitude lower, if laser power is available. As discussed by Gordon and Ashkin [123]*, this reduces spontaneous heating processes and increases the lifetime of atoms in the trap to many minutes or longer. Sections 17.1, 17.2, 17.5, and Chapter 20 also treat this topic. As an example of low saturation, we list in Fig. 3.4 the case of $P_o = 2W$, $(v - v_o) = 3.2 \times 10^6 \gamma_n$, $U = 10 h\gamma_n$, with $p = 6.3 \times 10^{-6}$. For these conditions, the stimulated absorption time of a single photon is $\tau_n/f = 2\tau_n/p = 5.2\,\text{ms}$, where τ_n is the natural lifetime. This is more than 10^6 times smaller than at $p = 1$. Traps such as these with reduced spontaneous absorption events are important in experiments, where it is necessary to maintain atomic coherence over long periods of time.

3.1.5. *Optimum potential p for a given laser power*

Section 3.1.4 considered the saturation of the gradient force and ways to increase the potential U by tuning far from resonance. It is interesting to ask, what is the value of p that maximizes U for a Gaussian beam with a given power P_o and w_o? This value can be derived using the following (unpublished) calculation.

Starting with $U = (h/2)\gamma_n(v_o - v)\ln(1 + p)$, we simply find the value of U for which $dU/dp = 0$. Since $(v - v_o)$ is also a function of p by virtue of Eq. (3.5) we find, using the short hand $\Delta v = v - v_o$,

$$\frac{dU}{dp} = \left(\frac{h}{2}\right)\gamma_n \Delta v \frac{d\ln(1+p)}{dp} + \left(\frac{h}{2}\right)\gamma_n \ln(1+p)\frac{d}{dp}(\Delta v) = 0. \tag{3.9}$$

From Eq. (3.5) we have:

$$\Delta v = \left[\frac{I_o}{I_{\text{sat}}}\frac{1}{p} + \frac{\gamma_n^2}{4}\right]^{1/2}. \tag{3.10}$$

Therefore

$$\frac{dU}{dp} = \left[\frac{I_o}{I_{\text{sat}}}\cdot\frac{1}{p} + \frac{\gamma_n^2}{4}\right]^{1/2}\frac{1}{(1+p)} + \ln(1+p)\frac{1}{\left[\frac{I_o}{I_{\text{sat}}}\cdot\frac{1}{p} - \frac{\gamma_n^2}{4}\right]^{1/2}}\cdot\left(-\frac{I_o}{I_{\text{sat}}}\cdot\frac{1}{2p^2}\right) \tag{3.11}$$

and

$$1 - p\frac{I_{\text{sat}}}{I_o} = \frac{(1+p)\ln(1+p)}{2p}. \tag{3.12}$$

For a case where $I_o \gg I_{\text{sat}}$ and p is modest, one can neglect the term pI_{sat}/I_o relative to 1 and then solve Eq. (3.12) for the optimum p which makes U a maximum. This is a transcendental equation that can be solved numerically by calculating for various values of p. One sees that $p = 4$ solves the equation, since $5(\ln 5)/8 \cong 1.006$.

One also sees from the general formula for the scattering force F_{scat} that the value $p = 4$ implies a fair amount of saturation. Thus

$$F_{\text{scat}} = \frac{h}{\lambda}\frac{1}{2\tau_n}\left(\frac{p}{1+p}\right) = \frac{F_{\text{sat}}p}{1+p} = F_{\text{sat}}\left(\frac{4}{5}\right) = F_{\text{sat}}(0.8)$$

and we are in the saturated regime where F_{scat} changes only slowly with p, or power. In the unsaturated regime, where p is small compared to one, $1 + p \cong 1$, $F_{\text{scat}} \cong F_{\text{scat}}p$ and the scattering force is proportional to p, or power.

The above considerations about the deepest gradient force trap can be applied to another interesting question having to do with molasses. Letokhov and Minogin [126] proposed using the gradient force potential of molasses beams to trap atoms cooled to the Doppler limit of $h\gamma_n/2$. Cooling to the Doppler limit with molasses was first observed by Chu *et al.* [27]* in three dimensions using six Gaussian beams of 10 mW each, with $w_o = 0.36$ cm. We can now ask, what is the deepest potential possible with these beams? Using $p = 4$, one finds $U_{max} \cong 0.66\, h\gamma_n$ with $(v - v_o) = 0.82\, \gamma_n$. However, $p = 4$ corresponds to some saturation, as we just saw, and $(v - v_o) \cong 0.82\, \gamma_n$ is not the optimum detuning. This in turn implies $kT_{atom} > h\gamma_n/2$. Thus the Boltzmann factor U_{max}/kT_{atom} is less than $(0.66\, h\gamma_n)/(h\gamma_n/2) = 1.3$, and we again conclude there is no significant trapping with molasses beams.

3.1.6. *Conservative and nonconservative properties of the radiation pressure force components*

It was clear from our early work with micrometer-sized lossless macroscopic particles that the gradient force component F_{grad} was conservative, whereas the scattering force component F_{scat} was not. This implies that the total force $\mathbf{F}_{tot} = \mathbf{F}_{grad} + \mathbf{F}_{scat}$ was not conservative. For a force to be conservative, it is necessary that the path integral of the work done in following an arbitrary path in space is zero; i.e., the integral around a closed curve $(\int_C)$ of $\mathbf{F} \cdot \mathbf{ds} = 0$.

For example, consider a sphere placed in a focused beam with a Gaussian cross section as shown in Fig. 3.5. Consider a closed path, following a ray close to the beam axis from O to Q, along the phase front from Q to R, along a ray R to S along the fringes of the beam and then back to O along a phase front. The integral is greater than zero, since the contribution of the section OQ exceeds that of R to S, and the contributions from the two sections along the phase fronts are zero.

To show that the gradient force is conservative, we place a sphere in a beam, as shown in Fig. 3.6, with a number of rays passing through it. Consider a typical ray "a" striking the sphere and an arbitrary path C. To calculate the path integral of the gradient force as we traverse the path C, we take note of the fact that the total force, due to refraction and reflectance, of a single ray hitting the sphere, as shown in Fig. 3.6, can be resolved into two components acting at the center of the sphere

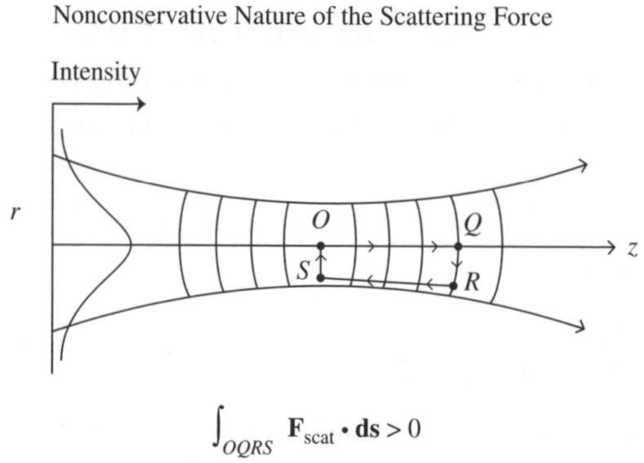

Nonconservative Nature of the Scattering Force

$$\int_{OQRS} \mathbf{F}_{scat} \cdot \mathbf{ds} > 0$$

Fig. 3.5. Nonconservative nature of the scattering force on a macroscopic sphere in a Gaussian TEM$_{00}$ mode beam. For closed paths such as shown, the $\int_{OQRS} \mathbf{F}_{scat} \cdot \mathbf{ds} > 0$.

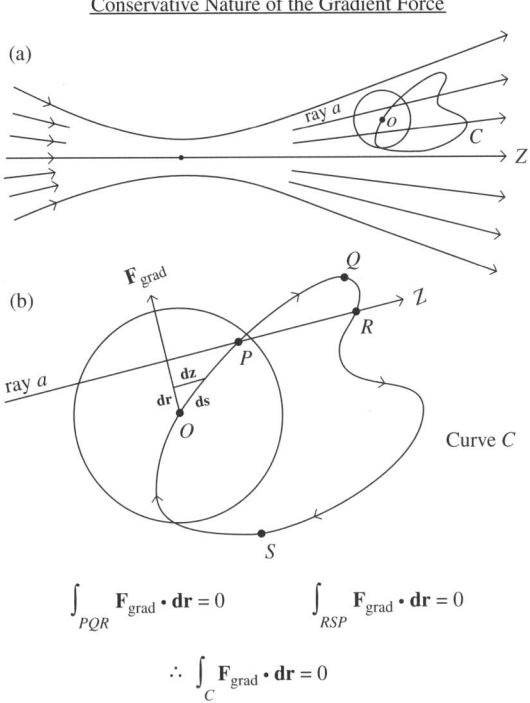

Conservative Nature of the Gradient Force

$$\int_{PQR} \mathbf{F}_{grad} \cdot \mathbf{dr} = 0 \qquad \int_{RSP} \mathbf{F}_{grad} \cdot \mathbf{dr} = 0$$

$$\therefore \int_{C} \mathbf{F}_{grad} \cdot \mathbf{dr} = 0$$

Fig. 3.6. Conservative nature of the gradient force on macroscopic sphere. (a) A sphere located at O is moved around an arbitrary path C. (b) A typical ray "a" has a gradient component F_{grad} perpendicular to ray "a" and a scattering force component F_{scat} in the direction along ray "a". The element of arc $\mathbf{ds}$ is resolved into $\mathbf{ds} = \mathbf{dr} + \mathbf{dz}$. Since F_{grad} is perpendicular to the ray, $\int_c \mathbf{F}_{grad} \cdot \mathbf{ds} = \int_c \mathbf{F}_{grad} \cdot \mathbf{dr} = 0$.

with the gradient force perpendicular to the direction of the incident ray "a" and the scattering force in the same direction as the incident ray [1]*, [49]*.

If one resolves $\mathbf{ds}$ into $\mathbf{dr} + \mathbf{dz}$ as shown, one sees that only the radial component of the integral is operational. That is, $\int_C \mathbf{F}_{grad} \cdot \mathbf{ds} = \int_C \mathbf{F}_{grad} \cdot \mathbf{dr}$, since $\mathbf{F}_{grad}$ is perpendicular to $\mathbf{dz}$. But the $\int_C \mathbf{F}_{grad} \cdot \mathbf{dr}$ is zero, since the contribution along lengths PQR and RSP separately cancel to zero. The same is true for all rays striking the sphere. Therefore, the gradient force is conservative.

The above results for a lossless sphere remain the same if we include some absorption loss. The absorption of some of the incident momentum simply becomes an additional contribution to the scattering force in the direction of the incident light.

Consider now Rayleigh sub-micrometer dielectric spheres, small compared to the wavelength. We can again resolve the total radiation pressure force into scattering and gradient forces in the direction of the incident light and in the direction of the intensity gradient. In this case, however, the light is scattered by Rayleigh scattering, rather than by refraction and reflection, as in micrometer-sized spheres. Since Rayleigh scattering occurs in a symmetric pattern, the scattering force F_{scat}, which is the momentum per second being given to the particle, is

$$F_{scat} = \frac{P_{Rayl}}{c} = \frac{I_o}{c} \frac{128}{3} \pi^5 r^6 \left(\frac{n-1}{n+2} \right)^2 ,$$

where P_{Rayl} is the standard Rayleigh scattered power, given in terms of the incident intensity I_o for a sphere of radius r and an index of refraction n. See p. 37 of Ref. [133]. The gradient force on the submicrometer Rayleigh sphere can be written, as discussed above, as $F_{grad} = 1/2\alpha_o \nabla E^2$, where α_o is the polarizability of the sphere. For a small sphere, one can simply use α_o, the electrostatic polarizability of the sphere, since it is capable of following the light frequency:

$$\alpha_o = n_m^2 \left(\frac{m^2 - 1}{m^2 + 2} \right) r^3 ,$$

where m is the relative index of refraction equal to the index of the particle divided by the index of the medium n_m. See p. 32 of Ref. [133] and p. 205 of Ref. [134].

For Rayleigh particles one can show that the scattering force is nonconservative and the gradient force is conservative, using essentially the same arguments as above.

Finally, for an atom, which one can consider as the limiting case of a submicrometer Rayleigh particle, one expects the same force conservation considerations to apply. For the conservative gradient force $(1/2\alpha_o \nabla E^2)$, one must, of course, use the proper atomic polarizability. As mentioned earlier, in connection with Eq. (3.3), one can calculate this polarizability using a simple classical harmonic oscillator model for an atom that is just a damped optically driven single bound electron. The nonconservative scattering force component, as given in the equation $F_{scat} = (h/\lambda)(1/t_N)f$ discussed earlier (see Sec. 1.7), arises from the symmetric spontaneous emission from the excited state of the optically driven atom. This is in analogy with the scattering force on a Rayleigh particle, which arises from the symmetric Rayleigh scattering pattern of a sub-micrometer dielectric sphere. It should be clear from the above discussion, that the nonconservative nature of the scattering force on atoms has nothing to do with the fact that "the reverse of spontaneous emission is not possible and therefore the action of the force cannot be reversed", as Metcalf and van der Straten say in their book [135]. Neither does the conservative nature of the gradient force have anything to do with the reversibility of stimulated emission processes, as implied in their book. As we saw above in the cases of classical light scattering, it is the resolution of the momentum of the scattered light into different directions that determines the differences between the two force components and their conservative properties.

3.1.7. *Two-beam optical dipole traps for atoms*

One of the main purposes of my 1978 trapping paper [3]* was to devise trapping geometries that would make deep traps using the dipole force with focused far-off-resonance beams. How this can be done is not obvious, since one also has the scattering force to contend with. The paper concentrates mostly on a two-beam atom trap that is the atomic analog of the first two-beam macroscopic particle trap [1]*. In the two-beam trap the scattering force is zero at the equilibrium point. A major problem, however, is how to get enough atoms into the relatively small volumes of such a focused beam trap. Figure 3.7 shows the design of the two-beam trap with a beam width, at the equilibrium point E, of $13\,w_o = 156\,\mu m$, a power of $P = 200\,mW$ in each beam, and a detuning of $\nu - \nu_o = 50\,\gamma_n$. It was calculated that atoms entering into the core of the trap on axis through a small hole in a mirror would be damped by the pair of opposing beams, as in molasses, and collect near the equilibrium

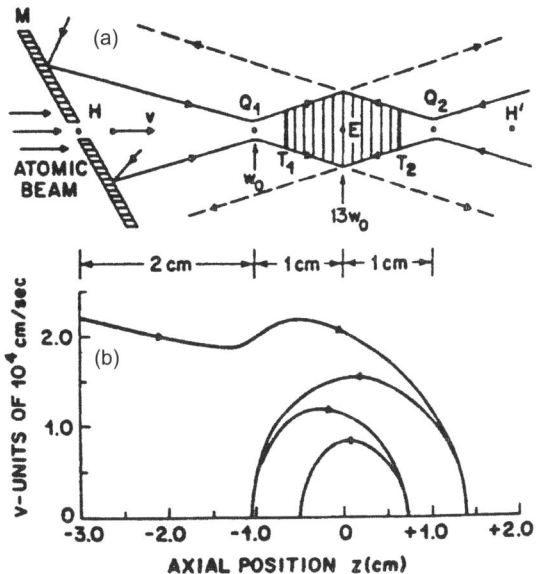

Fig. 3.7. (a) Proposed two-beam optical dipole trap for atoms; $w_o \cong 12\,\mu m$. (b) Calculated trajectory of an atom injected through H with $v = 2 \times 10^4\,\mathrm{cm/s}$.

point E, as shown in Fig. 3.7. The damping of an atom injected at $\sim 2 \times 10^4\,\mathrm{cm/s}$ is shown. This trap geometry should not only slow a reasonable number of atoms axially, due to the scattering force, but should also confine all atoms with transverse velocity less than $500\,\mathrm{cm/s}$, due to the dipole force from the combined power of the two beams. This combined two-beam atom trapping and one-beam slowing arrangement, shown in Fig. 3.7, has never been tried. It was necessary for the paper to have some plausible scheme for introducing atoms into the trap. I believe, to this day, that this simple idea has a good chance of working. Phillips and colleagues demonstrated the viability of the two-beam trapping geometry in 1988, but in conjunction with their Doppler slowing techniques and molasses cooling.

Not much serious attention was paid to the minimum temperature achievable in this two-beam trap. Without any real justification I stated at that time [3]* that one might eventually cool the atoms to the velocity of a single atomic recoil, or $\sim 3\,\mathrm{cm/s}$ for sodium. This was even lower than the Doppler limit of $h\gamma_n/2$ calculated by Letokhov and Minogin. This was just wishful thinking. However, I did propose that, if needed, one could add two additional pairs of cooling beams to cool the transverse components of the trapped atoms. As will be discussed later, use of additional cooling beams is essential for making viable traps [136]*.

3.1.8. *Single-beam optical dipole trap, or tweezer trap, for atoms*

Another important consequence of the ability to achieve strong dipole forces on atoms was the invention of the single-beam dipole trap, or as it was later known, the optical tweezer trap. This is the simplest of all possible traps, consisting of a single focused Gaussian beam. Figure 3.8 illustrates how it works.

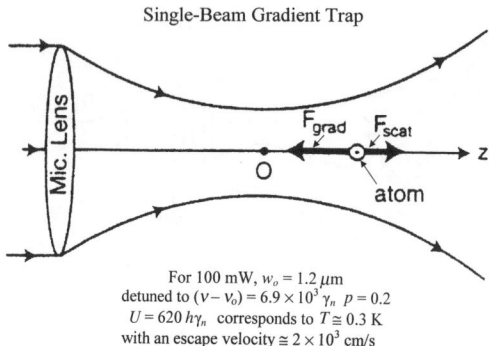

Fig. 3.8. Proposed single-beam gradient trap showing its important parameters.

Suppose one places an atom on the axis of a focused Gaussian beam, downstream from the focus. One might at first suspect that this atom would simply be blown out of the beam by the action of the scattering force. However, due to the focusing, there is also an axial gradient force at work pulling the atom back toward the high intensity beam focus, which can be made quite strong by detuning below resonance. If this backward gradient force exceeds the scattering force, the atom should be pulled back to the focal region and come to rest at an equilibrium point E, where the backward gradient force and the forward scattering force balance. This would clearly be a point of stable equilibrium. One has as a criterion for the existence of such a gradient force trap the statement that the backward gradient force must exceed the scattering force at the point of maximum intensity gradient on the Gaussian beam axis. One can show that the point of maximum intensity gradient lies at a distance of $Z = \pm \pi w_o^2 / \sqrt{3}\lambda$. As shown in Ref. [3]*, the above criterion of backward gradient force greater than the maximum scattering force is readily met and atom gradient force traps exist. The main problem with such single-beam gradient traps is their small volume and the difficulty of injecting atoms directly into them using slowed atomic beams or velocity selectors as sources. Nevertheless, as we shall see, the single-beam gradient traps became the first all-optical atom trap ever to work [4]*.

It is interesting, as a historical note, that this paper [3]* had considerable difficulty in getting published. I submitted it to Physical Review Letters, where it was rejected. One reviewer liked it and approved, but the second one had much to find fault with. He would not relent, despite my attempt to satisfy him. After a final rejection I appealed to the editor who said he had "some old guy" who volunteered to act as a reviewer of last resort in such disputed cases. This old fellow decided that I had indeed responded satisfactorily to all the second reviewer's points except for one that said atoms could not stand up to such high laser intensities as I proposed, without ionizing. I finally convinced him that this would not be a problem by referring to experiments by Steve Harris's group at Stanford and other nonlinear optic researchers, who studied photoionization. He insisted that I add a reference, number 11 of Ref. [3]*, in which photoionization of atoms from the trap was discussed, before the paper would be accepted. Years later, in the early 1980s, Chu asked me why I had not placed more emphasis on the single-beam gradient trap in this paper. "Your treatment of the single-beam gradient was essentially a 'one-liner'". I told him that if he only knew how much trouble I had in getting the paper accepted featuring the much more conventional and much larger volume two-beam trap, he would not have asked that question.

The day after the publication of the 1978 trapping proposal [3]* I received a letter from E. M. Purcell expressing considerable interest in the work, but at the same time wondering about the atom cooling problem (a copy of this letter and my response are reproduced in Figs. 3.9 and 3.10). Purcell sketched out a very simple way of calculating the fluctuational heating, optical damping and minimum temperature for a free atom in a molasses-type optical field. He asked politely whether he was missing anything, since he calculated minimum temperatures that were higher than mine. In my response I thanked him for his clear analysis of the minimum temperature and told him that in my trapping paper cooling was not the primary consideration, but that Gordon and I were writing another paper in which the frequency dependence of the cooling was being carefully considered.

3.1.9. *Separate trapping and cooling beams and the Stark shift problem*

This other paper by Gordon and me [136]* on cooling and trapping by resonance radiation pressure appeared in 1979 and broke some new ground. The traps considered were those of Ref. [3]* and involved focused beams and detuning far from resonance to achieve potential well depths of 10^2–10^3 $h\gamma_n$ and a separate cooling beam. In this paper, we used the simple analysis of the equilibrium temperature suggested by Purcell, including the effects of mutual saturation of the trapping and cooling beams.

In the formula for the equilibrium temperature of Doppler-cooled atoms, as calculated in Ref. [136]*, the dependence on the frequency of the cooling light appears in the factor $(\Delta + 1/\Delta)$, where $\Delta = (\nu_o - \nu)$, the detuning from resonance. Optimal cooling, i.e., the minimum of $(\Delta + 1/\Delta)$, occurs for $\Delta = (1/2)\gamma_n$, a half linewidth below resonance. If one combines the tuning behavior of the Doppler cooling with the fact that the trap potential U increases by the same factor of $\Delta = (\nu_o - \nu)$ when detuned from resonance, one sees that for the same detuning Δ for trapping and cooling, the Boltzmann factor $U/kT_{\min}$ remains about unity, as found previously for the Letokhov–Minogin six-beam trap [126]. The far-off-resonance detuned trap, though deeper by a factor of $\sim 10^2$–10^3 is by itself still leaky because its cooling degrades by the same factor!

The solution to this dilemma suggested in Refs. [3]*, [136]* was to use additional molasses cooling beams, tuned optimally to $\gamma_n/2$ below resonance, in addition to deep detuned trapping beams, to get minimum temperature of $\sim h\gamma_n$, giving values of $U/kT_{\min} \cong 10^2$–10^3 (see also Sec. 3.1.4).

I mention in passing that Wineland and Itano [137] also made calculations on the cooling limit for free atoms, following the approach of Purcell (see Figs. 3.9 and 3.10) and applied the results to atoms bound in ion traps.

However, even this suggestion of separate cooling and trapping had its own problems. Gordon and I pointed out in Ref. [136]* that the strong optical fields of the deep trapping beam generated a large optical Stark shift [138], which shifted the resonance frequency ν_o of an atom at the trap focus by as much as hundreds of linewidths γ_n. In fact the magnitude of the Stark shift of the ground level is $\Delta U = (h/2)\,(\nu' - \nu)\ln(1 + p)$, where ν' is the Stark shifted resonance frequency [138] and p is the saturation parameter. This implies that the molasses beam, to be effective, should be optimally tuned a half linewidth with respect to ν' and not ν_o. This is easily done, but a problem that is harder to

⊃ UNIVERSITY

DEPARTMENT ⊃ ⸺ ⸺⸺

LYMAN LABORATORY OF PHYSICS
CAMBRIDGE, MASSACHUSETTS 02138
March 21, 1978

Dr. A. Ashkin
Bell Telephone Laboratories
Holmdel, New Jersey 07733

Dear Dr. Ashkin:

Your extraordinarily interesting letter in PRL opens up fascinating possibilities. Although I have by no means fully digested it, I want to raise one point that bothers me. It concerns the limiting "temperature" of the axial motion, which you say corresponds to a single photon's momentum. Surely the axial kinetic energy must correspond, not to $p = h\nu/c$, but to $\overline{p^2} = n(h\nu/c)^2$, where n is approximately the number of photons scattered in one axial damping time. Scattering of left- and right-moving photons being statistically independent, the atom's momentum p_x must grow by a random walk, of step-length $h\nu/c$, until limited by the velocity-proportional drag. This is just the old fluctuation-dissipation theorem at work, as in Brownian motion!

When I calculate the limiting temperature that way (see note enclosed) I get a different but still very interesting result. T_c no longer depends on the mass of the atom - as it would according to your statement, if I interpret it correctly - but <u>only</u> on the resonance width γ. Putting in a γ appropriate to a radiation-damped electron resonance of unit oscillator strength I get $T_c = 2.5 \times 10^{-12}/\lambda^2$. This predicts millidegrees, rather than microdegrees, in the case of an optical resonance. Have I missed a fundamental point?

A millidegree is still a very attractive prospect. But to come close to it, won't you have to suppress stringently any non-statistical noise in $I_L - I_R$, at frequencies in the vicinity of $1/\tau_D$? Maybe that is made feasible by the circumstance that I_R is I_L reflected less than τ_D earlier - if that is indeed the case.

Sincerely yours,

E. M. Purcell

EMP/osr

Fig. 3.9. Purcell's two-page letter on limiting temperature achieved by optical cooling of atoms.

correct remains, and that is the variation of the Stark shift ν' with position within the trap itself, due to the spatial intensity variation of the trapping beam. This means that an atom oscillating around within the trap has its damping degraded whenever it moves away from the exact focus. Reference [136]* has a suggested fix for this problem with yet another beam, a Stark-correcting beam, but

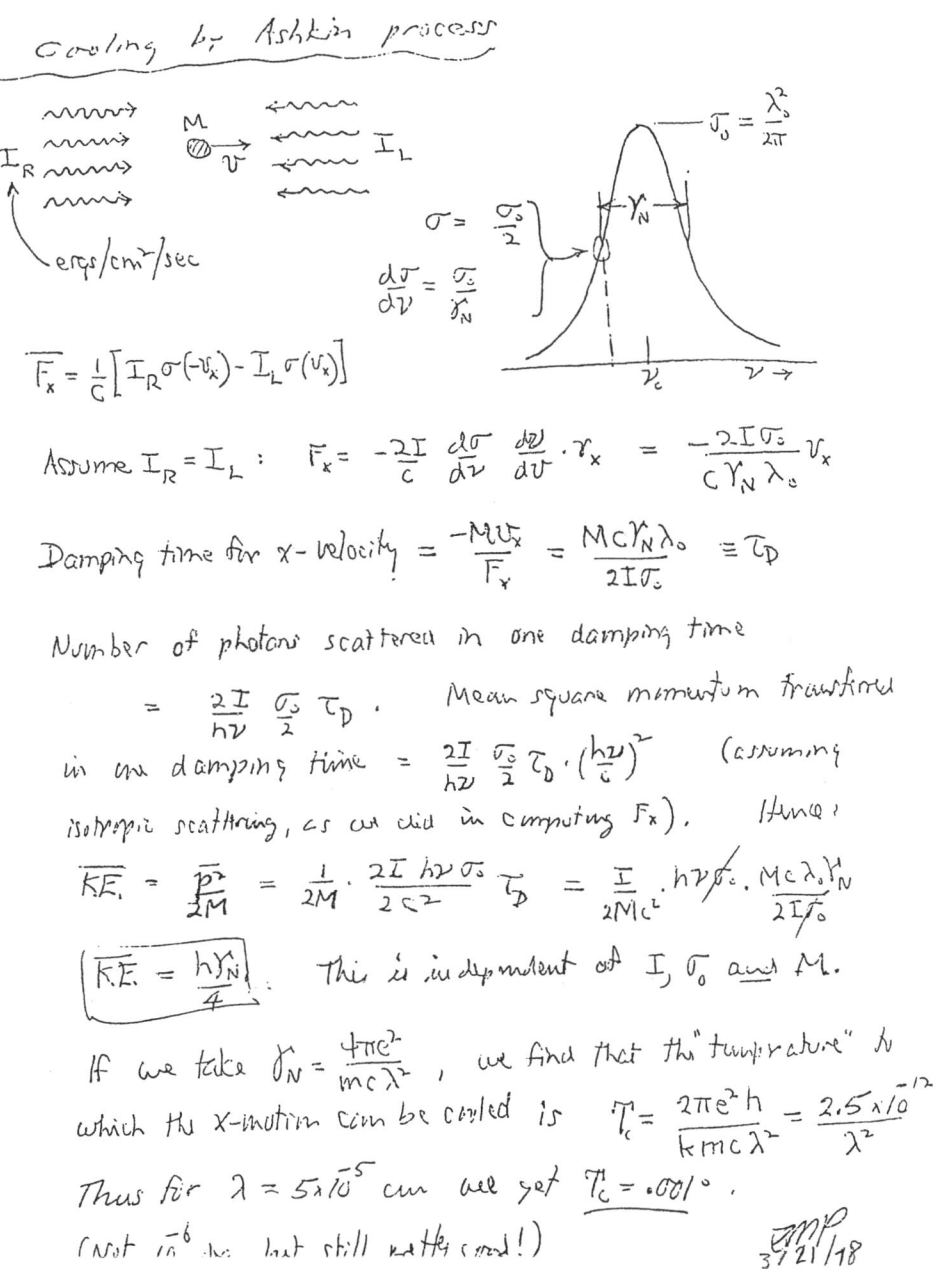

Fig. 3.9. (Continued).

things are becoming more complex. The Stark problem was the remaining outstanding problem in 1980 for achieving stable traps. As we shall see, there is a simple fix that was used in the first atom trapping experiment [4]*. The fact remains, however, that a single trap by itself cannot both stably trap and cool atoms at the same time. I will discuss this Stark shift further in connection with the

Bell Laboratories

Holmdel, New Jersey 07733
Phone (201) 949-3000

March 30, 1978

Professor E. M. Purcell
Department of Physics
Harvard University
Cambridge, Massachusetts   02138

Dear Prof. Purcell:

Thank you for your letter of March 21.   I don't
think you have missed any fundamental point. Fluctuations
and damping will set a lower limit on the temperature of
trapped atoms. In my PRL this was not specifically considered. I simply decided to leave it for another time.
J. P. Gordon and I, however, are now writing a follow-up
letter which treats these matters. We consider the fluctuations and the velocity dependence of the saturated
trapping forces for slowly moving atoms in the various
trap geometries, to get the damping. I do believe though
that the essence of the problem is correctly contained in
your very simple calculation. If I use your value of
$\overline{K.E.} = h\gamma_N/4$ and $kT = \overline{K.E.}$ with the actual value of γ_N for
sodium = 10.6 MHz, I get a value for $T_c \sim 10^{-4}\,°K$. This
corresponds to a velocity of $\sim$ 10 photon momenta.

Concerning your last point about nonstatistical
noise in $I_L - I_R$, it is our intention to derive both beams
from a single dye laser source and to keep the optical
paths substantially the same. This should suppress standing
wave fringe motions due to frequency wander and shift in
the equilibrium point E due to amplitude fluctuations.
The depth of the trap will fluctuate with the amplitude.
If need be, we can reduce amplitude fluctuations greatly
by feedback stabilization with an electro-optic modulator.

Sincerely yours,

A. Ashkin

A. Ashkin

Fig. 3.10. Ashkin's response to the Purcell letter.

first atom trapping experiments in 1987 and the experiment by Ido *et al.* in 2000 [139]*. See also Secs. 5.1.9, 17.8, and 17.9.

3.1.10. *First demonstration of the dipole force on atoms using detuned light*

The new detuning proposal for avoiding saturation and enhancing dipole forces was clearly a key concept impacting the future of atom trapping and manipulation, and it had to be checked experimentally. Bjorkholm, Freeman, Ashkin, and Pearson [43]*, [140]*, [141] devised atomic beam experiments to directly verify these effects. The experiments involved the observation of focusing, defocusing and steering of an atomic beam by the transverse dipole forces exerted by the radial intensity gradient of a superimposed co-propagating resonant cw light beam. Figure 3.11 shows the technique used to inject the atomic beam into the core of the laser beam and to detect the effects of the light on the atoms downstream.

In Ref. [43]* the detector was located in the far field of the beam (see Fig. 1 in Ref. [43]*). In Refs. [140]*, [141] the detector was at the focus of the laser beam. For tuning below resonance, we expect that atoms that would normally follow the dotted trajectory and leave the light beam should be pulled into the beam radially and execute an oscillatory motion, as shown in Fig. 3.12. Figure 3.13 shows particularly beautiful results from the data of Ref. [141], with the detector at the beam

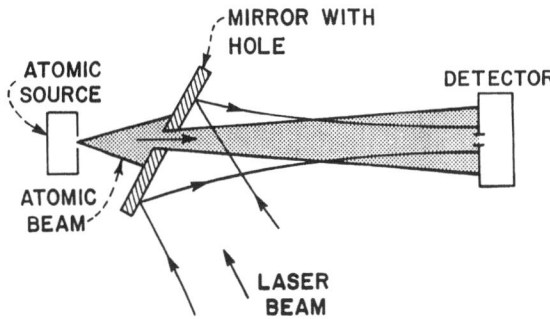

Fig. 3.11. Apparatus for observing focusing and defocusing of an atomic beam by the dipole force of an off-resonance laser beam with low saturation.

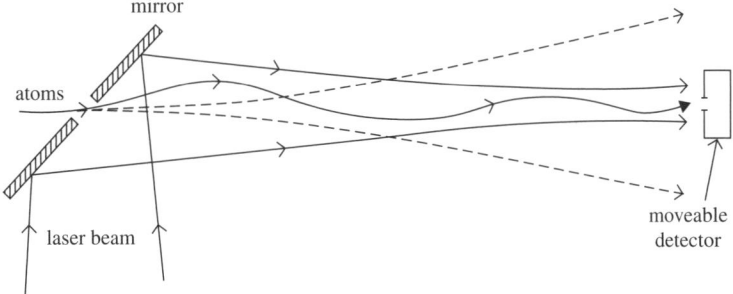

Fig. 3.12. For tuning below resonance, atoms that would spread out and leave the beam are expected to be confined by the transverse gradient force and focused to a small spot within the light beam, as shown.

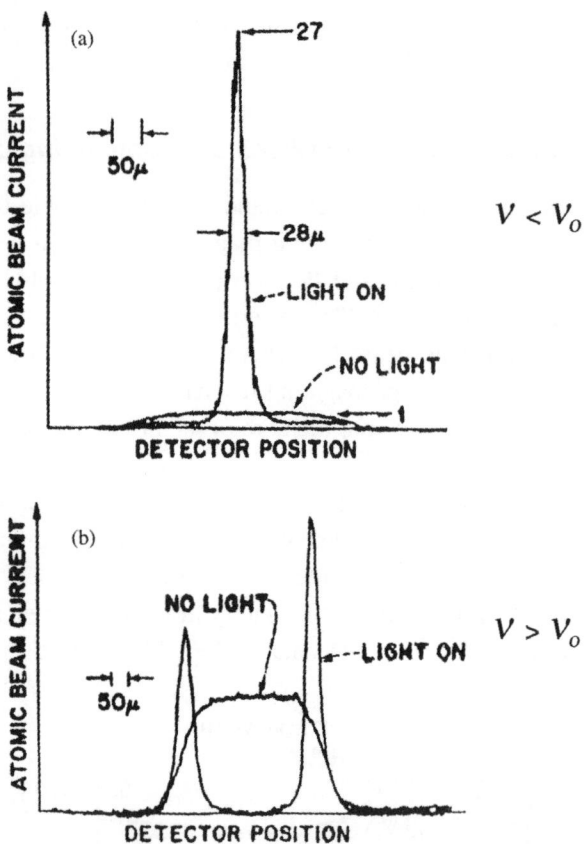

Fig. 3.13. (a) Focusing of the atomic beam at the detector by light tuned below resonance v_o. (b) Defocusing or expulsion of the atomic beam from the core of the light beam, for tuning $v > v_o$.

focus. For $v < v_o$, the atoms are strongly focused as the atomic beam current on axis rises 27 times. For $v > v_o$, atoms are expelled from the light beam, the on-axis current drops to zero, and two side lobes appear which arise from the ejected atoms. These results were the first ever demonstration of the dipole force on free atoms.

The main point of Ref. [141] was to demonstrate the limiting effect of quantum fluctuations on the focusing of neutral atoms by the dipole forces of resonance radiation pressure. The minimum spot diameter of the focused atomic beam at the optimum laser detuning from resonance, as shown in Fig. 3.13, was $\sim 28\,\mu$m at 1/2 height. The atomic beam intensity enhancement was 27 times, relative to the no-light intensity, as already noted. Based just on the geometry, we expect a focal spot diameter of $\sim 15\,\mu$m, which is about two times narrower. One can account for the observed $28\,\mu$m beam width using a simple model based on the fluctuation in the approximate number of photons scattered by the focused atoms. Figures 1 and 2 of Ref. [141], measuring beam diameter and optimum detuning from resonance, make a very convincing picture for the simple fluctuational model used. This measurement (based on the atomic beam focal spot size) was the first and possibly the only direct measurement of the quantum fluctuations of the light forces on atoms.

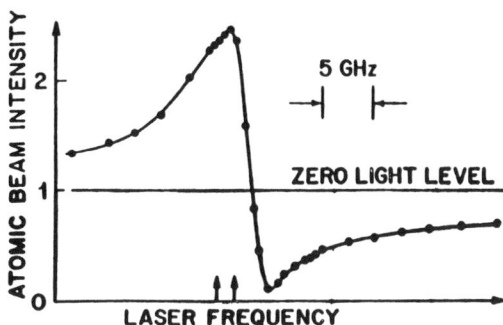

Fig. 3.14. Measurements of the on-axis atomic beam intensity as a function of frequency. This graphically displays the dispersive nature of the dipole force.

Figure 3.14, taken from Ref. [43]*, shows the on-axis atomic beam intensity at the detector as a function of laser frequency. These beautiful high-resolution data show the dispersion-like behavior of the dipole force very graphically. The two arrows mark the frequencies of the two resonance transitions for the sodium D line, the $3^2 S_{1/2}$ ($F = 2, 1$) $\rightarrow 3^2 P_{3/2}$ lines separated by $\sim 1.77 \, \mathrm{GH_Z}$. The shift of the curve crossing relative to these frequencies is consistent with the longitudinal Doppler shifts of the atoms.

3.1.11. *Origin of atom optics*

One sees the close analogy of light forces on atoms with light forces on macroscopic dielectric particles. The above observations on focusing, defocusing, and atomic beam steering can also be regarded as the origin of the field of "atom optics" in which light forces perform functions on atoms which are analogous to the manipulation of light by matter. In the guiding experiments of Ref. [43]*, the light beam was behaving very much as a standard GRIN lens. Optical analogies to thin positive and negative atom lenses are obvious. Applications of this new form of optics to atom epitaxy were suggested, as was the use of the TEM$_{01}$ or "donut" mode laser beams tuned to $\nu > \nu_o$ to confine atoms to the axis of the light beam where the intensity was a minimum. Axial cooling of confined atomic beams was suggested, as well as the use of additional transverse cooling beams, as a means of obtaining beams of ultra-slow atoms. It was clear that a world of opportunities was opened by these experiments. Perhaps, the most important application was to all-optical traps. It was evident from Refs. [3]*, [43]* that optical trapping in principle was achievable. In practice it required slow enough atoms to inject into traps and proper trapping geometry, such as the two-beam or single-beam gradient trap [3]*, or the Alternating Beam trap [142]*, [143]*. The atom optics field is alive and well and more will be said about it later.

The success of the two-dimensional guiding experiment using detuned gradient forces on atoms [43]* also implies that Letokhov's 1968 one-dimensional confinement proposal [56] could be made to work, if properly designed. This is obvious now, in the light of all that has happened. Furthermore, it does not imply that Letokhov had any inkling of the existence of stable three-dimensional traps or how to achieve them. It was only after the later demonstration of three-dimensional trapping of dielectric spheres by Ashkin and the proposal of atom trapping [1]* that Letokhov and Minogin started to work along such lines. Although they did contribute some useful ideas on atomic beam slowing [144] and the Doppler-limit of radiation pressure cooling of atoms [125], they never did come up with a stable trap using the dipole force, where the trap depth exceeded the temperature of the atoms. See discussion in Refs. [45]*, [123]*, [136]* for instance.

3.2. Theoretical Aspects of Optical Forces on Atoms

3.2.1. *Quantum theory of "The motion of atoms in a radiation trap"*

A major contribution was made to the study of laser trapping of atoms in 1980 by the paper entitled "Motion of atoms in a radiation trap" by Gordon and Ashkin [123]*. This theoretical paper has become one of the most referred to papers of the field. It uses quantum mechanics to calculate the optical forces, their first-order velocity dependence and their fluctuations *ab initio*, and applies the results to the problem of stable atom trapping. The quantum results showed that, starting from first principles including effects of saturation, Ashkin's earlier simple-minded, semi-classical rate equation approach to the derivation of the scattering force and gradient or dipole force components is correct [2]*, [3]*.

A new result, derived by Gordon and included in Ref. [123]*, is the complete calculation of the contributions of dipole force fluctuations to diffusional heating. This contribution to the heating is conceptually more difficult to understand than the scattering force fluctuations. It involves processes such as the interaction of the semi-classical dipole with zero-point field fluctuations. This gives rise to somewhat unexpected results, such as the fact that the diffusion constant, even at low values of the saturation parameter p, is independent of an atom's position in a standing wave field. There is heating even at standing wave minima where the field is zero. This is an example of an effect that cannot be obtained without the benefit of the quantum-electrodynamic analysis. This dipole force fluctuation effect was overlooked in the scattering force fluctuational analysis of Letokhov and Minogin, and of Purcell.

The use of a single trapping beam for both trapping and cooling atoms was addressed again more carefully. It was confirmed that such a single-beam atom trap is unstable for a simple two-level atom, unless of course one adds separate cooling beams whose sole purpose is to counteract fluctuational heating of the atoms. An interesting question addressed in this paper [123]* was how long an atom, initially at absolute zero, would last in an optimally tuned off-resonance trap in the absence of any cooling beams. Surprisingly, it was found that for a single-beam gradient or tweezer trap detuned by a typical $100 \gamma_n$ or greater, the lifetime due to fluctuational heating was many seconds. This time is comparable to the duration of atoms in many experiments using traps.

If more power is available with a dipole trap, one can either deepen the trap or keep the trap depth fixed and tune further off-resonance. Further detuning reduces absorption, which reduces the scattering force, and increases the lifetime of an uncooled atom in the trap. It is even possible, as will be seen later, to detune to the point where a Nd:YAG laser at $1.064 \, \mu m$ can trap a sodium atom resonant at 596 nm [145]* or trap a cesium atom resonant at 852 nm with $10.6 \, \mu m$ light [146]. In such detuned, traps the number of spontaneous emission events is negligible. In these cases other types of scattering, such as Raman scattering, can play a larger role. For instance, in a 20 W CO_2 laser beam at $10.6\,\mu$m, with a focal spot size of $\sim 100\,\mu$m, the calculated well depth is $115\,\mu$K and the time to scatter a single photon by Raman effect is 4400 s [146]. More will be said about such traps later.

A nice feature of a strongly detuned dipole trap is that the demands on the spectral purity of the trapping beam are much reduced and frequency-stabilizing techniques are no longer necessary. Not only is the experimental complexity reduced, but also one can use the full raw unregulated power of the laser for trapping, provided power fluctuations are not too severe [147].

These detuning capabilities just described are precisely those of ordinary tweezer traps for atoms. Perhaps less mindful of our earlier results, later workers in experiments taking advantage of the deep potential and low scattering force of off-resonance properties of tweezers saw fit to give it a new special name, the FORT, for far-off-resonance-trap [148]. This additional terminology, in my opinion, obscures history and just makes for some confusion. I propose that we should forget the additional jargon and simply call a "FORT" what it really is: an off-resonance dipole trap [3]*.

3.2.2. *Optical Stark shifts and dipole force traps for atoms*

Another unfortunate development, which obscures history and the logical development of the subject of atom trapping, is the use of the optical Stark shift as the primary explanation of the operation of tweezer traps and the optical potential of the dipole force. Gordon and I, in our 1979 and 1980 papers [123]*, [136]*, were the first to point out that the Stark shift of the ground level $\Delta U = (h/2)(\nu - \nu_o)\ln(1 + p)$, derived from semi-classical considerations and from Quantum Electrodynamics (QED), is exactly equal to the potential of atoms in a tweezer trap. This is fairly obvious after the fact has been pointed out. If you move an atom into an optical field and the energy drops (due to the Stark shift), this drop corresponds to the potential change, provided one can neglect the scattering forces. Actually, in the optical Stark shift for light tuned to a frequency $\nu < \nu_o$, the ground state energy is shifted down and the first excited state energy is shifted up an equal amount. Since an unsaturated atom spends most of its time in the ground state, the atom's average energy is lowered in the field of a trap. If the saturation parameter p is small, the level shift of a two-level atom becomes $\Delta U = (h/2)(\nu - \nu_o)p$, which agrees with the usual Stark shift formula. One should remember that associating the level shift with the trapping potential is a fairly sophisticated and incomplete description of the origin of the trap potential. Often the trap geometry plays a role, as we saw [3]*, depending on the value of p. For the single-beam gradient trap, if p is large enough to give a significant scattering force, one must use a sufficiently strongly focused beam to overcome the forward scattering force. I believe that anyone studying the optical Stark shift would find it hard to come up with the idea of a tweezer-type dipole trap, based on such an understanding alone, due to the role of the scattering force.

Cohen-Tannoudji was the first to observe low-intensity Stark shifts using lamp sources in his thesis in 1961 [149]. See also Ref. [150]. At the time they called the low-level Stark shifts "lamp shifts", since they were seen as a small disturbance. Clearly, optical traps were not on their minds at the time.

The optical Stark shift using lasers was fully confirmed experimentally in the early 1970s by the pioneering experimental work of Liao and Bjorkholm on Doppler-free two-photon spectroscopy in sodium [138], [151]. However, in spite of this work, no one in the Stark shift community at that time was motivated to think of traps.

Nowadays the Stark shift explanation for atom traps and the dipole force potential is often the only one mentioned by atomic physicists. While not wrong if p is small, as we just saw, chronologically it is backward. For example, Metcalf, in his recent book, starts his discussion of dipole traps with the optical Stark shift. Likewise, Cohen-Tannoudji, in his Nobel Lecture [37], gives a similar Stark shift explanation of optical gradient atom traps. This approach to the dipole or gradient forces on atoms overlooks the much more straightforward discovery process that led to deep dipole traps by tuning away from resonance to minimize saturation.

3.2.3. *Optical dipole forces*

In a slightly different context, my former colleague Steve Chu does often invoke dipoles to explain gradient forces on atoms, pretty much as I described earlier, in connection with the gradient force on Rayleigh particles and atoms. He then says, however, that dipoles are also at the root of trapping for macroscopic dielectric (micrometer-sized) spheres. This is an inadequate explanation of macroscopic particle trapping. It is not only far from obvious; it may also be misleading and incorrect without further qualification. Dick Slusher, a laser researcher and department head at Bell Labs., Murray Hill, once asked me how one understands trapping of micrometer-sized dielectric spheres. He had heard or read Chu's hand-waving dipole explanation of the forces and traps for micrometer-sized particles and decided he could not follow the reasoning. "What does it have to do with dipoles?" he asked. I gave him my simple ray picture of macroscopic particle trapping based on ray optics and conservation of light momentum as shown in Figs. 1.5–1.7, and he understood immediately. Then he asked, "Why does not Steve Chu use such simple pictures?" I told him how I had complained to Steve about this very point and got his answer, which was, "You give your talk your way, I give my talk my way".

If one wants to invoke dipoles to explain trapping of micrometer-sized particles, the argument is fairly complex. You first have to say dipoles give rise to the index of refraction and then use Fresnel's laws of refraction, which in turn govern the direction of light ray refracting through dielectrics. Another complexity is that the trapping force for macroscopic particles is shape dependent. A solid sphere acts as a positive lens and is pulled into the high intensity region of the beam, whereas a hollow sphere with the same number of dipoles acts as a negative lens and moves away from high intensity light. So I think the simple ray optic or geometric optic picture is preferable, and indeed this, as we saw, is how the optical force and optical trapping were discovered in the first place.

3.2.4. *Conservation of momentum in light scattering by atoms and sub-micrometer particles*

We saw how use of the principle of conservation of momentum led to an understanding of the radiation pressure forces on micrometer-sized dielectric particles. It was shown that the total light pressure force could be conveniently broken into two components, a so-called scattering force component in the direction of the incident light and a gradient force component in the direction of the gradient of the light intensity. It was discovered that stable optical traps could be made using just these optical forces. It was postulated that similar stable traps were possible using light pressure forces on atoms [1]*, [2]*. The properties of the scattering force component on atoms were directly deduced from simple conservation of momentum principles [2]*. The properties of the gradient force, however, were deduced, as discussed above, from the basic dipole force formula using the optical polarizability and the intensity gradient of the light [3]*. This semi-classical approach to the gradient or dipole force on atoms was quite successful and, subsequently, fully confirmed by the quantum mechanical calculation of Gordon and Ashkin [123]* for atoms. The quantum theory, of course, has the conservation of momentum and energy built into it from the start. Nevertheless, it still leaves open the question of how one can demonstrate the dipole force formula directly from light scattering considerations and the conservation of momentum, in direct analogy with the force on micrometer-sized particles. Such a demonstration should also apply to submicrometer Rayleigh particles, for which we use the same gradient force formula, only using Rayleigh polarizability, instead of the atomic polarizability.

I always considered that the direct light-scattering approach gave the most insight into the origin of the light forces. Indeed, I had deduced the force on micrometer particles by "following the momentum". Could one somehow treat atoms or Rayleigh particles as little refracting lenses that directed the scattered light? I discussed this problem with Jim Gordon. He pointed out that an atom or Rayleigh particles could only scatter an incident light beam in a symmetrical spherical wave pattern. He thought the solution to the problem of how to calculate the redistribution of the light involved the interference of the spherical scattered wave with the incident Gaussian beam in the very-far field. Depending on where the scattering particle was located axially or transversely within the incident beam, there would be a different field distribution and phase shift between the scattered wave and the incident wave in the very-far field. Thus the total intensity in the very-far field $E_{tot} = (E_{scat} + E_{inc})^2$. The E_{scat}^2 term corresponds to the standard Rayleigh scattering term in the absence of E_{inc}. This gives the scattering force. The E_{inc}^2 term gives the distribution of the incident beam in the absence of E_{scat}. The interference term $2E_{scat}E_{inc}$ gives the distribution of redirected light due to the presence of both fields in the very-far field. This is the gradient force contribution.

Gordon subsequently told me that this approach did give the correct forces. It was a bit tricky, since the Gaussian beam description of the incident light is only correct in the limit of small beam angles and transverse fields. The beauty of this approach, however, is that the result is general for any Rayleigh-type particle, either classical, such as a submicrometer dielectric particle, or quantum mechanical such as an atom. It confirms the simple dipole force picture and shows that one does not have to delve into the complexities of the optical Stark shift to understand the basics of radiation pressure forces on atoms.

I thought this result gave some valuable insights into light forces, and I urged Gordon to write it up for publication, but he never did. Other people, in the early literature on masers and lasers, have considered the similar problem of directionality of the emitted radiation from an atom, using classical Maxwell scattering theory. See the fascinating review paper by Lamb, Schleich, Scully, and Townes on Laser Physics [152]. Figure 3.15, taken from this reference, is a reproduction of their discussion on "Stimulated emission: Einstein and Dirac versus Maxwell". Two remarks are especially

Stimulated emission: Einstein and Dirac versus Maxwell

In his derivation of the Planck radiation formula in 1917 Albert Einstein introduced the A coefficient for the rate of spontaneous emission by atoms and the B coefficient for their absorption of radiation. He also introduced the new process of stimulated emission of radiation and found that the B coefficient determined its rate. Ten years later the quantum electrodynamics (QED) of P. A. M. Dirac provided the deeper foundation.

However, Einstein's result is perfectly natural when we disregard, for a moment, Maxwell's electromagnetic theory and, instead, believe in the 1905 concept of photons and in the Bohr orbits. Then it is natural to have spontaneous emission and absorption of the light particles, and the new feature is, indeed, stimulated emission. However, we emphasize that Maxwell's theory also predicts these phenomena.

To bring this out most clearly, we consider a charged particle oscillating back and forth in an electromagnetic wave. We recall that a particle of charge q moving with velocity v in an electric field E, gains or loses energy depending on the algebraic sign of the product qEv. An increase of the energy of the charge implies a loss of energy in the field. This is equivalent to the process of absorption of radiation. Likewise, if the charge is losing energy, the electromagnetic field must be gaining energy. This is equivalent to stimulated emission of radiation. The relative direction of the velocity and the electric-field vectors determines the direction of the energy flow between field and matter (Lamb, 1960). Moreover, the fact that an accelerated charge radiates corresponds to the process of spontaneous emission.

How does this translate into the language of QED? To answer this question, we consider the change of the electromagnetic field due to the transition of an excited atom into its ground state. We assume that initially only one mode is occupied by n-quanta and all the other modes are empty. The atomic transition creates one quantum of field excitation in any field mode. However, due to the property

$$\hat{a}^\dagger |n\rangle = \sqrt{n+1}\,|n+1\rangle$$

of the creation operator, the mode with n-quanta already present has a higher probability compared to the vacuum modes where $n = 0$. Hence the amplification, which is stimulated emission, is preferentially in the mode of the incident radiation.

But how can we use Maxwell's theory to explain the directionality of the emitted radiation, which is so obvious in the QED formulation? On first sight this seems to be impossible: A dipole does not radiate in the direction in which it is driven. However, when we calculate the energy flow, that is, the Poynting vector of the total field consisting of the incident and the radiated electromagnetic field, the interference term between the two provides the directionality. Indeed, this term is rapidly oscillating in space except along a narrow cone along the axis of propagation of the incident radiation (Sargent *et al.*, 1974).

We conclude this section by briefly alluding to one more feature of stimulated emission. Stimulated emission is said to be in phase with the incident radiation. We can understand this feature when we recall that the induced dipole is a driven oscillator. Therefore it is in phase and has the same frequency as the incident light—there is no way to see this easily from QED!

Fig. 3.15. Taken from review article by Lamb, Schleich, Scully and Townes in Ref. [152].

relevant: "But how can we use Maxwell's theory to explain the directionality of the emitted radiation, which is obvious in the QED (Quantum electrodynamics) formulation? On first sight it seems to be impossible ..." [153] and the second remark concerning the fact that stimulated emission is in phase with the incident radiation is easy to see with a classically driven induced dipole, but not easily seen from QED.

CHAPTER

4

Summary of the First Decade's Work on Optical Trapping and Manipulation of Particles

My 1980 review paper in Science, [109]*, entitled "Applications of Laser Radiation Pressure" gives a fairly complete summary of our understanding of the subject of optically trapping and manipulating of macroscopic particles and atoms after the first decade. It shows how use of lasers has revolutionized the study of radiation pressure. The basic forces on macroscopic particles and atoms were explained and experimentally observed. Stable laser trapping was discovered. The 1980 review summarizes the contributions of laser trapping and manipulation techniques to the fields of light scattering, atomic physics, and high-resolution spectroscopy.

By 1980 the techniques of optical levitation and macroscopic particle manipulation were quite mature. Sophisticated force measurements were made on small macroscopic particles using optical feedback techniques with many useful applications. Also, after ten years of theoretical and experimental work elucidating the nature of radiation pressure forces on atoms, the prospects for realizing stable atom trapping and cooling were excellent. Relatively deep dipole potentials were demonstrated and various trapping geometries, such as two-beam traps and the single-beam dipole traps (tweezers), were shown theoretically to be stable in the Boltzmann sense, if separate trapping and cooling beams were used. The understanding of the Doppler limit of the Hänsch and Schawlow cooling scheme was well in hand. Radiation pressure slowing of atomic beams was already being considered by Ashkin [3]* in the context of two-beam traps and by Balykin, Letokhov and Mishin, using chirped slowing beams [144].

Summary of the First Decade's Work on Optical Trapping and Manipulation of Particles

III
1980–1990

CHAPTER

5

Trapping of Atoms and Biological Particles in the 1980–1990 Decade

The new decade 1980–1990 was primarily a decade of fulfillment. The promises of the previous decade were realized and all facets of optical trapping and manipulation were improved on. Cooling and trapping of atoms was finally demonstrated, with a few surprises in both cooling and trapping techniques. Cooling well below the Doppler limit to about the temperature corresponding to a single photon recoil was accomplished. The single-beam gradient or dipole trap emerged as a surprise winner in the race for the first optical atom trap, only to be replaced shortly thereafter by the so-called "work-horse" of atom trapping, the magneto-optic trap (MOT). The MOT was not an all-optical trap but a hybrid trap that required a quadrupole magnetic field.

The single-beam tweezer trap had many virtues that the MOT lacked, such as simplicity, maneuverability, and the capability of trapping all atoms, not just low-field-seeking atoms, at high densities in volumes a fraction of a cubic wavelength. There are other virtues that were only recognized in the 1990s, as we shall see. The perceived disadvantages of single-beam tweezer or dipole traps were the weaker potential and the inability to trap large numbers of atoms. However, to my disappointment, Chu, who participated in the first dipole trapping experiment, did not pursue any of the many possible variants or improvements of the dipole trap after leaving Bell Labs. in 1987. In my opinion he missed an important opportunity there. He and most others continued work with the MOT for many more years. This lack of interest in dipole traps, as we shall see, extended well into the 1990s.

Applications of laser-cooled atoms using magneto-optic traps proliferated in this decade. The tweezer trap, although neglected by atomic physicists, emerged as the optical trap, which started the new field of laser trapping in biology. By the end of the decade one was able to use tweezers to manipulate living cells, with no attendant optical damage. One could move entire cells and also manipulate the cytoplasm and organelles within the interior of living cells. The potential for the future uses of tweezers in biology was well established in the latter half of the 1980–1990 decade. This was as much a surprise to me as to anyone. I had given casual lip service to this possibility, but never thought it possible. Nature at last forced this possibility upon us in a remarkable example of serendipity. This chance revelation by Nature is reminiscent of the earlier piece of good fortune when Nature revealed the presence of strong gradient forces on particles in beams. As we saw, this led to the concept of stable optical trapping.

After this brief introduction to the new decade, let us continue our discussion of optical trapping and cooling of atoms where we left off, at the close of the first decade.

5.1. Optical Trapping and Cooling of Neutral Atoms in the Decade 1980–1990

I recently read a "Foreword" written by Bill Phillips in 1999 for Harold Metcalf and Peter van der Straten's book, "Laser Cooling and Trapping" [135]. I was greatly surprised by a statement made in the very first few sentences. Phillips writes, "When Hal Metcalf and I began work on laser cooling of neutral atoms in about 1979, we found ourselves in a field that was nearly unoccupied by other researchers, or by any real understanding of what the problems or possibilities were. While the study of laser cooling of trapped ions was well underway, only two other groups had ventured into laser manipulation of neutral atoms, one in Moscow and one at Bell Labs. (although the latter had temporarily dropped this line of research). Today, laser cooling and its applications represent one of the major subfields of atomic, molecular, and optical physics, with over one hundred active groups around the world..."

Phillips's appraisal of the state of the field of optical manipulation of neutral atoms in 1979 is almost totally at odds with the picture I just painted of the state of the field at the end of the first decade. Why is this? Phillips defines the field of research very narrowly, as "laser cooling of neutral atoms", whereas I spoke of all aspects of laser trapping and manipulation of neutral atoms. It is true that at that time the number of researchers in this field was small, but the Bell Labs.' contributions to cooling and trapping in the years 1979 and 1980 were major both in concept and in experiment, as I just explained. In the years 1970–1980, we at Bell Labs. invented optical trapping of neutral macroscopic particles, extended the concepts to neutral atoms, and essentially guided its course through the decade. In fact, the years 1978–1980 were, in my opinion, exceptionally productive years. I personally wrote or co-authored 10 papers, seven of which were on atom trapping and manipulation [3]*, [43]*, [123]*, [136]*, [140]*, [141], [154]*; two others [155], [156] were on light forces on macroscopic particles; and one [109]* was a review in Science on atoms and macroscopic particles. It was true that Rick Freeman had left Holmdel, where all of the trapping work had been done, for Murray Hill after being promoted to department head, but only after he, John Bjorkholm and I had completed our exciting experiments on the dipole force in 1978–1980 [43]*, [140]*, [141]. After these experiments and the theoretical paper of Gordon and Ashkin in 1980 [123]*, John Bjorkholm and I were convinced that laser trapping of atoms was experimentally possible.

Concerning Phillips's remarks about the so-called Bell Labs. group, it should be pointed out that we never had a formal group at Bell Labs. A few people joined in as collaborators, as their interests dictated. After 1980, without Rick Freeman as our expert in atomic beams, and without much in the way of resources or encouragement from a new generation of managers, experiments in atom trapping did slow down. John Bjorkholm drifted off to other projects. I started some new experiments with

Peter Smith on using dipole forces on small Rayleigh particles in liquid as a new type of "artificial nonlinear medium". I will describe that work later.

Although my main focus remained on all things having to do with radiation pressure forces, I never considered this "my group". In fact, I never had a post-doc to assist me in any of my work on radiation pressure, although I was a department head. I knew that my bosses, from the early 1980s on, did not think that highly of the subject, and I always gave the privilege of having a post-doc to other senior members of my department.

Later, when Steve Chu, John Bjorkholm and I joined forces in our push to demonstrate atom trapping, Chu took great pains to make it clear that he was not joining "my group". By the same token, when Phillips, in his Nobel Lecture, and others speak of "Steve Chu's group at Bell Labs.", or "How Steve Chu assembled a group at Bell Labs.", John Bjorkholm and I find this irritating, since he did not lead us in any conceptual sense, whereas, initially, we had to contribute to his education what we learned in the 1970s and early 1980s about atom trapping. See Chu's Nobel Lecture [36].

5.1.1. *Slowing of atomic beams by the scattering force*

The immediate need for atom trapping was for a source of slow or cooled atoms to fill traps. This was easy in concept, but technically difficult in practice. My first two papers [1]*, [2]* made it clear what had to be done. My only immediate alternative, with no resources, was to adopt a wait-and-see attitude. The Russian group under Letokhov had addressed the problem of slowing atoms in 1979 [144], using a frequency chirped slowing beam to compensate for the Doppler shift, without notable success. In 1979, Phillips and Metcalf also started work on cooling of an atomic beam by chirping [38], as I have just mentioned. They realized that optical pumping was a significant problem. Later, Prodan and Phillips [120] obtained the first indications of strong deceleration and cooling of atoms by "chirp-cooling" in atomic beams (see references in Ref. [38]). These chirping experiments did not bring the atoms to rest. This was finally accomplished a few years later, by Ertmer, Blatt, Hall, and Zhu [122]. For more details on Phillips's work on laser cooling and trapping of neutral atoms, see his Nobel lecture [38]. In 1982, Phillips and Metcalf [157] also used another approach to the Doppler problem, namely the Zeeman cooling technique. They used a tapered solenoid in which the shifting Zeeman resonance frequency kept pace with the slowing atoms. This method succeeded in slowing atoms to zero velocity in one dimension. In 1985, Phillips and collaborators [121] finally produced a sample of atoms totally at rest in one dimension in the laboratory frame at an overall temperature of ~ 0.1 K and at a density of about 10^5 atoms/cm^3.

5.1.2. *Scattering force traps and the optical Earnshaw theorem*

In 1982, two new proposals were published for trapping atoms, using just the scattering force, one by Minogin [158] and the other by Minogin and Javanainan [159]. These were meant to be large volume, very deep traps and were supposed to overcome the problems of small-volume dipole traps. I saw at once that these traps were flawed. I had rejected traps like these, based solely on the scattering force, while working on my very first paper in 1970, because no matter how you configure such a trap, it leaks in some direction. I suspected in 1970 that what I called an optical Earnshaw theorem was at work; but I never proved it. It was a bit surprising, however, that the flaws in these two purely scattering force traps were not noticed by the authors or their colleagues in the Russian group at this late date.

Shortly after these two scattering force traps were published, Jim Gordon, John Bjorkholm, and I received an invitation from Bill Phillips in 1983 to attend a short conference at NBS Gaithersberg on "Laser-cooled and Trapped Atoms". Gordon was asked to speak on dipole heating and the Doppler limit for cooling from a quantum mechanical point of view. Phillips, of course, gave us his latest results on slowing atomic beams. Later at this meeting Phillips informed us that he intended to make the first optical trap for atoms. He said they would use the Minogin-Javaneinan trap, and that they were already building it. When we returned to Holmdel I told Jim Gordon that Phillips had chosen a flawed trap. Phillips obviously could not choose either my two-beam trap or my single-beam gradient trap [3]* because his slowed atoms were still too hot (~ 0.1 K at best). I mentioned to Jim Gordon my suspicion that an "optical Earnshaw theorem" (OET) was at work on all scattering force traps. I told him that I had never succeeded in devising a stable scattering force trap for atoms despite many tries in 1970, at the time of my first paper on trapping [1]*. He confirmed the calculations I had made on the Minogin traps and we set out to prove an OET in analogy with the electrostatic Earnshaw theorem for the stability of electric charges in electrostatic fields.

The scattering force in the far field falls off as $1/r^2$, where r is the distance from the beam focus, and always points in the direction of the radius vector of the incident light, just as the electrostatic force between two charged particles. Jim Gordon showed, in all generality for a Gaussian beam, that F_S, the scattering force of radiation pressure, satisfies the equation $\nabla \cdot F_S = 0$ just as $\nabla \cdot E = 0$ for the electrostatic case. This says that if lines of force of either F_S or E enter an arbitrary closed surface about a point P, they must also leave somewhere through the surface, which implies no purely scattering force trap is possible. For a brief moment we entertained the idea of not publishing our results immediately, leaving Phillips and collaborators to follow a lost cause. Inasmuch as Bell Labs. at that moment was doing nothing experimentally on the atom-trapping front, we decided otherwise. Jim Gordon sent a version of the OET to the workshop proceedings [160] and we sent a paper to Optics Letters [45]*. I was more or less resigned to staying on the sidelines and seeing Phillips, Metcalf, and the NBS group eventually succeed in making the first optical trap. They were well ahead of the Russians and us experimentally. I cared very little about magnetic traps, but I considered optical traps for neutral atoms to be my invention. Phillips's group, after presumably giving up on the Minogin trap, did succeed in 1985 in demonstrating magnetic trapping of neutral atoms, but at low densities of about 1000 atoms/cm^3 [161]. Trapping was restricted to atoms in the low-field-seeking hyperfine states of the ground level. So, after five or six years of work on slowing and cooling, Phillips's group finally had a large diffuse magnetic atom trap, but not an optical trap.

5.1.3. *Arrival of Steve Chu at the Holmdel Laboratory*

In late 1983, however, a fortunate thing happened in our laboratory: Steve Chu showed up, as a newly promoted fellow department head. He came from the Murray Hill branch of our laboratories with the reputation of being a great young experimentalist. He had done some famous experiments with Alan Mills on the spectroscopy of positronium. Chu was much better connected with the atomic physics community than I was. He reported, "People on the outside are getting interested in optical trapping of atoms". He told us of a workshop organized by the atomic physicist David Pritchard at MIT on trapping of ions and atoms, to which we at the Holmdel Labs. were not invited, although almost the entire prior optical atom trapping work had been done by us. Chu said they sat around and asked, "What is this gradient force on atoms all about?" This amused me greatly, since we had shown what the gradient force was all about in 1978 in my Phys. Rev. Letters [3]* and in our subsequent experimental demonstrations in 1979–1980, published in Phys. Rev. Letters [43]*, in Optics Letters [141], and in Applied Physics Letters [140]*. This confirmed my suspicion of atomic physicists in academia at the time, as being unresponsive to work done elsewhere by "applied types" at industrial labs. Chu also said there had been a semi-private international meeting in Finland, organized by Stig Stenholm, on laser cooling in 1984. Phillips mentions this meeting in his Nobel Lecture [38]. Phillips mentions that only a few of the attendees at the Stenholm meeting were even active in the field at the time. So, finally, the sleeping giants of atomic physics were slowly waking up.

The final big news was that Chu said he and his new post-doc, Leo Hollberg, wanted to optically trap atoms, in spite of not knowing much about traps at that time. Nothing could have made me happier. I had already mentally conceded the optical trapping of atoms to guys like Bill Phillips who had produced slow atoms at ~ 0.1 K and thus presumably had a big lead. This was a very bold step for Chu in many ways. First, he had to do a lot of quick learning about light forces on atoms. See Chu's Nobel Lecture [36]. Second, he had to build an apparatus from scratch. John Bjorkholm opted to join the project when he heard of it. He volunteered his skills, his high-resolution tunable dye laser, and the rest of his equipment. Between what equipment we had at Holmdel and what Chu brought from Murray Hill, we had the physical resources to start an experiment very quickly. This was very important because our mutual director, Chuck Shank, told Chu that he could do what he wanted now that he was at Holmdel, but if he was thinking of trapping atoms, he should forget it, because it would never work. Chu, however, recognized the full importance of the experiment and decided to try it. What was even a worse situation was that Bjorkholm, a star researcher in my department with considerable first-hand experience with forces on atoms, was told by the same director that he was forbidden to work on atom trapping. This all happened without my knowledge. Bjorkholm never mentioned this to anyone, but, to his credit, just went ahead and worked on the experiment with us. He was older than the director and considered it his right as a Bell Labs. researcher to do as he pleased. I only learned of this a few years ago from John Bjorkholm after Steve Chu's Nobel Prize, when we were reminiscing about "the old days".

5.1.4. *Planning for the first atom trapping experiment*

As for the experiment, the big question was, "What exactly do we do?" We, of course, chose to use separate trapping and cooling beams, as proposed by Ashkin and Gordon [136]*, in order to get $U/kT_{min} \gg 1$. This made sense. For cooling we decided on six-beam molasses *ala* Hänsch and Schawlow [26]*. To make molasses work we would need Doppler slowing by the scattering force to totally stop an atomic beam of sodium. This would give atoms of ~ 0.1 K, as shown by Phillips. This temperature is the lowest one can get by such a single-beam slowing process, for many reasons. For one, not all atoms slow to zero axial velocity simultaneously. Also, the transverse velocities are not slowed by the slowing beam and, in fact, are heated by the fluctuational heating process. Actually, we decided against Phillips's Zeeman slowing because, in the interim, the simpler Doppler chirping technique had been perfected by Ertmer *et al.* [122]. The slowed beam was considered necessary to feed the six molasses cooling beams, which in turn were supposed to Doppler cool the atoms to the Doppler limit of $\sim 240\,\mu$K. The final step was to turn on some sufficiently deep trap that would then confine the molasses cooled atoms. Incidentally, although Ertmer *et al.*'s experiment gave us a very useful technique for slowing atoms to 0.1 K, they themselves were not enthusiastic about some of "the wild ideas" being proposed for trapping. They suggested using their slowed atoms in a so-called "bounced trap" of the "Zackerias type" for making precision frequency standards. See Sec. 15.1.

5.1.5. *Stable alternating beam scattering force atom traps*

The question facing us was what trap do we use. Earlier in 1984 I had proposed a new type of neutral atom trap based solely on the scattering force [142]*, which did not violate the optical Earnshaw theorem. This was an alternating beam trap conceived of in direct analogy with quadrupole electrodynamic or alternating current (ac) traps for atomic ions or charged macroscopic particles in the micrometer-size range [72], [74], [162]. Since the fields are not static, the OET does not apply. Deep trap depths greater than 1 K were calculated. These alternating beam traps can be large volume traps of about 1–100 cm^3 tuned close to resonance with low saturation parameters. This means that the Stark shift $\Delta U = (h/2)(\nu - \nu_o)p$ is small and therefore not a problem. I estimated, for a 1 cm^3 trap, that when cooled, the atom cloud should shrink to dimensions of hundreds of μm, giving densities approaching one atom per 10 μm^3 or a total of about 10^7 atoms trapped. This alternating beam trap seemed very attractive and I decided to see if one could test the principles of that new trap using micrometer-sized dielectric spheres.

A test of principle was important, since this trap was a candidate for the planned atom trapping experiment. Joe Dziedzic and I succeeded very quickly in demonstrating an alternating beam trap using micrometer-sized spheres, under conditions where stable cw trapping was not possible [143]*. The experiment was a veritable *tour de force* in trapping. We started with a levitated 9 μm silicone oil drop and manipulated it into the waiting beams of a cw optical two-beam quadrupole trap. This trap was not a large ~ 1 cm^3 trap as envisioned for the atoms, but a compact, tightly focused trap as sketched in Fig. 5.1. With this fairly tightly focused geometry we were able to trap the oil drop at O in a two-beam cw trap that used the transverse gradient forces for radial confinement [1]*. Our test of alternating beam trapping stability was made in the axial direction only. The initial particle transfer from the levitation beam to the forward cw two-beam trap was done slowly so that we could sensitively adjust the alignment of the two-beam trapping geometry initially at low power before completing the transfer. The beam reversal was accomplished by a mechanical chopper, which operated over a wide range of frequencies. The trap demonstrated impressive stability under alternating beam conditions. Quite serendipitously, it turned out that if a particle escaped from the alternating trap at O, it did not escape totally, but ended up being trapped at either of the foci f_A or f_B (see Fig. 5.1). This was not anticipated and was a pleasant surprise. This trap at f_A and f_B was clearly a new type of two-beam gradient force trap (see Fig. 5.2). The presence of this new trap allowed us to transfer an escaped oil drop back into the vertical levitating beam and start another

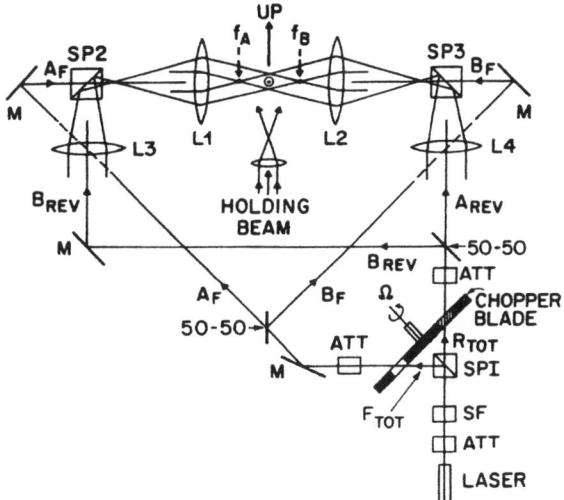

Fig. 5.1. Diagram of experimental setup used to observe radiation pressure trapping of particles by alternating light beams.

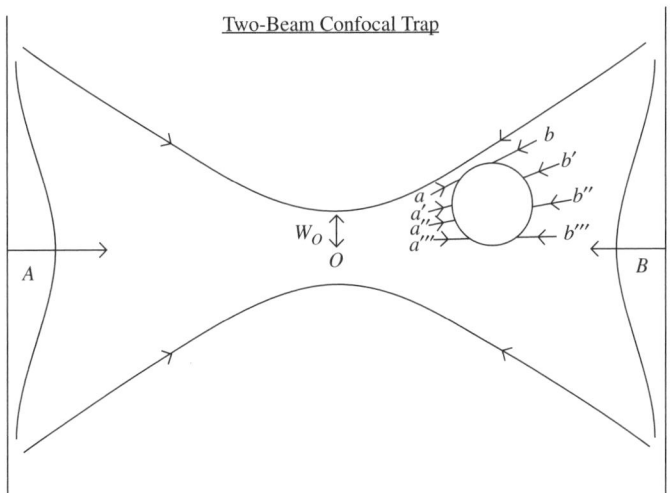

Fig. 5.2. Diagram of two-beam confocal gradient force trap.

run under different experimental conditions. In this way we quickly mapped out the full range of frequencies and other trapping parameters as described in the paper [143]*.

The new two-beam gradient force trap, of course, also worked equally well under cw conditions, as expected. It was thus a new kind of all-optical, two-beam gradient trap with many useful potential applications. With this two-beam geometry, one could completely cancel the effects of the scattering force. In the course of a five-hour series of experiments, with the help of the new two-beam gradient trap, the oil drop was transferred a total of around 50 times, without totally losing the particle as we juggled it, between five different types of optical traps. This illustrates the remarkable manipulative ability of all-optical traps for micrometer-sized spheres.

The above experiment suggested that large-volume, deep-potential, alternating beam traps might be our best choice for an atom trap. The only other possible choices were Ashkin's two-beam small-volume trap [3]*, or his single-beam smaller-volume gradient trap [3]* or possibly a trap proposed by William Wing in 1980 [163], [164]. The Wing trap was an electrostatic trap that got around the electrostatic Earnshaw theorem by using atoms in excited Rydberg states. Such atoms are low-electric-field seekers. This was complex and not attractive to us, since we were laser people. The near resonance standing wave trap of Letokhov and Minogin [126] was not deep enough and the scattering force traps of Minogin [158] and Minogin and Javanainen [159] were not stable, since they violated the OET [45]*. The choice at that point seemed clear, and we chose the alternating beam trap.

5.1.6. *First demonstration of optical molasses and early work on an optical trap*

The apparatus was quite complex and was designed, built, and assembled in early 1984, starting from a bare optical table. See Steve Chu's Nobel Lecture [36]. After numerous attempts, it was evident that, for some reason, we were not getting a trap. The situation was getting bleaker all the time. At this point Chu said, "Maybe we are trying too much at once". He suggested we just try to make molasses, which was a relatively quick, simple experiment. This indeed worked beautifully, but with some pleasant surprises. The slowed atoms collected in a roughly $0.2 \, \text{cm}^3$ volume at quite high densities of about 10^6 atoms/cm^3 [27]*. Also, they were so slow and the damping so strong that it took about 0.1 s for the cloud to decay by diffusion to half density. The diffusing cloud of atoms was clearly visible as a glowing ball, in the center of the vacuum chamber. We were able to measure the exponential decay of the atoms in molasses and compare it with the theoretical exponential decay expression for this geometry [27]*. We got a very nice fit. I was as surprised as Steve Chu and John Bjorkholm, by the slow decay, but I was also quite prepared and adept at that time at handling diffusion problems such as this. I had just finished studying the diffusional properties of liquid suspensions of submicrometer Rayleigh particles in experiments with Peter Smith on artificial Kerr nonlinear media, based on the gradient forces of radiation pressure [165], [166]*, [167], [168]. Fortunately, I was familiar with that great book entitled "Aerosol Mechanics", by N. A. Fuchs, which actually treats a diffusion problem equivalent to the molasses case [169]. See Eq. (1) of Ref. [27]*. Another pleasant feature, at the time, was that the temperature was about $240 \, \mu\text{K}$, very close to the calculated Doppler limit for molasses of $h\gamma_n/2$. This corresponded to an average atomic velocity of $\sim 60 \, \text{cm/s}$. Recall that the recoil momentum for a sodium atom scattering a single photon is $\sim 3 \, \text{cm/s}$. To put this further into perspective, recall that the temperature of the one-dimensionally slowed and stopped atoms of Phillips *et al.* [121] and of Ertmer *et al.* [122] was $\sim 0.1 \, \text{K}$. Our $240 \, \mu\text{K}$ temperature was also significantly lower than the cooled ions held in electromagnetic traps, where ion temperatures between 5 and 100 mK have been observed [170]. The achievement of molasses changed our view of how one fills traps. Cold atoms in molasses were free to simply diffuse into the waiting arms of an optical potential U. The actual demonstration of molasses, with its remarkable properties, was a major step toward atom trapping and was clearly useful for all sorts of new experiments with cold atoms.

Our immediate problem was still the demonstration of a stable optical trap. Using our newly demonstrated molasses, we again attempted to achieve trapping with the alternating beam atom trap, but without success. At this point Bjorkholm suggested using the tweezer or single-beam gradient trap [3]*. Chu, in his Nobel Lecture [36] says, in spite of his initial rejection of the suggestion to use the small-volume tweezer trap, that after a day or two he convinced himself and also John Bjorkholm and me that filling would take place by diffusion. He may have convinced himself, but he did not need to convince John or me at that point. I suspect that this remark reflects his own learning process at the time.

Indeed, filling the atom trap by diffusion is exactly analogous to filling the first optical trap by diffusion of macroscopic latex spheres in the first demonstration of optical trapping in 1970 [1]*. You simply put the trap inside the diffusing cloud of particles and wait. Damping actually plays two roles here. Not only do the particles wander into the trap by diffusion, but also the damping of the particle motion within the trap prevents the escape of particles over the potential barrier.

5.1.7. *Cooling below the Doppler limit of molasses and below the recoil limit*

In the meantime, before embarking on this new attempt at optical trapping, it was natural to wonder if one could cool to even lower temperatures than the Doppler limit of molasses. There was no obvious reason for not being able to do so. Since atoms are quantum-mechanical in nature and emit photons one at a time, one might imagine that cooling below the temperature corresponding to a single photon recoil was impossible. A little thought showed that even the single photon recoil was no real physical barrier. Indeed, in 1985 we proposed several schemes to achieve arbitrarily low atom temperatures [171]*.

One of these proposals is sketched in Fig. 5.3. It starts with an imagined deep gradient force trap filled with trapped atoms harmonically confined in a volume of $\sim (1\text{--}2)\,\mu$m in diameter, located

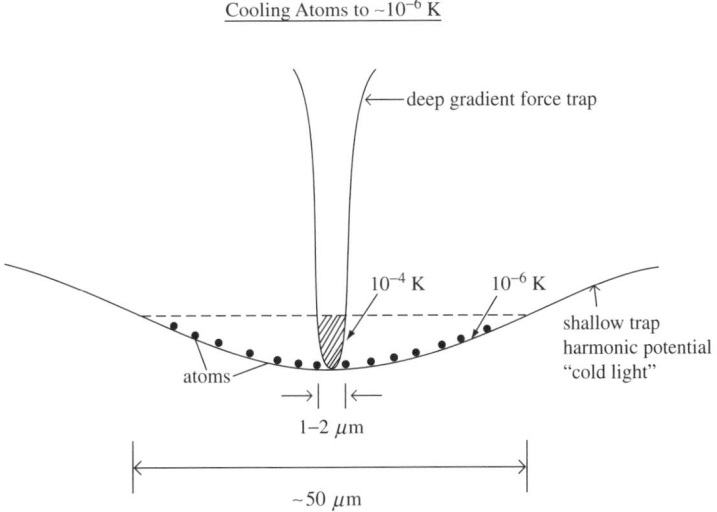

Fig. 5.3. Cooling of atoms from $\sim 10^{-4}$ K to $\sim 10^{-6}$ K by expansion of tightly confined atoms into a large shallow gradient force trapping potential.

within a shallow trap also having a harmonic potential. The initial temperature of the lightly confined atoms is $\sim 10^{-4}$ K. At $t = 0$, the deep potential is turned off and the atoms expand into the much larger volume of shallow potential. After one quarter of a period all the initial atoms, irrespective of their initial velocities, have come to rest at $v = 0$ with a spread in position out to $\sim 50\,\mu$m in diameter. The shallow potential is tuned very far from resonance, so it can be thought of as "cold light", which makes essentially no heating transitions in the quarter cycle. Since the atoms are not in fact all initially at $r = 0$, not all atoms actually come to $v = 0$ at one quarter cycle. The final residual temperature is calculated to be somewhat less than $\sim 10^{-6}$ K. This is close to the recoil limit.

This Optics Letter paper does not mention applications of atoms cooled to $\sim 10^{-6}$ K. However, in a longer version of this work published in Methods of Laser Spectroscopy [172], we mentioned, among several applications, the study of collective quantum effects, such as Bose–Einstein condensation. This proposed application of cold atoms was originally made in a talk given by Leo Hollberg in Israel in 1985 before we had even achieved atom trapping. See Figs. 5.4 and 5.5, taken from this paper, which are the title page and the page discussing possible applications of micro-Kelvin temperatures. Thus it is seen that, in late 1985, we at Bell Labs. were already thinking about micro-Kelvin temperatures, well before the cooling experts such as Phillips [38] and Cohen-Tannoudji [37] had even started their work on molasses and sub-Doppler cooling.

5.1.8. *Evaporative cooling from optical dipole traps*

Yet another even simpler scheme for cooling below 10^{-6} K, to temperatures below the recoil temperature, using dipole traps, was proposed in Ref. [171]*. This scheme was to just open a small-volume dipole trap for a short time and then close it. During the open time the fastest of the atoms originally at the Doppler limit of $240\,\mu$K leave the trap. If, for example, we allow all atoms faster than $v/10$ to escape, then the temperature of the remaining cold fraction of $0.022N_o$ atoms is less than $T/100 = 2.4\,\mu$K; where N_o is the original number of atoms. Thus, if we are willing to pay the price of reducing the number of atoms remaining in the trap, we can reduce the temperature of the remaining atoms to arbitrarily low values. This is basically an evaporative cooling technique.

At just about the same time Hess, working with Kleppner and Greytak, proposed evaporative cooling of atomic hydrogen held in magnetic traps as a possible way of cooling hydrogen gas to Bose–Einstein condensation [173]. Somewhat later, in 1987, Hess *et al.* demonstrated significant evaporative cooling of atomic hydrogen by this scheme [174].

As we shall see in some detail, BEC was achieved in rubidium and sodium in 1995 using evaporative cooling from magnetic traps [29]*, [175]*. The magnetic evaporative cooling technique considered in Ref. [173] and demonstrated in Refs. [29]*, [174], [175]* was more sophisticated than the simple scheme of Refs. [171]*, [172], in which one just opens and closes the trap. The Hess scheme involved the slow evaporation or coupling out of fast atoms. This leaves time for the remaining atoms to collide and constantly equilibrate to the gradually lowering temperature. This is much more efficient than a quick one shot venting of the hottest atoms. Eventually, cooling of hydrogen to a B–E condensate was achieved by Fried *et al.* in 1998 [176], about three years after the first achievement of BEC in alkali metal atoms. Of course, slow evaporative cooling from dipole traps is also possible, as we shall see in later experiments in the mid-1990s in Sec. 17.3.

From: METHODS OF LASER SPECTROSCOPY
      Edited by Yehiam Prior, Abraham Ben–Reuven
        and Michael Rosenbluh
      (Plenum Publishing Corporation, 1986)

COOLING AND TRAPPING OF ATOMS WITH

LASER LIGHT

Steven Chu, J. E. Bjorkholm
A. Ashkin, L. Hollberg*, and Alex Cable

AT&T Bell Laboratories
Holmdel, New Jersey 07733

Abstract:

Recent experiments have shown that it is now possible to cool and trap neutral atoms. We review the work done at Bell Laboratories to confine and cool sodium atoms to $T = 2.4 \times 10^{-4}K$ in a viscous molasses of photons tuned to the yellow "D-line" resonance of sodium. Potential optical traps are presented, and several ideas for cooling atoms to temperatures of $10^{-6}-10^{-10}K$ are suggested. Possible uses for ultra cold atoms are also mentioned.

The manipulation of atoms with laser light has seen major advances in recent years. Deflection [1] and focusing [2] of atomic beams was followed by the slowing and finally stopping [3] of a beam of sodium atoms by a laser beam directed opposite the atomic velocity. In the laser stopping experiments, the residual velocity spread corresponded to a temperature of 50-100mK, and such a sample of atoms led to the magnetic trapping of sodium atoms [4]. The confinement of sodium atoms in an "optical molasses" of light has also been demonstrated [5]. In addition to fairly long containment times, the optical confinement scheme simultaneously cools the atoms to temperatures of a fraction of a mK in a time much less than a millisecond. Because of the relative ease in obtaining a high density sample of cold atoms, optical molasses seems to be a good starting point for a variety of experiments.

After a brief review of our work on optical molasses, we will present schemes for laser trapping of atoms that we are trying. We also present several schemes for laser cooling atoms to temperatures of $10^{-6}K$ and possibly to $10^{-10}K$. Finally, the realization of ultra-cold atoms has opened up the potential for new experiments. We will briefly mention a few areas of research that can be addressed with these atoms.

This paper is not intended to be a review of the field of laser manipulation of atoms, but rather a snap-shot of the work currently underway at AT&T Bell Laboratories. In particular, we will not discuss the elegant magnetic trapping experiments of Phillips and his co-workers.[4] For a review of the earlier work in the field, the reader is referred to the earlier review articles of Ashkin [6] and Letokhov and Minogin [7]. More recent collections of articles in the field can be found in dedicated journal issues edited by Phillips [8] and Meystre and Stenholm [9].

---

* Present address: National Bureau of Standards, 325 Broadway, Boulder, Colorado 80303

Fig. 5.4. Title page of Leo Holberg's 1985 talk in Israel, taken from Ref. [172].

the cooling of magnetically confined spin aligned hydrogen experiments.[25] Adiabatic cooling, achieved by the slow weakening of an atom trap, is another method for achieving further cooling. We have suggested a cooling technique particularly well suited to light traps.[26] In this scheme, the atoms are first localized and cooled by a combination of optical molasses and a tightly confining gradient trap. Next, the deep, small trap is turned off and a broad harmonic trap is substituted in its place. The atoms, initially near the center of the trap will begin to oscillate. The trap light is turned off after approximately one quarter of a harmonic oscillator period - the time when most of the atoms will have reached their turning points in the oscillation. The reduction in atomic velocity is determined by Liouville's Theorem $\Delta v_i \Delta x_i = \Delta v_f \Delta x_f$, how closely a gradient trap can approximate a harmonic potential, and the probability of heating due to the trap. For realistic gradient traps and an initial temperature of $T \sim h\gamma/2 \sim 2.4 \times 10^{-4} K$, we have shown that temperatures below $10^{-6} K$ are possible.[26]

The above discussion treats the atoms as classical particles. If we consider sodium atoms in a trap with a potential depth $U \sim 4 \times 10^2 h\gamma$, 30 μm in diameter at a temperature $T = h\gamma/2$, the quantum mechanical harmonic oscillator energy $\hbar\omega$ corresponds to approximately $10^{-2} kT$. For lithium atoms, the lighter mass and narrower D-line resonance (2s -> 2p at 671 nm) will mean that only ~ 15 quantum states will be occupied. If the 2s -> 4p (323 nm) transition is used to cool the Li atoms, temperatures on the order of $10^{-5} K$ should be achieved. At this temperature, kT is comparable to $\hbar\omega$ If the atoms can be cooled into the ground state of the harmonic oscillator potential, the zero point "temperature" can be decreased by adiabatically enlarging the size or decreasing the depth of the trap. Since the oscillator frequency will initially be on the order of $\omega/2\pi \sim 1$ MHz, the adiabatic cooling time is short compared to the trap heating time. The limitation to atom cooling in this situation seems to be the size of the localization volume before the trap cannot support the weight mg of the atom. Atoms with harmonic oscillator wavefunctions on the order of $\lambda = 100$ μm should be accessible. The corresponding localization temperature which we define by $\lambda = \dfrac{h}{\sqrt{3KTm}}$, is $T \sim 10^{-10} K$.

IV. APPLICATIONS

Laser cooled and confined atoms open the possibility for a variety of novel experiments. Atoms cooled and confined in a mostly uniform magnetic field [27], or in free fall as in an atomic fountain [28], will have the long coherence times needed for precision spectroscopy. In addition to improved frequency standards, small frequency shifts due to feeble physics effects can be explored. For example, an electron dipole moment d_e will induce a linear Stark shift in a nondegenerate state of an atom. The present limit of $d_e \leq 2 \times 10^{-24} e - cm$ [29] corresponds to frequency shifts in strong electric fields in the millihertz range. It is possible that a two to four order of magnitude improvement can be obtained by using an intense beam or fountain of slow atoms.

Micro-Kelvin temperatures will allow us to examine atom-atom and atom-surface collisions in a new regime. In the gas phase, pure s-wave scattering, low energy resonances, and spin flip collisions can be studied.[30] In surface collisions where both the atom and surface temperatures approach zero, it is uncertain whether the atom will stick or bounce on the de-Broglie wavelength λ of the atom is significantly larger than the extent of the attractive potential. The results from ^4He atoms scattering from a liquid ^4He surface at grazing incidence agree surprisingly well with a model that neglects inelastic processes.[31] In these experiments, the atom momentum components perpendicular to the surface correspond to de-Broglie wavelengths as long as 150 Å. If the reflection probability goes to unity, cold atom scattering from surfaces would have important applications. One can imagine a box that could store the atoms in much the same way that ultra-cold neutrons are stored. In the limit of very long de Broglie wavelengths, the atom wavefunction should have a 180° phase shift in an elastic collision. The internal degrees of freedom, however may not be affected. Thus, the coherence in atomic transitions will be maintained, and the confinement box would be the ideal confinement volume for ultra-precise metrology.

The study of quantum collective effects such as Bose Condensation are also possible. For the case of ^7Li, temperatures of $10^{-5} K$ require densities of $\sim 10^{15}$ atoms/cm^3 before Bose condensation is expected. At these densities, three body recombination will create hot atoms,

Fig. 5.5. Copy of p. 47 from Holberg's Israel talk [172] discussing atomic fountains and Bose–Einstein condensation.

5.1.9. *First atom trapping experiment using the single-beam dipole trap*

Let us now return to the atom trapping experiment [4]*, which we fully expected to go well. We had our newly discovered cloud of slowly diffusing molasses at energies $\sim h\gamma_n/2$. We had a relatively deep single-beam gradient (tweezer) trap with a calculated potential $U \cong 20\, h\gamma_n$. We had also calculated that the diffusing cloud of slowly moving atoms would find its way into the tweezer trap, as small as it was. Finally, we took care of the Stark shift problem, which Gordon and Ashkin identified as due to the high intensity of the trapping light [123]*, [136]*. The large Stark shift caused the atoms to be out of resonance with the molasses beams and thus negated the cooling. We had proposed a physically sound but experimentally complex solution to this problem by adding an additional Stark-correcting beam [136]*. A simple solution to the Stark shift problem was proposed by Dalibard *et al.* [177]. This involved the complete decoupling of the trapping beam from the molasses cooling beams by turning off the trapping beam while the molasses beams were on, and *vice versa*. If the switching cycle was 50–50, the affected trap depth was halved while the equilibrium temperature doubled. Gordon and Ashkin in Ref. [123]* had previously discussed a similar switching scheme for avoiding the effect of the Stark shift on atoms during sensitive spectroscopic measurements. We opted for the simple switching scheme to take care of the Stark shift. We started the trapping experiment in 1986 and succeeded fairly easily after only a few false starts. I will discuss this brief period of difficulties later.

Figure 5.6 is a sketch of the trapping experiment and the principal results. The single-beam trap could be placed anywhere with respect to the molasses cloud, and it promptly filled with about 1000 atoms. The atoms were compressed in the trap to a volume of $\sim 10^{-9}\,\text{cm}^3$, which gives a density of $\sim 10^{11}\,\text{atoms/cm}^3$. This density was three orders of magnitude higher than the background density

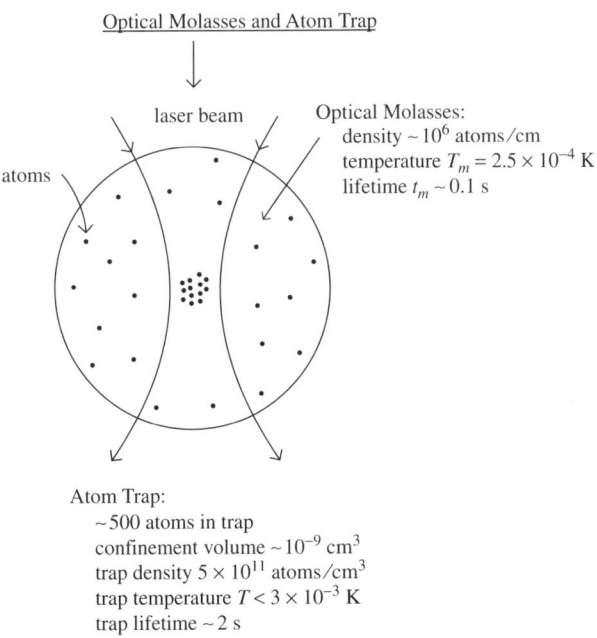

Fig. 5.6. Sketch of the single-beam gradient atom trap and list of principal results.

of the molasses. The high density of atoms in the trap was easily distinguished from the molasses as a bright point of fluorescing atoms, which remained for about a second after the molasses diffused away (in ~ 0.1 s). The trapped atoms could be manipulated easily in space, much as a single particle, by wiggling the lens of the trapping beam. This simple manipulation signaled the start of a new era of trapped atom manipulation. Although the temperature of trapped atoms was somewhat more difficult to measure, we estimated temperatures of $0.6 \leq kT/h\gamma_n \leq 0.8$. Considering the expected factor of two increase in temperature due to the switching, this is quite close to the expected $kT = h\gamma_n/2$.

The world's first laser trapping of atoms [4]* was considered big news at the time and was reported on the front page of the Sunday New York Times on July 13, 1986, and subsequently in all sorts of popular news media around the world. So, we had our moment of glory. My boss, Chuck Shank (the great skeptic), came around to see me shortly thereafter. He said, "Remember, Art, I never told you that you could not work on atom trapping".

Before leaving the discussion of the first molasses experiment and the first atom trapping experiment [4]*, [27]*, I should say something about the authorship of those papers. All of the authors of the molasses paper were aware of the fact that we were initially way behind the Phillips's group at NBS. We also all agreed that, if Chu had not offered to enter the field, we at Bell Labs. could have done nothing to compete with Phillips. It seemed a long shot at the time, as I explained before, but we decided to go all out to win the optical trapping race. It involved teaching Chu what we had learned over the last fifteen years or so on atom trapping, and combining all of our experimental equipment [36]. Chu's suggestion to limit the experiment to molasses, after some initial difficulties trying to both cool and trap simultaneously, was an important feature of the ultimate success of the experiment. It was unanimously agreed that he should be first author.

Things were less clear concerning the first atom trapping experiment with the single-beam dipole or tweezer trap. After early problems with my alternating beam trap, Bjorkholm's suggestion to use the single-beam tweezer-type trap was evidently the right choice. The strong viscous damping and long decay time of molasses made it clear that the tweezer trap would fill easily, despite its very small volume. Chu, with his great energy and unorthodox work schedule, probably put in more lab. time than Bjorkholm or anyone else. I spent the least time in the lab., even though I enjoy observing the data first hand. I was pretty much excluded, since Chu felt there were too many people already in the crowded lab. We all put in a lot of time in the write up of the results. I, in particular, insisted on writing a careful description of the basic forces, trapping principles, and the history of the development of the dipole trap. Bjorkholm did most of the calculations. At the end, Bjorkholm thought he should be first on this paper. Chu saw it differently. I said nothing. Finally, they decided to toss a coin. Chu won the toss. As for myself, I was quite content to be listed as a coauthor on these papers. My contributions to the planning of the experiments, the interpretation of the data, and the write-ups, I felt, had earned my inclusion. Any place in the list was fine with me. I treated it as a facetious remark when, in our earlier discussions on the molasses paper, Chu suggested that I leave my name off that paper entirely, because I was already "so famous" that I did not need to worry about credits. I did not agree to this suggestion. After struggling for 15 years to reach this point, I was amused by this remark about my fame, considering what our vice president of research, Arno Penzias, had said about what was acceptable work at one of his annual talks to the troops. He said, looking straight at me, that if you wrote papers in which you referenced yourself six times and then Einstein, you were well advised to drop the subject. In the early days of levitation and trapping, this was almost exactly what my papers were like. There was almost no one else to refer to, and I did reference Einstein in connection with the "Einstein *Rückstoß*", the Einstein recoil for atoms emitting a photon. I, of course, did not follow Penzias's advice to drop the subject; I just added some extraneous references to dilute the references to my work. In the course of the early work I did receive strong support from my immediate bosses, Rudy Kompfner and Herwig Kogelnik, in particular. Kogelnik told me that it was bad for a department head to compete directly with the work of his own men. He said it was best if he had a separate project of his own, which allowed him to show his own research capabilities. I thought, and he agreed, that studying radiation pressure qualified as such a project, especially for my laser research department.

5.1.10. *Proposal for stable spontaneous force light traps*

It turned out that a new atom trapping proposal based on the scattering force by Pritchard *et al.* [178] appeared in the very same 1986 issue of Phys. Rev. Letters as our paper on the first trapping experiment using the single-beam gradient trap [4]*. The thrust of this proposal was that by changing the spatial dependence of the scattering force with a spatially varying magnetic field, for example, one could obtain a spatially varying Zeeman shift in the resonant frequency of an atom, which could be made to give a stable trap without use of a gradient or dipole force. More will be said shortly about this trap. This type of trap is only one of a class of traps that change the internal degrees of freedom of atoms to make stable scattering force traps. Such traps, in effect, circumvent the optical Earnshaw theorem (OET). This is all true. The authors, however, concluded their paper with the remark that they hoped that their suggestion "will remove the optical Earnshaw theorem as a practical barrier to the design of spontaneous force light traps and will lead to the realization of successful traps of different types" [178]. This certainly puts the OET in a very negative light. The OET is no barrier, except to what is wrong.

These authors [178] seemed to forget that the OET had already identified two published scattering force traps as flawed and saved Bill Phillips from the big mistake of trying to make use of one of these traps at NBS. Gordon and I considered the OET as a big step forward in the understanding of traps, for which these authors should have been grateful.

The authors in this paper also introduced a new terminology in connection with the scattering force. They decided that the term "scattering force" should only be used for cases where the OET applies, and that one should use the term "spontaneous force" for cases where the OET does not. They say, "Note our use of the terms 'scattering' force to apply to light forces that obey the OET and 'spontaneous force' for the analogous forces in atoms (which do not). These terms have been used interchangeably by previous authors". I find this new terminology very ill advised. In all generality, as we have seen, one can resolve the radiation pressure force on particles into two force components: a scattering force and a gradient force. For atoms, the scattering force comes from the momentum of absorbed and spontaneous randomly emitted photons. This can be called correctly and equivalently the scattering force or the spontaneous force. For atoms, the gradient force on the optically induced dipole in the direction of the intensity gradient can be called either the gradient or dipole force, since it depends on optically induced dipoles in a gradient of the optical intensity. Also, calling it the stimulated force is acceptable. There is no place in this discussion for the OET. To impose such a connection causes gratuitous confusion.

After reading this nonsense about the OET being a barrier to progress and the new terminology, I suggested to Jim Gordon that we write a short note to the comments section of Phys. Rev. Letters about these confusing statements in an otherwise very nice paper. Jim said, "We stated the conditions under which the OET applies very clearly, so why repeat ourselves". We ended up doing nothing.

5.1.11. *Nature's comments on the first atom trapping experiment*

There was another surprising consequence of the Pritchard *et al.* paper on "Light Traps Using Spontaneous Forces" [178]. A few weeks after our first trapping paper [4]* and the Pritchard *et al.* paper appeared back to back in Phys. Rev. Letters, there was an editorial in Nature by the editor John Maddox [179] commenting on the achievement of optical trapping of atoms. He praised our accomplishment highly, but said that we had achieved it by taking advantage of the new understanding gained from the work of Pritchard *et al.* [178]. This was too much for me to bear, especially since it was from the editor of Nature, a theoretical physicist, no less. I sent a letter explaining there was no connection between Pritchard's proposal and our experiment and that we did not use or need a magnetic field for stability, but relied on the optical gradient force. This letter appeared in Nature [180] under the title "Caught in a Trap" and served as an acknowledgment of Maddox's error.

5.1.12. *The first experimental demonstration of a MOT*

Shortly thereafter, Pritchard proposed to Chu that they do a joint experiment to demonstrate the idea of a stable scattering force trap. This has come to be known as the magneto-optic trap, or MOT. Pritchard and his graduate student Eric Raab had a specific geometry in mind for implementing their proposal [178] and Bell Labs. was the only laboratory where molasses cooled atoms were readily available. I had no part in this experiment. The former collaborators on the first atom trapping experiment were drifting apart at this point. Chu was spending his time spreading the word on molasses and consulting with Pritchard and Wieman on the MOT. By that time I was fully involved in applying tweezer trapping to biology. Chu's new post-doc Mara Prentiss had arrived from MIT and he put her to work in 1987 with Raab on the new magnetic radiation trap [181]*. Figure 5.7 of Ref. [181]* illustrates the principle of the trap, somewhat simplified in one dimension.

Imagine a hypothetical two-level atom, with spin $s = 0$ ($m_s = 0$) ground state and a spin $s = 1$ excited state, having excited state levels ($m_s = -1, 0, +1$), which is placed in a magnetic field which is zero at a point O in the center of the trap and reverses linearly about O along the Z-axis, as shown. If this atom is irradiated by a pair of opposing beams, tuned below resonance, of opposite circular polarization, σ^+ and σ^- as shown, then the absorption becomes position-dependent along the Z-axis. As one moves away from O to the right, the right beam σ^- moves closer to the $m_s = -1$ resonance than to the $m_s = +1$ resonance, and sees an increasing restoring force. Similarly, if one moves to the left, increasing absorption by the $m_s = +1$ level gives a restoring force back to the origin O. The same argument can be generalized to give stability in three dimensions. A weak quadrupole

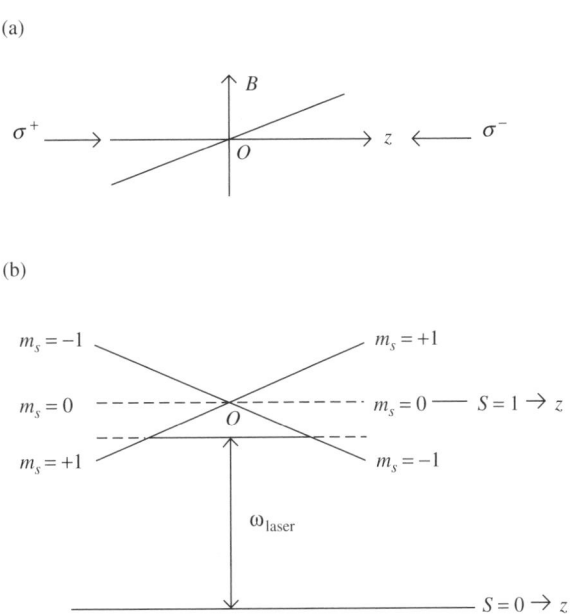

Fig. 5.7. Diagram of the operation of a MOT in one-dimension. (a) The atoms are in a linearly varying magnetic field and are irradiated by a pair of σ^- and σ^+ polarized light beams tuned below resonance in zero field. (b) The potential variation of the energy levels for atoms with spin 0 ground state and spin 1 excited state is shown. Atoms located at positive z are closer to resonance with the σ^- beam than with the σ^+ beam and are pushed toward the center of the trap at O. The reverse is true at negative z.

magnetic field, generated by a pair of coils with equal currents, gives the correct three-dimensional (3D) magnetic quadrupole field shape. In the MOT the three pairs of σ^+, σ^- beams also act as molasses cooling beams when tuned $\gamma_n/2$ below resonance. Thus the MOT beams simultaneously trap and cool the atoms.

This first MOT was very large, ~ 0.5 mm in diameter and captured $\sim 10^7$ atoms at densities of $\sim 10^{11}$ atoms/cm^3 at a temperature of ~ 0.4 mK. It is a deep trap, ~ 0.4 K and not sensitive to beam alignment or magnetic field. This was clearly a very versatile trap. However, it was reported by Raab *et al.* [181]* in the first MOT paper, that there was complex collective behavior of the atoms. The trapped atoms executed swirling motions and could collect at multiple points within the trap. This unexpected behavior was sensitive to small changes in alignment of the beams and the magnetic field.

5.1.13. *Radiation trapping in MOTs*

Later work by Walker, Sesko, and Wieman, in 1990 [182]*, studied the collective behavior of MOTs more carefully and identified stable rings and clumps of atoms executing complex orbital patterns. They identified a long-range force that acted across large distances in these large traps. They showed that this force, arising from radiation trapping, could account for much of the behavior, but there was still an array of unexplained behavior at higher atom densities. Nevertheless, it was clear that these repulsive radiation-trapping forces were limiting the density of atoms that could be trapped in a MOT. We will see later that there is a simple way to reduce the magnitude of the radiation trapping in MOTs, which results in several orders of magnitude increase in the density of trapped atoms. This device is the so-called dark spot MOT [183].

Apart from these few studies of trap behavior, the MOT, in spite of its complexity, has been used primarily as a source of cold atoms for other experiments. One other big advantage of the MOT is its ability to capture neutral atoms directly from an atomic beam without the need for radiation pressure slowing, as developed in the Phillips tapered magnetic slower [157] or the Ertmer *et al.* chirped laser slower [122]. This was first shown in 1990 by Cable, Prentice, and Bigelow [184] and somewhat later in 1990 by Walker *et al.* [182]*. This ability to stop and collect atoms with no slowing beams is due to the large dimensions of the trap.

Although, as we shall see later, the MOT has played a major role in neutral atom trapping, it lacks the easy maneuverability of a single-beam tweezer-type trap.

As we shall also see in later experiments, the magnetic field itself sometimes can cause complications, as in some Bose–Einstein condensation experiments. In this context I often wonder if the alternating-beam scattering force trap [142]*, which we originally used in our early attempts at trapping, would have worked. An alternating-beam trap might be preferable to the MOT. It has trap depths (~ 1 K) comparable to or deeper than a MOT and can be expected to hold comparable numbers of atoms with no complications from magnetic field and Zeeman shifts. Although it did not work initially, we never seriously returned to try it again after we got the molasses cooling technique to work. Recall that it worked beautifully for macroscopic particles [143]* and the basic physics has been tested many times. When I occasionally brought this matter up with Chu, he just said, "We tried it and it did not work". I personally think it will work and that it is important enough to try again, even now.

Finally, most people, including me, have characterized the MOT as the "workhorse trap". Pritchard, however, in one of his summer school lectures on cooling and trapping says of the MOT, it is "a trap whose importance cannot be overemphasized". This, I suspect, is a bit of hyperbole. If I were asked to characterize optical traps for atoms, I would say, "the importance of the MOT, which is not even an all-optical trap, is overemphasized, and that of the single-beam gradient trap or tweezer trap for atoms has been greatly underemphasized". The evidence supporting my characterization will be given in connection with more recent BEC experiments and other work in the late 1990s, where the magnetic field of MOTs and purely magnetic traps often is a source of problems.

5.1.14. *Atom cooling below the Doppler limit*

By 1987 groups in many physics laboratories had begun to use optical molasses for making cold atoms. Unexpected behavior was soon observed that was contrary to the accepted picture of molasses cooling. Lett *et al.*, in Phillips's group in 1988 [185], observed temperatures for sodium that were considerably lower than the expected minimum of $\sim 240 \, \mu$K. In addition, these low temperatures occurred at tunings quite a bit further from resonance than the theoretical value of half a linewidth. Figure 5.8 shows results of a measurement of fluorescence versus time-of-flight for atoms dropped into a detector located below the molasses, indicating a temperature of $\sim 40 \, \mu$K. The expected distribution for atoms at a temperature of $240 \, \mu$K is shown for reference. It became clear that simple two-level atoms could not account for the observed additional cooling.

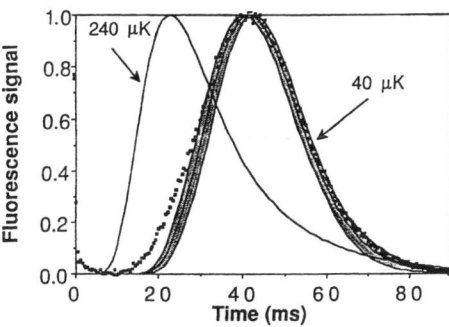

Fig. 5.8. Measured time of flight distribution of fluorescence from molasses-cooled sodium atoms, indicating a temperature of 40 μK. The curve for 240 μK is shown for reference. Figure from Ref. [185].

Measurements of the effects of the polarization of the molasses beams, the intensity and the effects of the ambient magnetic field helped to identify the factors responsible for the anomalous cooling [38]. The resolution of this problem was published in 1989 in papers by Dalibard and Cohen-Tannoudji of France [186] and Ungar *et al.* of the Stanford group [187] in a special issue of the Optical Society of America. See also the Nobel lectures of Cohen-Tannoudji [37] and Chu [36], and also [41]. The source of the additional cooling is a fortuitous combination of multilevel atoms, polarization gradients, dipole forces (light shifts), and optical pumping.

There are many possible polarization arrangements for the molasses beam that give cooling below the Doppler limit: three pairs of beams with linear polarization, or orthogonal polarization, or σ^+ and σ^- polarization (i.e., combinations of right and left circular polarizations). There is also the possibility of adding an external magnetic field.

A rough explanation for the additional damping can be found in the nonadiabatic response of the atom moving through the light field, the Zeeman structure of the ground state, which manifests itself in a spatial variation of the ground state Stark shifts, and position-dependent optical pumping rates. Because of time lags between the atom motion and the optical pumping, atoms end up having to climb up potential hills before they are pumped to higher levels, from which they decay to the lowest energy ground state with a net energy loss.

Figure 5.9, taken from Phillips [38], illustrates the physical basis of sub-Doppler "Sisyphus" cooling. It shows a particularly simple one-dimensional (1D) case of an atom with a $J = 1/2$ to 3/2 transition having two ground state Zeeman levels, $m = -1/2$ and $+1/2$, located in a pair of opposing orthogonally polarized cooling beams. As shown in Fig. 5.9(a), the polarization is spatially modulated, going from σ^+ to σ^- and back to σ^+ over a distance of $\lambda/2$. At intermediate positions one has points of elliptical and linear net polarization. Figure 5.9(b) shows the correlated spatial variations of the two ground state standing wave potentials. The light shifts of these two sublevels are changed sinusoidally in the regions of varying polarization, as shown. The atom in the σ^- ground state is shown moving uphill as it advances, after which it is optically pumped to the upper level. From there it preferably returns to the lower energy σ^+ level. The successive hill climbing followed by returns to the lower state drains kinetic energy from the atoms as they move along through the light field. Such cooling processes are called "Sisyphus" cooling, after Sisyphus of mythology, who

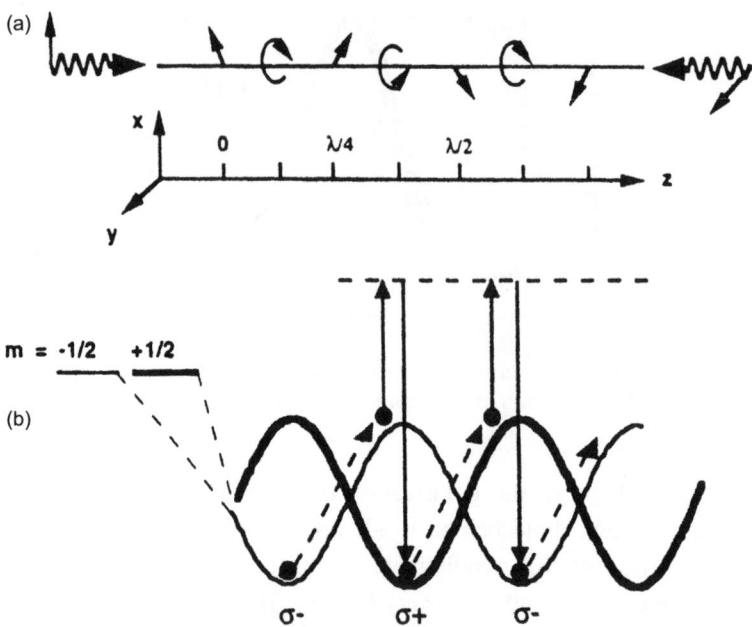

Fig. 5.9. (a) Interfering, counterpropagating beams having orthogonal, linear polarizations create a polarization gradient. (b) The different Zeeman sublevels are shifted differently in light fields with different polarizations; optical pumping tends to put atomic population on the lowest energy level, but nonadiabatic motion results in "Sisyphus" cooling. Figure 18 and caption from Ref. [38].

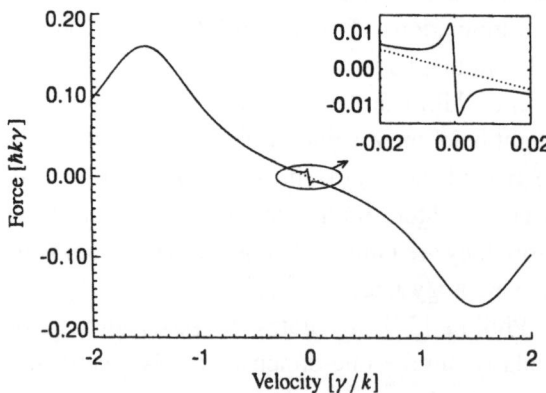

Fig. 5.10. The force as a function of velocity for an atom in a lin $\perp$ lin polarization gradient cooling configuration with $s_0 = 0.5$ and $\delta = 1.5\gamma$. The solid line is the combined force of Doppler and sub-Doppler cooling, whereas the dashed line represents the force for Doppler cooling only. The inset shows an enlargement of the curve around $v = 0$. Note, the strong increase in the damping rate over a very narrow velocity range that arises from the sub-Doppler process. Figure 8.2 and caption from Ref. [135].

was condemned to roll a huge stone to the top of a steep hill, only to have it slip from his grasp each time and roll down again.

It is instructive to look at a plot of force versus velocity for atoms moving through molasses, as shown in Fig. 5.10 for the same 1D case of mutually perpendicular polarizations as considered above in Fig. 5.9(a). Figure 5.10, taken from Fig. 8.2 of Metcalf and van der Straten's book [135], makes clear the separate damping force contributions due to Doppler cooling and the multilevel Sisyphus cooling contribution. Figure 5.10 is for low saturation and a detuning from resonance of $1.5\,\gamma_n$. The solid curve is the large reversible Doppler force contribution to the expression for the damping force $F = -\beta v$, requiring only a two-level atom. The much smaller dotted curve that is effective at much lower velocity is due to the multilevel Sisyphus contribution, i.e., from the Zeeman sublevels or hyperfine sublevels. As atoms cool under the combined influence of the separate damping features, it is the Sisyphus cooling that finally takes over at low velocities [135], [186].

The lower limit of the temperature one can achieve by Sisyphus-type cooling is difficult to compute exactly, because in three-dimensional molasses it is impossible to get simple arrangements of polarization. For example, one can never get the polarizations of the six beams of a 3D molasses to all point in the same direction. Theory says that the lowest achievable temperatures are of the order of the depth of the ground state potential or, equivalently, the Stark shift variations of the ground state. These are found to be effective down to temperatures of the order of several E_R/k, where E_R is the recoil energy of a single photon and k is the Boltzmann constant [37], [186]. See also Metcalf and van der Straten's book [135], where they have a detailed discussion of sub-Doppler cooling.

In extensive measurements of the minimum temperature of cesium molasses in 1990, Salomon *et al.* [188] observed temperatures as low as $\sim 3\,\mu$K, which is a factor of 40 below the Doppler temperature and a factor of 15 above the recoil temperature. As we saw for sodium atoms, Lett *et al.* [185] cooled to $40\,\mu$K, a factor of ~ 16 times the recoil temperature of $2.4\,\mu$K, and a factor of six times cooler than the originally measured Doppler temperature of $240\,\mu$K [27]*.

This work on sub-Doppler cooling in molasses was the final major achievement of atom manipulation in the 1980–1990 decade. It is satisfying to find that each of the basic force components, the scattering or spontaneous emission force due to the Doppler shift, and the gradient or dipole force due to the optical Stark shift variation, contributes significantly to the cooling process in optical molasses. In the next decade we will consider experiments on sub-recoil cooling of atoms.

The subjects of cold atomic collisions and atom optics were started shortly after the first trapping experiments, at the end of the 1980s. Since the work on these topics extended well into the decade of the 1990s, I will defer discussing these topics until later, when they will be considered in conjunction with the later work of the 1990s.

5.2. Trapping of Biological Particles

We will now show how the subject of trapping and manipulation of biological particles was suddenly discovered in 1986, sixteen years after the first macroscopic particle trapping experiments and after 16 years of work on trying to trap atoms. Since biological trapping was a by-product of this dual effort, it serves as an example of the basic interrelationships of the various sub-fields of the overall subject of laser trapping and manipulation of small neutral particles. As was mentioned above, there was a bad moment in the course of the first atom trapping experiment [4]* when it seemed that tweezer trapping was not going to work. This greatly worried me at the time. If the single-beam gradient or tweezer trap failed, it would be a second failure for one of my traps (after the failure of the alternating beam trap) and, more importantly, a failure in understanding.

John Bjorkholm, Steve Chu, and Alex Cable were all working away on the atom trapping apparatus, while I was just worrying. I decided at this point to try a simpler experiment, the trapping of submicrometer Rayleigh particles in my lab., with the help of Joe Dziedzic. Tweezers should work more easily on macroscopic particles than on atoms. The theory was much the same. One merely had to substitute the polarizability of a submicrometer Rayleigh particle for that of an atom in the basic formula $F_{grad} = F_{dip} = 1/2\alpha\nabla E^2$. Figure 5.11 gives the theoretical expressions for the gradient and scattering forces on submicrometer Rayleigh particles.

Forces on Submicrometer Rayleigh Particles

Gradient force:

Polarization vector: $\quad \vec{P} = \alpha \vec{E} \quad$ where α = polarizability

$$\vec{E} = \text{optical electric field}$$

$$\vec{F}_{grad} = (\vec{P} \cdot \vec{\nabla})\vec{E} = \tfrac{1}{2}\alpha \overrightarrow{\nabla E^2}$$

In free space: $\quad \alpha = \left(\dfrac{n^2 - 1}{n^2 + 2}\right)r^3 \qquad$ where $\quad r$ = radius

$\qquad\qquad\qquad\qquad\qquad\qquad\qquad\qquad\quad n$ = refractive index of particle

Within a dielectric medium:

$$\alpha = n_b\left(\frac{m^2 - 1}{m^2 + 2}\right)r^3 \qquad \text{where } m = \frac{n_a \;\; \text{(index of sphere)}}{n_b \;\; \text{(index of medium)}}$$

In general: $\qquad \vec{F}_{grad} = \tfrac{1}{2}\,n_b\left(\dfrac{m^2 - 1}{m^2 + 2}\right)\overrightarrow{\nabla E^2}$

Scattering force:

In free space: $\quad F_{scat} = \dfrac{P_{scat}}{c} \qquad$ where P = Rayleigh scattered power

i.e. $\quad \vec{F}_{scat} = \dfrac{\vec{I}_0}{c}\,\dfrac{128\,\pi^5 r^6}{3\lambda^4}\left(\dfrac{n^2 - 1}{n^2 + 2}\right)^2 \qquad$ where $\vec{I}_0$ = incident beam intensity

Within a dielectric medium of index n_b:

In general: $\quad \vec{F}_{scat} = \dfrac{\vec{I}_0}{c}\,\dfrac{128\,\pi^5 r^6}{3\,\lambda^4}\,n_b\left(\dfrac{m^2 - 1}{m^2 + 2}\right)^2$

Fig. 5.11. Theoretical expressions for the gradient and scattering forces of radiation pressure on submicron Rayleigh particles.

5.2.1. *Artificial nonlinear media*

As background for our discussion of trapping of biological particles, I must mention some work Peter Smith and I did on what we called "artificial nonlinear optical media". This work had just been completed in the period from 1981 to 1984 [165], [166]*, [167], [168]. In these experiments we observed very beautiful four-wave-mixing and self-focusing effects in liquid suspensions of submicrometer silica spheres, which acted as a new type of high-nonlinearity optical Kerr medium. In a Kerr nonlinear medium, the index of refraction of the medium becomes intensity dependent. For the case of a liquid suspension of high-index Rayleigh particles, the particles are physically pulled into regions of high light intensity by the gradient force. This increases the local density of particles and the average index of refraction. Since this new nonlinearity is based on the physical motion of the particles, it responds relatively slowly, compared with standard electronic nonlinearities. On the other hand, the magnitude of this new nonlinearity is about five orders of magnitude greater. This has made possible some quite remarkable optical self-focusing or soliton behavior [166]*. Figure 5.12 shows soliton behavior for the TEM_{01} mode in such a medium. The soliton community apparently

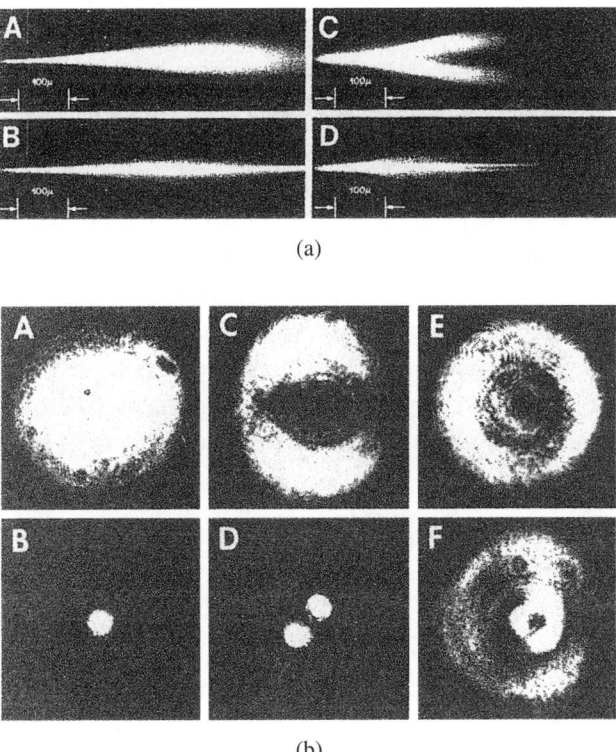

(a)

(b)

Fig. 5.12. (a) Beam-trajectory photographs of TEM_{00}- and TEM_{01}-modes in the nonlinear medium as seen in side scattering. A and C show normal diffractive spread at low power. B and D are taken above the critical power and show formation of self-trapped filaments. (b) Beam shapes at the output face of the nonlinear medium for TEM_{00}, TEM_{01}, and TEM_{01}^* modes. A, C, and E are taken at low power, where diffraction controls the spot size. B, D, and F are taken above the critical power and show the small, highly intense self-trapped filaments. Figures 2 and 3, and their captions from Ref. [166]*.

has overlooked these results in recent years. See reviews by Stegeman [189]. These earlier nonlinear optical experiments involved two-dimensional (2D) trapping of submicrometer particles and not the stable 3D trapping of a true optical trap. As we shall see later, this nonlinear behavior, based on the motion of individual particles, is quite relevant to the recently observed nonlinear behavior of Bose–Einstein condensates [190]*.

5.2.2. *Trapping of submicrometer Rayleigh particles*

The successful experiments on artificial nonlinearities bolstered our confidence in our understanding of the basic light forces on submicrometer Rayleigh particles and raised our hopes for the success of the planned tweezer trapping experiment on these particles. Figure 5.13 shows the new tweezer trapping apparatus we built in 1986 to trap submicrometer Rayleigh particles [5]*. It was very similar to the one used in the ongoing atom trapping experiment. A highly convergent TEM$_{00}$ mode Gaussian beam from a 5145 Å argon ion laser was focused into a glass cell filled with a dilute suspension of fairly mono-dispersed submicrometer colloidal silica particles. A low-power telescope looking from the side could be used for direct viewing of the beam or one could project an enlarged image of the focal region of the tweezer beam onto a screen for easy viewing or measurement with a photodetector in a darkened room. We could easily see the illuminated region because of the Rayleigh scattering from the particles. The Brownian motion of these submicrometer particles was particularly beautiful. Whenever a particle diffused close to the beam focus, it would quickly brighten up as it was pulled into the focus and trapped. Particle after particle was trapped in this way as the focal region became ever brighter. The trap worked instantaneously from the very first. We were able to capture large numbers of particles at a rate depending on the density of the colloid in the liquid suspension. Although silica is quite dense, one does not see any settling out of the particles, due to the large Brownian motion. This geometry with the diffusing particles and fixed tweezer trap is almost an exact analog of the atom tweezer trap in molasses and is a particularly graphic example of loading a small-volume tweezer trap by diffusion. The minimum size particle we were able to trap was as small as ~ 250 Å. The size of the trapped particle could be determined readily by comparing its Rayleigh scattered power with that from a reference polystyrene latex sphere of known diameter ($\sim 0.1\,\mu$m).

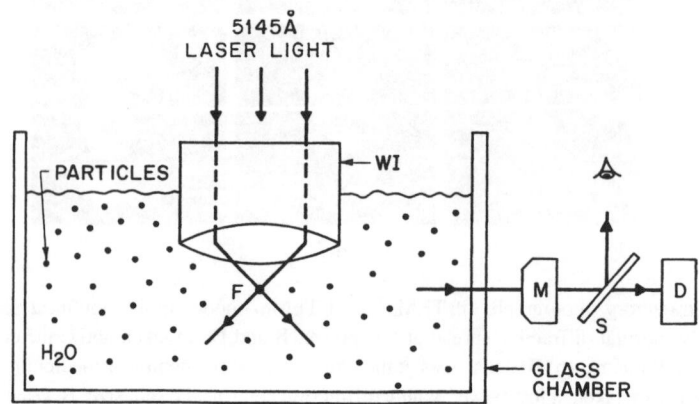

Fig. 5.13. Diagram of the first tweezer trapping apparatus for submicron Rayleigh particles.

An interesting aspect of these tweezer experiments was the observation that the high intensity light of the strongly focused argon laser beam trap apparently caused a form of optical damage to the small latex spheres [5]*. For example, our test latex spheres of diameter $\cong 0.109\,\mu$m shrunk in size (as judged by a drop in intensity of the Rayleigh scattered light) and disappeared in about a half minute. This was still long enough to give a good measurement of the Rayleigh scattered light. Larger-sized spheres of $0.173\,\mu$m survived minutes before being lost. The silica spheres were very much more rugged and smaller particles suffered only small shrinkage over longer times [5]*. This damage, which we surmised was photochemical in nature, made it clear that light of these intensities need not be benign. However, see Ref. [44]*, where we found how to stop the photodamage to small silica colloidal particles. The success of this experiment on trapping small Rayleigh particles [5]* boded well for our atom trapping experiment [4]*. Indeed, the first observation of tweezer trapping of atoms occurred shortly thereafter, as described above.

5.2.3. *Tweezer trapping of micrometer-sized dielectric spheres*

Having seen successful trapping of Rayleigh particles by tweezers, it was natural to look for tweezer trapping of micrometer-sized dielectric spheres. Figure 5.14, taken from Fig. 1(b) of Ref. [5]*, shows the scattered light from a trapped $10\,\mu$m diameter sphere in water, as seen in the red fluorescence of water through a filter blocking the incident argon laser light. The Mie-type interference rings appear as stria in Fig. 5.14. I use the term "Mie-type interference rings" because, strictly speaking, the term "Mie rings" applies only for the case of a plane wave striking a macroscopic sphere $> \lambda$. The highly focused beam used in the tweezer trap is about as different from a plane wave as one can get. Nevertheless, one expects to observe a Mie-like interference pattern due to interference of adjacent rays following slightly different paths through the sphere. The usual ray optic calculation of force on a sphere does not take into account the presence of these rings. However, exclusion of these rings causes no significant error in the total force calculation. Figure 5.15, taken from Refs. [5]*,

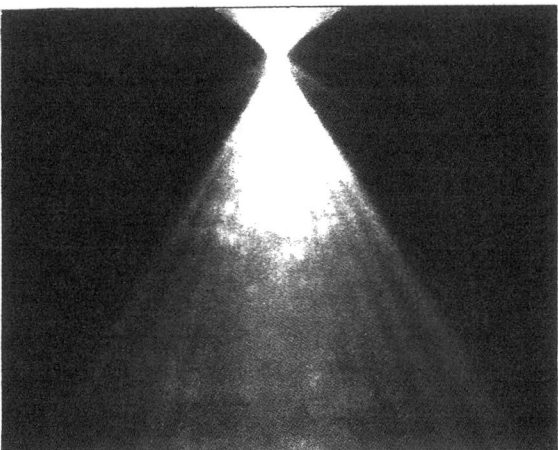

Fig. 5.14. Photograph, taken in red fluorescence, of the tweezer trapping of a $10\,\mu$m sphere in water. The paths of the incident and scattered light rays are seen. Evidence of Mie-type interference fringes is discernable as stria in the scattered light.

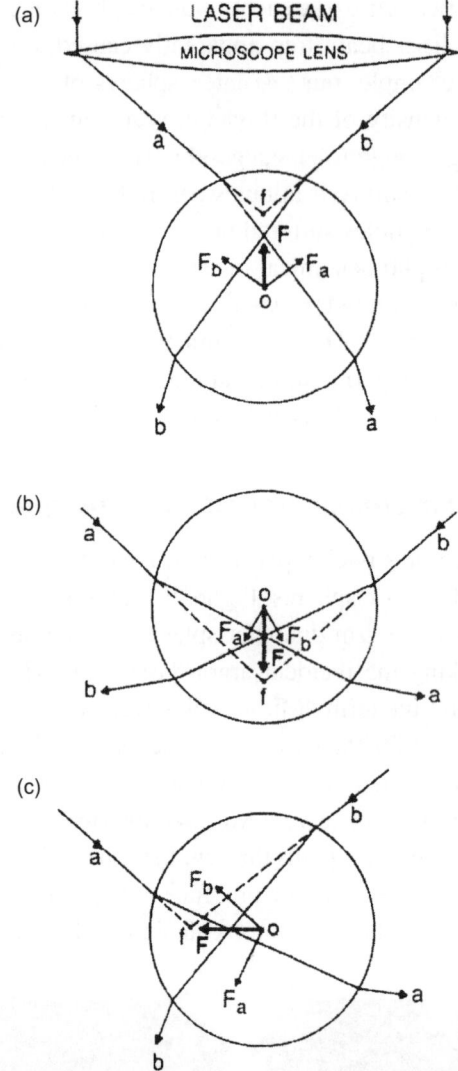

Fig. 5.15. Ray optic explanation of the stability of a tweezer trap for spherical particles large compared to λ, for different displacements of the sphere relative to the beam focus at *f*. Note the origin of the so-called "backward radiation pressure" in (a).

[49]*, gives a simple ray-optics or geometrical optics explanation of the stability of the tweezer trap for spherical particles large compared to λ. One sees that, for sufficiently high beam convergence angles, one expects a strong backward gradient component of the radiation pressure force, just as found experimentally for submicrometer Rayleigh particles. Although a backward radiation pressure force for a single beam hitting a particle seems impossible at first sight, it is not unphysical and is a direct consequence of the ray-optics picture as explained in Ref. [1]* and in the introduction to this book. The principal advantage of the tweezer trap is its easy maneuverability and its considerable well depth. Compared to a levitation trap, where the depth is always equal to mg, the tweezer trap

can have depths as high as 10^6 mg. As pointed out earlier, the levitation trap is a hybrid optical trap that depends on gravity, whereas the tweezer trap is an all-optical trap and independent of gravity.

5.2.4. *Optical trapping and manipulation of viruses and bacteria*

The year 1987, in a real sense, marks the beginning of the use of optical trapping and manipulation techniques in biology. Even though over the years I had often included a remark about the possible use of laser trapping in biology in many of my papers [1]*, [109]*, I never really believed it would happen except under unusual circumstances, and I never acted on it. I thought the damage would be a very severe problem. None of this was on my mind, however, when Joe Dziedzic and I started working on trapping tobacco mosaic virus (TMV) in early 1987 [44]*. I obtained our samples of this rod-like virus from Joe Zasadzinski of our Murray Hill Lab., after reading in progress reports about his X-ray studies on the spontaneous ordering of dense colloids of TMV in water [191]. Since TMV is a single molecule, it should be perfectly mono-disperse. It has a rod-like shape and is of a size we could easily trap. The dimensions of TMV are shown in Fig. 5.16. It does not qualify as a Rayleigh particle, because of its long dimension of ~ 3200 Å. At first I did not think of TMV as live entities. After all, they are just chunks of protein with a high index of refraction $n \cong 1.57$ that can crystallize like other inanimate particles. Joe Zasadzinski said that they only lasted a finite time before they began to deteriorate. He disposed of them by pouring them down the drain. He was happy to send his old samples to me and I could later pour them down my drain. I wondered about infecting all the tobacco plants in New Jersey, but he assured me there were no tobacco plants in this state to be infected.

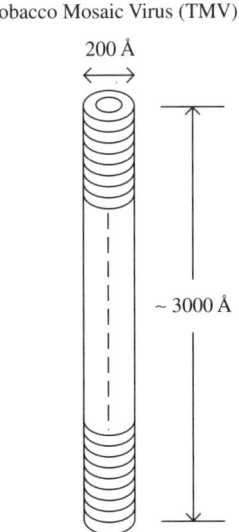

Fig. 5.16. Dimensions of the tobacco mosaic virus (TMV).

5.2.5. *Optical alignment of tobacco mosaic viruses*

We placed TMV in the same apparatus as shown in Fig. 5.13. It trapped very solidly, using about 100–300 mW. The scattered light from the TMV particles undergoing Brownian motion was unusual in that it flickered in intensity as the viruses jiggled about in space. This was due to the varying polarizability of the virus as it tumbled in rotational Brownian motion. There is about an order of magnitude difference in polarizability between the two extremes of polarizability, depending on whether the electric field of the light is perpendicular or parallel to the cylindrical axis of the virus. Once the TMV particles were captured in the trap, the flickering stopped and the scattering became very bright. Apparently, the trap stopped both the spatial and rotational Brownian motions. This suggested that the TMV was totally aligned by the optical electric field. The partial alignment of molecules of CS_2, for example, is the origin of its nonlinear optical Kerr coefficient. This is the first example of 100% alignment in an optical field that I am aware of. We were observing scattering in a direction perpendicular to the electric field E. If E is fully aligned along the long axis of the TMV, then it radiates perpendicular to the axis of the virus to the far field with the entire molecule in-phase, just as if it were a Rayleigh particle. With this assumption we were able to compute the length of the virus, based on a comparison of its scattering power to that of a standard $0.109\,\mu$m polystyrene latex reference Rayleigh particle. This method of length measurement gave $L = 3100 \pm 700\,\text{Å}$, in good agreement with expectations. This confirms that the molecular axis is fully aligned with E and the far-field radiation to the detector is in phase.

5.2.6. *Fixed particle arrays of tobacco mosaic viruses*

Another interesting observation was that the scattered light was constant on the scale of tens of minutes, indicating that TMV was quite damage free in $5145\,\text{Å}$ light at these laser intensities. It was, in fact, more rugged than silica particles in this regard (see Ref. [44]*). We finally did succeed at a later time in getting silica to be damage-free by either adding potassium silicate or by adjusting the pH of the surrounding solution. This all points to photochemical damage at the silica–water interface as the source of the damage for silica [44]*.

There was a puzzling aspect to the data that Joe and I noticed almost immediately. It had to do with the magnitude of the far-field Rayleigh scattering as successive particles entered the trap. The scattering usually increased, but not in uniform steps. Also, one occasionally saw the total scattered intensity drop as a new particle entered the trap. Was this a case of a new particle ejecting an old particle, or what? If a second particle the same size as the first entered the trap and simply fused with it to form a new particle of twice the volume, we would expect the scattering to increase proportional to V^2, or a factor of four times larger, according to the Rayleigh formula. A third particle fusing should give nine times the scattering. This did not happen for TMV or the silica or latex spheres of Ref. [44]*. These were all charged colloids that should repel one another. They must then form into a cloud with a fixed orientation governed by the equilibrium of the repulsive forces and optical forces of the tweezer trap. The total scattered signal for a cloud should then be the result of the interference of the scattered fields from the individual particles. To get a better picture of what was actually happening, we added a second far-field detector located at a different angle with respect to the cloud of particles, as shown in Fig. 5.17. To our surprise we found that the intensity changes seen

Multiple Particles — Fixed Array

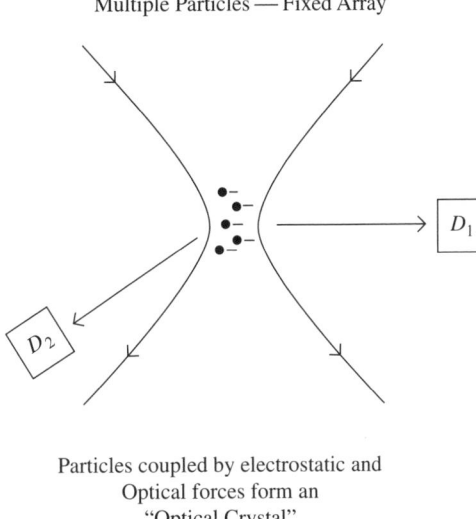

Particles coupled by electrostatic and
Optical forces form an
"Optical Crystal"

Fig. 5.17. Observation of scattering from multiple particles arranged in fixed arrays in the trap using a pair of detectors D_1 and D_2.

at the two angles, as new particles entered the trap, were very different. We even saw cases where the entry of a new particle caused the total intensity to drop in one detector and rise in another! This made it clear that a drop in intensity did not indicate particles were being ejected from the trap, but rather the lower signal was the result of destructive interference in one direction and constructive interference in the other direction. The particles evidently were forming fixed arrays coupled by a combination of electrostatic and optical forces. The addition of a new particle could conceivably cause a total rearrangement of the entire cloud. This was very reminiscent of the "optical crystals" we had seen with levitated liquid drops [68]*, only now we were using TMV particles in a tight tweezers trap. We found that the same type of particle arrays could be observed with submicrometer, latex, or silica Rayleigh spheres.

These measurements suggest that it is possible to deduce the relative positions of a pair of identical trapped TMV or other Rayleigh particles by measuring the detailed far-field angular distribution of scattered light. This would be a complete solution to the inverse light scattering problem of deducing the geometry of the scattering particles from measurement of the scattered light. Tweezer traps are great tools for studying the light scattering because, starting with two separate traps and two particles, for example, one can study each particle individually and then combine them in known ways. One can advance then to more particles. One does not necessarily have to look at the scattering from the trapping light beams, but can alternatively use separate low power plane wave scattering beams of a different wavelength, for instance, to simplify the process.

5.2.7. *Tweezer trapping of bacteria and "opticution"*

It was during our studies on trapping and scattering of TMV in 1987 that we made a truly serendipitous discovery [44]*. Since the various batches of TMV were often quite variable (some had few intact

Strange New Particles

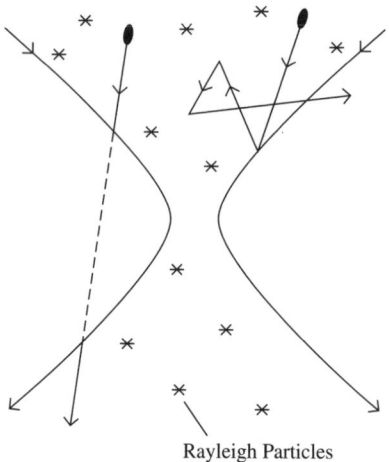

Rayleigh Particles

Fig. 5.18. The beam focal region as seen using the side-viewing microscope of Fig. 5.13. Strange new particles appeared with weird trajectories traversing the light beam, as shown.

viruses), we tended to use our good samples for longer periods of time than usual. We also left our samples open to the air, as shown in Fig. 5.13. When we looked into the sample with our side-viewing microscope, we occasionally saw some very strange moving particles. They were much larger than the TMV particles and moved about quite erratically (see Fig. 5.18). When they came to a boundary between light and dark, they would sometimes reverse their motion. They were not driven by light, as were runners and bouncers [48] since they were able to follow trajectories that took them through regions of darkness with no loss of propulsion. They could stop and then start moving in other directions. The number of these strange particles grew with time. If left overnight, there were many more present by morning. We even had a few move right into the beam focus, where they became trapped. We saw some really wild scattering as they seemed to tumble and twist. If we turned off the trap momentarily and turned it on again, they were gone. Joe and I started to call these particles "bugs". And they were bugs: live bacteria. The open sample had become contaminated from the air and the bugs apparently found enough nourishment in the solution to multiply. Joe searched for them using a medium power microscope and managed to catch glimpses of them as they swam about at remarkably high speeds, up to $\sim 500\,\mu\mathrm{m/s}$. We decided to flood a large volume of the sample near the tweezer focus with a red laser beam, to give more visibility. We could then clearly see the bugs swimming about and even enter into the trap. There was little doubt we had motile bacteria. If a bug was caught in the trap with only about 5 or 10 mW, it could stay alive for minutes (see Fig. 5.19). If we turned up the power to 100 mW, we got a brief burst of bright scattering and then almost nothing, just a small remnant of steady low-level scattering. We concluded this was a case of "optication", or death by light. The bright green 5745 Å light caused the bug to explode, spew forth its contents and leave an inert carcass in the tweezer trap. This explanation of the data seemed fairly certain.

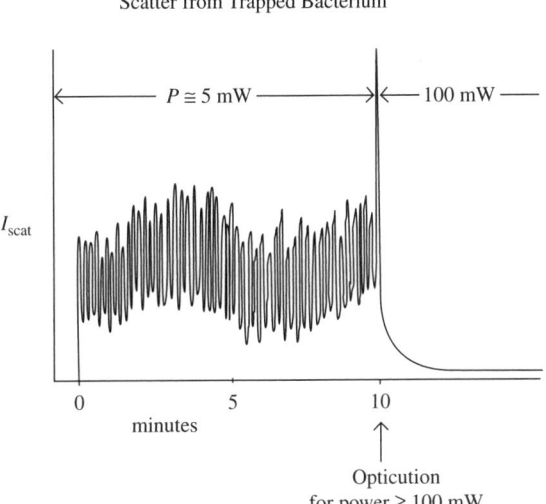

Scatter from Trapped Bacterium

Fig. 5.19. Diagram of the side-scattered light signal from a trapped bacterium at a low power of ~ 5 mW, and its subsequent "optication" as the power was raised to ~ 100 mW.

5.2.8. *Tweezer trapping of bacteria in a high-resolution microscope*

To be completely certain, we decided to purposely grow our own bacteria and examine them right in a high-resolution Zeiss universal microscope. We introduced our green trapping beam through the camera port of the microscope and focused the high numerical aperture tweezer trapping beam right into the viewing plane of the sample, as shown in Fig. 5.20. With a laser-blocking filter in front of the eyepiece we could observe the trapping of the bacteria by eye or with a video camera with the high resolution and convenience of a high-power microscope. The tweezer trap and microscope are completely compatible. To get our own bacteria, we simply grew them overnight from a piece of Joe's ham sandwich added to some water. This gave us a plethora of strange bugs, which had to be diluted down to make reasonable samples. We confirmed all of our low-power observations on trapping and optication. We found two modes of trapping. One was the usual 3D trapping, where we manipulated bacteria trapped at the focus freely in space, and the other in which the bugs were impaled against the bottom surface of the sample cell. This was observed by moving the focus below the bottom surface. This was a form of 2D optical trapping, combined with a 1D mechanical wall force trapping.

Of course one does not need a high numerical aperture beam to impale particles against a surface. This type of 2D trap was first observed by chance at the cell wall in the very first trapping experiment [1]*. Some early trappers in biology inadvertently used this type of wall trap before they realized how high to make their high numerical aperture microscope objectives to achieve a true 3D all-optical tweezer trap based on the "backward radiation pressure" force components. This was the case for the early experiments on sperm trapping [192].

Another interesting observation we made was that some of the motile bacteria we grew would occasionally attach themselves to the bottom surface of the sample by their flagella and then rotate

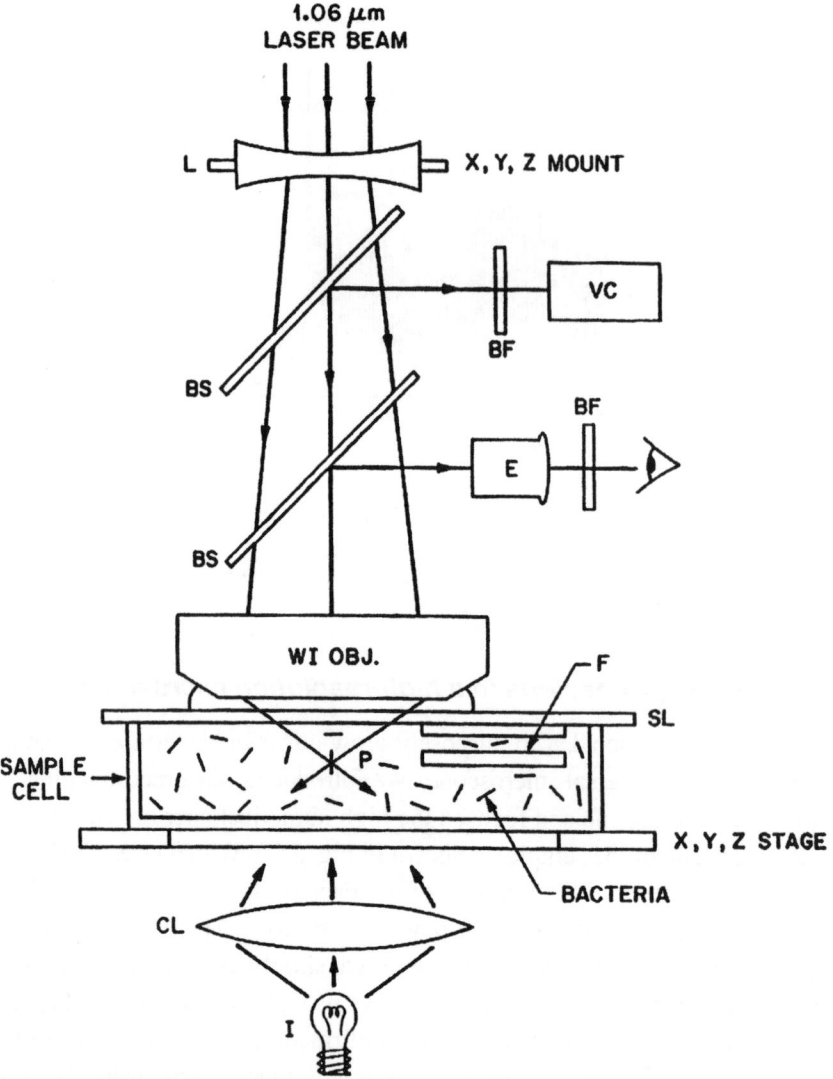

Fig. 5.20. Experimental setup for introducing a laser beam into a high-powered microscope for the simultaneous trapping and viewing of motile bacteria.

about their tethers. Under these circumstances, it was still possible to optically manipulate the bacterium in a circuit about its tether and observe the action of the optical trapping forces on the flagellum (see Fig. 5.21). This motion was achieved using the microscope stage to drag the bacterium around its tether along a square path. These early tweezer experiments on optical trapping and manipulation of viruses and bacteria in 1987 marked our accidental entry into the field of biology. As we shall see, Block *et al.* [193] subsequently used a tethered bacterium to measure the torsional compliance of the flagellum. Steve Block was one of the early pioneers in the use of tweezers in biology. He told me that he and Howard Berg (two experts on the motion of bacteria) read our paper [44]* in 1987 and decided that they would not use tweezer traps in their work because of the optical

Manipulation of Tethered Bacterium

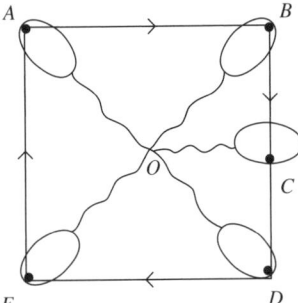

Fig. 5.21. Using tweezers to drag a tethered bacterium about a point O along a square path by manipulating the microscope stage around the path A, B, C, D, E, A.

damage problem. This was fortunate for us, because it gave us time to consider the problem without the competition of professional biologists. With regard to professional biologists, when our first version of the paper on trapping viruses and bacteria was sent to the biophysics area of Bell Labs. for internal review, the reviewer was not overly enthusiastic about the work, but did make a very helpful suggestion. He said that professional biologists would not like to read about bacteria grown from a ham sandwich. They would want to know the exact type of bacteria we used. He suggested we contact Tets Yamani in the biophysics area, who grew *E. coli* bacteria and who might give us some to experiment with. Yamani was very enthusiastic about the project. He taught Joe and me about the care and feeding of his *E. coli*. He explained that, sadly, many modern biologists were so focused on molecular biology that they barely knew anything about whole live microorganisms. In the final version of our paper [44]*, thanks to Yamani, we included work on *E. coli* optical damage using the 5145 Å line of the argon-ion laser.

5.2.9. *Optical tweezers using infrared light from a Nd:YAG laser*

Using infrared optical tweezers, one can perform the experiments described below under damage free conditions.

5.2.9.1. *Damage-free trapping of living cells*

Our hope for avoiding optical damage was to find some other laser wavelength where the damage would be less. In 1987, on a hunch, we tried using the 1.06μm line of the Nd:YAG laser [20]*. If one looks at the absorption coefficient of important organic molecules, such as hemoglobin (Hb) and oxyhemoglobin (HbO$_2$) one sees that there is a large drop as one moves to longer wavelengths in the infrared region. Similar behavior also occurs for chlorophyll (not shown here). The absorption of water, on the other hand, as seen in Fig. 5.22, increases as one goes to the infrared. At 1.06μm the water absorption is not too severe. So we settled on the 1.06μm YAG laser. Joe rebuilt one of the original Bell Labs. homemade YAG lasers for this purpose. The use of YAG laser light immediately gave a large improvement. We could hold a single *E. coli* in the trap with 50 mW of power for long

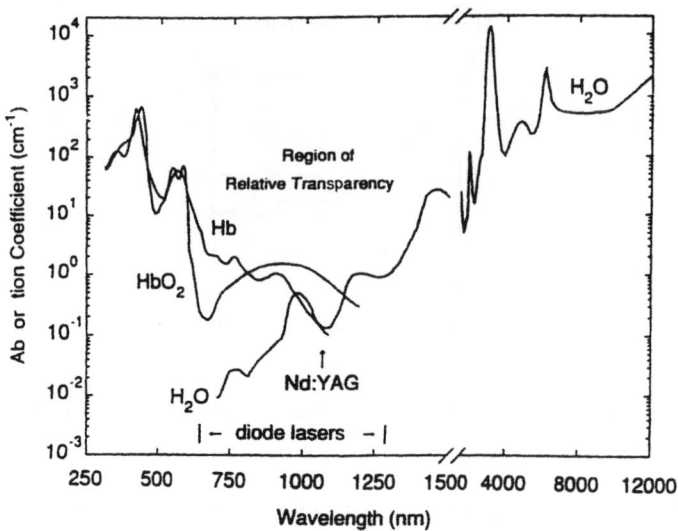

Fig. 5.22. Plot of the optical absorption coefficients of hemoglobin (Hb), oxyhemoglobin (HbO$_2$) and water versus the wavelength.

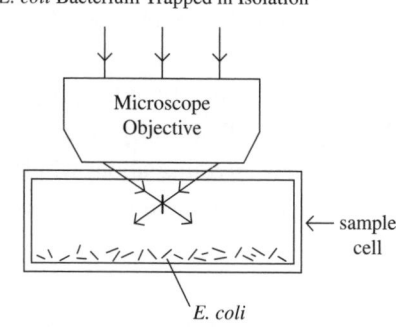

Fig. 5.23. Single *E. coli* bacterium held in a tweezer trap in isolation from other untrapped bacteria.

periods of time, with no evidence of damage. Dziedzic and I decided to cook a single *E. coli* and see how long it took to degrade. *E. coli* are motile bacteria, but the type we used did not move that fast, so we were able to lift a single bacterium higher in the sample cell, in complete isolation from the other bacteria (see Fig. 5.23). After an hour or so we released it for a good look and found, if anything, that it had grown somewhat. After a time, to our surprise and delight, the single bacterium turned into two. We put the two into the trap and waited an hour or so, and saw the two turn into four. Clearly there was no optical damage occurring if the *E. coli* bacteria could reproduce right in the high-intensity focus of the trap! Yamani suggested we also try yeast cells, which he was growing. He provided all the cells and nutrients and instructed us to trap single cells initially, because the clumps that grow up are really fairly independent and if trapping one of a clump caused damage, the others might still be able to grow by a process called budding. These were beautifully transparent ellipsoidal lens-like cells. They trapped easily and quickly grew into clumps of four or five cells that we could

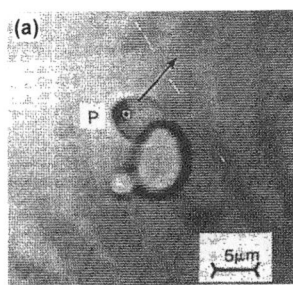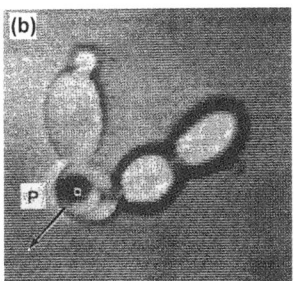

Fig. 5.24. Photograph of a trapped clump of yeast cells as they are being manipulated by infrared optical tweezers. The location of the trap is indicated.

drag around with the trap, even though the trap focus was localized to a small fraction of a single cell (see Fig. 5.24). Although one cannot see the beam focus when operating in the infrared, one can locate the focal region on the face of the video monitor by initially trapping a small macroscopic reference sphere and marking the spot with a dot or a cross.

Once we realized that it was possible to greatly reduce the damage to bacteria and yeast cells, Joe Dziedzic and I decided to survey as many types of cells as we could and look for damage. We tried red blood cells, plant cells, and the huge number of different types of protozoa, diatoms, and single cells of algae one can find in pond water. We obtained cell samples from the old farm pond on the Bell Labs. Holmdel property and from the surrounding streams and golf course ponds. It was a pleasant spring and summer for Joe Dziedzic and me. Not only were the cell types quite varied, but also their sizes and shapes. Shape and optical properties of particles are crucial to the trapping process. It became abundantly clear that tweezer-type traps are much more tolerant of shape variation than levitation-type traps. One can trap almost every type of cell. Only a few of the cells and biological particles that we studied were reported in the Nature report on infrared tweezer traps [20]*. Figure 5.25 gives some examples of different shapes that are easily trapped by tweezers. Spheres, spheroids and ellipsoids are also trappable by levitation traps [156].

In general, there was greatly reduced damage using the YAG laser light, compared with argon laser light at 5145 Å. This was particularly evident for colored cells, such as red blood cells and green cells with chlorophyll. Colored cells simply exploded at very low power of 5145 Å light.

5.2.9.2. *Internal cell trapping and manipulation*

With larger cells, such as scallion cells and *spirogyra*, we found later, in 1988 and 1989, that we were able to manipulate small particles within a living cell [19]*. We could collect small micrometer-sized vesicles and other particles and probe the geometry of the chloroplasts and central vacuole by dragging the particles about, moving them deep in the cell from the bottom to the top, under and between cell structures. The tweezer trap is robust enough to tolerate partial shading of the beam by intervening structures. In scallion cells there is strong cytoplasmic streaming occurring along channels. We could trap collections of the streaming particles, release them and watch them resume their journey along the channels (see Fig. 5.26). It was clear that traps could play a big role in studying internal cell processes [19]*, [20]*.

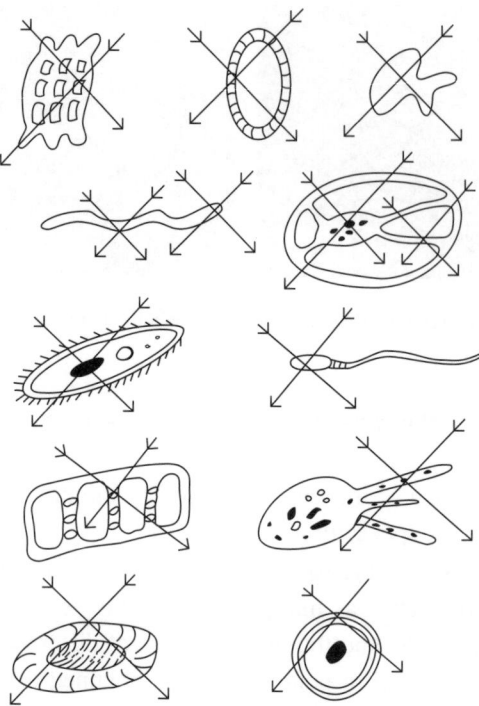

Fig. 5.25. Trappable micrometer-sized dielectric objects of various shapes. Crossed-rays indicate location of forces of tweezer trap. Reading left to right, top to bottom, the objects are: silicon diatom, dielectric ring, amoeba, snake-shaped particle, cell nucleus and support strands, protozoan with food vacuole, sperm cell, plant cell with chloroplasts, growth processes with moving mitochondria, red blood cell, and egg cell.

Cytoplasmic Streaming

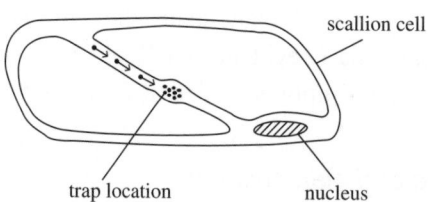

Fig. 5.26. Internal cell manipulation. Collection of particles and a blob of cytoplasm trapped within a streaming channel of cytoplasm inside a living scallion cell. When released, they simply move on.

5.2.9.3. *Separation of bacteria using tweezers*

We demonstrated the ability of tweezer traps to separate a single selected bacterium from a "*gemisch*" or collection of bacteria [20]*. In Fig. 5.20 we show an approximately $15\,\mu$m inner diameter hollow fiber F attached to the top of the sample chamber. We were able to trap one or more bacteria and manipulate them into the core of the hollow fiber. Despite the rather severe optical distortions at the input of the fiber, we could still maneuver the bacteria into the fiber core without losing them. To complete the separation we removed the chamber lid, washed it, rinsed it, and dried it to remove any possible bacteria clinging to the lid, and then with a gentle air stream blew the liquid contents of the fiber into another water filled vessel. The fiber acted as a very convenient storage vessel for the separated bacteria. The ability to separate bacteria and other biological cells is a very simple but important capability of tweezer techniques.

It was also shown that one could introduce two traps into the microscope [20]*. One can then grab a bacterium or cell at its ends and orient it in space or rotate it at will by moving one trap relative to the other. One can stretch a cell, for example, and observe its mechanical properties. A red blood cell is so pliable that simply running a tweezer trap over its surface makes a very noticeable distortion. As big as they are, one can squeeze many blood cells into a single trap. The elastic behavior of the cytoplasm was evident in almost all of the cells we tried.

5.2.9.4. *Elastic properties of the cytoplasm*

We made more careful observations on the elastic properties of the cytoplasm in subsequent work entitled "Internal cell manipulation using infrared traps" [19]*. Using scallion cells we could generate what we called "artificial cytoplasmic filaments" pulled by the trap from the surfaces of most internal cell organelles. Figure 5.27 shows such a filament pulled from the surface of the nucleus of the cell into the central vacuole of the cell. The filament stretches from the original location of the trap at A to its final location at B. If one quickly turns off the trap at B, the filament snaps right back to A. If one waits a minute or so at B, the filament snaps back, but more slowly. The longer one waits at

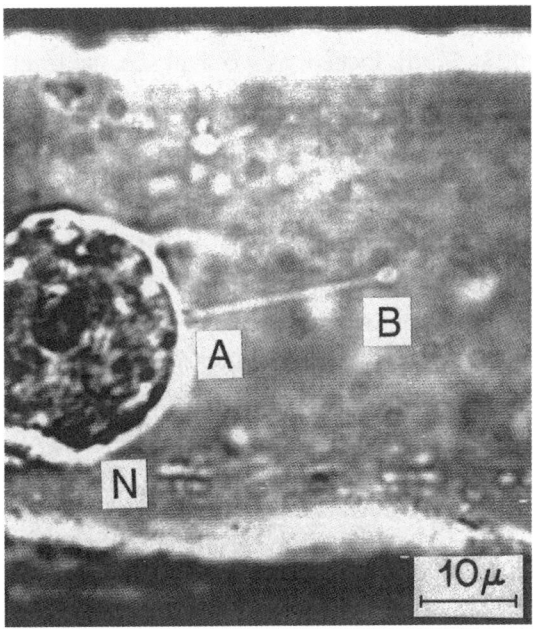

Fig. 5.27. Artificial filament made by manipulating a tweezer trap from point A to point B.

B, the more slowly the filament returns to A. At longer times the filament only partially retracts and one is left with a sagging remnant of cytoplasm. This is classic viscoelastic behavior. Cytoplasm is viscoelastic like the toy "silly putty". For quick distortions it is highly elastic. For slow distortions or long times, the material flows and sags and is relatively weak.

Evidence for viscoelasticity is everywhere within cells, and we used it to perform a new kind of internal cell surgery. If one tries to pull quickly with tweezers on some fairly large structure within a cell, such as the spiral chloroplast of a *spirogyra* cell, it looks quite rigid and it barely moves before it slips out of the trap. If, however, one pulls slowly on the chloroplast and continues to apply force after it moves somewhat, one finds that the tension slowly relaxes and that further motion is possible. This can be repeated and in time one can detach the entire chloroplast right off the cell wall and pull it into the fluid of the central vacuole, where it is quite free. One can thus make gross changes in cell structure. If a detached chloroplast is placed against the side wall of the vacuole, it does not adhere to the inner membrane initially. If one uses the trap to continue to hold it against the wall, its membrane fuses again with the wall membrane, completing the operation. The nucleus of the cell and its cytoplasmic supports can be similarly manipulated. At the time of this tweezer experiment there had been very few experiments showing the viscoelasticity of live cells. In 1950, the physicist Crick, of Watson and Crick fame, tried to study viscoelasticity by getting cells to ingest small magnetic particles, which he proceeded to move with a large electromagnet [194]. After much work, he barely observed any effect, probably due to the fairly rapid relaxation of the tension on the particles. Later a similar experiment with magnetic particles showed a somewhat larger effect [195].

With tweezers we observed more viscoelastic behavior of the cytoplasm within a cell in a few minutes than had been seen from the time of the Crick experiments up to this point. With tweezers we were able to probe the viscoelasticity of the different parts of the living cell at will. Tweezers has also been used recently to explore the viscoelastic behavior of single strands of polymeric molecules, such as DNA, in many contexts [23], [196]. Indeed, as we shall see, the level of sophistication of recent experiments is such that the compliance of motor molecules, or even parts of motor molecules, is an important consideration in the attempt to fully understand the stepping motion of motor molecules [197]. With tweezers one can interact strongly with the cytoplasmic streaming that one sees so readily in scallion and other plant cells. As mentioned above in Fig. 5.26, if the trap is placed in some streaming channels, one can capture the moving particles of the stream and then release them and see them move on. At higher power, one can totally stop the entire stream, particles and all. Subsequent particles are stalled or can even turn around and move backwards. One can probe the relative viscosities of the moving streams and of the other more liquid regions of the cell. By rupturing the cell and manipulating the same particles in pure water, one can estimate the absolute viscosities of the cell contents. This type of information about cells is generally not known [195]. There is clearly much to be learned about the streaming process and many other cellular processes with trapping techniques. Other methods do not provide access to the study of such processes.

The three papers discussed here on damage-free trapping [19]*, [20]*, [44]* in a real sense mark the beginning of the new field of optical trapping in biology.

IV
1990–2006

I previously characterized 1980–1990 as the decade of fulfillment in which the basic concepts and proposals of the previous decade were demonstrated for macroscopic particles and atoms. The work on macroscopic particles had its surprises and led to the discovery of damage-free trapping of biological particles and cells [19]*, [20]*, [44]*. Also, the work on atoms fulfilled the promise of atom trapping and cooling [4]*, [27]*, with the surprising extension of cooling techniques to temperatures well below the Doppler limit [36], [38], [198].

The third and most recent period from 1990 to 2006 can be considered as one in which the applications of atom trapping and cooling blossomed. This period is divided by subject into four parts: IVA, which considers "Biological Applications"; IVB, "Other Recent Applications in Physics and Chemistry"; IVC, "Applications of Atom Trapping and Cooling"; and IVD, "Bose–Einstein Condensation and Related Developments".

The applications to biology in IVA were in large measure based on the realization that laser damage could be minimized and that internal cell manipulation was possible. As will also be seen, many of these applications are based on the simple but virtually unique ability of a tweezer trap to pick up a selected single cell or biological particle and manipulate it freely in space [6], [20]*, [44]*.

Chapter 6 describes the use of tweezers to isolate single motile bacteria and to study the mechanical properties of their flagella, pili, and their rotary motors. Isolated extremophile bacteria were cloned and sequenced in studies of their evolutionary history. They were also used in searches for new types of enzymes.

Single sperm, T-lymphocytes and red blood cells were isolated and studied. The mechanical properties of membranes were studied, using tethers. Light forces distorted whole cells. Tweezers guided neuronal growth and acted as an artificial gravity controlling the direction of growth of plant root cells. A practical test for malarial infection was discovered, based on the observation of self-rotation of infected red blood cells in an optical trap.

Chapter 7 discusses the use of optical tweezers to study single motor molecules, such as dynein and kinesin moving on microtubules, and on actin filaments. In these experiments the "handles technique" of Block and colleagues allowed individual molecules to be optically manipulated by attaching them to micrometer-sized transparent spheres. This work opened a whole new window on the study of motor molecules. Novel measurements could determine the force generated by single motor molecules, their processivity and stepping motion over the cytoskeleton. New feedback techniques, such as force clamps, were developed to study the step-size and single enzyme kinetics of motors under constant force. The remarkable force sensitivity (to a small fraction of a pN) and positional sensitivity (to a small fraction of a nanometer) of the optical trapping microscope make it an ideal tool for studying the properties of single biological molecules.

Chapters 8 and 9 describe the application of tweezer traps to the study of RNA and DNA and the many mechanoenzymes that interact with them. The tweezer trapping techniques provide new ways of studying these important molecules. The forces generated and the velocity of these molecules were deduced using force clamp methods. Other important results are discussed in these sections: force-extension measurements of "unprecedented accuracy" of double-stranded and single-stranded DNA; measurements of the persistence length under different conditions, using controlled stretching forces;

the response of DNA to applied torque and its behavior as a rotary motor; and the elastic response of RecA-DNA filaments and DNA-chromatin fibers to feedback-controlled stretching forces.

The sophistication of the optical trapping microscope has reached a level where detailed measurements of transcriptional processes in RNA and DNA are possible. Experiments showed backtracking and transcriptional pausing of RNA polymerase molecules. Thus, tweezers provides a new way to study transcription and the error correcting processes on DNA samples with known base-pair sequences.

In Chapter 10 Part IVB we consider the origin of the net backward force in tweezer traps. Chapter 11 discusses the role of tweezers in studying the pair interaction forces in charge-stabilized colloids. Chapter 12 shows how to make a variety of optically driven motors, using the linear momentum of light, orbital angular momentum of light and intrinsic angular momentum. Chapter 13 considers a number of applications of tweezers to microchemistry. Various interesting chemical reactions are studied that only occur in single microdroplets. Chapter 14 describes a number of newly devised applications to nanotechnology based on holographically produced arrays of up to ~ 1000 optical tweezer traps. Some remarkable fluidic particle and cell sorting devices were made using static and dynamic arrays of tweezer traps. Other ingenious devices have been fabricated for use in microfluidic analytic systems, such as optically driven gear pumps, peristaltic screw pumps, and various micro-valve types.

Chapter 15 of Part IVC describes work on atomic fountains [199] and their use in novel high-precision atomic clocks [36], [200].

New types of atom interferometers made possible improvements in the determination of fundamental constants and new atom-based instruments for the measurement of gravity [201]. The history of the subject of atom optics and magnetic guiding was also discussed. Work on atomic collisions with ultra-cold atoms has proliferated, giving new insights into molecular structures and loss processes of atoms in traps [202].

In Chapter 16 of Part IVD we discuss the very surprising achievement of Bose–Einstein condensation in atomic vapors in 1995, using evaporative cooling to ultralow temperatures [203], [204]. This was followed by the demonstration of various forms of pulsed atom lasers (see Ref. [203] and Sec. 16.10). There even have been proposals for cw lasers (see Sec. 20.19). Bose–Einstein condensates, in which essentially all atoms are in a single coherent atomic state, provide atomic physicists with an ideal tool for studying quantum and many-body physics.

The importance and possibility of BEC in atomic vapors had been recognized for some time [172], [206], [207], [208], but the expectation was that it would be very difficult to realize in the near term. A Nobel Prize was awarded in 2001 for this work, just a short six years after the first observation of Bose–Einstein condensation. Interest in this new state of matter continues to grow rapidly. See especially Chapters 17–20 and 23.

Chapter 17 describes experiments that led to the eventual revival of interest in all-optical traps in the mid to late 1990s. This in turn led to the first experimental demonstration of spinor condensates (Chapter 18) and Feshbach resonances (Chapter 19) using optical dipole traps, in 1998.

Chapter 20 considers more recent work on BECs from around 2000 to 2006. One sees the increased use of optical dipole traps in the later years.

Indeed, in Chapters 21–23 we see more applications of BEC based on optical trapping. Chapter 21 describes experiments with single atoms in cavity quantum electrodynamics (QED) using single atoms trapped in single high-finesse optical cavities. Chapter 22 describes experimental techniques for trapping of single atoms in optical dipole traps for use in QED experiments and quantum computers. Chapter 23 discusses observations of vortices and superfluidity in BE condensates.

Chapter 24 summarizes recent experiments, extending trapping and manipulation techniques to small molecules. Methods for cooling small molecules and creating molecules from atoms in BE condensates are described.

Having achieved BEC of bosonic atoms, a new exciting challenge arose, the race to cool fermionic atoms to Fermi degeneracy and the possible observation of superfluidity and Cooper pairing. The pace quickened after the first successful cooling of ^{40}K to degeneracy in 1999 (Sec. 25.6). This goal is generally agreed to be more difficult to reach than BEC of bosons. Progress has been surprisingly rapid, as can be seen in Chapter 25. In large measure this can be ascribed to the revived use of off-resonance optical dipole traps. Observations of long-lived molecular condensates of fermions and Cooper pairing of strongly interacting fermions have been made using Feshbach resonances and optical dipole traps. A total of seven different experimental signatures of strong pairing have been observed, but weakly coupled Cooper pairs have yet to be observed.

IVA
Biological Applications

CHAPTER

6

General Biological Applications

6.1. Application of Tweezers to the Study of Bacteria

6.1.1. *Bacteria flagella and bacterial motors*

One of the earliest applications of tweezers to biology was the study of bacterial flagella, which are responsible for most bacterial motility. In their early experiments on trapping of bacteria, Ashkin and Dziedzic [44]* observed the ability of tweezers to manipulate bacteria that had become tethered or attached to the glass surfaces of a chamber by their flagella. They found they could capture the body of the bacterium with a tweezer trap and drag it clockwise or counter clockwise about its tether as is shown in Fig. 5.21. This motion causes the flagellum to twist. Subsequently Block *et al.* [193] using a similar technique measured the torsional compliance of a single flagellum of a tethered *E. coli* bacterium that had its rotary motors "locked up" by treatment with protonophores or by fixation with glutaraldehyde. They found that the compliance of the flagellum exhibited a soft phase up to about 180° rotation followed by a torsionally rigid phase for higher rotation angles. The complete flagellum of an *E. coli* bacterium, for instance, consists of a motor region located in the bacterial membrane, followed by a short flexible hook region and then a long thin filament. See Fig. 6.1 for an electron micrograph and detailed diagram of an *E. coli* bacterium flagellum, taken from Berg's review article on "Motile Behavior of Bacteria" in Physics Today [222]. Block, Blair, and Berg [193] suggested that the soft compliance was essentially all in the hook region and the rigid behavior was due to the stiff filament. The authors confirmed this in 1991 [223] by applying optical torques to a mutant strain of *E. coli* with an abnormally long hook region, but no filaments. The measured torsional compliance was the same as previously found with the wild type of bacteria. This confirms the notion that the soft hook compliance acts as a flexible universal joint that makes it possible for the filaments to join in bundles while driving the cell by clockwise rotation. In counterclockwise rotation the filaments fly apart and the cell tumbles [222]. In other experiments Charon *et al.* [224] used optical tweezers in an experiment in which they immobilized spirochete bacteria and established that their periplactic flagella rotate.

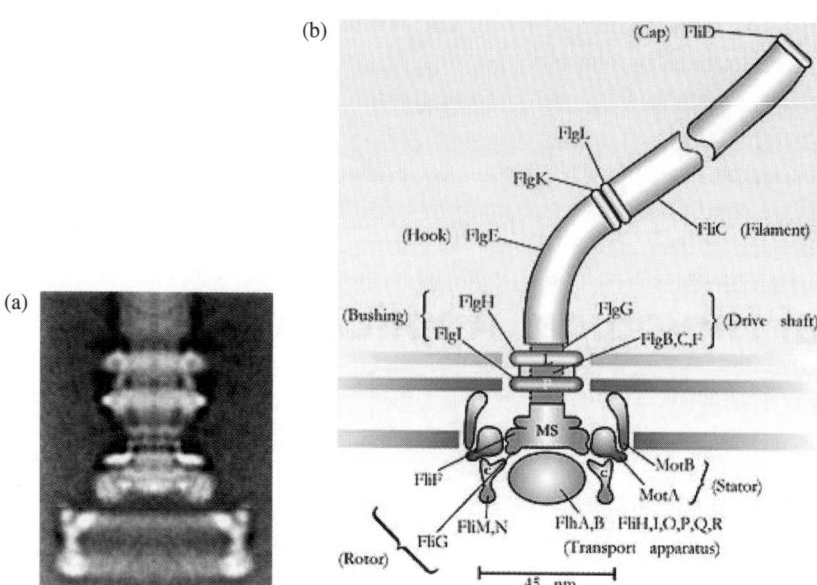

Fig. 6.1. Bacterial motor and drive chain. (a) Rotationally averaged reconstruction of electron micrographs of purified hook-basal bodies. The rings seen in the image and labeled in the schematic diagram (b) are the L ring, P ring, MS ring, and C ring (digital print courtesy of David DeRosier, Brandeis University).

Berg and Turner specifically studied the torque generated by the flagellar motor of *E. coli* bacteria [225]. More recently Berry and Berg [226] demonstrated the absence of a barrier to backward rotation of the bacterial flagellar motor using optical tweezers. They then proceeded to measure the torque generated by *E. coli* when driven backward [227]. Ryu, Berry, and Berg [228] have developed a new assay that permits them to measure and monitor the speed of the flagellar motor of *E. coli* under conditions of low viscous load using weak optical traps. They find that the limiting speed is independent of the number of torque generating units of MotA and MotB pairs driving the motors (see Fig. 6.1, taken from Fig. 2 of Berg's recent review article [222]). This result shows that each driving unit remains attached to the rotor for most of its mechanochemical cycle. This behavior is much more like kinesin motors that drive vesicles in cells than the myosin motors that power muscles, as we shall see. Use of tweezers has thus played a major role in understanding these powerful molecular bacterial motors, which Howard Berg calls a "nanotechnologist's dream" in his recent review [222].

6.1.2. *Optical manipulation of extremophilia*

6.1.2.1. *Archaea*

The ability to manipulate live bacteria, especially under physiological conditions, has had some surprising consequences. In recent years there has been growing interest in tracing the evolution of bacteria over geologic times as a clue to the origin of life on earth and possibly on other planets. For a period of about a billion and a half years in the early history of the earth, the only forms of life in existence were the so-called ancestral procaryotic cells that included bacteria. At some unknown point these were postulated to have divided into two separate branches, eubacteria and archaebacteria. Eubacteria eventually led to eucaryotic cells that make up multicelled plants and animals [229]. The archaea branch includes sulfur bacteria that live in very hot or acid conditions, extreme halophiles, which tolerate extreme salt conditions, and methanogens, which reduce CO_2 to methane. This early history and the important question of the origin of oxygen in geologic time is the subject of much recent research. There is a growing body of evidence of remarkable adaptation of bacteria to extreme conditions. These bacteria are often called extremophiles. The recent discoveries of anaerobic archaea in deep sea volcanic rifts (sulfur smokers) at temperatures as high as 113°C [230] and of anaerobic fresh water thermophiles in the hot springs of Yellowstone Park [231], in solid rock in deep mines [232], and in oil wells [233] are indications of this interest. Experiments are underway at Lake Vostok [234], a fresh water lake that exists under the Antarctic ice sheet. There apparently are bacteria that can survive and thrive in the presence of huge exposures to nuclear radiation [235]. Apart from the fascinating question of the early evolutionary history of life, there is an important practical aspect. The extremophiles, having adapted to extreme conditions, are often capable of producing unusual enzymes that catalyze unusual chemical processes. The hope is to find new enzymes possibly as valuable as Taq polymerase, which is widely used in the polymerase chain reaction. In the 1980s it was discovered that one of the hyperthermophilic bacteria in the hot springs of Yellowstone Park produces this unique high temperature enzyme. It became the basis for a multimillion-dollar industry [231], [236].

Huber *et al.*, in 1995, developed a new assay, using tweezers, for separating selected archaea under the high temperature anaerobic conditions of a hot obsidian pool in Yellowstone Park [6]. Using conventional dilution techniques one can grow pure cultures of only a few of the bacteria in a mixed sample, namely the predominant types. With tweezers, Huber and his co-workers were able to isolate and grow pure cultures of a new type of hyperthermophilic archaeum, which was outnumbered by other bacteria by at least four orders of magnitude. The isolated clones displayed DNA sequences indistinguishable from small quantities of the sequences detected in the natural environment. The author predicted this separation technique would become a powerful approach for studying microbial ecosystems [6]. Since an estimated 99.9% of the thermophiles are unknown, much remains to be explored. This work by Huber *et al.* has prompted the resource management chief of Yellowstone Park to seek ways in which the park can share in profits made by large drug companies using bacteria found in its hot springs [231].

6.1.2.2. *Sequencing of Thermotoga maritima, a eubacterium*

More recently, Huber used optical tweezers to isolate a single cell of a high-temperature eubacteria called *Thermotoga maritima* from marine sediments in Vulcano, Italy. He cloned it to obtain a pure culture. These eubacteria have been sequenced by Nelson *et al.* [237] and found to have the highest percentage (24%) of genes that are very similar to archaeal genes and are conserved in the same order as on the genome of Archaea. This suggests that lateral gene transfer may have occurred between thermophilic Eubacteria and Archaea. This bacterium, aside from its evolutionary significance, also has potential as an energy source, since it metabolizes carbohydrates to produce hydrogen gas and carbon. There also are vast numbers of unidentified water and soil bacteria that could be separated by similar tweezer techniques [238].

6.1.3. *Pilus retraction powers bacterial twitching motility*

A pioneering experiment on bacterial motility involving what is called the twitching or social gliding motion of bacteria along surfaces was performed in 2000 by Merz, So, and Sheetz [239]*, using optical tweezers. In this type of movement, bacteria generate, extend, attach, and retract their filaments, called type-IV pili, thereby pulling themselves hundreds of micrometers along on an inert surface. Although not fully understood even now, this process is widely used by many types of pathogenic bacteria that lack flagella. They crawl at velocities of about a μm/s over surfaces to form microcolonies and, at times, biofilms, and fruiting bodies.

N. Gonorrhea cells were first attached to $\sim 3\,\mu$m diameter latex beads and manipulated by optical tweezers operating in the infrared, close to a microscope slide where the pili could attach themselves. Once attached, the cell proceeded to ingest the pilus, thus generating a force on the cell-bead combination of $\sim 80\,$pN to draw it onto the slide. If the bead and its attached cell were placed anywhere near a microcolony of other cells, the cell proceeded to attach itself via its pili to the colony and was drawn in. It is clear that the cell was responding to some signal from the colony.

The authors performed a fascinating series of experiments involving the use of inhibiting chemicals to stop the ingestion process. It was shown that a particular pilus synthesizing protein, known as *PilT*, is essential for pilus-generated motility. *PilT* makes the pilus filament out of subunits of pilin protein. On retracting, it is thought that the retractile force is generated by filament disassembly into the inner part of the cell membrane. The filament itself is a single-stranded helix, about 60 Å in diameter, with five pilin subunits per turn and a 40 Å pitch. Retraction at a 1.2 μm/s rate requires the removal of 1500 pilin subunits per second from the fiber base.

There is evidence that the pilus retraction is used in many bacterial systems for a variety of other macromolecular translocation processes. These processes are implicated in adhesive plaque formation and virulence-related functions.

6.1.4. *Direct observation of extension and retraction of type-IV pili*

More recently, in 2001, Skerker and Berg [240] devised a dye labeling scheme that allows them to directly observe the extension and retraction of type-IV pili in bacteria without flagella that move over surfaces by the gliding or twitching motility process [241]. Type-IV pili have been seen under electron microscopes, but have never been visualized *in vivo*. The authors succeeded in labeling the pili with an amino-specific Cy3 fluorescent dye and observing them on a quartz slide using total internal reflection microscopy.

The extension of pili, their attachment to surfaces, and their subsequent motion by retraction could be directly observed by fluorescence. The rates of extension and retraction were closely equal to $\sim 0.5\,\mu$m/s. Lengths of pili of $\sim 15\,\mu$m broke off at times, with one end attached and the other floating free. The shape fluctuations of these pili were readily observed and gave a persistence length of $\sim 5\,\mu$m. Crudely, the persistence length is the length over which the flexible polymer maintains its directional integrity.

The authors clearly visualize bacterial motion by pili retraction. They expect to be able to use the technique to visualize motion of both phages and pili during infection or of cells and pili during conjugation. They see this as a useful technique, and when combined with the tweezer manipulation techniques [239]* it will be ideal for studying this previously overlooked important subject. There is an excellent set of references on pili in the above two papers [239]*, [240].

6.1.5. *A force-dependent switch reverses type-IV pilus retraction*

Maier *et al.* in Sheetz's group [242], in a recent experiment at Columbia, showed that the elongation and contraction of pilus polymers play an important role in many vital functions in a variety of prokaryotic cells. Among the functions controlled by this type of dynamics are the virulence of infectious strains of bacteria and the motility and DNA transfer through the cell membrane wall. Pilus dynamics is known to involve polymerization and depolymerization of pili in membranes. It depends on the presence of various retraction proteins such as *PilT* protein and its analogs. These are related to mechanoenzymes that hydrolyze ATP.

From previous work by Sheetz and collaborators, it is known that pili can generate force in excess of 100 pN. It is thought that control of these forces mediate infectivity of various bacteria. To study retraction and elongation of pili, the authors used a computer-controlled movable mirror optical tweezer system operating at 1.06 μm wavelength. An *N. gonorrhoeae* bacterium was immobilized on a coverslip. A single type-IV bacterium pilus attached to a latex bead was held taut by the single-beam tweezer trap. The authors measured the velocity–force relationship and the effects of *PilT* concentration. Surprisingly, they discovered that, during study of pilus retraction rates at low levels of *PilT* protein, increasing pilus force acted as a switch that reversed the retraction and initiated pilus elongation. If the force was turned off, the pili retracted immediately. This behavior at low levels of *PilT* sets important constraints on the possible mechanism of pilus retraction. It leads the authors to believe in a model of the cell membrane in which both external force and *PilT* concentrations dictate the direction of pilus dynamics.

6.1.6. *Characterization of photodamage to Escherichia coli in optical traps*

Neuman *et al.*, working with Steve Block at Princeton in 1999 [21]*, made a very important study of the photo damage to *E. coli* bacteria held in optical traps. The level of damage is perhaps the first question to be addressed in any optical tweezer experiment with living cells or organisms. Indeed, it was the discovery in 1986 of greatly reduced optical damage to living cells by 1.06 μm laser light in the infrared that marked the beginning of laser manipulation of biological particles *in vivo* and *in vitro*.

This detailed experiment by Neuman *et al.* used a rotating *E. coli* bacterial array with a tunable titanium sapphire laser to characterize the damage in the infrared part of the spectrum and attempted to identify the possible laser damage mechanism. *E. coli* bacteria were tethered by their flagella to a microscope slide while being constrained by an optical trap. They were periodically released in order to monitor their rotation rate about their tethers. *E. coli* are in many ways ideal organisms for such an experiment, as the authors explain. They can be grown either aerobically or anaerobically, and many mutant varieties are available.

The data covering the region from 790 to 1064 nm showed minimum damage at 830 and 970 nm. Also, damage was found to be negligible under anaerobic conditions. This indicates an important role for oxygen in the photodamage process. They also found that the aerobic damage was linear with optical intensity and is thus presumably a single photon process. This is contrary to some earlier work. See the extensive list of references for this paper. Notice also in Fig. 1.6 of this cited paper that the spectrum of damage in *E. coli*, as found here, was quite similar to that found previously in Chinese hamster ovary cells at considerably less accuracy. Nevertheless, this points to a possible common source of damage.

6.2. Use of UV Cutting Plus Tweezers to Study Cell Fusion and Chromosomes

Greulich and Berns were the first to use the tweezers technique in combination with the older "microbeam" technique of pulsed laser cutting (more recently called "laser scissors" or "scalpel") for cutting and moving cells and organelle. Greulich's early work involved pulsed ultraviolet (UV) laser cutting of pieces of chromosomes for gene isolation, and their subsequent manipulation by tweezers [243].

In other experiments tweezers was used to bring cells into contact with one another in order to effect cell fusion by UV cutting of the common wall [244]. If one of the cells is a cancer cell and the other has genes for producing desirable molecules, then one can get a self-reproducing strain of combined cells that continue to make the desired products indefinitely. Berns and his group initiated the use of tweezers, often combined with optical scissors, to manipulate chromosomes during cell division as a new way to study the complexities of mitosis [245], [246], [247].

6.3. Tweezer Manipulation of Live Sperm and Application to *In Vitro* Fertilization

Experiments were performed with tweezers to manipulate live sperm cells in three dimensions [192], [248] and to measure their swimming forces [249]. Applications of tweezers and scissors to all-optical *in vitro* fertilization are being considered [250]. UV drilling of channels in the zona pellucida of oocytes was performed and selected sperm were inserted into channels to effect fertilization [251], [252]. Experiments by Berns's group measured the effects of the wavelength on possible optical damage processes in sperm and in other contexts using tunable sapphire lasers [253].

6.4. Tweezer Study of the Immune Response of T-Lymphocytes

Recently an elegant experiment was performed with tweezers studying the contact conditions that must exist between T-lymphocyte cells and antigen-presenting cells to stimulate an immune response [254]. By using antibody coated beads, it was possible to determine the minimum number of receptors which must be engaged to stimulate a calcium-signaling response, which is an indication of the onset of the immune process. T-cells essentially count the number of receptors that are activated at the contact. By manipulating beads, it was possible to control the contact geometry, timing, and nature of the ligand, under physiological conditions. It was observed that crawling T-cells become polarized and are three times more sensitive at their leading edge than at their trailing edge. It was possible to determine a latency factor that indicated how many seconds of delay occurred between the moment of contact and a calcium fluorescent signal response from the dye-loaded T-cells. The authors foresee further more detailed experiments to learn more about gene expression and the effect of various types of ligand stimulating molecules.

6.5. Adhesion of Influenza Virus to Red Blood Cells Using OPTCOL Technique

Another assay to study the collision of two particles or cells under controlled biologically relevant conditions, called "OPTCOL", was developed with two tweezer traps by Mammer *et al.* [255]. The adhesion of influenza virus-covered spheres to erythrocytes during collision was studied with controlled velocities and controlled geometry in the presence of various attachment inhibitors. The new extremely sensitive technique has identified the most potent known inhibitor of the attachment process. The authors foresee wide usage of OPTCOL for studies of collisions of biological particles such as bacteria, viruses, T cells, ribosomes, liposomes, and even nonbiological objects.

6.6. Mechanical Properties of Membranes Studied by Tether Formation Using Tweezers

The ability of optical tweezers to probe the mechanical properties of living cells was demonstrated in the early experiments of Ashkin and Dziedzic [19]*. The use of tweezers to pull long filaments (or tethers) from cell membranes and to observe strong viscoelastic behavior was discussed above (see Fig. 5.27).

Sheetz and his coworkers have pursued the use of tweezers for the measurement of the mechanical properties of membranes in a variety of biological studies. Dai and Sheetz, in 1995 [256], studied neuronal growth cone membranes. In another experiment they measured the flow of axon membrane from the growth cone to the cell body [257]. They also observed the deformation and flow of membrane into tethers that were pulled from neuronal growth cones [258]. They point out some of the advantages of the tweezer technique for these applications. As opposed to the more primitive and restricted methods of probing membranes using micropipettes [259], tweezers can probe the external and internal membranes at any desired location on cells of virtually any shape. Although tethers can be formed with just the force of tweezers on the membrane surface itself, as Ashkin showed, it is often desirable to use a plastic of metal bead as a handle. The force one can exert in this way is stronger and more easily measured.

Measurements were made of apparent membrane viscosity. This is an important parameter of all membranes, which determine, for example, the ability of particles, such as integrin molecules, to diffuse on a membrane surface. Values in the range of 0.5×10^{-4} to 4×10^{-4} dyne·s/cm^2 were found for the chick dorsal root ganglion. By using cytoskeletal-perturbing drugs, the authors deduced that a major component of the viscosity is the viscous drag of the lipid across the adjacent cytoskeleton. Other basic parameters determined were the bending stiffness for a growth membrane of 2.7×10^{-19} Nm and the membrane tension. Remarkably, some of the radii of the tethers of the chick growth cone studied were as small as $0.2 \, \mu$m. There is a good review of "Cell Membrane Mechanics" by Dai and Sheetz in *Methods in Cell Biology* [260].

6.7. Deformation of Single Cells by Light Forces

There has been an interesting new application of optical traps in biology to the problem of deforming soft biological particles by Guck *et al.*, in 2000 [261]*. For this purpose they used a two-beam trap just like the very first optical trap of Ashkin in 1970 [1]*, rather than the more recently developed single-beam tweezers trap. They point out that for any trap the particle is subject to surface deforming forces, which are considerably higher than the trapping force on the particle. For large volume two-beam traps, the surface forces are distributed over a large area of the particle and do not cause local damage, as they might for soft biological samples trapped by tweezers. This is all true, but they overstate their case in the abstract when they say, "Radiation damage is avoided since a double-beam trap does not require focusing for stable trapping". It does not require strong focusing, as in tweezer traps, but it *is* the gradient force that accounts for transverse stability in the double-beam trap; otherwise, it violates the optical Earnshaw theorem.

The existence of surface deforming forces is no mystery and was studied by Ashkin and Dziedzic in 1973 [262]*. Since the velocity of light changes from c to c/n on entering a dielectric medium, or vice versa on leaving, one expects a surface force by conservation of momentum. One can crudely say that the momentum per photon changes from $h\nu/c$ to $h\nu/(c/n) = nh\nu/c$ on going from free space into the dielectric. Thus, for a particle of high index, this causes a stretching force to be applied on entering or leaving the particle. This, of course, cancels out, on a ray-by-ray basis, and is neglected in the determination of the net optical force on a rigid particle. The early experiment on surface forces by Ashkin and Dziedzic [262]* was motivated by a theoretical controversy on the nature of the momentum of light in a dielectric (see references in Ref. [262]*). Was it $nh\nu/c$, the so-called Minkowski momentum, or the Abraham momentum, $h\nu/cn$? The experiment involved shining a high-powered pulse YAG laser on a water surface and looking for the impulsive response of the surface. See Fig. 6.2,

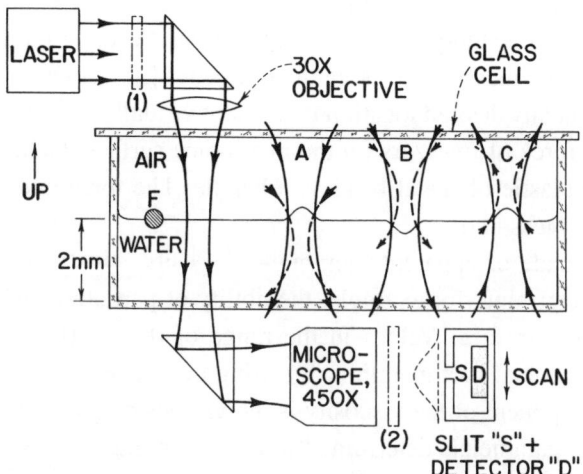

Fig. 6.2. Basic apparatus: (A) Beam shapes for low power (solid curve) and high power (dashed curve) for positive surface lens. (B) Shapes (for low and high power) for a negative surface lens. For (A) and (B) the beam is incident from above. (C) Beam shapes for low and high power for a positive surface lens with the beam incident from below.

taken from Ref. [262]*. If the photon momentum is $nh\nu/c$, one expects the surface to "pimple up", as in A or C of Fig. 6.2. If the photon momentum is $nh\nu/cn$, one expects the surface to "dimple in", as in B. We called our experiment, designed to settle this question, the "dimple-pimple" experiment. The results showed that a pulse of light entering or leaving the surface caused the surface to pimple out, as the intuitive Minkowski momentum implied. It would be interesting to perform this experiment on a cw basis. Perhaps a strongly dimpled surface would turn into a small liquid drop and the drop would be driven away from the surface by the scattering force. In this case, the light would act as a drop generator, where the drop diameter would be controlled by the light power. There are liquids where the surface tension is sufficiently small that one could hope to perform this experiment at reasonable powers. In some mixed liquids one can control the viscosity with temperature, near a phase transition. At the critical temperature, the viscosity approaches zero.

Subsequent theoretical work by Gordon in 1973 [263] showed that the complicated question of the momentum carried by the dielectric and the problem of the force in dielectric media could be correctly treated in terms of the conservation of a "pseudo-momentum" nP/c, where P is the beam power. The optical force of radiation pressure on a mirror in a liquid was subsequently remeasured very precisely by Jones [264] and found to be in excellent agreement with nP/c.

Guck *et al.* [261]* estimate that stretching forces as high as $400\,pN$ can be achieved with a pair of trapping beams, each with $500\,mW$ of power for a roughly $10\,\mu m$ particle of index $n_2 = 1.45$ in water of index $n_1 = 1.33$. There are, of course, surface reflection forces at the dielectric interfaces for every ray passing through the particle. These give rise to the usual scattering force used for trapping and amount to only about $20\,pN$.

The authors confirmed these simple calculations in experiments using soft red blood cells (RBCs) and so-called PC-12 cells. In their experiments the authors first converted the normal biconcave disk-shaped RBCs into a spherical shape by properly adjusting the osmolarity of the buffer. The index difference between the RBC ($n = 1.380$) and water ($n = 1.335$) is quite small. For powers up to $350\,mW$ the $3.32\,\mu m$ cell deformed by expanding axially by about $800\,nm$ and contracting radially by about $600\,nm$. At yet higher powers the cell's response was nonlinear until it finally ruptured at about $600\,mW$, after a deformation of about 160%. The results agree with a computation of the applied surface stress based on a ray-optic model.

This is not to say that the tweezer approach is not useful. See, for example, the early work on red blood cells by Ashkin and Dziedzic [20]*, the more recent experiments by Svoboda *et al.* in 1992 [265], and later experiments by Bronkhorst *et al.* in 1995 [209]. Guck *et al.* do not mention any of these previous papers. Ashkin *et al.* [20]* showed the ability of tweezers to distort the shape of RBCs and simultaneously trap many cells in a single trap. Svoboda and Block in 1992 [265] measured the elastic properties of isolated red blood cell membrane skeletons. Using three tweezer traps, Bronkhorst *et al.* in 1995 [209] developed a new assay to distort and sensitively measure the shape recovery time of single red blood cells, using physiologically relevant shapes and conditions. Significant differences in relaxation times were found for old and young cells. Measurements were made in blood plasma and gave markedly different results from previous assays using pipettes in buffer solution. With automation, this may be a powerful technique for the study of sub-populations of pathological cells. The three computer-controlled tweezer traps used a multiple scanning trap system developed by Visscher *et al.* in 1993 [266].

6.8. Artificial Gravity in Plants

A very elegant experiment was performed with the help of optical tweezers on the sensing of gravity by plant roots. In 1995, Leitz, Schnapf and Greulich Ref. [267] simulated the effect of gravity on the direction and growth of the root-like rhizoid cells of the alga *Chara*. The response to gravity of rhizoid cells while growing is called "gravitropism". Its detailed mechanism is not understood.

It is known, however, that the gravity sensors are barium sulfate microcrystals, 1–2 μm in diameter, surrounded by membranes. These are called statoliths and are distributed about in the rhizoid cells. It is the pull of gravity on these dense statoliths that is presumed to guide the cell growth downward.

Using tweezers one can move the focus about and collect anywhere from one to five of these statoliths from the central region of the cell and displace them transversely. This simulates transverse gravity on the root and causes the tip of the new root growth to bend in the direction of the transverse force. As one continues to move the trap and apply transverse force, the rhizoid tip continues to grow, and bend transversely, following the light force. See Fig. 6.3, taken from Ref. [267], which shows the effect of statolith transverse displacement on tip growth. The rate of tip growth was also found to correlate very well with statolith displacement. This method of applying artificial gravity is superior to simulated gravity in spacecraft experiments or in slowly rotating centrifuge microscopes. Optical tweezer simulation of gravity does not perturb different parts of the cell differently, as in a centrifuge. It is very simple and can be applied for many hours and the results are readily observable.

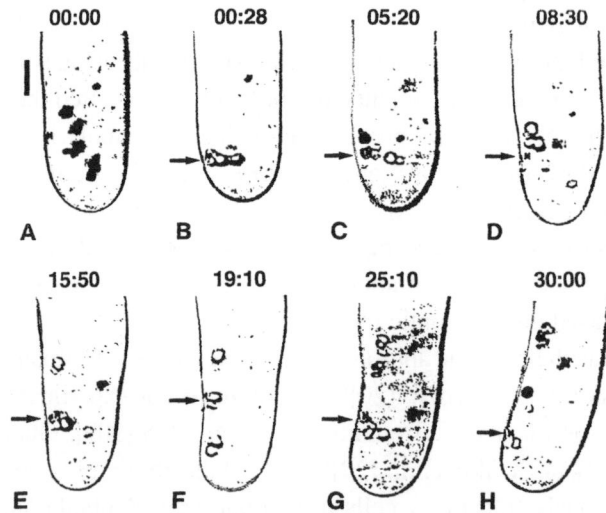

Fig. 6.3. Series of micrographs showing a change in the direction of rhizoid growth induced by continuous statolith displacement with optical tweezers (the exact position of the optical trap is marked by the small H-shaped cursor inside the cells). (A) Tip region of the rhizoid just before the optical-trapping experiment. (B) Lateral displacement of the statoliths into a stable position close to the side wall. The position of the optical trap remains unchanged during the experiment. The *arrow* points to the trapped statoliths in the laser focus. (C)–(H) The cell reacts by an extended growth of the opposite cell wall flank, resulting in a change in growth direction. (F)–(H) Three statoliths (*arrow*) have escaped the optical trap after having been held for a certain period of time. Note the two statoliths which still remain in the optical trap (just above the *arrow*). Time is given in minutes and seconds, starting after final reposition of the statoliths. X550; bar = 10 μm. Figures and captions from Ref. [267].

It was shown that the force sensing occurs due to the tension induced in actin filaments bound to the rhizoids. By treating the cells with cytochalasin B, an agent that destabilizes actin, one sees a sharp decrease in the force needed to displace the statoliths transversely or toward the tip. This new method of studying gravitropism can be used with many other types of plant root cells.

In this experiment, the authors applied the artificial gravity force by hand, so to say. As the *rhizona monadia* responded to the tweezer force, the tension on the statoliths was relieved and the tweezer trap had to be moved to restore the tension, so that sidewise growth could continue. The experiment could be updated using a recently developed feedback force clamp to apply a constant tension as the root grows. Such a setup would be the closest thing to artificial gravity.

6.9. Guiding of Neuronal Growth with Light

A method has been discovered for optically guiding the growth of actively spreading neuron cells adhering to a glass substrate [268]*. Control over neuronal growth is of interest for *in vitro* experiments involving neural networks interfaced with artificial external semiconductor probes and for *in vivo* attempts to achieve successful nerve regeneration. The authors found that optical tweezer forces were too weak to directly grab and move the leading edge of a growth cone of the cell. The optical gradient forces were, however, strong enough to guide the growth direction by biasing the actin polymerization process that drives the extending growth cone. Turns as sharp as 90° were possible by keeping the leading edge of the growth cone in the optical gradient. Using estimates of the optical potential for a power of $\sim 80 \, \text{mW}$ from a Ti:sapphire laser, they obtained a differential drift velocity for globular actin proteins of $\sim 26 \, \mu\text{m/h}$, close to the observed growth velocity of the neuron. There was no evidence of optical damage to the neuron at the power used.

6.10. Self-Rotation of Red Blood Cells in Optical Tweezers

In 2004 in India, Mohanty, Uppal and Gupta [269], [270]* developed a new method for detecting malarial infection based on rotation of red blood cells (RBCs) in optical tweezers. Each year, malaria affects 500 million people and kills 2.7 million of them. Therefore, a simple method for early detection of infection is of considerable interest. The authors point out that the membranes of malarial-infected RBCs are considerably more rigid than those of normal RBCs. Hence, measurement of RBC elasticity can be used for detection of infected RBCs.

The authors have discovered that normal RBCs, when placed in a hypertonic buffer and trapped by optical tweezers, rotate up to hundreds of rpm, whereas cells infected with malaria parasites do not rotate, or do so more slowly. They attribute the difference to the significant distortions in shape that healthy cells can undergo, while infected cells remain rigid. See Fig. 1 of Ref. [270]*, showing a schematic of the large distortions that occur due to the gradient force of the trap. The rotation is believed to be due to the effect of the linear momentum of the beam on the distorted cell. This gives a simple test for the presence of infection. The authors confirmed infection by acridine orange fluorescence staining. Another remarkable observation was that the other cells from an infected sample, but not containing the parasite, were found to rotate an order of magnitude slower than healthy cells. This further enhances the usefulness of the test. Flowing the liquid through the chamber can screen approximately 40 RBCs per minute. The high throughput and sensitivity make this diagnostic technique very useful in comparison with other techniques. Furthermore, one could increase the number of traps by a big factor with a holographic array of traps, giving an even higher throughput. The authors suggest that the technique can also be used for other diseases that change the elasticity of RBC membranes. The reader should refer to earlier relevant work in Secs. 6.6 and 6.7 on tether formation and whole cell stretching. See also Chapter 12 on "Rotation of particles by radiation pressure".

CHAPTER

7

Use of Optical Tweezers to Study Single Motor Molecules

One of the most distinctive characteristics displayed by living matter is motility, or the ability of a cell or organism to control its own motion. As already seen, an important application of tweezers is in the study of motility at the single cell level. It also has been realized that tweezers has the capability to study motility at the single molecule level. By the early 1980s many mechanoenzymes or motor molecules were identified which could interact with the microtubules and the actin filaments of the cytoskeleton of a cell to generate the forces driving flagella, muscle actin, amoeboid motion, cytoplasmic streaming, and organelle motion within the cell. The detailed mechanism for this motion at the single molecule level was not known. As we will see, tweezers, although not the only tool for studying these processes, has become the predominant research tool, contributing to the explosive growth in the understanding of motor molecules. In early work [210], [211], using what has become known as the handles technique, Stephen Block and collaborators attached single kinesin motor molecules to macroscopic dielectric spheres of a micrometer or so in diameter and then, using tweezers, placed them directly onto microtubules, where they could be activated in solution with ATP molecules. This new technique was a great improvement over earlier *in vitro* motility assays, which used spheres covered with many motors and relied on random particle diffusion for attachment of the spheres to the microtubules or filaments. Block was able to show with his tweezers single motor assay that kinesin was a highly processive motor molecule that, once attached and activated, could advance ~ 100 multiple rounds of activity without detaching from the microtubule polymer. Kinesin molecules, as many of the other motor molecules, are dimers with two identical heads attached to a helical stalk. See Fig. 7.1, taken from Fig. 1 of Ref. [197], which is a diagram of a kinesin molecule. This molecule, as we will see later [271], [272], is believed to advance along microtubules in a hand-over-hand fashion.

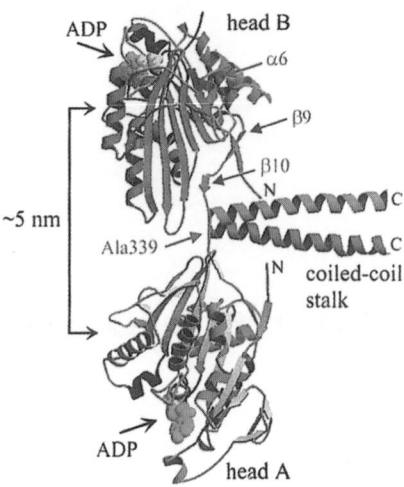

Fig. 7.1. The crystal structure of the dimeric Kinesin motor bound to ADP.

7.1. *In Vivo* Force Measurement of Dynein in Giant Amoeba *Reticulomyxa*

The Block *et al.* Nature paper [210] appeared in 1990, back to back with a paper by Ashkin *et al.* [17]* on "Force generation of organelle transport measured *in vivo* by an infrared trap". Ashkin *et al.*'s experiment was done in the giant amoeba called *Reticulomyxa*, which Schliwa and Eutenauer had previously identified as an interesting organism for studying motility processes [273]. There are literally hundreds of mitochondria moving through the cytoplasm of this organism. In this experiment it was determined by electron microscopy that anywhere from one to four cytoplasmic dynein motor molecules, with an average of about two, are attached to each mitochondrion in the amoeba. With tweezers it was possible to measure the force generated by these motors as they drew mitochondria along the microtubules of the peripheral network of growth processes of the amoeba. The mitochondria had a diameter somewhat larger than $0.3\,\mu$m and were moving along strands of $0.3\,\mu$m or less, so that they bulged the membrane of the strand somewhat as they moved. In such geometry, the motors are held in close contact with the microtubules by the elastic membrane and questions of processivity do not apply. The particles can move very long distances, greater than $30\,\mu$m or more, without stopping. See Fig. 7.2, taken from Fig. 2 of Ref. [17]*. With kinesin in Block *et al.*'s *in vitro* experiment, particles remained attached to microtubules for distances greater than a micrometer only for cases where two or more kinesin motors were attached to the sphere. For kinesin the particle velocity was $\sim 1\,\mu$m/s, whereas the dynein velocity was $\sim 10\,\mu$m/s.

Ashkin *et al.* measured the force of dynein by determining the power at which the particle pulled free from the trap. The value of force per dynein molecule was found to be $\sim 2.6 \times 10^{-7}$ dyne $= 2.6$ pN, which was similar but more closely defined than the force of ciliary dynein [274]. In our experiment we could almost resolve the forces due to individual motors. However, there were uncertainties in the size of the mitochondria. Also, we had uncertainties in measuring the laser power needed to stop mitochondria. With more care these problems could be reduced. This technique has the potential to measure the force per motor with increased accuracy and also to resolve the number

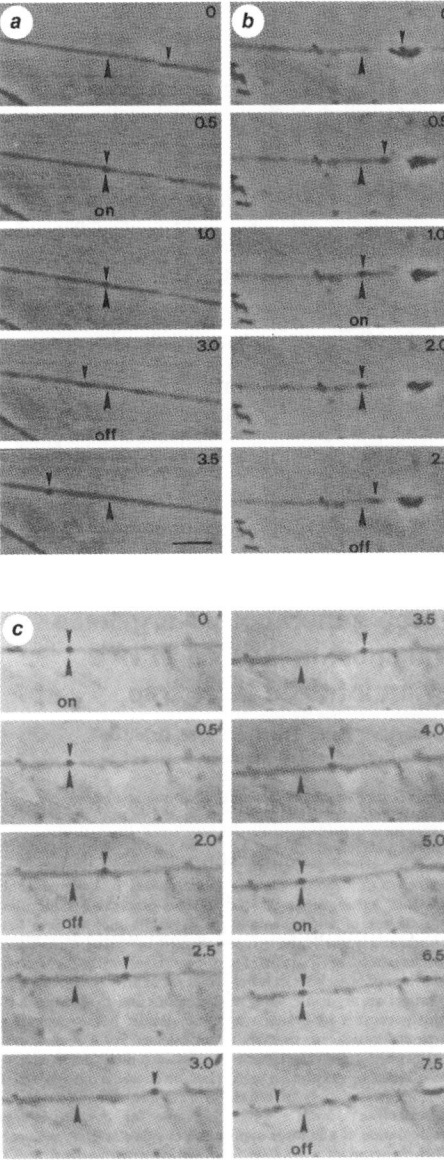

Fig. 7.2. Laser trapping of mitochondria in thin *Reticulomyxa* strands. Small arrowhead denotes the organelle; large arrowhead marks the position of the trap. Times in seconds are shown in the upper right-hand corner. "On" and "off" indicate the state of the laser trap. Scale bar, $5\,\mu$m. (a) A mitochondrion traveling from right to left is trapped (frame 2), held in the trap for ~ 2 s, and moves on in the same direction after the trap is turned off. Its speed is $12.5\,\mu$m/s before and $10.6\,\mu$m/s after trapping. (b) A mitochondrion reversing direction after being trapped for ~ 1 s (frames 3 and 4). Its speed is $8.7\,\mu$m/s before and $9.1\,\mu$m/s after trapping. (c) This mitochondrion entered the trap from the left (frame 1), was released after ~ 1.5 s and continued to move in the same direction (time 2–3 s). It then reversed its direction (between 3 and 5 s), and was trapped again at the same power (time 5–6.5 s). Figure and caption from Ref. [17]*.

of motors on individual mitochondria by observing the exact escape force. In 1998, Welte *et al.* in Block's group (see Secs. 7.20 and 7.21) found strong evidence for quantized force levels due to discrete numbers of motors in an *in vivo* experiment measuring force on individual vesicles in *Drosophila* embryos. Block *et al.*, in their initial experiment on kinesin, were only able to estimate the maximum force as somewhere in the broad range of 0.5–5 pN [210].

As an aside and in another context, I think *Reticulomyxa* could be an interesting vehicle for studying microtubule polymerization and membrane fusing. These processes, which are basic to many cell functions, are not fully understood. In unpublished work, Ashkin and Dziedzic observed that if one pulls out a filament, as in plant cells [19]*, and waits, the filament loses its elasticity and its core polymerizes into stiff rod-like microtubules that are readily observable on release of the filament from the trap. With tweezers one can also pull out these artificial filaments and make contact with nearby natural strands. One finds that the membranes immediately fuse. See Fig. 7.3, taken from Fig. 2 of Ref. [275]. After a few minutes the artificial connecting filament polymerizes and mitochondria are observed to move along the newly made connection.

In another experiment with *Reticulomyxa* we induced the amoeba to ingest latex spheres. These can act as convenient probes for measuring the viscosity of the fluids inside the cell. Reference [275], on *in vivo* manipulation of internal cell organelles, also describes some interesting work on cytoplasmic streaming in *spirogyra* plant cells, in which the chloroplasts were removed, making observation and trapping of mitochondria much simpler. This book chapter [275] has some quite useful references.

7.2. Measurement of the Force Produced by Kinesin

In a subsequent experiment on kinesin, Kuo and Sheetz [276] measured a maximum force of 1.9 pN, using a somewhat different geometry. They dispersed the kinesin motors on a glass substrate at low density and manipulated lengths of microtubule attached to a pair of trapped small dielectric beads. They measured the escape power of the spheres from the trap, which, after calibration against Stokes' law, converts directly into the maximum or so-called stall force for the molecule.

7.3. Resolution of the Stepping Motion of Kinesin on Microtubules by Interferometry

A major advance in the field was the resolution by Svoboda and Block *et al.* [217]* of the detailed motion of a single kinesin molecule into a sequence of 8 nm steps as it advances along a microtubule. Microtubules are made up of a modified helical structure consisting of two types of intertwined helices called α and β. It was known that kinesin bonds preferentially to the β helix. As a molecule advances axially along a β helix, the repeat distance of the β-helix is ~ 8 nm. See Fig. 7.4, taken from Fig. 2 of Block [277]. This first observation of the stepping motion used an optical trapping interferometric position monitor with sub-nanometer resolution [217]*. The interferometer makes use of standard differential interference contrast (DIC) microscope optics to interfere the two polarization components of an input focused laser trapping beams. The slight displacement of the two polarization components used in DIC microscopy does not affect the trapping and the combined beams function

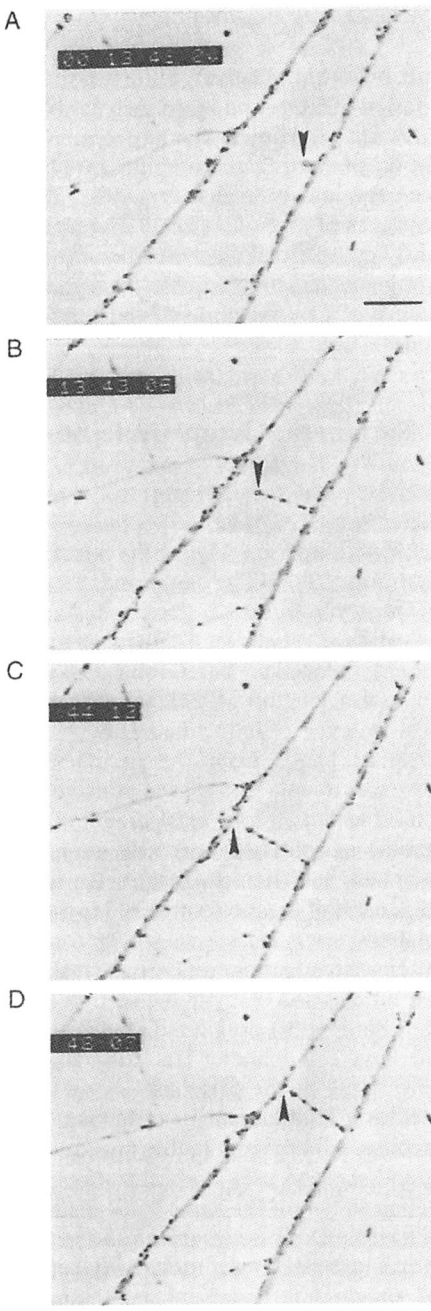

Fig. 7.3. Creating an artificial cell extension. By grabbing a particle in one strand of a reticulomyxa network (A), a new cell process (arrowhead) can be pulled out (B). Upon contact with a neighboring strand (C), the two membranes fuse instantaneously, forming a stable connection that slides along the two parallel strands (D). The entire sequence took place in 4 s. Bar, 5 μm. Figure 2 and caption from Ref. [275].

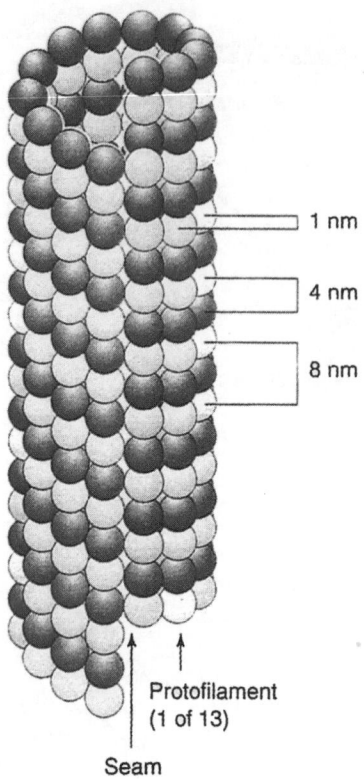

1 nm

4 nm

8 nm

Protofilament
(1 of 13)

Seam

Fig. 7.4. Subunit arrangements in a microtubule with the *B*-type lattice. Thirteen protofilaments are shown, each consist-ing of repeated tubulin heterodimers at 8 nm intervals. The α and β subunits of the heterodimer are indicated by different colors. Adjacent protofilaments are offset by ~ 1 nm axially, which introduces a beak in the helical symmetry, so that the phase of the lattice cannot return in exact register after a full turn. This results in a helical discontinuity, or seam, shown running down the front of the diagram, that is characteristic of the *B*-type lattice. Along the seam, the apposed protofilaments have the same symmetry (lattice contacts) as in the *A*-type lattice. Figure and caption from Ref. [277].

as a single trap (see Fig. 7.5). Displacement of a phase object from the center of the focal region of the trap results in a relative phase retardation between the two polarizations, which is detected as a difference between two photo detectors. This scheme has a measured position sensitivity of a fraction of a nanometer. Figure 7.5, taken from Ref. [217]* shows multiple 8 nm steps. In addition to having a trapping scheme with sufficient positional resolution to observe a stepping nature of ~ 8 nm, one needs to provide some form of damping to control the severe Brownian motion of the moving sphere. Fortunately, this is possible under close to stall conditions, where the trap is severely restraining the motion of the sphere [217], [277]. Statistical analysis of the displacement versus time under various loads and ATP concentrations yielded an average step size of ~ 8 nm. See Ref. [217]*. See also Fig. 7.9 in Sec. 7.14, showing 8 nm steps, taken with an advanced force clamp optical feedback technique.

Concerning questions of tweezer techniques, Block *et al.* calibrated the force constant of the trap from Stokes' law by measuring the velocity of fluid flow past a sphere of known diameter, as was previously done by Ashkin [17]*. Another convenient way of getting the force constant of the trap is to observe the thermal noise spectrum of a particle in a harmonic trap driven by thermal noise. This yields a Lorentzian power spectrum, whose so-called corner frequency α_0 is proportional to the trap

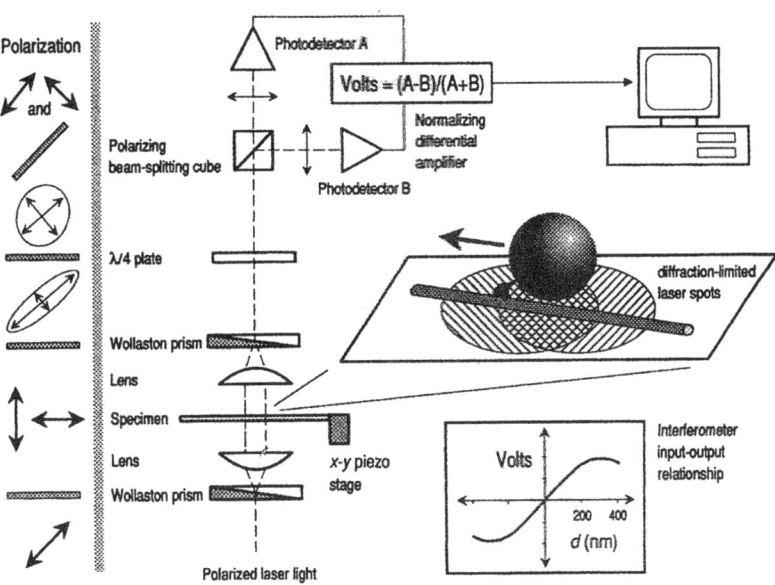

Fig. 7.5. The optical trapping interferometer. The diagram illustrates the polarization state of light as it passes through elements of the system, viewed along the optical axis. Polarized laser light passes through a Wollaston prism and is focused to two overlapping diffraction-limited spots ($\sim 1\,\mu$m diameter) with orthogonal polarization, separated by roughly $\sim 250\,\mu$m. After passage through the specimen (a bead propelled along a microtubule by a kinesin molecule), light recombines in the upper Wollaston and develops slightly elliptical polarization. Ellipticity is measured by a quarter waveplate, which produces nearly circularly polarized light, followed by a polarizing beam-splitting cube, which splits the light into two nearly equal components. The difference in intensity is detected by photodiodes and a normalizing differential amplifier. Signals were analyzed offline (Labview, National Instruments). Figure and caption from Ref. [217]*.

stiffness. The central region of the trap is found to be very linear, as might be expected. There is a very good early review on applications of tweezers in biology by Svoboda and Block, as of 1994 [211]. This covers topics such as trap design, force calibration, force stepping, position measurements, Brownian motion, and picotensiometers, which are often not treated in much detail in standard papers. Figure 7.6, for example, taken from Fig. 10 of Svoboda and Block [211], shows the force versus displacement calibration of an interferometric trap taken with modest powers of ~ 14 mW for a $\sim 0.6\,\mu$m diameter sphere. The observed trap sensitivity of $\sim 10^{-2}$ pN/nm shows that measurements on the sub-piconewton scale are readily made with tweezers. This type of sensitivity is not matched by other single particle force measuring techniques, such as microneedles [15] or AFM [13], [278], [279].

7.4. Observation of Single Stepwise Motion of Muscle Myosin-II Molecules on Actin Using Feedback and Tweezers

In 1994, Finer *et al.* [61]* introduced a new feedback enhanced tweezer trap [218] with a detection capability of sub-nanometers in position, sub-piconewtons in force and ms in time response. In 1994 Finer, Simmons, and Spudich [61]* applied this feedback-controlled trap to study the interaction of actin filaments with single muscle type myosin-II motor molecules. This is the interaction which gives rise to the usual muscle movement. There are many different classes and individual types of motor molecules, each with its own properties, adapted to its specific biological function. Despite

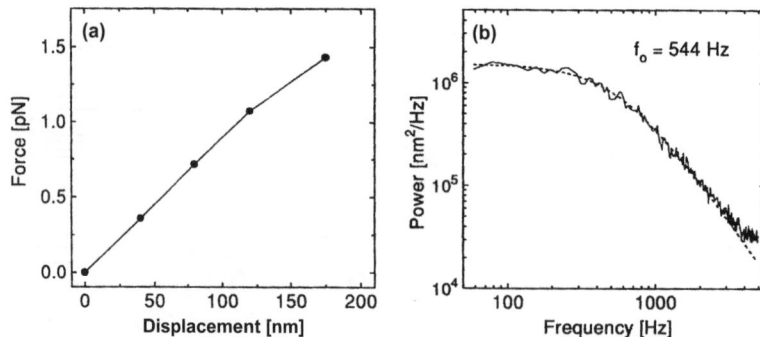

Fig. 7.6. Force calibration of an optical trap. (a) Force versus displacement for a trapped silica bead (diameter $\sim 0.6\,\mu$m; power $\sim 14\,$mW), calibrated according to Eqs. (14) and (15) ($\chi_0 = 2\,\mu$m; $\beta = 5.7 \times 10^{-6}$ pN s/nm). Stiffness is constant out to 150 nm ($\alpha = 9.0 \times 10^{-3}$ pN/nm), beyond which it decreases. (b) (solid line) The thermal noise spectrum of a trapped silica bead measured with the optical trapping interferometer (diameter $\sim 0.6\,\mu$m; power $\sim 28\,$mW). (Dotted line) The fit by a Lorentzian. The corner frequency of the Lorentzian (544 Hz) implies $\alpha = 1.9 \times 10^{-2}$ pN/nm. Figure and caption from Ref. [211].

some similarities in structure, muscle-type myosin-II motor molecules are only weakly bound or nonprocessive in nature. They essentially interact with actin filaments only once per catalytic cycle [280]. Muscle myosins are designed to operate in groups that rapidly move actin thin filaments and thus they behave quite differently from kinesins and dyneins.

To study this more complex interaction, Finer *et al.* [61]* used a dual trap scheme that suspended a taut actin filament between two spheres held by a pair of tweezer traps. The filament was allowed to interact with a small fixed (surface bound) silica sphere holding a single myosin motor molecule (see Fig. 7.7). With the feedback control they were able to observe single stepwise strokes of motion of about 11 nm in length, each with a maximum force of about 3–4 pN per molecule. Finer *et al.* used low ATP concentrations to minimize the magnitude of the effects of the considerable thermal diffusional noise compared to the step motion. Low ATP keeps the myosin locked to the actin filament for long times and thus minimizes diffusion. In the tweezer feedback system devised by Finer *et al.* [61]*, [218], the particle was locked to a fixed reference position and the entire tweezer beam, operating at fixed power, was servoed in space by an acousto-optic-modulator (AOM).

In the earlier levitation feedback scheme of Ashkin and Dziedzic [7]* the particle was locked to a fixed reference height by servoing the levitating power. The effect is equivalent. In the levitation

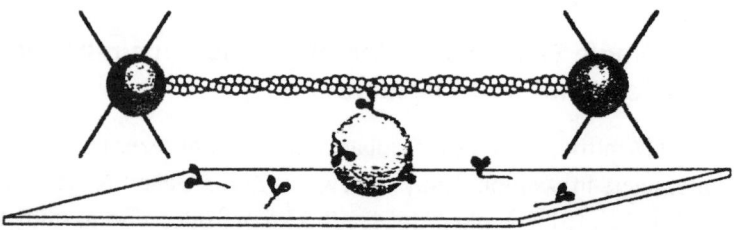

Fig. 7.7. Experimental geometry used to observe single molecules binding and pulling an actin filament. The filament is attached on either end to a trapped bead. These beads are used to stretch the filament taut and move it near surface-bound silica beads that were decorated sparsely with myosin molecules. Figure and caption from Ref. [61]*.

case the laser power is the measure of the applied force. In the tweezer case the voltage needed on the AOM to keep the particle fixed in space is the measure of externally applied force. The ability of feedback to compensate for laser power fluctuation gives a feedback scheme an advantage over interferometry in force measurement. As we shall see, the interferometric systems and feedback systems were the basis for many later tweezer experiments on kinesin and myosin motor molecules and other molecules. See Sec. 7.13 on optical feedback.

7.5. Measurement of Diffusional Motion and Stepping in Actin–Myosin Interactions

Malloy *et al.* [281] studied the interaction of myosin with a mutant form of *Drosophila* actin. They used low trap strengths to minimize the perturbing effects of the trap on the weakly coupled motor-actin filament system. The actin filament was suspended between two trapped spheres in a dumbbell arrangement. The authors made careful measurements of the combined diffusional displacement of the suspended dumbbell and the driven step size. The dumbbell diffusion was ~ 50 nm, which is large compared to the step size ~ 10 nm. With feedback they were able to resolve these two contributions. In other work Nishizaka *et al.* [282] measured the unbinding force of a single myosin molecule from an actin filament in the absence of ATP.

Ishijima *et al.*, in 1998 [283], using a similar tweezer dumbbell setup, showed that in the absence of ATP there was no step displacement. The distribution of diffusing bead movements was symmetrical about 0 nm. They also measured a decrease in diffusional amplitude as an indication of myosin attachment to the dumbbell.

7.6. Measurement of Myosin Step Size Using an Oriented Single-Headed Molecule

The question of step size for myosin is still not fully settled, since the orientation of the myosin heads relative to the actin filament, in general, is not known and various measurements give different results. A clever experiment by Tanaka *et al.* [284] used a synthetic single-headed myosin molecule with only the rod remaining of the missing head or neck region. They could orient the actin filament relative to the single myosin head and found a mean head displacement of 10 nm when aligned parallel to each other, 0 nm when perpendicular, and, strangely, 5 nm when opposite. This implies that measurements made with random myosin orientations may be underestimating the actual step size. This may account for some of the wide range in step sizes, from 4 to 15 nm, all measured by different groups using optical tweezer techniques.

7.7. Forces on Smooth Muscle Myosin and Use of Fluorescently Labeled ATP with Total Internal Reflection Microscopy

Guilford *et al.* [285] studied smooth muscle myosin with similar dumbbell techniques and observed increases of up to five times in the time spent in the bound state of the molecule. This probably reflects the ability of smooth muscle myosin to generate higher isometric force than regular muscle myosin.

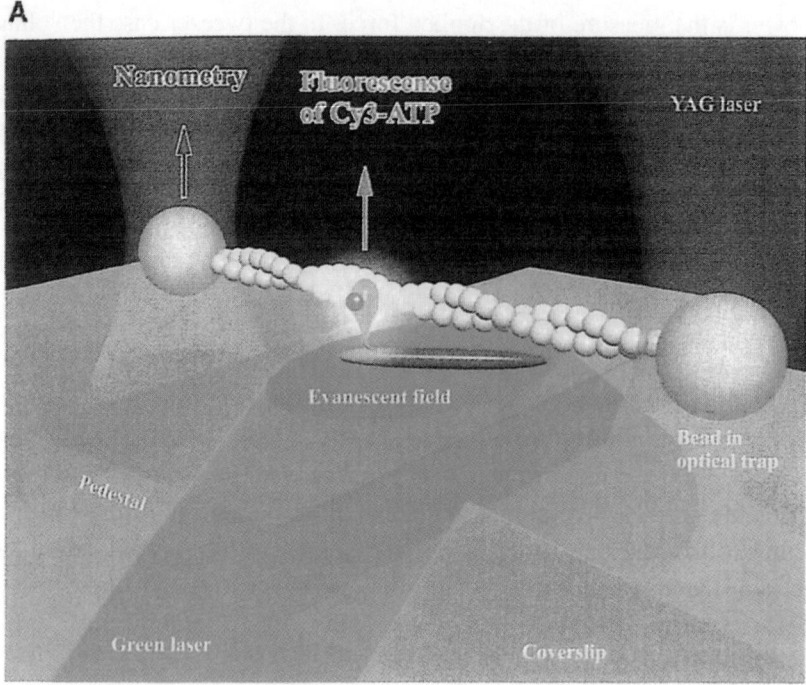

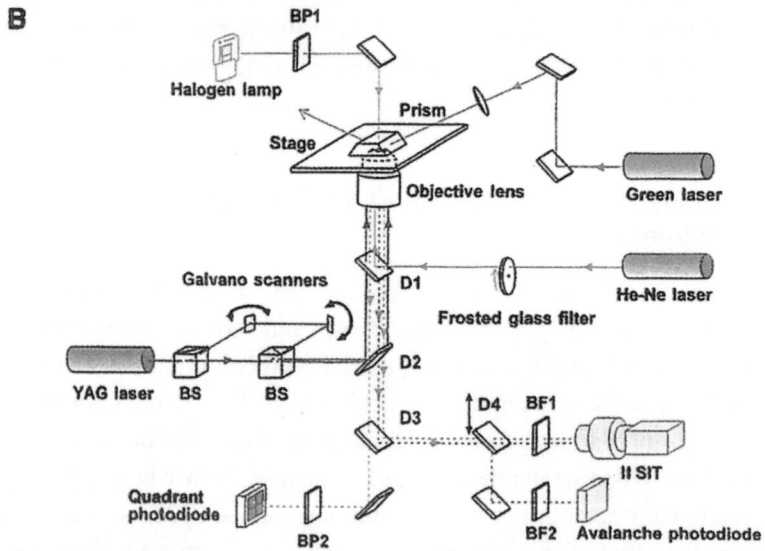

Fig. 7.8. Simultaneous Measurement of individual ATPase and Mechanical Reactions of Single One-Headed Myosin Molecules. (A) The experimental apparatus. A single actin filament with beads attached to both ends was suspended in solution by optical tweezers. The suspended actin filament was brought into contact with a single one-headed myosin molecule in a myosin-rod cofilament bound to the surface of a pedestal formed on a coverslip. Displacement or force due to actomyosin interactions was determined by measuring bead displacements with nanometer accuracy. Using total internal reflection fluorescence microscopy, individual ATPase reactions were monitored as changes in fluorescence intensity due to association-(hydrolysis)-dissociation events of fluorescent ATP analog, Cy3-ATP, with the myosin head. The diagram has been drawn upside down. (B) Optics. See Microscopy (in Experimental Procedures) for details. Figure 1 and caption from Ref. [283].

Ishijima *et al.* of Yanagida's group [283] further modified the force-measuring dumbbell assay to include techniques to follow fluorescent-labeled ATP during the actin–myosin interaction. This technique, called total internal reflection microscopy, detects ATP fluorescent-labeled fluorophores arriving at the surface of a totally internally reflecting prism. Figure 7.8, taken from Fig. 1 of Ref. [283] shows their setup. The measurements show the strong correlation between the onset of head displacement in the working stroke and the arrival of ATP (as predicted in the most popular strong coupling models of actin–myosin interaction).

At this point, general opinion among experts favors the strong coupling picture of interaction between actin and muscle myosin. However, despite all the accumulated new data, there is still controversy over the exact model for force production by muscle myosin. One such model invokes a one-way thermally driven diffusion mechanism called the "thermal ratchet". A paper by R. D. Astumian considers ratchet motion in considerable detail [286]. The basic principle of thermal diffusion to drive the motion of a single particle in one direction was very convincingly demonstrated experimentally by Faucheaux *et al.* of Libchaber's group in 1996 [287], using tweezers. They showed that diffusion could be unidirectional in a specially shaped ratchet-like optical potential. Astumian [286]. See also Sec. 7.17. Astumian also considers biased Brownian motion based on nonequilibrium chemical reactions.

Kitamura *et al.* presented evidence for loose coupling of actin and myosin in 1999 [15] using a microneedle manipulating technique. They found that hydrolysis of a single ATP molecule can give rise to a series of short-lasting 5.5 nm steps. This demonstrated a remarkable degree of control over a mechanical device. One advantage of the microneedle techniques is their ability to reduce the sizeable compliance between the actin filament and the dumbbell setup. Needles, however, are much more difficult to use and are not as adaptable as tweezers to rapid feedback control.

A lot of progress has been made and it is generally felt that further experiments using tweezer feedback systems [61]*, combined with fluorescent techniques [283] and with means to control myosin orientation, should further deepen present understanding of the actin–myosin force generating system at the single molecule level.

There was a recent conference on motor molecules in Japan, hosted by Prof. Yanagida, in which there was considerable discussion of the strong versus weak coupling issue of the actin–myosin interaction. See the report on this conference in Nature [12]*.

7.8. Observation of Two-Step Behavior of Myosin-I Using the Tweezer Dumbbell Technique

Other forms of myosin have been studied with markedly different properties from myosin-II, the muscle myosin. Veigel *et al.* in Malloy's group [288] have measured the working stroke of two forms of single-headed myosin-I: rat liver myosin-I and chicken intestine myosin-I. They found that the working stroke consists of two steps. The first step of about 6 nm occurs within 10 ms of actomyosin binding. They found another step of roughly 5.5 nm after a variable delay that depends on ATP concentration. No such second step could be found with single-headed sub-fragments of fast skeletal muscle myosin-II at the highest time resolution of about a millisecond. Presumably, the slow kinetics of myosin-I make the second step visible. The authors used the same dumbbell-type technique as

Finer *et al.* [61]* for mounting the actin filament over a roughly micrometer-sized sphere containing single myosin-I molecules. To improve the time resolution of their system the authors applied a 1 kHz oscillation to one of the two dumbbell tweezer beams and monitored the transmission signal of the second dumbbell tweezer beam on a quadrant detector. It is at present not known if the two-step behavior is peculiar to the two forms of myosin-I used in this experiment.

7.9. Study of Processive Class-V Myosins Using a Pair of Tweezer Traps

Mehta *et al.* [289] studied class-V myosin, one of 15 known classes of actin-based molecular motors involved in organelle transport. They used chick brain myosin-V, which is a double-headed molecule. Based on its function and biochemistry, it was suspected to be a processive motor. For example, myosin-V is known to transport vesicles to the endoplastic reticulum and in other regions where there are few motors and few actin filaments. This experiment was the first direct confirmation that myosin can act as a processive motor. They found that myosin-V can take at least 40–50 steps, on average, before releasing from the actin filament. The authors obtained some really impressive data using a modified two-beam dumbbell type geometry, in which they slowly (over a time span of ~3 s) displaced one of the two tweezer traps allowing the actin to go from a taut to a compliant state. Once the myosin-V attached, it went through a beautifully documented sequence of steps that demonstrated classic processive behavior. They measured resolved steps with a size distribution of 30–38 nm and a stall force of 3.0 ± 0.3 pN. The stall force was less than the 5–8 pN of kinesin. Also, the step size was much larger than the 4–17 nm range of kinesin.

7.10. Force versus Velocity Measurement on Kinesin Motor Molecules

Returning to the kinesin-microtubule story, after Svoboda and Block *et al.* observed the stepping motion of kinesin along microtubules [217]*, they measured the complete force–velocity relationship of single kinesin motors in considerable detail, as a function of ATP concentration, again using interferometry [290]. They found by applying calibrated pN retarding forces that the velocity decreased linearly with increased force up to a stall force of up to 5–6 pN. These measurements were made with varying ATP concentrations, which resulted in a greater than order of magnitude variation in velocity. It was observed during runs of kinesin displacement versus time, taken at different loads (powers) at saturating ATP concentrations, that at low loads (15 mW) the beads would escape the trap without detaching from the microtubules. At high loads (~60 mW) they would mostly pull off before leaving the trap. In either case the particles, once detached, would return to the center of the trap where they would reattach and the measurement could be repeated again and again on the same motor.

It is interesting to note that to measure the kinesin velocity it was necessary to make a correction for the flexibility of the kinesin molecule connecting the bead to the microtubule. They measured this flexibility directly in a simple experiment in which the microtubule plus motor are pulled by the piezoelectrically driven microscope stage relative to the fixed trap. Bead displacement versus time data taken at high loads and limiting ATP concentration often showed the 8 nm steps and even some examples of backward slippage of the motor by 8 nm.

7.11. Single Enzyme Kinetics of Kinesin

The abundance of new experimental observations on kinesin molecular motors has triggered work on detailed models of motion, the ATP hydrolysis cycle, and single enzyme kinetics. In 1994 Svoboda *et al.* [291] made a fluctuational analysis of motor protein movement and single enzyme kinetics. They experimentally studied the fluctuations in the displacement of silica beads driven by single kinesin motor molecules under low mechanical loads at saturating ATP concentrations. They found that the variance in particle position was quite low. A general fluctuation analysis of the data leads to quite general conclusions. For example, it suggests that single ATP hydrolysis with a cluster of several steps is unlikely and rules out the likelihood of thermal ratchet models. More will be said on both of these topics later.

A review of the molecular mechanics of kinesin by Block in 1995 [277] covers the early work on kinesin in the 1980s, the use of the gliding bead assay, the moving bead assay, the use of tweezers, step size, stepping force, force–velocity relationships, and possible molecular models of kinesin movement. See also commentary in Physics Today on molecular motors [292].

Articles such as this in Physics Today by B. G. Levi on biophysics serve an important function in alerting physicists to interesting problems.

7.12. Kinesin Hydrolyses One ATP Molecule per 8 nm Step

In 1997, Schnitzer and Block [293] performed an important experiment that indicates kinesin hydrolyses one ATP molecule for each 8 nm step it takes as it moves processively along a microtubule. This determination of the coupling ratio is crucial for calculating the correct model of force generation and motion. For the actomyosin system, as we saw above, this question of mechanochemical coupling is still not resolved. Knowledge of the stall force, step size, and number of ATP molecules hydrolyzed per step permits one to directly calculate the maximum efficiency of the motor. Using tweezers and their interferometric techniques to track motion with high spatial and temporal resolution, Schnitzer and Block use statistical analysis of fluctuations in motor speed as a function of ATP concentration to deduce the mechanochemical coupling. Measurements were made over three decades of ATP concentration, the motor velocity tracked linearly with ATP concentrations over two decades of ATP concentration until saturation of the velocity set in. The data taken at limiting ATP concentrations unambiguously support one ATP molecule per step and rule out two ATP molecules per step. The authors note that a problem in understanding still remains having to do with the 8 nm steps and the rather small size of the kinesin motor domain. The molecule appears a little too short to constitute a lever arm analogous to the structure of the myosin force generating heads. See Block's paper in 1998, entitled "Kinesin: What gives?" [197].

7.13. Feedback Control of Tweezers: Force Clamps and Position Clamps

The tweezer techniques used to study single molecules continue to evolve. In 1998, Visscher and Block [42] developed a new single molecule technique, which they call a molecular force clamp, for studying processive enzymes. It is based on optical feedback. Ashkin's original feedback control system for measuring force was based on locking a particle's position by servoing the laser intensity [7]*. Visscher and Block call this an "intensity-modulating position clamp". In their terminology, Finer *et al.*'s work on myosin force measurement used a "beam-deflecting position clamp" [61]* in which the trap was displaced by an acousto-optic modulator to keep position constant. The new technique of Visscher and Block was a "beam-deflecting force clamp" in which the trap followed a particle at a specified distance to maintain a constant force. They point out one can also have an "intensity-modulating force clamp" which keeps the force clamped by varying the intensity of a fixed trap.

The beauty of force clamps is that the force exerted by the trap on an object is dynamically maintained at constant tension, even if the particle is moving. This use of a force clamp solves the problem of the unknown "series elasticity" between the motor and microspheres by keeping the elasticity constant. This allows the true movements and forces to be measured with no need for elastic corrections. For the first time, using the moving force clamp trap, stepping data were acquired at high loads over distances as great as 200–300 nm. These long runs, which involve no elastic corrections, give the highest accuracy on kinesin step size of 8.3 ± 0.2 nm. In earlier work with fixed traps, corrections as large as 19% had to be applied to adjust for series compliance [294]. The article "Versatile Optical Traps with Feedback Control" [42] by Visscher and Block gives many practical details and references on the methods of feedback control, such as quadrant photodiodes, modulators, trap stiffness calibration and time-sharing of multiple traps. Equally useful for the experimentalist is the article "Construction of Multiple-Beam Optical Traps with Nanometer-Resolution Position Sensing" by Visscher *et al.* [295].

7.14. Study of Single Kinesin Molecules with a Force Clamp

Visscher, Schnitzer and Block used their beam deflecting molecular force clamp to take a huge amount of new data on kinesin dynamics over a wide range of conditions at different loads [219]. Knight and Malloy, in their "News and Views" article in Nature Cell Biology [296], characterized this work [42], [219] as "a technological break-through". Force clamps are much more difficult to achieve than position clamps used in actomyosin work. They require a computer system to take information on position and calculate the required acousto-optic modulator control signal in a time of a few microseconds. The computer also is used to store, display and analyze the data taken over wide ranges of both load and ATP concentrations. Constant loads, known to a fraction of a piconewton, are applied to the kinesin motor. See data in Fig. 7.9, taken from of Fig. 1 of Ref. [219], where the measured force is 6.5 ± 0.1 pN. Double logarithmic plots of velocity versus ATP concentration covering four decades of velocity and concentration at three different fixed loads are then fitted to the so-called "Michaelis–Menten" curves: $v = v_{max} [ATP]/([ATP] + K_m)$, where v_{max} is the maximum velocity, [ATP] is the ATP concentration, and K_m is the Michaelis–Menten constant. See Fig. 7.10, taken from Fig. 2 of Ref. [219]. It was known that the maximum velocity, v_{max}, at saturating ATP levels falls with increased load, but not how the Michaelis–Menten constant K_m varies with load. The new discovery was that K_m increased with load, that is, kinesin requires a higher ATP concentration to reach saturation at a higher load. The variation of K_m with load relates directly to the coupling of the chemical cycle to the stepping or mechanical cycle, and is quite model dependent. Measurements

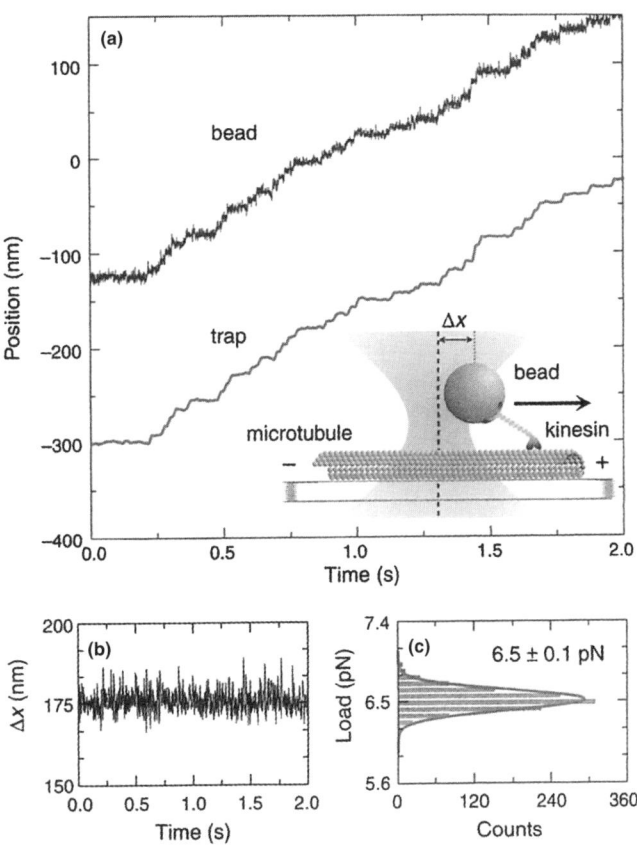

Fig. 7.9. Operation of the force clamp. (a) Sample record from the force clamp, showing kinesin-driven bead movement and corresponding optical trap displacement (2 mM ATP). Discrete steps of 8 nm are readily apparent. Inset, schematic representation of the motility assay used, showing the experimental geometry (not to scale). The separation between bead and trap was nominally fixed at $\Delta x = 175$ nm. Bead position was sampled at 20 kHz, filtered with a 12-ms boxcar window for the feedback on the trap deflection, and saved unfiltered at 2.0 kHz. (b) The measured bead-trap separation, Δx, for the record in (a), (c), Histogram of the displacements in (b), converted to force by multiplying by the trap stiffness (0.037 pN nm^{-1}). Solid red line is Gaussian fit to these data, yielding a load of 6.5 ± 0.1 pN (mean $\pm$ s.d.). Figure 1 and caption from Ref. [219].

were made of velocity versus load at a low 5 μM ATP concentration and high concentration of 2 mM ATP. See Fig. 7.11, taken from Fig. 3 of Ref. [219]. This is impressive data covering a range of loads from a few tenths of a piconewton out to the stall force values ~ 5.5 and 7 pN. Surprisingly, the high-resolution force clamp data are at odds with lower resolution data taken previously with fixed optical traps or microneedles (see Refs. 10–13 of Ref. [219]). The old data supported a loose coupling model of kinesin with a load-independent K_m. However, the above data argues for tight coupling of kinesin stepping to ATP hydrolysis over a wide range of forces with a single hydrolysis per mechanical advance. All these imply that kinesin operates at high efficiency at high-load conditions and low efficiency at low-load conditions. Furthermore, kinesin does not alter its coupling to improve its efficiency at low load. Tight coupling implies that thermal ratchet models of kinesin motion are ruled out [286], [287]. See also Ref. [12]* and the discussion in Sec. 7.17.

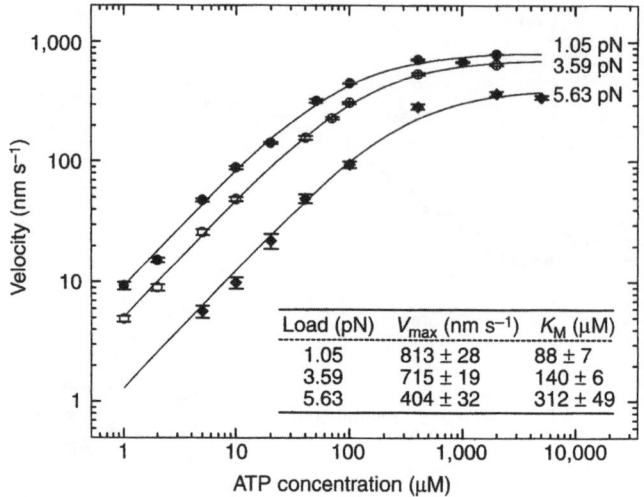

Load (pN)	V_{max} (nm s^{-1})	K_M (μM)
1.05	813 ± 28	88 ± 7
3.59	715 ± 19	140 ± 6
5.63	404 ± 32	312 ± 49

Fig. 7.10. Michaelis–Menten kinetics under load. Double logarithmic plot of the average bead velocity, v (mean ± SEM), versus ATP concentrataion for various loads (filled circles, 1.05 ± 0.01 pN, $N = 11$–102 runs; open circles, 3.59 ± 0.03 pN, $N = 8$–79 runs; diamonds, 5.63 ± 0.06 pN, $N = 19$–58 runs). Data were fitted to Michaelis–Menten curves (lines), $V = V_{max}[\text{ATP}]/([\text{ATP}]/K_m)$. Inset, fit parameters, V_{max} and K_m. Figure 2 and caption from Ref. [219].

7.15. Structural Measurements on Kinesin

As more and more detailed information about the motion of kinesin was accumulated, it became clear that there were problems in understanding on a structural basis how it worked. Its two heads seemed too small to span the two 8 nm steps that each motor head takes as it moves processively in one direction along microtubules from binding site to binding site [197]. This difficulty appears to be solved by new data on the detailed structure of the kinesin dimer as it undergoes conformational changes due to binding of ATP and its conversion to ADP. A series of structural measurements by a large group of workers from many institutions, Rice *et al.* in 1999 [272], found that when kinesin binds ATP to its front foot, it causes a 15-amino acid region called the neck linker to stretch out, make contact with the plus end of the microtubule, convert to ADP, and subsequently to swing the rear "foot" forward to start another cycle. Electron paramagnetic resonance (EPR), fluorescence resonance energy transfer (FRET), and cryo-electron microscopy (cryo-EM) were used to study the various steps in the cycle. Special chemical modifications of kinesin were made to focus on the neck region of kinesin for the EPR data. Dyes were added to the structure for the FRET measurements. Changes in conformation cause increases or decreases in the efficiency of energy transport between dye molecules. The position of the neck linker region was visualized directly in the cryo-EM experiments, using gold-labeled kinesin heads and difference mapping techniques to identify the various conformational states at the low temperature of 4°C. These structural measurements confirmed suspicions by Hancock and Howard that ATP hydrolysis of the leading kinesin head is needed to pull the lagging head off its old site and prompt forward movement [297] (see also Refs. [197], [271]). These two papers provide an excellent review of the state of kinesin knowledge as of 1998 leading up to the work of Rice *et al.* [272].

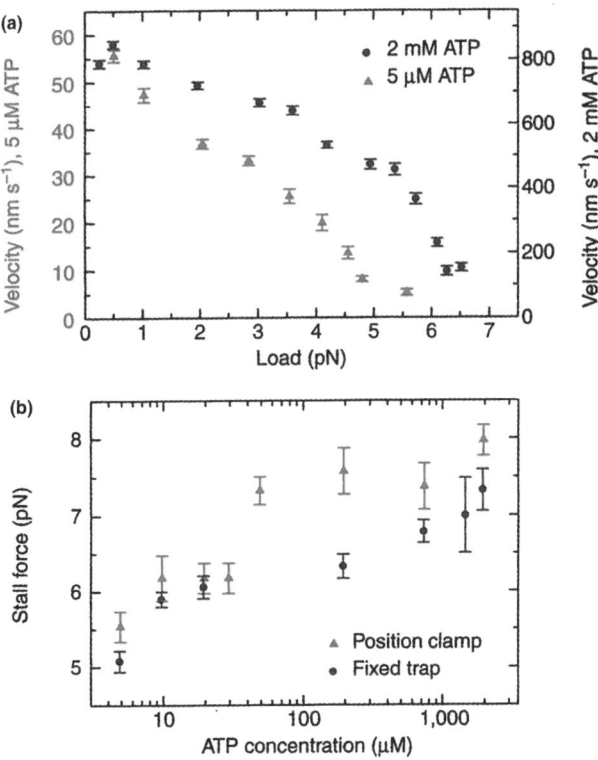

Fig. 7.11. Load dependence of motility. (a) Average bead velocity, v (mean $\pm$ SEM), versus applied load for fixed ATP concentrations (red triangles, left axis, $5\,\mu$M ATP, $N = 19$–57; blue circles, right axis, 2 mM ATP, $N = 37$–87). The velocity point at (5.6 pN, 5 μM ATP) is likely to represent an overestimate because beads which stalled completely ($v = 0$) were indistinguishable from beads lacking active motors, and so were not included in the analysis. (b) Stall force (mean $\pm$ SEM) versus ATP concentration, measured either with the position clamp (red triangles) or with a fixed optical trap (blue circles). Stalls had to last a minimum of 2 s to be included in the analysis. Data points represent an average of either 12–29 (position clamp) or 6–70 (fixed trap) stalls. Figure 3 and caption from Ref. [219].

7.16. Substeps within the 8 nm Step of the ATPase Cycle of Single Kinesin Molecules

In 2001, a Japanese group reported the resolution of the well-known 8 nm step of the processive kinesin motor molecule into fast and slow substeps, each corresponding to a displacement of ~ 4 nm [298]. To observe the finer details of the 8 nm stepping motion they developed a modified tweezer-trapping assay capable of resolving nanometer displacements with microsecond time resolution. This involved use of $\sim 0.2\,\mu$m diameter beads as handles to hold the single kinesin molecules. Such small beads have a faster response than the typical 1 μm beads used in previous work. This, however, makes it harder to resolve particle displacements. To overcome this problem, they observed the subwavelength beads using dark field illumination. The detector was a quadrant photodetector that only sees the scattered light from an off-axis red laser illuminating beam.

Data were taken by attaching beads to a microtubule and recording the displacement versus time as the motor pulled against an increasing optical force. In going a distance of ~ 100 nm in a time span of a few hundred ms, the restraining force rose from 0 to $\sim 8\,$pN. The data show the basic 8 nm steps very clearly, using a 20 kHz or a 0.5 kHz low-pass filter. It is necessary, however, to use quite a bit of statistical analysis to extract evidence of substeps hiding within the Brownian noise of the displacement versus time data.

Initially, the authors measured the response time of beads by spectral analysis of the bead fluctuations between 8 nm steps at different force levels. At zero force they found the response time was 15 μs. By analyzing data on 152 steps they then determined the initial rise time of an 8 nm step from a histogram plot of number of particles versus rise time measured with the 20 kHz low pass filter. They found that the rise time was more than two times longer than the free particle response time, indicating that the 8 nm steps do not occur abruptly.

As shown in Fig. 7.12(a), taken from Fig. 4(a) of Ref. [298], the 152 steps were divided into three groups, depending on rise time t. Group-I with $t < 50\,\mu$s, Group-II with t between 50 and 100 μs, and Group-III with $t > 100\,\mu$s. The average trace of the displacement versus time of the Group-I particles shows only a single fast ($t \neq 25\,\mu$s) rising 8 nm step. Group-II particles show a rapid ($t = 25\,\mu$s) first step of 4 nm, followed by a slow ($t = 40\,\mu$s) second step of 4 nm. Group-III particles had a similar fast 4 nm step, followed by a slower 4 nm second step.

These average displacement versus time curves are good evidence of substeps. Furthermore, as in Fig. 7.12(b), taken from Fig. 4(b) of Ref. [298], if one makes a histogram of the frequency of the distances between all pairs of particles in the average plots of distance versus time, one sees well-defined peaks at 4.2 ± 1.2 and $8.7 \pm 0.7\,$nm, again indicating that the 8 nm step is made up of two substeps of $\sim 4\,$nm each.

The authors speculate that the data showing 4 nm substeps could be experimental evidence of the conformational change that was deduced by Rice *et al.* [272] in their structural measurement of kinesin, as discussed above. Another possibility is real substeps along the 4 nm repeat distance of the tubulin molecule, which is made up of intertwined α and β helices, as shown in Fig. 6.3. More work must be done to further clarify these important new sub-step observations on kinesin.

7.17. Processivity of a Single-Headed Kinesin Construct C351 and the Brownian Ratchet

Although the understanding of the operation of the kinesin super-family of motor molecules on the molecular level seems to be converging rapidly, there still is plenty of room for surprises and controversy as the number of different types of motor molecules being studied increases. For example, recent work on a single-headed construct C351 of the kinesin molecule called KIF1A has shown that it behaves as a processive motor that can take more than 600 steps before detaching from a microtubule [299]. This exceeds the roughly 100 steps of kinesin II, the highly processive tightly bound two-headed motor. This is novel behavior since kinesin monomers produced by truncating one head are not processive but can give motion and stay attached to microtubules only if many monomer molecules are attached to a single test sphere (see Refs. 7–9 in Ref. [272]). Okada and Hirokawa describe the processive motion of the C351 construct as "very stochastic, it was a kind of biased Brownian movement". The question is what keeps the single-headed monomer from diffusing away after releasing from the microtubules. They believe that there is a mobile tether at work that holds

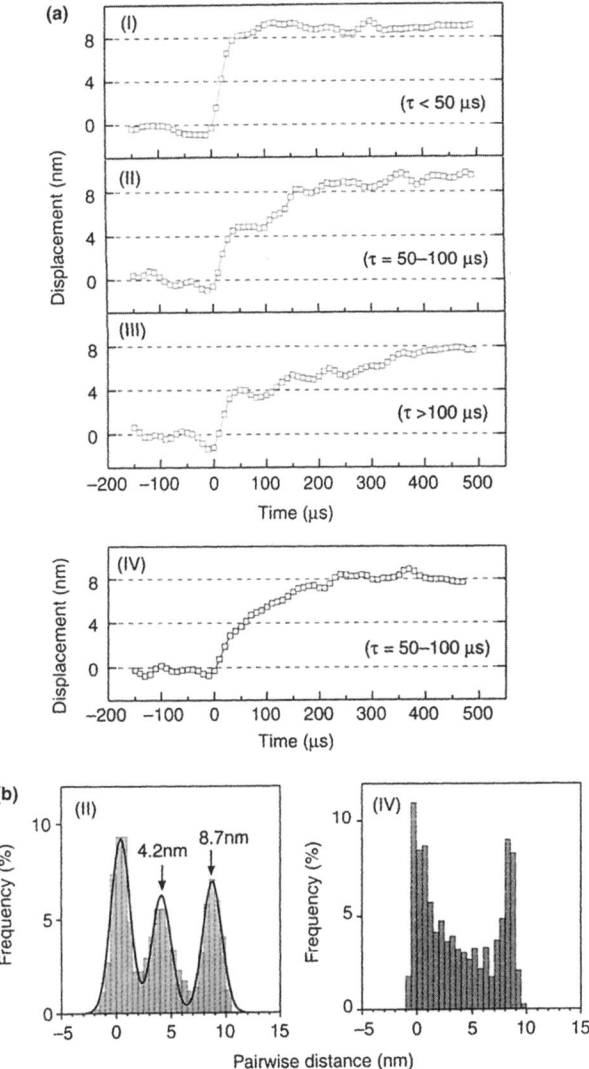

Fig. 7.12. Analysis of the rising phase of the 8 nm step. (a) Traces averaged over the rising phase of the 8-nm steps. Traces show the averaged rising phases of the 8-nm step, for which the rise times were 0–50 μs (group-I, $n = 90$), 50–100 μs (group-II, $n = 28$) and 100–400 μs (group-III, $n = 34$). Trace-IV shows an averaged trace of the rising phases of simulated 8-nm steps (see Method). Sampling time; 10 μs. (b) Histograms of the distances between all pairs of data points in the averaged traces of II and IV. The solid line is the sum of three Gaussian peaks fitted to the data by least squares. Figure 4 and caption from Ref. [298].

the C351 molecule close to the microtubule as it diffuses along. They cite considerable evidence that binding depends on a lysine-rich, highly positively charged motor domain (the K-loop) and a highly positively charged C-terminal region of tubulin (called the E-hook). They showed, by genetically engineering the number of lysines, that the binding is specifically affected. They also showed that addition of six extra lysines to the K-loop of conventional monomeric kinesin gives high processivity to a motor that previously lacked it. Okada and Hirokawa think the observations fit their model based on a Brownian ratchet mechanism [300]. See also Sec. 7.7. They state, however, that more experiments on the mobile tether are still needed.

It is generally felt that the new high-resolution single-molecule manipulation techniques (especially tweezers), when combined with new structural data, are making major improvements in our understanding of motor molecules [12]*, [271], [272]. The ability to engineer new single-molecular constructs and measure their behavior is a prime example of what is being called the new field of "molecular engineering".

7.18. Myosin-VI is a Processive Motor with a Large Step Size

In 2001, Rock, Spudich and their colleagues at Stanford [301]* extended their work on actin–myosin molecular motors with a study of the properties of myosin-VI motors. This very important motor is involved in intracellular vesicle and organelle transport in many different organisms and tissue types. It is unique among the myosin motor types in that it moves backward toward the pointed end of an actin filament, as opposed to the motion of other known myosins. Actin filaments are typically pointed away from the plasma membrane into the cell interior and this suggests that myosin-VI processes are involved in endocytosis. In this regard, myosin-VI is found to be localized in protrusions in the plasma membrane and in the trans-Golgi network. The highly processive motor myosin-V, which was studied by this group earlier, was found to move toward the "barbed end" of actin at the membrane surface. A motor molecule is processive if it advances through many catalytic cycles of motion before it releases from the guiding filament. In this study of myosin-VI it was found, using force feedback measurements at 1.7 pN values, that it, too, was highly processive. It advanced with an unexpectedly large irregular step size of $+30 \pm 12$ nm, based on the usual lever arm mechanism. It was also found that myosin-VI would take a few smaller steps backward of $\sim -13 \pm 8$ nm before proceeding on its forward path. The authors offer some speculations on the origin of the stepping mechanism. This work opens a whole new important area of motor molecule research and will very likely yield important information on the role of vesicles in cell processes.

7.19. Mapping the Actin Filament with Myosin

Steffen *et al.* [16], working at the Muscle and Cell Motility Unit at Kings College in London, have done an elegant experiment mapping out the "target zones", or sites, on an actin filament that are favorably oriented for attachment of myosin. Working with single S1 fragments of a myosin head attached to a 1.5 μm glass bead resting on a glass slide, the authors were able to locate the binding position along the length of an optically supported actin filament at which the myosin preferably binds. Figure 7.13, taken from Fig. 1 of Ref. [16], shows ten seconds of an experimental run measuring the position of the left-hand bead of the dumbbell while interacting with the myosin head. The data were taken using acousto-optics modulators to control the position of the dumbbell. For a loose trap with a stiffness of 0.04 pN/nm, one can see from the reduction in Brownian noise that the myosin head binds preferentially at sites more than 30 μm apart. The authors also moved the myosin head laterally along the actin filament with an acousto-optic modulator and recorded the frequency of attachments at different positions. The histogram of the distribution of the bound levelsdisplays the target zones

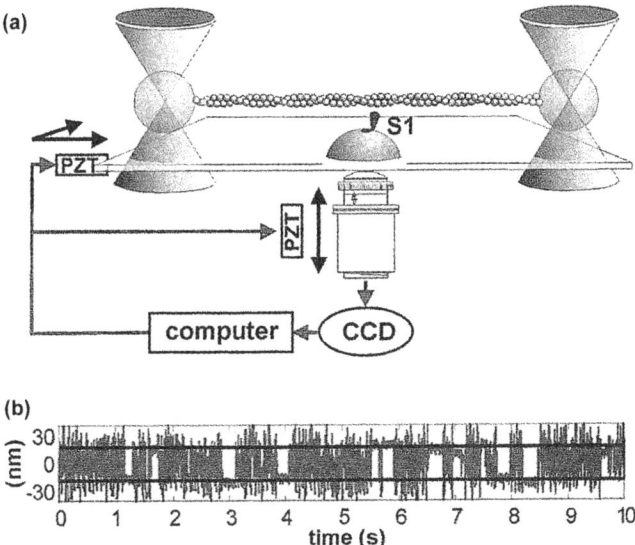

Fig. 7.13. (a) Schematic of the optical trap system. (b) The first 10 s of the position record of the left bead of the dumbbell while interacting with a myosin head, using traps with a combined stiffness of 0.04 PN/nm. A total of 28 binding/detachment events were detected (8), of which 12 are readily visible. A similar fraction of total to visible event counts applies to the whole 100-s record. Figure 1 and caption from Ref. [16].

very clearly and yields a value of ~ 36 nm for the helix repeat distances. See also Fig. 7.14, taken from Fig. 4 of Ref. [16], where the step sizes for myosins VI and V molecules are shown.

The ability to work at trap stiffnesses as low as 0.04 pN/nm and extract accurate positional data is impressive. This displays the unique ability of tweezers to work at sub-piconewton force levels. Traps with a sensitivity of $\sim 10^{-2}$ pN/nm were previously used by Block's group in their first measurement of the 8 nm kinesin stepping motion (see Sec. 7.3). Measurements at that level are not accessible with atomic force microscopy.

The authors point out that these results directly confirm that myosin interacts with a single strand and the actin is not free to rotate in this interaction. They also estimate that the interaction zone along the helix is made up of three monomers.

7.20. Development Regulation of Vesicle Transport in *Drosophila* Embryos: Forces and Kinetics

The ability to manipulate biological particles inside of living cells is quite unique to optical tweezers, as pointed out above (see Secs. 5.2.9 and 7.1). In the study of motor molecules with tweezers most measurements have been made *in vitro* using the handles technique, as we have seen. The reason for this is that, in general, the experimental conditions are most controlled *in vitro*. Ashkin *et al.*'s [17]* *in vivo* experiments with the amoeba *Reticulomyxa* were an exception. Ultimately, however, one wants to understand the more complex behavior of motor molecules as they develop and function within the more complex environment of the cell. In 1998, Welte *et al.* [18] studied the regulation of vesicle transport by molecular motors in *Drosophila* embryos as they grew and developed. *Drosophila*

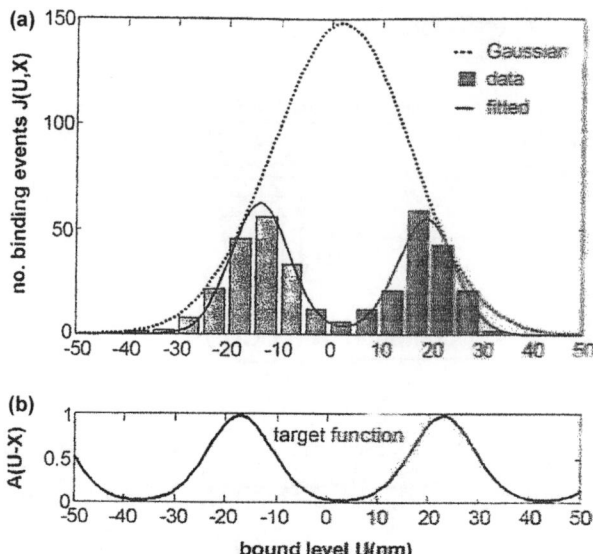

Fig. 7.14. The decomposition of a bimodal distribution of bound displacement levels into a free Gaussian and a target function. (a) The histogram of Fig. 3(a), shown with a Gaussian distribution of free displacements (Eq. (3)) and the fitted distribution based on the target function $A(U - X)$ (Eq. (4)), shown separately in (b). The fitted curve has $\alpha = 3.7$, period $b = 40$ nm, offset $X = 17.2$ nm, and a Gaussian with mean $h = 2.5$ nm and standard deviation $S = 13.5$ nm ($X^2/n = 0.23$, Levenberg–Marquadt method). For this distribution, the mean value is not well conditioned. For 0.04 pN/nm traps, the expected value of S is 10 nm; the excess width arises from residual drift and undetected events. Figure 4 and caption from Ref. [16].

embryos make an interesting model system for studying motor transport *in vivo* in part because of the many mutant strains available and in part because one can use squashed-mount embryo preparations that facilitate optical access to the samples. Using computer-controlled analysis, the authors studied the motion of storage organelles of neutral lipids, known as lipid droplets, $\sim 0.5\,\mu$m in diameter, with nanometer precision, as they moved forward and backward along oriented microtubule tracks. Forces were measured using the escape force method of Ashkin *et al.* [17]* to determine the stall force for the motors.

They compared the behavior of normal wild type embryos with embryos having the so-called *klarsicht* (clear view) mutation. The *klar* mutation embryos become more transparent due to accumulation of lipid vesicles during the various developmental phases. The authors found very clear changes in force and distribution of vesicles at the onset of gastrulation. For wild type embryos the stall force varied from 3 to 4 and 5.5 pN depending on the developmental stage, whereas for the *klar* embryos the stall force was only ~ 1.2 pN and did not change with the developmental stage. The force data are shown in Fig. 7.15, taken from Fig. 4 of Ref. [18]. The straight line shown in Fig. 4(C) leads the authors to speculate that the changes in stall force for vesicles in the various phases and in the *klar* mutation are solely due to changes in the number of motors attached to the vesicle in the various phases. Thus, they believe the *klar* mutation seems to act by affecting the coordination of the activity of molecular motors, which seriously affects embryo development in many aspects. The motor types responsible for forward and backward motion cannot be specifically identified, but it

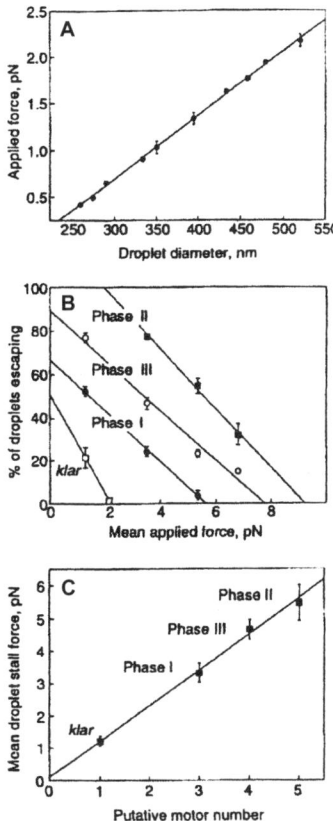

Fig. 7.15. Stall forces change throughout development. (A) Calibration. Droplet diameter versus escape force from the optical trap. Each point represents an average of four lipid droplets of similar size; the data are fit to a linear relationship (solid line). (B) Embryo stall force measurements. Percentage of droplets escaping the optical trap as a function of a variable trapping force, achieved by adjusting the laser power. Data points summarize measurements from 5 to 10 embryos, with 30–50 droplets tracked per embryo. The x-intercept for each fit line represents the extrapolated critical force, F_c, beyond which no droplets escape. (C) Mean droplet stall force, F_s, plotted against the presumed number of active motors. F_s values were computed from the F_c determinations in (B) by applying corrections, as described in the Experimental Procedures. The slope of the weighted line fit (solid line) corresponds to 1.10 ± 0.04 pN/motor; note that the unconstrained fit passes nearly through the origin (y-intercept 0.10 ± 0.09 pN). Figure 4 and caption from Ref. [18].

is suspected that the backward motion is due to cytoplasmic dynein, a motor with about the proper force to give the observed backward particle velocity.

7.21. Dynein-Mediated Cargo Transport *In Vivo*: A Switch Controls Travel Distance

In vivo experiments on *Drosophila* embryos reported on above showed that the transport of vesicles was governed by changes in the number of motors active at different phases of embryo development. This conclusion was based on studies of the effects of the *klar* mutation on vesicle transport. The authors also surmised that dynein was the active motor molecule driving vesicles in the backward direction.

In this more recent extension of the above work, the Princeton group of Gross *et al.* [302] showed that cytoplasmic dynein is indeed the active motor in backward lipid vesicle transport. The authors further deduced, based on the known dynein processivity, that a control mechanism must be at work controlling the travel distance of the vesicle cargo.

The conclusion that dynein is the active motor was based on a study of the effect of the *Dhc64C* mutation, which affects the function of cytoplasmic dynein. Analysis of force measurements on wild type embryos and *Dhc64C* mutant embryos in different phases of development gave a value for the force that a single cytoplasmic dynein motor exerts *in vivo* as 1.1 pN.

A detailed computer analysis of the statistics of lipid drop transport was made. Statistics were accumulated on the distance traveled in forward and backward directions, on the number and length of pauses and reversals in direction. From this the authors concluded that the particles do not stop because they detach from the microtubules, but rather a process exists that regulates the length and direction of the run. They believe the challenge for the future is to identify the molecular components of this switching process.

7.22. Kinesin Moves by an Asymmetric Hand-Over-Hand Mechanism

As techniques for studying single motor molecules advance and knowledge of their behavior increases, more detailed questions arise about the mechanism of their operation. For example, in kinesin the question of whether the 8 nm processive steps of these two-headed dimer molecules advance along microtubules in a hand-over-hand motion or by an inchworm type of motion, where

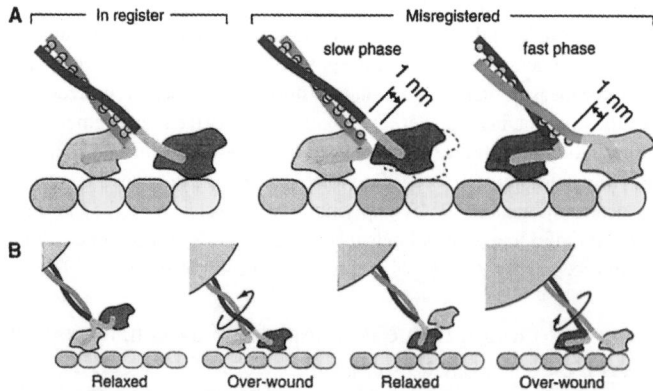

Fig. 7.16. Candidate models for limping in kinesin. (A) The misregistration model. Left (in register): The two-heads-bound state for native, non-limping kinesin, shown as it moves to the right along the microtubule lattice (tubulin α–β subunits, spaced pairwise by 8 nm, are indicated in gray and yellow). The coils of both heavy chains (blue and red ribbons) are in correct register. Middle and right (misregistered): A shift in registration of the coiled-coil by a single heptad changes the relative lengths of the head–neck linker regions by up to 1 nm. On alternate steps, this shift places one head (dark gray) farther from its binding site on the microtubule, reducing the stepping rate when this head attempts to take the lead (middle). When the partner head (light gray) steps, the shift is accommodated by slack in the longer tether (right). (B) The winding model. When one head (dark gray) steps forward, the neck coiled-coil (red and blue) is overwound relative to the relaxed state. When the partner head (light gray) leads, the coiled-coil is underwound. Asymmetry in torsional compliance slows the stepping associated with overwinding.

one head always stays ahead of the other. To help resolve this question Asbury *et al.* in Block's group [303] measured the stepwise motion of native and other kinesin constructs with varying molecular stalk lengths at high spatial and time resolution with their force-clamp apparatus. Surprisingly, they found that kinesin advances with a "limping" type of motion, in which the stepping rates can alternate between two different values at sequential steps. This type of asymmetric hand-over-hand limping motion rules out inchworm and symmetric hand-over-hand mechanisms. Different kinesin constructs with varying stalk lengths showed that the shorter the stalk length, the larger the ratio of dwell times of successive steps in the limping motion. See Fig. 7.16, taken from Fig. 4 of Ref. [303]. The authors proposed different candidate models for limping in kinesin, based on stretching of the head–neck linker region or possibly by a varying tension in a winding-model, involving two stalk coils. More experiments are needed to decide on the exact mechanism. Nevertheless, discovery of significant limping motion may require reexamination of some earlier chemical kinetic experiments.

CHAPTER

8

Applications to RNA and DNA

8.1. Observation of the Force of an RNA Polymerase Molecule as it Transcribes DNA

In 1995, Yin *et al.* of Block's group [212]* made another major advance in the study of motor molecules by extending tweezer force measuring techniques to new types of motors called the nucleic acid enzymes. These mechanoenzymes move along DNA molecules and carry out a variety of specialized functions. The authors studied one of the most important of these enzymes, RNA polymerase (RNAP), which synthesizes an RNA messenger copy of a DNA template. Remarkably, the movement of RNAP is powered by the free energy liberated by the nucleotide condensation process itself [305], [306]. Single RNAP enzyme movement along DNA previously had been observed *in vitro* by binding RNAP enzymes to a surface and watching them move DNA strands attached to beads [307], [308]. Yin *et al.* [212]* extended the *in vitro* technique by trapping the moving bead by tweezers and measuring the force generated by the RNAP as it pulled on the DNA being replicated. They found that RNAP behaves as an extremely powerful motor with a stall force >14 pN that progresses along DNA at speeds >10 nucleotides per second. Thus, RNAP is a true motor, like myosin, kinesin, and dyein.

8.2. Force and Velocity Measured for Single Molecules of RNA Polymerase

To improve on and extend these first measurements, a tweezer feedback position clamp was introduced in 1998 by Wang *et al.* [309] and force versus velocity measurements were made on *E. coli* RNAP. The improved setup increased the estimated stall force to ~21–27 pN. See Fig. 8.1, taken from Fig. 3 of Ref. [309]. This stall force exceeds the 5–7 pN of the kinesin processive motor. The data showed that the velocity of the RNAP motor is nearly independent of load, out to the stall load. In this respect, it is reminiscent of the behavior seen by Ashkin *et al.* [17]* of cytoplasmic dynein motors in their *in vivo* experiments done with *Reticulomyxa*. Ashkin *et al.* pointed out that constant speed at

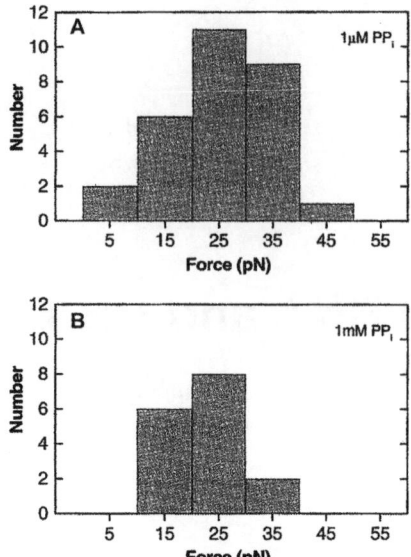

Fig. 8.1. Histograms of RNAP stall force. (A) Stall force distribution for 1 mM NTP, 1 μM PP$_i$. Average force was 25.0 $\pm$ 1.7 pN (mean $\pm$ SEM; $n = 29$). (B) Stall force distribution for 1 mM NTP, 1 mM PP$_i$. Average force was 22.9 $\pm$ 1.9 pN ($n = 16$). The difference in values is not significant (t text, $P = 0.44$). Figure 3 and caption from Ref. [309].

variable load is similar to the behavior of synchronous alternating current (AC) in electrical motors. Wang *et al.*'s data imply, with regard to the kinetics, that the enzyme carries out one condensation reaction on the growing RNA chain per base pair moved along the DNA. Thus RNAP is a tightly coupled motor.

Two theoretical models have been proposed [310], [311] that fit the general features of the experimental data. In one, conformational changes take place within a flexible RNAP molecule, which allow it to deform and remain in register with the DNA molecule over a variable distance of 0–8 base pairs [311]. In the other, the end of the RNA molecules undergoes thermal fluctuations against the physical barrier of the enzyme's catalytic site. This acts as a "Brownian ratchet" driven by the free energy of the nucleotide condensation that can generate considerable force [310]. These studies open a new way of studying the DNA transcription process [309]. Both models give load-independent velocities up to stall. Wang *et al.* [309] are particularly interested in looking for the sequence specific aspects of RNAP behavior, such as transcriptional pausing, stalling, and arrest, etc. Behavior of this sort has been observed. They anticipate that "single molecule experiments will supply new insights into this fascinating molecule".

8.3. Measurement of the Mechanical Properties of DNA Polymer Strands

We have seen that tweezers is especially useful for the study of the mechanical properties of single molecular motors and single mechanoenzymes. There is also a close relation between these molecules and the single-molecule biopolymers that act as the rails along which these motors move. These molecular pairs, each consisting of a motor molecule and a biopolymer, have evolved together and can only be fully understood together. However, much of the early work on biopolymers focused more on their physical properties as polymers than on their biological properties because biopolymers, such as DNA in solution, are such convenient models of single molecule polymers. See, for example, work by Smith, Finzi, and Bustamante *et al.* [312] on stretching experiments using magnetic beads and fluid flow. Perkins, Smith, and Chu [23] first applied tweezers to study DNA polymers in an experiment demonstrating the tube-like motion of a single, extended, fluorescently labeled DNA strand as it relaxed through a dense entangled polymer solution. This was done by attaching a sphere to the free end of a fluorescent length of DNA and dragging it through the dense intertwined mass of polymer to "write" the capital letter "R" with the tweezers. When the trap was turned off and the sphere released, it relaxed following the exact R-shaped path in reverse through the spaghetti-like mix of DNA polymer. This behavior lends strong support of the reptation model of de Gennes. This model helps to explain the viscoelastic behavior of many biological materials [19]*. Perkins *et al.* [313] also performed experiments on relaxation of a model DNA polymer strand in a dilute aqueous solution and compared the results with theories of dynamic scaling.

8.4. Measurement of Flexural Rigidity of Microtubule Fibers and Torsional Rigidity of Microtubules and Actin Filaments

Measurements were also made of the flexural rigidity of microtubules by Kurachi *et al.* [214] by attaching polystyrene beads to the two ends of microtubules and bending them with tweezers. Felgner *et al.* [314] studied the rigidity by directly bending free-floating single microtubules with tweezers. The torsional rigidity of a single actin filament was deduced by Yanagida *et al.* [215] from a measurement of the rotational Brownian motion of a single filament suspended from a freely rotating sphere held in a tweezers trap.

8.5. Measurement of the Stretching of Double- and Single-Stranded DNA

In 1992, Smith, Cui, and Bustamante [22] used tweezers to extend some of their earlier work done with magnetic beads [312] on the stretching of single-stranded DNA. They found that at 70 pN of force, there was a reversible transformation of the double-stranded DNA (dsDNA) to a single-stranded unraveled form of DNA. They measured the elastic response of the individual dsDNA and single-stranded DNA (ssDNA) molecules. In other work in 1966, Cluzel *et al.* [315] studied DNA stretching, using a glass microneedle technique. Finally, in 1997, Wang *et al.* in Block's group [24], following the work on RNAP [212]*, made force–extension (FX) measurements with single molecules of DNA under a wide range of buffer conditions, using an optical trapping interferometer modified to incorporate feedback control. With this technique, FX measurements could be made on lengths of DNA as short as $\sim 1 \, \mu m$, "with unprecedented accuracy, subjected to both low (~ 0.1 pN) and high (~ 50 pN) loads". Complete data sets were acquired in a minute. The $\sim 1 \, \mu m$ DNA sample lengths used were about 2000 base-pairs long, which is more than an order of magnitude less than previously used. Wang *et al.* determined the elastic compliance of DNA in order to determine the correct RNA polymerase translocations in their previous RNAP mechanoenzyme experiment [212]*. At low and moderate forces the molecule is purely elastic or entropic and behaves as an ideal spring characterized by a polymer persistence length, L_p. Roughly speaking, the persistence length is the length over which a linear polymer persists without bending. At higher forces the molecule continues to stretch into a region that is called enthalpic, in which irreversible molecular changes occur due to breakage of bonds. This is characterized by an elastic modulus K_0. Recently, theories of elasticity have been extended to account for both entropic and enthalpic contributions to stretching [316], [317]. By fitting the measured data to these theories over the full range of extensions, Wang *et al.* [24] could simultaneously determine the persistence length and the elastic modulus to high accuracy. The authors also observed a significant effect of the ionic strength on DNA stiffness. This is due to the electrostatic shielding of DNA's surface charge by the surrounding electrolyte. The experimental results found are remarkably close to the theoretical expectations. The authors expect techniques similar to those used here should be broadly applicable to the rapid and accurate determination of micromechanical properties for a variety of biopolymers.

8.6. Polymerization of RecA Protein on Individual dsDNA Molecules

There are many other molecules associated with the complex functioning of DNA machinery that can be studied with optical tweezers. In 1999, Shivashankar *et al.* in Libchaber's group [318] studied the polymerization of the protein RecA on individual dsDNA molecules. RecA is known to have a strong structural effect on DNA. It is widely found in most cells and is known to play an important role in DNA repair. During genetic processes the structural modification of DNA is a key step in sequence recognition. See Refs. 1–14 in Ref. [318]. The group measured the elongation of DNA by assembly of RecA, driven *in vitro* by ATP, and obtained values for the persistence length of the RecA-DNA complex by observing the force–extension relationship and hydrodynamic recoil. They found that the overall polymer length has a transient dynamics controlled by ATP hydrolysis. For complete polymerization of DNA by RecA the persistence length is about four times larger than for the naked molecule. The authors followed the detailed kinetics of the growth process by making measurements as a function of time, using ATP and a modification of ATP called ATP(γs). They also derived an assay based on the use of ^{32}P isotope in ATP. This allowed them to monitor the [ATP]/[ADP] ratio throughout the experiment.

8.7. Study of Elasticity of RecA-DNA Filaments with Constant Tension Feedback

Almost simultaneously with Shivashankar *et al.* [318], Hegner, Smith, and Bustamante [25] also studied the polymerization and mechanical properties of single RecA-DNA filaments using a force-measuring laser trap with feedback in real time. These authors also point out that the polymerization of RecA protein plays an important role in many of the important processes in the functioning of DNA machinery of eucaryotic species as diverse as yeast and humans. Among these, DNA processes are homologous recombination, in which two DNA sequences are compared over a distance of several base pairs, DNA repair, and chromosomal aggregation during cell division. Their basic technique, used previously, was to stretch the DNA fiber between two polystyrene beads, one bead held by suction to a glass micropipet and the other held by a two-beam laser trap. RecA is known to interact with

ADP or ATP, both of which affect the stability and structure of the RecA-DNA helical complex. Upon binding ATP or a nonhydrolyzable variety of ATP called ATP(γs), RecA undergoes a transformation into a high-DNA-affinity binding form. The filament forms a right-handed helical filament that is active in DNA processes. The complex can also polymerize in the presence of ADP in a low-DNA-affinity form, which will not initiate DNA processes. These filaments have shorter pitches and are called inactive filaments. The authors studied the longitudinal stiffness and bending rigidity of the many complex helical filaments as a clue to understand many functions of this important molecule and its homologs in many organisms.

Taking advantage of their feedback scheme, Bustamante *et al.* measured polymerization speeds as a function of time under constant 60 pN force conditions, using a feedback apparatus, in which the fiber position was moved to maintain constant tension. With feedback, the polymerization speed remained high until polymerization was complete. Measurements were also made without feedback. Without feedback, the polymerization speed fell steadily as the fiber increased its length and the force dropped from 60 to 0 pN. These data indicate that stretching the filament increased the polymerization speed and that there is a significant barrier for nucleation of RecA in ds DNA, which can be overcome by strongly stretching the helical filament. The authors also suggest that because the polymerization continues even as the tension drops, the presence of RecA molecules must catalyze the addition of subsequent RecA monomers on the DNA.

Data were fit into the inextensible worm-like-chain (WLC) model. See Refs. [316], [319], [320] taken from their paper on this subject. At low forces, the fiber behaves as an inextensible WLC. This yields the entropic elasticity of the chain. From the fit of the data near 5 pN, where the fiber approaches the limit of its elastic stretch, one can deduce the contour length and the persistence length. For forces beyond 10 pN, the elastic behavior deviates strongly from the WLC model and one is in the regime of enthalpic elastic behavior. Measurements were made for dsDNA, ssDNA, as well as dsDNA-RecA-ATP and ssDNA-RecA-ATP fibers. The data show a striking difference in the persistence lengths for dsDNA of 53 nm versus 0.75 nm for ssDNA. When decorated with RecA in the presence of ATP, the persistence length dsDNA-RecA-ATP = 936 $\pm$ 120 nm and ssDNA-RecA-ATP = 860 $\pm$ 130 nm. These increases show that the addition of RecA-ATP makes a more than order of magnitude increase in rigidity. Also, the persistence length of the filament is insensitive to whether one or two DNA chains are present at its core. Additionally, using ATP(γs) instead of ATP makes essentially no difference in persistence length. Measurements were also made with ADP, which showed that overstretching is required for the binding of RecA dsDNA. These detailed mechanical experiments on DNA-RecA fibers are providing insight into the functions of this complex molecule in DNA. The authors hope to learn more about the homologous recombination processes that occur in DNA and also learn more about conversion of active filaments to inactive filaments.

8.8. Possible Role of Tweezers in DNA Sequencing

Given the ability of optical tweezers to manipulate single strands of DNA, it is natural to ask whether tweezers can play any role in the current rush to sequence the genomes of various organisms. This question was addressed in a "News" article in Science referred to above, called "Watching DNA at Work" by R. F. Service [321], in which he reported on the 1999 Biophysical Society meeting. After interviewing many of the currently active researchers studying DNA, Service concludes, "the key to single molecule manipulation of DNA was development of optical traps". He says that a number of groups working with DNA think that use of optical traps could give rise to new approaches to sequencing based on decoding of just single copies of DNA, rather than using Taq polymerase machines that create thousands of copies of DNA fragments, as in current techniques. The advantage would be to be able to use DNA segments as long as 50 000 base pairs, rather than cutting short segments, each about 1000 base pairs long, which must be pieced together in the correct order by complex computer programs. This would simplify and speed up the sequencing process. The technique based on single long DNA segments relies on the protein exonuclease, which can chop off tens of thousands of individual bases one at a time from a long strand of DNA. Of course, one must be able to detect the single bases with some sort of laser-based detector. There are four, or possibly five, laboratories trying to detect single bases reliably one at a time by attaching different fluorescent molecules to DNA for each base pair or by using native DNA with no fluorescent tabs. Schemes without fluorescence are faster, but the signals are weaker. Single bases have been successfully detected, but there are still some problems with background fluorescence [320]. Taq polymerase machines have won the race, but it remains to be seen if tweezer techniques will play any useful role in the future.

8.9. Study of the Structure of DNA and Chromatin Fibers by Stretching with Light Forces

The packing of a cell's long DNA molecule making up its genome into the compact nucleus of a cell is done with the help of chromatin fibers. In 2000, Cui and Bustamante [322]* studied chromatin and measured the forces that maintain the various structures of single chromatin fibers as a function of ionic strength. Chromatin fiber is made up of DNA and several types of histone proteins. It exists in a number of discrete forms. There is the simplest form called the "beads on a string" form, in which the DNA wraps itself twice around a nucleosome core, made up of eight histones having a diameter of about 10 nm, and then strings on to the next nucleosome, and so on, as in a string of beads. At higher ionic strengths there is a transformation to a thicker, more condensed fiber arrangement in which the DNA and nucleosomes are packed more tightly into a 30 nm spiral or solenoidal fiber arrangement with six nucleosomes per turn. See, for example, Fig. 8.2, taken from Figs. 9–16 of "Structure of Chromatin" on pp. 330–333 of Ref. [323]. On a still larger scale these thicker fibers are packed into the arms of the chromosome.

As pointed out by Cui and Bustamante [322]*, DNA and the organization of chromatin have been studied intensely for about 25 years. Although the structures of the various forms of chromatin are known, little is known about the magnitude and the origin of forces that maintain and stabilize the various compact forms during the cell cycle. The dynamics of this transformation is related to the transcription process, since only part of the DNA in chromatin is available for transcription during the folding and unfolding at a given time.

The experiment was performed with a length of chromatin fiber connected at its ends to two avidin-coated polystyrene beads. One bead was held in a force measuring two-beam laser trap and the other end to a moveable glass micropipette. Measurements were taken of force versus extension by moving

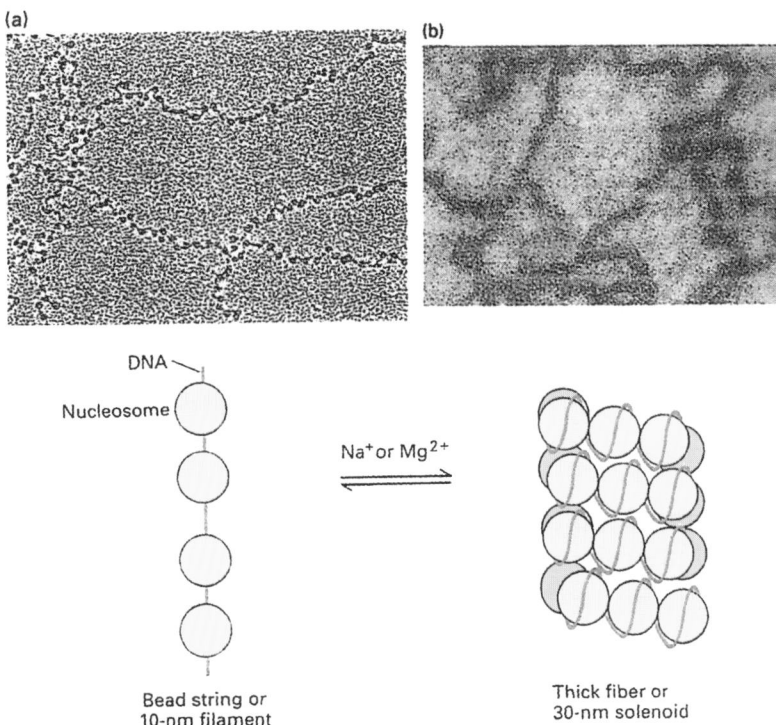

Fig. 8.2. Electron micrographs and schematic diagrams of chromatin fibers from the nuclei of *Drosophila melanogaster* cells. (a) Chromatin shows characteristic beads-on-a-string appearance at low ionic strength. The "beads" are nucleosomes, which are about 10 nm in diameter. (b) Chromatin extracted at high ionic strength (or following addition of divalent cations) consists of thick, 30-nm-diameter fibers; the nucleosomes are in a solenoidal arrangement. Figure 9.16 and caption from Ref. [323].

the pipette by small $0.05\,\mu$m steps until the stretching force rose to various predetermined levels. Stretch-release cycles were also determined at various ionic strengths. At the low ionic strength and a low force limit of 7 pN, the stretch release curves were completely reversible and repeatable. In the intermediate force regime the relaxation curve no longer coincided with the stretch curve and one had hysteresis. The process was still repeatable, however. At high force above 20 pN, the FX curves were neither reversible nor repeatable. For each successive stretch, it took less force for the same stretch, indicating an unraveling of the chromatin fiber. At higher ionic strength it took more force to achieve the same stretch, indicating the fibers are more condensed at high salt levels.

The low ionic FX data were fit to a theory using the extensible worm-like chain, or beads on a string model of Marko [324] (see also Fig. 8.1) and yielded a value of the persistence length of 30 nm for forces below 20 pN. As the fiber was pulled at physiological ionic strength there was a continuous transition from the 35 nm diameter form to the 10 nm diameter beads on a string form; that is, from the closed to the open form. Once the fiber was converted entirely to the stretched state, further stretching caused the spiral or solenoidal pitch angle to open up. The maximum length of the fiber at full extension was found to be 3.05 times the original length of $1\,\mu$m.

The authors were also able to see evidence of plateau regions in the FX curves indicating transitions between the two forms of chromatin fiber states. They even obtained an estimate of the internucleosome attraction energy as the separation between nucleosomes changed during the stretch cycle. See Fig. 8.2 for an electron micrograph showing the beads on a string form of chromatin. This paper shows how, with the help of tweezers, one can get quantitative data and new insight into the functioning of a quite complex biomolecule.

I should point out that Cui and Bustamante call their two-beam trap a "tweezer trap". This is incorrect terminology, since tweezers specifically refer to the single-beam gradient trap. The two-beam trap used by them has advantages over the single-beam tweezer trap in that it uses longer working distance, lower numerical aperture lenses to form the two-beam trap at the common focus. It is much less maneuverable than the single-beam tweezer trap, but where it is used in a fixed force measuring application it is quite convenient. Occasionally one sees the use of the expression "tweezering" to indicate any type of manipulation by any kind of optical trap. I hope this usage does not spread.

8.10. Condensation and Decondensation of the Same DNA Molecule by Protamine and Arginine Molecules

An elegant experiment by Brewer, Corzett, and Balhorn [325] was performed on condensation and decondensation of the same DNA molecule by protamine and arginine molecules in solution. The DNA in sperm and certain viruses is condensed by protamine and an associated set of arginine binding domains into toroidal subunits; a form of packaging that inactivates the entire genome until fertilization or reproduction occurs. Using tweezers, the authors manipulate $1\,\mu$m fluorescent beads with attached individual strands of DNA stained with YoYo + dye in a specially designed flow cell. They were able to trap a flowing bead and its attached DNA strand and transfer it into a stream of moving protamine molecules or arginine molecules and watch as the stretched DNA strand condensed into a tightly bound, compact helical coil. As with many other DNA structures, the coiled torus structure has been extensively studied by light scattering, electron and atomic force microscopy and by elasticity measurements (see references in Ref. [325]). However, there were no direct observations on the toroidal formation process or of the kinetics of the condensation and decondensation induced by protamine and arginine. By fluorescence they could see the shrinking of the DNA length and the formation of the coils when the DNA was inserted into the protamine or arginine. They determined rates of DNA length change and the corresponding protamine and arginine on and off rates. Because of the flow geometry, there were none of the problems due to competing aggregation reactions. The kinetic data bear on the activation of sperm DNA once it enters an egg cell. The authors suggest that the tweezer techniques developed here can be used to study the kinetics and biophysics of other protein-DNA interactions.

8.11. Non-Mendelian Inheritance of Chloroplast DNA in Living Algal Cells Using Tweezers

A very interesting experiment was performed by Nishimura *et al.* [326]* on the non-Mendelian inheritance of chloroplast DNA in the living single-celled green alga *Chlamydomanas reinhardtii* using tweezers. In this work the active digestion of uniparental chloroplast DNA was observed in a single isolated zygote using fluorescence and Polymerase Chain Reaction (PCR) techniques. The genetic material of each algal cell consists of a nucleus with its DNA and several chloroplast nucleoids containing chloroplast DNA and several mitochondrial nucleoids with mitochondrial DNA. Usually the alga reproduces asexually by splitting into two identical cells. When stressed, however, it can reproduce sexually. In this case, the cell divides into two same size (isogamus) gamete cells, denoted by mating type positive (mt^+) and mating type negative (mt^-). The mt^+ gamete has mt^+ chloroplast (cp) nucleoids and mt^+ mitrochondrial (mt) nucleoids, whereas the mt^- gamete has mt^- chloroplast nucleoids and mt^- mitochondrial nucleoids. The mt^+ and mt^- gametes become free swimming, each with its own type of DNA molecules. Subsequently, when two opposite gametes meet and fuse, the result is a zygote with double the number of DNA molecules. The zygote and its two gametes are visible under a microscope as separate entities within the zygote for about an hour. They only fuse totally after an hour or so. In the meantime, one can observe the nuclei and nucleoids by fluorescence. It was known that the fluorescence from the mt^- cp nucleoids disappeared preferentially by a digestive process in the zygote within an hour, but the details of this disappearance were not observed on a molecular basis using single zygotes. With tweezers the authors were able to isolate a single living zygote and observe changes in time of the cpDNA molecules during digestion. Separation of a single zygote or even organelle from a heterogeneous suspension of cells was accomplished using a special isolation chamber. Within a time of only 5–10 min the cell or organelle was cut off from the suspension, dropped into a PCR tube for 1 min of centrifugation and DNA analysis. It was demonstrated that the cpDNA and mtDNA could be detected from a single gamete after only two rounds of PCR amplifications. Gene sequences from a single chloroplast (or mitochondrion) could be detected.

The results show that the mt^- cpDNA molecules are digested during the 10 min that the fluorescent mt^- cp nucleoids disappear. The mt^+ cpDNA and mtDNA are somehow protected during the digestion of mt^- cp nucleoids. There is an evidence that gene expression or protein synthesis by mt^+ is important in the digestive process. All these activities occur before the gametes fuse within the zygote. Much remains to be learned about the process and the specific nucleases responsible for digestion.

The authors point out: "preferential disappearance of fluorescent chloroplast or mitochondrial nucleoids of uniparental origin also occurs in ferns and higher plants". They believe that the ability to study the chloroplast and mitochondrial genomes of single chloroplasts and mitochondria is of increasing importance. It is a way of avoiding the confusion of conventional biochemical analysis, which is based on an analysis of a mixed population of cells. This is clearly a pioneering paper and the details of the authors' technique are worth studying.

8.12. Measurement of the Force and Mechanical Properties of DNA Polymerase with Optical Tweezers

Wuite *et al.* in Bustamante's group started work in 2000 [213]* on using tweezers to study the mechanical properties of T7 DNA-polymerase. Just as RNA-polymerase functions as a molecular motor, so, too, does DNA-polymerase function as a molecular motor, converting chemical energy into mechanical energy. It is one of the cell's important mechanoenzymes. It can catalyze DNA replication and also has an exonuclease (nucleotide removing) activity. The rate-limiting step for replication is believed to involve a conformational change between ssDNA and dsDNA. An optical

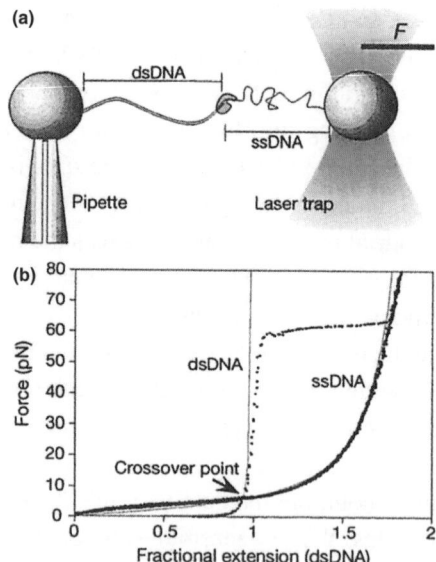

Fig. 8.3. Optical trap setup. (a) A 10 416-base pair plasmid DNA fragment was prepared as described and attached between two beads, one held on the tip of a glass pipette, the other in an optical trap. Single-stranded DNA was obtained by using the force-induced exonuclease activity of T7 DNAp to remove any desired length of the nontemplate strand. The end-to-end length of the DNA was obtained by video imaging of the bead positions, and the force (F) was measured using the change in light momentum which exits the dual-beam trap. (b) Force–extension data for dsDNA and ssDNA (dotted lines), compared with the wormlike chain model using ssDNA and dsDNA persistence lengths of 0.7 and 53 nm, respectively (solid lines). Difference between ssDNA and WLC curves at low tension is due to partial hairpin formation and disappears if magnesium is removed from buffer. Figure 1 and caption from Ref. [213]*.

trap setup was devised, in which a roughly 10 000 base-pair DNA fragment was attached between two beads, one end held on a glass pipette and the other in a tweezer trap. It was shown in an assay based on the differential elasticity of ssDNA and dsDNA that the motor can work against a maximum tension of about 34 pN. See Fig. 8.3, taken from Fig. 1 of this paper, showing the optical setup and the FX data for dsDNA and ssDNA. It was found that there was a force-induced 100-fold increase in exonucleolysis above 40 pN. Estimates of the mechanical work done by the enzyme show that T7 DNA polymerase polymerizes two bases per catalytic cycle. Data on the replication rate in bases per second were taken at different tensions using constant force-feedback by holding the strand at constant end-to-end length. These data gave the limiting tension, or stall force (at which replication ceased), of 34 ± 8 pN. The authors state that, "Future single-molecule experiments should elucidate the mechanism of the force-induced exonucleolysis and provide additional insight into the mechanochemistry of DNA polymerase".

8.13. Reversible Unfolding of Single RNA Molecules by Mechanical Force

Liphardt *et al.* in 2001, working with Bustamante at Berkeley [327], have made some significant advances in the understanding of molecular folding and unfolding, using three representative types of single RNA molecules and optical tweezers.

The first molecule was a simple double-helix hairpin type RNA, the second molecule with a triple-helix structure containing a triple-helix junction, and somewhat more complex molecule containing additional bulge-like structures. The RNA molecules were attached between a pair of $2\,\mu$m beads with the help of RNA–DNA handles. One bead was held in a force measuring optical trap and the other was attached to a micropipette connected to a piezoelectric drive. All three molecules could be unfolded by slowly applying force to their ends. Surprisingly, the molecules did not unzip in stages. When the force reach ~ 14 pN, this length suddenly increased by ~ 18 nm, consistent with complete unfolding, in an all-or-none transition from all folded to all unfolded. These length changes occurred in a time less than the 10 ms experimental time resolution. The molecules were also observed to be bi-stable within some very narrow critical force ranges and to hop back and forth between folded and unfolded states.

The authors show that the process follows simple two-state statistics with an activation energy ΔG, which they can determine with a fraction of a percent precision. These experiments show the advantages of tweezers for studying single molecules with piconewton and sub-piconewton force levels.

In commentary on this paper by Fernandez, Chu, and Oberhauser in Science [328], they point to the advantages of using simple "one-dimensional (1D) structures", such as RNA hairpins, where length is the only significant variable, to study folding and unfolding of proteins. They are particularly struck by the so-called Markovian all-or-none kinetics with steep force dependence of the observed dwell times. Such behavior is strikingly similar to the behavior of single-ion channel kinetics. See Fig. 8.4, taken from Ref. [328], showing single-ion channel kinetics and RNA hairpin unfolding kinetics. They speculate as to why there is an all-or-none kinetics for the RNA hairpins. The part of the answer may lie in thermal fluctuations. They point out that a trapped bead with a spring constant of 0.1–0.03 pN/nm will fluctuate on average ~ 6–12 nm. This is a considerable fraction of the entire extension of the hairpin. These commentators look forward to similar experiments on other isolated protein hairpins, in order to gain further physical insight into the protein-folding problem.

8.14. Grafting of Single DNA Molecules to AFM Cantilevers Using Optical Tweezers

Shivashankar and Libchaber [329] have developed a method of grafting and manipulating a single preselected DNA molecule using a combined atomic force microscope (AFM) and optical tweezers. In a demonstration of this technique they first tethered one end of a DNA molecule to a cover slip and then attached the other end to a latex bead of $\sim 3.0\,\mu$m diameter. They then manipulated the latex bead with its single attached DNA molecule under an optical microscope until the bead was in contact with the AFM silicon cantilever. The heat of the tweezer laser beam on the cantilever served to graft the bead to the cantilever. Pushing the bead against the cover slip and popping the bead off could release the grafting. To complete the demonstration the authors measured the force versus extension of the single DNA molecules using AFM. The results are very much like those obtained with tweezers (see, e.g., Ref. [24]), but with a much reduced force sensitivity because of the stiffer cantilever spring. The authors think the combined tweezer-AFM technique will be useful in DNA sensors and bioelectronic devices. They visualize uses in studying protein–DNA interactions with

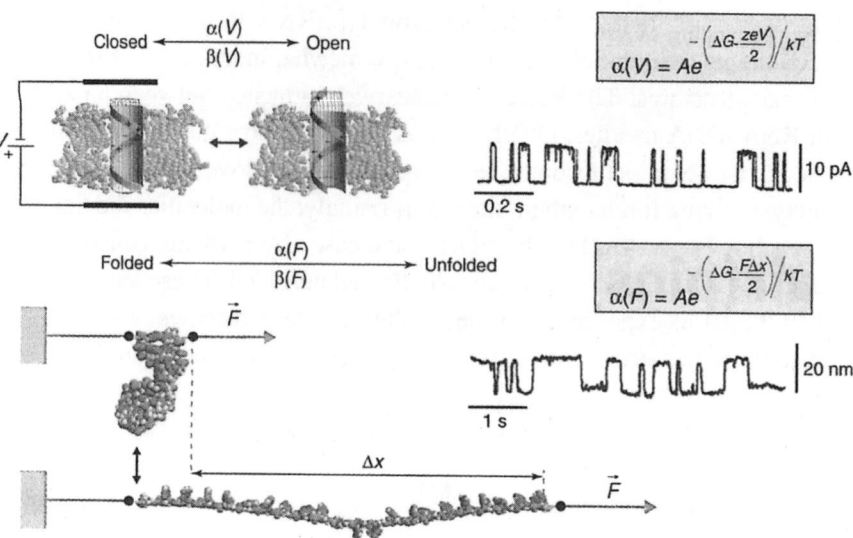

Fig. 8.4. Single-ion channel and-RNA hairpin unfolding kinetics. The opening and closure of an ion channel by an electric field is a highly cooperative event, leading to all-or-none fluctuations between the closed and open states (upper panel). The equation in the inset relates the opening rate constant α to the energy required to open the channel, where ΔG is the height of the activation energy barrier at zero voltage, e is the elementary charge, and z is the number of gating charges that move in the electric field. The trace on the right illustrates a typical recording of the activity of a single ion channel. The all-or-none unfolding of an RNA hairpin (lower panel) can be triggered by a mechanical stretching force, F, applied to the 3' and 5' ends. A mechanical force increases the probability of unfolding by exponentially speeding up the rate of unfolding α and decreasing the refolding rate β. The trace on the right shows a recording of the all-or-none changes in length from a single RNA hairpin. Figure and caption from Ref. [328].

known DNA sequences grafted to silicon substrates. Of course, other molecules could be attached to the cantilever using this combined tweezer-AFM technique.

8.15. Structural Transition and Elasticity from Torque Measurements on DNA

Extensive measurements have been made on the structural properties of DNA and RNA using optical tweezers and other techniques to apply controlled forces along their lengths. Less is known about the responses of these molecules to torque. Bustamante and colleagues have devised an ingenious apparatus for directly measuring the effects of torque on individual strands of DNA molecules, with the help of tweezer techniques [330]*.

Single DNA molecules 22 kilobase (kb) pairs long were stretched between two antibody-coated beads. One was attached to a pipette and the other was held in a two-beam force-measuring trap. A smaller third bead was attached at a specific point 14.8 kb away from the pipette and acted as a rotator. Torsional stress was built up in the 14.8 kb section through locking the rotor in place by applying a fluid flow while rotating the pipette. An optical force feedback system was used to measure positions and apply constant tension and prevent any twisting of the lower portion of the DNA. When the flow was stopped, torque caused the rotor bead to spin about the axis. The authors could observe the angular velocity in revolutions per second, the change in twist in turns and the change in length in nanometers. Torque was obtained from the angular velocity and rotational drag of the rotor. The torque and the angular twist of the rod-like molecule at low twist gave the torsional elasticity. Values found were $\sim 40\%$ higher than generally accepted.

As the number of turns increased to ~ 75 turns, the torque increased linearly until a transition was reached in which part of the molecule converted from the normal B-DNA state to an overwound state designated P-DNA. See Figs. 2 and 4 of Ref. [330]*. As the number of turns was further increased, while maintaining a 7 pN load, the fraction of the molecule in the overwound state continued to increase. The total length at constant torque decreased. A plot of the constant torque values versus force for both over-stretching and under-stretching is called the phase diagram of the molecule. The experimentally observed diagram was in general agreement with a simple theoretical model.

Thus, starting in the fully overwound P state, the molecule behaved as a constant-torque wind-up motor capable of repeatedly producing thousands of rotations. The overstretched molecule acted as a force–torque converter in the region where the P state relaxed to the B state. For the sample considered, the generation of about 35 pN of constant torque during 275 turns was possible over a period of about 400 s.

It occurs to me that a direct analogy exists between DNA motors of this type and the common child's wind-up rubber band motors for toy model airplanes. The authors pointed out that such motors can be useful for studying the properties of molecules or, more generally, for applying constant torque in nanotechnology.

8.16. Backtracking by Single RNA Polymerase Molecules Observed at Near-Base-Pair Resolution

Block and his colleagues, using optical tweezer techniques, were the first to extend the study of the motion of single molecules to other mechanoenzymes, such as RNA polymerase [331]*. In 2003, this group, working with improved apparatus capable of near-base-pair resolution, was the first to be able to observe the backtracking of *E. coli* RNA polymerase during RNA synthesis. Backtracking is thought to be a crucial step in the proofreading and error correcting mechanism of this mechanoenzyme. Backtracking of typically five base-pairs is associated with the so-called "long pauses" of varying lengths of time from 20 s to more than 30 min, during which time the molecule can cleave off and discard sections of RNA containing flawed bases, prior to resuming transcription. See Fig. 8.5 taken from Fig. 1 of Ref. [331]*, illustrating RNA polymerase transcription and proofreading. Short pauses were also observed to occur from 1 to 20 s without backtracking and presumably were not associated with error correction. These experiments were reported elsewhere [332].

In the backtracking experiment a DNA strand containing the rpoB gene of *E. coli* and its promoter sequence was suspended between two independent feedback-controlled beads in optical traps that could monitor position, with low drift. Base-pair resolutions of < 0.3 nm were achieved in averages of multiple scans. After long pauses detailed measurements were made of the position of the RNA polymerase molecule versus time (in seconds) at each of three distinct phases of motion (i.e., backtracking, pausing, and recovery).

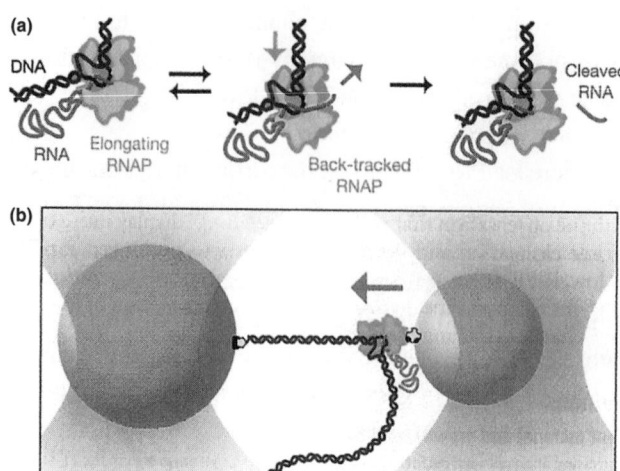

Fig. 8.5. RNA polymerase transcription and proofreading studied by optical trapping. (a) During normal elongation, RNAP (green) moves forward (downstream) on the DNA (blue) as it elongates the nascent RNA (red). At each position along the template, RNAP may slide backward along the template, causing transcription to cease temporarily. From the backtracked state, polymerase can either slide forward again, returning to its earlier state (left) or cleave the nascent RNA (right) and resume transcriptlonal elongation. (b) Cartoon of the experimental geometry employed for opposing force experiments (not to scale). Two beads (blue) are held in separate optical traps (red) in a force-clamp arrangement. The smaller bead (right) is bound to a single moleculeof RNAP, while the larger bead (left) is bound to the downstream end of the DNA by noncovalent linkages (yellow). During transcriptional elongation, the beads are pulled together. Nearly all the motion appears as a displacement of the right bead (green arrow), which is held in a comparatively weaker trap.

Data were also taken using constant force feedback, with force either opposing or assisting transcription. Such a force had a strong effect on the duration and frequency of backtracking pauses. Measurements were made on the effect on pauses of the ribonucleotide analog inosine triphosphate (ITP), which mimics guanosine and forms weak Watson–Crick pairs. The authors also studied the effect of the transcription factors GreA and GreB, which stimulate the cleavage of short segments of backtracked RNA. The observed changes in both the frequency and duration of long pauses were consistent with the proofreading hypothesis.

The authors think that the error correcting mechanism being studied with RNA polymerase using this new high-resolution technique may also apply to many polymerase systems, including both prokaryotes and eukaryotes.

The experiments reported here on long pauses and the subsequent experiments on short pauses described later include some of the most sophisticated measurements in the topic of transcription and error correction.

8.17. Ubiquitous Transcriptional Pausing is Independent of RNA Polymerase Backtracking

In 2003, Steve Block's laboratory built and made use of the highly stabilized optical tweezer trap to make extensive measurements of the pausing of RNA polymerase molecules during transcription of a gene of *E. coli* DNA [332]. The segment of the rpoB gene chosen for this study contains no regulatory pauses. The transcriptional pausing is a highly variable process. The majority of the ~ 2400 pauses studied were brief, lasting from about 1 to 6 s and occurred with equal probability along the gene segment. The frequency of occurrence of short pauses was ~ 0.7 per 100 base pairs. Much less frequent were longer pauses of about 20 s to more that 30 min. These long pauses were associated with backtracking events and were strongly affected by either hindering or assisting forces that were applied. These long pauses are associated with the process of transcriptional error correction. See the discussion in Sec. 8.16.

The short pauses, however, were found to be unaffected by applied forces ranging from -37 pN hindering forces to 27 pN assisting forces. Using a simple single barrier free energy model for backtracking-induced pausing, the authors showed that the absence of dependence load implies the absence of any backtracking motion, even by as much as a single base pair. Thus, short pauses arise from a different mechanism. The authors propose that small structural rearrangements within the polymerase enzyme itself are responsible for the observed short pauses. Much remains to be learned about the short pauses. The high-resolution tweezer technique developed in this work offers a new approach to the study of transcriptional problems in general.

8.18. RNA Polymerase can Track a DNA Groove During Promoter Search

The 2004 paper in PNAS [333] concerns itself with the process by which various RNA polymerase enzymes bind to DNA and search for the promoter sequence that initiates transcription. Various mechanisms have been proposed for the motion of the polymerase molecule, involving hopping between adjacent DNA segments or a sliding motion that follows a helical groove in the DNA molecule. Such a groove-tracking motion would impart a torque to the DNA that could then be experimentally detected.

Groove-tracking test apparatus was built by the authors, in which a 13 kb fragment of DNA containing no strong promoters was stretched between a ~ 910 nm diameter sphere held in a tweezer trap and an RNA polymerase molecule that was attached rigidly to a glass plate. When the glass plate was moved with a piezoelectric ceramic drive, dragging the DNA along a nearby polymerase molecule, a torque was transmitted to the tweezer-held sphere. This was detected optically by following the motion of a small ~ 90 nm fluorescent sphere permanently attached to the bead. Tracking was observed over lengths of about 80 Å for a time of about 1 s. This corresponds to a DNA rotation of ~ 2.3 turns. Because of the elasticity of DNA, only a portion of the twist is transmitted to the bead. The detection involves following the trapped sphere's motion on a statistical basis over many molecules, using a video camera to record the motion. The homogeneity of the sample was sufficient to allow such statistical averaging.

However, the motion could be clearly distinguished from normal Brownian motion of the sphere in the trap, in the absence of DNA, or the motion of the bead attached to the DNA in the absence of the dragging motion. These results directly show the existence of groove tracking. The authors think that such tracking could be a universal search mechanism for specific sequences.

8.19. The Bacterial Condensin MukBEF Compacts DNA into a Repetitive, Stable Structure

Bustamante and coworkers [334] have used optical tweezers to study the mechanism by which condensing molecules compact DNA in chromosomes. This process is crucial for cell division because compacted DNA is easier to split between daughter cells than its expanded form. The authors used an assay in which the DNA molecule was suspended between two spheres attached to a micropipette on one end and a tweezer trap on the other end. An *E. coli* condensin called MukBEF was used to tuck up and thus compact lengths of DNA, as shown in Fig. 8.6, taken from the commentary of X. Zhuang in Science [335]. When the tweezers pulled with a constant force of $\sim 20\,\text{pN}$, a transition occurred and the molecule underwent a series of decondensations and was stretched out, as shown in the figure.

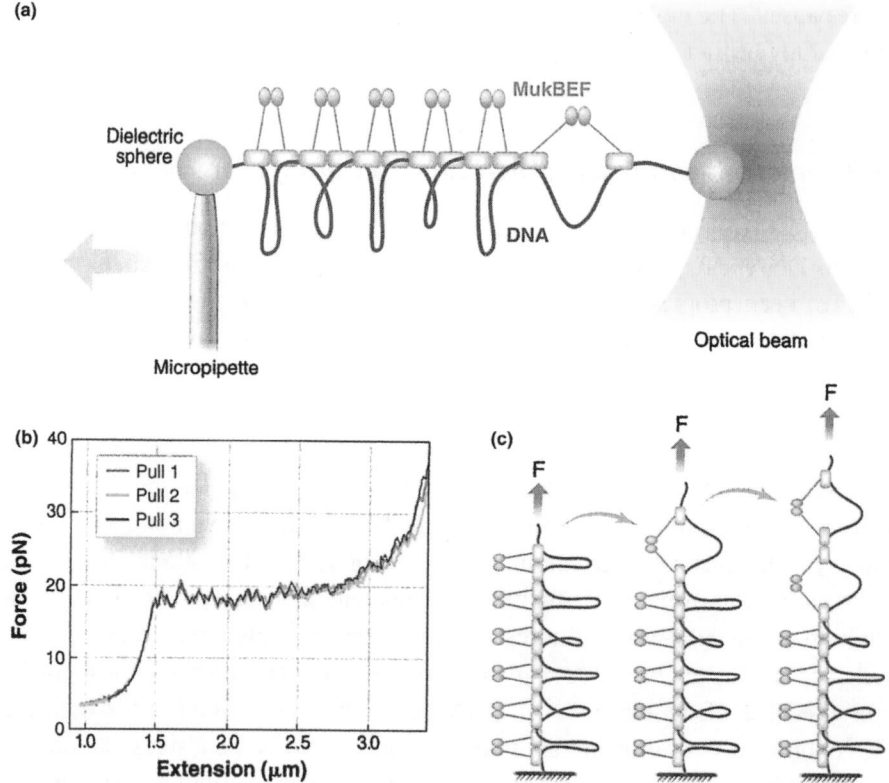

Fig. 8.6. A nip and tuck for DNA. Unraveling a DNA-MukBEF complex with optical tweezers. (a) In the experimental setup of Case *et al.*, polystyrene beads (beige spheres) were attached to the two ends of a piece of double-stranded DNA (purple). One bead was held by an optical trap (orange) and the other by a micropipette (gray handle). By moving the micropipette away from the optical beam, the investigators obtained force-versus-extension curves for DNA bound to the bacterial condensin MukBEF (blue). Here, only the SMC dimers (MukB) are shown; the non-SMC subunits, MukE and MukF, are believed to bind to the head domains of MukB (blue rectangles). (b) Force-versus-extension curves obtained for three consecutive pulling runs of the DNA-MukBEF complex. (c) A model of force-induced decondensation of the DNA-MukBEF complex.

The results show that ATP plays an important regulatory role for MukBEF condensation. It was observed that supercooled DNA was trapped in a condensed structure to which it returned with no addition of energy. The authors propose a new model for the condensation mechanism.

Zhuang was very impressed by the many elegant experiments performed with biological molecules using optical tweezers. In her commentary she summarizes a number of recent tweezer experiments with biological molecules. She also refers to other impressive results using other single-molecule force spectroscopy techniques.

8.20. Forward and Reverse Motion of RecBCD Molecules on DNA

S. Block and his colleagues [336] have initiated a study of single processive DNA-based motor enzyme RecBCD molecules, using new high-resolution tweezer trapping methods. RecBCD is an enzyme complex of *E. coli* bacteria that exhibits both helicase (unwinding) activity and nuclease (strand cutting and repair) activity. The individual RecBCD molecules were attached to the microscope cover glass. Enzyme molecules bind to the blunt end of a double-stranded length of DNA ~ 7.1 kb long, prepared by PCR techniques. The far end of the DNA was attached to a $\sim 0.5\,\mu$m diameter polystyrene sphere that was held in an optical tweezer trap. The trapping instrument incorporated a force clamp feedback capability that maintained a preset force up to ~ 8 pN.

They observed fine-scale uniform motion, smooth down to the detection limit of about 2 nm, indicating a motor stepping size below six base-pairs. Motion with constant velocity over hundreds to thousands of base-pairs occurred, often interrupted by quick switches in velocity, or pauses in motion of mean length of 3 s. The molecule was observed to occasionally reverse direction and slide backward. The fact that the velocity varied differently at different times to load changes was interpreted as due to changes in the functional state of the enzyme-DNA complex. Observations indicated that backsliding is not simply a load-induced reversal of RecBCD helicase activity. Elasticity measurements made during backsliding indicated persistence lengths considerably shorter than for dsDNA, revealing the presence of a ssDNA loop. Also, forward moving molecules proceed against a 2 pN force for long periods of time, while 2 pN of force applied to backsliding complexes had little effect on slippage. Lowering the applied external force could reduce backsliding.

The authors believe that backsliding is evidence for a different structural and functional state. They also speculate that the ability to backslide and then restart may be a way of overcoming mechanical obstacles, such as tangles, and allow the cellular function of DNA replication, repair, and degradation to continue.

The authors conclude that RecBCD-DNA complexes can exist in a number of functionally different states that persist for many catalytic turnovers. Much still remains to be explored. For example, one could also incorporate a χ-sequence into the DNA that acts as a helicase, cutting one strand, at which point a RecA enzyme can bind and perform yet another function. For background reading see "*E. coli* RecA and RecBCD Proteins Promote Recombination", pp. 481–485 in Ref. [323].

8.21. Direct Observation of Base-Pair Stepping by RNA Polymersase

Block and his group have made a major advance in technique with the development of an ultra-stable optical trapping system that has a positional resolution accurate to one angstrom. With this technique they have tracked the motion of an RNA polymerase molecule as it steps along a double-helix DNA molecule and fabricates a single RNA replica [717]*.

Two factors were responsible for the increase in resolution. The first was the replacement of the air surrounding the external optics by low index of refraction helium gas. This gave an order of magnitude reduction in the positional noise of the dumbbell pair of lasers supporting the DNA strand. The second factor was the replacement of the usual electronic feedback force clamp by a novel passive force clamp. In the passive clamp the power of one of the supporting lasers is decreased so that the bead sits on the edge of its trap near the peak of the force, where $dF/dx = 0$. At this point the force stays essentially constant for small displacements. This occurs on a real-time basis without the use of electrons, as in the past. See Sec. 7.13.

High-resolution transcription data taken with this setup clearly showed for the first time that RNA polymerase copies DNA one base-pair at a time with an average step size of 3.7 ± 0.6 Å. The authors also measured pausing, backstepping under assisted loads, and backtracking under hindering loads. Such observations were made previously, but with lower resolution feedback force clamps. See Secs. 8.16 and 8.17.

High-resolution measurements were made to study the driving force mechanism of RNA polymerase. Two types of models were considered: a power stroke model, in which the motor acts as a coiled spring that is periodically released; and a Brownian ratchet model, in which RNA polymerase fluctuates by Brownian motion and gets periodically locked in the forward position.

The authors took data on the force versus elongation velocity in base-pairs per second during transcription. They did this over a wide range of assisting and hindering loads at saturating and sub-saturating nucleotide concentrations. The analysis of the data strongly supports the Brownian ratchet model. See also Secs. 7.7 and 7.17.

A second paper describing the passive all-optical force clamp in detail was published in Physical Review Letters [718]. The authors applied the force clamp to study the important problem of molecular folding. They did experiments with a single DNA hairpin, in which a single DNA strand pairs with itself. They were able to measure the folding and unfolding transition at the angstrom level. It seems clear that the ability to follow single molecules with one-angstrom resolution as they move through water in real time opens a host of new possibilities.

CHAPTER

Study of the Mechanical Properties of Other Macromolecules with Optical Tweezers

9.1. Stretching and Relaxation of the Giant Molecule Titin

Titin is another important biopolymer that plays a major role in muscle function. It is a gigantic elastic protein about 1 μm long that serves to preserve the structural integrity and elasticity of relaxed muscle by flexibly attaching the myosin thick filaments to the Z-disk structures of the elementary sarcomeres. Titin keeps the actin thin filaments and myosin thick filaments in proper alignment. Figure 9.1 is a copy of Figs. 22–24 from p. 878 of Ref. [323] showing the role of titin in organizing the sarcomere in striated muscles. A number of experiments in 1997 used tweezers [14], [216], [337] to pull and reversibly unfold the single immunoglobulin and fibronectin domains that make up a single titin molecule. Part (A) of Fig. 9.2, taken from Fig. 5(A) of the review of Mehta *et al.* on "Single-Molecule Biomechanics with Optical Methods" [196], sketches the experimental setup of Ref. [216] to pull single titin molecules. The molecule is shown under tension with several of its domains unfolded. Part (B) of Fig. 9.2, taken from Ref. [337], shows the hysteretic behavior of the force-extensive curve, above 50 pN. The force levels off as the domains start to unfold. During relaxation the extended domains only refold themselves at low force, due to the low elasticity of the unfolded domains. Curve 5B does not resolve the individual unfolding and folding events. Tskhovrebova *et al.* in Simmons's group in Ref. [216] applied a jump in tension by making a sudden step in trap position. They then were able to watch the fiber tension relax in time, step by step, as the domains unfolded one at a time, as seen in Fig. 9.2(C). In Fig. 9.2(D), we see results of an AFM experiment by Rief *et al.* [14] in which they pull on a short length of titin eight domains long. Due to the high stiffness of the AFM cantilever, one observes large drops in force as each of the eight domains unfolds. An initial force of about 200–350 pN drops by a factor of about five as each successive domain opens.

Analysis shows that the unfolding rate constant should depend on the applied force [338], [339]. Also, the probability of unfolding depends on both the rate constant and the time allowed. However,

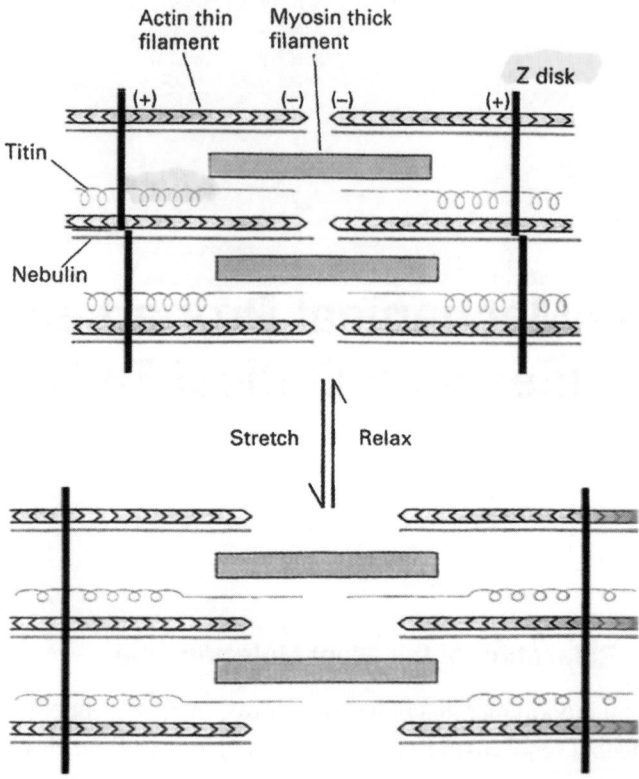

Fig. 9.1. Proposed model of role of titin and nebulin in organizing the sarcomere in striated muscle. Titin, a gigantic, elastic protein, links myosin thick filaments to Z disks. When the muscle is stretched, the elasticity of the titin filament keeps thick filaments centered in the sarcomere. Nebulin (MW 600 000–800 000) forms inextensible filaments, anchored at Z disks, which may act as organizing elements for the actin filaments.

even for the same rate constants, the different pulling speeds can account for the different unfolding forces. Rief *et al.* observed this experimentally [340], [341]. Such behavior of titin is quite reminiscent of Ashkin's early work on the viscoelastic behavior of cytoplasm in artificial filaments and in internal cell surgery in living *spirogyra* cells [19]*. The slow yielding of biopolymers even at rather low forces is one of the most characteristic features of most of the cell's cytoplasm *in vivo* and *in vitro*. Tskhovrebova *et al.* point out that the general behavior of mechanical unfolding of titin may well explain thermally induced or denaturant induced unfolding. They also remark that there is much to be gained from studies of this type, using other molecules and mutated structures.

9.2. Cell Motility of Adherent Cells over an Extra-Cellular Matrix

There is another important area of application of tweezers that has not yet been discussed. It has to do with the cell motility of adherent cells over an extra-cellular matrix (ECM). There are many cellular processes that depend on the ability of cells to migrate and spread along regulated pathways, such

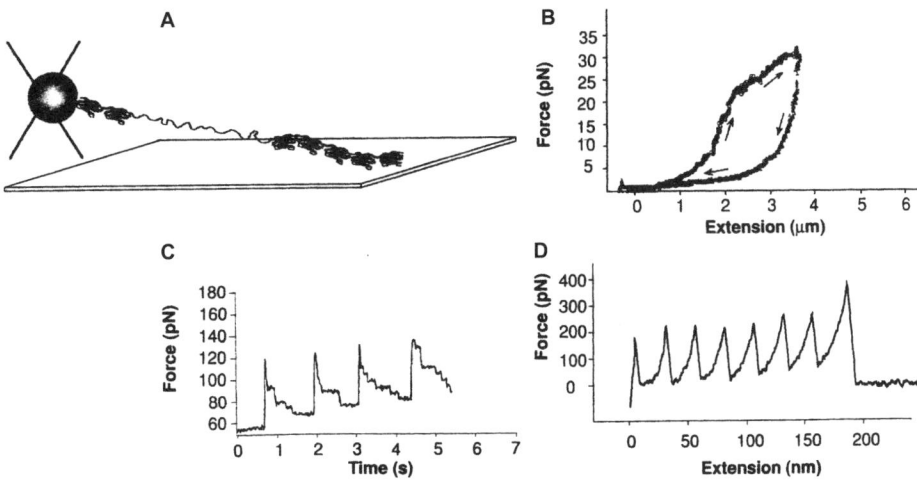

Fig. 9.2. (A) Experimental geometry used by Tskhovrebova *et al.* [216] to pull single titin molecules. A bead attached by an antibody to one end of a titin molecules was trapped and pulled; the other end was fixed to the microscope coverslip. Kellermayer *et al.* [337] trapped a bead on one end of the molecule and attached the other to a micropipette. Rief *et al.* [14] fixed one end to a coverslip surface and pulled the other using a stiff AFM cantilever. (B) Force–extension curve generated by Kellermayer *et al.* [337] with a compliant trap. Reprinted from Ref. [337]. (C) Force transients generated by Tskhovrebova *et al.* [216] by applying transient tension jumps to the molecule and then watching it relax incrementally. Reprinted from Ref. [216]. (D) Force extension curve generated by Rief *et al.* [14] by pulling a titin fragment with a stiff cantilever. Reprinted from Ref. [114].

as in wound healing, lymphocyte function, tissue regeneration, cell embryo development, metastasis of cancer cells, etc. Although a great deal is known about these processes at the cellular and molecular level [342], use of tweezers has shown that it has unique capabilities that can be used to elucidate many of the complex problems of force generation and selective regulation at the molecular level.

It has been postulated that a class of long thin adhesion receptor molecules called integrins, which protrude through the cell membrane and attach, via ligand binding, to EMC molecules such as fibronectin, for example, moderates cell motility on the ECM. The other end of the integrin establishes a linkage with the actin filament of the cytoskeleton via myosin molecular motors. This linkage between the molecule of the ECM and the actin–myosin system is the origin of the force slowly dragging cells along their paths.

The detailed motion over the ECM involves extension of a stiff, thin, ectoplasmic lamellipodium from a large endoplasmic reservoir at the back of the cell. The ectoplasm attaches to the ECM surface by means of the integral fibronectin linkages, and force is generated as the outer skin of ectoplasm is pushed back to the endoplasm. Once the material returns to the endoplasm, the integrin links dissolve, the back of the cell releases from the surface and the cycle continues. In the last half dozen years Sheetz and coworkers have performed a number of experiments with tweezers, which have helped confirm and extend the above force-generating model.

Choquet *et al.* in 1997 [343] showed directly that ligand-coated beads bound to integrins are pulled out of the laser trap by the forces of the cytoskeleton, when the beads are held at the leading edge of the cell. The authors also demonstrated a remarkable mechanotaxic effect, in which the integrins make stronger connections with the more rigid sites of the cytoskeleton than they do with the more pliable sites, and pull harder on these sites, as well. This effect could act as a mechanism for guiding migrating cells *in vivo*.

One sees here very graphically the ability of tweezers to observe the motile force-generating process in a living cell. The cell does not distinguish between a fibronectin molecule of the ECM attached at the experimentally inaccessible position at the bottom surface of the lamellipodium and a fibronectin molecule attached to an optically manipulated bead on the upper surface. It pushes on both.

In a more recent experiment in 1999, Felsenfeld *et al.* [344] studied the ability of the tyrosine inhibitor kinase Src to regulate integrin–cytoskeleton interactions, thus, in effect, controlling cell spreading and migration by internal means. By pushing a ligand-coated bead into the cell surface, they showed that tyrosine kinase Src affects the integrin–cytoskeleton interaction, but not the integrin–fibronectin bonding. By probing the surface attachment forces on the bead at different positions on the cell, they deduced that a process of integrin aggregation takes place once the bead attaches to the cell. They observe that the force necessary to return a moving bead back to the leading edge of a lamellipodium increases with time, indicating integrin aggregation and increasing binding to the cytoskeleton via the myosin motors. A Nature Cell Biology "News and Views" comments on the work entitled "Connections count in cell migration", by K. Vuori and E. Ruoslahti [345] and discusses the new results, showing the regulation of cell motion over surfaces by the interaction between Src and the integrin family of cell-adhesion receptors. The reviewers point out the importance of the results and mention that there are various other proteins that can mediate the effects of Src on integrin–cytoskeleton interactions. These compounds could also be used to reverse the effects of loss of Src on cytoskeleton functions.

In 2000, Nishizaka, Shi, and Sheetz [346]* sought to further clarify the position-dependent linkages of fibronectin to integrin and integrin to the cytoskeleton, using motile 3T3 fibroblast cells. Polystyrene beads coated with low concentrations of fibronectin were found three to four times more likely to bind to integrin when placed within $0.5 \, \mu$m of the leading edge of the moving cell, rather than further back. The authors deduced that this was not the result of the integrins being more concentrated at the leading edge, since anti-integrin coated beads showed no preferential binding at the front edge of the cell. The preferential binding at the front edge correlated with increased attachment to the cytoskeleton. The velocity of the rearward movement decreased and diffusive Brownian motion increased as the beads moved backward to the ectoplasm–endoplasm boundary. Release of the fibronectin-coated beads occurred close to this boundary, with the release of the integrins into the endoplasm. These observations are consistent with the hypothesis that the binding of the cytoskeleton to liganded integrins increases the strength of binding to the ECM complexes. Figure 9.3, taken from Fig. 5 of Ref. [346]* is a schematic illustration of the positional variation of the cytoskeletal binding of integrin–fibronectin complexes along a lamellipodium. The papers [344], [346]* contain a fairly complete set of recent references.

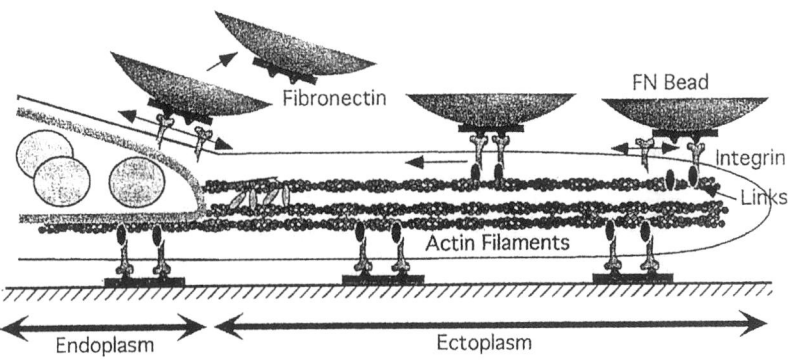

Fig. 9.3. Schematic illustration showing that cytoskeleton binding of integrin–fibronectin complexes at the leading edge could stabilize them. A fibronectin-coated bead (FN bead) attaches to the dorsal surface of the leading edge and recruits a second integrin, which recruits a second link to the cytoskeleton. Because the two bound integrins are attached to a rigid cytoskeleton, they cannot diffuse away if one should release from the fibronectin. Therefore, bead binding is stabilized until the actin cytoskeleton depolymerizes which is often seen at the endoplasm–ectoplasm boundary. Upon release from the cytoskeleton, the integrins could diffuse away leading to FN-bead release. On the ventral surface, additional components could stabilize the integrin cytoskeleton complex perhaps in a force-dependent process (1). Such a position-dependent binding and release cyclin could aid cell migration. Figure 5 and caption from Ref. [346]*.

9.3. Study of Forces that Regulate the Movement of Plasma Membrane Proteins

The outer plasma membranes of cells are multifaceted structures. As the window of the cell to the outside world, they are the site of many complex and vital processes associated with cell motility and sensing, as discussed above [344], [346]*, transport of fluids in and out of cells via vesicles in exocytosis and endocytosis, and transport of various ions through special single molecule ion channel gates. These processes have been the subjects of many separate studies using a variety of specialized techniques, including tweezers.

In a work by Sako and Kusumi [347] studying diffusional barriers (termed "fences") on membranes, the membrane as a whole was studied in its interaction with surface proteins. These experiments give insight into the mechanics and organization of the membrane itself. The procedures used involve the coating of small latex or gold beads with particular macromolecules that are trapped by tweezers and manipulated at the membrane surface. Dragging forces as small as fractions of a piconewton are involved. Such forces are in a much lower range than typically used in atomic force microscopy. The motion of particles released on the membrane surface can be tracked with nanometer precision using enhanced contrast microscopy [348], [349]. Hitherto unknown compartmentalized structures in the membrane were revealed by observing the untrapped motion of the particles or by dragging experiments using trapping forces. The dragging study used 210 nm latex particles containing transferring receptor molecules, whose behavior suggested the presence of barriers in the dragging path. By starting with strong trapping forces and gradually reducing the trapping forces, they could detect the presence of localized obstacles in their path that eventually pulled the particles out of the trap. The escaped particles were observed to rebound due to the elastic properties

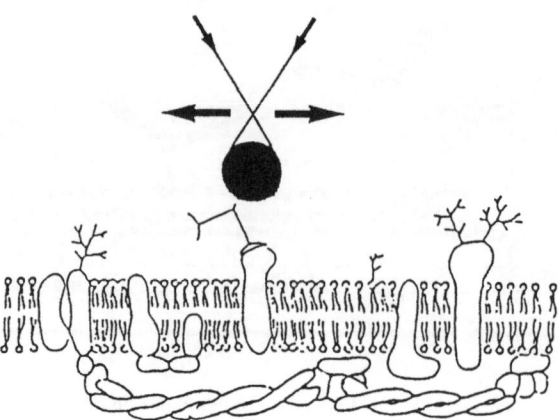

Fig. 9.4. Conceptual cartoon showing an experiment using laser tweezers to drag a membrane protein molecule along the plasma membrane. Figure 1 and caption from Ref. [350].

of the boundaries. The effective elastic constant of the boundary was found to be in the range of 1–$10\,\mathrm{pN}/\mu\mathrm{m}$ [347].

In other experiments using a variety of cell types, such as normal rat kidney fibroblasts, epithelial cells, etc., they found by tracking the diffusive motion of single particles over long periods of time, that the plasma membrane is compartmentalized into regions 0.1–$1\,\mu\mathrm{m}$. It was also observed that many membrane proteins undergo inter-compartmental hop diffusion between adjacent regions.

There is a very clear review article by Kusumi *et al.* discussing the use of tweezers to study the "Fences and tethers of the membrane skeleton that regulate the movement of plasma membrane proteins" in *Methods in Cell Biology* [350]. Figures 9.4 and 9.5, taken from Figs. 1 and 8 of Ref. [350], show diagrams of tweezer dragging experiments and the geometry of the membrane–skeleton fence model that gives rise to the compartments with elastic boundaries, as discussed above.

Although some of the results attained in these experiments are quite qualitative, it is hard to see how one could gain comparable insights into membrane mechanics and structure by other techniques.

I am struck by the resemblance between the problems being addressed here on forces and mechanics of single particles on membrane surfaces and the problems of measuring electrostatic and entropic forces on macroscopic particles in the field of colloidal research. As we will see later in Chapter 11, many of the methods for measuring weak forces on diffusing particles are quite similar. Indeed, techniques used in one field may prove useful in the other.

9.4. Membrane Tube Formation from Giant Vesicles by Dynamic Association of Motor Proteins

Liquid bilayer membranes play a crucial role in the functioning of cells. Membranes not only define the outer boundaries of cells but also form the complex tubular morphology of the endoplastic reticulum, Golgi apparatus, and intracellular traffic apparatus. The formation of these tubular membrane networks is dynamic in nature. Tubes continuously form and disappear during cell function. The

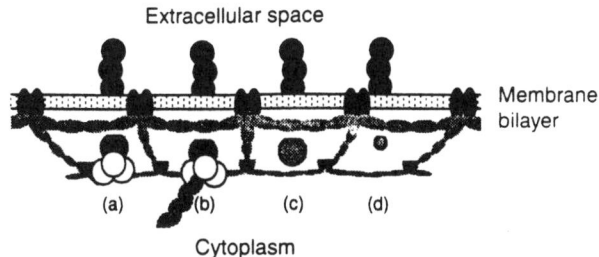

Fig. 9.5. Membrane–skeleton fence model. In this figure, the plasma membrane is viewed from inside a cell. The bilayer portion of the membrane, the transmembrane proteins (light shading) and those connecting the membrane skeleton to the membrane (darker shading), the cytoplasmic proteins bound to membrane proteins (open circles, in (a) and (b)), and the binding of this complex to the cytoskeleton (b) are shown. The membrane skeleton is in close proximity to the cytoplasmic surface of the plasma membrane. The cytoplasmic domain of the membrane protein (or its complex with a cytoplasmic protein) collides with the membrane skeleton and cannot readily move to an adjacent compartment ((a), (c), (d)). If the cytoplasmic domain is smaller (d), the protein moves to an adjacent compartment more readily. Binding of membrane proteins to the membrane skeleton ((b), but direct binding of transmembrane proteins to the membrane skeleton also occurs) has been found for a variety of proteins. Binding can be detected by using laser tweezers to drag the membrane protein. In many cases, the membrane skeleton itself is undergoing macroscopic diffusion as well as oscillative motion without any real displacement. Some membrane skeletons show directed, active transport-type movements, and the transmembrane proteins bound to such skeletons also undergo similar directed-type movements. Figure 8 and caption from Ref. [350].

dynamic nature of the process and the observation of the localization of these structures near microtubules suggest the involvement of motor proteins in tubular membrane formation. Although there are hypotheses, the precise mechanism by which forces are exerted on membranes and how they are regulated are not known (see references in Ref. [351]).

In this work, the authors have devised an *in vitro* assay for studying tubular membrane formation under known conditions involving purified kinesin motor molecules, synthetic lipid membranes in the form of giant vesicles of varying tension, and microtubules of varying rigidity. First, the properties of single tubules were studied. Optical tweezers was used to manipulate vesicles and to attach them to coated polystyrene beads for force measurements. Changing the separation between the bead and the vesicle formed single bilayer tubes of varying radius and tension.

In the assay, a random network of microtubules was fixed to the bottom surface of a chamber. Motor molecules were attached to giant vesicles and allowed to sediment onto the microtubules. The motors quickly attached themselves to the nanotubules and advanced along the microtubule network pulling membrane tubes with them, forming a quasi-static visible tubular structure. The authors believe the structure stops growing when the number of motors pulling the tube stabilizes, due to a dynamic balance of the number of new motors arriving on the microtubule from the vesicle and those motors leaving in the usual processive motion along the microtubules. In the future, the authors intend to use tweezers to quantify the force during the evolution of the network of tubes to its final steady state.

IVB
Other Recent Applications in Physics and Chemistry

CHAPTER

10

Origin of Tweezer Forces on Macroscopic Particles Using Highly Focused Beams

As we have seen, since the late 1980s optical tweezers has had a major impact on studies of single cells and biological particles. The "handles technique" of attaching a wide variety of biological materials, ranging from molecules to viruses to small macroscopic spheres, has greatly enhanced the usefulness of tweezer traps. At the same time, there has also been an abundance of other experiments using tweezers and optical manipulation, which broadly fall into the category of physics and chemistry. These new experiments with tweezers have extended the applications of optical levitation and optical manipulation of macroscopic particles.

10.1. Origin of the Net Backward Radiation Pressure Force in Tweezer Traps

An important aspect of this work, in the last 10 years, has been the characterization of the optical forces in tweezer traps for micrometer and sub-micrometer particles for beams of various shapes. An essential element in the conception of tweezer traps was the existence of a net backward radiation pressure force in strongly focused beams. Ashkin discovered this net backward force in 1978 in connection with tweezer traps for atoms [3]*. The trapping experiments by Ashkin *et al.* in 1986 on sub-micrometer Rayleigh-sized particles [5]* and the subsequent atom trapping experiment on atoms by Chu *et al.* [4]* confirmed the validity of this concept for particles small compared with the wavelength λ. The trapping of macroscopic dielectric spheres large compared with λ was also demonstrated and explained with a simple ray-optic model in the Ashkin *et al.* 1986 paper [5]*. It was somewhat surprising that in the early 1990s, even after the successful application of tweezers to biology [19]*, [20]*, [44]*, a number of authors saw fit to re-examine the origins of the so-called "backward radiation pressure" [352], [353]. They independently used a combined Gaussian beam wave-optics and ray-optics approach to calculating the axial force on large macroscopic particles in a highly focused Gaussian beam. They treated beam propagation according to the usual Gaussian beam propagation formula and used the normal to the phase fronts to give the direction of the energy flow and ray direction at the particle surface. This approach sounds correct at first sight, but in fact it can be seriously in error. The Gaussian beam formulae and the scalar wave equation from which they are derived are correct only in the paraxial approximation. Beams with numerical apertures of 1.2 or beam angles of 70° or more, typical of tweezer traps, are grossly different from the Gaussian beams approximation of beam angle $\theta \approx \lambda/\pi w_0$ approaching zero [354]. Thus, the papers using the combined Gaussian beam and ray-optics approach only added confusion and error to the simple purely ray-optics picture [5]*.

10.2. Light Propagation at the Focus of a High Numerical Aperture Beam

Light propagation in the near field about the focus of high numerical aperture (NA) beams does not follow the phase fronts. The electric fields of high NA beams are no longer transverse and have very large axial components. The proper description of these fields is quite complex and requires the full Maxwell vector wave equation. It is much simpler and more physically correct to use a geometrical or ray-optics approach to the problem. Figure 10.1 compares the approaches of Refs. [352], [353] with the strict ray-optics approach as described in Refs. [5]*, [49]*.

In the ray-optics picture shown in Fig. 10.1(A_1) we see an initially parallel beam focused by an assumed ideal high numerical aperture lens of NA $\cong$ 1.25 to an ideal point focus at maximum beam angle of $\theta_b \cong 70°$. The outer beam rays contribute a longitudinal electrical field component of $E \sin \theta_b \cong 0.94 E$ to the electric field at the focus F, as seen in Fig. 10.1(A_2). Adding the contribution of all rays, we expect a very large net axial field. If the beam focus is placed close to the particle surface, we expect strong refraction and a large backward radiation pressure component, as described in Refs. [5]*, [49]*.

In Figs. 10.1(B_1) and 10.1(B_2), we show the Gaussian beam propagation model. Figure 9.3(B_1) shows the same highly convergent beam as Fig. 10.1(A_1). The phase fronts and the postulated rays perpendicular to the phase fronts are shown at the particle surface. If the particle surface is located at the beam focus F, as shown in Fig. 9.3(B_2), the rays are taken to be essentially parallel to the axis and pass straight through the particle, giving very little refraction and a very small force. This minimum of force is just the opposite of the very large force of the ray-optics model. Also, there are no transverse field components at the focus F.

We trace the path of two comparable beam rays through the focal region out to the far field, in the absence of a particle for A and B. The trajectory for B is the exact opposite of the ray-optic trajectory in A, where the energy passes to the opposite side of the beam. It is, of course, unphysical and violates the conservation of momentum for a noninteracting light ray to follow a curved trajectory in free space. One can easily convince oneself of the correct ray path by blocking a piece of the incoming beam. Its shadow then appears on the *opposite* side after passing through the lens, as in the ray-optic picture, not on the *same* side, as in the Gaussian beam picture. Another way to see the constancy of the light momentum and ray direction in free space is to resolve the focused light field into an equivalent angular distribution of converging planes waves [355]. Each of these plane waves propagates undeviated in free space with its appropriate amplitude and phase, and interferes in space

Ray-Optic Picture

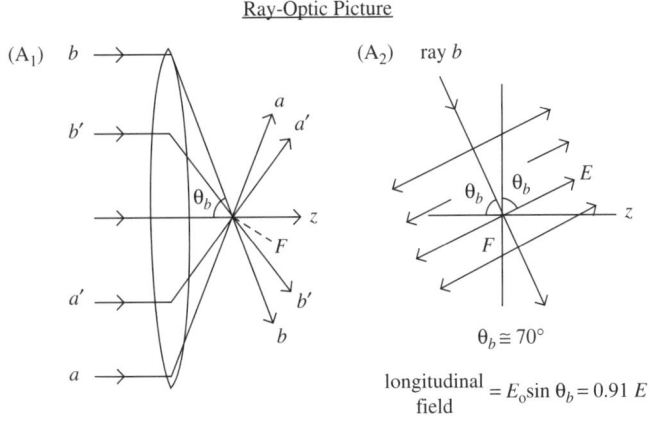

Gaussian Beam Picture

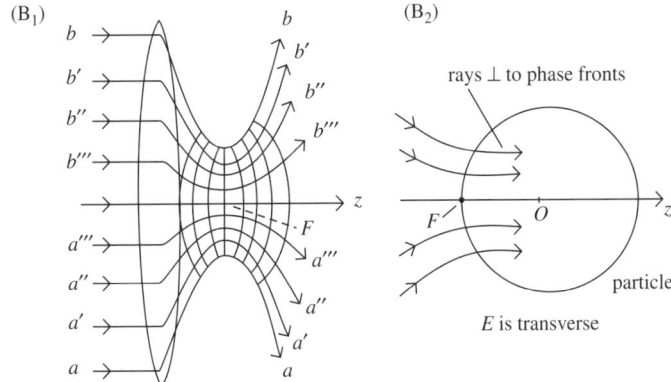

Fig. 10.1. (A_1) Ray-optics picture of the focusing of an initially parallel beam by a high numerical aperture lens of NA = 1.25. The rays cross at the focus F at beam angles θ_b up to 70°. (A_2) Rays at the beam angle of 70° contribute to a large axial electric field component $E_0 \sin \theta_b = 0.94\, E$. ($B_1$) A Gaussian beam optics picture for the same highly convergent beam. The phase fronts and directions of the rays, assumed to be perpendicular to the phase front are shown. (B_2) A spherical particle with the focus F at the particle surface. The rays, which are essentially perpendicular to the sphere's surface, enter nearly parallel to the beam axis and are hardly refracted, giving a very small force. A particle similarly placed in the ray optic picture experiences maximum force.

to give the correct beam shape at the focus and elsewhere. These errors are most serious close to the beam focus, in the near field, where particles are trapped.

10.3. Calculation of the Tweezer Forces on Dielectric Spheres in the Ray-Optics Regime

In 1992, I decided to make a complete computation of the forces on spheres using a geometrical optics model [49]*, as described in Fig. 5.15(a). This calculation gave the light force on spheres large compared with λ, not only for the beam focus located along the z-axis, but at all points inside and outside the sphere. This involved dividing the total force of a single ray into a scattering force component in the direction of the ray and a gradient force component perpendicular to the ray, and then integrating over all incident rays striking the sphere. This was done for beams of different numerical apertures and different intensity distributions over the cross section, including TEM$_{01}$* donut modes, and for particles of different indices of refraction.

The major intent of the paper [49]* containing these calculations was to correct the errors of previous treatments which combined a wave description of the incident Gaussian beam with a geometric or ray-optic description of the scattering from the sphere. In these treatments ray directions were assumed to be normal to the phase fronts and to follow the Poynting vector. As just discussed, this leads to serious errors at the beam focus. As we shall see, this incorrect description of the focus persisted in a substantial number of subsequent papers during the late 1990s, resulting in some very strange predictions [356], [357].

10.4. Corrections to Paraxial Ray Approximation for Strongly Focused Gaussian Beams

As extensive as this paper was, it did not cover the very important cases of particles with sizes of a few micrometers, which are outside the limits of geometric optics. Calculation of the forces for these intermediate cases requires a more complete description of the incident light fields and their scattering, based on the full vector wave equation or approximations to it, as we shall see. An important paper in this regard by Davis in 1979 [358] gives an approximate method of correction of all components of the Gaussian beam fields for cases of tighter beam focusing, where the paraxial ray approximation is no longer valid. He expanded the vector potential and corresponding fields to first order in the parameter $s = \lambda/2\pi w_0$. He also presented a procedure for developing higher-order corrected Gaussian beams descriptions. Barton et al. in 1988 [359] used the first-order corrected Gaussian beam to compute the internal and external fields of a sphere for a Gaussian beam focused to a beam waist $w_0 \cong 2\,\mu$m, using the first-order Davis correction to the fields. The results were significantly different from plane wave excitation of a moderate sized sphere of $\sim 5\,\mu$m in diameter.

Barton et al., in a companion paper [360]*, computed the internal fields of a spherical particle illuminated by a fairly tightly focused Gaussian beam, showing the effects of different positions of the focus of the input beam relative to the sphere. They showed that when the beam focus was located on the axis of the sphere there was very little excitation of the internal resonances of the sphere. Strong resonances were excited when the beam was focused on the edges of the sphere. This was in complete agreement with the experiments of Ashkin and Dziedzic [92]*, [97]*. This behavior supports the surface wave or whispering gallery mode picture of these high-Q resonances. If light does not hit the sphere at nearly tangential incidence, one does not observe resonances.

10.5. Fifth-Order Corrected Electromagnetic Field Components for a Focused Fundamental Gaussian Beam

Barton and Alexander in 1989 [361], following the prescription of Davis [358], extended the correction of electromagnetic field components for a tightly focused fundamental Gaussian beam to fifth order. These higher-order fields are applicable to even smaller values of the focused beam

s =		0.02	0.05	0.10	0.20	0.30	0.40
s^0	avg%	0.817	2.10	4.37	9.47	15.3	21.8
	max%	.3.07	7.94	16.8	37.0	60.8	88.0
s^1	avg%	1.73×10^{-2}	0.111	0.457	1.90	4.33	7.74
	max%	9.28×10^{-2}	0.603	2.51	10.3	22.6	38.4
s^2	avg%	6.43×10^{-4}	1.05×10^{-2}	8.85×10^{-2}	0.757	2.56	5.89
	max%	8.23×10^{-3}	0.133	1.14	10.7	31.9	49.3
s^3	avg%	2.36×10^{-5}	9.58×10^{-4}	1.61×10^{-2}	0.277	1.44	4.25
	max%	1.97×10^{-4}	8.26×10^{-3}	0.144	2.51	19.1	36.0
s^4	avg%	1.15×10^{-6}	1.19×10^{-4}	4.10×10^{-3}	0.148	1.13	3.85
	max%	2.46×10^{-5}	2.52×10^{-3}	8.85×10^{-2}	3.99	38.2	54.0
s^5	avg%	5.13×10^{-8}	1.27×10^{-5}	8.69×10^{-4}	6.19×10^{-2}	0.725	3.34
	max%	7.58×10^{-7}	1.99×10^{-4}	1.40×10^{-2}	1.19	22.2	36.6

Fig. 10.2. Average percent error and maximum percent error of solution to Maxwell's equations for zeroth- to fifth-order Gaussian beam descriptions versus s. Percent error calculated for 216 points consists of all combinations of $\xi, \eta, \zeta = 0.0$, 0.1, 0.2, 0.5, 1.0, and 1.5. Table I and caption from Ref. [361].

radius w_0. Figure 10.2, taken from Barton and Alexander's Table I, shows the average and maximum percent error to be expected for the various representations of the solutions to Maxwell's equations for zeroth- to fifth-order Gaussian beam descriptions in terms of the parameter $s = \lambda/2\pi w_0$. These values are obtained by direct substitution of the higher-order field components into Maxwell's equations. We see, for example, that a maximum error of $\sim 1.2\%$ is expected for $s = 0.20$. An s value of 0.2 corresponds to a beam waist to wavelength ratio of about 0.8 and a Gaussian beam angle $\theta_b \cong \lambda/\pi w_0 \cong 23°$. If $s = 0.4$, then $w_0/\lambda \cong 0.4$, the maximum error is about 37% and $\theta_b \cong 46°$.

10.6. Computation of Net Force and Torque for a Spherical Particle Illuminated by a Focused Laser Beam

Barton, Alexander and Schaub in 1989 [362] used the fifth-order corrected Gaussian beam to compute the net radiation force and torque for a spherical particle illuminated by a focused laser beam. They calculated for a 5 μm water droplet levitated by a $\lambda = 5145$ Å beam with a 2 μm beam waist diameter. The force was calculated by finding the correct incident, internal, and scattered electromagnetic field and integrating the normal component of the Maxwell stress tensor $\mathbf{T}$ over a surface enclosing the particle. To get the torque, one integrates the normal component of $\mathbf{T} \times \mathbf{r}$ over a surface enclosing the particle. The results of this procedure, using the highly corrected field vectors, should be quite exact. The results of the torque calculations using circularly polarized light, on a sphere in this size regime, show the same general characteristics of earlier calculations by Marston and Crichton [131], [363], Chang and Lee [132], and Gouesbet et al. in 1988 [364]. The induced torque for circular polarized light depends entirely on the absorption of the incident light. If the imaginary part of the refractive index is zero (i.e., there is no absorption), the induced torque is zero, regardless of the orientation of the beam on the sphere. It should be repeated that there are no torques on spheres from the linear momentum of the beam, since the force on the sphere due to any ray passes through the center of the sphere. As useful as these exact calculations using the stress tensor are for checking against experimental results, they do not give very much in the way of physical insight into the force mechanisms, as was the case with the simple ray-optics approach of the early work. Finally, the fifth-order correction process, as useful as it is for exact levitation problems where the beam angle is not too large, is of limited use for tweezer traps, since we are dealing with beam angles $\theta \cong 70°$, typical of the highest numerical aperture objectives.

10.7. Measurements of the Forces on Microspheres Held by Optical Tweezers

Returning to the tweezer trap, there was considerably more work in the 1990s on the backward radiation pressure force, extending Ashkin *et al.*'s treatment of Rayleigh particles [49]* and his geometric optics description of the tweezer trap [5]*. A pair of papers by Wright, Sonek and Berns in 1993 [365] and in 1994 [366] reexamined the trapping forces of tweezers more carefully. They considered the theory for axial and transverse tweezer trapping and compared the ray-optics of Ashkin [49]* and the electromagnetic field model of Barton *et al.* [362] with quite extensive measurements covering a fairly wide range of beam and particle parameters.

In their 1993 paper, Wright *et al.* [365] present a curve showing the maximum force of a tweezer trap vs. the particle radius. In the small-particle or Rayleigh limit, gradient forces predominate and increase in size proportional to r^3, whereas in the large-particle or geometric optics limit the force is independent of particle radius. In the intermediate size regime Wright *et al.* calculated forces, using the fifth-order electromagnetic field corrections, for beam angles $\theta_b \cong 60°$. This angle is rather large for the approximation, so one expects some error according to Fig. 10.2. Nevertheless, the general trend is correct and the overall curve is fairly plausible.

Wright *et al.* subsequently, in 1994 [366], made a much more extensive comparison of experimental results and calculations of the axial and transverse forces for particles covering the Rayleigh to geometric-optic-size regimes. In the ray-optic regime with $\lambda = 1.06$ and $20\,\mu$m spheres, they found quite good agreement with the strict ray-optic model of Ashkin [49]* over a range of beam angles from 60° to $\sim 30°$. In the electromagnetic model regime, following Barton *et al.* [361], they calculated the fields to fifth order for $1\,\mu$m spheres and integrated the stress tensor over a closed surface surrounding the sphere to get the force. This calculated force agrees quite well for beams with beam angles of $\sim 30°$. The calculated results for the electromagnetic model disagree with measured values by factors as high as six for cone angles of 60°, as expected.

A very useful aspect of this paper is the discussion of the experimental details of the measurements. The authors discuss a laser spot size measuring technique based on a moving knife edge, corrections to the viscous drag force measurement by Stokes' law due to the proximity to the microscope cover slip, choice of specific high numerical aperture objectives for achieving the smallest focal spot size, and the use of gravity and buoyancy as a force measuring method. There is an appendix that outlines the electromagnetic algorithm and a useful collection of references.

10.8. Generalized Lorenz–Mie Theory for Convergent Gaussian Beams

The interaction of focused light beams with macroscopic spherical particles has also been studied by Gouesbet and his colleagues ever since the 1970s. The usual Lorenz–Mie theory (LMT) is a plane wave theory and is not able to describe interactions in shaped beams, such as highly convergent Gaussian beams. Gouesbet *et al.* have developed what they call a generalized Lorenz–Mie theory

(GLMT) to study the interactions between arbitrary shaped beams and a sphere arbitrarily located within it. A summary of the early development of GLMT was given in Gouesbet *et al.* in 1988 [364].

More recently, in 1994, Gouesbet and colleagues applied GLMT to radiation pressure to study the electromagnetic resonances of focused beams [367]. They express the radiation pressure force components in terms of three radiation pressure force cross sections: $C_{pr,z}$ (for axial propagation along the z-axis), and $C_{pr,x}$ and $C_{pr,y}$ (for the transverse cross sections). GLMT introduced two infinite sets of constant beam shape coefficients to describe the incident beam shape. The authors claim that the beam description of Barton *et al.* [359], using their so-called arbitrary beam theory (ABT), is strictly equivalent to the beam description of GLMT, using a localized approximation (LA) [368]. Gouesbet, however, claims that their BSC method of calculation is more accurate than Barton *et al.*'s [362]. The results of GLMT show that the resonances of spherical particles occur at the same values of size parameter as for LMT (i.e., the plane wave case) but the peaks are much broader. GLMT, on the other hand, is not capable of giving the detailed internal field distribution, which Barton *et al.*'s method does [360]*.

Gouesbet and his colleagues showed that electromagnetic resonances are not excited for a Gaussian beam tightly focused on the particle, in agreement with the calculations of Barton *et al.* [360]* and the experiments of Ashkin and Dziedzic [92]*, [97]*. They, too, found that the electric and magnetic resonances could be strongly excited by moving the beam focus of a polarized beam to the appropriate edge of the sphere.

10.9. Computation of Backward Radiation Pressure Using GLMT

In 1996, Ren, Gréhan, and Gouesbet applied the GLMT procedure to the problem of predicting reverse radiation pressure in extremely focused Gaussian Beams [357]. It is this feature of strongly focused beams that makes possible the trapping of particles ranging in size from sub-micrometer Rayleigh particles to macroscopic particles large compared with the wavelength. The higher-order localized approximations of focused beams, LA^1 to LA^5, are equivalent to using the higher-order Davis approximations [361].

For sub-micrometer Rayleigh particles, results of computation with the LA^1 first order approximation to a converging beam are close to those obtained using the uncorrected Gaussian beam, as in Ref. [5]*. However, based on Fig. 109 of Ref. [361], we expect even the LA^5 fifth-order approximation to be inadequate for the high numerical aperture beams with beam angles $\sim 70°$ typical of experimental conditions. As we shall see shortly, one can compute the incident vectorial fields exactly, which should make more accurate Rayleigh calculations possible.

The results found using GLMT for intermediate-size particles with diameter $2a$ comparable to λ and for large so-called Mie particles with diameters $\ll \lambda$ are questionable. They find that negative radiation pressure, and therefore trapping, only exists for particle radii greater than $a = 1.78\,\mu$m. This is contrary to extensive experimental data in this size range. According to most theoretical expectations, all lossless dielectric spheres are, in principle, trappable by tweezer-type traps.

Harada and Asakura, in 1996 [369], made an extensive reexamination of the radiation forces on a dielectric sphere in the Rayleigh scattering regime. They calculated radiation forces using the GLMT of focused Gaussian beams. According to Kerker [133], Rayleigh scattering theory should apply for small particles of diameter less than about $\lambda/20$. Harada and Asakura used GLMT to examine the sizes at which deviations from Rayleigh scattering occur. The authors also recalculated the upper particle size limit of Rayleigh particles for which the tweezer criterion that the backward gradient force exceeds the forward scattering force is met. They calculated this to be 62 nm, as compared with Ashkin *et al.*'s original value of 95 nm. This discrepancy is accounted for by the factor of $2n_2^2$ in Ashkin's expression for the ratio R of the maximum backward gradient force to the forward scattering force [5]*. The factor 2 comes from a failure to use the time average of E^2 given by $\langle E^2(t)\rangle_t = 1/2|E^2|$ in $F_{\text{grad}} = 1/2\alpha\nabla E^2$. The n_2^2 factor comes from the erroneous inclusion of n_2^2 into the polarizability α. These errors amount to a size error of $(2n_2^2)^{1/3} = 2^{1/3} n_2^{2/3} = 1.26(1.2) = 1.5$ in the maximum particle diameter. This number, however, is quite academic since we know that tweezer trapping applies in principle for all particle sizes with a sufficiently small beam focal spot size w_0. It should also be pointed out that the GLMT used by the authors is not valid for the high numerical aperture incident beams typically used for tweezer traps. Another quantity recalculated in this paper was the minimum size Rayleigh particle that can be trapped for a given laser power and spot size w_0. They found, using the parameters of Ashkin *et al.* [5]*, namely $P = 1.4$ W, $2w_0 = 1.5\lambda$, $\lambda = 5145$ Å, that the minimum size particle that could be trapped was 18.9 nm in diameter, compared with 14 nm diameter as estimated by Ashkin. In practice, the input beam parameters are not well known. Therefore, an empirical approach to the problem of characterizing a tweezer trap is best.

10.10. Single-Beam Trapping of Rayleigh and Macroscopic Particles Using Exact Diffraction Theory

There are two interesting and much referenced papers by Visscher and Brakenhoff from 1992 on a "Theoretical study of optically induced forces on spherical particles in a single-beam trap I: Rayleigh scatterers" [370] and "II: Mie scatterers" [356]. The authors, who are experts in high resolution confocal microscopy, use a method devised by Richards and Wolf [371] for calculating the electromagnetic field near the focus of a high numerical aperture lens based on diffraction theory using the vector equivalent of the Kirchoff–Fresnell integral. This method expands the diffracted field into a superposition of plane waves whose wave vectors all lie within the converging geometrical light cone. This, they assert, should be accurate near the focal plane. This gives formulae for the vector field components and components of the Poynting vector near the focus. In paper I [370] the radiation pressure force, i.e., the scattering force, is deduced from the Rayleigh scattering cross section and the Poynting vector components. The gradient force can be calculated from the intensity gradients of E^2 using the force formula $F_{\text{gr}} = n_m(1/2)\alpha\nabla E^2$ as mentioned above in Sec. 5.2. Numerical results, however, are given only for the z component of the Poynting vector, from which one can deduce the parameters of the axial potential well for Rayleigh particles, under the reasonable assumption of negligibly small scattering forces. They calculated for 25 nm particles, typical for viruses, semi-aperture angles of $60°$ and $30°$, different wavelengths, and different indices of refraction. For example, at a power of 4 W and a semiaperture angle of $60°$ one gets a well depth of about $10 kT$. These, in principle, are the most reliable results for high numerical aperture tweezer traps for submicrometer Rayleigh particles. Calculations based on higher-order corrections [357], [365] are not adequate for these large beam angles and small beam spot sizes.

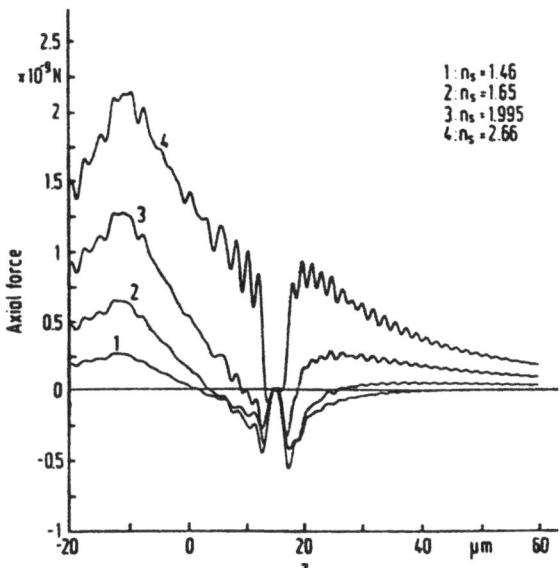

Fig. 10.3. Axial force exerted on a dielectric sphere with a radius of $15\,\mu$m in water. The refractive index of the sphere is 1:1.46, 2:1.65, 3:1.995 and 4:2.66, respectively and it is embedded in water. The wavelength and the power of the laser are 1064 nm and 1 W, respectively. The semi aperture angle of the focusing lens is 60°. The focal position (without sphere) is at $z = 0$. Along the horizontal axis the z position of the center of the sphere is plotted. The gravitational force on a polystyrene sphere of the given dimensions in water is about 8.6 pN, on such a glass sphere it is about 0.25 nN. Figure 3 and caption from Ref. [356].

In paper II Visscher and Brakenhoff [356] again used the same vector diffraction theory for a highly focused tweezer trapping beam to calculate the optically induced forces on Mie particles, which are $\gg \lambda$. They adopted a combined wave-optic and ray-optic method of computing the forces. First, they calculated the Poynting vector from the electromagnetic fields and then took the directions of the beam rays at the input surface of the sphere to be that of the local Poynting vector. As pointed out by Ashkin [49]*, this is correct except in the vicinity of the beam focus, where the spot size w_0 is less than λ. They assumed all the rays at the focus are parallel to the beam axis, which again gives essentially zero force when the focus is placed in contact with the input face of the sphere. Although they compute the exact field components when considering ray directions at the focus of tweezer beams, Visscher and Brakenhoff seem to have made the same error that Wright *et al.* [352] and Gussgard *et al.* [353] previously made.

Figure 10.3, adapted from Visscher and Brakenhoff [356], illustrates the error in their computation of the axial variation of the computed force for various values of n_m for a sphere large compared with λ. It shows very strikingly the zero in computed force when the focus is at the sphere surface. The authors remarked on this feature with some puzzlement.

10.11. Optical Gradient Forces of Strongly Localized Fields

In 1998, Tlusty, Meller and Bar-Ziv [372] published a very interesting paper on the "optical gradient forces of strongly focused beams on dielectric particles". They showed that when the beam is focused to a diffraction-limited spot, one can use a dipole approximation that is valid for particles of any size. This approach involved neither ray-optics where the particle size $R \gg \lambda/n$ nor the complications of the exact electromagnetic (EM) approach, based on the stress tensor and the decomposition of the incident focused beam into a set of Fourier plane wave components [49]*, [133]. For submicrometer particles of $R \ll \lambda$, the EM approach reduces to Rayleigh theory. When R is comparable to λ or larger than λ and the spot size is less than λ, interference effects are greatly reduced and, as the authors showed, simple formulae can be derived for the force versus radial and axial coordinates, and the radial and axial force constants. They calculated the force versus radial displacement and the radial force constants for $R = 0.5 \, \mu m$ and showed excellent agreements with data from Ref. [218]. They also showed very good agreement between data and the normalized maximum force versus normalized particle size, from R equal to a fraction of w_0 out to $R \cong 6 \, w_0$. The interest of these results is that they apply for the very beam sizes and particle sizes that are typical of the very high NA objectives used in tweezers, and are not well described by GLMT, even in the fifth-order approximation.

One of the advantages of this dipole approximation is its simple intuitive appeal. In the dipole limit even the normally complex integral of the stress tensor over the particle surface reduces to a very simple form.

10.12. Exact Theory of Optical Tweezers for Macroscopic Dielectric Spheres

Finally, in 2000, Maia Neto and Nussenzweig [373]* took up the challenge and developed what should be the exact theory of optical tweezers based on a partial-wave (Mie) expansion of the axial force on a transparent sphere by a highly focused beam from a high numerical aperture objective. Calculations were made for the simplest case of the force distribution along the beam axis. Previously, people were put off by the presumed magnitude of using an exact approach. In stating their results, the authors give no indication of how many plane-wave Mie contributions were needed to calculate each point of the force computation.

The authors showed that, in the limit of size parameter $\beta = 2\pi a/\lambda \ll 1$, one recovers the force formula $F = (\alpha/2)\nabla E^2$, where α is the static polarizability. In the opposite limit $\beta \gg 1$, the force is reduced to the geometric optical expression of the force as a sum over incident rays, as used by Ashkin [49]* in his ray-optics calculation of tweezer forces and previously by Roosen and Imbert [50], [51]. In the interesting size range inaccessible to either limit of β the authors plotted results for sphere diameters of about 1.4 and 2.2 μm. The high numerical aperture beam angle was taken as 78°, well beyond the range of validity of the higher-order corrected Gaussian beams of Barton and collaborators [359], [360]*, [361] and of the GLMT approach of Ren, Gréhan, and Gouesbet [357], [367]. The highly focused beam was also assumed to follow the Abbe sine condition typical of corrected high numerical aperture objectives. This is a better representation of the simplistic input beam used by Ashkin [49]* in his ray-optics calculations. The results of their exact calculations show departures from the geometric or ray-optics result, when the beam focus is located near the sphere surface and also when near the sphere center. The exact maximum backward and forward axial forces are larger than the corresponding geometric optics result and occur at somewhat larger distances from the sphere surface. At the sphere center, the net forces arise only from the normal surface scattering force. This gives rise to low Q in Fabry-Perot type resonances due to reflections from opposite sides of the sphere.

This interference, first identified here, implies a sinusoidal variation of this force as a function of sphere diameter. The authors show how axial stiffness about the trap equilibrium point is affected by this resonance behavior. At some specific values of particle radius or trapping wavelength the axial stiffness can get quite small, due to interference. The approach of Tlusty *et al.* [372], which is based on avoidance of all interference effects, does not show any such oscillatory behavior, as the particle size is varied as expected. The authors suggest that values of low axial force may be of interest for applications of optical tweezer to scanning force microscopy [374], [375]. See the discussion on scanning force microscopy just below in Sec. 10.13. The numerical values obtained from these computations are quite close to those of Simmons *et al.* [218]. The results are of theoretical value, but, in practice, most experimenters will still choose to calibrate their traps directly, using their specific beam profiles and particles. More will be said later about the experimental techniques for calibrating traps.

10.13. Use of Optical Tweezers as a Stylus Support for Scanning Force Microscopy

Friese *et al.* [374] investigated the potential of using single-beam gradient type tweezer traps as a way of holding a stylus for scanning force microscopy. In scanning force microscopy, a sharp stylus is mounted on a soft cantilever spring and is scanned over a surface to interact with and generate an image of the surface features of the sample. For sensitive performance it is desirable to have low spring constants and high resonant frequency of the tip-cantilever system. In this work the force constants and resonant frequencies of a tweezer trap holding 1–4 μm diameter polystyrene spheres were measured as a model force microscopy setup. The authors' work improved on earlier work by Ghislain *et al.* [375] on this concept. They developed a new sensitive way of measuring force constants and monitoring particle motions, based on backscattering from a trapped particle.

Use of backscattered light for atomic force applications implies that one is not restricted to transparent samples, as was the case for Ghislain *et al.* [375]. Friese *et al.* also studied the effect of using holographically generated donut mode beams as a way of enhancing performance. Spring constants as low as 10^{-6} and 4×10^{-6} N/m were measured with powers of 1–4 mW. Note that 10^{-6} N/m $= 10^{-3}$ pN/nm. This is 40 times more sensitive than the tweezer trap used in the experiment of Sec. 7.19. Resonant frequencies between 1 and 10 khz were measured with best performance occurring for the donut mode. These power levels are very modest. The spring constants are about three orders of magnitude less than typical scanning force microscopes [376], with comparable resonant frequency response. It may be that spring constants of $\sim 10^{-6}$ N/m are too sensitive in many applications, but one can easily stiffen up the trap to 10^{-5} or 10^{-4} and still be 10 times more sensitive than mechanical cantilevers. This is one more example of the advantage of using optical springs over mechanical springs. However, as mentioned earlier, one cannot expect atomic transverse resolutions using optical forces. Transverse resolution of a fraction of a nanometer has been achieved in other contexts (see Secs. 7.13 and 7.16).

10.14. Localized Dynamic Light Scattering

Bar Ziv *et al.*, in 1997 [377], developed a new technique, termed "localized dynamic light scattering" (LDLS), for probing single particle dynamics on the scale of nanometers. It involves scattering of a focused probe laser beam from a single trapped particle held in a tweezer trap under a microscope and measuring its fluctuations via the temporal autocorrelation function of the scattered intensity.

In conventional "dynamic light scattering" (DLS) one samples scattered light from a large number of particles illuminated by an essentially uniform beam. In such a case random thermal motion of the particle produces phase shifts, causing intensity fluctuations in the scattered light.

In LDLS, there are intensity fluctuations due to the motion of a single point-like scatterer within the highly nonuniform scattering beam. A separate trapping laser at $5145\,\text{Å}$ and a probe laser at $6328\,\text{Å}$ were used in order to be able to vary the trapping forces while probing with a constant scattering beam, for example. They measured the intensity autocorrelation function $\langle I_s(o)_s(t)\rangle$ of the scattered light. When measured over a decade in time, it clearly showed two well-separated decay times due to rapid transverse motion and longer time axial motion. The results fit well with a simple theoretically derived formula. From the data, one can deduce the transverse and axial force constants as functions of trapping intensity, as well as the anisotropy of the trap in the transverse and axial directions. For a numerical aperture of 1.4, the ratio of $w_x/w_z \cong 0.4$.

The shortest time measured in the experiment corresponds to a mean-square displacement of $10\,\text{nm}$; but this can be much reduced with small particles. The particle size used for these experiments was $\sim 1.0\,\mu\text{m}$, which is very interesting for many tweezer applications. The authors suggest that using DLS from single particles with more structures, such as cell membranes and organelles, may provide dynamical information down to the nanometer scale. This localized version of dynamic light scattering is potentially a very important technique.

10.15. Thermal Ratchet Motors

More should be said about the subject of optical thermal ratchets. As mentioned above in Secs. 7.7 and 7.17, some people think thermal ratchets play an essential role in molecular motor motion. For amusement and edification, one should read Feynman's treatment of thermal ratchets showing how they act as one-way thermally driven motors [378].

The experiment of Faucheux *et al.* in Libchaber's group in 1995 [287], using optical potentials, stands out as a beautiful and unambiguous demonstration of thermal ratchet motors. By rapidly rotating a tweezer trap around a horizontal circle, using a pair of oscillating mirrors, the authors formed a closed circular trap almost $10\,\mu\text{m}$ in diameter. Particles could be trapped at rest on the circle when the rotational frequency was high enough that the particles felt only the average light intensity. The authors then introduced a periodic asymmetric circumferential saw tooth spatial intensity modulation by synchronizing a rotating neutral density filter with the high-speed rotating tweezer trap. This drew particles to the deep points of the asymmetric potential wells. They then turned the saw tooth intensity modulation on and off periodically. In the off periods the particle diffused forward and backward with equal probability. In the on position it tended to be carried forward by the asymmetric light potential (see Fig. 10.4). This caused the randomly diffusing particles to spend most of their on

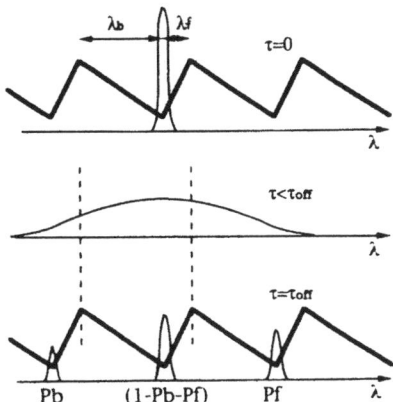

Fig. 10.4. The asymmetric potential is drawn as the thick line. The forward and backward length scales defining the asymmetry are λ_f and λ_b. The particle probability densities are drawn as thin lines. At time $\tau = 0$, the particle is localized and the probability density is sharply peaked. For times $\tau \leq \tau_{\text{off}}$, the potential is off and the particle diffuses freely. At time $\tau = \tau_{\text{off}}$, the potential is back on and the particle is forced to the forward and backward minimum with probabilities P_f and P_b. Figure 1 and caption from Ref. [287].

time running down the gradual slope of the asymmetric circumferential potential, thereby rotating on average in only one direction around the circle. The average circumferential velocity was fractions of a μm/s.

The authors allude to the possible connection of thermal ratchet motion and molecular motors. See the reference in Ref. [287] and the discussion in Sec. 1.4. They suggest that, in principle, this mechanism can be used to sort particles of different sizes. They also link this use of thermal noise to do mechanical work with considerations of entropy and Maxwell's demon.

10.16. Experimental Test of Kramers' Theory of Thermally Driven Transition Rates

McCann, Dykman, and Golding, in 1999 [379]*, made what is probably the best quantitative test of Kramers' theory of thermally driven transition rates for a particle confined in a metastable potential well. See Ref. 8 of Ref. [379]* for a review of the Kramers' problem. As the authors point out, escape from a thermally activated metastable state underlies many physical, chemical and biological processes. They used a bi-stable dual optical tweezer trap to make a well-controlled model system for studying thermally activated systems. With two tweezer beams focused through a single high numerical aperture microscope objective and adjusting the separation between the two potential wells by fractions of a micrometer, one gets a double-well potential with stable particle centers r_1 and r_2, and with a single saddle point r_s connecting them. By repeatedly sampling the particle coordinates over as many as 4×10^6 camera frames, one can deduce the entire potential shape to high accuracy and record as many as 94 000 internal transitions between the potential wells. The authors compared the experimentally measured transition rates with the theoretically computed rates from each of the potential wells, using the measured curvatures of the potential wells, and got remarkable agreement with Kramers' theory for a nearly three decade variation in transition rates.

The implications of this experiment are more than just academic. One, of course, can also probe the diffusional escape over the top of the potential barrier of a single potential well. The authors point out that the technique can be applied to other optically trapped particle types, i.e., not just spheres. The authors do not emphasize it, but this technique is capable of giving the absolute magnitude and detailed potential shape for tweezer traps acting on a variety of spherical particles, from the Rayleigh size regime through the intermediate range, where the diameter is comparable to the wavelength, and out to the geometrical optics limit. Looking at results of the present study for a particle of size that about equals a wavelength, it is experimental proof that the many force calculations and potentials for spheres along the optic axis are in error, as alluded to above. This technique can give the best experimental check of Maia Neto and Nussenzweig's exact force calculations using a Mie plane-wave expansion of the vector field. Of course, one also can experimentally probe and calibrate the shape of other specially designed optical potentials.

I think this technique is an example of the truth in one of Yogi Berra's famous expressions, "You can observe a lot by watching". [380].

CHAPTER

Study of Charge-Stabilized Colloidal Suspensions

One important area of research where optical tweezers and optical manipulation have played a major role is charge-stabilized colloids. Colloids have, from the very first experiment on radiation pressure and optical trapping, provided the highly mono-dispersive particle samples that have played so large a part in the optical manipulation field [1]*, [8], [109]*. These colloidal particles of a wide variety of materials are available from sizes in the nanometer to many tens of micrometer-size regimes. It was, in a sense, inevitable that optical trapping and manipulation techniques would shift from using existing single colloids as test particles to becoming a means of studying the colloids themselves and their interactions. This has become quite evident in the last 15 years or so.

Dense suspensions of charge-stabilized colloids have been studied for almost a century and are often regarded as a distinct state of matter [381]. Colloids play an important practical role in biological and physical chemistry [382]. They have been used as model systems for studying crystallization and phase transitions [383], [384]. As we shall see, the optical tweezers technique, with its ability to manipulate single colloidal particles and measure minute forces, has been playing an increasingly important role in this field, as Crocker and Grier show [385]*. Optical techniques have made it possible to directly check the almost fifty-year-old standard Derjaguin–Landau–Verwey–Overbeck (DLVO) theory. This theory has been the basis of most theoretical predictions of colloidal crystal behavior. The validity of this theory is a subject of considerable recent interest.

11.1. Optically Induced Colloidal Crystals

An interesting early work making use of the interaction of light forces and colloids was performed in 1985 by Chowdhury, Ackerson, and Clark on what they called "Laser-Induced Freezing" [69]. This experiment involved the optical alignment of particles in a two-dimensional (2D) colloidal liquid by a pair of laser beams, to produce a freezing transition to a solid-like structure at a sufficiently high light potential. A few years earlier Ashkin and colleagues [165], [167] had used a pair of beams to form a three-dimensional (3D) colloidal standing wave grating. This was a four-wave mixing type of nonlinear optics experiment in which a third beam was used to probe the particle index grating. In these earlier experiments the particle interactions played no essential role. In this 2D freezing experiment [69], the strong particle–particle interactions were crucial to the transition to the frozen state. The authors formulated a Landau-type theory to account for the phase transition to the solid-like structure. Data were taken for the self-scattered light for the first and second-order scattering peaks as a function of input light intensity for weakly interacting particles at high solution ionic strengths and for more strongly interacting particles at low solution ionic strengths. The agreement with Landau theory was quite good. The authors suggest that an interesting possibility for this type of experiment would be to produce other types of more complex periodic structures, even quasi-crystals, with multiple cross beams.

11.2. Optical Matter: Crystallization and Binding of Particles in Intense Laser Fields

Subsequent experiments published in 1990 in an article by Burns, Fournier, and Golovchenko, entitled "Optical Matter: Crystallization and Binding in Intense Laser Fields", extend the work of Chowdhury *et al.* on "Laser-Induced Freezing" [386]. The first part of their work showed how to use two to five external beams to form standing wave patterns that act as templates for forming dielectric crystals from uniform colloidal or other types of particles. With five equiangular coherent beams they formed the 2D standing wave pattern of a quasi-crystal located at the upper surface of the $200 \, \mu m$ thick sample cell and observed its diffraction pattern. They could directly observe the assembly of the quasi-crystal spheres, a few particles at a time, as the polystyrene latex spheres were lifted off the bottom of the cell by the radiation pressure of the incident beams and fell into the quasi-crystalline array at the upper surface.

Burns *et al.*, in 1989 [387], also observed some unexpected behavior, in the absence of any externally imposed potential, while watching successive particles driven by a single vertical beam rise to the upper surface, where they formed a closely packed crystal array. As each new particle appeared, it became clear that the particle's motion was being influenced not only by the incident beam, but also by the beam scattered from the central crystal. This behavior was attributed to a new type of interaction between two coherent dipoles placed in close proximity to each other, in a single incident beam. An analysis of this behavior showed that interaction energy exists between a pair of dipoles, depending on the inverse power of their separation, multiplied by an oscillatory factor.

This optical potential gives rise to bound states located at separations of nearly a wavelength. The forces arising from the scattered light of the two dipoles are similar to those arising in a standing wave made up of two opposing beams. To test for optical binding, the authors fabricated a narrow line-type trap with two cylindrical lenses, having dimensions of about $3 \, \mu m$ in width and $50 \, \mu m$ in length. First, they added a buffer to the water to screen out the electrostatic interactions of the charged spheres. They then captured two $1.43 \, \mu m$ spheres near the center of the line trap at the upper surface and followed their motions with a video camera and a diffraction screen as the spheres executed Brownian motion. A histogram of the separations of the spheres and a plot of their time course show

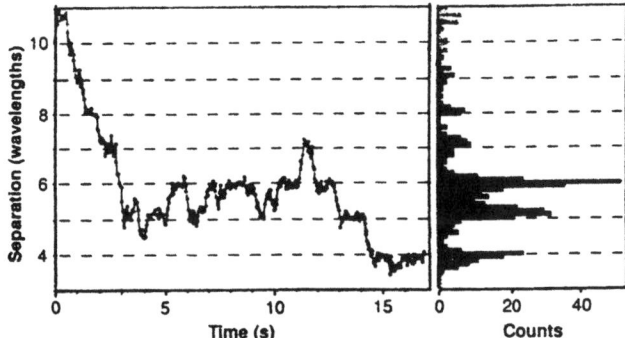

Fig. 11.1. Relative separation of two 1.43-μm-diameter spheres measured in units of wavelength of illuminating light in water. The plot on the left shows the time course of the separation, sampled at 1/30 second intervals, and the plot on the right is the corresponding histogram. Figure 8 and caption from Ref. [386].

that the particles strongly collect at positions separated by a wavelength, as predicted by the simple dipole approximation. See Fig. 11.1, taken from Fig. 8 of Ref. [386].

At higher powers in the beam, one expects the particle to lock into a fixed separation. Although the experimental particles are quite far from single Rayleigh dipoles, one gets an idea of how the near-field light intensity varies by using the Mie theory scattering intensity for the sphere sized used. The authors show results for this intensity variation that are quite close to the experimental data [387]. These optical binding effects are large enough that they must be considered in many optical interactions involving many particles.

The authors use the term "optical matter" for particles bound by the exchange of photons, in analogy with other physical forces, such as ordinary "electronic matter", which can be viewed as being held together by exchange of electrons. They also have in mind many of the other ordered structures mediated by light, such as colloidal and atomic crystals. For the future, they visualize using light as a template for many other types of complex structures, such as optical band gaps for periodic dielectrics. They even suggest *in situ* freezing or hardening of a fluid medium as a fabrication technique for producing permanent structures. See the discussion and references in Ref. [386].

11.3. Microscopic Measurement of the Pair Interaction of Charge-Stabilized Colloids Using Tweezers

Crocker and Grier, in their 1994 paper entitled "Microscopic Measurement of the Pair Interaction Potential of Charge-Stabilized Colloids" [385]* explicitly checked the standard theory of colloidal interaction formulated about 50 years ago by Derjaguin, Landau, Verwey, and Overbeck (DLVO), using video microscopy and optical trapping techniques. A pair of particles of about 32 nm radii were localized in space at a fixed separation of about 1.5 μm, and then released. The subsequent motions were recorded with a computer controlled video recorder. Each particle's transverse motions could be resolved to better than 25 nm (about 1/5 of a pixel) in the focal plane and to about 150 nm in depth, as estimated from its apparent size as it moved up and down. By analyzing some 6500 independent samples of the pair separations, they deduced the interaction potential as a function of separation.

From the experimental fit of the data they also obtained the effective charge of each particle and the Debye–Hückel screening length due to the presence of electrolyte in the water sample. They found that the agreement of their measurement with DLVO theory was excellent. The measurements made here corresponded to a force resolution of the order of 10^{-15} N. The authors concluded that the relative simplicity and versatility of the technique allow one to make measurements as a function of temperature, sphere size, and dielectric constant, which would be useful in understanding the bulk properties of bidisperse systems.

Now comes a surprise. Although DLVO theory predicts that an isolated pair of highly charged colloidal spheres feels only a purely repulsive screened coulomb force, experimental evidence began to accumulate pointing to the possibility of a long-range attractive component. The evidence came from unaccounted-for observations in colloidal crystals and in dilute suspensions. See references in Ref. [388]*. Also, direct measurements were made in which the potential was estimated using a digital video microscopy (DVM) technique, which allows direct visual observation of a suspension of colloidal particles and measures and digitizes all the particle separations in a single field of view.

Based on this type of data, Kepler and Fraden [389] calculated a pair correlation function by processing 500–5000 images ($90 \times 70\,\mu$m), depending on particle density. From this data they deduced that an attractive component to the potential exists between confined colloids at low ionic strength. Similar results indicating an attractive component to the potential were found by Carbajal-Tinoco *et al.* [390].

In the 1996 experiment by Crocker and Grier [388]*, they applied their tweezer technique to attacking the problem of "When Like-Charges Attract: The Effect of Geometric Confinement in Long Range Colloidal Interactions". They devised a special sample chamber in which the thin glass cover slip could be bowed by pressure to vary the separation of the glass chamber walls from about 2.6–6.5 μm. At 6.5 μm, the walls were so far apart that the particles diffused around, as in free space. For separations $d < 5\,\mu$m, the spheres were confined to the chamber's mid-plane by electrostatic interactions with the charged walls.

The authors used tweezers to isolate a pair of colloidal particles and to position them at a fixed separation. By repeatedly "blinking" the tweezers on and off, and by tracking the particles' motion to within 50 nm with DVM during the trap off time, they could sample and numerically solve for the potential at that point, with energy resolution of $0.1\,k_BT$, where k_B is the Boltzmann constant and T is room temperature.

They found for particles of 0.97 μm in diameter that a wall separation of 6.5 μm or greater resulted in a repulsive potential in agreement with their earlier experiment in 1994 [385]*. At $d = 4.0$ and 3.5 μm, under confined conditions, they found a strong attractive potential. However, at $d = 3.0\,\mu$m the potential reverts to repulsive! This indicates that the conditions for measuring attraction are not understood. Proximity to a wall is not sufficient to guarantee an attractive force. Measurements made with a larger sphere diameter of 1.53 μm gave an even stronger attractive long-range potential of about $1.0\,k_BT$.

The experiment was a real *tour de force* and shows some of the advantages of tweezer manipulation. Measurements were made with three different sphere sizes in the same electrolyte, in free space and also in confined conditions. They measured effective charges on the particles and the screening lengths. They discovered a small virtual leak of ions on the time scale of hours, which affected the screening lengths. This had to be corrected for.

The authors were able to show that the data taken with different sphere sizes were inconsistent with the alternative Sogami-Isa theory for the colloidal potential. Certainly DLVO theory is not formulated for such conditions and cannot explain the attractive effects.

11.4. Theoretical Approaches to the Understanding of Pair Interactions of Charge-Stabilized Colloids

The new data discussed above on long-range attractive potentials have raised a challenge to theorists to explain the effect. In 1998, Bowen and Sharif [391] published the results of a numerical calculation showing attraction, using the nonlinear Poisson–Boltzmann equation, which presumably accounts for the multibody interactions and the 2D confinement of the particles in question. They concluded there is "no need to revise the established concepts underlying theories of colloidal interaction". David Grier, in a "News and Views" comment in Nature on the work [392], gives a very nice summary of the importance of colloidal science and the state of knowledge on colloidal behavior. He points out the rancorous nature of the debate triggered by the surprising recent discovery of attractive forces. He also points out that numerical calculations of the type carried out by Bowens and Sharif for a particular case give no insights into the behavior under different circumstances.

In a continuing series of surprises, Neu in 1999 [393] reported an analytic proof based on the Poisson–Boltzmann model, showing that this model quite generally cannot predict attraction. This is contrary to the specific numerical calculations of Bowen and Sharif. Neu concluded that a new model is needed to explain the experimental results.

Later, an attempt was made by Gray, Chiang, and Bonnecaze [394] to directly reproduce the calculation of Bowen and Sharif [391]. They found, using a somewhat different, more rapidly convergent approach to the solution of the Poisson–Boltzmann equation, that there is no evidence for an attractive force. They therefore agree with Neu [393] that a new physical theory has to be developed to explain the anomalous attractive forces.

In 2001, a new feature was introduced into the investigation of the origin of anomalous attractive forces between a pair of like-charged colloids. Two theorists, Squires of Harvard and Brenner of MIT [395] proposed that hydrodynamics plays a significant role in explaining the data of Crocker and Grier [388]*. For example, they find that, for a sphere moving away from a wall by electrostatic repulsion, a fluid motion is created that drags a nearby sphere toward the first sphere, simulating an attractive force. A detailed calculation incorporating the hydrodynamic coupling of a pair of particles reproduces the experimental data almost exactly. See Fig. 11.2, taken from Fig. 4 of Ref. [395] Physics Today article on this subject by Richard Fitzgerald [396]. According to Squires, the hydrodynamic coupling is the only new feature that needs to be added to the standard DLVO screened potential to explain the effect. In other words, the force between the spheres, in their view, is still repulsive.

Thus, there are still questions to be answered to understand the force between charged colloidal particles. It seems clear that tweezers will play a big role in future experiments in this field.

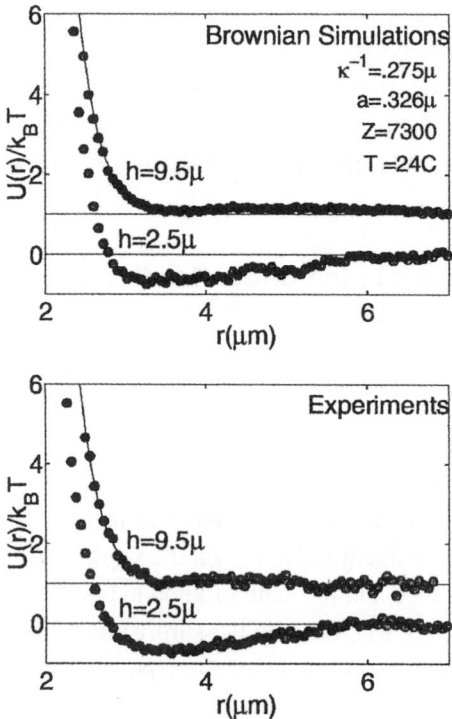

Fig. 11.2. Comparison between Brownian dynamics simulations and experiments for the effective potential between two collodial charged spheres near a wall. Two situations are presented: spheres close to the wall ($h = 2.5\,\mu$m) and far from the wall ($h = 9.5\,\mu$m). These are offset by $1\,k_BT$ for clarity. The simulations were carried out using standard methods, taking all parameters for the DLVO potential as those measured in the experiments. The simulations were analyzed using the same techniques used in the experiments. The only parameter that is not precisely measured is the charge density on the wall, which we take to be $\sigma_g = 0.4\,\sigma_p$. Figure and caption from Ref. [396].

11.5. Confinement-Induced Colloidal Attractions in Equilibrium

In 2003 Han and Grier made some careful experiments to help resolve the problem of the attraction of like-charged polystyrene (PS) colloidal spheres at large separations in thin confined cells in water [397]. This unexpected result contradicts the prediction of the Poisson–Boltzmann mean-field theory for electrostatic interactions. Is the theory wrong, or are there experimental errors or artifacts that can explain the data? (See Secs. 11.3 and 11.4).

The new measurements were taken with $1.58\,\mu$m diameter weakly charged colloidal silica spheres having a density twice that of polystyrene and one-fifth the surface charge. Such silica particles sedimented into a monolayer $\sim 0.9\,\mu$m above the lower cell wall. The cell was completely sealed after filling and left to equilibrate before taking measurements. It was found that attractive force also existed with an asymmetric geometry for cells of total thickness of 3–20 μm, but not for cells up to 200 μm thick. Figure 11.3, taken from Fig. 2 of Ref. [397] shows high precision data taken with PS spheres in a 1.3 μm-thick cell, as well as data taken with silica spheres in cells 195 and 9 μm thick. This showed that the upper confining wall, even when $\sim 20\,\mu$m away, still influenced the forces.

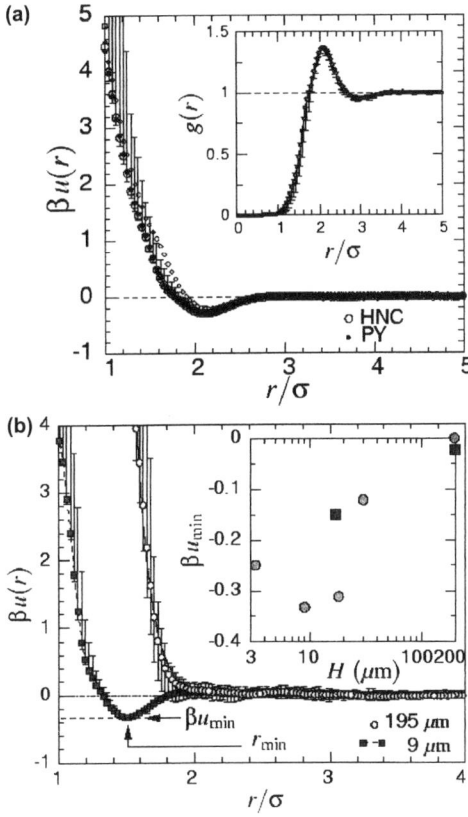

Fig. 11.3. Colloidal pair potentials. (a) PS spheres $\sigma = 0.652\,\mu$m diameter at $H = 1.3\pm0.1\,\mu$m and $h = H/2$. Circles: $n\sigma^2 = 0.056$; diamonds: $n\sigma^2 = 0.020$. Inset: Pair correlation function at $n\sigma^2 = 0.056$. (b) Silica spheres $\sigma = 1.58\,\mu$m diameter at $H = 195\,\mu$m (circles) and $9\,\mu$m (squares). The solid curve is a fit to Eq. (6). Inset: The minimum's dependence on H for $\sigma = 1.58\,\mu$m (circles) and $2.2\,\mu$m (squares).

The authors conclude that the existence of attractive forces in violation of mean-field theory is real and that explanations based on kinetic effects, such as hydrodynamic coupling, are excluded.

11.6. Entropic Forces in Binary Colloids

There are other types of forces on colloidal particles, called entropic forces, which are due to the presence of smaller particles in the background solvent. These forces play a role in many colloidal processes of technological importance.

When two large spheres approach each other, at some point they get close enough to exclude the small spheres. This causes an asymmetry in the Brownian bombardment of small particles on the rest of the spheres, which gives rise to an attractive force between the large spheres, or between a sphere and a surface. These entropic effects are sometimes called excluded volume effects and play a role in emulsions and in phase-separation phenomena.

11.7. Entropic Control of Particle Motion Using Passive Surface Microstructures

Dinsmore, Yodh, and Pine, in 1996 [398], show that these entropic forces for a mixture of large and small spheres correspond to an energy decrease near a surface or an edge that can be as large as about $k_B T$, the mean thermal energy. This results in an attractive force of about 0.04 pN that can operate over distances of tens or hundreds of nanometers. They suggest that passive structures etched into the walls of a container can create localized entropic forces, which can trap, repel, or induce particles to move in a controlled drift. This could be useful for making highly ordered arrays of particles, such as microelectronic masks, photonic band gap structures, or materials for clinical assays. See the figures and references in Ref. [398].

Using an optical tweezer trap with about 10 mW of power, the authors were able to place a single large sphere of diameter 0.45 μm very precisely on the surface of a glass cover slip. This large sphere was mixed with 0.083 μm small spheres in volume fractions of 10^{-5} and 0.30, respectively. A solution of NaCl (0.01 M) was added to screen the electrostatic interactions over a distance of ~ 5 nm, to give nearly hard surface interactions. The spheres were placed a distance of $0.56 \pm 0.06\,\mu$m away from the edge of the cover slip and then released. The subsequent motion was followed using a CCD (charged-coupled device) camera every 0.1 s with an accuracy of 0.04 μm. They repeated the process about 1400 times to obtain the average behavior of the particles, starting from this precise location. By following the 2D trajectories of the large spheres, they deduced the Helmholtz free energy F from the probability of finding a particle at different locations along the slide. The force on the sphere is given by the gradient of F in the direction of decreasing F. They found a distinct repulsive barrier of $\sim 2\,k_B T$ at the edge of the slide. They calculated the free energy F and the forces for a variety of cases, such as for spheres moving perpendicular to a plane surface, for spheres near the corners of a closed box, or for spheres in the neighborhood of a trench-like groove in a surface. The authors also showed that there is a directed entropic drift force along a wall of changing curvature.

11.8. Entropic Attraction and Repulsion in Binary Colloids Probed with a Line Optical Tweezer

The history of entropic forces in binary colloids goes back at least 40 years. At low concentrations of the small species the forces on the larger sphere of a binary mixture are usually described by the depletion model of Asakura and Oosawa [399]. This model predicts a monotonically attractive potential. However, recent theories for more concentrated smaller particles often predict a repulsive and even an oscillatory component. See Ref. 3 in the paper by Crocker *et al.* [400].

The paper of Crocker *et al.* makes the first direct measurement of these effects between two colloidal particles in suspension and shows the increasing level of sophistication reached in this field with the use of optical tweezer measurement techniques. They used a rapidly scanning focused laser beam to form a 1D rod-like optical trap $\sim 10\,\mu$m long on which they confined a pair of essentially freely diffusing spheres. By video taping the single pair of large spheres (1100 nm in diameter) in a background of concentrated smaller spheres (83 nm in diameter) they were able to follow a pair of diffusing particles for an hour or so, during which time they made $\sim 2 \times 10^5$ separation measurements. This detailed measurement yielded the first direct measurement of entropic attraction and repulsion forces, and even oscillatory behavior. Surprisingly, even at higher concentrations the pair of spheres was observed to coalesce only rarely. This anomalous slowing of relative Brownian motion was completely unexpected. The authors believe it may require new theoretical insights to explain it. Nevertheless, it suggests a new means for stabilizing suspensions, using the entropic force alone.

CHAPTER

Rotation of Particles by Radiation Pressure

There has been a recent resurgence of interest in the subject of optically induced rotation of macroscopic particles. This probably stems in part from the interest in the numerous types of small biological motors as well as interest in small mechanical motors for possible application in microtechnology. Optical manipulation techniques based on the linear light momentum have, as already seen, played a major role in making traps for studying small biological motors. It is natural to expect that these forces can also be used to power optically driven motors. In addition to its linear momentum, light also carries angular momentum of $h/2\pi$ per photon. In classical optics the angular momentum of light is found in circularly polarized light. Linear polarized light, which has no angular momentum, is a superposition of right and left circularly polarized light waves.

The angular momentum of light and some of its possible uses for spinning particles were discussed in Ashkin's early review paper in Science 1980 [109]*. Marston and Crichton in 1984 [131], [363] were probably the first to calculate the radiation torque on a sphere caused by circularly polarized light using Mie theory. They showed that the torque on a transparent sphere is zero. One gets net torque only for the case of absorbing spheres. Similar results were found by Chang and Lee [132], Gouesbet *et al.* [364], Marston and Crichton [131], [363], using different types of calculations.

12.1. Optically Induced Rotation of an Anisotropic Micro-Particle Fabricated by Surface Micromachining

In 1994, Higurashi *et al.* [130]* devised a very clever way of generating optical rotation of a fabricated asymmetric transparent particle held in a single-beam tweezer trap. It was, of course, well known from the first demonstrations of the linear momentum of light by Nichols and Hull [78] that a properly designed vaned object could generate torque. In their work, Higurashi *et al.*, starting with a $10\,\mu$m thick layer of SiO_2 deposited on a substrate, used reactive-ion etching to form a cylindrical particle with a cross sectional shape having four vanes, each of which lacked bilateral symmetry. The particles were subsequently released by dissolving the substrate. These cylindrical particles have

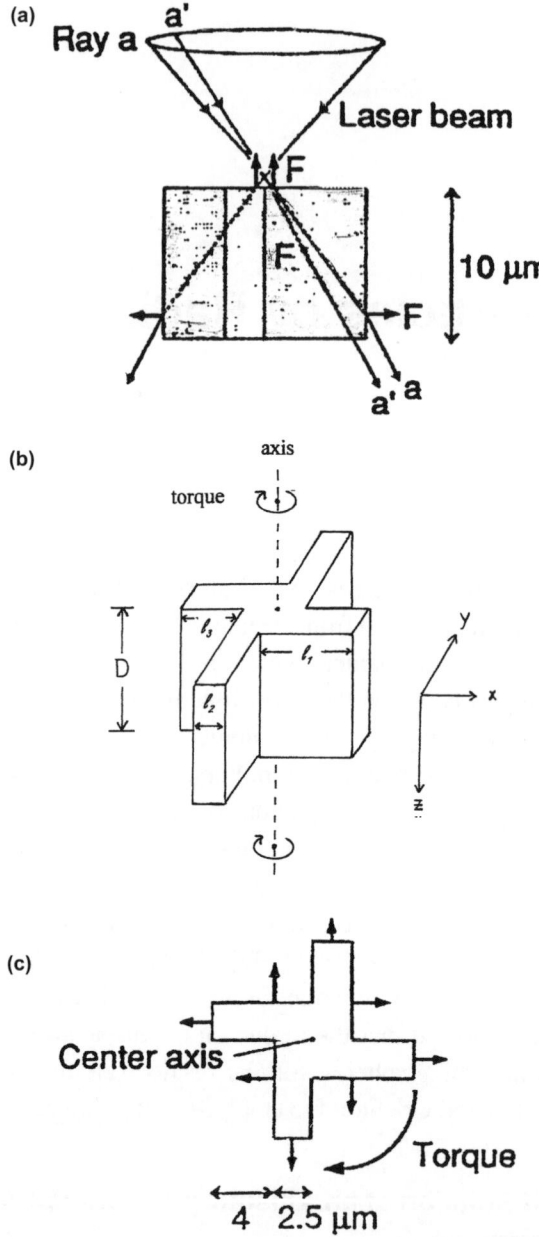

Fig. 12.1. Micro-object designed to rotate about the laser beam axis. (a) Side view. (b) Perspective diagram of micro-object. (c) Forces on the side faces of the micro-object, which lead to clockwise rotation. Figure derived from Ref. [130]*.

rotational symmetry and are readily trapped co-axially by the highly convergent tweezers beam. They spin vigorously and continuously, driven by the linear momentum. The torque arises basically from the lack of bilateral symmetry of the vanes. Figure 12.1, taken from Fig. 2(C) of Ref. [130]*, is a sketch of the cross section of the vaned particle, showing the direction of the forces on the faces of the vertical vanes by the levitating TEM_{00} mode beam. It is clear that there is a large torque about the axis.

The ray picture also shows that the particle has transverse stability. Once again, we see that a simple ray-optics model of the light scattering is sufficient to explain the origin of the torque and the trapping.

Rotation rates of about 22 rpm for the four-vaned cylinder were observed with a laser power of 80 mW in water. This is probably the smallest fabricated motor. It is 10 times smaller than electrostatic motors. See Ref. 7 of the authors' paper.

It would be exciting to try to observe spinning of a particle such as this in high vacuum, possibly to the point of rupture, as suggested by Ashkin in Ref. [109]*. Recall also the discussion in Sec. 3.1.3 on the rupture of small asymmetric dust particles in interplanetary space driven by solar radiation pressure.

It would also be interesting to see how long one can stably trap shaped particles, such as in Higurashi *et al.*'s experiment, with tweezers in vacuum without using feedback damping. Tweezer traps are much stronger that levitation traps and should be more tolerant of random fluctuations in laser power.

Following the experiment of Higurashi *et al.* [130]*, Gauthier in 1995 [401] made a more complete analysis of the torque, using the ray-optic model. Gauthier in 1996 [402], and also Higurashi *et al.* in 1997 [403] reported some improvements in the shape of the spinning particle, which should yield higher torques.

12.2. Optically Induced Rotation of a Trapped Micro-Object about an Axis Perpendicular to the Laser Beam Axis

In 1997, Gauthier [404] theoretically and experimentally demonstrated the tweezer trapping of a transparent thin ring. This of itself was not totally surprising, in view of the ability of tweezers to trap complicated biological organisms by grabbing localized parts of their anatomy [19]*, [20]* (see also Sec. 5.2.7). In 1998, however, Higurashi *et al.* [405]* made a 2.9 μm thick ring-like or disc-like particle rotate about an axis perpendicular to the beam axis by shaping the interior of the disc into an asymmetric tooth-like structure. Figure 12.2, taken from Fig. 1 of Ref. [405]*, shows the geometry of the 12 μm diameter disc and the direction of the optically induced spin. Figure 12.3, taken from Fig. 2 of Ref. [405]*, shows the origin of the torque on the tooth-shaped inner wall. The disc was fabricated by reactive-ion etching of a fluorinated polyimide film. Rotation speeds were much higher than previous [130]*, reaching about 150 rpm at 80 mW. The authors suggest that these motors can be used as probes of microfluid dynamics. In biology they think these microdiscs could serve as handles for twisting biological specimens, such as DNA strands, for example.

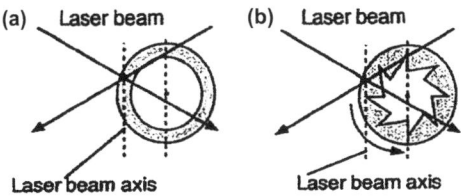

Fig. 12.2. Geometry of optical trapping of micro-objects (side view); (a) ring-shaped micro-object (microdisk having an opening), (b) micro-object having an opening and shape anisotropy in its interior. Figure 1 and caption from Ref. [405]*.

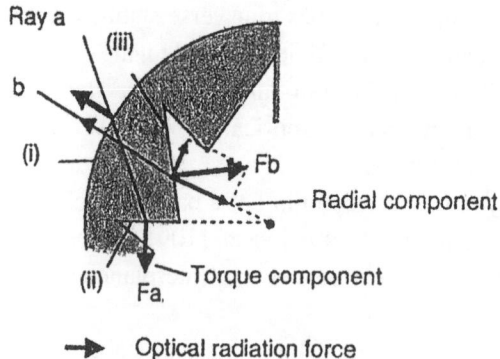

Fig. 12.3. Forces of radiation pressure exerted on the inner walls of the object in Fig. 1(b); showing origin of the torque on the tooth-shaped inner wall. Figure 2 and caption from Ref. [405]*.

12.3. Optical Microrotors

Galajda and Ormas, in 2001, fabricated a novel microrotor for use in optical tweezer rotation experiments [406]. Various rotor shapes, several micrometers in size, resembling windmills, propellers, and helices were fabricated by a two-photon polymerization technique using a resinous material. By using computer controlled motion of a piezo-controlled microscope stage, the authors could reproducibly irradiate almost arbitrarily complex-shaped structures with a resolution of $\sim 0.5\,\mu$m. They used a focused 5.14 nm argon-ion laser beam of 20 mW. The unpolymerized material, which was not irradiated, was subsequently dissolved away. Hundreds of identical particles were produced in this way. Since it is difficult to predict the optimum shape of a rotor, they empirically tested many different rotor shapes. A very efficient rotor is shown in Fig. 12.4, taken from Fig. 1 of Ref. [406]. The use of a control shaft on the particle axis adds greatly to the particle's stability. There are two stable orientations of the particle in the tweezer trap, in which its axis points either up or down along the beam direction. They both rotate well, but at somewhat different rates.

Remarkably, these particles also act as tiny gears or cogwheels that can be oriented and intermeshed on a surface, with their axes pointing upward. The authors were then able to engage a pair of these gears with a smaller optically levitated gear-shaped particle and optically drive the entire assembly. See Fig. 12.5, taken from Fig. 3 of Ref. [406]. This qualifies as a true micromotor and, I believe, would have won Feynman's prize for the smallest man-made motor. See the Afterword. As I recall, he had to reluctantly award his prize for a miniaturized electric motor. The authors suggest that one might convert the particles into a drill or mill to perform machining functions.

In addition to applying torques to miniature mechanical devices, the authors anticipate application in the study of the mechanical properties of protein or DNA molecules by attaching one end to a surface and the other end to a rotor and applying an optical torque.

12.4. Orbital Angular Momentum

We have already discussed the intrinsic linear and angular momentum of single photons and their conservation in interactions with particles of various types. Recently, a useful new concept has been

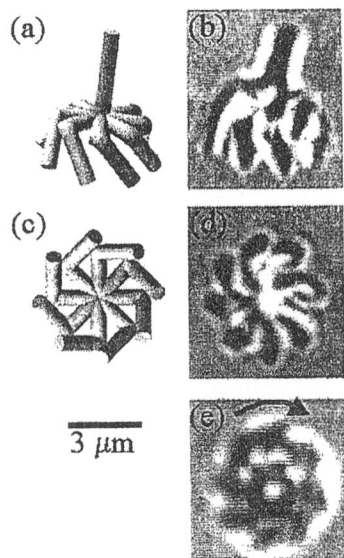

Fig. 12.4. Geometry and picture of an efficient microscopic light driven rotor. (a) Explanatory drawing and (b) photograph of the rotor in an arbitrary position when it is tumbling freely, viewed from identical directions. (c) Drawing and (d) equivalent photograph of the rotor when it is trapped in focus but held against the cover glass thereby preventing rotation and yielding a sharp image. (e) Photograph of the spinning rotor trapped in focus and rotated by the light. Figure 1 and caption from Ref. [406].

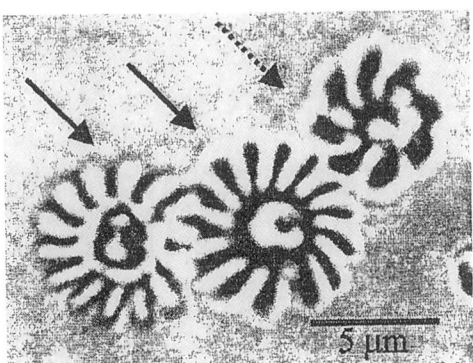

Fig. 12.5. Complex micromachine built by the two-photon technique: two engaged cogwheels are rotated by a light driven rotor. The solid arrows point to the cogwheels rotating on axes fixed to the glass surface. The dashed arrow points to the rotor similar to that in Fig. 1. The rotor is held and rotated by the laser tweezers and the rotating propeller drives the system. Figure 3 and caption from Ref. [406].

introduced, namely that of the orbital angular momentum of light, for a paraxial light beam, especially Laguerre–Gaussian (LG) mode beams. Laguerre–Gaussian beam modes are characterized by an $e^{i\ell\varphi}$ azimuthal dependence. Such LG modes have helical wave fronts and carry angular momentum for linearly polarized modes of $\ell h/2\pi$ per equivalent photon (see Refs. [407], [408], [409]). This orbital angular momentum is distinct from the $\pm h/2\pi$ angular momentum associated with spin in circularly polarized light. Thus, a circularly polarized LG mode carries $(\ell \pm 1)h/2\pi$ total angular momentum per photon. If an LG mode beam interacts with an object that converts it to another LG mode with a different angular momentum, one expects, by conservation of total angular momentum, that the change in angular momentum of the two modes will give rise to a torque on the converter. Most of us are more familiar with the Hermite–Gaussian (HG) laser modes, especially the TEM_{00} HG mode, and even occasionally use the low order HG modes. These modes, with their simpler phase fronts and absence of helicity, have no orbital angular momentum.

In Ref. [407], the authors give a design of a mode converter that transforms an HG mode of arbitrary high order to an LG mode and vice versa. It consists of two cylindrical lenses and depends on the appropriate use of the Gouy phase.

The Gouy phase refers to the phase shift of light when going through a beam waist compared to that of a plane wave. For isotropic or nonastigmatic HG modes there is the usual single phase shift along the beam axis. For LG modes that are astigmatic in character, there is a different phase shift and beam parameters along the two orthogonal xz and yz transverse planes. The authors show that with an appropriate pair of cylindrical lenses, one can convert an HG mode to a beam with different waist and phase shift parameters in the xz and yz planes, where z is the direction of beam propagations. If the phase shift in one of the two orthogonal planes is $\pi/2$ relative to the other, then we have a $\pi/2$ converter. This is like a birefringent $\pi/2$ phase plate. It converts an HG_{nm} mode with indices n and m into an LG_{nm} mode beam with the same indices. See Fig. 12.6 taken from Fig. 5 of Beijersbergen *et al.* of Woerdman's group [407]. Woerdman and collaborators [410] show that one can use a special spiral phase plate to accomplish conversions of a paraxial TEM_{00} laser beam into a paraxial LG mode. The spiral phase plate shown in Fig. 12.7 introduces the characteristic phase singularity on axis and gives the beam its spiral character. The amount of the azimuthal phase change

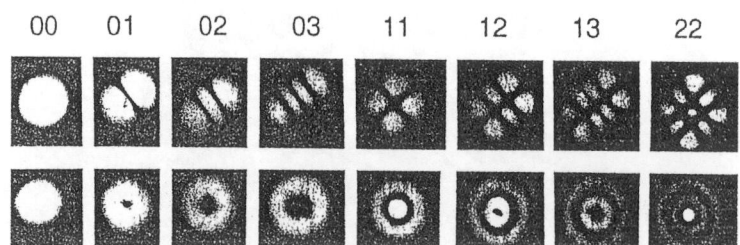

Fig. 12.6. Experimental results obtained with the $\pi/2$ converter. The top row shows the input HG_{nm} mode; the bottom the output LG_{nm} mode. where n, m is indicated above the modes. Figure 5 and caption from Ref. [407].

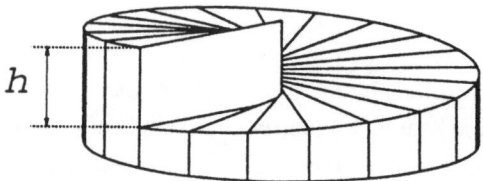

Fig. 12.7. Spiral phase plate to convert a TEM_{00} mode into a paraxial LG mode.

depends on the thickness (h) of the one turn "staircase". In a mode converter of this sort we would, of course, expect a torque on the phase plate in order to conserve angular momentum.

12.5. Observation of Transfer of Angular Momentum to Absorptive Particles from a Laser Beam with a Phase Singularity

There have been a number of experiments demonstrating the transfer of angular momentum from an LG mode beam with a phase singularity to an absorbing particle. He *et al.*, in 1995 [411], used absorbing particles, ~ 1–$2\,\mu$m in diameter, from a ceramic superconducting powder dispersed in kerosene, as well as CuO particles dispersed in water. Figure 12.8, taken from Ref. [411], gives a snapshot of the irradiance of a $TEM_{01}*$ mode beam containing a first-order phase singularity. This gives a graphic picture of the spiral nature of the LG beam modes. The beam used in their experiment consisted of a 7 mW He–Ne LG_{03} linearly polarized beam converted from a TEM_{00} mode beam by a blazed holographic grating. Such a higher-order LG beam has an equivalent angular momentum "ℓ" of $3h/2\pi$ per photon. It was focused by a high numerical aperture lens on the particles and trapped them in two dimensions against the glass output face of the sample cell, where they were observed to spin in the same sense as the LG mode. Reversing the holographic filter reversed the sense of rotation. Most particles appeared to be spherical under an electron microscope. The absorption was quite large. It was estimated to be about 25% for a particle with a 1 μm radius, based on a rotation rate of about 4 Hz. Similar reversals in the direction of rotation of CuO particles were observed when the sense of spiraling of the beam was reversed. Other sources of rotation due to radiometric forces or reflective forces seemed unlikely. The authors concluded that the source of the spin was clearly the absorption of "orbital" angular momentum associated with the helical wave front with its central phase singularity.

Friese *et al.*, in 1996 [412], extended the work of He *et al.* [411] on transfer of optical angular momentum to trapped absorbing particles. They used absorbing CuO particles in the 1–5 μm range as

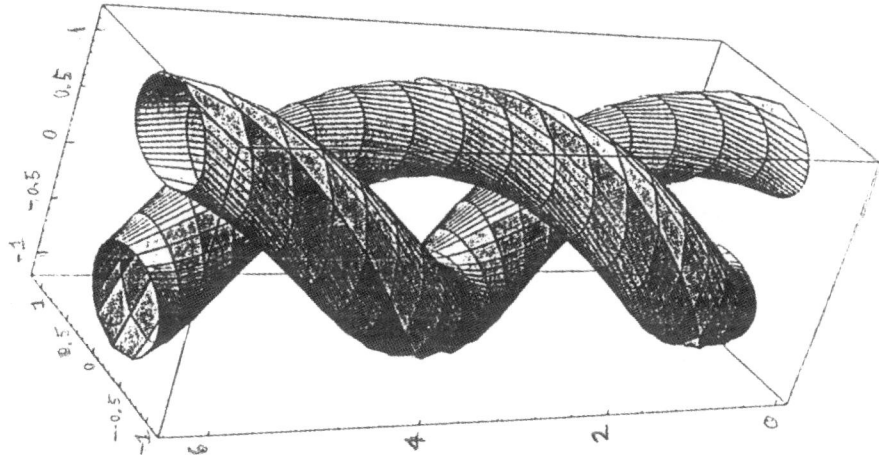

Fig. 12.8. Snapshot of the irradiance structure of a $TEM_{01}* [(r/\omega)e^{-r^2/\omega^2}e^{i\theta}e^{ikz}]$ beam containing a first-order phase singularity. The surface represented is that where the irradiance has half its peak value and we take $k = 1$, $\omega = 1$ for simplicity. Figure 1 and caption from Ref. [411].

they did earlier, and once again used the LG_{03} "donut" mode. The new aspect of this later experiment was to see the effect of right and left circularly polarized light on the particle rotation speeds. The total angular momentum carried by a photon of an LG mode beam is, in the paraxial beam approximation $(l + \sigma_z)h/2\pi$ as pointed out earlier, where $z = \pm 1$ for circularly polarized light and $z = 0$ for linearly polarized light. This divides the total angular momentum into separate spin and orbital parts. As one varies the polarization, one might expect the effective spin per photon to vary as 2:3:4 and cause the particle spin rate to vary in the same ratio.

There is, however, a complication. The highly focused, high NA beam used by the authors was hardly paraxial. In 1994, Barnett and Allen [413] developed a general nonparaxial theory using a mode that is a general LG mode in the paraxial limit. In the general theory there is no longer a clear separation between orbital and spin angular momentum. They derived an expression in which an additional cross-term appears, proportional to σ_z, which depends on both spin and angular momentum and varies as $1/z_R$, the inverse of z_R, where z_R is a length term that is the paraxial limit associated with the Rayleigh range (the Rayleigh range is one half the usual confocal distance). For $\sigma_z = 0$, the cross-term disappears and, despite the light beam convergence, the torque depends only on ℓ, the orbital angular momentum, as for the paraxial beam case! This explains why the experiment of He *et al.* [411] worked so well. Friese *et al.* calculated that the contribution of the cross term to the torque for the value of z_R used is only about 4%, which is negligible.

The results of the experiment showed that the rotation frequency increased and decreased by roughly equal amounts in switching from left to right circular polarization, as expected, but that the linear polarization rotation rate was too slow. The authors attribute this to static friction between the particle and the slide.

This is a nice experiment, but the authors' use of the word "trap" in both the experiments of Refs. [411], [412] is confusing. The so-called "trap" only optically confines the absorbing particles in two dimensions; i.e., transversely, using a donut-type mode. The particles are driven against the transparent wall in the z or axial direction. I believe the term "trap" should be reserved for stable three-dimensional (3D) traps. This same sort of confusion about two-dimensional (2D) and 3D trapping exists in many of the early papers in biology.

12.6. Mechanical Equivalence of Spin and Orbital Angular Momentum of Light: An Optical Spanner

A true 3D tweezer-type trap was demonstrated by Simpson *et al.* in 1997 [414]*, for mildly absorbing particles having a size larger than the focal spot size of a tightly focused LG mode beam. Since the particle can be made to rotate in such a trap, based on the angular momentum of the light, they call this an optical *spanner* or wrench. The major distinction between this experiment and the previous experiments [411], [412] is the use of a true 3D tweezer-type trap. Tweezer traps of this sort, based on the donut mode, were first described theoretically by Ashkin in 1992 [49]*. In the Simpson *et al.* experiment they showed that the spin angular momentum, $\pm h/2\pi$ associated with circularly polarized light can add to or subtract from the orbital angular momentum of an $\ell = 1$ LG mode. The results show that the $\ell = 1$ LG_{01} mode, which is the equivalent of the $TEM_{01}*$ mode, has well-defined momentum of $h/2\pi$ per photon.

12.7. Controlled Rotation of Optically Trapped Microscopic Particles

A very clever technique for controlling the rotation of optically trapped transparent micrometer or sub-micrometer particles was developed in 2001 by Paterson, *et al.* [415]*. It is based on the controlled rotation of the spiral interference pattern of LG mode beams [411], [412], [414] with a plane wave beam. The interference patterns of a higher-order ring-shaped LG mode beam and a co-propagating plane wave or TEM_{00} mode beam has a number of spiral arms, fixed in space, depending on the order of the LG mode. A tweezer trap based on such a combined LG mode and TEM_{00} mode beam has a multilobed pattern at the focus, which can trap a variety of small particles on the high intensity regions of the interference pattern. See Fig. 12.9, taken from Fig. 1 of Ref. [415]*. Changing the relative phase of the two beams causes the pattern and any trapped particles to rotate in space. One can control the sense and rate or angle of rotation with the relative phase shift. This technique does not depend on absorption or any other intrinsic properties other than index of refraction. The authors illustrated the technique by rotating three $5\,\mu$m sized glass spheres at $5\,$Hz rate with an order $\ell = 3$ LG mode beam and also a cylindrical glass rod $5\,\mu$m long in an $\ell = 2$ LG mode beam. The latter can serve as a controlled stirring rod. They also demonstrated the rotation of Chinese hamster chromosomes in their infrared Nd:YVO$_4$ laser trap. Rotation of objects within the interior of cells is of course possible.

This is a very practical scheme. The LG modes were formed from holographically manufactured elements. Phase shifting was produced crudely by rotating a glass plate, but phase modulators can

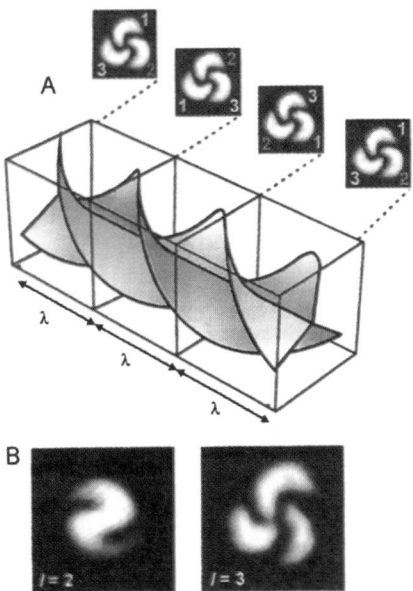

Fig. 12.9. (A) The phase fronts of an LG beam of azimuthal index $l = 3$ (helical structure) and intensity pattern when interfered with a plane wave. The phase fronts describe a triple start intertwined helix that repeats its shape every λ but only rotates fully after $l\lambda$. In (B), we can see the experimental forms of the interference patterns of LG beams of index $l = 2$ and 3 with plane waves used in our experiments. Figure 1 and caption from Ref. [415]*.

be used for higher rotation rates. As useful as this technique is, it should be pointed out that equivalent rotational performance could probably be demonstrated with the computer-controlled scanning techniques of Visscher, *et al.* [266], Sasaki *et al.* [112], and others.

12.8. Optical Torque Wrench: Angular Trapping, Rotation, and Torque Detection of Quartz Microparticles

A variety of techniques have been devised to optically rotate different types of particles, using the linear and angular momentum of light beams. Both absorbent and transparent particles have been previously studied under trapped and untrapped conditions (see Chapter 12).

There have been some additional experiments involving torque from polarized light on birefringent particles. The birefringence can arise from departures from sphericity, called shape birefringence, or from crystal structure birefringence. Bishop *et al.* [416] in 2004, in Australia, measured torques on micrometer-sized glass rods using polarized trapping light. Their torque measurements for rods of different sizes agreed well with their calculations based on mathematical models.

La Porta and Wang [417]* in 2004, at Cornell, generated torque using polarized light applied to crystal quartz particles. Using a pair of acousto-optic modulators, they were able to combine the phases of two orthogonal polarizations to orient or apply torque to the trapped quartz particles. They measured the torque by measuring the imbalance in right and left circularly polarized components of the light beam transmitted through the trapped particle. Measurements were made of applied torque versus the rotation rate. For example, with 22 mW they generated about 1300 pN nm of torque at a rotation rate of about 30 Hz.

The authors believe that the demonstrated control over torque and angular motion of small quartz particles can be used as the equivalent of the "handles" technique for study of biological rotary motors.

Bishop *et al.*, in an additional paper in 2004, entitled "Optical Microrheology Using Rotating Laser Trapped Particles" [418] also developed a technique for measuring the torque exerted by a tweezer trapping beam on a rotating birefringent particle. They say they can use this system to measure viscosity and surface effects on micrometer-sized particles in liquid media. An accurate measurement of viscosity was demonstrated inside a prototype cellular structure.

CHAPTER

13

Microchemistry

There were a number of innovations in laser trapping and manipulation techniques introduced in the early 1990s as part of a research project called "Microchemistry by laser microfabrication techniques" sponsored by the Exploratory Research for Advanced Technology Research Corporation of Japan (ERATO).

An interesting paper by Sasaki *et al.* [112] describes a computer-controlled beam scanning system developed to generate multiple optical traps (see also Ref. [419]). Figure 13.1, taken from Ref. [112], shows a schematic diagram of such a scanning micromanipulation system. The pattern of traps formed in Fig. 13.1 is of the Chinese character for light. This aspect of their work is very similar to the scanning trap system developed by Visscher *et al.* in 1993 [266], which was later used in their assay to distort red blood cells, as discussed above [209].

In addition, Sasaki *et al.* showed that one could trap low-index particles, absorbing particles, and metal particles, all of which are pushed out of high-intensity regions of a light beam by making a three-dimensional (3D) "light cage". This was created by scanning a highly focused tweezer beam around a small circle in space. This gave a region of low light intensity at the center of the circle, which acted as a trap for low-intensity seeking particles. Figure 13.2, taken from Ref. [112], shows the scheme used to form an optical trap. Trapping of micrometer-sized iron, aluminum, and carbon black particles in various liquid solvents was demonstrated experimentally. The authors think these new scanning techniques will prove useful in studying fine particles, in making microelectronic devices, and for fabricating micromachines.

Misawa *et al.* [420], [421] performed some quite interesting experiments on laser ablation and laser fusing of polymer latex particles using optical manipulation techniques. The authors found that they were able to drill small holes in small, levitated PMMA (polymethyl methacrylate) latex particles in water by the well-known process of ultraviolet (355 nm) ablative photodecomposition of polymers [422], [423]. These are the first observations of ablative decomposition in water, however, and the threshold energy is much higher than that observed in PMMA in air or vacuum. Remarkably, with a single shot, they could drill a quite precise $1\,\mu$m hole in a $10\,\mu$m sphere held in a tweezer

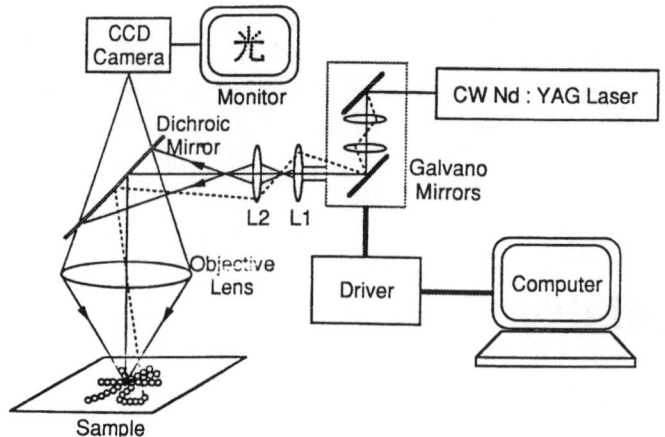

Fig. 13.1. Computer-controlled scanning system to generate multiple traps. The traps are shown forming the Chinese character for light. Figure from Ref. [112].

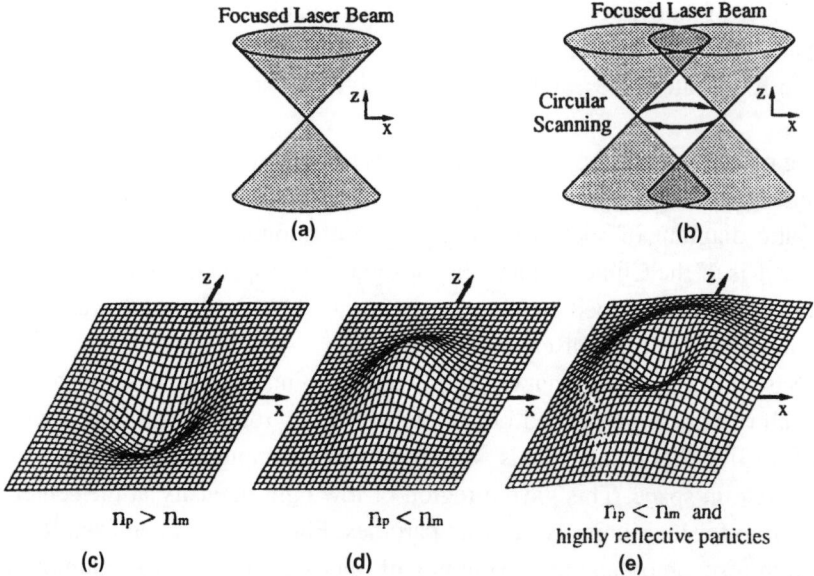

Fig. 13.2. Three-dimensional light "cage" or low intensity dark spot formed by scanning a highly focused tweezer beam in a small circle in space. Figure 9 from Ref. [112].

trap. Without tweezers, particles hit by a UV pulse were driven out of the field of view. It was found that even sub-micrometer diameter holes could be drilled in particles as large as $6\,\mu$m in diameter. The small hole diameters were not expected and might be an indication of an optical self-focusing effect driving a multiphoton nonlinear absorption.

Another potentially useful manipulation by Misawa *et al.* [424] was the assembly of strings of latex particles held in a line by tweezers and then photopolymerized by UV light. Light focused at the contact points of the particles caused fusion, but no noticeable distortions. The same approaches are expected to make it possible to create various microstructures comprised of other particle types, such as micrometer-sized particles (iron particles, carbon black particles, etc.). Particle structures comprised of three fused polymer particles were manipulated after their assembly with a pair of tweezer traps and then rotated in space. One beam was held fixed and the other was spun about it at rotation rates of $120°/s$, in the clockwise or counterclockwise direction.

13.1. Laser Trapping, Electrochemistry, and Photochemistry of a Single Microdroplet

The ERATO group performed a number of very innovative experiments in microchemical reaction systems, involving mass transfer chemical reactions across a micrometer-sized droplet/solution interface. The authors successfully combined laser trapping with electrochemistry and spectroscopy in microparticles. As an example, Misawa *et al.* in 1991 [425], [426] studied the electrolysis of ferrocene (FeCp) in a single nitrobenzene droplet in water. Drops of various sizes (4–$10\,\mu$m) of nitrobenzene were manipulated by tweezers and placed in physical contact with gold microelectrodes and were thus immobilized. The electrochemical measurements, such as current versus voltage and the total electron charge versus time, were made without light. The authors succeeded in getting quantitative results on the transfer of $FeCp/FeCp^+$ across the interface for the first time. They assert that such studies are only possible using the laser trapping-electrochemistry techniques.

In addition to perturbing the distribution equilibrium of FeCp between the oil droplet and the water phase by electrolysis in the dark as demonstrated above, Nakatani *et al.* in 1993 [426] extended the method to control a photoinduced electron transfer reaction of 9,10-diphenylanthracene (DPA) on a droplet by mass transfer of FeCp across the droplet/water interface. It was virtually impossible to control photoinduced transfer between FeCp and DPA with mm-sized droplets. However, for μm-sized droplets, FeCp was electrolyzed rapidly, so that the mass transfer of $FeCp/FeCp^+$ could be controlled directly by the redistribution of the equilibriums inside and outside the droplet. The authors believe that simultaneous poly(N-isopropylacrylamide) (PNIPAM) microparticle formation will be of use for laser-controlled micromanipulations, drug delivery systems, and other microdevices. Another possibility is the study of chemical reactions on a small scale by the laser manipulation and fusion of reactants localized in small droplets.

13.2. Control of Dye Formation Inside a Single Laser-Positioned Droplet by Electrolysis

Nakatani *et al.*, also in 1993 [427], further showed how to control a dye formation reaction in a single droplet by varying the droplet–electrode distance. They achieved control not only of the dye-forming reaction, but also the mass-transfer rate in the single microdroplet.

They started with an oil in water emulsion in which the oil droplets contained di-n-butylphthalate derivative C–Cp. These droplets were dispersed in a buffer solution containing molecules of PPD (4-N,N-diethyl-2-methylphenylenediamine). An oxidation reduction reaction was triggered at the microelectrode, converting PPD into QDI (4-N,N-diethyl-2-methylquinonediimine). If QDI molecules diffuse to the droplet–water interface, they react with C–Cp to form cyan dye in the oil droplet. This reaction is known from color developing processes in photographic emulsions [428]. There are, however, other side reactions occurring which compete for QDI, and it disappears at a rapid rate. This is the reason it becomes possible to control the dye formation in μm dimensions but not at mm dimensions, for example. Measurements show that the dye formation rate falls rapidly over distances of a few micrometers and approaches zero at distances greater than about $10\,\mu$m. This was demonstrated in experiments in which tweezers was used to position oil drops at varying distances from the microelectrode and then the absorbency of the created dye was observed. This was all done under a microscope.

The authors view this dye formation experiment as an example of the unique ability to control chemical reactions at micrometer dimensions. They believe the results show the potential of their laser trapping-spectroscopy–electrochemistry technique to induce, monitor, and control reactions in minute volumes.

13.3. Laser-Controlled Phase Transitions in PNIPAM and Reversible Formation of Liquid Drops

Ishikawa *et al.* in 1993 [429] studied laser-controlled photothermal phase transitions of aqueous polyacrylamide solutions in the micrometer-size regime. In particular, they studied PNIPAM solutions. Thermal phase transitions of PNIPAM have been studied in the past as an interesting stimulus responsive material. The material is clear and transparent up to a critical temperature $T_c \cong 31°$C, above which point sub-micrometer PNIPAM particles precipitate and the solution becomes turbid. The hydrophobic nature and coil structure of the polymer change markedly above T_c. The authors point out that much remains unknown about the nature of the phase transition.

Quite remarkably, however, they find that by using an infrared beam from a $1.06\,\mu$m YAG laser, strongly focused to $\sim 1\,\mu$m, it is possible to locally heat a sample of PNIPAM liquid a few degrees above T_c to the point where a single microparticle with a diameter of several micrometers forms and grows rapidly to a fixed size and is held in the trap. When the light is turned off the particle disappears rapidly in what is a highly reversible reaction. Particle size could be sensitively controlled between 2 and 9 μm by varying the laser power. Fluorescence measurements were made from the single trapped particle, which indicated changes in the micropolarity of the polymer chain due to changes in the hydrophilic and hydrophobic nature of the polymer below and above T_c. The authors believe that simultaneous PNIPAM microparticle formation and laser manipulation will be of use in laser-controlled microactuator and other microdevices.

Many of the above results of the microphotoconversion ERATO project, and more, are contained in a very useful book entitled "Microchemistry, Spectroscopy, and Chemistry in Small Domains", edited by H. Masuhara *et al.* [430].

This book has a very useful article entitled "Photophysics and photochemistry of individual microparticles in solution" by Koshioka *et al.* [431]. It describes various optical measurements made on single polymeric particles in various solutions to help to characterize their structures, dynamics, and nature. The authors used a dynamic microspectroscopy system designed to give information on dynamics in micrometer-sized particles. It combines the optical microscope, picosecond pulsed lasers, and fast response detectors for time-resolved fluorescence and absorption spectroscopy [432],

[433]. Polymeric microspheres of different types modified with different surface molecules are much used in applications of optical manipulation in physical, chemical, and biological fields, as we have discussed earlier. The authors study the polystyrene microsphere surface in particular because of its wide usage in research and the fact that it can be modified easily with various amine molecules. These in turn can be further modified for specific purposes. A table is presented of different surface modifications of polystyrene microspheres and their effects on the fluorescent spectra of a surface probe molecule called ANS (8-aniline-i-naphthaline sulfonic acid). ANS is known to be sensitive to the micropolarity and flexibility of the surface.

Studies were also made of microcapsules and oil droplets. Microcapsules have a unique geometry with a solvent being encapsulated inside a thin polymer resin wall. They are widely used for industrial applications. Excimer formation dynamics, solute concentration distributions, and the internal viscosity of single microcapsules and oil droplets were also probed using nanosecond and picosecond optical pulses [434]. The single-particle behavior differs considerably from average multiparticle behavior and emphasizes the importance of tweezer-type measurements in this field.

The book on microchemistry [430] also contains a very instructive article by M. Irie entitled "Stimuli-responsive polymer gels: an approach to microactivators" [435]. He considers gel systems that undergo isothermal phase transitions by external stimulation, such as by photons or chemicals. Reversible conformational changes in the various physical and chemical properties of gels, such as size, can occur. These changes are especially large near phase transitions. These changes are possibly useful for making microactivators or artificial muscles when implemented on a micrometer-size scale. Changes in size as large as a factor of 10 were shown in PNIPAM gel. See discussion above on reversible formation and trapping of PNIPAM droplets using the change in temperature of water due to absorption of YAG laser irradiation.

CHAPTER

14

Holographic Optical Tweezers and Fluidic Sorting

One can design holograms that convert a single laser light beam into an array of planar or three-dimensional (3D) traps in space. By continuously varying the phase shifts of the hologram one can manipulate the trap array in a controlled way in space and time. In the following sections, we discuss the device potential of this new technique. The holographic devices for controlling macroscopic particles in space strongly resemble the optical lattice-type devices developed to control ultracold atoms. This is another example of where interchanges of techniques for atoms and macroscopic particles could be useful.

14.1. Nanofabrication with Holographic Tweezers

This short paper by Grier and collaborators [436] deals with a problem that arises when one tries to fill a dense array of optical tweezer traps. Particles tend to occupy the outer regions of the array first, thereby blocking access to the inner traps. Figure 14.1, taken from Fig. 1 of Ref. [436], shows a diagram of the holographic tweezer apparatus. Within the input telescope formed by lenses L1 and L2 one has an enlarged conjugate plane (OP*) that exactly mimics the trap foci in the objective plane (OP). The solution to the filling problem was to use a moveable knife-edge to initially block most of the traps of the array. This allowed one side to fill. By retracting the knife-edge, one could gradually fill the remainder of the array.

14.2. Dynamic Holographic Tweezers

The 2002 paper by Grier and colleagues [437]* introduced new methods of creating large numbers of dynamic holographic tweezer traps in three dimensions. For example, they used a computer-controlled Hamamatsu nematic spatial light modulator (SLM). They found that one could shape the phase front of individual traps to make arrays of different light modes to form optical vortices,

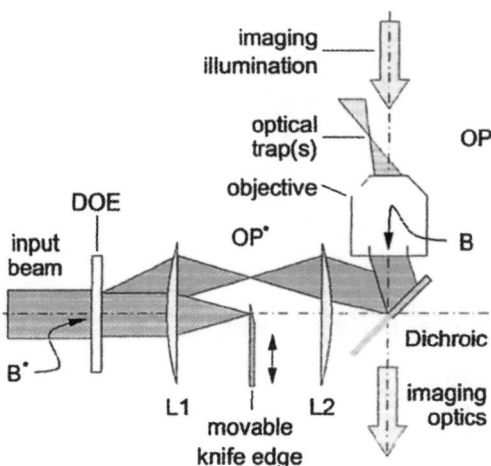

Fig. 14.1. Schematic diagram of holographic optical tweezers. A diffractive optical element (DOE) splits a collimated laser beam into several beams, each of which is transferred to the back aperture (B) of an objective lens by the telescope formed by lenses L1 and L2. The objective lens focuses each beam into a separate optical trap in the object plane (OP). Corresponding focii occur at the conjugate to the optical plane (OP*). The dichroic mirror separates imaging illumination from trapping light, allowing images to be formed of the trapped particles.

axial line traps, optical bottles, and optical rotators. Arrays of hundreds of traps with different characteristics could be formed. Particles could be manipulated individually or moved as a group through the array.

Optical traps can be formed by reshaping the phase profile of an ordinary Gaussian beam with the Hamamatsu X7550 nematic SLM. With the SLM one can adjust each of the $40\,\mu$m wide pixels of a 480×480 array to any of 150 different phase shifts between 0 and 2π with a 5 Hz refresh rate. This was used to generate an array of 400 functional optical tweezer traps, which exceeded by far the previous number of controllable traps. Previous work produced only trap distributions in two dimensions. By adding an additional phase term that effectively adjusted the divergence of the input beams, the authors were able to shift the focus up or down, thereby achieving 3D operation for the first time. Particles could be moved about to form different spatial patterns in three dimensions by gradually shifting the traps in a number of discrete small steps.

Starting with other input beam modes, they were able to demonstrate other functionalities. By modifying the phase profile of an ordinary Gaussian beam around the polar coordinates in multiples of $2\pi\ell$, where ℓ is an integer called the topological charge, they converted the Gaussian beam into a ring-shaped Laguerre–Gausssian (LG) mode beam and its corresponding tweezer trap into an optical vortex trap. Previously, vortex trapping with $\ell < 5$ was reported. The authors gave an example of trapping with $\ell = 30$, which formed the most highly charged vortices so far reported. Vortices with helical phase fronts carry orbital angular momentum and thus exert tangential or rotational forces on trapped particles. See Chapter 12 and especially Sec. 12.4.

The rotating ring of particles entrains fluid that can induce motion of neighboring vortices and acts as a pump for use in microfluidics and the so-called "lab on a chip" applications. Dynamic holograms can be combined with computer-vision techniques to make fully automated systems.

Among the applications that the authors foresee are massively parallel, high-throughput cell screening and macromolecular sorting.

14.3. Sorting by Periodic Potential Energy Landscapes: Optical Fractionation

Grier's group, working at the University of Chicago, has devised new ways of separating damped particles by flowing them through an optically periodic potential energy landscape [438], [439]. As a model system they used a stream of colloidal particles passing through a holographically generated square array of tweezer traps. They demonstrated the phenomenon of kinetic locking, in which moving particles become locked into specific directions as they pass through the array. This mechanism deflects the particles away from the driving direction, effectively resulting in separation.

In an initial study of the process of kinetic locking with optical tweezers [439], they observed that spheres entering the array in the [10] direction are pulled transversely into the [10] rows of the trap array and follow them to the end of the array. This behavior is similar in many respects to the channeling of high-speed ions through real crystals in preferred directions. If the array is rotated with respect to the incident flow by $\sim 9°$, the particles still closely follow the array's [10] rows of traps. Thus the channeling effectively deflects the particles from the flow and leaves a shadow behind it. At 28° rotation no channeling is seen. At 45° another channeling state is observed with the particles following the array's diagonal $[\bar{1}\,\bar{1}]$ rows.

Subsequently, Grier's group showed particle separation by optical fractionation using the same experimental setup [438]. Figure 14.2, taken from Fig. 1 of Ref. [438], illustrates optical fractionation for the case of two different types of incident particles flowing in liquid at an angle θ to the incident driving force. Results of calculation using experimental parameters show, as indicated in the figure, that the larger particles of radius $a_1 = 0.79\,\mu$m incident at an angle θ with respect to the array are kinetically locked-in and deflected, while the smaller particles with $a_2 = 0.5\,\mu$m are not. In

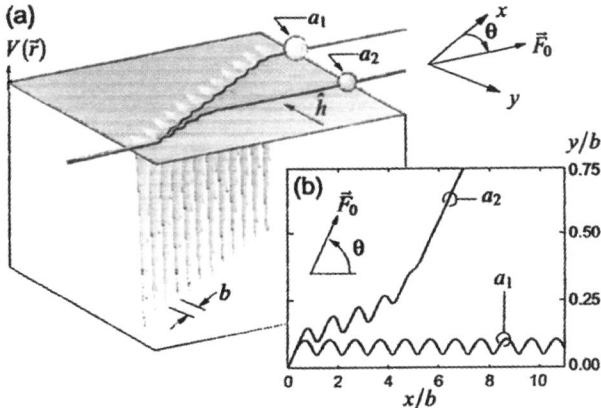

Fig. 14.2. Principle of optical fractionation. (a) Different types of particles are driven by external force, $\vec{F}_0$ through an array of optical traps inclined at angle θ with respect to $\hat{F}_0$. Strongly interacting particles (a_1) become kinetically locked-in to the array and deflected, while the others (a_2) are not. (b) Trajectories for large ($a_1 = 0.79\,\mu$m) and small ($a_2 = 0.5\,\mu$m) spheres calculated with Eq. (8) for experimental conditions described in the text.

such an experiment large and small particles can be collected separately on a continuous basis. Note, however, that if one increases the optical power to the array too much, one can enter the regime where the particles are simply trapped in the individual tweezer traps and cannot flow through the system. Theoretical calculations show that the threshold for a particle to become kinetically locked to the array depends exponentially on the particle size, for particle sizes larger than ~ 100 nm. Also, separation based on size is possible for smaller objects in the Rayleigh regime less than the optical wavelength. This regime requires much more optical power and gives only algebraic sensitivity to size. For increased selectivity one can feed the collected output sample on to additional stages of fractionation.

The authors believe that this technique will have wide application for particle separation and serve as a powerful research and practical tool. The American Institute of Physics (AIP) cited optical fractionation as one of the most noteworthy physics achievements of 2003. Follow their web link to story 627.

14.4. Optical Peristalsis

In 2003, Koss and Grier [440] devised a new way to controllably move and rearrange particles in a 3D space using a process they call "optical peristalsis", in analogy with physical peristaltic pumps. They started with particles located in a computer-generated holographic pattern of discrete symmetric tweezer traps. Replacing the original trap-forming hologram with an updated hologram that made a slightly shifted pattern of traps, they transferred the particles on to the shifted pattern of traps. Continuation of this process transported particles long distance along prescribed paths.

Particles of $\sim 1.6 \mu$m in diameter were transported at speeds of 7μm/s with $\sim 500 \mu$W of optical power, giving a typical well depth of $\sim 10 k_B T$. In addition to demonstrating linear displacement, the authors were able to create and maintain a void within a dense heterogeneous suspension and generate controlled rotation of particles in a number of circular rings about a common axis. Note that with peristaltic motion the particles are optically driven through a quiescent system, which, of course, is different from kinetic locking where flowing the liquid medium and particles through an array of traps causes the separation.

As pointed out by the authors, peristaltic motion, such as demonstrated in this experiment, has analogies with thermal ratchets [286], [287] in which one relies on asymmetric trapping potentials and thermal diffusion. With peristalsis the traps are symmetric, but the motion and directionality is due to temporal asymmetries.

14.5. Microfluidic Sorting in an Optical Lattice

The recent experiments by McDonald, Spalding, and Dholakia [441] at St. Andrews University on sorting or fractionation of particles in fluids passing through a computer-generated 3D optical lattice made significant advances over the previously discussed work [438], [439]. The authors demonstrated sorting by size and by refractive index. The sorting efficiency approached 100% even for concentrated samples with throughputs exceeding those reported for competing fluorescence-activated cell sorting (FACS) techniques.

Particle sorters operating over long lengths with arrays of discrete isotropic traps tend to jam up. By reshaping the traps to provide weak guidance between adjacent traps, the authors were able to free up the motion between traps and deflect incident particles at angles up to 45° through the lattice. They thus achieved fractionation with reduced path lengths at flows in excess of $35\,\mu m/s$, giving high throughput of ~ 35 particles per second. Experiments were also carried out, separating red blood cells from lymphocytes dispersed in a culture medium, with $1.06\,\mu m$ laser light to show the potential of the technique for biological applications.

14.6. Microfluidic Control Using Colloidal Devices

One aspect of microtechnology is the control of fluids on the micrometer scale within customized channels. The aim is to create complex, highly integrated, microscale, total analysis systems. To implement such systems, special means must be devised to pump and direct fluid flow in microchannels.

By rotating two optically trapped colloidal dumbbells (each formed from two $3\,\mu m$ silica spheres) in opposite directions within a microchannel, the authors simulated a two-lobed gear pump. This pumped fluid at a rate of up to $4\,\mu m/s$ in a $6\,\mu m$ channel. Such a pump, however, may be destructive of concentrated cellular suspensions. In that case, another alternative is the use of a peristaltic design, in which six spheres, each $3\,\mu m$ in diameter, are moved by an optical trap to form a forward or reverse traveling sine wave filling the $6\,\mu m$ channel. The encased volume of liquid is carried along in a manner that simulates a two-dimensional (2D) analog of a 3D screw pump. See Figs. 14.3 and 14.4 from Figs. 1 and 3 of Ref. [442]. These ingenious pumps take advantage of the low Reynolds number of fluids in micrometer channels.

Two types of valves were demonstrated. In the first type, the valve consisted of a $3\,\mu m$ sphere photopolymerized to a number of smaller $0.64\,\mu m$ spheres to form a linear structure. This structure was positioned in a straight channel with the $3\,\mu m$ sphere placed against one wall where it acted as a freely rotating pivot. For fluid motion in one direction, the linear valve structure lay flat against the wall and allowed free passage of a mixture of 3 and $1.5\,\mu m$ particles. For flow in the opposite

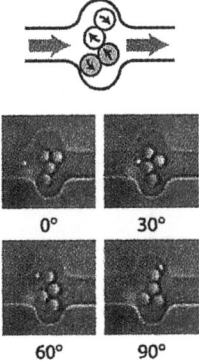

Fig. 14.3. Pump design illustrating lobe movement (the top pair rotate clockwise, the bottom counterclockwise). Also shown is 3-μm colloidal silica undergoing rotation at 2 Hz within a 6-μm channel. Frames are separated by two cylces to show movement of the 1.5 μm colloidal silica tracer particles.

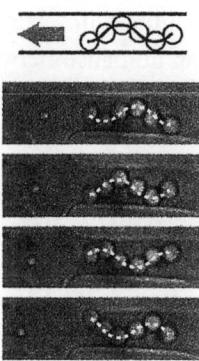

Fig. 14.4. Three-micrometer colloidal silica used as a peristaltic pump, operating at 2 Hz, to induce flow from right to left within a 6-μm channel. Frames are separated by four cycles to show movement of the 1.5-μm tracer particle.

direction, the valve rotated across the channel, where it blocked the flow of the 3 μm particles but allowed the passage of the 1.5 μm particles. For guides in a T-arrangement, one can locate the linear valve to rotate either up or down, so as to direct the fluid to either arm of the T.

   The authors believe that the ability to manipulate individual particles in a fluidic device roughly the size of a single red blood cell, using silica microspheres as the active control elements, avoids many of the previous complexities. They anticipate that many useful microfluidic analysis systems will emerge from this technology.

IVC
Applications of Atom Trapping
and Cooling

CHAPTER

15

Uses of Slow Atoms

15.1. Atomic Clocks Using Slow Atoms

One of the first applications based on the availability of relatively high densities of slow atoms was to high-precision measurements, such as atomic clocks. The idea of using slow atoms to make high-precision atomic clocks predates the invention of the laser. In the 1950s Zacharias tried to use the slow atoms in a vertically directed thermal atomic beam or "fountain" to increase the measurement time of atoms in a resonant cavity [443], [444]. Unfortunately, the number of slow atoms in the tail of the Boltzmann distribution emanating from a hole is severely depleted because of collisions with fast moving atoms. The scheme failed for that reason. The achievement of optical molasses [27]* with about 10^7 atoms cooled to temperatures of $\sim 50 \, \mu$K [445] in a MOT dramatically changed the prospects for use of atomic fountains for precision spectroscopy. In 1989, Kasevich *et al.* [199] built a sodium atomic fountain using initially cooled and trapped atoms.

Once cooled and trapped, the atoms could be launched upward by a pulse of resonant laser light into an rf cavity. Figure 15.1, taken from Fig. 1 of Ref. [199], shows the basic experimental setup. With this setup, they were able to measure the hyperfine splitting of atoms lofted into the cavity with a linewidth of 2 Hz. This was the result of Ramsey-type measurement using separated oscillatory fields [443] having an effective measurement time of 1/4 second. This linewidth was eleven times narrower than the then current NBS VI standard.

See Ref. [199] for details of this first experiment and ideas for possible uses for an atomic clock of increased accuracy. These results were not at the limit of atomic time measurements. Improvement by factors of a thousand were still considered possible [446], [447], especially for clocks based on cold atoms in the microgravity of space stations [448]. Work has proceeded on fountain-type atomic clocks to the point where, in March 2000, the first laser-cooled atomic fountain clock, called NIST-F-1, became the primary atomic time standard for the US [447]. It surpasses by a factor of three the performance of the NIST-7 atomic clock introduced in 1993.

In 1991, Kasevich *et al.* [449] made a significant improvement in technique. They launched the fountain out of the trap by changing the frequencies of the molasses beams, so that the atoms were not only pushed upward, but they also were cooled continuously by polarization gradient cooling in the moving frame of reference of the fountain. Such a scheme totally avoids the heating from single-beam pulses. Gibble and Chu in 1993 [447] constructed a fountain that exceeded the stability of the primary frequency standard. See p. 194 of Metcalf and van der Straten [135]. Figure 15.2, taken from Fig. 3 of Ref. [447], shows the central Ramsey fringes of the microwave clock transition of cesium.

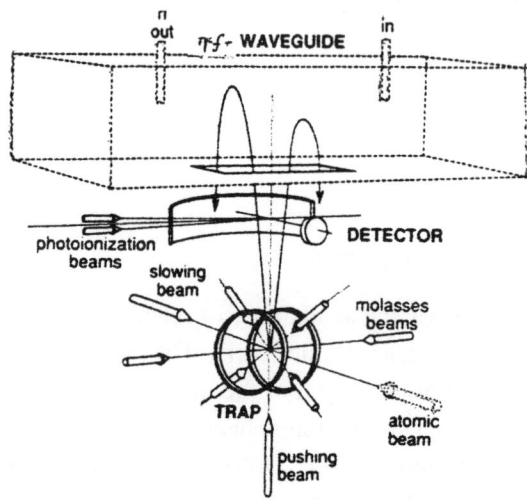

Fig. 15.1. Perspective view of the experimental setup for the atomic fountain. The atoms, initially confined to a small volume in the trap region, follow a ballistic trajectory through the waveguide and back to the detection region. A curved metal shield electrostatically focuses the photoionized atoms onto the detector. The waveguide is impedance matched to 50 Ω in both ends (Ref. 13). The distances from the trap to the detector is 14 and 31 cm to the top of the waveguide. The figure is roughly to scale. Figure 1 and caption from Ref. [199].

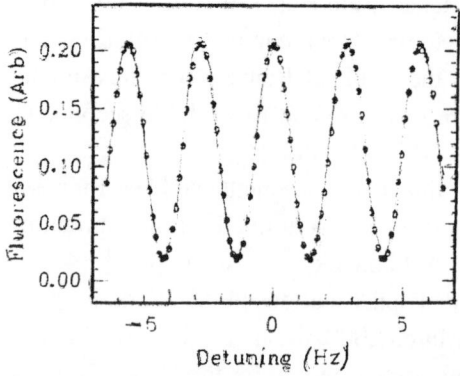

Fig. 15.2. The central Ramsey fringes of the $F = 3, m_F = 0 \rightarrow F = 4, m_F = 0$ transition. The FWHM linewidth is 1.41 Hz for a fountain height of 15 cm. The large circles are the data (1 launch per point, 0.933 s per launch) and the small circles and line represent the fit to the data. Figure 3 and caption from Ref. [447].

Patrick Gill, in a 2001 "Perspectives" article on atomic clocks in Science [451] reported on some of the papers presented at the Sixth Symposium on Frequency Standards and Metrology. The latest results on fountain clocks were given as well as a discussion of "optical clocks" that rely on optical rather than microwave transitions. He briefly mentioned applications to measuring the fine structure constant α and to atom interferometers as sensors for rotation, acceleration, and measurement of gravity gradients. He also mentioned limits due to gravity and discussed Salomon's projected experiment on the international space station. He looked forward to progress in the next decade toward clock performance with uncertainties of the order of one part in 10^{18}. The article contains a good set of references.

15.2. Atom Optics

15.2.1. *The first guiding and focusing of atoms using dipole forces*

The ability to optically manipulate atomic beams has given rise to the sub-field of atom optics. It refers loosely to the optical manipulation of neutral atoms in ways similar to the manipulation of light by conventional optical elements, such as light guides, lenses, mirrors, beam splitters, gratings, and interferometers. At times this specifically involves the wave properties of atoms. As mentioned earlier (Sec. 3.1.11) our first experiment demonstrating the effect of the gradient force on atoms in 1978 [43]* showed the ability of long thin laser beams to guide atoms in what was the analog of the optical GRIN lens (Gradient Index Lens). The observed focusing and defocusing of these atomic beams mark the beginning of atom optics.

15.2.2. *Lenses based on the scattering force*

Other types of thinner lenses for atoms were developed by Balykin *et al.* [450], based on the scattering force. As shown in Fig. 15.3(a), taken from Fig. 20 of Ref. [450], the lens consists of two pairs of opposing cylindrical beams tuned to the atomic resonance [2]*. For atoms entering the lens at exact right angles to the light, there are no Doppler shifts and the gradient force is strictly zero. Even for other atoms entering the lens at angles that depart slightly from 90° and are Doppler shifted away from resonance, the beams are sufficiently large in the vicinity of the axis that the gradient forces are negligible. In this predominantly cylindrical two-dimensional (2D) geometry the scattering forces are essentially radially inward. As atoms depart from the beam axis, axial or z components of the scattering force appear, which give rise to aberrations in focusing. The principal process limiting the spatial resolution, however, is momentum diffusion arising from spontaneous emission. This is a strong effect limiting the spot radius to a minimum of about $60 \, \mu$m [450]. Incidentally, we also know, by virtue of the optical Earnshaw theorem [45]*, that this lens geometry cannot function as a stable trap. This field configuration is stable in the xy plane, but leaks along the z-axis. A much better lens was proposed by Balykin *et al.* in 1987 [452] based on a blue-detuned TEM_{01}* donut mode gradient force configuration, as shown in Fig. 15.3(b), taken from Fig. 23 of Ref. [453]. The authors calculate a minimum focal spot size of $\sim 10 \, \text{Å}$, including spherical and chromatic aberrations.

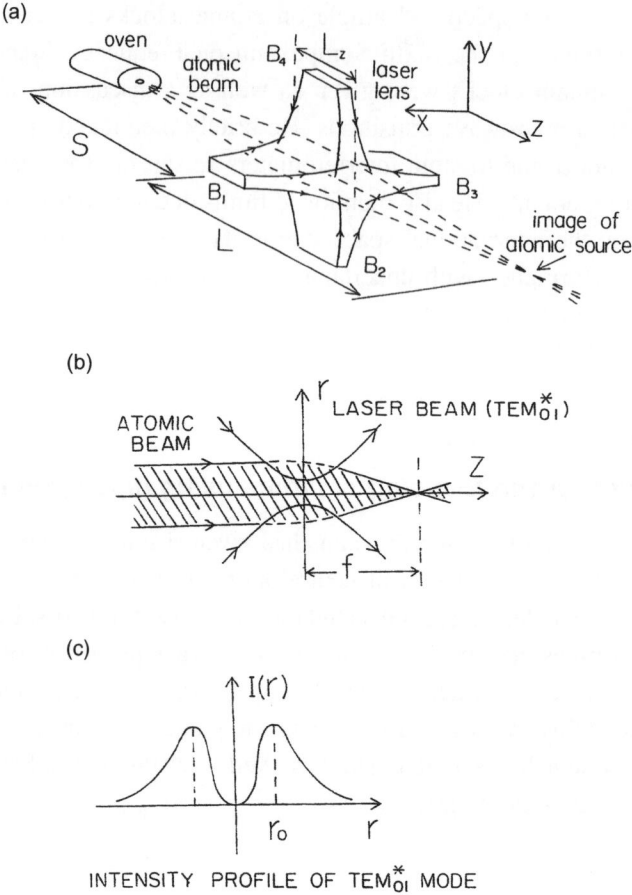

Fig. 15.3. (a) Four divergent Gaussian laser beams used to form a scattering force lens. Figure 20 from Ref. [452]. (b) TEM^*_{01} beam used to form a gradient force lens. (c) Cross-sectional intensity profile in the TEM^*_{01} laser mode field. Figure 23 from Ref. [453].

15.2.3. *Magneto-optic waveguide and atomic beam brightness*

Readers may recall that in optics there is a "brightness theorem", which states that in a lossless optical lens system, the brightness of an image equals the brightness of the source, where the brightness is defined as the luminous flux density in a pencil of rays in lumens per unit solid angle [454].

In atom optics there is an analogous quantity known as the atomic beam brightness, or radiance, that depends on the particle current, or flux density per unit solid angle. The beam brightness is closely related to a quantity known as the phase-space density of classical statistical mechanics [455]. See also Ref. [135] pp. 68–70 and 186–190, for a discussion of atomic beam brightness.

In statistical mechanics there is Liouville's theorem, which states that the phase density of a system of particles moving in six-dimensional (6D) phase space, made of three position coordinates and three corresponding momentum coordinates, is a constant. This applies for a conservative system. This implies, for a system of atomic particles moving in a beam subjected to conservative forces, such as dipole forces, that the beam brightness or phase-space density cannot be changed. On the other hand, it implies that the nonconservative forces, such as the scattering force, can be used to change the beam brightness, or phase-space density, by damping and extracting energy from the atoms. Beam brightness is clearly an important parameter that plays an important role in many atomic beam applications.

At the turn of the decade, in 1989 and 1990, a number of experiments were performed using transverse molasses cooling to increase the density and brightness of atomic beams. Ertmer and colleagues, who had pioneered the chirp cooling of atomic beams in 1989 [122], also did some of the earliest beam brightening experiments [456], [457]. Later in 1990, Nellessen, Werner, and Ertmer [458] used a distributed 2D MOT configuration to transversely cool and compress an atomic beam. The MOT field was produced by permanent magnets arranged to give a tapered quadrupole field. Starting with an input beam of several millimeters, the diverging beam was cooled and compressed to about $40\,\mu$m with about a thousand-fold increase in density, to roughly 10^9 atoms/cm^3.

Subsequently, in 1990, Riis *et al.* of Chu's group [459], also using a transverse MOT configuration, made some changes in technique to produce a slow atomic beam of high density, having a narrow velocity distribution. They produced a beam with a density of 10^8 atoms/cm^3 at a temperature of $\sim 200\,\mu$K, at a velocity of 270 cm/s. Their experiment, shown in Fig. 15.4, taken from their paper, involved the injection of a chirp-slowed atomic beam into a tapered funnel of four current-carrying hairpin wires at a small angle of about $7°$. The atoms were captured by the quadrupole MOT fields and were guided into a six-beam molasses region located close to the exit end of the MOT field, where the magnetic fields were diminished. This avoided the interference of the cooling process by magnetic fields. The final design feature was the use of traveling molasses in the axial cooling direction [449]. By using a frequency shift in one of the axial beams, they generated a moving axial standing wave. This was arranged to cool the axial beam velocity to 270 cm/s. The resulting beam

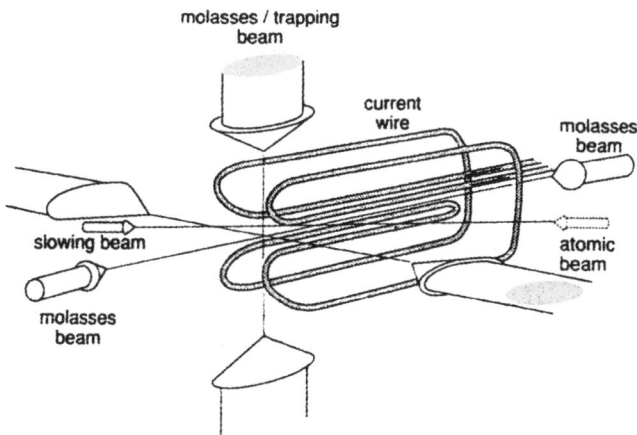

Fig. 15.4. Atoms from a thermal beam are slowed and funneled along the axis of the quadrupole field created by the four current-carrying "hairpin" wires. The square formed by the inner wires is 1.75 cm on a side. The upstream (right-hand side of figure) halves of the elliptical molasses-trapping beams are circularly polarized to trap the atoms transversely. The linearly polarized downstream (left) halves of these beams and the axial beams provide cooling. The axial beams also impose a drift velocity to the left. Figure 1 and caption from Ref. [459].

flux was about 40 times greater and 400 times colder than the flux one could achieve with just chirp cooling. The funneled beam is also about 3×10^3 times brighter than the original beam.

It is useful to point out that, for atoms and sub-micrometer Rayleigh particles, there is a close interrelationship of the physics of Liouville's theorem, in connection with beam brightness, and the previously discussed optical Earnshaw theorem [45]*, which tells us which radiation pressure force components can be used for making stable optical traps. Also, the conservative and nonconservative natures of the force components (see Sec. 3.1.6) are crucial to understanding the application of these theorems.

Concerning the two force components of the radiation pressure force, Liouville's theorem tells us that one cannot make traps with just the dipole force, because one cannot change the phase-space density within a dipole trap. One needs the scattering force to provide the damping to fill the trap. Earnshaw's theorem tells us that we cannot make a trap using just the scattering force, because such traps are not stable. Taken together, this implies that one needs both force components to make stable traps and to fill them with cooled atoms.

Although the two theorems discuss somewhat different quantities, the derivation of the two theorems is quite similar. Both rely on the divergence theorem, or Gauss's law. Combining the results of each theorem with the conservative and nonconservative nature of the dipole force and scattering force, one can draw the same conclusion about the need for both force components to make stable traps and fill them with cooled atoms. So, any way that you look at it, we deduce that both force components are necessary.

Having said this, there are two exceptions to this general rule about the need for both force components. The first involves the optical Earnshaw theorem, if one allows the internal energy of the particles to change. As explained earlier in Sec. 5.1.12, one can make a stable MOT scattering force trap, where the internal energy or resonance frequency varies with position due to a Zeeman shifting magnetic field. The other exception is in connection with Liouville's theorem. If one initially has a system of trapped particles moving about with a constant phase-space density and then suddenly turns on within the first trap a deep conservative second dipole trap, having a potential much deeper than the kinetic energy of the particles, one can then trap those particles that were within the trap volume, when the walls are raised around them. This arrangement is often used to transfer atoms and particles from a large trap to another more compact trap.

Of course, none of these general theorems tells us how to design a stable optical trap and fill it with particles in the first place. This was described in some detail earlier in the discussion of the work on radiation pressure in the time period 1970–1980. We saw the gradual evolution of the ideas from the simplest of concepts and experiments to the more complex concepts and applications in many subfields.

Concerning the obviousness of stable optical trapping, the conclusions of the general theorems on trapping make it even clearer to me that people like Letokhov, working on the problem of high resolution in spectroscopy in 1968 [56], in his suggestion to use just the far-off-resonance dipole force, was unlikely to have thought of stable trapping of atoms using dipole and scattering forces. Indeed, it was only after Ashkin discovered stable 3D trapping in 1970 in a different context, after stable traps for atoms were proposed, and after Hänsch suggested Doppler cooling, that he showed any interest in the subject. The same sort of remarks applies to Askar'yan's work on 2D confinement for demonstration of self-focusing light using atomic beams or plasmas [55]. In my opinion, practitioners of the Stark shift, like Cohen-Tannoudji, in their early work, were nowhere near the idea of a stable trap. As I pointed out, using the optical Stark shift is a very useful concept, after the fact, but it is not the simplest or most likely way to discover traps. For these reasons, I do not think it is the way to introduce the subject of traps to students.

15.2.4. *Evanescent wave mirrors*

Cook and Hill [460] introduced the idea of using the evanescent waves of laser fields penetrating the free space of totally internally reflected light beams as atomic mirrors (see also Ref. [461]). When tuned below resonance, these evanescent fields, which protrude just a fraction of a wavelength into free space, can repel incident atoms. Plane and curved mirrors of this type were used as atomic trampolines for atoms bouncing up and down under the force of gravity [462], [463]. They have also been proposed for making resonant cavities for atoms [464].

15.2.5. *Atomic beam splitters*

Atomic beam splitters, based on the "phase grating" formed by the dipole force in optical standing waves, were first used by Martin *et al.* in Pritchard's group to diffract atoms [465]. Another type of a more practical beam splitter was devised by Prentiss and collaborators using a new type of "blazed grating" based on standing waves [466]. Yet another type of elegant beam splitter, employing adiabatic passage, transfers momentum to atoms without ever populating the excited states of atoms, using the "superposition dark state" [467]. Weitz *et al.* [468] applied this general technique to transfer 140 photon momenta to sodium and thereby made a precision measurement of the fine structure constant.

15.2.6. *Neutral atom lithography*

Atom optics techniques also have potential applications in technology. Standing wave lenses made from the gradients of pairs of opposing beams have focused sodium and chromium atoms, passing transversely through the interference pattern onto surfaces, making grating patterns [469], [470]. A paper by Behringer *et al.* [471], using this technique on sodium, reports the achievement of minimum line widths of 13 nm having a contrast ratio of 6:1. Behringer and collaborators have developed a theory for the resolution of such lens systems. Their theory predicts even higher resolutions should be possible. Sodium gratings are, of course, not practical. Chromium atoms make more realistic gratings [470]. Experiments with chromium atoms showed features 200 nm wide. Cooled beams could reduce this line width considerably. Neutral atom lithography, such as demonstrated above, has advantages over electron beam and X-ray lithography.

15.2.7. *Atom Interferometers*

Atom interferometers are an important class of devices for making precision measurements in which one detects fringe shifts due to phase changes in one of the interferometer arms. Kasevich and Chu

[472]* used their fountain and Raman techniques to make an interferometric measurement of the acceleration of gravity, with an accuracy of one part in 10^6. Increases in accuracy to 10^{-10} g are possible. The authors suggest applications to geology, the search for net charge on atoms, fifth force experiments, and the test of general relativity.

15.3. Atomic Waveguide Devices

15.3.1. *Optical guiding of atoms*

As the subject of atom optics developed, attempts were made to improve on the early experiments of Bjorkholm *et al.* [43]* on atom guiding. Rather than use free space Gaussian beams, which spread as they propagate, new types of atom guides were devised that depended on the guiding of light in hollow fibers. It is known that fairly large hollow fibers of low numerical aperture can act as grazing incidence waveguides for laser light [473]. Such internal reflection guides can support a variety of modes. However, guides like these are lossless only in the limit of incident beams with zero beam angle and infinite radius of curvature. This implies that guides have a finite loss, which, in turn, limits their length. The advantage of grazing incidence guides is that with this new arrangement both the laser light and the atoms are guided in the hollow core of the fiber. Renn and colleagues in 1995 [474] demonstrated atom guiding in fibers of this type, using Rb atoms. These guides are useful for atom interferometers and delivering atoms to surfaces for microfabrication, for example.

15.3.2. *Magnetic guiding of atoms near wires*

In 1999, Denschlag *et al.*, in Austria, demonstrated a new type of magnetic atom guiding based on a current-carrying wire [475]. In such a scheme, atoms can be magnetically guided in close proximity to a surface or macroscopic object. The history of this type of guiding is well referenced in this paper. There are two types of such wire guides, the Kepler guides and the side guide. In the Kepler guide, atoms spiral about a single current-carrying guide, as shown in Fig. 15.5, taken from Fig. 1(a) of Ref. [475]. The atoms in the high-field seeking state follow a $1/r$ potential, centered about the wire. Also in Fig. 15.5, taken from Fig. 1(b) of Ref. [475], the guiding takes place in a side guide formed by adding a transverse magnetic bias field. This forms a magnetic potential with a nonzero minimum displaced to one side of the current-carrying wire. This can guide atoms rotating about the minimum in their low-field seeking states.

Figure 15.6, taken from Fig. 5 of Ref. [475], shows the measured distribution of atoms from the Kepler and side guides after free expansion. For the Kepler guide, one clearly sees the donut-shaped beam of high-field seeking atoms and the Gaussian distribution of the side guide for low-field seeking atoms.

The authors consider the side-guiding technique to be very promising for atom optics because it can be used with surface mounted wires. With several wires, it lends itself to the fabrication of beam splitters, interferometers, and more complex matter wave networks. This technique is seen as a way of using nanofabrication methods to make quantum atom circuits similar to modern integrated electronics.

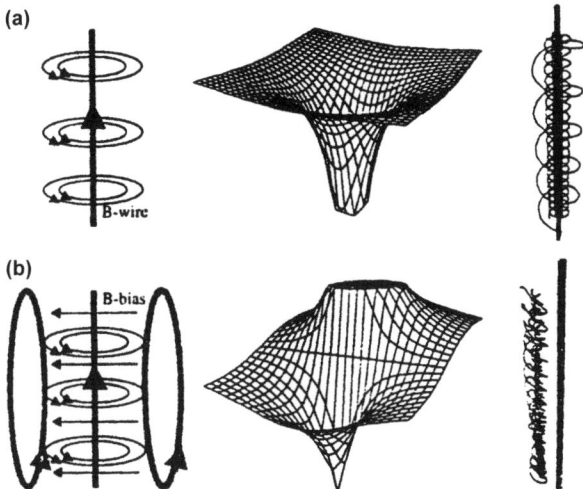

Fig. 15.5. Two configurations for guiding neutral atoms with a current carrying wire. The left graphs display the magnetic field configurations, the middle shows the corresponding guiding potentials, and on the right-hand side typical trajectories of guided atoms are drawn. (a) Guiding atoms in their *high-field-seeking* state that circle in Kepler orbits *around* the wire. (b) Guiding atoms in their *low-field-seeking* state along a line of magnetic field minimum *on the side* of the wire. Figure 1 and caption from Ref. [475].

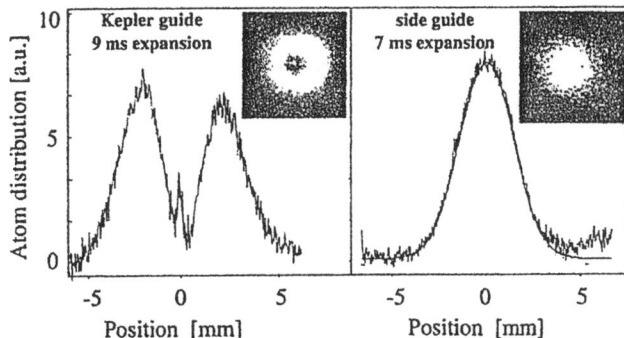

Fig. 15.6. Atomic distribution after free expansion of 9(7) ms for atoms that have been guided along the wire. The two profiles that are shown are cuts through the center of the respective CCD images in the right upper corner. For the high-field-seeker guide (left) the expanded cloud is doughnut-shaped due to the orbital motion of the atoms around the wire. For guiding on the side (right) we obtain a Gaussian distribution. Figure 5 and caption from Ref. [475].

15.3.3. *Magnetic guiding using atom chips*

In 2000, the Austrian group [476] demonstrated atom chips for controlling cold atoms close to nanofabricated surfaces. They took previous designs of wire devices, such as guides, beam splitters, and Z- or U-shaped 3D traps for free space operation and redesigned them for operation on chip surfaces. See references in Ref. [476]. To test their new chip design they first used U-shaped wires on the chip to capture about 10^8 atoms in an external MOT located several millimeters above the chip surface. In the process, the atoms were cooled to a temperature of $\sim 200\,\mu$K. By adjusting the bias field and lowering the current in the chip wires, they were able to lower the atoms to a few hundred micrometers, close to the surface, where the chip fields could take over and trap them. The atoms were finally transferred to a trap located just above a single, thin, $10\,\mu$m wire. The trap could be converted into a waveguide by adjusting the biases. As small as this trap was, the atoms still populated many of the energy levels. In the future, cold sources of atoms from a Bose–Einstein (BE) condensate would be needed to populate the ground state and reach conditions appropriate for experiments on quantum computation.

The experiments show, however, that the concept of an atom chip is feasible using standard microfabrication techniques. The hope is to reduce the current-carrying wire size to reach ground state trap sizes of ~ 10nm with trap frequencies in the MHz range. Direct loading of the surface MOT is possible. Other applications to atomic clocks, sensors, and quantum computing are considered possible.

15.3.4. *Magnetic guiding around curves*

Experiments were also reported by Müller *et al.*, at JILA [477], on guiding of ^{87}Rb neutral atoms around curves using lithographically patterned current-carrying wire. Using a current-carrying pair of parallel wires deposited on a surface, they guided a maximum flux of 2×10^6 atoms/s for a distance of 10cm, around a guide with a series of curves having a 15cm radius of curvature. Guiding took place along the zero of magnetic field, using low-field seeking atoms. The wire cross section was $100\,\mu$m $\times\, 100\,\mu$m. They were spaced $200\,\mu$m apart, from center to center. The guided atoms had a transverse temperature of $42\,\mu$K. The atoms were fed into the guide through a hole in one of the mirrors of an external MOT. The radiation pressure imbalance in the MOT pushed the atoms through the hole and into the guide. About 25% of the atoms were in the correct magnetic sublevel to be guided. The total flux exiting the MOT was $\sim 5 \times 10^8$ atoms/s. The guided atoms had a transverse temperature of $126\,\mu$K and were confined in a guiding potential of 1.1 mK. The current in the track wires was 2 A. Pulsed currents of 8 A are possible without excessive heating. This should reduce the curvature of the guiding track to ~ 3 or 4cm radius. This could then be used to guide the atoms around a full 360° bend. This would make it possible to create interferometers, provided that single transverse mode guiding is possible. In practice, this may require atoms loaded from a BE condensate.

The authors are very interested in applications of these guided structures to inertial and rotational sensors of extreme sensitivity, based on large enclosed-area atom interferometers [478].

15.3.5. *Magnetic guides as atomic beam splitters*

In an extension of the above guiding experiments, Müller *et al.*, in 2000, demonstrated a waveguide beam splitter for laser-cooled atoms [479]. By adding a transverse magnetic bias field, it was possible to guide and transfer atoms originally following the zero field of one wire guide over to a second wire guide in a coupling region, where the zero fields of the two wires merge. See Fig. 15.7, taken from Fig. 1 of Ref. [479]. As the two wire guides left the coupling region and separated, each carried off a fraction of the original atoms that depended on the relative strengths of the currents driving the two guides. Couplers with adjustable coupling from zero to 100% were observed with very little loss. Although nearly ideal 50/50 separation was achieved, examination of the separated beams indicated

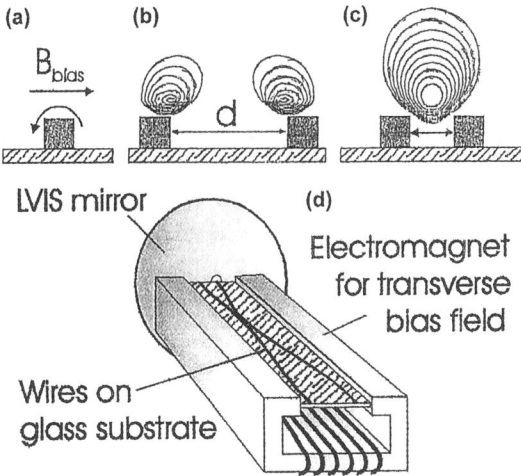

Fig. 15.7. Contour lines of magnetic-field potential and guide schematic. (a)–(c) We show a cross-sectional cut across the wires. When a bias field is applied transverse to the wires the magnetic field becomes zero just above the wire surface (a). For large track separations ($d = 300\,\mu$m) the magnetic field minima do not merge (b). As the track spacing is reduced ($d = 100\,\mu$m) the magnetic field minima merge to form one field minimum (c). The transverse bias field is generated with an electromagnet near the wire substrate (d). The LVIS mirror hole is aligned with one of the wires to couple the LVIS atoms into the guide. Figure and caption from Ref. [479].

an increase in transverse velocity from an initial $v_t = 10.0 \pm 1.5\,$cm/s to $v_t = 17.2 \pm 3.5\,$cm/s. The authors attribute this heating effect to limited nonadiabatic loading of the input atoms into the guide. This can be viewed as a mode mismatch problem. They also believe that the guide will have to merge more adiabatically to ensure single mode operation of the beam splitter.

I believe that magnetic guides and atom optics on chips hold the promise of interesting devices. The small size and strong magnetic field gradients possible near current-carrying wires make such magnetic devices more interesting than their larger free space analogs.

15.4. Cold Atom Collisions

Another major use of cooled and trapped atoms that developed very quickly after the early experiments was in the study of cold atomic collisions. This is a large, many faceted subject bearing on almost every aspect of optical trapping and manipulation of atoms. Cold collisions play a crucial role in determining the lifetime of trapped atoms at high densities. Collisions determine the re-thermalization time of perturbed atomic velocity distributions in traps. This affects the various cooling and heating processes that can take place within traps. Cold collisions also provide an opportunity to explore processes not seen at higher temperatures. Cold atoms behave very quantum mechanically. This is due to the long collision times and large deBroglie wavelengths of cold atoms. Use of cold atoms has revolutionized the subject of photoassociation spectroscopy, in which colliding atoms become coupled to each other in the presence of light to form bound-state excited molecules. Approaching atoms are initially attracted to each other by an R^{-6} van der Waals potential. When the separation of two ground state atoms approaches the Condon point at separation R_C, where the potential difference between the s ground state of the molecule and an excited p state $(h/2\pi)\omega_p$ exactly matches the photon energy, the potential of the atoms either becomes attractive or repulsive and proportional to $\pm R^{-3}$, depending on the energy of the coupling photon $(h/2\pi)\omega_1$ relative to $(h/2\pi)\omega_p$. If $(h/2\pi)\omega_1 < (h/2\pi)\omega_p$ the atoms approach with the attractive potential and end up as a stable but excited molecule. Na_2^* is an example. This stable but excited molecule will leave the trap via two possible processes. The first is photoassociation ionization, in which the excited molecule absorbs a second photon and converts to a molecular ion plus a photoelectron. The second is by fluorescent decay to a stable molecule in the ground state, which is, in general, not confined in the atom trap. If $(h/2\pi)\omega_1 > (h/2\pi)\omega_p$, the potential becomes repulsive and the two atoms simply separate. The repulsive case is an example of "optical shielding". However, the separated atoms have absorbed the photon energy and may, as a consequence, escape the trap if it is not deep enough.

If one scans the coupling laser frequency from the high side ω_p and passes through the free-bound resonance, one also generates a high-resolution fluorescence-loss spectrum as the newly excited molecules decay. Fortunately the fraction of atoms ending up as molecules is usually small [135], [202].

15.4.1. *Photoassociation vs. associative ionization*

Thorsheim *et al.* [480] in 1987 were the first to propose use of cold atoms for photoassociation measurements and to calculate free-bound photoassociation spectra at low temperatures, showing how one could clearly resolve individual rotational transitions in the free-bound absorption. The technique should also give accurate measurement of the s-wave scattering length. As we shall see, this parameter figures crucially in the formation of BE condensates.

Photoassociative ionization in collisions between cooled and trapped sodium atoms was first observed by Gould *et al.* in 1988 [481], using essentially the two-beam trap first described by Ashkin in 1978. They called this a hybrid trap because it used the scattering force for axial stability and the dipole or gradient force for transverse stability. See Ref. [38] for Phillips's comments in his Nobel Lecture on the use of the hybrid trap. He half apologizes for using it by saying that for him it represented a form of closure. This possibly is an oblique reference to the history of the NBS error on Minogin-type scattering force optical traps. However, the demonstration of the two-beam trap gives me a greater sense of closure. It was the first viable atom trap that I had conceived of. It led to the important concept of the single-beam gradient trap and the optical Earnshaw theorem and the subsequent developments discussed earlier. Phillips *et al.*'s experiment in 1988 was the only demonstration of the viability of this two-beam trap. Atoms captured in this trap had a density of $\sim 5 \times 10^9\,\mathrm{cm}^{-3}$ compared to the molasses density of $\sim 10^7\,\mathrm{cm}^{-3}$ and a temperature of $\sim 750\,\mu\mathrm{K}$.

Gould *et al.*, in their first experiment in 1998, looking at ions generated in photoassociative collisions, were able to identify the generated ions as Na_2^+, as opposed to Na^+ ions, using a time-of-flight technique. They interpreted the ions as due to collisions between excited atoms, i.e. the associative process, even though this gave problems in the magnitude of the cross section for this process. Their measured cross section, however, was more than an order of magnitude larger than the computed value. The associative process can be summarized by the two sequential steps:

$$Na + (h/2\pi)\omega \rightarrow Na^*,$$
$$Na^* + Na^* \rightarrow Na_2^+ + e^-.$$

Julienne in 1988 [482] identified the source of the problem. For slow atoms in molasses, which are red-detuned by only about a linewidth, the excitation occurs at a very long range, $\sim 1800\,a_0$, where a_0 is the Bohr radius, and most of the excited state population decays to the ground state before reaching the autoionization region. For atoms excited by the trapping light, which is detuned by ~ 70 linewidths, the excitation occurs at much closer internuclear distances and one expects that these excited atoms will contribute more than atoms excited in molasses.

Lett *et al.*, in 1991 [483], used the same hybrid trap to check the ideas put forward by Julienne in 1988. They measured separately the ions produced during the molasses cycle and the trapping cycle and found increased ion production during the trapping cycle by a factor of about 40–100. Also, changing the trap detuning from $800\,\mu$Hz to more than $4\,$GHz to the red continued to produce ions at rates comparable to those closer to the atomic resonance. If the origin of ions were due to collisional ionization, as postulated earlier, such a large detuning would have caused a reduction in the excited state atomic population to reduce the ion rate by over four orders of magnitude.

Julienne and Heather, in 1991 [484], analyzed the results of the Lett *et al.* experiment [483] and arrived at an interpretation of photoionization that incorporates all of the important features of the experiments. They identified the principal path by which atoms can approach one another, be excited, and reach the region of small internuclear separation necessary for ionization to occur. In their 1993 paper, Heather and Julienne [485] introduce the term "photoassociative ionization" to distinguish the process of optical excitation of the quasimolecule from the conventional associative ionization collision between excited atomic states. The photoassociative ionization process can be summarized by the following sequence of steps:

$$Na + Na + (h/2\pi)\omega_1 \rightarrow Na_2^* + (h/2\pi)\omega_2 \rightarrow Na_2^* \rightarrow Na_2^+ + e^-,$$

where the principal ionization route to sodium is through doubly excited autoionization. Figure 15.8, taken from Fig. 2 of Ref. [483], shows that excited atoms are not involved in the associative ionizing collisions.

As the technique advanced, experimenters introduced scanning probe fields to gain access to new regions of photoassociation spectra beyond the previously studied regions near the hyperfine-coupling zone. This greatly extended the utility of the technique and made possible a new domain of high-precision spectroscopy not previously achieved by other methods. These new scanning techniques have given rise to an extensive literature probing detailed molecular potentials, atomic lifetimes, and scattering lengths for a wide selection of trappable atoms, such as sodium, rubidium, lithium, and potassium [202]. These techniques were applied to the usual MOTs, as in Lett *et al.* in 1993, [486], and later to "dark spot" MOTs, which increased ionization signals by about a hundred times. This is the result of an approximate 100 times increased atom density in a dark spot MOT.

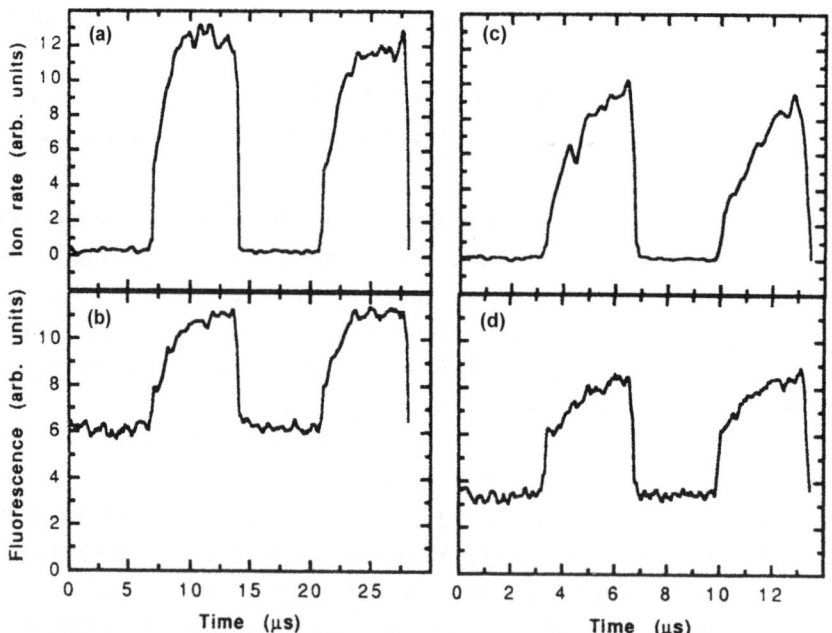

Fig. 15.8. (a) Ion signal versus time for $\Delta_{\text{molasses}} = -8.5\,\text{MHz}$, and cycle periods of $7.14\,\mu\text{s}$. (b) Fluorescence versus time for the same conditions as (a). (c) Ion signal versus time for $\Delta_{\text{molasses}} = -15\,\text{MHz}$, and cycle periods of $3.33\,\mu\text{s}$. (d) Fluorescence versus time for the same conditions as (c). The time-of-flight of the ions to the detector has been taken out to align the fluorescence and ion signals from the same trap or cooling phase. The trap laser is on during the periods of larger ion or fluorescence signal. The vertical scales in (a) and (c) and in (b) and (d), although arbitrary, are kept the same. Figure 2 and caption from Ref. [483].

15.4.2. *Dark spot MOTs*

Dark spot MOTs were introduced by Ketterle *et al.* in 1993 [183]* to overcome the "photoinduced repulsion effect" that one observes in ordinary MOTs. This repulsion arises when spontaneous emission from atoms trapped in the central regions of the MOT is intercepted and absorbed by atoms nearer the edge of the trap before it can escape to the outside. The kick of the absorbed spontaneous emission can cause atom loss, which limits the density of atoms in the trap [182]*. In the usual MOT one uses an auxiliary "repumper" light beam, co-propagating with the trapping beams, tuned to neighboring hyperfine levels, to prevent optical pumping to the other hyperfine levels. In the dark spot MOT, one places an obstruction in the center of the re-pumping beam, which gives rise to a central dark spot region where re-pumping does not occur and atoms in this sheltered region can accumulate in the dark hyperfine level. The dark central region, of course, gives no outward light pressure, which increases the maximum trap densities up to two orders of magnitude, as remarked above in connection with photoassociation.

15.4.3. *Scanning photoassociative spectroscopy using far-off-resonance dipole traps*

The scanning methods alluded to above using MOTs were also pioneered by Heinzen's group. He and his collaborators made use of the purely optical, far-off-resonance single-beam gradient trap, or FORT trap, as they called it [148]. Miller *et al.*, in 1993 [487], using rubidium, measured photoassociation spectra by observing the fluorescence as they scanned the trapping frequency. Up to that time most work on collisions used the MOT. The FORT trap has the following advantages over the MOT: (i) simplicity, (ii) absence of a magnetic field, (iii) very low excited state population, which greatly decreases optical heating, (iv) high atomic density of $\sim 10^{12}/cm^3$, and (v) a well-defined optical polarization axis for aligning the atoms. Its disadvantage, as seen by the authors at this time, was the relatively low number of trapped atoms, $\sim 10^4$. Much more will be said about off-resonance dipole-like tweezer traps later.

These FORT traps were loaded by transferring atoms from MOTs. In subsequent scanning experiments, the trapping frequency was held fixed and a tunable probe laser was used to scan through the resonances [488], [489]. In these measurements the authors avoided perturbations of the trapped atoms by Stark shift from the trapping beam by chopping the trapping and probing beams. In this way they were able to probe undisturbed cold atoms, as originally suggested by Gordon and Ashkin in 1980 [123]*. Figure 15.9, taken from Fig. 2 of Ref. [487], gives an example of some of the high-resolution spectra obtained in this way. As a further example, the technique allowed the value of the C_3/R^3 long-range potential of excited states of rubidium to be determined with an accuracy of C_3 $(1\,g) = (14.29 \pm 0.7)$ a.u. [488], [489]. Tsai *et al.*, in 1997 [490], used two-color photoassociation

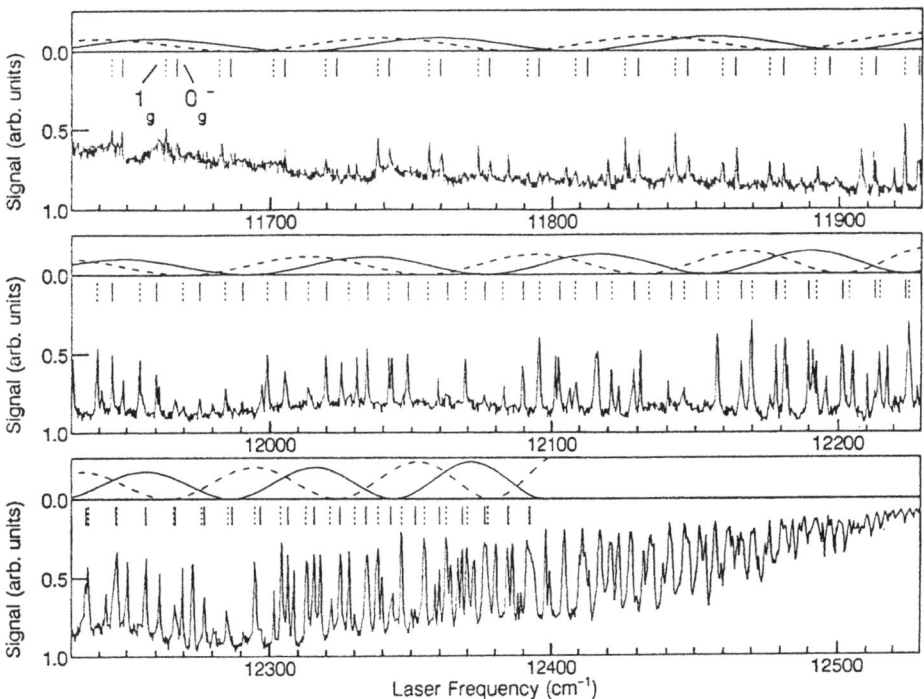

Fig. 15.9. Photoassociation spectrum of laser-cooled Rb atoms. The vertical axis corresponds to increased photoassociative loss of atoms from a far-off resonance optical dipole trap. The atomic $D1$ resonance occurs at $12578.9\,cm^{-1}$. Two vibrational series, indicated by the dashed and solid vertical lines, are clearly visible and are associated with the vibrational levels of the $^3\Sigma_g^+$ states illustrated in Fig. 1. The inset curves show the approximate variations in the Franck-Condon factors for these two series. Figure 2 and caption from Ref. [487].

spectroscopy in which the second color couples initially excited states to the higher vibrational levels of the $^{85}Rb_2$ ground state. This gives highly precise values for the collisional parameter of ultra cold rubidium, including the scattering length and the position of Feshbach resonances, as will be explained later in connection with BE condensation in rubidium.

The photoassociation ionization technique should apply to many trappable atomic species, but requires the use of dark spot MOT or FORT traps to compete with fluorescence trap-loss spectroscopic techniques.

15.4.4. *Optical shielding or suppression of trap loss*

The other side of the coin of photoassociation is optical shielding or suppression of the trap-loss process. Indeed, if one understands the origin of photoassociative trap loss, one can often devise schemes to use light to divert atoms away from the loss channels, thus shielding or greatly reducing trap loss. An early experiment manipulating the collision process was by Bali *et al.* in 1994 [491], using ^{85}Rb atoms held in a MOT. In their experiment they tuned an auxiliary control laser over a band from 5 to 20 GHz to the blue of the cooling frequency. This blue-detuned light transferred atoms approaching each other in the ground state to an excited repulsive state. They found, however, that at tunings beyond 10 GHz the extra kinetic energy gained by atoms in the repulsive potential caused them to escape the trap. This trap loss was, of course, a function of the intensity of the control laser. Sanches-Villicana in 1995 [492] studied a closely similar process and observed the effects of hyperfine-changing collisions by varying the trap depth.

Muniz *et al.* in 1997 [493] used a clever trick to study blue-detuned optical fields used to suppress hyperfine-changing collisions in sodium. They first captured 10^7–10^8 atoms in a deep MOT tuned near the D_2 line of sodium. They then transferred the cold atoms to a very much shallower MOT operating near the D_1 line. In this trap, atom loss is due to hyperfine-changing collisions. They then suppressed loss from this trap with a suppressor laser. Figure 15.10, taken from Fig. 3 of Ref. [493], is a plot of the fractional suppression vs. the suppressor laser intensity.

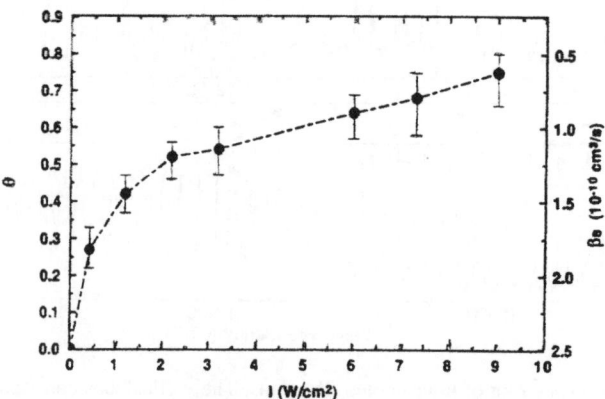

Fig. 15.10. Suppression factor versus suppressor laser intensity for a $D1$ trapping laser intensity of 40 mW/cm^2. Figure 3 and caption from Ref. [493].

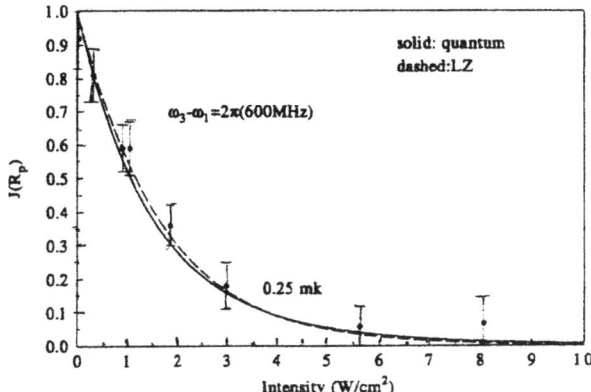

Fig. 15.11. Fractional photoassociative ionization rate, $J(R_p)$, as a function of suppression intensity. Points are measured data with statistical error bars; curves are calculations of one-dimensional, two-level models. The dashed curve is a one-dimensional (1D) Landau–Zener (LZ) calculation. The solid curve is a 1D close-coupling calculation. Both calculations assumed a MOT temperature at 0.25 mK. Figure 4 and caption from Ref. [495].

Marcassa *et al.* made careful studies in sodium in 1994 [494] of the important process of optical shielding in a MOT, where externally applied light intervenes and suppresses trap loss by photoionization. In a collision of two ground states, the optical fields transfer the atoms to a repulsive curve near the Condon point. In this way the atoms become shielded from ionization and the collision becomes essentially elastic. They studied this two-step process in detail, using a Monte Carlo wave function simulator and a Landau–Zener model of the interaction. They obtained agreement with their data to within about 15%. In a subsequent experiment in 1995 [495] the authors measured the suppression factor as a function of laser intensity, up to about 8 W/cm^2. The agreement with theory was impressive and is shown in Fig. 15.11, taken from Fig. 4 of Ref. [495]. Zilio *et al.* in 1996 [496] and Napolitano *et al.* in 1997 [497] compared experiment and theory in optical suppression in sodium using circular or linear polarization of the probe beam. Circular polarization is much more effective than linear. Measurements of optical shielding have been made in xenon, krypton, and in ^{85}Rb and ^{87}Rb collisions using MOTs. All these experimental works have stimulated much work on the theory of optical shielding by Suominen *et al.* in 1995 [498], Suominen *et al.* in 1996 [499], and Napolitano *et al.* in 1997 [497].

IVD
Bose–Einstein Condensation and Related Developments

CHAPTER

16

Introduction to Bose–Einstein Condensation

The big surprise of the last time period of 1990 to 2006 in the field of optical trapping and manipulation of neutral atoms was the achievement of Bose–Einstein condensation (BEC) in atomic vapors. It was clear in 1985, just after the very first demonstration of optical molasses, that cooling and subsequent optical trapping that BEC was a possibility [171]*, [172]. This was in part stimulated by attempts to achieve BEC in atomic hydrogen gas, using standard cryogenic techniques and later by evaporative cooling techniques [173], [174]. At this time there were already a number of theoretical works on BEC in gases. See the review of Dalfovo *et al.* [500]. The experiments on optical cooling and trapping in the late 1980s and early 1990s showed that, starting with conventional atomic beams at $\sim 1000\,$K, one could cool atoms by about six orders of magnitude to hundreds of microkelvin [4]* and later to tens of microkelvin [198] at densities as high as 10^{11} atoms/cm^3. This was a giant step for atomic physics at the time, but still a long way from achieving conditions where one could observe BEC.

According to modern quantum mechanics, such a condensation should occur when identical bosonic atoms (having integral spins) are cooled to the point where the deBroglie wavelengths λ_{db} of adjacent particles overlap and couple. This criterion for BEC, according to statistical mechanics [501], is met when the phase-space density $\rho_{ps} = n(\lambda_{db})^3 > 2.612$, where n is the number of atoms/cm^3. At this point the identical atoms should rapidly condense into a single coherent state very much like the photons in a coherent laser beam. According to the Pauli exclusion principle, only particles of integral spin with symmetric wave functions can occupy the same energy state of a system at the same time. Atoms such as sodium, rubidium and cesium, for example, fall into this category and are candidates for BEC.

The phase-space condition for onset of BEC is a very stringent condition, requiring another reduction in temperature of about three or four orders of magnitude and an increase in density of about two or three orders of magnitude, starting from a typical MOT [181]*, [183]*. The original idea of getting a BEC in a cryogenically cooled spin polarized hydrogen gas goes back to the late 1970s. The hydrogen experiments were ultimately successful [502], but only two years after the achievement of BEC in optically cooled atomic vapors. Bose–Einstein (BE) condensates in dilute gaseous media are expected to be quite different from the well-known condensates in strongly interacting systems such as liquid helium. Of course, there are also fermionic particles of half-integral spins, such as atomic vapors of ^{6}Li and ^{40}K, which can only condense into a Fermi gas with one atom per quantum state of the trap. As will be seen, cooling of ^{40}K to the point of degeneracy below the Fermi temperature was observed in 1999 by DeMarco and Jin [35]*.

Interactions occur in an atomic gas via collisions. The most frequent common collisions are two-particle collisions, which are purely elastic and result in no net loss of atoms. This can be seen from conservation of energy and momentum. Three-particle collisions are much less frequent, but can result in the formation of molecules, especially at very low temperatures. This possibility represents a loss channel for trapped atoms. In this sense, BE condensates are metastable even in perfect vacuum. Eventually three-particle collisions will deplete the condensate. In the presence of light there can be additional loss processes, such as formation of molecular ions, as considered above in some detail. Condensates do live long enough, however, to perform many interesting experiments, as will be discussed.

16.1. First Demonstration of BEC, Using the TOP Magnetic Trap

The first experiment to achieve BEC in atomic vapors using optical methods was performed by Anderson *et al.* in 1995 at JILA [29]*. Let us describe their experimental procedure. The authors started by collecting $\sim 10^7$ atoms ^{87}Rb in the dark spot MOT directly from a vapor of atoms at room temperature. These atoms were then cooled by molasses and polarization gradient cooling to $\sim 20\,\mu$K [198]. Then, using what they called a hybrid approach, they transferred the atoms into a purely magnetic trap for evaporative cooling. This transfer was effected with the help of a small bias field and a short pulse of circularly polarized light, which optically pumped atoms into the $F = 2$, $m_F = 2$ angular momentum ground state of rubidium. The MOT was then turned off and the optically pumped atoms were transferred to the purely magnetic quadrupole trap [503]. In such a trap the optically pumped atoms are low-field seekers and are attracted to the zero field point of the magnetic quadrupole. At this point they had $\sim 4 \times 10^6$ atoms magnetically trapped at about $90\,\mu$K. They used the maximum magnetic field available, which compressed the atom cloud in space and increased the elastic collision rate as much as possible. This was needed to allow the atoms to rethermalize as the evaporative cooling process began. Evaporative cooling was discussed above in connection with sub-recoil cooling. See Sec. 5.1.8 and Refs. [173], [174], [208].

However, a serious problem arises if one tries to evaporatively cool using the standard quadrupole magnetic trap. As atoms cool and spend more time in the vicinity of zero magnetic field, an increasing number of atoms undergo what are known as Majorana or spin-flipping transitions to the unbound state of the atoms. The rate of these nonadiabatic spin-flipping processes can be estimated in terms of the Larmor precession rate $\omega_Z = \mu B/(h/2\pi)$ relative to the angular frequency of atoms orbiting about the zero field point [504]. At really low temperatures, where the atoms spiral in close to zero field point, they see rapidly varying magnetic fields and the loss rate becomes prohibitive.

To solve this problem the authors devised what they call the TOP trap. In such a trap the zero of the field is moved around in space so rapidly, with the help of additional magnetic coils, that the atoms are not able to follow it. They respond only to the average field, which has no zero point, and the spin-flipping process is inhibited (see Refs. [29]*, [504]).

Evaporative cooling is controlled in such a trap by varying the frequency of an rf spin-flipping field. This causes transitions to an unbound spin state. The frequency was initially set to flip the spins of the higher-energy atoms that penetrate out to where the quadrupole magnetic fields are greatest. Then the frequency was gradually reduced, thereby continuing to skim off the high-energy atoms. The rf frequency reduction was made slow enough to give the atoms time to rethermalize. It is indeed fortunate for evaporative cooling that the elastic collision rate, as measured by Gardner *et al.* [505]*, with a far-off-resonance dipole trap, is about 200 times faster than the elastic loss rate in ^{85}Rb. In this way the average temperature falls as the fast atoms essentially continue to boil off. This process not only cools the atom cloud but also increases its density as it shrinks in size. This is a powerful process that increases the phase-space density by orders of magnitude. The authors finally achieved a BEC at a temperature of about $170\,$nK and a number density of about $2.5 \times 10^{12}\,$atoms/cm^3. The final number of atoms in the condensate was about 2×10^4, down by a factor of ~ 500 from an initial $\sim 10^7$ atoms captured in the MOT.

The very rapid onset of the condensation was observed very clearly by measuring the atomic velocity distribution of the cloud within the trap at various stages in the evaporation process. The velocity distribution was determined by turning off the trapping fields and waiting a period of time before making a spatial image of the atoms. The image was observed in silhouette, using a short pulse of resonant light. This procedure essentially gave a time-of-flight measurement. Figure 16.1, taken from Fig. 2 of Ref. [29]*, shows the appearance of a small compact core of condensate atoms within a broader background of thermal atoms, which rapidly grew to a full-fledged condensate containing essentially all of the atoms. The condensed fraction of atoms was $\sim 80\%$ in early experiments and rose to close to 100% in later work [204]. This condensed vapor condensate is a new form of matter, quite different from previously observed Bose–Einstein condensates in superfluids and solids, where only a small fraction of the atoms are in the condensed phase. For helium, only about 10% of the atoms are in the condensed superfluid phase [204]. The complexities of these denser superfluid systems make it difficult to investigate the properties of condensates in detail. In this sense, BE condensates in atomic vapors form an ideal quantum system.

The authors pointed out that it was possible to use their newly formed condensates to carry out a wide range of quantitative studies previously only dreamt of, such as: comparing the spectroscopy of coherent and incoherent matter [506]; studying condensates with positive and negative scattering lengths from atoms of ^{87}Rb and of ^{85}Rb [505]*, for example; studying the specific heats at the transition boundary of a condensate [501]; searching for analogies to the classic experiments on superfluid helium [501], [507]; and studying the change in scattering length at magnetic scattering resonances [508]. As we can see from the above references, many of these possibilities and many more have been realized and elaborated on in just a few of years.

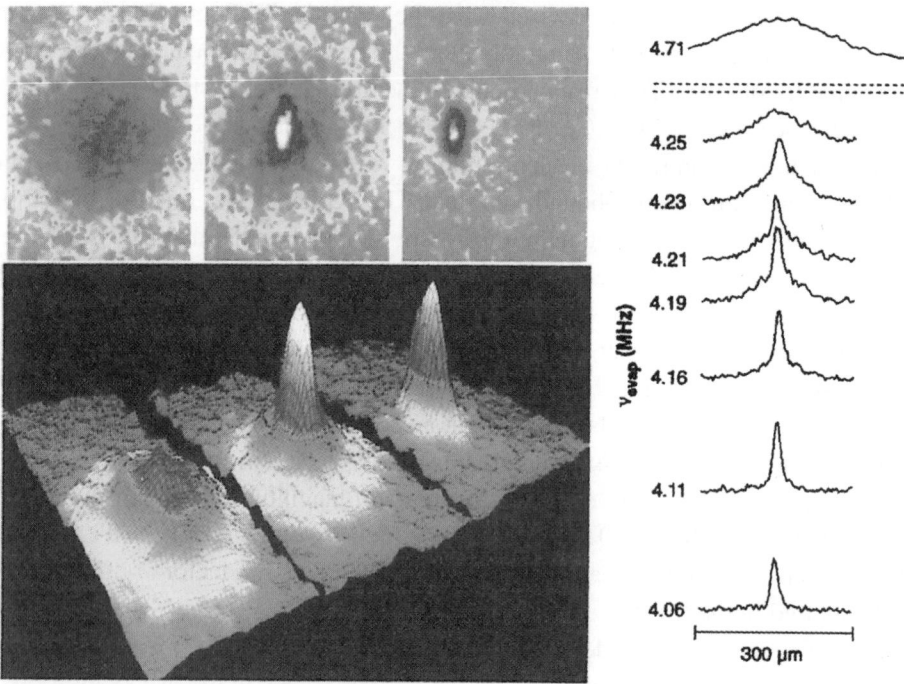

Fig. 16.1. Observation of BEC in rubidium by the JILA group. The upper left sequence of pictures shows the shadow created by absorption in the expanding atomic cloud released from the trap. Below, the same data are shown in another representation, where the distribution of the atoms in the cloud is depicted. In the first frame to the left, we see the situation just before the condensation sets in, in the middle a condensate peak with a thermal background is observed, whereas the third figure shows the situation where almost all atoms participate in the condensate. The thermal cloud is seen as a spherically symmetric broad background, whereas the sharp peak describing the condensate displays the squeezed shape expected in an asymmetric trap. The diagram to the right cuts through the atomic cloud when it is cooled by more and more atoms being evaporated. Figure 2 and caption from Ref. [29]*.

Shortly after the publication of the first successful BEC experiment [29]*, the editor of Nature, Maddox, in an editorial comment, made one of his less prescient remarks: "Remember the TOP trap because it is going to be around for a long while" (see Sec. 5.1.11).

16.2. Bose–Einstein Condensation Using an Optically Plugged Magnetic Trap

Within a few months, however, another group at MIT produced a BE condensate in sodium vapor without a TOP trap [175]*. In their trap they plugged the leak at zero field in their magnetic trap by using the repulsive dipole force of a blue-detuned off-resonance optical beam, focused at the zero of magnetic field. This gave improved results in several ways. They increased the density of their condensate to $\sim 1.5 \times 10^{14}$ atoms/cm^3 and reached the critical temperature at $\sim 2\,\mu$K. The number of atoms in the condensate increased by a factor of ~ 250 to $\sim 5 \times 10^5$ atoms. This was due, in part, to the fact that they chose to form their molasses from a slowed atomic beam rather than from a room temperature vapor, as at JILA. The optically plugged magnetic trap was much more compact than the TOP trap and cooled to the critical temperature in about nine seconds, compared with 6 min previously. The MIT group observed their condensate nondestructively by a type of phase contrast imaging that uses the dispersive properties of the atom cloud and is much more detailed than absorptive imaging.

16.3. Bose–Einstein Condensation Using the "Cloverleaf" Magnetic Trap

One of the properties of the optically plugged magnetic trap was that it had a double minimum, causing two separate condensates to appear as evaporation proceeded. About six months later the MIT group achieved BEC in a novel "cloverleaf" magnetic trap [509]. This trap modifies the usual quadrupole trap by addition of extra cloverleaf-like coils and results in a trap that is tight in the two transverse dimensions and weaker in the axial direction. It has a simple harmonic potential with a nonzero magnetic field and needs no optical plugging beam. This cloverleaf type of trap is derived from an older Ioffe-Pritchard magnetic trap [510], [511], which stems from the early days of fusion research. This new cloverleaf trap gave improved results. The number of atoms in the condensate increased by almost 10 times, but the particle density and the critical temperature remained about the same. The lifetime of the condensate increased by more than an order of magnitude. Cloverleaf traps or other modifications of the Ioffe-Pritchard trap are in general use at the present time, although modifications of the TOP trap are still used.

16.4. Bose–Einstein Condensation in ^{7}Li

In 1995, a group at Rice University [512] reported experiments on a gas of ^{7}Li atoms, which have a negative scattering length and thus have a mutually attractive interaction. Condensation experiments in this case are more difficult. Condensed atoms can only collect at very low densities in the energy levels of the trap. In free space such a gas is unstable against collapse. Therefore, some people questioned if such a condensate of was even possible (see Refs. [500], [204]). It was only in 1997 that the group under Hulet [513]* showed that condensates with attractive interactions indeed existed, but were stable only with a limited number of trapped atoms totaling about 1400. Within a trap it is the zero point energy of the ground state that keeps atoms having an attractive interaction from collapsing. It is the balance of this zero point energy and the attractive energy of the collapsing atoms that sets the critical number of atoms N_{cr} that can be contained in the trap. Condensates with negative scattering lengths are still the subject of much theoretical discussion [204]. Most of the subsequent work on BEC, however, proceeded with atoms having positive scattering lengths for which the mutual interactions are weakly repulsive. For such atoms there is no limitation on the ultimate number of atoms that can be condensed. As will be seen, this can be greater than 10^8 atoms.

16.5. Expanding Bose–Einstein Condensates

Figure 16.2, taken from a review by Townsend *et al.* in 1996 [514], summarizes the results of all the early condensation experiments as of 1996. Some of the earliest types of experiments on BE condensates involved the observation of the effect of fairly gross perturbations on the condensates by

	JILA 95 [12]	Rice 95 [14]	MIT 95 [13]	MIT 96 [16]
Atom	^{87}Rb	^{7}Li	Na	Na
Scattering length [nm]	+6	−1.5	+3	+3
Trap	TOP	permanent magnetic trap	optically-plugged magnetic trap	cloverleaf magnetic trap
$B''_x B''_y B''_z$ [(10^3 Gcm^{-2})^3]	27	1	4×10^6	100
First BEC	June 95	(July 95)	September 95	March 96
Evidence	TOF	(Halo)	TOF	TOF, in-situ image
N_C	2×10^4	2×10^5	2×10^6	15×10^6
T_C [μK]	0.1	0.4	2	1.5
n_C [cm^{-3}]	2×10^{12}	2×10^{12}	1.5×10^{14}	10^{14}
N_0	2,000		5×10^5	5×10^6
Cooling time	6 min.	5 min.	9 s	30 s
BEC atoms/s	6		60,000	200,000
Lifetime [s]	≈ 15	≈ 20	≈ 1	≈ 20

Fig. 16.2. Comparison of BEC experiments reported thus far. The relevant figure of merit of a trap for evaporative cooling is the product of the three effective magnetic field curvatures, which are defined to be the curvatures of the trapping potential divided by the magnetic moment. N_c, T_c and n_c are atom number, temperature and number density at the phase transition respectively. N_0 is the number of condensate atoms. Recently the Boulder group reported reaching $N_0 = 1.5 \times 10^6$ in a purely magnetic trap. Table 1 and caption from Ref. [514].

manipulating the trapping magnetic trap itself. As pointed out above, the principal signature of BEC is the sudden appearance of a compact, dense core of slow atoms from within the wider distribution of thermal atoms at temperatures higher than T_C. This was clearly observed when the magnetic trap was turned off and the condensate was allowed to expand by itself. The spatial development of this expanding condensate, driven by the weak repulsive inter-atomic forces, was compared with the predictions of the Gross-Pitaevskii (GP) nonlinear Schrödinger equation. This work by Holland and Cooper in 1996 [515] gave agreement to about 5%, with no adjustable parameters.

16.6. Gross-Pitaevskii Mean Field Theory

The Gross-Pitaevskii (GP) mean field theory is the main theoretical tool for treating BEC in dilute gases. It was originally developed to treat weakly interacting bosons in connection with vortex states of superfluids [516], [517]. Mean field theory was formulated to understand the role of interactions between particles. These interactions are capable of strongly modifying the behavior of static and dynamic properties of even weakly interacting condensates. GP theory can give density profiles, ground state configurations, and the dynamics of expansion. As we shall see later, it is also useful for understanding collective oscillations, thermodynamic properties such as the critical temperature, and other complex BEC phenomena in dilute condensates, such as superfluidity, quantized vortices and Josephson-type effects.

In the GP nonlinear wave equation the Hamiltonian includes the trap potential V_{trap} and a term $\mu\psi$, where μ is the chemical potential, which is equal to the energy required to add one more atom to the condensate. The atom–atom interaction energy U_0 is specifically included in the nonlinear term $NU_0|\psi|^2$, where N is the number of particles in the condensate and $U_0 = [4\pi(h/2\pi)^2/m]a_0$ (see Refs. [204], [500]). Thus U_0 depends only on a_0, the s-wave scattering length, and on m, the total mass of the N particles.

16.7. Collective Excitation of a Bose–Einstein Condensate

Another of the measurements on BE condensates was that of the excitation spectrum of the condensate when perturbed by sinusoidally varying the trapping potential of the magnetic trap. This experiment in 1996 by Jin *et al.* [518], using condensates of ~ 104 atoms, observed that the condensates oscillated resonantly in various modes at frequencies of about 100 Hz and then decayed away as the excitation source was removed. These oscillations in shape agreed very well, within about 2%, with the results of the zero temperature GP nonlinear Schrödinger equation. The theoretical treatment is different, however, depending on whether the number of atoms is $\sim 10^4$ or $\sim 10^6$, as in the experiments of Mewes *et al.* [519]. Stringari, in 1996 [520], found an analytic solution for the GP equation for the large N case, in which the frequencies are independent of N and the scattering length. This agreed with experiment to within a few percent. At a temperature $T = 0$, one expects no damping of the oscillations. In the Jin *et al.* and Mewes *et al.* experiments [518], [519] there were frequency shifts and increased damping effects as the temperature of the condensate rose. These effects at present are unaccounted for by the theory. Complex phenomena, such as correlations between atoms, which are related to particle pairing in the Bardeen, Cooper, Schrieffer (BCS) theory of superconductivity, may be involved. See Dalfovo [500] and Burnett [204]. Experiments by Stamper-Kurn *et al.* in Ketterle's group in 1999 [521] extended the study of the bulk properties of perturbed condensates to much higher frequencies, using a light scattering technique with an optical standing wave excitation. This two-beam technique was equivalent to extending the measurements to the phonon regime in quantum gases. Use of two beams also served to increase the signal detection sensitivity in these tenuous gases.

16.8. Coherence of Bose–Einstein Condensates

16.8.1. *Interference between two condensates*

A number of experiments were performed that directly showed the coherence of the condensate. Andrews *et al.*, in 1997 [31]*, used their cloverleaf magnetic trap to first make a cigar-shaped condensate. Figure 16.3, taken from Fig. 2 of Ref. [514], shows the cloverleaf geometry. By adjusting the cloverleaf electromagnets they could control all three dimensions of the condensate shape. The condensate was then split into two condensates by introducing a blue-detuned repulsive far-off-resonance sheet-like light beam perpendicular to the axis of a cigar-shaped condensate. This beam served not only to plug the trap leak at zero magnetic field, but also to cut the condensate into two parts. The repulsive sheet beam was from an argon ion laser, at 514 nm. It was so far from the sodium resonance at 589 nm that heating from spontaneous emission was negligible. By varying the power of the repulsive cutting beam they could vary the coupling and separation between the two condensates. They then turned off the magnetic trap and the repulsive sheet beam and let the two condensates fall, expand and overlap. The combining condensates were examined by absorption and dispersion imaging. Beautiful, high-contrast interference fringes across the entire overlap region were observed with a fringe spacing that varied with the separation of the two condensates. See Fig. 16.4, taken from Fig. 2 of Ref. [31]*. This pattern is in agreement with a simple picture of coherent interference from two point-like sources at varying distances apart. This not only proved that each condensate was completely coherent and could be described by a single wave function, but that it is possible

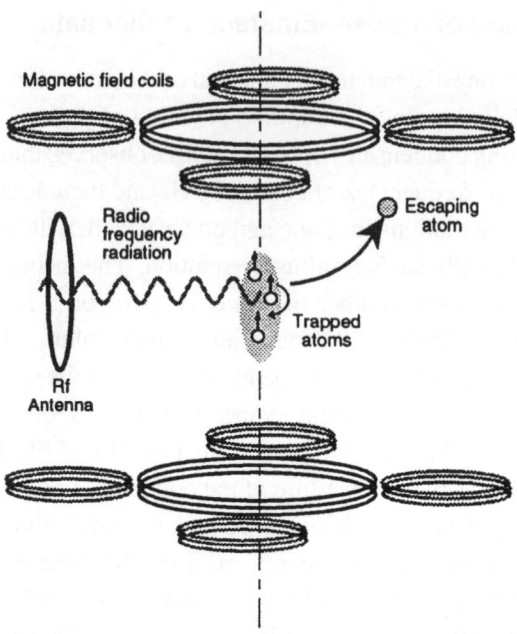

Fig. 16.3. Cloverleaf configuration of trapping coils used in experiments at MIT. The central coils provide axial confinement, the outer coils (the "cloverleaves") provide tight radial confinement. The rf radiation induces spinflips of hot atoms, leading to evaporative cooling. Figure 2 and caption from Ref. [514].

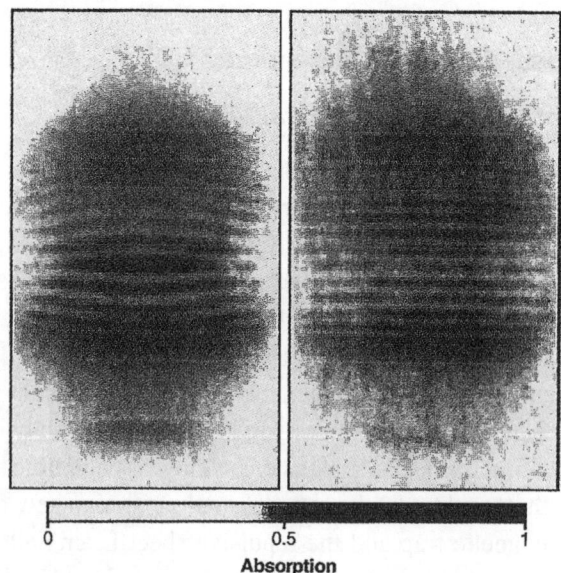

Fig. 16.4. Interference pattern of two expanding condensates observed after 40 ms time-of-flight, for two different powers of the argon ion laser-light sheet (raw-data images). The fringe periods were 20 and 15 μm, the powers were 3 and 5 mW, and the maximum absorptions were 90 and 50%, respectively, for the left and right images. The fields of view are 1.1 mm horizontally by 0.5 mm vertically. The horizontal widths are compressed fourfold, which enhances the effect of fringe curvature. For the determination of fringe spacing, the dark central fringe on the left was excluded. Figure 2 and caption from Ref. [31]*.

for the case of a high-power cutting beam and total separation to observe the relative phase of two totally distinct condensates. The authors further conclude that these results imply that a condensate with an output coupler is in fact an atom laser.

16.8.2. *Measurement of Δp and the uncertainty principle*

Another experiment showing the coherence of condensates by Stenger *et al.* in 1999 [522] involved irradiating condensates with a pair of standing wave beams. If an atom of the condensate absorbs a photon from one beam and is stimulated to emit the photon by the other beam, the atom gains a net momentum of $2h/\lambda$ and is ejected from the condensate. The number of photons ejected shows a sharp resonance as the frequency difference between the two beams is varied. The resonant width is an accurate measure of Δp, the momentum distribution of the atoms in the condensate. The ejected photons were detected free of background after a 20 ms ballistic expansion, which carried them free of the condensate. The width of the condensate is a measure of its uncertainty in position. It was found that $\Delta p \approx (h/2\pi)/\Delta x$ for a number of different sizes of condensates, Δx varying by more than a factor of two. This shows, in a very elegant way, that the coherence length of the condensate corresponds to its physical size and thus can be described by a single coherent wave.

16.8.3. *Coherence and interference in the time domain*

A coherence experiment analogous to the above experiment [522], but only in the time domain, was performed at NIST by Hagley *et al.* [523]. The experiment started with a BE condensate in a TOP trap and proceeded to check the overall coherence, using a self-interference method, which essentially compared the phase of the condensate at two different times. This was done by applying two short standing wave pulses, separated in time by Δt, to the condensate, which ejected two copies of the condensate. The ejection process was one of absorption of a photon from a beam in one direction, followed by stimulated emission of a photon in the opposite direction. This kicked or diffracted atoms out of the condensate with an impulse corresponding to two-photon momenta, $2(h/2\pi)k$. Each of these ejecting pulses was detuned from resonance by 600 MHz to keep the spontaneous emission rate low. Only a small fraction of atoms ($\sim 2\%$) was ejected into each copy of the condensate. The ejected copies moved in space with a velocity arising from the momentum impulse of $2(h/2\pi)k$ and from the conversion of the mean field potential of the condensate into kinetic energy. After a time, the copies separated from the original condensate and moved into a region of space free of the condensate, where it was possible to observe their interference. Sinusoidal interference fringes occurred in the region of spatial overlap of the two copies. This overlap depended on Δt, the time separation between the two pulses.

The decay of the sinusoidal oscillations with time depended on the momentum spread in the original condensate. If the phase varies over the condensate, then the decay should be faster due to dephasing. The authors measured a fringe decay time to one-half amplitude $t_{1/2} = 225 \pm 40\,\mu s$, compared to a calculated value from the GP equation of $t_{1/2} = 275 \pm 6\,\mu s$. The theoretical decay is mainly due to geometrical overlap with a small contribution from the mean field effect.

They also measured the decay of interference for the case where the condensate was released from the trap before the diffraction or copy pulses were applied. The measured decay time was

$t_{1/2} = 65 \pm 10\,\mu s$, compared with the theoretical value of $82 \pm 3\,\mu s$. The agreement between theory and experiment implied a uniform phase over the entire condensate. The authors suggest uses for this coherence technique, such as to study the development of a uniform phase in a trap, or to detect the effect of an uncondensed fraction of atoms, or to measure condensate temperature. They have used it to show that the NIST version of an atom laser was in phase. See the later discussion on atom lasers, in Sec. 16.10.

16.8.4. *Coherence of atoms tunneling out of arrays*

Anderson and Kasevich at Yale performed a different type of coherence experiment in 1998 [524]. They observed macroscopic quantum interference from BEC atoms localized on the array of standing wave intensity maxima of a vertically directed two-beam optical dipole trap. Tunneling out of the macroscopic array of coherent local traps was induced by gravity. As the atoms fell they interfered constructively at certain heights to form a train of falling pulses of atoms. The pulse frequency was determined by the gravitational potential energy difference between adjacent potential wells. This effect is closely related to the AC Josephson effect in superconducting electronic systems. The results imply that the output of the many beamlets from each source is coherent and properly phased. The constant time interval between pulses makes this the equivalent of a mode-locked atomic laser in direct analogy with mode-locked optical lasers. The authors also see the possibility of using this technique to measure weak forces, such as the Casimir force, van der Waals force, or gravity, with high accuracy, much as with the AC Josephson effect. Without too much effort they can now measure the acceleration of gravity to a resolution of $\delta g \approx 10^{-4}\,g$. Resolutions of $\delta g \approx 10^{-13}\,g$ are conceivable. See Ref. 26 of Anderson and Kasevich. I omitted any discussion of how Anderson and Kasevich managed to transfer their condensate from a TOP trap into a vertically directed two-beam trap. Transfers of BE condensates to dipole traps will be considered later in some detail. More will be said about the use of this apparatus as a pulsed atom laser in the later discussion on the various types of atom lasers.

16.8.5. *Spatial coherence of atoms ejected from a trap*

Bloch *et al.*, in 2000 [525], performed a related type of experiment showing the spatial coherence of atoms originating from two distinct regions of a condensate. Starting with a BE condensate of ^{87}Rb atoms in a magnetic trap, they used two different radio frequency waves, ω and ω', applied simultaneously to the condensate, to transfer atoms from the $m_F = -1$ trapped atomic state into the $m_F = 0$ untrapped atomic state.

Because of the differences in photon energy, the emerging untrapped atoms had different energies and therefore different deBroglie wavelengths. The condensate used had a cigar-type shape, with the long axis perpendicular to the vertical direction. In such a condensate cloud of atoms, the regions of constant magnetic field B are located in roughly horizontal slices through the condensate. One can see, by virtue of the spin resonance condition $h\omega = (1/2)\mu_0 B$, where μ_0 is the Bohr magneton, that different frequencies couple out at different B's located at different heights. Gravity collimates the two sets of atoms and pulls them downward to form two overlapping beams, which can then interfere. Using absorption imaging, one can measure the contrast of the interference pattern and deduce the degree of coherence of the two sources. Results obtained from different combinations of ω and ω' show that the condensate is highly coherent over its full extent.

Measurements were taken at temperatures above and below the critical temperature and for different effective slit separations at heights above and below the center of the magnetic trap. The spatial coherence demonstrated in this experiment has obvious connections with the previous experiment of Anderson and Kasevich [524] where the falling pulses of atoms emerged from the trapped condensate by the many tunneling beamlets, as opposed to the rf stimulated emissions of Bloch *et al.*

16.9. Condensate Formation by Bose Stimulation

A topic of considerable interest is how the atoms achieve coherence in the process of BEC. If all the atoms are in the lowest energy state of a trap at high enough density so that the wave packets of adjacent atoms overlap, it is intuitively clear that the collection of atoms will be indistinguishable and in a coherent state. Detailed observations on the sudden formation of the condensate at a critical temperature distinguish the Bose-stimulation of the condensate from the simple process of thermal relaxation. This atom stimulation process is the essence of the atom gain process and atom laser concept. New atoms are not generated; rather, thermal atoms are stimulated into the same coherent state. It is directly analogous to the gain mechanism of stimulated emission and gain in optical lasers, in which the presence of an incident beam of photons stimulates the emission of more photons into that mode.

In 1998, Miesner *et al.* at Ketterle's lab studied the formation of a BE condensate in considerable detail [526]. They cooled a condensate close to T_c, waited until the atoms came to thermal equilibrium, and then suddenly turned the condensation on by a quick burst of radio frequency cooling. They then watched its subsequent dynamics. Using nondestructive phase contrast imaging, they could observe, step by step, every 5 or 10 ms, the transition of thermal atoms into a growing condensate, in real time. Equilibrium was reached after about 400 ms. The authors showed very clearly that the number of condensate atoms grew as expected for a Bose-stimulated process. This growth history is considerably different from that of a purely thermal process. There are, however, very substantial differences between experiment and theoretical calculations of growth rates, especially for the early exponential part of the stimulation process. Experimental results for the growth parameter γ are factors of 3–15 times faster than predicted by the theory [527]. This paper by Gardiner *et al.* is a fully quantum mechanical theory of the formation process. Miesner *et al.* suggest that improvements in their measurements of the early exponential growth of the condensate should be made; but they especially believe that more detailed theoretical studies are needed. See also Ref. [500] for the theory of BE condensate formation in dilute gases. Irrespective of theory, strong experimental proof of the achievement of BEC comes from the many demonstrations of the coherence of the condensates, as discussed earlier.

16.10. Atom Lasers

The demonstration of coherence implies that atom lasers are possible. To obtain an atom laser, all one has to do is start with an ideal condensate in a trap where all atoms are in the same coherent single particle wave function and then devise a way of coupling out some of these coherent atoms. As we saw above, weakly interacting BE condensates are essentially 99% in a single coherent state and thus can serve as sources for atom lasers. Indeed, some of the experimental configurations described above qualify as atom lasers.

16.10.1. *Pulsed sodium atom laser*

The first coupling out experiment of this kind was performed at MIT by Mewes *et al.* in 1997 [32]. They started with 5×10^6 atoms of sodium in a condensate, confined in a cloverleaf magnetic trap and cooled to the point where there was no measurable normal fraction of atoms. They proceeded to couple out varying numbers of atoms using radio frequency pulses, which flipped the spins of the atoms to an untrapped state. The trapped atoms originally in the $F = 1$, $m_F = -1$ trapped low-field seeking state were transferred to the $m_F = 0$ untrapped state and to the $m_F = 1$, the high-field seeking untrapped state. Since the trap is in the inhomogeneous field of the magnetic trap, its rf resonant frequency varies over the condensate. This was no problem with the short pulses used. These pulses were either fixed frequency pulses of varying amplitude or chirped pulses varying at a constant rate from 0 to a high frequency, well above the resonant frequency. In either case, the spectral width was sufficient to couple uniformly over the full condensate.

The pulse of atoms released in the $m_F = 1$ high-field seeking state is expelled away from the trapping region and does not form a usable beam. The $m_F = 0$ atoms, however, expand somewhat due to their mutual repulsion and fall downward due to gravity as a series of pulses in a well-defined atom laser beam. Measurements made with fixed-frequency or with swept-frequency pulses of the variations of trapped fractions of atoms with rf amplitude agreed extremely well with theoretical expectations. Measurements confirmed the coherence of the coupled-out fraction of the condensate. The authors point out that continuous wave coupling out of BEC is possible. Since the strength of the trapping magnetic field varies by about $10\,\text{mG}$ over the condensate, one needs extremely high stability to control the output of the coupler. With such a coupler, one could, for example, couple just to the outer surface of the condensate, where the energy of the mean field repulsion is low. The authors further point out that rf output couplers are the equivalent of beam splitters or partially reflective mirrors.

16.10.2. *cw atom laser*

Esslinger, Bloch and Hänsch developed the first cw atom laser in 1998 [528] and later, in 1999 [33]*. It was cw in the sense that it was an extremely high brightness laser that ran for $100\,\text{ms}$ before draining the source condensate. The basic idea was similar to that of Mewes *et al.* [32], but it had some new features. They developed yet another modification of the quadrupole magnetic trap, called a QUIC trap, which incorporated an Ioffe-type trapping geometry into the quadrupole geometry. It was simpler than the many-coiled cloverleaf trap and ran with magnets of much lower power. It contained two quadrupole coils plus an Ioffe coil off to one side [528]. As the Ioffe coil was turned on, atoms held in the linear geometry of the standard quadrupole trap transferred to the tighter parabolic geometry of the Ioffe trap, resulting in tight confinement with lower power dissipation. Continuous wave rf-induced evaporation from the Ioffe trap, which had a minimum magnetic field of $\sim 2.5\,\text{G}$, cooled the transition temperature to form a dense BE condensate. The trap was compact and was put inside a μ-metal shielding box that reduced environmental magnetic noise to a low level. The low-power coils were all run in series with low thermal drift, all of which contributes to a very high overall magnetic stability, which makes cw operation of an atomic laser possible. Another innovation was the use of a dual trap configuration, in which they used a standard MOT to initially capture and pre-cool the atoms in a relatively high-pressure environment. They then transferred these atoms into the QUIC trap, situated below it, through a tube in a very high-vacuum environment.

The high magnetic stability made it possible to precisely control the location of the region of the trap where the atoms were coupled out. Since the magnetic field variation over the BE condensate in the QUIC trap was known, one could tune the cw radio frequency coupling to any desired magnetic field contour within the condensate, via the resonance condition $(1/2)\mu_B |B(r)|$, where μ_B is the Bohr magneton. The authors estimate a spatial resolution of about $1\,\mu$m. Of course, gravity caused the location of the equilibrium position of the trapped cloud to shift downward. Furthermore, one could predict the trajectories of the ejected atoms precisely. Atoms in the high-field seeking state ($m_F = 1$) were dispersed and those in the $m_F = 0$ untrapped state felt no magnetic force and simply fell by gravity. Figure 16.5, taken from Fig. 1 of Ref. [33]*, shows the atom laser output. As the atoms fall and their velocity increases, their deBroglie wavelength shrinks. After the atoms fell 1 mm, their deBroglie wavelength has shrunk to about 30 nm. See Ref. [529] for a commentary on atom lasers.

The future of cw atom lasers is promising, especially for atom optics. The authors foresee focusing atom beams to much less than 1 nm. The coherence length of cw atomic laser beams is truly macroscopic, extending over the entire beam, limited ultimately only by stability considerations, such as mechanical instabilities and magnetic field fluctuations. Hänsch estimates the present laser is limited to a coherence time of 1 ms [529]. Nevertheless, one expects large improvements in the precision of

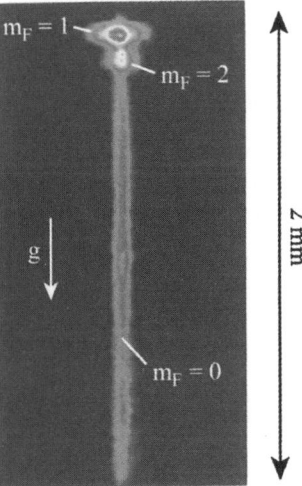

Fig. 16.5. Atom laser output: A collimated atomic beam is derived from a BE condensate over a 15 ms period of continuous output coupling. A fraction of condensed atoms has remained in the magnetically trapped $|F = 2, m_F = 2\rangle$ and $|F = 2, m_F = 1\rangle$ state. The magnetic trap has its weakly confining axis in the horizontal direction. Figure 1 and caption from Ref. [33]*.

interferometric experiments with cw atom lasers. Eventually, one should be able to achieve a truly cw atom laser by devising a scheme for providing a continuous supply of atoms. The present laser starts with $\sim 5 \times 10^5$ atoms and depletes them in 100 ms. This time of emission can be extended using weak rf coupling fields. See Secs. 19.6–19.11 for further discussion of true cw atom lasers.

16.10.3. *Quasi-continuous atom laser by Raman ejection*

Another significant advance in atoms lasers was achieved at NIST by Hagley *et al.*, also in 1999 [34]*. This was a quasi-continuous laser, which did not depend on gravity for ejection from the condensate, but ejected its atoms at a specific velocity. The authors used stimulated Raman transitions between magnetic sublevels to give the atoms a momentum kick in a specified direction. The direction and energetics of the Raman emission was specified by two nearly oppositely directed Raman beams having a frequency difference $\delta = \omega_2 - \omega_1$, which coupled atoms from the $m = -1$ trapped hyperfine state of the sodium condensate to the $m = 0$ untrapped state. This imparted the difference in momentum of the $m = -1$ and $m = 0$ states of $2(h/2\pi)k$, where k is the wave number $2\pi/\lambda$, to the untrapped atoms. Both beams were detuned from the excited state by -1.85 GHz to suppress significant spontaneous emission from the excited state. If the angle between the oppositely directed beams θ is varied somewhat, one can tune the output momentum anywhere from 0 to $2(h/2\pi)k$. If one tunes to the $m = +1$ untrapped high-field seeking state, one can then vary the output momentum from 0 to $(h/2\pi)k$. The $m = +1$ state is undesirable, however, as we saw earlier, because atoms exiting the trap in this state are distorted by the varying magnetic fields of the quadrupole magnetic trap.

Another advantage of Raman coupling is that it gives a highly directed beam with reduced transverse momentum spread. The initial spread of momentum in the condensate is very small, $\sim 0.09(h/2\pi)k$. Then the output beam suddenly acquires its $\sim 2(h/2\pi)k$ velocity of 6 cm/s without any of the slow isotropic mean spreading that occurs with gravity coupling.

A disadvantage of this atom laser, which uses the TOP trap, is that the magnetic field is constantly changing. This means that cw operation is impossible. In order to get output coupling at a resonance frequency at the same magnetic field, the authors had to synchronize their output Raman pulses with the rotating field. This gave a series of output pulses. These atom pulses emerging from the 50-μm-sized condensate advanced only $\sim 3\,\mu$m per pulse and appeared almost continuous. A continuous Raman laser needs a fixed field magnetic trap or a cw dipole trap. Figure 16.6, taken from Figs. 3 and 4 of Ref. [34]*, shows the output from one, three, and six 6-μs Raman pulses, and, finally, the production of a quasi-continuous atomic beam with the firing of 1-μs Raman pulses at the full repetition rate.

16.10.4. *Coherent beams by Bragg scattering*

Another way of manipulating coherent atoms in a condensate is by Bragg scattering [530]. Kozuma *et al.*, in 1999 [531]*, using a BE condensate, demonstrated the process of Bragg reflection of atoms by a moving standing wave potential. With this effect, one can selectively transfer a fixed momentum to an arbitrary fraction of a condensate without changing the magnetic sublevel. Hagley *et al.* [34]*, in their Raman coupler, changed $m = -1$ atoms into $m = 0$ atoms. The moving standing waves

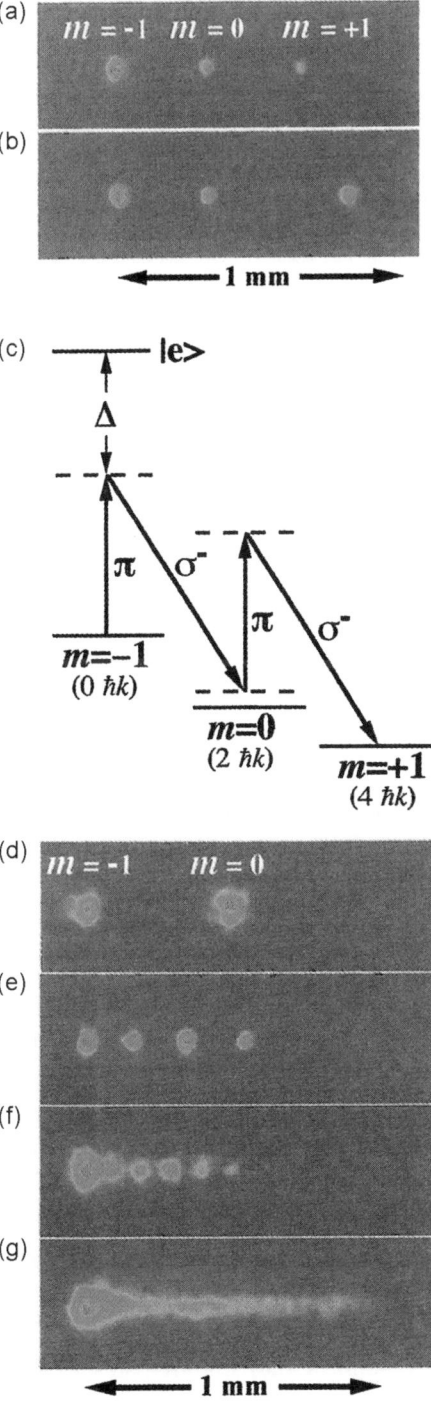

Fig. 16.6. Atoms are coupled to both the $m = 0$ and $m = +1$ and magnetic sublevels using a single Raman pulse. (a) Magnetic trap is switched off immediately after the Raman pulse. (b) Magnetic trap is held on for 4 ms after the Raman pulse. (c) Transition used and laser polarizations. (d) to (f) One, three, and six 6-μs Raman pulses, respectively, were applied to the condensate. (g) Firing 1-μs Raman pulses at the full repetition rate of about 20 kHz imposed by the frequency of the rotating bias field (140 pulses in 7 ms) produces a quasi-continuous atomic beam. Figures 3 and 4 and their captions from Ref. [34]*.

were formed from two beams with a frequency difference of δ intersecting at some angle θ. This Bragg geometry specifies a direction in space along which incident atoms scatter coherently from the optical standing waves. By controlling the length of the optical pulse or its intensity, one can transfer a controlled number, from 0 to 100%, of the incident atoms to the new momentum state. By varying θ from 0 to π, one can control the magnitude of the momentum imparted to the scattered atoms from 0 to $2(h/2\pi)k$.

The authors demonstrated Bragg scattering from free condensates and from condensates held within the trap. The scattering process preserves the coherence of the scattered atoms. The technique provides a way to study the coherence of the condensate in free space and within the trap. It also can be used to coherently couple atoms out of a trap. The authors point out that the Bragg process, under the conditions used above, is the equivalent of a stimulated Raman transition between two momentum states (see Ref. [34]*).

16.10.5. *Coherent beam generation by four-wave mixing*

A particularly interesting application of BEC derives from the ability to observe nonlinear mixing of coherent matter waves in condensates, in direct analogy with the mixing of light waves in nonlinear materials. Just as intense light waves can change the local index of refraction of an optical nonlinear material and generate coherent waves of different frequencies, so, too, can coherent matter waves interact with a condensate and cause density changes to generate new matter waves of different momentum (or wavelength).

Deng *et al.*, in 1999 [190]*, did a clever four-wave mixing experiment, in which they used pulsed Bragg diffraction from a moving standing wave to generate three moving matter wave packets within a condensate, which, in turn, mixed and generated a fourth matter wave (see also Ref. [531]*). For mixing to occur, the wave packets have to satisfy conservation of energy, momentum, and particle number. The four waves were observed by simply letting them emerge into free space after a delay of about 6 ms. The final location of the atoms and their velocities were consistent with the time-of-flight and momenta of the four beams.

The initial condensate of $\sim 2 \times 10^6$ atoms was formed in the NIST TOP trap and expanded adiabatically to reduce the momentum of the original condensate to $(0.14 \pm 0.02)(h/2\pi)k$. The trap was turned off and $600\,\mu s$ later, after the decay of the magnetic field, the four-wave mixing experiment was performed in free space. The Bragg pulses were detuned by about 13 GHz from resonance to avoid spontaneous emission. The three matter waves to be mixed were formed almost simultaneously by two pairs of light pulses. The pulses of the first pair were at right angles to each other in the x and y directions, with a frequency difference of 50 kHz. The pulses of the second pair were opposite each other in the x and $-x$ directions, with a frequency difference of 100 kHz. The first pair of pulses put one third of all atoms in the $\mathbf{P_2}$ momentum state. The second pair of pulses put half of the remaining atoms into the $\mathbf{P_1}$ and $\mathbf{P_3}$ momentum states. Four-wave mixing in the condensate generated the fourth beam $\mathbf{P_4}$. Figure 16.7, taken from Fig. 2 of Ref. [190]*, shows the calculated and measured distribution of the atoms in space. The theory based on the nonlinear wave equation predicts the intensity or number of atoms in the $\mathbf{P_4}$ to good accuracy. The authors foresee studies of many other nonlinear processes using condensates and matter waves. This four-wave mixing process gives another way of generating a coherent atomic beam in a controlled direction with a controlled

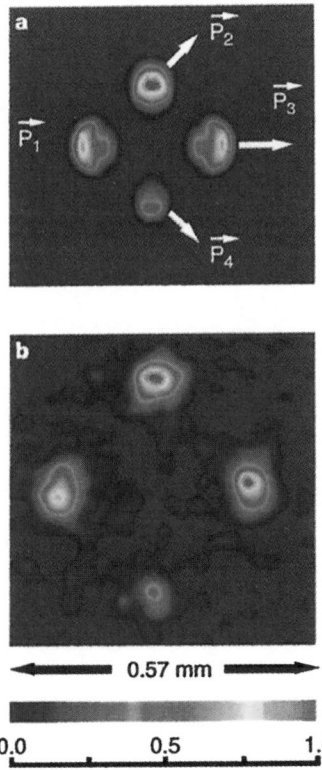

Fig. 16.7. Numerical simulation and experimental results for 4 WM. (a) Calculated 2D atomic distribution after 1.8 ms, showing the 4 WM. The calculations were performed only until the wavepackets completely separated due to constraints on the simulation grid-size. The momenta are those of Fig. 1(a). The field of view is 0.23 × 0.26 mm. We note that atoms are removed primarily from the back-end of the wavepackets because these regions overlap for the longest time. (b) A false-color image of the experimental atomic distribution showing the fourth (small) wavepacket generated by the 4 WM process. The four wavepackets form a square measuring 0.26 × 0.26 mm, corresponding to the distance of 0.25 mm calculated using the experimental time of flight of 6.1 ms and the wavepacket momenta. We have verified that if we make initial wavepackets such that energy and momentum conservation cannot be simultaneously satisfied, no 4 WM signal is observed. For instance, if we change the sign of the frequency difference between the two laser beams that comprise the second Bragg pulse, we will create a component with momentum $\mathbf{P_3} = -2\hbar k\hat{\mathbf{x}}$ instead of $\mathbf{P_3} = 2\hbar k\hat{\mathbf{x}}$. In this case there is no 4 WM signal. Figure 2 and caption from Ref. [190]*.

momentum. Its properties can be compared with the other atom laser beams discussed above [32], [33]*, [34]*.

The fact that the nonlinearity needed for observation of nonlinear optical effects with matter waves requires no special atom wave nonlinear material other than a uniform condensate itself is not as unusual as it first appears. Recall the so-called "artificial optical nonlinear materials" studied by Ashkin and colleagues in the early 1980s, which simply consisted of a uniform suspension of sub-micrometer colloidal spheres (see Sec. 5.2.1). We, in fact, studied degenerate four-wave mixing in this material, in very close analogy to the experiment of Deng *et al.* [190]*. Using gradient light forces from three input beams, two of which were called pump waves and the third an input signal wave, we generated a fourth optical wave, which grew nonlinearly in a direction opposite to the signal

Index Grating in Standing Wave

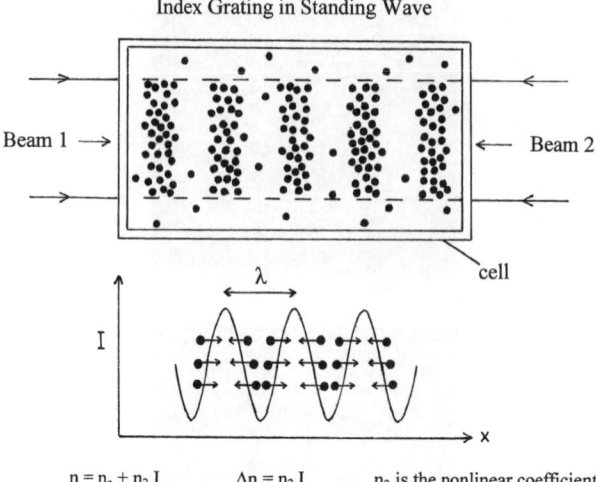

$n = n_o + n_2\,I$ $\qquad$ $\Delta n = n_2\,I$ $\qquad$ n_2 is the nonlinear coefficient

Fig. 16.8. Formation of an index of refraction grating in an artificial nonlinear medium made up of a colloidal suspension of submicron particles using an optical intensity standing wave. High index colloidal particles are pulled into the optical standing wave intensity maxima, causing the average index of refraction of the medium to increase. Such a medium can be characterized by an intensity-dependent nonlinear index of refraction $n = n_0 + n_2 I$, where n_2 is the nonlinear coefficient.

wave [165], [167]. In our degenerate four-wave mixing case, light waves interacted with standing wave index gratings, fixed in space and made up of small spheres. Figure 16.8 shows the gratings generated by the incoming signal wave and the pump waves. The reason for this is that a modulated spatial density grating of submicrometer spheres behaves as a spatial index grating. In the Deng experiment, the matter waves interacted with a moving standing wave made up of atoms of the condensate. Another important difference was the response time of the nonlinear medium. For the degenerate four-wave mixing work in liquid, the particles were heavily damped and the build-up and decay of the fixed grating was quite slow, whereas for condensates the damping was enormously less, making for a very rapid response.

One of the major applications of degenerate four-wave mixing with light is in producing a phase-conjugating mirror in which a light beam passing through a distorting medium can be reflected back on itself ray by ray, thus undoing the distortion [532]. It occurs to me that it should be quite interesting to demonstrate a phase-conjugate mirror by degenerate four-wave mixing with matter waves. Such a mirror would be a useful element in the new field of atom wave-optics with coherent atom laser beams and could be the basis of some quite useful devices.

16.10.6. *Commentary on Bose–Einstein condensates and nonlinear matter waves*

The BE condensates and the recent demonstrations of coherent nonlinear matter wave behavior have been the subjects of recent commentary. See Andrews "Bose gases and their Fermi cousins", in Nature [533] and Lubkin, "Bose Condensates Produce Coherent Nonlinear Behavior", in Physics Today [534], both appearing in 1999.

16.10.7. *Two-component Bose–Einstein condensates in ^{87}Rb and sympathetic cooling*

As the field of BEC developed using atoms trapped with specific low-field seeking spin states in a magnetic field, the possibility of simultaneously trapping atoms in different spin states in a single magnetic trap was considered. At JILA, in 1997, Myatt *et al.* [535]* discovered a two-component condensate that exists in both the upper and lower spin states of ^{87}Rb.

The new system makes use of what they call a double MOT [536] to collect large numbers of atoms, at higher pressures, that are later transferred through a long (40 cm × 1 cm diameter) tube to an Ioffe type baseball trap at pressures of less than 10^{-11} Torr. With this apparatus, the authors discovered they could trap atoms in either of the $|F = 2, m = 2\rangle$ or $|F = 1, m = -1\rangle$ ground states of about 2×10^6 atoms.

They found that they could also create mixtures of the two condensates, for the first time, in which the $|2, 2\rangle$ state was cooled only by thermal contact with the $|1, -1\rangle$ evaporatively cooled state. This established that the scattering length of the $|1, -1\rangle$ state is positive. The fact that both condensates form together and overlap is also an indication that the scattering lengths accidentally happen to be close to each other, so that the trapping forces are almost identical, and the two condensates overlap in space. There is an offset between their centers due to gravity. Each condensate can be examined independently by absorption imaging. They have different sizes, as they should, and decay from the trap at different rates. Interestingly, interactions of the two condensates can be seen as they are simultaneously released from the trap. There is clear evidence of a mutual repulsion between the two expanding condensate clouds.

The authors were able to measure the densities, temperatures and loss rates of the condensates. The loss rates are consistent with being due to binary spin flipping from inelastic collisions between species. In general, the authors foresee applications of the sympathetic cooling technique to situations where the bosonic atoms act as a cooling fluid to sympathetically cool a fermionic gas or other species to form a condensate where inelastic processes make conventional evaporative cooling impossible.

16.10.8. *Dynamics of two-component Bose–Einstein condensates in ^{87}Rb*

The JILA group made use of their new binary mixture of BE condensates to study the dynamics of the two interpenetrating bosonic quantum fluids (see Ref. [537]). Starting with all the atoms in the $|1\rangle$ state, they converted $\sim 50\%$ of the atoms to the $|2\rangle$ state with a 400 μs pulse of microwaves plus radio frequency photons, and then watched the two populations evolve. The authors were able to get the two populations to overlap, with no relative "sag", by adjusting the TOP trap parameters. After $\sim 400 \mu$s they found that a crater had formed in the $|1\rangle$ condensate that was filled in by a shell of $|2\rangle$ atoms. This was consistent with a theory by Pu and Bigelow [538], which stated that it is energetically favorable for the atoms with the larger scattering lengths to form a low-density ring about the atoms with the smaller scattering lengths. Introducing a small "sag", or offset, of the condensates induces quite violent oscillations in the relative centers of mass of the condensates, which are not visible in the combined distribution of atoms. These violent oscillations initially give rise to almost complete separation of the two components, but then quickly damp down after about 100 ms. The theory of

Pu and Bigelow, using a pair of coupled GP equations, predicts this oscillation, which is driven by the repulsion of the two components, but does not account for the rapid damping.

16.10.9. *Phase memory in ^{87}Rb two-component Bose–Einstein condensates*

Hall *et al.*, in their 1998 paper [537], also examined the evolution of the relative phase in their two component BE condensates in ^{87}Rb atoms. The two condensates were formed initially by a $\pi/2$ pulse. This transformed 50% of the atoms from the $|1\rangle$ spin state to the $|2\rangle$ spin state with a particular relative phase. After interacting with each other and possibly with the environment, the two components were examined in phase again in about 45 ms, when their relative oscillations had damped down. The comparison in phase was made by applying a second $\pi/2$ pulse and examining the relative densities of the two states in their region of overlap. If the phases had evolved similarly, to give the same densities after the passage of the same time, then one knows that the phase evolution was coherent. Indeed, even after experiencing heavy damping, the relative phases were the same, out to about 100 ms. This definite phase memory in the face of essentially complete damping is somewhat surprising. The authors foresee experiments on possible phase locking of the condensates and the study of analogs of superconducting Josephson junctions. See references in Hall *et al.*'s paper [537].

17

Role of All-Optical Traps and MOTs in Atomic Physics

As exciting as the BEC work was, one of the disappointing aspects of the field for me, in its initial years from 1995 to 1998, was the fact that optical dipole trapping played an increasingly minor role. As seen earlier, the final cooling of atoms to the critical temperature T_c and below was carried out by evaporation in purely magnetic traps of various sorts. They included (i) the TOP quadrupole trap and its variations used at JILA [29]* and NIST [523]; (ii) the optically plugged quadrupole at MIT [175]*; (iii) the cloverleaf trap, a variation of the Ioffe-Pritchard trap at MIT [509]; (iv) the quadrupole-Ioffe-configuration trap of the Munich group [528]; and (v) the Ioffe-type baseball trap at JILA [536]. Optics and lasers were gone at this time. Also, most of the early experimental work revolved around the use of these magnetic traps. Magnetic traps are large pieces of apparatus with many coils often taking up to kilowatts of power to produce high fields, at times in the 2T range.

Let me briefly recall the history of optical gradient force traps. Gradient force traps were invented in 1978 by Ashkin, initially for atoms. The first atom trap was achieved in 1986 in our experiments with Chu *et al.* [4]* and used the simplest of all traps, a single focused Gaussian beam. Using molasses, the trapped atoms were cooled to a temperature of $\sim 240\,\mu\text{K}$ [27]*. It also was shown, in 1986, that the single beam gradient trap could also trap micrometer and sub-micrometer particles [5]*. Later, it was discovered that this same basic trap, called "tweezers" for microscopic particles, could also manipulate living biological particles and cells [19]*, [20]*, [44]*. Biophysicists and biologists quickly embraced it and have performed some quite unique experiments with it on biological particles, including manipulating single molecules, as we described earlier.

Unfortunately, however, these gradient or dipole traps for atoms were not similarly embraced by atomic physicists. They favored the magneto-optic trap, or MOT, invented in 1986, [178] and demonstrated in 1987 by Raab, Prentiss, Cable, Chu, and Pritchard [181]*. This was a hybrid trap, part optical and part magnetic. It was based on magnetic tuning of the scattering force. With a few minor exceptions, dipole traps were neglected. The principal attraction of the MOT was its large volume, of about $1\,cm^3$, and its considerable depth, of about $100\,mK$. The first MOT held about 2×10^7 atoms at a density of about 2×10^{11} atoms/cm^3, as compared with about 500 atoms at a density of about 5×10^{11} atoms/cm^3 for the first dipole atom trap [4]*. A large fixed MOT with its magnetic coils lacks the manipulative ability in space and time of an all-optical dipole trap.

From 1987 up to the present, the MOT has been used in essentially all experiments in the initial collection of atoms cooled to molasses temperatures. It has undergone some improvements, such as the development of the dark spot MOT, which increased the density and total number of atoms that could be trapped [183]*. In dark spot traps, the central portion of the trap is blocked from the usual repumper light, which is used to keep the atoms in the correct hyperfine level to interact with the trapping beams. This reduces trap loss due to repulsions in rescattered light and results in about a hundred-fold increase, at best, in the density and number of trapped atoms [183]*, which is important for feeding the purely magnetic quadrupole traps used for evaporative cooling in BEC experiments [29]*, [175]*. See earlier discussions. Of course a similar "dark" feature is easily possible with dipole traps (see Secs. 19.5, 19.6–19.10).

17.1. Far-Off-Resonance Optical Traps for ^{85}Rb

In the continuing history of optical trapping from around 1998, we will see that the dominance of the MOT and purely magnetic traps began to erode, starting with the experiment of Miller *et al.* in Heinzen's lab. in 1993 [148], discussed above. With time there was an increasing appreciation of the considerable advantages of all-optical trapping. This manifested itself in new types of experiments involving the transfer of Bose–Einstein (BE) condensates into optical traps and the optical trapping of ground state atoms in any spin state, including high-field-seeker states. This, in turn, led to attempts to directly cool atoms to BEC without using purely magnetic traps. Miller *et al.* pointed out that confining low scattering atoms at high densities to subwavelength dimensions could be advantageous in many quantum optics experiments.

The experiment of Heinzen's group differed from the original sodium dipole trap experiment of Chu *et al.* [4]* in that the ^{85}Rb atoms were loaded into the dipole trap from a MOT with $\sim 10^6$ atoms at a density of $\sim 10^{10}$ atoms/cm^3. Also, their dipole trap detuning from resonance was much higher than for sodium. The peak atom scattering rate of the ^{85}Rb trapping beam at the focus was only $\sim 4 \times 10^2$ photons/s. This is quite low and gives a recoil heating lifetime of $\sim 43\,s$. About 1300 atoms were trapped in their optical dipole at a density of 8×10^{11} atoms/cm^3.

17.2. Far-Off-Resonance Traps for Cesium Using CO_2 Lasers

In 1995, Takekoshi *et al.* [539]* proposed trapping of neutral atoms using a high-power extremely far-off-resonance infrared laser. In particular, they considered the case of cesium atoms trapped at the focus of a CO_2 laser beam at $\lambda = 10.6\,\mu m$. For this case the laser frequency is far below any of the cesium resonances and the usual polarizability computation has to be modified somewhat. The atom can no longer be considered a simple two-level system. It behaves quasi-electrostatically, so they chose to call this a Quasi-Electrostatic trap, or QUEST. The scattering rate, and therefore optical heating, is really negligible, of the order of years when operating so far from resonance, if one

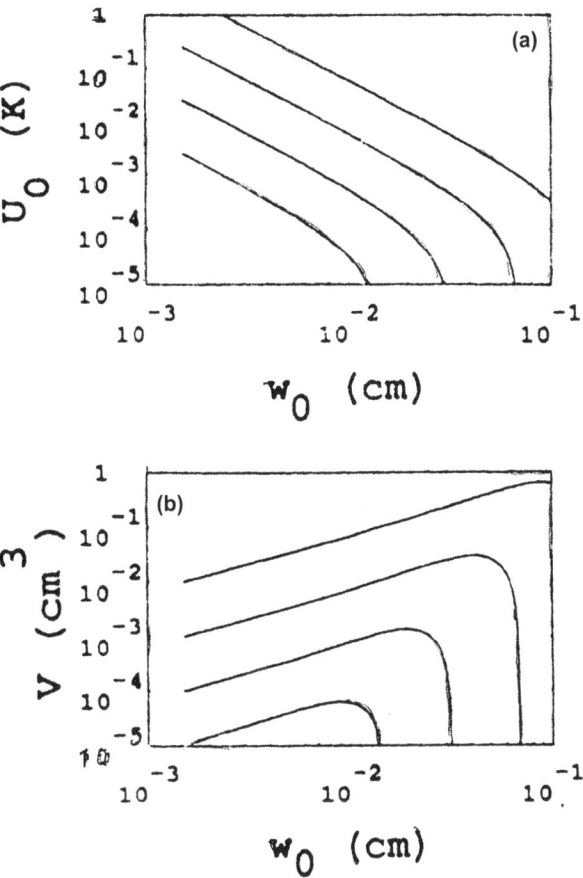

Fig. 17.1. (a) Trap depth U_0/k_b (including gravity) versus waist radius w_0 for cesium atoms. (b) Trap volume V versus waist radius w_0 for cesium atoms. The curves (from top to bottom) correspond to laser powers of 10 kW, 1 kW, 100 W and 10 W. Figure 1 and caption from Ref. [539]*.

excludes background collisions. In fact, the main optical scattering process that occurs is Rayleigh scattering from the atoms. Figure 17.1, taken from Fig. 1 of Ref. [539]*, shows the trap depth and volume of a CO_2 laser trap for different beam waist radii and different laser powers from 10 kW down to 10 W. The authors point out that CO_2 laser traps have the advantage that they could also be used to trap several atomic species and molecules simultaneously in one trap. Another possibility is to use such traps in search for the BE condensation of a cold atomic vapor.

In 1996, Takekoshi and Knize demonstrated a CO_2 laser trap for cesium atoms [146]. Starting with a dark spot MOT holding up to 10^8 cesium atoms, they transferred more than 10^6 atoms into an 18 W CO_2 laser trap with a beam waist $w_0 = 100\,\mu$m. Cesium densities of $\sim 10^8$ atoms/cm^3 were achieved in the dipole trap. The lifetime of the atoms in the trap was 10 s, consistent with their background gas pressure. This is quite impressive behavior. Powers much in excess of 18 W are possible with CO_2 lasers. For example, a 100 W CO_2 laser has an internal circulating power of about 2.5 kW; so even this very high power is experimentally conceivable.

17.3. Evaporative Cooling of Sodium Atoms from an Optical Dipole Trap

In the few years prior to the success of the first TOP trap a number of approaches tried to attain high phase-space density and low temperatures with dipole traps. One such attempt was an experiment in 1995 by Adams *et al.*, in Chu's lab. [145]*, on evaporative cooling in a crossed dipole trap. Sodium atoms were trapped in an optical trap formed at the intersection of two $1.06\,\mu$m laser beams. Initially they loaded densities as high as 4×10^{12} atoms/cm^3 at a temperature of $\sim 140\,\mu$K into the dipole trap from a MOT. By reducing the trap depth slowly over a time period of 2 s, they successfully evaporatively cooled the atoms to a final temperature of $\sim 4\,\mu$K, at a density of 6×10^{11} atoms/cm^3. This increased the phase-space density by a factor of 28. Cooling essentially stopped at this point, due to falling density. They estimated that the atom re-thermalization time changed from $\sim 10^{-1}$ to ~ 1 s over the 2 s evaporative cooling span. During the evaporation process the number of atoms in the trap dropped from ~ 5000 to ~ 500. This experiment, however, was the first demonstration of evaporative cooling from dipole traps by lowering the optical potential.

17.4. Raman Cooling of Trapped Atoms in a Dipole Trap

Work on dipole traps continued even after the achievement of BEC in magnetic quadrupole traps. At Stanford, in 1996, Lee *et al.* [540] experimented with a different version of a dipole trap. This consisted of four repulsive sheet beams formed into a large inverted pyramid that relied on gravity for vertical confinement. This "dark" large-volume optical levitation trap was cooled by the Raman cooling technique, developed earlier by Chu's group [541]. About 5×10^5 atoms were cooled to $\sim 1.0\,\mu$K at a density of $\sim 10^{11}$ atoms/cm^3 in this way, but the results were still short of BEC because the density was too low. The phase-space density reached in this experiment was $\sim 1/400$ of that required for BEC.

17.5. Laser Noise Heating in Far-Off-Resonance Optical Dipole Traps

In 1997, a paper by Savard, O'Hara and Thomas at Duke University [542] gave some important insights into the nature of laser-noise-induced heating in far-off-resonance traps. In dipole traps, one expects very low spontaneous scattering rates and therefore low heating rates. Low heating rates should increase the atomic density by reducing repulsive scattering forces [182]* and reduce excited state trap loss collisions [543]. Reduction in noise should make it possible to achieve the very long atom storage times inherent in these detuned optical traps.

In this paper, Thomas's group calculated other possible sources of laser heating from laser intensity fluctuations and beam pointing fluctuations. They found that achieving long trapping lifetimes beyond 10 s imposes stringent requirements on laser noise power spectra for either red- or blue-detuned laser trapping beams. A follow-up study by Gehm *et al.* of Thomas's group in 1998 [544] analyzed the dynamics of noise-induced heating in atom traps, using a simple harmonic oscillator potential model. They compared both optical traps and magnetic traps, and showed that the stability requirements for optical traps are more stringent than the magnetic field stability requirements for magnetic traps.

17.6. Sisyphus Cooling of Cesium in Far-Off-Resonance Optical Dipole Traps

A group in France in 1998 [545]* studied cold and dense clouds of cesium atoms trapped in far-detuned dipole traps. They used $1.06\,\mu$m light from YAG lasers to achieve nearly nondissipating potentials and very low spontaneous emission rates ($= 1\,\mathrm{s}^{-1}$). A new feature of their experiment was the use of very efficient blue-detuned Sisyphus cooling (see references in text) to get temperatures as low as $0.5\,\mu$K. Their densities vary, depending on conditions, from 10^{12} atoms/cm^3 at $2\,\mu$K to 10^{13} atoms/cm^3 at somewhat higher temperatures. The authors also imaged a rod-like hexagonal lattice structure in the atom cloud and trapped as many as 10^4 atoms on each of the lattice sites. A total of about 10^6 atoms were trapped in this manner. They believe these results are a big step toward reaching quantum degeneracy by purely optical methods. They point out that evaporative cooling from their dipole trap is also possible, as demonstrated at Stanford [145]* and is an additional conceivable path to BEC.

17.7. Raman Cooling of Cesium in Far-Off-Resonance Optical Dipole Traps

Another attempt was made in 1998 by Chu's group at Stanford to cool cesium atoms to high phase-space densities by optical means in far-detuned dipole traps. This is described in Ref. [546]*. At this point in time, they considered achieving BEC by optical means an important goal, despite the success of evaporative cooling in magnetic traps. Raman optical cooling techniques are much faster than evaporative cooling techniques and are also a way of avoiding the high loss of atoms associated with evaporative cooling. They hope to achieve BEC at higher condensate temperatures, which means working at higher densities. This may be possible with the high densities typical of optical traps. They used a Raman sideband technique, but at densities of about 10^{13} atoms/cm^3, two orders of magnitude higher than previously used for sodium in free space [541]. The optical trap was a two-beam vertical standing wave, made with a TEM$_{00}$ mode beam from a $1.06\,\mu$m Nd:YAG laser focused to a beam width of $260\,\mu$m. About 10^7 atoms collected in a cigar-shaped cloud about 2.5 mm long, made up of 4700 individual pancake shaped traps with an aspect ratio of 1000, spaced by 532 nm. The atoms were cooled in three dimensions, without loss of atoms, to a mean temperature of $2.8\,\mu$K and a density of 1.4×10^{13} atoms/cm^3. This corresponds to a phase-space density of $1/180$. These results are somewhat better than those of Boiron *et al.* [545]*, as discussed above in Sec. 17.6. Vuletić *et al.* in Ref. [546]* also say one could conceivably start optical evaporative cooling at this point, to try to reach BEC.

By 1998, it was finally becoming clear that traps involving a magnetic field, namely the MOT-type trap used to collect large numbers of laser-cooled atoms and the variants of the magnetic quadrupole traps used to form BE condensates by evaporatively cooling, had some significant drawbacks when it came to studying ultracold atoms and the properties of BE condensates (see Refs. [4]*, [146], [539]*, [545]*, [546]*).

17.8. Two-Step Narrow-Line Cooling of Strontium in Optical Dipole Traps

In 1999, Katori and his collaborators at the Cooperative Excitation Project of the ERATO group in Japan [547] made a very significant advance in atom cooling and trapping. Their stated interest was in extending BEC techniques to other atomic species, to fermion gases and degenerate mixtures of boson and fermions. They also believe that formation and manipulation of condensates by all-optical means represents the ultimate form of atom manipulation. The authors, in pursuit of these aims, report on an experiment using atoms of ^{88}Sr, an alkaline earth metal. They made use of an extremely narrow spin-forbidden intercombination transition to Doppler cool a sample of atoms held in a MOT down to the photon recoil temperature of 400 nK. Figure 17.2, taken from Fig. 1 of Ido et al. [139]*, shows the energy levels of ^{88}Sr. In the process they demonstrated an atomic density of over 10^{12} atoms/cm^3, corresponding to a phase-space density of 10^{-2}.

This striking result was achieved by using a two-step technique. First, they precooled the atoms using the broad $^1S_0$–$^3P_1$ transition at 461 nm, having a line width of 32 MHz. They then switched the cooling to the 689 nm spin-forbidden transition, having a linewidth of 7.6 kHz and a saturation intensity of 3 μW/cm^2. This cooling light was generated using frequency-stabilized diode lasers. To capture and cool as many as possible of the first stage cooled atoms, they sinusoidally swept the narrow-line second step light over the 1.5 MHz to cover the full Doppler linewidth of the atoms cooled in the first step. The sweeping was subsequently turned off and the intensity adjusted for highest phase-space density.

The very low temperatures achieved imply that radiation trapping effects due to spontaneous emission [182]* were greatly suppressed due to the very narrow linewidth of the second stage cooling. The phase-space density, using the narrow line cooling of width $\gamma \sim E_r/(h/2\pi)$, where E_r is the recoil energy, is about three orders of magnitude larger than conventional MOTs using relatively broad transition linewidths. The authors say that these cooled atoms could be easily loaded into a far-off-resonance dipole trap and be evaporatively cooled or further compressed.

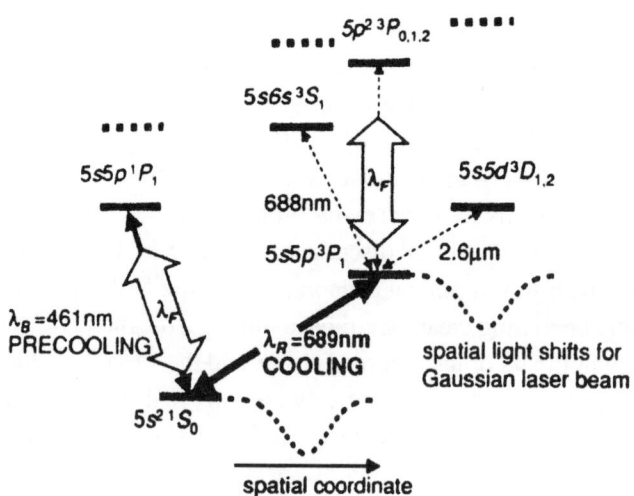

Fig. 17.2. The energy levels of ^{88}Sr. A FORT laser at $\lambda_F = 800$ nm couples the cooling ground state $^1S_0$ and the excited state $^3P_1$ to the upper singlet and triplet states, respectively, generating the same amount of Stark shifts. Spatial light shifts for these two states are schematically depicted by the dotted lines assuming the FORT laser with Gaussian profile. Because the atomic resonance frequency is unchanged in space, the optical confinement $^1S_0$ can be compatible with Doppler cooling on the intercombination line. Figure 1 and caption from Ref. [139]*.

17.9. Continuous Doppler Cooling of Strontium Atoms in an Optical Dipole Trap

Very recently, Ido *et al.* of the Japanese ERATO project [139]* improved their previous results [547] on optical-dipole trapping of strontium atoms at high phase-space density. As before, they used the narrow spin-forbidden $^1S_0$–$^3P_1$ transition to Doppler cool the atoms to the strontium recoil temperature of $\sim 440\,$nK in a MOT. They then transferred them into a specially designed far-off-resonance optical dipole trap that allowed simultaneous trapping and Doppler cooling. With this trap, they both cooled and compressed the atom cloud to a density of $\sim 10^{13}\,$atoms/cm^3. The total number of atoms in the trap varied from 4×10^4 to 3×10^5, which was about 20% of the original number in the MOT. They were able to achieve a phase-space density somewhat greater than 0.1. This was quite close to the threshold for BEC.

The problem with simultaneous dipole trapping and Doppler cooling is that the high-intensity dipole or gradient trap suffers from a position-dependent ground state Stark shift, which is incompatible with fixed frequency molasses beams. In our first demonstration of dipole trapping with sodium, in 1986, we avoided this problem by alternately switching between the trapping beam and the molasses cooling beams at a fairly rapid rate [4]*. The solution to this problem devised by Ido *et al.* involved choosing the frequency of the far-off-resonance trapping light beam such that the decrease in Stark energy of the $^1S_0$ ground state was exactly matched by a similar decrease in the Stark energy of the $^3P_1$ first excited state. Since this matching is independent of the intensity of the trapping beam, there was no net positional variation of the $^1S_0$–$^3P_1$ energy difference and the atoms in the trap could be cooled just as effectively as in free space. The value of λ_F for which the Stark shifts are equal in $^1S_0$ and $^3P_1$ is $\lambda_F = 800\,$nm. The authors say this is calculable by considering the detailed coupling of the $^3P_0$ level to the next three higher excited states. By using $\lambda_F = 800\,$nm it was possible to Doppler cool the strontium atoms to the recoil temperature of $\sim 440\,$nK with $\lambda = 689\,$nm molasses light. This was exactly the same temperature reached as in the authors' previous experiments on cooling in MOTs [547]. Refer to Fig. 17.2 to see the energy levels of ^{88}Sr and the Stark shift-correcting scheme.

I find this first demonstration of simultaneous Doppler cooling and dipole trapping of atoms quite exciting. Indeed, Ashkin and Gordon in the late 1970s [123]*, [136]*, when faced with the same problem of the positional dependence of the Stark shift, proposed a solution based on an additional Stark-correcting beam having a properly chosen intensity and the same spot size as the trapping beam, to give a positionally independent Stark shift. This Stark shift-corrected trap could, in principle, be cooled with conventional molasses to its minimum allowable temperature. This was never tried, although it avoids the factor of two loss in the minimum temperature one suffers with the 50–50 chopping of the trapping and molasses beams [4]*. Ido *et al.*'s solution to the Stark problem in very-far-off-resonance traps in strontium is more elegant than the Ashkin and Gordon Stark solution of the sodium Stark problem, since their solution involves beams of two wavelengths, λ_F, the far-off-resonance wavelength, and the molasses wavelength beams chosen slightly to the red of the resonance line, versus our solution involving three beams at different wavelengths: the off-resonance trapping beam, the Stark correcting beam and the molasses beams.

Ido *et al.* remark, finally, that increases in phase-space density beyond 0.1 should be possible using evaporative cooling.

17.10. Three-Dimensional (3D) Raman Sideband Cooling of Cesium in Optical Dipole Traps

Almost simultaneously with Ido's *et al.*'s work in Japan, Han *et al.* in Weiss's group at Berkeley, California [70] cooled a total of 5×10^7 cesium atoms by 3D Raman sideband cooling and adiabatic release in a 3D far-off-resonance optical lattice dipole trap, to a high density of 1.5×10^{12} atoms/cm^3. The lowest temperature achieved was about 300 nK, which gives a phase-space density of 1/30.

In previous work this group, in 1999, achieved essentially unity occupation of 3D far-off-resonance lattice sites with about 30% population occupancy of the ground state [548] using polarization gradient cooling. To improve on these previous results the authors implemented 3D Raman sideband cooling. Free space Raman cooling and one-dimensional (1D), two-dimensional (2D), and 3D Raman sideband cooling in far-off-resonance optical dipole traps have all been previously demonstrated. See references in the text of Ref. [70]. The authors point out that far-off-resonance lattice (FORL) traps typically can confine atoms at two orders of magnitude higher density than non-FORL traps; but that heating from population buildup due to scatter into one magnetic sublevel eventually becomes a barrier to further increase in phase-space density. In their experiment they designed a FORL trap that traps all magnetic sublevels identically, avoiding Raman scattering between different sublevels and allowing the cooling process to proceed, as in free-space Raman cooling.

The 3D FORL trap was made by combining a 1D vertical standing wave FORL trap with a horizontal 2D FORL standing wave trap. Each of the FORL standing wave beams had a power of about 300 mW in a 400 μm radius beam. A 180 MHz shift was introduced between the 1D standing wave beams and the 2D phase-locked beams, so that linear polarization was maintained in the 3D FORL trap. The trap depth at the center of the FORL trap was 160 μK vertically and 320 μK horizontally, but the lattice spacing vertically is $\sqrt{2}$ greater than horizontally, giving a trap with equal vibrational frequencies in all directions.

Starting with 9×10^7 atoms in the 3D FORL trap, the authors first cooled the atoms to a temperature of 8.5 μK using polarization gradient cooling molasses. They then spatially compressed and recooled the atoms, as in Ref. [548], and ended up with a density of 1.5×10^{12} atoms/cm^3 at an average temperature of 9 μK. This corresponds to a FORL trap lattice occupancy of 23%. The authors then added a 250 mG bias magnetic field and optically pumped 95% of the atoms into the $6S_{1/2}$, $F = 3$, $m_F = 3$ state, ending up with an additional 6 μK increase in temperature. They then used three traveling wave Raman beams to further cool each component of motion at the symmetric lattice sites. The final temperatures in the horizontal and vertical directions varied from $\sim 6 \mu$K and 5 μK at lower densities of $\sim 7 \times 10^{11}$ atoms/cm^3 to 8 and 7 μK at higher densities of $\sim 1.5 \times 10^{12}$ atoms/cm^3.

At the final cooling stage they adiabatically cooled [549] the atoms by slowly reducing the FORL trap intensity by a factor of $\sim 10^3$ until the atoms were released from the trap. The final release temperature was 250 ± 30 and 380 ± 10 nK in the horizontal and vertical directions, respectively, corresponding to a free space-phase density of $n_A \lambda_{DB}^3 = 1/30$. The authors feel that these results could be further improved and demonstrate that resolved sideband Raman cooling in a FORL trap could produce high-density spin-polarized laser-cooled samples, even with more than 10^7 atoms. Such samples are already of interest for studying quantum phase transitions in periodic potentials [550] and for studying quantum computers [551], [552], [553].

The methods demonstrated in the work of Han *et al.* [70] and of Ido *et al.* [139]* have a very good chance of achieving BEC by all-optical methods in far-off-resonance laser traps. As we can see, the field at this point was quite competitive, with different approaches achieving phase-space densities quite close to BEC. As we shall see in Sec. 20.5, BEC in all-optical traps was finally achieved in 2001.

17.11. Blue-Detuned Optical Dark Traps for Achieving High Atomic Density

As we know, dipole traps based on far-off-resonance red-detuned light can effectively trap atoms in relatively strong dipole potentials and reduce spontaneous emission [3]*, [123]*. The first demonstration of these forces showed focusing and defocusing of atoms detuned to the red or the blue side of the atomic resonance. The reduction of repulsive spontaneous emission forces between atoms due to detuning is desirable for trapping [182]*. This detuning makes it possible to avoid saturations and achieve high atom densities in dipole traps [4]*, [148], [205]*. Another powerful way of reducing atom–atom interactions inherent in dipole trapping is to use a blue-detuned optical dark trap. In such a trap the atoms are repelled from high-intensity beams. The atoms, in this case, are trapped at a spatial minimum of the optical intensity, which greatly reduces excitation. This type of trap is the equivalent of the MOT dark spot trap and is achieved much more simply. Davidson *et al.* in 1995 [554] and Ozeri *et al.* in 1999 [555] demonstrated traps of this sort using a holographic interference method to form a 3D intensity minimum in space. Ozeri obtained long spin relaxation times in this type of single-beam blue-detuned optical trap. Another way of forming a dark spot surrounded by higher-intensity light is to rotate a focused Gaussian beam about a radius that is several times larger than the beam waist [556]. If the rotation rate is rapid enough, the slow moving particles feel the force of only the average light potential. Similar computer-controlled rotating traps, as we have seen, were used in experiments by Sasaki *et al.* [419] on macroscopic metal particles and absorbing particles in tweezer traps. Visscher *et al.* [266] developed similar types of rotating traps. Davidson and colleagues demonstrated trapping of $> 10^6$ ^{85}Rb atoms in such rotating blue-detuned traps. They then proceeded to slowly decrease the radius of rotation from 70 to $27\,\mu$m and adiabatically compress and increase the atomic density by a factor of 350 to $\sim 5 \times 10^{13}$ atoms/cm^3. At this density they observed an elastic collision rate of $\sim 100\,\mathrm{s}^{-1}$ in the dark trap. They could get a further fourfold adiabatic increase in density by changing the shape of the trapping potential. The authors see this technique as a possible route to reaching BEC.

Earlier in 1997, Kuga *et al.* [557] demonstrated a somewhat different optical dipole trap based on a Laguerre–Gaussian (LG) (donut) mode beam. The LG modes were discussed in some detail in the text earlier in connection with beam angular momentum (see Refs. [408], [411], [412]). Kuga *et al.*'s trap consisted of a blue-detuned LG$_{03}$ mode that has a deep dark spot enclosed by a bright ring of $r_0 = 600\,\mu$m, plugged with two blue-detuned end cap beams. With a detuning of $60\,$GHz and a power of $600\,$mW they were able to trap 10^8 rubidium atoms from an original 3×10^8 atoms cooled to about $40\,\mu$K in a MOT trap. The authors suggest that the dark spot trap is useful for making precision measurements on unperturbed atoms or for further cooling and adiabatic compression to achieve BEC. LG mode beams are also ideal candidates for making light guides for atoms. In this context see also Ref. [558].

17.12. Transfer of Bose–Einstein Condensates into Optical Dipole Traps

Much of the interest in trapping and cooling atoms in dipole traps stems from the possibility of directly making BE condensates in such traps without resorting to magnetic traps.

Rather than pursue the development of sub-recoil techniques for achieving BEC in dipole traps by Raman cooling or evaporative cooling, as discussed above, Ketterle's group at MIT in 1998 [559]*, [560] took a different approach and simply transferred their existing condensate produced in magnetic traps into an off-resonance optical dipole trap. This avoided all of the difficulties of cooling the condensate. The temperature of the condensate was so low that only about 4 mW of off-resonance infrared power were needed for optical trapping, thereby reducing all power-dependent loss processes in optical dipole traps.

As the papers point out, this achievement eliminates the restrictions of magnetic traps for further studies of atom lasers and BE condensates. For atom lasers it eliminates the need to couple beams out of traps through an inhomogeneous field that exposes the atom laser beams to varying Zeeman shifts. Also, magnetic traps confine only the weak-field seeking atoms. Since the ground atomic state is always strong-field seeking, any inelastic dipole-relaxing scattering into the ground state results in heating and trap loss. This is a problem in studying multicomponent condensates, as discussed earlier [535]*. The ability of all-optical traps to trap the $m_F = 0$ hyperfine state is important, since this is the preferred state for atomic clocks and for precision metrology.

By putting the focused dipole trap into the center of the magnetically produced condensate and ramping up the infrared power in about 125 ms, they were able to transfer about 85% of the condensate into the optical trap. In this way, more than 5×10^6 condensed atoms were transferred at densities up to a remarkable 3×10^{15} atoms/cm^3. Densities were determined by turning off the light and letting the condensate expand in free space. It was also possible to transfer uncondensed cold atoms at a lower density into the optical trap and observe a condensate form as the atoms compress to a higher density in the tightly focused dipole trapping beam. By varying the optical power from 4 to 2 mW, one could vary the optical trap depth and, therefore, the density of the condensate by an order of magnitude. At 2 mW, the atoms began to spill out of the condensate, indicating a mean field energy equal to the optical well depth of 200 nK at a density of 3×10^{14} atoms/cm^3.

The ability of the dipole trap to confine in all three $F_s = 1$ spin states was confirmed in a beautifully simple Stern–Gerlach type experiment, in which one physically separates different spin states with an applied magnetic field gradient, as will be explained later. This makes possible many multicomponent spin condensate experiments. It was generally recognized that this achievement opened a whole new range of possible experiments on ultracold atoms and condensates.

Since the spins are no longer constrained, the trapping study of spin waves may be possible in a BE condensate. See references in Stamper-Kurn *et al.* [559]* Feshbach resonance in strong-field-seeking states also might be possible. See also a thorough review of "Optically Confined BE Condensates" [560]. The shallow, well-controlled depth of optical traps should make it possible to devise new output couplers for atom lasers. Finally, the authors mention the possibility of using the trap much as a conventional optical tweezers, for moving condensates in space and for placing them into microwave cavities or close to surfaces.

I found this turn of events very gratifying. This work confirmed my long-held opinion that optical traps such as the dipole trap, using all-optical forces, would prove to be superior to magnetic traps for most atom manipulation. I consider the magnetic quadrupole traps used for evaporative cooling the be real "kluges". Great pains were taken to make them stable for low-field seeking atoms by avoiding the zero field "hole", at considerable sacrifice in trap depth. The much-used MOT-type trap is a hybrid trap, part optical and part magnetic. I suspect that it could be replaced by an equally deep alternating beam optical trap, as suggested by Ashkin in 1984 and 1985 [142]*, [143]*. See also the discussion on alternating beam traps in Sec. 5.1.5. The single-beam optical dipole trap, which some people thought was limited to trapping ~ 1000 atoms, is capable of initially trapping $\sim 10^8$ atoms, as we now know. Also, as we recently learned, BEC has been achieved in far-off-resonance all-optical traps (see Sec. 20.5). Although the number of condensate atoms produced is considerably less than with magnetic traps, I believe that condensates and atom lasers ultimately will be based on optical traps and not on magnetic traps. Indeed, we shall see, in the following discussions of the most recent work, that the number of manipulations in atomic physics and BE condensates using optical traps is continuing to increase sharply. It remains to be seen if MOT-type or magnetic traps will continue to play any important role in the generation and use of BE condensates.

CHAPTER

18

Spinor Condensates in Optical Dipole Traps

The demonstration of the ability of dipole traps to confine Bose–Einstein (BE) condensates free of any constraints on the magnetic field has led to the observation of some remarkable phenomena using ground state spin domains. The MIT group calls condensates in which the spin is free to assume any orientation "spinor condensates".

18.1. Dynamics of Formation

Stenger *et al.* in 1998 [561]* were able to observe the dynamics of formation of the three possible ground state spin domains of sodium, over times of many seconds, in a dipole trap placed in an applied field of ~ 40 G. Arbitrary distributions of atoms in the three hyperfine states of an $F = 1$ angular momentum atom were prepared by sweeping through the rf transitions. These populations were then followed for times as long as 30 s, until they reached their final equilibrium distribution. Starting, for example, with the atoms all in the $m_F = 0$, or in equal fractions of $m_F = 1$ and -1, the same final distribution was produced. The distributions were observed in space and as a function of time by letting the condensate expand in space in the presence of an applied Stern–Gerlach field gradient, which spatially separates the spin components. Because the original condensate was 60 times as long as it was wide, the expansion was essentially radial and could be resolved into a one-dimensional (1D) distribution. Data taken at different applied fields led to different equilibrium distributions of states with different degrees of spatial overlap or miscibility. These equilibrium distributions, showing the amount of miscibility, can be calculated, as the authors show, by optimizing the free energy of the system under conditions of conservation of total spin.

18.2. Metastable Excited Spin States

In another study of spinor BE condensates, Miesner *et al.* of Ketterle's group [562] observed two new types of long-lived metastable excited spin states. In one such state, a two-component system

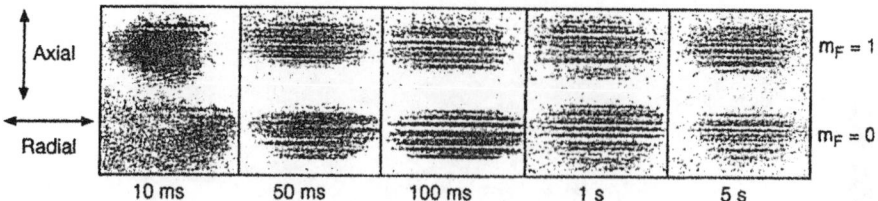

Fig. 18.1. Spontaneous formation of metastable states in a 15 G bias field. Condensates were probed at various times after an overlapping $|m_F = 1, 0\rangle$ mixed condensate was prepared. The $|m_F = 1\rangle$ (top) and $|m_F = 0\rangle$ (bottom) states were separated during their free expansion. Striations reflect the presence of alternating $\approx 40 \, \mu$m axial spin domains which form 50 ms after state preparation. Only a 200 μm slice of the cloud was imaged, by using a narrow optical pumping beam, to improve resolution. Arrows indicate the axial and radial directions in the optical trap. The height of each image is 1.3 mm. Figure and caption from Ref. [562].

involving $m_F = 1$ and 0 condensates, the hyperfine states were stable in spin composition due to immiscibility, but spontaneously formed a metastable array of spatial striations or domains with the $m_F = +1$ and 0 populations. See Fig. 18.1, taken from Fig. 1 of Ref. [562]. The figure shows the metastable arrangement of alternating spin domains in the two-component condensate, which they call a "quantum bubble", and how it separates under the influence of a magnetic field gradient applied after expanding in free space. In a second metastable state, a single component condensate initially in the $m_F = 0$ state was stable in space, but was metastable in its hyperfine spin composition against decomposing into $m_F = +1$ and -1 ground spin composition.

In both cases, the barriers to relaxation of these metastable states was only ~ 0.1 nK, much less than the condensate temperature of 100 nK. This would suggest a rapid thermal relaxation, but the authors ascribe the observed long lifetimes of at least 20 s for these metastable states to the low number of available uncondensed thermal atoms to drive this relaxation. The fraction of uncondensed atoms in dipole traps is very much lower (less than 1%) than in magnetic quadrupole traps. In this sense, the dipole condensates are much purer than magnetically held condensates, a significant advantage in many experimental situations. This comes from the fact that optical dipole traps have a relatively small volume. The authors find that the metastable condensates persist even when they apply external magnetic gradients that energetically favored their rearrangement. There is evidence that quantum tunneling plays a role in these condensates. See an excellent review of spinor condensates by Stamper-Kurn and Ketterle [563] for more details.

18.3. Optical Tunneling of Trapped Spinor States

In a separate experiment reported by Stamper-Kurn *et al.* in 1999 [564], the MIT group studied quantum tunneling in their optically trapped spinor condensates. The condensates in optical dipole traps are long and thin ($\sim 4 \, \mu$m in diameter by $\sim 100 \, \mu$m in length) and behave essentially as 1D systems.

The group used chirped pulses to set up an $F = 1$ spinor condensate consisting of equal populations of $|m_F = 0\rangle$ and $|m_F = 1\rangle$ states. The two components were separated into two domains by the application of a strong magnetic gradient along the axis on top of a fixed bias field. The condensates were then placed in a metastable state by applying a magnetic field gradient in the opposite direction to

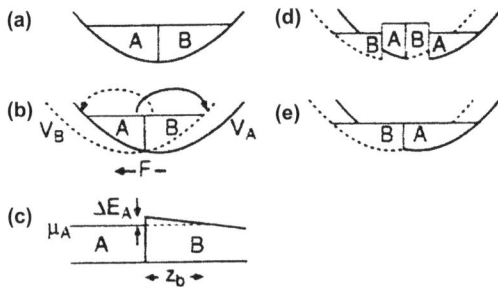

Fig. 18.2. Metastable spin domains and the energy barrier for decay. (a) The ground state of a two-component condensate consists of two phase-separated domains. (b) A state-selective force F displaces the trap potential V_B from V_A, creating metastable spin domains. Atoms tunnel form the metastable spin domains (direction of arrows) through an energy barrier (c) of maximum height ΔE_A and width $z_b = \Delta E_A/F$ (shown for component A). (d) Tunneling proceeds from the metastable domains (inner) to the ground state domains (outer) until (e) the condensate has completely relaxed to the ground state. Figure 1 and caption from Ref. [564].

the initially separated components. Figure 18.2, taken from Fig. 1 of Ref. [564], shows the separated condensate and the effects of the magnetic field gradients on the system. Part (c) of the figure shows the potential barrier through which the condensates must tunnel to return to the original two phase-separated domains. The condensates were probed by time-of-flight absorption imaging combined with Stern–Gerlach spin separation, as described above. All the variables were adjusted, such as the width of the energy barrier, the height of the barrier, the temperature, and the density of the condensates, leaving little doubt that the observed decay of metastable spin systems is due to quantum tunneling of a two-component system. The experimental results agree only qualitatively with simple two-state tunneling models. At low values of magnetic bias field, some new behaviors were observed. These were dramatic increases in the tunneling rate that were inconsistent with a simple two-component system. The authors believe this rapid relaxation was due to the formation of $m_F = -1$ atoms at the spin domain boundary, where its production is energetically favored at low fields of less than 1 G. Presence of small fractions of $m_F = -1$ atoms would weaken the repulsion between the domains at the domain walls. A description of this experiment requires a theory based on more than just a pair of coupled Gross–Pitaevskii equations, using just two different internal states.

I remind the reader and these authors that the tunneling of macroscopic particles studied with a pair of optical tweezer traps gave excellent agreement with Kramers' tunneling theory (see Sec. 10.16).

CHAPTER

19

Feshbach Resonances

As we discussed earlier in connection with the work on cold atomic collisions and Bose–Einstein (BE) condensates in gases, all the essential properties of these ultracold atomic gases are determined by a single parameter, the scattering length "a", used to describe the strength of the atomic interactions. For atom–atom interactions or collisions that are repulsive the scattering length "a" is positive. If they are attractive, then "a" is negative. If $a = 0$, the interaction is zero. With condensates, all the basic properties, such as the formation dynamics, the collective modes of oscillation, behavior of two-component systems, amplification of matter waves, nonlinear four-wave mixing behavior, etc., depend on the strength of the interaction. If one can modify the interaction by applying an external magnetic field, light, or rf fields, one can modify all of these properties. However, strong variations in the scattering length of atoms are only large near "Feshbach" resonances, where a quasi-bound molecular state formed by two colliding atoms has nearly zero energy (at the so-called Condon point) and can couple strongly to the free state of atoms. Theoretical calculations suggested that magnetic field modifications of the interaction, by affecting the hyperfine levels, were most likely to succeed. There are difficulties in studying Feshbach resonances in magnetic traps, since only weak-field-seeking hyperfine states can be trapped in the first place and any variation in magnetic field strength also affects the trapping potential.

19.1. Magnetic Tuning of the Scattering Length in a Dipole Trap

Inouye *et al.* in Ketterle's group, in 1998 [565]*, avoided these problems with magnetic traps by using condensates confined in optical dipole traps where one can apply an arbitrary magnetic field. Indeed, they succeeded in optically trapping two strong-field-seeking states, which, of course, cannot be trapped in quadrupole magnetic traps, and then tuned the scattering length by a factor of 10 as they scanned across the resonance field. This work was the first experimental observation of Feshbach resonances.

Starting with BE condensates in the $|F = 1, m_F = -1\rangle$ state as before in a dc magnetic trap [559]*, the condensates were transferred into an optical dipole trap. Once in the dipole trap, atoms were then spin flipped with rf into the $m_F = +1$ strong-field-seeking state. Since the authors used a bias magnetic field with a large gradient that gave a repulsive magnetic force, they added a pair of far-off-resonance blue-detuned end cap beams to prevent trap loss from the dipole trap. This gave an approximately W-shaped axial potential. The condensate formed adjacent to one of the end caps, as observed by nondestructive phase-contrast imaging. Feshbach resonances were then located by sweeping the magnetic field and looking for an enhanced rate of inelastic collisions. This could be caused by atom loss due to atom collapse in the region of negative scattering length or an enhanced rate of three-body inelastic collisions. Measurements of scattering length and number of atoms in the condensate were made by suddenly turning off the trap and allowing the stored kinetic energy of the condensate to convert into a freely expanding cloud. Using time-of-flight observations, one can determine both the interaction energy, which is proportional to the scattering length "a", and also the number of atoms N in the condensate. Figure 19.1, taken from Ref. [565]*, shows the results for

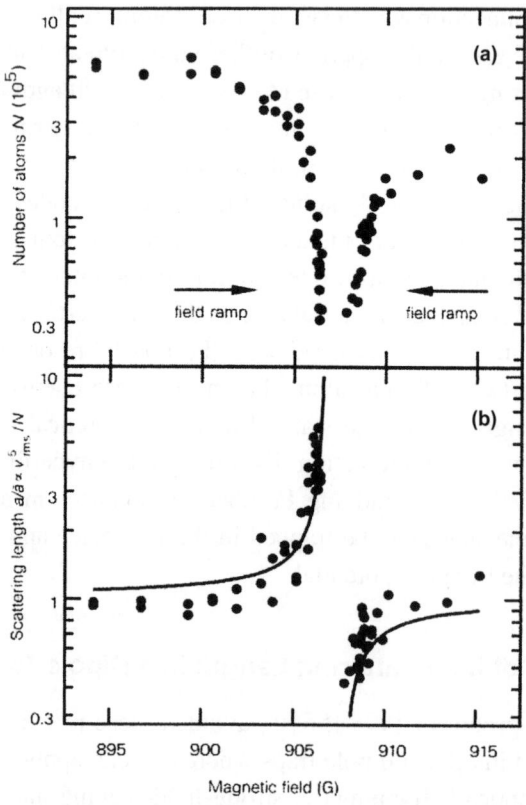

Fig. 19.1. Observation of the Feshbach resonance at 907 G using time-of-flight absorption imaging. (a) Number of atoms in the condensate versus magnetic field. Field values above the resonance were reached by quickly crossing the resonance from below and then slowly approaching from above. (b) The normalized scattering length $a/\bar{a} \propto V_{rms}^5/N$ calculated from the released energy, together with the predicted shape (Eq. (1), solid line). The values of the magnetic field in the upper scan relative to the lower one have an uncertainty of < 0.5 G. Figure 2 and caption from Ref. [565]*.

a resonance near 907 G in sodium. The resonant onset of trap loss and the dispersive variation of the scattering length are clearly seen in this figure. Some of the assumptions of the data analysis were not totally valid, as the magnetic field was ramped up and down during measurement. Therefore, more quantitative analysis is necessary in later work. Also, to work close to resonance will require more precise control of the bias field.

The authors concluded that tuning the scattering length will become an important tool for designing condensates with novel properties and for studying new phenomena, such as: (i) the collapse of condensates in the Fermion regime or negative scattering length regime; (ii) creating low-interacting condensates near zero scattering length; (iii) varying the interaction between different species and thereby controlling the phase diagram of multicomponent condensates; (iv) controlling nonlinear coefficients in atom optics using coherent beams of atoms.

19.2. Magnetic Tuning in Photoassociative Spectroscopy

At just about the same time, in 1998, Heinzen's group at the University of Texas [566]* also observed Feshbach resonances by studying cold atom collisions using photoassociation spectroscopy with a magnetically tunable far-off-resonance optical dipole trap. Starting with about 10^4 ^{85}Rb atoms in the dipole trap, at a temperature of about $60\,\mu$K, the authors applied a magnetic field and optically pumped the atoms into the $f = 2$ hyperfine ground state. They then applied a tunable probe laser beam that induced photoassociative transitions to a quasi-bound state of the excited molecule. After another 1000 ms or so, they switched off all laser beams and the magnetic field and detected the number of atoms remaining in the trap by laser-induced fluorescence. Photoassociative absorption usually results in loss of atoms from the trap because the excited atoms return to the ground state by spontaneous emission, and the excess kinetic energy is too great to stay in the trap. One detects the Feshbach resonance by plotting the photoassociative signal versus magnetic field, as seen in Fig. 19.2, taken from Fig. 4 of Ref. [566]*. Figure 19.3, taken from Fig. 5 of Ref. [566]*, shows the calculated magnetic field dependence of the scattering length $a_{2,-2}$, for collisions of ^{85}Rb ($f = 2, m_f = -2$) atoms. It clearly displays the expected dispersive behavior near resonance. The ^{85}Rb ($f = 2, m_f = -2$) atoms are expected to exhibit a very low two-body elastic collision rate. The authors point out that this method of Zeeman associated photoassociative spectroscopy is useful in searching for resonances with any hyperfine levels. They also suggest the possibility of ^{85}Rb and ^{87}Rb mixed condensates with a positive cross-species scattering length and a tunable ^{85}Rb scattering length.

19.3. Feshbach Resonance of Ground State Cesium at Low Magnetic Field

Later in 1999, following the MIT work on tunable Feshbach resonances in sodium [565]* and that done by the Texas group under Heinzen on rubidium [566]*, Vuletić *et al.*, in Chu's group at Stanford, studied Feshbach resonances in optical traps in cesium [567]. Although one can obtain the lowest temperature optical molasses in cesium, at this point, using magnetic traps, it resisted BEC because of extraordinarily high inelastic collision rates [568], [569]. Vuletić *et al.* in this paper point out the advantages of using the lowest energy ground state $F = 3$, $m_F = 3$, which is stable against inelastic decay collisions. Using the magnetic tunability of Feshbach resonances with optical traps,

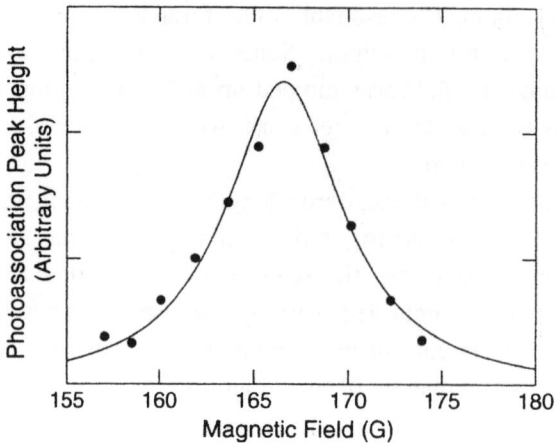

Fig. 19.2. Height of the photoassociation peaks shown in Fig. 3, as a function of magnetic field, showing clearly the Feshbach resonance. The solid curve shows a Lorentzian fit to the data. Figure 4 and caption from Ref. [566]*.

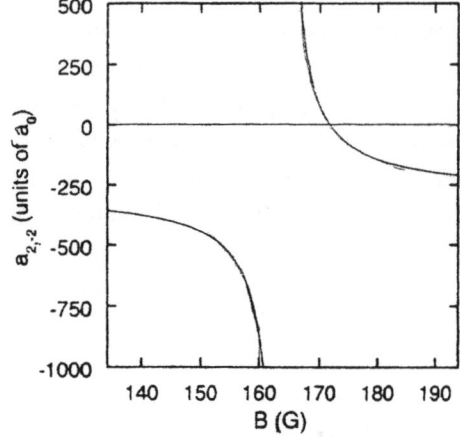

Fig. 19.3. Calculated field dependence of scattering length a_{2-2}, corresponding with the resonance field value and width observed in this experiment. Figure 5 and caption from Ref. [566]*.

they succeeded in locating a strong $F = 3$, $m_F = 3$ Feshbach resonance at the low magnetic field of 17 G. This resonance can presumably be magnetically tuned to a region of positive scattering length. Two more relatively low-field Feshbach resonances for $F = 3$, $m_F = -3$ were found by means of their strong increases in inelastic decay rate, one at 31 G and a weaker one at 34 G.

19.4. Elastic and Inelastic Collisions Near Feshbach Resonances in Sodium

A paper by the MIT group in 1999 by Stenger *et al.* [570] on the properties of BE condensates of sodium held in optical traps, showed that inelastic collisions are strongly enhanced near Feshbach resonances in magnetically tuned condensates. The rate of inelastic losses increased when tuned to positive or negative scattering lengths. This, they believe, will severely restrict experiments using the tunability of the scattering length of Feshbach resonances. The origin of these inelastic processes in high-density condensates is not fully understood and hints at new, many-body molecular events not accounted for by previous theory. It is a whole new regime that needs more careful study.

The authors used essentially the same technique as in their earlier work on the transfer of condensates from magnetic to optical traps [565]*. In the course of the work they discovered a new $F = 1$, $m_F = -1$ Feshbach resonance at 1195 G in sodium. This resonance may be useful for experiments with zero or negative scattering length, since it can be directly approached without crossing any other high-loss resonance. The losses in the high-density condensates can be as high as 70% of all the atoms in 1 μs in the strong 907 G resonance. In the 400 times weaker 853 G resonance, 70% of the atoms are lost in 400 μs. These measurements imply that one must reduce the atom density to below 10^{14} atoms/cm^3 to exploit the tunability of the scattering length near Feshbach resonances.

19.5. Suppression of Collision Loss in Cesium Near Feshbach Resonances

In 1999, Chu's group, Vuletić *et al.* [571] used a far-detuned optical dipole trap to make some interesting observations on suppression of atomic radiative collisions in cesium by magnetically tuning the ground state scattering length near a Feshbach resonance. Starting with a one-dimensional (1D) far-detuned optical lattice made from a 17 W Nd:YAG laser operating at 1.06 μm, they trapped 10^7 cesium atoms in 4000 1D pancake standing wave traps at a density of about 3×10^{12} atoms/cm^3, at a temperature of 4 μK (see also Ref. [546]*). The atoms were optically pumped to the lowest energy ground state $F = m_F = 3$ in 5 ms and an adjustable magnetic field was applied for tuning its s-wave scattering length. Radiative collisions were induced by applying an independently tunable loss laser of 100 mW to the trapped cloud. It was discovered, when the loss laser was detuned from the atomic resonance in a range from 60 GHz to 60 THz to the blue of the atomic resonance, that radiative trap loss was suppressed by factors up to 15 at the magnetic field near to a Feshbach resonance. The authors have devised a model that predicts suppressed collision loss when the blue-detuned laser is in resonance at the Condon point. With this method, they are able to find the magnitude and sign of the scattering length. The magnetic fields were determined where minimum loss occurs for different loss laser detuning, i.e., for different Condon points. See Fig. 19.4, which is taken from Fig. 2 of Ref. [571].

They also discovered, as seen in Fig. 19.4, a number of very narrow resonances with high radiative loss superimposed on the broader variations of loss near the Feshbach resonance. These are thought to arise from coupling to molecular states and give rise to the formation of molecules. If these sharp molecular states could be identified, it could lead to the determination of cesium ground state parameters with unprecedented accuracy.

Aside from the spectroscopic implications, the ability to control collisions could prove to be useful in the production of BE condensates by evaporative cooling, where elastic collisions are necessary for rapid thermalization. On the other hand, suppression of inelastic collisions is desirable for maintaining high-density ultracold samples.

19.6. Discovery of New Low-Field Feshbach Resonances by High-Resolution Spectroscopy

In 2000 an important advance in the study of Feshbach resonances in collisions between ultracold atoms was made by the Stanford group [572] on the high-resolution Feshbach spectroscopy of

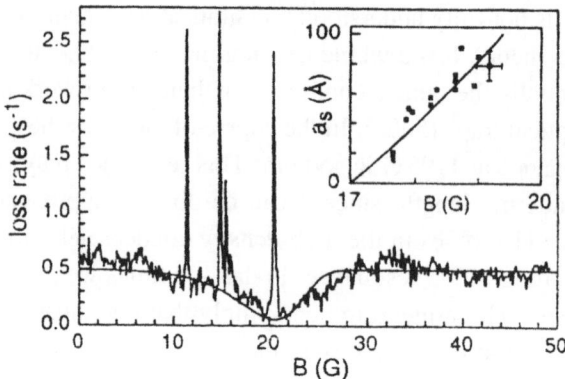

Fig. 19.4. Radiative loss rate for a detuning of $2\pi \times 32\,\text{GHz}$ and an average intensity of $26\,\text{mW/cm}^2$. The gas temperature and peak density are $5.0\,\mu\text{K}$ and $3 \times 10^{12}\,\text{cm}^{-3}$, respectively. The broad feature corresponds to the suppression of light-assisted collisions when a_s is positive and close to the Condon point R_C. Superimposed are very narrow resonances with high radiative loss. Figure 2 and caption from Ref. [571].

cesium. In a careful series of measurements, they observed the high-resolution resonances of more than 25 resonances in low magnetic field regions up to $230\,\text{G}$ with an accuracy of about $\pm 0.03\,\text{G}$. These cold scattering measurements are essentially a spectroscopic technique for measuring the energies of the weakly bound molecular states of cesium. By probing the collisions of atoms in different hyperfine and magnetic sublevels it is possible to determine for the first time a detailed picture of the long-range interaction parameters describing the cold collision properties of ^{133}Cs. Theorists at NIST, in an accompanying paper by Leo *et al.* [573], have used these data to extract singlet and triplet scattering lengths and the van der Waals potential coefficient, using what they call a coupled channel approach. The results differ significantly from earlier, less complete work in both sign and magnitude of some scattering lengths. A change of sign, of course, implies the difference between an attractive and a repulsive interaction. The authors point out that the inability to determine the sign of the scattering length stems from insufficient knowledge of the magnitude of the van der Waals force.

The authors [572] use their same basic experimental setup for their other recent experiments in Feshbach resonances and high density optical trapping [546]*, [567], [571]. They loaded up to 10^8 atoms into their 1D lattice, using $2\,\text{ms}$ long periods of three-dimensional (3D) degenerate Raman sideband cooling [574]. The axial and radial temperatures were typically between 2 and $6\,\mu\text{K}$. There was negligible heating from the trap, since the scattering rate was only 0.2 photons per second with the $1.06\,\mu\text{m}$ wavelength trap. Peak trap densities were nearly $10^{13}\,\text{atoms/cm}^3$.

The authors have developed a novel technique, based on evaporative cooling, for accurately measuring the scattering length of states connected to the lowest magnetic sublevel $F = 3, m_F = 3$. They set the trap depth at a low level, from which atoms can escape, and measure the fraction of atoms remaining, after a given cooling time, as a function of magnetic field B. At the resonance field the elastic scattering reaches a maximum, the fraction remaining reaches a minimum, and the final temperature reaches a minimum. In Fig. 19.5, taken from Fig. 1 of Ref. [572], showing this minimum, the temperature fell to $1.1\,\mu\text{K}$ at the narrow resonance field of $48.017 \pm 0.030 \pm 0.026\,\text{G}$.

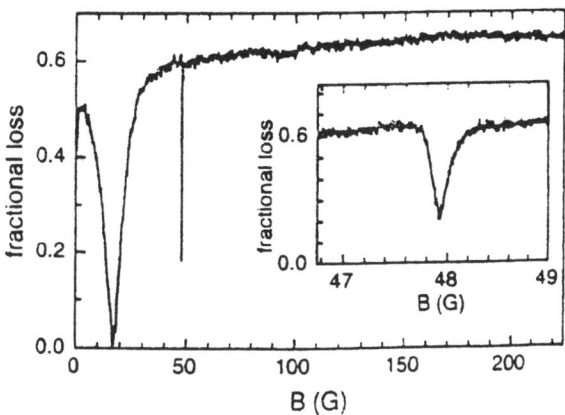

Fig. 19.5. Evaporative loss for atoms in the $|F = 3, m_F = 3\rangle$ state in 300 ms with a final temperature of 1.1 μK. The evaporative loss is smallest when the elastic cross section has a minimum. The inset shows an expanded view of the resonance near 48 G. Figure 1 and caption from Ref. [572].

The ± 0.030 is the estimated systematic error and the ± 0.026 is the statistical error. Statistical errors as low as ± 0.026 G have been observed with very narrow lines, using this technique.

The authors foresee applications of this work to the understanding of density-dependent frequency shifts in atomic clocks. Of even more interest is the possibility of achieving BEC in cesium. The analysis of Leo *et al.* [573] shows that the $F = 3, m_F = -3$ states near resonances at 70, 110, and 130 G are positive and inelastic collision losses are relatively small. They are thus candidates for successful evaporative cooling. There are possible problems with the ratio R of the thermalization rate to the inelastic rate. The good news is that the $F = 3, m_F = 3$ state, which is the lowest energy magnetic state that experiences no binary collision loss, is a good candidate for BEC with an optical dipole trap magnetically tuned near 17 G. The theorists state their conclusions about achieving BEC more conservatively by saying that, in contrast to previous work on cesium, BEC cannot be ruled out.

19.7. Observation of Optically Induced Feshbach Resonances in Collisions of Cold Atoms

In 2000, Fatemi *et al.* [575] succeeded in demonstrating optically induced Feshbach resonances in the collisions of cold sodium atoms. This first demonstration of this new technique offers a possibly useful alternative to magnetically tuned Feshbach resonances. Although optically induced resonances were originally proposed in 1996 (see references in Ref. [575]), it was thought that magnetically induced resonances were easier to implement. Fatemi *et al.* found it was possible to optically vary the scattering properties of atomic collisions by either changing the intensity or detuning of a laser near a photoassociative state of a dimer. The effects of varying the frequency and intensity of a strong applied laser field near the Condon point were monitored by looking at the changes in a probe ion signal induced by the combined application of a weak probe laser tuned slightly above the Condon point and a second ionizing laser. Very clear dispersive ion probe signals were measured, which could be fitted very well by adapting standard Feshbach resonance theory.

The authors point out that a potential advantage of using optical Feshbach resonances is the ability to turn on the inducing beam and adjust the frequency and intensity at will. This avoids the problems one can have with magnetic field tuning where one has to sweep the field through a lossy region to reach the "far side" of a desirable resonance [570]. Another novel capability of light is that one can vary the scattering length of different parts of a condensate by using nonuniform light intensity. Yet another advantage of optical resonances mentioned by the authors is that one can reduce spontaneous emission losses by detuning farther from resonance and, at the same time, increasing the intensity.

CHAPTER

Recent Work on Bose–Einstein Condensation

20.1. Diffraction of a Released Bose–Einstein Condensate by a Pulsed Standing Light Wave

Ovchinnikov *et al.*, in 1999, in a quite international collaboration in Phillips's group at NIST [576], performed the first experiment on the diffraction of coherent atoms from a Bose–Einstein (BE) condensate by a pulsed optical standing wave. This experiment culminates more than ten years of work on the interaction of atomic beams with continuous standing waves, dating back to experiments in Pritchard's group in the thin grating limit. For thin gratings the atoms make very small displacements during the time they traverse the standing wave at normal incidence (see Ref. 5 in Ref. [576]). This new condensate experiment, performed in the thick grating limit, has many advantages over past work, such as a very narrow high-brightness source, whose variation in incident momentum is much less than the recoil of a single photon. This makes it possible, for the first time, to experimentally resolve the momentum of the various orders of the atoms diffracted off the standing wave grating. In this regime the interaction time of the atoms with the optical grating can be longer than the vibrational period of the atoms in the standing waves. One is now able to see periodic growth and collapses of new momentum components in time, as the atoms oscillate inside the standing waves.

In previous related experiments on lithography using a similar standing wave geometry, as discussed previously in Sec. 15.2.6, Timp *et al.* [469] and McClelland *et al.* [470] had a much less uniform source. They therefore suffered considerable chromatic aberration, which degraded their resolution and made comparison of their results with theory very difficult [577].

Salomon *et al.* in 1987 [578] and Ovchinnikov and Letokhov in 1992 [579], in other earlier experiments with this geometry, were able to observe channeling along the one-dimensional (1D) standing waves, but were not able to resolve diffracted momentum peaks. In addition to chromatic aberration, these experiments suffered from dissipation due to spontaneous emission.

| 0.3 μs | 0.5 μs | 0.8 μs | 1.2 μs | 2.0 μs | 2.8 μs | 3.0 μs |

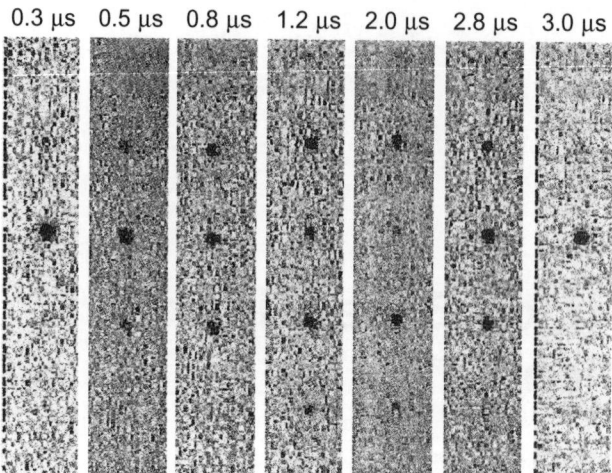

Fig. 20.1. Spatial distribution of the optical depth of the BEC diffracted by a standing wave pulse of various durations. The corresponding duration of the pulse is shown above each individual distribution. Figure 1 and caption from Ref. [576].

In the experiment of Ovchinnikov *et al.* [576] the experimental sample of about 10^6 sodium atoms was produced in a TOP magnetic trap in about 30 s. The spring constant of the trap was adiabatically reduced by expanding the condensate to about $60\,\mu$m in order to reduce the momentum spread of the condensate. The atoms were suddenly released and introduced into a red-detuned 1 mW standing wave pulsed beam with $w_0 \cong 2.8$ mm. The spatial distribution of the atoms was observed, after interacting with the light for varying lengths of time, using standard absorption imaging. Figure 20.1, taken from Fig. 1 of Ref. [576], shows diffraction into the $n = \pm 1$ orders, separated in momentum space by $4(h/2\pi)k$. The number of atoms diffracted out of the zero order into the $n = \pm 1$ grows with time, and after $3\,\mu$s returns again to the nearly undiffracted distribution. Oscillations with this period were seen out to $12\,\mu$s, with decaying amplitude. The authors explain this decay as due to continuous excitation to the excited state, with random decay by spontaneous emission.

It is suggested that the observed diffraction would be useful for atomic beam splitters, temporal atom interferometers, and examining the limits of atom lithography in the absence of chromatic aberration.

As an aside I mention that the 1D confinement of atoms, using the far-off-resonance optical dipole force in this standing wave geometry, was proposed by Letokhov as early as 1968 as a way of overcoming the Doppler shift. He essentially invoked Dicke narrowing [580] in one dimension. Unfortunately, in the experiment he proposed the number of slow atoms in the low velocity tail of a hot gas, entering at grazing incidence to the standing wave with low enough transverse velocity to be channeled in one dimension and execute more than one oscillation, is truly infinitesimal. He gave no indication of how one can improve the situation to make the proposal viable. This proposal, in my opinion, was just wishful thinking. Letokhov nevertheless has continued to imply over the years that this 1968 proposal marks the beginning of the field of optical atom trapping. The field of atom trapping and cooling really got its start with the independent discovery by Ashkin [1]*, [2]* of stable three-dimensional (3D) optical trapping and his proposal for stable atom trapping, using both scattering and dipole forces, and the proposal of Hänsch and Schawlow [26]* for molasses-type cooling.

20.2. Collective Collapse in a Bose–Einstein Condensate with Attractive Interactions

Interest in BE condensates of atoms with attractive interactions has grown steadily since the original work of Hulet's group at Rice University demonstrating the possibility of condensates with negative scattering lengths [204], [512]. Attractive interactions previously were thought to make a condensate unstable and prevent condensation. However, Hulet and collaborators showed that such condensates could exist metastably as long as the number of atoms N_0 remains small [513]*. The attractive interactions give the atom cloud a negative compressibility that drives it toward implosive collapse. In a trap this is opposed only by the zero-point kinetic energy. For ^{7}Li atoms with a scattering length $a = -1.46$ nm and a condensate volume of ℓ^3, where $\ell \approx 3\,\mu$m, one finds, by numerically solving the nonlinear Schrödinger equation, that the maximum numbers of atoms N_m in the trap at the stability limit is $N_m \approx 1250$ atoms Ref. [581].

These BE condensates with attractive interactions are expected to exhibit complex physical properties, such as solitons, macroscopic quantum tunneling, and complex dynamical behavior as the number of atoms N_0 approaches the stability limit (see Refs. 4–8 of Ref. [581]).

In the experiment performed in 1999 by Hulet's group [581], the authors examine in considerable detail the dynamics of the collective collapse in a BE condensate with attractive interactions in ^{7}Li atoms. The theory predicts that as the gas with a large number of atoms is cooled below the critical temperature for BEC, the number of condensed atoms in the trap N_0 will increase until N_m, the maximum number of atoms in the trap at the stability limit, is reached, and then collapse spontaneously to zero, ejecting the N_m atoms. The process repeats itself successively over time, only more slowly for as long as the number of atoms in the trap is greater than N_m. Finally, after some tens of seconds and about 15 such cycles, the gas comes to equilibrium with $N_0 < N_m$. The experimental setup was basically the same as before. About 10^8 laser-cooled ^{7}Li atoms were loaded into a permanent magnetic trap and evaporatively cooled, using a microwave frequency field, into the degenerate regime, giving a sample of $\sim 4 \times 10^5$ atoms at $T \approx 400$ nK. The number of atoms was reduced to $\sim 4 \times 10^4$ by cutting off the high-energy end of the energy distribution. This helped in distinguishing the condensate from the noncondensed cloud of atoms.

The authors experimentally probed the evolving condensate system after specific lengths of time t, varying from 5 to 90 s, by measuring the number of condensed atoms N_0 in the trap. A histogram showing the number of occurrences of atoms of different N_0 for different times was compared with the numerical solutions for N_0 vs time from the nonlinear Schrödinger equation. The results of this comparison were consistent with the model of condensate growth and collapse driven by an excess of noncondensed atoms.

20.3. ^{85}Rb Bose–Einstein Condensates with Magnetically Tunable Interactions

In 2000, Roberts *et al.* in Wieman's group [582] and Cornish *et al.*, working with Cornell and Wieman [583], studied the magnetic field dependence of ultracold inelastic collisions near a Feshbach resonance in ^{85}Rb. Roberts *et al.* systematically measured the inelastic collision rates in the $F = 2$, $m_F = -2$ state as a function of magnetic field, over a range of fields from ~ 100 to 250 G, and found the two- and three-body collisional loss rates. They found that as the magnetic field was reduced from 250 G toward a Feshbach resonance at 155 G the inelastic collisions decreased to a minimum and then increased dramatically, peaking at the resonance. The results agreed well with theory and are important for making a tunable BE condensate in ^{85}Rb. The measurements were extended into the negative scattering length region at the magnetic field below resonance. Order of magnitude changes in the two- and three-body inelastic rates were observed near resonance. They varied in detail with magnetic field near resonance, so that the total loss was a complicated mixture of the two processes.

Cornish *et al.* used the above data to devise a fairly complex evaporative cooling procedure to successfully reach BEC in ^{85}Rb. They successfully produced pure condensates up to 10^4 atoms with peak densities up to 1×10^{12} atoms/cm^3 at temperatures of 15 nK. The lifetime at 162 G was about ten seconds, which is quite long. This was the result of moving to high magnetic field, away from the Feshbach resonance at 155 G, where losses would be much higher.

Stenger *et al.*, working with sodium, had reported anomalously high inelastic loss rates when their condensate was swept rapidly through the Feshbach resonance [570] (see also Sec. 18.4). Cornish *et al.* [583] found they could sweep rapidly across the resonance and not lose many atoms. The situation with ^{85}Rb is thus much better. The authors could sweep controllably from the positive scattering length condensate with a repulsive interaction to a negative scattering length where the interaction was attractive. As they did so, they saw the condensate shrink in size and emit a burst of high-energy atoms. This was quite analogous to the behavior seen by Sackett *et al.* in Hulet's group, using ^{7}Li atoms. The Feshbach technique, however, has more reproducibility, making it easier to study the dynamic response of the collapse process. One is not cooling and collapsing simultaneously in the case of ^{85}Rb.

Using the magnetic field to control the magnitude of the repulsive scattering length "a", they were able to continuously control the width of the condensate by a factor of about four for a 10 G change.

There is an interesting discussion of this work in the "Search and Discovery" section of Physics Today, entitled "Researchers Can Now Vary the Atomic Interactions in BE Condensates", by B. Goss Levi [584]. In spite of the euphoric tone of the comments of some of the researchers in this commentary, based on work with magnetic traps, the ultimate control at a Feshbach resonance, of course, comes with optical trapping, where one can control beam size by controlling the trapping power independently of the magnetic field. As we know from the work of Ketterle's group in Refs. [559]*, [565]*, it is easily possible to transfer condensates directly into optical traps from magnetic traps.

20.4. Bose–Einstein Condensation in Metastable Helium Atoms

An important advance in the study of BEC in dilute gases occurred in 2001 with the observation of BEC in metastable helium gas by Robert *et al.* of Aspect's group in France [585], followed closely by a similar experiment by Dos Santos *et al.* of Cohen–Tannoudji's group in France [586]. Helium is normally inert when it is in its ground state. It becomes trappable when it is excited to its metastable state, where it can persist for approximately two hours. Robert *et al.* initially captured about 2×10^8 atoms in a cloverleaf-type MOT. After molasses cooling, optical pumping and magnetic compression, there were $\sim 2 \times 10^8$ atoms in the only trappable $m = +1$ hyperfine sublevel, at a temperature of 1 mK. The authors then applied a 60-second-long radio frequency evaporative cooling ramp to reach BEC, with $\sim 10^5$ atoms at a temperature of $\sim 0.7\,\mu$K. Each metastable helium atom carries about 20 eV of excitation energy above the electronic ground state. Any significant conversion of this internal energy to heat by Penning type of ionizing collisions could prevent condensation from occurring. This process was prevented by the almost total spin polarization of the trapped gas. The authors used the high internal energy of the metastable state to achieve essentially single atom detection with nanosecond time resolution, using a microchannel plate. They suggest that this opens the possibility of novel atom optics experiments, such as the detection of spatial interference between two independent condensates based on time and position detection of single atoms. By estimating the number of atoms in the condensate ($\sim 10^5$) and its size, they deduce the approximate scattering length "a". The result is quite a large number, $a = 20 \pm 10$ nm, which helps thermalization during the cooling process.

The authors hope to be able to cool ^{3}He fermion atoms to degeneracy by sympathetic cooling, using ^{4}He as a coolant gas. Finally, they point out that the upper-lying levels of triplet metastable He are radiatively coupled to the ground state. Laser excitation of these upper levels could lead to superradiance or possibly lasing at energies of more than 20 eV.

The experiment of Cohen–Tannoudji's group [586] proceeded in much the same manner as the one by Aspect's group. They achieved BEC with a somewhat higher number of 5×10^5 atoms and a temperature of $4.7\,\mu$K. Both groups deduced similar large s-wave scattering lengths of ~ 20 nm. With their higher density, their condensate is in the hydrodynamic regime where the collision rate is so high that it cannot be neglected, as in the usual condensates. In this regime, coupling can occur to the noncondensed thermal cloud. See commentary by Richard Fitzgerald on helium BEC in the "Search and Discovery" article in the May 2001 issue of Physics Today [587].

20.5. Observation of Bose–Einstein Condensation Using Optical Dipole Traps

Readers will recall that since the revival of interest in dipole traps for use in studies of cold atoms in atomic physics in the mid-1990s, there have been efforts to achieve BEC in optical dipole traps, thus totally avoiding use of magnetic traps [70], [139]*, [145]*, [146], [546]*. Some of these experiments were very sophisticated and yielded quite high phase-space densities, but they were still shy of BEC. Ketterle's group circumvented the problem of directly achieving BEC in dipole traps by simply transferring their BE condensates from magnetic traps into dipole traps [559]*. This yielded big results, such as the ability to make and study spinor condensates [561]*, [564] and the observation of the first Feshbach resonances [565]*.

Although I, personally, maintained my belief in the versatility of optical traps for manipulating neutral atoms, and felt it would only take some optimization to finally reach BEC, I always wondered whether there was some hidden flaw. Such a doubt can only be resolved by experimentation. Indeed, any doubts were relieved by the news of the success of experimenters in Chapman's group at Georgia Tech in achieving a BE condensate in ^{87}Rb in an optical dipole trap in May 2001.

In this experiment, Barrett *et al.* [205]* formed a dipole trap at the focus of two orthogonal very-far-off resonance CO_2 laser beams at $10.6\,\mu$m. This was placed at the center of a standard MOT, which, as usual, served to capture and precool the atoms. Turning off the MOT laser beams and the magnetic field transferred atoms to the dipole trap. This left the dipole trap filled and ready for evaporative cooling. Evaporation of hot atoms was accomplished by lowering the trapping power. It took only about 2.5 s of cooling to reach BEC at a temperature $T/T_c \cong 0.7$, where the critical temperature $T_c = 375$ nK. This was about 10 or more times faster than evaporative cooling from typical magnetic traps for ^{87}Rb. The final condensate contained $\sim 3.5 \times 10^4$ atoms, distributed among the $m_F = -1$, 0, +1 hyperfine states, thus demonstrating the insensitivity of optical traps to hyperfine states.

This experimental success was announced at the annual American Physical Society's meeting of the division of atomic, molecular and optical physics. Some researchers reacted to this announcement with excitement and wonderment. It is interesting to read the comments of various luminaries of the BEC field, as quoted by B. Goss Levi in her Physics Today account of the achievement of BEC in optical traps [588].

Carl Wieman remarked, concerning the optical technique, "It's very fast, it avoids a number of loss processes inherent in magnetic traps and — best of all — it looks downright easy". Wieman, who with colleagues at JILA and the University of Colorado made the first BE condensate, also in ^{87}Rb, took six minutes in their first experiment to evaporatively cool their sample, ending up with 2×10^4 in their TOP trap at their critical temperature T_c of ~ 170 nK, with a critical density of $\sim 2 \times 10^{12}$ atoms/cm^3. It seems that he was genuinely surprised by this new result.

B. Goss Levi also asks, "What is the secret?" of this success. The authors in their paper [205]* attribute their success in part to the very high densities of $\sim 2 \times 10^{14}$ atoms/cm^3 achieved initially in their dipole trap, before the onset of evaporation. This is the highest initial density achieved, by about a factor of 10, compared with previous dipole trap experiments, and very much more dense than what one can typically achieve in magnetic traps prior to evaporation. John Thomas of Duke University, who worked previously on dipole trapping with CO_2 lasers, attributed the high density achieved to the high ratio of trap depth to loading temperature and to the fact that CO_2 traps are dark traps that do not interact with the cooling beams of the MOT. David Weiss of Berkeley believed the use of ^{87}Rb was a very important factor, since it has a large elastic scattering rate favorable for evaporative cooling and a low inelastic scattering rate, which minimizes atom loss.

The "secret" could simply be a combination of all of these factors. I suspect, however, that the high density is the most important one. In previous attempts by Chu and colleagues to reach high phase-space density using evaporative and Raman cooling of atoms in optical traps [145]*, [540], it was basically low atom density that reduced the cooling rate to less than the trap loss rate. I also think the use by Barrett *et al.* of two crossed beams to form the dipole trap helps increase the density, because it makes a compact, nearly spherical trapping volume. In the future it seems likely that we will see improved performance in BEC formation using all-optical traps by simply increasing the density still further. If one increases the depth of the dipole trap by using higher CO_2 laser power, one would collect more of the MOT atoms and would further increase the initial density in the dipole trap. The cooling process would be improved because one then starts evaporative cooling from a higher energy. The fact that the first experiment yielded condensates with $\sim 2.5 \times 10^4$ atoms, as opposed to 2×10^7 condensate atoms from the best magnetic traps, does not present a fundamental difficulty. As Stenger *et al.* [559]* showed in their experiments transferring atoms from magnetic traps to dipole traps, dipole traps have no difficulty in holding as many as 10^7 atoms in a BE condensate.

There now seems little doubt that use of optical dipole traps for BEC represents a major simplification and improvement in technique over magnetic traps. Magnetic traps are currently fed by a complex double MOT arrangement consisting of an "upper MOT" that collects atoms in a high pressure cell and transfers them via a resonant light pulse, for example, to a second low pressure MOT, through a long, narrow guiding field with a hexapole magnetic guiding configuration. Differential pumping maintains the pressure difference between the two MOTs. It may take as many as 100 shots of atoms from the upper low pressure MOT to feed the second lower MOT. After further cooling in the low pressure MOT, the atoms are transferred to the purely magnetic trap, with some loss (~ 30–70%). The magnetic traps these days are mostly of the QUIC variety [33]* or of the cloverleaf variety, which are modifications of the Ioffe-Pritchard trap [509]. The QUIC trap, with its many compression and adjustment coils, and the original Ioffe-Pritchard trap use an array of as many as eight magnet coils to move, compress, and evaporatively cool the atomic cloud.

In 1998, after noting some of the many advantages of optical dipole traps for use in BEC studies, Stenger *et al.* [560] speculated on which trap, magnetic or optical, will prove to be the workhorse for studying nanokelvin atomic physics and BEC. Finally, in 2001, after the direct achievement of BEC in a dipole trap and after many experiments, where use of dipole traps was crucial to their success, it seemed clear to me that optical traps were emerging as superior, not only for forming condensates, but that they would also prove superior for making atom lasers, and in many other future applications.

20.6. Bose–Einstein Condensation of Potassium Atoms by Sympathetic Cooling

Modugno *et al.* in 2001, in Inguscio's laboratory at the University of Florence [589]*, reported the first experiments in which it was possible to observe the BEC of one atomic species (^{41}K, in this case) by sympathetic cooling with another evaporatively cooled species of atoms (^{87}Rb atoms). Sympathetic cooling of atoms in different hyperfine states of the same atoms has been observed previously [535]*, as it has been for two isotopes of the same species [590], [591], [592]. The authors used the conventional double MOT apparatus to trap both Rb and K atoms and transfer them into a quadruple Ioffe configuration (QUIC) [528]. After capturing Rb and K atoms in the first MOT they transferred $\sim 10^9$ Rb atoms over a 30 s period into the second MOT. This was followed by an eight second loading of $\sim 10^7$ atoms into the second MOT. Subsequent loading into the QUIC trap resulted in an initial population of 2×10^8 Rb atoms and 2×10^6 K atoms, with both at a temperature of $\sim 300\,\mu$K. Figure 20.2, taken from Fig. 1 of Ref. [589]*, shows the decline in temperature and number of atoms as the "microwave knife" frequency causing the evaporative cooling was lowered over a time span of more than 50 s. The final populations of $\sim 10^4$ atoms for Rb and K represent a loss of atoms by a factor of $\sim 2 \times 10^4$ for Rb and 2×10^2 for K, at a final temperature of about 200 nK. The authors think the large loss of K atoms during the cooling process was due to inelastic collisions within the K sample. It is possible to determine the various scattering lengths from the experimentally observed thermalization times.

The authors think that sympathetic cooling of the fermionic isotope ^{40}K is a natural extension of this work and can be a new way of studying the physics of atoms below the Fermi temperature

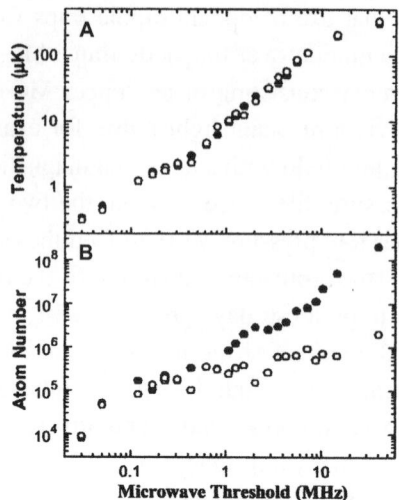

Fig. 20.2. Evolution of the number of atoms (A) and temperature (B) of the two atomic samples in the magnetic trap as a function of the microwave evaporation threshold of Rb. The solid circles correspond to ^{87}Rb and the open circles to ^{41}K. Figure 1 and caption from Ref. [589]*.

[35]*, [591]. The authors see implications for cooling molecules, and for atoms or molecules lacking a magnetic moment in their ground state (see references in Ref. [589]*).

Finally, use of sympathetic cooling is relevant in the context of the proposed cw optical atom laser, as discussed in Sec. 20.19.

20.7. Realization of Bose–Einstein Condensates in Lower Dimensions

Recent work with BE condensates has led to the realization that there are advantages to forming condensates with restricted dimensionality, commonly called quasi-condensates. Very long, thin, trapped condensates behave as 1D systems where motion is primarily restricted to one dimension. In flat disc-like condensates, the behavior is largely two dimensional (2D).

The first experimental realization of BEC in lower dimensions was reported by Görlitz et al. in Ketterle's group at MIT [593] with trapped condensates of sodium. Condensation in lower dimensions is impossible in homogeneous systems. The authors point out that to observe a 1D cigar-shaped quasi-condensate one needs $R_Z > \xi > R_\perp$, where R_Z and $R_\perp$ are the axial and transverse condensate half-length and radius, respectively. ξ is the so-called healing length and is equal to $(4\pi na)^{-1/2}$, where n is the condensate density and a is the scattering length. It is a measure of the strength of the interaction between atoms. For $R_\perp > \xi > R_Z$, one has a disc-shaped 2D quasi-condensate. We will encounter the healing length again in connection with vortices in BEC (see Sec. 23.4).

To observe 2D condensates, they started with $\sim 10^7$ atoms in a magnetic trap and successfully transferred about half of them into a $1.06\,\mu$m thin light sheet using $\sim 500\,$mW of laser power. The resulting trap had an axial trapping frequency of $790\,$Hz, compared to 30 and $10\,$Hz in the transverse directions. To observe the crossover from 3D to 2D behavior, they studied the aspect ratio and release energy as a function of the number of atoms N in the condensate after release from the trap. For a

3D trap one expects no change in aspect ratio of the condensate with N. For a 2D condensate, one expects a sudden change in aspect ratio of the ballistically expanding load. The number of atoms in the cloud was adjusted by exposing the condensate to a thermal atomic beam. Below $\sim 2 \times 10^5$ atoms there was indeed a rapid, approximately two- to three-fold increase in aspect ratio after 15 ms of expansion, indicating a transition to 2D behavior, as the density was decreased from the high density low temperature Fermi-Thomas limit. At the crossover to 2D behavior, the interaction energy per particle is roughly equal to the kinetic zero point energy of the trap.

Experiments in 1D were performed in magnetic traps with an aspect ratio of $\sim 100{:}1$. A similar but not so large change in aspect ratio at the crossover was observed in the 1D experiment.

The authors believe that 2D and 1D condensates will be much more stable than 3D condensates and will be useful in studying solitons and vortices. Vortices should not develop kink instabilities at lower dimensions, and collective oscillation of BE condensates should be simpler to study, as well. Another consideration is that lower dimensional condensates behave locally like ordinary condensates with a locally uniform phase, but do not have a globally uniform phase. Thus, study of phase fluctuations may be more advantageously studied in reduced dimensional condensates. A particularly interesting application of BEC in lower dimensions is mentioned in an earlier theoretical paper by Petrov et al. [594]. They assert that at temperatures well below T_c one can reach a quasi-2D regime, in which it is possible to switch the sign of the interaction from attractive to repulsive by changing the trap oscillation frequency in the tight confinement direction, and also control the rates of inelastic scattering loss. This is an important capability. Calculations are given for cesium, for example, which has a large attractive scattering length that they claim can be changed to a repulsive interaction.

There are extensive references in Ref. [593] on the earlier theoretical work in this field. For further remarks on this topic by W. Ketterle, W. D. Phillips, and R. Hulet, see the commentary by Charles Seife in Science [595].

20.8. Josephson Junction Arrays with Bose–Einstein Condensates

In 2001, Cataliotti et al. at the University of Florence [596]*, made a direct observation of an oscillating atomic current in a 1D array of Josephson junctions in an atomic BE condensate. The Josephson effect manifests itself as a tunneling current through a potential barrier between two superconductors or superfluids. In the case of BEC superfluids, it is a direct consequence of the phase coherence of the condensate. The ability to make a "weak link", or tunnel junction between adjacent BE condensates has led to experiments that are almost exact analogs of classical Josephson effect experiments in superconductors or liquid helium quantum fluids. The controlled environment of BEC in atomic vapors makes this an ideal environment to study Josephson effects in neutral fluids. Previous work by Kasevich's group at Yale [524], [599] has made the connection between the quantum coupling of 1D lattices and the Josephson effect.

Starting with $\sim 5 \times 10^8$ ^{87}Rb atoms in a magnetic trap, they evaporatively cooled to somewhat below the T_c for BEC and then superimposed a blue-detuned axial laser standing wave. They continued to evaporatively cool until no thermal component was seen. This split the condensate into an array of ~ 200 separate wells $\lambda/2$ apart, occupied by ~ 1000 atoms in each well, for a total of $\sim 2 \times 10^5$ atoms. By changing the laser power the barrier height between wells could be varied from zero to $\sim 5E_r$, where E_r is the recoil energy. The wave functions of the individual condensates, how-ever, were coupled, due to tunneling, and, if released from the combined trap, showed an interference pattern. The pattern gives information on the relative phase of the different condensates.

If they suddenly displaced the magnetic trap by a short distance of $\sim 30\,\mu$m relative to the $100\,\mu$m long optical lattice, they disturbed the equilibrium and the condensates started to move. Since the change in energy given to the atoms was less than the barrier height between condensates, the condensates moved along the axis by tunneling through the barriers. This, of course, imposed a relative phase difference between the adjacent condensates. This showed up as a collective oscillation in the position of the interferogram of the released expanded condensates. Figure 20.3, taken from Fig. 2 of Ref. [596]*, gives direct evidence of this oscillatory Josephson current. As seen in the figure, this manifested itself as the considerable difference in period from the sloshing back and forth motion of the atoms in a displaced magnetic trap, with no standing wave potential. As seen in Fig. 20.3(b), no motion was observed when only a thermal cloud of atoms was present. The static thermal cloud only served to damp the Josephson component.

Data on the frequency of the atomic current in the Josephson junction array, as a function of the interwell potential height, show a decreasing frequency in good agreement over a range of potentials from 1 to $5E_r$ with a discrete Josephson nonlinear Schrödinger equation. The authors point out that

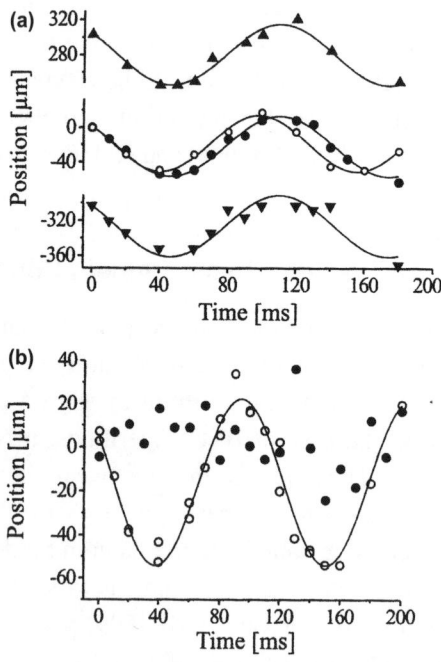

Fig. 20.3. (a) Center of mass positions of the three peaks in the interferogram of the expanded condensate as a function of the time spent in the combined trap after displacement of the magnetic field. Up and down triangles correspond to the first-order peaks; filled circles correspond to the central peak. Open circles show the center of mass position of the BEC in the absence of the optical lattice. The continuous lines are the fits to the data. (b) Center of mass positions of the thermal cloud as a function of time spent in the displaced magnetic trap with the standing wave turned on (filled circles) and off (open circles). Figure 2 and caption from Ref. [596]*.

the same equation applies to many nonlinear systems and could be of use in the study of solitons in BE condensates, formation of condensates, and low-dimensional systems.

20.9. Josephson Effects in Dilute Bose–Einstein Condensates

There was an earlier proposal by Giovanazzi *et al.* in 2000 [597] for an experiment demonstrating DC and AC Josephson effects in a simple geometry using two weakly linked BE condensates. It involved a time-dependent barrier moving adiabatically across the trapping potential. The phase dynamics of such a situation is governed by the same "driven-pendulum" equation as in current-driven superconducting Josephson junctions. The authors point out that at a critical velocity, proportional to the critical tunneling current, there is a sharp transition between the DC and AC regimes, whose signature is a sharp jump of a large fraction of the relative condensate population. The DC regime (the equivalent of the voltage $= 0$ regime), the weak link, is a moving blue-detuned laser sheet. This divides the condensate into two parts, which remain in equilibrium in their ground state because of the nonzero atom current flowing through the barrier. The barrier is sustained by a phase shift between the two condensates. The current is proportional to the velocity up to a critical current I_c at a critical velocity v_c. At this point the system switches to the AC mode.

Figure 20.4, taken from Fig. 1 of Ref. [597], shows the relative population of the two condensates $(N_1 - N_2)/N$, where N is the total number of atoms $N_1 + N_2$, as a function of the velocity of the moving barrier after 1 s of motion. We see a very significant fractional population imbalance of $\sim 30\%$ for a velocity of $\sim 0.42\,\mu\text{m/s}$, close to the critical velocity. At higher velocities the imbalance falls dramatically and one enters the AC regime. The small fraction of atoms in the AC regime undergoes plasma oscillations that are not seen on the scale of Fig. 20.4. Performing the experiment on a longer

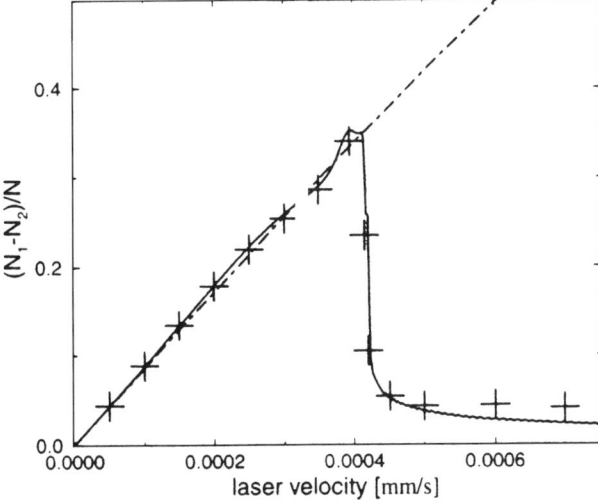

Fig. 20.4. Fractional population imbalance versus the velocity of the laser creating the weak link. A sharp transition between the "dc" and the "ac" branches occurs at a barrier critical velocity. The solid line and the crosses are the analytical and the numerical calculations, respectively. The dash-dotted line represents the static equilibrium value η_{eq} calculated with the center of the laser at vt_f. Figure 1 and caption from Ref. [597].

time scale would improve the observability of the above effects. The authors believe that the DC regime is readily observable, but that the AC oscillations are much harder to see. If one can make truly cw condensates and cw atom lasers, as discussed above in Secs. 19.8 and 19.10, it should be possible to do quite sophisticated Josephson type measurements on very long time scales, and study secondary quantum effects, such as collapsing and revival of population and phase oscillations between the two couple condensates [598].

20.10. Squeezed States in a Bose–Einstein Condensate

A potentially important application of BEC was devised by Orzel *et al.* in Kasevich's group at Yale in 2001 [599]. This involves the observation of an atomic squeezed state in an array of coupled 1D atom traps in an optical standing wave potential. The concept of squeezing has been developed as a way of manipulating the Heisenberg uncertainty principle connecting a pair of conjugate variables in the quantum theory of measurement. The uncertainty principle is fundamental to the quantum theory. It says, for example, that with conjugate variables of position and momentum, the uncertainty in the measurement of position Δx is related to the uncertainty in the measurement of its momentum Δp by $\Delta x \Delta p \geq h$. Similarly, for the conjugate pair of variable energy E and time t, we have $\Delta E \Delta t \geq h$.

The idea of squeezing is that if one is interested in determining only one of a conjugate pair of variables, one tries to measure that one variable with very high precision and not worry about the large uncertainty in the determination of the other variable. Squeezing has been successfully implemented with light, but with considerable difficulty. Kasevich's group has demonstrated squeezing with atoms in a BE condensate. This, they hope, will have advantages over optical squeezing, making possible new ways of achieving precision in atom-interferometer-based gyroscopes, gravimeters, and gravity gradiometers. They point out that such devices made with atoms already have accuracies and sensitivities comparable with more conventional devices.

In their experiment, they trapped about 10^3 condensed ^{87}Rb atoms in a linear array of about 12 wells formed at the antinodes of an off-resonance retroreflected standing wave beam, at $\lambda = 840$ nm, having a $1/e$ intensity radius of $50\,\mu$m. This tuning well on the long wavelength side of the resonance at $\lambda = 780$ nm gave trapping with negligible spontaneous emission. Trap depths up to $50\,E_R$ were possible, where E_R is the recoil energy from absorption of an 840 nm photon.

In this system the conjugate variables are the number of atoms and the phase of the wave function that describes them. If the number of atoms in the wells is very well known, then the relative phases of the atoms should be quite variable. The atomic system is probed by shutting off the quadrupole magnetic trap and the standing wave trap, which allows the released atoms to fall under gravity. The atoms in the standing wave lattice expand ballistically and overlap and interfere with atoms released from adjacent sites. The contrast of the interference pattern, measured by absorption imaging, is a measure of the phase coherence of the atoms.

Data were taken as a function of the trap depth, from values as low as $\sim 7E_R$ up to $44\,E_R$. At low E_R the atoms should easily tunnel between adjacent trapping sites, which means poor localization. One expects and finds small phase variations between adjacent sites and, therefore, deep contrasting interference fringes in the measured images. For deep traps, one expects almost complete localization, much reduced tunneling and thus a poorly defined phase relation between traps. With such traps one

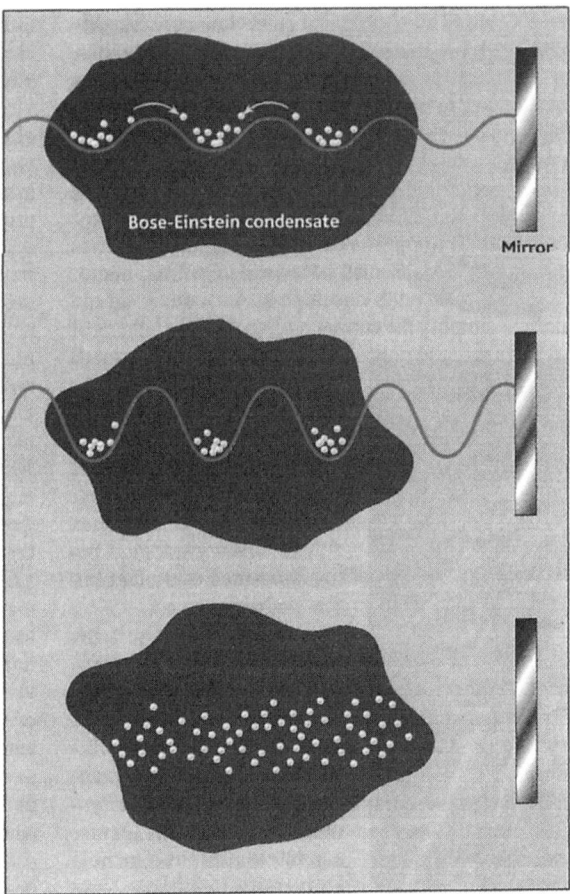

Fig. 20.5. Big squeeze. A laser beam bounced off a mirror creates a standing wave that traps atoms from a BEC (top). By hiking the intensity, physicists keep atoms from tunneling out of their pockets (middle). Lack of interference fringes when the laser is turned off (bottom) shows that the BEC's wave function was "squeezed". Figure and caption from Ref. [600].

finds very poor fringe contrast in the detection image. The simplicity of these measurements suggests that atom squeezing could be the basis for many novel measurements.

Figure 20.5, taken from "News Focus" commentary by David Voss in Science [600], illustrates the results of the squeezing measurements.

20.11. Quantum Phase Transition from a Superfluid to a Mott Insulator in a Gas of Ultracold Atoms

The atoms in a dilute ultracold BE condensate gas behave as a quantum superfluid with essentially no thermal energy, in close analogy with superfluid helium, for example. In such a system, quantum fluctuations are always present, due to the Heisenberg uncertainty principle, even at $T = 0$. They can cause phase transitions to a different state, such as the Mott transition from a superfluid-to-insulator in the case of helium.

Greiner *et al.* [601]*, in a beautiful experiment performed at the Max-Planck-Institute in Germany, demonstrated such a Mott transition using bosonic atoms with repulsive interactions hopping through an optical lattice potential. The BE condensate atoms, formed in a magnetic trap, were transferred into a cubic array of standing wave traps. If the lattice potential is fairly weak, there is negligible trapping and atoms are effectively spread over the entire lattice. The atomic wave function is phase coherent and the gas remains essentially in the superfluid phase. At high enough lattice potentials exact numbers of atoms become bound at the lattice sites and there is no phase coherence between atoms in the separate lattice sites. One is then in the insulating phase. The two phases are readily distinguishable experimentally. The authors simply turned off the confining potentials and let the expanding gas interfere. Atoms in the coherent state interfere to give the 3D pattern characteristic of a periodic array of phase-coherent sources. When in the insulator phase the interference pattern completely disappears due to incoherence of the localized atoms. The transition between phases occurs rather quickly. It is possible to reverse the transition by simply reducing the optical potential. The authors show that the Mott transition range can be closely predicted by theory. They probed the system with a tilted lattice potential, which allows tunneling to neighboring lattice sites. In that case, the excitation spectrum exhibits an energy gap Δ, which is characteristic of such a transition.

This experiment opens a new, highly controlled regime of ultracold gases, in which atom number fluctuations at the lattice sites are suppressed. The authors foresee futures experiments using Feshbach resonances with this system. The Mott insulator phase opens the possibility of quantum gates for quantum computing. See also the interesting commentary by Stoof [602] in the same issue of Nature.

20.12. Bose–Einstein Condensation on a Microelectronic Chip

The idea of using microscopic magnetic fields generated near fine current-carrying wires at the surface of chips for guiding and manipulation of cold atoms was developed in the late 1990s [603], [604], [605]. This technique overcomes some of the problems associated with magnetic trapping of cold atoms. By using small, closely spaced conductors on chips, for example, one can generate strongly localized magnetic field gradients and make traps with conductor dimensions $\sim 50\,\mu$m, which are much more compact than free space magnetic traps. This implies higher atom density, shorter collision times for evaporative cooling, and the ability to make more rapid adiabatic changes. These are some of the same advantages possessed by optical dipole traps.

In 2001 Hänsel *et al.* in Hänsch's group [606]* in Munich successfully demonstrated BEC on a microelectronic chip. They created a Ioffe-Pritchard magnetic potential with a nonzero field minimum using a "Z"-shaped conductor and a homogeneous external magnetic bias field. Increasing the bias field moved the trap center closer to the chip surface. Starting with a mirror-type MOT of their own design, they trapped $\sim 3 \times 10^6$ atoms at a point about a 1 mm from the chip surface. After transferring the atoms close to the surface, they had a temperature of $\sim 45\,\mu$K and a density of $\sim 5 \times 10^{10}$ atoms/cm^3. They then compressed the sample and evaporatively cooled for 100 ms, reducing the number of atoms to 5×10^5 atoms. From this point they could follow two routes to BEC, depending on the desired shape of the final condensate.

By the first route, they cooled to $\sim 6\,\mu$K with $\sim 7 \times 10^4$ atoms left at a density of 5×10^{13} atoms/cm^3 and a phase-space density of $\sim 10^{-2}$. After a decompression to avoid too high a three-body collision loss when the condensate approaches too close to the chip surface, they cooled to $T_c \approx 630$ nK with $\sim 11\,000$ atoms. The condensate lifetime was apparently short, ~ 500 ms, but could be sustained to ~ 1.3 s with continued radio frequency cooling below the transition temperature T_c, with typically ~ 3000 atoms in the final condensate. This is a rather poor yield. Part of the problem is a high measured heating rate of $\sim 2.7\,\mu$K/s in compressed traps, when held $\sim 100\,\mu$m from a hot surface. The authors say little is known about condensate-surface interactions at very small distances. For a $100\,\mu$m condensate to surface distance, they empirically arrived at a heating rate at $0.5 \pm 0.3\,\mu$K/s. The long, thin, cigar-shaped condensate found in the above manner is well within the regime of fluctuating phase 1D quasi-condensates, with a ratio of longitudinal to transverse trap frequencies of $\sim 200{:}1$. Lower-dimensional condensates were discussed in Sec. 19.7.

Following a different cooling regime, the authors arrived at a differently located condensate with a more spherical shape, with only a 12:1 ratio of the trap frequencies. The condensate yield was only slightly higher.

The total cycle from MOT loading to BEC was ~ 10 s. The evaporative part of the cycle was as short as 700 ms, which they point out is ~ 3 times shorter than the evaporative cooling time for the optical dipole BEC trap of Barrett *et al.* [205]*, as described above in Sec. 19.5. This is simply a question of trap size. If the dipole trap is reduced in size, one would expect even shorter evaporation times with less atom loss. A shorter evaporative cooling time implies less scattering loss with background atoms. It also allows operation at a higher pressure, with both the MOT and the BE condensate produced in the same chamber at 10^{-9} Torr. One might think this pressure is too low to vapor-feed the MOT, but the authors momentarily augmented the Rb atom vapor pressure during MOT loading by shining a 30 W halogen lamp into the chamber. This caused light-induced desorption of atoms from the chamber walls.

Once a BE condensate is formed, it can be coupled into a magnetic conveyor belt and magnetically transported along the surface of the chip a distance of 1.6 mm by modulating a pair of currents $90°$ out of phase. It is possible at any time to totally release the condensate from the inverted chip surface and watch it fall downward by gravity. The authors say these capabilities will be of great use in microchip experiments. See also the "News and Views" article on this work in the same issue of Nature [607].

20.13. Bose–Einstein Condensates Near a Microfabricated Surface

Although production of a BE condensate on a microchip was previously demonstrated, serious problems of large heating rates and short trap lifetimes remain for atom–surface separations in the $100\,\mu$m range. Several groups have also reported problems of condensate fragmentation near the surface of a microchip (see references in Ref. [608]).

In 2003 Ketterle, Pritchard, and colleagues made an experimental study of the behavior of condensates near a microfabricated surface [608]. Their technique involved a comparison of the behavior of a condensate loaded in either an optical tweezer trap or in a microfabricated Ioffe-Pritchard magnetic trap formed by a Z-shaped wire carrying current in the presence of an external magnetic bias field. As in previous experiments, the condensates confined near the surface in the magnetic trap were observed to fragment longitudinally as the condensate approached the surface. In contrast, condensates held in the optical trap under the same conditions remained intact. These results show that the fragmentation within the magnetic trap was somehow caused by the current in the trap. The authors suspect that small imperfections in the making of the microfabricated wires on the chip led to condensate fragmentation. Changes in the width or thickness of electroplating as small as an estimated 5%, departures from straightness, variation of resistivity over the cross section, or parasitic conductance to the substrate could all lead to irregular current flow. No fragmentation was observed in previous work using a macroscopic round copper wire trap.

In experiments on the condensate lifetime, it was found that the lifetime was ≥ 20 s for both magnetic and optical traps over distances from 70 to $500\,\mu$m from the surface. Thus, proximity to the surface is not a limiting factor over this range. This observation contrasts with earlier experiments showing loss of lifetime for magnetic surface traps. On the other hand, the authors found that the magnetic surface traps are very sensitive to spin-flip transitions driven by rf noise at the Zeeman splitting frequency. Spin-flips can be sensitively studied with the help of the optical trap configured as a Stern–Gerlach experiment. Small changes in the experimental configuration can lead to orders of magnitude increases in spin-flip rates.

The authors conclude that the extreme sensitivity to small static and dynamic electromagnetic fields represents a severe challenge to realizing condensate-on-a-chip devices, especially at distances less than $70\,\mu$m between condensates and chip surfaces.

It is also clear, I believe, that this work gives additional evidence that devices based on optical traps have distinct advantages over devices using magnetic traps, and are often to be preferred. See, for example, Sec. 20.19 on the design of an optical cw atom laser.

20.14. Tonks–Girardeau 1D Gas of Ultracold Atoms

Yet another novel state of matter, the Tonks–Girardeau (TG) gas, was observed in 2004 in an ultracold 1D condensate using optical confining potentials [609]. For such a system the repulsive interactions of bosonic atoms in a superimposed optical lattice dominated the interaction and atoms were prevented from occupying the same position in space. In this sense these bosonic atoms behaved as fermions. The system was formed from Rb atoms held in two strong orthogonal standing wave beams with an additional third optical lattice trap of variable strength along the long axis. To achieve a TG gas one must reach a large enough value of the parameter $\gamma = I/K$, representing the ratio of interaction energy, I, to kinetic energy, K. The authors indicate that this is a difficult criterion to achieve, calling for both a very low-temperature BE condensate and low atomic density. They used the variable axial standing wave potential to increase γ by localizing the atoms at low density to increase I and also reduced the number of atoms to about one per lattice site.

With ordinary 1D systems one achieves a value of $\gamma \cong 1$, or less. By adding the variable axial standing wave, one can increasingly localize the atoms and increase γ by one or two orders of magnitude, well into the TG regime, with densities of about one atom per lattice site. As proof of the existence of TG gas they showed that the momentum distribution of the atoms was in excellent agreement with a theory of fermionized trapped Bose gases. The authors think it may be possible to look for behavior similar to Cooper pairing in this system.

Very soon afterward, Weiss and his colleagues at Pennsylvania State University [610] also reported the observation of a 1D Tonks–Girardeau (TG) gas using a somewhat different technique, not involving a periodic potential along the 1D axis. They used a very clever system that involved the interplay between two types of optical traps; one red-detuned from resonance and the other blue-detuned.

They started with an all-optically produced BE condensate of $\sim 2 \times 10^5$ atoms that was produced every three seconds. The atoms were initially held in a far-off-resonance red-detuned horizontal crossbeam dipole YAG laser trap having $w_0 \cong 70\,\mu m$. They created an ensemble of parallel 1D traps by superimposing on the dipole trap two orthogonal sets of intersecting blue-detuned lattices from a Ti:sapphire laser. The blue-detuned beams were antitrapping in the axial direction but only slightly reduced the axial trapping of the red-detuned trap. As a result, the authors could widely vary the power of the blue-detuned beams and significantly vary the transverse confinement without affecting the axial confinement. With this arrangement the authors showed that a 1D condensate could be tuned from the BEC regime through the TG regime by varying the blue-detuned power. Measurements of the total 1D energy, which is proportional to T_{ID}, the 1D temperature, and the full root mean square length of the condensate cloud agreed with the exact 1D Bose theory, with no free parameters. The calculations showed that they achieved values of $\gamma_{av} = 5.5$, where wave function overlap was reduced and the atoms acted as if they were noninteracting fermions.

20.15. All-Optical Production of a Degenerate Fermi Gas

Shortly after the production of an all-optical BE condensate by Chapman's group, there was a related demonstration by Thomas's group at Duke of the first all-optical production of a degenerate Fermi gas, also using evaporative cooling from a CO_2 laser trap [611]*. This was part of an extensive effort by this group to produce degeneracy with fermions. Their technique involved cooling a mixture of two hyperfine states of ^6Li by direct evaporation of atoms in both states. They took advantage of the ability to use a bias field to magnetically tune the interaction strength with a Feshbach resonance. They achieved low temperatures of $\sim 580\,nK$, in agreement with scaling laws developed by the authors [612].

See Chapter 25 on "Trapped Fermi gases" for a more detailed discussion of work by Thomas's group.

20.16. Bose–Einstein Condensation of Cesium by Evaporative Cooling from Optical Dipole Traps

Of all the stable alkali atoms, only cesium has resisted BE condensation using magnetic techniques (see Sec. 19.3). In 2003, however, Weber et al., in Austria, successfully evaporated cesium to a BE condensate in a beautiful experiment using optical trapping techniques [613]*. They took advantage of the ability to optically trap the lowest energy state where inelastic two-body collisional loss is fully suppressed. They also made full use of the resonant behavior near a magnetic Feshbach resonance to tune through regions of attractive, repulsive, and null interaction strength, where imploding, exploding, and noninteracting behavior is seen. They were able to control the atomic density and evaporatively cool from a large-volume, low-density CO_2 laser trap and $1.06\,\mu m$ small-volume

YAG laser trap. The authors believe that the unique tunability of this important atom will make possible new applications to atomic clocks, formation of molecular condensates and MOT insulator transitions [601]*.

20.17. Optimized Production of a Cesium Bose–Einstein Condensate

In 2004, the Austrian group followed up its earlier demonstration of the production of a BE condensate in cesium, using optical traps with an updated experiment showing enhanced performance [614]. Using an improved trap loading and evaporation technique they increased the number of atoms in the condensate to more than 10^5 atoms. They also varied the scattering length with an independently controlled magnetic field and studied the expansion of the condensate. It was also possible to excite strong oscillation of the trapped condensate by varying the interaction strength. This work further shows the usefulness of optical trapping techniques for forming and manipulating BE condensates.

20.18. Cooling Bose–Einstein Condensates Below 500 pK

Leanhardt *et al.* at MIT recently succeeded in cooling condensates of atomic vapors to record low temperatures [615]. The intent was to gain access to new low energy phenomena at ever lower energy scales. Starting with partially condensed BE condensates of more than 10^7 ^{23}Na atoms in a magnetic trap, they used optical tweezers to transfer up to 3×10^6 atoms into a special auxiliary gravitomagnetic trap in an ultrahigh vacuum chamber. The atoms were stably trapped ~ 5 mm above a single magnetic coil ~ 1 cm in diameter, having trap frequencies $\omega_x = \omega_y - \omega_z \cong 2\pi \times 8$ Hz. After about five seconds in the trap, there were $\sim 5 \times 10^5$ atoms at a temperature T, such that $0.5 < T/Tc < 1.0$, where $T_c = 30$ nK is the BEC transition temperature. Further cooling took place by adiabatically decompressing the potential by turning on currents in two large external 10 cm coils while lowering the current in the single small coil in two five second stages with a five second delay for collisional damping. They ended up with $\varpi = 2\pi \times 1$ Hz and $\sim 2 \times 10^5$ atoms in the partial condensate at $T_c = 3$ nK. A final cooling stage occurred by holding the atoms for an additional 200 s during which the collisional cooling reduced the number of partially condensed atoms to ~ 3000, at a kinetic temperature below 1 nK.

The lowest 3D kinetic temperature achieved was 450 ± 80 pK for 2500 atoms at a peak condensate density of 5×10^{10} atoms/cm^3. These atoms had a thermal velocity of 1 mm/s. The authors believe that obtaining lower temperatures with these magnetic traps by further lowering the trap depth (or $\varpi/2\pi$ substantially below 1 Hz) is technologically challenging.

20.19. Design for an Optical cw Atom Laser

The achievement of a truly cw atom laser using ultracold atomic vapors remains a challenge for the field of BE condensation. As discussed in Sec. 16.10, several types of crude pulsed atom lasers were developed in early work. In later work, aimed at cw atom lasers, Ketterle's group, in 2002, demonstrated sustained BE condensates. But problems still remained [616], [617].

Another approach to reaching the goal of a cw coherent source of atoms was considered by a group at the Laboratoire Kastler Brossel in France [618]. The intent of this approach was to evaporatively cool an atomic beam to degeneracy as it proceeded along a magnetic atom waveguide. For this approach to succeed, the density had to be high enough to be well into the collisional regime where rapid atom thermalization could occur.

To test the basic method, the authors slowed $\sim 3 \times 10^9$ atoms in a Zeeman slower and collected them into a MOT. About 2×10^9 of these atoms, at a temperature of $\sim 400 \, \mu K$ were successfully launched into a long magnetic waveguide for further evaporative cooling. Two rf antennas were used; the first to flip the spins and allow the energetic atoms to leave the guide, and a second antenna to probe the beam four meters down stream to look for cooling. Cooling to a temperature of $\sim 280 \, \mu K$ was found, with a decrease in atomic flux of $\sim 27\%$, corresponding to an increased phase space density of around 1.9 times, which was very close to the calculated value. This experiment showed that transverse evaporation works in principle. However, the authors estimate that to finally achieve a coherent cw atom laser, they would have to increase the initial collision rate by one order of magnitude, which is quite challenging.

In 2004, Ashkin [46]* proposed a design for a viable cw atom laser using mainly optical techniques that he believes can operate at high intensity, high coherence, and very low temperature. It is based on a so-called "optical shepherd" technique, in which far-off-resonance blue-detuned swept shepherd sheet beams are used to make new types of high-density optical traps, repulsive wall box-like atom waveguides and components. Devices based on shepherd-enhanced traps should be superior to conventional MOTs and magnetic traps for producing an atom laser. A scheme is proposed in which shepherd beams capture and recycle essentially all of the escaped atoms in optical evaporative cooling, thereby making large increases in condensate output. Condensate atoms are stored in a shepherd trap shielded from absorbing light under effectively zero gravity conditions and are coupled out directly into an optical waveguide. The ability to optically cancel gravity uniformly throughout the waveguide volume is especially important at very low temperatures, where it makes it possible to maintain uniform density and avoid atom pile-up at the lower surface. The atom laser is designed so that the individual steps, from enhanced MOT atom collection and molasses cooling to evaporative cooling with feedback, can be optimized and tested individually. Achievement of a viable cw atom laser would be a major step forward for BEC research.

21

Trapping Single Atoms with Single Photons in Cavity Quantum Electrodynamics

As we discussed above, there has been a strong effort to achieve trapping of large numbers of ultra-cold atoms at high densities in optical dipole traps in order to make Bose–Einstein (BE) condensates and realize the many benefits of the technique of all-optical trapping, including perhaps a cw atom laser. On the other hand, there has been recent interest in studying the interaction of light with single atoms or small numbers of atoms in the context of Cavity Quantum Electrodynamics (CQED) and quantum computing. Although it is possible to trap single atoms in MOTs, the principal thrust of CQED research uses single atoms and far-off-resonance optical dipole traps. It should not come as a surprise that one can trap single atoms in such optical dipole traps, considering what was achieved in the high-density work described above. Detecting single atoms by resonance fluorescence is no great problem either. The big surprise, as we shall see, is that one can trap single atoms with single photons, if one uses very high-Q (quality factor) optical cavities and very cold atoms. When one thinks of how hard it was to devise optical traps for atoms with high intensity laser beams in the 1970s, it is quite remarkable to see an atom trapped by a single photon.

I will briefly trace the history of CQED and the work on single atoms.

Purcell observed, as early as 1946 [619], that the spontaneous emission rate for a two-level atomic system is significantly increased proportionate to Q for atoms placed in a cavity tuned to the resonance transition.

In 1981, Kleppner [620] proposed the inhibition of the spontaneous emission rate when an ideal cavity is mistuned and no mode is available into which the spontaneous emission can enter.

The topic of the interaction of a single atom with a single mode of the electromagnetic field initially received a lot of theoretical attention after the invention of the microwave maser.

21.1. The Simple One-Atom Maser

A major advance in experimental technique in CQED was the observation by Meschede *et al.* in 1985 [621] of a one-atom maser using laser techniques. The experiment used laser excitation of long-lived ^{85}Rb atoms placed in a superconducting microwave cavity with a quality factor $Q = 8 \times 10^8$ at a temperature of 2 K. The atoms were excited with a frequency-doubled dye laser to the upper end of the $63p_{3/2}$–$61d_{3/2}$ microwave maser transition. The incident flux was simply an atomic beam from an oven. Remarkably, maser oscillation was observed at $\sim 21\,506$ MHz with only a single photon in the cavity, with an average number of only 0.06 atoms in the cavity at any one time. At higher atom densities asymmetries in the line shape were observed, attributable to the optical Stark shift.

21.2. The Two-Photon Maser

An equally remarkable experiment was performed in 1987 by Brune *et al.* and colleagues of Haroche's group [622] on the realization of a two-photon maser oscillator. The authors used a technique very similar to that of Meschede *et al.* [621]. The level scheme used is shown in Fig. 21.1, taken from Fig. 1 of Ref. [622]. The close proximity of the virtual two-photon level to the $39p_{3/2}$ level helps to enhance the two-photon process. Oscillation was seen with only 10–20 atoms in the high-Q cavity. The authors thought that a moderate increase in cavity Q would make oscillation possible, with an average of one photon or less in the cavity.

21.3. Trapping Single Atoms in a MOT

To further implement experiments with single atoms for CQED or for spectroscopy with squeezed or other forms of nonclassical light, Hu and Kimble, in 1994 [623], trapped single cooled atoms in a magneto-optical trap. A cesium ampule kept at $\sim 35°$C served as a source of atoms for the trap. Atoms entering or leaving the MOT at low atomic densities were detected by fluorescence from the cesium atoms. Discrete steps were observed as a function of time as individual atoms entered and left the trap. Situations with zero, one, and two atoms in the trap could be readily distinguished and clearly represented the localization of single cold atoms. The authors also pointed out at this time that it is possible to use far-off-resonance dipole force traps for greatly improved atomic localization.

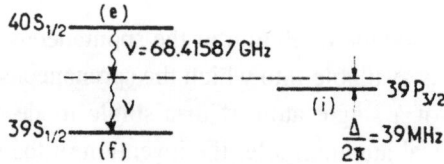

Fig. 21.1. Level scheme relevant to the Rb two-photon maser. Figure 1 and caption from Ref. [602].

e Optical Cavity

Tech. [624] considered use of single atoms
crucial elements for implementing quantum
nterest in quantum computing networks was
ork the photons are the quantum carriers of

ngle atom strongly coupled to a high-finesse
passing through the cavity under conditions
ss than one. They proposed using these phase
quantum logic. Optical phase shifts can be as
mp photons. Such a system can be viewed as

to a High-Finesse Optical Cavity

In a further experiment, Kimble and his colleagues [625] used the strong coupling of atoms and photons to detect the passage of cold atoms through a high-finesse optical cavity. Cold cesium atoms from a MOT were dropped from above through a cavity tuned to the resonant frequency of the atom and detected by a drop in the power of a resonant probe beam being transmitted along the axis of the optical resonator. This technique makes it possible to detect the presence of single atoms for CQED or quantum logic experiments on a real time basis. They suggested that one might be able to use feedback to raise the coupling of the atom to the cavity and achieve optical trapping of the atom within the cavity.

To illustrate how a quantum communication channel might possibly work, Cirac *et al.* [626] proposed a scheme to utilize photons for quantum transmission between atoms located at two different locations. It involves an optical transmission link between two cavities, each of which contains atoms strongly coupled to their respective cavity modes. Each cavity acts as a quantum system that stores quantum information in quantum bits and processes this information with quantum gates. To implement such a scheme one needs to be able to trap the atoms in the optical cavities. A system such as this opens possible applications to cryptography and teleportation, in addition to quantum computation (see references in Ref. [626]).

21.6. Cooling an Atom Strongly Coupled to a High-Q Standing Wave Cavity

The rising interest in the dynamics of single atoms strongly coupled to a standing wave cavity has led to a reexamination of the cooling of a two-level atom in the regime where the atom–field coupling dominates the atomic and cavity decay rates. Hechenblaikner *et al.* in Austria in 1998 [627] have found that an entirely new cooling mechanism appears, similar to Sisyphus cooling in free space, which can dominate the cooling process. They examined the overall cooling in "good and bad cavity" limits, where in the good cavity the internal atomic dynamics is much faster than the cavity dynamics, and in the bad cavity it is the reverse. In the bad cavity limit the friction coefficient is given by the usual free space Doppler friction force. In the good cavity case, where the strong atom–cavity coupling dominates the frictional force, the new Sisyphus-type cavity-induced cooling mechanism dominates. This gives rise to an equilibrium temperature of about an order of magnitude lower than the Doppler minimum temperature. This is, in essence, a one-dimensional (1D) calculation. The authors say that calculations in two- and three-dimensions show that long confinement times (>ms) in the optical trap are possible, sufficient for typical CQED experiments.

21.7. Real-Time CQED and Atom Channeling with Single Atoms

The work of Mabuchi *et al.* [625], discussed above, showed how one could detect single cold atoms falling into a high-finesse optical cavity. In 1998, Hood *et al.* [628] and colleagues in Kimble's group used single slow atoms to make real time CQED measurements with an intracavity photon number less than one. With a finesse $F = 1.8 \times 10^5$ they formed a cavity that was 10.1 μm between the mirrors. With these parameters they reached the highest atom–cavity coupling yet achieved. This was deduced from probe measurements near the atom–cavity resonance. This was the first experiment in which the interaction energy $(h/2\pi)g_0$ was greater than the atomic kinetic energy.

In measurements of the probe transmission signal versus probe detuning they found an asymmetry about zero detuning. They interpret this as due to light forces that channel atoms along the intensity maxima for tunings to the red side of resonance, or along the intensity minima for tunings to the blue side of resonance.

21.8. Formation of Giant Quasi-Bound Cold Diatoms by Strong Atom–Cavity Coupling

The achievement of strong coupling of cold atoms with high-finesse cavities in the "good cavity" regime has stimulated the imagination of theorists and experimentalists alike. Deb and Kurizki in 1999 [629] posed the question of what happens when two identical cold atoms exchange photons in the strong-coupling regime, during collisions and subsequent dissociation. They showed that giant quasi-bound diatoms could form by intracavity photoassociation for a photon tuned below resonance, and could live in a metastable state as long as the cavity holds the photon inside. This is a novel CQED optical binding effect. The authors explain this effect as atomic quasi-binding in a potential well formed by the competing effects of the resonant dipole–dipole interaction and strong atom–cavity coupling.

21.9. Single Atoms Trapped in Orbit by Single Photons

It was becoming increasingly evident from CQED experiments, especially by Hood *et al.* in 1998 [628], that the light forces on single cold atoms passing through high-finesse optical cavities were able to cause large changes in the particle dynamics, even at the single photon level. This has led to the achievement of optical trapping in the optical potential well of single photons. Such trapping was achieved by Hood *et al.* at the Cal. Tech. laboratory of Kimble in 2000 [63]*. They were able to confine single cesium atoms for times up to ~ 2 ms on one of the standing wave maxima of a

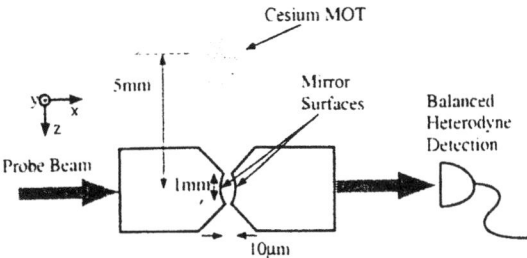

Fig. 21.2. Schematic of the experimental apparatus. Figure 1 and caption from Ref. [628].

retro-reflected Gaussian beam, having a beam waist $w_0 \approx 10 \,\mu$m. The atoms were dropped into the horizontal standing wave from a MOT located ~ 3 mm above the optical resonator. The dropped atoms were originally at a temperature of $\sim 20 \,\mu$K and had a velocity of ~ 4 cm/s. See Fig. 21.2, taken from Fig. 1 of the paper [628]. When they crossed the standing wave axis, they were moving with an average velocity of ~ 24 cm/s. To capture single atoms, the authors started with a low-intensity probe beam passing through the cavity to a sensitive heterodyne detector. When the detector signaled the presence of a single atom, the probe power was quickly increased, raising the trapping potential sufficiently to capture the atom in the cavity field with no more than a single photon in the cavity at any given time.

As the cavity field decayed in time, the input probe field was reactivated to allow another photon to enter the cavity and continue the trapping process. This reactivation was necessary because the cavity decay time was quite short compared with the oscillation period of the trap. The atoms ultimately left the trap due to heating. For the parameters used here, the heating time in the usual free-space trap was quite short, barely long enough for an atom to traverse the trap. However, the heating in the case of the atom coupled to a high-finesse cavity was strongly reduced by about a factor of 10 or more. If the cavity potential was deepened or the finesse increased (a factor of 10 is possible) then the trapping lifetime could be increased further.

It should be mentioned that optical cooling was also occurring in these cavity traps. The paper of Hechenblaikner *et al.* [627] showed that cooling, at least in one dimension, was quite a strong process. Thus, one certainly gains lifetime due to optical cooling.

21.10. The Atom Cavity Microscope

Another major aspect of this work [628] was the ability of Hood *et al.* to use the single atom cavity trapping as an atom cavity microscope (ACM), which permitted them to follow the individual atom trajectories in real time. The position of an atom along its orbit in the cavity was deduced from the large variations in the transmitted probe signal as the atom moved about in the cavity field. This position could be observed with a large signal to noise ratio. This allowed them to deduce the position in the cavity field with a $2 \,\mu$m spatial resolution within a $10 \,\mu$s measuring interval. This measurement sensitivity was close to the quantum limit for sensing atomic motion. Many trajectories were computed in this way, all localized in a single node of the standing wave field transverse plane.

Throughout this paper, the authors emphasized that the role of the cavity is not simply one of increasing the internal field for a given input power. For the same peak field there were changes in trap depth, atom heating rates, and atom cooling rates, along with a new way of sensing particle position. One was not dealing with an atom–field interaction, as in free space, but with an atom–field–cavity interaction. The field of one photon served as an intermediary, coupling the atom to the cavity. The system of one atom, one photon, and cavity acted, in a sense, as a type of single molecule.

The authors believe that implicit in the ability to sense particle position in real time, with high sensitivity via the transmitted probe signal, is the ability to apply feedback. With such feedback, one could damp the atomic motion to the bottom of the potential well. Such damping was, of course, first demonstrated by Ashkin and Dziedzic in their early work on levitation of dielectric spheres in vacuum in 1977 [7]*. Hood *et al.* suggest that operation of the ACM, even at levels of more than one photon in the cavity at a time, might be useful for monitoring of chemical and biological processes. Additionally, one could provide separate means of particle localization, using feedback control of optical dipole (tweezer-type) traps. Such traps are much more strongly focused than are ACM traps and therefore localize particles much more precisely.

They further suggest the possibility of indirectly controlling a molecule in an ACM that does not directly couple to the cavity field, through its interaction with an atom that does strongly couple to the cavity. In this context the proposal of Deb and Kurizki [629] to study the attractive forces between two atoms sharing one photon in the cavity, is quite relevant.

There are, of course, many other techniques for studying molecular motion in biology, chemistry, and physics, as discussed in detail above. The authors' suggestions [63]* on other applications of ACM are quite imaginative, but it remains to be seen if they can be implemented in practice. This article by Hood *et al.*, however, is very informative, with a useful collection of references, and is well worth studying.

21.11. Dynamics of Single Atom Motion in the Field of a Single Photon

Almost simultaneously with Hood *et al.*, Pinkse *et al.*, writing in Nature in 2000 [64], also observed the trapping of an atom with single photons. The essence of their experiment was very much the same as in the Hood *et al.* [63]* experiment just discussed, but differs in some details. They used ^{85}Rb atoms rather than cesium. They loaded their atoms from an atomic fountain, rather than by dropping them from above. With a feedback switch to keep one photon on average in the cavity, Pinkse *et al.* were able to hold single atoms in the trap for times of ~ 0.4 ms. This, however, was a bit marginal. By increasing the power input to the two- to three-photon level, the retention time of single atoms was increased to 1.5 ms, comparable with Hood *et al.*

Another important difference was that the trap used by Pinkse *et al.* had a well depth of 0.8 mK compared with 5.3 mK for Hood *et al.* The facts accounting for this were the following: (i) the mirror finesse was $\sim 4.3 \times 10^5$ compared with 1.8×10^5 for Hood *et al.* Higher finesse is, of course, better. (ii) The beam and cavity dimensions were considerably larger for Pinkse *et al.* Their Gaussian beam radius was $w_0 = 29\,\mu$m and cavity length $= 116\,\mu$m, compared with $w_0 \approx 10\,\mu$m with a cavity length of $\approx 10\,\mu$m in a spherical mirror geometry for Hood *et al.* A smaller cavity volume is better because it gives rise to higher fields for the same internal power. (iii) Overall, this implied that the atom–field coupling constant g_0 was smaller by the ratio of trap depths $5.3\,\text{mK}/0.8\,\text{mK} \cong 7$. The differences in g_0 and in the dimensions account for the better performance seen by Hood *et al.*

Indeed, Pinkse *et al.* found, by analyzing the power transmitted through their cavity, that the atoms made only about three or four orbits in the trap and wandered considerably along the axial direction of the cavity, hopping between various antinode maxima. Axial motion through the standing wave was deduced from amplitude oscillations in the transmitted light. Axial cooling was probably occurring, but eventually spontaneous emission kicks caused the atoms to leave the trap. Pinkse *et al.*, however, think that feedback cooling might prevail over spontaneous emission and make possible cooling to the quantum mechanical ground state of the atom in the trap. In general, they believe that trapping in high-finesse cavities has application to the rapidly growing field of quantum communications. Their paper also has many useful references.

There is an interesting commentary on "An Atom is Trapped by the Field of One Photon" by B. Goss Levi in Physics Today, July 2000 [630]. The authors and other experts comment on the recent work.

21.12. Commentary on CQED in Nature's "News and Views"

Another commentary on trapping of single atoms with single photons, entitled "Tricks with a single photon", by Peter Zoller, appeared in "News and Views" in Nature, March 23, 2000 [631]. Zoller considers how it is possible to trap an atom with the smallest quantity of light, namely one photon. He then discusses the significance of the experiments of Pinkse *et al.* of the Munich group [64] and of the Hood *et al.* group of Cal. Tech. [63]* on one-photon atom trapping in the context of quantum physics, CQED, and quantum information processing. He considers these one-photon experiments as important milestones along the road leading to many important goals in the fields of CQED and quantum communications.

21.13. Experimental Realization of a One-Atom Laser in the Regime of Strong Coupling

Kimble's group at Cal. Tech. made a significant advance in single atom research with the first demonstration of a truly one-atom laser in 2003 [632]. The authors took advantage of their ability to capture single atoms in a FORT (far-off-resonance trap) dipole trap and achieved very strong coupling of atoms to high Q resonators [624], [625], [629]. Initially, cesium atoms were held at a very low density in a MOT. They were dropped from above into a close-spaced TEM_{00} mode high-Q stabilized Fabry-Perot cavity with a high probability of catching a single atom. Trapped single atoms thus can survive 2–3 s if cooled by a pair of σ^+, σ^- beams. If, however, the transverse cooling beams are switched in intensity and frequency to an optical pumping configuration, the atom is observed to lase for times up to $200\,\text{ms}$ before the atom escapes from the FORT.

The laser output exhibits some remarkable properties. Measurement of output power vs. pump power shows essentially no evidence of a threshold. The output builds up and saturates in $\sim 10^{-7}\,\text{s}$, which is much shorter than the trap lifetime of $\sim 0.05\,\text{s}$. The output, however, is typically about 10 times stronger than the fluorescence power. Furthermore, the light output is much more stable than with conventional many-atom lasers and exhibits sub-Poissonian statistics. Also, the photons are "anti-bunched", i.e., they are spread out more uniformly over time. The authors stress that this behavior is due to the action of a "one-and-the-same" trapped atom tightly coupled to the high-Q cavity. This differs from previous so-called one atom masers or lasers where the lasing was built up from the action of many random single atoms entering and leaving the cavity randomly over a period of time (see Secs. 21.1 and 21.2). The authors expect that the realization of this much anticipated quantum nonclassical light source will have practical applications. For further commentary on single atom lasers, see Ref. [633].

21.14. Cavity Cooling of a Single Atom

Mauntz *et al.* [634] at the Max Planck Institute of Quantum Optics has recently demonstrated cooling of an optically trapped single atom that was coupled strongly to a high-finesse cavity. It has been predicted that a new cooling mechanism is possible in such a system, which is basically different from usual mechanisms based on repeated cycles of spontaneous emission. In cavity cooling, photons of a weak blue-detuned optical cooling beam are Doppler upshifted in frequency. For strong atom–cavity coupling the presence of a single atom changes the optical path between the cavity mirrors and thus shifts the resonance and internal intensity. An atom when near a node of the standing wave, however, does not couple strongly to the field. When at an antinode, the atom interacts strongly and blue shifts the cavity to a higher frequency. This finally leads to a blue shift of the probe photons escaping from the cavity, which translates into a net cooling of the atomic kinetic energy.

In the experiment, an atom was injected into the cavity from below into a weak far-red-detuned trapping beam. Once the probe cooling beam detected its presence, the dipole trap power was raised, trapping the atom for about 18 ms. Measurements made on the average storage time as a function of the probe cooling power gave a maximum storage time of about 36 ms at about 0.4 pW of probe light. The storage time falls at higher than 0.4 pW presumably due to the onset of the absorption and spontaneous emission heating process. Much longer storage times than 36 ms were expected based on theoretical estimates of heating due to light scattering (85 s) and quantum electronic dipole fluctuations (200 ms). The difference is presumably attributed to parametric heating coming from power fluctuations of the trapping laser, as described by Thomas *et al.* [542]. By making improvements in the stabilization of the Ti:sapphire trapping laser, the atom storage time was increased to 60 ms.

One thus sees clear evidence of extended storage time and improved localization of atoms due to cooling. Tuning the probe to the red gives heating and a decreased storage time below the dark cavity retention time. The authors estimate the cavity cooling rate to be at least a factor of five higher than for free space cooling methods at comparable excitation. They also point out that the technique of cooling without need for excitation could be applied to molecules in the ground state or to an atom with a stored qubit during quantum computation.

21.15. Deterministic Generation of Single Photons from One Atom Trapped in a Cavity

An essential step in the transmission of quantum information from one atom to another is the deterministic generation of controlled single photons. Kimble and his group [635] at Cal. Tech. have

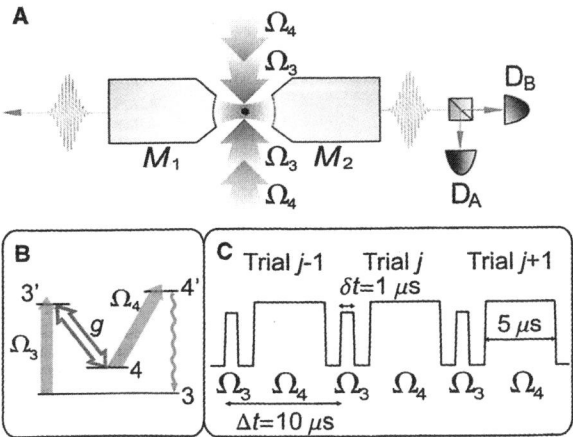

Fig. 21.3. Illustration of the generation of single photons by one atom trapped in an optical cavity. (**A**) A single Cs atom is trapped in a cavity formed by the reflective surfaces of mirrors (M_1, M_2) and is pumped by the external fields (Ω_3, Ω_4). (**B**) The relevant atomic levels of the Cs D_2 line at 852.4 nm. Strong coupling at rate g is achieved for the transition $F' = 3' \rightarrow F = 4$ near a cavity resonance, where $g = 2\pi \times 16$ MHz. Atom and cavity decay rates $(\gamma, \kappa)/2\pi = (2.6, 4.2 \text{ MHz})$. (**C**) The timing sequence for the generation of successive single photons by way of the $\Omega_{3,4}$ fields. Figure 1 and its caption from Ref. [635].

demonstrated the ability to generate a single photon on demand in a single mode of a Gaussian beam by externally controlled driving fields with close to unity probability. Starting with a cooled single cesium atom optically trapped in the regime of strong coupling to the mode of a high-finesse optical cavity, they illuminated the atom with a sequence of laser pulses $\Omega_3(t)$ and $\Omega_4(t)$, as illustrated in Fig. 21.3 taken from Fig. 1 of Ref. [635]. The control pulse $\Omega_3(t)$ transfers the trapped atom from the $F = 3$ level hyperfine ground state to the $F = 4$ level via a dark eigenstate of the coupled atom-cavity system. The $\Omega_4(t)$ pulse excites the atom to level $4'$ from which it decays back to the $F = 3$ ground state, generating a single photon flying qubit, that can transmit information over long distances. The pulse can then be repeated. Up to 10^4 or more photons can be generated for each trapped atom. The output pulses are directed by a beam splitter to two photon-counting detectors, D_A and D_B, as shown in Fig. 21.3.

22

Trapping of Single Atoms in an Off-Resonance Optical Dipole Trap

In spite of the demonstration of trapping of single atoms by single photons in a very high-Q resonant cavity and the observation of well-defined orbits of the confined atoms [63]*, [64], the fact remains that these confinement times, although long (many ms) on the scale of a single orbit, were quite short compared to the many seconds of confinement possible in more conventional MOTs and optical traps [636].

22.1. Single Atoms in an Optical Dipole Trap: Towards a Deterministic Source of Cold Atoms

Following the work on trapping of single atoms by photons, Frese *et al.*, in 2000 in Germany, developed a scheme based on off-resonance dipole traps for what they call a step towards "deterministic source of cold atoms" [636]. This involved improvements in the technique for capturing just a few individual atoms of cesium. They used a special very-small-volume MOT, from which they transferred the laser-cooled atoms with 100% efficiency to an off-resonance YAG diode trap operating at 1.06 μm wavelength. They point out the growing realization in the last few years that dipole traps are "an elegant and simple way of storing atoms". They cite the virtues of trapping atoms in all ground states and the subwavelength localization of an exactly known number of atoms. The authors measured a 51 second storage time for atoms in the off-resonance dark dipole trap. This is identical to the lifetime measured in a purely magnetic trap, showing that the background gas pressure limits the lifetime in both cases.

The center of the dipole trap, which had a beam waist of $\sim 5 \mu$m, was lined up with the center of the MOT by running both traps simultaneously and observing the fluorescence of the trapped atoms. Due to the Stark shift of the optical trapping beam, the best alignment corresponded to a minimum of the fluorescence.

The dipole trap potential depth was $\sim 16\,\text{mK}$ for a typical YAG laser power of 2.5 W. Since the temperature of the MOT-cooled atoms was $\sim 125\,\mu\text{K}$, there was no problem in capturing several atoms within the trapping volume when the trap was turned on. The exact number captured could be adjusted by varying the background cesium vapor pressure.

The authors foresee use of this new controlled source of atoms in experiments in cavity QED and information processing. This would be a useful adjunct to the experiments of Kimble's group [63]* and Rempe's group [64]. Use of this source would allow one to place atoms at precise locations in the high-Q cavity at precise times.

A second subsequent experiment was performed by the same group from the Institut für Angewandte Physik at Bonn in 2001 [637], which they characterize as "the realization of a deterministic source of single atoms". This involved capture of one or more cesium atoms in a $1.06\,\mu\text{m}$ standing wave dipole trap. The atoms were transferred with essentially 100% efficiency into the confocal region of two opposing $w_0 = 30\,\mu\text{m}$, 5 W Nd:YAG laser beams from the same small-volume MOT used above with the single-beam trap. Maximum trap depths of 2 mK were achieved and easily confined the atoms at $\sim 0.1\,\text{mK}$. Once captured, the atoms were transported along the common beam axis of the standing wave trap by changing the frequencies of the pair of standing wave beams, using a pair of acousto-optic modulators. They used the standing wave as a "single-atom moveable conveyor belt" to transport and relocate the trapped atoms at variable speeds to new locations, with submicrometer precision. Atoms were detected by resonance fluorescence excited by a weak on-resonance pulsed probe laser beam at $\lambda_{\text{res}} = 852\,\text{nm}$. The detection optics consisted of a fixed detector monitoring the center of the MOT and a moveable detector on a precision stage that followed the moving atoms on the conveyor belt. Fluorescence detection was limited to a narrow field of view of $\sim 40\,\mu\text{m}$.

Fixed numbers of atoms were moved out of the MOT source region at controlled speeds, stopped, returned, or launched into free space in a given direction at modest velocities by adjusting the frequency difference of the constituent standing wave beams. Figure 22.1, taken from Figs. 1–3 of Ref. [637] gives a clear picture of the loading of the standing wave trap, the apparatus, and the single-atom conveyor belt.

It is interesting to note that continued use of the resonance probe beam served to heat the atoms and to eventually cause the loss of the trapped atoms. Also, the movement of atoms away from the focus into regions of larger beam widths w_0 in the far field resulted in a lowering of the trapping potential. Both of these effects could cause a loss of atoms. Fortunately, as the authors point out, the lowering of the potential depth is slow enough to allow adiabatic cooling of the transported atoms to occur. As a result, atoms that would have ended up with a kinetic energy of half the potential depth in the far field were cooled enough to remain trapped over distances of the order of 5 mm.

A goal of this work on the atom conveyor belt is the feeding of atoms into a high-finesse cavity to form a string of atoms with well-defined separation to create an atomic shift register and quantum logic structures (see references in Ref. [637]).

22.2. Sub-Poissonian Loading of Single Atoms in a Microscopic Dipole Trap

There was another advance, in 2001, by Schlosser *et al.* of France, in the trapping and manipulation of single atoms in small optical dipole traps for use in connection with the processing of information at

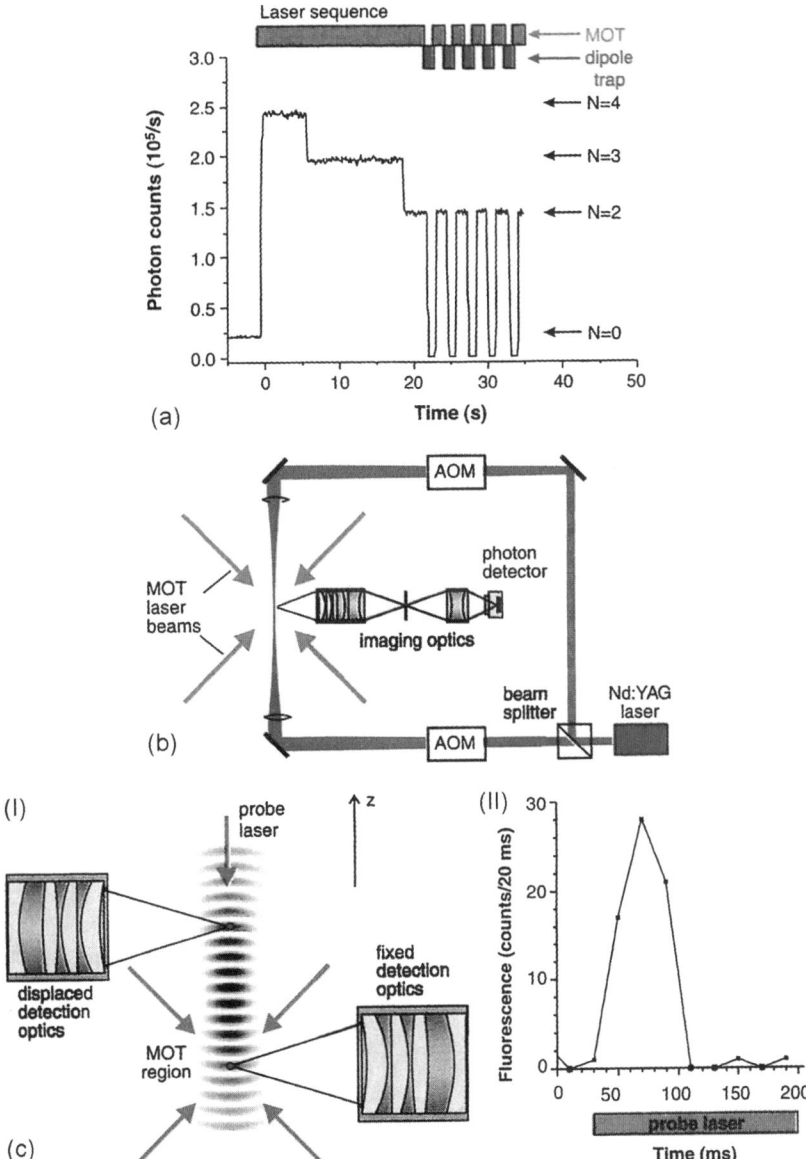

Fig. 22.1. (a) Loading the dipole trap with a desired number of atoms. Fluorescence from the MOT shows discrete signal levels and directly monitors the number of trapped atoms N. Here, the MOT is switched on at $t = 0$ and four atoms are captured from the background gas. A desired number of atoms (two, in this case) is transferred into the dipole trap by turning on the dipole trap laser a few milliseconds before the MOT lasers are turned off. After 1 s, the reverse procedure recaptures the atoms into the MOT, showing the same fluorescence level (i.e., the same number of atoms) as before. Transfer of atoms between the two traps can be repeated many times without losing atoms. (b) The MOT and dipole trap are overlapped in the center of a vacuum cell (not shown). AOMs are used to control the frequencies of the two laser beams that form the dipole trap. Fluorescence light is collected by imaging optics with spatial filtering apertures and is detected by a single photon counting detector. The photon count rate from the MOT is $5 \times 10^4 \, \text{s}^{-1}$ per atom (see Fig. (a)). (c) "Single-atom conveyor belt". (I) Moving the interference pattern transfers a trapped atom from the MOT region to a new position where the resonance fluorescence is collected by the displaced detection optics. (II) Detection of a single atom in the dipole trap. The graph shows a burst of fluorescence photons from one atom in the dipole trap displaced by $500 \, \mu$m. The probe laser is switched on with a time delay of 30 ms after transportation and produces no measurable contribution to stray light. Figures 1–3 and their captions from Ref. [637].

the quantum level [638]. In the previous experiments [636] single-atom dipole traps were developed to place one or more atoms into a small high-Q cavity electrodynamic system [63]*. In the present work, the object was to directly address individual atoms with high-spatial resolutions in extremely small trapping volumes. Using high-numerical aperture objectives (NA = 0.7) they successfully trapped single [87]Rb atoms using light detuned well below the resonance D1 line at 795 nm and the D2 line at 780 nm. The high NA objective lens was mounted within the vacuum and gave a focal spot $w_0 < 1\,\mu$m. Only single atoms could be trapped. These were observed by residual fluorescence that emerged from a diffraction-limited spot of $0.7\,\mu$m, which was the diffraction limit of the imaging system. With a dipole trap power of 1 mW, the atom scattered about 4000 photons per second, which were very easily detected. The authors attributed the inability to trap more than one atom at a time in the trap to the process of optically assisted collisions from the residual MOT light [182]*. They observed trapping times of single atoms that were as long as two seconds.

Although a single trap could hold only a single atom, they were able to introduce a second beam at a slight angle through the same optics, which trapped a second atom located at a controlled distance of about 1–$10\,\mu$m away from the first atom. This opens the possibility of studying atom–atom entanglements. The authors discuss the possibility of one-qubit and two-qubit operations. They think this setup fulfills the requirements for realizing various fast logic gates (see discussion and references in Ref. [638]).

23

Vortices and Frictionless Flow in Bose–Einstein Condensates

It was evident from the start of the subject of Bose–Einstein (BE) condensation of dilute vapors that novel properties could be expected, in analogy with the strange behavior of the denser helium superfluids, which were the first examples of BE condensate systems. The early experiments on the dynamics of BEC and the oscillatory behavior of perturbed condensates, as well as sound propagation, as discussed above, were all, in a sense, examples of simple superfluid behavior. One of the novelties of the later work with vortices and frictionless flow in BE condensates was the ability to compare the results with the relatively simple Gross–Pitaevskii theory [204], [500]. As mentioned earlier, this theory was formulated with dilute gaseous condensation in mind. The purity of the vapor condensates contrasts with the substantial fraction of noncondensed atoms in liquid helium condensation. The subject reached another level of sophistication with the more recent proposals and experiments on vortex and frictionless flow in gases.

23.1. Vortices in a Two-Component Bose–Einstein Condensate

In late 1999, Williams and Holland [639] proposed a clever scheme for observing superfluid vortex behavior in BE condensates. They showed how to use a two-component condensate to generate persistent vortex modes with specific angular momenta. They ascribe the stabilizing mechanism for such persistent modes to the macroscopic occupancy of a single quantized mode free of thermal relaxation.

The proposed vortex generating scheme involves two steps: first, the formation of two strongly coupled spin states from an original condensed spin state; followed by the mechanical rotation of the trapping potentials that confine the condensate. The rotation speed is such that a phase shift of $2n\pi$, where n is an integer, is introduced in one of the two coupled states. The other state remains as a fixed core in space and serves to pin the vortex. With $n = 1$, a single vortex is formed in the condensate with angular momentum $h/2\pi$. The authors analyze the dynamics of the vortex evolution for the two coupled states, using two coupled Gross–Pitaevskii nonlinear Schrödinger equations.

Important factors in the theory of vapor phase condensates are the weakness of the atom–atom coupling and the purity of the condensate. Condensates are very close to 100% condensate, compared to only about 10% condensate in liquid helium. Solutions for $n = 1$ to 4, corresponding to single, double, triple, and quadruple vortices, are given by Williams and Holland.

The quantum mechanical vortices predicted by the theory resemble classical vortices in classical fluids only qualitatively. The local velocity of a quantum fluid in a BE condensate is proportional to the rate of change of phase of the fluid and is greatest at the center. For a rotating vortex in a classical fluid the velocity is greatest at the outer edges and lowest near the axis. Thus, quantum vortices behave contrary to ordinary expectations.

23.2. Observation of Two-Component Vortices in a Bose–Einstein Condensate

Cornell, Wieman, and collaborators Matthews *et al.*, also at JILA [640] implemented the proposal of Williams and Holland [639] for generating two-component vortices. In their experiment, they started with a condensate with all atoms in a single hyperfine state $|1\rangle$. They then applied a red-detuned microwave field far from a second hyperfine state $|2\rangle$, so that essentially no transition occurred. Finally, they shined a focused spinning off-axis laser into the condensate, with intensity sufficient to Stark shift the condensate atoms into resonance with the microwaves, to make transitions from $|1\rangle$ to $|2\rangle$ possible.

By controlling the angular velocity of the rotating beam, the authors succeeded in introducing a $2\pi n$ rotational phase shift in a significant number of condensate atoms in the $|2\rangle$ state, as proposed by Williams and Holland [639]. The authors were able to view the vortex structures thus formed using nondestructive phase contrast imaging. This involved converting part of the fixed uniform condensate $|1\rangle$ into state $|2\rangle$ as a reference. The image of the combined $|2\rangle$ atoms then displays the relative phase as a cosine variation in azimuth. The vortices, thus formed with their inner core of $|1\rangle$ state atoms, were stable up to 600 ms, after which they dispersed.

Following the experiments of Matthews *et al.* [640] in 1999, demonstrating vortices in two species of condensates, García and Pérez-García, in 2000 [641], made a theoretical analysis of the stability of vortices in multi-component BE condensates. Using a model based on coupled Gross–Pitaevskii equations, they showed that a two-component vortex is stable only when it is placed in the component with the largest scattering length. Their results are in agreement with the experimental results of Matthews *et al.* If the roles of the two components are reversed and a vortex with a higher scattering length is formed within the lower scattering length condensate, the vortex is found to be unstable and tends to wander out of the condensate. The model used does not include dissipative effects, so that there is no mechanism to get rid of the vortex angular momentum. In the unstable cases, they start with the vortex on axis and then introduce a perturbation that sets off complex dynamics. Oscillations develop and the central hole breaks up and complex spirals rotate about in the condensate. The hollow core can transfer from one condensate to the other. For a larger perturbation the dynamics is even more complex. The authors describe it as spatiotemporal chaos.

23.3. Single-Component Vortices in Bose–Einstein Condensates

Another experiment on vortex behavior in a single-component BE condensate was published in 1999 by Raman *et al.* [642] of Ketterle's group, back to back with the letter describing the two-component experiment [640], in the same issue of Physical Review Letters. In this experiment the authors studied the frictional forces resulting from linear motion, rather than rotational motion of an "object" moving through a condensate. The moving "object" was not a physical object such as might be used with liquid helium condensates, but a focused laser beam. They used a beam tuned above the atomic resonance that served to repel atoms in the condensate, and so acted as a cold physical rod that displaces atoms. The authors oscillated their beam at constant velocity through the condensate, at varying frequencies, and observed the amount of energy deposited in the condensate after a period of oscillating back and forth by measuring the initial and final temperatures of the condensate and the fraction of noncondensed atoms. At high-beam velocities, they found heating proportional to beam velocity. At low-beam velocities, there was a threshold velocity of about 1.6 mm/s, below which no heating was observed. The authors believe it is significant that this velocity is close to the theoretically calculated critical velocity for the formation of vortices.

Some interesting commentary and relevant references can be found in October 1999 "News and Views" article in Nature by Rokhsar [643], as well as in a "Search and Discover" article on vortices and frictionless flow by Levi in the November 1999 issue of Physics Today [644].

23.4. Single-Component Vortices Generated by an Optical Stirring Spoon

At the beginning of 2000, Madison *et al.* of Dalibard's group [645] demonstrated single-component vortex formation in a particularly graphic manner by stirring a BE condensate using attractive far-off-resonance optical dipole potential as a "stirring spoon". This is the analog of vortex formation in a rough-walled rotating bucket of liquid. In this case the liquid was a cigar-shaped condensate in a Ioffe-Pritchard magnetic trap. The stirring spoon was formed from a focused laser beam directed parallel to the long axis of the condensate and swept rapidly back and forth along a line connecting points equidistant from the axis. This effectively formed a sheet beam potential that attracted atoms. The sheet beam was then rotated about the axis at a fixed angular velocity. The condensate was formed by evaporation while the spoon was stirring it. When the evaporation cycle was over, the condensate was allowed to reach thermal equilibrium in the "rotating bucket" for a time of 500 ms. The experimental procedure was to increase the angular stirring velocity gradually from zero up to ~ 250 Hz. As the rotational frequency increased, no vortices were observed until a critical rotational frequency was reached at ~ 147 Hz. Above this frequency single, multiple and then, finally, turbulent vortex patterns were successively observed. One of the advantages of these vortices was their macroscopic size. The empty vortex cores had diameters of $2\xi = 0.4\,\mu$m. Thus, the radius of the empty core was of the order of the healing length $\xi = (8\pi a \rho)^{-1/2}$. Here, ρ is the density of the condensate and "a" is the scattering length. See also Sec. 20.7, which discusses ξ in the context of BE condensation in lower dimensions. This size is about 10^3 times larger than the ^{4}He vortices. Once the vortices formed, the authors turned off the driving spoon potential and allowed the condensate to fall and expand to $\sim 250\,\mu$m in overall diameter, by virtue of its mean field energy. They were then able to photograph

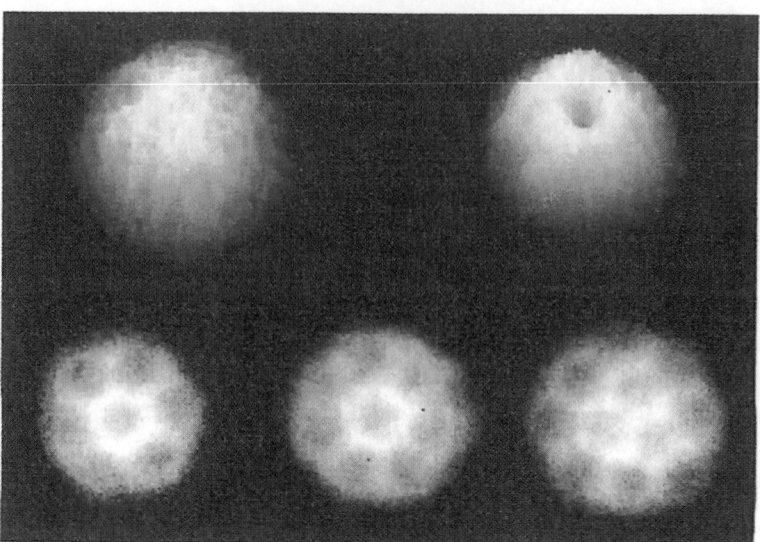

Fig. 23.1. Stirred BE Condensates show no vortices when the stirring frequency is below the critical frequency for vortex nucleation (top left). Once the critical frequency is crossed, a vortex appears at the center of the condensate (top right). Arrays of vortices are found for higher stirring frequencies. Shown below are images of 7, 8 and 11 vortices. Figure and caption from Ref. [646]*.

the condensate by absorption imaging and determine the number of atoms in the condensate and the temperature. The results were $N_0 \cong 1.4 \times 10^5$ atoms and $T \leq 80$ nK.

Figure 23.1, taken from commentary by Fitzgerald in Physics Today of August 2000 [646]* shows some excellent photographs of vortices from the work of Madison *et al.* [645]. Notice the uniform spacing of the multiple vortices on the outer edges of the condensate. As Madison *et al.* point out, this supports a prediction of a repulsive interaction between vortices of the same sign of rotation.

The authors believe that the ease of formation of these single-component vortices and their ready observability make this technique particularly useful for studying the dynamics of vortex nucleation and decay.

23.5. Scissors Mode Excitation of Superfluidity

There was yet another experiment giving strong evidence for superfluidity in a trapped BE condensed gas by Maragò *et al.* [647], working with Foot at Oxford, in 2000. These authors implemented a suggestion by Guery-Odelin and Stringari [648] to use a "scissors mode oscillation" to demonstrate superfluidity. This involved initiating the scissors oscillation of the condensate by a sudden rotation of an anisotropic harmonic trapping potential through a small angle. This type of scissors mode oscillation has been used previously in nuclear physics to demonstrate the superfluidity of neutron and proton clouds in deformed nuclei (see references in Ref. [647]). In this experiment, for small twist amplitudes, the cigar-shaped condensate was not deformed, but oscillated as a whole in the anisotropic trapping potential of a TOP trap. Figure 23.2, taken from Fig. 1 of Ref. [647], shows the method of excitation of the scissors mode.

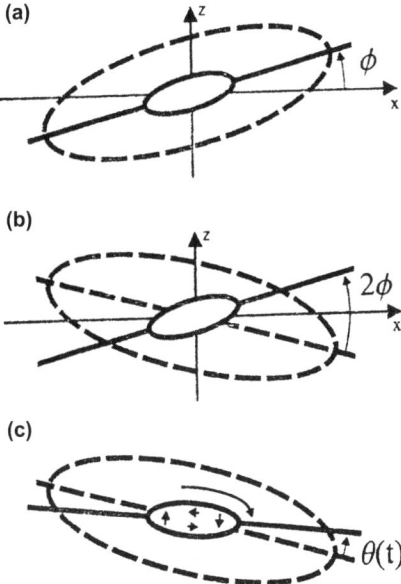

Fig. 23.2. The method of exciting the scissors mode by a sudden rotation of the trapping potential. The solid lines indicate the shape of the atomic cloud and its major axes. The dotted lines indicate the shape of the potential and its major axes. (a) The initial situation after adiabatically ramping on the field in the z direction, with cloud and potential aligned. (b) The configuration immediately after rotating the potential, with the cloud displaced from its equilibrium position. (c) The large arrow indicates the direction of the scissors mode oscillation and the smaller arrows show the expected quadrupolar flow pattern in the case of a BEC. The cloud is in the middle of an oscillation period (the angles have been exaggerated for clarity). Figure 1 and caption from Ref. [647].

In the thermal (normal) regime, at a temperature of $\sim 1\,\mu$K, which is about five times the critical temperature, the condensates oscillate in a combination of two damped frequencies, as shown in Fig. 23.3(a), taken from Fig. 3(a) of Ref. [647]. These frequencies were measured to be $\omega_1/2\pi = 338.5 \pm 0.8$ Hz and $\omega_2/2\pi = 159.1 \pm 0.8$ Hz, which is in good agreement with the prediction of theory [648] of 339 ± 3 and 159 ± 2 Hz. When the condensate was cooled well below the critical temperature where no thermal cloud is observable, the pure condensate was observed to oscillate at a new undamped frequency $\omega_c/2\pi = 265 \pm 2$ Hz, as shown in Fig. 23.3(b). This agrees with the value of 265 ± 2 Hz, as predicted by the theory in Ref. [648]. These observations give clear-cut evidence of superfluidity. The authors think that it would be informative to have measurements of the scissors mode frequency in the finite temperature regime, where a sizeable background thermal gas cloud is present in addition to the BE condensate. One expects a measurable damping [648]. The interaction of a condensate moving through an uncondensed component has apparently not been fully understood in the past [648].

23.6. Suppression and Enhancement of Impurity Scattering in a Bose–Einstein Condensate

Chikkatur *et al.* in Ketterle's group in 2000 [649] studied the interaction of impurity atoms (really fast moving untrapped atoms) with a stationary gaseous condensate. This new work closely relates

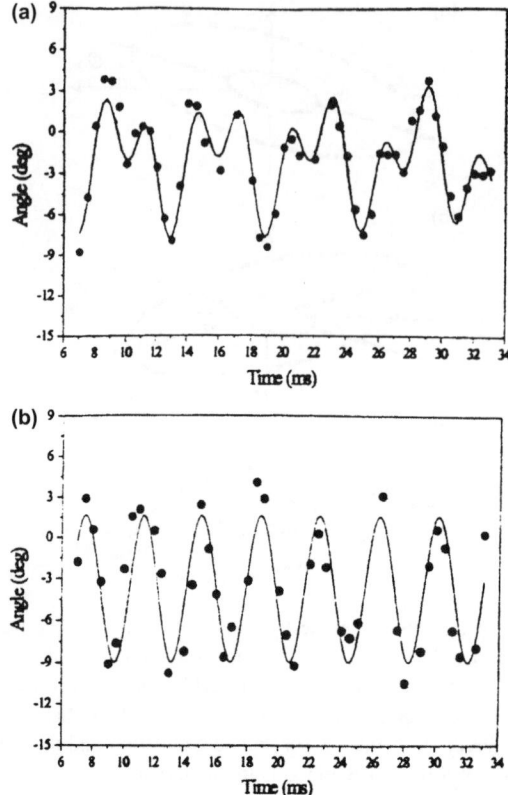

Fig. 23.3. (a) The evolution of the scissors mode oscillation with time for a thermal cloud. For a classical gas the scissors mode is characterized by two frequencies of oscillation. The temperature and density of our thermal cloud are such that there are few collisions, so no damping of the oscillations is visible. (b) The evolution of the scissors mode oscillation for the condensate on the same time scales at the data in (a). For the BEC there is an undamped oscillation at a single frequency ω_c. This frequency is not the same as either of the thermal cloud frequencies. Figures 3(a) and 3(b), and their captions from Ref. [647].

to the question of the interaction of a moving condensate with a background of uncondensed thermal atoms that Maragò *et al.* [647] referred to.

Chikkatur *et al.* discuss the propagation of a fast moving impurity atom using a stationary condensate in terms of the Landau critical velocity v_L, below which there is no dissipation. For the case of motion of a macroscopic object, such as an optical stirring rod, one observes the onset of dissipation at velocities much less than the critical Landau velocity [642]. This is believed to be due to turbulence and vortex generation. For microscopic objects, such as impurity atoms, no such turbulence is possible and one expects to observe dissipationless motion up to the critical Landau velocity. As Chikkatur *et al.* point out, the critical velocity for a gaseous condensate turns out to equal the velocity of sound, as one might suspect.

The impurity atoms were created using a stimulated Raman process that transferred a small number of condensate atoms from the $|F = 1, m_F = -1\rangle$ state into an untrapped hyperfine state $|F = 1, m_F = 0\rangle$ with a well defined initial velocity. As the impurity atoms pass through the condensate, they can interact and change their momentum distribution. When the velocity of the impurity atoms is reduced below the speed of sound, the probability of collision with the condensate atoms reduces drastically. This is strong evidence for the existence of superfluidity in BE condensates. For about 10^7 atoms in the condensate in a cigar-shaped magnetic trap, the velocity of sound was varied in the range from 0.55 to 1.0 cm/s.

The experimental techniques for observing the impurity atoms have grown quite sophisticated. The pulse of Raman-heated impurity atoms was allowed to scatter from the condensate cloud for ~ 4 ms, the time required to traverse the condensate, after which the magnetic trap was turned off. After another 5 ms, a Stern-Gerlach magnetic field gradient was pulsed on for 30 ms, spatially separating the $m_F = 0$ impurity atoms from the $m_F = -1$ condensate. All the atoms were then examined after another 60 ms of free flight, using resonant absorption imaging. This required optical pumping of all the atoms into the $|F = 2, m_F = 2\rangle$ state of sodium.

Figure 23.4, taken from Fig. 2 of Ref. [649], shows the suppression of collisions for slow impurity atoms traveling upward along the axis of the condensate at a velocity of 7 mm/s. In Fig. 2(a), which is above the critical velocity, the number of scattered impurity atoms is large. By comparison, in Fig. 2(b), below the critical velocity, the number of collisions is suppressed.

Using data such as Fig. 2(b), it was discovered that the fraction of impurity atoms that collided increased by about a factor of two as the number of impurity atoms increased. According to a simple perturbation picture, one expects the fraction of impurity atoms scattered to be a constant independent of impurity atom density. The authors interpret this result as due to a collective self-amplification of atomic scattering similar to the recently observed super-radiant amplification of light scattered from a BE condensate [650]. Collisions with impurity atoms transfer atoms from the macroscopically occupied initial state of the condensate to unoccupied higher momentum states. Atoms in these higher states can then stimulate further scattering by bosonic enhancement. This effect increases with density. Such collisional amplification is not directional and resembles recently observed superfluorescence [651].

The authors mention that they have also used Bragg transition [522] to make impurity atoms. They also observe that these impurity atom scattering results affect schemes for making Raman and Bragg

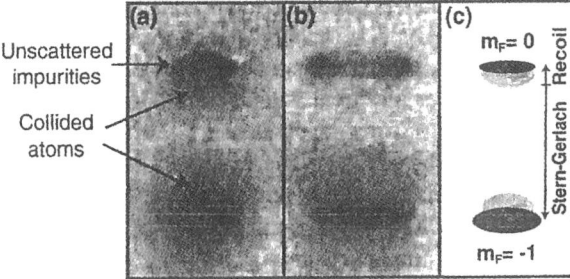

Fig. 23.4. Superfluid suppression of collisions. The impurity $m_F = 0$ atoms (top) traveled at 7 mm/s along the condensate axis (upward in image) and were separated from the condensate (bottom) by a magnetic field gradient applied during ballistic expansion [see (c)]. (a) Absorption image after 50 ms of time of flight shows the collisional products as indicated by the arrow. For this image, $v_g/c = 2.7$ (see text), and the collided fraction is about 20%. (b) Similar image as (a) with $v_g/c = 1.6$, where the collisions are suppressed. The outcoupled atoms (impurities) were distorted by mean-field repulsion. The images are 2.0 mm × 4.0 mm. Figures 2(a) and 2(b), and caption from Ref. [649].

output couplers for atom lasers [32], [33]*, [34]*, [520]. They suggest that to avoid the enhanced scattering losses, one can either reduce the density to where the stimulated effects are small, or increase the density to where the speed of sound is greater than that of the uncoupled atoms, thereby realizing a "superfluid output coupler". The latter solution is superior, since it gives a more intense output beam.

23.7. Hydrodynamic Flow in a Bose–Einstein Condensate Stirred by a Macroscopic Object

In 2000, Onofrio *et al.* in Ketterle's group [652] updated their previous sodium stirring experiment [642] in a BE condensate by using a blue-detuned 5145 Å laser beam. They studied the onset of dissipation with higher sensitivity and using nondestructive phase-sensitive imaging in very pure condensates with up to 5×10^7 sodium atoms, at densities up to 3.5×10^{14} atoms/cm^3. The macroscopic repulsive laser stirring rod had a beam size $2\omega_0 = 10 \,\mu$m. It was swept through an $\sim 200 \,\mu$m condensate with a constant velocity. At low-stirring velocities, in the superfluid regime, there was little effect on the condensate. At higher velocities the density distribution became asymmetric and one could see condensate piled up in front of the moving beam. Figure 23.5, taken from Fig. 2 of Onofrio *et al.* [652], shows the observed pressure difference across the laser stirring beam as the beam was swept to the right and left through the condensate. There was a critical velocity for the onset of a drag force of the laser on the condensate. The authors were able to determine the force from the asymmetric density and pressure difference across the moving laser stirring rod. By using different condensate densities, they were able to observe the density dependence of the critical velocity. They were able to improve on their previous calorimetric measurements and determine the energy being transferred into the condensate as a function of the stirring velocity. The calorimetric measurements

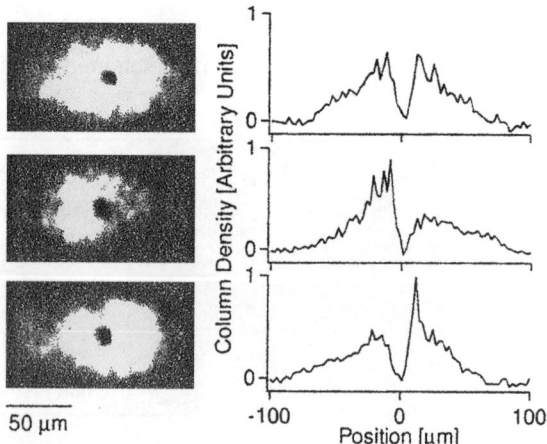

Fig. 23.5. Pressure difference across a laser beam moving through a condensate. On the left side *in situ* phase contrast images of the condensate are shown, strobed at each stirring half period: beam at rest (top), beam moving to the left (middle), and beam moving to the right (bottom). The profiles on the right are horizontal cuts through the center of the images. The stirring velocity and the maximum sound velocity were 3.0 and 6.5 mm/s, respectively. Figure 2 and caption from Ref. [652].

were in excellent agreement with the force versus stirring velocity measurements. The authors point out that there are now many different theoretical models for the critical velocity and that the experiment is now in a position to help resolve the contradictory predictions (see references to the theory in Ref. [652]).

23.8. Observation of Vortex Lattices in Bose–Einstein Condensates

A significant advance in the generation of vortices was made in 2001 by Abo-Shaeer *et al.* in Ketterle's group in experiments demonstrating the formation of large, highly ordered vortex lattices in a rotating BE condensed gas [653]*. Triangular lattices of as many as 130 extremely regular vortices were formed in large condensates containing up to 5×10^7 sodium atoms at a temperature of $310\,\text{nK}$. These lattices lasted for times of several seconds, while individual vortices could persist for as long as $40\,\text{s}$. They were formed by a pair of blue-detuned stirring beams rotating symmetrically about the long axis of cylindrical magnetic traps. Figure 23.6, taken from Fig. 1 of Ref. [653]*, shows examples of vortices crystallized in a triangular pattern.

The authors devised a clever way of imaging just the central slice of the condensate. This was done by optically pumping the slice from $F = 1$ to $F = 2$ hyperfine state. The condensate was allowed to expand and then only the $F = 2$ atoms were viewed by absorption imaging. This avoided distortions in the lattice due to surface effects. Moving a single laser beam once radially through the condensate generated vortex lattices that were essentially identical to those formed with the two rotating beams. The authors think that imperfect beam alignment is the origin of the torques given to the condensate in that case.

They suggest use of these lattices as a tool for studying the dynamics of lattice formation and decay. They may serve as a model system for understanding lattice formation in superconductors. The role of the uncondensed thermal component can also be investigated.

It occurs to me that it might be possible to use the pair of optical beams as traps for vortices for directly studying the forces known to exist between vortices. Such measurements may be possible in analogy with force measurements between spherical charged particles in colloidal suspensions [385]*, [388]* and between binary entropic colloids [400]. Some experimental details are lacking in Ref. [653]*, so it is hard to say how easily one can adapt the apparatus for this purpose.

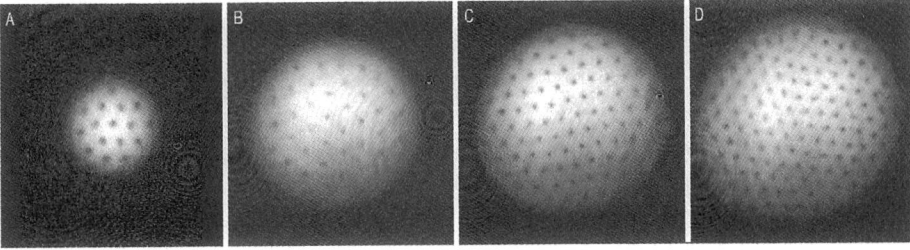

Fig. 23.6. Observation of vortex lattices. The examples shown contain approximately (A) 16, (B) 32, (C) 80, and (D) 130 vortices. The vortices have "crystallized" in a triangular pattern. The diameter of the cloud in (D) was 1 mm after ballistic expansion, which represents a magnification of 20. Slight asymmetries in the density distribution were due to absorption of the optical pumping light. Figure 1 and caption from Ref. [653]*.

23.9. Measurement of the Angular Momentum of a Rotating Bose–Einstein Condensate

Later in 2000, Chevy *et al.* of Dalibard's laboratory [654], following up on their previous experiment [645], succeeded in measuring the angular momentum per atom in a single vortex by observing the precession of the axes of the anisotropic transverse profile of an angularly stirred BE condensate. The stirring was done by rotating an attractive sheet-like beam that propagates parallel to the long, or z-axis of a cigar-shaped condensate. This acts as a rotating "optical spoon" [646]*. This was different from the stirring of Onofrio *et al.* [652], where they stirred a condensate with a repulsive "rod-like" beam moving at fixed velocity back and forth along the condensate, perpendicular to the z-axis of the beam. In Chevy *et al.*'s experiment vortices were formed with spin axes parallel to the z-axis. For Onofrio, they were spun off due to the "rod" pushing through the condensate, much as a when a solid rod moves through a viscous liquid. A single line vortex spinning parallel to the z-axis is much better controlled than the vortices formed from the moving rod.

The authors follow the theoretical treatment of angular momentum of a condensate harmonically trapped in a quasi-cylindrically symmetric potential. For such a case, the angular momentum L_z of the condensate along the symmetry axis can be expressed in terms of the frequencies of the excitation modes carrying angular momentum. In particular, as suggested by Zambelli and Stringari [655], they studied the two transverse quadrupole modes $m = \pm 2$ carrying angular momentum and having frequencies ω_+ and ω_-.

Chevy *et al.* started by generating a single vortex at beam center by rotating their optical spoon potential at an angular velocity Ω close to the critical angular velocity Ω_c. Once it was formed, they excited a quadrupolar oscillation using the optical dipole potential of the stirring laser, on a fixed basis, as a short (compared to the transverse oscillation period), intense pulse of 0.3 ms duration and having 10 times the intensity previously used for stirring. This excited the transverse quadrupole mode of the slightly asymmetric condensate and set it into angular precession. This quadrupole mode can be viewed as a linear superposition of the two quadrupole modes $m = \pm 2$ of frequency ω_+ and ω_-. For a condensate stirred below the critical frequency Ω_c these two modes are degenerate and the angular precession frequency θ, which is proportional to $(\omega_+ - \omega_-)$, is zero. For $\Omega > \Omega_c$, the degeneracy of the modes is broken and the condensate precesses. As the authors explain, to measure the total angular momentum per atom L_z, one must know the precession frequency $(\omega_+ - \omega_-)$, the average radius of the condensate r_{perp}, and the total number of atoms. By using the relation: $2L_z = M r_{\text{perp}}^2 (\omega_+ - \omega_-)$, where M is the atomic mass, the authors deduced L_z, the angular momentum per atom. Figure 23.7, taken from Fig. 1 of Ref. [654], shows how they were able to measure the transverse oscillations and find the precessive frequency of the stirred condensate using time-of-flight expansion and absorption imaging. The expansion factor for Fig. 23.7 was ~ 27. From these images one also finds the average condensate dimensions (r_{perp}) and the total number of atoms.

They found that $L_z/(h/2\pi) = 1.2 \pm 0.1$, which is quite close to the expected value of $L_z = h/2\pi$. The biggest source of error was in the measurement of the angular precession frequency θ.

Measurements were also made for stirring frequencies well above Ω_c, where one can nucleate up to five vortices. They found that the total angular momentum per atom L_z increased by steps considerably less than $h/2\pi$ as the number of vortices increased. Also, as the vortex number increased, they distributed themselves farther and farther apart from one another and moved toward the edges

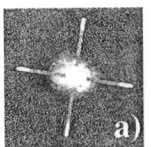 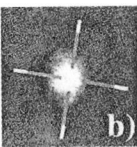 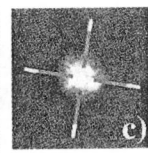   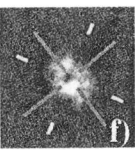

Fig. 23.7. Transverse oscillations of a stirred condensate with $N = 3.7\,(\pm1.1) \times 10^5$ atoms and $\omega_\perp/2\pi = 171\,\mathrm{Hz}$. For (a)–(c) the stirring frequency is $\Omega/2\pi = 114\,\mathrm{Hz}$, below the vortex nucleation threshold $\Omega_c/2\pi = 115\,\mathrm{Hz}$. For (d)–(f) $\Omega/2\pi = 120\,\mathrm{Hz}$. For (a), (d) $\tau = 1\,\mathrm{ms}$; (b), (e) $\tau = 3\,\mathrm{ms}$; (c), (f) $\tau = 5\,\mathrm{ms}$. The fixed axes indicate the excitation basis and the rotating ones indicate the condensate axes. A single vortex is visible at the center of the condensate in (d)–(f). Figure 1 and caption from Ref. [654].

of the condensate. These observations indicate that the individual vortices, which rotate in the same direction, repel one another and have less than $h/2\pi$ angular momentum per atom as they move away from the condensate axis toward the beam edge. This behavior of the condensate with multiple vortices is in agreement with the theoretical predictions of Ref. [656] (see Sec. 23.7).

23.10. Vortex Precession in Bose–Einstein Condensates: Observations with Filled and Empty Cores

One of the anticipated phenomena relating to the recent observations of single quantized vortices in dilute gas BE condensates is the observation of vortex precession about the condensate axis. Thinking of a vortex as a rotating top, if one exerts a force perpendicular to the spin axis, one expects the top to respond by precessing at right angles to the plane of the force and spin axis. Much the same happens with a quantized vortex that is located asymmetrically in a BE condensate. As the authors Anderson *et al.* from the JILA Laboratory of Wieman and Cornell pointed out in 2000 [657]*, a vortex core off-axis should feel a radial force towards lower condensate densities.

In this experiment the JILA group formed their vortices as previously [640] in a two-component condensate in ^{87}Rb. Such a system is formed by rotating a nonresonant optical beam about the axis of a stationary condensate, in a hyperfine ground state labeled $|1\rangle$, with an intensity sufficient to Stark shift the condensate into resonance with an rf field to convert some fraction of the $|1\rangle$ atoms to hyperfine $|2\rangle$ state atoms. They also introduced an automatic 2π phase shift in the generated $|2\rangle$ component, if the optical beam rotation frequency is made equal to the Rabi oscillation frequency between the states $|1\rangle$ and $|2\rangle$. This generates a simple stable on-axis vortex in a $|2\rangle$ state condensate in the presence of a stationary $|1\rangle$ state condensate.

Due to instabilities in the formation process, many of the vortices are created off-center. By using nondestructive phase contrast imaging it was possible to observe these off-center single vortices precess about the condensate axis, as shown in Fig. 23.8, taken from Fig. 1 of Ref. [657]*. Normally, one cannot resolve the core of a condensate without letting the condensate expand in an untrapped state. Here, the presence of stationary atoms in the core enlarged the diameter sufficiently to give the large-diameter core seen in the figure. There was some indication of a decrease in the diameter of the vortex core with time (Fig. 23.8). This may be due to decay of $|2\rangle$ fluid through inelastic collisional processes. The observed precession rates, as deduced from data such as in Fig. 23.8, are in qualitative agreement with numerical simulations for two-component condensates.

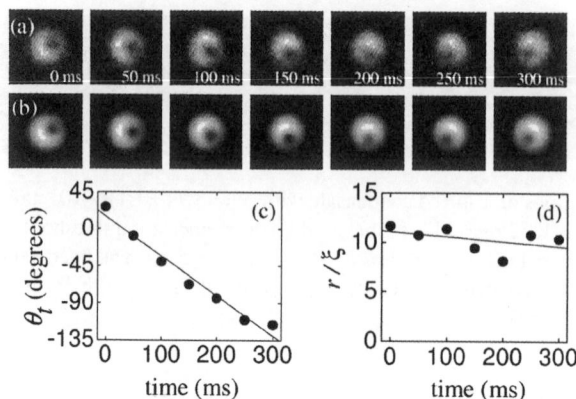

Fig. 23.8. (a) Seven successive images of a condensate with a vortex and (b) their corresponding fits. The 75 μm-square nondestructive images were taken at the times listed, referenced to the first image. The vortex core is visible as the dark region within the bright condensate image. (c) The azimuthal angle of the core is determined for each image, and is plotted versus time held in the trap. A linear fit to the data indicates a precession frequency of 1.3(1) Hz for this data set. (d) Core radius r in units of healing length ξ. The line shown is a linear fit to the data. Figure 1 and caption from Ref. [657]*.

Measurements were also made of the precession rates for bare-core vortices or single-component vortices. For these measurements the authors adiabatically removed the stationary $|1\rangle$ component completely, in order not to perturb the system. Since bare-core vortices are made visible only by a destructive type of measurement, in which the condensate is allowed to expand by releasing it from the trap, one has to observe the precession using a series of vortices and varying the hold time in the condensate. They determined a precession frequency of 1.8 Hz, which is slightly faster than the measured precession of filled condensates. Theoretical predictions for the precession rate vary from ~ 1.6 Hz for a solution of the Gross–Pitaevskii equation [658] to ~ 1.24 Hz [659] and to even lower values of ~ 0.8 Hz [660], using a two-dimensional model.

23.11. Generating Solitons by Phase Engineering of a Bose–Einstein Condensate

A very large number of researchers, including many from other institutions around the world, working in Phillips's group at NIST, contributed to a paper by Denschlag *et al.* in 2000 [661] on generating solitons in condensates. This paper introduced the notion of manipulating the phase of the quantum mechanical wave function of a condensate to control the behavior of the condensate. For example, by irradiating a condensate with an off-resonance blue-laser pulse having a controlled intensity distribution, one can imprint a desired phase variation on the condensate. This initial two-dimensional phase is short enough that it causes no motion, but simply changes the initial condensate wave function. The condensate then evolves under the influence of the imposed wave function. One of the theoretically expected results for an initial condensate with a repulsive atom–atom interaction at zero temperature and zero thermal atom content is the propagation of a dark soliton. In such a soliton, the atom distribution has a dark or empty region that propagates, without changing its shape, through the condensate at a velocity that can be considerably slower than the velocity of sound in the condensate.

The authors imprinted their sodium condensate using various shaped masks, such as a razor blade, that illuminated only one half the condensate, or a mask with a narrow slit, that irradiated the condensate with a thin stripe of light.

The authors measured the amount of phase change introduced with a Mach-Zender type of matter wave interferometer. They then followed the time development of the condensate over times up to $\sim 10\,$ms using absorption imaging. Figure 23.9, taken from Fig. 5 of Ref. [661], shows some of the impressive results obtained, which are in very good agreement with theory from the Gross–Pitaevskii equation. As the initial phase change increases, one sees in Fig. 23.9 that many solitons appear and propagate through the condensate.

The correspondence between the physics of atom solitons in condensates and optical solitons in optical fibers, for instance, is very close. In both cases, the soliton shape propagates without change due to a balancing of dispersive effects and nonlinear compression effects. Interestingly, if one has a condensate with an attractive atom–atom interaction, such as in ^{7}Li, one expects to find bright solitons, as found in optics, rather than dark solitons.

There is a close relation between solitons in BE condensates and vortices. For complex excitation in three-dimensions, for instance, one expects the solitons to eventually break up into vortices. It is also possible, with proper phase imprinting, to generate vortices directly. One can achieve the 2π phase winding of a vortex wave function by imaging a linear azimuthally varying intensity pattern.

For the future, the authors also suggest studying soliton–soliton collisions and other nonlinear dynamical effects, using solitons in condensates.

Another experiment, on "dark solitons" in BE condensates, was performed by Burger et al. at about the same time at the University of Hanover in Germany [662]. They used ^{87}Rb atoms in a long cigar-shaped condensate. The authors induced solitons by imprinting a local phase onto the BE condensate wave function. They then monitored the resulting dynamics of the wave function and the evolution of the induced density profile by releasing the condensate after times up to $\sim 15\,$ms,

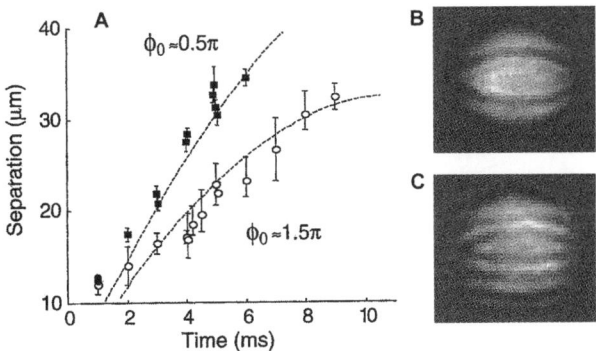

Fig. 23.9. (A) Plot of separation versus time for two oppositely propagating solitons after a phase imprint in the form of a stripe. For a small phase imprint ($\phi_0 \approx 0.5\pi$, squares), the solitons move at almost the local speed of sound. For a larger phase imprint ($\phi_0 \approx 1.5\pi$, circles), they are much slower. The dashed lines are from numerical simulations, from which we extract speeds for the corresponding solitons of 2.56 mm/s ($\phi_0 = 0.5\pi$) and 1.75 mm/s ($\phi_0 = 1.5\pi$) at 4 ms. (B) The condensate 6 ms after a stripe phase imprint of $\phi_0 \approx 1.5\pi$. (C) For a larger phase imprint of $\phi_0 \approx 2\pi$ many solitons appeared. Figure 5 and caption from Ref. [661].

and then taking absorption images of the expanded condensate. For the case of ^{87}Rb, dark solitons or "kink states" are expected to form that travel through the condensate at a velocity lower than the speed of sound, c_s. With a cigar-shaped condensate this approximates propagation in a simple one-dimensional problem.

Phase imprinting was done by masking off one half of the cigar-shaped condensate and illuminating the other half with a short pulse of blue-detuned far-off-resonance light of intensity $I \cong 20\,\text{W/mm}^2$. At this intensity one induces a π phase shift in the condensate wave function. Data that were taken with phase shifts of 0.5–1.5π show clear evidence of a dark soliton moving into the illuminated half of the condensate with a velocity measurably less than c_s. The authors also observed, after a short time, the formation of a second density wave, which was not a soliton, moving in the opposite direction, with a velocity very close to c_s. This sort of behavior is well predicted by the Gross–Pitaevskii equation for $T = 0$. The soliton is predicted to propagate indefinitely, with no shape change, by bouncing back and forth, whereas the oppositely directed density wave at the velocity of sound should dissipate by spreading. Unfortunately, there is an approximately 10% thermal atom component present with the condensed atoms that causes the soliton to dissipate after about 30 ms. So, even though conditions are not ideal, one clearly sees the generation of solitons.

The authors see this as an opportunity to study solitons in a dissipative environment. They also suggest use of an additional off-resonance blue-detuned hollow beam to presumably increase the initial atom density and slow the soliton velocity. The recently achieved all-optical BEC with its very high-density condensate and negligible population of thermal atoms would be ideal for soliton studies. An interesting avenue for further work might be the study of double and triple solitons generated by phase shifts of π, 2π, 3π, etc., using cleaner condensates. The two papers [661] and [662] discussed here have opened a new area of nonlinear atom optics.

24

Trapping and Manipulation of Small Molecules

Following the great success of optical trapping and manipulation techniques for controlling atoms, small dielectric particles and large molecules in physics and biology, as discussed above, the chemical physicists, physical chemists, and molecular physicists, who traditionally study small molecules, began to consider applying similar optical techniques to these particles.

Large macromolecules, such as TMV [44]*, and microtubules [314] can be directly trapped with optical tweezers, much as was done for sub-micrometer dielectric Rayleigh particles [5]*. In other experiments, large biological molecules, such as molecular motors, kinesin, dynein, and myosin, were manipulated by attaching them to small dielectric spheres that acted as "handles" (see Chapter 7). The handles technique was used extensively for applications involving the mechanoenzymes RNA polymerase and DNA polymerase, and for studying the properties of DNA and RNA and of their associated molecules protamine, arginine, chromatin, and titin (see Chapters 8 and 9).

Atoms have been readily trapped and manipulated since the mid-1980s by taking advantage of the relatively strong scattering and dipole forces that exist near strong atomic resonance transitions (see Sec. 5.1, Chapters 15–23 and 25).

There are problems, however, in applying similar forces to small molecules. Their more complex molecular spectroscopy, with many vibrational and rotational levels, greatly increases the lifetime of excited states. This increase greatly reduces the scattering force near molecular resonances. The result is a form of optical pumping. The longer lifetimes make it difficult to optically slow and cool molecular beams. There are, of course, optically induced dipole forces on molecules, as there are on atoms that depend on the molecular polarizability. Thus, optical trapping of small molecules via the gradient force in a tweezer-type trap [5]* is possible, if one can somehow cool them to low enough temperatures. If the molecules have a magnetic moment or an electric dipole moment, they can also be trapped in DC magnetic or electric quadrupole traps, as we will discuss later.

As was mentioned above, a great deal was learned about molecular spectroscopy in the studies on ultracold collisions [202], where formation of dimer molecules was one of the possible loss channels for cold atoms confined in traps.

Experimental and theoretical interest in trapping and manipulating small molecules has grown considerably since the mid-1990s. As we shall see, there was even at that time discussion of the possibility of achieving Bose–Einstein (BE) condensation of ultracold trapped molecules and molecular lasers. Molecular BE condensates were finally observed in 2003 (see Sec. 25.10).

In 1996 [663] and 1997 [664] Seideman analyzed the problem of the focusing of molecular beams by light in terms of high-field seeker molecules that are attracted to the high-intensity regions of a light beam. The trapping of sufficiently cool molecules in a high-intensity focus of a laser beam was considered by Friederich and Herschbach earlier in 1995 [665], [666].

24.1. Deflection of Neutral Molecules Using the Nonresonant Dipole Force

In 1997 Stapelfeldt *et al.* [220]*, in Corkum's group in Ottawa, in a proof of principle experiment, demonstrated the ability of the dipole force to focus molecules. They used Cs_2 molecules and a pulsed far-off resonance YAG laser beam to generate a dipole force sufficiently large that there was no need for laser cooling. This greatly simplifies the experiment and avoids the need to consider the details of the dense level structure of the molecules close to resonance. A pulsed Cs_2 beam was made translationally cold by expanding the molecules through a $250\,\mu$m diameter aperture into the authors' time-of-flight spectrometer. With this instrument they analyzed the position and velocity of molecules that had passed perpendicularly through the axis of the focused YAG laser beam ($w_0 \approx 7\,\mu$m). At a peak intensity of $\sim 9 \times 10^{11}\,$W/cm^2 they generated an optical potential well of about $7\,$meV in depth that was about four orders of magnitude greater than the transverse energy of their beam ($\sim 10^{-6}\,$eV). The light beam acted as a lens, strongly focusing the Cs_2 molecules. The measured transverse velocities were in reasonable agreement with the calculated deflections using the simple dipole force formula $F = (1/2)\alpha \nabla E^2$, where α is the average polarizability of the molecule.

The authors point out that the maximum intensity an atom or molecule can withstand without strong photoionization determines the maximum Stark shift that can be obtained. They calculate that with an ionization potential of $10.1\,$eV one expects a 1% probability of ionization during the $10\,$ns laser pulse.

It is interesting to note, as an aside, that the first observation of the optical Stark effect on vibrational and rotational levels of molecules was made in 1980 by Rahn *et al.* [667] with H_2 and N_2 molecules, using a pulsed YAG laser. The results were in agreement with polarizability computations for these molecules. The motivation for these measurements was the importance of the Stark shift on the ultimate sensitivity and spectral resolution of CARS spectroscopy and Raman gain techniques in nonlinear optics. No mention was made of the connection of Stark measurements to optical forces on molecules. This connection between Stark shift and dipole traps was made by Corkum's group, but only after the connection between optical traps and Stark shifts was made by Takekoshi and Knize [146] and Heinzen's group [148], who in turn learned it from Chu *et al.* [4]*, who got it from Gordon and Ashkin [123]*, [136]*.

24.2. Observation of Optically Trapped Cold Cesium Molecules

A big step forward in the optical manipulation of molecules was taken by Takekoshi, Patterson, and Knize in 1998 [668]* with the first observation of optically trapped cold neutral molecules. Direct optical cooling of molecules is difficult because the many vibrational and rotational degrees of freedom make it hard to create large scattering forces, as alluded to above. One way around this is to produce cold dimers from laser-cooled atoms. This was done recently for the case of cesium by Fioretti *et al.* in 1998 [669]. Takekoshi *et al.* independently verified this result in Ref. [668]* and additionally trapped the Cs_2 molecules in an optical dipole force trap formed at the focus of a 17 W CO_2 laser beam operating at $\lambda = 10.6\,\mu$m. The depth of the optical trap was about $350\,\mu$K for Cs_2 dimer molecules and about $200\,\mu$K for Cs atoms. Starting with a dark spot MOT for Cs atoms, they transferred the normally produced dimer Cs_2 molecules from the dense population of Cs atoms in the MOT into a dipole trap. The molecules were observed by a two-photon photoionization process. This produces either an ionized Cs_2^+ molecule plus an electron or else an ionized Cs^+ atom, a neutral Cs atom, and an electron. The ions were measured using time-of-flight spectroscopy with a channeltron for observing the ions.

In the transfer process the authors very cleverly removed unwanted Cs atoms from their trap with a pushing beam, presumably tuned to the Cs atomic resonance. Although the Cs_2 dimer production rate and collection rate were quite inefficient, nevertheless the authors were able to observe a molecular lifetime in the dipole trap of about half a second. This is the same as the atomic lifetime and is presumably limited by collisions with background gas. The number of molecules produced in this manner is quite small, but no effort to collect large numbers was made.

24.3. Magnetic Trapping of Calcium Monohydride Molecules at mK Temperatures

Weinstein *et al.* [670], in Doyle's group at Harvard in 1988, investigated the use of collisions with a cold buffer gas as a way to magnetically trap molecules of calcium monohydride (CaH). They hope that this nonoptical technique will also be applicable to many other paramagnetic molecules.

They used a superconducting magnet at liquid helium temperatures to make a magnetic quadrupole field up to 3 T, thereby making a magnetic trap having a depth of ~ 2 K. The CaH molecules were produced by laser ablation from a lump of solid CaH_2 in a chamber located within a dilution refrigerator. The CaH molecules were cooled by thermal contact with He gas within the chamber to a temperature ~ 300 mK. Of the molecules produced, only the low-field seekers were retained in the magnetic quadrupole trap. The high-field-seeking molecules were ejected. The quadrupole trap, of course, has a leak in the low-field region near the zero field point due to nonadiabatic spin-flipping Majorana transitions to the high-field-seeking states. At the temperature of ~ 300 mK this is not a large effect, as in MOTs for atoms. The authors observed the initial ejection of high-field seekers and retention of low-field seekers using fluorescence spectroscopy. They observed as many as 10^{10} ground state CaH trapped molecules. A necessary condition for trapping is a small spin-flip collision cross-section between the helium gas buffer and the trapped molecules. This allows kinetic thermalization to occur before spin relaxation. Removal of the buffer gas should lead to evaporative cooling, if this can be done rapidly enough. This is not the case now. With deeper traps and more rapid pumping it may be possible to evaporatively cool to quantum degeneracy.

A commentary on the magnetic trapping of molecules by Doyle's group appeared in Physics Today in November 1998, written by Gloria Lubkin [671]. There are interesting quotes from Doyle on the technique, its relation to other molecular trapping techniques, and its possible future applications. Lubkin also discusses other approaches to molecule trapping, involving optical dipole traps, multiline Raman cooling and observations of translationally cold alkali dimer molecules.

24.4. Stimulated Raman Molecule Production in Bose–Einstein Condensates

In a rather far-ranging collaboration, Julienne from NIST, Burnett from Oxford, Band from Ben-Gurion University, and Stwalley from the University of Connecticut, in 1999, proposed a Raman photoassociation process to make translationally cold molecules from a BE condensate [672]. Photoassociation, which involves the collision of two atoms in the presence of light to make bound molecules, has already been used to make and trap cold Cs_2 molecules in a magneto-optical trap [669] and in an optical dipole trap [668]*. In this proposal [672], the authors study photoassociation using two different laser frequencies and an initial atom condensate. They consider a total of five possible coupled channels to describe the collision of two incident ground state atoms giving rise to either trapped or untrapped molecules. Using model potentials they calculated molecular production possibilities and loss probability as a function of collision energies that might characterize an evaporatively cooled gas before reaching condensation. Tuning the Raman frequencies to a molecular resonance, they find that the probability for loss here was very small and molecules are produced efficiently. The results should apply for Cs_2 and Rb_2 molecule production.

Considering an atom condensate and using a conservative value for the loss rate, they estimate it takes a laser pulse of $\sim 100\,\mu s$ to convert a fraction $1/e$ of the atoms to molecules at the densities possible in an optical trap [559]*. This process is very similar to coherent out-coupling of atoms using a short rf pulse. Thus the molecules coherently coupled out of the trap will result in a "pulsed molecular laser". The authors further point out that if they tune their Raman pulses between two photoassociative molecular resonances, molecule production is inhibited, compared to the free atom scattering rate. Thus it is conceivable to switch between out-coupled atoms and out-coupled molecules by tuning the two Raman laser pulses.

24.5. Optical Centrifuge for Molecules

In 1999, Karczmarek *et al.* [673], in Canada, showed that it should be possible to make an optical centrifuge based on trapping of cold anisotropic molecules. They showed that optical torques applied to anisotropic molecules could be used to spin such molecules to very high angular momentum states, even sufficient to break the molecular bonds. An anisotropic molecule placed in a linearly polarized field develops a time-averaged induced dipole moment. The polarizability α has two components, α_{par} and α_{perp}, for field components parallel and perpendicular to the molecular axis, as we saw earlier for a cylindrical particle, such as TMV [44]* (see Secs. 5.25 and 5.26). The potential of the particle in a laser field is $-U_0 \cos^2 \theta$, where $U_0 = (1/4)(\alpha_{par} - \alpha_{perp})E^2$ and θ is the angle between

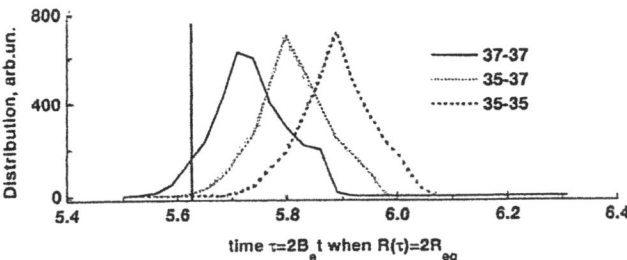

Fig. 24.1. Isotope separation for rotationally cold Cl_2. The pulse is the same as in Figs. 1 and 2. Probability distribution for time at which $R = 2R_{eq}$ is shown; 3200 trajectories were used for each plot. Figure 3 and caption from Ref. [673].

E and the molecular axis. If one slowly rotates the polarization of the infrared trapping field about a fixed axis, the molecule will follow and rotate with the same angular frequency in a controlled manner. The authors analyze the rotation process carefully and specify the rates at which the fields can be increased. They use two chirped circularly polarized fields to generate the rotating field. They describe the increasing rotational velocity as "rotational ladder climbing". They previously considered the process of using a chirped linearly polarized field for "vibrational ladder climbing" [674], [675].

Calculations were made for rotationally breaking up Cl_2 molecules. The authors pointed out that one can use this breaking process to separate the different chlorine isotopes. Figure 24.1 from Ref. [673] shows the distribution of breaking times (defined as the time when the atom separation is two times the equilibrium separation) for each of the three types of molecules shown. If the spin producing pulse stopped at the position of the vertical line, then 8.5% of the ^{37}Cl–^{37}Cl is broken compared to 0.6% of the ^{35}Cl–^{37}Cl and none of the ^{35}Cl–^{35}Cl. The ratio of ^{37}Cl–^{35}Cl in the "debris" is 5.6:1, compared with the natural abundance of 1:3. The proposed optical centrifuge can be used to selectively dissociate a given diatomic molecule from a mixture. For example, one can easily dissociate the 40% weaker bond of the I_2 molecule, without breaking the stronger bond of the Cl_2 molecule.

24.6. Cooling of Molecules by DC Electric Field Gradients

The inability to slow molecular beams using resonant optical forces, as in many atoms, was discussed above. This is the result of the complex vibrational and rotational spectroscopy, which inhibits the recycling of resonantly excited molecules and greatly reduces the scattering force. This stimulated the search for alternative cooling methods for molecules, such as the method developed by Doyle's group, based on collisions with cold helium buffer gas [670].

Another alternative for cooling molecules was developed in 1999 by Bethlem, Berden and Meijer of the Netherlands [676], and subsequently by Bethlem *et al.* in 2000 [677]. They showed how one could use an array of pulsed electric fields to decelerate a beam of neutral polar molecules possessing an electric dipole moment. Figure 24.2, taken from an article in Physics Today on "Hot Prospects for Ultracold Molecules" by B. Goss Levi [221], illustrates the scheme. A molecule entering the gradient of the electric field of a charged pair of wire electrodes with its dipole oriented opposite to the field is slowed down by the electric dipole force. When the molecule reaches the high-field,

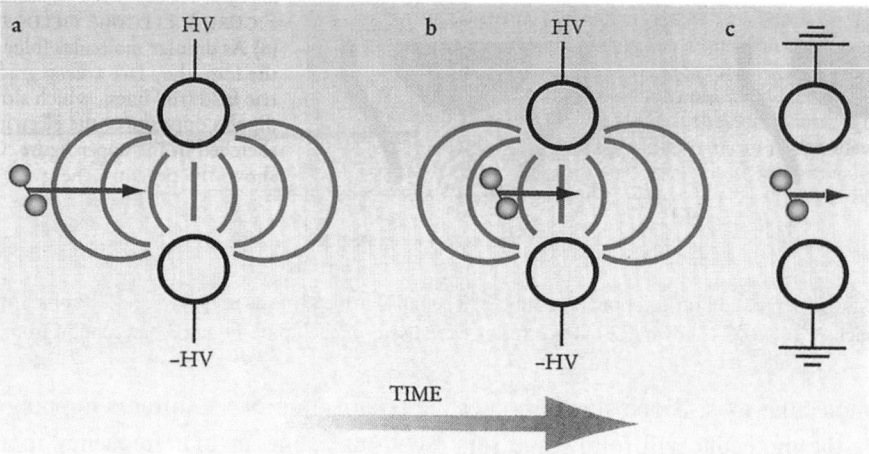

Fig. 24.2. Electric fields slow molecules. (a) A dipolar molecule (green) experiences a Stark shift as it enters the electric field (red lines) between electrodes. (b) As the molecules near the center, those with dipoles antiparallel to the field are decelerated, as shown by the shorter velocity vector (blue). (c) Electrodes are turned off when the dipole reaches the center, so that the molecule does not gain velocity as it leaves the field. Connections to a high voltage source are labeled by ±HV. Figure 1 and caption from Ref. [221].

low-gradient region of the electrode pair, the field is quickly reduced to zero and the slowed molecule coasts on into a field-free region. The process is then repeated through many such cooling stages. The molecules that are slowed in this way are aptly referred to as "low-field seekers" in the electric field. Molecules with their dipole parallel to the electric field are "high-field seekers" and are pushed toward high electric fields and are thus accelerated, heated up, and ejected, as they pass through the electric field gradient. This is in direct analogy with the case of atoms in magnetic field gradients, as we discussed. This molecular deceleration process can also be described as Stark deceleration. As the molecules enter into the high electric field, the Stark potential rises at the expense of the kinetic energy of the molecules. Looking at the process as a repulsive gradient force, however, is closer to the spirit of the original optical atom gradient force component of radiation pressure forces, as discussed above.

As an additional piece of insight, I mention that this cooling technique is quite analogous to an electron interacting with a microwave field in an interdigital resonant circuit, as shown in Fig. 24.3. Suppose we have an electron of velocity v_0 at point P_1 in the gap between two fingers of the interdigital circuit operating in the π-mode, where the charges at the peaks of the rf cycles on the fingers are alternately plus and minus, as shown. This electron feels a slowing force and loses kinetic energy. Imagine that the electron's velocity is synchronized with the microwaves, so that it travels to P_2, the center of the next gap, in one cycle of the rf. Notice that as the electron travels from P_1 to P_2, it arrives at point H_1 at the time that the fields in the gaps reverse. Point H_1 is a "hiding point" close to a metallic surface where, by continuity, the tangential electric field is zero. Thus, it fails to gain any energy during field reversal and arrives at P_2 in time to be slowed again, and so on, as it proceeds. For $d_1 \cong 1$ cm and $v_0 \cong c/20$, a typical velocity for a kilovolt electron beam, we calculate $\lambda = \sim 3$ cm. This puts us squarely in the microwave radar regime.

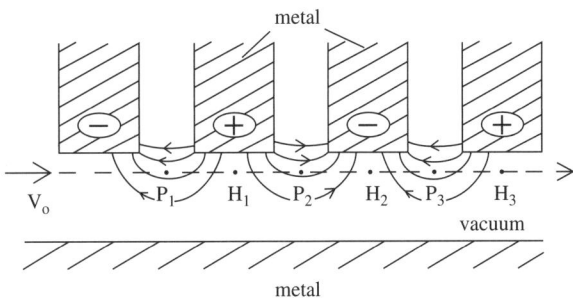

Fig. 24.3. Slowing of electrons by a π-mode electromagnetic field.

Electrons arriving at points P at the wrong time, in analogy with the high-field seekers, gain velocity as they move through the circuit. This picture results in a situation with no net rf gain. Old microwave hands know that if v_0 is tuned to lower frequencies, the beam interaction results in "axial bunching". These bunches slow as they interact and give net gain. This crudely describes the gain process of many microwave tubes, such as the traveling wave tubes, magnetrons, backward wave oscillators, etc. [678]. Again, we see the role of tuning above or below the synchronism or resonance velocity.

In the scheme of Meijer's group, the low-field-seeking CO molecules entered their 63 stages of slowing with a temperature spread of ~ 0.7 K and emerged with a minimum translational temperature of ~ 4 mK. These molecules, moving with an initial velocity of 275 m/s, are effectively confined within a quite deep traveling potential well of ~ 1.15 K.

Alternate cooling stages have their electrodes rotated 90°. This serves to reduce the buildup of transverse oscillations if the spacing of successive stages is correct. Old timers like myself recognize this as the "strong-focusing" principle of synchrotrons and linear accelerators (see references in Ref. [677]).

24.7. Cooling Molecules by Time-Varying Inhomogeneous Fields and Expansion from Nozzles

In a "News and Views" article in Nature entitled "Molecules are Cool", Doyle and Friedrich in 1999 [679]* discuss the complex energy level structure of molecules that makes it difficult to cool molecules by laser techniques. They discuss alternative molecular cooling methods based on switching time-varying inhomogeneous fields and using supersonic expansion from a spinning nozzle [676], [680], [681]. In Refs. [676], [680] electric fields are synchronized and switched so as to oppose the linear motion of pulses of molecules as they pass through regions of electric field gradient. After passing through many stages of linear cooling, the molecules might be caught in a deep molecular trap, such as Doyle's CaH trap [670]. The spinning nozzle concept has not yet been tried. It is based on canceling the linear velocity of molecules emerging from a supersonic nozzle by the velocity of the mechanical rotating of the nozzle [681]. These schemes may work for molecules, but there is no need for such techniques with atoms, where one can cool to lower temperatures by many orders of magnitude.

24.8. Electrostatic Trapping of Ammonia Molecules

The technique of adiabatic slowing of molecules by time-varying inhomogeneous electric fields was used by the Netherlands group of Bethlem *et al.* in 2000 to electrostatically trap ammonia molecules [682]. Figure 24.4, taken from this paper, shows a bunch of slow molecules coasting into the input orifice of the trap after having been slowed from a mean velocity of 260 m/s to a velocity sufficiently slow to climb the last potential hill into the center of the trap. They arrived with a final velocity spread of 2 m/s, which corresponds to a temperature of 2 mK. At this point, the input electrode was switched to 0 V, making an essentially symmetric DC quadrupole trap having a depth of 0.35 K. A density of 10^6 molecules/cm^3 was achieved in a volume of 0.25 cm^3. The final temperature was certainly less than the 0.35 K trap depth and possibly, depending on the loading conditions, as low as the 2 mK translational temperature. The authors actually used ND$_3$ rather than NH$_3$, because the heavier deuterated ammonia gives a lower mean initial velocity out of their expansion valve. By knowing the initial translational temperature of 0.8 K out of the nozzle, the authors can specify the only low-field-seeking state that can possibly be populated. Only an eighth of all molecules are in this state. The authors further say that these results indicate it may be possible to achieve BEC by subsequent evaporative cooling.

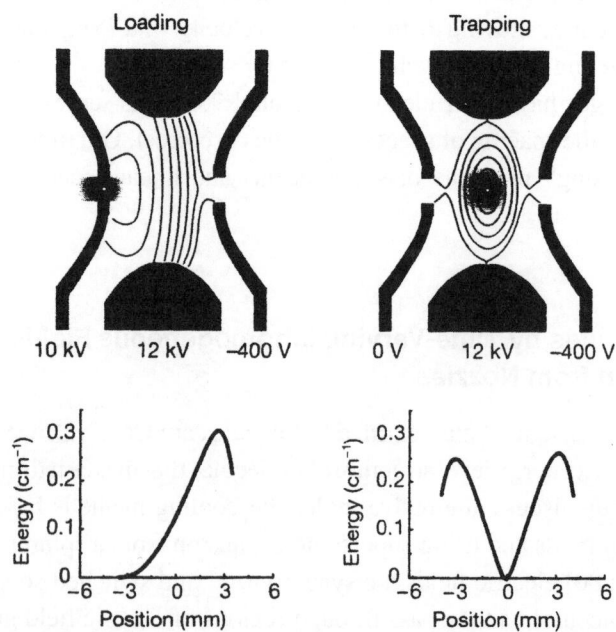

Fig. 24.4. Configuration of the trap with the voltages as applied during loading and trapping. In the trap, lines of equal electric field are indicated and the cloud of molecules is sketched. The potential energy along the molecular beam axis of the ND$_3$ molecules in the $|J\ K\rangle = |1\ 1\rangle$ state with positive Stark shift is shown for both field geometries. Figure 2 and caption from Ref. [682].

The criteria for trapping molecules in this way are that the molecules must have a large positive Stark shift (about $1 \, \text{cm}^{-1}$) that can be achieved with reasonable fields and also an initially slow enough pulsed beam velocity. A number of excellent candidates exist that fulfill the first criterion, among which are OH, NO, CO, H_2O, H_2CO, and CH_3F.

24.9. Creation of Molecules from Atoms in a Bose–Einstein Condensate

A significant advance in the production of ultra-cold molecules was made in 2000 by Wynar *et al.* in Heinzen's group [683], using photoassociative techniques. Molecules had previously been produced by photoassociation with a relatively large energy spread of $\sim 100 \, \mu\text{K}$ [669]. This new experiment involved use of a stimulated Raman process with two tunable laser fields to target specific states of the $^{87}\text{Rb}_2$ molecule, very much as suggested by Julienne *et al.* in 1998 [672]. Starting with a condensate of ^{87}Rb atoms in a magnetic TOP trap, they tuned a pair of laser fields so that their energy difference $(h/2\pi)(\omega_2 - \omega_1)$ was close to the binding energy E of the targeted level. Since the two beams propagate in the same direction, the momentum transferred to the molecule is exactly zero at resonance and the molecule is formed at rest. One of the important features of this two-beam Raman technique is that neither of the two beams is tuned to resonance with any excited state transition. This implies that no spontaneous emission can occur, thus avoiding decay to the many vibrational or rotational states of the stable molecules. This is the feature that makes it possible to target specific molecular levels.

The authors did not observe the molecules directly, but rather measured the excitation spectrum or number of atoms remaining in the condensate as they tuned the Raman difference frequency through resonance. Since Doppler shifts were essentially zero, they got an extremely narrow line, as small as 1.0 kHz. They studied the line shape and position of the resonance as a function of condensate density and laser intensity of the Raman beams. Remarkably, at intensities as low as $I_2 = 5.6 \, \text{W/cm}^2$ and $I_1 = 0.5 \, \text{W/cm}^2$, the net AC Stark shift due to the laser is $\Delta E_L = 95 \, \text{kHz}$ and the resonant width $\gamma_L/2\pi = 350 \, \text{Hz}$. However, a detailed study of the line shape and line shifts indicates that mean field interactions are responsible for most of the observed changes. From the analysis they deduced that the value of the molecule–atom scattering length is negative, which enhances the molecular stability in the condensate. The authors believe that they should try to directly image the molecules, rather than detect their absence and also determine the lifetime of the molecules in the magnetic trap. With their Raman technique they may be able to populate the ground state of the molecule and produce a molecular condensate.

Williams and Julienne of NIST, two experts on photoassociation spectroscopy of ultracold molecules, commented on the work of Wynar *et al.* [683] in a "Perspectives" article, entitled "Molecules at Rest", in the same issue of Science [684]*. They compared the various methods for producing ultracold molecules and emphasized the new features of Wynar *et al.*

24.10. Prospects for Trapping and Manipulating Ultracold Molecules

In the September 2000 issue of Physics Today, B. Goss Levi [221] compared the various techniques devised to trap and manipulate molecules at sub-millikelvin temperatures. Of particular interest are the comments by many of the authors on their recent work, and on their views of the status of the field and its future prospects. This article and the perspective article of Williams and Julienne alluded to above [684]* have an abundance of useful references.

24.11. Dynamics of Coupled Atomic and Molecular Bose–Einstein Condensates

Following the experimental demonstration of the production of molecules in a BE condensate by two-frequency photoassociation [683], Heinzen and Wynar, in collaboration with Drummond and Kheruntsyan, in 2000, analyzed the dynamics of a trapped atom condensate coupled to a diatomic molecular condensate by coherent Raman transitions [685]. They showed that a new type of giant collective oscillation occurs, transferring populations back and forth between the two types of condensates in a nonlinear process that they term "superchemistry". This effect is caused by stimulated Raman emission of bosonic atoms or molecules into their condensate phases.

The authors point out the close analogy between this process in condensates and nonlinear frequency conversion in coherent laser beams of different frequencies in nonlinear optical media. This analogy may have motivated the present suggestion.

Consider, for example, the case of a parametric amplifier in which a signal beam of frequency ω_S is coupled to a high frequency pump beam ω_P in a nonlinear medium and is amplified, while producing an idler beam at frequency $\omega_i = \omega_P - \omega_S$. This last equation expresses conservation of energy (i.e., $h\omega_P = h\omega_i + h\omega_S$). If the process is phase matched, then $k_P = k_S + k_i$. Since the wave vector $k = 2\pi/\lambda$ and the momentum per photon $p = h/\lambda$, phase matching implies conservation of momentum.

For a phase-matched parametric amplifier, one has complete depletion of the pump wave by the signal beam. If, however, one has a phase mismatch, then only a certain amount of conversion of the pump wave to signal wave occurs before the phase slips completely out of phase and the signal power reverts back into pump power.

This is closely what the authors predict for the conversion of atoms in a condensate into molecules for a slightly off-resonance two-photon Raman process. In that case, the atom condensate converts a fraction of its atoms to molecules and back again to atoms, in an oscillatory manner, depending on the energy mismatch and the original atom densities. Figure 24.5 taken from Ref. [685] illustrates the collective oscillation of the atomic and molecular condensates. One can therefore convert an arbitrary fraction of the atom condensate to a coherent molecule condensate. If the coupling beams, i.e., Raman gain beams, are turned off at the correct time, one is left with two coexisting condensates.

The authors ignore any effects due to inelastic collisions in their calculations, since they are not known at present. If one wants to vary the scattering lengths of the various components, it would be desirable to have a dipole trap rather than a magnetic trap, since the scattering lengths can then be independently varied with an external magnetic field without affecting the trap depth.

The theoretical treatment of the dynamics of coupled ultracold atomic-molecular condensate gases is still controversial. A new analysis of the problem has appeared by Góral *et al.* from Poland in 2001 [686]. This analysis claims that the oscillations predicted by Heinzen *et al.* [685] will not occur under the conditions postulated. Heinzen *et al.* used just a two-mode model of the condensates

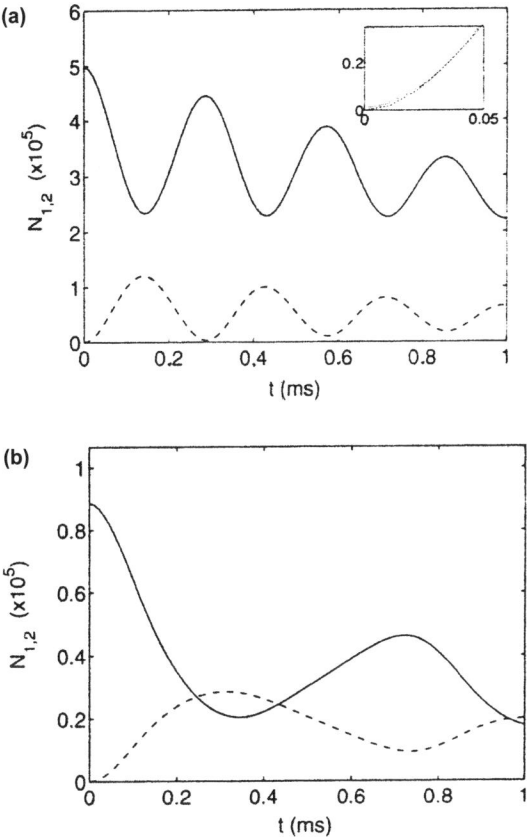

Fig. 24.5. (a) Occupation numbers $N_i = \int d\mathbf{x} |\phi_i(\mathbf{x}, t)|^2$ of the atomic (solid line) and molecular (dashed line) fields, as a function of time t, for the parameter values of Fig. 2; (b) same as in (a) but for the half the initial atomic density and the same initial effective detuning $\tilde{\delta}$. Figure 3 and caption from Ref. [685].

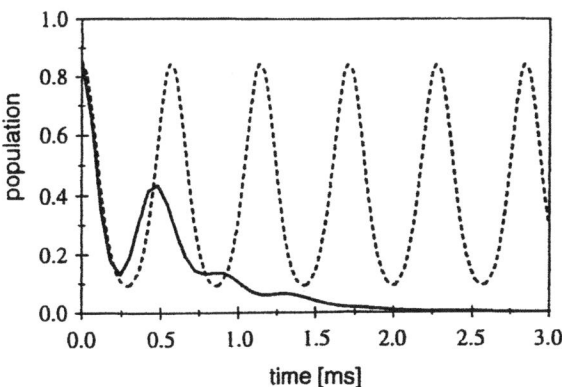

Fig. 24.6. 2-mode (dashed line) versus multimode (solid line) dynamics of an atomic condensate population. Figure 3 and caption from Ref. [686].

with a neglect of the coupling to the residual uncondensed 15% of the atoms. Góral *et al.* do a multi-mode calculation that, they assert, results in a rapid washout of the oscillations on a short time scale. Figure 24.6, taken from Ref. [686], shows the results of a two-mode versus a multimode dynamics calculation illustrating the washout of the predicted oscillations. The authors predict that one can recover the oscillation behavior if one reduces the volume of the condensate trap, assuming the same initial atom condensate density. This would have the effect of spreading the level spacing of the uncondensed 15% of the atoms and making it more difficult to convert these randomly phased atoms into molecules. They do not point it out, but, clearly, use of optical dipole traps for holding the original atom condensate would be a great advantage. In a typical compact dipole trap, not only is the level spacing greatly increased as desired, but the fraction of atoms in the untrapped cloud of thermal atoms is drastically reduced to a percent or less, as discussed by Ketterle's group [559]*, [565]. As Góral *et al.* point out, the results of their new analysis should be taken into account in the design of coupled atomic-molecular systems.

CHAPTER

25

Trapped Fermi Gases

The achievement in 1995 of a Bose–Einstein (BE) condensate in dilute gases of the atomic bosons ^{87}Rb [29]*, ^{7}Li [512], and, shortly thereafter, ^{23}Na [30]* served as a powerful stimulus to attempts to attain similar quantum degeneracy in dilute gases of atomic fermions. Attention has focused on vapors of the fermionic isotopes ^{6}Li and ^{40}K. Of the alkali metals these are the only two to have half-integral spin.

Fermions can be cooled and trapped in MOTs or dipole traps in exactly the same manner as bosons at temperatures higher than the critical temperature. Of course, for MOTs, one can trap only spin-polarized atoms in the low-field-seeking hyperfine states, whereas dipole traps can trap all hyperfine states.

25.1. Superfluid State of Atomic ^{6}Li in a Magnetic Trap

In early 1996, Stoof and Houbiers at Utrecht, working with Sackett and Hulet at Rice University, made an analysis of superfluidity in spin-polarized ^{6}Li fermions [687]. An expanded version of this work by Houbiers [688] appeared in 1997. As these papers pointed out, problems arise with fermions if one tries to cool the gas down to lower temperatures and reach quantum degeneracy by evaporative cooling. At low temperatures the statistics of the fermion gas begin to affect the cooling process. The principal cooling channel in evaporative cooling is through s-wave collisions. However, for fermions, which obey the Pauli exclusion principle, s-wave collisions are strongly inhibited at low temperatures. Thus, evaporative cooling essentially stops, preventing the temperature from reaching quantum degeneracy. There are really two transitions that occur with fermions as they are cooled. The first is the transition to quantum degeneracy, where one cools below the Fermi energy E_F for the cooled fermion gas and where quantum statistics is important. This is the Fermi gas phase. Indeed, as one cools the gas, the crossover to quantum degeneracy is gradual, compared with the abrupt phase transition to a BE condensate. At still lower temperatures there is a second possible transition, known as the Bardeen, Cooper, Schrieffer (BCS) transition, to a state where one hopes to find so-called Cooper pairs. This is the superfluid transition, analogous to what is seen in helium superfluids.

To solve the evaporative cooling problem alluded to above, the authors [687], [688] suggested magnetically trapping atoms in two hyperfine states, rather than having all atoms in a single hyperfine state. They recommended levels $|6\rangle = |m_s = 1/2, m_i = 1\rangle$ and $|5\rangle = |m_s = 1/2, m_i = 0\rangle$, the upper two states of the three weak-field-seeking hyperfine states of ^{6}Li. The advantage of having a two-component gas is that the two hyperfine levels can now interact strongly through thermalizing collisions and evaporatively cool one another without violating the Pauli exclusion principle. The interaction of atoms in these two states is characterized by the s-wave scattering length of the triplet potential. Fortunately, the scattering length for ^{6}Li is $a = -2160\, a_0$, which is anomalously large and negative. This large value of the s-wave scattering length contributes to the relatively high value of the $T_c \cong 11\, n$K, the transition temperature for formation of Cooper pairs. The results of a computation on T_c as a function of the number of particles in each of the two hyperfine components were presented. They also considered the lifetime of the two-component Fermi gas in phase–space and showed that, for densities of $\sim 10^{12}$ atoms/cm^3 within the stability regime, the lifetime was about a second, provided that a magnetic bias field of about 5 T or higher was applied.

An important experimental problem is how one detects a transition from a Fermi gas into a superfluid state. There is almost no change in the density of the gas in such a transition. The authors suggested that one might look for a change in the decay rates or a change in average energy at the critical temperature.

One of the main conclusions to be drawn from this work, however, is that a BCS-like transition to a superfluid state of fermions is possible at experimentally accessible temperatures and densities, in spite of some difficulties such as 5 T bias fields.

25.2. Elastic and Inelastic Collisions in ^{6}Li

In 1998, Houbiers *et al.* [689] considered elastic and inelastic collisions of ^{6}Li in magnetic and optical traps. This paper goes to the heart of the problem of the ^{6}Li superfluid transition of a two-component fermion gas as originally proposed by Stoof *et al.* [687] and by Houbiers *et al.* [688] for magnetic traps and now also for optical traps. The important question in cooling is the magnitude of the elastic and inelastic collisions for the pair of hyperfine states chosen to be cooled. In Stoof *et al.*'s (1996) paper and in Houbiers *et al.*'s (1997) paper, it was proposed to apply a high-bias field of about 5 T to enhance the elastic collisions and decrease the inelastic exchange rate collisions. The 1998 paper [689] considered these collisions in much greater detail and also at lower magnetic fields. Using an exact technique of calculation they considered the "good" or elastic collisions of all combinations of the magnetically trapped low-field seeker hyperfine states $|6\text{-}5\rangle$, $|6\text{-}4\rangle$, and $|5\text{-}4\rangle$. They also considered the very important case of $|2\text{-}1\rangle$ collisions in optical traps with an added magnetic bias. They used the full exact coupled-channels (CC) calculation and displayed the real part of the zero temperature s-wave scattering lengths as a function of magnetic field. In the calculation with the CC technique they used the most up-to-date singlet and triplet potentials to describe the radial potential. Figure 25.1 taken from Fig. 1 of Houbiers *et al.* [689] shows the results of this exact calculation. One sees that one must apply fields of ~ 10 T or more in order to reach the value of $a_{65} = -2160\, a_0$. This large negative

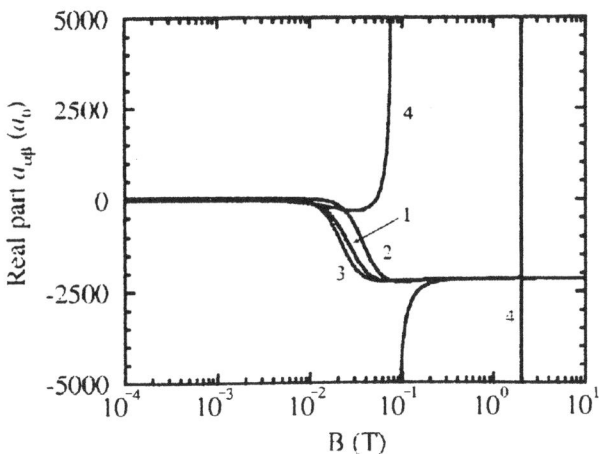

Fig. 25.1. Real part of zero-temperature s-wave scattering length for (1) 65, (2) 54, (3) 64, and (4) 21 collisions as a function of the magnetic field. Figure 1 and caption from Ref. [689].

value of triplet wave scattering length, as we saw, was the origin of the relatively high value of T_c for formation of a fermion superfluid. However, at magnetic fields below $\sim 10^{-2}$ T, the scattering length changes drastically in magnitude and sign to $a_{65} = 47\,a_0$, which should give drastically different behavior. Interestingly, as seen in Fig. 25.1, the scattering length a_{21} exhibits a Feshbach resonance at $B \approx 0.08$ and $B \approx 1.98$ T. Recall that Feshbach resonances arise when the Zeeman energy of the incoming wave function of the $|21\rangle$ state, which is almost purely triplet, coincides with a bound-state energy of the singlet part of the central potential. The existence of these Feshbach resonances makes the use of optical dipole trapping of ^{6}Li even more attractive.

The authors also calculated and displayed exchange decay rate processes for the various combinations of two-component inelastic or "bad" collisions. These results allow one to compare cooling prospects with different combinations of hyperfine levels at different magnetic fields. However, for the case of an all-optical far-off-resonance dipole trap, they did not need to calculate decay rates since the $|21\rangle$ combination cannot decay through collisions. This is, therefore, the most favorable choice experimentally. This new consideration of the possible use of optical far-off-resonance traps for trapping fermions stands in contrast to the authors' considerations of just a year or two earlier [687], [688]. It is a further indication of the rapid renewal of interest in all-optical traps for atoms.

25.3. Sympathetic Cooling of an Atomic Bose–Fermi Gas Mixture

Another possible solution to the problem of cooling of fermion gases to quantum degeneracy that was proposed was the use of a two-component gas made up of a mixture of fermion and boson gases in a single trap. In such a boson–fermion mixture, the bosons of the mixture can be used as a coolant to reduce the temperature of the fermions, provided the two components remain in collisional contact with one another. Such sympathetic cooling is necessary in cooling to Fermi gas degeneracy because of the suppression of s-wave scattering between identical fermions in spin-symmetric states as alluded to above. Cooling in two-component mixtures of bosons had been studied before in Ref. [535]* and later in experiments on cooling ^{40}K fermions (see Sec. 25.6).

Geist *et al.*, in 1999 [690], considered the sympathetic cooling of fermions in some details, using the solutions of the coupled Boltzmann equations for a confined gas mixture. The authors give results on equilibrium temperatures and relaxation dynamics and show that the Fermi gas can be cooled to the degenerate regime, where quantum statistics and mean field effects are important. They deduce final spatial distributions of the degenerate gas of fermions. Mean field effects are discussed only qualitatively. Many examples of sympathetic cooling are given for ^{40}K, ^{39}K Fermi–Bose isotope mixtures.

25.4. Cooper Pair Formation in Trapped Atomic Fermi Gases

Turning again to ^{6}Li atoms, Houbiers and Stoof [691] make a quantum mechanical computation on Cooper pair formation dynamics in magnetically trapped atomic Fermi gases as they are cooled to the superfluid state. They consider the mixture of two different hyperfine levels, the $|6\rangle$ and the $|5\rangle$ low-field seeker states, in order to take advantage of the very large triplet *s*-wave scattering that occurs in that case. This large scattering length is the source of the relatively high predicted critical temperatures $T_c = 37$ nK for $\sim 2.8 \times 10^5$ atoms in each spin state. A sample lifetime of 1 s is also possible, provided that a magnetic bias field of about 10 T is used [689].

In this more accurate calculation, the authors confirmed that the nucleation time of 1 ms for a Cooper pair is considerably less than the overall one second lifetime of the gas in the magnetic trap. If the magnetic field is reduced from ~ 10 T to about 0.7 T, then the trap lifetime decreases from about 1 s to 0.7 ms. So, a very high magnetic field is a requirement for magnetic trapping.

The authors, at this point of time (1999), suggested "a better option, therefore, appears to trap and cool the lowest two hyperfine states in an optical trap, which was recently shown to be possible for bosons" [559]*. This, of course, would be a great simplification, since the lowest two hyperfine levels $|2\rangle$ and $|1\rangle$, which are high-field seekers in magnetic traps, are now trappable in optical traps and cannot undergo inelastic collisions, since they are energetically the lowest two of the six hyperfine levels.

25.5. Collisional Relaxation in a Fermionic Gas

Ferrari, in 1999 [692], reviews the various suggestions for probing an atomic Fermi gas in the degenerate regime. There have been theoretical proposals to study modifications of refractive index, or the absorption coefficient of the gas, to look for reduction of the spontaneous emission of an excited atom within the cloud of fermions, or to measure the angular dependency of the radiation pattern. See Ferrari for specific references.

In this paper, Ferrari proposed studying the collisional relaxation of a test probe particle passing through the degenerate gas. It is shown that for a test particle with energy well below the Fermi energy there is a strong decrease in the relaxation rate as a consequence of Fermi–Dirac statistics. This method can be used to measure directly the temperature of the Fermi gas. The author proposed using ^{7}Li as a probe of a ^{6}Li Fermi gas, where the ^{7}Li atoms are excited by means of successive optical Raman transitions. This isotopic selectivity is possible due to the isotope shift of the Li resonance line.

Three different types of test particle "trajectories" were considered: (i) the particle is initially at rest in the center of the trap, (ii) the particle oscillates with a given energy and angular momentum, and (iii) the test particle trajectory consists of a circular orbit within the cloud. The calculations show evidence of resonance behavior in the damping, when the energy of the probe particle equals the Fermi energy. The authors state that the effect of Cooper pairing on the probe particle can be extended to even lower temperatures.

25.6. Observation of Fermi Degeneracy in Trapped ^{40}K Atomic Gas

Another milestone in the continuing study of quantum degeneracy in ultracold atomic vapors was achieved in 1999 by DeMarco and Jin [35]* in their demonstration of the "Onset of Fermi Degeneracy in a Trapped Atomic Gas". This was just about four years after the first demonstration of BEC in dilute atomic vapors of bosons [29]*, [30]*, [512]. Fermions are much more difficult to cool to quantum degeneracy, as already discussed, because of the difficulty of maintaining thermal contact with the cooling gas as it approaches the Fermi temperature. This is due to Fermi–Dirac statistics, that prohibits the s-wave collisions responsible for evaporative cooling.

To overcome this cooling limitation DeMarco and Jin resorted to one of the two-component Fermi gas strategies suggested earlier [535]*, [690], [693]. In their experiment using ^{40}K they trapped a mixture of atoms in two magnetic sublevels that could maintain thermal contact with each other by allowed s-wave collisions. In effect, they were able to simultaneously cool both atomic sublevels, using evaporative cooling.

As explained in the text, they used either a single radio frequency or separate radio frequencies to eject hot atoms from the magnetic trap at different points in the cooling cycle. Some clever tricks involving adiabatic cooling and magnetic field changes were needed to maintain nearly equal populations of the two components as they cooled, and to cool below $T = T_F$, the Fermi temperature.

After evaporative cooling, one of the two components was ejected from the trap, in what amounted to a final cooling step, and the remaining component was analyzed by resonant absorption imaging to determine the number of remaining atoms, the final temperature, and their momentum distribution. Remarkably, the final temperature of whatever remained was essentially limited to about $T_F/2$, irrespective of the cooling trajectory. The final Fermi temperatures T_F were in the range of 0.36– $1.0\,\mu$K and the final numbers of atoms in the condensate were in the range from 3.5×10^5 to 1.2×10^6.

Although there is no quantitative theory for this behavior, the authors ascribe it to the onset of the well-known quantum mechanical Fermi pressure, which arises from the Pauli exclusion principle when all the available states of a Fermi gas distribution are occupied. This is the same Fermi pressure that resists the gravitational collapse of white dwarfs and neutron stars in astronomy. In the case of the final temperature $F \cong 0.5\,T_F$, the occupancy of the lowest trap states is $\sim 60\%$ and the elastic collision rate is much inhibited.

A nice way to detect the emergence of quantum degeneracy is to measure the excess of the total energy of the trapped gas over the classical gas at the same temperature. One clearly sees a very large increase in the excess energy at $T = T_F$, as expected. Another indication of the onset of quantum degeneracy is in the change in the shape of the measured momentum distribution from the classical Gaussian momentum distribution at temperatures well above T_F to the very different distribution

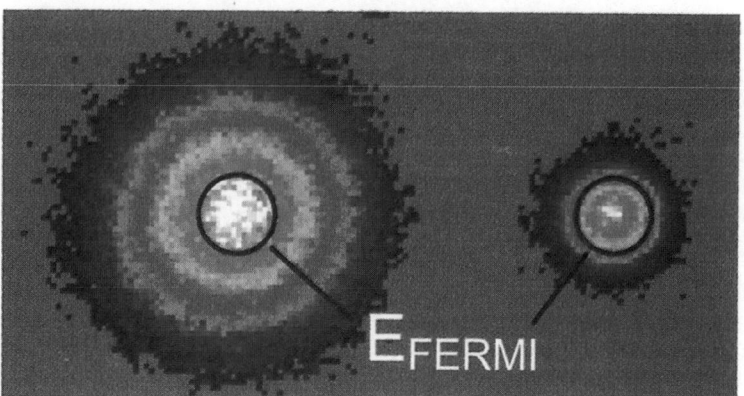

Fig. 25.2. Approach to the Fermi surface. The absorption images of an expanded ultracold Fermi gas show that more of the atoms lie within the Fermi surface (black circles) below the Fermi temperature (right) than above it (left) (in the false color images, white indicates the highest density; blue/black, the lowest). The images were taken 15 ms after release from a magnetic trap. The hotter cloud ($T = 2.4\,\mu$K) has 2.5 million atoms; the colder cloud ($T = 0.29\,\mu$K) has 0.78 million, having lost atoms through evaporative cooling. Figure and caption from Ref. [694].

expected at $T = 0.5\,T_F$. The transition also manifests itself in a clear change in the radial distribution of atoms in space. One clearly sees, from optical density measurements, the compact core of atoms within the Fermi surface of the distribution. See the photographs in Fig. 25.2, taken from B. Goss Levi's commentary on Fermi condensates in Ref. [694].

The authors foresee the use of this novel dilute Fermi gas as an ideal system for quantitative studies of various predicted phenomena, such as narrowing of spontaneous emission, the suppression of inelastic and elastic collisions, changes in damping rates, emergence of zero-sound at low temperatures, possible appearance of shell structure in degenerate Fermi gases, and, finally, the possibility of achieving the very low temperature phase change to a superfluid state of Cooper paired atoms, as discussed above. See the references in DeMarco and Jin.

Levi's "Search and Discovery" article in Physics Today mentioned above recapitulates the essence of the DeMarco and Jin article and gives comments by theorists on what new problems they would find interesting to solve, using this new atomic Fermi gas. There is also discussion about the approach of Hulet to cooling ^{6}Li and the problem of detecting Cooper pairs. Of special interest to me are remarks about the advantages of purely optical trapping for use in forming Cooper pairs and the statement attributed to Jin that she is considering transferring a quantum degenerate Fermi gas into an optic trap once it is formed. Levi also discusses the experiments of Thomas and his colleagues in the context of cooling ^{6}Li atoms to form Fermi condensates, using ultrastable CO_2 lasers.

25.7. Stable, Strongly Attractive Two-State Mixtures of ^{6}Li Fermions in an Optical Trap

In 2000, Thomas and his colleagues O'Hara *et al.* at Duke University, [695]*, made quite significant advances in the experimental trapping of stable, strongly attractive, two-state mixtures of ^{6}Li fermions in far-off-resonance optical traps. In earlier work with this system, they observed a trapping lifetime for ^{6}Li atoms of $300\,$s in a $0.4\,$mK deep trap, consistent with the background pressure of $10^{-11}\,$Torr. The stability demonstrated here relies heavily on the previous studies by Thomas's group on laser-noise-induced heating in far-off-resonance optical traps [542], [544]. See the earlier discussion in Sec. 16.5. This is the longest storage time achieved in all-optical traps, comparable to the best magnetic traps. The scattering time per photon for one atom is about $3400\,$s. Also, the Stark shifts of the ground and excited states are nearly identical, making it possible to optically cool this system further. The authors see the CO_2 laser trap as a good system for studying elastic and inelastic collisions between fermions in different hyperfine states. Such a study is usually not possible in a MOT or a purely magnetic trap.

Previously in Refs. [687], [688], it had been suggested that one could use the anomalously large triplet s-wave scattering length $a_T = -2160\,a_0$ between states $|6\rangle$ and $|5\rangle$ to evaporatively cool ^{6}Li atoms to the transition temperature in magnetic traps. To achieve stability of these magnetically trappable mixtures one needed a high magnetic field of at least $5\,$T. Houbiers $et\ al.$ in Hulet's group [689] proposed use of optical trapping of ^{6}Li, using mixtures of the lowest two hyperfine states $|2\rangle$ and $|1\rangle$, as a more favorable alternative, as discussed above.

In their paper, O'Hara $et\ al.$ [695]* point out that one still needs quite high magnetic fields, $B \geq 800\,$G, in the $|2\rangle$ and $|1\rangle$ mixture to get a large negative scattering length, or to be near a Feshbach resonance. They have discovered, however, that by using a mixture of the $|3\rangle$ and $|1\rangle$ states one can achieve very large negative scattering length at the low magnetic field of $B = 8.3\,$G. This mixture is also stable against inelastic spin exchange, as long as one provides a small bias magnetic field. They calculated that the a_{31} scattering length for collisions between these two states can be widely tuned at low-fields in the range from about $0.01\,$G up to about $100\,$G. See Fig. 2 of O'Hara $et\ al.$ [695]*, where we see that scattering length is expected to vary from a minimum of $\sim -480\,a_0$ at $10\,$G to values as large as $-1500\,a_0$ near 0 and $100\,$G. The calculation is based on the asymptotic boundary condition approximation of Houbiers $et\ al.$ [689] using the most up-to-date data on ^{6}Li scattering parameters.

Experiments were performed on the behavior of this two-state mixture in the authors' ultrastable CO_2 far-off-resonance dipole laser trap [542], [544]. Atoms initially trapped in $F = 3/2$ low-field-seeking state of a MOT were transferred into the optical trap and then were optically pumped into a 50/50 mixture of $|1\rangle$ and $|2\rangle$ states. By applying a π Raman pulse, the authors transformed the $|2\rangle$ state atoms into the $|3\rangle$ state. The CO_2 laser trap depth was $330\,\mu$K and the atoms were at 100–$200\,\mu$K. They then initiated optical evaporation of the mixture of the $|3\rangle$ and $|1\rangle$ states by lowering the trap depth with an acoustooptic modulator to $100\,\mu$K, in the presence of a bias magnetic field of $8.3\,$G. After an initial rapid decrease in the number of trapped atoms in the first 20–50 s, the evaporation stagnated and the number of atoms in the mixture decayed exponentially over a period of several hundred seconds. The data matched very closely with a numerically integrated Boltzmann-type theoretical evaporation curve, over their full time span. The calculation was done for an s-wave scattering length of $|a_{31}| = 540\,a_0$ at $8.3\,$G and a temperature of $46\,\mu$K. The calculation predicts a final temperature of $8.7\,\mu$K. To demonstrate further that evaporative cooling is occurring rather than just trap loss, the authors measured the final temperature and obtained $9.8 \pm 1\,\mu$K, which is very good agreement. This represents cooling by a factor of $\sim 46\,\mu$K$/9.8\,\mu$K ≈ 5.

In future work the authors will attempt to use continuous evaporation to obtain even lower temperatures and to take advantage of reduced magnetic field bias to obtain larger scattering lengths. They believe this system may be a good candidate for cooling close to superfluidity. This work, as we saw, led to the first demonstration of the all-optical production of a degenerate Fermi gas as discussed in Sec. 20.15.

Subsequent careful measurements of the magnetic field dependence of the scattering length accurately located the zero crossing field for the two lowest hyperfine states of ^{6}Li [696]. This was done by monitoring the magnetic field dependence of the temperature decrease and atom loss caused by evaporative cooling from a CO_2 laser trap. Such data are needed in designing an experimental search for a high-temperature superfluid transition near a Feshbach resonance.

25.8. Observation of Fermi Pressure in a Doubly Degenerate Gas of Fermions and Bosons

Following the pioneering work of DeMarco and Jin [35]* on the achievement of quantum degeneracy in a gas of ^{40}K fermions in 1999, Truscott et al. in 2001, working in Hulet's laboratory, succeeded in cooling ^{6}Li fermions to degeneracy [697]. Direct evaporative cooling of a Fermi gas to degeneracy is difficult because Fermi gases obey the Pauli exclusion principle, which forbids identical particles from occupying the same quantum state. Once temperatures near the Fermi temperature T_F are reached, thermalizing collisions between atoms are reduced and further evaporative cooling is inhibited, as we discussed above. DeMarco and Jin circumvented this problem by simultaneously trapping an equal mixture of different spin states, between which thermalizing collisions are allowed. They evaporatively cooled by ejecting roughly equal numbers of atoms in the two spin states from the trap. They succeeded in reaching temperatures of $T/T_F \cong 0.5$.

With ^{6}Li, Truscott et al. performed sympathetic cooling, using a mixed gas of ^{6}Li fermions and ^{7}Li bosons. They loaded $\sim 3 \times 10^{10}$ atoms of ^{7}Li and $\sim 10^7$ atoms of ^{6}Li into a MOT using properly chosen hyperfine levels. These atoms were transferred to a cloverleaf magnetic trap for evaporative cooling of the small sample of ^{6}Li fermions by the large sample of ^{7}Li bosons. This was highly successful and cooled the ^{6}Li sample to the low value of $T/T_F = 0.25$. The authors claim that for this experimental situation where the heat capacity of the bosons, C_B, is much greater than C_F, the heat capacity of the fermions that one expects theoretically is $T/T_F \geq 0.3$. This is close to the experimentally observed value. At the temperature of $T/T_F = 0.25$, the spatial size of the Fermi cloud is strongly affected by the Fermi pressure, resulting from the single level occupancy. This is clearly seen in Fig. 25.3, taken from Fig. 3 of Ref. [697]. One sees that at $T/T_F = 0.25$, the Fermi gas has compressed to about as small a radius as possible.

The authors see the success of this experiment as opening the possibility of observing the Cooper pairing phase transition to a superfluid state at yet lower temperatures.

K. M. O'Hara and J. E. Thomas, in a "Perspective: Quantum Physics" article in Science [591], discuss the Truscott et al. paper and the various approaches to cooling dilute Fermi gases to degeneracy. They point out that to reach the low temperatures and higher densities required for the Cooper pair phase transition, one must avoid magnetic field gradients that can destroy the pairing. These

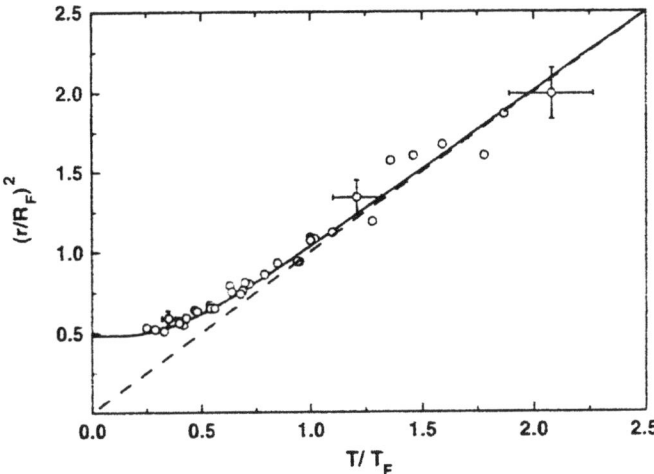

Fig. 25.3. Square of the $1/e$ axial radius, r, of the ^{6}Li clouds versus T/T_F. The radius is normalized by the Fermi radius, $R_F = (2k_B T_F / m\omega_a^2)^{1/2}$, where m is the atomic mass of ^{6}Li. The solid line is the prediction for an ideal Fermi gas, whereas the dashed line is calculated assuming classical statistics. The data are shown as open circles. The divergence of the data from the classical prediction is the result of Fermi pressure. Several representative error bars are shown. These result from the uncertainties in number and temperature as described in the legend to Fig. 1. In addition, we estimate an uncertainty of 3% in the determination of r. Figure 3 and caption from Ref. [697].

conditions seem to make the use of an all-optical dipole trap a necessity (Ref. 12 in Ref. [591] provides an Internet address for a comprehensive listing of articles on degenerate Fermi gases).

25.9. Observation of a Strongly Interacting Degenerate Fermi Gas of Atoms

In 2002, Thomas's group [698] performed experiments on a strongly interacting Fermion gas made up of a 50–50 mixture of the two lowest hyperfine states of ^{6}Li with a scattering length a_s tuned to $\sim -10^4\, a_0$, where a_0 is the Bohr radius. Resonance superfluidity should occur at a sufficiently low temperature. Both spin states were cooled by evaporative cooling from an optical trap at a desired magnetic field of 910 G. When abruptly released from the trap, the trapped cloud was observed to expand anisotropically, i.e., rapidly in the radial direction and only slightly in the axial direction. This contrasts with the isotropic expansion for the case of a noninteracting Fermi gas. The authors interpret such an anisotropic expansion dynamics in terms of a collisionless superfluid and collisional hydrodynamics. Data taken at the lowest achievable temperature is in disagreement with collisional hydrodynamics, but is in plausible agreement with superfluidity.

The difficult problem of finding an unequivocal signature for the presence of superfluidity and Cooper pairing was discussed in some interesting commentary in Science [711] in 2003. This problem has been resolved in the last five years or so, as we shall see in Secs. 25.10–25.16.

25.10. Emergence of a Molecular Bose–Einstein Condensate from a Fermi Gas

It has been predicted theoretically that in a fermionic superfluid, cooled far below the Fermi temperature, T_F, one could tune the system continuously by varying the interaction strength between two limits: a Bardeen–Cooper–Schrieffer (BCS)-type superfluid involving atom pairing and a molecular BEC.

Jin and colleagues have recently experimentally demonstrated the production of molecules [699] by sweeping the magnetic field of a very cold degenerate Fermi gas from above the Feshbach resonance, where the scattering length $a > 0$ (repulsive) to $a < 0$ (attractive).

Using a mixture of two different spin states of ^{40}K, they first evaporatively cooled the Fermi gas in a magnetic trap. As a final step in cooling, they loaded the atoms into a $1.06\,\mu$ far-off-resonance optical dipole trap for evaporative cooling to temperatures between $0.36\,T_F$ and $0.04\,T_F$. By adiabatically lowering the magnetic field, they adiabatically converted 78–80% of the atoms into highly excited molecules that stayed trapped in the dipole trap. They lasted for about 10 ms, which was sufficient time for the molecules to thermalize.

Grimm's group in Austria also observed the formation of a molecular condensate using a somewhat different method [700]. They worked with a 50:50 mixture of the lowest spin states of ^{6}Li fermions in an optical dipole trap and a fixed magnetic field close to a Feshbach resonance, giving them $a \approx +3500\,a_0$, where a_0 is the Bohr radius. Forced evaporative cooling led directly to the formation of weakly bound dimer molecules by three-body recombination, rather than to a degenerate Fermi gas, as found with "a" negative. Molecular condensates having as many as 1.5×10^5 molecules and temperatures as low as ~ 50 nK were observed. Condensate lifetimes as long as 20 s were obtained. This is to be compared with the ~ 10 ms molecular condensate lifetimes measured by Jin *et al.* [Jin, Nature December 2003]. This illustrates the remarkable stability of dimers made from fermions. Interesting experiments were performed measuring the maximum number of atoms and molecules that could be trapped in dipole traps on opposite sides of a Feshbach resonance as the potential and temperatures were reduced.

25.11. Observation of Resonance Condensation of Fermionic Atom Pairs

Regal, Greiner and Jin [701]* have made a major advance in the study of fermions near a magnetic Feshbach resonance. They present strong evidence showing the condensation of fermionic atom pairs for detuning on the low field or attractive side of the BCS–BEC resonance. In this regime $a < 0$, where a is the s-wave scattering length. Previously they and others have observed the formation of molecular BE condensates on the $a > 0$ or repulsive side of resonance [699], [700], [709], [710].

The main problem to be solved in the experiment was how to detect the presence of fermionic pairs. The usual method of BE condensate detection involving observation of time-of-flight images fails because the weakly bound pairs depend on many-body effects and would not survive during the free expansion of the gas.

In their experiment the authors started by cooling ^{40}K fermions in a magnetic trap to well below degeneracy, at $T/T_F = 0.07$, where T_F is the Fermi temperature. These atoms were transferred into a far-off-resonance optical trap at a magnetic field far from resonance on the fermionic side of resonance. The authors then induced pairing of the cold fermions by slowly tuning the field close to resonance and waiting about 2 to 30 ms. To search for condensation of fermion pairs, they used a technique that projects the fermionic atoms onto molecules. This was accomplished by turning off the trap and quickly lowering the field well into the BE condensate side or resonance, where it is weakly interacting. This converted the fermion gas into a molecular gas that was then allowed to expand freely. Time-of-flight measurements indicated that about 60–80% of the atoms were converted into molecules with close to zero momentum. Measurements of the fraction of molecules near zero momentum versus T/T_F and versus detuning showed a distinct threshold. This was interpreted as the onset of a pre-existing condensate of fermion pairs.

The authors have studied the BCS–BEC crossover regime and observed a smooth variation in properties, as had been predicted. Interestingly, it is found that the lifetime of the condensed state in the crossover regime is significantly longer than in the BEC limit. The authors remark that one expects the resonance fermion condensate observed in these experiments should give rise to superfluidity, as for BEC.

A distinction was made by the authors between the "condensation of fermionic pairs" observed in these experiments and the more commonly termed condensation of Cooper pairs. For fermionic condensates one works much closer to the resonance, where the interaction strength, as determined by a, is much larger than that of the weaker Cooper pair condensate.

The authors believe the ability to detect fermion pairing in the BCS–BEC crossover region will open twenty years of theoretical analysis to experimental scrutiny.

25.12. Evidence for Superfluidity in a Resonantly Interacting Fermi Gas

Thomas and his colleagues in their earlier work, in 2002, on strongly interacting degenerate fermionic gases [698] observed asymmetric expansion of the Fermi gas released from a trap. Such behavior is consistent with pair formation and superfluidity, but cannot be explained by collisional hydrodynamics. As seen above, the experiment of Jin and colleagues in 2004 gave the first direct evidence for pair formation in the region of strong interactions near a Feshbach resonance [701]*.

In this experiment [702]* in 2004, the Thomas group presented strong evidence for superfluidity in a resonantly interacting ^{6}Li Fermi gas held in an optical trap, at a magnetic field just above a Feshbach resonance, where molecules are forbidden. Collective oscillations of the degenerate gas were started by turning off the laser trap for a short time, allowing the gas to expand, and then turning the trap back on. They measured the vibration frequency of the cloud and its lifetime as a function of temperature. They found that lifetime continued to increase as the temperature fell. This is the opposite of an ordinary hydrodynamic gas, where the damping increases at lower temperatures. The frequency of the radial breathing mode of the oscillation was measured to be 2830 Hz, in excellent agreement with a theoretical prediction of 2830 Hz for a hydrodynamic Fermi gas. Figure 25.4, taken from Ref. [702]* shows one cycle of the oscillating motion of the atom cloud in the optical trap. Both of the above measurements give direct evidence of superfluid behavior of paired fermions in the resonantly interacting gas.

25.13. Collective Excitations of a Degenerate Gas at the BEC–BCS Crossover

In 2004, Bartenstein *et al.*, in Grimm's group [703], performed experiments on collective oscillations of ^{6}Li gases in the BEC–BCS crossover region, which are quite similar to those of the Thomas group discussed above in Sec. 25.12. Measurements made on the axial compression mode of the

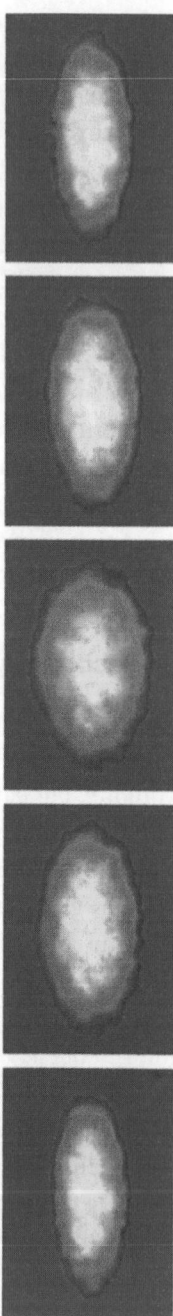

Fig. 25.4. By observing the way their soup of lithium-6 atoms jiggled at ultracold temperatures, NASA-funded researchers hit upon the first direct evidence for a frictionless fluid made up of fermion atoms. This series of pictures shows the "fermion atom superfluid" oscillating inside a laser beam trap.

cigar-shaped trap close to a Feshbach resonance confirmed theoretical expectations. They found, however, that the radial compression mode shows surprising features. When in the strongly interacting molecular BEC regime, they observed a negative frequency shift with increasing coupling strength. In the regime of a strongly interacting Fermi gas, an abrupt change in the collective oscillation frequency occurred, which may be a signature for a transition to a collisionless phase.

25.14. Observation of the Pairing Gap in a Strongly Interacting Fermi Gas

Chin *et al.*, in Grimm's group in Austria in 2004 [704] experimentally demonstrated the existence of pairing of fermionic atoms in the strongly interacting region of an ultracold Fermi gas by observing a pairing energy gap in the rf excitation spectrum. The existence of such a pairing gap has long been considered a signature of superfluidity and superconductivity in fermionic systems in cryogenic liquids and metals. The pairing gap is the main parameter used to characterize the pairing process.

In their experiment, the authors prepared an ultracold sample of ^{6}Li atoms in a balanced mixture of sub-levels $|1\rangle$ and $|2\rangle$, the lowest two sub-levels of the triplet ground state. About 4×10^5 atoms were placed in a far-off-resonance optical trap, where it was possible to independently control the coupling strength with an external magnetic field, the temperature by varying the evaporative cooling, and the Fermi energy by adiabatic recompression of the gas after evaporation via the laser power.

To spectroscopically search for a pairing gap, they drove an rf transition from state $|2\rangle$ to the empty state $|3\rangle$ at $\sim 80\,\mu$Hz and monitored the loss of atoms in $|2\rangle$ with weak excitation, after passage of about 1 s, using state selective absorption imaging. They then took spectra showing the fractional loss in state $|2\rangle$ versus the frequency offset in kHz relative to the $|2\rangle \rightarrow |3\rangle$ atomic transition for different degrees of cooling in various coupling regimes. See Fig. 25.5, taken from Fig. 1 of Ref. [704].

Spectral signatures of pairing have been considered theoretically [705]. A clear signal of pairing is the appearance of a double-peaked structure showing the coexistence of unpaired and paired atoms. The pair-related peak is located at a higher frequency than the peak of unpaired atoms because of the extra energy required for pair breaking. To make a comparison with theory, one must make a detailed analysis of both the homogeneous line shape and the inhomogeneous line shape broadening due to the density distribution in the harmonic trap.

The spectra of Fig. 25.5 below resonance at 720 G shows a sharp peak of the unpaired atoms at zero offset and the dissociation signal of the dimer molecules. They determined the molecular binding energy from the measured frequency shift. On the low field side of the resonance regime at 822 G, where the scattering length "a" is greater than zero, they saw the double peaked response of unpaired atoms and dimer molecules, formed by only two-body interactions. At 837 G on the high field side of resonance, where $a < 0$, bound molecules are forbidden, and BCS pairing can only occur because of many-body effects. At low enough temperatures one clearly sees double peaked spectra showing both the unpaired atoms at zero rf offset and the shifted BCS pairing signal. This appears at higher frequency due to the extra energy required for pair breaking. Finally at 875 G, well beyond resonance, the data show a pure BCS pairing signal at a temperature of $T/T_F < 0.2$. This corresponds to the many-body regime where the pair size exceeds the inter-particle spacing.

Data were also taken to show the effect of temperature on the measured pairing gap, keeping all other parameters constant. This was done using a method of controlled heating. The authors started

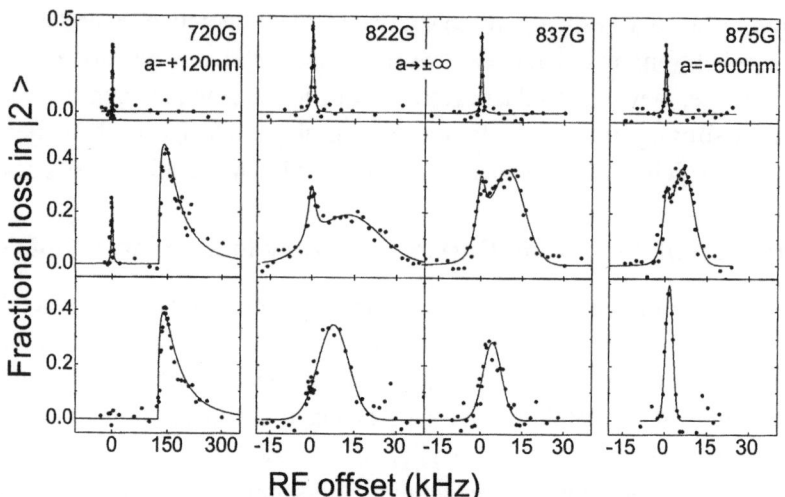

Fig. 25.5. RF spectra for various magnetic fields and different degrees of evaporative cooling. The RF offset ($k_B \times 1\,\mu$k $\cong$ $h \times 20.8$ kHz) is given relative to the atomic transition $|2\rangle \rightarrow |3\rangle$. The molecular limit is realized for $B = 720$ G (first column). The resonance regime is studied for $B = 822$ G and $B = 837$ G (second and third columns). The data at 875 G (fourth column) explore the crossover on the BCS side. Top row, signals of upaired atoms at $T' \approx 6T_F$ ($T_F = 15\,\mu K$); middle row, signals for a mixture of unpaired and paired atoms at $T' = 0.5\,T_F$ ($T_F = 3.4\,\mu K$); bottom row, signals for paired atoms at $T' < 0.2\,T_F$ ($T_F = 1.2\,\mu K$). The true temperature T of the atomic Fermi gas is below the temperature T', which we measured in the BEC limit. The solid lines are introduced to guide the eye.

with a trapped condensate adiabatically adjusted to $B = 837$ G and $T/T_F < 0.2$. They then increased the trapping laser power by different amounts using exponential ramps of different durations. The fast ramp gives a nonadiabatic recompression of the atom cloud and an increase in temperature. By variation of the ramp time they explored a range from their lowest temperature up to $T/T_F = 0.8$. The emergence of the gap with decreasing temperature is clearly seen in the spectra of Fig. 3 of Ref. [704].

The spectra calculated by Kinnunen *et al.* [705] are in excellent qualitative agreement with the data of Fig. 1 of Chin *et al.* [704]. Data on the shift of the pair peak with temperature agree quantitatively with the measurements of Grimm's group. The theory also clarifies the role of the so-called "pseudo-gap" regime in which pairs begin to be formed even above the critical temperature before superfluidity is reached. The experimentally measured gap is the sum of the pseudo-gap and the true superfluid-pairing gap. As the temperature decreases the pseudo-gap contribution totally disappears, leaving just the pairing gap.

The conclusion of Chin *et al.* [704] is that their observation provides a strong case for superfluidity based on a many-body effect in resonantly interacting trapped Fermi gases.

Kinunnen *et al.*, in their conclusion, point to the agreement of their theory with the measurements of Chin *et al.* They also emphasize their theoretical contributions to the understanding of the role of the pseudo-gap to the total pairing gap.

25.15. Heat Capacity of a Strongly Interacting Fermi Gas

Kinast, Turlapov, and Thomas of Duke University and Chen, Stajic, and Levin of the University of Chicago have collaborated on an experiment and analysis of the heat capacity of a strongly interacting

Fermi gas in the crossover region of a Feshbach resonance in ^{6}Li [706]. The data show a clear transition in the heat capacity of the ultracold gas that is in excellent agreement with theory and indicates the onset of superfluidity at the observed transition point. Measurements of heat capacity versus temperature have long been used in studying superfluidity and superconductivity in fermionic systems.

The experiment started with a 50:50 mixture of $\sim 2 \times 10^5$ atoms in the lowest two spin states, held in a deep optical dipole trap magnetically tuned close to a Feshbach resonance. The atoms were initially evaporatively cooled to a temperature of $\sim 0.04\,T_F$, where $T_F \cong 2.5\,\mu$K, well into the degenerate region. The optical trap depth was $\sim 35\,\mu$K. Energy was added to the trapped gas in small precise amounts by releasing the atoms from the trap for short times ($t_{\text{heat}} \leq 460\,\mu$s), after which time the trap was turned back on and the atoms were recaptured. The gas was allowed to thermalize for 0.1 s before being released from the trap for imaging purposes after one millisecond of expansion. The imaging measurements gave the number of atoms and their temperature. For each value of heating time, the authors determined the total energy or heat capacity of the gas. Data were taken for various heating times at two magnetic fields: at 526 G, far from resonance in the noninteracting regime, where the scattering length a is essentially zero, and at 840 G in the so-called strongly interacting unitary regime, where fermions are strongly coupled.

The data on heat capacity versus temperature in the two regimes were compared with theory. When the strong interaction data were plotted on an expanded log-log scale, a striking result appeared. A clear transition in slope was evident at $T/T_F = 0.27 \pm 0.02$, clearly suggestive of the onset of superfluidity. This value is close to the theoretical value for the onset of pairing at $T/T_F = 0.29$.

The authors point out an important feature of the strong coupling theory: pair formation can occur at a temperature, T, that is higher than T_c, where pairs condense. In the range from $T_c < T < T^*$ the fermion attraction is strong enough to form quasi-bound or preformed pairs that affect the thermodynamics. A finite energy, known as the pseudogap T^*, is needed to create such single quasi-bound fermion pairs. They find it interesting that in the regime of strong-coupling T^* and T_c are large and approach T_F, resulting in high temperature pair formation and very high temperature superfluidity. These heat capacity experiments give yet another signature for Cooper pairing of fermions in the strong-interaction regime near a Feshbach resonance.

Theorists believe that superfluidity due to pairing in all strongly interacting systems will behave similarly. This is called the universality principle. On this basis one might expect to obtain new high-temperature superconductors in which the transition temperature is a significant fraction of the Fermi temperature of electrons in solids.

25.16. Commentary on the Search for Superfluidity in Fermi Gases

As seen in Chapter 25, it has taken five short years to progress from the first achievement of degeneracy in fermionic gases to the observation of the BEC of fermionic molecules and of fermionic pairing in the high interaction region near a Feshbach resonance. These results have been established using a total of six different experimental signatures. The search for superfluidity in fermionic gases has elicited many interesting comments, during the course of this work, in major scientific magazines and journals.

B. Goss Levi, in her 1999 article in Physics Today [694], discussed DeMarco and Jin's first demonstration of cooling ^{40}K fermions to degeneracy [35]* (see Sec. 25.6). Levi points out the difficulty of cooling below T_F due to Pauli blocking and of its solution using a two-component mixture of two spin states held in a magnetic trap. Levi also discusses the possible use of Feshbach resonance in a search for Cooper pairing. Although Jin suggested a possible switch to optical traps in the future, Levi stressed the "daunting task" of directly cooling a two-state mixture within an optical trap.

Later in 2002, Pitaevski and Stringari [707] commented on the work of Thomas's group on the "first experimental realization of the strongly interacting resonance regime of ^{6}Li" [698]. They point out that Thomas's experimental conditions placed them close to the unitary limit where universal behavior is expected. In this regime one expects superfluid behavior and anisotropic expansion of the gas when released from the trap. This is exactly what Thomas's group observed. However, there might be other explanations for anisotropic expansion, and other signatures for superfluidity should also be looked for. Pitaevskii and Stringari suggest looking for quantized vortices. In superfluid Fermi systems vortices are characterized by quanta of circulation that are multiples of $\pi\hbar$, in contrast to multiples of $2\pi\hbar$ for bosons. See Sec. 25.9 and Ref. [645] for measurement of the boson angular momentum per atom.

B. Goss Levi commented again in 2003 in Search and Discovery [708] on the experiments showing the creation of bosonic molecular BE condensates from fermionic atoms in the resonance region. This is the region of universal behavior for all strongly interacting particles. Theorist Jason Ho of Ohio State University stated that experiments with atomic vapors in this regime would help shed light on some of the most interesting problems encountered in all strongly interacting systems. In addition to the work of Jin's group at Boulder [699] using ^{40}K, four other groups, Grimm *et al.* at Innsbruck [700], Salomon *et al.* in Paris, Hulet *et al.* at Rice University [709], and Ketterle *et al.* [710] have all reported observation of molecules from degenerate ^{6}Li Fermi gas near a Feshbach resonance. The lifetimes of the molecules in the ^{6}Li system are a few seconds compared to the millisecond lifetime of Jin's ^{40}K molecules. They pointed out that an interesting aspect of molecular condensation from fermions is that it is a lot easier and more efficient than is the production of molecules from bosons. This probably is the result of being on the molecular side of the Feshbach resonance where condensation is favored. Levi reports that Hulet believes that people may have already seen tightly bound fermionic superfluids close to resonance, but have not recognized them. He also thinks it may take several different pieces of evidence to confirm pairing: observation of a superconductivity-like energy gap, for example, plus the detection of resistance-free flow.

Adrian Cho, commenting on the subject of "Ultracold Atoms Spark a Hot Race", [711] discussed the race to achieve Cooper pairing of fermions. At that time he identified six research teams as contenders in this race: Jin and colleagues using fermionic ^{40}K in two different spin states to cool each other; Hulet and his team who refrigerate fermionic ^{6}Li with bosonic ^{7}Li atoms; Salomon and his group also using a combination of ^{6}Li and ^{7}Li atoms; Thomas's group cooling with two different spin states of ^{6}Li; Ketterle and colleagues who cool ^{6}Li with bosonic ^{23}Na; and Inguscio and collaborators who cool ^{40}K with bosonic ^{87}Rb. Cho, commenting as late as 2003, foresaw a long trek ahead to reach Cooper pairing starting from degeneracy.

Many people agree with statements by Hulet and Jin that the ultracold fermionic system will ultimately have a bigger impact than ultracold bosonic systems. Thomas thinks that the study of ultracold fermionic vapors might be the perfect tool for discovery of the principles that could unify seemingly disparate phenomena.

Tin-Lun Ho, commenting on "The Arrival of the Fermion Superfluid" [712], discusses the paper of Grimm's group in Austria and the paper by Kinnunan and Törmä in Finland on the observation and theory of the pairing gap in a strongly interacting Fermi gas [704], [705]. Ho makes specific note of the large size of Cooper pairs in BCS conductivity and of the newfound ability to vary this size continuously in atomic vapor systems near a Feshbach resonance. He discusses the origin of the pairing gap that represents the energy of the bound pairs and the variation of the gap with magnetic field tuning near resonance. One detects the presence of the gap using rf spectroscopy to carefully measure the transition frequency from level 2 to level 3 of the three-level-ground state of ^{6}Li. Ho points out that the discovery of the energy gap and pseudo-gap in the regime of universal behavior strengthens the connection to other fields, such as nuclear physics and information theory.

At the same time, Charles Seife, in July 2004, commented in Science [713] on the achievement of superfluidity by Thomas's group and Grimm's group, in which they observed very long-lived vibrations of the gas cloud in a perturbed optical trap. Grimm said that the damping rate was so low as to be incompatible with normal hydrodynamic theories. Grimm also mentions the observation of another signature of pairing: the measurement of the variation of the binding energy with temperature.

B. Goss Levi commented yet again, in Physics Today of March 2004 [714], on the work of Regal, Greiner and Jin [701]* on the first experiments showing pair formation in the region of strong interactions near a Feshbach resonance. Pair formation occurs only when tuned on the side of resonance where the scattering length a_0 is negative. Levi describes the novel technique devised by Jin and colleagues for detecting coupled ^{40}K pairs. This involved first making pairs in an optical trap with a_0 negative and then rapidly switching the magnetic field well into the BEC side of resonance. There the pairs split apart and formed molecules with essentially zero momentum. To determine the number of pairs in the original condensate, they measured the fraction of atoms having zero momentum in the usual manner by turning off the trap and letting the atoms expand freely. One clearly sees the growth of a sharp zero momentum peak as the magnetic field is tuned closer to resonance. Pairs were detected in the resonance region but not in the weakly interacting region. Weakly interacting pairs are yet to be detected. Researchers Grimm and Ketterle pointed out the strongly interacting crossover region is a very complicated many-body regime where a clear distinction between fermionic pairs and fermionic molecules is not possible. They remark that at present no complete theoretical description is available for this region. Nevertheless Jin feels that "the strongly interacting region is more interesting than the weakly interacting region of low temperature superconductors".

The above extensive commentaries on the search for Cooper pairing give a clear picture of the history of the subject, starting in 1999, at the time of the first observations of cooling of fermionic atoms to degeneracy, up to 2005, after observation of the final six signatures of fermion pairing. We see that in about six short years we have gone from a period of doubt about the possibility of success to one of elation over the demonstrations of pairing in the strong interaction regime near Feshbach resonances. It is now widely believed that this achievement opens many new opportunities for future research.

Levi's commentary in 1999, at the time of the first cooling to degeneracy in a Pritchard–Ioffe magnetic trap, focused on the doubts and difficulties of the quest for superfluidity for fermions and her worries about the daunting problems associated with using optical traps for cooling purposes. According to Adrian Cho, in Science [711], these doubts persisted into 2003, even after six different groups achieved degeneracy with fermions. Judging from the quotes, there was a general belief that the road to Cooper pairing would be long and hard.

However, amid all the final excitement about the arrival of the molecular superfluid [708], [711] and the many observations of fermionic pairing superfluid [712], [714] there was an absence of commentary on the virtues of the optical trapping technique and the key role it played in achieving these successes. Optical traps were used in the fermionic molecule experiments and in all six of the experiments displaying signatures of fermionic pairing. Without the use of optical trapping, I doubt if optical pairing could have been identified. Of all the race contestants, Thomas was the first to recognize the unique advantages of all-optical trapping for the manipulation of fermions (see Secs. 17.5 and 20.5). Many of these advantages have already manifested themselves in other work, as well (see Chapters 17 and 20).

A final comment worth making is that the simplest and probably best way of detecting Cooper pairing is to use cw samples. One could then prepare fixed samples with which to do diagnostic experiments using weak non-perturbing probes, without resorting to the usual imaging of transient expanding samples. This would, however, involve the use of a cw fermion atom laser, which may possibly be designed along the lines of the optical cw atom lasers suggested by Ashkin for bosons [46]* in Sec. 20.19.

25.17. Vortices and Superfluidity in a Strongly Interacting Fermi Gas

Ketterle's group, in mid-2005 made the first observation of vortices and superfluidity in a strongly interacting Fermi gas [719]. Experiments inferring Bose–Einstein condensation of fermionic gases, and therefore superfluidity, had previously been reported [698], [701]*, [702]*, [703], [704], [705], [706]. See also Sec. 25.16. However, this experiment was the first direct evidence of condensation of fermion pairs in dilute gases under controlled conditions. They made observations in a far-off-resonance optical dipole trap by tuning an external magnetic field near a Feshbach resonance. They observed vortex lattices on both the BEC side of resonance and the BCS side of resonance, where a Bardeen-Cooper superfluid of loosely bound pairs formed.

This was a difficult experiment. The authors found it necessary to use very round optical trapping beams and to optimize the setup by first generating vortex lattices in ^{23}Na before being able to generate lattices in ^{6}Li. It was also necessary to strictly maintain the cylindrical symmetry of the optical trap, optical stirring rod, magnetic field, and gravity.

The authors see this work as a big step for studying superfluid dynamics. The cited reference [720]* by R. Grimm entitled "A Quantum Revolution" provides an excellent summary of this work, and its significance for the future of Fermi systems in many-body quantum physics.

25.18. Fermion Pairing in a Gas with Unequal Spin Populations

As has been pointed out, this is an important problem to consider that applies to many physical situations, such as in superconductors in the presence of a high magnetic field, pairing of quarks in the cold, dense cores of neutron stars, and in the superfluidity of quarks in the early universe. These phenomena are inaccessible to direct experimental study.

Two groups used their previously developed techniques to study fermion pairing in ^{6}Li gases with unequal spin populations. The MIT group cooled sympathetically by collision with ^{23}Na atoms in a magnetic trap and then transferred them to an optical trap [721]. The Rice University group first created the desired spin mixture by adjusting the rf power connecting the two spin states [722]. The mixture was evaporatively cooled by reducing the optical trapping power. The magnetic field was shifted adiabatically to the desired value. This gives very precise control of the spin population imbalance.

Although standard theory forbids superfluidity with unbalanced spin populations, both groups found evidence of a superfluid core, surrounded by a halo of unpaired atoms. Imbalances as large as 70% are possible. The Rice group also observed that unpaired fermions could actually exist within the superfluid state, up to a small imbalance in numbers. If true, this coexistence represents a new type of superfluidity. These first results on imbalanced superfluids have evoked extravagant praise from theorists. Theorist Frank Wilczek, the 2004 Nobel Prize Winner, who works with quark superfluids in neutron stars, speaks of these experiments, "as a gift from heaven" [723].

Afterword

In this book, the history of optical trapping and manipulation of small neutral particles has been followed over the last 35 years or so, from its origin in 1970 up to the present. The scope of the subject continues to widen, encompassing more types of small particles, new phenomena, and new applications. In spite of attempts by some practitioners to treat their specialties as independent subjects, I have maintained that the subject is broader than the specialties and that each of the specialists can benefit from the knowledge of what has been achieved in other areas. After all, the basic light forces and techniques are the same, no matter how they are applied. I believe this broad applicability and the interrelationships have been amply demonstrated by the history, as recounted here.

I am particularly intrigued by the history of atom trapping and cooling in the last 20 years, since the success of the first experiments with optical dipole atom traps in 1986 [4]*. I fully expected that atomic physicists would appreciate the simplicity and virtues of this trap and proceed to make full use of it. This did not happen. Instead, for seven or more years attention was almost exclusively devoted to the magnetooptic or MOT trap and purely magnetic traps such as the Ioffe-Pritchard trap or other variants of purely magnetic traps. The MOT was touted as the trap that overcame the limitations of the dipole trap and the limitations of the optical Earnshaw theorem [178], [181]*. In its first manifestation in 1987, the MOT confined $\sim 10^7$ atoms, compared with the 1500 atoms of the first dipole trap [4]*. This was by virtue of its large volume, up to $(1.5\,\text{cm})^3$ or more, compared to the $10\,\mu\text{m}$ beam radius of the first optical dipole trap. MOTs combine trapping and molasses cooling. They use the scattering force in combination with a magnetic quadrupole field and can only trap atoms in a low-field-seeking hyperfine level. By contrast, dipole traps confine atoms in all hyperfine levels without any need for magnetic fields.

MOTs were found to be limited by reabsorption of spontaneous scattered light [182]*. With the introduction of the so-called "dark spot MOTs" the performance was improved to the point where $\sim 10^{10}$ atoms could be trapped at densities approaching 10^{12} atoms/cm^3 [183]*. This was an almost two orders of magnitude improvement over the ordinary MOT.

However, optical dipole traps with larger volumes and dark spot geometries are possible [545]*, [555] that can confine much larger numbers of atoms at higher densities than the earlier small-volume optical traps [4]*, [148]. Trapping of $\sim 10^8$ atoms has been reported in a large-volume blue-detuned trap at a density of $\sim 10^{11}$ atoms/cm^3 [557]. More than 10^6 atoms were trapped at densities of 5×10^{13} atoms/cm^3 in a dark spot optical trap [545]*. In other experiments $\sim 5 \times 10^6$ atoms were transferred from a magnetic trap into a red-detuned dipole trap at the high density of $\sim 3 \times 10^{15}$ atoms/cm^3. Ashkin, in his proposal for an optical cw atom laser [46]*, discussed the possibility of even higher densities for short periods of time, using optical shepherd-type traps.

It became increasingly clear, following the achievement of BEC in 1995, using evaporative cooling from magnetic traps, that optical traps had distinct advantages for BEC research (see Chapter 17). This led to a revival of interest in optical traps. One also sees from Chapter 18 that the use of optical dipole traps to confine all the hyperfine levels made the study of spinor condensates possible. Phenomena such as metastable excited spin states [555], optical tunneling and Stern–Gerlach separation of spin states [557] were observed.

Dipole traps, as has been discussed, are readily capable of being cooled to Fermi degeneracy using various two-component cooling schemes [689], [691], [694], [695]*.

Far-off-resonance dipole forces tuned below and also above resonance have been used in the study of superfluid vortex behavior in condensates to make "optical stirring rods" [640], [642], [652] or "optical spoons" [645] to generate these vortices.

Optical dipole traps were used to achieve the first observations of Feshbach resonance and the tunability of the atomic interaction strength (see Chapter 19). This ability to magnetically tune the interaction strength and reach very high values of the scattering length without affecting the trap depth or other parameters of the system is one of the most important assets of optical dipole traps.

A number of attempts were made to use optical dipole traps rather than purely magnetic traps to reach the critical phase-space density for BEC in the period from 1995 to 2000. The highest value reached was ~ 0.1, or about an order of magnitude shy of BEC [70], [139]*. Finally, in 2001, Chapman's group achieved BEC using optical dipole traps [205]*, although the final condensate contained only $\sim 3.5 \times 10^4$ atoms. Later, Thomas and colleagues, in their work on fermions, obtained excellent results in 2002, when they cooled more than 10^5 fermions to ~ 600 nK, well below degeneracy [611]*.

One of the most surprising developments using optical dipole trapping techniques involved the trapping of single atoms by single photons. This occurs within the context of cavity-quantum-electrodynamics (CQED) [63]*, [64]. In these experiments a single atom injected into a short ultra-high-finesse optical cavity forms a novel atom–field–cavity complex that can trap the atom with only a single photon circulating in the cavity. Considering how much power is required to trap an atom in the atom–field interaction of the simple optical dipole trap, it is remarkable that a single photon can stably bind a single atom. The atom–field–cavity complex exhibits other remarkable properties, such as strong optical cooling, ability to generate single photons on demand, and even serve as a one-atom laser (see Chapter 20). It still remains to be seen, however, if trapped single atoms ever become the basis of a practical quantum computer.

An area of application of optical dipole trapping is in the field of trapping of small-sized molecules. Deflection of cesium dimer molecules by dipole gradient forces was observed [220]*, as well as trapping of cesium molecules [668]*. A technique to make cold molecules using a Raman pho-toassociation process in a BE condensate of cesium or rubidium atoms was proposed in dipole or magnetic traps [672], and subsequently demonstrated [683]. This method can be used as a pulsed molecular laser. There are, however, other molecular traps operating at higher temperatures that use magnetic fields [670] and electrostatic fields [682].

Probably the most exciting recent events in atomic and molecular physics have been the race to cool fermionic atoms to degeneracy and to reach the Cooper pair regime at even lower temperatures. In the Cooper pair regime one should observe superfluidity of weakly interacting fermions in the regime of many-body interactions. This search began in earnest in 1999, after DeMarco and Jin observed the first cooling to Fermi degeneracy [35]*. As explained in Chapter 25, and especially in the commentary in Sec. 25.16, the dipole trap played a dominant role in the observation of a long-lived dimer molecular BE condensate from an ultracold Fermi gas [700] when tuned to large positive scattering length "a" near a Feshbach resonance. Strongly interacting fermionic pairs were observed on the other side of resonance when "a" was negative. The optical trap also played a crucial role in a total of the seven different experiments showing clear evidence of pairing and superfluidity in the strong interaction region of the pairing regime. This is the so-called unitary regime where one expects universal behavior for all strongly interacting superfluid systems. This explains the strong interest of theorists in this work. They hope to learn more about analogous systems, such as in neutron stars, in nuclei, and possibly even in high temperature superconductors in a simple, well-controlled pure condensate. This work also opens the possibility of cw fermionic atom lasers, as well as cw bosonic atom lasers, and experiments on fermionic solitons and vortices. Thus far, no one has reached the weakly interacting Cooper pair regime where many-body effects play a large role. Optical dipole traps will continue to be widely used in the trapping and manipulation of small ultracold molecules.

In the light of its past neglect and recent impressive performance, I like to think of the all-optical dipole trap as the "Phoenix Trap", because it rose from the ashes, as did the Phoenix of mythology. This trap is certain to play a unique role in future work with fermion superfluids, which many researchers consider more interesting than BEC of bosons.

One wonders why it took so long for the virtues of the Phoenix trap to be recognized. I suspect there was an element of the Not Invented Here (NIH) syndrome among atomic physicists of the early and mid-1980s. It must have been hard on their egos to have to learn about the forces of trapping and manipulation of atoms from some outsider, not even a specialist in atomic physics (those interested in my scientific background can read my brief biography at the end of the book).

In 1997 a Nobel Prize in physics was awarded for "techniques for cooling and trapping atoms". At that time there was, however, no indication that the Nobel Committee in 1997 had any appreciation of the advantages of the optical dipole trap. The committee, in their press releases on the Nobel Prize actually placed very little emphasis on trapping, but rather concentrated on cooling. See the releases A and B in Figs. 187 and 188. This neglect of optical atom trapping was strange, since optical cooling techniques were a direct outgrowth of the early work on the trapping forces. The only mention of optical trapping in press release A was to say, "The limitation of the dipole trap was overcome by the advantages of the MOT". In view of what has happened, it is clear that the Nobel Committee and most atomic physicists, following the demonstration of the optical dipole trap in 1986, grossly misjudged its capabilities. Indeed, it now appears as if the opposite is true. It was necessary to revive the "Phoenix trap" to overcome the difficulties of the MOT and other magnetic traps and realize the full potential of bosonic and fermionic BE condensates (see Chapters 17–20 and 25).

In press release *B*, on the background of the Prize, in a discussion of optical traps, the committee got the origins of tweezer traps wrong. They say, "The tweezer trap was developed earlier by Letokhov and Ashkin". As we know from his papers, Letokhov had nothing to do with tweezer traps or the first conception of stable optical trapping. One wonders how much of the literature on optical trapping the Nobel Committee actually read first-hand. I suspect the committee was relying heavily on the accounts submitted to them by the atomic physicists from the home institutions of the Prizewinners, who were pushing for their own candidates.

These press releases upset me and I decided to send the Nobel Committee on physics a letter protesting their "poor scholarship". Not wanting to embarrass Bell Labs. I sent Bill Brinkman, our then vice president of research, a copy of my proposed letter and a copy of my Proceedings of the National Academy of Sciences inaugural review paper, entitled "Optical trapping and manipulation of neutral particles using lasers", written earlier in 1997 [8]. Brinkman responded that he was convinced I was the discoverer of optical trapping and optical tweezers, but that the Nobel Committee would never admit to "poor scholarship". He suggested to me that I avoid such remarks. On the other hand, he said that since the press releases were historical documents, I had every right to complain about factual errors in these releases. I rewrote my letter along these lines, but, in the end, never sent it, having decided there was nothing to be gained by complaining, other than the resentment of the committee. The fact is that the historical record, consisting of my many published papers and the acknowledgements of my work before and after the granting of the prize by the many other scientists using optical trapping and manipulation, is there for anyone to see. The work, I believe, speaks for itself.

It is interesting to note that there was a Russian complaint about this award that was reported on the Internet just days after the announcement of the prize. See Addendum C, which is a copy of the news release. It says that a team led by Letokhov carried out the same research at the Russian Academy of Sciences in 1986. They quote V. Minogin, a close collaborator of Letokhov, as saying that a translation of this work, published in English in the US in 1987 "gave a schema for the experimental apparatus used by the Nobel Laureates". This is apparently the paper by Minogin and Rozhdestvenskii entitled "Stable Localization of Atoms in a Standing Light Wave Field" [715].

I find this paper is quite amazing. Reference [715] first explains why their early proposals in the late 1970s [126] for stable localization of atoms in a standing wave would not work and how the problems of trapping were solved by Ashkin and Gordon [45]*, [123]* through their suggested use of separate trapping and cooling beams and tuning the trap far off resonance. The Russian paper points out that Chu *et al.* at Bell Labs. demonstrated the first stable localization (trapping). None of this was their work. They go on to state the obvious conclusion, after the fact, that stable localization (trapping) of atoms on the periodic potential of a standing wave is possible. Minogin and Rozhdestvenskii overlook the problem of Stark shift in Ref. [715]. They claim Letokhov suggested the possibility of stable localization in his 1968 paper [56]. As I pointed out in Sec. 20.1, the 1968 proposal was *not* a proposal of stable trapping but of one-dimensional localization. The potential of the standing wave was so weak it required ultracold atoms. Letokhov gave no indication that stable trapping was possible, nor any indication of how to proceed to get the cold atoms necessary for such an experiment. He and his group can hardly claim any priority based on any of the above. In the course of the subsequent work in the mid-1970s they did suggest slowing of atoms by chirped beams [144] and they were the first to estimate the Doppler cooling limit of molasses [125] (see also Secs. 5.1.1 and 3.1.2). The Nobel Committee considered the Russian criticism and responded that their committee "of extremely competent experts" found the work "not as developed as that of the prizewinners". I have to agree, not about the competency of the experts, but with their conclusion about the Russian work.

Note that the information given in the Nando Internet release mistakenly states that the three Nobel Prize winners worked at AT&T Bell Labs. when they began their research. Phillips and Cohen Tannoudji never were at Bell Labs. Of the three, only Steve Chu worked there.

In 2001 a Nobel Prize in physics was awarded for BE condensation; specifically "for the achievement of BE condensation in dilute gases of alkali atoms, and for early fundamental studies of the properties of the condensates". This is a well-deserved award. Work on BE condensates is discussed in (Chapters 16, 18–20, 23, and 25 of the text). The ability to condense bosonic atoms into a single atomic energy level of a trap at nanokelvin temperatures and also to condense fermionic atoms to quantum degeneracy has extended the revolution in atomic physics wrought by the work leading up to the 1997 Nobel Prize on optical cooling and trapping techniques.

From 1995, the time of the first observation of BEC, to 2001, the role of optical forces in this achievement was further reduced, as we have seen, by the widespread use of purely magnetic traps for the final evaporative cooling step leading to subcritical temperatures. However, in May of 2001, as part of the "phoenix revolution", BEC was achieved with evaporative cooling from an all-optical trap, in a remarkably simple and straightforward way [205]*, [588], [611]*, [706]. This achievement and the recent successes of optical trapping techniques with bosons and fermions has prompted me to think of how to take the next big step in BEC technique, namely, the achievement of cw atom lasers [46]*. This paper introduces the concept of sweeping shepherd beams made from blue-detuned focused laser beams to fabricate novel optical traps, waveguides and other all-optical structures. It also describes ways of recirculating evaporatively cooled atoms and of canceling gravity over the entire volume of an atom trap. An important need at this time is for experimental tests of these designs. Although success is not assured, I believe that the shepherding principles will ultimately lead to viable cw atom lasers for bosons and fermions. It appears to me that cw atom lasers based on magnetic traps are much more difficult to implement and will have inferior performances.

For the future, I predict that work on ultracold atoms, BE condensates, and probably cw atom lasers using optical techniques will continue to dominate the atomic physics scene. Many of the interesting specific applications and topics for study were already mentioned above in connection with the phoenix trap. Other topics for study are: improvements in atom clocks using neutral atoms; atom interferometry, including the measurement of gravity and angular rotation by coherent atom gyroscopes; atom lithography using coherent atomic beams; coherent atomic microscopy; applications of cavity quantum electrodynamics, including optical computing and cryptography; superfluid properties of BE condensates, including Josephson effects; study of solitons in BE condensates; nonlinear optical interactions using coherent atomic beams; study of BEC of molecules; sympathetic cooling of other atomic species, possibly even solids and molecules in BE condensates; study of cooling methods at high atom densities in optical dipole traps; and the study of optical cw atom laser designs. All these topics are discussed in the text.

Taken together, these applications of optical trapping, cooling, and manipulation of atoms by laser light cover a remarkable range of topics. It is safe to predict that laser manipulation will continue to play a major role in atomic and molecular physics in the years ahead. The total number of atomic physics papers produced since the first experiment on optical trapping in 1986 and optical molasses cooling in 1985 must be in the thousands. In 2005, at the time of the tenth anniversary of the demonstration of Bose–Einstein condensation, Carl Wieman remarked that a total of fifteen groups worldwide were working with condensates.

The possibility has been raised of a third Nobel Prize in the area of atomic and molecular physics based on the realization of superfluidity in fermionic systems [712]. Some people think the prospects are good, but Jin remarks, "more things are worth the Nobel Prize than it is given out for" [711].

The history of optical trapping and manipulation in biology has been much more straightforward. In the 19 years since the first tweezer trapping of live bacteria and viruses the scope of applications has grown greatly (see Sec. 5.2 and Chapters 6–9 of the text). Optical micromanipulation techniques have unique advantages not shared by other micromanipulative techniques. This has made tweezers the predominant micromanipulation technique in many areas of biological research.

The ability to noninvasively move, separate, and study living cells, viruses, and extremophile bacteria under physiological conditions is quite unique. This new capability makes it possible to establish monocultures, study their genomes, and search for novel enzymes of technological interest (see Chapter 6).

One of the most basic characteristics of living cells and organisms is their ability to move, respond, and reproduce itself in a controlled manner. The use of optical tweezers has played a major role in the recent revolution in understanding cell locomotion, the motion of organelles within the cell, and the molecular motors and mechanoenzymes that make this motion possible. With tweezers, feedback, and the handles technique one can measure the forces of single motor molecules down to a fraction of a piconewton. The spatial and temporal resolution of tweezers have made it possible to observe the nanometer steps and even substeps of molecular motors as they move over the microtubules and actin filaments of the cell's cytoskeleton. With mechanoenzymes such as RNA and DNA polymerase, researchers can measure the forces and detailed motion of these molecules as they read, copy, pause, and repair DNA molecules, using feedback schemes. Forces in the $<100\,\text{fN}$ and distances down to the one angstrom range [717]* can be resolved. Use of tweezers in this context gives a totally new way of studying these DNA and RNA processes, and will certainly continue well into the future (see Chapters 7 and 8, in particular).

Another important capability of optical tweezer techniques in biology is the ability to measure the mechanical properties of the cytoplasm of cells. Single tweezer traps or pairs of tweezers can be used to twist bacterial flagella and measure their rotational compliances; to distort entire cells; to study the elastic and viscoelastic properties of cytoplasm and of internal cell membranes by generating artificial filaments or tethers; and to measure the flexural rigidity of microtubules and actin filaments. Stretching measurements on single- and double-stranded DNA molecules, and on

RecA-DNA complexes with force clamps or constant tension feedback techniques have been made that determine the entropic or elastic and enthalpic or bond-breaking contributions to the elastic modulus and persistence length. These techniques are broadly applicable to the study of a variety of biopolymers (see Chapter 8 of the text). I again remind the reader of the heavy reliance of the work on optical manipulation in biology on the optical trapping techniques originally developed for macroscopic particles in the 1970s and 1980s.

The successful application of tweezers to biology gives me, personally, much satisfaction. I look back now with amusement at the response of our then vice-president of research and Nobel Prize winner, Arno Penzias, to my suggestion in 1986 at a demonstration of optical tweezer manipulation of live bacteria, that tweezer manipulation might some day be important in biology. His response was, "Ashkin, don't exaggerate".

As we saw, the early discussion of optical forces and stable optical trapping involved single macroscopic dielectric particles (see Chapter 2). More recently, tweezer trapping of macroscopic colloidal particles focused on measurements of the interparticle potential of a pair of charge-stabilized colloidal particles, with very controversial results (see Chapter 11). Another important application of optical forces to colloidal and solid state physics is the formation of optically induced colloidal crystals in a process known as "laser-induced freezing". Even two-dimensional quasi-crystals can be formed in this way using multiple beams. The capabilities of tweezers for these colloidal applications are virtually unique. Measurement of entropic forces in binary colloids can also be probed with tweezers (see Chapter 11 of the text for a discussion of these matters). Experimental studies of this important and unusual state of matter are likely to continue for some time into the future.

Beautiful studies were made of optically induced rotation of particles by the spin and orbital angular momentum of light using optical traps. Controlled rotation of various vane-shaped particles by the linear momentum of light was also demonstrated. These complex-shaped particles were fabricated using lithography and a two-photon polymerization technique. In view of the current interest in microtechnology involving small motors and devices, it seems likely that work on these clever optically driven motors will continue for biological and physical applications (see Chapter 12).

The demonstration of small optically driven motors on the scale of a few micrometers brings to mind the story about the "Feynman Prize". Feynman gave a famous lecture at the American Physical Society meeting in 1956 called, "There's Plenty of Room at the Bottom", in which he considered the possibility of devices on the micron or even atomic scale. This talk is often referred to as the start of the field of nanotechnology. Hoping to stimulate some new physics, he offered a US$1000 prize for the demonstration of a rotating motor that would fit into a 1/64th inch cube, not counting the leads. To his disappointment an engineer quickly built a standard motor of the required size. Feynman paid up, but remarked that he should have specified a smaller size. I wonder what he would have thought of the optically driven micrometer-sized rotary motors described in Chapter 12.

As we saw in Chapter 13, microchemists and physical chemists have already made significant advances using microchemical techniques involving chemical reactions in small optically manipulated liquid drops at room temperature. As discussed in Chapter 24, direct optical cooling of small molecules is ruled out, due to the long lifetime of the excited rotational and vibrational states. However, as seen in Chapter 24, trapping of small numbers of alkali dimers has been observed and the prospects for cooling large numbers of molecules by sympathetic cooling by atoms are good. A whole host of new applications exists based on BE condensation of molecules and even molecular matter-wave lasers at ultracold temperatures (see Chapter 25 on making molecules from fermions).

In recent years there have been important advances in optical trapping and manipulation techniques based on the use of dynamic holographic tweezers. Using computer controlled nematic spatial modulators, it is possible to generate arrays of many hundreds of functioning tweezer traps in two and three dimensions. Chapter 14 shows how these arrays can be used to make novel fluidic particle sorting devices based on "optical fractionation". Microparticle and cell sorters have been designed using optical fractionation where particles are driven through static arrays of traps. Other types of devices use controlled fluid flow to separate and analyze particles. Ingenious optical gear pumps and optically driven valves for controlling fluid motion in microchannels were demonstrated for use in integrated "lab-on-a chip" applications. The American Institute of Physics cited optical fractionation as one of the most noteworthy physics achievements of 2003.

A new technical conference, covering optical sorting of particles, "lab-on-a-chip" applications, and other related optical trapping and micromanipulation applications, was sponsored by SPIE, the International Society of Optical Engineering, in 2004. This was a highly successful conference. The conference proceedings contain an excellent collection of around seventy papers, with a large number of useful references. It is available from the Proceedings of SPIE as Volume 5514, with the title, "Optical Trapping and Optical Micromanipulation", K. Dholakia and G. C. Spalding, editors. The conference was repeated in 2005 and is available as Volume 5930.

Another very relevant conference held annually at Aspen, Colorado and run by S. M. Block is the "Single Molecule Workshop" (see http://andy.bu.edu/aspen/winterworkshops.html). Steve Block says that in 2004 they had many papers that reported use of optical tweezers.

An especially valuable review article, called "Optical Trapping", by Keir Newman and Steve Block appeared in 2004 [716], focusing particularly on the instrument design considerations, position detection schemes, and force calibration techniques. Block's laboratory has played a pioneering role in developing optical trapping techniques to the point where sub-pN forces and a DNA base pair positional resolution of one angstrom is possible. This review concentrates principally on recent advances in technique and applications. It contains close to 200 references covering both recent and earlier work on instrument design.

There is a very useful website containing "Recent Publications Related to Optical Tweezers" by David McGloin of St. Andrews University, http://www/st-andrews.ac.uk/~atomtrap/tweezers/tweezer-papers.ktm, which presently contains 23 pages of references starting in 2002. Although not meant to be comprehensive, its roughly 400 publications give a measure of the recent work using optical tweezers.

Finally, just a quick glance at the table of contents of this book is sufficient, I believe, to convince one that even after 35 years of progress, the subject of optical trapping and manipulation of small neutral particles is still in a growing phase. This results from optical trapping techniques having features that other techniques lack. As optical lasers improve, so, too, does the scope of applications. The role of optical manipulation techniques should further increase as we enter the "age of nanotechnology". One can conclude that the future of optical trapping and manipulation of small particles with lasers continues to look bright.

Addendum A

Press Release: The 1997 Nobel Prize in Physics

October 15, 1997

The Royal Swedish Academy of Sciences has decided to award **the 1997 Nobel Prize in Physics** jointly to

Professor **Steven Chu**, Stanford University, Stanford, California, USA,

Professor **Claude Cohen-Tannoudji**, Collège de France and École Normale Supérieure, Paris, France, and

Dr. **William D. Phillips**, National Institute of Standards and Technology, Gaithersburg, Maryland, USA,

for development of methods to cool and trap atoms with laser light.

A.1. Atoms Floating in Optical Molasses

At room temperature the atoms and molecules of which the air consists move in different directions at a speed of about 4000 km/h. It is hard to study these atoms and molecules because they disappear all too quickly from the area being observed. By lowering the temperature one can reduce the speed, but the problem is that when gases are cooled down they normally first condense into liquids and then freeze into a solid form. In liquids and solid bodies, study is made more difficult by the fact that single atoms and molecules get too close to one another. If, however, the process takes place in a vacuum the density can be kept low enough to avoid condensation and freezing. But even a temperature as low as $-270°$C involves speeds of about 400 km/h. Only as one approaches absolute zero $(-273°$C$)$ does the speed fall greatly. When the temperature is one-millionth of a degree from this point (termed $1\,\mu$K, mK) free hydrogen atoms, for example, move at speeds of less than 1 km/h $(= 25\,\text{cm/s})$.

Steven Chu, Claude Cohen-Tannoudji, and William D. Phillips have developed methods of using laser light to cool gases to the μK temperature range and keeping the chilled atoms floating or captured in different kinds of "atom traps". The laser light functions as a thick liquid, dubbed optical molasses, in which the atoms are slowed down. Individual atoms can be studied there with very great accuracy and their inner structure can be determined. As more and more atoms are captured in the same volume a thin gas forms, and its properties can be studied in detail. The new methods of investigation that the Nobel Laureates have developed have contributed greatly to increasing our knowledge of the interplay between radiation and matter. In particular, they have opened the way to a deeper understanding of the quantum-physical behavior of gases at low temperatures. The methods may lead to the design of more precise atomic clocks for use in, e.g., space navigation and accurate determination of position. A start has also been made on the design of atomic interferometers with which, e.g., very precise measurements of gravitational forces can be made, and atomic lasers, which may be used in the future to manufacture very small electronic components.

A.2. Slowing Down Atoms with Photons

Light may be described as a stream of particles, photons. Photons have no mass in the normal sense but, just like a curling stone sliding along the ice they have a certain momentum. A curling stone that collides with an identical stone can transfer all its momentum (mass × velocity) to that stone and itself become stationary. Similarly, a photon that collides with an atom can transfer all its momentum to that atom. For this to happen the photon must have the right energy, which is the same as saying that the light must have the right frequency, or color. This is because the energy of the photon is proportional to the frequency of the light, which in turn determines the latter's color. Thus red light consists of photons with lower energy than those of blue light.

What determines the right energy for photons to be able to affect atoms is the inner structure (energy levels) of the atoms. If an atom moves, the conditions change because of what is termed the Doppler effect — the same effect that gives a train whistle a higher pitch when the train is approaching than when it is standing still. If the atom is moving towards the light, the light must have a lower frequency than that required for a stationary atom if it is to be "heard" by the atom. Assume that the atom is moving in the opposite direction of the light at a considerable speed and is struck by a stream of photons. If the photons have the right energy the atom will be able to absorb one of them and take over its energy and its momentum. The atom will then be slowed down somewhat. After an extremely short time, normally around a hundred-millionth of a second, the retarded atom emits a photon. The atom can now immediately absorb a new photon from the oncoming stream. The emitted photon also has a momentum, which gives the atom a certain small recoil velocity. But the direction of the recoil varies at random, so that after many absorptions and emissions the speed of the atom has diminished considerably. To slow down an atom an intensive laser beam is needed. Under the right conditions effects can be achieved with a strength corresponding to what would be seen if a ball was thrown upwards from the surface of a planet with a gravity 100 000 times the Earth's.

A.3. Doppler Cooling and Optical Molasses

The slowing down effect described above forms the basis for a powerful method of cooling atoms with laser light. The method was developed around 1985 by *Steven Chu* and his co-workers at the Bell Laboratories in Holmdel, New Jersey. They used six laser beams opposed in pairs and arranged in three directions at right angles to each other. Sodium atoms from a beam in vacuum were first stopped by an opposed laser beam and then conducted to the intersection of the six cooling laser beams. The light in all six laser beams was slightly red-shifted compared with the characteristic color absorbed by a stationary sodium atom. The effect was that whichever direction the sodium atoms tried to move they were met by photons of the right energy and pushed back into the area where the six laser beams intersected. At that point there formed what to the naked eye looked like a glowing cloud the size of a pea, consisting of about a million chilled atoms. This type of cooling was named Doppler cooling.

At the intersection of the laser beams, atoms move as in thick liquid, and the name optical molasses was coined. To calculate the temperature of the atoms cooled in the optical molasses the lasers were switched off. It was found that the temperature was about 240μK. This corresponds to a sodium atom speed of about 30 cm/s, and agreed very well with a theoretically calculated temperature — the Doppler limit — then considered the lowest temperature that could be reached with Doppler cooling.

The atoms in the above experiment are cooled, but not captured. Gravity causes them to fall out of the optical molasses in about 1 s. *To really capture atoms, a trap is required, and a highly efficient one was constructed in 1987. It was called a magneto-optical trap (MOT).* It uses six laser beams in the same sort of array as in the experiment described above, but has in addition two magnetic coils that give a slightly varying magnetic field with a minimum in the area where the beams intersect. Since the magnetic field affects the atoms' characteristic energy levels (the Zeeman effect) a force will develop which is greater than gravity and which therefore draws the atoms in to the middle of the trap. *The atoms are now really caught*, and can be studied or used for experiments.

A.4. Doppler Limit Broken

Magnetic fields had already been used at the beginning of the 1980s by *William D. Phillips* and his co-workers in a method of slowing down and completely stopping atoms in slow atomic beams. Phillips had developed what was termed a Zeeman slower, a coil with a varying magnetic field, along the axis of which atoms could be retarded by an opposed laser beam. With his device Phillips had in 1985 stopped and captured sodium atoms in a purely magnetic trap. Enclosure in this trap, however, is relatively weak, for which reason the atoms within it must be extremely cold to remain inside. When Chu managed to cool atoms in optical molasses Phillips designed a similar experiment and started a systematic study of the temperature of the atoms in the molasses. He developed several new methods of measuring the temperature, including one in which the atoms are allowed to fall under the influence of gravity, the curve of their fall being determined with the help of a measuring laser.

Phillips found in 1988 that a temperature as low as 40μK could be attained. This value was six times lower than the theoretically calculated Doppler limit! It turned out that the Doppler limit had been calculated for a simplified model atom that had previously been considered sufficiently realistic. However, *Claude Cohen-Tannoudji* and his co-workers at the École Normale Supérieure in Paris had already in theoretical works studied more complicated cooling schemes. The explanation of Phillips' result lay in the structure of the lowest energy levels of the sodium atom. What happens can be likened to Sisyphus' endlessly rolling his stone up the slope, but in this case finding that the slope beyond the crest is also an uphill one. The comparison has led to the process being termed Sisyphus cooling.

The recoil velocity an atom gains when it emits a single photon corresponds to a temperature termed the recoil limit. For sodium atoms the recoil limit is $2.4\,\mu$K and for the somewhat heavier cesium atoms about $0.2\,\mu$K. In collaboration with Cohen-Tannoudji and his Paris colleagues Phillips showed that cesium atoms could be cooled in optical molasses to about 10 times the recoil limit, i.e., to about $2\,\mu$K. It first appeared that in optical molasses it was generally possible to reach temperatures only about 10 times higher than the recoil limit. In a later development both Phillips and the Paris group have showed that with suitable laser settings it is possible to trap the atoms so that they group at regular intervals in space, forming what is termed an optical lattice. The atom groupings in the lattice occur at distances of one light wavelength from each other. Atoms in an optical lattice can, as has been shown, be cooled to about five times higher temperature than the recoil limit.

A.5. Recoil Limit also Broken

The reason why the recoil velocity an atom obtains from a single photon sets a limit to both Doppler cooling and Sisyphus cooling is that even the slowest atoms are continually being forced to absorb and emit photons. These processes give the atom a small but not negligible speed and hence the gas has a temperature. If the slowest atoms could be made to neglect all the photons in the optical molasses, perhaps lower temperatures could be reached. One mechanism through which a stationary atom can be caused to assume a "dark" state in which it does not absorb photons, was known. But a difficulty was to combine this method with laser cooling.

Claude Cohen-Tannoudji and his group between 1988 and 1995 developed a method based on use of the Doppler effect and which converts the slowest atoms to a dark state. He and his colleagues showed that the method functions in one, two and three dimensions. All his experiments use helium atoms, for which the recoil limit is $4\,\mu$K. In the first experiment two opposed laser beams were used and a one-dimensional velocity distribution was achieved which corresponded to half the recoil limit temperature. With four laser beams a two-dimensional velocity distribution was achieved, corresponding to a temperature of $0.25\,\mu$K, 16 times lower than the recoil limit. Finally with six laser beams a state was attained in which the whole velocity distribution corresponded to a temperature of $0.18\,\mu$K. Under these conditions helium atoms crawl along at a speed of only about $2\,$cm/s!

A.6. Applications Just Round the Corner

Intensive development is in progress concerning laser cooling and the capture of neutral atoms. Among other things, Chu has constructed an atomic fountain, in which laser-cooled atoms are sprayed up from a trap like jets of water. When the atoms turn at the top of their trajectory and start falling again, they are almost stationary. There they are exposed to microwave pulses that sense the atoms' inner structure. With this technique it is believed that it will be possible to build atomic clocks with a hundredfold greater precision than at present. The technique rewarded this year also forms the basis for the discovery of Bose–Einstein condensation in atomic gases, a phenomenon that has attracted great interest.

Further Reading

Additional background material on the Nobel Prize in Physics 1997, The Royal Swedish Academy of Sciences.

Cooling and Trapping Atoms, by W. D. Phillips and H. J. Metcalf, Scientific American, March 1987, p. 36.

New Mechanisms for Laser Cooling, by C. N. Cohen-Tannoudji and W. D. Phillips, Physics Today, October 1990, p. 33.

Laser Trapping of Neutral Particles, by S. Chu, Scientific American, February 1992, p. 71.

Experimenters Cool Helium below Single-Photon Recoil Limit in Three Dimensions, by G.B. Lubkin, Physics Today, January 1996, p. 22.

Steven Chu was born 1948 in St. Louis, Missouri, USA. American citizen. Doctoral degree in physics 1976 at the University of California, Berkeley. Theodore and Frances Geballe Professor of Humanities and Sciences at Stanford University 1990. Among other awards Chu received the 1993 King Faisal International Prize for Science (Physics) for development of the technique of laser-cooling and trapping atoms.

Professor **Steven Chu**
Physics Department
Stanford University
Stanford, CA 94305
USA

Claude Cohen-Tannoudji was born 1933 in Constantine, Algeria. French citizen. Doctoral degree in physics 1962 at the École Normale Supérieure in Paris. Professor at the Collège de France 1973. Member of, among other institutions, the Acadèmie des Sciences (Paris). Among many prizes and distinctions Cohen-Tannoudji received the 1996 Quantum Electronics Prize (European Physical Society) for, among other things, his pioneering experiments on laser cooling and the trapping of atoms.

Professor **Claude Cohen-Tannoudji**
Laboratoire de Physique de École Normale Supérieure
24, Rue Lhomond
F-75231 Paris Cedex 05
France

William D. Phillips was born 1948 in Wilkes-Barre, Pennsylvania, USA. American citizen. Doctoral degree in physics in 1976 at the Massachusetts Institute of Technology, Cambridge, USA. Among other awards Phillips has received the 1996 Albert A. Michelson Medal (Franklin Institute) for his experimental demonstrations of laser cooling and atom trapping.

Dr. William D. Phillips
National Institute of Standards and Technology
Gaithersburg, MD 20899 USA

Addendum B

Additional background material on the Nobel Prize in Physics 1997

October 15, 1997

The Royal Swedish Academy of Sciences has decided to award **the 1997 Nobel Prize in Physics** jointly to

Professor **Steven Chu**, Stanford University, Stanford, California, USA,

Professor **Claude Cohen-Tannoudji**, Collège de France and École Normale Supérieure, Paris, France, and

Dr. **William D. Phillips**, National Institute of Standards and Technology, Gaithersburg, Maryland, USA,

for development of methods to cool and trap atoms with laser light.

This additional *background material* is written mainly *for physicists*.

The work of Steven Chu, Claude Cohen-Tannoudji and William D. Phillips in the field of laser cooling and trapping has meant a breakthrough for both theory and experiment within the field and has led to a deeper understanding of the interaction between light and matter. It has also led to an intense world-wide activity within the atomic, molecular and optical physics community and has, in particular, opened up new roads toward the study of the quantum behavior of dilute atomic vapors at very low temperatures. The techniques of laser cooling and trapping are used in fundamental high resolution spectroscopy and the study of ultracold collisions. They also find application in the construction of atomic clocks, atomic interferometers and atom lasers, and in the development of instruments for atom optics and atomic lithography. Recent applications related to the 1997 Nobel Prize in Physics are the first observation of Bose–Einstein condensation in a dilute atomic gas and the development of the first rudimentary atom laser.

B.1. Historical Background

Johannes Kepler, in an attempt to explain why the tails of comets entering our solar system always point away from the sun, suggested already in 1619 that light can have a mechanical effect. Important contributions to the theory of the so called "light pressure" were given by James Clerk Maxwell in 1873 and by Albert Einstein in 1917. Einstein showed, in particular, that absorption and emission of photons by atoms modifies the linear momentum of the atoms. The first process in which photon momentum played an important role was the Compton effect, i.e., the scattering of X-rays against electrons. The first observation of recoiling electrons was made in 1923 by C. T. R. Wilson in his cloud chamber. The first experimental observation of recoiling atoms was made by O. R. Frisch in 1933. With the invention of the tunable dye laser by P. P. Sorokin and F. P. Schäfer in 1966, an eminent tool for the further exploration of what has become known as "the mechanical properties of light" had been created.

Important early theoretical and experimental work on the action of photons on neutral atoms was made in the 1970s by *V. S. Letokhov* and other physicists in the USSR *and* in the group of *A. Ashkin at Bell Laboratories*, Holmdel, NJ, in the USA. Among other things, *they suggested bending and focusing atomic beams and trapping atoms in focused laser beams. This early work has, for example, led to the development of "optical tweezers" that can manipulate living cells and other small objects.*

The first proposal to cool neutral atoms in counter-propagating laser beams was made by T. W. Hänsch and A. L. Schawlow in 1975. At the same time, a similar proposal was put forward by D. J. Wineland and H. G. Dehmelt, to be used for ions in ion traps. Hänsch and Schawlow proposed to cool neutral atoms in pairs of counter-propagating laser beams detuned slightly below a resonance transition of the atoms. Since the Doppler effect will tend to tune atoms moving toward one of the laser beams into resonance, these atoms will systematically be slowed down by the absorption of photons coming from the same direction and having energy below the resonance energy. Since the emission occurs in random directions it only leads to a small isotropic velocity distribution. Thus, the atoms moving toward the laser will lose velocity and effectively cool. Other atoms, with velocity components toward the other lasers, will cool in the same manner. In a gas of ideal two-level atoms one can easily calculate the limiting temperature, the so-called Doppler limit, which for the resonance transition in sodium would amount to 0.24 mK.

An important part in any attempt to cool and trap neutral atoms is to produce a beam of atoms slow enough to stay in the photon–atom interaction region for a sufficiently long time. Several attempts to slow down atomic beams by photons were made around 1980. A difficulty is that as the free atoms slow down, the frequency of the laser light has to follow the Doppler shifted resonance frequency, a technique proposed by Letokhov and called "frequency chirping". W. D. Phillips and co-workers at the National Institute of Standards and Technology, NIST, Gaithersburg, USA, instead developed a scheme where the atomic beam propagates along the axis of a varying solenoidal magnetic field so that the Doppler and Zeeman shifts compensate and the resonance transition frequency is constant ("Zeeman slower"). Phillips used this technique in 1985 and was able to stop an atomic beam and to trap the atoms in a magnetic trap (of a kind proposed by D. E. Pritchard at MIT in 1983). Also in 1985, the full stopping of a neutral atomic beam by the frequency chirping method was demonstrated by J. L. Hall and co-workers at NIST in Boulder.

B.2. Optical Molasses

In 1984, S. Chu and co-workers (among them Ashkin and J. E. Bjorkholm) at Bell Laboratories, Holmdel, NJ, set out to realize the Doppler cooling proposal of Hänsch and Schawlow. They used a beam of sodium atoms which first was slowed down by a frequency-chirped pre-cooling laser. After this laser had been turned off, the sodium atoms drifted into the intersection of six pairwise orthogonal counter-propagating laser beams (see Fig. B.1).

In 1985, the group reported cooling a dilute vapor of about 10^5 neutral sodium atoms in a volume of $0.2 \, cm^3$ to a temperature of about $0.2 \, mK$.

The motion of the atoms in the intersection region is similar to the movement in a hypothetical viscous medium, given the name "optical molasses" in the original publication (Phys. Rev. Lett. 55, 48 (1985)). Since the atoms are not trapped but fall slowly in the gravitational field, the cloud of cold atoms has a finite lifetime which in the original experiment was 0.1 s. This effect was actually used to estimate the temperature of the atoms in the intersection region by monitoring the decay of the fluorescence as the cooling lasers were turned off for a variable time interval. The result of this temperature measurement was consistent with the theoretical Doppler limit for sodium of $0.24 \, mK$. Further developments of the optical molasses experiment eventually led to an increase in the density of atoms to $10^9 \, cm^{-3}$ and to observation times up to 1 s.

For the laser-cooled neutral atoms moving in optical molasses to become really useful one needed a trap. This trap had to be deeper than the magnetic trap used by Phillips in 1985 *or the focused laser beam trap proposed by Letokhov and Ashkin and realized in 1986 by Chu and co-workers in optical molasses experiments.* In 1987 Pritchard and Chu, following a suggestion by J. Dalibard, *developed the workhorse of later experiments, the magneto-optical trap (MOT)* . This also uses the three pairs

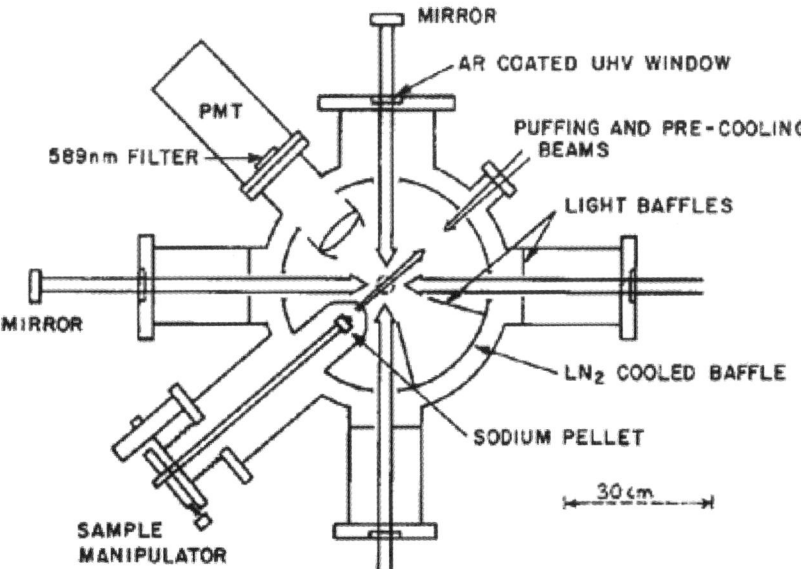

Fig. B.1. Schematic drawing of the vacuum chamber, intersecting laser beams and atomic beam used for the Doppler cooling experiment. The laser beams enter the UHV windows vertically and horizontally.

of counter-propagating laser beams, but now with circular polarization and in conjunction with a weak magnetic field.

Following suggestions by J. R. Zacharias and Hänsch, Chu also developed an atomic fountain for high precision spectroscopy. Cooled and trapped atoms are there launched upwards in the gravitational field, in trajectories with turning points inside a microwave cavity, in which the slow atoms are resonantly excited by a succession of two microwave pulses. The technique of using two excitation regions for atomic beams was pioneered by N. F. Ramsey and is used in the most precise atomic clocks. Today these clocks have a precision of 10^{-14}, while a new design based on atomic fountains is predicted to improve the precision by a factor 100.

B.3. Sub-Doppler Cooling

Meanwhile, W. D. Phillips and his group at NIST studied a cold cloud of neutral sodium atoms slowly moving in optical molasses. Prompted by small disagreements between theory and experiment, which were also noted by Chu, they developed methods to measure the temperature of the cloud under varying cooling conditions in a more precise manner. In particular, they adopted a technique to determine the time-of-flight of falling atoms to reach a set of probe laser beams below the optical molasses region. In early 1988, they discovered that the atoms had a temperature of about $40\,\mu\mathrm{K}$, much below the predicted Doppler limit of $240\,\mu\mathrm{K}$ (Phys. Rev. Lett. 61, 169 (1988)) (Fig. B.2). They also found that the lowest temperatures were reached under conditions that contradicted those of the theoretical Doppler limit.

Experiments by the groups of Chu, who now had moved to Stanford University, and of C. Cohen-Tannoudji at École Normale Supérieure in Paris, soon confirmed that the discovery of Phillips was real. The explanation of the discrepancy came almost immediately from J. Dalibard and

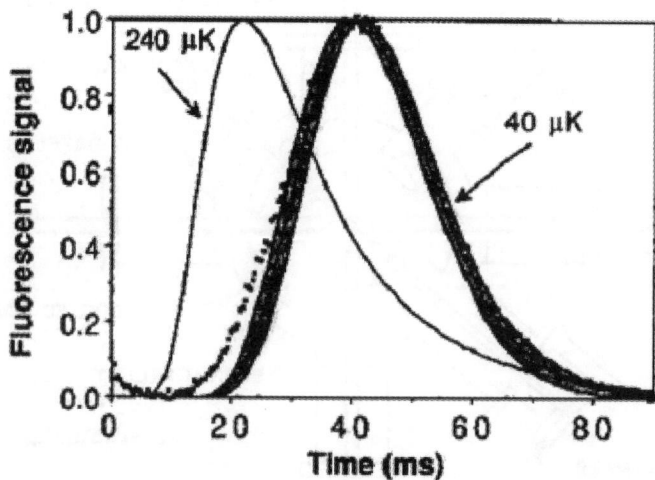

Fig. B.2. Time-of-flight data compared to calculations for assemblies of sodium atoms at two different temperatures. The shaded area indicates the range of error in the calculation. Evidently, the cloud has reached a temperature of $40\,\mu\mathrm{K}$, which is six times below the Doppler limit.

Cohen-Tannoudji in Paris and from the Stanford group. The theory of Doppler cooling and the Doppler limit assumes an atom with a simple two-level energy spectrum. Real sodium atoms, on the other hand, have several Zeeman sublevels, both in the ground state and in the excited state. The ground state sublevels can become optically pumped, i.e., laser light can transfer sodium atoms into different distributions of sublevel populations and give rise to new cooling mechanisms. Details of the population distributions depend on the laser polarization, which is changing rapidly over a distance of one optical wavelength within the molasses. For this reason the new cooling mechanisms go under the general name of polarization gradient cooling. The particular mechanism originally discovered by Phillips has been given the name Sisyphus cooling, in analogy to the character in Greek mythology, who was condemned to push a heavy stone uphill only to find after having reached the crest that the stone rolled down and he had to start all over again. The atoms are always losing kinetic energy as if moving uphill, being optically pumped back to a valley by the laser field, and start uphill again.

During a visit to Paris in 1989, Phillips collaborated with the group at École Normale Supérieure. They showed that neutral cesium atoms could be cooled to $2.5\,\mu$K. As for Doppler cooling, there seems to exist a fundamental lower limit also for other kinds of laser cooling. This so-called recoil limit would correspond to the temperature of a cloud of atoms moving with velocities of the order of the velocity of the recoil from a single photon. For sodium atoms the recoil temperature is $2.4\,\mu$K while for cesium atoms it is as low as $0.2\,\mu$K. The experimental results quoted above thus seemed to indicate that it is possible to reach a temperature of about 10 times the recoil limit using polarization gradient cooling of a disordered cloud of atoms. In a recent development one has been able to trap the cooled atoms in the sites of what is called an optical lattice. Such a lattice has a spacing of the order of an optical wavelength and it can be adjusted by changing the laser beam configuration. Since the atoms are more effectively cooled at the lattice sites than at arbitrary positions, a temperature about half of what can be reached in the disordered state can be achieved. For example, $1.1\,\mu$K has been reached for cesium.

B.4. Sub-Recoil Cooling

The reason that the recoil energy of a single photon sets a limit to both Doppler cooling and polarization gradient cooling is that in both cooling schemes there is a continous cycle of absorption and emission processes taking place. Each process gives the atom a small but non-negligible recoil energy. If atoms almost at rest could be exempted from the absorption–emission cycle, one might in principle reach sub-recoil limit temperatures in dilute atomic vapors. Already in the 1970s a mechanism in which atoms in an intense laser field can be optically pumped into a nonabsorbing coherent superposition of states, a so called dark state, was discovered at the University of Pisa. C. Cohen-Tannoudji, together with several co-workers at the École Normale Supérieure in Paris, among them E. Arimondo (from Pisa) and A. Aspect, has shown in a series of experiments how the Doppler effect can be used to ensure that only the coldest atoms end up in the dark state. This so-called velocity selective coherent population trapping (VSCPT) method was first applied in 1988 in one dimension (Phys. Rev. Lett. 61, 826 (1988)), in 1994 in two dimensions (Phys. Rev. Lett. 73, 1915 (1994)) and in 1995 in three dimensions (Phys. Rev. Lett. 75, 4194 (1995)).

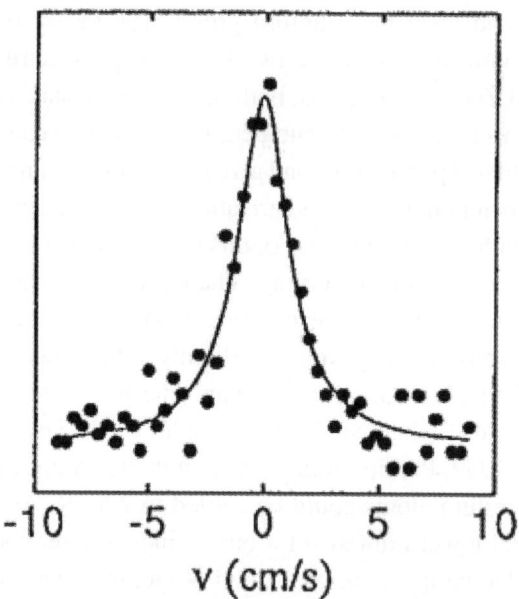

Fig. B.3. Profile of the atomic velocity distribution in three-dimensional sub-recoil cooling. The single photon recoil velocity is 9.2 cm/s while the spread of the cooled atoms is about 2 cm/s.

All three experiments used metastably excited helium atoms He (1s2s^{3}S) for which the Doppler limit is 23 μK while the recoil limit is 4 μK. Already in 1988, Cohen-Tannoudji and co-workers used two counter-propagating laser beams and could show cooling in one dimension to a temperature of 2 μK, a factor of two below the recoil limit. The experiment was developed to two-dimensional cooling in the beginning of the 1990s. In 1994, together with Aspect and an otherwise new set of co-workers and using two pairs of mutually orthogonal and counter-propagating laser beams, Cohen-Tannoudji could show cooling in two dimensions to 250 nK, about 16 times below the recoil limit temperature. Finally, in 1995, the experiment was developed to include three sets of laser beams and cooling in three dimensions was shown. The minimum temperature now became 180 nK, 22 times below the recoil limit (see Fig. B.3). Even though other groups have participated in the developments of sub-recoil cooling, notably Chu and collaborators, it is the work of Cohen-Tannoudji which has opened the new area of sub-recoil laser cooling.

B.5. Applications

There has been a tremendous progress in the field of laser cooling and trapping of neutral atoms over the last decade. The three Nobel Laureates and their collaborators have, together with many other scientists, laid the groundwork for exciting developments within a number of subfields of physics. The development of methods to cool dilute vapors of trapped atoms to temperatures hitherto only reached in isolated parts of condensed matter systems, has made it possible to construct atomic clocks useful for precise timekeeping, e.g., in connection with navigation in space and the exploration of the solar system. Another application using laser cooling is the development of atomic interferometers in which the de Broglie wavelength of slow atoms is used for interferometric measurements with ultrahigh precision, e.g., of the acceleration of gravity. Refined instruments for atom optics have also opened the road towards atomic lithography. The atomic beams may be used to form nanometer structures on surfaces, e.g., for electronic components. The recent observation of a Bose–Einstein condensation in a dilute atomic gas is also related to the groundbreaking work on laser cooling and trapping described in this short background material.

Further Reading

"Cooling and Trapping Atoms", by William D. Phillips and Harold J. Metcalf, Scientific American, March 1987, p. 36.

"New Mechanisms for Laser Cooling", by Claude N. Cohen-Tannoudji and William D. Phillips, Physics Today, October 1990, p. 33.

"Laser Trapping of Neutral Particles", by Steven Chu, Scientific American, February 1992, p. 71.

"Le Refroidissement des Atomes par Laser", by A. Aspect and J. Dalibard, La Recherche, January 1994, p. 30.

"Experimenters Cool Helium below Single-Photon Recoil Limit in Three Dimensions", by Gloria B. Lubkin, Physics Today, January 1996, p. 22.

Addendum C

Uproar over Nobel Physics Prize

Copyright © 1997 Nando.net

Copyright © 1997 Agence France-Presse

STOCKHOLM (October 18, 1997 4:05 p.m. EDT http://www.nando.net) — The Nobel physics prize, awarded to a Frenchman and two US nationals, was plunged into controversy on Saturday when Russian scientists said they had carried out the same research more than a decade ago.

Claude Cohen-Tannoudj and US scientists Steven Chu and William Phillips were rewarded on Wednesday for developing techniques for superfreezing gases to slow down their molecules for study.

But a business newspaper in Moscow, Kommersant, said on Saturday that a team lead by Vladilen Letokhov carried out the same research at the Russian academy of sciences in 1986.

"We obtained the same results", a member of the team, Vladimir Minoguine, told the newspaper. He said the results were first published in 1986 and were translated into English and published in the United States a year later.

He said the work "gave a schema for the experimental apparatus used by the Nobel laureates".

Kommersant quoted other senior Russian researchers who complained that the Nobel committee systematically favored Westerners at the expense of Russians.

But the Nobel committee in Stockholm rejected criticism of its choice for the prize.

"We were aware of the Russian work, but it was not as developed as that of the ptize-winners", Bengt Nagel, president of the science committee, told the Swedish agency TT.

He added: "I understand the Russians' protests. They currently have enormous problems and a Nobel prize would perhaps help them to continue their research.

"However, the work submitted to us is studied by a committee of extremely competent experts, and it is wrong to say we favor Westerners over Russians".

Cohen-Tannoudji, 64, is a professor at the prestigious Ecole Normale Superieure in Paris. Chu is a Standford University physicist while William Phillips National works at the Institute of Standards and Technologies in Gaithersburg, Maryland.

The three men worked together at AT&T Bell Laboratories when they began their research into ways to slow atoms down to a standstill.

Their award means that 11 French scientists and 65 Americans have become Nobel Physics Laureates. Researchers from the ex-Soviet Union and Russia have won the prize seven times.

References

[1]* A. Ashkin, Acceleration and trapping of particles by radiation pressure, *Phys. Rev. Lett.* **24**, 156–159 (1970).

[2]* A. Ashkin, Atomic-beam deflection by resonance-radiation pressure, *Phys. Rev. Lett.* **25**, 1321–1324 (1970).

[3]* A. Ashkin, Trapping of atoms by resonance radiation pressure, *Phys. Rev. Lett.* **40**, 729–732 (1978).

[4]* S. Chu, J. E. Bjorkholm, A. Ashkin and A. Cable, Experimental observation of optically trapped atoms, *Phys. Rev. Lett.* **57**, 314–317 (1986).

[5]* A. Ashkin, J. M. Dziedzic, J. E. Bjorkholm and S. Chu, Observation of a single-beam gradient force optical trap for dielectric particles, *Opt. Lett.* **11**, 288–290 (1986).

[6] R. Huber, S. Burggraf, T. Mayer, S. M. Barns, P. Rossnagel and K. O. Stetter, Isolation of a hyperthermophilic archaeum predicted by *in situ* RNA analysis, *Nature* **376**, 57–58 (1995).

[7]* A. Ashkin and J. M. Dziedzic, Feedback stabilization of optically levitated particles, *Appl. Phys. Lett.* **30**, 202–204 (1977).

[8] A. Ashkin, Optical trapping and manipulation of neutral particles using lasers, *Proc. Natl. Acad. Sci. USA* **94**, 4853–4860 (1997).

[9] A. Ashkin, Optical trapping and manipulation of neutral particles using lasers, *Opt. Photonic News* **10**, 41–46 (1999).

[10]* A. Ashkin, History of optical trapping and manipulation of small neutral particle, atoms, and molecules, *IEEE J. Sel. Top. Quantum Electron.* **6**, 841–855 (2000).

[11] A. Kishino and T. Yanagida, Force measurements by micromanipulation of a single actin filament by glass needles, *Nature* **334**, 74–76 (1988).

[12]* D. Cyranoski, Swimming against the tide, *Nature* **408**, 764–766 (2000).

[13] M. Grandbois, M. Beyer, M. Rief, H. Clausen-Schaumann and H. E. Gaub, How strong is a covalent bond? *Science* **283**, 1727–1730 (1999).

[14] M. Rief, M. Gautel, F. Oesterhelt, J. M. Fernandez and H. E. Gaub, Reversible unfolding of individual titin immunoglobin domains by AFM, *Science* **276**, 1109–1112 (1997).

[15] K. Kitamura, M. Tokunaga, A. H. Iwane and T. Yanagida, A single myosin head moves along an actin filament with regular steps of 5.3 nanometers, *Nature* **397**, 129–134 (1999).

[16] W. Steffen, D. Smith, R. Simmons and J. Sleep, Mapping the actin filament with myosin, *Proc. Natl. Acad. Sci. USA* **98**, 14949–14954 (2001).

[17]* A. Ashkin, K. Schütze, J. M. Dziedzic, U. Euteneuer and M. Schliwa, Force generation of organelle transport measured *in vivo* by an infrared laser trap, *Nature* **348**, 346–348 (1990).

[18] M. A. Welte, S. P. Gross, M. Postner, S. M. Block and E. F. Wieschaus, Developmental regulation of vesicle transport in *Drosophila* embryos: Forces and kinetics, *Cell* **92**, 547–557 (1998).

[19]* A. Ashkin and J. M. Dziedzic, Internal cell manipulation using infrared laser traps, *Proc. Natl. Acad. Sci. USA* **86**, 7914–7918 (1989).

[20]* A. Ashkin, J. M. Dziedzic and T. Yamane, Optical trapping and manipulation of single cells using infrared laser beams, *Nature* **330**, 769–771 (1987).

[21]* K. C. Neuman, E. H. Chadd, G. F. Liou, K. Bergman and S. M. Block, Characterization of photodamage to *Escherichia coli* in optical traps, *Biophys. J.* **5**, 2856–2863 (1999).

[22] S. B. Smith, Y. Cui and C. Bustamante, Overstretching B-DNA: The elastic response of individual double-stranded and single-stranded molecules, *Science* **258**, 1122–1126 (1992).

[23] T. T. Perkins, D. E. Smith and S. Chu, Direct observation of tube-like motion of a single polymer chain, *Science* **264**, 819–822 (1994).

[24] M. D. Wang, H. Yin, R. Landick, J. Gelles and S. M. Block, Stretching DNA with optical tweezers, *Biophys. J.* **72**, 1335–1346 (1997).

[25] M. Hegner, S. B. Smith and C. Bustamante, Polymerization and mechanical properties of single RecA-DNA filaments, *Proc. Natl. Acad. Sci. USA* **96**, 10109–10114 (1999).

[26]* T. W. Hänsch and A. L. Schawlow, Cooling of gases by laser radiation, *Opt. Commun.* **13**, 68–69 (1975).

[27]* S. Chu, L. Hollberg, J. E. Bjorkholm, A. Cable and A. Ashkin, Three-dimensional viscous confinement and cooling of atoms by resonance radiation pressure, *Phys. Rev. Lett.* **55**, 48–51 (1985).

[28] R. Frisch, Experimenteller Nachweis des Einsteinschen Strahlungsrückstoßes, *Z. Phys.* **86**, 42–48 (1933).

[29]* M. H. Anderson, J. R. Ensher, M. R. Matthews, C. E. Wieman and E. A. Cornell, Observation of Bose-Einstein condensation in a dilute atomic vapor, *Science* **269**, 198–201 (1995).

[30]* K. B. Davis, M.-O. Mewes, M. R. Andrews, N. J. van Druten, D. S. Durfee, D. M. Kurn and W. Ketterle, Bose-Einstein condensation in a gas of sodium atoms, *Phys. Rev. Lett.* **75**, 3969–3973 (1995).

[31]* M. R. Andrews, C. G. Townsend, H.-J. Miesner, D. S. Durfee, D. M. Kurn and W. Ketterle, Observation of interference between two Bose condensates, *Science* **275**, 637–641 (1997).

[32] M.-O. Mewes, M. R. Andrews, D. M. Kurn, D. S. Durfee, C. G. Townsend and W. Ketterle, Output coupler for Bose-Einstein condensed atoms, *Phys. Rev. Lett.* **78**, 582–585 (1997).

[33]* I. Bloch, T. W. Hänsch and T. Esslinger, Atom laser with a cw output coupler, *Phys. Rev. Lett.* **82**, 3008–3011 (1999).

[34]* E. W. Hagley L. Deng, M. Kozuma, J. Wen, K. Helmerson, S. L. Rolston and W. D. Phillips, A well-collimated quasi-continuous atom laser, *Science* **283**, 1706–1709 (1999).

[35]* B. DeMarco and D. S. Jin, Onset of fermi degeneracy in a trapped atomic gas, *Science* **285**, 1703–1706 (1999).

[36] S. Chu, The manipulation of neutral particles, *Rev. Mod. Phys.* **70**, 685–706 (1998).

[37] C. N. Cohen-Tannoudji, Manipulating atoms with photons, *Rev. Mod. Phys.* **70**, 707–719 (1998).

[38] W. D. Phillips, Laser cooling and trapping of neutral atoms, *Rev. Mod. Phys.* **70**, 721–741 (1998).

[39] B. G. Levi, Cornell, Ketterle and Wieman, share nobel prize for Bose-Einstein condensates, *Phys. Today* **54**, 14–16 (2000).

[40] N. R. Newbury and C. E. Wieman, Trapping of neutral atoms, *Am. J. Phys.* **64**, 1–129 (1996).

[41] C. Cohen-Tannoudji and W. D. Phillips, New mechanisms of laser cooling, *Phys. Today* **43**, 33–40 (1990).

[42] K. Visscher and S. M. Block, Versatile optical traps with feedback control, *Methods Enzymol.* **298**, 460–489 (1998).

[43]* J. E. Bjorkholm. R. R. Freeman, A. Ashkin and D. B. Pearson, Observation of focusing of neutral atoms by the dipole forces of resonance-radiation pressure, *Phys. Rev. Lett.* **41**, 1361–1364 (1978).

[44]* A. Ashkin and J. M. Dziedzic, Optical trapping and manipulation of viruses and bacteria, *Science* **235**, 1517–1520 (1987).

[45]* A. Ashkin and J. P. Gordon, Stability of radiation-pressure traps: An optical Earnshaw theorem, *Opt. Lett.* **8**, 511–513 (1983).

[46]* A. Ashkin, Design for an optical cw atom laser, *Proc. Natl. Acad. Sci. USA* **101**, 12108–12113 (2004).

[47] *Microwave Magnetrons*, ed. G. B. Collins, in *MIT Radiation Series,* Vol. 6 (McGraw Hill, 1948), Sec. 19.13.

[48] E. G. Rawson and A. D. May, Propulsion and angular stabilization of dust particles in a laser cavity, *App. Phys. Lett.* **8**, 93–95 (1966).

[49]* A. Ashkin, Forces of a single-beam gradient laser trap on a dielectric sphere in the ray optics regime, *Biophys. J.* **61**, 569–582 (1992).

[50] G. Roosen, Optical levitation of spheres, *Can. J. Phys.* **57**, 1260–1279 (1970).

[51] G. Roosen and C. Imbert, Optical levitation by means of two horizontal laser beams: Theoretical and experimental study, *Phys. Lett. A* **59**, 6–8 (1976).

[52] N. A. Fuchs, *The Mechanics of Aerosols* (Pergamon Press, 1964), Chapt. II.

[53] A. Einstein, Zur Quanten Theorie der Strahlung, *Phys. Zeit.* **18**, 121–128 (1917).

[54] A. Yariv, Introduction to optical electronics, in *Interaction of Radiation and Atomic Systems*, Holt, Rinehart and Winston (1971), Chapt. 5.

[55] G. A. Askar'yan, Effect of the gradient of a strong electromagnetic beam on electrons and atoms, *Zh. Eksp. Teor. Fiz.* **42**, 1567–1569 (1962) [*Sov. Phys. JET* **15**, 1088–1090 (1962)].

[56] V. S. Letokhov, Narrowing of the Doppler width in a standing wave, *Pis'ma Zh. Eksp. Teor. Fiz.* **7**, 348 (1968) [*JETP Lett.* **7**, 272–275 (1968)].

[57] H. G. Kuhn, Atomic spectra, in *The Radiation of Atoms* (Academic Press, 1969), pp. 59–84.

[58]* A. Ashkin and J. M. Dziedzic, Optical levitation by radiation pressure, *Appl. Phys. Lett.* **19**, 283–285 (1971).

[59] A. Ashkin, After image, *Opt. Photonics News* **10** (1999).

[60] A. Ashkin, The pressure of laser light, *Sci. Am.* **226**, 63–71 (1972).

[61]* J. T. Finer, R. M. Simmons and J. A. Spudich, Single myosin molecule mechanics: Piconewton forces and nanometre steps, *Nature* **368**, 113–119 (1994).

[62] M. D. Wang, H. Yin, R. Landick, J. Gelles and S. M. Block, Stretching DNA with optical tweezers, *Biophys. J.* **72**, 1335–1346 (1997).

[63]* C. J. Hood, T. W. Lynn, A. C. Doherty, A. S. Parkins and H. J. Kimble, The atom-cavity microscope: Single atoms bound in orbit by single photons, *Science* **287**, 1447–1453 (2000).

[64] P. W. H. Pinkse, T. Fischer, P. Maunz and G. Rempe, Trapping an atom with single photons, *Nature* **404**, 365–368 (2000).

[65]* A. Ashkin and J. M. Dziedzic, Optical levitation in high vacuum, *Appl. Phys. Lett.* **28**, 333–335 (1976).

[66]* A. Ashkin and J. M. Dziedzic, Stability of optical levitation by radiation pressure, *Appl. Phys. Lett.* **24**, 586–588 (1974).

[67] N. G. Borisenko, V. S. Bushuev, A. I. Gromov, V. M. Dorogotovtsev, A. I. Isakov, E. R. Koresheva, Yu. A. Merkul'ev, A. I. Nikitenko and S. M. Tolokonnikov, Laser target technology at the Lebedev Physics Institute, *Sov. J. Quantum Electron* **19**, 1221–1224 (1989).

[68]* A. Ashkin and J. M. Dziedzic, Optical levitation of liquid drops by radiation pressure, *Science* **187**, 1073–1075 (1975).

[69] A. Chowdhury, B. J. Ackerson and N. A. Clark, Laser-induced freezing, *Phys. Rev. Lett.* **55**, 833–836 (1985).

[70] D.-J. Han, S. Wolf, S. Oliver, C. McCormick, M. T. DePue and D. Weiss, 3D Raman sideband cooling of cesium atoms at high density, *Phys. Rev. Lett.* **85**, 724–727 (2000).

[71] G. Birkl, M. Gatzke, I. H. Deutsch, S. L. Rolston and W. D. Phillips, Bragg scattering from atoms in optical lattices, *Phys. Rev. Lett.* **75**, 2823–2826 (1995).

[72] R. F. Weurker, F. H. Shelton and R. V. Langmuir, Electrodynamic containment of charged particles, *J. Appl. Phys.* **30**, 342–349 (1959).

[73] A. Ashkin and J. M. Dziedzic, unpublished.

[74] H. G. Dehmelt, Radiofrequency spectroscopy of stored ions, *Adv. At. Mol. Phys.* **3**, 53–72 (1967).

[75] D. J. Berkeland, J. D. Miller, J. D. Bergquist, W. M. Itano and D. J. Wineland, Laser-cooled mercury ion frequency standard, *Phys. Rev. Lett.* **80**, 2089–2092 (1998).

[76] S. Arnold, Determination of particle mass and charge by one electron differentials, *J. Aerosol. Sci.* **10**, 49–53 (1979).

[77] *Aerosol Science,* Chapts. V and VI., ed. C. N. Davies (Academic Press, 1966).

[78] E. F. Nichols and G. F. Hull, A preliminary communication on the pressure of heat and light radiation, *Phys. Rev. (Series I)* **13**, 307–320 (1901).

[79] G. F. Hull, The elimination of gas action in experiments on light pressure, *Phys. Rev. (Series I)* **20**, 292–299 (1905).

[80] R. W. Wood, *Physical Opt.*, Chapt. XXIII (Dover Publications, 1967), pp. 793–797.

[81] P. Lebedev, Untersuchungen über die Druckkräfte des Lichtes, *Ann. der Physik, (Leipzig)* **6**, 433–458 (1901).

[82] J. H. Poynting, *Cambridge Meeting of the British Association*, 1904.

[83] R. W. Ditchburn, *Light* (John Wiley, 1962), pp. 665–669.

[84] *Van Nostrand's Scientific encyclopedia* 4th edn. (D. Van Nostrand Co., Princeton, 1968), p. 1403.

[85] O. Preining, *Aerosol Science*, in *Photophoresis*, ed. C. N. Davies (Academic Press, 1966).

[86] J. Tyndall, *Proc. R. Instn. Gt. Br.* **6** (1870).

[87] I. Waldermann, Aerosol Science, *Thermophoresis and Diffusiophoresis of Aerosols*, ed. C. N. Davies (Academic Press, 1966).

[88] F. Ehrenhaft, On the physics of millionth of centimeters, *Phys. Z.* **18**, 352–368 (1917).

[89] M. Reiss, Die Photophorese, Elektro- und Magnetophorese, *Phys. Z.* **33**, 185–202 (1932).

[90] F. Ehrenhaft, Photophoresis and its interpretation by electric and magnetic ions, *J. Franklin Inst.* **233**, 235–255 (1942).

[91] M. Lewittes, S. Arnold and G. Oster, Radiometric levitation of micron sized spheres, *Appl. Phys. Lett.* **40**, 455–457 (1982).

[92]* A. Ashkin and J. M. Dziedzic, Observation of resonances in radiation pressure on dielectric spheres, *Phys. Rev. Lett.* **38**, 1351–1353 (1977).

[93] W. M. Irvine, Light scattering by spherical particles: Radiation pressure asymmetry factor and extinction cross section, *J. Opt. Soc. Am.* **55**, 16–21 (1965).

[94] H. C. Van de Hulst, *Light Scattering by Small Particles*, Chapt. 17 (Dover, 1981).

[95] D. Q. Chowdhury, D. H. Leach and R. K. Chang, Effect of the Goos-Hänchen shift on the geometrical-optics model for spherical-cavity mode spacing, *J. Opt. Soc. Am. A* **11**, 1110–1116 (1994).

[96] P. Chylek, J. T. Kiehl and M. K. W. Ko, Optical levitation and partial wave resonance, *Phys. Rev. A* **18**, 2229–2232 (1978).

[97]* A. Ashkin and J. M. Dziedzic, Observation of optical resonances of dielectric spheres by light scattering, *Appl. Opt.* **20**, 1803–1814 (1981).

[98] P. Chylek, V. Ramaswamy, A. Ashkin and J. M. Dziedzic, Simultaneous determination of refractive index and size of spherical dielectric particles from light scattering data, *Appl. Opt.* **22**, 2302–2307 (1983).

[99] *Optical Effects Associated with Small Particles*, eds. P. W. Barber and R. K. Chang (World Scientific, 1988).

[100] *Optical Processes in Microcavities*, eds. R. K. Chang and A. J. Campillo (World Scientific, 1996).

[101] S. Haroche, Cavity quantum electrodynamics, in *Fundamental Systems in Quantum Optics, Les Houches Summer School, Session LIII*, eds. J. Dalibard, J. M. Raimond and J. Zinn-Justen (North Holland, 1992).

[102] V. Lefévre-Seguin, J. C. Knight, V. Sandoghdar, D. S. Weiss, J. Hare, J. M. Raimond and S. Haroche, *Very High Q Whispering-Gallery Modes in Silica Microspheres for Cavity-QED Experiments*, in reference [93].

[103] S. Suzuki, K. Shuto and Y. Hibino, Integrated-optics ring resonators with 2 stacked layers of silica waveguide on Si, *IEEE Photonics Tech. Lett.* **4**, 1256–1258 (1992).

[104] B. E. Little, J. S. Foresi, G. Steinmeyer, E. R. Thoen, S. T. Chu, H. A. Haus, E. P. Ippen, L. C. Kimerling and W. Greene, Ultra-compact Si-SiO$_2$ microring resonator optical channel dropping filters, *IEEE Photonics Tech. Lett.* **10**, 149–151 (1998).

[105] B. E. Little, S. T. Chu, W. Pan, D. Ripin, T. Kaneko, Y. Kokubun and E. P. Ippen, Vertically coupled glass microring resonator channel dropping filters, *IEEE Photonics Tech. Lett.* **11**, 215–217 (1999).

[106] S. L. McCall, A. F. J. Levi, R. E. Slusher, S. J. Pearton and R. A. Logan, Whispering-gallery mode microdisk lasers, *Appl. Phys. Lett.* **60**, 289–291 (1992).

[107] C. Gmachl, F. Capasso, E. E. Narimanov, J. U. Nöckel, A. D. Stone, J. Faist, D. L. Sivio and A. Y. Cho, High-power directional emission from microlasers with chaotic resonators, *Science* **280**, 1556–1564 (1998).

[108] R. A. Millikan, *The Electron*, ed. J. W. M. DuMond (University of Chicago Press, 1917).

[109]* A. Ashkin, Applications of laser radiation pressure, *Science* **210**, 1081–1088 (1980).

[110] G. S. La Rue, W. M. Fairbank and A. F. Hebard, Evidence for the existence of fractional charge on matter, *Phys. Rev. Lett.* **38**, 1011–1014 (1977).

[111] J. W. M. DuMond and R. A. Millikan, *The Electron* (University of Chicago Press, 1917), p. xvii.

[112] K. Sasaki, M. Koshioka, H. Misawa, N. Kitamura and H. Masuhara, Optical trapping of a metal-particle and a water droplet by a scanning laser beam, *Appl. Phys. Lett.* **60**, 807–809 (1992).

[113] A. Peters, K. Y. Chung and S. Chu, Measurement of gravitational acceleration by dropping atoms, *Nature* **400**, 849–852 (1999).

[114] C. P. Collier, E. W. Wong, M. Belohradsky, F. M. Raymo, J. F. Stoddart, P. J. Kuekes, R. S. Williams and J. R. Heath, Electronically configurable molecular-based logic gates, *Science* **285**, 391–394 (1999).

[115] C. G. B. Garrett, W. Kaiser and W. L. Long, Stimulated emission into optical whispering modes of spheres. *Phys. Rev.* **124**, 1807–1809 (1961).

[116] R. Schieder, H. Walther and L. Wöste, Atomic beam deflection by the light of a tunable dye laser, *Opt. Commun.* **5**, 337–340 (1972).

[117] P. Jacquinot, D. Liberman, J.-L. Picqué and J. Pinard, High resolution spectroscopic application of atomic beam deflection by resonant light, *Opt. Commun.* **8**, 163–165 (1973).

[118] A. F. Bernhardt, Isotope separation by laser deflection of an atomic beam, *Appl. Phys.* **9**, 19–34 (1976).

[119] A. F. Bernhardt, D. E. Duerre, J. R. Simpson and L. L. Wood, High resolution spectroscopy using photodeflection, *Opt. Commun.* **16**, 166–168 (1976).

[120] J. V. Prodan and W. D. Phillips, Chirping the light-fantastic? Recent NBS atom cooling experiments, *Prog. Quantum Electron.* **8**, 231 (1984).

[121] J. Prodan, A. Migdall, W. D. Phillips, I. So, H. Metcalf and J. Dalibard, Stopping atoms with laser light, *Phys. Rev. Lett.* **54**, 992–995 (1985).

[122] W. R. Ertmer, R. Blatt, J. L. Hall and M. Zhu, Laser manipulation of atomic beam velocities: Demonstration of stopped atoms and velocity reversal, *Phys. Rev. Lett.* **54**, 996–999 (1985).

[123]* J. P. Gordon and A. Ashkin, Motion of atoms in a radiation trap, *Phys. Rev. A* **21**, 1606–1617 (1980).

[124] S. Stenholm, The semiclassical theory of laser cooling, *Rev. Mod. Phys.* **58**, 699–739 (1986).

[125] V. S. Letokhov, V. G. Minogin and B. D. Pavlik, Cooling and capture of atoms and molecules by a resonant light field, *Zh. Eksp. Teor. Fiz.* **72**, 1328–1341 (1977) [*Sov. Phys. JETP* **45**, 698–705 (1977)].

[126] V. S. Letokhov and V. G. Minogin, Trapping and storage of atoms in a laser field, *Appl. Phys.* **17**, 99–103 (1978).

[127] H. J. Habing, C. Dominik, M. Jourdain de Muizon, M. F. Kessler, R. J. Laureijs, K. Leech, L. Metcalfe, A. Salama, R. Siebenmorgen and N. Trams, Disappearance of stellar debris disks around main-sequence stars after 400 million years, *Nature* **401**, 456–458 (1999).

[128] M. R. Showalter, Neptune's misbehaving rings, news and views, *Nature* **400**, 709–710 (1999).

[129] N. Y. Misconi, Rotational bursting of interplanetary dust particles, *Geophys. Res. Lett.* **3**, 585–588 (1976).

[130]* E. Higurashi, H. Ukita, H. Tanaka and O. Ohguchi, Optically induced rotation of anisotropic micro-objects fabricated by surface micromachining, *Appl. Phys. Lett.* **64**, 2209–2210 (1994).

[131] P. L. Marston and J. H. Crichton, Radiation torque on a sphere caused by a circularly-polarized electromagnetic wave, *Phys. Rev. A* **30**, 2508–2516 (1984).

[132] S. Chang and S. S. Lee, Optical torque exerted on a homogeneous sphere levitated in the circularly polarized fundamental-mode laser beam, *J. Opt. Soc. Am. B* **2**, 1853–1860 (1985).

[133] M. Kerker, *The Scattering of Light and Other Electromagnetic Radiation*, in Chapt. III (Academic Press, 1969).

[134] J. A. Stratton, *Electromagnetic Theory* (McGraw-Hill, 1941).

[135] H. J. Metcalf and P. van der Straten, *Laser Cooling and Trapping* (Springer-Verlag, 1999).

[136]* A. Ashkin and J. P. Gordon, Cooling and trapping of atoms by resonance radiation pressure, *Opt. Lett.* **4**, 161–163 (1979).

[137] D. J. Wineland and W. M. Itano, Laser cooling of atoms, *Phys. Rev. A* **20**, 1521–1540 (1979).

[138] P. F. Liao and J. E. Bjorkholm, Direct observation of atomic energy level shifts in two-photon absorption, *Phys. Rev. Lett.* **34**, 1–4 (1975).

[139]* T. Ido, Y. Isoya and H. Katori, Optical-dipole trapping of Sr atoms at a high phase-space density, *Phys. Rev. A* **61**, 061403–061404 (2000).

[140]* D. B. Pearson, R. R. Freeman, J. E. Bjorkholm and A. Ashkin, Focusing and defocusing of neutral atomic beams using resonance-radiation pressure, *Appl. Phys. Lett.* **36**, 99–101 (1980).

[141] J. E. Bjorkholm, R. R. Freeman, A. Ashkin and D. B. Pearson, Experimental observation of the influence of the quantum fluctuations of resonance-radiation pressure, *Opt. Lett.* **5**, 111–113 (1980).

[142]* A. Ashkin, Stable radiation-pressure particle traps using alternating light beams, *Opt. Lett.* **9**, 454–456 (1984).

[143]* A. Ashkin and J. M. Dziedzic, Observation of radiation pressure trapping of particles by alternating light beams, *Phys. Rev. Lett.* **54**, 1245–1248 (1985).

[144] V. I. Balykin, V. S. Letokhov and V. I. Mishin, Observation of the cooling of free sodium atoms in a resonance laser field with a scanning frequency, *Pisma Zh. Eksp. Teor. Fiz.* **29**, (1979) [*JETP Lett.* **29**, 560–564 (1979)].

[145]* C. S. Adams, H. J. Lee, N. Davidson, M. Kasevich and S. Chu, Evaporative cooling in a crossed dipole trap, *Phys. Rev. Lett.* **74**, 3577–3580 (1995).

[146] T. Takekoshi and R. J. Knize, CO_2 laser trap for cesium atoms, *Opt. Lett.* **21**, 77–79 (1996).

[147] K. M. O'Hara, S. R. Granade, M. E. Gehm, T. A. Savard, S. Bali, C. Freed and J. E. Thomas, Ultrastable CO_2 laser trapping of lithium fermions, *Phys. Rev. Lett.* **82**, 4204–4207 (1999).

[148] J. D. Miller, R. A. Cline and D. J. Heinzen, Far-off-resonance optical trapping of atoms, *Phys. Rev. A* **47**, R4567–R4570 (1993).

[149] C. Cohen-Tannoudji, Théorie quantique du cycle pompage optíque. Vérification expérimentale des nouveaux effets prévus, *Ann. Phys. (Paris), 13*, 423–495 (1962).

[150] C. Cohen-Tannoudji, Spectroscopie Hertzienne — Observation d'un déplacement de raie de resonance magnétique cause par l'excitation optique, *Comp. Rend. Des séances de l'Acad. des Scie.* **252**, 394–396 (1961).

[151] P. F. Liao and J. E. Bjorkholm, Optically-induced energy level shifts for intermediate intensities, *Opt. Commun.* **16**, 392–395 (1976).

[152] W. E. Lamb, W. P. Schleich, M. O. Scully and C. H. Townes, Laser physics: Quantum controversy in action, in *More Things in Heaven and Earth, a Celebration of Physics at the Millenium*, ed. Benjamin Bederson (Springer Verlag, 1999), pp. 442–459. See also *Rev. Mod. Phys.* **71** (1999).

[153] M. Sargent, M. O. Scully and W. E. Lamb, *Laser Physics* (Addison-Wesley, 1974).

[154]* J. E. Bjorkholm, R. R. Freeman, A. Ashkin and D. B. Pearson, Transverse resonance radiation pressure on atomic beams and the influence of fluctuations, in *Laser Spectroscopy IV, Proc. 4th Int. Conf. Rottach-Egern*, Springer Series in Optical Sciences (Springer Verlag, 1979), pp. 49–55.

[155] A. Ashkin and J. M. Dziedzic, Observation of light scattering from oriented non-spherical particles using optical levitation, in *Light Scattering by Irregularly Shaped Particles, Proc. of the Workshop* (Plenum Publishing Corp., 1979).

[156] A. Ashkin and J. M. Dziedzic, Observation of light scattering from nonspherical particles using optical levitation, *Appl. Opt.* **19**, 660–668 (1980).

[157] W. Phillips and H. Metcalf, Laser deceleration of an atomic beam, *Phys. Rev. Lett.* **48**, 596–599 (1982).

[158] V. G. Minogin, Kvantovaya Elecktron. (Moscow) **9**, 505 (1982). [Theory of a radiative atomic trap, *Sov. J. Quantum Electron.* **12**, 299–303 (1982).]

[159] V. G. Minogin and J. Javanainen, A tetrahedral light pressure trap for atoms, *Opt. Commun.* **43**, 119–122 (1982).

[160] W. Phillips, ed., Laser cooled and trapped atoms, in *U.S. National Bureau of Standards Special Publications No. 653* (U.S.G.P.O., 1983) or *Prog. Quantum Electron.* **8**, 119–127 (1984).

[161] A. L. Migdall, J. V. Proden, W. D. Phillips, T. H. Bergeman and H. J. Metcalf, First observation of magnetically trapped neutral atoms, *Phys. Rev. Lett.* **54**, 2596–2599 (1985).

[162] D. J. Wineland, Trapped ions, laser cooling and better clocks, *Science* **226**, 395–400 (1984).

[163] W. Wing, On neutral particle trapping in quasistatic electromagnetic field, *Prog. Quant. Elect.* **8**, 181–199 (1984).

[164] W. H. Wing, Electrostatic trapping of neutral atomic particles, *Phys. Rev. Lett.* **45**, 631–634 (1980).

[165] P. W. Smith, A. Ashkin and W. J. Tomlinson, Four-wave mixing in an artificial Kerr medium, *Opt. Lett.* **6**, 284–286 (1981).

[166]* A. Ashkin, J. M. Dziedzic and P. W. Smith, Continuous-wave self-focusing and self-trapping of light in artificial Kerr media, *Opt. Lett.* **7**, 276–278 (1982).

[167] P. W. Smith, P. J. Maloney and A. Ashkin, Use of a liquid suspension of dielectric spheres as an artificial Kerr medium, *Opt. Lett.* **7**, 347–349 (1982).

[168] P. W. Smith, A. Ashkin, J. E. Bjorkholm and D. J. Eilenberger, Studies of self-focusing bistable devices using liquid suspensions of dielectric particles, *Opt. Lett.* **10**, 131–133 (1984).

[169] N. A. Fuchs, *The Mechanics of Aerosols* (Pergamon Press, 1964), pp. 193–200.

[170] W. Neuhauser, M. Hohenstatt, P. E. Toschek and H. Dehmelt, Localized visible Ba^+ mono-ion oscillator, *Phys. Rev. A* **22**, 1137–1140 (1980).

[171]* S. Chu, J. E. Bjorkholm, A. Ashkin, J. P. Gordon and L. W. Hollberg, Proposal for optically cooling atoms to temperatures of the order of 10^{-6} K, *Opt. Lett.* **11**, 73–75 (1986).

[172] S. Chu, J. E. Bjorkholm, A. Ashkin, L. Hollberg and A. Cable, Cooling and trapping of atoms with laser light, in *Methods of Laser Spectroscopy*, eds. Y. Prior, A. Ben-Reuven and M. Rosenbluh (Plenum Publishing Corp., 1986), pp. 41–49.

[173] H. F. Hess, Magnetic trapping and cooling of atomic hydrogen, *Bull. Am. Phys. Soc.* **30**, 854 (1985).

[174] H. F. Hess, G. P. Kochanski, J. M. Doyle, N. Masuhara, D. Kleppner and T. J. Greytak, Magnetic trapping of spin-polarized atomic hydrogen, *Phys. Rev. Lett.* **59**, 672–675 (1987).

[175]* K. B. Davis, M.-O. Mewes, M. R. Andrews, N. J. van Druten, D. S. Durfee, D. M. Kurn and W. Ketterle, Bose-Einstein condensation in a gas of sodium atoms, *Phys. Rev. Lett.* **75**, 3969–3973 (1995).

[176] D. Fried, T. Killian, L. Willman, D. Landhuis, S. Moss, D. Kleppner and T. Greytak, Bose-Einstein condensation of atomic hydrogen, *Phys. Rev. Lett.* **81**, 3811–3814 (1998).

[177] J. Dalibard, S. Reynaud and C. Cohen-Tannoudji, Proposals of stable optical traps for neutral atoms, *Opt. Commun.* **47**, 395–398 (1983).

[178] D. Pritchard, E. L. Raab, V. Bagnato, C. E. Wieman and R. N. Watts, Light traps using spontaneous forces, *Phys. Rev. Lett.* **57**, 310–313 (1986).

[179] J. Maddox, Catching atoms in beams of light, news and views, *Nature* **322**, 403 (1986).

[180] A. Ashkin, J. E. Bjorkholm and S. Chu, Caught in a trap, *Nature* **323**, 585 (1986).

[181]* E. L. Raab, M. Prentiss, A. Cable, S. Chu and D. E. Pritchard, Trapping of neutral sodium atoms with radiation pressure, *Phys. Rev. Lett.* **59**, 2631–2634 (1987).

[182]* T. Walker, D. Sesko and C. Wieman, Collective behavior of optically trapped neutral atoms, *Phys. Rev. Lett.* **64**, 408–411 (1990).

[183]* W. Ketterle, K. B. Davis, M. A. Joffe, A. Martin and D. Pritchard, High densities of cold atoms in a dark spontaneous force optical trap, *Phys. Rev. Lett.* **70**, 2253–2256 (1993).

[184] A. Cable, M. Prentiss and N. Bigelow, Observations of sodium atoms in a magnetic molasses trap loaded by a continuous uncooled source, *Opt. Lett.* **15**, 507–509 (1990).

[185] P. D. Lett, R. N. Watts, C. I. Westbrook, W. D. Phillips, P. L. Gould and H. J. Metcalf, Observation of atoms laser cooled below the Doppler limit, *Phys. Rev. Lett.* **61**, 169–172 (1988).

[186] J. Dalibard and C. Cohen-Tannoudji, Laser cooling below the Doppler limit by polarization gradients: Simple theoretical models, *J. Opt. Soc. Am. B* **6**, 2023–2045 (1989).

[187] P. J. Ungar, D. S. Weiss, E. Riis and S. Chu, Optical molasses and multilevel atoms: Theory, *J. Opt. Soc. Am. B* **6**, 2058–2071 (1989).

[188] C. Salomon, J. Dalibard, W. D. Phillips, A. Clairon and S. Guellati, Laser cooling of cesium atoms below 3 μK, *Europhys. Lett.* **12**, 683–688 (1990).

[189] G. I. Stegeman and M. Segev, Optical spatial solitons and their interactions: Universality and diversity, *Science* **286**, 1518–1523 (1999).

[190]* L. Deng, E. W. Hagley, J. Wen, M. Trippenbach, Y. Band, P. S. Julienne, J. E. Simsarian, K. Helmerson, S. L. Rolston and W. D. Phillips, Four-wave mixing with matter waves, *Nature* **398**, 218–220 (1999).

[191] J. A. N. Zasadzinski and R. B. Meyer, Molecular imaging of tobacco mosaic virus lyotropic nematic phases, *Phys. Rev. Lett.* **56**, 636–638 (1986).

[192] Y. Tadir, W. H. Wright, O. Vafa, T. Ord, R. Asch and M. W. Berns, Micromanipulation of sperm by a laser generated optical trap, *Fertil. Steril.* **52**, 870–873 (1989).

[193] S. M. Block, D. F. Blair and H. C. Berg, Compliance of bacterial flagella measured with optical tweezers, *Nature* **338**, 514–517 (1989).

[194] F. H. C. Crick and A. F. W. Hughes, The physical properties of cytoplasm, *Exp. Cell Res.* **1**, 37–80 (1950).

[195] R. D. Allen, Amoeboid movement in *Chaos carolinensis*, in *The Application of Laser Light Scattering to the Study of Biological Motion*, eds. J. C. Earnshaw and M. W. Steer (Plenum Press, 1982), pp. 519–528.

[196] A. D. Mehta, M. Rief, J. A. Spudich, D. A. Smith and R. M. Simmons, Single-molecule biomechanics with optical methods, *Science* **283**, 1689–1695 (1999).

[197] S. M. Block, Kinesin: What gives? *Cell* **93**, 5–8 (1998).

[198] P. D. Lett, R. N. Watts, C. I. Westbrook, W. D. Phillips, P. L. Gould and H. Metcalf, Observation of atoms cooled below the doppler limit, *Phys. Rev. Lett.* **61**, 169–172 (1988).

[199] M. A. Kasevich, E. Riis, S. Chu and R. G. DeVoe, rf Spectroscopy in an atomic fountain, *Phys. Rev. Lett.* **63**, 612–615 (1989).

[200] Scientists seek further improvements to quantum measurements and standards, *APS News* **5**, 4 (1996).

[201] M. A. Kasevich and S. Chu, Atomic interferometry using stimulated Raman transitions, *Phys. Rev. Lett.* **67**, 181–184 (1991).

[202] J. Weiner, V. S. Bagnato, S. Zilio and P. S. Julienne, Experiments and theory in cold and ultracold collisions, *Rev. Mod. Phys.* **71**, 1–85 (1999).

[203] W. Ketterle, Experimental studies of Bose-Einstein condensation, *Phys. Today* **52**, 30–35 (1999).

[204] K. Burnett, M. Edwards and C. W. Clark, The theory of Bose-Einstein condensation of dilute gases, *Phys. Today* **52**, 37–42 (1999).

[205]* M. D. Barrett, J. A. Sauer and M. S. Chapman, All-optical formation of an atomic Bose-Einstein condensate, *Phys. Rev. Lett.* **87**, 010404-1-010404-4 (2001).

[206] T. J. Greytak and D. Kleppner, New trends in atomic physics, in *Las Houches Summer School*, eds. G. Greenberg and R. Stors (North Holland, 1984).

[207] R. V. E. Lovelace, C. Mehanian, T. J. Tommila and D. M. Lee, Magnetic confinement of a neutral gas, *Nature* **318**, 30–36 (1985).

[208] H. F. Hess, Evaporative cooling of magnetically trapped and compressed spin-polarized hydrogen, *Phys. Rev. B* **34**, 3476–3479 (1986).

[209] P. J. H. Bronkhorst, G. J. Streekstra, J. Grimbergen, E. J. Nijhof, J. J. Sixma and G. J. Brakenhoff, A new method to study shape recovery of red blood cells using multiple optical trapping, *Biophys. J.* **69**, 1666–1673 (1995).

[210] S. M. Block, L. S. B. Goldstein and B. J. Schnapp, Bead movement by single kinesin molecules studied with optical tweezers, *Nature* **348**, 348–352 (1990).

[211] K. Svoboda and S. M. Block, Biological applications of optical forces, *Ann. Rev. Biophys. Biomol. Struct.*, 247–285 (1994).

[212]* H. Yin, M. D. Wang, K. Svoboda, R. Landick, S. M. Block and J. Gelles, Transcription against an applied force, *Science* **270**, 1653–1657 (1995).

[213]* G. J. L. Wuite, S. B. Smith, M. Young, D. Keller and C. Bustamante, Single-molecule studies of the effect of template tension on T7 DNA polymerase activity, *Nature* **404**, 103–106 (2000).

[214] M. Kurachi, M. Hoshi and H. Tashiro, Buckling of a single microtubule by optical trapping forces — direct measurement of microtubule rigidity, *Cell Motil. Cytoskeleton* **30**, 221–228 (1995).

[215] Y. Tsuda, H. Yasutake, A. Ishijima and T. Yanagida, Torsional rigidity of single actin filaments and actin–actin bond breaking force under torsion measured directly by *in vitro* micromanipulation, *Proc. Natl. Acad. Sci.* **93**, 12937–12947 (1996).

[216] L. Tskhovrebova, J. Trinick, J. A. Sleep and R. M. Simmons, Elasticity and unfolding of single molecules of the giant muscle protein titin, *Nature* **387**, 308–312 (1997).

[217]* K. Svoboda, C. F. Schmidt, B. J. Schnapp and S. M. Block, Direct observation of kinesin stepping by optical trapping interferometry, *Nature* **365**, 721–727 (1993).

[218] R. M. Simmons, J. T. Finer, S. Chu and J. A. Spudich, Quantitative measurements of force and displacement using an optical trap, *Biophys. J.* **70**, 1813–1822 (1996).

[219] K. Visscher, M. J. Schnitzer and S. M. Block, Single kinesin molecules studied with a molecular force clamp, *Nature* **400**, 184–189 (1999).

[220]* H. Stapelfeldt, H. Sakai, E. Constant and P. B. Corkum, Deflection of neutral molecules using the nonresonant dipole force, *Phys. Rev. Lett.* **79**, 2787–2790 (1997).

[221] B. G. Levi, Hot prospects for ultracold molecules, *Phys. Today* **53**, 46–50 (2000).

[222] H. Berg, Motile behavior of bacteria, *Phys. Today* **53**, 24–29 (2000).

[223] S. M. Block, D. F. Blair and H. C. Berg, Compliance of bacterial polyhooks measured with optical tweezers, *Cytometry* **12**, 492–496 (1991).

[224] N. W. Charon, S. F. Goldstein, S. M. Block, K. Curci and J. D. Ruby, Morphology and dynamics of protruding spirochete, eriplastic spirochete flagella, *J. Bacteriol.* **174**, 832–840 (1992).

[225] H. C. Berg and L. Turner, Torque generated by the flagella of *Escherichia coli*, *Biophys J.* **65**, 2201–2216 (1993).

[226] R. M. Berry and H. C. Berg, Absence of a barrier to backward rotation of the bacterial flagellar motor demonstrated with optical tweezers, *Proc. Natl. Acad. Sci. USA* **94**, 14433–14437 (1997).

[227] R. M. Berry and H. C. Berg, Torque generated by the flagellar motor of *Escherichia coli* while driven backward, *Biophys. J.* **76**, 580–587 (1999).

[228] W. S. Ryu, R. M. Berry and H. C. Berg, Torque generating units of the flagellar motor of *Escherichia coli* have a high duty ratio, *Nature* **403**, 444–447 (2000).

[229] B. Alberts, D. Bray, J. Lewis, M. Ruff, K. Roberts and J. D. Watson, *Molecular Biology of the Cell*, 2nd edn. (Garland Publishing, 1989), pp. 10–11.

[230] D. Lloyd, Microbial ecology — how to avoid oxygen, *Science* **286**, 249 (1999).

[231] M. Milstein, Yellowstone managers stake a claim on hot-springs microbes, *Science* **270**, 226 (1995).

[232] R. Monastersky, Deep dwellers — microbes thrive far below ground, *Sci. News* **151**, 192–193 (1997).

[233] R. J. Parkes, Oiling the wheels of controversy, a review of *The Deep Hot Biosphere,* by Thomas Gold, *Nature* **401**, 644 (1999).

[234] R. Stone, Permafrost comes alive for Siberian researchers, *Science* **286**, 36–37 (1999).

[235] M. W. Browns, Odd microbe survives vast doses of radiation, N. Y. Times, p. C1, *Science Times*, Tues. Nov. 28, 1995.

[236] S. Barker, MITI turns up heat on research into thermophile genes, *Nature* **381**, 455 (1996).

[237] K. E. Nelson, R. A. Clayton, S. R. Gill, M. L. Gwinn, R. J. Dodson, D. H. Haft, E. K. Hickey, J. D. Peterson, W. C. Nelson, K. A. Ketchum, L. McDonald, T. R. Utterback, J. A. Malek, K. D. Linher, M. M. Garrett, A. M. Stewart, M. D. Cotton, M. S. Pratt, C. A. Phillips, D. Richardson, J. Heidelberg, G. G. Sutton, R. D. Fleischmann, J. A. Eisen, O. White, S. L. Salzberg, H. O. Smith, J. C. Venter and C. M. Fraser, Evidence for lateral gene transfer between Archaea and Bacteria from genome sequence of *Thermotoga maritima*, *Nature* **399**, 323–329 (1999).

[238] J. G. Mitchell, R. Weller, M. Beconi, J. Sell and J. Holland, A practical optical trap for manipulating and isolating bacteria from complex microbial communities, *Microb. Ecol.* **25**, 113–119 (1993).

[239]* A. J. Merz, M. So and M. P. Sheetz, Pilus retraction powers bacterial twitching motility, *Nature* **407**, 98–102 (2000).

[240] J. M. Skerker and H. C. Berg, Direct observation of extension and retraction of type IV pili, *Proc. Natl. Acad. Sci. USA* **98**, 6901–6904 (2001).

[241] D. Wall and D. Kaiser, Type IV pili and cell motility, *Mol. Microbiol.* **32**, 1–10 (1999).

[242] B. Maier, M. Koomey and M. P. Sheetz, A force dependent switch reverse type IV pilus retraction, *PNAS (USA)* **101**, 10961–10966 (2004).

[243] S. Seegar, S. Manojembaski, K. J. Hutter, G. Futterman, J. Wolfrum and K. O. Greulich, Application of laser optical tweezers in immunology and molecular genetics, *Cytometry* **12**, 497–504 (1991).

[244] R. W. Steubing, S. Chang. W. H. Wright, Y. Namajiri and M. W. Berns, Laser induced cell fusion in combination with optical tweezers: The laser cell fusion trap, *Cytometry* **12**, 505–510 (1991).

[245] H. Liang, W. H. Wright, W. He and M. W. Berns, Micromanipulation of mitotic chromosomes in PTK-2 cells using laser-induced optical forces ("optical tweezers"), *Exp. Cell Res.* **197**, 21–35 (1991).

[246] H. Liang, W. H. Wright, W. He and M. W. Berns, Micromanipulation of chromosomes in PTK-2 cells using laser microsurgery (optical scalpel) in combination with laser-induced optical force (optical tweezers), *Exp. Cell Res.* **204**, 110–120 (1993).

[247] M. W. Berns, Laser scissors and tweezers, *Sci. Am.* **278**, 62–67 (1998).

[248] J. M. Colón, P. Sarosi, P. G. McGovern, A. Ashkin, J. M. Dziedzic, J. Skurnick, G. Weiss and E. M. Bonder, Controlled micromanipulation of human sperm in three dimensions with an infrared laser optical trap: Effect on sperm velocity, *Fertil. Steril.* **57**, 695–698 (1992).

[249] E. M. Bonder, J. M. Colón, J. M. Dziedzic and A. Ashkin, Force production by swimming sperm: Analysis using optical tweezers, *J. Cell Biol.* **111**, (No. 5, Pt.2) 421a (1990) [Abstract No. 2349].

[250] Y. Tadir, W. H. Wright, O. Vafa, L. H. Liaw, R. Asch and M. W. Berns, Micromanipulation of gametes using laser microbeams, *Hum. Reprod.* **6**, 1011–1016 (1991).

[251] K. Schütze, A. Clement-Sengewald and A. Ashkin, Zone drilling and sperm insertion with combined laser microbeam and optical tweezers, *Fertil. Steril.* **61**, 783–786 (1994).

[252] A. Clement-Sengewald, K. Schütze, A. Ashkin, G. A. Palma, G. Kerlen and G. Brem, Fertilization of bovine oocytes induced solely with combined laser microbeam and optical tweezers, *J. Assist. Reprod. Genet.* **13**, 259–265 (1996).

[253] H. Liang, K. T. Vu, P. Krishnan, T. C. Trang, D. Shin, S. Kimel and M. W. Berns, Wavelength dependence of cell cloning efficiency after optical trapping, *Biophys. J.* **70**, 1529–1533 (1996).

[254] X. Wei, B. J. Tromberg and M. D. Cahalan, Mapping the sensitivity of T cells with an optical trap: Polarity and minimal number of receptors for Ca^{2+} signaling, *Proc. Natl. Acad. Sci. USA* **96**, 8471–8476 (1999).

[255] M. Mammen, K. Helmerson, R. Kishore, S.-K. Choi, W. D. Phillips and G. M. Whitesides, Optically controlled collisions of biological objects to evaluate potent polyvalent inhibitors of virus-cell adhesion, *Chem. Biol.* **3**, 757–762 (1996).

[256] J. Dai and M. P. Sheetz, Mechanical properties of neuronal growth cone membranes studied by tether formation with optical tweezers, *Biophys. J.* **68**, 988–996 (1995).

[257] J. Dai and M. P. Sheetz, Axon membrane flows from the growth cone to the cell body, *Cell* **83**, 693–701 (1985).

[258] R. M. Hochmuth, J. Shao, J. Dai and M. P. Sheetz, Deformation and flow of membrane into tethers extracted from neuronal growth cones, *Biophys. J.* **70**, 358–369 (1996).

[259] J. M. Mitchison and M. M. Swann, The mechanical properties of the cell surface: I. The cell elastimeter, *J. Exp. Biol.* **31**, 443–460 (1954).

[260] J. Dai and M. P. Sheetz, Cell membrane mechanics, in *Methods in Cell Biology*, Vol. 55, Chap. 9, ed. M. P. Sheetz (Academic Press, 1998).

[261]* J. Guck, R. Ananthakrishnan, T. J. Moon, C. C. Cunningham and J. Käs, Optical deformability of soft biological dielectrics, *Phys. Rev. Lett.* **84**, 5451–5454 (2000).

[262]* A. Ashkin and J. M. Dziedzic, Radiation pressure on a free liquid surface, *Phys. Rev. Lett.* **30**, 139–142 (1973).

[263] J. P. Gordon, Radiation forces and momenta in dielectric media, *Phys. Rev. A* **8**, 14–21 (1973).

[264] R. V. Jones, Radiation pressure of light in a dispersive medium, *Proc. R. Soc. Lond. A* **360**, 365–371 (1978).

[265] K. Svoboda, C. F. Schmidt, D. Branton and S. M. Block, Conformation and elasticity of isolated red blood cell membrane skeleton, *Biophys. J.* **63**, 784–793 (1992).

[266] K. Visscher, G. J. Brakenhoff and J. J. Krol, Micromanipulation by 'multiple' optical traps created by a single fast scanning trap integrated with the bilateral confocal scanning microscope, *Cytometry* **14**, 105–114 (1993).

[267] G. Leitz, E. Schnepf and K. O. Greulich, Micromanipulation of statoliths in gravity sensing *Chara* rhizoids by optical tweezers, *Planta* **197**, 278–288 (1995).

[268]* A. Ehrlicher, T. Betz, B. Stuhrmann, D. Koch, V. Milner, M. G. Raizen and J. Käs, Guiding neuronal growth with light, *Proc. Natl. Acad. Sci. USA* **99**, 16024–16028 (2002).

[269] S. K. Mohanty, A. Uppal and P. K. Gupta, Self-rotation of red blood cells in optical tweezers: Prospects for high throughput malaria diagnosis, *Biotechnol. Lett.* **26**, 971–974 (2004).

[270]* S. K. Mohanty, A. Uppal and P. K. Gupta, Self-rotation of red blood cells in optical tweezers: Prospects for high throughput malaria diagnosis, *OPN* **15**, No. 12, p. 19 (2004).

[271] S. M. Block, Leading the procession: New insights into Kinesin motors, *J. Cell Biol.* **140**, 1281–1284 (1998).

[272] S. Rice, A. W. Lin, D. Safer, C. L. Hart, N. Naber, B. O. Carragher, S, M. Cain, E. Pechatnikova, E. M. Wilson-Kubalek, M. Whittaker, E. Pate, R. Cooke, E. W. Taylor, R. A. Milligan and R. D. Vale, A structural change in the kinesin motor protein that drives motility, *Nature* **402**, 778–784 (1999).

[273] U. Euteneuer, K. Johnson, M. P. Koonce, K. L. McDonald, J. Tong and M. Schliwa, *In vitro* analysis of cytoplasmic organelle transport, in *Cell Movement*, Vol. 2, eds. F. D. Warner and J. R. McIntosh (Liss, 1989), pp. 155–167.

[274] K. Oiwa and K. Takahashi, The force–velocity relationship for microtubule sliding in demembranated sperm flagella of the sea urchin, *Cell Struct. Funct.* **13**, 193–205 (1988).

[275] H. Felner, F. Grolig, O. Müller and M. Schliwa, *In vivo* manipulation of internal cell organelles, in *Laser Tweezers in Biology*, Vol. 55, ed. M. Sheetz (Academic Press, 1998).

[276] S. C. Kuo and M. P. Sheetz, Force of single kinesin molecules measured with optical tweezers, *Science* **260**, 232–234 (1993).

[277] S. M. Block, Nanometres and piconewtons: The macromolecular mechanics of kinesin, *Trends Cell Biol.* **5**, 169–175 (1995).

[278] E.-L. Florin, V. T. Moy and H. E. Gaub, Adhesion forces between individual ligand-receptor pairs, *Science* **264**, 415–417 (1996).

[279] P. Hinterdorfer, W. Baumgartner, H. J. Gruber, K. Schilcher and H. Schindler, Detection and localization of individual antibody-antigen recognition events by atomic force microscopy, *Proc. Natl. Acad. Sci. USA* **93**, 3477–3481 (1996).

[280] J. Howard, Molecular motors: Structural adaptations to cellular functions, *Nature* **389**, 561–567 (1997).

[281] J. E. Molloy, J. E. Burns, J. Kendrick-Jones, R. T. Tregear and D. C. S. White, Force and movement produced by a single myosin head, *Nature* **378**, 209–212 (1995).

[282] T. Nishizaka, H. Miyata, H. Yoshikawa, S. Ishiwata and K. Kinosita Jr., Unbinding force of a single motor molecule of muscle measured using optical tweezers, *Nature* **377**, 251–254 (1995).

[283] A. Ishijima, H. Kojima, T. Funatsu, M. Tokunaga, H. Higuchi, H. Tanaka and T. Yanagida, Simultaneous observation of individual ATPase and mechanical events by a single myosin molecule during interaction with actin, *Cell* **92**, 161–171 (1998).

[284] H. Tanaka, A. Ishijima, M. Honda, K. Saito and T. Yanagida, Orientation dependence of displacement by a single one-headed myosin relative to the actin filament, *Biophys. J.* **75**, 1886–1894 (1998).

[285] W. H. Guilford, D. E. Dupuis, G. Kennedy, J. Wu, J. B. Patlak and D. M. Warshaw, Smooth muscle and skeletal muscle myosins produce similar unitary forces and displacements in the laser trap, *Biophys. J.* **72**, 1006–1021 (1997).

[286] R. D. Astumian, Thermodynamics and kinetics of a brownian motor, *Science* **276**, 917–922 (1997).

[287] L. P. Faucheaux, L. S. Bourdieu, P. D. Kaplan and A. J. Libchaber, Optical thermal ratchet, *Phys. Rev. Lett.* **74**, 1504–1507 (1995).

[288] C. Veigel, L. M. Coluccio, J. D. Jontes, J. C. Sparrow, R. A. Milligan and J. E. Molloy, The motor protein myosin-I produces its working stroke in two steps, *Nature* **398**, 530–533 (1999).

[289] A. D. Mehta, R. S. Rock, M. Rief, J. A. Spudich, M. S. Mooseker and R. E. Cheney, Myosin-V is a processive actin-based motor, *Nature* **400**, 590–593 (1999).

[290] K. Svoboda and S. M. Block, Force and velocity measured for single kinesin molecules, *Cell* **77**, 773–784 (1994).

[291] K. Svoboda, P. P. Mitra and S. M. Block, Fluctuation analysis of motor protein movement and single enzyme kinetics, *Proc. Nat. Acad. Sci. USA* **91**, 11782–11786 (1994).

[292] B. G. Levi, Measured steps advance the understanding of molecular motors, *Phys. Today* **48**, 17–19 (1995).

[293] M. J. Schnitzer and S. M. Block, Kinesin hydrolyses one ATP per 8-nm step, *Nature* **388**, 386–390 (1997).

[294] S. M. Block and K. Svoboda, Analysis of high resolution recordings of motor movement, *Biophys. J.* **68**, 230–241 (1995).

[295] K. Visscher, S. P. Gross and S. M. Block, Construction of multiple-beam optical traps with nanometer-resolution position sensing, *IEEE J. Sel. Top. Quant. Electr.* **2**, 1066–1076 (1996).

[296] A. E. Knight and J. E. Molloy, Coupling ATP hydrolysis to mechanical work, *Nat. Cell Biol.* **1**, E87–E89 (1999).

[297] W. O. Hancock and J. Howard, Kinesin's processivity results from mechanical and chemical coordination between the ATP hydrolysis cycle of the two motor domains, *Proc. Natl. Acad. Sci. USA* **96**, 13147–13152 (1999).

[298] M. Nishiyama, E. Muto, Y. Inoue, T. Yanagida and H. Higuchi, Substeps within the 8-nm step of the ATPase cycle of single kinesin molecules, *Nat. Cell Biol.* **3**, 425–428 (2001).

[299] Y. Okada and N. Hirokawa, Mechanism of the single-headed processivity: Diffusional anchoring between K-loop of kinesin and the C terminus of tubulin, *Proc. Natl. Acad. Sci. USA* **97**, 640–645 (2000).

[300] Y. Okada and N. Hirokawa, A processive single-headed motor: Kinesin superfamily protein KIF1A, *Science* **283**, 1152–1157 (1999).

[301]* R. S. Rock, S. E. Rice, A. L. Wells, T. J. Purcell, J. A. Spudich and H. Lee Sweeney, Myosin VI is a processive motor with a large step size, *Proc. Natl. Acad. Sci. USA* **98**, 13655–13659 (2001).

[302] S. P. Gross, M. A. Welte, S. M. Block and E. F. Wieschaus, Dynein-mediated cargo transport *in vivo*: A switch controls travel distance, *J. Cell Biol.* **148**, 945–955 (2000).

[303] C. L. Asbury, A. N. Fehr and S. M. Block, Kinesin moves by an asymmetric hand-over-hand mechanism, *Science* **302**, 2130–2134 (2003).

[304] L. Nugent-Glandorf and T. T. Perkins, Measuring 0.1-nm motion in 1 ms in an optical microscope with differential back-focal-plane detection, *Opt. Lett.* **29**, 2611–2613 (2004).

[305] S. M. Uptain, C. M. Kane and M. J. Chamberlin, Basic mechanisms of transcript elongation and its regulation, *Ann. Rev. Biochem.* **66**, 117–172 (1997).

[306] R. A. Mooney, I. Artsimovitch and R. Landick, Information processing by RNA polymerase: Recognition of regulatory signals during RNA chain elongation, *J. Bacteriol.* **180**, 3265–3275 (1998).

[307] D. A. Schafer, J. Gelles, M. P. Sheetz and R. Landick, Transcription by single molecules of RNA-polymerase observed by light-microscopy, *Nature* **352**, 444–448 (1991).

[308] H. Yin, R. Landick and J. Gelles, Tethered particles motion method for studying transcript elongation by a single RNA-polymerase molecule, *Biophys. J.* **67**, 2468–2478 (1994).

[309] M. D. Wang, M. J. Schnitzer, H. Yin, R. Landick, J. Gelles and S. M. Block, Force and velocity measured for single molecules of RNA polymerase, *Science* **282**, 902–907 (1998).

[310] H. Y. Wang, T. Elston, A. Mogilner and G. Oster, Force generation in RNA polymerase, *Biophys. J.* **74**, 1186–1202 (1998).

[311] F. Jülicher and R. Bruinsma, Motion of RNA polymerase along DNA: A stochastic model, *Biophys. J.* **74**, 1169–1185 (1998).

[312] S. B. Smith, L. Finzi and C. Bustamante, Direct mechanical measurement of the elasticity of single DNA molecules by using magnetic heads, *Science* **258**, 122–126 (1992).

[313] T. T. Perkins, D. E. Smith and S. Chu, Relaxation of a single DNA molecule observed by optical microscopy, *Science* **264**, 822–826 (1994).

[314] H. Felgner, R. Frank and M. Schliwa, Flexural rigidity of microtubules measured with use of optical tweezers, *J. Cell Sci.* **109**, 509–516 (1996).

[315] P. Cluzel, A. Lebrun, C. Heller, R. Lavery, J. L. Viovy, D. Chatenay and F. Caron, DNA: An extensible molecule, *Science* **271**, 792–794 (1996).

[316] J. F. Marko and E. D. Siggia, Stretching DNA, *Macromolecules* **28**, 8759–8770 (1995).

[317] T. Odijk, Stiff chains and filaments under tension, *Macromolecules* **28**, 7016–7018 (1995).

[318] G. V. Shivashankar, M. Feingold, O. Krichevsky and A. Libchaber, RecA polymerization on double-stranded DNA by using single-molecule manipulation: The role of ATP hydrolysis, *Proc. Natl. Acad. Sci. USA* **96**, 7916–7921 (1999).

[319] C. Bustamante, J. F. Marko, E. D. Siggia and S. Smith, Entropic elasticity of λ-phage DNA, *Science* **265**, 1599–1600 (1994).

[320] R. F. Service, Deconstructing DNA for faster sequencing, *Science* **283**, 1669 (1999).

[321] R. F. Service, Watching DNA at work, *Science* **283**, 1668–1669 (1999).

[322]* Y. Cui and C. Bustamante, Pulling a single chromatin fiber reveals the forces that maintain its higher-order structure, *Proc. Natl. Acad. Sci. USA* **97**, 127–132 (2000).

[323] J. Darnell, H. Lodish and D. Baltimore, Structure of chromatin, 2nd edn., in *Molecular Cell Biology* (Scientific American Books, 1990), pp. 330–338.

[324] J. F. Marko, DNA under high tension: Overstretching, undertwisting, and relaxation dynamics, *Phys. Rev. E* **57**, 2134–2149 (1998).

[325] L. R. Brewer, M. Corzett and R. Balhorn, Protamine-induced condensation and decondensation of the same DNA molecule, *Science* **286**, 120–123 (1999).

[326]* Y. Nishimura, O. Misumi, S. Matsunaga, T. Higashiyama, A. Yokota and T. Kuroiwa, The active digestion of uniparental chloroplast DNA in a single zygote of *Chlamydomonas reinhardtii* is revealed by using the optical tweezer, *Proc. Natl. Acad. Sci. USA* **96**, 12577–12582 (1999).

[327] J. Liphardt, B. Onoa, S. B. Smith, I. Tinoco, Jr. and C. Bustamante, Reversible unfolding of single RNA molecules by mechanical force, *Science* **292**, 733–737 (2001).

[328] J. M. Fernandez, S. Chu and A. F. Oberhauser, Pulling on hair(pins), *Science* **292**, 653–654 (2001).

[329] G. V. Shivashankar and A. Libchaber, Single DNA molecule grafting and manipulation using a combined atomic force microscope and an optical tweezer, *Appl. Phys. Lett.* **71**, 3727–3729 (1997).

[330]* Z. Bryant, M. D. Stone, J. Gore, S. B. Smith, N. P. Cozzarelli and C. Bustamante, Structural transitions and elasticity from torque measurements on DNA, *Nature* **424**, 338–341 (2003).

[331]* J. W. Shaevitz, E. A. Abbondanzieri, R. Landick and S. M. Block, Backtracking by single RNA polymerase molecules observed at near-base-pair resolution, *Nature* **426**, 684–687 (2003).

[332] K. C. Neuman, E. A. Abbondanzieri, R. Landick, J. Gelles and S. M. Block, Ubiquitous transcriptional pausing is independent of RNA polymerase backtracking, *Cell* **115**, 437–447 (2003).

[333] K. Sakata-Sogawa and N. Shimamoto, RNA polymerase can track a DNA groove during promoter search, *PNAS (USA)* **101**, 14731–14735 (2004).

[334] R. B. Case, Y.-P. Chang, S. B. Smith, J. Gore, N. R. Cozzarelli and C. Bustamante, The bacterial condensin muk BEF compacts DNA into a repetitive, stable, structure, *Science* **505**, 222–227 (2004).

[335] Perspectives: Zhuang on Unraveling DNA condensates with optical tweezers, *Science* **505**, 188–191 (2004).

[336] T. T. Perkins, H.-W. Li, R. V. Dalal, J. Gelles and S. M. Block, Forward and reverse motion of single RecBCD molecules on DNA, *Biophys. J.* **86**, 1640–1648 (2004).

[337] M. S. Z. Kellermayer, S. B. Smith, H. L. Granzier and E. Bustamante Folding-unfolding transitions in single titin molecules characterized with laser tweezers, *Science* **276**, 1112–1116 (1997).

[338] G. I. Bell, Models for the specific adhesion of cells to cells, *Science* **200**, 618–627 (1978).

[339] E. Evans and K. Ritchie, Dynamic strength of molecular adhesion bonds, *Biophys. J.* **72**, 1541–1555 (1997).

[340] M. Rief, M. Gautel, A. Schemmel and H. E. Gaub, The mechanical stability of immunoglobulin and fibronectin III domains in the muscle protein titin measured by atomic force microscopy, *Biophys. J.* **75**, 3008–3014 (1998).

[341] M. Rief, J. M. Fernandez and H. E. Gaub, Electrically coupled two-level systems as a model for biopolymer extensibility, *Phys. Rev. Lett.* **81**, 4764–4767 (1998).

[342] S. Alberts, D. Bray, L. Lewis, M. Raff, K. Roberts and J. D. Watson, Cell adhesion, cell junctions, and the extracellular matrix, 2nd edn., in *Molecular Biology of the Cell* (Garland Publishing, 1989).

[343] D. Choquet, D. P. Felsenfeld and M. P. Sheetz, Extracellular matrix rigidity causes strengthening of integrin-cytoskeleton linkages, *Cell* **88**, 39–48 (1997).

[344] D. P. Felsenfeld, P. L. Schwartzberg, A. Venegas, R. Tse and M. P. Sheetz, Selective regulation of integrin-cytoskeleton interactions by the tyrosine kinase Src, *Nat. Cell Biol.* **1**, 200–206 (1999).

[345] K. Vuori and E. Ruoslahti, Connections count in cell migration, *Nat. Cell Biol.* **1**, E85–E87 (1999).

[346]* T. Nishizaka, Q. Shi and M. P. Sheetz, Position-dependent linkages of fibronectin-integrin-cytoskeleton, *Proc. Natl. Acad. Sci. USA* **97**, 692–697 (2000).

[347] Y. Sako and A. Kusumi, Barriers for lateral diffusion of transferring receptor in the plasma membrane as characterized by receptor dragging by laser tweezers: Fence vs. tether, *J. Cell Biol.* **129**, 1559–1574 (1995).

[348] M. J. Saxton, Single-particle tracking: Effects of corrals, *Biophys. J.* **69**, 389–398 (1995).

[349] M. J. Saxton, Single particle tracking: New methods of data analysis, *Biophys. J.* **70**, A334 (1996).

[350] A. Kusumi, Y. Sako, T. Fujiwara and M. Tomishiga, Application of laser tweezers to studies of the fences and tethers of the membrane skeleton that regulate the movements of plasma membrane proteins, in *Methods in Cell Biology*, Vol. 55, ed. M. P. Sheetz (Academic Press, 1998).

[351] G. Koster, M. VanDuijn, B. Hofs and M. Dogterom, Membrane tube formation from giant vesicles by dynamic association of motor proteins, *PNAS (USA)* **100**, 15583–15588 (2003).

[352] W. H. Wright, G. J. Sonek, Y. Tadir and M. W. Berns, Laser trapping in cell biology, *IEEE J. Quant. Elect.* **26**, 2148–2157 (1990).

[353] R. Gussgard, T. Lindmo and I. Brevik, Calculation of the trapping force in a strongly focused laser beam, *J. Opt. Soc. Am. B* **9**, 1922–1930 (1992).

[354] A. Yariv, Quantum electronics, in *The Propagation of Optical Beams in Homogeneous and Lenslike Media*, (Wiley, 1975).

[355] M. Born and E. Wolf, *Principles of Opt.*, 5th edn. (Pergamon Press, 1975), Sec. 11.4.2.

[356] K. Visscher and G. J. Brakenhoff, Theoretical study of optically induced forces on spherical particles in a single beam trap II: Mie scatterers, *Optik* **90**, 57–60 (1992).

[357] K. F. Ren, G. Gréhan and G. Gouesbet, Prediction of reverse radiation pressure by generalized Lorenz-Mie theory, *Appl. Opt.* **35**, 2702–2710 (1996).

[358] L. W. Davis, Theory of electromagnetic beams, *Phys. Rev. A*, **19**, 1177–1179 (1979).

[359] J. P. Barton, D. R. Alexander and S. A. Schaub, Internal and near-surface electromagnetic fields for a spherical particle irradiated by a focused laser beam, *J. Appl. Phys.* **64**, 1632–1639 (1988).

[360]* J. P. Barton, D. R. Alexander and S. A. Schaub, Internal fields of a spherical particle illuminated by a tightly focused laser beam: Focal point positioning effects at resonance, *J. Appl. Phys.* **65**, 2900–2906 (1989).

[361] J. P. Barton and D. R. Alexander, Fifth-order corrected electromagnetic field components for a fundamental Gaussian beam, *J. Appl. Phys.* **66**, 2800–2802 (1989).

[362] J. P. Barton, D. R. Alexander and S. A. Schaub, Theoretical determination of net radiation force and torque for a spherical particle illuminated by a focused laser beam, *J. Appl. Phys.* **66**, 4594–4602 (1989).

[363] P. L. Marston and J. H. Crichton, Radiation torque on a sphere illuminated with circularly polarized light, *J. Opt. Soc. Am. B* **1**, 528–529 (1984).

[364] G. Gouesbet, B. Maheu and G. Gréhan, Light scattering from a sphere arbitrarily located in a Gaussian beam, using a Bromwich formulation, *J. Opt. Soc. Am. A* **5**, 1427–1443 (1988).

[365] W. H. Wright, G. T. Sonek and M. W. Berns, Radiation trapping forces on microspheres with optical tweezers, *Appl. Phys. Lett.* **63**, 715–717 (1993).

[366] W. H. Wright, G. T. Sonek and M. W. Berns, Parametric study of the forces on microspheres held by optical tweezers, *Appl. Opt.* **33**, 1735–1748 (1994).

[367] K. F. Ren, G. Gréhan and G. Gouesbet, Radiation pressure forces exerted on a particle arbitrarily located in a Gaussian beam by using the generalized Lorentz-Mie theory, and associated resonance effects, *Opt. Commun.* **108**, 343–354 (1994).

[368] G. Gouesbet, G. Gréhan and B. Maheu, Localized interpretation to compute all the coefficients $g_n{}^m$ in the generalized Lorenz-Mie theory, *J. Opt. Soc. Am. A* **7**, 998–1007 (1990).

[369] Y. Harada and T. Asakura, Radiation forces on a dielectric sphere in the Rayleigh scattering regime, *Opt. Commun.* **124**, 529–541 (1996).

[370] K. Visscher and G. J. Brakenhoff, Theoretical study of optically induced forces on spherical particles in a single beam trap I: Rayleigh scatterers, *Optik* **89**, 174–180 (1992).

[371] B. Richards and E. Wolf, Electromagnetic diffraction in optical systems II: Structure of the image field in an aplanatic system, *Proc. R. Soc.* **253**, 358–379 (1959).

[372] T. Tlusty, A. Meller and R. Bar-Ziv, Optical gradient forces of strongly localized fields, *Phys. Rev. Lett.* **81**, 1738–1741 (1998).

[373]* P. A. M. Neto and H. M. Nussenzweig, Theory of optical tweezers, *Europhys. Lett.* **50**, 702–708 (2000).

[374] M. E. J. Friese, H. Rubinsztein-Dunlop, N. R. Heckenberg and E. W. Dearden, Determination of the force constant of a single-beam gradient trap by measurement of backscattered light, *Appl. Opt.* **35**, 7112–7116 (1996).

[375] L. P. Ghislain, N. A. Switz and W. W. Webb, Measurement of small forces using an optical trap, *Rev. Sci. Instrum.* **65**, 2762–2768 (1994).

[376] G. Y. Chen, R. J. Warmack, T. Thundat and D. P. Allison, Resonant response of scanning force microscopy cantilevers, *Rev. Sci. Instrum.* **65**, 2532–2537 (1944).

[377] R. Bar-Ziv, A. Meller, T. Tlusty, E. Moses, J. Stavans and S. A. Safran, Localized dynamic light scattering: Probing single particle dynamics at the nanoscale, *Phys. Rev. Lett.* **78**, 154–157 (1997).

[378] R. P. Feynman, The ratchet and pawl, in *The Feynman Lectures on Physics*, Vol. III, eds. R. P. Feynman, R. B. Leighton and M. Sands (Addison Wesley, 1963).

[379]* L. I. McCann, M. Dykman and B. Golding, Thermally activated transitions in a bistable three-dimensional optical trap, *Nature* **402**, 785–787 (1999).

[380] Yogi Berra, *The Yogi Book* (Workman Publishing, 1998), p. 95.

[381] W. B. Russel, D. A. Saville and W. R. Schwalte, *Colloidal Dispersions* (Cambridge University Press, 1989).

[382] P. A. Rundquist, P. Photinos, S. Jagannathan and S. A. Asher, Dynamical bragg diffraction from crystalline colloidal arrays, *J. Chem. Phys.* **91**, 4932–4941 (1989).

[383] P. Pieranski, L. Strzelecki and B. Pansu, Thin colloid crystals, *Phys. Rev. Lett.* **50**, 900–903 (1983).

[384] C. A. Murray, in *Bond-Orientational Order in Condensed Matter Systems*, ed. K. J. Strandber (Springer Verlag, 1991).

[385]* J. C. Crocker and D. G. Grier, Microscopic measurement of the pair interaction potential of charge-stabilized colloids, *Phys. Rev. Lett.* **73**, 352–355 (1994).

[386] M. M. Burns, J.-M. Fournier and J. A. Golovchenko, Optical matter: Crystallization and binding in intense optical fields, *Science* **249**, 749–754 (1990).

[387] M. M. Burns, J.-M. Fournier and J. A. Golovchenko, Optical binding, *Phys. Rev. Lett.* **63**, 1233–1236 (1989).

[388]* J. C. Crocker and D. G. Grier, When like charges attract: The effects of geometrical confinement on long-range colloidal interactions, *Phys. Rev. Lett.* **77**, 1897–1900 (1996).

[389] G. M. Kepler and S. Fraden, Attractive potential between confined colloids at low ionic strength, *Phys. Rev. Lett.* **73**, 356–359 (1994).

[390] M. D. Carbajal-Tinoco, F. Castro-Román and J. L. Arauz-Lara, Static properties of confined colloidal suspensions, *Phys. Rev. E* **53**, 3745–3749 (1996).

[391] W. R. Bowen and A. O. Sharif, Long-range electrostatic attraction between like-charge spheres in a charged pore, *Nature* **393**, 663–665 (1998).

[392] D. G. Grier, A surprisingly attractive couple, *News Views Nat.* **393**, 621–623 (1998).

[393] J. C. Neu, Well-mediated forces between like-charged bodies in an electrolyte, *Phys. Rev. Lett.* **82**, 1072–1074 (1999).

[394] J. J. Gray, B. Chiang and R. T. Bonnecaze, Origin of anomalous multibody interactions, *Nature* **402**, 750 (1999).

[395] T. M. Squires and M. P. Brenner, Like-charge attraction and hydrodynamic interaction, *Phys. Rev. Lett.* **85**, 4976–4979 (2001).

[396] R. Fitzgerald, Hydrodynamics may explain like-charge colloid attraction, *Phys. Today* **54**, 18–20 (2001).

[397] Y. Han and D. G. Grier, Confinement-induced colloidal attractions in equilibrium, *Phys. Rev. Lett.* **91**, 038302 (2003).

[398] A. D. Dinsmore, A. G. Yodh and D. J. Pine, Entropic control of particle motion using passive surface microstructures, *Nature* **383**, 239–242 (1996).

[399] S. Asakura and F. Oosawa, Interaction between particles suspended in solutions of macromolecules, *J. Polym. Sci.* **33**, 183–192 (1958).

[400] L. C. Crocker, J. A. Matteo, A. D. Dinsmore and A. G. Yodh, Entropic attraction and repulsion in binary colloids probed with a line optical tweezer, *Phys. Rev. Lett.* **82**, 4352–4355 (1999).

[401] R. C. Gauthier, Ray optics model and numerical computations for the radiation pressure micromotor, *Appl. Phys. Lett.* **67**, 2269–2271 (1995).

[402] R. C. Gauthier, Theoretical model for an improved radiation pressure micromotor, *Appl. Phys. Lett.* **69**, 2015–2017 (1996).

[403] E. Higurashi, O. Ohguchi, T. Tamamura, H. Ukita and R. Sawada, Optically induced rotation of dissymmetrically shaped fluorinated polyimide micro-objects in optical traps, *J. Appl. Phys.* **82**, 2773–2779 (1997).

[404] R. C. Gauthier, Theoretical investigation of the optical trapping force and torque on cylindrical micro-objects, *J. Opt. Soc. Am. B* **14**, 3323–3333 (1997).

[405]* E. Higurashi, R. Sawada and T. Ito, Optically induced rotation of a trapped micro-object about an axis perpendicular to the laser beam axis, *Appl. Phys. Lett.* **72**, 2951–2953 (1998).

[406] P. Galajda and P. Ormos, Complex micromachines produced and driven by light, *Appl. Phys. Lett.* **78**, 249–251 (2001).

[407] M. W. Beijersbergen, L. Allen, H. E. L. O. van der Veen and J. P. Woerdman, Astigmatic laser mode converters and transfer of orbital angular momentum, *Opt. Commun.* **96**, 123–132 (1993).

[408] L. Allen, M. W. Beijersbergen, R. J. C. Spreeuw and J. P. Woerdman, Orbital angular momentum of light and the transformation of Laguerre-Gaussian laser modes, *Phys. Rev. A* **45**, 8185–8189 (1992).

[409] S. J. van Enk and G. Nienhuis, Eigenfunction description of laser beams and orbital angular momentum of light, *Opt. Commun.* **94**, 147–158 (1992).

[410] M. W. Beijersbergen, R. P. C. Coerwinkel, M. Kristensen and J. P. Woerdman, Helical-wavefront laser beams produced with a spiral phaseplate, *Opt. Commun.* **112**, 321–327 (1994).

[411] H. He, M. E. J. Friese, N. R. Heckenberg and H. Rubinsztein-Dunlop, Direct observation of transfer of angular momentum to absorptive particles from a laser beam with a phase singularity, *Phys. Rev. Lett.* **75**, 826–829 (1995).

[412] M. E. J. Friese, J. Engen, H. Rubinsztein-Dunlop and N. R. Heckenberg, Optical angular-momentum transfer to trapped absorbing particles, *Phys. Rev. A* **54**, 1593–1596 (1996).

[413] S. M. Barnett and L. Allen, Orbital angular momentum and nonparaxial light beams, *Opt. Commun.* **110**, 670–678 (1994).

[414]* N. B. Simpson, K. Dholakia, L. Allen and M. J. Padgett, Mechanical equivalence of spin and orbital angular momentum of light: An optical spanner, *Opt. Lett.* **22**, 52–54 (1997).

[415]* L. Paterson, M. P. MacDonald, J. Arlt, W. Sibbett, P. E. Bryant and K. Dholakia, Controlled rotation of optically trapped microscopic particles, *Science* **292**, 912–914 (2001).

[416] A. I. Bishop, T. A. Nieminen, N. R. Heckenberg and H. Rubinsztein-Dunlop, Optical application and measurement of torque on microparticles of isotropic nonabsorbing material, *Phys. Rev. A* **68**, 033802 (2003).

[417]* A. La Porta and M. D. Wang, Optical torque wrench: Angular trapping, rotation, and torque detection of quartz microparticles, *Phys. Rev. Lett.* **92**, 190801 (2004).

[418] A. I. Bishop, T. A. Nieminen, N. R. Heckenberg and H. Rubinsztein-Dunlop, Optical microrheology using rotating laser-trapped particles, *Phys. Rev. Lett.* **92**, 198104 (2004).

[419] K. Sasaki, M. Koshioka, H. Misawa, N. Kitamura and H. Masuhara, Laser-scanning micromanipulation and spatial patterning of fine particles, *Jpn. J. Appl. Phys.* **30**, L907–L909 (1991).

[420] H. Misawa, M. Koshioka, K. Sasaki, N. Kitamura and H. Masuhara, 3-dimensional optical trapping and laser ablation of a single polymer latex particle in water, *J. Appl. Phys.* **70**, 3829–3836 (1991).

[421] H. Misawa, K. Sasaki, M. Koshioka, N. Kitamura and H. Masuhara, Multibeam laser manipulation and fixation of microparticles, *Appl. Phys. Lett.* **60**, 310–312 (1992).

[422] R. Srinivasan, B. Braren, R. W. Drefus, L. Hadel and D. E. Seeger, Mechanism of the ultraviolet laser ablation of polymethyl methacrylate at 193 and 248 nm: Laser-induced fluorescence analysis, chemical analysis, and doping studies, *J. Opt. Soc. Am. B* **3**, 785–791 (1986).

[423] R. Srinivasan and B. Braren, Ultraviolet laser ablation of organic polymers, *Chem. Rev.* **89**, 1303 (1989).

[424] H. Misawa, K. Sasaki, M. Koshioka, N. Kitamura and H. Masuhara, Laser manipulation and assembling of polymer latex-particles in solution, *Macromolecules* **26**, 282–286 (1993).

[425] H. Misawa, N. Kitamura and H. Masuhara, Laser manipulation and ablation of a single microcapsule in water, *J. Am. Chem. Soc.* **113**, 7859–7863 (1991).

[426] K. Nakatani, T. Uchida, H. Misawa, N. Kitamura and H. Masuhara, Electrochemistry and fluorescence spectroscopy of a single, laser-trapped oil droplet in water — mass-transfer across microdroplet water interface, *J. Phys. Chem. US* **97**, 5197–5199 (1993).

[427] K. Nakatani, T. Uchida, S. Funakura, H. Misawa, N. Kitamura and H. Masuhara, Control of a dye formation reaction in a single micrometer-sized oil-droplet by laser trapping and microelectrochemical methods, *Chem. Lett.* **4**, 717–720 (1993).

[428] T. H. James, in *The theory of the photographic process* (Macmillan, 1977).

[429] M. Ishikawa, H. Misawa, N. Kitamura and H. Masuhara, Poly(n-isopropylacrylamide) microparticle formation in water by infrared laser-induced photothermal phase-transition, *Chem. Lett.* **3**, 481–484 (1993).

[430] *Microchemistry, Spectroscopy, and Chemistry in Small Domains*, eds. H. Masuhara, F. C. DeSchryver, N. Kitamura, N. Tamai (North Holland, 1994).

[431] M. Koshioka, U. Pfeifer-Fukumura, S. Funakura, K. Nakaani and H. Masuhara, Photophysics and photochemistry of individual microparticles in solution, in *Microchemistry, Spectroscopy, and Chemistry in Small Domains*, eds. H. Masuhara, et al. (North Holland, 1994), pp. 349–362.

[432] K. Sasaki, M. Koshioka and H. Masuhara, 3-dimensional space-resolved and time-resolved fluorescence spectroscopy, *Appl. Spectrosc.* **45**, 1041–1045 (1991).

[433] N. Tamai, T. Asahi and H. Masuhara, Femtosecond transient absorption microspectrophotometer combined with optical trapping technique, *Rev. Sci. Instrum.* **64**, 2496–2503 (1993).

[434] M. Koshioka, H. Misawa, K. Sasaki, N. Kitamura and H. Masuhara, Pyrene excimer formation dynamics in a single microcapsule by space-resolved and time-resolved fluorescence spectroscopy, *J. Phys. Chem.* **96**, 2909–2914 (1992).

[435] M. Irie, Stimuli-responsive polymer gels: An approach to micro actuators, in *Microchemistry, Spectroscopy, Chemistry of Small Domains*, eds. H. Masuhara, et al. (North Holland, 1994), pp. 363–371.

[436] P. Korda, G. C. Spalding, E. R. Dufresne and D. G. Grier, Nanofabrication with holographic optical tweezers, *Rev. Sci. Instr.* **73**, 1956–1957 (2002).

[437]* J. E. Curtis, B. A. Koss and D. G. Grier, Dynamic holographic optical tweezers, *Opt. Commun.* **207**, 169–175 (2002).

[438] K. Ladavac, K. Kasza and D. G. Grier, Sorting by periodic energy landscapes: Optical fractionation, *Phys. Rev. Lett.*

[439] P. T. Korda, M. B. Taylor and D. G. Grier, Kinetically locked-in colloidal transport in an array of optical tweezers, *Phys. Rev. Lett.* **89**, 128301 (2002).

[440] B. A. Koss and D. G. Grier, Optical peristalsis, *Appl. Phys. Lett.* **82**, 3985–3987 (2003).

[441] M. P. MacDonald, G. C. Spalding and K. Dholakia, Microfluidic sorting in an optical lattice, *Nature* **426**, 421–424 (2003).

[442] A. Terray, J. Oakley and D. W. M. Marr, Microfluidic control using colloidal devices, *Science* **296**, 1841–1844 (2002).

[443] N. F. Ramsay, *Molecular Beams* (Oxford University Press, 1956).

[444] J. R. Zacharias, Precision measurements with molecular beams, *Phys. Rev.* **94**, 751 (1954).

[445] P. D. Lett, N. Watts, C. I. Westbrook, W. D. Phillips, P. L. Gould and H. Metcalf, Observation of atoms laser cooled below the doppler limit, *Phys. Rev. Lett.* **61**, 169–172 (1988).

[446] K. Gibble and S. Chu, Future slow-atom frequency standards, *Metrologie* **29**, 201–212 (1992).

[447] K. Gibble and S. Chu, Laser-cooled cs frequency standard and a measurement of the frequency shift due to ultracold collisions, *Phys. Rev. Lett.* **70**, 1771–1774 (1993).

[448] P. Laurent, P. Lemonde, E. Simon, G. Santarelli, A. Clairon, N. Dimarcq, P. Petit, C. Audoin and C. Salomon, A cold atom clock in absence of gravity, *Eur. Phys. J. D* **3**, 201–204 (1998).

[449] M. Kasevich, D. S. Weiss, E. Riis, K. Moler, S. Kasapi and S. Chu, Atomic velocity selection using stimulated Raman transitions, *Phys. Rev. Lett.* **66**, 2297–2300 (1991).

[450] V. I. Balykin, V. S. Letokhov, A. I. Sidorov and Yu. B. Ovchinnikov, Focusing of an atomic beam and imaging of atomic sources by means of a laser lens based on resonance radiation pressure, *J. Mod. Opt.* **35**, 17–34 (1988).

[451] P. Gill, Raising the standards, *Science* **294**, 1666–1668 (2001).

[452] V. I. Balykin and V. S. Letokhov, The possibility of deep focusing of an atomic beam into the Å-region, *Opt. Commun.* **64**, 151–156 (1987).

[453] V. I. Balykin and V. S. Letokhov, *Atom Optics with Laser Light* (Harwood Academic Publishers, 1995).

[454] F. A. Jenkins and H. E. White, *Fundamentals of Optics* (McGraw Hill, 1959), pp. 108–112.

[455] E. H. Kennard, *Kinetic Theory of Gases* (McGraw-Hill, 1938), pp. 339–348.

[456] J. Nellessen, J. H. Muller, K. Sengstock and W. Ertmer, Laser preparation of a monoenergetic sodium beam, *Europhys. Lett.* **9**, 133–138 (1989).

[457] J. Nellessen, J. H. Muller, K. Sengstock and W. Ertmer, Large-angle beam deflection of a laser-cooled sodium beam, *J. Opt. Soc. Am. B* **6**, 2149–2154 (1989).

[458] J. Nellessen, J. Werner and W. Ertmer, Magnetooptical compression of a monoenergetic sodium atomic-beam, *Opt. Commun.* **78**, 300–308 (1990).

[459] E. Riis, D. S. Weiss, K. A. Moler and S. Chu, Atom funnel for the production of a slow, high-density atomic beam, *Phys. Rev. Lett.* **64**, 1658–1661 (1990).

[460] R. J. Cook and R. K. Hill, An electromagnetic mirror for neutral atoms, *Opt. Commun.* **43**, 258–260 (1982).

[461] R. Kaiser, Y. Lévy, N. Vansteenkiste, A. Aspect, W. Seifert, D. Leipold and J. Mlynek, Resonant enhancement of evanescent waves with a thin dielectric waveguide, *Opt. Commun.* **104**, 234–240 (1994).

[462] M. A. Kasevich, D. S. Weiss and S. Chu, Normal-incidence reflection of slow atoms from an optical evanescent wave, *Opt. Lett.* **15**, 607–609 (1990).

[463] C. G. Aminoff, A. M. Steane, P. Bouyer, P. Desbiolles, J. Dalibard and C. Cohen-Tannoudji, Cesium atoms bouncing in a stable gravitational cavity, *Phys. Rev. Lett.* **71**, 3083–3086 (1993).

[464] V. I. Balykin, V. S. Letokhov, Atomic cavity with light-induced mirrors, *Appl. Phys. B* **48**, 517–523 (1989).

[465] P. J. Martin, B. G. Oldaker, A. H. Miklich and D. E. Pritchard, Diffraction of atoms moving through a standing wave, *Phys. Rev. Lett.* **60**, 515–518 (1988).

[466] K. S. Johnson, A. Chu, T. W. Lynn, K. K. Berggren, M. S. Shahriar and M. Prentiss, Demonstration of a nonmagnetic blazed-grating atomic beam splitter, *Opt. Lett.* **20**, 1310–1312 (1995).

[467] J. Lawall and M. Prentiss, Demonstration of a novel atomic beam splitter, *Phys. Rev. Lett.* **72**, 993–996 (1994).

[468] M. Weitz, B. C. Young and S. Chu, Atomic interferometer based on adiabatic population transfer, *Phys. Rev. Lett.* **73**, 2563–2566 (1994).

[469] G. Timp, R. E. Behringer, D. M. Tennant, J. E. Cunningham, M. Prentiss and K. Berggren, Using light as a lens for submicron, neutral-atom lithography, *Phys. Rev. Lett.* **69**, 1636–1639 (1992).

[470] J. J. McClelland, R. E. Scholten, E. C. Palm and R. J. Celotta, Laser-focused atomic deposition, *Science* **262**, 877–880 (1993).

[471] R. E. Behringer, V. Natarajan, G. Timp and D. M. Tennant, Limit of resolution of a standing wave atom optical lens, *J. Vac. Sci. Technol. B* **14**, 4072–4075 (1996).

[472]* M. Kasevich and S. Chu, Atom interferometry using stimulated Raman transitions, *Phys. Rev. Lett.* **67**, 181–184 (1991).

[473] E. A. J. Marcatili and R. A. Schmeltzer, Hollow metallic and dielectric wave-guides for long distance optical transmission and lasers, *Bell Syst. Tech. J.* **43**, 1783–1808 (1964).

[474] M. J. Renn, D. Montgomery, O. Vdovin, D. Z. Anderson, C. E. Wieman and E. A. Cornell, Laser-guided atoms in hollow-core optical fibers, *Phys. Rev. Lett.* **75**, 3253–3256 (1995).

[475] J. Denschlag, D. Cassettari and J. Schmiedmayer, Guiding neutral atoms with a wire, *Phys. Rev. Lett.* **82**, 2014–2017 (1999).

[476] R. Folman, P. Krüger, D. Cassettari, B. Hessmo, T. Maier and J. Schmiedmayer, Controlling cold atoms using nanofabricated surfaces: Atom chips, *Phys. Rev. Lett.* **84**, 4749–4752 (2000).

[477] D. Müller, D. Z. Anderson, R. J. Grow, P. D. D. Schwindt and E. A. Cornell, Guiding neutral atoms around curves with lithographically patterned current-carrying wires, *Phys. Rev. Lett.* **83**, 5194–5197 (1999).

[478] J. Schmiedmayer, M. S. Chapman, C. R. Ekstrom, T. D. Hammond, D. A. Kokorowski, A. Lenef, R. A. Rubinstein, E. T. Smith and D. E. Pritchard, Optics and interferometry with atoms and molecules, in *Atom Interferometry*, ed. P. R. Berman (Academic Press, 1997), pp. 155–167.

[479] D. Müller, E. A. Cornell, M. Prevedelli, P. D. D. Schwindt, A. Zozulya and D. Z. Anderson, Waveguide atom beam splitter for laser-cooled neutral atoms, *Opt. Lett.* **25**, 1382–1388 (2000).

[480] H. R. Thorsheim, J. Weiner and P. Julienne, Laser-induced photoassociation of ultracold sodium atoms", *Phys. Rev. Lett.* **58**, 2420–2423 (1987).

[481] P. L. Gould, P. D. Lett, P. S. Julienne, W. D. Phillips, H. R. Thorsheim and J. Weiner, Observation of associative ionization of ultracold laser-trapped sodium atoms, *Phys. Rev. Lett.* **60**, 788–791 (1988).

[482] P. S. Julienne, Laser modification of ultracold atomic collisions in optical traps, *Phys. Rev. Lett.* **61**, 698–701 (1988).

[483] P. D. Lett, P. S. Jessen, W. D. Phillips, S. L. Rolston, C. I. Westerbrook and P. L. Gould, Laser modification of ultracold collisions: Experiment, *Phys. Rev. Lett.* **67**, 2139–2142 (1991).

[484] P. S. Julienne and R. Heather, Laser modification of ultracold atomic collisions: Theory, *Phys. Rev. Lett.* **67**, 2135–2138 (1991).

[485] R. W. Heather and P. S. Julienne, Theory of laser induced associative ionization of ultracold Na, *Phys. Rev. A* **47**, 1887–1906 (1993).

[486] P. D. Lett, K. Helmerson, W. D. Phillips, L. P. Ratcliff, S. L. Ralston and M. E. Wagshul, Spectroscopy of Na_2 by photoassociation of laser-cooled Na, *Phys. Rev. Lett.* **71**, 2200–2203 (1993).

[487] J. D. Miller, R. A. Cline and D. J. Heinzen, Photoassociation spectrum of ultracold rb atoms, *Phys. Rev. Lett.* **71**, 2204–2207 (1993).

[488] R. A. Cline, J. D. Miller and D. J. Heinzen, Study of rb_2 long-range states by high-resolution photoassociation spectroscopy, *Phys. Rev. Lett.* **73**, 632–635 (1994).

[489] R. A. Cline, J. D. Miller and D. J. Heinzen, Study of rb_2 long-range states by high-resolution photoassociation spectroscopy, *Phys. Rev. Lett.* **73**, 2636–2639 (1994).

[490] C. C. Tsai, R. S. Freeland, J. M. Vogels, M. M. J. M. Boesten, B. J. Verhaar and D. J. Heinzen, Two-color photoassociation spectroscopy of ground state rb_2, *Phys. Rev. Lett.* **79**, 1245–1248 (1997).

[491] S. Bali, D. Hoffmann and T. Walker, Novel intensity dependence of ultracold collisions involving repulsive states, *Europhys. Lett.* **27**, 273–277 (1994).

[492] V. Sanchez-Villicana, S. D. Gensemer, K. Y. N. Tan, A. Kumarakrishnan, T. P. Dinneen, W. Süptitz and P. L. Gould, Suppression of ultracold ground-state hyperfine-changing collisions with laser light, *Phys. Rev. Lett.* **74**, 4619–4622 (1995).

[493] S. R. Muniz, L. G. Marcassa, R. Napolitano, G. D. Telles, J. Weiner, S. C. Zilio and V. S. Bagnato, Optical suppression of hyperfine-changing collisions in a sample of ultracold sodium atoms, *Phys. Rev. A* **55**, 4407–4411 (1997).

[494] L. Marcassa, S. Muniz, E. de Queiroz, S. Zilio, V. Bagnato, J. Weiner, P. S. Julienne and K.-A. Suominen, Optical suppression of photoassociative ionization in a magneto-optical trap, *Phys. Rev. Lett.* **73**, 1911–1914 (1994).

[495] L. Marcassa, R. Horowicz, S. Zilio, V. Bagnato and J. Weiner, Intensity dependence of optical suppression in photoassociative ionization collisions in a sodium magneto-optic trap, *Phys. Rev. A* **52**, R913–R916 (1995).

[496] S. C. Zilio, L. Marcassa, S. Muniz, R. Horowicz, V. Bagnato, R. N. Napolitano, J. Weiner and P. S. Julienne, Polarization dependence of optical suppression in photoassociative ionization collisions in a sodium magneto-optic trap, *Phys. Rev. Lett.* **76**, 2033–2036 (1996).

[497] R. Napolitano, J. Weiner and P. S. Julienne, Theory of optical suppression of ultracold-collision rates by polarized light, *Phys. Rev. A* **55**, 1191–1197 (1997).

[498] K.-A. Suominen, M. J. Holland, K. Burnett and P. Julienne, Optical shielding of cold collisions, *Phys. Rev. A* **51**, 1446–1457 (1995).

[499] K.-A. Suominen, Theories for cold atomic collisions in light fields, *J. Phys. B* **29**, 5981–6007 (1996).

[500] F. Dalfovo, S. Giorgini, L. P. Pitaevskii and S. Stringari, Theory of Bose-Einstein condensation in trapped gases, *Rev. Mod. Phys.* **71**, 463–512 (1999).

[501] K. Huang, *Statistical Mechanics* (John Wiley and Sons, 1987).

[502] D. G. Fried, T. C. Killian, L. Willmann, D. Landhuis, S. C. Moss, D. Kleppner and T. J. Greytak, Bose-Einstein condensation of atomic hydrogen, *Phys. Rev. Lett.* **81**, 3811–3814 (1998).

[503] A. L. Migdall, J. V. Prodan, W. D. Phillips, T. H. Bergeman and H. J. Metcalf, First observation of magnetically trapped neutral atoms, *Phys. Rev. Lett.* **54**, 2596–2599 (1985).

[504] W. Petrich, M. H. Anderson, J. R. Ensher and E. A. Cornell, Stable, tightly confining magnetic trap for evaporative cooling of neutral atoms, *Phys. Rev. Lett.* **74**, 3352–3355 (1995).

[505]* J. R. Gardner, R. A. Cline, J. D. Miller, D. J. Heinzen, H. M. J. M. Boesten and B. J. Verhaar, Collisions of doublyspin-polarized, ultracold ^{85}rb atoms, *Phys. Rev. Lett.* **74**, 3764–3767 (1995).

[506] L. You, M. Lewenstein and J. Cooper, Quantum-field theory of atoms interacting with photons: Scattering of short laser-pulses from trapped bosonic atoms, *Phys. Rev. A* **51**, 4712–4727 (1995).

[507] P. A. Ruprecht, M. J. Holland, K. Burnett and M. Edwards, Time-dependent solution of the nonlinear Schrödinger-equation for Bose-condensed trapped neutral atoms, *Phys. Rev. A* **51**, 4704–4711 (1995).

[508] E. Tiesinga, A. J. Moerdijk, B. J. Verhaar and H. T. C. Stoof, Conditions for Bose-Einstein condensation in magnetically trapped atomic cesium, *Phys. Rev. A* **46**, R1167–R1170 (1992).

[509] M.-O. Mewes, M. R. Andrews, N. J. van Druten, D. M. Kurn, D. S. Durfee and W. Ketterle, Bose-Einstein condensation in a tightly confining DC magnetic trap, *Phys. Rev. Lett.* **77**, 416–419 (1996).

[510] Y. V. Gott, M. S. Ioffe and V. G. Telkovsky, in *Nuclear Fusion, 1962 Suppl. Pt. 3* (International Atomic Energy Agency, Vienna, 1962), p. 1045.

[511] D. E. Pritchard, Cooling neutral atoms in a magnetic trap for precision spectroscopy, *Phys. Rev. Lett.* **51**, 1336–1339 (1983).

[512] C. C. Bradley, C. A. Sackett, J. J. Tollett and R. G. Hulet, Evidence of Bose-Einstein condensation in an atomic gas with attractive interactions, *Phys. Rev. Lett.* **75**, 1687–1690 (1995).

[513]* C. C. Bradley, C. A. Sackett and R. G. Hulet, Bose-Einstein condensation of lithium: Observation of limited condensate number, *Phys. Rev. Lett.* **78**, 985–989 (1997).

[514] C. G. Townsend, N. J. van Druten, M. R. Andrews, D. S. Durfee, D. M. Kurn, M.-O. Mewes and W. Ketterle, Bose–Einstein condensation of a weakly-interacting gas, OSA TOPS on *Ultracold Atoms and BEC, 1996 Vol. 7*, ed. Keith Burnett (1996), pp. 2–13.

[515] M. Holland and J. Cooper, Expansion of a Bose-Einstein condensate in a harmonic potential, *Phys. Rev. A* **53**, R1954–R1957 (1996).

[516] E. P. Gross, Structure of a quantized vortex in Boson systems, *Nuovo Cimento* **20**, 454–477 (1961).

[517] L. P. Pitaevskii, Vortex lines in an imperfect Bose gas, *Zh. Eksp. Teor. Fiz.* **40**, 646 [Sov. Phys. JETP **13**, 451–454] (1961).

[518] D. S. Jin, J. R. Ensher, M. R. Matthew, C. E. Wieman and E. A. Cornell, Collective excitations of a Bose-Einstein condensate in a dilute gas, *Phys. Rev. Lett.* **77**, 420–423 (1996).

[519] M.-O. Mewes, M. R. Andrews, N. J. van Druten, D. M. Kurn, D. S. Durfee, C. G. Townsend and W. Ketterle, Collective excitations of a Bose-Einstein condensate in a magnetic trap, *Phys. Rev. Lett.* **77**, 988–991 (1996).

[520] S. Stringari, Collective excitations of a trapped Bose condensed gas, *Phys. Rev. Lett.* **77**, 2360–2363 (1996).

[521] D. M. Stamper-Kurn, A. P. Chikkatur, A. Görlitz, S. Inouye, S. Gupta, D. E. Pritchard and W. Ketterle, Excitation of phonons in a Bose-Einstein condensate by light scattering, *Phys. Rev. Lett.* **83**, 2876–2879 (1999).

[522] J. Stenger, S. Inouye, A. P. Chikkatur, D. M. Stamper-Kurn, D. E. Pritchard and W. Ketterle, Bragg spectroscopy of a Bose-Einstein condensate, *Phys. Rev. Lett.* **82**, 4569–4573 (1999).

[523] E. W. Hagley, L. Deng, M. Kozuma, M. Trippenbach, Y. B. Band, M. Edwards, M. Doery, P. S. Julienne, K. Helmerson, S. L. Rolston and W. D. Phillips, Measurement of the coherence of a Bose-Einstein condensate, *Phys. Rev. Lett.* **83**, 3112–3115 (1999).

[524] B. P. Anderson and M. A. Kasevich, Macroscopic quantum interference from atomic tunnel arrays, *Science* **282**, 1686–1689 (1998).

[525] I. Bloch, T. W. Hänsch and T. Esslinger, Measurement of the spatial coherence of a trapped Bose gas at the phase transition, *Nature* **403**, 166–170 (2000).

[526] H.-J. Miesner, D. M. Stamper-Kurn, M. R. Andrews, D. S. Durfee, S. Inouye and W. Ketterle, Bosonic stimulation in the formation of a Bose-Einstein condensate, *Science* **279**, 1005–1007 (1998).

[527] C. W. Gardiner, P. Zoller, R. J. Ballagh and M. J. Davis, Kinetics of Bose-Einstein condensation in a trap, *Phys. Rev. Lett.* **79**, 1793–1796 (1997).

[528] T. Esslinger, I. Bloch and T. W. Hänsch, Bose-Einstein condensation in a quadrupole-Ioffe-configuration trap, *Phys. Rev. A* **58**, R2664–R2667 (1998).

[529] G. B. Lubkin, New atom lasers eject atoms or run CW, *Search Discov. Phys. Today* **52**, 17–18 (1999).

[530] D. M. Giltner, R. W. McGowan and S. A. Lee, Atom interferometer based on bragg scattering from standing light waves, *Phys. Rev. Lett.* **75**, 2638–2641 (1995).

[531]* M. Kozuma, L. Deng, E. W. Hagley, J. Wen, R. Lutwak, K. Helmerson, S. L. Rolston and W. D. Phillips, Coherent splitting of Bose-Einstein condensed atoms with optically induced bragg diffraction, *Phys. Rev. Lett.* **82**, 871–875 (1999).

[532] Y. R. Shen, Phase conjugation by four-wave mixing, in *The Principles of Nonlinear Optics* (John Wiley, 1984).

[533] M. R. Andrews, Bose gases and their Fermi cousins, *Nature* **398**, 195–198 (1999).

[534] G. B. Lubkin, Bose condensates produce coherent nonlinear behavior, *Search Discov. Phys. Today* **52**, 17–19 (1999).

[535]* C. J. Myatt, E. A. Burt, R. W. Ghrist, E. A. Cornell and C. E. Wieman, Production of two overlapping Bose-Einstein condensates by sympathetic cooling, *Phys. Rev. Lett.* **78**, 586–589 (1997).

[536] C. J. Myatt, N. R. Newbury, R. W. Ghrist, S. Loutzenhiser and C. E. Wieman, Multiply loaded magneto-optical trap, *Opt. Lett.* **21**, 290–292 (1996).

[537] D. S. Hall, M. R. Mathews, C. E. Wieman and E. A. Cornell, Measurements of relative phase in two-component Bose-Einstein condensates, *Phys. Rev. Lett.* **81**, 1543–1546 (1998).

[538] H. Pu and N. P. Bigelow, Properties of two-species Bose condensates, *Phys. Rev. Lett.* **80**, 1130–1133 (1998).

[539]* T. Takekoshi, J. R. Yeh and R. J. Knize, Quasi-electrostatic trap for neutral atoms, *Opt. Commun.* **114**, 421–424 (1995).

[540] H. J. Lee, C. S. Adams, M. Kasevich and S. Chu, Raman cooling of atoms in an optical dipole trap, *Phys. Rev. Lett.* **76**, 2658–2661 (1996).

[541] M. Kasevich and S. Chu, Laser cooling below a photon recoil with three-level atoms, *Phys. Rev. Lett.* **69**, 1741–1744 (1992).

[542] T. A. Savard, K. M. O'Hara and J. E. Thomas, Laser-noise-induced heating in far-off-resonance optical traps, *Phys. Rev. A* **56**, R1095–R1098 (1997).

[543] M. Prentiss, A. Cable, J. E. Bjorkholm, S. Chu, E. L. Raab and D. E. Pritchard, Atomic-density-density losses in an optical trap, *Opt. Lett.* **13**, 452 (1988).

[544] M. E. Gehm, K. M. O'Hara, T. A. Savard and J. E. Thomas, Dynamics of noise-induced heating in atom traps, *Phys. Rev. A* **58**, 3914–3921 (1998).

[545]* D. Boiron, A. Michaud, J. M. Fournier, L. Simard, M. Sprenger, G. Grynberg and C. Salomon, Cold and dense cesium clouds in far-detuned dipole traps, *Phys. Rev. A* **57**, R4106–R4109 (1998).

[546]* V. Vuletić, C. Chin, A. J. Kerman and S. Chu, Degenerate Raman sideband cooling of trapped cesium atoms at very high atomic densities, *Phys. Rev. Lett.* **81**, 5768–5771 (1998).

[547] H. Katori, T. Ido, Y. Isoya and M. Kuwata-Gonokami, Magneto-optical trapping and cooling of strontium atoms down to the photon recoil temperature, *Phys. Rev. Lett.* **82**, 1116–1119 (1999).

[548] M. T. DePue, C. McCormick, S. L. Winoto, S. Oliver and D. S. Weiss, Unity occupation of sites in a 3D-optical lattice, *Phys. Rev. Lett.* **82**, 2262–2265 (1999).

[549] A. Kastberg, W. D. Phillips, S. L. Rolston and R. J. C. Spreeuw, Adiabatic cooling of cesium to 700 nK in an optical lattice, *Phys. Rev. Lett.* **74**, 1542–1545 (1995).

[550] D. Jaksch, C. Bruder, J. I. Cirac, C. W. Gardiner and P. Zoller, Cold bosonic atoms in optical lattices, *Phys. Rev. Lett.* **81**, 3108–3111 (1998).

[551] D. Jaksch, H.-J. Briegel, J. I. Cirac, C. W. Gardiner and P. Zoller, Entanglement of atoms via cold controlled collisions, *Phys. Rev. Lett.* **82**, 1975–1978 (1999).

[552] G. K. Brennen, C. M. Caves, P. S. Jessen and I. H. Deutsch, Quantum logic gates in optical lattices, *Phys. Rev. Lett.* **82**, 1060–1063 (1999).

[553] A. Sørensen and K. Mølmer, Spin–spin interaction and spin squeezing in an optical lattice, *Phys. Rev. Lett.* **83**, 2274–2277 (1999).

[554] N. Davidson, H. J. Lee, C. S. Adams, M. Kasevich and S. Chu, Long atomic coherence times in an optical dipole trap, *Phys. Rev. Lett.* **74**, 1311–1314 (1995).

[555] R. Ozeri, L. Khaykovich and N. Davidson, Long spin relaxation times in a single-beam blue-detuned optical trap, *Phys. Rev. A* **59**, R1750–R1753 (1999).

[556] N. Friedman, L. Khaykovich, R. Ozeri and N. Davidson, Single-beam dark optical traps for cold atoms, *Opt. Photonics News* **10**, 36–37 (1999).

[557] T. Kuga, Y. Torii, N. Shiokawa and T. Hirano, Novel optical trap of atoms with a doughnut beam, *Phys. Rev. Lett.* **78**, 4713–4716 (1997).

[558] S. Kuppens, M. Rauner, M. Schiffer, G. Wokurka, T. Slawinski, M. Zinner, K. Sengstock and W. Ertmer, Atom guiding in a blue-detuned donut mode, ultra-cold atoms and Bose–Einstein condensates, OSA Vol. 7, ed. Keith Burnett, papers from EQEC '96, *European Quantum Electronics Conference*, September 1996, pp. 102–107.

[559]* D. M. Stamper-Kurn, M. R. Andrews, A. P. Chikkatur, S. Inouye, H.-J. Miesner, J. Stenger and W. Ketterle, Optical confinement of a Bose-Einstein condensate, *Phys. Rev. Lett.* **80**, 2027–2030 (1998).

[560] J. Stenger, D. M. Stamper-Kurn, M. R. Andrews, A. P. Chikkatur, S. Inouye, H.-J. Miesner and W. Ketterle, Optically confined Bose-Einstein condensates, *J. Low Temp. Phys.* **113**, 167–188 (1998).

[561]* J. Stenger, S. Inouye, D. M. Stamper-Kurn, H.-J. Miesner, A. P. Chikkatur and W. Ketterle, Spin domains in ground-state Bose-Einstein condensates, *Nature* **396**, 345–348 (1998).

[562] H.-J. Miesner, D. M. Stamper-Kurn, J. Stenger, S. Inouye, A. P. Chikkatur and W. Ketterle, Observation of metastable states in spinor Bose-Einstein condensates, *Phys. Rev. Lett.* **82**, 2228–2231 (1999).

[563] D. M. Stamper-Kurn and W. Ketterle, *Spinor condensates and light scattering from Bose–Einstein condensates*, Les Houches 1999 Summer School, Session LXXII.

[564] D. M. Stamper-Kurn, H.-J. Miesner, A. P. Chikkatur, S. Inouye, J. Stenger and W. Ketterle, Quantum tunneling across spin domains in a Bose-Einstein condensate, *Phys. Rev. Lett.* **83**, 661–665 (1999).

[565]* S. Inouye, M. R. Andrews, J. Stenger, H.-J. Miesner, D. M. Stamper-Kurn and W. Ketterle, Observation of feshbach resonances in a Bose-Einstein condensate, *Nature* **392**, 151–154 (1998).

[566]* P. Courteille, R. S. Freeland and D. J. Heinzen, Observation of a feshbach resonance in cold atom scattering, *Phys. Rev. Lett.* **81**, 69–72 (1998).

[567] V. Vuletić, A. J. Kerman, C. Chin and S. Chu, Observation of low-field feshbach resonances in collisions of cesium atoms, *Phys. Rev. Lett.* **82**, 1406–1409 (1999).

[568] J. Söding, D. Guéry-Odelin, P. Desbiolles, G. Ferrari and J. Dalibard, Giant spin relaxation of an ultracold cesium gas, *Phys. Rev. Lett.* **80**, 1869–1872 (1998).

[569] D. Guéry-Odelin, J. Söding, P. Desbiolles and J. Dalibard, Strong evaporative cooling of a trapped cesium gas, *Opt. Express* **2**, 323–329 (1998).

[570] J. Stenger, S. Inouye, M. R. Andrews, H.-J. Miesner, D. M. Stamper-Kurn and W. Ketterle, Strongly enhanced inelastic collisions in a Bose-Einstein condensate near feshbach resonances, *Phys. Rev. Lett.* **82**, 2422–2425 (1999).

[571] V. Vuletić, C. Chin, A. J. Kerman and S. Chu, Suppression of atomic radiative collisions by tuning the ground state scattering length, *Phys. Rev. Lett.* **83**, 943–946 (1999).

[572] C. Chin, V. Vuletić, A. J. Kerman and S. Chu, High resolution feshbach spectroscopy of cesium, *Phys. Rev. Lett.* **85**, 2717–2720 (2000).

[573] P. J. Leo, C. J. Williams and P. S. Julienne, Collision properties of ultracold ^{133}Cs atoms, *Phys. Rev. Lett.* **85**, 2721–2724 (2000).

[574] A. J. Kerman, V. Vuletić, C. Chin and S. Chu, Beyond optical molasses: 3D Raman sideband cooling of atomic cesium to high phase-space density, *Phys. Rev. Lett.* **84**, 439–442 (2000).

[575] F. K. Fatemi, K. M. Jones and P. D. Lett, Observation of optically induced feshbach resonances in collisions of cold atoms, *Phys. Rev. Lett.* **85**, 4462–4465 (2000).

[576] Yu. B. Ovchinnikov, J. H. Müller, M. R. Doery, E. J. D. Vredenbregt, K. Helmerson, S. L. Rolston and W. D. Phillips, Diffraction of a released Bose-Einstein condensate by a pulsed standing light wave, *Phys. Rev. Lett.* **83**, 284–287 (1999).

[577] J. J. McClelland and M. P. Scheinfein, Laser focusing of atoms: A particle-optics approach, *J. Opt. Soc. Am. B* **8**, 1974–1986 (1991).

[578] C. Salomon, J. Dalibard, A. Aspect, H. Metcalf and C. Cohen-Tannoudji, Channeling atoms in a laser standing wave, *Phys. Rev. Lett.* **59**, 1659–1662 (1987).

[579] Yu. B. Ovchinnikov and V. S. Letokhov, Channeling of atoms in a standing laser light wave, *Comments At. Mol. Phys.* **27**, 185–201 (1992).

[580] R. H. Dicke, The effect of collisions upon the Doppler width of spectral lines, *Phys. Rev.* **89**, 472–473 (1953).

[581] C. A. Sackett, J. M. Gerton, M. Welling and R. G. Hulet, Measurements of collective collapse in a Bose-Einstein condensate with attractive interactions, *Phys. Rev. Lett.* **82**, 876–879 (1999).

[582] J. L. Roberts, N. R. Claussen, S. L. Cornish and C. E. Wieman, Magnetic field dependence of ultracold inelastic collisions near a feshbach resonance, *Phys. Rev. Lett.* **85**, 728–731 (2000).

[583] S. L. Cornish, N. R. Claussen, J. L. Roberts, E. A. Cornell and C. E. Wieman, Stable ^{85}Rb Bose-Einstein condensates with widely tunable interactions, *Phys. Rev. Lett.* **85**, 1795–1798 (2000).

[584] B. G. Levi, Researchers can now vary the atomic interactions in a Bose-Einstein condensate, *Phys. Today* **53**, 17–18 (2000).

[585] A. Robert, O. Sirjean, A. Browaeys, J. Poupard, S. Nowak, D. Boiron, C. I. Westbrook and A. Aspect, A Bose-Einstein condensate of metastable atoms, *Science* **292**, 461–464 (2001).

[586] F. P. Dos Santos, J. Léonard, J. Wang, C. J. Barrelet, F. Perales, E. Rasel, C. S. Unnikrishnan, M. Leduc and C. Cohen-Tannoudji, Bose-Einstein condensation of metastable helium, *Phys. Rev. Lett.* **86**, 3459–3462 (2001).

[587] R. Fitzgerald, Helium joins family of gaseous Bose-Einstein condensates, *Phys. Today* **54**, 13–14 (2001).

[588] B. G. Levi, Magnetic forces need not apply: Bose-Einstein condensates can be made in an optical trap, *Phys. Today* **54**, 20–22 (2001).

[589]* G. Modugno, G. Ferrari, G. Roati, R. J. Brecha, A. Simoni and M. Inguscio, Bose-Einstein condensation of potassium atoms by sympathetic cooling, *Science* **294**, 1320–1322 (2001).

[590] F. Schreck, G. Ferrari, K. L. Corwin, J. Cubizolles, L. Khaykovich, M.-O. Mewes and C. Salomon, Sympathetic cooling of bosonic and fermionic lithium gases towards quantum degeneracy, *Phys. Rev. A* **87**, 1402R–1405R (2001).

[591] K. M. O'Hara and J. E. Thomas, Standing room only at the quantum scale, *Science* **291**, 2556–2557 (2001).

[592] I. Bloch, M. Greiner, O. Mandel, T. W. Hänsch and T. Esslinger, Sympathetic cooling of ^{85}Rb and ^{87}Rb, *Phys. Rev. A* **64**, 1402R–1405R (2001).

[593] A. Görlitz, J. M. Vogels, A. E. Leanhardt, C. Raman, T. L. Gustavson, J. R. Abo-Shaeer, A. P. Chikkatur, S. Gupta, S. Inouye, T. Rosenband and W. Ketterle, Realization of Bose-Einstein condensates in lower dimensions, *Phys. Rev. Lett.* **87**, 130402-1–130402-4 (2001).

[594] D. S. Petrov, M. Holzmann and G. V. Shlyapnikov, Bose-Einstein condensation in quasi-2D trapped gases, *Phys. Rev. Lett.* **84**, 2551–2555 (2000).

[595] C. Seife, Quantum condensate gets a fresh squeeze, *Science* **293**, 2368 (2001).

[596]* F. S. Cataliotti, S. Burger, C. Fort, P. Maddaloni, F. Minardi, A. Trombettoni, A. Smerzi and M. Inguscio, Josephson junction arrays with Bose-Einstein condensates, *Science* **293**, 843–846 (2001).

[597] S. Giovanazzi, A. Smerzi and S. Fantoni, Josephson effects in dilute Bose-Einstein condensates, *Phys. Rev. Lett.* **84**, 4521–4524 (2000).

[598] A. Smerzi, Classical and quantum Josephson effects with Bose-Einstein condensates, in *Bose-Einstein Condensates and Atom Lasers*, eds. Martelucci *et al.*

[599] C. Orzel, A. K. Tuchman, M. L. Fenselau, M. Yasuda and M. A. Kasevich, Squeezed states in a Bose-Einstein condensate, *Science* **291**, 2386–2389 (2001).

[600] D. Voss, Doing the Bose Nova with your main squeeze, *News Focus Sci.* **291**, 2301–2303 (2001).

[601]* 1M. Greiner, O. Mandel, T. Esslinger, T. W. Hänsch and I. Block, Quantum phase transition from a superfluid to a Mott insulator in a gas of ultracold atoms, *Nature* **415**, 39–44 (2002).

[602] H. T. C. Stoof, Breaking up a superfluid, *Nature* **415**, 25–26 (2002).

[603] J. D. Weinstein and K. G. Libbrecht, Microscopic magnetic traps for neutral atoms, *Phys. Rev. A* **52**, 4004–4009 (1995).

[604] J. Reichel, W. Hänsel and T. W. Hänsch, Atomic micromanipulation with magnetic surface traps, *Phys. Rev. Lett.* **83**, 3398–3401 (1999).

[605] D. Cassettari, B. Hessmo, R. Folman, T. Maier and J. Schmiedmayer, Beam splitter for guided atoms, *Phys. Rev. Lett.* **85**, 6483–5487 (2000).

[606]* W. Hänsel, P. Hommelhoff, T. W. Hänsch and J. Reichel, Bose-Einstein condensation on a microelectronic chip, *Nature* **413**, 498–501 (2001).

[607] R. Folman and J. Schmiedmayer, Mastering the language of atoms, *Nature* **413**, 466–467 (2001). (Kluwer Academic/Plenum Publishers, 2000), pp. 249–263.

[608] A. E. Leanhardt, Y. Shin, A. P. Chikkatur, D. Kielpinski, W. Ketterle and D. E. Pritchard, Bose-Einstein condensates near a microfabricated surface, *Phys. Rev. Lett.* **90**, 100404 (2000).

[609] B. Paredes, A. Widera, V. Murg, O. Mandel, S. Fölling, I. Cirac, G. V. Shlyapnikov, T. W. Hänsch and I. Bloch, Tonks-girardeau gas of ultracold atoms in an optical lattice, *Nature* **429**, 277–281 (2004).

[610] T. Kinoshita, T. Wenger and D. S. Weiss, Observation of a one-dimensional tonks-girardeau gas, *Sciencexpress*/www.sciencexpress.org/29 July 2004/ 1–10 (2004).

[611]* S. R. Grenade, M. E. Gehm, K. M. O'Hara and J. E. Thomas, All optical production of a degenerate fermi gas, *Phys. Rev. Lett.* **88**, 120465-1 (2002).

[612] K. M. O'Hara, M. E. Gehm, S. R. Granada and J. E. Thomas, Scaling laws for evaporative cooling in time-dependent optical traps, *Phys. Rev. A* **64**, 051403 (2001).

[613]* T. Weber, J. Herbig, M. Mark, H.-C. Nägerl and R. Grimm, Bose-Einstein condensation of cesium, *Science* **299**, 232–235 (2003).

[614] T. Kraemer, J. Herbig, M. Mark, T. Webver, C. Chin, H.-C. Nägerl and R. Grimm, Optimized production of Bose-Einstein condensate, *App. Phys. B Lasers Opt.* **79**, 1013–1019 (2004).

[615] A. E. Leanhardt, T. A. Pasquini, M. Saba, A. Schirotzek, Y. Shin, D. Kielpinski, D. E. Pritchard and W. Ketterle, Cooling Bose-Einstein condensates below 500 picokelvin, *Science* **301**, 1513–1515 (2003).

[616] A. P. Chikkatur, Y. Shin, A. E. Leanhardt, D. Kielpinski, E. Tsikata, T. L. Gustavson, D. E. Pritchard and W. Ketterle, A continuous source of Bose-Einstein condensed atoms, *Science* **296**, 2193–2195 (2002).

[617] T. L. Gustavson, A. P. Chikkatur, A. E. Leanhardt, A Görlitz, S. Gupta, D. E. Pritchard and W. Ketterle, Transport of Bose–Einstein condensates with optical tweezers, *Phys. Rev. Lett.* **88**, 020401 (2002).

[618] T. Lahaye, J. M. Vogels, K. J. Günter, Z. Wang, J. Dalibard and D. Guéry-Odelin, Realization of a magnetically guided atomic beam in the collisional regime, *Phys. Rev. Lett.* **93**, 093003 (2004).

[619] E. M. Purcell, Spontaneous emission probabilities at radio frequencies, *Phys. Rev.* **69**, 681 (1946).

[620] D. Kleppner, Inhibited spontaneous emission, *Phys. Rev. Lett.* **47**, 233–236 (1981).

[621] D. Meschede, H. Walther and G. Müller, One-atom maser, *Phys. Rev. Lett.* **54**, 551–554 (1985).

[622] M. Brune, J. M. Raimond, P. Goy, L. Davidovich and S. Haroche, Realization of a two-photon maser oscillator, *Phys. Rev. Lett.* **59**, 1899–1902 (1987).

[623] Z. Hu and H. J. Kimble, Observation of a single atom in a magneto-optical trap, *Opt. Lett.* **19**, 1888–1890 (1994).

[624] Q. A. Turchette, C. J. Hood, W. Lange, H. Mabuchi and H. J. Kimble, Measurement of conditional phase shifts for quantum logic, *Phys. Rev. Lett.* **75**, 4710–4713 (1995).

[625] H. Mabuchi, Q. A. Turchette, M. S. Chapman and H. J. Kimble, Real-time detection of individual atoms falling through a high-finesse optical cavity, *Opt. Lett.* **21**, 1393–1395 (1996).

[626] J. I. Cirac, P. Zoller, H. J. Kimble and H. Mabuchi, Quantum state transfer and entanglement distribution among distant nodes in a quantum network, *Phys. Rev. Lett.* **78**, 3221–3224 (1997).

[627] G. Hechenblaikner, M. Gangl, P. Horak and H. Ritsch, Cooling an atom in a weakly driven high-Q cavity, *Phys. Rev. A* **58**, 3030–3042 (1998).

[628] C. J. Hood, M. S. Chapman, T. W. Lynn and H. J. Kimble, Real-time cavity QED with single atoms, *Phys. Rev. Lett.* **80**, 4157–4160 (1998).

[629] B. Deb and G. Kurizki, Formation of giant quasibound cold diatoms by strong atom-cavity coupling, *Phys. Rev. Lett.* **83**, 714–717 (1999).

[630] B. G. Levi, An atom is trapped by the field of just one photon, *Phys. Today* **53**, 19–22 (2000).

[631] P. Zoller, Tricks with a single photon, *News Views Nat.* **404**, 340–341 (2000).

[632] J. McKeever, A. Boca, A. D. Boozer, J. R. Buck and H. J. Kimble, Experimental realization of a one-atom laser in the regime of strong couplings, *Nature* **425**, 268–271 (2003).

[633] H. Carmichael and L. A. Orozco, Single atom laser orderly light, *Nature* **425**, 246 (2003).

[634] P. Maunz, T. Puppe, I. Schuster, N. Syassen, P. W. H. Pinkse and G. Rempe, Cavity cooling of a single atom, *Nature* **428**, 50–52 (2004).

[635] J. McKeever, A. Boca, A. D. Boozer, R. Miller, J. R. Buck, A. Kuzmich and H. J. Kimble, Deterministic generation of single photons from one atom trapped in a cavity, *Science* **303**, 1992–1994 (2004).

[636] D. Frese, B. Ueberholz, S. Kuhr, W. Alt, D. Schrader, V. Gomer and D. Meschede, Single atoms in an optical dipole trap: Towards a deterministic source of cold atoms, *Phys. Rev. Lett.* **85**, 3777–3780 (2000).

[637] S. Kuhr, W. Alt, D. Schrader, M. Müller, V. Gomer and D. Meschede, Deterministic delivery of a single atom, *Science* **293**, 278–280 (2001).

[638] N. Schlosser, G. Reymond, I. Protsenko and P. Grangler, Sub-poissonian loading of single atoms in a microscopic dipole trap, *Nature* **411**, 1024–1026 (2001).

[639] J. E. Williams and M. J. Holland, Preparing topological states of a Bose-Einstein condensate, *Nature* **401**, 568–572 (1999).

[640] M. R. Matthews, B. P. Anderson, P. C. Haljan, D. S. Hall, C. E. Wieman and E. A. Cornell, Vortices in a Bose-Einstein condensate, *Phys. Rev. Lett.* **83**, 2498–2501 (1999).

[641] J. J. García-Ripoll and V. M. Pérez-García, Stable and unstable vortices in multicomponent Bose-Einstein condensates, *Phys. Rev. Lett.* **84**, 4264–4267 (2000).

[642] C. Raman, M. Kohl, R. Onofrio, D. S. Durfee, C. E. Kuklewicz, Z. Hadzibabic and W. Ketterle, Evidence for a critical velocity in a Bose-Einstein condensed gas, *Phys. Rev. Lett.* **83**, 2502–2505 (1999).

[643] D. S. Rokhsar, Condensates in a twist, *News Views Nat.* **401**, 533–534 (1999).

[644] B. G. Levi, Researchers put a new spin on Bose-Einstein condensates, search and discovery, *Phys. Today* **52**, 17–18 (1999).

[645] K. W. Madison, F. Chevy, W. Wohlleben and J. Dalibard, Vortex formation in a stirred Bose-Einstein condensate, *Phys. Rev. Lett.* **84**, 806–809 (2000).

[646]* R. Fitzgerald, An optical spoon stirs up vortices in a Bose-Einstein condensate, *Phys. Today* **53**, 19–21 (2000).

[647] O. M. Maragò, S. A. Hopkins, J. Arlt, E. Hodby, G. Hechenblaikner and C. J. Foot, Observation of the scissors mode and evidence for superfluidity of a trapped Bose-Einstein condensed gas, *Phys. Rev. Lett.* **84**, 2056–2059 (2000).

[648] D. Guery-Odelin and S. Stringari, Scissors mode and superfluidity of a trapped Bose-Einstein condensed gas, *Phys. Rev. Lett.* **83**, 4452–4455 (1999).

[649] A. P. Chikkatur, A. Görlitz, D. M. Stamper-Kurn, S. Inouye, S. Gupta and W. Ketterle, Suppression and enhancement of impurity scattering in a Bose-Einstein condensate, *Phys. Rev. Lett.* **85**, 483–486 (2000).

[650] S. Inouye, A. P. Chikkatur, D. M. Stamper-Kurn, J. Stenger, D. E. Pritchard and W. Ketterle, Superradiant Rayleigh scattering from a Bose-Einstein condensate, *Science* **285**, 571–574 (1999).

[651] A. I. Lvovsky, S. R. Hartmann and F. Moshary, Omnidirectional superfluorescence, *Phys. Rev. Lett.* **82**, 4420–4423 (1999).

[652] R. Onofrio, C. Raman, J. M. Vogels, J. R. Abo-Shaeer, A. P. Chikkatur and W. Ketterle, Observation of superfluid flow in a Bose-Einstein condensed gas, *Phys. Rev. Lett.* **85**, 2228–2231 (2000).

[653]* J. R. Abo-Shaeer, C. Raman, J. M. Vogels and W. Ketterle, Observation of vortex lattices in Bose-Einstein condensates, *Science* **292**, 476–479 (2001).

[654] F. Chevy, K. W. Madison and J. Dalibard, Measurement of the angular momentum of a rotating Bose-Einstein condensate, *Phys. Rev. Lett.* **85**, 2223–2227 (2000).

[655] F. Zambelli and S. Stringari, Quantized vortices and collective oscillations of a trapped Bose-Einstein condensate, *Phys. Rev. Lett.* **81**, 1754–1757 (1998).

[656] D. Butts and D. Rokhsar, Predicted signatures of rotating Bose-Einstein condensates, *Nature* **397**, 327–329 (1999).

[657]* B. P. Anderson, P. C. Haljan, C. E. Wieman and E. A. Cornell, Vortex precession in Bose-Einstein condensates: Observations with filled and empty cores, *Phys. Rev. Lett.* **85**, 2857–2860 (2000).

[658] A. A. Svidzinsky and A. L. Fetter, Stability of a vortex in a trapped Bose-Einstein condensate, *Phys. Rev. Lett.* **84**, 5919–5923 (2000).

[659] J. Tempere and J. T. Devreese, Vortex dynamics in a parabolically confined Bose-Einstein condensate, *Solid State Commun.* **113**, 471–474 (2000).

[660] E. Lundh and P. Ao, Hydrodynamic approach to vortex lifetimes in trapped Bose condensates, *Phys. Rev. A* **61**, 3612–3618 (2000).

[661] J. Denschlag, J. E. Simsarian, D. L. Feder, C. W. Clark, L. A. Collins, J. Cubizolles, L. Deng, E. W. Hagley, K. Helmerson, W. P. Reinhardt, S. L. Rolston, B. I. Schneider and W. D. Phillips, Generating solitons by phase engineering of a Bose-Einstein condensate, *Science* **287**, 97–101 (2000).

[662] S. Burger, K. Bongs, S. Dettmer, W. Ertmer, K. Sengstock, A. Sanpera, G. V. Shlyapnikov and M. Lewenstein, Dark solitons in Bose-Einstein condensates, *Phys. Rev. Lett.* **83**, 5198–5201 (1999).

[663] T. Seideman, Manipulating external degrees of freedom with intense light: Laser focusing and trapping of molecules, *J. Chem. Phys.* **106**, 2881–2892 (1997).

[664] T. Seideman, Shaping molecular beams with intense light, *J. Chem. Phys.* **107**, 10420–10429 (1997).

[665] B. Friedrich and D. Herschbach, Alignment and trapping of molecules in intense laser fields, *Phys. Rev. Lett.* **74**, 4623–4626 (1995).

[666] B. Friedrich and D. Herschbach, Polarization of molecules induced by intense nonresonant laser fields, *J. Phys. Chem.* **99**, 15686–15693 (1995).

[667] L. A. Rahn, R. L. Farrow, M. L. Koszykowski and P. L. Mattern, Observation of an optical stark effect on vibrational and rotational transitions, *Phys. Rev. Lett.* **45**, 620–623 (1980).

[668]* T. Takekoshi, B. M. Patterson and R. J. Knize, Observation of optically trapped cold cesium molecules, *Phys. Rev. Lett.* **81**, 5105–5108 (1998).

[669] A. Fioretti, D. Comparat, A. Crubellier, O. Dulieu, F. Masnou-Seeuws and P. Pillet, Formation of cold cs_2 molecules through photoassociation, *Phys. Rev. Lett.* **80**, 4402–4405 (1998).

[670] J. D. Weinstein, R. deCarvalho, T. Guillet, B. Friedrich and J. M. Doyle, Magnetic trapping of calcium monohydride molecules at millikelvin temperatures, *Nature* **395**, 148–150 (1998).

[671] G. B. Lubkin, Molecules are magnetically trapped, *Phys. Today* **51**, 19–20 (1998).

[672] P. S. Julienne, K. Burnett, Y. B. Band and W. C. Stwalley, Stimulated Raman molecule production in Bose-Einstein condensates, *Phys. Rev. A* **58**, R797–R800 (1998).

[673] J. Karczmarek, J. Wright, P. Corkum and M. Ivanov, Optical centrifuge for molecules, *Phys. Rev. Lett.* **82**, 3420–3423 (1999).

[674] S. Chelkowski, A. D. Bandrauk and P. B. Corkum, Efficient molecular dissociation by a chirped ultrashort infrared laser pulse, *Phys. Rev. Lett.* **65**, 2355–2358 (1990).

[675] W.-K. Liu, B. Wu and J.-M. Yuan, Nonlinear dynamics of chirped pulse excitation and dissociation of diatomic molecules, *Phys. Rev. Lett.* **75**, 1292–1295 (1995).

[676] H. L. Bethlem, G. Berden and G. Meijer, Decelerating neutral dipolar molecules, *Phys. Rev. Lett.* **83**, 1558–1561 (1999).

[677] H. L. Bethlem, G. Berden, A. J. A. van Roij, F. M. H. Crompvoets and G. Meijer, Trapping neutral molecules in a traveling potential well, *Phys. Rev. Lett.* **84**, 5744–5747 (2000).

[678] W. J. Kleen, *Electronics of Microwave Tubes*, Chapt. 18 (Academic Press, 1958).

[679]* J. M. Doyle and B. Friedrich, Molecules are cool, *Nature* **401**, 749–751 (1999).

[680] J. A. Maddi, T. P. Dinneen and H. Gould, Slowing and cooling molecules and neutral atoms by time-varying electric-field gradients, *Phys. Rev. A* **60**, 3882–3891 (1999).

[681] D. Herschbach, Chemical physics: Molecular clouds, clusters, and corrals, *Rev. Mod. Phys.* **71**, S411–S418 (1999).

[682] H. L. Bethlem, G. Berden, F. M. H. Crompvoets, R. T. Jongma, A. J. A. van Roij and G. Meijer, Electrostatic trapping of ammonia molecules, *Nature* **406**, 491–494 (2000).

[683] R. Wynar, R. S. Freeland, D. J. Han, C. Ryu and D. J. Heinzen, Molecules in a Bose-Einstein condensate, *Science* **287**, 1016–1019 (2000).

[684]* C. J. Williams and P. S. Julienne, Molecules at rest, *Science* **287**, 986–987 (2000).

[685] D. J. Heinzen, R. Wynar, P. D. Drummond and K. V. Kheruntsyan, Superchemistry: Dynamics of coupled atomic and molecular Bose-Einstein condensates, *Phys. Rev. Lett.* **84**, 5029–5033 (2000).

[686] K. Góral, M. Gajda and K. Rzażewski, Multimode dynamics of a coupled ultracold atomic-molecular system, *Phys. Rev. Lett.* **86**, 1397–1401 (2001).

[687] H. T. C. Stoof, M. Houbiers, C. A. Sackett and R. G. Hulet, Superfluidity of spin-polarized ^{6}Li, *Phys. Rev. Lett.* **76**, 10–13 (1996).

[688] M. Houbiers, R. Ferwerda, H. T. C. Stoof, W. I. McAlexander, C. A. Sackett and R. G. Hulet, Superfluid state of atomic ^{6}Li in a magnetic trap, *Phys. Rev. A* **56**, 4864–4878 (1997).

[689] M. Houbiers, H. T. C. Stoof, W. I. McAlexander and R. G. Hulet, Elastic and inelastic collisions of ^{6}Li atoms in magnetic and optical traps, *Phys. Rev. A* **57**, R1497–R1500 (1998).

[690] W. Geist, L. You and T. A. B. Kennedy, Sympathetic cooling of an atomic Bose-Fermi gas mixture, *Phys. Rev. A* **59**, 1500–1508 (1999).

[691] H. Houbiers and H. T. C. Stoof, Cooper-pair formation in trapped atomic fermi gases, *Phys. Rev. A* **59**, 1556–1561 (1999).

[692] G. Ferrari, Collisional relaxation in a fermionic gas, *Phys. Rev. A* **59**, R4125–R4128 (1999).

[693] E. Timmermans and R. Côté, Superfluidity in sympathetic cooling with atomic Bose-Einstein condensates, *Phys. Rev. Lett.* **80**, 3419–3423 (1998).

[694] B. G. Levi, The fermionic cousin of a Bose-Einstein condensate makes its debut, *Search Discov. Phys. Today* **52**, 17–18 (1999).

[695]* K. M. O'Hara, M. E. Gehm, S. R. Granade, S. Bali and J. E. Thomas, Stable, strongly attractive, two-state mixture of lithium fermions in an optical trap, *Phys. Rev. Lett.* **85**, 2092–2095 (2000).

[696] K. M. O'Hara, S. L. Hemmer, S. R. Granada, M. E. Gehm and J. E. Thomas, Measurement of the zero crossing in a Feshbach resonance of fermionic ^{6}Li, *Phys. Rev. A* **66**, 041401 (2002).

[697] A. G. Truscott, K. E. Strecker, W. I. McAlexander, G. B. Partridge and R. G. Hulet, Observation of fermi pressure in a gas of trapped atoms, *Science* **291**, 2570–2572 (2001).

[698] K. M. O'Hara, S. L. Hemmer, M. E. Gehm, S. R. Granada and J. E. Thomas, Observation of a strongly interacting degenerate fermi gas of atoms, *Science* **298**, 2179–2182 (2002).

[699] M. Greiner, C. A. Regal and D. S. Jin, Emergence of a molecular Bose-Einstein condensate from a fermi gas, *Nature* **426**, 537–540 (2003).

[700] S. Johim, M. Bartenstein, A. Altmeyer, G. Hendl, S. Richl, C. Chin, J. Hecker Denschlag and R. Grimm, Bose-Einstein condensation of molecules, *Science* **302**, 2101–2103 (2003).

[701]* C. A. Regal, M. Greiner and D. S. Jin, Observation of resonance condensation of fermionic atom pairs, *Phys. Rev. Lett.* **92**, 040403 (2004).

[702]* J. Kinast, S. L. Hemmer, M. E. Gehm, A. Turlapov and J. E. Thomas, Evidence for superfluidity in a resonantly interacting fermi gas, *Phys. Rev. Lett.* **92**, 150402 (2004).

[703] M. Bartenstein, A. Altmeyer, S. Riedl, S. Jochim, C. Chin, J. Hecker Denschlag and R. Grimm, Collective excitations of a degenerate gas at the BEC-BCS crossover, *Phys. Rev. Lett.* **92**, 203201 (2004).

[704] C. Chin, M. Bartenstein, A. Altmeyer, S. Reidl, S. Jochim, J. Hecker Denschlag and R. Grimm, Observation of the pairing gap in a strongly interacting fermi gas, *Science* **305**, 1128–1130 (2004).

[705] J. Kinnunen, M. Rodriquez, P. Törmä, Pairing gap and in-gap excitations in trapped fermionic superfluids, *Science* **305**, 1131–1133 (2004).

[706] J. Kinast, A. Turlapov, J. E. Thomas, Q. Chen, J. Stajic and K. Levin, Heat capacity of a strongly-interacting fermi gas, *Science* **27**, 0502087 (2005).

[707] L. Pitaevskii and S. Strigari, The quest for superfluidity in fermi gases, *Science* **298**, 2144–2146 (2002).

[708] B. Goss Levi, Ultracold fermionic atoms team up as molecules: Can they form cooper pairs as well? *Phys. Today* 18–20 (2003).

[709] K. E. Strecker, G. B. Partridge and R. G. Hulet, Conversion of an atomic fermi gas to a long-lived molecular Bose gas, *Phys. Rev. Lett.* **91**, 080406 (2003).

[710] M. W. Zwierlein, C. A. Stan, C. H. Schunk, S. M. F. Raupach, S. Gupta, Z. Hadzibabic and W. Ketterle, Observation of Bose-Einstein condensation of molecules, *Phys. Rev. Lett.* **91**, 250401 (2003).

[711] A. Cho, Ultracold atoms spark a hot race, *Science* **301**, 750–752 (2003).

[712] T.-L. Ho, Arrival of the fermion superfluid, *Science* **305**, 1114–1115 (2004).

[713] C. Seife, Energy curve confirms paired-up fermi condensate, *Science* **305**, 459–460 (2004).

[714] B. G. Levi, Fermionic atoms appear to pair up much as electrons do in a superconductor, *Phys. Today* 21–23 (2004).

[715] V. G. Minogin and Yu. V. Rozhdestvenskii, Stable localization of atoms in a standing wave field, *Opt. Commun.* **64**, 172–174 (1987).

[716] K. C. Neuman and S. M. Block, Optical trapping, *Rev. Sci. Inst.* **75**, 2787–2809 (2004).

[717]* E. A. Abbondanzieri, W. J. Greenleaf, J. W. Shaevitz, R. Landick and S. M. Block, Direct observation of base-pair stepping by RNA polymerase, *Nature* **438**, 460–465 (2005).

[718] W. J. Greenleaf, M. T. Woodside, E. A. Abbondanzieri and S. M. Block, Passive all-optical force clamp for high-resolution laser trapping, *Phys. Rev. Lett.* **95**, 208102 (2005).

[719] M. W. Zwierlein, J. R. Abo-Shaeer, A. Schirotzek, C. H. Schunk and W. Ketterle, Vortices and superfluidity in a strongly interacting Fermi gas, *Nature* **435**, 1047–1051 (2005).

[720]* R. Grimm, A quantum revolution, *Nature* **435**, 1035–1036 (2005).

[721] M. W. Zwierlein, A. Schirotzek, C. H. Schunk and W. Ketterle, Fermionic superfluidity with imbalanced spin populations, *Science* **311**, 492–496 (2006).

[722] G. B. Partridge, W. Li, R. I. Kamar, Y.-A. Liao and R. G. Hulet, Pairing and phase separation in a polarized Fermi gas, *Science* **311**, 503–505 (2006).

[723] A. Cho, Mismatched cold atoms hint at a Stellar new superfluid, *Science* **310**, 1892 (2005).

Acknowledgments

I would especially like to thank my wife, Aline, and our three children for putting up with an often-preoccupied husband and father all these years. As far as this book is concerned, it could never have made it to publication without my wife's assistance. Aline not only read and typed every word, she helped greatly to clarify the presentation. It is a pleasure to dedicate this book to her.

I would like to particularly acknowledge and thank two close colleagues, John Bjorkholm and Jim Gordon, for their help during the crucial discovery years of atom trapping in the 1970–1980 decade. John's experimental skills and Jim's theoretical insights contributed substantially to the founding of the field.

Joe Dziedzic deserves special thanks for his roughly 35 years of experimental collaboration with me, even predating the work on optical trapping and manipulation of particles. He started his career, as a navy-trained technician. In mid-career, he took courses and became an electrical engineer. By the time he retired he had risen to be a member of technical staff and was a colleague in all our joint experiments on trapping of macroscopic and biological particles. Without his help I never would have achieved as much as I did. Together we shared the Rank Prize for the experiments on living cells with optical tweezers.

I would also like to thank Herwig Kogelnik, one of my former directors who was particularly supportive of this work. I also value the memory of Rudi Kompfner, who, as my director in the early days of trapping, enthusiastically encouraged my efforts. I would also like to express my gratitude for the support of the late Chapin Cutler, yet another former director who also endorsed and encouraged my endeavors.

Biography of Arthur Ashkin

Arthur Ashkin was born in Brooklyn, New York, on September 2, 1922. He received an A. B. degree in physics from Columbia College in 1947. During World War II he served three years in the army, mostly at the Columbia Radiation Laboratory. This laboratory, founded by Prof. I. I. Rabi from a nucleus of his former atomic beam students J. M. B. Kellogg, S. Millman, P. Kusch, and the then instructor W. E. Lamb to work on short wavelength microwave magnetrons. Ashkin, as a college student, was their first technician. He ended up as a staff member responsible for the design of the AX9 megawatt 3 cm Rising Sun magnetron. After the war, in 1947, he went to Cornell as a graduate student at the Nuclear Studies Laboratory. His doctoral thesis, with Dr. W. M. Woodward, involved experiments on positron-electron scattering, in which relativistic effects were first seen. He was fortunate to have taken courses from a group of young faculty members, including such now famous scientists as Richard Feynman, Hans Bethe, Robert Wilson, Philip Morrison, and others formerly from the Los Alamos project.

In 1952 Ashkin joined the technical staff of AT&T Bell Laboratories, where he worked for about 10 years on microwave tubes and electron beam and low noise cyclotron-wave parametric amplifiers. At that time this laboratory, led by luminaries such as Sid Millman, John Pierce, Rudi Kompfner, Cal Quate, and Chapin Cutler, was arguably the premier microwave research laboratory in the world.

In 1963, after the invention of the laser, Ashkin became head of the Laser Science Research Department. His personal research at the time centered on nonlinear optics. With Gary Boyd, he demonstrated the first cw second harmonic and parametric amplification. In 1966, in the course of this work they also discovered optically induced changes in the refractive index of electrooptic crystals such as $LiNbO_3$, now known as the photorefractive effect. This discovery led to literally thousands of papers on applications, such as: holographic storage, soliton generation, four-wave nonlinear mixing and amplification, and optical phase conjugation. In the early 1970s, Ashkin participated with his departmental colleagues, Erich Ippen and Rogers Stolen, to found the field of nonlinear optics in fibers. To this day, fiber nonlinear optics is a very active field of research, due to the major role it plays in understanding the residual noise in optical communications fibers and in optical soliton generation in high capacity communications systems.

In 1970 Ashkin discovered how to use the forces of radiation pressure to stably trap dielectric particles. He was the first to suggest stable atom trapping and, in the course of the next ten years, experimentally demonstrated, with John Bjorkholm and Richard Freeman, large radiation pressure forces on atoms. He proposed the use of just a single focused Gaussian beam as an atom trap. This was the single-beam gradient or dipole trap. Atom cooling and single-beam dipole trapping was accomplished in 1985 and 1986 with Steve Chu and others at Bell Labs. Trapping and manipulation of living cells was demonstrated with Joe Dziedzic in 1987. The subfields of atom trapping and cooling and laser trapping and manipulation of biological particles and living cells have grown enormously in recent years and are the principal subjects of this book. The unique features of manipulation by optical forces have led to the remarkable range of applications discussed in the text.

Ashkin has published a total of 105 papers and was granted a total of 47 patents. Of these, 61 papers and six patents are on radiation pressure on small particles.

Ashkin has been granted a number of awards and honors for his work, as listed below:

As a student at Columbia College, he was elected to Phi Beta Kappa in his junior year, and he was awarded the Michaelis Prize in Physics on graduation.

He is a Fellow of the American Physical Society, the American Optical Society, the Institute of Electrical and Electronic Engineers, and the American Association for the Advancement of Science.

He was elected a member of the National Academy of Engineering in 1984 and of the National Academy of Science in 1996.

He received the Quantum Electronics Award of the Laser and Electro-Optic Society of the IEEE in 1987, the Charles Hard Townes Award of the Optical Society of America in 1988, the Rank Prize in Opto-Electronics in 1993, the Frederic Ives Medal/Jarus W. Quinn Award of the Optical Society of America in 1998, the Keithley Award of the American Physical Society in 2003, and the Harvey Prize of Technion — the Israel Institute of Technology in 2004.

The citations, or excerpts of the citations, for these awards are as follows:
The citation for the 1987 Quantum Electronics Award reads:

> For his seminal experimental and theoretical work which initiated worldwide study of laser radiation pressure and for his continuing exceptional contributions to the development of the field.

The citation for the 1988 Townes Award is almost the same as above:

> …in recognition of his original, creative, experimental and theoretical work which initiated the study of laser radiation pressure…

The citation for the 1993 Rank Prize in Opto-Electronics reads, in part, as follows:

> …ability to trap and manipulate living micro-organisms, non-destructively and without making potentially damaging physical contact with them… This is achieved using a new family of opto-electronic instruments, aptly named "optical tweezers…

Optical tweezers, deriving directly from this work, have also recently been used, by others, to trap and manipulate single strands of DNA. This extends the applications of these opto-electronic instruments into the field of molecular biology and illustrates their considerable further potential to influence future developments in biology and medical science for the well being of mankind.

The citation for the 1998 Ives Medal/Quinn Award reads, in part, as follows:

For his pioneering work on the manipulation of particles with light, including the invention of the "optical tweezers" trap and his studies of radiation forces on atoms...

The citation for the 2003 Keithley Award reads as follows:

For theoretical and experimental contributions to the understanding of laser cooling and trapping of atoms and particles, for demonstrating the optical gradient forces on atoms and the trapping of atoms with light, and for inventing optical tweezers and showing how they can be used to measure the physical forces generated by biological molecular motors.

The citation for the 2004 Harvey Prize reads as follows:

In recognition of his pioneering research on manipulation of particles by laser light forces, including the invention of optical tweezers, which revolutionized atomic and biological physics, and for his basic contributions to nonlinear optics.

List of Reprints

[1]*   Ashkin, see p. 441

[2]*   Ashkin, see p. 445

[3]*   Ashkin, see p. 449

[4]*   Chu, et al., see p. 453

[5]*   Ashkin, et al., see p. 457

[7]*   Ashkin and Dziedzic, see p. 460

[10]*   Ashkin, see p. 463

[12]*   Cyranoski, see p. 479

[17]*   Ashkin, et al., see p. 482

[19]*   Ashkin and Dziedzic, see p. 485

[20]*   Ashkin, et al., see p. 490

[21]*   Neuman, et al., see p. 493

[26]*   Hänsch and Schawlow, see p. 501

[27]*   Chu, et al., see p. 503

[29]*   Anderson, et al., see p. 507

[30]*   Davis, et al., see p. 511

[31]*   Andrews, et al., see p. 516

[33]*   Bloch, et al., see p. 521

[34]*   Hagley, et al., see p. 525

[35]*   DeMarco and Jin, see p. 529

[43]*   Bjorkholm, et al., see p. 533

[44]*   Ashkin and Dziedzic, see p. 537

[45]*   Ashkin and Gordon, see p. 541

[46]*   Ashkin, see p. 544

[49]*   Ashkin, see p. 550

[58]*   Ashkin and Dziedzic, see p. 564

[61]*   Finer, et al., see p. 567

[63]*   Hood, et al., see p. 574

[65]*   Ashkin and Dziedzic, see p. 581

[66]*   Ashkin and Dziedzic, see p. 584

[68]*   Ashkin and Dziedzic, see p. 587

[92]*   Ashkin and Dziedzic, see p. 591

[97]*   Ashkin and Dziedzic, see p. 595

[109]*   Ashkin, see p. 607

[123]*   Gordon and Ashkin, see p. 615

[130]*   Higurashi, et al., see p. 627

[136]*   Ashkin and Gordon, see p. 629

[139]*   Ido, et al., see p. 632

[140]*   Pearson, et al., see p. 636

[142]*   Ashkin, see p. 639

[143]*   Ashkin and Dziedzic, see p. 642

[145]*   Adams, et al., see p. 646

[154]*   Bjorkholm, et al., see p. 650

[166]*   Ashkin, et al., see p. 657

[171]*   Chu, et al., see p. 660

[175]*   Davis, et al., see p. 663

[181]*   Raab, et al., see p. 668

[182]*   Walker, et al., see p. 672

[183]*   Ketterle, et al., see p. 676

[190]*   Deng, et al., see p. 680

[205]*   Barrett, et al., see p. 683

[212]*   Yin, et al., see p. 687

Volume 24, Number 4   PHYSICAL REVIEW LETTERS   26 January 1970

ACCELERATION AND TRAPPING OF PARTICLES BY RADIATION PRESSURE

A. Ashkin

Bell Telephone Laboratories, Holmdel, New Jersey 07733

(Received 3 December 1969)

Micron-sized particles have been accelerated and trapped in stable optical potential wells using only the force of radiation pressure from a continuous laser. It is hypothesized that similar accelerations and trapping are possible with atoms and molecules using laser light tuned to specific optical transitions. The implications for isotope separation and other applications of physical interest are discussed.

This Letter reports the first observation of acceleration of freely suspended particles by the forces of radiation pressure from cw visible laser light. The experiments, performed on micron-sized particles in liquids and gas, have yielded new insights into the nature of radiation pressure and have led to the discovery of stable optical potential wells in which particles were trapped by radiation pressure alone. The ideas can be extended to atoms and molecules where one can predict that radiation pressure from tunable lasers will selectively accelerate, trap, or separate the atoms or molecules of gases because of their large effective cross sections at specific resonances. The author's interest in radiation pressure from lasers stems from a realization of the large magnitude of the force, and the observation that it could be utilized in a way which avoids disturbing thermal effects. For instance a power $P = 1$ W of cw argon laser light at $\lambda = 0.5145$ μm focused on a lossless dielectric sphere of radius $r = \lambda$ and density $= 1$ gm/cc gives a radiation pressure force $F_{rad} = 2qP/c = 6.6 \times 10^{-5}$ dyn, where q, the fraction of light effectively reflected back, is assumed to be of order 0.1. The acceleration $= 1.2 \times 10^{8}$ cm/sec^2 $\cong 10^5$ times the acceleration of gravity.

Historically,[1,2] the main problem in studying radiation pressure in the laboratory has been the obscuring effects of thermal forces. These are caused by temperature gradients in the medium surrounding an object and, in general, are termed radiometric forces.[3] When the gradients are caused by light, and the entire particle moves, the effect is called photophoresis.[3,4] These forces are usually orders of magnitude larger than radiation pressure. Even with lasers, photophoresis usually completely obscures radiation pressure.[5] In our work, radiometric effects were avoided by suspending relatively transparent particles in relatively transparent media. We operated free of thermal effects at 10^3 times the power densities of Ref. 5.

The first experiment used transparent latex spheres[6] of 0.59-, 1.31-, and 2.68-μm diam freely suspended in water. A TEM$_{00}$-mode beam of an argon laser of radius $w_0 = 6.2$ μm and $\lambda = 0.5145$ μm was focused horizontally through a glass cell 120 μm thick and manipulated to focus on single particles. See Fig. 1(a). Results were observed with a microscope. If a beam with milliwatts of power hits a 2.68-μm sphere off center, the sphere is simultaneously <u>drawn in to the beam axis</u> and <u>accelerated in the direction of the light</u>. It moves with a limiting velocity of microns per second until it hits the front surface of the glass cell where it remains <u>trapped</u> in the beam. If the beam is blocked, the sphere wanders off by Brownian motion. Similar effects occur with the other sphere sizes but more power is required for comparable velocities. When mixed, one can accelerate 2.68-μm spheres and leave 0.585-μm spheres behind. The particle velocities and the trapping on the beam axis can

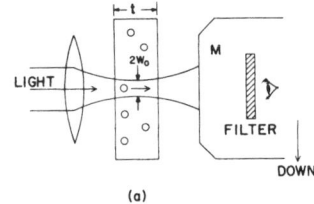

(a)

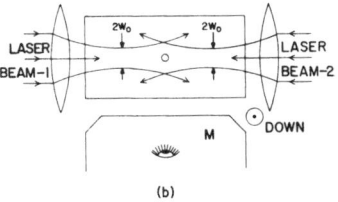

(b)

FIG. 1. (a) Geometry of glass cell, $t = 120$ μm, for observing micron particle motions in a focused laser beam with a microscope M. (b) The trapping of a high-index particle in a stable optical well. Note position of the TEM$_{00}$-mode beam waists.

VOLUME 24, NUMBER 4   PHYSICAL REVIEW LETTERS   26 JANUARY 1970

be understood as follows (see Fig. 2): The sphere of high index $n_H = 1.58$ is situated off the beam axis, in water of lower index $n_L = 1.33$. Consider a typical pair of rays symmetrically situated about the sphere axis B. The stronger ray (a) undergoes Fresnel reflection and refraction (called a deflection here) at the input and output faces. These result in radiation pressure forces F_R^i, F_R^o (the input and output reflection forces), and F_D^i, F_D^o (the input and output deflection forces), directed as shown. Although the magnitudes of the forces vary considerably with angle Φ, qualitatively the results are alike for all Φ. The radial (r) components of F_D^i, F_D^o are much larger than F_R^i and F_R^o (by ~10 at $\Phi = 25°$). All forces give accelerations in the $+z$ direction. F_R^i and F_R^o cancel radially to first order. F_D^i and F_D^o add radially in the $-r$ direction, thus the net radial force for the stronger ray is <u>inward</u> toward higher light intensity. Similarly the symmetrical weak ray (b) gives a net force along $+z$ and a net <u>outward but weaker</u> radial force. Thus the sphere as a whole is accelerated <u>inward</u> and <u>forward</u> as observed. To compute the z component of the force for a sphere on axis, one integrates the perpendicular (s) and parallel (p) components of the plane-polarized beam over the sphere. This yields an effective $q = 0.062$. This geometric optic result (neglecting diffraction) is identical with the asymptotic limit of a wave analysis by Debye[2] for an incident plane wave. He finds $q = 0.06$. From the force we get the limiting velocity v in a viscous medium using Stokes's law. For $r \ll w_0$,

$$v = 2qPr/3c\pi w_0^2 \eta, \qquad (1)$$

where η is the viscosity. For $P = 19$ mW, $w_0 = 6.2$ μm, and a sphere of $r = 1.34$ μm in water ($\eta = 1 \times 10^{-2}$ P), one computes $v = 29$ μm/sec. We measured $v = 26 \pm 5$ μm/sec which is good agreement. In the above, the sphere acts as a focusing lens. If one reverses the relative magnitudes of the indices of the media, the sphere becomes a diverging lens, the sign of the radial deflection forces reverse, and the sphere should be pushed <u>out</u> of the beam. This prediction was checked experimentally in an extreme case of a low-index sphere in a high-index medium, namely an air bubble. Bubbles, of ~8-μm diam, were generated by shaking a high-viscosity medium consisting of an 80% by weight mixture of glycerol in water. It was found that the bubbles were <u>always pushed out</u> of the light beam as they were accelerated along, as expected. In the same

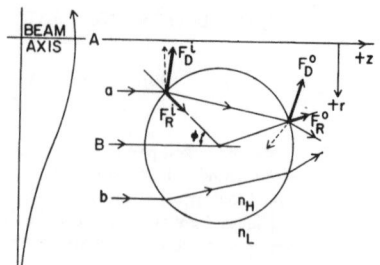

FIG. 2. A dielectric sphere situated off the axis A of a TEM_{00}-mode beam and a pair of symmetric rays a and b. The forces due to a are shown for $n_H > n_L$. The sphere moves toward $+z$ and $-r$.

medium of $n = 1.44$, the 2.68-μm spheres of $n = 1.58$ were still focusing. At higher powers the bubbles are expected to deform. This would result in a deformation contribution to the radiation pressure force as postulated by Askaryon.[7] Our observation of the attraction of high-index spheres into regions of high light intensity is related to the deformation of a liquid surface postulated by Kats and Kantorovich.[8]

The experimentally observed radial inward force on the high-index spheres suggest a means of constructing a true <u>optical potential well</u> or "optical bottle" based on radiation pressure alone. If one has two opposing equal TEM_{00} Gaussian beams with beam waists located as shown in Fig. 1(b), then a sphere of high index will be in stable equilibrium at the symmetry point as shown (i.e., any displacement gives a restoring force). <u>Such trapping was observed experimentally</u> in an open cell filled with 2.68-μm spheres in water as sketched in Fig. 1(b). Here the entire beam is viewed at once. Particles are observed by their brilliant scattered light. With 128 mW in only one of the beams, a maximum particle velocity of ~220 μm/sec was observed as particles traversed the entire near field. The calculated velocity is 195 μm/sec. For trapping, the two opposing beams were introduced. Particles that drift near either beam are drawn in, accelerated to the stable equilibrium point, and stop. To check for stability one can interrupt one beam for a moment. This causes the particle to accelerate rapidly in the remaining beam. When the opposing beam is turned on again the particle returns to its equilibrium point, only more slowly since it is now acted on by the differential force. Interrupting the other beam reverses the behavior. In other experiments, ~5-μm-diam water droplets from an atomizer were accelerated in air with a single beam. At 50 mW, velocities ~0.25 cm/sec were observed. Such motions could be seen with the naked eye. The behavior of the droplets was in qualitative agreement with expectation.

In our experiments it is clear that we have discriminated against radiometric forces. These forces push more strongly on hot surfaces and would push high-index spheres and bubbles out of the beam; whereas our high-index spheres were drawn into the beam. Even the observed direction of acceleration along the beam axis is the opposite of the radiometric prediction. A moderately absorbing focusing sphere concentrates more heat on the downstream side of both the ball and the medium and should move upstream into the light (negative photophoresis).[9] For water drops in air we can invoke the well-confirmed formula of Hettner[10] and compute the temperature gradient needed across a 5-μm droplet to account radiometrically for the observed velocity of 0.25 cm/sec. From Stokes's formula, $F = 2.1 \times 10^{-7}$ dyn. Hettner's formula then requires a gradient of 0.5°C across the droplet. No such gradients are possible with the 50 mW used. For water and glycerol the gradients are also very low.

The extension to vacuum of the present experiments on particle trapping in potential wells would be of interest since then any motions are frictionless. Uniform angular acceleration of trapped particles based on optical absorption of circular polarized light or use of birefringent particles is possible. Only destruction by mechanical failure should limit the rotational speed. In vacuum, particles will heat until they are cooled by thermal radiation or vaporize. With the minimum power needed for levitation, micron spheres will assume temperatures of hundreds to thousands of degrees depending on the loss. The ability to heat in vacuum without contaminating containing vessels is of interest. Acceleration of neutral spheres to velocities ~10^6-10^7 cm/sec is readily possible using powers that avoid vaporization. In this regard one could attempt to observe and use the resonances in radiation pressure predicted by Debye[2] for spheres with specific radii. The separation of micron- or submicron-sized particles by radiation pressure based on radius as demonstrated experimentally could also be useful [see Eq. (1)].

Finally, the extension of the ideas of radiation pressure from laser beams to atoms and molecules opens new possibilities. In general, atoms

and molecules are quite transparent. However, if one uses light tuned to a particular transition, the interaction cross section can be much larger than geometric. For example, an atom of sodium has $\pi r^2 = 1.1 \times 10^{-15}$ cm² whereas, from the absorption coefficient,[11] the cross section σ_T at temperature T for the D_2 resonance line at $\lambda = 0.5890$ μm is $\sigma_T = 1.6 \times 10^{-9}$ cm² $= 0.5\lambda^2$ for $T < 40°K$ (region of negligible Doppler broadening). The absorption and isotropic reradiation by spontaneous emission of resonance radiation striking an atom results in an average driving force or pressure in the direction of the incident light. We shall attempt to show that radiation pressure from a laser beam on resonance can work as an actual optical gas pump and operate against significant gas pressures. Figure 3(a) shows a schematic version of such a pump. Imagine two chambers initially filled with sodium vapor, for example. A transparent pump tube of radius w_0 is uniformly filled with laser light tuned to the D_2 line of Na from the left. Let the total optical power P and the pressure p_0 be low enough to neglect light depletion and absorption saturation. Most atoms are in the ground state. The average force on an atom is $P\sigma_T/c\pi w_0^2$ and is constant along the pump. Call x_{cr} the critical distance. It is the distance traveled by an atom in losing its average kinetic energy $\frac{1}{2}mv_{av}^2$. That is, $Fx_{cr} = \frac{1}{2}mv_{av}^2 \cong kT$. The variation of pressure in a gas

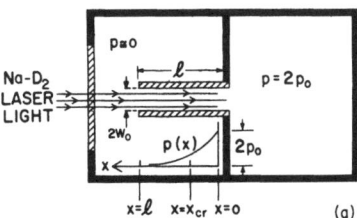

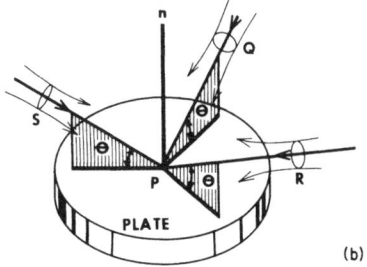

FIG. 3. (a) Schematic optical gas pump and graph of Na pressure $p(x)$. (b) Geometry of gas confinement about point P of a plane surface.

VOLUME 24, NUMBER 4   PHYSICAL REVIEW LETTERS   26 JANUARY 1970

with a constant force is exponential at equilibrium. Thus

$$p(x) = p_0 e^{-Fx/kT} = p_0 e^{-x/x_{cr}}, \qquad (2)$$

$$x_{cr} = \pi w_0^2 ckT / P\sigma_T. \qquad (3)$$

Next, consider higher power. Saturation sets in. Population equalization occurs between upper and lower levels for those atoms of the Doppler-broadened line of width $\Delta\nu_D$, within the natural width $\Delta\nu_n$ of line center. A "hole" is burned in the absorption line and the power penetrates more deeply into the gas. But there is a net absorption, even when saturated, due to the ever-present spontaneous emission from the upper energy level. The average force per atom also saturates and is constant along the tube. Its value is $(h/\tau_n\lambda)(\Delta\nu_n/\Delta\nu_D)$, where τ_n is the upper level natural lifetime.[12] Lastly, we consider the effect of collision broadening due to a buffer gas on the force per atom. With collision one replaces $1/\tau_n$ by $(1/\tau_n + 1/\tau_L)$ and $\Delta\nu_n$ by $(\Delta\nu_n + \Delta\nu_L)$ in the average saturated force, where $\Delta\nu_L = \frac{1}{2}\pi\tau_L$ is the Lorentz width. This enhances the force greatly. Then

$$x_{cr} = \frac{kT\lambda}{h}\left(\frac{\tau_n\tau_L}{\tau_n + \tau_L}\right)\left(\frac{\Delta\nu_D}{\Delta\nu_n + \Delta\nu_L}\right). \qquad (4)$$

As an example, consider Na vapor at $p_0 = 10^{-3}$ Torr ($n_0 = 3.4 \times 10^{13}$ atoms/cc and $T = 510°$K), buffered by helium at 30 Torr. Take a tube of $l = 20$ cm with diameter $2w_0 = 10^{-2}$ cm. For $\tau_n = 1.48 \times 10^{-8}$ sec, $\Delta\nu_D = 155\Delta\nu_n$ (at $T = 510°$K), and $\Delta\nu_L \cong 30\Delta\nu_n$,[13] one finds $x_{cr} = 1.5$ cm and $l = 20$ cm $= 13.3 x_{cr}$. Thus $p(l) = 2p_0 e^{-13.3} = 2 \times 10^{-3} \times 1.7 \times 10^{-6} = 3.4 \times 10^{-9}$ Torr. Essentially complete separation has occurred. This requires a total number of photons per second of $2\pi w_0^2 x_{cr} n_0/(1/\tau_n + 1/\tau_L) \cong 1.7 \times 10^{19} \cong 6$ W. Under saturated conditions there is little radiation trapping of the scattered light. Almost all the incident energy leaves the gas without generating heat. The technique applies for any combination of gases. Even different isotopes of the same atom or molecule could be separated by virtue of the isotope shift of the resonance lines. The possibilities for forming atomic or molecular beams with specific energy states and for studying chemical reaction kinetics are clear. The possibility of obtaining significant population inversions by resonant gas pumping remains to be evaluated. One can also show that gas can be optically trapped at the surface of a transparent plate. For example [see Fig. 3(b)], three equal TEM$_{00}$-mode beams with waists at points Q, R, and S directed equilaterally at point P, at some angle θ, result in a restoring force for displacements of an atom about P. Gas trapped about P could serve as a windowless gas target in many experimental situations. The perfection of accurately controlled frequency-tunable lasers is crucial for this work.

It is a pleasure to acknowledge stimulating conversations with many colleagues; in particular, J. G. Bergman, E. P. Ippen, J. E. Bjorkholm, J. P. Gordon, R. Kompfner, and P. A. Wolff. I thank J. M. Dziedzic for making his equipment and skill available.

[1] E. F. Nichols and G. F. Hull, Phys. Rev. 17, 26, 91 (1903).

[2] P. Debye, Ann. Physik 30, 57 (1909).

[3] N. A. Fuchs, The Mechanics of Aerosols (The Macmillan Company, New York, 1964).

[4] F. Ehrenhaft and E. Reeger, Compt. Rend 232, 1922 (1951).

[5] A. D. May, E. G. Rawson, and E. H. Hara, J Appl. Phys. 38, 5290 (1967); E. G. Rawson and E. H. May, Appl. Phys. Letters 8, 93 (1966).

[6] Available from the Dow Chemical Company.

[7] G. A. Askar'yan, Zh. Eksperim. i Teor. Fiz.—Pis'ma Redakt. 9, 404 (1969) [translation: JETP Letters 9, 241 (1969)].

[8] A. V. Kats and V. M. Kantorovich, Zh. Eksperim. i Teor. Fiz.—Pis'ma Redakt. 9, 192 (1969) [translation: JETP Letters 9, 112 (1969)].

[9] See Ref. 3, p. 60.

[10] G. Z. Hettner, Physics 37, 179 (1926); Ref. 3, p. 57.

[11] A. C. G. Mitchell and M. W. Zemansky, Resonance Radiation and Excited Atoms (Cambridge University Press, New York, 1969), p. 100.

[12] J. P. Gordon notes that power broadening occurs at still higher powers. This increases the hole width and the average force $\sim\sqrt{P}$.

[13] See Ref. 11, p. 166.

VOLUME 25, NUMBER 19   **PHYSICAL REVIEW LETTERS**   9 NOVEMBER 1970

Atomic-Beam Deflection by Resonance-Radiation Pressure

A. Ashkin

Bell Telephone Laboratories, Holmdel, New Jersey 07733
(Received 14 September 1970)

It is proposed to use the saturated value of the radiation pressure force on neutral atoms to produce a constant central force field to deflect atoms in circular orbits and make a high-resolution velocity analyzer. This is useful for studying the interaction of atoms with high-intensity monochromatic light, and to separate, velocity analyze, or trap neutral atoms of specific isotopic species or hyperfine level.

This Letter proposes a new method of studying the interaction of single atoms with intense light. It involves the use of the resonance radiation pressure of laser light[1] to give a force for deflecting neutral atoms out of an atomic beam. At high light intensity this force saturates to an essentially constant value, independent of intensity variations. If the force is always applied perpendicular to an atom's velocity, one can produce a constant central force field in which atoms follow circular orbits. An atom of specific velocity in such an orbit maintains strict resonance (Doppler effects are avoided) and we have in essence a neutral-atom velocity analyzer of high resolution. The deflection technique to be described affects the linear momentum of the atom. If circularly polarized light is used, angular momentum can simultaneously be imparted to the atom and optical orientation in the sense of Kastler[2] should also be achieved.

This Letter treats the saturation of the radiation pressure force at high light intensities phenomenologically using the Einstein A and B coefficients. The absorption cross section and linewidths used are those derived by perturbation theory[3] which is inherently a low-level theory. If modifications are needed at high values of saturation,[4] it should be possible to study them with the proposed velocity analyzer.

It has been suggested[1] that light pressure can exert on gases a sizable pressure which can be used to separate atomic or isotopic species due to the selective nature of this force. With the present technique specific isotopic species can also be selected out of an atomic beam containing many atomic species. Here the simplicity of the geometry makes the calculation of the power required to separate a given mass straightforward. The problem of trapping neutral atoms by light is also considered. We show that an extension of the velocity-analyzer technique makes it possible to trap atoms stably in circular orbits.

If we irradiate an atom with a beam of resonance radiation, connecting the ground state with an excited state, a radiation pressure force F is exerted on the atom, given by

$$F = (h/\lambda)\tau_N{}^{-1}f, \tag{1}$$

where τ_N is the natural lifetime of the excited state and f is the fraction of time the atom spends in the upper state. Equation (1) describes the linear momentum per second scattered isotropically out of an incident beam due to resonance fluoresence from the random spontaneous emission from the upper atomic state. To get f we compute the equilibrium population distribution of N atoms between the ground state and the excited state in the presence of resonance radiation of energy density $w(\nu)$. $(n_1 + n_2 = N)$. Using the well known Einstein A and B coefficients one has

$$f = \frac{n_2}{N} = \frac{x}{1 + Ax/BW(\nu)}, \tag{2}$$

where $x = (1 + g_1/g_2)^{-1}$. Saturation occurs if the stimulated emission rate $Bw(\nu)$ is much higher than the spontaneous emission rate Ax. Then $f \cong x$. g_1 and g_2 are degeneracy factors for the lower and upper states. Absorption followed by stimulated emission, by itself, contributes negligibly to the motion of the atom. It results only in a small net drift velocity in the direction of the incident light equal, on the average, to one-half the velocity corresponding to the absorption of a single photon.

Consider an atomic beam emerging from an oven into a vacuum. If irradiated with saturating resonance radiation transverse to the initial velocity of the beam, acceleration occurs until the transverse velocity is such that atoms are Doppler shifted out of resonance. This restricts the maximum deflection angle obtainable and thus reduces the usefulness of this geometry.[5]

As an alternative to the strictly transversely directed resonance radiation we consider now a radially directed cylindric light beam as shown

VOLUME 25, NUMBER 19   PHYSICAL REVIEW LETTERS   9 NOVEMBER 1970

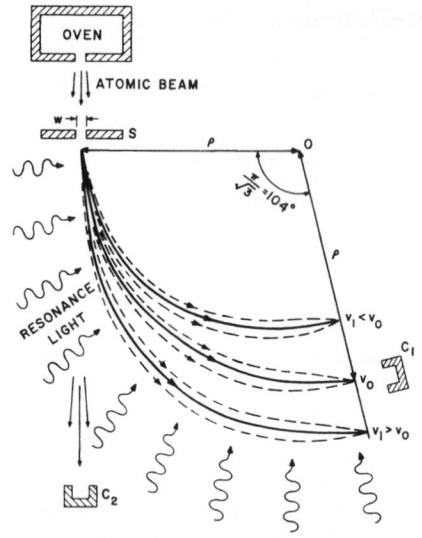

FIG. 1. The 104° atomic velocity analyzer. Resonance light directed toward the cylindric axis O is perpendicular to the equilibrium circular orbit labeled v_0. Collector C_1 detects refocused particles where atoms of different velocity arrive at different radii. Nonresonant species go to C_2. For Na atoms of velocity $v_0 = 2 \times 10^4$ cm/sec, deflected by Na D_2 resonance light, $\rho = 4.0$ cm.

in Fig. 1. If the light intensity is high enough a resonant atom will experience a constant radially directed central force given by (1) throughout the illuminated volume. This is true as long as the transverse component of the atom's velocity is less than the velocity required to Doppler shift out of resonance. For such a force field, atoms of velocity v_0 satisfying the equation

$$F_{\text{sat}} = (h/\lambda)x/\tau_N = mv_0^2/\rho \qquad (3)$$

will follow a circular orbit of radius ρ and thus can experience large deflection from the original direction. Since v_0 is constant and perpendicular to the force no Doppler shift occurs.

This force field has focusing properties in direct analogy with other central-force fields, as for instance the cylindrical E field of electron velocity analyzers.[6,7] From the differential equation for a particle in a central force field one can derive the differential equation for δ, the departure from the equilibrium orbit of particles entering with the correct velocity v_0 but an incorrect direction (i.e., not tangential to the

equilibrium circle ρ):

$$-(3v_0^2/\rho^2)\delta = d^2\delta/dt^2. \qquad (4)$$

Equation (4) has solutions sin and $\cos(\sqrt{3}v_0 t/\rho)$. Thus particles of the correct velocity but wrong injection angle are refocused at $r = \rho$ after transversing an angle of $\pi/\sqrt{3} = 104°$ (see Fig. 1). This is in analogy with the focusing angle of $\pi/\sqrt{2} = 127°$ for electrons in a field $\sim 1/r$ and an angle of $\pi/\sqrt{1} = 180°$ for planetary orbits where the field $\sim 1/r^2$. Particles with the correct injection angle (i.e., tangential, for which $dr/dt = 0$) but wrong velocity v_1, oscillate radially with the same period and return to $dr/dt = 0$ after $\pi/\sqrt{3}$ rad but at a new radius r_1. This radius is found from the equation for the apsides of the motion,

$$\left(\frac{dr}{dt}\right)^2 = v_1\left(1 - \frac{\rho^2}{r^2}\right) + 2v_0\left(1 - \frac{r}{\rho}\right) = 0, \qquad (5)$$

which is derived by integrating the equation of motion or simply writing the law of conservation of energy subject to the boundary condition $dr/dt = 0$ at $r = \rho$. From (5) we find that atoms with $v_1 < v_0$ arrive at the collector C_1 with $r_1 < \rho$ and those with $v_1 > v_0$ arrive with $r_1 > \rho$. For example, if $v_1 = 1.05v_0$, $r_1 = 1.07\rho$ and if $v_1 = 0.95v_0$, $r_1 = 0.93\rho$. Atoms injected with $v = v_1$ but at different angles with respect to the equilibrium orbit at the input are refocused at r_1 at the output (see Fig. 1). Thus we have a true velocity analyzer.

Consider now the power required to saturate the force. We rewrite the stimulated absorption (or emission) rate in more detail in terms of the on-resonance absorption cross section $[(\lambda_0^2/2\pi) g_2/g_1]$, the incident intensity $I(\nu)$ of monochromatic light, and the Lorentzian line shape $S(\nu)$[3,8]:

$$Bw(\nu) = \frac{\lambda_0^2}{2\pi} \frac{g_2}{g_1} \frac{I(\nu)}{h\nu} \frac{S(\nu)}{4\tau_N},$$

where

$$S(\nu) = \frac{\gamma_N}{2\pi[(\nu - \nu_0)^2 + \gamma_N^2/4]}; \qquad (6)$$

γ_N is the natural width. ($\gamma_N = 1/2\pi\tau_N$ and $1/\tau_N = A$). Referring to (2) we define a saturation parameter $\varphi(\nu)$:

$$Bw(\nu) = \varphi(\nu)Ax \qquad (7)$$

which specifies the amount of saturation at frequency ν. From $\varphi(\nu)$ we get f and also therefore the value of the saturated force F. [$F \sim (1 + 1/\varphi)^{-1}$ from Eq. (1)]. $\varphi(\nu)$ also gives the force on atoms entering the field at different angles with respect to the equilibrium orbit, since by virtue of the

VOLUME 25, NUMBER 19   PHYSICAL REVIEW LETTERS   9 NOVEMBER 1970

Doppler shift such atoms absorb at a shifted frequency. Explicitly,

$$\wp(\nu) = \frac{\lambda_0^2(1 + g_2/g_1)I(\nu)S(\nu)}{8\pi h\nu} = \wp(0)\frac{S(\nu)}{4\tau_N}, \qquad (8)$$

where $\wp(0)$ is the degree of saturation achieved with intensity $I(\nu) = I_0$ at line center.

As an example, consider Na atoms irradiated with Na D_2 resonance light with $\lambda_0 = 5890$ Å and $\gamma_N = 10.7$ MHz. Due to the nuclear spin of Na^{23} $(I = \frac{3}{2})$, the $^2S_{1/2}$ ground level is actually split into two levels with $F = 1, 2$ whereas the $^2P_{3/2}$ level is split into four closely spaced hyperfine levels with $F = 0, 1, 2, 3$. The selection rule $\Delta F = \pm 1, 0$ permits one, in principle, to interact with either level of the split ground state without coupling to the other. However, to avoid the possibility of cross coupling due to the close spacing of the upper F levels, it is advantageous to use light with circular polarization (σ^+) connecting the degenerate magnetic sublevel $m_F = 2$ of the $F = 2$ ground state with the $m_F = 3$ magnetic sublevel of the $F = 3$ upper level. Decay from the $m_F = 3$ excited sublevel to any other degenerate sublevels of either $F = 2$ or $F = 1$ is prohibited by the $\Delta m_F = \pm 1$ selection rule for σ components. Statistically one-eighth of all ground-state atoms emerging from the source will be in the $F = 2$, $m_F = 2$ magnetic sublevel. This fraction will be increased by the σ^+ optical pumping[2,9] from the other m_F sublevels of $F = 2$. We have thus, in effect, achieved an ideal two-level system of the type considered above in which, in addition, all the deflected atoms have completely oriented spins. For the case considered $g_2/g_1 = 1$. This example suggests that the deflection technique can be used to supplement existing atomic beam techniques for studying hyperfine structure, nuclear mangetic moments, and atomic orientation by optical pumping.

Applying Eq. (8) to the Na D_2 line, one finds

$$\wp(0) = \frac{I_0(\text{W/cm}^2)}{2.1 \times 10^{-2}}. \qquad (9)$$

If incoming atoms are restricted by slits of width w and height h parallel to the cylindric axis of the analyzer, then we must provide the saturating light intensity I_0 over an area of $\pi\rho h/\sqrt{3}$. If the source temperature is $T = 510°$, then the Na pressure $= 10^{-3}$ Torr, the density $n_0 = 3.4 \times 10^{13}$ atoms/cm^3, the mean free path $L = 30$ cm, and the average atomic velocity $v_{av} = (2kT/m)^{1/2} = 6.1 \times 10^4$ cm/sec. If we consider atoms with velocity $v_0 = v_{av}/3 \cong 2 \times 10^4$ cm/sec, then from Eq. (3) the

equilibrium orbit radius $\rho = 4.0$ cm. Taking $h = 0.1$ cm, and $\wp(0) = 10^2$, we must therefore have a total power of $2.1 \times 10^{-2} \times 10^2 \times \pi(4)(0.1)/\sqrt{3} = 1.5$ W of resonance power in the incident beam. This situation is appropriate for a cw experiment. To calculate an acceptance angle for incoming particles we specify that $\wp(\nu)$ vary from a minimum of 10 to a maximum of 10^2 within the acceptance angle. This yields a range of $\pm 2.6°$. An atom making an angle of 2.6° with respect to the equilibrium orbit is absorbing at a Doppler-shifted frequency of $(\nu - \nu_0) = 1.5\gamma_N$ with $\wp(\nu) = 10$. The number of atoms emerging from the source within this angular tolerance with velocity $v = v_{av}/3 \pm \frac{1}{2}\%$ is $\sim 2.2 \times 10^{11}$ atoms/cm^2 sec. If we use slits of width $w = \rho/100 = 0.04$ cm and $h = 0.1$ we have a flux of $\sim 10^8$ atoms/sec which is adequate for most experiments.

Since the transit time for an atom through the analyzer is $\sim 4 \times 10^{-4}$ sec, a pulsed experiment is possible with much higher peak power and higher saturation. Thus, if the peak power is 10^4 W, then $\wp(0) \cong 7 \times 10^5$. The angular tolerance based on Doppler shift is so large that other considerations dictate the acceptance angle. The main advantage of high power is the relaxation of the requirement on the frequency control of the laser. Thus one can scatter power with $\wp(\nu) = 10^2$ or larger within the band $(\nu - \nu_0) = \pm 42\gamma_N = \pm 450$ MHz as seen from (8). In some cases one might take advantage of the wide bandwidth of a mode-locked laser to ease the frequency control problem, provided the power per mode is enough to saturate the force.

A useful feature of the analyzer is its insensitivity to light intensity variations when strongly saturated $[\wp(\nu) \gtrsim 10]$. Thus the incoming light could be part of a TEM$_{00}$ Gaussian-mode beam, with the atoms following a phase front. Since a far-field diffraction angle of $\pi/\sqrt{3}$ represents rather tight focusing one could break the incoming beam into separate beams without much difficulty. Although the total optical power required is fairly modest, it can be further reduced by using a scheme where the unscattered light is recirculated through the analyzer. A further modification of interest is to depart from strict cylindric geometry and use spherical beams in analogy to the Purcell velocity analyzer for electrons.[10] This would give vertical focusing as well as radial focusing and one could in principle trap atoms in a stable circular orbit in a geometry in which the flux is incident over 2π rad. Figure 2 shows such a trapping apparatus

VOLUME 25, NUMBER 19   PHYSICAL REVIEW LETTERS   9 NOVEMBER 1970

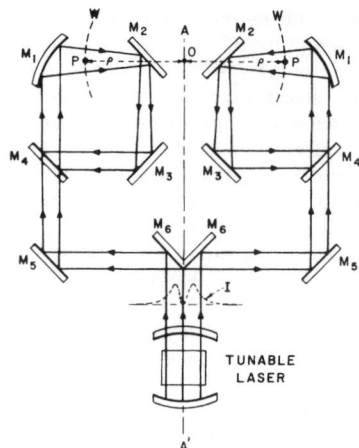

FIG. 2. Apparatus for trapping atoms in a circular orbit. AA' is the axis of a figure of revolution. Mirrors M_1, M_2, M_3, and M_4 recirculate light converging with spherical wavefronts w on the circular orbit P of radius ρ centered at O. A tunable laser with an intensity profile I (i.e., a "do-nut" or $\text{TEM}_{01}{}^*$ mode) feeds the recirculating resonator M_1, M_2, M_3, M_4 via M_5 and M_6.

including means for recirculating the light.

The analyzer can also serve as an isotope selector. Only resonant atoms are deflected to collector C_1 in Fig. 1. Nonresonant species proceed undeflected to C_2. For particle selectors of this type it is simple to calculate the optical power needed to separate atoms. An atom of velocity v_0 spends a time $= (\pi/\sqrt{3})\rho/v_0$ in the light field and scatters $2t_N{}^{-1}(\pi/\sqrt{3})\rho/v_0)$ photons which by (1) and (2) equals $(\pi/\sqrt{3})mv_0(h/\lambda)^{-1}$. If we use $qP \times 10^7/h\nu$ photons/sec, where qP is the fraction of the incident power in watts scattered, the number of atoms/sec N we collect is

$$N = \frac{\sqrt{3}}{\pi}\frac{qP \times 10^7}{mv_0 c}. \tag{9}$$

Equation (9) shows that the number of atoms/sec collected depends on P and the incident velocity

v_0 but not on the force F_{sat}. F_{sat} and v_0 do, however, determine the size of the orbit radius as seen in (3). Using $v_0 = 2 \times 10^4$ cm/sec one finds that it takes 1.2×10^4 photons per atom collected. Assuming $q \sim 1$, $N = 2.4 \times 10^{14}$ atoms/sec W collected. With the expenditure of 1 kW h we can collect a mass of 30 mg.

In conclusion this Letter points out the use of resonance radiation pressure to separate, trap, and velocity analyze neutral atoms. It should also prove useful in studying the details of the basic interaction of atoms with high-intensity monochromatic light.

It has come to the author's attention that Frisch,[11] in the last paper of the thirty "Untersuchungen zur Molekular Strahlmethode," has observed the recoil of sodium atoms on emitting a single photon of resonance radiation.

It is a pleasure to thank my colleagues A. G. Fox, J. P. Gordon, and R. Kompfner for many stimulating discussions.

[1]A. Ashkin, Phys. Rev. Lett. **24**, 156 (1970).

[2]A. Kastler, J. Opt. Soc. Amer. **47**, 460 (1957).

[3]W. Heitler, *The Quantum Theory of Radiation* (Oxford Univ., London, 1954), 3rd. ed., Chap. V.

[4]C. R. Stroud, Jr. and E. T. Jaynes, Phys. Rev. A **1**, 1 (1970); J. P. Gordon, private communication.

[5]In astronomy rectilinear acceleration of various ions to high velocities (3×10^8 cm/sec) by resonance radiation pressure has been observed in the planetary nebulae around hot stars. The ions are not Doppler shifted out of resonance due to the continuous nature of black-body radiation. The forces involved are much less than the saturated values postulated here. See L. B. Lucy and P. M. Solomon, Astrophys. J. **159**, 879 (1970).

[6]A. Hughes and V. Rojansky, Phys. Rev. **34**, 286 (1929).

[7]D. Roy and J. D. Carette, Appl. Phys. Lett. **16**, 413 (1970).

[8]A. Yariv, *Quantum Electronics* (Wiley, New York, 1967), chap. 13.

[9]W. B. Hawkins, Phys. Rev. **98**, 478 (1955).

[10]E. M. Purcell, Phys. Rev. **54**, 818 (1938).

[11]O. R. Frisch, Z. Phys. **86**, 42 (1933).

PHYSICAL REVIEW LETTERS

Ref in
Cavity Optomech
Aspelmeyer

VOLUME 40   20 MARCH 1978   NUMBER 12

Trapping of Atoms by Resonance Radiation Pressure

A. Ashkin

Bell Telephone Laboratories, Holmdel, New Jersey 07733
(Received 17 October 1977)

A method of stably trapping, cooling, and manipulating atoms on a continuous-wave basis is proposed using resonance radiation pressure forces. Use of highly focused laser beams and atomic beam injection should give a very deep trap for confining single atoms or gases at temperatures $\sim 10^{-6}$ °K. An analysis of the saturation properties of radiation pressure forces is given.

A method of optically trapping and cooling atoms on a continuous-wave (cw) basis is proposed based on radiation pressure forces. The new trap geometry provides stable confinement, optical damping, and means for optical manipulation of trapped atoms. Injection into the trap is from an atomic beam. The radiation pressure trapping forces used are the scattering force due to spontaneous emission[1-4] and the ponderomotive force[5-9] which exists on the induced atomic dipole in an optical field gradient. It is known that the scattering force can increase,[1,2,4] decrease,[3] or deflect[2,10] atomic velocities. Dipole gradient forces can be attractive or repulsive giving optical self-defocusing or self-focusing[5] as well as novel beam interaction forces[6] and a possible means of accelerating atoms.[8] Proposals exist for optically trapping atoms dynamically[2] and statically.[9] This proposal, based on a new treatment of the saturation of these forces and a new geometry, results in a trap with remarkable properties. The trapping energy is more than two orders of magnitude greater than previous proposals,[9] it can accept $\sim 10^7$ atoms, cool them to about a single photon momentum ($\sim 10^{-6}$ °K), and hold them indefinitely even as single atoms. The technique should have wide application in experiments in atomic physics.

Consider the behavior of the proposed trap qualitatively. Light from two opposing TEM_{00} mode beams is focused at points Q_1 and Q_2 located symmetrically about point E [see Fig. 1(a)]. The beams grow in radius from w_0 to $13w_0$ in going

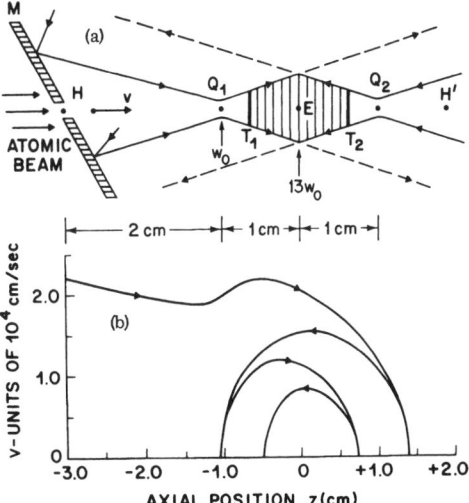

FIG. 1. (a) Sketch of the proposed optical trap for atoms; $w_0 \cong 12 \ \mu\text{m}$. (b) Calculated trajectory of an atom injected through H with $v = 2 \times 10^4$ cm/sec.

VOLUME 40, NUMBER 12   PHYSICAL REVIEW LETTERS   20 MARCH 1978

1 cm from the foci to E ($w_0 \simeq 12$ μm for $\lambda \simeq 5900$ Å). Each beam is tuned $50\gamma_N$ (the natural width) below the sodium D resonance frequency, for example, and has a cw power of 200 mW. E is a point of stable equilibrium since any displacement of an atom from E results in a restoring force. There is an axial restoring force from scattering due to intensity imbalance and a radial restoring dipole force due to radial field gradients when tuned below resonance. To trap atoms one needs damping. Damping due to the Doppler shift occurs when tuned below resonance since moving atoms interact more strongly with the opposing beam. An atomic beam with average velocity $v_{av} = 6 \times 10^4$ cm/sec is injected into the trap through a hole H in mirror M. Atoms traversing the trap with transverse velocity $v_T < 5 \times 10^2$ cm/sec are confined radially by dipole forces. Those with proper axial velocities damp down and stop in the trap along the axis HH'. For example, an atom entering with 2×10^4 cm/sec stops at a point ~ 4 mm beyond Q_2. It then recoils and executes a damped oscillation about E. Several cycles of this motion are shown in Fig. 1(b).

Next consider the effect of the standing-wave fringes which exist with varying depths within the trap. Because of the dipole force from the axial field gradient, atoms are attracted to the peaks of the fringes (i.e., the standing wave loops) with decreasing strength away from E. Beyond planes T_1 and T_2 this attractive force becomes less than the scattering force toward E. Thus atoms executing a damped oscillation about E can only be trapped on a loop if they come to rest inside the region from T_1 to T_2. Once trapped, atoms continue to be damped due to Doppler shift down to about a single photon momentum. Atoms trapped on loops can be manipulated by slowly moving the loops toward one of the T planes by changing the optical path of one of the beams. This drags the atoms along and deposits them on just a few loops near T. These can then be retracted to E where they are axially most stable and held indefinitely.

Consider the trapping forces following Ashkin,[2] who used rate equations to describe the saturation of the scattering force F_{scat} of a single beam acting on a two-level atom. F_{scat} depends on the fraction of time f an atom spends in the upper state

$$F_{scat} = (h/\lambda\tau_N)f, \tag{1}$$

where h/λ is the photon momentum, τ_N the natu-

al lifetime, and

$$f = \tfrac{1}{2}[1 + 1/p(\nu)]^{-1}. \tag{2}$$

$p(\nu)$ is a saturation parameter given by $BW(\nu)/A$, the ratio of the (Einstein B coefficient) stimulated absorption rate to the (Einstein A coefficient) spontaneous emission rate, multiplied by $W(\nu)$ the energy density:

$$p(\nu) = \frac{\lambda_0^2}{2\pi} \frac{I(\nu)}{h\nu} \frac{S(\nu)}{4\tau_N} \left(\frac{1}{\tau_N} \right)^{-1} = \frac{I}{I_{sat}} \frac{S(\nu)}{4\tau_N}, \tag{3}$$

where $\lambda_0^2/2\pi$ is the absorption cross section at the resonance frequency ν_0, $I(\nu)$ the intensity at frequency ν, and $S(\nu)/4\tau_N$ is a line-shape factor

$$\frac{S(\nu)}{4\tau_N} = \frac{\gamma_N^2/4}{(\nu - \nu_0)^2 + \gamma_N^2/4}, \tag{4}$$

$\gamma_N = \frac{1}{2\pi\tau_N}$ is the natural width. For high intensities $p(\nu)$ is large, $f \simeq \tfrac{1}{2}$, and $F_{scat}^{max} \simeq h/2\lambda\tau_N$.

Suppose an atom at rest is irradiated by two opposing beams of different intensities and frequencies I_R, ν_R and I_L, ν_L. For $|\nu_R - \nu_L| > \tau_N^{-1}$, the time-averaged force on the atom is determined by the average number of photons absorbed from each beam [$\sim p(\nu)$] and the interference effects of the two opposite beams can essentially be ignored. Thus the f value for the atom is derived from $p_{tot} = p(\nu_R) + p(\nu_L)$, which determines the total force $F_{tot} = |F_R| + |F_L|$. The ratio of the $|F_R|/|F_L|$ is the ratio of the number of photons absorbed from each beam $p(\nu_R)/p(\nu_L)$, which gives

$$|F_R| = \frac{h}{\lambda} \frac{1}{2\tau_N} \frac{p(\nu_R)}{1 + p_{tot}}, \tag{5}$$

$$|F_L| = \frac{h}{\lambda} \frac{1}{2\tau_N} \frac{p(\nu_L)}{1 + p_{tot}}. \tag{6}$$

Atoms moving in the trap see different ν_R and ν_L because of the Doppler shift, and different I_R and I_L because of trap geometry. The trap is detuned below resonance by $\nu_0 - \nu = q\gamma_N$. Therefore $\nu_0 - \nu_R = (q - b)\gamma_N$ and $\nu_0 - \nu_L = (q + b)\gamma_N$, where $b = v/\lambda\gamma_N$ is the Doppler shift in units of γ_N. Equations (5) and (6) accurately describe the scattering forces acting on an atom injected into the trap over most of its velocity range ($v > 2 \times 10^3$ cm/sec) and were thus used to compute the damping curves of Fig. 1(b) by calculating the forces and velocity changes on an incremental basis. This damping calculation neglected dipole forces which would only cause a slight modulation of the computed velocities.

Consider next the dipole or ponderomotive force F_{dip} acting on an optically produced atomic dipole of moment αE placed in a region of electric field gradient. α is the atomic polarizability. As shown by Gordon,[7] for example, $F_{dip} = \frac{1}{2}\alpha \nabla E^2$. The magnitude of α, however, depends on the field strength E. If α_0 is the zero-field polarizability, at higher field there is a contribution to the dispersion of $-\alpha_0 f$ from the population of the upper state and $\alpha_0(1-f)$ from the ground state giving

$$\alpha = \alpha_0(1-2f) = \alpha_0/[1+p(\nu)], \qquad (7)$$

$$\alpha_0 = \frac{-\lambda_0^3}{16\pi^3}\frac{\gamma_N}{2}\frac{\nu - \nu_0}{(\nu - \nu_0)^2 + \gamma_N^2/4}. \qquad (8)$$

Thus α saturates to zero as $1/E^2$ at high field where we can neglect unity compared to $p(\nu)$ in (7). For the specific case of a Gaussian beam $E^2 = E_0^2 \exp(-2r^2/w_0^2)$, the radial dipole force at saturation is

$$F_{dip}^{sat} = -\frac{1}{2}\alpha_0 E_{sat}^2\frac{4\tau_N}{S(\nu)}\frac{4r}{w_0^2} = h(\nu - \nu_0)\frac{4r}{w_0^2}, \qquad (9)$$

where we have used $I_{sat} = cE_{sat}^2/4\pi$ in (3). In general we have the exact expression

$$F_{dip} = \frac{\alpha_0/2}{1+p(\nu)}\nabla\left(\frac{E_{sat}^2}{S(\nu)/4\tau_N}p(\nu)\right) \equiv -\nabla U, \qquad (10)$$

where U is a potential having the value

$$U = h(\nu - \nu_0)\ln[1+p(\nu)]. \qquad (11)$$

The factor $4r/w_0^2$ in (9) is a shape factor arising from the radial variation of α and the gradient of E^2. Also the force (9) is independent of E_0^2 as long as $p(\nu) \gg 1$. To calculate the effectiveness of this radial confining force, assume saturation out to $r = R$ [i.e., $p(\nu) = 1$ at $r = R$] and neglect the force for $r > R$. Then the trapping energy

$$\left|\int_0^R F_{dip}^{sat}dr\right| = -h(\nu - \nu_0)2R^2/w_0^2.$$

Thus the trapping energy continues to rise with increasing power. If, for example, $R = 2w_0$ the trapping energy is $\sim [-8h(\nu - \nu_0)]$. This implies $p(\nu) = e^8$ at $r = 0$ and therefore $|U| \simeq -8h(\nu - \nu_0)$ at $r = 0$. Thus the approximate trapping energy based on (9) and the potential energy (11) agree for large $p(\nu)$. Note that the trapping energy increases in proportion to the detuning $(\nu - \nu_0)$.

Next consider the axial or z component of F_{dip} for an atom at rest in a simple standing wave $E = 2E_0\cos\omega t\cos kz$ due to two equal opposing beams.

The same procedure gives $F_{dip}^{sat}(z) = h(\nu - \nu_0)2k \times \sin kz/\cos kz$. This disagrees with Ref. 9, which apparently neglects the variation of α with z. More generally for two unequal opposing beams there is a standing wave plus a running wave. α is determined by p_{tot} due to absorption from the standing wave S and the running wave R. $p_{tot} = p_S(\nu) + p_R(\nu)$ and

$$F_{dip} = \frac{1}{2}\alpha_{tot}\nabla E_S^2$$
$$= \frac{1}{2}\alpha_0[1 + p_S(\nu) + p_R(\nu)]^{-1}\nabla E_S^2. \qquad (12)$$

If now $p_R(\nu) \gg 1$ and $p_R(\nu) \gg p_S(\nu)$,

$$F_{dip}(z) = (4E_0^2/E_R^2)h(\nu - \nu_0)\sin 2kz. \qquad (13)$$

Thus the trapping force of the standing wave is reduced by the addition of a strong saturating running wave. At the foci Q_1 and Q_2 of the trap $F_{scat} = 1.7 F_{dip}^{max}(z)$; so no trapping on loops is possible. At the plane T_1 and T_2 the two forces just balance.

Knowing the potential U at point E we get the maximum transverse velocity of a trapped atom using $\frac{1}{2}m(v_T^{max})^2 = U$. From $w_0 = 12$ μm and the total intensity we find $p_{tot}(\nu) \simeq 2.5$ at E. Thus $U = 62.5h\gamma_N$ and $v_T^{max} = 480$ cm/sec. Using v_T^{max} and the range of axial velocities captured by the trap (0 cm/sec to $\sim 3 \times 10^4$ cm/sec) we estimate a trapping rate of $\sim 10^6$ atoms/sec and a trap capacity of $\sim 3 \times 10^7$ atoms. With the trap filled and the source off the 3×10^7 atoms cool axially to a velocity of ~ 3 cm/sec corresponding to a single photon momentum. In time, collisions among trapped atoms should thermalize all velocity components to ~ 3 cm/sec or $T \simeq 10^{-6}$ °K. If needed one can damp the transverse components by adding two pairs of opposing beams tuned below resonance but weak enough not to reduce the trapping by additional saturation. This saturation effect, which was neglected in Ref. 9, reduces their trapping energy by a factor of 3. As T approaches 10^{-6} °K, the gas density approaches that of a solid and atoms may be lost by condensation. However, one should easily trap and observe low densities or even single atoms since a trapped atom scatters $\sim 10^8$ photons/sec and can be observed free of background gas. Once cooled to a few centimeters per second, the trapping light can be shut off for times $\sim 4 \times 10^{-4}$ sec without atoms drifting out of the trap. This time is adequate to perform even long lifetime spectroscopic measurements under strictly Doppler-free conditions. One can apply strong external

VOLUME 40, NUMBER 12   PHYSICAL REVIEW LETTERS   20 MARCH 1978

fields on trapped atoms. Also one can study re-actions between different types of cooled atoms by manipulating two traps so that they overlap.

Consider now modified trap geometries. Shifting the foci Q_1 and Q_2 into coincidence at E gives a confocal-type trap 4.8 times deeper in energy but lacking the features which allow coalescence of atoms trapped on different loops. This trap geometry can also be made using an optical resonator with much reduced power. Finally there is perhaps the conceptually simplest trap: a single highly focused Gaussian beam tuned well below resonance. Such a TEM_{00} mode beam has radially inward dipole forces as discussed and also strong axial dipole forces $F_{\mathrm{dip}}'(z)$ directed toward the focus due to the axial intensity gradient. (This axial contribution was negligible in previous traps where the focusing was weaker.) Further, there is the saturated axial scattering force $F_{\mathrm{scat}}(z)$ in the direction of the light. If $F_{\mathrm{dip}}'(z)$ ever exceeds $F_{\mathrm{scat}}(z)$, there is a barrier to the escape of atoms from the focal region and a stable trap exists. This condition, i.e., $R = F_{\mathrm{dip}}'(z)/F_{\mathrm{sat}}(z) > 1$, is easily met for a saturated beam with tight focusing. At $z = \pi w_0^2/\sqrt{3}\,\lambda$, the position of maximum axial gradient, $R = \sqrt{3}\,q\lambda^2/2\pi^2 w_0^2$. For a power of 25 mW, $q = 400$, and $w_0 = 2.5\ \mu\mathrm{m}$, then $p(\nu) = 10$, $R = 2.0$ and strong trapping exists. However, atomic beam injection into such a small trap is difficult. It can only damp and trap velocities $\sim 10^3$ cm/sec within narrow limits. The flux of such atoms is low, ~ 1–10 atoms/sec, or even less if low velocities are depleted. One can, however, transfer cooled atoms into a single beam trap from a two-beam-type trap.

In the above Na was treated as an ideal two-level atom. In fact, the ground state is split into two hyperfine components. To avoid problems due to transfer of atoms to the other hyperfine component by optical pumping, one can use two laser frequencies, one tuned below each of the two hyperfine components. In geometries where the intensity is high enough a single frequency can saturate both components. One can also couple the two components with rf fields and thus avoid loss of atoms from the trap.[11]

Thus, based on a new analysis of the forces we propose use of high-intensity, strictly cw, highly focused beams tuned well below resonance to give strong damping and transverse gradient forces for trapping and cooling individual atoms to $\sim 10^{-6}$ °K. These traps are remarkably similar in both geometry and general behavior to those used to trap and levitate macroscopic dielectric spheres by radiation pressure.[1,12] With atomic beam injection, one gets background-free operation; there is no need to cool large volumes[3,9] or to shift the light frequency adiabatically[9] to resonance with its loss of trap depth. The proposed trap should be useful not only for novel experiments on cooled atoms but also for studies on the resonance radiation pressure forces themselves using, for example, monoenergetic atoms injected into the trap by an atomic beam velocity selector.

The author acknowledges helpful discussions with J. Ashkin, J. E. Bjorkholm, and J. P. Gordon.

[1]A. Ashkin, Phys. Rev. Lett. **24**, 156 (1970).

[2]A. Ashkin, Phys. Rev. Lett. **25**, 1321 (1970).

[3]T. W. Hänsch and A. L. Schawlow, Opt. Commun. **13**, 68 (1975).

[4]J. E. Bjorkholm, A. Ashkin, and D. B. Pearson, Appl. Phys. Lett. **27**, 534 (1975).

[5]G. A. Askar'yan, Zh. Eksp. Teor. Fiz. **42**, 1567 (1962) [Sov. Phys. JETP **15**, 1088 (1962)]; D. Grischkowsky, Phys. Rev. Lett. **24**, 866 (1970); J. E. Bjorkholm and A. Ashkin, Phys. Rev. Lett. **32**, 129 (1974).

[6]A. C. Tam and W. Happer, Phys. Rev. Lett. **38**, 278 (1977).

[7]J. P. Gordon, Phys. Rev. A **8**, 14 (1973).

[8]A. P. Kazantsev, Zh. Eksp. Teor. Fiz. **63**, 1628 (1972) [Sov. Phys. JETP **36**, 861 (1973)].

[9]V. S. Letokhov, V. G. Minogin, and B. D. Pavlik, Opt. Commun. **19**, 72 (1976).

[10]R. Schieder, H. Walther, and L. Woste, Opt. Commun. **5**, 337 (1972); J. L. Picque and J. L. Vialle, Opt. Commun. **5**, 402 (1972).

[11]Loss of atoms due to photoionization out of the strongly excited upper level of the D transition is a two-photon process and should be negligible at the intensities considered. See T. B. Lucatorto and T. J. McIlrath, Phys. Rev. Lett. **37**, 428 (1976), Ref. 17.

[12]A. Ashkin and J. M. Dziedzic, Appl. Phys. Lett. **19**, 283 (1971), and Science **187**, 1073 (1975), and Appl. Phys. Lett. **28**, 333 (1976).

VOLUME 57, NUMBER 3   PHYSICAL REVIEW LETTERS   21 JULY 1986

Experimental Observation of Optically Trapped Atoms

Steven Chu, J. E. Bjorkholm, A. Ashkin, and A. Cable

AT&T Bell Laboratories, Holmdel, New Jersey 07733
(Received 14 April 1986)

We report the first observation of optically trapped atoms. Sodium atoms cooled below 10^{-3} K in "optical molasses" are captured by a dipole-force optical trap created by a single, strongly focused, Gaussian laser beam tuned several hundred gigahertz below the D_1 resonance transition. We estimate that about 500 atoms are confined in a volume of about 10^3 μm^3 at a density of 10^{11}–10^{12} cm^{-3}. Trap lifetimes are limited by background pressure to several seconds. The observed trapping behavior is in good quantitative agreement with theoretical expectations.

PACS numbers: 32.80.Pj

We report the optical trapping of neutral atoms by the forces of resonance-radiation pressure in a single-beam optical trap. At the time of the first demonstration of stable optical trapping and manipulation of small dielectric particles[1] it was predicted that similar effects were possible with atoms. Since then there have been extensive studies of the basic forces of laser light on neutral particles and atoms.[2-8] The trapping and manipulation of neutral macroscopic dielectric particles from ~ 30 μm to ~ 25 nm has been demonstrated.[5,9-11] With atoms, resonance-radiation forces have been used for atomic deflection,[2,12-14] the guiding and focusing of atomic beams,[15] the slowing and stopping of atomic beams,[16,17] and the three-dimensional cooling of atoms to a temperature of 2.4×10^{-4} K.[18,19] While there have been many proposals of optical traps for atoms in the last sixteen years,[1,2,5,6,8] none has been demonstrated previously.

Optical atom traps are difficult to achieve for several reasons. (i) Their potential wells are shallow, typically 10^{-1}–10^{-2} K. (ii) Their volumes, with the exception of alternating-beam scattering-force traps,[8,20] are quite small. (iii) Once confined within a trap, atoms are heated by the random fluctuations of the light forces and will "boil" out of the trap in a fraction of a second.[4] The single-beam trap demonstrated here has a well depth ~ 5 mK and a volume $\sim 10^{-7}$ cm^3. Use of the recently demonstrated "optical molasses"[18,19] was crucial to the present experiment. Optical molasses (OM) is formed by the intersection of three pairs of counterpropagating, mutually orthogonal laser beams tuned half a linewidth below a resonance transition. The OM forms a highly viscous medium of photons capable of cooling the sodium atoms used in this experiment to 2.4×10^{-4} K. The dense collection of atoms confined within OM provides an excellent source with which to load an optical trap. The random-walk motion of the atoms and their long storage time allows atoms to continuously diffuse to the trap surface and be captured. OM was also used to cool atoms within the trap.

The optical trap demonstrated here is the single-beam gradient-force trap proposed in 1978.[3] It con-

sists of a single strongly focused Gaussian laser beam tuned about 10^4 natural linewidths below resonance. Conceptually it is the simplest of the proposed traps and offers many advantages. It has no standing waves and consequently minimal dipole-force heating.[4] It is capable of giving deep optical potential wells, localized to a few optical wavelengths, and is ideal for achieving high atomic densities and for optical manipulation of atoms.

The properties of the single-beam trap result from the dominance of the dipole force over the scattering force in a focused Gaussian beam. The scattering force[2] is due to the spontaneous scattering of photons. Below saturation, it is proportional to the light intensity and points in the direction of the incident light. Quantitatively, $F_{\text{scat}} = (h/\lambda)(1/2\tau_N)[p/(1+p)]$ where τ_N is the natural lifetime and p is the saturation parameter. Also, $p = (I/I_s)[\gamma^2/4(\Delta\nu^2 + \gamma^2/4)]$ where I_s is the saturation intensity (20 mW/cm^2 for the D_2 line), $I = I(r,z)$ is the Gaussian beam intensity, $\Delta\nu$ is the detuning from resonance, and $\gamma = 1/2\pi\tau_N$ is the natural linewidth. The dipole (or gradient) force,[3] which arises from the atomic dipole moment induced by the laser field, is proportional to and points in the direction of the intensity gradient. It can be derived from a conservative potential[3,4] U where $U = (h\Delta\nu/2)\ln(1+p)$. For tunings below resonance ($\Delta\nu < 0$), the dipole force pushes atoms toward the high-intensity region of the light. This assures the radial stability of the trap. In the axial direction one has the additional complication of the scattering force. However, axial stability can be achieved since the axial gradient force can dominate over the destabilizing influence of the scattering force for tunings well below resonance and for strongly focused beams. The above considerations on the scattering- and dipole-force components on atoms are fully consistent with conservation of momentum in the optical-scattering process.[1,4,11] The dipole potential U can also be understood in terms of the optical Stark shift of an atom. A laser tuned below resonance lowers the ground-state energy of an atom by an amount U and raises the energy of the excited state by the same

VOLUME 57, NUMBER 3   PHYSICAL REVIEW LETTERS   21 JULY 1986

amount. Since an atom spends most of its time in the ground state its total energy is minimized in the region of the maximum laser intensity. Figure 1(a) shows the potential well along the z axis for three laser detunings. Figure 1(b) shows the axial well depth versus detuning below the sodium D_2 transition. It is the presence of the axial scattering force which accounts for both the asymmetry of the axial potential well and the fact that the depth of the axial potential is always shallower than the corresponding transverse well depth. The basic principles of single-beam gradient traps fed by visously confined particles were demonstrated[11] in experiments on the trapping of Rayleigh particles as small as 25 nm in water.

Damping of the atomic motion in the trap is accomplished by periodically turning the trap off and applying OM by use of square-wave modulation.[21] This avoids any interference with the damping process due to the optical Stark shift of the atomic resonance caused by the trap beam,[4] or interference with the trapping process due to the large fraction of excited atoms caused by the OM beams. If the chopping cycle is short compared with one-half of the oscillation period of an atom in the potential well ($\sim 10 \mu s$),

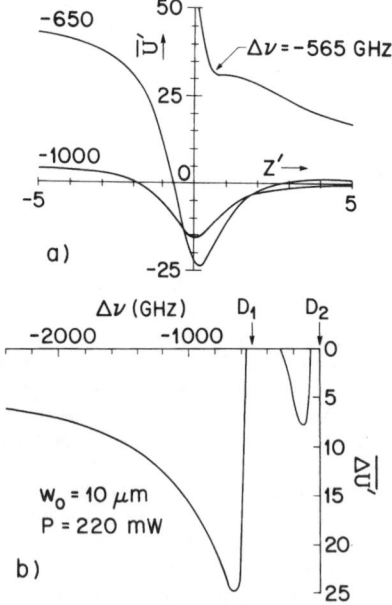

a)

b)

FIG. 1. (a) The normalized, time-averaged axial potential $\bar{U}' = 2\bar{U}/h\gamma$ as a function of the axial coordinate $z' = z\lambda/\pi w_0^2$ for $w_0 = 10 \mu$m, $P = 220$ mW, and for three values of trap-laser detuning below the D_2 resonance line. The calculation uses $I_s(D_1) = 2I_s(D_2) = 40$ mW/cm². (b) The normalized, time-averaged, potential well depth versus $\Delta\nu$ for $w_0 = 10 \mu$m and $P = 220$ mW.

then an atom in the trap does not move much during the chopping cycle and its motion is roughly that of a damped harmonic oscillator with an average well depth of one-half the cw well depth. In our trap, the heating rate due to the trapping forces is so small that the temperature of the atoms in the trap remains close to the cw OM temperature.

Our experimental apparatus is similar to the apparatus used earlier to demonstrate OM[18,19] with the addition of a focused trapping beam nearly parallel to one of the OM beams. Pulses of sodium atoms were created every 0.1–10 s by evaporating sodium metal with a 10-ns, pulsed, yttrium-aluminum-garnet laser beam. Atoms with a velocity of about 2×10^4 cm/s were slowed over a distance of 8 cm to about 2×10^3 cm/s by the radiation pressure of a frequency-chirped laser beam counterpropagating with the atomic beam. An electro-optic modulator at 856.2 MHz is used to create sidebands 1712.4 MHz apart to prevent optical pumping.[18] The slowed atoms drift into the OM region where three-dimensional cooling and viscous confinement occurs. The OM-beam spot size is about 1 cm and the peak intensity is about 20 mW/cm² in the relevant sidebands. The optimum tuning of the OM beams ($\sim \gamma/2$ below resonance) is determined empirically by tuning for maximum storage time. We routinely achieve OM lifetimes of 0.5 s and densities of 10^6 atoms/cm³. Temperatures as low as 0.24×10^{-3} K were previously measured.[18] The trap laser beam, obtained from a second dye laser, enters the vacuum chamber nearly parallel to one of the OM beams. The linearly polarized trapping beam has a power of about 220 mW and is focused to a diffraction-limited spot radius $w_0 = 10 \mu$m.

The optically trapped atoms were detected visually, by video camera, and photographically. Figure 2(a) is a 0.5-s exposure that shows the fluorescence from atoms in the initial slowing beam, the subsequently formed OM cloud, and atoms collecting in the trap. Figure 2(b) taken later shows the bright spot from trapped atoms which remain in the trap after most of the surrounding OM atoms have diffused away. The brightness of the fluorescence from the trapped atoms indicates a density much higher than the surrounding cloud of cold atoms. Virtually all of the fluorescence is caused by the OM beams since our trap is tuned far from the atomic resonance.

Several tests were made to confirm trapping: (i) The bright oval-shaped spot coming from trapped atoms occurs only for tunings within the range expected for axial trapping. For example, strong trapping was observed between −570 to at least −1300 GHz below the D_2 resonance with the deepest traps occurring at −650 ± 25 GHz. This agrees well with calculations based on Fig. 1(b). (ii) We observed visually that the lifetime of trapped atoms was longer than for

315

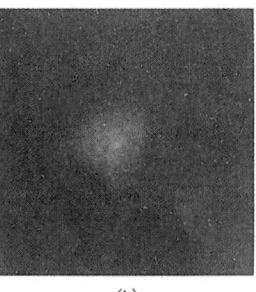

(a)   (b)

FIG. 2. (a) Photo showing the collimating nozzle, atomic beam, and atoms confined in OM. The distance from the nozzle to the OM region is 5 cm. (b) Photo taken after the atomic source and the slowing laser beam have been turned off, showing trapped atoms.

confined OM atoms. To quantify these observations we analyzed video tapes of the trap decay using a video waveform monitor. Figure 3(a) shows the signal of one horizontal video line through the trap center. The curve S_1 in Fig. 3(b) shows that atoms confined within only OM decay with a lifetime of about 0.5 s. Curve S_2 is for trapped atoms plus atoms in OM. The fact that S_2 is nonzero for more than a second after S_1 has fully decayed indicates that trapped atoms are longer lived. We surmise that background pressure is limiting the trap decay lifetimes to ~ 1 s. Since background gas atoms are at ~ 300 K and trapped atoms will be ejected by collisions transferring $< 10^{-2}$ K, we expect a large cross section for ejection of $\sim 10^{-13}$ cm². (iii) The shape of the trap fluorescence varied with tuning and trap-laser intensity as expected from the calculated axial potential profiles shown in Fig. 1(a). As an example, for strong trapping at -650 GHz we see a bright oval-shaped spot ~ 0.5 mm long, but not fully resolved. As the power is lowered, the oval lengthens and finally a long weak fluorescent streak appears only on the $+z$ side of the trap due to atoms confined transversely by the stronger radial potential well. (iv) Measurements were made on the ef-

fect of our varying the chopping period for the trapping and OM beams. At 220-mW and -650-GHz tuning, good trapping was obtained for chopping periods between 0.4 and 10.0 μs. Computer simulations of the radial motion by use of average optical forces shows that the atomic motion becomes unstable for periods longer than 8 μs. We believe that trapping fails at fast chopping rates as a result of the generation of frequency sidebands, which is detrimental to optical cooling.

A measurement of the trap heating time was made by turning off the OM while keeping the trap on for varying times (0, 5, 10, 15 ms). At times of 5 ms the brightness of the trapped atoms decreases by about a factor of 2. This agrees with our calculated trap heating time[4] of 4 ms.

We can determine the temperature of trapped atoms from the observation that at -750 GHz strong traps were obtained for 220 and 110 mW while no trapping was seen for 90 mW. This is done by calculating the intrinsic leakage rate of the trap as a function of power and temperature. The leakage rate is given by the escape-attempt frequency (twice the oscillation frequency of an atom in the trap) times the escape probability, $\exp(-\Delta U/kT)$. To agree with observation, we require intrinsic trap lifetimes of 1 s or greater for 110 mW and much less than 1 s for 90 mW. The result is, $0.6 \leqslant kT/h\gamma \leqslant 0.8$; thus the trapped-atom temperature is remarkably close to the quantum limit for atoms in OM,[18] $kT = \frac{1}{2}h\gamma = 240$ μK. The intrinsic trap lifetimes corresponding to $kT = 0.7h\gamma$ are 0.12 s for 90 mW, 1.0 s for 110 mW, and 3.7×10^4 s for 220 mW. The inferred temperature for the trapped atoms allows us to calculate the fraction of the trapping volume in which the atoms reside. For a detuning of -650 GHz, a trap beam power of 220 mW, and an atomic temperature of $0.7h\gamma$, the atoms are confined within a cylinder of length 210 μm and diameter of 2.5 μm, giving a volume of 1×10^{-9} cm³.

We deduced the number of atoms in the trap and

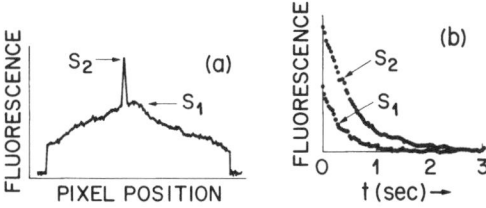

FIG. 3. (a) Video signal on a single horizontal video line through the trap showing scattered light, camera noise, atoms in OM, and a sharp spike due to the fluorescence of the trapped atoms. Data were quantified by use of boxcar integrators to measure the time dependence of the levels S_2 and S_1. (b) Time dependence of S_2 (trap plus OM fluorescence) and S_1 (OM fluorescence).

VOLUME 57, NUMBER 3   PHYSICAL REVIEW LETTERS   21 JULY 1986

their density in several independent ways. From an absolute calibration of the detection system (zoom lens, video camera, and waveform monitor combination), we conclude that our best trap signals are due to about 500 atoms. Based on the confinement volume of 1×10^{-9} cm^3, the corresponding trapped-atom density is 5×10^{11} cm^{-3}. An accurate estimate of the trapped-atom density is also obtained by noting that the signal from the trapped atoms is roughly equal to the signal from the atoms in OM. Using the f-stop number of the video-camera lens and assuming a OM radius of 0.5 cm, we estimate that the average trap density is 1.5×10^6 times the OM density. A OM density of 5×10^5 cm^{-3} yields a trap density of 8×10^{11} cm^{-3}, reasonably consistent with the previous estimate.

Finally we note that the collection of trapped atoms could be moved easily at speeds on the order of 1 cm/s by manually scanning the location of the trap focal spot.

In summary, optical trapping of neutral atoms has been observed in good quantitative agreement with theoretical expectations. We have achieved orders-of-magnitude higher densities of trapped atoms and lower temperatures than reported previously using magnetic neutral-atom traps[22] and electrodynamic ion traps.[23,24] It may be possible to increase the density by $\geq 10^4$ and to decrease the temperature to $\leq 10^{-6}$ K. This would involve tighter beam focusing, additional cooling of OM with narrower linewidth transitions, or possibly beam-expansion schemes.[25] Confinement and spatial manipulation of cold trapped atoms to dimensions less than the optical wavelength are also possible. These capabilities should prove useful in studies of Bose and Fermi gas statistics and other interactions at high atomic density, atom-atom collisions, atom-surface interactions, atom diffraction and tunneling, molecular formation using individual atoms, modifications of atomic spontaneous-emission lifetimes and other collective effects, spectroscopy of the trap energy levels, and, possibly, high-resolution spectroscopy and atomic time standards.

The authors thank J. P. Gordon for valuable discussions.

[1]A. Ashkin, Phys. Rev. Lett. **24**, 156 (1970).

[2]A. Ashkin, Phys. Rev. Lett. **25**, 1321 (1970).

[3]A. Ashkin, Phys. Rev. Lett. **40**, 729 1978).

[4]J. P. Gordon and A. Ashkin, Phys. Rev. A **21**, 1606 (1980).

[5]A. Ashkin, Science **210**, 1081 (1980).

[6]V. Letokhov and V. Minogin, Physics Rep. **73**, 3 (1981).

[7]*Laser-Cooled and Trapped Atoms*, edited by W. D. Phillips, Progress in Quantum Electronics Vol. 8 (Pergamon, New York, 1984).

[8]A. Ashkin, Opt. Lett. **9**, 454 (1984).

[9]G. Roosen, Can. J. Phys. **57**, 1260 (1979).

[10]A. Ashkin and J. M. Dziedzic, Phys. Rev. Lett. **54**, 1245 (1985).

[11]A. Ashkin, J. M. Dziedzic, J. E. Bjorkholm, and Steven Chu, Opt. Lett. **11**, 288 (1986).

[12]R. Schieder, H. Walther, and L. Wöste, Opt. Commun. **5**, 337 (1972).

[13]P. Jacquinot, S. Liberman, J. L. Picque, and J. Pinard, Opt. Commun. **8**, 163 (1973).

[14]J. E. Bjorkholm, R. R. Freeman, and D. B. Pearson, Phys. Rev. A **23**, 491 (1981).

[15]J. E. Bjorkholm, R. R. Freeman, A. Ashkin, and D. B. Pearson, Phys. Rev. Lett. **41**, 1361 (1978).

[16]J. Prodan, A. Migdall, W. D. Phillips, I. So, and H. Dalibard, Phys. Rev. Lett. **54**, 992 (1985).

[17]E. Ertmer, R. Blatt, J. L. Hall, and M. Zhu, Phys. Rev. Lett. **54**, 996 (1985).

[18]Steven Chu, L. Hollberg, J. E. Bjorkholm, Alex Cable and A. Ashkin, Phys. Rev. Lett. **55**, 48 (1985).

[19]Steven Chu, J. E. Bjorkholm, A. Ashkin, L. Hollberg, and Alex Cable, "Methods of Laser Spectroscopy" (Plenum, New York, to be published).

[20]We have recently learned of a newly proposed scattering-force optical trap that can have volumes and well depths comparable with alternating-beam scattering-force traps: D. E. Pritchard, private communication, and D.E. Pritchard *et al.*, preceeding Letter [Phys. Rev. Lett. **57**, 310 (1986)].

[21]J. Dalibard, S. Reynaud, and C. Cohen-Tannoudji, Opt. Commun. **47**, 395 (1983).

[22]A. L. Migdall, J. V. Prodan, W. D. Phillips, T. H. Bergeman, and H. J. Metcalf, Phys. Rev. Lett. **54**, 2596 (1985).

[23]D. J. Wineland, W. M. Itano, J. C. Bergquist, J. J. Bollinger, and H. Hemmate, Prog. Quantum Electron. **8**, 139 (1984).

[24]W. Nagourney, G. Janik, and H. Dehmelt, Proc. Natl. Acad. Sci. U.S.A. **80**, 643 (1983).

[25]S. Chu, J. E. Bjorkholm, A. Ashkin, J. P. Gordon, and L. W. Hollberg, Opt. Lett. **11**, 73 (1986).

Reprinted from Optics Letters, Vol. *11*, page 288, May, 1986.
Copyright © 1986 by the Optical Society of America and reprinted by permission of the copyright owner.

Observation of a single-beam gradient force optical trap for dielectric particles

A. Ashkin, J. M. Dziedzic, J. E. Bjorkholm, and Steven Chu

AT&T Bell Laboratories, Holmdel, New Jersey 07733

Received December 23, 1985; accepted March 4, 1986

Optical trapping of dielectric particles by a single-beam gradient force trap was demonstrated for the first reported time. This confirms the concept of negative light pressure due to the gradient force. Trapping was observed over the entire range of particle size from 10 μm to ~25 nm in water. Use of the new trap extends the size range of macroscopic particles accessible to optical trapping and manipulation well into the Rayleigh size regime. Application of this trapping principle to atom trapping is considered.

We report the first experimental observation to our knowledge of a single-beam gradient force radiation-pressure particle trap.[1] With such traps dielectric particles in the size range from 10 μm down to ~25 nm were stably trapped in water solution. These results confirm the principles of the single-beam gradient force trap and in essence demonstrate the existence of negative radiation pressure, or a backward force component, that is due to an axial intensity gradient. They also open a new size regime to optical trapping encompassing macromolecules, colloids, small aerosols, and possibly biological particles. The results are of relevance to proposals for the trapping and cooling of atoms by resonance radiation pressure.

A wide variety of optical traps based on the basic scattering and gradient forces of radiation pressure have been demonstrated or proposed for the trapping of neutral dielectric particles and atoms.[2–4] The scattering force is proportional to the optical intensity and points in the direction of the incident light. The gradient force is proportional to the gradient of intensity and points in the direction of the intensity gradient. The single-beam gradient force trap is conceptually and practically one of the simplest radiation-pressure traps. Although it was originally proposed as an atom trap,[1] we show that its uses also cover the full spectrum of Mie and Rayleigh particles.

It is distinguished by the feature that it is the only all-optical single-beam trap. It uses only a single strongly focused beam in which the axial gradient force is so large that it dominates the axial stability. In the only previous single-beam trap, the so-called optical levitation trap, the axial stability relies on the balance of the scattering force and gravity.[5] In that trap the axial gradient force is small, and if one turns off or reverses the direction of gravity the particle is driven out of the trap by the axial scattering force.

There were also relevant experiments using gradient forces on Rayleigh particles that did not strictly involve traps, in which liquid suspensions of submicrometer particles acted as an artificial nonlinear optical Kerr medium.[6]

The physical origin of the backward gradient force in single-beam gradient force traps is most obvious for particles in the Mie size regime, where the diameter is large compared with λ. Here one can use simply ray optics to describe the scattering and optical momentum transfer to the particle.[7,8] In Fig. 1a) we show the scattering of a typical pair of rays A of a highly focused beam incident upon a 10-μm lossless dielectric sphere, for example. The principal part of the momentum transfer from the incident light to the particle is due to the emergent rays A', which are refracted by the particle. Successive surface reflections, such as R_1 and R_2, contribute a lesser scattering. For a glass particle in water the effective index m, equal to the index of the particle divided by the index of the medium, is about 1.1 to 1.2, and the sphere acts as a weak positive lens. If we consider the direction of the resulting forces F_A on the particle that are due to refraction of rays A in the weak-lens regime, we see as shown in Fig. 1a) that there is a substantial net backward trapping-force component toward the beam focus.

Figure 2 sketches the apparatus used for trapping Mie or Rayleigh particles. Spatially filtered argon-laser light at 514.5 nm is incident upon a high-numerical-aperture (N.A. 1.25) water-immersion microscope objective, which focuses a strongly convergent downward-directed beam into a water-filled glass cell. Glass Mie particles are introduced into the trap by an auxiliary vertically directed holding beam,[5] which lifts particles off the bottom of the cell and manipulates them to the focus. Rayleigh particles are simply dispersed in water solution at reasonable concentrations and enter the trapping volume by Brownian diffusion. A microscope M is used to view the trapped particles visually off a beam splitter S or by recording the 90° scatter with a detector D.

Figure 1b) is a photograph of a 10-μm glass sphere of index about 1.6 trapped and levitated just below the beam focus of a ~100-mW beam. The picture was taken through a green-blocking filter using the red fluorescence of the argon laser beam in water in order to make the trajectories of the incident and scattered

May 1986 / Vol. 11, No. 5 / OPTICS LETTERS **289**

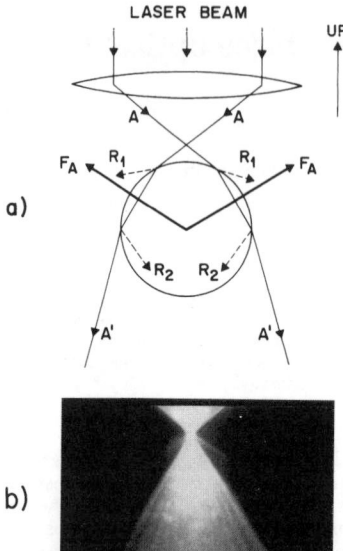

Fig. 1. a) Diagram showing the ray optics of a spherical Mie particle trapped in water by the highly convergent light of a single-beam gradient force trap. b) Photograph, taken in fluorescence, of a 10-μm sphere trapped in water, showing the paths of the incident and scattered light rays.

beams visible. The sizable decrease in beam angle of the scattered light, which gives rise to the backward force, is clearly seen. The stria in the forward-scattered light arise from the usual Mie-scattering ring pattern.

Next consider the possibility of single-beam trapping of submicrometer Rayleigh particles whose diameter $2r$ is much less that λ. Although we are now in the wave-optic regime, we will again see the role of the strong axial gradient in producing a net backward axial force component. For Rayleigh particles in a medium of index n_b the scattering force in the direction of the incident power is $F_{scat} = n_b P_{scat}/c$, where P_{scat} is the power scattered.[9] In terms of the intensity I_0 and effective index m

$$F_{scat} = \frac{I_0}{c} \frac{128\pi^5 r^6}{3\lambda^4} \left(\frac{m^2-1}{m^2+2}\right)^2 n_b. \quad (1)$$

The gradient force F_{grad} in the direction of the intensity gradient for a spherical Rayleigh particle of polarizability α is[6]

$$F_{grad} = -\frac{n_b}{2}\alpha\nabla E^2 = -\frac{n_b^3 r^3}{2}\left(\frac{m^2-1}{m^2-2}\right)\nabla E^2. \quad (2)$$

This Rayleigh force component, in analogy with the gradient force for Mie particles, can be related to the lenslike properties of the scatterer.

As for atoms,[1] the criterion for axial stability of a

single-beam trap is that R, the ratio of the backward axial gradient force to the forward-scattering force, be greater than unity at the position of maximum axial intensity gradient. For a Gaussian beam of focal spot size w_0 this occurs at an axial position $z = \pi w_0^2/\sqrt{3}\,\lambda$, and we find that

$$R = \frac{F_{grad}}{F_{scat}} = \frac{3\sqrt{3}}{64\pi^5}\frac{n_b^2}{\left(\frac{m^2-1}{m^2+2}\right)}\frac{\lambda^5}{r^3 w_0^2} \geq 1, \quad (3)$$

where λ is the wavelength in the medium. This condition applies only in the Rayleigh regime where the particle diameter $2r \lesssim 0.2\lambda \cong 80$ nm. In practice we require R to be larger than unity. For example, for polystyrene latex spheres in water with $m = 1.65/1.33 = 1.24$ and $2w_0 = 1.5\lambda = 0.58$ μm we find for $R \geq 3$ that $2r \leq 95$ nm. Thus with this choice of spot size we meet the stability criterion over the full Rayleigh regime. The fact that $R < 3$ for $2r > 95$ nm does not necessarily imply a lack of stability for such larger particles since we are beyond the range of validity of the formula. Indeed, as we enter the transition region to Mie scattering we expect the ray-optic forward-scattering picture to be increasingly valid. As will be seen experimentally we have stability from the Rayleigh regime, through the transition region, into the full Mie regime. For silica particles in water with $m = 1.46/1.33 = 1.10$ and $2w_0 = 0.58$ μm we find for $R \geq 3$ that $2r \leq 126$ nm. For high-index particles with $m \equiv 3.0/1.33 = 2.3$ we find that $2r \leq 61$ nm.

The stability condition on the dominance of the backward axial gradient force is independent of power and is therefore a necessary but not sufficient condition for Rayleigh trapping. As an additional sufficient trapping condition we have the requirement[1] that the Boltzmann factor $\exp(-U/kT) \ll 1$, where $U = n_b\alpha E^2/2$ is the potential of the gradient force. As was previously pointed out,[6] this is equivalent to requiring that the time to pull a particle into the trap be much less than the time for the particle to diffuse out of the trap by Brownian motion. If we set $U/kT \gtrsim 10$,

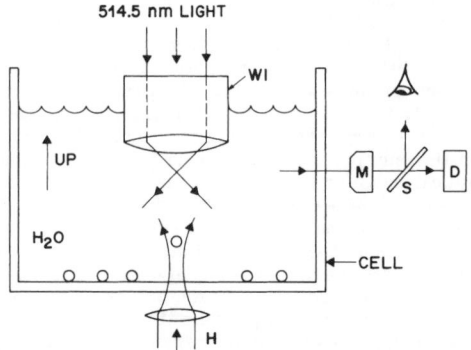

Fig. 2. Sketch of the basic apparatus used for the optical trapping of Mie and Rayleigh particles in water by means of a single-beam gradient force radiation-pressure trap.

for example, and use a power of ~1.5 W focused close to the limiting spot diameter of 0.58 μm $\cong$ 1.5λ, we find for silica that the minimum theoretical particle size that satisfies this condition is $2r$ = 19 nm. For polystyrene latex, the minimum particle size that can be trapped under these conditions is $2r$ = 14 nm. With a high-index particle of m = 3/1.33 = 2.3 the theoretical minimum size is $2r \cong$ 9 nm.

Additional experiments were performed on individual colloidal polystyrene latex particles in water. Unfortunately the particles exhibit a form of optical damage at high optical intensities. For 1.0-μm spheres with a trapping power of a fraction of a milliwatt, particles survived for tens of minutes and then shrank in size and disappeared. Spheres of 0.173 μm were trapped for several minutes with a power of a milliwatt before being lost. Particles of 0.109-μm diameter required about 12–15 mW and survived about 25 sec. With 85- and 38-nm latex particles the damage was so rapid that it was difficult to observe the scattering reliably. It was nevertheless clear that trapping occurred over full size range from Mie to Rayleigh particles.

The remarkable uniformity of latex particles was evident from the small variation of ±15% in the 90° scatter of 0.109-μm particles. Since the scattering is closely Rayleigh this corresponds to a diameter variation of ±2.4%. Subsequently we determined the size of unknown silica Rayleigh particles by comparing their scatter with the scatter from the 0.109-μm particles taken as a standard, using Eq. (1). Although the 0.109-μm particles are not strictly Rayleigh, one can make a modest theoretical correction[10] of ~1.06 to the effective particle size.

Trapping of nominally spherical colloidal silica particles was observed by using commercially available Nalco and Ludox samples[11] diluted with distilled water. With a high concentration we quickly collect many particles in the trap and observe a correspondingly large scattering. At reduced concentration we can observe single particles trapped for extended time. Once a particle is captured at the beam focus we observe an apparent cessation of all Brownian motion and a large increase in particle scattering.

With silica samples we always observe a wide distribution of particle sizes as evidenced by the more than an order-of-magnitude difference in scattering from particles trapped with a given laser power. Particle damage by the light was not a serious problem with silica particles. The smaller particles of the distribution showed only slight changes of scattering over times of minutes. The larger particles would often decay by factors up to 3 in comparable times.

Measurements were made on a Nalco 1060 sample with a nominal size of about 60 nm and an initial concentration of silica of 50% by weight diluted to one part in 10^5–10^6 by volume. Trapping powers of 100–400 mW were used. The absolute size of the Nalco 1060 particles as determined by comparison with the 0.109-μm latex standard varied from ~50 to 90 nm with many at ~75 nm. We also studied smaller silica particles using a Ludox TM sample with nominal particle diameter of ~21 nm and a Nalco 1030 sample with nominal distribution of 11 to 16 nm. Dilutions of ~10^6–10^7 and powers of ~500 mW to 1.4 W were used. With both samples we were limited by laser power in the minimum-size particle that could be trapped. With 1.4 W of power the smallest particle trapped had a scattering that was a factor of ~3 × 10^4 less than from the 0.109-μm standard. This gives a minimum particle size of 26 nm assuming a single spherical scatterer. The measured minimum size of 26 nm compares with the theoretically estimated minimum size of ~19 nm for this power, based on U/kT = 10 and spot size $2w_0$ = 0.58 μm. This difference could be resolved by assuming a spot size $2w_0$ = 0.74 μm = 1.28 (0.58 μm) $\cong$ 1.9 λ.

Experimentally we found that we could introduce a significant drift of the fluid relative to the trapped particle by moving the entire cell transversely relative to the fixed microscope objective. This technique gives a direct method of measuring the maximum trapping force. It also implies the ability to separate a single trapped particle from surrounding untrapped particles by a simple flushing technique.

Our observation of trapping of a 26-nm silica particle with 1.4 W implies, by simple scaling, the ability to trap a 19.5-nm particle with m = 1.6/1.33 = 1.20 and a 12.5-nm particle of m = 3.0/1.33 = 2.26 at the same power. These results suggest the use of the single-beam gradient force traps for other colloidal systems, macromolecules, polymers, and biological particles such as viruses. In addition to lossless particles with real m there is the possibility of trapping Rayleigh particles with complex m for which one can in principle achieve resonantly large values of the polarizability α. Finally, we expect that these single-beam traps will work for trapping atoms[1] as well as for macroscopic Rayleigh particles since atoms can be viewed as Rayleigh particles with a different polarizability.

References

1. A. Ashkin, Phys. Rev. Lett. **40**, 729 (1978).
2. A. Ashkin, Science **210**, 1081 (1980); V. S. Letokhov and V. G. Minogin, Phys. Rep. **73**, 1 (1981).
3. A. Ashkin and J. P. Gordon, Opt. Lett. **8**, 511 (1983).
4. A. Ashkin and J. M. Dziedzic, Phys. Rev. Lett. **54**, 1245 (1985).
5. A. Ashkin and J.M. Dziedzic, Appl. Phys. Lett. **19**, 283 (1971).
6. P. W. Smith, A. Ashkin, and W. J. Tomlinson, Opt. Lett. **6**, 284 (1981); and A. Ashkin, J. M. Dziedzic, and P. W. Smith, Opt. Lett. **7**, 276 (1982).
7. A. Ashkin, Phys. Rev. Lett. **24**, 146 (1970).
8. G. Roosen, Can. J. Phys. **57**, 1260 (1979).
9. See, for example, M. Kerker, *The Scattering of Light* (Academic, New York, 1969), p. 37.
10. W. Heller, J. Chem. Phys. **42**, 1609 (1965).
11. Nalco Chemical Company, Chicago, Illinois; Ludox colloidal silica by DuPont Corporation, Wilmington, Delaware.

Feedback stabilization of optically levitated particles

A. Ashkin and J. M. Dziedzic

Bell Telephone Laboratories, Holmdel, New Jersey 07733
(Received 5 November 1976)

We demonstrate the locking of an optically levitated sphere to an external reference using an electronic
feedback system. This provides a new external source of damping for the stabilization and manipulation of
particles in vacuum and at atmospheric pressure. The method permits accurate and continuous monitoring
of applied forces. Numerous applications are suggested.

PACS numbers: 42.80.−f, 42.60.−v, 06.60.Sx

We report on experiments demonstrating the use of an electronic feedback system to provide damping, stabilization, and manipulation of optically levitated particles in high vacuum and air. The feedback method also gives an accurate means for continuously measuring external forces applied to levitated particles and other useful features. Optical levitation[1] is basically a technique for supporting, manipulating, and studying small transparent particles using the forces of radiation pressure from laser beams. It can be used to probe optical, gravity, and electrical forces[2,3] applied to small particles. Optically levitated particles are, in principle, stable and should therefore remain fixed in the laser beam at the point of stable equilibrium. This is largely true in air[2] where strong viscous forces rapidly damp out motions due to any perturbations. Recently optical levitation was extended to the high-vacuum regime[4] where viscous damping can be made arbitrarily small. In vacuum, particle oscillations never totally cease but grow and decay due to residual random laser and mechanical perturbations. Also, in the earlier experiments, all manipulations in vacuum had to be performed adiabatically to prevent loss of the particle. These difficulties can in part be overcome passively by minimizing the perturbations. We have chosen however to apply active electronic feedback to damp the oscillations, both vertical and horizontal, and lock the particle's height to a fixed reference. This method also provides a means of manipulating particles in vacuum by manipulating the external reference. With the external stability provided by the feedback system it is now possible to levitate in those regions where

optical levitation by itself is inherently unstable. Although particles are locked to a fixed reference they are still free to rotate. Thus optical levitation with feedback has many potential applications, for example, in laser fusion experiments,[2] construction of ultracentrifuges, study of photoemission,[5] and measurement of gravity forces. This feedback scheme has features in common with the one used by Beams[6] to magnetically levitate and rotate steel spheres in vacuum.

Figure 1 illustrates the basics of the feedback scheme. The particle height is sensed optically by projecting an enlarged image of the levitated sphere, illuminated by the support beam, on a split photodiode detector $D_1 D_2$.[7] This image, as viewed at 90° by the microscope M is actually made up of two bright spots[8] but we use only the brighter upper one. By centering the upper spot on the photodetector and masking the lower spot we derive an error signal E from the difference of the signals from the two diodes $(D_1 - D_2)$ which is proportional to the displacement from equilibrium. A vertical displacement of a fraction of a micron results in a significant error signal E. This is amplified and processed to give a feedback voltage V which is applied to the KDP electro-optic modulator. The modulator is biased so that a change in transmitted optical power is proportional to the applied voltage V.

The simplest feedback circuit is one in which V is proportional to E. Such a scheme effectively increases the restoring force of the levitated particle but cannot by itself extract energy from particle motion. This nevertheless works well for levitation in air since the system has internal viscous gas damping which reduces any oscillation and locks the particle to the reference. In vacuum where internal damping is lacking, the particle simply oscillates about its equilibrium position at a frequency of several hundred hertz which is higher than the natural frequency of ~20 Hz due to the larger restoring force. If, however, a second component proportional to dE/dt is added to the feedback voltage V, a 90° phase lead is introduced which changes the light power in the proper phase to slow all vertical motions. This is a true damping in which a light force proportional to velocity opposes any particle motion. Such external or circuit damping was used by Beams[6] and is known in feedback theory as error rate damping.[9] Thus for vacuum operation we added the derivative voltage dE/dt to the error voltage E using a separate amplifier and adder circuit and then varied the magnitude of the dE/dt voltage until vertical oscillations were damped

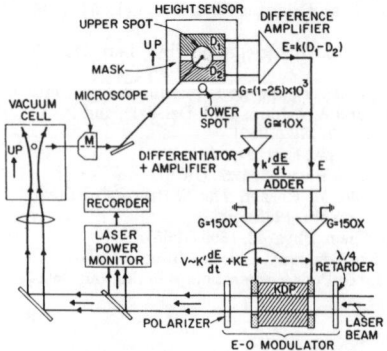

FIG. 1. Sketch of the feedback stabilization apparatus.

out. It was discovered that the vertical damping mechanism can also damp horizontal oscillations. In levitation a horizontal displacement of a particle, besides giving a radial restoring force, results in a vertical displacement as well, because of the variation of intensity with horizontal position. This coupling due to beam shape implies that any horizontal oscillation has a vertical component which can be damped by the feedback circuit. Thus by supporting the particle in a beam with a tight focus we can increase the coupling sufficiently to lock the particle to the reference with no detectable vertical or horizontal motion to within a fraction of a micron in vacuum. The amplified error signal E meanwhile displays all the fluctuations to which the particle is subject within its response bandwidth. This is from dc to 10 kHz. The total amplifier gain used varied with the detector output voltage. For typical particle sizes this was 10^5 to 10^7 times. The circuit is capable of correction voltage swings of ± 300 V without saturating. Within these limits the system corrects for all fluctuations including changes in input laser power.

An important aspect of the feedback scheme is that it, in effect, acts as an accelerometer or force measuring device. Indeed, the smoothed light power at the output of the modulator is a sensitive measure of the force needed to levitate the particle at the reference height. If one applies any additional force, as for example a downward electric force on a charged sphere, the system automatically readjusts the power to keep the particle at the reference height. The added light power is then a direct measure of the applied electric force. From this force one can get the charge on the sphere. This method of measuring charge is more accurate and faster than the previous particle deflection technique used in photoemission experiments.[5]

The ability of feedback to provide external stability, a means of manipulation, and continuous monitoring of forces applied to levitated particles is illustrated in the next experiment. Here we manipulate a particle

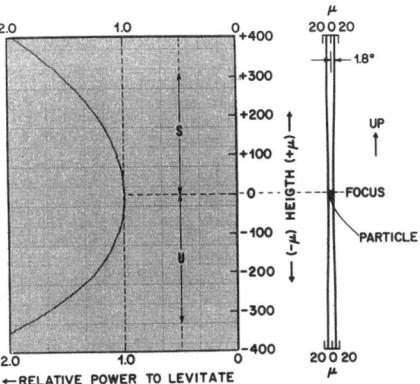

FIG. 3. Photograph of the chart recorder trace shows the relative stabilized power at $\lambda \cong 0.60\ \mu$ needed to levitate an 8.7-μ-diam sphere over a range of heights about the focus of a beam having a 12.2-μ-diam minimum spot size. S and U indicate the regions where the sphere is stable or unstable with no feedback. The beam geometry is shown to the right.

up and down within the levitating beam and continuously measure the levitating force at different heights. This is done by moving the reference height sensor vertically relative to the beam with a motor drive. The particle is automatically forced to follow the reference and the stabilized laser power is a measure of the relative light force at each height. In Fig. 2 the 10.8-μ-diam particle is larger than the 4.1-μ-diam beam waist and the force, which is proportional to the inverse of the power, displays the expected minimum at the focus.[2] In Fig. 3 the 8.7-μ-diam particle is smaller than the 12.2-μ-diam beam waist and the force is a maximum at the focus.[1] Comparing with results of previous force measurements taken point by point without feedback,[1,2] we see that the feedback data is continuous and much more accurate (a few tenths of a percent). Also, because of the added stability of the feedback, we can hold particles and measure force in the previously inaccessible regions labeled U where particles are inherently unstable in the absence of feedback. This technique of feedback manipulation by moving the reference detector relative to the beam should be generally useful especially for vacuum experiments, for example, in the introduction and exact location of laser fusion targets[2] in the target area.

For manipulation where the feedback circuit is called upon to change the power level beyond its range of control, we simply make an occasional change in the overall power level by hand with an attenuator to restore the average modulator voltage V close to zero. Even this occasional manual adjustment can be avoided with a circuit capable of larger voltage swings.

The beam configuration with a tight focus and highly divergent beam, as shown in Fig. 2, was originally chosen[2] because it results in a strong optical trap in which the particle height is insensitive to power or force changes. With a feedback we can now easily levitate in beams with weak focusing (nearly parallel beams) and try to take advantage of the opposite situa-

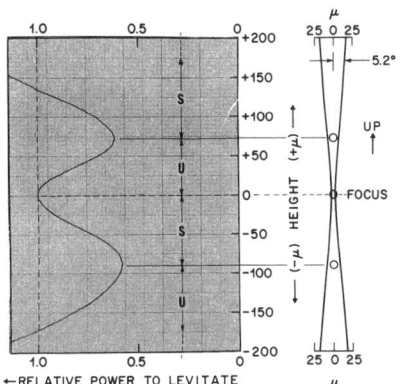

FIG. 2. Photograph of the chart recorder trace showing the relative stabilized power at $\lambda \cong 0.60\ \mu$ needed to levitate a 10.8-μ-diam sphere over a range of heights about the focus of a beam having a 4.1-μ-diam minimum spot size. S and U indicate the regions where the sphere is stable or unstable with no feedback. The beam geometry is shown to the right.

A. Ashkin and J.M. Dziedzic   203

tion, namely, a weak trap with high sensitivity to small force changes. This is an extreme case of Fig. 3 with a much larger beam diameter. For example a particle levitated at the focus of a beam having a $70\text{-}\mu\text{-focal-}$ spot diameter moves $10\ \mu$ in height for a force change or gravity change of about one part in 10^6 of the particle's weight. If laser perturbations and mechanical fluctuations can be reduced or separately stabilized, then the feedback system should not only hold a particle close to the focus but be able to record force changes of one part in 10^6. The ultimate limitation on sensitivity will come from the inability to supply enough power to very wide beams or the inability to reduce the background fluctuations.

The force measuring technique is proving to be useful in the study of radiometric forces. We can continuously record the expected[4] variation of the radiometric force from its low value at atmospheric pressure through its maximum and on down to negligible values as the pressure is reduced. In addition, however, with fused silica spheres we often observe a steady increase in the radiometric force with time at a fixed pressure. The effect is clearly due to an increase in absorption and is reminiscent of the more drastic effect observed qualitatively in silicone oil spheres.[4] If severe enough it prevents us from pumping to low pressure without losing the silica particle. Our only present solution to this difficulty is to either pump down rapidly before the absorption becomes too high or to hunt for particularly resistant particles. So far cleaning attempts only made things worse.

Another potential application of feedback stabilization is to experiments based on rotation in vacuum. Although all linear motions are damped, rotational motion re-

mains undamped. Experiments based on the angular momentum of light or experiments involving high rotational speeds can therefore use this technique. Beams[6] in his classic rotational experiments using 0.4-mm-diam magnetically levitated steel spheres, stabilized by feedback, achieved a 0.8-MHz rotational frequency and a centrifugal acceleration of 5×10^8 g. The maximum rotational frequency at particle rupture scales inversely with the radius of the sphere and directly with the tensile strength of the material and its density. Beam's minimum sphere size was dictated by his method of providing transverse stability. With optical levitation we can readily levitate $4\text{-}\mu\text{-diam}$ fused silica spheres whose strength can be five times higher than steel and whose density is less by a factor of 3.6. This scales to a limiting rotational frequency of ~1500 MHz with a centrifugal acceleration of ~10^{12} g.

[1]A. Ashkin and J.M. Dziedzic, Appl. Phys. Lett. **19**, 283 (1971).
[2]A. Ashkin and J.M. Dziedzic, Appl. Phys. Lett. **24**, 586 (1974).
[3]A. Ashkin and J.M. Dziedzic, Science **187**, 1073 (1975).
[4]A. Ashkin and J.M. Dziedzic, Appl. Phys. Lett. **28**, 333 (1976).
[5]A. Ashkin and J.M. Dziedzic, Phys. Rev. Lett. **36**, 267 (1976).
[6]J.W. Beams, Science **120**, 619 (1954); Rev. Sci. Instrum. **21**, 182 (1950).
[7]The split photodiode is a United Detector Technology PIN Spot 2D detector.
[8]A. Ashkin, Sci. Am. **226**, 63 (1972).
[9]H. Lauer, R.N. Lesnick, and L.E. Matson, *Servomechanism Fundamentals* (McGraw-Hill, New York, 1960), Chap. 5.

IEEE JOURNAL ON SELECTED TOPICS IN QUANTUM ELECTRONICS, VOL. 6, NO. 6, NOVEMBER/DECEMBER 2000   841

History of Optical Trapping and Manipulation of Small-Neutral Particle, Atoms, and Molecules

A. Ashkin, *Life Fellow, IEEE*

Invited Paper

Abstract—This paper reviews the history of optical trapping and manipulation of small-neutral particles, from the time of its origin in 1970 up to the present. As we shall see, the unique characteristics of this technique are having a major impact on the many subfields of physics, chemistry, and biology where small particles play a role.

I. INTRODUCTION

I WILL review the history of optical trapping and manipulation of small neutral particles, with particular emphasis on the origins of the field. This subject, which did not even exist before the advent of lasers, now plays a major role in single particle studies in physics, chemistry, and biology. It was known from physics and the early history of optics that light had linear and angular momentum, and, therefore, could exert radiation pressure and torques on physical objects. These effects were so small, however, that they were not easily detected. To quote J. H. Poynting's presidential address to the British Physical Society in 1905, concerning radiation pressure forces, "A very short experience in attempting to measure these forces is sufficient to make one realize their extreme minuteness—a minuteness which appears to put them beyond consideration in terrestrial affairs ..." The study of radiation pressure was considered exciting physics, but not very practical at the turn of the previous century when Nichols and Hull [1] and Lebedev [2] first succeeded in experimentally detecting radiation pressure on macroscopic objects and absorbing gases. The subject essentially dropped into obscurity until the invention of the laser in 1960 [3].

In 1970, Ashkin [4] showed that one could use the forces of radiation pressure from focused laser beams to significantly affect the dynamics of small transparent micrometer sized neutral particles. Two basic light pressure forces were identified: a scattering force in the direction of the incident light beam, and a gradient force in the direction of the intensity gradient of the beam. It was shown experimentally that, using just these forces, one could accelerate, decelerate, and even stably trap small μm sized neutral particles using mildly focused laser beams. It was not known previously that one could use radiation pressure forces to make a stable 3-dimensional optical trap. Such a trap has an

equilibrium point in space, with the property that any displacement of a particle away from this point results in a restoring force.

Over the years, these newly found laser trapping and manipulation techniques were found to apply over a wide range of particle types, including particles as diverse as atoms, molecules, submicron particles, and macroscopic dielectric particles hundreds of micrometers in size. Even living biological cells and organelles within cells can be trapped and manipulated free of optical damage.

The unique capabilities of these techniques have had a revolutionary impact in various sub-fields of these same sciences, where single particles play a role. In the field of light scattering, it has led to the highest resolution studies of Mie scattering, the first high resolution observations of the resonant behavior of macroscopic spherical particles, and the use of these resonances in many applications in linear and nonlinear optics and lasers. The highest Q optical resonances ever observed have been found in these so-called Mie resonance or "whispering gallery modes." In atomic physics, laser trapping and cooling techniques have led to the optical trapping of individual atoms, to atom cooling down to the lowest kinetic temperatures in the universe, to Bose–Einstein condensation, and, more recently, to atom lasers. Practical advances in atomic clocks and the measurement of gravitational forces have also been made. In the biological sciences and chemistry, use of laser techniques has led to the trapping and manipulation of single living cells, organelles within cells, single biological molecules, and the measurement of mechanical forces and elastic properties of cells and molecules.

The use of optical tweezer techniques has led to an explosion of new understanding of the mechanics, force generation, and kinetics of a wide variety of motor molecules and mechanoenzymes. Other studies have been made of the elastic properties, folding and unfolding of DNA, titan, and other large biopolymers, as well as the elasticity of membranes and entire cells. With optics, we have fine control over forces from as low as hundredths of a piconewton up to several hundreds of piconewtons.

The unique ability of tweezers to manipulate small macroscopic particles has been used in colloidal science to detect anomalous new effects, such as the attraction of like charged colloidal particles. Another is the anomalous prevention of the entropic coalescence of large particles in the presence of a high concentration of small particles.

Manuscript received August 21, 2000.
The author was with Bell Laboratories, Lucent Technologies, Holmdel, NJ 07733 (e-mail: aashkinshome@aol.com).
Publisher Item Identifier S 1077-260X(00)11555-2.

As we shall see, one needs to use only the simplest of concepts, such as momentum conservation, ray optics, and semiclassical rate equations, to understand the basic forces and optical trapping. Indeed, it was only with the help of such simple concepts, simple experiments, and a little luck that trapping of particles was discovered in the first place.

I feel qualified to survey this history since I discovered optical trapping of small neutral particles and proposed optical trapping of atoms. I invented optical tweezers and did pioneering work in all the major areas of optical manipulation. I will trace the development of the subject chronologically, starting in 1969, with the simple considerations leading to the discovery of the first trap, and ending with some of the more complex work in Bose–Einstein condensation (BEC) and applications of tweezers to the study of single-motor molecules, and DNA folding and sequencing in the year 2000.

II. BASIC FORCES AND THE FIRST TRAP

My interest in the subject was aroused in 1969 by the following "back of the envelope" calculation of the magnitude of the radiation pressure force of light on a totally reflecting mirror. As we know, the momentum of a single photon is $h\nu/c$. If we have an incident power of P, we have $P/h\nu$ photons striking the mirror per second. Since all these photons are assumed to reflect straight back, the total change in momentum of the light per second is $(2P/h\nu)(h\nu/c) = 2P/c$. By conservation of momentum, this implies that the mirror acquires an equal momentum/sec, or force, in the direction of the light. That is, $F_{\text{mirror}} = 2P/c$. For $P = 1$ watt $= 10^7$ ergs/sec, one gets $F_{\text{mirror}} \cong 10^{-3}$ dynes $= 10$ nanonewtons, which is quite a small force in absolute terms. It represents the maximum force one can extract from the light momentum at a power of one watt. Suppose we have a laser and we focus our one watt to a small spot size of about a wavelength $\cong 1$ μm, and let it hit a particle of diameter also of ~ 1 μm. Treating the particle as a 100% reflecting mirror of density $\cong 1$ gm/cm^3, we get an acceleration of the small particle $= A = F/m = 10^{-3}$ dynes/10^{-12} gm $= 10^9$ cm/sec^2. Thus, $A \cong 10^6\, g$, where $g \cong 10^3$ cm/sec, the acceleration of gravity. This is quite large and should give readily observable effects, so I tried a simple experiment [4] to look for particle motion by laser radiation pressure. I used a sample of transparent latex spheres of density ~ 1, in water, to avoid any problems with heating or so called radiometric effects. With just milliwatts of power, particle motion was observed in the direction of a mildly focused Gaussian beam. The particle velocity was in approximate agreement with our crude-force estimates, suggesting that this was indeed a radiation pressure effect. However, an additional unanticipated force component was soon discovered which strongly pulled particles located in the fringes of the beam into the high intensity region on the beam axis. Once on axis, particles stayed there and moved forward, even if the entire beam was slued back and forth within the chamber. Particles were being guided by the light! They finally collected in a clump at the output face of the chamber. When the light was turned off, they wandered toward the fringes of the beam. When the light was turned on again, they were quickly pulled to the beam axis. Was this transverse force component light pressure, too?

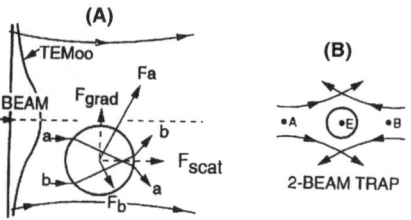

Fig. 1. (a) Origin of F_{scat} and F_{grad} for high index sphere displaced from TEM$_{00}$ beam axis. (b) Geometry of 2-beam trap.

Fig. 1(a) shows that both these force components do indeed originate from radiation pressure. Imagine a high index of refraction sphere, many wavelengths in diameter, placed off axis in a mildly focused Gaussian beam. Consider a typical pair of rays "a" and "b" striking the sphere symmetrically about its center. Neglecting relatively minor surface reflections, most of the rays refract through the particle, giving rise to forces F_a and F_b in the direction of the momentum change. Since the intensity of ray "a" is higher than ray "b," the force F_a is greater than F_b. Adding all such symmetrical pairs of rays striking the sphere, one sees that the net force can be resolved into two components F_{scat} called the scattering force component pointing in the direction of the incident light, and F_{grad}, a gradient component arising from the gradient in light intensity and pointing transversely toward the high intensity region of the beam. For a particle on axis or in a plane wave, $F_a = F_b$ and there is no net gradient force component. A more detailed calculation of the sum of the forces of all the rays striking the sphere gave a net force in excellent agreement with the observed velocity. For a low index particle placed off-axis, the refraction through the particle reverses, F_a is less than F_b and such a particle should be pushed out of the beam. This behavior was seen in μm-sized air bubbles in glycerine. One also observes by mixing large and small diameter spheres in the same sample that the large spheres move faster and pass right by the smaller spheres as they proceed along the beam. This is a form of particle separation and is expected from the simple ray-optic calculations.

The understanding of the magnitude and properties of these two basic force components made it possible to devise the first stable three-dimensional (3-D) optical trap for single neutral particles [4]. The trap consisted of two opposing moderately diverging Gaussian beams focused at points A and B as shown in Fig. 1(b). The predominant effect in any axial displacement of a particle from the equilibrium point E is a net opposing scattering force. Any radial displacement is opposed by the gradient force of both beams. The trap was filled by capture of randomly diffusing small particles which wandered into the trap. The viscous damping of the liquid serves to dissipate all the kinetic energy gained from the trapping potential and particles come to rest at the trap center. If one blocks one beam, the particle is driven forward and guided by the second beam. If one restores the first beam the particle is pushed back to the equilibrium point E, as expected. It is surprising that this simple first experiment [4], intended only to show forward motion due to laser radiation pressure, ended up demonstrating not only this force but the existence of the transverse force component, particle guiding, particle separation, and stable 3-D particle trapping.

The success of these experiments on macroscopic particles prompted the hypothesis in reference [4] "that similar acceleration and trapping are possible with atoms and molecules using laser light tuned to specific optical transitions." It was shown that a scattering force should exist for atoms in the direction of the incident light due to the process of absorption and subsequent isotropic spontaneous emission of resonant photons. The low intensity absorption cross-section of an atom is huge, about $\lambda^2/2$. This says that almost every photon of a strongly focused resonant beam hitting an atom is absorbed and subsequently scattered by the atom. Since the atomic mass is small, this implies a huge acceleration. For a sodium atom, one gets a kick of ~ 3 cm/sec per photon scattered. This process, however, is strongly limited by saturation, even at very modest light intensities of hundreds of watts per square centimeter. The problem of saturation of the scattering force was treated phenomenologically using the so called Einstein A and B coefficients to calculate the fraction of time f an atom spends in the excited state. The scattering force is given by the rate of scattering momentum $F_{\text{scat}} = hf/\lambda t$, where t is the spontaneous emission lifetime [5]. At high saturating intensities the population of a 2-level atom equalizes and $f = 1/2$. The magnitude of this saturated force is sufficient, however, to turn an atomic beam of sodium of average thermal velocity of $\sim 10^5$ cm/sec through a radius of curvature $\rho \cong 20$ cm, if applied continuously at right angles to the velocity to avoid any Doppler shifts. If one applies the saturated force in opposition to the atomic motion, one can stop atoms at the average velocity in a distance of $\rho/2 \cong 10$ cm, assuming one compensates for the large Doppler shift of the atomic resonance. It was suggested that one could use the scattering force to make an atomic beam velocity selector or an isotope separator [5]. A scheme for exerting significant optical pressure on a gas of atoms was also proposed [4].

No consideration was given in [5] to the gradient component of the force on atoms inasmuch as I did not understand how to treat the saturation of this force. The classical formula for the gradient force of an electromagnetic wave on a neutral atom, considered as a simple dipole, is the dipole force formula $(1/2)\alpha\nabla E^2$ where α is the optically induced polarizability of the atom or particle. The polarizability α can be calculated using a simple harmonic oscillator model of an atom. For atoms, the polarizability is dispersive and changes signs above and below resonance in analogy with the change in sign of the gradient force on high and low index particles. This gradient force formula, applied to atoms, electrons, and plasmas, was considered previously by Askar'yan [6], [188] using lasers in a two-dimensional (2-D) geometry in connection with self focusing. Letokhov [7] also considered very weak off-resonant one-dimensional (1-D) confinement of atoms in laser standing waves for spectroscopic purposes. Neither work discusses the possibility of stable 3-D trapping of atoms.

III. OPTICAL LEVITATION AND APPLICATIONS

The next advance in optical trapping and manipulation was the demonstration of an optical levitation trap in air, under con-

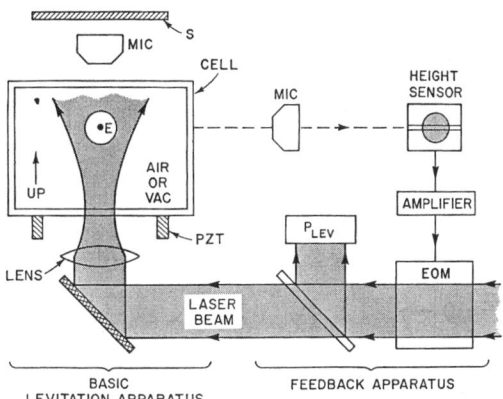

Fig. 2. Basic apparatus for optically levitating dielectric spheres and feedback stabilization apparatus for levitating in vacuum and measuring forces. PZT is a piezoelectric ceramic shaker; EOM is an electrooptic modulator.

ditions where gravity plays a significant role [8]. In the levitation trap, as shown in Fig. 2, a single vertical beam confines a macroscopic particle at a point E where gravity and the upward scattering force balance. The equilibrium is stable because of the increase in axial scattering force with decreasing height near E and the transverse confinement of the gradient force. Once aloft, levitated particles can be freely manipulated by simply moving the beam. With a pair of movable beams one can combine the two beams and thus collide a pair of levitated particles and assemble compound particles. In this way, one can fuse oil droplets and also form spheroids, teardrops, spherical doublets, triplets, etc. [9]. These complex particles align themselves in the beam and make ideal test particles for light scattering experiments. Levitation of hollow glass spheres, which are sometimes used as laser fusion targets, is also possible [10]. One uses TEM_{01}^* or donut mode beams having a hole in their center, since these hollow particles behave as air bubbles and are pushed out of the high intensity region of the light. Levitation in high vacuum is also possible [11]. However, feedback is needed to damp particle oscillations caused by random beam fluctuations [12] (see Fig. 2). The feedback scheme both locks the particle to a split photodiode height sensor and also varies the levitating power in proportion to the negative of the velocity to give strong optical damping. Importantly, feedback locking in either vacuum or air provides a means of automatically measuring forces on particles, since the change in power needed to hold the particle at a fixed height is a direct measure of an externally applied force [13].

The feedback force measuring technique was used to measure the electric force on oil drops as they accumulated single-electron charges in a modern version of the Millikan oil drop experiment [13], where we could easily resolve the changes in levitating power associated with changes in the charge on a sphere by single-electron amounts (see Fig. 3). We could also measure viscous drag forces on small particles as the velocity of a fluid varies, changes in radiometric forces with pressure, and changes in the optical scattering force with axial position in the light beam [12]. We also used feedback

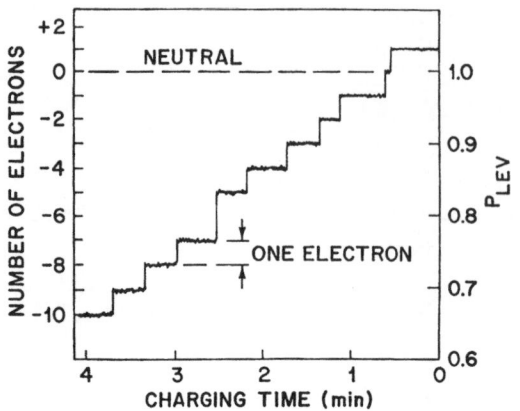

Fig. 3. Changes in the optical levitating power caused by the automatic feedback system as the charge on the sphere in an electric field increases by single-electron amounts.

to measure the wavelength dependence of the levitating force with a tunable dye laser [14]. This led to the experimental discovery of a complex series of high Q resonances in the force as shown in [13, Fig. 4]. At first, we wondered if they were even real. According to the simple ray picture of the forces as I described it, one should not expect much of a change in force with wavelength. These resonances are in fact real and are predicted by Mie–Debye Theory [14]. They manifest themselves as dips in the radiation pressure force and peaks in the light scattering of a trapped spherical particle. Applications of these resonances, variously called surface-wave resonances, morphology-dependent resonances, or whispering-gallery resonances, have had a great impact on light scattering studies. Precision measurements of the resonant backscattered spectrum or "glory scattering" from silicone oil drops probably give the best experimental test of Mie–Debye theory of electromagnetic scattering [15, Fig. 5]. High Q resonances give a two–three order of magnitude improvement in absolute and relative size and index of refraction measurement of spheres [14], [16]. More recently, drops have served as extremely high Q dye laser and Raman laser resonators and as a medium for studying and enhancing a wide range of linear and nonlinear optical interactions [17]. Mirrorless optical resonators based on these modes have the highest Qs of any optical cavity ($>2 \times 10^9$) [18] and have recently been used for low threshold semiconductor lasers [19], [20] and for experiments in cavity electrodynamics [21].

IV. ORIGINS OF OPTICAL ATOM TRAPPING

Following the early work on light forces on atoms [4], [5], experiments were performed demonstrating atomic beam deflection [22], [23] and isotope separation [24] using the scattering force. In 1975, Hänsch and Schawlow made the important suggestion that it was possible to use the strong velocity dependence of the scattering force due to Doppler shift for the optical cooling or damping of atomic motions [25]. For example, in one dimension with a pair of identical opposing beams tuned below

resonance, any atomic motion along the axis meets a net opposing force due to the strong Doppler shifts of the absorption. Three pairs of such opposing beams should damp all degrees of freedom. Does this imply the ability to cool atoms to absolute zero? Certainly not. Due to quantum fluctuations there is randomness in the directions of emission of successive photons and also in the times of emission, which corresponds to a constant heating process. The equilibrium temperature finally achieved is a balance of the optical cooling rate and the quantum heating rate. Letokhov and Minogin were the first to estimate the equilibrium temperature based on the fluctuations of the scattering force [26], [27]. For a tuning $\gamma_n/2$ below resonance, which gives the optimum cooling rate, they estimate an equilibrium energy of $\sim h\gamma_n$. They also proposed at this time that one could use the same 6-beam cooling geometry for stably trapping atoms on the intensity maxima of the 3-D standing wave pattern by virtue of the gradient force. Unfortunately, they estimated that the trap depth was also $\sim h\gamma_n$ which implies a very leaky trap.

In order to see if one could succeed in getting a deep enough trap, I decided in 1978 to address the problem of saturation of the gradient force using the same semiclassical rate-equation approach used earlier for understanding saturation of the scattering force [5]. The key point was the realization that the classical value of the polarizability α in the formula $(1/2)\alpha\nabla E^2$ applies to an atom in its ground state, and that an atom in its excited state contributes polarizability of the opposite sign in proportion to the fraction of time f it spends in the excited state. With this approach one finds for the potential U of the gradient force for arbitrary tuning and arbitrary light intensity, $U = (h/2)(\nu - \nu_o)\ln(1 + p)$ where p is an intensity dependent saturation parameter [28]. It is readily seen from the expression for U that one can greatly increase the potential depth U by factors of 10^2 or more for a given intensity by keeping the saturation modest ($p \cong 1$) and greatly increasing the detuning ($\nu - \nu_o$) to values of about $10^2\gamma_n$ or more. The value of $p = 1$ implies that the stimulated emission rate equals the spontaneous emission rate. Use of detuning to reduce saturation made it possible, for the first time, to devise trapping geometries for atom traps which were stable in the Boltzmann sense, i.e., $U/kT \gg 1$. A 2-beam trap was proposed at this time in analogy with the first macroscopic particle trap. Also suggested was the simplest of all traps, the optical tweezers trap [28] consisting of a single strongly-focused Gaussian beam (see Fig. 6). Although the tweezer trap at first sight seems counter-intuitive, it is axially stable because of the dominance of the backward axial gradient force over the forward-scattering force. With these new strongly detuned atom traps, one cannot use the trapping beams to also provide cooling. This requires the use of optimally tuned auxiliary cooling beams to keep the atom temperature at $\sim h\gamma_n$ [29].

An experiment was performed to demonstrate the new concept of large gradient forces with detuned light [30]. An atomic beam was injected into the core of a Gaussian beam, as shown in Fig. 7. Depending on the tuning, either below or above resonance, the atoms are strongly focused or defocused by the transverse gradient forces of the laser beam, as seen in [30], [31], and in [13, Fig. 8]. This work was the first experimental demonstration of the gradient force on atoms. It also represents a demonstration of 2-D trapping of atoms using light forces. In addition,

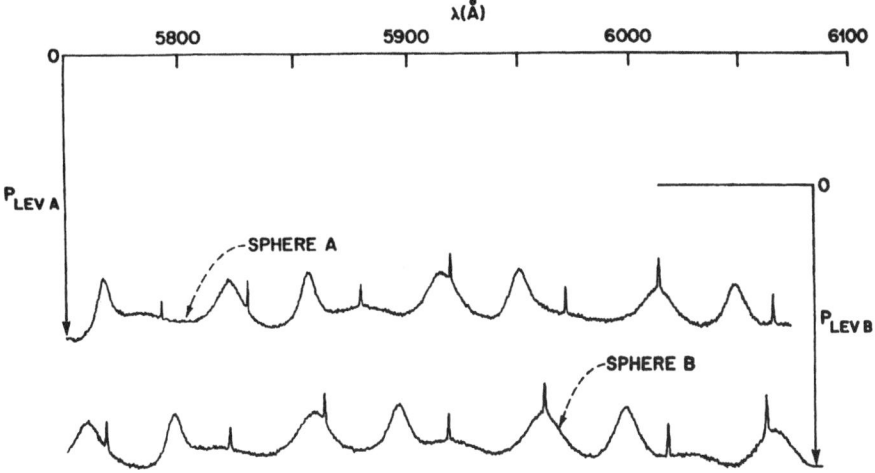

Fig. 4. Resonant behavior of light forces on dielectric spheres. The spectra show the variation with wavelength of $P_{\mathrm{lev A}}$ and $P_{\mathrm{lev B}}$, the power needed to levitate oil drops A and B, which have index of refraction $n = 1.47$ and slightly different diameters (~ 10 μm). The resonances of sphere A are shifted ~ 50 Å higher in wavelength than the corresponding ones in sphere B.

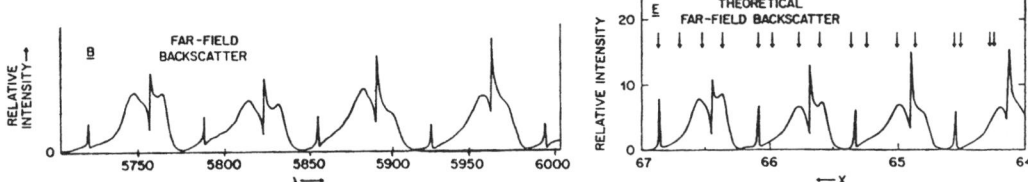

Fig. 5. Comparison of experimental far-field backscatter versus λ (curve B) with the theoretical far-field backscatter (curve E) plotted versus $x = 2\pi a/\lambda$.

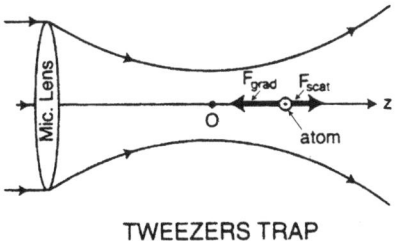

TWEEZERS TRAP

Fig. 6. Tweezer trap for atoms. $F_{\mathrm{grad}} > F_{\mathrm{scat}}$ giving a net backward force toward E.

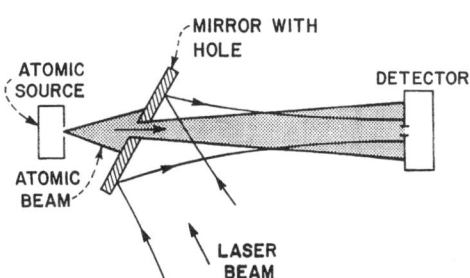

Fig. 7. Apparatus for observing focusing and defocusing of an atomic beam by the dipole force of a nearly resonant laser beam.

it marks the beginning of the so called field of atom optics in which atoms are guided by the light. The Gaussian beam, in effect, acts as a gradient index or GRIN lens. Additional work studying the variation of the atomic beam focal spot size with light intensity gave the first evidence of quantum heating of atoms by light [32].

Prospects for optical atom trapping were bolstered by a theoretical analysis by Gordon *et al.* [33]. This work derived the basic optical forces on atoms, their saturation, and their fluctuations from first principles, using a fully quantal theory, and applied the results to traps. It confirmed the correctness of the

scattering and gradient force components which I deduced from a combination of experiment, intuition, and semiclassical analysis. A new result of [33] was the derivation of the fluctuations of the gradient or dipole force. This is conceptually more difficult to understand than the scattering force fluctuations, but it contributes equally with the scattering force fluctuations to the quantum heating rate and the equilibrium temperature. This paper has become a basic reference on questions about optical forces on atoms.

846   IEEE JOURNAL ON SELECTED TOPICS IN QUANTUM ELECTRONICS, VOL. 6, NO. 6, NOVEMBER/DECEMBER 2000

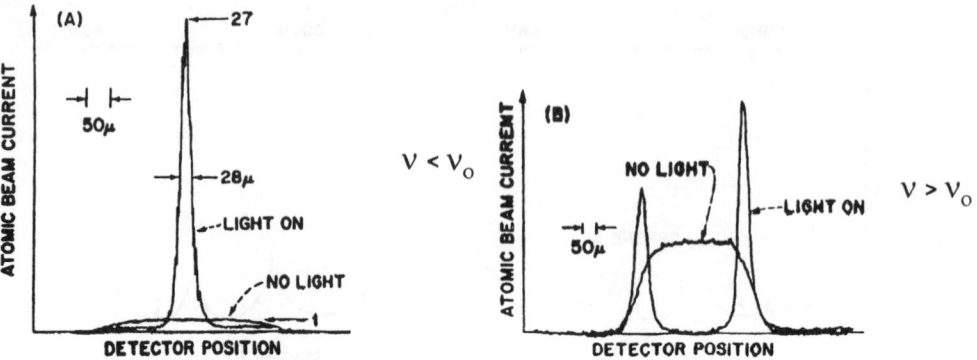

Fig. 8.   (a) Focusing of an atomic beam by light tuned below resonance. (b) Defocusing of an atomic beam by light tuned above resonance.

A further big experimental step on the way to atom trapping was the gross slowing of atomic beams using the scattering force of an opposing laser beam by Phillips *et al.* [34], [35] (see also [36]). The major problem here was to compensate for the large Doppler shifts that occur as the atoms are slowed. This was done by magnetically tuning the resonant frequency of the atoms with a properly tapered magnetic field to keep the peak of the distribution of slowing atoms in resonance with the light. Chirping the light frequency was also suggested by Letokhov *et al.* [37], and subsequently demonstrated [38]. Although these 1-D techniques could slow the peak of the axial velocity distribution to zero, there was no transverse cooling and the lowest average temperature achieved was about 0.1 °K. At this temperature, relatively few atoms are available for filling small volume atom traps. One solution to the difficulty actively being pursued at that time at the National Bureau of Standards was a different type of trap [39] in which atoms were confined in a relatively large volume deep trap solely by the scattering force from mildly diverging beams. Unfortunately, this proposal was flawed. A theorem called the Optical-Earnshaw Theorem was proven by Ashkin and Gordon [40], showing that any trap based solely on scattering forces, which are strictly proportional to the light intensity, is inherently unstable. This was proven in analogy with the Earnshaw Theorem in electrostatics.

In 1984, an experiment was started at Bell Laboratories, Holmdel, NJ, USA, on optical trapping of atoms. This was stimulated by the arrival of Steve Chu as a new Fellow Department Head in the Electronics Research Laboratory with intentions of trapping atoms (see [41]). Our initial plan was to combine slowing, cooling, and trapping in a single experiment. Chu argued for a simpler first step, to first study the 3-D cooling scheme using the Doppler-cooling technique [25] now referred to as "optical molasses." This was wise since molasses cooling succeeded so well, it affected our subsequent choice of traps. The molasses experiment [42] produced a roughly 1 cm^3 volume of atomic vapor at a density of 10^9 atoms/cm^3, viscously confined at a temperature of about 250 μK, close to the Doppler limit [26], [27], [29], [33], which persisted for times up to a second before diffusing away. Indeed, with this remarkable sample of cooled atoms it became possible to demonstrate the first 3-D stable atom trap [43] using the very

simple tweezer trap consisting of just a single strongly focused Gaussian beam. Despite its small volume, the tweezer trap, placed anywhere within the sample of cold atoms, proceeded to fill up to densities of about 10^{11} atoms/cm^3 by diffusion from the surrounding vapors, in analogy to the filling of the first particle trap by diffusion from the surrounding latex spheres [4]. Trapped atoms persisted in the trap and could be freely manipulated in space after the surrounding molasses atoms diffused away. The success of these cooling and trapping experiments marked the beginning of a new era of experimentation which has revolutionized experimental atomic physics.

V. RECENT WORK ON ATOM TRAPPING AND MANIPULATION

In 1986, the achievement of optical cooling and trapping of a dense cloud of atoms greatly stimulated interest in optical manipulation techniques. A new large volume magneto-optical trap (MOT) was developed using the scattering force [44]. In this trap, an external magnetic quadrupole Zeeman splitting field was used which made the resonance frequency, and α, the polarizability, position dependent. This results in a stable scattering force trap that doesn't violate Earnshaw's Theorem. This robust large volume deep trap is widely used as a workhorse trap in spite of some poorly understood behavior [45].

Although the initial molasses temperature of 240 μK was close to the Doppler cooling limit for a 2-level atom [42], disagreements soon arose. Unexpectedly, temperatures almost ten times less than the Doppler limit were observed [46], [36]. Explanations of this increased cooling are based on the multilevel nature of the cooling transition used [47], [48] (also refer to [36]).

One might think the minimum possible temperature of cooled atoms would be T_r, which is the temperature due to the recoil of a single photon. For sodium, the recoil velocity is about 3 cm/sec with a temperature T_r of about 2 μK. However, cooling below even T_r can be achieved. In 1986, the Holmdel group [49] was the first to propose that one could reduce the temperature of atoms held in an optical trap at the molasses temperature to temperatures of $\sim 10^{-6}$ K by momentarily turning off the trap and then turning it on again. This allows fast atoms to escape and is the equivalent of evaporative cooling, as proposed by Hess for cooling hydrogen gas in magnetic traps [50]. Evaporative cooling techniques have the disadvantage that well over 90% of the atoms originally in the trap are lost. It was also proposed that one could cool, with no loss of atoms, by simply letting tightly confined atoms expand and cool into a larger volume harmonic trap. Subrecoil temperatures were also calculated. This idea has never been implemented.

It was proposed, and later demonstrated, that cooling below T_r also occurs using velocity selective coherent population trapping (VSCPT). In this technique, atoms randomly scatter photons until they fall into a superposition ground state with close to zero velocity where they are decoupled from the light [51], [52] (see also [53]). Another practical scheme for cooling below T_r involves selective Raman cooling [54], [55].

An early use of cold atoms was in the achievement of a practical "atomic fountain" [56]. Chu *et al.* showed that atoms optically launched vertically from a MOT trap could interact with a microwave cavity for long times to improve the accuracy of atomic clocks [57], [41]. Another growing use of cold atoms is the study of ultra-cold atomic collisions. With ultra-cold collisions one can explore processes not seen at higher temperature [58]–[61].

A subfield of optical manipulation has developed called "atom optics." It loosely refers to the optical manipulation of atoms in ways similar to manipulation of light by conventional optical elements like lenses, mirrors, beam splitters, gratings, and interferometers. At times, it makes use of the wave properties of atoms. The first experiments showed guiding and focusing of atoms using the distributed lens action of the gradient force in a long thin laser beam [30]. Other single-optical lenses were demonstrated [62], [63], but all suffer from chromatic aberration. Nevertheless, focusing to spot sizes of 20 Å has been seen [64]. Optical mirrors for atoms have been developed using reflection from the dipole force of evanescent waves of laser fields [65], [66]. Atomic beam splitters have been extensively studied, based on a variety of interactions [67]–[70]. Atom optics techniques also have potential uses in technology for neutral atom lithography [71], [72]. See a review of atom optics in [73]. Atom interferometers have been developed which are an important class of devices for making precision measurements. Interferometric measurements of the acceleration of gravity were made with an accuracy of one part in 10^6. Increases in accuracy to $10^{-10}\,g$ are anticipated [74]. Applications to geology, a search for the net charge on atoms, fifth force experiments, and a test of general relativity were suggested.

The most important recent development in the field has been the final achievement of Bose–Einstein condensation (BEC) of atomic vapors in atom traps. This was made possible by the realization of a combination of sufficiently cold and dense vapors of atomic bosons where the deBroglie wavelength of the atoms becomes large enough so that individual atomic wave functions overlap and become coherent in a single ground state extending over the sample. BEC has previously been observed in superconducting solids and in superfluid liquids, but never in atomic vapor. In Bose–Einstein condensates in the vapor phase, the likelihood of three body collisions, which might result in the formation of molecules or atomic clusters, is much reduced. In

this sense, the vapor condensate is metastable for times of seconds or minutes. However, once formed, the atomic vapor condensate is the purest macroscopic quantum system yet achieved. Almost all atoms are in the condensed state.

Bose–Einstein condensation was achieved by using evaporative cooling and specially designed magnetic traps [75], [76], which have no point of zero magnetic field. Since its discovery, there has been an ever growing number of theoretical and experimental studies probing the novel properties of this new state of matter. The collective oscillations of the condensate have been observed [77]. Mechanical measurements also have been made on the propagation of sound waves and phonons [78] in condensates. Ketterle *et al.* showed the coherence of the BEC by splitting it into two parts with a far off-resonance laser beam and observing macroscopic atom wave interference when they recombine [79]. More recently, measurements of the coherence length of a condensate were made by deducing the momentum uncertainty in the Heisenberg uncertainty principle [80]. The results showed the entire condensate was a coherent matter wave.

Techniques were devised for expelling atoms from condensates in traps to give an external beam of coherent atoms which is in fact a form of atom laser. For example, in 1997, Ketterle and colleagues flipped the spins of a condensate in a magnetic trap, causing condensed atoms to fall downward due to gravity [81]. Hänsch and collaborators used a continuous radio frequency signal to make a continuous output coupler. Phillips and his group used optical Raman transitions in which the photon recoil ejected atoms from the trap to give a directional output coupler [82]. These atom lasers make possible a new form of atom nonlinear optics in which the atoms of the condensate itself act as a nonlinear medium. This has resulted in the observation by Phillips *et al.* [83] of 4-wave mixing with matter waves. In recent experiments, Ketterle and a combined NIST and Japanese group have observed real gain in the number of atoms in a coherent atom laser beam by using a pump laser to convert condensate atoms into the coherent incident signal beam [84], [85]. Coherent atom laser beams may some day play an important role in atom lithography. The principal difficulty with use of present atom lasers for lithography is the lack of sufficient beam intensity. In time, we expect that all the atom optics components that have been developed for incoherent atomic beams will have their coherent atom laser beam analogs.

One of the interesting recent developments in the BEC field is the realization of the importance of the simple-single beam dipole trap, or tweezer trap [86], [87]. This is essentially the same trap used in the first atom trap in 1986 and, subsequently, used in biology and elsewhere for trapping and manipulating macroscopic particles and Rayleigh particles. These single beam tweezer type traps, which have been largely neglected by the Bose–Einstein community, overcome many of the drawbacks of magnetic traps. Magnetic traps are large, fixed in space and limited to particular hyperfine levels. One can load more than 5×10^6 condensed atoms at densities as high as 3×10^{15} atoms/cm^3 into single beam dipole traps from magnetic traps [86] and trap several hyperfine states using adjustable external magnetic fields. With dipole traps, one can magnetically tune a uniform external field to so called Feshbach magnetic resonances without affecting the trap operation [87]–[89]. Experiments also have been done to measure scattering lengths [89], observe spin domains [90], and measure quantum tunneling across spin domains [91]. Reversible formation of condensates has been studied using optical traps [87]. Heinzen and collaborators [92] showed how one can use a tweezers trap tuned by a magnetic field to locate a zero energy magnetic Feshback resonance in trapped ^{85}Rb atoms with a photoassociation spectroscopic detection method. At such a resonance the atoms change from being attractive to being repulsive on going through the exact resonance field.

Recently, the Boulder group [93], [94], in a complex experiment, managed to achieve BEC at the same ^{85}Rb resonance in a MOT type trap. They observed a dramatic change from Bosonic behavior to Fermion behavior, with ejection of atoms from the trap as they passed to the Fermion repulsive side of resonance.

Another area of excitement in the BEC community stems from the recent observations on superfluidity and vortices in condensates [95]–[97]. Various optical force stirring methods were used for imparting angular momentum to the condensates [95], [98]. The field of BEC is growing rapidly. The number of papers per year on the subject is fast approaching a thousand [99].

A new area of application is the use of optical trapping, cooling, and manipulation techniques for producing cold molecules. Although in its infancy, this area of application has high potential. People are thinking of the possibility of trapped molecular condensates and even coherent molecular lasers. See [100] and references therein.

VI. ORIGINS OF OPTICAL TRAPPING IN BIOLOGY

Although the optical tweezers trap was originally designed as an atom trap and was used in the first optical trapping experiment [43], that experiment did not represent the first use of tweezers. During the atom trapping experiment, at a time of temporary difficulty, it was decided to try the tweezer trap on simpler Rayleigh dipole particles, such as submicron silica spheres. Using the known polarizability of a submicron Rayleigh sphere, one can show that it is possible to satisfy the criterion for a tweezer trap that the backward gradient force exceeds the forward scattering force at the point of maximum axial gradient of a strongly focused Gaussian beam [101]. Submicron colloidal silica particles were placed in a simple water filled chamber and irradiated with a strongly focused 5145 Å argon laser beam, as shown in Fig. 9. Scattered light at 90° was viewed with a low power microscope, either visually or with a photodetector. Individual particles were easily seen in the cone of the focused laser light and displayed beautiful Brownian motion. Whenever a particle wandered close to the beam focus, it was immediately pulled into the trap. It brightened up, and all Brownian motion ceased. Particle after particle entered and we showed that they didn't just fuse into a single lump, but formed a fixed array in the focus, held apart by their colloidal charge. Particles as small as ~250 Å were trapped. We also showed that tweezers could trap micron sized spheres large compared with the wavelength. This extended the notion of a backward gradient force to large particles as well. The origin of the backward light force for tweezers

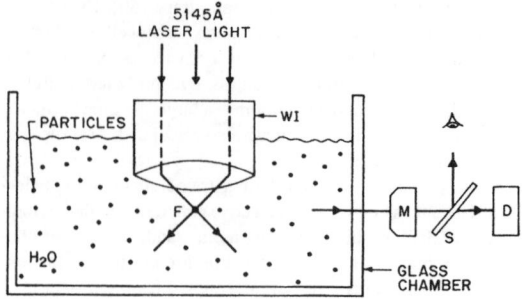

Fig. 9. Apparatus for trapping and observing submicron colloidal particles in water.

in the ray optic regime [102] is shown in Fig. 10. For many applications with macroscopic particles, tweezers is superior to the levitation trap. Levitation traps depend on gravity and have forces of $\sim mg$, where m is the mass and g is the acceleration of gravity. Tweezers, however, is an all optical trap and can have forces of thousands of times mg, limited only by the optical power. This is useful for confining submicron particles in situations where gravity plays a minor role and Brownian motion dominates. The compact tweezer trap is also more tolerant of particle shape irregularities than the levitation trap. As we shall see, tweezer trapping of macroscopic particles has turned out to be widely useful in biology and the physical sciences.

Our next experiments involved trapping of colloidal tobacco mosaic virus [103]. Tobacco mosaic virus (TMV) is a rugged rod-like protein ~ 200 Å in diameter and ~ 3000 Å long. It is easily trapped by ~ 100 mW and it, too, can form arrays of aligned viruses in the beam focus. This may well be the first rod-like molecule to be fully aligned by an optical electric field. While we were doing these experiments a very serendipitous event occurred. We tended to keep these TMV samples a long time in a container open to the ambient atmosphere. With time we noticed the appearance of increasing numbers of strange, relatively large, apparently self-propelled particles. Some, occasionally, were trapped when they neared the focus, where they gave rise to a wild display of scattering. We called these new particles "bugs" and, it turned out, they were bugs. Bacteria, that is, which presumably contaminated the chamber from the air. At low powers of ~ 5 mW, they continued scattering for many minutes. If we raised the power to ~ 100 mW, there was a huge burst of scattering and then nothing but a weak steady scattering. We interpreted this as "optication," death by light. The weak scattering was from the empty carcasses of the bugs. To check this we introduced the trap into a high resolution microscope as shown in Fig. 11. We could then trap, observe, and manipulate bacteria which we grew from bits of Joe Dziedzic's ham sandwich. We readily confirmed our hypothesis. Our paper in Science [103] on laser trapping of viruses and bacteria was the first report of optical manipulation of living cells, although optical damage to bacteria cells was apparent.

We decided to try other laser wavelengths that might be less damaging to living cells [104]. Since the absorption of molecules like chlorophyll and hemoglobin falls rapidly in the

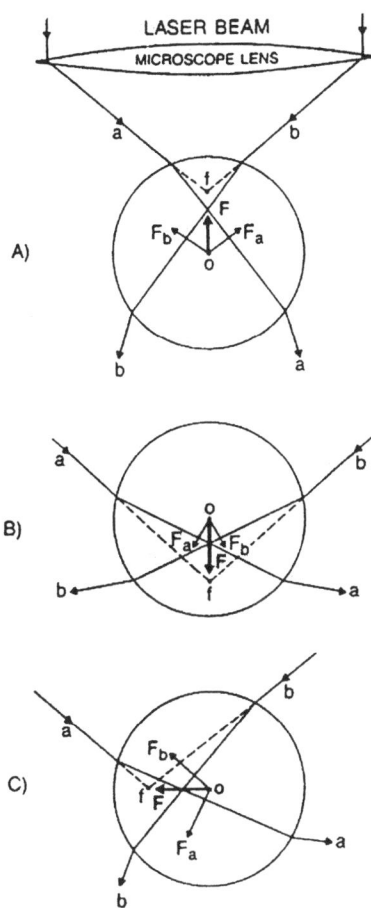

Fig. 10. Simple ray-optics picture of the stability of tweezer trap. Any displacement of a macroscopic sphere away from the focus, f, either axially as in (a) or (b), or transversely as in (c), results in a net restoring force.

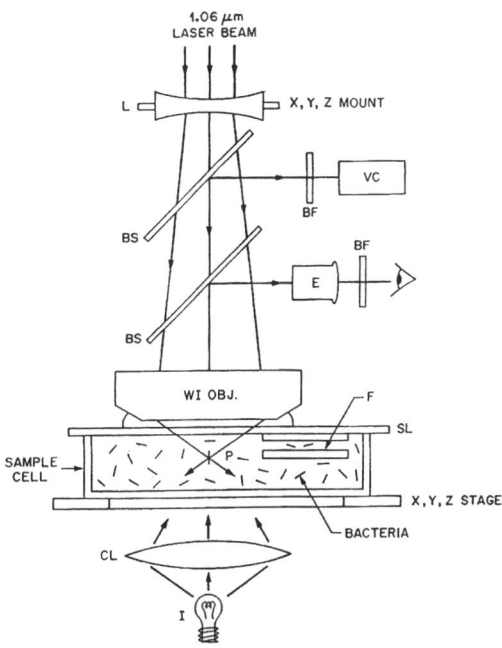

Fig. 11. Combined high-resolution optical microscope and 1.06 μm infrared laser trap for observing, manipulating, and separating bacteria and other organizms.

near infrared, we decided to try trapping with a Nd : YAG laser at 1.06 μm. Also, the absorption of water at 1.06 μm is quite small. Using 1.06-μm YAG laser, we immediately found a large decrease in damage to bacteria. We could collect large numbers of motile bacteria within a single trap for many minutes and see them swim away, apparently undamaged, with powers of 50 mW or more. As a test, we obtained E. coli bacteria and trapped one in the middle of the microscope chamber for a long time, far from other untrapped cells resting on the bottom surface. To our pleasant surprise we saw it grow in size and finally divide into two cells. We retrapped these two cells and after another hour or so we had four cells. E. coli was reproducing right in the high intensity trap focus! Clearly there was no serious optical damage occurring. We then showed that single yeast cells could also reproduce by budding right in the trap. We grew large clumps of cells this way, which we could freely manipulate within the chamber. Joe Dziedzic and I proceeded to trap all

sorts of cells: pigmented red blood cells, green algae, diatoms, amoebas, and other protozoans. In general, there was greatly reduced damage using the YAG laser light. In contrast, red blood cells and green cells with chlorophyll simply exploded with argon laser light at 5145 Å. With larger cells, such as scallion cells, we were able to manipulate small particles within the living cell. We could collect small μm-sized vesicles and other particles and probe the geometry of the chloroplasts and central vacuole by dragging the particles about, moving them deep in the cell from the bottom to the top, under and between cell structures. The tweezer trap is robust enough to tolerate partial shading of the beam by intervening structures.

We demonstrated the ability of tweezer traps to separate a single selected bacterium from a "gemisch" or collection of bacteria. In Fig. 11, we show an approximately 15 μm-inner diameter fiber F attached to the top of the sample chamber. We were able to trap one or more bacteria and manipulated them into the core of the hollow fiber. Despite the rather severe optical distortions at the input of the fiber, one could still maneuver the bacteria into the fiber core without losing them. To complete the separation we removed the chamber lid, washed it, dried it, and then with a gentle air stream blew the liquid contents of the fiber into another water filled vessel. The ability to separate bacteria and other biological cells is a very simple but important capability of tweezer techniques. It was also shown that one could introduce two traps into the microscope. One can then grab a bacterium or cell at its ends and orient it in space or rotate it at

850   IEEE JOURNAL ON SELECTED TOPICS IN QUANTUM ELECTRONICS, VOL. 6, NO. 6, NOVEMBER/DECEMBER 2000

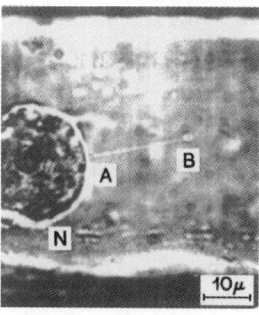

Fig. 12. Artificial cytoplasmic filaments in a scallion cell. The laser trap originally located on the surface of the nucleus (N) at A is moved to B, pulling out the viscoelastic filament AB into the central vacuole.

will by moving one trap relative to the other. One can stretch it and observe its mechanical properties. A red blood cell is so pliable that simply running a tweezer trap over its surface makes a very noticeable distortion. As big as they are, one can squeeze many blood cells into a single trap. The elastic behavior of the cytoplasm is evident in almost all cells.

We made more careful observations on the elastic properties of the cytoplasm in subsequent work [105]. Using scallion cells, we could generate what we called "artificial cytoplasmic filaments" pulled by the trap from the surfaces of most internal cell organelles. Fig. 12 shows such a filament pulled from the surface of the nucleus of the cell into the central vacuole of the cell. The filament stretches from the original location of the trap at A to its final location at B. If one quickly turns off the trap at B, the filament snaps right back to A. If one waits a minute or so at B, the filament snaps back, but more slowly. The longer one waits at B, the more slowly the filament returns to A. At longer times the filament only partially retracts and we are left with a sagging remnant of cytoplasm. This is classic viscoelastic behavior. Cytoplasm is like the toy "silly putty." For quick distortions, it is highly elastic. For slow distortions or long times, the material flows and sags and is relatively weak. Using tweezers, evidence for viscoelasticity is seen everywhere within cells. We used tweezers to perform a new kind of internal cell surgery. If one tries to pull quickly with tweezers on some fairly large structure within a cell, such as the spiral chloroplast of a spirogyra cell, it looks quite rigid and it barely moves before it slips out of the trap. If, however, one pulls slowly on the chloroplast and continues to apply force after it moves somewhat, one finds that the tension slowly relaxes and that further motion is possible. This can be repeated and in time one can pull the entire chloroplast right off the cell wall and into the fluid of the central vacuole, where it is quite free. One can, thus, make gross changes in cell structure. If the chloroplast is placed against the side wall of the vacuole, it doesn't adhere to the inner membrane initially. If one uses the trap to continue to hold it against the wall, its membrane fuses with the wall membrane, completing the operation. The nucleus of the cell and its cytoplasmic supports can be similarly manipulated. At the time of these tweezer experiments there had been very few experiments showing the viscoelasticity of live cells. In 1950, the physicist Crick, of Watson and Crick fame, tried to study viscoelasticity by getting cells

to ingest small magnetic particles, which he proceeded to move with a large electromagnet [106]. After much work, he barely observed any effect, probably due to the fairly rapid relaxation of the tension on the particles. Later, a similar experiment with magnetic particles showed a somewhat larger effect [107].

With tweezers, one can observe more viscoelastic behavior of the cytoplasm within a cell in a few minutes than has been seen from the time of the Crick experiments to the present. With tweezers we are able to probe the viscoelasticity of the different parts of the living cell. Tweezers has also been used recently to explore the viscoelastic behavior of single strands of polymeric molecules, such as DNA in many contexts [108], [109]. With tweezers, one can interact strongly with the cytoplasmic streaming that one sees so readily in scallion and other plant cells. If the trap is placed in some streaming channels, one can capture the moving particles of the stream and then release them and see them move on. Depending on the power, we can stop the entire stream, particles and all. Subsequent particles are stalled or can even turn around and move backward. One can probe the relative viscosities of the moving streams and of the other more liquid regions of the cell. By rupturing the cell and manipulating the same particles in pure water, one can estimate the absolute viscosities of the cell contents. This type of information about cells was generally not known [110]. There is clearly much to be learned about the streaming process with trapping techniques.

The three papers discussed here on damage free trapping [103]–[105] in a real sense mark the beginning of the new field of optical trapping in biology.

VII. RECENT WORK ON OPTICAL TRAPPING AND MANIPULATION IN BIOLOGY

An early application of this new manipulative technique of tweezers in biology involved the measurement of the torsional compliance of bacterial flagella by twisting a bacterium about a tethered flagellum [111]. It was shown that this compliance was located within the bacterial motor itself [112]. Tweezers helped show that the flagella of spirochete bacteria also work by the rotary action of their motors [113]. Greulich and Berns were the first to use the tweezers technique in combination with the so called "microbeam" technique of pulsed laser cutting (sometimes called "laser scissors" or "scalpel") for cutting and moving cells and organelles. Greulich's early work involved ultraviolet cutting and tweezer manipulation of pieces of chromosomes for gene isolation [114]. Tweezers were also used to bring cells into contact with one another in order to effect cell fusion by cutting the common wall [115]. Recently, the spatial and temporal contact needed between antigen-presenting cells and T lymphocyte cells to initiate the activation process was studied with a new optical tweezer based assay involving intracellular calcium signaling [116]. Berns and his group used tweezers, often combined with optical scissors, to manipulate chromosomes during cell division [117] as a new way to study the complexities of mitosis.

Experiments were performed with tweezers to manipulate live sperm cells in 3-D [118], [119] and to measure their swimming forces [120]. Applications of tweezers and scissors to all-optical *in vitro* fertilization are being considered [121]. Ultraviolet drilling of channels in the zona pellucida of oocytes was performed and selected sperm were inserted into channels to effect fertilization [122], [123]. Experiments by Berns' group measured the effects of the wave length on possible optical damage processes in sperm and in other contexts using tunable Ti sapphire lasers [124].

One of the most important biological applications of tweezers is in the study of molecular motors. These mechanoenzymes interact with the microtubules or actin filaments of the cell to generate the forces responsible for cell motility, muscle action, cell locomotion, and organelle movement within cells. In early work using the "handles technique," Block *et al.* [125], [126] attached single kinesin motor molecules to spheres and placed them directly onto microtubules where they could be activated by ATP. This new technique greatly improved on earlier *in vitro* motility assays which used many motors and relied on random diffusion for attachment to filaments. Ashkin and colleagues [127], using a related *in vivo* technique, estimated the force generated by a few dynein motors attached to mitochondria as they moved along microtubules in the giant amoeba reticulomyxa. Kuo and Sheetz [128], working *in vitro* with tweezers and handles, attached to a microtubule filament estimated the force generated by a single kinesin molecule.

A major advance in the field was the resolution by Svoboda *et al.* [129] of the detailed motion of a single kinesin molecule into a sequence of 8-nm steps as it advanced along a microtubule. The first observation of this previously postulated stepping motion used an optical trapping interferometric position monitor with subnanometer resolution [129]. Proper damping of the Brownian motion of the sphere by the trap was also needed to see the steps [130]. Later, Svoboda and Block [131] measured the complete force-velocity relationship of single kinesin motors as a function of ATP concentrations. A maximum force of ∼5–6 pN was observed. Finer *et al.* [132], [133] shortly thereafter introduced a new feedback enhanced tweezer trap with a detection capability of subnanometers in position, piconewtons in force, and ms in time response. They studied the interaction of actin with myosin in a dual trap scheme which suspended the actin filament over a single myosin molecule. They observed stepwise motion of about 11 nm and forces of about 3–4 pN. Malloy *et al.* [134] also used feedback to study the interaction of myosin with mutant drosophila actins. The unbinding force of a single myosin molecule and actin filaments in the absence of ATP was measured with tweezers by Nishizaka *et al.* [135]. Another form of myosin, called myosin-V, was recently studied with a pair of tweezer traps. Myosin-V was shown to be a processive actin-based motor that can move in large steps approximating the 36-nm pseudo-repeat distance of actin filaments [136], [137]. Single motor molecule experiments have triggered work on detailed models of motion, the ATP hydrolysis cycle, and single enzyme kinetics [130], [138]–[141]. Recently, Visscher, Schnitzer, and Block have developed a novel feedback technique called a molecular force clamp which maintains constant force on a single kinesin molecule as it moves along microtubules [142]–[144]. New data show that one molecule of ATP is hydrolyzed per 8-nm step of kinesin over wide ranges of

the force. The new data on tight coupling between ATP hydrolysis and mechanical stepping seem to rule out many theoretical models for force generation by kinesin, including the so called thermal ratchet model [145], [146].

A recent exciting advance in the field was the extension of tweezer force measuring techniques to a new class of motors, nucleic acid motor enzymes. Using a handles technique the force generated by a single RNA polymerase enzyme was measured as it pulled itself along a DNA molecule while synthesizing an RNA transcript [147]. The motion is slow, but the motor is surprisingly powerful. It was observed to stall reversibly at 14 pN. Force versus velocity measurements were recently obtained for single RNA polymerase molecules using an optical feedback technique in which the position of the molecule was held fixed as the laser power was varied. Novel behavior was seen which can be modeled as a double potential well [148], [149]. Use of tweezer techniques opens a new way of studying the transcription process of RNA polymerase [150]. Increasingly, tweezers is becoming the technique of choice for the study by many groups of the mechanics of the many types of motor molecules [146], [151].

Work has also been done with tweezers to examine the mechanical properties of microtubules, actin filaments, and DNA bio-polymers. Kurachi *et al.* [152] measured the flexural rigidity of microtubles by attaching polystyrene beads and bending them with tweezers. Feigner *et al.* [53] studied the rigidity directly by manipulating free floating single microtubules. The torsional rigidity of actin was deduced from a measurement of the rotational Brownian motion of a single actin filament suspended from a freely rotating sphere held in a tweezers trap [154]. Chu *et al.* [155] made the first direct observation of the tube-like motion of a single extended fluorescently labeled DNA polymer strand as it relaxed through a dense entangled polymer solution. The behavior supports the reptation model of deGennes. The model explains the observed viscoelastic behavior of many biological materials [104], [105]. The stretching of double stranded DNA was studied with optical forces. At 70 pN of force a reversible transition to a single stranded unraveled form of DNA was seen [156]. Libchaber *et al.* [157] have studied the role of ATP hydrolysis in RecA polymerization on double-stranded DNA using single-molecule manipulation with optical tweezers. RecA protein plays a complex role in DNA growth and DNA repair. Block and collaborators stretched DNA molecules with optical tweezers using feedback control of position. They achieved great improvements in accuracy for sample lengths as short as ∼1 μm with forces ranging from ∼0.1 pN to ∼50 pN. An accurate value of the DNA persistence length of 40 nm was obtained [158]. There have been recent reports in Science that tweezers can possibly play an important role in DNA sequencing [159]. Fürst and Gast have used a pair of tweezer traps with tethers as a powerful quantitative technique to study the micromechanics of dipole chains made from superparamagnetic particles in liquids. Rupture forces and yield stresses were measured [160]. A beautiful experimental *in vivo* study of vesicle transport was performed in drosophila embryos. Knowledge of the regulation of forces and kinetics within a cell is a key to understanding cell and embryo development [161].

The ability of tweezers to separate single bacteria from a mixed sample in a chamber was recently used for separating selected archaea bacteria under high temperature anaerobic conditions for cloning purposes [162]. This new technique has already yielded a new species of hyperthermophilic archaeum from the hot springs of Yellowstone Park. A high temperature marine bacterium *Thermotoga maritima* has recently been isolated by tweezers and sequenced [163]. This has evolutionary significance. It is also a potential energy source, since it metabolizes carbohydrates to produce hydrogen gas and carbon. The hope is to find new high temperature enzymes, possibly as valuable as Tac polymerase used in PCR [164], [165]. There are vast numbers of unidentified water and soil bacteria which could be separated by similar tweezer techniques [166].

Burkhart *et al.* [167] used tweezers in a study to identify the mechanisms within killer cells and T lymphocytes by which so-called lytic particles move to attack target cells. They developed an *in vitro* assay that showed kinesin dependent motility of these particles on microtubles. A study of the cell-substrate adhesive process was made by Sheetz *et al.* [168] using the ability of tweezer manipulated coated microspheres to stick to the surface of moving fibroblast cells. They identified increased integrin-cytoskeleton adhesive interactions at the front of moving cells and increased deformability of the cell membrane at the rear of such cells. Measurement of changes in plasma membrane lipid structure and viscoelasticity during hypoxia were made by Kuo *et al.* [169] with tweezers. Results showed a transition to a more rigid state and the loss of membrane viscoelasticity during hypoxia. Sheetz *et al.* [170] made the first study of the mechanical properties of membranes on the leading edges of migrating neuronal growth cones by pulling out membrane tethers with tweezers. The force to extend the membrane and the membrane surface viscosity were determined.

The early work of Ashkin *et al.* [104] showed the ability of tweezers to distort the shape of red blood cells and confine many cells in a single trap. Svoboda and Block [171] measured the elastic properties of isolated red blood cell membrane skeletons. Using three tweezer traps, Brakenhoff *et al.* [172] developed a new assay to sensitively measure the shape recovery time of single red blood cell, using physiologically relevant shapes and conditions. Significant differences in relaxation times were found for old and young cells. Measurements were made in blood plasma and gave markedly different results from previous assays using pipettes in buffer solution. With automation, this may be a powerful technique for study of subpopulations of pathological cells. The three computer controlled tweezer traps used a multiple scanning trap system developed by Visscher *et al.* [173].

Greulich *et al.* have used tweezers to simulate the effect of gravity on the growing tips of algal cells [174]. Dragging the statoliths or gravity sensors of the cell to one side can induce the cell to reorient its growth in that direction.

An assay to study the collision of two particles or cells under controlled biologically relevant conditions, called "OPTCOL," was developed with two tweezer traps [175]. The adhesion of influenza virus covered spheres to erythrocytes during collision with controlled velocities and controlled geometry was studied in the presence of various attachment inhibitors. The new extremely sensitive technique has identified the most potent known inhibitor of this process. The authors foresee wide usage of OPTCOL for studies of collisions of biological particles such as bacteria, viruses, T cells, ribosomes, liposomes, and even nonbiological objects.

VIII. Other Recent Work on Optical Trapping and Manipulation in Physics and Chemistry

Interesting applications of optical manipulation techniques exist in other diverse areas of physics and chemistry.

In the field of statistical physics and nonlinear dynamics, Simon and Libchaber [176] used stochastic resonance to synchronize the escape of a Brownian particle from a pair of coupled tweezer traps.

Ackerson *et al.* [177] studied phase transitions and crystallization of a random 2-D colloidal suspension to a colloidal crystal using the optical forces of a standing wave beam.

Higurashi *et al.* [178] have observed optically induced torques and rotations of micromachined μm-sized anisotropic particles held in a tweezers trap.

Svoboda and Block [179] have shown that small metallic Rayleigh particles have polarizabilities larger than dielectric particles and can be trapped by tweezers.

Ghislain and Webb [180] have built a novel scanning force microscope based on a tweezer trapped stylus particle having a much lower spring constant than a mechanical cantilever. Applications to imaging soft samples in water are anticipated.

Tweezers were used to help measure the entropic forces of about 40 femtonewtons which control motion of colloidal particles at passive surface microstructures [181]. Entropic attraction and repulsive forces were directly observed with tweezers in binary colloids depending on the concentration of the small spheres [182].

Extensive use of optical trapping techniques has been made in the field of microchemistry, which studies the spectroscopy and chemistry of small μm-sized domains. Experiments combining trapping with fluorescence, absorption spectroscopy, photochemistry, and electrochemistry were performed. Polymerization, ablation, and other microfabrication techniques were demonstrated with micron samples. Beam scanning techniques were developed for trapping of μm-sized metal particles, low index particles, and moving of particle arrays in complex patterns. These experiments are by Masuhara *et al.* [183], summarizing the results of a five year ERATO project.

Bar-Zvi *et al.* [184], [185] have used tweezers to study the physical properties of membranes and vesicles. The local unbinding of pinched membranes [184] and pressurization and entropic expulsion of inner vesicles from large vesicles [185] was studied.

In the field of colloidal science, direct measurements using tweezers showed quite surprisingly that an attractive force can exist between like-charged particles in a colloidal suspension near a surface, contrary to theory [186]. Metastable colloidal crystals were made based on this attractive potential. This new work in colloidal science has importance on theoretical and possibly practical grounds [187].

IX. THE FUTURE

Looking ahead to the early years of the new century, it seems fair to predict that use of optical manipulation techniques will continue to grow at an increasingly rapid pace in the many subfields of physics, chemistry, and biology involving small particles. We are entering an era of increasing emphasis on the small for applications and for basic science. Microtechnology, small machines, small motors, motor molecules, gene sequencing, genetic engineering, and biological computers are already familiar terms. The role of laser tweezers and manipulation in basic sciences has been truly revolutionary. Atomic physics is once again growing in vitality. The cooling of atoms to the lowest temperature yet observed, the achievement of BEC, superfluid behavior of condensates, and atom lasers give new ways of studying quantum effects. The impact of laser technology on the biological sciences may prove to be equally revolutionary.

REFERENCES

[1] E. F. Nichols and G. F. Hull, *Phys. Rev.*, vol. 13, p. 293, 1901.
[2] P. N. Lebedev, "Untersuchungen über die Druckkräfte des Lichtes," *Annalen der Physik*, vol. 6, p. 433, 1901.
[3] C. H. Townes, *How The Laser Happened*. Oxford, U.K.: Oxford Univ. Press, 1999.
[4] A. Ashkin, "acceleration and trapping of particles by Radiation Pressure," *Phys. Rev. Lett.*, vol. 24, p. 156, 1970.
[5] ——, "Atomic Beam Deflection by Resonance-Radiation Pressure," *Phys. Rev. Lett.*, vol. 25, p. 1321, 1970.
[6] V. S. Askar'yan, "Effects of the Gradient of a Strong Electromagnetic Beam on Electrons and Atoms," *Zh. Eksp. Teor. Fiz*, vol. 42, p. 1567, 1962.
[7] V. L. Letokhov, "Narrowing of the Doppler Width in a Standing Wave," *Pis'maZh. Eksp. Teor. Fiz.*, vol. 7, p. 348, 1968.
[8] A. Ashkin and J. M. Dziedzic, "Optical Levitation by Radiation Pressure," *Appl. Phys. Lett.*, vol. 19, p. 283, 1971.
[9] ——, "Observation of Light Scattering from Nanospherical Particles Using Optical Levitation," *Appl. Optics*, vol. 19, p. 660, 1980.
[10] ——, "Stability of Optical Levitation by Radiation Pressure," *Appl. Phys. Lett.*, vol. 24, p. 586, 1974.
[11] ——, "Optical Levitation in High Vacuum," *Appl. Phys. Lett.*, vol. 28, p. 333, 1976.
[12] ——, "Feedback Stabilization of Optically Leviated Particles," *Appl. Phys. Lett.*, vol. 30, p. 202, 1977.
[13] A. Ashkin, "Applications of Laser Radiation Pressure," *Science*, vol. 210, p. 1081, 1980.
[14] A. Ashkin and J. M. Dziedzic, "Observation of Resonance in the Radiation Pressure on Dielectric spheres," *Phys. Rev. Lett.*, vol. 38, p. 1351, 1977.
[15] ——, "Observation of Optical Resonances of Diectric Spheres by Light Scattering," *Appl. Optics*, vol. 20, p. 1803, 1981.
[16] P. Chylek, V. Ramaswamy, A. Ashkin, and J. M. Dziedzic, "Simultaneous Determination of Refractive Index and Size of Spherical Dielectric Particles from Light Scattering Data," *Appl. Optics*, vol. 22, p. 2303, 1983.
[17] S. C. Hill and R. E. Benner, "Morphology-dependent resonances," in *Optical Effects Associated with Small Particles*, P. W. Barber and R. K. Chang, Eds. Singapore: World Scientific, 1988, pp. 3–61.
[18] V. Sandoghder, F. Treussart, J. Hare, V. Lefever-Seguin, J.-M. Raimond, and S. Haroche, "Very low threshold whispering-gallery mode microsphere laser," *Phys. Rev. A*, vol. 54, p. 1777, 1996.
[19] Y. Yamamoto and R. E. Slusher, "," *Physics Today*, vol. 46, p. 66, 1993.
[20] C. Gmachl, F. Capasso, E. E. Narimanov, J. U. Nöckel, A. D. Stone, J. Faist, D. L. Sivco, and A. Y. Cho, "High-Power Directional Emission from Microlasers with Chaotic Resonators," *Science*, vol. 280, p. 1556, 1998.
[21] S. Haroche, "Cavity electrodynamics," in *Proc. Les Houches Summer School of Theoretical Physics, Session LIII, 1990*, J. Dalibard, J.-M. Raimond, and J. Zinn-Justin, Eds. Amsterdam, The Netherlands: North-Holland, 1992.

[22] R. Schieder, H. Walther, and L. Woste, "Atomic Beam Deflection by the Light of a Tunable Dye Laser," *Opt. Comm.*, vol. 5, p. 337, 1972.
[23] P. Jacquinot, D. Liberman, J. L. Pigne, and J. Pinard, "High Resolution Spectroscopic Application of Atomic Beam Deflection by Resonant Light," *Opt. Comm.*, vol. 8, p. 163, 1973.
[24] A. F. Bernhardt, *Appl. Phys.*, vol. 9, p. 19, 1976.
[25] T. W. Hänsch and A. L. Schawlow, "Cooling of Gases by Laser Radiation," *Opt. Comm.*, vol. 13, p. 68, 1975.
[26] V. L. Letokhov, V. G. Minogin, and B. D. Pavlik, *Zh. Eksp. Teor. Fiz.*, vol. 72, p. 1328, 1977.
[27] V. L. Letokhov and V. G. Minogin, *Appl. Phys.*, vol. 17, p. 99, 1978.
[28] A. Ashkin, "Trapping af Atoms by Resonance Radiation Pressure," *Phys. Rev. Lett.*, vol. 40, p. 729, 1978.
[29] A. Ashkin and J. P. Gordon, "Cooling and Trapping of Atoms by Resonance Radiation Pressure," *Opt. Lett.*, vol. 4, p. 161, 1979.
[30] J. E. Bjorkholm, R. R. Freeman, A. Ashkin, and D. B. Pearson, "Observation of Focusing of Neutral Atoms by the Dipole Forces of Resonance Radiation Pressure," *Phys. Rev. Lett.*, vol. 41, p. 1361, 1978.
[31] D. B. Pearson, R. R. Freeman, J. E. Bjorkholm, and A. Ashkin, "Focusing and Defocusing of Neutral Atomic Beams Using Resonance Radiation Pressure," *Appl. Phys. Lett.*, vol. 36, p. 99, 1980.
[32] J. E. Bjorkholm, R. R. Freeman, A. Ashkin, and D. B. Pearson, "Experimental Observation of influence of the Quantum Fluctuations of Resonance Radiation Pressure," *Opt. Lett.*, vol. 5, p. 111, 1980.
[33] J. P. Gordon and A. Ashkin, "Motion of Atoms in Radiation Trap," *Phys. Rev. A*, vol. 21, p. 606, 1980.
[34] W. D. Phillips and H. Metcalf, "Laser Declaration of an Atomic Beam," *Phys. Rev. Lett.*, vol. 48, p. 596, 1982.
[35] P. Prodan, A. Migdall, W. D. Phillips, I. So, H. Metcalf, and J. Dalibard, "Stopping Atoms with Laser Light," *Phys. Rev. Lett.*, vol. 54, p. 992, 1985.
[36] W. D. Phillips, "Laser cooling and trapping of neutral atoms," *Rev. Mod. Phys.*, vol. 70, p. 21, 1998.
[37] V. S. Letokhov and V. G. Minogin, *Phys. Rep.*, vol. 73, p. 1, 1981.
[38] W. Ertmer, R. Blatt, J. L. Hall, and M. Zhu, "Laser Manipulation of Atomic Beam Velocities: Demonstration of stopped Atoms and Velocity Reversal," *Phys. Rev. Lett.*, vol. 54, p. 996, 1985.
[39] V. G. Minogin and J. Javanainen, "A Tetrahedral Light Pressure Trap for Atoms," *Opt. Commun.*, vol. 43, p. 119, 1982.
[40] A. Ashkin and J. P. Gordon, "Stability of Radiation-Pressure Traps: An Optical Earnshaw Theorem," *Opt. Lett.*, vol. 8, p. 511, 1983.
[41] S. Chu, "The manipulation of neutral particles," *Rev. Mod. Phys.*, vol. 70, p. 685, 1998.
[42] S. Chu, L. Holberg, J. E. Bjorkholm, A. Cable, and A. Ashkin, "Three-Dimensional Viscous Confinement and Cooling of Atoms by Resonance Radiation Pressure," *Phys. Rev. Lett.*, vol. 55, p. 48, 1985.
[43] S. Chu, J. E. Bjorkholm, A. Cable, and A. Ashkin, "Experimental Observation of Optically Trapped Atoms," *Phys. Rev. Lett.*, vol. 57, p. 314, 1986.
[44] E. L. Raab, M. Raab, A. Cable, S. Chu, and D. E. Pritchard, "Trapping of Neutral Sodium Atoms with Radiation Pressure," *Phys. Rev. Lett.*, vol. 59, p. 2631, 1987.
[45] T. Walker, D. Sasko, and C. Wieman, "Collective Behavior of Optically Trapped Neutral Atoms," *Phys. Rev. Lett.*, vol. 64, p. 408, 1990.
[46] P. D. Lett, R. N. Watts, C. I. Westbrook, W. D. Phillips, P. L. Gould, and H. J. Metcalf, "Observation of atoms laser cooled below the Doppler limit," *Phys. Rev. Lett.*, vol. 61, p. 169, 1988.
[47] P. J. Ungar, D. S. Weiss, E. Riis, and S. Chu, "Optical molasses and multilevel atoms: Theory," *Opt. Soc. Am. B*, vol. 6, p. 2058, 1989.
[48] J. Dalibard and C. Cohen-Tannoudji, "Laser Cooling Below the Doppler Limit by Polarization Gradients-Simple Theoretical Models," *Opt. Soc. Am. B*, vol. 6, p. 2023, 1989.
[49] S. Chu, J. E. Bjorkholm, A. Ashkin, J. P. Gordon, and L. Holberg, "Proposal for Optically Cooling Atoms to Temperatures of the Order of 10^{-6} K," *Opt. Lett.*, vol. 11, p. 73, 1986.
[50] H. Hess, G. P. Kochenski, D. Kleppner, and T. J. Greytak, "Magnetic trapping of spin-polarized atomic hydrogen," *Phys. Rev. Lett.*, vol. 59, p. 672, 1987.
[51] A. Aspect, E. Arimondo, R. Kaisor, N. Vansteenkiste, and C. Cohen-Tannoudji, "Laser cooling below the One-Photon Recoil Energy by Velocity Selective Coherent Population Trapping," *Phys. Rev. Lett.*, vol. 61, p. 826, 1988.
[52] J. Lawall, S. Kulin, B. Saubamea, N. Bigelow, M. Leduc, and C. Cohen-Tannoudji, "Three-Dimensional Laser Cooling of Helium Beyond the Single-Photon Recoil limit," *Phys. Rev. Lett.*, vol. 75, p. 4194, 1995.
[53] C. N. Cohen-Tannoudji, "Manipulation of atoms with photons," *Rev. Mod. Phys.*, vol. 70, p. 707, 1998.

[54] M. Kasevich and S. Chu, "Laser cooling below a Photon Recoil with Three-Level Atoms," *Phys. Rev. Lett.*, vol. 69, p. 1741, 1992.
[55] H. J. Lee, C. S. Adams, M. Kasevich, and S. Chu, "Raman Cooling of Atoms in an Optical Dipole Trap," *Phys. Rev. Lett.*, vol. 76, p. 2658, 1996.
[56] M. A. Kasevich, E. Riis, S. Chu, and R. G. DeVoe, "rf Spectroscopy in an Atomic Fountain," *Phys. Rev. Lett.*, vol. 63, p. 612, 1989.
[57] R. Drullinger, *APS News*, vol. 5, no. 6, p. 4, 1996.
[58] M. Prentiss, A. Cable, J. E. Bjorkholm, S. Chu, E. Raab, and D. E. Pritchard, *Opt. Lett.*, vol. 13, p. 452, 1988.
[59] D. Sesko, T. Walker, C. Monroe, A. Gallagher, and C. Wieman, "Collisional Losses From a Light-Force Trap," *Phys. Rev. Lett.*, vol. 63, p. 961, 1989.
[60] H. R. Thorsheim, J. Weiner, and P. S. Julienne, "Laser-Induced Photoassociation of Ultracold Sodium Atoms," *Phys. Rev. Lett.*, vol. 58, p. 2420, 1987.
[61] P. S. Julienne and F. H. Mies, *J. Opt. Soc. Am. B*, vol. 6, p. 2257, 1989.
[62] T. Sleator, T. Pfau, V. Balykin, and J. Mlynek, *Appl. Phys. B*, vol. 54, p. 375, 1992.
[63] V. Balykin *et al.*, *J. Mod. Opt.*, vol. 35, p. 17, 1988.
[64] G. M. Gallatin and P. L. Gould, "Laser focusing of atomic beams," *J. Opt. Soc. Amer. B*, vol. 8, p. 502, 1991.
[65] R. J. Cook and R. K. Hill, "An Electromagnetic Mirror for Neutral Atoms," *Opt. Comm.*, vol. 43, p. 258, 1982.
[66] R. Kaiser *et al.*, "Resonant enhancement of evanescent waves with a thin dielectric waveguide," *Opt. Comm.*, vol. 104, p. 234, 1993.
[67] P. J. Martin, B. G. Oldaker, A. H. Miklich, and D. E. Pritchard, "Bragg Scattering of Atoms from a standing Light Wave," *Phys. Rev. Lett.*, vol. 60, p. 515, 1988.
[68] K. S. Johnson, A. Chu, T. W. Lynn, K. K. Berggren, M. S. Shahriar, and M. Prentiss, "Demonstration of a Novel Atomic Beam Splitter," *Opt. Lett.*, vol. 20, p. 1310, 1995.
[69] J. Lawall and M. G. Prentiss, *Phys. Rev. Lett.*, vol. 72, p. 993, 1994.
[70] M. Weitz, B. C. Young, and S. Chu, *Phys. Rev. Lett.*, vol. 72, p. 2563, 1994.
[71] G. Timp, R. E. Behringer, D. M. Tennant, J. E. Cunningham, M. Prentiss, and K. Berggren, "Using Light as a Lens for Submicron, Neutral Atom Lithography," *Phys. Rev. Lett.*, vol. 69, p. 1636, 1992.
[72] J. J. McClelland, R. E. Scholten, E. C. Palm, and R. J. Celotta, "Laser-Focused Atomic Deposition," *Science*, vol. 262, p. 877, 1993.
[73] V. I. Balykin and V. S. Letokhov, "Laser Optics of Neutral Atomic Beams," *Physics Today*, vol. 42, p. 23, 1989.
[74] M. Kasevich and S. Chu, "Atomic Interferometry Using Stimulated Raman Transitions," *Phys. Rev. Lett.*, vol. 67, p. 181, 1991.
[75] M. H. Anderson, J. R. Ensher, M. R. Mathews, C. E. Wieman, and E. A. Cornell, "Observation Of Bose-Einstein Condensation in a dilute vapor below 200 nanokelvin," *Science*, vol. 269, p. 198, 1995.
[76] K. B. Davis, M.-O. Mewes, M. R. Andrews, N. J. vanDruten, D. S. Durfee, D. M. Kurn, and W. Ketterle, "Bose-Einstein condensation in a gas off sodium atoms," *Phys. Rev. Lett.*, vol. 75, p. 3969, 1996.
[77] D. S. Jin, M. R. Mathews, J. R. Ensher, C. E. Wieman, and E. A. Cornell, "Temperature dependent Damping and Frequency shifts in collective Excitations of a Dilute Bose-Einstein Condensate," *Phys. Rev. Lett.*, vol. 78, p. 764, 1997.
[78] D. M. Stamper-Kurn *et al.*, *Phys. Rev. Lett.*, vol. 83, p. 2876, 1999.
[79] M. R. Andrew, C. G. Townsend, H.-J. Miesner, D. S. Durfee, D. M. Kurn, and W. Ketterle, *Science*, vol. 275, p. 637, 1997.
[80] J. Stenger *et al.*, *Phys. Rev. Lett.*, vol. 82, p. 4569, 1999.
[81] M.-O. Mews *et al.*, "Output Coupler for Bose-Einstein Condensed Atoms," *Phys. Rev. Lett.*, vol. 78, p. 582, 1997.
[82] E. W. Hagley, L. Deng, M. Kozuma, J. Wen, K. Helmerson, S. L. Rolston, and W. D. Phillips, *Science*, vol. 283, p. 1706, 1999.
[83] L. Deng, E. W. Hagley, J. Wen, M. Trippenback, Y. Band, P. S. Julienne, J. E. Simsarian, K. Helmerson, S. L. Rolston, and W. D. Phillips, "Four-wave mixing with matter waves," *Nature*, vol. 398, p. 218, 1990.
[84] S. Inouye, T. Pfau, S. Gupta, A. P. Chikkatur, A. Görlitz, D. E. Pritchard, and W. Ketterle, *Nature*, vol. 402, p. 641, 1999.
[85] M. Kozumi, Y. Suzuki, Y. Torii, T. Sugiwa, T. Kuga, E. W. Hagley, and L. Deng, *Science*, vol. 286, p. 2309, 1999.
[86] D. M. Stamper-Kurn, M. R. Andrews, A. P. Chikkatur, S. Inouye, H.-J. Miesner, J. Stenger, and W. Ketterle, "Optical confinement of a Bose-Einstein Condensate," *Phys. Rev. Lett.*, vol. 80, p. 2027, 1998.
[87] J. Stenger, D. M. Stamper-Kurn, M. R. Andrews, A. P. Chikkatur, S. Inouye, H.-J. Miesner, and W. Ketterle, *J. LowTemp. Phys.*, vol. 113, p. 167, 1998.

[88] S. Inouye, M. R. Andrew, J. Stenger, H.-J. Miesner, D. M. Stamper-Kurn, A. P. Chikkatur, and W. Ketterle, "Resonances in a Bose-Einstein Condensate," *Nature*, vol. 392, p. 151, 1998.
[89] J. Stenger, S. Inouye, M. R. Andrews, H.-J. Miesner, D. M. Stamper-Kurn, and W. Ketterle, "Strongly Enhanced Inelastic Collisions in a Bose-Einstein Condensate near Feshbach Resonances," *Phys. Rev. Lett.*, vol. 82, p. 2422, 1999.
[90] J. Stenger, S. Inouye, D. M. Stamper-Kurn, H.-J. Miesner, A. P. Chikkatur, and W. Ketterle, *Nature*, vol. 396, p. 345, 1998.
[91] D. M. Stamper-Kurn, H.-J. Miesner, A. P. Chikkatur, S. Inouye, J. Stenger, and W. Ketterle, *Phys. Rev. Lett.*, vol. 83, p. 661, 1999.
[92] P. Courteille, R. S. Freeland, and D. J. Heinzen, "Observation of a Feshbach Resonance in Cold Atom Scattering," *Phys. Rev. Lett.*, vol. 81, p. 69, 1998.
[93] J. L. Roberts, N. R. Claussen, S. L. Cornish, and C. E. Wieman, "Magnetic field Dependence of Ultracold Inelastic Collisions near a Feshbach Resonance," *Phys. Rev. Lett.*, vol. 85, p. 728, 2000.
[94] B. Goss Levi, "Researchers Can now Vary the Atomic Interactions in a Bose-Einstein Condensate," *Physics Today*, vol. 53, no. 8, p. 17, August 2000.
[95] M. R. Mathews, B. P. Anderson, P. C. Haljan, D. S. Hall, C. E. Wieman, and E. A. Cornell, *Phys. Rev. Lett.*, vol. 33, p. 2498, 1999.
[96] C. Raman, M. Köhl, R. Onofrio, D. S. Durfee, C. E. Kuklewicz, Z. Hadzibabic, and W. Ketterle, *Phys. Rev. Lett.*, vol. 83, p. 2502, 1999.
[97] O. M. Marago, S. A. Hopkins, J. Arlt, E. Hodby, G. Hechenblaikner, and C. J. Foot, *Phys. Rev. Lett.*, vol. 84, p. 2056, 2000.
[98] K. W. Madison, F. Chevy, W. Wohlleben, and J. Dalibard, "Vortex Formation in a Stirred Bose-Einstein Condensate," *Phys. Rev. Lett.*, vol. 84, p. 806, 2000.
[99] W. Ketterle, "Experimental Studies of Bose-Einstein Condensation," *Phys. Today*, vol. 52, no. 11, p. 30, Dec. 1999.
[100] C. Gabbanini, A. Fioretti, A. Luchesini, S. Gozzini, and M. Mazzoni, "Cold Rubidium Molecules formed in a Magneto-Optical Trap," *Phys. Rev. Lett.*, vol. 84, p. 2814, 2000.
[101] A. Ashkin, J. M. Dziedzic, J. E. Bjorkholm, and S. Chu, "Observation of a Single Beam Gradient Force Optical Trap for Dielectric Particles," *Opt. Lett.*, vol. 11, p. 288, 1986.
[102] A. Ashkin, "Forces of a Single Beam Gradient Laser Trap on a Dielectric Sphere in the Ray Optics Regime," *Biophys. J.*, vol. 61, p. 569, 1992.
[103] A. Ashkin and J. M. Dziedzic, "Optical trapping and Manipulation of Viruses and Bacteria," *Science*, vol. 235, p. 1517, 1987.
[104] A. Ashkin, J. M. Dziedzic, and T. Yamane, "Optical trapping and Manipulation of Single Cells Using Infrared Laser Beams," *Nature*, vol. 330, p. 769, 1987.
[105] A. Ashkin and J. M. Dziedzic, "internal Cell Manipulation Using Infrared Laser Traps," *Proc. Natl. Acad. Sci. USA*, vol. 86, p. 7914, 1989.
[106] F. H. C. Crick and A. F. W. Hughes, *Exp. Cell Res.*, vol. 1, p. 37, 1950.
[107] M. Sato, T. Z. Wong, and R. D. Allen, *J. Cell Biol.*, vol. 97, p. 1089, 1983.
[108] T. T. Perkins, D. E. Smith, and S. Chu, "Direct Observation of Tube-Like Motion of a Single Polymer Chain," *Science*, vol. 264, p. 822, 1994.
[109] S. B. Smith, Y. Cui, and C. Bustamante, "Overstretching B-DNA: The Elastic Response of Individual Double-Stranded and Single-Stranded DNA Molecules," *Science*, vol. 271, p. 795, 1996.
[110] M. Sato, T. Z. Wong, and R. D. Allen, "A preliminary investigation of living physarum endoplasm," in *The Application of Light Scattering to the Study of Biological Motion*, J. C. Earnshaw and M. W. Stear, Eds. New York: Plenum, 1983.
[111] S. M. Block, D. F. Blair, and H. C. Berg, "Compliance of bacterial flagella measured with optical tweezers," *Nature*, vol. 338, p. 514, 1989.
[112] ——, "Compliance of Compliance of bacterial Polyhooks Measured with optical Tweezers," *Cytometry*, vol. 12, p. 492, 1991.
[113] N. W. Charon, S. F. Goldstein, S. M. Block, K. Curci, and J. D. Ruby, *J. Bacteriol.*, vol. 174, p. 832, 1992.
[114] S. Seeger, S. Manojembaski, K. J. Hutter, G. Futterman, J. Wolfrum, and K. O. Greulich, "Application of laser tweezers in immunology and molecular genetics," *Cytometry*, vol. 12, p. 497, 1991.
[115] R. W. Steubing, S. Cheng, W. H. Wright, Y. Namajiri, and M. W. Berns, "Laser induced fusion in combination with optical tweezers: the laser cell fusion trap," *Cytometry*, vol. 12, p. 505, 1991.
[116] X. Wei, B. J. Tromberg, and M. D. Cahalan, "Mapping the sensivity of T celle with an optical trap: Polarity and minimum number of receptors for Ca^{2+} signaling," *Proc. Natl. Acad. Sci. USA*, vol. 96, p. 18 471, 1999.
[117] H. Liang, W. H. Wright, W. He, and M. W. Berns, *Exp. Cell Res.*, vol. 204, p. 110, 1993.

[118] Y. Tadir, W. H. Wright, O. Vafa, T. Ord, R. H. Asch, and M. W. Berns, "Micromanipulation of sperm by a laser generated optical trap," *Fertility & Sterility*, vol. 52, p. 870, 1989.

[119] J. M. Colon, P. Sarosi, P. G. McGovern, A. Ashkin, J. M. Dziedzic, J. Skurnick, G. Weiss, and E. M. Bonder, "Controlled micromanipulation of human spermatozoa in three dimensions with a infrared laser optical trap: effect on sperm velocity," *Fertility & Sterility*, vol. 57, p. 695, 1992.

[120] E. M. Bonder, J. M. Colon, J. M. Dziedzic, and A. Ashkin, "Force production by swimming sperm using optical tweezers," *J. Cell Biol.*, vol. 111, p. 421a, 1990.

[121] Y. Tadir, W. H. Wright, O. Vafa, L. H. Liaw, R. Asch, and M. W. Berns, "Micromanipulation of gametes using laser microbeams," *Human Reproduct.*, vol. 6, p. 1011, 1991.

[122] K. Schütze, A. Clement-Sengewald, and A. Ashkin, "Zona drilling and sperm insertion with combined laser microbeam and optical tweezers," *Fertility&Sterility*, vol. 61, p. 783, 1994.

[123] A. Clement-Sengewald, K. Schütze, G. A. Palma, G. Kerlen, and B. Brem, *J. Assisted Reprod. & Genetics*, vol. 13, p. 259, 1996.

[124] H. Liang, K. T. Vu, P. Krishnan, T. C. Trang, D. Shin, S. Kimel, and M. W. Berns, "Wavelength dependence of cell cloning efficiency after optical trapping," *Biophys. J.*, vol. 70, p. 1529, 1996.

[125] S. M. Block, L. S. B. Goldstein, and B. J. Schnapp, "Bead movement by single kinesin molecules studied with optical tweezers," *Nature*, vol. 348, p. 348, 1990.

[126] K. Svoboda and S. M. Block, "Biological Applications of optical Forces," *Ann. Rev. Biophys; Biomol. Struct.*, vol. 23, p. 247, 1994.

[127] A. Ashkin, K. Schütze, J. M. Dziedzic, U. Eutenauer, and M. Schliwa, "Force Generation of Organelle Transport Measured *in vivo* by an Infrared Laser Trap," *Nature*, vol. 348, p. 346, 1990.

[128] S. C. Kuo and M. P. Sheetz, "Force of Single Kinesin Molecules Measured with Optical Tweezers," *Science*, vol. 260, p. 232, 1993.

[129] K. Svoboda, C. F. Schmidt, B. J. Schnapp, and S. M. Block, "Direct observation of kinesin stepping by optical trapping interferometry," *Nature*, vol. 365, p. 721, 1993.

[130] S. M. Block, "Nanometers and piconewtons: the macromolecular mechanics of kinesin," *Trends in Cell Biology*, vol. 5, p. 169, 1995.

[131] K. Svoboda and S. M. Block, "Force and Velocity Measured for Single Kinesin Molecules," *Cell*, vol. 77, p. 773, 1994.

[132] J. T. Finer, R. M. Simmons, and J. A. Spudich, "Single myosin molecule mechanics: piconewton forces and nanometer steps," *Nature*, vol. 368, p. 113, 1994.

[133] R. M. Simmons, J. T. Finer, S. Chu, and A. Spudich, "Quantitative measurements of forces and displacement using an optical trap," *Biophys. J.*, vol. 70, p. 1813, 1996.

[134] J. E. Molloy, J. E. Burns, J. C. Sparrow, R. T. Tregear, J. Kendrick-Jones, and D. C. White, *Biophys. J.*, vol. 68, p. 2985, 1995.

[135] T. Nishizaka, H. Miyata, H. Yoshikawa, S. Ishiwata, and K. Kinosita Jr., "Unbinding force of a single motor molecule of muscle measured using optical tweezers," *Nature*, vol. 377, p. 251, 1995.

[136] A. D. Mehta, R. S. Rock, M. Rief, J. A. Spudick, M. S. Moosekor, and R. E. Cheney, "Myosin-V is a processive actin-based motor," *Nature*, vol. 400, p. 590, 1999.

[137] H. Sakakibara, H. Kojima, Y. Sakai, E. Katayana, and K. Oiwa, "Inner-arm dynein C of chlamydomanas flagella is a single-headed processive motor," *Nature*, vol. 400, p. 586, 1999.

[138] H. Kojima, E. Muto, H. Higuichi, and T. Yanagida, "Mechanics of Single-Kinesin Molecules Measured by Optical Trapping Nanometry," *Biophys. J.*, vol. 73, p. 2012, 1997.

[139] B. G. Levi, *Phys. Today*, vol. 48, no. 4, p. 17, 1995.

[140] K. Svoboda, P. P. Mitra, and S. M. Block, "Fluctuation analysis of motor Protein movement and single enzyme kinetics," *Proc. Natl. Acad. Sci. USA*, vol. 91, p. 11 782, 1994.

[141] M. J. Schnitzer and S. M. Block, "Kinesin Hydrolysis one ATP per 8–nm step," *Nature*, vol. 388, p. 386, 1997.

[142] A. E. Knight and J. E. Malloy, *Nature Cell Biol.*, vol. 1, p. E87, 1999.

[143] K. Visscher, M. J. Schnitzer, and S. M. Block, "Single Kinesin molecule studied with a molecular force clamp," *Nature*, vol. 400, p. 184, 1999.

[144] K. Visscher and S. M. Block, "Versatile Optical Traps with Feedback Controls," *Meth. Enzymol.*, vol. 298, p. 460, 1998.

[145] R. D. Asturmian and M. Bier, *Phys. Rev. Lett.*, vol. 72, p. 1766, 1994.

[146] L. P. Faucheux, L. S. Bourdieu, P. D. Kaplan, and A. D. Libchaber, "Optical Thermal Ratchet," *Phys. Rev. Lett.*, vol. 74, p. 1504, 1995.

[147] H. Yin, M. D. Wang, K. Svoboda, R. Landick, S. M. Block, and J. Gelles, "Transcription Against an Applied Force," *Science*, vol. 270, p. 1653, 1995.

[148] M. D. Wang, M. J. Schnitzer, H. Yin, R. Landrick, J. Gelles, and S. M. Block, "Force and Velocity Measurements for Single Molecule of RNA Polymerase," *Science*, vol. 282, p. 902, 1998.

[149] R. F. Service, "Watching DNA at work," *Science*, vol. 283, p. 1668, 1999.

[150] C. O'Brien, "RNA Polymerase Gets Very Pushy," *Science*, vol. 270, p. 1668, 1995.

[151] A. D. Mehta, M. Rief, J. A. Spudick, D. A. Smith, and R. H. Simmons, "Molecule Biomechanics with Optical Methods," *Science*, vol. 283, p. 1689, 1999.

[152] M. Kurachi, M. Hoshi, and H. Tashiro, *Cell Motil. Cytoskel.*, vol. 30, p. 221, 1995.

[153] H. Feigner, R. Frank, and M. Schliwa, *J. Cell Sci.*, vol. 109, p. 509, 1996.

[154] Y. Tsuda, H. Yasutake, A. Ishijima, and T. Yanagida, "Torsional Rigidity of Single Actin Filaments and Actin-Actin Bond breaking Force under Torsion Measured Directly by *in vitro* Micromanipulation," *Proc. Natl. Acad. Sci. USA*, vol. 93, p. 12 937, 1996.

[155] T. T. Perkins, D. E. Smith, and S. Chu, *Science*, vol. 264, p. 822, 1994.

[156] S. B. Smith, Y. Cui, and C. Bustamante, "Overstretching B-DNA: the elastic response of individual double-stranded DNA molecules," *Science*, vol. 271, p. 795, 1996.

[157] G. V. Shivashankar, M. Feingold, O. Krichevsky, and A. Libchaber, "Rec A polymerization on double-stranded DNA by using single-molecule manipulation: the role of ATP hydrolysis," *Proc. Natl. Acad. Sci. USA*, vol. 96, p. 7916, 1999.

[158] M. D. Wang, H. Yin, R. Landrick, J. Gelles, and S. M. Block, "Stretching DNA with Optical Tweezers," *Biophys. J.*, vol. 72, p. 1335, 1997.

[159] R. F. Service, *Science*, vol. 283, p. 1669, 1999.

[160] E. M. Furst and A. P. Gast, "Micromechanics of Dipolar Chains Using Optical Tweezers," *Phys. Rev. Lett.*, vol. 83, p. 4130, 1999.

[161] M. A. Welte, S. P. Gross, M. Postner, S. M. Block, and E. F. Wieschaus, *Cell*, vol. 92, p. 547, 1998.

[162] R. Huber, S. Burggraf, T. Mayer, S. M. Barns, P. Rossnagel, and K. O. Stetter, "Isolation of a hyperthermophilic archaeum predicted by *in situ* RNA analysis," *Nature*, vol. 376, p. 57, 1995.

[163] K. E. Nelson *et al.*, *Nature*, vol. 399, p. 323, 1999.

[164] M. Milstein, "Yellowstone managers strike claim on hot-springs microbes," *Science*, vol. 270, p. 226, 1995.

[165] S. Barker, *Nature*, vol. 381, p. 455, 1996.

[166] J. G. Mitchell, R. Weller, M. Beconi, J. Sell, and J. Holland, *Microbial Ecology*, vol. 25, p. 113, 1993.

[167] J. K. Burkhardt, J. M. McIlvain Jr., M. P. Sheetz, and Y. Argon, *J. Cell Sciences*, vol. 104, p. 151, 1993.

[168] C. E. Schmidt, A. F. Horwitz, D. A. Lauffenburger, and M. P. Sheetz, *J. Cell Biol.*, vol. 123, p. 977, 1993.

[169] X. F. Wang, J. J. Lemaster, B. Herman, and S. Kuo, *Opt. Engin.*, vol. 32, p. 284, 1993.

[170] J. Dai and M. P. Sheetz, *Biophys. J.*, vol. 68, p. 988, 1995.

[171] K. Svoboda, C. F. Schmidt, D. Branton, and S. M. Block, "Conformation and elasticity of the isolated red blood cell membrane skeleton," *Biophys. J.*, vol. 63, p. 784, 1992.

[172] P. J. H. Bronkhorst, G. J. Streekstra, J. Grimbergen, E. J. Nijhof, J. J. Sixma, and G. J. Brakenhoff, *Biophys. J.*, vol. 69, p. 1666, 1995.

[173] K. Visscher, G. J. Brakenhoff, and J. Krol, "Micromanipulation by 'multiple' optical traps created by a single fast scanning trap integrated with the bilateral confocal scanning microscope," *Cytometry*, vol. 14, p. 105, 1993.

[174] G. Leitz, E. Schnepf, and K. O. Greulich, "Micromanipulation of statoliths in gravity-sensing chara rhizoids by optical tweezers," *Planta*, vol. 197, p. 278, 1995.

[175] M. Mammer, K. Helmerson, R. Kishore, C. Seok-Ki, W. D. Phillips, and G. M. Whitesides, "optically controlled collisions of biological objects to evaluate potent polyvalent inhibitors of virus-cell adhesion," *Chem. & Biology*, vol. 3, p. 757, 1996.

[176] A. Simon and A. Libchaber, "Escape and Synchronization of a Brownian Particle," *Phys. Rev. Lett.*, vol. 68, p. 3375, 1992.

[177] A. Choudhury, B. J. Ackerson, and N. A. Clark, "Laser-Induced Freezing," *Phys. Rev. Lett.*, vol. 55, p. 833, 1985.

[178] E. Higurashi, H. Ukita, H. Tanaka, and O. Ohguchi, "Optically induced rotation of anisotropic micro-objects fabricated by surface micromachining," *Appl. Phys. Lett.*, vol. 64, p. 2209, 1994.

[179] K. Svoboda and S. M. Block, "Optically trapping of metallic Rayleigh particles," *Opt. Lett.*, vol. 19, p. 930, 1994.

[180] L. P. Ghislain and W. W. Webb, "Scanning force microscope using an optical trap," *Opt. Lett.*, vol. 18, p. 1678, 1993.

[181] A. D. Dinsmore, A. G. Yodh, and D. J. Pine, "Entropic control of particle motion using passive surface microstructures," *Nature*, vol. 383, p. 239, 1996.

[182] J. C. Crocker, J. A. Mattes, A. D. Dinsmore, and A. G. Yodh, "Entropic Attraction and Repulsion in Binary Colloids Probed with a Line Optical Tweezer," *Phys. Rev. Lett.*, vol. 82, p. 4352, 1999.

[183] H. Masuhara, F. C. deSchryver, N. Kitamura, and N. Tamai, *Microchemistry-Spectroscopy and Chemistry in Small Domains*: North Holland, 1994.

[184] R. Bar-Ziv, R. Menes, E. Moses, and S. A. Safran, "Local unbinding of pinched membranes," *Phys. Rev. Lett.*, vol. 75, p. 3356, 1995.

[185] R. Bar-Ziv, T. Frisch, and E. Moses, "Entropic expulsion in vesicles," *Phys. Rev. Lett.*, vol. 75, p. 3481, 1995.

[186] J. C. Crocker and D. G. Grier, "When like charges attract: the effects of geometrical confinement on long-range colloidal interactions," *Phys. Rev. Lett.*, vol. 77, p. 1897, 1996.

[187] C. A. Murray, "When like charges attract," *Nature*, vol. 385, p. 203, 1997.

[188] V. S. Askar'yan, *Sov. Phys. JETP Lett.*, vol. 15, p. 1088, 1962.

A. Ashkin (M'61–SM'75–F'76–LF'92), photograph and biography not available at the time of publication.

news feature

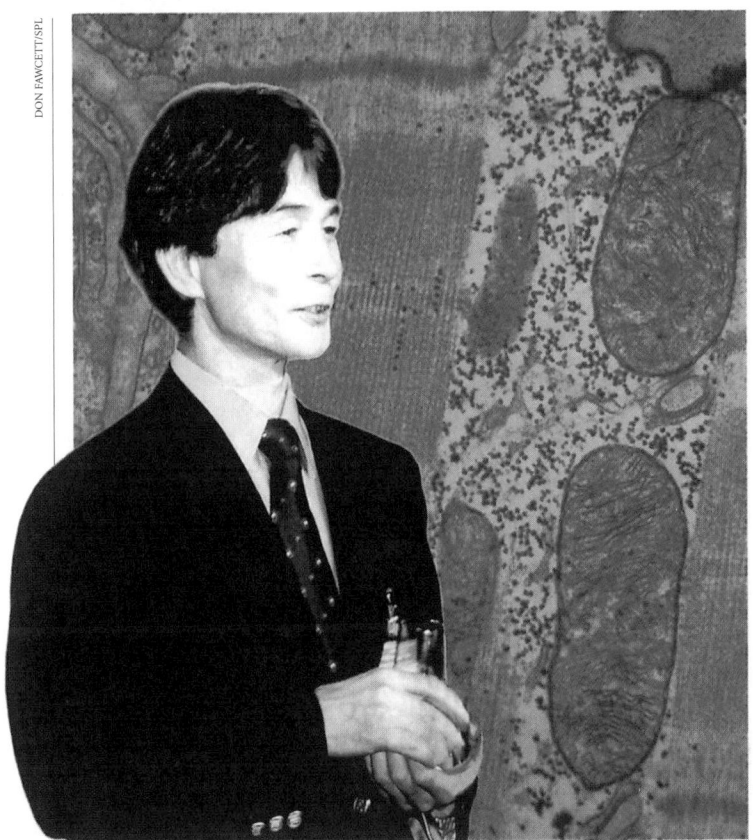

Swimming against the tide

Toshio Yanagida rejects the conventional biophysical explanation of muscle contraction. No one doubts his technical genius, but could the debate he started ultimately hold back the field? David Cyranoski investigates.

Iconoclast, radical, technical wizard — these are just some of the descriptions that have been applied to Toshio Yanagida, host of last month's Frontiers in Molecular Motors Research symposium in Japan.

Yanagida's technological brilliance has been crucial in developing methods to study what drives muscle contraction. But for 15 years, he has been at odds with the majority of muscle biophysicists over the precise mechanisms involved.

Held on the island of Awaji, just across the water from Yanagida's laboratories in Osaka, last month's meeting was a lively affair. Both sides of the debate presented their latest results. And these served to underline the field's impressive achievements in manipulating and visualizing the individual protein molecules involved in muscle movement.

But from similar experiments, the participants drew very different conclusions. And this continuing trend for both camps to bolster their arguments without finding common ground leaves many in the field perplexed. The debate is where it was ten years ago, says Clive Bagshaw, a biophysicist at the University of Leicester. "It's like nothing has changed."

The argument centres on two proteins:

myosin II and actin. Myosin II has two 'heads' joined by 'neck' regions to a coiled 'tail'. In muscle cells, the tails of myosin II molecules clump together to form 'thick filaments' with the necks and heads jutting out from the sides. Actin, on the other hand, forms helical 'thin filaments', which line up alongside their myosin counterparts. Muscles contract when the thick and thin filaments slide past one another. The puzzle for biophysicists is how the myosin II heads interact with the actin filaments to bring about this movement.

Head start

Most researchers back the lever-arm theory. In this, myosin II heads latch onto a nearby actin thin filament. By flipping in a lever-like motion, each head propels the actin in a 'power stroke' (see figure, opposite). The myosin head then lets go of the actin, the neck cocks back, and the head reattaches at a new point on the thin filament, starting the process again.

Muscle contraction is driven by adenosine triphosphate (ATP), the energy 'currency' of the cell, which binds to the myosin head. In the lever-arm model, a single cycle of actin attachment and power stroke for one myosin

II head involves the hydrolysis of one ATP molecule. Because of this straightforward relationship between energy input and mechanical action, the lever-arm theory is also called the tight-coupling model.

Yanagida proposes that, for each ATP molecule consumed, a myosin II head moves several steps, bumping along the actin filament like a railway wagon rattling down a rickety track. But the details of his theory are frustratingly diffuse. "My opponents are always saying: 'give more interpretation of your results'," says Yanagida, "but I think it's better not to."

An input of ATP is still crucial, although its precise role in this 'loose-coupling' model is unclear. One possibility, says Yanagida, is that myosin somehow absorbs the energy from ATP hydrolysis, releasing this energy over a series of steps to drive its movement. But the role of ATP might simply be to cause a change in myosin's shape that allows movement to begin. In this case, much of the energy needed for myosin to move against actin would come from the Brownian motion of the molecules in the surrounding fluid. Brownian motion is the random movement of molecules, driven by thermal forces and characterized by collisions between the molecules. Yanagida believes that the structures of actin and myosin might bias this motion in one direction — although he does not have a full explanation of how this could work.

Seeds of doubt

Yanagida made his mark in 1984, when his group was the first to image a single actin filament in solution[1]. But it was in the following year that he sowed the seeds of the debate that continues to this day. By relating the 'sliding velocity' of an actin filament to the consumption of ATP, Yanagida estimated that a single molecule of ATP causes a myosin head to advance some 60 nanometres down an actin filament[2]. This posed a problem for the lever-arm theory, as the dimensions of the myosin head and neck mean that a single ATP molecule should cause a displacement of only around 6 nm. Japanese newspapers immediately hailed Yanagida as a scientific folk hero, claiming that he had overturned a popular 'Western' theory.

Yanagida was working with his mentor Fumio Oosawa at Osaka University, even though Oosawa could only employ him as a technician. After Oosawa retired, the university took the unusual step in 1988 of promoting Yanagida straight to full professor. He has since enjoyed high levels of government funding. His current grant, for his Single Molecule Processes project, is worth more than US$9.3 million over five years to 2002.

With this generous financial support, Yanagida has assembled a team of 15 scientists in a renovated warehouse. Most have a strong physics or engineering background, and have built their own equipment more or less from scratch. Their technical accomplishments, it is widely agreed by their peers, represent a *tour de force*.

Image is everything

In the 1980s, biophysicists realized that solving the mysteries of muscle contraction would require better technology. They wanted to trap and manipulate individual protein molecules tagged with fluorescent markers. And they needed sophisticated microscopes to watch these proteins interact.

Yanagida helped to overcome many of these hurdles, building the necessary tools in his lab — and, in doing so, spurred others to higher levels of technological sophistication. "He has pushed the envelope," says Ron Vale, a biophysicist at the University of California at San Francisco.

In 1995, for instance, Yanagida's group revealed the first images of a single myosin II molecule binding to an ATP molecule in an aqueous solution[3]. Imaging fluorescently tagged proteins in solution using conventional microscopes is difficult because the laser that makes the molecules visible also causes scattering and luminescence throughout the solution. This problem was solved by

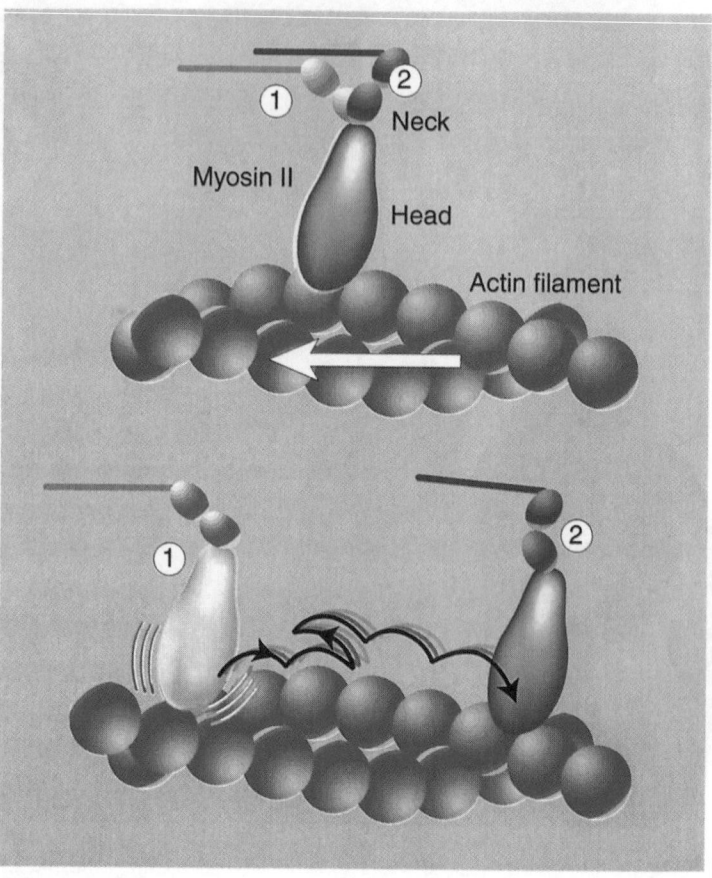

Flipping neck: the motion of actin and myosin II filaments relative to each other causes muscle contraction. In the lever-arm theory (top), myosin heads repeatedly attach to the actin fibre and then flip forward, propelling the actin along. But according to Yanagida's loose-coupling theory (bottom), the myosin runs along the actin fibre like a railway wagon on a rickety track.

a protégé of Yanagida's, Takashi Funatsu. He refined a technique called total internal reflection fluorescence so that he could limit to 150 nm the depth to which the laser's light penetrated the solution. This set up an 'evanescent' light field, which illuminated the target molecules but reduced the background luminescence more than 2,000-fold.

Other experiments had a more direct bearing on Yanagida's loose-coupling theory. In 1998, Yanagida and his colleagues reported that they had simultaneously observed the binding of ATP and recorded the displacement of a single myosin molecule along an actin filament[4]. The experiment combined Funatsu's evanescent field with another key tool — 'optical tweezers'. These were first applied to muscle biophysics by researchers working in the lab of James Spudich at Stanford University in California[5]. By fixing microscopic beads to either end of an actin filament, the filament can be held taut by directing laser beams — the tweezers — at the beads. From his experiments, Yanagida con-

cluded that the motion of a myosin head down an actin filament induced by one ATP molecule did not occur immediately. He observed a delay of several tenths of a second between the release of adenosine diphosphate, the product of ATP hydrolysis, and the myosin's motion. This delay is incompatible with the tight-coupling model.

Last year, Yanagida and his colleagues refined their estimate of 60 nm for the myosin displacement induced by a single ATP molecule to between 11 and 30 nm (ref. 6). This revision resulted from measurements taken during experiments in which individual myosin II heads were attached to glass microneedles. By measuring the deflection of the needles and their stiffness, the researchers recorded both the displacement of the myosin heads and whether or not they were attached to the actin filament.

They concluded that a single molecule of ATP causes a myosin head to move down the filament in a series of distinct steps, each about 5 nm long. Yanagida cannot rule out ▶

news feature

that the myosin head detaches from the filament momentarily during these steps, but he argues that if this happens it is too fast for the myosin to hydrolyse another ATP molecule, and that the net effect is one of loosely coupled movement.

Necks on the block

The problem is that most of the other groups working in the field, using similar imaging approaches to those used by Yanagida, have consistently recorded results that fit the lever-arm theory. They remain convinced that there is a tight coupling between ATP and myosin displacement, and measure this displacement at less than 10 nm (refs 5,7,8). Further evidence for their case comes from structural studies, which indicate that myosin can bend its neck as proposed by the lever-arm model[9–12].

At the Osaka symposium, both sides presented fresh results. A key discussion focused on myosin's neck region. According to the lever-arm theory, changing the length of the molecule's neck should alter myosin's power stroke, and hence the displacement induced by a single ATP molecule. But Yanagida predicts that neck length will make little difference.

David Warshaw of the University of Vermont described experiments published last month[13] in which his group halved the length of the neck of myosin II. Sure enough, this decreased by 40% the average displacement induced by one ATP molecule — and increasing the neck length had the opposite effect. But Yanagida's colleagues described unpublished experiments in which they removed the neck of myosin II altogether, and found no difference in displacement.

Every which way but loose?

Reconciling conflicting results in this field is extremely difficult — especially given the huge investment of time and money needed to create the experimental set-ups. Although everyone in the field acknowledges these constraints, some of Yanagida's competitors feel that his approach makes resolving the debate almost impossible.

"The problem with the loose-coupled thermal ratchet model is that it is difficult to devise experiments to specifically exclude it," says Justin Molloy of the University of York. "The tightly coupled lever-arm idea is simple, predictive and inherently testable because of its more restrictive nature."

Yanagida's glass microneedle experiment, for instance, is controversial. Is the myosin head really positioned on the glass needle as described in the paper? Might there be two myosin heads attached, the fluorescent one, plus one that has lost its fluorescence? If so, this might help explain the multiple steps that Yanagida sees.

But anyone wanting to repeat the experiment faces a formidable challenge. The microneedles were produced by Kazuo Kita-

Field project: Takashi Funatsu's experimental set-up allows muscle proteins to be seen in action.

mura, who says it took him six months of practice before he perfected their manufacture. "No one wants to repeat this experiment in Japan — not even in our own group," says Yoshiharu Ishii, group leader for Yanagida's Single Molecule Processes project.

Yanagida admits that his loose-coupled model can appear a little vague. "But just because I can't explain the whole system doesn't mean my data are wrong," he insists.

In explaining the continuing difference of opinions, Yanagida hints at cultural rifts. In Eastern thought, "it is necessary to pile experimental fact upon fact before asserting anything to be true", he says. This makes Yanagida and his supporters — most of whom are Japanese — comfortable with the absence of a simple, complete model.

But to most muscle biophysicists, this is no excuse for Yanagida's failure to provide a better description of his theory. And they find it difficult to accept his cheerful admission that he rejects the tight-coupling model on intuitive and aesthetic grounds. "It's so boring," he says.

Yanagida invokes his original training as a semiconductor physicist. Transistors, he explains, are fast and simple. "Biological molecules are not that simple," he says, holding his elbow with one hand and pumping it from side to side in imitation of a lever. "They are too soft, too flexible." He also questions the significance of the lever-arm camp's structural studies. "They cannot tell us about function," he argues.

Confusion reigns

Given the difficulty of repeating Yanagida's experiments, many of his competitors complain that they are left to grapple with unassailable data shrouded in a ghost-like theory. One researcher at the Osaka meeting likened him to the US Second World War general George Patton, who gained a reputation for making relentless advances, leaving his col-

leagues to sort out the mess left behind in his wake. "Sometimes I wonder if he's really looking for answers or just out to cause trouble," observes another biophysicist.

Despite the exasperation felt by many in the field, the arguments do not seem to have become tainted with personal animosity. Yanagida is amiable and generally well-liked. And at the Osaka meeting, he maintained a light-hearted playfulness even while disagreeing strongly with some of his peers. Nevertheless, his steely determination becomes clear when asked why he continues to swim against the tide of scientific opinion. "Because I'm right," he responds.

Most biophysicists agree that Yanagida has so far been a force for good, his charisma and technical skills helping draw money into the field. But some researchers fear that the field is becoming mired in an ultimately unproductive debate.

Perhaps the best summing up of the current state of play came in the Osaka symposium's closing speech. "I came here confused about actin and myosin," said Nobel laureate Andrew Huxley of the University of Cambridge. "Now, I am still confused, but at a higher level." ∎

David Cyranoski is *Nature*'s Asian-Pacific correspondent.

1. Yanagida, T., Nakase, M., Nishiyama, K. & Oosawa, F. *Nature* **307**, 58–60 (1984).
2. Yanagida, T., Arata, T. & Oosawa, F. *Nature* **316**, 366–369 (1985).
3. Funatsu, T., Harada, Y., Tokunaga, M., Saito, K. & Yanagida, T. *Nature* **374**, 555–559 (1995).
4. Ishijima, A. *et al. Cell* **92**, 161–171 (1998).
5. Finer, J. T., Simmons, R. M. & Spudich, J. A. *Nature* **368**, 113–119 (1994).
6. Kitamura, K., Tokunaga, M., Iwane, A. H. & Yanagida, T. *Nature* **397**, 129–134 (1999).
7. Molloy, J. E., Burns, J. E., Kendrick-Jones, J., Treager, R. T. & White, D. C. S. *Nature* **378**, 209–212 (1995).
8. Mehta, A. D., Finer, J. T. & Spudich, J. A. *Proc. Natl Acad. Sci. USA* **94**, 7927–7931 (1997).
9. Raymont, I. *et al. Science* **261**, 58–65 (1993).
10. Dominguez, R., Freyzon, Y., Trybus, K. M. & Cohen, C. *Cell* **94**, 559–571 (1998).
11. Corrie, J. E. T. *et al. Nature* **400**, 425–430 (1999).
12. Irving, M. *et al. Nature Struct. Biol.* **7**, 482–485 (2000).
13. Warshaw, D. W. *et al. J. Biol. Chem.* **275**, 37167–37172 (2000).

Reprinted from Nature, Vol. 348, No. 6299, pp. 346–348, 22nd November, 1990
© *Macmillan Magazines Ltd.*, 1990

Force generation of organelle transport measured *in vivo* by an infrared laser trap

A. Ashkin*, Karin Schütze†, J. M. Dziedzic*, Ursula Euteneuer† & Manfred Schliwa†

* Laser Science Research Department, AT&T Bell Laboratories, Holmdel, New Jersey 07733, USA
† Department of Molecular and Cell Biology, University of California, Berkeley, California 94720, USA

ORGANELLE transport along microtubules is believed to be mediated by organelle-associated force-generating molecules[1]. Two classes of microtubule-based organelle motors have been identified: kinesin[2-7] and cytoplasmic dynein[8-12]. To correlate the mechanochemical basis of force generation with the *in vivo* behaviour of organelles, it is important to quantify the force needed to propel an organelle along microtubules and to determine the force generated by a single motor molecule. Measurements of force generation are possible under selected conditions *in vitro* (for example, see refs 13 and 14), but are much more difficult using intact or reactivated cells. Here we combine a useful model system for the study of organelle transport, the giant amoeba *Reticulomyxa*[15], with a novel technique for the non-invasive manipulation of and force application to subcellular components, which is based on a gradient-force optical trap, also referred to as 'optical tweezers'[16-19]. We demonstrate the feasibility of using controlled manipulation of actively translocating organelles to measure direct force. We have determined the force driving a single organelle along microtubules, allowing us to estimate the force generated by a single motor to be 2.6×10^{-7} dynes.

We measured the driving force of mitochondria moving along microtubules within fine strands of the peripheral network of *Reticulomyxa*. These mitochondria are spherical, with a uniform diameter of 320 ± 70 nm ($n = 50$) by electron microscopy; they have a high refractive index and move in large numbers at relatively high speed within strands, being clearly distinguishable by light microscopy from other particles. The number of motor molecules per mitochondrion is small, between 1 and 4. This is based on electron microscopy of microtubule-mitochondria complexes showing an average of 2.4 ± 1 ($n = 77$) cross-bridges and a maximum of four crossbridges per organelle (Fig. 1). The motors in *Reticulomyxa* are thought to be cytoplasmic dynein[11,20] and to be capable of transport in opposite directions along microtubules[11,20-22].

To trap a mitochondrion moving along a strand requires a maximum trapping force greater than the maximum driving force, irrespective of the viscosity of the medium. A conceptually simple way of measuring the force is to trap a particle at high power and gradually reduce it until the particle just escapes the trap. This 'escape' power is a measure of the force. In practice this is difficult because of the heavy traffic of particles in active strands, so we used another approach based on limits. The procedure was to catch any actively moving single mitochondrion at an initially high power, and then to reduce the power

TABLE 1 Particle behaviour in 'trap and escape' experiments

Power range at escape (mW)	N	Average speed before trapping ($\mu m\ s^{-1}$)	Average speed after trapping ($\mu m\ s^{-1}$)
30–80	13	14.9	11.6
80–110	17	10.7	8.8

N is the total number of mitochondria observed in escape experiments and analysed for speed before trapping and after escape. All particles were trapped at 220 mW. The power was then dropped rapidly to a preset lower value by rotating the attenuator against a mechanical stop. Observations are from at least 12 strands in 3 amoeba samples.

quickly to a preset lower value to see whether it could escape. The force of any escaping particle has a value (equating, for the moment, power and force) between the upper and lower limits fixed by the stopping and escape powers.

Even at high power, some particles can elude the trap owing to the small size of the trap[23] relative to the size of the strand. Trapping is best in narrow, freely suspended strands 0.3 μm or less in diameter, when the trap captures the strand itself, which then guides mitochondria to the trap. In such narrow strands, mitochondria ride on the outside of a bundle of 1-6 microtubules, bulging the plasma membrane as they travel[24]. Figure 2 shows some simple experiments involving trapping and release of mitochondria moving along strands. These experiments illustrate the ability to trap fast-moving organelles and show that there are no obvious differences in behaviour of organelles moving in opposite directions along strands. Trap and release experiments can also roughly locate the relevant power (force) range corresponding to molecular motors. As we lower the power, the number of mitochondria trapped begins to drop at ~220 mW and is essentially zero at 30 mW, indicating that most mitochondria have forces between these power extremes. Table 1 summarizes the 'trap and escape' force measurements on 30 particles, for which the upper limit is taken at 220 mW and the lower limit varies from 30 to 110 mW, well into the relevant range. We list the number of particles measured in each of two power groups and give the average speeds measured before trapping and after escape. The fact that there is no significant change in speed after trapping indicates that the motors are not being damaged by the laser light. For a given particle that was, for example, initially trapped at 220 mW and later escaped at 100 mW, the force is taken as $(220 + 100)/2 = 160$ mW, the value midway between the upper and lower limits. Averaging over the 30 particles of Table 1 gives a mean driving force of ~150 mW per mitochondrion. Assuming 2.4 motors per mitochondrion (the value determined from electron microscopy), an average driving force for a single motor of ~63 mW is obtained.

The force of a trap on a particle depends on its size and relative index of refraction[23]. To calibrate the force inside the amoeba, we compared the viscous drag forces on mitochondria in water with those in the amoeba. Free mitochondria were obtained by gentle cell homogenization in 100 mM HEPES buffer. From the drag velocity at which mitochondria just escape, we deduce (using Stoke's law) that 1.0 mW gives a maximum

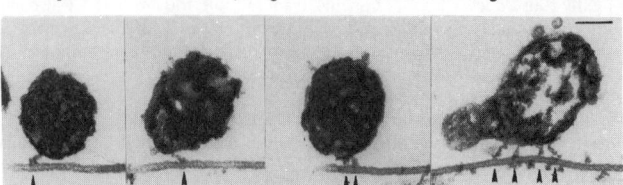

FIG. 1 Examples of mitochondria with 1, 2 and 4 crossbridges to a microtubule, as seen by electron microscopy. Bar, 0.1 μm. To avoid confusion of *bona fide* crossbridges with other components of the ground cytoplasm, we used cells lysed with Brij 58 in a stabilization buffer[32]. After this gentle lysis, movement of mitochondria along microtubules can be reactivated by ATP (ref. 20), indicating that the mitochondria have active crossbridges to microtubules.
METHODS. Networks spread on coverslips were lysed in a buffer consisting of 50% PHEM (ref. 32; 60 mM PIPES, 25 mM HEPES, 10 mM EGTA, 2 mM $MgCl_2$) 0.2% Brij 58, 1 mM vanadate and 5% hexylene glycol[20]. After 1-2 min they were fixed with 0.5% glutaraldehyde

in 50% PHEM buffer and processed for electron microscopy[33]. Sections were cut on a Porter-Blum MT2b microtome and viewed in a JEOL 1200 CX electron microscope.

2

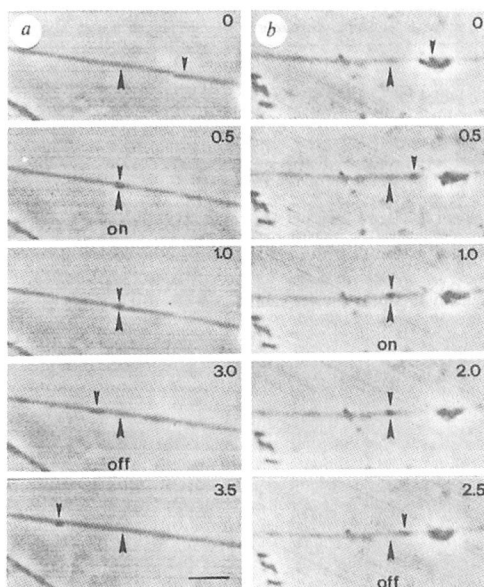

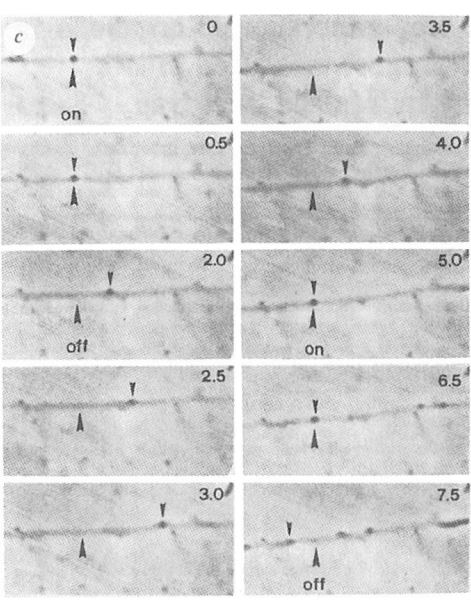

FIG. 2 Laser trapping of mitochondria in thin *Reticulomyxa* strands. Small arrowhead denotes the organelle; large arrowhead marks the position of the trap. Times in seconds are shown in the upper right-hand corner. 'On' and 'off' indicate the state of the laser trap. Scale bar, 5 μm. *a,* A mitochondrion travelling from right to left is trapped (frame 2), held in the trap for ~2 s, and moves on in the same direction after the trap is turned off. Its speed is 12.5 μm s⁻¹ before and 10.7 μm s⁻¹ after trapping. *b,* A mitochondrion reversing direction after being trapped for ~1 s (frames 3 and 4). Its speed is 8.7 μm s⁻¹ before and 9.1 μm s⁻¹ after trapping. *c,* This mitochondrion entered the trap from the left (frame 1), was released after ~1.5 s

and continued to move in the same direction (time 2–3 s). It then reversed its direction (between 3 and 5 s), and was trapped again at the same power (time 5–6.5 s).

METHODS. *Reticulomyxa* was grown and prepared for light microscopy as described[34]. The laser trap system was the same as before[17,18], except that phase contrast optics were used. All experiments were recorded on video tape and analysed frame-by-frame. Sequences were photographed off the monitor using a Pentax 35 mm camera equipped with an automatic motor drive.

trapping force of 5.8×10^{-9} dynes ($n = 21$) in water. In a comparable experiment within large low-viscosity amoeba droplets, it takes about twice the power for the same maximum drag velocity. We attribute this factor of two to either decreased force (due to a presumed lower relative index of refraction) or increased viscosity in the amoeba. By assuming a force $\sqrt{2}$ smaller and viscosity $\sqrt{2}$ higher, we make a maximum error of $\sqrt{2}$ in these quantities. On this basis, 1.0 mW gives a force of 4.1×10^{-9} dynes inside the amoeba. The value of 63 mW for the force of one motor translates into 2.6×10^{-7} dynes for the force generated by a single motor molecule, with an overall accuracy of a factor of 2–3 (combining errors in particle size distribution, force calibration, and power meter accuracy).

Our *in vivo* force measurement of $\sim 2.6 \times 10^{-7}$ dynes for a single (presumably dynein-like) motor is similar to, but more closely defined than, *in vitro* measurements for ciliary dynein of 1×10^{-7} dynes[25] and 0.5–10×10^{-7} dynes[26], and indirect estimates of 2–9×10^{-7}dynes[27,28]. For kinesin motors driving microtubules across a glass surface, a force $>6 \times 10^{-9}$ dynes was estimated[14], whereas Block *et al.*[29], using kinesin-covered beads and sea-urchin axonemes, report a force of $>(1$–$50) \times 10^{-9}$ dynes. Kishino and Yanagida[30] estimate a force of $>2 \times 10^{-8}$ dynes for a single myosin head. This compares with a force per head of $\sim 10^{-7}$ dynes for myosin in muscle during isometric contraction[31].

Inherent in our *in vivo* technique is the potential to see the effects on motility due to differences in viscosity and in the number of motors in mitochondria. Estimates of viscosity or, more strictly, viscous resistance[18] were made by dragging trapped mitochondria back and forth within strands. Values ranged from about 1,000 in stiff strands, to twice the viscous

drag of water in loose amoeba droplets.

A striking feature of the data in Table 1 is the uniformity of the velocity (or stepping rate of the molecular motors) of mitochondria over many samples with different viscosities and also presumably with different numbers of motors. This constancy of speed at ~ 10 μm s⁻¹ indicates that individual active motors can run at narrowly fixed rates against a range of resistive loads (due to varying viscous drag for example). Similar behaviour was observed with kinesin driving microtubules *in vitro*[14]. This general behaviour is reminiscent of electrical synchronous motors. We expect such a rate-controlled motor to produce an increasing force as the load increases up to a point of overload, beyond which its behaviour should become erratic.

A Stoke's law calculation shows that mitochondria with one motor having a force of 2.6×10^{-7} dynes can run without overloading at a full velocity of 10 μm s⁻¹ for all viscosities up to ~ 90 times that of water. Mitochondria with four motors can operate at 10 μm s⁻¹ up to viscosities of ~ 360 times that of water without overload. These values seem reasonable in light of the approximate range of viscous resistances found in different amoeba samples.

There is qualitative evidence supporting the overload picture from the experiments measuring the ability of the trap to stop active mitochondria at gradually decreasing powers. Trapping ability is gauged from the approximate time it takes to capture a particle. We found that in more viscous strands that the range of possible stopping powers was restricted to high values, from ~ 220 to ~ 150 mW. Only in less viscous samples could trapping proceed down to values as low as ~ 55 mW. This is an indication that at high viscosity, mitochondria with more motors (3 and 4) are fully active and those with fewer (1 and 2) are in overload.

The particles in overload are presumed to be those particles in viscous strands that are seen to move erratically at velocities markedly lower than $10 \, \mu m \, s^{-1}$. The data also suggest that at low viscosity, when all particles should be active, we are observing the trapping of particles with 1 or 2 motors at powers as low as 55 mW. This is reasonably consistent with our previously determined force for a single molecular motor of ~63 mW. In future more detailed *in vivo* studies of the effects of the numbers of motors and viscosity on the motility of individual organelles will probably make use of some type of more rapid automated force-measuring scheme.   □

Received 23 April; accepted 2 October 1990.

1. Warner, F. D. & McIntosh, J. R. (eds) in *Cell Movement* Vol. 2 (Liss, New York, 1989).
2. Brady, S. T. *Nature* **317**, 73–75 (1985).
3. Vale, R. D., Reese, T. S. & Sheetz, M. P. *Cell* **42**, 39–50 (1985).
4. Scholey, J. M., Porter, M. E., Grissom, P. M. & McIntosh, J. R. *Nature* **318**, 483–486 (1985).
5. Schroer, T. A., Schnapp, B. J., Reese, T. S. & Scheetz, M. P. *J. Cell Biol.* **107**, 1785–1792 (1988).
6. Pfister, K. K., Wagner, M. C., Stenoien, D. L., Brady, S. T. & Bloom, G. S. *J. Cell Biol.* **108**, 1453–1463 (1988).
7. Brady, S. T., Pfister, K. K. & Bloom, G. S. *Proc. natn Acad. Sci. U.S.A.* **87**, 1061–1065 (1990).
8. Paschal, B. M., Shpetner, H. S. & Vallee, R. B. *J. Cell Biol.* **105**, 1273–1282 (1987).
9. Paschal, B. M. & Vallee, R. B. *Nature* **330**, 181–183 (1988).
10. Lye, R. J., Porter, M. E., Scholey, J. M. & McIntosh, J. R. *Cell* **51**, 309–318 (1988).
11. Euteneuer, U., Koonce, M. P., Pfister, K. K. & Schliwa, M. *Nature* **332**, 176–178 (1988).
12. Shroer, T. A., Steuer, E. & Sheetz, M. P. *Cell* **56**, 937–946 (1989).
13. Kishino, A. & Yanagida, T. *Nature* **334**, 74–76 (1988).
14. Howard, J., Hudspeth, A. J. & Vale, R. D. *Nature* **342**, 154–158 (1989).
15. Euteneuer, U. *et al.* in *Cell Movement* Vol. 2 (eds Warner, F. D. & McIntosh, J. R.) 155–167 (Liss, New York, 1989).
16. Ashkin, A. & Dziedzic, J. M. *Science* **235**, 1517–1520 (1987).
17. Ashkin, A. & Dziedzic, J. M. & Yamane, T. *Nature* **330**, 769–771 (1987).
18. Ashkin, A. & Dziedzic, J. M. *Proc. natn. Acad. Sci. U.S.A.* **86**, 7914–7918 (1989).
19. Block, S. M., Blair, D. F. & Berg, H. C. *Nature* **338**, 514–517 (1989).
20. Koonce, M. P. & Schliwa, M. *J. Cell Biol.* **103**, 605–612 (1986).
21. Euteneuer, U., Johnson, K. B. & Schliwa, M. *Eur. J. Cell Biol.* **50**, 34–40 (1989).
22. Schliwa, M., Shimizu, T., Vale, R. D. & Euteneuer, U. *J. Cell Biol.* (in the press).
23. Ashkin, A., Dziedzic, J. M., Bjorkholm, J. E. & Chu, S. *Optics Lett.* **11**, 288–290 (1986).
24. Koonce, M. P. & Schliwa, M. *J. Cell Biol.* **100**, 322–326 (1985).
25. Kamimura, S. & Takahashi, K. *Nature* **293**, 566–568 (1981).
26. Oiwa, K. & Takahashi, K. *Cell Struct. Funct.* **13**, 193–205 (1988).
27. Hiramoto, Y. in *Cilia and Flagella* (ed. Sleigh, M. A.) 177–196 (Academic, London, 1974).
28. Brokaw, C. J. in *Molecules and Cell Movement* (eds Inoue, S. & Stephens, R. E.) 165–179 (Raven, New York, 1975).
29. Block, S., Goldstein, L. S. B. & Schnapp, B. J. *J. Cell Biol.* **109**, 81a (1989).
30. Kishino, A. & Yanagida, T. *Nature* **334**, 74–76 (1988).
31. Oosawa, F. *Biorheology* **14**, 11–19 (1977).
32. Schliwa, M. & van Blerkom, J. *J. Cell Biol.* **90**, 222–235 (1981).
33. Euteneur, U., Haimo, L. T. & Schliwa, M. *Eur. J. Cell Biol.* **49**, 373–376 (1989).
34. Koonce, M. P., Euteneuer, V., McDonald, K. L., Menzel, D. & Schiwa, M. *Cell Motil. Cytoskel.* **6**, 521–533 (1986).

ACKNOWLEDGEMENTS. We thank R. Wiegand-Steubing for collaboration in preliminary experiments. Supported by a grant from NSF to M.S. K.S. is a Feodor-Lynen Fellow of the Alexander von Humboldt Foundation.

Proc. Natl. Acad. Sci. USA
Vol. 86, pp. 7914–7918, October 1989
Cell Biology

Internal cell manipulation using infrared laser traps

(laser cell surgery/optical tweezers/viscoelasticity/mechanical properties/cytoplasmic streaming)

A. ASHKIN AND J. M. DZIEDZIC

Laser Science Research Department, AT&T Bell Laboratories, Holmdel, NJ 07733-1988

Communicated by James P. Gordon, July 5, 1989

ABSTRACT   The ability of infrared laser traps to apply controlled forces inside of living cells is utilized in a study of the mechanical properties of the cytoplasm of plant cells. It was discovered that infrared traps are capable of plucking out long filaments of cytoplasm inside cells. These filaments exhibit the viscoelastic properties of plastic flow, necking, stress relaxation, and set, thus providing a unique way to probe the local rheological properties of essentially unperturbed living cells. A form of internal cell surgery was devised that is capable of making gross changes in location of such relatively large organelles as chloroplasts and nuclei. The utility of this technique for the study of cytoplasmic streaming, internal cell membranes, and organelle attachment was demonstrated.

Infrared laser traps have recently been developed using the forces of radiation pressure that offer a way of micromanipulating entire living cells and even organelles within cells (1, 2). By using these traps, preliminary observations were made of the elastic behavior of the cytoplasm when small particles were optically displaced inside living cells (1). Measurements have also recently been made on the compliance of bacterial flagella using laser traps (3). One of the unique features of this manipulative technique, sometimes referred to as the "optical tweezers" technique, is the ability to apply controlled manipulative forces inside of cells while leaving the cell wall intact. We show here that application of highly local internal forces makes it possible to study the mechanical properties of living cytoplasm with minimal damage. There is essentially no precedent for such a capability in the extensive literature on micromanipulation techniques and measurement of mechanical properties of biological materials. Standard micromanipulation techniques based on micrugy (4–6), although capable of removing organelles from within cells, are not easily controlled and are fairly destructive of the cell. Weak variable magnetic forces have been applied to magnetic particles ingested into living cells for studying the viscoelastic properties of the more fluid parts of cells (7–9). Extensive measurements exist external to cells on the mechanical response of individual muscle fibers isolated from muscles (10–13). The mechanical properties of organic gels of actin and other organic polymers have been much studied (14–16) using standard rheological instruments (17, 18). External forces applied to living cells using a "cell-poker" have been used to probe gross mechanical properties of living cells (19). A related technique measures the local pressure needed to aspirate cell membrane into a micropipette (20). The principal limitation on the optical tweezers technique for internal cell manipulation is the restricted magnitude of the force consistent with minimal optical absorption damage to the cell.

In this largely qualitative study, we explore areas of internal cell manipulation in plant cells where optical tweezers can make significant changes in internal cell structure. We have discovered the ability of traps to pull out long thin

filaments of cytoplasm from different parts of the cell. These so-called artificial filaments are observed to exhibit, in striking fashion, the viscoelastic properties of plastic flow, necking, stress relaxation, and set, in close analogy with the known physical behavior of polymeric fiber materials (21, 22) and the more limited measurements on biological samples. By using filaments, one has the unique capability of probing point to point and temporal variations of the mechanical properties of cytoplasm in essentially unperturbed cells. As pointed out by Allen (23) and Allen (24), the local rheological properties of plant cells and amoeboid cells are not well characterized and are needed for a full understanding of cytoplasmic streaming and amoeboid movement. We also show that an appreciation of viscoelastic properties enables gross changes in the internal structure of the large organelles of the cell. This amounts to a limited form of optical cell surgery. Effects due to internal membranes are observed in experiments on making connections between cytoplasmic filaments and between optically displaced organelles. The displacement and reconnection of large organelles give rise to novel cell structures not normally found in nature. In other experiments, we see and strongly affect the phenomenon of cytoplasmic streaming and particle transport. Streaming motions represent the response of the cytoplasm to its own internally generated forces of motility (23–27). Traps also offer a unique approach to the study of these processes. We believe that the optical methods demonstrated here will enhance the prospects for use of laser traps for micromanipulation in cell biology.

MATERIALS AND METHODS

The single-beam infrared laser trap consists of a highly convergent $1.06\text{-}\mu\text{m}$ laser beam focused within the viewing plane of a standard optical microscope (1). It can trap and manipulate cells and particles from tens of microns down to submicron sizes (1, 2, 28). For large cells, where the beam focal volume of a few μm^3 is smaller than the cell, the trap acts much as a laser tweezers, which exerts strong forces near the boundaries of the cell, where the cell curvature and light refraction are largest. For small (submicron) organisms and particles, it exerts maximum force when the particle is at the point of maximum intensity gradient within the beam focal region. Traps of this type are capable of trapping a remarkable range of cell shapes and particle types (1, 2).

Manipulation is accomplished by fixing the trap in space and manipulating the stage and sample chamber in x, y, and z directions with micropositioners. For internal cell manipulation, we often use piezoelectric controls on the coordinate micrometers. Samples of *Spirogyra* and water-net (*Hydrodicton*) algae cells were collected from the back pond of the AT&T Bell Laboratories at Holmdel. Scallion (*Allium cepa*) cell samples came from a local supermarket.

In this work, we discovered a further capability of traps—namely, the ability to pull out a cytoplasmic strand or filament from an essentially planar internal interface. This occurs in the absence of any particles in the trap. When we

Proc. Natl. Acad. Sci. USA 86 (1989) 7915

move the trap across a cytoplasmic boundary such as exists, for example, between the higher index of refraction layer of cytoplasm surrounding the nucleus of a scallion cell and the lower index material of the central vacuole, the trap pulls out a long thin strand of high index cytoplasm into the lower index material. We can understand the initial surface forces causing this effect as due to the sum of the forces of the individual rays of the trap converging on the planar index discontinuity. For any ray, irrespective of its incident direction, there are two main contributions to the net surface force pointing out from the high to the low index medium: simple refraction and the change in P/cn, the effective momentum per second of the light within dielectric media (29, 30), where P is the power, n is the refractive index, and c is the velocity of light. There is also a normal force due to surface reflection but this is in general negligible. Surprisingly, this gives a substantial outward normal surface force, irrespective of the orientation of the planar interface even when the plane is vertical and rays are converging on the surface from both sides. Fig. 1 is a photograph of a so-called "artificial" elastic filament pulled out from the cytoplasm surrounding a scallion nucleus into the central vacuole. The trap initially at A creates a normal surface force that pulls the cytoplasm into a hemispherical dome and finally a long filament as the trap moves from A to B.

RESULTS AND DISCUSSION

Artificial Filaments. Artificial filaments or strands provide a way to observe the viscoelastic properties of cell cytoplasm. Filaments of different elastic properties were pulled from different regions of plant cells—for example, from the thin layer of cytoplasm adjacent to the cell wall into the central vacuole, from the edges of organelles such as chloroplasts or nuclei into the nearby cytoplasm, or from the boundaries of the many naturally occurring streaming cytoplasmic channels into surrounding areas. Strands pulled into

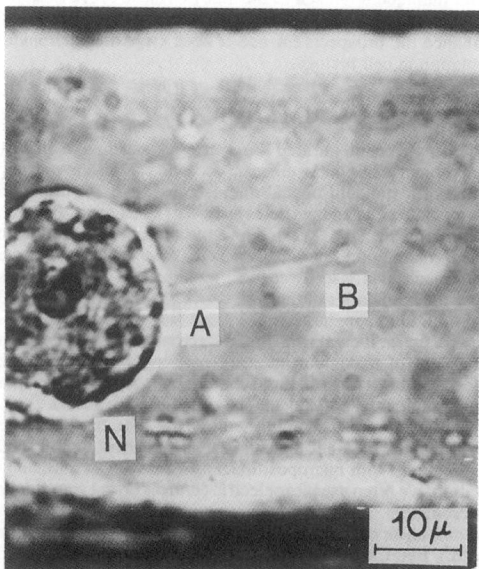

FIG. 1. Artificial cytoplasmic filaments in a scallion cell. The laser trap originally located on the surface of the nucleus (N) at A is moved to B, pulling out the viscoelastic filament AB into the central vacuole.

the central vacuole can often be hundreds of microns long and span almost the entire cell. Filament diameters vary and often are submicron. When a filament is released from the trap, it snaps back to its point of origin with speeds from about 100 μm/sec to a few μm/sec depending on its point of origin and whether it is released quickly or after being held for tens of seconds. After release, a small blob of a few cubic μm at the origin is the only remnant of the filament. If one rapidly pulls out a strand at a lower optical power level, one often loses it after a finite displacement due to increasing tension, and it snaps back. If, however, one stops and waits after shorter displacements, the extension can be essentially arbitrarily large.

If a filament is held extended for tens of seconds and one then rapidly moves the trap back toward the origin, one finds that at some point the filament no longer contracts but begins to sag due to loss of tension. After a few seconds, the strand tightens up again as it apparently continues to shrink slowly under relaxed conditions. Insight into how the filament extends itself when pulled can be gained by observing a pattern of fixed particles attached to the filament that act as markers. Although some filament expansion does occur, the principal gain in length comes from the feeding of new material from a short "neck-like" region that protrudes out a few microns from the point of origin.

Another striking property of filaments is their ability to "walk" over cytoplasmic surfaces. If one pulls out a filament normal to a surface and then displaces the trap parallel to the surfaces, the point of origin of the filament simply "walks" over the surface, following the trap. This serves to minimize the length and tension of the filament. "Walking" can occur over the surfaces of complicated shaped objects like the nucleus or its cytoplasmic supporting strands.

Remarkably, the whole range of rather complex mechanical effects observed above in living cytoplasmic filaments is closely analogous to the physical properties of viscoelastic fiber materials as seen in polymer science (21, 22). At their simplest, physical viscoelastic polymeric fibers and solids closely follow a Hooke's law stress–strain relation at low forces or stresses, up to a so-called yield point. Beyond this point, the solid starts to exhibit plastic flow. Large displacements (strain) are now possible with little increase in stress, and the material extends out principally at the expense of a "necked-out" region. If the extension is too rapid for plastic flow to occur, there is a rapid rise of stress and the fiber breaks. At fixed extension (strain) beyond the yield point, the stress relaxes steadily with time as the material continues plastic deformation. This in turn allows further extension. Upon removal of stress, a plastic fiber does not return to its original configuration but ends up with a "permanent set." There is also a transient set component due to relaxation of the deformed plastic material.

The fact that the mechanical properties of living cytoplasm so closely mimic those of viscoelastic polymers is not entirely surprising. Indeed, many polymeric materials and fibers are of organic origin, such as rubber, lattices, keratin, wool, etc. At the cellular level, the viscoelasticity of cytoplasm arises from the polymeric properties of actin and microtubule filaments that make up the basic support matrix and transport system of cells (14, 25, 31). Rapid transitions in cytoplasmic properties (ectoplasm–endoplasm transitions and possible sol–gel transitions) are considered important processes in cell motility (25, 31, 32) and contribute to local variations in mechanical properties. Only limited measurements exist on viscoelastic properties of living cytoplasm in prior work. Evidence of yield, stress relaxation, and shear-rate dependent viscosities was seen in experiments using ingested magnetic particles (7–9). In "poking" experiments, hysteresis was seen in the force needed to indent living mouse fibroblast cells (20). Measurements external to cells on mus-

7916   Cell Biology: Ashkin and Dziedzic   *Proc. Natl. Acad. Sci. USA 86 (1989)*

cle fibers show evidence of stress relaxation, crosslinking, and set (11, 12). Extensive viscoelastic measurements have been made on artificial actin gels that serve as models of cytoplasm of the variation of stress, compliance, and viscosity with shear rate (14–16). Strong non-Newtonian and shear thinning behavior was seen.

A significant factor in our ability to use traps for probing viscoelastic properties of living cells is the relatively low values of yield point of much of the cell's cytoplasm. For plant cells, we find yield points from as low as $\approx 10^{-8}$ dynes/μm^2 for the more fluid streaming parts of the cytoplasm up to values of 10^{-6}–10^{-4} dynes/μm^2 for the stiffer cytoplasm of artificial filaments. The maximum surface stress available using light traps is about 2×10^{-4} dynes/μm^2 (corresponding to ≈ 0.8 W applied at a focal spot of ≈ 0.5 μm in diameter). This greatly exceeds the stresses of approximately 10^{-8}–10^{-7} dynes/μm^2 used in magnetic particle experiments (9). It also exceeds the measured yield stress values of pure actin gels (32) of about 0.5–3×10^{-8} dynes/μm^2 and the stiffer shear-enhanced values of stress of about 10^{-5} dynes/μm^2 for gels of actin plus α-actinin crosslinking protein (15). The maximum stress using traps is also closely equal to the value of pressure needed to indent cells in the poking experiments. The pressure to distort the surface of red blood cells (erythrocytes) was measured to be about 50 times less ($\approx 4 \times 10^{-6}$ dynes/μm^2) using the micropipette technique. In unpublished results using optical tweezers, we distorted the surface of erythrocytes with tens of milliwatts of power, which corresponds to $\approx 2 \times 10^{-6}$ dynes/μm^2. Another factor in our filament experiments is the apparently low surface tension of internal interfaces. We attribute this to the presence of internal membranes at these cytoplasmic interfaces. If we ignore the viscoelastic stress, then the surface force required to pull out a filament of radius r into the initial shape of a hemispheric dome is at least $2\pi r S$, where S is the surface tensions (8). The fact that we can pull out filaments from a focal spot about 0.5 μm in diameter using powers of 0.1 W or less implies a surface tension of a cytoplasmic interface about 10^3 times less than an air–water interface.

The advantages of this technique for studying viscoelastic properties are its simplicity and ability to probe much of cell's

cytoplasm on a point to point basis and as a function of time and treatment with minimal disturbance. By inducing an alternating motion in the microscope stage, one can hope to make dynamic stress and viscosity measurements using artificial fibers and small trapped organelles in analogy with rheological techniques used external to the cell.

Manipulation of Large Organelles. An understanding of the viscoelastic properties of cytoplasm also makes possible the micromanipulation of the larger organelles of the cell. Initial attempts to move larger cell structures such as chloroplasts or nuclei, which are held in place by thick cytoplasmic connections, only resulted in plucking out of filaments from the surrounding cytoplasm. If, however, we apply force to a large organelle slowly, and stop after observing a slight displacement, we can then wait for plastic flow and force relaxation to occur. Assuming we are beyond the yield point, it then becomes possible to advance slowly with increasing displacements, until large changes are made in the position of the organelle. During this process, the thick cytoplasm holding the organelle gradually pulls out, "necks down," and ends up as, at most, a few weak strands, leaving the organelle essentially free. If one stops the process at an intermediate displacement where the cytoplasm is not too severely "necked down," the organelles creep back close to their original locations. Fig. 2 shows some of the drastic changes that can be made in the position of the spiral band of chloroplast of a *Spirogyra* cell. Fig. 2 *a* and *b* are views of the undisturbed cell close to the surface and the mid-plane. In Fig. 2*c*, the left end of the chloroplast spiral was pulled free from the outer cytoplasmic skin into the central vacuole close to the nucleus. In Fig. 2*d*, the loosened chloroplast is lifted close to the upper surface of the cell. Similarly, we have pulled off the central section of the chloroplast spiral and the fairly rigid nuclear support structure from their original location at points P and Q of Fig. 3*a* and moved them into the central vacuole to points R and S of Fig. 3*b*. This frees the nucleus from the cell wall and rotates it by about 45°. We also find that the cell has the ability to heal itself. If we "park" detached chloroplasts or the detached nuclear support structure against the cytoplasm at the cell wall and wait for tens of minutes, we observe the more fluid parts of the cytoplasm reattach the loose structures to the cell wall. The reattached

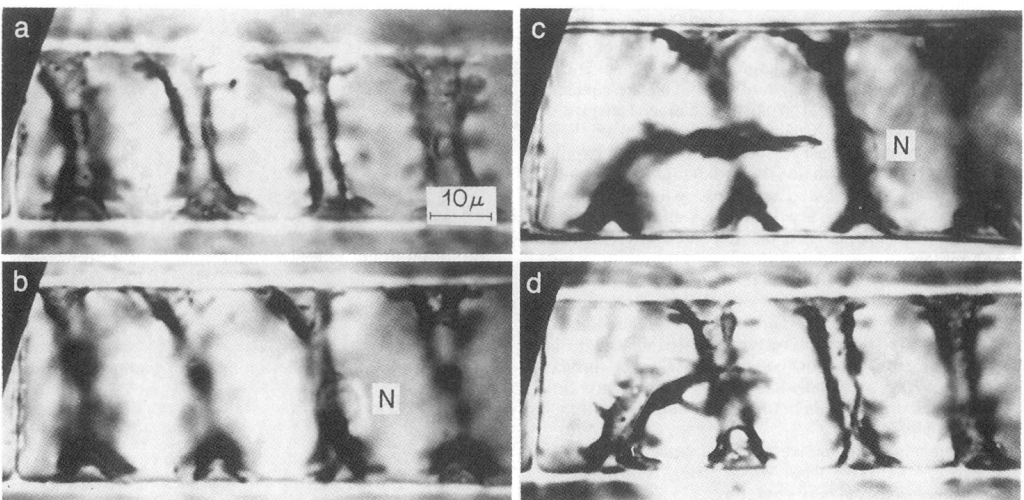

FIG. 2.   Manipulation of spiral chloroplast of *Spirogyra* from its undisturbed position (*a* and *b*) to grossly new locations (*c* and *d*). In *c*, the lefthand piece of chloroplast was moved into the vacuole near the nucleus (N). In *d*, it is raised to the cell surface.

Cell Biology: Ashkin and Dziedzic   *Proc. Natl. Acad. Sci. USA* 86 (1989)   7917

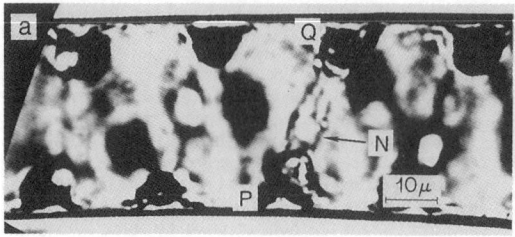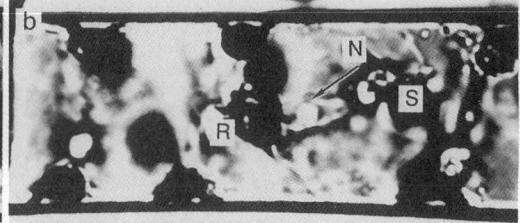

FIG. 3. Gross relocation of chloroplast and nucleus of *Spirogyra*. Material originally near the cell wall at P and Q (*a*), was manipulated into the central vacuole at R and S (*b*), thereby detaching a section of chloroplast and rotating the nucleus (N).

organelles appear to continue to function. With *Spirogyra*, we cannot pull the nucleus free of its support structure, though we can distort it, and move the nucleolus and small particles within the nucleus. With scallion cells it is occasionally possible, when the cytoplasm is not too viscous, to slowly drag the entire nucleus about within the cell along with its network of feeding cytoplasmic channels.

The power level used for manipulation of the larger organelles was often raised to 800 mW at the sample, which is 5–10 times higher than previously used (1). At this power, we still see no obvious thermal damage to plant cells but destruction of some protozoa can occur. In plants, use of high power accelerates the gross manipulation of large organelles. Manipulation occurs more slowly at powers of 100–200 mW. This technique for manipulation of large organelles using the plastic properties of the cytoplasm represents a limited form of cell surgery with which one can effect fairly gross changes in cell structure with minimum disruption to the cell. We believe this will prove useful for the study of cell function. This tweezers technique differs from the previous microsurgical techniques (33, 34) using pulsed high-powered lasers for microdissection and microirradiation.

Particle Motion and Cytoplasmic Streaming. Laser traps are able to affect internal particle motion and cytoplasmic streaming. In scallion cells we can trap and store up collections of the small particles usually seen moving in the well-defined channels of the cell interior or the more diffuse channels near the cell wall. When particles are trapped within a channel, a clump of cytoplasm often collects around them, which locally disturbs the flow. When released, the particles and clump disperse and move on along the channel. If the disturbance to the channels is large, some of the advancing particles will turn and go backward. Artificial side channels or filaments, pulled out from an existing channel, can grow or shrink as cytoplasm and particles flow in or out of them. When the flowing cytoplasm is very fluid, one can capture and transfer particles, with no attached cytoplasmic strands, from one stream to another. The powers required to manipulate streaming particles vary from a few tenths of a watt to a few milliwatts depending on the velocity and viscosity of the cytoplasm.

It is possible using artificial filaments to observe and possibly measure the velocity of cytoplasmic streaming independently of any particle motion within the stream. We find under circumstances of very rapid streaming that filaments, which under low-flow conditions walk over surfaces and come to rest perpendicular to the surface, are now strongly pulled downstream by the current and come to rest at angles as high as 45°.

Measurements of the motive force of cytoplasmic streaming made on the giant alga *Nitella* using centrifugation, external squeezing, and perfusion through opened cells gave values of $1.0–2.0 \times 10^{-8}$ dynes/μm^2 (25). Since we can generate forces about 3 orders of magnitude larger, we can easily affect the streaming process. However, as pointed out

by Allen and Kamiya (24, 25), the inability to measure local viscosity results in an inability to know if changes in streaming velocity are due to changes in viscosity or changes in driving force. Relative viscosity measurements are now possible with tweezers using a small organelle as a moveable test particle. Absolute measurements require a knowledge of the relative index of the particle. Measurements on particles and cytoplasmic drops vented into the surrounding liquid could be made. Effects of temperature, pH, and reagents on streaming have been reported (25, 35). Measurement of local viscosity and forces in such systems with tweezers would be useful.

Cell Membranes and Fusing of Cytoplasm. Another aspect of cytoplasm that affects the ability to manipulate it is the presence of membranes surrounding the outer surfaces of artificial filaments, natural filaments, and organelles. Attempts to fuse or join one filament with another or to organelles or cell surfaces are generally unsuccessful if the cytoplasmic surfaces are held in contact for times up to tens of seconds. Cytoplasmic joints are however possible if one waits considerably longer, presumably allowing the intervening membranes to break down. One unusual structure thus formed (see Fig. 2) consisted of a pair of chloroplasts (C1 and C2) stretched across the central vacuole of a water-net algae and held in place by elastic strands of cytoplasm. The chloroplasts, each with a residual connecting strand, were pulled out from opposite cell walls and held in contact within the same trap long enough for a connecting strand of cytoplasm to form. The positions of C1 and C2 shift as the tension of the connecting filaments varies. In Fig. 4b, the trap is also

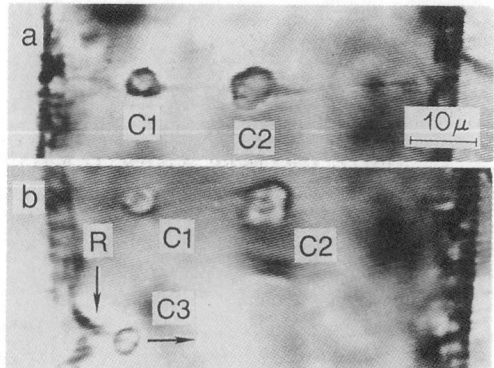

FIG. 4. Unique cell structure made in water-net algae: Chloroplasts C1 and C2 stretched across the vacuole by thin cytoplasmic filaments. The chloroplasts move as the tension of the filaments varies. In *b*, the trap is also pulling chloroplast C3 into the vacuole forming a necked-out region of cytoplasm (R).

Proc. Natl. Acad. Sci. USA 86 (1989)

seen pulling chloroplast C3 into the vacuole forming a necked-out region of cytoplasm.

An interesting organism to study with traps, in this context, is the giant amoeba *Reticulomyxa*, which has unusual membranes that fuse instantly on contact. It can therefore spontaneously reassemble itself from a suspension of separated pieces (36).

1. Ashkin, A. & Dziedzic, J. M. (1987) *Nature (London)* **330**, 769–771.
2. Ashkin, A. & Dziedzic, J. M. (1987) *Science* **235**, 1517–1520.
3. Block, S. M., Blair, D. F. & Berg, H. C. (1989) *Nature (London)* **338**, 514–518.
4. Crossway, A., Hauptli, H., Houck, C. M., Irvine, J. M., Oakes, J. V. & Perani, L. A. (1986) *BioTechniques* **4**, 320–333.
5. Miller, S. D. (1980) *Am. Biotechnol. Lab.* **71** (1), 30–34.
6. El-Badry, H. M. (1964) *Micromanipulators and Micromanipulation* (Academic, New York).
7. Crick, F. H. C. & Hughes, A. F. W. (1950) *Exp. Cell Res.* **1**, 37–80.
8. Yagi, K. (1961) *Comp. Biochem. Physiol.* **3**, 73–91.
9. Sato, M., Wong, T. Z. & Allen, R. D. (1983) *J. Cell Biol.* **97**, 1089–1097.
10. Huxley, H. E. (1969) *Science* **164**, 1356–1366.
11. Huxley, A. F. & Simmons, R. M. (1972) *Cold Spring Harbor Symp. Quant. Biol.* **37**, 669–680.
12. Cecchi, G., Griffiths, P. J. & Taylor, S. (1986) *Biophys. J.* **49**, 437–451.
13. Haskell, R. C. & Carlson, F. D. (1981) *Biophys. J.* **33**, 39–62.
14. Stossel, T. P. (1978) *Annu. Rev. Med.* **29**, 427–457.
15. Sato, M., Schwarz, W. H. & Pollard, T. D. (1987) *Nature (London)* **325**, 828–830.
16. Zaner, K. S. & Hartwig, J. H. (1988) *J. Biol. Chem.* **263**, 4532–4536.
17. Ferry, J. (1980) *Viscoelastic Properties of Polymers* (Wiley, New York).
18. Zaner, K. S., Fotland, R. & Stossel, T. P. (1981) *Rev. Sci. Instrum.* **52**, 85–87.
19. Petersen, N. O., McConnaughey, W. B. & Elson, E. L. (1982) *Proc. Natl. Acad. Sci. USA* **79**, 5327–5331.
20. Evans, E. A. & Hochmuth, R. M. (1978) *Curr. Top. Membr. Transp.* **10**, 1–64.
21. Nielsen, L. E. (1962) *Mechanical Properties of Polymers* (Rheinhold, New York).
22. Hadley, D. W. & Ward, I. M. (1986) *Encyclopedia of Polymer Science and Engineering* (Wiley, New York), 2nd Ed., Vol. 9, pp. 379–466.
23. Allen, N. S. (1983) in *The Application of Laser Light Scattering to the Study of Biological Motion*, eds. Earnshaw, J. C. & Steer, M. W. (Plenum, New York), pp. 529–543.
24. Allen, R. D. (1983) in *The Application of Laser Light Scattering to the Study of Biological Motion*, eds. Earnshaw, J. C. & Steer, M. W. (Plenum, New York), pp. 519–527.
25. Kamiya, N. (1981) *Annu. Rev. Plant Physiol.* **32**, 205–236.
26. Allen, R. D. & Allen, N. S. (1978) *Annu. Rev. Biophys. Bioeng.* **7**, 497–526.
27. Schliwa, M. (1984) in *Cell and Muscle Motility*, ed. Shay, J. (Plenum, New York), Vol. 5, pp. 1–85.
28. Ashkin, A., Dziedzic, J. M., Bjorkholm, J. E. & Chu, S. (1986) *Opt. Lett.* **11**, 288–290.
29. Ashkin, A. & Dziedzic, J. M. (1973) *Phys. Rev. Lett.* **30**, 139–142.
30. Gordon, J. P. (1973) *Phys. Rev. A* **8**, 14–21.
31. Taylor, D. L. & Condeelis, J. S. (1979) *Int. Rev. Cytol.* **56**, 57–144.
32. Brotschi, E. A., Hartwig, J. H. & Stossel, T. P. (1978) *J. Biol. Chem.* **253**, 8988–8993.
33. Berns, M. W. (1972) *Nature (London)* **240**, 483–485.
34. Berns, M. W. (1974) *Science* **186**, 700–705.
35. Shimmen, T. & Tazawa, M. (1985) *Protoplasma* **127**, 93–100.
36. Koonce, M. P., Eutenauer, V., McDonald, K. L., Menzel, D. & Schliwa, M. (1986) *Cell Motil. Cytoskeleton* **6**, 521–533.

Reprinted from Nature, Vol. 330, No. 6150, pp. 769-771, 24 December 1987
© *Macmillan Magazines Ltd., 1987*

Optical trapping and manipulation of single cells using infrared laser beams

A. Ashkin*, J. M. Dziedzic* & T. Yamane†

* AT&T Bell Laboratories, Holmdel, New Jersey 07733, USA
† AT&T Bell Laboratories, Murray Hill, New Jersey 07974, USA

Use of optical traps for the manipulation of biological particles was recently proposed, and initial observations of laser trapping of bacteria and viruses with visible argon-laser light were reported[1]. We report here the use of infrared (IR) light to make much improved laser traps with significantly less optical damage to a variety of living cells. Using IR light we have observed the reproduction of *Escherichia coli* within optical traps at power levels sufficient to give manipulation at velocities up to ~500 μm s[-1]. Reproduction of yeast cells by budding was also achieved in IR traps capable of manipulating individual cells and clumps of cells at velocities of ~100 μm s[-1]. Damage-free trapping and manipulation of suspensions of red blood cells of humans and of organelles located within individual living cells of spirogyra was also achieved, largely as a result of the reduced absorption of haemoglobin and chlorophyll in the IR. Trapping of many types of small protozoa and manipulation of organelles within protozoa is also possible. The manipulative capabilities of optical techniques were exploited in experiments showing separation of individual bacteria from one sample and their introduction into another sample. Optical orientation of individual bacterial cells in space was also achieved using a pair of laser-beam traps. These new manipulative techniques using IR light are capable of producing large forces under damage-free conditions and improve the prospects for wider use of optical manipulation techniques in microbiology.

The trapping and manipulation of dielectric particles by laser light is based on forces due to radiation pressure[2,3]. In the so-called single-beam gradient force traps used here[1,4,5] the dominant force component is the gradient force which pushes dielectric particles into the high-intensity region of a highly focused light beam. For physical particles such traps have been shown to operate over a range of particle sizes from tens of microns down to hundreds of Å[5] and also for individual atoms[6]. For operation with micron-sized biological particles the gradient trap was introduced into a high-resolution optical microscope where it can be used to manipulate particles within a sample while being viewed under high magnification[1].

In these experiments a 1.06 μm Nd:YAG (neodymium-doped yttrium aluminium garnet) laser trapping beam was coupled into a conventional microscope as shown in Fig. 1, with a beam splitter, BS, and focused with an adjustable external lens, L, into the viewing plane at point P. The high-numerical-aperture water-immersion objective (WI-OBJ) of the microscope (numerical aperture = 1.25) functions in the IR to form the single-beam gradient trap at P and in the visible with the high-numerical-aperture condenser lens, CL, and illuminator I to give high-resolution viewing. IR-blocking filters, BF placed in front of the eyepiece, E, and video camera, VC, serve to isolate the viewing optics from back reflections from the sample cell. The optical laser trap can be moved about within the 1 cm × 3 cm 100 μm thick sample cell by moving either the external lens on its X, Y, Z mount or by moving the X, Y, Z microscope stage.

To observe the reproduction of *E. coli* within the laser trap, an individual bacterium was captured and lifted up from among a collection of bacteria that had settled at the bottom of the sample to a clear region, where it was monitored continuously. As individual rod-like bacteria line themselves up vertically along the trapping-beam axis, the bacteria were released momentarily every 10 min to determine better their size. In a run of almost 5 h about 2.5 life cycles were observed with all four of

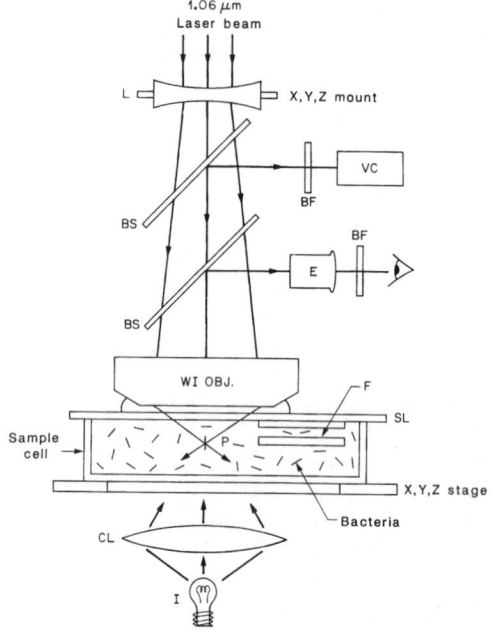

Fig. 1 Combined high-resolution optical microscope and 1.06 μm infra-red laser trap for observing, manipulating and separating bacteria and other organisms.

the resulting offspring remaining in the trap. The ability to reproduce within the laser trap clearly demonstrates that the bacterium was not damaged by the IR beam.

Separation of bacteria from one sample and their introduction into another was accomplished by first manipulating them into a liquid-filled hollow-glass fibre, F, with an inside diameter of 14 μm and a length of 5 mm, which was attached to the lid, SL, of the sample cell. The lid and fibre were then removed, washed externally and blown dry, without disturbing the bacteria inside the small core of the liquid-filled fibre. The lid and fibre were later placed on top of another sample cell into which the bacteria could be transferred using the optical trap. Reproduction of *E. coli* while being held in the light trap inside the confines of the hollow fibre, was also observed.

In the above reproduction experiments no evidence of damage to *E. coli* was seen up to the maximum power available at 1.06 μm of 80 mW, as measured at the sample. Other more motile types of bacteria were also captured at 80 mW with no evidence of optical damage as indicated by the absence of changes in

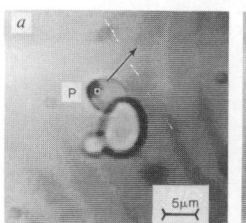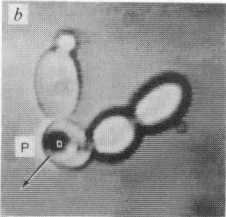

Fig. 2 The division of yeast cells in the IR trap. An original clump of two cells increases to four cells and then to six cells as shown in *a* and *b* after a total elapsed time of about 3 h.

2

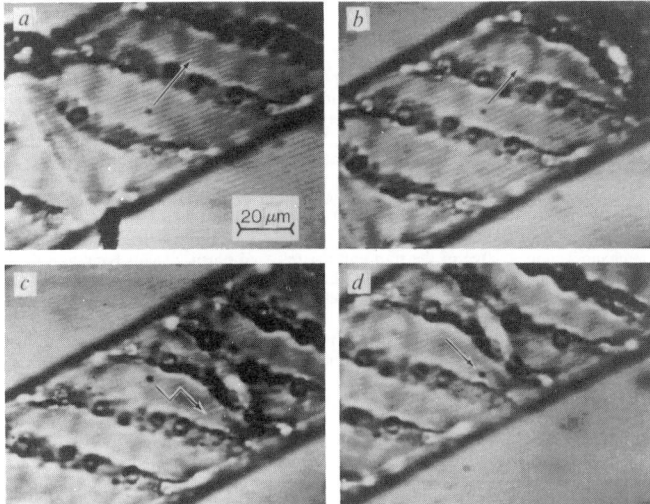

Fig. 3 *a–d*, The manipulation of a clump of particles within the cytoplasm of a cell of a spirogyra along the path of the arrows, by moving the microscope stage.

motility after many minutes of trapping. At this IR-power level it was possible to drag trapped bacteria at velocities up to ~500 μm s^{-1} through the fluid. In previous experiments[1] with 0.51 μm visible laser light the threshold for damage to *E. coli* and other more motile bacteria occurred at considerably lower intensities and lower manipulation velocities. The manipulation velocity at the damage threshold is about an order of magnitude higher in the 1.06 μm IR traps than in the 0.51 μm visible traps. The trapping force corresponding to a 500 μm s^{-1} drag velocity is about 800 mg assuming a bacterium of 1 μm diameter and unit density. This is sufficient to capture most motile bacteria.

To demonstrate the ability to orient biological particles in space we introduced a second independently adjustable trapping beam into the sample cell. This is possible through another microscope objective at the bottom of the sample cell which functions as both an IR trapping lens and a visible-condenser lens or through the same trapping and viewing objective at the top using beam splitters.[7] With two beams we were able to hold a rod-like bacterium, for example, at each of its ends and orient it at will.

Optical trapping of yeast cells (*Saccharomyces cerevisiae*), was achieved over a range of powers of about 5–80 mW. Individual yeast cells held in the laser trap were observed to bud into clumps of as many as eight connected cells; each achieving a size up to ~5–10 μm in diameter over a time span of about 5 h. It was possible to manipulate individual cells at velocities up to ~100 μm s^{-1} with a power of 80 mW. Manipulation at velocities of ~30 μm s^{-1} was also possible holding any one of the cells of a single large clump. By comparison severe optical damage to yeast was seen after a few minutes using visible-light traps at 0.5145 μm at lower force manipulation velocities and lower power levels (~20 mW). Figure 2 illustrates the reproduction of yeast cells in the IR trap. The first photograph (*a*), taken after about a half-hour in the trap, shows the increase of an original clump of two yeast cells to a clump of three with an incipient fourth bud appearing near the location of the trap at point P. The clump is being dragged along through the water of the cell in the direction of the arrow, by moving the microscope stage. This motion creates viscous drag that causes the clump, which at rest hangs vertically, to tip up into the viewing plane for better visibility. After ~3 h the clump has grown to six cells as shown in the second photograph (*b*).

Red blood cells (erythrocytes) of humans were trapped and

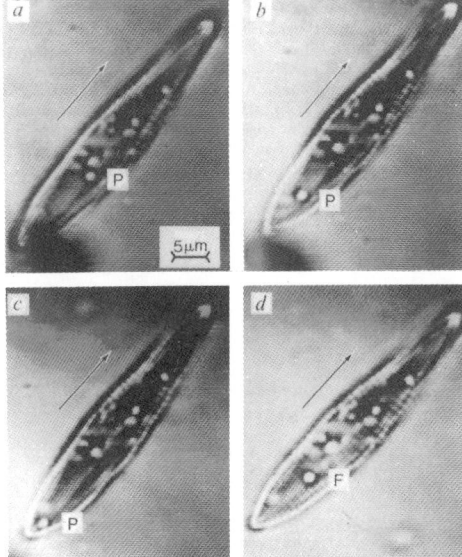

Fig. 4 *a–d*, The manipulation of an optically trapped organelle within the interior of a protozoan as it moves along freely on a glass surface in the direction of the arrow. In photographs *a–c*, an organelle, originally trapped at P (*a*), is being dragged by the advancing protozoan to the rear of the cell (*c*). In (*d*) the organelle pulls free of the trap and snaps back to its final position at F.

manipulated with no apparent change in flexibility or appearance with powers of 4–40 mW in the IR. The extreme flexibility of these cells was evident by the ease with which their shape was distorted by optically generated forces. Trapping occurs most readily on the thick doughnut-shaped rim of the cell. Manipulation at a velocity of ~100 μm s^{-1} was observed at a power of 40 mW. At powers of 80 mW some loss of flexibility was seen after trapping for tens of minutes. With visible argon-laser light catastrophic damage occurred with only a few milliwatts of power and low force levels.

In experiments on individual living cells of spirogyra, we were able to scan a focused 80 mW beam over the entire cell with no sign of damage. When the chloroplasts or pyrenoid bodies were irradiated there were no changes in appearance or changes in the nearby cytoplasmic streaming. We were, however, able to trap individual particles or collections of particles of about a micron or less, which are characteristically seen streaming along in well-defined channels in the cytoplasm near the cell walls. If manipulated into nearby interior regions of the cell vacuole, for example, the particles appear to drag part of the channels and cytoplasm with them and when released snap back as if on elastic bands. If some particles are freed totally from the channel, they can be manipulated about anywhere within the cell and used as probes for tracing out the internal structure of the cell. The cell appears to be sufficiently transparent and the trap strong enough to allow manipulation of particles under and around localized obstacles within the cell. Figure 3 shows a sequence of photographs illustrating manipulation, within the cytoplasmic layer of a cell of a spirogyra, of a small clump of captured steaming particles. The particles were moved along the path indicated by the arrows in Fig. 3a-d, taking them past the pyrenoid bodies and chloroplasts situated near the top cell wall of the spirogyra. Damage to spirogyra occurred within the chloroplasts using a visible-argon-laser light trap at powers of a few mW, probably due to chlorophyll absorption. The ability to manipulate organelles within the interior of a living cell without damaging the cell wall is probably unique to the optical-manipulation technique.

Experiments were also performed on the trapping of protozoa and of organelles within protozoa. Little evidence of damage due to the light was seen at powers up to ~160 mW, as measured at the sample. Trapping was observed for free-swimming small motile protozoa and even varieties of surface-adhering protozoa that were occasionally pried free of a surface. In one instance an amoeba was successfully trapped free of any contacting surface and manoeuvered about as it continuously changed its shape. It was finally redeposited on a glass surface apparently undamaged. Figure 4 shows an example of manipulation of an organelle within the interior of a fairly large protozoan as the organism moves along freely on the bottom glass surface of the sample cell in the direction of the arrow. In Fig. 4a, the laser trap has just been manipulated to point P where it captures the prominent circular particle located at that position within the interior of the protozoan. As the protozoan advances, the trapped particle at P is pulled back toward the rear of the moving cell as shown in Fig. 4b. In Fig. 4c the restrained particle is seen colliding with the cell wall at the rear of the advancing protozoan. Subsequently, as seen in Fig. 4d the protozoan pulls the particle free of the trap and the released particle snaps back to a final position F, part of the way back to its original position in Fig. 4a. Observations such as these are clearly giving information on the viscosity and elastic properties of the cytoplasm in the region of the trapped organelle. It was also possible to manipulate particles within stationary protozoa by simply moving the trap.

Apart from possible reductions in absorption of the particles, IR traps have the further advantage over visible traps of a fourfold reduction in intensity due to the larger focal spot size, without a reduction in force, provided the local particle curvature does not vary significantly over the focal spot. Another factor in use of IR for trapping, is the temperature rise due to absorption of the water of the sample cell itself. At a power of 80 mW, assuming an absorption coefficient a of ~0.1 cm^{-1}, the temperature rise is estimated to be several degrees Centigrade. At higher powers this effect will eventually cause damage to the trapped particle and be a limiting factor even if additional particle absorption does not occur.

Received 21 July; accepted 5 November 1987.

1. Ashkin, A. & Dziedzic, J. M. *Science* **235**, 1517–1520 (1987).
2. Ashkin, A. *Phys. Rev. Lett.* **24**, 156–159 (1970).
3. Ashkin, A. *Science* **210**, 1081–1088 (1980).
4. Ashkin, A. *Phys. Rev. Lett.* **40**, 729–732 (1978).
5. Ashkin, A., Dziedzic, J. M., Bjorkholm, J. E. & Chu, S. *Opt. Lett.* **11**, 288–290 (1986).
6. Chu, S., Bjorkholm, J. E., Ashkin, A. & Cable, A. E. *Phys. Rev. Lett.* **57**, 314–317 (1986).
7. Ashkin, A. & Dziedzic, J. M. *Appl. Opt.* **19**, 660–668 (1980).

2856

Biophysical Journal   Volume 77   November 1999   2856–2863

Characterization of Photodamage to *Escherichia coli* in Optical Traps

Keir C. Neuman,*[#][||] Edmund H. Chadd,[#] Grace F. Liou,[§] Keren Bergman,[¶][||] and Steven M. Block*[#][||]

Departments of *Physics, [#]Molecular Biology, [§]Chemical Engineering, and [¶]Electrical Engineering, and [||]Princeton Materials Institute, Princeton University, Princeton, New Jersey 08544 USA

ABSTRACT   Optical tweezers (infrared laser-based optical traps) have emerged as a powerful tool in molecular and cell biology. However, their usefulness has been limited, particularly in vivo, by the potential for damage to specimens resulting from the trapping laser. Relatively little is known about the origin of this phenomenon. Here we employed a wavelength-tunable optical trap in which the microscope objective transmission was fully characterized throughout the near infrared, in conjunction with a sensitive, rotating bacterial cell assay. Single cells of *Escherichia coli* were tethered to a glass coverslip by means of a single flagellum: such cells rotate at rates proportional to their transmembrane proton potential (Manson et al., 1980. *J. Mol. Biol.* 138:541–561). Monitoring the rotation rates of cells subjected to laser illumination permits a rapid and quantitative measure of their metabolic state. Employing this assay, we characterized photodamage throughout the near-infrared region favored for optical trapping (790–1064 nm). The action spectrum for photodamage exhibits minima at 830 and 970 nm, and maxima at 870 and 930 nm. Damage was reduced to background levels under anaerobic conditions, implicating oxygen in the photodamage pathway. The intensity dependence for photodamage was linear, supporting a single-photon process. These findings may help guide the selection of lasers and experimental protocols best suited for optical trapping work.

INTRODUCTION

"Optical tweezers," or optical traps, provide a unique means of manipulating and controlling biological objects (Svoboda and Block, 1994). Since the first demonstration of optical trapping by Ashkin (1978, 1986), a host of applications have arisen in biology, both in vivo and in vitro. A drawback of optical trapping has been the damage induced by the intense trapping light. In practice, such damage limits the exposure time for trapped specimens and has proved to be a significant problem for some optical trapping studies, particularly those in vivo. Indeed, Ashkin first encountered this problem and coined the colorful term "opticution" to describe the laser-induced death of specimens (Ashkin and Dziedzic, 1989). The potential for damage is readily appreciated by computing the light level at the diffraction-limited focus of a typical trapping laser: for a power of just 100 mW, the intensity is 10^7 W/cm^2, with an associated flux of 10^{26} photons/s·cm^2 (traps used in cell biology are generally based on lasers producing from 25 mW to 2 W in the specimen plane). Proposed mechanisms for photodamage include transient local heating (Liu et al., 1996), two-photon absorption (Berns, 1976; König et al., 1995, 1996a; Liu et al., 1996), and photochemical processes leading to the creation of reactive chemical species (Calmettes and Berns,

1983; Block, 1990; Svoboda and Block, 1994; Liu et al., 1996).

Some practical progress has been made toward decreasing photodamage in optical trapping systems, primarily through the choice of trapping lasers with wavelengths in the near-infrared region (Ashkin et al., 1987). This corresponds to a waveband that is comparatively transparent to biological material, situated between the absorption bands of many biological chromophores in the visible, and the increasing absorption of water toward longer wavelengths (Svoboda and Block, 1994). The most common source used in optical traps is the continuous-wave (CW) diode-pumped Nd:YAG laser (1064 nm) or its close relatives, Nd:YLF (1047 nm) and Nd:YVO$_4$ (1064 nm). These represent the most economical choices for achieving the requisite power (1–10 W) and output stability. But other sources suitable for optical trapping exist. Recent years have seen the emergence of high-intensity, single-mode diode lasers, available in the wavelength region from 700-1500 nm, with powers up to ~1 W. Diode lasers possess exceptional amplitude stability and are more economical than Nd-based lasers. Another option is the CW Ti:sapphire laser, which affords continuous tuning through much of the near-infrared region (700–1000 nm), along with high output power. However, it requires a separate pump source, typically suffers reduced amplitude stability, and is far and away more costly than the alternatives. For now, Nd-based lasers continue to dominate the optical trapping field, but sources at other wavelengths may represent more advantageous choices for reducing photodamage.

Berns and co-workers pioneered investigations of photodamage in optical traps, using a variety of biological assays. Their work with temperature-sensitive fluorescent dye reporters in Chinese hamster ovary (CHO) cells and lipo-

Received for publication 13 May 1999 and in final form 30 July 1999.

Address reprint requests to Dr. Steven M. Block, Department of Biological Sciences, Gilbert Building, Room 109, 371 Serra Mall, Stanford University, Stanford, CA 94305-5020. Tel.: 650-724-4046; fax: 650-723-6132; E-mail: sblock@stanford.edu.

Mr. Neuman's and Dr. Block's present address is Department of Biological Sciences, Stanford University, Stanford, CA 94305.

Ms. Liou's present address is Department of Chemical Engineering, Stanford University, Stanford, CA 94305.

somes confirmed the prediction that local heating of micron-sized specimens is negligible from a tightly focused CW laser source, thereby ruling out direct heating as a source of damage (Block, 1990; Liu et al., 1995a, 1996). Additional studies, based on assays of the rates of chromosome bridge formation in rat kangaroo cells (Vorobjev et al., 1993) or cloning efficiency in CHO cells (Liang et al., 1996), established rough action spectra for damage over portions of the near-infrared region. Following this work, additional studies, scoring either CHO cell-cloning efficiency or loss of viability in human spermatozoa, led to the suggestion that damage is generated by a two-photon process (König et al., 1995, 1996a,b; Liu et al., 1996). In addition, work with fluorescent probes demonstrated no changes in the intracellular pH of trapped cells and no detectable changes in DNA structure following CW laser illumination (as opposed to pulsed lasers, which do produce changes in acridine orange staining) (Liu et al., 1996).

While such experiments provide important clues to the photodamage process, the bioassays upon which they are based have certain intrinsic limitations. Chromosome bridge formation is largely qualitative and difficult to score. Cloning efficiency and sperm viability essentially provide a binary output (alive or dead), necessitating many measurements to gain adequate statistics. The assays are indirect, complex, and time consuming, requiring long incubation and/or growth periods, together with sensitive fluorescence-measuring capabilities. Furthermore, they do not readily lend themselves to the continuous monitoring of photodamage during experimental exposure.

To address these limitations, we employed a rotating bacterial cell assay that provides a quantitative, real-time measure of the metabolic state of the cell. The assay is based on attaching *Escherichia coli* cells to a glass coverslip by a single flagellum (Block et al., 1982, 1989). When the tethered cell turns its flagellar motor, the cell body is driven into rotation about its point of attachment, typically ~0–15 Hz, depending upon the cell size (and therefore on the load posed by viscous rotational drag). Motors of tethered cells spin at rates proportional to the transmembrane proton potential (Manson et al., 1980).

Although based on a prokaryote, this assay has some advantages over the eukaryotic systems employed previously. *E. coli* are robust and well-characterized organisms, which can be grown either aerobically or anaerobically, permitting evaluation of the role of oxygen in photodamage. Moreover, an enormous variety of mutants is available.

Using this assay, in conjunction with a broadly tunable optical trapping system, we determined the action spectrum for photodamage from 790 to 1064 nm. This spectrum shows a roughly sevenfold variation in damage across this range, with two pronounced maxima at 870 and 930 nm. The least damaging wavelength was found to be 970 nm, followed closely by 830 nm. By growing and trapping cells in the absence of oxygen (or by removing oxygen after growth with a chemical scavenging system), we tested the effect of oxygen on the lifetime of cells. There was a

significant increase in lifetime under anaerobic conditions: in fact, damage was reduced to nearly background levels. Determining photodamage as a function of laser power (at two different wavelengths, 870 and 1064 nm), we found that the sensitivity of cells (defined as the reciprocal of the lifetime) was linearly related to the intensity. These results suggest that photodamage in optical traps is mediated by oxygen, and that it involves a one-photon process.

MATERIALS AND METHODS

Optics

The optical trap (schematic shown in Fig. 1) was based on three separate lasers: a Ti:sapphire ring laser tunable between 780 and 970 nm (model 899; Coherent, Santa Clara, CA), a MOPA diode laser at 991 nm (model 5762-A6; SDL, San Jose, CA), and a Nd:YAG laser at 1064 nm (model BL-106C; Spectra-Physics Lasers, Mountain View, CA). The Ti:sapphire laser was pumped with all lines from a large-frame argon ion laser (Innova 400; Coherent). To ensure true continuous-wave output from the Ti: sapphire laser, we incorporated an intercavity etalon (model 895; Coherent), which reduces the bandwidth and prevents temporal mode beating and partial modelocking (König et al., 1996a). The laser output was monitored in both temporal and frequency domains to check for pulses, which are indicative of temporal mode beating. Without the etalon, pulses were observed at a repetition rate of 186 MHz, corresponding to the round-trip time in the cavity. With the etalon in place, all mode beating ceased. The spatial mode of the Ti:sapphire laser and of the YAG laser was TEM$_{00}$, while the mode from the MOPA was slightly elliptical (ellipticity = 1.3). The output from the laser was expanded to slightly overfill the back pupil of the microscope objective (63×/1.2 numerical aperture (NA) Plan NeoFluar, water/glycerol immersion, model 461832; Carl Zeiss, Oberkochen, Germany) and brought into an inverted microscope (Diaphot TMD; Nikon, Tokyo, Japan) via the epiillumination port. The optical path

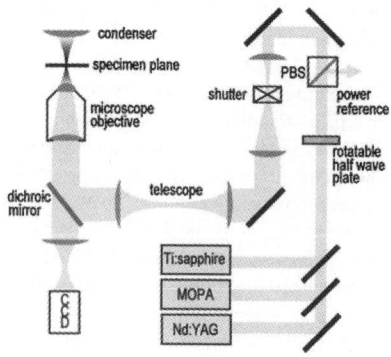

FIGURE 1 Simplified schematic of the tunable optical trap (not to scale). To cover the near-infrared spectrum, one of three separate lasers was selected: the Ti:sapphire ring laser allows continuous tuning from 790 nm to 970 nm; the MOPA laser is at 991 nm; the Nd:YAG is at 1064 nm. The laser power is controlled via a rotatable halfwave plate and polarizing beam splitter (PBS). After the polarizer, the beam is expanded to slightly overfill the back pupil of the microscope objective. In the middle of the beam expander is a computer-controlled shutter. The laser is then directed into the epiillumination path of an inverted microscope and reflected by a dichroic mirror into the microscope objective. The objective focuses the laser light to form a trap in the specimen plane and collects visible light from the condenser to form an image. The visible light passes through the dichroic mirror to a video camera (CCD).

2858   Biophysical Journal   Volume 77   November 1999

included a computer-driven shutter (model 845; Newport Corp., Irvine, CA) controlling the laser trap. A dichroic mirror (model 635DCSPX; Chroma Technology Corp., Brattleboro, VT) in the microscope directed the laser into the objective while permitting the visible light, imaged by the objective, to pass through. Blue light artifacts induced by the microscope illumination source (50-W, 12-V DC halogen bulb) were minimized by placing a green interference filter (Nikon) in the illumination pathway.

Rotating, tethered cells were imaged on a CCD camera (model V-1056SX CCD; Video Runner, Culver City, CA). A time code generator (model TRG-50; Horita Co., Mission Viejo, CA) added a time stamp to the video signal, which was displayed on a B/W monitor (model PVM-97; Sony Corp., Montvale, NJ) and recorded by VCR (model AG-1980; Panasonic Co., Secaucus, NJ). In most cases, rotation rates of cells were simultaneously analyzed using a custom-built video cursor box placed in the video chain, which delivered a TTL pulse to a computer whenever the position of a rotating cell crossed a user-defined cursor position (Block and Berg, 1984). The same cursor box could also be used off-line with video-taped records of cells.

Microscope objective transmission calibration

To determine accurately the power delivered to the specimen plane, the transmission of the microscope objective must be characterized. Because of the high NA and short working distance of objectives used for optical trapping work, transmission cannot be measured by simply passing a beam of light through the lens and collecting it with an ordinary photodetector. Instead, the objective transmission as a function of wavelength was measured using a dual-objective technique (Misawa et al., 1991), as described by Svoboda and Block (1994). Measured transmission curves for several candidate objectives are displayed in Fig. 2.

Calibration of power in the specimen plane

The power in the specimen plane was determined by a multistep procedure. First, the microscope objective used for optical trapping was replaced by a low-NA objective with a known transmittance (20×/0.4 NA, model M-20X; Newport Corp.; transmittance determined separately). A pyroelectric optical power meter (model LM-10; Coherent) was placed in front of this objective, at (or near) the specimen position, to record the intensity of light passing through. The power at this position, $P_{\rm m}$, is related to the power delivered to the specimen plane in an actual experiment using a high-NA objective, $P_{\rm a}$, by $P_{\rm a} = P_{\rm m} \cdot T_1(\lambda)/T_2(\lambda)$, where $T_1(\lambda)$ is the

measured transmission of the high-NA objective and $T_2(\lambda)$ is the measured transmission of the low-NA objective. (The entrance pupils of the low-NA objective and the high-NA objectives have the same diameter.) Next, to set the power at the specimen for any given wavelength, λ', the half-wave plate in front of the polarizing beam splitter (PBS) was adjusted to obtain a reading of $P_{\rm m} = P_{\rm a} \cdot T_2(\lambda')/T_1(\lambda')$ on the optical power meter. The low-NA objective was then replaced with the high-NA trapping objective. Once the power was established in this way, any drift in power could be monitored via the second PBS port and corrected during an experiment. Power measurements as just described were performed before trapping in each experiment and after each change in the wavelength.

Bacterial assay

We employed a tethered cell assay (Block et al., 1982, 1989) based on a strain of *E. coli* that carries two useful mutations (KAF95, a gift of Karen Fahrner, Harvard University; Berg and Turner, 1993). The first mutation is a deletion of the *cheY* gene. CheY-P protein induces clockwise rotation of the flagellar motor; in its absence, cells rotate smoothly in the counter-clockwise direction (Parkinson, 1978; Parkinson et al., 1983), facilitating measurements of rotation rates. The second mutation affects the flagellar protein flagellin. In KAF95, the *fliC* gene encoding flagellin has an internal deletion leading to a nonspecific binding interaction between flagella and the negative surface charge on the coverglass (Kuwajima, 1988). Cells carrying both of these mutations spontaneously tether themselves and rotate continuously in the counterclockwise direction.

Cells of *E. coli* strain KAF95 were grown as described by Block et al. (1982), except that cultures were grown in T-broth (10 mg ml^{-1} Bacto-Tryptone, Difco Laboratories, Detroit, MI; 5 mg ml^{-1} NaCl, Sigma, St. Louis, MO), supplemented with 100 μg ml^{-1} ampicillin (Sigma) at 30°C, and the motility medium was that described by Block et al. (1983). Cells were loaded into a flow cell consisting of a coverslip attached to a microscope slide by two pieces of double-sided tape. Cells were allowed to tether for 10–15 min, after which time the flow cell was washed with 900–1200 μl of motility medium to remove untethered cells.

The experimental procedure was modified slightly to study cells under reduced oxygen tension. To ensure anaerobic conditions, mineral oil (Fisher Scientific, Pittsburgh, PA) was layered over the surface of the growth medium before incubation to prevent oxygen from entering the test tube (cells consume any residual oxygen during the early stages of growth). The entire shearing and tethering process was carried out under nitrogen inside a glove bag, and the flow cell was sealed all around with vacuum grease (Apiezon M; M&I Materials, Manchester, England) before exposure to air. In other experiments, anaerobic conditions were achieved by introducing an oxygen-scavenging system into the flow cell after tethering but before trapping (250 μg ml^{-1} glucose oxidase, 30 μg ml^{-1} catalase, 4.5 mg ml^{-1} glucose; Sigma). We estimate the time required to deplete the remaining oxygen in the flow cell under these conditions to be less than 1 s.

Tethered cells were held by the optical trap and periodically released to monitor their rotation rates (Fig. 3). In a typical experiment, once a suitably tethered cell was identified (initial frequency of 5–12 Hz), between 30 and 100 s of data was collected before the trap was turned on. Thereafter, during each successive 10-s interval, the cell was held for 8 s by the trap and then released for 2 s. The rotation rate was determined from the timing of pulses generated by the video cursor box corresponding to full rotations (above). Pulses were captured by a data acquisition board (model AT-MIO-16E-10; National Instruments, Austin, TX), using a Labview program (Labview 4; National Instruments), which was also used to control data acquisition and analyze rotation rates. Rotational data were further analyzed with Igor software (Igor Pro; Wavemetrics, Lake Oswego, OR). The data were smoothed, the start time (corresponding to when the trap was first turned on) was established, and the LD$_{50}$ time, operationally defined as the time at which the rotation rate decreased to 50% of its initial value, was determined (see Fig. 4). Control data were obtained in a similar manner, but with cells exposed only to the microscope illumination. Experiments were performed at 25–27°C. A typical flow cell had one or two well-tethered cells per field of view (200 μm²). After data were

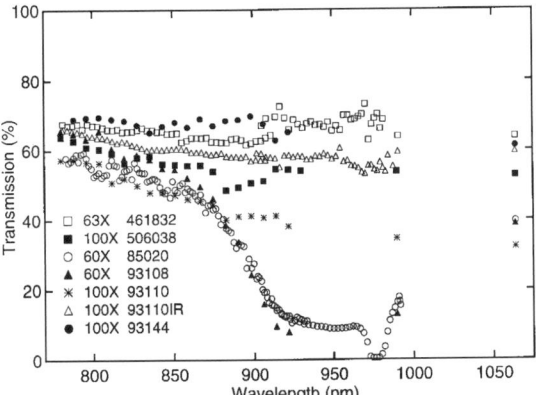

FIGURE 2 Microscope objective transmission curves. Transmission measurements were made by a dual-objective method (see Materials and Methods). Part numbers are cross-referenced in Table 1. The uncertainty associated with a measurement at any wavelength is ~5%.

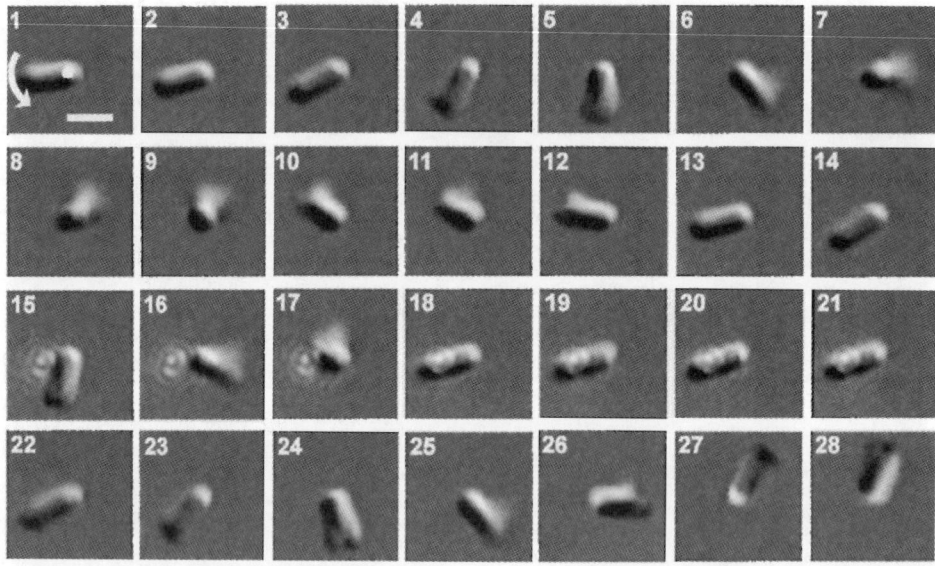

FIGURE 3 Images of a single rotating cell of *E. coli* tethered to a glass coverslip and imaged with DIC microscopy: successive video fields are displayed (time interval, 33 ms/field). The scale bar is 1 μm; the curved arrow indicates the direction of rotation, and the white dot shows the approximate center of rotation. The first 14 frames show the cell spinning freely (at ~2.5 Hz). The next 14 frames were taken with the trap on (the diffraction-limited laser focus is seen as concentric rings in frame 15). The cell rotates into, and is held by, the trap for four frames (18–21), before release (frame 22), after which it continues to rotate (frames 23–28).

acquired from a cell, the next cell was chosen at least 400 μm away from the first. No more than two flow cells were made from a single culture. To mitigate the effect of systematic variation in cell behavior from day to day, data for each point were collected from a minimum of three preparations over 2 days, with each point representing the average of 6–23 individual LD_{50} determinations. There was no correlation between initial rotation rate and LD_{50} time (correlation coefficient $r = 0.1$). We defined sensitivity as the inverse of the LD_{50} time. Data are presented as mean ± SEM.

RESULTS

Microscope objective transmission calibration

Measured transmission data for seven high-NA microscope objectives from three manufacturers are presented in Fig. 2 and Table 1. Overall transmission for the group varied from 1% to 73%. All objectives showed acceptable transmission in the short-wavelength region of the infrared spectrum (~45–65%, ~790–830 nm). Beyond 850 nm, the transmission of most Plan Apo objectives fell dramatically, in certain cases to levels unacceptable for optical trapping work. However, objectives designed primarily for fluorescence work (Plan NeoFluar, Zeiss; Plan Fluor, Nikon) or explicitly for work in the near IR (93110IR; Nikon) had improved transmission characteristics in the longer wavelength region.

Wavelength-dependent damage

Control cells exposed to light from the microscope lamp, but not from the trapping laser, had an average LD_{50} time of 3300 ± 400 s, with a corresponding sensitivity of 3.1 ×

$10^{-4} \pm 0.4 \times 10^{-4}$ s^{-1}. The action spectrum (i.e., the wavelength-dependent sensitivity) for *E. coli* trapped at 100 mW of laser power (determined in the specimen plane) is presented in Fig. 5. There was a roughly sevenfold difference between the most damaging wavelength (930 nm) and the least (970 nm). A direct comparison between the photodamage spectrum measured for *E. coli* and that reported

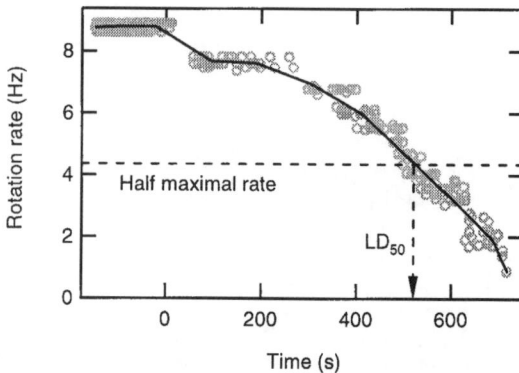

FIGURE 4 Rotation rate as a function of time for a single cell (*open gray circles*, experimental data; *solid line*, a linear spline fit to the data). The trap is turned on at $t = 0$, after which there is a gradual decrease in rotation rate. The LD_{50} is depicted on the graph as the time at which the rotation rate decreased to 50% of the initial value. The appearance of layer lines in the data is a consequence of the video detection scheme, which requires an integral number of frames.

2860 Biophysical Journal Volume 77 November 1999

TABLE 1 Transmission of microscope objectives, cross-referenced with Fig. 2

Part no.	Manufacturer	Magnification/tube length (mm)/numerical aperture	Type designation	Transmission ($\pm 5\%$)			
				830 nm	850 nm	990 nm	1064 nm
461832	Zeiss	63/160/1.2 water	Plan NeoFluar	66	65	64	64
506038	Leica	100/∞/1.4–0.7 oil	Plan Apo	58	56	54	53
85020	Nikon	60/160/1.4 oil	Plan Apo	54	51	17	40
93108	Nikon	60/∞/1.4 oil	Plan Apo CFI	59	54	13	39
93110	Nikon	100/∞/1.4 oil	Plan Apo CFI	50	47	35	32
93110IR	Nikon	100/∞/1.4 oil	Plan Apo IR CFI	61	60	59	59
93144	Nikon	100/∞/1.3 oil	Plan Fluor CFI	67	68	—	61

by Liang et al. (1996), based on cell cloning efficiency, is displayed in Fig. 6.

Oxygen-dependent damage

A comparison between cells trapped under either aerobic or anaerobic conditions at two different wavelengths is presented in Fig. 7. Anaerobic conditions were achieved either by growing and maintaining cells in an oxygen-free environment or by introducing an oxygen-scavenging system just before trapping. The experimental results were statistically identical in the two cases. The effect on photodamage of removing oxygen was dramatic, resulting in a three- to sixfold increase in LD_{50}. Notably, trapping lifetimes under anaerobic conditions were the same as for the controls.

Intensity dependence of photodamage

Clues to the photochemical process underlying optical damage can be gained from the study of its intensity dependence. A simplified model for photodamage takes the form $S(P) = A + BP^n$, where S is the sensitivity, A is the control sensitivity, B is the wavelength-dependent sensitivity, and P is the power. For a single photon-based process, n should be

1, while for a two-photon process, n should be 2. A double-logarithmic plot of the reduced sensitivity, S-A, as a function of power at 870 and 1064 nm is plotted in Fig. 8. Data sets for each wavelength were fit to lines. At 1064 nm, the slope was 1.14 ± 0.03 (reduced $\chi^2 = 4.2$), while at 870 nm the slope was 0.91 ± 0.06 (reduced $\chi^2 = 2.5$). Taken together, the average slope is 1.06 ± 0.07, consistent with a linear, one-photon process.

Temporal dependence of photodamage

A distinct attribute of the rotating cell assay is an ability to obtain quantitative data from a single cell in real time (Fig. 4). Averaged single-cell curves for data taken at 870 nm with 100 mW are plotted in Fig. 9. To compute this average, individual curves were first normalized by their initial rotation rates, and then the time was normalized by the measured LD_{50}. While there was considerable variation among individual curves, the average behavior displays an approximately linear decrease in rotation speed with time.

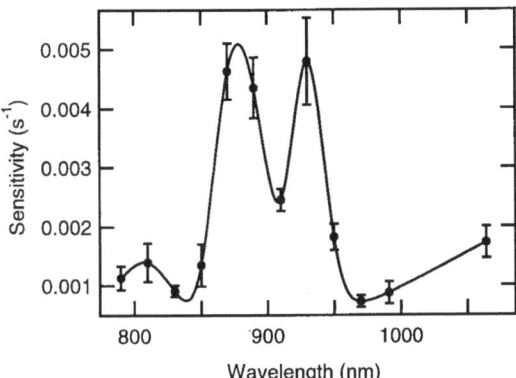

FIGURE 5 The action spectrum for *E. coli* trapped with 100 mW. Sensitivity is defined as the reciprocal of the average LD_{50}. (*solid circles*, experimental data; *solid line*, a cubic spline fit to the data). Each point represents an average of 12–23 determinations, with the errors shown ($\pm$SEM).

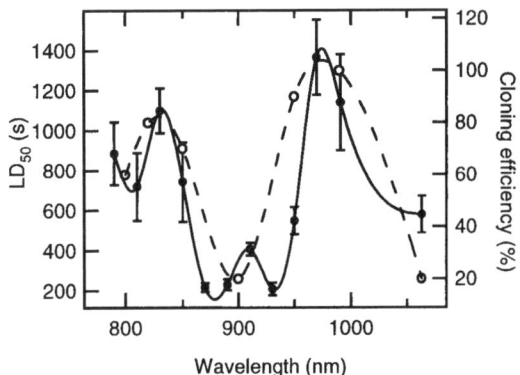

FIGURE 6 The wavelength dependence of photodamage in *E. coli* compared to CHO cells (*solid circles and solid line, left axis*, data replotted from Fig. 5 as LD_{50}; *open circles and dashed line, right axis*, cloning efficiency determined by Liang et al., 1996 (used with permission)). Lines represent cubic spline fits to the data. The cloning efficiency in CHO cells was determined after 5 min of trapping at 88 mW in the specimen plane (error bars unavailable), selected to closely match to our experimental conditions (100 mW in the specimen plane; errors are shown as $\pm$SEM).

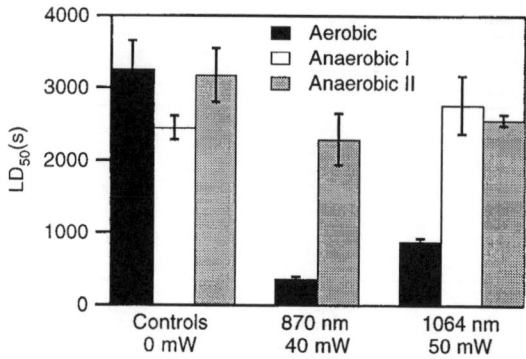

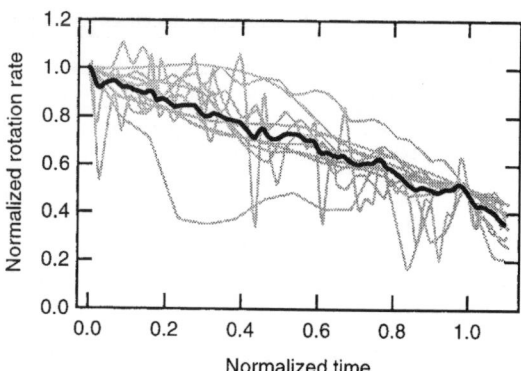

FIGURE 7   The oxygen dependence of photodamage. Comparison between *E. coli* cells trapped under aerobic and anaerobic conditions (*solid bars*, cells grown and maintained aerobically; *open bars*, cells trapped in the presence of an oxygen scavenging system; *gray bars*, cells that were grown and maintained anaerobically). Each point represents an average of 6–12 determinations, with the errors shown (± SEM).

FIGURE 9   Photodamage as a function of time at 870 nm and 100-mW trapping power. *Gray lines*, normalized rotation rates versus normalized time. Rates were normalized according to their initial values, determined over an interval of ~30–100 s before the start of trapping at $t = 0$. Time was normalized according to the individual LD_{50} time determined for each cell, and typically ranged from ~200 to 300 s. *Solid line*, the unweighted average of these curves.

DISCUSSION

The prominent features exhibited by the photodamage action spectrum (Fig. 5) are not easily understood. For example, the spectrum does not bear any superficial resemblance to the absorption spectrum of suspensions of *E. coli* cells, to water absorption (Palmer and Williams, 1974), or to the absorption of molecular oxygen (Krupenie, 1972). The relatively sharp spectral features suggest that light is absorbed by one or more specific photopigments. However, our effort to match the observed spectrum with known chromophores was hampered by a dearth of spectral data for biological molecules in the near-infrared region (most published spectra do not extend beyond ~750 nm). One noteworthy characteristic is the rough similarity between the wavelength dependence of photodamage seen in *E. coli* and in CHO

cells (Fig. 6). This may indicate a common basis for damage in both prokaryotic and eukaryotic systems, possibly involving a ubiquitous intracellular chromophore, and suggests that it may be possible to generalize the present results, with caveats, to other organisms.

The dramatic increase in LD_{50} under anaerobic conditions (Fig. 7) implies a critical role for oxygen in the damage pathway. In its absence, trapped cells display a LD_{50} comparable to that of control cells. Whether oxygen is directly responsible, through the formation of a reactive oxygen species (the primary candidate being singlet molecular oxygen), or simply mediates the process remains to be determined.

The nearly linear relationship between sensitivity and power strongly suggests that a single-photon mechanism leads to photodamage (Fig. 8). This implies a direct absorption by some molecule (or molecules) in the infrared region, as opposed to a two-photon excitation mechanism in the visible (or UV) by unidentified fluorophores. This conclusion is at variance with previous reports implying a role for a two-photon process (König et al., 1995, 1996a; Liu et al., 1996), which were based on the finding that photodamage depended on the peak intensity, and not the average intensity, when short-pulse laser irradiation was used (pulsed lasers are not normally used for optical trapping work). However, the clearest signature for a two-photon process is a quadratic dependence of damage on laser intensity, which was not explicitly established. One possible resolution of the discrepancy may be that there are two regimes for photodamage: at the extremely high peak intensities generated by mode-locked and Q-switched lasers (GW/cm²; König et al., 1996a), photodamage may be dominated by some two-photon process, while at the lower intensities encountered in CW optical traps (operating at MW/cm²), the single-photon mechanism prevails. An alternative ex-

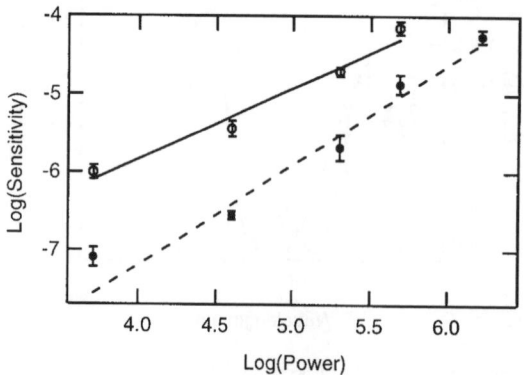

FIGURE 8   The intensity dependence of photodamage. Double logarithmic plot of the reduced sensitivity, *S-A*, versus power in the specimen plane at 870 nm (*open circles, solid line*) and at 1064 nm (*filled circles, dashed line*). The data are fit to lines, the slope of which gives the apparent order of the photodamage process. Fitted slope at 870 nm, 0.911 ± 0.06 (reduced $\chi^2 = 2.5$); slope at 1064 nm, 1.14 ± 0.03 (reduced $\chi^2 = 4.2$).

Biophysical Journal Volume 77 November 1999

planation for the increased damage seen with pulsed lasers may be the onset of optoacoustic shock waves (Hu, 1969; Bushanam and Barnes, 1975; Patel and Tam, 1981), which are pressure waves generated from high-intensity light pulses focused into a liquid medium. The overpressures produced can amount to several atmospheres and may have deleterious effects. Optoacoustic damage has been studied in bulk tissues (Yashima et al., 1990, 1991; Lustmann et al., 1992) but not in single cells.

The ability to continuously monitor single cells in the optical trap reveals the progress of the damage process. The nearly linear decline in rotation rate displayed by Fig. 9 was found for all wavelengths and laser powers investigated. Photodamage therefore seems to be a gradual process, not a catastrophic one. A damage threshold did not appear to exist. Even at the lowest power investigated, the rotation rate started to decrease immediately after trapping began.

A source of photodamage consistent with our data is the production of excited-state (singlet) oxygen, mediated by a sensitizer molecule (Calmettes and Berns, 1983; Block, 1990; Svoboda and Block, 1994). Singlet oxygen is a long-lived, highly reactive species with well-established toxicity (Pryor, 1986; Dahl et al., 1987). While it is possible to produce singlet oxygen directly with laser illumination (Rosenthal, 1985), transitions from the ground state of molecular oxygen to the low-lying excited states are forbidden (Krupenie, 1972). Moreover, the absorption spectrum for molecular oxygen does not resemble the action spectrum for *E. coli*. Singlet oxygen may also be produced indirectly by exciting the triplet state of some sensitizer molecule, which in turn excites oxygen (Foote, 1976). It is conceivable, therefore, that the action spectrum for *E. coli* matches the spectrum of an unidentified sensitizer. This conjecture is consistent with the observed reduction in damage when oxygen is removed from the sample, and by the relationship between intensity and damage. The lack of a damage threshold and its linear time course suggests that the toxic species may have a short lifetime (a longer-lived species that accumulated would be expected to produce damage at a rate that increased with time). Other possibilities exist. For example, the absorbing species could itself directly damage cells, independent of oxygen per se, but be present in concentrations that depended indirectly on the oxygen tension.

This work was motivated, in part, by a search for the most favorable wavelength for optical trapping in biological work. Based on these data, some general conclusions can be reached concerning the design of optical tweezers. Spectral transmission characteristics suggest that microscope objectives designed for fluorescence are better suited to optical trapping work than the (more costly) highly corrected objectives designed for general high NA use. The large variation in throughput across the near-infrared portion of the spectrum means that careful consideration should be given to transmission characteristics before any objective for trapping work is selected. We also note that our measurements of transmission for most of the objectives tested differed from the test data supplied by various manufacturers, with

our figures invariably being lower by 10–30%. This difference may be attributable to their use of integrating spheres to measure transmission through high-NA objectives, rather than the dual-objective method employed here. Integrating spheres do not distinguish between scattered and refracted light and therefore count scattered rays, which do not contribute usefully to trapping.

The action spectrum (Figs. 5 and 6) suggests that the region between 870 and 910 nm is particularly damaging and should be avoided, especially for work in vivo. The least harmful wavelengths are 830 and 970 nm, which are about a factor of 2 less destructive than the 1064 nm Nd:YAG wavelength in common use. Currently, single-mode diode lasers are available at all the favorable wavelengths, but only at relatively low power (typically, ~50–1000 mW). Continuing developments in diode laser technology may improve this situation, but there has been little increase in peak powers over the last 4 years. The fact that 970 nm is near the wavelength favored for pumping erbium fiber lasers in the communications industry (980 nm) augurs well for the development of economical, hybrid diode-based designs that may eventually reach higher powers.

The dramatic increase in lifetime promoted by the removal of oxygen suggests that where possible, scavengers or other means should be employed to reduce the oxygen tension in trapping experiments. While this strategy works well for in vitro protein assays and anaerobic organisms, it is obviously untenable for work with most eukaryotes. For the latter, a useful approach may involve adding quenchers of singlet oxygen to media. These include simple amino acids (e.g., histidine, methionine, or tryptophan) and powerful antioxidant compounds such as β-carotene, DABCO (diazabicyclo [2,2,2]octane), or α-tocopherol (vitamin E). The trapped-and-tethered cell assay presented here should provide a ready means for testing the protective potential of such compounds.

We thank Prof. Steven Lyon for generously providing lab space, equipment, and technical advice. We thank Prof. Howard Berg for the generous loan of the video cursor box, and Dr. Karen Fahrner for the generous gift of strain KAF95. We thank the Princeton University Department of Chemical Engineering teaching lab for the use of their incubator. We are indebted to Drs. Lisa Satterwhite, Koen Visscher, and Mark Schnitzer for helpful discussions, Jason Hsu for preliminary work on this project, Anja Brau for assistance with the anaerobic data collection, and Jeff Lehrman for assistance with LabView programming. We thank Neil Barlow of Micron Optics for the loan of Nikon microscope objectives and Geoff Daniels of Leica America for the loan of Leica objectives for transmission measurements.

KCN was supported by a training grant from the National Institutes of Health. SMB acknowledges support from grants from the National Science Foundation, the National Institutes of Health, and the W. M. Keck Foundation.

REFERENCES

Ashkin, A. 1974. Trapping of atoms by resonance radiation pressure. *Appl. Phys. Lett.* 19:283–285.

Ashkin, A., J. M. Dziedzic, J. E. Bjorkholm, and S. Chu. 1986. Observation of a single beam gradient force optical trap for dielectric particles. *Opt. Lett.* 11:288–290.

Ashkin, A., and J. M. Dziedzic. 1989. Optical trapping and manipulation of single living cells using infra-red laser beams. *Ber. Bunsenges. Phys. Chem.* 93:254–260.

Ashkin, A., J. M. Dziedzic, and T. Yamane. 1987. Optical trapping and manipulation of single cells using infrared laser beams. *Nature.* 330: 769–771.

Berg, H. C., and L. Turner. 1993. Torque generated by the flagellar motor of *Escherichia coli*. *Biophys. J.* 65:2201–2216.

Berns, M. W. 1976. A possible two-photon effect in vitro using a focused laser beam. *Biophys. J.* 16:973–977.

Block, S. M. 1990. Optical tweezers: a new tool for biophysics. *In* Noninvasive Techniques in Cell Biology. Modern Review of Cell Biology, Vol. 9. J. K. Foskett and S. Grinstein, editors. Wiley-Liss, New York. 375–402.

Block, S. M., and H. C. Berg. 1984. Successive incorporation of force-generating units in the bacterial rotary motor. *Nature.* 309:470–472.

Block, S. M., D. F. Blair, and H. C. Berg. 1989. Compliance of bacterial flagella measured with optical tweezers. *Nature.* 338:514–518.

Block, S. M., J. E. Segall, and H. C. Berg. 1982. Impulse responses in bacterial chemotaxis. *Cell.* 31:215–226.

Block, S. M., J. E. Segall, and H. C. Berg. 1983. Adaptation kinetics in bacterial chemotaxis. *J. Bacteriol.* 154:312–323.

Bushanam, G. S., and F. S. Barnes. 1975. Laser-generated thermoelastic shock wave in liquids. *J. Appl. Phys.* 46:2074–2082.

Calmettes, P. P., and M. W. Berns. 1983. Laser induced multiphoton processes in living cells. *Proc. Natl. Acad. Sci. USA.* 80:7197–7199.

Dahl, T. A., R. A. Midden, and P. E. Hartman. 1987. Pure singlet oxygen cytotoxicity for bacteria. *Photochem. Photobiol.* 46:345–352.

Foote, C. S. 1976. Photosensitized oxidation and singlet oxygen: consequences in biological systems. *In* Free Radicals in Biology, Vol. II. W. A. Pryor, editor. Academic Press, New York. 85–133.

Hu, C. 1969. Spherical model of an acoustical wave generated by rapid laser heating in a liquid. *J. Acoust. Soc. Am.* 46:728–735.

König, K., H. Liang, M. W. Berns, and B. J. Tromberg. 1995. Cell damage by near-IR microbeams. *Nature.* 377:20–21.

König, K., H. Liang, M. W. Berns, and B. J. Tromberg. 1996a. Cell damage in near-infrared multimode optical traps as a result of multiphoton absorption. *Opt. Lett.* 21:1090–1092.

König, K., Y. Tadir, P. Patrizio, M. W. Berns, and B. J. Tromberg. 1996b. Effects of ultraviolet exposure and near infrared laser tweezers on human spermatozoa. *Hum. Reprod.* 11:2162–2164.

Krupenie, P. H. 1972. The spectrum of molecular oxygen. *J. Phys. Chem. Ref. Data.* 1:423–520.

Kuwajima, G. 1988. Construction of a minimum-size functional flagellin of *Escherichia coli*. *J. Bacteriol.* 170:3305–3309.

Liang, H., K. T. Vu, P. Krishnan, T. C. Trang, D. Shin, S. Kimel, and M. W. Berns. 1996. Wavelength dependence of cell cloning efficiency after optical trapping. *Biophys. J.* 70:1529–1533.

Liu, Y., D. K. Cheng, G. J. Sonek, M. W. Berns, C. F. Chapman, and B. J. Tromberg. 1995a. Evidence for localized cell heating induced by infrared optical tweezers. *Biophys. J.* 68:2137–2144.

Liu, Y., G. J. Sonek, M. W. Berns, K. König, and B. J. Tromberg. 1995b. Two-photon fluorescence excitation in continuous-wave infrared optical tweezers. *Opt. Lett.* 20:2246–2248.

Liu, Y., G. J. Sonek, M. W. Berns, and B. J. Tromberg. 1996. Physiological monitoring of optically trapped cells: assessing the effects of confinement by 1064-nm laser tweezers using microfluorometry. *Biophys. J.* 71:2158–2167.

Lustmann, J., M. Ulmansky, A. Fuxbrunner, and A. Lewis. 1992. Photo-acoustic injury and bone healing following 193 nm excimer laser ablation. *Lasers Surg. Med.* 12:390–396.

Manson, M. D., P. M. Tedesco, and H. C. Berg. 1980. Energetics of flagellar rotation in bacteria. *J. Mol. Biol.* 138:541–561.

Misawa, H., M. Koshioka, K. Sasak, N. Kitamura, and H. Masuhara. 1991. Three dimensional optical trapping and laser ablation of a single polymer latex in water. *J. Appl. Phys.* 70:3829–3836.

Palmer, K. F., and D. Williams. 1974. Optical properties of water in the near infrared. *J. Opt. Soc. Am.* 64:1107–1110.

Parkinson, J. S. 1978. Complementation analysis and deletion mapping of *Escherichia coli* mutants defective in chemotaxis. *J. Bacteriol.* 135: 45–53.

Parkinson, J. S., S. R. Parker, P. B. Talbert, and S. E. Houts. 1983. Interactions between chemotaxis genes and flagellar genes in *Escherichia coli*. *J. Bacteriol.* 155:265–274.

Patel, C. K. N., and A. C. Tam. 1981. Pulsed optoacoustic spectroscopy of condensed matter. *Rev. Mod. Phys.* 53:517–550.

Pryor, W. A. 1986. Oxy-radicals and related species: their formation, lifetimes, and reactions. *Annu. Rev. Physiol.* 48:657–667.

Rosenthal, I. 1985. Chemical and physical sources of singlet oxygen. *In* Singlet O_2, Vol. I, Physical and Chemical Aspects. A. A. Frimer, editor. CRC Press, Boca Raton, FL. 13–38.

Svoboda, K., and S. M. Block. 1994. Biological applications of optical forces. *Annu. Rev. Biomol. Struct.* 23:247–285.

Vorobjev, I. A., H. Liang, W. H. Wright, and M. W. Berns. 1993. Optical trapping for chromosome manipulation: a wavelength dependence of induced chromosome bridges. *Biophys. J.* 64:533–538.

Yashima, Y., D. J. McAuliffe, and T. J. Flotte. 1990. Cell selectivity laser induced photoacoustic injury of skin. *Lasers Surg. Med.* 10:280–283.

Yashima, Y., D. J. McAuliffe, S. L. Jacques, and T. J. Flotte. 1991. Laser-induced photoacoustic injury of skin: effect of inertial confinement. *Lasers Surg. Med.* 11:62–68.

Volume 13, number 1   OPTICS COMMUNICATIONS   January 1975

COOLING OF GASES BY LASER RADIATION[1*]

T.W. HÄNSCH[2†] and A.L. SCHAWLOW

Department of Physics, Stanford University, Stanford, California 94305, USA

Received 20 October 1974

It is shown that a low-density gas can be cooled by illuminating it with intense, quasi-monochromatic light confined to the lower-frequency half of a resonance line's Doppler width. Translational kinetic energy can be transferred from the gas to the scattered light, until the atomic velocity is reduced by the ratio of the Doppler width to the natural line width.

It is well known that light exerts a radiation pressure on any substance which reflects or scatters it. It is also known that the scattering cross section of an atom can be quite large when the light frequency is in resonance with a sharp absorption line [1]. Thus the radiation pressure of laser light has been used to selectively deflect atoms of a chosen isotopic species from a beam.

We wish to point out that if the laser radiation is essentially isotropic, but confined to frequencies on the lower half of the Doppler-broadened absorption line of an atomic vapor, the gas can be cooled. That is, the average translational kinetic energy of the atoms can be reduced.

To understand why this occurs let us first consider the irradiation of the vapor by a single directed laser beam. Only those atoms which are moving towards the laser source will find the light Doppler-shifted up in frequency so as to have a large scattering cross section. To atoms moving away from the laser, the frequency of the light will appear lowered out of resonance with the scattering transition. Thus if the laser light is confined to the lower half of the Doppler line width, the atoms can only lose energy and momentum by scattering of the laser light, and never gain.

If the light comes from all directions, atoms will lose energy by scattering the oncoming light, while the Doppler-shift will detune any light wave traveling in the same direction as the atoms. In this way the translational temperature of the atoms can be reduced until ultimately the Doppler line width is as small as the natural line width[2*].

To estimate the size of the effect, consider a gas of magnesium atoms at a temperature of 600 K, illuminated by intense light on the low-frequency side of the singlet resonance line at 2852.1 Å. The Doppler width at this temperature is 3.8×10^9 Hz full width at half maximum, and the natural line width, determined by the 2 nsec radiative lifetime of the upper state, is 8×10^7 Hz. Thus radiation cooling could reduce the average atomic velocity by a factor of about 50, which is equivalent to a reduction in temperature to $600/(50)^2 = 0.24$ K.

Cooling of this order could be achieved quickly. When a photon of momentum $h\nu/c$ is scattered by an atom of mass M, moving towards it with a velocity v, the average change in velocity is

$$\Delta V = \frac{\Delta(Mv)}{M} - \frac{h\nu}{Mc}.$$

Thus at each scattering, the velocity of an atom will decrease, on the average, by about 6 cm/sec. Since the r.m.s. velocity $v_0 = (3kT_0/M)^{1/2}$ is, at $T_0 = 600$ K,

---

[1*] Work supported by the National Science Foundation under Grant NSF MPS74-14786 A01.

[2†] Alfred P. Sloan Fellow 1973–75.

[2*] A different mechanism of "cooling" with quasi-monochromatic light, the depletion of low, thermally populated molecular energy levels via optical pumping, has been discussed elsewhere [2].

Volume 13, number 1   OPTICS COMMUNICATIONS   January 1975

initially 80,000 cm/sec, about $v_0/\Delta v \approx 13,000$ scattering events will substantially complete the velocity reduction. If the light intensity is comparable to that needed for saturation of the atomic absorption, a photon can be scattered essentially every $\tau = 2$ nsec, and the cooling process need only take about $t_0 = \tau v_0/\Delta v \approx 3 \times 10^{-5}$ sec. During this time, the average atom will move a path length

$$l_0 = \tfrac{1}{2} v_0 t_0 = 3kT_0 c\tau/2h\nu,$$

independent of the atomic mass. For the present example, this path is on the order of 1 cm, i.e. it is not necessary to illuminate a large volume, if the radiative lifetime τ is sufficiently short, and/or the light frequency ν is sufficiently high.

To saturate the Doppler-broadened magnesium resonance line requires a flux of about 1000 W/cm^2. Half of this would be needed to irradiate only the lower half of the Doppler profile. To irradiate a 1 cm cubic volume with six such beams directed along the six perpendicular directions would require a total power of 3000 W. Since the power would need to be applied for about 3×10^{-5} sec for complete cooling, a pulse energy of 0.1 J would be required. Such a pulse could be generated by harmonic generation from a flashlamp-pumped dye laser. The required power can be substantially reduced, if the six rays are generated by multiple reflections of the same laser beam. The limiting effect of power broadening of the atomic resonance can be avoided by using a lower light intensity towards the end of the cooling pulse.

Radiation cooling could be applied to provide slow-moving atoms which would remain for a long time in interaction with a weak optical or radiofrequency field. A particularly interesting case might be hydrogen, where experiments are under way to study the $1s \rightarrow 2s$ two-photon absorption. The upper state has a lifetime of 1/7 sec, so that fractional line widths of 10^{-15} might eventually be obtained. However, at 300 K the r.m.s. velocity of hydrogen atoms is nearly 3×10^5 cm/sec. Thus transit time is likely to be an important source of line broadening unless the atomic velocity is reduced. Another possible application might be to improve the collimation of an atomic beam by reducing the transverse velocities through two-dimensional radiation cooling.

It is possible that radiation cooling might occur naturally in some astronomical objects. A continuous-spectrum light source will have little influence on atoms, except for the narrow band within the Doppler width of a resonance line. If just those frequencies on the upper side of the Doppler line are removed, cooling will occur. Such removal might come about by absorption from a volume of gas moving toward the region being cooled.

The process of cooling by narrow-band light requires that each atom scatters many photons. Thus the atom must have a high probability of returning to the original lower state so that it can scatter repeatedly. If strong monochromatic sources of X-rays become available, cooling of even complex atoms and molecules could occur by just a few scattering events, although the residual recoil momentum may then become an important limiting factor. It should also be noted that the cooled region should be optically thin enough so that most of the scattered light can escape to a distant absorber.

The possibility of cooling by nearly monochromatic light illustrates that such radiation is equivalent, in a thermodynamic sense, to mechanical work or electricity rather than heat energy [2], even though this particular process is envisioned as operating far from equilibrium. When monochromatic light of low entropy is scattered by a moving atom, the frequency of the scattered light and so the energy of the scattered photon, is higher by an amount depending on the recoil direction. Thus the bandwidth and so the disorder of the light is increased in the cooling process.

A.L. Schawlow wishes to thank Professor D.J. Bradley for his hospitality at Imperial College, London and the Rank Prize Fund for support during the time when this manuscript was prepared.

References

[1] A. Ashkin, Phys. Rev. Lett. 25 (1970) 1321.
[2] W.H. Christiansen and A. Hertzberg, Proc. IEEE 61 (1973) 1060.

VOLUME 55, NUMBER 1   PHYSICAL REVIEW LETTERS   1 JULY 1985

Three-Dimensional Viscous Confinement and Cooling of Atoms by Resonance Radiation Pressure

Steven Chu, L. Hollberg, J. E. Bjorkholm, Alex Cable, and A. Ashkin

AT&T Bell Laboratories, Holmdel, New Jersey 07733
(Received 25 April 1985)

We report the viscous confinement and cooling of neutral sodium atoms in three dimensions via the radiation pressure of counterpropagating laser beams. These atoms have a density of about $\sim 10^6$ cm^{-3} and a temperature of ~ 240 μK corresponding to a rms velocity of ~ 60 cm/sec. This temperature is approximately the quantum limit for this atomic transition. The decay time for half the atoms to escape a ~ 0.2-cm^3 confinement volume is ~ 0.1 sec.

PACS numbers: 32.80.Pj

The deflection of atoms by light resonant with an atomic transition was observed as early as 1933 by Frisch.[1] Much later Ashkin[2] pointed out that laser light can exert a substantial scattering force on an atom. He raised the possibility that this force could be used to trap atoms, and subsequently there have been numerous proposals to cool and ultimately trap neutral atoms.[3] Various experiments have been proposed that would utilize trapped atoms; these generally require long observation times and/or low atomic velocities. We report the demonstration of a confinement scheme based on the damping of atomic velocities. This scheme is not a trap, but can confine atoms in a small region in space for times on the order of 0.1 sec, and cool them to ~ 240 μK, the quantum limit for our experimental conditions. For comparison, we note that two-dimensional radiative cooling has reduced the temperatures transverse to the motion of an atomic beam from 40 to 3.5 mK,[4] and atomic beams stopped by light have resulted in three-dimensional temperatures of 50–100 mK.[5] Laser cooling of electromagnetically trapped ions has resulted in ion temperatures between 5 and 100 mK.[6]

The basic physics of the viscous damping scheme is briefly outlined. Consider an atom irradiated by a laser beam tuned near a resonance line. For each photon absorbed, an atom receives a net change of momentum $\Delta p = h/\lambda$, where λ is the wavelength. Since the subsequent reemission of the photon has no preferred direction, an average of many scattering events gives a net scattering force along the direction of the light.[2] Hänsch and Schawlow[7] noted that if counterpropagating beams were tuned to the low-frequency side of the absorption line, there would always be a net force opposing the velocity of an atom. For example, an atom moving with velocity $+v_x$ will blue shift into resonance with a laser beam propagating towards $-\hat{x}$ and red shift out of resonance with the laser beam propagating towards $+\hat{x}$. Thus, the atom is more likely to absorb photons going towards $-\hat{x}$. With the use of six beams along $\pm\hat{x}$, $\pm\hat{y}$, and $\pm\hat{z}$ and an averaging over many absorptions, the net effect is a viscous damping

force $\mathbf{F} = -\alpha\mathbf{v}$ opposite the velocity of the atom. The cooling rate is $\mathbf{v} \cdot \mathbf{F} = -\alpha v^2$. The expression for the damping force which includes standing waves and saturation has been previously derived.[8]

In addition to the average force, statistical fluctuations must be considered.[9] These fluctuations lead to heating. For a simple picture of the fluctuations, consider the momentum impulses on an atom due to the absorption and emission of photons. In the absence of damping, the atoms will execute a random walk in velocity, and although $\langle v \rangle = 0$, $\langle v^2 \rangle$ will increase linearly with the total number scattered photons. Increasing $\langle v^2 \rangle$ corresponds to heating, as first observed by Bjorkholm et al.[10] If we equate the heating and cooling terms, the steady-state kinetic temperature is obtained. In the absence of stimulated processes, the minimum kinetic energy for a two-level atom is given by $kT = \frac{1}{4}h\gamma$, where γ is the width (FWHM) of the absorption line. If one includes stimulated processes, the minimum temperature is increased by a factor of 2.[8] For sodium, $\gamma = 10$ MHz and $T_{\min} = 240$ μK.

An estimate of the confinement time can be obtained by the observation that the motion of atoms in a viscous fluid of photons ("optical molasses") is analogous to diffusion in classical Brownian motion. The diffusion constant D is given by the Einstein relation $D = kT/\alpha$, and for an infinite medium, $D = \langle x^2 \rangle / 2t$, where $\langle x^2 \rangle$ is the mean square displacement after time t. However, an analysis based on an infinite medium overestimates the storage time. A more appropriate model is a viscous fluid surrounded by a spherical boundary (defined by the extent of the laser beams) such that the atoms that reach the boundary escape. If we assume an initial uniform concentration of atoms n_0, the average concentration $\bar{n}$ has been shown[11] to vary as

$$\bar{n} = n_0 \frac{6}{\pi^2} \sum_{v=1}^{\infty} \frac{1}{v^2} e^{-D\pi^2 v^2 t/R^2}. \qquad (1)$$

The spherical-boundary modification of the random-walk analysis reduces the storage time for our experimental conditions by a factor of 3.1.

48

VOLUME 55, NUMBER 1   PHYSICAL REVIEW LETTERS   1 JULY 1985

The experimental apparatus is schematically shown in Fig. 1. We use a pulsed atomic beam source in order to simplify diagnostics. The beam of sodium atoms is produced by irradiation of a pellet of sodium metal with a ~ 10-nsec pulse from a frequency-doubled Nd-YAlG laser (~ 30-mJ pulse focused to $\sim 5 \times 10^{-2}$ cm^2). A hot plasma is formed which produces sodium ions, suprathermal atoms with average energy of ~ 3.5 eV, and a small fraction of thermal atoms at ~ 1000 K. The source of atoms is apertured to produce a directed 0.6-cm-diam atomic beam at the confinement region shown in Fig. 1.

Since our "optical molasses" can only capture atoms moving with velocities $\lesssim 3 \times 10^3$ cm/sec, we slow some of the atoms in the atomic beam by using a counterpropagating laser beam. Following Ertmer et al.,[5] we use an electro-optic modulator to generate a frequency-shifted sideband which can be swept in frequency to stay in resonance with the changing Doppler shift of the atomic resonance frequency as the atoms slow down. The laser intensity used to slow (precool) the atoms is ~ 120 mW/cm^2. Atoms with initial velocities of $\sim 2 \times 10^4$ cm/sec and less are slowed down in less than 5 cm to velocities $\sim 2 \times 10^3$ cm/sec. After slowing for 0.5 msec, the precooling beam is shut off by an acoustic-optic modulator, and the atoms then drift into the region defined by the six intersecting laser beams where they are cooled and viscously confined. The precooling laser beam and the confining beams are obtained from the combined beams of two cw actively stabilized, ring dye lasers. The lasers are operated at frequencies differing by ~ 1.7 GHz to prevent optical pumping of the sodium ground state.

The power of each confining beam is between 4 and 20 mW, and the beam radius is $w_0 = 0.36$ cm. Despite the fact that the vast majority of atoms are not slowed down, the pulsed beam is sufficiently intense that confinement densities on the order of 10^6 atoms/cm^3 are obtained. At these densities, the cloud of confined atoms is clearly visible by eye.

An averaged fluorescence signal (sixteen pulses) as a function of time is shown in Fig. 2. Pulse-to-pulse amplitude fluctuations are $\sim 30\%$. The initial abrupt spike is due to fast atoms passing rapidly through the interaction region. The baseline is the scattered light level obtained by blockage of the precooling beam. If any of the confining dimensions is blocked, a fluorescence pulse $\lesssim 1$ msec long is seen as a result of the passage of the slowed atoms through the interaction region. In addition, the laser frequencies must be tuned to the low-frequency side of the absorption lines and be critically tuned with respect to each other. In Fig. 2, we plot the number of atoms remaining in the observation region as a function of Dt/R^2, where D is the diffusion constant, t is the time, and R is an effective radius to the spherical boundary. We extract a value for D by scaling the horizontal axis of the experimental data to match to the theoretical decay given by Eq. (1). If we take $R = 0.4$ cm (where the intensity of the laser beam is $\simeq 0.1$ of the peak intensity), the decay curves give us an effective diffusion constant, D_{eff}. Using a computed value of $\alpha = \alpha_{max} = 5.8 \times 10^{-18}$ g/sec,[8] we obtain an upper limit on the temperature of $T_{max} = D_{eff}\alpha_{max}/k = 1.9$ mK. The actual temperature is expected to be lower since we have ignored drift velocities due to beam intensity imbalance, intensity

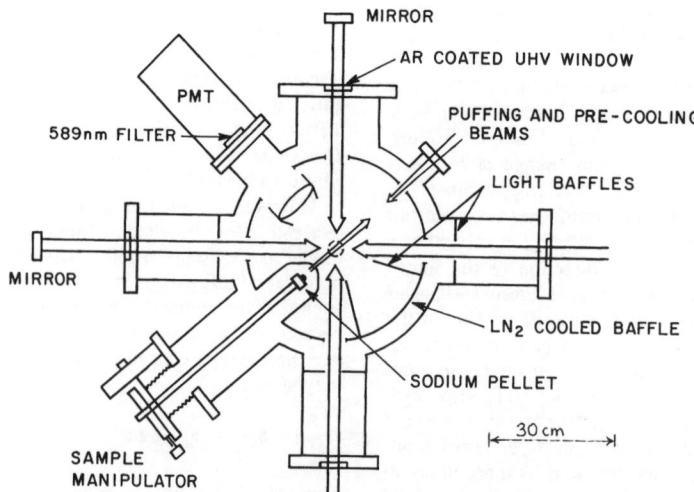

FIG. 1. Schematic of the vacuum chamber and intersecting laser beams and atomic beam. The vertical confining beam is indicated by the dashed circle. The "puffing" beam is from the pulsed YAlG laser.

49

VOLUME 55, NUMBER 1   PHYSICAL REVIEW LETTERS   1 JULY 1985

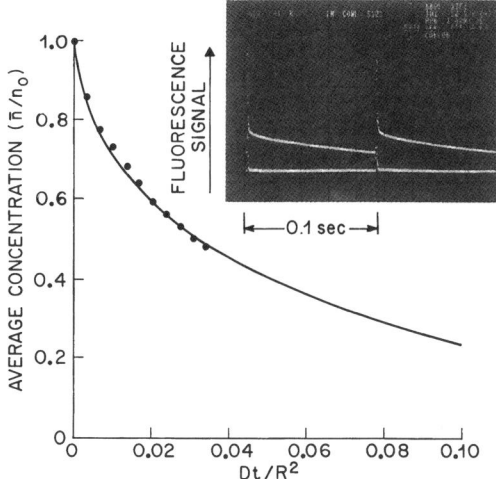

FIG. 2. Fluorescence signal as a function of time is shown in the inset. The baseline shows the scattered light level. The confinement region is loaded every 0.1 sec, governed by the repetition rate of the YAlG laser. The fraction of atoms remaining in the observation region is plotted as a function of Dt/R^2. The solid line is the theoretical calculation based on the modified random walk.

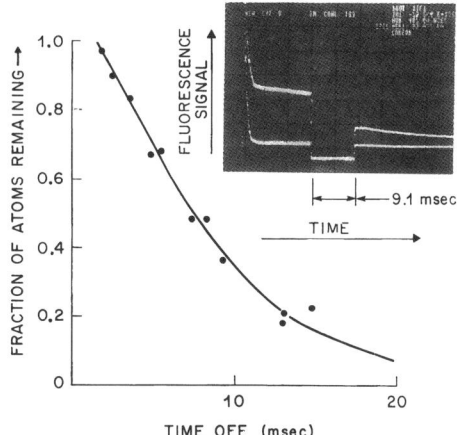

FIG. 3. Inset shows the effect of blocking the confining laser beams 15 msec after the YAlG laser fires. The horizontal trace is the scattered light level. The fraction of atoms remaining is plotted as a function of the time-off period. The solid line is the theoretical curve.

hot spots, and beam misalignment, all of which increase D_{eff} but do not substantially increase the temperature of the atoms. For example, if the counterpropagating beams are imbalanced by 2%, the drift velocity is $v_{drift} = 3$ cm/sec, and the storage time will be $R/v_{drift} \sim 0.13$ sec.

A direct measurement of the temperature of the cooled atoms is obtained by a time-of-flight technique. After a 15-msec cooling and confinement time, all six beams are turned off in ~ 0.1 msec and left off for a variable time. During that time, the atoms will leave the observation region ballistically with their instantaneous velocities. We show in the inset of Fig. 3 an example time-of-flight measurement for a 9.1-msec light-off time. Given an initial uniform spherical distribution of atoms with a Maxwell-Boltzmann distribution of velocities, we calculate the fraction of atoms remaining in the observation region as a function of the light-off time. To fit the data with calculated decay curve, we introduce a 1.4-msec time shift in the theoretical curve to account for the fact that the actual atomic distribution will be depleted near the edge of the sphere by Brownian diffusion to the escape boundary.[11] The measured fraction of atoms remaining and the theoretical curve are plotted in Fig. 3. Additional uncertainty in the temperature measurement arises from the fact that our Gaussian-type beam profiles must be convoluted with saturation effects and the acceptance volume of our phototube. Note that the tem-

perature depends on R_{eff}^2, where R_{eff} is an effective observation radius. Using a fairly conservative value of $R_{eff} = 0.4$ cm, we obtain $T = 240^{+200}_{-60}$ μK.

In summary, we have confined sodium atoms in a ~ 0.2-cm^3 volume for times on the order of 0.1 sec and cooled them to temperatures approaching the quantum limit. These kinetic temperatures are significantly lower than the electromagnetically trapped ions[6] or stopped atoms[5] previously reported. The low velocities and long confinement times that we have achieved can increase the observation and/or coherence time for atoms, opening up new possibilities in areas such as collisions, quantum statistics, and precision spectroscopy. In addition, the low temperatures make possible the efficient loading of atom traps. Such traps will help in reaching higher densities and possibly temperatures as low as 10^{-6} K.[3] At these temperatures, the de Broglie wavelength of the atoms is on the order of 1 μm. Even modest densities of these ultracold atoms (which can be either fermions or bosons) should reveal interesting physics.

We acknowledge helpful discussions with J. P. Gordon and assistance from L. Buhl in the preparation of the LiTaO$_3$ crystal used in the electro-optic modulator.

[1]O. R. Frisch, Z. Phys. **86**, 42 (1933).
[2]A. Ashkin, Phys. Rev. Lett. **24**, 156 (1970), and **25**, 1321 (1970).
[3]See, for example, A. Ashkin, Science **210**, 1081 (1980), and Prog. Quantum Electron. **8**, 204 (1984); for recent proposals, see D. E. Pritchard, Phys. Rev. Lett. **51**, 1336

VOLUME 55, NUMBER 1   PHYSICAL REVIEW LETTERS   1 JULY 1985

(1983); A. Ashkin, Opt. Lett. **9**, 454 (1984); J. Dalibard, S. Reynaud, and C. Cohen-Tannoudji, J. Phys. B **17**, 4577 (1984).

[4]V. I. Balykin, V. S. Letokhov, and A. I. Sidorov, Pis'ma Zh. Eksp. Teor. Fiz. **40**, 251 (1984) [JETP Lett. **40**, 1026 (1984)].

[5]J. Prodan, A. Migdall, W. D. Phillips, I. So, H. Metcalf, and J. Dalibard, Phys. Rev. Lett. **54**, 992 (1985); W. Ertmer, R. Blatt, J. L. Hall, and M. Zhu, Phys. Rev. Lett. **54**, 996 (1985).

[6]W. Neuhauser, M. Hohenstatt, P. E. Toschek, and H. Dehmelt, Phys. Rev. A **22**, 1137 (1980); D. J. Wineland and W. M. Itano, Phys. Lett. **82A**, 75 (1981); W. Nagourney, G. Janik, and H. Dehmelt, Proc. Natl. Acad. Sci. U.S.A. **80**, 643 (1983).

[7]T. W. Hänsch and A. L. Schawlow, Opt. Commun. **13**, 68 (1975).

[8]J. P. Gordon and A. Ashkin, Phys. Rev. A **21**, 1606 (1980); R. J. Cook, Phys. Rev. A **20**, 224 (1979), and Phys. Rev. Lett. **44**, 976 (1980).

[9]A. Yu. Pusep, Zh. Eksp. Teor. Fiz. **70**, 851 (1976) [Sov. Phys. JETP **43**, 441 (1976)]; V. S. Letokhov, V. G. Minogin, and B. D. Pavlik, Zh. Eksp. Teor. Fiz. **72**, 1328 (1977) [Sov. Phys. JETP **45**, 698 (1977)]; J. L. Picqué, Phys. Rev. A **19**, 1622 (1979); D. J. Wineland and W. M. Itano, Phys. Rev. A **25**, 35 (1982).

[10]J. E. Bjorkholm, R. R. Freeman, A. Ashkin, and D. B. Pearson, Opt. Lett. **5**, 111 (1980).

[11]N. A. Fuchs, *The Mechanics of Aerosols* (Pergamon, Oxford, 1964), pp. 193–200.

Observation of Bose-Einstein Condensation in a Dilute Atomic Vapor

M. H. Anderson, J. R. Ensher, M. R. Matthews, C. E. Wieman,* E. A. Cornell

A Bose-Einstein condensate was produced in a vapor of rubidium-87 atoms that was confined by magnetic fields and evaporatively cooled. The condensate fraction first appeared near a temperature of 170 nanokelvin and a number density of 2.5×10^{12} per cubic centimeter and could be preserved for more than 15 seconds. Three primary signatures of Bose-Einstein condensation were seen. (i) On top of a broad thermal velocity distribution, a narrow peak appeared that was centered at zero velocity. (ii) The fraction of the atoms that were in this low-velocity peak increased abruptly as the sample temperature was lowered. (iii) The peak exhibited a nonthermal, anisotropic velocity distribution expected of the minimum-energy quantum state of the magnetic trap in contrast to the isotropic, thermal velocity distribution observed in the broad uncondensed fraction.

On the microscopic quantum level, there are profound differences between fermions (particles with half integer spin) and bosons (particles with integer spin). Every statistical mechanics text discusses how these differences should affect the behavior of atomic gas samples. Thus, it is ironic that the quantum statistics of atoms has never made any observable difference to the collective macroscopic properties of real gas samples. Certainly the most striking difference is the prediction, originally by Einstein, that a gas of noninteracting bosonic atoms will, below a certain temperature, suddenly develop a macroscopic population in the lowest energy quantum mechanical state (1, 2). However, this phenomenon of Bose-Einstein condensation (BEC) requires a sample so cold that the thermal deBroglie wavelength, λ_{db}, becomes larger than the mean spacing between particles (3). More precisely, the dimensionless phase-space density, $\rho_{ps} = n(\lambda_{db})^3$, must be greater than 2.612 (2, 4), where n is the number density. Fulfilling this stringent requirement has eluded physicists for decades. Certain well-known physical systems do display characteristics of quantum degeneracy, in particular superfluidity in helium and superconductivity in metals. These systems exhibit counterintuitive behavior associated with macroscopic quantum states and have been the subject of extensive study. However, in these systems the bosons are so closely packed that they can be understood only as strongly interacting systems. These strong interactions have made it difficult to understand

the detailed properties of the macroscopic quantum state and allow only a small fraction of the particles to occupy the Bose condensed state. Recently, evidence of Bose condensation in a gas of excitons in a semiconductor host has been reported (5). The interactions in these systems are weak but poorly understood, and it is difficult to extract information about the exciton gas from the experimental data. Here, we report evidence of BEC in a dilute, and hence weakly interacting, atomic vapor. Because condensation at low densities is achievable only at very low temperatures, we evaporatively cooled a dilute, magnetically trapped sample to well below 170 nK.

About 15 years ago, several groups began to pursue BEC in a vapor of spin-polarized hydrogen (6). The primary motivation was that in such a dilute atomic system one might be able to produce a weakly interacting condensate state that is much closer to the original concept of Bose and Einstein and would allow the properties of the condensate to be well understood in terms of basic interatomic interactions. In the course of this work, 1000-fold increases in phase-space density have been demonstrated with the technique of evaporative cooling of a magnetically trapped hydrogen sample (7); recently, the phase-space density has approached BEC levels. Progress has been slowed, however, by the existence of inelastic interatomic collisions, which cause trap loss and heating, and by the lack of good diagnostics for the cooled samples.

The search for BEC in a dilute sample of laser-cooled alkali atoms has a somewhat shorter history. Developments in laser trapping and cooling over the past decade made it possible to increase the phase-space density of a vapor of heavy alkali atoms by more than 15 orders of magnitude. However, several processes involving the scattered pho-

tons were found to limit the achievable temperatures (8) and densities (9), so that the resulting value for ρ_{ps} was 10^5 to 10^6 times too low for BEC. We began to pursue BEC in an alkali vapor by using a hybrid approach to overcome these limitations (10, 11). This hybrid approach involves loading a laser-cooled and trapped sample into a magnetic trap where it is subsequently cooled by evaporation. This approach is particularly well suited to heavy alkali atoms because they are readily cooled and trapped with laser light, and the elastic scattering cross sections are very large (12), which facilitates evaporative cooling.

There are three other attractive features of alkali atoms for BEC. (i) By exciting the easily accessible resonance lines, one can use light scattering to sensitively characterize the density and energy of a cloud of such atoms as a function of both position and time. This technique provides significantly more detailed information about the sample than is possible from any other macroscopic quantum system. (ii) As in hydrogen, the atom-atom interactions are weak [the S-wave scattering length a_0 is about 10^{-6} cm, whereas at the required densities the interparticle spacing (x) is about 10^{-4} cm] and well understood. (iii) These interactions can be varied in a controlled manner through the choice of spin state, density, atomic and isotopic species, and the application of external fields. The primary experimental challenge to evaporatively cooling an alkali vapor to BEC has been the achievement of sufficiently high densities in the magnetic trap. The evaporative cooling can be maintained to very low temperatures only if the initial density is high enough that the atoms undergo many (~100) elastic collisions during the time they remain in the trap. Using a combination of techniques to enhance the density in the optical trap, and a type of magnetic trap that provides long trap holding times and tight confinement, has allowed us to evaporatively cool to BEC.

A schematic of the apparatus is shown in Fig. 1. The optical components and magnetic coils are all located outside the ultra-high-vacuum glass cell, which allows for easy access and modification. Rubidium atoms from the background vapor were optically precooled and trapped, loaded into a magnetic trap, then further cooled by evaporation. The TOP (time orbiting potential) magnetic trap (13) we used is a superposition of a large spherical quadrupole field and a small uniform transverse field that rotates at 7.5 kHz. This arrangement results in an effective average potential that is an axially symmetric, three-dimensional (3D) harmonic potential providing tight and stable confinement during evaporation. The evaporative cooling works by selectively re-

M. H. Anderson, J. R. Ensher, M. R. Matthews, C. E. Wieman, JILA, National Institute of Standards and Technology (NIST), and University of Colorado, and Department of Physics, University of Colorado, Boulder, CO 80309, USA.
E. A. Cornell, Quantum Physics Division, NIST, JILA-NIST, and University of Colorado, and Department of Physics, University of Colorado, Boulder, CO 80309, USA.
*To whom correspondence should be addressed.

leasing the higher energy atoms from the trap; the remaining atoms then rethermalize to a colder temperature.

We accomplished this release with a radio frequency (rf) magnetic field (14). Because the higher energy atoms sample the

Fig. 1. Schematic of the apparatus. Six laser beams intersect in a glass cell, creating a magneto-optical trap (MOT). The cell is 2.5 cm square by 12 cm long, and the beams are 1.5 cm in diameter. The coils generating the fixed quadrupole and rotating transverse components of the TOP trap magnetic fields are shown in green and blue, respectively. The glass cell hangs down from a steel chamber (not shown) containing a vacuum pump and rubidium source. Also not shown are coils for injecting the rf magnetic field for evaporation and the additional laser beams for imaging and optically pumping the trapped atom sample.

trap regions with higher magnetic field, their spin-flip transition frequencies are shifted as a result of the Zeeman effect. We set the frequency of the rf field to selectively drive these atoms into an untrapped spin state. For optimum cooling, the rf frequency was ramped slowly downward, causing the central density and collision rate to increase and temperature to decrease. The final temperature and phase-space density of the sample depends on the final value of the rf frequency (ν_{evap}).

A typical data cycle during which atoms are cooled from 300 K to a few hundred nanokelvin is as follows: (i) For 300 s the optical forces from a magneto-optical trap (15) (MOT) collect atoms from a room temperature, $\sim 10^{-11}$ torr vapor (10) of ^{87}Rb atoms; we used a so-called dark MOT (16) to reduce the loss mechanisms of an ordinary MOT, enabling the collection of a large number (10^7) of atoms even under our unusually low pressure conditions (17). (ii) The atom cloud is then quickly compressed and cooled to 20 μK by adjustment of the field gradient and laser frequency (18). (iii) A small magnetic bias field is applied, and a short pulse of circularly polarized laser light optically pumps the magnetic moments of all the atoms so they are parallel with the magnetic field (the $F = 2$, $m_F = 2$ angular momentum state.) (19). (iv) All laser light is removed and a TOP trap is constructed in place around the atoms, the necessary quadrupole and rotating fields being turned on in 1 ms. (v) The quadrupole field component of the TOP trap is then adiabatically ramped up to its maximum value, thereby

increasing the elastic collision rate by a factor of 5.

At this point, we had about 4×10^6 atoms with a temperature of about 90 μK in the trap. The trap has an axial oscillation frequency of about 120 Hz and a cylindrically symmetric radial frequency smaller by a factor of $\sqrt{8}$. The number density, averaged over the entire cloud, is 2×10^{10} cm^{-3}. The elastic collision rate (19) is approximately three per second, which is 200 times greater than the one per 70 s loss rate from the trap.

The sample was then evaporatively cooled for 70 s, during which time both the rf frequency and the magnitude of the rotating field were ramped down, as described (13, 20). The choice of the value of ν_{evap} for the cycle determines the depth of the rf cut and the temperature of the remaining atoms. If ν_{evap} is 3.6 MHz, the rf "scalpel" will have cut all the way into the center of the trap and no atoms will remain. At the end of the rf ramp, we allowed the sample to equilibrate for 2 s (21) and then expanded the cloud to measure the velocity distribution. For technical reasons, this expansion was done in two stages. The trap spring constants were first adiabatically reduced by a factor of 75 and then suddenly reduced to nearly zero so that the atoms essentially expanded ballistically. A field gradient remains that supports the atoms against gravity to allow longer expansion times. Although this approach provides small transverse restoring forces, these are easily taken into account in the analysis. After a 60-ms expansion, the spatial distribution of the

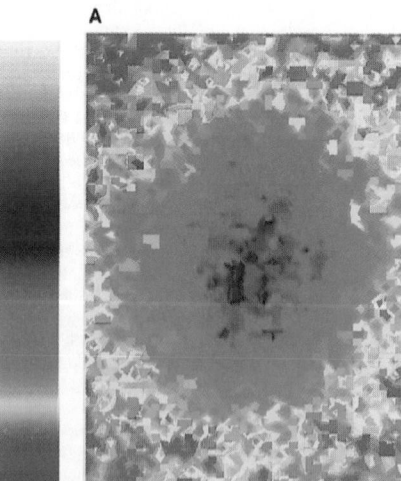
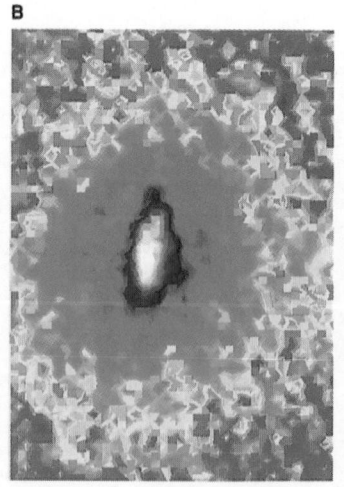
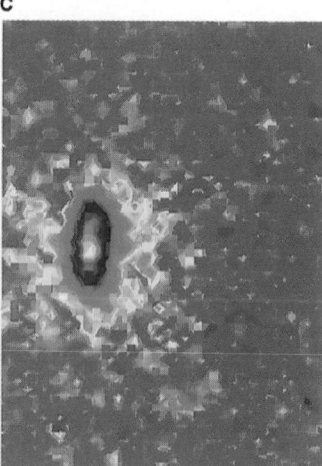

Fig. 2. False-color images display the velocity distribution of the cloud (**A**) just before the appearance of the condensate, (**B**) just after the appearance of the condensate, and (**C**) after further evaporation has left a sample of nearly pure condensate. The circular pattern of the noncondensate fraction (mostly yellow and green) is an indication that the velocity distribution is isotropic, consistent with thermal equilibrium. The condensate fraction (mostly blue and white) is elliptical, indicative that it is a highly nonthermal distribution. The elliptical pattern is in fact an image of a single, macroscopically occupied quantum wave function. The field of view of each image is 200 μm by 270 μm. The observed horizontal width of the condensate is broadened by the experimental resolution.

cloud was determined from the absorption of a 20-μs, circularly polarized laser pulse resonant with the $5S_{1/2}$, $F = 2$ to $5P_{3/2}$, $F = 3$ transition. The shadow of the cloud was imaged onto a charge-coupled device array, digitized, and stored for analysis.

This shadow image (Fig. 2) contains a large amount of easily interpreted information. Basically, we did a 2D time-of-flight measurement of the velocity distribution. At each point in the image, the optical density we observed is proportional to the column density of atoms at the corresponding part of the expanded cloud. Thus, the recorded image is the initial velocity distribution projected onto the plane of the image. For all harmonic confining potentials, including the TOP trap, the spatial distribution is identical to the velocity distribution, if each axis is linearly scaled by the harmonic oscillator frequency for that dimension (22). Thus, from the single image we obtained both the velocity and coordinate-space distributions, and from these we extracted the temperature and central density, in addi-

tion to characterizing any deviations from thermal equilibrium. The measurement process destroys the sample, but the entire load-evaporate-probe cycle can be repeated. Our data represent a sequence of evaporative cycles performed under identical conditions except for decreasing values of ν_{evap}, which gives a corresponding decrease in the sample temperature and an increase in phase-space density.

The discontinuous behavior of thermodynamic quantities or their derivatives is always a strong indication of a phase transition. In Fig. 3, we see a sharp increase in the peak density at a value of ν_{evap} of 4.23 MHz. This increase is expected at the BEC transition. As cooling proceeds below the transition temperature, atoms rapidly accumulate in the lowest energy state of the 3D harmonic trapping potential (23). For an ideal gas, this state would be as near to a singularity in velocity and coordinate space as the uncertainty principle permits.

Thus, below the transition we expect a two-component cloud, with a dense central condensate surrounded by a diffuse, non-condensate fraction. This behavior is clearly displayed in sections taken horizontally through the center of the distributions, as shown in Fig. 4. For values of ν_{evap} above 4.23 MHz, the sections show a single, smooth, Gaussian-like distribution. At 4.23 MHz, a sharp central peak in the distribution begins to appear. At frequencies below 4.23 MHz, two distinct components to the cloud are visible, the smooth broad curve and a narrow central peak, which we identify as the noncondensate and condensate

fractions, respectively. (Figs. 2B and 4). As the cooling progresses (Fig. 4), the noncondensate fraction is reduced until, at a value of ν_{evap} of 4.1 MHz, little remains but a pure condensate containing 2000 atoms.

The condensate first appears at an rf frequency between 4.25 and 4.23 MHz. The 4.25 MHz cloud is a sample of 2×10^4 atoms at a number density of 2.6×10^{12} cm^{-3} and a temperature of 170 nK. This represents a phase-space density ρ_{ps} of 0.3, which is well below the expected value of 2.612. The phase-space density scales as the sixth power of the linear size of the cloud. Thus, modest errors in our size calibration could explain much of this difference. Below the transition, one can estimate an effective phase-space density by simply dividing the number of atoms by the observed volume they occupy in coordinate and velocity space. The result is several hundred, which is much greater than 2.6 and is consistent with a large occupation number of a single state. The temperatures and densities quoted here were calculated for the sample in the unexpanded trap. However, after the adiabatic expansion stage, the atoms are still in good thermal equilibrium, but the temperatures and densities are greatly reduced. The 170 nK temperature is reduced to 20 nK, and the number density is reduced from 2.6×10^{12} cm^{-3} to 1×10^{11} cm^{-3}. There is no obstacle to adiabatically cooling and expanding the cloud further when it is desirable to reduce the atom-atom interactions, as discussed below (24).

A striking feature evident in the images shown in Fig. 2 is the differing axial-to-radial aspect ratios for the two components of the cloud. In the clouds with no condensate ($\nu_{evap} > 4.23$ MHz) and in the noncondensate fraction of the colder clouds, the velocity distribution is isotropic (as evidenced by the circular shape of the yellow to green contour lines in Fig. 2, A and B). But the condensate fraction clearly has a larger velocity spread in the axial direction than in the radial direction (Fig. 2, B and C). This difference in aspect ratios is readily explained and in fact is strong evidence in support of the interpretation that the central peak is a Bose-Einstein condensate. The noncondensate atoms represent a thermal distribution across many quantum wave functions. In thermal equilibrium, velocity distributions of a gas are always isotropic regardless of the shape of the confining potential. The condensate atoms, however, are all described by the same wave function, which will have an anisotropy reflecting that of the confining potential. The velocity spread of the ground-state wave function for a noninteracting Bose gas should be 1.7 ($8^{1/4}$) times larger in the axial direction than in the radial direction. Our observations are in qualitative agreement with this

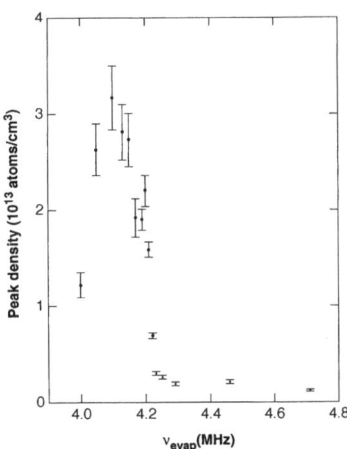

Fig. 3. Peak density at the center of the sample as a function of the final depth of the evaporative cut, ν_{evap}. As evaporation progresses to smaller values of ν_{evap}, the cloud shrinks and cools, causing a modest increase in peak density until ν_{evap} reaches 4.23 MHz. The discontinuity at 4.23 MHz indicates the first appearance of the high-density condensate fraction as the cloud undergoes a phase transition. When a value for ν_{evap} of 4.1 Mhz is reached, nearly all the remaining atoms are in the condensate fraction. Below 4.1 MHz, the central density decreases, as the evaporative "rf scalpel" begins to cut into the condensate itself. Each data point is the average of several evaporative cycles, and the error bars shown reflect only the scatter in the data. The temperature of the cloud is a complicated but monotonic function of ν_{evap}. At $\nu_{evap} = 4.7$ MHz, $T = 1.6$ μK, and for $\nu_{evap} = 4.25$ MHz, $T = 180$ nK.

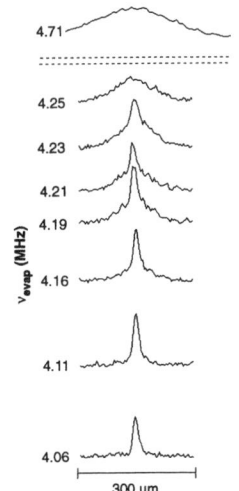

Fig. 4. Horizontal sections taken through the velocity distribution at progressively lower values of ν_{evap} show the appearance of the condensate fraction.

simple picture. This anisotropy rules out the possibility that the narrow peak we see is a result of the enhanced population of all the very low energy quantum states, rather than the single lowest state.

A more quantitative treatment of the observed shape of the condensate shows that the noninteracting gas picture is not completely adequate. We find that the axial width is about a factor of 2 larger than that calculated for a noninteracting ground state and the ratio of the axial to radial velocity spread is at least 50% larger than calculated. However, the real condensate has a self-interaction energy in the mean-field picture of $4\pi n a_0 \hbar^2/m$, which is comparable to the separation between energy levels in the trap. Simple energy arguments indicate that this interaction energy will tend to increase both the size and the aspect ratio to values more in line with what we observed.

Although an atomic vapor of rubidium can only exist as a metastable state at these temperatures, the condensate survives in the unexpanded trap for about 15 s, which is long enough to carry out a wide variety of experiments. The loss rate is probably a result of three-body recombination (25, 26), which could be greatly reduced by adiabatically expanding the condensate after it has formed.

Much of the appeal of our work is that it permits quantitative calculations of microscopic behavior, heuristic understanding of macroscopic behavior, and experimental verification of both. The technique and apparatus described here are well suited for a range of experiments. The basic glass cell design provides flexibility in manipulating and probing the atoms. In addition, it is not difficult to substantially improve it in several ways. First, our position and velocity resolution can be improved with minor changes in optics and in expansion procedures. Second, a double MOT technique that spatially separates the capture and storage of atoms will increase our number of atoms by more than 100 (27). Third, with improvement in measurement sensitivity it should be possible to probe the cloud without destroying it, in order to watch the dynamics in real time.

An abbreviated list of future experiments includes (i) performing optical spectroscopy, including higher order correlation measurements, on the condensate in situ to study how light interacts differently with coherent matter and incoherent matter (28); (ii) comparing the behaviors of ^{87}Rb with ^{85}Rb, which is known to have a negative scattering length (19), potentially making the condensate unstable; (iii) studying time-dependent behavior of the phase transition including the

stability of the supersaturated state; (iv) exploring the specific heat of the sample as it goes through the transition boundary (2) by measuring how condensate and nonconensate fractions evolve during cooling; (v) studying critical opalescence and other fluctuation-driven behavior near the transition temperature; and (vi) carrying out experiments analogous to many of the classic experiments on superfluid helium (2, 29). There is a prediction that the scattering length of heavy alkalis can be modified, and even be made to change sign, by tuning the ambient magnetic field through a scattering resonance (26). Directly modifying the scattering length would provide the ultimate control, but whether or not this is practical, one can still study the properties of the condensate as functions of the strength of the residual interactions because we now have the ability to cross the phase-transition curve over a large range of densities. Thus, it will be possible to observe, and to compare with theoretical prediction, the emergence of nonideal behavior such as singularities in the specific heat and many other phenomena, including those mentioned above.

REFERENCES AND NOTES

1. S. N. Bose, *Z. Phys.* **26**, 178 (1924); A. Einstein, *Sitzungsber. Kgl. Preuss. Akad. Wiss.* **1924**, 261 (1924); *ibid.* **1925**, 3 (1925); A. Griffin, D. W. Snoke, A. Stringari, Eds., *Bose Einstein Condensation* (Cambridge Univ. Press, Cambridge, 1995).
2. K. Huang, *Statistical Mechanics 2nd Edition* (Wiley, New York, 1987).
3. $\lambda_{db} = h/(2\pi mkT)^{1/2}$, where h is Planck's constant, m is the mass of the atom, k is Boltzmann's constant, and T is the temperature.
4. V. Bagnato, D. E. Pritchard, D. Kleppner, *Phys. Rev. A* **35**, 4354 (1987).
5. J.-L. Lin and J. P. Wolfe, *Phys. Rev. Lett.* **71**, 1222 (1993).
6. For reviews of the hydrogen work, see T. J. Greytak and D. Kleppner, in *New Trends in Atomic Physics, Proceedings of the Les Houches Summer School, Session XXXVIII*, Les Houches, France, 2 to 28 June 1993, G. Greenberg and R. Stora, Eds. (North-Holland, Amsterdam, Netherlands, 1984), pp. 1127–1158; I. F. Silvera and J. T. M. Walraven, in *Progress in Low Temperature Physics*, D. Brewer, Ed. (North-Holland, Amsterdam, Netherlands, 1986), vol. 10, pp. 139–173; T. J. Greytak, in *Bose Einstein Condensation*, A. Griffin, D. W. Snoke, A. Stringari, Eds. (Cambridge Univ. Press, Cambridge, 1995), pp. 131–159.
7. N. Masuhara *et al.*, *Phys. Rev. Lett.* **61**, 935 (1988); O. J. Luiten *et al.*, *ibid.* **70**, 544 (1993); H. F. Hess, *Phys. Rev. B* **34**, 3476 (1986).
8. C. Wieman and S. Chu, Eds., *J. Opt. Soc. Am. B* **6** (no. 11) (1989) (special issue on laser cooling and trapping of atoms; in particular, see the "Optical Molasses" section).
9. D. Sesko, T. Walker, C. Monroe, A. Gallagher, C. Wieman, *Phys. Rev. Lett.* **63**, 961 (1989); T. Walker, D. Sesko, C. Wieman, *ibid.* **64**, 408 (1990); D. Sesko, T. Walker, C. Wieman, *J. Opt. Soc. Am. B* **8**, 946 (1991).
10. C. Monroe, W. Swann, H. Robinson, C. Wieman, *Phys. Rev. Lett.* **65**, 1571 (1990).
11. C. Monroe, E. Cornell, C. Wieman, in proceedings of the Enrico Fermi International Summer School on Laser Manipulation of Atoms and Ions, Varenna, Italy, 7 to 21 July 1991, E. Arimondo, W. Phillips, F. Strumia, Eds. (North-Holland, Amsterdam, Netherlands, 1992), pp. 361–377.
12. C. Monroe, E. Cornell, C. Sackett, C. Myatt, C. Wieman, *Phys. Rev. Lett.* **70**, 414 (1993); N. Newbury, C. Myatt, C. Wieman, *Phys. Rev. A* **51**, R2680 (1995).
13. W. Petrich, M. H. Anderson, J. R. Ensher, E. A. Cornell, *Phys. Rev. Lett.* **74**, 3352 (1995).
14. D. Pritchard *et al.*, in *Proceedings of the 11th International Conference on Atomic Physics*, S. Haroche, J. C. Gay, G. Grynberg, Eds. (World Scientific, Singapore, 1989), pp. 619–621. The orbiting zero-field point in the TOP trap supplements the effect of the rf by removing some high-energy atoms by Majorrana transitions. Two other groups have evaporatively cooled alkali atoms [C. S. Adams, H. J. Lee, N. Davidson, M. Kasevich, S. Chu, *Phys. Rev. Lett.* **74**, 3577 (1995); K. B. Davis, M-O. Mewes, M. A. Joffe, M. R. Andres, W. Ketterle, *ibid.*, p. 5202].
15. E. Raab, M. Prentiss, A. Cable, S. Chu, D. E. Pritchard, *Phys. Rev. Lett.* **59**, 2631 (1987).
16. W. Ketterle, K. B. Davis, M. A. Joffe, A. Martin, D. E. Pritchard, *ibid.* **70**, 2253 (1993).
17. M. H. Anderson, W. Peterich, J. R. Ensher, E. A. Cornell, *Phys. Rev. A* **50**, R3597 (1994).
18. W. Petrich, M. H. Anderson, J. R. Ensher, E. A. Cornell, *J. Opt. Soc. Am. B* **11**, 1332 (1994).
19. J. R. Gardner *et al.* [*Phys. Rev. Lett.* **74**, 3764 (1995)] determined the ground-state triplet scattering lengths and found that they are positive for ^{87}Rb and negative for ^{85}Rb. It is believed that a positive scattering length is necessary for the stability of large samples of condensate. The $F = 2, m_F = 2$ state also has the advantage that, of the Rb $5S$ states, it is the spin state with the maximum magnetic trapping force.
20. After the rotating field has been reduced to one-third its initial value, which increases the spring constant by a factor of 3, it is held fixed (at 5 G) and the final cooling is done only with the rf ramp.
21. After the sample is cooled to just below the transition temperature, the condensate peak does not appear immediately after the ramp ends but instead grows during this 2-s delay.
22. This exact correspondence between velocity and coordinate-space distributions requires that the particles be an ideal gas, which is an excellent approximation in our system, except in the condensate itself. It also requires that sinusoidal trajectories of the atoms have random initial phases. This is much less restrictive than requiring thermal equilibrium.
23. Below the transition temperature the fraction of the atoms that go into the condensate is basically set by the requirement that the phase-space density of the noncondensate fraction not exceed 2.612 (for an ideal gas). As the cloud is further cooled or compressed, the excess atoms are squeezed into the condensate (2).
24. The temperature of a classical gas that would correspond to the kinetic energy of the pure condensate cloud (v_{evap} = 4.11 MHz), after adiabatic expansion, is only 2 nK, and during the near-ballistic expansion it becomes substantially lower.
25. E. Tiesinga, A. J. Moerdijk, B. J. Verhaar, H. T. C. Stoof, *Phys. Rev. A* **46**, R1167 (1992).
26. E. Tiesinga, B. J. Verhaar, H. T. C. Stoof, *ibid.* **47**, 4114 (1993).
27. C. Myatt, N. Newbury, C. Wieman, personal communication.
28. L. You, M. Lewenstein, J. Cooper, *Phys. Rev A* **51**, 4712 (1995) and references therein; O. Morice, Y. Castin, J. Dalibard, *ibid.*, p. 3896; J. Javanainen, *Phys. Rev. Lett.* **72**, 2375 (1994); B. V. Svistunov and G. V. Shylapnikov, *JETP* **71**, 71 (1990).
29. P. A. Ruprecht, M. J. Holland, K. Burnett, M. Edwards, *Phys. Rev. A.* **51**, 4704 (1995).
30. During the adiabatic stage of expansion, we already routinely changed the sample density by a factor of 25.
31. We thank K. Coakley, J. Cooper, M. Dowell, J. Doyle, S. Gilbert, C. Greene, M. Holland, D. Kleppner, C. Myatt, N. Newbury, W. Petrich, and B. Verhaar for valuable discussions. This work was supported by National Science Foundation, National Institute of Standards and Technology, and the Office of Naval Research.

26 June 1995; accepted 29 June 1995

PHYSICAL REVIEW
LETTERS

VOLUME 75   27 NOVEMBER 1995   NUMBER 22

Bose-Einstein Condensation in a Gas of Sodium Atoms

K. B. Davis, M.-O. Mewes, M. R. Andrews, N. J. van Druten, D. S. Durfee, D. M. Kurn, and W. Ketterle

Department of Physics and Research Laboratory of Electronics, Massachusetts Institute of Technology,
Cambridge, Massachusetts 02139
(Received 17 October 1995)

We have observed Bose-Einstein condensation of sodium atoms. The atoms were trapped in a novel trap that employed both magnetic and optical forces. Evaporative cooling increased the phase-space density by 6 orders of magnitude within seven seconds. Condensates contained up to 5×10^5 atoms at densities exceeding 10^{14} cm^{-3}. The striking signature of Bose condensation was the sudden appearance of a bimodal velocity distribution below the critical temperature of $\sim 2~\mu$K. The distribution consisted of an isotropic thermal distribution and an elliptical core attributed to the expansion of a dense condensate.

PACS numbers: 03.75.Fi, 05.30.Jp, 32.80.Pj, 64.60.–i

Bose-Einstein condensation (BEC) is a ubiquitous phenomenon which plays significant roles in condensed matter, atomic, nuclear, and elementary particle physics, as well as in astrophysics [1]. Its most striking feature is a macroscopic population of the ground state of the system at finite temperature [2]. The study of BEC in weakly interacting systems holds the promise of revealing new macroscopic quantum phenomena that can be understood from first principles, and may also advance our understanding of superconductivity and superfluidity in more complex systems.

During the past decade, work towards BEC in weakly interacting systems has been carried forward with excitons in semiconductors and cold trapped atoms. BEC has been observed in excitonic systems, but a complete theoretical treatment is lacking [1,3]. The pioneering work towards BEC in atomic gases was performed with spin-polarized atomic hydrogen [4,5]. Following the development of evaporative cooling [6], the transition was approached within a factor of 3 in temperature [7]. Laser cooling provides an alternative approach towards very low temperatures, but has so far been limited to phase-space densities typically 10^5 times lower than required for BEC. The combination of laser cooling with evaporative cooling [8–10] was the prerequisite for obtaining BEC in alkali atoms. This year, within a few months, three independent and different approaches succeeded in creating BEC in

rubidium [11], lithium [12], and, as reported in this paper, in sodium. Our results are distinguished by a production rate of Bose-condensed atoms which is 3 orders of magnitude larger than in the two previous experiments. Furthermore, we report a novel atom trap that offers a superior combination of tight confinement and capture volume and the attainment of unprecedented densities of cold atomic gases.

Evaporative cooling requires an atom trap which is tightly confining and stable. So far, magnetic traps and optical dipole traps have been used. Optical dipole traps provide tight confinement, but have only a very small trapping volume (10^{-8} cm^3). The tightest confinement in a magnetic trap is achieved with a spherical quadrupole potential (linear confinement); however, atoms are lost from this trap due to nonadiabatic spin flips as the atoms pass near the center, where the field rapidly changes direction. This region constitutes a "hole" in the trap of micrometer dimension. The recently demonstrated "TOP" trap suppresses this trap loss, but at the cost of lower confinement [8].

We suppressed the trap loss by adding a repulsive potential around the zero of the magnetic field, literally "plugging" the hole. This was accomplished by tightly focusing an intense blue-detuned laser that generated a repulsive optical dipole force. The optical plug was created by an Ar$^+$-laser beam (514 nm) of 3.5 W focused

0031-9007/95/75(22)/3969(5)$06.00

VOLUME 75, NUMBER 22   PHYSICAL REVIEW LETTERS   27 NOVEMBER 1995

to a beam waist of 30 μm. This caused 7 MHz (350 μK) of light shift potential at the origin. Heating due to photon scattering was suppressed by using far-off-resonant light, and by the fact that the atoms are repelled from the region where the laser intensity is highest.

The experimental setup was similar to that described in our previous work [9]. Typically, within 2 s 10^9 atoms in the $F = 1$, $m_F = -1$ state were loaded into a magnetic trap with a field gradient of 130 G/cm; the peak density was $\sim 10^{11}$ cm^{-3}, the temperature ~ 200 μK, and the phase-space density 10^6 times lower than required for BEC. The lifetime of the trapped atoms was $\sim$ 30 s, probably limited by background gas scattering at a pressure of $\sim 1 \times 10^{-11}$ mbar.

The magnetically trapped atoms were further cooled by rf-induced evaporation [8,9,13]. rf-induced spin flips were used to selectively remove the higher-energy atoms from the trap resulting in a decrease in temperature for the remaining atoms. The total (dressed-atom) potential is a combination of the magnetic quadrupole trapping potential, the repulsive potential of the plug, and the effective energy shifts due to the rf. At the point where atoms are in resonance with the rf, the trapped state undergoes an avoided crossing with the untrapped states (corresponding to a spin flip), and the trapping potential bends over. As a result, the height of the potential barrier varies linearly with the rf frequency. The total potential is depicted in Fig. 1. Over 7 s, the rf frequency was swept from 30 MHz to the final value around 1 MHz, while the field gradient was first increased to 550 G/cm (to enhance the initial elastic-collision rate) and then lowered to 180 G/cm (to avoid the losses due to inelastic processes at the final high densities).

Temperature and total number of atoms were determined using absorption imaging. The atom cloud was imaged either while it was trapped or following a sudden switch-off of the trap and a delay time of 6 ms. Such time-of-flight images displayed the velocity distribution of the trapped cloud. For probing, the atoms were first pumped to the $F = 2$ state by switching on a 10 mW/cm^2 laser beam in resonance with the $F = 1 \rightarrow F = 2$ transition. 10 μs later the atoms were concurrently exposed to a 100 μs, 0.25 mW/cm^2 probe laser pulse in resonance with the $F = 2 \rightarrow F = 3$ transition, propagating along the trap's y direction. This probe laser beam was imaged onto a charge-coupled device sensor with a lens system having a resolution of 8 μm. Up to 100 photons per atom were absorbed without blurring the image due to heating.

At temperatures above 15 μK the observed trapped clouds were elliptical with an aspect ratio of 2:1 due to the symmetry of the quadrupole field. At the position of the optical plug they had a hole, which was used for fine alignment. A misalignment of the optical plug by ~ 20 μm resulted in increased trap loss and prevented us from cooling below 50 μK. This is evidence that the Majorana spin flips are localized in a very small region around the center of the trap. At temperatures below 15 μK, the cloud separated into two pockets at the two minima in the potential of Fig. 1. The bottom of the potential can be approximated as a three-dimensional anisotropic harmonic oscillator potential with frequencies $\omega_y^2 = \mu B'/(2mx_0)$, $\omega_z^2 = 3\omega_y^2$, $\omega_x^2 = \omega_y^2[(4x_0^2/w_0^2) - 1]$, where μ is the atom's magnetic moment, m the mass, B' the axial field gradient, w_0 the Gaussian beam waist parameter ($1/e^2$ radius) of the optical plug, and x_0 the distance of the potential minimum from the trap center. x_0 was directly measured to be 50 μm by imaging the trapped cloud, w_0 (30 μm) was determined from x_0, the laser power (3.5 W), and B' (180 G/cm). With these values the oscillation frequencies are 235, 410, and 745 Hz in the y, z, and x directions, respectively.

When the final rf frequency ν_{rf} was lowered below 0.7 MHz, a distinctive change in the symmetry of the velocity distribution was observed [Figs. 2(a) and 2(b)]. Above this frequency the distribution was perfectly spherical as expected for a thermal uncondensed cloud [14]. Below the critical frequency, the velocity distribution contained an elliptical core which increased in intensity when

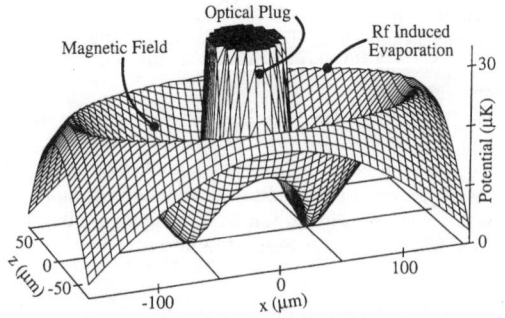

FIG. 1. Adiabatic potential due to the magnetic quadrupole field, the optical plug, and the rf. This cut of the three-dimensional potential is orthogonal to the propagation direction (y) of the blue-detuned laser. The symmetry axis of the quadrupole field is the z axis.

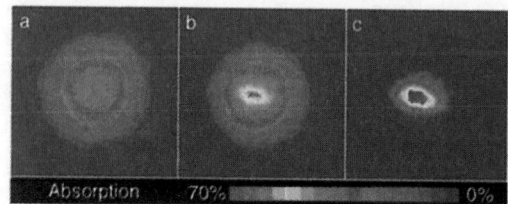

FIG. 2 (color). Two-dimensional probe absorption images, after 6 ms time of flight, showing evidence for BEC. (a) is the velocity distribution of a cloud cooled to just above the transition point, (b) just after the condensate appeared, and (c) after further evaporative cooling has left an almost pure condensate. (b) shows the difference between the isotropic thermal distribution and an elliptical core attributed to the expansion of a dense condensate. The width of the images is 870 μm. Gravitational acceleration during the probe delay displaces the cloud by only 0.2 mm along the z axis.

VOLUME 75, NUMBER 22   PHYSICAL REVIEW LETTERS   27 NOVEMBER 1995

the rf was further swept down, whereas the spherical cloud became less intense. We interpret the elliptical cloud as due to the Bose condensate, and the spherical cloud as due to the normal fraction.

In the region just below the transition frequency one expects a bimodal velocity distribution: a broad distribution due to the normal gas and a narrow distribution due to the condensate. The cross sections of the time-of-flight images (Fig. 3) indeed show such bimodal distributions in this region. Figure 4 shows how suddenly the time-of-flight image changes below $\nu_{rf} = 0.7$ MHz. The effective area of the observed cloud becomes very small [Fig. 4(a)], while the velocity distribution is no longer Gaussian [Fig. 4(b)] and requires different widths for the condensate and the normal fraction [Fig. 4(c)].

At the critical frequency, a temperature of (2.0 ± 0.5) μK was derived from the time-of-flight image. An independent, though less accurate estimate of the temperature T is obtained from the dynamics of evaporative cooling. Efficient evaporation leads to a temperature which is about 10 times smaller than the depth of the trapping potential [15]. Since the speed of evaporation depends exponentially on the ratio of potential depth to temperature, we expect this estimate of $T = 2$ μK to be accurate to within a factor of 2.

The critical number of atoms N_c to achieve BEC is determined by the condition that the number of atoms per cubic thermal de Broglie wavelength exceeds 2.612 at the bottom of the potential [2]. For a harmonic oscillator potential this is equivalent to $N_c = 1.202(k_B T)^3/\hbar^3 \omega_x \omega_y \omega_z$ [16]. For our trap and 2.0 μK, $N_c = 2 \times (1.2 \times 10^6)$, where the factor of 2 accounts for the two separated clouds. The predicted value for N_c depends on the sixth power of the width of a time-of-flight image and is only accurate to within a factor of 3. We determined the number of atoms by integrating over the absorption image. At the transition point, the measured number of 7×10^5 agrees with the prediction for N_c. The critical peak density n_c at 2.0 μK is 1.5×10^{14} cm^{-3}. Such a high density appears to be out of reach for laser cooling, and demonstrates that evaporative cooling is a powerful technique to obtain not only ultralow temperatures, but also extremely high densities.

An ideal Bose condensate shows a macroscopic population of the ground state of the trapping potential. This picture is modified for a weakly interacting Bose gas. The mean-field interaction energy is given by $n\tilde{U}$, where n is the density and $\tilde{U}$ is proportional to the scattering length a: $\tilde{U} = 4\pi\hbar^2 a/m$ [2]. Using our recent experimental result $a = 4.9$ nm [9], $\tilde{U}/k_B = 1.3 \times 10^{-21}$ K cm^3. At the transition point, $n_c\tilde{U}/k_B T_c = 0.10$. Consequently, above the transition point, the kinetic energy dominates over the interaction energy, and the velocity

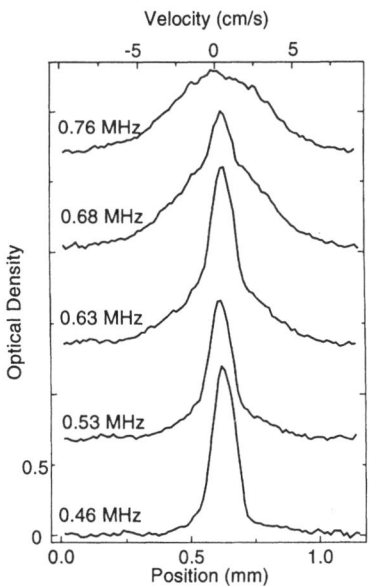

FIG. 3. Optical density as a function of position along the z axis for progressively lower values of the final rf frequency. These are vertical cuts through time-of-flight images like those in Fig. 2. For $\nu_{rf} < 0.7$ MHz, they show the bimodal velocity distributions characteristic of the coexistence of a condensed and uncondensed fraction. The top four plots have been offset vertically for clarity.

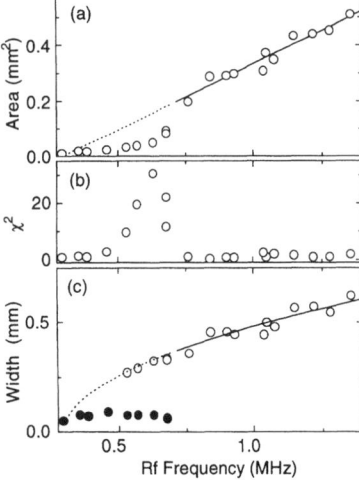

FIG. 4. Further evidence for a phase transition is provided by the sudden change of observed quantities as the final rf frequency ν_{rf} is varied. (a) Area of the cloud in the time-of-flight image versus ν_{rf}. The area was obtained as the ratio of the integrated optical density and the peak optical density. The area changes suddenly at $\nu_{rf} = 0.7$ MHz. Below the same frequency, the velocity distributions (Fig. 3) cannot be represented by a single Gaussian, as demonstrated by the χ^2 for a single Gaussian fit (b), and required different widths for the condensate (full circles) and noncondensate fraction (c). In (a) and (c) the lines reflect the behavior of a classical gas with a temperature proportional to the trap depth.

VOLUME 75, NUMBER 22   PHYSICAL REVIEW LETTERS   27 NOVEMBER 1995

distribution after sudden switch-off of the trap is isotropic. For the condensate, however, the situation is reversed. As we will confirm below, the kinetic energy of the condensate is negligible compared to its interaction energy [17]. Furthermore, well below the transition point, the interaction with the noncondensate fraction can be neglected. In such a situation, the solution of the nonlinear Schrödinger equation reveals that the condensate density $n_0(\mathbf{r})$ is a mirror image of the trapping potential $V(\mathbf{r})$: $n_0(\mathbf{r}) = n_0(\mathbf{0}) - V(\mathbf{r})/\tilde{U}$, as long as this expression is positive, otherwise $n_0(\mathbf{r})$ vanishes (see, e.g., Refs. [5,18]). For a harmonic potential, one obtains the peak density $n_0(\mathbf{0})$ for N_0 atoms in the condensate $n_0(\mathbf{0}) = 0.118(N_0 m^3 \omega_x \omega_y \omega_z / \hbar^3 a^{3/2})^{2/5}$.

Typically, we could cool one-fourth of the atoms at the transition point into a pure condensate. For an observed $N_0 = 1.5 \times 10^5$, we expect the condensate to be 2 times more dense than the thermal cloud at the transition point, and about 6 times larger than the ground-state wave function. The kinetic energy within the condensate is $\sim \hbar^2/(2mR^2)$, where R is the size of the condensate [18], while the internal energy is $2n_0\tilde{U}/7$. Thus the kinetic energy of the condensed atoms is around 1 nK, much less than the zero-point energy of our trap (35 nK) and the calculated internal energy of 120 nK. This estimate is consistent with our initial assumption that the kinetic energy can be neglected compared to the interaction energy.

The internal energy is ~ 25 times smaller than the thermal energy $(3/2)k_BT_c$ at T_c. Consequently, the width of the time-of-flight image of the condensate is expected to be about 5 times smaller than at the transition point. This is close to the observed reduction in the width shown in Fig. 4(c). This agreement might be fortuitous because we have so far neglected the anisotropy of the expansion, but it indicates that we have observed the correct magnitude of changes which are predicted to occur at the BEC transition of a weakly interacting gas. In several cooling cycles, as many as 5×10^5 condensed atoms were observed; we estimate the number density in these condensates to be 4×10^{14} cm^{-3}.

A striking feature of the condensate is the nonisotropic velocity distribution [11,19]. This is caused by the "explosion" of the condensate due to repulsive forces which are proportional to the density gradient. The initial acceleration is therefore inversely proportional to the width of the condensate resulting in an aspect ratio of the velocity distribution, which is inverted compared to the spatial distribution. When we misaligned the optical plug vertically, the shape of the cloud changed from two vertical crescents to a single elongated horizontal crescent. The aspect ratio of the time-of-flight image of the condensate correspondingly changed from horizontal to vertical elongation. In contrast, just above the transition point, the velocity distribution was found to be spherical and insensitive to the alignment of the plug. However, we cannot account quantitatively for the observed distributions because we have two separated condensates which overlap in the time-of-flight images, and also because of some residual horizontal acceleration due to the switch-off of the trap, which is negligible for the thermal cloud, but not for the condensate [20].

The lifetime of the condensate was about 1 s. This lifetime is probably determined either by three-body recombination [21] or by the heating rate of 300 nK/s, which was observed for a thermal cloud just above T_c. This heating rate is much higher than the estimated 8 nK/s for the off-resonant scattering of green light and may be due to residual beam jitter of the optical plug.

In conclusion, we were able to Bose-condense 5×10^5 sodium atoms within a total loading and cooling cycle of 9 s. During evaporative cooling, the elastic collision rate increased from 30 Hz to 2 kHz resulting in a mean free path comparable to the dimensions of the sample. Such collisionally dense samples are the prerequisite for studying various transport processes in dense ultracold matter. Furthermore, we have reached densities in excess of 10^{14} cm^{-3}, which opens up new possibilities for studying decay processes like dipolar relaxation and three-body recombination, and for studying a weakly interacting Bose gas over a broad range of densities and therefore strengths of interaction.

We are grateful to E. Huang and C. Sestok for important experimental contributions, to M. Raizen for the loan of a beam pointing stabilizer, and to D. Kleppner for helpful discussions. We are particularly grateful to D.E. Pritchard, who not only contributed many seminal ideas to the field of cold atoms, but provided major inspiration and equipment to W.K. This work was supported by ONR, NSF, JSEP, and the Sloan Foundation. M.-O.M., K.B.D., and D.M.K. would like to acknowledge support from Studienstiftung des Deutschen Volkes, MIT Physics Department Lester Wolfe fellowship, and NSF Graduate Research Fellowship, respectively, and N.J.v.D. from "Nederlandse Organisatie voor Wetenschappelijk Onderzoek (NWO)" and NACEE (Fulbright fellowship).

[1] A. Griffin, D.W. Snoke, and S. Stringari, *Bose-Einstein Condensation* (Cambridge University Press, Cambridge, 1995).

[2] K. Huang, *Statistical Mechanics* (Wiley, New York, 1987), 2nd ed.

[3] J.L. Lin and J.P. Wolfe, Phys. Rev. Lett. **71**, 1222 (1993).

[4] I.F. Silvera and M. Reynolds, J. Low Temp. Phys. **87**, 343 (1992); J.T.M. Walraven and T.W. Hijmans, Physica (Amsterdam) **197B**, 417 (1994).

[5] T. Greytak, in Ref. [1], p. 131.

[6] N. Masuhara *et al.*, Phys. Rev. Lett. **61**, 935 (1988).

[7] J. Doyle *et al.*, Phys. Rev. Lett. **67**, 603 (1991).

[8] W. Petrich, M.H. Anderson, J.R. Ensher, and E.A. Cornell, Phys. Rev. Lett. **74**, 3352 (1995).

VOLUME 75, NUMBER 22   PHYSICAL REVIEW LETTERS   27 NOVEMBER 1995

[9] K. B. Davis *et al.*, Phys. Rev. Lett. **74**, 5202 (1995).

[10] C. S. Adams *et al.*, Phys. Rev. Lett. **74**, 3577 (1995).

[11] M. H. Anderson *et al.*, Science **269**, 198 (1995).

[12] C. C. Bradley, C. A. Sackett, J. J. Tollett, and R. G. Hulet, Phys. Rev. Lett. **75**, 1687 (1995).

[13] D. E. Pritchard, K. Helmerson, and A. G. Martin, in *Atomic Physics 11,* edited by S. Haroche, J. C. Gay, and G. Grynberg (World Scientific, Singapore, 1989), p. 179.

[14] The measured $1/e$ decay time for the magnet current is 100 μs, shorter than the ω^{-1} of the fastest oscillation in the trap (210 μs). We therefore regard the switch-off as sudden. Any adiabatic cooling of the cloud during the switch-offtibs would result in a nonspherical velocity distribution due to the anisotropy of the potential.

[15] K. B. Davis, M.-O. Mewes, and W. Ketterle, Appl. Phys. B **60**, 155 (1995).

[16] V. Bagnato, D. E. Pritchard, and D. Kleppner, Phys. Rev. A **35**, 4354 (1987). This formula is derived assuming $k_B T_c \gg \hbar \omega_{x,y,z}$, which is the case in our experiment.

[17] Note that already for about 200 atoms in the ground state, the interaction energy in the center of the condensate equals the zero-point energy.

[18] G. Baym and C. Pethick (to be published).

[19] M. Holland and J. Cooper (to be published).

[20] These effects do not affect the vertical velocity distributions shown in Fig. 3.

[21] A. J. Moerdijk, H. M. J. M. Boesten, and B. J. Verhaar, Phys. Rev. A (to be published).

RESEARCH ARTICLE

RESEARCH ARTICLE

Observation of Interference Between Two Bose Condensates

M. R. Andrews, C. G. Townsend, H.-J. Miesner, D. S. Durfee, D. M. Kurn, W. Ketterle

Interference between two freely expanding Bose-Einstein condensates has been observed. Two condensates separated by ~40 micrometers were created by evaporatively cooling sodium atoms in a double-well potential formed by magnetic and optical forces. High-contrast matter-wave interference fringes with a period of ~15 micrometers were observed after switching off the potential and letting the condensates expand for 40 milliseconds and overlap. This demonstrates that Bose condensed atoms are "laser-like"; that is, they are coherent and show long-range correlations. These results have direct implications for the atom laser and the Josephson effect for atoms.

The realization of Bose-Einstein condensation (BEC) in dilute atomic gases has created great interest in this new form of matter. One of its striking features is a macroscopic population of the quantum-mechanical ground state of the system at finite temperature. The Bose condensate is characterized by the absence of thermal excitation; its kinetic energy is solely the result of zero-point motion in the trapping potential (in general, modified by the repulsive interaction between atoms). This is the property that has been used to detect and study the Bose condensate in previous experiments. The Bose-Einstein phase transition was observed by the sudden appearance of a "peak" of ultracold atoms, either in images of ballistically expanding clouds (time-of-flight pictures) (1–3) or as a dense core inside the magnetic trap (4, 5). The anisotropic expansion of the cloud (1–3) and the appearance of collective excitations at frequencies different from multiples of the trapping frequencies (6, 7) were found to be in quantitative agreement with the predictions of the mean-field theory for a weakly interacting Bose gas (8–11). However, similar anisotropic expansion and excitation frequencies have been predicted for a dense classical gas in the hydrodynamic regime (12, 13) and are therefore not distinctive features of BEC. Indeed, the nonlinear Schrödinger equation is equivalent to a hydrodynamic equation for superfluid flow, which, in many situations, is very similar to a classical hydrodynamic equation (9, 13, 14). Previous BEC studies have mainly concerned the "very cold" nature of the Bose condensate but have not revealed properties that directly reflect its coherent nature, such as its phase, order parameter (macroscopic wave function), or long-range order. In superconductors, the phase of the order parameter was directly observed through the Josephson effect, whereas in superfluid helium the observation of the motion of quantized vortices (15) provided indirect evidence.

The coherence of a Bose condensate has been the subject of many theoretical studies. Kagan and collaborators predicted that the Bose condensate will form first as a quasi-condensate consisting of very cold atoms but lacking long-range order, which is only established on a much longer time

The authors are in the Department of Physics and Research Laboratory of Electronics, Massachusetts Institute of Technology, Cambridge, MA 02139, USA.

scale (16). Stoof predicted that a coherent condensate would form immediately (17). Several groups discussed interference experiments and quantum tunneling for condensates (18–29). If the condensate is initially in a state of well-defined atom number, its order parameter, which is the macroscopic wave function, vanishes. However, the quantum measurement process should still lead to quantum interference and "create" the phase of the condensate (20, 23–25, 27, 28), thus breaking the global gauge invariance that reflects particle number conservation (30). This is analogous to Anderson's famous gedanken experiment, testing whether two initially separated buckets of superfluid helium would show a fixed value of the relative phase—and therefore a Josephson current—once they are connected (31).

Arguments for and against such a fixed relative phase have been given (31, 32). Even if this phase exists, there has been some doubt as to whether it can be directly measured, because it was predicted to be affected by collisions during ballistic expansion (12, 26) or by phase diffusion resulting from the mean field of Bose condensed atoms (21, 25, 27, 33). Additionally, the phase of the condensate plays a crucial role in discussions of an atom laser, a source of coherent matter waves (34–37).

The phase of a condensate is the argument of a complex number (the macroscopic wave function) and is not an observable. Only the relative phase between two condensates can be measured. Here, we report on the observation of high-contrast interference between two atomic Bose condensates, which is clear evidence for coherence in such systems.

The experimental setup. Two Bose condensates were produced using a modification of our previous setup (3, 7). Sodium atoms were optically cooled and trapped and were then transferred into a double-well potential. The atoms were further cooled by radio frequency (rf)–induced evaporation (38). The condensates were confined in a cloverleaf magnetic trap (3), with the trapping potential determined by the axial curvature of the magnetic field B'' = 94 G cm^{-2}, the radial gradient B' = 120 G cm^{-1}, and the bias field B_0 = 0.75 G. The atom clouds were cigar-shaped, with the long axis horizontal. A double-well potential was created by focusing blue-detuned far-off-resonant laser light into the center of the magnetic trap, generating a repulsive optical dipole force. Because of the far detuning of the argon ion laser line at 514 nm relative to the sodium resonance at 589 nm, heating from spontaneous emission was negligible. This laser beam was focused into a light sheet with a cross section of 12 μm

by 67 μm ($1/e^2$ radii), with its long axis perpendicular to the long axes of the condensates. The argon ion laser beam propagated nearly collinearly with the vertical probe beam. We aligned the light sheet by imaging the focused argon ion laser beam with the same camera used to image the condensates.

Evaporative cooling was extended well below the transition temperature to obtain condensates without a discernible normal fraction. Condensates containing 5×10^6 sodium atoms in the $F = 1$, $m_F = -1$ ground state were produced within 30 s. The presence of the laser-light sheet neither changed the number of condensed atoms from our previous work (3) nor required a modification of the evaporation path; hence, problems with heating encountered earlier with an optically plugged magnetic trap (2) were purely technical. In the present application, the argon ion laser beam was not needed to avoid a loss process, and thus we had complete freedom in the choice of laser power and focal parameters.

The double condensate was directly observed by nondestructive phase-contrast imaging (Fig. 1A). This technique is an extension of our previous work on dispersive imaging (4) and greatly improved the signal-to-noise ratio. The probe light frequency was far detuned from a resonant transition (1.77 GHz to the red), and thus absorption was negligible. Images were formed by photons scattered coherently in the forward direction. The phase modulation caused by the condensate was transformed into an intensity modulation at the

camera by retarding the transmitted probe beam by a quarter-wave with a phase plate in the Fourier plane. Previously, the transmitted probe beam was blocked by a thin wire (dark-ground imaging).

Interference between the condensates was observed by simultaneously switching off the magnetic trap and the argon ion laser-light sheet. The two expanding condensates overlapped and were observed by absorption imaging. After 40 ms time-of-flight, an optical pumping beam transferred the atoms from the $F = 1$ hyperfine state to the $F = 2$ state. With a 10-μs delay, the atoms were exposed to a short (50 μs) circularly polarized probe beam resonant with the $F = 2 \rightarrow F' = 3$ transition and absorbed ~20 photons each. Under these conditions, the atoms moved ~5 μm horizontally during the exposure.

Absorption imaging usually integrates along the line of sight and therefore has only two-dimensional spatial resolution. Because the depth of field for 15-μm fringes is comparable to the size of an expanded cloud, and because the fringes are in general not parallel to the axis of the probe light, line-of-sight integration would cause considerable blurring. We avoided this problem and achieved three-dimensional resolution by restricting absorption of the probe light to a thin horizontal slice of the cloud. The optical pumping beam was focused into a light sheet of adjustable thickness (typically 100 μm) and a width of a few millimeters; this pumping beam propagated perpendicularly to the probe light and parallel to the long axis of the trap (39). As a result, the

Fig. 1. (A) Phase-contrast images of a single Bose condensate (left) and double Bose condensates, taken in the trap. The distance between the two condensates was varied by changing the power of the argon ion laser-light sheet from 7 to 43 mW. **(B)** Phase-contrast image of an originally double condensate, with the lower condensate eliminated.

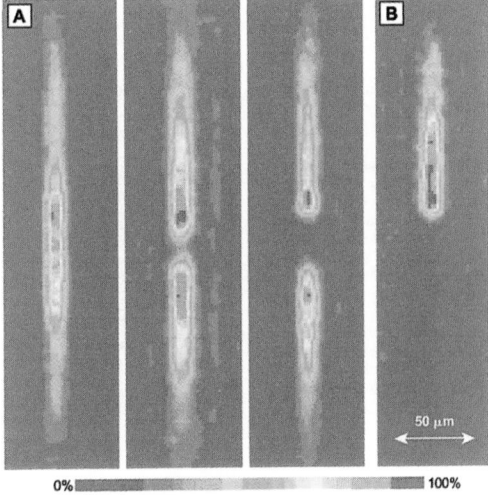

0%   Intensity (arbitrary units)   100%

probe light was only absorbed by a thin slice of the cloud where the atoms were optically pumped. Because high spatial resolution was required from only the fraction of atoms residing in the slice, a good signal-to-noise ratio required condensates with millions of atoms.

Interference between two Bose condensates. In general, the pattern of interference fringes differs for continuous and pulsed sources. Two point-like monochromatic continuous sources would produce curved (hyperbolic) interference fringes. In contrast, two point-like pulsed sources show straight interference fringes; if d is the separation between two point-like condensates, then their relative speed at any point in space is d/t, where t is the delay between pulsing on the source (switching off the trap) and observation. The fringe period is the de Broglie wavelength λ associated with the relative motion of atoms with mass m,

$$\lambda = \frac{ht}{md} \tag{1}$$

where h is Planck's constant. The amplitude and contrast of the interference pattern depends on the overlap between the two condensates.

The interference pattern of two condensates after 40 ms time-of-flight is shown in Fig. 2. A series of measurements with fringe spacings of ~15 μm showed a contrast varying between 20 and 40%. When the imaging system was calibrated with a standard optical test pattern, we found ~40% contrast at the same spatial frequency. Hence, the contrast of the atomic interference was between 50 and 100%. Because the condensates are much larger than the observed fringe spacing, they must have a high degree of spatial coherence.

We observed that the fringe period became smaller for larger powers of the argon ion laser-light sheet (Fig. 3A). Larger power increased the distance between the two condensates (Fig. 1A). From phase-contrast images, we determined the distance d between the density maxima of the two condensates versus argon ion laser power. The fringe period versus maxima separation (Fig. 3B) is in reasonable agreement with the prediction of Eq. 1, although this equation strictly applies only to two point sources. Wallis et al. (26) calculated the interference pattern for two extended condensates in a harmonic potential with a Gaussian barrier. They concluded that Eq. 1 remains valid for the central fringes if d is replaced by the geometric mean of the separation of the centers of mass and the distance between the density maxima of the two condensates. This prediction is also shown in Fig. 3B. The agreement is satisfactory given our experimental uncertainties in the determination of the maxima separations (~3 μm) and of the center-of-mass separations (~20%). We conclude that the numerical simulations for extended interacting condensates (26) are consistent with the observed fringe periods.

We performed a series of tests to support our interpretation of matter-wave interference. To demonstrate that the fringe pattern was caused by two condensates, we compared it with the pattern from a single condensate (this is equivalent to performing a double-slit experiment and covering one of the slits). One condensate was illuminated with a focused beam of weak resonant light 20 ms before release, causing it to disappear almost completely as a result of optical pumping to untrapped states and evaporation after heating by photon recoil (Fig. 1B).

The resulting time-of-flight image did not exhibit interference, and the profile of a single expanded condensate matched one side of the profile of a double condensate (Fig. 4). The profile of a single expanded condensate showed some coarse structure, which most likely resulted from the nonparabolic shape of the confining potential. We found that the structure became more pronounced when the focus of the argon ion laser had some weak secondary intensity maxima. In addition, the interference between two condensates disappeared when the argon ion laser-light sheet was left on for

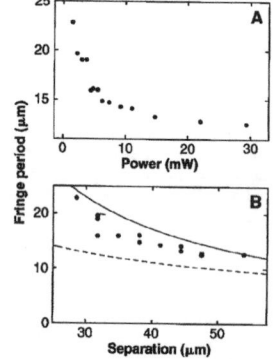

Fig. 3. (**A**) Fringe period versus power in the argon ion laser-light sheet. (**B**) Fringe period versus observed spacing between the density maxima of the two condensates. The solid line is the dependence given by Eq. 1, and the dashed line is the theoretical prediction of (26) incorporating a constant center-of-mass separation of 96 μm, neglecting the small variation (±10%) with laser power.

Fig. 2. Interference pattern of two expanding condensates observed after 40 ms time-of-flight, for two different powers of the argon ion laser-light sheet (raw-data images). The fringe periods were 20 and 15 μm, the powers were 3 and 5 mW, and the maximum absorptions were 90 and 50%, respectively, for the left and right images. The fields of view are 1.1 mm horizontally by 0.5 mm vertically. The horizontal widths are compressed fourfold, which enhances the effect of fringe curvature. For the determination of fringe spacing, the dark central fringe on the left was excluded.

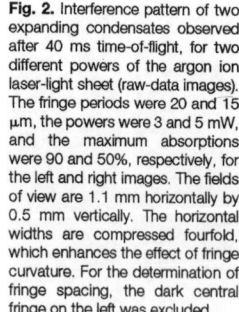

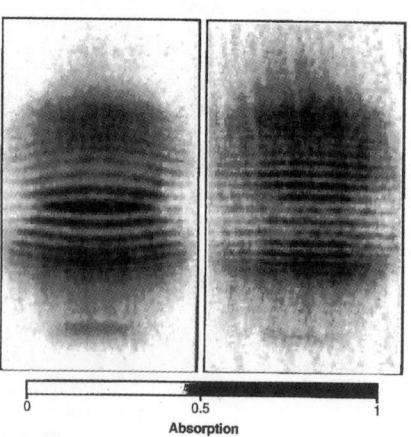

Fig. 4. Comparison between time-of-flight images for a single and double condensate, showing vertical profiles through time-of-flight pictures similar to Fig. 2. The solid line is a profile of two interfering condensates, and the dotted line is the profile of a single condensate, both released from the same double-well potential (argon ion laser power, 14 mW; fringe period, 13 μm; time of flight, 40 ms). The profiles were horizontally integrated over 450 μm. The dashed profile was multiplied by a factor of 1.5 to account for fewer atoms in the single condensate, most likely the result of loss during elimination of the second half.

2 ms after the magnetic trap was switched off. The absorption images showed that the two condensates were pushed apart and did not subsequently overlap.

Another test confirmed that the fringes were not attributable to density waves of two colliding condensates. Because the interference pattern depends on the phases of the condensates, the fringes should be sensitive to perturbations that strongly affect the phase but weakly affect the motion. Applying resonant rf radiation during the expansion of the two condensates caused a reduction of the fringe contrast by up to a factor of 4. The greatest reduction in contrast was found when the rf was swept 25 times between 0 and 300 kHz at 1 kHz. When a single condensate was exposed to the same rf radiation, no clearly discernible differences in the time-of-flight pictures were found. A possible explanation for the reduced fringe contrast is that frequent sweeps through the resonance in slightly inhomogeneous dc and rf magnetic fields created atoms in different superpositions of hyperfine states that only partially interfered.

The visibility of the fringes depended critically on several imaging parameters, as expected for the observation of such a finely striated structure. The fringes became almost invisible when the thickness of the optical pumping sheet was increased to 800 μm, whereas the focus of the imaging system could be varied over a wider range of up to ±1 mm without losing contrast. This implies that the fringes were at a small angle (~20 mrad) with respect to the probe beam.

The interference was remarkably robust. The fringes were very regular, although no attempt was made to control residual magnetic fields during the expansion. The high contrast implies that neither phase diffusion during expansion nor collisions with normal atoms were important. The latter aspect was studied in more detail when the rf evaporation was stopped at higher temperatures. We still observed fringes of identical contrast (40), but with decreasing amplitude because of the smaller number of condensed atoms. At the transition temperature, the fringes and the condensate disappeared.

We now consider whether the two condensates were truly independent. When the power of the argon ion laser was varied, we realized both well-separated and connected condensates. The chemical potential of the Bose condensates was ~4 kHz. The height of the barrier created by the argon ion laser was estimated to be ~2 kHz per milliwatt of power. At 100-mW laser power, the barrier height was 10 μK, resulting in a cloud that was already split well above the phase transition temperature of ~2 μK. The tunneling time of well-separated condensates was estimated to be greater than the age of the

universe (19), and thus our experiment should be equivalent to Anderson's gedanken experiment ("What is the relative phase of two buckets of liquid helium?") (31) and also to an interference experiment between two independent lasers (41). Two independent condensates will show high-contrast interference fringes with a phase that varies between experiments (20, 23, 24, 27). In our experiment, however, even a fixed relative phase would have been detected as being random because of mechanical instabilities on a 10-μm scale. Once it becomes possible to distinguish between fixed and random phases, we should be able to investigate how phase coherence is established and lost. One possible experiment would be to adiabatically switch on the argon ion laser after condensation, thus splitting a single condensate, and to study how a definite phase becomes random as a function of time.

For argon ion laser powers below 4 mW, the interference pattern was slightly curved and symmetric about a central fringe that was always dark (Fig. 2). We conjecture that for small separations, the two condensates overlap very early during the expansion and interactions between them are not negligible. When the power of the laser-light sheet was lowered further, the number of fringes decreased, while the central dark feature persisted and eventually lost contrast. For such low powers we were in the regime where the condensates were not fully separated.

The observation of matter-wave interference with a 15-μm period required sources of atoms with a matter wavelength of 30 μm, corresponding to a kinetic energy of 0.5 nK or 1/2600th of the single-photon recoil energy. This energy is much smaller than the mean-field energy of Bose condensates in our trap (~100 nK) and also much less than the zero-point energy (~15 nK). Fortunately, the extremely anisotropic expansion of the condensates released from the cloverleaf trap yields atoms with very long de Broglie wavelengths in the axial direction.

Outlook. The techniques of condensate cutting and three-dimensional absorption imaging described above open up possibilities for further investigations. We have switched off the trap and observed the existence of the relative phase of two condensates. The next logical step is to combine this technique with our recently demonstrated output coupler for a Bose condensate (42). In that case, recording the interference pattern for the first output pulse creates a coherent state of the trapped condensate through the quantum measurement process. Subsequent output pulses could be used to study the time evolution of the

phase and the loss of coherence resulting from phase diffusion (21, 27, 33).

By using a thinner barrier (~1 μm) between the two condensates, it should be possible to reliably establish a weak link and study quantum tunneling, or the Josephson effect, for atoms (18, 19, 29). For superconductors, the Josephson effect is the usual way of detecting the phase of the order parameter. For atomic Bose condensates, we observed a relative phase directly. This is an example of the complementary physics that can be explored with Bose condensation in dilute atomic gases. Moreover, we have shown the technical feasibility of manipulating magnetically trapped Bose condensates with far-off-resonant laser beams. Hence, it is possible to perform "microsurgery" of Bose condensates, such as shaping the trapping potential or creating localized excitations (for example, using such a laser beam as a "paddle wheel" to excite rotational motion).

The observation of high-contrast interference fringes is clear evidence for spatial coherence over the extent of the condensates (43). In theoretical treatments, coherence (off-diagonal long-range order) has been used as the defining criterion for BEC (30). Our results also demonstrate that a Bose condensate consists of "laser-like" atoms, or atoms that interfere without any further selection by collimating apertures. This opens up the field of coherent atomic beams. Our recent work on an output coupler for a Bose condensate (42) already contained all the elements of an atom laser (44), because it created multiple pulses that should have a coherence length exceeding the size of a single condensate. Although this has been described as the first realization of an atom laser (45), we felt the demonstration that Bose condensed atoms have a measurable phase was a crucial missing feature. The present work addresses this issue and demonstrates that a Bose condensate with an output coupler is an atom laser.

Note added in proof: We have recently combined the rf output coupler (42) with the observation of interference between two condensates. The output pulse from a split condensate showed high-contrast interference that was very similar to the results discussed above (46). This proves that the rf output coupler preserves the coherence of the condensates.

REFERENCES AND NOTES

1. M. H. Anderson, J. R. Ensher, M. R. Matthews, C. E. Wieman, E. A. Cornell, *Science* **269**, 198 (1995).
2. K. B. Davis *et al.*, *Phys. Rev. Lett.* **75**, 3969 (1995).
3. M.-O. Mewes *et al.*, *ibid.* **77**, 416 (1996).
4. M. R. Andrews *et al.*, *Science* **273**, 84 (1996).
5. C. C. Bradley, C. A. Sackett, R. G. Hulet, in preparation; see also C. C. Bradley *et al.*, *Phys. Rev. Lett.* **75**, 1687 (1995).

6. D. S. Jin, J. R. Ensher, M. R. Matthews, C. E. Wieman, E. A. Cornell, *Phys. Rev. Lett.* **77**, 420 (1996).
7. M.-O. Mewes *et al.*, *ibid.*, p. 988.
8. M. Edwards, P. A. Ruprecht, K. Burnett, R. J. Dodd, C. W. Clark, *ibid.*, p. 1671.
9. S. Stringari, *ibid.*, p. 2350.
10. M. Holland and J. Cooper, *Phys. Rev. A* **53**, R1954 (1996).
11. Y. Castin and R. Dum, *Phys. Rev. Lett.* **77**, 5315 (1996).
12. Y. Kagan, E. L. Surkov, G. V. Shlyapnikov, *Phys. Rev. A* **54**, R1753 (1996).
13. A. Griffin, W.-C. Wu, S. Stringari, in preparation.
14. P. Nozières and D. Pines, *The Theory of Quantum Liquids* (Addison-Wesley, Reading, MA, 1990), vol. 2.
15. P. W. Anderson, *Rev. Mod. Phys.* **38**, 298 (1966).
16. Y. Kagan, in *Bose-Einstein Condensation*, A. Griffin, D. Snoke, S. Stringari, Eds. (Cambridge Univ. Press, Cambridge, 1995), pp. 202–225.
17. H. T. C. Stoof, *ibid.*, pp. 226–245.
18. J. Javanainen, *Phys. Rev. Lett.* **57**, 3164 (1986).
19. F. Dalfovo, L. Pitaevskii, S. Stringari, *Phys. Rev. A* **54**, 4213 (1996).
20. J. Javanainen and S. M. Yoo, *Phys. Rev. Lett.* **76**, 161 (1996).
21. E. M. Wright, D. F. Walls, J. C. Garrison, *ibid.* **77**, 2158 (1996).
22. W. Hoston and L. You, *Phys. Rev. A* **53**, 4254 (1996).
23. M. Naraschewski, H. Wallis, A. Schenzle, J. I. Cirac, *ibid.* **54**, 2185 (1996).
24. J. I. Cirac, C. W. Gardiner, M. Naraschewski, P. Zoller, *ibid.*, p. R3714.
25. T. Wong, M. J. Collett, D. F. Walls, *ibid.*, p. R3718.
26. H. Wallis, A. Röhrl, M. Naraschewski, A. Schenzle, *ibid.*, in press.
27. Y. Castin and J. Dalibard, in preparation.
28. E. M. Wright, T. Wong, M. J. Collett, S. M. Tan, D. F. Walls, in preparation.
29. M. W. Jack, M. J. Collett, D. F. Walls, *Phys. Rev. A* **54**, R4625 (1996).
30. K. Huang, *Statistical Mechanics* (Wiley, New York, ed. 2, 1987).
31. P. W. Anderson, in *The Lesson of Quantum Theory*, J. D. Boer, E. Dal, O. Ulfbeck, Eds. (Elsevier, Amsterdam, 1986), pp. 23–33.
32. A. J. Leggett and F. Sols, *Found. Physics* **21**, 353 (1991).
33. M. Lewenstein and L. You, *Phys. Rev. Lett.* **77**, 3489 (1996).
34. C. J. Bordé, *Phys. Lett. A* **204**, 217 (1995).
35. R. J. C. Spreeuw, T. Pfau, U. Janicke, M. Wilkens, *Europhys. Lett.* **32**, 469 (1996).
36. M. Holland, K. Burnett, C. Gardiner, J. I. Cirac, P. Zoller, *Phys. Rev. A* **54**, R1757 (1996).
37. M. Olshanii, Y. Castin, J. Dalibard, in *Laser Spectroscopy XII*, M. Ignuscio, M. Allegrini, A. Sasso, Eds. (World Scientific, Singapore, 1996), pp. 7–12.
38. W. Ketterle and N. J. van Druten, in *Advances in Atomic, Molecular, and Optical Physics*, B. Bederson and H. Walther, Eds. (Academic Press, San Diego, CA, 1996), vol. 37, pp. 181–236, and references therein.
39. Inhomogeneities in the pumping sheet caused weak striations that were perpendicular to the observed fringes and could therefore be clearly distinguished.
40. The thermal cloud had expanded so much that it contributed negligible background.
41. R. L. Pfleegor and L. Mandel, *Phys. Rev.* **159**, 1084 (1967).
42. M.-O. Mewes *et al.*, *Phys. Rev. Lett.* **78**, 582 (1997).
43. We are not distinguishing here between different aspects of coherence that are expressed by expectation values of products of one, two, or four field operators.
44. E. Cornell, *J. Res. Natl. Inst. Stand. Technol.* **101**, 419 (1996).
45. K. Burnett, *Physics World*, 18 (October 1996).
46. In these experiments, we transferred ~50% of the atoms into the $F = 1$, $m_F = 0$ state, immediately turned off the argon ion laser-light sheet to allow the two out-coupled condensates to overlap, and switched off the magnetic trap 2 ms later to avoid acceleration by quadratic Zeeman shifts.
47. We thank M. Naraschewski and H. Wallis for enlightening discussions; their theoretical simulations (26) were helpful in selecting the final parameters for the experiment. We also thank M.-O. Mewes for essential contributions during the early phase of the experiment, S. Inouye for experimental assistance, and D. Kleppner and D. Pritchard for valuable discussions. Supported by the Office of Naval Research, NSF, Joint Services Electronics Program, and the Packard Foundation. D.M.K. was supported by a NSF Graduate Research Fellowship, C.G.T. by a North Atlantic Treaty Organization (NATO) Science Fellowship, and H.-J.M. by Deutscher Akademischer Austauschdienst (NATO Science Fellowship).

11 December 1996; accepted 19 December 1996

VOLUME 82, NUMBER 15   PHYSICAL REVIEW LETTERS   12 APRIL 1999

Atom Laser with a cw Output Coupler

Immanuel Bloch, Theodor W. Hänsch, and Tilman Esslinger

Sektion Physik, Ludwig-Maximilians-Universität, Schellingstrasse 4/III, D-80799 Munich, Germany
and Max-Planck-Institut für Quantenoptik, D-85748 Garching, Germany
(Received 3 December 1998)

We demonstrate a continuous output coupler for magnetically trapped atoms. Over a period of up to 100 ms, a collimated and monoenergetic beam of atoms is continuously extracted from a Bose-Einstein condensate. The intensity and kinetic energy of the output beam of this atom laser are controlled by a weak rf field that induces spin flips between trapped and untrapped states. Furthermore, the output coupler is used to perform a spectroscopic measurement of the condensate, which reveals the spatial distribution of the magnetically trapped condensate and allows manipulation of the condensate on a micrometer scale. [S0031-9007(99)08914-0]

PACS numbers: 03.75.Fi, 05.30.Jp, 32.80.Pj, 42.55.–f

Four decades ago the first optical lasers [1] were demonstrated [2,3], marking a scientific breakthrough: Ultimate control over frequency, intensity, and direction of optical waves had been achieved. Since then, lasers have found innumerable applications, for both scientific and general use. It may now be possible to control matter waves in a similar way, as Bose-Einstein condensation (BEC) has been attained in a dilute gas of trapped atoms [4–6]. In a Bose-Einstein condensate a macroscopic number of bosonic atoms occupy the ground state of the system, which can be described by a single wave function. A pulsed output coupler which coherently extracts atoms from a condensate was demonstrated recently [7,8]. Because of its properties, this source is often referred to as an atom laser.

In this Letter we report on the successful demonstration of an atom laser with a continuous output. The duration of the output is limited only by the number of atoms in the condensate. The Bose-Einstein condensate is produced in a novel magnetic trap [9] which provides an extremely stable trapping potential. A weak rf field is used to extract the atoms from the condensate over a period of up to 100 ms, thereby forming an atomic beam of unprecedented brightness. The output coupling mechanism can be visualized as a small, spatially localized leak in the trapping potential. The condensate wave function passes through the leak and forms a collimated atomic beam. This continuous output coupler allows the condensate wave function to be studied and manipulated with high spatial resolution.

Let us consider a simple model for a cw output coupler [10–12] and assume that atoms which are in a magnetically trapped state are transferred into an untrapped state by a monochromatic resonant rf field. In the untrapped state the atoms experience the repulsive mean-field potential of the condensate, which can be approximated by the parabolic form of a Thomas-Fermi distribution. When the atoms leave the trap the corresponding potential energy of the atoms is transformed into kinetic energy. The velocity of the atoms leaving the trap can be adjusted by the frequency

of the rf field, because it determines the spatial region where the atoms are transferred into the untrapped state.

So far, only pulsed output couplers have been demonstrated [7,13,14]. If a short and, hence, broadband rf pulse is applied, the output pulse has a correspondingly large energy spread. Each portion of the condensate that leaves the trap has a kinetic energy distribution with a width comparable to the mean-field energy of the condensate. For this case the output coupling process is no longer spatially selective and therefore largely insensitive to fluctuations in the magnetic field.

Continuous wave output coupling, as done in this paper, can be achieved only if the magnetic field fluctuations that the trapped atoms experience are minimized. The level of fluctuations in the magnetic field has to be much less than the change of the magnetic trapping field over the spatial size of the condensate.

In the experiment, we use a novel magnetic trap [9], the quadrupole and the Ioffe configuration (QUIC) trap, which is particularly compact and operates at a current of just 25 A. The compactness of the trap allowed us to place it inside a μ-metal box, which reduces the magnetic field of the environment and its fluctuations by a factor of approximately 100. In combination with an extremely stable current supply ($\Delta I/I < 10^{-4}$) we are able to reduce the residual fluctuations in the magnetic field to a level below 0.1 mG. The rf field for the output coupler is produced by a synthesizer (HP-33120A) and is radiated from the same coil as is used for evaporative cooling. The coil has 10 windings, a diameter of 25 mm, and is mounted 30 mm away from the trap center. The magnetic field vector of the rf is oriented in the horizontal plane, perpendicular to the magnetic bias field of the trap.

To obtain Bose-Einstein condensates we use the same setup and experimental procedure as in our previous paper [9]. Typically, 10^9 rubidium atoms are trapped and cooled in a magneto-optical trap. Then the atoms are transferred into the QUIC trap where they are further cooled by rf-induced evaporation. The QUIC trap is operated with

0031-9007/99/82(15)/3008(4)$15.00

VOLUME 82, NUMBER 15   PHYSICAL REVIEW LETTERS   12 APRIL 1999

trapping frequencies of $\omega_\perp = 2\pi \times 180$ Hz in the radial and $\omega_y = 2\pi \times 19$ Hz in the axial direction. In this configuration the trap has a magnetic field of 2.5 G at its minimum.

After the creation of the Bose-Einstein condensate, the rf field used for evaporative cooling is switched off, and 50 ms later the radio frequency of the output coupler is switched on for a time of 15 ms in a typical experiment. The field of the output coupler is ramped up to an amplitude of $B_{rf} = 2.6$ mG within 0.1 ms. Its frequency follows a linear ramp from 1.752 to 1.750 MHz, to account for the shrinking size of the condensate, as discussed below. Over this period, atoms are extracted from the condensate and are accelerated by gravity. Subsequently, the magnetic trapping field is switched off, and 3.5 ms later the atomic distribution is measured by absorption imaging. The atom laser output is shown in Fig. 1. The beam contains 2×10^5 atoms and its divergence in the plane of observation is below our experimental resolution limit of 3.5 mrad. We obtain an output beam over a longer period of time when we reduce the magnetic field amplitude B_{rf} of the rf field. With absorption imaging we are able to directly image the

continuous output from the atom laser over 40 ms, with $B_{rf} = 1.2$ mG. A more sensitive method is to measure the number of atoms that remain in the condensate after a certain period of time. This enables us to monitor the output coupling process over up to 100 ms, with $B_{rf} = 0.2$ mG. The magnetic field amplitudes have been calibrated with an accuracy of 20%.

It is instructive to estimate the brightness of the beam produced by our atom laser. Defining the brightness as the integrated flux of atoms per source size divided by the velocity spreads in each dimension $\Delta v_x \, \Delta v_y \, \Delta v_z$ [15], we find that the brightness of our beam has to be at least 2×10^{24} atoms s^2 m^{-5}. To obtain this lower limit for the brightness, we estimate that the atomic flux is 5×10^6 atoms/s and that the longitudinal velocity spread is given by $\Delta v_z = 3$ mm/s. We further assume a velocity spread $\Delta v_x = 5$ mm/s for the strongly confining axis, which corresponds to the chemical potential of the condensate. Our measurements show that the velocity spread along the weakly confining axis is less than $\Delta v_y = 0.3$ mm/s.

Assuming a Fourier-limited longitudinal velocity width of $\Delta v_z = 0.3$ mm/s and diffraction-limited transverse velocity spreads, a brightness of 4×10^{28} atoms s^2 m^{-5} can be reached. Both numbers show that continuous output coupling from a condensate creates an atomic beam with a brightness that is orders of magnitude higher than that of a state-of-the-art Zeeman, slower [18] with a brightness of 2.9×10^{18} atoms s^2 m^{-5}, or an atomic source derived from a magneto-optical trap [19], with a brightness of 8.5×10^{16} atoms s^2 m^{-5}.

Let us now consider the geometry of the trap with respect to the output coupling mechanism in more detail. The magnetic field $B(\mathbf{r})$ gives rise to a harmonic trapping potential which confines the condensate in the shape of a cigar, with its long axis oriented perpendicular to the gravitational force. The rf field of frequency ν_{rf} induces transitions from the magnetically trapped $|F = 2, m_F = 2\rangle$ state to the untrapped $|F = 2, m_F = 0\rangle$ state via the $|F = 2, m_F = 1\rangle$ state. Here F denotes the total angular momentum and m_F is the magnetic quantum number. The resonance condition $\frac{1}{2}\mu_B |B(\mathbf{r})| = h\nu_{rf}$, where μ_B is the Bohr magneton, is satisfied on the surface of an ellipsoid which is centered at the minimum in the magnetic trapping field [20]. Without gravity the condensate would have the same center, so that an undirected output could be expected [11]. The frequency range in which significant output coupling occurs would then be determined by the magnetic field minimum B_{off} and by the chemical potential μ of the condensate: $\frac{1}{2}\mu_B B_{off} \leq h\nu_{rf} \leq \frac{1}{2}(\mu_B B_{off} + \mu)$.

Because of gravity, the minimum of the trapping potential is displaced relative to the minimum of the magnetic field. With g being the gravitational acceleration, this displacement is given by $g/\omega_\perp^2$, which is 7.67 μm for our trapping parameters. The confinement of the trap

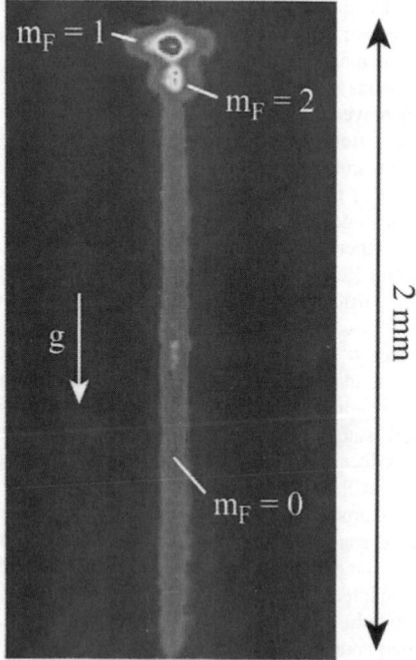

FIG. 1(color). Atom laser output: A collimated atomic beam is derived from a Bose-Einstein condensate over a 15 ms period of continuous output coupling. A fraction of condensed atoms has remained in the magnetically trapped $|F = 2, m_F = 2\rangle$ and $|F = 2, m_F = 1\rangle$ state. The magnetic trap has its weakly confining axis in the horizontal direction.

VOLUME 82, NUMBER 15   PHYSICAL REVIEW LETTERS   12 APRIL 1999

and hence the spatial size of the condensate remain the same. In this geometry, which is illustrated in Fig. 2, output coupling occurs only at the intersection of the displaced condensate with the ellipsoid that is determined by the resonance condition. Atoms leaving the condensate therefore experience a directed force which is dominated by gravity in our experiment and gives rise to a collimated output beam. The frequency range over which output coupling can be achieved is larger than without gravity, because the condensate is shifted into a region of an increasingly stronger magnetic field gradient. The frequency interval $\Delta \nu = g \sqrt{2 \ \mu \text{m}}/(\hbar \omega_\perp)$, where m is the atomic mass, gives the difference in frequency between an rf field that is resonant with the upper edge and an rf field that is resonant with the lower edge of the condensate, assuming a Thomas-Fermi distribution. For our trapping parameters and 7×10^5 rubidium atoms in the condensate this frequency interval is $\Delta \nu = 10.2$ kHz.

We investigate the condensate spectroscopically by measuring the number of atoms which remains in the condensate after 20 ms of output coupling, for various amplitudes and frequencies of the rf field. The number of atoms in the condensate is measured by absorption imaging. Before the absorption pulse is applied, the condensate is released from the magnetic trap and expands ballistically for 10.5 ms. Because of an inhomogeneous magnetic field, which is applied during the switch-off period, atoms in the $m_F = 2$ and $m_F = 1$ magnetic sublevel are spatially separated from each other, allowing us to determine the population in each of the sublevels independently (see also Fig. 1). The experimental parameters are carefully controlled for each series of measurements, so that the number of atoms in the condensate does not fluctuate by more than 5%.

In Fig. 3 the number of atoms in the condensate is plotted versus the square of the Rabi frequency which

is induced by the rf field, for a fixed frequency of 1.736 MHz. With increasing rf power the population N in the condensate decreases until it approaches a constant level N_0. This is to be expected, since the size of the condensate shrinks with its depopulation until the overlap with the resonance ellipsoid vanishes. Our measurements can therefore be explained by the simple rate equation $\frac{dN(t)}{dt} = -\Gamma[N(t) - N_0]$, in which the depopulation rate Γ is proportional to the square of the Rabi frequency $\Omega = g_F \mu_B B/\hbar$, with g_F being the Landé factor. From a fit to our experimental data we obtain $\Gamma/\Omega^2 = 1.2(2) \times 10^{-5}$ s.

Figure 4 shows the dependency of the condensate population on the frequency of the rf field, for $\Omega = 2\pi \times 0.7$ kHz. The measured distribution has a width of 13.1(5) kHz (10% values) and stretches over a larger frequency range than the 10.2 kHz estimated (see above). This is to be expected, since the Thomas-Fermi distribution does not properly take into account the decay of the wave function near the outer edge of the condensate. The condensate wave function extends beyond the classical radius R, determined by the Thomas-Fermi distribution, and falls off exponentially on a characteristic length scale, which is given by $\delta = R/2(\hbar \omega_\perp/\mu)^{2/3}$ [21,22]. For our parameters we obtain $\delta/R = 0.074$. Our measurements are in good agreement with these considerations. The slight asymmetry of the measured curve is caused by the asymmetry in the number of atoms per resonance ellipsoid.

The results clearly show the high spectral and spatial resolution of our output coupler. From an estimated spectral resolution of 1 kHz we obtain a spatial resolution of about 1 μm. A more precise comparison of our measurements with a theoretical model [11,12] should be possible if the magnetic structure of the $F = 2$ spin state is taken into account and if the output coupling process is studied beyond the Thomas-Fermi approximation.

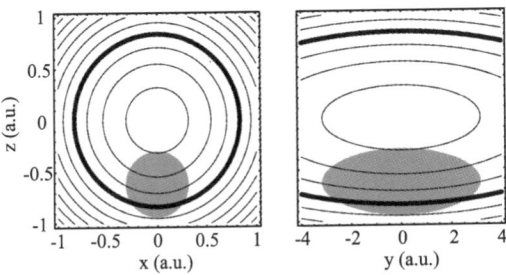

FIG. 2. Continuous output coupling from a Bose-Einstein condensate. The contour lines represent the absolute value of the magnetic trapping field. The thick line indicates the region where the rf field transfers atoms from the magnetically trapped state into an untrapped state. Because of gravity, the condensate is trapped 7.67 μm below the minimum in the magnetic field. In the untrapped state the atoms experience a directed force caused by gravity and the mean field of the condensate. This results in a collimated output beam.

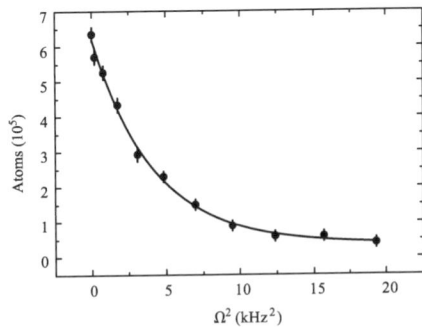

FIG. 3. Condensate population of the $|F = 2, m_F = 2\rangle$ state after a 20 ms period of output coupling versus the square of the Rabi frequency induced by the rf field. The solid line represents a fit based on a simple rate equation model (see text).

VOLUME 82, NUMBER 15   PHYSICAL REVIEW LETTERS   12 APRIL 1999

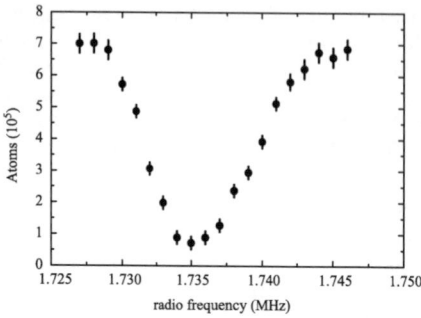

FIG. 4. Spectroscopy of the Bose-Einstein condensate. The condensate population of the $|F = 2, m_F = 2\rangle$ state is shown for different radio frequencies after a 20 ms period of output coupling. Assuming a Thomas-Fermi distribution for the condensate density we expect output coupling to occur within a frequency interval of 10.2 kHz (see text). This does not take into account the exponential decay of the condensate wave function at the surface of the condensate.

The method reported in this paper is a starting point to systematically analyze the spectroscopic properties of alkali condensates [23]. Our experiments also demonstrate that Bose-Einstein condensates can be manipulated on a micrometer scale with rf fields. A similar scheme to our output coupler could be used to realize a weak link between two condensates which are trapped in different internal states [24]. This would provide the means to observe Josephson-type oscillations, as discussed in a recent theoretical paper [25].

Studies on the output properties of the atom laser will provide a fascinating field of research. It has already been shown that the rf output coupling process preserves first order coherence [8], but more quantitative and detailed measurements are required. One of the crucial properties of an optical laser is its second order coherence [26]. A direct measurement of the second order coherence of an atom laser would reveal the intriguing consequences of quantum statistics.

There is a wide range of applications for atom lasers in the field of atom optics. It is now conceivable to produce diffraction-limited atomic beams which could be focused down to a spot size of much less than 1 nm. Atom lasers will also revolutionize atom interferometers. Highly collimated and slow beams of atoms, as demonstrated in this paper, make it possible to create atom interferometers with large enclosed areas and a superior signal-to-noise ratio, which are ideally suited for precision measurements.

We would like to thank Jens Schneider for stimulating discussions.

[1] A. L. Schawlow and C. H. Townes, Phys. Rev. **112**, 1940 (1958).
[2] T. H. Maiman, Nature (London) **187**, 493 (1960).
[3] A. Javan, W. B. Bennett, Jr., and D. R. Herriott, Phys. Rev. Lett. **6**, 106 (1961).
[4] M. H. Anderson et al., Science **269**, 198 (1995).
[5] C. C. Bradley, C. A. Sackett, J. J. Tollett, and R. G. Hulet, Phys. Rev. Lett. **75**, 1687 (1995); C. C. Bradley, C. A. Sackett, and R. G. Hulet, Phys. Rev. Lett. **78**, 985 (1997).
[6] K. B. Davis et al., Phys. Rev. Lett. **75**, 3969 (1995).
[7] M.-O. Mewes et al., Phys. Rev. Lett. **78**, 582 (1997).
[8] M. R. Andrews et al., Science **275**, 637 (1997).
[9] T. Esslinger, I. Bloch, and T. W. Hänsch, Phys. Rev. A **58**, R2664 (1998).
[10] R. J. Ballagh, K. Burnett, and T. F. Scott, Phys. Rev. Lett. **78**, 1607 (1997).
[11] M. Naraschewski, A. Schenzle, and H. Wallis, Phys. Rev. A **56**, 603 (1997).
[12] H. Steck, M. Naraschewski, and H. Wallis, Phys. Rev. Lett. **80**, 1 (1998).
[13] B. P. Anderson and M. A. Kasevich, Science **282**, 1686 (1998).
[14] K. Helmerson et al. (private communication).
[15] In this, we follow the definition of [16] and additionally include the source size of the beam, as is usual for the definition of the brightness of a light source in optics [17].
[16] E. Riis, D. S. Weiss, K. A. Moler, and S. Chu, Phys. Rev. Lett. **64**, 1658 (1990).
[17] L. Mandel and E. Wolf, *Optical Coherence and Quantum Optics* (Cambridge University Press, Cambridge, England, 1995).
[18] D. Meschede (private communication).
[19] Z. T. Lu et al., Phys. Rev. Lett. **77**, 3331 (1996).
[20] Using the Breit-Rabi formula, we equate that the magnetic splitting between the $m_F = 0$ and $m_F = 1$ state is 900 Hz larger than the splitting between the $m_F = 1$ and $m_F = 2$ state.
[21] F. Dalfovo, L. Pitaevskii, and S. Stringari, Phys. Rev. A **54**, 4213 (1996).
[22] E. Lundh, C. J. Pethick, and H. Smith, Phys. Rev. A **55**, 2126 (1997).
[23] Two-photon optical spectroscopy was recently used to observe BEC in atomic hydrogen. See T. C. Killian et al., Phys. Rev. Lett. **81**, 3807 (1998); D. G. Fried et al., Phys. Rev. Lett. **81**, 3811 (1998).
[24] D. S. Hall et al., Phys. Rev. Lett. **81**, 1539 (1998); D. S. Hall, M. R. Matthews, C. E. Wieman, and E. A. Cornell, Phys. Rev. Lett. **81**, 1543 (1998); M. R. Matthews et al., Phys. Rev. Lett. **81**, 243 (1998).
[25] J. Williams, R. Walser, J. Cooper, E. Cornell, and M. Holland, Phys. Rev. A **59**, R31 (1999).
[26] F. T. Arecchi, E. Gatti, and A. Sona, Phys. Lett. **20**, 27 (1966).

A Well-Collimated Quasi-Continuous Atom Laser

E. W. Hagley,[1] L. Deng,[1,2] M. Kozuma,[3] J. Wen,[1] K. Helmerson,[1] S. L. Rolston,[1] W. D. Phillips[1]

Extraction of sodium atoms from a trapped Bose-Einstein condensate (BEC) by a coherent, stimulated Raman process is demonstrated. Optical Raman pulses drive transitions between trapped and untrapped magnetic sublevels, giving the output-coupled BEC fraction a well-defined momentum. The pulsed output coupling can be run at such a rate that the extracted atomic wave packets strongly overlap, forming a highly directional, quasi-continuous matter wave.

The occupation of a single quantum state by a large number of identical bosons (1–5) is a matter-wave analog to the storage of photons in a single mode of a laser cavity. Just as one extracts a coherent, directed beam of photons from a laser cavity by using a partially transmitting mirror as an output coupler, one can analogously extract directed matter waves from a condensate. Such a source of matter waves, or "atom laser," is important in the field of atom optics (6), the manipulation of atoms analogous to the manipulation of light. Its development is providing atom sources that are as different from ordinary atomic beams as lasers are from light bulbs.

The first demonstration of a BEC output coupler was reported in 1997 (7) where coherent, radiofrequency (rf)-induced transitions were used to change the internal state (magnetic sublevel) of the atoms from a trapped to an untrapped state. This method, however, did not allow the direction of the output-coupled atoms to be chosen. The extracted atoms fell under the influence of gravity and expanded because of their intrinsic repulsion. We demonstrate a highly directional method to optically couple out a variable fraction of a condensate and apply this method to produce a well-collimated, quasi-continuous beam of atoms, an important step toward a truly continuous wave (cw) atom laser (8).

The output coupling is based on stimulated Raman transitions between magnetic sublevels (9, 10). The sublevel into which the

atoms are transferred is unaffected by the trapping potential, and the process imparts a well-defined momentum to the output-coupled condensate fraction. In contrast, previous work on Bragg diffraction (11) transferred momentum without changing the internal state of the atom. A single Raman pulse can couple out any desired fraction of the condensate. By changing the angle between the wave vectors k of the Raman lasers ($k = 2\pi/\lambda$, $\lambda = 589$ nm) and using higher order ($2n$-photon) Raman transitions, it is possible to impart any momentum of magnitude 0 to $2n\hbar k$ to the atoms (for sodium, $2\hbar k$ corresponds to a velocity of 6 cm/s). In this way it is possible to choose the energy of the extracted deBroglie wave, producing a widely tunable atom laser.

In this experiment, we used a hybrid evaporation technique with a time orbiting potential (TOP) trap (12, 13) to form a sodium condensate (11). We typically obtain a condensate, without a discernible normal fraction, with about 10^6 atoms in the $3S_{1/2}$, $F = 1$, $m = -1$ state (14). Once the condensate is formed we adiabatically expand the trapping potential in 0.5 s, reducing the trapping frequencies to $\omega_x/2\pi = 18$ Hz, $\omega_y/2\pi = 25$ Hz, and $\omega_z/2\pi = 35$ Hz. We have measured (15) that our adiabatic cooling reduces the asymptotic root mean square (rms) momentum width of the released condensate to $0.09(1)\hbar k$ (16).

In Raman output coupling (Fig. 1), a moving standing wave, composed of two nearly counterpropagating laser beams with frequency difference $\delta = \omega_2 - \omega_1$ (17), is

[1]National Institute of Standards and Technology, Gaithersburg, MD 20899, USA. [2]Georgia Southern University, Statesboro, GA 30460, USA. [3]Institute of Physics, University of Tokyo, Tokyo 153-8902, Japan.

REPORTS

applied to the condensate for a short period of time. These beams propagate nearly along the $\hat{z}$ axis of the trap (gravity is along $\hat{x}$) and each beam is detuned from the $3S_{1/2}$, $F = 1 \rightarrow 3P_{3/2}$, $F' = 2$ transition by $\Delta/2\pi = -1.85$ GHz to suppress spontaneous emission. A stimulated Raman transition occurs when an atom changes its state by coherently exchanging photons between the two laser beams (absorption from ω_1 and stimulated emission into ω_2). The atom acquires momentum $P = \hbar(k_1 - k_2) = P\hat{z}$ (in our case) with $P = 2\hbar k \sin(\theta/2)$ (18). Therefore, an atom initially at rest acquires kinetic energy $\hbar\delta_{Recoil} = P^2/(2M)$, where M is the atomic mass. The stimulated Raman process can change the internal energy state of an atom by driving $\Delta m = 1$ or even $\Delta m = 2$ transitions (19). The energy difference between the absorbed and emitted photons must account for both the change in kinetic energy and any change in the internal (magnetic) energy level of the atom: $\hbar\delta = \hbar\delta_{Zeeman} - \hbar\delta_{recoil}$. By changing the internal state of a trapped atom to $m = 0$, a state that feels no trapping forces, we release the atoms and impart a momentum that kicks them away.

Our Raman output coupling scheme dramatically reduces the transverse momentum width of the extracted atoms compared with other methods such as rf output coupling (7). The $0.09(1)\hbar k$ rms momentum width discussed previously corresponds to the average release energy of $2/7\mu$ (the chemical potential $\mu \propto N^{2/5}$, where N is the number of atoms in the condensate) per atom caused by the intrinsic repulsion between the atoms. If, however, only a small number of atoms are coupled out of the condensate into the $m = 0$ state, the average energy per extracted atom is two times larger (20). For atoms coupled out of a spherically symmetric trap without an initial momentum kick, this release energy causes an isotropic momentum spread. In our case, where significant momentum is imparted to the atoms, the release energy is primarily channeled into the forward direction. This dramatically reduces the transverse rms momentum width, resulting in a highly collimated output beam. The transverse momentum width (21) is reduced by roughly the ratio of the characteristic time it takes the output-coupled atoms to leave the still-trapped condensate divided by the time scale over which the mean field repulsion acts on the freely expanding condensate (22) and will therefore be extremely small compared with its longitudinal momentum of $2\hbar k$. This has been confirmed by numerical calculations, which estimate the mean field component of the transverse rms momentum width to be $0.004\hbar k$.

The uncertainty principle imposes an additional transverse momentum width of $0.002\hbar k$ due to the finite size of the condensate. The predicted mean-field momentum width is close to this lower limit and corresponds to a divergence of a few milliradians, comparable to that of a typical commercial optical laser. As this momentum width is so small, it is difficult to measure experimentally (during the typical 7-ms duration of the experiment the radius of the atomic cloud would expand only 1 μm).

To implement this Raman output coupling scheme requires that special attention be paid to the time-varying magnetic field in a TOP trap. In the presence of gravity the atoms sag (in this case about 0.8 mm) away from the center of the trap to a position where the magnitude of the magnetic field changes as the bias field rotates (23). The time-dependent detuning for the two-photon transition is

$$\delta(t) = \frac{\Delta m \mu_B}{2\hbar} |\mathbf{B}(t)| - \frac{2\hbar k^2 \sin^2(\theta/2)}{M} \quad (1)$$

where μ_B is the Bohr magneton and $\Delta m = 0, 1$, or 2. The frequency difference between $m = -1$ and $m = 0$ sublevels changes by nearly 4 MHz as the TOP field rotates. This frequency difference is large compared with the effective width of the transition, the inverse of the Raman pulse length (typically 6 μs). Therefore, the Raman pulse is synchronized to the maximum value of $\delta(t)$ to minimize variations in the resonance frequency during the transition. Compared with continuous output coupling, this pulsed Raman output coupling results in less spontaneous emission for the same percentage of output-coupled atoms.

When the Raman pulses were applied, the magnetic field direction was along the $\hat{x}$ axis (vertical), as was the polarization of one of the two Raman lasers [which drove only $\Delta m = 0$ (π) transitions]. This laser had an intensity of 300 mW/cm^2 (24). The second Raman laser drove $\Delta m = 1$ (σ) transitions when its polarization was along $\hat{y}$ and it had an intensity of 600 mW/cm^2, but only half this power was useful because of selection rules. For $\Delta/2\pi = -1.85$ GHz, this intensity corresponds to an average time per atom between spontaneous emission events of 70 μs.

Directional output coupling is observed by imaging the atoms several milliseconds after the Raman pulse. Figure 2 shows optical depth images obtained by first optically pumping the atoms to the $3S_{1/2}$, $F = 2$ state and then absorption-imaging (2) on the $3S_{1/2}$, $F = 2 \rightarrow 3P_{3/2}$, $F' = 3$ transition. The sequence (Fig. 2, A to C) shows a BEC ($F =$

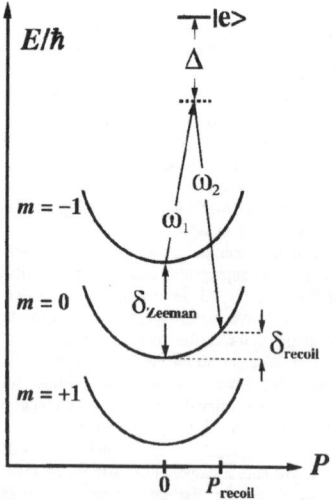

Fig. 1. Principle of the Raman output coupler. Energy conservation requires a relative detuning, $\delta = \omega_2 - \omega_1$, between the Raman lasers. Total energy as a function of atomic momentum is plotted, where the parabolas correspond to kinetic energy $P^2/2M$.

Fig. 2. (A) Condensate before the application of a Raman pulse. **(B)** $\Delta m = +1$ transition (19) from a 6-μs pulse with $\delta/2\pi = 6.4$ MHz. This detuning is chosen to be slightly larger than the 6.27-MHz resonance frequency to suppress four-photon coupling to the $m = +1$, $4\hbar k \hat{z}$ state (25). **(C)** $\Delta m = 0$ transition (26) after a 14-μs Raman pulse with equal laser intensities of 25 mW/cm^2. The relative detuning was $\delta/2\pi = -98$ kHz and the polarizations of both lasers were aligned with $\hat{x}$. The diagrams to the right of (B) and (C) show the polarization of the lasers with respect to the local magnetic field. We verified that no transitions occurred when incorrect polarizations were used. **(D)** The rotating magnetic field zero (circle of death) results in Majorana transitions of an output-coupled condensate fraction in the $m = 0$ state. Arrow denotes physical location of the rotating bias field zero. This is a graphic depiction of Majorana transitions.

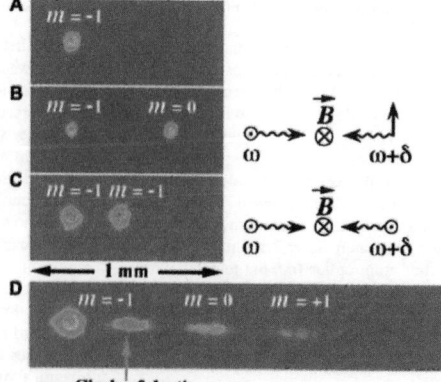

Circle of death

1, $m = -1$) followed by two BECs after application of a single Raman pulse. The TOP trap confining fields were held on for 7 ms after application of the Raman pulse before being switched off. The system was then allowed to evolve freely for 1.6 ms before being imaged. Note that the position of the output-coupled atoms in Fig. 2B is different from that in Fig. 2C. In the former, the atoms that have undergone the Raman transition are in the state $m = 0$ and therefore no longer feel the trapping potential, whereas in Fig. 2C the atoms are still trapped. The position of the $m = 0$ atoms corresponds to free flight with momentum $2\hbar k\hat{z}$ during the entire 8.6 ms, whereas the position of the atoms in Fig. 2C corresponds to their classical turning point in the trap (7 ms is about one-quarter of the 28.6-ms oscillation period along $\hat{z}$).

In the case where the detuning of the lasers from the excited state is large compared with the excited-state hyperfine structure splitting, it is not possible to drive $\Delta m = 2$ transitions directly with two photons because the ground state looks like a spin 1/2 system for which there are only two states. Instead, we can couple to the $m = +1$ state by combining the Raman

process with Majorana transitions due to the TOP rotating magnetic field zero (13). Atoms were first output-coupled to the $m = 0$ state and imaged 10.6 ms later after having crossed the orbit of the zero of the magnetic field (27), known as the "circle of death," which orbits in the $\hat{x}$-$\hat{z}$ plane (Fig. 2D). For this image the rotating bias field was reduced by a factor of 3, which reduced the distance to the orbiting magnetic field zero to 0.3 mm. As the atoms crossed this orbit they lost their quantization axis and were repeatedly projected to all three magnetic sublevels at the 20-kHz TOP frequency. The atoms in $m = -1$, 0, and $+1$ states were, respectively, retarded, unimpeded, and ejected by the trapping potential, giving rise to three spatially separated stripes of atoms. At the time of imaging, the atoms that ended in the $m = -1$ state have already been pulled back to the circle of death by the trap.

Although it is not possible to drive the $\Delta m = 2$ transition with two photons, it is possible with four by using the $m = -1 \rightarrow m = 0 \rightarrow m = +1$ transition scheme (Fig. 3C). A single 6-μs Raman pulse with $\delta/2\pi = 6.15$ MHz and the same intensities used for Fig. 2B was applied. In Fig. 3A, the TOP was switched off immediately after the Raman pulse and the atoms were imaged 5.6 ms later. The $m = +1$ atoms received $4\hbar k\hat{z}$ of momentum from four photons, and the $m = 0$ atoms received only $2\hbar k\hat{z}$ from two photons; thus the $m = +1$ atoms moved twice as far as the $m = 0$ atoms. With the TOP switched off 4 ms after the Raman pulse and the system imaged 1.6 ms later (Fig. 3B), the atoms in the $m = +1$ antitrapped state

are accelerated away from the trap, causing them to move further.

To produce quasi-continuous output coupling, we used multiple Raman pulses. The laser intensities were reduced and the detuning was again chosen to be $\delta/2\pi = 6.4$ MHz to primarily couple to the $m = 0$ state. For the optical depth images of the condensate after one, three, and six Raman pulses (Fig. 4, A to C), the TOP was held on for a 9-ms window during which time 6-μs Raman pulses were fired at a subharmonic of the rotating bias frequency. The magnetic fields were then extinguished and the atoms were imaged 1.6 ms later. The intensity of the laser whose polarization was aligned with $\hat{x}$ was 300, 150, and 100 mW/cm^2, respectively, to couple out different fractions of the condensate (the intensity of the second Raman laser was twice that of the first).

In Fig. 4D the TOP trap was held on during a 7-ms window during which time 140 Raman pulses were fired at the 20-kHz frequency of the rotating bias field and the distribution of atoms was imaged 1.6 ms later. The Raman pulse duration and intensity were reduced to 1 μs and 40 mW/cm^2 for Fig. 4D to ensure that the total integrated pulse time of 140 μs was much less than the spontaneous emission time of about 500 μs. The phase of the output-coupled matter wave evolves at about 100 kHz with respect to the condensate itself because of the kinetic energy imparted by the two-photon Raman transition (28). Because 100 kHz is an integer multiple of the ~20-kHz output coupling repetition rate, the interference of successive pulses is almost completely constructive (29). In the time between two Raman pulses each output-coupled wave packet moves only 2.9 μm, much less than the ~50-μm size of the condensate, so the output-coupled atoms form a quasi-continuous coherent matter wave. By varying the delay between pulses, the interference between wave packets can be used to investigate the coherence properties of the condensate.

It is apparent that there is also coupling to the $m = +1$ state in Fig. 4D because some output-coupled atoms have moved the distance that an atom with momentum $4\hbar k\hat{z}$ moves in 8.6 ms. Such coupling to the $m = +1$ state occurs in this case because the spectral width of the 1-μs Raman pulse is sufficiently broad (30) to drive a transition from the state $m = 0$ (momentum $2\hbar k\hat{z}$) to the state $m = +1$ (momentum $4\hbar k\hat{z}$). In our experiment the trajectories of the two output-coupled beams ($m = 0$ and $m = +1$) are spatially separated because the direction of momentum transfer is orthogonal to gravity. These two beams appear to overlap in Fig. 4D because the camera views them from above. Coupling to the $m = +1$ state could be suppressed by using a larger bias magnetic field and a larger detuning Δ to exploit the second-order Zeeman shift and to reduce the pulse bandwidth without excessive spontaneous emission. To completely suppress

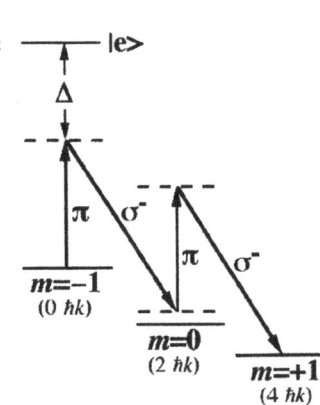

Fig. 3. Atoms are coupled to both the $m = 0$ and $m = +1$ and magnetic sublevels using a single Raman pulse. (**A**) Magnetic trap is switched off immediately after the Raman pulse. (**B**) Magnetic trap is held on for 4 ms after the Raman pulse. (**C**) Transition used and laser polarizations.

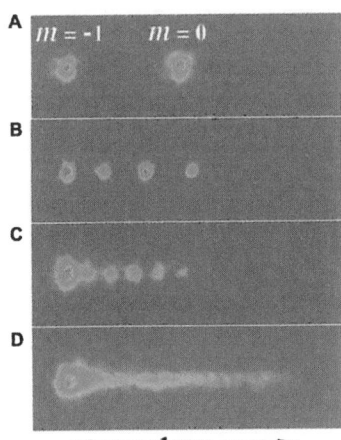

Fig. 4. (**A to C**) One, three, and six 6-μs Raman pulses, respectively, were applied to the condensate. (**D**) Firing 1-μs Raman pulses at the full repetition rate of about 20 kHz imposed by the frequency of the rotating bias field (140 pulses in 7 ms) produces a quasi-continuous atomic beam.

REPORTS

coupling to unwanted antitrapped states a Raman transition to the $F = 2$, $m = 0$ ground state of Na could be used.

An important property of the condensate, and any output-coupled fraction, is its coherence. Coherence effects between two condensates have already been observed by dropping them and allowing them to interfere (30). Because we use a stimulated Raman process, our output beam should be fully coherent. The effect of the mean field on the atoms as they leave the BEC will be to distort the outgoing wave without resulting in any true loss of coherence. In a separate experiment we observed matter-wave interference due to the 100-kHz phase evolution discussed above and we are using it to measure the coherence properties of the condensate.

References and Notes

1. A. Einstein, *Sitzungsbe. Kgl. Preuss. Akad. Wiss.* (1924), p. 261; *ibid.* (1925), p. 3.
2. M. H. Anderson, J. R. Ensher, M. R. Matthews, C. E. Wieman, E. A. Cornell, *Science* **269**, 198 (1995).
3. K. B. Davis et al., *Phys. Rev. Lett.* **75**, 3969 (1995).
4. C. C. Bradley, C. A. Sackett, R. G. Hulet, *ibid.* **78**, 985 (1997); see also C. C. Bradley, C. A. Sackett, J. J. Tollett, R. G. Hulet, *ibid.* **75**, 1687 (1995).
5. D. G. Fried et al., *ibid.* **81**, 3807 (1998).
6. *J. Physique* **4**, 11 (1994); *Appl. Phys. B.* **54**, 321 (1992); *J. Phys. Rep.* 240 (1994) (special issues on optics and interferometry with atoms).
7. M.-O. Mewes et al., *Phys. Rev. Lett.* **78**, 582 (1997).
8. A truly cw atom laser produces a continuous, coherent matter wave output while being continuously replenished with new atoms, in direct analogy with a cw optical laser. The coherence length of such a laser would be longer than the size of the trapped condensate just as the coherence length of a cw optical laser is longer than the laser cavity.
9. G. Moy, J. Hope, C. Savage, *Phys. Rev. A* **55**, 3631 (1997).
10. M. A. Edwards, C. W. Clark, K. Burnett, S. L. Rolston, W. D. Phillips, *J. Phys. B*, in press.
11. M. Kozuma et al., *Phys. Rev. Lett.* **82**, 871 (1999).
12. Our TOP trap is different from previous TOP traps (13) because the rotating bias field orbits in a plane that includes the quadrupole axis. The field gradient along the quadrupole axis ($\hat{z}$) is 9.2 T/m, and the rotating bias field is 1.0 mT. The time-averaged magnetic field forms a trap with harmonic frequencies $\omega_x/2\pi = 180$ Hz, $\omega_y/2\pi = 250$ Hz, and $\omega_z/2\pi = 360$ Hz.
13. W. Petrich, M. H. Anderson, J. R. Ensher, E. A. Cornell, *Phys. Rev.* **74**, 3352 (1995).
14. Atoms in the state $m = -1$ are trapped by the magnetic fields, whereas those in state $m = +1$ are antitrapped. The state $m = 0$ does not feel the confining potential of the magnetic trap.
15. This was done by switching off the trap and measuring the rate of the mean-field-driven ballistic expansion of the condensate at long times (>10 ms).
16. All uncertainties reported in this paper are 1-SD combined statistical and systematic uncertainties.
17. For frequency stability, both beams are derived from a single dye laser with the frequency difference controlled by two acousto-optical modulators.
18. θ, which equals 166° in this case, is the angle between k_1 and k_2, and $|k_1| \approx |k_2| = k$. Therefore, $P = 1.99\hbar k \approx 2\hbar k$.
19. We define $\Delta m = m_{final} - m_{initial}$; $m_{initial} = -1$ is the only magnetically trapped state.
20. Assuming that the scattering lengths among all m states are the same, this can be derived from expressions found in [F. Dalfovo, S. Giorgini, L. P. Pitaevskii, S. Stringari, *Rev. Mod. Phys.* **71**, 2 (1999)]. The energy needed to add one atom is μ, which has a magnetic contribution of 3/7 μ and a mean-field contribution of 4/7 μ. If a small number of atoms are ouput-coupled to $m = 0$, a state that is not magnetically trapped, their

release energy will simply be 4/7 μ, or twice the average release energy of 2/7 μ for the whole condensate.
21. In addition, the longitudinal momentum width is reduced by about the same factor because of kinematic compression.
22. The characteristic time during which the mean field potential energy turns into kinetic energy in the released BEC is $1/\bar{\omega}$ (in our case about 6 ms), where $\bar{\omega}$ is the geometric mean of the three trapping frequencies. For our two-photon Raman transition the characteristic time scale for leaving the region of the condensate is 300 μs.
23. This is because our TOP field rotates in $\hat{x}$-$\hat{z}$ plane, which includes the direction of gravity.
24. The power quoted was the average over a 3-mm-diameter aperture in the center of a somewhat inhomogeneous 7-mm beam. These powers were empirically chosen to produce good output coupling.
25. The resonance frequency, for the $\Delta m = 2$ four-photon transition discussed later, was found to be 6.15(5) MHz, in good agreement with the calculated value of 6.0(2) MHz based on measurements of the trapping magnetic fields. This additional detuning of 2×250 kHz = 500 kHz from the four-photon resonance frequency is large compared with the Fourier width of the Raman pulse and results in a suppression of coupling to the $4\hbar k\hat{z}$, $m = +1$ state.
26. A stimulated Raman transition that changes the momentum state of an atom but does not change the internal energy state can be viewed as Bragg diffraction

(11); see also P. J. Martin, B. G. Oldaker, A. H. Miklich, D. E. Pritchard, *Phys. Rev. Lett.* **60**, 515 (1988).
27. Because of our choice of applying the Raman beams along the quadrupole axis of the trap ($\hat{z}$), the trajectory of the output-coupled atoms (initially along $\hat{z}$) lies in the $\hat{x}$-$\hat{z}$ plane because gravity is along $\hat{x}$. This is the plane of the rotating magnetic field zero and so the atoms will, at some point in time, cross this circle of death.
28. In the case of θ = 180° the recoil momentum from a first-order Raman transition is exactly $2\hbar k\hat{z}$, which corresponds to a frequency of 100.1 kHz.
29. This was confirmed in a separate experiment, which looked at the interference of two clouds of atoms diffracted out of the condensate.
30. If the output coupling process were made continuous, by using an optical dipole or magnetic trap with no time-dependent magnetic fields, such coupling would not occur because the Fourier width of the light pulse could be made arbitrarily small. It would therefore be a simple matter to make a continuous Raman output coupler in such a case.
31. M. R. Andrews et al., *Science* **273**, 637 (1997).
32. Supported in part by the Office of Naval Research and NASA. M.K. acknowledges the support of the Japanese Society for the Promotion of Science for Young Scientists. We thank C. W. Clark, M. A. Edwards, and P. S. Julienne for their valuable comments and suggestions.

24 November 1998; accepted 3 February 1999

Onset of Fermi Degeneracy in a Trapped Atomic Gas

B. DeMarco and D. S. Jin*†

An evaporative cooling strategy that uses a two-component Fermi gas was employed to cool a magnetically trapped gas of 7×10^5 ^{40}K atoms to 0.5 of the Fermi temperature T_F. In this temperature regime, where the state occupation at the lowest energies has increased from essentially zero at high temperatures to nearly 60 percent, quantum degeneracy was observed as a barrier to evaporative cooling and as a modification of the thermodynamics. Measurements of the momentum distribution and the total energy of the confined Fermi gas directly revealed the quantum statistics.

Fermions, such as electrons, protons, and neutrons, compose all of the matter around us, and phenomena derived from the quantum degeneracy of Fermions are ubiquitous in nature. Fermi-Dirac (FD) statistics governs the structure and behavior of such diverse systems as atoms, nuclei, electrons in metals, and white dwarf and neutron stars. However, Fermi systems are generally dense and strongly interacting. The only realization of a low-density Fermi system up to the present has been a dilute solution of liquid ^{3}He dissolved in superfluid ^{4}He (1). We report here the creation of a nearly ideal Fermi gas composed of atoms cooled to the regime where effects of quantum statistics can be observed.

This ultracold atomic gas constitutes a dilute system in which the interparticle interactions are weak and readily treated theoretically. Furthermore, fundamental control over the interactions is available, for example, through the recently realized magnetic field Feshbach resonances (2). In addition, the inhomogeneous trapping potential leads to a spatial separation of high- and low-energy atoms, giving rise to a Fermi surface that is manifest in position as well as in momentum. Features comparable with those listed above have already been demonstrated and exploited in studies of dilute-gas Bose-Einstein condensates (3). Hence, an ultracold atom gas of Fermions is an ideal system for quantitative study of quantum statistical effects in a controlled environment. Novel phenomena predicted for this system include shell structure (4), linewidth narrowing in spontaneous emission (5, 6), suppression of inelastic and elastic collisions (7, 8), changes in the excitation spectrum and damping rates (9, 10), and the emergence of a zero-sound mode at

low temperature (11). Another prospect in this system is the possibility of a phase transition at very low temperature to a superfluid state of Cooper-paired atoms (12).

We magnetically confined and evaporatively cooled a gas of Fermionic atoms, ^{40}K, to temperatures T below 300 nK. The gas enters a regime where T is less than the Fermi temperature T_F and quantum statistical effects become significant. We monitored the evaporative cooling efficiency by measuring the scaled temperature T/T_F as a function of the number of atoms N and probed the thermodynamics of the Fermi gas with time-of-flight optical imaging. The quantum statistics are directly observable in the momentum distribution and in the total energy of the ultracold, trapped Fermionic atom gas.

The ^{40}K atoms are collected and precooled optically in a double magneto-optic trap (MOT) apparatus (Fig. 1). In the first MOT, atoms are captured from a room-temperature vapor provided by a potassium atom source enriched in the Fermionic isotope ^{40}K (13). A series of light pulses pushes the atoms to the second MOT located in a higher vacuum section of the apparatus (14) where the gas is Doppler-cooled to 150 μK (15) before being loaded into a purely magnetic trap. A Ioffe-Pritchard–type magnetic trap (16) provides a cylindrically symmetric, harmonic potential with an axial frequency of $\omega_z = 2\pi \times 19.5$ Hz and a variable radial frequen-

cy. The radial frequency, as well as the minimum magnetic field, can be smoothly varied from $\omega_r = 2\pi \times 44$ Hz to $2\pi \times 370$ Hz by changing the current in a pair of Helmholtz bias coils. The lifetime of atoms in the magnetic trap has an exponential time constant of 300 s limited by collisions with residual room-temperature atoms. This gives ample time for forced evaporation.

Forced evaporative cooling of Fermionic atoms to quantum degeneracy presents particular challenges because of the FD statistics. Evaporative cooling in the magnetic trap (17) relies on binary elastic collisions to rethermalize the gas after selective removal of the most energetic atoms. At the temperatures of interest (below 100 μK), these collisions are primarily s-wave in character and are prohibited between spin-polarized identical Fermions. Various schemes to circumvent this limit to evaporative cooling have been proposed, including sympathetic cooling between Fermionic and Bosonic species (18, 19) or enhancement of the p-wave collision rate by means of an applied dc electric field (20). Our strategy for cooling ^{40}K involves magnetically trapping two spin states of the Fermionic species (21). In effect, we simultaneously cool two gases that maintain thermal equilibrium only through thermal contact with each other by means of allowed s-wave collisions. Runaway evaporation, in which the collision rate in the gas increases as the temperature decreases, thus requires a nearly equal mixture of the two spin states and balanced removal of energy from each component. These constraints necessitate selective removal of atoms from both spin state gases during evaporation.

For the experiment presented here we trapped a mixture of atoms in two magnetic sublevels, $|F = 9/2, m_F = 9/2\rangle$ and $|9/2, 7/2\rangle$, of the hyperfine ground state having total atomic spin $F = 9/2$ (m_F is the magnetic quantum number) (Fig. 2A). This particular mixture of spin states is metastable against m_F changing collisions at low T so that the number of atoms in each state is separately conserved. To selectively remove atoms in either spin state for evaporation we drive

Fig. 1. Schematic of the double-MOT apparatus used to trap and cool ^{40}K atoms. The transfer tube permits differential pressure between first MOT (left) and second MOT (right) and hexapole magnetic confinement for atoms in transit between the two glass cells. Atoms in the second MOT are transferred into a Ioffe-Prit-

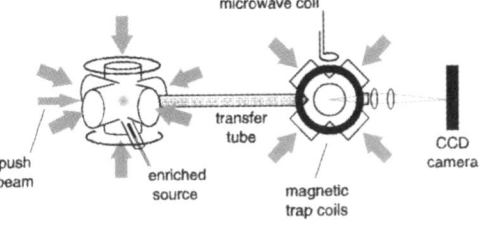

chard–type magnetic trap where they are evaporatively cooled by a microwave field delivered by a small coil. After the atoms are released from the trap, the shadow that the atom cloud casts onto a resonant probe beam is imaged onto a charge-coupled device (CCD) array.

JILA, National Institute of Standards and Technology, and Physics Department, University of Colorado, Boulder, CO 80309-0440, USA.

*Quantum Physics Division, National Institute of Standards and Technology, University of Colorado, Boulder, CO 80309-0440, USA.
†To whom correspondence should be addressed. E-mail: jin@jilau1.colorado.edu

microwave transitions with frequencies $\nu_{9/2}$ and $\nu_{7/2}$ (~1.3 GHz) to an untrapped spin state in the $F = 7/2$ hyperfine level. Zeeman shifts, due to the interaction of the atomic magnetic moment μ with the spatially varying magnetic field of the trap, cause these transitions to be nondegenerate and allow selective removal of the high-energy atoms present in either gas.

As illustrated in Fig. 2B, this type of evaporation can occur in three regimes set by the energy scales of the gas temperature compared with the trap's minimum magnetic field B_0. The kinetic energy scale $k_B T$ sets the width of the cloud in the magnetic potential and hence the spread of microwave frequencies $\delta\nu \sim k_B T/h$ that will remove atoms from the trap. Here k_B is Boltzmann's constant and h is Planck's constant. The trap bias field B_0 sets the Zeeman shift $\Delta\nu = \nu_{9/2} - \nu_{7/2}$ between the two microwave lines at the trap center. At relatively high T (case 1 in Fig. 2B) where $\delta\nu \gg \Delta\nu$, a single frequency removes atoms nearly equally from both spin states. At low T (case 3 in Fig. 2B) where $\delta\nu \ll \Delta\nu$, the microwave lines are distinct and two microwave frequencies are needed to cool both components of the gas in parallel. In the intermediate case (case 2, Fig. 2B) where $\delta\nu \approx \Delta\nu$, the application of any relevant microwave frequency will remove unequal numbers of atoms from each species. It is therefore impossible to efficiently cool through this regime with a microwave field.

To overcome this problem we controlled the evaporation regime through adiabatic changes in the magnetic trap strength, which affect both T and B_0. After ~10^8 atoms are loaded into a relatively weak magnetic trap, an adiabatic ramp to a high-ω_r trap increases the collision rate for evaporation. The evap-

oration begins in this $B_0 = 1.0$ gauss trap with a 60%/40% mixture of the $m_F = 9/2$ and $m_F = 7/2$ states at $T \approx 1$ mK ($T/T_F \approx 240$). In the first stage of evaporation, a single-frequency microwave field is applied to evaporatively cool the gas as in case 1 (Fig. 2B). When the condition $\delta\nu \approx \Delta\nu$ is reached (at $T/T_F \approx 1$), a second stage begins with an adiabatic ramp of the bias field to $B_0 = 5.0$ gauss, resulting in a radial trap frequency $\omega_r = 2\pi \times 137$ Hz. In this trap, evaporation uses a two-frequency microwave field (case 3, Fig. 2B) with a frequency difference that keeps the spin mixture constant to within 5%.

After evaporation an analysis of resonant absorption images, similar to that used to investigate Bose-Einstein condensates (22), was used to determine N, T, and the momentum distribution of the Fermi gas. To simplify interpretation of these images, we produced a single-component gas of $m_F = 9/2$ atoms by removing the $m_F = 7/2$ atoms with a microwave sweep (23). To keep the remaining $m_F = 9/2$ atoms in thermal equilibrium, we chose a time scale for the sweep that was slow compared with the collision rate in the sample. Indeed, this sweep provides the final evaporative cooling. The time-of-flight image was taken by suddenly switching off the current that provides the magnetic trapping field (always releasing from the $B_0 = 5.0$ gauss trap) was switched off, which allowed the gas to expand freely for 15 to 20 ms. The absorption shadow, generated by illumination of the expanded gas with a 24-μs pulse of light resonant with the $4S_{1/2}$, $F = 9/2$ to $4P_{3/2}$, $F = 11/2$ transition, was imaged onto a charge-coupled device array, and the optical depth was calculated. The probe beam, which travels along the double-MOT axis (Fig. 1), is circularly polarized and has a uniform inten-

sity profile with $I/I_{\text{saturation}} = 0.05$. The imaging system uses $f/5$ optics and has a resolution better than 15 μm, much smaller than the 270-μm size of the smallest expanded atom cloud.

The quantum statistics dramatically influences the evaporative cooling of Fermions, as illustrated in a plot of the observed evaporation trajectory (Fig. 3). The trajectory is shown as T/T_F versus N, where $T_F = \hbar/k_B (6\omega_z\omega_r^2 N)^{1/3}$ (Fig. 3) ($\hbar = h/2\pi$) (24, 25). The value of T_F, which changes as N decreases during evaporation, is $T_F = 0.6$ μK for a million atoms in the $\omega_r = 2\pi \times 137$ Hz trap. A measure of the efficiency of evaporation is the slope of this T/T_F versus N curve. Although T/T_F follows an efficient trend with N for the bulk of evaporation, the behavior changes drastically as T/T_F nears 0.5. For $T/T_F \lesssim 0.5$, the forced evaporation becomes grossly inefficient in reducing T/T_F, and many more atoms are removed to accomplish the same change in T.

This plunge in cooling efficiency does not coincide with any observed change in the atoms' loss or heating rate and is robust against changes in the details of evaporation. The behavior survives variation in the initial number of atoms and temperature of the sample, changes in the duration of the microwave sweep that removes $m_F = 7/2$ atoms, changes in the evaporation timing, and even replacement of the second stage of evaporation with continued single-frequency removal of atoms in a lower B_0 trap (as in case 1, Fig. 2B). In addition to the data presented in Fig. 3 obtained with the evaporation procedure described above, we observed that the reduction in evaporation efficiency always occurs near $T/T_F \approx 0.5$, even when we varied N and ω_r. At the point where the evaporation effi-

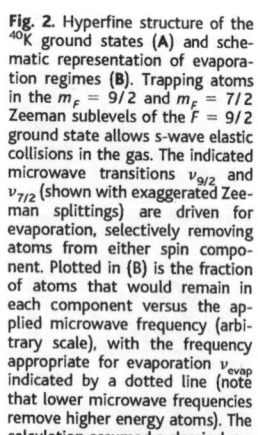

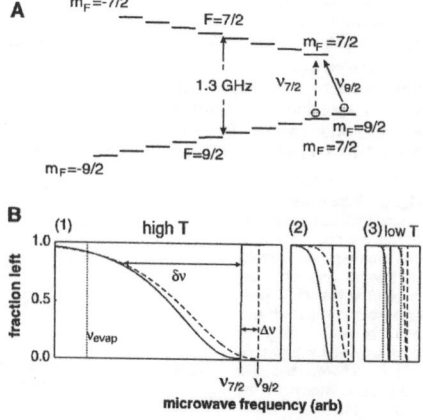

Fig. 2. Hyperfine structure of the ^{40}K ground states (**A**) and schematic representation of evaporation regimes (**B**). Trapping atoms in the $m_F = 9/2$ and $m_F = 7/2$ Zeeman sublevels of the $F = 9/2$ ground state allows s-wave elastic collisions in the gas. The indicated microwave transitions $\nu_{9/2}$ and $\nu_{7/2}$ (shown with exaggerated Zeeman splittings) are driven for evaporation, selectively removing atoms from either spin component. Plotted in (B) is the fraction of atoms that would remain in each component versus the applied microwave frequency (arbitrary scale), with the frequency appropriate for evaporation ν_{evap} indicated by a dotted line (note that lower microwave frequencies remove higher energy atoms). The calculation assumed a classical gas in which the microwave field removes all atoms above a particular energy. The high T (case 1), low T (case 3), and intermediate regimes (case 2) are distinguished by the spread of frequencies $\delta\nu$ resonant with atoms in the trap compared with the Zeeman shift $\Delta\nu$ at the trap center.

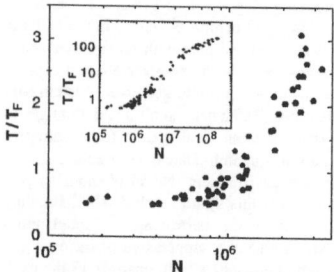

Fig. 3. Evaporation trajectory. A plot T/T_F versus N shows the result of evaporation; the inset displays the entire trajectory, starting at $T/T_F \approx 240$ and $N = 10^8$ atoms, and the main figure shows the low-temperature region. The bulk of the evaporation is very efficient, as seen in the large slope of the T/T_F versus N curve. However, the cooling process becomes limited at $T/T_F \approx 0.5$, where effects of FD statistics are observed in the momentum distribution of the gas.

ciency falls, N varied from 3.5×10^5 to 1.2×10^6 and ω_r from 127 to 373 Hz, corresponding to a T_F of 0.36 to 1.0 μK.

Although no quantitative theory exists yet, two general arguments (7) suggest that the onset of quantum behavior should lead to marked changes in the evaporation process. One effect of the quantum statistics is to alter the equilibrium size of the cloud in the magnetic trap by means of the Fermi pressure. This quantum mechanical pressure arises from the Pauli exclusion principle and causes a Fermi gas to resist compression, for example stabilizing white dwarf and neutron stars against gravitational collapse (26). In the magnetic trap, the confined gas approaches a fixed size as T approaches 0, with a root-mean-square radius of $0.6R_F$ where $R_F = \sqrt{2k_BT_F/mw^2}$ (24, 25); at $T/T_F = 0.5$, the radius of the gas is only $0.9R_F$. However, efficient "runaway" evaporation depends on the continual decrease in the size of the trapped cloud as T decreases and the accompanying increase in collision rate. Thus, the Fermi pressure will eventually negatively affect the evaporation.

A second effect of the quantum statistics on evaporation is Pauli blocking, which alters the dynamics of rethermalization after removal of high-energy atoms. At low T/T_F the atoms begin to form a Fermi sea arrangement, with the mean occupancy per state of the low-lying motional states approaching unity. At $T/T_F = 0.5$, for example, the occupancy of the lowest trap states is already 60%. The Pauli exclusion principle will begin to block the elastic collisions (8) essential to evaporation because

the collisions will involve occupied low-energy final states. A quantitative understanding is lacking for how and when these effects become important in the forced evaporative cooling of Fermionic atoms.

In addition to witnessing a decrease in the evaporation efficiency, we observed the emerging quantum degeneracy in measurements of the total energy and the momentum distribution of the trapped Fermi gas. Classically, at high T, the gas has total energy $U_{cl} = 3Nk_BT$ and a gaussian momentum distribution. At $T = 0$, however, the atoms occupy the energy levels of the harmonic confining potential in a Fermi sea arrangement with $U_{FD} = \frac{3}{4} Nk_BT_F$, and the Fermi pressure results in a parabolic momentum distribution (24). We measured the extra energy due to FD statistics and observed the transition between these two distributions by analyzing the optical depth images of expanded clouds.

A deviation from classical thermodynamics is exposed in a measurement of the total energy U of the trapped gas (Fig. 4). The total energy was obtained from the second moment calculated directly from absorption images of expanded clouds (27). This moment analysis of time-of-flight images is independent of any assumption of the exact statistical distribution. In addition, the effect of interactions on the expansion can be neglected because the mean field interaction energy is more than two orders of magnitude smaller than the trap potential energy for our single-component Fermi gas (10). In contrast to a dilute-gas Bose-Einstein condensate, where expansions typically exhibit a large effect due to interactions (28), the free expansion of the Fermi gas arises entirely from the kinetic energy of the sample.

The difference $\delta U = U - U_{cl}$ between the measured energy and the classical energy at the same T is plotted in Fig. 4. The temperature T is determined from a fit to the periphery of the absorption image where the effects of the quantum statistics are reduced

because of the low mean occupancy at these high-momentum states (29). The emergence of excess energy in the gas due to FD statistics coincides with the change in evaporation efficiency and shows quantitative agreement with thermodynamic theory for a noninteracting Fermi gas (Fig. 4).

The excess energy due to FD statistics should be accompanied by a characteristic nongaussian momentum distribution. To explore the shape of the momentum distribution (6), we fit the cloud optical depth, $OD(\rho)$, to the following functional form that varies smoothly between the correct low- and high-T limits:

$$OD(\rho) = \begin{cases} A\left(1 - \dfrac{\rho^2}{R^2}\right)^2 & ,1 - \dfrac{\rho^2}{R^2} \geq L \\ Be^{-\frac{\rho^2}{2}} & , \text{otherwise} \end{cases} \quad (1)$$

Here, ρ is a scaled distance $\rho = \sqrt{x^2/\sigma_x^2 + z^2/\sigma_z^2}$ from the peak of the distribution, and A, σ_x, σ_z, and L are fit parameters. The requirement of continuity of the function and its first derivative at the boundary of the inner quartic and the outer gaussian form fixes the parameters B and R. The parameter L characterizes the deviation from the classical gaussian profile with $L = 1$ at $T/T_F \gg 1$ and $L = 0$ at $T/T_F = 0$.

The data in Fig. 5 show the onset of a clear deviation from a gaussian momentum distribution ($L = 1$) as T decreases below T_F. At low T/T_F the best fit value of the parameter L decreases from the classical value $L = 1$ as the momentum distribution becomes nongaussian. This nongaussian character of the time-of-flight images can also seen in an analysis of fit residuals that uses azimuthally averaged data. When low T/T_F images are fit to the classical gaussian distribution, a pattern appears in the fit residuals as a function of scaled radius ρ (Fig. 5, inset). At these low temperatures, fits to the function in Eq. 1 typically give a factor of 3 improvement in the reduced χ^2 compared with the simple gaussian fit. The deformation of the momen-

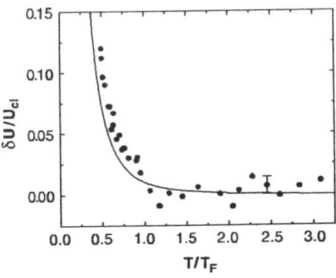

Fig. 4. Emergence of quantum degeneracy as seen in the energy of the trapped Fermi gas. A moment analysis was used to extract the energy of the gas from time-of-flight absorption images. The excess energy $\delta U = U - U_{cl}$ is shown versus T/T_F, where U is the measured energy and $U_{cl} = 3Nk_BT$ is the energy of a classical gas at the same temperature. Each point represents the average of two points from the evaporation trajectory shown in the main part of Fig. 3, and the single error bar shows the typical statistical uncertainty. The measured excess energy at low T/T_F agrees well with thermodynamic theory for a noninteracting Fermi gas (line).

Fig. 5. Emergence of quantum degeneracy as seen in the shape of the momentum distribution. Surface fits to the same absorption images used for Fig. 4 reveal the nongaussian character of the momentum distribution at low T/T_F. For a particularly low noise image at $T/T_F = 0.5$ the inset shows fit residuals normalized by the peak OD versus the scaled cloud radius ρ. A classical gaussian fit (solid triangles) is contrasted with the nongaussian fitting function from Eq. 1 (open circles). The main figure shows the fit parameter L of the nongaussian form versus T/T_F. For a classical gas $L = 1$; for a Fermi gas at $T/T_F = 0$, $L = 0$. The data compare well to theory (line), in which interactions are neglected and the distribution is calculated with the Thomas-Fermi approximation (24).

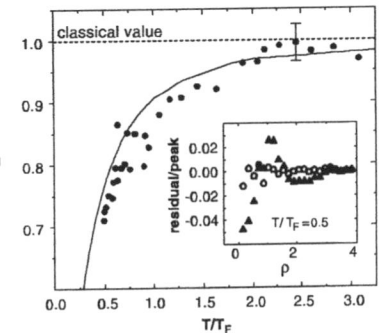

REPORTS

tum distribution observed in the nongaussian fits signals the onset of FD quantum degeneracy, as evidenced by the good agreement between the data and the theory line presented in Fig. 5. In fact, the comparison with theory suggests that our lowest temperatures are actually closer to $T/T_F = 0.4$, consistent with our 20% systematic uncertainty in T/T_F (primarily due to uncertainty in N) (30).

We detected the emergence of quantum degeneracy in a trapped gas of Fermionic atoms and observed a barrier to the evaporative cooling process in a two-component Fermi gas below $0.5 \; T_F$. We observed a nonclassical momentum distribution and found that the total energy of the gas is larger than the classical expectation. This excess energy is a manifestation of the Pauli exclusion principle that gives rise to an expanded momentum distribution at low T/T_F by forcing the atoms to fill higher motional states of the harmonic trapping potential. Even as T approaches zero, an ideal Fermi gas still has $\frac{3}{4} k_B T_F$ energy per particle; indeed, at our lowest $T/T_F \approx 0.5$ we measured an energy that is only 2.2 times this $T = 0$ limit. Reaching this quantum regime in the dilute Fermi gas extends the field of quantum degenerate gases and sets the stage for further experimental probes of a Fermi sea of atoms.

References and Notes

1. See, for example, E. P. Bashkin and A. E. Meyerovich, *J. Phys. Colloq. France* **41**, C7-61 (1980).
2. S. Inouye *et al.*, *Nature* **392**, 151 (1998); Ph. Courteille, R. S. Freeland, D. J. Heinzen, F. A. van Abeelen, B. J. Verhaar, *Phys. Rev. Lett.* **81**, 69 (1998); J. L. Roberts *et al.*, *ibid.*, p. 5109; V. Vuletic, A. J. Kerman, C. Chin, S. Chu, *ibid.* **82**, 1406 (1999).
3. For recent reviews, see E. A. Cornell, J. R. Ensher, C. E. Wieman, online abstract available at http://xxx.lanl.gov/abs/cond-mat/9903109; W. Ketterle, D. S. Durfee, D. M. Stamper-Kurn, online abstract available at http://xxx.lanl.gov/abs/cond-mat/9904034
4. J. Schneider and H. Wallis, *Phys. Rev. A* **57**, 1253 (1998); G. M. Bruun and K. Burnett, *ibid.* **58**, 2427 (1998).
5. K. Helmerson, M. Xiao, D. Pritchard, International Quantum Electronics Conference 1990, book of abstracts (IEEE, New York, 1990), abstr. QTHH4; Th. Busch, J. R. Anglin, J. I. Cirac, P. Zoller, *Europhys. Lett.* **44**, 1 (1998); J. Ruostekoski and J. Javanainen, *Phys. Rev. Lett.* **82**, 4741 (1999).
6. B. DeMarco and D. S. Jin, *Phys. Rev. A* **58**, R4267 (1998).
7. J. M. K. V. A. Koelman, H. T. C. Stoof, B. J. Verhaar, J. T. M. Walraven, *Phys. Rev. Lett.* **59**, 676 (1987).
8. G. Ferrari, *Phys. Rev. A* **59**, R4125 (1999).
9. G. Bruun and C. Clark, online abstract available at http://xxx.lanl.gov/abs/cond-mat/9905263
10. L. Vichi and S. Stringari, online abstract available at http://xxx.lanl.gov/abs/cond-mat/9905154
11. S. K. Yip and T. L. Ho, *Phys. Rev. A* **59**, 4653 (1999).
12. H. T. C. Stoof, M. Houbiers, C. A. Sackett, R. G. Hulet, *Phys. Rev. Lett.* **76**, 10 (1996); M. A. Baranov and D. S. Petrov, *Phys. Rev. A* **58**, R801 (1998); M. Houbiers and H. T. C. Stoof, *ibid.* **59**, 1556 (1999); G. Bruun, Y. Castin, R. Dum, K. Burnett, online abstract available at http://xxx.lanl.gov/abs/cond-mat/9810013.
13. B. DeMarco, H. Rohner, D. S. Jin, *Rev. Sci. Instrum.* **70**, 1967 (1999).
14. C. J. Myatt, N. R. Newbury, R. W. Ghrist, S. Loutzenhiser, C. E. Wieman, *Opt. Lett.* **21**, 290 (1996).
15. In our experiment we saw no evidence for sub-Doppler cooling processes that are used to bring some alkali gases to much colder temperatures. However, sub-Doppler cooling of ^{40}K is reported in G. Modugno, C. Benko, P. Hannaford, G. Roati, M. Inguscio, online abstract available at http://xxx.lanl.gov/abs/cond-mat/9908102
16. Y. V. Gott, M. S. Ioffe, V. G. Tel'kovski, *Nucl. Fusion* (1962 suppl.), 1045 (1962); *ibid.*, p. 1284; D. E. Pritchard, *Phys. Rev. Lett.* **51**, 1336 (1983).
17. H. F. Hess, *Phys. Rev. B* **34**, 3476 (1986); H. F. Hess *et al.*, *Phys. Rev. Lett.* **59**, 672 (1987).
18. C. J. Myatt, E. A. Burt, R. W. Ghrist, E. A. Cornell, C. E. Wieman, *Phys. Rev. Lett.* **78**, 586 (1997).
19. W. Geist, L. You, T. A. B. Kennedy, *Phys. Rev. A* **59**, 1500 (1999); E. Timmermans and R. Côté, *Phys. Rev. Lett.* **80**, 3419 (1998).
20. W. Geist, A. Idrizbegovic, M. Marinescu, T. A. B. Kennedy, L. You, online abstract available at http://xxx.lanl.gov/abs/cond-mat/9907222
21. B. DeMarco, J. L. Bohn, J. P. Burke Jr., M. Holland, D. S. Jin, *Phys. Rev. Lett.* **82**, 4208 (1999).
22. J. R. Ensher, D. S. Jin, M. R. Matthews, C. E. Wieman, E. A. Cornell, *ibid.* **77**, 4984 (1996).
23. The remaining gas was composed of 99% $m_f = 9/2$ atoms, determined from a measurement of the cross-dimensional relaxation rate (21).
24. D. A. Butts and D. S. Rokhsar, *Phys. Rev. A* **55**, 4346 (1997).
25. I. F. Silvera and J. T. M. Walraven, *J. Appl. Phys.* **52**, 2304 (1981); J. Oliva, *Phys. Rev. B* **39**, 4204 (1989).
26. W. Greiner, L. Neise, H. Stöcker, *Thermodynamics and Statistical Mechanics* (Springer-Verlag, New York, 1995), pp. 359–362.
27. The kinetic energy extracted from the time-of-flight absorption images equals half the total energy of the harmonically confined gas (from the equipartition theorem).
28. M. O.-Mewes *et al.*, *Phys. Rev. Lett.* **77**, 416 (1996); M. J. Holland, D. S. Jin, M. L. Chiofalo, J. Cooper, *ibid.* **78**, 3801 (1997).
29. T is obtained from the widths of the outer gaussian, σ_x and σ_z, in fits of the form given in Eq. 1. Although the effects of the FD statistics are less severe on the outer edges of the momentum distribution, these fits become less accurate as T/T_F decreases. We made a correction to T that is at most 7% based on the measured T/T_F and the results of identical fits to calculated (semiclassical) momentum distributions for an ideal Fermi gas.
30. The number of atoms N is calibrated by a florescence measurement, which has an uncertainty of $\pm 50\%$ because of intensity variations across the laser beams. The trap frequencies are determined to better than $\pm 5\%$ from center-of-mass oscillations of the trapped gas.
31. Supported by the National Institute of Standards and Technology, the NSF, and the Office of Naval Research. We thank C. Wieman, E. Cornell, and the other members of the JILA BEC group for useful discussions.

19 July 1999; accepted 16 August 1999

VOLUME 41, NUMBER 20   PHYSICAL REVIEW LETTERS   13 NOVEMBER 1978

Observation of Focusing of Neutral Atoms by the Dipole Forces
of Resonance-Radiation Pressure

J. E. Bjorkholm, R. R. Freeman, A. Ashkin, and D. B. Pearson
Bell Telephone Laboratories, Holmdel, New Jersey 07733
(Received 21 July 1978)

For sodium atoms in an atomic beam, we demonstrate focusing, defocusing, and steering caused by the transverse dipole forces exerted by the radial intensity gradient of a superimposed and co-propagating resonant cw light beam. Dipole radiation-pressure forces differ from the forces due to spontaneous emission and are needed to achieve optical traps for neutral atoms.

We have observed that a cw laser beam superimposed upon and co-propagating with a beam of neutral atoms can cause substantial changes in the atomic trajectories when the light frequency is tuned near an atomic resonance. The atoms can be confined, ejected, or steered by the light beam. This new effect, the focusing of atoms by light, results from the same physical mechanism (momentum exchange) responsible for self-focusing of light in atomic vapors.[1] These deflections are caused by the transverse *dipole* resonance-radiation-pressure forces exerted on an induced dipole by an electric field gradient. Deflection of neutral atoms by dc field gradients is well known[2] and the deflection of neutral molecules by gradients of resonant microwave fields has been observed.[3] The analogous effects in atoms caused by resonant fields have not previously been observed, but they have been discussed lately in applications of light pressure.[4,5] Indeed, transverse dipole forces are important in proposed optical traps for neutral atoms.[5] Since the effects we observe are quite strong, other applications will also be apparent.

Dipole resonance-radiation pressure arises from stimulated light-scattering processes and exists only in optical field gradients; it thus differs fundamentally from spontaneous resonance-radiation pressure[6] which arises from spontaneous light scattering and which exists even in uniform resonant light fields. Spontaneous forces have been observed and discussed in many situations, for example, deflection of atoms,[6,7] cooling of atomic vapors,[8] induced density gradients in a vapor,[9] and isotope separation.[10] Recently, they have been used to cool ions contained in ion traps.[11] Both pressures, of course, derive from light momentum. However, the dipole force can be made the larger of the two forces.

A diagram of our experiment is shown in Fig. 1. Light from a continuously tunable, single-mode cw dye laser was superimposed upon an effusive atomic beam of neutral sodium using a 3-mm-thick dielectric-coated mirror with a 230-μm-diam hole in it. The light was focused by a 75-cm lens to a focal spot size $w_0 = 100\ \mu$m situated 25 cm from the mirror. The laser spot size on the mirror was 500 μm and the confocal parameter of the beam was 10 cm. Because the mirror was in the far field of the light, the dark spot in

VOLUME 41, NUMBER 20   **PHYSICAL REVIEW LETTERS**   13 NOVEMBER 1978

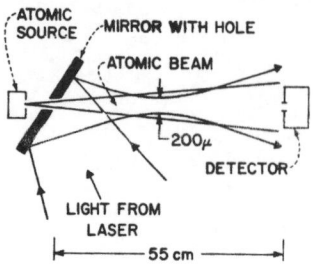

FIG. 1. Schematic diagram of experimental setup. Note the scale differences between the longitudinal and transverse directions.

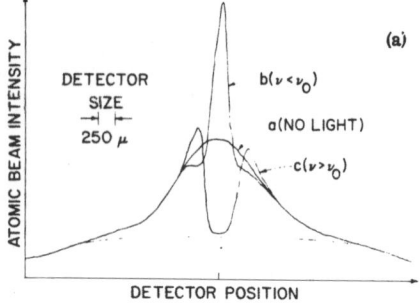

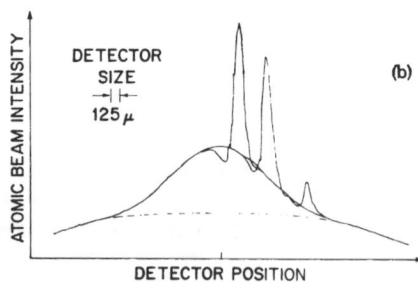

FIG. 2. (a) The atomic-beam current at the detector as a function of transverse detector position. The light dashed line shows the approximate background intensity. Peak beam current is approximately 2×10^8 atoms/sec. (b) The atomic-beam current as a function of transverse detector position for laser frequencies $\nu < \nu_0$, and various degrees of misalignment of the laser beam relative to the atomic beam.

the center of the reflected laser beam caused by the hole in the mirror was totally washed out in the near field of the light. Thus the laser intensity distribution was nearly Gaussian in the central 20-cm region where the interaction between the atoms and light predominantly took place. The laser was tuned near 5890 Å in order to excite the $3\,^2S_{1/2} \rightarrow 3\,^2P_{3/2}$ resonance transition. The atomic-beam profile was measured by a movable hot-wire detector. We found the detector to be insensitive to the incident light.

The time-averaged transverse dipole force exerted on a two-level atom can be obtained from the radial potential energy $U(r)$, given by[5]

$$U(r) = \tfrac{1}{2} h (\nu - \nu_0) \ln(1 + p), \tag{1}$$

where the atomic saturation parameter p is given by

$$p = \frac{\lambda^3}{\pi^3 hc\Delta\nu_N} \frac{\Delta\nu_N^2/4}{(\nu - \nu_0)^2 + \Delta\nu_N^2/4} \frac{P}{w^2} \exp(-2r^2/w^2).$$

Here ν and λ are the frequency and wavelength, P is the optical power, w is the Gaussian beam spot size, ν_0 is the atomic resonance frequency, and $\Delta\nu_N$ is the natural linewidth. For $\nu < \nu_0$, $U(r)$ is negative and atoms are pulled into regions of high light intensity; for $\nu > \nu_0$, atoms tend to be expelled from that region. The transverse forces can be appreciable; for $\nu < \nu_0$ and parameters typical of our experiment, $\lambda = 5890$ Å, $\Delta\nu_N = 10^7$ Hz, $w = 100\ \mu$m, $P = 50$ mW, and $(\nu - \nu_0) = -2$ GHz, we find $p = 0.1$ and $U(0) = 6.9 \times 10^{-19}$ erg. This means that a sodium atom with a transverse velocity as great as 190 cm/sec can be confined transversely within the laser beam. While sodium is not a two-level atom, our experiment demonstrates that possible deleterious effects caused by optical pumping of the ground state are not important.

The atomic-beam profile at the detector is shown in Fig. 2(a). The spatial resolution was set by a 250-μm-diam aperture on the detector; the narrow, intense peak in the atomic-beam intensity induced for $\nu < \nu_0$ and the dip induced for $\nu > \nu_0$ are almost fully resolved. As indicated by the above calculation and as demonstrated in Fig. 2(b), one can guide or steer the atoms by slightly changing the alignment of the laser beam relative to the atomic beam. Calculations show that for $\nu > \nu_0$ nearly all atoms should be expelled from the light beam. This is not observed in Fig. 2 because of background atoms superimposed upon the direct atomic beam; the approximate inferred background level is shown by the dashed lines. Atoms in the background apparently do not interact much, if at all, with the light. To remove the background and force any atom which reaches the detector to interact with the light we placed a 250-μm-diam aperture in the atomic beam, located at the focus of the laser beam. Results for

VOLUME 41, NUMBER 20   PHYSICAL REVIEW LETTERS   13 NOVEMBER 1978

these conditions, and with a 125-μm-diam aperture on the detector, are shown in Fig. 3. As expected, the atoms are almost totally expelled from the laser beam for $\nu > \nu_0$. (In a separate ion-counting experiment we have not been able to detect any photoionization of the atoms caused by the cw light.)

The on-axis atomic-beam intensity as a function of the laser tuning is shown in Fig. 4. The dispersionlike shape of this curve is in accord with Eq. (1). Within several GHz of resonance, the longitudinal force due to spontaneous light pressure is also important. This spontaneous force accelerates the atoms, which reduces the atomic-beam divergence and increases the on-axis atomic-beam intensity. Note the strong effects obtained for tunings more than 15 GHz away from resonance; only the transverse dipole forces are significant in these regions.

The height and width of the peak induced in the atomic-beam intensity distribution for $\nu < \nu_0$ have been roughly explained by approximating the potential of Eq. (1) to be harmonic. Ignoring the longitudinal acceleration due to spontaneous forces, the atomic trajectories in the interaction region are then sinusoidal with a typical oscillation frequency of 4 kHz. Thus, even atoms with a large longitudinal velocity execute more than a full oscillation cycle, and the interaction region can be considered as a distributed strong lens which guides and focuses the atoms by varying amounts depending on initial conditions. A simple calculation along these lines agrees qualitatively with our observations. This analysis predicts a strongly non-Maxwellian velocity distribution on axis. Spontaneous forces will also alter the velocity distribution. We have observed velocity-distribution changes in preliminary experiments using a slotted-disk velocity selector

placed in front of the detector.

By observing changes of atomic trajectories in an atomic beam, we have directly observed the effects of transverse dipole resonance-radiation pressure. Dipole forces may cause important indirect effects in other types of experiments. For instance, the inclusion of transverse dipole forces seem to explain anomalies seen in experiments which observed pressure changes induced by radiation pressure in a long capillary filled with an atomic vapor.[9] It has also been suggested[12] that the density changes induced in a vapor by the dipole forces could cause distortions of line shapes observed in saturation spectroscopy. Whenever nonlinear optical effects are important, the effects of dipole forces may be observable.[13]

Finally, we discuss some possible applications of transverse dipole forces. First, applications to isotope separation are obvious; purification of materials by elimination of trace elements would also seem possible. Using the dipole forces, light beams can be used to confine and guide atoms; in conjunction with spontaneous forces, they can be used to construct optical traps and "bottles" for neutral atoms. An interesting situation would be the use of a TEM_{01} or "doughnut"-mode laser beam tuned to the high-frequency side of resonance. In this case, the atoms would be confined to the axis of the light beam where the intensity is a minimum. The use of a counter-propagating TEM_{00} laser beam would allow one to decelerate the atoms; this will increase the atomic beam divergence and will tend to destroy the beam. The use of transverse dipole forces for confinement (plus possible transverse cooling

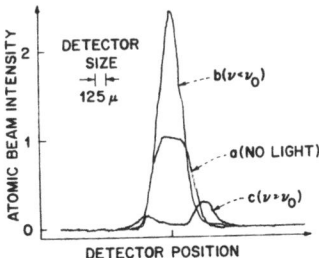

FIG. 3. Same sweep as in Fig. 2(a) only with a 250-μm-diam aperture placed in the atomic beam, located at the focus of the laser beam.

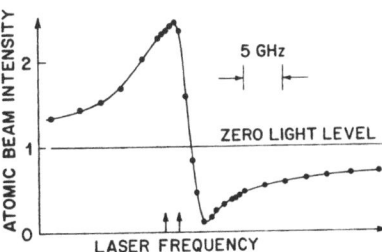

FIG. 4. The on-axis atomic-beam intensity as a function of laser frequency. The arrows mark the frequencies of the two resonance transitions $3^2S_{1/2}(F=2,1)$ $\rightarrow 3^2P_{3/2}$ and are separated by approximately 1.77 GHz. The shift of the curve zero crossing relative to these frequencies is consistent with the longitudinal Doppler shifts of the atoms.

VOLUME 41, NUMBER 20   PHYSICAL REVIEW LETTERS   13 NOVEMBER 1978

using additional light beams) might make it possible to create beams of ultraslow atoms. There may even by applications in molecular-beam epitaxy in which light beams are used to guide the atoms being deposited to desired locations. Of course, the discussion presented here is not restricted to neutral atoms; dipole forces are also exerted on neutral molecules and on ions.

We acknowledge useful discussions with J. P. Gordon and thank D. E. Pritchard for the velocity selector.

[1]G. A. Askar'yan, Zh. Eksp. Teor. Fiz. **42**, 1567 (1962) [Sov. Phys. JETP **15**, 1088 (1962)]; D. Grischkowsky, Phys. Rev. Lett. **24**, 866 (1970); J. E. Bjorkholm and A. Ashkin, Phys. Rev. Lett. **32**, 129 (1974); A. C. Tam and W. Happer, Phys. Rev. Lett. **38**, 278 (1977).

[2]N. F. Ramsey, *Molecular Beams* (Clarendon Press, Oxford, 1956), Chap. X.

[3]R. M. Hill and T. F. Gallagher, Phys. Rev. A **12**, 451 (1975).

[4]V. S. Letokhov, V. G. Minogin, and B. D. Pavlik, Opt. Commun. **19**, 72 (1976).

[5]A. Ashkin, Phys. Rev. Lett. **40**, 729 (1978).

[6]A. Ashkin, Phys. Rev. Lett. **25**, 1321 (1970).

[7]R. Schieder, H. Walther, and L. Wöste, Opt. Commun. **5**, 337 (1972); J. L. Picque and J. L. Vialle, Opt. Commun. **5**, 402 (1972).

[8]T. W. Hänsch and A. L. Schawlow, Opt. Commun. **13**, 68 (1975).

[9]J. E. Bjorkholm, A. Ashkin, and D. B. Pearson, Appl. Phys. Lett. **27**, 534 (1975).

[10]A. F. Bernhardt, D. E. Duerre, J. R. Simpson, and L. L. Wood, Appl. Phys. Lett. **25**, 617 (1974).

[11]D. J. Wineland, R. E. Drullinger, and F. L. Walls, Phys. Rev. Lett. **40**, 1639 (1978); W. Neuhauser, M. Hohenstatt, P. E. Toschek, and H. Dehmelt, Phys. Rev. Lett. **41**, 233 (1978).

[12]P. L. Kelley, private communication.

[13]M. E. Marhic and L. I. Kwan, J. Opt. Soc. Am. **68**, 644 (1978).

Optical Trapping and Manipulation of Viruses and Bacteria

A. ASHKIN AND J. M. DZIEDZIC

Optical trapping and manipulation of viruses and bacteria by laser radiation pressure were demonstrated with single-beam gradient traps. Individual tobacco mosaic viruses and dense oriented arrays of viruses were trapped in aqueous solution with no apparent damage using ~120 milliwatts of argon laser power. Trapping and manipulation of single live motile bacteria and *Escherichia coli* bacteria were also demonstrated in a high-resolution microscope at powers of a few milliwatts.

WE REPORT THE EXPERIMENTAL demonstration of optical trapping and manipulation of individual viruses and bacteria in aqueous solution by laser light using single-beam gradient force traps. Individual tobacco mosaic viruses (TMV) and oriented arrays of viruses were optically confined within volumes of a few cubic micrometers, without obvious damage, and manipulated in space with the precision of the optical wavelength. The ability of the same basic optical trap to confine and manipulate motile bacteria was also demonstrated. We have used the trap as an "optical tweezers" for moving live single and multiple bacteria while being viewed under a high-resolution optical microscope. These results suggest that the techniques of optical trapping and manipulation, which have been used to advantage with particles in physical systems, are also applicable to biological particles. Optical trapping and manipulation of small dielectric particles and atoms by the forces of radiation pressure have been studied since 1970 (1–4). These are forces arising from the momentum of the light itself. Early demonstrations of optical trapping (1) and optical levitation (5) involved micrometer-size transparent dielectric spheres in the Mie regime (where the dimensions *d* are large compared to the

wavelength λ). More recently optical trapping of submicrometer dielectric particles was demonstrated in the Rayleigh regime (where $d << \lambda$). Single dielectric particles as small as ~260 Å (6) and even individual atoms (7) were trapped with single-beam gradient traps (8).

Single-beam gradient traps are conceptually and practically the simplest. They consist of only a single strongly focused Gaussian laser beam having a Gaussian transverse intensity profile. In such traps the basic scattering forces and gradient force components of radiation pressure (1, 3, 4, 8) are configured to give a point of stable equilibrium located close to the beam focus. The scattering force is proportional to the optical intensity and points in the direction of the incident light. The gradient force is proportional to the gradient of intensity and points in the direction of the intensity gradient. Particles in a single-beam gradient trap are confined transverse to the beam axis by the radial component of the gradient force. Stability in the axial direction is achieved by making the beam focusing so strong that the axial gradient force component, pointing toward the beam focus, dominates over the scattering force trying to push the particle out of the trap. Thus one has a stable trap based solely on optical forces, where gravity plays no essential role as was the case for the levitation trap (5). It works over a particle

size range of 10^5, from ~10 μm down to a few angstroms, which includes both Mie- and Rayleigh-size particles.

The sensitivity of laser trap effectiveness to optical absorption and particle shape is of particular importance for the trapping of biological particles. Absorption can cause an excessive temperature rise or additional thermally generated (radiometric) forces as a result of temperature gradients within a particle (9). In general, the smaller the particle size the less the temperature rise and the less the thermal gradients for a given absorption coefficient (10). Particle shape plays a larger role in the trapping of Mie particles than Rayleigh particles. For Mie particles both the magnitude and direction of the forces depend on the particle shape (3). This restricts trapping to fairly simple overall shapes such as spheres, ellipsoids, or particles whose optical scattering varies slowly with orientation in the beam. In the Rayleigh regime, however, the particle acts as a dipole (6) and the direction of the force is independent of particle shape; only the magnitude of the force varies with orientation. A significant conclusion of this work is that important types of biological particles in both the Mie and Rayleigh regimes have optical absorptions and shapes that fall within the scope of single-beam gradient traps.

As a first test for trapping of small biological particles we tried TMV, a much studied virus that can be prepared in monodisperse colloidal suspension in water at high concentrations (11, 12). Its basic shape is cylindrical with a diameter of 200 Å and a length of 3200 Å (13). Although its volume of about 470 Å^3 is typical of a Rayleigh particle, its length is comparable to the wavelength in the medium and we expect some Mie-like behavior in its light scattering. TMV particles have a negative charge in solution (14) and an index of refraction (15) of about 1.57. Our virus samples were prepared from the same batches used in experi-

AT&T Bell Laboratories, Holmdel, NJ 07733.

ments on the self-alignment of TMV in parallel arrays in dense aqueous suspensions (*13, 16*). These samples, suitably diluted, were studied in essentially the same apparatus (*6*) previously used for trapping of silica and polystyrene latex colloidal suspensions and Mie-size dielectric spheres (see Fig. 1).

When one visually observes the 90° scattering from untrapped viruses in the vicinity of the focus, one sees the random-walk motion and intensity fluctuations, or flickering, characteristic of Brownian diffusion in position and orientation. These rotational intensity fluctuations of about an order of magnitude result from the large changes in polarizability of a cylindrical particle with orientation in the light beam (*17*). At laser power levels of about 100 to 300 mW we begin to see trapping. The capture of a virus manifests itself as a sudden increase in the 90° scattering, as shown in Fig. 2. As more viruses are captured in the trap, we see further abrupt changes in scattering. If we block the beam momentarily (at points B) the trap empties and the trapping sequence repeats. Not only does all apparent positional Brownian motion disappear when viruses are captured but the intensity fluctuations characteristic of a freely rotating TMV particle are also greatly reduced (Fig. 2). This strongly indicates the angular alignment of the individual and multiple viruses within the trap. In previous work on trapped silica colloids we deduced that the successive particles entering the trap do not coalesce into a single particle but form a dense fixed array of separated particles (*18*), presumably held apart by electrostatic repulsion. It is likely therefore that charged viruses in the trap also form a dense array of separate particles. We further suspect that the trapped viruses are oriented parallel to one another as in dense oriented arrays of free TMV suspensions (*13, 16*).

The size of an unknown captured particle can be obtained by comparing the magnitude of its 90° scattering with that from a polystyrene latex calibrating sphere of known size, in the same trap. This comparison technique was used in our previous work on colloidal silica (*6*). It gives high accuracy for spherical Rayleigh particles because of the r^6 dependence of Rayleigh scattering. In the Rayleigh limit, all parts of the particle radiate in phase as a single dipole. Since TMV is not strictly a Rayleigh particle, we might expect interference effects due to optical path differences from different parts of the same particle to reduce the scattering below the full Rayleigh value. However, if we compare the 90° scatter of the average of 72 single particles from one sample, measured over a period of several days, with the 90° scatter from the calibrat-

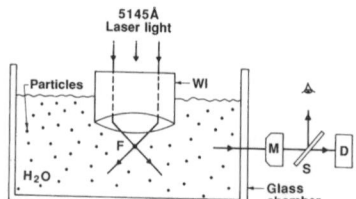

Fig. 1. Apparatus used for optical trapping of TMV particles and mobile bacteria. Spatially filtered argon laser light at 5145 Å is focused to a spot diameter of about 0.6 μm in the water-filled chamber by the high numerical aperture (1.25) water-immersion microscope objective (WI) forming a single-beam gradient trap near the beam focus (F). The 90° scattering from trapped particles can be viewed visually through a beam splitter (S) with a microscope (M) or recorded using a photodetector (D).

ing sphere, using the Rayleigh formula, we find an effective volume of about $(450 \text{ Å})^3$. This volume corresponds to a cylinder 200 Å in diameter and about 3100 Å long, which is quite close to the volume of TMV. We conclude from this that we are looking at single TMV particles, and further that the axis of the TMV particle in the trap is oriented closely perpendicular to the beam axis along the optical electric field, since only then can all parts of the cylindrical virus radiate in phase at 90° as a Rayleigh particle. Looking at the size uniformity of our 72 particles we find, assuming that all differences in scatter are due to changes in the length of the viruses, that 75% of all particles lie in a length range within about 20% of the average. Thus we find that this particular sample is quite monodisperse and the measured length is $L = 3100 \pm 700$ Å.

Not much size information can be deduced from the observed changes in 90° scatter as additional virus particles enter a trap and form an aligned array, since the combined scattering field in such cases is the result of interference of the fields from each

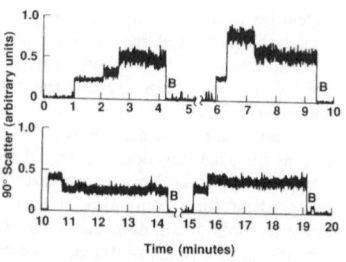

Fig. 2. Scattered light observed at 90° as successive TMV viruses enter the optical trap. At times labeled "B" the trapping beam is momentarily blocked, releasing the viruses. The trap subsequently refills with new virus particles.

particle of the array. Not only are the positions and phases of the fields of the various particles unknown, but they probably change as additional particles enter the trap. Examples of destructive interference and a decrease in total 90° scattering as additional particles enter a trap are seen at times of 7.4 and 10.8 minutes in Fig. 2. In a related experiment (*18*) on the angular distribution of scattered light from trapped colloidal silica particles we found that when a similar decrease in 90° scattering occurred on entry of an additional particle, it was possible to find another direction where the phases added constructively and actually gave an increase in scattering.

In the above discussion we deduced the optical alignment of TMV along the optical field *E* on purely experimental grounds. This also makes sense energetically since this optical alignment results in the maximum polarizability α of the cylindrical virus. Indeed, the TMV, when aligned, not only feels the deepest trapping potential $\alpha E^2/2$ but also experiences strong realigning torques when rotated, due to the angular dependence of $\alpha E^2/2$. More detailed information on the angular orientation of TMV within the trap could be obtained from scattering experiments with an additional low-power probe laser beam of different wavelength and varying directions of polarization relative to the virus axis.

Another conclusion that can be drawn from the data of Fig. 2 is that it is possible to trap viral material without any gross optical damage. Indeed, single viruses have been trapped for tens of minutes with no apparent changes in size as indicated by the constant magnitude of the scattering. The viability of the virus after trapping was not examined. In previous experiments optical damage was a serious problem, which limited the trapping of small Rayleigh-sized polystyrene latex particles and even to some extent silica particles. For silica this damage was subsequently eliminated (*18*) by pH changes or additions of potassium silicate to the solution, which points to a surface photochemical reaction as the damage mechanism. For virus particles the fact that the strong optical absorptions are in the ultraviolet probably contributes to their optical stability in the visible light range. As with silica colloids (*6*) we were able to manipulate captured TMV particles within their environment by moving either the light beam or the entire chamber. This implies the ability to separate trapped viruses by means of a simple flushing technique, for example.

The major problem encountered in the experiment was one of reproducibility. Thus far we have had two batches of dense virus (10 and 50% by weight) and had successful trapping runs with samples diluted from each batch which lasted several days. However, with other samples from the same batches we trapped many fewer TMV-sized particles. Instead we trapped mainly larger-sized clumps and smaller-sized single particles. These larger trapped clumps usually decrease in size (damage) in just a few minutes. Large clumps, which also decayed in time, were previously trapped in our silica experiments (6). The smaller-sized trapped particles are stable in size and are probably just smaller pieces of virus. On this assumption, the lengths of some of these smaller viral pieces that we observed at trapping powers of ~1.5 W were ~270 Å. If these small particles are typical of the index of refraction of other proteins, then this observation implies an ability to trap proteins with molecular weight $M \geq 3 \times 10^6$ at a power of 1.5 W.

The lack of consistency in TMV trapping could be the result of either the dilution process or the trap geometry. With TMV samples, proper pH and low ionic content are needed to avoid polydispersity. At pH > 9 the TMV falls apart and at pH < 6 it aggregates. We attempted to maintain the pH of our diluted samples close to 7. To do this we used deionized or distilled water with small volumes of buffer solution added to adjust the pH, with no improvement in the consistency of our results. Regarding trap geometry, we often found laser beam wander to be a problem and usually checked our overall geometry by looking for good trapping of 600-Å silica test particles. It is possible, however, that the trap is much more tolerant of aberrations for the spherical 600-Å Rayleigh particles than for 3200-Å TMV. If the TMV sits transverse to the beam axis, the virus would extend to the edges of the beam in regions not felt by a silica particle. Although it requires more power, a trap with a larger diameter focal spot might be more favorable for a particle of this shape. Another obvious experiment is to study the trapping of more spherical viruses such as tomato bushy-stunt virus (19) with a diameter of ~450 Å.

In most of our experiments with silica colloids or TMV in water, we noticed the appearance of some strange new particles in diluted samples that had been kept around for several days. They were quite large compared to Rayleigh particles, on the basis of their scattering of light, and were apparently self-propelled. They were clearly observed moving through the distribution of smaller slowly diffusing Rayleigh-sized colloidal particles at speeds as high as hundreds of micrometers per second. They could stop, start up again, and frequently reversed their direction of motion at the boundaries of the

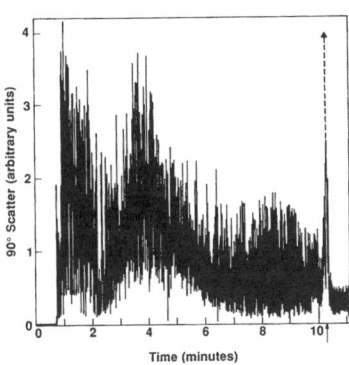

Fig. 3. Scattered light at 90° from a live bacterium trapped by ~5 mW of laser power. At about 10.3 minutes (indicated by arrow) the power was increased to 100 mW. The bacterium was killed and apparently loses much of its cell contents.

trapping beam, when they encountered a dark region, indicating some sort of attraction toward the light. Their numbers increased rapidly as time went by. When examined under 800× magnification in an optical microscope, they were clearly identifiable as rod-like motile bacteria, propelled by rotating tails. There were at least two types of bacteria, about 0.5 and 1.5 μm in length. Optically they resembled small, transparent Mie particles with an index of refraction close to unity.

When one of these bacteria wandered into or was possibly attracted into the trap, it was captured. It was observable through microscope M either by eye or on the photo detector, as a fluctuating signal, as it struggled unsuccessfully to escape from the trap. The far-field forward scatter from the bacteria could also be seen on a screen placed outside the cell. To help capture the bacteria we initially set the laser power at ~50 mW. Once the bacterium had been captured, we quickly lowered the power to ~5 mW to reduce the possibility of optically damaging the bacterium. Figure 3 shows the recorded 90° scatter as a function of time, taken at 5 mW, as a bacterium propels itself about in the trap. After about 10 minutes we raised and maintained the power at ~100 mW. This, as is seen, was sufficient to kill the trapped bacterium. The light scattering stopped fluctuating and decreased to quite a low value as the bacterium apparently vented some of it contents into the surroundings. The remains of the cell could be held in the trap with laser power as low as ~0.5 mW. It was reported that similar venting of a cell's contents occurred in experiments on the puncture of blood cells by pulsed laser beams (20).

In other experiments we illuminated the entire trapping region with a wide low-

power auxiliary red laser beam directed transverse to the trap axis. We then viewed the scene through a red-pass filter with microscope M, either visually or with a video camera and recorder. We could observe the capture of free-swimming bacteria and their subsequent release as the trapping beam was blocked. Several bacteria were occasionally trapped simultaneously. We were also able to demonstrate optical micromanipulation of single trapped bacteria, within the liquid, by moving either the trapping beam or the entire chamber and its liquid.

In the above experiment the low resolution of the side-viewing microscope M and the presence of red laser interference rings made it difficult to resolve details of the trapped bacteria. An obvious extension of this viewing technique is to introduce the trapping laser beam directly into a high-resolution microscope through a beam splitter. By using a water-immersion microscope objective with a high numerical aperture for both laser trapping and viewing through a filter, we were able to simultaneously trap, manipulate, and observe bacteria or other particles with high resolution. For convenience we placed our water samples containing bacteria under a cover slip and used a water-immersion objective designed for operation through the cover slip. An additional lens mounted outside the microscope on an *xyz* mechanical micromanipulator was used to focus the laser trapping beam within the field of view of the microscope and move it about transversely without any need to touch the microscope. We observed the scene either by eye or with a video camera. At power levels as low as 3 to 6 mW we were able to move the beam about and capture a free-swimming bacterium anywhere in the field of view. Once the bacterium had been captured, we could rapidly move it transversely and continue to catch more bacteria until we had a half dozen or more within the trap. Rapid transverse motion without loss of bacteria is possible because of the strong transverse gradient trapping forces. The trapping forces in the axial or *z* direction are stronger in the forward direction of the light rather than in the backward direction (6, 7). Thus any rapid upward motion of the focus can result in escape of particles.

At lower power levels, from 1 to 3 mW, and probably less, we discovered another trapping mode in which bacteria were trapped against the bottom surface of the slide. In this low-power surface mode of operation the mechanical surface provides the backward trapping force needed to prevent the escape of the particle out the bottom of the trap in the direction of the weakest trapping force. It is still possible to move particles about transversely over the surface in this mode because the transverse forces remain quite strong even at the lower power. Bacteria captured by either mode of trapping with powers in the range from 1 to 6 mW have survived for hours in the laser light with no apparent damage.

We performed subsequent experiments using the high-resolution microscope with *Escherichia coli* bacteria. These bacteria are much less motile and could be captured and manipulated rapidly at surfaces and in the bulk fluid with powers as low as a fraction of a milliwatt with no apparent change in behavior or appearance. At powers of 100 mW or more it was possible to observe a shrinkage in the size of the *E. coli* as they become optically damaged in a time of about a minute. With yet another sample of highly motile bacteria we observed a gradual loss in motility of trapped bacteria in about a half-minute with powers as low as 10 to 20 mW. In all cases where optical damage was observed with the 5145-Å green argon laser line it might be advantageous to use other laser wavelengths.

One advantage offered by the high-resolution microscope was the ability to study the trapping forces on bacteria in some detail. For example a bacterium, while being manipulated close to the surface of the slide,

would occasionally manage to attach itself to the surface with its tail and remain tethered. Under these conditions it was still possible to optically manipulate the particle in a circle around its tether and observe the action of the optical forces. Although we do not have a complete description of the trapping forces for complex shaped particles like bacteria, it is clear from these experiments that the same qualitative features, based on simple ray-optics and refraction, that account for trapping of Mie-sized spheres (3, 4, 6) and spheroids (21) still apply here. For example when the beam center is moved toward the edges of a bacterium where refraction is large and asymmetric, we generate large transverse gradient forces which in effect drag the particle with the beam in the direction which recenters the particle on the beam. This is the same basic effect that accounts for the centering of a small Mie particle at the position of maximum light intensity of a large Gaussian beam. However, for the case of a large particle and a small beam, as we have here, it is clearly the local shape of particle that dominates the net force.

REFERENCES AND NOTES

1. A. Ashkin, *Phys. Rev. Lett.* **24**, 156 (1970).
2. _____, *ibid.* **25**, 1321 (1970).
3. _____, *Science* **210**, 1081 (1980).
4. G. Roosen, *Can. J. Phys.* **57**, 1260 (1979).
5. A. Ashkin and J. M. Dziedzic, *Appl. Phys. Lett.* **19**, 283 (1971).
6. _____, J. E. Bjorkholm, S. Chu, *Opt. Lett.* **11**, 288 (1986).
7. S. Chu, J. E. Bjorkholm, A. Ashkin, A. Cable, *Phys. Rev. Lett.* **57**, 314 (1986).
8. A. Ashkin, *ibid.* **40**, 729 (1978).
9. _____ and J. M. Dziedzic, *Appl. Phys. Lett.* **28**, 333 (1976).
10. P. W. Dusel, M. Kerker, D. D. Cooke, *J. Opt. Soc. Am.* **69**, 55 (1979).
11. S. Fraden, A. J. Hurd, R. B. Meyer, M. Cahoon, D. L. D. Caspar, *J. Phys. (Paris) Colloq.* **46**, C3-85 (1985).
12. H. Boedtkèr and N. S. Simmons, *J. Am. Chem. Soc.* **80**, 2550 (1958).
13. J. A. N. Zasadzinski and R. B. Meyer, *Phys. Rev. Lett.* **56**, 636 (1986).
14. V. A. Parsegian and S. L. Brenner, *Nature (London)* **259**, 632 (1976).
15. M. A. Lauffer, *J. Phys. Chem.* **42**, 935 (1938).
16. J. A. N. Zasadzinski, M. J. Sammon, R. B. Meyer, in *Proceedings of the 43rd Annual Meeting of the Electron Microscopy Society of America*, G. Bailey, Ed. (San Francisco Press, San Francisco, 1985), p. 524.
17. H. C. van de Hulst, *Light Scattering by Small Particles* (Dover, New York, 1981), pp. 70–73.
18. A. Ashkin and J. M. Dziedzic, unpublished data.
19. C. Tanford, *Physical Chemistry of Macromolecules* (Wiley, New York, 1961), chap. 2.
20. K. O. Greulich *et al.*, paper Tu GG-11 presented at the International Quantum Electronics Conference, San Francisco, June 1986.
21. A. Ashkin and J. M. Dziedzic, *Appl. Opt.* **19**, 660 (1980).
22. We thank J. A. N. Zasadzinski and R. B. Meyer for providing samples of TMV colloidal solution, and T. Yamane for *E. coli* samples. Helpful discussions with J. E. Bjorkholm and S. Chu are gratefully acknowledged.

29 September 1986; accepted 30 January 1987

October 1983 / Vol. 8, No. 10 / OPTICS LETTERS   **511**

Stability of radiation-pressure particle traps: an optical Earnshaw theorem

A. Ashkin and J. P. Gordon

Bell Laboratories, Holmdel, New Jersey 07733

Received June 6, 1983

We prove an optical radiation Earnshaw theorem: A small dielectric particle cannot be trapped by using only the scattering force of optical radiation pressure. A corollary is that the gradient or dipole force is necessary to any successful optical trap. We discuss the implications of the theorem for recent proposals for the optical trapping of neutral atoms.

We derive an optical Earnshaw theorem that states that it is impossible to trap a small dielectric particle at a point of stable equilibrium in free space by using only the scattering force of radiation pressure. The theorem is analogous to Earnshaw's theorem in electrostatics,[1] which states that it is impossible to trap a charged particle by using only electrostatic forces. We show explicitly that two recently proposed traps for neutral atoms using the scattering force are in fact unstable, in agreement with our general theorem. The implications of the theorem for the stable trapping and cooling of atoms are discussed.

It is well known that optical trapping of dielectric particles is possible by using the forces of radiation pressure from lasers.[2,3] These forces arise from the scattering of light momentum by the particles.[4] Micrometer-sized particles (in the Mie-scattering regime, $d \gtrsim \lambda$) have been optically levitated[5] by a single laser beam and trapped stably by two beams.[2,6]

Here the principal contribution to the light scattering comes from refraction of the incident light rays passing through the particle.[2] For a spherical Mie particle in a plane-wave beam, the scattering is symmetric and the force is in the direction of the incident Poynting vector. If the Mie particle is in a beam with a transverse gradient, the scattering is no longer symmetric and the force has an additional component transverse to the Poynting vector. With submicrometer-sized particles (in the Rayleigh-scattering regime, $d \ll \lambda$; this includes atoms) it is again convenient and useful to divide the total force into two components. One is called the scattering force.[2,7] It is proportional to the scattering cross section of the particle and for paraxial beams and scalar polarizability is in the direction of the Poynting vector. It is a nonconservative force resulting from the removal of momentum from the incident beam. The other is called the gradient force.[8–11] It arises from the interference of the scattered field with the incident field and is proportional to the in-phase component of the particle's polarizability. It attracts particles with positive polarizability to regions of high electric-field strength. The gradient force is a conservative force whose potential is the free energy[12] of the particle.

Thus it can also be regarded as the electrostrictive force on the optically induced dipole of the particle in a field-intensity gradient.[13] For this reason the gradient force on Rayleigh particles is also referred to as the dipole force. Stable trapping of submicrometer dielectric particles has been observed in the standing-wave field of two beams.[14] Trapping of neutral atoms is more difficult and has not yet been accomplished, although the basic forces have been experimentally demonstrated.[15] In this case resonance can be used to increase the magnitude of the forces, but saturation effects limit the trap depth, and the heating effects of quantum fluctuations become important.

It is our thesis that the gradient force is essential to any trap for small particles. The strong velocity dependence of the scattering force has led to the useful concept of optical cooling[16] of atoms by light tuned below resonance, but the scattering force by itself cannot form a trap.

Before the recent proposals of Minogin[17] and Minogin and Javanainen[18] for atom traps based on the scattering force, all the experimentally demonstrated or proposed optical traps for neutral particles involved use of the gradient force. The earliest proposal for trapping atoms about a point of stable equilibrium[10] involved confining atoms to the intensity maxima of standing-wave fields. The light was tuned below resonance by $\gamma_N/2$ (half of the natural line width) so that the same beams would give trapping and optical cooling. This proposed trap is stable, in that the gradient force is restoring for arbitrary displacements from a point of maximum electric energy density. However, saturation of the atomic resonance limits the depth of this trap to an energy of about $h\gamma_N$, which is the same as the minimum kinetic energy to which the atoms can be cooled. Thus this trap is leaky, and in fact the same result seems to occur for any trap in which the same beam is used for both trapping and cooling.

To overcome these difficulties another class of trap was proposed[11] in which the trapping field was tuned much farther below resonance. This inhibits saturation and with the help of beam focusing allows the trap depth to be increased, in proportion to the detuning, by

0146-9592/83/100511-03$1.00/0

512   OPTICS LETTERS / Vol. 8, No. 10 / October 1983

as much as 10^4 times. Since the detuning similarly decreases the optical cooling, the use of auxiliary damping beams[19,20] tuned closer to resonance was proposed for additional cooling. The dynamic Stark shift of the atomic resonance caused by the trapping field substantially reduces the effectiveness of the auxiliary damping beams, however, so this proposal has its own difficulties.

The purpose of the most recent trap proposals of Minogin[17] and of Minogin and Javanainen[18] was to circumvent all these difficulties by relying solely on the scattering force for stability. These workers suggested the use of four or six beams to provide simultaneous trapping and cooling. However, as we now show specifically for these proposed traps and in general for any trap, directions exist at which the scattering force points away from the trap, making it unstable. As Earnshaw's theorem in electrostatics is a direct result of div($\mathbf{E}$) = 0 in vacuum, so our theorem is a direct result of div(scattering force) = 0, a relation that applies so long as saturation effects are neglected.

These recent trap proposals make use of the scattering force in the far field of Gaussian laser beams, so let us first describe this force. Figure 1 shows the geometry. In the far field ($z \gg \pi w_0^2/\lambda$), the scattering force is proportional to the optical power hitting the particle and is directed normal to the spherical phase fronts, whose centers of curvature are at the beam waist. In the paraxial approximation, with $z \gg x$ or $z \gg y$, we can express the nonzero components of the scattering force $\mathbf{F}$ in cylindrical coordinates as

$$F_z = Kw^{-2}\exp(-2r^2/w^2), \qquad (1)$$

$$F_r = F_z(r/z), \qquad (2)$$

where the spot size $w = \lambda z/\pi w_0$ and K is a constant proportional to the particle's scattering cross section and to the total power in the light beam. Only the exponential factor distinguishes this force from the electrostatic force on a test charge that is due to a point charge located on the axis at the beam waist at $z = 0$. Otherwise the force is similarly directed and has the same $1/R^2$ dependence on position.

Consider now the four-beam trap of Minogen and Javanainan,[18] as illustrated in Fig. 2. Here four beams are tetrahedrally arranged in an attempt to form a trap

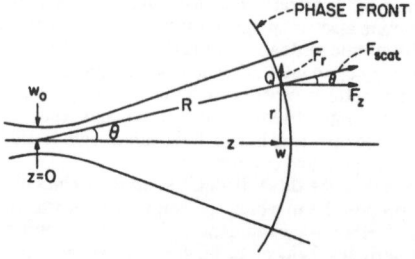

Fig. 1. Geometry of the scattering force on a particle placed at Q in the far field of a Gaussian beam with waist w_0 at $z = 0$.

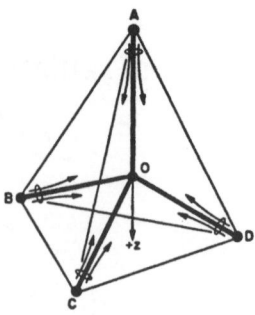

Fig. 2. Geometry of a proposed tetrahedral far-field scattering force atom trap, which we show to be unstable. The four beams waists are are at A, B, C, and D.

at their intersection O. Now Earnshaw's theorem says that the analogous electrostatic trap with a tetrahedral arrangement of four charges located at A, B, C, and D is unstable. In the optical trap, as one proceeds along the z axis, for example, away from the intersection point O, the z-directed beam maintains its electrostatic analogy but the other beams, making an angle with the z axis, lose effectiveness because of the exponential falloff. Hence, as a particle proceeds out of the trap along the $+z$ axis, the restoring force that is due to the other beams is less than in the electrostatic case, which is already unstable. Indeed, if one adds the forces that are due to the four beams, one finds along the z axis in the neighborhood of the intersection ($x = y = z = 0$) the result that $F_z = K(4z/3w^2)^2 + \ldots$. Not only is there no linear restoring force, but the force always points in the direction of positive z.

Similar considerations apply to Minogin's six-beam trap.[17] Here the beam waists are located along the x, y, and z coordinate axes at a distance l from the origin, and all six beams shine inward toward the trap situated at the origin. Again one finds that there is no linear restoring force. There is a cubic restoring force along the axes given by $F_i = -8Kx_i^3/lw^4$ (for the x_i axis), but the particle can escape along any of the [1, 1, 1] directions that make equal angles with the coordinate axes. If s is the distance away from the origin in one of these directions, one finds that $F_s = 16Ks^3/3lw^4$. Again the trap is unstable.

These specific results can be generalized to any similar trap that uses the far fields of Gaussian beams by noting from Eqs. (1) and (2) that div($\mathbf{F}$) = 0. Such a divergenceless force can be represented by continuous lines, which must leave any volume that they enter, thus necessarily providing escape routes for the particles. Any sum of such forces has the same property.

We can proceed further to prove in general that the scattering force has zero divergence for small dipolar particles of arbitrary shape and properties and for optical fields of arbitrary geometry, provided only that the particle's dipole is linearly related to the field. We do not consider saturation effects, but they are detrimental to the formation and stability of any trap. The nonrelativistic Lorentz force exerted on a small neutral

October 1983 / Vol. 8, No. 10 / OPTICS LETTERS **513**

particle by the light field in the electric-dipole approximation is[20] $\mathbf{F}_l = (1/2)\mathrm{Re}\{\hat{x}_j\mu^* \cdot \partial\mathbf{E}/\partial x_j\}$, where μ is the particle's dipole moment, $\mathbf{E}$ is the electric field of the incident wave, $\hat{x}_j$ are the Cartesian-coordinate unit vectors, and the sum over j is implied. A single frequency $[\mathbf{E} \propto \exp(-i\omega t)]$ is assumed. This force is the sum of the electric force on the dipole moment of the particle and the magnetic force on the associated current. For quantum systems such as atoms, the same expression applies[20] for the expectation value of the force if the light field is coherent and μ is the expectation value for the atomic dipole. The trapping-force field is this force evaluated for a stationary atom.

If μ and $\mathbf{E}$ are related linearly, then $\mu = \chi \cdot \mathbf{E}$, where χ is an arbitrary constant polarizability tensor. Now any such tensor can be developed as $\chi = \chi' + i\chi''$, where χ' and χ'' are Hermitian. In the absence of a dc magnetic field χ' and χ'' are symmetric and therefore real, but that is not necessary to our argument. On inserting this development for μ into the expression for $\mathbf{F}_l$, we obtain $\mathbf{F}_l = \mathbf{F}_g + \mathbf{F}_s$, where

$$\mathbf{F}_g = (1/4)\mathrm{grad}(\mathbf{E}^* \cdot \chi' \cdot \mathbf{E}), \qquad (3)$$

$$\mathbf{F}_s = (1/2)\mathrm{Im}\{\hat{x}_j\mathbf{E}^* \cdot \chi'' \cdot \partial\mathbf{E}/\partial x_j\}. \qquad (4)$$

These are the general expressions for the gradient force and the scattering force, respectively. One can see that the potential for the gradient force is just the ac analog of the electrostatic free energy $[-(1/2)\mu \cdot \mathbf{E}]$ of a dipole.[12] The scattering force is intimately related to the rate of work done on the dipole by the incident field $[(\omega/2)\mathbf{E}^* \cdot \chi'' \cdot \mathbf{E}]$ and therefore to the absorption plus scattering cross section of the particle. In the case of scalar polarizability and paraxial radiation the scattering force is proportional to Poynting's vector. Using the wave equation $\nabla^2\mathbf{E} = -(\omega/c)^2\mathbf{E}$, one can quickly show from Eq. (4) that $\mathrm{div}(\mathbf{F}_s) = 0$. The optical Earnshaw theorem is thus proved in considerable generality. In contrast to the electrostatic case, we do not have $\mathrm{curl}(\mathbf{F}_s) = 0$, so some lines of the scattering force may circulate within some volume. Such vortex behavior has been shown in the similar case of the Poynting vector.[21] A trap, however, must have some volume where the force is inward over its whole surface, and this is impossible for the scattering force since $\mathrm{div}(\mathbf{F}_s) = 0$.

The major implication of the optical Earnshaw theorem and its corollary that gradient forces are necessary for traps is that to produce deep traps one must maximize the gradient contributions to the force. Traps that rely on maximizing the scattering forces to the neglect of gradient forces are necessarily flawed, as we have proven. The actual scattering-force traps of Minogen and Javanainen were tuned $\gamma_N/2$ below resonance for cooling purposes. This introduces a small inward gradient force, which for their geometry cannot compensate for the gross instability that is due to the scattering force in the unstable directions.

Finally, the traps based on maximizing the gradient force[11,19,20] present the best prospects for experimentally achieving all optical neutral-atom traps. For optical powers of ~1 W, trap depths ~$10^4 h\gamma_N$ are achievable for sodium atoms, which confine atoms of

velocities $\lesssim 2 \times 10^3$ cm/sec. Optical traps are therefore relatively shallow compared with ion traps.[22] However, recent experiments on optical cooling of sodium atomic beams using the scattering force[23] have produced slow atoms with velocity ~4×10^3 cm/sec. Such slowing techniques may thus ultimately provide a means of filling atom traps. Cooling difficulties that are due to a dynamic Stark shift remain for highly detuned gradient traps. The suggested use of an additional Stark-shift-canceling beam[19,20] is still a possible solution to the cooling problem. However, as has been pointed out,[20] quantum heating can be reduced enough to give a 1-sec retention time for cold atoms in such traps in the absence of any cooling.

References

1. J. C. Maxwell, *Treatise on Electricity and Magnetism* (Clarendon, Oxford, 1904; Dover, New York, 1962); W. T. Scott, Am. J. Phys. **27**, 418 (1959).
2. A. Ashkin, Phys. Rev. Lett. **24**, 156 (1970).
3. A. Ashkin, Science **210**, 1081 (1980).
4. Absorption of light by the particles may occur and is included in the general theory outlined below.
5. A. Ashkin and J. M. Dziedzic, Appl. Phys. Lett. **19**, 283 (1971).
6. G. Roosen, Can. J. Phys. **57**, 1260 (1979).
7. A. Ashkin, Phys. Rev. Lett. **25**, 1321 (1970).
8. G. A. Askar'yan, Zh. Eksp. Teor. Fiz. **42**, 1567 (1962) [Sov. Phys. JETP **15**, 1088 (1962)].
9. A. P. Kazantsev, Zh, Eksp. Teor. Fiz. **63**, 1628 (1972) [Sov. Phys. JETP **36**, 861 (1973)].
10. V. S. Letokhov, V. G. Minogin, and B. D. Pavlik, Zh. Eksp. Teor. Fiz. **72**, 1328 (1977) [Sov. Phys. JETP **45**, 698 (1977)]; V. S. Letokhov and V. G. Minogin, Appl. Phys. **17**, 99 (1978).
11. A. Ashkin, Phys. Rev. Lett. **40**, 729 (1978).
12. L. D. Landau and E. M. Lifshitz, *Electrodynamics of Continuous Media* (Pergamon, London, 1960), p. 54.
13. J. P. Gordon, Phys. Rev. A **8**, 14 (1973).
14. A. Ashkin, unpublished experiments. See related work demonstrating gradient forces on submicrometer particles: P. W. Smith, A. Ashkin, and W. J. Tomlinson, Opt. Lett. **6**, 284 (1981); P. W. Smith, P. J. Maloney, and A. Ashkin, Opt. Lett. **7**, 347 (1982).
15. See, for example, J. E. Bjorkholm, R. R. Freeman, and D. B. Pearson, Phys. Rev. A **23**, 491 (1981); J. E. Bjorkholm, R. R. Freeman, A. Ashkin, and D. B. Pearson, Phys. Rev. Lett. **41**, 1361 (1978); A. Ashkin, Science **210**, 1081 (1980).
16. T. W. Hänsch and A. L. Schawlow, Opt. Commun. **13**, 68 (1975).
17. V. G. Minogin, Kvantovaya Elektron. (Moscow) **9**, 505 (1982) [Sov. J. Quantum Electron. **12**, 299 (1982)].
18. V. G. Minogin and J. Javainen, Opt. Commun. **43**, 119 (1982).
19. A. Ashkin and J. P. Gordon, Opt. Lett. **4**, 161 (1979).
20. J. P. Gordon and A. Ashkin, Phys. Rev. A **21**, 1606 (1980).
21. A. Boivin, J. Dow, and E. Wolf, J. Opt. Soc. Am. **57**, 1171 (1967).
22. D. J. Wineland, R. E. Drullinger, and F. L. Walls, Phys. Rev. Lett. **40**, 1639 (1978); W. Neuhauser, M. Hohenstatt, P. Toschek, and H. Dehmelt, Phys. Rev. Lett. **41**, 223 (1978).
23. W. D. Phillips and H. Metcalf, Phys. Rev. Lett. **48**, 596 (1982); J. V. Prodan, W. D. Phillips, and H. Metcalf, Phys. Rev. Lett. **49**, 1149 (1982).

Design for an optical cw atom laser

Arthur Ashkin[†]

Bell Laboratories, Lucent Technologies (Retired), Holmdel, NJ 07733-3030

Contributed by Arthur Ashkin, June 24, 2004

A new type of optical cw atom laser design is proposed that should operate at high intensity and high coherence and possibly record low temperatures. It is based on an "optical-shepherd" technique, in which far-off-resonance blue-detuned swept sheet laser beams are used to make new types of high-density traps, atom waveguides, and other components for achieving very efficient Bose–Einstein condensation and cw atom laser operation. A shepherd-enhanced trap is proposed that should be superior to conventional magneto-optic traps for the initial collection of molasses-cooled atoms. A type of dark-spot optical trap is devised that can cool large numbers of atoms to polarization-gradient temperatures at densities limited only by three-body collisional loss. A scheme is designed to use shepherd beams to capture and recycle essentially all of the escaped atoms in evaporative cooling, thereby increasing the condensate output by several orders of magnitude. Condensate atoms are stored in a shepherd trap, protected from absorbing light, under effectively zero-gravity conditions, and coupled out directly into an optical waveguide. Many experiments and devices may be possible with this cw atom laser.

One of the remaining challenges of ultralow-temperature research with atomic vapors is the achievement of a truly cw atom laser. This problem has resisted solution since the time of the first demonstration of Bose–Einstein condensation (BEC) in 1995 with magneto-optic traps (MOTs) and evaporative cooling from purely magnetic traps (1, 2). Crude pulsed atom lasers were made shortly thereafter by using similar magnetic techniques (3–5).

Recent attempts to achieve cw atom lasers by combining magnetic techniques of BEC formation with optical tweezer techniques for transporting and combining condensates succeeded in demonstrating sustained B–E condensates (6, 7). Serious problems still remain to be overcome, mainly having to do with insufficient atoms, before a useful cw atom laser is achieved. Proposed here is a design of a viable cw atom laser with mainly optical techniques. This design uses new types of optical traps and waveguides based on so-called "optical–shepherd" beams, which should be superior to the currently used standard MOTs and magnetic traps. "Shepherding" involves use of thin reflective blue-detuned sheet beams to make box-like atom traps and waveguide structures to confine and move atoms at controlled velocities. The sheet beams are made by cyclically sweeping Gaussian beams in space at a sufficiently rapid rate so that the beams act as fixed repulsive walls for slow-moving atoms. The sweeping shepherd beams can be controlled electronically by using well known beam-scanning technology, as discussed later. With shepherd beams one can make complex optical structures with strikingly new capabilities, such as high-density dark-spot optical traps and the ability to capture and recycle evaporatively cooled atoms, essentially canceling the effect of gravity on atoms. This capability leads to the design of a powerful, highly coherent, optical cw atom laser operating at very low temperatures.

Advantages of Optical Trapping and Cooling Techniques

Insight into the problems of making a cw atom laser and how to solve them can be gained by examining the history of optical trapping and comparing the relative advantages of optical and magnetic trapping and cooling techniques.

The discovery of stable optical trapping of neutral particles (8) and the understanding of the properties of the basic optical

scattering and dipole forces on neutral particles and atoms (9–11) date back to the 1970s. By 1980 the basic principles of optical atom trapping and cooling were well understood (11–13), and the foundations of "atom optics" were laid (14). A key concept in proposed optical dipole force traps and shepherding was detuning far from resonance to avoid saturation and scattering force heating (11, 13).

The first demonstration of optical molasses cooling of atoms was made in 1985 (15), followed by the first optical dipole trapping of ≈1,000 sodium atoms in 1986 (16). Soon thereafter the first MOTs were introduced. These hybrid magnetic scattering force traps initially confined >10^7 atoms (17, 18). At this point work on optical dipole traps essentially ceased while MOT-type traps and, later, dark-spot MOTs became the workhorse traps for collecting and cooling large numbers of atoms to Doppler and subDoppler temperatures (19–21).

In the late 1980s many new applications of ultracold atoms were pursued and searches began for methods to reach the high densities and low temperatures needed to achieve BEC (20, 22).

Achievement of BEC in 1995 (1, 23) led to an explosion of interest in the physics of these novel quantum systems and their applications (24, 25). Experiments with B–E condensates also brought a further appreciation of some of the many advantages of optical dipole traps and optical manipulation techniques over magnetic techniques. With optical traps, one can confine all hyperfine states, both low- and high-field seekers (26). This ability made sophisticated experiments in spintronics possible (27) in dipole traps using condensates transferred from larger-volume magnetic traps (28). With dipole traps the trapping parameters are independent of externally applied magnetic fields. This property made possible the first observations of Feshbach resonances (29, 30). It was shown that evaporative cooling was possible in dipole traps (31) and efforts were made to reach BEC in all-optical traps (31), but the original densities were too low. Finally, in 2001, Barrett *et al.* (32) succeeded in demonstrating BEC in all-optical traps, although the final number of condensate atoms was small. More recently, Granada *et al.* (33) used an all-optical technique to produce a degenerate Fermi gas.

The optical shepherding techniques described here have evolved from the original work in atom optics on the focusing and defocusing of atoms by laser beams (14), and, more recent experiments, on atom guiding in TEM$_{01}$* mode beams (34), trapping of atoms with a donut-mode beam (35), and experiments with "atom-optics billiards" (36). Relevant experiments were also performed involving the trapping of macroscopic particles in spatial arrays and in a "light-cage" by scanning a pair of computer-controlled galvo mirrors (37, 38).

Fig. 1 shows a cross-sectional sketch of the proposed cw atom laser in ^{87}Rb, fabricated from shepherd beams. The first step in achieving BEC is the collection of atoms in a molasses-cooled trap. This usually involves capturing atoms from an atomic vapor source in a MOT, which simultaneously traps and molasses cools atoms. Here, we enhance performance of the MOT by the

Abbreviations: BEC, Bose–Einstein condensation; MOT, magneto-optic trap; PGC, polarization gradient cooling.

[†]E-mail: aashkinshome@aol.com.

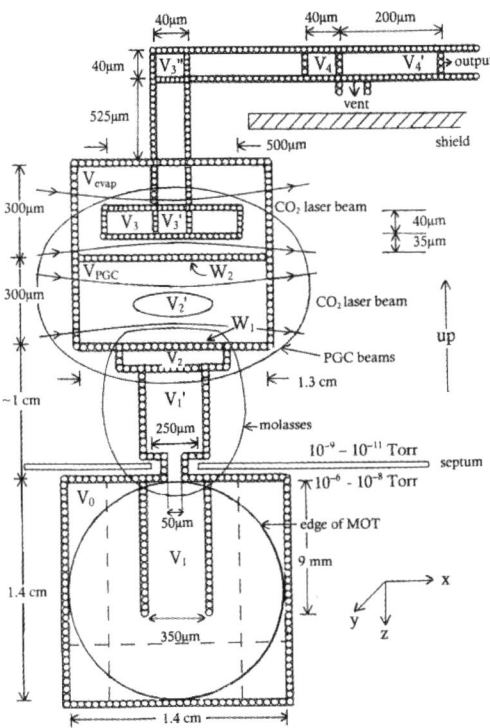

Fig. 1. Improved cw optical atom laser design. This design features a double-vacuum chamber separated by a septum, a separate molasses and PGC region, and a recirculating evaporative cooling chamber; all in an optical waveguide.

addition of a cubic, blue-detuned shepherd trap $\approx(1.4 \text{ cm})^3$ surrounding the spherical MOT volume, as shown. Atoms in a standard MOT feel a linearly decreasing trapping force as the radius decreases, whereas the average shepherding force is highly localized and can be larger than the MOT force and actually increases with decreasing distance from the origin because of the increase in light intensity as the shepherd beams shrink. These considerations imply an ability to collect all the atoms of the MOT and compress them at higher speeds to higher densities by shrinking the repulsive walls of the shepherd-type trap. One concludes, therefore, that shepherd enhanced MOTs should be superior to conventional MOTs as a source of cold atoms.

Looking at the number of Rb atoms typically collected in a MOT, one conservatively expects, by using shepherding, to surround $\approx 3 \times 10^7$ atoms cooled to a temperature approaching $T_D \cong 140 \mu$K in a volume of $\approx(1.4 \text{ cm})^3$, in ≈ 2 s or less, at $\approx 10^{-6}$ Torr (35). Next, the shepherd beams surrounding V_0 are rapidly collapsed, as indicated by dotted lines, compressing the molasses-cooled atoms into the volume V_1. For a volume of $V_1 \cong 350 \mu$m $\times$ 350 μm $\times$ 9 mm and 3×10^7 atoms, we have a density of 2.7×10^{10} atoms per cm^3.

To take full advantage of the shepherd beam's ability to rapidly compress atoms in the MOT volume we have to prevent atom pileup as the shepherd pushes the slow-moving atoms diffusing in molasses. By chopping the MOT molasses beams as the shepherd beams advance, one can periodically undamp the

atoms and keep the density uniform. Compression times of a few tenths of a second are anticipated to reach volume V_1.

It should be stressed that the figure of merit for an atom source is not just the maximum number of atoms that can be collected, but rather the maximum number of atoms per s that can be collected.

Because the conditions are not optimal for producing BEC at the typical vapor-source pressures of $\approx 10^{-5}$–10^{-8} Torr, it is desirable to adopt a double-dipole trap arrangement that separates and optimizes the initial process of trapping thermal-source atoms from the final condensation process at pressures of $\approx 10^{-9}$–10^{-11} Torr, or even lower. With this scheme atoms are transported along the 350-μm-square waveguide of Fig. 1, as a unit, in a fraction of a second, by using a pair of pusher shepherd beams. This transport is analogous to the standard double-MOT procedure (39, 40). However, instead of the long, lossy magnetic atom guides commonly used to isolate high- and low-vacuum chambers, one can use a simple, ≈ 250-μm pinhole in a thin antireflection-coated septum between the chambers and the shepherd waveguide to reach high-vacuum conditions at a distance of only ≈ 1 cm into the high-vacuum region, with no loss of atoms.

Cooling Stages Leading to cw Atom Laser in High Vacuum

Once in the high-vacuum chamber, cooling follows several distinct stages leading to BEC and a highly efficient optical cw atom laser: (i) molasses cooling of input atoms moving from V_1' to V_2; (ii) polarization gradient cooling (PGC) of atoms in volume, V_{PGC}, and their compression into V_2'; (iii) preevaporative cooling from V_2' into the V_{PGC} chamber, followed by evaporative cooling from V_3 into the V_{evap} chamber to form a B–E condensate in V_3' (included in the cooling step are atoms fed back to V_2' from the previous evaporation cycle); and (iv) the feeding of condensed atoms from V_3', through V_3'' and V_4 to the laser storage volume V_4' and the coupling out of cw laser atoms.

In the first stage (i) of high-vacuum cooling, the $\approx 3 \times 10^7$ atoms collected in V_1 from the vapor source are pushed through the molasses volume V_1' at a constant density of $\approx 10^{11}$ atoms per cm^3 and cooled to a temperature approaching the Doppler limit temperature of Rb, $T_D \cong 143 \mu$K.

In the second stage (ii), the $\approx 3 \times 10^7$ atoms are advanced, maintaining the same density of $\approx 10^{11}$ atoms per cm^3, into volume V_2 located at the entrance of the V_{PGC} all-optical-shepherd trap. The V_{PGC} trap is used to cool the atoms in the PGC volume V_{PGC} to temperatures of ≈ 10–20μK, as seen by Barrett *et al.* (32) and Granada *et al.* (33) and compress them to densities of $\approx 10^{14}$ atoms per cm^3, in preparation for subsequent evaporative cooling. Included in this second cooling stage, however, is also a large number of atoms, $\approx 10^8$–10^9, which have been collected in V_{PGC} and V_{evap} and fed back to V_2' from the previous evaporation cycle. The handling of such a large number of atoms presents a problem. Experience with MOT cooling at densities in excess of $\approx 10^{11}$ atoms per ml shows that difficulties arise from reabsorption of spontaneously emitted fluorescence (19, 40, 41). Use of a "dark-spot" MOT (19), with a dark core of atoms, reduced absorption, and increased the density of cooled atoms by an order of magnitude.

In this proposal, as shown in Fig. 1, one starts with six PGC beams surrounding the 300 μm $\times$ 1.3 cm $\times$ 1.3 cm V_{PGC} volume. An optical dark-spot trap V_2' is formed by sweeping a focused $w_0 = 55$-μm-diameter red-detuned CO_2 laser beam transversely over a width of 330 μm inside the V_{PGC} volume. Atoms in V_{PGC} are combined with atoms from V_{evap} and V_2 for cooling and compression by turning off the trap wall W_1 of V_2 and gradually lowering the potential of wall W_2 of V_{evap}. This process is done while keeping the average atomic density in the cooling volume at $\approx 10^{11}$ atoms per cm^3. The atoms in V_{PGC} are quickly cooled to their minimum PGC temperature and should start to diffuse into the deep-red-detuned dipole storage trap V_2', where they collect at high density. A final stage in this filling process involves shrinking the surrounding shepherd walls of V_{evap} and V_{PGC} to collect all of the remaining atoms and feed them into the red-detuned dipole trap V_2'. Although not driven by shepherd beams, similar filling behavior into a red-detuned optical trap was observed in the experiments of Barrett *et al.* (32) and Thomas and collaborators (33). Barrett *et al.* (32) were able to transfer PGC-cooled Rb atoms into their dipole trap, achieving the high density of $\approx 10^{14}$ atoms per cm^3. They offer several explanations for this successful behavior: the damping of atoms in the tails of the red-detuned dipole potential; the formation of an effective dark-spot MOT due to the Stark shift of the red-detuned trap; and, possibly, the existence of some blue-detuned Sisyphus cooling (42). An important aspect of the buildup of high density in the red-detuned trap is the thermalization of atoms captured in the fringes of the trap by two-body collisions. Any atoms that gain energy in such collisions and leave the trap are recooled in the PGC volume by the PGC molasses beams and eventually returned to the red trap.

Here, as an additional aid to the CO_2 trap-filling process, one may resort to chopping of the cw PGC beams at a rapid rate. With the trapping beam turned on, and the damping beams turned off, atoms move freely and are drawn into the CO_2 trap and move more rapidly toward the focus, where they can thermalize. With the cooling on and the trap on, the atoms are damped down to PGC temperature, where they only can move diffusively everywhere except near the red-trap focus, where they are Stark-shifted out of resonance. As the process continues, the atoms follow this alternating free motion toward the focus and the diffusive damping motion. Finally, all the atoms collect in the red trap, with the help of the shrinking shepherd beams. This chopping technique is somewhat reminiscent of the first optical trapping experiment (16).

The principal reason optical dark-spot cooling is superior to dark-spot cooling in MOTs is that the optical dark-spot is a true trap that prevents PGC-cooled atoms from reentering the PGC volume.

Stamper-Kurn *et al.* (28) were also able to transfer a sodium condensate directly from a large-volume, low-density magnetic trap into a high-density, compact red-detuned dipole trap. Densities as high as 2×10^{15} atoms per cm^3 were observed.

The dimensions of the red-detuned CO_2 trap were selected to give a final density in the 10^{13}–10^{14} range. With an assumed trap $U/kT_{atoms} = 160 \, \mu K/20 \, \mu K = 8$, it is estimated that $\approx 7.5 \times 10^8$ atoms are compressed in V_2' to a density of 7×10^{13} atoms per cm^3. Compression times of a second or less should be sufficient to reach this density at PGC temperatures. At this point the PGC molasses beams are turned off.

The third stage (*iii*) in producing a cw atom laser, the evaporative cooling stage, involves a large departure from conventional practice. In previous work on evaporative cooling to BEC with magnetic traps and optical traps, one evaporated the high-energy component of trapped atoms into free space, thereby cooling the remaining atoms in the trap (1, 2). For example, for Rb atoms in magnetic traps, many groups have started with $\approx 2 \times 10^8$ to 5×10^9 atoms and ended up with condensates having 10^4 to 2×10^5 atoms at temperatures in ranging from 100 to 500 nK after cooling times of 20–45 s. Typically, 100–1,000 or more atoms are evaporated away and lost for every cold condensate atom remaining in the magnetic trap. This is a very inefficient process.

In this proposal involving atom feedback with shepherd beams, starting with a modest source of 3×10^7 atoms at the input from V_2', assuming a moderate feedback ratio $r = 0.99$, corresponding to 99 atoms fed back for every condensed atom that remains in V_3', one expects to collect and recirculate $\approx 3 \times 10^9$ atoms internally. The feedback effectively increases the

source by a factor of 100. Under equilibrium conditions, this result implies an output yield of $\approx 3 \times 10^7$ condensate atoms in the final volume V_3' per evaporation cycle, assuming no other losses. This calculation may not be totally realistic with high densities of $\approx 10^{13}$–10^{14} atoms per cm^3 in V_2', V_3, and V_3', because of some loss from three-body recombinations.

One can analyze the feedback and buildup of internally circulating atoms, N_m, in V_2' after m cycles as a geometric series, $N_m = \Sigma \, ar^n$ from $n = 0$ to $n = m - 1$. To see the correctness of this formula, one can rewrite N_m as $N_m = a + r \, \Sigma \, ar^n$ from $n = 1$ to $n = m - 1$. This equation says that the number of atoms in the mth cycle is made up of "a" from the input plus the fraction r times the number in the $(m - 1)$ cycle. After a large number of cycles, N_m approaches $N_\infty = a/(1 - r)$, whereas a number of atoms $(1 - r) \, N_\infty = a$ is effectively lost to the output and recombination. This formally accounts for all of the atoms. However, it cannot determine the division of the so-called "lost" atoms between the output of condensed atoms and recombination, without additional data.

To recapitulate numerically, for a case with no recombination loss, having an input of $\mathbf{a} = 3 \times 10^7$ atoms and $r = 0.99$, one gets $N_\infty = 100 \, a = 3 \times 10^9$ atoms and an equilibrium condensate output of 3×10^7 atoms. For a case including recombination loss, $a = 3 \times 10^7$ and an assumed $r = 0.96$, one gets recirculating atoms $N_\infty = a/(1 - r) = 25a = 7.5 \times 10^8$ atoms and a total of 3×10^7 atoms divided between condensate atoms and atoms lost to recombination. Experimental results for evaporative cooling from a magnetic trap of fixed volume show that the temperature falls approximately linearly with the number of atoms evaporated (41, 43). Thus, if one wants a final temperature $T = 0.15$ μK, starting from 7.5×10^8 atoms at 20 μK, this calls for a final number of atoms $(7.5 \times 10^8)/(20 \, \mu K/0.15 \, \mu K) = 5.6 \times 10^6$. This, in turn, implies a loss of atoms due to three-body recombination of $3 \times 10^7 - 5.6 \times 10^6 \approx 2.4 \times 10^7$ atoms.

With optical cooling from a shepherd trap, where one can optimally adjust the density and trap proportions during evaporation, one expects to achieve even lower temperatures. Thomas and colleagues previously indicated that evaporation from optical traps has advantages over that from magnetic traps (33). Considering the experimental results of Burnett *et al.* (32), one anticipates reaching temperatures of 0.10 μK or less after approximately 2 s of evaporation time. Ultimately, it is the loss of atoms that limits the lowest achievable temperatures by using evaporative cooling. One simply runs out of atoms.

In practice, one performs the optical evaporative cooling in two stages. In the first stage one preevaporates from the V_2' red-detuned CO_2 laser trap back into V_{PGC}, starting at $\approx 7.5 \times 10^8$ atoms and $T_{atoms} \approx 20 \, \mu K$. Conservatively, assuming performance comparable with magnetic evaporation, one should be able to cool down to $T_{atoms} \approx 2.7 \, \mu K$, ending up with $\approx 1 \times 10^8$ atoms. As one reduces the V_2' potential in preevaporation, one reduces the transverse sweep of the CO_2 trapping beam to help maintain optimum density. The escaping atoms leaving V_2' trampoline over the lower surface of V_{PGC} and very few return to interfere with the evaporative process.

Next, the potential and width of the CO_2 trap are increased and the remaining $\approx 1 \times 10^8$ atoms are lifted into the V_{evap} chamber, where they are surrounded by V_3, a box-like 4880 Å blue-shepherd trap in preparation for the second step of forced evaporative cooling down to BEC. The dimensions and potential of the box-like V_3 trap are chosen to enclose essentially all the 1×10^8 atoms at the same density and, therefore, temperature, as in V_2' after preevaporation.

Gravitational forces often play a considerable role in the dynamics of ultracold atoms. With a box-like blue-detuned shepherd trap, the possibility exists of buildup of excessively high densities at the lower repulsive wall. One can avoid such difficulties by simply canceling gravity within the 4880 Å shepherd trap by fabricating a blue shepherd intensity ramp with a constant upward gradient force equaling gravity. The gravity ramp keeps the density uniform within V_3 during evaporative cooling and subsequent condensate manipulation. Gravity ramps are likely to find other important applications, using ultralow-temperature atoms and condensates.

In the second stage of forced evaporative cooling, starting from the 4880 Å blue-detuned V_3 trap with 1×10^8 atoms and $T \approx 2.7$ μK, one ends up with a condensate of $\approx 5.6 \times 10^6$ atoms at a temperature of $T = 0.15$ μK or less. The dimensions of V_{evap} are chosen to be large enough so that very few of the evaporating atoms return to V_3 and V_3' and interfere with the production of the condensate. This is due to the small volume of V_3 relative to V_{evap} and also the long time of flight for evaporated atoms to reach the outer walls and return.

A matter of further interest is the time it takes for the cw atom laser to reach full output after being turned on. The analysis above shows that for $r = 0.96$ it takes ≈ 64 feedback cycles to reach 93% of N_∞, the equilibrium output. If each cycle is ≈ 5 s long, this implies a startup time of 320 s or ≈ 5 min.

The last step (iv) in making a cw atom laser involves guiding the condensate a distance <2 mm from its source in V_3' into the final storage volume V_4' from which it is continuously coupled out. This, however, must be done under effectively dark conditions, free of any destructive resonant or near-resonant light from other parts of the optical structure. The guiding of the condensate can be done with appropriate $+y$ and $+x$ 4880 Å light beams (as seen in Fig. 4B) or, alternatively, guides made from $+y$ and $-z$ beams. An opaque light shield protects the condensate in V_4 and V_4' from resonance light coming from the molasses and PGC volumes during subsequent cycles.

One sees, in step iv, that use of shepherding with all-optical traps gives a solution to the problem of effectively sustaining a condensate in the dark that is superior to the one used by Chikkatur *et al.* (6).

If the final storage volume V_4' contains five times the number of atoms of V_4, for example, then the maximum change in the number of atoms in V_4' during any cycle will be much less than 20%. One can essentially eliminate all fluctuations with a simple feedback setup. By sensing the fluctuations in output and controlling the number of atoms being vented from the storage volume by a separate venting port, one can stabilize the output at the expense of a small loss in output. A stabilized condensate such as this in V_4', with minimal perturbations, should have high spatial and temporal coherence (6). Atoms can be coupled out of V_4' by adjusting the output sheet beam potential. They can then go directly into the shepherd waveguide at a rate equal to the average rate of atoms being fed in from V_4. For the case of 5.6×10^6 atoms entering V_4' per cycle and a total estimated cycle time of ≈ 5 s one expects a cw output flux of $\approx 5.6 \times 10^6$ atoms every 5 s, or $\approx 1.1 \times 10^6$ atoms every s. The breakdown of the total ≈ 5-s cycle time is: ≈ 2 s to collect source atoms and transfer them to V_2, plus ≈ 1 s for preevaporation and transfer to V_3, and ≈ 2 s of evaporation time to reach V_3' and V_4. One anticipates a final temperature of 0.1 μK or less, as indicated above. Picokelvin temperatures are conceivable if one accepts lower output flux.

It has been pointed out for red-detuned optical trapping that an additional heating source exists because of fluctuations in the power and pointing direction of the trapping beam (44). Beam fluctuational heating should be much reduced for blue box-like shepherd traps, as in V_4, because atoms in such traps interact with the repulsive light walls at much lower light intensities and for only a fraction of the time. Laser noise introduced by mechanically driven moving lenses (7) should be much reduced by using electronically driven shepherd beams.

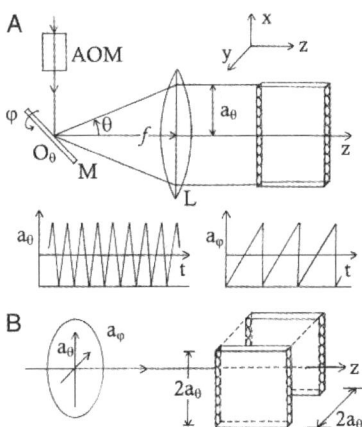

Fig. 2. Formation of scanned sheet mirror. (*A*) Scheme for making a thin repulsive light mirror from a single blue-detuned laser beam by using a computer-controlled, gimbaled mirror rotating in θ with a saw-tooth amplitude, a_θ. (*B*) Scheme to advance the mirror forward at a linear rate and back again by means along y by means of a ramp voltage controlling a_φ.

Possible Applications

A viable cw atom laser would result in novel designs for precision interferometers and devices, such as gyroscopes, gravitometers, and high-precision atomic clocks. In addition, they could be used as a superior source for the sympathetic cooling of other atomic vapors and in two-component Fermi mixtures (43, 44). Observation of the very-low-temperature Cooper-pairing transition (33, 44) may also be possible with the proposed very-low-temperature cw optical laser.

It becomes possible to study the Josephson effect in waveguide-confined B–E condensates that are the exact analog of the DC and AC Josephson effect (45). Applications are also conceivable to optical computing involving arrays of atoms in optical lattices in atom waveguides.

Finally, the atom laser itself could serve as an ideal experimental tool for detailed studies of the processes of atom collection from the vapor, PGC rates at high density with dark-spot traps, and controlled optical evaporative cooling at high densities and very low temperatures.

Apparatus

Fig. 2*A* shows the scheme for making a thin, moveable, repulsive light mirror from a single blue-detuned laser beam. A blue-detuned beam strikes a computer-controlled gimbaled mirror, M, for example, located at the focus, O_θ, of a thin lens, L, and is rotated at a uniform angular rate by a saw-tooth drive in voltage having an amplitude, a_θ, thus forming a scanned sheet mirror beam. If the mirror M is moved downward off axis in the focal plane, the rays of the sheet beam are then tipped upward. One can repeatedly translate the sheet mirror in a direction perpendicular to the plane of the mirror by rotating the mirror M about an orthogonal φ axis with an a_φ voltage waveform, as shown in Fig. 2*B*. Using two orthogonal pairs of scanned sheet mirror beams, we can form an optical waveguide for atoms having an adjustable rectangular cross section, as shown in Fig. 3. Samples of atoms can be translated along such an optical waveguide by advancing a pair of sheet mirror beams while the beams are held at constant separation, as mentioned above.

The beam-sweeping elements just described can be fabricated by using either microelectro-mechanical mirror-type devices

PNAS | **August 17, 2004** | vol. 101 | no. 33 | 12111

PHYSICS

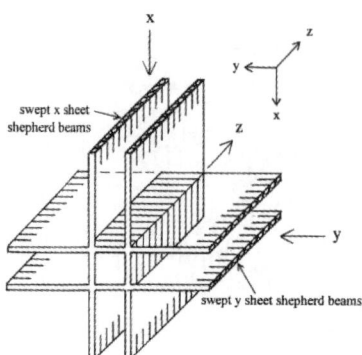

Fig. 3. A blue-detuned repulsive optical waveguide formed from two pairs of sheet mirror beams.

(MEMs), or acousto-optic deflectors (AODs), or galvo mirrors. Microelectro-mechanical mirror-type devices and acousto-optic deflectors act as electronically controlled gimbaled mirrors and can deflect in θ and φ directions. Galvo mirrors are single deflectors and must be used in pairs in conjunction with cylindrical lenses to give deflections in orthogonal directions. The relevant beam-deflection design parameters are the angular amplitude and the frequency responses of the different scanners. Beam-steering techniques are well developed.

Implementation of the atom laser of Fig. 1 should be fairly straightforward with these elements. Fig. 4A shows how one feeds an essentially parallel Gaussian beam into lens L_x to generate a focused beam of diameter, $2w_0$, in the x direction, which is then swept. Fig. 4B is a perspective sketch of the entire atom laser apparatus, showing the location of all the most important L_x, L_y, and L_z lenses of varying aperture and focal length. With the L_x lenses pointing in the x direction, one can project the xz waveguide surfaces and xy sheet beam surfaces of the structure. The L_y lenses pointing in the y direction form the yz waveguide surfaces and also yx sheet beam surfaces.

The laser beam parameters needed to fabricate different parts of the apparatus vary according to the local geometry and

temperature of the guided atoms. The following basic equations can be used to determine the optical potential, U, the saturation parameter, p, and the fraction of time that an atom spends in the excited state.

$$U = (h/2)(\nu - \nu_0)\ln(1 + p)$$

$$p(\nu) = (I_0/I_{sat})[(\gamma_n^2/4)/(\nu - \nu_0)^2]$$

$$f = \frac{1}{2}[p/(1 + p)]$$

For a single Gaussian beam of the form $I(r) = I_0 \exp(-2r^2/w_0^2)$, one determines these parameters in terms of the total power, P_0, the spot size, w_0, and the detuning from resonance $(\nu - \nu_0)$. The intensity I_0 on the beam axis is $2P_0/\pi w_0^2$. I_{sat} and γ_n are the saturation intensity and line width of the atomic transition. See refs. 9, 11, and 13.

For a shepherd Gaussian beam swept uniformly over a total distance, L_{tot}, large compared with w_0, the peak intensity $I_0 = (2P_0/\pi w_0^2)[(\pi/2)^{1/2} w_0/L_{tot}]$.

Initially, atoms are collected from the vapor in the relatively high-pressure regions V_0 and V_1 of a $(1.4 \text{ cm})^3$ shepherd-enhanced MOT, using a pair of swept 250-mW Ti:sapphire laser beams with $w_0 = 50\ \mu m$ and saturation parameter $p = 0.1$, giving a blue-detuned peak wall potential of $U_0 \cong 14\ h\gamma_n/2$. The same shepherd beams guide the molasses-cooled atoms through volumes V_1' and into V_2 in the low-pressure region.

One sees in Fig. 1 that an atom leak may occur because of the shadow cast by the thin septum dividing the high- and low-pressure chambers. This can be avoided by launching two pairs of additional L_x and L_y beams (not shown in Fig. 4B) at an angle into the shadow region to bridge the waveguide gap.

Resonance fluorescence from atoms in the V_0, V_1, V_1', and V_2 can cause heating of previously evaporated atoms being held in V_{PGC} and V_{evap} by far-off-resonance shepherd beams from the earlier cooling cycle. To prevent such heating and possible atom loss, once a new collection cycle is started, one switches from the far-off-resonance V_{PGC} and V_{evap} shepherd traps to PGC-cooled near-resonance traps, having the same shepherd beam parameters as used for V_0, V_1, V_1', and V_2.

The CO_2 red-detuned trap V_2 used to collect PGC-cooled atoms at $\approx 20\ \mu K$ is formed from a 125-W beam with $w_0 = 55\ \mu m$, swept over a width of $\approx 330\ \mu m$, giving a trap depth of $U_0 = 8 \times 20\ \mu K =$

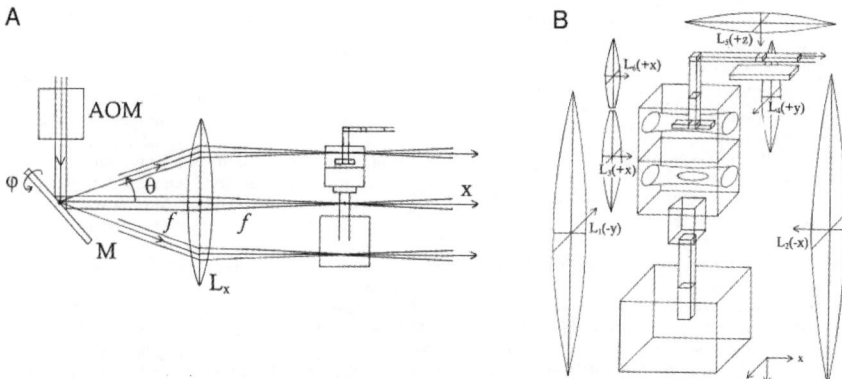

Fig. 4. Lens geometry for implementing cw atom laser. (A) Input optics for focusing an incident beam to a desired spot size w_0 at the laser structure. By feeding the gimbaled mirror M with either of two polarizations, one can switch between either of two input beams. (B) A perspective view showing the principal lenses needed to make the optical cw atom laser.

160 μK. This trap confines all 7.5×10^8 atoms up to the average velocity within dimensions of 30 μm $\times$ 360 μm $\times$ 990 μm at an average density of $\approx 7.0 \times 10^{13}$ atoms per cm^3.

During the subsequent preevaporation step, atoms escaping from V_2' trampoline over the lower wall of V_{PGC}, after falling an effective distance of ≈ 75 μm. This wall is made from a 1.5-W far-off-resonance Ti:sapphire beam having $w_0 = 67$ μm with an effective $p_{swept} \cong 2 \times 10^{-5}$ and $U_0 = 160$ μK. The power needed for the side walls of V_{PGC} is only a fraction of a watt, because the perpendicular component of velocity at the side wall is quite small.

After preevaporation the power and width of the CO_2 shepherd trap are then readjusted for transfer to V_3 and subsequent forced evaporative cooling. A CO_2 power of ≈ 20 W and a swept width of ≈ 100 μm give a V_3' trap depth of $U_0 = 90$ μK and an average density of $\approx 3.4 \times 10^{13}$ atoms per ml. The CO_2 trap is then raised into V_{evap} and placed ≈ 75 μm above the wall dividing V_{PGC} and V_{evap}. As a way of reducing the 4880 Å power requirements for the V_3 blue trap and gravity ramp, one can first reduce the axial length of the atom cloud in the CO_2 trap from $\approx 1,000$ μm to ≈ 500 μm by using the yz sheet beams before fully forming the 4880 Å V_3 trap and ramp. It takes ≈ 4 W of power divided between +y and +z beams with $w_0 = 3.9$ μm to get an initial V_3 trap potential $U_0 = 8 \times 2.7$ μK $\cong 22$ μK. Atoms in V_3 are maintained at uniform density by an antigravity ramp made from a +y swept 4880 Å beam of ≈ 750 μW with $w_0 = 3.9$ μW.

Forced evaporation and volume compression to optimum density results in a BEC in V_3', of dimensions 40 μm $\times$ 100 μm $\times$ 38 μm, with an estimated 5.6×10^6 atoms at a temperature of ≈ 0.1 μK. The evaporated atoms leaving V_3 trampoline over the lower wall of V_{evap}, and, as in V_{PGC}, are stored for later feedback to V_2' in the next cycle. Only minor heating of the trampolined atoms held in V_{PGC} and V_{evap} occurs for later feedback, in part, because of the small fraction of time the atoms interact with the walls of the repulsive box-like shepherd traps.

The newly formed condensate in V_3' is stabilized against further evaporation by increasing the wall potential to ≈ 1.2 μK

by using a power of 7.5 mW. It can then can be raised to V_3'' by transferring it to an off-axis trap, made from a +y sheet beam from L(+y) and a downward-angled beam launched off axis from the L(+x) lens. Otherwise, one can tip the L(+x) axis downward ≈ 4–$5°$ to reach the V_3'' volume. Next, the +y and +z beams are used to guide the condensate from V_3'' to V_4, V_4', and the output.

The shepherd power needed for the storage volume V_4 is ≈ 40 mW.

The final density of the condensate in V_4 and V_4' is $\approx 3.7 \times 10^{13}$ atoms per cm^3. With a cw output of $\approx 1.1 \times 10^6$ atoms per s, uniformly distributed over the guide area of 40 μm $\times$ 100 μm = 4,000 μm^2, one has an output intensity of $\approx 2.8 \times 10^2$ atoms per μm^2 s.

A single time-shared antigravity beam of total power ≈ 1.1 W can be used to make antigravity ramps for all of the manipulations involving condensate atoms. Besides the 750 mW needed for V_3, one needs ≈ 60 mW to follow the atoms from V_3' to V_1'' and on to V_4. A continuous power of 5×60 mW is required for the storage volume V_4'.

The detailed geometry of the MOT- and molasses-cooling and PGC beams was not specifically shown in Figs. 1 and 4B. The MOT magnetic field axis and one beam pair can be located 45° to the xy coordinate axes. The remaining two orthogonal beam pairs lie in a plane perpendicular to the MOT axis at 45° to the z axis. The three mutually orthogonal molasses beam pairs can be conveniently oriented at the so-called $\langle 1,1,1 \rangle$ angle of $\approx 35°$ to the z axis and rotated in azimuth to avoid other obstructing lenses. The same considerations apply to the PGC beams.

Conclusion

The discussion above indicates that a cw atom laser, as proposed here, should be possible by using present-day technology. Such a laser should have high spatial and temporal coherence, high intensity, and low temperature. The experimental achievement of such a cw atom laser would be a truly revolutionary development in BEC research and application.

1. Anderson, M. H., Ensher, J. R., Matthews, M. R., Wieman, C. E. & Cornell, E. A. O. (1995) *Science* **269**, 198–201.
2. Davis, K. B., Mewes, M.-O., Andrews, M. R., van Druten, N. J., Durfee, D. S., Kurn, D. M. & Ketterle, W. (1995) *Phys. Rev. Lett.* **75**, 3969–3973.
3. Mewes, M.-O., Andrews, M. R., Kurn, D. M., Durfee, D. S., Townsend, C. G. & Ketterle, W. (1997) *Phys. Rev. Lett.* **78**, 582–585.
4. Hagley, E. W., Deng, L., Kozuma, M., Wen, J., Helmerson, K., Rolston, S. L. & Phillips, W. D. (1999) *Science* **283**, 1706–1709.
5. Bloch, I., Hänsch, T. W. & Esslinger, T. (1999) *Phys. Rev. Lett.* **82**, 3008–3011.
6. Chikkatur, A. P., Shin, Y., Leanhardt, A. E., Kielpinski, D., Tsikata, E., Gustavson, T. L., Pritchard, D. E. & Ketterle, W. (2002) *Science* **296**, 2193–2195.
7. Gustavson, T. L., Chikkatur, A. P., Leanhardt, A. E., Görlitz, A., Gupta, S., Pritchard, D. E. & Ketterle, W. (2002) *Phys. Rev. Lett.* **88**, 020401(4).
8. Ashkin, A. (1970) *Phys. Rev. Lett.* **24**, 156–159.
9. Ashkin, A. (1970) *Phys. Rev. Lett.* **25**, 1321–1324.
10. Hänsch, T. W. & Schawlow, A. L. (1975) *Opt. Commun.* **13**, 68–69.
11. Ashkin, A. (1978) *Phys. Rev. Lett.* **40**, 729–732.
12. Ashkin, A. (1980) *Science* **210**, 1081–1088.
13. Gordon, J. P. & Ashkin, A. (1980) *Phys. Rev. A.* **21**, 1606–1617.
14. Bjorkholm, J. E., Freeman, R. R., Ashkin, A. & Pearson, D. B. (1978) *Phys. Rev. Lett.* **41**, 1361–1364.
15. Chu, S., Hollberg, L., Bjorkholm, J. E., Cable A. & Ashkin, A. (1985) *Phys. Rev. Lett.* **55**, 48–51.
16. Chu, S., Bjorkholm, J. E., Ashkin A. & Cable, A. (1986) *Phys. Rev. Lett.* **57**, 314–317.
17. Pritchard, D., Raab, E. L., Bagnato, V., Wieman, C. E & Watts, R. N. (1986) *Phys. Rev. Lett.* **57**, 310–313.
18. Raab, E. L., Prentiss, M., Cable, A., Chu, S. & Pritchard, D. E. (1987) *Phys. Rev. Lett.* **59**, 2631–2634.
19. Ketterle, W., Davis, K. B., Joffe, M. A., Martin, A. & Pritchard, D. (1993) *Phys. Rev. Lett.* **70**, 2253–2256.
20. Chu, S., Cohen-Tannoudji, C. N., Phillips, W. D. (1998) *Rev. Mod. Phys.* **70**, 685–741.
21. Ashkin, A. (1997) *Proc. Natl. Acad. Sci. USA* **94**, 4853–4860.
22. Dalfovo, F., Giorgini, S., Pitaevskii, L. P. & Stringari, S. (1999) *Rev. Mod. Phys.* **71**, 463–512.
23. Davis, K. B., Mewes, M.-O., Andrews, M. R., van Druten, N. J., Durfee, D. S., Kurn, D. M. & Ketterle, W. (1995) *Phys. Rev. Lett.* **75**, 1687–1690.
24. Ketterle, W. (1999) *Phys. Today* **52**, 30–35.
25. Martellucci, S., Chester, A. N., Aspect, A. & Inguscio, M, eds. (2000) *Bose-Einstein Condensates and Atom Lasers* (Kluwer Academic/Plenum Publishers, New York).
26. Miller, J. D., Cline, R. A. & Heinzen, D. J. (1993) *Phys. Rev. A* **47**, R4567–R4570.
27. Stenger, J., Inouye, S., Stamper-Kurn, D. M., Miesner, H.-J., Chikkatur, A. P. & Ketterle, W. (1998) *Nature* **396**, 345–348.
28. Stamper-Kurn, D. M., Andrews, M. R., Chikkatur, A. P., Inouye, S., Miesner, H.-J., Stenger, J. & Ketterle, W. (1998) *Phys. Rev. Lett.* **80**, 2027–2030.
29. Inouye, S., Andrews, M. R., Stenger, J., Miesner, H.-J., Stamper-Kurn, D. M. & Ketterle, W. (1998) *Nature* **392**, 151–154.
30. Courteille, P., Freeland, R. S. & Heinzen, D. J. (1998) *Phys. Rev. Lett.* **81**, 69–72.
31. Adams, C. S., Lee, H. J., Davidson, N., Kasevich, M. & Chu, S. (1995) *Phys. Rev. Lett.* **74**, 3577–3580.
32. Barrett, M. D., Sauer, J. A. & Chapman, M. S. (2001) *Phys. Rev. Lett.* **87**, 010404-1–010404-4.
33. Granada, S. R., Gehm, M. E., O'Hara, K. M. & Thomas, J. E. (2002) *Phys. Rev. Lett.* **88**, 120405 (4).
34. Renn, M. J., Montgomery, D., Vdovin, O., Anderson, D. Z., Wieman, C. E. & Cornell, E. A. (1995) *Phys. Rev. Lett.* **75**, 3253–3256.
35. Kuga, T., Torii, Y., Shiokawa, N. & Hirano, T. (1997) *Phys. Rev. Lett.* **78**, 4713–4716.
36. Friedman, N., Kaplan, A., Carasso, D, & Davidson, N. (2001) *Phys. Rev. Lett.* **86**, 1518 (4).
37. Visscher, K., Brakenhoff, G. J. & Krol, J. J. (1993) *Cytometry* **14**, 105–114.
38. Masuhara, H., DeSchryver, F. C., Kitamura, N. & Tamai, N., eds. (1994) *Microchemistry, Spectroscopy, and Chemistry in Small Domains* (North–Holland, Amsterdam).
39. Myatt, C. J., Burt, E. A., Ghrist, R. W., Cornell, E. A. & Wieman, C. E. (1997) *Phys. Rev. Lett.* **78**, 586–589.
40. Walker, T., Sesko, D. & Wieman, C. (1990) *Phys. Rev. Lett.* **64**, 408–411.
41. Metcalf, H. J. & van der Straten, P. (1999) *Laser Cooling and Trapping* (Springer, New York).
42. Boiron, D., Michaud, A., Fournier, J. M., Simard, L., Sprenger, M., Grynberg, G. & Salomon, C. (1998) *Phys. Rev. A* **57**, R4106–R4109.
43. Modugno, G., Ferrari, G., Roati, G., Brecha, R. J., Simoni, A. & Inguscio, M. (2001) *Science* **294**, 1320–1322.
44. O'Hara, K. M., Granada, S. R., Gehm, M. E., Savard, T. A., Bali, S., Freed, C. & Thomas, J. E. (1999) *Phys. Rev. Lett.* **82**, 4204–4207.
45. Giovanazzi, S., Smerzi, A. & Fantoni, S. (2000) *Phys. Rev. Lett.* **84**, 4521–4524.

PHYSICS

Forces of a single-beam gradient laser trap on a dielectric sphere in the ray optics regime

A. Ashkin

AT&T Bell Laboratories, Holmdel, New Jersey 07733

ABSTRACT   We calculate the forces of single-beam gradient radiation pressure laser traps, also called "optical tweezers," on micron-sized dielectric spheres in the ray optics regime. This serves as a simple model system for describing laser trapping and manipulation of living cells and organelles within cells. The gradient and scattering forces are defined for beams of complex shape in the ray-optics limit. Forces are calculated over the entire cross-section of the sphere using TEM_{00} and TEM_{01}^* mode input intensity profiles and spheres of varying index of refraction. Strong uniform traps are possible with force variations less than a factor of 2 over the sphere cross-section. For a laser power of 10 mW and a relative index of refraction of 1.2 we compute trapping forces as high as $\sim 1.2 \times 10^{-6}$ dynes in the weakest (backward) direction of the gradient trap. It is shown that good trapping requires high convergence beams from a high numerical aperture objective. A comparison is given of traps made using bright field or differential interference contrast optics and phase contrast optics.

INTRODUCTION

This paper gives a detailed description of the trapping of micron-sized dielectric spheres by a so-called single-beam gradient optical trap. Such dielectric spheres can serve as first simple models of living cells in biological trapping experiments and also as basic particles in physical trapping experiments. Optical trapping of small particles by the forces of laser radiation pressure has been used for about 20 years in the physical sciences for the manipulation and study of micron and submicron dielectric particles and even individual atoms (1–7). These techniques have also been extended more recently to biological particles (8–18).

The basic forces of radiation pressure acting on dielectric particles and atoms are known (1, 2, 19–21). For dielectric spheres large compared with the wavelength, one is in the geometric optics regime and can thus use simple ray optics in the derivation of the radiation pressure force from the scattering of incident light momentum. This approach was used to calculate the forces for the original trapping experiments on micron-sized dielectric spheres (1, 22). These early traps were either all optical two-beam traps (1) or single beam levitation traps which required gravity or electrostatic forces for their stability (23, 24). For particles in the Rayleigh regime where the size is much less than the wavelength λ the particle acts as a simple dipole. The force on a dipole divides itself naturally into two components: a so-called scattering force component pointing in the direction of the incident light and a gradient component pointing in the direction of the intensity gradient of the light (19, 21).

The single-beam gradient trap, sometimes referred to as "optical tweezers," was originally designed for Rayleigh particles (20). It consists of a single strongly focused laser beam. Conceptually and practically it is one of the simplest laser traps. Its stability in the Rayleigh regime is the result of the dominance of the gradient force pulling particles toward the high focus of the beam over the scattering force trying to push particles away from the focus in the direction of the incident light. Subsequently it was found experimentally that single-beam gradient traps could also trap and manipulate micron-sized (25) and a variety of biological particles, including living cells and organelles within living cells (8, 10). Best results were obtained using infrared trapping beams to reduced optical damage. The trap in these biological applications was built into a standard high resolution microscope in which one uses the same high numerical aperture (NA) microscope objective for both trapping and viewing. The micromanipulative abilities of single-beam gradient traps are finding use in a variety of experiments in the biological sciences. Experiments have been performed in the trapping of viruses and bacteria (8); the manipulation of yeast cells, blood cells, protozoa, and various algae and plant cells (10); the measurement of the compliance of bacterial flagella (11); internal cell surgery (13); manipulation of chromosomes (12); trapping and force measurement on sperm cells (14, 15); and recently, observations on the force of motor molecules driving mitochondrion and latex spheres along microtubules (16, 17). Optical techniques have also been used for cell sorting (9).

Qualitative descriptions of the operation of the single-beam gradient trap in the ray optics regime have already been given (25, 26). In Fig. 1 taken from reference 26, the action of the trap on a dielectric sphere is described in terms of the total force due to a typical pair of rays a and b of the converging beam, under the simplifying assumption of zero surface reflection. In this approximation the forces F_a and F_b are entirely due to refraction and are shown pointing in the direction of the momentum change. One sees that for arbitrary displacements of the sphere origin O from the focus f that the vector sum of F_a and F_b gives a net restoring force F directed back to the focus, and the trap is stable. In this paper we quantify the above qualitative picture of the trap. We show how to define the gradient and scttering force on a sphere $\gg \lambda$ in a natural way for beams of arbitrary shape. One can then describe trapping in the ray optics regime in the same terms as in the Rayleigh regime.

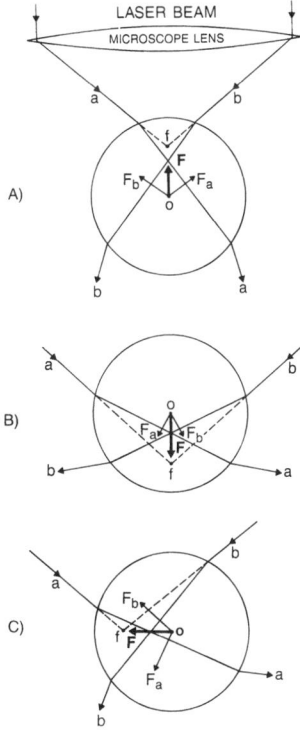

FIGURE 1 Qualitative view of the trapping of dielectric spheres. The refraction of a typical pair of rays a and b of the trapping beam gives forces F_a and F_b whose vector sum F is always restoring for axial and transverse displacements of the sphere from the trap focus f.

Results are given for the trapping forces over the entire cross-section of the sphere. The forces are calculated for input beams with various TEM_{00} and TEM_{01}^* mode intensity profiles at the input aperture of a high numerical aperture trapping objective of NA = 1.25. The results confirm the qualitative observation that good trapping requires the input aperture to be well enough filled by the incident beam to give rise to a trapping beam with high convergence angle. One can design traps in which the trapping forces vary at most by a factor of ~ 1.8 over the cross-section of the sphere with trapping forces as high as $Q = 0.30$ where the force F is given in terms of the dimensionless factor Q in the expression $F = Q(n_1 P/c)$. P is the incident power and $n_1 P/c$ is the incident momentum per second in a medium of index of refraction n_1. There has been a previous calculation of single-beam gradient trapping forces on spheres in the geometrical optics limit by Wright et al. (27), over a limited portion of the sphere, which gives much poorer results. They find trapping forces of $Q = 0.055$ in the above units which vary over the sphere cross-section by more than an order of magnitude.

LIGHT FORCES IN THE RAY OPTICS REGIME

In the ray optics or geometrical optics regime one decomposes the total light beam into individual rays, each with appropriate intensity, direction, and state of polarization, which propagate in straight lines in media of uniform refractive index. Each ray has the characteristics of a plane wave of zero wavelength which can change directions when it reflects, refracts, and changes polarization at dielectric interfaces according to the usual Fresnel formulas. In this regime diffractive effects are neglected (see Chapter III of reference 28).

The simple ray optics model of the single-beam gradient trap used here for calculating the trapping forces on a sphere of diameter $\gg \lambda$ is illustrated in Fig. 2. The trap consists of an incident parallel beam of arbitrary mode structure and polarization which enters a high NA microscope objective and is focused ray-by-ray to a dimensionless focal point f. Fig. 2 shows the case where f is located along the Z axis of the sphere. The maximum convergence angle for rays at the edge of the input aperture of a high NA objective lens such as the Leitz PL APO 1.25W (E. Leitz, Inc., Wetzlar, Germany) or the Zeiss PLAN NEOFLUAR 63/1.2W water immersion objectives (Carl Zeiss, Inc., Thornwood, NY), for example, is $\phi_{max} \cong 70°$. Computation of the total force on the sphere consists of summing the contributions of each beam ray entering the aperture at radius r with respect to the beam axis and angle β with respect the Y axis. The effect of neglecting the finite size of the actual beam focus, which can approach the limit of $\lambda/2n_1$ (see reference 29), is negligible for spheres much larger than λ. The point focus description of the convergent beam in which the ray directions and momentum continue in straight lines through the focus gives the correct incident polarization and momentum for each ray. The rays then reflect and refract at the surface of the sphere giving rise to the light forces.

The model of Wright et al. (27) tries to describe the single-beam gradient trap in terms of both wave and ray optics. It uses the TEM_{00} Gaussian mode beam propagation formula to describe the focused trapping beam and takes the ray directions of the individual rays to be

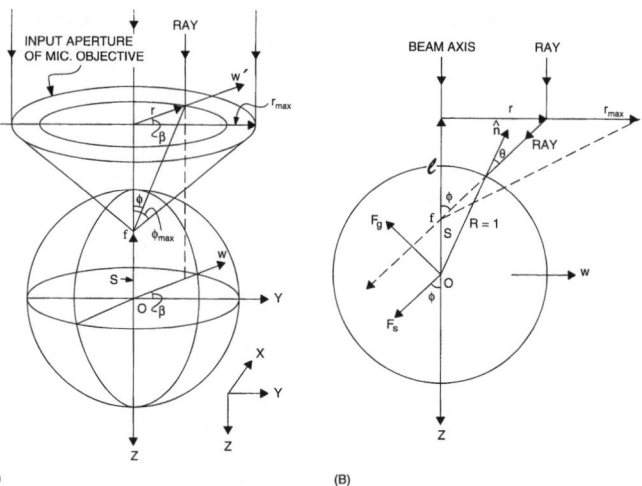

FIGURE 2 (A) Single beam gradient force trap in the ray optics model with beam focus f located along the Z axis of the sphere. (B) Geometry of an incident ray giving rise to gradient and scattering force contributions F_g and F_s.

perpendicular to the Gaussian beam phase fronts. Since the curvature of the phase fronts vary considerably along the beam, the ray directions also change, from values as high as 30° or more with respect to the beam axis in the far-field, to 0° at the beam focus. This is physically incorrect. It implies that rays can change their direction in a uniform medium, which is contrary to geometrical optics. It also implies that the momentum of the beam can change in a uniform medium without interacting with a material object, which violates the conservation of light momentum. The constancy of the light momentum and ray direction for a Gaussian beam can be seen in another way. If one resolves a Gaussian beam into an equivalent angular distribution of plane waves (see Section 11.4.2 of reference 28) one sees that these plane waves can propagate with no momentum or direction changes right through the focus. Another important point is that the Gaussian beam propagation formula is strictly correct only for transversely polarized beams in the limit of small far-field diffraction angles θ', where $\theta' = \lambda/\pi w_o$ (w_o being the focal spot radius). This formula therefore provides a poor description of the high convergence beams used in good traps. The proper wave description of a highly convergent beam is much more complex than the Gaussian beam formula. It involves strong axial electric field components at the focus (from the edge rays) and requires use of the vector wave equation as opposed to the scalar wave equation used for Gaussian beams (30).

Apart from the major differences near the focus, the model of Wright et al. (27) should be fairly close to the ray optics model used here in the far-field of the trapping beam. The principal distinction between the two calculations, however, is the use by Wright et al. of beams with relatively small convergence angle. They calculate forces for beams with spot sizes $w_o = 0.5$, 0.6, and 0.7 μm, which implies values of θ' of ~29, 24, and 21°, respectively. Therefore, these are beams having relatively small convergence angles compared with convergence angles of $\phi_{max} \cong 70°$ which are available from a high NA objective.

Consider first the force due to a single ray of power P hitting a dielectric sphere at an angle of incidence θ with incident momentum

per second of $n_1 P/c$ (see Fig. 3). The total force on the sphere is the sum of contributions due to the reflected ray of power PR and the infinite number of emergent refracted rays of successively decreasing power PT^2, PT^2R, ... PT^2R^n, ... The quantities R and T are the Fresnel reflection and transmission coefficients of the surface at θ. The net force acting through the origin O can be broken into F_Z and F_Y components as given by Roosen and co-workers (3, 22) (see Appendix I for a sketch of the derivation).

$$F_Z = F_S = \frac{n_1 P}{c}$$
$$\left\{ 1 + R \cos 2\theta - \frac{T^2[\cos(2\theta - 2r) + R \cos 2\theta]}{1 + R^2 + 2R \cos 2r} \right\} \quad (1)$$

$$F_Y = F_g = \frac{n_1 P}{c}$$
$$\left\{ R \sin 2\theta - \frac{T^2[\sin(2\theta - 2r) + R \sin 2\theta]}{1 + R^2 + 2R \cos 2r} \right\} \quad (2)$$

where θ and r are the angles of incidence and refraction. These formulas sum over all scattered rays and are therefore exact. The forces are polarization dependent since R and T are different for rays polarized perpendicular or parallel to the plane of incidence.

In Eq. 1 we denote the F_Z component pointing in the direction of the incident ray as the scattering force component F_S for this single ray. Similarly, in Eq. 2 we denote the F_Y component pointing in the direction perpendicular to the ray as the gradient force component F_g for the ray. For beams of complex shape such as the highly convergent beams used in the single-beam gradient trap, we define the scattering and gradient forces of the beam as the vector sums of the scattering and gradient force contributions of the individual rays of the beam. Fig. 2 B depicts the direction of the scattering force component and gradient force component of a single ray of the convergent beam

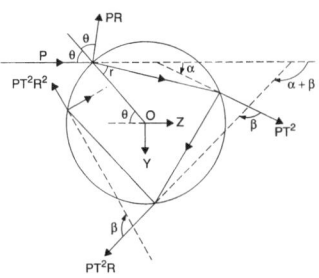

FIGURE 3 Geometry for calculating the force due to the scattering of a single incident ray of power P by a dielectric sphere, showing the reflected ray PR and infinite set of refracted rays PT^2R^n.

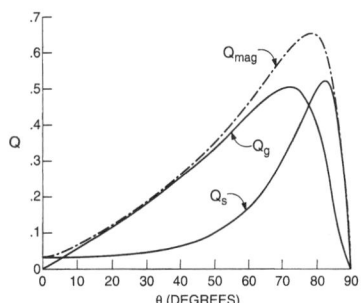

FIGURE 4 Values of the scattering force Q_s, gradient force Q_g, and magnitude of the total force Q_{mag} for a single ray hitting a dielectric sphere of index of refraction $n = 1.2$ at an angle θ.

striking the sphere at angle θ. One can show that the gradient force as defined above is conservative. This follows from the fact that F_g, the gradient force for a ray, can be expressed solely as a function of ρ, the radial distance from the ray to the particle. This implies that the integral of the work done on a particle in going around an arbitrary closed path can be expressed as an integral of $F_g(\rho)d\rho$ which is clearly zero. If the gradient force for a single ray is conservative, then the gradient force for an arbitrary collection of rays is conservative. Thus the conservative property of the gradient force as defined in the geometric optics regime is the same as in the Rayleigh regime. The work done by the scattering force, however, is always path dependent and is not conservative in any regime. As will be seen, these new definitions of gradient and scattering force for beams of more complex shape allow us to describe the operation of the gradient trap in the same manner in both the geometrical optics and Rayleigh regimes.

To get a feeling for the magnitudes of the forces, we calculate the scattering force F_s, the gradient force F_g, and the absolute magnitude of the total force $F_{mag} = (F_s^2 + F_g^2)^{1/2}$ as a function of the angle of incidence θ using Eqs. 1 and 2. We consider as a typical example the case of a circularly polarized ray hitting a sphere of effective index of refraction $n = 1.2$. The force for such a circularly polarized ray is the average of the forces for rays polarized perpendicular and parallel to the plane of incidence. The effective index of a particle is defined as the index of the particle n_2 divided by the index of the surrounding medium n_1; that is, $n = n_2/n_1$. A polystyrene sphere in water has $n = 1.6/1.33 \cong 1.2$. Fig. 4 shows the results for the forces F_s, F_g, and F_{mag} versus θ expressed in terms of the dimensionless factors Q_s, Q_g, and $Q_{mag} = (Q_s^2 + Q_g^2)^{1/2}$, where

$$F = Q\frac{n_1 P}{c}. \tag{3}$$

The quantity $n_1 P/c$ is the incident momentum per second of a ray of power P in a medium of index of refraction n_1 (19, 31). Recall that the maximum radiation pressure force derivable from a ray of momentum per second $n_1 P/c$ corresponds to $Q = 2$ for the case of a ray reflected perpendicularly from a totally reflecting mirror. One sees that for $n = 1.2$ a maximum gradient force of Q_{gmax} as high as ~ 0.5 is generated for rays at angles of $\theta \cong 70°$. Table I shows the effect of an index of refraction n on the maximum value of gradient force Q_{gmax} occurring at angle of incidence θ_{gmax}. The corresponding value of scattering force Q_s at θ_{gmax} is also listed. The fact that Q_s continues to grow relative to Q_{gmax} as n increases indicates potential difficulties in achieving good gradient traps at high n.

FORCE OF THE GRADIENT TRAP ON SPHERES

Trap focus along Z axis

Consider the computation of the force of a gradient trap on a sphere when the focus f of the trapping beam is located along the Z axis at a distance S above the center of the sphere at O, as shown in Fig. 2. The total force on the sphere, for an axially-symmetric plane-polarized input trapping beam, is clearly independent of the direction of polarization by symmetry considerations. It can therefore be assumed for convenience that the input beam is circularly polarized with half the power in each of two orthogonally oriented polarization components. We find the force for a ray entering the input aperture of the microscope objective at an arbitrary radius r and angle β and then integrate numerically over the distribution of input rays using an AT&T 1600 PLUS personal computer. As seen in Fig. 2, the vertical plane ZW which is rotated by β from the ZY plane contains both the incident ray and the normal to the sphere $\hat{n}$. It is thus the plane of incidence. We can compute the angle of

TABLE 1 **For a single ray. Effect of index of refraction n on maximum gradient force Q_{gmax} and scattering force Q_s occurring at angle of incidence θ_{gmax}**

n	Q_{gmax}	Q_s	θ_{gmax}
1.1	−0.429	0.262	79°
1.2	−0.506	0.341	72°
1.4	−0.566	0.448	64°
1.6	−0.570	0.535	60°
1.8	−0.547	0.625	59°
2.0	−0.510	0.698	59°
2.5	−0.405	0.837	64°

incidence θ from the geometric relation $R \sin \theta = S \sin \phi$, where R is the radius of the sphere. We take $R = 1$ since the resultant forces in the geometric optics limit are independent of R. Knowing θ we can find F_g and F_s for the circularly polarized ray by first computing F_g and F_s for each of the two polarization components parallel and perpendicular to the plane of incidence using Eqs. 1 and 2 and adding the results. It is obvious by symmetry that the net force is axial. Thus for S above the origin O the contribution of each ray to the net force consists of a negative Z component $F_{gz} = -F_g \sin \phi$ and a positive Z component $F_{sz} = F_s \cos \phi$ as seen from Fig. 2 B. For S below O the gradient force component changes sign and the scattering force component remains positive. We integrate out to a maximum radius r_{max} for which $\phi = \phi_{max} = 70°$, the maximum convergence angle for a water immersion objective of NA = 1.25, for example. Consider first the case of a sphere of index of refraction $n = 1.2$ and an input beam which uniformly fills the input aperture. Fig. 5 shows the magnitude of the antisymmetric gradient force component, the symmetric scattering force component, and the total force, expressed as Q_g, Q_s, and Q_t, for values of S above and $(-S)$ below the center of the sphere. The sphere outline is shown in Fig. 5 for reference. It is seen that the trapping forces are largely confined within the spherical particle. The stable equilibrium point S_E of the trap is located just above the

center of the sphere at $S \cong 0.06$, where the backward gradient force just balances the weak forward scattering force. Away from the equilibrium point the gradient force dominates over the scattering force and Q_t reaches its maximum value very close to the sphere edges at $S \cong 1.01$ and $(-S) \cong 1.02$. The large values of net restoring force near the sphere edges are due to the significant fraction of all incident rays which have both large values of θ, near the optimum value of $70°$, and large convergence angle ϕ. This assures a large backward gradient force contribution from the component $F_g \sin \phi$ and also a much-reduced scattering force contribution from the component $F_s \cos \phi$.

Trap along Y axis

We next examine the trapping forces for the case where the focus f of the trapping beam is located transversely along the $-Y$ axis of the sphere as shown in Fig. 6. The details of the force computation are discussed in Appendix II. Fig. 7 plots the gradient force, scattering force, and total force in terms of Q_g, Q_s, and Q_t as a function of the distance S' of the trap focus from the origin along the $-Y$ axis for the same conditions as in III A. For this case the gradient force has only a $-Y$ component. The scattering force is orthogonal to it along the $+Z$ axis. The total force again maximizes at a value $Q_t \cong 0.31$ near the sphere edge at $S' \cong 0.98$ and makes a small angle $\phi = \arctan F_g/F_s \cong 18.5°$ with respect to the Y axis. The Y force is, of course, symmetric about the center of the sphere at O.

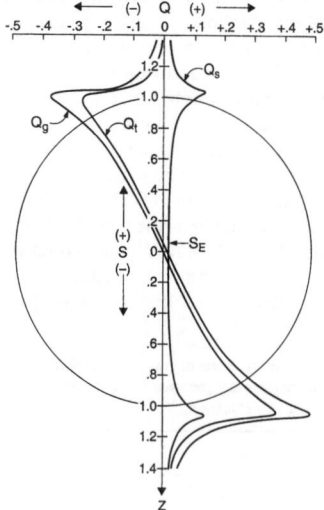

FIGURE 5 Values of the scattering force, gradient force, and total force Q_s, Q_g, and Q_t exerted on a sphere of index of refraction $n = 1.2$ by a trap with a uniformly filled input aperture which is focused along the Z axis at positions $+s$ above and $-s$ below the center of the sphere.

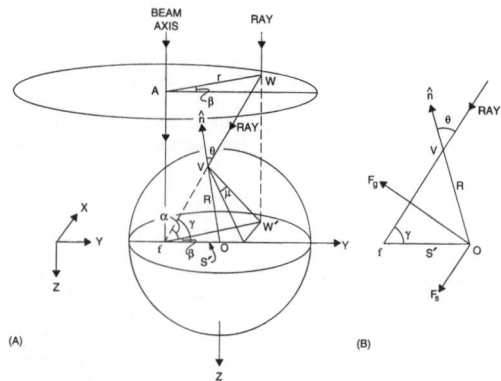

FIGURE 6 (A) Trap geometry with the beam focus f located transversely along the $-Y$ axis at a distance S' from the origin. (B) Geometry of the plane of incidence showing the directions of the gradient and scattering forces F_g and F_s for the input ray.

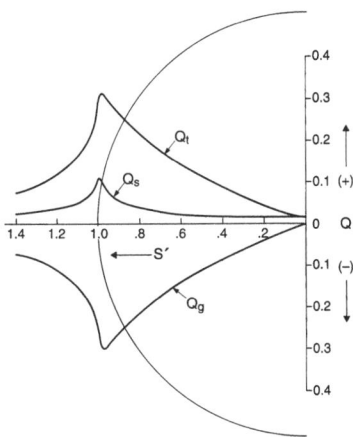

FIGURE 7   Plot of the gradient force, scattering force, and total force Q_g, Q_s, and Q_t as a function of the distance S' of trap focus from the origin along the $-Y$ axis for a circularly polarized trapping beam uniformly filling the aperture and a sphere of index of refraction $n = 1.2$.

General case: arbitrary trap location

Consider finally the most general case where the focus f is situated arbitrarily in the vertical plane through the Z axis at the distance S' from the sphere origin O in the direction of the $-Y$ axis and a distance S'' in the direction of the $-Z$ axis as shown in Fig. 8. Appendix III summarizes the method of force computation for this case.

Fig. 10 shows the magnitude and direction of the gradient force Q_g, the scattering force Q_s, and the total force Q_t as functions of the position of the focus f over the left half of the YZ plane, and by mirror image symmetry about the Y axis, over the entire cross-section of the sphere. This is again calculated for a circularly polarized beam uniformly filling the aperture and for $n = 1.2$. Although the force vectors are drawn at the point of focus f, it must be understood that the actual forces always act through the center of the sphere. This is true for all rays and therefore also for the full beam. It is an indication that no radiation pressure torques are possible on a sphere from the linear momentum of light. We see in Fig. 10 A that the gradient force which is exactly radial along the Z and Y axes is also very closely radial (within an average of $\sim 2°$ over the rest of the sphere. This stems from the closely radially uniform distribution of the incident light in the upper hemisphere. The considerably smaller scattering force is shown in Fig. 10 B (note the change in scale). It is strictly

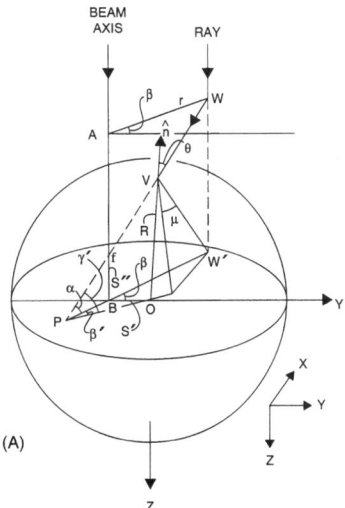

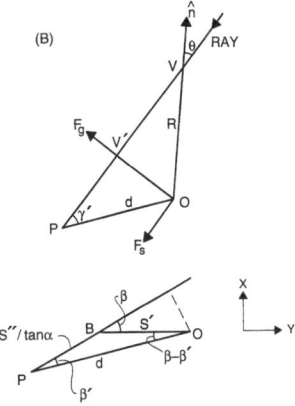

FIGURE 8   (A) Trap geometry with the beam focus located at a distance S' from the origin in the $-Y$ direction and a distance S'' in the $-Z$ direction. (B) Geometry of the plane of incidence POV showing the direction of gradient and scattering forces F_g and F_s for the ray. Geometry of triangle POB in the XY plane for finding β' and d.

axial only along the Z and Y axes and remains predominantly axial elsewhere except for the regions farthest from the Z and Y axes. It is the dominance of the gradient force over the scattering force that accounts for the overall radial character of the total force in Fig. 10 C. The rapid changes in direction of the force that occur when the focus is well outside the sphere are mostly due to the rapid changes in effective beam direction as parts of the input beam start to miss the

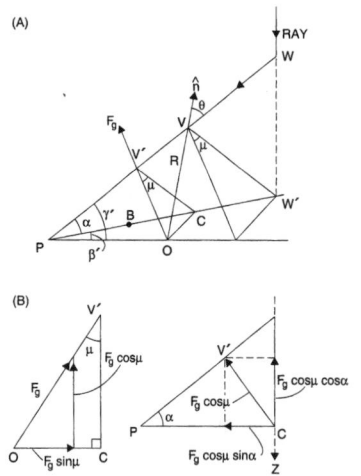

FIGURE 9 Another view of Fig. 8 *A* containing the angle μ between the plane of incidence POV' and the vertical plane WW'P for resolving force components along the coordinate axis.

sphere. We note that the magnitude of the total force Q_t maximizes very close to the edge of the sphere as we proceed radially outward in all directions, as does the gradient and scattering forces. The value of maximum restoring forces varies smoothly around the edge of the sphere from a minimum of $Q_t = 0.28$ in the axially backward direction to a maximum of $Q_t = 0.49$ in the forward direction. Thus, for these conditions the maximum trapping force achieved varies quite moderately over the sphere by a factor of $0.49/0.28 = 1.78$ and conforms closely to the edges of the sphere.

The line EE' marked on Fig. 10 *C* represents the locus of points for which the Z component of the force is zero; i.e., the net force is purely horizontal. If one starts initially at point E, the equilibrium point of the trap with no externally applied forces, and then applies a +Y-directed Stokes force by flowing liquid past the sphere to the right, for example, the equilibrium position will shift to a new equilibrium point along EE' where the horizontal light force just balances the viscous force. With increasing viscous force the focus finally moves to E', the point of maximum transverse force, after which the sphere escapes the trap. Notice that there is a net z displacement of the sphere as the equilibrium point moves from E to E'. We have observed this effect in experiments with micron-sized polystyrene spheres. Sato et al. (18) have recently reported also seeing this displacement.

EFFECT OF MODE PROFILES AND INDEX OF REFRACTION ON TRAPPING FORCES

To achieve a uniformly filled aperture in practice requires an input TEM_{00} mode Gaussian beam with very large spot size, which is wasteful of laser power. We therefore consider the behavior of the trap for other cases of TEM_{00} mode input beam profiles with smaller spot sizes, as well as TEM_{01}^* "do-nut" mode beam profiles which preferentially concentrate input light intensity at large input angles φ.

TEM_{00} mode profile

Table II compares the performance of traps with $n = 1.2$ having different TEM_{00} mode intensity profiles of the form $I(r) = I_o \exp(-2r^2/w_o^2$ at the input aperture of the microscope objective. The quantity a is the ratio of the TEM_{00} mode beam radius w_o to the full lens aperture r_{max}. A is the fraction of total beam power that enters the lens aperture. A decreases as a increases. In the limit of a uniform input intensity distribution $A = 0$ and $a = \infty$. For $w_o \leq r_{max}$ we define the convergence angle of the input beam as θ' where $\tan \theta' = w_o/\ell$. ℓ is the distance from the lens to the focus f as shown in Fig. 2 *B*. For $w_o > r_{max}$ the convergence angle is set by the full lens aperture and we use $\theta' = \phi_{max}$, where $\tan \phi_{max} = r_{max}/\ell$. For a NA = 1.25 water immersion objective $\phi_{max} = 70°$. The quality of the trap can be characterized by the maximum strength of the restoring forces as one proceeds radially outward for the sphere origin O in three representative directions taken along the Z and Y axes. We thus list Q_{1max}, the value of the maximum restoring force along the −Z axis, and S_{max}, the radial distance from the origin at which it occurs. Similarly listed are Q_{2max} occurring at S'_{max} along the −Y axis and Q_{3max} occurring at $(−S)_{max}$ along the +Z axis (see Figs. 2, 5, and 6 for a reminder on the definitions of S, $−S$, and S'). S_E in Table II gives the location of the equilibrium point of the trap along the −Z axis as noted in Fig. 5.

One sees from Table II that the weakest of the three representative maximum restoring forces is Q_{1max} occurring in the −Z direction. Furthermore, of all the traps the $a = \infty$ trap with a uniformly filled aperture has the largest Q_{1max} force and is therefore the strongest of all the TEM_{00} mode traps. One can also define the "escape force" of a given trap as the lowest force that can pull the particle free of the trap in any direction. In this context the $a = \infty$ trap has the largest magnitude of escape force of $Q_{1max} = 0.276$. One also sees that the $a = \infty$ trap is the most uniform trap since it has the smallest fractional variation in the extreme values of the restoring forces Q_{1max} and Q_{3max}. If, however, we reduce a to 1.7

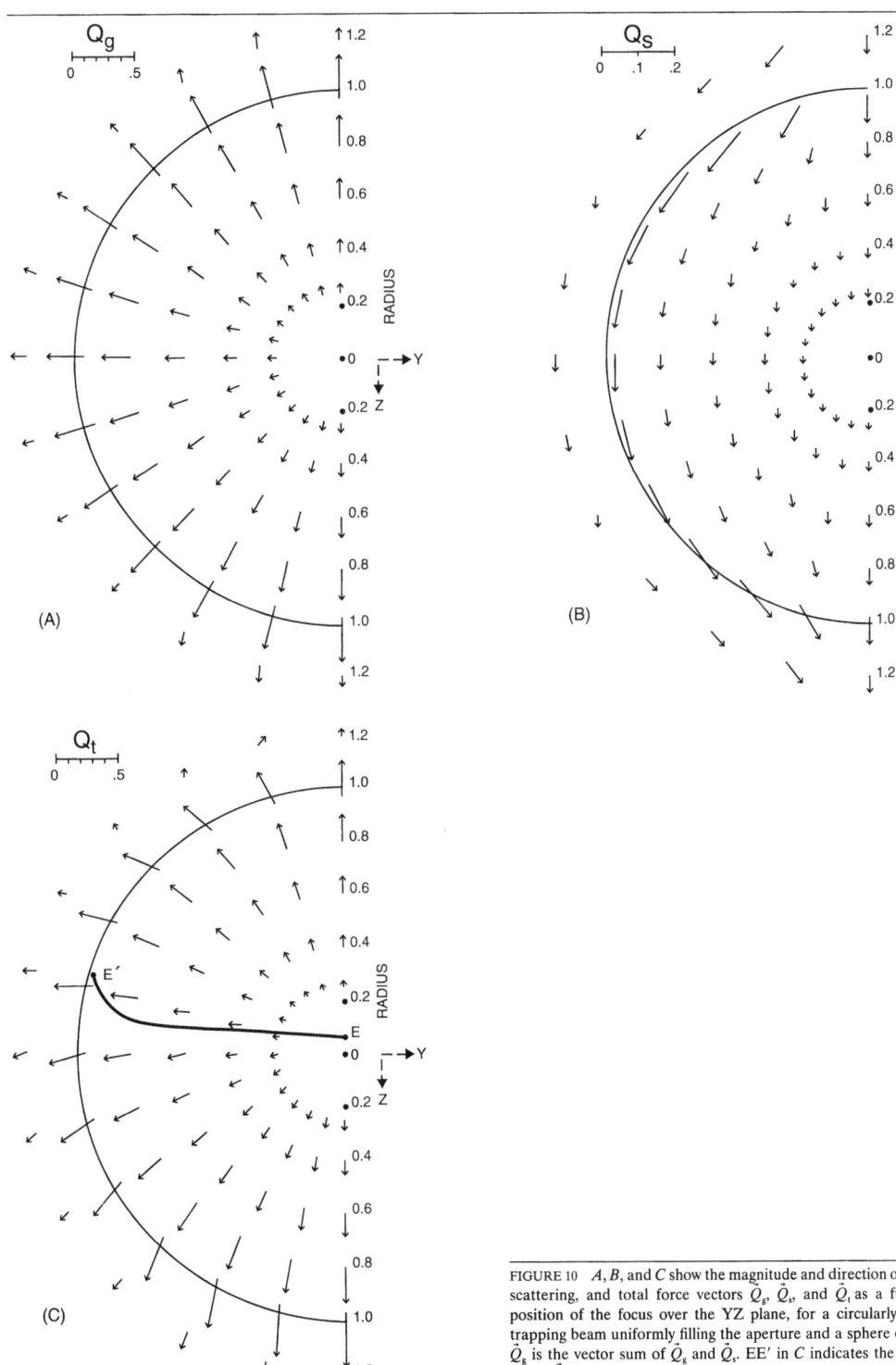

FIGURE 10 $A, B,$ and C show the magnitude and direction of gradient, scattering, and total force vectors $\vec{Q}_g$, $\vec{Q}_s$, and $\vec{Q}_t$ as a function of position of the focus over the YZ plane, for a circularly polarized trapping beam uniformly filling the aperture and a sphere of $n = 1.2$. $\vec{Q}_g$ is the vector sum of $\vec{Q}_g$ and $\vec{Q}_s$. EE' in C indicates the line along which $\vec{Q}_t$ is purely horizontal.

TABLE 2 Performance of TEM_∞ mode tapes with $n = 1.2$ having different intensity profiles at the input of the microscope objective

a	A	$[Q_{1max},$	$S_{max}]$	$[Q_{2max},$	$S'_{max}]$	$[Q_{3max},$	$(-S)_{max}]$	S_E	θ'
∞	0	-0.276	1.01	0.313	0.98	0.490	1.05	0.06	70°
1.7	0.5	-0.259	1.01	0.326	0.98	0.464	1.05	0.08	70°
1.0	0.87	-0.225	1.02	0.349	0.98	0.412	1.05	0.10	70°
0.727	0.98	-0.184	1.03	0.383	0.98	0.350	1.06	0.13	63°
0.364	1.0	-0.077	1.15	0.498	0.98	0.214	1.3	0.32	45°
0.202	1.0	-0.019	1.4	0.604	0.98	0.147	1.9	0.80	29°

or even 1.0, where the fraction of input power entering the aperture is reasonably high (~ 0.50 or 0.87), one can still get performance close to that of the uniformly filled aperture. Trap performance, however, rapidly degrades for cases of underfilled input aperture and decreasing beam convergence angle. For example, in the trap with $a = 0.202$ and $\theta' \cong 29°$ the value of Q_{1max} has dropped more than an order of magnitude to $Q_{1max} = -0.019$. The maximum restoring forces Q_{1max} and Q_{2max} occur well outside the sphere and the equilibrium position has moved away from the origin to $S_E = 0.8$. This trap with $\theta' \cong 29°$ roughly corresponds to the best of the traps described by Wright et al. (27) (for the case of $w_o = 0.5$ μm). They find for $w_o = 0.5$ μm that the trap has an equilibrium position outside of the sphere and a maximum trapping force equivalent to $Q_{1max} = -0.055$. Any more direct comparision of our results with those of Wright et al. is not possible since they use an approximate force calculation which overestimates the forces somewhat. They do not calculate forces for the beam focus inside the sphere and there are other artifacts associated with their use of Gaussian beam phase fronts to give the incident ray directions near the beam focus.

TEM_{01}^* "do-nut" mode profile

Table III compares the performance of several traps based on the TEM_{01}^* mode, the so-called "do-nut" mode, which has an intensity distribution of the form $I(r) = I_o (r/w_o')^2 \exp(-2r^2/w_o')^2$. The quantity a is now the ratio

of w_o', the spot size of the do-nut mode, to the full lens aperture r_{max}. All other items in the table are the same as in Table II. For $a = 0.76 \sim 87\%$ of the total beam power enters the input aperture r_{max} and one obtains performance that is almost identical to that of the trap with uniformly filled aperture as listed in Table II. For larger values of a the absolute magnitude of Q_{1max} increases, the magnitude of Q_{2max} decreases, and the fraction of power entering the aperture decreases. Optimal trapping, corresponding to the highest value of escape force, is achieved at values of $a \cong 1.0$ where the magnitudes $Q_{1max} \cong Q_{2max} \cong 0.30$. This performance is somewhat better than achieved with TEM_{00} mode traps.

It is informative to compare the performance of do-nut mode traps with that of a so-called "ring trap" having all its power concentrated in a ring 95–100% of the full beam aperture, for which $\phi \cong \phi_{max} = 70°$. When the ring trap is focused at $S \cong 1.0$ essentially all of the rays hit the sphere at an angle of incidence very close to $\theta_{gmax} = 72°$, the angle that makes Q_g a maximum for $n = 1.2$ (see Table I). Thus the resulting backward total force of $Q_{1max} = 0.366$ at $S = 0.99$, as listed in Table III, closely represents the highest possible backward force on a sphere of $n = 1.2$. The ring trap, however, has a reduced force $Q_{2max} = 0.254$ at $S'_{max} = 0.95$ in the $-Y$ direction since many rays at this point are far from optimal. If we imagine adding an axial beam to the ring beam then we optimally increase the gradient contribution to the force in the $-Y$ direction near $S' = 1.0$ and decrease the

TABLE 3 Performance of TEM_{01}^* mode traps with $n = 1.2$ having different intensity profiles at the input of the microscope objective

a	A	$[Q_{1max},$	$S_{max}]$	$[Q_{2max},$	$S'_{max}]$	$[Q_{3max},$	$(-S)_{max}]$	S_E
				TEM_{01}^* do-nut mode traps				
1.21	0.40	-0.310	1.0	0.290	0.98	0.544	1.05	0.06
1.0	0.59	-0.300	1.01	0.296	0.98	0.531	1.05	0.06
0.938	0.66	-0.296	1.01	0.298	0.98	0.525	1.05	0.07
0.756	0.87	-0.275	1.01	0.311	0.98	0.494	1.06	0.10
				Ring beam with $\phi = 70°$				
		-0.366	0.99	0.254	0.95	0.601	1.03	
				Ring beam plus axial beam				
		-0.31	0.99	0.31	0.95	0.51	1.03	

Comparison data on a Ring Beam having $\phi = 70°$ and a Ring Beam plus an Axial Beam containing 18% of the power.

overall force in the $-Z$ direction. With 18% of the power in the axial beam one gets $Q_{1max} = Q_{2max} \cong 0.31$. This performance is now close to that of the optimal do-nut mode trap. It is possible to design gradient traps that approximate the performance of a ring trap using a finite number of individual beams (for example, four, three or two beams) located symmetrically about the circumference of the ring and converging to a common focal point at angles of $\phi \cong 70°$. Recent reports (32, 33) at the CLEO-'91 conference presented observations on a trap with two individual beams converging to a focus with $\phi \cong 65°$ and also on a single beam gradient trap using the TEM_{01}^* mode.

Knowledge of the forces produced by ring beams allows one to compare the forces generated by bright field microscope objectives, as have thus far been considered, with the forces from phase contrast objectives of the same NA. For example, assume a phase contrast objective having an 80% absorbing phase ring located between radii of 0.35 and 0.55 of the full input lens aperture. For the case of an input beam uniformly filling the aperture with $n = 1.2$, one finds that the bright field escape force of $Q_{1max} = 0.276$ (see Table II) increases by $\sim 4\%$ to $Q_{1max} = 0.287$ in going to the phase contrast objective. With a TEM_{00} mode Gaussian beam input having $A = 0.87$ and $n = 1.2$, the bright field escape force magnitude of $Q_{1max} = 0.225$ increases by $\sim 2\%$ to $Q_{1max} = 0.230$ for a phase contrast objective. The reason for these slight improvements is that the force contribution of rays at the ring corresponds to $Q_{1max} \cong 0.204$, which is less than the average force for bright field. Thus any removal of power at the ring radius improves the overall force per unit transmitted power. Differential interference contrast optics can make use of the full input lens aperture and thus gives equivalent trapping forces to bright field optics.

Index of refraction effects

Consider, finally, the role of the effective index of refraction of the particle $n = n_1/n_2$ on the forces of a single-beam gradient trap. In Table IV we vary n for two types of trap, one with a uniformly filled input aperture, and the other having a do-nut input beam with $a = 1.0$, for which the fraction of total power feeding the input aperture is 59%. For the case of the uniformly filled aperture we get good performance over the range $n = 1.05$ to $n \cong 1.5$, which covers the regime of interest for most biological samples. At higher index Q_{1max} falls to a value of -0.097 at $n = 2$. This poorer performance is due to the increasing scattering force relative to the maximum gradient force as n increases (see Table I). Also the angle of incidence for maximum gradient force falls for higher n. At $n = 2$ (which corresponds roughly to a particle of index ~ 2.7 in water of index 1.33), the do-nut mode trap is clearly better than the uniform beam trap.

CONCLUDING REMARKS

It has been shown how to define the gradient and scattering forces acting on dielectric spheres in the ray optics regime for beams of complex shape. One can then describe the operation of single beam gradient force traps for spheres of diameter $\gg \lambda$ in terms of the dominance of an essentially radial gradient force over the predominantly axial scattering force. This is analogous to the previous description of the operation of this

TABLE 4 **Effect of index of refraction n on the performance of a trap with a uniformly filled aperture ($a = \infty$) and a do-nut trap with $a = 1.0$**

n	$[Q_{1max},$	$S_{max}]$	$[Q_{2max},$	$S'_{max}]$	$[Q_{3max},$	$(-S)_{max}]$	S_E
			Trap with uniformly filled aperture				
1.05	−0.171	1.06	0.137	1.00	0.219	1.06	0.02
1.1	−0.231	1.05	0.221	0.99	0.347	1.06	0.04
1.2	−0.276	1.01	0.313	0.98	0.490	1.05	0.06
1.3	−0.288	0.96	0.368	0.97	0.573	1.04	0.11
1.4	−0.282	0.93	0.403	0.96	0.628	1.02	0.15
1.6	−0.237	0.89	0.443	0.94	0.693	1.00	0.25
1.8	−0.171	0.88	0.461	0.94	0.723	0.99	0.37
2.0	−0.097	0.88	0.469	0.94	0.733	0.99	0.53
			TEM_{01}^* do-nut mode trap with $a = 1.0$				
1.05	−0.185	1.06	0.134	1.00	0.238	1.06	0.02
1.1	−0.250	1.05	0.208	0.99	0.379	1.06	0.03
1.2	−0.300	1.01	0.296	0.98	0.531	1.05	0.06
1.4	−0.309	0.93	0.382	0.95	0.667	1.02	0.13
1.8	−0.204	0.88	0.434	0.94	0.748	0.99	0.32
2.0	−0.132	0.88	0.439	0.94	0.752	0.99	0.42

trap in the Rayleigh regime, where the diameter $\ll \lambda$. Quite strong uniform traps are possible for $n = 1.2$ using the TEM_{01}^* do-nut mode in which the trapping forces vary over the sphere cross-section from a Q value of -0.30 in the $-Z$ direction to 0.53 in the $+Z$ direction. The magnitude of trapping force of 0.30 in the weakest trapping direction gives the escape force which a spherically shaped motile living organism, for example, must exert in order to escape the trap. For a laser power of 10 mW the minimum trapping force or escape force of $Q = 0.30$ is equivalent to 1.2×10^{-6} dynes. This implies that a motile organism 10 μm in diameter which is capable of propelling itself through water at a speed of 128 μm/s will be just able to escape the trap in its weakest direction along the $-Z$ axis. The only possible drawback to using the do-nut mode in practice is the difficulty of generating that mode in the laser. With the simpler TEM_{00} mode beams one can achieve traps with Q's as high as 0.23, for example, with 87% of the laser power entering the aperture of the microscope objective.

The calculation confirms the importance of using beams with large convergence angles θ' as high as $\sim 70°$ for achieving strong traps, especially with particles having lower indices of refraction typical of biological samples. At small convergence angles, less than $\sim 30°$, the scattering force dominates over the gradient force and single beam trapping is either marginal or not possible. One can, however, make a two-beam gradient force trap using smaller convergence angles based on two confocal, oppositely directed beams of equal power in which each ray of the converging beam is exactly matched by an oppositely directed ray. Then the scattering forces cancel and the gradient forces add, giving quite a good trap. Gradient traps of this type have been previously observed in experiments on alternating beam traps (34). The advantage of lower beam convergence is the ability to use longer working distances.

This work using ray optics extends the quantitative description of the single beam gradient trap for spheres to the size regime where the diameter is $\gg \lambda$. In this regime the force is independent of particle radius r. In the Rayleigh regime the force varies as r^3. At present there is no quantitative calculation for the intermediate size regime where the diameter is $\approx \lambda$, in which we expect force variations between r^0 and r^3. This is a more difficult scattering problem and involves an extension of Mie theory (35) or vector methods (36) to the case of highly convergent beams. Experimentally, however, this intermediate regime presents no problems. One can often directly calibrate the magnitude of the trapping force using Stokes dragging forces and thus successfully perform experiments with biological particles of size $\approx \lambda$ (16).

One can get a good idea of the range of validity of the

trapping forces as computed in the ray optics regime from a comparison of the scattering of a plane wave by a large dielectric sphere in the ray optics regime with the exact scattering, including all diffraction effects, as given by Mie theory. It suffices to consider plane waves since complex beams can be decomposed into a sum of plane waves. It was shown by van de Hulst in Chapter 12 of his book (35) that ray optics gives a reasonable approximation to the exact angular intensity distribution of Mie theory (except in a few special directions) for sphere size parameters $2\pi r/\lambda = 10$ or 20. The special directions are the forward direction, where a large diffraction peak appears which contributes nothing to the radiation pressure, and the so-called glory and rainbow directions, where ray optics never works. Since these directions contribute only slightly to the total force, we expect ray optics to give fair results down to diameters of approximately six wavelengths or ~ 5 μm for a 1.06-μm laser beam in water. The validity of the approximation should improve rapidly at larger sphere diameters. A similar result was also derived by van de Hulst (35) using Fresnel zones to estimate diffractive effects.

One of the advantages of a reliable theoretical value for the trapping force is that it can serve as a reference for comparison with experiment. If discrepancies appear in such a comparison one can then look for the presence of other forces. For traps using infrared beams there could be significant thermal (radiometric) force contributions due to absorptive heating of the particle or surrounding medium whose magnitude could then be inferred. Detailed knowledge of the variation of trapping force with position within the sphere is also proving useful in measurements of the force of swimming sperm (15).

Received for publication 19 June 1991 and in final form 16 August 1991.

APPENDIX I

Force of a ray on a dielectric sphere

A ray of power P hits a sphere at an angle θ where it partially reflects and partially refracts, giving rise to a series of scattered rays of power $PR, PT^2, PT^2R, \ldots, PT^2R^n, \ldots$. As seen in Fig. 3, these scattered rays make angles relative to the incident forward ray direction of $\pi + 2\theta$, α, $\alpha + \beta, \ldots, \alpha + n\beta \ldots$, respectively. The total force in the Z direction is the net change in momentum per second in the Z direction due to the scattered rays. Thus:

$$F_z = \frac{n_1 P}{c}$$
$$- \left[\frac{n_1 PR}{c} \cos(\pi + 2\theta) + \sum_{n=0}^{\infty} \frac{n_1 P}{c} T^2 R^n \cos(\alpha + n\beta) \right], \quad (A1)$$

where $n_1 P/c$ is the incident momentum per second in the Z direction. Similarly for the Y direction, where the incident momentum per second is zero, one has:

$$F_Y = 0$$

$$-\left[\frac{n_1 P R}{c} \sin(\pi + 2\theta) - \sum_{n=0}^{\infty} \frac{n_1 P}{c} T^2 R^n \sin(\alpha + \beta)\right]. \quad (A2)$$

As pointed out by van de Hulst in Chapter 12 of reference 35 and by Roosen (22), one can sum over the rays scattered by a sphere by considering the total force in the complex plane, $F_{tot} = F_Z + iF_Y$. Thus:

$$F_{tot} = \frac{n_1 P}{c}[1 + R \cos 2\theta] + i\frac{n_1 P}{c} R \sin 2\theta$$

$$- \frac{n_1 P}{c} T^2 \sum_{n=0}^{\infty} R^n e^{i(\alpha+n\beta)}. \quad (A3)$$

The sum over n is a simple geometric series which can be summed to give:

$$F_{tot} = \frac{n_1 P}{c}[1 + R \cos 2\theta]$$

$$+ i\frac{n_1 P}{c} R \sin 2\theta - \frac{n_1 P}{c} T^2 e^{i\alpha}\left[\frac{1}{1 - Re^{i\beta}}\right]. \quad (A4)$$

If one rationalizes the complex denominator and takes the real and imaginary parts of F_{tot}, one gets the force expressions A1 and A2 for F_Z and F_Y using the geometric relations $\alpha = 2\theta - 2r$ and $\beta = \pi - 2r$, where θ and r are the angles of incidence and refraction of the ray.

APPENDIX II

Force on a sphere for trap focus along Y axis

We treat the case of the beam focus located along the $-Y$ axis at a distance S' from the origin O (see Fig. 6). We first calculate the angle of incidence θ for an arbitrary ray entering the input lens aperture vertically at a radius r and azimuthal angle β in the first quadrant. On leaving the lens the ray stays in the vertical plane AWW' f and heads in the direction towards f, striking the sphere at V. The forward projection of the ray makes an angle α with respect to the horizontal (X, Y) plane. The plane of incidence, containing both the input ray and the normal to the sphere OV, is the so-called γ plane fOV which meets the horizontal and vertical planes at f. Knowing α and β, we find γ from the geometrical relation $\cos \gamma = \cos \alpha \cos \beta$. Referring to the γ plane we can now find the angle of incidence θ from $R \sin \theta = S' \sin \gamma$ putting $R = 1$.

In contrast to the focus along the Z axis, the net force now depends on the choice of input polarization. For the case of an incident beam polarized perpendicular to the Y axis, for example, one first resolves the polarized electric field E into components E $\cos \beta$ and E $\sin \beta$ perpendicular and parallel to the vertical plane containing the ray. Each of these components can be further resolved into the so-called p and s components parallel and perpendicular to the plane of incidence in terms of these angle μ between the vertical plane and the plane of incidence. By geometry, $\cos \mu = \tan \alpha/\tan \gamma$. This resolution yields fractions of the input power in the p and s components given by:

$$f_p = (\cos \beta \sin \mu - \sin \beta \cos \mu)^2 \quad (A5)$$

$$f_s = (\cos \beta \cos \mu + \sin \beta \sin \mu)^2. \quad (A6)$$

If the incident polarization is parallel to the Y axis, then f_p and f_s reverse. Knowing θ, f_p, and f_s, one computes the gradient and scattering force components for p and s separately using Eqs. A5 and A6 and adds the results.

The net gradient and scattering force contribution of the ray thus computed must now be resolved into components along the coordinate axes (see Fig. 6 B). However, comparing the force contributions of the quartet of rays made up of the ray in the first quadrant and its mirror image rays in the other quadrants we see that the magnitudes of the forces are identical for each of the rays of the quartet. Furthermore, the scattering and gradient forces of the quartet are directly symmetrically about the Z and Y axes, respectively. This symmetry implies that the entire beam can only give rise to a net Z scattering force coming from the integral of the $F_s \cos \phi$ component and a net Y gradient force coming from the $F_g \sin \gamma$ component. In practice we need only integrate these components over the first quadrant and multiply the results by 4 to get the net force. The differences in force that result from the choice of input polarization perpendicular or parallel to the Y axis are not large. For the conditions of Fig. 7 the maximum force difference is $\sim 14\%$ near $S' \cong 1.0$. We have therefore made calculations using a circularly polarized input beam with $f_p = f_s = \frac{1}{2}$, which yields values of net force that are close to the average of the forces for the two orthogonally polarized beams.

APPENDIX III

Force on a sphere for an arbitrarily located trap focus

We now treat the case where the trapping beam is focused arbitrarily in the XY plane at a point f located at a distance S' from the origin in the $-Y$ direction and a distance S'' in the $-Z$ direction (see Fig. 8). To calculate the force for a given ray we again need to find the angle of incidence θ and the fraction of the ray's power incident on the sphere in the s and p polarizations. Consider a ray of the incident beam entering the input aperture of the lens vertically at a radius r and azimuthal angle β in the first quadrant. The ray on leaving the lens stays in the vertical plane AWW'B and heads toward f, hitting the sphere at V. The extension of the incident ray to f and beyond intersects the XY plane at point P at an angle α. The plane of incidence for this ray is the so-called γ' plane POV which contains both the incident ray and the normal to the sphere OV. Referring to the planar figure in Fig. 8 B one can find the angle β' by simple geometry in terms of S', S'', and the known angles α and β from the relation

$$\tan \beta' = \frac{S' \sin \beta}{S' \cos \beta + S''/\tan \alpha}. \quad (A7)$$

We get γ' from $\cos \gamma' = \cos \alpha \cos \beta'$. Referring to the γ' plane in Fig. 8 B we get the angle of incidence θ for the ray from $R \sin \theta = d \sin \gamma'$, putting $R = 1$. The distance d is deduced from the geometric relation:

$$d = \frac{S'' \cos \beta'}{\tan \alpha} + S' \cos(\beta - \beta'). \quad (A8)$$

As in Appendix II, we compute f_p and f_s, the fraction of the ray's power in the p and s polarizations, in terms of the angle μ between the vertical plane W'VP and the plane of incidence POV. We use Eqs. A5 and A6 for the case of a ray polarized perpendicular to the Y axis and the same expressions with f_p and f_s reversed for a ray polarized parallel to the Y axis. To find μ we use $\cos \mu = \tan \alpha / \tan \gamma'$. As in Appendix II we can put $f_p = f_s = \frac{1}{2}$ and get the force for a circularly polarized ray, which is the average of the force for the cases of two orthogonally polarized rays.

The geometry for resolving the net gradient and scattering force contribution of each ray of the beam into components along the axes is now more complex. The scattering force F_s is directed parallel to the incident ray in the VP direction of Fig. 8. It has components $F_s \sin \alpha$ in the $+Z$ direction and $F_s \cos \alpha$ pointing in the BP direction in the XY plane. $F_s \cos \alpha$ is then resolved with the help of Fig. 8 B into $F_s \cos \alpha \cos \beta$ in the $-Y$ direction and $F_s \cos \alpha \sin \beta$ in the $-X$ direction. The gradient force F_g points in the direction OV' perpendicular to the incident ray direction VP in the plane of incidence OPV. This is shown in Fig. 8 and also Fig. 9, which gives yet another view of the geometry. In Fig. 9 we consider the plane V'OC, which is taken perpendicular to the γ' plane POV and the vertical plane WW'P. This defines the angle OV'C as μ, the angle between the planes, and also makes the angles OCV', OCP, and CV'P right angles. As an aid to visualization one can construct a true three-dimensional model out of cardboard of the geometric figure for the general case as shown in Figs. 8 and 9. Such a model will make it easy to verify that the above stated angles are indeed right angles, and to see other details of the geometry. We can now resolve F_g into components along the X, Y, and Z axes with the help of right triangles OV'C and CV'P as shown in Fig. 9 B. In summary, the net contribution of a ray in the first quadrant to the force is:

$$F(Z) = F_s \sin \alpha + F_g \cos \mu \cos \alpha \qquad \text{(A9)}$$

$$F(Y) = -F_s \cos \alpha \cos \beta$$
$$+ F_g \cos \mu \sin \alpha \cos \beta + F_g \sin \mu \sin \beta \qquad \text{(A10)}$$

$$F(X) = -F_s \cos \alpha \sin \beta$$
$$+ F_g \cos \mu \sin \alpha \sin \beta - F_g \sin \mu \cos \beta. \qquad \text{(A11)}$$

The force equations A9–A11 are seen to have the correct signs since F_s and F_g are, respectively, positive and negative as calculated from Eqs. 1 and 2.

For the general case under consideration we lose all symmetry between first and second quadrant forces and we must extend the force integrals into the second quadrant. All the above formulas which were derived for rays of the first quadrant are equally correct in the second quadrant using the appropriate values of the angles β, β', γ', and μ. For example, in the second quadrant β' can be obtuse. This gives obtuse γ' and obtuse μ. Obtuse μ implies that the γ' plane has rotated its position beyond the perpendicular to the vertical plane AWW'. In this orientation the gradient force direction tips below the XY plane and reverses its Z component as indicated by the sign change in the F_g $\cos \mu \cos \alpha$ term.

There are, however, some symmetry relations in the force contributions of rays of the input beam which still apply. For example, there is symmetry about the Y axis; i.e., rays of the third and fourth quadrants give the same contribution to the Z and Y forces as rays of the first and second quadrants, whereas their X contributions exactly cancel. To find the net force we need only integrate the Y and Z components of first and second quadrants and double the result.

If we make S'' negative in all formulas, we obtain the correct magnitudes and directions of the forces for the case of the focus below the XY plane. Although we find different total force values for S'' positive and S'' negative, i.e., symmetrical beam focus points above and below the XY plane, there still are symmetry relations that apply to the scattering and gradient forces separately. Thus we find that the

Z components of the scattering force are the same above and below but the Y component reverse. For the gradient force the Z components reverse above and below and the Y components are the same. This is seen to be true in Fig. 10. It is also consistent with Fig. 5 showing the forces along the Z axis. This type of symmetry behavior arises from the fact that the angle of incidence for rays entering the first quadrant from above the XY plane (S'' positive) is the same as for symmetrical rays entering in the second quadrant below the XY plane (S'' negative). Likewise the angles of incidence are the same for the second quadrant above and the first quadrant below. These results permit one to directly deduce the force below the XY plane from the values computed above the XY plane. The results derived here for the focus placed at an arbitrary point within the YZ plane are perfectly general since one can always choose to calculate the force in the cross-sectional plane through the Z axis that contains the focus f.

As a check on the calculations one can show that the results putting $S'' = 0$ in the general case are identical with those from the simpler Y axis integrals derived in Appendix II. Also in the limit $S' \to 0$ one gets the same results as are given by the simpler Z axis integral discussed above.

REFERENCES

1. Ashkin, A. 1970. Acceleration and trapping of particles by radiation pressure. *Phys. Rev. Lett.* 24:156–159.

2. Ashkin, A. 1970. Atomic-beam deflection by resonance-radiation pressure. *Phys. Rev. Lett.* 24:1321–1324.

3. Roosen, G. 1979. Optical levitation of spheres. *Can. J. Phys.* 57:1260–1279.

4. Ashkin, A. 1980. Applications of laser radiation pressure. *Science (Wash. DC)* 210:1081–1088.

5. Chu, S., J. E. Bjorkholm, A. Ashkin, and A. Cable. 1986. Experimental observation of optically trapped atoms. *Phys. Rev. Lett.* 57:314–317.

6. Chu, S., and C. Wieman. 1989. Feature editors, special edition, laser cooling and trapping of atoms. *J. Opt. Soc. Am.* B6:2020–2278.

7. Misawa, H., M. Koshioka, K. Sasaki, N. Kitamura, and H. Masuhara. 1990. Laser trapping, spectroscopy, and ablation of a single latex particle in water. *Chem. Lett.* 8:1479–1482.

8. Ashkin, A., and J. M. Dziedzic. 1987. Optical trapping and manipulation of viruses and bacteria. *Science (Wash. DC)* 235:1517–1520.

9. Buican, T., M. J. Smith, H. A. Crissman, G. C. Salzman, C. C. Stewart, and J. C. Martin. 1987. Automated single-cell manipulation and sorting by light trapping. *Appl. Opt.* 26:5311–5316.

10. Ashkin, A., J. M. Dziedzic, and T. Yamane. 1987. Optical trapping and manipulation of single cells using infrared laser beams. *Nature (Lond.).* 330:769–771.

11. Block, S. M., D. F. Blair, and H. C. Berg. 1989. Compliance of bacterial flagella measured with optical tweezers. *Nature (Lond.).* 338:514–518.

12. Berns, M. W., W. H. Wright, B. J. Tromberg, G. A. Profeta, J. J. Andrews, and R. J. Walter. 1989. Use of a laser-induced force trap to study chromosome movement on the mitotic spindle. *Proc. Natl. Acad. Sci. USA.* 86:4539–4543.

13. Ashkin, A., and J. M. Dziedzic. 1989. Internal call manipulation using infrared laser traps. *Proc. Natl. Acad. Sci. USA.* 86:7914–7918.

14. Tadir, Y., W. H. Wright, O. Vafa, T. Ord, R. H. Asch, and M. W.

Berns. 1989. Micromanipulation of sperm by a laser generated optical trap. *Fertil Steril.* 52:870–873.

15. Bonder, E. M., J. Colon, J. M. Dziedzic, and A. Ashkin. 1990. Force production by swimming sperm-analysis using optical tweezers. *J. Cell Biol.* 111:421A.

16. Ashkin, A., K. Schütze, J. M. Dziedzic, U. Euteneuer, and M. Schliwa. 1990. Force generation of organelle transport measured in vivo by an infrared laser trap. *Nature (Lond.).* 348:346–352.

17. Block, S. M., L. S. B. Goldstein, and B. J. Schnapp. 1990. Bead movement by single kinesin molecules studied with optical tweezers. *Nature (Lond.).* 348:348–352.

18. Sato, S., M. Ohyumi, H. Shibata, and H. Inaba. 1991. Optical trapping of small particles using 1.3 μm compact InGaAsP diode laser. *Optics Lett.* 16:282–284.

19. Gordon, J. P. 1973. Radiation forces and momenta in dielectric media. *Phys. Rev. A.* 8:14–21.

20. Ashkin, A. 1978. Trapping of atoms by resonance radiation pressure. *Phys. Rev Lett.* 40:729–732.

21. Gordon, J. P., and A. Ashkin. 1980. Motion of atoms in a radiation trap. *Phys. Rev. A.* 21:1606–1617.

22. Roosen, G., and C. Imbert. 1976. Optical levitation by means of 2 horizontal laser beams–theoretical and experimental study. *Physics. Lett.* 59A:6–8.

23. Ashkin, A., and J. M. Dziedzic. 1971. Optical levitation by radiation pressure. *Appl. Phys. Lett.* 19:283–285.

24. Ashkin, A., and J. M. Dziedzic. 1975. Optical levitation of liquid drops by radiation pressure. *Science (Wash. DC).* 187:1073–1075.

25. Ashkin, A., J. M. Dziedzic, J. E. Bjorkholm and S. Chu. 1986. Observation of a single-beam gradient force optical trap for dielectric particles. *Optics Lett.* 11:288–290.

26. Ashkin, A., and J. M. Dziedzic. 1989. Optical trapping and manipulation of single living cells using infra-red laser beams. *Ber. Bunsen-Ges. Phys. Chem.* 98:254–260.

27. Wright, W. H., G. J. Sonek, Y. Tadir, and M. W. Berns. 1990. Laser trapping in cell biology. *IEEE (Inst. Electr. Electron. Eng.) J. Quant. Elect.* 26:2148–2157.

28. Born, M., and E. Wolf. 1975. Principles of Optics. 5th ed. Pergamon Press, Oxford. 109–132.

29. Mansfield, S. M., and G. Kino. 1990. Solid immersion microscope. *Appl. Phys. Lett.* 57:2615–2616.

30. Richards, B., and E. Wolf. 1959. Electromagnetic diffraction in optical systems. II. Structure of the image field in an aplanatic system. *Proc. R. Soc. London. A.* 253:358–379.

31. Ashkin, A., and J. M. Dziedzic. 1973. Radiation pressure on a free liquid surface. *Phys. Rev. Lett.* 30:139–142.

32. Hori, M., S. Sato, S. Yamaguchi, and H. Inaba. 1991. Two-crossing laser beam trapping of dielectric particles using compact laser diodes. Conference on Lasers and Electro-Optics, 1991 (Optical Society of America, Washington, D.C.). *Technical Digest.* 10:280–282.

33. Sato, S., M. Ishigure, and H. Inaba. 1991. Application of higher-order-mode Nd:YAG laser beam for manipulation and rotation of biological cells. Conference on Lasers and Electro-Optics, 1991 (Optical Society of America, Washington, D.C.). *Technical Digest.* 10:280–281.

34. Ashkin, A., and J. M. Dziedzic. 1985. Observation of radiation pressure trapping of particles using alternating light beams. *Phys. Rev. Lett.* 54:1245–1248.

35. van de Hulst, H. C. 1981. Light Scattering by Small Particles. Dover Press, New York. 114–227.

36. Kim, J. S., and S. S. Lee. 1983. Scattering of laser beams and the optical potential well for a homogeneous sphere. *J. Opt. Soc. Am.* 73:303–312.

APPLIED PHYSICS LETTERS   VOLUME 19, NUMBER 8   15 OCTOBER 1971

Optical Levitation by Radiation Pressure

A. Ashkin and J. M. Dziedzic

Bell Telephone Laboratories, Holmdel, New Jersey 07733
(Received 14 June 1971; in final form 13 August 1971)

The stable levitation of small transparent glass spheres by the forces of radiation pressure has been demonstrated experimentally in air and vacuum down to pressures ~1 Torr. A single vertically directed focused TEM_{00}-mode cw laser beam of ~250 mW is sufficient to support stably a ~20-μ glass sphere. The restoring forces acting on a particle trapped in an optical potential well were probed optically by a second laser beam. At low pressures, effects arising from residual radiometric forces were seen. Possible applications are mentioned.

This letter reports the observation of stable optical levitation of transparent glass spheres in air and vacuum by the forces of radiation pressure from laser light. The technique used involves properties of radiation pressure previously deduced from experiments on small transparent micron-sized spheres in liquid.[1] It was shown that a light beam striking a sphere of a high index of refraction in a position of transverse gradient of light intensity not only exerts a force directed along the light beam, but also has a transverse component of force which pushes the particle toward the region of maximum light intensity. This fact makes stable optical potential wells possible.[1] In contrast to magnetic[2] and electrostatic[3] feedback levitation, or electrodynamic levitation,[4] optical levitation based on the potential well provided by radiation pressure is truly stable in a dc sense with the particle at rest at the equilibrium point. In this regard, magnetically levitated superconductors[5] are similarly stable at rest.

In our experiment a single vertically directed focused cw laser beam was used to lift a glass sphere off a glass plate and stably levitate it. Figure 1 shows the basic apparatus. About 100—500 mW of 5145-Å laser light in the TEM_{00} mode is focused by a 5-cm lens and directed vertically on a selected sphere of ~15—25 μ in diameter, initially at rest in position A on the glass plate. The power is such that a force of several g is applied at A with the particle at the beam waist ($2w_0 \cong 25 \mu$). This force is directly calculable from the index of refraction of the sphere ($n = 1.65$) as indicated in Ref. 1. This, by itself, is insufficient to break the strong van der Waals attraction to the supporting plate, which for a 20-μ sphere is ~$10^4 g$. This bond can, however, be broken acoustically by setting up a vibration with a piezoelectric ceramic cylinder cemented to the glass plate. Tuning the driving frequency rapidly through a mechanical resonance momentarily shakes the particle loose and it begins to rise up into the diverging Gaussian beam. It comes to equilibrium at position B about 1 mm above the beam waist where radiation pressure and gravity balance. A glass enclosure over the plate serves to minimize air currents. By moving the lens, the beam and hence the particle can easily be moved anywhere within the enclosure. It can even be deposited on the roof for careful subsequent examination. Figure 2 shows a 20-μ particle levitated in air and photographed by

its scattered light. The particle is extremely stable and can remain aloft for hours.

Operation at reduced pressure was accomplished with a simple vacuum cell and an adjustable leak valve. Levitation down to pressures as low as 1 Torr was observed before the particles were lost. It is felt that residual radiometric forces coupled with reduced viscous air damping account for this loss. As the pressure was reduced, the equilibrium position B gradually dropped down toward the beam focus indicating the onset of an additional downward force, thought to be radiometric in origin. Due to residual optical absorption and the lenslike character of the sphere, the top of the sphere is slightly hotter than the bottom. This gives a downward radiometric force (negative photophoresis) which initially increases as the pressure is reduced. The thermal force F_{th} is proportional to $1/p$ down to ~10 Torr, where the mean free path is comparable with the sphere diameter.[6] Also, the viscosity and thermal conductivity of the gas are roughly constant for pressures down to ~10 Torr. Below this pressure, radiometric forces, viscosity, and thermal conductivity begin to decrease. Experimentally, at low pressure the particles begin to become less stable

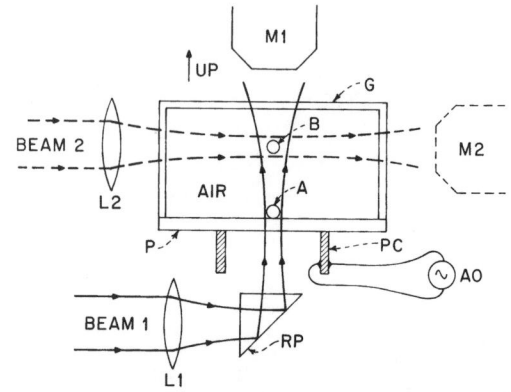

FIG. 1. Levitation apparatus. Particle at A is shaken loose acoustically and lifted to B by TEM_{00}-mode beam 1. TEM_{00}-mode beam 2 is introduced later as a probe beam to study the strength of the trapping forces. L1 and L2 are lenses, P is a glass plate, G is a glass enclosure about 1.5 cm high, RP is a reflecting prism, PC is a piezoelectric ceramic cylinder driven by audio-oscillator AO, and M1 and M2 are microscopes.

A . ASHKIN AND J.M. DZIEDZIC

FIG. 2. Photograph of a ∼20-μ transparent glass particle being levitated about 1 cm above a glass plate by a 250-mW vertically directed laser beam (shown as beam 1 in Fig. 1). The bright spot is the particle (vastly overexposed) photographed by its own scattered light. The ∼90° Mie scattering from the particle is seen on a screen placed at the rear and side of the glass enclosure.

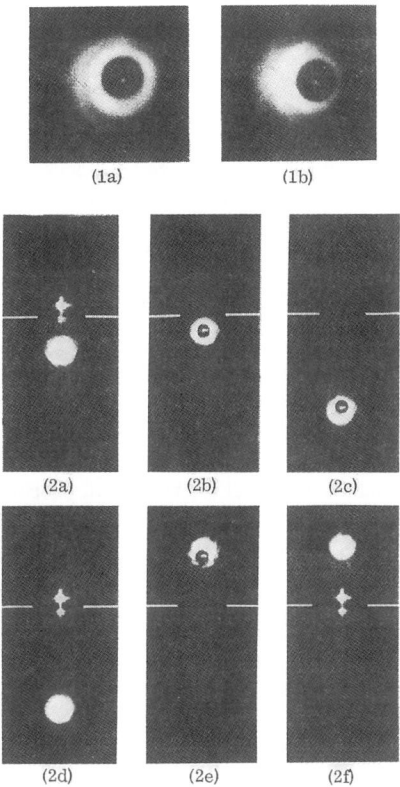

FIG. 3. Probing the horizontal and vertical trapping forces with auxiliary transverse beam 2. Top views 1(a) and 1(b) from microscope 1 show the horizontal displacement of the particle in beam 1. Side views 2(a) and 2(f) from microscope 2 show the vertical position of the particle in beam 1 as it is hit by transverse beam 2.

horizontally and vertically and even begin to spin at an increasing rate about a vertical axis prior to breaking loose from the optical trapping forces. Thus we see that even radiometric forces which are one or two orders of magnitude less than radiation pressure at atmospheric pressure can cause trouble at reduced pressure. A calculation of optical absorption in the particle, based on the estimated radiometric forces, indicates an absorption loss $\alpha \cong 5 \times 10^{-2}$ cm^{-1} which is quite high. The glass spheres were made commercially by the Flexolite Corp. of St. Louis, Mo. for reflectors. Since a loss of $\alpha \cong 10^{-4}-10^{-5}$ cm^{-1} is possible in glass, this thermal limitation is not fundamental.

It can be estimated that the horizontal restoring forces are much stronger than the vertical restoring forces about the equilibrium point of the optical trap. This is directly observable in an auxiliary experiment in which a particle levitated in a 250-mW vertical beam 1 is illuminated by a horizontal beam 2 of adjustable power (see Fig. 1). Microscopes 1 and 2 are used to project enlarged views of the beams and particles on viewing screens. In photographs 1(a) and 1(b) of Fig. 3, taken with microscope 1, we observe the horizontal displacement of the particle in beam 1 as the power in beam 2 is increased to ∼125 mW. This power almost pushes the particle out of beam 1. Thus we find a maximum transverse trapping acceleration of ∼$\frac{1}{2}g$ for beam 1. This is very large and accounts for the high degree of horizontal stability observed. The much weaker vertical stability manifests itself in microscope 2 in the sensitivity of the vertical equilibrium level to minor power fluctuations. It is interesting to note that the

probe beam 2 itself can be used to stabilize the particle vertically. View 2(a) shows the particle, seen by its own scattered light as two bright sources located near the top and bottom of the particle,[7] sitting at its equilibrium level marked by the white line. Just below it is beam 2, adjusted to about 40 mW. In view 2(b), beam 2 is raised so its fringe field hits the particle, whereupon it draws the particle down into it, close to its axis. In this position the transverse stability of only 40 mW in beam 2 is sufficient to hold the particle essentially fixed vertically as the power fluctuates. In fact, it is now possible to lower beam 2 and drag the particle down many particle diameters into the region of slowly increasing vertical force [view 2(c)] before it breaks free and returns to its equilibrium position [view 2(d)]. The particle can also be lifted up, as shown in view 2(e), before it breaks free and drops back to the equilibrium level [view 2(f)]. In situations 2(d) and 2(f), where the particle returns to equilibrium, we can conveniently observe the effects of viscous damping inasmuch as the particle returns without undergoing any vertical oscillations.

OPTICAL LEVITATION BY RADIATION PRESSURE   285

Many other trapping configurations having even greater stability based on several beams are possible. For instance, a third beam, opposing beam 2, and of the same power would keep the equilibrium point on the axis of beam 1 while adding to the vertical stability. Other laser modes such as TEM_{01} and TEM_{01}^* (the doughnut mode) have been successfully used, although the TEM_{00} mode is optimum. Other particle shapes, if not too extreme, are useable, possibly even partly hollow particles. With a different launching technique, it certainly should be possible to levitate very much smaller particles, even perhaps in the submicron range. It should be noted, however, that the principles of optical levitation should not be expected to provide any significant trapping of atoms, even though the pressure of resonance radiation on atoms is quite a significant effect.[8]

The technique of optical levitation will probably have use in applications where the precision micromanipulation of small particles, free from any supports, is important such as in light scattering from single small particles (Mie scattering) or in laser-initiated fusion experiments. If the viscous damping can be further reduced, applications to inertial devices such as gyroscopes and accelerometers become possible. Measurement of low optical absorptions, absolute optical power measurement, and pressure measurement are also likely areas of application. Levitation may also provide an interesting adjunct to Millikan-type experiments on charged particles. The extreme simplicity of the technique and its remarkable stability recommend its use.

[1]A. Ashkin, Phys. Rev. Letters 24, 156 (1970).

[2]J.W. Beams, Science 120, 619 (1954).

[3]C.B. Strang of the Martin-Marietta Corp. (private communication); also Martin-Marietta, Report No. OR9638, 1968 (unpublished).

[4]R.F. Wuerker, H. Shelton, and R.V. Langmuir, J. Appl. Phys. 30, 342 (1959).

[5]I. Simon, J. Appl. Phys. 24, 19 (1953).

[6]N.A. Fuchs, *The Mechanics of Aerosols* (Macmillan, New York, 1964).

[7]The interference observed in the 90° Mie scattering of Fig. 2 can be ascribed to the interference from these two bright sources. One can determine the particle diameter d by measuring its distance from the observing screen and the fringe spacing, since the separation between the two near-field sources is $\cong (1+\sqrt{2})\,d/2$.

[8]A. Ashkin, Phys. Rev. Letters 25, 1321 (1971).

ARTICLES

Single myosin molecule mechanics: piconewton forces and nanometre steps

Jeffrey T. Finer[*], Robert M. Simmons[†] & James A. Spudich[‡*]

* Departments of Biochemistry and Developmental Biology, Beckman Center, Stanford University School of Medicine, Stanford, California 94305, USA
† MRC Muscle and Cell Motility Unit, Randall Institute, King's College London, 26–29 Drury Lane, London WC2B 5RL, UK

A new *in vitro* assay using a feedback enhanced laser trap system allows direct measurement of force and displacement that results from the interaction of a single myosin molecule with a single suspended actin filament. Discrete stepwise movements averaging 11 nm were seen under conditions of low load, and single force transients averaging 3–4 pN were measured under isometric conditions. The magnitudes of the single forces and displacements are consistent with predictions of the conventional swinging-crossbridge model of muscle contraction.

DURING muscle contraction, chemical energy from ATP hydrolysis is converted to the relative sliding of actin and myosin filaments and force production. Although the actomyosin system has been extensively studied, the mechanism underlying its mechanochemical energy transduction remains unknown. The conventional swinging-crossbridge theory[1,2] proposes that for each ATP hydrolysis, myosin binds to actin and undergoes a conformational change or power stroke before subsequently detaching. This theory assumes that the myosin step size, or movement of actin relative to myosin for each ATP hydrolysis, is less than 40 nm, a value limited by the physical dimensions of the myosin head. The solution of the crystal structure of the myosin head led to the suggestion that a long α-helical domain might serve as a lever-arm capable of producing a movement of about 6 nm by a conformational change[3]. Early mechanical studies on muscle fibres estimated the myosin step size to be well within the structural constraints of the conventional theory[4,5]. There have been conflicting estimates of the myosin step size from *in vitro* motility assays, however: some measurements[6-8] are consistent with the conventional theory, whereas others[9,10] suggest much larger step sizes which are incompatible with the theory. There is also evidence from mechanical studies on muscle fibres that challenges the conventional theory by suggesting that myosin may be capable of undergoing multiple power strokes for each ATP hydrolysis[11-15].

These inconsistencies arise because each estimate of the step size is based upon a number of uncertain assumptions, particularly concerning the number of myosin heads involved in a given measurement. The duty ratio, or the fraction of the myosin ATPase cycle that is spent in a force-generating or strongly bound state, has been estimated to be as low as 5 per cent[7] or as high as 60–90 per cent[10,13] under sliding conditions in *in vitro* motility assays. As a result, the estimated number of myosin molecules involved in a measurement could easily differ by an order of magnitude. Varying estimates have also been obtained of the number of myosin heads bound to actin or producing force in isometrically contracting muscle fibres as compared with rigor. Under physiological conditions, stiffness measurements suggest that 75 per cent of the myosin molecules are bound to actin[16]; X-ray diffraction intensity measurements give values as high as 90 per cent[17]; and electron spin resonance measurements suggest that 24 per cent of the myosin heads are bound to actin in a rotationally disordered state[18] and 12 per cent in a rigor-like state[19]. It has therefore become clear that in order to resolve these issues, it is necessary to develop new techniques with the resolution necessary to probe the mechanical properties of myosin at the level of single molecular events.

‡ To whom correspondence should be addressed.

Towards this goal, actomyosin *in vitro* motility assays have become increasingly sophisticated. Unlike kinesin–microtubule assays, in which single kinesin molecules can support continuous movement of microtubules[20], actomyosin motility assays are complicated by the comparatively large fraction of time in the ATPase cycle that myosin spends detached from actin. During this time, the actin filament will diffuse away from a myosin molecule unless it attaches to another molecule in the vicinity. This difficulty has been overcome by including in the assay buffer the inert polymer methylcellulose to prevent actin from diffusing away[7], which made it possible to measure the velocity of actin filaments moving on only a few myosin molecules[8]. It is also possible now to measure the force on single actin filaments moving on a myosin-coated surface either by measuring the bending of a glass microneedle attached to one end of an actin filament[13,21] or by attaching a latex bead to the actin filament and measuring its movement in a single-beam gradient optical trap[22]. The optical trap technique has been used to measure single molecule events by kinesin molecules[23-25] and it offers advantages over microneedles, such as ease of use and the ability to change the stiffness of the trap in the middle of an experiment.

Here we extend the use of the optical trap technique in motility assays to measure nanometre movements and piconewton forces at millisecond rates[26]. By adding a second optical trap an actin filament can be held and manipulated through beads attached to each end of the filament (Fig. 1a), which prevents the filament from diffusing away from surfaces sparsely coated with myosin. By placing myosin molecules on a support above the coverglass surface, interactions of either the actin filament or the latex beads with the surface are minimized. By using an electronic feedback system, it is possible to measure forces under approximately isometric conditions.

Single displacements

The basic design of the experiment involved the firm attachment of silica beads to a microscope coverslip to provide docking platforms for myosin molecules (Fig. 1a). This coverslip was used to construct a flow cell as in the myosin-coated surface *in vitro* motility assay[27]. The coverslip was then coated with skeletal muscle heavy meromyosin (HMM) at a density that was insufficient to support continuous movement of actin filaments. Polystyrene beads coated with *N*-ethylmaleimide (NEM)-treated HMM were attached to actin filaments and applied to the flow cell in the presence of ATP. An actin filament with a bead attached near each end was caught and held in mid-solution with two optical traps. The brightfield image of one of the beads was projected onto a quadrant photodiode detector for high resolution position detection. For displacements of the bead up to ~200 nm away from the centre of the trap, the bead position

ARTICLES

is proportional to an applied force acting on the bead, so that it is possible to measure force as well as displacement[26]. The actin filament was pulled taut by moving the second bead until the force on the actin filament was ~2 pN, and it was then brought close to the surface of the coverslip so that it could interact with one or a few HMM molecules on the silica bead support. The feedback enhanced laser trap is shown schematically in Fig. 1b.

To study myosin movement under conditions of low load, it was necessary to use relatively compliant optical traps. The stiffness of each trap (0.02 pN per nm; total stiffness opposing motion, 0.04 pN per nm) was chosen to be as large as possible to decrease the amplitude of the brownian motion of the beads while not exceeding a force that would impede a myosin

molecule from producing its full displacement. When the actin filament was brought in contact with a sparsely coated surface on the silica bead support, the trapped bead attached to the actin filament showed rapid transient movements in the direction along the actin filament (Fig. 2a, top trace), but not in the direction perpendicular to the actin filament (Fig. 2a, bottom trace). These displacements were almost without exception in one direction, presumably corresponding to the polarity of the actin filament. The leading and falling edges of the displacements were typically as fast as the response time of the measuring system (a few milliseconds).

At saturating ATP concentration (2 mM), the average size and duration of the singe displacements were 12 nm and $\leqslant 7$ ms, respectively (Table 1). The duration was close to the resolution

FIG. 1 a, Schematic diagram (not drawn to scale) illustrating the use of two optical traps that are focused on beads attached to a single actin filament, which is held near a single HMM molecule. The filament is pulled taut and lowered onto a silica bead that is firmly fixed to a microscope coverslip and sparsely coated with HMM. b, Schematic diagram of the feedback enhanced laser trap. The thick solid lines represent the optical trap paths; the dashed lines represent imaging paths. Optics include a half-wave plate ($\lambda/2$), polarizing beam splitters (PBS), mirrors (M), lenses (L), dichroic filters (D1-3), microscope objective (O), condenser (C), filter (F), and beam-splitting mirror (BS). Two orthogonal acousto–optic modulators (AOMs) driven by voltage-controlled oscillators (VCO) are used for rapidly deflecting the laser beam in two dimensions. The bead image is projected onto a quadrant photodiode detector (QD) and displacements of the bead in two dimensions, Δx_B and Δy_B, are obtained from the differential outputs of the photodiode elements. The dotted line represents a feedback loop which is closed in order to make measurements under isometric conditions, in which case the quadrant detector outputs are fed through amplifiers to the drivers for the AOMs. A substage consisting of piezo-electric-transducers (PZT) is used for calibrating force (Fig. 4). This system is described in more detail in ref. 26.

METHODS. Silica beads (1 μm diameter; Bangs Laboratories) were firmly fixed to a microscope coverslip by suspending them in 0.05% Triton X-100 and spreading them onto the coverslip, which was then air-dried. The surface was then either coated with nitrocellulose[41] or siliconized by treatment with 0.2% dichlorodimethylsilane (Dow Corning, Z1219) in chloroform. The coverslip was then used to construct a flow cell as described[27]. Rabbit skeletal muscle HMM (1–5 μg ml⁻¹)

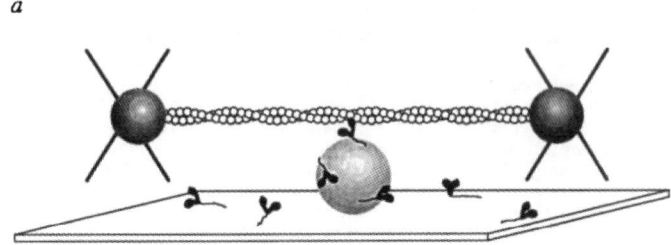

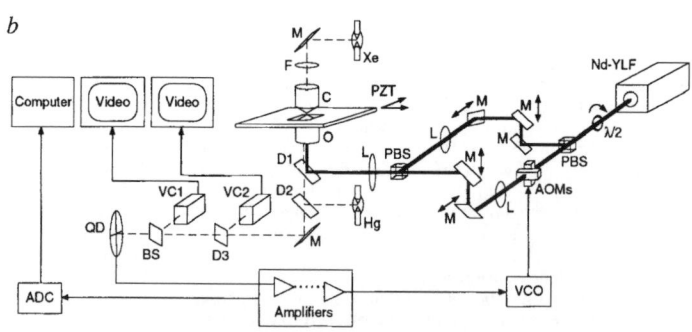

was applied to the flow cell for 2 min. Polystyrene beads (1.0 μm MX Covaspheres, Duke Scientific), labelled with NEM–HMM as described[42], were mixed with rhodamine–phalloidin-labelled actin filaments at a ratio of ~1 bead to 10 μm F-actin before being applied to the flow cell in motility buffer (25 mM KCl, 4 mM MgCl₂, 1 mM EGTA, 10 mM DTT, 25 mM imidazole, pH 7.4) at 21 °C. Two beads were placed on an actin filament near its ends (typically 5–10 μm apart), with each bead held in an optical trap. The most successful technique was to use one optical trap to hold a bead with a filament already attached. The actin filament was then straightened by a solution flow produced by the motorized stage controls, and a second bead manoeuvred by the second optical trap was attached to the free end of the filament. The filament was then pulled taut by moving the second trap with a motorized mirror under computer control, and the filament was lowered onto the silica bead using the fine focus of the microscope objective. The optical trap design used a Nd:YLF diode-pumped laser (Spectra-Physics, TFR; 1.047 μm) and a custom-built inverted microscope with a high numerical aperture objective (Zeiss, 63× Planapochromat, 1.4 NA, DIC grade).

Two traps were produced by splitting the laser beam before the AOMs using a half-wave plate ($\lambda/2$) followed by a polarizing beam splitter (PBS). The traps were the same strength (12 mW, measured before the objective). One trap position was controlled for small, fast movements (resolution < 1 nm; response time about 10 μs) using the AOMs (Isomet, 1206C), and both trap positions were controlled for larger, slower movements using d.c. motors (Newport, 860A-1-HS) to move mirrors. Specimens were observed simultaneously by brightfield and epifluorescence using video cameras (VC1-2) using a 75 W xenon arc lamp (Xe) and a 100 W mercury arc lamp (Hg), respectively. Dichroic filters (D1-3) were used to separate the laser light, the brightfield illumination, and the epifluorescence. (Transmission: D1, 450–1,000 nm; D2, >565 nm; D3, 700–1,000 nm; F, >700 nm; BS, 95% 700–1,000 nm.) The position of the bead in the trap equipped with AOMs was monitored by the quadrant photodiode detector (QD) (Hamamatsu, S1557). Δx_B and Δy_B were obtained by appropriate additions and subtractions of the four quadrant outputs. The bandwidth was 100 Hz, limited by the 200 MΩ resistors used in the current-to-voltage converters and stray capacitance. Data was sent through an analog-digital converter (ADC) and stored on a PC computer (see Fig. 2). The quadrant detector was mounted on micrometers, and these were used to determine the sensitivity of the detector and the AOMs.

TABLE 1 Average single myosin molecule measurements

ATP concentrations (mM)	Single displacements (nm)	Displacement durations (ms)	Single forces (pN)	Force durations (ms)
2	12 ± 2.0 (22)	≤ 7 (22)	3.4 ± 1.2 (85)	18 ± 6 (85)
0.01	11 ± 2.6 (36)	72 ± 29 (36)	3.5 ± 1.3 (43)	25 ± 9 (43)
0.001	11 ± 2.5 (21)	260 ± 140 (21)	3.4 ± 1.4 (50)	190 ± 150 (50)

Values are mean ± standard deviation (n). The differences between force and displacement durations were examined for statistical significance, using error estimates obtained from the distribution of the means of experimental runs. At 2 mM and 10 μM ATP these differences were highly significant ($P < 0.01$), but at 1 μM ATP the difference was not significant ($P > 0.05$).

of measurement, and we therefore lowered the ATP concentration to 1 μM or 10 μM to delay the dissociation of myosin from actin. These concentrations are well below the apparent K_m (~50 μM) for the sliding velocity[27]. The HMM density was also decreased, to the point where many of the actin filaments tested showed no transient displacements, so that when interactions were detected they most likely involved only one or a very few HMM molecules. Single displacements were detected above the noise at 10 μM ATP and 1 μM ATP (Fig. 2b, c). The size of the steps, 11 ± 2.4 nm (mean ± s.d.), was the same at both high and

low ATP concentrations, but the average duration of the displacements increased as the ATP concentration decreased (Table 1). The peak amplitude distribution was independent of total trap stiffness over the range 0.014–0.08 pN per nm, implying that the step size is nearly independent of load in this range. This experiment was done using siliconized surfaces and nitro-cellulose-coated surfaces, with similar results (Fig. 2d). It was noticeable in many records that the brownian noise was markedly reduced during the steps, presumably because of the increased stiffness associated with the HMM–actin link[25]. In some records it appeared that the rise time of the displacements (Fig. 2b, c) was relatively slow, but examination of all of the records failed to show a consistency in this pattern. This point requires further investigation.

When the HMM density on the surface was increased to levels just below that which supported continuous movement of actin over larger distances, the trapped bead no longer returned to its baseline position after each displacement. Instead the bead moved for distances larger than a single step before abruptly returning to the baseline (Fig. 3a). At the trap stiffness used in these measurements, a free bead returns to its equilibrium position with a time constant of <1 ms[26], so that each rapid return to the baseline in these experiments probably corresponded to

FIG. 2 Single displacements at low load with a, 2 mM; b, 10 μM; and c, 1 μM ATP. The upper trace in each record shows the movement of the trapped bead in the direction along the actin filament. The lower trace in each shows the movement of the bead in the perpendicular direction. The HMM density was highest in a and lowest in c. d, Distribution of single displacement amplitudes obtained from data at 2 mM, 10 μM and 1 μM ATP. The distribution shows data obtained from both siliconized surfaces (white) and nitrocellulose surfaces (shaded). Each trap: laser power 12 mW, stiffness 0.02 pN/nm.
METHODS. The HMM concentrations applied to the flow cell were: 5 μg ml^{-1} (a), 2 μg ml^{-1} (b) and 1 μg ml^{-1} (c). ATP concentrations of 1 μM or 10 μM were maintained by the presence of 1 mM phosphocreatine and 0.1 mg ml^{-1} creatine phosphokinase. The laser trap stiffness was calibrated by applying viscous forces to a trapped bead as described in Fig. 4. Signals were sampled at 4 kHz (R. C. Electronics, ISC-16), recorded on a computer, and subsequently filtered during analysis using up to 8-point averaging, and a 2–8-point Hanning filter. Essentially the same results were achieved using steep-cut Bessel filters. Single displacement amplitudes were measured, using the minimum filtering possible, by fitting lines through the baseline noise on either side of an event, and (where possible) through the noise at the plateau of the event. Not all of the displacements shown here were scored for inclusion in the distribution shown in d because many did not meet the following criteria: the displacements had to be isolated events; the baseline brownian motion had to be at the same level on either side of the peak; and the displacements could not appear similar in character to the multiple displacements shown in Fig. 3. An example of a displacement that was not scored for the latter reason is shown in b, right panel. Displacements smaller than or equal to the level of the brownian noise (equivalent to the perpendicular motion shown in the lower trace) were not scored, and as a result the distribution shown in d may be truncated below ~5 nm. Compliance of the filament was ignored, as it was estimated to be ≪ 0.3 nm per pN, assuming a Young's modulus of ≫ 6.8 × 10^8 N m^{-2}, a filament length of ≤ 5 μm, and a cross-sectional area of 24 nm^2 (refs 43, 44).

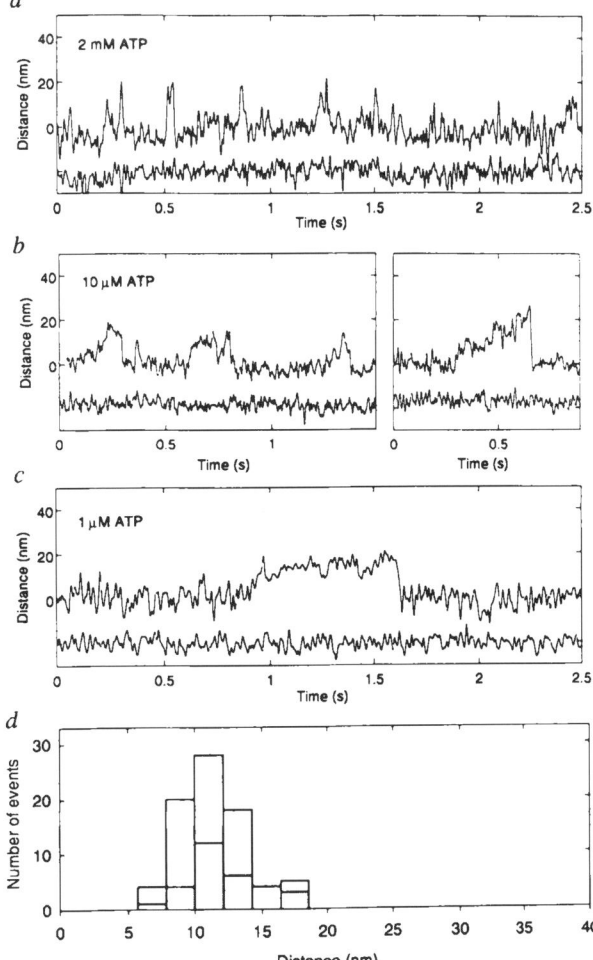

ARTICLES

a period when there were no HMM molecules bound to the actin filament. This would be expected to occur frequently at the subsaturating HMM densities used in these measurements. In many instances the bead clearly moved in a few discrete steps (Fig. 3b, left panel) before returning to the baseline. This feature is quantified in Fig. 3b (right panel), which is a histogram that shows the distribution of distances away from the baseline as a function of time spent at each distance. The bead spent long periods of time at discrete levels and moved quickly between levels. The distance between levels averaged 11 ± 3.0 nm (mean $\pm$ s.d.) (Fig. 3c), which is consistent with the size of the single movements shown in Fig. 2 and Table 1. As such multiple steps were observed infrequently at very low surface HMM densities, it is likely that they correspond to a small number of HMM molecules attaching to a filament and moving it sequentially.

It would be expected that at high, saturating surface density of HMM molecules, the velocity of filament sliding would be at least as great as the average unitary step divided by its duration. We measured these velocities in a motility assay at the three ATP concentrations, and found them to be 3.0 μm s^{-1} (2 mM ATP), 0.26 μm s^{-1} (10 μM) and 0.08 μm s^{-1} (1 μM). These compare favourably with the calculated velocities from the single displacements of ≥ 2, 0.15 and 0.04 μm s^{-1}, respectively. The approximate agreement between the values of velocity observed at high

HMM density and the values calculated from the size and duration of single displacements, strongly suggests that these single displacements are the underlying basis of filament sliding, although precisely what happens during overlapping interactions remains to be investigated.

Single forces

To measure the force produced by HMM molecules, a much stiffer trap was needed than was used to measure displacements. The trap stiffness was increased to 6 pN per nm by means of a feedback system in which the quadrant detector output signals were fed into the driver circuits for acousto–optic modulators which were used for making rapid trap displacements[22] (Fig. 1b). When an external force in the form of a viscous drag was applied to a trapped bead without feedback, the bead was displaced from the centre of the trap (Fig. 4a); this displacement was directly proportional to the force applied (Fig. 4c). In contrast, when the feedback loop was closed, the centre of the trap moved by an amount proportional to the applied force and the bead was essentially prevented from moving (Fig. 4b, c). Thus, the displacement of the trap position could be used as a measure of force on the trapped bead under approximately isometric conditions.

The procedure for measuring single forces was as follows. When an actin filament was brought in contact with HMM molecules and single displacements were observed, the feedback loop was closed and force fluctuations were then measured. At sufficiently low HMM densities, single force transients were observed (Fig. 5a–c). Just as for displacements at low load, forces were nearly always found to be in one direction along a particular actin filament. The magnitude of the forces covered a broad distribution, ranging from 1 to 7 pN, and averaged 3.4 ± 1.2 pN (mean $\pm$ s.d.) (Fig. 5d). The distribution of forces was the same at the different ATP concentrations (Table 1). As observed for the single displacements at low load, the duration of the single force transients increased at low ATP concentrations (Table 1). At these low HMM densities, the force transients usually appeared as isolated events, but at slightly higher HMM densities the events occurred very close together and often were additive (data not shown). This is the corresponding process to multiple steps at low load, and presumably is due to the overlapping action of several HMM molecules.

Discussion

The ability to measure the mechanical properties of single myosin molecules provides us with direct evidence that under conditions of low load, myosin undergoes stepwise displacements with an average step of about 11 nm, and under isometric conditions, myosin produces an average force of 3–4 pN. In discussing our results we assume that the single forces and displacements were derived from interactions of actin with single myosin heads, as there is evidence that the two heads on a given myosin molecule act independently[28,29]. However, we cannot at present discount the possibility that some of the interactions we have observed involve both heads of a myosin molecule. One complication that probably affects the values of the force and displacement in our experiments is that the HMM molecules are randomly oriented on the surface, which is likely to reduce the efficacy of force production or displacement. However, even myosin heads that are in the completely wrong orientation still move actin filaments in vitro at one-tenth the normal velocity[30,31]. It seems likely, therefore, that the true average force and average displacements produced by correctly oriented myosin heads are not higher than the upper end of our range of values. These values are well within the range of those allowed by the swinging-crossbridge model[2], and the average value of 11 nm agrees with the value of 14 nm found by Ford et al.[5] for the maximum range of rapid recovery after a shortening step. However, the largest displacements that we observed may not fit easily into the specific model suggested by Rayment et al.[3],

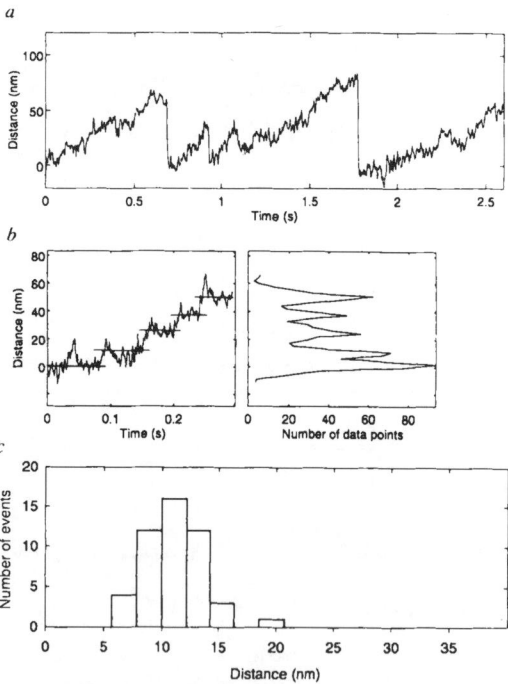

FIG. 3 a and b (left panel), Multiple displacement records at higher HMM density than for Fig. 2, 10 μm ATP. b (right panel), Distribution of distances away from the baseline for the records shown in the left panel. The abscissa is proportional to the time the bead spent at each position along the ordinate. c, Distribution of distances between levels for multiple displacement records at 10 μM ATP and 1 μM ATP. Each trap: laser power 12 mW, stiffness 0.02 pN per nm.

METHODS. As described in Fig. 2 legend, but 5 μg ml^{-1} HMM was applied to the flow cell. The distribution in b (right panel) had a bin size of 2.2 nm. The distribution of steps in c was formed by fitting lines through adjacent plateaus and measuring the distance between them.

FIG. 4 Example of optical trap stiffness calibration with and without feedback control. *a*, Without feedback: a viscous force that alternated in direction was applied to a trapped bead by applying a triangular wave (lower trace) to the microscope stage position; the bead position showed an approximately square wave response. *b*, With feedback: as in *a*, but in this case the trap position showed an approximately square wave response, and the bead remained essentially stationary. *c*, Displacement versus force curve. ●, Bead movement without feedback; ○, trap movement with feedback. The trap stiffness under feedback control increased by ~300-fold from 0.05 pN per nm without feedback to 15 pN per nm with feedback.

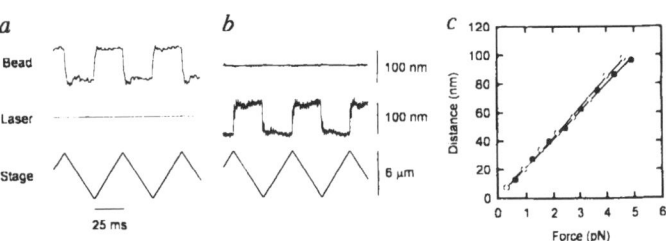

METHODS. A viscous force was applied to a trapped bead 5 μm from the coverslip surface by moving a substage with piezo-electric transducers (Physik Instrumente, P771) to produce a solution flow. Stokes' law was used to calculate the force, which is directly proportional to the velocity of the solution flow[26]. The small amount of creep in the response was due to hysteresis in the PZT response (not shown). The average value of displacement was used and shown to be directly proportional to the measured average stage velocity. Under feedback control the bead movement, including brownian motion, was less than 2 nm peak-to-peak. As the displacement was small, the trap stiffness was obtained by averaging 50 traces of the bead response (*b*, top trace). The bandwidth of the feedback system, measured from the response to a small input perturbation, was 830 Hz. The increase in bandwidth compared to that of the detector (100 Hz, single pole) resulted from the high feedback gain.

FIG. 5 Single force transients near isometric conditions with *a*, 2 mM; *b*, 10 μM; and *c*, 1 μM ATP concentrations. The upper trace in each record shows the forces in the direction along the actin filament. The lower trace in each shows forces in the perpendicular direction. *d*, Distribution of single force amplitudes obtained from data using 2 mM, 10 μM and 1 μM ATP concentrations. The distribution shows data obtained from both siliconized surfaces (white) and nitrocellulose surfaces (shaded). Each trap: laser power 12 mW. Stiffness of trap with feedback, 6 pN per nm; stiffness of trap without feedback, 0.02 pN per nm.

METHODS. Single actin filaments were pulled taut as described in Fig. 1, and feedback control was applied to the bead imaged on the quadrant detector. The HMM concentrations applied to the flow were: 2 μg ml⁻¹ (*a*) and 1 μg ml⁻¹ (*b*, *c*). The criteria used to select single force transients were similar to those used for single displacements (Fig. 2 legend). Force transients were not scored if they were in the direction that caused the filament to slacken. Calibration for force was made as for Fig. 4, but with the bead centred 1 μm from the surface (equal to the distance of the actin-bound beads from the surface in this experiment). It was found by two methods that the trap stiffness was the same at 5 μm and 1 μm from the surface: First, the same calibration procedure gave 1.4 times greater bead displacement at 1 μm, corresponding well to the raised viscous force near a surface[45]; second, we compared the amplitude of brownian noise at the two depths (as r.m.s. brownian motion is inversely proportional to trap stiffness) and found that the two were identical. Errors in the force calibration were estimated to be less than 10%.

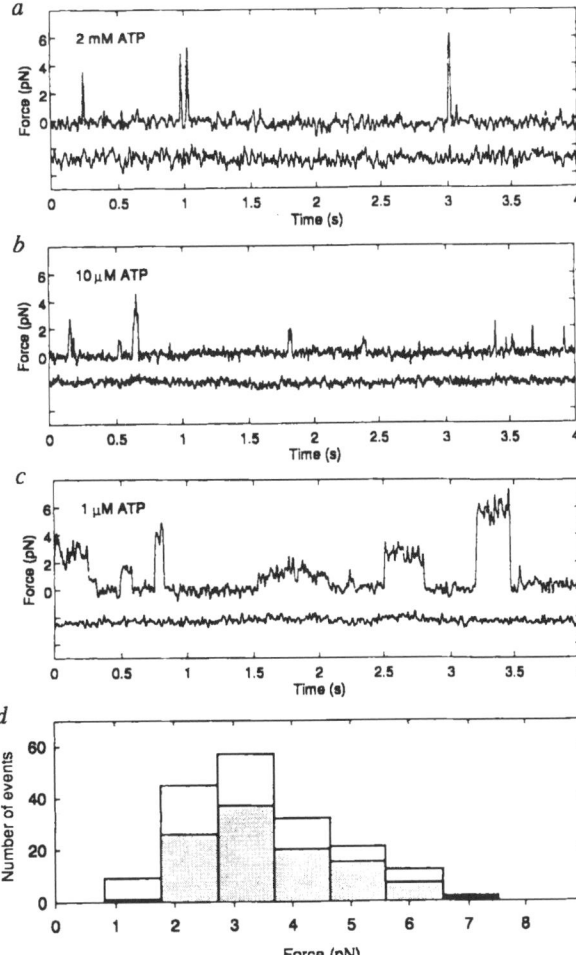

ARTICLES

which is based on the S1 crystal structure. The values of the displacements reported here do not support models in which myosin heads are able to generate movement over very large distances[9-11]. Models in which a myosin molecule makes multiple interactions with an actin filament for one ATP hydrolysis[12-15] would also be unlikely, because the single displacements and single forces that we observed at very low HMM density did not appear as a rule to be clustered[25]. Furthermore, our data show that the durations of the transient displacements and forces depend on ATP concentration so that, at least at low ATP, each interaction requires the completion of the ATP hydrolysis cycle. In addition, as discussed in more detail below, the durations are very close to those that would be expected for single ATP hydrolysis cycles. Models proposing that a myosin molecule makes multiple interactions with an actin filament for one ATP hydrolysis would predict that, at low ATP concentrations, multiple interactions would occur before the final step of product release, and they would presumably be similar in duration to the interactions seen at high ATP. Such models would therefore require a marked preponderance of events of very short duration at low ATP, and this was not observed in our preliminary analysis of the data.

Our results for the ATP dependence of single forces and displacements give some preliminary information about the biochemical steps involved and the effects of mechanical strain. The reciprocal of the mean value of duration of single interactions gives the rate constant for the process which terminates the event. For both single forces and displacements, the variation of the rate constant with ATP concentration is consistent with Michaelis–Menten kinetics, with an apparent K_m for ATP of the same order of magnitude as for sliding velocity in a motility assay, $50 \mu M$[27]. However, we found significant differences between the durations for single forces and those for single displacements (that is, high and low loads) in the case of $10 \mu M$ and 2 mM ATP, which would be consistent with a lower saturating rate and apparent K_m at high loads, In turn, such a difference would imply that the second-order rate constant for ATP binding is larger at high loads. The values for the second-order rate constant have to be calculated from concentrations of substrate well below the K_m, and at $1 \mu M$ ATP the values calculated from $1/(Ct)$, where C is the ATP concentration and t is the average duration, are $3.8 \pm 0.4 \times 10^6 M^{-1} s^{-1}$ (mean $\pm$ s.e.m.) for displacement and $5.3 \pm 1.6 \times 10^6 M^{-1} s^{-1}$ for force. Although the difference between these values is not statistically significant, it is consistent with our argument. These values are within range of the value of $4.0 \times 10^6 M^{-1} s^{-1}$ for the acto-S1 ATPase in solution[32]. A detailed study will be required to confirm these results, but the difference in the values for the rate of ATP binding at low and high load is probably due to a strain-dependent increase in this rate, which is to be expected for a strong-to-weak transition[33]. At saturating ATP the situation is reversed, and single forces lasted longer than single displacements, implying that the rate of detachment rises during filament sliding[34]. The duration of $\leqslant 7$ ms for single displacements is equivalent to a rate constant of $\geqslant 140 s^{-1}$. This value can be compared directly with the rates that affect detachment in the acto-S1 ATPase, and it probably corresponds to an isomerization between two AM.ADP states preceding ADP release[35]. The difference between the durations for force and displacement at high ATP can be explained by a strain-dependence of this transition.

If the force exerted by a correctly oriented myosin head is 3–7 pN, how does this compare with the average force per myosin head produced by a contracting muscle fibre under isometric conditions? It is difficult to make exact comparisons because of differences in the conditions used, but rabbit skinned muscle fibres produce an isometric force of $0.22 N mm^{-2}$ (ionic strength, 200 mM, pH 7.1, 20 °C; M. G. Bell, J. A. Dantzig and Y. E. Goldman, personal communication). Assuming a filament lattice spacing $(d_{1,0})$ during activation of 38 nm (ionic strength 170 mM, pH 7.0, 5 °C)[36], the average force is 1.3 pN per myosin

head[37]. A similar value was found for frog intact muscle near 0 °C (force, $0.25 N mm^{-2}$ (ref. 38); $d_{1,0}$, 35 nm (ref. 39)), and our results would then suggest that only 20–40% of the heads develop force at any one time, that is, the duty ratio is 0.2–0.4. Earlier estimates of the fraction of myosin crossbridges attached at any moment in an isometric contraction were obtained using several techniques by comparison with the rigor state, in which it is thought that all the crossbridges are bound to actin, and with the relaxed state, in which the crossbridges are either not bound or only weakly bound. Measurements on frog intact muscle of fibre stiffness[16] and X-ray diffraction equatorial intensities[17] suggest values for the fraction of myosin heads attached at any moment of 0.75 and 0.9, respectively, though these estimates may include bound crossbridges not producing force. On the other hand, electron spin resonance measurements on activated, crosslinked myofibrils show that at low ionic strength (45 mM, pH 7.0, 25 °C) a fraction of the heads (0.37) are bound to actin in a rotationally disordered state[18], with 0.12 bound in a rigor-like state[19]. Measurements of the intensity of the X-ray layer lines also suggest a low fraction of ordered heads, 0.2–0.3 (ref. 17). Our results support these estimates.

The duty ratio under low load can be calculated from the durations of the single displacements measured here and the k_{cat} value for the actin-activated HMM ATPase under similar conditions. At low load and 2 mM ATP, the average duration of single displacements was $\leqslant 7$ ms. There is no direct measurement of the ATPase rate under the single molecule conditions of our experiments (this presents a challenge for the future), but under comparable solution and temperature conditions, the k_{cat} for the ATPase would be about 10 per second per head[40]. Thus under these conditions, the upper limit of the duty ratio under low load is 0.07. This value is consistent with that obtained by Uyeda et al.[7]. Therefore under both low- and high-load conditions, the results reported here are consistent with small duty ratios, and are compatible with the predicted large duty ratios[10,13] only if the ATPase rate in the motility experiments is 100 per second per head, which seems unlikely.

The picture that emerges from this study is of a simple cycle of force production followed by detachment, in which the isometric force per interaction of a correctly oriented myosin head is 3–7 pN and the displacement at low load is 8–17 nm. *Note added in proof*: The number of interactions per ATP hydrolysed can be obtained by dividing the useful energy per ATP $(0.5 \times 10^{-19} J)$[37] by the energy per interaction (estimated as $0.5 \times 3.4 \times 11 \times 10^{-21} J$), giving 2.6. As our values of force and displacement are underestimates, the correct value may be close to unity. □

Received 27 October 1993; accepted 2 February 1994.

1. Reedy, M. K., Holmes, K. C. & Tregear, R. T. Nature 207, 1276–1280 (1965).
2. Huxley, H. E. Science 164, 1356–1366 (1969).
3. Rayment, I. et al. Science 261, 50–58 (1993).
4. Huxley, A. F. & Simmons, R. M. Nature 233, 533–538 (1971).
5. Ford, L. E., Huxley, A. F. & Simmons, R. M. J. Physiol. 269, 441–515 (1977).
6. Toyoshima, Y. Y., Kron, S. J. & Spudich, J. A. Proc. natn. Acad. Sci. U.S.A. 87, 7130–7134 (1990).
7. Uyeda, T. Q. P., Kron, S. J. & Spudich, J. A. J. molec. Biol. 214, 699–710 (1990).
8. Uyeda, T. Q. P., Warrick, H. M., Kron, S. J. & Spudich, J. A. Nature 352, 307–311 (1991).
9. Harada, Y. & Yanagida, T. Cell Motil. Cytoskel. 10, 71–76 (1988).
10. Harada, Y., Sakurada, K., Aoki, T., Thomas, D. D. & Yanagida, T. J. molec. Biol. 216, 49–68 (1990).
11. Yanagida, T., Arata, T. & Oosawa, F. Nature 316, 366–369 (1985).
12. Higuchi, H. & Goldman, Y. E. Nature 352, 352–354 (1991).
13. Ishijima, A., Doi, T., Sakurada, K. & Yanagida, T. Nature 352, 301–306 (1991).
14. Brenner, B. Proc. natn. Acad. Sci. U.S.A. 88, 10490–10494 (1991).
15. Lombardi, V., Piazzesi, G. & Linari, M. Nature 355, 638–641 (1992).
16. Goldman, Y. E. & Simmons, R. M. J. Physiol. 269, 55p–57p (1977).
17. Huxley, H. E. & Kress, M. J. Muscle Res. Cell Motil. 6, 153–161 (1985).
18. Berger, C. L. & Thomas, D. D. Biochemistry 32, 3812–3821 (1993).
19. Fajer, P. G., Fajer, E. A. & Thomas, D. D. Proc. natn. Acad. Sci. U.S.A. 87, 5538–5542 (1990).
20. Howard, J., Hudspeth, A. J. & Vale, R. D. Nature 342, 154–158 (1989).
21. Kishino, A. & Yanagida, T. Nature 334, 74–76 (1988).
22. Simmons, R. M. et al. in Mechanism of Myofilament Sliding in Muscle (eds Sugi, H. & Pollack, G. H. J.) (Plenum, New York and London, 1993).
23. Block, S. M., Goldstein, L. S. B. & Schnapp, B. J. Nature 348, 348–352 (1990).
24. Kuo, S. C. & Sheetz, M. P. Science 260, 232–234 (1993).
25. Svoboda, K., Schmidt, C. F., Schnapp, B. J. & Block, S. M. Nature 365, 721–727 (1993).

ARTICLES

26. Simmons, R. M., Finer, J. T., Chu, S. & Spudich, J. A. *Biophys. J.* (submitted).
27. Kron, S. J. & Spudich, J. A. *Proc. natn. Acad. Sci. U.S.A.* **83,** 6272–6276 (1986).
28. Taylor, E. W. *Crit. Rev. Biochem.* **6,** 103–164 (1979).
29. Cooke, R. *Crit. Rev. Biochem.* **21,** 53–118 (1986).
30. Sellers, J. R. & Kachar, B. *Science* **249,** 406–408 (1990).
31. Yamada, A., Ishii, N. & Takahashi, K. *J. Biochem.* **108,** 341–343 (1990).
32. White, H. D. & Taylor, E. W. *Biochemistry* **15,** 5818–5826 (1976).
33. Goldman, Y. E., Hibberd, M. G. & Trentham, D. R. *J. Physiol.* **354,** 577–604 (1984).
34. Huxley, A. F. *Prog. Biophys.* **7,** 225–318 (1957).
35. Taylor, E. W. *J. biol. Chem.* **266,** 294–302 (1991).
36. Brenner, B. & Yu, L. C. *J. Physiol.* **441,** 703–718 (1991).
37. Bagshaw, C. R. *Muscle Contraction*, 98 (Chapman & Hall, London, 1993).
38. Woledge, R. C., Curtin, N. A. & Homsher, E. *Energetic Aspects of Muscle Contraction*, 103 (Academic, London, 1985).
39. Haselgrove, J. C. & Huxley, H. E. *J. molec. Biol.* **77,** 549–568 (1973).
40. Margossian, S. S. & Lowey, S. *Meth. Enzym.* **85,** 55–71 (1982).
41. Toyoshima, Y. Y. *et al. Nature* **328,** 536–539 (1987).
42. Warrick, H. M. *et al.* in *Methods in Cell Biology* (eds Scholey, J. M.) 1–21 (Academic, San Diego, 1993).
43. Yoshino, S., Umazume, Y., Natori, R., Fujime, S. & Chiba, S. *Biophys. Chem.* **8,** 317–326 (1978).
44. Ford, L. E., Huxley, A. F. & Simmons, R. M. *J. Physiol.* **311,** 219–249 (1981).
45. Svoboda, K. & Block, S. M. *A. Rev. Biophys. biomol. Str.* (in the press).

ACKNOWLEDGEMENTS. We thank S. Chu for help with the design of the prototype instrument that led to this work, and J. Sleep, T. Uyeda, C. Coppin, A. Friedman and Y. Goldman for comments on the manuscript. This work was supported in part by the NIH Medical Scientist Training Program (J.T.F.), the MRC, the Wellcome Trust, the Fulbright Commission, and NATO (R.M.S.), and a grant from the NIH (J.A.S.).

RESEARCH ARTICLES

The Atom-Cavity Microscope: Single Atoms Bound in Orbit by Single Photons

C. J. Hood,[1] T. W. Lynn,[1] A. C. Doherty,[2] A. S. Parkins,[2]
H. J. Kimble[1*]

The motion of individual cesium atoms trapped inside an optical resonator is revealed with the atom-cavity microscope (ACM). A single atom moving within the resonator generates large variations in the transmission of a weak probe laser, which are recorded in real time. An inversion algorithm then allows individual atom trajectories to be reconstructed from the record of cavity transmission and reveals single atoms bound in orbit by the mechanical forces associated with single photons. In these initial experiments, the ACM yields 2-micrometer spatial resolution in a 10-microsecond time interval. Over the duration of the observation, the sensitivity is near the standard quantum limit for sensing the motion of a cesium atom.

We report a type of measurement capability that achieves continuous position measurement by using an optical cavity to enhance the sensitivity for atomic detection while achieving high spatial resolution. In this case, the signal-to-noise (S/N) ratio R_c for atomic detection within the cavity becomes $R_c \sim R_0 \sqrt{F}$, where R_0 is the S/N ratio for sensing the presence of the atom with absorption cross section σ within the resolution area A, $R_0 \sim \sqrt{\sigma \Delta t / A\tau}$ (Δt is the measurement time and τ is the minimum allowed interval between successive absorption events) (1, 2), and F is the cavity finesse (roughly the number of intracavity photon round trips during the cavity decay time) (3, 4). With low-loss dielectric coatings deposited on superpolished substrates, the cavity finesse F can be quite large, with the record value for a Fabry-Perot cavity being $F = 1.9 \times 10^6$ (5), thereby suggesting potentially large gains in sensitivity for sensing motion within the cavity (6).

Improving sensitivity by placing a sample inside a high-quality optical cavity is in and of itself a well-known technique, with implementations ranging from multipass absorption cells to high-finesse optical cavities (7–9). However, these experiments most often involve a concomitant loss in spatial resolution [with (resolution $\delta r = \sqrt{A}$) ~ (cavity waist w_0) $\gg$ (wavelength λ)] and have usually detected changes of cavity transmission caused by ensembles of atoms or molecules.

In contrast, real-time modifications in cavity transmission wrought by single atoms within an optical cavity have been observed within the

setting of cavity quantum electrodynamics (cavity QED), beginning in 1996 (10) and in several subsequent experiments (11–15). In fact, cavity transmission modified by a factor of 10^2 associated with the 80-μs passage of a single atom through a Fabry-Perot cavity was reported in (11). To translate this high sensitivity for atomic detection into a capability for atomic microscopy requires achieving $\delta r < w_0$ for motion within the cavity, attainable by trading back a fraction of the gain associated with large F for increased spatial resolution.

Toward this end, consider a cavity of length l driven by an input probe laser, with the transmitted light detected to generate a photocurrent, as shown in Fig. 1. The intracavity field $\vec{E}(\vec{r}')$ is $E_0\psi(\vec{r}')\hat{\epsilon}$, with cavity polarization vector $\hat{\epsilon}$ and spatially varying mode function $\psi(\vec{r}') = \cos(2\pi x/\lambda)\exp[-(y^2 + z^2)/w_0^2]$ for $-(l/2) < x < l/2$. An atom falling into the cavity modifies E_0; examples of the resulting variation are displayed in Fig. 2 for $\bar{m} \equiv |E_0|^2 \leq 1$ photon mean field strength. Because the large changes in $\bar{m}(t)$ evident in Fig. 2 are caused by the motion $\vec{r}(t)$ of a single atom within the cavity

mode, spatial resolution $\delta r < w_0$ can be achieved if the association between $\bar{m}(t)$ and $\vec{r}(t)$ can be quantified.

In fact, the quantum master equation (16) describing the radiative interaction between atom and cavity field allows E_0 to be calculated for any atomic position. Knowledge of E_0 in turn enables deduction of the total field transmitted by the cavity and thence of the photocurrent generated by measuring this transmitted field. We developed an algorithm that inverts this chain of deduction—namely, we infer the position $\vec{r}$ of a single atom within the cavity mode from the recorded photocurrent, albeit with some caveats. We can then use the cavity field as a microscope to track atomic motion in real time, with spatial resolution $\delta r \simeq 2$ μm attained in time $\delta t \simeq 10$ μs. These capabilities realize a form of time-resolved microscopy—the atom-cavity microscope (ACM).

Cavity quantum electrodynamics. Our work was carried out within the setting of cavity QED for which a single atom is strongly coupled to the electromagnetic field of a high-finesse (optical or microwave) cavity (16, 17). Here, the interaction energy between atom and cavity field is given by $\hbar g(\vec{r})$, where $g(\vec{r}) = g_0\psi(\vec{r})$ and $\hbar$ is Planck's constant divided by 2π. In a regime of strong coupling, the rate g_0 that characterizes the interaction of an atom with the cavity field for a single photon can dominate the dissipative rates for atomic spontaneous emission γ and cavity decay κ. Explicitly, $2g_0$ is the Rabi frequency for the oscillatory exchange of a single quantum between atom and cavity field, with $g_0\tau_0 = (V_0/V_C)^{1/2}$, where $\tau_0 = 1/2\gamma$ is the atomic lifetime, V_C is the cavity mode volume $V_C = (\pi/4)w_0^2 l$, and V_0 is the "radiative" volume $V_0 = \sigma c\tau_0$ (where c is the speed of light). For strong coupling ($g_0 \gg \kappa, \gamma$), the number of photons required to saturate an intracavity atom $n_0 \sim \gamma^2/g_0^2 \ll 1$ and the number of atoms required to have an appreciable effect on the intracavity field $N_0 \sim \kappa\gamma/g_0^2 \ll 1$, thereby enabling the observations of Fig. 2.

These observations should be viewed within the context of important laboratory advances

[1]Norman Bridge Laboratory of Physics 12-33, California Institute of Technology, Pasadena, CA 91125, USA. [2]Physics Department, The University of Auckland, Private Bag 92019, Auckland, New Zealand.

*To whom correspondence should be addressed. E-mail: hjkimble@cco.caltech.edu

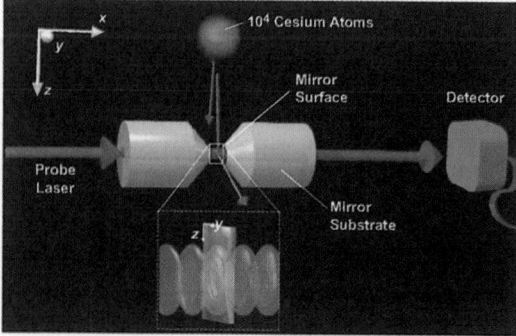

Fig. 1. Experimental schematic. Cesium atoms are captured in a magneto-optical trap (MOT) and dropped through a high-finesse optical cavity. A single atom (green arrow) transiting the cavity mode substantially alters the measured transmission of a probe beam through the cavity.

RESEARCH ARTICLES

that demonstrate the effect of strong coupling on the internal degrees of freedom of an atomic dipole coupled to the quantized cavity field, including the realization of diverse new phenomena such as the creation of nonclassical states of the radiation field (18–20). However, until now, the consequences of strong coupling for the external, atomic center-of-mass (CM) motion with kinetic energy E_k have remained largely unexplored experimentally (11, 12, 14). The seminal work of (21, 22) and numerous analyses since then (23–27) have made it clear that a rich set of phenomena should arise from the interaction of atoms with a quantized light field. In the regime of strong coupling for both the internal and external degrees of freedom, $g_0 \gg (E_k/\hbar, \gamma, \kappa)$, a single quantum is sufficient to profoundly alter the atomic CM motion, as shown, for example, in (27).

Following this theme, our experiment represents the observation of a single atom trapped by an intracavity field with $\bar{m} \simeq 1$ photon mean field strength. Such trapping is possible because the coherent coupling energy $\hbar g_0 \simeq 5.3$ mK is larger than the atomic kinetic energy $E_k \simeq 0.46$ mK for the cold atoms that fall into the cavity (Fig. 1).

Moreover, beyond providing single-quantum forces sufficient to trap atoms, strong coupling also enables real-time detection by way of the light emerging from the cavity (10–15), although actual atomic trajectories have not

been previously extracted. Stated more quantitatively, the ability to sense atomic motion within an optical cavity by way of the transmitted field can be characterized by the optical information $I = \alpha(g_0^2 \Delta t/\kappa) \sim \alpha R_C^2$, which roughly speaking is the number of photons collected as signal in time Δt with efficiency α as an atom transits between a region of optimal coupling g_0 and one with $g(\vec{r}) \ll g_0$. When $I \simeq 3 \times 10^4$ for $\Delta t = 30$ μs as in Fig. 2, atomic motion through the spatially varying cavity mode leads to variations in the transmitted field that can be recorded with high S/N ratio.

Atom trapping at the single-photon level. Relative to earlier work in cavity QED with cold atoms, we demonstrate a mechanism for trapping an atom within the cavity (12), rather than settling for a single transit through the cavity mode (10, 11, 13–15). However, we emphasize at the outset that the operation of the ACM is not restricted to this particular trapping mechanism. The functions of trapping and sensing within the cavity mode can be separated, both in theory and in practice, as, for example, by way of the dipole-force trap of (15).

The conceptual basis for our scheme is illustrated in Fig. 3A and involves trapping with single quanta in cavity QED. Displayed is the energy-level diagram for the eigenstates of the coupled atom-cavity system (that is, the Jaynes-Cummings ladder of dressed states). We focus first on the spatial dependence of the energies

$\hbar\beta_{\pm}(\rho)$ for the first excited states $|\pm\rangle$ of the atom-cavity system along the radial direction $\rho = \sqrt{y^2 + z^2}$, for optimal standing-wave position x_0 [such that $\cos(2\pi x_0/\lambda) = 1$] and neglecting dissipation. The ground state of the atom-cavity system is $|a, 0\rangle$; the atom is in its ground state a and there are no photons in the cavity. For weak or no coupling, the first two excited states are that of one photon in the cavity and the atom in the ground state, $|a, 1\rangle$, and of the atom in the excited state e with no photons in the cavity, $|e, 0\rangle$. These two states are separated by an energy $\hbar\Delta_{ac}$, where $\Delta_{ac} \equiv \omega_{cavity} - \omega_{atom}$ is the detuning between the "bare" (uncoupled) atom and cavity resonances. As an atom enters the cavity along ρ, it encounters the spatially varying mode of the cavity field and hence a spatially varying interaction energy $\hbar g(\vec{r})$. The bare states map via this coupling to the dressed states $|\pm\rangle$ shown in Fig. 3A, with energies $\beta_{\pm} = (\omega_{atom} + \omega_{cavity})/2 \pm [g(\vec{r})^2 + (\Delta_{ac}^2/4)]^{1/2}$. Our interest is in the state $|-\rangle$. The spatial dependence of the energy $\hbar\beta_{-}(\vec{r})$ represents a pseudopotential well that can be selectively populated by our choice of the strength and frequency $\omega_{probe} = \omega_{atom} + \Delta_{probe}$ of an external driving field, thereby enabling an atom with kinetic energy $E_k \ll \hbar g_0$ to be trapped.

As discussed in more detail in the theory section below, Fig. 3B shows one example of the effective potential $U(\vec{r})$ that results from this trapping mechanism, which involves contributions from the higher lying levels shown in Fig. 3A, as well as the state $|-\rangle$. Displayed are both the radial and axial dependencies $U(\rho, x_0)$ and $U(0, x)$ (that is, perpendicular to and along the cavity axis, where x_0 is an antinode of the standing wave). The depth of the potential $U_0 \simeq 2.3$ mK is greater than the initial kinetic energy of atoms in our experiment, $E_k \simeq 0.46$ mK, thereby enabling an atom to be trapped within the cavity mode. The perturbing effect of gravity on this potential is negligible.

Also shown in Fig. 3B are the heating rates (mean rates of energy increase) $dE(\rho, x_0)/dt$ and $dE(0, x)/dt$ along the radial and axial directions, with dE/dt related to the momentum diffusion coefficient D by $dE/dt = D/m$, where m is the atomic mass of cesium. Near a field antinode (for example, $x = 0$), the random or diffusive component of the motion arising from dE/dt on experimental time scales of ~50 μs is on the whole much smaller than that associated with conservative motion in the potential U. Thus, we expect a predominantly orbital motion within the cavity mode with a smaller (but nonnegligible) diffusive component.

For comparison with the well-established theory of laser cooling and trapping in free space (28), Fig. 3B also displays in dashed lines the corresponding potential $V(\vec{r})$ and heating rate $d\mathcal{E}(\vec{r})/dt$ derived in the absence of the cavity, but for the same beam geometry and the same peak field strength (29). Al-

Fig. 2. (A and B) Examples of atom transits, that is, cavity transmission as a function of time as an atom passes through the cavity field. Red traces show atoms trapped with the triggering method described, with $\bar{m} \simeq 1$ photon mean field strength and with the dashed line at the level $\bar{n}_t$. For comparison, an untriggered (untrapped) atom transit is shown in black. For these traces, $\Delta_{probe}/2\pi = -125$ MHz and $\Delta_{ac}/2\pi = -47$ MHz. **(C and D)** Theoretical simulation of atom transits for the same Δ_{probe} and Δ_{ac}. Shot noise and technical noise were added to the transmission signals, shown in red. Other traces show the 3D motion of the atom. Motion along x, the standing-wave direction, was multiplied by 10 to be visible on the plot. The atom is very tightly confined in x until rapid heating in this direction causes the atom to escape.

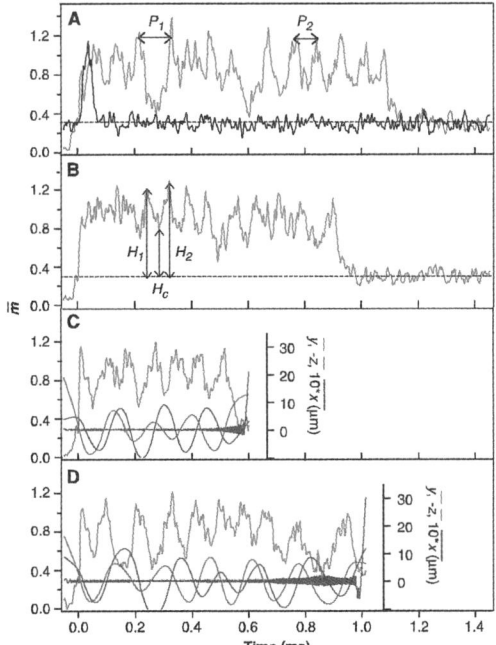

though the free-space potential $V(\bar{r})$ is similar to the cavity QED potential $U(\bar{r})$, suggesting that trapping could be achieved without the cavity [as, for example, in the pioneering experiments with optical lattices (30–32)], in fact, in the axial direction, the free-space heating rate $d\mathcal{E}/dt$ is much greater than the corresponding cavity QED quantity dE/dt. Indeed, the trapping time for an atom in the free-space setting would be more than 10-fold less than the observations of Fig. 2, so short that the atom would not even make one orbit before being heated out of the potential well.

We emphasize that the comparison in Fig. 3 is made for the same peak electric field— the cavity is not simply a convenient means for increasing the electric field for a given incident drive strength. Rather, there are profound differences between the standard theory of laser cooling and trapping and its extension into the domain of cavity QED in a regime of strong coupling. At root is the distinction between the nonlinear response of an atom in free space and one strongly coupled to an optical cavity. In the latter case, it is the composite response of the atom-cavity system illustrated in Fig. 3A that must be considered, as is described by the corresponding one-atom master equation in cavity QED. That this full quantum treatment of the atom-cavity system is required has been experimentally confirmed by way of measurements of the nonlinear susceptibility for the coupled system in a setting close to that used here (11–13).

A second and critically important point of distinction between the current work and traditional laser cooling and trapping in free space (28) relates to the ability to sense atomic motion in real time with high S/N ratios. We stress that this is not simply a matter of a practical advantage, but a fundamental improvement beyond what is possible by way of alternate detection strategies demonstrated to date [such as absorption (1, 2) or fluorescence (33–35) for single atoms and molecules]. An estimate of this enhanced capability is given by the ratio $R_c/R_0 \gg 1$, or alternatively by way of the optical information rate $I/\Delta t \sim 10^9/s$, which in the current work is the largest value yet achieved in optical physics.

Apparatus and protocol. A cloud of cesium atoms was collected in a magneto-optical trap [MOT (28)], cooled to a temperature of $\simeq 20\ \mu K$ and then released, all in a vacuum chamber at 10^{-8} torr (Fig. 1). With initial mean velocity $\bar{v} \simeq 4$ cm/s, the cold atoms then fell 3 mm toward an optical resonator (cavity) (36) and reached velocity $v \simeq 24$ cm/s. Even with 10^4 atoms initially, only one or two atoms crossed the standing-wave mode of the cavity each time the MOT was dropped (37) (green arrows, Fig. 1).

We trapped an atom by driving the cavity with a weak circularly polarized probe laser at a frequency $\omega_{probe} \simeq \beta_-(0)$ [corresponding to

$\beta(\bar{r})$ for maximum coupling, $g(\bar{r}) = g_0$] and intracavity photon number $\bar{n}_p = 0.05$ to provide small, off-resonant excitation of the empty cavity. With reasonable probability, a falling atom will be channeled by the resulting (shallow) potential $U_p(\bar{r})$ toward regions of high coupling, resulting in a corresponding increase in probe transmission as β_- comes into resonance with ω_{probe} in the fashion illustrated in Fig. 3A (11). When $g(\bar{r})$ exceeded some predetermined threshold g_t, we switched the probe power up to a level $\bar{n}_t = 0.3 \pm 0.05$ intracavity photons to create a deep confining potential $U(\bar{r})$ around the atom, thus trapping it (12). (We denote by $\bar{n}$ the photon number for the empty cavity and by $\bar{m}$ the corresponding quantity with an atom present; these quantities are directly proportional to the detected transmission signal.) The transmission was measured by heterodyne detection at 100-kHz bandwidth and digitized at 1 MHz, with an overall efficiency $\alpha = 25\%$ to detect an intracavity photon.

The probe transmission recorded by way of this protocol for two individual atom transits is displayed in Fig. 2, A and B. At time = 0, atom detection triggered the increase $\bar{n}_p \to \bar{n}_t$ to catch the atom. The cavity transmission was highest (with $\bar{m} \simeq 1$) when the atom was near the center of the cavity. The observed oscillations in $\bar{m}$ resulted from modifications in cavity transmission as the atom moved within the cavity mode. We emphasize that the corresponding quantum state is a bound state of atom and cavity. The situation is analogous to a molecule for which two atoms share an electron to form a bound state with a lower energy than two free atoms. Here a "molecule" of one atom and the cavity field is formed through the sharing of one photon excitation on average, thereby binding the

atomic CM motion. Our atom-cavity molecule only exists while an excitation is present, with decay set by κ, because $\kappa > \gamma$. To compensate for this decay-induced destruction of the atom-cavity molecule, the probe field continuously drove the cavity to repeatedly recreate the bound state before the atom had a chance to escape. When the atom eventually did leave the cavity mode, transmission returned to $\bar{n}_t$.

To demonstrate the strong effect of the triggering-trapping strategy, Fig. 2A also shows an atom transit (black trace) recorded with a drive strength of $\bar{n} = 0.3$ and no triggering. In this case, atoms fell through $U(\bar{r})$ with an average transit time of 74 μs. In contrast, the triggering protocol described above extended this average to 340 μs, in good agreement with theoretical simulations for both the mean and distribution of trap times. Indeed, many individual atom transits lasted much longer (the maximum observed time was 1.9 ms). Single atoms have recently been trapped with a lifetime of 28 ms in a regime of strong coupling by way of a classical dipole-force trap (15) but not with a quantum field at the single-photon level (38).

A striking feature of the traces in Fig. 2 are the oscillations in atom-cavity transmission. As illustrated in Fig. 2, C and D, our numerical simulations show that these oscillations arise from elliptical atomic orbits in planes perpendicular to the cavity axis. From the simulations discussed below, we find that motion along the cavity axis x is tightly confined to a region $\delta x \simeq \pm 50$ nm due to the steepness of $U(x)$ (39).

Theory and numerical simulations. Beyond the intuitive picture of trapping with the lower components of the dressed states as discussed in connection with Fig. 3, we car-

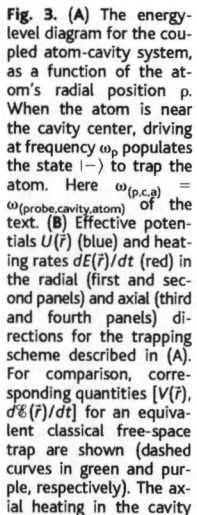
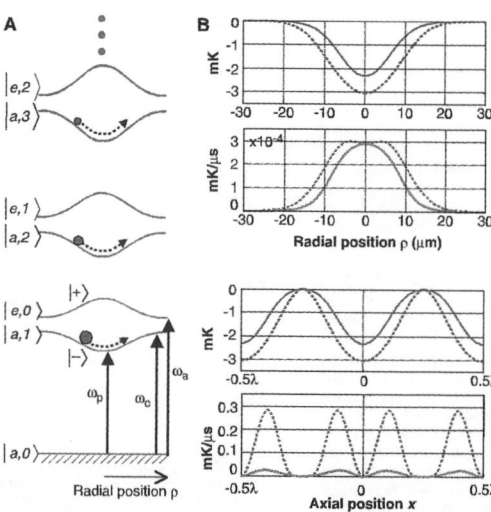

Fig. 3. (A) The energy-level diagram for the coupled atom-cavity system, as a function of the atom's radial position ρ. When the atom is near the cavity center, driving at frequency ω_p populates the state $|-\rangle$ to trap the atom. Here $\omega_{(p,c,a)} = \omega_{(probe,cavity,atom)}$ of the text. **(B)** Effective potentials $U(\bar{r})$ (blue) and heating rates $dE(\bar{r})/dt$ (red) in the radial (first and second panels) and axial (third and fourth panels) directions for the trapping scheme described in (A). For comparison, corresponding quantities [$V(\bar{r})$, $d\mathcal{E}(\bar{r})/dt$] for an equivalent classical free-space trap are shown (dashed curves in green and purple, respectively). The axial heating in the cavity trap is 10-fold smaller, greatly enhancing the trap lifetime. All quantities are calculated for $\Delta_{probe}/2\pi = -145$ MHz and $\Delta_{ac}/2\pi = -74$ MHz, with empty cavity photon number $\bar{n} = 0.3$.

ried out extensive analytical and numerical simulations of atomic motion for the parameters of our experiment, following the basic model of (23). As is the case for motion of an atom in a free-space standing wave (40, 41), there is a separation of time scales between the internal (atomic dipole + cavity field) degrees of freedom and the external atomic CM motion. The effective potential $U(\vec{r})$ presented in Fig. 3 is determined by integration of the expectation value of the force operator:

$$\hat{\mathbf{F}}(\vec{r}) = -\hbar\nabla g(\vec{r})(\hat{a}^\dagger\hat{\sigma}_- + \hat{\sigma}_+\hat{a}) \quad (1)$$

with $(\hat{a}, \hat{a}^\dagger)$ as the annihilation and creation operators for photons in the cavity field and $\hat{\sigma}_\pm$ as the raising and lowering operators for atomic excitation. There are also nonconservative (velocity-dependent) and random (diffusive) forces that act on the atom and are described by matrices $\alpha(\vec{r})$ and $\mathscr{D}(\vec{r})$, respectively. The nonconservative forces may include cooling such as the Sisyphus cooling mechanism discussed in (24). The diffusive forces have a component due to "recoil kicks" from spontaneous emission that is the dominant contribution for the radial motion and a reactive component due to fluctuations of the atomic dipole that is substantial for motion along the direction of the optical standing wave. All of the contributions to the atomic CM motion are strongly position dependent. For example, the reactive component of $\mathscr{D}(\vec{r})$ depends on the square of the gradient of the coupling $g(\vec{r})$ as well as the atomic internal state at $\vec{r}$, so that this contribution to $\mathscr{D}(\vec{r})$ is strongly suppressed around the antinodes of the standing wave. The separation of time scales in the problem means that all of the quantities $\{U(\vec{r}), \alpha(\vec{r}), \mathscr{D}(\vec{r})\}$ may be evaluated by solving the steady-state quantum master equation for the internal degrees of freedom alone (42, 43). Hence, the local atom-field coupling $g(\vec{r})$, probe

parameters ($\mathscr{E}_{probe}$, Δ_{probe}), and detuning Δ_{ac} suffice to determine the various forces on the atom at $\vec{r}$.

The motion of an atom in the cavity may be simulated by means of a system of Langevin equations for the position and momentum $\vec{p}$ of the atom:

$$\frac{d}{dt}\vec{r} = \frac{\vec{p}}{m} \quad (2)$$

$$\frac{d}{dt}\vec{p} = \langle\hat{\mathbf{F}}(\vec{r})\rangle - \alpha(\vec{r})\frac{\vec{p}}{m} + B(\vec{r})\vec{e}(t) \quad (3)$$

where B is such that $\mathscr{D}(\vec{r}) = B(\vec{r})B^T(\vec{r})/2$ and the vector $\vec{e}(t)$ is made up of noises of zero mean that are delta-correlated in time (44). These equations allow us to investigate the statistics of the length of time atoms spend in the cavity and the characteristics of atomic oscillation in the optical potential as discussed below, as well as the characteristics of the heating processes. The ensemble of these trajectories provides information about the correlation between the motional dynamics and the cavity field state (38), which in turn forms the basis of the reconstruction algorithm for the atomic motion discussed below. In the experimental regime, spontaneous emission will lead to a coherence length of the quantum mechanical state of motion that is small compared with the length scales of the variation of the coupling $g(\vec{r})$, and as a result, individual trajectories of such simulations may be tentatively identified with the random motion of the mean position and momentum of a localized atomic wave packet.

The simulations as well as observations (39) indicate that the motion along the cavity axis x is tightly confined (for example, to a region $\delta x \approx \pm 50$ nm from the simulations) because of the steepness of $U(x)$. However, as shown in

Fig. 2, C and D, ultimately the atom does escape because of a "burst" of heating along the cavity axis that occurs over a time less than the orbital period. This dominant loss mechanism appears repeatedly in the simulations over a wide range of operating parameters. The mechanism for this heating is the very steep growth of the diffusion constant away from the antinode. Once an atom is heated sufficiently to leave the antinode to which it was initially confined, it is very rarely recaptured in another antinode but rather escapes the cavity altogether, because the Sisyphus-type mechanisms (24) for cooling are ineffective in the current setting (as confirmed in our simulations).

Validation of $U(\rho)$. Restricting our attention then to motion in transverse $(y, z) \to (\rho, \theta)$ planes, we can investigate the validity of our model for the effective potential $U(\rho)$ by comparing the predicted and observed oscillation frequencies. Oscillations with a short period (P_2 in Fig. 2A) have a smaller amplitude than those of longer period (P_1 in Fig. 2A) because of the anharmonicity of our approximately Gaussian-shaped potential $U(\rho)$; large-amplitude oscillations are expected to have a longer period than nearly harmonic oscillations at the bottom of the well. The data in Fig. 4A reveal this anharmonicity. Plotted is the period P versus the amplitude A for individual oscillations, where $A \equiv 2[(H_1 + H_2)/2 - H_c]/(H_1 + H_2)$, with parameters $\{H_1, H_2, H_c\}$ indicated in Fig. 2B. The blue curve is calculated for motion in the effective potential $U(\rho)$ shown in the inset to Fig. 4A; the comparison is absolute with no adjustable parameters.

We also present in Fig. 4B similar results for A versus P from our numerical simulations (for the same parameters as Fig. 4A). This plot reveals the relative importance of different mechanisms that cause deviations from the one-dimensional (1D), conservative-force model. To this end, we select from the simulation points corresponding to atoms with low angular momentum about the center of the cavity, that is, those that pass close to the center of the potential ($\rho = 0$) and therefore have close to a 1D trajectory. As expected, these points (shown in blue in Fig. 4B) fall closest to the curve given by the 1D potential $U(\rho)$. The green points in Fig. 4B have larger angular momentum, corresponding to atoms in more circular orbits. The presence of this separation by angular momentum in the simulation indicates that friction and momentum diffusion, which tend to invalidate the conservative-force model, have a relatively small effect on the motion, as is evident from the plots of $U(\vec{r})$ and $dE(\vec{r})/dt$ in Fig. 3B. The spread in observed angular momenta is constrained by our triggering conditions—the potential is switched up only when an atom reaches a position near the center of the cavity mode, so that the measured trajectories tend to be in a regime of tight binding. The wider spread in the data of Fig. 4A relative to Fig. 4B comes from

Fig. 4. Oscillation period as a function of amplitude from (A) experimental and (B) simulated atom transits, for the parameters of Fig. 3B. Calculated 1D oscillation in the anharmonic effective potential (inset) is shown by the blue curve, with no adjustable parameters. In simulated data, note the separation of data points by angular momentum; lowest angular momentum transits (blue) most closely follow the 1D model.

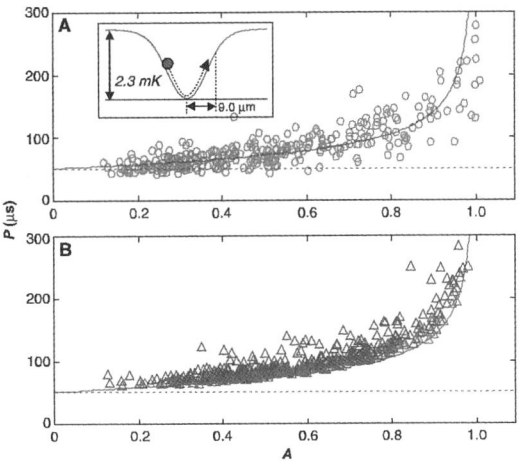

experimental noise (present in Fig. 4A but not added to Fig. 4B), with both shot noise and technical noise contributing substantially. We made comparisons as in Fig. 4 for several data sets with varying values of $\{(\mathscr{E}_{probe}, \Delta_{probe}), \Delta_{ac}, \bar{n}_t\}$ with the same conclusions.

Reconstruction of atomic orbits. Our understanding of atomic motion in the effective potential $U(\rho)$ (including confirmation that motion in the standing-wave direction is minimal) together with a knowledge of the mapping between atom position and probe beam transmission via the master equation enables accurate reconstructions of 2D trajectories for the individual atom transits of Fig. 2. The reconstruction algorithm consists first of digitally filtering the transmission data of Fig. 2 with a 20-kHz low-pass Butterworth filter to reduce noise unrelated to the atomic motion. The smoothed transmission is then mapped to atomic radial position to obtain $\rho(t)$. To infer a 2D trajectory for the atom, it is necessary to determine $\theta(t)$ as well. If we consider that the atom orbits in a known central potential, the solution to this problem becomes apparent. The angular momentum of such an orbit can be calculated from the maximum and minimum radius it attains, via

$$L = \rho_{max}\rho_{min}\sqrt{\frac{2m[U(\rho_{max}) - U(\rho_{min})]}{(\rho_{max}^2 - \rho_{min}^2)}}$$
(4)

where m is the atomic mass. An atom in our cavity does not orbit in a strictly conservative potential, so its angular momentum and orbit change over time because of the velocity-dependent and random forces discussed above. However, if the angular momentum changes by a small fractional amount in the course of a single orbit, we may use successive maximum and minimum radial positions $\rho_{max}^i, \rho_{min}^i$ to estimate a piecewise angular momentum L^i for each half-orbital period. A smooth interpolation in L can then be made along the segments from L^i to L^{i+1}. Knowledge of $\rho(t)$ and $L(t)$ allows determination of $\theta(t)$ via $\dot{\theta} = L/m\rho^2$. We stress that trajectories derived from this algorithm contain three fundamental ambiguities: the initial angle of entry, the overall sign of the angular momentum, and the specific antinode in which the orbit is confined. These initial conditions are not given by the reconstruction algorithm and may be considered degrees of freedom in the final result. In the trajectories of Fig. 5, A and B, the initial angle was chosen to display the atoms falling into the cavity from above, as is physically appropriate. In Fig. 5, C and D, the initial conditions were chosen to provide best agreement with the corresponding actual (simulated) trajectories.

We validated this inversion algorithm by analyzing a series of the simulated atom transits and associated transmissions (as in Fig. 2, C and D). Atomic trajectories are reconstructed from the simulated transmission (including fundamental and technical noise) via our algorithm

and compared with the actual positions from the simulation. Particular results for the simulations of Fig. 2, C and D, are shown in Fig. 5, C and D, respectively, where the actual trajectory is traced in gray, with the reconstruction in green. The quality of these reconstructions is typical of the results for most atom trajectories. In general, reconstructions exhibit good agreement until the very end of the trajectory, where our algorithm fails because (i) the angular momentum cannot be estimated once the atom has escaped from a bound orbit and (ii) the reconstruction ignores x axis motion, which becomes nonnegligible at the end of the trajectory (see Fig. 2, C and D).

For a small fraction of atom transits, our reconstruction method cannot be applied reliably. These are the atoms with nearly linear orbits that pass near the origin of the potential. Reconstructions fail in this case because these atoms have very low angular momentum that changes by a large fraction in the course of a single orbit and may even change sign from one orbit to the next. Such cases are characterized by distinct oscillations in the cavity transmission that repeatedly reach the maximum allowed value of $\bar{m}$ for the known probe parameters ($\mathscr{E}_{probe}, \Delta_{probe}$) and detuning Δ_{ac}. Reconstructions were not attempted in such cases and,

indeed, when attempted tended to produce reconstructed trajectories with sharp corners and unphysical kinks near the origin (45).

On the basis of this ability to reconstruct trajectories in the simulations (with the associated caveats), we applied the same technique to the actual experimental data. In this way, the two individual atom transits of Fig. 2, A and B, were translated into the trajectories of Fig. 5, A and B, respectively. We now see directly that the transmission changes of Fig. 2, A and B, relate to elongated orbits, with time-varying distance to the cavity center. The size of the green dot at the start of each trajectory indicates the typical error in the estimate of the atomic location, from comparisons as in Fig. 5, C and D (46).

Extensions of the ACM. Subject to the caveats concerning the reconstruction algorithm, the results of Fig. 5 represent a capability for tracking the position of a single atom with about 2-μm resolution achieved on a 10-μs time scale. Stated in terms of a sensitivity S_ρ for tracking atomic motion in the radial plane, these numbers translate to $S_\rho \simeq 2 \times 10^{-8}$ m/$\sqrt{\text{Hz}}$, as set by (among other things) the slope of the cavity mode in the radial direction, $|d\psi(\vec{r})/d\rho|_{max} \sim w_0^{-1}$. Increasing this slope would lead directly into

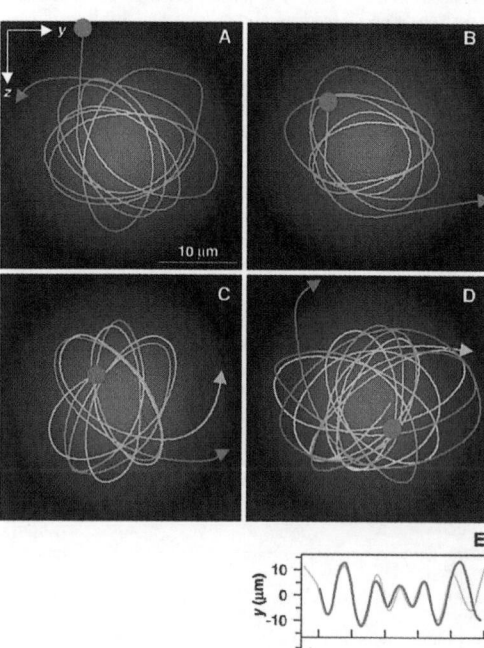

Fig. 5. Atom trajectories lie in a plane perpendicular to the cavity axis, as illustrated by the schematic in Fig. 1. (**A** and **B**) Reconstructed atomic trajectories for the transits of Fig. 2, A and B. (**C** and **D**) Trajectories reconstructed from the simulated transits of Fig. 2, C and D, with the actual trajectories shown in gray for comparison. (**E**) Positions $y(t)$ and $z(t)$ for (D) are shown for clarification. The green dot at the start of each reconstructed trajectory indicates an estimated error in the reconstruction. Animated versions of these orbits can be viewed at www.its.caltech.edu/~qoptics/atomorbits/.

RESEARCH ARTICLES

improvements in sensitivity, both through the explicit increase in the rate of change of the coupling coefficient with displacement $dg(\vec{r})/d\rho$ as well as through the implicit increase in g_0 with reduced cavity volume V_C [and hence also decreases in the critical parameters (n_0, N_0)].

Although the axial motion was not observed in our current experiments, we can nonetheless make an estimate of the sensitivity S_x for detecting atomic position along the standing wave direction through the simple relation $|d\psi/dx|_{max}/|d\psi/d\rho|_{max} \sim 10^2$, leading to $S_x \simeq 2 \times 10^{-10}$ m/$\sqrt{\text{Hz}}$. This estimate should be compared with that of (13), namely $S_x \simeq 1 \times 10^{-10}$ m/$\sqrt{\text{Hz}}$, which was, however, obtained by operating in a dispersive regime and detecting the full optical phase to optimize sensitivity. Given that there is a large separation in time scales associated with motion in the radial and axial dimensions ($\sim |d\psi/d\rho|_{max}/|d\psi/dx|_{max}$), a possible strategy for full 3D reconstruction of atomic motion within the cavity would be to split the detected photocurrent into two components—one with a low-pass filter relating to the radial motion and another with a high-pass filter for the axial motion. Additionally, ambiguities in the initial angle and the sign of the angular momentum may in principle be overcome by strategies that break the radial symmetry, such as the use of external field gradients or transverse cavity modes.

With respect to fundamental quantum limits of the ACM, we estimate that the sensitivity S_ρ together with the time of our observations brings us close to the standard quantum limit (SQL) for position measurement. The SQL is the limit at which measurement-induced back action on the momentum of the particle becomes an important component in the noise budget for position sensing and can limit further improvements in sensitivity (47). Initial estimates based on the theoretical analysis of (48) indicate that the current experiment is perhaps a factor of five above the SQL, which is again consistent with the estimate of (13). Straightforward improvements to the experiment [such as enhanced detector quantum efficiency, a single-sided versus the current two-sided cavity, and reduced technical noise along the lines of (13)] should improve both the spatial and temporal resolution of our ACM for monitoring atomic motion.

Implicit in this ability to sense with high sensitivity and bandwidth is the possibility for control of a single atomic trajectory by quantum feedback. By implementing our inversion algorithm in real time, a suitable error signal can be derived to modulate the effective potential $U(\vec{r}) \rightarrow U(\vec{r}, t)$ in a fashion that damps atomic motion to the bottom of the well. Indeed, with generalized strategies for active control, it should be possible to surpass the SQL and to synthesize novel nonclassical states of motion (49).

Even without reaching such fundamental limits and extremely low levels of incident light ($\bar{m} \lesssim 1$ photon in the cavity), we suggest that the type of real-time microscopy represented by the ACM might be more broadly applicable to monitoring of chemical and biological processes at the single-molecule scale. In this setting, a key feature of the ACM would be the ability to sense changes of the optical properties of an intracavity medium with high bandwidth and sensitivity. The function of localization within the cavity mode would be provided by a separate means other than the single-photon trapping used in the current work (as, for example, by an optical dipole-force trap or indeed by in vitro diffusion).

Certainly optical techniques already exist with single-molecule resolution (34, 35, 50–53). However, a potentially powerful aspect of the ACM would be the ability to track molecular dynamics in real time for a single molecule within a resolution volume within the cavity. Implicit in realizing such a capability would be a detailed understanding of the nature of the radiative interaction between molecule and cavity field, as well as of the detection mechanism, thereby allowing the development of an inversion algorithm such as that leading to the results of Fig. 5, where now the "trajectory" could be in a space such as, for example, molecular conformations (54).

Although it might seem at first sight hopeless to accomplish this for complex chemical or biological species, in fact, the situation can be considerably simpler than in the full quantum case presented here. In many situations, knowledge only of the linear susceptibility of the particle in question should be sufficient for the purpose of realizing an ACM. A rather extensive theory of the input-output characteristics for cavities containing such linear (or indeed nonlinear) media exists within the context of the literature on optical bistability (55). Within this setting, the key parameters become the so-called cooperativity parameter C_1 for a single particle ($\sim 1/N_0$ for our experiment) and the saturation intensity I_s for the intracavity field [$\sim n_0 c/(\hbar\omega V_C)$], both of which can be determined by traditional means with bulk samples. Of particular interest might be detection of dispersive shifts of the cavity resonance by a target molecule, thereby potentially avoiding photobleaching. What is required is an extension of this literature directed toward the development of suitable inversion algorithms, as has been carried out in one particular case in this research article.

An alternative to such a case-specific approach involving the direct modification of the cavity field by an intracavity molecule is to exploit a detailed knowledge of the atom-cavity interaction to sense molecular dynamics indirectly. Consider an atom (e.g., cesium as here) trapped within the cavity mode but subject to an additional interaction with a molecule that has

negligibly small direct coupling to the cavity field. The interaction energy of the (sensing) cavity atom and (sensed) molecule leads to changes in the level structure of the cavity atom as well as to a force that shifts the equilibrium atomic position within the cavity. In either event, the amplitude and phase of the transmitted field are modified, from which an inference of the molecular interaction (such as dipole-dipole coupling) can be drawn. We envision a geometry that would allow the atom-cavity system to be scanned spatially, thereby combining the very high quality factors available from the atom-cavity interaction with more conventional scanning probe microscopies.

References and Notes
1. D. J. Wineland, W. M. Itano, J. C. Bergquist, *Opt. Lett.* **12**, 389 (1987).
2. W. E. Moerner and L. Kador, *Phys. Rev. Lett.* **62**, 2535 (1989).
3. The finesse is assumed to be large compared with the cavity Fresnel number to ensure rapid diffractive mixing within the cavity lifetime.
4. For a discussion, see L. A. Orozco et al., *Phys. Rev.* **39**, 1235 (1989).
5. G. Rempe, R. J. Thompson, H. J. Kimble, R. Lalezari, *Opt. Lett.* **17**, 363 (1992).
6. G. Rempe, *Appl. Phys. B* **60**, 233 (1995).
7. P. Cerez, A. Brillet, C. N. Man-Pichot, R. Felder, *IEEE Trans. Instrum. Meas.* **29**, 352 (1980).
8. M. de Labachlelerie, K. Nakagawa, M. Ohtsu, *Opt. Lett.* **19**, 840 (1994).
9. J. Ye, L.-S. Ma, J. L. Hall, *Opt. Lett.* **21**, 1000 (1996).
10. H. Mabuchi, Q. A. Turchette, M. S. Chapman, H. J. Kimble, *Opt. Lett.* **21**, 1393 (1996).
11. C. J. Hood, M. S. Chapman, T. W. Lynn, H. J. Kimble, *Phys. Rev. Lett.* **80**, 4157 (1998).
12. J. Ye et al., *IEEE Trans. Instrum. Meas.* **48**, 608 (1999).
13. H. Mabuchi, J. Ye, H. J. Kimble, *Appl. Phys. B* **68**, 1095 (1999).
14. P. Münstermann et al., *Phys. Rev. Lett.* **82**, 3791 (1999).
15. J. Ye, D. W. Vernooy, H. J. Kimble, *Phys. Rev. Lett.* **83**, 4987 (1999).
16. P. Berman, Ed., *Cavity Quantum Electrodynamics* (Academic Press, San Diego, CA 1994).
17. For a more recent review, see contributions in the Special Issue of *Physica Scripta* T76, (1998).
18. G. Rempe, F. Schmidt-Kaler, H. Walther, *Phys. Rev. Lett.* **64**, 2783 (1990).
19. G. Rempe, R. J. Thompson, R. J. Brecha, W. D. Lee, H. J. Kimble, *Phys. Rev. Lett.* **67**, 1727 (1991).
20. M. Brune et al., *Phys. Rev. Lett.* **77**, 4887 (1996).
21. S. Haroche, M. Brune, J. M. Raimond, *Europhys. Lett.* **14**, 19 (1991).
22. B. G. Englert, J. Schwinger, A. O. Barut, M. O. Scully, *Europhys. Lett.* **14**, 25 (1991).
23. A. C. Doherty, A. S. Parkins, S. M. Tan, D. F. Walls, *Phys. Rev. A* **56**, 833 (1997).
24. P. Horak et al., *Phys. Rev. Lett.* **79**, 4974 (1997).
25. M. O. Scully et al., *Phys. Rev. Lett.* **76**, 4144 (1996).
26. W. Ren and H. J. Carmichael, *Phys. Rev. A* **51**, 752 (1995).
27. D. W. Vernooy and H. J. Kimble, *Phys. Rev. A* **56**, 4287 (1997).
28. For an overview of atom trapping techniques, see the Nobel Lectures by S. Chu [*Rev. Mod. Phys.* **70**, 685 (1998)], C. Cohen-Tannoudji [*Rev. Mod. Phys.* **70**, 707 (1998)], and W. D. Phillips [*Rev. Mod. Phys.* **70**, 721 (1998)].
29. C. Cohen-Tannoudji, in *Fundamental Systems in Quantum Optics, Les Houches, Session LIII, 1990*, J. Dalibard, J. M. Raimond, J. Zinn-Justin, Eds. (Elsevier Science, Amsterdam, 1992), pp. 21–52.
30. P. Verkerk et al., *Phys. Rev. Lett.* **68**, 3861 (1992).
31. P. S. Jessen et al., *Phys. Rev. Lett.* **69**, 49 (1992).
32. P. Marte, R. Dum, R. Taieb, P. D. Lett, P. Zoller, *Phys. Rev. Lett.* **71**, 1335 (1993).
33. R. J. Cook, in *Progress in Optics*, vol. XXVIII, E. Wolf, Ed. (North Holland, Amsterdam, 1990), p. 361.

RESEARCH ARTICLES

34. M. Oritz, J. Bernard, R. Personov, *Phys. Rev. Lett.* **65**, 2716 (1990).

35. W. P. Ambrose and W. E. Moerner, *Nature* **349**, 225 (1991).

36. The optical cavity is formed by two 1-mm diameter, 10-cm radius of curvature mirrors, located on the tapered end of 4 mm by 3 mm glass substrates. The multilayer dielectric mirror coatings have a transmission of 4.5×10^{-6} and absorption/scatter losses of 2.0×10^{-6}, giving rise to a cavity finesse $F = 480,000$. For the measured cavity length $l = 10.9\ \mu m$ and waist $w_0 = 14\ \mu m$, we have parameters $(g_0, \kappa, \gamma) = 2\pi(110, 14.2, 2.6)$ MHz, where these rates refer to the transition $|a\rangle \equiv 6S_{1/2}, F = 4, m_F = 4\rangle \rightarrow |e\rangle \equiv 6P_{3/2}, F = 5, m_F = 5\rangle$ at $\lambda = 852$ nm in atomic cesium.

37. To interact with light in the cavity mode, an atom must fall through the 9-μm gap between the edges of the mirrors and also through the waist $w_0 = 14\ \mu m$ of the Gaussian mode of the resonator.

38. Our numerical simulations indicate that the intracavity field exhibits manifestly quantum or nonclassical characteristics for the parameters used in the experiment. For example, the normalized two-time intensity correlation function $\simeq 0.9$ for short time delays, where a value < 1 indicates nonclassicality.

39. Experimentally, we can set a limit on atomic excursions along x by examining the amplitude of the observed oscillations. Because the oscillation frequency $\nu_x \simeq 1.5$ MHz is well above our detection bandwidth of 100 kHz, large excursions in x would lead to substantial reduc-

tions in amplitude for the observed oscillations in transmission, from which we deduce a bound $\delta x^e \lesssim 70$ nm for $\rho \ll w_0$. Inference of axial localization without such bandwidth limitations can be found in (*13*).

40. J. Dalibard and C. Cohen-Tannoudji, *J. Phys. B* **18**, 1661 (1985).

41. J. P. Gordon and A. Ashkin, *Phys. Rev. A* **21**, 1606 (1980).

42. The approximations used in some of the calculations of (*23*) and also the weak driving approximation of (*24*) are not appropriate in the regime of the current work. All quantities are calculated numerically from the full quantum master equation for the internal degrees of freedom.

43. This part of the calculation was performed with a code based on Sze Tan's "Quantum Optics Toolbox," available at www.phy.auckland.ac.nz/Staff/smt/qotoolbox/download.html

44. These are technically stochastic differential equations written in their Ito form.

45. On first inspection, it might seem that the reconstruction algorithm should fail for circular orbits as well, because in the absence of large transmission oscillations the algorithm will rely on small differences between ρ'_{max} and ρ'_{min}, which are uncertain because of technical and detection noise. However, the resulting estimated angular momentum closely approximates that of a circular orbit at radius $\rho'_{max} \simeq \rho'_{min}$ and is therefore the correct angular momentum for the orbit.

46. A more extensive gallery of reconstructed orbits can be

viewed at www.its.caltech.edu/~qoptics/atomorbits/gallery.html.

47. V. B. Braginsky and F. Ya. Khalili, *Quantum Measurement* (Cambridge Univ. Press, Cambridge, 1992).

48. H. Mabuchi, *Phys. Rev. A* **58**, 123 (1998).

49. J. A. Dunningham, H. M. Wiseman, D. F. Walls, *Phys. Rev. A* **55**, 1398 (1997).

50. E. Betzig and R. J. Chichester, *Science* **262**, 1422 (1993).

51. X. S. Xie and J. K. Trautman, *Annu. Rev. Phys. Chem.* **49**, 441 (1998).

52. W. E. Moerner and M. Orrit, *Science* **283**, 1670 (1999).

53. S. Weiss, *Science* **283**, 1676 (1999).

54. H. Mabuchi *et al.*, in preparation.

55. H. M. Gibbs, *Optical Bistability: Controlling Light with Light* (Academic Press, Orlando, FL, 1995).

56. We gratefully acknowledge the contributions of K. Birnbaum, M. S. Chapman, H. Mabuchi, J. Ye, and S. Tan to the current research. This work is supported by the NSF, by the Office of Naval Research, by the Defense Advanced Research Projects Agency via the Quantum Information and Computation Institute administered by Army Research Office, and by Hewlett-Packard Research Labs. Work at the University of Auckland is supported by the Marsden Fund of the Royal Society of New Zealand.

25 October 1999; accepted 18 January 2000

Optical levitation in high vacuum

A. Ashkin and J. M. Dziedzic

Bell Telephone Laboratories, Holmdel, New Jersey 07733
(Received 17 November 1975)

Optical levitation of highly transparent particles has been observed in the high-vacuum regime where viscous damping and thermal conductivity are small, the particle is cooled only by thermal radiation, and radiometric forces are negligible. The effects of an impulse and adiabatic manipulation on the dynamics of a sphere were studied from atmospheric pressure down to $\sim 10^{-6}$ Torr. The calculated time for an oscillating particle to decay to half-amplitude due to the intrinsic optical damping at zero pressure is ~ 0.7 years.

PACS numbers: 42.60.Q, 06.60.S, 42.80.

We report on experiments demonstrating optical levitation of highly transparent particles under high-vacuum conditions. Optical levitation[1-3] is a technique for stably supporting and manipulating transparent particles by the forces of radiation pressure from cw laser beams. It results from the ability of light to not only push axially on transparent particles but also to exert transverse forces[4] which can either push the particles into or out of the high-light-intensity regions of the beam depending on their light scattering properties. Levitation of glass, plastic, and liquid spheres as well as spherical shells has been demonstrated. The technique is of interest for support and manipulation of laser fusion targets,[2,5] studies of light scattering from oriented particles,[3] studies of drop-drop interactions in cloud-physics experiments,[3] and as a sensitive detector of emitted photoelectrons in studies of nonlinear photoelectric effects[6] in transparent materials.

For many of these applications and for other new ones, levitation under high-vacuum conditions is an advantage or even a necessity. In earlier work[1] levitation of glass spheres down to pressures $\sim 1-10$ Torr was reported at which pressures the particles were lost. This loss was attributed to residual radiometric (thermal) forces which, although negligible at atmospheric pressure, grow in magnitude as the pressure is reduced to the point where they compare with or exceed radiation pressure. The temperature gradients responsible for the destabilizing radiometric forces arise from residual optical absorption in the particle (photophoresis).[7] In our present experiments using considerably lower absorption particles (silicone oil drops and fused quartz spheres) we have achieved levitation down to pressures $\sim 10^{-6}$ Torr. This pressure is well into the high-vacuum regime. This we define as the low-pressure regime where radiometric forces are completely negligible, viscosity is low, thermal conductivity of the gas is negligible, and the particle is cooled solely by thermal radiation. Once in this regime there is no further barrier to pumping to arbitrarily low pressures where the viscosity is vanishingly small apart from possible problems arising from external vibrations and beam fluctuations.

In Fig. 1, we show schematically the expected pressure variation of the radiation pressure force F_{rad}, the radiometric (thermal) force F_T,[7] the viscosity η,[8] the thermal conductivity Λ,[8] and the particle temperature T, for a 20-μ-diam sphere levitated in a 5145-Å beam of ~ 150 mW. F_{rad} is independent of pressure. The other quantities vary according to kinetic theory in a way which depends on the quantity p_0, the pressure for which the mean free path of a gas molecule equals the particle diameter. $p_0 \simeq 2.5$ Torr for a 20-μ-diam sphere. The straight lines indicate the asymptotic behavior of the curves away from p_0. Curve F_T^a of Fig. 1(a) is calculated for a loss $\alpha \simeq 2 \times 10^{-2}$ cm^{-1} and an estimated temperature gradient across the particle of 0.13 °C per 20 μ using the Hettner formula for F_T.[7] It closely represents

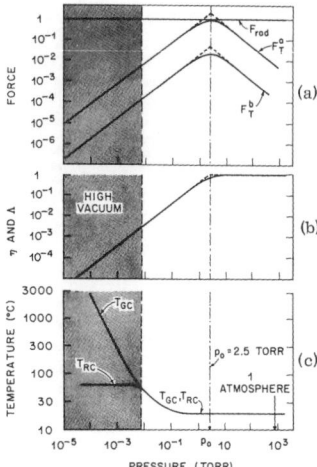

FIG. 1. (a) Dependence of the radiation pressure force F_{rad} and the radiometric (thermal) force F_T on gas pressure, for a 20-μ particle. F_T^a is for $\alpha \cong 2 \times 10^{-2}$ cm^{-1} and F_T^b is for $\alpha \cong 5 \times 10^{-4}$ cm^{-1}. (b) Thermal conductivity Λ and viscosity η vs gas pressure. (c) Particle temperature vs gas pressure for a 20-μ particle with $\alpha \cong 5 \times 10^{-4}$ cm^{-1}. T_{GC} is only for gas cooling, T_{RC} is for gas cooling plus radiation cooling.

the situation with previous experiments[1] where an essentially negligible value of F_T of ~1% of F_{rad} at a pressure of 1 atm grew to a value comparable with F_{rad} and caused loss of levitation at a lower pressure p which was $\cong p_0$. Using a clear silicone oil we reduced the loss to $\alpha \cong 5 \times 10^{-4}$ cm^{-1}.[9] The force F_T^b for this case is reduced relative to F_{rad} and F_T^a by a factor of $\sim 2 \times 10^{-2}/5 \times 10^{-4} = 40$. F_T^b is thus only a few percent of F_{rad} at $p = p_0$ and with these oil drops we were able to pump down through the radiometric maximum to pressures ~10^{-6} Torr, the limit of our present vacuum system, with no problems. The oil drops were introduced into the cell with a spraying technique[3] and the cell sealed for pumping. At 10^{-6} Torr the thermal force $F_T^b \cong 4 \times 10^{-8}$ F_{rad} and is quite negligible compared to F_{rad}.

Since the thermal conductivity Λ of the gas is constant down to p_0 as seen in Fig. 1(b), and the power absorbed by the levitated drop is small (~3×10^{-8} W), we expect the average temperature of the particle to be constant at essentially $T_{room} \cong 20\,^\circ$C down to p_0. At pressures below p_0, Λ falls as p and we expect T to rise linearly as $1/p$. Curve T_{GC} of Fig. 1(c) shows this behavior and is based on the known coefficient of molecular conductivity[8] $\alpha \Lambda_0 = 8.3 \times 10^{-6}$ W/cm^2 deg μ. T_{GC} assumes only gas cooling and extrapolates to a temperature ~$2.7 \times 10^5\,^\circ$C at 10^{-6} Torr. This is clearly unreasonable since DC-200 silicone oil becomes highly volatile at ~300 °C. What actually happens we believe is that once the temperature rises to ~60 °C as p falls to ~10^{-2} Torr, radiation cooling proportional to ϵT^4 takes over and the particle maintains this temperature down to arbitrarily low vacuum [curve T_{RC} of Fig. 1(c)]. The value of 60 °C is based on the power absorbed and an estimated value of emissivity ϵ of ~0.1. For all pressures below ~10^{-2} Torr we are then in the high-vacuum regime as defined above since as seen from the curve

for η in Fig. 1(b) we are also in the low-viscosity regime. η varies exactly as Λ since η/Λ is a constant. Thus η is independent of pressure down to p_0 and then falls asymptotically as p at lower pressure. At 10^{-6} Torr η is about 6 orders of magnitude less than at p_0.

The effects of changes of viscosity on the particle dynamics are quite dramatic and were studied experimentally by displacing the particle from its equilibrium position with an impulse and observing its subsequent return to equilibrium at different pressures. We recorded the data photographically by sweeping an enlarged image of the particle across a screen placed in front of a camera while passing a wand through the beam to give the particle a vertical displacement. Figure 2 shows the results. The particle is seen by its own 90° scattered light and appears as two dots[1] which are drawn out into two separate lines as the image is swept across the screen. From atmospheric pressure down to $p \cong p_0 = 2.5$ Torr, the particles return to equilibrium with no overshoot, due to strong viscous damping [Fig. 2(a)]. Below p_0 an overshoot develops, [2(b)], and then finally the characteristic behavior of a damped harmonic oscillator [2(c)−2(h)]. Below ~3×10^{-1} Torr the half-amplitude decay time increases as $1/p$ down to ~3×10^{-3} Torr as expected from the viscosity curve of Fig. 1(b). Below 3×10^{-3} Torr random perturbations make it increasingly difficult to observe long clean decay curves. At 10^{-6} Torr these random fluctuations, entering with random phases, cause vertical and horizontal oscillations which rise and fall with an average period of seconds to minutes which is short compared to the extrapolated half-amplitude decay time of ~4.5 h. Thus the particle's response to perturbations is not much affected by the damping at 10^{-6} Torr. This suggests that pumping to still lower pressures should present no new difficulties. As seen in Fig. 2 we cannot

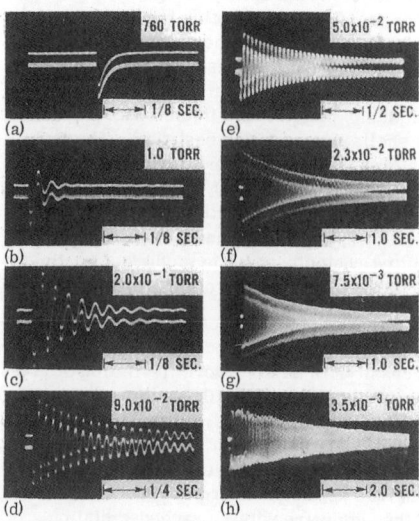

FIG. 2. Photographs of the swept image of a levitated particle following a downward impulse showing the subsequent return of the particle to equilibrium as a function of time at various pressures. The vertical axis is height, the horizontal is time. For $p < p_0 \cong 2.5$ Torr the damping begins to decrease and the decay time increases as $1/p$.

make rapid changes in particle position under low-damping conditions without introducing strong oscillations. If, however, we changed conditions adiabatically (that is slow compared to the oscillation period of $\sim \frac{1}{20}$ sec) we were able to perform all the usual particle manipulation without introducing oscillations.

Silicone drops with their low loss, low vapor pressure, and perfect shape are almost ideal particles for low-pressure work except for one unfortunate problem. After about 10 min in the beam at pressures below p_0 the oil apparently undergoes a photochemical reaction causing the losses to rise to a point where the drops are heated and ejected. We therefore made low-loss fused quartz spheres starting from high-quality natural quartz sand.[10] With these particles levitation was achieved from atmospheric pressure down to 10^{-6} Torr with no changes in properties. Particles were held for periods up to $\sim \frac{1}{2}$ h at 10^{-6} Torr and again we believe we could go to lower pressure with a better pump. The fused silica spheres do not pass through the radiometric maximum at p_0 quite so smoothly as oil drops. This may indicate somewhat higher loss. On the other hand, liquid drops probably do not sustain as large a temperature gradient even for the same loss because of convection within the drop. Lower-loss silica spheres should be achievable based on observed losses of $\sim 5 \times 10^{-5}$ cm^{-1} in high-purity silica optical fibers.[11] Our fused quartz spheres were launched into the beam from the base plate of the cell using a piezoelectric ceramic shaker as previously described.[1] Because of the low loss of the quartz we could launch and successfully trap spheres at pressures from 1 atm down to a $\sim 2 \times 10^{-1}$ Torr. At lower pressures the damping was insufficient to permit trapping. If launched at 1 atm we could pump to ~ 1 Torr in a time as short as 1 min without the gas dragging the particle out of the beam. The particle does, however, displace noticeably in the direction of a gas flow while pumping at this rate, suggesting use of a levitated sphere as an essentially nonperturbing directional flow meter. By manipulating the position of the sphere one should be able to trace out complex gas flow patterns. Once below p_0, where η is small, one can pump as fast as possible without gas currents disturbing the particle.

It is of interest to consider the intrinsic damping of an oscillating levitated particle in the limit of total vacuum, i.e., zero viscous damping. Consider first a static particle in neutral equilibrium between the forces of gravity and radiation pressure from a plane wave. In this case $F_{rad} = N\Delta(h\nu/c) = mg$, where N is the photon flux hitting the particle, $\Delta(h\nu/c)$ is the momentum change of the photons hitting the particle, and mg is the force of gravity. If now the particle moves with a velocity v toward or away from the light, F_{rad} changes to F'_{rad} due to changes in N and $\Delta(h\nu/c)$, causing a velocity-dependent damping term. Thus the motion alters the incident flux N by a factor $1 \pm v/c$. $\Delta(h\nu/c)$ is also changed by the same factor $1 \pm v/c$ due to Doppler shift. Therefore

$$F'_{rad} = F_{rad}(1 \pm 2v/c) = F_{rad} \pm mg(2v/c) \equiv F_{rad} \pm \gamma v. \quad (1)$$

From (1) $\gamma/m = 2g/c$, where γ is the damping constant.

This damping factor slows the particle velocity according to $v = v_0 \exp(-2gt/c)$. Using this same γ in a simple harmonic oscillator equation which adequately describes a levitated sphere stably trapped by light and gravity, we deduce that an oscillation of amplitude A_0 decays as

$$A = A_0 \exp(-\gamma t/2m) = A_0 \exp(-gt/c). \quad (2)$$

Therefore, the time for an oscillation to fall to half-amplitude is $\sim 2.2 \times 10^7$ sec or ~ 0.7 years. This is, therefore, one of the lowest-loss mechanical oscillators known. Extrapolating the measured damping times to low pressure (using $1/p$) we find that a pressure less than $\sim 10^{-9}$ Torr is needed to reduce viscous damping below the intrinsic optical damping. As far as we know only magnetically levitated superconducting particles have lower damping. Theoretically it is zero. In practice, experiments on superconducting gyroscopes show that flux trapping problems lead to considerable damping.[12]

In conclusion we have observed optical levitation under high-vacuum conditions down to pressures of 10^{-6} Torr where radiometric forces are negligible, particles are cooled solely by radiation, and the viscosity is greatly reduced. We see no barriers to levitation at arbitrarily low pressures. We still have the ability to move particles about in high vacuum, provided it is done adiabatically, which should be useful in laser fusion and photoelectric emission experiments. Applications of levitation based on low gas damping and low intrinsic damping can now be studied, for example, the generation of high angular velocities using the torque from absorbed circularly polarized light. Finally, the observed levitation under high-vacuum conditions where radiometric forces are less than 10^{-7} of the radiation pressure force shows unequivocally that particles can be stably trapped solely by the forces of radiation pressure.

The authors thank J. Ashkin and J.P. Gordon for helpful discussions and J.W. Fleming, K. Nassau, A.D. Pearson, J.W. Shiever, and J.C. Williams for making the low-loss fused quartz spheres.

[1]A. Ashkin and J.M. Dziedzic, Appl. Phys. Lett. **19**, 283 (1971).
[2]A. Ashkin and J.M. Dziedzic, Appl. Phys. Lett. **24**, 586 (1974).
[3]A. Ashkin and J.M. Dziedzic, Science **187**, 1073 (1975).
[4]A. Ashkin, Phys. Rev. Lett. **24**, 156 (1970).
[5]J.H. Nuckolls, Bull. Am. Phys. Soc. **20**, 1226 (1975).
[6]A. Ashkin and J.M. Dziedzic, Phys. Rev. Lett. **36**, 267 (1976).
[7]N.A. Fuchs, *The Mechanics of Aerosols* (Macmillan, New York, 1964), Chap. II; G. Hettner, Z. Phys. **37**, 179 (1926).
[8]S. Dushman, *Scientific Foundations of Vacuum Technique* (Wiley, New York, 1962), Chap. 1.
[9]The oil was Dow Corning-200 silicone oil. The loss was measured directly with 5145-Å light and a 1-m-long cell.
[10]Berkeley high-purity silica ($\sim 0.0005\%$ Fe$_2$O$_3$) from Pennsylvania Glass Sand Corp.
[11]P. Kaiser and H.W. Astle, Bell Syst. Tech. J. **53**, 1021 (1974).
[12]V.L. Newhouse, *Applied Superconductivity* (Wiley, New York, 1964), Chap. 5; I. Simon, J. Appl. Phys. **24**, 19 (1953); J.T. Harding and R.H. Tuffias, in *Advances in Cryogenic Engineering*, Vol. 6, edited by K.D. Timmerhous (Plenum, New York, 1961), p. 95.

Stability of optical levitation by radiation pressure

A. Ashkin and J. M. Dziedzic

Bell Telephone Laboratories, Holmdel, New Jersey 07733
(Received 4 March 1974)

Stable optical levitation of transparent hollow dielectric spheres has been demonstrated using TEM_{01} mode laser beams. The levitation of solid dielectric spheres has been made much more stable using highly convergent TEM_{00} mode beams. We have discovered the existence of two distinct stable regimes of levitation for solid spheres, one located above the beam focus, the other below it. A particle can be switched back and forth between these regimes. Three separate stable regimes are also possible.

The stable optical leviation of particles by the forces of radiation pressure has been recently demonstrated for a class of particles which are not only pushed axially by a laser light beam but are also attracted to the high-intensity region of the beam.[1] Transparent solid glass spheres were trapped at a point above the focus of a vertically directed TEM_{00} mode laser beam.[2] We demonstrate here the stable leviation of a second class of particles which are pushed out of the high-intensity region of a light beam as well as being pushed axially along the beam. This is shown by supporting a hollow glass sphere at a point above the focus of a vertically directed TEM_{01}^* mode (do-nut mode) laser beam. We have also greatly improved the positional stability of the leviated particles using strongly focused beams and discovered a new regime of stable levitation for solid glass spheres in which the particles are trapped *below* the focus of a TEM_{00} mode laser beam. Particles can be made to switch back and forth between upper and lower stable regimes demonstrating bistable operation. These results are important for the precise micromanipulation of small particles.

A simple example of a particle that is pushed out of the high-intensity regions of a beam as well as being pushed along the beam is the perfectly reflecting sphere. This follows directly from the fact that the net transverse and axial components of the force come from inwardly directed radial surface reflection forces. In our experiments, to avoid losses, we used transparent thin-walled hollow dielectric spheres which are only *partially* reflecting. In this case in addition to the radially inward surface reflection forces F_{RI} we have,

due to the transmitted light, outwardly directed radial reflection forces F_{R0} and radially outward deflection forces F_D due to refraction [see Fig. 1(a)]. The radially inward surface reflection forces, however, dominate and the hollow dielectric spheres behave as perfectly reflecting spheres in their net transverse behavior.

To give hollow spheres transverse stability in a levitation experiment, we placed the $30-60-\mu$-diam hollow spheres[3] (wall ~ 1.5 μ) on the axis of a vertically directed TEM_{01}^* mode (do-nut mode) laser beam ($\lambda = 4880$ Å). This mode has an intensity *minimum* on axis. Thus any radial displacement of the particle results in a restoring force. The spheres were launched into the beam by lifting them off an acoustically agitated glass plate and into the air as described in Ref. 2 for solid spheres. Levitation of a $45-\mu$ hollow sphere was achieved at a point above the focus of a 1.2-W beam using a 5-cm-focal-length lens. The relation of the particle's size to the actual radial beam intensity distribution is shown in Fig. 1(b). The distribution is predominantly TEM_{01}^* with a residual of TEM_{00} which is not enough to prevent stable levitation. Figure 1(c) is a photograph of the levitated particle in the beam. Figure 1(d) shows the beam itself. More power is needed to levitate hollow spheres than solid spheres of the same weight. For example, with an almost perfect do-nut mode a minimum of ~ 400 mW was needed to levitate a $40-\mu$ hollow sphere, which is ~ 3 times more than for the equivalent $24-\mu$ solid sphere.

In most levitation experiments high positional stability is desirable. We define the positional stability as

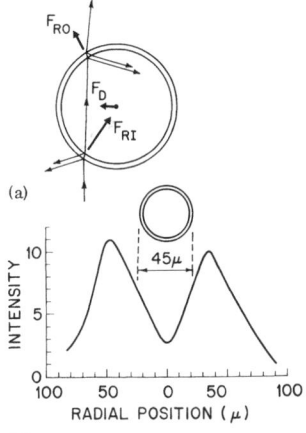

(a)

(b)

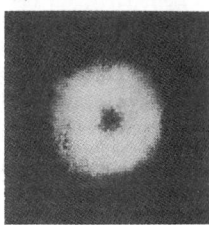

(c)

(d)

FIG. 1. (a) Reflection and refraction of a ray passing through a hollow glass sphere and the resulting forces. (b) Position of the levitated sphere relative to the measured intensity distribution. (c) Photo of the hollow sphere in the beam. (d) The beam itself (do-nut mode).

the distance the equilibrium position moves in microns per unit of perturbing force equal to a percent of the particle's weight. For *solid spheres* and gentle focusing (i.e., small far-field diffraction angle θ), the vertical stability is much less than the transverse stability. Thus for conditions approximately those of Ref. 2, where a $20\text{-}\mu$ *solid sphere* was levitated with $\theta = \lambda/\pi w_0 = 1.2°$ using a lens of focal length $f = 5$ cm, the vertical stability was ~ 10 $\mu/(\%$ of weight) whereas the average transverse stability was ~ 0.4 $\mu/(\%$ of weight). Since this value of vertical stability can cause jitter problems due to power fluctuations, we attempted to improve the vertical stability. To do this we increased the vertical gradient of the light intensity of using a short-focal-length lens ($f = 0.5$ cm and $\theta = 12.5°$) with a long work-

ing distance. This indeed increased the vertical stability as will be seen in Fig. 2(b). There we plot the total beam power P versus the equilibrium height h relative to the beam focus of a solid sphere levitated in a 5145-Å TEM_{00} mode beam. The height was measured by two microscopes, one viewing the particle at 90° to the vertical by its own scattered light, and the other viewing the particle and beam from above. Enlarged images from the microscopes were projected directly on screens for ease of observation. The experimental uncertainty in the position of the focus is $\sim \pm 10$ μ. Since force and power are proportional, the vertical stability in $\mu/(\%$ of weight) is identical with the slope of the curve in Fig. 2(b) in $\mu/(\%$ power change). Referring for the moment in Fig. 2(b) to the upper stable regime labeled S_U which corresponds to levitation *above* the beam focus, the stability at point S is 1.2 $\mu/(\%$ power change) $\equiv 1.2$ $\mu/(\%$ of weight). This represents an improvement of about 10 times for about a 10 times increase in θ.

A further consequence of the short-focal-length lens is a new feature, namely the existence of a lower stable regime labeled S_L in Fig. 2, located *below* the focus. This arises as we shall see from the fact that for the short-focal-length lens used, the beam waist (~ 2 μ) becomes considerably less than the particle diameter (~ 20 μ). The criterion for vertical stability is that the vertical force increase as the particle is displaced downward. If the beam waist is larger than the particle, the vertical force is a maximum at the focus where the intensity is a maximum and decreases above and below the focus. In this case only one regime of stable levitation is possible above the focus. If, however, the beam waist is considerably smaller than the particle as in the present experiment [see Fig. 2(a)], then the vertical force has a *minimum* in the region of the focus. At the focus the beam is concentrated near the axis of the sphere and the force comes mainly from surface reflections at near normal incidence. Above and below the focus, however, the beam spreads transversely over the sphere where surface reflection and refraction increase, giving a larger force. At points above and below the

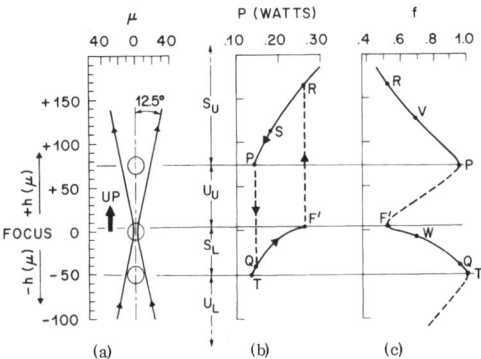

(a)   (b)   (c)

FIG. 2. (a) Relation of TEM_{00} mode beam ($\theta = 12.5°$) to a levitated $22\text{-}\mu$-diam solid glass sphere at the boundaries between stable and unstable regimes. (b) Laser power P vs equilibrium height h in microns. (c) Relative force f for a fixed power normalized to T vs h in microns.

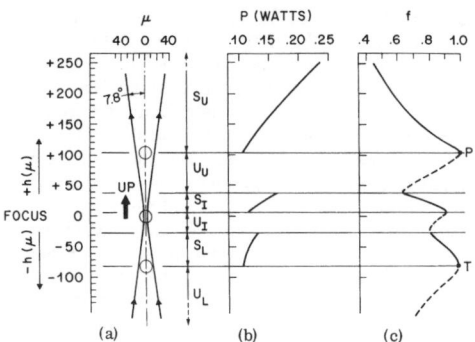

FIG. 3. (a) Relation of TEM_{00} mode beam ($\theta = 7.8°$) to a levitated 19-μ-diam solid glass sphere at various heights. (b) Laser power P vs equilibrium height h in microns. (c) Relative force f for a fixed power vs h in microns.

focus where the beam fills the particle the force should reach a maximum and then decrease again as the beam size exceeds the particle size. This behavior is seen in Fig. 2(c) where we plot the normalized vertical force f for a fixed power versus h as derived from the experimental data of Fig. 2(b) ($f \alpha 1/P$). We do not expect perfect symmetry about the focus due to the shift from beam convergence to divergence in going through the focus. This general type of force curve with two maxima gives rise to the two stable regimes S_U (stable upper) and S_L (stable lower) and the two unstable regimes U_U (unstable upper) and U_L (unstable lower). In Fig. 2(b) only the stable portions are accessible experimentally and, therefore, the values of f in the unstable regions are merely sketched in qualitatively.

In practice we are able to inject particles into the lower stable regime S_L in three ways. First, we can acoustically launch spheres directly into S_L from their resting place on the glass plate by placing the beam focus above the spheres at launch time. This is about 10—20% successful. Second, starting with a particle in S_U we can reduce the power until it reaches the boundary with U_U at point P and falls. If the lower force maximum T is greater than the upper at P, the particle will fall through the focus and be caught in S_L at Q. Raising the power lifts the particle smoothly to the point F' on the boundary between S_L and U_U, whereupon it jumps discontinuously to the point R in S_U. This hysteresis cycle can be repeated indefinitely. We thus have bistable operation with two possible stable positions for a given power, one *above* the focus in S_U, and the other *below* in S_L. For some particles the lower force maximum T is slightly less than the upper force maximum P. In such cases a particle dropping from S_U at the maximum P will fall through S_L and not be caught. Small variations in particles or the beam shape can perhaps account for the variation in behavior. Finally, we found a simple,

virtually flawless method of entering the S_L regime by briefly interrupting the beam when in S_U with the power set at a value corresponding to stable levitation in S_U *and* S_L. V is such a point for example. If the power is turned off just long enough for the particle to fall anywhere in the S_L regime, it gets caught at W when the power comes back on. This is done simply by waving a "wand" through the beam while watching the particle's position on the screen. By varying the power we can trace out the full extent of S_L and the relevant parts of S_U. We see that F' corresponds to the beam focus F within experimental error. Levitation in S_L is interesting in part because it is more stable than in S_U. Thus the vertical stability at W is ~ 0.2 μ/(% power change), a factor of 3 better than in S_U. Also the transverse stability in S_L is better than in S_U since in S_L most of the light passes through the particle and a slight transverse displacement results in a large restoring force. Other incidental consequences of this light distribution in S_L are increased difficulty in clearly observing the particle with a microscope from above and a large reduction in the 90° scattering and Mie rings. Levitation with a $f = 0.8$ cm lens ($\theta = 7.8°$) gave similar behavior except for the frequent appearance of a *third* intermediate stable regime S_I in the region of the beam focus. This is illustrated in Fig. 3. We believe this may be associated with residual beam astigmatism which causes the beam waist to occur at two different heights for two orthogonal transverse directions, giving rise to a double minimum in the force curve near the focus. This is supported by the fact that adjustments that increase the astigmatism enhance the extent of S_I. Similarly, for the $f = 0.5$ cm lens where the focus is sharper and the separation of the minima is smaller, we only occasionally see the S_I regime.

In conclusion, levitation techniques have been extended to make it possible to design highly stable optical levitators for particles which move either into or out of the high-intensity regions of a beam. These techniques should be useful for experiments on Mie scattering, studies of the lens characteristics of small transparent spheres for fiber optics work, and for holding either solid or hollow[4] dielectric targets for laser fusion experiments.

The authors thank E.P. Ippen for helpful discussions.

[1]A. Ashkin, Phys. Rev. Lett. **24**, 156 (1970).
[2]A. Ashkin and J.M. Dziedzic, Appl. Phys. Lett **19**, 283 (1971).
[3]The hollow glass spheres, called eccospheres, were obtained from Emerson and Cuming Inc., Canton, Mass.
[4]John Nuckolls, John Emmett, and Lowell Wood, Phys. Today **26**, 46 (1973).

Reprinted from
21 March 1975, Volume 187, pp. 1073-1075

SCIENCE

Optical Levitation of Liquid Drops by Radiation Pressure

A. Ashkin and J. M. Dziedzic

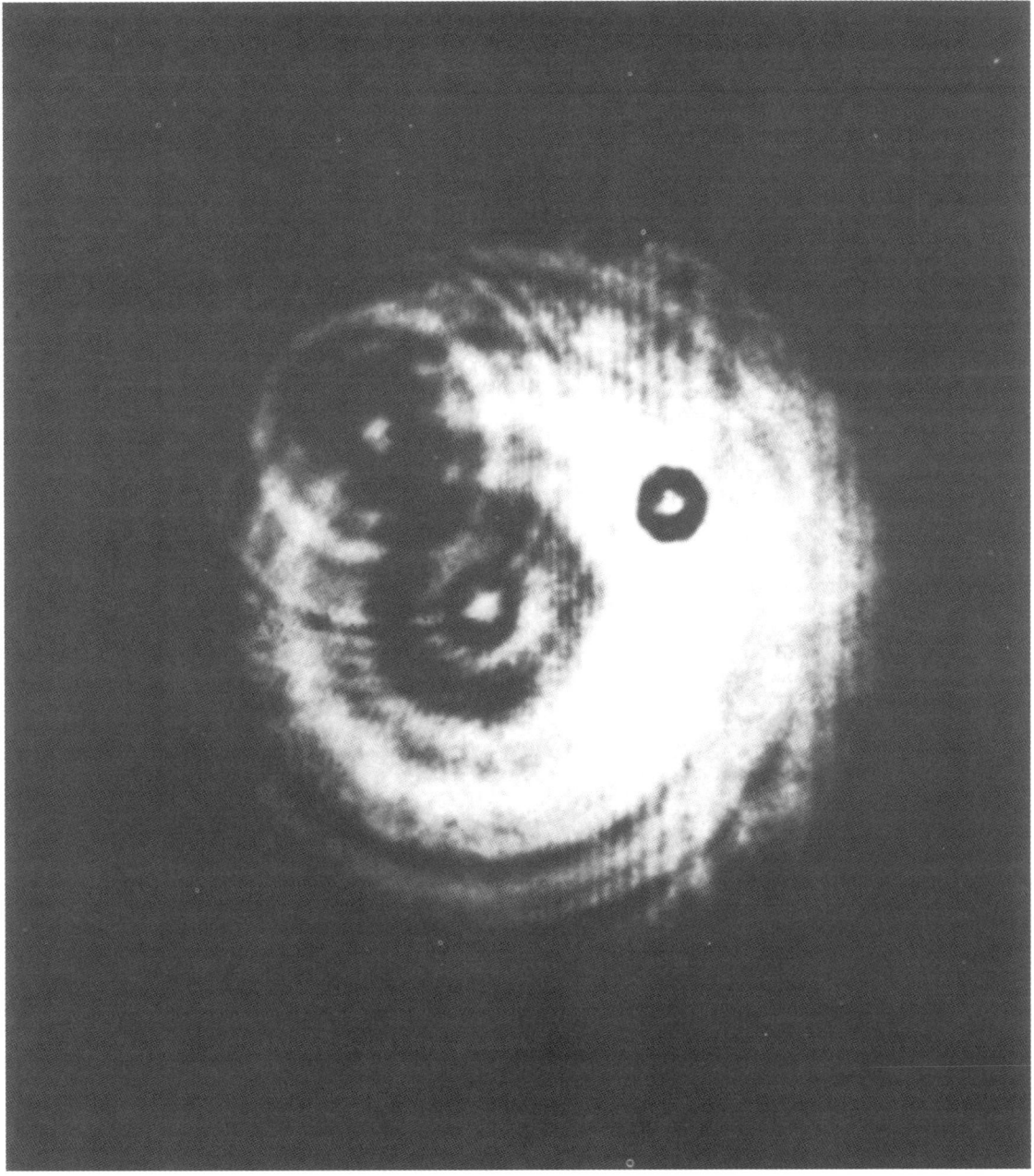

Optical Levitation of Liquid Drops by Radiation Pressure

Abstract. *Charged and neutral liquid drops in the diameter range from 1 to 40 microns can be stably levitated and manipulated with laser beams. The levitation technique has been extended toward smaller particles (about 1 micron), lower laser power (less than 1 milliwatt), and deeper traps (greater than ten times the particle's weight). The techniques developed here have particular importance in cloud physics, aerosol science, fluid dynamics, and optics. The interactions of the drops with light, the electric field, the surrounding gas, and one another can be observed with high precision.*

We report here on a study of the optical levitation of charged and neutral liquid drops with laser beams. The techniques developed here have particular importance in cloud physics, aerosol science, fluid dynamics, and optics. Optical levitation is based on the ability of light to stably trap nonabsorbing particles by the force of radiation pressure (*1*). In this technique a continuous-wave (cw) vertically directed focused TEM$_{00}$ Gaussian-mode laser beam not only supports the particle's weight but pulls the particle transversely into the region of high light intensity on the beam axis. Once the particle is trapped, one can manipulate it by moving the beam about. This levitation technique has been demonstrated for glass spheres and other nonabsorbing particles (*1, 2*). Applications of levitation and radiation pressure have been discussed (*3*).

The proposed application of levitation to cloud physics and aerosol science is a natural extension of the technique, since progress in these fields has hinged crucially on studies of the interactions of single particles (*4*). A comparison of optical levitation with other techniques that have been used in these sciences for manipulating particles, such as electrodynamic suspension, acoustic suspension, the use of wind tunnels, and suspension on fine wires, indicates that only optical levitation has the combined ability to handle the drops (~ 1 to $40 \ \mu$) found in natural clouds with such simplicity, high positional stability, ease of manipulation,

and ease of observation. The advantage of this technique results from the depth of the optical trap, its highly localized nature approximating the particle size, and the ease with which the beam can be moved in space. We will discuss specifically how we generate charged and neutral drops in the 1- to 40-μ diameter range, trap them singly and in small clouds in the light beams, move them about, measure their charge, and observe their interactions with the electric field, light, one another, and the surrounding gas.

Figure 1 is a sketch of the basic apparatus. A vertically directed cw laser beam is focused by a lens L and introduced into a glass box B from be-

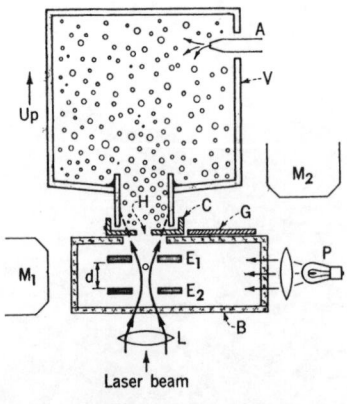

Fig. 1. Sketch of the experimental apparatus. Dimension d is ~ 0.6 cm.

low. The beam is aligned with a hole H, about 0.5 mm in diameter, located in a sliding roof cover C. A liquid droplet cloud is sprayed into a large storage vessel V with an atomizer A where it can settle slowly by gravity. Some drops fall through the hole and enter the light beam where, if their sizes are in the correct range, they can be trapped and levitated. With microscope M_1 one can view the levitated drops from the side. The drops are not only seen by scattered light from the levitating beam but are visible in silhouette against the projection lamp P. The enlarged view from M_1 is projected on a screen for ease of observation and height measurement. Two plane electrodes E_1 and E_2 with a narrow slit, which allows the beam to pass through, can be used to apply an essentially uniform electric field to the particles. Once drops have been collected by the beam, one can stop the rain of particles by pushing the sliding roof cover C aside with the sliding glass plate G. Thus, without opening the box to significant disturbing air currents, the source can be removed and replaced by a transparent plate for viewing with microscope M_2 from above.

The levitated drops in general arrange themselves in order of size, with the largest closest to the beam focus. Figure 2 shows various situations. With multiple drops the upper drops are arranged in a light intensity distribution modified by the lower drops. Thus in Fig. 2b' the upper drop is located outside the shadow cast by the lower drop and sits off-axis in the high-intensity light ring which diffracts past the lower drop. With many drops (Fig. 2, c and d) the particles become more closely coupled as a result of their effects on the light distribution and also as a result of electrostatic attractions and repulsions. Up to 20 or so drops can collect in a fixed array which undergoes rearrangement when significantly disturbed. Drops were made from various liquids. Drops of pure water in the 10- to 30-μ range evaporate so rapidly in room air that they are held only for about 30 seconds. Instead of adding water vapor to the surrounding air to increase drop life, we made drops from liquids with low vapor pressure such as silicone oil or water-glycerol mixtures with composition ratios varying from ten parts water and one part glycerol to pure glycerol. Silicone oil drops persist in this apparatus almost indefinitely. However, a pure glycerol drop with a diameter of $\sim 12~\mu$ was observed to evaporate to $\sim 1~\mu$ in 3 hours. We gradually reduced the power of the argon ion laser operating at 5145 Å from ~ 40 to ~ 0.2 mw to keep the drop from rising too high in the levitating beam as its mass decreased. The impressively low power of 0.2 mw indicates how little power is required for the levitation of particles approaching 1 μ in diameter. The 35-μ particle of Fig. 2a is held by ~ 400 mw.

The electrical charge carried by drops can be determined from a comparison of the applied electric forces with the light forces. Thus, if we displace a drop upward with an electric field, then the percentage reduction in light needed to restore the drop to its original position represents the fraction of the particle's weight supported by the field. From the weight and the applied field, one can determine the total charge. Water-glycerol mixtures are polar liquids and give drops with various charges (plus, minus, or neutral) as expected. The charges are often high enough so that a force several hundreds of times the weight of the particle, mg (where m is the mass and g is the gravitational acceleration), can be applied with reasonable fields. Nonpolar silicone oil gives drops with much less charge.

The magnitude of the restoring forces acting on levitated particles corresponds to $\sim mg$ (2). For larger particles ($\sim 40~\mu$) the magnitude of the restoring forces gives rise to a very stable trap. Thus it would take an air current of ~ 5 cm/sec to eject a 40-μ particle from the beam. Because of increased viscous drag, small particles ($\sim 1~\mu$) are stable up to an air current of only $\sim 3 \times 10^{-3}$ cm/sec. This makes manipulation of 1-μ drops difficult. However, it is possible to deepen these traps by orders of magnitude. By increasing the strength of the upward supporting light and balancing it with a downward electric or optical force, one can maintain levitation and increase the trap depth in proportion to the increase in light power. Fortunately this method is most useful with small drops where the levitating power is initially quite low. In practice, we have deepened traps by more than an order of magnitude by using a downward electric force. This scheme may permit the levitation of even smaller particles, possibly in the submicron range. It also has the important advantage of operating in the zero-gravity environment of space experiments. The opposite procedure of applying an upward electric force and decreasing the supporting light beam results ultimately in a transition to the Millikan type of support where uniform electric forces alone support a particle in neutral equilibrium. Although easily done for small particles (~ 1 to $4~\mu$), this procedure has less practical interest since we lose the advantages of stability. For larger particles where light power is a consideration and damping is not severe, partial electric support has merit. With the field uniformity of our plates we have easily supported 90 percent of a particle's weight electrically.

Potential applications of the levitation technique for single trapped drops include measurements of the rates of evaporation, condensation, charging, and neutralization. The effects of changes in the ion content and flow rate of the surrounding air and also changes in the drop itself, such as the effects of the addition of dissolved impurities (salts) and various solid nuclei, can be studied. In addition to making possible the direct observation of the drops in a high-power microscope while levitated, the levitation technique can also be used to deposit drops elsewhere for measurements.

Concerning solid nuclei, the smallest solids that we have levitated thus far are ~ 4-μ dried latex spheres (Dow Corning). Because latex is slightly absorbing in 5145-Å light, we used ~ 1 mw of 6328-Å light to levitate the spheres. We also expect that one can grow a small crystal within a levitated drop by partial evaporation of a drop containing dissolved salts. Possibly we can evaporate the solvent completely and levitate the crystal itself, if its shape is not too extreme. The most irregularly shaped objects that we have levitated thus far are random clumps of two to five glass spheres. These invariably orient themselves in the beam. These objects, incidentally, could be

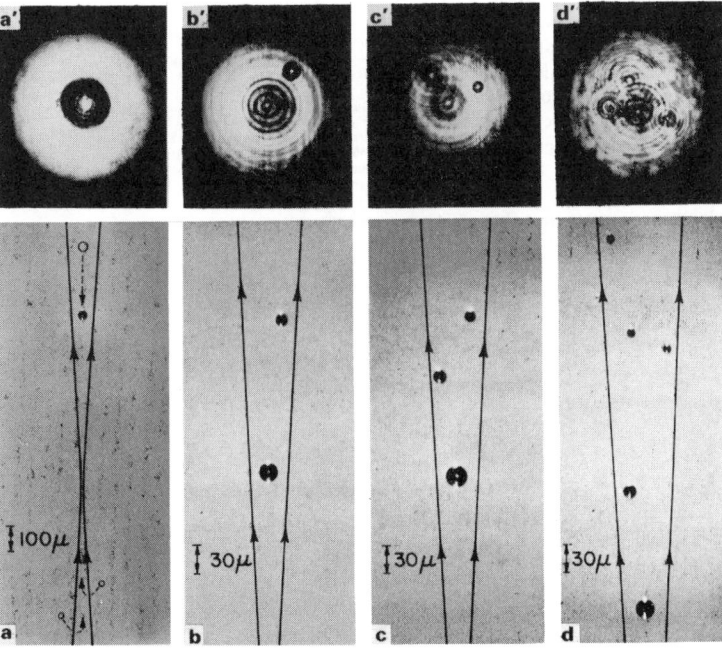

Fig. 2. Photos of optically levitated drops. (a–d) Side views taken with microscope M_1. Beam shapes are shown for reference; (a) also shows the trajectories of drops colliding with the levitated drop. (a'–d') Corresponding top views taken with microscope M_2; (a'–c') focus on the highest drop, and (d') focuses lower in the beam, showing the diffraction rings from the four lowest drops.

used for studying Mie scattering from oriented irregular particles. In addition, our observations of levitation in complex light intensity distributions, such as in Fig. 2d, as well as interesting cases of levitation of small drops in the irregularities of the light beam occurring just above splattered fallen drops lying on the cell floor, indicate that levitation does not require an extreme uniformity of geometry. Finally, concerning ice crystals, we believe that fairly regular crystals such as prisms *should* levitate and that more complex shapes *may* levitate. Possibly an electric field would orient the more plate-like crystals and thus aid in levitation.

Levitation permits various drop interactions to be observed. Thus, if two levitated drops (Fig. 2b) have opposite charges, we can force them closer and closer together by applying an external field until they finally coalesce. The fused drop remains levitated at a new height with the combined mass and charge. We have also directly observed drop-drop collisions. Often a levitated drop is struck from above by a heavier drop that is drawn into the beam as it falls. Alternatively, a levitated drop can be hit from below by lighter drops that wander into the lower regions of the beam where they are drawn in, are driven upward through the beam focus, and then encounter the levitated drop (see Fig. 2a). In these encounters we have often observed drop coalescence when the partners have opposite charges and misses when the drops have like charges. Drops can grow by a factor of 4 or 5 in diameter by successive collisions.

The simple experiments described here indicate the potential of the technique for studying the important problems of droplet growth by accretion in Langmuir type collisions with the use of cloud size droplets. In more sophisticated experiments prepared target and incident drops of known size and charge, each initially held in its own optical trap, could be used. They could subsequently be placed in the same beam and be brought together at varying speeds, with an electric field or light being used as the driving mechanism. The parameters of the collision, the particle trajectories, and the detailed fluid dynamics of the drop coalescence could be observed with high-speed movie cameras viewing from different angles. Finally we have made observations on the question of the coalescence efficiency of drops making physical contact. Occasionally we observed two drops roughly equal in size come side by side in the beam and seemingly touch for seconds prior to coalescence. This result should be checked with the use of movie cameras, as suggested above.

A. Ashkin
J. M. Dziedzic

Bell Laboratories,
Holmdel, New Jersey 07733

References and Notes

1. A. Ashkin, *Phys. Rev. Lett.* **24**, 156 (1970).
2. —— and J. M. Dziedzic, *Appl. Phys. Lett.* **19**, 283 (1971); *ibid.* **24**, 586 (1974).
3. A. Ashkin, *Sci. Am.* **226**, 63 (February 1972).
4. B. J. Mason, *The Physics of Clouds* (Oxford Univ. Press, London, 1971); C. N. Davies, Ed., *Aerosol Science* (Academic Press, London, 1966); Proceedings of the International Colloquium on Drops and Bubbles, California Institute of Technology, Pasadena, 28–30 August 1974.

14 November 1974

COVER

Three liquid drops levitated in a laser beam. The microscope is focused on the upper 10-micron drop located at a position of high light intensity in the diffraction rings formed about the two lower drops. See page 1073. [A. Ashkin, Bell Laboratories, Holmdel, New Jersey]

VOLUME 38, NUMBER 23   PHYSICAL REVIEW LETTERS   6 JUNE 1977

Observation of Resonances in the Radiation Pressure on Dielectric Spheres

A. Ashkin and J. M. Dziedzic

Bell Telephone Laboratories, Holmdel, New Jersey 07733

(Received 20 April 1977)

We report an experimental check of the Mie-Debye theory for the variation of radiation pressure on dielectric spheres with wavelength and size using optical-levitation techniques. Sharp resonances are observed which are shown to be related to dielectric surface waves. They permit particle-size measurement to a precision of 1 part in 10^5 to 10^6.

We report the first observation of the variation of the radiation-pressure force on transparent dielectric spheres with wavelength and size. We use a technique based on optical levitation which we call force spectroscopy. The measured force shows a regular series of sharp optical resonances which are in excellent qualitative agreement with the limited force data presently available from Mie-Debye theory. This resonant behavior can provide the most precise check on Mie scattering theory and also a way of measuring sizes of spheres to an accuracy exceeding that of present far-field scattering techniques by at least two orders of magnitude. These resonances are thought to be due to dielectric surface waves. This view is strongly supported by the observation of the scattered-light distribution in the near field. Both the resonant coupling of light striking the sphere edge and its subsequent isotropic tangential scattering are seen. These measurements should stimulate more precise calculation of radiation pressure and of the scattered-light distributions in the near field. Resonant effects are also of interest for the interpretation of particle scattering processes using optical models.

The Mie theory[1] for scattering of light by a sphere large compared to a wavelength is the best understood and most carefully checked example of scattering of waves by a particle. It is the basis of a vast theoretical and experimental literature and is widely used for particle-size measurement.[2] Using Mie theory, Debye[3] calculated expressions for the radiation-pressure force on a sphere as a function of the size parameter $X = 2\pi a/\lambda$, where a is the radius and λ the wavelength. Recently Irvine[4] made the first computer evaluation of the radiation pressure and extended computations of the total scattering cross section to large values of X using relatively high resolution (up to $\Delta X/X = 10^{-4}$). At low X, calculations of the total scattering cross section for low-loss spheres show the well-known "ripple structure"[5] as X is varied. Ripples are experimentally observed in measurements of the total scattering cross section for small x,[6] in far-field radar backscatter,[7] and in 90° scattering.[8] They are attributed to dielectric surface waves as originally proposed by Van de Hulst.[5] Irvine[4] shows that the ripple structure is larger on the radiation pressure than on the total scattering cross section and that at large X and high index n it sharpens dramatically and eventually becomes an unresolved sequence of resonances. Unfortunately he did not increase the resolution of his calculation further, since recent advances in the study of radiation pressure have now made possible the observation of Mie-Debye resonances with a resolution exceeding these existing calculations.

Indeed, with focused laser beams of modest powers one can use radiation-pressure forces to

VOLUME 38, NUMBER 23   PHYSICAL REVIEW LETTERS   6 JUNE 1977

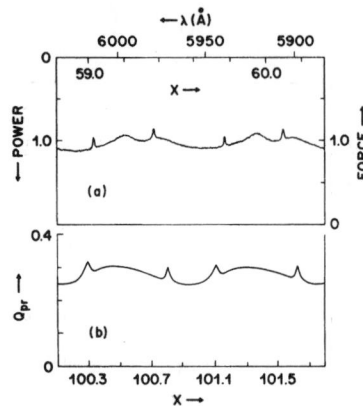

FIG. 1. (a) Measured power for levitation and approximate radiation-pressure force vs λ and X for a drop of $n = 1.40$. (b) Theoretical force Q_{pr} vs X, $n = 1.33$.

stably trap, levitate, and manipulate micronsized dielectric spheres.[9] Particles levitated against the forces of gravity act as sensitive probes for various applied forces,[10] e.g., electrical, gravitational, and optical. Recently an electronic feedback scheme[11] was devised to lock a levitated particle's height to an external reference by controlling the laser power. The laser power needed to hold the height constant is a continuous measure of any variations of force on the particle. Thus to measure the wavelength variation of the radiation-pressure force we use the feedback stabilizer with a tunable dye laser as the levitating laser. We then simply record the light power needed to hold the sphere at fixed height as λ is continuously tuned by a motor-driven birefringent plate rotating inside the dye-laser cavity. We call this technique of force measurement radiation-pressure–force spectroscopy. For most experiments we used liquid drops as test spheres because of their remarkable perfection. Drops were made from highly transparent, low-vapor-pressure silicone oil[10] having $n = 1.40$ and somewhat more volatile, in-

dex-matching oils with indices up to 1.53. To levitate drops in the size range of 4 to 30 μm required dye-laser powers in the range of 1 to 400 mW. Since the theory is calculated for plane waves, particles were held at a height in the beam where the intensity variation of the Gaussian beam across the particle was very small.

Figure 1(a) is a tracing from the chart recording of the levitating power vs λ for an 11.3-μm-diam silicone oil drop, shown upside down (with zero power on the top). Since the radiation-pressure force per unit power is proportional to the inverse of the levitating power this view gives a good approximate representation of the force variation vs λ for comparison with the theoretical computations of Irvine. The X scale is derived from the λ scale using an approximate value of a ($\pm 5\%$) as measured with a microscope. Figure 1(b) shows Irvine's theoretical force per unit power, Q_{pr}, versus X for $n = 1.33$ calculated with a resolution $\Delta X/X \cong 10^{-4}$. The experimental resolution $\Delta\lambda/\lambda$ is determined by the dye-laser linewidth of $\frac{1}{4}$ Å. Thus $\Delta\lambda/\lambda \cong 4 \times 10^{-5} = \Delta X/X$. Although the data were taken for $n \cong 1.40$ and $X \cong 60$, the resemblance to the theory is striking. We find much sharper resonances and more structure to the force variation for other size ranges of silicone drops (i.e., different X) and for drops of higher index of refraction. Figure 2 shows a section of one such curve for a slightly volatile index oil of $n = 1.47$. Since a scan over the laser tuning range takes $\sim \frac{1}{2}$ h and it takes minutes to change drops we can easily map out the variation of force with X and n. A range of X from ~ 10 to 200 is readily accessible. We believe the measured force to be highly accurate. The response of the laser-power monitor is constant with λ. Other than weak interference ripples from optics in the beam path there are no artifacts of the tunable laser in the data. Since the force depends solely on X for a given n we can, as a check, obtain identical force curves, appropriately shifted in λ, for different drop

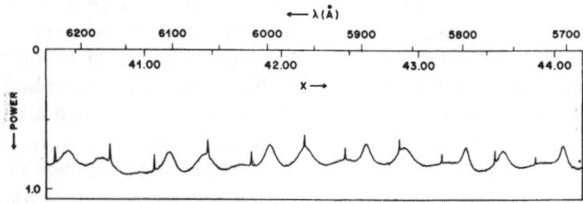

FIG. 2. Measured power for levitation vs λ and X for an approximately 8.0-μm-diam drop of $n = 1.47$. The diameter decreased $\sim 0.75\%$ by evaporation as the wavelength was increased during the scan.

VOLUME 38, NUMBER 23   PHYSICAL REVIEW LETTERS   6 JUNE 1977

sizes. We can also get force-vs-X data using a fixed λ and a continuously varying radius a by measuring the levitating power as drops of the more volatile index oils slowly evaporate. These data are completely free of interference ripples from the optics. With evaporating drops we can rerun parts of the force curve by decreasing λ and letting X repeat itself as a decreases. With evaporation we can view the force variation over wide ranges of X and observe the evolution of successive groups of interleaved resonances of gradually diminishing sharpness as X decreases. This variation in sharpness is in part seen in Fig. 2. Irvine noted similar behavior in low-resolution calculations (see his Figs. 2 and 3).

Experiments were performed to understand the nature of the observed resonances. By simultaneously observing the force and the 90° scattered light in the near field from two orthogonal directions, the so-called parallel and perpendicular polarization directions, it is clear that the resonances fall into two interleaved sets depending on incident polarization. Next, if these resonances are due to Van de Hulst–type surface waves they should be excited by light striking the sphere edge tangentially. To check this we manipulated the drop with the feedback stabilizer away from its usual plane-wave position to a location at the focus of a strongly convergent beam where the beam completely misses the edges of the sphere and the sphere is supported solely by light passing through its axis.[11] In this case all sharp resonances disappear and we are left with a broad force variation of period $\Delta X \cong \pi/2n$ due to interference from the weak front- and back-surface reflections of the sphere. This is strong evidence for the surface-wave picture. The fact that edge illumination can cause strong surface-wave resonances at first sight violates one's intuition based on the high reflectivity of rays striking the sphere at nearly tangential incidence. However, viewing the sphere as a resonant, low-loss optical cavity, it is clear that the buildup of internally circulating light can modify the reflectivity sufficiently to give strong coupling into the sphere. Light thus coupled into the sphere should then be scattered isotropically, emerging tangentially. This suggests looking for direct visual evidence of surface-wave coupling and tangential scattering by near-field observations. Indeed tangential scatter is clearly seen in the backward direction. Not only do the edges of the sphere brighten at resonance but we can distinguish perpendicular and parallel resonances. When view-

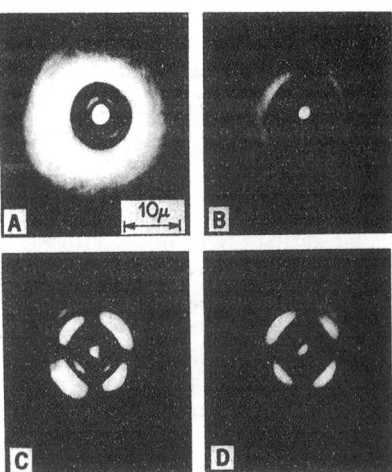

FIG. 3. Near-field views in the forward direction of an 11.3-μm-diam levitated drop: (a) no polarizer (exposure $E = \frac{1}{16}$ sec); (b) through a crossed polarizer, off resonance ($E = \frac{1}{4}$ sec); (c) through a crossed polarizer, on resonance for parallel polarization ($E = \frac{1}{8}$ sec); (d) through a crossed polarizer on resonance for perpendicular polarization ($E = \frac{1}{8}$ sec).

ing in the forward direction [Fig. 3(a)] we can see the coupling process by using a crossed polarizer placed in front of the near-field image of the sphere. This blocks almost all transmission when off resonance [see Fig. 3(b)]. When tuned on resonance for one of the polarization components, the sphere by absorbing the resonant component of the light striking the edge and passing the orthogonal nonresonant polarization, in effect acts as a rotated polarizer inserted between two crossed polarizers and gives transmission. Since the effective angle of the sphere "polarizer" varies with azimuth and the absorption occurs only near the sphere edge, we expect four arcs of transmission with the maximum intensity at 45°. This is what is observed; see Figs. 3(c) and 3(d).

Consider the accuracy of size measurement and the related question of the accuracy of Mie-Debye theory. Assuming that we had a high-resolution curve of theoretical force vs X for the given n, we could then determine a by identifying each of a sequence of measured resonances with its corresponding X value. This comparison is unambiguous because of the distinctiveness of the shape variations of the force curve (see Fig. 2). From the theoretical X for each resonance and the measured λ we get a by using $a = X\lambda/2\pi$. The accuracy of each radius determination, $\delta a/a$, is

VOLUME 38, NUMBER 23   PHYSICAL REVIEW LETTERS   6 JUNE 1977

equal to $\delta\lambda/\lambda$, the accuracy of the wavelength determination. Since we have observed resonant widths of $\sim\frac{1}{4}$ Å and can determine the resonance peak to $\delta\lambda\cong\frac{1}{5}$ of full width, we expect an accuracy $\delta a/a\cong(\frac{1}{20}$ Å$)/(6000$ Å$)\cong 1:10^5$. The extent to which a values from different resonances are the same tests the internal consistency of Mie-Debye theory. In the absence of theoretical curves we determined the ratio of radii of two silicone drops roughly 13.1 μm in diameter by comparing a sequence of ten sharp resonances. These data give the same ratio 1.007 22 for all ten resonances to a rms accuracy of ± 0.000 03. The precision of 3 parts in 10^5 can be improved. It still shows the high accuracy and internal consistency of the resonance method. Even higher accuracies are possible in measuring size changes of a single particle. By tuning to a point of steep slope on the side of a sharp resonance we have easily detected changes of $\sim 10\%$ of the resonance height as a volatile drop evaporated. This corresponds to a change $\delta X\cong\frac{1}{10}$ of the resonance width which gives a sensitivity $\delta a/a = \delta X/X$ of 1 part in 5×10^5. If $a = 5$ μm, then $\delta a\cong 0.1$ Å. Since δa is considerably less than a monolayer this is interpreted as an average radius change. For comparison the accuracy of standard Mie-theory radius measurements[12] based on the far-field angular distribution of scattered light is only $\sim 1\%$ using Mie fringe-counting techniques. By comparing intensities with theoretical distributions this can be improved to $\sim 1:10^3$.

The high precision of force spectroscopy is useful for measurement of evaporation and condensation; study of the deposition of monolayers; optical detection of drop oscillations and minute drop distortions; and precise determination of electronic charge in a Millikan experiment. At higher values of n we expect even narrower resonances. Eventually drop distortion due to the radiation pressure itself should broaden the resonances. Perhaps we can then use highly precise glass spheres. An understanding of Mie-Debye resonances is clearly necessary for accurate measurement of radiation pressure and for proper formulation of optical analogs[13] of molecular, nuclear, and atomic scattering.

In conclusion we believe that radiation-pressure–force spectroscopy can provide the most exacting test of Mie theory. We have visually observed dielectric surface waves and demonstrated measurement of sphere sizes to an accuracy of 1 part in 10^5 to 10^6.

---

[1]G. Mie, Ann. Phys. (Leipzig) 25, 377 (1908).

[2]H. C. van de Hulst, *Light Scattering by Small Particles* (Wiley, New York, 1957); M. Kerker, *The Scattering of Light and Other Electromagnetic Radiation* (Academic, New York, 1969).

[3]P. Debye, Ann. Phys. (Leipzig) 30, 57 (1909).

[4]W. M. Irvine, J. Opt. Soc. Am. 55, 16 (1965).

[5]See van de Hulst, Ref. 2, Chap. 17; and Kerker, Ref. 2, Chap. 4.

[6]W. Heller, in *Electromagnetic Scattering*, edited by M. Kerker (MacMillan Co., New York, 1963); and Kerker, Ref. 2, Chap. 7.

[7]P. S. Ray and J. J. Stephens, Radio Sci. 9, 43 (1974).

[8]R. A. Dobbins and T. I. Eklund, Appl. Opt. 16, 281 (1977).

[9]A. Ashkin, Phys. Rev. Lett. 24, 156 (1970); A. Ashkin and J. M. Dziedzic, Appl. Phys. Lett. 19, 283 (1971), and Science 187, 1073 (1975).

[10]A. Ashkin and J. M. Dziedzic, Phys. Rev. Lett. 36, 267 (1976), and Appl. Phys. Lett. 28, 333 (1976).

[11]A. Ashkin and J. M. Dziedzic, Appl. Phys. Lett. 30, 202 (1977).

[12]H. H. Blau, Jr., D. J. McCleese, and D. Watson, Appl. Opt. 9, 2522 (1970).

[13]V. Khare and H. M. Nussenzveig, Phys. Rev. Lett. 33, 976 (1974); U. Buck, Rev. Mod. Phys. 46, 369 (1974); H. C. Bryant and N. Jarmie, Ann. Phys. (N.Y.) 47, 127 (1968).

Reprinted from **Applied Optics,** Vol. *20*, page 1803, May 15, 1981.
Copyright © 1981 by the Optical Society of America and reprinted by permission of the copyright owner.

Observation of optical resonances of dielectric spheres by light scattering

A. Ashkin and J. M. Dziedzic

Use of the wavelength and size dependence of light scattering from optically levitated liquid drops is demonstrated as a sensitive means of detecting optical resonances of dielectric spheres. High resolution spectra are presented of the radiation pressure, far- and near-field backscatter, and 90° scatter. Excellent agreement is found between experimental spectra and high resolution Mie calculations of Chylek *et al.* Strong evidence supporting the van de Hulst dielectric surface-wave model for these resonances is presented. Use of resonances for high precision measurement of sphere size and sphere distortion, index of refraction, temperature, and vapor pressure is discussed.

I. Introduction

This paper describes new high resolution measurements of the resonant behavior of light scattering from dielectric spheres. We show that scattered light can be used as a sensitive means of detecting dielectric surface-wave resonances of spherical particles. Interest in high resolution measurement of the resonant behavior of spheres has been stimulated in part by the recent realization that dielectric spheres have many sharp resonances with a distinctive spectroscopy,[1] which makes possible relative and absolute size measurements to new precision, and in part by its importance to the basic understanding of light scattering from this fundamental scattering particle. In that earlier resonance work[1] measurements were made of the radiation pressure and scattering using single optically levitated dielectric spheres having diameters of ~10 μm. This is large compared with the wavelength and therefore in the Mie-Debye scattering regime. Sharp optical resonances were demonstrated using a technique called force spectroscopy in which the radiation pressure force needed to support a particle at a given height in a Gaussian laser beam is recorded as the wavelength of the light is continuously varied. Data were taken with a resolution exceeding any theoretical computation of

Mie scattering for spheres of this approximate size at that time. The data revealed a complex spectroscopy of readily identifiable resonances whose wavelength varied with sphere size. At their narrowest these resonances had halfwidths $\Delta\lambda$ so that $\Delta\lambda/\lambda \cong 10^{-5}$. The remarkable discovery, made experimentally from these data, was that the spectrum of these resonances was sufficiently distinctive, in spite of apparent periodicities, to distinguish one resonance from another. This makes it possible to use the sharpest of the resonances to measure relative and absolute sizes of spheres to a precision of 1 part in 10^5–10^6. Absolute size measurement involves a comparison of experiment with theory. Relative measurements simply involve the wavelength ratios of the same resonance for different size spheres. The accuracy of the resonance technique is 2–3 orders higher than conventional size measuring techniques based on the angular distribution of Mie-scattered light at a fixed wavelength. The observed sharp resonances were qualitatively attributed to the so-called van de Hulst dielectric surface waves,[2] which couple in and out of spheres tangentially and can circulate around with very low losses. These waves propagate partly on the surface by diffraction and partly on so-called shortcuts traversing the sphere at the critical angle. When these waves close on themselves, in phase, resonance occurs. Experimental evidence of resonant coupling at the sphere edges was found. Such behavior must, of course, be inherently contained in the complete Mie-Debye electromagnetic theory for scattering from spherical particles. Indeed a recent high resolution Mie-Debye computation[3] of the radiation pressure force spectrum using increments in size parameter Δx so that $\Delta x/x = \Delta\lambda/\lambda \cong 10^{-7}$ (where $x \equiv 2\pi r/\lambda$ and r is the sphere radius) gave remarkable agreement with all the experi-

The authors are with Bell Laboratories, Holmdel, New Jersey 07733.

Received 20 December 1980.

0003-6935/81/101803-12$00.50/0.

© 1981 Optical Society of America.

mentally observed resonances. From a theoretical point of view each of the experimentally observed resonances could be ascribed to a resonance in the partial wave expansion of the scattering amplitude. In addition the computation showed another set of still narrower resonances of width $\Delta\lambda/\lambda \cong 10^{-7}$ due to the first-order partial wave resonances which were not observed experimentally, implying the possibility of even higher precision size measurements.

Historically van de Hulst introduced the notion of dielectric surface waves as a physical explanation of what is termed the ripple structure in the variation of the extinction cross section with size parameter. He also invoked surface waves propagation to explain features of the backscatter[2] of light not interpretable by means of simple ray optics. Surface wave concepts were given additional validity by analyses of Walstra[4] and Irvine[5] on the ripple structure and by Bryant and Cox[6] and Khare and Nussenzveig[7] on the relation of the intuitive concept of edge coupling of dielectric surface waves to Mie theory. Experiments on optical backscatter from approximately millimeter-sized hanging water droplets[8] showed resonant behavior as they evaporated, which was definitely correlated with a full Mie theory backscatter calculation. There were experimental problems with drop distortion and mechanical vibrations. These are also some microwave radar backscatter measurements and theoretical analyses,[9] but the resolution of the data and the computations was rather coarse. Attempts were made to interpret the microwave results in terms of surface waves.[10] Additional evidence for the van de Hulst surface-wave picture comes from the largely theoretical work on backscatter of short pulses from dielectric spheres,[11] where one can in principle observe time-resolved responses from different surface-wave modes.

From an experimental point of view, the introduction of optical levitation techniques[12–15] for supporting individual dielectric spheres at rest in a light beam has greatly improved the capability for making precise light-scattering measurements on spherical particles. With levitation, particles are held free of interfering obstacles where they can be observed in detail in the near and far field. Highly transparent liquid drops in the ~(1–100)-μm size range are almost ideal spherical test particles. Surface tension is so strong that distortions in shape are minimized.

In this work we explicitly study various ways of detecting the resonant behavior of spheres. Clearly a resonance in the radiation pressure on a sphere implies a resonant change in the light scattered from the sphere, since the optical force on a particle arises directly from the scattering process.[1] This is made clear in simultaneous measurements of the wavelength dependence of the levitating force, the far-field backscattered light, and the far-field 90° scattered light perpendicular and parallel to the polarization of the incident light. As will be seen the magnitude of the resonances relative to the background is by far the greatest as observed in the backscattered light. Comparison of the observed

backscattered light to preliminary computations[16] carried out with high resolution using Mie-Debye theory shows remarkable agreement. This agreement not only serves as a check on both the experiment and the computation but also confirms our previous notions about the perfection of liquid drops in this micron size range.[1] By looking at the near-field light distribution emerging directly from the sphere we are able, by aperturing, to identify light from the sphere edges as the most resonant part of the scattered light. This is an aid in the observation of weaker resonances and also gives additional support for the surface-wave resonance pictures. Further support for this picture comes from data which also identify the edge as the point where resonant light is coupled into the sphere. Data on the resonant behavior of radiation pressure and scattered light are also presented using a laser of fixed wavelength in which the diameter of a levitated liquid drop is allowed to change continuously by evaporation.

Study of surface-wave resonances we believe opens a new and important field of application of Mie-Debye scattering. The sensitivity of sharp surface-wave resonances to small changes makes them one of the most sensitive probes for determining the parameters of spheres. For a fixed particle one can in principle measure absolute size, index of refraction, and dispersion by varying the wavelength of the input light and determining the spectrum of the resonances. This assumes the existence of a sufficiently complete compendium of high resolution theoretical scattering resonance data for varying size parameter and index of refraction with which to compare experimental results. The incentive for compiling such a compendium we believe now exists. Relative size measurement involves detection of shifts in the resonance wavelength of spheres. A sensitivity to radius change of a fraction of an angstrom can be achieved.[1] A possible application is to the study of the deposition of monomolecular layers on liquid drops. Another interesting use of resonances is for measurement of temperature. Shifts in resonant wavelength with temperature[17] result from the combined effect of the temperature coefficient of expansion and change of index of refraction. If temperature changes are produced by absorption of the incident laser light, we get a measure of the optical absorption coefficient of the particle. Also, for liquid drops in equilibrium with a surrounding vapor one can hope to use shifts in particle size as detected by shifts in the resonances as a measure of vapor pressure. In recent preliminary experiments[17] shifts in resonances were used to monitor minute distortions of liquid drops produced by externally applied electric fields. This gives a measure of the surface tension of the liquid. Surface-wave resonances can also play a role in inelastic scattering from dielectric spheres. For example, in observations of dye fluorescence or Raman effect from spheres, we expect sphere resonances to modify the wavelength dependence of the scattered light. Resonance effects have recently been observed using dye impregnated spheres.[17,18] At sufficiently high incident light intensities one can conceivably obtain lasing on the

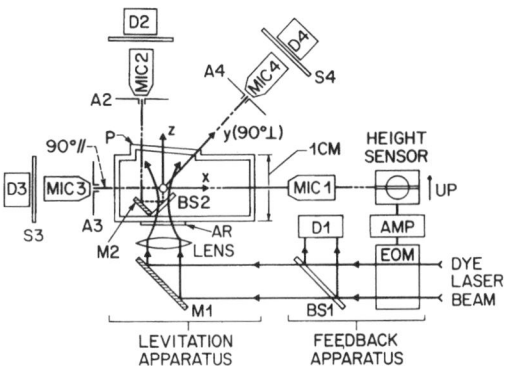

Fig. 1. Sketch of levitation and feedback apparatus.

high Q resonances of such active particles as dye impregnated spheres.

II. Experimental Apparatus and Techniques

Figure 1 is a sketch of the experimental apparatus used for levitating liquid drops and for measuring the levitating force and scattered light as a function of wavelength. Drops are introduced into the glass box having 2.5×7.5 (1×3-in.) $\times 1$-cm dimensions through a pinhole in the top with a spraying technique[13] as described earlier. The vertically directed levitating beam was focused typically by an ~6-cm focal length lens. Levitation of an ~10-μm oil drop requires ~15-mW of power. Drops, once captured by the light, are moved to a clean area of the base plate which is free of debris from the pinhole.

Measurements of the spectrum of the force of radiation pressure on a drop are performed using the automatic feedback system[14] developed previously. In this technique the power needed to stabilize a given levitated particle at a fixed height P_{LEV} is measured as the wavelength of the levitating dye laser beam is continuously scanned. This power is inversely proportional to the levitating force. A microscope labeled $Mic1$ projects an image of the particle on a split photodiode height sensor which generates an error signal if the particle tries to move. The error signal is appropriately amplified and used to drive an electrooptic modulator (EOM) which controls the input laser power to maintain levitation at the reference height. The levitating power P_{LEV} is monitored by broadband detector $D1$ and recorded on a time swept chart recorder. The lower surface of the box has an antireflection coated plate (AR), attached to it with index-matching fluid to avoid Fabry-Perot ripples on the data due to the base plate acting as an etalon located beyond the power monitor $D1$.

Backscattered light and 90° scattered light from levitated drops can be measured in the near and far field with the help of microscopes $Mic2$, $Mic3$, and $Mic4$. For incident light polarized along the x axis, $Mic4$ viewing along the y axis detects only x-polarized scattered light. We call this the 90° parallel ($\parallel$) direction.

$Mic3$, viewing along the $-x$, axis, detects only z-polarized light. This is called the 90° perpendicular ($\perp$) direction. $Mic2$ looks in the $-z$ direction at the backscattered light, which has its polarization maintained in the x direction. The combination of beam splitter $BS2$ and mirror $M2$ was to reduce the working distance from the drop to $Mic2$. The top plate P of the box in the neighborhood of the drop was tilted to prevent backscatter from the plate from reaching $Mic2$. By focusing the microscopes right on the sphere and projecting the enlarged images directly on screens $S2$, $S3$, and $S4$, one can obtain a complete near-field view of all the light emerging from the sphere within the acceptance angle of the microscopes. The images are bright enough to be easily observed visually or photographically. To obtain high magnification and good near-field spatial resolution, we typically use microscope objectives of $\sim32 \times$ with a full acceptance angles of $\sim35°$.

Far-field intensity measurements can be obtained by removing the screens and using detectors $D2$, $D3$, or $D4$ to measure all the light entering the aperture of the microscope. To improve the angular resolution of the far-field measurements, we introduced apertures $A2$, $A3$, and $A4$ in front of the microscope objectives to reduce the acceptance angle of the light. This, of course, degrades the near-field resolution. The small acceptance angle is especially important for far-field backscatter measurements where we found that the intensity is extremely sensitive to angle. This sensitivity makes it difficult to line up the detection apparatus on the exact backward direction. This was done by first visually observing the high resolution near-field backscatter pattern and sensitively adjusting the direction for the most symmetrical shape. Aperture $A2$ was then inserted for the high resolution far-field measurements.

With this apparatus one can simultaneously measure the spectrum of the levitating force, the far-field 90° $\parallel$ and 90° $\perp$ scattered light, and the far-field backscattered light as the incident dye laser beam is mechanically scanned with a motor driven birefringent plate. A comfortable scanning rate of ~15 Å/min is used. Spectral data were also recorded of the near-field light. In this type of measurement we first visually observe the full near-field light distribution and then sample selected portions of this light pattern by placing detectors $D2$, $D3$, and $D4$ behind appropriately located small pinholes.

Measurement of resonance effects by evaporation was performed on drops made up of slightly volatile liquids such as glycerol and water mixtures and some index-matching oils. In this technique we hold the drop in the feedback stabilizer with a fixed laser wavelength (such as 5145 Å from an argon-ion laser) and record the change in levitating force and scattered light as the size parameter varies continuously by evaporation.

III. Observations and Discussion

Figure 2 shows simultaneous recordings for a 11.4-μm silicone oil drop of the wavelength dependence of the power to levitate at a fixed height P_{LEV}, the far-field

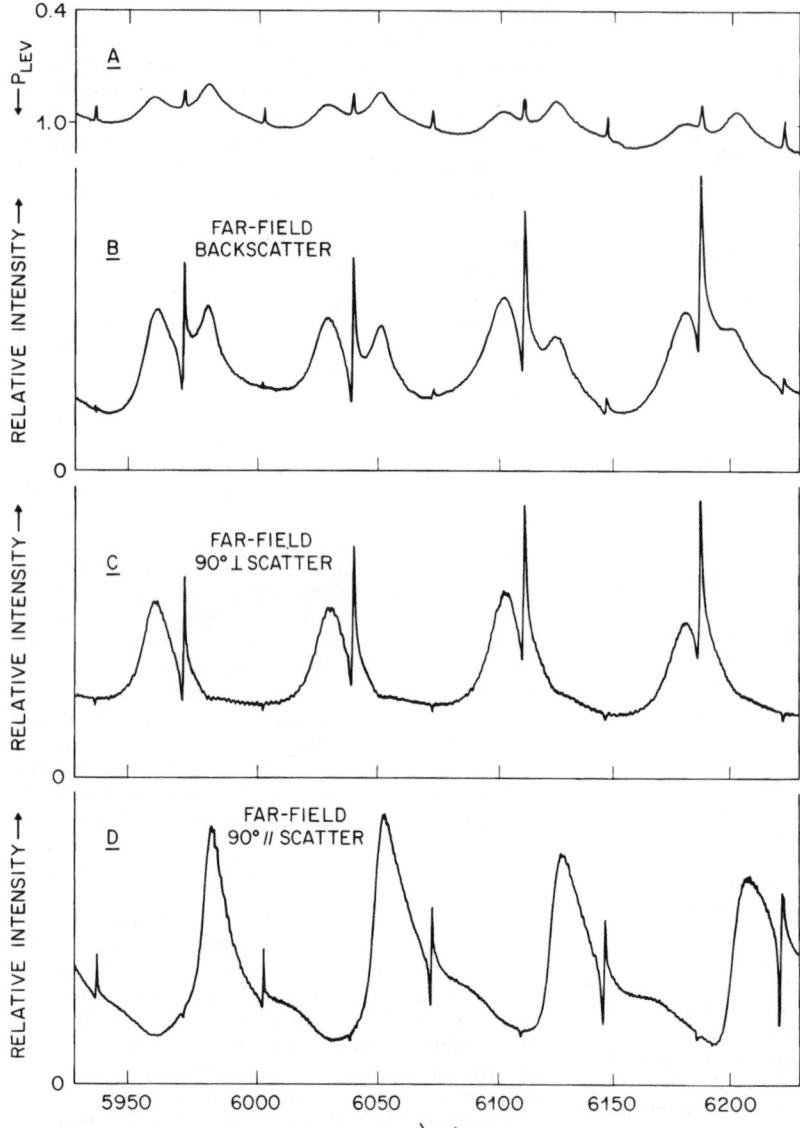

Fig. 2. Wavelength dependence of (A) P_{LEV}, (B) far-field backscatter, (C) far-field 90° ⊥ scatter, and (D) far-field 90° ∥ scatter.

backscatter, the far-field 90° ⊥ scatter, and the far-field 90° ∥ scatter. The scattering data were taken with ~1 × microscope objectives having an acceptance angle of ~4°. Comparison of Figs. (A) and (B) shows that each characteristic evolving sequence of moderate and high Q resonances observable in the power to levitate or the force[1,3] is also detectable in the far-field backscatter. Most of the resonances, as observed in backscatter, are detected with greater contrast relative to the background. The data from the two orthogonal 90° directions (C) and (D) make clear that the resonances fall into two relatively distinct polarization classes, the so-called perpendicular and parallel resonances.

A preliminary comparison[16] was recently made of the experimentally determined P_{LEV} and backscatter spectrum of silicone oil as shown in curves (A) and (B) of Fig. 2 with high resolution Mie computations. Curves (D) and (E) of Fig. 3 show the theoretical computation of P_{LEV} and backscatter for silicone oil in this range of size parameter x. The results of this comparison leave no doubt about the proper correspondence between particular experimental and theoretical resonances. The agreement between experiment and theory for P_{LEV} is excellent. The experimental backscatter in Fig. 2(B), however, exhibits an anomalously high background level of light not evident in the theory

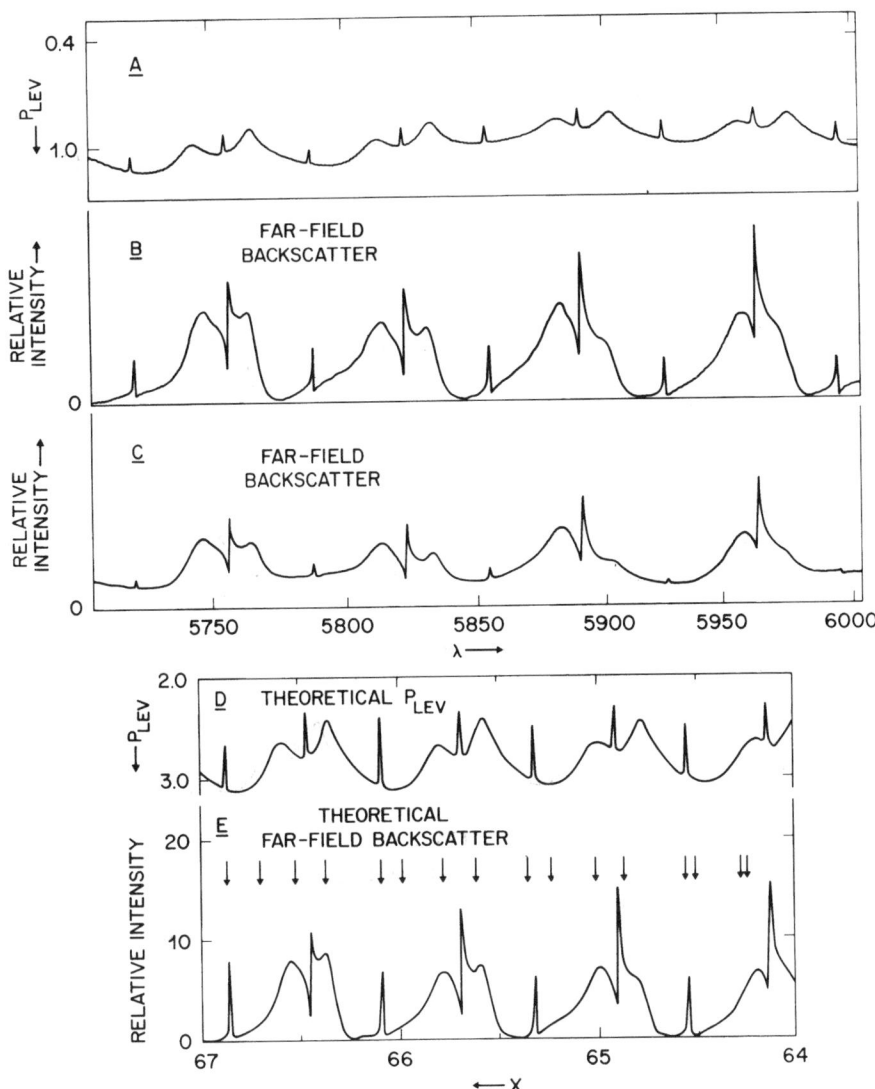

Fig. 3. (A) P_{LEV} spectrum, (B) backscatter with an ~1.5° acceptance angle, (C) backscatter with an ~4.0° acceptance angle, (D) theoretical P_{LEV}, and (E) theoretical backscatter.

[Fig. 3(E)]. The origin of this discrepancy was discovered to lie in the size of the acceptance angle of the backscatter detection apparatus. Figures 3(A), (B), and (C) show data on the P_{LEV} spectrum and the backscatter taken on a slightly larger sized silicone oil drop using two different acceptance angles, ~1.5° for curve (B) in Fig. 3 and ~4.0° for curve (C) in Fig. 3. The narrower acceptance angle brings the background light down almost to zero and makes the smaller of the high Q resonances much more visible. Compare, for example, the 5855- and 5925-Å resonances in Figs. 3(B) and (C). Further reductions in the acceptance angle below ~1.5° results in little change in shape. The significant differences in shape between curves (B) and (C) in Fig. 3 are an indication of the great sensitivity of the backscatter to angle. The agreement between the experimental data of Fig. 3(B) and the theory in Fig. 3(E) is now excellent. In comparing the detailed shapes of curves (B) and (E) in Fig. 3, we should note that the theoretical backscatter curve (E) in Fig. 3 is for constant power, while the experimental backscatter curve (B) in Fig. 3 is for constant force. One must therefore correct the vertical scale of (E) in Fig. 3 slightly to reflect the power variation in curve D of Fig. 3 before comparing with experiment. This small correction to Fig. 3(E) makes agreement with Fig. 3(B) even closer.

The very small differences in shape that remain in the comparison of the high resolution backscatter data with theory could be due to slight residual misalignments or more likely to small departures of the scattering beam from an ideal plane wave. Indeed the actual focused Gaussian scattering beam has some phase front curvature and intensity variation across the particle. The phase front curvature can most easily be reduced by decreasing the far-field diffraction angle. For the 6.3-cm focal length lens typically used, and our dye laser beam, the far-field diffraction half angle $\cong 1.7°$. In practice we reduce the intensity variation across the particle a great deal by keeping the levitation beam large compared with the particle size.

The 90° side-scatter data in Figs. 2(C) and (D) also show an unexpected feature. This is the weak appearance in the 90° $\perp$ scatter spectrum of each of the sharp 90° $\parallel$ resonances as, for example, at ~ 6003, ~ 6072, and ~ 6145 Å. Similarly the sharp 90° $\perp$ resonances at ~ 6040, ~ 6111, ~ 6188 Å, etc. show up weakly in the 90° $\parallel$ spectrum. This is unexpected from symmetry considerations and indeed did not appear in preliminary theoretical computations.[19] The presence of these weak resonant features is apparently not related to the size of the acceptance aperture or to deviations from the exact 90° direction, since the 90° patterns were found experimentally to be quite insensitive to aperture size and angle about the 90° direction. Drop distortion due to the supporting levitation force could in principle result in departures from symmetry; however, we believe this effect to be rather small. Deviation of the scattering beam from an ideal plane wave could be the source of the aberrations in the 90° spectra.

To see if small deviations from a plane wave beam are important, one can conceive of a refinement of the levitation technique used above. One could support a height stabilized sphere with a fixed wavelength beam and use a superimposed weak, power stabilized, tunable plane wave dye laser probe for scattering measurements. Such a two-beam technique would be generally applicable to all scattering experiments where very careful comparisons are to be made with plane wave theory. On the other hand, there also exists a simplified less precise version of our scattering technique involving only a single-tuned dye laser levitating beam with no feedback stabilization apparatus. As the incident power varies with tuning, the sphere simply moves up and down in the levitating beam, always coming to rest where the incident power gives a constant force equal to the particle's weight. If the incident light intensity distribution at the particle continues to be reasonably approximated by a plane wave as the height changes, and if the far-field detection optics is made insensitive to the vertical displacements of the particle, the resulting spectrum will be closely the same as with the automatic feedback height stabilizer.

An important point in the comparison of Figs. 3(A) and (B) with Figs. 3(D) and (E) is that, as previously found,[3,16] the theoretical computation shows the existence of even sharper resonances which were undetected experimentally. Previously at $n = 1.47$ there was one set of undetected resonances[3] with computed widths of less than the $\sim 1/4$-Å resolution of the dye laser used. Now there are two such undetected sets for $n \cong 1.40$ and this range of size parameter. The positions of these resonances are simply marked by arrows on the theoretical curve of Fig. 3(E). They apply to Fig. 3(D) as well. More will be said about the undetected resonances later.

Figure 4 shows the P_{LEV} (or force) spectrum and backscatter spectrum of a slightly volatile oil drop of $n \cong 1.47$ and diameter $\cong 11.4$ μm for comparison with the silicone oil data at $n \cong 1.40$ in Fig. 2. The larger n gives a richer spectrum. As in the earlier data[1] there is probably only one set of narrower undetected resonances. Figure 4 gives enough of the spectrum to show clearly the evolution of the well-defined characteristic sequences of resonances that appear out of the background, grow, and gradually decay as the wavelength or size parameter changes. The data were taken with an $\sim 4.0°$ far-field acceptance angle.

We now discuss observations on backscatter and 90° scatter in the near field. Figures 5(A) and (B) show near-field photographs of an ~ 12.0-μm diam silicone oil drop as seen in backscatter at wavelengths corresponding to a sharp parallel resonance and perpendicular resonance, respectively. This was taken with an acceptance angle of $\sim 35°$. Figures 6(A) and (B) gives sketches identifying various features of the backscattered light as seen in Fig. 5. The opposing pairs of arcs labeled $\perp RES$ and $\parallel RES$ in Fig. 6(A) are interpreted as arising from resonant surface wave light emitted tangentially at the sphere edges as indicated by rays R in Fig. 6(B). The existence of the arcs is consistent with the fact that the $\perp$ and $\parallel$ surface waves couple strongly to the sphere at those points of the surface where the polarization of the incident light is, respectively, $\perp$ or $\parallel$ to the surface. This view of the coupling-in process was also used to interpret the related arclike patterns observed in the forward direction through a crossed polarizer as discussed in Ref. 1. The central spot, labeled Axial Reflection in Fig. 6(A), comes from the combined retroreflection from input and output surfaces of the drop at close to normal incidence [see schematically drawn rays labeled a and b in Fig. 6(B)]. In Fig. 5 we see that tuning to either a $\parallel$ or $\perp$ resonance results in enhancement of the appropriate pair of surface wave arcs relative to the remaining features. By placing apertures in the image plane of the backscattered light as sketched in Fig. 6(A) at locations $L1$, $L2$, and $L3$, one can sample light and measure spectra from the specific features mentioned above to the almost total exclusion of the others.

Figure 7 shows data for a 16.6-μm diam silicone oil drop comparing the P_{LEV} spectrum (A) with spectra (B) and (C) due to backscattered light coming from the two pairs of arcs at apertures $L1$ and $L2$. Whereas the P_{LEV} spectrum contains all resonances, the spectrum at $L1$ consists of only $\parallel$ resonances, and the spectrum at $L2$ consists of only $\perp$ resonances. The fact that spectra (B) and (C) do not fall to zero at wavelengths between major resonances can again be attributed to the large

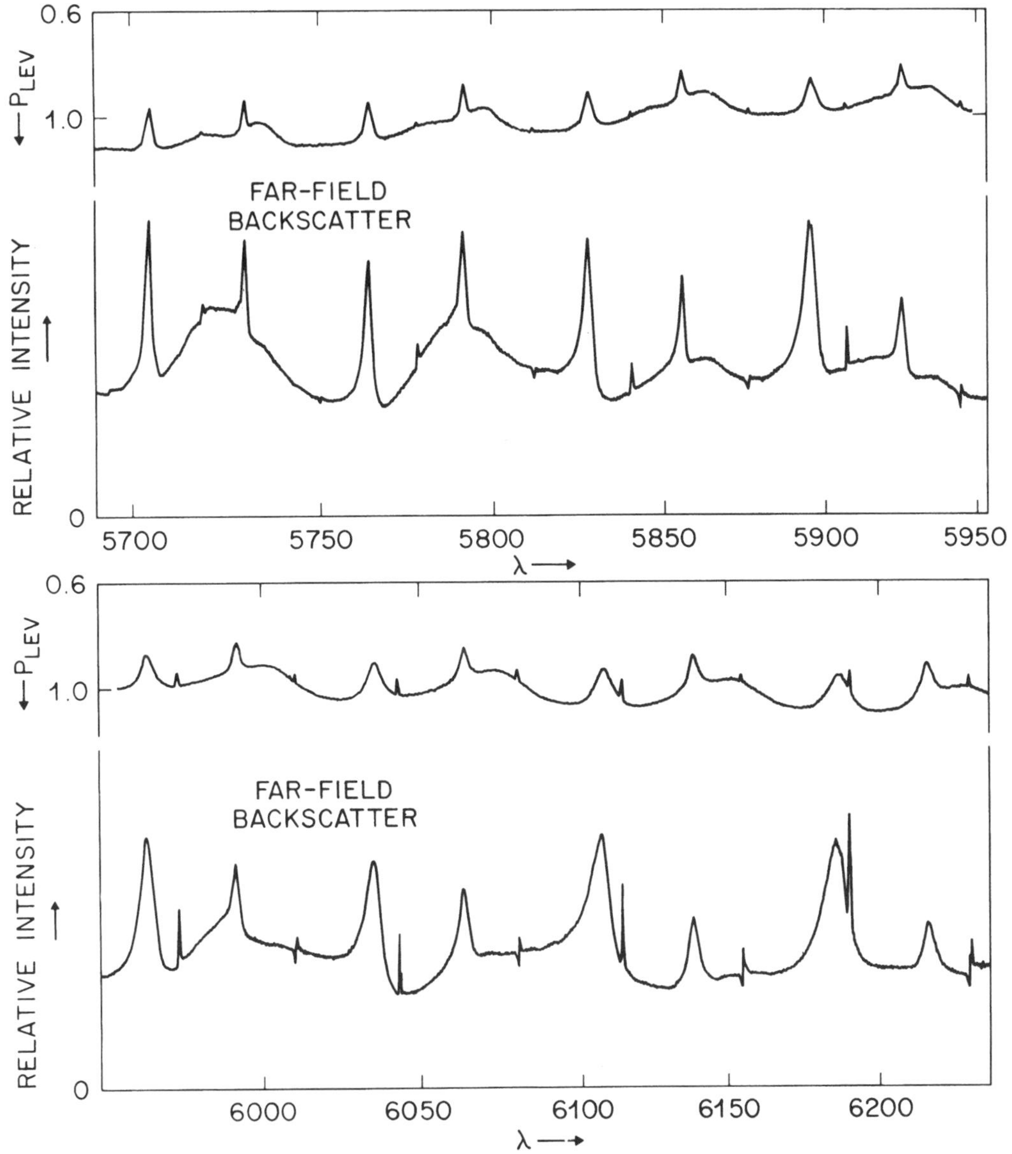

Fig. 4. P_{LEV} and backscatter spectrum for $\cong 11.4$-μm diam oil drop of $n \cong 1.47$.

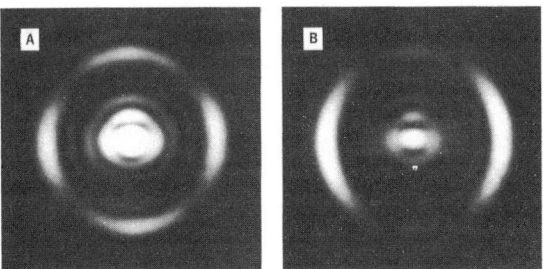

Fig. 5.   Near-field backscatter photographs of a 12-μm diam silicone oil drop (A) at a sharp parallel resonance (B) at a sharp perpendicular resonance.

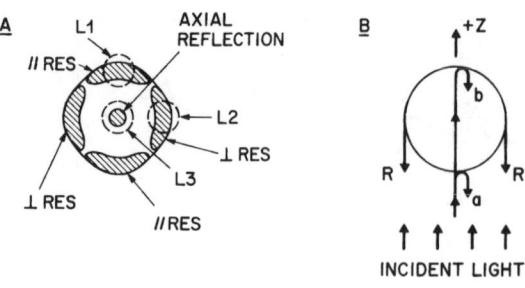

Fig. 6. (A) Sketch identifying various features of backscatter light in Fig. 5; (B) sketch of the origin of the axial reflection and resonant surface-wave light.

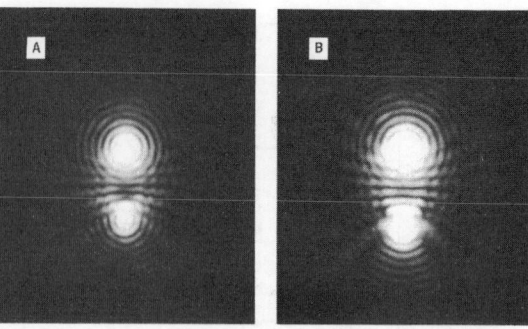

Fig. 8. Near-field photographs of 90° perpendicular light (A) when tuned off resonance and (B) when tuned onto a sharp resonance.

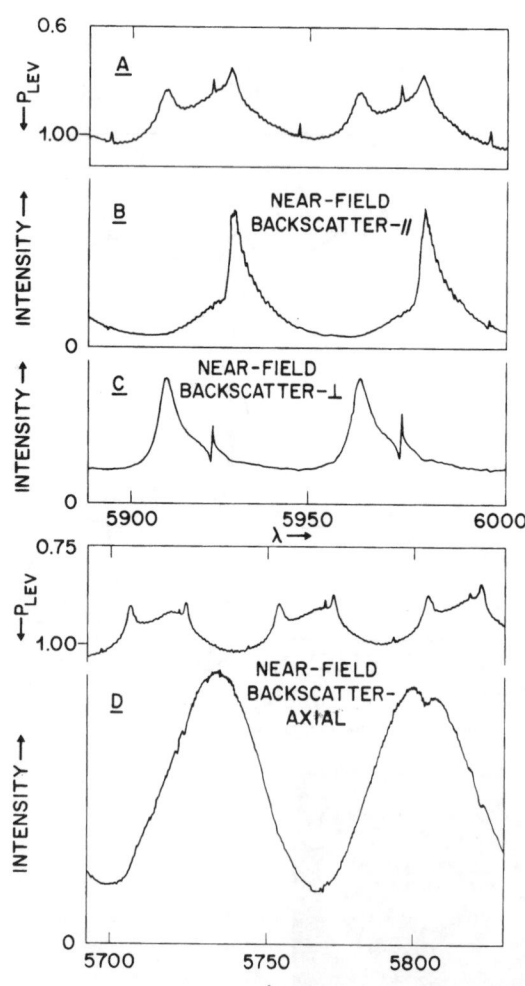

Fig. 7. (A) $P_{\rm LEV}$ spectrum; (B) parallel resonances in backscattered light as seen at aperture $L1$; (C) perpendicular resonances in backscattered light as seen at aperture $L2$; and (D) $P_{\rm LEV}$ spectrum and axially reflected light as seen at aperture $L3$.

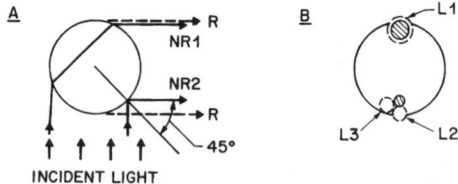

Fig. 9. (A) Sketch of 90° scattered light and (B) location of apertures for recording near-field light.

acceptance angle of the microscope. As seen from Fig. 5, the spatial resolution of the images is very high and could be considerably reduced and still resolve the ⊥ and ∥ arcs. Such a reduction should enhance the contrast of the sharp resonances of (B) and (C) as we observed in the far-field backscatter data. Spectrum (D) from aperture $L3$, sampling the central spot, barely shows any evidence of surface-wave resonances. The broad Fabry-Perot-type pattern that appears is just the low Q interference from input and output surface reflection a and b shown in Fig. 6(B).

Figure 8 shows near-field photographs of an ~13.5-μm silicone oil drop taken in the 90° ⊥ direction when tuned off resonance [Fig. 8(A)] and when tuned onto a sharp resonance [Fig. 8(B)]. The general features of Figs. 8(A) and (B) are the same, consisting of two main spots of light. Tuning from Figs. 8(A) and (B) results in some changes in the detailed structure of the two spots and in the appearance of a weak fan of light principally below and to the side of the lower spot. Figure 9(A) identifies the two main spots as an upper 90° nonresonant refracted ray ($NR1$) and a lower nonresonant 90° reflected ray ($NR2$). Resonant surface-wave emission is expected to appear tangentially from the top and bottom edges as indicated by dashed lines R in Fig. 9(A). Also indicated in Fig. 9 are the locations $L1$, $L2$, and $L3$ where apertures were placed for recording scattering spectra from restricted portions of the 90° near-field light. Figure 10 shows 90° ⊥ scattering spectra taken on a 14.1-μm diam silicone oil drop.

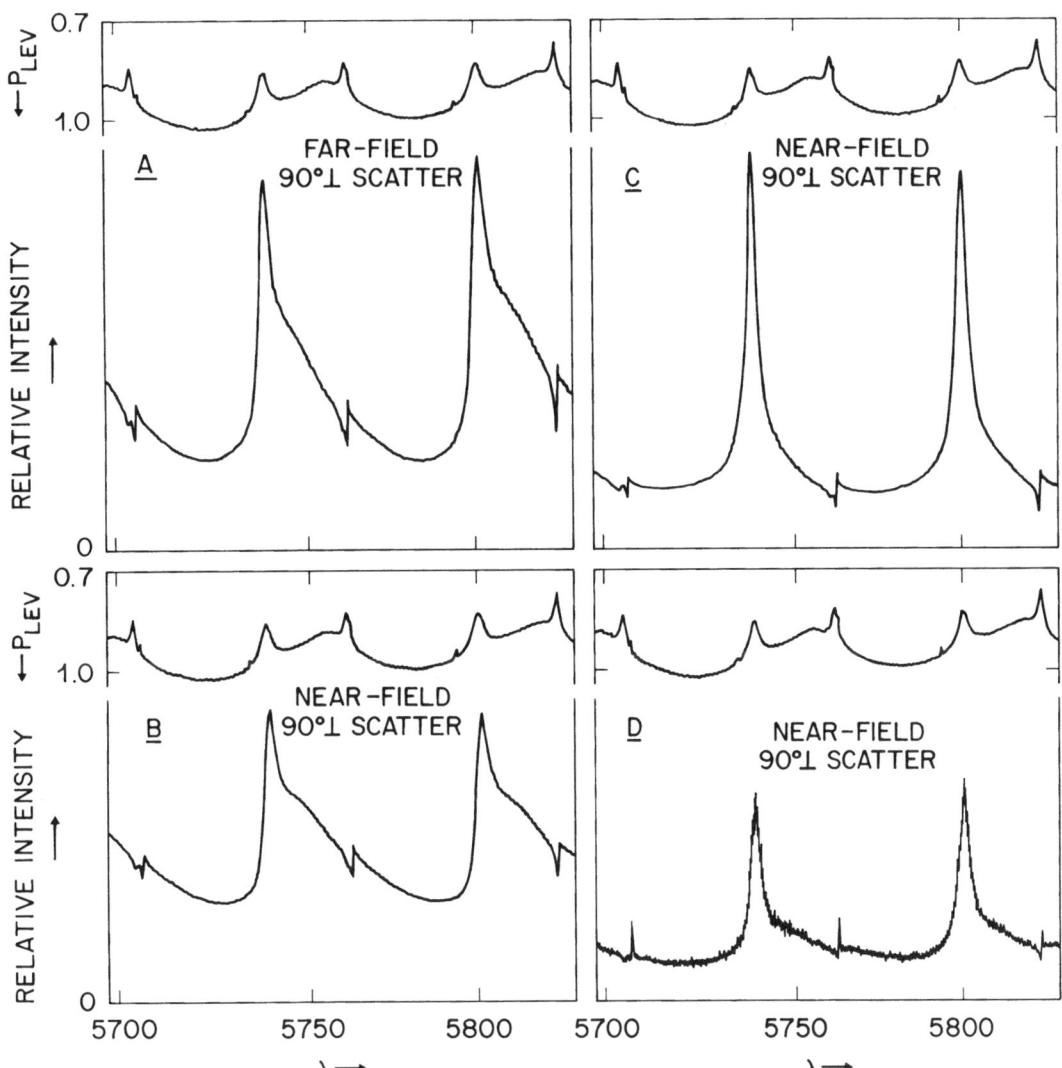

Fig. 10. Segment of the spectra of (A) P_{LEV} and total 90° perpendicular scattered light, (B) 90° perpendicular scattered light from aperture $L1$, (C) 90° perpendicular scattered light from aperture $L2$, and (D) 90° perpendicular scattered light from aperture $L3$.

In Fig. 10(A) we show a segment of the P_{LEV} or force spectrum and the total 90° ⊥ scattering spectrum which was obtained by collecting all the light with no aperturing as in Fig. 2(C), for example. This spectrum of the total light is in essence the far-field spectrum taken with an acceptance angle of ∼35° as compared with the 4° acceptance angle for Fig. 2(C). P_{LEV} is sensitive to all resonances, whereas the 90° ⊥ scattering responds strongly only to the ⊥ resonances. Aperturing around the upper spot at $L1$ results in 90° ⊥ resonances with a larger background of nonresonant light [Fig. 10(B)]. Aperturing at $L2$ just below the lower

spot reduces the background and increases the detectability of the low and high Q resonances as seen in Fig. 10(C). Aperturing at $L3$ where we visually observed the fan of light when tuned to a sharp resonance results in a further enhancement of the sharp resonance relative to other features [Fig. 10(D)]. Although the total amount of light is reduced at a position such as $L3$, it is an excellent place to look for small sharp resonances with reduced background. The width of the sharp resonance at ∼5710 Å is probably resolution limited.

Figure 11 sketches an experiment alluded to previously[1] which helps identify how resonant light is cou-

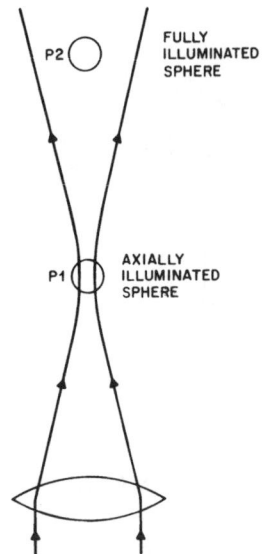

Fig. 11.   Sketch of a sphere levitated in the beam focus at $P1$ and above the beam focus at $P2$.

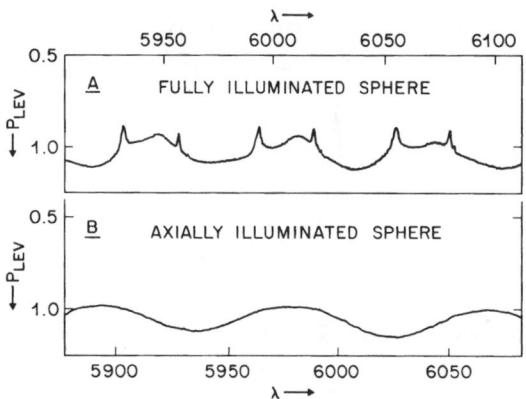

Fig. 12.   (A) P_{LEV} spectrum for sphere at position $P2$ of Fig. 11 and (B) P_{LEV} spectrum for the same sphere at position $P1$ of Fig. 11.

pled into a dielectric sphere. With the use of the feedback stabilizer[14] we can locate an ~13.1-μm diam silicone oil drop either at a point $P1$ at the beamwaist of a strongly focused Gaussian beam, where the sphere diameter is greater than the beam size, or at a point $P2$ well above the focus where the beam is much larger than the sphere diameter. In this experiment the incident dye laser beam was focused by a 2.3-cm focal length lens. For our beam size this gives a spot diameter of ~5.6 μm with a far-field diffraction half angle of 4.6°. Figure 12(A) shows the force spectrum taken at $P2$, the fully illuminated position, and Fig. 12(B) shows the force

spectrum taken at $P1$, the axially illuminated position, for the same sphere. It is seen that Fig. 12(A) exhibits well-defined surface-wave resonances, whereas in spectrum (B) of Fig. 12 they are totally absent. This indicates that the resonances of Fig. 12(A) are excited by the part of the incident beam striking the outer edges of the sphere at nearly tangential incidence in agreement with the surface-wave model. The residual broad ripples in P_{LEV} seen in Fig. 12(B), where the beam is confined principally to the axis of the sphere, are simply the low Q Fabry-Perot interference fringes due to reflections from the front and back surfaces of the sphere.

Finally we illustrate use of continuous size changes by evaporation as another way of observing the variation of scattering from a sphere as a function of size parameter. Figure 13 shows the variation of the power to levitate a somewhat volatile Cargille index-matching oil mixture of index 1.51 at a fixed height in a 5145-Å argon laser beam as a function of time for selected time intervals. The resonant behavior is clearly seen. The sequences of surface-wave resonances with gradually increasing time separation indicates a continuously decreasing evaporation rate. During the 64-min observation time, the drop size decreased from ~11.9 to 9.8 μm as measured with the microscope. Initially the evaporation rate may be too fast for the recorder to follow on the sharper resonances. This is no problem, however, at later times as evaporation slows. In the evaporation of a mixture of several fluids one expects the more volatile fluids to evaporate more rapidly. This, of course, results in a varying index of refraction as well as a varying size. The change of levitating power with decreasing drop size observed in Fig. 13 is consistent with such a model.

IV.   Concluding Remarks

From the above data, it is evident that measurement of scattered light as a continuous function of the size parameter, by changing wavelength or sphere size, is a powerful way of detecting surface-wave resonances. All the resonant features seen in force spectroscopy appear in the spectra of the scattered light. Separation of the resonances into perpendicular and parallel polarization components is possible in 90° scattering and apertured backscattering. Very high resolution detection of sharp surface wave resonances with low background light levels was demonstrated in far-field backscatter measurements with small acceptance angles and also in near-field measurements of light emitted from the edges of spheres.

The detection of sharp resonances with low background should be useful in experimentally searching for the undetected high Q surface-wave resonances. Some possible reasons for the absence of these resonances are insufficient spectral resolution, sphere distortions, surface roughness, and excessive optical absorption. Increased spectral resolution is possible with dye lasers, but it often is not compatible with continuous tunability. Use of a narrowband fixed wavelength laser and

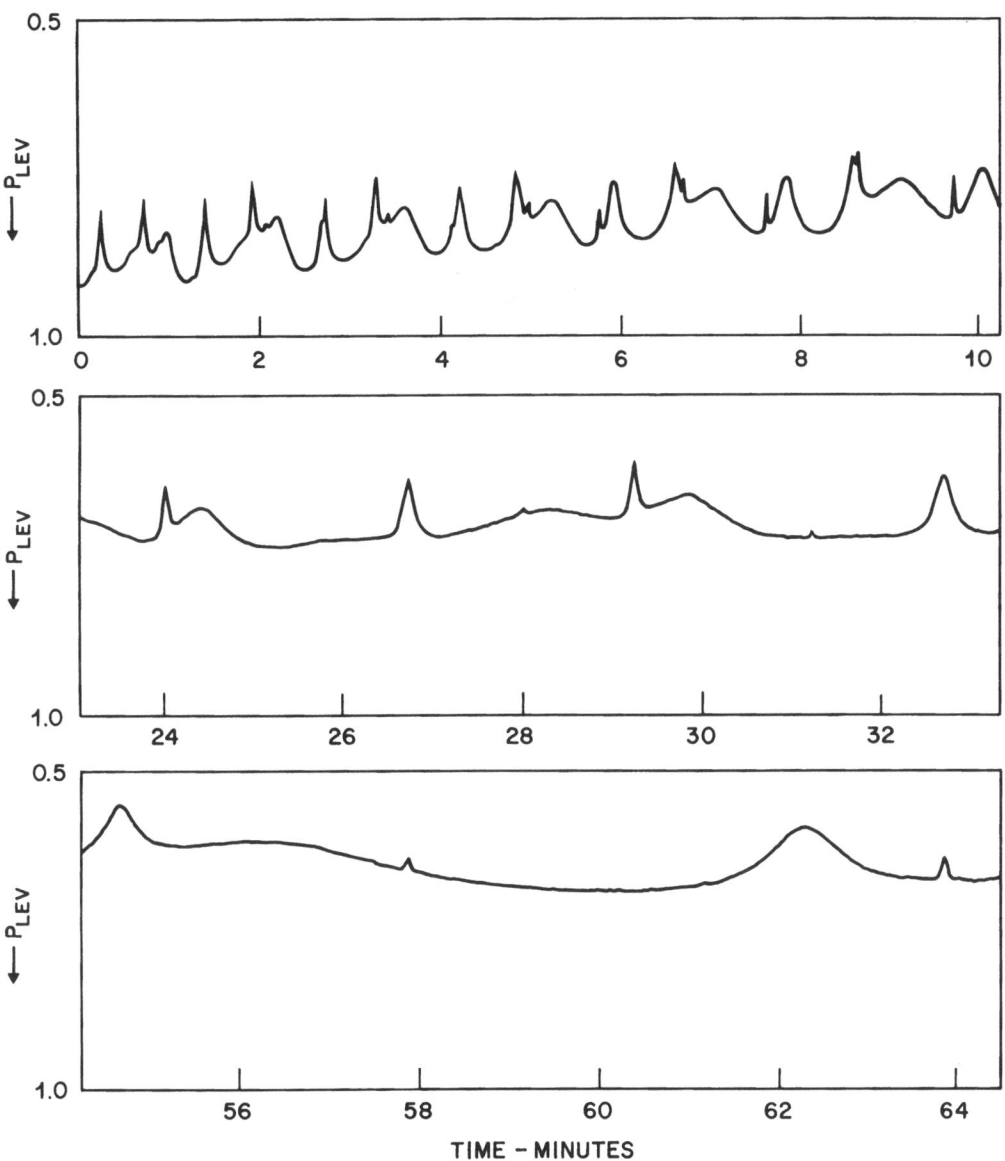

Fig. 13. P_{LEV} of an evaporating oil vs time in minutes.

a varying sphere size as demonstrated above is a possible solution. Minor distortions of a sphere shift the surface-wave resonances but apparently do not destroy them.[17] High Q resonances are probably sensitive to surface distortions, but we believe that liquid surfaces can be almost atomically smooth. The sensitivity of the high Q resonances to absorption[3] is probably the most serious experimental problem.

The experiments probing the near-field light distribution and spectrum of light emitted from various points on the face of the sphere give some of the strongest evidence presently available for the tangential coupling of van de Hulst surface waves in and out of dielectric spheres at the edges.

The near- and far-field scattering data presented here indicate a need for further theoretical calculations. There is the need for high resolution far-field scattering spectra as a function of size parameter and index of refraction at various scattering angles as mentioned above. There is the further need for high resolution near-field calculations of the light intensity being emitted from various points over the face of the sphere, within a given acceptance angle. The near-field light distribution is strongly wavelength dependent, as observed, and contains a great deal of information about the detailed physical processes occurring in Mie-Debye scattering. Besides learning about surface waves, one can hope to learn more about rainbow effects[20] in the near field.

Surface waves give a means for measuring sphere size, index of refraction, and dispersion with potentially the highest precision of any light-scattering method. This high precision opens the possibility of using spheres as probes for the sensitive measurement of temperature, humidity, or vapor pressure of a surrounding gas. Study of surface waves is also important from a fundamental point of view. Surface-wave resonances can provide probably the most precise check on the predictions of Mie-Debye-scattering theory. Scattering of electromagnetic waves is theoretically and experimentally the best understood scattering process and often serves as a physical model for other wave-scattering processes.

The measurements presented here indicate that optical levitation is a nearly ideal experimental tool for the precise measurement of light scattering from transparent dielectric spheres or other well-characterized nonspherical dielectric particles.[21]

References

1. A. Ashkin and J. M. Dziedzic, Phys. Rev. Lett. **38,** 1351 (1977).
2. H. C. van de Hulst, *Light Scattering by Small Particles* (Wiley, New York, 1957).
3. P. Chylek, J. T. Kiehl, and M. K. W. Ko, Phys. Rev. A: **18,** 2229 (1978); Appl. Opt. **17,** 3019 (1978).
4. P. Walstra, Proc. K. Ned. Akad. Wet. **B67,** 491 (1964).
5. W. M. Irvine, J. Opt. Soc. Am. **55,** 16 (1965).
6. H. C. Bryant and A. J. Cox, J. Opt. Soc. Am. **56,** 1529 (1966).
7. V. Khare and H. M. Nussenzveig, Phys. Rev. Lett. **33,** 976 (1974); **38,** 1279 (1977).
8. T. S. Fahlen and H. C. Bryant, J. Opt. Soc. Am. **58,** 304 (1968).
9. D. Atlas, L. J. Battan, W. G. Harper, B. M. Herman, M. Kerker, and E. Matijevio, IEEE Trans. Antennas Propag. **AP-11,** 68 (1963).
10. J. R. Probert-Jones, *Electromagnetic Scattering* (Macmillan, New York, 1963), p. 237.
11. J. Rheinstein, IEEE Trans. Antennas Propag. **AP-16,** 89 (1968); J. J. Stephens, P. S. Ray, and R. J. Kurzeja, J. Atmos. Sci. **28,** 785 (1971); H. Inada, Appl. Opt. **13,** 1928 (1974); P. S. Ray, J. S. Stephens, and T. W. Kitterman, Appl. Opt. **14,** 2492 (1975).
12. A. Ashkin and J. M. Dziedzic, Appl. Phys. Lett. **19,** 283 (1971).
13. A. Ashkin and J. M. Dziedzic, Science **187,** 1073 (1975).
14. A. Ashkin and J. M. Dziedzic, Appl. Phys. Lett. **30,** 202 (1977).
15. A. Ashkin, Science **210,** 1081 (1980).
16. P. Chýlek, J. T. Kiehl, M. K. W. Ko, and A. Ashkin, in *Light Scattering by Irregularly Shaped Particles*, D.W. Scheuerman, Ed. (Plenum, New York, 1980), p. 153.
17. A. Ashkin and J. M. Dziedzic; unpublished results.
18. R. E. Benner, P. W. Barber, J. F. Owen, and R. K. Chang, Phys. Rev. Lett. **44,** 475 (1980).
19. P. Chýlek, private communication.
20. H. M. Nussenzveig, J. Opt. Soc. Am. **69,** 1068 (1979); Sci. Am. **236,** 116 (1977).
21. A. Ashkin and J. M. Dziedzic, Appl. Opt. **19,** 680 (1980).

5 December 1980, Volume 210, Number 4474

SCIENCE

Applications of Laser Radiation Pressure

A. Ashkin

Historically, the idea that light carries momentum and therefore can exert forces on electrically neutral objects goes back to Kepler and Newton. It was confirmed by Maxwell. However, Maxwell found the momentum to be small, implying small forces when light from conventional sources is absorbed or re-flected by macroscopic objects. Indeed, it was only after the turn of the century, when the high-vacuum pump was invented and experiments were performed with mirrors suspended on fine torsion fibers, that researchers were able to eliminate disturbing thermal or radiometric forces from the reflection of light, in agreement with Maxwell's theory (1).

Nothing in this early history suggested that there would be practical applications for these light forces. Only in astronomy, where light intensities and distances were huge, did radiation pressure play a significant role in moving matter. It took another invention, that of the laser, to radically alter this situation and make radiation pressure a useful laboratory tool. The optical forces arising from the momentum of laser light are capable of strongly affecting the dynamics of small neutral particles ranging from micrometer-sized macroscopic particles down to molecules and atoms. This new capability permits one to stably trap small particles, levitate them against gravity, manipulate them singly, combine them in pairs, channel them selec-tively along laser beams, and use them as sensitive probes for measuring optical, electric, magnetic, radiometric, viscous drag, and gravity forces. These techniques based on light pressure have present and potential applications in a wide variety of subjects such as light scatter-ing, cloud physics, aerosol science, planetary physics, laser fusion, atomic and molecular physics, quantum optics, iso-tope separation, and high-resolution spectroscopy.

Qualitatively, one can see how the properties of laser light, namely its high degree of spatial coherence and spectral purity, have resulted in large light forces. For example, by focusing a laser beam of modest power, about 1 watt, to a spot size of about a wavelength, λ, one can subject a dielectric sphere 1 micrometer in diameter to the very high light in-tensity of about 10^8 watts per square cen-timeter. Assuming the light is reflected from the sphere with an average reflec-tivity of 10 percent one achieves an ac-celeration of approximately $10^6 g$, where g is the acceleration due to gravity (2). This is huge by any previous terrestrial standard. Transparent dielectric parti-cles are considered in this example in or-der to avoid thermal problems such as melting and radiometric forces.

For large forces to be exerted on atoms (2) it is necessary to use light of high spectral purity, which will interact strongly with narrow atomic resonance lines. Under strong excitation (or satu-rated conditions), an atom with a typical spontaneous emission lifetime of $\sim 10^{-8}$ second can absorb and spontaneously emit a maximum of about 10^8 photons per second. Since the velocity of atoms is changed by a few centimeters per sec-ond per photon absorbed or emitted, this gives forces that can significantly affect atoms moving with velocities typical of thermal atomic beams. To achieve satu-ration requires an intensity of only about 10^{-2} W/cm² because the absorption cross section of atoms for resonant light is very large, $\sim \lambda^2$. This intensity can be ex-ceeded by factors of around 10^8 with dye lasers, which has important conse-quences for achieving large forces on atoms. Another feature of focused laser light not shared by any other light source is the existence of high-intensity gradi-ents. This leads to large gradient forces (2-4) with the properties that make pos-sible stable optical trapping and manipu-lation of particles on the scale of the opti-cal wavelength.

In this article the discussion of radia-tion pressure will be restricted, as is tra-ditional, to neutral particles, since it is here that one can best describe the forces as arising from absorption and scattering of light momentum. Forces on charged particles such as electrons can, of course, be similarly described. How-ever, forces on electrons are most con-veniently considered in terms of conven-tional electric and magnetic fields.

Summary. Use of lasers has revolutionized the study and applications of radiation pressure. Light forces have been achieved which strongly affect the dynamics of indi-vidual small particles. It is now possible to optically accelerate, slow, stably trap, and manipulate micrometer-sized dielectric particles and atoms. This leads to a diversity of new scientific and practical applications in fields where small particles play a role, such as light scattering, cloud physics, aerosol science, atomic physics, quantum optics, and high-resolution spectroscopy.

The author is head of the Physical Optics and Electronics Research Department at Bell Labora-tories, Holmdel, New Jersey 07733.

Forces on Macroscopic Particles

Consider first the force of laser light on transparent uniform dielectric spheres (2, 5-7) in the size range 1 to 100 μm, for example. Let a sphere with an index of refraction higher than that of the surrounding medium be placed off-axis at position Q in the field of a nearly parallel TEM_{00}-mode Gaussian beam, as shown in Fig. 1. In a simple ray optics picture, the principal contribution to the force comes from rays such as a and b, which are predominantly refracted through the sphere, giving rise to forces F_a and F_b in the direction of the momentum change. Surface reflections are negligible. Since there is more light at a than at b, $F_a > F_b$, and one sees that there are two components to the net force: one in the axial direction of the light, denoted by F_{ax}, that would exist even for a plane wave, and the other a transverse gradient force, F_{tr}, pulling the sphere into the high-intensity region of the light. The gradient force can be comparable with the axial force for typical beam diameters, and at achievable light intensities both can be made as much as thousands of times the particle weight mg, where m is the particle mass.

The existence of these light forces was first demonstrated in experiments with transparent micrometer-sized spheres in liquids, where viscosity is high and gravity plays a minor role. Spheres were guided along laser beams and driven by the light with velocities proportional to their radii. Stable optical trapping of individual particles was first observed with spheres in liquids in a trap geometry consisting of a pair of opposing focused beams (2). Optical trapping is also pos-sible in less dense media, such as air, where one can stably trap or levitate particles against the force of gravity with a single, vertically directed laser beam (5).

Figure 2 shows the basic optical levitation apparatus. A uniform dielectric sphere is located at equilibrium point E above the focus of a TEM_{00}-mode Gaussian beam, where gravity and the axial light force in the upward direction balance. The equilibrium is stable since any vertical displacement from E results in a restoring force due to the change in light intensity caused by the beam divergence, and any transverse displacement results in a restoring force due to the transverse gradient force. Levitation is done in a glass cell to avoid disturbing air currents. Solid particles are introduced into the beam by lifting them off the transparent base plate. This requires an impulse from a piezoelectric ceramic shaker to break the van der Waals bond with the plate. Liquid drops can be captured by allowing drops produced by an atomizer to fall through a small hole in the roof of the cell and enter the light beam. If they are in the proper size range, they are captured. Inherent in the process of capturing a particle and bringing it to rest in a stable trap is the need for damping. In air, the viscosity of the gas prevents the escape of particles that enter the essentially conservative optical potential well from the outside. Once captured, a particle can be moved anywhere in the cell by simply moving the beam; the particle is constrained to follow. The power required to levitate uniform solid or liquid dielectric spheres in the size range 1/2 to 100 μm varies from microwatts to several watts cw (continuous wave).

For Gaussian beams with focal spots smaller than the particle diameter, there is another stable region of levitation below the focus of the beam (6). With a pair of opposing horizontal beams with focal spots less than the particle diameter there are several stable regions of levitation (8). One can also capture and hold assemblages of more than a dozen spheres in a single beam, locked in stable rigid arrays (9). Each sphere becomes trapped at a local intensity maximum in the optical diffraction pattern of all particles located below it. Drastic rearrangements occur only when one of the lower particles is displaced from its local trap. Nonspherical particles such as spheroids, teardrops, and spherical doublets, triplets, and quadruplets have also been levitated. Such particles were assembled in the air by an extension of the levitation technique, using two beams to make various solid and liquid spheres collide and combine (10). Such nonspherical particles orient themselves stably at fixed angles with respect to the levitating beam.

There is another class of particles for which the transverse gradient force is reversed and the particles are pushed out of the high-intensity region of a beam. Included in this class are transparent hollow dielectric spheres (6) and highly reflecting metals (11), where the transverse force is dominated by reflected rays, and low-index particles in a high-index medium, where refraction through the particles reverses (2). Such particles can be levitated or trapped by using the TEM_{01}^* laser beam mode, which has an intensity minimum on the beam axis.

An important aspect of levitation is the remarkable visibility it affords. Particles

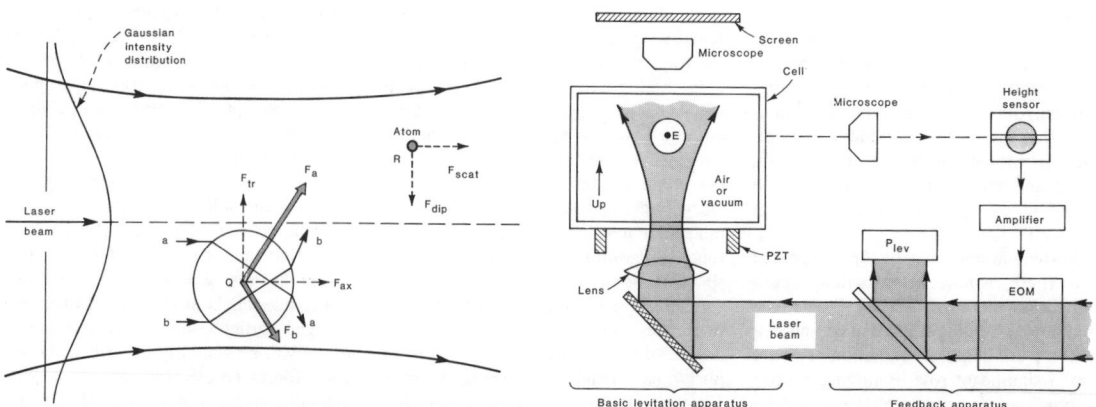

Fig. 1 (left). Optical forces on a dielectric sphere and an atom in a TEM_{00} mode Gaussian laser beam.   **Fig. 2 (right).** Basic apparatus for optically levitating dielectric spheres and feedback stabilization apparatus for levitating in vacuum and measuring forces. Abbreviations: *PZT*, piezoelectric ceramic shaker; *EOM*, electrooptic modulator.

sit at rest in high-intensity light beams, free of other scattering sources. One can thus focus microscopes on the particles, as shown in Fig. 2, and project enlarged images on screens for convenient visual observation, measurement, or photography.

Levitation is also possible in high vacuum (*12*). This was done by capturing a particle in air, where the viscosity is high, and then slowly pumping out the air. If the particle's optical absorption is low enough, there are no serious disturbances from residual thermal or radiometric forces, which reach a maximum during the evacuation process. In high vacuum, viscous damping and radiation damping are weak and only radiation pressure forces are operative. One complication that then arises is the spontaneous buildup of random oscillations due to fluctuations in the levitating beam itself. This can be regarded as a form of kinetic heating, and it can result in escape of the particle.

The problem of kinetic heating was overcome by introducing feedback optical damping (*12*). A feedback system, as sketched in Fig. 2, senses the height and vertical velocity of the particle and electronically feeds back to an electrooptic modulator which controls the levitating light power. The system can "lock" the average position of the particle to a fixed height and change the optical power in proportion to the velocity to give strong optical damping. It can, in fact, damp both vertical and horizontal oscillations due to a coupling of these two motions. In practice, it holds particles motionless in high vacuum. Feedback control of levitated particles provides another important capability, namely automatic force measurement. By monitoring the laser power needed to levitate a particle at a fixed height, P_{lev}, in the presence of an external force, one automatically obtains a direct measure of the magnitude of the applied force relative to the particle's weight. This technique is useful in many applications of levitation in air as well as in vacuum.

Applications with Macroscopic Particles

One of the major applications of radiation pressure is to the study of light scattering. This is natural since the basic light pressure forces themselves arise from light scattering. Indeed, measurement of the wavelength dependence of radiation pressure forces on dielectric spheres by levitation techniques (*13*) revealed for the first time the existence of a

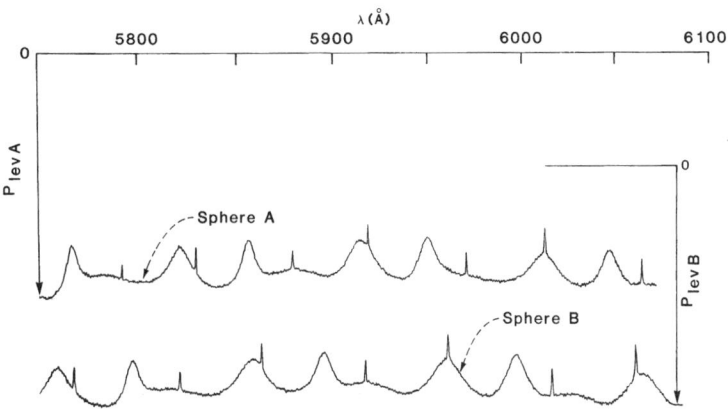

Fig. 3. Resonant behavior of light forces on dielectric spheres. The spectra show the variation with wavelength of P_{levA} and P_{levB}, the power needed to levitate oil drops A and B, which have index of refraction $n = 1.47$ and slightly different diameters ($\sim 10\ \mu$m). The resonances of sphere A are shifted ~ 50 Å higher in wavelength than the corresponding ones in sphere B.

complex spectroscopy of very sharp resonances. This is illustrated in Fig. 3, which shows the variation of P_{lev} for two spherical oil drops with slightly different diameters as the wavelength of a levitating dye laser is varied. The underlying spectroscopy of the sphere is shown by the fact that the sequences of force resonances for sphere A are identical with those for sphere B, only shifted in wavelength. The resonances were attributed (*13*) to the so-called van de Hulst dielectric surface waves (*14*), which couple into and out of spheres at the edges and can run around spheres and resonantly close on themselves. At resonance, the force increases and P_{lev} drops. This process involves diffraction and cannot be understood by simple ray optics.

Use of these surface wave resonances provides a new method of relative and absolute size measurement that is two to three orders of magnitude more accurate than previous methods based on Mie-Debye scattering theory (*14*) for spheres. Relative measurement of size is based on a comparison of the wavelength of a particular resonance for different spheres and can have an accuracy of about 1 part in 10^5, determined by the width of the sharp resonances. Absolute measurement of size and also index of refraction requires a comparison of experiment with theory. Recent high-resolution computer calculations of the force spectrum (*15*) from Mie-Debye theory gave impressive agreement with experiment but also predicted sharper resonances, which were unresolved experimentally. This implies that even more precise measurements may be possible.

Another sensitive way to detect surface wave resonances is by direct obser-

vation of near- and far-field light scattering from levitated spheres (*13, 16, 17*). Near-field observations, in addition, give information on the internal light fields and the origins of surface wave resonances. For example, in the photograph shown in Fig. 4 of the near-field backscatter from a levitated oil drop, one sees that surface wave emission from the edges of the sphere dominates the backscatter. The existence of sharp surface wave resonances can also affect the wavelength dependence of inelastic light scattering from small particles. Such particle resonance effects were recently observed in the fluorescence from dye-impregnated spheres (*17, 18*). Minute distortions of liquid drops—as small as 1 part in 10^5—caused by electric fields have been observed by shifts of surface wave resonances (*17*). Such distortions give a new way of measuring the surface tension of liquids. In retrospect, it is remarkable that it has taken so long to recognize the usefulness of the basic spectroscopy of the fundamental spherical scattering particle.

Extension of the levitation technique to nonspherical particles (*10*) should make possible equally detailed scattering and force measurements on these particles. For example, one can hope to measure surface wave resonances of spheroids. In levitation experiments with oriented particles of more complex shape and internal structure, the ability to directly observe near-field transmission patterns and correlate them with the full far-field scattering patterns provides a new way to understand complex scattering without need for difficult mathematics (*10*).

Application of radiation pressure to

cloud physics and aerosols is based on the ability to manipulate and sensitively observe individual cloud-size liquid drops (9). One can study such fundamental processes as drop evaporation or condensation, drop-drop collision, interaction of charged drops, supersaturation of drops, and their crystallization. Evaporation or condensation can be observed directly with microscopes or, more sensitively, by measuring size changes with surface wave resonances (13). For example, by observing the scattering from a drop with a fixed-wavelength laser as the drop diameter varies continuously with time, one obtains the full surface wave spectrum of the sphere. By tuning a laser to the steep wings of a narrow resonance, one can observe changes in the average diameter of a drop of 1 part in 10^6 or about 0.1 angstrom for a 10-μm drop—a sensitivity of a fraction of a monolayer. Collisions between two drops of known size and charge can be seen by using the two-beam levitation technique (10). Fusion of drops of opposite electric charge can be induced by an externally applied electric field (9). Supersaturation and crystallization were observed in levitation experiments with drops of various salt solutions in environments of varying humidity (17). Use of radiation pressure has also been proposed for manipulation of water droplets in the near-zero-gravity environment of the Atmospheric Cloud Physics Lab payload being designed for the Shuttle Spacelab (19).

Use of optically levitated particles for the measurement of applied forces relies on direct observation of the displacement of particles from equilibrium or on detection of changes in the stabilized levitating power in the feedback force measuring technique. Measurement of optical forces was discussed above.

Measurement of the electric force on charged levitated particles gives a sensitive determination of electric charge. Indeed, this led to the discovery of photoemission rates as low as a few electrons per minute from optically levitated fused silica spheres due to a new type of three-photon nonlinear photoelectric effect in dielectrics (20). Using the optical levitation feedback technique to measure electric force, one can detect changes of the electric charge of single-electron units (21). Figure 5 illustrates the steplike decreases in optical levitating power caused by feedback in response to increased electric force as the charge on a sphere changes by units, from -1 electron up to 10 electrons, due to charging by ultraviolet light. Transition to full electric field or Millikan-type sup-

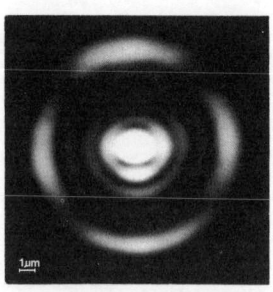

Fig. 4. Near-field photograph of $\sim$ 12-μm dielectric sphere as seen in backscatter, showing the predominance of dielectric surface wave scatter from the sphere edges.

port can be made with increased charge. Combining Millikan support of an oil drop with optical measurement of size, and therefore mass, makes possible an absolute measurement of the charge of the electron accurate to about 1 part in 10^4 to 10^5. It is also seen in Fig. 5 that one has a sensitivity of measurement of a fraction of an electronic charge with oil drops in this general size range, which makes a search for quarks possible. Static electric charges deposited on surfaces can be measured with a sensitivity corresponding to a few electrons and a spatial resolution of micrometers by scanning a charged levitated particle over the surface and detecting the electrostatic force. It is hard to conceive of measuring surface charge with comparable sensitivity by other methods.

Diamagnetic forces on optically levitated particles have also been observed by manipulating the particles into regions with static magnetic field gradients (17). In this way one can sensitively measure relative diamagnetic susceptibilities. It is well known that stable diamagnetic levitation is possible at a magnetic field minimum (22). Surprisingly, one can apply sufficiently strong magnetic gradients to such weakly diamagnetic particles as fused silica spheres and glycerol drops to achieve stable diamagnetic levitation. These particles were transferred from optical levitation traps into small magnetic traps.

Another basic force on small particles that can be measured by levitation with the feedback force measuring technique is the radiometric force. This is caused by thermal gradients and is called photophoresis when the origin of the heat is optical absorption in the particle. The variation of the radiometric force with pressure can be observed (12) from its value at atmospheric pressure through its maximum and down to negligible values in high vacuum. The radiometric force is a sensitive measure of optical ab-

sorption in particles. Levitation under high vacuum, where radiometric forces are less than 10^{-7} of the radiation pressure forces, proves unequivocally that particles can be trapped solely by the forces of radiation pressure.

Whenever gas surrounding a fixed levitated particle flows, it subjects the particle to viscous drag forces by Stokes' law. Measurement of the viscous forces as a function of position with a maneuverable levitated particle gives a means of mapping gas flow patterns in chambers of varying shape. Flows as low as micrometers per second are readily detectable, for example, when vacuum chambers are slowly evacuated (12).

An important potential application of radiation pressure is to high-speed mechanical rotation of micrometer-sized particles in vacuum (12). This is possible with the feedback stabilization technique, which damps all linear motion while leaving rotational motion undamped. If one calculates, for example, the highest rotational frequency and centrifugal acceleration possible with a 4-μm-diameter silica sphere, at the point of rupture, one obtains $\sim$ 1500 megahertz with an acceleration of $\sim 10^{12}g$. This exceeds by a factor of about 10^3 the record values achieved by Beams (23) in his classic ultracentrifuge experiment on 0.4-millimeter-diameter magnetically levitated steel spheres. Angular acceleration can be achieved with torques based on the angular momentum of light. Measurable torques in reasonable agreement with the angular momentum of circularly polarized light have been observed on optically levitated particles at atmospheric pressure (17). In planetary science the lifetime of moderate-sized interplanetary dust grains is thought to be governed by just such a rotational bursting process driven by solar radiation pressure (24). The optical torques, however, are due to a windmill-type effect caused by asymmetries in particle shape. Attempts to study such torques are being made with lasers and levitation techniques (25).

It has been suggested that optical levitation could be useful for the support of targets in laser fusion experiments (6, 26). This is based on actual levitation of hollow quartz Microballoons that are typical of fusion targets and on the ability to stabilize the position of particles in high vacuum by optical feedback. Levitation may also be useful for manipulation and fabrication of more complex fusion target structures (10).

Another potential application of radiation pressure is to the separation and manipulation of biological particles such as cells and viruses in liquids (2). Lasers can do this with transparent dielectric spheres in liquids, where the spheres range in size from about 0.1 to 50 μm. With real biological particles care must be taken to avoid thermal absorption in the particles and the liquid. In order to understand optical forces on particles smaller than the wavelength of light ($\sim$ 0.5 μm), one has to abandon simple ray optical pictures (2) and invoke Rayleigh scattering (14). Indeed, the trapping of particles in the 0.1-μm region is more akin to the trapping of atoms in the off-resonance regime.

Radiation pressure from lasers has been used to help resolve some fundamental questions about the momentum of light in dielectric media. In a well-known gedankenexperiment, the sign of the light force on a free liquid surface is used to distinguish between the proposals of Minkowski and Abraham for the momentum in dielectrics (27). A pulsed laser experiment of this sort was performed on a water surface and sufficient force was achieved to overcome surface tension and generate a readily observable surface distortion (28). The result favored the Minkowski momentum. Analysis of this and other force measurements in liquids with conventional light sources has given a better picture of momentum in dielectrics (29, 30).

Radiation Pressure on Atoms

The basic forces and applications of radiation pressure on atoms are conceptually similar to those already discussed for macroscopic particles. For example, there is a scattering force that drives atoms in the direction of the light and a gradient force that pulls atoms into or out of regions of high light intensity. Use of these forces leads to transverse confinement of atomic beams within an optical beam and to the possibility of stable trapping of individual atoms in single Guassian beams, in analogy to levitation. The concept of heating of trapped macroscopic particles due to laser beam fluctuations and the idea of optical damping have their direct counterparts in atomic behavior. Applications to isotope separation are analogous to macroscopic particle sorting. There is a similar capability for particle manipulation on the scale of the light wavelength. Potential applications to high-resolution atomic spectroscopy are reminiscent of the demonstrated spectroscopic applications to macroscopic dielectric spheres. The details of radiation pressure on atoms must, however, be different because of the considerable difference in particle

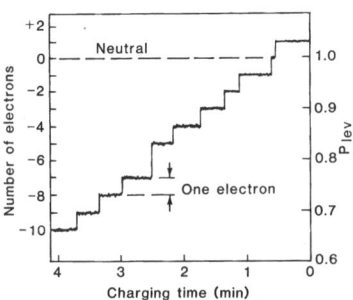

Fig. 5. Changes in the optical levitating power caused by the automatic feedback system as the charge on a sphere in an electric field increases by single-electron amounts.

size and the quantum nature of the interaction. Indeed, light forces on atoms are large only in the vicinity of a resonance transition. Their effects have therefore been termed resonance radiation pressure.

Figure 1 shows the resolution of the average forces on an atom into axial and transverse gradient components when the atom is off-axis at position R in an essentially parallel, nearly resonant Gaussian beam. The axial force is called the spontaneous scattering force and is denoted by F_{scat}. It is the average driving force in the direction of the incident light arising from the scattering process involving absorption of photons and their subsequent, on the average symmetric, spontaneous emission. This scattering force exists even for incident plane wave light. The idea of atomic recoil, which is the basis of this force, goes back to Einstein. Recoil of an atom due to absorption and emission of a single photon (the Einstein Rückstoss) was observed (31) with resonance lamps in prelaser times. At laser intensities the scattering force (32) increases to a maximum value set by saturation of the atomic transition. The force is a maximum at exact resonance. Its saturated magnitude is sufficient to stop a sodium atom moving at thermal velocities of $\sim$ 5 $\times$ 10^4 cm/sec in about 10 cm if continuously applied.

The scattering force has been directly observed in atomic beam deflection experiments with lasers (33-35). In these experiments resonant light striking a collimated atomic beam at right angles to its direction of motion causes sizable transverse deflections. The frequency sensitivity of this process has been used in high-resolution atomic beam deflection spectroscopy (35). Such deflections have also been used as a method of isotope separation (34) based on the finite isotope shift of resonance frequencies. A scheme for a continuously operating

atomic beam velocity selector has been proposed (32) that uses the scattering force. The scattering force has also been observed to exert a significant pressure on a resonant atomic vapor; light-induced pressure increases of about 50 percent were seen (36). The strong velocity dependence of the scattering force due to the Doppler shift led to the important concept of optical cooling or damping of atomic motions (37). A pair of oppositely directed plane wave beams of equal intensity, tuned to a frequency below an atomic resonance, should damp any atomic motion along the direction of the beam pair since the Doppler shift always increases the scattering force of the opposing beam. Cooling of all components of atomic motion in an atomic vapor and transverse cooling of an atomic beam were proposed as uses for this technique. More will be said about this in connection with atom traps.

The transverse component of the average force on an atom in the Gaussian beam of Fig. 1 is due to the so-called dipole force (3, 4, 38, 39) and is designated by F_{dip}. This force arises in general from a gradient of the light intensity. The dipole force can be considered as the force on an optically induced atomic dipole in the gradient of the optical electric field. It is directed along the intensity gradient and is dispersive in nature because of its dependence on the atomic polarizability. Atoms are pulled either into or out of the region of high light intensity, depending on whether the light frequency is tuned below or above resonance. The dipole force is zero, on the average, right at resonance. It is the near-resonance form of electrostriction in gases. The dipole force arises from stimulated emission processes and can be equivalently understood in terms of the net light scattering coming from the interference of dipole radiation from the atom and the stimulating incident beam (40). To obtain large dipole forces one has to balance the effect of detuning from exact resonance and the amount of saturation for a given amount of available power (3). It was shown that the dipole force can be represented as the negative gradient of a potential which, for $\sim$ 1 W of power and sodium atoms, for example, can be as deep as $\sim$ 10^4 $h\gamma_N \cong$ 10^{-4} electron volt, where h is Planck's constant and γ_N is the natural width of the resonance line. A potential of this depth is capable of confining atoms moving with a velocity of $\sim$ 2 $\times$ 10^3 cm/sec.

Observations of this dipole force of resonance radiation pressure at optical frequencies were first made in a Gaussian laser beam where the transverse component of the dipole force was used to transversely confine and focus a co-propagating neutral sodium atomic beam to a small spot size (*41, 42*). Figure 6 shows the experimental setup used to inject the atomic beam into the core of a tunable Gaussian dye laser beam. Figure 7A shows the shape of the atomic beam at the focal plane of the optical beam in the presence of light. When light tuned below resonance is turned on, atoms are transversely trapped within the light beam and focused or concentrated to an intense spot as shown. For light tuned above resonance (Fig. 7B), atoms are ejected or defocused from the beam, as expected from the disperse character of the dipole force. One can in principle also trap atomic beams within light beams with light tuned above resonance by confining them on the axis of the TEM_{01}* mode of the laser beam, where the light intensity is a minimum. This is analogous to levitation of hollow dielectric spheres in air or of bubbles in liquids with the TEM_{01}* laser mode.

The fundamental experiment described above not only shows the essential properties of the dipole force but suggests several applications. For example, one should be able to perform isotope separation with an atomic beam containing two isotopes and a laser beam tuned between their resonance frequencies. The light beam should confine one isotope and eject the other, thereby achieving single-step separation ratios that can approach 10^3 (*41*). Optical steering of atomic beams was experimentally demonstrated by moving the guiding light beam (*41*). Other possibilities are the cleaning up of "dirty" atomic beams by confining only a desired species of atom. The ability to increase atomic beam intensities in small focal spots or in a long apparatus by the focusing and confining action of the dipole force is also useful.

Other proposals for use of the dipole force involve the transient behavior of the force when the interaction of the atoms with the light is short compared to a natural lifetime (*43, 44*). Transient acceleration of atoms to high velocity in a rapidly moving standing wave field has been analyzed theoretically (*38*). Use of the transient dipole force to deflect atoms passing transversely through a standing wave field has been considered for isotope separation (*43, 45*). Atomic beam deflection has been observed experimentally under these circumstances (*46*).

We now turn to the important topic of the fluctuations of the light forces acting on atoms due basically to the quantum nature of the interaction. Indeed, the

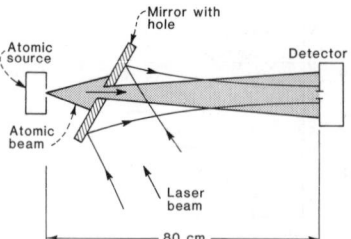

Fig. 6. Apparatus for observing focusing and defocusing of an atomic beam by the dipole force of a nearly resonant laser beam.

forces described above, for cases where the atom interacts with the field for times long compared to a natural lifetime, were only the average forces that the light exerts on atoms. Fluctuations in spontaneous scattering are intuitively understood as arising from the temporal and spatial randomness of the absorption and spontaneous emission processes (*39, 47–49*). Less obvious is the origin of fluctuations in the dipole force (*48–51*). Force fluctuations add randomness to the prediction of the dynamics of motion of atoms based on the average force and can be considered as a constant source of heat, which is added to the initially cold orbits predicted by the average force. The limiting effects of the fluctuations of the scattering force were observed directly, for the first time, in the atomic beam focusing experiment, where the size of the atomic focal spot as calculated for a

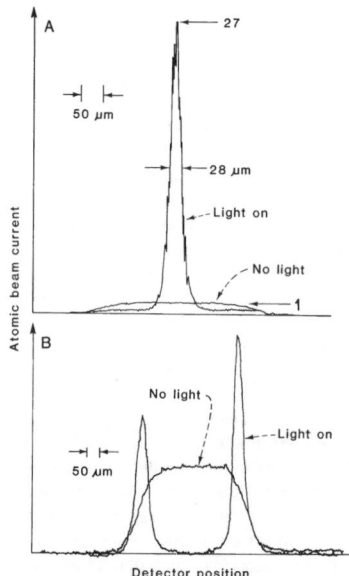

Fig. 7. (A) Focusing of an atomic beam by light tuned below resonance. (B) Defocusing of an atomic beam by light tuned above resonance.

transversely cold beam was found to increase monotonically with the amount of fluctuation heating of the transverse motion of the atoms (*52*). This fluctuation heating, in the absence of any additional cooling mechanism, must eventually lead to escape of all the initially cold atoms from the light beam.

Consider now ideas of how to make optical traps, using the above information on light forces. The observed confinement of atomic beams within the core of light beams (*41, 42, 52*) is a form of atom trapping, but only in two dimensions. Trapping of atoms in stable circular orbits was also considered in early proposals for use of the saturated scattering force (*32*). More recent concepts consider localized traps (*3, 39, 48, 49*), with the atoms held as nearly at rest as possible. Thus the basic optical trap consists of an optical field configuration with a point of stable equilibrium such that any displacement of an atom from this point results in an average restoring force. The maximum kinetic energy of an atom that can be confined in a trap is defined as the well depth. If one places an atom with zero velocity at the equilibrium point of a trap, one expects that the average energy of the atom will increase at a steady rate due to heating from the quantum fluctuations until the atom eventually escapes from the trap in a time called the retention time. If the fluctuation heating of atoms can be counteracted by a sufficiently strong damping or cooling mechanism, such as optical cooling (*37*), then presumably atoms can be retained indefinitely. Indeed, the equilibrium kinetic energy of an atom in a trap results from a balance of the heating and cooling rates (*39, 48, 49*). The effectiveness of a trap in containing an atom is determined by the Boltzmann factor, given by the ratio of the well depth to the equilibrium kinetic energy. A large Boltzmann factor implies a low probability of thermal excitation over the top of the trap barrier and a very low probability of quantum mechanical tunneling through the barrier for typical trap dimensions. In practice, therefore, we seek both a low average kinetic energy and a large Boltzmann factor.

It has been proposed (*39*) that atoms of an atomic vapor can be trapped in relatively large volumes on the many local intensity maxima of three orthogonal pairs of standing wave beams tuned $\sim \gamma_N/2$ below the atomic resonance. This tuning was chosen to provide trapping and optimal damping. The resulting well depth of the trap is $\sim h\gamma_N$. Unfortunately, it was recently shown (*48, 49*) that damping for a standing wave field varies as a function of position, at low atomic velocities, and is actually zero at the maximum of the standing wave, where atoms are expected to collect. A further problem with traps tuned for maximum damping is that one obtains an equilibrium kinetic energy of $\sim h\gamma_N$, giving a Boltzmann factor of about unity. Such traps are therefore very leaky.

Traps proposed more recently have wells that are two to three orders of magnitude deeper (3) than the standing wave traps considered above. These traps are based on strongly focused Gaussian beams tuned to produce essentially maximum depth for a given laser power. In geometry, they resemble the macroscopic particle traps described above. Perhaps the simplest such atom trap is a single strongly focused Gaussian beam. In this trap stability is achieved by making the axial gradient strong enough that the backward dipole force on an atom exceeds the scattering force tending to drive it away from the focus. For a tuning of $\sim 10^2$ to $10^3 \gamma_N$ below resonance, trap depths of $\sim 10^2$ to $10^3 h\gamma_N$ are possible, but with much reduced damping. In fact, the Boltzmann factor is again ~ 1. However, use of an additional optimally tuned plane wave damping beam has been considered as a means of cooling atoms in such a trap to a minimum kinetic energy of $\sim h\gamma_N$. This corresponds to a large Boltzmann factor and a minimum temperature of $\sim 10^{-3}$ to 10^{-4} K. A possible problem with this damping scheme is related to the optical Stark shift of the atomic resonance of atoms moving in the trapping beam. Solutions to this problem have been proposed (45). It has been shown, however, that single-beam traps with no additional damping beams can be designed with sufficiently low dipole and scattering force heating that the retention time of initially cold atoms can be many seconds (49). This implies that an experimental demonstration of localized atom trapping is not contingent on damping. The effects of adding cooling beams can then be studied subsequently. Other versions of cooled traps based on a pair of opposing focused Gaussian beams having deep transverse and deep standing wave potentials are possible (3, 48).

Injection of slow or cool atoms into traps will probably involve radiation pressure slowing of atomic beams with the scattering force (3, 53). This is necessary because of the well-known absence of slow atoms in the low-energy tail of the Maxwellian distribution of atomic beams. Slow atoms are generally useful for other devices such as atomic clocks. Atoms caught in optical traps should be

directly observable by their scattered fluorescence. Indeed, it should be possible to see even single atoms in this fashion, since scattering rates of about 10^6 to 10^8 photons per second are expected. Single atoms have been experimentally detected by their resonance fluorescence in other contexts (54, 55). The ability to observe the fluorescence of a single atom also leads to the interesting possibility of sensing the position and velocity of an atom in a trap and then using electronic feedback on the trapping laser power to damp out fluctuations, just as was done for dielectric spheres in vacuum (12).

In the well-developed field of ion trapping there have recently been some highly relevant experiments on the optical cooling of ions (56, 57). The concepts of optical cooling of ions held in electromagnetic traps were developed independently of those for radiation pressure cooling of neutral atoms, although in essence they are equivalent (58). The experiments have demonstrated significant optical cooling of ions by laser light. This success with ion traps bolsters hopes for optical cooling of neutral atoms in the more complex environment of an optical trap.

One immediate application of optical atom traps is to the study of basic atomic behavior under high optical excitation: the radiation forces, their fluctuations, the optical Stark shifts and other nonlinearities, and optical cooling mechanisms. High-resolution spectroscopy is an application common to optical neutral atom traps and ion traps. With very slow confined atoms one can make observations for long periods of time under conditions where first- and second-order Doppler effects are small. For many spectroscopic studies with optical traps the presence of the high-intensity light with its optical Stark shift would be undesirable. This problem could be overcome by turning the trapping beam off periodically for times long enough to study the high-resolution spectroscopy of the cold unexcited atoms but short enough to prevent escape of the slow atoms.

Some of the fascination with optical atom traps comes from their manipulative possibilities. For example, one can conceive of doing chemistry with individual atoms by combining separately trapped neutral atoms to form molecules, in analogy with the experiments on combining levitated macroscopic particles (10). With trapped cold neutral atoms one could study forces between atoms such as the weak van der Waals forces. There is also the possibility of ob-

serving for the first time coupling forces that exist between pairs of atoms under strong optical excitation when they approach one another in the near field of their radiation patterns (59). Furthermore, with atoms closely coupled to one another in their near field, or similarly an atom coupled to itself through its image in closely spaced mirrors, one has the prospect of observing modifications of the natural lifetime and radiation patterns of free atoms (60). One can also imagine arrays of atoms. The closest possible distance of approach of individually localized optically trapped cold atoms is a quarter of the optical wavelength of the trapping beam. This is possible in standing wave traps based on a pair of focused Gaussian beams (3). Tunneling of atoms between the successive field maxima of a standing trap has been considered (39). Tunneling between shallow standing wave maxima can be conveniently studied in traps with strong overall confinement in regions of focused-beam standing wave traps (48). The periodic potential of an optical standing wave has been considered for use as a stop and passband filter for the transmission of atoms of different velocities (39).

Many of the applications of optical traps mentioned above should be realizable with single-beam traps, in which the retention time for cooled atoms can be seconds. Successful addition of optical cooling, giving minimum kinetic energies of 10^{-3} to 10^{-4} K and much longer retention times, would greatly increase the utility of the technique. Whether this additional cooling can be achieved is the biggest unresolved question in optical trapping of atoms.

Although little direct work has been done on resonance radiation pressure on molecules, the same basic principles should apply. The scattering forces in general will be weaker because of the longer lifetimes, but strong dipole forces and trapping should be possible near molecular resonances.

References and Notes

1. P. Lebedew, *Ann. Phys. (Leipzig)* **6**, 433 (1901); *Astrophys. J.* **31**, 385 (1910); E. F. Nichols and G. F. Hull, *Phys. Rev.* **13**, 307 (1901); *ibid.* **17**, 26 (1903).
2. A. Ashkin, *Phys. Rev. Let.* **24**, 156 (1970).
3. _____, *ibid.* **40**, 729 (1978).
4. G. A. Askar'yan, *Zh. Eksp. Teor. Fiz.* **42**, 1567 (1962) [*Sov. Phys. JETP* **15**, 1088 (1962)].
5. A. Ashkin and J. M. Dziedzic, *Appl. Phys. Lett.* **19**, 283 (1971); A. Ashkin, *Sci. Am.* **226**, 63 (February 1972).
6. A. Ashkin and J. M. Dziedzic, *Appl. Phys. Lett.* **24**, 586 (1974).
7. G. Roosen, *Can. J. Phys.* **57**, 1260 (1979).
8. _____, *Opt. Commun.* **21**, 189 (1977).
9. A. Ashkin and J. M. Dziedzic, *Science* **187**, 1073 (1975).
10. _____, *Appl. Opt.* **19**, 660 (1980).
11. G. Roosen and C. Imbert, *Opt. Commun.* **26**, 432 (1978).

12. A. Ashkin and J. M. Dziedzic, *Appl. Phys. Lett.* **28**, 333 (1976); *ibid.* **30**, 202 (1977).
13. _____, *Phys. Rev. Lett.* **38**, 1351 (1977).
14. H. C. van de Hulst, *Light Scattering by Small Particles* (Wiley, New York, 1957); M. Kerker, *The Scattering of Light and Other Electromagnetic Radiation* (Academic Press, New York, 1969).
15. P. Chýlek, J. T. Kiehl, M. K. W. Ko, *Phys. Rev. A* **18**, 2229 (1978); *Appl. Opt.* **17**, 3019 (1978).
16. _____, A. Ashkin, in *Light Scattering by Irregularly Shaped Particles*, D. W. Scheuerman, Ed. (Plenum, New York, 1980), p. 153.
17. A. Ashkin and J. M. Dziedzic, unpublished results.
18. R. E. Benner, P. W. Barber, J. F. Owen, R. K. Chang, *Phys. Rev. Lett.* **44**, 475 (1980).
19. L. R. Eaton and S. L. Neste, *AIAA J.* **17**, 261 (1979).
20. A. Ashkin and J. M. Dziedzic, *Phys. Rev. Lett.* **36**, 267 (1976).
21. _____, unpublished results. Such a single electron sensitivity with a feedback scheme was recently demonstrated in an all-electrical modified Millikan support technique [S. Arnold, *J. Aerosol. Sci.* **10**, 49 (1979)].
22. G. Morpurgo, G. Gallinaro, G. Palmieri, *Nucl. Instrum. Methods* **79**, 95 (1970).
23. J. W. Beams, *Science* **120**, 619 (1954); *Rev. Sci. Instrum.* **21**, 182 (1950).
24. N. Y. Misconi, *Geophys. Res. Lett.* **3**, 585 (1976); S. J. Paddock and J. W. Rhee, *ibid.* **2**, 365 (1975).
25. N. Y. Misconi, S. J. Paddock, K. Ratcliff, private communication.
26. G. Roosen, B. G. de Saint Louvent, S. Slansky, *Opt. Commun.* **24**, 116 (1978).
27. M. G. Burt and R. Peirls, *Proc. R. Soc. London Ser A* **333**, 149 (1973).
28. A. Ashkin and J. M. Dziedzic, *Phys. Rev. Lett.* **30**, 139 (1973).
29. J. P. Gordon, *Phys. Rev. A* **8**, 14 (1973).
30. I. Brevik, *Phys. Rep.* **52**, 133 (1979).
31. O. R. Frisch, *Z. Phys.* **86**, 42 (1933).
32. A. Ashkin, *Phys. Rev. Lett.* **25**, 1321 (1970).
33. R. Schieder, H. Walther, L. Woste, *Opt. Commun.* **5**, 402 (1972).
34. A. F. Bernhardt, *Appl. Phys.* **9**, 19 (1976).
35. P. Jacquinot, S. Liberman, J. L. Pique, J. Pinard, *Opt. Commun.* **8**, 163 (1973); A. F. Bernhardt, D. E. Duerre, J. R. Simpson, L. L. Wood, *ibid.* **16**, 166 (1976).
36. J. E. Bjorkholm, A. Ashkin, D. B. Pearson, *Appl. Phys. Lett.* **27**, 534 (1975).
37. T. W. Hänsch and A. L. Schawlow, *Opt. Commun.* **13**, 68 (1975).
38. A. P. Kazantsev, *Zh. Eksp. Teor. Fiz.* **63**, 1628 (1972) [*Sov. Phys. JETP* **36**, 861 (1973)]; *Zh. Eksp. Teor. Fiz.* **66**, 1599 (1974) [*Sov. Phys. JETP* **39**, 783 (1974)].
39. V. S. Letokhov, V. G. Minogin, B. D. Pavlik, *Zh. Eksp. Teor. Fiz.* **72**, 1328 (1977) [*Sov. Phys. JETP* **45**, 698 (1977)]; V. S. Letokhov and V. G. Minogin, *Appl. Phys.* **17**, 99 (1978).
40. J. P. Gordon, private communication.
41. J. E. Bjorkholm, R. R. Freeman, A. Ashkin, D. B. Pearson, *Phys. Rev. Lett.* **41**, 1361 (1978).
42. D. B. Pearson, R. R. Freeman, J. E. Bjorkholm, A. Ashkin, *Appl. Phys. Lett.* **36**, 99 (1980).
43. A. P. Kazantsev, *Usp. Fiz. Nauk* **124**, 113 (1978) [*Sov. Phys. Usp.* **21** (No. 1), 56 (1978)].
44. R. J. Cook, *Phys. Rev. Lett.* **41**, 1788 (1978).
45. _____ and A. F. Bernhardt, *Phys. Rev. A* **18**, 2533 (1978).
46. E. Arimondo, H. Lew, T. Oka. *Phys. Rev. Lett.* **43**, 753 (1979).
47. A. Yu. Pusep, *Zh. Eksp. Teor. Fiz.* **70**, 851 (1976) [*Sov. Phys. JETP* **43**, 441 (1976)].
48. A. Ashkin and J. P. Gordon, *Opt. Lett.* **4**, 161 (1979).
49. J. P. Gordon and A. Ashkin, *Phys. Rev. A* **21**, 1606 (1980).
50. A. P. Botin and A. P. Kazantsev, *Zh. Eksp. Teor. Fiz.* **68**, 2075 (1975) [*Sov. Phys. JETP* **41**, 1038 (1975)].
51. R. J. Cook, *Phys. Rev. Lett.* **44**, 976 (1980).
52. J. E. Bjorkholm, R. R. Freeman, A. Ashkin, D. B. Pearson, *Opt. Lett.* **5**, 111 (1980).
53. V. I. Balikin, V. S. Letokhov, V. I. Mishin, *Pisma Zh. Eksp. Teor. Fiz.* **29**, 614 (1979) [*JETP Lett.* **29**, 560 (1979)].
54. W. Neuhauser, M. Hohenstatt, P. Toschek, H. Dehmelt, *Appl. Phys.* **17**, 123 (1978).
55. W. M. Fairbank, Jr., and C. Y. She, *Opt. News* **5** (No. 2), 4 (1979).
56. D. J. Wineland, R. E. Drullinger, F. L. Walls, *Phys. Rev. Lett.* **40**, 1639 (1978).
57. W. Neuhauser, M. Hohenstatt, P. Toschek, H. Dehmelt, *ibid.* **41**, 233 (1978).
58. D. J. Wineland and W. Itano, *Phys. Rev. A* **20**, 1521 (1979).
59. N. I. Zhokova, A. P. Kazantsev, E. F. Kazentsev, V. P. Sokolov, *Zh. Eksp. Teor. Fiz.* **76**, 896 (1979) [*Sov. Phys. JETP* **49**, 452 (1979)].
60. K. H. Drexhage, *Prog. Opt.* **12**, 165 (1974); W. Lukosz and R. E. Kunz, *J. Opt. Soc. Am.* **67**, 1607 (1977).

PHYSICAL REVIEW A   VOLUME 21, NUMBER 5   MAY 1980

Motion of atoms in a radiation trap

J. P. Gordon and A. Ashkin

Bell Telephone Laboratories, Holmdel, New Jersey 07733

(Received 4 December 1979)

The force exerted by optical-frequency radiation on neutral atoms can be quite substantial, particularly in the neighborhood of an atomic resonance line. In this paper we derive from quantum theory the optical force, its first-order velocity dependence, and its fluctuations for arbitrary light intensity, and apply the results to the problem of creating a stable optical trap for sodium atoms. New results include the position dependence of the velocity-dependent force, a complete expression for the momentum diffusion constant including the substantial contribution from fluctuations of the dipole force, and an estimate of trapping times in excess of 1 sec even in the absence of effective damping. The paper concludes with a discussion of the prospects and difficulties in providing sufficient damping to stabilize such a trap.

INTRODUCTION

The force exerted by optical-frequency radiation on neutral atoms can be quite substantial, particularly in the neighborhood of an atomic resonance line. This force can be viewed, equivalently, as the Lorentz force exerted by the field on the optical-frequency atomic dipoles, or as the consequence of momentum conservation in the absorption and reemission of light by the atoms. The total force has three distinguishable components, resulting, respectively, from absorption, spontaneous emission, and induced emission. The first two of these[1,2] have together been called the scattering force, and are the only forces present for plane-wave radiation. Spontaneous emission, by the symmetry of its angular distribution, contributes only to the fluctuations of this force component. The third force component[3-6] is dispersive in nature, and, below resonance, results in an attraction of the atom toward regions of higher field intensity. It has been called the dipole force, and is the near-resonance form of electrostriction in a gas.

The term resonance radiation pressure has been used, heuristically, to denote these forces. Among early proposals for their use were suggestions[3,2,7] for confining atoms to restricted regions of space. The concept of optically cooling atomic motions by resonance radiation forces[8] led to proposals[5,6,9] of spatial traps for atoms in which their temperature could be cooled to ~$(10^{-3}$–$10^{-4})$ K. Such a trap would render the atoms available for long times for spectroscopic studies or other novel experiments.

In this work we consider the motion of atoms situated in optical resonance radiation traps. We derive the radiation pressure forces, their velocity dependence, and their quantum fluctuations in a consistent way starting from fundamentals. The

calculations yield new results on both the velocity dependence and quantum fluctuations of the force and lead to some new insights into the overall trapping problem. The average scattering force, by conservation of momentum, is in the direction of the incident wave vector. It has been detected in atomic-beam-deflection experiments[10,11] and by the generation of significant pressure differences in atomic vapors.[12] The strong velocity dependence on this force has led to the concept of optical cooling or damping.[8] If the frequency of the light is tuned below resonance, then any atomic motion either toward or away from the light beam results in a sizable Doppler shift leading to an incremental force opposing the motion. The idea of optical cooling was proposed independently within the context of ion trapping.[13] Experimental cooling of ions held in electromagnetic traps has recently been observed.[14,15] The average dipole force is directed along the gradient of the optical field intensity and is dispersive in character, being in the direction of the gradient when below resonance, and in the opposite direction when above resonance. Dipole forces have recently been observed experimentally in atomic beam focusing and defocusing experiments.[16] In a somewhat different context, nonresonant dipole forces have been invoked[17] to explain observations on the distortion of liquid surfaces by radiation pressure.[18] For the resonance radiation forces on atoms the effect of saturation and tuning must be considered[2,6] to obtain the greatest force for a given power.

Many uses have been suggested for radiative forces on atoms, such as velocity selection,[2] isotope separation[2,11,19,16] production of transversely cooled[8] or axially slowed and confined atomic beams,[16] and acceleration of neutral atoms to high velocity.[4] Transient dipole forces also exist for light tuned to exact resonance if the atom interacts with the field for times short compared to the natural lifetime.[19-21] Atomic-beam deflection has recently been observed under these circumstances.[22] Analogous effects have been seen in the deflection of molecular beams by resonant microwave electric fields.[23] We shall, however, restrict our present discussion principally to the initially mentioned application of optical trapping and cooling.

Basically an optical trap consists of an optical field configuration[5,6,9] with a point of stable equilibrium such that any displacement of an atom from this point results in an average restoring force. The maximum kinetic energy (at the equilibrium point) of an atom that can be confined in a trap is defined as the well depth. An important aspect of such traps is the fact that the trapping forces have fluctuations due to the quantum nature of the interactions. These fluctuations constitute a source of heat. In the absence of cooling this will result in a finite retention time for atoms introduced into the trap. The fluctuations due to the scattering force can be readily appreciated on intuitive grounds.[24,5,9,14] Recently evidence on heating due to the scattering force was observed experimentally.[25] Fluctuations in the dipole force[26,9] are conceptually more difficult to understand and have not until now been adequately considered in the context of traps.

The heating of atoms by fluctuations must be counteracted by cooling if atoms are to be retained in the trap. Indeed the equilibrium kinetic energy of an atom in a trap results from a balance of the fluctuation heating and the degree of optical cooling that exists. The effectiveness of a trap in containing an atom is determined by the Boltzmann factor given by the ratio of the well depth to the equilibrium kinetic energy. In practice we seek a low average kinetic energy for trapped atoms and a large Boltzmann factor.

Letokhov et al.[5] proposed trapping atoms in large volumes on the standing wave maxima of 3 orthogonal pairs of standing waves tuned $\sim \Gamma/2$ below the atomic resonance, where Γ is the radiative decay rate of the atom, the reciprocal of the radiative lifetime τ. This tuning gives optimum damping and results in a minimum kinetic energy $\sim \hbar \Gamma$. Unfortunately this trap has a well depth which is also $\sim \hbar \Gamma$ giving a Boltzmann factor of about unity. Thus thermal excitation out of this trap is probable and the trap is very leaky. The traps proposed by Ashkin et al.[6,9] are based on strongly focused Gaussian beams tuned $\sim(10^2-10^3)\Gamma$ below resonance. In this way traps of depth $\sim(10^2-10^3)\hbar\Gamma$ are achieved but at a price of much reduced damping. In fact, the Boltzmann factor is again ~ 1. However, it has been proposed[9] to use additional optimally tuned damping

beams with these focused beam traps with the intent of simultaneously obtaining trap depths of $(10^2-10^3)\hbar\Gamma$ and minimum temperatures $\sim \hbar\Gamma$ to give highly stable traps with Boltzmann factors of 100 or more.

Currently the outstanding conceptual problems remaining for the understanding of optical traps for neutral atoms are the questions of the magnitude of the fluctuations of the dipole force and the viability of the concept of separate trapping and damping beams. In the earlier estimates[5,9] of minimum kinetic energy the contributions of the dipole force fluctuations were neglected. One of the principal results of this work is a quantum-mechanical calculation of the dipole fluctuations exact to arbitrary field strength. The results show that these fluctuations can often be large. In standing waves dipole force fluctuations make a contribution which in the absence of saturation effects is the same size as the scattering force fluctuations. Interestingly, their spatial variation is such that when added to the scattering fluctuations, there results a velocity diffusion constant which is independent of position in the standing wave. In Gaussian beam traps and at high powers dipole fluctuations can exceed the spontaneous scattering force fluctuations. Fortunately, however, conditions exist in the Gaussian trap of Refs. 6 and 9 where the dipole force fluctuation can be neglected relative to the scattering force fluctuations. This paper draws no new conclusions about the problem of using separate trapping and damping beams. However, we develop the whole subject of cooling and heating in a unified way and give some new insights. For example, a new calculation is given of the damping for a standing-wave field as a function of position, correct to all intensities, for low atomic velocities. An important aspect of this result is that the damping varies with position and is zero at the maximum of the standing-wave fields where atoms are expected to collect. This represents another complication for single-frequency standing-wave traps as proposed in Ref. 5. Although an atom can be viewed as a simple harmonic oscillator (SHO) only when saturation effects are absent, it is shown that a single beam trap is capable of stably trapping a SHO. This result at least shows that nature does not necessarily abhor an optical trap. Finally, our estimate of the magnitudes of the dipole and spontaneous force fluctuations point out a new experimental possibility. This involves an estimate of the retention lifetime of an atom put into a single beam trap in the absence of significant cooling. It is shown that by working at sufficiently low saturation and with low dipole fluctuations, lifetimes of many seconds are possible for deep traps using reasonable optical powers. This implies that an experimental demonstration of trapping is not contingent on cooling. Cooling can then be studied subsequently by the addition of separate damping beams, for example. Parenthetically, one further important use of traps may be as an experimental probe for the study of the fluctuations of the radiation pressure forces themselves.

The approach used in our calculation of the force, its velocity dependence, and the atom's momentum diffusion is to treat the interaction of the optical radiation field with the atom's momentum quantum mechanically. The quantum treatment of field and momentum is only necessary in finding the force fluctuations. That, however, being a needed result, the quantum treatment is used throughout. Two approximations are used. First, we treat the atom's position as a classical variable. In the case of sodium, or any comparably heavy atom, this approximation is justified by its small de Broglie wavelength, which amounts to only 0.03 μm for sodium atom whose kinetic energy is of the order of the natural width $\hbar\Gamma$ of its resonance level. For sodium, $\Gamma = (16.1 \text{ nsec})^{-1}$. Alternatively one can appeal to Heisenberg's uncertainty principle, which implies that if an atom's position is defined to $\sim \lambda/2\pi$, its momentum uncertainty is equal to that occasioned by the random scattering of one photon. As we are concerned here with the cumulative effects of the scattering of many photons, the neglect of an equivalent few more should have no major consequence. Second, we treat the velocity as small, retaining effects only to first order in v. In context, the reason for this is that atoms caught in an optical trap must be moving quite slowly. Again using numbers typical of the sodium resonance line at 590-nm wavelength, the trap depth might be of the order $10^3\hbar\Gamma$. A trapped atom with, say, one tenth of that energy would have an effective kinetic temperature of about 0.02 K, and would travel, at most, about 0.1 μm in one lifetime. This being a small fraction of an optical wavelength, it is appropriate, indeed desirable, to treat the atom as moving slowly. A symptom of such slowness is that the Doppler shift caused by the atom's motion is less than the natural width.

CALCULATION OF THE FORCES AND MOMENTUM DIFFUSION

In the electric-dipole approximation the force of radiation on a neutral, slowly moving atom is given by

$$f_i = \vec{\mu} \cdot \partial \vec{E}/\partial x_i , \qquad (1)$$

where $\vec{\mu}$ is the atomic dipole moment, $\vec{E}$ is the electric field, and f_i is the ith Cartesian compo-

nent of the force. Expression (1) includes both the force the electric field exerts on the dipole and that the magnetic field exerts of the associated current.

The force (1), and all the other quantities we need may be derived from the dipole Hamiltonian function,

$$\mathcal{K} = H_{\text{field}} + H_{\text{atom}} - \vec{\mu} \cdot \vec{E}(\vec{x}) , \qquad (2)$$

where

$$H_{\text{field}} = \frac{1}{8\pi} \int (E^2 + H^2)dV$$

and

$$H_{\text{atom}} = P^2/2M + \hbar\omega_0 \sigma_{22} .$$

In (2) $\vec{\mu}$ is the atomic dipole moment, $\vec{E}$ is the electric field of the radiation, $\vec{H}$ the magnetic field, and σ_{22} is the projection operator for the upper atom level. Also $\vec{P}$ is the atom's momentum, $\vec{x}$ its position, and M its mass.

If we deal with sharply resonant two-level atoms, we can approximate the dipolar energy term by

$$\vec{\mu} \cdot \vec{E} \approx (\vec{\mu}_{12} \cdot \vec{E}^\dagger)\sigma + \sigma^\dagger (\vec{\mu}_{21} \cdot \vec{E}) . \qquad (3)$$

Here the total field $\vec{E}$ has been expanded according to

$$\vec{E} = \vec{E}e^{-i\omega t} + \vec{E}^\dagger e^{i\omega t} ,$$

where the new $\vec{E}$ contains the "positive"-frequency part of the field, and represents an energy-lowering operator, while $\vec{\mu}$ has been similarly expanded according to

$$\vec{\mu} = \vec{\mu}_{12}\sigma e^{-i\omega t} + \vec{\mu}_{21}\sigma^\dagger e^{i\omega t} ,$$

where $\vec{\mu}_{12}$ is the dipole matrix element connecting the two pertinent atomic levels, and σ is the lowering operator for the atom. The operators $E^\dagger$ and $\sigma^\dagger$ are the Hermitian-conjugate energy-raising operators. In the dipolar energy (3), we have kept only the secular, or energy-conserving, terms, and have arranged them in normal order (lowering operators to the right). The explicit carrier frequency ω is added for later convenience. For a monochromatic applied field, ω will be chosen as the field frequency.

We now wish to find expressions for the mean force on the atom, the first-order velocity dependence of that force, and finally the two-time autocorrelation function of the force, which determines the momentum diffusion constant. The first two of these require only semiclassical theory, but the last requires that we stay with the quantum theory. We shall be working in the Heisenberg picture, where the operators are time dependent, satisfying the equations of motion

$$dO/dt = -(i/\hbar)[\,O, \mathcal{K}\,]\,. \tag{4}$$

Here the square bracket represents the commutator, and the (constant) quantum state of the system is prescribed at some appropriate initial time. Note that, e.g., $\sigma e^{-i\omega t}$ represents a Heisenberg operator. Using (2), (3), and (4) we can immediately write down the equation for the force on the atom, namely

$$\vec{f} \equiv d\vec{P}/dt = -\vec{\text{grad}}(\mathcal{K})$$
$$= \sigma^{\dagger}\,\vec{\text{grad}}[\,\vec{\mu}_{21} \cdot \vec{E}(x)] + \text{H.c.}, \tag{5}$$

where H.c. represents the Hermitian-conjugate operator. Note that (5) has the same form as the classical equation (1).

The next step, solving for the field at the atom, is plagued by the usual difficulties of quantum electrodynamics. However, if we can approximate that field by the sum of the external field and the radiation reaction field, assuming that the effect of the divergent local dipole field can be suitably renormalized into the excitation energy of the atom, our purpose is served. Thus we assume that $\vec{E}(\vec{x})$ can be developed as

$$\vec{E}(\vec{x}) = \vec{E}^{0}(\vec{x}) + i\,\tfrac{2}{3}\,k^{3}\vec{\mu}_{12}\sigma, \tag{6}$$

where $E^{0}(\vec{x})$ is the free external incident field, and the other term is the local reaction field, with $k = \omega/c$, where ω is the frequency of the atomic dipole. Recalling that the reaction field has no gradient at the atom's position, we note that it does not contribute to the force (5). Hence we can write

$$\vec{f} = -i\sigma^{\dagger}\,\vec{\text{grad}}G + \text{H.c.}, \tag{7}$$

where

$$G \equiv i\vec{\mu}_{21} \cdot \vec{E}^{0}(\vec{x}, t)/\hbar\,.$$

We now turn to the equations of motion of the atomic operators. Let σ_{11} be the projection operator for the lower atomic level, and D be the population difference

$$D \equiv \sigma_{11} - \sigma_{22}\,.$$

(Note $\sigma_{11} + \sigma_{22} = 1$.) Using the operator relations

$$\sigma_{ij}\sigma_{hl} = \sigma_{il}\delta(j,k), \tag{8}$$

where $\sigma_{12} \equiv \sigma$, $\sigma_{21} \equiv \sigma^{\dagger}$, and $\delta(j,k)$ is the Kronecker delta function, we gain from (4) the usual equations of motion

$$\dot{\sigma} - i\Omega\sigma = (i/\hbar)D(\vec{\mu}_{21} \cdot \vec{E})\,,$$

and

$$\dot{D} = (2i/\hbar)[\,(\vec{\mu}_{12} \cdot E^{\dagger})\sigma - \sigma^{\dagger}(\vec{\mu}_{21} \cdot E)]\,,$$

where $\Omega \equiv \omega - \omega_{0}$ is the detuning of the atomic resonance frequency ω_{0} from the chosen ω. Next

we make the reaction field approximation (6), to obtain [using (8)]

$$\dot{\sigma} + (\Gamma/2 - i\Omega)\sigma = DG$$

and $\tag{9}$

$$\dot{D} + \Gamma D = \Gamma - 2(G^{\dagger}\sigma + \sigma^{\dagger}G)\,,$$

where $\Gamma = (4/3\hbar)k^{3}|\mu_{12}|^{2}$ is the usual expression for the natural radiative decay rate of the atom. Note that the operator order in Eqs. (9) has become important, for while $\vec{E}(\vec{x})$ commutes with $\sigma^{\dagger}$ at the same time, for example, as they represent different physical entities, it is evident from (6) that the external field $\vec{E}^{0}(\vec{x})$ (and hence G) does not. The reason for choosing normal ordering will surface in the next paragraph.

Let us now consider the quantum expectation value of (7) and (9). We may represent the initial state of the system by a Dirac ket $|S\rangle$, or sometimes more simply just by $\rangle$. The quantum expectations of (7) and (9) are then just those equations surrounded by angular brackets, as in $abc \rightarrow \langle abc\rangle$. If the initial field is the coherent state $|E'(\vec{r})\rangle$ we can apply the well-known result that

$$\vec{E}^{0}(\vec{r}, t)|\vec{E}'(r)\rangle = |\vec{E}'(\vec{r})\rangle\vec{E}'(\vec{r}, t)\,, \tag{10}$$

where $\vec{E}'(\vec{r}, t)e^{-i\omega t}$ represents the classical field satisfying the vacuum Maxwell equations and having $\vec{E}'(\vec{r})$ as its initial value. We use $\vec{r}$ to represent any point in space, and $\vec{x}$ to represent the atom's position. Thus, because of the normal ordering of Eqs. (7) and (9), if we assume the above initial coherent field state, we achieve the following equations

$$\langle \vec{f}\rangle = -i\hbar[\langle \sigma\rangle^{*}\vec{\text{grad}}g - \langle \sigma\rangle\,\vec{\text{grad}}g^{*}]\,,$$

$$\langle \dot{\sigma}\rangle + (\Gamma/2 - i\Omega)\langle \sigma\rangle = \langle D\rangle\,g\,, \tag{11}$$

$$\langle \dot{D}\rangle + \Gamma\langle D\rangle = \Gamma - 2(g^{*}\langle \sigma\rangle + g\langle \sigma\rangle^{*})\,,$$

where

$$g = i\vec{\mu}_{21} \cdot \vec{E}'(\vec{x}, t)/\hbar = \langle G\rangle\,.$$

Here the complex conjugate (*) has replaced the Hermitian conjugate ($\dagger$) of (7) and (9), and we have with a minimum of complexity obtained the appropriate semiclassical equations with damping.

If we represent $\vec{\text{grad}}g$ by

$$\vec{\text{grad}}g = (\vec{\alpha} + i\vec{\beta})g\,, \tag{12}$$

where $\vec{\alpha}$ and $\vec{\beta}$ are real (Note: if $g = ue^{i\phi}$ with u and ϕ real, then $\vec{\alpha} = \text{grad}\ln u$ and $\vec{\beta} = \text{grad}\phi$) then the force equation expands to

$$\langle \vec{f}\rangle = \vec{\alpha}[\,i\hbar(g^{*}\langle \sigma\rangle - g\langle \sigma\rangle^{*})]$$
$$+ \hbar\vec{\beta}(g^{*}\langle \sigma\rangle + g\langle \sigma\rangle^{*})\,. \tag{13}$$

In (13), the coefficient of $\vec{\alpha}$ is the negative of the expectation value of the dipolar interaction energy; i.e., $\langle \vec{\mu} \cdot E^0 \rangle$ if we use (3). This is the dipole force familiar also for dc fields. The coefficient of $\hbar \vec{\beta}$ is the absorption rate, as may be seen from the third of Eqs. (11). This is the force component usually called the scattering force.

If the external field $E'(\vec{x}, t)e^{-i\omega t}$ is monochromatic, of frequency ω, and the atom is motionless, then $\vec{E}'$ and hence g is independent of time, and the stationary solution of (11) is

$$\langle \sigma \rangle = g/[\gamma(1+p)] ,$$
$$\langle D \rangle = (1+p)^{-1} , \tag{14}$$

where

$$\gamma \equiv \Gamma/2 - i\Omega , \quad p \equiv 2|g|^2/|\gamma|^2 ,$$

and the force, from (13), is

$$\langle \vec{f} \rangle = \hbar p (1+p)^{-1}(-\Omega \vec{\alpha} + \Gamma \vec{\beta}/2) . \tag{15}$$

The quantity p is called the saturation parameter.

From (15) we can demonstrate that the scattering force is associated with spontaneous emission, the dipole force with a coherent redistribution of the incident field due to stimulated emission. The stationary upper-state probability $\langle \sigma_{22} \rangle = p/[2(1+p)]$. Hence one may express the scattering force component as

$$\langle \vec{f} \rangle_{\text{scat}} = \Gamma \langle \sigma_{22} \rangle \hbar \vec{\beta} .$$

This evidently may be regarded as the result of removing quanta of average momentum $\hbar \vec{\beta}$ from the incident external field at the same rate that the atom is undergoing spontaneous decay. The dipole force by contrast depends on the detuning Ω as well as on the excitation of the atom. It must therefore depend on the phase relationship between the mean dipole $\vec{\mu}_{12}\langle \sigma \rangle$ and the external field $\vec{E}'$. It may be regarded as the result of the redistribution of field momentum caused by coherent interference between the emitted field of the dipole and the outgoing waves of the incident field. For example, an atom below resonance (negative Ω) sitting in a focused Gaussian beam is pulled toward the beam focus by the dipole force because the atom acts rather like a weak positive lens. The "light scattering" involved in generating this force is the intrabeam coherent forward scattering.

Now we examine the first-order velocity dependence of the force. A moving atom experiences a modified field since

$$\frac{d\vec{E}'(\vec{x}, t)}{dt} = \frac{\partial \vec{E}'(\vec{x}, t)}{\partial t} + (\vec{v} \cdot \text{grad})\vec{E}'(\vec{x}, t) .$$

For the same monochromatic field as above, we have, using (12),

$$\dot{g} = \vec{v} \cdot (\vec{\alpha} + i\vec{\beta})g . \tag{16}$$

This modifies the solution of (11). We can obtain an expression for the force accurate to first order in the velocity by taking the time derivative of the zero-order solutions (14), using (16), and then using these first-order results for $\langle \dot{\sigma} \rangle$ and $\langle \dot{D} \rangle$ to re-solve Eqs. (11) to first order in $\vec{v}$. We find thus

$$\langle \dot{D} \rangle \cong -\frac{2p}{1+p}(\vec{v} \cdot \vec{\alpha})\langle D \rangle ,$$

$$\langle \dot{\sigma} \rangle \cong \left((\vec{v} \cdot \vec{\sigma})\frac{1-p}{1+p} + i(\vec{v} \cdot \vec{\beta})\right)\langle \sigma \rangle ,$$

and it is then straightforward to solve Eqs. (11) again to find the modified force. In the simple case of a plane wave of wave vector $\vec{k}$, we have $\vec{\alpha} = 0$, $\vec{\beta} = \vec{k}$, so that in this case the only change in Eq. (15) is the Doppler shift $\Omega \to \Omega - \vec{v} \cdot \vec{k}$. This change modifies p. Then to first order we find

$$\langle \vec{f} \rangle = \hbar p (1+p)^{-1}(\Gamma/2)\vec{k}\left(1 + \frac{2\Omega (\vec{v} \cdot \vec{k})}{|\gamma|^2(1+p)}\right) . \tag{17}$$

For negative Ω, this velocity dependence damps the motion of the atom along $\vec{k}$, a necessity for a stable trap. Note that the damping force is maximized for $p = 1$ and $\Omega = -\Gamma/2$. Another relatively simple case that has been examined in the literature is the case of a pure standing wave. For this case we have in (12), $g = 2g_0 \cos(\vec{k} \cdot \vec{x})$; whence $\vec{\beta} = 0$, $\vec{\alpha} = -\vec{k}\tan(\vec{k} \cdot \vec{x})$, and we obtain after some algebra the result

$$\langle \vec{f} \rangle = \frac{\hbar p}{1+p}[\Omega \vec{k} \tan(\vec{k} \cdot \vec{x})]$$
$$\times \left(1 + \frac{\Gamma^2(1-p) - 2p^2|\gamma|^2}{\Gamma|\gamma|^2(1+p)^2}(\vec{v} \cdot \vec{k})\tan(\vec{k} \cdot \vec{x})\right). \tag{18}$$

Here p represents the local value of the saturation parameter, or $p = 4p_0 \cos^2(\vec{k} \cdot \vec{x})$, where p_0 is the saturation parameter corresponding to one of the two oppositely directed traveling waves that comprise the standing wave. For small values of p_0, the force reduces to

$$\langle \vec{f} \rangle = 2\hbar \vec{k} p_0 \Omega \{ \sin(2\vec{k} \cdot \vec{x})$$
$$+ (\Gamma/|\gamma|^2)\vec{v} \cdot \vec{k}[1 - \cos(2\vec{k} \cdot \vec{x})]\} .$$

Our result (18) is consistent with a derivation[27] of the first few spatial Fourier components of the force, which were derived for all velocities. In particular, we checked that the spatial averages were the same to first order in the velocity at all intensities. Two things are noteworthy about (18). First, the velocity-dependent force vanishes at the standing-wave maxima (e.g., at $\vec{k} \cdot \vec{x} = 0$), exactly where one might expect trapped atoms to accumulate. In addition, there is a sign reversal of this force when $p^2/(1-p) = \Gamma^2/2|\gamma|^2$, which, for large detuning $|\Omega| \gg \Gamma$, can have a quite small value. Thus, for negative Ω, which yields traps at the standing-wave maxima (a desirable feature since this also makes a trap for the directions perpendicular to $\vec{k}$), the damping which exists at low intensities in the neighborhood of the maxima reverses sign and transforms into heating at rather small values of the saturation parameter p. These features complicate the conception of trapping atoms in a standing wave, as in Ref. 5.

We turn now to the investigation of the force fluctuations, which ultimately determine how long an atom will stay in the trap. We assume a mono-chromatic field, negligible velocity, and quasi-stationary conditions. We seek the value of the momentum diffusion constant $2D_p$ due to the quantum fluctuations of the force. It is given by

$$2D_p = (d/dt)(\langle \vec{P} \cdot \vec{P} \rangle - \langle \vec{P} \rangle \cdot \langle \vec{P} \rangle)$$
$$= 2\operatorname{Re}(\langle \vec{P} \cdot \vec{f} \rangle - \langle \vec{P} \rangle \cdot \langle \vec{f} \rangle)$$
$$= 2\operatorname{Re} \int_{-\infty}^{0} dt \left[\langle \vec{f}(t) \cdot \vec{f}(0) \rangle - \langle \vec{f}(t) \rangle \cdot \langle \vec{f}(0) \rangle \right],$$
$$(19)$$

where the second line results from $\vec{f} = d\vec{P}/dt$, and the third from expressing $\vec{P}$ as the time integral of the force. Note that the product of two Hermitian operators is Hermitian only if they commute, which $\vec{P}$ and $\vec{f}$ do not. The time zero is arbitrary, and the time minus infinity is an exaggeration, since the autocorrelation of the force lasts only for the order of the atomic lifetime Γ^{-1}. For quasistationary conditions, we can advance the time arguments of the integrand by $|t|$ obtaining equivalently,

$$D_p = \operatorname{Re} \int_0^\infty dt \left[\langle \vec{f}(0) \cdot \vec{f}(t) \rangle - \langle f \rangle^2 \right]. \qquad (20)$$

A quantity written without a specific time argument is assumed to have its stationary equilibrium value; and $\langle f \rangle^2 \equiv \langle \vec{f} \rangle \cdot \langle \vec{f} \rangle$. Inserting (7) for the forces, and using (10) we obtain the result

$$\langle \vec{f}(0) \cdot \vec{f}(t) \rangle = \hbar^2 \{ \langle \sigma^\dagger(0)\sigma(t) + \sigma(0)\sigma^\dagger(t) \rangle |(\operatorname{grad}g)^2|$$
$$- \langle \sigma^\dagger(0)\sigma^\dagger(t) \rangle (\operatorname{grad}g)^2$$
$$- \langle \sigma(0)\sigma(t) \rangle (\operatorname{grad}g^*)^2$$
$$+ \langle \sigma^\dagger(0)\operatorname{com}(0,t)\sigma(t) \rangle \}, \qquad (21)$$

where

$$\operatorname{com}(0,t) \equiv \vec{\operatorname{grad}}G(0) \cdot \vec{\operatorname{grad}}G^\dagger(t)$$
$$- \vec{\operatorname{grad}}G^\dagger(t) \cdot \vec{\operatorname{grad}}G(0).$$

The last term of (21) arises upon rearranging the field operators into normal order so that (10) may

be used. One can show that the field *gradient* operators commute with the atomic operators at all times.

Consider first the final term of (21). It is the only term which depends on the quantum fluctuations of the field (i.e., on a field commutator). One would thus expect it to yield the effects of spontaneous emission. The quantity $\operatorname{com}(0,t)$ is a free field commutator; its value, casting out a possible high-frequency divergence as usual, is

$$\operatorname{com}(0,t) = k^2 \Gamma \delta(t).$$

The last term of (21) thus reduces to

$$(\hbar k)^2 \Gamma \langle \sigma_{22} \rangle \delta(t), \qquad (22)$$

and indeed this result can be modeled by the random instantaneous emission of quanta of momentum $\hbar k$ at an average rate $\Gamma \langle \sigma_{22} \rangle$, as one might expect.

The remaining terms of (21) encompass the effects of the external field gradient interacting with the atomic dipole fluctuations, and constitute the same result one would obtain from a semiclassical theory. The autocorrelation times involved are, as we have mentioned above, of the order of the atomic lifetime, and their modeling in terms of the emission and absorption of quanta is not at all obvious. To proceed with the evaluation, we see that we need quantities such as

$$u \equiv \int_0^\infty dt \left[\langle \sigma^\dagger(0)\sigma(t) \rangle - \langle \sigma^\dagger \rangle \langle \sigma \rangle \right]. \qquad (23)$$

Suppose we have solved the set of first-moment equations in (11) for *arbitrary* initial conditions to yield a result of the form

$$\langle \sigma(t) \rangle = a_0(t) + a_1(t)\langle \sigma(0) \rangle$$
$$+ a_2(t)\langle \sigma^\dagger(0) \rangle + a_3(t)\langle D(0) \rangle. \qquad (24)$$

One can show by study of its equation of motion[28] that the quantity $\langle \sigma^\dagger(0)\sigma(t) \rangle$ has the similar solution

$$\langle \sigma^\dagger(0)\sigma(t) \rangle = a_0(t)\langle \sigma^\dagger(0) \rangle + a_1(t)\langle \sigma^\dagger(0)\sigma(0) \rangle$$
$$+ a_2(t)\langle \sigma^\dagger(0)\sigma^\dagger(0) \rangle + a_3(t)\langle \sigma^\dagger(0)D(0) \rangle$$
$$= [a_0(t) + a_3(t)]\langle \sigma^\dagger(0) \rangle + a_1(t)\langle \sigma_{22}(0) \rangle. \qquad (25)$$

The solution for u that we seek follows from insertion of (25) into (23), yielding

$$u = (A_0 + A_3)\langle \sigma^\dagger \rangle + A_1\langle \sigma_{22} \rangle, \qquad (26)$$

where

$$A_i \equiv \int_0^\infty dt \left[a_i(t) - \langle \sigma \rangle \delta(i,0) \right], \quad i = 0\text{-}3. \qquad (27)$$

The other terms of (20) may be treated similarly;

we obtain thus

$$2D_p = 2\hbar^2 \operatorname{Re}[(2A_0\langle \sigma^\dagger \rangle + A_1)|(\operatorname{grad} g)^2|$$

$$+ (2A_0\langle \sigma \rangle + A_2)(\operatorname{grad} g^*)^2]$$

$$+ (\hbar k)^2 \Gamma \langle \sigma_{22} \rangle. \tag{28}$$

The quantities A_i may be conveniently found by taking and solving the Laplace transform of Eqs. (11). One finds thus

$$A_0 = \frac{g}{|\gamma|^2(1+p)^2}\left[\frac{\gamma}{\Gamma} - \frac{\Gamma}{\gamma} - i2p\left(\frac{\Omega}{\Gamma}\right) \right],$$

$$A_1 = \frac{\Gamma + p\gamma}{\gamma\,\Gamma(1+p)}, \quad A_2 = -\frac{2g^2}{|\gamma|^2\,\Gamma(1+p)}, \tag{29}$$

$$A_3 = \frac{g}{\gamma\,\Gamma(1+p)}.$$

Finally, using these results, along with the stationary values (14) of the atomic variables, and using (12) for $\operatorname{grad} g$, we obtain the final result

$$2D_p = \hbar^2\alpha^2\Gamma\,\frac{p}{2(1+p)^3}\left[1 + \left(\frac{\Gamma^2}{|\gamma|} - 1 \right)p + 3p^2 + \frac{4|\gamma|^2}{\Gamma^2}p^3 \right] + \hbar^2\beta^2\Gamma\,\frac{p}{2(1+p)^3}\left[1 + \left(3 - \frac{\Gamma^2}{|\gamma|^2} \right)p + p^2 \right]$$

$$+ 2\hbar^2(\vec{\alpha}\cdot\vec{\beta})\Omega\,\frac{p^2}{(1+p)^3}\left[\frac{\Gamma^2}{|\gamma|^2} + p \right] + (\hbar k)^2\Gamma\,\frac{p}{2(1+p)}. \tag{30}$$

We now have in hand the quantities necessary for discussion of trap stability.

DISCUSSION OF THE DIFFUSION CONSTANT

The above result for the diffusion constant has some properties which are at first sight somewhat surprising; hence it merits some discussion. For small excitation of the atom (to first order in the saturation parameter p) we find

$$2D_p \approx \tfrac{1}{2}\hbar^2\Gamma p(k^2 + \alpha^2 + \beta^2)$$

$$\approx \hbar^2\Gamma\langle\sigma_{22}\rangle\,(k^2 + \alpha^2 + \beta^2). \tag{31}$$

We see here three terms, each associated with one of the elementary processes, absorption (β^2), induced emission (α^2), and spontaneous emission (k^2). The spontaneous-emission term can be alternatively associated with the interaction of the semiclassical dipole (note that for small p, $\langle \sigma_{22} \rangle \approx |\langle \sigma \rangle|^2$) with the zero-point field fluctuation, while the other two terms may be similarly associated with interaction of the semiclassical field gradient (recall that $\alpha^2 + \beta^2 = |g^{-1}\operatorname{grad} g|^2$) with the zero-point dipole fluctuation. In this regard we note that

$$\langle \vec{\mu} \cdot \vec{\mu} \rangle = |\mu_{12}|^2 \langle \sigma^\dagger\sigma + \sigma\sigma^\dagger \rangle = |\mu_{12}|^2,$$

independent of the state of the atom. Thus an atom even in its ground state has a substantial random dipole moment which gives rise to a random force in interaction with the external classical field gradient. Another view may be had by noting that a weakly excited two-level system is indistinguishable from a one-dimensional harmonic oscillator, whose zero-point fluctuation can interact with the external field gradient. In passing, we remark that calculation of the diffusion constant for a one-dimensional harmonic oscillator yields exactly (31), with $\langle \sigma_{22} \rangle$ replaced by $\langle n \rangle$, the mean excitation number of the oscillator.

An interesting and perhaps somewhat unexpected aspect of (31) occurs when the atom is in the presence of several beams of radiation of the same frequency but different directions. In particular, consider the standing-wave case

$$g = 2g_0 \cos(\vec{k}\cdot\vec{x}).$$

Then

$$\vec{\beta} = 0, \quad \vec{\alpha} = -\vec{k}\tan(\vec{k}\cdot\vec{x}),$$

$$D_p = (\hbar k)^2\Gamma p_0,$$

where $p_0 = 2|g_0|^2/|\gamma|^2$ represents the saturation parameter corresponding to a single one of the two associated traveling waves. We see that the diffusion constant is independent of the atom's position in the standing wave, even though the field strength, excitation of the atom, and mean force are strongly position dependent. The explanation of this curious behavior we have noted above; that is, in the field minima the diffusion results from the interaction of the zero-point dipole fluctuation with the large gradient of the external field amplitude. Further, we see that in this approximation, the diffusion depends only on p_0; hence if we hold p_0 fixed while increasing $|\Omega|$, we can increase the depth of the sinusoidal potential while the diffusion constant remains unchanged. This is encouraging with regard to trapping atoms, but fortunately does not ensure a stable trap, because the damping is reduced as $|\Omega|$ increases.

The other particularly interesting feature of (30) is the term proportional to $\alpha^2 p^4(1+p)^{-3}$, which becomes dominant at large p. It is the only term which does not saturate or decrease for large p. Thus, if $p \gg 1$, and $\vec{\alpha} \neq 0$ we find,

$$2D_p \to 2\hbar^2\alpha^2 p|\gamma|^2/\Gamma = 4\hbar^2\alpha^2|g|^2/\Gamma. \tag{32}$$

It is noteworthy that this term can contribute significantly to the diffusion constant even for small p when the detuning Ω is large. Thus it merits careful consideration in constructing a trap. Its proportionality to α^2 shows it to be associated with the dipole force.

At large p we can understand (32) by a not-too-complex argument. In this limit, the atom is strongly coupled to the externally excited field mode, and it is appropriate to consider the atom and that one field mode as a single quantum system, according to the picture of the "dressed" atom.[29] This system has two quasistationary states, in each of which the atom has an equal mixture of upper and lower levels, with its dipole $\vec{\mu}_{12}\langle\sigma\rangle$, respectively, in phase with and in opposition to the field, giving them equal and opposite interaction energies $\pm|\vec{\mu}_{12}\cdot\vec{E}'|$ and thus equal and opposite dipole forces. Each time this system spontaneously emits a photon into one of the unexcited field modes, the atom of necessity finds itself in its lower state, and thus in an equal mixture of the above two quasistationary states. Thus, after each spontaneous emission event the atom is forced randomly in either direction by the field gradient, the same steady force persisting until the next decay. Finally, the spontaneous decay rate of each quasistationary state is $\Gamma/2$, since in each such state the upper atomic level probability is $\frac{1}{2}$ and the spontaneous emission rate is always proportional to the upper level probability. On this basis one can calculate the resulting diffusion constant. The interaction energies are

$$\pm\hbar|g|,$$

hence the forces $\vec{f}$ are $\pm\hbar\,\vec{\text{grad}}\,|g| = \pm\hbar\vec{\alpha}|g|$. The diffusion constant from (19) is

$$2D_p = 2\int_{-\infty}^{0} dt\,\langle\vec{f}(t)\cdot\vec{f}(0)\rangle,$$

where we have used the fact that the forces are real. To make the required average we observe that the force is constant between decays and takes a new random direction at each decay. Thus $2D_p$ reduces to

$$2D_p = 2f^2\langle t_d\rangle,$$

where $-t_d$ is the time of the last spontaneous decay prior to time zero. Since the spontaneous decay rate is $\Gamma/2$, we have $\langle t_d\rangle = 2/\Gamma$, and hence

$$2D_p = 4f^2/\Gamma = 4\hbar^2\alpha^2|g|^2/\Gamma,$$

in exact agreement with (32).

In conclusion of this section, we will risk some remarks concerning the photon concept. It is most precise, and often useful, to think of the momentum exchange between atom and field as oc-curring in quantum units $\hbar\vec{k}$. In the present case one can nicely understand the scattering force and its associated fluctuations in such terms. However, the dipole force and its associated fluctuations cannot be simply understood on this basis; in particular, our heuristic picture of the fluctuations in the high saturation limit invokes a *steady* force giving many $\hbar\vec{k}$ of momentum to the atom, *interrupted* by the spontaneous-emission events. The photon concept does *not* seem particularly helpful in understanding this part of the force on the atom.

IMPLICATIONS FOR TRAPPING

We are now prepared to consider the problem of trapping. First we note that the mean dipole force [from Eq. (15)] may be written as the negative gradient of a potential U.[6] Since $\vec{\text{grad}}\,p = 2\vec{\alpha}p$, we have

$$\langle\vec{f}\rangle_{\text{dip}} = -\vec{\text{grad}}[\,(\hbar\Omega/2)\ln(1+p)\,], \qquad (33)$$

so that

$$U = (\hbar\Omega/2)\ln(1+p).$$

This is called the trap potential. The mean scattering force

$$\langle\vec{f}\rangle_{\text{scat}} = \hbar\Gamma p(1+p)^{-1}\vec{\beta}/2 \qquad (34)$$

is nonconservative and must be offset by the dipole force. As we shall see, the trap parameters may be chosen so that the mean scattering force is negligibly small. Thus for negative detuning Ω (i.e., below resonance) atoms might be expected to collect at the positions of the field amplitude maxima.

The important question is then how long a trapped atom, subject to the force fluctuations, will remain in the trap. If sufficient damping could be obtained, a stable trap would ensue, but as we shall see, any simple single-frequency trap is unstable for a two-level atom. We shall return to the discussion of damping below. Neglecting the effects of damping altogether, it turns out that it should be possible to keep an atom in the trap for a considerable time, of the order of seconds. Define the trap depth U_0 as the maximum of $|U|$, i.e.,

$$U_0 = |U|_{\text{max}}.$$

Also, let W be the energy of the atom relative to the bottom of the trap; i.e.,

$$W = U_0 + U + P^2/2M.$$

A trapped atom gains energy due to the force fluctuations, so that, in the absence of damping,

$$dW/dt = D_p/M.$$

The residence time T of an atom in the trap is thus of the order of

$$T \approx U_0 (dW/dt)^{-1} \approx MU_0/(D_p)_{max} . \quad (35)$$

The use of $(D_p)_{max}$ here, rather than some more accurate average, makes this a conservative estimate. It turns out that the optimum value of p is always very much smaller than unity, and the optimum detuning is always very much larger than Γ. If we define a normalized detuning parameter by

$$q \equiv -\Omega/\Gamma , \quad (36)$$

then we can here approximate $p \ll 1$ and $q \gg 1$. In the small p approximation, we find for the trap depth,

$$U_0/\hbar \Gamma = q p_{max}/2 . \quad (37)$$

We want this quantity to be reasonably large compared to unity, which shows immediately that for small p, the detuning q must be large. We now need a suitable approximation for D_p. For small p, we need keep from (30) only the terms linear in p, and the potentially troublesome $\alpha^2 p^4$ term. Thus, when $q \gg 1$, $p \ll 1$,

$$2D_p \approx (\hbar^2 \Gamma p/2)(k^2 + \alpha^2 + \beta^2 + 4\alpha^2 q^2 p^3) . \quad (38)$$

It is helpful to note how the residence time depends on the parameters describing the trap, particularly the light intensity, trap depth, and trap dimension. If we define intensity by

$$I \equiv (c/4\pi)\overline{E}^2 = (c/2\pi)|E'|^2 ,$$

which is the same as the Poynting vector for a plane wave, then one can demonstrate for a free atom the relation

$$p(1 + 4q^2) = I/I_s , \quad (39)$$

where

$$I_s = \hbar \omega \Gamma k^2/12\pi = 6.29 \text{ mW/cm}^2 ,$$

where the evaluation[30] is for the case of sodium. For $q \gg 1$ we have

$$4pq^2 = I/I_s . \quad (40)$$

Using expressions (37) and (40), we can express p_{max} and q in terms of the trap depth and the light intensity I_{max} at the bottom of the trap. In particular,

$$q = (I_{max}/4I_s)/(2U_0/\hbar \Gamma) ,$$
$$p_{max} = (2U_0/\hbar \Gamma)^2/(I_{max}/4I_s) ,$$

and $\quad (41)$

$$q^2 p_{max}^4 = (2U_0/\hbar \Gamma)^6/(I_{max}/4I_s)^2 .$$

One can observe, then, from (35), (38), and (41),

that so long as the $\alpha^2 q^2 p^4$ term in the diffusion constant remains small, the trap lifetime is proportional to (I_{max}/U_0), whereas if the $\alpha^2 q^2 p^4$ term becomes dominant, the proportionality changes to (I_{max}^2/U_0^5) and the making of a deeper trap becomes very costly. Recalling that $\vec{grad} p = 2\vec{\alpha}p$, we observe that α is inversely proportional to the trap dimension; hence smaller traps are limited to smaller depths.

To get an idea of the numbers involved, we will look at two types of traps that have been proposed, namely a standing-wave trap and a traveling-wave Gaussian beam trap. For the former, we consider only longitudinal trapping in the standing-wave maxima. Recalling that here the atom field coupling has the form

$$g = 2g_0 \cos kx ,$$

where g_0 is the magnitude of the coupling constant corresponding to a single traveling plane wave, we have

$$p = 4p_0 \cos^2 kx , \quad \vec{\alpha} = -kx \tan kx , \quad \beta = 0 ,$$

where $\hat{x}$ is the x directed unit vector, so that (38) becomes

$$2D_p = 2(\hbar k)^2 \Gamma p_0(1 + 256q^2 p_0^3 \sin^2 kx \cos^6 kx) . \quad (42)$$

The maximum value of $\sin^2\theta \cos^6\theta$ is $\frac{27}{256}$, occurring when $\cos^2\theta = \frac{3}{4}$. Thus the p_0^4 term will begin to be in evidence when [using $p_{max} = 4p_0$, $I_{max} = 4I_0$, in (41)]

$$27q^2 p_0^3 = 27(U_0/2\hbar \Gamma)^4/(I_0/I_s) \gtrsim 1 ,$$

or $\quad (43)$

$$\tfrac{27}{16}(U_0/\hbar \Gamma)^4 \gtrsim I_0/I_s ,$$

where I_0 is now the intensity of one of the traveling waves that comprise the standing wave. Thus for $(U_0/\hbar \Gamma) = 100$, we need $I_0 \gtrsim 1$ MW/cm^2 for sodium to avoid the effects of the troublesome dipole force fluctuation term, but we then have [using (36), (41), and (42)]

$$q \approx 8 \times 10^5 , \quad p_0 \approx 6 \times 10^{-5} ,$$
$$T \approx 2q(M/\hbar k^2) \approx 5 \text{ sec} . \quad (44)$$

The detuning required is about 270 cm^{-1}, or about 1.6% of the $3p$ state energy. For this trap there is no average scattering force, so its negligibility is ensured.

The other pertinent example is the traveling-wave Gaussian beam trap. Assuming the beam is focused at the origin, and travels in the x direction, we have now[31]

$$g = g_0(b/is) \exp[ik(x + r^2/2s)] , \quad (45)$$

where

$$s = x - ib, \quad r^2 = y^2 + z^2,$$

and b is the confocal length. Then, one has

$$p = p_0 [b^2/(b^2 + x^2)] \exp[-kbr^2/(x^2 + b^2)], \quad (46)$$

and if $kb \gg 1$, then to good approximation,

$$\vec{\beta} \approx k\hat{x}, \quad \vec{\alpha} \approx -(x\hat{x} + kbr\hat{r})/(x^2 + b^2). \quad (47)$$

In evaluating the diffusion coefficient (38) for this case we can ignore $\alpha^2 p$ with respect to the other terms, leaving

$$2D_p \approx (\hbar k)^2 \Gamma p (1 + 2\alpha^2 k^{-2} q^2 p^3). \quad (48)$$

The quantity $2\alpha^2 k^{-2} q^2 p^3$ of (48) maximizes at $x = 0$, $r^2 = b/3k$, where it equals $2q^2 p_0^3/3ekb$. Thus corresponding to (43), the p^4 term will in this case be negligible if

$$(2/3ekb)(2U_0/\hbar\Gamma)^4 \lesssim I_0/4I_s,$$

or, approximately

$$(16/kb)(U_0/\hbar\Gamma)^4 \lesssim I_0/I_s. \quad (49)$$

From this expression it is evident that somewhat deeper traps may be obtained if the trap is less sharply focused. Recalling from (43) that the $1/e$ beam radius w_0 at the beam focus is related to b by

$$kb = (kw_0)^2.$$

we see that we can gain trap depth in proportion to the square root of the beam radius, but of course only at the expense of a less localized trap. For comparable light intensity and fairly tight focus, one sees that the trap depth and residence time are comparable to the standing-wave case. In the present case

$$T \approx (m/\hbar k^2)q, \quad (50)$$

a factor of two less than for the standing-wave case for the same detuning.

We can check that the mean scattering force is negligible. From (33), (34), and (47), one finds the ratio of the mean x-directed dipole force to the mean scattering force to be

$$\langle f_{\text{dip}} \rangle_x / \langle f_{\text{scat}} \rangle_x = -2qx/k(x^2 + b^2).$$

This ratio maximizes at $x = b$, where it is $-2q/kb$. If we are thinking of values of q near 10^6, then so long as $kb \lesssim 10^4$, the mean scattering force will not be important.

The interesting point here is that the traps we have considered, with light intensities of 1 MW/cm² or greater and with depths of $100\hbar\Gamma$ or greater, can hold atoms for periods of several seconds even in the absence of effective damping.

Finally, consider the question of damping. The component of the mean scattering force which is proportional to velocity damps the atom's motion. If the components of the damping force may be expressed in terms of a damping tensor γ_{ij} by

$$f_i = -M \sum \gamma_{ij} v_j,$$

then the equation of motion of the energy W may be expanded from its from (35) to read now

$$dW/dt \approx U_0/T - M \sum \gamma_{ij} v_i v_j.$$

If the trap can be so structured that the velocity distribution remains nearly isotropic, then we may average over directions to obtain

$$dW/dt \approx U_0/T - 2\bar{\gamma}(E_k), \quad (51)$$

where $\bar{\gamma} = \sum \gamma_{ii}/3$, and (E_k) is the kinetic energy. If we further assume that $(E_k) \approx W/2$, as it is for a harmonic potential well, then we find the relation

$$dW/dt \approx U_0/T - \bar{\gamma}W. \quad (52)$$

If steady-state conditions come about, then

$$W/U_0 = (\bar{\gamma}T)^{-1} = D_p/M\bar{\gamma}U_0. \quad (53)$$

Thus, the trap will be stable if $(\bar{\gamma}T) > 1$, and unstable otherwise.

It is quickly evident that a single beam trap is unstable. Consider the Gaussian beam trap. The damping is very nearly that for a plane wave; hence, for an x-directed beam, picking out the velocity-dependent term from (17), we find (for $p \ll 1$)

$$\gamma_{xx} = \frac{\hbar k^2}{M} \frac{pq}{q^2 + \frac{1}{4}}.$$

Using (37) and (48), there results

$$\bar{\gamma}T = \tfrac{1}{3} p(1 + 1/4q^2)^{-1} \ll 1, \quad p \ll 1.$$

Hence in traps with $p \ll 1$, as in the example discussed above, the damping is ineffectual. An attempt to increase $\bar{\gamma}T$ by increasing p (with $U_0 \propto pq = \text{constant}$) fails, for while γ_{xx} then varies as p^2, so does D_p (and hence T^{-1}) after its $q^2 p^4$ term becomes dominant. Similar arguments pertain to the standing-wave trap. Thus while single-beam (really single-frequency) traps can contain an atom for a sufficiently long time to envisage experiments, they are essentially unstable.

We have proposed[9] the idea of stabilizing the trap by using one or more additional light beams, tuned closer to resonance, whose sole purpose is to damp the atomic motion. It turns out that nature has put an obstacle in the way of this solution, namely the dynamic Stark shift of the damping beam's resonance as the atom moves around in the trap.

To examine this idea, we suppose that in addition to a Gaussian trapping beam, whose param-

J. P. GORDON AND A. ASHKIN

eters will now be labeled by the subscript t, there is also a damping beam (subscript d) tuned closer to resonance. If both saturation parameters are small, and the damping beam is not strongly focused, then the damping due to the trapping beam and the trapping due to the damping beam can both be safely neglected. The only significant interaction between the two beams is the dynamic Stark shift.

The important parameters of the two-beam problem are thus

$$U_0 = \hbar \Gamma q_t p_t / 2 \,,$$

$$\bar{\gamma} = \hbar k^2 p_d q_d / 3M(q_d^2 + \tfrac{1}{4}) \,,$$

$$2D_p = (\hbar k)^2 \Gamma (p_t + p_d) \,,$$

where we must add the two beams' contributions to the momentum diffusion, and the detuning parameter q_d of the damping beam is subject to the Stark shift. With these values, we find for the equilibrium energy [see (53)]

$$W/\hbar \Gamma = D_p/M\bar{\gamma}\hbar \Gamma = 3(p_t + p_d)(q_d^2 + \tfrac{1}{4})/2p_d q_d \,.$$

If $p_d \gg p_t$, this expression reduces to

$$W/\hbar \Gamma = \tfrac{3}{2}\left(q_d + \frac{1}{4q_d}\right). \tag{54}$$

Without the effects of Stark shift we could quickly minimize this by setting $q_d = \tfrac{1}{2}$, thus maximizing the damping and obtaining an equilibrium energy of the order of $\hbar \Gamma$. The Stark shift, however, has the following effect. As an atom of energy W (with respect to the bottom of the trap), moves in the trap, it encounters changes in the potential equal to W. Now for small p_t, the Stark shift of the resonance as seen by the damping beam is just twice the potential U; that is

$$q_d = q_{d0} - 2U/\hbar \Gamma \,,$$

where q_{d0} is the unperturbed detuning parameter. Thus for a change in potential equal to W, the change in q_d is

$$\delta q_d = -2W/\hbar \Gamma = -3\left(q_d + \frac{1}{4q_d}\right),$$

where we have used (54). One sees that the change in q_d occasioned by the motion of the atom in the trap is larger than q_d itself, and hence the effect of the damping beam is not at all simple. We have shown that if only one dimension is considered, the trap is still stable, because the damping is most effective at the bottom of the trap, where the atom's momentum is largest. For three dimensions, that conclusion may not hold.

Lest one think that nature somehow will not allow an optical trap to be stable, we remark that a simple-harmonic-oscillator dipole (SHO) can be trapped even by a single beam. For the case of the linear SHO one takes the small p limit of the theory, and then replaces

$$p \to 2\langle n \rangle \,,$$

where $\langle n \rangle$ is the mean excitation number for the SHO. The results so derived are valid for any $\langle n \rangle$. For the Gaussian beam trap, we then have, for $q \gg 1$,

$$\bar{\gamma}T = \tfrac{2}{3}\langle n \rangle \,,$$

and the trap is stable for large $\langle n \rangle$. For a real atom, however, we don't have the privilege of large $\langle n \rangle$, and some clever methods are called for. If the two-beam trap also turns out to be unstable in three dimensions, then there are several possible ways to proceed. One can reduce the Stark shift by a factor of 2 by using different upper levels for the trapping and damping resonances, so that only the lower level of the damping resonance is shifted. In addition, or alternatively, one might use a third beam tuned near a resonance of the upper damping level to cancel the Stark shift of the damping resonance.

Thus it would seem in principle possible to form a stable trap for atoms. Obtaining the optimum form of damping may take some experimentation, but since the traps can contain slow atoms for long times anyway, such experimentation would seem feasible and worthwhile.

[1]O. R. Frisch, Z. Phys. 86, 42 (1933).

[2]A. Ashkin, Phys. Rev. Lett. 25, 1321 (1970).

[3]G. A. Askar'yan, Zh. Eksp. Teor. Fiz. 42, 1567 (1962) [Sov. Phys.—JETP 15, 1088 (1962)].

[4]A. P. Kazantsev, Zh. Eksp. Teor. Fiz. 63, 1628 (1972) [Sov. Phys.—JETP 36, 861 (1973)]; ibid. 66, 1599 (1974) [ibid. 39, 784 (1974)].

[5]V. S. Letokhov, V. G. Minogin, and B. D. Pavlik, Zh. Eksp. Teor. Fiz. 72, 1328 (1977) [Sov. Phys.—JETP 45, 698 (1977)]; V. S. Letokhov and V. G. Minogin, Appl. Phys. 17, 99 (1978).

[6]A. Ashkin, Phys. Rev. Lett. 40, 729 (1978).

[7]A. Ashkin, Phys. Rev. Lett. 24, 156 (1970).

[8]T. W. Hansch and A. L. Schawlow, Opt. Commun. 13, 68 (1975).

[9]A. Ashkin and J. P. Gordon, Opt. Lett. 4, 161 (1979).

[10]R. Schieder, H. Walther, and L. Woste, Opt. Commun. 5, 337 (1972); J. L. Picque and J. L. Vialle, ibid. 5, 402 (1972).

[11]A. F. Bernhardt, D. E. Duerre, J. R. Simpson, and L. L. Wood, Appl. Phys. Lett. 25, 617 (1974).

[12]J. E. Bjorkholm, A. Ashkin, and D. B. Pearson, Appl. Phys. Lett. 27, 534 (1975).

[13]D. J. Wineland and H. Dehmelt, Bull. Am. Phys. Soc.

<u>20</u>, 637 (1975).

[14]D. J. Wineland, R. E. Drullinger, and F. L. Walls, Phys. Rev. Lett. <u>40</u>, 1639 (1978).

[15]W. Neuhauser, M. Hohenstatt, P. Toschek, and H. Dehmelt, Phys. Rev. Lett. <u>41</u>, 233 (1978).

[16]J. E. Bjorkholm, R. R. Freeman, A. Ashkin, and D. B. Pearson, Phys. Rev. Lett. <u>41</u>, 1361 (1978).

[17]J. P. Gordon, Phys. Rev. A <u>8</u>, 14 (1973).

[18]A. Ashkin and J. M. Dziedzic, Phys. Rev. Lett. <u>30</u>, 139 (1973).

[19]A. P. Kazantsev, Usp. Fiz. Nauk <u>124</u>, 113 (1978) [Sov. Phys.—Usp. <u>21(1)</u>, 56 (1978)].

[20]R. J. Cook, Phys. Rev. Lett. <u>41</u>, 1788 (1978).

[21]R. J. Cook and A. F. Bernhardt, Phys. Rev. A <u>18</u>, 2533 (1978).

[22]E. Arimondo, H. Lew, and T. Oka, Phys. Rev. Lett. <u>43</u>, 753 (1979).

[23]R. M. Hill and T. F. Gallagher, Phys. Rev. A <u>12</u>, 451 (1975).

[24]A. Yu. Pusep, Zh. Eksp. Teor. Fiz. <u>70</u>, 851 (1976) [Sov. Phys.—JETP <u>43</u>, 441 (1976)].

[25]J. E. Bjorkholm, R. R. Freeman, A. Ashkin, and D. B. Pearson, in *Laser Spectroscopy IV*, *Proceedings of the Fourth International Conference, Rottach-Egern, Germany, 1979*, edited by H. Walther and K. W. Rothe (Springer, Berlin, 1979).

[26]A. P. Botin and A. P. Kazantsev, Zh. Eksp. Teor. Fiz. <u>68</u>, 2075 (1975) [Sov. Phys.—JETP <u>41</u>, 1038 (1975)].

[27]V. G. Minogin and O. T. Serimoa, Opt. Commun. <u>30</u>, 373 (1979).

[28]M. Lax, Phys. Rev. <u>129</u>, 2343 (1963). The theory of such two-time autocorrelation functions for driven atoms was worked out in his study of resonance fluorescence by B. R. Mollow, Phys. Rev. <u>188</u>, 1969 (1969). See also H. J. Kimble and L. Mandel, Phys. Rev. A <u>13</u>, 2123 (1976), and references therein.

[29]C. Cohen-Tamoudji and S. Reynaud, J. Phys. B <u>10</u>, 345 (1977).

[30]To get this number we assume that the sodium atom is behaving like an oriented two-level system. Optical pumping may in fact bring this about. Other cases involve further complication, which we have not taken into account.

[31]H. Kogelnik, Appl. Opt. <u>4</u>, 1562 (1965).

Optically induced rotation of anisotropic micro-objects fabricated by surface micromachining

E. Higurashi, H. Ukita, H. Tanaka, and O. Ohguchi

NTT Interdisciplinary Research Laboratories, 3-9-11, Midori-cho, Musashino-shi, Tokyo 180, Japan

(Received 26 August 1993; accepted for publication 6 February 1994)

Optical trapping and directional high-speed rotation by radiation pressure are demonstrated for anisotropic micro-objects fabricated by reactive ion-beam etching. These micro-objects, which have shape dissymmetry (not bilateral symmetry but rotational symmetry) in the horizontal cross section, rotate about the laser beam axis in the designed direction in a liquid medium (e.g., water or alcohol). The rotation speed is almost proportional to the input laser power.

Radiation pressure from a strongly focused laser beam is known to trap and manipulate micrometer-sized transparent particles. The phenomenon of optical trapping (optical tweezers) was first demonstrated by Ashkin and co-workers.[1,2] They initially used their technique to study biological physics: manipulating various biological cells[3] and measuring the motor force of molecules.[4] Recently, it has also been used to study microchemistry: measuring the fluorescence spectra of individual trapped particles and extracting the internal solution from trapped microcapsules by laser ablation.[5] These studies used the force of radiation pressure to position and move micro-objects.

In this letter, we use radiation pressure for microdynamics: directional high-speed rotation of anisotropic micro-objects in a liquid. Two methods of rotating micro-objects by using only a single laser beam have been reported. One used a circularly polarized laser beam[6] and the other used the rotating nonuniform input intensity profile of a higher-order mode (TEM_{0n}) laser beam.[7] However, the rotation speeds of both methods are very slow, about 6.7×10^{-1}–6.7×10^{-2} rpm[6] and 6 rpm.[7]

We propose a new directional high-speed rotation method that uses a single laser beam. This method is useful for more applications. Giving the micro-objects an anisotropic geometry produces a torque as a result of the net radiation pressure on their surfaces from a symmetrical input intensity profile. We used micro-objects that were flat on the top and bottom, but had shape dissymmetry (rotational symmetry) on the side. The dissymmetrically shaped micro-objects were made by reactive ion-beam etching (RIBE)[8] of a 10-μm-thick silicon dioxide (SiO_2) layer.

The apparatus used in this study for optical trapping and rotational driving is shown in Fig. 1. The light source was a Gaussian mode Nd:YAG laser with a wavelength of 1.06 μm. The beam was strongly focused by a microscope objective lens [NCF Plan 100×, numerical aperture (NA)=1.25] onto the micro-object. The laser power on the micro-object was varied from 1 to 100 mW for the experiments. The experiments were performed on micro-objects dispersed in ethanol (refractive index: $\sim$1.36), where gravity plays a minor role. Micro-objects made of SiO_2 (refractive index $\sim$1.5) and ethanol are both transparent to the YAG laser wavelength. Trapping and rotation were recorded by a charged-coupled device (CCD) camera to measure the rotation speed of micro-objects.

Figure 2 shows the origin of the optically induced rotation for the designed micro-object. The forces of radiation pressure arise from the momentum change of photons due to the refraction and reflection of light.[1] We ignored the reflection because the difference in refractive indices between the micro-object (SiO_2) and surrounding ethanol was small in our experiments. Radiation pressure forces are exerted on the surfaces where light enters and exits the micro-object as shown in Fig. 2(a). They act perpendicular to the object surface because the tangential momentum of light on the surface is preserved. When incident light is refracted at the flat top surface, the upward radiation pressure F pulls the micro-object but does not produce a torque because the pressure force is parallel to the laser beam axis. Since the laser beam is strongly focused through an objective with high NA, the incident laser light is emitted from the side surfaces. When the output light is refracted at these side surfaces as shown in Fig. 2(b), the force of radiation pressure F is again exerted on side surfaces (i) and (ii) in the direction opposite to the momentum change ΔP for these rays. Side surface (iii) does not feel F because it is parallel to the radial direction and does not refract the laser beam. The net radiation pressure, therefore, gives rise to a clockwise torque as shown in Fig.

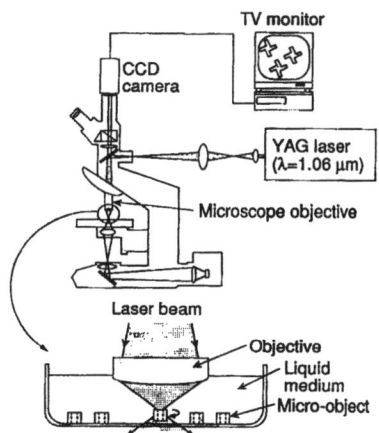

FIG. 1. Block diagram of the experimental apparatus used for the optically induced rotation of micro-objects. Micro-objects were optically trapped near the focus of the laser beam and rotated by the same laser beam.

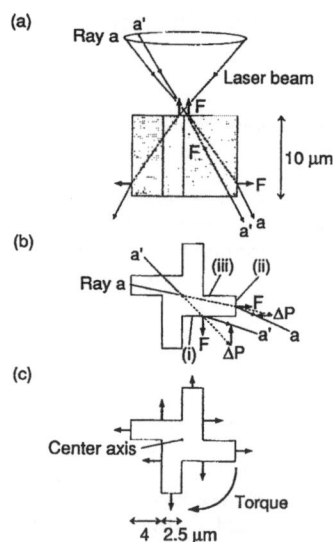

FIG. 2. Origin of the optically induced rotation for the designed micro-object. (a) Side view. (b) Top view. The force of radiation pressure F arises from momentum change ΔP due to ray refraction. (c) Forces on the side faces of the designed micro-object, which lead to clockwise rotation.

2(c). Thus, optically induced rotation requires a cross section that is dissymmetric, i.e., it does not have bilateral symmetry but has rotational symmetry.

We fabricated these micro-objects by the following process: (i) a SiO_2 layer (10 μm thick) was deposited on a gallium arsenide (GaAs) substrate by radio-frequency (rf) sputtering, (ii) the SiO_2 layer was shaped by RIBE,[8] and (iii) the micro-objects were separated from the GaAs substrate by dissolving the substrate in a selective etching liquid. Typical micro-objects were 10–25 μm in diameter and 10 μm thick.

Fabricated micro-objects rotated about the laser beam axis in the expected direction as soon as they were trapped in the vicinity of the focal point of the laser beam, as shown in Fig. 3(a). Rotation stopped immediately when the laser was

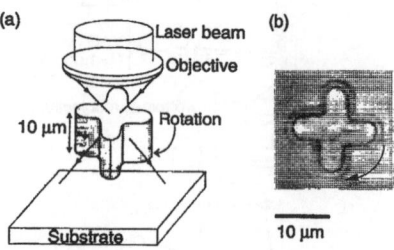

FIG. 3. Optically induced rotation of an anisotropic SiO_2 micro-object. (a) Schematic representation of laser-driven rotation of a micro-fabricated SiO_2 object, 13 μm wide and 10 μm high. (b) Photomicrograph of the micro-object rotating in ethanol. It rotates clockwise at 22 rpm with a laser power of 80 mW. The laser beam axis is perpendicular to the plane of the photograph.

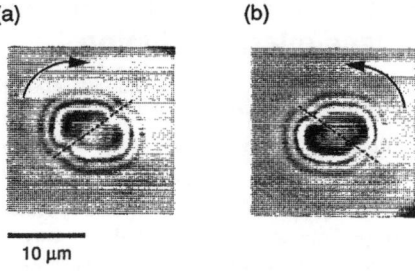

FIG. 4. Rotational direction control by dissymmetric geometry of the micro-objects. (a) Clockwise rotation. (b) Counter-clockwise rotation.

switched off. Figure 3(b) shows a top view photomicrograph of a microfabricated SiO_2 object. The fabricated shape is not exactly as designed, especially surface (iii), but it nevertheless rotated clockwise at 22 rpm with a laser power of 80 mW. This is because the radiation pressure on surface (i) is greater than that on surface (iii). The rotation speed was almost proportional to the incident laser power. We also found that the larger the NA, the faster the micro-object rotated, because more light was emitted from the side surfaces.

The rotational direction can be chosen by designing an appropriate dissymmetric geometry. The object in Fig. 4(a) rotates clockwise and the one in Fig. 4(b) rotates counter clockwise.

In conclusion, directional and high-speed rotation were demonstrated for microfabricated objects having anisotropic geometry that is not bilateral symmetry but rotational symmetry on the horizontal cross section. Microfabricated electrostatic motors[9] (typical dimensions of 100 μm) usually have short lifetimes because of surface friction between the rotor and the substrate. This optically induced rotation, on the other hand, requires no mechanical contact and no electric wires, so it can be expected to play a significant role in photomicrodynamical systems[10] such as remotely driven motors and actuators.

The authors would like to thank Tohru Maruno of NTT Interdisciplinary Research Laboratories for offering microsamples and many useful discussions and Eduardo S. Vera of NTT Advanced Technology Corporation for helpful discussions.

[1] A. Ashkin, Phys. Rev. Lett. **24**, 156 (1970).
[2] A. Ashkin, J. M. Dziedzic, J. E. Bjorkholm, and S. Chu, Opt. Lett. **11**, 288 (1986).
[3] A. Ashkin and J. M. Dziedzic, Science **235**, 1517 (1987).
[4] S. C. Kuo and M. P. Sheetz, Science **260**, 232 (1993).
[5] H. Misawa, N. Kitamura, and H. Masuhara, J. Am. Chem. Soc. **113**, 7859 (1991).
[6] T. Sugiura, S. Kawata, and S. Minami, J. Spectrosc. Soc. Jpn. **39**, 342 (1990) (in Japanese).
[7] S. Sato, M. Ishigure, and H. Inaba, Electron. Lett. **27**, 1831 (1991).
[8] J. Shimada, O. Ohguchi, and R. Sawada, J. Lightwave Technol. **9**, 571 (1991).
[9] R. S. Muller, Sensors and Actuators A **21-23**, 1 (1990).
[10] H. Ukita, Y. Uenishi, and H. Tanaka, Science **260**, 786 (1993).

Reprinted from **OPTICS LETTERS**, Vol. 4, page 161, June, 1979
Copyright © 1979 by the Optical Society of America and reprinted by permission of the copyright owner.

Cooling and trapping of atoms by resonance radiation pressure

A. Ashkin and J. P. Gordon

Bell Laboratories, Holmdel, New Jersey 07733

Received February 16, 1979

The combined use of trapping and cooling laser beams for optical trapping and cooling of neutral atoms by the forces of resonance radiation pressure is examined. Calculations show that atoms can be held in traps as deep as 10^{-4} eV at temperatures of $\sim 10^{-3}$ K, close to the minimum set by quantum fluctuations. Spatial confinement of atoms to a region a fraction of a wavelength in length should be possible.

Recently a new method was proposed for optically trapping and cooling neutral atoms based on resonance radiation pressure forces.[1] The technique is potentially useful for high-resolution spectroscopy and novel experiments on a few or even possibly single atoms. The new trap geometry, based on strongly focused beams, provides deep cw optical potential wells with means for damping atoms injected from atomic beams. This Letter considers the damping of atoms in these traps in more detail and shows that, with use of optimally tuned auxiliary damping beams, trapped atoms can be cooled to minimum temperatures of $(10^{-3}-10^{-4})$ K in potential wells as deep as $10^3 h\gamma_N$, where γ_N is the natural linewidth. For sodium this well depth is $\sim 10^{-4}$ eV. We compare our trap with other proposals[2] for trapping atoms by light pressure and with recent experiments on trapping and radiation pressure cooling of ions using electromagnetic traps.[3,4]

The resonance radiation pressure forces used for optical trapping are of two types. First, there is the spontaneous scattering force based on resonant optical absorption and isotropic scatter of spontaneous emission.[5] This force has been observed in atomic-beam-deflection experiments.[6] The scattering force is also the basis of Hänsch and Schawlow's proposal for cooling or damping of free atoms.[7] Atoms moving in optical beams tuned below resonance experience either an increase or decrease in scattering force, depending on whether they move toward or away from the light, because of Doppler shift. This velocity-dependent increment in force causes damping. The second radiation pressure force is the so-called dipole force that exists on optically induced atomic dipoles when placed in an optical field gradient. This force is dispersive in character. That is, because of the sign change of the atomic polarizability on either side of resonance, it pulls atoms either into or out of the high-intensity regions of the light beam. This effect is responsible for self-focusing or self-defocusing of near-resonance light beams in atomic vapors.[8] Dipole forces on atoms have recently been directly observed in an atomic-beam-focusing experiment.[9]

The trap proposed by Ashkin[1] is based on the focusing of strictly cw light beams tuned well below resonance [typically by $\sim(10^2-10^4)\gamma_N$]. One gets deep traps and

strong damping of fast atoms injected into the trap from thermally produced atomic beams. Unfortunately, slow atoms held in these traps are not optimally damped because of the large detuning. Optimal damping corresponds to $(\nu_0 - \nu_d) = \gamma_N/2$ (see below). It was suggested[1] that one could get additional damping by adding auxiliary damping beams that are optimally tuned. This suggestion must be examined carefully since, because of mutual saturation, presence of a damping beam decreases the trapping force and presence of a trapping beam decreases the damping force. In addition, the trapping beam causes an optical Stark shift of the atomic resonance, which further complicates the damping process.

The minimum temperature of an atom in a trap results from the balance of the damping force and any fluctuations in the trapping environment. Assuming stable lasers, there is still a constant noise source arising from the quantum fluctuations in the absorption and spontaneous emission processes inherently responsible for the scattering force and from fluctuations in the dipole force. The fluctuations in the scattering force were discussed by Letokhov *et al.*[2] and Wineland *et al.*[3] in the context of their traps. The fluctuations in the dipole force are discussed in a more-complete quantum treatment of the radiation field trap.[10] They are important for standing-wave traps and in cases of high field gradients and strong saturation. They may be ignored for the simple trap discussed below. To demonstrate the basic principles of cooling an atom in an optical trap with addition of an auxiliary damping beam, consider the conceptually simplest atom trap, namely, a single focused Gaussian beam tuned below resonance[1] and a single counterpropagating plane-wave damping beam, also tuned below resonance (see Fig. 1). We calculate first the effect of the damping beam on the radial trapping potential, for example. Denoting by p_t and ν_t the saturation parameter and frequency of the trapping beam and similarly by p_d and ν_d, the damping beam, one can write the radial potential as[1]

$$U(r) = \frac{1}{2} h(\nu_t - \nu_0)\ln[1 + p_{tot}(r)], \qquad (1)$$

where $p_{tot}(r) = p_t(r) + p_d$, since the atom is absorbing from both trapping and damping beams. We neglect

162   OPTICS LETTERS / Vol. 4, No. 6 / June 1979

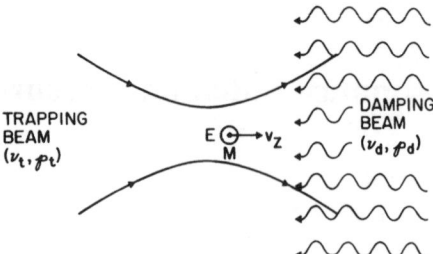

Fig. 1. Atom of mass M and velocity v_z located at the equilibrium point E of a Gaussian-beam trap with a plane-wave damping beam. The axial location of E depends on the magnitudes of p_t and p_d.

any high-frequency interference terms in Eq. (1). The radial trap depth ΔU is the difference between the potential on axis and the edges of the trapping beam.

$$\Delta U = \frac{1}{2} h(\nu_t - \nu_0) \ln\left[\frac{1 + p_t(r=0) + p_d}{1 + p_d}\right]. \quad (2)$$

For p_t and $p_d \ll 1$, the depth ΔU is the same as in the absence of damping. For p_t and p_d of order unity, the fractional change in trap depth that is due to damping is rather small [i.e., $\ln(3/2)/\ln(2) = 0.58$]. This modest reduction can be compensated for by an increase in power.

Next consider the effect of the trapping beam on the damping process. We calculate the minimum axial temperature of an atom in the trap of Fig. 1 first for a fixed tuning of the damping beam below the atomic resonance. We follow a highly physical method for calculating fluctuations and damping for a free atom pointed out by Purcell.[11] For an atom at rest at the equilibrium position E, there is a balance of the axial dipole force and the axial scattering force of the trapping and damping beams. An atom moving with velocity v_z at E experiences an additional damping force ΔF_z proportional to v_z because of Doppler shift of the scattering force. There is negligible change in the axial dipole force for small v. Thus, following the notation of Ref. (1),

$$\Delta F_{z_{\text{scatt}}} = \frac{h}{\lambda}\frac{1}{2\tau_N}\left[\left(\frac{p_t - p_d}{1 + p_{\text{tot}}}\right)_{v=v} - \left(\frac{p_t - p_d}{1 + p_{\text{tot}}}\right)_{v=0}\right], \quad (3)$$

where

$$p_{\text{tot}} = p_t + p_d \text{ and } p = \frac{I}{I_{\text{sat}}}\frac{\gamma_N^2/4}{(\nu - \nu_0)^2 + \gamma_N^2/4}.$$

Denote the static tuning of the lasers by $(\nu_0 - \nu_t) \equiv q_t\gamma_N$ and $(\nu_0 - \nu_d) \equiv q_d\gamma_N$. For atoms moving with velocity v_z to the right, one has $(\nu_0 - \nu_t) = (q_t + b)\gamma_N$ and $(\nu_0 - \nu_d) = (q_d - b)\gamma_N$, where $b = v_z/\lambda\gamma_N$ is the Doppler shift in units of γ_N. We find to first order in b that

$$\Delta F_{z_{\text{scatt}}} = -\frac{h}{\lambda}\frac{1}{2\tau_N}\frac{1}{(1 + p_d + p_t)^2}$$
$$\times \left(\frac{8p_d q_d}{4q_d^2 + 1} + \frac{8p_t q_t}{4q_t^2 + 1}\right) b. \quad (4)$$

Define a velocity relaxation time

$$\tau_D \equiv \frac{-Mv_z}{\Delta F_{z_{\text{scatt}}}}, \quad (5)$$

where M is the mass of the atom. The kinetic energy $(\propto v^2)$ decays in a time $\tau_D/2$.[12] One can then compute the number of photons N scattered by the atom in the decay time $\tau_D/2$ from

$$N = \frac{1}{\tau_N}f\frac{\tau_D}{2}. \quad (6)$$

τ_N is the natural lifetime and f is the fraction of time an atom spends in the excited state, $f = p_{\text{tot}}/2(1 + p_{\text{tot}})$. Since the N photons are both absorbed and emitted randomly, the atom momentum fluctuates with a mean-square value given by

$$\overline{p^2} = 2N(h/\lambda)^2. \quad (7)$$

Thus the average kinetic energy of an atom in the trap $\overline{KE} = \overline{p^2}/2M$ can be deduced combining Eqs. (4)–(7). In the limit $q_t \gg 1$ and $p_t/q_t \ll 1$, the contribution to the damping from the trapping beam in Eq. (4) becomes negligible and

$$\overline{KE} = \frac{h\gamma_N}{8}\left(2q_d + \frac{1}{2q_d}\right)\left(1 + \frac{p_t}{p_d}\right)(1 + p_t + p_d). \quad (8)$$

$\overline{KE}$ is a minimum for $q_d = \frac{1}{2}$. For the case in which $p_t \ll p_d$ and p_t and $p_d \ll 1$, one has $\overline{KE}_{\text{min}} = h\gamma_N/4$. This corresponds to a minimum temperature of $\sim 10^{-4}$ K for sodium, where $\gamma_N = 10.6$ MHz. Approximately this value was found by Letokhov et al.[2] and by Wineland for optically cooled ions.[3,4] For p_d and $p_t \approx 1$, one finds from Eq. (8) that $\overline{KE}_{\text{min}} = 3 h\gamma_N/2$.

The above considerations on damping must be modified in one important respect. Because of the relatively large detuning of the trapping beam needed for achieving deep traps, there is a sizable optical Stark shift of the resonant frequency ν_0 of the atom to a new frequency ν_0'. This shift,[13]

$$\nu_0' - \nu_0 = (\nu_t - \nu_0)[(1 + 2p_t)^{1/2} - 1], \quad (9)$$

is typically $\sim(10^2–10^3)\gamma_N$.

To obtain adequate damping, one must therefore tune the damping-beam frequency ν_d with respect to ν_0'. There is, however, the additional aspect that the resonance frequency ν_0' now varies with position in the trap. This is a minor effect for cool atoms, where the orbital excursions are small. For more-energetic atoms, the excursions can be sufficient to shift ν_0' so that $\nu_d > \nu_0'$ and one has heating over part of the atomic orbit. Assuming a harmonic potential of infinite height, one can estimate the effect of the variation of ν_0' by integrating the energy loss that is due to the damping force F_d from Eq. (4) over the orbit, $\int F_d dz$, and comparing with the net heating occurring over the orbit using Eq. (7). One finds an equilibrium value of $\overline{KE}$ approximately 4 times higher than that of Eq. (8). Also, all atoms of energy greater than $\overline{KE}$ experience net damping, although less than in the Stark-free case considered above. One can, however, halve the Stark shift relative to the damping beam by using separate transitions for trapping and damping purposes, such as the D_1 and D_2 lines of sodium. This reduces the value of $\overline{KE}$ by 2. Finally, one can totally avoid the positional

June 1979 / Vol. 4, No. 6 / OPTICS LETTERS   163

variation of the $3s$–$3p$ resonance frequency in the trap that is due to the Stark shift of the trapping beam by adding a Stark correction beam tuned below the $3p$–$4d$ transition frequency to lower the energy of the $3p$ level sufficiently to make $\nu_0' - \nu_0 = 0$. If the correction beam has the same spatial variation as the trapping beam, then $\nu_0' - \nu_0 = 0$ everywhere within the trap.

Thus far only z-directed damping was considered. The damping can be made three dimensional by using a single beam having equal projections on the x, y, and z axes. If the vibrational frequencies in the three directions are sufficiently different, the atom is then equally damped in all directions, giving a value of $\overline{KE}_{\min} = 3\,(h\gamma_N/4)$.

Our derivation has extended the simple rate equations of Refs. 1 and 5, which correctly describe the time-averaged scattering and dipole forces to the two-beam case of combined trapping and damping beams with the phenomenological addition of the optical Stark shift. A time-dependent semiclassical analysis of the two-beam problem confirms the validity of using rate equations plus the Stark shift for low values of p. For p's approaching unity, one still expects damping but with enough complexity from nonlinear mixing effects to make literal use of Eq. (8) questionable. Another interesting result of the time-dependent analysis for two beams is the case of a standing-wave-damping field $E_d{}^2 \cos^2 kz$. One finds[10] that the damping force is a function of position. At low p, $F_{\text{damp(st w)}} = F_{\text{damp(pl w)}} \, 4\sin^2 kz$, and the damping force is zero where the field is a maximum. At higher p the behavior is complex and can even give heating at certain positions. The average of the damping force over the standing wave, however, equals the plane-wave damping. Thus, use of a single standing-wave field for trapping and damping as described in Ref. 2 is unfavorable since atoms are trapped about a point of zero damping. In free-particle experiments, as in Ref. 7, the average damping is unaffected by the positional dependence. Finally, for an atom localized in a deep well at the maximum of a standing-wave-trapping beam (see Ref. 1 and below), one can position the standing-wave-damping beam to gain a factor of 2 in minimum temperature, giving $\overline{KE}_{\min} = h\gamma_N/8$ for $q_d = \frac{1}{2}$.

The effectiveness of the trap in preventing escape of atoms over the top of the potential barrier depends on the ratio of well depth to the average temperature. For a typical depth $\Delta U \cong 10^2 h\gamma_N$ and an average temperature of $\sim 10^{-3}$ K, $\Delta U/kT \cong 100$. Thus, from the Maxwell–Boltzmann velocity distribution, the fraction of atoms having an energy greater than kT is $\sim\exp(-100)$, which is negligible. One can also show that quantum-mechanical tunneling into the potential barrier is minute on the scale of the optical wavelength and trap dimensions. An atom of energy E penetrating a barrier of height V has an exponential wavefunction of the form $\psi = A\exp(-z/b)$, where $b = h/[2m(V - E)]^{1/2}$. For $V - E \cong 10^2 h\gamma_N$, one finds that $b \cong \lambda/1000$ and the penetration is negligible.

These considerations apply also to the two-beam traps proposed by Ashkin (see Fig. 1 of Ref. 1), which have deep-axial potential wells with superposed standing-wave ripples of varying depth. Atoms of $\overline{KE} \cong h\gamma_N$ cannot tunnel through or pass over the top of the barrier to a neighboring maximum. Atoms thus confined reside in the lowest-few vibrational-energy levels of a closely parabolic trap and remain confined to a dimension of $\sim\lambda/35$. This contrasts with the traps described by Letokhov et al.,[2] in which atoms of $\overline{KE} \cong h\gamma_N/2$ are moving through standing-wave ripples of depth $\sim h\gamma_N$. In that case, the tunneling probability is high and thermal excitation over the top of the barrier is likely. Their picture of an atom moving in a periodic potential is appropriate. Their configuration also has problems with weak transverse confinement. One can, however, study tunneling with strong transverse confinement in the two-beam geometry of Ref. 1, Fig. 1, by observing atoms in the region of shallower standing-wave traps near planes T_1 and T_2.

In conclusion, calculations show that combined use of optimally tuned trapping and damping beams results in traps capable of confining atoms at temperatures as low as $(10^{-3}\text{–}10^{-4})$K in optical potential wells as deep as 10^{-4} eV, within a dimension as small as $(\lambda/35)$, with negligible probability of escape by tunneling through or by thermal excitation over the top of the potential barrier.

We thank J. E. Bjorkholm and R. R. Freeman for helpful discussions.

References

1. A. Ashkin, Phys. Rev. Lett. **40**, 729 (1978).
2. V. S. Letokhov, V. G. Minogin, and B. D. Pavlik, Sov. Phys. JETP **45**, 698 (1977); V. S. Letokhov and V. G. Minogin, Appl. Phys. **17**, 99 (1978), and references therein.
3. D. J. Wineland, R. E. Drullinger, and F. L. Walls, Phys. Rev. Lett. **40**, 1639 (1978).
4. W. Neuhauser, M. Hohenstatt, R. Toschek, and H. Dehmelt, Phys. Rev. Lett. **41**, 233 (1978).
5. A. Ashkin, Phys. Rev. Lett. **25**, 1321 (1970).
6. R. Schieder, H. Walther, and L. Wöste, Opt. Commun. **5**, 337 (1972); J. L. Picque and J. L. Vialle, Opt. Commun. **5**, 402 (1972).
7. T. W. Hänsch and A. L. Schawlow, Opt. Commun. **13**, 68 (1975).
8. G. A. Askar'yan, Zh. Eksp. Teor. Fiz. **42**, 1567 (1962) [Sov. Phys. JETP **15**, 1088 (1962)]; D. Grischkowsky, Phys. Rev. Lett. **24**, 866 (1970); J. E. Bjorkholm and A. Ashkin, Phys. Rev. Lett. **32**, 129 (1974).
9. J. E. Bjorkholm, R. R. Freeman, A. Ashkin, and D. B. Pearson, Phys. Rev. Lett. **41**, 1361 (1978).
10. A unified theory of the scattering and dipole forces and their quantum fluctuations has been developed by J. P. Gordon using quantum electrodynamics (to be submitted for publication).
11. E. M. Purcell, Department of Physics, Harvard University, Cambridge, Massachusetts 02138, personal communication.
12. This kinetic-energy decay time is actually valid only for a free particle. Addition of a conservative potential U, however, gives the same equilibrium $\overline{KE}$.
13. P. F. Liao and J. E. Bjorkholm, Opt. Commun. **16**, 392 (1976), and references therein.

RAPID COMMUNICATIONS

PHYSICAL REVIEW A, VOLUME 61, 061403(R)

Optical-dipole trapping of Sr atoms at a high phase-space density

Tetsuya Ido,[1] Yoshitomo Isoya,[1] and Hidetoshi Katori[1,2]

[1]*Cooperative Excitation Project, ERATO, Japan Science and Technology Corporation (JST),
KSP D-842, 3-2-1 Sakado Takatsu-ku, Kawasaki, 213-0012, Japan*
[2]*Engineering Research Institute, University of Tokyo, Bunkyo-ku, Tokyo 113-8656, Japan*

(Received 9 August 1999; published 2 May 2000)

Employing a far-off resonance optical-dipole trap (FORT), we attained a phase-space density exceeding 0.1, or an order to quantum degeneracy. Strontium atoms were magneto-optically cooled and trapped using the spin-forbidden $^1S_0$-$^3P_1$ transition and then compressed into a FORT that was designed to allow simultaneous Doppler cooling. We discussed that the phase-space density was finally limited by the light-assisted collisions occurring in the optical cooling.

PACS number(s): 32.80.Pj

The attainment of quantum degenerate atomic gases [1] has been one of the driving forces promoting laser-cooling techniques. Toward this goal, various kinds of cooling techniques have been developed, allowing us to reach down to subrecoil temperatures. Even for ultracold atoms that are laser-cooled down to the photon recoil momentum h/λ, an atom density of $n_R \sim \lambda^{-3}$ is necessary to reach a quantum degenerate regime. For atom densities such as n_R, since the average atom spacing is on the order of λ, strong atom-atom interactions via the near resonant photon, i.e., the radiation trapping [2] and the light-assisted collisions [3], manifest themselves and drastically disturb the cooling dynamics.

To overcome the difficulties inherent in optical-cooling schemes at such high atom densities, evaporative cooling in magnetic traps has been successfully employed to create Bose-Einstein condensation (BEC) in alkali-metal atomic gases [1]. Optical cooling and the trapping schemes, however, still attract strong interest because they are expected to realize rapid creation of degenerate atoms as well as their better handling. Moreover, optical schemes will give unique access to some different atom species not condensed thus far: polarized fermions, some kinds of bosons with unfavorable scattering length, or spinless particles [4], to which thermalization due to elastic collisions or magnetic trapping cannot be applied. Among them, alkaline-earth atoms [5] can be important for future applications. Using narrow-line cooling [4], fermionic isotopes such as ^{87}Sr can be Doppler-cooled down to a submicrokelvin regime, where degenerate fermion gases show its distinctive features. Spinless bosonic isotopes, on the other hand, have potential importance as an atom interferometer because of their insensitivity to magnetic fields, and as candidates for optical standards based on intercombination lines [5].

This Rapid Communication demonstrates the Doppler cooling of ^{88}Sr atoms in a far-off resonance optical-dipole trap (FORT) [6] to achieve a phase-space density higher than 0.1, which is, to our knowledge, the highest ever achieved by purely optical means. This alternative FORT design, which allows simultaneous Doppler cooling, enables high loading efficiency of magneto-optically trapped atoms. We have shown that the maximum phase-space density is finally limited by light-assisted inelastic collisions occurring in optical cooling. The developed scheme can be widely applied to

alkaline earth and other species such as Yb [7], in which an intercombination transition is used for cooling.

The light-induced interactions between atoms have been the main obstacle to the creation of cold and dense atomic gas in optical cooling. The radiation-trapping effect [2], which limits the density and temperature of laser-cooled atoms, has been moderated by applying Raman (sideband) cooling [8–10], gray molasses [11], or narrow-line [4] cooling, which minimize the photon scattering or reabsorption in the course of laser cooling. Now, the last obstacle that limits the attainable atom density is light-assisted collisions [3], in which the excitations of the attractive quasimolecular potential by the red-detuned cooling photon cause inelastic two-body collisions. These light-assisted collisions with a loss rate β determine the steady-state atom density as $\sqrt{\phi/(\beta V)}$, assuming an atom flux ϕ into a trap volume V. This formula infers a simple guiding principle for high-density trapping: for the given β, the atom flux per volume ϕ/V needs to be increased.

In order to realize the idea, we have developed an optical-dipole trap that can be combined with Doppler cooling, and thus enhanced ϕ/V by efficiently transferring the atoms into a small conservative trap with the help of optical cooling. In laser cooling that employs an intercombination transition, since the cooling ground state and the excited state are mainly coupled to the respective spin states by an applied laser field, arbitrary light shift potentials for these states can be generated by tuning the laser parameters. Especially, by adjusting the light shifts of both cooling states so that they are equal, simultaneous Doppler cooling is expected.

We employed magneto-optically cooled and trapped ^{88}Sr atoms using the spin-forbidden $^1S_0$-$^3P_1$ transition at $\lambda = 689$ nm. The narrow linewidth of the transition $\gamma/2\pi = 7.6$ kHz, which is less than the photon recoil shift of $\hbar k^2/m = 9.5$ kHz, cooled atoms down to the photon recoil temperature of 440 nK and enabled high-density trapping by efficiently suppressing the radiation-trapping effects, leading to a phase-space density of $\rho \sim 0.01$ [4]. To further increase the phase-space density, we applied the FORT described above. Figure 1 shows the energy levels of ^{88}Sr. The FORT laser couples the cooling ground state $5s^2 \, ^1S_0$ to the upper singlet series of $5snp \, ^1P_1$, while the cooling upper state $5s5p \, ^3P_1$ is coupled to the triplet series of $5sns \, ^3S_1$,

TETSUYA IDO, YOSHITOMO ISOYA, AND HIDETOSHI KATORI

PHYSICAL REVIEW A **61** 061403(R)

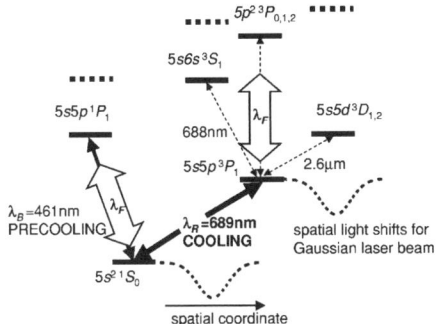

FIG. 1. The energy levels of ^{88}Sr. A FORT laser at λ_F = 800 nm couples the cooling ground state $^1S_0$ and the excited state $^3P_1$ to the upper singlet and triplet states, respectively, generating the same amount of Stark shifts. Spatial light shifts for these two states are schematically depicted by the dotted lines assuming the FORT laser with gaussian profile. Because the atomic resonance frequency is unchanged in space, the optical confinement in $^1S_0$ can be compatible with Doppler cooling on the intercombination line.

$5p^2\,^3P_{0,1,2}$, and $5snd\,^3D_{1,2}$. Our calculation and preliminary experiment [12] showed that the same negative light shifts for both states can be realized by tuning the laser wavelength to $\lambda_F = 800$ nm, at which the Stark shift coefficient for both $^1S_0$ and $^3P_1$ is $U/h = I_F \times 1.45$ mHz/(W/m^2), where I_F is the laser power density. This same amount of negative light shift guaranteed that the FORT is compatible with the Doppler cooling or the magneto-optical trapping because the atomic resonance frequency is virtually unchanged in space. In addition, as the FORT laser at λ_F is 340-nm far-off resonant from the $^1S_0$-$^1P_1$ transition, the photon-scattering rate is below 10^{-1} s for a typical laser power of $I_F = 18$ kW/cm^2; thus, recoil heating is negligible.

A crossed dipole trap formed by two horizontal beams propagating along the x and y axes was applied to provide three-dimensional tight confinement of atoms. The actual shape of the trap potential, determined by the combination of the Stark shift potentials and the gravitational potential mgz along the vertical direction, is written as

$$U(x,y,z) = -U_0\{e^{-(x^2+z^2)/a^2} + e^{-(y^2+z^2)/a^2}\} + mgz, \quad (1)$$

where a is the e^{-1} radius of the laser beam and U_0 is the depth of the Stark shift potential given by a single beam. The effective depth of the trap is set by the lowest potential barrier in either direction. $U(x,y,0)$ provides the potential depth of $u_{xy} = U_0$ on the x-y plane, and $U(0,0,z)$ gives the depth of u_z along the z axis, which varies $0 < u_z < 2U_0$ depending on the competition between gravity and the optical dipole force. For high laser intensity, where the Stark shift potential dominates the gravitational one and thus $u_{xy} = U_0 < u_z < 2U_0$ holds, energetic atoms leak along each beam axis in the horizontal plane. On the other hand, for lower intensity, since $u_z < u_{xy} = U_0$, the atoms leak along the z direction. When

decreasing the laser intensity further, the local minimum of the potential disappears at $U_0 = mga\sqrt{e/2}$, below which atoms cannot be trapped.

Ultracold strontium atoms with a few photon-recoil energies were prepared by two-stage magneto-optical cooling and trapping [4]. First, strontium atoms were decelerated in a Zeeman slower and loaded into a magneto-optical trap using the $^1S_0$-$^1P_1$ transition to precool atoms. Second, we switched the laser to excite the spin-forbidden $^1S_0$-$^3P_1$ transition for further cooling and magneto-optical trapping (MOT). In order to cover the whole Doppler width (a few MHz) of the precooled atoms, the second-stage cooling laser was frequency modulated for a duration of 70 ms at the beginning, by which 30% of the precooled atoms were recaptured. The modulation of the cooling laser was then turned off, and the detuning δ_L and the total intensity I_L were set to -200 kHz and 160 μW/cm^2, respectively. Simultaneously, two laser beams, generated by the titanium-sapphire laser and passed through optical fibers, were introduced to start the loading of atoms. In order to form the FORT, these laser beams perpendicularly crossed one another at their waists almost in the center of the MOT. The laser beams had orthogonal linear polarization to avoid causing interference at the intersection. The atom transfer into the FORT continued for 35 ms in the presence of the MOT laser. After that, the MOT laser was turned off to operate the FORT alone for 10 ms, in which period the atoms not transferred into the FORT were separated.

Typically, 20% of the atoms recaptured in the MOT were transferred into the FORT, while the atom temperature remained constant. The various loading conditions were experimentally adjusted to optimize the final phase-space density. The laser detuning δ_L and the magnetic field gradient of the MOT (typically ~ 5 G/cm) set the size of the MOT, while the intensity I_L set the atom temperature, as indicated in Ref. [4]. The optimum parameters were determined by the competition among the loading efficiency, the loss of atoms due to inelastic collisions, and the temperature of the atoms. For example, reducing the trap volume improved the loading efficiency but enhanced the collision losses. Furthermore, the FORT loading time with the MOT laser present was optimized by considering the balance between the atom loading and the collision losses. The loading atom flux ϕ decreased continuously because the available atom number in the MOT was limited, while the collision loss βn^2 increased quadratically as the FORT density n increased.

The trapped atoms were observed using an absorption imaging technique. To image the atoms in the FORT, a probe laser with an intensity of 120 μW/cm^2 tuned to 17 MHz above the $^1S_0$-$^1P_1$ resonance was sent through the atoms for 20 μs after 10 ms of FORT operation. The shadow of the atoms was imaged on a charge-coupled device (CCD) through a microscope objective lens. The magnification of the imaging system was 2.3, corresponding to the rescaled CCD-pixel size of 5.2 μm. The resolution was measured to be 6.6 μm. The image is shown in Fig. 2(a). Since the axis of the probe beam was directed to $\mathbf{e}_x + \mathbf{e}_y$ in the notation of Eq. (1), the trapped atoms were observed as a faint line and a dense ellipsoid on it, corresponding to the atoms confined

RAPID COMMUNICATIONS

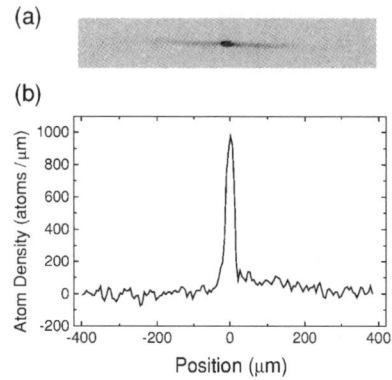

FIG. 2. (a) Absorption image of strontium atoms trapped in a crossed dipole trap. (b) Corresponding atom distribution along the horizontal direction. Over half of the atoms were trapped in the crossed region.

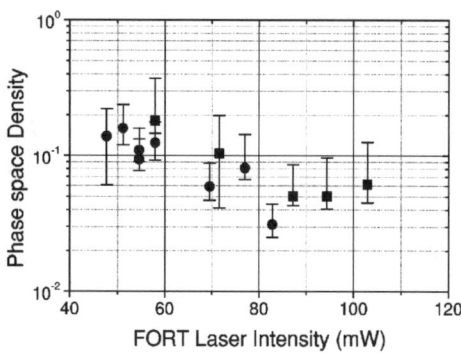

FIG. 4. Phase-space density of the atoms trapped in the crossed FORT as a function of the single FORT laser intensity. Different symbols correspond to different runs. Although the uncertainties in the estimation of the atom cloud size introduced relatively large error bars, as shown, a phase-space density exceeding 0.1 was obtained.

in each FORT laser axis and in the intersection, respectively. The typical e^{-1} radius of the ellipsoid was $\sigma_v = 6$ μm vertically and $\sigma_h = 12$ μm horizontally. The atom-number distribution corresponding to Fig. 2(a) horizontally is shown in Fig. 2(b), which indicates that over half of the atoms captured by the FORT are confined in the crossed region.

The temperature T_F and the number of atoms in the FORT were determined by time-of-flight (TOF) measurements. At 6 ms after turning off the FORT laser, i.e., 16 ms after turning off the MOT laser, the probe laser was flashed for 50 μs to image the two expanded atom clouds, corresponding to the atoms transferred into the FORT and those not transferred. These two atom clouds indicated the temperature of the atoms in the FORT and the MOT, respec-

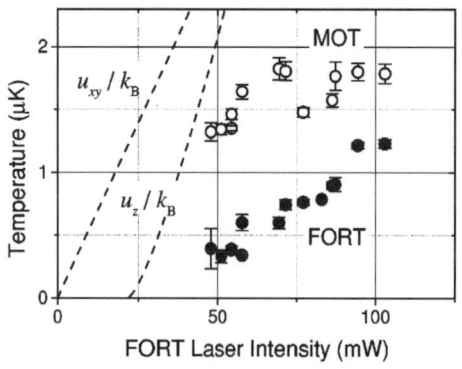

FIG. 3. Temperature of the atoms T_F in the crossed FORT (●) and the MOT (○) as a function of the single FORT laser intensity. The dashed lines u_z and u_{xy} are the calculated FORT potential along the vertical direction and in the horizontal plane, respectively. These barriers truncated the kinetic energy of atoms in the FORT, forcing the T_F to decrease proportionately to these barrier heights at the expense of trapped atoms.

tively. Both temperatures are summarized in Fig. 3 as a function of the FORT laser intensity. While the temperature of atoms in the MOT was constant, T_F linearly depended on the laser intensity because the effective depth of the FORT potential u_{xy} or u_z, as shown by the dashed lines in Fig. 3, truncated the kinetic energy of trapped atoms. In the calculation of the trap depth, we assumed a beam waist of $a = 20$ μm that reasonably agreed with the measured beam radius.

We carefully determined the number of atoms in the crossed FORT by combining the FORT image and the TOF image, because the high atom density ($n > 10^{12}$ cm^{-3}) in the crossed region modified the absorption rate of the probe laser due to radiation-trapping effects. We made the TOF measurement when the atoms not transferred into the FORT had been spatially separated. The total number of atoms N_t finally detected by the TOF was that of atoms trapped anywhere in the FORT laser beams, i.e., outside and inside of the beam intersection. With the help of the FORT image, we could precisely determine the number of atoms outside of the beam intersection N_{out} because of its low density. Hence the atom number inside the intersection N_{in} was inferred from both numbers, i.e., $N_{in} = N_t - N_{out}$, which ranged from 4 $\times 10^4$ to 3×10^5 depending on the FORT laser intensity.

We estimated the size of the atom cloud in the crossed region by a FORT image assuming a gaussian distribution of atoms. It should be noted that these fits could overestimate the actual trap size because the radiation-trapping effects could distort the atom density profiles by reducing the probe absorption, especially in the trap center. In order to estimate the lower bound of the atom density or the phase-space density, we took the volume as $V = \pi^{3/2} \sigma_v \sigma_h^2$, where σ_v and σ_h are the vertical and horizontal e^{-1} radii obtained by the fitting. The phase-space density is thus estimated by $\rho = (N/V)\lambda_{dB}^3$ with $\lambda_{dB} = h/\sqrt{2\pi m k_B T_F}$, where m is the mass of an atom. Results are summarized in Fig. 4 as a function of FORT laser intensity. The asymmetric error bars shown in

RAPID COMMUNICATIONS

TETSUYA IDO, YOSHITOMO ISOYA, AND HIDETOSHI KATORI

PHYSICAL REVIEW A **61** 061403(R)

the figure are mainly due to the uncertainties in the volume measurements: The radiation-trapping effects described above may result in the reduction of volume by 60% or the increase in ρ by 150%, while the other statistical uncertainties were on the 20% level. In spite of these relatively large error bars, it is clear that the phase-space density well exceeds 0.1, with slightly increasing tendency as the FORT depth decreases [13].

The obtained phase-space density was mainly limited by the achievable atom density in the presence of light-assisted collisions occurring during atom transfer into the FORT. To examine the influence, we measured the fluorescence decay of the MOT and thus obtained the binary loss rate β with the help of atom-density measurements. The light-assisted collision loss rate was found to be $5 \times 10^{-12} < \beta < 1.5 \times 10^{-11}$ (cm^3/s), rather insensitive to the MOT laser parameters that were used in the FORT loading experiment [14]. With the loading flux ϕ into the FORT volume V, the change of atom density n obeys the rate equation $dn/dt = \phi/V - \beta n^2$. We define the rising time constant as $\tau = 1.5(\beta \phi/V)^{-1/2} = 1.5(\beta n_f)^{-1}$, in which the density reaches 90% of the equilibrium density n_f [15]. The measured rising time $\tau = 15$ ms and loss rate $\beta = 10^{-11}$ cm^3/s yielded the at-

tainable density $n_f = 10^{13}$ cm^{-3}, which moderately agreed with the atom density we experimentally attained in the FORT. Since the density n_f is proportional to $\sqrt{\phi/(\beta V)}$, two orders of enhancement in $\phi/(\beta V)$ are necessary to reach quantum degeneracy.

In summary, we have demonstrated a rapid creation of nearly degenerate strontium atoms by employing an alternative FORT scheme, which is based on the successful combination of properly designed dipole force trapping and Doppler-cooling on a narrow intercombination line. We observed that the phase-space density was finally limited by the light-assisted inelastic collisions occurring in laser cooling. Therefore the reduction of these inelastic collisions is crucial to increase the density by another order. A laser suppression of light-assisted collisions or evaporative cooling [1] in an optical trap may overcome these difficulties. Even in the latter case, evaporative cooling time, which is reported typically tens of seconds for alkali-metal atoms [1], can be significantly reduced because of the high phase-space density we start with.

We would like to thank M. Gonokami for helpful discussion and M. Daimon for his advice on focusing optics.

[1] M. H. Anderson *et al.*, Science **269**, 198 (1995); K. B. Davis *et al.*, Phys. Rev. Lett. **75**, 3969 (1995); C. C. Bradley, C. A. Sackett, and R. G. Hulet, *ibid.* **78**, 985 (1997).

[2] T. Walker, D. Sesko, and C. Wieman, Phys. Rev. Lett. **64**, 408 (1990).

[3] A. Gallagher and D. E. Pritchard, Phys. Rev. Lett. **63**, 957 (1989).

[4] H. Katori, T. Ido, Y. Isoya, and M. K-Gonokami, Phys. Rev. Lett. **82**, 1116 (1999).

[5] J. L. Hall, M. Zhu, and P. Buch, J. Opt. Soc. Am. B **6**, 2194 (1989).

[6] J. D. Miller, R. A. Cline, and D. J. Heinzen, Phys. Rev. A **47**, R4567 (1993).

[7] T. Kuwamoto, K. Honda, Y. Takahashi, and T. Yabuzaki, Phys. Rev. A **60**, R745 (1999).

[8] H. J. Lee, C. S. Adams, M. Kasevich, and S. Chu, Phys. Rev. Lett. **76**, 2658 (1996).

[9] V. Vuletić, C. Chin, A. J. Kerman, and S. Chu, Phys. Rev. Lett. **81**, 5768 (1998).

[10] S. E. Hamann *et al.*, Phys. Rev. Lett. **80**, 4149 (1998).

[11] D. Boiron *et al.*, Phys. Rev. A **57**, R4106 (1998).

[12] H. Katori, T. Ido, and M. K-Gonokami, J. Phys. Soc. Jpn. **68**, 2479 (1999).

[13] In decreasing the trap depth, the atom temperature decreased monotonically while the atom density remained almost constant, leading to a net increase in phase-space density. This fact may infer the manifestation of evaporative cooling in 10 ms of optical trapping.

[14] In spite of three orders of magnitude less excitation rate than that for typical alkali-metal atom experiments, the inelastic loss rate was even larger by an order of magnitude. These results may be attributed to the much longer lifetime of quasi-molecules formed by two atoms in $^1S_0$ and $^3P_1$ states as well as the reduced trap depth.

[15] We assumed ϕ to be constant to simplify the discussion because the change of ϕ was estimated to be within 50% in the loading process.

Focusing and defocusing of neutral atomic beams using resonance-radiation pressure

D. B. Pearson, R. R. Freeman, J. E. Bjorkholm, and A. Ashkin

Bell Telephone Laboratories, Holmdel, New Jersey 07733

(Received 1 October 1979; accepted for publication 22 October 1979)

We demonstrate strong focusing and defocusing of a sodium atomic beam using the transverse resonance-radiation pressure of a superimposed cw dye laser beam tuned near the atomic resonance. Focal spot diameters of about 60 μm and a 30-fold increase of the on-axis atomic beam intensity have been obtained. For defocusing, the on-axis intensity can be reduced to less than 10^{-2} of its original value.

PACS numbers: 82.40.Dm

In recent experiments we demonstrated focusing and defocusing forces exerted on a beam of neutral sodium atoms by the radiation pressure of a superimposed and copropagating light beam from a cw dye laser.[1] The forces and the associated effects were made large by tuning the frequency of the laser to be near, but not equal to, the frequency of the sodium resonance transition. In Ref. 1, the modifications to the atomic beam profile caused by the light were determined in the far field of the laser beam. The work discussed here uses the same basic technique to study much stronger focusing or defocusing of the atomic beam. Specifically, we measured the atomic beam profile in the focus of the laser beam where the transverse confinement of the atoms is strongest. We found that the atomic beam can be focused to a spot diameter of about 60 μm and that its on-axis intensity can be increased more than 30-fold using 200 mW of light. Under defocusing conditions, the on-axis atomic beam intensity can be reduced to less than 10^{-2} of its value in the absence of light. These effects are strong enough that significant applications can be envisioned.

The forces of resonance-radiation pressure can be classified into two types. The focusing effects are caused by *dipole* resonance-radiation pressure.[2,3] This force arises from stimulated scattering of the light by the atoms, and it is the same as the force exerted on an induced dipole situated in an electric-field gradient. Thus it is proportional to and in the direction of the optical intensity gradient. For a weakly focused TEM_{00} (Gaussian) mode laser beam the force is essentially transverse to the direction of propagation of the light. When the light is tuned below the atomic resonance frequency, the sign of the force is such that the atom tends to be pulled into the high intensity regions of the light beam; for tuning above resonance, the atoms tend to be expelled from the light. *Spontaneous* resonance-radiation pressure arises from spontaneous scattering of the light by the atoms and it exists even in uniform plane waves.[4] The average spontaneous force is in the direction of light propagation. In our experiments the resulting acceleration of the atoms along the light beam is rather unimportant to an understanding of the results.

A schematic diagram of the experimental setup is shown in Fig. 1. Light from a continuously tunable, single-mode cw ring dye laser (Spectra-Physics Model 380A) was superimposed upon an effusive beam of neutral sodium atoms ($T \simeq 500$ °C) by reflection off a 3-mm-thick dielectric-coated mirror having a 230-μm-diam hole in it. The atomic beam divergence was determined by this hole, the 500-μm-diam hole in the atomic source, and the 55-cm separation between them. Typical laser powers superimposed on the atoms ranged between 20 and 200 mW. The light was linearly polarized and no precautions to prevent optical pumping of the sodium atoms were taken. A 75-cm focal length lens focused the light to a spot size $w_0 \approx 100$ μm at a nominal distance of 25 cm from the mirror. A moveable hot-wire detector was used to measure the atomic beam profile in the focus of the laser beam. The detector was insensitive to light incident upon it and its resolution was determined by an aperture of 30 μm diameter mounted in front of the hot iridium wire. We also had the option of using a velocity selector to narrow the velocity distribution of the atomic beam. However, since we observed no strong dependence upon atomic velocity in these experiments, we present results obtained without velocity selection.

Dramatic changes in the atomic beam profile were obtained when the laser was tuned to within several GHz of the sodium $3^2S_{1/2} \rightarrow 3^2P_{3/2}$ resonance transition at 5890 Å. Examples of the atomic beam profiles measured when the laser was tuned below resonance for focusing are shown in Fig. 2. Curve a shows the measured atomic beam current as a function of detector position in the absence of light; the peak beam current was approximately 10^{-11} A and is normalized to 1. Curves b and c show the beam profiles obtained with 200 and 25 mW, respectively; for each power level the tuning

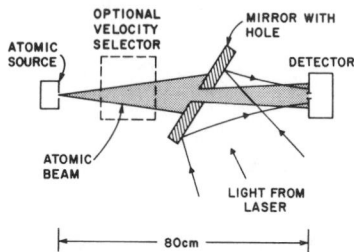

FIG. 1. Schematic diagram of the experimental setup. The transverse scale is greatly magnified relative to the longitudinal scale; for instance, the diameter of the hole in the mirror is 230 μm.

     Appl. Phys. Lett. **36**(1), 1 January 1980   0003-6951/80/010099-03$00.50

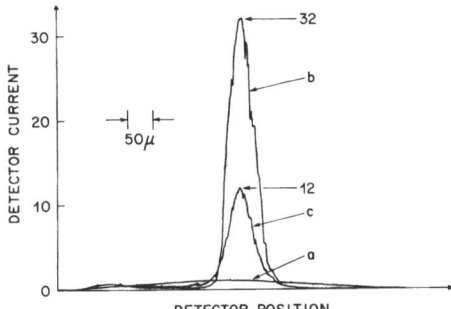

FIG. 2. Atomic beam current measured by the hot-wire detector as a function of its transverse position. The laser has been tuned to maximize the peak atomic beam intensity for each laser power: (a) no light, (b) 250 mW (c) 25 mW. The peak beam current corresponds to roughly 2×10^9 atoms/sec.

of the laser was chosen to maximize the on-axis atomic beam current. For profiles b and c the laser frequency was approximately 4.5 and 1 GHz, respectively, below the $3S_{1/2}(F=2) \rightarrow 3P_{3/2}$ transition resonant frequency for atoms at the peak of the velocity distribution (9×10^4 cm/sec). For both cases the measured diameter of the atomic beam profile was 65 μm; in the absence of light the diameter is 390 μm. The corresponding on-axis atomic beam intensities are increased by factors of 32 and 12 for profiles b and c, respectively. These focusing effects are much stronger than those previously reported in Ref. 1 where the atomic beam profile was measured 25 cm beyond the focus of the laser beam. The measured atomic beam profiles represent a convolution of the actual atomic beam intensity distribution with the 30-μm aperture on the detector; consequently, the actual atomic beam focal spot size is somewhat smaller, and the peak intensities are correspondingly higher than the values directly measured. Deconvolution is not easily accomplished since the aperture is roughly the same size as the atomic beam and both have circular cross sections. In principle the atomic beam profile is cylindrically symmetric; however, the symmetry of the measured profiles is critically dependent upon precise overlap and alignment of the laser and atomic beam. Effects due to the input laser beam not being perfectly Gaussian and due to distortion of the laser mode caused by the hole in the mirror may also cause asymmetries in the atomic beam profiles, such as the observed secondary peak on the wings of the profiles in Fig. 2. Because the transverse position of the small focal spot is strongly dependent upon alignment, our measurements are very sensitive to small movements of the atomic beam apparatus relative to the laser beam; this is the main reason for the lack of smoothness in the recorded beam profiles. Enhancements of the on-axis atomic beam intensity as large as 40 have been observed in real time.

In Fig. 3 we display a typical atomic beam profile measured for the laser tuned above the atomic resonance. In this case the laser was tuned to minimize the on-axis atomic beam intensity. For Fig. 3 the laser power was 160 mW and the laser was tuned about 2 GHz above resonance; the exact tuning for this case, however, is not at all critical. It is seen

that the atoms are essentially expelled from the central region of the atomic beam profile, a region having a diameter of 210 μm. For this defocusing case the measured on-axis detector current is primarily background noise; the atomic flux was, at most, less than 10^{-2} of the on-axis flux measured with no light present. Thus in going from the defocusing case to the focusing case, the on-axis atomic beam intensity changes by more than a factor of 3×10^3.

As demonstrated by Fig. 2, with the present experimental setup the spot diameter of the focused atomic beam is independent of the laser power over the power range we investigated and for laser tunings which maximize the on-axis atomic beam intensity. In spite of the constant spot size, we observe that the peak beam intensity does increase with increasing power; this occurs because more of the atoms in the initial atomic beam spatial distribution are pulled into the focal region at high powers than at low powers. For the simple case in which all the atoms are pulled into the focal region, such as for an atomic beam with a much smaller initial divergence than used here, the spot size of the focused atomic beam is determined by the interplay of the initial atomic beam divergence, the strength of the focusing forces, and the transverse heating of the atoms by the fluctuations inherent in spontaneous scattering.[5] In general, this leads to a power dependence of the minimum focal spot size, with higher powers giving smaller spots; an investigation of this situation is currently in progress. We presently have no quantitative description for the case studied here in which the initial beam divergence is so large that only a fraction of the atoms are confined in the focal volume, and that fraction is power dependent.

In conclusion, we have demonstrated that strong focusing and defocusing of neutral atomic beams can be readily obtained using cw dye lasers and the transverse dipole force of resonance radiation pressure. Atomic beam focal spot diameters of 60 μm and increases of the on-axis atomic beam intensity of a factor of 30 have been demonstrated. Of course, the effects observed here are not limited to neutral atomic beams; similar effects are possible with neutral molecular beams and with ion beams.[6] In the addition to the focusing or defocusing of a single-component atomic beam, the ability to selectively focus or defocus a given species in a mixed atomic beam offers interesting possibilities. For instance, a "dirty"

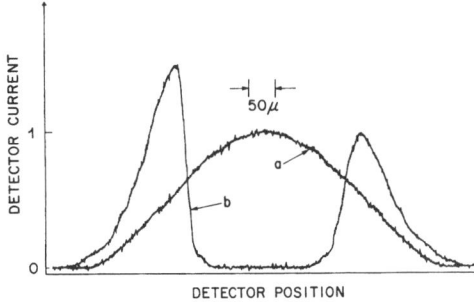

FIG. 3. Atomic beam curent measured by the hot-wire detector as a function of its transverse position. The laser has been tuned to minimize the on-axis atomic beam intensity; profile (a) no light, profile (b) 160 mW.

atomic beam can be cleaned by selectively focusing the desired species through an aperture, by selectively defocusing the undesired species, or by using both techniques at the same time. In isotope separation, the simultaneous focusing of one species and defocusing of the other could yield large single-step separation ratios.

[1]J.E. Bjorkholm, R.R. Freeman, A. Ashkin, and D.B. Pearson, Phys. Rev. Rev. Lett. **41**, 1361 (1978).

[2]G.A. Askar'yan, Zh. Eksp. Teor. Fiz. **42**, 1567 (1962) [Sov. Phys.-JETP **15**, 1088 (1962)]; V.S. Letokhov, V.G. Minogin, and B.D. Pavlik, Opt. Commun. **19**, 72 (1976).

[3]A. Ashkin, Phys. Rev. Lett. **40**, 729 (1978).

[4]A. Ashkin, Phys. Rev. Lett. **25**, 1321 (1970).

[5]J.E. Bjorkholm, R.R. Freeman, A. Ashkin, and D.B. Pearson, in *Laser Spectroscopy IV, Proc. 4th International Conference, Rottach-Egern*, 1979, *Springer Series in Optical Sciences*, edited by H. Walther and K.W. Rothe (Springer-Verlag, Berlin, 1979).

[6]G.A. Askar'yan, Ref. 2, part 1.

Reprinted from Optics Letters, Vol. *9*, page 454, October, 1984
Copyright © 1984 by the Optical Society of America and reprinted by permission of the copyright owner.

Stable radiation-pressure particle traps using alternating light beams

A. Ashkin

AT&T Bell Laboratories, Holmdel, New Jersey 07733

Received June 1, 1984; accepted July 5, 1984

A new type of stable alternating-beam light trap is proposed for confinement of neutral atoms and macroscopic dielectric particles. This trap, based only on the scattering force of radiation pressure, overcomes the limitations of the optical Earnshaw theorem. Trapping of $\sim 10^7$ sodium atoms in large volumes (100 cm^3) seems possible with well depths >1 K and with optical cooling close to the Purcell limit of $\sim 10^{-4}$ K. Trapping at points of zero light intensity is considered.

It is proposed here that a new kind of stable optical trap is possible for neutral dielectric particles and atoms that uses only the scattering force of radiation pressure from alternating light beams. These are beams whose direction of propagation reverses periodically. It is well known that stable optical trapping of neutral particles is possible with cw laser beams.[1–3] There are two types of radiation-pressure forces (the so-called scattering force and the gradient force); these cw traps rely solely on the action of the gradient force for their stability. Indeed, a recent theorem, called the optical Earnshaw theorem, has been proved; it states that stable cw traps based only on the scattering force are not possible.[3] The optical Earnshaw theorem was proved in analogy to Earnshaw's theorem in electrostatics, which states that stable traps for charged particles that use only electrostatic forces are not possible. For charged particles, however, it is well known that one way to circumvent Earnshaw's theorem is to make stable electrodynamic traps[4–6] with alternating electric fields. Indeed, the alternating light trap proposed here extends the analogy between charged-particle traps and radiation-pressure traps for neutral particles one step further, to include the case of stable alternating light traps based on the scattering force.

As we shall see, the alternating light traps can be at least as deep as cw gradient-force traps, extend over much larger volumes, and use much lower optical intensities. These features of the new trap considerably enhance the possibility of achieving optical trapping and cooling of neutral atoms.

Figure 1a shows in cross section the geometry of a typical quadrupole electrodynamic trap. A hyperbolically shaped ring electrode R of minimum radius r_0 is capped by a pair of hyperbolically shaped cap electrodes C_1 and C_2 placed at distance Z_0 from the origin O. Alternating voltage $V_0 \sin \Omega t$ is applied between the ring electrode and the pair of commonly connected cap electrodes, generating a hyperbolically shaped quadrupole field about the symmetry point O of the trap. If, as shown, a positively charged particle is placed at a point P along the Z axis when the caps are at a positive

voltage with respect to the ring, it feels a stable restoring force toward the origin O of the form $-4Az$, where A is a constant; whereas at a point Q along a radius from the origin it feels an unstable force away from O given by $2Ar$.[5] When the alternating voltage V_0 reverses, the force on P is outward (unstable) and at Q is inward (stable). The effect of this alternating voltage and inhomogeneous field on a charged particle is to give a net stability. The actual particle motion consists of an oscillatory component, with an average drift toward the weaker fields at the origin. This behavior is described by Mathieu's equation, which gives the effective well depth and range of frequencies Ω over which the motion is stable.[4,5]

Figure 1b shows the geometry of the optical-scattering-force analog of this electrodynamic trap. Consider two identical opposing coaxial TEM_{00}-mode Gaussian beams A and B with foci f_A and f_B displaced from the symmetry point O or the origin of the trap by a distance

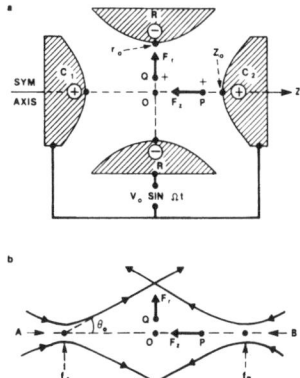

Fig. 1. a, Geometry of a typical quadrupole electrodynamic trap for charged particles. b, Geometry of the alternating-beam optical-scattering-force trap for neutral particles that is analogous to a.

October 1984 / Vol. 9, No. 10 / OPTICS LETTERS **455**

z_0. O is assumed to be in the far field of A and B. The net scattering force[3] at any point P displaced axially from the origin is toward O (stable). At a point Q displaced radially from O, the net scattering force is radially outward (unstable). Close to O these forces take the form $F_z = 4Kz/z_0^3$ and $F_r = 2Kr/z_0^3$, where K is proportional to the total light power. The optical fields are thus hyperbolic near the zero-field point O, in exact analogy with the electrodynamic trap of Fig. 1a. In addition to the outward (unstable) radial scattering force there is, of course, a net inward (stable) gradient force that is due to beams A and B, which can be the basis of a stable cw light trap.[1-3] However, this inward gradient force is negligible compared with the outward scattering force in this instance, since, as postulated, we are well into the far field of beams A and B. If, as now proposed, we exactly reverse the direction of propagation of Gaussian light beams A and B (ray for ray), the scattering force on a particle at any position P or Q will exactly reverse. The force at P will then be away from O (unstable) and at Q toward O (stable). If this beam reversal is accomplished sinusoidally at frequency Ω we expect to find the same net trapping effect on a neutral particle in this light field as on charged particles in electrodynamic traps.

Figure 2 shows one proposed method for accomplishing the sinusoidal beam reversal necessary for the alternating-beam light trap. The TEM_{00}-mode power P from laser source S, feeding a polarizing beam splitter SP, has its polarization switched 90° by application of a square-wave voltage of frequency Ω to electro-optic polarization modulator EOM_s. This effectively switches the full laser power back and forth between the two beam paths, which are labeled F_{tot} for total forward light and R_{tot} for total reverse light. Power in the forward-light path is further divided equally by a 50–50 beam splitter into beams A_F and B_F, which are injected axially into the trap by 45° mirrors M_1 and M_2 and focused by lenses L_1 and L_2 to focal points f_A and f_B. The focused beams subsequently spread to form the two-beam far-field trapping geometry described in Fig. 1b. The forward beams are refocused by lenses L_2 and L_1 as they leave the trap and emerge through small pinholes in the 45° mirrors M_1 and M_2, as shown. The total reverse power is split equally into reverse beams A_{rev} and B_{rev} by a 50–50 beam splitter and enter the trap axially through the pinholes in mirrors M_1 and M_2. By proper adjustment of lens combinations L_3, L_2 and L_4, L_1, the reverse beams A_{rev} and B_{rev} are focused to the same foci and beam spot sizes at points f_A and f_B as the forward beams. This ensures that the backward-directed beams follow the same light path (ray for ray) as the forward beams. Distortion of the forward beams because of the pinholes in the 45° mirrors is negligible for the spot sizes being considered. To achieve true sinusoidal modulation of the light force in the trap, we can simply sinusoidally amplitude modulate each of the components F_{tot} and R_{tot} of the switched beam with amplitude modulators EOM_F and EOM_R driven 180° out of phase and synchronized with the beam-switching modulator EOM_L.

We now estimate the achievable well depths of optical alternating-beam traps for neutral atoms. This is most

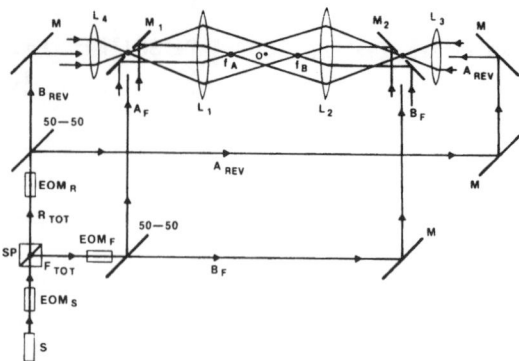

Fig. 2. Schematic diagram of apparatus for the sinusoidal reversal of the beams A and B feeding the alternating-beam light trap centered about the equilibrium point O.

easily done by comparison with the known well depths of quadrupole electric-field traps. For quadrupole traps the effective electrodynamic well depth is smaller than the full applied potential V_0 by the factor $\sqrt{2}\Omega/\overline{\omega}_z$, where Ω (the applied frequency) is the frequency of the micromotion of the particle and $\overline{\omega}_z$ is the average oscillation frequency of the so-called secular motion of the particle about the equilibrium point O of the effective harmonic well.[5,6] Theoretically, the trap is stable for ratios of $\Omega/\overline{\omega}_z$ from infinity down to all values greater than 2. Indeed, Wuerker et al.[4] experimentally demonstrated trapping down to a ratio of 2. At 2, or below, parametric instabilities occur. Thus well depths of $V_0\overline{\omega}_z/\sqrt{2}\Omega = V_0/2\sqrt{2} \cong V_0/3$ are achievable. For ion traps, for which well depth is not a factor, larger ratios of $\Omega/\overline{\omega}_z$, the so-called stability parameter, are often used with well depths of about $V_0/10$.[5,6] This helps to avoid the buildup of instabilities if the fields are not strictly quadrupole.

Consider the case of an alternating-beam optical trap for sodium atoms. We use large-numerical-aperture beams of N.A. = $\sin \theta_0 = 0.6$, for example, in order to maximize the radial components of the force and get a deep radial potential. Suppose that we feed two Gaussian beams A and B of the form $I_0 \exp(-2r^2/w^2)$, with a total incident power $P_0 \sin \Omega t$, where $P_0 = 0.14$ W. By choosing a large spot size at the origin, $w = 3$ cm, we can avoid excessive saturation. For these parameters the maximum light intensity at the origin is $I_0 = 2P_0/\pi w^2 = 10^{-2}$ W $= I_{sat}$, the saturation intensity of sodium atoms. We first evaluate the radial and axial cw (dc) potentials U_r^{cw} and U_z^{cw}. For U_r^{cw} we integrate $F_r dr = 2F_{scat}(r)\sin\theta dr$ from zero to infinity. For U_z^{cw} we integrate $F_z dz = F_{scat}(z)dz$ from zero to $z = w$ since this represents the approximate axial limit of the trap for our choice of beam geometry. One finds that $U_r^{cw} = 1.3 \times 10^{-15}$ erg and $U_z^{cw} = 3.9 \times 10^{-15}$ erg. To get a feeling for these magnitudes one can rewrite them as $U_r^{cw} = F_{sat}(0.34$ cm$)$ and $U_z^{cw} = F_{sat}(1.0$ cm$)$, where $F_{sat} = h/\lambda 2\tau = 3.7 \times 10^{-15}$ dyne. By analogy with the quadrupole alternating electric-field traps, we expect an alternating light well depth U^{alt}, which is in the range

456 OPTICS LETTERS / Vol. 9, No. 10 / October 1984

from $U^{cw}/10$ to $U^{cw}/3$. Since for our trap $U_r{}^{alt} < U_z{}^{alt}$, the actual trap depth is given by $U_r{}^{alt}$. This falls in the range from 1.3×10^{-16} to 4.2×10^{-16} erg. By setting $mv_{max}{}^2/2 = U_r{}^{alt}$, one finds the maximum velocity v_{max} of an atom in a trap. This falls in the range from 2.6×10^3 to 4.7×10^3 cm/sec. In terms of temperature, $U_r{}^{alt}$ falls in the range from 0.6 to 2.1 K. For these well depths and trap dimensions, the required driving frequency Ω is of the order of 10 kHz, assuming harmonic potential wells. One has the following options to deepen these traps. If the trap is scaled up with the same intensity profile, the well depth will increase linearly and the total power quadratically with trap dimensions. A factor of 2.5 in depth seems possible. By analogy with quadrupole traps one can add a cw light component to the reverse beams and deepen the radial depth by a factor of ~ 1.5 at the expense of the axial depth until both potentials are equal. Overall, one hopes to achieve trap depths in the range 2.3–7.6 K. Thus we achieve stable alternating-beam traps at least as deep as cw gradient-force traps but with much larger volumes (by a factor of 10^7), which greatly eases the problem of filling the trap with slow atoms.

A crucial consideration for all optical atom traps is the effectiveness of optical cooling. For cw gradient-force traps, it is known to be difficult to cool to the Purcell limit of $\sim \hbar\Gamma$ that is characteristic of plane-wave-beam optical cooling by using only the scattering force.[7,8] Use of the same beams for both trapping and cooling of gradient-force traps results in leaky traps.[7,8] If one resorts to separate trapping and cooling beams, problems arise because of optical Stark shifts.[7–9] Contributions from gradient-force heating can often be severe.

However, with alternating-beam traps tuned close to resonance it is anticipated that cooling problems will be much relieved. Since we use large-diameter far-field Gaussian beams, there should be negligible contributions to the heating from the radial gradient forces. If the saturation parameter $p = I/I_{sat} \leqslant 1$, one has negligible nonlinear problems. Low saturation also ensures that heating from the standing-wave gradients is small.[3] It is thus expected that simply adding an additional cooling beam optimally tuned one half linewidth $\Gamma/2$ below resonance will result in effective cooling to an equilibrium temperature within about a factor of 2 of the Purcell limit.[7] A possibly preferable cooling method is to tune the alternating beams $\Gamma/2$ below resonance and to use these beams for both trapping and cooling. The alternating beams clearly cool the low-frequency secular motion of atoms in the trap, which simultaneously reduces the high-frequency micromotion although this component may not be directly cooled.

Assuming a source of slow atoms with an average velocity $v_{av} \sim 3 \times 10^3$ cm/sec and a density of $\sim 10^5$ atoms/cm^3, which now seems achievable,[10] we estimate that a total of 10^7 atoms can be trapped in the ~ 100-cm^3 volume. If the trapped cloud is cooled to the Purcell limit of $3\hbar\Gamma/2 \cong 30$ cm/sec, assuming a harmonic potential, its radius reduces from ~ 3 cm to 300 μm, with an increase in density of 10^6 to about 1 atom per 10 μm^3.

This confined dense-atom cloud would be useful in a variety of slow-atom-interaction experiments and for spectroscopy. One interesting possibility is to transfer some of the cooled atoms to the much smaller and more tightly confined cw gradient-force traps for further study. If desired, one can make alternating-beam atom traps in which the equilibrium point O is also a point of zero light intensity. Use of TEM$_{01}$* doughnut-shaped mode beams for the A and B beams gives radial stability but no z directed force. Addition of a second orthogonal pair of TEM$_{01}$* beams gives z stability and results in a zero-field trap. Such a trap can be cooled ideally with an additional cooling beam. The absence of optical perturbations makes this trap geometry nearly ideal for spectroscopy and frequency standards[11] that use neutral atoms. Use of these traps with macroscopic particles might also be useful if large numbers of particles need to be trapped in deep wells.

In conclusion, the use of alternating-beam far-field radiation-pressure traps is proposed to circumvent Earnshaw's theorem and achieve stable traps with significant advantages for trapping and cooling neutral atoms and other neutral dielectric particles.

I would like to thank J. E. Bjorkholm and J. P. Gordon for helpful discussions.

References

1. A. Ashkin, Phys. Rev. Lett. **24**, 156 (1970).
2. A. Ashkin, Science **210**, 1081 (1980).
3. A. Ashkin and J. P. Gordon, Opt. Lett. **8**, 511 (1983).
4. R. F. Wuerker, H. Shelton, and R. V. Langmuir, J. Appl. Phys. **30**, 342 (1959).
5. H. G. Dehmelt, Adv. At. Mol. Phys. **3**, 53 (1967).
6. W. Neuhauser, M. Hohenstatt, P. E. Toschek, and H. G. Dehmelt, Appl. Phys. **17**, 123 (1978).
7. A. Ashkin and J. P. Gordon, Opt. Lett. **4**, 161 (1979).
8. J. P. Gordon and A. Ashkin, Phys. Rev. A **21**, 1606 (1980).
9. J. Dalibard, S. Reynaud, and C. Cohen-Tannoudji, Opt. Commun. **47**, 395 (1983).
10. W. D. Phillips and H. Metcalf, Phys. Rev. Lett. **48**, 596 (1982); J. V. Prodan, W. D. Phillips, and H. Metcalf, Phys. Rev. Lett. **49**, 1149 (1982).
11. D. J. Wineland, W. M. Itano, J. C. Berquist, J. J. Bollinger, and H. Hemmati, in *Laser Cooled and Trapped Atoms*, W. D. Phillips, ed., Nat. Bur. Stand. (U.S.) Spec. Publ. **653** (1983), p. 19.

VOLUME 54, NUMBER 12 PHYSICAL REVIEW LETTERS 25 MARCH 1985

Observation of Radiation-Pressure Trapping of Particles by Alternating Light Beams

A. Ashkin and J. M. Dziedzic

AT&T Bell Laboratories, Holmdel, New Jersey 07733

(Received 14 November 1984)

The new principle of radiation-pressure trapping of neutral dielectric particles by means of alternating light beams was demonstrated in an optical levitation experiment. Stable trapping of micron-sized spheres was observed due to alternating scattering-force fields under conditions where stable cw trapping was not possible. Application of this principle to neutral-atom trapping as a means of circumvention of the optical Earnshaw theorem is suggested. A new stable cw optical trap for spheres was also discovered.

PACS numbers: 32.80.−t, 42.60.−v

We demonstrate the first example of a stable alternating-beam radiation-pressure particle trap for neutral dielectric particles in an optical levitation experiment using micron-sized dielectric spheres. The newly proposed principle of alternating-beam trapping of neutral dielectric particles, including atoms, by radiation pressure[1] is based on a dynamic effect and uses only the scattering-force component of the radiation pressure. Such scattering-force traps were recently conceived as a means of overcoming the limitations of the optical Earnshaw theorem.[2] This theorem states that stable cw trapping of dielectric particles using only the scattering force is not possible. The scattering force is the simpler of the two radiation-pressure force components[3-6] acting on a neutral particle. It acts in the direction of the incident light beam and is directly proportional to the light intensity. The second force component, the so-called gradient force, points in the direction of the optical intensity gradient and is proportional to the strength of the optical gradient. It is well known that stable cw trapping of dielectric particles is possible with use of the gradient force.[3-6] Indeed, it is a corollary of the optical Earnshaw theorem that all stable cw traps must rely on the gradient force. Stable alternating-beam traps of the general type demonstrated here are expected to have distinct advantages[1] over cw traps for trapping of neutral atoms and macroscopic Rayleigh-size particles. For atoms, large-volume traps (~ 1 cm^3) are predicted with greatly reduced optical cooling problems and well depths of ~ 1 K. The well depth and volume of these new traps are thus quite compatible with the energy and density of atomic beams slowed by the scattering force of radiation pressure.[7]

The originally proposed alternating-light-beam trap[1] using only the time-varying scattering force is shown schematically in Fig. 1. It consists of a pair of identical coaxial TEM$_{00}$-mode Gaussian beams A and B with foci f_A and f_B located symmetrically about the trap equilibrium point O. Action by the gradient force is assumed to be negligible. This can be achieved by placing O well into the far field of the beams A and B, or, for atoms, by tuning the beams to exact resonance.

This removes the possibility of cw trapping. The light intensity in A and B is further assumed to vary sinusoidally in time as $I_{A,B} = |\mathbf{I}_{A,B}|\sin\Omega t$, where $|\mathbf{I}_{A,B}|$ is the peak amplitude of A or B and Ω is the frequency. Thus the beams switch periodically from the situation of Fig. 1(a) where A and B are in the so-called forward direction to that of Fig. 1(b), a half-cycle of Ω later, where A and B are exactly reversed. This causes a corresponding periodic reversal of the direction of the net scattering force as shown at representative points P and Q of Figs. 1(a) and 1(b). Trapping based on alternating scattering-force fields in this geometry was proposed in direct analogy with ac electrodynamic ion traps[8-10] where the dynamic stability is governed by the Mathieu equation. Indeed the shape of the scattering-force fields near the origin O in Fig. 1 are closely hyperbolic,[1] as are the electric fields of quadrupole ion traps. The detailed particle motion in alternating-field traps consists of the so-called oscillatory micromotion at the driving frequency Ω superimposed on a slower oscillatory macromotion in response to the average stable trapping potential that exists about the zero-field equilibrium point O.

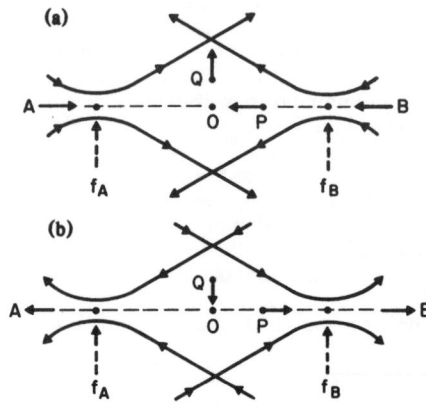

FIG. 1. Geometry of alternating-beam light trap. (a) Beams A and B in the so-called forward direction. (b) Beams A and B a half-cycle later in the reverse direction.

VOLUME 54, NUMBER 12   PHYSICAL REVIEW LETTERS   25 MARCH 1985

Figure 2 shows how we accomplish the periodic beam reversal necessary to give alternating light fields with the beam configuration of Fig. 1. Argon laser light at 5145 Å is spatially filtered by SF and split equally by polarizing beam splitter SPl into two orthogonally polarized beams, labeled forward total F_{tot} and reverse total R_{tot}. A chopper blade rotating at frequency Ω is poisitoned to switch alternately between the F_{tot} and R_{tot} beams. The beam F_{tot} is split into the two equal forward beams A_F and B_F by a 50-50 beam splitter. The beams A_F and B_F are injected into the trap through polarizing beam splitters SP2 and SP3 and focused by a pair of microscope objectives L1 and L2 to beam foci f_A and f_B. This gives rise to the beam configuration of Fig. 1(a). The beam R_{tot} is also split into two equal reverse beams A_{rev} and B_{rev} by a 50-50 beam splitter. They are focused by lenses L3 and L4 and injected into the trap through SP2 and SP3 lenses L1 and L2. The beams B_{rev} and A_{rev} are focused to the same focal points f_B and f_A with essentially the same spot sizes as the forward beams. This effectively reverses the beam configuration of Fig. 1(a) ray for ray, giving the so-called reverse-beam geometry of Fig. 1(b). In this experiment we accomplished the beam reversal in square-wave fashion with a chopper rather than sinusoidally as originally proposed.[1]

An additional movable vertical levitating beam was used for capturing, holding, stabilizing, or transporting the silicone-oil-drop particle used in this experiment. The optical absorption of the high-purity silicone oil is so low that radiometric forces[11] are negligible in these experiments. Work was performed at atmospheric pressure within a small glass box to avoid air currents. Top- and side-viewing microscopes projected enlarged views of the particle on screens for viewing and locating the particle positions in space. Attenuators (ATT)

and power monitors were used to adjust the beam powers.

The most direct demonstration of alternating-beam trapping using the scattering force would involve an ideal trap as in Fig. 1 where all gradient forces are negligible, cw trapping is ruled out, and only alternating-beam trapping is possible. In the levitation experiment actually performed it was difficult to reduce the radial gradient forces to a negligible value since that requires large-diameter beams with large numerical aperture and high powers. We therefore used an alternating-beam trap of the same general configuration as in Fig. 1 but with dimensions such that we still had radial stability from gradient forces. Our proof of alternating-beam stability in this version of the trap rests solely on demonstration of axial stability due to alternating scattering-force fields. Indeed, we show alternating-beam axial stability under conditions where cw axial stability is shown to be impossible.

In our experiment lenses L1 and L2 had focal lengths of ~ 3.7 mm which focused the forward beams A_F and B_F to spot radii $w_{OF} \cong 1.2$ μm at f_A and f_B. The reverse beams were focused by 14-cm focal-length lenses L3 and L4 and then by L1 and L2 to the same foci f_B and f_A with closely the same spot radius $w_{Orev} \cong 1.0$ μm. The separation between f_A and f_B was ~ 170 μm. This results in a spot radius $w \cong 13.0$ μm at O which puts the particle at O in the far field of the beams. Since the beam diameter is comparable to the particle diameter of ~ 9.0 μm we have strong radial confinement from the gradient forces of A and B when operating with either cw or alternating-beam conditions. As we shall see, the existence of radial stability was actually a great experimental advantage. It prevented the particle from ever totally escaping from the light beams during the course of the experiment and made it possible to perform a series of stability tests with the same oil drop. The rather precise beam alignment needed for this experiment was accomplished as follows. A silicone oil drop, 9 μm in diameter, was captured in the vertical levitating beam of ~ 30 mW power by use of a spraying technique[11] and locked in a fixed position by use of feedback stabilization.[5] The four component beams of the trap were then positioned with use of the drop fixed in space as a reference.

To demonstrate alternating-beam stability we start by first determining the range of conditions within which the trap is axially stable under cw conditions. This is done by removing the stabilizer and chopper blade and transferring the particle at O to the cw trap consisting initially of the pair of equal-power forward cw beams A_F and B_F. This transfer is accomplished by gradually turning up the power of the combined forward beam pair to about 140 mW as the vertical levitating beam power is reduced to 0. We note in passing

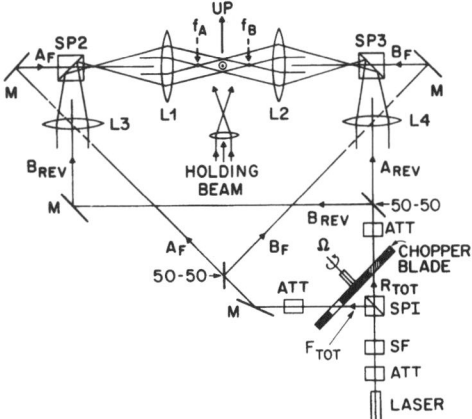

FIG. 2. Schematic diagram of the experimental setup.

VOLUME 54, NUMBER 12   PHYSICAL REVIEW LETTERS   25 MARCH 1985

that the first stable optical trap for macroscopic particles[3] and the first proposed version of a focused-beam atom trap[12] were based on such a cw two-beam geometry. We next introduce the pair of equal-power reverse beams A_{rev} and B_{rev} and probe the range of cw stability of the trap by gradually increasing the power of the reverse beams with the forward beams still on. This adds an outward or destabilizing contribution to the net scattering force at each point P on the beam axis as shown in Fig. 1(b) while adding additional inward or stabilizing radial gradient force. At some ratio of reverse power to forward power, which in the case of ideal alignment and equal spot sizes is unity, the particle is no longer cw stable at O and escapes axially. In practice this ratio is close to unity, indicating no large deviations from ideal geometry. Fortunately the escaped particle is not totally lost but is guided along the axis by the radial gradient forces and quickly trapped at one of the strong cw traps which exist at points f_A and f_B on either side of O. The trap geometry at each of these foci consists essentially of two strongly focused opposing beams of equal power with a common beam focus whose diameter (~ 2 μm) is considerably less than the particle diameter (~ 9 μm). Although traps of this type can be understood with the simple ray model[3] they were never explicitly described in previous studies of related trapping geometries.[3,13,14] With the particle safely trapped at f_A or f_B we convert to alternating-beam operation. We transfer the particle to the vertical holding beam, insert the chopper blade, and transport the particle back to position O. With a chopping frequency Ω of ~ 500 Hz, we then gradually raise the overall alternating-beam trapping power while maintaining the measured ratio of forward-to-reverse beam powers which previously resulted in the onset of cw axial instability. In this way we successfully transferred the particle from levitation in the vertical holding beam to levitation in the alternating-beam trap at powers of ~ 140 mW in each beam pair. This observation demonstrates the existence of stability in the axial direction in an alternating-beam trap which was shown to be unstable with the very same beams under cw conditions.

It is now possible to estimate experimentally the depth of the axial potential well of the alternating trap. By further increasing the power of the reverse beam pair relative to the forward beam pair we add an additional axial destabilizing force component to the trap. One can increase the reverse power by $\sim 34\%$ before the alternating-beam trap goes unstable and the particle escapes axially. This is equivalent to adding a 17% reverse cw beam. Since well depth is proportional to power, this implies an alternating-beam well depth of $\sim 17\%$ of the stable cw well depth in the axial direction. It was found that the escaped particle again becomes trapped at one of the foci. Indeed, with alternating beams an average trapping force still occurs at f_A or f_B as a result of the strong viscous damping at atmospheric pressure. We were thus able to retrieve the particle and repeat all measurements three more times under cw and chopped conditions, with essentially identical results. Our experimentally observed well depth of $\sim 17\%$ of the cw well depth can be compared with ion traps. For ion traps a maximum theoretical depth of $\sim 33\%$ of the dc well depth is predicted for strictly sinusoidal harmonic traps when operating at the edge of the stability region.[1,10] In practice, operation with depths of $\sim 10\%$ is more usual.

Finally we studied the stability and alternating-beam well depth as a function of chopping frequency Ω. Stability existed from the highest frequency used (1000 Hz) down to a minimum of ~ 100 Hz. The well depth which theoretically continues to increase down to the lowest stable frequency was observed to increase down to ~ 500 Hz, where the above extensive measurements were made, and slowly decrease at lower frequencies. This decrease may result from residual beam misalignments.

We observe that particles in our alternating-beam traps remain essentially motionless when trapped. No micromotion or macromotion is seen since all particle motion is heavily overdamped at atmospheric pressure. At low pressure oscillatory effects should be seen as in electrodynamic traps for macroscopic particles[9] and in cw levitation traps.[15] The experiment described above is an excellent example of optical micromanipulation of small particles and use of levitated particles as sensitive probes of optical and other forces. Indeed, in this experiment, the single 9-μm oil drop was flawlessly transferred a total of 50 times between five different types of traps located in a region of space about $25 \times 25 \times 200$ μm^3. This occurred over a 5-h period during which the particle position was directly observed with micrometer resolution.

In conclusion, we have made the first demonstration of the principle of alternating-beam trapping of neutral particles by radiation pressure. Using alternating scattering-force fields we achieved an optical levitation trap with a considerable well depth and range of stability. This demonstration strongly suggests further use of the alternating-beam technique for trapping and cooling of neutral atoms and possibly Rayleigh-sized particles under conditions where gradient forces are totally negligible.[1] Finally, a new and useful type of strong cw optical trap was discovered at the common focus of a pair of strongly focused opposing beams of equal power.

---

[1]A. Ashkin, Opt. Lett. **9**, 454 (1984).

VOLUME 54, NUMBER 12   PHYSICAL REVIEW LETTERS   25 MARCH 1985

[2]A. Ashkin and J. P. Gordon, Opt. Lett. **8**, 511 (1983).

[13]A. Ashkin, Phys. Rev. Lett. **24**, 156 (1970), and **25**, 1321 (1970).

[4]J. P. Gordon and A. Ashkin, Phys. Rev. A **21**, 1606 (1980).

[5]A. Ashkin, Science **210**, 1081 (1980).

[6]V. S. Letokhov and V. G. Minogin, Phys. Rep. **73**, 1 (1981).

[7]W. D. Phillips and H. Metcalf, Phys. Rev. Lett. **48**, 596 (1982)1 J. V. Prodan, W. D. Phillips, and H. Metcalf, Phys. Rev. Lett. **49**, 1149 (1982).

[8]D. J. Wineland, Science **226**, 395 (1984).

[9]R. F. Wuerker, H. Shelton, and R. V. Langmuir, J. Appl. Phys. **30**, 342 (1959).

[10]H. G. Dehmelt, Adv. At. Mol. Phys. **3**, 53 (1967).

[11]A. Ashkin and J. M. Dziedzic, Science **187**, 1073 (1975).

[12]A. Ashkin, Phys. Rev. Lett. **40**, 729 (1978).

[13]A. Ashkin and J. M. Dziedzic, Appl. Phys. Lett. **24**, 586 (1974).

[14]G. Roosen, Opt. Commun. **26**, 432 (1978).

[15]A. Ashkin and J. M. Dziedzic, Appl. Phys. Lett. **28**, 333 (1976).

VOLUME 74, NUMBER 18   PHYSICAL REVIEW LETTERS   1 MAY 1995

Evaporative Cooling in a Crossed Dipole Trap

Charles S. Adams, Heun Jin Lee, Nir Davidson, Mark Kasevich, and Steven Chu

Department of Physics, Stanford University, Stanford, California 94301
(Received 17 November 1994)

Laser cooled sodium atoms are trapped in an optical dipole force trap formed by the intersection of two 1.06 μm laser beams. Densities as high as 4×10^{12} atoms/cm^3 at a temperature of $\sim$140 μK have been obtained in a $\sim$900 μK deep trap. By reducing the trap depth over a 2 s interval, we have evaporatively cooled the atoms to a final temperature of $\sim$4 μK at a density of 6×10^{11} atoms/cm^3. This corresponds to a factor of 28 increase in atomic phase-space density.

PACS numbers: 32.80.Pj

Laser cooling and trapping techniques are capable of yielding dramatic increases in the phase-space density of dilute atomic vapors. For example, atoms from thermal atomic sources are routinely cooled and compressed in a magneto-optic trap (MOT) to densities of $\sim$10^{11}/cm^3 and temperatures of $\sim 20 T_{\rm rec}$, where $k_B T_{\rm rec} = (\hbar k)^2/2m$ is the photon recoil limit, k is the wave vector of the cooling laser, and m is the atomic mass [1]. Even higher phase-space densities are needed for the observation of quantum many body effects, such as Bose-Einstein condensation, and also for precision experiments requiring large fluxes of ultracold atoms.

Phase-space density in magneto-optic traps is constrained in temperature by the standard limits of polarization gradient cooling [2,3] and in density by radiative repulsion [4], light-assisted collisions [5], and the trap spring constant. Recent increases in MOT phase-space density have resulted from the development of improved techniques for minimizing the effects of radiative repulsion and light-assisted collisions, and the use of stronger magnetic trapping fields [6,7]. In this Letter, we report on the use of related techniques to achieve high densities at low temperatures in an optical dipole force trap, where we expect similar limiting mechanisms to be present.

We also report on the use of evaporative cooling to further increase the phase-space density in the trap [8]. Evaporative cooling was first demonstrated with magnetically trapped hydrogen [9] and has recently been observed with laser cooled, magnetically trapped alkali atoms [10,11]. Evaporative cooling is an attractive final stage cooling technique since, unlike laser cooling methods, it does not involve density limiting interactions with light. Successful evaporative cooling requires that the atom-atom elastic collision rate, which sets the time scale for thermalization in the trap dominate inelastic collision channels that can cause unwanted trap loss or heating. Two channels which are relevant to evaporation of sodium in optical dipole force traps are heating due to Rayleigh scattering of the trapping light and photoassociative two-body inelastic collisions [12,13]. In the work presented below, the use of far-detuned 1.06 μm radiation for trapping minimizes both of these potentially adverse interactions.

Evaporative cooling in dipole traps offers flexibility in designing trap potentials, especially those with dynamically changing trap volume and depth. Crucially, such flexibility allows for optimization of the evaporative cooling path. In this proof-of-principle work, we cross two tightly focused laser beams to produce the trapping potential illustrated in Fig. 1. Previous dipole trapping work has been restricted to the use of a single tightly focused beam [14–16]. Our new geometry was chosen to give the trap a strong, nearly isotropic, spring constant and a relatively large volume. Trap depth, and to some extent trap volume, was changed by adjusting the intensity of the trapping beams.

We have recently demonstrated the efficacy of Raman cooling techniques for atoms confined in a single focus dipole force trap [17]. In this work we cooled the axial motion (parallel to the propagation vector of the dipole trapping beam) to an effective temperature of $0.7 T_{\rm rec}$. In the crossed dipole trap geometry, where the oscillation frequency of trapped atoms is balanced in all directions, we expect to be able to cool all three degrees of freedom. The use of a Raman cooling stage to precool an atomic sample in conjunction with a final evaporative cooling stage may provide access to even higher phase space densities.

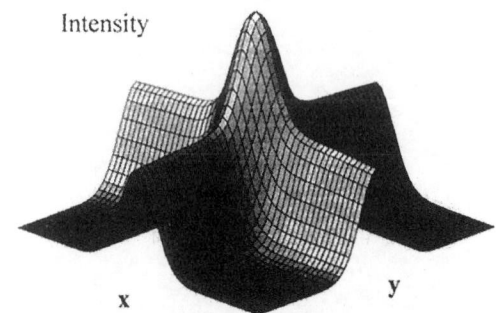

FIG. 1. A cross section of the dipole trap intensity distribution. The section plane is defined by the propagation axes of the two trapping beams, which are taken to be parallel **x** and **y** axes.

0031-9007/95/74(18)/3577(4)$06.00

The experimental details are as follows [18]. Sodium atoms from a thermal atomic source were initially trapped and cooled in a MOT, which was then used to feed atoms into the crossed dipole trap. A 10 W single mode Nd:YAG laser generated the dipole trapping light. The laser output passed through an acousto-optic modulator, which provided electronic control of the intensity, before being divided into two independent beams. The trap was formed by intersecting these two beams at a 90° angle in the center of the MOT. Each beam had a measured 15 μm $1/e^2$ radius at the point of intersection and a maximum power of 4 W. The polarizations of the beams were chosen to be orthogonal in the crossed region in order to suppress standing wave effects. Under these conditions the maximum attainable trap depth was ~900 μK, much deeper than the 30 μK temperatures which are routinely achieved with polarization gradient molasses [2].

The dipole trap was loaded by leaving the dipole beams on at full power while loading atoms into the MOT. After a fixed loading interval (typically 2 s), the MOT beams and magnetic field were turned off, leaving atoms confined solely by the dipole trapping beams. The alignment of the trapping beams was optimized by maximizing the number loaded into the trap. The total number of atoms in the trap was measured by a probing the ensemble with laser light tuned to the $3S_{1/2} \rightarrow 3P_{3/2}$ transitions: After a fixed time following the shutoff of the MOT, the dipole beams were turned off and the light used to form the MOT was turned back on. Resonance fluorescence from the diffusively expanding ensemble of atoms was focused into a photomultiplier tube. Under the above static loading conditions, the number loaded into the trap was maximized for MOT beams of ~3 mW/cm^2 peak intensity (intensity refers to total intensity in one of the six trapping beams), tuned -15 MHz below the $3S_{1/2}, F = 2 \rightarrow 3P_{3/2}, F = 3$ transition and a magnetic field gradient of 1 G/mm. To avoid optical pumping, a resonant electro-optic modulator imposed 1.73 GHz frequency sidebands on the MOT beams. One of these sidebands, whose power was 10% of that in the carrier, was nearly resonant with the $3S_{1/2}, F = 1 \rightarrow 3P_{3/2}, F = 2$ repumping transition. The number of atoms loaded into the dipole trap saturated for loading times ~2 s, which also corresponds to the time necessary to saturate the number of atoms loaded into the MOT.

An order of magnitude more atoms were loaded into the trap by reducing the intensity of the repumping sideband by a factor of 16 during the final 20 ms of the trap loading interval. The overall effect of reducing the sideband intensity is to optically pump a greater fraction of the trapped atoms into the $F = 1$ ground state hyperfine level and reduce the excitation rate. Although we have not extensively studied the mechanism behind this increase in number, we conjecture that it results from a reduction in three density limiting processes: radiative repulsion forces [5], photo-associative

collisions, and ground state hyperfine changing collisions [19]. The first two processes are suppressed by the reduction of population in the $3P_{3/2}$ manifold. Although the 1.77 GHz energy exchange associated with the final process is not enough to eject atoms from the MOT, it will eject atoms from the dipole trap. Optical pumping into the $F = 1$ level minimizes population in the $F = 2$ level and therefore reduces this loss.

We measured the trap size by imaging fluorescence from the detection pulse onto a charge coupled device array. The imaging system had a 4.5× magnification and a measured resolution of ~3 μm. The optical axis of the imaging system was normal to the plane containing both trapping beams as is illustrated in Fig. 2. For a total dipole trap power of 8 W, the measured $1/e$ half-width of a cross section passing through the center of the trap, at a 45° angle with respect to the trapping beams, was 7 ± 2 μm. For these measurements, the detection light had a 10 mW/cm^2 peak intensity per beam and was detuned -25 MHz from the cooling transitions. The diffusive spreading of the atomic cloud during the detection time was corrected for by measuring the size for 5 and 10 μs detection pulses. No corrections have been made for possible broadening of the imaging arising from the optical thickness of the sample.

The density in the dipole trap was determined by measuring the absorption of a weak probe beam. This beam was aligned to pass through the center of the crossed region as is illustrated in Fig. 2. The linearly polarized beam had a 7 μm $1/e^2$ waist and a power of ~1 nW to avoid saturation of the transition. The probe light was pulsed on for a 2 μs measurement interval following the extinction of the dipole trapping beams. This was short enough so that ballistic expansion of the atomic distribution during the measurement could be neglected. A plot of transmission against frequency is shown in Fig. 3. The three dips correspond to the $3S_{1/2}, F = 1$ to $3P_{3/2}$ $F = 0$, 1, and 2 transitions, respectively. The

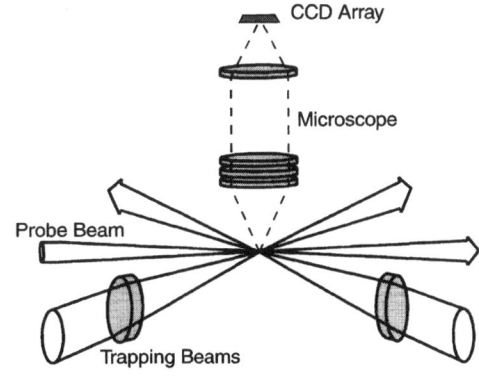

FIG. 2.   Schematic illustration of the apparatus.

VOLUME 74, NUMBER 18   PHYSICAL REVIEW LETTERS   1 MAY 1995

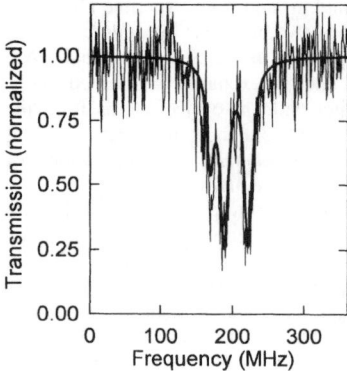

FIG. 3. Transmitted intensity as a function of probe laser frequency for a weak probe beam passing through the crossed region of the dipole trap. The solid curve is a fit of the modeled line shape to the data.

equal absorption on the $F = 1$ and $F = 2$ transitions indicates that the intensity and duration of the probe pulse were sufficiently small to avoid optical pumping of atoms into the $F = 2$ state. From the peak absorption of 75% and assuming a Gaussian atomic distribution $n = n_0 \exp(-r^2/r_0^2)$, where $r_0 = 7 \pm 2$ μm was the measured $1/e$ radius of the atomic distribution, we calculate the peak density n_0 to be $4^{+6}_{-2} \times 10^{12}$ atoms/cm^3 [20]. This measurement was also used to calibrate the sensitivity and collection efficiency of the video imaging system. Subsequent absolute density measurements were made with the video camera alone by inferring the number of trapped atoms from the integrated video signal and trap volume from the size measurement described above. The relative density in the crossed region was found to be an order of magnitude higher than the peak density in the trap wings.

The lifetime of the trap was measured by monitoring the number of atoms in the crossed region as a function of time. The lifetime data were fitted well by a single exponential decay, and were found to vary inversely with the trapping intensity and to be independent of the initial atomic number and density. At present we have no consistent explanation for the observed intensity and density dependence of the trap loss. The exponential decay time constant was 0.8 s for 8 W total trapping power and increased to 2.7 s for 2 W total power. For intensities less than 1 W, trap decay is dominated by background gas collisions. In the low intensity limit, the extrapolated lifetime is 14 s at our operating pressure of $\sim 6 \times 10^{-10}$ torr.

The temperature of atoms trapped in the crossed region was measured by a time-of-flight technique using the video imaging system. Images of the ballistically expanding ensemble were taken a fixed time delay after the dipole

trap was turned off. The delay was typically chosen to be long enough to allow the size of the ensemble to expand by roughly an order of magnitude over its initial size [21]. The effective temperature for atoms loaded into a ~ 900 μK deep trap (corresponding to 8 W trapping power) was 140 μK. From this measured temperature, we can use the virial theorem to estimate the expected spatial distribution of atoms in the trap. Modeling the trap potential in the central region as $V(x, y, z) = V_0\{\exp[-2(x^2 + y^2)/w_0^2] + \exp[-2(y^2 + z^2)/w_0^2]\}$ and the atomic density as $n(x, y, z) = n_0 \exp[-(x^2 + y^2 + z^2)/r_0^2]$ we calculate the expected $1/e$ radius to be 5.3 μm, in rough agreement with the measured radius of 7 ± 2 μm.

We measured the heating rate due to off-resonant scattering of trap photons by observing the evolution of trap temperature with time in the deep (8 W) trap. The temperature in the crossed region was observed to increase at a rate corresponding to scattering one 1.06 μm photon every 0.8 s. This rate is consistent with the estimated photon scattering rate.

The atoms in the trap were evaporatively cooled by lowering the trapping potential. The Nd:YAG laser power was exponentially ramped down from 8 to 0.4 W in 2 s with a 0.7 s time constant. Examples of the video signal depicting the cooling sequence are shown in Fig. 4. Figure 4(a), which illustrates the change in trap density, was recorded with a 10 μs detection pulse applied immediately after the dipole trap was turned off. After the ramp the atomic density decreased by a factor of about $\sim 7 \times 10^{11}$ to $6^{+6}_{-2} \times 10^{11}$ atoms/cm^3. Figure 4(b), which illustrates the change in trap temperature, shows the width of the atomic distribution following a 200 μs ballistic expansion interval. The half-width of the final velocity distribution $1/e$, was found to be 5.6 cm/s, which corresponds to a temperature of 4 μK. Compared to the initial conditions, the ramp produced a 28-fold increase in the phase-space density of atoms in the crossed region of the trap, while the number of trapped atoms decreased from ~ 5000 to ~ 500 atoms.

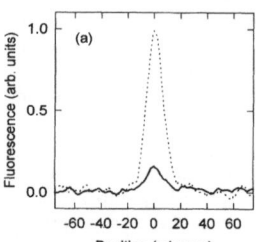
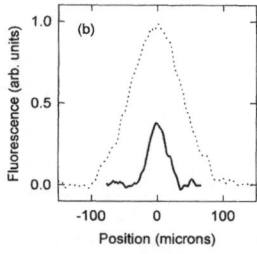

FIG. 4. (a) Image of trap size before (thin line) and after (thick line) the intensity ramp. (b) Time-of-flight measurement of trap temperature before (thin line) and after (thick line) the intensity ramp.

A fraction of the atoms which escape from the central region of the trap will remain confined by the wings of the dipole region. These atoms can return, in principle, to the central region of the trap, where they will collisionally heat the ensemble. However, a numerical simulation of single particle trajectories in the trap, which included both the light induced potential and the gravitational potential, indicated the probability of return to be less than 2% during the potential ramp. In addition, atoms which do return spend a relatively brief time in the central region before again escaping to the wings, further reducing the possibility of trap heating.

Recent experiments with magnetically trapped Na indicate an elastic collision cross section $\sim 10^{-12}$ cm^2 or larger for a sample in the $F = 1$, $m_f = +1$ state at a temperature of ~ 50 μK [22]. For our initial trap density of 4×10^{12} atoms/cm^3 and temperature of ~ 140 μK, and assuming this cross section is a reasonable estimate for our unpolarized atomic sample, the collision rate is ~ 100 s^{-1}, giving an estimated rethermalization time $\sim 10^{-1}$ s [23,24]. At the end of the ramp, this time is ~ 1 s. Experimentally, the 0.7 s ramp time constant gave optimal cooling results, and is roughly consistent with this estimate.

Although this represents a clear demonstration of evaporative cooling, the process is far from optimized. The rapid evaporation and rethermalization which occurs during the initial stages of the ramp soon shut off due to the loss of density as the potential relaxes. Efficient evaporative cooling would require simultaneous evaporation and compression, which we plan to implement in subsequent experiments.

We wish to thank E. Cornell, J. Doyle, W. Ketterle, and C. Wieman for useful discussions, and M. Perry for his loan of a Nd:YAG laser. C. S. A. acknowledges the support of the English Speaking Union. This work was supported in part by grants from the NSF and the AFOSR.

[1] E. Raab, M. Prentiss, A. Cable, S. Chu, and D. Pritchard, Phys. Rev. Lett. **59**, 2631 (1987).
[2] For discussions of polarization gradient cooling limits, see Special issue on *Laser cooling and trapping*, edited by S. Chu and C. E. Wieman [J. Opt. Soc. Am. B **6**, (1989)].
[3] A. Steane and C. Foot, Europhys. Lett. **14**, 231 (1991).
[4] T. Walker, D. Sesko, and C. Wieman, Phys. Rev. Lett. **63**, 957 (1989); D. Sesko, T. Walker, and C. Wieman, J. Opt. Soc. Am. B **8**, 946 (1991).
[5] D. Sesko, T. Walker, C. Monroe, A. Gallagher, and C. Wieman, Phys. Rev. Lett. **63**, 961 (1989).
[6] W. Petrich *et al.*, J. Opt. Soc. Am. B **11**, 1332 (1994).
[7] W. Ketterle, K. Davis, M. Joffe, A. Martin, and D. Pritchard, Phys. Rev. Lett. **70**, 2253 (1993).
[8] H. Hess, Phys. Rev. B **34**, 3476 (1986).
[9] N. Masuhara *et al.*, Phys. Rev. Lett. **61**, 935 (1988).
[10] K. Davis, M. Mewes, M. Joffe, and W. Ketterle, in *Atomic Physics 14*, edited by C. Wieman and D. Wineland (AIP, New York, 1994).
[11] W. Petrich, M. Anderson, J. Ensher, and E. Cornell, in *Atomic Physics 14* (Ref. [10]).
[12] P. Lett *et al.*, Phys. Rev. Lett. **71**, 2200 (1993).
[13] J. Miller, R. Cline, and D. Heinzen, Phys. Rev. Lett. **71**, 2204 (1993).
[14] S. Chu, J. Bjorkholm, A. Ashkin, and A. Cable, Phys. Rev. Lett. **57**, 314 (1986).
[15] J. Miller, R. Cline, and D. Heinzen, Phys. Rev. A **47**, R4567 (1994).
[16] S. Rolston *et al.*, Proc. SPIE **1726**, 205 (1992).
[17] H. J. Lee, C. S. Adams, N. Davidson, B. Young, M. Weitz, M. Kasevich, and S. Chu, in *Atomic Physics 14* (Ref. [10]).
[18] A detailed description of our trapping apparatus can be found in M. Kasevich and S. Chu, Appl. Phys. B **54**, 321 (1992).
[19] See, for example, P. Julienne, A. Smith, and K. Burnett, in *Advances in Atomic, Molecular, and Optical Physics*, edited by B. Bederson and H. Walther (Academic Press, San Diego, 1992); T. Walker and P. Feng, in *Advances in Atomic Molecular, and Optical Physics, ibid.*
[20] The density was calculated by integrating over the finite size of the probe beam and using the absorption cross section for linearly polarized light resonant with the $3S_{1/2}$ $F = 1$ to $3P_{3/2}$ $F = 2$ transition, $\frac{5}{6}\lambda^2/2\pi$, where λ is the transition wavelength.
[21] Atoms from the wings of the distribution contribute to the ballistic expansion signal. We have made a Monte Carlo simulation of the atomic distribution following the ballistic expansion interval in order to assess the impact of the contribution on the measured temperature. We find that the wing atoms increase the width of the final distribution by $\sim 5\%$.
[22] W. Ketterle (private communication).
[23] This estimate assumes the collision cross section for the $m_f = 0$ state to be comparable to that of the $m_f = \pm 1$ state.
[24] Results for the rethermalization of ultracold Cs appeared in C. Monroe, E. Cornell, C. Sackett, C. Myatt, and C. Wieman, Phys. Rev. Lett. **70**, 414 (1993).

Transverse Resonance-Radiation Pressure on Atomic Beams and the Influence of Fluctuations

J.E. Bjorkholm, R.R. Freeman, A. Ashkin, and D.B. Pearson

Bell Telephone Laboratories, Holmdel, NJ 07733, USA

The development of tunable laser sources has led to a resurgence of interest in various effects caused by resonance-radiation pressure on atoms and ions. Our immediate interest in this subject was initiated by the independent proposals of LETOKHOV [1] and of ASHKIN [2] for using radiation pressure to construct optical traps for neutral atoms. The intent of both proposals is to confine and cool neutral atoms in these optical traps. Recently it was demonstrated that ions stored in electromagnetic traps can be cooled using the radiation pressure of nearly-resonant light [3]. We are currently investigating the transverse dipole force of resonance-radiation pressure; an understanding of this type of force is crucial to the construction of optical traps.

Recently we made the first direct observation of the effects of the optical dipole force on free atoms [4]. Those experiments demonstrated focusing and defocusing forces exerted on a beam of neutral sodium atoms by the transverse radiation pressure associated with a superimposed and copropaga- ting light beam from a cw dye laser. The forces and the resulting effects were made large by tuning the frequency of the laser to be near the sodium resonance transition. The work presented here was aimed at determining the degree to which the atomic beam could be focused using this basic technique. The atomic beam focal spot size is, in some instances, determined by input characteristics of the atomic beam such as longitudinal velocity distribu- tion and initial divergence. However, an important role can also be played by fluctuations of the radiation-pressure forces which occur because of the quantized nature of light. These fluctuations serve to heat the atoms; this heating can also limit the atomic beam focal-spot sizes that can be obtained. Fluctuations of the radiation pressure can also be a limiting factor in the design of optical traps [5,6]. In general, fluctuations can severely limit the applicability of resonance-radiation pressure if they are not properly compensated for.

A diagram of the basic situation we consider is shown in Fig.1. An atom having a velocity v along the z-axis (the longitudinal direction) is illuminated by a Gaussian laser beam of frequency ν, also propagating along the z-axis. The transverse intensity distribution of the light is

$$I(r) = I_0 e^{-2r^2/w^2}$$

where w is the spot size of the laser beam. The forces of resonance- radiation pressure can be classified into two types and, for a weakly

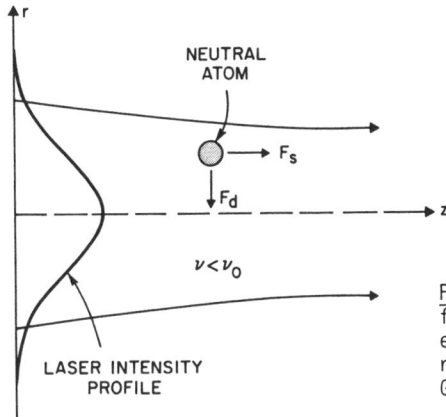

Fig.1 The transverse and longitudinal forces exerted on a neutral atom situated in the field of a weakly-focused, nearly-resonant laser beam having a Gaussian intensity profile

focused laser beam as being considered here, they are mainly transverse and longitudinal in character, as shown in Fig.1.

The focusing effects arise primarily from the transverse dipole resonance-radiation pressure [7,2]. This force arises from stimulated scattering of the light by the atoms and it is the same as the force exerted on an induced dipole situated in an electric-field gradient. Thus it is proportional to, and in the direction of, the optical intensity gradient. The dipole force is written as

$$\vec{F}_d = -\frac{1}{2} \alpha \vec{\nabla} |E|^2, \tag{1}$$

where $\vec{E}$ is the optical electric field and α is the atomic polarizability which, for a simple two-level atom, is

$$\alpha = \frac{\lambda^3 \Delta\nu_n}{32\pi^3} \frac{\Delta\nu}{\Delta\nu^2 + \Delta\nu_n^2/4} \frac{1}{1+p} . \tag{2}$$

In (2), $\Delta\nu = (\nu' - \nu_0)$ where $\nu' = (1-\nu/c)c/\lambda$ is the Doppler-shifted frequency of the optical field, ν_0 and $\Delta\nu_n$ are the frequency and natural linewidth of the resonance transition of the atom, and p is a saturation parameter. Explicitly, p is given by

$$p = \frac{I}{I_s} \frac{\Delta\nu_n^2/4}{\Delta\nu^2 + \Delta\nu_n^2/4} \tag{3}$$

where $I = |E|^2 c/8\pi$ is the optical intensity and $I_s = 2\pi^2 \Delta\nu_n h\nu/\lambda^2$ is the saturation intensity for the two-level atom. In the limit $p \gg 1$, the dipole force reduces to $h\Delta\nu \vec{\nabla}I/I$; this shows that, with sufficient laser power to insure that p remains large, the dipole force can be arbitrarily increased by increasing $|\Delta\nu|$. For a weakly focused TEM_{00} (Gaussian) mode laser beam the force is essentially transverse to the direction of propagation of the light. When the light is tuned below the atomic resonance frequency, the atoms are attracted to the high intensity regions of the light beam, while for tunings above resonance the atoms tend to be expelled from the light.

Spontaneous resonance-radiation pressure arises from spontaneous scattering of the light by the atoms and it exists even in uniform plane-waves [8]. The average spontaneous force is in the direction of the light propagation and its magnitude is given by

$$F_s = \frac{h}{2\lambda\tau} \quad \frac{p}{1+p} \; , \tag{4}$$

where $\tau = 1/2\pi\Delta\nu_n$ is the natural lifetime of the excited state of the atom. The average spontaneous force is simply the rate at which photon momentum is absorbed by the atom; for $p \gg 1$ it saturates to a maximum value of $h/2\lambda\tau$. Thus in many situations the dipole force can greatly exceed the spontaneous force. The effects of the spontaneous force have been observed in numerous experiments [9]. The role of spontaneous scattering in our experiment is twofold. First, the average spontaneous force accelerates the atoms in the direction of light propagation; this acceleration is rather unimportant to an understanding of our results. Secondly, spontaneous scattering occurs in a discrete and random fashion. The recoil experienced by an atom in a single scattering event can be in any direction; only the average recoil is in the direction of the light propagation. This random scattering serves to "heat" the atoms in the transverse and longitudinal directions, as recently discussed elsewhere [5,6,10]. We will present a discussion indicating that, under some conditions, it is transverse heating due to spontaneous scattering which determines the degree to which an atomic beam can be focused using resonance-radiation pressure. Spontaneous decay of excited atoms can also lead to fluctuations in the dipole force and consequently to another source of heating for the atoms. For our experimental conditions, however, the heating due to the random nature of the spontaneous force was dominant [11]. It should be emphasized that heating caused by fluctuations of the radiation pressure forces is to be distinguished from the heating or cooling due to the velocity dependence of the average spontaneous force resulting from Doppler shifts of moving atoms [12,3].

A schematic diagram of our experimental setup is shown in Fig.2. Light from a continuously-tunable, single-mode cw dye laser was superimposed upon an effusive beam of neutral sodium atoms ($T \sim 500°C$) by reflection off a 3 mm thick dielectric-coated mirror having a 230 μ diameter hole in it. Typical laser power superimposed on the atoms was 40 mw and the light was linearly polarized. A 75 cm focal-length lens focused the light to a spot size $w_0 = 75 \mu$ at a nominal distance of 25 cm from the mirror. A moveable hot-wire detector was used to measure the atomic beam profile in the focus of the laser beam. The detector was insensitive to light incident upon it and its resolution was determined by an aperture of 50 μ diameter mounted in front of the hot iridium wire. Strong focusing or defocusing effects were obtained when the laser was tuned to within several GHz of the sodium D_2 resonance transition at 5890A° for which $\Delta\nu_n = 10$ MHz. The corresponding value for I_s is about 19 mW/cm^2. As a cautionary note for the application of the various formulas given here, it must be realized that under most situations the sodium atom is not a good approximation to the idealized two-level atom.

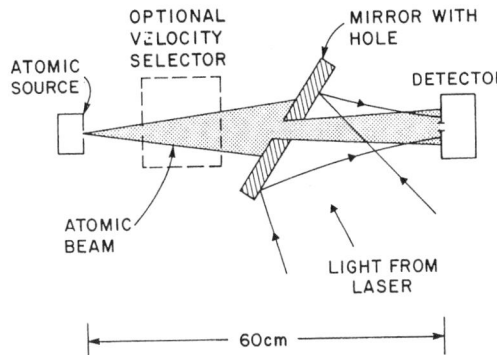

Fig.2 Schematic diagram of the
experimental setup. The transverse scale is greatly magnified
relative to the longitudinal
scale; for instance, the diameter of the hole in the mirror
is 250μ

Examples of the atomic beam profiles we measured under different conditions
are shown in Fig.3. Curve (a) shows the beam profile in the absence of
light; the peak beam current is normalized to 1. For profile (b) the laser
is tuned about 1 GHz below resonance for atoms at the peak of the velocity
distribution (~9 x 10⁴ cm/sec); the on-axis beam intensity is increased to
a normalized value of 5.8 and the diameter of the focused atomic beam (FWHM)
is 68 μ, a factor a 6 reduction. Clearly the actual atomic beam diameter is
smaller and the peak intensity higher as the experimental result is a convolution of the actual profile with the 50 μ aperture on the detector.
Profile (c) shows the results obtained when the laser was tuned about 1 GHz
above resonance. In this case the atoms are strongly expelled from the
light. In principle the beam profiles are cylindrically symmetric; it
isn't immediately obvious that the areas under the three curves represent
the same total number of atoms in the beam. Numerical integration verifies
that the numbers are in fact the same to within 10 percent. The symmetry of
the beam profiles we measured with the light applied was crucially dependent
upon precise overlap and alignment of the laser and atomic beams. Effects
due to the input laser beam not being perfectly Gaussian and due to
distortion of the laser mode caused by the hole in the mirror also may
cause asymmetries in the atomic beam profiles.

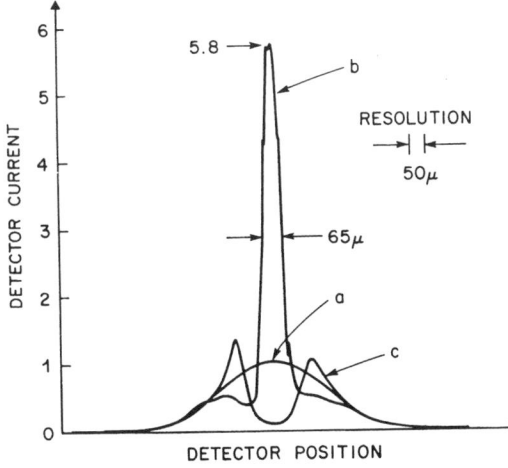

Fig.3 Atomic beam current measured by detector as a function
of its transverse position; a,
no light; b, laser tuned ~1GHz
below resonance for focusing; c,
laser tuned ~1GHz above resonance for defocusing. Peak beam
current is approximately 5 x
10⁸ atoms/sec

A simple calculation carried out for a two-level atom illustrates that heating caused by the fluctuations of spontaneous scattering can limit the spot size of the focused atomic beam. The transverse dipole force experienced by the atoms can be obtained from a transverse potential energy given by [2]

$$U(r) = \frac{1}{2} h \Delta\nu \ln[1+p(r)] \tag{5}$$

where $p(r) = p_0 \exp(-2r^2/w^2)$ and p_0 in the on-axis value of the saturation parameter. For our experimental parameters the well depth in the focus of the laser beam is 1.6×10^{-18} ergs, meaning that an atom with a maximum transverse velocity of 280 cm/sec can be confined in the well. The divergence of our atomic beam is about 3×10^{-4} rad, which corresponds to a maximum initial transverse energy for an atom emerging from the hole in the mirror of about 1.4×10^{-20} ergs. Assuming no change of transverse energy as the atoms propagate along the light, confinement in the transverse potential well implies that the spot size of the atomic beam in the focus of the light would be about 11 μ in diameter, much smaller than we observe.

Now consider the additional transverse energy acquired by the atoms due to the heating caused by spontaneous scattering. Assuming that the spontaneous scattering is isotropic (for simplicity), a random walk analysis indicates that, for an atom having zero initial transverse velocity, the probability density that its transverse velocity is v_t after N scattering events is

$$P_N(v_t) = \frac{m\lambda}{h} \left[\frac{3}{2\pi N} \right]^{1/2} \exp \left[-v_t^2 \frac{3m^2\lambda^2}{2h^2 N} \right] .$$

The corresponding rms transverse velocity is

$$v_t^{rms} \ (N) = \frac{h}{m\lambda}(N/3)^{1/2} \sim 1.7(N)^{1/2} \ \text{cm/sec.}$$

For our experimental parameters and for the most probable atomic velocity, we estimate that an atom undergoes $N \sim 1200$ scattering events in travelling along the laser beam. At the focus, then, we have $v_t^{rms} \sim 60$ cm/sec which corresponds to a transverse energy of 6.6×10^{-20} ergs. This is about a factor of 5 more than the initial transverse energy of the beam and it corresponds to an atomic beam focal spot size of about 25 μ. This simple analysis indicates that the effects of heating by spontaneous scattering are significant.

To more rigorously verify the importance of fluctuations we used a computer to numerically calculate atomic trajectories and the related beam profiles; a detector resolution of 50 μ was included in the computation. In the absence of light the atomic trajectories were taken to be straight lines emanating from a point source; the various rays were weighted to closely approximate the actual beam profile. The average dipole and spontaneous forces were exactly accounted for. Figures 4a and 4b show the beam profiles calculated for a laser power of 40 mW and $\Delta\nu = -1$ GHz. For Figure 4a the fluctuations in the spontaneous force were not included and the calculated

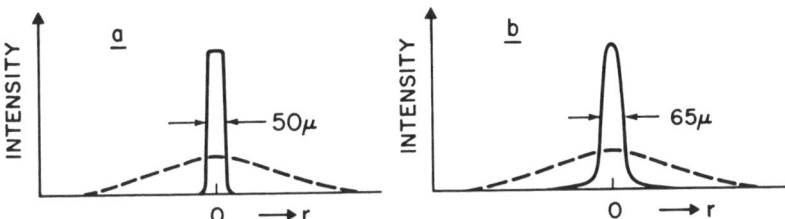

Fig.4 Computed atomic beam profiles for 40 mW of laser power and Δν = -1GHz; a, spontaneous heating not included; b, spontaneous heating included. The dashed curves show the beam profile in the absence of light. The vertical scales for all curves are arbitrary and comparisons between the various amplitudes should not be made.

atomic beam diameter is clearly much smaller than the detector resolution, in disagreement with our measurements. For Figure 4b the fluctuations of the spontaneous force were modeled using a random number generator and by keeping track of time and the rate of spontaneous scattering. The beam profile shown is the result of averaging over 40 such calculations. Clearly the atomic beam spot size has been greatly increased by the inclusion of spontaneous scattering; the computed beam profile is similar to the experimental profile b in Fig.3. This computation demonstrates that the random nature of spontaneous scattering can limit the degree of atomic beam focusing.

Assuming spontaneous scattering to be an important limitation, smaller spot sizes could be obtained if it were possible to reduce N while keeping the dipole focusing forces constant. With increased laser power this can, in fact, be done. For instance, in the limit p>>1, we have $F_d \propto$ power/Δν and $N \propto$ power/Δν². Consequently, if the laser power and Δν are increased by the same factor m, the dipole forces remain approximately constant while N is reduced by the factor 1/m. Consider 1W of laser power and Δν = -25GHz; for this case N∿100 and the additional transverse energy imparted by spontaneous heating is less than the initial transverse energy in the beam. The resulting atomic beam spot diameter indicated by the simple model is about 11 μ (as compared with the approximately 50 μ presently achieved); further improvements could be made by decreasing the initial atomic beam divergence. A computer beam profile for this case verifies the conclusions of this simple analysis. Thus it appears that atomic beam focal spot sizes much smaller than obtained in our present experiments should be readily attainable; we hope to verify this shortly using the increased power available from ring-type cw dye lasers.

In conclusion, we have experimentally demonstrated that a beam of neutral sodium atoms can be focused to a spot diameter of approximately 50 μ using the transverse dipole resonance-radiation pressure exerted by a 40 mW laser beam. Simple analysis shows that in some cases the spot sizes are limited by the random fluctuations of the spontaneous radiation pressure; with 1W of laser power, spot sizes less than 10 μ should be attainable. Our discussion makes it clear that the effects of heating by spontaneous scattering can have important detrimental effects in other applications of resonance-radiation pressure on atoms, such as the slowing or guiding of atoms. As discussed recently [6], consideration of heating effects is of paramount importance in the design of optical traps for neutral atoms.

Acknowledgement

The authors are indebted to J. P. Gordon for many useful discussions.

References

1. V. S. Letokhov, V. G. Minogin, and B. D. Pavlik, Opt. Commun. <u>19</u>, 72 (1976).

2. A. Ashkin, Phys. Rev. Lett. <u>40</u>, 729 (1978).

3. D. J. Wineland, R. E. Drullinger, and F. L. Walls, Phys. Rev. Lett. 40, 1639 (1978); W. Neuhauser, M. Hohenstatt, P. E. Toschek, and H. Dehmelt, Phys. Rev. Lett. <u>41</u>, 233 (1978).

4. J. E. Bjorkholm, R. R. Freeman, A. Ashkin, and D. B. Pearson, Phys. Rev. Lett. <u>41</u>, 1361 (1978).

5. V. S. Letokhov, V. G. Minogin, and B. D. Pavlik, Zh. Eksp. Teor. Fiz. <u>72</u>, 1328 (1977) [Sov. Phys. JETP <u>45</u>, 698 (1977)].

6. A. Ashkin and J. P. Gordon, Opt. Lett. <u>4</u>, 161 (1979).

7. G. A. Askar'yan, Zh. Eksp. Teor. Fiz. <u>42</u>, 1567 (1962) [Sov. Phys. - JETP <u>15</u>, 1088 (1962)].

8. A. Ashkin, Phys. Rev. Lett. <u>25</u>, 1321 (1970).

9. See, for example; O. R. Frisch, Z. Phys. <u>86</u>, 42 (1933); R. Schieder, H. Walther, and L. Woste, Opt. Commun. 5, 337 (1972); J. L. Picque and J. L. Vialle, Opt. Commun. 5, 402 (1972); A. F. Bernhardt, D. E. Duerre, J. R. Simpson, and L. L. Wood, Appl. Phys. Lett. 25, 617 (1974); J. E. Bjorkholm, A. Ashkin, and D. B. Pearson, Appl. Phys. Lett. <u>27</u>, 534 (1975).

10. J. L. Picque, Phys. Rev. A <u>19</u>, 1622 (1979).

11. J. P. Gordon, private communication.

12. T. W. Hansch and A. L. Schawlow, Opt. Commun. <u>13</u>, 68 (1975).

276   OPTICS LETTERS / Vol. 7, No. 6 / June 1982

Continuous-wave self-focusing and self-trapping of light in artificial Kerr media

A. Ashkin, J. M. Dziedzic, and P. W. Smith

Bell Laboratories, Holmdel, New Jersey 07733

Received February 2, 1982

Artificial Kerr media made from liquid suspensions of submicrometer particles were used as a new type of nonlinear medium for observing cw self-focusing and self-trapping of laser beams. Self-trapping of TEM_{00}-mode beams and higher-order TEM_{01}- and $TEM_{01}*$-mode beams were investigated. Saturation-free operation down to filament diameters of ~ 2 μm was observed. The independence of the critical power for self-trapping on the beam diameter in the unsaturated regime was confirmed for the first time to our knowledge. Values of the nonlinear coefficient were determined for a range of particle diameters from 0.038 to 0.234 μm.

We report the observation of cw self-focusing and self-trapping of laser beams in artificial Kerr media consisting of liquid suspensions of submicrometer dielectric particles. These experiments provide a simple way to determine the nonlinear coefficients of this new, highly nonlinear optical Kerr medium and add insights into the self-trapping phenomenon. We have observed the smallest-diameter cw self-trapped filaments yet reported and the first self-trapping of higher-order Gaussian modes.

The first observation of nonlinearities in artificial Kerr media[1] involved degenerate four-wave mixing. In these experiments optically induced index gratings were formed by the physical motion of submicrometer particles that is due to light pressure forces that exist in the field gradients of an optical standing wave. Such forces pull higher-index particles into regions of high optical intensity,[2] thus increasing the local index of refraction. The effective nonlinear coefficient[1] n_2 for a medium of 0.234-μm-diameter transparent polystyrene spheres in water at a concentration $N = 6.5 \times 10^{10}/cm^3$ was large ($\sim 10^5$ that of CS_2). Associated with the large nonlinearity was a slow response time of ~ 100 msec and a scattering loss of ~ 15 cm^{-1}. Another consequence of this large nonlinear increase of index of refraction with intensity is the production of intensity-dependent lenses in Gaussian beams, which cause cw self-focusing and self-trapping. Although we use quite different particles, this application of the artificial dielectric medium resembles the original proposal of Askaryan[3] on self-trapping by particles.

Self-focusing[4] has been widely studied on a pulsed or transient basis in many media. Continuous-wave self-trapping was previously seen in solids as a result of thermally produced index changes[4,5] and in sodium vapor.[6,7] The simple theory of self-focusing and self-trapping[4,8] shows that there is a critical power P_{crit} at which the nonlinearly generated lens just balances diffraction and the beam propagates without spreading. This balance should occur for a beam of any diameter at the same power, P_{crit}. At powers above P_{crit} the beam would collapse to a point, except for the onset of

saturation effects[9] or other nonlinear processes[4,10] that limit the collapsing beam to some finite equilibrium diameter. In sodium vapor,[6,11] saturation limits the trapped filament diameter to ~ 70 μm. For thermal filaments a minimum diameter of ~ 50 μm was observed.[5] For our media we see no saturation down to beam diameters ~ 2 μm. We also show the independence of P_{crit} on beam diameter in the saturation-free region.

The basic apparatus is sketched in Fig. 1. A 5145-Å cw laser beam is focused near the input face of the liquid-filled cell. Highly uniform submicrometer polystyrene latex spheres were used in water suspension, covering a size range from 0.038 to 0.234 μm. The beam trajectory in the nonlinear medium can be easily observed by looking at the 90° Rayleigh-scattered light from the particle suspension with microscope M1 and projecting a magnified image of the beam onto the screen S1. With increasing beam power we see the unperturbed diffracted Gaussian beam (rays a) shrink because of the onset of self-focusing (rays b) and finally propagate as a self-trapped filament (rays c) when the beam is at the critical power or above. The beam diameter far into the cell is observed to grow somewhat,

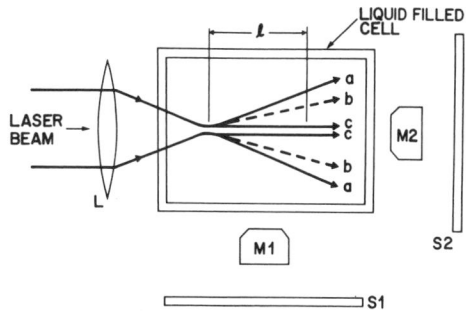

Fig. 1. Sketch of basic self-focusing and self-trapping experiment. The nonlinear medium filling the cell is a liquid suspension of submicrometer particles.

June 1982 / Vol. 7, No. 6 / OPTICS LETTERS **277**

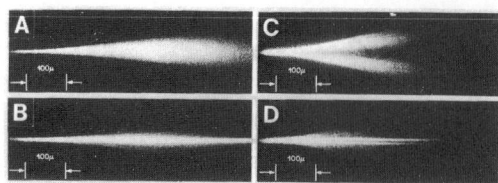

Fig. 2. Beam-trajectory photographs of TEM_{00}- and TEM_{01}-modes in the nonlinear medium as seen in side scattering. A and C show normal diffractive spread at low power. B and D are taken above the critical power and show formation of self-trapped filaments.

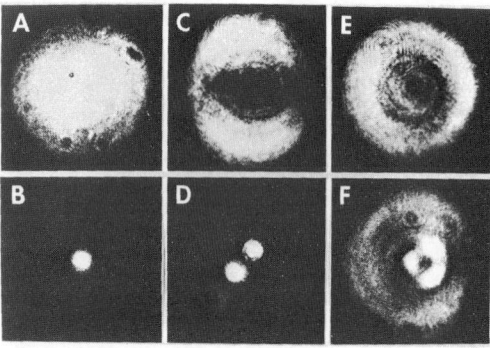

Fig. 3. Beam shapes at the output face of the nonlinear medium for TEM_{00}, TEM_{01}, and TEM_{01}^* modes. A, C, and E are taken at low power, where diffraction controls the spot size. B, D, and F are taken above the critical power and show the small, highly intense self-trapped filaments.

presumably because of the significant attenuation coming from scattering. At higher power the diameter maintains itself farther into the cell. Self-focusing and self-trapping can also be seen by observing the beam cross section at the output face of the cell with microscope M2 and screen S2.

Figure 2A is a photograph of the low-power-beam trajectory of a TEM_{00}-mode beam focused to a 3.2-μm diameter in a cell containing 0.038-μm-diameter polystyrene latex spheres having a particle concentration of ~3 × 10^{15}/cm³ and a scattering loss $\alpha_0 \cong 5$ cm⁻¹. The beam focus and diffractive spread are clearly visible. Figure 2B, taken with a power above P_{crit}, shows a long self-trapped filament propagating without significant beam spread for a distance of ~600 μm $\cong$ 15 near-field distances. Also evident is a feature observed in cw thermal self-trapping,[5] namely, the presence of a significant untrapped diffracting portion of light emanating from the focal region. This untrapped component is small (~25%) for input powers close to P_{crit} and increases at higher power. Figure 3A shows the beam spot of a 3.2-μm beam photographed at low power at the output face of the cell after traversing a cell length of ~250 μm filled with a suspension of 0.085-μm spheres having $\alpha_0 \cong 45$ cm⁻¹. In Fig. 3B, taken above P_{crit}, we see the focusing of most of the light into a small

self-trapped filament. We also performed for the first time cw self-trapping experiments using higher-order Gaussian-beam modes. In Fig. 2C we show the diffractive spread at low power of a TEM_{01}-mode beam tightly focused by a 0.5-cm focal-length lens in a medium of 0.085-μm particles with $\alpha_0 \cong 45$ cm⁻¹. Figure 2D, taken above P_{crit}, shows two separate filaments emanating from the focus, one from each lobe of the TEM_{01} mode. Comparing the light distributions of Figs. 3C and 3D at the output face of a cell at low and high power, we see an angular misorientation of the filament pair with respect to the low-power beam. We believe that this is due to distortions in the shapes of the beam lobes of the input TEM_{01}-mode beam, which shift the location of the self-trapped filaments. Figures 3E and 3F show an attempt to generate a self-trapped ring- or doughnut-shaped filament from a TEM_{01}^* "doughnut"-mode beam. Since our doughnut mode was imperfect, as is seen in Fig. 3E, we made only an imperfect ring filament (see Fig. 3F). At still higher power this imperfect ring broke up into three separate filaments. We expect that clean doughnut filaments can be formed with more-perfect beam shapes.

To determine the nonlinear coefficient n_2 we measured the critical power P_{crit} and applied simple theory relating n_2 to P_{crit}. As is shown in Fig. 1, we used a focused beam and observed the diameter at a distance l of ~200 μm, corresponding to many confocal distances. Values of scattering loss α_0 were chosen for which the power variation over this distance was small. The input power was increased, and P_{crit}, the power required to propagate to l as a filament with no noticeable diameter increase, was then recorded. Table 1 gives results on P_{crit} for three TEM_{00} input-beam diameters and five particle sizes. Distances of 2 μm represent the minimum beam diameter conveniently reached with high-quality short-focal-length lenses. Although the actual α_0's used varied from 5 to 8 cm⁻¹, we normalized values of P_{crit} in Table 1 to $\alpha_0 = 5$ cm⁻¹ on the assumption that P_{crit} varies as $1/\alpha_0$. Table 1 shows that for each particle size P_{crit} is independent of beam diameter within the experimental error of ± 0.05 W. This result also indicates that no significant saturation occurs over the range of beam diameters used. The observed independence of P_{crit} on beam diameter probably represents the first experimental verification of this basic feature of the simple theory in the absence of saturation. Saturation-free operation down to diameters as small as 2 μm is in contrast to other cw self-focusing media that saturate at more than a 10-times-larger filament diameter. It is not known how much smaller one can go without encountering saturation. Theory indicates that an

Table 1. Critical Power in Watts versus Particle Diameter and Beam Diameter for $\alpha_0 = 5$ cm⁻¹

Beam Diameter (μm)	Particle Diameter (μm)				
	0.038	0.085	0.091	0.109	0.234
2.0	0.77	0.77	0.66	0.95	—
3.2	0.68	0.78	0.64	0.85	0.76
4.8	0.80	0.84	0.69	0.89	0.68

278   OPTICS LETTERS / Vol. 7, No. 6 / June 1982

initially unsaturated beam for $P > P_{crit}$ should compress and finally oscillate about some equilibrium diameter set by saturation. In practice it is difficult to observe further beam compression below 2 μm by using our side-viewing technique. Only with a larger-diameter beam, ~5 μm, were we able to see beam compression for powers above P_{crit}. A further effect observed with 2-μm filaments at powers above P_{crit} was the onset of beam breakup at some localized point well into the cell. This occurs most strongly with smaller particles, in which the particle concentration is high. The effect may be due to particle damage or clumping occurring at the high intensities of a compressed or saturated region of the beam. Particle damage and clumping often occur at high particle concentrations when the beam is focused at the cell input face.

The blank value in Table 1 indicates that we were unable to see a clean self-trapped filament in that case. We attribute this to the coarse-grainedness of the medium for this combination of small beam diameter and large particle size. In general, the quality of the filament and the transmitted beam improves as the number of particles in the beam increases. One also sees from Table 1 that P_{crit} for the values listed is essentially independent of particle diameter, for the same scattering loss α_0. Since $P_{crit} \sim 1/n_2$, this implies that n_2 is independent of diameter for constant α_0. This important observation agrees with a simple theory based on the gradient force, which shows that n_2/α_0 = constant, independently of particle diameter.[13]

Although both self-focusing and four-wave mixing effects are due to gradient forces, these two types of experiment differ in one important respect. Self-focusing involves only a single beam that subjects all particles to an axially directed scattering force, which is absent in the opposing-beam four-wave mixing case. This scattering force causes strong axial motion of the individual particles in the direction of the light, especially for the 0.234-μm particles. This is best seen at low concentrations by viewing from the side. The axial motion is reduced for 0.085-μm particles and unobservable for 0.038-μm particles. This reduction of scattering force effects for particles in which self-focusing still occurs is due to the n^6 dependence of Rayleigh scattering versus the n^3 dependence of the gradient force. We also conclude from the independence of P_{crit} from particle size that the axial motion, where it exists, does not strongly affect n_2.

To find the nonlinear coefficient n_2 of the intensity-dependent refractive index (defined by $n = n_0 + n_2 I$), from a knowledge of P_{crit}, we use the relation

$$P_{crit} = \frac{5.763}{4\pi^2} \frac{\lambda^2}{n_0 n_2} , \quad (1)$$

where n_0 is the index of the medium at low intensity. This relation was derived from the computer solution of P_{crit} for the lowest-order mode of a self-trapped filament in terms of the nonlinear dielectric constant ϵ_2, where $\epsilon = \epsilon_0 + \epsilon_2 E^2$. See, for example, Eq. (50) of Ref. 4, p. 1171. To derive Eq. (1) from Eq. (50) of Ref. 4, first correct the misprint in the factor π to read π^3 and then use $\epsilon = n^2$ and $I = n_0 c E^2/8\pi$.

Using the average P_{crit} from Table 1 of 0.77 W, one finds from Eq. (1) that $n_2 = 0.38 \times 10^{-9}$ cm^2/W, independently of particle diameter for a scattering loss α_0 = 5 cm^{-1}. This compares with the four-wave mixing value[1] of $n_2 = 1.2 \times 10^{-9}$ cm^2/W for 0.234-μm latex spheres scaled to a loss of 5 cm^{-1} using the proportionality of n_2 with α_0. The four-wave mixing n_2 assumed no contribution from the so-called fine grating[1] for this relatively large particle size. New results[13] show that the fine grating does contribute, even for 0.234-μm particles, which reduces the four-wave mixing value by 2 to $n_2 = 0.6 \times 10^{-9}$ cm^2/W, in close agreement with the self-trapping value. We can, of course, scale n_2 for α_0 = 5 cm^{-1} to other values of α_0 by using particles of the same refractive index. Thus, in Figs. 2C and 2D, where $\alpha_0 \cong 45$ cm^{-1}, n_2 increases by 45/5 = 9 to $n_2 \cong 3.4 \times 10^{-9}$ cm^2/W, and P_{crit} drops to 0.77 W/9 = 85 mW.

We thank J. E. Bjorkholm for helpful discussion.

References

1. P. W. Smith, A. Ashkin, and W. J. Tomlinson, Opt. Lett. **6**, 284 (1981).
2. A. Ashkin, Science **210**, 1081 (1980).
3. G. A. Askar'yan, Zh. Eksp. Teor. Fiz. **42**, 1567 (1962) [Sov. Phys. JETP **15**, 1088 (1962)].
4. S. A. Akhmanov, R. V. Khokhlov, and A. P. Sukhorukov in *Laser Handbook*, F. T. Arecchi and E. O. Schulz-Dubois, eds. (North-Holland, Amsterdam, 1972), Vol. 2, part E2.
5. F. W. Dabby and J. R. Whinnery, Appl. Phys. Lett. **13**, 284 (1968).
6. J. E. Bjorkholm and A. Ashkin, Phys. Rev. Lett. **32**, 129 (1974).
7. A. C. Tam and W. Happer, Phys. Rev. Lett. **38**, 278 (1977).
8. R. Y. Chiao, E. Garmire, and C. H. Townes, Phys. Rev. Lett. **13**, 479 (1964).
9. W. G. Wagner, H. A. Haus, and J. H. Marburger, Phys. Rev. **172**, 256 (1968).
10. C. C. Wang, Phys. Rev. Lett. **16**, 344 (1966).
11. J. E. Bjorkholm, P. W. Smith, W. J. Tomlinson, and A. E. Kaplan, Opt. Lett. **6**, 345 (1981).
12. Dow Chemical Company uniform polystyrene latex particles. The refractive index of the spheres is ~1.59, and that of the surrounding liquid is ~1.33.
13. P. W. Smith, P. J. Maloney, and A. Askin, Opt. Lett. (to be published).

Reprinted from Optics Letters, Vol. *11*, page 73, February 1986
Copyright © 1986 by the Optical Society of America and reprinted by permission of the copyright owner.

Proposal for optically cooling atoms to temperatures of the order of 10^{-6} K

S. Chu, J. E. Bjorkholm, A. Ashkin, J. P. Gordon, and L. W. Hollberg

AT&T Bell Laboratories, Holmdel, New Jersey 07733

Received October 15, 1985; accepted November 18, 1985

We propose a technique for cooling optically trapped atoms to microkelvin temperatures, and lower, by using the dipole force of resonance-radiation pressure.

Substantial progress has been made recently in the laser cooling of atoms and ions using the *spontaneous force* of resonance-radiation pressure. For example, sodium atoms have been cooled to a temperature of 240 μK by using laser light tuned within 10 MHz of the sodium resonance transition.[1] The temperature achieved there was the so-called "quantum limit" for that experimental configuration,[2,3] $kT = \frac{1}{2}h\gamma$, where γ is the natural linewidth (FWHM) of the atomic transition. This limit is obtained in equilibrium when the rate of quantum heating of the atoms due to the random spontaneous and stimulated scattering of the photons[4] equals the rate of cooling due to the average spontaneous force. Importantly, the quantum limit is not an absolute limit to the temperature. In fact, several nonoptical schemes have been proposed for cooling atoms contained in magnetic traps to temperatures below $h\gamma/2k$.[5,6] The recent demonstration of magnetically trapped sodium atoms[7] presents the possibility of trying those schemes.

In this Letter we propose an optical method for the three-dimensional cooling of atoms to temperatures considerably less then $h\gamma/2k$. For sodium atoms temperatures of the order of 10^{-6} K should be possible. The technique that we propose uses the *dipole force* of resonance-radiation pressure to bring about cooling in a new way. The method that we propose is straightforward in principle, but it anticipates the availability of dipole-force optical traps[8] capable of localizing a collection of atoms cooled to the quantum limit within small regions of space (of the order of several micrometers). Such traps may be demonstrated in the near future.

The basic idea behind our scheme is understood as follows. Consider an ensemble of atoms laser cooled to a temperature of $kT = h\gamma/2$ and tightly confined within a small region of space by an optical trap having small dimensions. At time $t = 0$ the small, deep trap used to localize the atoms initially is turned off and is replaced by a shallow trap of much larger dimensions; the atoms then expand into that much larger volume. The atoms can be loosely thought of as initially being centered in the potential minimum of the new trap; as time proceeds, they start to execute oscillatory motion in that trap. Assuming that the trap is an ideal har-

monic potential well and that all the atoms start from the exact center of the trap, all the atoms reach their turning points (where $v = 0$) at the same time, equal to one quarter of the harmonic oscillator period, *irrespective of their initial velocities*. At this time the trapping light is turned off and all the atoms are at rest ($T = 0$)!

The above discussion is overly simplistic, and there are several factors that place limits on the ultimate temperature that can be achieved by this scheme. First, spontaneous scattering of the photons by the atoms as they expand into the large trap must be minimized. As an example, the recoil velocity experienced by a sodium atom absorbing or emitting a single resonance photon is 3 cm/sec, corresponding to an energy of about 0.8 μK. This constraint requires the use of a dipole-force trap with the light frequency tuned far away from resonance. Second, at $t = 0$ the atoms will not all be at the exact center of the large trap. Consequently, the atoms will not simultaneously come to rest after a quarter period. Finally, practical traps have anharmonic components to the potential, and this also prevents the atoms from simultaneously coming to rest. This constraint is minimized by confining the atoms to the central portion of the trap where the potential is nearly harmonic. Given a fixed amount of laser power, the above factors combine to determine the lowest achievable temperature. In what follows we evaluate the constraints imposed by each of these factors and we determine the limiting temperature.

The dipole-force optical trap that we consider is formed using three intersecting and mutually perpendicular laser beams; other trap geometries could also be used. To minimize heating of the atoms by spontaneous scattering we use TEM_{01}*-mode laser beams; their transverse intensity distributions are given by $I(r) = (4P/\pi w^2)(r^2/w^2)\exp(-2r^2/w^2)$, where r is the transverse coordinate, P is the power of the beam, and w is its spot size. When these beams are used, the optical trap formed is nearly radially symmetric for displacements small compared with w, and the light intensity has a local minimum of zero at the origin. For the trap geometry and $r \ll w$ the light intensity is approximately $I(r) = (4P/\pi w^2)(2r^2/w^2)$, where r is now

0146-9592/86/020073-03$2.00/0

74   OPTICS LETTERS / Vol. 11, No. 2 / February 1986

the displacement from the origin and P is the power in each of the beams. When the light frequency is tuned above the frequency of the atomic resonance the direction of the dipole force is such that the atoms are confined in the vicinity of the origin, where the light intensity is low and spontaneous scattering is minimized. The optical potential well is given by $U(r) = h\Delta\nu p(r)/2$ when p, the saturation parameter, is small and where $\Delta\nu$ is the laser detuning above resonance. Also, when $\Delta\nu \gg \gamma$, we have

$$p(r) = \frac{4P}{\pi w^2}\frac{1}{I_{\mathrm{sat}}}\frac{\gamma^2}{4\Delta\nu^2}\frac{2r^2}{w^2},\qquad(1)$$

where I_{sat} is the saturation intensity for the atomic transition. An atom trapped in the harmonic potential executes harmonic motion with an angular resonance frequency ω given by

$$\omega^2 = \frac{2h\gamma^2}{\pi m I_{\mathrm{sat}}}\frac{P}{\Delta\nu w^4},\qquad(2)$$

where m is the atomic mass.

We use the following model to carry out our calculation. For times $t < 0$ we assume that the atoms are initially confined in a deep, tight potential well having resonance frequency ω_1 and that the initial atomic temperature T_1 is given by $kT_1 = h\gamma/2$. At time $t = 0$ the initial trap is replaced by a broad shallow trap characterized by resonance frequency ω_2 and spot size w_2. For $t > 0$ the motion of an atom located at r_1 and having radial velocity v_1 at $t = 0$ is given by

$$r(t) = \frac{v_1}{\omega_2}\sin\omega_2 t + x_1\cos\omega_2 t.\qquad(3)$$

A quarter period after $t = 0$, at time $t_2 = \pi/2\omega_2$, the second trap is turned off. At this time the atom is located at $r_2 = v_1/\omega_2$ and its velocity is $v_2 = -\omega_2 x_1$. The temperature of the atoms at this time is evaluated by calculating their rms velocity at time t_2 using the above model. For simplicity the calculations are carried out in one dimension. For the actual three-dimensional problem the principles involved are exactly the same and the numbers obtained are only slightly different.

First we consider the effect of the spread of initial atomic positions on the final atomic temperature. Consider the v_0 group of atoms, those having total energy of $mv_0^2/2$ in the initial trap. At time $t = 0$ these atoms are distributed between $x_1 = -v_0/\omega_1$ and v_0/ω_1 and the probability of finding an atom at the position x within that range is $(1/\pi)(v_0^2/\omega_1^2 - x^2)^{1/2}$. Thus at t_2 the rms velocity for the v_0 group of atoms is

$$(\overline{v_2^2})_0 = \frac{1}{2}\frac{\omega_2^2}{\omega_1^2}v_0^2,$$

where the initial rms velocity of this group was $\tfrac{1}{2}v_0^2$. The rms velocity of the entire atomic velocity distribution is obtained by integrating over the Maxwellian distribution of velocities at $t = 0$. The result is $v_2^2 = \tfrac{1}{2}(\omega_2^2/\omega_1^2)(kT_1/m)$. Since the velocity distribution at t_2 is also Maxwellian we have $m\overline{v_2^2} = kT_2$, where T_2 is the temperature of the atoms at t_2. Thus the tem-

perature T_2 determined by the initial spread of atomic positions is

$$\left(\frac{T_2}{T_1}\right)_{\mathrm{spread}} = \frac{1}{2}\frac{\omega_2^2}{\omega_1^2}.\qquad(4)$$

This result indicates that we want ω_2 to be as small as possible to achieve the most cooling. Note that Eq. (4) can also be obtained by applying Liouville's theorem, which states that the quantity $v_{\mathrm{rms}}x_{\mathrm{rms}}$ is a constant of the motion for a conservative potential. Thus $\overline{x_1^2 v_1^2} = \overline{x_2^2 v_2^2}$ and

$$\frac{T_2}{(\tfrac{1}{2}T_1)} = \frac{\overline{v_2^2}}{\overline{v_1^2}} = \frac{\overline{x_1^2}}{\overline{x_2^2}} = \frac{(v_0/\omega_1)^2}{(v_0/\omega_2)^2} = \frac{\omega_2^2}{\omega_1^2},$$

where the factor of $\tfrac{1}{2}$ associated with T_1 arises because at $t = 0$ only one half of the mean energy is in the form of kinetic energy.

A limit to how small ω_2 can be made is imposed by spontaneous scattering of photons by the atom, which must be avoided. The spontaneous scattering rate for an atom is $\pi\gamma p(x)$, when $p \ll 1$. We calculate N, the average number of photons scattered by an atom, by integrating the scattering rate over the atomic trajectory from $t = 0$ to $t = t_2$. Consider the v_0 group of atoms; we find that N depends on x_1, with the most scattering occurring for atoms starting from $x_1 = 0$. For the v_0 group, $N \leq (\pi\gamma^3 P/2I_{\mathrm{sat}}w_2^4\Delta\nu^2\omega_2^3)v_0^2$. At $t = 0$ the majority of atoms have $mv_0^2 < 4kT_1 = 2h\gamma$, and we arrive at the following estimate for N:

$$N < \frac{\pi^2\gamma^2}{2\Delta\nu\omega_2} \equiv N_{\max}.\qquad(5)$$

The majority of atoms scatter far fewer than $N_{\max}$ photons. Requiring that $N_{\max} = 1$ will ensure that heating of the atoms due to spontaneous scattering is small; this places a lower limit on ω_2.

Finally we account for the effects of anharmonicity of the potential well. To the next higher order in x/w_2, the optical potential along the x axis is given by

$$U(x) = \frac{h\Delta\nu}{2}\left(\frac{4P}{\pi w_2^2}\frac{1}{I_{\mathrm{sat}}}\frac{\gamma^2}{4\Delta\nu^2}\right)\frac{2x^2}{w_2^2}\left(1 - \frac{2x^2}{w_2^2}\right).$$

To first order in the expansion parameter $\epsilon = v_1^2/w_2^2\omega_2^2 \ll 1$, the motion for an atom starting at $x_1 = 0$ at $t = 0$ is given by

$$x(t) = \frac{v_1}{\omega}\left(1 + \frac{9}{8}\epsilon\right)\sin\omega t + \frac{1}{8}\frac{v_1}{\omega_2}\epsilon\sin 3\omega t,$$

where $\omega^2 = \omega_2^2(1 - 3\epsilon)$. Notice that ω is now a function of v_1; because of this, all atoms starting from zero do not simultaneously reach their turning points, giving a spread of velocities when the expansion trap is turned off. The v_0 group atoms come to rest at the time $\tau_0 = (1 + 3v_0^2/2w_2^2\omega_2^2)t_2$, and the residual velocity for the v_1 group of atoms is

$$[v^2(\tau_0)]_1 = \frac{9\pi^2}{16}\frac{(v_0^2 - v_1^2)^2}{w_2^4\omega_2^4}v_1^2.$$

We calculate $\overline{v^2(\tau_0)}$ by integrating the above expression over the Maxwellian distribution of v_1's. We find

February 1986 / Vol. 11, No. 2 / OPTICS LETTERS   **75**

that $\overline{v^2(\tau_0)}$ is minimized by choosing $v_0^2 = 3kT_1/2m$; the result is $\overline{v^2} = 189\pi^2 k^3 T_1^3/64m^3 w_2^4 \omega_2^4$. Rearranging terms, we arrive at the final temperature as limited by anharmonicity of the potential well:

$$\left(\frac{T_2}{T_1}\right)_{\text{anharm}} = \frac{189\pi^3}{512}\frac{hI_{\text{sat}}\Delta\nu}{mP\omega_2^2}. \qquad (6)$$

In contrast with Eq. (4), this expression gives lowest T_2 when ω_2 is large.

We now solve for the minimum T_2/T_1 consistent with Eqs. (4)–(6). We do this by setting $(T_2/T_1)_{\text{spread}} = (T_2/T_1)_{\text{anharm}} \equiv (T_2/T_1)_0$ and solving for ω_2. Using Eq. (5), we eliminate ω_2 and find the optimum detuning $(\Delta\nu)_0$ in terms of N_{max}, P, and $\overline{x_1^2} = h\gamma/4m\omega_1^2$, the mean-square displacement in the initial trap. The result is

$$(\Delta\nu)_0 = \left(\frac{64\pi^5 m^2\gamma^7}{189h^2 I_{\text{sat}}}\right)^{1/5}\left(\frac{P\overline{x_1^2}}{N_{\text{max}}^4}\right)^{1/5}. \qquad (7)$$

For sodium atoms the result is $(\Delta\nu)_0 \cong 1.12 \times 10^{12}$ $(P\overline{x_1^2}/N_{\text{max}}^4)^{1/5}$ Hz, where P is in watts, $\overline{x_1^2}$ is in square centimeters, and we have used $I_{\text{sat}} = 2 \times 10^{-2}$ W/cm².[9] Using Eq. (7) to solve for (T_2/T_1), we obtain

$$\left(\frac{T_2}{T_1}\right)_0 = \frac{\pi^2}{8}\left(\frac{189}{2}\right)^{2/5}\left(\frac{m\gamma I_{\text{sat}}^2}{h}\right)^{1/5}\left[\frac{(\overline{x_1^2})^3}{P^2 N_{\text{max}}^2}\right]^{1/5}. \qquad (8)$$

For sodium, this yields $(T_2/T_1)_0 \approx 226$ $[(\overline{x_1^2})^3/P^2 N_{\text{max}}^2]^{1/5}$. We would expect the actual atomic temperature to be roughly twice $(T_2/T_1)_0$. Likewise we can solve for w_2; we find that

$$w_2 = \frac{2}{\pi}\left(\frac{2}{189}\right)^{1/20}\left(\frac{h}{m\gamma I_{\text{sat}}^2}\right)^{3/20}(\overline{x_1^2}N_{\text{max}}^6 P^6)^{1/20}. \qquad (9)$$

For sodium atoms this is $w_2 \approx 3.98 \times 10^{-2}$ $(\overline{x_1^2}N_{\text{max}}^6 P^6)^{1/20}$ cm.

As an example of what Eqs. (7)–(9) predict, consider using 0.4 W per beam and choose $N_{\text{max}} = 1$. Further, take $(\overline{x_1^2})^{1/2} = 0.5$ μm; such a choice is a reasonable one and can be achieved by using the following parameters for the initial trap: $P = 0.2$ W, $w_1 = 20$ μm, and $\Delta\nu = 40$ GHz. Under these conditions the heating effects of spontaneous and stimulated scattering are entirely negligible.[3] With these choices for the initial trap, the predictions of Eqs. (7)–(9) are $\Delta\nu = 17.8$ GHz, $w_2 = 112$ μm, and $(T_2/T_1)_0 = 2.25 \times 10^{-3}$, which implies that $T_2 \approx 1.1$ μK [twice $(T_2)_0$]. Further examples are given in Table 1.

There are other closely related ways of using the principles discussed here to cool trapped atoms. For instance, consider a version in which at $t = 0$ the initial trap is turned off and the atoms are allowed to expand into free space. At a later time the larger trap is pulsed on for a short period of time. Since the pulsed trap is harmonic, the impulse delivered to an atom at position r is proportional to r, as is the atom's velocity. Thus when the pulse energy is chosen correctly all atoms can be brought nearly to rest.

Once atoms are cooled to temperatures of about 10^{-6} K gravity plays a major role in determining the atomic motion. Dipole optical traps can be devised

Table 1. Final Atomic Temperature $(T_2)_0$, Required Laser Spot Size, and Required Laser Detuning as a Function of the Laser Power P [a]

P (W)	$(T_2)_0$ (μK)	w_2 (μm)	$\Delta\nu_2$ (GHz)
0.05	1.24	60	11.7
0.1	0.94	74	13.5
0.2	0.71	91	15.5
0.4	0.54	112	17.8
0.6	0.46	127	19.3

[a] We have taken $N_{\text{max}} = 1$ and $(\overline{x_1^2})^{1/2} = 0.5$ μm; this value for $(\overline{x_1^2})^{1/2}$ can be achieved with initial trap parameters of $P_1 = 0.2$ W, $w_1 = 20$ μm, and $\Delta\nu_1 = 40$ GHz.

that counteract the force of gravity and that confine the cold atoms for times of about 60 sec without any heating due to spontaneous or stimulated transitions. This is accomplished by using a smaller volume trap than the expansion trap and by tuning the light farther from resonance. The spatially varying optical Stark shifts in such gravity traps are several tens of kilohertz. Further reduction of this Stark shift using an additional Stark-correcting beam is conceivable.[2]

In summary, we have proposed a straightforward, albeit currently technologically challenging, optical technique for cooling of optically trapped atoms to temperatures of the order of 1 μK. It is expected that temperatures below those predicted here can be achieved by appropriate modifications such as by designing more nearly harmonic optical traps or by perturbing the initial velocity distribution, for instance by letting the fast atoms escape from the initial trap at times before $t = 0$. The ideas discussed here can also be applied in two dimensions, for instance in the collimation (transverse cooling) of atomic beams.[10]

References

1. S. Chu, L. Hollberg, J. E. Bjorkholm, A. Cable, and A. Ashkin, Phys. Rev. Lett. **55**, 48 (1985).
2. A. Ashkin and J. P. Gordon, Opt. Lett. **4**, 161 (1979); D. J. Wineland and W. M. Itano, Phys. Rev. A **20**, 1521 (1979).
3. J. P. Gordon and A. Ashkin, Phys. Rev. A **21**, 1606 (1980).
4. For an earlier experimental observation of the effects of spontaneous heating see J. E. Bjorkholm, R. R. Freeman, A. Ashkin, and D. B. Pearson, Opt. Lett. **5**, 111 (1980).
5. D. E. Pritchard, Phys. Rev. Lett. **51**, 1336 (1983).
6. H. F. Hess, Bull. Am. Phys. Soc. **30**, 854 (1985).
7. A. L. Migdall, J. V. Prodan, W. D. Phillips, T. H. Bergeman, and H. J. Metcalf, Phys. Rev. Lett. **54**, 2596 (1985).
8. A. Ashkin, Phys. Rev. Lett. **40**, 729 (1978); for a review of optical traps, see A. Ashkin, Science **210**, 1081 (1980); for more recent discussions, see references in Refs. 1 and 6.
9. This value seems appropriate for the actual combination of levels used in sodium; see J. E. Bjorkholm, R. R. Freeman, and D. B. Pearson, Phys. Rev. A **23**, 491 (1981).
10. J. E. Bjorkholm, R. R. Freeman, A. Ashkin, and D. B. Pearson, Phys. Rev. Lett. **41**, 1361 (1978); J. E. Bjorkholm and R. R. Freeman, Comments Atom. Mol. Phys. **10**, 31 (1980).

PHYSICAL REVIEW
LETTERS

VOLUME 75   27 NOVEMBER 1995   NUMBER 22

Bose-Einstein Condensation in a Gas of Sodium Atoms

K. B. Davis, M.-O. Mewes, M. R. Andrews, N. J. van Druten, D. S. Durfee, D. M. Kurn, and W. Ketterle

Department of Physics and Research Laboratory of Electronics, Massachusetts Institute of Technology,
Cambridge, Massachusetts 02139
(Received 17 October 1995)

We have observed Bose-Einstein condensation of sodium atoms. The atoms were trapped in a novel trap that employed both magnetic and optical forces. Evaporative cooling increased the phase-space density by 6 orders of magnitude within seven seconds. Condensates contained up to 5×10^5 atoms at densities exceeding 10^{14} cm^{-3}. The striking signature of Bose condensation was the sudden appearance of a bimodal velocity distribution below the critical temperature of ~ 2 μK. The distribution consisted of an isotropic thermal distribution and an elliptical core attributed to the expansion of a dense condensate.

PACS numbers: 03.75.Fi, 05.30.Jp, 32.80.Pj, 64.60.–i

Bose-Einstein condensation (BEC) is a ubiquitous phenomenon which plays significant roles in condensed matter, atomic, nuclear, and elementary particle physics, as well as in astrophysics [1]. Its most striking feature is a macroscopic population of the ground state of the system at finite temperature [2]. The study of BEC in weakly interacting systems holds the promise of revealing new macroscopic quantum phenomena that can be understood from first principles, and may also advance our understanding of superconductivity and superfluidity in more complex systems.

During the past decade, work towards BEC in weakly interacting systems has been carried forward with excitons in semiconductors and cold trapped atoms. BEC has been observed in excitonic systems, but a complete theoretical treatment is lacking [1,3]. The pioneering work towards BEC in atomic gases was performed with spin-polarized atomic hydrogen [4,5]. Following the development of evaporative cooling [6], the transition was approached within a factor of 3 in temperature [7]. Laser cooling provides an alternative approach towards very low temperatures, but has so far been limited to phase-space densities typically 10^5 times lower than required for BEC. The combination of laser cooling with evaporative cooling [8–10] was the prerequisite for obtaining BEC in alkali atoms. This year, within a few months, three independent and different approaches succeeded in creating BEC in

rubidium [11], lithium [12], and, as reported in this paper, in sodium. Our results are distinguished by a production rate of Bose-condensed atoms which is 3 orders of magnitude larger than in the two previous experiments. Furthermore, we report a novel atom trap that offers a superior combination of tight confinement and capture volume and the attainment of unprecedented densities of cold atomic gases.

Evaporative cooling requires an atom trap which is tightly confining and stable. So far, magnetic traps and optical dipole traps have been used. Optical dipole traps provide tight confinement, but have only a very small trapping volume (10^{-8} cm^3). The tightest confinement in a magnetic trap is achieved with a spherical quadrupole potential (linear confinement); however, atoms are lost from this trap due to nonadiabatic spin flips as the atoms pass near the center, where the field rapidly changes direction. This region constitutes a "hole" in the trap of micrometer dimension. The recently demonstrated "TOP" trap suppresses this trap loss, but at the cost of lower confinement [8].

We suppressed the trap loss by adding a repulsive potential around the zero of the magnetic field, literally "plugging" the hole. This was accomplished by tightly focusing an intense blue-detuned laser that generated a repulsive optical dipole force. The optical plug was created by an Ar$^+$-laser beam (514 nm) of 3.5 W focused

0031-9007/95/75(22)/3969(5)$06.00   © 1995 The American Physical Society   3969

VOLUME 75, NUMBER 22   PHYSICAL REVIEW LETTERS   27 NOVEMBER 1995

to a beam waist of 30 μm. This caused 7 MHz (350 μK) of light shift potential at the origin. Heating due to photon scattering was suppressed by using far-off-resonant light, and by the fact that the atoms are repelled from the region where the laser intensity is highest.

The experimental setup was similar to that described in our previous work [9]. Typically, within 2 s 10^9 atoms in the $F = 1$, $m_F = -1$ state were loaded into a magnetic trap with a field gradient of 130 G/cm; the peak density was $\sim 10^{11}$ cm^{-3}, the temperature ~ 200 μK, and the phase-space density 10^6 times lower than required for BEC. The lifetime of the trapped atoms was $\sim$ 30 s, probably limited by background gas scattering at a pressure of $\sim 1 \times 10^{-11}$ mbar.

The magnetically trapped atoms were further cooled by rf-induced evaporation [8,9,13]. rf-induced spin flips were used to selectively remove the higher-energy atoms from the trap resulting in a decrease in temperature for the remaining atoms. The total (dressed-atom) potential is a combination of the magnetic quadrupole trapping potential, the repulsive potential of the plug, and the effective energy shifts due to the rf. At the point where atoms are in resonance with the rf, the trapped state undergoes an avoided crossing with the untrapped states (corresponding to a spin flip), and the trapping potential bends over. As a result, the height of the potential barrier varies linearly with the rf frequency. The total potential is depicted in Fig. 1. Over 7 s, the rf frequency was swept from 30 MHz to the final value around 1 MHz, while the field gradient was first increased to 550 G/cm (to enhance the initial elastic-collision rate) and then lowered to 180 G/cm (to avoid the losses due to inelastic processes at the final high densities).

Temperature and total number of atoms were determined using absorption imaging. The atom cloud was imaged either while it was trapped or following a sudden switch-off of the trap and a delay time of 6 ms. Such time-of-flight images displayed the velocity distribution of the trapped cloud. For probing, the atoms

were first pumped to the $F = 2$ state by switching on a 10 mW/cm^2 laser beam in resonance with the $F = 1 \rightarrow F = 2$ transition. 10 μs later the atoms were concurrently exposed to a 100 μs, 0.25 mW/cm^2 probe laser pulse in resonance with the $F = 2 \rightarrow F = 3$ transition, propagating along the trap's y direction. This probe laser beam was imaged onto a charge-coupled device sensor with a lens system having a resolution of 8 μm. Up to 100 photons per atom were absorbed without blurring the image due to heating.

At temperatures above 15 μK the observed trapped clouds were elliptical with an aspect ratio of 2:1 due to the symmetry of the quadrupole field. At the position of the optical plug they had a hole, which was used for fine alignment. A misalignment of the optical plug by $\sim$ 20 μm resulted in increased trap loss and prevented us from cooling below 50 μK. This is evidence that the Majorana spin flips are localized in a very small region around the center of the trap. At temperatures below 15 μK, the cloud separated into two pockets at the two minima in the potential of Fig. 1. The bottom of the potential can be approximated as a three-dimensional anisotropic harmonic oscillator potential with frequencies $\omega_y^2 = \mu B'/(2mx_0)$, $\omega_z^2 = 3\omega_y^2$, $\omega_x^2 = \omega_y^2[(4x_0^2/w_0^2) - 1]$, where μ is the atom's magnetic moment, m the mass, B' the axial field gradient, w_0 the Gaussian beam waist parameter ($1/e^2$ radius) of the optical plug, and x_0 the distance of the potential minimum from the trap center. x_0 was directly measured to be 50 μm by imaging the trapped cloud, w_0 (30 μm) was determined from x_0, the laser power (3.5 W), and B' (180 G/cm). With these values the oscillation frequencies are 235, 410, and 745 Hz in the y, z, and x directions, respectively.

When the final rf frequency ν_{rf} was lowered below 0.7 MHz, a distinctive change in the symmetry of the velocity distribution was observed [Figs. 2(a) and 2(b)]. Above this frequency the distribution was perfectly spherical as expected for a thermal uncondensed cloud [14]. Below the critical frequency, the velocity distribution contained an elliptical core which increased in intensity when

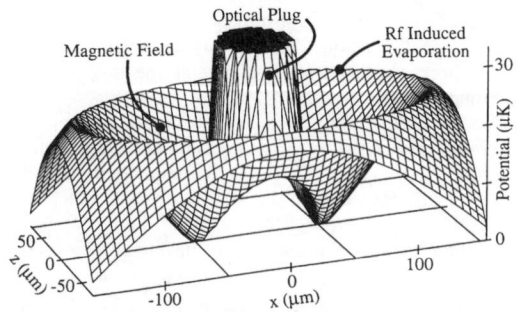

FIG. 1. Adiabatic potential due to the magnetic quadrupole field, the optical plug, and the rf. This cut of the three-dimensional potential is orthogonal to the propagation direction (y) of the blue-detuned laser. The symmetry axis of the quadrupole field is the z axis.

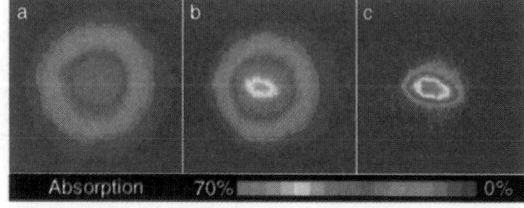

FIG. 2 (color). Two-dimensional probe absorption images, after 6 ms time of flight, showing evidence for BEC. (a) is the velocity distribution of a cloud cooled to just above the transition point, (b) just after the condensate appeared, and (c) after further evaporative cooling has left an almost pure condensate. (b) shows the difference between the isotropic thermal distribution and an elliptical core attributed to the expansion of a dense condensate. The width of the images is 870 μm. Gravitational acceleration during the probe delay displaces the cloud by only 0.2 mm along the z axis.

VOLUME 75, NUMBER 22   P H Y S I C A L   R E V I E W   L E T T E R S   27 NOVEMBER 1995

the rf was further swept down, whereas the spherical cloud became less intense. We interpret the elliptical cloud as due to the Bose condensate, and the spherical cloud as due to the normal fraction.

In the region just below the transition frequency one expects a bimodal velocity distribution: a broad distribution due to the normal gas and a narrow distribution due to the condensate. The cross sections of the time-of-flight images (Fig. 3) indeed show such bimodal distributions in this region. Figure 4 shows how suddenly the time-of-flight image changes below $\nu_{\rm rf} = 0.7$ MHz. The effective area of the observed cloud becomes very small [Fig. 4(a)], while the velocity distribution is no longer Gaussian [Fig. 4(b)] and requires different widths for the condensate and the normal fraction [Fig. 4(c)].

At the critical frequency, a temperature of (2.0 ± 0.5) μK was derived from the time-of-flight image. An independent, though less accurate estimate of the temperature T is obtained from the dynamics of evaporative cooling. Efficient evaporation leads to a temperature which is about 10 times smaller than the depth of the trapping potential [15]. Since the speed of evaporation depends exponentially on the ratio of potential depth to temperature, we expect this estimate of $T = 2$ μK to be accurate to within a factor of 2.

The critical number of atoms N_c to achieve BEC is determined by the condition that the number of atoms per

cubic thermal de Broglie wavelength exceeds 2.612 at the bottom of the potential [2]. For a harmonic oscillator potential this is equivalent to $N_c = 1.202(k_B T)^3/\hbar^3 \omega_x \omega_y \omega_z$ [16]. For our trap and 2.0 μK, $N_c = 2 \times (1.2 \times 10^6)$, where the factor of 2 accounts for the two separated clouds. The predicted value for N_c depends on the sixth power of the width of a time-of-flight image and is only accurate to within a factor of 3. We determined the number of atoms by integrating over the absorption image. At the transition point, the measured number of 7×10^5 agrees with the prediction for N_c. The critical peak density n_c at 2.0 μK is 1.5×10^{14} cm^{-3}. Such a high density appears to be out of reach for laser cooling, and demonstrates that evaporative cooling is a powerful technique to obtain not only ultralow temperatures, but also extremely high densities.

An ideal Bose condensate shows a macroscopic population of the ground state of the trapping potential. This picture is modified for a weakly interacting Bose gas. The mean-field interaction energy is given by $n\tilde{U}$, where n is the density and $\tilde{U}$ is proportional to the scattering length a: $\tilde{U} = 4\pi\hbar^2 a/m$ [2]. Using our recent experimental result $a = 4.9$ nm [9], $\tilde{U}/k_B = 1.3 \times 10^{-21}$ K cm^3. At the transition point, $n_c\tilde{U}/k_B T_c = 0.10$. Consequently, above the transition point, the kinetic energy dominates over the interaction energy, and the velocity

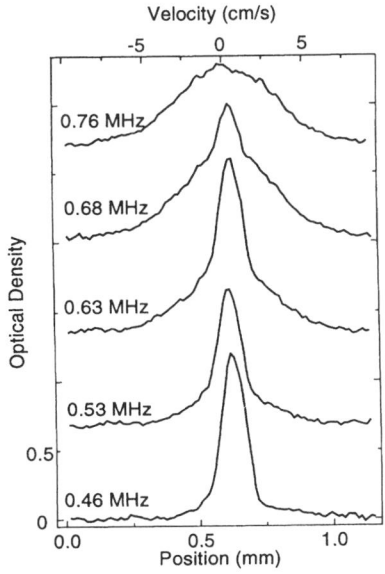

FIG. 3. Optical density as a function of position along the z axis for progressively lower values of the final rf frequency. These are vertical cuts through time-of-flight images like those in Fig. 2. For $\nu_{\rm rf} < 0.7$ MHz, they show the bimodal velocity distributions characteristic of the coexistence of a condensed and uncondensed fraction. The top four plots have been offset vertically for clarity.

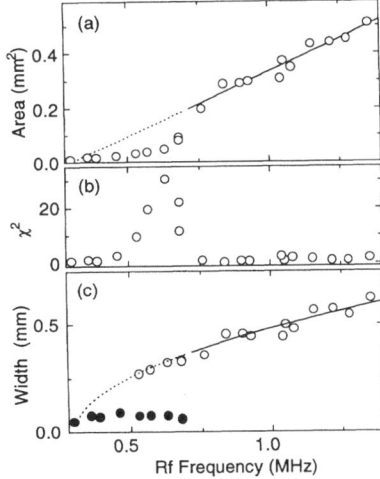

FIG. 4. Further evidence for a phase transition is provided by the sudden change of observed quantities as the final rf frequency $\nu_{\rm rf}$ is varied. (a) Area of the cloud in the time-of-flight image versus $\nu_{\rm rf}$. The area was obtained as the ratio of the integrated optical density and the peak optical density. The area changes suddenly at $\nu_{\rm rf} = 0.7$ MHz. Below the same frequency, the velocity distributions (Fig. 3) cannot be represented by a single Gaussian, as demonstrated by the χ^2 for a single Gaussian fit (b), and required different widths for the condensate (full circles) and noncondensate fraction (c). In (a) and (c) the lines reflect the behavior of a classical gas with a temperature proportional to the trap depth.

VOLUME 75, NUMBER 22   PHYSICAL REVIEW LETTERS   27 NOVEMBER 1995

distribution after sudden switch-off of the trap is isotropic. For the condensate, however, the situation is reversed. As we will confirm below, the kinetic energy of the condensate is negligible compared to its interaction energy [17]. Furthermore, well below the transition point, the interaction with the noncondensate fraction can be neglected. In such a situation, the solution of the nonlinear Schrödinger equation reveals that the condensate density $n_0(\mathbf{r})$ is a mirror image of the trapping potential $V(\mathbf{r})$: $n_0(\mathbf{r}) = n_0(\mathbf{0}) - V(\mathbf{r})/\tilde{U}$, as long as this expression is positive, otherwise $n_0(\mathbf{r})$ vanishes (see, e.g., Refs. [5,18]). For a harmonic potential, one obtains the peak density $n_0(\mathbf{0})$ for N_0 atoms in the condensate $n_0(\mathbf{0}) = 0.118(N_0 m^3 \omega_x \omega_y \omega_z/\hbar^3 a^{3/2})^{2/5}$.

Typically, we could cool one-fourth of the atoms at the transition point into a pure condensate. For an observed $N_0 = 1.5 \times 10^5$, we expect the condensate to be 2 times more dense than the thermal cloud at the transition point, and about 6 times larger than the ground-state wave function. The kinetic energy within the condensate is $\sim \hbar^2/(2mR^2)$, where R is the size of the condensate [18], while the internal energy is $2n_0\tilde{U}/7$. Thus the kinetic energy of the condensed atoms is around 1 nK, much less than the zero-point energy of our trap (35 nK) and the calculated internal energy of 120 nK. This estimate is consistent with our initial assumption that the kinetic energy can be neglected compared to the interaction energy.

The internal energy is ~ 25 times smaller than the thermal energy $(3/2)k_B T_c$ at T_c. Consequently, the width of the time-of-flight image of the condensate is expected to be about 5 times smaller than at the transition point. This is close to the observed reduction in the width shown in Fig. 4(c). This agreement might be fortuitous because we have so far neglected the anisotropy of the expansion, but it indicates that we have observed the correct magnitude of changes which are predicted to occur at the BEC transition of a weakly interacting gas. In several cooling cycles, as many as 5×10^5 condensed atoms were observed; we estimate the number density in these condensates to be 4×10^{14} cm^{-3}.

A striking feature of the condensate is the nonisotropic velocity distribution [11,19]. This is caused by the "explosion" of the condensate due to repulsive forces which are proportional to the density gradient. The initial acceleration is therefore inversely proportional to the width of the condensate resulting in an aspect ratio of the velocity distribution, which is inverted compared to the spatial distribution. When we misaligned the optical plug vertically, the shape of the cloud changed from two vertical crescents to a single elongated horizontal crescent. The aspect ratio of the time-of-flight image of the condensate correspondingly changed from horizontal to vertical elongation. In contrast, just above the transition point, the velocity distribution was found to be spherical and insensitive to the alignment of the plug. However, we cannot account quantitatively for the observed distributions because we have two separated condensates which overlap in the time-of-flight images, and also because of some residual horizontal acceleration due to the switch-off of the trap, which is negligible for the thermal cloud, but not for the condensate [20].

The lifetime of the condensate was about 1 s. This lifetime is probably determined either by three-body recombination [21] or by the heating rate of 300 nK/s, which was observed for a thermal cloud just above T_c. This heating rate is much higher than the estimated 8 nK/s for the off-resonant scattering of green light and may be due to residual beam jitter of the optical plug.

In conclusion, we were able to Bose-condense 5×10^5 sodium atoms within a total loading and cooling cycle of 9 s. During evaporative cooling, the elastic collision rate increased from 30 Hz to 2 kHz resulting in a mean free path comparable to the dimensions of the sample. Such collisionally dense samples are the prerequisite for studying various transport processes in dense ultracold matter. Furthermore, we have reached densities in excess of 10^{14} cm^{-3}, which opens up new possibilities for studying decay processes like dipolar relaxation and three-body recombination, and for studying a weakly interacting Bose gas over a broad range of densities and therefore strengths of interaction.

We are grateful to E. Huang and C. Sestok for important experimental contributions, to M. Raizen for the loan of a beam pointing stabilizer, and to D. Kleppner for helpful discussions. We are particularly grateful to D. E. Pritchard, who not only contributed many seminal ideas to the field of cold atoms, but provided major inspiration and equipment to W. K. This work was supported by ONR, NSF, JSEP, and the Sloan Foundation. M.-O. M., K. B. D., and D. M. K. would like to acknowledge support from Studienstiftung des Deutschen Volkes, MIT Physics Department Lester Wolfe fellowship, and NSF Graduate Research Fellowship, respectively, and N. J. v. D. from "Nederlandse Organisatie voor Wetenschappelijk Onderzoek (NWO)" and NACEE (Fulbright fellowship).

[1] A. Griffin, D. W. Snoke, and S. Stringari, *Bose-Einstein Condensation* (Cambridge University Press, Cambridge, 1995).

[2] K. Huang, *Statistical Mechanics* (Wiley, New York, 1987), 2nd ed.

[3] J. L. Lin and J. P. Wolfe, Phys. Rev. Lett. **71**, 1222 (1993).

[4] I. F. Silvera and M. Reynolds, J. Low Temp. Phys. **87**, 343 (1992); J. T. M. Walraven and T. W. Hijmans, Physica (Amsterdam) **197B**, 417 (1994).

[5] T. Greytak, in Ref. [1], p. 131.

[6] N. Masuhara *et al.,* Phys. Rev. Lett. **61**, 935 (1988).

[7] J. Doyle *et al.,* Phys. Rev. Lett. **67**, 603 (1991).

[8] W. Petrich, M. H. Anderson, J. R. Ensher, and E. A. Cornell, Phys. Rev. Lett. **74**, 3352 (1995).

VOLUME 75, NUMBER 22　　PHYSICAL REVIEW LETTERS　　27 NOVEMBER 1995

[9] K. B. Davis *et al.*, Phys. Rev. Lett. **74**, 5202 (1995).

[10] C. S. Adams *et al.*, Phys. Rev. Lett. **74**, 3577 (1995).

[11] M. H. Anderson *et al.*, Science **269**, 198 (1995).

[12] C. C. Bradley, C. A. Sackett, J. J. Tollett, and R. G. Hulet, Phys. Rev. Lett. **75**, 1687 (1995).

[13] D. E. Pritchard, K. Helmerson, and A. G. Martin, in *Atomic Physics 11,* edited by S. Haroche, J. C. Gay, and G. Grynberg (World Scientific, Singapore, 1989), p. 179.

[14] The measured $1/e$ decay time for the magnet current is 100 μs, shorter than the ω^{-1} of the fastest oscillation in the trap (210 μs). We therefore regard the switch-off as sudden. Any adiabatic cooling of the cloud during the switch-offtibs would result in a nonspherical velocity distribution due to the anisotropy of the potential.

[15] K. B. Davis, M.-O. Mewes, and W. Ketterle, Appl. Phys. B **60**, 155 (1995).

[16] V. Bagnato, D. E. Pritchard, and D. Kleppner, Phys. Rev. A **35**, 4354 (1987). This formula is derived assuming $k_B T_c \gg \hbar\omega_{x,y,z}$, which is the case in our experiment.

[17] Note that already for about 200 atoms in the ground state, the interaction energy in the center of the condensate equals the zero-point energy.

[18] G. Baym and C. Pethick (to be published).

[19] M. Holland and J. Cooper (to be published).

[20] These effects do not affect the vertical velocity distributions shown in Fig. 3.

[21] A. J. Moerdijk, H. M. J. M. Boesten, and B. J. Verhaar, Phys. Rev. A (to be published).

3973

VOLUME 59, NUMBER 23   PHYSICAL REVIEW LETTERS   7 DECEMBER 1987

Trapping of Neutral Sodium Atoms with Radiation Pressure

E. L. Raab,[a] M. Prentiss, Alex Cable, Steven Chu,[b] and D. E. Pritchard[a]

AT&T Bell Laboratories, Holmdel, New Jersey 07733

(Received 16 July 1987)

We report the confinement and cooling of an optically dense cloud of neutral sodium atoms by radiation pressure. The trapping and damping forces were provided by three retroreflected laser beams propagating along orthogonal axes, with a weak magnetic field used to distinguish between the beams. We have trapped as many as 10^7 atoms for 2 min at densities exceeding 10^{11} atoms cm^{-3}. The trap was $\simeq 0.4$ K deep and the atoms, once trapped, were cooled to less than a millikelvin and compacted into a region less than 0.5 mm in diameter.

PACS numbers: 32.80.Pj

The ability to cool and trap neutral atoms has recently been demonstrated by several groups.[1-3] Their traps utilized the intrinsic atomic magnetic dipole moment or the induced oscillating electric dipole moment to confine sodium atoms about a local-field strength extremum. We report the first optical trap which relies on near-resonant radiation pressure (also called *spontaneous* light force, in contrast to *induced* light forces[4]) to both confine and cool the atoms. The trap has an effective depth of about 0.4 K, about 10 times deeper than the deepest traps previously reported.[3] It is the first trap which exploits an atom's internal structure to induce a greater absorption probability for light moving toward the center of confinement.[4,5]

The basic principle of the trap can be illustrated by considering a hypothetical atom with a spin $S=0$ ($m_s=0$) ground state and a spin $S=1$ ($m_s=-1,0,+1$) excited state. In a weak inhomogeneous magnetic field $B_z(z)=bz$, the energy levels are Zeeman split by an amount $\Delta E = \mu m_s B = \mu b m_s z$ [Fig. 1(a)]. Now illuminate the atom with weak, collimated σ^- light propagating in the $-\hat{z}$ direction and σ^+ light propagating towards $+\hat{z}$. If the laser is tuned below the $B=0$ resonance frequency, the atom at $z>0$ will absorb more σ^- photons than σ^+ photons (since the laser frequency is closer to the $\Delta m = -1$ transition frequency) and consequently will feel a net time-averaged force toward the origin. For an atom at $z<0$, the Zeeman shift is reversed, and the force will again be directed to $z=0$. Tuning the low-intensity laser to the red of resonance also provides damping, as in the "optical molasses" demonstrated previously.[6]

The scheme is readily extended to three dimensions by adding counterpropagating beams along the x and y axes, and a "spherical quadrupole" magnetic field as shown in Fig. 1(b). The field is of the type used by Migdall *et al.* to confine spin polarized atoms magnetically,[1] though the field magnitudes in the light trap are about 100 times smaller and contribute negligibly to the confining force. If the x and y axis beams are polarized as shown, the conditions for confinement will be satisfied independently along each of the three axes.

The method can also work for atoms with a more complicated hyperfine structure. In the case of the sodium $3S_{1/2}\text{-}3P_{3/2}$ transition, e.g., the ground states have total

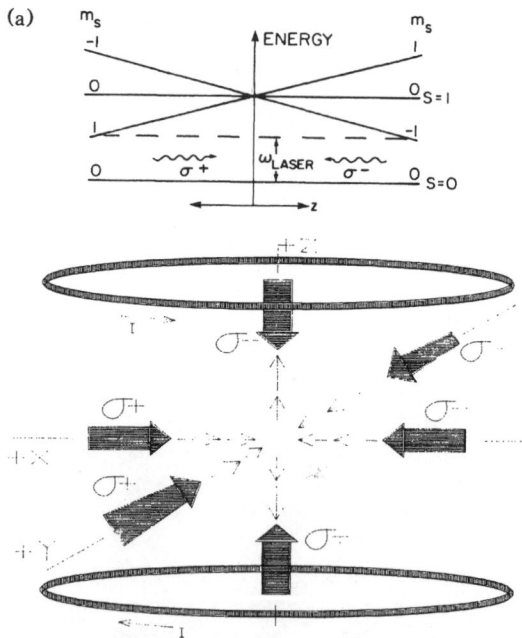

(a)

FIG. 1. (a) Energy-level diagram of hypothetical atom having spin $S=0$ ground state and spin $S=1$ excited state, immersed in a magnetic field $B_z(z)=bz$. The frequency and polarization of the counterpropagating laser are chosen to produce damping and restoring forces for the atom's z-axis motion. (b) Trapping scheme in three dimensions. The "spherical quadrupole" field is generated by two coils of opposing current placed along the z axis approximately as shown. The field along the axes, indicated by the light arrows, is parallel to its respective axis. Laser light, indicated by the heavy arrows, counterpropagates along x, y, and z, and is polarized as shown with respect to the axis of propagation.

VOLUME 59, NUMBER 23   PHYSICAL REVIEW LETTERS   7 DECEMBER 1987

angular momentum $F = 2,1$ and the excited states have $F' = 3,2,1,0$. Figure 2 shows the results of a computer model of the trap where the average force in one dimension was calculated for several magnetic field strengths and atomic velocities. Two laser frequencies are used in the model to avoid optically pumping the atom into an inaccessible ground hyperfine state. The results show the presence of both restoring and damping elements of the force. If we consider small displacements and velocities, we may write $\mathbf{F} \approx -\kappa \mathbf{x} - \alpha \dot{\mathbf{x}}$; the model in this regime gives $\kappa = (dB/dz)2.67 \times 10^{-16}$ dynes cm^{-1}, and $\alpha = 6.07 \times 10^{-18}$ dynes/cm s^{-1}. The equation of motion for small oscillations about the origin is simply that of a damped harmonic oscillator: $\ddot{x} + 2\beta\dot{x} + \omega_0^2 x = 0$, where $\omega_0^2 \equiv \kappa/m$, $\beta \equiv \alpha/2m$, and m is the mass of the atom. If we assume a nominal field gradient of 5 G/cm, we obtain $\omega_0 \approx 6 \times 10^3$ s^{-1}, and $\beta \approx 8 \times 10^4$ s^{-1}. Thus, the motion is strongly overdamped; the relaxation to the origin (at this gradient) is governed by a time constant $\tau_{simul} \approx 2\beta/\omega_0^2 \approx 4$ ms. The model was also used to examine the case when the atom ventures off from the principal axes, where the additional complication of $\Delta m = 0$ transitions arises. It was found that the trap was indeed restoring for small displacements in any direction.[7]

The spontaneous–light-force trap described above was demonstrated in the same apparatus previously used to generate optical molasses,[6] with only a few modifications. Six antireflection-coated quarter wave plates were placed adjacent to the six windows to generate circularly polarized light and to reverse the polarization of the reflected beams. Also, a pair of coils with opposing current was positioned within the vacuum chamber to

generate the desired magnetic field. Each 5-cm-diam coil consisted of three turns of $\frac{1}{8}$-in.-o.d. copper refrigerator tubing sheathed in fiberglass insulation. Water was passed through the coils for cooling.

Improvements were made in the remaining apparatus to increase its reliability and repeatability. The ring dye laser was actively locked to a crossover resonance of the Na $3S_{1/2}$-$3P_{3/2}$ transition in a saturated absorption cell. This allowed us to determine the laser frequency accurately and to provide the frequency stability necessary to observe atoms that remain in the trap for half an hour or more. A 5-m single-mode optical fiber was used as a spatial filter for the trapping beams; this also improved their pointing stability (and day to day alignment) since spatial drift of the dye laser output could be compensated by minor changes in the input coupling of the fiber. We used a larger $1 \times 1 \times 20$-mm^3 LiTaO$_3$ crystal as an electro-optic modulator to provide optical sidebands with a minimum of beam distortion. The crystal was driven by a resonant circuit tuned to 856 MHz with $Q \approx 100$, allowing us to maximize the fraction of light in the two first-order sidebands ($\approx 70\%$ total) with an rf drive power of less than 0.5 W.

The trap was loaded with atoms evaporated by a pulsed yttrium-aluminum-garnet laser and cooled by a frequency-swept laser beam as previously described.[6] Once slowed to velocities less than 2×10^3 cm s^{-1}, the atoms drifted into the molasses region and were trapped. The combination of restoring and damping forces compressed the trapped atoms into a small bright ball. Since the storage time of the atoms in the trap was considerably longer than the 10-Hz rate of the pulsed atomic beam, many (≈ 100) pulses of atoms could be injected into the trap before an equilibrium density was achieved.

Trapping was observed over a wide range of conditions. With a fixed 1712.4-MHz optical sideband splitting, the laser frequencies were tuned to the red of the $F = 2 \rightarrow 3$ and $1 \rightarrow 2$ transitions or the $F = 2 \rightarrow 2$ and $1 \rightarrow 0$ transitions. The former tuning produced a much more compact ball of atoms than the latter. For a fixed light intensity and a variety of laser detunings and magnetic field strengths, the size of the atom cloud, measured with a video camera and video wave-form analyzer, is shown in Fig. 3. The resolution of the camera system was better than 200 μm. We found that the diameter of the trapped atom cloud varies inversely as the square root of the current, consistent with our model of a harmonic trap potential and an atomic temperature independent of the current. Using a independent measure of the temperature (described below), we obtain a force constant of $8.8B' \times 10^{-17}$ dynes cm^{-1}; When $B' = 5$ G cm^{-1}, this implies an oscillation frequency $\omega_0 = 3.4 \times 10^3$ s^{-1} when trapping with the stronger transition. The results are thus in accord with the predictions of the model.

The restoring and damping forces were studied by

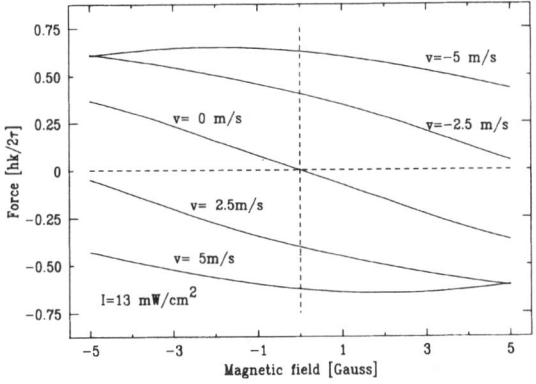

FIG. 2. Result of computer model for the forces felt by a sodium atom along the z axis in the light trap. The lasers are tuned 10 MHz to the red of the $F = 2$ to 3 and 1 to 2 transitions in the $D2$ line, with an intensity of 13 mW/cm^2 per sideband. $k = 2\pi/\lambda$, $\tau = 16$ ns is the natural lifetime for the sodium $D2$ line; $\hbar k/2\tau = F_{max} = 3.5 \times 10^{-15}$ dynes, the maximum theoretical spontaneous force attainable with this transition.

VOLUME 59, NUMBER 23   PHYSICAL REVIEW LETTERS   7 DECEMBER 1987

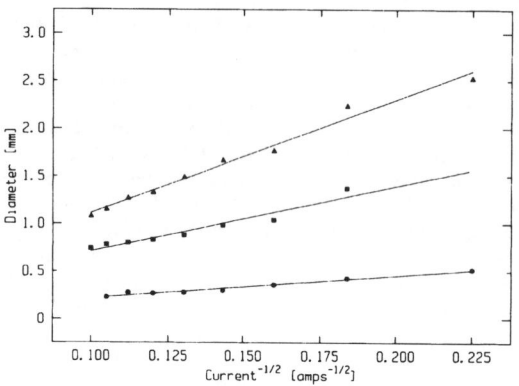

FIG. 3. The trap size (FWHM) as a function of the current through the magnets, for various laser tunings. The circles represent tuning 8 MHz to the red of the $F=2$ to 3 and 1 to 2 transitions, with an intensity of 13 mW/cm^2 per sideband. The squares and triangles represent tuning 8 and 20 MHz to the red of the $F=2$ to 2 and 1 to 0 transitions, respectively, with an intensity of 15 mW/cm^2.

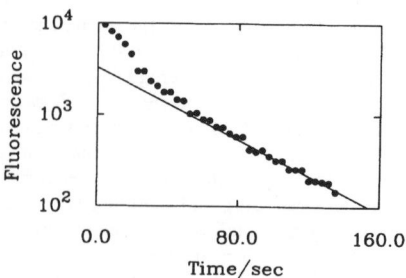

FIG. 4. Decay of atoms from the trap. The magnetic field gradient was $\simeq 12$ G cm^{-1}, and the laser intensity was $\simeq 20$ mW/cm^2 per sideband. The pressure was $\simeq 5 \times 10^{-11}$ Torr. The line indicates an exponential decay having a time constant of 95 s, which we observed at lower densities.

pushing the atoms from the center of the trap, then releasing them and allowing them to reequilibrate. Their relaxation to the origin was detected by placing an aperture around the image of the trap at equilibrium. Light passing through the aperture was detected by a photomultiplier tube; when the atoms were displaced from equilibrium, the signal from the photomultiplier tube would decrease. Two methods were used to displace the atoms: Either a neutral density filter was placed in one of the retroreflected beams (the trap was stable for a total attenuation of 0.6), or a cw probe laser was introduced (stability was destroyed for $I > 0.1I_{sat}$). The maximum displacement was $\simeq 1$ mm. With a field gradient of 5 G cm^{-1}, both measurements yielded a restoring force of $\simeq 10^{-14}$ dynes cm^{-1} when $I=10$ mW/cm^2 per beam (in each sideband) and $\delta = -10$ MHz, giving an oscillation frequency of $\omega_0 = 1.5 \times 10^4$ s^{-1}. This is approximately 2 times larger than ω_0 obtained from the model results at the higher intensity, which is fair agreement considering the experimental uncertainty and theoretical simplifications.

Once the trap was fully loaded, the yttrium-aluminum-garnet laser and chirped laser pulses were turned off, and the decay of the trap fluorescence was recorded as shown in Fig. 4. The longest lifetimes ($1/e \simeq 2$ min) were recorded at pressures of 5×10^{-11} Torr, achieved by cooling a cooper shroud inside the vacuum chamber to liquid-nitrogen temperatures. For low densities or later times ($t > 5$ s), the decay curve approached a simple exponential whose time constant decreased with increasing background pressure. This indicates that atoms were being expelled from the trap by collisions with the background gas. The initial decay for higher

densities was dominated by a loss which can be fitted to the square of the density of the trapped atoms. A detailed study of this loss mechanism will be presented in a future publication.

By measuring the absorption of a weak, resonant probe beam with the trapping light both on and off, we determined the ratio of the average excited-state population to the average ground-state population under various experimental conditions. With a magnet current of 50 A (corresponding to $B' \approx 10$ g cm^{-1} along z) and tuning to the red of the stronger ($F=2 \rightarrow 3$, $1 \rightarrow 2$) transitions, e.g., we found this ratio to be $\approx 3:4$. Measurements of the absolute fluorescence from the trapped atoms with a calibrated photomultiplier tube and lens assembly then implied that 3×10^6 atoms could be confined to a region $\simeq 320$ μm in diameter (FWHM). Thus the atomic density is 1.8×10^{11} atoms cm^{-3}. When the laser was tuned to the red of the $F=2 \rightarrow 2$ transition, the number of atoms trapped was $\simeq 1.2 \times 10^7$ with a trap diameter of 1 mm, giving a density of $\simeq 2 \times 10^{10}$ atoms cm^{-3}. The absorption of the probe beam was also used as an independent measure of the density. The peak absorption observed was 80% through $\simeq 300$-μm path which, with the assumption of a simple two-level atom, corresponds to a density of roughly 5×10^{10} atoms cm^{-3}. It is important to note that the rapid nonexponential loss mechanism seen at higher densities is responsible for keeping the atomic densities below 10^{12} atoms cm^{-3} under the present loading conditions. By adjustment of experimental parameters to limit the density of atoms (misalignment of the trapping beams, the tuning to the weaker transition, and the use of weak light intensities and field gradients), up to 10^8 atoms have been contained.

The effective depth of the trap was measured by giving the trapped atoms an impulse from an additional beam while the trapping lasers were momentarily off. We find that with an intensity per side band of 30 mW/cm^2, a

2633

VOLUME 59, NUMBER 23　　　PHYSICAL REVIEW LETTERS　　　7 DECEMBER 1987

light pulse of 18 μs was necessary to eject $\simeq 80\%$ of the atoms from the trap. Thus, atoms at the center of the trap require a velocity of $\simeq 1600$ cm s^{-1} to escape, implying a trap depth of $\simeq 0.4$ K.

We have also measured the fraction of atoms that remain in the trap after the molasses beams have been shut off for various times. At early times, the loss is dominated by atoms that leave the trapping region ballistically before the light is turned back on. We can estimate the mean atomic velocity $\bar{v}$ by observing that half the atoms are lost in the first 15 ms of darkness; if we assume a capture radius of 1 cm, we obtain a $\bar{v}$ of 45 to 85 cm s^{-1}, corresponding to a temperature of 300 to 1000 μK. Trapping with the $2 \rightarrow 2$ transition produced a gas 2 orders of magnitude hotter than trapping on the $2 \rightarrow 3$ transition.

The trap is very robust and does not critically depend on balanced light beams, purity of the circular polarization, or laser frequency (the trap worked over a 25-MHz tuning range). Trapping was observed for peak magnetic fields as low as 5 G and laser intensities ranging from 30 to 0.4 mW/cm^2. We were able to load the trap without the use of the chirped slowing laser by capturing atoms in the slow velocity tail of the pulsed atomic beam. When the laser beams were slightly misaligned, the potential well was no longer simply harmonic, causing the atoms to swirl around in rings or form irregular shapes. The atoms would sometimes settle into one of several local potential minima, and could be made to oscillate between them.

To summarize, we have trapped over 10^7 neutral atoms for over 2 min. We utilized a magnetic field to tune the atomic resonance, enabling radiation pressure to provide both cooling and damping forces. The confinement volume is several cubic centimeters and the effective depth is $\simeq 0.4$ K. The density of atoms is $\simeq 2 \times 10^{11}$ atoms cm^{-3} at a temperature of $\simeq 600$ μK.

We are extremely grateful to Jean Dalibard for giving us the seminal idea for this trapping scheme. This work was partially supported by the U.S. Office of Naval Research, Grant No. N00014-83-K-0695.

[a]Permanent address: Department of Physics and Research Laboratory of Electronics, Massachusetts Institute of Technology, Cambridge, MA 02139.

[b]Current address: Physics Department, Stanford University, Palo Alto, CA 94305.

[1]A. Migdall, J. Prodan, W. Phillips, T. Bergeman, and H. Metcalf, Phys. Rev. Lett. **54**, 2596 (1985).

[2]S. Chu, J. Bjorkholm, A. Ashkin, and A. Cable, Phys. Rev. Lett. **57**, 314 (1986).

[3]V. Bagnato, G. Lafyatis, A. Martin, E. Raab, and D. Pritchard, Phys. Rev. Lett. **58**, 2194 (1987).

[4]D. E. Pritchard, E. L. Raab, V. Bagnato, C. E. Wieman, and R. N. Watts, Phys. Rev. Lett. **57**, 310 (1986).

[5]S. Chu, M. Prentiss, J. Bjorkholm, and A. Cable, in *Laser Spectroscopy VIII*, edited by S. Swanberg and W. Pearson (Springer-Verlag, Berlin, 1987).

[6]S. Chu, L. Hollberg, J. Bjorkholm, A. Cable, and A. Ashkin, Phys. Rev. Lett. **55**, 48 (1985).

[7]Proceedings of International Laser Science Conference II, October, 1986 (to be published).

VOLUME 64, NUMBER 4   PHYSICAL REVIEW LETTERS   22 JANUARY 1990

Collective Behavior of Optically Trapped Neutral Atoms

Thad Walker, David Sesko, and Carl Wieman

*Joint Institute for Laboratory Astrophysics, University of Colorado and National Institute of Standards and Technology,
and Department of Physics, University of Colorado, Boulder, Colorado 80309-0440*

(Received 5 October 1989)

We describe experiments that show collective behavior in clouds of optically trapped neutral atoms.
This collective behavior is demonstrated in a variety of observed spatial distributions with abrupt bistable
transitions between them. These distributions include stable rings of atoms around a small core and
clumps of atoms rotating about the core. The size of the cloud grows rapidly as more atoms are loaded
into it, implying a strong long-range repulsive force between the atoms. We show that a force arising
from radiation trapping can explain much of this behavior.

PACS numbers: 32.80.Pj, 42.50.Vk

In the past few years there have been major advances
in the trapping and cooling of neutral atoms using laser
light. It has recently become possible to hold relatively
large samples of atoms for minutes at a time and cool
them to a fraction of a mK using a spontaneous-force optical trap.[1,2] One would expect these atoms to behave
like any other sample of neutral atoms, namely they
would move independently as in an ideal gas, except
when they undergo short-range collisions with other
atoms.

In this paper we report on experiments which show
that contrary to these expectations, optically trapped
atoms behave in a highly collective fashion under most
conditions. The spatial distribution of the cloud of atoms
is profoundly different from that of an ideal gas, and we
observe dramatic dynamic behavior in the clouds. We
believe that this unusual behavior arises from a long-
range repulsive force between the atoms which is the result of multiple scattering of photons by the trapped
atoms. This force is sensitive to the modification of the
emission and absorption profiles of the atom in the laser
field. In the past, this force has been neglected in any
system smaller and colder than stars. However, because
of the extremely low temperature and relatively high
densities of optically trapped atoms, it becomes very important.

The apparatus for this experiment has been described
elsewhere.[3,4] The trap was a Zeeman-shift spontaneous-force trap,[2] and was formed by the intersection
of three orthogonal retro-reflected laser beams. The
beams came from a diode laser tuned 5–10 MHz below
the $6S_{1/2}$ $F=4$ to $6P_{3/2}$ $F=5$ transition frequency of
cesium. The beams were circularly polarized with the
retro-reflected beams possessing the opposite circular polarization. A magnetic field was applied that was zero at
the center of trap and had a vertical field gradient (5–20
G/cm) 2 times larger than the horizontal field gradient.
Bunches of slowed cesium atoms from an atomic beam
were loaded into the trap by chirping a counterpropagating "stopping" laser at 20 times a second.[4] We were
able to load about 2×10^6 atoms per cooling chirp. The
number of atoms in the trap was determined by the load-

ing rate and the collisional loss[3] rate. Two other lasers
were used to deplete the $F=3$ hyperfine ground state in
both the trapped atoms and the atomic beam. All the
lasers were diode lasers with long-term stabilities and
linewidths well under 1 MHz.

The trapped atoms were observed by a photodiode,
which monitored the total fluorescence, and a charge-
coupled-device (CCD) television camera, which showed
the size and shape of the cloud of atoms. The optical
thickness of the cloud was determined by measuring the
absorption of a weak probe beam 1 ms after the trap
light was turned off. To measure the temperature of the
atoms, the probe beam was moved a few mm to the side
of the trap and the time-of-flight spectrum was observed
after the trap light was rapidly turned off, as described in
Ref. 5. By pushing the atoms with an additional laser,
as was done in Ref. 2, we determined that the trapping
potential was harmonic out to a radius of 1 mm with a
spring constant of 6 K/cm^2.

In this experiment, care was taken to obtain a symmetric trapping potential. The Earth's magnetic field
was zeroed to 0.01 G, and the trapping laser beams were
spatially filtered, collimated, and carefully aligned with
respect to the magnetic field zero. Before these precautions were taken we observed a variety of random shapes
and large density variations within the trap.

Three separate well-defined modes of the trapped
atoms were observed, depending on the number of atoms
in the trap and on how we aligned the trapping beams.
The "ideal-gas" mode occurred when the number of
trapped atoms was less than 40 000. The atoms formed a
small sphere with a constant diameter of approximately
0.2 mm. In this regime, the density of the sphere had a
Gaussian distribution, as expected for a damped harmonic potential, and increased linearly in proportion to the
number of atoms. The atoms in this case behaved just as
one would expect for an ideal gas with the measured
temperature and trap potential.

When the number of atoms was increased past 40 000,
the behavior deviated dramatically from an ideal gas,
and strong long-range repulsions between the atoms became apparent. In this "static" mode the diameter of

VOLUME 64, NUMBER 4   PHYSICAL REVIEW LETTERS   22 JANUARY 1990

the cloud smoothly increased with increasing numbers of atoms. Instead of a Gaussian distribution, the atoms in the trap were distributed fairly uniformly, as shown in Fig. 1(a). To study the growth of the cloud we measured the number of atoms versus the diameter of the cloud using a calibrated CCD camera. The results are shown in Fig. 2. To fully understand the growth mechanism the temperature as a function of diameter was also measured. For a detuning of 7.5 MHz, the temperature increased from the asymptotic value of 0.3 up to 1.0 mK as the cloud grew to a 3 mm diam. If the cloud were an ideal gas, the temperature increase would need to be orders of magnitude higher to explain this expansion. One distinctive feature of this regime is that the density did not increase as we added more atoms to the trap.

When all the trapping beams were reflected exactly back on themselves we obtained a maximum of 3.0×10^8 atoms in this distribution. In contrast, if the beams were slightly misaligned in the horizontal plane, we observed unexpected and dramatic changes in the distribution of the atoms. The atoms would collectively and abruptly jump to "orbital" modes when the cloud contained approximately 10^8 atoms. The number of atoms needed for a transition depended on the degree of misalignment.

The shapes of these orbital modes are illustrated in Fig. 1 [1(b), 1(c), and 1(d) show the top view, and 1(e) shows the view from the side]. We first observed the ring around a center shown in Fig. 1(b) and then, by strobing the camera, discovered it was actually a clump of atoms orbiting counterclockwise about a central ball

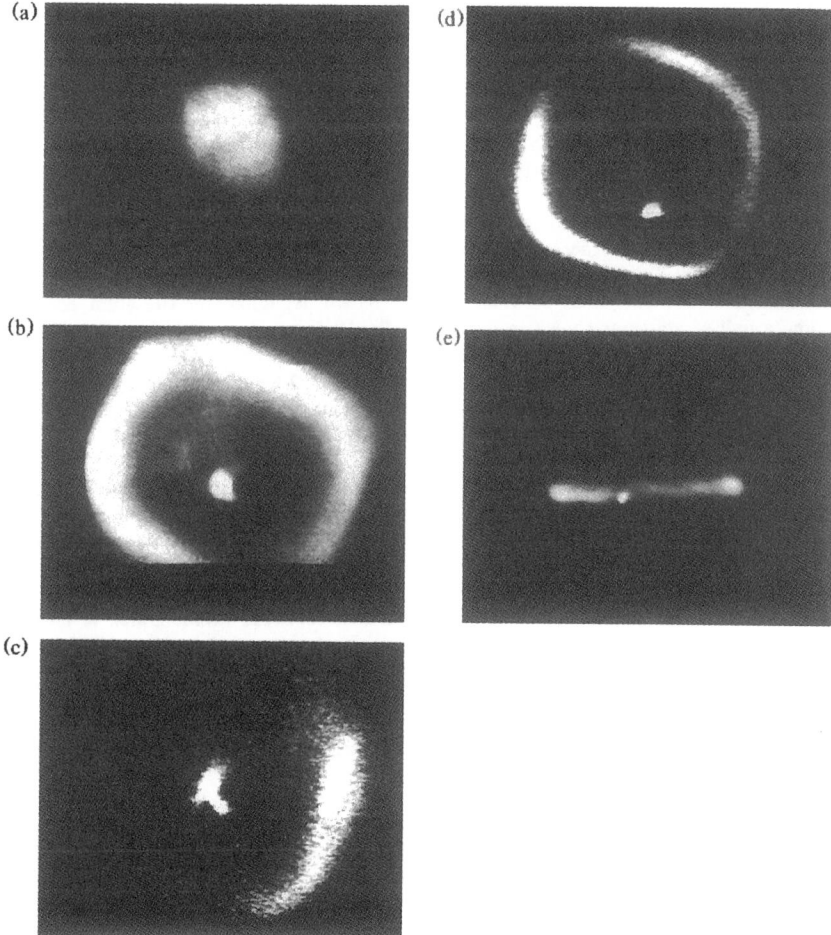

FIG. 1. Spatial distributions of trapped atoms. (a) Below 10^8 atoms the cloud forms a uniform density sphere. (b) Top view of rotating clump of atoms without strobing. (c) Top view of (b) with the camera strobed at 110 Hz. (d) Top view of a continuous ring. (e) Side view of (d). Horizontal full scale for (a), (d), and (e) is 1.0 cm; for (b) and (c) it is 0.8 cm.

VOLUME 64, NUMBER 4   **PHYSICAL REVIEW LETTERS**   22 JANUARY 1990

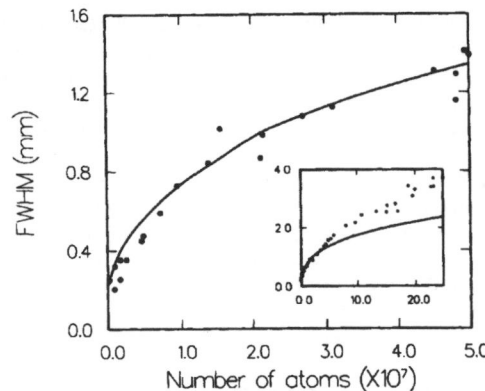

FIG. 2. Plot of the diameter (FWHM) of the cloud of atoms as a function of the number of atoms contained in the cloud. For the full figure the magnetic field gradient is 9 G/cm and 16.5 G/cm for the inset. The laser detuning is -7.5 MHz, and the total laser intensity is 12 mW/cm^2. The solid lines show the predictions of the model described in the text.

as shown in Fig. 1(c). Note that there is a tenuous connection between the clump and the center, and that the core has an asymmetry which rotates around with the clump. As more atoms were slowly added to the trap there were several abrupt increases in the radius of these orbits. The clumps orbited at well-defined frequencies between 80 and 130 Hz. We have also observed a continuous ring encircling a ball of atoms as shown in Fig. 1(d). In this case we assume that the atoms are still orbiting, but no clumping of the atoms was seen and there was no connection between the ring and the central ball. We observed only one stable radius (2.5 mm) for this ring. The ring is shaped by the intersection geometry of the laser beams with the corners of the ring corresponding to the centers of the beams. The atoms lie in a horizontal plane about 0.5 mm thick [Fig. 1(e)].

As mentioned above, the formation of these stable orbital modes depended on the alignment of the trapping beams in the horizontal plane. The beams had a Gaussian width of 6 mm and the return beams were misaligned horizontally by 1–2 mm (4–8 mrad) at the trap region. It was observed that the direction of the rotation corresponded to the torque produced by the misalignment. The degree of the misalignment affected which of the orbital modes would occur. Typically, the atoms would switch into the rotating clump for misalignments of 1–1.5 mm and into the continuous ring for 1.5–2 mm. If the light beams were misaligned even further, the cloud would be in the ring mode for any number of atoms.

The rotational frequencies of the atoms were measured by strobing the image viewed by the camera or by observing a sinusoidal modulation of the fluorescence of the atoms on a photodiode. The frequency depended on the detuning of the trapping laser, the magnetic field gradient, and the average intensity of the trapping light.

410

For each of these, the frequency of the rotation increased as the spring constant of the trap was increased. We also discovered we could induce the atoms to jump into rotating clumps by modulating the magnetic field gradient at a frequency within 8 Hz of the rotational frequency.

The transition between the static and orbital modes showed pronounced hysteresis. When the number of atoms was slowly increased until it reached 10^8, the cloud would jump in <20 msec from the static mode to the rotating clump and the trap would then lose approximately half its atoms in the next 50–100 msec. This orbital mode would remain stable until there was a small fluctuation in the loading rate. Then the cloud would suddenly switch back to the static mode without losing any atoms. The number of atoms would have to build up to 10^8 again before it would jump back into the orbital mode. We have also observed hysteresis between the static and orbital modes when the trap would jump into the continuous ring. The trap started with 10^8 atoms in the static mode and then 80% of these atoms would jump into a continuous ring with the remaining 20% in the central ball. For certain loading rates the cloud would repeatedly switch back and forth between the two modes.

The expansion of the cloud with increasing numbers of atoms and the various collective rotational behaviors clearly show that strong, long-range forces dominate the behavior of the atomic cloud. Since normal interatomic forces are negligible for the atom densities in the trap $(10^{10}$–10^{11} cm$^{-3})$, other forces must be at work. In particular, since the optical depth for the trap lasers is on the order of 0.1 for our atom clouds (3 for the probe absorption at the resonance peak), forces resulting from attenuation of the lasers or radiation trapping can be important. The attenuation force is due to the intensity gradients produced by absorption of the trapping lasers, and has been discussed by Dalibard in the context of optical molasses.[6] This force compresses the atomic cloud and so cannot explain our observations. In the following, we demonstrate that radiation trapping produces a repulsive force between atoms which is larger than this attenuation force and thus causes the atomic cloud to expand. We will also show how radiation trapping leads to the rotational orbits we observe.

The attenuation force, $\mathbf{F}_A$, for small absorption, obeys the relation

$$\nabla \cdot \mathbf{F}_A = -6\sigma_L^2 I_\infty n/c. \tag{1}$$

The atom density is n, the cross section for absorption of the laser light is σ_L, and the incident intensity of a single laser beam is I_∞.

The absorbed photons must subsequently be reemitted and can then collide with other atoms. Although this was neglected in Ref. 6, this reabsorption of the light (radiation trapping) provides a repulsive force, $\mathbf{F}_R$, which is larger than $\mathbf{F}_A$. Two atoms separated a dis-

VOLUME 64, NUMBER 4   PHYSICAL REVIEW LETTERS   22 JANUARY 1990

tance d in a laser field of intensity I repel each other with a force

$$|\mathbf{F}_R| = \sigma_R \sigma_L I / 4\pi c d^2 , \qquad (2)$$

where for simplicity we have assumed isotropic radiation and unpolarized atoms. (The observed fluorescence is unpolarized.) The cross section σ_R for absorption of the scattered light is in general different than σ_L due to the differing polarization and frequency properties of the scattered light.[7] For a collection of atoms in an optical trap, the radiation trapping force obeys

$$\nabla \cdot \mathbf{F}_R = 6\sigma_L \sigma_R I_\infty n / c . \qquad (3)$$

Comparison of (1) and (3) shows that the net force between atoms due to laser attenuation and radiation trapping is repulsive when $\sigma_R > \sigma_L$. In the limit that the temperature can be neglected, $\mathbf{F}_R + \mathbf{F}_A$ must balance the trapping force $-k\mathbf{r}$. This implies a maximum achievable density in the trap of $n_{max} = ck / 2\sigma_L(\sigma_R - \sigma_L) I_\infty$.

This allows us to explain the behavior shown in Fig. 1. Once the density reaches n_{max}, increasing the number of atoms only produces an increase in the size of the atomic cloud. To compare this model to experiment we numerically solve the above equations with the finite temperature taken into account. The only free parameter in the calculation is $\sigma_R/\sigma_L - 1$. A value of $\sigma_R/\sigma_L - 1 = 0.3$ gives the solid curve shown in Fig. 2, which agrees well with experiment.

We have estimated σ_R/σ_L by convoluting the emission and absorption profiles for a two-level atom[7] in a 1D standing wave field. The intense laser light causes ac-Stark shifts, giving rise to the well-known Mollow triplet emission spectrum. The absorption profile is also modified by the laser light, being strongly peaked near the blue-shifted component of the Mollow triplet, causing the average absorption cross section for the emitted light to be greater than that for the laser light. We calculate $\sigma_R/\sigma_L - 1 = 0.2$ for our detuning and intensity ($I_\infty = 2$ mW/cm^2). While this is less than the value of 0.3 needed to match the data in Fig. 2, it is very reasonable that our twenty-level atom in a 3D field should differ from our simple estimate by this amount.

The inset of Fig. 2 shows that the expansion of the cloud deviates from this model for sizes greater than 1.5 mm. However, the precise behavior in this region is very sensitive to alignment and is not very reproducible. Other effects not included in the above model, such as magnetic field broadening, optical pumping, multiple levels of the atom, the spatial dependence of the laser fields, and multiple (> 2) scattering of the light may also be important in this region.

We can explain the rings of Fig. 1 by considering the motion of atoms in the x-y plane when the laser beams are misaligned. A first approximation to the force produced by this misalignment is $\mathbf{F}_I = k'\hat{\mathbf{z}} \times \mathbf{r}$. The atoms are also subject to a harmonic restoring force $-k\mathbf{r}$ and a

damping force $-\gamma d\mathbf{r}/dt$. If we neglect radiation trapping, we find circular orbits can exist with angular frequency $\omega = k'/\gamma$, if $k'/\gamma = (k/m)^{1/2}$. Since the orbits may have any radius, this clearly does not explain the formation of rings. If, however, we add a force $(\alpha N/r^2)\hat{\mathbf{r}}$ due to the radiation pressure from a cloud of N atoms within the orbit radius R, we find circular orbits exist only for

$$R = \left[\frac{\alpha N}{k - m\omega^2} \right]^{1/3} , \qquad (4)$$

if $\omega < (k/m)^{1/2}$. Thus the effect of the radiation from the inner cloud of atoms is to cause circular orbits at a particular radius (i.e., rings are formed), and to allow circular orbits for a large range of misalignments. This is in accord with our observations that the orbits always encompass a small ball of atoms (Fig. 1). We have done a more detailed calculation including the Gaussian profiles of the lasers which gives a rotational frequency of 130 Hz and an orbit diameter of 3.5 mm for the conditions of Fig. 1(d).

While the simple model of the radiation trapping force we have presented explains the expansion of the cloud and the existence of the circular rings, there are several interesting aspects of the observed phenomena which are not explained. These include the formation of rotating clumps of atoms and the dynamics of the transitions between different distributions.

We have produced dense cold samples of optically trapped atoms. We find that this unique new physical system shows a fascinating array of unexpected collective behavior at densities orders of magnitude below where such behavior was expected. Much of this collective behavior can be explained by the interaction between the atoms in the trap due to their radiation fields. However, the detailed distributions and the transition dynamics deserve considerable future study.

This work was supported by the Office of Naval Research and the National Science Foundation. We are pleased to acknowledge helpful discussions with T. Mossberg and the contributions made by C. Monroe and W. Swann to the experiment.

[1] D. Pritchard, E. Raab, V. Bagnato, C. Wieman, and R. Watts, Phys. Rev. Lett. 57, 310 (1986).

[2] E. L. Raab, M. G. Prentiss, A. E. Cable, S. Chu, and D. E. Pritchard, Phys. Rev. Lett. 59, 2631 (1987).

[3] D. Sesko, T. Walker, C. Monroe, A. Gallagher, and C. Wieman, Phys. Rev. Lett. 63, 961 (1989).

[4] D. Sesko, C. G. Fan, and C. Wieman, J. Opt. Soc. Am. B 5, 1225 (1988).

[5] P. D. Lett et al., Phys. Rev. Lett. 61, 169 (1988).

[6] J. Dalibard, Optics Commun. 68, 203 (1988).

[7] B. R. Mollow, Phys. Rev. 188, 1969 (1969); Phys. Rev. A 5, 2217 (1972); D. A. Holm, M. Sargent, III, and S. Stenholm, J. Opt. Soc. Am. B 2, 1456 (1985).

VOLUME 70, NUMBER 15   PHYSICAL REVIEW LETTERS   12 APRIL 1993

High Densities of Cold Atoms in a *Dark* Spontaneous-Force Optical Trap

Wolfgang Ketterle, Kendall B. Davis, Michael A. Joffe, Alex Martin, [a] and David E. Pritchard

Department of Physics and Research Laboratory of Electronics, Massachusetts Institute of Technology,
Cambridge, Massachusetts 02139
(Received 28 December 1992)

A new magneto-optical trap is demonstrated which confines atoms predominantly in a "dark" hyperfine level, that does not interact with the trapping light. This leads to much higher atomic densities as repulsive forces between atoms due to rescattered radiation are reduced and trap loss due to excited-state collisions is diminished. In such a trap, more than 10^{10} sodium atoms have been confined to densities approaching 10^{12} atoms cm^{-3}.

PACS numbers: 32.80.Pj

Although the original suggestion [1] that spontaneous light forces could be used to trap atoms included several general ways to do this, the development of the magneto-optical trap (MOT) [2-4] opened the way to the practical use of slow atoms in several different types of experiments involving cold collisions, quantum optics, and atom interferometers [5]. Recently, there has been a resurgence of interest in light traps which offer the possibility of containing polarized atoms [6] or higher density samples [7,8]. Overcoming the density limit of $\sim 10^{11}$ atoms/cm^3 in a MOT may open the way to study collective effects like Bose-Einstein condensation and spin waves, and free-bound transitions in long-range molecules.

The density limit is set by two processes: First, by collisions between ground- and excited-state atoms in which part of the excitation energy can be transformed into kinetic energy, resulting in a trap loss rate per atom βn with $\beta \approx (1-5) \times 10^{11}$ cm^3/s [9]. For densities n approaching 10^{11} atoms/cm^3, the loading time of the trap is limited to less than 1 s. The second limit is due to repulsive forces between the atoms caused by reabsorption of scattered photons (radiation trapping) [10]. At a certain atomic density, the outward radiation pressure of the fluorescence light balances the confining forces of the trapping laser beams. Further increase of the number of trapped atoms leads to larger atom clouds, but not to higher densities. As a practical matter, the power of the rescattered light sets a limit to the *number* of atoms which can be confined in a magneto-optical trap: 10^{11} atoms scatter about 100 mW of near-resonant laser light.

In this paper, we demonstrate a dark spontaneous-force optical trap ("dark SPOT"), in which all the above-mentioned limitations are mitigated by confining the atoms mainly in a ("dark") hyperfine ground state which does not interact with the trapping light. The key idea is that optimum confinement of atoms is not necessarily achieved with the maximum light force because of the limitations mentioned above. Light forces which are orders of magnitude smaller than the saturated scattering force are still strong enough to confine atoms tightly, e.g., 1 m/s sodium atoms (corresponding to a temperature of 1

mK) can be stopped in a distance of 100 μm at 1% saturation. All spontaneous-light-force traps realized so far have operated with close to saturated excitation, whereas our dark SPOT works at scattering rates 2 orders of magnitude smaller.

The simple model used to explain the density limit in a MOT [10] is readily generalized to include a "dark" and a "bright" hyperfine ground state. The trapping force is $\mathbf{F}_T = -kpr\hat{\mathbf{r}}$, where p denotes the probability that the atom is in the bright hyperfine state and k the spring constant of the normal MOT (i.e., for $p=1$). Attenuation of the trapping light and radiation trapping give rise to a density-dependent repulsive force which is quadratic in p because it involves two scattering events: $\mathbf{F}_R = k(n/n_0)p^2 r\hat{\mathbf{r}}$, where n_0 is a constant. From the stability criterion $|\mathbf{F}_T| > |\mathbf{F}_R|$, one obtains one limit for the maximum atom density in a MOT: $n < n_0/p$. For a very large number of atoms, the *column density* of atoms is limited by the fact that the atom cloud of diameter d has to be transparent for the trapping light [11], i.e., $ndp < b_0$, where b_0 is a constant. Substituting $d^3 = N/n$, one obtains a second limit for n: $n < (b_0/p)^{3/2} N^{-1/2}$. Finally, for small p, the density is limited by the fact that the volume of low-density gas at fixed temperature varies as $p^{-3/2}$, as the spring constant of confinement is proportional to p. This results in a third limit to the atomic density: $n < Np^{3/2}/d_0^3$ (d_0 is the cloud diameter in a standard MOT for low N). The constants n_0, b_0, and d_0 depend on experimental parameters and are typically 5×10^{10} cm^{-3}, 5×10^9 cm^{-2}, and 200 μm [10].

In a simplified model, the atom density in a MOT is the largest value compatible with the three limits as shown in Fig. 1. The value of p which maximizes density depends on N and is smaller for larger N. The Stanford group [12] and our group have recently succeeded in trapping more than 10^{10} atoms in a normal MOT. For such an N, the predicted optimum p of ~ 0.01 corresponds to a density increase of more than 2 orders of magnitude over the normal MOT (Fig. 1).

In the case of sodium, the bright and dark hyperfine states are the $F=2$ and $F=1$ hyperfine levels of the $3S_{1/2}$ ground state, respectively. Spontaneous light forces are

VOLUME 70, NUMBER 15   PHYSICAL REVIEW LETTERS   12 APRIL 1993

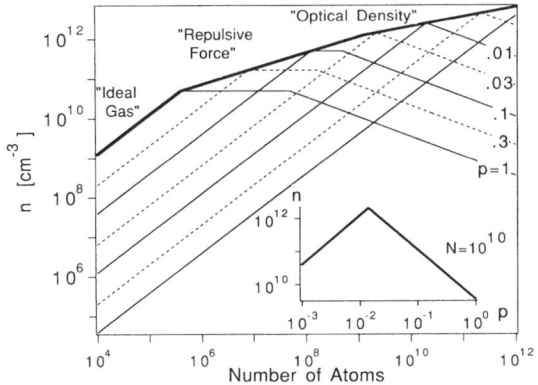

FIG. 1. Atomic densities n vs number N of trapped atoms for different values of the fractional population p of the "bright" hyperfine state. The thick line shows the highest density obtainable with an optimum value of p. For large N, this density is much higher than the one obtained in the normal MOT ($p=1$). For 10^{10} trapped atoms (inset), the optimum trap is 100 times "darker" than the normal MOT, resulting in more than 2 orders of magnitude increase in density.

applied using the cycling $F=2\rightarrow F'=3$ transition to the excited $3P_{3/2}$ state. As a result of nonresonant excitation of the $2\rightarrow 2$ transition, atoms are optically pumped into the $F=1$ ground state via a spontaneous Raman process. In all optical cooling and trapping experiments it has been necessary therefore to add repumping light resonant with the $1\rightarrow 2$ transition to pump atoms back to the $F=2$ state. Usually, the intensity of the repumping light has been high enough to keep the atoms mostly in $F=2$. In a dark SPOT with $p=0.01$, the atoms spend most of their time ($\sim99\%$) in $F=1$; this is accomplished by appropriately reducing the intensity of the repumping light.

Although a small excitation rate is superior for confining large numbers of atoms at high density, the maximum possible excitation rate is necessary to efficiently capture atoms from a thermal background or a slow atom beam, and load them into the trap. Therefore, a dark SPOT requires a "bright" capturing region which is separated from the dark trap spatially or temporally. In the bright region, the sodium atoms are mainly in the $F=2$ level and experience the maximum light force. The temporal separation is accomplished by loading atoms into a normal MOT and then switching to a dark trap, and will be discussed later. Spatial separation is superior since it allows continuous loading of atoms into a dark trap. This was accomplished by using a normal MOT and applying only weak or no repumping light to the center of the trap ("a MOT with a dark spot").

In our experimental setup, a crucial part was a slow-atom source employing an increasing-field Zeeman slower [13] capable of producing $>10^{12}$ sodium atoms/s at 100 m/s and $\sim10^{11}$ atoms/s at 30 m/s [14]. With this slower,

$\gtrsim10^{10}$ atoms/s could be loaded into our MOT. The trap consisted of three orthogonal retroreflected beams with diameters of ~3 cm and intensities of ~10 mW/cm^2 per beam. The frequency was tuned to the red of the $2\rightarrow3$ transition by 15–25 MHz. All beams were circularly polarized with helicities appropriate for magneto-optical trapping; they intersected at the center of a quadrupole magnetic field with a gradient of 10–15 G/cm. Repumping light close to resonance with the $1\rightarrow2$ transition was passed through a glass plate with a black dot, which was imaged into the trap center with an image size of ~10 mm. With a second similar repumping beam (diameter 3 cm, intensity 3 mW/cm^2) at an angle of $\sim20°$ to the first, the whole trapping region was efficiently repumped except for the center, where the dark regions of the two beams intersected.

Additional repumping light could be added to the trapping laser beams by means of an electro-optical modulator (EOM) operated at 1.71 GHz. With EOM sidebands of variable intensity, p could be smoothly varied between a value p_{min} and ~1. p_{min} was determined by two processes: (i) stray light from the repumping beams scattered by windows and the atomic beam, and (ii) spontaneous Raman transitions induced by the trapping light. With an estimated rate of $\sim10^3$ s^{-1}, the Raman process alone should cause $\sim0.1\%$ equilibrium population in $F=2$. As it turned out that p_{min} was close to the optimum value of p (see Fig. 1), most of the experiments were done without EOM sidebands.

In a normal MOT we observed a cloud of atoms ~1 cm in diameter containing roughly 10^{10} atoms. In a dark SPOT, one could clearly see a dark central region in the fluorescence of the thermal beam corresponding to the dark spots in the repumping beams. In the center was a compact ball of trapped atoms (2–4 mm in diameter) with an apparent brightness lower than the normal MOT, but still much brighter than the background fluorescence.

The density in the dark SPOT was determined by absorption spectroscopy using a weak probe beam with an intensity of ~1 μW/cm^2 and a photodiode. The probe laser beam was split off the trapping or repumping light and could be scanned by ±120 MHz with two acousto-optic modulators. Figure 2 shows an absorption spectrum of the trapped atoms. As at the highest densities the excited-state hyperfine structure could no longer be resolved, optical densities were deduced by fitting a theoretical spectrum to the one observed. The diameter of the cloud of trapped atoms was determined by imaging the fluorescence onto a charge-coupled-device camera or by recording spatially resolved absorption giving the same result.

An independent determination of the number of trapped atoms was performed by switching off the trapping and repumping beams and rapidly switching on a strong probe laser beam (diameter 10 mm, 0.5 mW/cm^2) close to resonance with $F=1$ atoms, optically pumping them into the $F=2$ state. From the transient absorption

VOLUME 70, NUMBER 15   PHYSICAL REVIEW LETTERS   12 APRIL 1993

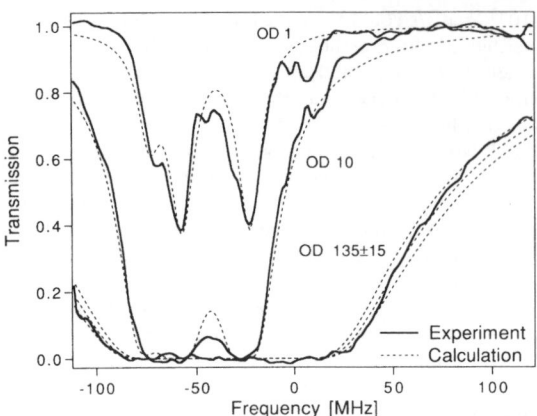

FIG. 2. Absorption spectrum of a 4 mm diam cloud of sodium atoms trapped in a dark SPOT. The best fit yields an optical density (OD) of 135 which corresponds to an atomic density of 7×10^{11} cm^{-3} and $\sim 5 \times 10^{10}$ trapped atoms. Traces with lower OD were recorded with a reduced number of atoms. The dashed lines are calculated spectra for OD = 1, 10, 120, 135, and 150, respectively.

signal, the number of absorbed photons was obtained. This number, divided by the number q of photons needed to optically pump one atom, gives directly the number of atoms in the trap. q was obtained from a knowledge of the matrix elements and a simple model of radiation trapping. The number of trapped atoms deduced should be accurate to within a factor of 2.

The highest density was observed for a 3.0 ± 0.5 mm cloud of atoms with an optical density of 110 ± 10 implying a density $n = (8 \pm 2) \times 10^{11}$ cm^{-3}. Another (less accurate) value for n derived from the number of atoms $N = 1.5 \times 10^{10}$ and the diameter agreed to within 30%. In slightly larger clouds, optical densities up to 160 and 5×10^{10} trapped atoms were observed. This combination of number and density is unprecedented in light traps: Slightly smaller densities have been reported to date only for at least 100 times fewer atoms [3,7,10,15]. A similar number of atoms has been trapped by light forces only at densities 20 times lower [12]. In some of the previous work, large detunings were used [12,15,16] to reduce the reabsorption of scattered photons. This and other ways to affect n_0, b_0, and d_0 may eventually be combined with our approach of reduced repumping to achieve even higher densities.

The maximum optical density (~ 2) observed when probing atoms in the bright state ($F = 2$) was only about 2 times smaller than the value obtained in a bright MOT for the same experimental parameters (except for the dark spot in the repumping light). This shows that eventually the dark SPOT has similar limitations as the normal MOT, but at much higher densities. The direct comparison between dark SPOT and bright MOT showed a

density ratio of ~ 100 in agreement with Fig. 1. The temperature of the trapped atoms was determined by switching off the trapping light and recording the decay of the absorption signal. The temperature of 1.2 ± 0.5 mK is in agreement with the prediction of ~ 0.8 mK for Doppler molasses [17] at the detuning and laser intensities used. Trap loading and decay times were 1 to 2 s and limited by collisions with the thermal atomic beam rather than with the residual gas at a pressure of 10^{-9} Torr. The long trapping time clearly demonstrates the reduced trap loss by excited-state collisions in a dark SPOT since extrapolating trap losses observed in a normal MOT [9] to our densities implies trapping times of only ~ 50 ms.

The repumping of the trapped atoms was mainly due to light scattered from the repumping beams into the "dark" trapping region. An absorption spectrum for $F = 2$ atoms showed that about 1% of the atoms were in $F = 2$, close to the optimum p predicted for $\sim 10^{10}$ trapped atoms. To vary the population in $F = 2$, we switched off the repumping light and switched on EOM generated sidebands of the trapping light with variable intensity. In agreement with the prediction of Fig. 1, a maximum in atomic density was found for a sideband intensity of about 0.1%. The traps with very weak repumping showed larger trap loss, probably due to the smaller potential well depth and therefore increased radiative escape rate [18]. [Since the trapped atoms are mainly in the lowest hyperfine state ($F = 1$), inelastic hyperfine changing collisions cannot account for the extra loss rate observed.] When atoms were loaded into a normal MOT and the intensity of the repumping light was rapidly decreased, a considerable increase in density was observed ("temporal" separation between bright capturing and dark trapping). However, the densities thus obtained were 30% lower than those achieved by loading directly into a dark trap. The probable reason is larger trap loss during the loading phase due to both collisions and leakage of atoms out of the trap because of imperfect beam profile (one could clearly see rays of atoms leading out of the bright trap). It should be noted that the trap works even without any repumping light at all—sufficient repumping is provided by spontaneous Raman scattering of the trapping light. This offers an alternative explanation for the trapping of Rb in a MOT without repumping light reported recently [19].

The realization of a dark SPOT for other alkali atoms seems rather straightforward. In Cs (or Rb) the atom cycles many times on the bright transition before it falls into the dark $F = 3$ state due to the very large hyperfine splittings. This cycling time could be shortened by using weak additional "depumping" light in resonance with the $4 \rightarrow 3$ or $4 \rightarrow 4$ transitions. The smaller rate of spontaneous Raman repumping might allow observation of trapping in a square well potential (*bounce trap*), where the atoms move freely in the central (nonrepumped) region of the trap (having $p \approx 0$) and are reflected at the boundary with the outer (repumped) region. For very weak

VOLUME 70, NUMBER 15   PHYSICAL REVIEW LETTERS   12 APRIL 1993

repumping the transition from the usual case of strong overdamping (damping rate $\alpha \gg \omega$, the oscillation frequency) to the oscillatory regime could be observed because $\alpha \propto p$ and $\omega \propto \sqrt{p}$. Generally, it appears that a larger hyperfine splitting is advantageous for the dark SPOT because off-resonant optical pumping processes are less important and the dwell times of the atom in the bright and dark states can be controlled independently by applying additional laser frequencies.

The high densities achieved in a dark SPOT are promising for the study of cold collisions and for the observation of evaporative cooling after transferring the atoms into a magnetic trap. At densities of 10^{12} cm^{-3}, the estimated elastic collision rate is already 100 s^{-1}, much larger than the trap loss rate due to collisions with the background gas.

The dark SPOT is the first cooling and trapping scheme in which the repumping light is intentionally reduced to "shelve" the atoms, i.e., cooling and trapping forces are only exerted on a small fraction of the atoms, while most of the atoms are kept in the dark, thus avoiding strong absorption of the cooling light. This concept should allow polarization-gradient cooling of trapped atoms below higher ultimate temperatures observed at high atomic densities [15]. Another possibility for realizing a dark trap would be repumping on the $1 \to 1$ transition of the D_1 line with elliptically polarized light. This transition has a coherent dark state only for magnetic fields $B = 0$ [20] which inhibits repumping in the center of the trap. Recently, a new scheme in velocity-selective coherent population trapping (VSCPT) has been suggested which, in addition to the momentum diffusion process, features a weak damping force towards low velocities [21]. A simple way of combining strong damping and VSCPT would be the use of polarization-gradient molasses acting on the bright hyperfine state together with a velocity-selective repumping scheme. Alternatively, polarized cold atoms could be obtained by using a repumping scheme which does not repump atoms from a certain m_F level (e.g., σ^+ light and a 1-1 transition). This could be implemented in the recently demonstrated vortex-force trap which confines polarized atoms at nonvanishing magnetic field [6]. Finally, a tapered two-dimensional version of the dark SPOT, a dark funnel [22], should allow the compression of intense slow atomic beams to unprecedented brightness.

In conclusion, we have demonstrated a dark spontaneous-force optical trap, which confines atoms predominantly ($\sim 99\%$) in a dark hyperfine ground state. In this way, limitations of the normal magneto-optical trap have been overcome and densities close to 10^{12} cm^{-3} for more than 10^{10} trapped atoms have been achieved.

We would like to acknowledge experimental assistance from M. Mewes. This work was supported by ONR and AFOSR through Contract No. N00014-90-J-1642, and by NSF Grant No. 8921769-PHY. W.K. and A.M. would like to acknowledge fellowships from the NATO Science Committee and DAAD, Germany, and from the DGICYT, Spain, respectively.

(a)On leave from Instituto de Optica, CSIC, Madrid, Spain.

[1] D. E. Pritchard et al., Phys. Rev. Lett. **57**, 310 (1986).

[2] D. E. Pritchard and E. L. Raab, in *Advances in Laser Science II*, edited by M. Lapp, W. C. Stwalley, and G. A. Kenney-Wallace (AIP, New York, 1987), p. 329.

[3] E. L. Raab et al., Phys. Rev. Lett. **59**, 2631 (1987).

[4] C. Monroe, W. Swann, H. Robinson, and C. Wieman, Phys. Rev. Lett. **65**, 1571 (1990).

[5] "Laser Manipulation of Atoms and Ions," Proceedings of the Varenna Summer School, edited by E. Arimondo and W. D. Phillips (North-Holland, Amsterdam, to be published).

[6] T. Walker, P. Feng, D. Hoffmann, and R. S. Williamson III, Phys. Rev. Lett. **69**, 2168 (1992).

[7] O. Emile, F. Bardou, and C. Salomon (to be published).

[8] W. D. Phillips, in "Laser Manipulation of Atoms and Ions" (Ref. [5]); D. J. Heinzen, J. D. Miller, and R. A. Cline, in The Thirteenth International Conference on Atomic Physics, Munich, 1992, Book of Abstracts, Paper C4.

[9] M. Prentiss et al., Opt. Lett. **13**, 452 (1988); L. Marcassa et al. (to be published).

[10] T. Walker, D. Sesko, and C. Wieman, Phys. Rev. Lett. **64**, 408 (1990).

[11] K. Lindquist, M. Stephens, and C. Wieman, Phys. Rev. A **46**, 4082 (1992).

[12] K. E. Gibble, S. Kasapi, and S. Chu, Opt. Lett. **17**, 526 (1992).

[13] T. E. Barrett, S. W. Dapore-Schwartz, M. D. Ray, and G. P. Lafyatis, Phys. Rev. Lett. **67**, 3483 (1991).

[14] M. A. Joffe, W. Ketterle, A. Martin, and D. E. Pritchard, in The Thirteenth International Conference on Atomic Physics, Munich, 1992, Book of Abstracts, Paper C9.

[15] A. Clairon et al., in Proceedings of the Sixth European Time and Frequency Forum, Noordwijk, Netherlands, 1992, edited by J. J. Hunt (to be published).

[16] E. A. Cornell and C. R. Monroe (private communication).

[17] P. D. Lett et al., J. Opt. Soc. Am. B **6**, 2084 (1989).

[18] P. S. Julienne and J. Vigué, Phys. Rev. A **44**, 4464 (1991).

[19] P. Kohns et al., in International Conference on Quantum Electronics, 1992, Technical Digest Series, Vol. 9, p. 258.

[20] A. M. Tumaikin and V. I. Yudin, Zh. Eksp. Teor. Fiz. **98**, 81 (1990) [Sov. Phys. JETP **71**, 43 (1990)].

[21] F. Mauri and E. Arimondo, Europhys. Lett. **16**, 717 (1991).

[22] E. Riis, D. S. Weiss, K. A. Moler, and S. Chu, Phys. Rev. Lett. **64**, 1658 (1990); J. Nellessen, J. Werner, and W. Ertmer, Opt. Commun. **78**, 300 (1990).

letters to nature

Four-wave mixing
with matter waves

**L. Deng*†, E. W. Hagley*, J. Wen*, M. Trippenbach*‡,
Y. Band*‡, P. S. Julienne*, J. E. Simsarian*, K. Helmerson*,
S. L. Rolston* & W. D. Phillips***

* *Atomic Physics Division, National Institute of Standards and Technology,
Gaithersburg, Maryland 20899, USA*
† *Department of Physics, Georgia Southern University, Statesboro, Georgia 30460,
USA*
‡ *Department of Chemistry and Physics, Ben-Gurion University of the Negev,
Beer-sheva 84105, Israel*

The advent of the laser as an intense source of coherent light gave
rise to nonlinear optics, which now plays an important role in
many areas of science and technology. One of the first applications
of nonlinear optics was the multi-wave mixing[1,2] of several optical
fields in a nonlinear medium (one in which the refractive index
depends on the intensity of the field) to produce coherent light of a
new frequency. The recent experimental realization of the matter-
wave 'laser'[3,4]—based on the extraction of coherent atoms from a
Bose–Einstein condensate[5]—opens the way for analogous experiments with intense sources of matter waves: nonlinear atom
optics[6]. Here we report coherent four-wave mixing in which
three sodium matter waves of differing momenta mix to produce,
by means of nonlinear atom–atom interactions, a fourth wave
with new momentum. We find a clear signature of a four-wave
mixing process in the dependence of the generated matter wave on
the densities of the input waves. Our results may ultimately
facilitate the production and investigation of quantum correlations
between matter waves.

The analogy between nonlinear optics with lasers and nonlinear
atom optics with Bose–Einstein condensates can be seen in the
similarities between the equations that govern each system. For a
condensate of interacting bosons, in a trapping potential V, the
macroscopic wavefunction Ψ satisfies a nonlinear Schrödinger
equation[7]

$$i\hbar\frac{\partial\Psi}{\partial t} = \left(-\frac{\hbar^2}{2M}\nabla^2 + V + U_0|\Psi|^2 \right)\Psi \qquad (1)$$

where M is the atomic mass, U_0 describes the strength of the atom–
atom interaction ($U_0 > 0$ for sodium atoms), and $|\Psi|^2$ is propor-
tional to atomic number density. The nonlinear term $U_0|\Psi|^2\Psi$ in
equation (1) is similar to the third order term $\chi^{(3)}|E|^2E$ in the wave
equation for the electric field E describing optical four-wave mixing
(4WM; where the susceptibility $\chi^{(3)}$ depends on the nonlinear
medium). We therefore expected 4WM with coherent matter
waves, analogous to optical 4WM. In contrast to optical 4WM,
the nonlinearity in matter-wave 4WM comes from atom–atom
interactions; there is no need for an additional nonlinear medium.

The first theoretical study of nonlinear atom optics was reported
in 1993[6], and the idea of 4WM using condensates prepared in
different electronic states to enhance the nonlinearity was discussed
in 1995[8]. A recent calculation[9] showed that the nonlinearity associated with the interaction between ground-state atoms is large
enough to observe 4WM with wavepackets created from existing
Bose–Einstein condensates. To produce matter-wave mixing, we
create three overlapping wavepackets with momenta $\mathbf{P}_n$ ($n = 1, 2, 3$)
and observe the creation of the 4WM wavepacket $\mathbf{P}_4$ that satisfies
energy, momentum and particle-number conservation (Fig. 1).

In our experiment, we use Bragg diffraction of atoms from a
moving optical standing wave[10] to create the necessary three
wavepackets, starting from a Bose–Einstein condensate. Briefly,
we first form a condensate of $\sim 2 \times 10^6$ sodium atoms in the

$3S_{1/2}$, $F = 1$, $m = -1$ state using a combination of laser cooling and radio-frequency-induced evaporative cooling in a TOP (time-orbiting-potential) trap[11], without a discernible non-condensed fraction. We then adiabatically expand the potential[10] in 4 s by simultaneously reducing the magnetic field gradient and increasing the rotating bias field. This reduces the trap frequencies in the $\hat{x}$, $\hat{y}$ and $\hat{z}$ directions to 84, 59 and 42 Hz, respectively. The asymptotic r.m.s. momentum width of the released condensate after adiabatic expansion is measured to be $0.14(\pm 0.02)\hbar k$ (all uncertainties reported here are one standard deviation combined statistical and systematic uncertainties). Because this is small compared to $\sqrt{2}\hbar k$, the smallest momentum imparted to the condensate with the Bragg diffraction, the wavepackets will spatially separate as the system evolves.

After adiabatic expansion, we switch off the trap, wait 600 μs so that the trapping magnetic fields decay away and then apply a sequence of two Bragg pulses. Each 30-μs pulse is composed of two linearly polarized laser beams detuned from the $3S_{1/2}$, $F = 1 \rightarrow 3P_{3/2}$, $F' = 2$ transition by $\Delta/2\pi = -2$ GHz to suppress spontaneous emission. This large detuning makes negligible Bragg scattering of the optical waves by the atoms, which could lead to a spurious scattering of atoms into $\mathbf{P}_4$. The frequency difference

between the two laser beams of a single Bragg pulse is chosen to fulfil a first-order Bragg diffraction condition that changes the momentum state of the atoms without changing their internal state[10]. The first Bragg pulse is composed of two mutually perpendicular laser beams of frequencies ν_1 and $\nu_2 = \nu_1 - 50$ kHz, and wavevectors $\mathbf{k}_1 = k\hat{x}$ and $\mathbf{k}_2 = -k\hat{y}$ ($k = 2\pi/\lambda$, $\lambda = 589$ nm). The maximum intensity of each beam is ~ 10 mW cm^{-2}. The intensity was chosen so that roughly 1/3 of the condensate atoms acquire momentum $\mathbf{P}_2 = \hbar(\mathbf{k}_1 - \mathbf{k}_2) = \hbar k(\hat{x} + \hat{y})$. The second Bragg pulse is applied 20 μs after the end of the first Bragg pulse (well before the wavepackets are separated). This second Bragg pulse is composed of two counter-propagating laser beams with frequencies ν_1 and $\nu_3 = \nu_1 - 100$ kHz, and wavevectors $\mathbf{k}_1 = k\hat{x}$ and $\mathbf{k}_3 = -k\hat{x}$. The intensities of these laser beams were chosen to cause half of the remaining atoms in the momentum state $\mathbf{P}_1 = 0$ to acquire a momentum $\mathbf{P}_3 = \hbar(\mathbf{k}_1 - \mathbf{k}_3) = 2\hbar k\hat{x}$ (atoms in $\mathbf{P}_2$ are not affected by this pulse because of the Doppler shift of the light). We chose this pulse sequence so that only $\mathbf{P}_2 = \hbar k(\hat{x} + \hat{y})$ and $\mathbf{P}_3 = 2\hbar k\hat{x}$ are produced from $\mathbf{P}_1 = 0$. Thus we create, nearly simultaneously, three overlapping wavepackets of the requisite momenta. Without the nonlinear term in equation (1), one would expect only to observe these three wavepackets after they have spatially separated. But as

a Lab frame:

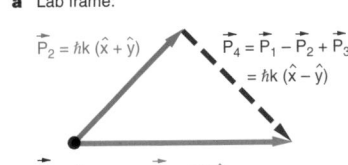

b Moving frame: $\vec{V} = \dfrac{\hbar k}{M}\hat{x}$

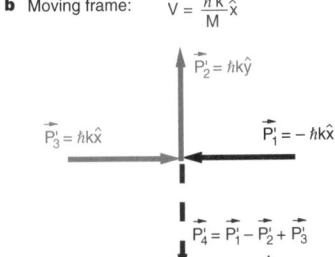

Figure 1 Momentum-energy conservation for 4WM and the bosonic stimulation viewpoint in a moving frame. **a**, Momentum conservation, $\mathbf{P}_4 = \mathbf{P}_1 - \mathbf{P}_2 + \mathbf{P}_3$ (equivalent to phase-matching in optical 4WM), in the laboratory frame. For clarity, over-arrows indicate vectors. Energy conservation requires $\mathbf{P}_4^2 = \mathbf{P}_1^2 - \mathbf{P}_2^2 + \mathbf{P}_3^2$. **b**, It is always possible to view matter 4WM in a frame moving with velocity **v** such that the three input momenta have the same magnitude, and two are counter-propagating. Then, in our case two atoms in momentum states $\mathbf{P}_1' = -\hbar k\hat{x}$ and $\mathbf{P}_3' = -\hbar k\hat{x}$ are bosonically stimulated by wavepacket $\mathbf{P}_2' = -\hbar k\hat{y}$ to scatter into momentum states $\mathbf{P}_2'$ and $\mathbf{P}_4' = -\mathbf{P}_2' = -\hbar k\hat{y}$. We note that the energy and momentum conditions are satisfied independent of the direction of $\mathbf{P}_2'$. The 4WM wavepacket is a consequence of energy, momentum and particle-number conservation when atoms are stimulated into the momentum state $\mathbf{P}_2'$. Thus 4WM can be viewed as the annihilation of momentum states $\mathbf{P}_1'$ and $\mathbf{P}_3'$, and the creation of momentum states $\mathbf{P}_2'$ and $\mathbf{P}_4'$ (the minus signs in the energy and momentum conditions are attached to the sate that gains atoms). It is this bosonic stimulation of scattering that mimics the stimulated emission of photons from an optical nonlinear medium. Alternatively, by choosing a frame of reference in which $\mathbf{P}_1' = -\mathbf{P}_2'$ (or $\mathbf{P}_2' = -\mathbf{P}_3'$), 4WM can also be viewed as matter-wave Bragg diffraction of $\mathbf{P}_3'$ ($\mathbf{P}_1'$) from the grating produced by the interference of two others.

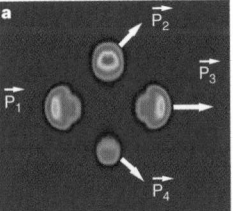

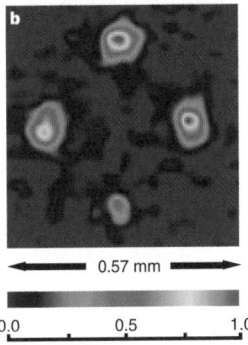

Figure 2 Numerical simulation and experimental results for 4WM. **a**, Calculated two-dimensional atomic distribution after 1.8 ms, showing the 4WM. The calculations were performed only until the wavepackets completely separated due to constraints on the simulation grid-size. The momenta are those of Fig. 1a. The field of view is 0.23×0.26 mm. We note that atoms are removed primarily from the back-end of the wavepackets because these regions overlap for the longest time. **b**, A false-colour image of the experimental atomic distribution showing the fourth (small) wavepacket generated by the 4WM process. The four wavepackets form a square measuring 0.26×0.26 mm, corresponding to the distance of 0.25 mm calculated using the experimental time of flight of 6.1 ms and the wavepacket momenta. We have verified that if we make initial wavepackets such that energy and momentum conservation cannot be simultaneously satisfied, no 4WM signal is observed. For instance, if we change the sign of the frequency difference between the two laser beams that comprise the second Bragg pulse, we will create a component with momentum $\mathbf{P}_3 = -2\hbar k\hat{x}$ instead of $\mathbf{P}_3 = 2\hbar k\hat{x}$. In this case there is no 4WM signal.

letters to nature

the three initial wavepackets separate, the nonlinear term will produce an additional wavepacket that satisfies the condition $\mathbf{P}_4 = \mathbf{P}_1 - \mathbf{P}_2 + \mathbf{P}_3 = \hbar k(\hat{\mathbf{x}} - \hat{\mathbf{y}})$, see Fig. 1a, as well as energy and particle-number conservation.

We have performed a two-dimensional numerical simulation of 4WM using equation (1) and the technique of ref. 9. The interaction energy (chemical potential) was chosen to be the same as it would be in three dimensions with a scattering length of 2.8 nm. The simulation releases 10^6 atoms from a trap with $\nu_x = 84$ Hz and $\nu_y = 59$ Hz. After 600 μs, the condensate was projected into the three initial momentum states. Figure 2a shows the atomic density 1.8 ms after this projection. The most important feature of Fig. 2a is the new wavepacket of atoms with momentum $\mathbf{P}_4 = \mathbf{P}_1 - \mathbf{P}_2 + \mathbf{P}_3$ generated by 4WM. The 4WM peak does not appear when the nonlinear term is absent.

Figure 2b is a false-colour image showing the results of the experiment. The atoms were imaged 6.1 ms after the second Bragg pulse by optically pumping the atoms to the $3S_{1/2}$, $F = 2$ state, and absorption-imaging[5] on the $3S_{1/2}$, $F = 2 \rightarrow 3P_{3/2}$, $F' = 3$ transition. The 4WM wavepacket is clearly visible. For this image, the numbers of atoms in each wavepacket were measured to be: $N_1 = 4.8(\pm 0.5) \times 10^5$, $N_2 = 5.3(\pm 0.5) \times 10^5$, $N_3 = 5.1(\pm 0.5) \times 10^5$ and $N_4 = 1.8(\pm 0.2) \times 10^5$, where the uncertainties are mainly due to uncertainties in background subtraction. The numbers of atoms in the three initial wavepackets N_1^0, N_2^0 and N_3^0 can be deduced using particle number conservation: $N_1^0 = N_1 + N_4$, $N_2^0 = N_2 - N_4$, and $N_3^0 = N_3 + N_4$. Defining the 4WM efficiency to be $\epsilon = N_4/N$, where $N = \Sigma_{j=1}^3 N_j^0 = \Sigma_{j=1}^4 N_j$, we obtain a conversion efficiency of 10.6(± 0.13)%. This is the best we have observed, although under similar conditions we have also observed conversion efficiencies of only 6%. This difference suggests the influence of some uncontrolled experimental conditions, such as laser beam inhomogeneities, or non-zero average velocity of the released condensate. By comparison, the calculation of Fig. 2a gives an efficiency of 10%, albeit for only 10^6 atoms.

Equation (1) can be used to make a simple prediction about the expected nonlinear dependence of the 4WM signal on the numbers of atoms in the initial wavepackets. Substituting $\Psi = \Sigma_{j=1}^4 \Psi_j$ (where Ψ_j correspond to the individual momentum components) into equation (1), we find the initial rate of growth of the 4WM amplitude, $\partial \Psi_4/\partial t \propto \Psi_1 \Psi_2^* \Psi_3$. We estimate the number of atoms in the fourth wave by multiplying this rate by a characteristic interaction time τ, proportional to the diameter of the condensate, squaring and integrating over space: $N_4 \propto n_1 n_2 n_3 V \tau^2$, where the density $n_j = N_j^0/V$. In the Thomas–Fermi limit[7], the volume of the condensate $V \propto N^{3/5}$, and $\tau \propto N^{1/5}$. Hence we expect $N_4/N \propto (N_1^0 N_2^0 N_3^0)N^{-9/5}$, a dependence which is supported by the numerical calculations. This nonlinear behaviour is clearly manifested in the initial linear growth seen in Fig. 3, where we vary the

number of atoms in the original BEC and measure the number of atoms in the respective wavepackets. The data also show saturation at high N, as does the corresponding theory, although the maximum theoretical efficiency is somewhat higher.

We now reconsider Fig. 1b. Here the process is seen as degenerate 4WM (where the magnitudes of all momenta are equal) in a geometry equivalent to phase-conjugation in optics[12]: indeed $\mathbf{P}_4'$ is the momentum conjugate of $\mathbf{P}_2'$. So this can also be considered as a demonstration of phase conjugation with matter waves. As in the case of optical phase conjugation, the process would work regardless of the angle between $\mathbf{P}_1'$ and $\mathbf{P}_2'$ (90° in the present case). If one alters the first Bragg pulse by changing only the angle of $\mathbf{k}_2$ (and appropriately changing ν_2) the magnitude and direction of $\mathbf{P}_2$ in the laboratory frame are changed, so that in the moving frame only the angle of $\mathbf{P}_2'$ is changed.

We emphasize that just as optical 4WM requires coherent light sources to coherently build up the generated wave, a condensate is also crucial for coherent generation of matter waves. If atoms are above the Bose–Einstein condensation temperature, the number density is necessarily low and the phase-matching condition is different for each velocity class. Both dramatically diminish the 4WM conversion efficiency.

In spite of the strong analogy between atom and optical 4WM, there are fundamental differences. In optical 4WM, the energy–momentum dispersion relation is $E = [c/n(k)]\hbar k$ (where $n(k)$ is the dispersive refractive index), whereas for massive particles (neglecting the matter-wave refractive index due to the atom–atom interaction energy) $E = P^2/2M$. Because atoms are neither created nor destroyed, the only 4WM processes allowed for matter waves conserve particle number. This is not the case for optical 4WM where, for example, in frequency tripling three photons are annihilated and one is created. Particle, energy and momentum conservation limit all matter 4WM processes to configurations that can be viewed as degenerate 4WM in an appropriate moving frame.

The present experiment used relatively large momenta. If we were to use momenta small enough to couple to phonons or other collective excitations of the condensate, we would be able to study these excitations and their nonlinear interactions with each other and with large-momentum excitations We could also change the internal states by using Raman transitions[4] to scatter atoms in one internal state from the matter-wave grating formed by atoms in a different internal state. It should even be possible to study 4WM between different isotopes or elements. Furthermore, just as nonlinear optics can create quantum correlations between photon beams, nonlinear atom optics may lead to the study of non-classical matter-wave fields. □

Received 29 December 1998; accepted 16 February 1999.

1. Franken, P. A., Hill, A. E., Peters, C. W. & Weinreich, G. Generation of optical harmonics. *Phys. Rev. Lett.* **7**, 118–119 (1961).
2. Maker, P. D. & Terhune, R. W. Study of optical effects due to an induced polarization third order in the electric field strength. *Phys. Rev.* **137A**, A801–A818 (1965).
3. Mewes, M.-O. *et al.* Output coupler for Bose-Einstein condensed atoms. *Phys. Rev. Lett.* **78**, 582–585 (1997).
4. Hagley, E. W. *et al.* A well collimated quasi-continuous atom laser. *Science* (in the press).
5. Mewes, M. H. *et al.* Observation of Bose-Einstein condensation in a dilute atomic vapor. *Science* **269**, 198–201 (1995).
6. Lens, G., Meystre, P. & Wright, E. W. Nonlinear atom optics. *Phys. Rev. Lett.* **71**, 3271–3274 (1993).
7. Dalfovo, F., Giorgini, S., Pitaevskii, L. P. & Stringari, S. Theory of trapped Bose-condensed gases. *Rev. Mod. Phys.* **71** (in the press).
8. Goldstein, E. V., Plättner, K. & Meystre, P. Atomic phase conjugation. *Quantum Semiclass. Opt.* **7**, 743–749 (1995).
9. Trippenbach, M., Band, Y. B. & Julienne, P. S. Four wave mixing in the scattering of Bose-Einstein condensates. *Opt. Express* **3**, 530–537 (1998).
10. Kozuma, M. *et al.* Coherent splitting of Bose-Einstein condensed atoms with optically induced Bragg diffraction. *Phys. Rev. Lett.* **82**, 871–875 (1999).
11. Petrich, W., Anderson, M. H., Ensher, J. R. & Cornell, E. A. Stable, tightly confining magnetic trap for evaporative cooling of neutral atoms. *Phys. Rev. Lett.* **74**, 3352–3355 (1995).
12. Shen, Y. R. *The Principles of Nonlinear Optics* 249–251 (Wiley, New York, 1984).

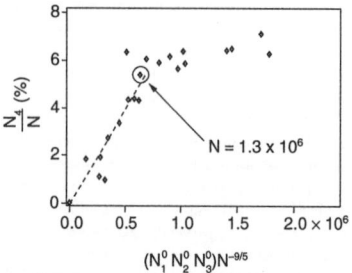

Figure 3 Measured conversion efficiency. Efficiency N_4/N is plotted as a function of $(N_1^0 N_2^0 N_3^0)N^{-9/5}$. The initial linear dependence is a signature of 4WM with matter waves. The dashed line is a fit to the first 12 points to guide the eye.

Acknowledgements. We thank K. Burnett, C. W. Clark, M. Kozuma and D. E. Pritchard for discussions. This work was supported in part by the US Office of Naval Research and NASA.

Correspondence and requests for materials should be addressed to L.D. (e-mail: lu.deng@nist.gov).

VOLUME 87, NUMBER 1   PHYSICAL REVIEW LETTERS   2 JULY 2001

All-Optical Formation of an Atomic Bose-Einstein Condensate

M. D. Barrett, J. A. Sauer, and M. S. Chapman

School of Physics, Georgia Institute of Technology, Atlanta, Georgia 30332-0430
(Received 10 May 2001; published 19 June 2001)

We have created a Bose-Einstein condensate (BEC) of ^{87}Rb atoms directly in an optical trap. We employ a quasielectrostatic dipole force trap formed by two crossed CO_2 laser beams. Loading directly from a sub-Doppler laser-cooled cloud of atoms results in initial phase space densities of $\sim 1/200$. Evaporatively cooling through the BEC transition is achieved by lowering the power in the trapping beams over ~ 2 s. The resulting condensates are $F = 1$ spinors with 3.5×10^4 atoms distributed between the $m_F = (-1, 0, 1)$ states.

DOI: 10.1103/PhysRevLett.87.010404   PACS numbers: 03.75.Fi, 03.67.–a, 32.80.Pj

The first observation of Bose-Einstein condensates (BEC) in dilute atomic vapors in a remarkable series of experiments in 1995 [1–3] has stimulated a tremendous volume of experimental and theoretical work in this field. Condensates are now routinely created in over 30 laboratories around the world, and the pace of theoretical progress is equally impressive [4]. The recipe for forming a BEC is by now well established [5,6]. The atomic vapor is first precooled, typically by laser cooling techniques, to sub-mK temperatures and then transferred to a magnetic trap. Further cooling to BEC is then achieved by evaporatively cooling the atoms in the magnetic trap using energetically selective spin transitions [7].

All-optical methods of reaching the BEC phase transition have been pursued since the early days of laser cooling. Despite many impressive developments beyond the limits set by Doppler cooling, including polarization gradient cooling [8], velocity selective coherent population trapping [9], Raman cooling [10–12], and evaporative cooling in optical dipole force traps [13–15], the best efforts to date produce atomic phase space densities $n\lambda_{dB}^3$ a factor of 10 away from the BEC transition [15]. The principal roadblocks have been attributed to density-dependent heating and losses in laser cooling techniques, residual heating in optical dipole force traps or the unfavorable starting conditions for evaporative cooling. Hence, optical traps have played only an ancillary role in BEC experiments. The MIT group used a magnetic trap with an "optical dimple" to reversibly condense a magnetically confined cloud of atoms evaporatively cooled to just above the phase transition [16]. Additionally, Bose condensates created in magnetic traps have been successfully transferred to shallow optical traps for further study [17–19]. In all these cases, however, magnetic traps provided the principle increase of phase space density (by factors up to $\sim 10^6$) to the BEC transition.

In this Letter, we present an experiment in which we have created a Bose condensate of ^{87}Rb atoms directly in an optical trap formed by tightly focused laser beams. Following initial loading from a laser cooled gas, evaporative cooling through the BEC transition is achieved by simply lowering the depth of the optical trap. Our success is due in part to a high initial phase space density realized in the loading of our optical dipole trap and in part to the tight confinement of the atoms that permits rapid evaporative cooling to the BEC transition in ~ 2 s. This fast evaporation relaxes considerably the requirement for extremely long trap lifetimes typical of magnetic trap BEC experiments. Additionally, in contrast to magnetic traps which confine only one or two magnetic spin projections [20], our technique is spin independent and the condensates that we form are $F = 1$ three-component spinors [18].

We utilize a crossed-beam optical dipole force trap employing tightly focused high-powered (12 W) infrared ($\lambda = 10.6 \ \mu$m) laser beams. For trapping fields at this wavelength, the trapping potential for the ground state atoms is very well approximated by $U(\mathbf{r}) = -\frac{1}{2}\alpha_g|E(\mathbf{r})|^2$ where α_g is the ground state dc polarizability of the atom (5.3×10^{-39} m^2 C/V for atomic rubidium) and $E(\mathbf{r})$ is the electric field amplitude. A significant feature of these "quasielectrostatic" traps (QUESTs) [21] is that heating due to spontaneous emission of the atoms is completely negligible. In contrast to previous QUESTs that have employed single focused beams [14,21] or standing wave configurations [22], we employ here a cross-beam geometry [13] to provide a balance of tight confinement in three dimensions (~ 1.5 kHz oscillation frequencies at full power) and a relatively large loading volume.

Our experiments begin with a standard vapor loaded magneto-optical trap (MOT). The trapping beams consist of three orthogonal retroreflected beams in the $\sigma^+ - \sigma^-$ configuration. They are tuned 15 MHz below the $5S_{1/2}\text{-}5P_{3/2}$ $F = 2 \rightarrow F' = 3$ transition of ^{87}Rb, and each of the three beams has a waist of 0.7 cm and a power of 25 mW. A repump beam tuned to the $F = 1 \rightarrow F' = 2$ transition is overlapped with one of the MOT trapping beams. The MOT is loaded for 5 s directly from the thermal vapor during which we collect 30×10^6 atoms. After loading the MOT, the trap configuration is changed to maximize the transfer of atoms to the optical trap. The repump intensity is first lowered to $\sim 10 \ \mu$W/cm^2 for a duration of 20 ms, and then the MOT trap beams are shifted

VOLUME 87, NUMBER 1   PHYSICAL REVIEW LETTERS   2 JULY 2001

to the red by 140 MHz for a duration of 40 ms. At this point the MOT beams are extinguished and the current in the MOT coils is turned off. In order to optically pump the atoms into the $F = 1$ hyperfine states, the repump light is shuttered off 1 ms before the trap beams are extinguished; we measure the efficiency of the optical pumping to the $F = 1$ state to be >95%. The CO_2 laser beams are left on at full power (12 W) throughout the MOT loading and dipole trap loading process.

The trapping beams are generated from a commercial CO_2 gas laser ($\lambda = 10.6 \ \mu m$). The beams are tightly focused and intersected at right angles; one beam is oriented in the horizontal direction and one beam is inclined at 45° from the vertical direction. Each beam passes through an acousto-optic modulator to provide independent control of the power in the two beams. Additionally, the beams are frequency shifted 80 MHz relative to each other so that any spatial interference patterns between the two beams are time averaged to zero [23]. Each beam has a maximum power of 12 W, and the beams are focused to a minimum waist $\lesssim 50 \ \mu m$ with $f = 38$ mm focal length ZeZn aspherical lenses inside the chamber.

Following standard techniques, the number of trapped atoms and their momentum distribution are observed using absorptive imaging of the released atoms. The trapped atoms are released by suddenly (<1 μs) switching off the trapping laser beams. Following a variable ballistic expansion time (typically 2–20 ms), the cloud is illuminated with a 50 μs pulse of $F = 1 \rightarrow F' = 2$ light applied to the atoms concurrent with a vertically oriented circularly polarized probe beam tuned to the $F = 2 \rightarrow F' = 3$ transition. The probe intensity is ~0.3 mW/cm^2, and each atom scatters up to 150 photons from the probe with no observable blurring of the cloud. The shadow of the atom cloud is imaged onto a slow-scan CCD camera. The measured spatial resolution of our imaging system is $\lesssim 10 \ \mu m$. From these images, the number of atoms and their temperature are determined. Together with the trap oscillation frequencies (which are measured directly using parametric excitation) the spatial density, elastic collision rate, and phase space density can be derived.

We first discuss the properties of the trap following loading from the MOT without employing forced evaporative cooling. The earliest that we observe the trapped atoms is 100 ms after loading to allow for the untrapped atoms to fall away. At this time, the trap contains 2×10^6 atoms at a temperature of 75 μK. Maintaining full power in the trap beams, we observe a rapid evaporation from the trap in the first 1.5 s, during which two-thirds of the trapped atoms are lost and the temperature falls to 38 μK. The relative phase space density increases by a factor of 3 during this stage. This rapid evaporation gives way to a much slower exponential trap decay with a time constant of 6(1) s. The temperature continues to fall gradually to a final value of 22 μK on a 10 s time scale.

The mean frequency of the trap at these powers is measured to be $1.5(^{+0}_{-5})$ kHz, from which we infer an initial peak density of the trap of $\sim 2 \times 10^{14}$ atoms/cm^3. The initial phase space density following the rapid free evaporation stage is calculated to be $1/200$ (or $1/600$ assuming equal distribution of the atoms in the three trapped internal states). Using the measured value for the three-body loss rate [24] yields an initial three-body loss rate constant of 1.3 s^{-1}. Hence it is possible that three-body inelastic collisions contribute to the rapid initial loss of atoms; however, we note that such collisions do not explain the observed cooling. The estimated elastic collision rate at this point is a remarkable 12×10^3 s^{-1}—higher than the trap oscillation frequency. Although the derived quantities must be taken with some caution, we note that densities of 3×10^{13} atoms/cm^3 and a phase space density of $1/300$ have been reported in a 1D CO_2 lattice trap with ^{85}Rb (albeit with much fewer atoms) [25], and we have measured similar results in our laboratory in 1D CO_2 lattice traps with ^{87}Rb. It is not clear what yields these extremely high initial densities in these traps at these low temperatures. A blue detuned Sisyphus process is suggested in [25] (densities of $\sim 10^{13}$ atoms/cm^3 have been observed in a dipole trap with blue detuned Sisyphus cooling [26]), but we suspect that there is an effective dark MOT [27] in the repump beam at the trap center induced by the ac Stark shift due to the dipole trap beams, which reduces density-limiting interactions with the MOT light fields. Using our trap parameters, we estimate that the repump scattering rate is reduced at the trap center by a factor of 40 relative to outside the trap. Our improved performance even relative to the 1D lattice traps may be due to more efficient loading from the "tails" of the crossed trap geometry. In any case, the net result for our trap is a very high initial phase space density combined with a large number of atoms and a fast thermalization time—together these conditions provide a very favorable starting point for evaporative cooling.

Efficient forced evaporation requires selectively ejecting the more energetic atoms from the trap such that the remaining atoms rethermalize at a lower temperature with a higher net phase space density. For optical traps, the simplest way to force evaporation is to lower the trap depth by decreasing the power in the trap beams. This technique was used in one of the first demonstrations of evaporative cooling in alkali atoms [13], where, starting with only 5000 atoms, a phase space density increase of ~30 was realized. The drawback to this method is that lowering the trap depth also lowers the trap oscillation frequency and the rethermalization rate. Hence the evaporation rate can slow down prohibitively. In our case, although the rethermalization rate falls by a factor of 50 by the end of the evaporation cycle, it nonetheless remains fast enough to allow us to reach the BEC transition in 2.5 s as described below.

Our procedure for forced evaporation is as follows: immediately after loading the trap, the power in both trap

VOLUME 87, NUMBER 1   PHYSICAL REVIEW LETTERS   2 JULY 2001

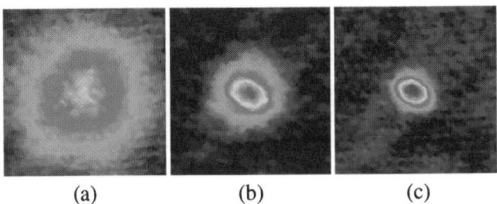

(a)   (b)   (c)

FIG. 1 (color). Absorptive images (false color) of atomic cloud after 10 ms free expansion for different final trap laser powers. (a) Thermal cloud above BEC transition ($P = 480$ mW), (b) thermal-condensate mixture ($P = 260$ mW), and (c) pure condensate ($P = 190$ mW). Field of view is 350 μm.

beams is ramped to 1 W in 1 s. The power in both beams is then ramped to a variable final power in 1 s and maintained at this low level for 0.5 s, after which the atoms are released from the trap and imaged as discussed above. Figure 1 shows three such images at different final powers for the trap lasers. The left image shows a cloud well above the BEC transition point, and reveals an isotropic Gaussian momentum distribution expected for a thermal cloud of atoms in equilibrium. As the evaporative cooling proceeds to lower powers, a bimodal momentum distribution appears with a central nonisotropic component characteristic of Bose condensates (Fig. 1, center). As the power is lowered further, the central, nonspherical peak of the distribution becomes more prominent, and the spherical pedestal diminishes. At a trap power of 190 mW (Fig. 1, right), the resulting cloud is almost a pure condensate and contains 3.5×10^4 atoms. For lower powers, the cloud rapidly diminishes as the trap can no longer support gravity. Line profiles of the images in Fig. 1 are shown in Fig. 2 for quantitative comparison. The profiles are taken along the orientation of the minor axis of the condensate. Also shown are Gaussian fits to the data in the wings that clearly show the bimodal nature of the momentum distributions near the BEC phase transition.

The critical condensation temperature T_c is a function of the number of atoms N and the mean trap frequencies $\bar{\omega} = (\omega_1 \omega_2 \omega_3)^{1/3}$ according to $k_B T_c = \hbar \bar{\omega} (N/1.202)^{1/3}$. Below the critical temperature, the condensate fraction should grow as $N_0/N = 1 - (T/T_c)^3$ [5,6]. We obtain the condensate fraction and the temperature of the normal component from 2D fits to the absorptive images using the methods described in [6]. In Fig. 3, the condensate fraction is plotted versus normalized temperature, $T/T_c(N, \bar{\omega})$ near the critical point, where we assume that $\bar{\omega}^2$ is proportional to the trap power. The agreement with the theoretical curve is reasonable given the scatter in the data. At the trap power of 350 mW, which is near the critical point, the trap contains 180 000 atoms at a temperature of 375 nK. Together with the measured trap frequencies at this power $\omega_1, \omega_2, \omega_3 = 2\pi(72, 175, 350)$ Hz, we infer a phase space density at this point of 1.4 (or 0.45 if the

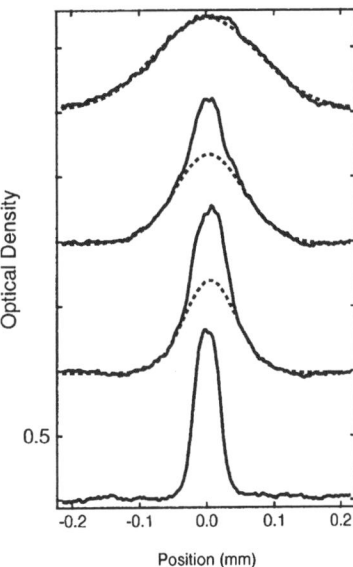

FIG. 2. One-dimensional profiles through the images of Fig. 1 (solid lines) along with Gaussian fits to the wings of the profile (dashed lines). Trap laser powers 480, 310, 260, and 190 mW, top to bottom, respectively.

three internal states are equally populated), a density of 4.8×10^{13} cm^{-3}, and an elastic collision rate of 300 s^{-1}.

We have measured the growth of the freely expanding condensate for expansion times from 5–20 ms. The measured aspect ratio of the cloud is 1.5(0.1) and shows a slight increase for longer expansion times. The measured size of the minor axis of the cloud grows linearly with time (within the limits of our imaging resolution) to a size of 100 μm (full width) at 15 ms. The observed 2D projection of the cloud is oriented at a 30° angle relative to the horizontal trap. Quantitative comparison of the observed condensates with theory requires knowledge of the trap oscillation frequencies as well as orientation of the principal axes of the trap. Although we can measure the frequencies,

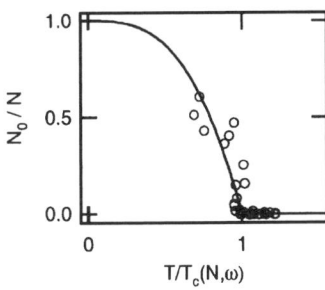

FIG. 3. (circles) Condensate fraction (ratio of condensate atoms to total number of atoms) vs scaled temperature, $T/T_c(N, \bar{\omega})$. Also shown is the theoretical prediction (solid curve).

VOLUME 87, NUMBER 1   PHYSICAL REVIEW LETTERS   2 JULY 2001

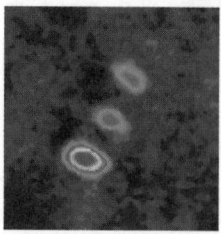

FIG. 4 (color). Absorptive image of atomic cloud after 10 ms free expansion in a Stern-Gerlach magnetic field gradient. Three distinct components are observed corresponding to $F = 1$, $m_F = (-1, 0, 1)$ spin projections from bottom to top, respectively.

in atoms or molecules lacking a suitable magnetic moment, perhaps by using sympathetic cooling to precool the sample. Finally, our technique offers considerable experimental simplicity and speed, easing the requirement for ultrahigh vacuum environments, and eliminating the need for strong magnetic trapping fields.

We acknowledge the technical assistance of D. Zhu and helpful discussions with T. A. B. Kennedy, C. Raman, S. Yi, and L. You. This work was supported by the National Security Agency (NSA) and Advanced Research and Development Activity (ARDA) under Army Research Office (ARO) Contract No. DAA55-98-1-0370.

the principal axes depend on the beam ellipticity and alignment. In our case, the trap beams are aberrated due to off-center propagation through the focusing lenses (required to overlap the beams), and the fact that the laser output beam is not a pure Gaussian TEM_{00} spatial mode. Nonetheless, using the measured trap frequencies, scaled for power, mean field theory (see, e.g., [4]) predicts aspect ratios of 1.2–1.7 for a 15 ms expansion and cloud sizes of 60–130 μm, consistent with our observations.

The $1/e$ lifetime of the condensate is measured to be 3.5(1) s, which is somewhat smaller than the 6(1) s lifetime observed in the trap. We observe no residual heating of the condensate for the lifetime of the condensate, although because the trap potential is quite shallow, it is possible that any heating would be offset by subsequent evaporation from the trap.

The trapped atoms are optically pumped during loading into the $F = 1$ ground state; however, no attempt is made to further optically pump the atoms into a single m_F state. Hence we expect the population to be a mixture of the $m_F = -1, 0, 1$ spin projections. To measure the spin content of our condensate, we apply a weak field gradient to the cloud after the cloud is released from the dipole trap [18]. An absorptive image of the cloud is shown in Fig. 4, and reveals that the condensate is composed of a mixture of spin states. Each cloud has a similar nonspherical momentum distribution characteristic of condensates. Determining the relative weights of the spin states is complicated by the differing matrix elements of the probe transitions. Nonetheless, if we assume that the atoms are all optically pumped by the circularly polarized probe and that all atoms scatter the same number of photons, we infer a 3:1:1 weighting of the $m_F = -1, 0, 1$ mixture, respectively, for the image shown in Fig. 4.

In summary, we have realized a Bose condensate of ^{87}Rb atoms directly in an optical trap. This technique seems promising for creating and studying more complex spinor condensates (e.g., ^{85}Rb, $F = 2$ [28]) as well as multiatom mixtures. Additionally, by eliminating the need for a magnetic trap, it may be possible to realize condensates

[1] M. H. Anderson *et al.*, Science **269**, 198 (1995).
[2] K. B. Davis *et al.*, Phys. Rev. Lett. **75**, 3969 (1995).
[3] C. C. Bradley *et al.*, Phys. Rev. Lett. **78**, 985 (1997); C. C. Bradley *et al.*, Phys. Rev. Lett. **75**, 1687 (1995).
[4] F. Dalfovo *et al.*, Rev. Mod. Phys. **71**, 463 (1999).
[5] E. A. Cornell *et al.*, in *Bose-Einstein Condensation in Atomic Gases*, Proceedings of the International School of Physics "Enrico Fermi," Course CXL, edited by M. Inguscio, S. Stringari, and C. Wieman (IOS Press, Amsterdam, 1999).
[6] W. Ketterle, D. S. Durfee, and D. M. Stamper-Kurn, in *Bose-Einstein Condensation in Atomic Gases* (Ref. [5]).
[7] For a recent review of evaporative cooling, see W. Ketterle and N. J. van Druten, in *Advances in Atomic, Molecular, and Optical Physics*, edited by B. Bederson and H. Walther (Academic Press, San Diego, 1996), Vol. 37, p. 181.
[8] P. D. Lett *et al.*, Phys. Rev. Lett. **61**, 169 (1988).
[9] A. Aspect *et al.*, Phys. Rev. Lett. **61**, 826 (1988).
[10] M. Kasevich and S. Chu, Phys. Rev. Lett. **69**, 1741 (1992).
[11] A. J. Kerman *et al.*, Phys. Rev. Lett. **84**, 439 (2000).
[12] D-J. Han *et al.*, Phys. Rev. Lett. **85**, 724 (2000).
[13] C. S. Adams *et al.*, Phys. Rev. Lett. **74**, 3577 (1995).
[14] K. M. O'Hara *et al.*, Phys. Rev. Lett. **85**, 2092 (2000); H. Engler *et al.*, Phys. Rev. A **62**, 031402(R) (2000).
[15] T. Ido, Y. Isoya, and H. Katori, Phys. Rev. A **61**, 061403(R) (2000); D. J. Han, M. T. DePue, and D. S. Weiss, Phys. Rev. A **63**, 023405 (2001).
[16] D. M. Stamper-Kurn *et al.*, Phys. Rev. Lett. **81**, 2194 (1998).
[17] D. M. Stamper-Kurn *et al.*, Phys. Rev. Lett. **80**, 2027 (1998).
[18] H.-J. Miesner *et al.*, Phys. Rev. Lett. **82**, 2228 (1999); D. M. Stamper-Kurn *et al.*, Phys. Rev. Lett. **83**, 661 (1999).
[19] B. P. Anderson and M. A. Kasevich, Science **282**, 1686 (1998).
[20] C. J. Myatt *et al.*, Phys. Rev. Lett. **78**, 586 (1997).
[21] T. Takakoshi and R. J. Knize, Opt. Lett. **21**, 77 (1996).
[22] S. Friebel *et al.*, Phys. Rev. A **57**, R20 (1998).
[23] M. T. DePue *et al.*, Phys. Rev. Lett. **82**, 2262 (1999).
[24] E. A. Burt *et al.*, Phys. Rev. Lett. **79**, 337 (1997).
[25] S. Friebel *et al.*, Appl. Phys. B **67**, 699 (1998).
[26] D. Boiron *et al.*, Phys. Rev. A **57**, R4106 (1998).
[27] W. Ketterle *et al.*, Phys. Rev. Lett. **70**, 2253 (1993).
[28] S. L. Cornish *et al.*, Phys. Rev. Lett. **85**, 1795 (2000).

Transcription Against an Applied Force

Hong Yin, Michelle D. Wang, Karel Svoboda, Robert Landick, Steven M. Block, Jeff Gelles*

The force produced by a single molecule of *Escherichia coli* RNA polymerase during transcription was measured optically. Polymerase immobilized on a surface was used to transcribe a DNA template attached to a polystyrene bead 0.5 micrometer in diameter. The bead position was measured by interferometry while a force opposing translocation of the polymerase along the DNA was applied with an optical trap. At saturating nucleoside triphosphate concentrations, polymerase molecules stalled reversibly at a mean applied force estimated to be 14 piconewtons. This force is substantially larger than those measured for the cytoskeletal motors kinesin and myosin and exceeds mechanical loads that are estimated to oppose transcriptional elongation in vivo. The data are consistent with efficient conversion of the free energy liberated by RNA synthesis into mechanical work.

RNA polymerases play a critical role in gene expression by synthesizing RNA transcripts containing genetic information copied from DNA templates. The nascent RNA chain is elongated in a chemical reaction during which appropriate ribonucleoside triphosphates (NTPs) are condensed with the RNA 3' end and pyrophosphate anions (PP_i) are released. RNA polymerases move along DNA while copying it, advancing on average a distance of 1 base pair (bp) (~0.34 nm along the DNA helix axis) for each nucleotide added to the transcript [(1) but see also (2, 3)]. In vivo, this movement performs mechanical work against hydrodynamic drag and other external forces (4, 5) and therefore requires a source of free energy. The energy is provided by the condensation reaction itself, which is energetically favorable at physiological NTP and PP_i concentrations (1). Thus, RNA polymerases may be viewed as molecular motors that catalyze a biosynthetic reaction while using a portion of the excess free energy from the reaction to perform mechanical work. Although translocation of RNA polymerase molecules along DNA has been detected by microscopy (6–8), little is known about the ability of these enzymes to move against opposing forces or about their energy conversion effi-

H. Yin and J. Gelles, Department of Biochemistry, Biophysics Program, and Center for Complex Systems, Brandeis University, Waltham, MA 02254, USA.
M. D. Wang and S. M. Block, Department of Molecular Biology and Princeton Materials Institute, Princeton University, Princeton, NJ 08544, USA.
K. Svoboda, Biological Computation Department, AT&T Bell Laboratories, Murray Hill, NJ 07974, USA.
R. Landick, Department of Bacteriology, University of Wisconsin, Madison, WI 53706, USA.

*To whom correspondence should be addressed.

ciencies. We report here observations of the movement of single *E. coli* RNA polymerase molecules along DNA templates. The measurements were made with a microscope-based optical trapping interferometer (9). Calibrated forces were applied in the direction opposing translocation by the polymerase to a small spherical particle (a polystyrene bead) attached to the DNA, while the position of the particle was simultaneously measured. When a sufficiently high force was applied, movement by the enzyme could be stalled, in many cases reversibly. Our results show that this nucleic acid polymerase is a powerful biological motor that can exert considerable force and may operate with energy conversion efficiencies comparable to those of prototypical mechanoenzymes, such as myosin and kinesin.

We conducted experiments using an assay developed (6, 7) to study transcriptional elongation by single molecules of *E. coli* RNA polymerase in vitro (10). Ternary transcription complexes consisting of single molecules of RNA polymerase associated with a DNA template and a nascent RNA transcript were assembled in solution, halted by NTP depletion, and adsorbed onto the cover glass surface of a microscope flow cell. Polystyrene beads (0.52 μm in diameter) were attached to the transcriptionally downstream ends of the DNA molecules so that each bead became tethered to the surface by its connection through the DNA and the polymerase. When supplied with NTPs, up to half of the immobilized transcription complexes are enzymatically active; bead-labeled complexes display elongation kinetics indistinguishable from those of unlabeled complexes in solution (6, 7). In a typical experiment, a transcription complex was first located by video-enhanced, differential interference contrast light microscopy. The complex was identified visually by the Brownian motion of the bead, which was constrained by the length of its DNA tether to a small region centered over the attachment position of the polymerase (6, 7). For the studies described here, the microscope was equipped with an optical trapping interferometer ("optical tweezers" plus a position sensor) that could exert calibrated forces up to ~100 pN on a bead while simultaneously measuring its displacement with subnanometer precision and millisecond time resolution (9, 11). With the laser light shuttered, the optical trap was moved to the region near the bead. The shutter was then opened to activate the trap, and the bead was captured. By adjusting the trap controls, we positioned the center of the bead directly over the polymerase, typically at a height of ~590 nm above the cover glass surface (this particular height was chosen to bring the bead as close as possible to the surface without

the risk of touching during subsequent measurements). We then repositioned the trap by moving it at a constant height along the direction of greatest detector sensitivity [the direction defined by the shear axis of the Wollaston prism; see (9, 12)] until the DNA straightened and was held under very light tension, with the bead displaced just 30 to 70 nm from the trap center (Fig. 1A).

We started transcriptional elongation of bead-labeled complexes by exchanging the buffer inside the flow cell with one containing NTPs. In most experiments, elongation was begun immediately before establishment of the initial configuration just described. During elongation, the template was pulled by the stationary polymerase molecule, developing further tension in the DNA between the bead and the polymerase and drawing the bead away from the trap center (Fig. 1B). The optical trap acts as a nearly linear spring of stiffness α_{trap} attached to a stationary reference frame and exerts a force F_{trap} on the bead (Fig. 1C). The bead adopts a position where F_{trap} is balanced by the force F_{tc} exerted by the polymerase, acting through the DNA. The series elasticity due to the DNA, polymerase, and associated linkages acts as a spring of stiffness α_{tc}. By calibrating the optical trap stiffness and determining the displacement of the bead, we could measure F_{trap} and thus F_{tc}. The force could be determined without knowledge of the stiffness α_{tc}, which depends on the length of DNA as well as the applied force (13).

During control experiments in which no NTPs were added, beads remained at an approximately fixed distance from the trap center. Slight changes seen in the interferometer signal (mean rate ~0.3 nm s^{-1} toward the center of the trap) were attributable to instrumental drift.

In elongation experiments using high concentrations of NTPs (1 mM each of adenosine triphosphate, guanosine triphosphate, cytidine 5'-triphosphate, and uridine 5'-triphosphate), 1 μM PP$_i$, and low trapping force (laser power at specimen, 25 mW; trap stiffness α_{trap}, ~0.03 pN nm^{-1}) (14), we observed continuous movement of beads out to the limit of the usable range of the trap, located roughly 200 nm from the trap center. Once a bead reached this limit, the trap was manually repositioned to bring the bead closer to the trap center so that observation could continue (Fig. 2A). In five of seven elongating complexes studied in this way, beads moved continuously to the limit without stopping, moving at similar velocities after the trap was repositioned. (One complex stopped and failed to restart; another stopped for ~18 s and then continued elongation to the trap limit.) The bead velocity, 4.3 ± 1.3 nm s^{-1} (mean ± SD) (15), was comparable to elongation rates measured under similar condi-

tions in solution (4.4 to 6.8 nm s^{-1}) (6) and in previous microscope experiments (4.2 ± 1.7 nm s^{-1}) (7).

Different behavior was observed when higher trapping forces were used (laser power at specimen, 82 to 107 mW; α_{trap}, ~0.09 to 0.12 pN nm^{-1}). In this regime, 66 of 77 elongating complexes (86%) stopped translocating (stalled) once the bead encountered the high-force region of the trap (Fig. 2, B and C) (16, 17). To test whether stalling was reversible, we maintained the bead at stall, typically for 10 to 15 s, then repositioned the trap center closer to the bead, reducing F_{trap}. After force reduction, 24 of the 66 stalled complexes resumed movement, and some complexes could even be stalled multiple times this way (Fig. 2C). After recovery from stall, the velocity of bead movement in the low-force region of the trap was similar to that

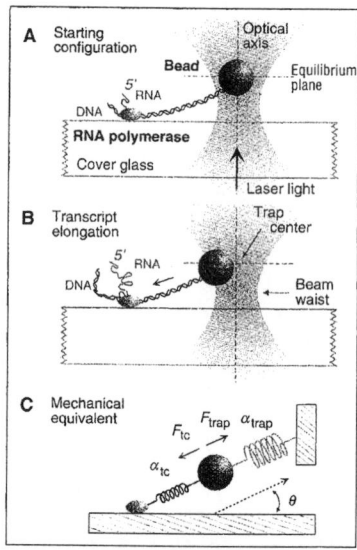

Fig. 1. A cartoon illustrating essential features of the experiment (not to scale). (**A**) The configuration at the start of transcriptional elongation. An RNA polymerase molecule (green ellipsoid) is attached to the cover glass surface of a flow cell. The polymerase is bound to template DNA (red) and has begun to synthesize a transcript RNA (green). A polystyrene bead (blue sphere) is attached to the downstream end of the DNA and is captured and held under slight tension by the light of the optical trap (pink). (**B**) The configuration during subsequent transcriptional elongation. The trap center is located on the optical axis but slightly above the narrow waist of the focused laser beam. The polymerase has proceeded for some distance along the DNA, shortening the segment between the bead and the polymerase. The bead is pulled away from the trap center, increasing the restoring force of the trap. (**C**) The mechanical equivalent of the experimental geometry shown in (A) and (B). All components lie centered in a vertical plane oriented parallel to the Wollaston shear axis.

before stall. The remaining fraction of RNA polymerase molecules stalled irreversibly in that they did not resume elongation when F_{trap} was reduced. Irreversible stalling may be attributable to one or more of several possible causes: The polymerase may have been directly inactivated by the mechanical load, have suffered photodamage from the laser light, or have spontaneously converted into an inactive species ("transcriptional arrest") similar to that formed during transcriptional stalling induced by NTP depletion (18, 19).

The simplest physical interpretation of reversible stalling is that it corresponds to the situation in which F_{trap} has increased to a level where it balances the maximal force that the polymerase can exert (F_{stall}), and no further progress is made. When F_{trap} is reduced, enzyme activity resumes (20). During both reversible and irreversible stalls, movement sometimes slowed gradually during the ap-

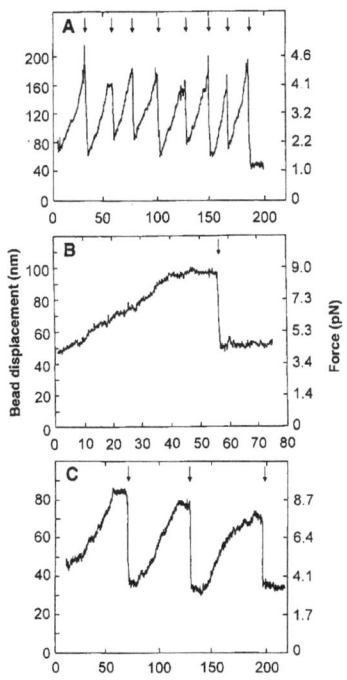

Fig. 2. Time courses for displacement from the trap center and the corresponding optical force applied to single beads driven by translocation of an RNA polymerase molecule along DNA. Force scales (right-hand axes) are nonlinear (14). The zero of the time axis is arbitrary. Reactions were conducted at 1 mM NTPs with 0.001 mM (**A**) or 1 mM (**B** and **C**) PP$_i$ at laser powers of 25 (A), 82 (B), or 99 (C) mW. Vertical arrows designate times at which the trap was repositioned to reduce the optical force. An irreversible stall is shown in (B), and reversible stalls (first two stalls) and an irreversible stall (third stall) are shown in (C).

proach to F_{stall} (for example, the third stall in Fig. 2C), whereas in other cases movement continued at nearly constant velocity, slowing abruptly close to F_{stall} (for example, the first and second stalls in Fig. 2C). Transcription complexes display differing biochemical properties depending on the nucleotide sequence of the DNA to which they are bound (21). Abrupt reversible stalling may correspond to the arrival of the enzyme at a template position for which F_{stall} is somewhat lower than that of the preceding positions, producing rapid arrest. This interpretation also could explain why F_{stall} values determined for multiple stalls of a single complex often differed by more than the experimental uncertainty in measurement (22). It is tempting to speculate that some sites of abrupt stalling might correspond to DNA sequences that trigger cycles of discontinuous elongation (2, 3) or "jumping." At such sequences, a portion of the RNA polymerase molecule may, during a single nucleotide addition cycle, move along the DNA by ~10 bp, an axial distance of ~3.4 nm (23). These large movements of the enzyme would be expected, all else being equal, to give stall forces significantly lower than those of single base pair movements, because the free energy available from nucleotide addition would be applied over a larger distance. Nucleotide sequence effects on behavior under mechanical load could be studied in future experiments through the use of DNA templates containing homopolymer tracts or direct repeats.

The distributions of F_{stall} values were obtained at 1 mM NTPs with 1 μM PP$_i$. The distributions of reversible and irreversible stall forces were statistically indistinguishable, with values of 13.0 ± 4.0 pN (mean ± SD, n = 8) and 12.6 ± 3.5 pN (n = 14), respectively. A possible explanation for the similarity of the distributions is that the reversibly stalled state is a precursor of the irreversibly stalled state. In this view, most or all of the stalled complexes are initially stalled reversibly, but a fraction of these are subsequently inactivated and therefore fail to resume elongation when F_{trap} is reduced.

F_{stall} was also determined at two higher concentrations of PP$_i$ (0.5 and 1 mM) with 1

mM NTPs. Increasing the PP$_i$ concentration to 1 mM slowed transcriptional elongation by about twofold (24), but F_{stall} remained nearly constant [at 0.5 mM PP$_i$, 13.0 ± 4.1 pN (n = 5) for reversible and 13.1 ± 2.6 pN (n = 10) for irreversible stalls; at 1.0 mM PP$_i$, 11.5 ± 2.9 pN (n = 11) for reversible and 9.9 ± 3.4 pN (n = 18) for irreversible stalls]; this observation may place important constraints on the RNA polymerase force-generation mechanism. In light of this result, stalls observed at all PP$_i$ concentrations were pooled (Fig. 3A) to generate a global distribution with mean F_{stall} = 12.3 ± 3.5 pN (n = 24) for all reversible stalls. We previously found that single immobilized transcription complexes exhibit a range of velocities (7). Such heterogeneity among complexes may also contribute to the width of F_{stall} distributions. The polymerase is fixed to the glass at random and presumably is not free to rotate. Different spatial orientations of the polymerase with respect to the direction of the applied force might also cause F_{stall} to vary from molecule to molecule. A small fraction of the complexes did not stall before reaching the limit of the usable range of the trap (Fig. 3B). The assumption that these complexes would stall at a force larger than the maximum measurable force at the laser power used yields a lower limit estimate of 13.6 pN for the mean reversible F_{stall} (25). Despite systematic instrumentation errors for these measurements estimated at ~30% (22), the F_{stall} values for single RNA polymerase molecules are clearly much larger than are forces previously measured for single molecules of other mechanoenzymes: up to 6 pN for kinesin (26) and 3 to 5 pN for myosin (27).

The efficiency of chemomechanical energy conversion in RNA polymerase may be defined as the fraction of the total free energy change of the chemical RNA polymerization reaction ($\Delta G'_{polym}$) that the enzyme expends to perform mechanical work against an external load. Thermodynamically reversible motors are most efficient as they approach the point of stalling. The free energy available from the chemical reaction varies with the PP$_i$-to-NTP concentration ratio because these are the product and

Fig. 3. Stall force distributions measured from pooled data obtained at 1 mM NTPs with 0.001, 0.5, and 1.0 mM PP$_i$. For beads exhibiting reversible stalls, only data from the first stall were included. Laser powers (and corresponding maximum measurable forces) of 82 mW (15 pN), 99 mW (18 pN), and 107 mW (20 pN) were used in 25, 33, and 19 measurements, respectively. (**A**) Stacked histogram of stall forces for irreversible (open portion of bars, n = 42) and reversible (solid portion of bars, n = 24) stalls. (**B**) Histogram of estimated lower bound of stall forces for complexes (n = 11) that did not stall before exiting the calibrated range of the trap. The number below each bar represents the maximum measurable force for the laser power used.

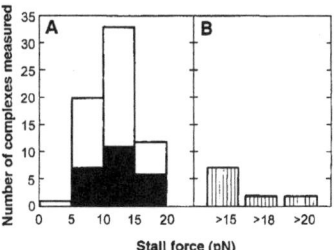

reactant species, respectively. For the conditions used, $\Delta G'_{\text{polym}}$ ranged from -7.2 kcal mol^{-1} (at 1 μM PP$_i$) to -3.1 kcal mol^{-1} (at 1 μM PP$_i$) (1). Under the assumption that each NTP consumed advances the polymerase 1 bp on average along the DNA (0.34 nm along the DNA axis), the mean F_{stall} estimate corresponds to maximal energy conversion efficiencies of 9, 19, and 22% at 0.001, 0.5, and 1 mM PP$_i$, respectively. These efficiencies are comparable to values obtained for biological motors such as kinesin [40 to 60% (11, 28)] or myosin [12 to 42% (27)] (29).

Nucleic acid polymerases carry out biosynthetic reactions and are not ordinarily classified as mechanoenzymes. Nevertheless, our results indicate that individual molecules of E. coli RNA polymerase exert forces and operate with energy conversion efficiencies that are similar to those of prototypical mechanoenzymes, whose specialized function is to generate biologically useful force and motion. Inside living cells, interactions between transcription complexes and cellular structures or DNA-bound proteins create substantial forces that oppose the translocation of polymerases relative to DNA (4, 5). For example, to function in opposition to loads imposed by transcription-induced supercoiling of plasmid DNA in vivo, E. coli RNA polymerase must generate forces estimated at $\sim$6 pN (30). Forces of this magnitude are sufficient, for example, to stall solitary molecules of kinesin or myosin, but are nonetheless smaller than the forces achieved by RNA polymerase molecules in vitro. We anticipate that further development of optical measurements on single transcription complexes will allow detailed characterization of the multiple mechanical processes (2, 3, 23) by which RNA polymerase moves along DNA.

REFERENCES AND NOTES

1. T. D. Yager and P. H. von Hippel, in Escherichia coli and Salmonella typhimurium: Cellular and Molecular Biology, F. C. Neidhardt et al., Eds. (American Society for Microbiology, Washington DC, 1987), vol. 2, pp. 1241–1275; D. A. Erie, T. D. Yager, P. H. von Hippel, Annu. Rev. Biophys. Biomol. Struct. 21, 379 (1992).
2. C. L. Chan and R. Landick, in Transcription: Mechanism and Regulation, R. Conaway and J. Conaway, Eds. (Raven, New York, 1994), pp. 297–320.
3. M. J. Chamberlin, in The Harvey Lectures (Wiley-Liss, New York, 1994), Ser. 88, pp. 1–21.
4. P. R. Cook, Bioessays 16, 425 (1994).
5. J. C. Wang, in Transcriptional Regulation, S. L. McKnight and K. R. Yamamoto, Eds. (Cold Spring Harbor Laboratory Press, Cold Spring Harbor, NY, 1992), pp. 1253–1269; D. N. Cook, D. Ma, J. E. Hearst, in Nucleic Acids and Molecular Biology, F. Eckstein and D. M. J. Lilley, Eds. (Springer-Verlag, Berlin, 1994), vol. 8, pp. 133–146.
6. D. A. Schafer, J. Gelles, M. P. Sheetz, R. Landick, Nature 352, 444 (1991).
7. H. Yin, R. Landick, J. Gelles, Biophys. J. 67, 2468 (1994).
8. H. Kabata et al., Science 262, 1561 (1993).
9. K. Svoboda, C. F. Schmidt, B. J. Schnapp, S. M. Block, Nature 365, 721 (1993).
10. Sample preparation methods, buffers, and reaction

conditions were the same as those used previously (7), except that disodium pyrophosphate (Sigma) was added at the specified concentration together with the NTPs, and 0.52-μm-diameter carboxylated polystyrene beads (Polysciences, Warrington, PA) were used in place of 0.226-μm beads. DNA template no. 5 (6), which contains the strong T7 A1 promoter followed by 3908 downstream base pairs, was used in all experiments. Transcription complex surface density was measured and controlled as in (7), so that each bead had a probability $\leq$0.20 of being attached to more than one transcription complex.
11. K. Svoboda and S. M. Block, Cell 77, 773 (1994).
12. _____, Annu. Rev. Biophys. Biomol. Struct. 23, 247 (1994).
13. S. B. Smith, L. Finzi, C. Bustamante, Science 258, 1122 (1992); C. Bustamante, J. F. Marko, E. D. Siggia, S. Smith, ibid. 265, 1599 (1994). More complete characterization of the series elastic spring constant α_{tc} (which includes contributions from the stiffness of the DNA as well as from the stiffness of other structures, such as the polymerase, through which the DNA is linked to the cover slip) could be used to simultaneously measure both the force and velocity generated by the polymerase, as has been done for kinesin (11). A complication of this approach is the fact that α_{tc} has a time-varying component, due to the changing length of the DNA tether.
14. We calibrated trap stiffness (α_{trap}) using free beads, either by measuring the power spectrum of the Brownian motion or by measuring the bead displacement from the trap center produced by viscous drag forces (9, 11, 12). Stiffnesses obtained with the use of these two methods differed by <10% for excursions below half the usable detector range (<100 nm). For larger excursions, the drag force method was used exclusively. Calibration was accomplished in two stages. First, the relation between digitized interferometer output voltage and true displacement (in nanometers) was established as described previously (9). Second, the relation between trapping force (in piconewtons) and displacement (in nanometers) was determined by the drag force method, with viscous forces being corrected for the proximity of the cover glass surface (12). The two data sets were then combined. In contrast to the case of silica particles (11), the force-displacement relation for polystyrene beads has significant nonlinearities in the outermost part of the trap; therefore, calibration data out to 150 nm were fit by a fifth-order polynomial to extrapolate interferometer signals into the region between 150 nm and $\sim$200 nm. This procedure may miscalculate the actual force produced in the region beyond 150 nm; however, exclusion from the analysis of data taken in this region did not significantly alter the mean stall forces reported here. The estimated force at 200 nm displacement was taken to be the largest measurable force at a given laser power.
15. A correction for the contribution of tether elasticity (spring constant α_{to}) has not been applied to the bead velocity measurement. Therefore, this bead velocity represents a lower bound on the polymerase elongation rate.
16. Interferometer output was low-pass–filtered at 1 kHz and digitized at 1-ms intervals. Before analysis, records were averaged in blocks of 100 points to give a final time resolution of 100 ms. Bead velocity (v_{bead}) was measured as the slope of the line fit to bead position versus time within a 10-s sliding window. Beads achieving $v_{bead} \geq 0.6$ nm s^{-1} continuously for $\geq$5 s (positive velocity being defined as movement away from the trap center) were judged to be attached to functional transcription complexes and were selected for further analysis. Any interval during which v_{bead} was <0.6 nm s^{-1} continuously for $\geq$5 s was taken to represent a stall. By this criterion, 13 of 77 complexes spontaneously underwent brief transient stalls that were apparently uncorrelated with the amount of force and were excluded from subsequent analyses. Such events may correspond to transcriptional pausing (1, 2) because movement resumed without the trap being repositioned.
17. Initially, 87 elongating complexes were observed at

high laser powers. However, in 10 of them the bead spontaneously detached from its tether before the trap could be repositioned. The site of breakage could not be determined. Data from such complexes were excluded from further analysis. Eleven of the remaining 77 complexes did not stall before reaching the limit of the usable range of the trap.
18. D. A. Erie, O. Hajiseyedjavadi, M. C. Young, P. H. von Hippel, Science 262, 867 (1993).
19. Although inactive enzyme species induced by chemical stalling in solution can be reactivated by GreB protein [S. Borukhov, V. Sagitov, A. Goldfarb, Cell 72, 459 (1993)], the fraction of complexes that were irreversibly stalled by mechanical load did not decrease when 100 nM GreB protein was included in the buffer (24), which suggests either that these complexes are in a different state than that produced in the chemical stalling experiments or that GreB cannot act on complexes adsorbed to cover slips.
20. It is unlikely that the reversible stalling events are due to radiation-induced damage to RNA polymerase, because enzyme photodamage is generally irreversible. The view that reversible stalls are due to applied force, not laser irradiation alone, is further supported by the observation that 95% of reversibly stalled complexes resumed bead movement shortly after the trap was repositioned. Repositioning the trap decreases F_{trap} but increases the laser intensity at the RNA polymerase because the trap is moved closer to the enzyme molecule. Control experiments [at laser powers 82 to 99 mW; all reported laser powers are estimates in the specimen plane determined by the method of (11)] in which the trap was repositioned before the stall (in 14 of 15 complexes) subjected the transcription complexes to higher average light intensities and lower average trap forces than those in the stalling experiments. In the controls, complexes were inactivated in 82 ± 58 s (mean ± SD, n = 15) exposure to the laser, whereas in the stalling experiments first stalls (reversible or irreversible) occurred significantly earlier, in 38 ± 16 s (mean ± SD, n = 66), confirming that the first stalls cannot be explained by photodamage alone.
21. B. Krummel and M. J. Chamberlin, J. Mol. Biol. 225, 221 (1992).
22. Stall force measurements are subject to systematic errors, which we attempted to estimate. The two primary sources of error have opposing effects and arise from the fact that the bead motion is not strictly parallel to the cover glass surface but has a vertical component. During elongation, beads experience a force directed downward at an angle θ, where θ is the angle between the DNA and the cover glass (Fig. 1C). This angle varies continuously during the course of an experiment as the DNA tether shortens. Neglect of the downward force component will cause underestimation of the actual force produced by the polymerase by a factor (cosθ). The minimal DNA tether length at stall was estimated by video analysis of the bead Brownian motion (7) with the trap shuttered. The tether length was >1915 bp (651 nm) in all experiments, corresponding to θ < 41°. This geometrical consideration alone would lead to a value for F_{stall} that underestimates the applied force by a variable amount up to $\sim$24%. A second source of systematic error, which has the opposite sign, arises from the fact that the interferometer is most sensitive to motion in the plane of the laser beam waist and along the Wollaston shear direction. Polystyrene spheres are initially trapped at a point slightly above the position of the true beam waist (Fig. 1B), and the downward force component acting along the DNA has the effect of carrying the bead into the region of greater detector sensitivity (in contrast, viscous drag calibration experiments, in which purely horizontal forces are generated, are not subject to this artifact). This leads to an overestimate of the distance moved, hence to an overestimate of the force F_{stall}. We estimated the magnitude of this effect at 26% or less, based on experiments in which beads fixed to a cover glass were moved in the specimen plane through the detector at various heights relative to the laser beam waist (24). Yet other uncertainties arise because the trap stiffness, hence the restoring force, varies with height, the trap being weaker in the vertical direction than in the horizontal direction. Given these various

opposing effects, it seems reasonable to estimate the overall systematic errors in force at ≤ ~30%.

23. B. Krummel and M. J. Chamberlin, *J. Mol. Biol.* **225**, 239 (1995); E. Nudler, A. Goldfarb, M. Kashlev, *Science* **265**, 793 (1994); D. Wang *et al.*, *Cell* **81**, 341 (1995); E. Nudler, M. Kashlev, V. Nikiforov, A. Goldfarb, *ibid.*, p. 351.

24. H. Yin *et al.*, data not shown.

25. This estimate is a lower limit because it includes data from complexes that did not stall before reaching the trap limit and because we cannot exclude the possibility that some events classified as stalls may in fact be lengthy transcriptional pauses (*16*).

26. S. M. Block, *Trends Cell Biol.* **5**, 169 (1995).

27. J. T. Finer, R. M. Simmons, J. A. Spudich, *Nature.* **368**, 113 (1994); J. T. Finer, A. D. Mehta, J. A. Spudich, *Biophys. J.* **68**, 291s (1995); A. Ishijima *et al.*, *Biochem. Biophys. Res. Commun.* **199**, 1057 (1994); H. Miyata *et al.*, *Biophys. J.* **68**, 286s (1995).

28. A. J. Hunt, F. Gittes, J. Howard, *Biophys. J.* **67**, 766 (1994).

29. Underestimation of F_{stall} (*25*) will cause underestimation of the maximum energy conversion efficiency. Moreover, our efficiency calculation assumes that stalling occurs when the free energy available to drive movement in one cycle of the chemical reaction balances the free energy required to translocate 1 bp along the DNA against the applied force. However, stalling could also occur because of structural alterations in the transcription complex (for example, stabilization of a catalytically inactive enzyme conformation by the applied force). In that case, the maximum efficiency calculated from F_{stall} would underestimate the true efficiency. In contrast, enzymatic consumption of NTPs uncoupled to translocation would reduce the energy conversion efficiency below that calculated from F_{stall}. Although little such consumption has been reported for transcription complexes in solution [for example see M. Chamberlin, R. L. Baldwin, P. Berg, *J. Mol. Biol.* **7**, 334 (1963)], we cannot exclude the possibility that such uncoupled reactions are catalyzed by immobilized enzyme molecules subjected to the high applied forces used in the stalling experiments. Estimates of energy conversion efficiency from the stall forces of single kinesin and myosin molecules carry analogous uncertainties. Recent evidence (*11*) suggests that kinesin is not tightly coupled near stall.

30. *Escherichia coli* plasmids that carry multiple transcription units can accumulate high densities of transcription-induced supercoils (*31*). This effect can be observed in the absence of DNA gyrase or topoisomerase I activities (which relax negative or positive supercoils, respectively) when one or more of the transcribed genes encodes a membrane-interacting protein (*5*). Specific linking differences (σ) of −0.013 in topoisomerase I mutants [D. N. Cook, D. Ma, N. G. Pon, J. E. Hearst, *Proc. Natl. Acad. Sci. U.S.A.* **89**, 10603 (1992)] and 0.024 in the presence of a DNA gyrase inhibitor [H. Y. Wu, S. H. Shyy, J. C. Wang, L. F. Liu, *Cell* **53**, 433 (1988)] have been reported. RNA polymerase transcription is accompanied by an obligatory rotation relative to the DNA helix. Therefore, the supercoiling torque τ_s [($|\sigma|$) 1.4×10^{-19} N m rad^{-1} (*31*)] is expected to exert a force opposing translocation that is estimated by $F_s = 2\pi\tau_s/h$, where h is the pitch of the DNA helix, 3.6 nm. $F_s = 3$ and 6 pN, respectively, in the two cited cases. However, this calculation may underestimate the force by a factor of ~2 because supercoils in vivo are thought to be confined to a limited portion of the plasmid DNA.

31. L. F. Liu and J. C. Wang, *Proc. Natl. Acad. Sci. U.S.A.* **84**, 7024 (1987).

32. Supported by grants from the National Institute of General Medical Sciences to J.G., R.L., and S.M.B. M.D.W. was supported by a Damon Runyon–Walter Winchell Cancer Research Fund postdoctoral fellowship. K.S. and S.M.B. thank the Rowland Institute for Science for support during the early stages of this work. A movie of the experiment shown in Fig. 2A can be viewed on the World Wide Web at http://www.rose.brandeis.edu/users/gelles/stall/.

11 August 1995; accepted 26 October 1995

letters to nature

Single-molecule studies of the effect of template tension on T7 DNA polymerase activity

Gijs J.L. Wuite*, Steven B. Smith*, Mark Young†, David Keller‡ & Carlos Bustamante*

* Department of Physics and Department of Molecular and Cell Biology, University of California, Berkeley, California 94720, USA
† Institute of Molecular Biology, University of Oregon, Eugene, Oregon 97403, USA
‡ Department of Chemistry, University of New Mexico, Albuquerque, New Mexico 87131, USA

T7 DNA polymerase[1,2] catalyses DNA replication *in vitro* at rates of more than 100 bases per second and has a $3' \rightarrow 5'$ exonuclease (nucleotide removing) activity at a separate active site. This enzyme possesses a 'right hand' shape which is common to most polymerases with fingers, palm and thumb domains[3,4]. The rate-limiting step for replication is thought to involve a conformational change between an 'open fingers' state in which the active site samples nucleotides, and a 'closed' state in which nucleotide incorporation occurs[3,5]. DNA polymerase must function as a molecular motor converting chemical energy into mechanical force as it moves over the template. Here we show, using a single-molecule assay based on the differential elasticity of single-stranded and double-stranded DNA, that mechanical force is generated during the rate-limiting step and that the motor can work against a maximum template tension of

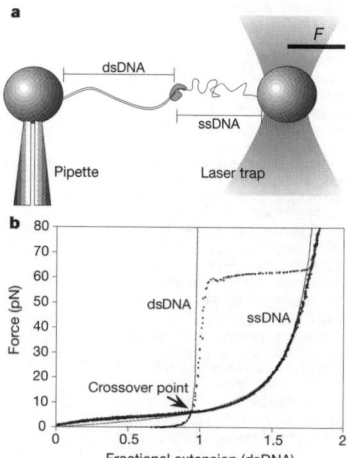

Figure 1 Optical trap setup. **a,** A 10,416-base pair plasmid DNA fragment was prepared as described[18] and attached between two beads, one held on the tip of a glass pipette, the other in an optical trap. Single-stranded DNA was obtained by using the force-induced exonuclease activity of T7 DNAp to remove any desired length of the non-template strand. The end-to-end length of the DNA was obtained by video imaging of the bead positions, and the force (F) was measured using the change in light momentum which exits the dual-beam trap[6]. **b,** Force–extension data for dsDNA and ssDNA (dotted lines), compared with the wormlike chain model using ssDNA and dsDNA persistence lengths of 0.7 nm and 53 nm respectively (solid lines)[7,8]. Difference between ssDNA and WLC curves at low tension is due to partial hairpin formation and disappears if magnesium is removed from buffer.

103

letters to nature

~34 pN. Estimates of the mechanical and entropic work done by the enzyme show that T7 DNA polymerase organizes two template bases in the polymerization site during each catalytic cycle. We also find a force-induced 100-fold increase in exonucleolysis above 40 pN.

We measured T7 DNA polymerase (DNAp) activity by using the optical-trap shown in Fig. 1a[6]. Because single-stranded (ss) and double-stranded (ds) DNA differ in length at any given tension[6–8] (Fig. 1b), conversion between these two forms changes the molecule's tension if the end-to-end distance of the template is held constant. Alternatively, the molecule's end-to-end distance changes if the tension is held constant. The end-to-end distance of ssDNA (Fig. 1b) is shorter than that of dsDNA for tensions below 6.5 pN ('crossover point') because ssDNA, despite having about twice the contour length of dsDNA, is more retractile owing to its greater flexibility. Above 6.5 pN, however, contour length predominates over entropy and ssDNA is longer than dsDNA. The force–extension curve of a molecule that is partly ssDNA and partly dsDNA can be fit to a linear combination of ssDNA and dsDNA stretching curves[6]. Thus, the progress of DNAp can be followed by the number of single-stranded bases, N_{ss}, remaining in the template at time t.

$$N_{ss}(t) = \frac{x_{meas}(F, t) - x_{ds}(F)}{x_{ss}(F) - x_{ds}(F)} * N_{tot} \quad (1)$$

where x_{meas} is the end-to-end distance of the molecule at force F; $x_{ds,ss}(F)$ are the end-to-end distances of fully double- or single-stranded DNA at that force, and N_{tot} is the total number of bases in the template.

Figure 2a (upper line) plots the ssDNA fraction remaining at time t as DNAp replicates against an applied tension of 20 pN. The instantaneous polymerization rate (lower line) determined from the time derivative of this curve shows bursts of activity. Two lines of evidence suggest that each burst corresponds to a DNAp molecule loading onto the 3′ end of the growing chain, replicating proces-

sively, and falling off. First, the mean width of a burst is force and concentration independent, and the rate corresponding to this width (0.13 ± 0.1 s^{-1}; $N = 62$) is near the off rate from the polymerization site measured in bulk[9]. Second, varying the DNAp concentration changed the width of the gaps between bursts. At 0.8 nM, gap widths average 7 ± 4 s, consistent with ~7 s calculated from the enzyme loading rates from solution (~180 s^{-1} μM^{-1}; Fig. 2b). Increasing the DNAp concentration to 8 nM, decreased gap size to 1.7 ± 0.8 s. At 80 nM and 880 nM, however, this trend reversed and the gaps increased to 7 ± 2 s and 50 ± 26 s, respectively. Notably, T7 DNAp can bind non-specifically along ssDNA with a dissociation constant of ~800 nM[10]. Therefore, such binding may block enzyme reloading at the 3′ end, increasing gap widths. Because the average burst height remained independent of enzyme concentration (data not shown), non-specific binding does not seem to impede translocation once an enzyme is specifically bound. Occasionally, replication would stop for extended periods (up to 30 min), indicating 'roadblocks' that the polymerase would not cross. Causes might include exogenous DNA hybridized to the template or bases missing from the template because of chemical or enzymatic damage. Template hairpins are an unlikely cause as blockage persists above 15 pN, where such structures should pull out[11].

Figure 2a shows surprising diversity in polymerization rates among individual DNAp molecules. Burst heights typically vary between 26 and 60 bases s^{-1} ($N = 39$, s.d. = 17), at a template tension of 20 pN. As the intrinsic rate of each DNAp molecule can be determined from analysis of burst heights (see Methods), such variations probably reflect differences in enzymatic activity among individual molecules. Similar differences have been reported for other enzymes[12,13].

The effect of template tension on the polymerization rate was determined either by holding the tension constant through force-feedback or by holding the template strand at constant end-to-end length. This length was chosen beyond the crossover point, so that polymerization increased template tension until the system halted itself, having sampled all intermediate tensions (Fig. 3). At the lowest tension, the replication rate of T7 DNAp was ~100 bases s^{-1}. Raising template tension increased the replication rate until a maximum of ~200 bases s^{-1} was reached at about 6 pN. Further increase in tension, however, caused the rate to decrease until polymerization stalled. For 12 constant-distance runs, the mean stall force was 34 ± 8 pN. On rare occasions, polymerization

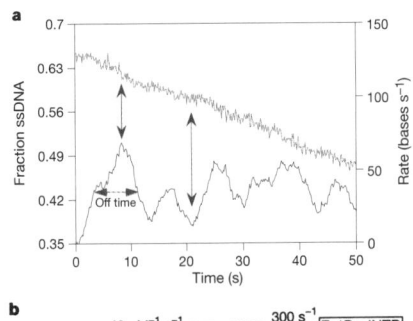

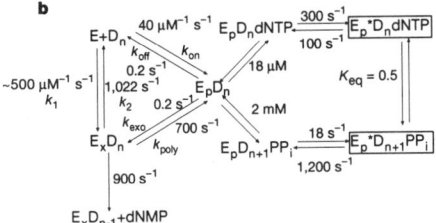

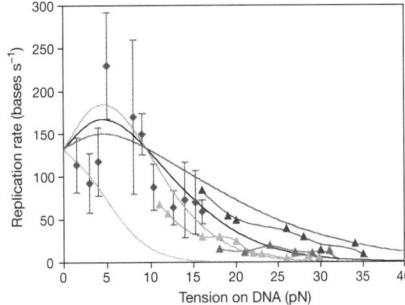

Figure 2 Polymerization kinetics. **a**, Replication of a ssDNA template under 20 pN tension using 8 nM DNA. Upper curve, conversion to dsDNA plotted as fraction of ssDNA left in the template versus time. Lower curve, polymerization rate obtained by differentiating upper curve after smoothing it with a moving-average filter (24 data points) to reduce Brownian noise. **b**, Induced fit kinetic pathway of T7 DNAp according to Patel et al.[9] with k_1 and k_2 taken from Donlin et al.[16]. States of the DNA–enzyme complex are shown as: E, enzyme in solution; E$_{p,x}$, DNA bound in polymerase active site or exo active site; D$_n$, DNA primer with length n; dNTP, deoxynucleotide; and PP$_i$, pyrophosphate.

Figure 3 Polymerization rate versus template tension. Diamonds represent 50 polymerization bursts taken at 11 different tensions (error bar, s.d.); triangles represent three traces fitted through a succession of replication bursts measured at constant end-to-end distances until a stalling force is reached; thick lines represent fits of equation (4) using $a = 0$, $k_0 = 130$ bases s^{-1}, and $n = 1$ (brown), $n = 2$ (black), $n = 3$ (grey) and $a = 1$ nm, $n = 2$ (purple). Noise increases near 0 pN and the crossover point, because the numerator and denominator of equation (1) become small while the brownian motion does not decrease.

proceeded briefly above 50 pN. Replication did not restart spontaneously at these high tensions even after several minutes, but replication always resumed after lowering the tension. The force would then rise once again until a new stalling force, usually different from the previous one, was reached, perhaps reflecting the effect of template sequence, or the stochastic nature of the stalling process itself.

The sensitivity of the polymerization rate even to low tensions indicates that the rate-limiting step is directly affected by force. This force dependence is consistent with the induced-fit model of Wong et al.[14], in which the enzyme changes conformation during the rate-limiting step. Crystal structures of the closed state show the fingers rotated ~40° to align the different components in the active site[3,4], and the ssDNA bending sharply in relation to the primer as it leaves this site.

If the tension, F, on the template can exert a torque on the fingers, the work done by the enzyme would include a term aF, where a is the distance moved by the fingers along the pulling direction of the template. Further, assume that closing the fingers organizes n adjacent sugar-phosphate units from single- to double-stranded geometry, $n-1$ of which are released when the fingers reopen. Then the work W to close the fingers is

$$W(F) = aF + nF(x_{ss}(F) - x_{ds}(F)) \qquad (2)$$

where $x_{ss}(F)$ and $x_{ds}(F)$ are the end-to-end distances per base of ssDNA and dsDNA at tension F. The second term in equation (2) changes sign at 6.5 pN, aiding or opposing replication below or above this force and causing the instrument to do work on the reaction or the reaction to do work on the instrument, respectively. Closing of the fingers also clamps the template strand reducing its degrees of freedom. Tensions applied to the template decrease its entropy and, consequently, the energetic cost of closing the fingers, thus speeding up the reaction. The entropy change, ΔS, to convert ssDNA into dsDNA at any given force can be obtained from the areas under the experimental force–extension curves (Fig. 1b), that is,

$$T\Delta S(F) = n\left[\int_0^{x_{ss}(F)} F_{ss}\,dx - \int_0^{x_{ds}(F)} F_{ds}\,dx \right] \qquad (3)$$

where $F_{ds,ss}$ are the experimental forces required to extend the chains by an amount x. If we assume that the terms in equations (2) and (3) contribute to the activation energy required to reach a transition state (for example, the complex with fingers half-closed) from the open state, then the rate coefficient (k) for this step is

$$k = k_0 e^{-((w - T\Delta S)/k_b T)} \qquad (4)$$

where k_0 is the rate coefficient at zero force; k_b is the Boltzman constant and T is the temperature. Figure 3 compares equation (4) to the data using various values of a and n. The best fit is obtained for $a = 0$ and $n = 2$, indicating that template tension exerts little torque opposing finger closure and that two adjacent sugar-phosphate units are organized by the fingers in this process. These results are supported by the structure of the closed complex which shows the template strand avoiding the finger tips and passing around the side of the fingers through a shallow cleft[3,4]. Moreover, two adjacent template bases appear immobilized in this structure. The first, opposite the incoming nucleotide, adopts a B-form structure, whereas the second is kinked outward, almost perpendicular to the pulling direction. The interphosphate distance corresponding to these two bases is close to that of dsDNA. Subsequent bases ($n > 2$) appear disorganized in the structure. Single-molecule studies of an exonuclease-deficient mutant of T7 DNAp (Sequenase) also suggest immobilization of two bases during finger closure (B. Maier, D. Bensimon and V. Croquette, personal communication).

When template tension was increased above 40 ± 3 pN ($N = 16$), a fast exonucleolysis (30 ± 11 bases s⁻¹) was initiated with or

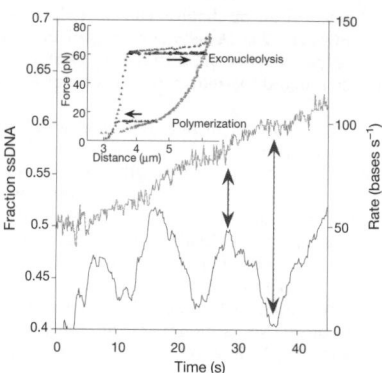

Figure 4 Exonuclease digestion of primer strand by T7 DNAp (at 8 nM) at template tension of 50 pN. Upper curve, ssDNA fraction of 10-kilobase template. Lower curve, exonuclease rate obtained by differentiating upper curve after application of a 3-s moving-average filter. Inset, force–extension curve for a dsDNA molecule (green points) before it was almost entirely converted to ssDNA by exonucleolysis at a constant force of 60 pN (upper horizontal points, right arrow). At the end of this process a force-extension curve for the ssDNA was obtained (red points). Finally, in the presence of dNTPs, the tension was decreased to 15 pN to allow T7 DNAp to reconvert ssDNA into dsDNA (lower horizontal points, left arrow).

without dNTPs present (Fig. 4). Decreasing tension below ~34 pN caused exonucleolysis to halt and polymerization to resume. Switching between these opposite activities (inset, Fig. 4) could be repeated many times on one template (data not shown). Exonucleolysis force dependence was measured either using constant force or constant end-to-end distance (Fig. 5). The exonucleolysis rate became force independent above 42 pN and is ~100 times faster than observed at zero tension on dsDNA, where the exo rate is limited by the escape of the 3′ end from the polymerization site (Fig. 2b). Force-induced exonucleolysis appears as bursts of activity (Fig. 4, lower line). Presumably, template tension shifts the equilibrium in favour of exonucleolysis either by increasing the escape rate from (k_{off} and k_{exo}) or decreasing the binding rate to (k_{on} and k_{poly}) the polymerization site, or both. However, if the escape rate from the polymerization site were increased by two orders of magnitude, such escape would still be rate limiting for exonucleolysis (0.2 s⁻¹ × 100) which would appear continuous at our temporal resolution. The gaps observed in the exonuclease rate are instead consistent with a 100-fold decrease in binding rate to the polymerization site. Because the enzyme bound through its exonuclease site associates/dissociates rapidly from solution to the 3′ end (Fig. 2b), many DNAps can bind, exonucleate and dissociate, before one of them moves back to the polymerase site. It then takes several seconds for this enzyme to escape the polymerase site ($k_{exo} = 0.2$ s⁻¹), resulting in gaps of activity as observed in Fig. 4. Thus, each exo-pause results from one DNAp molecule lingering in the polymerization active site, while each exo-burst, seeming continuous because of our temporal resolution, could result from the action of many individual DNAp molecules.

Although the gaps can be explained by a slow escape of the 3′ end from the poly site, the heights of the exo-bursts are inconsistent with published association/dissociation rates of the exo site from solution (k_1 and k_2, Fig. 2b). Unexpectedly, the forced-induced exonucleolysis rate remains relatively constant for DNAp concentration between 800 nM and 8 nM, dropping by 50% only when the concentration is lowered to 0.8 nM. Under zero tension, it is thought that several base pairs must be melted to allow binding of the DNA primer strand to the exonuclease site[5], and forces greater

letters to nature

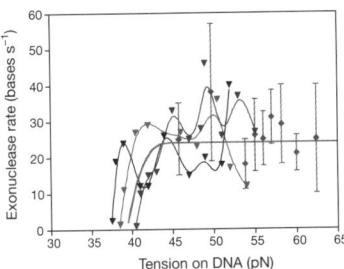

Figure 5 Force dependence of 3′→5′ exonuclease reaction. Diamonds represent average rates for 49 (total) exonuclease bursts measured at 9 different forces. Traces represent three lines fitted through successions of exonucleolysis burst heights (triangles), initiated at high tensions, on DNAs kept at constant end-to-end distances. Digestion lowers the tension in the DNA until the fast exonucleolysis stops. The upper limit for these experiments is determined by the overstretching force, 65 pN.[6,19]. Near the stalling force for polymerization, a competition was occasionally observed between exonucleolysis and polymerization which caused the template tension to bounce up and down every few seconds.

than 40 pN are known to promote fraying in dsDNA[15]. Perhaps the exonucleolysis rates simply reflect the rate of fraying of DNA at the tensions applied to the template and the known fast exo activity of T7 DNAp[16] on ssDNA. To test this idea, we investigated the effect of template tension on the activity of *Escherichia coli* exonuclease I. This enzyme attacks only ssDNA[17] at its 3′ end, so it should not digest the primer strand unless it frays. No activity was detected under a tension of 40 pN, but at 50 pN bases were removed at a rate of ~200 s[-1]. This fast exonucleolysis suggests that 3′ end fraying is not rate limiting during exonucleolysis by T7 DNAp. Thus, the force-induced exonucleolysis initiated at 40 pN is probably a specific property of T7 DNAp itself. For example, 40-pN tension may deform the dsDNA geometry at the 3′ end enough to trigger the enzyme's proof-reading function. But, as Fig. 5 shows, the fraying is rate limited in the presence of T7 DNAp. Perhaps the melting rate is mediated by some interaction with the enzyme which is independent of the tension. Future single-molecule experiments should elucidate the mechanism of force-induced exonucleolysis and provide additional insight into the mechanochemistry of DNA polymerase. □

Methods

Single-molecule assay

T7 DNAp (T7 gene 5 in a 1:1 complex with thioredoxin) was used in various concentrations (0.8–880 nM), where 1 nmol = ~230 activity units (Amersham). Replication buffer was 40 mM Tris pH 7.5, 5 mM MgCl$_2$, 50 mM NaCl, 50 μg ml^{-1} BSA, 0.1% NaN$_3$, 5 mM dithiothreitol and 0.6 mM (each) dNTPs. The end of one of the DNA chains was covalently coupled to a bead while the other end of the same chain was attached through a biotin/streptavidin linkage to the second bead[18](Fig. 1a). Exonuclease I (USB) was used at 50 units ml^{-1} in replication buffer altered to pH 8 and without dNTPs.

Temporal resolution

Data were collected at 8 Hz with a bandwidth of 60 Hz. The polymerization rate data was averaged over 3 s and, therefore, the height of bursts longer than 3 s should represent the intrinsic activity of individual molecules. For the expected average replication time (~5 s), most burst heights will be accurately determined with this temporal resolution. No correlation between burst height and width (measured at the half maximum) was found for bursts longer than 3 s.

Processivity

Because the replication rate varies with force (Fig. 3), whereas the dissociation rate (k_{off}) seems to remain constant, the processivity varies with force and is ~420 bases (60 bases s^{-1}, 0.13 s^{-1}) at 15 pN.

Received 3 September 1999; accepted 6 January 2000.

1. Modrich, P. & Richardson, C. C. Bacteriophage T7 deoxyribonucleic acid replication *in-vitro*. Bacteriophage T7 DNA polymerase: an enzyme composed of phage- and host-specific subunits. *J. Biol. Chem.* **250**, 5515–5522. (1975).
2. Tabor, S., Huber, H. E. & Richardson, C. C. *Escherichia coli* thioredoxin confers processivity on the DNA polymerase activity of the gene 5 protein of bacteriophage T7. *J. Biol. Chem.* **262**, 16212–16223 (1987).
3. Doublie, S. & Ellenberger, T. The mechanism of action of T7 DNA polymerase. *Curr. Opin. Struct. Biol.* **8**, 704–712 (1998).
4. Doublie, S. *et al.* Crystal structure of a bacteriophage T7 DNA replication complex at 2.2 Å resolution. *Nature* **391**, 251–258 (1998).
5. Johnson, K. A. conformational coupling in DNA polymerase fidelity. *Annu. Rev. Biochem.* **62**, 685–713 (1993).
6. Smith, S. B., Cui, Y. & Bustamante, C. Overstretching B-DNA: the elastic response of individual double-stranded and single-stranded DNA molecules. *Science* **271**, 795–799 (1996).
7. Marko, J. F. & Siggia, E. D. Stretching DNA. *Macromolecules* **28**, 8759–8770 (1995).
8. Bustamante, C., Marko, J. F., Siggia, E. D. & Smith, S. B. Entropic elasticity of lambda-phage DNA. *Science* **265**, 1599–1600 (1994).
9. Patel, S. S., Wong, I. & Johnson, K. A. Pre-steady-state kinetic analysis of processive DNA replication including complete characterization of an exonuclease-deficient mutant. *Biochemistry* **30**, 511–525 (1991).
10. Huber, H. E., Tabor, S. & Richardson, C. C. *Escherichia coli* thioredoxin stabilizes complexes of bacteriophage T7 DNA polymerase and primed templates. *J. Biol. Chem.* **262**, 16224–16232 (1987).
11. Essevaz-Roulet, B., Bockelmann, U. & Heslot, F. Mechanical separation of the complementary strands of DNA. *Proc. Natl Acad. Sci. USA* **94**, 11935–11940 (1997).
12. Xue, Q. & Yeung, E. Differences in the chemical reactivity of individual molecules of an enzyme. *Nature* **373**, 681–683 (1995).
13. Lu, H. P., Xun, L. & Xie, X. S. Single-molecule enzymatic dynamics. *Science* **282**, 1877–1882 (1998).
14. Wong, I., Patel, S. S. & Johnson, K. A. An induced-fit kinetic mechanism for DNA replication fidelity: direct measurement by single-turnover kinetics. *Biochemistry* **30**, 526–537 (1991).
15. Gurrieri, S., Smith, S. B. & Bustamante, C. Trapping of megabase-sized DNA molecules during agarose gel electrophoresis. *Proc. Natl Acad. Sci. USA* **96**, 453–458 (1999).
16. Donlin, M. J., Patel, S. S. & Johnson, K. A. Kinetic partitioning between the exonuclease and polymerase sites in DNA error correction. *Biochemistry* **30**, 538–546 (1991).
17. Lehman, I. R. & Nussbaum, A. L. On the specificity of E-coli exonuclease I. *J. Biol. Chem.* **239**, 2628–2636 (1964).
18. Hegner, M., Smith, S. B. & Bustamante, C. Polymerization and mechanical properties of single RecA-DNA filaments. *Proc Natl Acad. Sci. USA* **96**, 10109–10114 (1999).
19. Cluzel, P. *et al.* DNA: an extensible molecule. *Science* **271**, 792–794 (1996).

Acknowledgements

We thank J. Davenport and M. Hegner for their help with the template and suggestions, and D. Bensimon, B. Maier and V. Croquette for sharing their results before publication.

Correspondence and requests for materials should be addressed to C.B. (e-mail: carlos@alice.berkeley.edu).

Direct observation of kinesin stepping by optical trapping interferometry

Karel Svoboda[*][†], Christoph F. Schmidt[*][‡], Bruce J. Schnapp[§]
& Steven M. Block[*][||]

* Rowland Institute for Science, 100 Edwin Land Boulevard, Cambridge, Massachusetts 02142, USA
† Committee on Biophysics, Harvard University, Cambridge, Massachusetts 02138, USA
§ Department of Cell Biology, Harvard Medical School, Boston, Massachusetts 02115, USA

Do biological motors move with regular steps? To address this question, we constructed instrumentation with the spatial and temporal sensitivity to resolve movement on a molecular scale. We deposited silica beads carrying single molecules of the motor protein kinesin on microtubules using optical tweezers and analysed their motion under controlled loads by interferometry. We find that kinesin moves with 8-nm steps.

ENZYMES such as myosin, kinesin, dynein and their relatives are linear motors converting the energy of ATP hydrolysis into mechanical work, moving along polymer substrates: myosin along actin filaments in muscle and other cells; kinesin and dynein along microtubules. Motion derives from a mechanochemical cycle during which the motor protein binds to successive sites along the substrate, in such a way as to move forward on average[1-3]. Whether this cycle is accomplished through a swinging crossbridge, and how cycles of advancement are coupled to ATP hydrolysis, have been the subject of considerable debate[4-6]. In vitro assays for motility[7,8], using purified components interacting in well-defined experimental geometries, permit, in principle, measurement of speeds, forces, displacements, cycle timing and other physical properties of individual molecules, using native or mutant proteins[9-14].

Are there steps?

Do motor proteins make characteristic steps? That is, do they move forward in a discontinuous fashion, dwelling for times at

‡ Present address: Department of Physics, University of Michigan, Ann Arbor, Michigan 48109, USA.
|| To whom correspondence should be addressed.

well-defined positions on the substrate, interspersed with periods of advancement? We define the step size, which may be invariant or represent a distribution of values, as the distance moved forwards between dwell states. (Here we use 'step size' in its physical sense, and not to mean the average distance moved per molecule of ATP hydrolysed, also termed the 'sliding distance'[4,9–11,15–18]. The latter corresponds to the physical step size, or an integral multiple thereof, in models in which ATP hydrolysis and stepping are tightly coupled.) If stepping occurs, the distributions of step sizes and dwell times will place constraints on possible mechanisms for movement.

The motor protein kinesin has advantages over myosin for physical studies, despite a comparative dearth of biochemical data. It can be made to move slowly, permitting better time-averaging of position, and it remains attached to the substrate for a substantial fraction of the kinetic cycle[12,13,19], reducing the magnitude of brownian excursions. Movement by single molecules of kinesin has been demonstrated[12,13], and kinesin can transport small beads, which provide high-contrast markers for motor position in the microscope[8]. Unlike actin filaments, microtubules are relatively rigid[20], and can be visualized by video-enhanced differential interference-contrast microscopy[21] (DIC). Finally, recombinant kinesin expressed in bacteria has been shown to move *in vitro*, paving the way for future study[14].

Significant technical difficulties nevertheless exist in measuring movements of single molecules. The motions occur on length scales of ångströms to nanometres and on timescales of milliseconds and less. To obtain the high spatial and temporal sensitivity required, we combined optical tweezers[22,23] with a dual-beam interferometer[24] to produce an 'optical trapping interferometer'. A laser, focused through a microscope objective of high numerical aperture, provides position detection and trapping functions simultaneously, and can produce controlled, calibratable forces in the piconewton range. Using this device, we captured silica beads with kinesin molecules bound to their surface out of a suspension and deposited them onto microtubules immobilized on a coverslip. We then observed the fine structure of the motion as beads developed load by moving away from the centre of the trap.

Our data provide evidence for steps under three sets of conditions. At moderate levels of ATP and low loads, kinesin movement was load-independent, and the relatively high speed, in combination with brownian motion, precluded direct visualization of steps. A statistical analysis of the trajectories, however, revealed a stepwise character to the motion. At saturating levels of ATP and high loads, or at low levels of ATP and low loads, (when movement is slowed mechanically or chemically), it was possible to see the abrupt transitions directly. We estimate the step size to be 8 nm, a distance that corresponds closely to the spacing between adjacent α–β tubulin dimers in the protofilament of the microtubule[25], suggesting that the elementary step spans one dimer.

Optical trapping interferometer

Polarized laser light is introduced at a point just below the objective Wollaston prism into a microscope equipped with DIC optics (Fig. 1a). The prism splits the light into two beams with orthogonal polarization: these are focused to overlapping, diffraction-limited spots at the specimen plane, and together they function as a single optical trap[23]. A phase object in the specimen plane inside the region illuminated by the two spots (the detector zone) introduces a relative retardation between the beams, so that when they recombine and interfere in the condenser Wollaston prism, elliptically polarized light results (Fig. 1a, left). The degree of ellipticity is measured by additional optics[24] and provides a sensitive measure of retardation, which changes during movement, passing through zero when the object exactly straddles the two spots (for example, a bead moved along a microtubule, as shown in Fig. 1a, right). For small excursions

(out to $\sim$150 nm), the output of the detector system is linear with displacement (Fig. 1a, inset).

Over most of its bandwidth, detector noise is at or below 1 Å/$\sqrt{\text{Hz}}$ (Fig. 1b). The response to a 100 Hz sinusoidal calibration signal of 1 nm amplitude (Fig. 1b, inset) shows this ångström-level noise. An optical trapping interferometer has advantages over conventional split-photodiode systems[26,27]. Because it is a non-imaging device, it is relatively insensitive to vibrations of the photodetector. Laser light levels ensure that detectors do not become shot-noise-limited. The detector zone can be repositioned rapidly within the microscope field of view. Because trapping and position-sensing functions are provided by the same laser beam, the two are intrinsically aligned. Finally, the arrangement does not interfere with the simultaneous use of conventional DIC imaging.

Trapping force can be calibrated in a number of ways[28–31]. Two independent methods were used here. A rapid, convenient method applicable for small excursions from the trap centre is to measure the thermal motion of an unbound, trapped bead (Fig. 1c). The optical trap behaves like a linear spring, so that dynamics correspond to brownian motion in a harmonic potential, which has a lorentzian power spectrum. Experimental spectra are well fitted by lorentzians, and the corner frequency provides the ratio of the trap spring constant to the frictional drag coefficient of the bead[32]. The latter is obtained from Stokes' law, after correction for the proximity of the coverslip surface[33]. This approach permits forces on individual beads to be characterized *in situ*, just before or after their use in motility assays. A second approach, useful for larger excursions, is to move the stage in a sinusoidal motion while recording bead displacement from the fixed trap position (details are given in Fig. 1 legend). Both methods give results agreeing within 10%.

Low-load regime

Silica beads were incubated with small amounts of kinesin, such that they carried fewer than one active molecule, on average, and were deposited on microtubules with the optical trap. The power was set to $\sim$17 mW at the specimen plane, providing a nominal trapping force that varied linearly from 0 pN at the centre to $\sim$1.5 pN at the edge of the detector zone ($\sim$200 nm). The ATP concentration was fixed at 10 μM. Mean bead velocity in the outermost region of the trap, corresponding to the greatest force, was within 10% of that measured near the trap centre (51 nm s^{-1} versus 54 nm s^{-1}). As observed previously[13], beads frequently released from the microtubule after a variable period of progress (runs) and were drawn rapidly back to the trap centre (within $\leqslant$2 ms), whereupon they rebound and began to move again (Fig. 2a). Multiple cycles of attachment, movement and release were observed, with the same bead passing repeatedly through the detector region (as many as 20 times). The points of release, hence the corresponding force levels, were not the same at each pass. Measurements were terminated when a bead travelled out of the trap altogether or became stuck.

Bead displacement during one run is shown in Fig. 2b. The thermal noise in the displacement signal decreased as the bead developed load and walked toward the edge of the trap (Fig. 2c), indicating a nonlinear elasticity. This can be explained as follows. The bead's position is determined by its linkage to two springs. The first is the optical trap, with spring constant k_{trap}. The second is the linkage connecting the bead to the microtubule, acting through the motor, with spring constant k_{motor}. Both springs are extended as the motor pulls forward. If these springs were linear, thermal motion would be independent of the equilibrium position, with a mean-square displacement given by $\langle x^2 \rangle = k_B T/(k_{\text{trap}} + k_{\text{motor}})$, where k_B is Boltzmann's constant and T is the temperature. Therefore, the data of Fig. 2c imply that k_{motor} gets stiffer with increased extension. It is this diminution of noise, in part, that makes it possible to see steps: when the amplitude of the thermal noise is substantially larger than the step size, steps cannot be detected by any of the methods used here (our unpublished computer simulations). This under-scores the need to keep the bead-to-motor linkage stiff in experiments designed to look for molecular-scale motion, and also for imposing a stiff external tension. Considerable variability was observed in the mean noise level from bead to bead in these experiments, as well as in the degree of noise reduction under tension (by factors of 1–3). For this reason, and because of the relatively large extension (tens of nanometres), we consider it unlikely that stretching of the kinesin molecule itself was chiefly responsible for nonlinear behaviour. One reasonable explanation

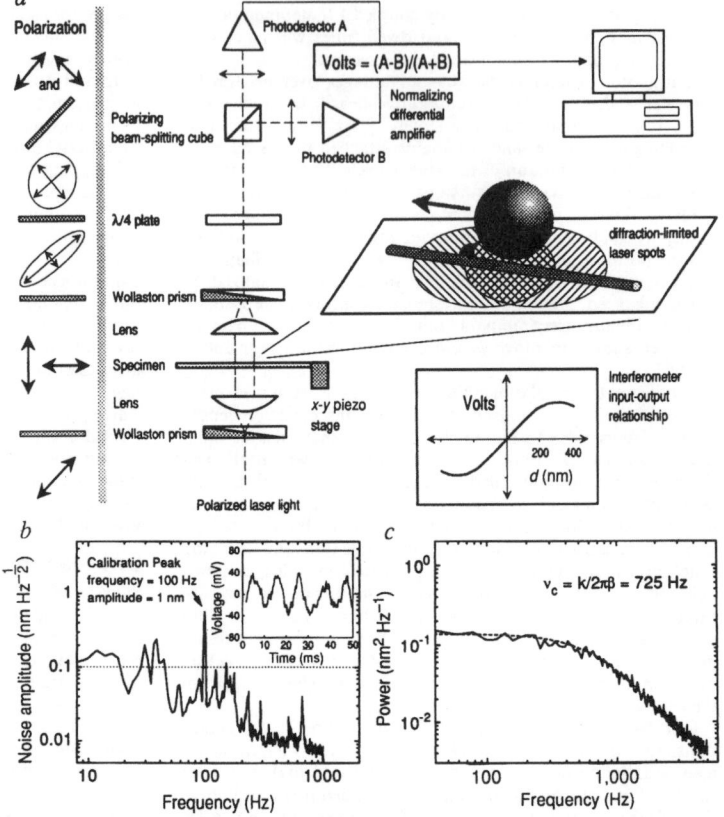

FIG. 1 The optical trapping interferometer. a (left), The diagram illustrates the polarization state of light as it passes through elements of the system, viewed along the optical axis. a (right), A schematic of the instrument. Polarized laser light passes through a Wollaston prism and is focused to two overlapping diffraction-limited spots ($\sim$1 μm diameter) with orthogonal polarization, separated by roughly $\sim$250 nm. After passage through the specimen (a bead propelled along a microtubule by a kinesin molecule), light recombines in the upper Wollaston and develops slightly elliptical polarization (see text). Ellipticity is measured by a quarter waveplate, which produces nearly circularly polarized light, followed by a polarizing beam-splitting cube, which splits the light into two nearly equal components. The difference in intensity is detected by photodiodes and a normalizing differential amplifier. Signals were analysed offline (LabView, National Instruments). b, Sensitivity of the interferometer. The graph shows the spectral noise density of the interferometer responding to a 100-Hz calibration signal of 1-nm amplitude. The large peak (arrow) corresponds to the signal. The detector voltage output (inset) shows both signal and noise. c, Force calibration of the optical trap. The thermal noise spectrum of a bead trapped using 58 mW of laser power is shown (solid line). The spectrum is fitted by a lorentzian (dashed line).

METHODS. We used a modified inverted microscope (Axiovert 35, Carl Zeiss) equipped with Nomarski DIC optics (Plan Neofluar 100×/1.3NA oil objective) fixed to a vibration isolation table (TMC Corp.). Light from a Nd:YLF laser (CW, 3W TEM_{00}, λ=1,047 nm; Spectra Physics) was coupled by an optical fibre. Beam-steering was accomplished with a telescope arrangement[13,23]. An x–y piezo stage (Physik Instrumente) allowed positioning of the specimen under computer control. To measure instrument noise, a bead was embedded in polyacrylamide to suppress brownian motion and introduced into the detector. The stage was moved in a sinusoidal motion along the x axis to provide a calibration signal while recording voltage, and the power spectrum was computed. The power spectrum was scaled by computing the integral under the calibration peak and setting this equal to the mean-square value of the sinusoidal displacement amplitude. To calibrate the trap stiffness, and thereby the force, a bead in solution was trapped (typically $\sim$2 μm above the coverslip) and its brownian motion recorded, from which a power spectrum was computed. The corner frequency provides the ratio of trap stiffness to the viscous drag of the bead; drag was calculated from the bead's diameter and corrected for proximity to the coverslip. One-sided power spectra were normalized such that $\langle x^2 \rangle = \int_0^\infty S(f)\,df$, where $\langle x^2 \rangle$ is the mean-square displacement and $S(f)$ is the power at a given frequency. Trapping force changed by less than 10% as the distance from the coverslip was reduced from 2 to 1 μm (data not shown). The force profile of the trap was also mapped over a larger range of displacements by moving the stage in sinusoidal fashion (amplitude A = 2.5 μm at frequency f) while monitoring the peak displacement, y, of a bead. At low frequencies, the force exerted by the fluid is $F = 2\pi A\beta f = k_{trap} y$, where $\beta = 5.7 \times 10^{-6}$ pN s nm^{-1} is the drag coefficient of the bead. The proportionality between y and f gave the mean stiffness, $(4.3 \pm 0.3) \times 10^{-4}$ pN nm^{-1} mW^{-1}, which was approximately constant out to displacements of $\pm$200 nm (data not shown). The foregoing stiffness was used to compute mean forces in the outermost regions of the trap. Power at the specimen plane was estimated by measuring the external power with a meter and applying an attenuation factor of

58%, corresponding to the transmittance of the objective at 1,064 nm, determined separately (data not shown). Kinesin was purified from squid optic lobe by microtubule affinity[47] and tubulin from bovine brain[48]. Experiments were done at room temperature. Silica beads (0.6 μm diameter, 6×10^{-6} w/v final concentration; Bangs Labs) were incubated for at least 5 min in buffer (80 mM PIPES, 1 mM $MgCl_2$, 1 mM EGTA, pH 6.9, 50 mM KCl, 0.5 mM dithiothreitol, 50 μg ml^{-1} filtered casein, 1–500 μM ATP, 0.5 μg ml^{-1} phosphocreatine kinase, 2 mM phosphocreatine; in some experiments, the last three reagents were replaced by 2 mM AMP-PNP). Beads were incubated for >1 h with kinesin diluted $\sim$1:10,000 from stock ($\sim$50 μg ml^{-1} kinesin heavy chain). Taxol-stabilized microtubules were introduced into a flow chamber in which the coverslip had been treated with 4-aminobutyl-dimethylmethoxysilane (Huls America): microtubules bound tightly to this surface. The chamber was then incubated with 1 mg ml^{-1} casein (in 80 mM PIPES, 1 mM $MgCl_2$, 1 mM EGTA, pH 6.9) for 10 min, rinsed with buffer, and kinesin-coated beads introduced. Beads were captured from solution with the trap and deposited on a microtubule selected with its long axis parallel to the Wollaston shear direction, the direction of detector sensitivity. Fewer than half the beads bound and moved when placed on a microtubule. Under these conditions, beads carry Poisson-distributed numbers of functional motors[13]. The bead size in these studies was such that the chance that any bead carried two motors in sufficient proximity to interact simultaneously with the microtubule was less than 2%, assuming, generously, that kinesin heads can reach 100 nm from their points of attachment.

for the nonlinearity is that the bead–motor linkage behaves like an entropic spring, with the swivelling motion of the bead at the end of its tether contributing degrees of freedom. This linkage becomes taut under load, such that subsequent displacements of the motor are communicated sharply to the bead.

Beads under the conditions shown in Fig. 2a, b moved close to 50 nm s^{-1}. The speed and the thermal noise in the data made a statistical analysis necessary to detect periodic structure. We used the following approach. First, reduced-noise segments of all runs in an experiment were selected, corresponding to movements under load, and filtered to further reduce noise (see below). Then, a histogram of all pairwise differences in displacement was computed for each record[34]. This 'pairwise distance distribution function' (PDF) shows spatial periodicities in stepping motion, independent of times at which steps occur. Next, PDFs were averaged. Finally, the power spectrum of the average PDF was computed: this produces a peak at the mean spatial frequency of the stepper.

For a noiseless, stochastic stepper (equidistant advances after exponentially distributed time intervals), the PDF gives a set of evenly spaced peaks at multiples of the step distance. For noisy steppers, peaks become broadened, disappearing altogether when their peak widths (r.m.s. amplitudes of brownian motion) exceed the step size. The detection of steps in the presence of noise can be improved by low-pass filtering the records with an appropriate filter frequency so as to preserve stepwise character. For this purpose, we used a nonlinear median filter[35]. Computer simulations showed that the peak stepping signal recovered from a background of gaussian noise by an appropriately chosen median filter was approximately twice the amplitude of an equivalent, linear Bessel filter (our unpublished data).

The PDF and associated power spectrum for a single run along a microtubule are shown in Fig. 3a, b. Even in individual runs, there was often an indication of periodicity in the PDF, with a spacing from 6–8 nm, although there was considerable variation from run to run. Averaged data (Fig. 4a, b), gave peaks corresponding to an uncorrected spacing of 6.7 ± 0.2 nm (mean $\pm$ s.e.). Essentially identical data were obtained for beads moving on axonemes (data not shown). Note that backward movements, corresponding to negative distances in the PDF, were practically non-existent. Can our instrument, in combination with the analysis, reliably detect nanometre-sized steps in a noisy background? To test this, we attached beads carrying single motors to microtubules with AMP-PNP, a non-hydrolysable ATP analogue. Beads bound in rigor by AMP-PNP exhibited brownian motion but did not translocate. The piezo stage of the microscope was then stepped stochastically to move the entire specimen (bead and microtubule) through the detector zone, parallel to the long axis of the microtubule, in a series of 8-nm increments, at the same speed as the kinesin-based movement of Fig. 4a. The PDF and power spectrum for this experiment gave peaks at nearly the same positions (Fig. 4c, d). (Similarly, experiments with a stage moved in 4-nm increments gave peaks at half this spacing; data not shown.) Smaller, secondary peaks (measured relative to the sloping background) at subharmonics of the main spatial frequency are usually seen: these arise mainly from the variation in peak heights in the PDF. The measured periodicity was 6.5 ± 0.2 nm, a distance that is 19% smaller than the anticipated value (8 nm). The discrepancy is a consequence of stretching in the elastic bead–motor linkage under load, causing beads to move only a fraction of the motor displacement. (The fraction is $k_{motor}/(k_{trap}+k_{motor})$ for linear springs.) This interpretation was confirmed by an experiment with beads bound directly to the coverslip that had dramatically reduced brownian motion. When the stage was again stepped stochastically by 8 nm, we obtained a strong peak at 8.0 ± 0.2 nm (Fig. 4e, f). Therefore, applying a 19% correction to the spectrum of Fig. 4b yields an adjusted estimate of 8.3 ± 0.2 nm for the mean periodicity of kinesin-based movement.

Could the filtering and/or statistical analysis produce

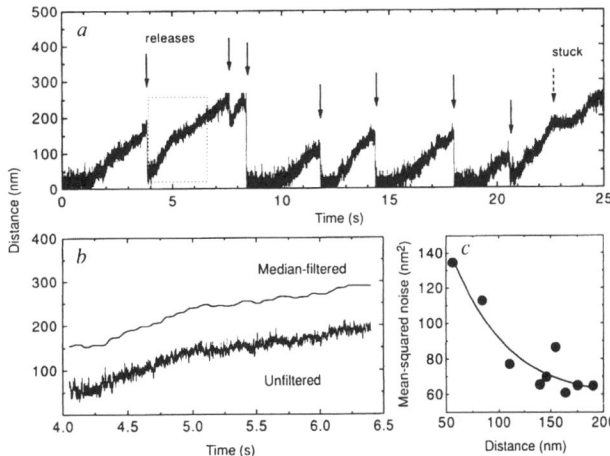

FIG. 2 a, Distance of a bead from the trap centre over 25 s. Multiple cycles of movement, release (solid arrows) and reattachment can be seen. The bead stuck briefly (<1 s) at 23 s (dashed arrow), then continued. The apparent peak between seconds 8 and 9 (third arrow) is a consequence of interferometer nonlinearity: the bead actually passed the turnover point in voltage at ~280 nm (Fig. 1a, inset) and continued its forward movement briefly before releasing. Portions of records beyond 200 nm were not analysed. b, Detail from the dashed box in a. The raw data (lower trace) were median-filtered for subsequent analysis (upper trace). c, The mean-square noise of the track in b as a function of distance from the trap centre. The noise decreases as the bead is placed under tension by the trap.

METHODS. Data were acquired at 1 kHz. A cubic polynomial was fitted to the response function (Fig. 1a, inset), and used to convert voltage to distance, a procedure that extended the usable instrument range to ±200 nm with a ±5% error (data not shown). The algorithm computes the absolute value of displacement, so that the distance from the centre of the trap is rectified. Record segments corresponding to movement near the trap centre were not used for analysis. The response function was determined by tracking the motion of beads stuck tightly to the coverslip surface and moved with computer-controlled voltage waveforms supplied to the piezo stage. The piezo stage was calibrated with nanometre-scale precision using video-based methods[38] against a 10-μm diamond-ruled grating (Donsanto Corp.). A calibration procedure was implemented that allowed beads of slightly varying size to be used, even though these scatter different amounts of light. This was made possible by the observation that the turnover point in the response function (Fig. 1a, inset) occurs at a fixed distance from the trap centre, independent of the bead size, and that the response function scales with voltage at that point. By measuring the turnover voltage, it was possible to establish an absolute correspondence between voltage and distance. Only runs longer than 100 nm were analysed for steps, as follows. A line was fitted to the run, and the slope used as a first estimate of velocity, v'. A set of non-overlapping segments of duration $2d/v'$ was then fitted to the run, where d is an assumed step size, and an improved estimate of velocity, $\langle v \rangle$, was computed from these segments. The filter frequency was chosen such that $f_c \approx 2\langle v \rangle/d$. Data were filtered with a Bessel filter at $1.2f_c$ and then by a median filter[35] of rank r, chosen such that $f_c = 1/(2r+1)$. This procedure was determined by computer simulations to provide reliable results with stochastic steppers subject to gaussian white noise. The exact choice of d is not critical: for the final data analysis, $d=8$ nm was assumed, but other values (from 4 to 16 nm) gave essentially identical results. The noise in c is the mean-square deviation of unfiltered data from successive line segments.

artefactual results? To answer this question, we again bound beads to microtubules with AMP-PNP, but moved the specimen smoothly through the trap at the same mean speed. The PDF and associated power spectrum (Fig. 4g, h) had similar baselines, but no peaks corresponding to spatial periodicities. The variation in several such spectra provided a means of estimating the statistical significance of the peak in Fig. 4b: its likelihood of random occurrence is $P < 0.00001$.

High-load regime

The laser power was set to ~58 mW at the specimen plane, a level that provided a nominal force that varied linearly from 0 pN at the centre to ~5 pN at the edge of the detector zone. The ATP concentration was raised to 500 μM (saturation level). When beads carrying single kinesin molecules were placed on microtubules, their motion was more erratic than at lower powers. Beads near the trap centre, experiencing low loads, moved rapidly at 300–500 nm s^{-1}. Under higher loads towards the edge of the trap, beads markedly slowed (or became stuck), although some were still able to escape altogether, opposing forces up to 5 pN. Other beads slowed or stuck before reaching the trap's edge, then continued forward at increased speed. Shuttering the trap briefly (<0.5 s) during a slow (or stuck) episode enabled some beads, but not all, to regain forward motion. Multiple transitions between fast- and slow-moving phases were seen in individual records, with beads entering the slow phase over a range of positions (loads) in the trap. Beads still underwent

multiple cycles of movement, release and reattachment, similar to the behaviour observed at lower loads.

In the slow-moving phase, the reduced speed permitted visualization of steps. Figure 5a, b shows two records at high load, each containing roughly ten abrupt displacements forwards. The heights of these steps varied from ~5–18 nm, with the smaller transitions averaging 8 ± 2 nm (mean ± s.d.; $n = 16$; Fig. 5c-f). Beads also advanced through what appeared to be 'double steps' (~17 nm), but it was not possible, given the bandwidth and noise in these signals, to determine whether such jumps represented two steps in rapid succession or one single step. Assuming that motors are stochastic steppers, one expects an exponential distribution of dwell times with numerous short steps. Records

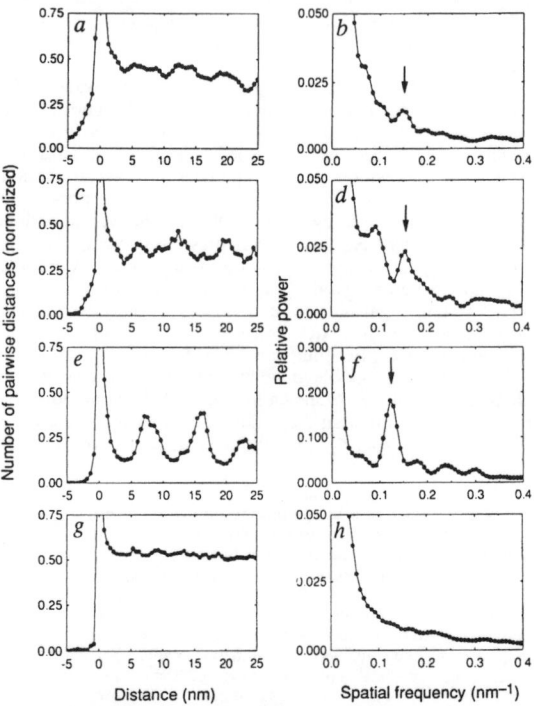

FIG. 4 a, The averaged pairwise distance distribution function (PDF) for kinesin movement, taken from 17 records of 10 different beads. b, The power spectrum of the data in a, with a peak at 0.149 ± 0.004 nm^{-1} (arrow), corresponding to a periodicity of 6.7 ± 0.2 nm. c, The averaged PDF for beads attached to the microtubule using AMP-PNP, with an 8-nm stochastically stepped stage. d. The power spectrum of the data in c, with a main peak at 0.154 ± 0.004 nm^{-1} (arrow), corresponding to a periodicity of 6.5 ± 0.2 nm. e, The averaged PDF for beads attached directly to the coverslip, with an 8-nm stochastically stepped stage. f, The spectrum of the data in e, with a peak at 0.125 ± 0.003 nm^{-1} (arrow), corresponding to a periodicity of 8.0 ± 0.2 nm. g, The averaged PDF for beads attached to the microtubule using AMP-PNP with a smoothly moved stage. h, The power spectrum of g. The average amplitude fluctuation about the mean for records of smooth movement was 0.002 units over the range of spatial frequencies 0.1–0.2; that is, there are no statistically significant peaks in this trace. All PDFs and power spectra were normalized to unity. Note different scale in e.

METHODS. PDFs in each panel represent averages of 17 records. Peak positions in power spectra and corresponding error estimates were determined by fitting gaussians. To estimate the statistical significance of the peak in b, the standard deviation per point, σ_k, for 17 spectra of individual runs of a smoothly stepped stage was computed. The likelihood of the peak was $\langle (S_k - \bar{N}_k)^2 \rangle^{1/2} / \langle \sigma_k \rangle$, where S_k is the average spectrum of kinesin-driven movement in b, and $\bar{N}_k$ is the average spectrum of smooth movement in h; averages were taken over the four spatial frequencies comprising the peak.

FIG. 3 a, The pairwise distance distribution function (PDF) for a single run. The density of interpoint distances in a median-filtered record is shown: multiple peaks with regular spacing can be identified. b, The normalized power spectrum of the data in a, showing a prominent peak at ~0.130 nm^{-1} (arrow), corresponding to periodicity of 7.7 nm.

METHODS. PDFs were computed by binning distance differences $(x_j - x_i)$ for all $(j > i)$ in a histogram, with bin width 0.5 nm. The PDF was normalized to unity at the first bin and smoothed with a 3-point moving window. The one-sided power spectrum, $S(k)$, was computed from the PDF by FFT and normalized to unity at $S(0)$. The PDF is identical to the autocorrelation function of the density of distances from the trap centre. $S(k)$ is analogous to the square of the structure factor used in X-ray scattering, with the density of distances being analogous to electron density[49].

of the type shown in Fig. 5a, b are relatively hard to obtain: additional records must be collected before a more complete statistical analysis, including measurements of dwell-time distributions, will be possible. In parts of the record, the distance from the centre of the trap appeared to shorten briefly (~ 10 ms) just before a step (Fig. 5e, f), but this was not seen in all cases (Fig. 5c, d). We do not know if this behaviour reflects an intrinsic property of the motor, but the shortening is not due to kinesin unbinding because substrate release for periods as short as a millisecond would result in the bead being pulled off completely (Fig. 2).

Low-ATP regime

Reasoning that steps might also be resolved under low loads if beads moved slowly enough we restored the power to 17 mW and lowered the ATP concentration to 1 or 2 μM. When beads carrying single kinesin molecules were placed on a microtubule, their speeds were slow, about 5–15 nm s^{-1}, and single steps could again be seen in records that were selected for low noise (and therefore, presumably, stiffer bead–motor linkages) (Fig. 6a–e). A statistical analysis of these, like that performed in the low-load

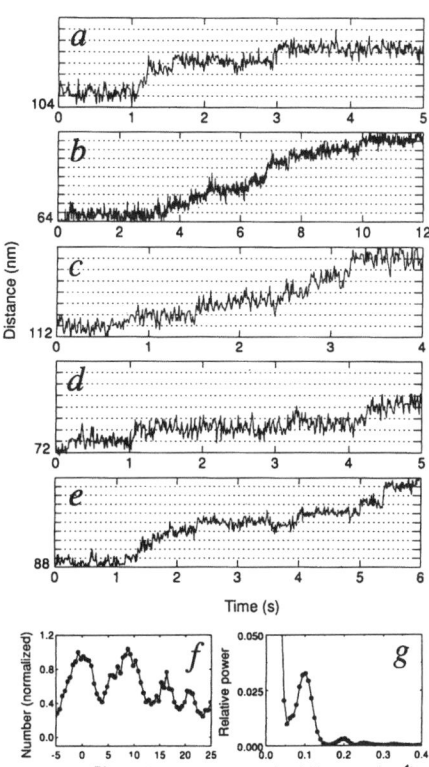

FIG. 6 a, b, Multiple steps in the displacement records of two different beads moving at an ATP concentration of 2 μM. c–e, Multiple steps in the displacement records of three different beads moving at an ATP concentration of 1 μM. Horizontal gridlines (dotted) have been drawn at a spacing of 8 nm. A 10-point jumping average of data taken at 1 kHz is plotted for an effective rate of 100 Hz. f, g, The PDF and associated power spectrum of the record in e, showing a periodicity of 8.8 ± 0.5 nm. Mean step size and s.d. for records in a–e were computed from peak positions determined from PDFs. A correction for the extension in k_{motor} has not been applied to these data, which were selected for their low noise and which presumably reflect stiffer linkages.

regime, showed that steps measured 8.2 ± 1.1 nm (mean ± s.d., as determined from peak positions in power spectra for all five panels), the same size as those seen in the high-load regime. Figure 6f, g shows an example of the PDF and associated power spectrum for the data of Fig. 6e, where the mean periodicity was 8.8 ± 0.5 nm.

Is it possible that kinesin motors travel a longer, possibly variable, distance with each ATP hydrolysis, moving by some integral multiple of this spacing? That is, does a single hydrolysis result in a sequence of physical steps, rather than just one, as has been proposed for myosin[4,5,16,36]? If this were the case for kinesin, then for extremely low concentrations of ATP (the diffusion limit), one would expect a motor to advance in a rapid burst of steps (through ≫ 8 nm) before stopping to wait for the next ATP molecule. This would produce motion characterized by periods of zero advancement interspersed with clusters of steps. The mean rate of advance during a step series ought to be high, close to the speed for saturating levels of ATP (~ 500 nm s^{-1}). We saw no evidence for step clusters in any of our records at 1 or 2 μM ATP: overall rates of advancement were quite uniform.

Discussion

The finding of discrete stepping behaviour should allow direct measurement of kinetic parameters of the mechanochemical cycle, by determining the timing of various phases of the motion. For example, releases of the type shown in Fig. 2a permit an estimate of the cycle off-time. Once a bead near the edge of the trap detaches under low load conditions, it is returned rapidly to the centre through a distance, x, of ~200 nm within an average time $\tau = 1.8 ± 0.4$ ms (mean ± s.e.). To make net forward progress, the kinesin off-time, τ_{off}, must be less than the time required for the trap to pull the bead back by one step, d, that is, $\tau_{off} < d/v = d\tau/x$, where v is the return speed at the edge of the trap. Using 8 nm for d gives $\tau_{off} \leqslant 72$ μs. This time is an upper limit, because velocity is practically unaffected by trapping forces. Such a short off-time signifies that kinesin somehow remains bound, sustaining load throughout most of its activity, perhaps moving hand-over-hand[12,37]. Estimates of other kinetic parameters are ongoing subjects of investigation.

It is known that kinesin motors do not wander over the surface lattice of a microtubule, but move instead along straight paths parallel to a protofilament[38–40]. Kinesin head fragments cloned from squid[41] or Drosophila[42] and expressed in bacteria have been used to decorate microtubules. Such fragments are non-functional in motility assays, but they bind microtubules and have microtubule-activated ATPase activity. Saturation binding experiments indicate a stoichiometry of one kinesin heavy-chain fragment to each tubulin dimer[42], and crosslinking studies demonstrate that it is the β-subunit of tubulin that is bound[41]. Electron microscopy of decorated microtubules shows an 8-nm repeat arising from the head spacing[41,42]. Taken together with our observations, this suggests that the two heads of a kinesin molecule walk along a single protofilament—or walk side-by-side on two adjacent protofilaments—stepping ~8 nm at a time, making one step per hydrolysis (or perhaps fewer, requiring multiple hydrolyses per step). If the foregoing model is correct, then during movement of single molecules, the ATP hydrolysis rate, r_{ATP}, should be related to the step size, d, and speed, v, through $r_{ATP} = v/d$. Movement at saturating levels of ATP ($v \approx 500$ nm s^{-1} and even higher[20]) implies $r_{ATP} \geqslant 60$ s^{-1} mol^{-1}, 6–10-fold higher than reported[43–45]. The hydrolysis rates for motility assays in vitro must therefore be substantially higher than values inferred from solution biochemistry, or motors must make several steps per ATP hydrolysed. The latter is difficult to reconcile with our data, particularly at low concentrations of ATP.

Occasionally jumps have been seen[38] in selected records of kinesin movement during video tracking, measuring 3.7 ± 1.7 nm, although most of the movement was subjectively

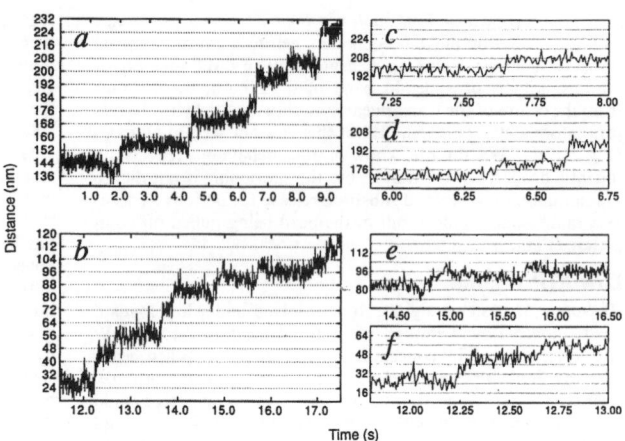

FIG. 5 *a*, *b*, Multiple steps in the displacement records of two different beads in the high-load regime. *c*, *d*, Details of selected steps in *a*. *e*, *f*, Details of selected steps in *b*. Horizontal gridlines (dotted) have been drawn at a spacing of 8 nm. A 5-point jumping average of data taken at 1 kHz is plotted for an effective rate of 200 Hz. The mean step size for these records was computed from averaged values of distance determined during segments of traces before and after transitions, identified by eye. Records at high loads were not corrected for stretching in k_{motor} because the load presumably extends that linkage to be taut.

smooth. Our instrument resolves distances of this size, but we did not find evidence for significant periodicities near ∼4 nm. The earlier work, however, was done before the development of single-motor assays, and beads carried unknown numbers of motors, probably several. It is possible that the shorter spacing, if real, might reflect an effect of multiple heads stepping along neighbouring protofilaments.

Our data from moving beads subjected to trap-induced tensions suggest that speeds are largely unaffected by forces of 1.5 pN, and that single molecules of kinesin can still transport beads against loads up to ∼5 pN. These findings seem incompatible with the conclusions of Kuo and Sheetz[30], who reported measuring the 'isometric' force generated by single kinesin molecules at 1.9 ± 0.4 pN. In their experiments, nominally arrested beads continued to move; that is, the isometric condition was not truly fulfilled, but this difficulty alone is probably insufficient to explain the discrepancy.

Finding displacement steps in the movement of kinesin molecules is in many respects analogous to detecting current steps in single-channel recordings of neurons. In both cases, steps reflect underlying motions of individual proteins, and allow one to make meaningful measurements at the level of single molecules. Before the advent of single-channel recording, Johnson (thermal) noise due to the leakage resistance of electrodes prevented the resolution of picoampere-sized current steps. Improved methods for increasing this resistance provided conditions that led directly to step detection[46]. As a result, single-channel recording has produced many insights in molecular neuroscience during the past decade. For mechanoenzymes, brownian (thermal) noise of objects pulled by motors prevented the resolution of nanometre-sized displacement steps. Additional stiffness, coming in our case from the optical trap and its effect on the bead linkage, provides the required reduction in noise, permitting visualization of steps. □

Received 16 July; accepted 27 September 1993.

1. Huxley, H. E. *Science* **164**, 1356–1366 (1969).
2. Squire, J. *The Structural Basis of Muscle Contraction* (Plenum, London, 1981).
3. Bagshaw, C. R. *Muscle Contraction* (Chapman & Hall, London, 1982).
4. Burton, K. *J. Musc. Res. Cell Motil.* **13**, 590–607 (1992).
5. Simmons, R. M. *Curr. Biol.* **2**, 373–375 (1992).
6. Cooke, R. *CRC Crit. Rev. Biochem.* **21**, 53–117 (1986).
7. Kron, S. J. & Spudich, J. A. *Proc. natn. Acad. Sci. U.S.A.* **83**, 6262–6276 (1986).
8. Vale, R. D., Schnapp, B. J., Reese, T. S. & Sheetz, M. P. *Cell* **40**, 559–569 (1985).
9. Ishijima, A., Doi, T., Sakurada, K. & Yanagida, T. *Nature* **352**, 301–306 (1991).
10. Uyeda, T. Q. P., Kron, S. J. & Spudich, J. A. *J. molec. Biol.* **214**, 699–714 (1990).
11. Uyeda, T. Q. P., Warrick, H. M., Kron, S. J. & Spudich, J. A. *Nature* **352**, 307–311 (1991).
12. Howard, J., Hudspeth, A. J. & Vale, R. D. *Nature* **342**, 154–158 (1989).
13. Block, S. M., Goldstein, L. S. B. & Schnapp, B. J. *Nature* **348**, 348–352 (1990).
14. Yang, J. T., Saxton, W. M., Stewart, R. J., Raff, E. C. & Goldstein, L. S. B. *Science* **249**, 42–47 (1990).
15. Toyoshima, Y., Kron, S. J. & Spudich, J. A. *Proc. natn. Acad. Sci. U.S.A.* **87**, 7130–7134 (1990).
16. Yanagida, T., Arata, T. & Oosawa, F. *Nature* **316**, 366–369 (1985).
17. Harada, Y., Sakurada, K., Aoki, T., Thomas, D. & Yanagida, T. *J. molec. Biol.* **216**, 49–68 (1990).
18. Higuchi, H. & Goldman, Y. *Nature* **352**, 352–354 (1991).
19. Spudich, J. A. *Nature* **348**, 284–285 (1990).
20. Gittes, F., Mickey, B., Nettleton, J. & Howard, J. *J. Cell Biol.* **120**, 923–934 (1993).
21. Schnapp, B. J. *Meth. Enzym.* **134**, 561–573 (1986).
22. Ashkin, A., Dziedzic, J. M., Bjorkholm, J. E. & Chu, S. *Optics Lett.* **11**, 288–290 (1986).
23. Block, S. M. in *Noninvasive Techniques in Cell Biology* 375–401 (Wiley-Liss, New York, 1990).
24. Denk, W. & Webb, W. W. *Appl. Optics* **29**, 2382–2390 (1990).
25. Amos, L. & Klug, A. *J. Cell Sci.* **14**, 523–549 (1974).
26. Kamimura, S. *Appl. Optics* **26**, 3425–3427 (1987).
27. Kamimura, S. & Kamiya, R. *J. Cell Biol.* **116**, 1443–1454 (1992).
28. Block, S. M., Blair, D. F. & Berg, H. C. *Nature* **338**, 514–517 (1989).
29. Ashkin, A., Schuetze, K., Dziedzic, J. M., Euteneuer, U. & Schliwa, M. *Nature* **348**, 346–348 (1990).
30. Kuo, S. C. & Sheetz, M. P. *Science* **260**, 232–234 (1993).
31. Simmons, R. M. *et al.* in *Mechanisms of Myofilament Sliding in Muscle* (eds Sugi, H. & Pollack, G.) 331–336 (Plenum, New York, 1993).
32. Wang, C. W. & Uhlenbeck, G. E. in *Selected Papers on Noise and Stochastic Processes* (ed. Wax, N.) 113–132 (Dover, New York, 1954).
33. Happel, J. & Brenner, H. *Low Reynolds Number Hydrodynamics* 322–331 (Kluwer Academic, Dordecht, 1991).
34. Kuo, S. C., Gelles, J., Steuer, E. & Sheetz, M. P. *J. Cell Sci.* suppl. **14**, 135–138 (1991).
35. Gallagher, N. C. Jr & Wise, G. L. *IEEE Trans. Acoust. Speech Sign. Proc.* **29**, 1136–1141 (1981).
36. Oosawa, F. & Hayashi, S. *Adv. Biophys.* **22**, 151–183 (1986).
37. Schnapp, B. J., Crise, B., Sheetz, M. P., Reese, T. S. & Khan, S. *Proc. natn. Acad. Sci. U.S.A.* **87**, 10053–10057 (1990).
38. Gelles, J., Schnapp, B. J. & Sheetz, B. J. *Nature* **331**, 450–453 (1988).
39. Kamimura, S. & Mandelkow, E. J. *Cell Biol.* **188**, 865–875 (1992).
40. Ray, S., Meyhoefer, E., Milligan, R. A. & Howard, J. A. *J. Cell Biol.* **121**, 1083–1093 (1993).
41. Song, Y.-H. & Mandelkow, E. *Proc. natn. Acad. Sci. U.S.A.* **90**, 1671–1675 (1993).
42. Harrison, B. C. *et al. Nature* **362**, 73–75 (1993).
43. Kusnetsov, S. A. & Gelfand, V. I. *Proc. natn. Acad. Sci. U.S.A.* **83**, 8530–8534 (1986).
44. Hackney, D. *Proc. natn. Acad. Sci. U.S.A.* **85**, 6314–6318 (1988).
45. Gilbert, S. P. & Johnson, K. A. *Biochemistry* **32**, 4677–4684 (1993).
46. Sakmann, B. & Neher, E. *Single-Channel Recording* (Plenum, New York, 1983).
47. Weingarten, M. D., Suter, M. M., Littman, D. R. & Kirschner, M. W. *Biochemistry* **13**, 5529–5537 (1974).
48. Schnapp, B. J. & Reese, T. S. *Proc. natn. Acad. Sci. U.S.A.* **86**, 1548–1552 (1989).
49. Glatter, O. & Kratky, H. C. *Small Angle X-Ray Scattering* (Academic, New York, 1982).

ACKNOWLEDGEMENTS. We thank W. Denk, C. Godek, M. Meister, P. Mitra and R. Stewart for discussions and advice, W. Hill for electronic design, and R. Stewart for help with motility assays and protein purification. This work was supported by the Rowland Institute for Science (K.S., C.F.S., S.M.B.), and partial support from the NIH (K.S. and B.J.S.), the Lucille P. Markey Charitable Trust (B.J.S.), and the University of Michigan (C.F.S.).

VOLUME 79, NUMBER 15   P H Y S I C A L   R E V I E W   L E T T E R S   13 OCTOBER 1997

Deflection of Neutral Molecules using the Nonresonant Dipole Force

H. Stapelfeldt,[1,*] Hirofumi Sakai,[1,2] E. Constant,[1,3] and P. B. Corkum[1]

[1]*Steacie Institute for Molecular Sciences, National Research Council of Canada, Ottawa, Ontario, Canada K1A 0R6*
[2]*Electrotechnical Laboratory, 1-1-4, Umezono, Tsukuba, Ibaraki 305, Japan*
[3]*Département de Physique, Faculté des Sciences, Université de Sherbrooke, Sherbrooke, Québec, Canada J1K 2R1*
(Received 3 April 1997)

The ac Stark shift produced by nonresonant radiation creates a potential minimum for a ground state molecule at the position where the laser intensity is maximum. The gradient of this potential exerts a force on the molecule. We experimentally observe this force when a beam of CS_2 molecules is redirected by sending it through the intensity gradient near the focus of a laser beam. We trace the direction of the molecules in the molecular beam, showing that the molecules that pass near the center of the high intensity laser beam will focus. [S0031-9007(97)04209-9]

PACS numbers: 42.50.Vk, 33.80.Ps, 51.70.+f

The use of laser light to manipulate and control the position and velocity of electrons [1], atoms [2], and microscopic particles [3,4] is a subject of intense activity in physics and biology. The manipulation of atoms in the gas phase is based on either the scattering force, exploited in most laser cooling techniques [5], or the induced dipole force. In the latter case a dipole moment is induced in the atom using a near-resonant continuous-wave laser beam. The polarized atoms experience a force that is proportional to the spatial gradient of the laser intensity, which can be used, for instance, to focus [6] or trap them [5].

The powerful laser cooling and trapping techniques developed for atoms are not readily applicable to molecules, due to their complicated level structure and their weak transition moments for any given rovibrational transition. Consequently, laser manipulation and control of neutral particles beyond the atomic case has been restricted to biological molecules in solution [4] and small dielectric particles [3]. Manipulation and control of molecules in the gas phase, however, remains a topic of great interest [7,8].

We use the nonresonant molecular polarizability to exert an induced dipole force that modifies the trajectory of gas phase molecules. By sending a beam of molecules through the focus of a laser beam oriented perpendicularly to their direction of propagation, we observe a change of the transverse velocity of the molecules. The velocity change is proportional to the spatial gradient of the laser intensity. We demonstrate that an intense laser beam can be used as a lens to focus the molecular beam.

Using nonresonant polarizability and intense laser fields, we solve two problems that have impeded the development of molecular optics. First, we remove the restrictions caused by the dense level structure of the molecule by using far-off resonance radiation to induce a dipole force. In this way, there is no critical dependence on the particular level structure of the molecule. Therefore, our approach is applicable to all molecules (or atoms [9]). Second, by employing the intense field from a pulsed laser to induce the molecular dipole moments, we obtain light-induced forces that are many orders of magnitude larger

than the forces induced on atoms by standard continuous-wave laser beams. This allows manipulation of molecules without laser cooling [10].

For experimental convenience, we use a molecular beam to produce molecules that are translationally cold in the direction perpendicular to the jet axis. Our experiment is designed to observe changes in the transverse velocity (deflection) against this very cold background. We will show potential well depths of about 7.0 meV, approximately 4 orders of magnitude greater than the transverse beam energy (10^{-6} eV).

The schematic of the experimental layout is shown in Fig. 1. A pulsed beam of CS_2 molecules is formed by expanding a CS_2 gas at ~25 Torr, either buffered with 1 atm of neon or without the neon buffer, through a 250-μm-diameter nozzle into a time-of-flight (TOF) spectrometer with the molecular beam axis (the x axis) perpendicular to the TOF axis (the y axis). The molecules travel freely for ~8 cm, after which they are crossed at 90° by the focused beam from a pulsed (10 Hz) Nd:YAG laser ($\lambda = 1.06$ μm) propagating along the z axis. The duration of the YAG pulse is 14 ns (FWHM). We use

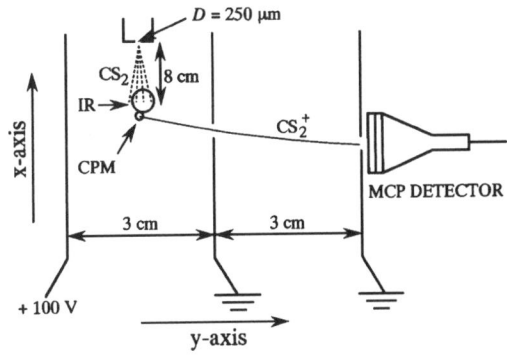

FIG. 1. Schematic of the experimental arrangement of the target chamber viewed in the direction of the laser beam (z axis).

VOLUME 79, NUMBER 15   PHYSICAL REVIEW LETTERS   13 OCTOBER 1997

laser pulses with energy in the range of 10 mJ. The laser beam is focused with an on-axis parabolic mirror to an $\omega_0 = 7$ μm Gaussian focal spot giving a peak intensity of $I_0 \sim 9 \times 10^{11}$ W/cm² at 10 mJ.

This laser field exerts an induced dipole force on the CS_2 molecules, which causes a change in the velocity of the molecules. We measure the velocity change [11] along the TOF axis—i.e., perpendicular to the molecular velocity before the interaction. To do this we multiphoton ionize the neutral molecules using a tightly focused femtosecond laser beam ($\lambda = 625$ nm) and observe the distribution of flight times of the CS_2^+ ions from the interaction region to the microchannel plate detector (illustrated in Figs. 1 and 2). The TOF mass spectrometer [11] consisted of an acceleration region defined by two plates and an equal length field-free drift region (Fig. 1). With this design, molecular ions with their transverse velocity component towards the detector at the time of ionization arrive before zero-transverse velocity ions. Those initially deflected away from the detector arrive after zero-transverse velocity ions.

The 80 femtosecond duration pulses, originating from a 10 Hz amplified colliding pulse mode-locked (CPM) laser, contain 0.66 μJ per pulse and are focused to a spot size $\omega_0 \sim 2.5$ μm that corresponds to a peak intensity of 8×10^{13} W/cm². To ensure that ionization occurred without the strong infrared pulse present, a 25 ns delay is introduced between the YAG pulse and the CPM pulse (requiring a corresponding 10 μm offset in the direction of the molecular beam velocity in the position of their focal spots for pure CS_2, 20 μm for CS_2 buffered with neon). The nonlinearity of multiphoton ionization allows

us to restrict the probe volume to a region smaller than the YAG focal volume by adjusting the focal spot size as well as the intensity of the CPM laser beam. Therefore, we can use the CPM laser to probe the spatial dependence of the induced dipole force simply by moving its focus with respect to the larger YAG focus.

Examples of the relevant portions of three TOF spectra are shown in Fig. 2. Each spectrum is the average of 1000 shots. The central (full) curve, recorded without the YAG laser present, gives information on the initial velocity distribution along the y axis of the molecular beam. Since our velocity measurement is restricted to those molecules that pass through the very small focal spot of the CPM laser, we expect the distribution to be very narrow. In fact, simple line-of-sight arguments show that the initial velocity spread along the y axis, Δv_y^{init}, is determined by $\Delta v_y^{\text{init}} = D/(l/v_{CS_2})$, where $D = 250$ μm is the diameter of the nozzle, $l = 8$ cm is the distance from the nozzle to the laser focus, and v_{CS_2} is the longitudinal velocity of the CS_2 molecules. For the unbuffered (buffered) expansion we measured $v_{CS_2} = 450$ (800 m/s). At this velocity, the CS_2 molecules therefore have a transverse velocity spread of ~ 1.4 m/s (~ 2.5 m/s) equivalent to a kinetic energy of 0.8×10^{-6} eV (2.5×10^{-6} eV). In Fig. 2 the CS_2 molecules are not buffered with neon during the expansion. The FWHM of the full peak (~ 3.4 ns) corresponds to a y-axis velocity spread of 7.2 m/s, which is significantly larger than the expected value of ~ 1.4 m/s. Thus, it provides a calibration of our ability to measure transverse velocities with the TOF spectrometer.

The remaining two spectra in Fig. 2 were obtained for molecules that transmit the focus of the YAG laser (peak intensity $= 1 \times 10^{12}$ W/cm², linear polarization) prior to the measurement of their velocity by the CPM laser. They were obtained for molecules that pass approximately 3.5 μm to the right of the center of the YAG focus (dotted curve in Fig. 2) and to the left of the center of the YAG focus (dashed curve in Fig. 2), respectively. The molecules arriving earlier at the detector (dashes) have acquired a y-axis velocity towards the microchannel plate detector [11]. Similarly, the molecules arriving later at the detector (dots) have acquired a negative y-axis velocity component from the YAG laser pulse. As expected, the laser-induced dipole force provides a central potential deflecting the molecules towards the high intensity region.

By recording TOF spectra of molecules for several positions of the CPM focus with respect to the YAG focus, we can measure the detailed y-coordinate dependence of the velocity change [11], and therefore of the dipole force. This is implemented by scanning the YAG focus over a range of ~ 30 μm while keeping the CPM focus fixed. In this way, the ions produced by the CPM laser always originate from the same spatial location, ensuring that the observed changes in the TOF spectra are not caused by the position of the ionizing laser inside the TOF spectrometer. The results are displayed in Fig. 3, where the velocity shift

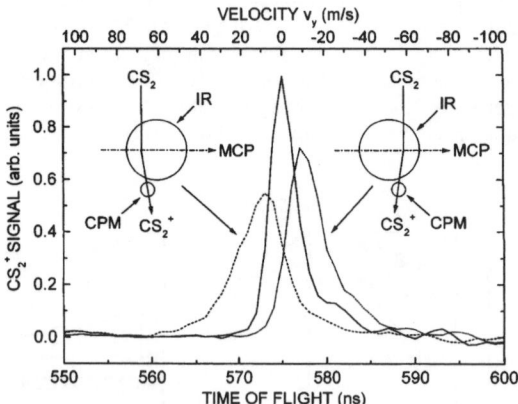

FIG. 2. Portion of the time-of-flight spectrum showing the time of arrival of the undeflected beam (solid curve) and the deflected beams (dashed and dotted curves). A schematic shows the relative placement of the focus of the deflecting laser (YAG) and the measurement laser (CPM). Deviations of the arrival time of a deflected molecule from the arrival time of a zero transverse velocity molecule allow the transverse velocity to be measured. The horizontal scale shows both the flight time and the transverse velocity v_y.

VOLUME 79, NUMBER 15   PHYSICAL REVIEW LETTERS   13 OCTOBER 1997

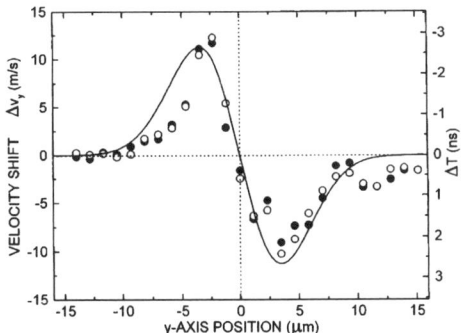

FIG. 3. The shift in arrival time ΔT induced by the dipole force and the associated transverse velocity Δv_y of the molecular beam plotted as a function of the position of the molecule that is probed in the focus of the deflecting laser beam. Data obtained with the deflection laser linearly polarized is represented by the symbol ●; data obtained with circular polarization is represented by ○. In both cases, the expansion of CS_2 molecules was buffered with neon. The derivative of the measured focal distribution of the deflecting laser beam is also shown by the solid curve.

of the center of the half maximum $CS_2{}^+$ signal is plotted as a function of the y coordinate of the YAG laser focus for both linear and circular polarization. The dispersionlike shape shows that the force causing the velocity change [Eq. (2)] is proportional to the y derivative of the YAG laser intensity profile (solid curve). (Deviations on the right of Fig. 3 experimental data and the y derivative of the intensity profile are due to space charge caused by ionization of CS_2 by the YAG beam [12].)

Those molecules that pass near the center of the focus (within $\pm\omega_0/2$) experience a displacement that causes them to meet at a common position; that is, the beam will focus. The focal length of our molecular lens is approximately 230 μm, corresponding to a redirection of the CS_2 molecules through an angle of up to 0.9°. The lens is thin since the size of the YAG beam is much smaller than the focal length.

We can estimate the magnitude of the potential well produced by the laser field for the CS_2 molecules. When a molecule enters the YAG laser focus, a dipole moment is induced. This results in a Stark shift U of the ground state of the molecule given by [13]

$$U(x, y, z, t) = -\tfrac{1}{4}\alpha E^2(x, y, z, t), \quad (1)$$

where $E(x, y, z, t)$ is the space and time dependent pulse envelope. Equation (1) includes a time average over one optical period and neglects alignment of the molecule by averaging over all angles [$\alpha = (\alpha_\parallel + 2\alpha_\perp)/3$, where $\alpha_{\parallel,\perp}$ is the polarizability parallel and perpendicular to the molecular axis]. Using the static polarizability [14], Eq. (1) predicts a Stark shift $U_0 = 10$ meV at the center of the laser focus and at the peak of a laser pulse with $I = 9 \times 10^{11}$ W/cm^2. In comparison, typical potential

depths reached in continuous-wave studies on atoms are about 1 μeV [2,8].

We can approximately neglect alignment if we use either circularly polarized light (where the molecule can align only to a plane) or, linearly polarized light when the orientational well depth [$U_{al} = \tfrac{1}{4}(\alpha_\parallel - \alpha_\perp)E^2$] is smaller than the rotational energy. Although U_{al} should be much greater than the estimated rotational temperature (≈ 5 K) [15] of our jet cooled molecules, alignment does not play a significant role in our experiment. We reach this conclusion by comparing the deflection with linearly and circularly polarized light. If the molecules substantially align with the field, the deflection should be greater with linearly polarized light since then $U = -\tfrac{1}{4}\alpha_\parallel E_0^2$ and $\alpha_\parallel$ is always larger than the averaged α. Figure 3 compares the results obtained with linearly (solid circles) and circularly (open circles) polarized light. The lack of alignment may be due to our multilongitudinal mode YAG laser. Mode beating can produce transient spikes as short as 30 ps which is comparable to or even shorter than the rotational period τ_{rot} of the CS_2 molecules ($\tau_{rot} \sim 34$ ps for a rotational quantum number $J = 4$). Under these conditions alignment is not efficient [7]. By contrast, laser-induced alignment has been reported [16] when temporally smooth nanosecond laser pulses are used.

We now determine the maximum Stark shift $|U_0|$ from the experimental deflection data. Since the induced dipole force F exerted on a molecule is equal to $-\nabla U$, Eq. (1) shows that F is proportional to the gradient of the intensity. The YAG beam can be approximated by a Gaussian intensity distribution in both space and time: $I(x, y, z, t) = I_0 \exp[-2(x^2 + y^2)/\omega_0^2] \exp(-4 \ln 2 t^2/\tau^2)$ (we probe near $z = 0$). Therefore, the y component of the dipole force F_y is given by $F_y = -4(U_0/\omega_0^2)y \exp \times \{-2[(v_{CS_2}t)^2 + y^2]/\omega_0^2\} \exp[-4 \ln(2t^2/\tau^2)]$, where x has been replaced with $-v_{CS_2}t$ to describe the motion of the molecules along the x axis. The experimentally observed velocity shift Δv_y is related to F_y by $\Delta v_y = \tfrac{1}{m} \int_{-\infty}^{\infty} F_y(t)\, dt$ yielding

$$\Delta v_y = -4 \frac{\sqrt{\pi}\, U_0}{\sqrt{2}\, m\omega_0 v_{CS_2}}$$

$$\times \frac{1}{\sqrt{1 + 2\ln 2 (\frac{\omega_0/v_{CS_2}}{\tau})^2}}\, y \exp(-2y^2/\omega_0^2), \quad (2)$$

where m is the mass of the deflected molecule. The dispersionlike shape of the data points in Fig. 3 is consistent with the above expression for Δv_y. For reference, we plot the derivative of the measured spatial profile of the YAG beam on the same figure (determined by measuring a 20 times magnified image of the focus).

The maximum Stark shift U_0 can be estimated from the experimental measurements. Solving Eq. (2) for U_0 at $y = \omega_0/2$, where the measured $\Delta v_y = 12$ m/s, yields

VOLUME 79, NUMBER 15   PHYSICAL REVIEW LETTERS   13 OCTOBER 1997

$U_0 = 6.2$ meV. This agrees with the calculated value ($U_0 = 10$ meV) within experimental uncertainty in measuring the pulse duration, energy, focal properties, and the longitudinal velocity of the CS_2 molecules.

The maximum intensity that an atom or molecule can withstand without substantial multiphoton ionization determines the maximum Stark shift that can be obtained. At sufficiently long wavelength and high intensity light, ionization can be accurately calculated as tunneling [17]. Applying tunnel ionization formulas [17] to CS_2, with its ionization potential of 10.1 eV, we find 1% probability of ionization during a 10 ns pulse (ionization rate $\rho = 10^6$ s^{-1}) at an intensity of 8×10^{12} W/cm^2 ($U_0 = 90$ meV). Since multiphoton ionization is a very nonlinear function of the intensity, 8×10^{12} W/cm^2 is an effective threshold intensity for CS_2.

In our experiment, ionization becomes important at a much lower peak laser intensity for two reasons. First, 1.06 μm irradiation of CS_2 does not satisfy the conditions for tunnel ionization [17]. Typically, nontunneling processes raise the ionization rate and therefore lower the peak intensity at which significant ionization is reached. Second, the high-intensity spikes of the YAG pulse limit the average intensity that can be employed before multiphoton ionization occurs as compared to the situation with a temporally smooth pulse.

Assuming tunneling, the maximum Stark shift on any neutral atom or small molecule is approximately constant, although the polarizability can vary greatly. As the polarizability α decreases, the ionization potential increases and the maximum intensity ($\propto E^2$) that the molecule can withstand without ionization [17] roughly compensates for the decreasing polarizability [see Eq. (1)]. For example, the Stark shift of H_2, a very unpolarizable molecule, is 50 meV at an intensity where the tunnel ionization rate reaches 10^6 s^{-1}.

In conclusion, through nonresonant molecular polarizability, large forces can be applied to molecules. Controlling the laser radiation means controlling the external molecular coordinates. Our results point to a method for trapping, focusing, wave guiding, accelerating, or decelerating molecules. In short, many aspects of atomic optics can be transferred to molecular optics by using nonresonant intense fields. It appears that manipulation of molecules without significant reduction of their thermal temperature is within reach.

We acknowledge the technical assistance of B. Avery, D. Joines, and D. Roth. Discussions with M. Yu. Ivanov, T. Seideman, and A. Stolow are also acknowledged as well as the careful reading of the manuscript by M. Drewsen.

---

*Present address: Department of Chemistry, Aarhus University, Langelandsgade 140, DK-8000 Aarhus C, Denmark.

[1] P. H. Bucksbaum, D. Schumacher, and M. Bashkansky, Phys. Rev. Lett. **61**, 1182 (1988); J. Chaloupka *et al.* (to be published).

[2] See, for example, D. J. Wineland, C. E. Weiman, and S. J. Smith, in *Atomic Physics 14*, AIP Conf. Proc. No. 323 (AIP, New York, 1994).

[3] A. Ashkin, J. M. Dziedzic, J. E. Bjorkholm, and S. Chu, Opt. Lett. **11**, 288 (1986).

[4] S. K. Svoboda and S. Block, Annu. Rev. Biophys. Biomol. Struct. **23**, 247 (1994); A. Ashkin and J. M. Dziedzic, Science **235**, 1517 (1987); T. T. Perkins *et al.*, Science **264**, 822 (1994).

[5] For a review, see C. S. Adams and E. Riis, Prog. Quantum Electron. **21**, 1 (1997).

[6] J. J. McClelland *et al.*, Science **262**, 877 (1993); C. Kurtsiefer *et al.*, in *Atomic Interferometry*, edited by P. R. Berman (Academic, New York, 1997), p. 177.

[7] B. Friedrich and D. Herschbach, Phys. Rev. Lett. **74**, 4623 (1995); T. Seideman, J. Chem. Phys. **106**, 2881 (1997).

[8] J. T. Bahns, W. C. Stwalley, and P. L. Gould, J. Chem. Phys. **104**, 9689 (1996).

[9] J. Takekoshi and R. J. Kinze, Opt. Lett. **21**, 77 (1996); J. D. Miller, R. A. Cline, and D. J. Heinzen, Phys. Rev. A **47**, 4567 (1993).

[10] Neutral molecules can also be manipulated with static fields, but the forces are much weaker than those obtained here. See N. F. Ramsey, *Molecular Beams* (Clarendon, Oxford, 1956).

[11] H. Stapelfeldt, E. Constant, and P. B. Corkum, Phys. Rev. Lett. **74**, 3780 (1995).

[12] H. Sakai *et al.* (to be published).

[13] A. A. Radtzig and B. M. Smirnov, *Reference Data on Atoms, Molecules, and Ions* (Springer-Verlag, New York, 1985).

[14] K. J. Miller, J. Am. Chem. Soc. **112**, 8543 (1990).

[15] We did not measure the rotational temperature T_{rot}, but previous measurements under essentially identical experimental conditions yielded $T_{rot} \sim 5$ K; see D. T. Cramb, H. Bitto, and J. R. Huber, Chem. Phys. **96**, 8761 (1992).

[16] W. Kim and P. M. Felker, J. Chem. Phys. **104**, 1147 (1996).

[17] M. V. Ammosov, N. B. Delone, and V. P. Krainov, Sov. Phys. JETP **64**, 1191 (1986); P. Dietrich and P. B. Corkum, J. Chem. Phys. **97**, 3187 (1992).

letters to nature

Pilus retraction powers bacterial twitching motility

Alexey J. Merz*†, Magdalene So* & Michael P. Sheetz†‡

*Department of Molecular Microbiology and Immunology,
Oregon Health Sciences University, Portland, Oregon 97201-3098, USA
‡ Department of Cell Biology, Duke University Medical School, Durham,
North Carolina 27705, USA

Twitching and social gliding motility allow many Gram negative bacteria to crawl along surfaces, and are implicated in a wide range of biological functions[1]. Type IV pili (Tfp) are required for twitching and social gliding, but the mechanism by which these filaments promote motility has remained enigmatic[1-4]. Here we use laser tweezers[5] to show that Tfp forcefully retract. *Neisseria gonorrhoeae* cells that produce Tfp actively crawl on a glass surface and form adherent microcolonies. When laser tweezers are used to place and hold cells near a microcolony, retractile forces pull the cells toward the microcolony. In quantitative experiments, the Tfp of immobilized bacteria bind to latex beads and retract, pulling beads from the tweezers at forces that can exceed 80 pN. Episodes of retraction terminate with release or breakage of the Tfp tether. Both motility and retraction mediated by Tfp occur at about 1 μm s^{-1} and require protein synthesis and function of the PilT protein. Our experiments establish that Tfp filaments retract, generate substantial force and directly mediate cell movement.

Type IV pili are implicated in motility[1-3,6], biofilm formation[7], virulence[8-11] and all three modes of prokaryotic horizontal genetic transfer (transformation[12,13], conjugation[14] and transduction[15,16]). The Tfp fibre, a helical polymer of the pilin protein, is 6 nm in diameter and up to several micrometres in length[17,18]. Type IV pilus biosynthesis occurs through the type II protein translocation pathway and requires several accessory proteins in addition to the pilin subunit[1]. The PilT protein is dispensable for Tfp biosynthesis but is essential for both Tfp-mediated motility[3,19,20] and for the DNA uptake step of genetic transformation[13,20]. PilT belongs to a highly conserved family of presumed ATPases, which partition to the inner membrane and cytosol and are thought to energize type II and IV protein translocation systems[1,13,19,21]. Two proteins of this family form hexameric rings strikingly similar to the rings formed by many AAA-type ATP-dependent chaperones and proteases[22]. Electron-microscopic studies of *pilT* mutants led to the hypothesis that Tfp promote cellular motility by retracting[3,15], perhaps through PilT-mediated filament disassembly[1,4,19]. Indirect evidence further suggests that other bacterial surface filaments might retract, including the conjugative F pili of *Escherichia coli*[23] and a type III export filament of *Salmonella enteriditis* associated with cell contact[24]. However, filament retraction has been neither observed directly nor proven to power motility in any prokaryotic system.

N. gonorrhoeae, the causative agent of gonorrhoea, provides an excellent model for studies of twitching motility. It lacks rotary flagella and components for type III export, and it produces only one known fimbrial structure, the Tfp. When we suspended Tfp-producing *N. gonorrhoeae* cells at low density in liquid medium, they attached to and crawled over the surface of a glass coverslip (Fig. 1). Under optimal conditions (see Methods), over half of the bacteria on the coverslip were motile. Figure 1b depicts the path of a crawling diplococcus (a joined pair of cells is the neisserial functional unit). Although most movements were short and directional changes occurred frequently, many directed movements of 2–5 μm were observed. Cells crawled at ~1 μm s^{-1} (Fig. 1c, d). This motility was not due to passive diffusion. Motile cells consistently crawled

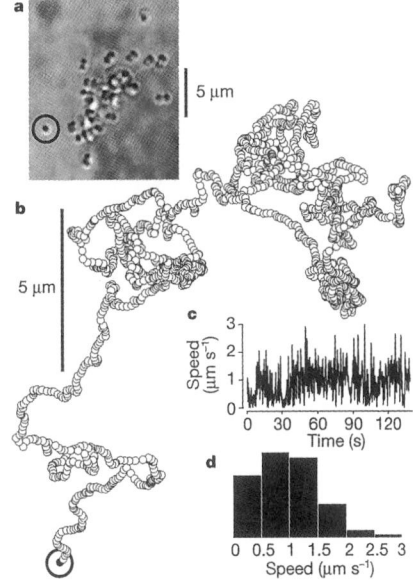

Figure 1 Piliated *N. gonorrhoeae* cells crawl on an inert surface. **a**, Cells crawling on a glass coverslip. This is the first frame of the sequence analysed in **b–d**, and the tracked diplococcus is circled. Note that most cells are present as diplococci. **b**, Tracking of the diplococcus indicated in **a** during an interval of 140 s. Small circles show the position of the tracked diplococcus at intervals of 67 ms. Large circle indicates the start point. **c**, Plot of velocity against time for the same track. **d**, Histogram of the velocities shown in **c**.

† Present addresses: Department of Biochemistry, Dartmouth Medical School, Hanover, New Hampshire 03755-3844, USA (A.J.M.); Department of Biological Sciences, Columbia University, New York, New York 10027, USA (M.P.S.).

out of a laser-tweezers trap strong enough to firmly arrest suspended cells or 1-μm latex beads (20 pN at 100 nm displacement). Cells crawled at temperatures from 20 to 42 °C. Motility ceased ~10 min after the addition of chloramphenicol or tetracycline and resumed upon drug washout, indicating a requirement for protein synthesis (Table 1). Similarly, bacteria were non-motile and unable to grow in defined medium lacking L-glutamine and pyruvate, and became motile upon addition of these nutrients. As expected, non-piliated *pilE* and *pilF* mutants were completely non-motile. Three piliated *pilT* mutants also failed to undergo large movements (>1 μm) and never crawled out of the laser trap (Table 1). *N. gonorrhoeae* cells thus can crawl over surfaces at rates of ~1 μm s⁻¹ by an active process requiring protein synthesis, Tfp biogenesis and PilT[3,19,20].

Tfp-producing *N. gonorrhoeae* cells not only crawl but aggregate into microcolonies that contain twitching or writhing cells. Cells within 1–5 μm of a microcolony often move into the colony, whereas cells within a microcolony move out of the colony[2,6–8]. To determine whether retractile forces would pull dispersed cells together, we used laser tweezers to position isolated cells 1–5 μm (1–2 pilus lengths) away from microcolonies attached to a coverslip (Fig. 2). The trapped cells were repeatedly pulled from the laser trap toward the microcolonies, directly showing that there are retractile forces between cells (Fig. 2a). The cells were pulled towards the microcolonies at speeds of ~1 μm s⁻¹, the same rate at which cells crawled on coverslips (Figs 1 and 2a). Cells pulled from the laser trap sometimes bound irreversibly to the microcolonies, but more often were released back into the laser trap (Fig. 2a). Upon release, the tethers between cells were broken, as trapped cells could be moved freely away from the microcolony using the laser tweezers. When released cells were moved ~10 μm away and then placed near the opposite side of the same microcolony, they were again pulled out of the trap toward the microcolony (Fig. 2b). Retraction and tethering are therefore transient and can be re-established. In contrast, when we carried out identical manipulations using piliated, non-motile *pilT* mutants, cells were never pulled from the trap (Table 1), although nonretractile static tethers between the trapped cells and the microcolonies sometimes formed (data not shown). Together these experiments show that episodes of PilT-dependent Tfp retraction pull bacterial cells towards one another over distances of up to 5 μm at speeds of ~1 μm s⁻¹.

Quantitative analyses of Tfp retraction are impeded by variations in cell morphology and by experimental geometries in which two groups of cells pull toward one another. We therefore developed a bead-based assay for Tfp retraction (Fig. 3a). Individual diplococci were immobilized on 3-μm latex beads that had first been coated with anti-*N. gonorrhoeae* antiserum and anchored to the coverslip. Smaller 1-μm beads, coated with a monoclonal antibody that recognizes a surface-exposed epitope on the Tfp fibre, were introduced into the sample chamber. Laser tweezers were then used to hold the 1-μm beads near the immobilized diplococci so that the beads could interact with Tfp.

When anti-pilus 1-μm beads were placed within 1–3 μm of immobilized cells, the beads became dynamically tethered and were repeatedly pulled towards the immobilized cells (Fig. 3b). The mean speed of retraction was $1.17 \pm 0.49\ \mu m\ s^{-1}$ (mean ± s.d; n = 713), and was independent of the distance traversed by the beads (Fig. 3c). Retraction events were separated by intervals of 1–20 s. One-micrometre beads lacking the anti-pilin monoclonal antibody did not bind to pili as frequently as anti-pilin beads; but when they bound, they also were pulled toward immobilized bacteria. Type IV pili therefore exert retractile force through both nonspecific and specific binding interactions, consistent with our finding that Tfp facilitate bacterial motility on inert substrates (Fig. 1). Retraction ceased ~10 min after the addition of chloramphenicol or tetracycline and resumed after drug washout, again indicating a requirement for protein synthesis.

The restoring force of the laser tweezers increases with radial displacement from the trap centre, and can be calibrated for homogeneous particles using laminar flow (see Methods). Figure 3c shows a trace of displacement against time, with force shown on the displacement axis. Isolated, immobilized diplococci were often able to exert forces of 80 pN or more on trapped 1-μm beads. By comparison, ~20 pN is the force needed to extract an integral membrane protein from a lipid bilayer[25], and ~30 pN of tensile force is sufficient to elongate a microvillus on a host cell[26].

Retraction usually terminated with release or breakage of the pilus tether and bead movement back into the laser trap (Fig. 3b, c). Transitions from retraction to release usually occurred in less than 67 ms. The velocities of bead movement back into the trap (Fig. 3c) were much more variable than retraction speeds and were proportional to the distance displaced before release ($4.34 \pm 3.48\ \mu m\ s^{-1}$;

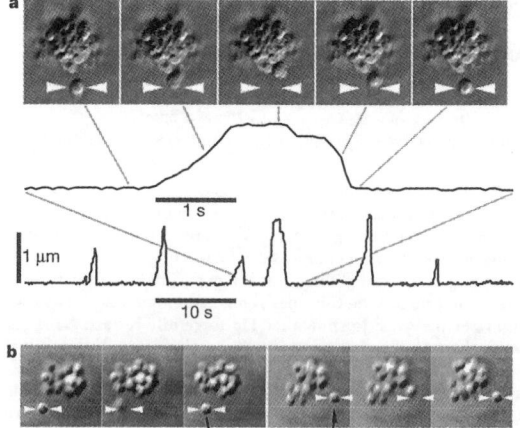

Figure 2 Optical tweezers reveal retractile forces between piliated *N. gonorrhoeae* cells. **a**, Cell–microcolony assay. A diplococcus held in the laser-tweezers trap (indicated by arrowheads) was repeatedly pulled toward a microcolony attached to the coverslip. Micrographs (separated by ~1-s intervals) show one retraction event from the sequence used to generate the trace. Traces indicate radial displacement of the trapped diplococcus from the centre of the laser trap. The top trace shows a time-magnified portion of the bottom trace. Note that retraction occurs at ~1 μm s⁻¹. **b**, Retraction and tethering are transient and can be re-established. A diplococcus was trapped and positioned next to a microcolony as in **a**. The diplococcus was pulled from the trap towards the microcolony and then released back into the trap (first three panels). Next, the trap and trapped diplococcus were moved more than 10 μm away from the microcolony, then repositioned on the opposite side of the same microcolony, where the diplococcus was again pulled from the trap towards the microcolony (last three panels).

Table 1 Motility and retraction phenotypes of *N. gonorrhoeae* strains

	Crawling	Cell–cell retraction	Cell–bead retraction
MS11 N400	Yes	Yes	n.d.
MS11 GT102 *pilT*$_{\Delta QSL}$	No	No	n.d.
MS11 GT103 *pilT*::mTnEGNS	No	No	n.d.
MS11 AM92	Yes	Yes	Yes
MS11 AM92*T1 pilT*::mTnEGNS	No	No	No
MS11 AM92			
+ Tetracycline	No	n.d.	No
+ Chloramphenicol	No	n.d.	No
−L-glutamine and pyruvate	No	n.d.	n.d.
20 °C (cell-division block)	Yes	n.d.	Yes

n.d., not determined.

letters to nature

$n = 171$). Release events could result from force-regulated release of the fibre at a threshold tension level, from breakage of the retracting fibre or the basal body, or from dissociation of the retracting fibre from the 1-μm bead. More detailed analyses of many traces made at laser power levels of 400 and 800 mW (Fig. 3e) indicated that bead release is dependent on force, and is not solely a function of linear displacement. Histograms of the forces reached before bead release show small peaks at about 40, 60 and 90 pN (Fig. 3e, arrows). Studies of other motor-dependent movements have also revealed peaks in displacement histograms. These peaks have been interpreted as multiples of the critical force generated by the motors[27]. The peaks observed in our study may reflect the presence of multiple retracting fibres or multiple motor units on a single fibre; however, we stress that additional, more refined measurements will be required to establish the unit critical force of the Tfp retraction machinery.

As in the cell–microcolony assays, *pilT* mutant cells were unable to generate retractile forces but could sometimes form static tethers to laser-trapped 1-μm beads. When a tethered bead was held in a stationary laser-tweezers trap, movement of the microscope stage

(and hence the immobilized bacterium) within a defined radius did not displace the tethered 1-μm bead out of the trap, but movement outside this radius displaced the tethered bead from the trap centre (Fig. 3f). The static character of the tether was tested by placing the tethered bead under tension, 0.2 μm from the centre of the trap. No further displacements of the bead (> 25 nm) occurred over of a period of more than 120 s, indicating that the tether was static (Fig. 3f). These results confirm that PilT is essential for Tfp retraction.

What is the mechanism of Tfp retraction? The available evidence is most consistent with molecular ratchet models[28] in which retractile force is generated by filament disassembly into the inner membrane. Genetic analyses suggest that PilT promotes pilus disassembly, pilin degradation, or both[4,19]. PilT, like the closely related ATPases thought to energize Tfp assembly and type II and IV secretion, localizes to the cytoplasm and inner membrane[1,13,19,21]. Unassembled pilin is a type II integral inner membrane protein[29]. Upon assembly, the membrane-spanning domain of pilin moves into a hydrophobic coiled-coil at the core of the polymeric fibre[17,18]. These observations support models in which cytoplasmic ATPases

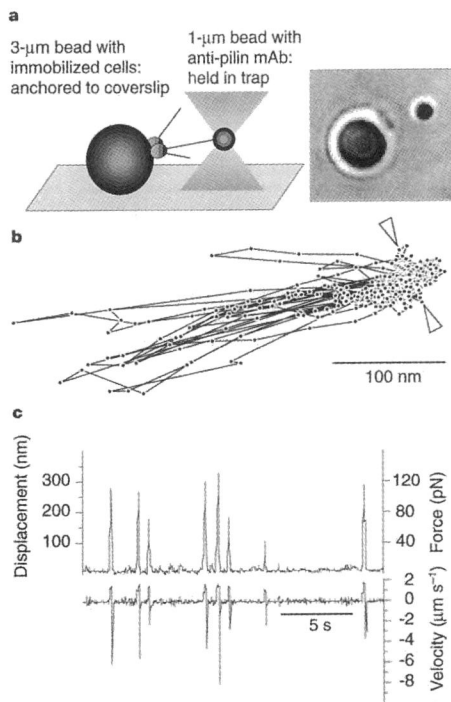

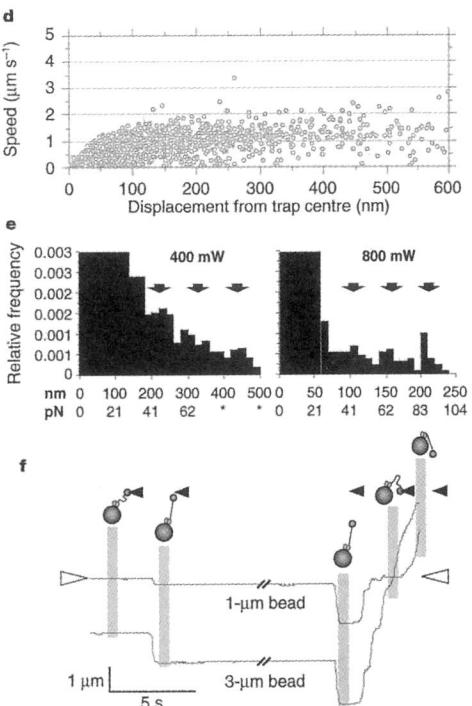

Figure 3 Quantitative type IV pili retraction assay. **a**, Experimental geometry. The cartoon shows a side view; the micrograph shows the assay in progress from above, with scale indicated by the beads (1 and 3 μm). **b**, Direction of displacements. Trace from a recording of a single immobilized diplococcus that shows the locations of the 1-μm bead centroid in the plane parallel to the coverslip surface. The centre of the laser trap is indicated (arrowheads). The immobilized diplococcus is in the same position relative to the laser trap as in the micrograph in **a**. **c**, Velocity, timing and force of retraction. Top trace, radial displacement of the 1-μm bead versus time, and calibrated restoring force imposed by the laser-tweezers trap. Bottom trace, velocities of retraction and release from the same recording. Positive velocities denote displacement away from the trap centre (toward the immobilized cells), and negative velocities denote release into the trap centre. **d**, Plot of retraction velocity versus displacement, showing pooled data from recordings of

14 different immobilized diplococci. Note that for displacements of 100 nm or more, retraction velocities are relatively constant at ~1 μm s⁻¹. **e**, Histograms of forces reached prior to bead release at 400 and 800 mW input laser power. **f**, Static tethering of 1-μm beads by immobilized *pilT* mutant cells. Cartoons show interpretation of the traces. The Tfp could not be seen, but their state (slack or under tension) was inferred from movements of the 1-μm bead relative to the trap. Arrowheads denote the position of the laser trap (filled for the cartoons, open for the trace), which was stationary throughout the experiment. The 3-μm bead was attached to the coverslip on the microscope stage, and was moved using a stage motor. Traces show the displacements of the 1-μm and 3-μm beads relative to the laser trap plotted against time. Breaks in the traces denote an interval of more than 120 s in which there were no movements more than 25 nm of the 1-μm bead.

letters to nature

catalyse pilus extension and retraction from the base of the pilus fibre, within the plane of the inner membrane. The Tfp filament is a single-stranded helix with five pilin subunits per turn and a 40 Å pitch[17,18]. Assembly and disassembly are therefore predicted to occur in 8 Å steps, consistent with the high levels of retractile force observed (Figs. 3c, e). Retraction mediated by disassembly at the observed rate of $1.2 \mu m \, s^{-1}$ (Fig. 3d) would entail the removal of 1,500 pilin subunits per second from the fibre base.

Pilus retraction is thought to occur in several bacterial systems[3,4,15,19,23,24]. To our knowledge our experiments provide the first direct observations of pilus retraction and may provide a model for understanding related retractile processes. In *N. gonorrhoeae*, DNA transformation facilitates pilin antigenic variation[12] and other horizontal genetic transfer functions. Like twitching motility, DNA uptake requires Tfp biosynthesis and PilT function[13,20]. Moreover, many bacteria use PilT or PilT orthologues for DNA uptake, including species that lack Tfp. Consistent with the hypothesis that DNA uptake and pilus retraction use a common machinery, the rate of processive DNA uptake in *Haemophilus influenzae* ($\sim 0.17 \mu m \, s^{-1}$; ref. 13) is comparable to the speed of Tfp retraction in our experiments. A broadly conserved mechanism may therefore power not only pilus retraction, but also other macromolecular translocation processes. Finally, PilT and twitching motility are implicated in virulence-related functions[8-10]. Eukaryotic cells sense and respond to mechanical forces[30], and *N. gonorrhoeae* Tfp and PilT function together to elicit adhesive plaque formation in epithelial cells[11]; we therefore propose that the mechanical forces generated by retractile pili are important and previously overlooked signals between pathogenic bacteria and host cells. □

Methods

Bacterial strains and media

N. gonorrhoeae were maintained on GCB agar with supplements (Difco). Antibiotics (Sigma) were used at the following concentrations (in mg l⁻¹): kanamycin, 100; erythromycin, 4; tetracycline, 50; chloramphenicol, 40. We carried out motility assays in phenol red-free DMEM (Gibco BRL), supplemented with L-glutamine, pyruvate and 1% (w/v) bovine serum albumin (BSA) fraction V (Sigma). BSA limited the adsorption of bacteria onto the glass coverslips and was essential for optimal motility. MS11 N400 and its isogenic derivatives GT102 (*pilTΔQSL*) and GT103 (*pilT5*::mTnEGNS) were gifts from M. Koomey[20]. MS11 AM92 is an MS11A derivative that produces a pilin variant recognized by monoclonal antibody 20D910 (ref. 10). The *pilT*::mTnEGNS allele from GT103 was moved into MS11 AM92 by DNA transformation, then backcrossed against MS11 AM92 three times to generate AM92T1. AM92 and AM92T1 are piliated as judged by immunofluorescent staining with monoclonal antibody 20D9 and by immunoblotting with monoclonal antibodies 20D9 and 10H5 (anti-SM1 epitope). As expected, AM92 is piliated, motile, competent for DNA transformation, and elicits cortical plaques in epithelial cells, whereas AM92T1 is piliated, non-motile, transformation deficient and exhibits defective cortical plaque induction[11,12,20].

Bead-based retraction assay

Three-micrometre latex beads (Bangs Laboratories) were coated with rabbit anti-*N. gonorrhoeae* antiserum by passive adsorption as recommended by the supplier. The beads were washed thoroughly in PBS (phosphate-buffered saline, pH 7.4) and adsorbed onto clean coverslips in the presence of PBS. Coverslips were washed in PBS to remove unbound beads and blocked in DMEM with 1% BSA at 25 °C for 15 min or more before use. One-micrometre beads were coated with biotinylated monoclonal antibody 20D9 by a sandwich method[30]. One-micrometre beads (Polysciences) were covalently coupled to biotinylated casein (Sigma). Biotinylated monoclonal antibody 20D9 (prepared with sulpho-NHS biotin; Pierce) was then linked to the biotinylated casein by an avidin bridge; control beads lacked monoclonal antibody 20D9. The beads were blocked with biotinylated BSA (biotinamido caproyl BSA; Sigma) and washed thoroughly. A dot blot assay indicated that roughly 5×10^4 molecules of monoclonal antibody 20D9 bound per bead. A serial perfusion method greatly reduced aggregate formation among the beads and cells. Coverslips with attached 3-μm beads were used to form the bottom surface of a perfusion chamber, which was placed on the microscope stage. A dilute, vortex-dispersed suspension of *N. gonorrhoeae* cells in DMEM with 1% BSA was perfused into the chamber. We monitored bacterial binding to the anchored 3-μm beads by microscope. When 10–20% of the 3-μm beads had bound 1 or 2 diplococci, unbound bacteria were washed out. One-micrometre beads were suspended in DMEM with 1% BSA and dispersed by bath sonication, then perfused into the sample chamber, and manipulated using the laser tweezers. Except where indicated, we carried out assays at 35 °C.

Laser trapping and single particle tracking

The optical tweezers workstation was configured as described[5] and included a Zeiss Axiovert microscope, a Nd-YAG laser (1,064-nm wavelength) directed into the bottom port of the microscope, a piezoelectric stage controller, and a Nuvicon camera with analogue signal processor and S-VHS video recorder for data capture. We carried out video digitization and data analysis on a SGI workstation running Isee particle tracking software (Inovision). Speeds of crawling bacteria (Fig. 1c, d) were obtained by calculating centroid position using a moving average function with a three-frame window to minimize artefacts arising from the irregular shapes of the cells. In the bead-based assay, velocities were calculated frame by frame. Positive velocities were assigned to displacements away from the trap centre (retraction) and negative velocities were assigned to movement toward the trap centre (release). Values less than three times the r.m.s. retraction or release velocities were discarded as noise. The remaining values were used to calculate the mean retraction and release velocities shown in the text. Results obtained by this method agreed with results obtained by hand-fitting the slopes of displacement versus time plots for individual retraction events. The trap force calibration was done as described[5].

Received 28 April; accepted 20 June 2000.

1. Wall, D. & Kaiser, D. Type IV pili and cell motility. *Mol. Microbiol.* **32**, 1–10 (1999).
2. Henrichsen, J. Twitching motility. *Annu. Rev. Microbiol.* **37**, 81–93 (1983).
3. Bradley, D. E. A function of *Pseudomonas aeruginosa* PAO polar pili: twitching motility. *Can. J. Microbiol.* **26**, 146–154 (1980).
4. Wolfgang, M., Park, H. S., Hayes, S. F., van Putten, J. P. M. & Koomey, M. Suppression of an absolute defect in type IV pilus biogenesis by loss-of-function mutations in *pilT*, a twitching motility gene in *Neisseria gonorrhoeae*. *Proc. Natl Acad. Sci. USA* **95**, 14973–14978 (1998).
5. Sheetz, M. P. (ed.) *Laser Tweezers in Cell Biology* (Academic, New York, 1997).
6. Swanson, J. Studies on gonococcus infection. XII. Colony color and opacity variants of gonococci. *Infect. Immun.* **19**, 320–331 (1978).
7. O'Toole, G. A. & Kolter, R. Flagellar and twitching motility are necessary for *Pseudomonas aeruginosa* biofilm development. *Mol. Microbiol.* **30**, 295–304 (1998).
8. Bieber, D. *et al.* Type IV pili, transient bacterial aggregates, and virulence of enteropathogenic *Escherichia coli*. *Science* **280**, 2114–2118 (1998).
9. Comolli, J. C. *et al. Pseudomonas aeruginosa* gene products PilT and PilU are required for cytotoxicity in vitro and virulence in a mouse model of acute pneumonia. *Infect. Immun.* **67**, 3625–3630 (1999).
10. Pujol, C., Eugene, E., Marceau, M. & Nassif, X. The meningococcal PilT protein is required for induction of intimate attachment to epithelial cells following pilus-mediated adhesion. *Proc. Natl Acad. Sci. USA* **96**, 4017–4022 (1999).
11. Merz, A. J., Enns, C. A. & So, M. Type IV pili of pathogenic *Neisseriae* elicit cortical plaque formation in epithelial cells. *Mol. Microbiol.* **32**, 1316–1332 (1999).
12. Seifert, H. S., Ajioka, R. S., Marchal, C., Sparling, P. F. & So, M. DNA transformation leads to pilin antigenic variation in *Neisseria gonorrhoeae*. *Nature* **336**, 392–395 (1988).
13. Dubnau, D. DNA uptake in bacteria. *Annu. Rev. Microbiol.* **53**, 217–244 (1999).
14. Yoshida, T., Kim, S. R. & Komano, T. Twelve *pil* genes are required for biogenesis of the R64 thin pilus. *J. Bacteriol.* **181**, 2038–2043 (1999).
15. Bradley, D. E. Evidence for the retraction of *Pseudomonas aeruginosa* RNA phage pili. *Biochem. Biophys. Res. Commun.* **47**, 142–149 (1972).
16. Karaolis, D. K., Somara, S., Maneval, D. R. Jr., Johnson, J. A. & Kaper, J. B. A bacteriophage encoding a pathogenicity island, a type-IV pilus and a phage receptor in cholera bacteria. *Nature* **399**, 375–379 (1999).
17. Parge, H. E. *et al.* Structure of the fibre-forming protein pilin at 2.6 Å resolution. *Nature* **378**, 32–38 (1995).
18. Forest, K. T. & Tainer, J. A. Type-4 pilus structure: outside to inside and top to bottom—a minireview. *Gene* **192**, 165–169 (1997).
19. Whitchurch, C. B., Hobbs, M., Livingston, S. P., Krishnapillai, V. & Mattick, J. S. Characterisation of a *Pseudomonas aeruginosa* twitching motility gene and evidence for a specialised protein export system widespread in eubacteria. *Gene* **101**, 33–44 (1991).
20. Wolfgang, M. *et al. pilT* mutations lead to simultaneous defects in competence for natural transformation and twitching motility in piliated *Neisseria gonorrhoeae*. *Mol. Microbiol.* **29**, 321–330 (1998).
21. Brossay, L., Paradis, G., Fox, R., Koomey, M. & Hebert, J. Identification, localization, and distribution of the PilT protein in *Neisseria gonorrhoeae*. *Infect. Immun.* **62**, 2302–2308 (1994).
22. Krause, S. *et al.* Sequence-related protein export NTPases encoded by the conjugative transfer region of RP4 and by the *cag* pathogenicity island of *Helicobacter pylori* share similar hexameric ring structures. *Proc. Natl Acad. Sci. USA* **97**, 3067–3072 (2000).
23. Novotny, C. P. & Fives-Taylor, P. Retraction of F pili. *J. Bacteriol.* **117**, 1306–1311 (1974).
24. Ginocchio, C. C., Olmsted, S. B., Wells, C. L. & Galan, J. E. Contact with epithelial cells induces the formation of surface appendages on *Salmonella typhimurium*. *Cell* **76**, 717–724 (1994).
25. Evans, E., Berk, D. & Leung, A. Detachment of agglutinin-bonded red blood cells. I. Forces to rupture molecular-point attachments. *Biophys. J.* **59**, 838–848 (1991).
26. Shao, J. Y., Ting-Beall, H. P. & Hochmuth, R. M. Static and dynamic lengths of neutrophil microvilli. *Proc. Natl Acad. Sci. USA* **95**, 6797–6802 (1998).
27. Coppin, C. M., Finer, J. T., Spudich, J. A. & Vale, R. D. Detection of sub-8-nm movements of kinesin by high-resolution optical-trap microscopy. *Proc. Natl Acad. Sci. USA* **93**, 1913–1917 (1996).
28. Mahadevan, L. & Matsudaira, P. Motility powered by supramolecular springs and ratchets. *Science* **288**, 95–100 (2000).
29. Dupuy, B., Taha, M. K., Pugsley, A. P. & Marchal, C. *Neisseria gonorrhoeae* prepilin export studied in *Escherichia coli*. *J. Bacteriol.* **173**, 7589–7598 (1991).
30. Felsenfeld, D. P., Schwartzberg, P. L., Venegas, A., Tse, R. & Sheetz, M. P. Selective regulation of integrin-cytoskeleton interactions by the tyrosine kinase Src. *Nature Cell Biol.* **1**, 200–206 (1999).

Acknowledgements
We thank our colleagues in the Sheetz and So labs for invaluable technical assistance and

letters to nature

stimulating discussions; E. Barklis and L. Kenney for critical comments on the manuscript; and M. Koomey for providing bacterial strains. This work was supported by NIH grants to M.S. and M.P.S. A.J.M. received pre-doctoral support from an NIH NRSA grant and postdoctoral support from the Cancer Research Fund of the Damon Runyan-Walter Winchell Foundation.

Correspondence and requests for materials should be addressed to M.P.S. (e-mail: ms2001@columbia.edu).

VOLUME 84, NUMBER 23   PHYSICAL REVIEW LETTERS   5 JUNE 2000

Optical Deformability of Soft Biological Dielectrics

J. Guck,[1] R. Ananthakrishnan,[1] T. J. Moon,[3] C. C. Cunningham,[4] and J. Käs[1,2]

[1]*Center for Nonlinear Dynamics, Department of Physics, University of Texas at Austin, Austin, Texas 78712*
[2]*Institute for Molecular and Cellular Biology, University of Texas at Austin, Austin, Texas 78712*
[3]*Department of Mechanical Engineering, University of Texas at Austin, Austin, Texas 78712*
[4]*Physicians Reliance Network, Dallas, Texas 75246*
(Received 21 October 1999)

Two counterpropagating laser beams were used to significantly stretch soft dielectrics such as cells. The deforming forces act on the surface between the object and the surrounding medium and are considerably higher than the trapping forces on the object. Radiation damage is avoided since a double-beam trap does not require focusing for stable trapping. Ray optics was used to describe the stress profile on the surface of the trapped object. Measuring the total forces and deformations of well-defined elastic objects validated this approach.

PACS numbers: 87.80.Cc, 87.15.La, 87.16.Ka

Because of the small momentum of photons, radiation pressure can be neglected in our immediate environment. In the microscopic world, however, the effects of the interaction of light with matter can be significant. Lasers are used to manipulate objects ranging from atoms to micron-sized beads or biological cells [1–3]. The total momentum transfer from a laser beam to a transparent object results in a propulsive force in the direction of the light propagation (scattering force) and an attractive force along the intensity gradient perpendicular to the laser axis (gradient force) [4]. One-beam gradient traps, called optical tweezers, have been used for a variety of biological experiments in which cells, organelles, or beads, attached to biological objects as tiny handles, have been trapped and moved [5,6]. However, attempts to deform whole cells by pulling on two handles have been limited by the small holding forces on the handles. Even soft red blood cells could be deformed only by 15% of the original cell size with this method [7].

In spite of this, laser beams can be used for the deformation of cells. While scattering and gradient forces are due to the total momentum transferred to the particle's center of gravity, the transfer actually occurs on the particle's surface. Our study shows that the resulting local surface forces are much larger than the total forces. This effect can be observed in a two-beam trap [1,8,9], where two slightly divergent, counterpropagating laser beams are used to trap single cells. Intuitively, one might expect that the scattering forces from the two beams compress the cell. In contrast, exactly the opposite occurs: The cell is stretched out along the beam axis. This optical deformability of soft dielectrics can be motivated by a simple *gedankenexperiment*.

The momentum p_1 of a ray of light with energy E traveling in water is $p_1 = n_1 E/c$ (n_1: refractive index of water, ≈ 1.33; c: speed of light in vacuum) [10,11]. Let such a ray hit the surface of a dielectric transparent cube with length $l = 10 \ \mu$m and a refractive index $n_2 = 1.45$, which is typical for biological materials. At normal incidence (incident angle $\alpha = 0°$) only a fraction $R = 0.2\%$ of the light

is reflected. Almost all the light enters the cube and gains momentum due to the higher index of refraction. Upon exiting the cube, the same fraction of light, R, is reflected and the exiting light loses momentum. The conserving momenta transferred to the two surfaces per second, i.e., the forces experienced by the two surfaces, are

$$F_{\text{front}} = [n_1 - (1 - R)n_2 + Rn_1]P/c \qquad (1a)$$

and

$$F_{\text{back}} = [n_2 - (1 - R)n_1 + Rn_2](1 - R)P/c, \qquad (1b)$$

where P is the total light power. The forces F_{front} (≈ 190 pN for $P = 500$ mW) and F_{back} (≈ 210 pN) point in opposite directions—away from the cube. The total force acting on the cube is the difference between those two surface forces $F_{\text{total}} = F_{\text{back}} - F_{\text{front}} \approx 20$ pN. This total force is in essence the scattering force. In addition, the surface forces stretch the cube with $(F_{\text{back}} + F_{\text{front}})/2 \approx 200$ pN, which is 10 times greater than the scattering force. If an identical ray hits the cube from the opposite side, there is no total force acting on the center of the cube. However, the forces stretching the cube are now twice as large as before (≈ 400 pN). The cube experiences a deforming stress $\sigma = 400$ pN/$(10 \ \mu$m$)^2 \approx 4$ N m^{-2} that results in a deformation of $\Delta l = l\sigma/E \approx 400$ nm for a Young's modulus $E = 100$ N m^{-2}. Any soft dielectric material can be stretched in this fashion as long as its refractive index is larger than the refractive index of the surrounding medium. Thus, we termed this two-beam setup the *optical stretcher.*

An essential benefit of using a two-beam trap for cells is the possibility to work with higher laser powers than in a one-beam trap. Radiation damage is avoided because there is no focusing required for the trap's stability. In fact, the laser beams must be divergent. The trap is stable as long as the radii w of the divergent laser beams at the position of the cell are larger than the cell size ($w \approx 10 \ \mu$m) [8]. In optical tweezers there are also surface forces present that distort the cell shape. These deformations are very small because the light power is limited to $P < 20$–250 mW

VOLUME 84, NUMBER 23   PHYSICAL REVIEW LETTERS   5 JUNE 2000

depending on the cell type and the wavelength used [3,4]. The reason is that the laser beam has to be highly focused for the trap to be stable, which can lead to opticution of living biological cells. Typical beam sizes at the focal point of optical tweezers are on the order of $w \approx \lambda/2 \approx 500$ nm. The light power, i.e., the deforming stresses, in the optical stretcher can be 400 times greater than in optical tweezers before similar light intensities in the cell are reached. Thus, the stresses accessible for cell elasticity measurements with an optical stretcher range between the highest stresses possible with optical tweezers and the lowest stresses exerted by an atomic force microscope.

In our experiments, even sensitive eukaryotic cells, such as PC12 cells, were trapped in the optical stretcher without any sign of radiation damage with up to 700 mW of light power in each beam [12]. Red blood cells (RBCs) were used to test the concept of optical deformability because they are well-defined mechanical objects, intensively studied, and easy to handle. The use of oil drops was disregarded. Their shape is determined by surface tension that is sensitive to miniscule temperature changes due to light absorption. Giant vesicles did not seem to be a good choice because laser beams induce instabilities in their often multilamellar membranes [13].

The experimental setup consists of a cw-Ti:sapphire laser emitting at $\lambda = 785$ nm, an acousto-optic modulator as light power control, and an inverted microscope equipped with a CCD camera. Images of the trapping and stretching of cells are captured and analyzed with a computer. Similar to the method described in [9], two single mode optical fibers are used to deliver the light to the microscope for ease of use and as spatial filters for the Gaussian beam profile. The maximum light power output from each optical fiber is $P \approx 700$ mW.

Ray optics (RO) was used to calculate the deforming stress acting on the surface of a cell trapped in the optical stretcher. The cells studied can be well approximated by nonabsorbing spheres with an isotropic index of refraction. This is justified because the eukaryotic cells assumed a spherical shape in suspension and the RBCs were osmotically swollen into a sphere. Cells are almost transparent in the near infrared, so absorption can be neglected. The relative refractive index is $n = n_2/n_1 = 1.05–1.15$ for biological objects, where n_1 and n_2 are the refractive indices of the medium and the object, respectively. The value of n can be determined by index matching in phase contrast microscopy [14]. The size of cells is on the order of tens of microns (typical radius of an eukaryotic cell, $\rho = 8–15 \ \mu$m; radius of a spherical RBC, $\rho = 3.0–3.4 \ \mu$m). Thus, RO can be used to describe their interaction with $\lambda = 785$ nm light [15]. This approach is commonly used for laser traps [1,4–6,8]. The deforming stress calculation proceeds similar to the simple estimate above, with the rays intersecting the cell surface in general at angles $\alpha \neq 0$. The direction of the transmitted ray is given by Snell's law. The fraction of reflected light, R, varies depending on α and the state of polarization and can

be calculated from the Fresnel formulas. To simplify the calculation and to maintain symmetry with respect to the laser axis, R is taken to be the average of the coefficients for perpendicular and parallel polarization relative to the plane of incidence. This is a negligible deviation from the true situation [16].

With this simple RO model, the forces acting on any surface element, i.e., the stress on the cell surface, was calculated. One result is that all surface forces act normal to the surface. Figure 1 shows the stress profiles for one laser beam shining on the cell. The profile has rotational symmetry around the beam axis (z axis). The cell acts as a lens and focuses the beam towards the axis. The resulting asymmetry between front and back profile leads to a total propulsive force in the positive z direction. This is the origin of the scattering force. The exact shape of the stress distribution and the magnitude of the total force F_{total} depend on the ratio between the cell radius ρ and the beam radius w, which in turn depends on the distance d from the tip of the optical fiber.

In order to test the RO calculations, we measured the total force acting on different objects. After trapping silica beads, polystyrene beads, or cells in the optical stretcher, one of the beams was blocked. The total force from the other beam then accelerated the object in the direction of the light propagation. The resulting velocities were measured and the total force on the object was estimated to equal the Stokes drag force. Figure 2 shows the agreement between the measured and calculated total force. The fact

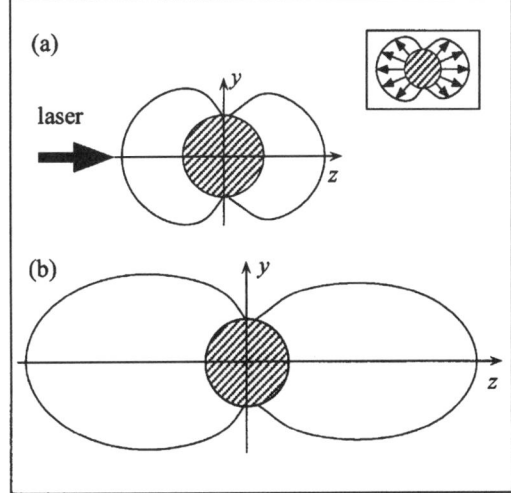

FIG. 1. Stress profiles on the surface of a sphere ($n_1 = 1.33$, $n_2 = 1.45$) due to one laser beam with Gaussian profile and total power $P = 500$ mW; (a) for a ratio of $w/\rho = 2.0$ the stresses along the z axis are $\sigma_{\text{front}} = 2.8$ N m^{-2} and $\sigma_{\text{back}} = 3.1$ N m^{-2} resulting in a total force $F_{\text{total}} = 25$ pN and (b) for $w/\rho = 1.1$, $\sigma_{\text{front}} = 9.0$ N m^{-2} and $\sigma_{\text{back}} = 9.8$ N m^{-2} resulting in $F_{\text{total}} = 38$ pN. The inset illustrates the direction of the surface forces.

VOLUME 84, NUMBER 23   PHYSICAL REVIEW LETTERS   5 JUNE 2000

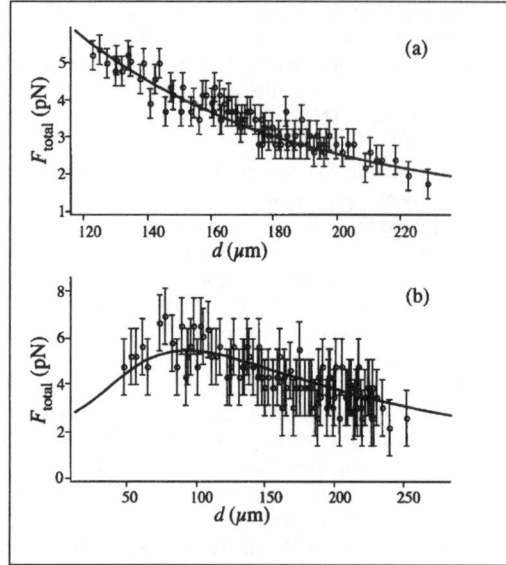

FIG. 2. Calculated and measured total force F_{total} as a function of the distance d between fiber tip and particle for (a) silica beads ($\rho = 2.50 \pm 0.02$ μm, $n_2 = 1.43 \pm 0.01$, $P = 350$ mW) and (b) PC12 cells ($\rho = 7.7 \pm 0.2$ μm, $n_2 = 1.38 \pm 0.01$, $P = 300$ mW) in aqueous solution.

that the RO model works as well for spherical beads with homogeneous index of refraction (silica beads and polystyrene beads) as for cells justifies the assumptions about the physical properties of biological cells. The magnitude and the d dependence of the total force also agree well with previous results [8].

In the optical stretcher the cell was trapped between two identical, counterpropagating laser beams. Thus, the total force on the cell was zero, and the cell was stably trapped if $w/\rho > 1$. However, when the stress on the cell surface was large enough, the cell was stretched out along the beam axis. To verify this optical deformability, osmotically swollen spherical RBCs were investigated because their elastic properties are very well characterized [17] and, due to their softness, deformations are easily quantified. Figure 3(a) shows a typical stress profile calculated for a RBC. It can be approximated by $\sigma(\alpha) = \sigma_0 \cos^2(\alpha)$, σ_0 being the peak stress along the z axis.

RBCs consist mainly of a thin elastic shell with a ratio of radius ρ to thickness h, $\rho/h \approx 100$. They are filled with hemoglobin, which leads to a homogeneous index of refraction of $n_2 = 1.380$ (for spherical shape) [18]. In contrast to eukaryotic cells, RBCs do not have a three-dimensional polymer scaffold throughout the cytoplasm, which makes them much softer. The buffer for the RBCs with osmolarity of 270 mOsm was adapted from [19]. Under physiological conditions, they have a biconcave, disk-like shape. However, the osmolarity of the buffer was adjusted to ≈ 130 mOsm ($n_1 = 1.335$) and the RBCs assumed a spherical shape prior to the trapping and stretch-

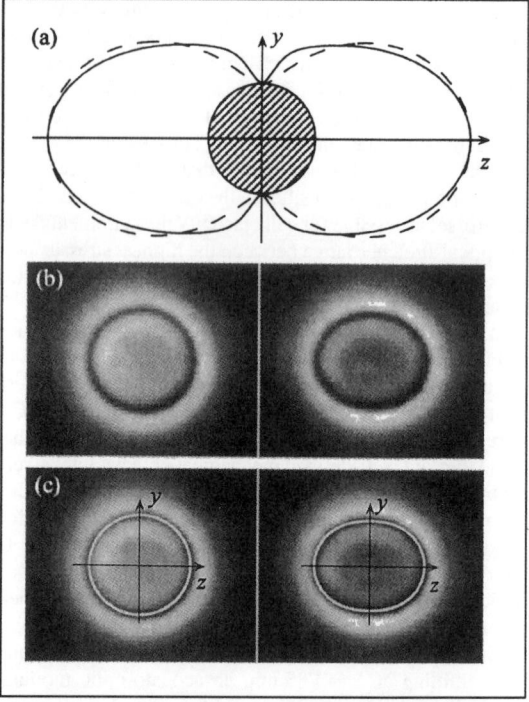

FIG. 3. (a) Stress profile for a spherical RBC with radius $\rho = 3.32 \pm 0.02$ μm ($n_1 = 1.335$, $n_2 = 1.380$) in between two 86 mW laser beams ($w/\rho \approx 1.1$). The peak stress along the z axis is $\sigma_0 = 1.02$ N m^{-2}. The solid line is for the calculated stress; the dashed line is for the $\sigma(\alpha) = \sigma_0 \cos^2(\alpha)$ approximation. (b) Phase-contrast images of a RBC in the optical stretcher at 5 mW and at 86 mW light power in each laser beam. (c) The white line shows shapes expected from linear membrane theory due to the stress shown in (a).

ing. Prepared this way, they perfectly resembled the model cell: They were spherical, had an isotropic index of refraction, and had virtually no absorption at the wavelength used ($\lambda = 785$ nm).

Single RBCs were trapped in the optical stretcher at low light powers (≈ 5 mW). The distance d between cell and fibers was adjusted so that $w/\rho \approx 1.1$–1.2. The light power was then increased in steps up to 350 mW and the resulting deformation of the cell was recorded [see Fig. 3(b)]. In the linear regime the maximum expansions in the z direction were ≈ 800 nm at 350 mW, while in the y direction the cells contracted ≈ -600 nm. At light powers higher than 350 mW the elastic response of the RBCs became nonlinear, deformations reached values up to 160% of the original cell size at ≈ 600 mW, and finally the cells ruptured. For each step, the stress distribution on the cell was calculated using the power P, the radius ρ, and the distance d measured. Figure 4 shows the relative deformations in the linear regime along the beam axis (positive values) and perpendicular to it (negative values) as a function of the peak stress σ_0 along the z axis.

VOLUME 84, NUMBER 23   PHYSICAL REVIEW LETTERS   5 JUNE 2000

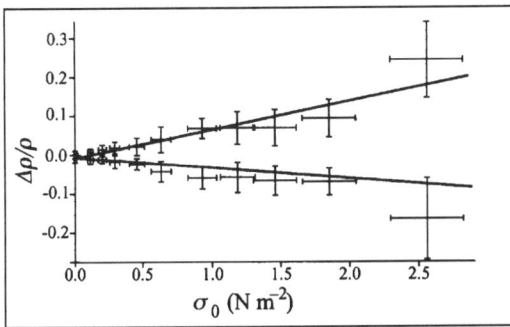

FIG. 4. Relative deformation of RBCs measured along (positive) and perpendicular to the laser axis (negative) as a function of the peak stress σ_0 on each cell. Solid lines show the prediction from linear membrane theory.

To verify our stress profile calculations we related the observed deformations of the RBCs to the known material constants of their membranes by using linear membrane theory [20]. The stress on a thin shell with isotropic Young's modulus E, thickness h, and radius ρ leads to tensions in the shell that result in displacements of surface elements. The rotational symmetry makes spherical coordinates a natural choice, which are oriented so that the incident angle α is identical to the polar angle. For a thin shell the bending energy is negligible compared to the membrane (stretching) energy [21]. Because of the rotational symmetry of the stress profile, the radial displacements $\Delta\rho(\alpha)$ of the surface elements depend only on the polar angle,

$$\Delta\rho(\alpha) = \frac{\rho^2 \sigma_0}{4Eh} [(5 + \nu)\cos^2(\alpha) - \nu - 1], \quad (2)$$

where the Poisson ratio $\nu \approx 0.5$ for biological membranes. These displacements result in an ellipsoid with major axis $a = \rho + \Delta\rho(0)$ and minor axes $b = \rho + \Delta\rho(\pi/2)$. Figure 3(c) shows the agreement between the theoretical and the observed shape of a RBC in the optical stretcher. The relative deformations along the z axis $[\Delta\rho(0)/\rho]$ and the y axis $[\Delta\rho(\pi/2)/\rho]$ are both linearly proportional to the peak stress and the material properties ρ/Eh. Plotting the relative deformation for $Eh \approx 3.9 \times 10^{-5}$ N m^{-1} (see Fig. 4) shows the consistency between experiment and theory. This value for Eh is in agreement with literature values for RBC membranes [22].

In conclusion, we have demonstrated the possibility of stretching soft biological dielectrics in a two-beam laser trap. A RO approach is sufficient to explain this stretching. The momentum transferred from the light to the surface of the trapped object results in forces on the object's surface. The surface forces lead to stretching of an elastic object [23]. The deforming forces can exceed the total trapping forces. As illustrated with the *gedankenexperiment*, the stretching forces of two counterpropagating

500 mW laser beams on an object with a relative refractive index of $n \approx 1.1$ are $F \approx 400$ pN. They could be even greater for higher indices of refraction and higher light powers. The optical stretcher can be used to measure the elasticity of biological cells. The advantages are that optical deformation does not require any kind of mechanical contact and covers a stress range previously inaccessible to cell elasticity measurements.

We thank M. Raizen, C. Schmidt, S. Kuo, J. Black, and H. Swinney for their support. This work was supported by Grant No. MCB-9808849 from the National Science Foundation.

[1]   A. Ashkin, Phys. Rev. Lett. **24**, 156 (1970).
[2]   S. Chu, Science **253**, 861 (1991).
[3]   A. Ashkin, J. M. Dziedzic, and T. Yamane, Nature (London) **330**, 769 (1987).
[4]   S. C. Kuo and M. P. Sheetz, Trends Cell Biol. **2**, 116 (1992).
[5]   A. Ashkin et al., Opt. Lett. **11**, 288 (1986).
[6]   K. Svoboda and S. M. Block, Annu. Rev. Biophys. Struct. **23**, 147 (1994).
[7]   S. Hénon et al., Biophys. J. **76**, 1145 (1999).
[8]   G. Roosen, Opt. Commun. **21**, 189 (1977).
[9]   A. Constable et al., Opt. Lett. **18**, 1867 (1993).
[10] I. Brevik, Phys. Rep. **52**, 133 (1979).
[11] A. Ashkin and J. M. Dziedzic, Phys. Rev. Lett. **30**, 139 (1973).
[12] The viability of PC12 cells in the optical stretcher was addressed in three different ways: They had a normal appearance, were able to prevent the vital stain Trypan Blue from entering their interior, and their growth rate after trapping was the same as for control cells.
[13] R. Bar-Ziv, E. Moses, and P. Nelson, Biophys. J. **75**, 294 (1998).
[14] R. Barer and S. Joseph, Q. J. Microsc. Sci. **95**, 399 (1954).
[15] Ray optics can be used if $2\pi\rho/\lambda \approx 25$–$130 \gg 1$. H. C. van de Hulst, *Light Scattering by Small Particles* (Dover Publications, New York, 1981), p. 174.
[16] The error in the stress introduced by this simplification is smaller than 2% for $n_2 = 1.45$ and smaller than 0.5% for $n_2 = 1.38$.
[17] N. Mohandas and E. Evans, Annu. Rev. Biophys. Biomol. Struct. **23**, 787 (1994).
[18] E. Evans and Y. C. Fung, Microvasc. Res. **4**, 335 (1972).
[19] K. Zeman, Ph.D. thesis, Technische Universität München, Germany, 1989.
[20] E. Zbigniew and R. T. N. Mazurkiewicz, *Shells of Revolution* (Elsevier Science, New York, 1991).
[21] The ratio between the two energies for a $\sigma_0 \cos^2(\alpha)$ distribution of stress on the surface is proportional to $4h^2/3\rho \approx 10^{-4}$ for RBCs.
[22] E. A. Evans, Biophys. J. **13**, 941 (1973).
[23] While the ray optics approach explains the deformation by momentum transfer and forces acting on the surface it is equivalent to think in terms of the minimization of energy when the dielectric object deforms its shape so that more of its volume is located in the higher field along the laser axis.

VOLUME 30, NUMBER 4   PHYSICAL REVIEW LETTERS   22 JANUARY 1973

Radiation Pressure on a Free Liquid Surface

A. Ashkin and J. M. Dziedzic

Bell Telephone Laboratories, Holmdel, New Jersey 07733

(Received 10 November 1972)

The force of radiation pressure on the free surface of a transparent liquid dielectric has been observed using focused pulsed laser light. It is shown that light on either entering or leaving the liquid exerts a net outward force at the liquid surface. This force causes strong surface lens effects, surface scattering, and nonlinear absorption. The data relate to the understanding of the momentum of light in dielectrics.

We report the observation of the forces of radiation pressure from a focused pulsed laser beam on the free surface of a lossless liquid dielectric medium. We find that light on either entering or leaving the liquid exerts a net outward force at the surface of the dielectric medium. This force causes surface motion which results in strong surface lens effects, strong surface scattering, and nonlinear absorption. This result has bearing on the momentum of light in dielectric media, the ponderomotive force of electromagnetic waves in dielectric media, and the self-focusing of laser light in liquids. Light on entering a dielectric from free space has its momentum changed because of Fresnel reflection and by interaction with the medium. If p_0 is the momentum in free space and p the momentum in the medium of refractive index n, the net change in momentum is $p_0(1+R) - p(1-R)$, where R is the Fresnel reflection coefficient. This difference must be balanced by a mechanical force on the medium. On this basis, J. J. Thomson and Poynting[1] concluded that light on entering a dielectric exerts a net outward force at the surface. For the momentum in the medium they used $p = Un/c$, where U is the energy. This simply replaces c by c/n in the free-space momentum $p_0 = U/c$. Over the years there has been controversy over the proper form of the momentum of light in dielectrics.[2] Apart from Un/c, which is the so-called Minkowski value, there is U/cn proposed by Abraham. Recently, Burt and Peierls[3] gave arguments in favor of Abraham's value in agreement with other recent work.[4] Using $p = U/cn$ they predict that light should exert a net inward force on a dielectric interface. They also fail to understand the measurement of Jones and Richards[5] of the light force on a metal vane in liquid, which agrees with $p = Un/c$. The existence of these disparate views on the direction of the surface force prompted the present experiment.[6] Also, radiation pressure on a dielectric discontinuity has served as the basis of numerous other *Gedankenexperimente*.[7] Kats and Kontorovich[8] considered the effect of laser light on liquid surfaces in connection with nonlinear effects and suggest that surface lenses can be made comparable with the nonlinear lenses generated in self-focusing experiments.

In our experiment we have generated surface lenses free of background thermal or nonlinear index changes. This was accomplished using 20 pulses per second of single transverse-mode doubled neodymium:yttrium-aluminum-garnet radiation at $\lambda = 0.53$ μm having a peak power of 1–4 kW and a width τ of 60 nsec focused to a spot diameter of $2w_0 = 4.2$ μm on the surface of water. Water should give low thermal lens effects[9] for $\lambda = 0.53$ μm since absorption is low ($\alpha \cong 3 \times 10^{-4}$ cm^{-1} and two-photon absorption is negligible), dn/dT is lower than most liquids, and finally the light pulse width is less than the thermal time constant which reduces thermal lensing by ~70. Since our power of 4 kW is well below the threshold for self-focusing[10] (~1 MW), nonlinear lens effects should be minimal. With 4 kW and a 4.2-μm spot diameter we expect a force sufficient to overcome surface tension and give surface motion. For the surface tension force we use $2\pi r S \times \sin\theta$, where r is the beam radius to the half-power point, S the surface tension, and θ the angle of the surface normal and the beam. This has a maximum value ~5×10^{-2} dyn for $\theta = 90°$ using $r = 1.2$ μm. With 4 kW and using either Un/c or U/cn for p, the radiation pressure force is ~4×10^{-1} dyn which exceeds surface tension by a factor of 8. This force is also 10^3–10^4 larger than needed to move or support micrometer-size free particles.[11]

Figure 1 shows the apparatus for studying laser-induced lenses. Light is focused on the water surface from above and viewed from below with a microscope and scanning slit-detector

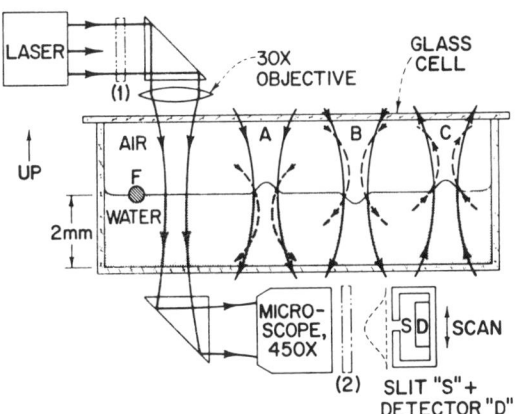

FIG. 1. Basic apparatus: *A*, Beam shapes for low power (solid curve) and high power (dashed curve) for positive surface lens; *B*, shapes (for low and high power) for a negative surface lens. For *A* and *B* the beam is incident from above. *C*, beam shapes for low and high power for a positive surface lens with the beam incident from below.

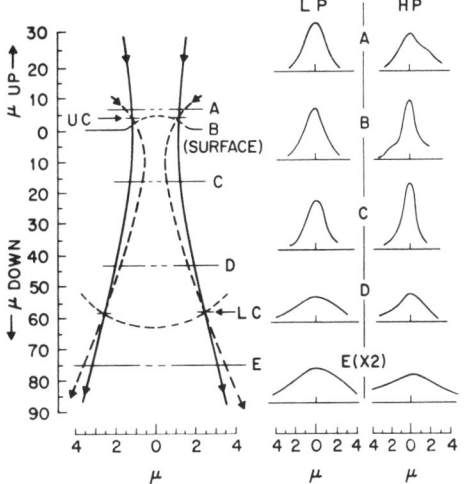

FIG. 2. Left, observed beam shapes at the half-power points, at low power (LP) (solid curve) and high power (HP) (dashed curve). A positive lens is dashed in at the upper crossover UC. A negative lens is dashed in at the lower crossover LC. Right, detailed scans of shapes, i.e., power versus position, at the indicated planes.

combination. To identify the surface with the microscope we floated a glass fiber *F*, 2–4 μm in diameter, as a local reference. Damage to the optics was avoided with long–working-distance objectives. The beam shapes were recorded at various planes above and below the surface at low and high power. To change the power level an attenuator was moved from position (1) in front of the laser to position (2) in front of the detector. Using a boxcar integrator with a 10-nsec gate we could measure the beam shapes at different times during the light pulse and thus follow the time development of the lens. If the force is outward on entering the liquid, the surface should lift up and form a positive or focusing lens which changes the beam shape at high power as shown at *A* in Fig. 1. If the force is inward, the surface should depress forming a negative lens as at *B* in Fig. 1. The beam shapes observed at high power will be the real beam shapes when viewed below the surface lens and virtual shapes when viewed above the lens. Figure 2 shows results for the beam shapes taken with 3 kW of power when measured ~250 nsec after the peak of the light pulse. The data show strong lens effects at high power. The data suggest a positive lens at the upper crossover as at *A* in Fig. 1. From data taken at other times we get the time development of the lens shown in Fig. 3(b). The light pulse shape is shown in Fig. 3(a). In Fig. 3(b) the lens strength is taken as

the reciprocal of the focal length of the ideal positive lens which best fits the observed beam shape. We see that the lens develops strongly hundreds of nanoseconds after the peak of the light pulse. At these times the light intensity, though weak, is adequate to study the lens shape. We show that the time development of the lens and the surface displacement can only be reason-

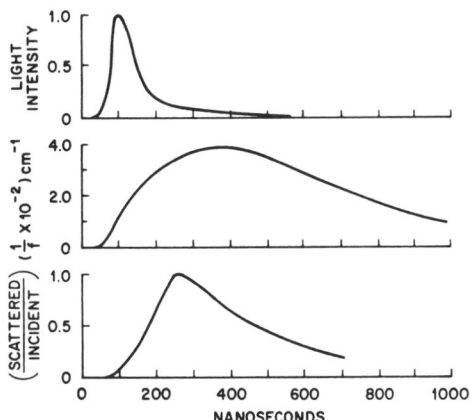

FIG. 3. Time development of (a) light pulse, (b) lens strength in units of $1/f$ (cm^{-1}), where f is the focal length, and (c) normalized scattering, i.e., (scattered light)/(incident light).

VOLUME 30, NUMBER 4   PHYSICAL REVIEW LETTERS   22 JANUARY 1973

ably understood in terms of a positive lens located ~ 4 ± 2 μm above the original surface at the upper crossover UC of Fig. 2. Assuming the force is on for only $t = \tau = 60$ nsec, the liquid receives an impulse $\int F \, dt \cong 1.8 \times 10^{-8}$ dyn sec. If we take a mass $= 1.5 \times 10^{-11}$ g, which is reasonable for a lens at the upper crossover, this results in a velocity $v \cong 1.2 \times 10^3$ cm/sec and a surface displacement $d \cong 0.4$ μm after time τ. Subsequently, the moving mass coasts upward to a stop restrained only by surface tension (i.e., $\int 2\pi rS \, dt = \int F \, dt$). Thus, after a time of 390 nsec the lens has moved up to $d \cong 2.5$ μm and is fully developed. The lens should then relax as a result of the continuing pull of surface tension after another 660 nsec, i.e., at $t \cong 1000$ nsec. Thus, these semiquantitative considerations agree with a positive lens in Fig. 2 and the time dependence of Fig. 3(b). Even the strength of the lens is reasonable. To get the 20-μm average focal length observed in Fig. 2, we need a lens with an average radius of curvature ~ 5 μm. If we try now to interpret the data of Fig. 2 in terms of a negative lens at the lower crossover, then because of the larger mass involved we get discrepancies between the calculated and observed displacements of ~ 10^4. We also experimentally ruled out a negative lens at the lower crossover. If we raise the power to ~ 4 kW this increases the lens strength, and we find that the upper and lower crossovers both move up. This is only consistent with the surface located at the upper crossover. Similarly for weaker lenses both the upper and lower crossovers move down.

We have thus far attributed the observed lenses to surface deformation and given reasons for expecting negligible thermal and nonlinear lens effects. We also showed this experimentally by looking for lenses when the laser was focused ~ 500 μm into the depth of the water. Here thermal and nonlinear lens effects, if present, should still occur, whereas surface lens effects should disappear. We found no detectable lenses.

Finally, we irradiated the free surface from below, i.e., from within the liquid. We find that the surface has again lifted up to form a positive lens as indicated at C in Fig. 1. This is as expected since if light on entering a dielectric medium exerts an outward force on the entering surface, it must also exert an outward force on the exiting surface to maintain the momentum balance.

Another feature accompanying strong lens development is the appearance of strong surface light scattering. This is seen at all angles, but mostly as a broad halo in the forward direction. Scattering is also evident in Fig. 2 where the integrated intensity at high and low powers can often differ by 20%. This lost power is the scattered light. We expect some scattering and beam-shape distortions from a smooth nonideal lens of small radius of curvature. However, we attribute most of the large-angle scattering at higher powers to finer scale ripples or possibly some necking in of the lens due to incipient droplet formation. Figure 3(c) shows the time development of the scattering. This differs from the lens development indicating a different origin. Surface scattering provides another method of determining the position of the lens. Since the scattering source must lie at the surface within the core of the beam, we located with the microscope the depth at which the scattered halo shrinks into the beam core. This occurs as expected at the upper crossover and *not* at the lower crossover. Also, by viewing the scattering at grazing incidence above and below the surface, we observe that the intensity falls almost to zero as the angle below the surface decreases to a few degrees, whereas the intensity increases somewhat as the grazing angle decreases from above. This is understood from the reflecting and transmitting properties of a surface only if the scattering source is located *above* the background surface. Thus, this gives an independent determination of the position of the surface at high power.

The presence of surface motion implies the existence of a nonlinear optical energy loss for transparent dielectrics since the kinetic energy of the moving liquid is eventually lost to heat. In our experiment it represents a fractional energy loss of ~ 10^{-8}. This varies as the impulse squared. At higher impulses we eventually expect droplets a few micrometers in diameter will be ejected from the surface at considerable velocity. With lower surface tension all effects should occur more strongly. Indeed some of our strongest lenses occurred with detergent added to the water.

In conclusion, the observed direction of the net force at the surface due to radiation pressure agrees with the predictions of the Minkowski momentum. In this sense it is in agreement with the experiment of Jones and Richards.[5] This disagrees with the expectations of Burt and Peierls[3] based solely on the Abraham momentum. More detailed study of the liquid surface dynamics is needed to get more quantitative results on the magnitude of the observed force. A recent analysis by Gordon,[12] prompted by this experiment, shows that the Minkowski momentum is a pseudo-momentum that gives the correct value of the observed forces on a liquid surface or a metal vane[5] and yet does not invalidate the true Abraham momentum.

We acknowledge helpful discussions with J. A. Arnaud, E. I. Blount, J. P. Gordon, A. G. Fox, C. K. N. Patel, and R. Kompfner.

[1]J. H. Poynting, Phil. Mag. S6 **9**, 393 (1905).

[2]W. Pauli, *Theory of Relativity* (Pergamon, New York, 1958), pp. 106-111 and p. 216; C. Møller, *The Theory of Relativity* (Oxford Univ. Press, London, 1962), pp. 202-206.

[3]M. G. Burt and R. Peierls, to be published.

[4]W. Shockley and R. P. James, Phys. Rev. Lett. **18**, 876 (1967).

[5]R. V. Jones and J. C. S. Richards, Proc. Roy. Soc., Ser. A **221**, 480 (1954).

[6]The authors thank Professor N. Rosenzweig and Professor R. Peierls for bringing these problems to their attention.

[7]See I. Brevik, Kgl. Dan. Vidensk. Selsk., Mat.-Fys. Medd. **37**, No. 13 (1970), for a discussion and additional references.

[8]A. V. Kats and V. M. Kontorovich, Pis'ma Zh. Eksp. Teor. Fiz. **9**, 192 (1969) [JETP Lett. **9**, 404 (1969)].

[9]J. P. Gordon *et al.*, J. Appl. Phys. **36**, 3 (1965).

[10]P. L. Kelley, Phys. Rev. Lett. **15**, 1005 (1965).

[11]A. Ashkin, Appl. Phys. Lett. **19**, 283 (1971).

[12]J. P. Gordon, to be published.

Guiding neuronal growth with light

A. Ehrlicher*†‡§, T. Betz*†‡, B. Stuhrmann*†, D. Koch*†, V. Milner*, M. G. Raizen*‡, and J. Käs*†‡

*Center for Nonlinear Dynamics, Department of Physics, University of Texas, Austin, TX 78712; and †Lehrstuhl für die Physik Weicher Materie, Fakultät für Physik und Geowissenschaften, Universität Leipzig, Linnéstrasse 5, D-04103 Leipzig, Germany

Communicated by Harry L. Swinney, University of Texas, Austin, TX, October 17, 2002 (received for review May 6, 2002)

Control over neuronal growth is a fundamental objective in neuroscience, cell biology, developmental biology, biophysics, and biomedicine and is particularly important for the formation of neural circuits *in vitro*, as well as nerve regeneration *in vivo* [Zeck, G. & Fromherz, P. (2001) *Proc. Natl. Acad. Sci. USA* 98, 10457–10462]. We have shown experimentally that we can use weak optical forces to guide the direction taken by the leading edge, or growth cone, of a nerve cell. In actively extending growth cones, a laser spot is placed in front of a specific area of the nerve's leading edge, enhancing growth into the beam focus and resulting in guided neuronal turns as well as enhanced growth. The power of our laser is chosen so that the resulting gradient forces are sufficiently powerful to bias the actin polymerization-driven lamellipodia extension, but too weak to hold and move the growth cone. We are therefore using light to control a natural biological process, in sharp contrast to the established technique of optical tweezers [Ashkin, A. (1970) *Phys. Rev. Lett.* 24, 156–159; Ashkin, A. & Dziedzic, J. M. (1987) *Science* 235, 1517–1520], which uses large optical forces to manipulate entire structures. Our results therefore open an avenue to controlling neuronal growth *in vitro* and *in vivo* with a simple, noncontact technique.

Many of the chemical cues that control neuronal growth during the development of an organism have been identified, although their interplay is rather complex (1–4). These biochemical signals eventually address the actin cytoskeleton, which advances the leading edge of a growth cone, known as the lamellipodium, through polymerization of new actin filaments and interactions with molecular motors (5–7). This understanding of neuronal growth is the basis for ongoing efforts to form defined neuronal network architectures, which can be interfaced with artificial structures such as semiconductors (8, 9), and to achieve successful nerve regeneration. In both cases, alternative approaches to complex guidance cues have been explored. On artificial substrates, such as silicon wafers, nerves have been directed by topographically structured surfaces (10) or by selectively patterning the substrate with materials that act as adhesives for nerves (11, 12). Damaging tensions, however, often rip apart the neuronal structures formed on the substrate due to the tendency of axons to straighten and stiffen with time. Thus, only small random neuronal networks have been successfully built in contact with semiconductor structures (13). There also have been reports of guiding neurons with electrodes (14, 15), but the specific impact of induced electrophoresis effects is not well understood (16, 17). *In vivo*, various types of guidance channels (18, 19) are used to repair neuronal damage of the peripheral nerve system. Fiber-based optical guidance may offer an alternative for aided regeneration of peripheral nerves. For the more complex situation of spinal cord injuries, stem cell approaches may be more viable (20). With this background, it is clear that optical nerve guidance is an important alternative to existing methods of nerve guidance.

Over the past 30 years, lasers have been used to manipulate objects ranging from atoms to μm-sized beads or biological cells (21–24). One-beam gradient traps, also known as optical tweezers, have been used successfully for a variety of biological experiments in which cells, organelles, or beads, attached to biological objects as tiny handles, have been held and moved (25–28). Single polymers have been micromanipulated with light

(29, 30) and polymer solutions have been patterned with a laser (31). The results reported in this article show that lasers can be used to actually control a natural biological process and can address the long-standing goal of controlling neuron growth.

Cell Culture

We used PC12 cells, a rat neuron precursor cell line, stimulated to spread with neuronal growth factor (5×10^{-5} mg/ml), and NG108 cells, an immortalized mouse neuroblastoma rat glioma hybrid cell line. Cells were cultured in medium at 37°C (PC12: 85% RPMI-1640, 10% horse serum, 5% FBS; NG108: 90% DMEM, 10% FBS) (32, 33). Cells and media were purchased from the American Type Culture Collection. Cells under optical direction were grown for up to 4 h on plain cover glasses for NG108 cells and either plain or laminin-coated cover glasses for PC12 cells. Cells adhered to these cover glasses were integrated into a temperature-controlled sample chamber to assure continuing cell viability. Because no differences were observed between experiments with normal cell medium and cell medium that was pH-stabilized through addition of 10 mM Hepes we concluded that pH effects can be neglected although the sample's pH was not actively controlled. This finding is not surprising because the measured changes in medium without Hepes were <5% during the observation time.

Optical Guidance Methods

Control over the extending growth cone was achieved with a Ti:sapphire laser ($\lambda = 800$ nm). The laser light was guided and focused through the beam pathway of an inverted confocal microscope (Zeiss LSM 410) equipped with a ×63 objective (Zeiss Plan Neofluar, Ph3, numerical aperture = 1.25), which allowed us simultaneously to image the growth cone in phase contrast (Fig. 1). No particular efforts were made to achieve a tightly focused beam as is typically required for optical tweezers (34). We varied the spot size diameter in the plane of the lamellipodium between ≈2 and 16 μm and directly measured the power after the microscope objective from 20 to 120 mW. Whereas the variation in beam diameters <4 μm was achieved by optically defocusing the beam, the larger spot sizes were simulated by scanning the beam at a frequency of ≈0.1 Hz along the desired area of the growth cone's leading edge. Scanning at a frequency faster than intracellular processes generated an illuminated band along the lamellipodial outline. This process is beneficial for guiding larger areas of the lamellipodium, because it combines the sharper optical gradient of a smaller spot with an artificially larger spatial coverage. To achieve optical guidance, the beam was interactively steered with the confocal scanning microscope or by moving the microscope stage relative to the sample with an xy-piezo stage (Physik Instrumente, Berlin). We placed approximately one-fourth to one-half of the beam image on the lamellipodium, with the remainder of the spot placed ahead of the leading edge.

Heating effects caused by absorption of the laser beam, which could influence the cell's biochemical reactions, play a minor role because the steady-state temperature increase is estimated

---

‡A.E., T.B., M.G.R., and J.K. contributed equally to this work.

§To whom correspondence should be addressed. E-mail: allen@physik.uni-leipzig.de.

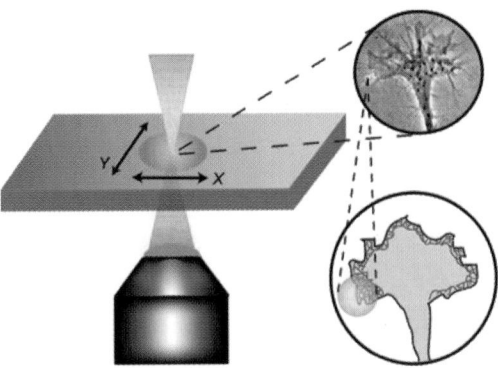

Fig. 1. Experimental setup for the optical guidance of growing neurons. A laser spot ($\varnothing$ = 2–16 μm, power = 20–120 mW, λ = 800 nm) was placed with partial overlap in front of an actively extending growth cone. The overlap area was chosen in the direction of the preferred growth and to cover the actin cortex, which directly underlies the plasma membrane and drives the advancement of the leading edge of the nerve.

to be only a few degrees. The rate of actin nucleation is relatively insensitive to small temperature changes $\approx 37°$C, although the rate can drastically change at lower temperatures (35, 36). Although small heating enhances actin polymerization, it is not restricted to the area of the laser spot, due to thermal conductivity. We can therefore exclude heating as the cause of controlled optical guidance. Because the neuronal cells maintained their viability and growth after guiding, we also conclude that no significant radiation damage occurred.

Results

The observed optical guidance of neurons is a robust effect, successful for both rat and mouse neuronal cell lines and a broad range of laser powers. For our experiments, we chose actively ruffling lamellipodia, which were vigorously extending and sufficiently separated from the cell body and other axons to be clearly defined. Successfully biased growth is characterized by a lamellipodia extension in the radius of influence (approximately three laser beam radii) that grows toward the beam center. Subsequent extensions into the radius of influence are not considered to be new events here if no beam parameters (size, power, position, direction angle) have changed.

Our laser beam, asymmetrically positioned to the left or right of the leading edge of a growth cone, clearly affects advancing nerves (Fig. 2). The laser spot determines the active area of lamellipodia extension. We observe that the extending lamellipodia grow into the focus of the laser beam in 35 of 44 experiments. This is in 79.5% of all cases. Experiments where normally advancing growth cones are visible in the same field of view as optically guided neurons indicate that the laser increases the rate at which the nerve's leading edge advances (Fig. 3). Because growth cones extend discontinuously and two well-defined advancing growth cones in the field of view are rarely observed, a precise quantification of the effect was not possible. Nevertheless, from the time series in Fig. 3 we are able to estimate the trend that the lamellipodia extension rate increases from 7 ± 3 μm/h to 37.5 ± 22.5 μm/h.

The biased growth events are transient, lasting a few minutes. However, their cumulative effect leads to permanent results. Continuous optical guidance of the spreading growth cone results in controlled turns. We define a turn as a persistent change of growth direction that tracks the path of the laser spot

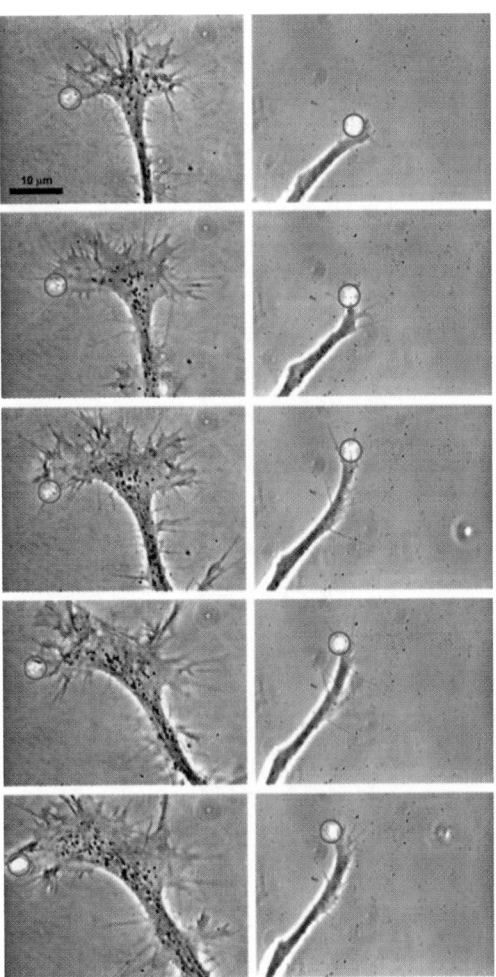

Fig. 2. Time sequences of optically guided turns of neurons and optically enhanced neuronal growth. Optically induced turns are shown for a time period of 40 min (*Left*) and 20 min (*Right*). The time interval between successive pictures is 10 min (*Left*) and 5 min (*Right*). The power of the laser spot is 100 mW (*Left*) and 60 mW (*Right*), and a red circle indicates the position of the laser spot (see Movies 1 and 2, which are published as supporting information on the PNAS web site, www.pnas.org). Optical control was achieved for extensive flat growth cones (*Left*) as well as for small, tube-like growth cones (*Right*). Before the laser altered the direction of the growth cone, the nerve was growing upward (*Left*) or to the right side (*Right*). The growth direction changes on the order of 90° under optical guidance. Note that the apparent change growth direction appears to be smaller because the axon straightens into the new direction.

for at least three radii of influence. The change in direction was chosen to be between 30° and 90° from the initial direction of growth. In these experiments, the laser spot is continuously positioned sufficiently in front of the leading edge to ensure uninterrupted extension and to prevent the leading edge from filling the entire laser spot. The laser guidance speed is limited by the extension speed of the cone. We observe successfully

Ehrlicher *et al.*

BIOPHYSICS

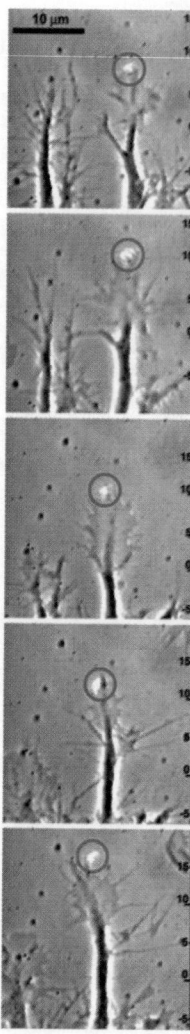

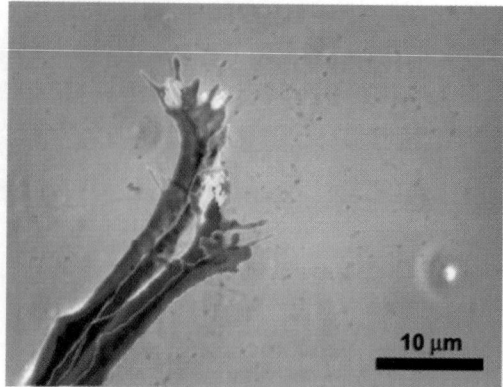

Fig. 4. Superposition of the shapes of an actively extending growth cone under optical guidance. The same growth cone is shown in Fig. 2 *Right*. The progressing time is coded by the following series of colors: yellow (*t* = 0 min), red (*t* = 5 min), green (*t* = 10 min), and blue (*t* = 20 min). Two effects are clearly visible: stimulated by the laser, the small growth cone extends to the top of the picture in a pronounced lamellipodia structure with noticeable filopodia; the axonal stump, which does not actively participate in the growth process, changes its orientation by pointing more toward the top of the picture. Thus, the laser spot is able to induce significant changes in growth direction if the following axon can relax and avoid bends.

Fig. 3. The parallel extension of two growth cones, one optically guided (right) and one normally growing, illustrates the increase in the speed of growth cone extension in the presence of a 20-mW laser spot (red circle) (see Movie 3, which is published as supporting information on the PNAS web site). The time interval between successive pictures is 10 min. The reference marks on the left are spaced in 5-μm steps.

guided turns as successions of controlled lamellipodia extensions in 17 of 20 experiments. This finding means that optically guided turns were achieved with a success rate of 85.0% whereas in control experiments with an imaginary laser spot the success rate for apparent optically guided turns is only 20% in 10 experiments. The control experiments use fictitious radiation times and time intervals between trials that both agree with real optically guided turns. Our results are clearly statistically significant. The Fisher–Yates test, which is used for experiments with a small number of tests and controls, gives a P value of 0.097% for our

optically guided turn experiments (a value below the standard confidence level of 5% is considered to be statistically significant). The P value denotes the probability to get an 85% success rate by chance in control experiments.

An overlay of the shapes of an optically guided neuron reveals the direct impact of the laser beam on lamellipodia structure as well as an indirect, temporally delayed turn of the axonal stump (Fig. 4). The area radiated by the beam promptly responds with the extension of a lamellipodium and with an accumulation of filopodia in the beam area. The induced turn does not appear as large as the true change in direction because the axonal stump behind the growth cone slowly straightens in the new direction of lamellipodia extension. The directional change of the leading edge can be on the order of 90°, as can be seen in Figs. 2 and 4. Thus, optically guided growth cones are able to drastically change their direction of movement if the axon behind the lamellipodia is allowed to turn in the new direction of growth cone extension. Because the high rigidity of the axon favors low curvatures (i.e., no curves and bends), it is preferable to guide a neuron in a fashion that allows the axonal stump to straighten, relaxing high curvatures. Optical guidance avoids high tensions due to bending, yielding a substantial but gradual turn.

In some cases where the angle of directed growth is large with respect to the original growth direction, the laser beam induced a bifurcation of the growth cone (Fig. 5). This allows us to compare the underlying structure of the actin cytoskeleton in optically guided and normally developed lamellipodia. The narrower shape of the optically guided growth cone is not significant because we observe small and large growth cones during both normal, as well as optically guided, growth. For both growth cones, we find prominent structures of filamentous actin, which are absent in the main body of the growth cone, when visualized by rhodamine-phalloidin staining (which only binds to actin filaments). Actin bundles, originating from the filopodia and the filamentous actin cortex directly underlying the plasma membrane, are also visible. Because these two structures are key

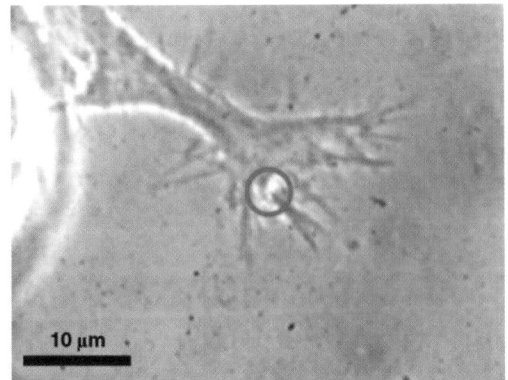

10 µm

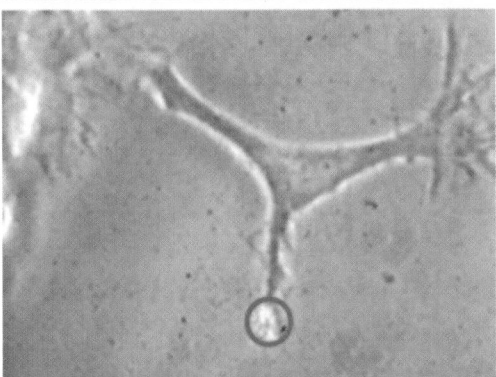

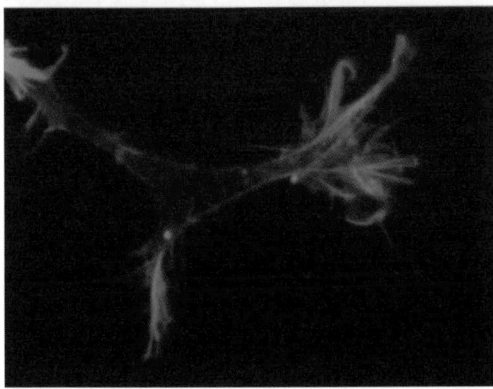

Fig. 5. Optically induced bifurcation of a growth cone. A growth cone, which is growing to the upper right, sprouts off an extension to the lower right under the influence of the beam marked by a circle. The last picture displays the distribution of actin filaments by rhodamine-phalloidin staining. Actin filaments are clearly accumulated at the areas of lamellipodia extension.

to lamellipodia extension, an essential prerequisite for the expansion of the leading edge is the transport of actin monomers (G-actin) to the front of the cell to provide for the polymerization of new actin filaments (37).

Discussion

At the chosen power levels of the laser, we are not able to detach the growth cone from the substrate and move it around by steering the focus of the laser beam. Therefore, we do not use the gradient forces of our optical trap in the conventional fashion, whereby the tweezing effect of the trap is used to move an entire structure. We conclude that our laser spot influences the general actin-based processes for growth cone extension when we optically guide turns. Because these intracellular processes are not completely understood, it is difficult to fully explain the molecular basis of optical guidance. Nevertheless, the optical forces interact with the proteins of the cytoplasm. Because the larger filamentous structures of the cytoplasm are crosslinked and coupled to the plasma membrane by focal adhesions, the optical forces can only exert a tension on these structures, but cannot move them. However, globular proteins and small oligomeric structures are free to move in the cytoplasm, and thus the optical forces can impact their diffusive behavior. The smallest forces are felt by globular proteins and are a lower bound for this effect.

The laser beam, placed just outside the growth cone, creates an intensity gradient for the globular proteins of the cytoplasm, such as actin monomers, and they feel a weak optical dipole force. This force is directed toward the center of the laser spot. In the electromagnetic field of the laser E, the force on an induced electric dipole $p = \alpha E$ is given by (38)

$$F(r)_{dipole} = \langle p|\nabla|E(r,t)|\rangle_{timeaverage} = \frac{1}{4}\alpha\nabla|E(r)|^2 = \frac{\alpha\nabla I(r)}{2n_{cyt}\varepsilon_0 c},$$

$$[1]$$

where α is the polarizability of the protein experiencing the dipole force, $n_{cyt} = 1.37$ is the refractive index of the cytoplasm (39), ε_0 is the dielectric constant, and c is the speed of light. $I(r)$ represents the Gaussian intensity profile of the laser beam. The corresponding dipole potential (i.e., the attractive energy toward the center of the laser beam) is

$$U(r) = \frac{\alpha I(r)}{2n_{cyt}\varepsilon_0 c}.$$

$$[2]$$

We estimated α for G-actin, which we needed to calculate F_{dipole} and U, using the Clausius–Mosotti equation (40)

$$\alpha = 4\pi\varepsilon_0 r_{act}^3 n_{water}^2 (m^2 - 1)/(m^2 + 2)$$

$$[3]$$

($r_{act} = 3$ nm is the radius of an actin monomer, n_{water} is the refractive index of water, $m = n_{act}/n_{water}$). With a refractive index for actin of $n_{act} = 1.59$ (41–43) we obtain for G-actin (as well as other proteins of similar size within the cell) a polarizability of $\approx 6.5 \times 10^{-37}$ Cm²/V.

With a laser power of 60 mW focused to a beam radius of 1.5 µm, Eq. **2** yields a potential well depth of $U = 1.5 \times 10^{-24}$ J $= 3.6 \times 10^{-4} k_B T$ (k_B Boltzmann's constant, $T = 310$ K the cell temperature, $k_B T$ thermal energy). Because polarizability increases with molecular size, small actin oligomers (fragments of actin filaments) will experience a deeper potential well than actin monomers. These fragments originate in the depolymerization of the actin cortex toward the interior of the growth cone. Nevertheless, this potential well remains smaller than the thermal energy $k_B T$ and thus is much too weak to function as optical tweezers because Brownian diffusion can overcome its trapping power.

A significant portion of the cytoplasmic proteins, including G-actin, undergo Brownian motion within the cell, and their diffusive transport to the leading edge of a lamellipodium is essential for cell motility. By approximating the dipole force F_{dipole} with a spatially constant force equaling the dipole force at half the radius of the laser spot, a drift velocity caused by the dipole force toward the center of the laser beam can be estimated. This drift velocity biases diffusion of cytoplasmic proteins toward the laser. The drift velocity, v_{Drift}, satisfies

$$v_{Drift} = F_{dipole}/\xi \qquad [4]$$

with $\xi = k_B T/D$ (D is the diffusion coefficient). The diffusion coefficient D of a globular macromolecule of the size of G-actin in the cytoplasm is $\approx 2.5 \times 10^{-11}$ m^2/s (44). Based on these parameters and Eqs. 1 and 4, we predict a drift velocity of 26 μm/h for actin monomers, which is very close to growth rates of an active growth cone observed in the experiments. Dipole forces create even higher drift velocities for fragments of actin filaments. We therefore propose that the optical gradient enhances actin polymerization at the leading edge by pooling actin monomers and providing nucleation seeds in the form of actin filament fragments. The resulting locally enhanced actin polymerization promotes a turn of the lamellipodium in the direction of optically increased actin density. Optical forces that distend the membrane might also allow actin monomers to flow into new extensions. Furthermore, although they cannot move the crosslinked actin cortex, the optical forces do pull on this actin network and relieve pressure at the rearward cortex-microtubule

junction, thus enhancing pressure-dependent microtubule polymerization (45). In summary, we conjecture that weak optical dipole forces could be used to control biological growth processes such as growth cone extension. The molecular basis of this optical guidance is complex and may rely on a combination of all of the optically induced processes discussed above.

Conclusion

We have shown that weak optical dipole forces can be used to bias the molecular process of cell motility so that we can guide neuronal growth. Because we influence lamellipodia extension, a process essential for all motile cells, optical guidance may be extended as a general cell guidance method and provide an investigative tool to understand the underlying processes of cell motility. Moreover, we hope that optical guidance will allow us to form controlled neuronal structures *in vitro*, and we can also imagine that optical guidance could find applications in nerve repair *in vivo*.

We thank S. A. Moore and D. Martin for help in editing the manuscript and M. Löffler, A. Reichenbach, W. Thompson, K. Franze, C. Duggan, J. Shear, M. Goegler, F. Käs, D. Sanchez, and P. Fromherz for helpful discussions. The Alexander von Humboldt Foundation through the Wolfgang Paul Prize has supported J.K. in this work. The research of M.G.R. was supported by the Sid W. Richardson Foundation.

1. Bomze, H. M., Bulsara, K. R., Iskandar, B. J., Caroni, P. & Skene, J. H. (2001) *Nat. Neurosci.* **4**, 38–43.
2. Tessier-Lavigne, M. & Goodman, C. S. (1996) *Science* **274**, 1123–1133.
3. Mueller, B. K. (1999) *Annu. Rev. Neurosci.* **22**, 351–388.
4. Hong, K., Nishiyama, M., Henley, J., Tessier-Lavigne, M. & Poo, M. (2000) *Nature* **403**, 93–98.
5. Pantaloni, D., Le Clainche, C. & Carlier, M.-F. (2001) *Science* **292**, 1502–1506.
6. Suter, D. M. & Forscher, P. (2000) *J. Neurobiol.* **44**, 97–113.
7. Borisy, G. G. & Svitkina, T. M. (2000) *Curr. Opin. Cell Biol.* **12**, 104–112.
8. Weis, R., Müller, B. & Fromherz, P. (1996) *Phys. Rev. Lett.* **76**, 327–330.
9. Vassanelli, S. & Fromherz, P. (1997) *Appl. Phys. A* **65**, 85–88.
10. Duncan, A. C., Weisbuch, F., Rousais, F., Lazore, S. & Baquez, C. (2002) *Biosensors Bioelectronics* **17**, 413–426.
11. Fromherz, P., Schaden, H. & Vetter, T. (1991) *Neurosci. Lett.* **129**, 77–80.
12. Prinz, A. A. & Fromherz, P. (2000) *Biol. Cybernet.* **82**, L1–L5.
13. Zeck, G. & Fromherz, P. (2001) *Proc. Natl. Acad. Sci. USA* **98**, 10457–10462.
14. Marsh, G. & Beams, H. W. (1946) *J. Cell. Comp. Physiol.* **27**, 139–157.
15. Patel, N. & Poo, M. (1982) *J. Neurosci.* **2**, 483–496.
16. Poo, M., Lam, J. W., Orida, N. & Chao, A. W. (1979) *Biophys. J.* **26**, 1–21.
17. Ming, G., Henley, J., Tessier-Lavigne, M., Song, H. & Poo, M. (2001) *Neuron* **29**, 441–452.
18. Valentini, R. F. (1995) in *The Biomedical Engineering Handbook*, ed. Bronzino, J. D. (CRC, Boca Raton, FL), pp. 1985–1996.
19. Rivers, T. J., Hudson, T. W. & Schmidt, C. E. (2002) *Adv. Funct. Mater.* **12**, 33–37.
20. Gage, F. H. (2000) *Science* **287**, 1433–1438.
21. Ashkin, A. (1970) *Phys. Rev. Lett.* **24**, 156–159.
22. Ashkin, A. & Dziedzic, J. M. (1987) *Science* **235**, 1517–1520.
23. Chu, S. (1991) *Science* **253**, 861–866.
24. Kuo, S. C. & Sheetz, M. P. (1992) *Trends Cell Biol.* **2**, 116–118.
25. Ashkin, A., Dziedzic, J. M., Bjorkholm, J. E. & Chu, S. (1986) *Opt. Lett.* **11**, 288–290.
26. Svoboda, K. & Block, S. M. (1994) *Annu. Rev. Biophys. Struct.* **23**, 147.
27. Frohlich, J. & Konig, H. (2000) *FEMS Microbiol. Rev.* **24**, 567–572.
28. Holm, A. & Sundqvist, T. (1999) *Med. Biol. Eng. Comput.* **37**, 410–412.
29. Bustamante, C., Macosko, J. C. & Wuite, G. J. L. (2000) *Nat. Rev. Mol. Cell Biol.* **1**, 130–136.
30. Fujii, T., Sun, Y. L., An, K. N. & Luo, Z. P. (2002) *J. Biomech.* **35**, 527–531.
31. Sigel, R., Fytas, G., Vainos, N., Pispas, S. & Hadjichristidis, N. (2002) *Science* **297**, 67–70.
32. Greene, L. A. & Tischler, A. S. (1976) *Proc. Natl. Acad. Sci. USA* **73**, 2424–2428.
33. Hamprecht, B. (1977) *Int. Rev. Cytol.* **49**, 99–170.
34. Block, S. (1995) in *Noninvasive Techniques in Cell Biology*, eds. Foskett, K. & Grinstein, M. (Wiley, New York), pp. 375–401.
35. Niranjan, P. S., Forbes, J. G., Greer, S. C., Dudowicz, J., Freed, K. F. & Douglas, J. F. (2001) *J. Chem. Phys.* **114**, 10573–10576.
36. Zimmerle, C. T. & Frieden, C. (1986) *Biochemistry* **25**, 6432–6438.
37. Lin, C. H., Espreafico, E. M., Mooseker, M. S. & Forscher, P. (1996) *Neuron* **16**, 769–782.
38. Harada, Y. & Toshimitsu, A. (1996) *Optics Commun.* **124**, 529–541.
39. Drezek, R., Dunn, A. & Richards-Kortum, R. (1999) *Appl. Optics* **38**, 3651–3661.
40. Rohrbach, R. & Stelzer, E. (2001) *Opt. Soc. Am. A* **18**, 839–853.
41. Kratochvíl, P. (1987) in *Classical Light Scattering from Polymer Solutions*, ed. Jenkins, A. D. (Elsevier, Amsterdam), pp. 79–83.
42. Huglin, M. B. (1972) in *Light Scattering from Polymer Solutions*, ed. Huglin, M. B. (Academic, London), p. 204.
43. Suzuki, N., Tamura, Y. & Mihashi, K. (1996) *Biochim. Biophys. Acta* **1292**, 265–272.
44. Popov, S. & Poo, M. (1992) *J. Neurosci.* **12**, 77–85.
45. Dennerll, T., Joshi, H., Steel, V., Buxbaum, R. & Heidemann, S. (1988) *J. Cell Biol.* **107**, 665–674.

Biophotonics

Self-Rotation of Red Blood Cells in Optical Tweezers: Prospects for High Throughput Malaria Diagnosis

Samarendra Kumar Mohanty, Abha Uppal and Pradeep Kumar Gupta

Malaria affects 500 million people a year and kills 2.7 million of them,[1] which makes methods for the detection of malarial infection of considerable interest. The membrane of malaria-infected red blood cells (RBCs) is more rigid than the membrane of normal RBCs.[2] A measurement of membrane elasticity of a RBC can, therefore, be used for the detection of malaria-infected RBCs.[3] More interestingly, it has been determined[4] that, in a hypertonic buffer medium (> 800-mOsm/kg osmolarity), malaria-infected RBCs either do not rotate at all or rotate at a significantly slower speed because of their rigid membrane. A normal RBC, in contrast, rotates by itself when placed in a laser optical trap at the same trap beam power. This difference in rotational speed has been exploited for the detection of malaria-infected cells.[4]

In a hypertonic buffer, a meniscus-shaped normal RBC rotates when it is optically trapped with trap beam power beyond ~40 mW (Fig. 1) because of the torque that is generated on the cell by the transfer of linear momentum from the trapping beam. For a given osmolarity, the rotational speed was observed to increase superlinearly with an increase in trap beam power, which we believe is the result of deformation of the RBC caused by the radiation pressure of the trap beam. In contrast, the infected cells having malaria parasite (as confirmed by acridine orange fluorescence staining) suspended in the same hypertonic buffer were not observed to rotate even up to 240-mW trap beam power. Even more significant is the fact that the rotational speed of other RBCs from a malaria-infected blood sample (which did not show Acridine orange fluorescence) was an order of magnitude smaller and increased much slower, with an increase in trap beam power in comparison with normal cells. We could screen approximately 40 RBCs/min by making the cells flow through the trapping point at a velocity of ~10 μm/s. When a flowing RBC struck a trapped RBC, the trapped RBC was thrown out by the collision and the other RBC would get trapped and—if normal—begin to rotate. Higher-screening rates can be achieved by increasing the flow rate of RBCs and by increasing the number of traps in an array orthogonal to the flow direction. This approach can provide higher throughput and sensitivity of detection of malaria in comparison with the current front-line approaches and can also be used for the diagnosis of other diseases, such as leukemia, that change the elasticity of a RBC membrane.

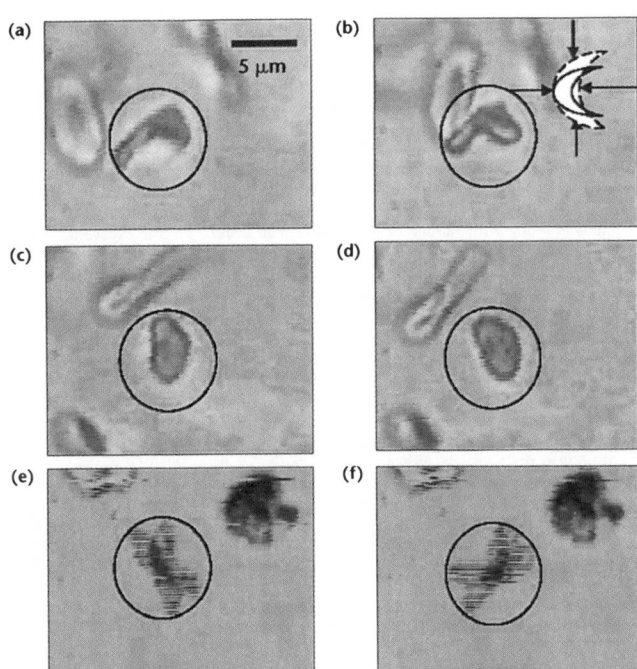

Figure 1. Rotation of a normal RBC trapped by optical tweezers. The cell (*encircled*) was suspended in a hypertonic buffer and trapped at power levels that varied from (a) 40 mW to (b) 200 mW. Figure 1(b) inset shows a schematic of the deformation observed in the horizontal cross section of a RBC structure at a higher trap beam power (*solid curve*). The dotted curve corresponds to the shape observed without the trapping beam. Arrows illustrate the transverse gradient force of optical tweezers. Figures 1 (c) and (d) show time-lapsed digitized video images of RBC rotation at a buffer osmolarity of 1,000 mOsm/kg; images in (e) and (f) represent an osmolarity of 1,250 mOsm/kg. The trap power was 80 mW and the time lapse between consecutive frames was 80 ms. From the sequence of digitized frames, the speed of rotation was estimated to be 25 rpm at 1,000 mOsm/kg and 200 rpm at 1,250 mOsm/kg.

Samarendra Kumar Mohanty, Abha Uppal and Pradeep Kumar Gupta (pkgupta@cat.ernet.in) are with the Biomedical Applications Section, Centre for Advanced Technology, Indore, India.

References
1. U. Weiss, Nature **415**, 669–715 (2002).
2. F. K. Glenister et al., Blood **99**, 1060–3 (2002).
3. S. K. Mohanty et al., presented at the Sixth International Conference on Optoelectronics, Fiber Optics, and Photonics, Mumbai, India, 16–18 Dec. 2002.
4. S. K. Mohanty et al., Biotechnol. Lett. **26**, 971–4 (2004).

Myosin VI is a processive motor with a large step size

Ronald S. Rock*, Sarah E. Rice*, Amber L. Wells†, Thomas J. Purcell*, James A. Spudich*‡, and H. Lee Sweeney†

*Department of Biochemistry, Stanford University School of Medicine, Stanford, CA 94305; and †Department of Physiology, University of Pennsylvania School of Medicine, 3700 Hamilton Walk, Philadelphia, PA 19104-6085

Contributed by James A. Spudich, September 27, 2001

Myosin VI is a molecular motor involved in intracellular vesicle and organelle transport. To carry out its cellular functions myosin VI moves toward the pointed end of actin, backward in relation to all other characterized myosins. Myosin V, a motor that moves toward the barbed end of actin, is processive, undergoing multiple catalytic cycles and mechanical advances before it releases from actin. Here we show that myosin VI is also processive by using single molecule motility and optical trapping experiments. Remarkably, myosin VI takes much larger steps than expected, based on a simple lever-arm mechanism, for a myosin with only one light chain in the lever-arm domain. Unlike other characterized myosins, myosin VI stepping is highly irregular with a broad distribution of step sizes.

Myosin VI is a molecular motor that is ubiquitously expressed across organisms and tissue types and is involved in a variety of functions (1–3). It is the motor that is defective in *Snell's waltzer* mice, characterized by deafness and coordination problems, suggesting that myosin VI is involved in stereocilia function in cochlear hair cells (4). Additionally, myosin VI is believed to be a vesicle transporter in other cell types. Immunocytochemistry has shown that myosin VI is directly associated with vesicles in *Drosophila* embryos (5), and imaging of green fluorescent protein (GFP) fusions indicates that myosin VI is broadly localized in the trans-Golgi network and in protrusions in the plasma membrane (6). Unique among characterized myosin motors, myosin VI moves toward the pointed ends of actin filaments (7). Actin is typically oriented with the barbed end toward the plasma membrane and the pointed end toward the cell interior. This fact, coupled with the observation that myosin VI colocalizes with clathrin-coated pits (8), suggests that myosin VI is involved in endocytosis. Kinetic characterization of single-headed constructs shows that myosin VI is a high-duty-ratio motor, meaning that it spends much of its ATPase cycle strongly bound to actin. Furthermore, myosin VI is kinetically processive, meaning that after a diffusional encounter with actin, it hydrolyzes multiple ATPs before completely releasing again (9). Together, these kinetic and functional characteristics led to the expectation that myosin VI may be capable of transporting cargo at the single molecule level. In this study, we show that myosin VI is a processive motor with an unusually large step size.

Methods

Protein Constructs and Expression. To create the double-headed myosin VI/GFP construct, the porcine myosin VI cDNA was truncated at Arg-994 to include 20 native heptad repeats of predicted coiled-coil and was followed by a leucine zipper (GCN4) to ensure dimerization (10). This was then followed by the cDNA for enchanced GFP (EGFP; CLONTECH), and then a Flag tag (encoding GDYKDDDDK) at the C terminus to facilitate purification (11). This cDNA was used to generate a recombinant baculovirus that was used for coexpression of the myosin VI/GFP with chicken calmodulin. To create the double-headed myosin V/GFP construct, the chicken myosin V cDNA was truncated at Glu-1099 and was followed by a leucine zipper (GCN4) to ensure dimerization (10). This was then followed by the cDNA for EGFP, and then a Flag tag (encoding GDYKD-DDDK) at the C terminus to facilitate purification (11). This cDNA was used to generate a recombinant baculovirus that was used for coexpression of the myosin V/GFP with separate

recombinant baculoviruses coding for calmodulin and two essential light chains, LC1-sa and LC23. The generation of recombinant baculovirus, expression in SF9 cells, and protein purification followed published procedures (9, 11). Human fascin protein and cDNA were the generous gift of Steve Almo (Albert Einstein College of Medicine, Bronx, NY). Bacterial expression and purification of fascin followed published procedures (12).

Motility Assays. In all assays described, assay buffers included 25 mM imidazole HCl (pH 7.4), 25 mM KCl, 5 μM calmodulin, 1 mM EGTA, 10 mM DTT, 4 mM MgCl$_2$, 2 mM ATP, and an oxygen scavenging system to retard photobleaching (25 μg·ml^{-1} glucose oxidase, 45 μg·ml^{-1} catalase, and 1% glucose). Assays were performed at 30°C, with the exception of the optical trapping assays, which were performed at 23°C. Actin bundles were made by polymerizing G-actin in the presence of a 10-fold molar excess of purified fascin. Loose bundles appeared after >1 week at 4°C. Single-molecule fluorescence assays were performed on an evanescent field microscope patterned after the design of Yanagida and coworkers (13) (R.S.R. and J.A.S., unpublished work). Only spots that moved >300 nm and for >1 s were tabulated. Landing and continuous movement assays were performed as previously described (14). Flow cells were pretreated with anti-GFP monoclonal antibodies (Quantum, Durham, NC, 0.05 mg·ml^{-1}), followed by 1 mg·ml^{-1} BSA in assay buffers, before addition of the GFP-tagged myosin VI. All surface myosin VI densities given assume that every molecule entered the flow cell, none were denatured, and half were adsorbed to each surface of the flow cell. Fits to the continuous movement assay were described by

$$P(> length) = \frac{P(2n)}{P(n)} = \frac{1 - \sum_{i=0}^{2n-1} \frac{(\rho/\rho_0)^i}{i!} e^{-\rho/\rho_0}}{1 - \sum_{i=0}^{n-1} \frac{(\rho/\rho_0)^i}{i!} e^{-\rho/\rho_0}}$$

where n is the number of motor units required for motility, ρ is the surface density, and ρ_0 is a fit parameter. All errors are SD unless otherwise indicated.

Optical Trap. A single actin filament attached at both ends to optically trapped 1-μm diameter polystyrene beads was stretched to tension, then relaxed slightly. This actin dumbbell was moved near surface-bound silica spheres that served as platforms, which were decorated sparsely with myosin VI molecules. Both polystyrene beads were tracked with nm and ms resolution. Force feedback was performed as described (15), with feedback control on one of the two trapped beads. The actin filament was arranged so that the motor protein would pull the bead under feedback control out of its trap. Because the actin filament between the bound motor and the second bead was slack (see Fig. 2A), the second bead did not move over the entire

---

Abbreviation: GFP, green fluorescent protein.

†To whom reprint requests should be addressed. E-mail: jspudich@cmgm.stanford.edu.

The publication costs of this article were defrayed in part by page charge payment. This article must therefore be hereby marked "*advertisement*" in accordance with 18 U.S.C. §1734 solely to indicate this fact.

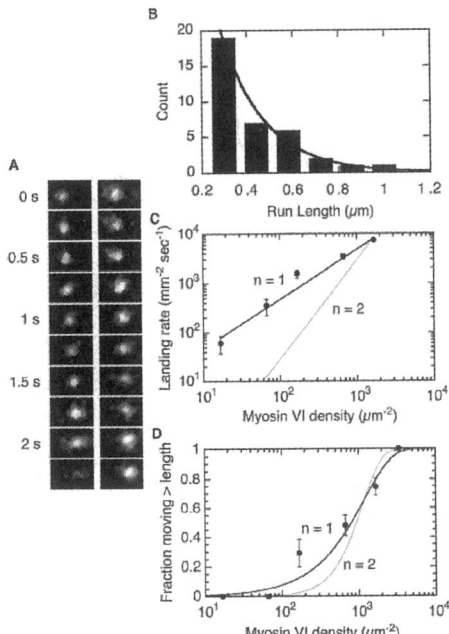

Fig. 1. Processivity assays. (A) Motility of GFP-tagged myosin VI on fascin-actin bundles as seen by total internal reflection fluorescence. Two separate runs are shown. Frame width, 1.5 μm. (B) Run-length distribution for single, fluorescently labeled myosin VI molecules. Exponential curves were fit to the data by using only runs that lasted longer than 1 s, all of which were ≥0.3 μm. n = 36 measurements, histogram bin widths are 0.14 μm. (C) Actin filament landing rates as a function of myosin VI density. First-power (reduced χ^2 = 2.45) and second-power (reduced χ^2 = 17.4) fits are shown. (D) Fraction of filaments that moved more than their length before dissociating, as a function of motor density. Fits describing single-molecule motility (reduced χ^2 = 2.2) and double molecule motility (reduced χ^2 = 8.9) are shown. Error bars are SE obtained from counting statistics.

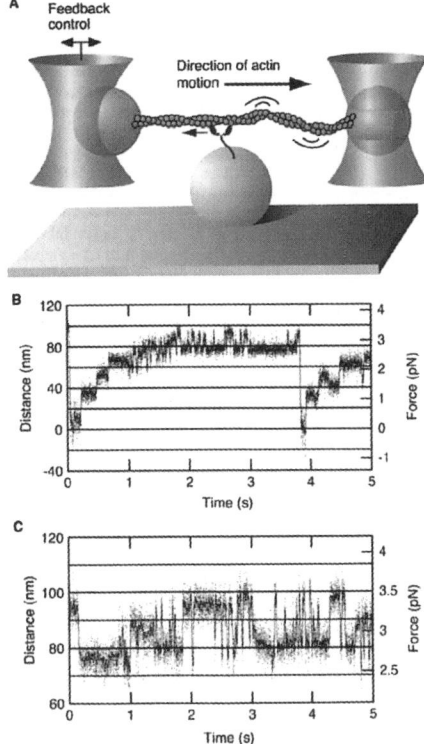

Fig. 2. Processive stepping by myosin VI observed by the dual-beam optical trap. (A) Dual-bead optical trap scheme. An actin filament is held between two 1-μm polystyrene beads. Bead positions are recorded with single nm resolution over 10 kHz. For fixed trap measurements (B and C) the optical load increases with displacement. (B) Sample trace of stepping behavior at 2 mM ATP with a fixed trap. Dwell periods at low load (<1.75 pN) fit single exponential statistics, with a mean stepping rate of 9.1 ± 0.6 s^{-1} (n = 234). This rate is similar to the 8- to 9-s^{-1} steady-state ATPase rate for single-headed constructs (9). (C) Oscillatory behavior observed at 2 mM ATP and high load (>2.5 pN).

feedback range. As a result, the second trap did not contribute to the optical load experienced by myosin VI, and constant load could be maintained through feedback control of a single trap. Steps in the optical trapping records were tabulated manually, and all steps within the feedback range were tabulated.

Results and Discussion

Processive movement of myosin VI was observed by *in vitro* motility assays using total internal reflection fluorescence microscopy to track fluorescently labeled molecules (baculovirus expressed truncated myosin VI with GFP engineered at the C terminus; ref. 16). Although no processive motility was seen on single actin filaments attached to the glass surface, myosin VI did move processively on loosely bundled, unidirectional actin filaments created by crosslinking with fascin (Fig. 1A) (12). The actin bundles were attached to the glass surface. Myosin VI molecules in solution were observed to move as single fluorescent spots along filaments that were frayed out of the bundles at their ends, perhaps tracking around the actin helix as they moved. The velocity of moving myosin VI spots in these assays was 291 ± 77 nm·s^{-1} (n = 36) similar to that observed in sliding filament assays (312 ± 17 nm·s^{-1}, n = 212). Moving fluorescent spots disappeared in at most two events, as expected for the sequential photobleaching of each of the two GFP dyes fused to a single two-headed myosin VI. The average processive run length of myosin VI in these assays is 226 nm (n = 36, r^2 = 0.96, single exponential, Fig. 1B). This run length may be an underestimate because of steric hindrance from the fascin or interaction of filaments with the glass coverslip.

Independent evidence for processivity came from myosin VI motility observed at a variety of surface densities. At high density (1,600 μm^{-2}), myosin VI exhibits smooth and continuous actin filament movement. Motility was observed at low densities (16 μm^{-2}) as well, where actin appeared to thread through and swivel about isolated surface attachments, as observed for other processive cytoskeletal motors (17–20). The actin velocity did not fall as the myosin VI surface density was decreased, consistent with a high duty ratio. Such continuous motility may be caused by the coincident colocalization of several nonprocessive motors on the surface. To exclude this possibility, actin filament landing rates and distances moved were measured over a range of surface densities. There was a first-power dependence of the landing rate on the surface density, as well as a gradual transition

Rock *et al.*

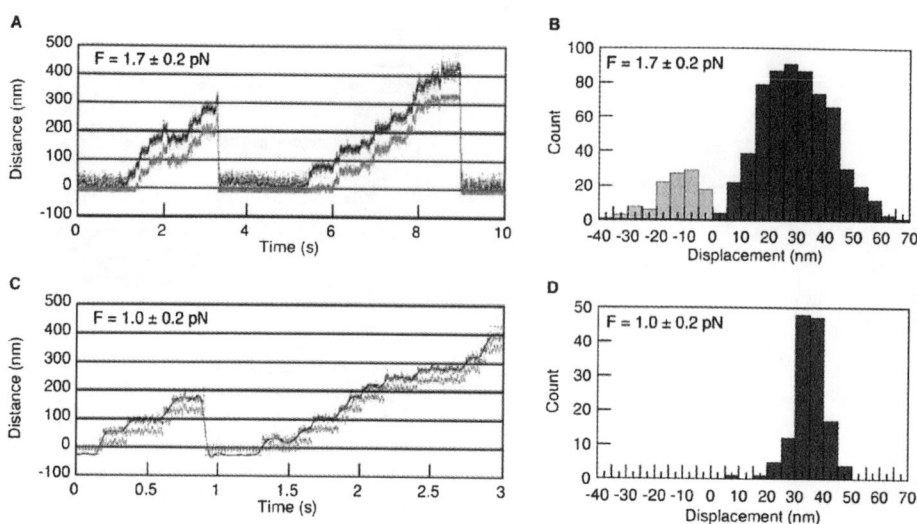

Fig. 3. Force feedback measurements. Constant separation between the bead center and the trap center was maintained by a feedback loop for the left-hand bead (see Fig. 2A). Forward steps indicate positive displacements of the left-hand bead out of its trap, and do not reflect the polarity of the actin filament. (A) Sample force-feedback trace of myosin VI stepping under 1.7 pN of load, and 2 mM ATP. Black trace, bead position; red trace, trap position. (B) Histogram of measured myosin VI step sizes. Forward steps were 30 ± 12 nm (n = 615), and backsteps were −13 ± 8 nm (n = 114). Both the |mean| and variances of the forward and backward steps differed significantly (t probability and F probability < 0.0001). Therefore the backsteps observed here arise from a different process than the oscillations observed at high load in Fig. 2 B and C (where backward and forward transitions are of equal magnitude). (C) Sample force-feedback trace of myosin V stepping under 1 pN of load and 2 mM ATP, using the same dual-bead trapping setup described for myosin VI (see text). (D) Step size histogram of myosin V. The narrow distribution of myosin V step sizes near 36 nm (35 ± 6 nm, n = 131) is in agreement with published results (15). This indicates that the broad distribution of myosin VI steps around 30 nm is not due to the geometry used, and the myosin VI step size is significantly less than 36 nm (t probability < 0.0001).

(from 0 to 1) in the fraction of filaments that moved greater than their length before detaching (Fig. 1 C and D). Both of these results are expected for processive motors, where encounters with functional motor units are processes that are first-order in motor density and differ markedly from the behavior of non-processive motors (21).

Dual bead optical trapping (Fig. 2A) was used to examine the stepping process of myosin VI in detail (22, 23). Of 177 platforms tested, 14 (8%) exhibited processive binding events with multiple steps as in Fig. 2B, whereas only 4 platforms (2%) yielded single-step binding events such as seen for myosin II. Based on the observed incidence of binding (10% of platforms tested), 95% of the binding events detected should have been produced by single molecules (24). The small fraction of platforms that give rise to single-step binding events are likely due to motors that have been damaged by surface adsorption.

In a typical processive event, the actin filament and its attached bead were pulled for 3–5 steps before reaching a stall point, detaching from the surface, and then returning to the baseline position (Fig. 2B). Myosin VI stepped against the increasing load from the optical trap until it reached a stall force of 2.8 ± 0.3 pN (n = 34), comparable to the 3 pN seen for myosin V (18). Unlike myosin V, which slows its stepping rate and occasionally steps backward under high load, myosin VI exhibited remarkable oscillatory behavior at high load with rapid transitions between two bound positions, as previously observed for RNA-folding transitions (25). Either forward steps or detachment events followed all backsteps. Because the rate of forward (and reverse) stepping in such oscillations against high load exceeds the ATPase rate of unloaded myosin VI, these oscillations reflect a reversible or off-pathway process in the chemomechanical cycle of myosin VI.

To enable an accurate measurement of myosin VI step size, a force-feedback system was used (Fig. 2A) (15). This system maintains a constant load on myosin VI by moving one of the optical traps to follow the trapped bead. This system enables myosin VI molecules to take a large number of steps (≈10) and eliminates errors in step size measurements because of compliance of the protein-bead linkages in the system. Large forward steps were interrupted with less frequent, smaller backward steps (Fig. 3A). Application of a moderate load (1.7 ± 0.2 pN, below the level that produced the oscillatory behavior in Fig. 2 B and C) yielded a broadly distributed step-size histogram (Fig. 3B). Two clear lobes were observed, one from forward steps (30 ± 12 nm, modal value 27 nm, n = 615) and a smaller one from backsteps (−13 ± 8 nm, modal value −11 nm, n = 114). In each of these lobes, the spread of the step size distribution greatly exceeds the 1.4-nm measurement precision. Therefore, it appears that myosin VI binds actin in a promiscuous manner, attaching to within ± ≈3 sites of the preferred binding site along the helical actin filament. This spread is significantly greater than the ±6 nm observed for myosin V in both single-bead (15) and dual-bead trap geometries (Fig. 3 C and D). Myosin V has been shown by electron microscopy to predominantly bind within one site of the preferred binding site 36 nm away (26) (Fig. 4 *Lower*, green monomers).

Myosin VI has only a single calmodulin light chain in the lever-arm domain, which, according to a simple lever-arm model, implies a maximum working stroke of ≈5 nm. However, there is a unique 53-residue insert of unknown function between the converter domain and the light chain-binding domain (7, 27). This insert (and perhaps adjacent structures) may adopt an extended conformation to allow myosin VI to take the observed 30-nm steps. Myosin VI likely operates by the combination of a

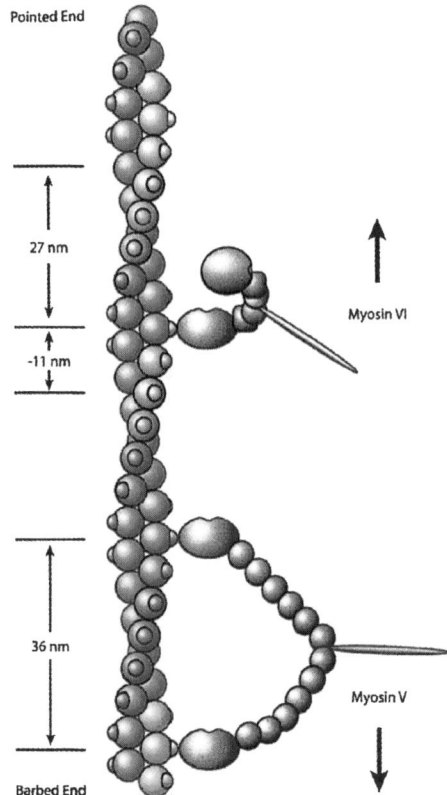

Pointed End

27 nm

-11 nm

Myosin VI

36 nm

Myosin V

Barbed End

Fig. 4. Model of myosin VI stepping. Myosin V and VI molecules are shown in gray, and the actin filament is shown in color. Knobs on the actin filament indicate stereospecific myosin-binding sites (although the actual binding site is located between two actin monomers, binding is shown on only one monomer for simplicity). Both motors are bound to actin subunits facing directly right. Myosin V (*Lower*), having six light chains, spans the 36-nm actin pseudorepeat to within ± 1 binding site (green and blue-green). Myosin VI (*Upper*), having only one light chain and the unique insert (square), cannot. Instead, the bound myosin VI head swings the unbound head to the left side of the actin filament as suggested by the transition from the ADP-bound state to the rigor state in electron microscopy structures (7). The actin subunits shown in red indicate the preferred binding sites of the unbound myosin VI head toward both the pointed end (≈27 nm) and the barbed end (≈11 nm) of actin, corresponding to the modal values seen in the optical trapping experiments. The spread of the color from red to blue reflects the step size histogram. Therefore, nearby actin subunits shown in purple are somewhat less accessible, but steps to these subunits can occur. Subunits shown in blue are relatively inaccessible because their myosin VI-binding sites are on either the right side or the underside of the actin filament. The measured step sizes of myosin VI toward both the pointed end (modal value, 27 nm) and the barbed end (modal value, −11 nm) of actin are a direct result of the accessibility of binding sites on the left side of the actin filament to the unbound myosin VI head.

short powerstroke in the strongly bound head, coupled with a significant conformational change that allows the free head to extend to the next available binding site. It is formally possible that the 30-nm step of myosin VI involves a conformational change of the actin filament, or that the motor takes two rapid, tightly coupled 15-nm steps in succession to create an apparent 30-nm forward step. However, no motor has been shown to move by either of these two mechanisms. If myosin VI is indeed taking single 30-nm steps without altering the actin track, it is not operating by a canonical lever-arm mechanism. Alternate stepping models that do not involve a rotation of the lever arm have been proposed for other myosin classes (28, 29).

To explain the step size histogram in Fig. 3*B*, a mechanism that biases the step 20–30 nm toward the pointed end of actin is required. A model for the stepping of myosin VI is outlined in Fig. 4. The actin filament shown has two right-handed long-pitch helices with a pseudorepeat of 36 nm, and the pointed end is at the top of the figure. In this model (in which a state with one bound head and one free head is shown), the bound head undergoes a powerstroke parallel to the long axis of actin, toward the pointed end, but in addition has a perpendicular component that involves a left-handed rotation around the actin filament. This rotation positions the free head to preferably bind the red monomers shown in Fig. 4. Modal values for steps are indicated (27 nm and −11 nm, respectively). Binding sites within 1 SD of these modal values are located ≈5–15 actin subunits (14–41 nm) toward the pointed end and ≈2–6 actin subunits (5–17 nm) toward the barbed end of the filament, and are shown in red and purple. In this manner, the step size is determined largely by the periodicity of the actin filament. This model is supported by two observations: (*i*) electron microscopy studies of single headed myosin VI have shown a left-handed rotation of the lever arm domain in the transition from the ADP state to the rigor state (7); and (*ii*) the distance between the two lobes of the step size distribution is ≈38 nm, or roughly one pseudohelical repeat of actin.

Myosin VI has evolved features that make it unique among motor proteins. Not only is it processive, it is also backward directed (7), making myosin the first family of motor proteins shown to move processively in both directions. This property of myosins greatly facilitates steady-state cellular trafficking, because separate tracks for trafficking in each direction need not be assembled. The processive nature of myosin VI demonstrates that a long light chain-binding region like that of myosin V is not required for myosin processivity (18, 30). The stepping behavior of myosin VI is also quite revealing. Unlike all other characterized motors, myosin VI rapidly oscillates at its stall load and has different forward and backward step sizes at lower loads. It is possible that the insert between the converter and light chain-binding domains, along with adjacent structures, undergoes a major conformational change to extend the reach of myosin VI, and that the geometry of the actin filament determines the myosin-VI binding sites.

We thank A. Mehta, J. Dawson, A. Sääf, and D. Robinson for their comments on the manuscript and E. Landahl for stimulating discussions. R.S.R. is a Helen Hay Whitney postdoctoral fellow. S.E.R. is supported by the Tumor Biology Training Program, awarded by the National Cancer Institute. J.A.S. and H.L.S. are supported by grants from the National Institutes of Health.

1. Kellerman, K. A. & Miller, K. G. (1992) *J. Cell Biol.* **119**, 823–834.
2. Titus, M. A. (2000) *Curr. Biol.* **10**, R294–R297.
3. Rodriguez, O. C. & Cheney, R. E. (2000) *Trends Cell Biol.* **10**, 307–311.
4. Avraham, K. B., Hasson, T., Steel, K. P., Kingsley, D. M., Russell, L. B., Mooseker, M. S., Copeland, N. G. & Jenkins, N. A. (1995) *Nat. Genet.* **11**, 369–375.
5. Mermall, V., McNally, J. G. & Miller, K. G. (1994) *Nature (London)* **369**, 560–562.
6. Buss, F., Kendrick-Jones, J., Lionne, C., Knight, A. E., Cote, G. P. & Paul Luzio, J. (1998) *J. Cell. Biol.* **143**, 1535–1545.
7. Wells, A. L., Lin, A. W., Chen, L. Q., Safer, D., Cain, S. M., Hasson, T., Carragher, B. O., Milligan, R. A. & Sweeney, H. L. (1999) *Nature (London)* **401**, 505–508.

8. Buss, F., Arden, S. D., Lindsay, M., Luzio, J. P. & Kendrick-Jones, J. (2001) *EMBO J.* **20,** 3676–3684.
9. De La Cruz, E. M., Ostap, E. M. & Sweeney, H. L. (2001) *J. Biol. Chem.* **276,** 32373–32381.
10. Trybus, K. M., Freyzon, Y., Faust, L. Z. & Sweeney, H. L. (1997) *Proc. Natl. Acad. Sci. USA* **94,** 48–52.
11. Sweeney, H. L., Rosenfeld, S. S., Brown, F., Faust, L., Smith, J., Xing, J., Stein, L. A. & Sellers, J. R. (1998) *J. Biol. Chem.* **273,** 6262–6270.
12. Tseng, Y., Fedorov, E., McCaffery, J. M., Almo, S. C. & Wirtz, D. (2001) *J. Mol. Biol.* **310,** 351–366.
13. Tokunaga, M., Kitamura, K., Saito, K., Iwane, A. H. & Yanagida, T. (1997) *Biochem. Biophys. Res. Commun.* **235,** 47–53.
14. Rock, R. S., Rief, M., Mehta, A. D. & Spudich, J. A. (2000) *Methods* **22,** 373–381.
15. Rief, M., Rock, R. S., Mehta, A. D., Mooseker, M. S., Cheney, R. E. & Spudich, J. A. (2000) *Proc. Natl. Acad. Sci. USA* **97,** 9482–9486.
16. Sakamoto, T., Amitani, I., Yokota, E. & Ando, T. (2000) *Biochem. Biophys. Res. Commun.* **272,** 586–590.
17. Howard, J., Hudspeth, A. J. & Vale, R. D. (1989) *Nature (London)* **342,** 154–158.
18. Mehta, A. D., Rock, R. S., Rief, M., Spudich, J. A., Mooseker, M. S. & Cheney, R. E. (1999) *Nature (London)* **400,** 590–593.
19. Sakakibara, H., Kojima, H., Sakai, Y., Katayama, E. & Oiwa, K. (1999) *Nature (London)* **400,** 586–590.
20. Mehta, A. (2001) *J. Cell Sci.* **114,** 1981–1998.
21. Hancock, W. O. & Howard, J. (1998) *J. Cell Biol.* **140,** 1395–1405.
22. Finer, J. T., Simmons, R. M. & Spudich, J. A. (1994) *Nature (London)* **368,** 113–119.
23. Mehta, A. D., Finer, J. T. & Spudich, J. A. (1998) *Methods Enzymol.* **298,** 436–459.
24. Block, S. M., Goldstein, L. S. & Schnapp, B. J. (1990) *Nature (London)* **348,** 348–352.
25. Liphardt, J., Onoa, B., Smith, S. B., Tinoco, I. J. & Bustamante, C. (2001) *Science* **292,** 733–737.
26. Walker, M. L., Burgess, S. A., Sellers, J. R., Wang, F., Hammer, J. A., 3rd, Trinick, J. & Knight, P. J. (2000) *Nature (London)* **405,** 804–807.
27. Homma, K., Yoshimura, M., Saito, J., Ikebe, R. & Ikebe, M. (2001) *Nature (London)* **412,** 831–834.
28. Kitamura, K., Tokunaga, M., Iwane, A. H. & Yanagida, T. (1999) *Nature (London)* **397,** 129–134.
29. Yanagida, T. & Iwane, A. H. (2000) *Proc. Natl. Acad. Sci. USA* **97,** 9357–9359.
30. Howard, J. (1997) *Nature (London)* **389,** 561–567.

BIOPHYSICS

Pulling a single chromatin fiber reveals the forces that maintain its higher-order structure

Yujia Cui* and Carlos Bustamante*[†‡§]

Departments of *Molecular and Cell Biology, †Physics, University of California, Berkeley, CA 94720; and ‡Physical Biosciences Division, Lawrence Berkeley Laboratory, Berkeley, CA 94720

Communicated by Ignacio Tinoco, Jr., University of California, Berkeley, CA, November 12, 1999 (received for review August 6, 1999)

Single chicken erythrocyte chromatin fibers were stretched and released at room temperature with force-measuring laser tweezers. In low ionic strength, the stretch-release curves reveal a process of continuous deformation with little or no internucleosomal attraction. A persistence length of 30 nm and a stretch modulus of ≈5 pN is determined for the fibers. At forces of 20 pN and higher, the fibers are modified irreversibly, probably through the mechanical removal of the histone cores from native chromatin. In 40–150 mM NaCl, a distinctive condensation-decondensation transition appears between 5 and 6 pN, corresponding to an internucleosomal attraction energy of ≈2.0 kcal/mol per nucleosome. Thus, in physiological ionic strength the fibers possess a dynamic process in which the fiber locally interconverting between "open" and "closed" states because of thermal fluctuations.

The DNA of all eukaryotic cells is organized in the form of chromatin and its structure has been the subject of intense research during the last 25 years. These studies have shown that the basic structural unit of chromatin is the nucleosome comprising the core particle and linker DNA. The core particle contains two of each of four core histones H2A, H2B, H3, and H4, and 146 bp of DNA wrapped around this core. The chromatosome (1, 2) includes the core particle and an additional 20 bp of linker DNA associated to a linker histone (H1 or H5). Although there is still some controversy about the position of the linker histone (3, 4) and the location of H3 and H4 histone tails in the chromatosome, many details have been revealed by the crystal structure of the nucleosome core particle (5–7).

Much less is known about the next level of chromatin structure, i.e., the spatial organization of chromatosomes interspersed by linker DNA (8, 9). Many models involving the regular, three-dimensional organization of nucleosomes into chromatin fibers have been proposed (10). Recently, new methods of direct visualization such as scanning force microscopy (11–13) and cryo-electron microscopy (14–16), as well as reinterpretation of older data, indicate that, at low ionic strength at least, nucleosomes in the so-called 30-nm fiber are organized in an irregular three-dimensional zigzag.

Chromatin undergoes a process of condensation and decondensation during the cell cycle *in vivo*. Higher-order structures occur in transcriptionally inactive regions, whereas regions of decondensed nucleosomal arrays often are associated with active chromatin (10). Because in most cell types only a small percentage of the total chromatin content is active at any given time, dynamic changes in the folding state of local chromatin domains must occur to modify the accessibility of the transcription machinery to these domains. These structural transitions may involve H1 removal, histone modifications (acetylation, phosphorylation, or methylation), and changes in the nonhistone protein complement (10, 17). Despite the role played by changes in fiber compaction in the regulation of gene expression, little is known about the magnitude and the origin of the forces that maintain and stabilize these variously compacted forms of the fiber or how these forces are modified during the cell cycle. A simple way to address these questions is to grab a single chromatin fiber by its ends and pull it to determine its mechan-

ical behavior. The response of the fiber to a range of forces thus can provide insights into the nature, range, and magnitude of the interactions holding together its three-dimensional structure. *In vitro*, chromatin fibers also can become compacted or extended in high and low ionic strength, respectively. By carrying out experiments in different ionic strengths, it may be possible to determine how these conditions selectively modulate the strength of some interactions over others to bring about the condensation-decondensation transition of the fiber.

Single molecule manipulation methods recently have been used to investigate the mechanical responses of DNA and proteins. In particular, the entropic elasticity of double-stranded DNA (dsDNA) has been studied with magnetic beads (18), laser tweezers (19), and hydrodynamic drag (20). The response of DNA to torsional force also has been characterized (21). Single molecules of the muscle protein titin (22–24) and tenascin (25) have been mechanically unfolded. In the present study, single, intact fibers of chicken erythrocyte chromatin are extended between two polystyrene beads by using a force-measuring, dual-beam, laser tweezers apparatus (Fig. 1) under various ionic strength conditions.

Materials and Methods

Nuclei. Fresh adult chicken blood mixed with equal volume of 25 mM EDTA, 75 mM NaCl (EDTA saline) was purchased from Lampire Biological Laboratory, Pipersville, PA. All operations were carried out at 4°C unless stated otherwise. Red blood cells were washed three times in EDTA saline at 4,000 × g and resuspended in 10 mM MgCl₂, 250 mM sucrose, 0.2 mM PMSF (nuclei washing buffer) back to the original volume of blood. Then, 1% Triton X-100 was added slowly, and the solution was stirred for 1 hr before the nuclei were spun at 5,000 × g for 1 hr. Nuclei then were washed repeatedly in nuclei washing buffer until clean at 4,000 × g for 20 min each time and stored at −80°C.

Chromatin. Nuclei were thawed at 37°C and washed twice in 10 mM Tris, pH 8.0/1 mM CaCl₂/0.2 mM PMSF/250 mM sucrose (isolation buffer) at 800 × g for 10 min each time and resuspended in the same buffer. At A_{260} = 50, nuclei were digested with 0.014 unit/ml micrococcal nuclease (Sigma) for 6 min at 34°C, and the digestion was stopped by adding 2 mM EGTA, pH 8.0. Nuclei were spun at 3,000 × g for 10 min, and the pellet was resuspended and dialyzed against 10 mM Tris, pH 8.0/0.5 mM EGTA/0.2 mM PMSF (dialysis buffer) overnight at 4°C. Nuclear debris was removed by spinning at 8,000 × g for 15 min, and the chromatin solution was stored on ice.

All fibers used in this study were previously fractionated by sucrose gradients. About 2 ml of chromatin solution was carefully loaded on top of a linear 10–40% sucrose gradient in dialysis buffer (≈34 ml), and the gradient was centrifuged in a

Abbreviation: dsDNA, double-stranded DNA.

§To whom reprint requests should be addressed. E-mail: carlos@alice.berkeley.edu.

The publication costs of this article were defrayed in part by page charge payment. This article must therefore be hereby marked "advertisement" in accordance with 18 U.S.C. §1734 solely to indicate this fact.

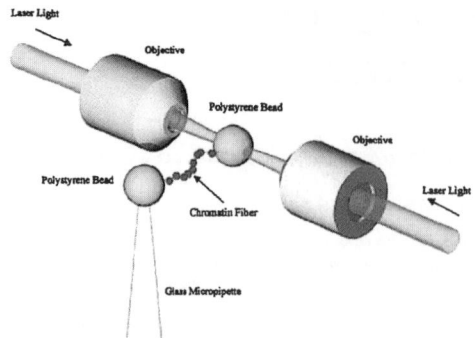

Fig. 1. Schematic drawing of a chromatin fiber pulled between two beads by laser tweezers (19) formed by two laser beams counter-propagating through the objectives with a common focus, not to scale. The stiffness of the trap is ≈28 pN/μm. The Brownian noise in the force ~0.34 pN, which is also close to the thermal drift.

Beckman LM-8 μltracentrifuge by using a SW28 rotor at 27,000 rpm for 6 hr. Fractions of about 1.8 ml each were collected. The DNA lengths of different fractions were determined by pulse-field gel electrophoresis. Samples were incubated in 1% SDS at 37°C for 30 min before being loaded on 1% Fastlane agarose (FMC) gel. Fractions that contained the longest fibers were selected for biotinylation.

Chromatin was washed once in Centricon Plus 100 (Amicon) with 10 mM Tris, 0.1 mM EDTA, pH 7.5, then in 25 mM Na phosphate, 0.5 mM MgCl$_2$, pH 7.5. Three units of T4 DNA polymerase (NEB, Beverly, MA) was added to about 30 μg of DNA at 37°C for 8 min to create 3′ single-stranded overhangs. Six micromolar each of biotin-14-dATP, biotin-14-dCTP, dGTP, and dTTP (GIBCO/BRL) were added, and the reaction was incubated for another 15 min before being stopped by adding 10 mM EDTA, pH 8.0. The chromatin then was washed repeatedly in 10 mM Tris, 2 mM EDTA, pH 7.5 through a Centricon Plus 100 filter to remove free biotinylated nucleotides and stored at 4°C. The ratio of linker histones to core histones of the modified fibers is not changed by the labeling process judged by SDS/PAGE.

Fiber Assembly. A single biotinylated chromatin fiber was connected by its ends between two avidin-coated polystyrene beads (2.54 μm in diameter, Bangs Laboratory, Carmel, IN) inside a flow chamber. One bead was held in a force-measuring dual beam laser tweezers (19); the other was held on top of a movable glass micropipette (Fig. 1). A diluted chromatin solution was passed through the two beads to prevent multiple fiber connections. The tethering of a single fiber also can be confirmed by stripping the fiber of histones by using SDS, and recording the elastic behavior of the DNA, at the end of the experiments. After the connection was made, a saturated solution of calf thymus nucleosomes (Worthington) was passed through the chamber to prevent nonspecific interactions. Throughout this process, care was exercised to prevent permanent mechanical damage to the fiber.

Force-Extension Curves. During one stretch/release cycle, the fiber was first stretched by moving the glass micropipette with steps of ≈0.05 μm and increasing the distance between the beads until a predetermined force was reached. The end-to-end distance of the fiber (extension) was determined from the distance between the centers of the two beads, by using video microscopy. The force was measured directly from the change in momentum of the light in the trap as described (19). The extension and the corresponding force were averaged typically over a period of 0.25 s. The fiber then was relaxed by gradually decreasing the distance. The normal rate of data acquisition was 2 points/s. The experiments were done at room temperature, in buffers ranging from 5 to 150 mM NaCl with 2 mg/ml BSA to prevent nonspecific interactions.

Results

The stretch/release curves of chromatin fibers display more variability than those observed for DNA (18, 19). This finding is expected given the greater heterogeneity and complexity of the chromatin samples. Some variability may have been introduced by the interaction of the fibers with the bead or with the glass micropipette during the assembly of the fibers (such as nonspecific binding and mechanical damage to the fiber). A few very long fibers with force-extension characteristics similar to that of naked dsDNA (19) sometimes were encountered. These fibers, presumably to have lost either part or most of their linker histones and/or histone octamers, were segregated in a different class. Only curves corresponding to intact fibers are presented here. In general, these fibers displayed more consistent, less variable force-extension curves. The examples illustrated are typical of this class of fibers.

Fiber Elasticity in 5 mM NaCl. At low ionic strengths (5 mM NaCl), three distinct force regimes can be recognized in the elastic response of native chromatin fibers. Between 0 and 7 pN, the stretch-relaxation cycles yield reversible force-extension curves (Fig. 2A), i.e., the curve obtained during relaxation coincides with that obtained during stretching, indicating that the process

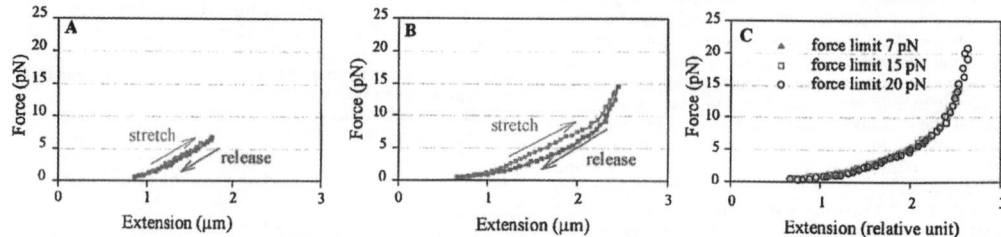

Fig. 2. Force-extension curves of chicken erythrocyte chromatin fibers at low and intermediate force regimes in 5 mM NaCl, 10 mM Tris (pH 7.5), 2 mM EDTA, 2 mg/ml BSA. (A) In the low force regime, below 6 pN, the stretch release curves are repeatable and reversible with a positive curvature. (B) Between 6 and 20 pN the stretch and release curves no longer coincide and hysteresis is evident in the cycle. (C) Shown are the relaxation curves, adjusted to the same length by a constant factor, corresponding to a fiber stretched in three cycles with different maximum forces.

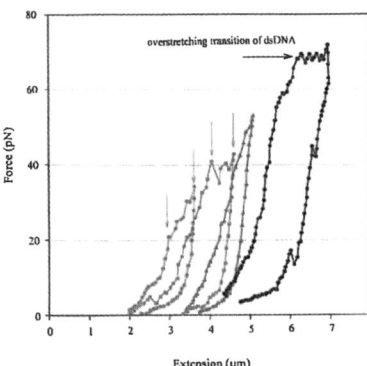

Fig. 3. Force-induced irreversible changes in chromatin fibers successively extended to increasing forces well above 20 pN in 5 mM NaCl, 10 mM Tris (pH 7.5), 2 mM EDTA, 2 mg/ml BSA. Hysteresis between the stretch and release part of the curve is evident in all curves. During the first stretch-release cycle (green) the fiber undergoes a transition that appears between 20 and 35 pN (between the green vertical arrows). In the successive cycles this transition occurs at higher forces (red curve and red vertical arrows) and eventually become less noticeable (blue curve). When the fiber is stretched up to ≈65 pN, a plateau corresponding to the overstretching of dsDNA can be seen (black curve and horizontal arrow).

is reversible and the fiber is at equilibrium throughout the cycle. In this force regime, the curves are also repeatable, namely, the same fiber can be cycled over and over, yielding each time the same force-extension curves. More significantly, the force-extension curves are monotonic and display a slightly positive curvature.

In the intermediate force regime (7–20 pN), the relaxation curve no longer coincides with the stretching curve (hysteresis) and rejoins it only below 2 pN (Fig. 2B). But the stretch and relaxation halves of the cycle are still repeatable during successive cycles. Another feature is observed between cycles that reach different maximum stretching force. The stretch curves of a fiber extended by different amounts in successive cycles will approximately coincide in the shared force interval, whereas the release curves differ each time only by a constant length factor (Fig. 2C), indicating that fibers subjected to successively higher forces become proportionately longer without changing their elastic properties. If a cycle is repeated with the same force limit the stretch and release curves also repeat.

The hysteresis observed above 7 pN indicates that the rate at which the fiber is stretched or released exceeds the rate of extension and contraction of the fiber at equilibrium at that extension. Chromatin fibers pulled 4–6 times slower than those showed in Fig. 2 (data not shown) did not display a reduced hysteresis.

Above 20 pN, the force-extension curves are neither reversible nor repeatable, i.e., hysteresis appears between each stretch and release curve, and for each successive stretch it takes less force to obtain an equivalent extension (Fig. 3). Once the fiber has been pulled above 20 pN, successive stretch/release curves of this fiber in the low force regime no longer reproduce the curves of fibers never exposed to high force. Between 20 and 35 pN, a transition appears in the stretching curve the first time the fiber is subjected to these forces (Fig. 3, green curve, between the vertical arrows). In successive stretching curves the transition becomes less evident, occurring at higher forces (Fig. 3, red curve, between the vertical arrows) until it finally disappears altogether (Fig. 3, blue curve). The relaxation curves are re-

peatable, as long as they start from the same fiber extension. If the maximum force applied to the fiber is the same between successive cycles, the fiber gets longer with each cycle. The change experienced by the fiber appears to be permanent, as the fibers never regain their initial properties regardless of the waiting time elapsed between cycles. These observations suggest that extension beyond 20 pN is accompanied each time by an irreversible change in the fiber, possibly involving the loss of linker histones and histone octamers to the histone-free surrounding buffer. When the tension reaches ≈65 pN, the over-stretching transition characteristic of dsDNA is observed (19) (Fig. 3 horizontal arrow). These observations taken together are consistent with the interpretation that forces beyond 20 pN lead to the mechanical detachment of the histone octamers from the DNA, a process that may continue at higher forces. Not all the histone octamers are eliminated when the fiber is stretched to 65 pN, as revealed by analysis of the low force region of irreversibly modified fibers. The persistence length of these fibers treated as inextensible worm-like chains (estimated from many low force data points) is significantly smaller than that of dsDNA (53 nm; ref. 26). This reduced persistence length is most likely caused by the residual histones still present on the DNA that bend and thus decrease its apparent persistence length (27, 28). The reduced persistence length also may have resulted from the highly positively charged tails of the residual histones, which decrease the net negative charge of the phosphate backbone. Only after the fiber is washed by SDS does its elastic response become indistinguishable from that of naked dsDNA (data not shown).

Fiber Elasticity in 40 and 150 mM NaCl. To determine the effect of ionic strength on the elastic response of chromatin, individual fibers were stretched in 40 and 150 mM NaCl buffer. Force-extension curves of chromatin fibers obtained in these conditions show features not present at the low ionic strengths. At high ionic strength, it takes higher forces to arrive at the same extension during the stretching half of the cycle than at low ionic strength, consistent with the fiber being more condensed in high salt (2, 8, 29). At very low force (0–4 pN), the stretch and release curves of fibers pulled in 40 mM NaCl are both repeatable and reversible, as observed in low salt, but their curvature is negative (Fig. 4A). Between 4 and 6 pN, the fiber starts to get longer with little increase in tension and give rise to a plateau in the force-extension curve (Fig. 4, horizontal arrow). Once the fiber has been stretched beyond the plateau at 6 pN, the relaxation curves are repeatable but no longer coincide with the stretching curves, as they regain the positive curvature characteristic of curves obtained in 5 mM NaCl in the same force regime.

If the fiber is stretched between 10 and 20 pN, the stretching curves behave again as curves obtained in 5 mM NaCl but with higher slope and the stretch/release cycle displays larger hysteresis (Fig. 4C). The stretch and release curves are not reversible but repeatable, and the stretching curve still shows the plateau each time between 4 and 6 pN, an indication that, even after being stretched up to 20 pN, the fiber still can attain its condensed state. As in 5 mM NaCl, no histone dissociation occurred in this force regime, because the curves within 20 pN are repeatable in the histone-free buffer. As in the previous regime, chromatin fibers are more compact than in low salt, judged by the consistently higher forces that are needed to stretch the fibers to the same extension. The relaxation curves display similar characteristics as those obtained in this force regime at low ionic strength. In particular, fibers extended to higher forces yield curves that are related to those of fibers extended to lower forces by a constant length factor. However, a region of negative curvature also appears below 5 pN (Fig. 4D). Above 20 pN, the stretch/release curves are indistinguishable from those in low-salt buffer. In this high force regime the fiber again is modified irreversibly by the force. Curves corresponding

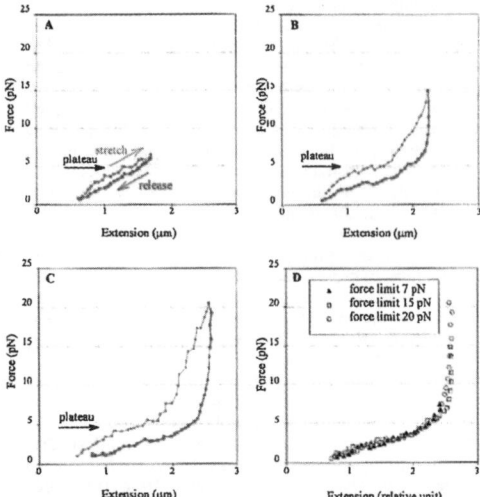

Fig. 4. Force-extension curves for chicken erythrocyte chromatin fibers pulled below 20 pN in 40 mM NaCl, 10 mM Tris (pH 7.5), 2 mM EDTA, 2 mg/ml BSA. (*A*) In low force regime, the stretch and release halves of the cycle nearly coincide. (*B* and *C*) Intermediate force regime. The stretch and release curves no longer coincide and the process displays hysteresis. The plateau corresponding to the condensation-decondensation transition is indicated by the horizontal arrows. (*D*) Three relaxation curves, adjusted to the same length by using a constant factor, corresponding to a fiber stretched successively to different maximum forces. The negative curvature and the plateau although less obvious, are still present.

to fibers stretched in 150 mM NaCl show similar characteristics to those observed in 40 mM (data not shown).

Discussion

Hysteresis. The mechanical force applied to the fiber should act mainly on the entry and exit angle of the DNA around the core particle, fixed by the interaction of the linker histones H1 and H5 with the DNA. The hysteresis observed in the intermediate force regime (7–20 pN) thus may reflect the mechanically induced modification of these interactions. Such modification could involve the detachment of the linker histone from binding sites on the DNA and the core particle, and even its partial denaturation, but not its release into the solution. During the stretch half of the cycle, the force increases until it is high enough to break these interactions. Therefore, to extend the fiber at the experimental pulling rate, the force must increase above its equilibrium value. During the release part of the cycle, the tension in the fiber may prevent the immediate reformation of the linker histone-DNA contacts. The force then must be lowered below its equilibrium value for the contacts to be re-established, giving rise to the observed hysteresis. This scenario could account for the apparent "lengthening" of the fiber seen in the release curves: removal of the angle constraint imposed on the DNA at its entry and exit point around the core particle would lead to additional extension of the fiber. However, this model will not easily explain the invariance of the elastic property of the fiber as they get longer (Figs. 2*C* and 4*D*).

Such invariance suggests a different type of hysteresis model. Here it is required that distant, nonadjacent nucleosomes along the fiber interact with each other at low force. During the stretch half of the cycle, the force increases until it breaks those interactions. During the release part of the cycle, the tension in

the fiber aligns the nucleosomes and prevents them from making the nonadjacent contacts. The force then must be lowered below its equilibrium value for the contacts and the loops to be reformed. In this model, the apparent "lengthening" of the fiber occurs as the sections of chromatin delimited by the broken interactions become incorporated in the extendable length of the fiber. Different relaxation curves thus are obtained when the fiber is extended to different maximum forces, but these curves can be related to one another each time by a constant length factor. Consistent with this interpretation is the observation that hysteresis is reduced if the fiber is repeatedly extended and released in the range between 5 and 20 pN, presumably because in these conditions the force is never low enough to permit the full reformation of re-entrant structures. That the extension and release curves are still repeatable in this regime indicates that the changes involved in the extension of the fiber can be reversed at low tensions. It is not clear what the nature of these interactions would be. They do not appear to be electrostatic, because they persist over 2 orders of magnitude variation in ionic strength. One possibility is nucleosomal interactions via H3 and H4 tails, another is that the nucleosomes in the zigzag become re-entrant (tangled) so, as the fiber is stretched, the nucleosomes may become hooked on each other forming local loops (30). Here, the tension effectively tightens the tangles at the re-entrant points. This process continues until the tension deforms the fiber, allowing one nucleosome to slip pass the other to unhook the tangle. At present, it is not possible to discriminate between any of these models.

Force Extension in Low Salt. The positive curvature of the force-extension curves at forces between 0 and 20 pN suggests that, in this regime of ionic strength and forces, the extension of the fiber corresponds to a continuous deformation process. The elasticity presumably is dominated by the alignment and local straightening of the fiber, possibly through the deformation of the intrinsically straight linker DNA whose entry-exit angle appears to be maintained by steric (14, 31, 32) (globular domain of the linker histones H1 and H5) and electrostatic (linker repulsion) interactions (Fig. 5). Thus, a continuous, extensible worm-like chain model (33) can be used to attempt a first-order description of the release half of the force-extension cycle of chromatin fibers in low ionic strength buffer and between 0 and 20 pN. The release part of the cycle is used for the model, because the stretch part is more likely to be dominated by the nonequilibrium processes responsible for the observed hysteresis (see previous section). These processes, which involve the disruption of the nonlocal, nonadjacent contacts formed in the fiber at low or zero force, lead to the overestimation of the forces in the stretching part of the cycle. In contrast, during release, particularly between 20 pN and 5 pN (see above), no re-entrant interactions are expected to reform and the elastic behavior of the fiber mainly is controlled by the nature of its local interactions, such as the entry-exit angle of the DNA around the nucleosome maintained by the linker histones. In this regime, the release curves can be reasonably well described by an equilibrium model.

The effective energy of an extensible worm-like chain can be expressed as (33):

$$\frac{E}{kT} = \int_0^{L_0} d\xi \left[\frac{A}{2} \left(v \frac{dt}{d\xi} \right)^2 + \frac{\gamma}{2} u^2 + V(u) - \frac{F}{kT} v t \cdot z \right], \quad [1]$$

where L_0 is the contour length of the chain in the absence of thermal fluctuations, ξ is the chemical distance along the chain. The axial strain u is defined as $u = v - 1$, where $dr/dx = vt$, and r is the position of point ξ along the chain, and t is the tangent to the chain at that point. The first term in this equation describes the bending energy of an inextensible worm-like chain, where A

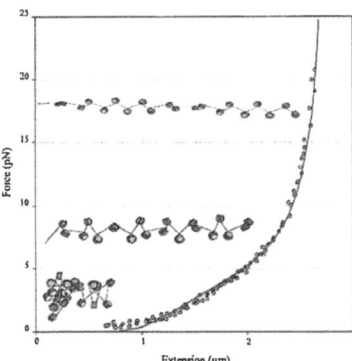

Fig. 5. Fit of the release part of the force-extension curve of a chromatin fiber at low ionic strength by using an extensible worm-like chain model. Fitting of the data gives a persistence length of 30 nm, a stretch modulus of 5 pN, and a maximum length of 3.05 times the original fiber length, estimated to be 1 μm based on the x-intercept of the linear part of the curve. The fiber attains its maximum length when the linkers become completely aligned in the direction of the applied force. For reference, the chromatin models, schematic drawing depicting qualitatively the continuous deformation of the chromatin fiber as it is subjected to increasing tension at low ionic strengths, have been superimposed on the relaxation curve of a fiber pulled under those conditions.

is the persistence length. The second term describes the energy associated with the stretching of the chain, where γ is the stretch modulus. The third term is the potential energy that defines the maximum extension of the chain, and the last term describes the free energy of the chain under a force F acting in z direction. The model assumes that the fiber can rotate freely between its points of attachment and no torsional energy is stored in the fiber as it is being extended. The equilibrium extension is calculated from the conformation of the chain that minimizes the effective energy at any given force. In this model, the mechanical properties of the fiber are determined by its persistence length, its stretch modulus (defined here as the force at which the fiber doubles its contour length at zero force), and its maximum extension. When using this model for a chromatin fiber, the stretch modulus is obtained directly from the measurements, whereas the persistence length and the maximum fiber extension are treated as fitting parameters. The fit of the data to this model (Fig. 5) reveals some of the unique mechanical properties of the chromatin fiber at low ionic strengths, i.e., a bending rigidity comparable to that of dsDNA and an extremely low stretch modulus. The low value of the stretch modulus of chromatin compared with dsDNA ($\approx$1,000 pN; ref. 19) probably reflects the tertiary nature of the interactions disrupted during the mechanical extension of the fiber at low ionic strength, such as the straightening of the linker zigzags during stretching. A more elaborate model of the force-extension characteristics of chromatin at low ionic strength will be presented elsewhere (30).

Mechanical Dissociation of the Histone Core Particle. A force of 20 pN or higher appears to be required to induce the mechanical removal of the core particles from native chromatin in these experiments. This value is $\approx$10 times larger than recently predicted (34) by using an equilibrium model in which the DNA-core binding energy (20 kT) was dissipated over the complete DNA length (146 bp). However, in our experiments the dissociation process between histone octamer and DNA does not occur at equilibrium because (*i*) pulling is too rapid and (*ii*) once

detached, the histone core can never rebind the DNA because it is washed away in the histone-free buffer. Further, the equilibrium model is equivalent to peeling the DNA off the core, but tension in a chromatin fiber actually may wrap the DNA tighter around the particle, rather than peeling it off. Could such tightening deform the protein core and extrude it out the ends of the DNA helix? An estimate of the pressure generated on the core particle by pulling the linker with 20 pN yields a pressure of $\approx 2 \times 10^6$ Pa, which is $\approx$1,000 times less than the Young's modulus of a typical protein, effectively ruling out the core-extrusion hypothesis. In fact, the external force probably induces core dissociation by twisting the core particle so that the force acquires a component perpendicular to the plane in which the DNA wraps around the histones. Then the force can act to peel the DNA wraps off the ends of the two-turn helix.

Assume some sizable twist, say 45°, is necessary on the core particle before the 2 pN peeling force (34) can act. Further assume fiber tension acts through the torque generated on a 4-nm offset at the entry and exit points of the DNA to bend the linker DNA and enable twist. Then a fiber tension of $\approx$6 pN is required to achieve such a twist. If greater twist angle is required for peeling, or if the bending rigidity of linker DNA is increased by its interaction with histone tails, then this force estimate will increase and may account for the observed value (20 pN).

Internucleosomal Interaction. Electron microscopy and scanning force microscopy images of chromatin in high ionic strength ($\geq$ 40 mM NaCl) reveal chromatin fibers condensed into compact rod-like structures with a reduced average distance between nucleosomes (29) and smaller entry-exit angles (2). This structure may be stabilized by the additional neutralization of the linker DNA charge (effectively reducing the entry-exit angle of the linker DNA and bringing the nucleosomes closer to each other), and by an increased internucleosomal attraction that may be mediated by the N-terminal tails of the core histones (31). Correspondingly, the stretch/release curves obtained in high ionic strength show a distinct, mechanically induced transition between 5 and 6 pN in which the fiber becomes longer within a relatively small force range. This transition is identified here with the disruption of nucleosomal-nucleosomal attraction. The negative curvature observed in this force regime and leading to the plateau in Fig. 4, is, in fact, the signature of a continuous transition between two distinct forms of the fiber characterized by different end-to-end extensions. According to this interpretation, the short form corresponds to the condensed structure in which nucleosomes interact with each other through an attractive potential, whereas the long form corresponds to the extended structure with little or no residual internucleosomal interactions (30). Throughout this range, the fiber thus can be thought of as possessing bistability and displaying the coexistence of two phases corresponding to condensed and decondensed structures. At high ionic strength and below 6 pN, the fiber is in its condensed state and the force extension curve simply corresponds to the energy required to orient and straighten the condensed fiber against thermal fluctuations. At some critical tension (5–6 pN), however, it becomes energetically more favorable to decondense the fiber, resulting in the observed transition (34).

The average energy required to pull apart the nucleosomes in high ionic strength can be estimated directly from the stretch/release curves at low forces, assuming that the plateau observed at low force (5–6 pN) corresponds to the interaction of the fiber between condensed and decondensed states. An equilibrium calculation is justified in this case, because at high ionic strength the curves in the regime between 0 and 6 pN (Fig. 4*A*) are very nearly equilibrium curves, with minimal hysteresis, which is not the case for forces above this range (Fig. 4 *B* and *C*). However, the low force/low extension regime is of interest here as in this region the short-ranged internucleosomal interactions are likely to play a role in the elasticity of the fiber. The total length of DNA in the fiber shown in Fig. 4C is about 20 μm, as estimated from fibers treated with SDS at the end of the experiments. Using 210 bp per nucleosome for chicken erythrocyte chromatin, this length corresponds to $\approx$280 nucleosomes. A force of 5 pN acting over a distance of 0.6 μm (the extent of the plateau) during the decondensation (Fig. 4C) yields a binding energy of $\approx$2.6 kT per nucleosome. However, because of the re-entrant nature of the fiber, only $\approx$78% of the fiber participates in the extension in the low force regime comparing the extensions of the fiber in the stretch and release parts of the curve at $\approx$6 pN. Taking this figure into account, a binding energy per nucleosome of $\approx$3.4 kT is obtained.

The force-extension curve of a condensed fiber in high ionic strength also can be fitted to a simplified two-state model (34) (Fig. 6). In this model, the fiber can interconvert between a short (condensed) and a long (decondensed) states. For simplicity, both states are assumed to behave as inextensible worm-like chains. The transition occurs when the work done by the external force equals the difference of the chemical potential between these two states. This model reproduces the magnitude and the main features of the transition region reasonably well, when the internucleosomal attractive energy is set to be 3.8 kT, a figure comparable to that estimated from the experiment directly. The critical force required to convert the short into the long state is given by the ratio of the free energy difference to the difference in end-to-end distance between the two states at that force (34). A critical force of 5 pN gives the change in length upon decondensation of 3 nm, which is two times the contour length of the condensed form in 40 mM NaCl. The distance between the projections of two adjacent nucleosomes on the polymer axis of the decondensed chromatin in 40 mM NaCl therefore is estimated to be 4.5 nm. This value yields, in turn, an entry-exit angle of linker DNA of $\approx$40°, consistent with the value estimated at these ionic strengths from cryo-electron microscopy studies. Notice that, as expected, the model is good only up to $\approx$6 pN.

The value of the internucleosomal attractive energy estimated here from the force-extension curves at physiological ionic strength ($\approx$3.4 kT) indicates that the compact, inactive form of the fiber is a dynamic structure that can locally interconvert

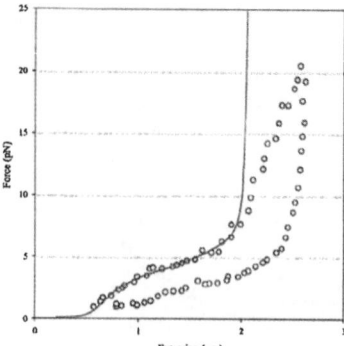

Fig. 6. Fit of chromatin decondensation data by a two-state model. A persistence length of 30 nm is assumed for both before and after the transition (either side of the plateau). A contour length of 0.7 μm was used for the condensed fiber (short form), whereas the contour length for the decondensed fiber (long form) is assumed to be two times longer. The energy to convert the short form to long form is found to be 3.8 kT from the fit. The divergence between the experimental values (O) and the calculated values (solid line) at high forces results from the inextensible nature of the model.

between an "open" and a "closed" state through thermal fluctuations. This interaction energy indicates that, on the average, two adjacent nucleosomes should be found in the open state about 4% of the time. The local dynamics of the fiber structure thus may play an essential role in the regulation of gene expression by providing local access to trans-acting factors such as acetyltransferases, deacetylases, and phosphorylases (17, 35), that can modify the fiber to stabilize its active or inactive conformations.

We thank S. Smith and V. Katritch for many valuable discussions. We also thank P. Yau and C. Castro for developing the method to isolate long chromatin fibers. We are grateful to J. Marko who did the fitting of data to his models. This work was supported by grants from the National Institute of Health (GM-32543) and the National Science Foundation (MBC-9118482 and DBI-9732140).

1. Simpson, R. T. (1978) *Biochemistry* **17**, 5524–5531.
2. Bednar, J., Horowitz, R. A., Dubochet, J. & Woodcock, C. L. (1995) *J. Cell Biol.* **131**, 1365–1376.
3. Pruss, D., Bartholomew, B., Persinger, J., Hayes, J., Arents, G., Moudrianakis, E. N. & Wolffe, A. P. (1996) *Science* **274**, 614–617.
4. Hamiche, A., Schultz, P., Ramakrishnan, V., Oudet, P. & Prunell, A. (1996) *J. Mol. Biol.* **257**, 30–42.
5. Richmond, T. J., Finch, J. T., Rushton, B., Rhodes, D. & Klug, A. (1984) *Nature (London)* **311**, 532–537.
6. Arents, G., Burlingame, R. W., Wang, B. C., Love, W. E. & Moudrianakis, E. N. (1991) *Proc. Natl. Acad. Sci. USA* **88**, 10148–10152.
7. Luger, K., Mader, A. W., Richmond, R. K., Sargent, D. F. & Richmond, T. J. (1997) *Nature (London)* **389**, 251–260.
8. van Holde, K. & J. Zlatanova, J. (1996) *Proc. Natl. Acad. Sci. USA* **93**, 10548–10555.
9. Butler, P. J. & Thomas, J. O. (1998) *J. Mol. Biol.* **281**, 401–407.
10. van Holde, K. E. (1988) in *Chromatin*, ed. Rich, A. (Springer, New York), pp. 111–408.
11. Leuba, S. H., Yang, G., Robert, C., Samori, B., van Holde, K., Zlatanova, J. & Bustamante, C. (1994) *Proc. Natl. Acad. Sci. USA* **91**, 11621–11625.
12. Bustamante, C., Zuccheri, G., Leuba, S. H., Yang, G. & Samori, B. (1997) *Methods Companion Methods Enzymol.* **12**, 73–83.
13. Yang, G., Leuba, S. H., Bustamante, C., Zlatanova, J. & van Holde, K (1994) *Nat. Struct. Biol.* **1**, 761–763.
14. Horowitz, R. A., Agard, D. A., Sedat, J. W. & Woodcock, C. L (1994) *J. Cell Biol.* **125**, 1–10.
15. Woodcock, C. L & Horowitz, R. A. (1998) *Methods Cell Biol.* **53**, 167–186.
16. Furrer, P., Bednar, J., Dubochet, J., Hamiche, A. & Prunell, A. (1995) *J. Struct. Biol.* **114**, 177–183.

17. Wolffe, A. P. & Hayes, J. J. (1999) *Nucleic Acids Res.* **27**, 711–720.
18. Smith, S. B., Finzi, L. & Bustamante, C. (1992) *Science* **258**, 1122–1126.
19. Smith, S. B., Cui, Y. & Bustamante, C. (1996) *Science* **271**, 795–799.
20. Perkins, T. T., Smith D. E. & Chu, S. (1997) *Science* **276**, 2016–2021.
21. Strick, T. R., Allemand, J. F., Bensimon, D. & Croquette, V. (1998) *Biophys. J.* **74**, 2016–2028.
22. Kellermayer, M. S. Z., Smith, S. B., Granzier, H. L. & Bustamante, C. (1997) *Science* **276**, 1112–1116.
23. Tskhovrebova, L., Trinick, J., Sleep, J. A. & Simmons, R. M. (1997) *Nature (London)* **387**, 308–312.
24. Rief, M., Gautel, M., Oesterhelt, F., Fernandez, J. M. & Gaub, H. (1997) *Science* **276**, 1109–1112.
25. Oberhauser, A. F., Marszalek, P. E., Erickson, H. P. & Fernandez, J. M. (1998) *Nature (London)* **393**, 181–185.
26. Bustamante, C., Marko, J. F., Siggia, E. D. & Smith, S. (1994) *Science* **265**, 1599–1600.
27. Trifonov, E. N. (1985) *CRC Crit. Rev. Biochem.* **19**, 89–106.
28. Schellman, J. A. & Harvey, S. C. (1995) *Biophys. Chem.* **55**, 95–114.
29. Zlatanova, J., Leuba, S. H., Yang, G., Bustamante, C. & van Holde, K. E. (1994) *Proc. Natl. Acad. Sci. USA* **91**, 5277–5280.
30. Katritch, V., Bustamante, C. & Olson, W. K. (1999) *J. Mol. Biol.*, in press.
31. Leuba, S. H., Bustamante, C., Zlatanova, J. & van Holde, K. E. (1998) *Biophys. J.* **74**, 2823–2829.
32. Bednar, J., Horowitz, R. A., Grigoryev, S. A., Carruthers, L. M., Hansen, J. C., Koster, A. J. & Woodcock, C. L. (1998) *Proc. Natl. Acad. Sci. USA* **95**, 14173–14178.
33. Marko, J. F. (1998) *Phys. Rev. E.* **57**, 2134–2149.
34. Marko, J. F. & Siggia, E. D. (1997) *Biophys. J.* **73**, 2173–2178.
35. Widom, J. (1998) *Annu. Rev. Biophys. Biomol. Struct.* **27**, 285–327.

The active digestion of uniparental chloroplast DNA in a single zygote of *Chlamydomonas reinhardtii* is revealed by using the optical tweezer

Yoshiki Nishimura*[†], Osami Misumi*, Sachihiro Matsunaga[‡], Tetsuya Higashiyama*, Akiho Yokota[§], and Tsuneyoshi Kuroiwa*

*Department of Biological Sciences, Graduate School of Science, University of Tokyo, Hongo, Tokyo 113-0033, Japan; [‡]Department of Integrated Biosciences, Graduate School of Frontier Sciences, University of Tokyo, Hongo, Tokyo 113-0033, Japan; and [§]Research Institute of Innovative Technology for the Earth (RITE), Kidu, Kyoto 619-02, Japan

Edited by Diter von Wettstein, Washington State University, Pullman, WA, and approved September 8, 1999 (received for review July 19, 1999)

The non-Mendelian inheritance of organelle genes is a phenomenon common to almost all eukaryotes, and in the isogamous alga *Chlamydomonas reinhardtii*, chloroplast (cp) genes are transmitted from the mating type positive (*mt*⁺) parent. In this study, the preferential disappearance of the fluorescent cp nucleoids of the mating type negative (*mt*⁻) parent was observed in living young zygotes. To study the change in cpDNA molecules during the preferential disappearance, the cpDNA of *mt*⁺ or *mt*⁻ origin was labeled separately with bacterial *aadA* gene sequences. Then, a single zygote with or without cp nucleoids was isolated under direct observation by using optical tweezers and investigated by nested PCR analysis of the *aadA* sequences. This demonstrated that cpDNA molecules are digested completely during the preferential disappearance of *mt*⁻ cp nucleoids within 10 min, whereas *mt*⁺ cpDNA and mitochondrial DNA are protected from the digestion. These results indicate that the non-Mendelian transmission pattern of organelle genes is determined immediately after zygote formation.

In nearly all eukaryotes, chloroplast DNA (cpDNA) and mtDNA are inherited from only one parent (1–3). The non-Mendelian inheritance of organelle DNA is a phenomenon common to diverse taxa of plants and animals, but its actual molecular mechanism remains a mystery.

In the unicellular alga *Chlamydomonas reinhardtii*, the chloroplast cpDNA is transmitted only from the mating type positive (*mt*⁺) parent (4), whereas mitochondrial genes are believed to be inherited from the mating type negative (*mt*⁻) parent (5). The *mt*⁺ and *mt*⁻ gametes are the same size (isogamous) and contribute the same copy numbers of cp and mtDNA molecules to the zygote.

There are approximately 80 copies of the 196-kb cpDNA molecule per cell (6), and they are organized into 5–8 cpDNA–protein complexes (cp nucleoids). The nucleoids can be visualized under a fluorescent microscope by using DNA-specific fluorochrome 4′,6-diamidino-2-phenylindole (7). This technique permits direct observation of the number, distribution, and behavior of cp nucleoids in gametes and zygotes of *C. reinhardtii*. In 1982, Kuroiwa *et al.* (8) discovered that the cp nucleoids derived from the *mt*⁻ parent disappear preferentially within 60 min after zygote formation, whereas the cp nucleoids of the *mt*⁺ parent remained unaffected.

The preferential disappearance of fluorescent *mt*⁻ cp nucleoids occurs well before the DNA digestion is detected by biochemical techniques 6 hr after zygote formation (4). Therefore, the changes in cpDNA molecules that cause the preferential disappearance of fluorescent *mt*⁻ cp nucleoids remain controversial. Several possibilities have been proposed. One is that the dispersal of cpDNA molecules, presumably caused by the digestion or relaxation of the proteins that form the cp nucleoids, might lead to the disappearance of fluorescent cp nucleoids (9, 10). Another is that the rapid digestion of cpDNA molecules might lead to the disappearance of cpDNA nucleoids (1, 11, 12).

To understand the molecular mechanism that causes the preferential disappearance of fluorescent *mt*⁻ cp nucleoids, more precise cytological observation and further molecular analysis of the phenomenon are indispensable. In this study, the process of preferential disappearance is visualized in a living zygote, and the changes in cpDNA molecules during preferential digestion are investigated further by using optical tweezers.

The use of optical tweezers is a novel technique for manipulating living cells or organelles in suspension without physically touching or damaging them (13–15). Using this technique, we isolated a single zygote under direct observation, according to the presence or absence of fluorescent *mt*⁻ cp nucleoids. The *mt*⁺ and *mt*⁻ cpDNA molecules were labeled separately by using bacterial *aadA* (aminoglycoside adenyl transferase) sequences (16, 17). Single zygotes obtained in this manner were subjected to highly sensitive nested-PCR analysis. Consequently, it was found that the rapid digestion of *mt*⁻ cpDNA molecules causes the disappearance of fluorescent *mt*⁻ cp nucleoids, whereas *mt*⁺ cpDNA and mtDNA molecules are protected from the digestion.

Materials and Methods

Cell Strain and Culture. The wild-type strain, *mt*⁺ and *mt*⁻, was derived from strain *137c* of *C. reinhardtii*. Cells were grown separately on agar plates [1.2% agar in Snell's medium (18, 19)] at 22°C under 12 h of light followed by 12 h of darkness to synchronize cell division, which occurs in the middle of the dark period. The light intensity was approximately 6,600 lux at the surface of the flat culture container.

The chloroplast genome of *L03c* was transformed with plasmid pUCCEB*aadA* 53(NN), which is derived from the plasmid constructed by Goldschmidt-Clermont (16) and contains an *aadA* cassette driven by the *atpA* promoter and the *aadA* coding region and the *rbcL* 3′ region downstream from the coding region. The bacterial *aadA* gene encodes aminoglycoside adenyl transferase and is expressed in chloroplasts conferring resistance to spectinomycin and streptomycin. Chloroplast transformation was carried out with a particle gun as described in a previous report (17). The *aadA* cassette was inserted into the noncoding region downstream from the *rbcL* gene in the wild-type chloroplast genome by homologous recombination. This mutant strain was maintained on Snell's medium supplemented with 100 μg/ml spectinomycin (Sigma). The mating reaction was induced by incubating the vegetative cells in N-free medium (Tsubo mating buffer: 0.6 mM MgSO₄/1.2 mM Hepes·NaOH, pH 6.8) according to the previously described method (19). For UV

This paper was submitted directly (Track II) to the PNAS office.

Abbreviations: cpDNA, chloroplast DNA; *mt*⁺ and *mt*⁻, mating type positive and negative, respectively.

[†]To whom reprint requests should be addressed. E-mail: yoshiki@biol.s.u-tokyo.ac.jp.

The publication costs of this article were defrayed in part by page charge payment. This article must therefore be hereby marked "*advertisement*" in accordance with 18 U.S.C. §1734 solely to indicate this fact.

irradiation, each gamete culture (mt^+ and mt^-) was exposed to UV (2,270 erg s^{-1}·cm^{-1}; 1 erg = 0.1 μJ) for 5 min immediately before mixing.

SYBR Green I Staining. To stain the DNA in live gametes and zygotes, SYBR Green I nucleic acid stain (Molecular Probes) was added to the sample to give a final dilution of 1:1,000 (20). In all the experiments, the mixture was kept under the same conditions as the original culture and either illuminated or placed in the dark. The cells were observed under blue (B) excitation with a fluorescence microscope, and the squashed, motionless, live cells were photographed.

Optical Tweezers. The optical tweezers used were similar to those described by Ashkin and colleagues (13–15). The main components of the optical tweezers device are a laser and a microscope system. The laser was a diode-pumped neodymium-YAG (yttrium/aluminum/garnet) laser ADLAS DPY 421 (Adlas, Lubeck, Germany) that emits continuous infrared (IR) light at 1,064 nm with a maximum power of 2,000 mW. The IR laser beam is deflected into a Zeiss photomicroscope (Axiovert 135; Zeiss) equipped with a PALM laser interface system (PALM, Brnrid, Germany; ref. 21) and focused with a Neofluar ×100 objective into the optical field. The laser power in the 1- to 2-μm focal spot was varied from 10 to 2,000 mW. In this study, optical manipulations were conducted at a power of 1,000 mW. The optics of the optical tweezers were improved by adding filters to allow observation of the fluorescent images of cells or organelles during the manipulation.

Immobilization of Cells. To manipulate *C. reinhardtii* cells with the optical tweezers, the cells were immobilized by deflagellation or formaldehyde fixation, because optical tweezers cannot trap the swimming gametes or zygotes. The flagella were removed by following the method described by Witman *et al.* (22). For some experiments, the cells were fixed with formaldehyde. Formaldehyde solution in methanol was added to the cell suspension to give a final concentration of 4%. The suspension was incubated for 10 min at room temperature. Then the cells were harvested by centrifugation at 3,000 rpm for 1 min and resuspended in cold Tsubo mating buffer.

Nested-PCR Analysis. The nested-PCR method was used to detect *rbcL* (ribulose 1,5-biphosphate carboxylase oxygenase large subunit: cpDNA), *cox I* (cytochrome *c* oxidase subunit I: mtDNA), and *aadA* (bacterial gene introduced into cpDNA of *L03c* strain) genes in single cells of *C. reinhardtii*. Nested-PCR primer sets specific for the *rbcL*, *cox I*, and *aadA* regions were generated. The primer sequences were as follows: (*i*) *rbcL*_F1: 5'-CATGGAC-TACAGTATGGACTGACGG-3', *rbcL*_R1: 5'-GTACAAGCT-TCAAGAGCTACACGGT-3'; (*ii*) *rbcL*_F2: 5'-TGCTTACGT-TAAAACATTCGTAGGT-3', *rbcL*_R2: 5'-ATACGTGAAT-ACCGCCTGAAGCAAC-3'; (*iii*) *cox I*_F1: 5'-GCCTT-CTTTGGCGGTTTGCTAGGTA-3', *cox I*_R1: 5'-TACCATA-GCACGACCCTCGTGGTAA-3'; (*iv*) *cox I*_F2: 5'-TTGCTA-CCAATCATGATCGGTGCCC-3', *cox I*_R2: 5'-GGCACCCA-TAGCGCAAATCATACCA-3'; (*v*) *aadA*_F1: 5'-CTCTAGCT-AACTTAGTATAC-3', *aadA*_R1: 5'-GAAGTATCGACT-CAACTATC-3'; (*vi*) *aadA*_F2: 5'-CTATCAGAGGTAGTTG-GCGTCATCG-3', *aadA*_R2: 5'-GCACTACATTTCGCTCAT-CGCCAGC-3'.

The expected sizes of the fragment amplified with primer pairs 1–6 were 1.09, 0.73, 1.16, 0.56, 1.40, and 0.52 kb, respectively. PCR was performed in a 50-μl reaction volume that contained 1× *Ex Taq* buffer (Takara Shuzo, Otsu, Japan), 200 μmol dNTPs, 25 pmol of each primer pair, and 1 unit of *Ex Taq* polymerase (Takara). For the first amplification, the outer primer pairs (1, 3, and 5) were used for the *rbcL*, *cox I*, and *aadA*

sequences, respectively. Thirty-five cycles of PCR (*rbcL* and *cox I*: 1 min at 94°C, 1 min at 64°C, and 1 min at 72°C; *aadA*: 1 min at 94°C, 2 min at 55°C, and 3 min at 72°C) were carried out. One microliter of the first PCR product was used as the template for the second PCR. The second PCR was performed with the inner primer pairs (2, 4, and 6) for *rbcL*, *cox I*, and *aadA*, respectively, under the same conditions as the first PCR.

Results

Serial Observation of the Preferential Disappearance of *mt⁻* Chloroplast Nucleoids in a Living Zygote. The conventional DNA-specific fluorochrome 4'-6-diamidino-2-phenylindole is a very sensitive dye that is used to detect DNA molecules in cells, but it cannot penetrate living cells or detect mitochondrial nucleoids. The use of SYBR Green I eliminates these two problems and permits simultaneous observation of the cell nuclei and cp and mitochondrial nucleoids in living *C. reinhardtii* (20).

In this study, SYBR Green I staining was used to visualize the preferential digestion of fluorescent *mt⁻* cp nucleoids in living zygotes. The preferential disappearance commenced about 40 min after zygote formation and was completed within 10 min (Fig. 1). During the preferential disappearance, all the cp nucleoids in *mt⁻*-derived chloroplasts simultaneously became smaller and disappeared completely, without swelling, whereas the *mt⁺* cp nucleoids and mitochondrial nucleoids of both parents remained unchanged. The two chloroplasts remained at a distance from each other during the preferential disappearance of the *mt⁻* cp nucleoids.

Labeling of *mt⁺* and *mt⁻* cpDNA Molecules by Transformation with Bacterial *aadA* Gene Sequences by Using a Particle Gun. To investigate the changes in cpDNA molecules during the preferential disappearance of *mt⁻* cp nucleoids, it was necessary to label the *mt⁺* and *mt⁻* cpDNA molecules in individual zygotes separately. Transformed strain *L03c* was used for this purpose. The cpDNA of *L03c* strain was transformed with the bacterial *aadA* gene, by using a particle gun (17). The behaviors of *mt⁺* and *mt⁻* cpDNA were monitored separately by PCR amplification of the *aadA* sequences by using zygotes resulting from the cross between *L03c* gametes and wild-type *137c* gametes of the opposite mating type.

To determine whether the preferential disappearance of fluorescent *mt⁻* cp nucleoids occurred normally, crosses between *L03c* and wild-type gametes were examined by using SYBR Green I fluorescent microscopy (Fig. 2). The insertion of *aadA* sequences into cpDNA apparently had no effect on cell morphology or the activity of *L03c* gametes (Fig. 2 *a–h*). Moreover, the preferential disappearance of the fluorescent *mt⁻* cp nucleoids occurred 60–90 min after mating, whereas the fluorescent *mt⁺* nucleoids were preserved. The two cell nuclei fused 120 min after zygote formation, just as in the cross between wild-type gametes (Fig. 2 *i– x*). These results suggested that the cross between *L03c* and wild-type gametes reflects the events that occur in the cross between wild-type gametes precisely.

Optical Isolation and Molecular Analysis of a Single Gamete or Zygote. The use of optical tweezers is a novel technique for manipulating living cells or organelles in suspension without physical contact or damage. Fig. 3 summarizes the optical isolation method developed in this research, which can harvest cells or organelles of interest from a heterogeneous suspension within 5–10 min. Harvesting takes place under the direct microscopic visualization of cells or organelles. During the process, the cells or organelles can be observed in detail by phase-contrast or fluorescent microscopy with UV, B, or G excitation. Therefore, it is possible to distinguish between gametes and zygotes before and after the preferential disappearance of *mt⁻* cp nucleoids.

To investigate how many cells are required to detect cpDNA

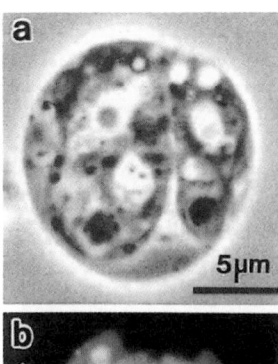

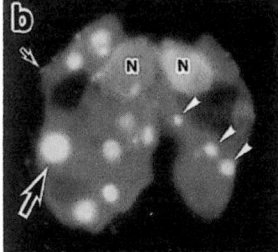

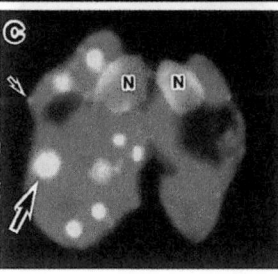

Fig. 1. Preferential disappearance of *mt⁻* cp nucleoids visualized in a living zygote of *C. reinhardtii*. The living zygote was stained with SYBR Green I, and the preferential disappearance was observed under B irradiation. Phase-contrast (a) and fluorescent images of identical zygotes before (b) and after (c) the preferential disappearance are shown. To distinguish the *mt⁺* and *mt⁻* chloroplasts, small *mt⁻* gametes were used. (b and c) The *mt⁺* (*Left*) and *mt⁻* (*Right*) chloroplasts are emitting red autofluorescence. Cell nucleus (N), cp nucleoids (big fluorescent spots; large arrow), and mitochondrial nucleoids (small fluorescent spots; small arrow) all are visualized with SYBR Green I staining. The cp nucleoids in the *mt⁻* chloroplasts (b, white arrowheads) disappeared completely within 10 min (b and c).

and mtDNA molecules, *rbcL* (cpDNA), *cox I* (mtDNA), and *aadA* gene sequences were amplified by PCR by using one, three, and five gametes isolated with the optical tweezers (Fig. 4). The first PCR was insufficient to detect the genes from 1–5 gametes, and a visible band was obtained only by using 10^5 gametes. When the second PCR was performed, all three genes could be detected as visible bands, even when only one gamete was used.

Detecting the Nuclease Activity That Causes Uniparental Inheritance of cpDNA by Using a Single Zygote. All the bands shown in Fig. 5 were detected from a single gamete or zygote harvested with optical tweezers. The *rbcL* (cpDNA), *cox I* (mtDNA), and *aadA* sequences were amplified, and each experiment was repeated five times to ensure its reproducibility. Fig. 5*A* shows the results obtained from wild-type and *L03c* gametes. The *rbcL* and *cox I*

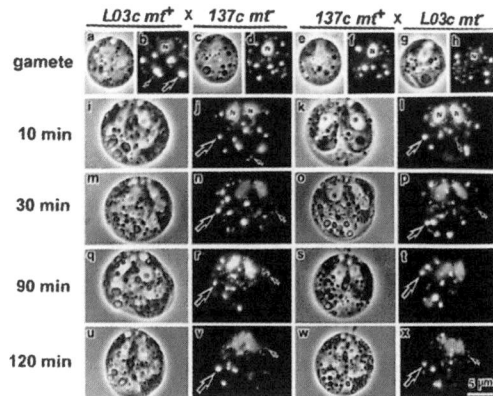

Fig. 2. Phase-contrast (a, c, e, g, i, k, m, o, q, s, u, and w) and SYBR Green I fluorescent (b, d, f, h, j, l, n, p, r, t, v, and x) images of living gametes (a–h) and zygotes (i–x). The cross between *L03c* (harboring bacterial gene *aadA* in cpDNA) *mt⁺* and *137c* (wild type) *mt⁻* and the reciprocal cross were examined. Cell nuclei (N), mitochondrial nucleoids (small arrow), and cp nucleoids (large arrow) are visible in the gametes [*137c mt⁺* (a and b) and *mt⁻* (c and d), *L03c mt⁺* (e and f), and *mt⁻* (g and h)] and zygotes 10 (i–l), 30 (m–p), 90 (q–t), and 120 (u–x) min after mating. The fluorescent *mt⁻* cp nucleoids were preferentially digested within 60–90 min, whereas the mitochondrial nucleoids were preserved biparentally in both crosses.

sequences, which were used as control genes, were amplified from all of the single gametes. The *aadA* sequence was not detected in any of the wild-type gametes, whereas it was clearly detected in *mt⁺* and *mt⁻ L03c* gametes in all cases.

Next, single zygotes were isolated from cell cultures 10, 30, 90, and 120 min after zygote formation and subjected to nested-PCR analysis. From each cell culture, single zygotes were isolated optically and analyzed. At 10 min, the zygote had equal-sized *mt⁺* and *mt⁻* cp nucleoids. At 30 min, the zygote had normal *mt⁺* cp nucleoids and slightly smaller *mt⁻* cp nucleoids. At 90 min, the zygote had only *mt⁺* cp nucleoids. At 120 min, the cell nuclei in the zygote had fused. When *L03c mt⁺* gametes were crossed with wild-type *mt⁻* gametes, *rbcL*, *cox I*, and *aadA* gene sequences were detected in all the zygotes examined. On the contrary, when the *L03c mt⁻* gametes were crossed with wild-type gametes, the *aadA* sequences were amplified only in younger zygotes (10 and 30 min after zygote formation). After the fluorescent *mt⁻* cp nucleoids disappeared, the *aadA* sequences were no longer detected in the zygotes (90 and 120 min after zygote formation).

UV irradiation of *mt⁺* gametes just before zygote formation significantly inhibited the preferential disappearance of *mt⁻* cp nucleoids whereas the irradiation of *mt⁻* gametes had almost no effect, which is consistent with the previous reports (1, 9). When the zygotes whose *mt⁺* parents were pretreated with UV irradiation were analyzed, the *aadA* sequences in *mt⁻* cpDNA were clearly amplified even 120 min after zygote formation (data not shown).

Discussion

The preferential disappearance of fluorescent cp nucleoids in *C. reinhardtii* zygotes was reported first in 1982 (8). However, the molecular mechanism for the disappearance of the *mt⁻* cp nucleoids remains controversial (9, 10). In this study, we demonstrated that *mt⁻* cpDNA molecules are digested rapidly during the preferential disappearance of fluorescent *mt⁻* cp nucleoids. This suggests that highly effective nucleases are activated to digest the *mt⁻* cpDNA molecules just after zygote formation.

Isolation Chamber

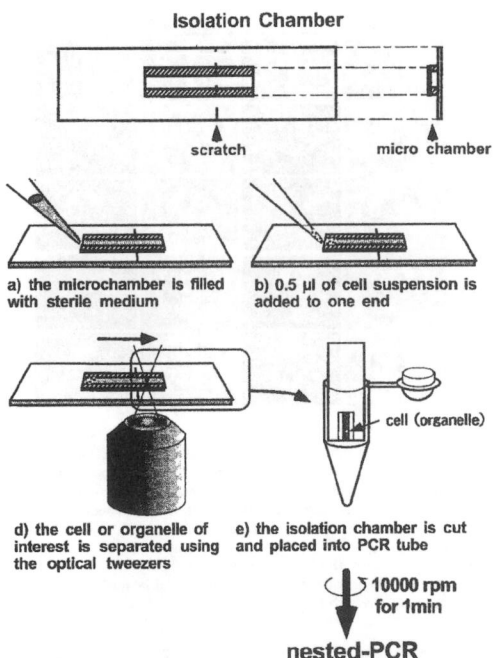

Fig. 3. The optical isolation process and the "isolation microchamber." The microchamber is formed by separating a large (5 × 30 × 0.15 mm, length × width × thickness) and a small (9 × 3 × 0.15mm) coverslip with two thin strips of adhesive tape (9 × 1 × 0.1 mm). The inside dimensions of the chamber were 9 × 1 × 0.1 mm. A small scratch was made on the chamber with a diamond knife. The bottoms of 5-cm diameter Petri dishes were cut out with a knife and replaced with thin, plastic film (≈100-μm thick). Filter paper was cut and placed in the dishes, and roughly 200 μl of distilled water was added dropwise to keep the inside of the dish moist. Then, the chamber was placed inside the dish and attached to the plastic film by using one drop of distilled water. First, the chamber was filled with sterile buffer containing 1.5% sucrose and 0.1% BSA (a). The BSA was added to prevent cells from adhering to the glass chamber. Then, 0.5 μl of cell suspension was carefully applied to one end of the chamber (b). The cells were observed with a microscope, and a single cell was trapped with the optical tweezers and transferred to the opposite end of the chamber (c). The transfer was processed automatically with a microscopic stage control system (MCU26 X, Y, Z-Axes Motor Control; Zeiss) at a velocity of 1–60 μm/sec. When the cell moved past the scratch in the chamber, the chamber was cut immediately, and the piece containing the cell of interest was placed in a PCR tube (d). The PCR tube was centrifuged for 5 sec to drop the cell into the tube. With this technique, it was possible to procure a living cell or intact organelles in a suspension within only 5–10 min.

The transmission pattern of organelle genes in *C. reinhardtii* appears to be determined at an extremely early stage of zygote maturation.

Fig. 1 shows serial observations of the preferential disappearance of *mt⁻* cp nucleoids in a living zygote. During the process, all the *mt⁻* cp nucleoids simultaneously became smaller approximately 40 min after zygote formation (Fig. 1b) and disappeared completely within 10 min (Fig. 1c). These results suggest that at least one enzyme is activated or synthesized only in the *mt⁻* chloroplast, resulting in the disappearance of *mt⁻* cp nucleoids. The two chloroplasts remained separated during the process. This suggests a possible role of the chloroplast membrane in localizing the high activity of the putative enzyme(s) that causes the disappearance of the cp nucleoids.

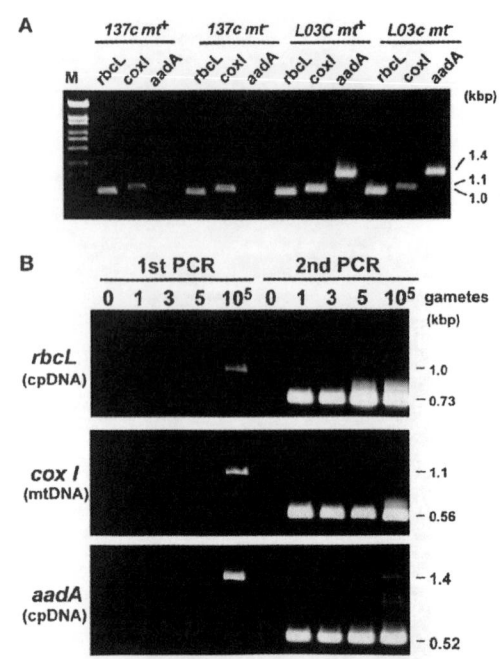

Fig. 4. PCR amplification of *rbcL* (cpDNA), *cox I* (mtDNA), and *aadA* (bacterial sequence introduced into the cpDNA of strain *L03c*) genes from 10^5 gametes of *137c* (wild type) *mt⁺*, *137c mt⁻*, L03c *mt⁻*, and *L03c mt⁻* gametes with the outer primer pairs (A). Nested PCR amplification of *rbcL*, *cox I*, and *aadA* sequences from one, three, or five gametes obtained optically and immobilized by deflagellation (B). The requisite number of gametes was isolated from the cell culture with the optical tweezers and subjected directly to nested-PCR analysis in which the *rbcL*, *cox I*, and *aadA* sequences were amplified by two rounds of PCR by using nested primers.

To learn more about the changes in cpDNA molecules during the preferential disappearance of fluorescent *mt⁻* cp nucleoids, we labeled the cpDNA molecules with bacterial *aadA* gene sequences by particle gun transformation (refs. 16 and 17; Fig. 2). Then we procured gametes and zygotes with or without *mt⁻* cp nucleoids, using the newly developed optical isolation method (Fig. 3). The result, shown in Fig. 4, demonstrates that cpDNA and mtDNA molecules can be detected clearly by using only a single gamete after two rounds of PCR amplification. Because each *Chlamydomonas* gamete contains only one chloroplast, this indicates that the gene sequences of a single chloroplast can be detected.

With this technique, we collected zygotes with and without *mt⁻* cp nucleoids and showed that the *aadA* sequences in *mt⁻* cpDNA molecules were no longer detected in zygotes lacking fluorescent *mt⁻* cp nucleoids (90 and 120 min after zygote formation) (Fig. 5). From this, we concluded that at least one highly effective nuclease is activated in the *mt⁻* chloroplast just after zygote formation, and this determines the non-Mendelian transmission pattern of cpDNA. The *mt⁻* cpDNA molecules are digested completely during the 10 min in which the fluorescent *mt⁻* cp nucleoids disappear. There is a discrepancy in the timing of DNA digestion observed in biochemical [6 h after zygote formation (4)] and cytological [60 min after zygote formation (8)] studies. This might result from the fact that the biochemical studies examined a cell population (including unmated gametes

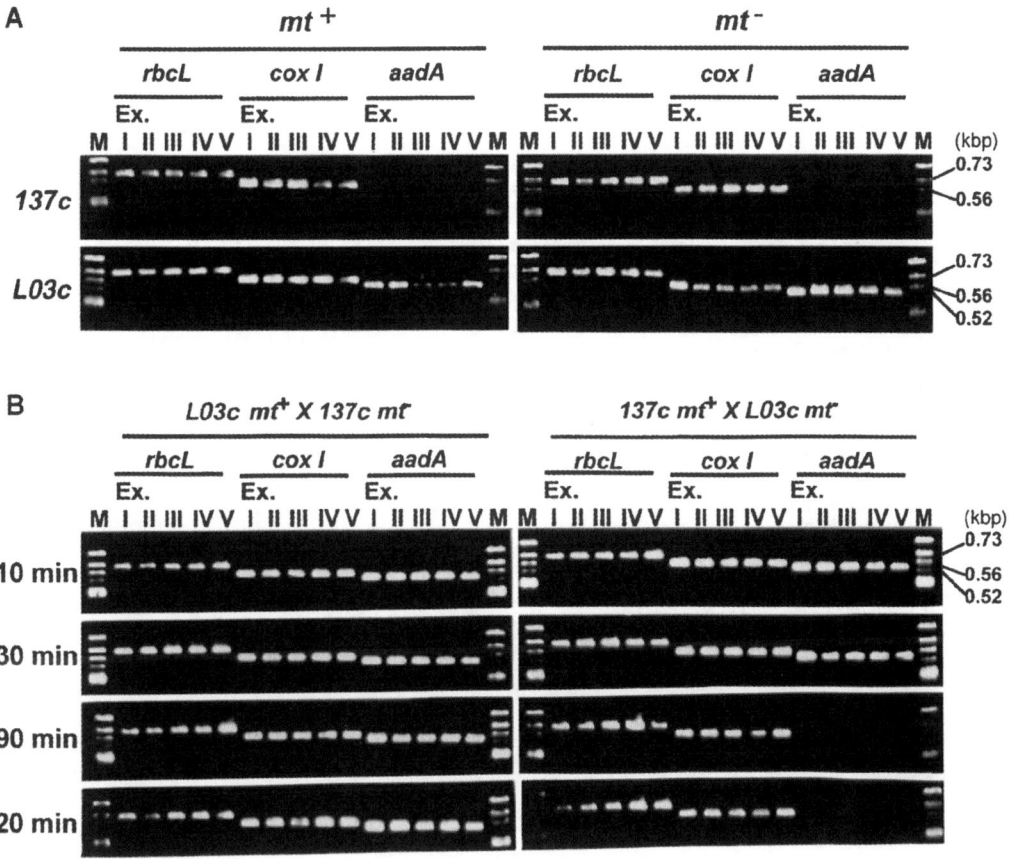

Fig. 5. Nested-PCR amplification of *rbcL* (cpDNA), *cox I* (mtDNA), and *aadA* sequences from one optically isolated gamete or zygote 10, 30, 90, and 120 min after zygote formation. One gamete of *137c mt⁺*, *137c mt⁻*, *L03c mt⁺*, or *L03c mt⁻* (A) or one zygote resulting from the crosses *L03c mt⁺ × 137c mt⁻* or *137c mt⁺ × L03c mt⁻* (B) was isolated by using the optical tweezers and subjected immediately to nested PCR analysis. One typical zygote was isolated from each of the cell cultures 10, 30, 90, and 120 min after mating. To ensure reproducibility, each experiment was repeated five times (*I– V*). The lanes marked M are loaded with marker.

and zygote with and without *mt⁻* cp nucleoids) whereas the cytological study dealt with individual cells.

UV irradiation of *mt⁺* gametes just before zygote formation significantly inhibited the digestion of the cpDNA whereas the irradiation of *mt⁻* gamete had almost no effect. This result suggests that the gene expression or protein synthesis by *mt⁺* parent is important in the digestion of *mt⁻* cpDNA. At the time of writing, reports indicated that at least 200 polypeptides are synthesized *de novo* just after zygote formation. Six of these [94(α), 94(β), 94(γ), 52, 50, and 38 kDa] appear to be essential for the preferential digestion of *mt⁻* cpDNA (23). Ezy-1 polypeptide, which is thought to be the 52-kDa polypeptide, localizes in the chloroplast (24). These polypeptides might be involved in the activation or synthesis of the unknown nuclease(s).

The nuclease activity is likely to be nonspecific, because it digests exogenous DNA sequences such as *aadA*. A previous attempt to purify nucleases in *C. reinhardtii* cells identified nuclease C, a nonspecific nuclease that requires Ca²⁺ for full activation (11, 12). The preferential disappearance of *mt⁻* cp nucleoids is also effectively inhibited by the addition of EGTA to the cell culture (25). We think that further investigation of nuclease C will be an important key to understanding the molecular mechanism for maternal inheritance.

On the other hand, the amplification of *mt⁺* cpDNA molecules and mtDNA was not affected during the preferential digestion of *mt⁻* cpDNA, indicating that these molecules are protected from the digestion by nuclease(s). The molecular mechanism that is responsible for the selective protection of *mt⁺* cpDNA and mtDNA is still an open question for further study. It is possible that methylation of DNA molecules might play an important role in the protection, in a way that is analogous to the bacterial restriction-methylation system (26). Indeed, increased methylation of cpDNA molecules is detected during zygote maturation (26–28). In our experiment, however, treatment of zygotes with the methylation inhibitor 5-azaCyd had no effect on the preferential disappearance of fluorescent *mt⁻* cp nucleoids or on the preferential digestion of cpDNA molecules (data not shown).

This result combined with the results of genetic studies using 5-azaCyd (29) and the hypermethylation mutant *me-1* (30) suggest that methylation of cpDNA molecules is unlikely to be involved in the protection. Instead, because the preferential digestion of cp nucleoids is complete before the fusion of the mt^+ and mt^- chloroplasts, it is reasonable to assume that some modification to the chloroplast membrane might play an important role in protecting mt^+ cpDNA and mtDNA of both mating types.

The preferential disappearance of fluorescent chloroplast or mitochondrial nucleoids of uniparental origin is not restricted to *C. reinhardtii* zygotes; it also occurs in ferns and higher plants. In the fern *Pteris vittata*, the cp nucleoids of male origin disappear during maturation of the sperm, whereas the plastid without cp nucleoids remains visible until the final stage of sperm development (31). In higher plants, such as *Triticum aestivum*, *Lillium longiform*, and *Nicotiana tabacum*, the fluorescent cp and mitochondrial nucleoids in generative cells disappear during pollen mitosis (32). Perhaps ferns and higher plants also have a similar nuclease to digest uniparental cp or mtDNA molecules.

Molecular analysis of a cell procured under direct observation is becoming increasingly important in various clinical and bio-logical studies when it is difficult to prepare a pure cell population (21, 33–35). Conventional biochemical analysis of a complex cell population would merely produce confusion. The technique introduced in this report allows the researcher to obtain a single living cell, or even one chloroplast or mitochondria, from a population under direct visualization. Recently, much more attention has been paid to heterogeneity in the chloroplast (36, 37) and mitochondrial genomes (38) of higher plants. With this technique, it might be possible to explore the complexity of the chloroplast and mitochondrial genome in individual organelles. In the future, this technique combined with reverse transcription–PCR and microassays of enzyme activity will be very useful in various molecular studies.

We thank Dr. M. Goldschmidt-Clermont for generous permission to use plasmid pUCEBaadA 53(NN) for the chloroplast transformation. This work was supported by a research fellowship to Y.N. (5024) from the Japan Society for the Promotion of Science by Young Scientists and a Grant-in-Aid of Specially Promoted Research (06101002) and Scientific Research in Priority Areas (1163206) to T.K. from the Japanese Ministry of Education, Science and Culture.

1. Kuroiwa, T. (1991) *Int. Rev. Cytol.* **128**, 1–62.
2. Gillham, N. W. (1994) in *Organelle Genes and Genomes* (Oxford Univ. Press, New York), pp. 149–152.
3. Birky, C. W., Jr. (1995) *Proc. Natl. Acad. Sci. USA* **92**, 11331–11338.
4. Sager, R. & Lane, D. (1972) *Proc. Natl. Acad. Sci. USA* **69**, 2410–2413.
5. Boynton, J. E., Harris, E. H., Burkhart, B. D., Lamerson, P. M. & Gillham, N. W. (1987) *Proc. Natl. Acad. Sci. USA* **84**, 2391–2395.
6. Gillham, N. W. (1978) in *Organelle Heredity* (Raven, New York).
7. Kuroiwa, T., Suzuki, T., Ogawa, K. & Kawano, S. (1981) *Plant Cell Physiol.* **22**, 381–396.
8. Kuroiwa, T., Kawano, S., Nishibayashi, S. & Sato, C. (1982) *Nature (London)* **298**, 481–483.
9. Harris, E. H. (1989) in *Chlamydomonas Sourcebook* (Academic, San Diego).
10. Boynton, J. E., Gillham, N. W. & Harris, E. H. (1990) *Advances in Plant Gene Research* (Springer, New York).
11. Ogawa, K. & Kuroiwa, T. (1985) *Plant Cell Physiol.* **26**, 481–491.
12. Ogawa, K. & Kuroiwa, T. (1985) *Plant Cell Physiol.* **26**, 493–503.
13. Ashkin, A. & Dziedzic, J. M. (1987) *Science* **235**, 1517–1520.
14. Ashkin, A., Shütze, K., Dziedzic, J. M., Euteneuer, U. & Schliwa, M. (1990) *Nature (London)* **348**, 346–348.
15. Shütze, K., Clement-Segewald, A. & Ashkin, A. (1994) *Fertil. Steril.* **61**, 783–786.
16. Goldschmidt-Clermont, M. (1991) *Nucleic Acids Res.* **19**, 4083–4089.
17. Ishikura, K., Takaoka, Y., Sekine, M., Yoshida, K. & Shinmyo, A. (1999) *J. Biosci. Bioeng.* **87**, 307–314.
18. Snell, W. (1982) *J. Mol. Biol.* **25**, 47–66.
19. Nakamura, S., Itoh, S. & Kuroiwa, T. (1986) *Plant Cell Physiol.* **27**, 775–784.
20. Nishimura, Y., Higashiyama, T., Suzuki, L., Misumi, O. & Kuroiwa, T. (1998)

Eur. J. Cell Biol. **77**, 124–133.
21. Richert, J., Krantz, E., Lörz, H. & Dresselhaus, T. (1996) *Plant Sci.* **114**, 93–99.
22. Witman, G. B., Carlson, K., Berliner, J. & Rosenbaum, J. L. (1972) *J. Cell Biol.* **54**, 507–539.
23. Nakamura, S., Sato, C. & Kuroiwa, T. (1988) *Plant Sci.* **56**, 129–136.
24. Ferris, P. J. & Goodenough, U. W. (1993) *Cell* **74**, 801–811.
25. Kuroiwa, T. (1985) *Microbiol. Sci.* **2**, 267–272.
26. Burton, W. G., Gravowy, C. T. & Sager, R. (1979) *Proc. Natl. Acad. Sci. USA* **76**, 1390–1394.
27. Royer, H. D. & Sager, R. (1979) *Proc. Natl. Acad. Sci. USA* **76**, 5794–5798.
28. Sano, H., Gravowy, C. T. & Sager, R. (1981) *Proc. Natl. Acad. Sci. USA* **78**, 3118–3122.
29. Feng, T.-Y. & Chiang, K.-S. (1984) *Proc. Natl. Acad. Sci. USA* **81**, 3438–3442.
30. Bolen, B. L., Grant, D. M., Swinton, D., Boynton, J. E. & Gillham, N. W. (1982) *Cell* **28**, 335–343.
31. Kuroiwa, H., Sugai, M. & Kuroiwa, T. (1988) *Protoplasma* **146**, 89–100.
32. Miyamura, S., Kuroiwa, T. & Nagata, T. (1987) *Protoplasma* **141**, 149–159.
33. Karrer, E. E., Lincoln, J. E., Hogenhout, S., Bennet, A. B., Bostock, R. M., Martineau, B. Lucas, W. J., Gilchrist, D. G. & Alexander, D. (1995) *Proc. Natl. Acad. Sci. USA* **92**, 3814–3818.
34. Emmert-Buck, M. R., Bonner, R. F., Smith, P. D., Chuaqui, R. F., Zhuang, Zhengping, Goldstein, S. R., Weiss, R. A. & Liotta, L. A. (1996) *Science* **274**, 998–1001.
35. Simone, N. L., Bonner, R. F., Gillespie, J. W., Emmert-Buck, M. R. & Liotta, L. A. (1998) *Trends Genet.* **14**, 272–276.
36. Furg, J. E. (1998) *Nature (London)* **398**, 115–116.
37. Suzuki, H., Ingersoll, J., Stern, D. B. & Kindle, K. L. (1997) *Plant J.* **11**, 635–648.
38. Backert, S., Nielsen, B. L. & Börner, T. (1997) *Trends Plant Sci.* **2**, 477–483.

letters to nature

•••

Structural transitions and elasticity from torque measurements on DNA

Zev Bryant*, Michael D. Stone*, Jeff Gore†, Steven B. Smith†‡, Nicholas R. Cozzarelli* & Carlos Bustamante*†‡§

* *Department of Molecular and Cell Biology,* † *Department of Physics,* ‡ *Howard Hughes Medical Institute, and* § *Physical Biosciences Division of Lawrence Berkeley National Laboratory, University of California, Berkeley, California 94720, USA*

Knowledge of the elastic properties of DNA is required to understand the structural dynamics of cellular processes such as replication and transcription. Measurements of force and extension on single molecules of DNA[1–3] have allowed direct determination of the molecule's mechanical properties, provided rigorous tests of theories of polymer elasticity[4], revealed unforeseen structural transitions induced by mechanical stresses[3,5–7], and established an experimental and conceptual framework for mechanical assays of enzymes that act on DNA[8]. However, a complete description of DNA mechanics must also consider the effects of torque, a quantity that has hitherto not been directly measured in micromanipulation experiments. We have measured torque as a function of twist for stretched DNA—torsional strain in over- or underwound molecules was used to power the rotation of submicrometre beads serving as calibrated loads. Here we report tests of the linearity of DNA's twist elasticity, direct measurements of the torsional modulus (finding a value ~40% higher than generally accepted), characterization of torque-induced structural transitions, and the establishment of a framework for future assays of torque and twist generation by DNA-dependent enzymes. We also show that cooperative structural transitions in DNA can be exploited to construct constant-torque wind-up motors and force–torque converters.

Previous investigations of the force–extension behaviour of supercoiled DNA have found that, under low tension, DNA behaves like an isotropic flexible rod[2,9]: as turns are added to the molecule, its extension remains nearly constant until a critical twist density is

reached and the molecule buckles to form plectonemic (interwound) structures; thereafter, additional turns cause a rapid decrease in extension as twist is traded for writhe. At higher tensions, the behaviour of over- and underwound molecules differ. In each case, DNA undergoes a structural change before the twist density necessary for buckling is reached, and the molecule lengthens (or contracts only gradually) as additional turns are introduced[6,7]. These results have been interpreted to reflect cooperative torque-induced transitions in DNA structure[6,7], and a theoretical force–torque phase diagram has been proposed that explains the major features of force–extension curves for torsionally constrained molecules over a large range of forces and supercoiling densities[10]. Our work makes direct tests of both the isotropic rod model and the proposed force–torque phase diagram.

To perform dynamic torque measurements on single DNA molecules, we generated molecular constructs shown diagrammatically in Fig. 1a. The use of three distinct chemical modifications of DNA allows oriented tethering of the ends of the molecule, and attachment of a small bead (the 'rotor') to an internal position. A site-specific nick engineered below the rotor attachment point serves as a swivel.

We begin each assay by assembling the DNA molecule and rotor between two antibody-coated beads held in a micropipette and a force-measuring optical trap[11] (Fig. 1b and Supplementary Movie 1). Typically, we build up torsional stress in the molecule by holding the rotor bead stationary using fluid flow, and rotating the micropipette by ≥300 turns. When the flow is released, torque stored in the upper DNA segment causes the central bead to rotate continuously about its edge (Fig. 1c–e, and Supplementary Movies 2 and 3) until the torsional stress has been relieved, after which the bead rotates slightly back and forth under the influence of thermal fluctuations. Constant tension is maintained using force feedback[11], preventing buckling of the molecule[2] during the experiment, so that the observed dynamics reflect changes in twist and not writhe. At low Reynolds numbers, the magnitude of the torque can be determined as the product of the angular velocity (ω) of the rotor and its rotational drag (γ).

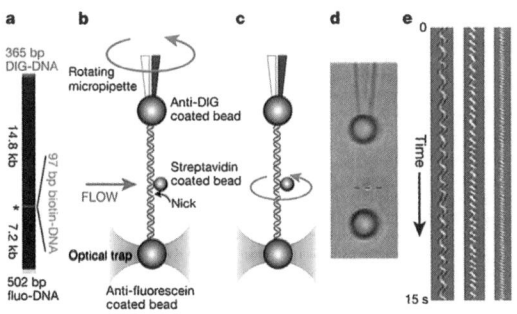

Figure 1 Experimental design. **a**, The molecular construct contains three distinct attachment sites and a site-specific nick (asterisk), which acts as a swivel. **b**, Each molecule was stretched between two antibody-coated beads using a dual-beam optical trap[11]. A rotor bead was then attached to the central biotinylated patch (see Supplementary Movie 1). The rotor was held fixed by applying a fluid flow, and the micropipette was twisted to build up torsional strain in the upper segment of the molecule. **c**, Upon releasing the flow, the central bead rotated to relieve the torsional strain. **d**, Video of the rotating bead (Supplementary Movies 2 and 3) was analysed to track cumulative changes in angle. Horizontal sections (red dashed line) of successive video frames can be stacked (**e**) to allow visualization of the helical path of the bead in space and time. Left to right, traces of 920 nm, 760 nm and 520 nm rotor beads.

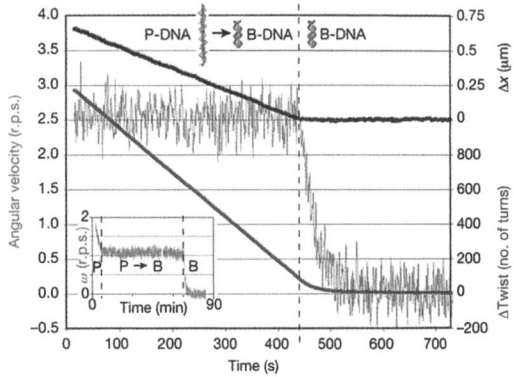

Figure 2 Analysis of bead rotation. A tether was overwound by 1,200 turns before releasing the 760 nm rotor bead. Red, cumulative rotor angle (Δtwist); green, angular velocity; and blue, molecular extension (Δx) versus time under constant tension (45 pN). The data support a model in which overwinding the molecule converts a fraction of the DNA into P-form[6,7]. P-DNA converts back into B-DNA at constant torque, after which the molecule regains its B-form extension and the remaining torque decays to zero. Inset, a molecule was completely converted to P-DNA by introducing 4,800 turns before releasing the 920 nm rotor bead. Torsional relaxation of hyperwound P-DNA preceded the long (~4,000 turns) P–B transition. During the gap at ~50 min, rotation was paused by turning on flow.

In highly overwound DNA held at tensions >7 pN, a portion of the molecule is converted from standard B-DNA into an over-extended, high-helicity form called P-DNA[6,7]. If the B → P transition is highly cooperative, it should occur at a constant torque. Confirming this prediction, the angular velocity of the rotor remained constant for much of the torsional relaxation of over-wound molecules (Fig. 2). P to B conversion was reflected in a progressive reduction in extension. When all of the molecule had been converted to B-DNA, the extension ceased to change and the angular velocity of the rotor decayed to zero as the remaining torsional strain in the B-DNA was removed. Complete conversion of the 14.8 kilobase (kb) molecule into P-DNA (Fig. 2, inset) required the introduction of $\sim 4.0 \times 10^3$ turns, implying a helical repeat of ~2.7 base pairs (bp) per turn for P-DNA, close to previous estimates[6,7] of ~2.6 bp per turn. From the changes in extension at 45 pN, we conclude that P-DNA is 50% longer than B-DNA under our conditions (slightly shorter than previous estimates[6,7] of 60–75% longer than B-form).

To quantify the torques (τ) in our experiments, the angular velocity of the rotor during the P → B transition ($\omega_{P \to B}$) was measured using rotor beads of several different radii (r) (Fig. 3a, inset). The intended rotor geometry (a sphere rotating about its edge) was confirmed by video analysis, which showed that the orbital radius was equal to the bead radius (within a 2% error). This geometry predicts $\tau = \gamma\omega = 14\pi r^3 \eta\omega$, where η is the viscosity of the solution. The appearance of the expected relationship $\omega \propto r^{-3}$ allowed us to determine the critical torque of the P → B transition, $\tau_{\text{crit},+} = 34 \pm 2$ pN nm (at 45 pN tension). This value was thereafter used as a calibration factor: instantaneous angular velocities of individual rotor beads were converted into torques using $\tau = \tau_{\text{crit},+}(\omega/\omega_{P \to B})$.

To precisely determine the twist elasticity of DNA, numerous molecules were repeatedly over- or underwound and allowed to relax. Plots of torque as a function of twist show considerable run-to-run variation due to random thermal fluctuations (Fig. 3a). Therefore, torque–twist data from 103 runs were averaged together (Fig. 3b). The torque–twist relationship for DNA shows two constant-torque regions, reflecting structural transitions at $\tau_{\text{crit},+} = +34$ pN nm and $\tau_{\text{crit},-} = -9.6$ pN nm, and a nearly linear region reflecting the twist elasticity of B-DNA. Our direct measurements of $\tau_{\text{crit},+}$ and $\tau_{\text{crit},-}$ are in good agreement with previous estimates from force–extension analysis[6,7,9,12]. Entry into the 34 pN nm plateau can now be seen to be very sharp, demonstrating the cooperativity of the B–P transition. (Using Ising/Zimm-Bragg theory, we find that a free-energy penalty of at least $4k_B T$ at P–B domain boundaries is required to explain the data.) The transition at −9.6 pN nm is somewhat less cooperative (showing a shallow slope in the initial portion of the plateau), and probably reflects the formation of denatured DNA[6] and some combination of non-canonical structures (for example, Z-DNA[7,10]). Because the DNA has a net left-handed twist (~13 bp per turn; data not shown) upon completion of this transition[7,8,10], we have collectively designated the underwound states 'L-DNA'[8].

The twist elasticity of B-DNA is often assumed to be that of an isotropic rod: torque builds up linearly with twist according to $\tau = (C/L)(\theta - \theta_0)$, where L is the rod length, θ is the twisting angle, θ_0 is the equilibrium twist (35° per bp), and the torsional modulus C is constant. However, the asymmetry of twisting a chiral molecule suggests that this approximation may break down. Indeed, a study of fluorescence polarization anisotropy decay (FPA) reported a linear dependence of C on supercoiling density[13]. The anharmonic model proposed from FPA[13] fits our data better than a simple harmonic model (Fig. 3b). However, the deviation from linearity is small, accounting for a <10% reduction in torque near entry into the P–B transition.

The torsional modulus C of DNA (near $\theta - \theta_0 = 0$) has been

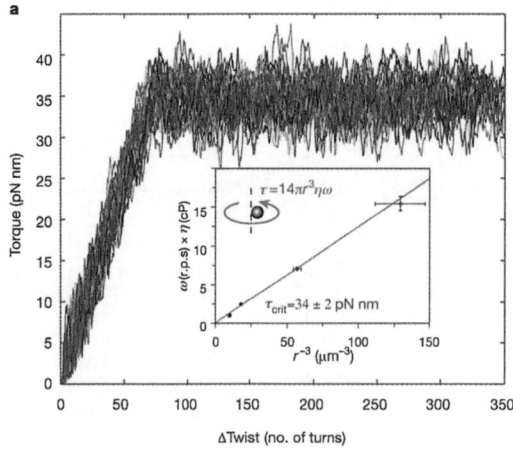

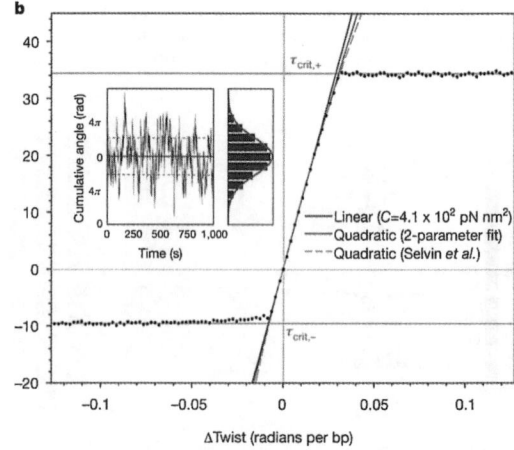

Figure 3 Torque calibration and twist elasticity of DNA. **a**, Torque versus twist for 37 runs from 16 molecules (520 nm rotor beads; 45 pN constant tension). Inset, calibration of $\tau_{\text{crit},+}$. Angular velocities (ω) of beads with mean diameters of 920 nm ($n = 5$), 760 nm ($n = 6$) or 520 nm ($n = 16$) were measured during the P → B transition at 45 pN (filled circles, ±s.d.). Viscosity-corrected ω is proportional to r^{-3}; the best-fit slope (red) gives $\tau_{\text{crit},+} = 34 \pm 2$ pN nm. Open circle, extrapolated P → B angular velocity for 400 nm beads ($n = 3$; see Methods). **b**, Averaged twist elasticity data. Negative torques: average of 39 runs at 15 pN. Positive torques: 37 runs at 45 pN and 27 runs at 15 pN gave very similar traces and were averaged together. Green lines, constant-torque structural transitions. Blue, linear fit to the data points falling within ±8 pN nm. Anharmonic models

$(C(\Delta\theta) = a - b\Delta\theta/N)$, where $N = 14,795$ bp) give superior fits to the data over the full range of B-DNA stability. Red, two-parameter anharmonic fit ($b/a = 4.5$); dashed purple, anharmonic fit constraining $b/a = 8.16$ to agree with FPA data[13]. Blue and red fits give $C(0) = 4.1 \times 10^2$ pN nm²; purple fit gives $C(0) = 4.3 \times 10^2$ pN nm². Fits to 45 pN or 15 pN positive torque data alone (not shown) give $C(0) = 4.1 \times 10^2$ or 4.3×10^2 pN nm², respectively. Inset, independent measure of $C(0)$ using the equipartition theorem $<\Delta\theta^2> = k_B TL/C$, where $L = 5.03$ μm. Twist fluctuations of five molecules (520 nm rotor beads, 15 pN tension) were tracked at 50 Hz (example shown). Angular variance (red gaussian plot over blue data histogram) gives $C(0) = (4.4 \pm 0.4) \times 10^2$ pN nm².

letters to nature

estimated from FPA[14,15], equilibrium topoisomer distribution of ligated circles[16,17], and circularization kinetics[18]. These methods give a wide range of values for C. FPA typically gives $C \approx 200$ pN nm^2, although elevated values have been reported for small circular DNA[15]. Topoisomer distribution analysis yields $C = 300$–400 pN nm^2, and best fits as high as 480 pN nm^2 have been reported from circularization kinetics[18]. Theoretical treatments of force and extension data from single supercoiled DNA molecules[9,12,19,20] have yielded $C = 300$ pN nm^2 (ref. 19), 350 pN nm^2 (refs 9, 12) and $C = 450$ pN nm^2 (ref. 20), depending on the analytical method. For all of the methods above, analytical theories or simulations are needed to estimate the relative contributions of twisting and bending or writhing to the experimental results. Topoisomer distributions of small circles (yielding $C \approx 300$ pN nm^2) have generally been considered the most reliable measurements, since the writhing contribution is thought to be small.

Our measurements provide a direct determination of C in the absence of writhe. A simple linear fit to the torque–twist data near $\theta - \theta_0 = 0$ gives $C = 410 \pm 30$ pN nm^2 (Fig. 3b). The uncertainty in this value depends heavily on imprecision in the absolute torque calibration from viscous drag (Fig. 2a, inset), since the relative scatter in the averaged data is small. To provide a drag-independent measure of C and check the calibration of our data, we used our experimental system to observe the amplitude of thermal fluctuations in twist (around $\theta - \theta_0 = 0$) and applied the equipartition theorem, as was previously done for single actin filaments[21]. This analysis (Fig. 3b, inset) yields $C = 440 \pm 40$ pN nm^2, in good agreement with our dynamic torque measurements.

These direct measurements of the torsional modulus of stretched DNA agree with the largest estimates of C from other methods, and are 40–50% higher than the widely used value of ~300 pN nm^2. The difference may be due to the presence of tension in our experiments. Stretching may induce structural changes that raise the torsional modulus, analogous to the proposed role of bending strain[15], or it may raise the effective C value owing to elastic coupling between

twisting, bending and stretching (A. Matsumoto and W. Olson, personal communication). No increase in rigidity was observed between 15 pN and 45 pN, so any such effect must be saturated at these forces.

It is also possible that previous studies have not sufficiently accounted for writhing, and consequently underestimated the torsional rigidity. In this case, the only important role of tension is suppression of writhe fluctuations, which would otherwise lead to torsional softening[20]. Our measurement should then accurately reflect the local torsional modulus of the molecule.

We extended our analysis of torque and twist to investigate the overstretching transition of DNA (Fig. 4a). When DNA is pulled to high forces, it undergoes a cooperative transition at ~65 pN to a form ('S-DNA') that is 70% longer than B-form[3,5]. The structure of S-DNA remains the subject of debate[22], but it is known to be substantially underwound relative to B-DNA[7,10]. Therefore, untwisting must accompany overstretching. When our molecule-rotor assemblies are held at tensions above 65 pN, torque is generated that spins the rotor as the B → S transition progresses (Fig. 4a and Supplementary Movie 4). The ratio of changes in extension and twist ($\Delta x / \Delta\theta = 3.7$ nm per turn) corresponds to a helical repeat of ~33 bp per turn for S-DNA, slightly smaller than the previously reported value of 37.5 bp per turn[7,10]. Relaxation of the molecule below 65 pN allows complete rewinding to B-form twist. Higher tensions produce higher unwinding torques, following a Clausius–Clapeyron-like force–torque coexistence line for the B–S transition. Together with tension-dependent critical torque data for the P–B and L–B transitions, these measurements validate the proposed force–torque phase diagram for stretched and twisted DNA[10] (Fig. 4b).

We have shown that an over- or underwound DNA molecule behaves as a constant-torque wind-up motor capable of repeatedly producing thousands of rotations, and that an overstretched molecule acts as a force–torque converter. (During overstretching at forces above $F_{crit} \approx 65$ pN, the efficiency, ε, of converting force–extension work into rotational work is given by the Carnot-like expression $\varepsilon = 1 - F_{crit}/F$.) The production of continuous directed rotation by molecular devices has potential applications in the construction of nanomechanical systems. Previous approaches to this problem have included co-opting the biological motor F1 ATP synthase[23], which produces torques comparable to the P–B transition[23,24]. DNA-based devices that produce continuous rotation, according to the principles demonstrated here, might be integrated with existing DNA nanotechnology[25] using the intrinsic self-assembly properties of DNA.

An important future application of the methods developed here is the measurement of torque generation and changes in twist induced by DNA-associated enzymes. A previous measurement of torque generation by RNA polymerase[26] was complicated by buckling of the DNA to form plectonemes. Here, we have established a method for torque measurement that decouples twisting from writhing, with the potential for broad application to biological motors. □

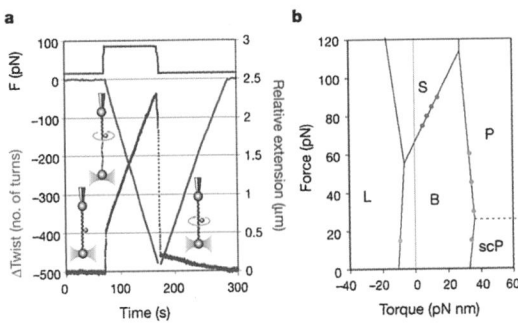

Figure 4 Unwinding during overstretching, and the global force–torque phase diagram of DNA. **a**, Unwinding during overstretching. Constructs were pulled to forces (black trace) above 65 pN. Cumulative rotations (red) of a 400 nm rotor bead showed continuous constant-torque untwisting while tension was maintained at 85 pN (Supplementary Movie 4). Concurrent decrease in twist (red) and increase in extension (blue) reflect S-DNA formation. When the force was relaxed to 15 pN, the molecule rewound, returning to B-form twist and extension. **b**, The global force–torque phase diagram of DNA. A single molecule was unwound by pulling to successively higher forces (relaxing at 15 pN between each run) to measure the torque of the B–S transition (judged from $\omega_B \rightarrow$ S) as a function of force (red). The critical torque of the P–B transition was also measured as a function of force by relaxation of overwound tethers under varying tensions (green), and all critical torque data (orange: $\tau_{crit,-} = 9.6$ pN nm) were found to agree with a simple model[10] in which DNA can access five distinct structural states differing only in extension, twist, and free energy. scP, P-DNA that has been shortened by supercoiling[6,7].

Methods

Beads and molecular constructs

Beads (2.8 μm) coated with Protein G (Spherotech) were crosslinked to rabbit anti-fluorescein (Molecular Probes) or sheep anti-digoxigenin (DIG) (Roche). Streptavidin-coated rotor beads were purchased from Spherotech (760 nm) or Bangs (all others), and their diameters were measured from transmission electron micrographs using diffraction grating replicas as standards (400 nm, $n = 55$; 520 nm, $n = 81$; 760 nm, $n = 48$; 920 nm, $n = 66$); all beads appeared round.

Molecular constructs were generated by serial ligation of purified restriction fragments of chemically modified and unmodified DNA. Hapten-modified polymerase chain reaction (PCR) fragments (966 bp) were generated by amplification of a multiple cloning site using 0.2 mM dATP, dCTP and dGTP, 0.13 mM dTTP, and 0.07 mM biotin-dUTP, DIG-dUTP or fluorescein-dUTP (Roche). PCR fragments were digested and gel-purified to recover a 374 bp biotin-modified KpnI fragment, a 365 bp DIG-modified BamHI fragment, and a 502 bp fluorescein-modified ClaI fragment. pSV8 (a plasmid generated by

. insertion of the 5.5 kb BamHI fragment of λ phage into pSV01[27]) was linearized with KpnI, treated with alkaline phosphatase (AP), and ligated to an excess of the biotinylated KpnI fragment. The product (containing nicked ligation junctions owing to the absence of plasmid-contributed phosphates) was digested with ClaI and XhoI. The resulting purified 7.3 kb fragment, designated protoSM, has the structure ClaI:7.2 kb:nick:97 bp(biotin-modified):XhoI. The torque-bearing DNA segment is the 14.8 kb BamHI:SalI fragment of pPIA6[28]. The final construct (SM1) was generated in a four-way ligation of the fluorescein-modified fragment, protoSM, torque-bearing segment, and DIG-modified fragment. Full-length products were selected by sequential binding to anti-fluorescein- and anti-DIG-coated beads in the optical tweezers. SM1 (Fig. 1a) was used in all reported experiments, with the following exceptions: SM2 (identical except DIG-DNA section extended to 601 bp for added stability) was used for the run shown in Fig. 2. SM3 (torque-bearing segment replaced with the 8.4 kb BglII-SalI fragment of pSV8; fluorescein-DNA section extended to 4 kb) was used for the force–torque analysis of the B–S transition (red data points in Fig. 4b).

Experimental assembly and data collection

Anti-fluorescein beads were incubated with DNA and introduced into the flow chamber. Anti-DIG beads were introduced via a separate channel, and a molecular tether was assembled by keeping an anti-DIG bead on the micropipette by suction, and 'fishing' near a DNA:anti-fluorescein bead held in the laser trap. The trapped bead was then released into flow, and a streptavidin-coated 'rotor' bead was trapped and brought to the vicinity of the biotinylated portion of the molecule, where it became attached laterally to the DNA (Supplementary Movie 1). The micropipette was rotated using a computer-controlled electric motor (LEGO Mindstorms) while the rotor bead was held fixed by flowing buffer at ~0.5 mm s^{-1}.

All experiments were performed in 100 mM NaCl and 40 mM Tris-HCl (pH 8.2). EDTA was typically present at 1 mM; omission caused no perceptible changes. Ambient temperature (23 ± 1 °C) was recorded prior to each experiment for use in viscosity calculations. Drag was also corrected for hydrodynamic coupling with the outer beads[29]; correction factors for the different rotor diameters were 1.005 (400 nm), 1.01 (520 nm), 1.02 (760 nm) and 1.03 (920 nm). Video was digitized at 30 Hz unless otherwise indicated, and the instantaneous angle of the rotor was extracted from the x-position and brightness (indicative of focal depth) of the bead. Angular velocities were obtained by numerical differentiation of the cumulative bead angle over a 1 s (Fig. 3) or 2 s (Fig. 2) window. The extrapolated P → B velocity of 400 nm beads (open circle in Fig. 3a inset) was obtained by measuring the velocity at large negative twists and scaling by $\tau_{\text{crit},+}/\tau_{\text{crit},-}$, since P → B rotation was too fast to track. During data collection, constant tension was maintained using stage-based force feedback[11]. During the exceptionally long run shown in Fig. 2 inset, force feedback (45 pN) was inoperative (out of actuator range) prior to $t = 23$ min, but $F > 30$ pN throughout.

Phase diagram

In the 'zero-temperature' approximation, the five-state structural model[10] leads to force–torque coexistence lines with constant slopes $\delta F/\delta \tau = -\Delta\theta/\Delta x$, where $\Delta\theta$ and Δx are the changes in twist and extension, respectively, for a particular structural transition. The slopes of the boundaries shown (Fig. 4b) were taken from experimental measurements of $\Delta\theta/\Delta x$, and predict the trends of the force–torque measurements. The intercepts of the boundaries were varied to fit the data. No torque measurements were made at the S–L or S–P boundaries, so these predicted slopes remain to be confirmed.

Received 6 February; accepted 28 May 2003; doi:10.1038/nature01810.

1. Smith, S. B., Finzi, L. & Bustamante, C. Direct mechanical measurements of the elasticity of single DNA molecules by using magnetic beads. *Science* **258**, 1122–1126 (1992).
2. Strick, T. R., Allemand, J. F., Bensimon, D., Bensimon, A. & Croquette, V. The elasticity of a single supercoiled DNA molecule. *Science* **271**, 1835–1837 (1996).
3. Smith, S. B., Cui, Y. & Bustamante, C. Overstretching B-DNA: The elastic response of individual double-stranded and single-stranded DNA molecules. *Science* **271**, 795–799 (1996).
4. Bustamante, C., Marko, J. F., Siggia, E. D. & Smith, S. Entropic elasticity of lambda-phage DNA. *Science* **265**, 1599–1600 (1994).
5. Cluzel, P. et al. DNA: An extensible molecule. *Science* **271**, 792–794 (1996).
6. Allemand, J. F., Bensimon, D., Lavery, R. & Croquette, V. Stretched and overwound DNA forms a Pauling-like structure with exposed bases. *Proc. Natl Acad. Sci. USA* **95**, 14152–14157 (1998).
7. Leger, J. F. et al. Structural transitions of a twisted and stretched DNA molecule. *Phys. Rev. Lett.* **83**, 1066–1069 (1999).
8. Bustamante, C., Bryant, Z. & Smith, S. B. Ten years of tension: Single-molecule DNA mechanics. *Nature* **421**, 423–427 (2003).
9. Bouchiat, C. & Mezard, M. Elasticity model of a supercoiled DNA molecule. *Phys. Rev. Lett.* **80**, 1556–1559 (1998).
10. Sarkar, A., Leger, J. F., Chatenay, D. & Marko, J. F. Structural transitions in DNA driven by external force and torque. *Phys. Rev. E* **63**, 051903 (2001).
11. Smith, S. B., Cui, Y. & Bustamante, C. Optical-trap force transducer that operates by direct measurement of light momentum. *Methods Enzymol.* **361**, 134–162 (2003).
12. Strick, T. R., Bensimon, D. & Croquette, V. Micro-mechanical measurement of the torsional modulus of DNA. *Genetica* **106**, 57–62 (1999).
13. Selvin, P. R. et al. Torsional rigidity of positively and negatively supercoiled DNA. *Science* **255**, 82–85 (1992).
14. Millar, D. P., Robbins, R. J. & Zewail, A. H. Direct observation of the torsional dynamics of DNA and RNA by picosecond spectroscopy. *Proc. Natl Acad. Sci. USA* **77**, 5593–5597 (1980).
15. Heath, P. J., Clendenning, J. B., Fujimoto, B. S. & Schurr, J. M. Effect of bending strain on the torsion elastic constant of DNA. *J. Mol. Biol.* **260**, 718–730 (1996).
16. Horowitz, D. S. & Wang, J. C. Torsional rigidity of DNA and length dependence of the free energy of DNA supercoiling. *J. Mol. Biol.* **173**, 75–91 (1984).
17. Shore, D. & Baldwin, R. L. Energetics of DNA twisting. II. Topoisomer analysis. *J. Mol. Biol.* **170**, 983–1007 (1983).
18. Crothers, D. M., Drak, J., Kahn, J. D. & Levene, S. D. DNA bending, flexibility, and helical repeat by cyclization kinetics. *Methods Enzymol.* **212**, 3–29 (1992).
19. Vologodskii, A. V. & Marko, J. F. Extension of torsionally stressed DNA by external force. *Biophys. J.* **73**, 123–132 (1997).
20. Moroz, J. D. & Nelson, P. Entropic elasticity of twist-storing polymers. *Macromolecules* **31**, 6333–6347 (1998).
21. Yasuda, R., Miyata, H. & Kinosita, K. Jr Direct measurement of the torsional rigidity of single actin filaments. *J. Mol. Biol.* **263**, 227–236 (1996).
22. Williams, M. C., Rouzina, I. & Bloomfield, V. A. Thermodynamics of DNA interactions from single molecule stretching experiments. *Acc. Chem. Res.* **35**, 159–166 (2002).
23. Soong, R. K. et al. Powering an inorganic nanodevice with a biomolecular motor. *Science* **290**, 1555–1558 (2000).
24. Yasuda, R., Noji, H., Kinosita, K. Jr & Yoshida, M. F1-ATPase is a highly efficient molecular motor that rotates with discrete 120 degree steps. *Cell* **93**, 1117–1124 (1998).
25. Seeman, N. C. DNA in a material world. *Nature* **421**, 427–431 (2003).
26. Harada, Y. et al. Direct observation of DNA rotation during transcription by *Escherichia coli* RNA polymerase. *Nature* **409**, 113–115 (2001).
27. Wobbe, C. R., Dean, F., Weissbach, L. & Hurwitz, J. In vitro replication of duplex circular DNA containing the simian virus 40 DNA origin site. *Proc. Natl Acad. Sci. USA* **82**, 5710–5714 (1985).
28. Davenport, R. J., Wuite, G. J., Landick, R. & Bustamante, C. Single-molecule study of transcriptional pausing and arrest by E. coli RNA polymerase. *Science* **287**, 2497–2500 (2000).
29. Davis, M. H. The slow translation and rotation of two unequal spheres in a viscous fluid. *Chem. Eng. Sci.* **24**, 1769–1776 (1969).

Supplementary Information accompanies the paper on **www.nature.com/nature**.

Acknowledgements We thank E. Watson and Y. Inclán for technical assistance, E. Nogales for microscope time, and A. Vologodskii, V. Croquette, D. Bensimon, D. Collin, N. Pokala and Y. Chemla for critical readings of the manuscript and/or discussions. Z.B. is an HHMI predoctoral fellow, M.D.S. is supported by a PMMB training grant, and J.G. holds a fellowship from the Hertz Foundation. This work was supported by the NIH and DOE.

Competing interests statement The authors declare that they have no competing financial interests.

Correspondence and requests for materials should be addressed to C.B. (carlos@alice.berkeley.edu).

letters to nature

● ●

Backtracking by single RNA polymerase molecules observed at near-base-pair resolution

Joshua W. Shaevitz[1]*, **Elio A. Abbondanzieri**[2]*, **Robert Landick**[4] **& Steven M. Block**[2,3]

[1]*Department of Physics,* [2]*Department of Applied Physics, and* [3]*Department of Biological Sciences, Stanford University, Stanford, California 94305, USA*
[4]*Department of Bacteriology, University of Wisconsin, Madison, Wisconsin 53706, USA*

* These authors contributed equally to this work

Escherichia coli RNA polymerase (RNAP) synthesizes RNA with remarkable fidelity *in vivo*[1]. Its low error rate may be achieved by means of a 'proofreading' mechanism comprised of two sequential events. The first event (backtracking) involves a transcriptionally upstream motion of RNAP through several base pairs, which carries the 3′ end of the nascent RNA transcript away from the enzyme active site. The second event (endonucleolytic cleavage) occurs after a variable delay and results in the scission and release of the most recently incorporated ribonucleotides, freeing up the active site. Here, by combining ultrastable optical trapping apparatus with a novel two-bead assay to monitor transcriptional elongation with near-base-pair precision, we observed backtracking and recovery by single molecules of RNAP. Backtracking events (~5 bp) occurred infrequently at locations throughout the DNA template and were associated with pauses lasting 20 s to >30 min. Inosine triphosphate increased the frequency of backtracking pauses, whereas the accessory proteins GreA and GreB, which stimulate the cleavage of nascent RNA, decreased the duration of such pauses.

Recent studies have implicated the nucleolytic activity of RNA polymerase as part of a proofreading mechanism[2–4], similar to that found in DNA polymerases[5]. A key feature of this proofreading mechanism is a short backtracking motion of the enzyme along the DNA template (directed upstream, opposite to the normal direction of transcriptional elongation). Similar rearward movements are thought to accompany the processes of transcriptional pausing[6–8], arrest[9,10], and transcription-coupled DNA repair[11]. During backtracking, the transcription bubble shifts and the DNA–RNA hybrid duplex remains in register, while the 3′ end of the RNA transcript moves away from the active site, and may even protrude into the secondary channel (nucleotide entrance pore) of the enzyme[6,7,9], blocking the arrival of ribonucleoside triphosphates (NTPs). In its backtracked state, RNAP is able to cleave off and discard the most recently added base(s) by endonucleolysis, generating a fresh 3′ end at the active site for subsequent polymerization onto the nascent RNA chain. In this fashion, short RNA segments carrying misincorporated bases can be replaced, leading to the correction of transcriptional errors (Fig. 1a). Accessory proteins have been identified that increase transcriptional fidelity by preferentially stimulating the cleavage of misincorporated nucleotides: GreA and GreB for *E. coli* RNA polymerase[4] and SII/TFIIS for eukaryotic RNA polymerase II[2,3].

We studied transcription by RNAP at physiological nucleotide concentrations using a new single-molecule assay together with improved optical trapping instrumentation. In combination, these achieve subnanometre resolution along with extremely low positional drift. Our current system is capable of near-base-pair resolution in individual records of RNAP displacements, and achieves base-pair resolution (<0.3 nm) in averages of multiple records. During an experiment, two beads are optically trapped in buffer above a microscope coverglass by independently steered laser traps.

A recombinant derivative of *E. coli* RNAP is bound specifically via a biotin–avidin linkage to the smaller of two polystyrene beads, while the transcriptionally downstream end of the DNA template (or the upstream end, in the case of assisting forces) is bound to the larger bead via a digoxygenin-antibody linkage, forming a bead–RNAP–DNA–bead 'dumbbell' (Fig. 1b).

The tension in the DNA was kept nearly constant (8.4 ± 0.8 pN), for loads both opposing and assisting transcription, by feedback control of the position of the optical trap holding the larger bead. A force of this magnitude has a negligible effect on transcription rates, and is well below the stall force for RNAP[12]. An opposing load was applied in all experiments, except where noted. Transcriptional elongation was observed by measuring the position of the smaller bead as the polymerase moved (Fig. 2a). We chose to make the trap holding the larger bead an order of magnitude stiffer than that holding the smaller bead so that all motion appeared in the latter (see Methods). None of the components of the assay were attached to the coverglass surface: this isolates the system from drift of the microscope stage relative to the objective and other optics, which represented a major source of low-frequency noise in previous single-molecule studies[12–17]. Measured drift rates during our experiments were typically below 5 nm h⁻¹ (data not shown). We recorded the transcriptional motion of over 150 individual RNAP molecules at 1 mM NTPs moving on a DNA template derived from the *E. coli rpoB* gene sequence. As previously noted[12,13,15,17], RNAP activity consists of periods of continuous motion interrupted by distinct pauses of variable duration (Fig. 2a). The velocity during the continuous-motion phase averaged ~15 bp s⁻¹, but varied among molecules, consistent with earlier reports[12,13,15].

Computer analysis of RNAP records identified transcriptional pauses ranging from 1 s (our detection threshold) to more than 30 min. Only intervals where transcriptional elongation ceased and subsequently recovered were scored as pauses. Pausing events could

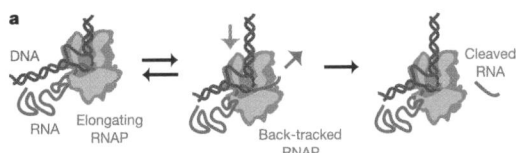

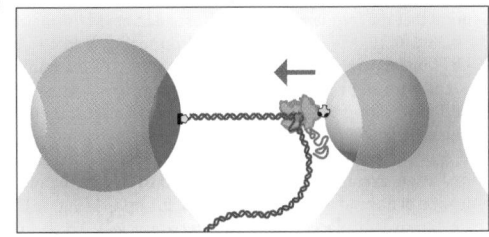

Figure 1 RNA polymerase transcription and proofreading studied by optical trapping. **a**, During normal elongation, RNAP (green) moves forward (downstream) on the DNA (blue) as it elongates the nascent RNA (red). At each position along the template, RNAP may slide backward along the template, causing transcription to cease temporarily. From the backtracked state, polymerase can either slide forward again, returning to its earlier state (left) or cleave the nascent RNA (right) and resume transcriptional elongation. **b**, Cartoon of the experimental geometry employed for opposing force experiments (not to scale). Two beads (blue) are held in separate optical traps (red) in a force-clamp arrangement. The smaller bead (right) is bound to a single molecule of RNAP, while the larger bead (left) is bound to the downstream end of the DNA by non-covalent linkages (yellow). During transcriptional elongation, the beads are pulled together. Nearly all the motion appears as a displacement of the right bead (green arrow), which is held in a comparatively weaker trap.

NATURE | VOL 426 | 11 DECEMBER 2003 | www.nature.com/nature

letters to nature

be broken up into two broad categories: 95% of events were 'short,' with lifetimes drawn from a double-exponential distribution with time constants of 1.5 s and 6.5 s, similar to our previous findings[12]. The remaining 5% of events were 'long,' with lifetimes >20 s and a broad, non-exponential temporal distribution. Long pauses occurred at positions randomly distributed along the DNA template, rather than at stereotyped locations, and appeared to be sequence-independent within the resolution of these experiments. On average, long pauses occurred with a frequency of $0.95 \pm 0.21 \, \text{kb}^{-1}$, a value that corresponds closely to ribonucleotide misincorporation rates during RNA synthesis *in vitro*[1], suggesting a possible role for such pauses in proofreading.

Operationally, we defined the duration of a 'pause' as the interval between the cessation of forward transcriptional motion and its subsequent recovery. At high spatial resolution, however, long pauses were found to consist of three distinct phases of motion that could be discerned in some individual records (Fig. 2b), as well as in averages of multiple records (see below). After abruptly stopping forward transcription, the enzyme underwent a slow rearward movement (phase 1, backtracking), typically lasting from 1–5 s, before stopping altogether for a variable interval (phase 2, pause). At the end of phase 2, rather than immediately resuming transcription at normal rates, RNAP moved forward gradually, typically for 3–10 s, transitioning to elongation mode only after a significant fraction of the initial backtracking distance had been retraced (phase 3, recovery). In contrast, neither the backtracking nor the recovery phases were evident in records of short pauses, where transitions both to and from normal elongation were abrupt (Fig. 2c).

To analyse the mean behaviour during phases 1 and 3, we averaged records of long pauses obtained under opposing loads after placing these in register along their rising edges, immediately before the cessation or the resumption of elongation, respectively (Fig. 3a). This procedure allows one to probe details of the motion that would otherwise be obscured by noise in individual records: similar trace-averaging techniques have been successfully employed to detect nanoscale steps in motor proteins such as myosin[18] and NCD[19], as well as to look for fast transients within the 8-nm step of kinesin[20]. The average backtracking displacement during phase 1 of long pauses was 4.7 ± 0.8 bp, and could be fitted by a decaying exponential with a time constant of 1.2 ± 0.1 s. Both the duration and frequency of backtracking pauses are expected to display a strong force dependence due to the underlying motions involved. We found that the frequency of long pauses decreased dramatically from $0.95 \pm 0.21 \, \text{kb}^{-1}$ under an ~8 pN opposing load to below $0.03 \, \text{kb}^{-1}$ under an ~8 pN assisting load (Table 1). This finding is consistent with a previous report of force dependence in the duration of very long pauses (~90 s) using a low-resolution optical trapping assay[13].

Averages of records during phase 3 displayed a gradual forward motion, at an average velocity of $0.29 \pm 0.01 \, \text{bp s}^{-1}$, before the resumption of normal elongation at $13.2 \pm 0.1 \, \text{bp s}^{-1}$ (Fig. 3a). The average forward displacement during recovery was 2.5 ± 1.0 bp, that is, about half of the initial backtracking distance. This reduced distance may reflect a mixed population of records, some of which exited from the pause more abruptly than others. However, the difference might also reflect the trace-alignment procedure. The exit from phase 2 is far less distinct than the entry into phase 1, and is therefore harder to pinpoint: minor registration errors tend to alter the magnitude of motions in averaged traces.

For comparison, we aligned and averaged an identical number of short pause records (Fig. 3d). In contrast to the long pause average, the short pause average displayed sharp transitions both into and out of the pause (that is, no phase 1 or phase 3 motions), with no associated movement greater than a base pair. The average of a much larger population of short pauses (>250) showed the same behaviour (data not shown). The absence of backtracking in short

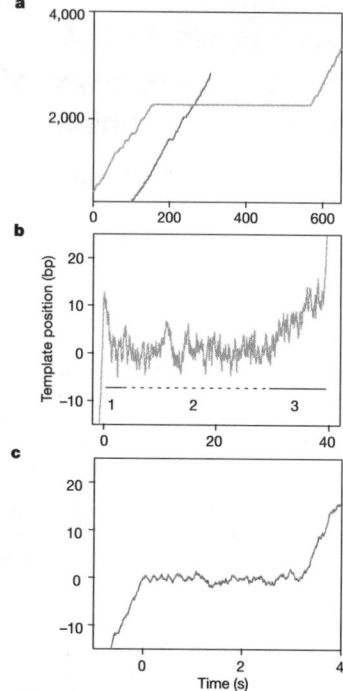

Figure 2 Backtracking occurs upon entry into long, but not short, pauses. **a**, Transcription records of two individual RNAP molecules are shown, each moving over the same template sequence. Both traces contain multiple short pauses (most are too short to be seen on this timescale); one includes a very long pause (410 s, red trace). **b**, In some records of long pauses, backtracking could be seen by eye: a representative record is shown. The three phases of motion are indicated below the trace: phase 1 (backtracking, solid line), phase 2 (pause, dotted line), and phase 3 (recovery, solid line). **c**, A representative record of a short pause (3 s); such pauses do not exhibit backtracking. Data were recorded at 2 kHz and boxcar-filtered at 100 ms for display.

pauses directly confirms and extends the conclusions of a recent study that examined the frequency and duration of short pauses, and found that these were independent of external load. This lack of force dependence implies an absence of backtracking motion, even as small as a single base pair[12].

We performed parallel experiments in the presence of the ribonucleotide analogue inosine triphosphate (ITP). Inosine mimics guanosine, forming a weak Watson–Crick pair that is slightly more stable than some measured mispairings[21,22]. ITP incorporation inhibits next-nucleotide addition in human polymerase II by an amount similar to that of a mismatched base[3]. In elongating complexes, inosine incorporation decreases the stability of the RNA–DNA hybrid, changing the relative stability of the backtracked and non-backtracked states, and also decreases the stability of secondary structures in the nascent RNA formed behind the complex. The addition of 200 μM ITP to the standard transcription buffer (containing 1 mM levels of GTP, CTP, ATP and UTP) increased both the frequency and duration of long pauses (Fig. 3e, f; Table 1; Supplementary Information). Both phase 1 and phase 3 of long pauses were quantitatively similar to those observed in the absence of ITP (Fig. 3b). However, ITP did not affect either the frequency or the duration of short pauses, nor the average transcriptional velocity between pauses.

To assay the effects of transcript cleavage on long pauses, we added the *E. coli* transcription factors GreA and GreB. Addition of

letters to nature

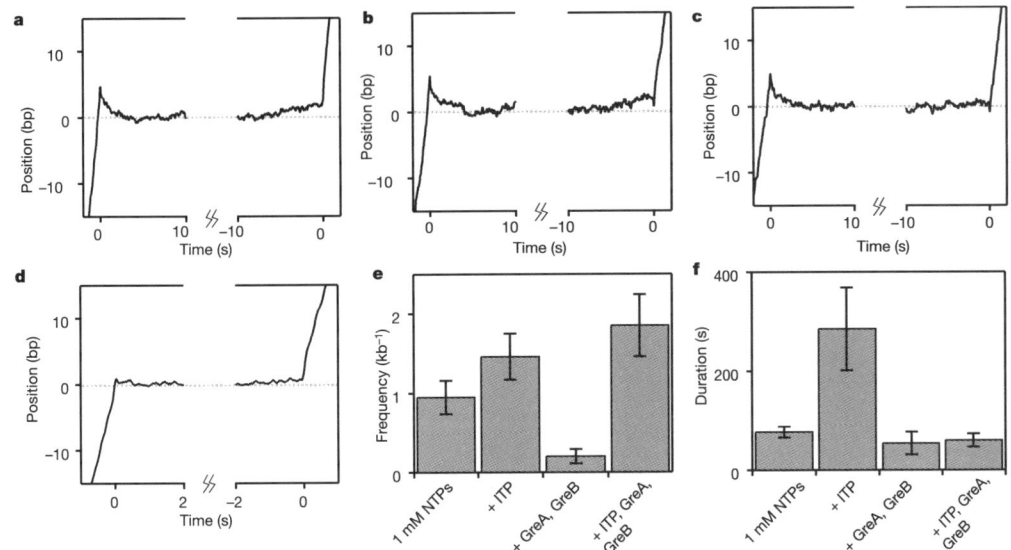

Figure 3 Averages of aligned long-pause records reveal details of backtracking and transcript cleavage events. **a**, Long-pause average of records at 1 mM NTPs ($N = 56$) displays a backtracking motion of ~5 bp (phase 1). Recovery (phase 3) is gradual, lasting ~5 s, before the resumption of normal elongation speed. **b**, Addition of ITP increases the frequency of long pauses that are indistinguishable from those in **a** ($N = 26$). **c**, Addition of GreA and GreB reduced the duration of long pauses. Recovery from these pauses (phase 3) was abrupt ($N = 22$), distinct from **a** and **b**. **d**, The short pause average ($N = 56$) displays no backtracking motion. Average records were smoothed with a 100-ms boxcar filter for display. ITP and the accessory proteins GreA and GreB affect both the frequency (**e**) and duration (**f**) of long pauses. See text.

either 2 μM GreA or 1 μM GreB decreased the frequency of long pauses. Moreover, GreB significantly decreased the duration of long pauses, whereas the effect of GreA on duration was negligible (Fig. 3e, f; Table 1). The average distance backtracked in the presence of GreA increased to 6.6 ± 0.7 bp, whereas GreB appeared to abolish backtracking pauses altogether, yielding a mean backtracking distance close to zero for the few remaining long pauses, 0.5 ± 0.8 bp (Supplementary Information). These findings are consistent with the known properties of the two factors. GreA stimulates the cleavage of short, dinucleotide segments of backtracked RNA, and so its addition should relieve only those pauses associated with short backtracking motions, leaving pauses involving larger displacements. Conversely, GreB accelerates the removal of larger fragments to such a degree that the duration of backtracking pauses falls below the discrimination threshold of 20 s[23].

In the presence of both GreA and GreB, the duration (phase 2) of ITP-induced long pauses decreased dramatically, from 285 ± 83 s to 60 ± 13 s, while the average backtracking distance (phase 1) remained the same, 5.0 ± 1.0 bp (Table 1; Fig. 3f). Significantly, in the presence of both these transcription factors, the pause recovery (phase 3) became abrupt, and not gradual. We interpret this difference as being caused by Gre-stimulated cleavage of the RNA

blocking the secondary channel after backtracking. Such cleavage would lead to the prompt removal of oligonucleotides containing a potential mismatch, thereby reducing the lifetime of the backtracked, paused state (phase 2), and restoring the new 3′ end of the nascent RNA to a position adjacent to the enzyme active site, ready for immediate polymerization.

In the absence of Gre-stimulated cleavage, RNAP must recover during phase 3 in a less direct fashion. In one mechanism, RNAP relies on its slow endogenous endonucleolytic activity to cleave the RNA at the enzyme active site, leading to long pause lifetimes but still allowing the possibility of error correction. In an alternative mechanism, thermal motions within the stalled complex may reverse the backtracking motion in a random walk process, carrying the original 3′ end of the RNA back to the enzyme active site, once again leading to longer lifetimes, but without concomitant error correction. The gradual recovery seen during phase 3 in the absence of Gre factors may reflect these processes. Random fluctuations of polymerase motion during phase 2 would not be apparent in averaged records, which only show ensemble behaviour.

Taken together, our high-resolution, single-molecule experiments are consistent with a proofreading mechanism in *E. coli* RNA polymerase involving entry into an initial backtracked state of

Table 1 Long pause statistics under different experimental conditions

Condition*	Mean pause frequency (kb⁻¹)	Mean pause duration (s)	Mean backtrack distance (bp)
1 mM ATP, UTP, GTP, CTP	0.95 ± 0.21 ($N = 56$)	77 ± 11	4.7 ± 0.8
1 mM NTPs, assisting force	<0.03 ($N = 1$)	–	–
+200 μM ITP	1.46 ± 0.29 ($N = 26$)	285 ± 83	5.5 ± 1.1
+2 μM GreA	0.14 ± 0.08 ($N = 3$)	56 ± 12	6.6 ± 0.7
+1 μM GreB	0.24 ± 0.09 ($N = 8$)	36 ± 7	0.5 ± 0.8
+2 μM GreA, 1 μM GreB	0.20 ± 0.09 ($N = 5$)	54 ± 23	5.8 ± 0.8
+200 μM ITP, 2 μM GreA, 1 μM GreB	1.85 ± 0.39 ($N = 22$)	60 ± 13	5.0 ± 1.0
+200 μM ITP, assisting force	0.15 ± 0.10 ($N = 2$)	–	–

* The applied force opposes transcription, unless otherwise noted.

NATURE | VOL 426 | 11 DECEMBER 2003 | www.nature.com/nature

the enzyme on the DNA template, followed by cleavage of the most recently polymerized RNA (1–10 bp) and enzymatic recovery. Under the conditions explored here (including an opposing load of ~8 pN), RNAP appeared to enter into long, backtracking pauses spontaneously at a rate of roughly once per kilobase: this rate was sensitive to transcription errors, and enhanced at least twofold by the addition of a nucleotide analogue. Incorporation of inosine leading to the backtracked state was relieved quantitatively by the action of transcription factors GreA and GreB, which are known to stimulate transcript cleavage. This simple editing mechanism may function, in principle, in many polymerase systems, including both prokaryotes and eukaryotes. □

Methods

Transcription assays

A bead–RNAP–DNA–bead dumbbell was constructed by binding a small 0.5-μm diameter polystyrene bead to a biotin tag located on the β′ subunit of a stalled *E. coli* RNAP transcription elongation complex, and a larger 0.7-μm-diameter bead to the downstream end of the DNA template using a digoxygenin antibody (Fig. 1b). Stalled complexes and avidin-coated 0.5-μm-diameter polystyrene beads were prepared as described previously[12]. Polyclonal anti-digoxigenin antibody was covalently attached to carboxylated 0.7-μm diameter polystyrene beads (Bangs Labs) via an EDC/Sulfo–NHS-coupled reaction. RNAP was stalled 29 base pairs after the T7A1 promoter on a template derived from the *rpoB* gene of *E. coli*[16]. Each of the two beads of the dumbbell was held in a separate optical trap ~1 μm above the coverglass surface. Transcription along the DNA template was recorded by monitoring the position of the smaller bead (see below). All experiments were performed in transcription buffer (50 mM HEPES, pH 8.0, 130 mM KCl, 4 mM $MgCl_2$, 0.1 mM EDTA, 0.1 mM DTT, 20 mg ml^{-1} heparin) in the presence of 1 mM NTPs and an oxygen scavenging system[24] at 22 ± 5 °C. ITP was purchased from Sigma-Aldrich Company. GreA and GreB proteins were purified as described[25].

Optical trapping

The salient aspects of the instrument used in these experiments have been described previously[26]. Briefly, the apparatus is based on an inverted microscope (Nikon) modified for exceptional mechanical stability and the incorporation of two lasers for trapping and position detection. To form two optical traps, the trapping laser was split into two orthogonally polarized beams that can be steered independently. Software written in LabView 6i (National Instruments) controlled the position of the trapped 0.7-μm-diameter bead via acousto-optical deflectors (IntraAction). The 0.5-μm-diameter bead was illuminated by the detection laser: scattered light was projected onto a position-sensitive detector (Pacific Silicon Sensors) placed in a plane conjugate to the back focal plane of the microscope condenser. Bead position signals were smoothed at 1 kHz by an eight-pole low-pass Bessel filter (Krohn-Hite) and digitized at 2 kHz.

During an experiment under opposing loads, transcriptional elongation by RNAP shortens the DNA tether and leads to an increase in tension between the two beads. The laser power in each trap was high to reduce brownian noise in position, which is inversely proportional to the trap stiffness[27]. The relative power in the two traps was fixed so that their ratio of stiffnesses was at least 1:10, which ensured that the majority of RNAP motion appeared as a change in the position of the smaller bead, held in the weaker trap. The resting tension in the DNA was maintained at 8.4 ± 0.8 pN (mean ± s.d.) by moving the 0.7-μm-diameter bead in discrete 50-nm increments whenever the tension on the DNA exceeded 10 pN.

In our measurements, zero-mean, brownian fluctuations of the smaller bead represent the dominant source of noise. Position records of pauses along short pieces of DNA, that is, from enzymes that have already transcribed a substantial portion of the template, exhibit reduced noise because the amplitude of the brownian fluctuations is inversely proportional to the compliance of the linkage holding the bead. Additional sources of noise include tiny departures from sphericity in the beads (slightly non-spherical beads generate low-frequency noise through brownian rotation) and submicroscopic particle contaminants in the buffer that may fall into the optical trap (modulating the level of scattered light and thereby altering the apparent position). The combined effect of such sources causes the root mean square (r.m.s.) noise level to vary slightly from trace to trace: typically, noise levels were ± 5 bp at a 1-kHz bandwidth.

Data analysis

The contour length of the downstream DNA was computed from the measured position of the 0.5-μm-diameter bead and the series elastic compliances of the optical traps and the DNA (using the nonlinear modified Marko–Siggia force–extension relation[28]). Template position was determined by subtracting the initial tether length (4,226 bp for experiments using forces opposed to transcriptional elongation, and 1,406 bp for experiments using forces assisting transcriptional elongation) from the computed DNA contour length. We estimate our absolute error in determining the template position at ±90 bp, due mainly to minor variations among individual bead diameters.

Transcriptional pauses in individual recordings were scored by eye and also by a custom computer algorithm (similar to ref. 12, but combining the numerical derivative and filtering operations into a single step). This algorithm recovers >99% of all pauses with lifetimes more than 1.5 s. Pauses occurred with a wide range of lifetimes, from seconds to many minutes. Broadly, events could be placed into two categories: 95% of all events belonged to a population that was satisfactorily fitted by a double-exponential relation with time constants of 1.5 s (65% of the normalized amplitude) and 6.5 s (35% of the normalized amplitude). The remaining 5% of all events could not be so represented, and belonged to a much longer-lived population with a broader, non-exponential distribution. To better separate these populations for statistical analysis, we operationally defined 'short' pauses as having durations shorter than 5 s and 'long' pauses as having durations longer than 20 s, thereby excluding from analysis all pauses with lifetimes $5 < t < 20$. Individual pause records were aligned along their initial and final rising edges[18–20] and averaged to reduce measurement noise. Analysis was performed using Igor Pro 4.0 (Wavemetrics). All errors are reported as (mean ± s.e.m), except for backtracking distances and durations, which were estimated using a bootstrap resampling analysis, and represent 68% confidence intervals.

Received 14 October; accepted 5 November 2003; doi:10.1038/nature02191.
Published online 23 November 2003.

1. Erie, D. A., Yager, T. D. & von Hippel, P. H. The single-nucleotide addition cycle in transcription: a biophysical and biochemical perspective. *Annu. Rev. Biophys. Biomol. Struct.* **21**, 379–415 (1992).
2. Jeon, C. & Agarwal, K. Fidelity of RNA polymerase II transcription controlled by elongation factor TFIIS. *Proc. Natl Acad. Sci. USA* **93**, 13677–13682 (1996).
3. Thomas, M. J., Platas, A. A. & Hawley, D. K. Transcriptional fidelity and proofreading by RNA polymerase II. *Cell* **93**, 627–637 (1998).
4. Erie, D. A., Hajiseyedjavadi, O., Young, M. C. & von Hippel, P. H. Multiple RNA polymerase conformations and GreA: control of the fidelity of transcription. *Science* **262**, 867–873 (1993).
5. Kunkel, T. A. & Bebenek, K. DNA replication fidelity. *Annu. Rev. Biochem.* **69**, 497–529 (2000).
6. Komissarova, N. & Kashlev, M. RNA polymerase switches between inactivated and activated states by translocating back and forth along the DNA and the RNA. *J. Biol. Chem.* **272**, 15329–15338 (1997).
7. Nudler, E., Mustaev, A., Lukhtanov, E. & Goldfarb, A. The RNA-DNA hybrid maintains the register of transcription by preventing backtracking of RNA polymerase. *Cell* **89**, 33–41 (1997).
8. Marr, M. T. & Roberts, J. W. Function of transcription cleavage factors GreA and GreB at a regulatory pause site. *Mol. Cell* **6**, 1275–1285 (2000).
9. Komissarova, N. & Kashlev, M. Transcriptional arrest: *Escherichia coli* RNA polymerase translocates backward, leaving the 3′ end of the RNA intact and extruded. *Proc. Natl Acad. Sci. USA* **94**, 1755–1760 (1997).
10. Reeder, T. C. & Hawley, D. K. Promoter proximal sequences modulate RNA polymerase II elongation by a novel mechanism. *Cell* **87**, 767–777 (1996).
11. Tornaletti, S., Reines, D. & Hanawalt, P. C. Structural characterization of RNA polymerase II complexes arrested by a cyclobutane pyrimidine dimer in the transcribed strand of template DNA. *J. Biol. Chem.* **274**, 24124–24130 (1999).
12. Neuman, K. C., Abbondanzieri, E. A., Landick, R., Gelles, J. & Block, S. M. Ubiquitous transcriptional pausing is independent of RNA polymerase backtracking. *Cell* **115**, 437–447 (2003).
13. Forde, N. R., Izhaky, D., Woodcock, G. R., Wuite, G. J. & Bustamante, C. Using mechanical force to probe the mechanism of pausing and arrest during continuous elongation by *Escherichia coli* RNA polymerase. *Proc. Natl Acad. Sci. USA* **99**, 11682–11687 (2002).
14. Schafer, D. A., Gelles, J., Sheetz, M. P. & Landick, R. Transcription by single molecules of RNA polymerase observed by light microscopy. *Nature* **352**, 444–448 (1991).
15. Wang, M. D. *et al.* Force and velocity measured for single molecules of RNA polymerase. *Science* **282**, 902–907 (1998).
16. Yin, H., Landick, R. & Gelles, J. Tethered particle motion method for studying transcript elongation by a single RNA polymerase molecule. *Biophys. J.* **67**, 2468–2478 (1994).
17. Adelman, K. *et al.* Single molecule analysis of RNA polymerase elongation reveals uniform kinetic behavior. *Proc. Natl Acad. Sci. USA* **99**, 13538–13543 (2002).
18. Veigel, C. *et al.* The motor protein myosin-I produces its working stroke in two steps. *Nature* **398**, 530–533 (1999).
19. deCastro, M. J., Fondecave, R. M., Clarke, L. A., Schmidt, C. F. & Stewart, R. J. Working strokes by single molecules of the kinesin-related microtubule motor ncd. *Nature Cell Biol.* **2**, 724–729 (2000).
20. Nishiyama, M., Muto, E., Inoue, Y., Yanagida, T. & Higuchi, H. Substeps within the 8-nm step of the ATPase cycle of single kinesin molecules. *Nature Cell Biol.* **3**, 425–428 (2001).
21. Aboul-ela, F., Koh, D., Tinoco, I. Jr & Martin, F. H. Base-base mismatches. Thermodynamics of double helix formation for dCA3XA3G+dCT3YT3G (X, Y=A,C,G,T). *Nucleic Acids Res.* **13**, 4811–4824 (1985).
22. Martin, F. H., Castro, M. M., Aboul-ela, F. & Tinoco, I. Jr Base pairing involving deoxyinosine: implications for probe design. *Nucleic Acids Res.* **13**, 8927–8938 (1985).
23. Borukhov, S., Sagitov, V. & Goldfarb, A. Transcript cleavage factors from *E. coli*. *Cell* **72**, 459–466 (1993).
24. Yildiz, A. *et al.* Myosin V walks hand-over-hand: single fluorophore imaging with 1.5-nm localization. *Science* **300**, 2061–2065 (2003).
25. Feng, G. H., Lee, D. N., Wang, D., Chan, C. L. & Landick, R. GreA-induced transcript cleavage in transcription complexes containing *Escherichia coli* RNA polymerase is controlled by multiple factors, including nascent transcript location and structure. *J. Biol. Chem.* **269**, 22282–22294 (1994).
26. Lang, M. J., Asbury, C. L., Shaevitz, J. W. & Block, S. M. An automated two-dimensional optical force clamp for single molecule studies. *Biophys. J.* **83**, 491–501 (2002).
27. Svoboda, K. & Block, S. M. Biological applications of optical forces. *Annu. Rev. Biophys. Biomol. Struct.* **23**, 247–285 (1994).
28. Wang, M. D., Yin, H., Landick, R., Gelles, J. & Block, S. M. Stretching DNA with optical tweezers. *Biophys. J.* **72**, 1335–1346 (1997).

Supplementary Information accompanies the paper on **www.nature.com/nature**.

Acknowledgements We acknowledge intellectual contributions from J. Gelles, and we thank the entire Block Laboratory, especially K. Neuman, for support and discussions. We also thank A. Meyer for reading of the original manuscript. This work was supported by grants from the NIGMS.

Competing interests statement The authors declare that they have no competing financial interests.

Correspondence and requests for materials should be addressed to S.M.B. (sblock@stanford.edu).

Position-dependent linkages of fibronectin–integrin–cytoskeleton

Takayuki Nishizaka*†‡, Qing Shi*, and Michael P. Sheetz*§

*Department of Cell Biology, Duke University Medical Center, Durham, NC 27710; and †Department of Physics, School of Science and Engineering, Waseda University, 3-4-1 Okubo, Shinjuku-ku, Tokyo169-8555, Japan

Edited by Thomas D. Pollard, The Salk Institute for Biological Studies, La Jolla, CA, and approved November 1, 1999 (received for review February 24, 1999)

Position-dependent cycling of integrin interactions with both the cytoskeleton and extracellular matrix (ECM) is essential for cell spreading, migration, and wound healing. Whether there are regional changes in integrin concentration, ligand affinity or cytoskeleton crosslinking of liganded integrins has been unclear. Here, we directly demonstrate a position-dependent binding and release cycle of fibronectin–integrin–cytoskeleton interactions with preferential binding at the front of motile 3T3 fibroblasts and release at the endoplasm–ectoplasm boundary. Polystyrene beads coated with low concentrations of an integrin-binding fragment of fibronectin (fibronectin type III domains 7–10) were 3–4 times more likely to bind to integrins when placed within 0.5 microns vs. 0.5–3 microns from the leading edge. Integrins were not concentrated at the leading edge, nor did anti-integrin antibody-coated beads bind preferentially at the leading edge. However, diffusing liganded integrins attached to the cytoskeleton preferentially at the leading edge. Cytochalasin inhibited edge binding, which suggested that cytoskeleton binding to the integrins could alter the avidity for ligand beads. Further, at the ectoplasm–endoplasm boundary, the velocity of bead movement decreased, diffusive motion increased, and approximately one-third of the beads were released into the medium. We suggest that cytoskeleton linkage of liganded integrins stabilizes integrin-ECM bonds at the front whereas release of cytoskeleton-integrin links weakens integrin-ECM bonds at the back of lamellipodia.

Motility of adherent cells is critical for many biological functions such as wound healing, lymphocyte function, and development. Cells apply force to the matrix through the integrin-cytoskeleton linkage (1, 2), and the integrins appear to participate in multiple cycles of binding to and release from ECM in cell migration on specific substrata (3). Models of cell migration have described several steps in the process, including extension to new regions, attachment to extracellular matrix (ECM) (adhesion), force generation, and release from ECM to allow further movement and recycling (4). An additional requirement for directed migration is that there must be position dependence of attachment to and release from substrate-bound ECM.

ECM contacts are initiated at the newly extended edge of the cell because that is the first region to contact exposed ECM molecules. Experiments in fish keratocytes have shown that the leading edge is also the domain where crosslinked glycoproteins are rapidly attached to the cytoskeleton (5). An implication of those studies is that the edge region is specialized for the binding of cross-linked membrane glycoproteins to the cytoskeleton. One explanation for the attachment in that region is that membrane molecules involved in attachment are concentrated there. Indeed, integrins are concentrated at the leading edge of fish keratocytes (C. G. Galbraith and M.P.S., unpublished results). Alternatively, cytoskeletal attachment proteins are concentrated at the edge such as those which catalyze actin filament assembly in the keratocyte (6). Such edge specificity has not been reported for crosslinked integrin–cytoskeleton interactions related to cell migration.

Once the matrix has moved rearward, the integrin must release from the matrix molecule. Three possible mechanisms for integrin release from ECM-binding sites are mechanical release caused by high forces at the back of the cell (2, 7, 8), calpain-mediated enzymatic cleavage of integrin/cytoskeleton linkages (9, 10), or biochemical release. Phosphatase-dependent release has been suggested for vitronectin receptors, in the case of calcineurin-dependent $\alpha v\beta 3$ integrin release (11), but not for the major fibronectin-binding site, the $\alpha 5\beta 1$ integrin. A fourth mechanism could involve the loss of cytoskeletal attachment to the ECM-crosslinked integrins. Unbound integrins could then diffuse away from the ECM molecules before rebinding. Such an avidity mechanism would not necessarily involve alterations in integrin-ECM affinity but would rely primarily on position-dependent cytoskeleton assembly and disassembly (explained further in Fig. 5).

Here, we examined the position dependence of fibronectin bead binding and release by using optical tweezers manipulation on 3T3 cells. We found a strikingly narrow region of preferential binding at the leading edge that correlated with the region of increased attachment to the cytoskeleton. Release of fibronectin beads occurred preferentially at the ectoplasm–endoplasm boundary after apparent detachment from the cytoskeleton. Both observations are consistent with the hypothesis that the cytoskeleton binding of liganded integrins increases avidity of binding for multimeric ECM complexes.

Materials and Methods

Bead Preparation. The fragment of fibronectin (integrin-binding domain of fibronectin type III, FNIII7–10; ref. 12) was used to avoid an aggregation of ECM and beads. Carboxylated polystyrene (1 μm) beads (Polysciences, Warrington, PA) were pre-coated with the mixture of 97% native BSA (Sigma) and 3% biotinamidocaproyl-labeled BSA (biotinylated BSA; Sigma) using a standard procedure (13). Then 2 mg/ml Neutra Lite avidin (Molecular Probes) was added and incubated overnight on ice. The beads were washed and mixed with the complex of biotinylated BSA and biotinylated FNIII7–10 to avoid the local aggregation of ligand on single beads. We used biotinylated BSA:FNIII7–10 (1:1 in weight ratio) for all experiments, except 1:0 for the control experiment in Fig. 2a. Antibody for chicken $\beta 1$ integrin chain, biotinylated ES66 mAb (refs. 14 and 15), was used for the antibody-coated bead.

Cell Preparation and Experiments. NIH 3T3 mouse fibroblasts transfected with chick $\beta 1$ integrin (7, 14) were seeded on

This paper was submitted directly (Track II) to the PNAS office.

Abbreviations: ECM, extracellular matrix; FNIII7–10, fibronectin type III domains 7–10; RGD, Arg-Gly-Asp peptide.

†Present address: Core Research for Evolutional Science and Technology (CREST) Genetic Programming Team 13, 907 Nogawa, Miyamae-ku, Kawasaki, Kanagawa 216-0001, Japan.

§To whom correspondence and reprint requests should be addressed. E-mail m.sheetz@cellbio.duke.edu.

The publication costs of this article were defrayed in part by page charge payment. This article must therefore be hereby marked "advertisement" in accordance with 18 U.S.C. §1734 solely to indicate this fact.

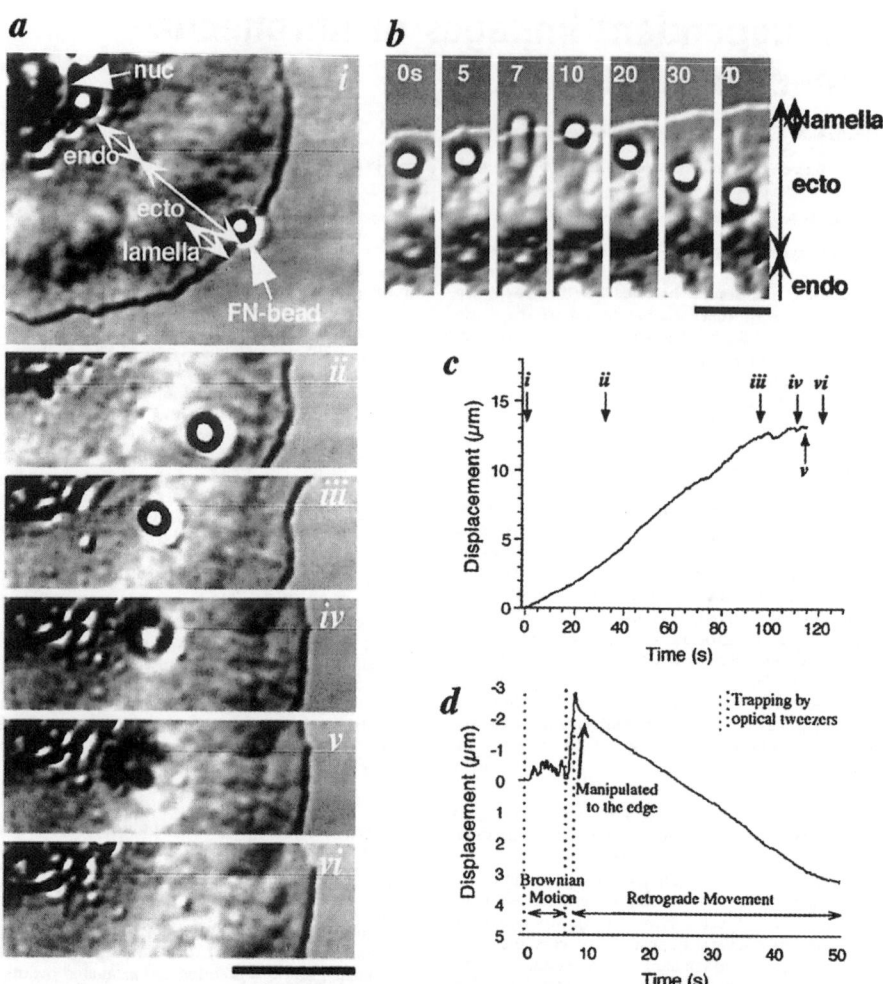

Fig. 1. Behavior of a bead coated with a low concentration of FNIII7–10. (a) Video-enhanced differential interference contrast micrographs show binding and release of a FNIII7–10 bead. The bead was carried to the edge of the lamellipodium with an optical tweezers and released from the trap (i). It immediately moved retrogradely across the lamellipodium (ii), and the velocity decreased at the boundary between ectoplasm and endoplasm (iii). The bead then detached from the cell surface (iv) and diffused out of focus into the medium (v and vi). (b) Video micrographs show attachment of diffusing bead and rearward movement after being brought to edge. (c) The trajectory of the bead along the direction of the displacement in a. i–vi Correspond to the micrographs in a. (d) The bead trajectory of b. [Scale bars, 10 μm (a) and 5 μm (b)]. Nuc, nuclear; endo, endoplasm; ecto, ectoplasm; and lamella, lamellipodium.

laminin-coated coverslips after trypsin treatment (1, 14, 16). DMEM with 10% serum (GIBCO/BRL) and Hanks' balanced salt solution with 1 mM Ca^{2+} were used as media for standard experiments and the Mn^{2+} experiment (Fig. 2d), respectively. After >4 h at 37°C, we chose trigonal cells with an actively spreading lamellipodium (7, 14) and checked the edge specificity with 7–47 beads on each cell in Fig. 2a. For the Arg-Gly-Asp peptide (RGD) experiment (Fig. 2c), Gly-Arg-Gly-Asp-Ser (GRGDS) peptide (Sigma) was used. For the fluorescence analysis cells were fixed with 4% paraformaldehyde for 10 min, washed three times, and then exposed to the medium containing 20 μg/ml biotinylated FNIII7–10 and 5 mg/ml BSA for 1 hr. Cells

were fixed again, and finally, a rhodamine antibody for biotin (Sigma) was infused. Control experiments were performed in the presence of 1 mg/ml GRGDS peptide in FNIII7–10 treatment procedure.

Optical Tweezers System and Data Analysis. We used the same optical tweezers system as described before (1) except that it was equipped with a movable mirror with dc servomotors (opt-mike-e; Sigma Koki, Hidaka, Japan) for moving the trap center at a constant rate (Fig. 1 b and d). The bead was trapped with 3.7–4.9 mW of laser power after the objective, which produced a maximal force of ≈5 pN for displacements in the xy-direction (1). High laser powers caused nonspecific cytoskeletal attachment of BSA beads. The probability of the cytoskeletal attachment increased with the laser power, and finally all of the beads showed retrograde movement after being pushed into the cell membrane for 3 s with 60 mW of laser power (data are not shown). The position of the bead on the cell surface was determined with nanometer precision (17, 18). The velocity of forward and of retrograde movements were determined from the displacement of the nucleus for 30–140 s and from the linear portion of the track of the bead on the ectoplasm (e.g., 0–90 s in Fig. 1c), respectively. In Fig. 4a, the one exceptional bead that detached at 20 μm was neglected in calculating the correlation coefficient.

Bead Attachments. Beads were trapped in the medium, manipulated to a cell surface, held there for 2.5–3.5 s, and then released. The bead behavior after release from the tweezers was classified in four ways: unattached, indirectly attached, membrane attached, and cytoskeletally attached (retrograde movement) (14). To confirm whether the bead was "indirectly attached" or "membrane attached," we retrapped all beads showing two-dimensional Brownian motion on a cell surface and tried to move them off the cell edge. Beads attached directly to a membrane protein always stopped at the edge of a cell, whereas beads attached to the membrane indirectly through a fibrous component went beyond the cell edge and on release diffused toward the cell center. When we used 1-μm diameter beads coated with BSA as the control experiment for ligand-coated beads, $\approx$50% of the beads were indirectly attached. Thus, we judged that the beads that were indirectly attached were nonspecifically bound and grouped the beads that were indirectly attached and "unattached" in the same class, "no membrane attachment" in the histograms of Figs. 2 and 3.

Definition of Terms. Ectoplasm is defined as the thin, rigid structure in the front part of the cell, which is rich in actin and depleted of membranous vesicles. Endoplasm is defined as the region just behind ectoplasm where movements of many membranous vesicles were observed. The boundary between them is often structurally visible as in Fig. 1 a and b and functionally clear because the release of fibronectin beads preferentially occurred there (see *Results*, *Discussion*, and Fig. 4).

Results

Position Dependence of Fibronectin-Coated Bead Binding. Fig. 1a shows a typical example of cytoskeletal attachment of a bead coated with a low concentration of FNIII7–10. The bead immediately showed steady directed movement on the dorsal surface of the cell (*i–iii*), i. e., retrograde movement, after turning off the trap (1, 16). Although the majority of the beads that bound to the membrane moved rearward (82%), some beads showed two-dimensional Brownian motion on the cell surface instead of retrograde movement (0–6 s in Fig. 1b). When we recaptured diffusing beads, brought them to the edge (7 s in Fig. 1b), and released them, the Brownian motion did not resume but retrograde movement immediately started (10–40 s in Fig. 1b).

Fig. 1 c and d show the trajectories of the bead in Fig. 1 a and b, respectively. Beads move steadily rearward in *i–iii* in Fig. 1c, and for 8–50 s in Fig. 1d beads show retrograde flow from the edge of the cell toward the nucleus ($1.2 \pm 0.3 \times 10^{-1}$ μm/s, $n = 12$). The beads moved with the same speed as small structures on the cell surface, suggesting retrograde movement of the bead relates to the flow of cytoskeletal structures in ectoplasm. Note that these rearward movements are measured relative to the substratum. In other words, data do not include the relatively slow forward movement of each cell body ($2.2 \pm 0.7 \times 10^{-2}$ μm/s, $n = 10$).

In Fig. 1 b and d, the Brownian motion changed to retrograde movement when the bead was moved to the edge $\approx$7 s. This

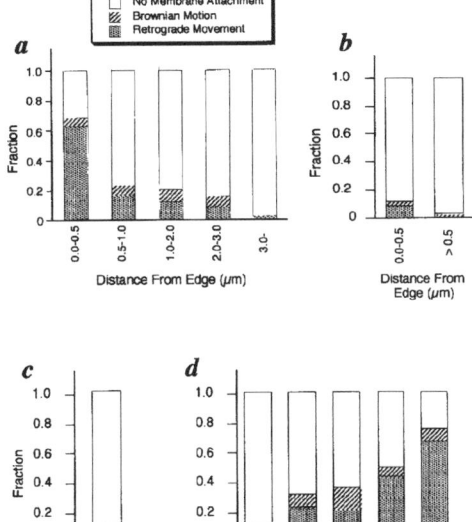

Fig. 2. Histograms are shown of the probability of bead binding. (a) The position dependence of beads coated with a low concentration of FNIII7–10 (total $n = 272$). At this concentration, we estimate that 300 FNIII7–10 molecules are bound per bead (4–10/bead-membrane contact area). (b) The control experiment for a. Beads were coated with 100% BSA (total $n = 75$) instead of FNIII7–10. (c) The same experiment as a in the presence of 2 mg/ml RGD peptide, which inhibits fibronectin binding to the β1 integrin (19, 20). This experiment was done within 0.5 μm from the edge (total $n = 94$). This result coincides with ref. 1. (d) Edge binding of FN beads is measured as a function of [Mn^{2+}], which is reported to change the affinity of integrin for fibronectin (20–22). The binding probability is dependent on the concentration of Mn^{2+} (total $n = 252$).

result supported the hypothesis that attachment to the cytoskeleton and retrograde movement occurred preferentially at the edge of the cell (5). To quantify binding, we placed beads at various positions on the lamellipodium for 3 s using optical tweezers (see Fig. 2a). Within 0.5 μm of the edge, 63% of the beads bound and showed retrograde movement. In contrast, the cytoskeletally attached population decreased below 20% when beads were held >0.5 μm from the edge using the same trapping force and holding time. We also found that a fraction of cells that did not show the edge specificity (20%, 5 cells in 25). In these cells, the binding probability was above 60% at any position, independent of the distance from the edge (data from these cells are not included in Fig. 2). The shape of these cells was morphologically the same as the other cells that showed edge specificity, so that we could not determine the origin of this difference so far. It may relate to the motility state because cells that were not motile also did not have higher binding at the edge or rearward movement. In the majority of motile cell lamellipodia, fibronectin beads bound preferentially at the very leading edge.

To confirm that the cytoskeletal attachment of the beads was caused by the interaction between FNIII7–10 and integrin, we performed four different control experiments. First, we used

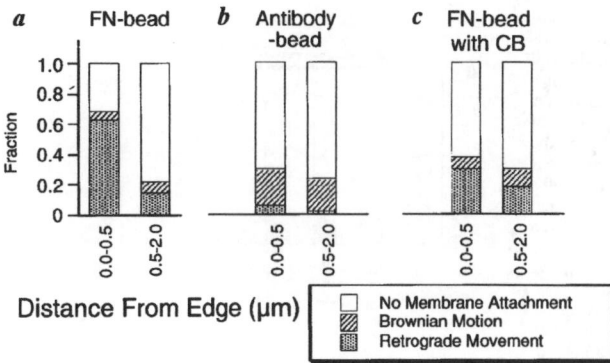

Fig. 3. (*a, b,* and *c*) Histograms of the bead binding behavior at the front part of migrating fibroblasts. (*a*) Fibronectin-coated bead. This histogram is based on the same set of data for Fig. 2*a*. (*b*) ES66 antibody-coated beads. (*c*) Fibronectin-coated beads in the presence of 100 nM cytochalasin B. For more details, see *Materials and Methods*.

beads coated with BSA and with no FNIII7–10 (Fig. 2*b*). The probability of the membrane attachment was greatly decreased at all positions on the lamellipodium. Second, we added RGD to compete for fibronectin binding to integrin (19, 20) and found that only 10% of the beads bound (Fig. 2*c*). These results indicate that the majority of the beads that we studied were attached in a ligand-specific manner. Third, we changed the concentration of divalent cations in the medium because the fibronectin–integrin interaction is divalent cation dependent (20–22). In low concentrations of manganese ion, spreading will occur on laminin but fibronectin–integrin binding is blocked (20). We measured the binding of fibronectin beads to cells spread on laminin, as a function of manganese ion. As shown in Fig. 2*d*, the membrane attachment of the fibronectin beads depends on the concentration of manganese (apparent K_d: 210 μM). A similar manganese ion concentration dependence has been reported for $\alpha 5\beta 1$ integrin–fibronectin binding (21). To further test whether or not $\alpha 5\beta 1$ integrin was the major fibronectin-binding site, we added an inhibiting anti-$\alpha 5$ antibody before measuring fibronectin bead binding. Anti-$\alpha 5$ antibody caused a $70 \pm 5\%$ decrease in binding both at the edge and back from the edge, whereas an anti-αv antibody only caused $\approx 15 \pm 4\%$ inhibition of fibronectin bead binding. Thus, we suggest that binding of the FNIII7–10 beads is primarily to $\alpha 5\beta 1$ integrin.

Distribution of $\beta 1$ Integrin on the Cell Surface and the Contribution of Cytoskeleton to the Edge-Specific Binding. To determine whether the edge specificity simply reflects the concentration of the fibronectin receptor on the cell surface, we checked whether or not there was any evidence of a concentration of $\beta 1$ integrin at the edge, using fluorescent antibody staining or antibody-coated beads. As was observed previously (23), we found that fluorescent anti-$\beta 1$ antibody staining of the membranes did not show any edge concentration of $\beta 1$. Surface carbohydrates could interfere with fibronectin bead as well as antibody bead binding but may not interfere with antibody binding alone. Thus, the distribution of antibody-coated bead binding controls for functional differences in bead binding as well as the distribution of integrin. Beads were coated with a nonperturbing antibody for chicken $\beta 1$ integrin chain, ES66. The binding probability of the beads was checked on NIH 3T3 mouse fibroblasts transfected with chick $\beta 1$ (7, 14). The lamellipodium was separated into two areas, within 0.5 μm and from 0.5 μm to 2.0 μm back from the leading edge, and compared with the results of fibronectin-coated bead binding to the equivalent cells (Fig. 3 *a* and *b*). There was no detectable position-dependent binding of ES66 beads (the difference in Fig. 3*b* is almost the same as the difference of the control experiment in Fig. 2*b*), and the probability of

retrograde movement was decreased to the level of the control experiment (compare Fig. 2*b* with Fig. 3*b*). Only 3% of control antibody beads bound under these conditions and ES66 beads bound to the parent 3T3 cells (without chicken $\beta 1$ integrin) at the control level (3%). In another experiment, a higher level of bead binding (43% at edge and 39% back from the edge) was obtained but again no positional difference was observed. These results indicate that the edge specific binding does not originate from the asymmetric distribution of $\alpha 5\beta 1$ integrin.

The contribution of cytoskeleton to the edge specific binding was studied by altering actin dynamics with cytochalasin B (5, 24). More than 0.5 μM of cytochalasin caused migrating fibroblasts to immediately become round and detach. This result suggested that the rigid cytoskeletal structure, which depended on receptor binding to ECM, is needed to keep the cells attached to the substrate. With moderate cytochalasin treatment (100 nM), the cells were still attached to the substrate. We then examined the edge specificity of FNIII7–10 bead binding (Fig. 3*c*) with and without cytochalasin. The edge-specific binding was decreased, even though retrograde movement did not disappear completely. No remarkable differences were observed in actin filament distribution as visualized by rhodamine-phalloidin with or without 100 nM cytochalasin (data are not shown). Thus, the edge specificity of binding is altered by altering actin dynamics.

Position Dependence of Release of Fibronectin-Coated Beads from the Cell Surface. After beads left the leading edge, we followed them until they stopped moving and in many cases ($n = 22$) released from the cell surface. Cessation of movement and release were position and not time dependent. The velocity of retrograde movement decreased and diffusive movement increased when the beads reached the endoplasm–ectoplasm boundary (at 95 s in Fig. 1*c* and >50% of beads started diffusing). Many of the diffusing beads (32% of total, $n = 69$) released completely (Fig. 1*a, iv–vi*). Nonspecific binding between the beads and the membrane may hinder the complete detachment of some of the beads, or multiple weak interactions could keep the beads attached for the observation period (1–2 min). BSA beads that were nonspecifically attached did not detach (data are not shown), indicating that the release of the bead involved specific integrin binding.

Two models for bead release were quantitatively tested, either time-dependent or position-dependent release. In Fig. 1*a*, the bead detached from the cell at the boundary between ectoplasm and endoplasm (see *Materials and Methods* for definition), which indicated that release was position specific. Alternatively, release could occur in a stochastic manner. If release was stochastic, the frequency of the release events should follow a single exponen-

CELL BIOLOGY

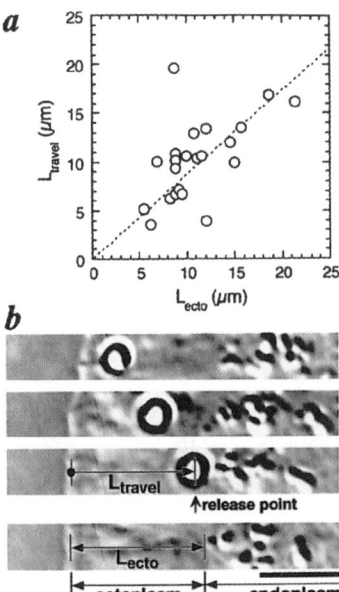

Fig. 4. (a) The position dependence of the unbinding of the beads from cell surfaces. The position where the unbinding of the bead occurred, indicated by the length of bead travel before unbinding (L_{travel}), is related to the position of the ectoplasm–endoplasm boundary, indicated by the length of ectoplasm (L_{ecto}). The dashed line is a linear approximation and the slope is 0.87. (b) Video micrographs show the definition of L_{travel} and L_{ecto} in a. The bead ceased moving rearward in the third micrograph and diffused away in the fourth micrograph. In micrographs (1–3), intervals are 21 s. (Scale bar, 5 μm).

tial decay (18, 25). However, the time dependence of release did not follow a single exponential decay (correlation coefficient $r = 0.31$, data are not shown). On the other hand, the position where the release occurred was related to the boundary between ectoplasm and endoplasm ($r = 0.80$, Fig. 4a). From these data, we conclude that release of the beads was not stochastic but was dependent on the position on the cell surface.

Discussion

We have found a position-dependent cycle of fibronectin–integrin–cytoskeleton binding, movement, and release. Fibronectin-coated beads bound preferentially within 0.5 microns

of the leading edge to motile but not cytochalasin B-treated fibroblasts. When the beads were coated with a low density of fibronectin to reduce but not eliminate cooperative binding by multiple ligands, release of bead binding was found at the ectoplasm–endoplasm boundary. We hypothesize that this cycle of fibronectin binding, movement, and release is involved in cell motility.

The preferential edge binding is not explained by a simple binding mechanism. There is no preferential binding of anti-integrin antibody beads to the leading edge (Fig. 3b), and fluorescent anti-integrin antibody distribution shows no edge concentration. We have previously observed the preferential trapping of $\beta 1$ integrin at the leading edge of these fibroblasts but that did not result in a concentration there, possibly because of the other factors causing depletion of integrins from the leading edge (14). In the fish keratocyte system, we have found an edge concentration of the $\beta 1$ integrin that can explain some of the edge concentration of binding in that system. In the 3T3 fibroblast system, however, neither a concentration of active integrins at the leading edge nor a masking of bead binding back from the leading edge could explain edge specificity.

Differences in the avidity of binding caused by cytoskeletal linkage could explain the preference for binding at the leading edge. When ligands bind to receptors with low affinity, the presentation of multiple ligands on a bead greatly increases the avidity. A further increase in cooperativity may occur when the receptors are crosslinked by the cytoskeleton after ligand binding. Fibronectin binds weakly to integrins and an off-rate of 3 s has been measured with fibronectin-coated gold particles and soluble FNIII7–10 (D. Choquet and D. P. Felsenfeld, unpublished observations). Thus, multiple fibronectin-integrin bonds are needed for beads to remain bound for several minutes as observed and cytoskeletal crosslinking of the liganded integrins would greatly decrease the rate of release of beads from the membrane (Fig. 5).

Preferential binding to the cytoskeleton does occur at the edge of fish keratocytes (5) in the case of concanavalin A-coated beads. In the case of fibronectin-coated beads on 3T3 cells, attachment to the cytoskeleton occurs preferentially at the edge. The integrins appear unattached to cytoskeleton before ligand binding (1, 14). At intermediate anti-$\beta 1$ antibody concentrations on the bead surface, the beads often diffused but would attach to the cytoskeleton if taken to the leading edge (D. Choquet and D. P. Felsenfeld, unpublished results). Fibronectin–integrin binding often produces cytoskeletal attachment of integrin (1, 16), but a larger fraction of beads diffused away when bound away from the edge. In these experiments conditions were chosen to give a low level of binding on the lamellipodium ($<30\%$). This level of binding enabled the definition of the most optimal binding location. An increased rate of cytoskeletal attachment of

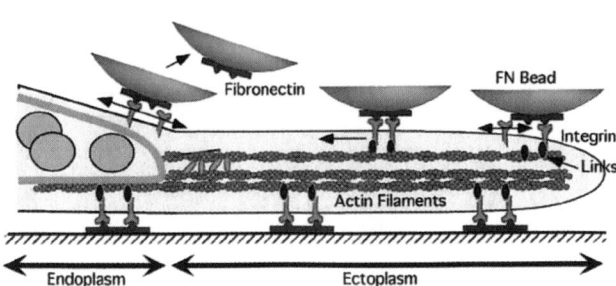

Fig. 5. Schematic illustration showing that cytoskeleton binding of integrin-fibronectin complexes at the leading edge could stabilize them. A fibronectin-coated bead (FN-bead) attaches to the dorsal surface of the leading edge and recruits a second integrin, which recruits a second link to the cytoskeleton. Because the two bound integrins are attached to a rigid cytoskeleton, they cannot diffuse away if one should release from the fibronectin. Therefore, bead binding is stabilized until the actin cytoskeleton depolymerizes, which is often seen at the endoplasm–ectoplasm boundary. Upon release from the cytoskeleton, the integrins could diffuse away leading to FN-bead release. On the ventral surface, additional components could stabilize the integrin-cytoskeleton complex perhaps in a force-dependent process (1). Such a position-dependent binding and release cycle could aid cell migration.

liganded integrins at the leading edge is consistent with other findings and could explain increased bead binding.

If crosslinking of liganded receptors caused increased edge binding, then the loss of cytoskeletal attachment at the ectoplasm–endoplasm boundary could explain the release of the beads at that point. All beads that released from the membrane started to diffuse laterally before they released. In other words, detachment of integrins from the cytoskeleton preceded the fibronectin bead release. There are three major alternative explanations for detachment of integrins from the cytoskeleton: i) a reversible biochemical process decreases the affinity of individual integrins for the cytoskeleton, ii) proteolysis of the integrins leads to detachment, or iii) depolymerization of actin filaments occurs in that region. The second possibility is less likely than others because of previous observations of integrin recycling (3, 11, 26, 27). Depolymerization of actin does occur at the ectoplasm–endoplasm boundary, and modification of the cytoplasmic tails of the integrins does modify their interactions with the forward transport system and with the cytoskeleton (14, 16). Both mechanisms, i and iii, could be involved in the dissociation of the liganded integrins from the cytoskeleton at the ectoplasm–endoplasm boundary.

Release of the beads from the membrane is potentially analogous to the release of a portion of the integrins from fibronectin surfaces. Both biochemical release and mechanical release have been postulated previously. In the case of the $\alpha v \beta 3$ integrin, there is evidence for cytoplasmic regulation of the integrin-matrix binding through the action of a calcium-dependent phosphatase (11). Mechanical dissociation of matrix linkages has been reported at the rear of cells (7, 8). If the binding of ligand stabilizes the conformation of an integrin, which has high affinity to the cytoskeleton, it is also possible that binding of the cytoskeleton stabilizes a high affinity conformation. Thus, both the bead release from the cytoskeleton and the release from the membrane can be coupled.

Here, we observed that the binding and some release of fibronectin molecules occurred in the front and the back of the endoplasmic region of the cell, respectively. In considering the bottom of the cell, the ventral integrins will perhaps show different position dependence because of reinforced integrin-fibronectin complexes (1) or the persistence of actin filaments at the lower surface. Some large aggregates persist from the front to the rear of the cell (7, 8). However, there is a major decrease in the area of contact of fibroblasts with the substratum from the front to the rear of the cell (trigonal shape) (7, 14). Furthermore,

the force on ventral contacts switches direction near the nucleus or back from the endoplasm–ectoplasm boundary (2). These observations are consistent with the hypothesis that many integrin-ECM contacts are released near the nucleus of fibroblasts.

Models of cell migration have included a force generation step at the front of the cell that can be rate-limiting under certain conditions (see reviews in refs. 4, 28, and 29). Linkage between the force generating cytoskeleton and the fibronectin matrix could occur through cytoskeleton attachment of liganded integrins preferentially at the leading edge (see diagram in Fig. 5a). The rearward movement of the cytoskeleton would draw the cell forward. The release process could begin at the ectoplasm–endoplasm boundary as these findings show but may persist to the tail of the cell. Preferential attachment at the leading edge would enable the cell to pull for the maximal distance on the new ECM ligands that it encounters. In fan-shaped fish keratocytes, there is a very large lamellipodium that generates pulling forces on substrate contacts (C. G. Galbraith and M.P.S., unpublished observations; ref. 30), and the preferential attachment of crosslinked glycoproteins to the cytoskeleton at the edge has been observed as well (5). The finding of some release at the ectoplasm–endoplasm boundary indicates that a portion of integrins could recycle in front of the nucleus. Unliganded integrins can diffuse (16, 31) and single integrins and small aggregates move to the leading edge in the absence of bound ligand (14, 32). Such a surface transport mechanism could recycle integrins to the front of the lamellipodium. The alternative mechanism of endocytosis and transport back to the front of the cell by intracellular vesicles (3, 11, 26, 27) is unlikely because there are few intracellular microtubules in the lamellipodium. Thus, we hypothesize that cytoskeleton binding of liganded integrins stabilizes integrin–ligand complexes at the leading edge. Integrin release from the cytoskeleton would then shorten the lifetime of integrin–ligand bonds by allowing integrins to diffuse away from ECM ligands upon unbinding.

We thank H. P. Erickson for providing the recombinant FNIII7–10 and advice. The software and system for the bead trajectory analysis were developed by K. Kinosita, Jr. We thank D. Choquet, D. P. Felsenfeld, C. G. Galbraith, and D. Raucher for helpful comments on this work in progress. We also thank A. F. Horwitz for the cell line and K. M. Yamada for the hybridoma, ES66. T.N. was a recipient of a Japan Society for the Promotion of Science Fellowship for Japanese Junior Scientists during this work. This work was partly supported by a National Institutes of Health grant to M.P.S.

1. Choquet, D., Felsenfeld, D. P. & Sheetz, M. P. (1997) *Cell* **88**, 39–48.
2. Galbraith, C. G. & Sheetz, M. P. (1997) *Proc. Natl. Acad. Sci. USA* **94**, 9114–9118.
3. Bretscher, M. S. (1992) *EMBO J.* **11**, 405–410.
4. Sheetz, M. P., Felsenfeld, D. P. & Galbraith, C. G. (1998) *Trends Cell Biol.* **8**, 51–54.
5. Kucik, D. F., Kuo, S. C., Elson, E. L. & Sheetz, M. P. (1991) *J. Cell Biol.* **114**, 1029–1036.
6. Symons, M. H. & Mitchison, T. J. (1991) *J. Cell Biol.* **114**, 503–513.
7. Regen, C. M. & Horwitz, A. F. (1992) *J. Cell Biol.* **119**, 1347–1359.
8. Palecek, S. P., Schmidt, C. E., Lauffenburger, D. A. & Horwitz, A. F. (1996) *J. Cell Sci.* **109**, 941–952.
9. Palecek, S. P., Huttenlocher, A., Horwitz, A. F. & Lauffenburger, D. A. (1998) *J. Cell Sci.* **111**, 929–940.
10. Huttenlocher, A., Palecek, S. P., Lu, Q., Zhang, W., Mellgren, R. L., Lauffenburger, D. A., Ginsberg, M. H. & Horwitz, A. F. (1997) *J. Biol. Chem.* **272**, 32719–32722.
11. Lawson, M. A. & Maxfield, F. R. (1995) *Nature (London)* **377**, 75–79.
12. Leahy, D. J., Aukhil, I. & Erickson, H. P. (1996) *Cell* **84**, 155–164.
13. Suzuki, N., Miyata, H., Ishiwata, S. & Kinosita, K., Jr. (1996) *Biophys. J.* **70**, 401–408.
14. Schmidt, C. E., Horwitz, A. F., Lauffenburger, D. A. & Sheetz, M. P. (1993) *J. Cell Biol.* **123**, 977–991.
15. Reszka, A. A., Hayashi, Y. & Horwitz, A. F. (1992) *J. Cell Biol.* **117**, 1321–1330.
16. Felsenfeld, D. P., Choquet, D. & Sheetz, M. P. (1996) *Nature (London)* **383**, 438–440.
17. Gelles, J., Schnapp, B. J. & Sheetz, M. P. (1988) *Nature (London)* **331**, 450–453.
18. Nishizaka, T., Miyata, H., Yoshikawa, H., Ishiwata, S. & Kinosita, K., Jr. (1995) *Nature (London)* **377**, 251–254.
19. Gehlsen, K. R., Argraves, W. S., Pierschbacher, M. D. & Ruoslahti, E. (1988) *J. Cell Biol.* **106**, 925–930.
20. Elices, M. J., Urry, L. A. & Hemler, M. E. (1991) *J. Cell Biol.* **112**, 169–181.
21. Gailit, J. & Ruoslahti, E. (1988) *J. Biol. Chem.* **263**, 12927–12932.
22. Lange, T. S., Bielinsky, A. K., Kirchberg, K., Bank, I., Herrmann, K., Krieg, T. & Scharffetter-Kochanek, K. (1994) *Exp. Cell Res.* **214**, 381–388.
23. Felsenfeld, D. P., Schwartzberg, P. L., Venegas, A., Tse, R. & Sheetz, M. P. (1999) *Nat. Cell Biol.* **1**, 200–206.
24. Forscher, P. & Smith, S. J. (1988) *J. Cell Biol.* **107**, 1505–1516.
25. Erickson, H. P. (1994) *Proc. Natl. Acad. Sci. USA* **91**, 10114–10118.
26. Bretscher, M. S. (1989) *EMBO J.* **8**, 1341–1348.
27. Szekan, M. M. & Juliano, R. L. (1990) *J. Cell. Physiol.* **142**, 574–580.
28. Lauffenburger, D. A. & Horwitz, A. F. (1996) *Cell* **84**, 359–369.
29. Mitchison, T. J. (1996) *Cell* **84**, 371–379.
30. Oliver, T., Dembo, M. & Jacobson, K. (1999) *J. Cell Biol.* **145**, 589–604.
31. Yauch, R. L., Felsenfeld, D. P., Kraeft, S. K., Chen, L. B., Sheetz, M. P. & Hemler, M. E. (1997) *J. Exp. Med.* **186**, 1347–1355.
32. Kucik, D. F., Elson, E. L. & Sheetz, M. P. (1989) *Nature (London)* **340**, 315–317.

Internal fields of a spherical particle illuminated by a tightly focused laser beam: Focal point positioning effects at resonance

J. P. Barton, D. R. Alexander, and S. A. Schaub

Center for Electro-Optics, College of Engineering, University of Nebraska-Lincoln, Lincoln, Nebraska 68588-0525

(Received 16 September 1988; accepted for publication 16 December 1988)

The spherical particle/arbitrary beam interaction theory developed in an earlier paper is used to investigate the dependence of structural resonance behavior on focal point positioning for a spherical particle illuminated by a tightly focused (beam diameter less than sphere diameter), linearly polarized, Gaussian-profiled laser beam. Calculations of absorption efficiency and distributions of normalized source function (electric field magnitude) are presented as a function of focal point positioning for a particle with a complex relative index of refraction of $\bar{n} = 1.33 + 5.0 \times 10^{-6} i$ and a size parameter of $\alpha \approx 29.5$ at both nonresonance and resonance conditions. The results of the calculations indicate that structural resonances are not excited during the on-center focal point positioning of such a tightly focused beam but structural resonances can be excited by proper on-edge focal point positioning. Electric wave resonances were found to be excited by moving the focal point from on-center towards the edge of the sphere *parallel* to the direction of the incident beam electric field polarization. Magnetic wave resonances were found to be excited by moving the focal point from on-center towards the edge of the sphere *perpendicular* to the direction of the incident beam electric field polarization.

I. INTRODUCTION

In an earlier paper,[1] theoretical expressions for the internal and external electromagnetic fields of a homogeneous spherical particle illuminated by an arbitrarily defined beam were derived. In particular, calculations of absorption efficiency and distributions of normalized source function (electric field magnitude) were presented for a spherical particle illuminated by a tightly focused (beam diameter less than sphere diameter), linearly polarized, Gaussian-profiled, monochromatic beam. This situation corresponds to the important experimental arrangement of focusing a high-quality TEM_{00} mode laser beam upon a small liquid droplet in air (or upon any suspended spherical particle). In this paper, additional calculations are presented in which the effect of beam focal point positioning on the absorption efficiency and internal normalized source function distribution is investigated for spherical particles at structural resonance conditions.

A spherical particle has a series of associated structural resonances which can be excited by incident radiation. For transparent (weakly absorbing) particles, resonance excitation is exhibited by an accompanying significant increase in the absorption and scattering of the incident radiation by the particle. Resonance effects have been experimentally observed through measurements of elastic scattering,[2,3] inelastic scattering,[4,5,6] and radiation pressure.[3,7]

Plane-wave Lorenz–Mie theory has been used to understand and predict resonance behavior, and excellent agreement with experimental measurements has been obtained.[2-7] However, plane-wave theory is not applicable for experiments utilizing tightly focused laser beams where the local beam diameter is less than the spherical particle diameter. Indeed, Ashkin and Dziedzic[3,7] have reported that resonances are not excited when such a tightly focused beam is aligned through the center of the spherical particle. But

Baer[8] and Zhang, Leach, and Chang[9] have observed *increased* resonance inelastic scattering when a tightly focused beam is aligned along the *edge* of a spherical particle.

In this paper, the spherical particle/arbitrary beam interaction theory developed in our earlier paper[1] is used to investigate the dependence of structural resonance behavior on focal point positioning when the beam diameter is tightly focused to less than the particle diameter. Corresponding plane-wave calculations are also presented for comparison. These calculations provide insight into understanding the experimental observations stated in the previous paragraph and also provide new observations which could be tested as part of future experiments.

II. INCIDENT PLANE-WAVE CALCULATIONS

Previous investigators have used plane-wave Lorenz–Mie theory to analyze the structural resonance behavior of spherical particles and a brief review of this work, including calculations corresponding to the incident beam conditions of the next section, is useful here. In Lorenz–Mie theory a plane wave of transverse polarized electromagnetic radiation propagating within an infinite homogeneous dielectric medium is incident upon a homogeneous spherical particle of complex relative index of refraction $\bar{n}$. The spatial coordinates are referenced to the center of the spherical particle, as shown in Fig. 1. The incident plane wave is assumed to propagate in the $+z$-axis direction with electric field polarization in the x-axis direction.

Following the development approach of the Lorenz–Mie theory, the electromagnetic field components internal to and scattered by the spherical particle can each be mathematically expressed in the form of an infinite series of partial waves. The partial waves, in turn, are of two types: electric waves, which have no radial component of magnetic field, and magnetic waves, which have no radial component

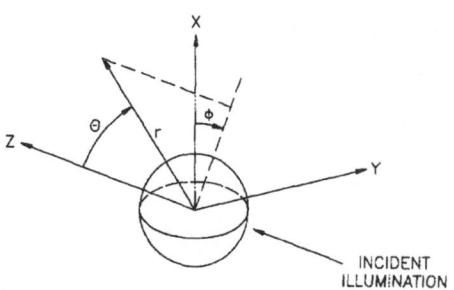

FIG. 1. Geometry for a plane-wave or focused Gaussian beam incident upon a spherical particle.

of electric field. Series expressions for the scattered electromagnetic fields, including a complete derivation, can be found in Born and Wolf,[10] and series expressions for the internal electromagnetic field components can be found in Kerker and Cooke.[11]

The coefficient for the l th electric wave term within the scattered field series is given by

$$a_l = \frac{\psi_l'(\bar{n}\alpha)\psi_l(\alpha) - \bar{n}\psi_l'(\alpha)\psi_l(\bar{n}\alpha)}{\xi_l^{(1)}(\alpha)\psi_l'(\bar{n}\alpha) - \bar{n}\xi_l^{(1)'}(\alpha)\psi_l(\bar{n}\alpha)}, \quad (1)$$

and the coefficient for the l th magnetic wave term within the scattered field series is given by

$$b_l = \frac{\bar{n}\psi_l(\alpha)\psi_l'(\bar{n}\alpha) - \psi_l'(\alpha)\psi_l(\bar{n}\alpha)}{\bar{n}\xi_l^{(1)}(\alpha)\psi_l'(\bar{n}\alpha) - \xi_l^{(1)'}(\alpha)\psi_l(\bar{n}\alpha)}, \quad (2)$$

where the prime refers to the derivative with respect to the argument, $\xi_l^{(1)} = \psi_l - i\chi_l$, ψ_l, and χ_l are the Riccati–Bessel functions, and α is the size parameter, $2\pi a/\lambda$, where a is the sphere radius and λ is the wavelength of the incident radiation.

Conditions for l th mode electric wave structural resonance can be determined by setting the denominator of a_l equal to zero and, likewise, conditions for l th mode magnetic wave structural resonance can be determined by setting the denominator of b_l equal to zero. (The c_l, d_l coefficients for the internal field have the same respective denominators as the a_l, b_l coefficients of the scattered field, thus the conditions for internal field structural resonance are identical to the conditions for the scattered field structural resonance.) For a specified value of $\bar{n}$, the solution of these equations requires a complex size parameter, the imaginary part of which describes the strength and width of the resonance (in electrical circuit analogy, the Q of the resonance) and the real part of which corresponds (approximately) to the actual sphere radius/incident wavelength combination at which the resonance occurs. There is an infinite sequence of roots for each equation, with the root having the smallest real part designated as the first-order resonance, the root having the next to smallest real part designated as the second-order resonance, and so on.[12,13,14]

To illustrate resonance behavior, calculations were performed for a spherical particle in air with $\bar{n} = 1.33 + 5.0 \times 10^{-6}i$ and α values of the order of 29.5. These parameters correspond to an approximately 10-μm-

diam water droplet with 1.06-μm (Nd:YAG laser) wavelength illumination. The absorption efficiency of the particle Q_{abs}, defined as the ratio of the total power absorbed by the particle to the power incident upon the projected area of the particle, can be expressed in terms of the incident plane-wave coefficients

$$Q_{abs} = \frac{2}{\alpha^2} \sum_{l=1}^{\infty} (2l+1)[\text{Re}(a_l + b_l)$$
$$- (|a_l|^2 + |b_l|^2)], \quad (3)$$

and is given in Fig. 2 for a range of size parameters from 28 to 32. Each peak in absorption efficiency seen in Fig. 2 can be directly related to the first-order resonance indicated. Higher-order resonances of lower modes are also present within this range of size parameters, but are weak and not observable. As discussed by Chylek, Kiehl, and Ko,[12,13] for larger size parameters the first-order resonances become narrow while the higher-order (second, third, etc.) resonances heighten and become dominant.

Plots of the normalized source function within the spherical particle provides additional information with regard to the formation of structural resonances. The normalized source function is defined as

$$\tilde{S} = |\mathbf{E}|^2 / E_0^2, \quad (4)$$

where $\mathbf{E}$ is the local electric field vector and E_0 is the electric field amplitude of the incident plane wave. The local volumetric heating rate is directly proportional to the normalized source function. In this paper, plots of the internal normalized source function distribution are presented either in the transverse (x-z) plane or the equatorial (y-z) plane. External sphere source function values are suppressed to zero in order to emphasize the internal sphere distribution and spatial coordinates are normalized by the sphere radius. (A tilde above a spatial quantity indicates that it has been normalized relative to the particle radius a.)

To compare internal sphere source function distributions at nonresonance, electric wave resonance, and magnetic wave resonance conditions, calculations were performed for the nonresonance case of $\alpha = 29.5$, the adjacent 34th mode, first-order electric wave resonance case of $\alpha = 29.753$, and the adjacent 34th mode, first-order magnetic wave resonance case of $\alpha = 29.365$ (refer to Fig. 2).

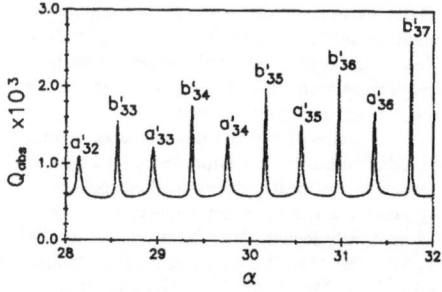

FIG. 2. Absorption efficiency vs size parameter for a plane wave incident upon a spherical particle. $\bar{n} = 1.33 + 5.0 \times 10^{-6}i$.

2901   J. Appl. Phys., Vol. 65, No. 8, 15 April 1989

Barton, Alexander, and Schaub   2901

Calculations for plane-wave illumination are shown in Figs. 3–6. Figure 3 is the normalized source function in the transverse $(x\text{-}z)$ plane for the nonresonance case of $\alpha = 29.5$. An equatorial $(y\text{-}z)$ plane plot for the same non-resonance case is given in Fig. 4. For the 34th mode, first-order electric wave resonance, it was found that, though there was insignificant difference in the equatorial plane normalized source function distribution in comparison with the nonresonance case (Fig. 4), there was a significant increase in the normalized source function in the transverse plane, as shown in Fig. 5. For the 34th mode, first-order magnetic wave resonance, little difference with the nonresonance case was observed in the transverse plane, but a significant increase in normalized source function was present in the equatorial plane, as shown in Fig. 6. In general, it was found that electric wave resonances were excited predominately in the transverse plane (parallel to the direction of incident electric field polarization) and magnetic wave resonances were excited predominately in the equatorial plane (perpendicular to the direction of incident electric field polarization) for plane-wave illumination.

The ring formation of the normalized source function just inside the surface of the sphere for the plane-wave illumination resonance cases, seen in Figs. 5 and 6, was reported by Chylek, Pendleton, and Pinnick,[15] who also observed that, in general, an lth mode resonance will have $2l$ peaks around the circumference, which is the case here. A physical explanation for the occurrence of structural resonances, which is consistent with the ring formation of the normalized source function just beneath the surface of the particle, is the "surface wave" description first proposed by van de Hulst[16] and later discussed by other investigators.[17,18] van de Hulst proposed that structural resonances occur when surface waves constructively interfere about the circumference of the particle.

III. INCIDENT BEAM CALCULATIONS

The theory of our earlier paper[1] was used to investigate structural resonance behavior for a spherical particle illuminated by a tightly focused laser beam as a function of focal point positioning. A focused, linearly polarized, Gaussian-profiled, monochromatic beam (a "focused Gaussian

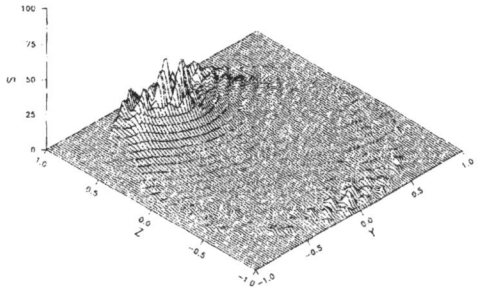

FIG. 4. Normalized source function in the equatorial $(y\text{-}z)$ plane for a transverse (x direction) polarized plane wave propagating in the $+z$-axis direction incident upon a sphere with $\bar{n} = 1.33 + 5.0 \times 10^{-6}i$ and $\alpha = 29.5$ (nonresonance).

beam") propagating in the $+z$-axis direction and linearly polarized in the x-axis direction (refer to Fig. 1) is assumed. The beam was expressed mathematically using the first-order corrected paraxial beam description of Davis.[1,19] The coordinates (x_0, y_0, z_0) are used to indicate the position of the center of the spherical particle relative to the focal point of the beam.

That an arbitrary beam can excite the same structural resonances as a plane wave is physically reasonable, and is also confirmed mathematically. As discussed in our earlier paper,[1] the derivation approach for the arbitrary beam incident upon a homogeneous spherical particle theory, similar to the Lorenz–Mie approach, is to express the internal and scattered electromagnetic fields of the spherical particle in the form of an infinite series of electric and magnetic waves. However, unlike the plane-wave Lorenz–Mie solution, the most general series form is chosen. The result is a nested series with each radial mode l having $(2l + 1)$ associated angular modes. (For each radial mode l the angular mode index m may have integer values from $-l$ to $+l$.) The coefficients of the scattered field electric wave and magnetic wave lth radial mode, mth angular mode terms, taken from our earlier paper,[1] are, respectively,

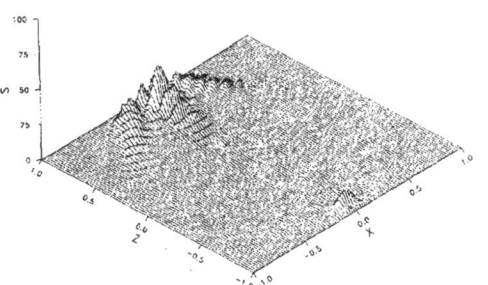

FIG. 3. Normalized source function in the transverse $(x\text{-}z)$ plane for a transverse (x direction) polarized plane wave propagating in the $+z$-axis direction incident upon a sphere with $\bar{n} = 1.33 + 5.0 \times 10^{-6}i$ and $\alpha = 29.5$ (nonresonance).

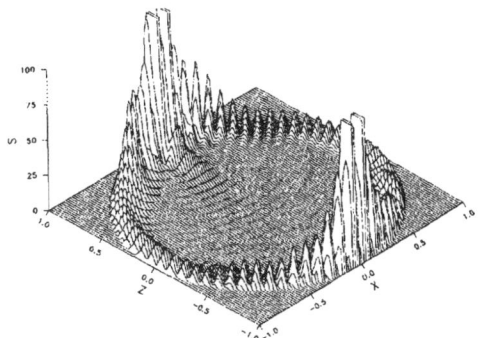

FIG. 5. Normalized source function in the transverse $(x\text{-}z)$ plane for a transverse (x direction) polarized plane wave propagating in the $+z$-axis direction incident upon a sphere with $\bar{n} = 1.33 + 5.0 \times 10^{-6}i$ and $\alpha = 29.753$ (34th mode, first-order electric wave resonance). (Note: Normalized source function truncated for values exceeding 100.)

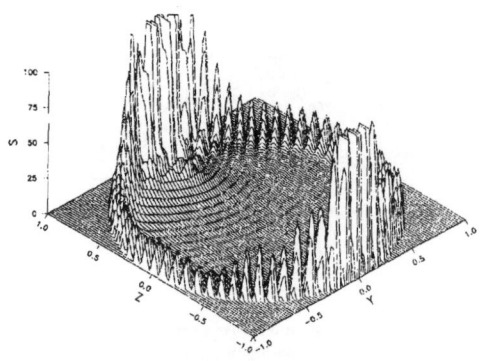

FIG. 6. Normalized source function in the equatorial (y-z) plane for a transverse (x direction) polarized plane wave propagating in the $+z$-axis direction incident upon a sphere with $\bar{n} = 1.33 + 5.0 \times 10^{-6}i$ and $\alpha = 29.365$ (34th mode, first-order magnetic wave resonance). (Note: Normalized source function truncated for values exceeding 100.)

$$a_{lm} = -a_l A_{lm} \qquad (5)$$

and

$$b_{lm} = -b_l B_{lm} , \qquad (6)$$

where A_{lm} and B_{lm} are incident field coefficients, the value of which are dependent upon the character of the incident beam, and a_l and b_l are the Lorenz–Mie plane-wave coefficients given in Eqs. (1) and (2). Since the scattered field electric wave and magnetic wave mode coefficients of the arbitrary beam theory are proportional to the respective plane-wave coefficients, structural resonance occurs at the same size parameters for arbitrary beam illumination as for plane-wave illumination. However, unlike the plane-wave resonance case, a particular beam radial resonance mode has associated with it $(2l + 1)$ angular modes. The relative excitation of each of these angular modes depends upon the character of the incident beam.

Beam calculations were performed for the same $\bar{n} = 1.33 + 5.0 \times 10^{-6}i$, $\alpha \approx 29.5$ conditions that were used for the plane-wave calculations of the previous section. A constant beam waist radius of $w_0 = 1.887\lambda$ was chosen. (For $\lambda = 1.06$-μm illumination, this would correspond to a 4-μm waist diameter beam incident upon an ≈ 10-μm-diam particle.)

To explore the general effect of beam focal point positioning on structural resonance excitation, the spherical particle absorption efficiency was calculated for focal point positioning along the x, y, and z axes of the spherical particle. A modified absorption efficiency $\tilde{Q}_{abs}$, defined as the ratio of the power absorbed by the spherical particle to the *total* power of the incident beam, was utilized. An expression for the power absorbed by the particle W_{abs} was given in our earlier paper.[1] Dividing this expression by the total power of the incident beam

$$(c/16)E_0^2 w_0^2 ,$$

where, here, E_0 is the electric field amplitude at the beam focal point, provides a series expression for the modified absorption efficiency

$$\tilde{Q}_{abs} = -\frac{2\alpha^2}{\pi\bar{w}_0^2 a^4 E_0^2} \sum_{l=1}^{\infty} \sum_{m=-l}^{m=l} l(l+1)\left[|a_{lm}|^2 + |b_{lm}|^2 \right.$$
$$\left. + \text{Re}(A_{lm}a_{lm}^* + B_{lm}b_{lm}^*) \right] . \qquad (7)$$

Figure 7 shows the modified absorption efficiency for the $\alpha = 29.5$ (nonresonance), $\alpha = 29.753$ (34th mode, first-order electric wave resonance), and $\alpha = 29.365$ (34th mode, first-order magnetic wave resonance) cases for incident beam focal point positioning on the particle's x axis ($0.0 \leqslant -\tilde{x}_0 \leqslant 2.0$, $\tilde{y}_0 = 0.0$, $\tilde{z}_0 = 0.0$). As can be observed, the modified absorption efficiency is approximately equal for all three cases for focal point positioning near the center of the particle, but as the focal point of the beam is moved along the x axis towards the surface of the particle, the modified absorption efficiency increases for the electric wave resonance case until it peaks at a focal point position just outside the particle surface before decreasing to zero as the beam is moved on away from the particle. In contrast, the modified absorption efficiency for the nonresonance and magnetic resonance cases decrease monotonically as the beam focal point position is moved away from the particle center.

Figure 8 presents the modified absorption efficiency for the same three cases presented in Fig. 7, but for beam focal point positioning along the y axis ($\tilde{x}_0 = 0.0$, $0.0 \leqslant -\tilde{y}_0 \leqslant 2.0$, $\tilde{z}_0 = 0.0$). In this arrangement, it is the magnetic wave resonance that is excited as the beam is moved from the center towards the particle surface, while the nonresonance and electric wave resonance cases decrease monotonically.

The lack of resonance excitation for on-center focal point positioning is consistent with the observation of Ashkin and Dziedzic[3,7] that resonances in radiation pressure did not appear for beams tightly focused through the center of a droplet. The excitation of resonance for beam focal point positioning near the surface of the spherical particle may explain the observations of Baer[8] and Zhang and co-workers[9] which indicated that increased resonance inelastic scattering can be obtained using such edge illumination. Apparently, to excite a structural resonance with a tightly focused beam it is necessary to position the focal point of the beam near the resonance ring formation that occurs for plane-wave illumination. According to van de Hulst's sur-

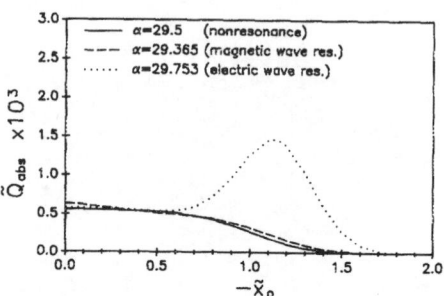

FIG. 7. Modified absorption efficiency vs focal point positioning along the x axis. Transverse (x direction) polarized focused Gaussian beam incident upon a sphere. $\bar{n} = 1.33 + 5.0 \times 10^{-6}i$, $\bar{w}_0\alpha = 11.86$, and $\alpha = 29.365$, 29.5, and 29.753.

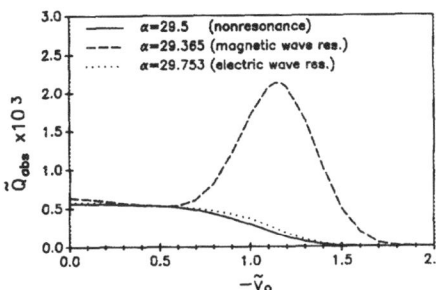

FIG. 8. Modified absorption efficiency vs focal point positioning along the y axis. Transverse (x direction) polarized focused Gaussian beam incident upon a sphere. $\bar{n} = 1.33 + 5.0 \times 10^{-6}i$, $\bar{w}_0\alpha = 11.86$, and $\alpha = 29.365, 29.5$, and 29.753.

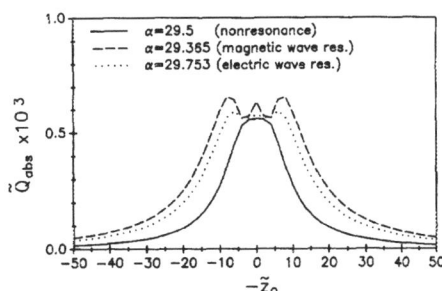

FIG. 9. Modified absorption efficiency vs focal point positioning along the z axis. Transverse (x direction) polarized focused Gaussian beam incident upon a sphere. $\bar{n} = 1.33 + 5.0 \times 10^{-6}i$, $\bar{w}_0\alpha = 11.86$, and $\alpha = 29.365, 29.5$, and 29.753.

face wave description, edge illumination is required to create the surface waves that constructively interfere to create the resonance.

In general, the calculations presented in Figs. 7 and 8 also suggest that only electric wave resonances are excited for edge illumination in the direction of incident electric field polarization and only magnetic wave resonances are excited for edge illumination perpendicular to the direction of incident electric field polarization. This observation, which was verified for the other resonance modes shown in Fig. 2, could be tested experimentally.

Another interesting observation is that the peak in structural resonance excitation occurs for beam focal point positioning outside the surface of the spherical particle. A possible physical explanation for this may lie within the "localization principle" discussed by van de Hulst[16] in which he indicates that, for radii much greater than a wavelength, the l th partial wave may be associated with a ray passing through a radial position

$$\bar{r} = (l + 1/2)/\alpha \qquad (8)$$

from the particle center. [Grehan and co-workers[20,21] have applied the localization principle to provide a mathematically simple, but approximate, determination of the interaction coefficients for their generalized Lorenz–Mie theory (GLMT) analysis.] Since for an l th mode structural resonance, it is the l th partial wave that is in resonance, a beam that is predominately characterized by that particular partial wave will excite the structural resonance. For the calculations considered here with $l = 34$, Eq. (7) gives $\bar{r} = 1.160$ for the $\alpha = 29.753$, electric wave resonance case, and $\bar{r} = 1.175$ for the $\alpha = 29.365$, magnetic wave resonance case. The peaks in the curves of Figs. 7 and 8 correspond approximately with these respective values. It appears then that van de Hulst's localization principle as expressed in Eq. (6) can be used to predict the approximate beam focal point positioning for maximum structural resonance excitation for an arbitrary l th mode resonance. This observation was also verified for several other complex relative index of refraction and resonance mode number combinations.

The modified absorption efficiency for focal point positioning along the z axis (the incident beam propagation axis) for the nonresonance, electric wave resonance, and magnetic

wave resonance cases is given in Fig. 9. For the nonresonance case, the modified absorption efficiency decreases monotonically as the beam focal point is moved away from the spherical particle due to the spreading of the beam incident upon the particle. However, for the resonance cases, moving the focal point of the tightly focused beam a relatively short distance away from the center of the sphere can actually increase the modified absorption efficiency when the beam spreads so as to provide illumination along the edge of the sphere that excites the structural resonances.

In general it was found, as was the case for plane-wave illumination, that electric wave resonances were predominately excited in the transverse plane (parallel to the direction of incident polarization) and magnetic wave resonances were predominately excited in the equatorial plane (perpendicular to the direction of incident polarization). The normalized source function distribution in the transverse (x-z) plane for the focused Gaussian beam positioned at the top edge of the particle ($\bar{x}_0 = -1.0$, $\bar{y}_0 = 0.0$, $\bar{z}_0 = 0.0$) is shown in Fig. 10 for the nonresonance case and in Fig. 11 for the electric wave resonance case. The normalized source function distribution in the equatorial (y-z) plane for the focused Gaussian beam positioned at the right edge of the particle ($\bar{x}_0 = 0.0$, $\bar{y}_0 = -1.0$, $\bar{z}_0 = 0.0$) is shown in Fig. 12 for the nonresonance case and in Fig. 13 for the magnetic

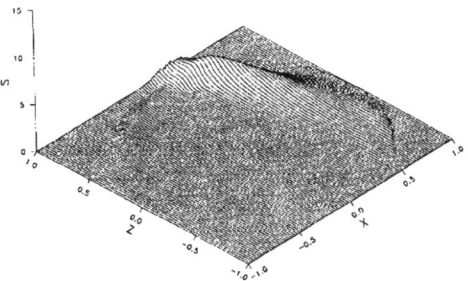

FIG. 10. Normalized source function in the transverse (x-z) plane for a transverse (x direction) polarized focused Gaussian beam incident upon a sphere at nonresonance. $\bar{n} = 1.33 + 5.0 \times 10^{-6}i$, $\bar{w}_0 = 0.402$, $\bar{x}_0 = -1.0$, $\bar{y}_0 = 0.0$, $\bar{z}_0 = 0.0$, and $\alpha = 29.5$.

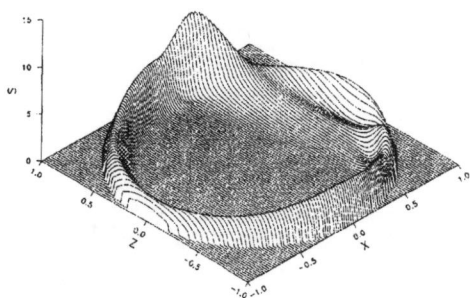

FIG. 11. Normalized source function in the transverse (x-z) plane for a transverse (x direction) polarized focused Gaussian beam incident upon a sphere at electric wave resonance. $\bar{n} = 1.33 + 5.0 \times 10^{-6}i$, $\bar{w}_0 = 0.399$, $\bar{x}_0 = -1.0$, $\bar{y}_0 = 0.0$, $\bar{z}_0 = 0.0$, and $\alpha = 29.753$.

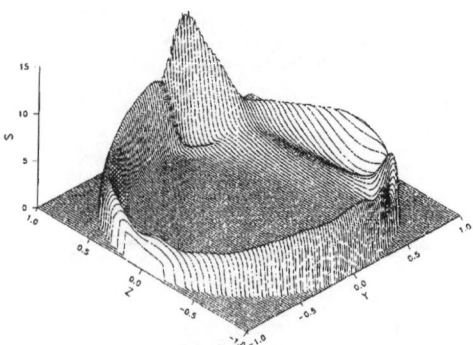

FIG. 13. Normalized source function in the equatorial (y-z) plane for a transverse (x direction) polarized focused Gaussian beam incident upon a sphere at magnetic wave resonance. $\bar{n} = 1.33 + 5.0 \times 10^{-6}i$, $\bar{w}_0 = 0.404$, $\bar{x}_0 = 0.0$, $\bar{y}_0 = -1.0$, $\bar{z}_0 = 0.0$, and $\alpha = 29.365$.

wave resonance case. For the resonance cases of Figs. 11 and 13 the beam excites a structural resonance producing a ring of increased normalized source function just inside the surface of the particle somewhat similar to the ring formation observed for plane-wave resonance illumination (Figs. 5 and 6). However, the ring formation for resonance beam illumination does not exhibit the 2*l* circumferential peaks that were observed for resonance plane-wave illumination but consists more of a solid ring. The difference in ring structure between the resonance plane wave and resonance beam cases is a consequence of the fact that for the plane-wave resonance only a single angular mode is "excited" while for beam resonance a series of angular modes ($-l \leqslant m \leqslant +l$) can be excited.

IV. SUMMARY

The theory of our earlier paper[1] has been used to investigate the structural resonance behavior of a spherical particle illuminated by a tightly focused Gaussian beam. For the $\bar{n} = 1.33 + 5.0 \times 10^{-6}i$, $\alpha \approx 29.5$, and $w_0 = 1.887\lambda$ conditions considered here, it appears that:

(1) Electric wave resonances are excited predominately in the transverse plane (parallel to the direction of incident electric field polarization) and magnetic wave resonances

are excited predominately in the equatorial plane (perpendicular to the direction of incident electric field polarization).

(2) Structural resonances are not excited for on-center focal point positioning, but can be excited by on-edge focal point positioning.

(3) Electric wave resonances are excited for on-edge illumination in the direction of incident electric field polarization and magnetic wave resonances are excited by on-edge illumination perpendicular to the direction of incident electric field polarization.

(4) van de Hulst's "localization principle" can be used to predict the approximate radial location for focal point positioning that will provide maximum excitation of an arbitrary *l* th mode structural resonance.

(5) The 2*l* circumferential peaks in the ring formation of normalized source function that occurs for plane-wave resonance are not exhibited for on-edge beam resonance excitation.

Observation (2) has already been verified by the experiments of Ashkin and Dziedzic.[3,7] Observations (1), (3), (4), and (5) could be tested as part of future experiments.

ACKNOWLEDGMENT

This work was supported by the Army Chemical Research and Development Center under Contract Nos. DAAA15-85-K-0001 and DAAL03-87-K-0138.

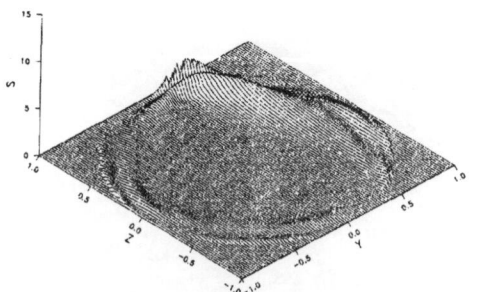

FIG. 12. Normalized source function in the equatorial (y-z) plane for a transverse (x direction) polarized focused Gaussian beam incident upon a sphere at nonresonance. $\bar{n} = 1.33 + 5.0 \times 10^{-6}i$, $\bar{w}_0 = 0.402$, $\bar{x}_0 = 0.0$, $\bar{y}_0 = -1.0$, $\bar{z}_0 = 0.0$, and $\alpha = 29.5$.

[1]J. P. Barton, D. R. Alexander, and S. A. Schaub, J. Appl. Phys. **64**, 1632 (1988).
[2]P. Affolter and B. Eliasson, IEEE Trans. Microwave Theory Tech. **MTT-21**, 573 (1978).
[3]A. Ashkin and J. M. Dziedzic, Appl. Opt. **20**, 1803 (1981).
[4]R. E. Benner, P. W. Barber, J. F. Owen, and R. K. Chang, Phys. Rev. Lett. **44**, 475 (1980).
[5]R. Thurn and W. Kiefer, Appl. Opt. **24**, 1515 (1985).
[6]J. B. Snow, S-X. Qian, and R. K. Chang, Opt. Lett. **10**, 37 (1985).
[7]A. Ashkin and J. M. Dziedzic, Phys. Rev. Lett. **38**, 1351 (1977).
[8]T. Baer, Opt. Lett. **12**, 392 (1987).
[9]J-Z. Zhang, D. H. Leach, and R. K. Chang, Opt. Lett. **13**, 270 (1988).
[10]M. Born and E. Wolf, *Principles of Optics* (Pergamon, Oxford, 1970).
[11]M. Kerker and D. D. Cooke, Appl. Opt. **12**, 1378 (1973).

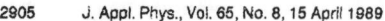

[12]P. Chylek, J. T. Kiehl, and M. K. W. Ko, Appl. Opt. **17**, 3019 (1978).

[13]P. Chylek, J. T. Kiehl, and M. K. W. Ko, Phys. Rev. A **18**, 2229 (1978).

[14]P. R. Conwell, P. W. Barber, and C. K. Rushforth, J. Opt. Soc. Am. A **1**, 62 (1984).

[15]P. Chylek, J. D. Pendleton, and R. G. Pinnick, Appl. Opt. **24**, 3940 (1985).

[16]H. C. van de Hulst, *Light Scattering by Small Particles* (Dover, New York, 1981).

[17]A. B. Pluchino, Appl. Opt. **20**, 2987 (1981).

[18]J. D. Murphy, P. J. Moser, A. Nagel, and H. Uberall, IEEE Trans. Antennas Propag. **AP-28**, 924 (1980).

[19]L. W. Davis, Phys. Rev. A **19**, 1177 (1979).

[20]G. Grehan, B. Maheu, and G. Gouesbet, Appl. Opt. **25**, 3539 (1986).

[21]B. Maheu, G. Grehan, and G. Gouesbet, Appl. Opt. **26**, 23 (1987).

EUROPHYSICS LETTERS

1 June 2000

Europhys. Lett., **50** (5), pp. 702–708 (2000)

Theory of optical tweezers

P. A. Maia Neto and H. M. Nussenzveig

Instituto de Física, Universidade Federal do Rio de Janeiro - Caixa Postal 68528
21945-970 Rio de Janeiro, Rio de Janeiro, Brazil

(received 2 November 1999; accepted in final form 22 March 2000)

PACS. 87.80.Cc – Optical trapping.
PACS. 42.50.Vk – Mechanical effects of light on atoms, molecules, electrons, and ions.
PACS. 42.25.Fx – Diffraction and scattering.

Abstract. – We derive an exact partial-wave (Mie) expansion of the axial force exerted on a transparent sphere by a laser beam focused through a high numerical aperture objective. The results hold throughout the range of interest for practical applications, as well as in the Rayleigh and geometrical optics limits. They allow, in principle, an absolute calibration of optical tweezers. Starting from the Mie result, we derive a closed analytic representation for the size-averaged short-wavelength limit that takes into account the Abbe sine condition. Numerical plots show large deviations from geometrical optics near the focal region and around the edge of the sphere, and oscillatory behavior of the force as a function of the size parameter. The oscillations are explained in terms of a simple interferometer picture derived from the Mie expansion. The few existing experimental data look more consistent with the present model than with previous ones.

Optical tweezers are single-beam laser traps for neutral particles that have a wide range of applications in physics and biology [1]. Dielectric microspheres are trapped and employed as handles in most of the quantitative applications. The gradient trapping force is applied by bringing the laser beam to a diffraction-limited focal spot through a large numerical aperture microscope objective. Measurements employ indirect force calibrations against Stokes' law, under complicated boundary conditions for the flow of the fluid past the sphere. Clearly, a reliable derivation of the optical trapping force from first principles is important for the numerous quantitative applications of optical tweezers, since it allows to perform absolute calibrations and to unravel the effects of optical aberrations and boundary perturbations.

Typical size parameters $\beta = ka$ ($a =$ microsphere radius, $k =$ laser wave number) range in order of magnitude from values < 1 to a few times 10^1. A theory of the trapping force based on geometrical optics (GO) [2] should not work in this range. Other proposals (cf. [1]), based on Mie theory, have employed unrealistic near-paraxial models for the transverse laser beam structure near the focus, incompatible with its large angular aperture, leading to discrepancies with experiments by factors of 3-5 or higher [3].

In this letter, we derive, from first principles, exact results for the trapping force. We take for the incident beam *before* the objective, propagating along the positive z-axis, the usual Gaussian (TEM)$_{00}$ transverse laser mode profile, with beam waist w_0 at the input aperture.

where $kw_0 \gg 1$. We employ the Richards and Wolf [4] representation for the corresponding strongly focused beam *beyond* the objective, with a large opening angle θ_0 (no paraxial assumption), taking due account of the Abbe sine condition. This Debye-type integral representation is well substantiated and is thus taken to provide a realistic model.

The microsphere, with real refractive index n_2 (we neglect absorption), is immersed in a homogeneous medium with refractive index n_1. We consider here the most symmetric situation, in which the sphere center is aligned with the laser beam axis, so that we evaluate the *axial* trapping force. With origin at the sphere center, we denote by $r = -q\hat{z}$ the focal point position. The fraction A of total beam power that enters the lens aperture is

$$A = 1 - \exp[-2\gamma^2 \sin^2 \theta_0]. \tag{1}$$

where γ is the ratio of the objective focal length to the beam waist w_0.

By axial symmetry, the trapping force in this situation is independent of input beam polarization: we take circular polarization. The electric field of the strongly focused beam (we omit the time factor $\exp[-i\omega t]$) is given by [4]

$$E_0(r) = E_0 \int_0^{2\pi} d\phi \int_0^{\theta_0} d\theta \sin\theta \sqrt{\cos\theta} \exp\left[-\gamma^2 \sin^2\theta\right] \exp\left[i\mathbf{k} \cdot (\mathbf{r} + q\hat{z})\right] \hat{\epsilon}(\theta, \phi), \tag{2}$$

where $k = |\mathbf{k}(\theta, \phi)| = n_1 \omega/c$, $\hat{\epsilon}(\theta, \phi) = \hat{x}' + i\hat{y}'$, and the unit vectors $\hat{x}'$ and $\hat{y}'$ are obtained from $\hat{x}$ and $\hat{y}$, respectively, by rotation with Euler angles $\alpha = \phi, \beta = \theta, \gamma = -\phi$. The factor $\sqrt{\cos\theta}$ arises from the Abbe sine condition.

For each plane wave $\exp[i\mathbf{k} \cdot \mathbf{r}]$ in the superposition (2), the corresponding scattered field is given by the well-known Mie partial-wave series [5], in terms of the Mie coefficients a_l, b_l, that are functions of the size parameter β and the relative refractive index $n = n_2/n_1$. By substitution into (2), we obtain the total scattered field $E_s(r)$.

The trapping force is found by replacing the total field $E = E_0 + E_s$ (likewise for B) into the Maxwell stress tensor and integrating over the surface of the sphere. The resulting axial force F is proportional to the focused laser beam power P,

$$F = (n_1/c)PQ, \tag{3}$$

where Q is the (dimensionless) axial trapping efficiency [1].

We denote by Q_e the contribution from terms in $E_0 E_s$ and $B_0 B_s$ (that also give rise to the extinction efficiency) and by Q_s the remaining terms, so that $Q = Q_e + Q_s$. We find

$$Q_e = \frac{4\gamma^2}{A} \text{Re} \sum_{l=1}^{\infty} (2l+1)(a_l + b_l) G_l G_l'^*, \tag{4}$$

$$Q_s = \frac{-8\gamma^2}{A} \text{Re}\left[\sum_{l=1}^{\infty} \frac{l(l+2)}{l+1}(a_l a_{l+1}^* + b_l b_{l+1}^*) G_l G_{l+1}^* + \frac{(2l+1)}{l(l+1)} a_l b_l^* G_l G_l^*\right], \tag{5}$$

where G_l and G_l' are multipole coefficients for the focused beam,

$$G_l = \int_0^{\theta_0} d\theta \sin\theta \sqrt{\cos\theta} \exp[-\gamma^2 \sin^2\theta] \exp[i\delta \cos\theta] d_{1,1}^l(\theta), \tag{6}$$

$$G_l' = -i\partial G_l/\partial\delta, \tag{7}$$

with $\delta = kq$. In (6), $d_{1,1}^l(\theta)$ are the matrix elements of finite rotations [6], that can be expressed as

$$d_{1,1}^l(\theta) = [p_l(\cos\theta) + t_l(\cos\theta)] / (2l + 1),\qquad(8)$$

in terms of the Mie angular functions [7] p_l and t_l. The results (4) and (5), apart from converging beam effects, have the same structure as the radiation pressure efficiency [7], with which they are closely related, as will be seen below.

In the Rayleigh limit, $\beta \ll 1$, Q is dominated by the electric dipole Mie term a_1, and the trapping force (3) becomes $F = (\alpha/2)\nabla E^2$, where α is the static polarizability of the sphere [1]. In the opposite limit $\beta \gg 1$, the connection with geometrical optics is established by applying to (4) and (5) the following steps [7,8]. i) In (6), substitute p_l and t_l by their (non-uniform) asymptotic expansions for large l, and approximate G_l and G_l' by the method of stationary phase [9]. ii) Compute the average $\langle Q \rangle$ over a size parameter range associated with a quasiperiod of the Mie coefficients. The result is

$$\langle Q \rangle_{GO} = \frac{4\gamma^2}{A} \int_0^{\theta_0} d\theta \sin\theta \cos\theta \exp[-2\gamma^2 \sin^2\theta] \times$$

$$\times \left\{ \cos\theta + \frac{1}{2}\sum_{j=1}^2 r_j \cos(2\theta_1 - \theta) - \frac{1}{2}\mathrm{Re}\sum_{j=1}^2 (1 - r_j)^2 \frac{e^{i[2(\theta_1 - \theta_2) - \theta]}}{1 + r_j e^{-2i\theta_2}} \right\}.\qquad(9)$$

In (9), $\theta_1 = \arcsin(q \sin\theta/a)$, $\theta_2 = \arcsin(\sin\theta_1/n)$ are the angles of incidence and refraction (defined so as to be negative if $q < 0$) at the sphere surface associated with a component in the direction θ of the focused beam (2). The corresponding Fresnel reflectivity for polarization j ($\|$, $\perp$) is r_j. Equation (9) may also be derived in the framework of GO. Thus, the expression within curly brackets agrees with the GO result for the force exerted by each component ray as first obtained in [10]. The remaining pre-factors in (9), not accounted for previously, represent the intensity distribution of the focused beam as implied by the sine condition and the transverse profile of the laser beam at the input aperture of the objective.

In fig. 1, Q is plotted as a function of q/a, the center offset from the focus in units of the sphere radius [11]. The numerical values chosen correspond to the experiment of Friese *et al.* [12]: $n_1 = 1.33, n_2 = 1.57, A = 0.85, \theta_0 = 78^\circ$, which by (1) yield $\gamma^2 = 0.99$. The dotted curve represents the GO result (9). The other curves represent the exact Mie results (4) and (5) for two different β values, corresponding to microsphere radii employed in [12]: $1.42\,\mu m$ (dashed) and $2.16\,\mu m$ (solid), respectively, $\beta = 18.8$ and $\beta = 28.4$. The qualitative behavior of the GO curve has been explained [2] in terms of competition between radiation pressure (scattering force) and gradient force. However, the GO result for the maximum backward trapping efficiency Q_m is smaller (by a factor of the order of 2) [13] than the values obtained in ref. [2]. This is in line with the discrepancy between experimental and theoretical values noted in ref. [3].

The Mie theory provides values for Q_m (0.088 for $\beta = 18.8$ and 0.086 for $\beta = 28.4$ [14]) below the GO result $Q_m = 0.095$. The position at which the backward force is maximum lies beyond the corresponding GO value $q/a = 1.01$. The stiffness decreases as this point is approached from the focus, contradicting the GO prediction and in agreement with an experiment reported in ref. [3].

GO is also a poor approximation near the geometrical focus, as expected. In fact, fig. 1 shows that the exact values deviate substantially from GO near $q = 0$. The stable equilibrium position shows large positive as well as negative offsets from GO (further discussed below), and the linear Hooke's law range around the equilibrium position is narrower than predicted by GO. Because of the axial focusing effect [7], the (non-uniform) asymptotic approximations to

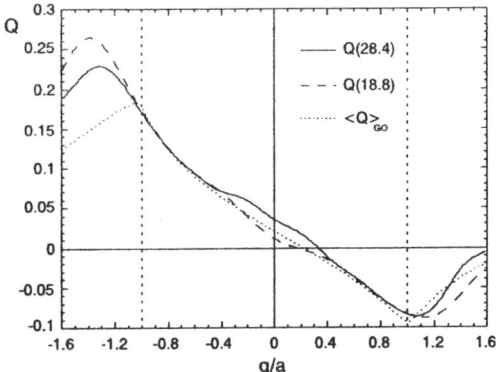

Fig. 1 – Normalized axial force *vs.* position (in units of the sphere radius). The dotted line is computed from ray optics theory, whereas the solid and dashed lines are calculated from the wave-optics theory with size parameters $\beta = 28.4$ and $\beta = 18.8$, respectively. The vertical dashed lines mark the microsphere boundaries.

the Mie angular functions employed in the derivation of (9) break down at $q = 0$, although (9) is continuous at this point, yielding, with $\theta_1 = \theta_2 = 0$,

$$\langle Q \rangle_{GO}(q = 0) = \frac{4r}{1+r} \langle \cos\theta \rangle, \tag{10}$$

where r is the Fresnel reflectivity for normal incidence and $\langle \cos\theta \rangle$ denotes an average over the intensity distribution of the focused beam (2). Since the incident rays are either backscattered or undeviated in this approximation, (10) represents pure radiation pressure in GO.

The region around $q = 0$ deserves special treatment, in view of its relevance to the evaluation of trap axial stiffness. For $\beta \gg 1$, the above discussion and the localization principle imply that the main contributions to (4) and (5) should arise from partial waves with $l \ll \beta$, so that we apply Hankel's asymptotic expansions to the spherical Bessel functions in the Mie coefficients. The results are independent of l, and the summations over multipole coefficients can then be carried out, resulting in

$$Q(q = 0) = \frac{8r \sin^2 \Delta/2}{1 + r^2 - 2r\cos\Delta} \langle \cos\theta \rangle, \tag{11}$$

where $\Delta = 4n_2\omega a/c$. This expression corresponds to the radiation pressure efficiency (twice the reflectivity) of an infinite set of parallel-plate interferometers (width $2a$, refractive index n_2, so that Δ is the round-trip phase), each one oriented at an angle θ with respect to the axis, traversed at normal incidence by the corresponding beam angular component. The GO result (10) follows from (11) by taking an incoherent average. Since $n - 1$ is small, we have $r \ll 1$, so that the interferometer reflectivity is nearly sinusoidal.

In fig. 2, for the same parameters as in fig. 1, we plot Q at $q = 0$ as a function of β. The Mie curve (full line) displays the expected near-sinusoidal oscillation as β increases, approaching the interferometer behavior (11) (shown as dotted line). The GO value (10) (dashed line) is approached in the average sense. The two points corresponding to the β values employed in fig. 1 are shown by circles. Since the radiation pressure at $\beta = 28.4$ is above the GO value, the Mie value for the equilibrium position q_{eq} is larger than the GO result, in agreement with fig. 1 (the opposite applies at $\beta = 18.8$). The values for q_{eq} are found by numerically solving

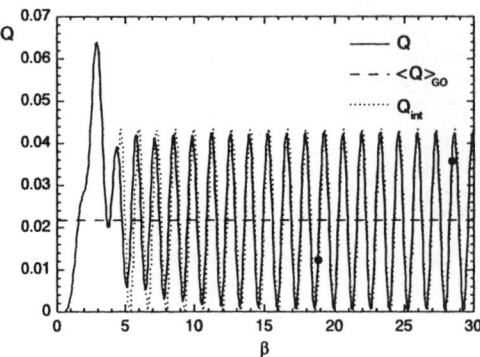

Fig. 2 – Normalized force at the geometrical focal point *vs.* size parameter β: exact (full line), interferometer model (dotted line) and geometrical optics (horizontal dashed line). The black circles indicate the two values of β used in fig. 1.

the equation $Q(q_{\text{eq}}) = 0$. In the limit $\beta \gg 1$, they are vanishingly small at β values that are minima of $Q(q = 0)$. For $\beta \gtrsim 5$, q_{eq}/a as a function of β oscillates in phase with the oscillations of $Q(q = 0)$, around the GO value $(q_{\text{eq}}/a)_{\text{GO}} = 0.217$, and with amplitude of the order of 0.17.

The trap axial stiffness is given by

$$\kappa = -\frac{n_1 P}{c} \left(\frac{\partial Q}{\partial q} \right)_{q=q_{\text{eq}}} \tag{12}$$

Within GO, κ decreases as $1/\beta$. This follows from scaling: Q_{GO} depends on q only through q/a. Hence, $\partial Q_{\text{GO}}/\partial q = Q'_{\text{GO}}(q/a)/a$, yielding

$$\kappa_{\text{GO}} = -\frac{n_1 P}{c} Q'_{\text{GO}}\left(\frac{q_{\text{eq}}}{a} \right) \frac{k}{\beta}. \tag{13}$$

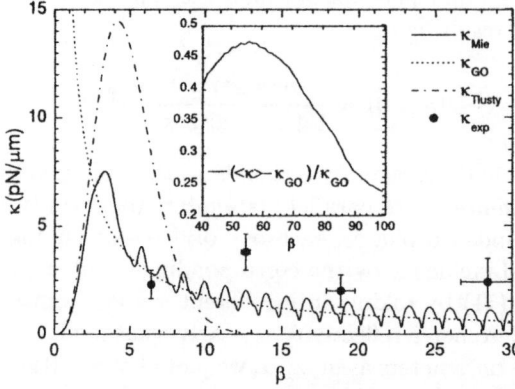

Fig. 3 – Axial stiffness κ of the optical tweezer as a function of β. Solid, dotted and dot-dashed lines correspond to the (exact) wave-optics theory, geometrical optics, and electrostatic theory [15], respectively. Also shown are the experimental data points of ref. [12], with corresponding error bars. In the inset, we plot the relative discrepancy between the average of the exact values and the geometrical optics results.

Again for the parameters of ref. [12] (power $P = 3\,\text{mW}$), we plot in fig. 3 the Mie values of κ (solid line) [16], the GO result $\kappa_{GO} = (18/\beta)(\text{pN}/\mu\text{m})$ calculated from eq. (13) (dotted line) and the experimental data points from ref. [12], with the respective error bars. We also show (dot-dashed line) the values predicted by the electrostatic model recently suggested by Tlusty *et al.* [15]. As could be expected, their approach may be applied only in the low-frequency (Rayleigh) limit, where it may be replaced by the simpler electric dipole approximation (neglecting the variation of the field over the sphere volume) already discussed above in connection with (4).

For $\beta \gtrsim 10$, the Mie values for κ oscillate around the GO curve with period $\Delta\beta = \pi/(2n)$, like the force $Q(q = 0)$ (cf. (11)) and the equilibrium position. This corresponds to a frequency interval $\Delta\nu = c/(4n_2 a)$, which is in the THz range for spheres with radii of a few microns. As shown in the inset in fig. 3, where we plot $(\langle\kappa\rangle - \kappa_{GO})/\kappa_{GO}$ as a function of β, the average of the Mie values, $\langle\kappa\rangle$, stands above the GO curve, but the relative difference decreases to zero as β increases beyond $\beta \approx 55$. For large sphere diameters, κ may become very small over short β intervals.

While it is difficult to draw definite conclusions from only four experimental points with sizable error bars, the following inferences may be drawn: i) The electrostatic model is not only inconsistent with the GO limit, but also with experiment. ii) The experimental results appear to be more compatible with oscillatory behavior than with the strictly monotonic fall-off predicted by geometrical optics. A fair comparison would require not only better statistics, but also disentangling the possible effects of additional aberrations, corrections to Stokes' law, radiometric forces [3] and back-reflected light [17]. Such experimental comparisons would be very valuable for providing an absolute calibration of optical tweezers.

In conclusion, starting from a Debye-type integral representation of the focused laser beam, we have derived an exact Mie expansion for the axial trapping efficiency whose validity covers the range of interest for applications. Our model is more realistic than previous ones. We have obtained analytic expressions for the multipole coefficients associated to the laser beam, which allowed us to compute the axial force by using well-tested numerical methods of Mie theory. The results are expressed in terms only of experimentally accessible parameters. The connection with the correct GO limit (taking into account the sine condition) has been derived from the Mie expansion by size averaging. The behavior near the focus has been obtained and interpreted in terms of an interferometer model, which also accounts for the equilibrium position and trap axial stiffness oscillations. The predicted oscillations should be accessible to experiment by scanning the laser beam frequency. Extension of the present theory to the computation of the transverse stiffness is under way.

$$* * *$$

We thank W. WISCOMBE for useful suggestions and programs for quadrature integration and for Mie scattering calculations, and CNPq for partial support. One of us (PAMN) acknowledges support by Programa de Núcleos de Excelência (PRONEX), grant 4.1.96.08880.00-035-1.

REFERENCES

[1] ASHKIN A., *Proc. Natl. Acad. Sci. USA*, **94** (1997) 4853, and references therein.
[2] ASHKIN A., *Biophys. J.*, **61** (1992) 569.
[3] SVOBODA K. and BLOCK S. M., *Annu. Rev. Biophys. Biomol. Struct.*, **23** (1994) 247.
[4] RICHARDS B. and WOLF E., *Proc. R. Soc. London, Ser. A*, **253** (1959) 358.

708   EUROPHYSICS LETTERS

[5]   BOHREN C. F. and HUFFMAN D. R., *Absorption and Scattering of Light by Small Particles*
      (Wiley, New York) 1983.

[6]   EDMONDS A. R., *Angular Momentum in Quantum Mechanics* (Princeton University Press) 1957.

[7]   NUSSENZVEIG H. M., *Diffraction Effects in Semiclassical Scattering* (Cambridge University
      Press) 1992.

[8]   NUSSENZVEIG H. M. and WISCOMBE W. J., *Phys. Rev. Lett.*, **45** (1980) 1490.

[9]   The stationary-phase point ($q = 0$ is excluded) is at $\bar{\theta} = \arcsin[(l + 1/2)/k|q|]$, as expected by
      the localization principle [7].

[10] ROOSEN G., *Can. J. Phys.*, **57** (1979) 1260.

[11] The numerical integrations in eqs. (6) and (9) were performed with the help of a Kronrod-
      Patterson adaptative Gaussian-type quadrature method.

[12] FRIESE M. E. J. *et al.*, *Appl. Opt.*, **35** (1996) 7112. These authors employ red laser light
      ($\lambda_0 = 2\pi c/\omega = 0.6328$ μm). Since microscope objectives are corrected for the visible, spherical
      aberration is probably smaller than in experiments using infrared light [3].

[13] The gradient force is overestimated in [2] as a consequence of neglecting the sine condition and
      the corresponding factor $\cos\theta$ in eq. (9), which diminishes the contribution of rays at large
      angles. By the same reason, the stable equilibrium positions as predicted by (9) are further from
      the focus than the values obtained in [2].

[14] This shows that, in contradiction with the near-paraxial results obtained by Wright *et al.* (*Appl.
      Phys. Lett.*, **63** (1993) 715), Q_m is not a monotonically increasing function of the sphere radius.

[15] TLUSTY T. *et al.*, *Phys. Rev. Lett.*, **81** (1998) 1738.

[16] The Mie values for κ are obtained by deriving from (4) and (5) the partial-wave expansion for
      $\partial_q Q$, and then replacing the results for q_{eq} into (12).

[17] SASAKI K., TSUKIMA M. and MASUHARA H., *Appl. Phys. Lett.*, **71** (1997) 37.

letters to nature

Thermally activated transitions in a bistable three-dimensional optical trap

Lowell I. McCann†, Mark Dykman & Brage Golding

Department of Physics and Astronomy, Michigan State University, East Lansing, Michigan 48824-1116, USA

Activated escape from a metastable state underlies many physical, chemical and biological processes: examples include diffusion in solids, switching in superconducting junctions[1,2], chemical reactions[3,4] and protein folding[5,6]. Kramers presented the first quantitative calculation[7] of thermally driven transition rates in 1940. Despite widespread acceptance of Kramers' theory[8], there have been few opportunities to test it quantitatively as a comprehensive knowledge of the system dynamics is required. A trapped brownian particle (relevant to our understanding of the kinetics, transport and mechanics of biological matter[9,10]) represents an ideal test system. Here we report a detailed experimental analysis of the brownian dynamics of a sub-micrometre sized dielectric particle confined in a double-well optical trap. We show how these dynamics can be used to directly measure the full three-dimensional confining potential—a technique that can also be applied to other optically trapped objects[11,12]. Excellent agreement is obtained between the predictions of Kramers' theory and the measured transition rates, with no adjustable or free parameters over a substantial range of barrier heights.

A mesoscopic particle, suspended in a liquid and confined within a metastable potential well provides an ideal representation of Kramers' ideas. The particle moves at random within the well until a large fluctuation propels it out of the well over an energy barrier. The potential can be created with a gradient optical trap—a technique widely used in biophysical studies[9]. Dual optical traps were introduced initially to study the synchronization of the interwell transitions by periodic forcing[13]. A particle in a dual optical trap is a well-controlled model system which can be used to address quantitatively the problem of transition rates provided the confining potential can be accurately determined.

The optically induced potential wells constructed in the present experiments are formed by focusing two parallel laser beams through a single objective lens. Each beam creates a stable three-dimensional trap as a result of electric field gradient forces exerted on a transparent dielectric spherical silica particle of diameter $2R = 0.6\,\mu m$ (Bangs Laboratories). Displaced by 0.25 to 0.45 μm, the beams create a double-well potential, with the stable positions of the particle centre at $\mathbf{r}_1$ and $\mathbf{r}_2$. Relatively infrequent random transitions between the potential wells occur through a saddle point at $\mathbf{r}_s$ as depicted in Fig. 1a. Both the depth of each potential well and the height of the intervening barrier are under experimental control.

The two HeNe lasers (17 mW, 633 nm) that create the traps are stabilized by Pockels cell electro-optic modulators and imaged into a sample cell by a 100x 1.4 NA PlanApo objective lens mounted in an Olympus IX-70 microscope. The beams are mutually incoherent and circularly polarized as they enter the microscope. A single trapped sphere is imaged onto a Dalsa CA-D1 CCD camera operated at 200 frames per second. The coordinates of the sphere's centre in the focal plane of the objective lens are found to within ±10 nm with a pattern-matching routine, and the coordinate in the beams' propagation direction is extracted by analysis of the image size as it moves above and below the focal plane. The sealed sample

† Present address: Department of Physics, University of Wisconsin-River Falls, River Falls, Wisconsin 54022, USA.

cell, constructed from two glass coverslips and epoxy resin, holds a dilute suspension of silica spheres in water at room temperature. The experimental output of this system is a time record of the coordinates $\mathbf{r}(t)$; as shown in Fig. 1b.

As a result of the short equilibration time of the sphere in water ($\gamma^{-1} = M/(6\pi\eta R) \sim 10^{-7}$ s, where η is the viscosity of water and M is the particle mass), the brownian particle relaxes to equilibrium on a timescale much shorter than the sampling time. The spatial probability density is therefore:

$$\rho(\mathbf{r}) = Z^{-1}\exp(-U(\mathbf{r})/k_B T) \qquad (1)$$

Here $U(\mathbf{r})$ is the potential energy as a function of particle position, and Z is a normalization constant. The probability density is found from the time series $\mathbf{r}(t)$, typically using 10^6 to 10^7 frames, for durations much longer than the mean interwell transition times. Equation (1) then allows us to obtain the potential directly from measurements.

Figure 2 shows $U(\mathbf{r})$ for a two-beam trap. We choose the x axis to be in the direction from one beam to the other and the z axis along the propagation direction of the beams. The potential minima, $\mathbf{r}_1$ and $\mathbf{r}_2$, lie in the symmetry plane $y = 0$ formed by the beam axes. Figure 2a shows a two-dimensional cross-section, at $y=0$, of the potential with energy contours at 1.0 $k_B T$ intervals distinguished by colour-coding, where k_B is the Boltzmann constant and T is temperature. If, for a given x, we find the minimum of $U(\mathbf{r})$ over y and z, we obtain the familiar one-dimensional representation of a double-well potential shown in Fig. 2b. Figure 2c shows the energy contours at $\mathbf{r}_2$ for a cross-section in the y–z plane with the corresponding energy profile in Fig. 2d. The elongated profile along z is expected, since the radial field gradient is determined by the transverse beam profile whereas the weaker axial gradient depends on the angular divergence of the focused beam near its diffraction-limited waist[14].

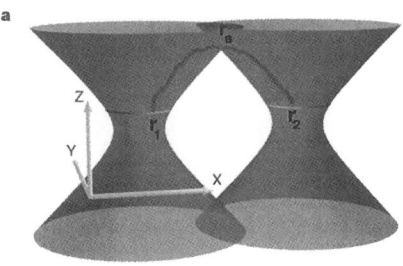

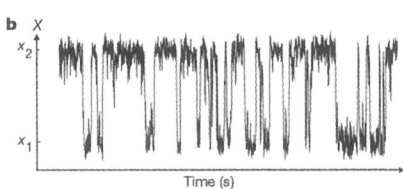

Figure 1 Interwell transitions in a dual optical trap. **a**, Two neighbouring focused beams in the absence of a perturbing trapped sphere are shown. The dark line illustrates the path of a trapped sphere in an interwell transition between $\mathbf{r}_1$ and $\mathbf{r}_2$ through the saddle at $\mathbf{r}_s$. The equilibrium positions $\mathbf{r}_1$ and $\mathbf{r}_2$, at the level of the rings around the beams, are displaced above the focal plane of the incident beams. **b**, Projection of particle trajectory on the x-axis perpendicular to the beams, where x_1, x_2 are components of $\mathbf{r}_1$, $\mathbf{r}_2$. The sampling interval is 5 ms and the total time duration of the record is approximately 8 s. During the acquisition time, a computer performs a pattern-matching routine that returns the three particle coordinates. The particle spends most of its time in the vicinity of the stable points $\mathbf{r}_1$ and $\mathbf{r}_2$ with infrequent transitions between them.

letters to nature

Table 1 Features of the optical potential in Fig. 2 and associated transition rates

	$|\omega^{(1)}|$	$\omega^{(2)}$	$\omega^{(3)}$	$\Delta U/k_BT$	W_0^K	W^K	W^{meas}
Well at r_1	16 ± 4	55 ± 14	6 ± 3	2.77 ± 0.05	80 ± 40	5 ± 2	6.57 ± 0.03
Saddle at r_s	7 ± 2	50 ± 25	6 ± 3				
Well at r_2	21 ± 5	56 ± 14	6 ± 3	3.51 ± 0.04	110 ± 50	3 ± 2	3.70 ± 0.02

Characteristic frequencies $\omega_i^{(j)}$ (in $10^4 \, s^{-1}$, with $j = 1, 2, 3$) associated with the extrema, r_i, of the potential in Fig. 2. Note that $\omega_s^{(1)2} < 0$. W_0^K is the Kramers prefactor, W^K is the Kramers rate, and W^{meas} is the experimentally determined rate in transitions per second, reported for the two potential wells at r_1 and r_2. $\Delta U/k_BT$ is the reduced barrier height. The errors reported for $\omega^{(3)}$ do not include the uncertainty in the z-position calibration since it does not influence the value of W_0^K.

The double-well potential in Fig. 2 has minima $r_{1,2}$ separated by $\delta x = 0.35 \, \mu m$, a single intervening saddle point at r_s, and $U(r_2) - U(r_1) = 0.8 \, k_BT$. We emphasize that the potential measured in this manner is the overall effective potential experienced by the particle in its environment.

A feature of the effective potential evident in Fig. 2a is the strong symmetry breaking about the focal plane, which is the symmetry plane of the beams, unperturbed by the particle. This symmetry breaking leads to the single saddle point in $U(r)$ instead of two saddle points as might be inferred from Fig. 1a. This aspect of the potential is not an artefact of specific experimental conditions, such as non-parallel optical beams, but is a consequence of the beam–particle interaction. The dielectric particle acts as a spherical lens to refocus the beam inside the sphere. When the particle is displaced in the +z direction above the focal plane, the electromagnetic field is most strongly 'squeezed' into the particle, thus minimizing the total free energy of the polarized particle in the field.

In the vicinity of r_1, r_2, and r_s, the potential $U(r)$ is quadratic in the displacements $\delta r = r - r_i$ with $i=1,2$, or s:

$$U(r) = U(r_i) + \frac{1}{2}M\sum_{\alpha,\beta}\Lambda_i^{\alpha\beta}\delta r_\alpha \delta r_\beta \qquad (2)$$

Here $\alpha, \beta = x, y, z$. In practice, we perform a least-squares fit of equation (2) to the data in the vicinity of r_i. As an example of the results and errors in this procedure, Table 1 shows the eigenvalues ω_i^2 of the matrix Λ_i for the potential shown in Fig. 2. The characteristic frequencies $|\omega_i|$ are small compared to the damping rate γ so the particle is overdamped.

We now consider the interwell dynamics of the particle. Specifically, we examine the rates W_{12} (W_{21}) of transitions $1 \rightarrow 2$ ($2 \rightarrow 1$), and their dependence on the energy barriers $\Delta U_1 = U(r_s) - U(r_1)$ and $\Delta U_2 = U(r_s) - U(r_2)$. We tune $\Delta U_i/k_BT$ up to 8.5 by adjusting the optical intensity in each beam and the beam separation. The presence of the energy asymmetry $\Delta U_{12} = U(r_2) - U(r_1)$ between 0.5–3 k_BT allows us to measure two independent rates with a single optical configuration. For each set of experimental conditions, we determine the full three-dimensional potential, similar to Fig. 2. To obtain satisfactory statistics, we accumulated between 2,400 and 94,000 interwell transitions that occurred over 5×10^5 to 10^7 frames.

A quantitative description of thermally activated escape from a one-dimensional metastable potential was given by Kramers[7] and subsequently extended to multidimensional potentials[15]. The subject has been extensively reviewed[8,16,17]. For an overdamped brownian particle in a potential $U(r)$, the Kramers transition rate is:

$$W^K = W_0^K \exp\left(-\frac{\Delta U}{k_BT}\right) \qquad (3)$$

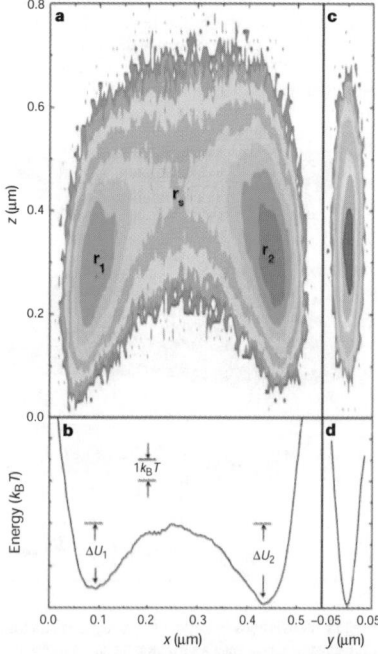

Figure 2 Experimentally determined potential energy of the particle in a double-well optical trap. **a**, Energy contours for a cross-section in the x–z plane that contains the stable points r_1 and r_2, and the saddle point r_s. Each colour indicates a 1.0 k_BT energy interval. **b**, The energy, minimized with respect to y and z, as a function of x as shown in **a**. ΔU_1 and ΔU_2 are the barrier heights for the corresponding wells. **c**, The energy contours in a y–z cross-section containing point r_2. **d**, The energy, minimized with respect to x and z, as a function of y as shown in **c**. The potential described in the figure was extracted from a data set of 4×10^6 camera frames containing 94,000 interwell transitions.

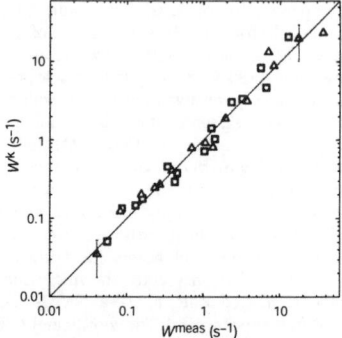

Figure 3 Experimental and theoretical transition rates. We compare the directly measured transition rates, W^{meas}, with the rates calculated from the three-dimensional Kramers theory, W_K, using the measured curvatures of the potential wells. The squares represent escapes from the well at r_1 and the triangles represent escapes from the well at r_2. The line of slope unity indicates the result expected if the data coincide with the Kramers theory.

NATURE | VOL 402 | 16 DECEMBER 1999 | www.nature.com

Here the prefactor is given by the following expression for a three-dimensional potential[15]:

$$W_0^K = \frac{|\omega_s^{(1)}| |\omega^{(1)}|}{2\pi\gamma} \frac{\omega^{(2)} \omega^{(3)}}{\omega_s^{(2)} \omega_s^{(3)}} \qquad (4)$$

Here $\omega^{(j)}$ and $\omega^{(j)}$ characterize, respectively, the curvatures of the potential at the saddle point and at the minimum from which the system escapes, with $(\omega_s^{(1)})^2 < 0$. Therefore, with knowledge of the potential, not only the exponential term, but also the prefactor can be explicitly computed, as shown in Table 1.

Figure 3 shows a plot of the Kramers rates, W^K, calculated from equations (3) and (4) as a function of the transition rates, W^{meas}, obtained from the mean dwell time in each state or by fitting an exponential to a histogram of dwell times, in accordance with a Poisson distribution, which yielded equivalent results. The systematic variation of the potential barrier ΔU by approximately 6 $k_B T$ is responsible for the nearly three-decade variation in transition rates. The solid line with unity slope denotes the coincidence of theory and experiment. The data fall remarkably close to the line, confirming the multidimensional Kramers theory of transition rates.

One of the major contributors to the uncertainty in calculating W^K is the error in the saddle point frequencies entering equation (4). This is primarily statistical, since the particle spends little time in the vicinity of r_s. This error is amplified for the highest barriers with the lowest transition rates. At low barriers, the statistical uncertainty is small since the transition rates are large, but the Kramers theory is not strictly valid here. If the thermal diffusion length exceeds the size of the parabolic region in the vicinities of the stationary states, then the transition rates are affected by the shape of the potential away from the stationary states. However, the corrections to the theory are small in the range of parameters studied here.

Detailed knowledge of the overall potential, as afforded by our experiments, should enable investigations of the escape rate of underdamped particles in the region in which escape occurs by diffusion over energy, as well as the Kramers turnover region[8,18]. This can be accomplished with present methods by reducing the viscous damping on the particle. Knowledge of the potential is also crucial for understanding strategies for control of escape. Escape occurs by large fluctuations[19] that move the system from a minimum of the potential to the barrier top along optimal paths. The results of this work make it possible to find these paths, enabling selective control of escape rates by external modulating fields[20].

Received 12 August ; accepted 7 October 1999.

1. Han, S., Lapoint, J. & Lukens, J. E. Effect of a two-dimensional potential on the rate of thermally induced escape over the potential barrier. *Phys. Rev. B* **46**, 6338–6345 (1992).
2. Devoret, M. H., Esteve, D., Martinis, J. M., Cleland, A. & Clarke, J. Resonant activation of a brownian particle out of a well: microwave-enhanced escape from the zero-voltage state of a Josephson junction. *Phys. Rev. B* **36**, 58–73 (1987).
3. van Kampen, N. G. *Stochastic Processes in Physics and Chemistry* (Elsevier, Amsterdam, 1992).
4. Gillespie, D. T. Exact stochastic simulation of coupled chemical reactions. *J. Chem. Phys.* **81**, 2340–2361 (1977).
5. Sali, A., Shakhnovich, E. & Karplus, M. How does a protein fold? *Nature* **369**, 248–251 (1994).
6. White, S. H. & Wimley, W. C. Membrane protein folding and stability: Physical principles. *Annu. Rev. Biophys. Biomol. Struct.* **28**, 319–365 (1999).
7. Kramers, H. A. Brownian motion in a field of force and the diffusion model of chemical reactions. *Physica* **7**, 284–304 (1940).
8. Melnikov, V. I. The Kramers problem: fifty years of development. *Phys. Rep.* **209**, 2–71 (1991).
9. Mehta, A. D., Reif, M., Spudich, J. A., Smith, D. A. & Simmons, R. M. Single molecule biomechanics with optical methods. *Science* **283**, 1689–1695 (1999).
10. Smith, D. E., Babcock, H. P. & Chu, S. Single-polymer dynamics in steady shear flow. *Science* **283**, 1724–1727 (1999).
11. Svoboda, K., Schmidt, C. F., Schnapp, B. J. & Block, S. M. Direct observation of kinesin stepping by optical trapping interferometry. *Nature* **365**, 721–727 (1993).
12. Tskhovrebova, L., Trinick, J., Sleep, J. A. & Simmons, R. M. Elasticity and unfolding of single molecules of the giant muscle protein titin. *Nature* **387**, 308–312 (1997).
13. Simon, A. & Libchaber, A. Escape and synchronization of a brownian particle. *Phys. Rev. Lett.* **68**, 3375–3378 (1992).
14. Ghislain, L. P., Switz, N. A. & Webb, W. W. Measurement of small forces using an optical trap. *Rev. Sci. Instrum.* **65**, 2762–2768 (1994).
15. Landauer, R. & Swanson, J. A. Frequency factors in the thermally activated process. *Phys. Rev.* **121**, 1668–1674 (1961).
16. Risken, H. *The Fokker-Planck Equation* (Springer, Berlin, 1989).
17. Hänggi, P., Talkner, P. & Borkovec, M. Reaction-rate theory: fifty years after Kramers. *Rev. Mod. Phys.* **62**, 251–341 (1990).
18. Linkwitz, S., Grabert, H., Turlot, E., Esteve, D. & Devoret, M. H. Escape rates in the region between the Kramers limits. *Phys. Rev. A* **45**, R3369–R3372 (1992).
19. Luchinsky, D. G. & McClintock, P. V. E. Irreversibility of classical fluctuations studied in analogue electrical circuits. *Nature* **389**, 463–466 (1997).
20. Smelyanskiy, V., Dykman, M. I. & Golding, B. Time oscillations of escape rates in periodically driven systems. *Phys. Rev. Lett.* **82**, 3193–3197 (1999).

Acknowledgements

We thank R. Kruse for his assistance with the experiment and graphics. Support from the Center for Fundamental Materials Research at Michigan State University and from the NSF Division of Physics, and Division of Materials Research is gratefully acknowledged.

Correspondence and requests for materials should be addressed to B.G. (email: golding@pa.msu.edu).

VOLUME 73, NUMBER 2   PHYSICAL REVIEW LETTERS   11 JULY 1994

Microscopic Measurement of the Pair Interaction Potential of Charge-Stabilized Colloid

John C. Crocker and David G. Grier

The James Franck Institute, The University of Chicago, 5640 South Ellis Avenue, Chicago, Illinois 60637

(Received 14 January 1994)

We present a microscopic measurement of the interaction potential between isolated pairs of charged colloidal spheres. The measured spatial dependence of the potential is consistent with the screened Coulomb repulsion expected from the Derjaguin-Landau-Verwey-Overbeek theory of colloidal interactions.

PACS numbers: 82.70.Dd

The past century of research on charge-stabilized colloidal suspensions has been inspired by their intrinsic interest as distinct states of matter [1], by their considerable technological value [2], and more recently by their utility as model condensed matter systems [3,4]. The standard theory for the interaction between charged colloidal spheres was formulated almost 50 years ago by Derjaguin-Landau-Verwey-Overbeek (DLVO) [5,6]. Calculations based on the DLVO theory [7–9] account at least qualitatively for the observed fluid-crystal and fcc-bcc phase boundaries [10,11] as well as for many of these phases' bulk properties [12,13]. Persistent quantitative discrepancies [10,11] and anomalous observations including the unaccounted for monodisperse glassy phase [11] suggest that the theory is not yet complete, however. Differences with the DLVO theory also have been raised on theoretical grounds [14–18]. In light of the fundamental importance of these considerations to the understanding of colloidal systems, we have measured the colloidal pair potential directly, utilizing digital video microscopy and optical trapping techniques. In particular, we extract the two-particle interaction from the dynamics of isolated pairs of particles moving away from artificially created initial configurations.

The colloid in our study consists of commercially available polystyrene sulfate spheres with radius $a = 32$ nm, monodisperse to within 1% (Duke Scientific, No. 5065A). Particles of this size undergo vigorous Brownian motion in water and yet are sufficiently large to be imaged with a conventional light microscope. When such spheres are dispersed in water, the ionic groups bonded to their surfaces dissociate and give rise to a screened electrostatic interaction. For sphere separations large enough that the Debye-Hückel approximation holds, the DLVO theory gives the potential:

$$U_{\text{DLVO}}(x) = \frac{Z^{*2}e^2}{\varepsilon}\left[\frac{e^{2\kappa a}}{(1+\kappa a)^2}\right]\frac{e^{-\kappa x}}{x}. \tag{1}$$

Here x is the center-to-center distance between identical spheres, Z^* is their effective surface charge, ε is the dielectric constant of the solution, and κ^{-1} is its Debye-Hückel screening length set by the counterion concentration. Nonlinear contributions near the spheres' surfaces cause the effective surface charge to be considerably

smaller than the fully dissociated surface charge. The spheres in our study, for example, have titratable charges of $(3.2 \pm 0.5) \times 10^5$ electron equivalents although their effective surface charge, Z^*, is more than 2 orders of magnitude smaller, as we will see below. For such highly charged spheres, the screened Coulomb interaction dominates van der Waals and hydrodynamic contributions which therefore do not appear in Eq. (1).

The possibility has been raised that a more comprehensive treatment of the counterion free energy might reveal a long-range attractive component [14–18] to the interaction potential. Such attractive regimes are known to occur in the parallel plate geometry [19]. Strong long-range attraction would account for such experimental observations as phase separation in monodisperse colloidal fluids [20,21] and stable multiparticle voids in colloidal crystals.

The majority of previous tests of the DLVO theory have relied on indirect probes such as elastic constant and phase boundary measurements. The calculated values for these bulk properties, however, are quite insensitive to the form of the local interaction potential, and the measurements also may reflect many-body contributions. Inversion of light scattering data is notoriously sensitive to noise and requires the assumption of a functional form for the potential [22]. The second virial coefficient of dilute suspensions extracted from light scattering data similarly relies for its interpretation on a model for the interaction [23]. The pioneering work of Takamura, Goldsmith, and Mason [24] sought the microscopic interaction's spatial dependence in the dynamics of particle pairs colliding in a shear flow. Deviations from paths predicted for noninteracting spheres provided estimates for parameters in their interaction model. Since their method relies on hydrodynamic calculations, it is applicable only in the absence of Brownian motion. Furthermore, as we will see below, the scale of forces which characterizes long-range colloidal interactions is small enough to challenge even atomic force microscopes. The method of the present study requires no assumptions for the form of the interaction beyond spherical symmetry, explicitly excludes many-body contributions, can probe weakly interacting systems with strong Brownian motion, and thus addresses colloidal systems of the greatest theoretical and practical interest.

0031-9007/94/73(2)/352(4)$06.00

VOLUME 73, NUMBER 2   **PHYSICAL REVIEW LETTERS**   11 JULY 1994

Unlike indirect probes of the potential, particle trajectories exactly represent their progenitor forces. However, extracting the pair potential from an enormous background of random thermal forces requires some care. Assuming the colloidal interaction is spherically symmetric, we cast the analysis into a one-dimensional form by considering only the particles' relative coordinate. This is conceptually equivalent to studying the dynamics of one particle (with correspondingly larger Brownian motion) moving in the potential $U(x)$ set up by another stationary particle. The probability density, $\rho(x,t)$, for finding a particle as it diffuses through a viscous fluid in a one-dimensional static potential, $U(x)$, satisfies the Smoluchowski equation [25],

$$\frac{\partial \rho(x,t)}{\partial t} = L_S \rho(x,t) , \tag{2}$$

where

$$L_S = \left[\frac{1}{m\gamma} \frac{\partial}{\partial x} U'(x) + \frac{k_B T}{m\gamma} \frac{\partial^2}{\partial x^2} \right] . \tag{3}$$

$U'(x)$ denotes the spatial derivative of $U(x)$ and $\gamma = 6\pi\eta a$ is the Stokes drag coefficient for a particle of mass m and radius a moving through a fluid of viscosity η. The probability density satisfying the Smoluchowski equation evolves in time according to

$$\rho(x,t+\tau) = \int P(x,t+\tau|x',t)\rho(x',t)dx' , \tag{4}$$

where

$$P(x,t+\tau|x',t) = e^{L_S \tau}\delta(x-x') . \tag{5}$$

Because the operator L_S does not include hydrodynamic coupling and $U(x)$ depends only on the instantaneous separation, the propagator, $P(x,t+\tau|x',t)$, depends only on the propagation time τ and not on t. The history independence of the propagator permits us to select as the propagation time $\tau = 1/30$ sec, the delay between consecutive video frames.

The equilibrium probability density, $\rho^{eq}(x)$, is the stationary solution of Eq. (4). Discretizing the spatial dependence in Eq. (4) then yields

$$\rho_i^{eq} = \sum_j P_{ij}\rho_j^{eq} , \tag{6}$$

where ρ_i^{eq} is the probability of finding the particle in the ith spatial bin and P_{ij} is the discrete approximation to the continuous transition probability, $P(x,\tau|x',0)$. Because the equilibrium probability density is related to the interaction potential through the Boltzmann distribution, $\rho_i^{eq} = \exp[-U(x_i)/k_B T]$, the experimental problem of determining the interaction potential is reduced to characterizing the transition probability matrix, P_{ij}.

The experimental arrangement is rendered in Fig. 1(a). We strove to minimize the concentration of ionic impurities and thereby maximize the range of the colloidal interaction. Our cell was constructed by hermetically seal-

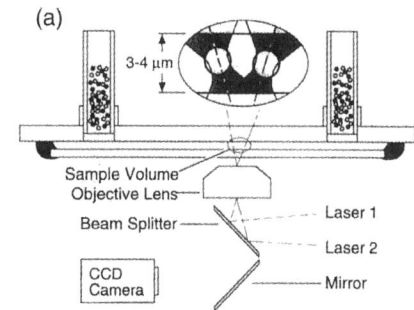

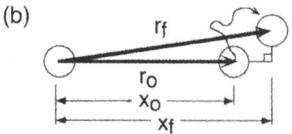

FIG. 1. (a) Schematic representation of the sample cell in cross section. The reservoir tubes are connected to gas transfer systems which are omitted for clarity. The inset shows a magnified view (not to scale) of two colloidal spheres trapped in a pair of optical tweezers between the glass walls of the sample cell. (b) Projection of the one-dimensional separations, x_0 and x_f.

ing the edges of a No. 1 microscope cover slip (150 μm thick) to the surface of a standard microscope slide. Holes drilled through the slide connect the thin sample volume to reservoir tubes filled with mixed-bed ion exchange resin. All glass surfaces were cleaned thoroughly before assembly. After filling with de-ionized colloidal suspension, the reservoir tubes are continuously flushed with water-saturated argon at a small overpressure to prevent contamination by airborne CO_2. Regulating the gas pressure in the reservoir tubes allows us to adjust the separation between slide and cover slip with submicron precision. When such a cell was filled with a dense suspension (volume fraction $\phi = 0.04$–0.05), colloidal crystals with fcc ordering formed rapidly and were observed to persist for months, indicating that such cells can remain relatively free of ionic impurities for long periods.

To perform interaction measurements, the cell is filled with a dilute suspension and mounted on an inverted (bright field) optical microscope with an attached video camera. We use a 100× numerical aperture 1.2 oil immersion objective and a total system magnification of 115 nm per pixel on our CCD camera. Two particles, selected at random, are held at a fixed separation (either 1.5 or 1.7 μm) with dual optical tweezers projected into the cell through the microscope objective [26,27]. The objective lens forms the tweezers by tightly focusing two laser beams, each of which localizes a dielectric sphere in three dimensions with optical gradient forces. A single 15 mW

VOLUME 73, NUMBER 2   PHYSICAL REVIEW LETTERS   11 JULY 1994

argon-ion laser operating at a wavelength of 488 nm provides enough power to form both traps. We release the spheres by interrupting the laser beam when no other particles are near. Their subsequent motions are captured with a computer-controlled SVHS video deck (NEC PC/VCR). In all, the dynamics of 796 pairs of spheres were recorded in this manner. Weak absorption of laser light at the experimental wavelength coupled with rapid thermal conduction by the water limit temperature increases in the trapped particles to less than 0.1 K. This small degree of heating will not measurably alter the interaction between the spheres. Thermal conduction also dissipates temperature gradients across the sample volume with a characteristic time shorter than 1 μsec.

Several steps are required to extract P_{ij} from the videotaped trajectories. First, the video tape is digitized with a precision frame grabber (Data Translation DT-3851A). The images are then corrected for geometric distortion and spatially filtered to suppress variations in illumination and camera noise. An individual sphere appears in an image as a maximum in the local brightness field. We refine its location to better than 25 nm (about 1/5 pixel) in the focal x-y plane by calculating the brightness-weighted centroid of the pixels it subtends. Because motion out of the focal plane (in the z direction) changes a sphere's appearance, its z coordinate can be estimated to within 150 nm by calibrating its apparent brightness and size as functions of its depth. Since the particles' separation vector lies approximately in the x-y plane, the measurement error in z adds only in quadrature; the mean error in the separation measurement is estimated to be 50 nm.

A single video frame is composed of two fields containing odd- and even-numbered lines, respectively, which are acquired 1/60 sec apart. We locate spheres in each video field separately to avoid location error due to their motion between fields. Having located a sphere in one video frame, its identification in subsequent frames is estimated with a maximum likelihood algorithm. To avoid possible systematic errors, we measure displacements between even fields and odd fields separately. The propagation time is still 1/30 sec and the four fields from two frames provide two independent samples of the particles motion. Details of the image processing and analysis will be presented elsewhere.

In consecutive video frames, a pair of spheres evolves from an initial vector separation, r_0, to a final separation, r_f. There are more ways for a pair of noninteracting spheres to wander apart than to come together in two or more dimensions. To avoid the resulting bias in our data set, we project the final separation vector onto the direction of the initial separation vector, as shown in Fig. 1(b). The transverse motion in one time step is sufficiently smaller than the center-to-center separation that the resulting error is smaller than our spatial resolution. Our data set also could be biased in the opposite direction if

we were to consider the initial conditions too close to either wall. In this case, the walls can prevent particles from wandering apart. To avoid this bias we restrict our data set to those trajectory steps with initial positions no further from the cell's midplane than $2\sqrt{2D\tau} = 422$ nm, where the self-diffusion coefficient $D = 0.67$ μm^2/sec for spheres of radius $a = 326$ nm.

Our 796 pair trajectories yield 6502 independent (x_0, x_f) samples of the transition probability. We distribute these measurements into the matrix elements of P_{ij} with weightings which are inversely proportional to the local density of initial separations. Weighting in this manner compensates for the nonuniform distribution of x_0 values. The bin size is half a pixel (57 nm). We approximate the errors $\delta\rho_i^{eq}$ in the elements of ρ_i^{eq} from the matrix δP_{ij} of standard deviations of P_{ij} as $(\delta\rho_i^{eq})^2 = \sum(\delta P_{ij}\rho_j^{eq})^2$.

In practice, we limit the range of separations considered to a value x_{max}. Those measurements (x_0, x_f) with $x_0 < x_{max}$ and $x_f > x_{max}$ are distributed into P_{ij} as $(x_0, 2x_{max} - x_f)$. This reflecting boundary condition compensates for the tendency of ρ_i^{eq} to vanish under free diffusion. Provided that x_{max} is larger than the range of the interaction, our results do not depend on x_{max}.

The pair interaction potential measured in this manner appears in Fig. 2. The solid line is a two-parameter fit to Eq. (1) which yields an effective charge of $Z^* = 1991 \pm 150$ electronic charges and a Debye-Hückel screening length of $\kappa^{-1} = 161 \pm 10$ nm. This value is considerably smaller than the maximum screening length of 960 nm in pure water and is consistent with a $3.6 \times 10^{-6}M$ concentration of a 1:1 electrolyte. This estimate for the surface charge should be considered a lower limit because the interaction curve is broadened slightly by measurement errors in the spheres' locations. The qualitative agreement between our measurement and the DLVO theory is excellent with no statistically significant evidence of a long-

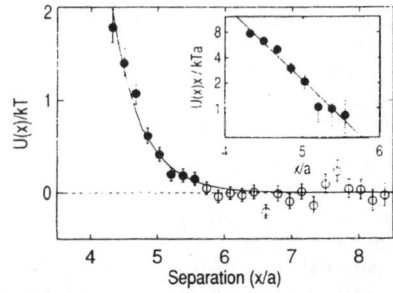

FIG. 2. Measured pair interaction potential for spheres of radius $a = 0.326$ μm as a function of center-to-center separation. The solid curve is a fit to Eq. (1) with $Z^* = 1991e$ and $\kappa^{-1} = 161$ nm. Inset: Semilogarithmic plot of the filled data points together with corresponding fit line. The line's slope provides an estimate for the Debye-Hückel screening length.

VOLUME 73, NUMBER 2   PHYSICAL REVIEW LETTERS   11 JULY 1994

range attractive interaction.

The estimated surface charge on our spheres is more than 2 orders of magnitude smaller than their titratable charge. Charge renormalization calculations [12] suggest that for very highly charged colloid the effective charge should saturate at $Z^* = C(a/\lambda_B)$, where the Bjerrum length is $\lambda_B = z^2 e^2/\epsilon k_B T = 0.715$ nm for a 1:1 electrolyte in water at 25 °C, and C is a constant around 10. A molecular dynamics simulation [28] of a spherically symmetric charged colloidal system finds $C = 7$ in reasonable agreement with measurements on micelles [29] which find $C = 6$. Our measured charge of $Z^* = 1991$ corresponds to $C = 4.4$.

While the present study has focused on a system with fairly large screening length and high surface charge, our technique also should apply to systems with weaker interactions. Such measurements hold the promise of systematizing the study of colloidal phases and their properties. It is noteworthy that the present measurements correspond to a force resolution on the order of 10^{-15} N. Our method should prove complementary to two other new techniques: That of Calderon et al. measures the highly energetic interactions of monodisperse systems driven into linear chains by external fields [30], and the direct imaging technique of Fraden derives information from extensive statistics on the instantaneous distribution of thermally equilibrated systems [31]. The relative simplicity and versatility of our technique suggests several immediate applications. Measurements of the effective charge as a function of temperature, sphere size, and dielectric constant will facilitate refinement of the charge renormalization theory. We can also test the linearity of the interaction by studying spheres with unequal radii. Such measurements will be crucial in understanding the bulk properties of bidisperse systems.

We would like to thank C. Murray, T. Witten, L. Kadanoff, M. Robbins, D. Weitz, and S. Fraden for valuable insights. This work was supported by the NSF Materials Research Laboratory at The University of Chicago through Grant No. DMR88-19860. Additional funding was provided by DOE.

[1] W. B. Russel, D. A. Saville, and W. R. Schowalter, *Colloidal Dispersons* (Cambridge University Press, Cambridge, 1989).
[2] P. A. Rundquist, P. Photinos, S. Jagannathan, and S. A. Asher, J. Chem. Phys. **91**, 4932 (1989).
[3] P. Pieranski, L. Strzlecki, and B. Pansu, Phys. Rev. Lett. **50**, 900 (1983).
[4] C. A. Murray, in *Bond-Orientational Order in Condensed Matter Systems*, edited by K. J. Strandburg (Springer-Verlag, Berlin, 1991).
[5] B. V. Derjaguin and L. Landau, Acta Physicochimica (USSR) **14**, 633 (1941).
[6] E. J. Verwey and J. T. G. Overbeek, *Theory of the Stability of Lyophobic Colloids* (Elsevier, Amsterdam, 1948).
[7] P. M. Chaikin, P. Pincus, S. Alexander, and D. Hone, J. Colloid Interface Sci. **89**, 555 (1982).
[8] D. Hone, S. Alexander, P. M. Chaikin, and P. Pincus, J. Chem. Phys. **79**, 1474 (1983).
[9] M. O. Robbins, K. Kremer, and G. S. Grest, J. Chem. Phys. **88**, 3286 (1988).
[10] Y. Monovoukas and A. P. Gast, J. Colloid Interface Sci. **128**, 533 (1989).
[11] E. B. Sirota, H. D. Ou-Yang, S. K. Sinha, P. M. Chaikin, J. D. Axe, and Y. Fujii, Phys. Rev. Lett. **62**, 1524 (1989).
[12] S. Alexander, P. M. Chaikin, P. Grant, G. J. Morales, P. Pincus, and D. Hone, J. Chem. Phys. **80**, 5776 (1984).
[13] H. M. Lindsay and P. M. Chaikin, J. Chem. Phys. **76**, 3774 (1982), and references therein.
[14] I. Sogami, Phys. Lett. **96A**, 199 (1983).
[15] I. Sogami and N. Ise, J. Chem. Phys. **81**, 6320 (1984).
[16] L. Guldbrand, B. Jönsson, H. Wennerström, and P. Linse, J. Chem. Phys. **80**, 2221 (1983).
[17] C. E. Woodward, J. Chem. Phys. **89**, 5140 (1988).
[18] C. E. Woodward, B. Jönsson, and T. Åkesson, J. Chem. Phys. **89**, 5145 (1988).
[19] J. Israelachvili, *Intermolecular and Surface Forces* (Academic, London, 1992).
[20] T. Okubo, J. Chem. Phys. **90**, 2408 (1989).
[21] B. V. R. Tata, M. Rajalakshmi, and A. K. Arora, Phys. Rev. Lett. **69**, 3778 (1992).
[22] R. Rajagopalan, in *The Structure, Dynamics, and Equilibrium Properties of Colloidal Systems*, edited by D. M. Bloor and E. Wyn-Jones (Kluwer Academic, Boston, 1990), p. 695.
[23] M. L. Gee, P. Tong, J. N. Israelachvili, and T. A. Witten, J. Chem. Phys. **93**, 6057 (1990).
[24] K. Takamura, H. L. Goldsmith, and S. G. Mason, J. Colloid Interface Sci. **82**, 175 (1981).
[25] H. Risken, *The Fokker-Planck Equation* (Springer-Verlag, Berlin, 1989).
[26] R. S. Afzal and E. B. Treacy, Rev. Sci. Instrum. **63**, 2157 (1992).
[27] A. Ashkin, J. M. Dziedzic, J. E. Bjorkholm, and S. Chu, Opt. Lett. **11**, 288 (1986).
[28] M. Robbins (private communication).
[29] S. Bucci, C. Fagotti, V. Degiorgio, and R. Piazza, Langmuir **7**, 824 (1991).
[30] F. L. Calderon, T. Stora, O. M. Monval, P. Poulin, and J. Bibette (to be published).
[31] S. Fraden (private communication).

VOLUME 77, NUMBER 9   PHYSICAL REVIEW LETTERS   26 AUGUST 1996

When Like Charges Attract: The Effects of Geometrical Confinement on Long-Range Colloidal Interactions

John C. Crocker and David G. Grier

The James Franck Institute and Department of Physics, The University of Chicago, 5640 S. Ellis Avenue, Chicago, Illinois 60637
(Received 6 May 1996)

High-resolution measurements of the interaction potential between pairs of charged colloidal microspheres suspended in water provide stringent tests for theories of colloidal interactions. The screened Coulomb repulsions we observe for isolated spheres agree quantitatively with predictions of the Derjaguin-Landau-Verwey-Overbeek (DLVO) theory. Confining the same spheres between charged glass walls, however, induces a strong long-range attractive interaction which is not accounted for by the DLVO theory. [S0031-9007(96)01030-7]

PACS numbers: 82.70.Dd

The Derjaguin-Landau-Verwey-Overbeek (DLVO) theory [1] predicts that an isolated pair of highly charged colloidal microspheres will experience a purely repulsive screened Coulomb interaction at large separations. This prediction is at odds with mounting evidence that the effective pair interaction in dense suspensions sometimes has a long ranged attractive component. This evidence includes observations of stable multiparticle voids in colloidal fluids and crystals [2], phase separation between fluid phases of different densities [3], and long-lived, metastable colloidal crystallites in dilute suspensions [4]. Recently, two measurements [5,6] have revealed a strong long-range attraction acting between colloidal spheres confined to a plane by charged glass walls, while the corresponding measurements on unconfined colloid have not found any such attraction [7–9]. The confusing state of the present experimental evidence raises a question of fundamental importance to colloid science: When do like-charged colloidal spheres attract each other?

We first describe direct measurements of the pairwise interaction potential between three different sizes of colloidal microspheres mixed together in the same dilute suspension at low ionic strength. Requiring consistency among the parameters describing the interactions of different sized spheres makes possible stringent tests of the DLVO theory and of an alternative theory due to Sogami and Ise [10]. A second series of measurements strives to resolve the apparent discrepancy between interactions measured with and without planar confinement. By performing a sequence of interaction measurements in the same electrolyte but at different wall separations, we find that the attraction seen in the confined geometry vanishes as the walls are drawn apart.

The DLVO theory provides approximate solutions to the Poisson-Boltzmann equation describing the nonlinear coupling between the electrostatic potential and the distribution of ions in a colloidal suspension. The resulting interaction between isolated pairs of well-separated spheres has the simple form [11]

$$\frac{U_{\mathrm{DLVO}}(r)}{k_B T} = Z_1^* Z_2^* \frac{e^{\kappa a_1}}{1 + \kappa a_1} \frac{e^{\kappa a_2}}{1 + \kappa a_2} \lambda_B \frac{e^{-\kappa r}}{r}$$
$$+ \frac{V(r)}{k_B T}, \qquad (1)$$

where r is the center-to-center separation between two spheres of radii a_i with effective charges Z_i^*, in an electrolyte with Debye-Hückel screening length κ^{-1} and where $\lambda_B = e^2/\epsilon k_B T$ is the Bjerrum length, equal to 0.714 nm in water at $T = 24\,°\mathrm{C}$. For an electrolyte containing concentrations n_j of z_j-valent ions,

$$\kappa^2 = 4\pi \lambda_B \sum_{j=0}^{N} n_j z_j^2, \qquad (2)$$

where N is the number of ionic species. $V(r)$ accounts for van der Waals attraction but is weaker than $0.01 k_B T$ for submicron-diameter latex spheres more than 100 nm apart [12] and so is neglected in the following analysis.

Our technique for measuring colloidal interactions is described in detail elsewhere [7,8]. We use a pair of optical tweezers [13] to position a pair of colloidal microspheres reproducibly at fixed separations. Repeatedly blinking the laser tweezers and tracking the particles' motions with digital video microscopy while the traps are off allow us to sample and numerically solve the master equation for the equilibrium pair distribution function $g(r)$ with 50 nm spatial resolution. The interaction potential $U(r)$ can then be calculated (up to an additive offset) from the Boltzmann distribution, $U(r) = -k_B T \ln[g(r)]$. Roughly 4×10^4 images of sphere pairs made over a range of tweezer separations are required to produce a single interaction curve with an energy resolution of $0.1 k_B T$. The data set for a typical potential curve is collected using 4 or 5 different pairs of nominally identical spheres at several different locations in the sample volume. Repeatability of our results and the continuity of individual curves suggest both that the populations of spheres are homogeneous and also that chemical conditions in the sample volume are uniform.

VOLUME 77, NUMBER 9   PHYSICAL REVIEW LETTERS   26 AUGUST 1996

We performed a series of such measurements on a mixture of polystyrene sulfate spheres of diameters 0.652 ± 0.005, 0.966 ± 0.012, and 1.53 ± 0.02 μm dispersed in water [14]. The suspension was contained in a $2 \times 1 \times 0.005$ cm^3 sample volume formed by hermetically sealing the edges of a glass microscope cover slip to the face of a glass microscope slide. All glass surfaces were stringently cleaned with an acid-peroxide wash and therefore developed a negative surface charge density on the order of one electron equivalent per 10 nm^2 in contact with water [15]. The suspension was in diffusive contact, via holes drilled in the glass slide, with reservoirs of mixed bed ion exchange resin flushed with humidified Ar to prevent contamination by atmospheric CO_2. Finally, the sample temperature was regulated at $T = 24.0 \pm 0.1\,°C$ to ensure reproducibility of our results.

Despite these precautions, glass surfaces act as a small virtual leak of ions. The screening length in the sample volume consequently decreases,

$$\kappa^{-1}(t) \approx \kappa^{-1}(0)(1 - \alpha t), \qquad (3)$$

at a rate α which we estimate by comparing identical measurements made over sufficiently long time intervals.

Figure 1 shows $U(r)$ measured for pairs of spheres from each of the three populations. The optical tweezers were set to maintain the spheres more than 8 μm away from the nearest glass wall throughout the measurements and in a region of the sample volume devoid of other spheres. Thus the data in Fig. 1 represent the pairwise interaction potentials in the limit of infinite dilution. The curves in Fig. 1 are fits by Eq. (1) for the screening length κ^{-1}, the effective charge Z^*, and an additive offset. The fit parameters appear in Table I. All three data sets were obtained in the same electrolyte during a period of 4 h. The fit values for the screening lengths are all consistent with $\kappa^{-1} = 280 \pm 15$ nm, corresponding to a $n = 1.2 \times 10^{-6}M$ concentration of 1:1 electrolyte.

To estimate α, we repeated the initial measurement on the 1.5 μm diameter spheres after the 5 h interval in which data for the other size spheres were obtained. Assuming constant effective charge, we fit both data sets to Eq. (1) with values of Z^* constrained to be equal and obtain $\alpha = 0.009 \pm 0.002$ h^{-1}. This suggests that the electrolytic strength increases by 2×10^{-8} M h^{-1}, which can be accounted for by a flux of 3 ions nm^{-2} yr^{-1} from the walls. Substituting this result into Eq. (3) with $\kappa^{-1}(0) = 280$ nm and refitting the data in Fig. 1 for the two remaining free parameters results in the solid curves in Fig. 1. The constrained fit parameters appear in Table II. The dashed curves in Fig. 1 represent the unconstrained fits. While the constrained and unconstrained fits are barely distinguishable, the constrained fits facilitate comparison between values of Z^* obtained from the different data sets. The error in the estimate for Z^* in an individual data set is roughly 25% because of the estimated

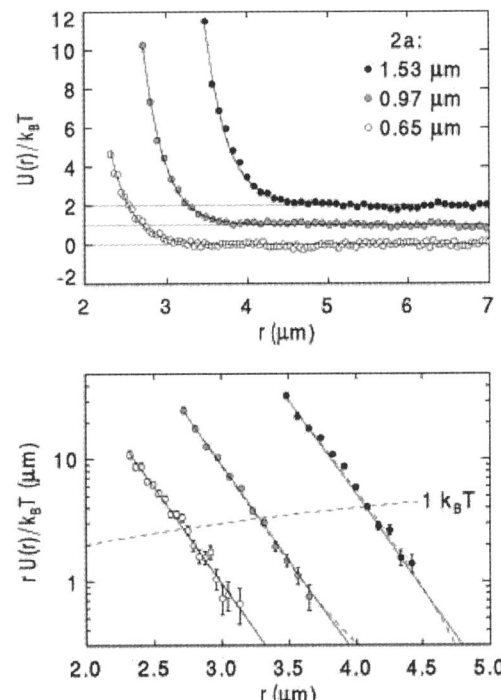

FIG. 1. (Top) Pairwise interaction potentials for three different populations of colloidal microspheres measured in the same electrolyte. The different data sets are offset by $1k_BT$ for clarity. Curves are fits by Eq. (1) with parameters given in Tables I and II. (Bottom) Data replotted to emphasize the screened Coulomb functional form of the interaction.

5% error in the value for $\kappa^{-1}(0)$, but the error in ratios of Z^* values from the constrained fits is only about 5%.

The data of Fig. 1 may also be used to test alternative theories of colloidal interactions, including the Sogami-Ise (SI) potential [10],

$$\frac{U_{SI}(r)}{k_B T} = Z_{SI}^{*2}\left(\frac{\sinh \kappa a}{\kappa a}\right)^2 \left(1 + \kappa a \coth \kappa a - \frac{\kappa r}{2}\right)$$
$$\times \lambda_B \frac{e^{-\kappa r}}{r}. \qquad (4)$$

Equation (4) predicts a deep potential minimum at large separations for some conditions, although there are regimes where this minimum is small or nonexistent.

TABLE I. Interaction parameters for isolated pairs of spheres obtained by fits of the data in Fig. 1 by Eq. (1) and by Eq. (4).

$2a$ (μm)	Z^*	κ^{-1} (nm)	$\zeta_0(mV)$	Z_{SI}^*	κ_{SI}^{-1} (nm)
1.53	22 793	289	-167	1767	960
0.97	13 796	268	-165	1525	730
0.65	5964	272	-145	777	670

VOLUME 77, NUMBER 9   PHYSICAL REVIEW LETTERS   26 AUGUST 1996

TABLE II. Interaction parameters obtained by fits of the data in Figs. 1 and 2 by Eq. (1) with κ^{-1} given by Eq. (3).

$2a_1$ (μm)	$2a_2$ (μm)	κ^{-1}	Z^*_{fit}	$\sqrt{Z^*_1 Z^*_2}$	Time (h)
1.53	1.53	280	26 136	$\cdots$	0
0.97	0.97	278	11 965	$\cdots$	0.9
0.65	0.65	275	5638	$\cdots$	1.7
1.53	0.97	270	17 401	17 684	4.2
1.53	0.65	266	12 393	12 139	5.7
0.97	0.65	265	8684	8213	6.4

Indeed, Eq. (4) provides satisfactory fits to the curves in Fig. 1. These fits, however, result in values for the screening length, κ^{-1}_{SI}, which vary systematically with a. Variation in κ^{-1}_{SI} is not likely to reflect drifts in the ionic strength since the control measurement on the 1.5 μm spheres yields a fit value for κ^{-1}_{SI} within 10% of the tabulated value. This inconsistency provides compelling evidence that the SI theory does not correctly describe the interactions between isolated pairs of charged colloidal spheres.

The effective charges Z^*_i can be related to the effective sphere surface potentials ζ_0 by [11]

$$Z^* = \left(\frac{e\zeta_0}{k_B T} \right) \frac{a}{\lambda_B} (1 + \kappa a). \quad (5)$$

In the limit of high surface charge density, both ζ_0 and Z^*_i should saturate at finite values due to incomplete dissociation of the surface groups [16] and strong nonlinear screening near the sphere surfaces neglected in the DLVO theory [17]. In this limit, Eq. (5) reproduces both the roughly quadratic dependence of Z^* on a seen in Table II and also the linear dependence predicted by the Poisson-Boltzmann cell model [18] in the limit $\kappa a \ll 1$.

The linear superposition approximation (LSA) used in calculating Eq. (1) requires each sphere's effective charge to be independent of the size and charge state of the other. We test the LSA's validity in our system by measuring interactions between dissimilar spheres. Figure 2 presents such potentials measured immediately after the like-sphere measurements of Fig. 1. As before, the measured interactions are fit well by Eq. (1) with κ^{-1} given by Eq. (3) and show no attractive component. The extracted charge numbers in Table II agree well with the geometric means of the individual sphere charges. Thus the DLVO theory successfully describes all six experimental curves with only five free parameters: $\kappa^{-1}(0)$, α, and the three Z^*_i.

The DLVO theory's quantitative agreement with measurements on isolated spheres needs to be reconciled with reports of attractive interactions when spheres are confined by glass walls [5,6]. We performed a series of measurements on a sample cell whose thin cover slip could be bowed inward by applying negative pressure. Figures 3(a)–3(d) show interaction curves for spheres of diameter $2a = 0.97$ μm measured in different regions of the bowed sample volume with wall separations varying

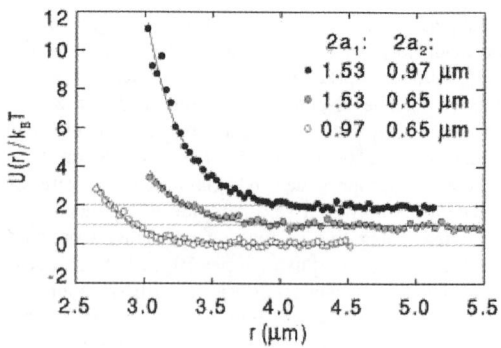

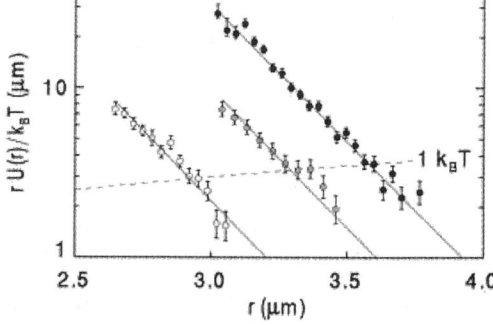

FIG. 2. (Top) Interaction potentials for isolated dissimilar pairs of spheres. Different data sets are offset by $1k_B T$ for clarity. Curves are constrained fits by Eq. (1) with parameters given in Table II. (Bottom) Data replotted to emphasize the functional form of the interaction.

from $d = 6.5 \pm 0.5$ down to 2.6 ± 0.3 μm. We measured d at each location by focusing the laser traps onto the glass-water interfaces and estimate the wedge angle to be less than 10^{-3} rad.

At the widest separation, spheres are free to roam in all three dimensions and the measured interaction follows the DLVO form with $\kappa^{-1} = 100 \pm 10$ nm. In regions where $d < 5$ μm, the spheres are confined to the cell's midplane by electrostatic interactions with the charged walls. Constancy of the spheres' images suggests they move out of the focal plane by less than 150 nm [7]. Under these conditions, an attractive minimum appears in the measured potential whose form is comparable to those previously reported [5,6]. Repeated measurements such as those in Fig. 3 suggest that the as yet unexplained attractive interaction is stronger and longer ranged for larger spheres.

When the wall separation is reduced to $d = 2.6 \pm 0.3$ μm, the interaction potential changes once again to a purely repulsive form. We interpret the data in Fig. 3(d) as resulting from the superposition of the DLVO repulsive core, the confinement-induced attraction (leading to the plateau in the curve), and an additional

VOLUME 77, NUMBER 9   PHYSICAL REVIEW LETTERS   26 AUGUST 1996

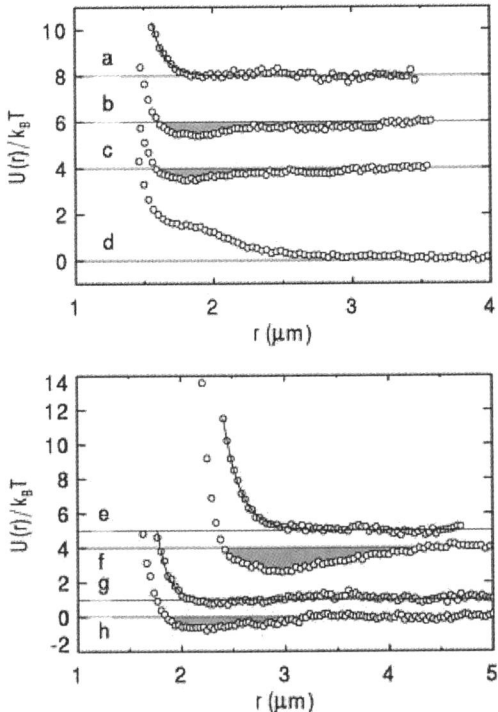

FIG. 3. (Top) Interaction potentials for $2a = 0.97$ μm diameter spheres measured in the same electrolyte between parallel glass walls separated by (a) $d = 6.5 \pm 0.5$ μm, (b) 4.0 ± 0.3 μm, (c) 3.5 ± 0.3 μm, (d) 2.6 ± 0.3 μm. (Bottom) Confinement-induced attraction for two different sphere size populations measured in the same electrolyte. $2a = 1.53$ μm: (e) unconfined, (f) $d = 3.0 \pm 0.5$ μm; $2a = 0.97$ μm: (g) unconfined, (h) $d = 3.5 \pm 0.5$ μm. Curves are offset for clarity.

dense colloidal suspensions [2–4]. Our measurements suggest that their origin is not to be found in the dilute-limit pair interaction, but do not rule out attractions mediated by many-body effects at finite volume fraction. Such an effect might be related to the unexplained attractions arising in the confined geometry, with the ensemble of spheres in a dense suspension playing a similar role to the charged walls.

We acknowledge valuable conversations with Tom Witten, Stuart Rice, Andy Marcus, and Seth Fraden. This work was supported by the National Science Foundation under Grant No. DMR-9320378.

long-range repulsive interaction. Such a repulsion can be mediated by unscreened electric fields propagating through the nearby glass walls [19].

Data such as those in Fig. 3(a) demonstrate that mere proximity of the charged walls is not sufficient to induce attractive interactions. Attractions are only seen when the spheres are rigidly confined, and not otherwise. This coincidence suggests that strong coupling between the counterion clouds of the spheres and the walls is necessary to produce the observed attraction. The DLVO theory is not formulated for such conditions, and its failure is not surprising. Regardless of their explanation, these observations indicate a need to reinterpret experiments on colloidal suspensions in porous media and confined colloidal monolayers, particularly in the context of two-dimensional melting.

Attractive pairwise interactions would provide a natural explanation for the anomalous phase behavior seen in

[1] B. V. Derjaguin and L. Landau, Acta Physicochimica (USSR) **14**, 633 (1941); E. J. Verwey and J. Th. G. Overbeek, *Theory of the Stability of Lyophobic Colloids* (Elsevier, Amsterdam, 1948).
[2] N. Ise and H. Matsuoka, Macromolecules **27**, 5218 (1994).
[3] B. V. R. Tata, M. Rajalakshmi, and A. Arora, Phys. Rev. Lett. **69**, 3778 (1992); T. Palberg and M. Würth, Phys. Rev. Lett. **72**, 786 (1994); B. V. R. Tata and A. K. Arora, Phys. Rev. Lett. **72**, 787 (1994).
[4] A. E. Larsen and D. G. Grier, Phys. Rev. Lett. **76**, 3862 (1996).
[5] G. M. Kepler and S. Fraden, Phys. Rev. Lett. **73**, 356 (1994).
[6] M. D. Carbajal-Tinoco, F. Castro-Román, and J. L. Arauz-Lara, Phys. Rev. E **53**, 3745 (1996).
[7] J. C. Crocker and D. G. Grier, J. Colloid Interface Sci. **179**, 298 (1996).
[8] J. C. Crocker and D. G. Grier, Phys. Rev. Lett. **73**, 352 (1994).
[9] K. Vondermassen, J. Bongers, A. Mueller, and H. Versmold, Langmuir **10**, 1351 (1994).
[10] I. Sogami and N. Ise, J. Chem. Phys. **81**, 6320 (1984).
[11] W. B. Russel, D. A. Saville, and W. R. Schowalter, *Colloidal Dispersions* (Cambridge University Press, Cambridge, 1989).
[12] B. A. Pailthorpe and W. B. Russel, J. Colloid Interface Sci. **89**, 563 (1982).
[13] A. Ashkin, J. M. Dziedzic, J. E. Bjorkholm, and S. Chu, Opt. Lett. **11**, 288 (1986).
[14] Duke Scientific Catalogs No. 5065A, No. 5095A, and No. 5153A.
[15] R. K. Iler, *The Chemistry of Silica* (Wiley, New York, 1972).
[16] T. Gisler, S. F. Schulz, M. Borkovec, H. Sticher, P. Schurtenberger, B. D'Aguanno, and R. Klein, J. Chem. Phys. **101**, 9924 (1994).
[17] H. Löwen, P. A. Madden, and J.-P. Hansen, Phys. Rev. Lett. **68**, 1081 (1992); H. Löwen, J.-P. Hansen, and P. A. Madden, J. Chem. Phys. **98**, 3275 (1993).
[18] S. Alexander, P. M. Chaikin, P. Grant, G. J. Morales, P. Pincus, and D. Hone, J. Chem. Phys. **80**, 5776 (1984).
[19] F. H. Stillinger, J. Chem. Phys. **35**, 1584 (1961).

APPLIED PHYSICS LETTERS   VOLUME 72, NUMBER 23   8 JUNE 1998

Optically induced rotation of a trapped micro-object about an axis perpendicular to the laser beam axis

E. Higurashi,[a)] R. Sawada, and T. Ito

NTT Opto-electronics Laboratories, 3-9-11, Midori-cho, Musashino-shi Tokyo 180-8585, Japan

(Received 2 February 1998; accepted for publication 7 April 1998)

A strongly focused Nd:YAG laser beam has been used experimentally to achieve optical trapping and simultaneous rotation of a micro-object about an axis perpendicular to the trapping laser beam axis. The micro-object (12 μm in diameter), which has shape anisotropy in its interior, can be produced by oxygen reactive ion etching of a fluorinated polyimide film (refractive index n = 1.53). This optically induced rotation of the micro-object in ethanol (n = 1.36) is accomplished by utilizing the net radiation pressure force arising from the momentum change of light scattered at the micro-object's inner walls. © *1998 American Institute of Physics.* [S0003-6951(98)03523-2]

The technique of optical manipulation[1,2] provides a unique means of controlling the microdynamics of small particles without physical contact. Several methods using radiation pressure have been reported for rotating microparticles by using a circularly (elliptically) polarized laser beam[3-5] or a laser beam with a special phase structure[6,7] for absorptive particles, and a rotating asymmetric beam[8] for transparent particles. Furthermore, one powerful approach using the anisotropic shape of the trapped particles has been reported for creating a remotely driven radiation pressure micromotor for micromechanical systems.[9] This optically induced rotation arises from the net radiation pressure exerted on the outside of a micro-object that has shape anisotropy. This torque has also been analyzed by a ray optics model.[10] Recently some studies improving the design of the shape of the micro-object have been reported.[11-14] Nevertheless, all these rotations occurred only around the laser beam axis.[3-14] There have been no reports yet on the possibility of both trapping by backward axial gradient force of radiation pressure and simultaneous rotation about an axis perpendicular to the laser beam axis, induced by the axial scattering force of radiation pressure.

In this letter, we report the experimental observation of rotation of a trapped micro-object about an axis perpendicular to the laser beam axis. To demonstrate this phenomenon, we used a novel type of micro-object having an opening and shape anisotropy in its interior. The scattering force of radiation pressure exerted on the inner walls of the micro-object creates a torque while keeping the micro-object in an optical trap by a strong axial gradient force.

Optical radiation forces arising from the transfer of optical momentum can be divided into two basic components: scattering force and gradient force. Scattering force is proportional to the light intensity and acts in the direction of propagation of the laser light. Gradient force is proportional to the gradient of the spatial light intensity and acts in the direction of the intensity gradient. If we use a strongly focused laser beam, the backward axial gradient force is large and overcomes the gravitational force. As a result, the micro-object is trapped at a stable equilibrium point. Furthermore, if the axial scattering force has a torque component, we can achieve both trapping and rotation of the micro-object about an axis perpendicular to the laser beam axis.

Since trapping and rotation tend to become unstable when the light incident surface (exterior) of the micro-object has an irregular shape, we designed the shape of the outside cross section of the micro-object as a smooth circle. A thin, ringlike micro-object (a microdisk having an opening) with a refractive index greater than the surrounding medium stood parallel to the laser beam axis when trapped in three dimensions and was offset with respect to the laser beam axis, as shown in Fig. 1(a). This is because trapping is stable when the high-refractive index part of the micro-object is located in the high-intensity region of the laser beam. This phenomenon of aligning along the laser beam axis was theoretically and experimentally studied by Gauthier for cylindrical micro-objects.[15] To induce the rotation, we gave the micro-object anisotropic geometry in its interior. This micro-object (outer diameter=12 μm; width=2.9 μm) has shape dissymmetry (rotational but not bilateral symmetry) in its internal vertical cross sections [Fig. 1(b)].

Figure 2 shows the radial and torque components of the forces of radiation pressure exerted on the inner walls of the micro-object. Typical rays a and b striking its inner walls give rise to radiation pressure forces in the direction of the surrounding medium and perpendicular to the object surface. Summing the forces of all the rays, one sees that the net radiation pressure generates an optical torque. This is because radiation pressure force on inner wall (ii) contributes

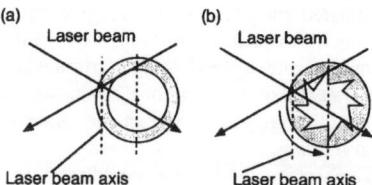

FIG. 1. Geometry of optical trapping of micro-objects (side view); (a) ring-shaped micro-object (microdisk having an opening), (b) micro-object having an opening and shape anisotropy in its interior.

[a)]Electronic mail: eiji@ilab.ntt.co.jp

2952   Appl. Phys. Lett., Vol. 72, No. 23, 8 June 1998

Higurashi, Sawada, and Ito

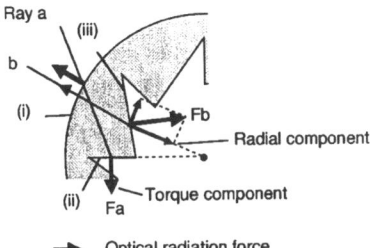

FIG. 2. Forces of radiation pressure exerted on the inner walls of the micro-object in Fig. 1(b).

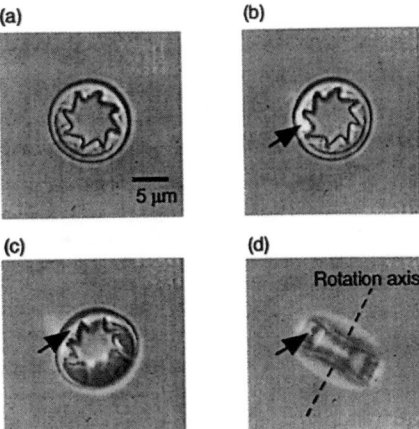

FIG. 4. A series of four frames from a video record demonstrating the trapping and rotation of the micro-object about an axis perpendicular to the laser beam axis in ethanol. The laser beam (power=50 mW) was irradiated perpendicular to the plane of the photograph. The position of the laser focal point is indicated by the arrow; (a) before laser irradiation, (b) the micro-object trapped two-dimensionally on the bottom of the sample cell, (c) when the microscope stage was moved downward, the micro-object followed the focus of the laser beam, (d) the three-dimensionally trapped micro-object standing parallel to the laser beam axis was rotated about an axis perpendicular to the laser axis.

only the torque component, while radiation pressure force on inner wall (iii) has a radial component as well as the torque component.

Transparent fluorinated polyimide (6FDA/TFDB, refractive index $n=1.53$)[16] was used as the micro-object material to avoid radiometric forces. Fluorinated polyimide micro-objects were fabricated by electron-beam lithography and oxygen reactive ion etching of a fluorinated polyimide film. Figure 3 shows a scanning electron micrograph (SEM) photograph of a fabricated micro-object. These micro-objects were freed from the substrate and dispersed in ethanol. The experimental setup was almost identical to that reported in our earlier papers.[9,14] A downward-directed Nd:YAG laser (cw, $\lambda=1064$ nm, TEM$_{00}$ mode) was focused with a water-immersion objective lens (Leitz, PL Fluotar, $\times100$, NA $=1.20$) onto a micro-object in ethanol ($n=1.36$). The maximum laser power incident on the micro-object was 100 mW (measured after the objective lens).

Figure 4 shows the trapping and rotation of the micro-object about an axis perpendicular to the laser beam axis in ethanol. The laser beam was irradiated perpendicular to the plane of the photograph. When the light was tuned on, the micro-object was quickly pulled to the focal point of the laser beam and trapped in two dimensions [Fig. 4(b)]. Once a micro-object was trapped two-dimensionally, the microscope stage was moved downwards. The micro-object followed the focal point of the laser beam [Fig. 4(c)] and stood parallel to the laser beam axis, which was consistent with the theoretical prediction.[15] Its center axis was offset with respect to the laser beam axis. As soon as the micro-object was

trapped in three dimensions, it rotated about an axis perpendicular to the laser beam axis [Fig. 4(d)].

The rotation speed was measured using recorded video images and light scattered from the micro-object.[17] A photodiode was positioned on part of the forward scattered pattern. Because the micro-object's cross section had eightfold rotational symmetry, one revolution produced eight peaks with equal intervals of intensity change. Because all the peaks had equal intensity during one revolution, we conclude that this optically induced rotation was stable, without eccentricity. Since the three-dimensionally trapped micro-object was not very stable at lower laser power (<30 mW), we measured the rotation characteristics of the two-dimensionally trapped micro-object (standing parallel to the laser beam axis) on the bottom of the sample cell. Figure 5 shows the measured rotation speed versus incident laser power. The rotation speed was approximately proportional to the incident laser power used for the trapping. It is determined by the balance between the externally applied optical torque and the torque arising from viscous drag.

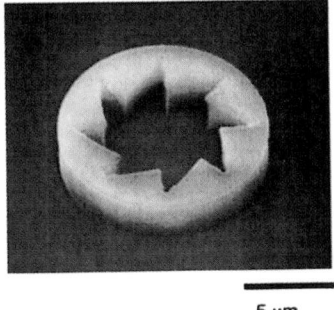

FIG. 3. SEM photograph showing the fabricated fluorinated polyimide micro-object (height=2.9 μm).

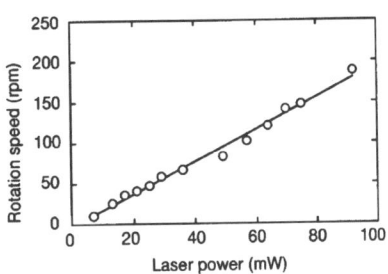

FIG. 5. Measured rotation speed vs incident laser power.

Appl. Phys. Lett., Vol. 72, No. 23, 8 June 1998

It is clear that this rotation was caused by the optical torque exerted on the micro-object's inner walls, because radiation pressure forces exerted on its outside act radially and do not contribute to optical torque. The stability of optical trapping depends on the balance between the gradient force and the scattering force. In our experiment, we used scattering forces, which try to push the micro-object away from the focus, to induce a torque. This tends to destabilize the trap. Therefore, a micro-object must be illuminated with a strongly focused laser beam (by a high-NA objective lens) to overcome the scattering force and achieve stable axial equilibrium. Because the scattering force changes during rotation in the light beam, we gave the micro-object manifold rotational symmetry which promotes stable rotation. Besides, in our experiment high viscous damping from the surrounding liquid (ethanol) further helped to confine the micro-object within an optical trap.

In conclusion, we have demonstrated three-dimensional optical trapping and simultaneous rotation of a micro-object about an axis perpendicular to the laser beam axis. This optically induced rotation requires no physical contact or electrical wires, so it has potential for use in remotely driven radiation pressure micromotors and sensitive microprobes for measuring the properties of microfluid dynamics. In biological applications of this technique, it should be possible to extend the manipulative capability, such as by using the micro-objects as handles to twist biological specimens (e.g., DNA).

[1] A. Ashkin, Phys. Rev. Lett. **24**, 156 (1970).
[2] A. Ashkin, J. M. Dziedzic, J. E. Bjorkholm, and S. Chu, Opt. Lett. **11**, 288 (1986).
[3] A. Ashkin and J. M. Dziedzic, Appl. Phys. Lett. **28**, 333 (1976).
[4] T. Sugiura, S. Kawata, and S. Minami, J. Spectrosc. Soc. Jpn. **39**, 342 (1990) [in Japanese].
[5] M. E. J. Friese, T. A. Nieminen, N. R. Heckenberg, and H. Rubinsztein-Dunlop, Opt. Lett. **23**, 1 (1998).
[6] H. He, M. E. J. Friese, N. R. Heckenberg, and H. Rubinsztein-Dunlop, Phys. Rev. Lett. **75**, 826 (1995).
[7] N. B. Simpson, K. Dholakia, L. Allen, and M. J. Padgett, Opt. Lett. **22**, 52 (1997).
[8] S. Sato, M. Ishigure, and H. Inaba, Electron. Lett. **27**, 1831 (1991).
[9] E. Higurashi, H. Ukita, H. Tanaka, and O. Ohguchi, Appl. Phys. Lett. **64**, 2209 (1994).
[10] R. C. Gauthier, Appl. Phys. Lett. **67**, 2269 (1995).
[11] E. Higurashi, O. Ohguchi, H. Ukita, and T. Tamamura, in Proceedings of International Symposium on Microsystems, Intelligent Materials and Robots, Sendai, Japan, Sept. 27–29, 1995, p. 63.
[12] R. C. Gauthier, Appl. Phys. Lett. **69**, 2015 (1996).
[13] H. Ukita and K. Nagatomi, Opt. Rev. **4**, 447 (1997).
[14] E. Higurashi, O. Ohguchi, T. Tamamura, H. Ukita, and R. Sawada, J. Appl. Phys. **82**, 2773 (1997).
[15] R. C. Gauthier, J. Opt. Soc. Am. B **14**, 3323 (1997).
[16] T. Matsuura, N. Yamada, S. Nishi, and Y. Hasuda, Macromolecules **26**, 419 (1993).
[17] A. Yamamoto and I. Yamaguchi, Jpn. J. Appl. Phys., Part 1 **34**, 3104 (1995).

52   OPTICS LETTERS / Vol. 22, No. 1 / January 1, 1997

Mechanical equivalence of spin and orbital angular momentum of light: an optical spanner

N. B. Simpson, K. Dholakia, L. Allen, and M. J. Padgett

J. F. Allen Physics Research Laboratories, Department of Physics and Astronomy, University of St. Andrews, North Haugh, St. Andrews, Fife KY16 9SS, Scotland

Received August 27, 1996

We use a Laguerre–Gaussian laser mode within an optical tweezers arrangement to demonstrate the transfer of the orbital angular momentum of a laser mode to a trapped particle. The particle is optically confined in three dimensions and can be made to rotate; thus the apparatus is an optical spanner. We show that the spin angular momentum of $\pm \hbar$ per photon associated with circularly polarized light can add to, or subtract from, the orbital angular momentum to give a total angular momentum. The observed cancellation of the spin and orbital angular momentum shows that, as predicted, a Laguerre–Gaussian mode with an azimuthal mode index $l = 1$ has a well-defined orbital angular momentum corresponding to $\hbar$ per photon. © 1997 Optical Society of America

The circularly symmetric Laguerre–Gaussian (LG) modes form a complete basis set for paraxial light beams.[1] Two indices identify a given mode, and the modes are normally denoted $LG_p{}^l$, where l is the number of 2π cycles in phase around the circumference and $(p + 1)$ is the number of radial nodes. In contrast to the planar wave fronts of the Hermite–Gaussian (HG) mode, for $l \neq 0$ the azimuthal phase term $\exp(il\phi)$ in the LG modes results in helical wave fronts.[2] These helical wave fronts are predicted by Allen *et al.* to give rise to an orbital angular momentum for linearly polarized light of $l\hbar$ per photon.[3] This orbital angular momentum is distinct from the angular momentum associated with the photon spin manifested in circularly polarized light. For a collimated beam of circularly polarized light the photon spin is well known to be $\pm\hbar$; however, for a tightly focused beam the polarization state is no longer well defined.[4] The total angular momentum of a beam, found by a rigorous solution of Maxwell's equations that reduces to a LG beam in the paraxial approximation, is given by[5]

$$\left[l + \sigma_z + \sigma_z \left(\frac{2kz_r}{2p + l + 1} + 1 \right)^{-1} \right] \hbar \qquad (1)$$

per photon, where $\sigma_z = 0$, ± 1 for linearly and circularly polarized light, respectively, and where z_r is the Rayleigh range and k is the wave number of the light. Note that for a collimated beam $kz_r \gg 1$, and expression (2) reduces to

$$(l + \sigma_z)\hbar \qquad (2)$$

per photon, whereas for linearly polarized light $l\hbar$ per photon is strictly correct even in the absence of any approximation.

Lasers have been made that operate in LG modes,[6] but it is easier to obtain them from the conversion of HG modes. Three different classes of mode converter have been demonstrated. Two of these, spiral phase plates[7] and computer-generated holographic converters,[8,9] introduce the azimuthal phase term to a $HG_{0,0}$ Gaussian beam to produce a LG mode with $p = 0$ and specific values of l. In all these devices a screw phase

dislocation, produced on axis, causes destructive interference, leading to the characteristic ring intensity pattern in the far field. The other class of converter is the cylindrical-lens mode converter,[10] which employs the change in Gouy phase in the region of an elliptically focused beam to convert higher-order HG modes, of indices m and n, into the corresponding LG modes (or vice versa) with 100% efficiency. The transformation is such that $l = |m - n|$ and $p = \min(m, n)$.

Consider the various macroscopic mechanisms whereby angular momentum can be transferred between a well-defined transverse laser mode and matter. One can transfer spin angular momentum by using a birefringent optical component, resulting in a change in the polarization state of the light. One can transfer orbital angular momentum by using a component that possesses an azimuthal dependence of its optical thickness, as with a cylindrical lens or a spiral phase plate. In contrast, absorption of the light allows both spin and orbital angular momentum to be transferred.

The prediction that LG laser modes possess well-defined orbital angular momentum has led to considerable research activity.[3,6–14] The transfer of orbital angular momentum from a light beam to small particles held at the focus has been modeled,[12] but to date only one group of researchers has demonstrated that these modes do indeed possess orbital angular momentum. He *et al.*[13] demonstrated that orbital angular momentum could be transferred from the laser mode to micrometer-sized ceramic and metal-oxide particles near the focus of a LG beam with an azimuthal mode index $l = 3$. More recently the same authors reported the use of a circularly polarized mode and the corresponding speeding up or slowing down of the rotation, depending on the relative sign of the spin and orbital angular-momentum terms.[14] In each case the particles were trapped with radiation pressure, first forcing the totally absorbing particle into the dark region on the beam axis (x–y trapping) and then forcing the particle against the microscope slide (z trapping). LG modes were recently also used in an optical tweezers

0146-9592/97/010052-03$10.00/0

January 1, 1997 / Vol. 22, No. 1 / OPTICS LETTERS   **53**

geometry for trapping hollow glass spheres[15]; however, no rotation was observed because these spheres were totally transparent.

Here we use weakly absorbing dielectric particles, larger than the dimensions of the tightly focused LG mode. The particle experiences a net force within the electric field gradient toward the focus of the beam and is thus held in an all-optical $x-y-z$ trap. This clearly distinguishes our experiments from previous ones,[13,14] which relied on mechanical restraint in the z direction. Such all-optical traps, using zero-order Gaussian modes, were demonstrated by Ashkin $et\ al.$[16] and are commonly referred to as optical tweezers. The transfer of angular momentum and the subsequent rotation of the trapped particle through the use of a LG mode leads us to refer to our modified optical geometry as an optical spanner.

Figure 1 shows our experimental arrangement. We use a diode-pumped Nd:YLF laser at 1047 nm operating in a high-order HG mode. This mode is converted into the corresponding LG mode by the use of the previously mentioned cylindrical-lens mode converter; a quarter-wave plate sets the polarization state. A telescope arrangement optimizes the mode size to fill the back aperture of the objective lens, and an adjustable mirror enables the focused beam to be translated within the field of view of the objective lens. A dichroic beam splitter couples the beam into the optical spanner. The optical spanner is based on a 1.3-N.A., 100× oil-immersion microscope objective. The associated tube lens forms an image of the trapped particle on a CCD array. The sample cell comprises a microscope slide, an $\approx$60-μm-thick vinyl spacer, and a cover slip and is backlit with a standard tungsten bulb. The LG mode is focused such that the trapped object is held in the focal plane of the objective lens, and a colored glass filter can be inserted before the CCD array to attenuate the reflected laser light. We independently confirmed the polarization state of the light in the plane of the trapped particle by monitoring the power transmitted through a linear polarizer at various orientations.

When the $l = 1$ LG mode was focused by the objective lens, its beam waist was approximately 0.8 μm, which corresponds to a high-intensity ring 1.1 μm in diameter. For dielectric particles smaller than $\sim$1 μm we observe that the gradient force results in the particles' being trapped off axis, centered within the high-intensity region of the ring itself. However, particles 1 μm or larger are trapped on the beam axis and for incident power above a few milliwatts are held stably in three dimensions. If the particle is partially absorbing, the transfer of angular momentum from the laser beam to the particle causes the particle to rotate. The strong z-trapping force enables us to lift the particle optically off the button of the sample cell, allowing it to rotate more freely. We have successfully rotated a number of a different particle types, including particles of absorbing glass (e.g., Schott BG38 glass) and Teflon spheres. The particles were dispersed in various fluids, including water, methanol, and ethanol.

For the linearly polarized $l = 1$ LG mode and an incident power of $\approx$25 mW, typical rotation speeds for the Teflon particles were $\sim$1 Hz. The relationship between applied torque τ and limiting angular velocity ω_{lim} of a sphere of radius r in a viscous medium of viscosity η is[17]

$$\omega_{lim} = \tau/(8\pi\eta r^3).\qquad(3)$$

This implies that the absorption of the Teflon particle is of the order of 2%.

The principal motivation of this study was to demonstrate that, with respect to absorption and the resulting mechanical rotation of a particle, the spin and the orbital angular momenta of light are equivalent. As discussed above, for the spin and orbital angular-momentum content of the laser mode to interact in an equivalent manner with the trapped particle it is essential that the dominant mechanism for transfer of the angular momentum be absorption. The birefringence or astigmatism of the trapped particle will preferentially transfer spin or orbital angular momentum, respectively. For comparison of spin and orbital angular momentum we selected Teflon particles, or amalgamated groups of Teflon particles, larger than the focused beam waist. Trapped on the beam axis, these particles interact uniformly with the whole of the beam, can be optically levitated off the bottom of the sample cell, and are observed to rotate smoothly with a constant angular velocity. This ability to lift the particles optically away from the cell boundary differentiates our research from that previously reported. Less regular particles could be lifted and made to rotate in the same sense as the more regular particles but often failed to stop when the spin and orbital angular-momentum terms were subtractive. This imperfect cancellation is due to additional transfer of orbital angular momentum owing to the asymmetry of the particle.

For our experimental configuration and an $l = 1$ mode, expression (1) shows that the total angular

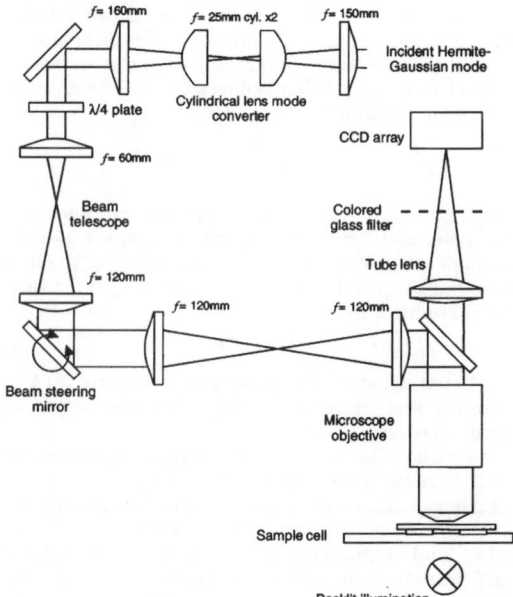

Fig. 1. Experimental configuration of the optical spanner.

54   OPTICS LETTERS / Vol. 22, No. 1 / January 1, 1997

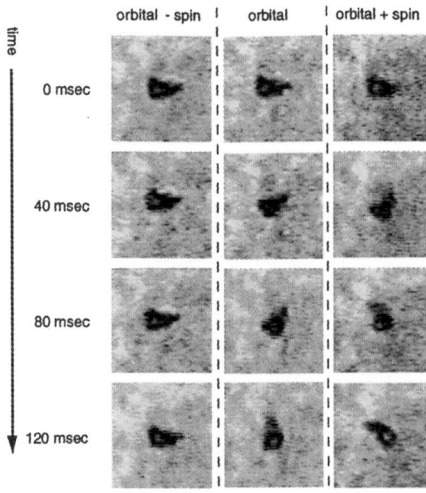

time

orbital - spin | orbital | orbital + spin

0 msec

40 msec

80 msec

120 msec

Fig. 2. Successive frames of the video image showing the stop–start behavior of a 2-μm-diameter Teflon particle held with the optical spanner.

momentum is $\sim 2.06\hbar$ per photon when the spin and orbital terms are additive and $\sim 0.06\hbar$ per photon when the spin and orbital terms are subtractive.

Figure 2 shows successive frames of the video image of a 2-μm-diameter Teflon particle held within the optical spanner, trapped with an $l = 1$ mode of various polarization states. The slight asymmetry in the particle geometry allows the particle rotation to be assessed. In more than 80% of cases, once a smoothly rotating particle had been selected, orienting the wave plate to circularly polarize the light would cause the particle to speed up or stop completely, depending on whether the spin and orbital angular-momentum terms were added or subtracted. In the other 20% of cases, although the particles would speed up or slow down, they could not be stopped completely. We believe that this was because these particles were insufficiently uniform, resulting in an unwanted mode transformation and giving an additional exchange of orbital angular momentum. With the well-behaved particles, when the spin and orbital terms are subtracted the low value of the torque, owing to a total angular momentum of no more than $0.06\hbar$ per absorbed photon, is insufficient to overcome the stiction present within the system, and the particle ceases to rotate. Stiction can arise from slight particle asymmetry coupled with residual astigmatism in the laser mode, which will favor the particles' being trapped in particular orientations. When the spin and orbital terms are additive the rotation speed increases significantly, but in only few cases is it seen to double. We attribute this to a nonlinear relationship between the applied torque and the terminal rotation speed, and the particle asymmetry. We confirmed this nonlinearity in the relationship between torque and rotation speed by deliberately changing the optical power while maintaining the same laser mode.

We observe, in agreement with our interpretation, that reversing the sense of the cylindrical-lens mode converter caused the particles to rotate in the opposite direction; similarly, by use of a $HG_{0,0}$ mode, particles could be rotated in either direction when circularly polarized light of the appropriate handedness was used.

Our experiment uses a LG mode within an all-optical $x-y-z$ trap to form an optical spanner for rotating micrometer-sized particles. By controlling the polarization state of a LG mode with $l = 1$ we can arrange for the angular momentum associated with the photon spin to add to the orbital angular momentum, giving a total angular momentum of $2\hbar$ per photon, in which case the Teflon particle spins more quickly. Alternatively, that momentum can subtract, resulting in no overall angular momentum, and the particle comes to a halt. We observe the cancellation of the spin and orbital angular momentum in a macroscopic system, which verifies that for an $l = 1$ LG mode the orbital angular momentum is indeed well defined and corresponds to $\hbar$ per photon.

This research is supported by Engineering and Physical Sciences Research Council grant GR/K11536. M. I. Padgett is a Royal Society Research Fellow.

References

1. A. E. Siegman, *Lasers* (University Science, Mill Valley, Calif., 1986), Sec. 17.5, pp. 685–695.
2. J. M. Vaughan and D. V. Willetts, Opt. Commun. **30**, 263 (1979).
3. L. Allen, M. W. Beijersbergen, R. J. C. Spreeuw, and J. P. Woerdman, Phys. Rev. A **45**, 8185 (1992).
4. D. N. Pattanayak and G. P. Agrawal, Phys. Rev. A **22**, 1159 (1980).
5. S. M. Barnett and L. Allen, Opt. Commun. **110**, 670 (1994).
6. M. Harris, C. A. Hill, and J. M. Vaughan, Opt. Commun. **106**, 161 (1994).
7. M. W. Beijersbergen, R. P. C. Coerwinkel, M. Kristensen, and J. P. Woerdman, Opt. Commun. **112**, 321 (1994).
8. N. R. Heckenberg, R. McDuff, C. P. Smith, and A. G. White, Opt. Lett. **17**, 221 (1992).
9. N. R. Heckenberg, R. McDuff, C. P. Smith, H. Rubinsztein-Dunlop, and M. J. Wegener, Opt. Quantum Electron. **24**, S951 (1992).
10. M. W. Beijersbergen, L. Allen, H. E. L. O. van der Veen, and J. P. Woerdman, Opt. Commun. **96**, 123 (1993).
11. M. Babiker, W. L. Power, and L. Allen, Phys. Rev. Lett. **73**, 1239 (1994); S. J. van Enk and G. Nienhuis, Opt. Commun. **94**, 147 (1992).
12. N. B. Simpson, L. Allen, and M. J. Padgett, J. Mod. Opt. **43**, 2485 (1996).
13. H. He, M. E. J. Friese, N. R. Heckenberg, and H. Rubinsztein-Dunlop, Phys. Rev. Lett. **75**, 826 (1995).
14. M. E. J. Friese, J. Enger, H. Rubinsztein-Dunlop, and N. R. Heckenberg, Phys. Rev. A **54**, 1593 (1996).
15. K. T. Gahagan and G. A. Swartzlander, Jr., Opt. Lett. **21**, 827 (1996).
16. A. Ashkin, J. M. Dziedzic, J. E. Bjorkholm, and S. Chu, Opt. Lett. **11**, 288 (1986).
17. S. Oka, in *Rheology*, F. R. Eirich, ed. (Academic, New York, 1960), Vol. 3.

R E P O R T S

Controlled Rotation of Optically Trapped Microscopic Particles

L. Paterson,[1] M. P. MacDonald,[1] J. Arlt,[1] W. Sibbett,[1]
P. E. Bryant,[2] K. Dholakia[1]*

We demonstrate controlled rotation of optically trapped objects in a spiral interference pattern. This pattern is generated by interfering an annular shaped laser beam with a reference beam. Objects are trapped in the spiral arms of the pattern. Changing the optical path length causes this pattern, and thus the trapped objects, to rotate. Structures of silica microspheres, microscopic glass rods, and chromosomes are set into rotation at rates in excess of 5 hertz. This technique does not depend on intrinsic properties of the trapped particle and thus offers important applications in optical and biological micromachines.

Optical forces have been used to trap and manipulate micrometer-sized particles for more than a decade (*1*). Since it was shown that a single tightly focused laser beam could be used to hold, in three dimensions, a microscopic particle near the focus of the beam, this optical tweezers technique has now become an established tool in biology, enabling a whole host of studies. They can be used to manipulate and study whole cells such as bacterial, fungal, plant, and animal cells (*2*) or intracellular structures such as chromosomes (*3*). Optical tweezers make use of the optical gradient force. For particles of higher refractive index than their surrounding medium, the laser beam induces a force attracting the trapped particle into the region of highest light intensity.

The ability to rotate objects offers a new degree of control for microobjects and has important applications in optical micromachines and biotechnology. Various schemes have, therefore, been investigated recently to induce rotation of trapped particles within optical tweezers. This could be used to realize biological machines that could function within living cells or optically driven cogs to drive micromachines.

[1]School of Physics and Astronomy, St. Andrews University, North Haugh, St. Andrews, Fife KY16 9SS, Scotland. [2]School of Biology, Bute Building, St. Andrews University, St. Andrews, Fife KY16 9TS, Scotland.

*To whom correspondence should be addressed. E-mail: kd1@st-and.ac.uk

Besides the use of specially fabricated microobjects (*4*), two major schemes have successfully enabled trapped microobjects to be set into rotation. The first scheme uses Laguerre-Gaussian (LG) light beams (*5–7*). These beams have an on-axis phase singularity and are characterized by helical phase fronts (Fig. 1A). The Poynting vector in such beams follows a corkscrewlike path as the beam propagates, and this gives rise to an orbital angular momentum component in the light beam (*8*). This angular momentum is distinct from any angular momentum due to the polarization state of the light and has a magnitude of $l\hbar$ per photon. Specifically, l refers to the number of complete cycles of phase ($2\pi l$) upon going around the beam circumference. However, to transfer orbital angular momentum to a trapped particle with such a beam, the particle must typically absorb some of the laser light yet still be transparent enough to enable tweezing to occur. This in turn restricts the range of particles to which this method can be applied, and it also further limits this technique because any heating that arises from this absorption could damage the rotating particle. Furthermore, as the particle absorption can be difficult to quantify, controlled rotation of trapped objects in such a beam is very difficult to realize.

The other technique for rotation makes use of the change in polarization state of light upon passage through a birefringent particle (*9, 10*). For example, circularly polarized light has spin angular momentum

that can be exchanged with a birefringent medium (e.g., calcite) upon propagation of the beam through the medium. This is analogous to Beth's famous experiment—where he measured the torque on a suspended half-wave plate as circularly polarized light passed through it (*11*)—but here we are working on a microscopic scale. This method has shown rotation rates of a few hundred hertz for irregular samples of crushed calcite, but it is difficult to control and is limited solely to birefringent media so it is not widely applicable. Although both of these methods have proven useful in specific applications, they do have serious shortcomings for general applications in rotating optical microcomponents and realizing optical micromachines.

We introduce a general scheme for rotating trapped microobjects. Specifically, we trap objects within the interference pattern of an LG beam and a plane wave (Fig. 1B) (*12*). By changing the path length of the interferometer, we are able to cause the spiral

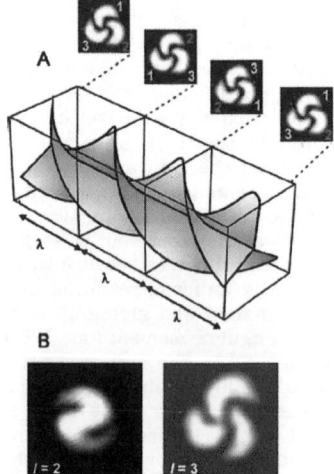

Fig. 1. (A) The phase fronts of an LG beam of azimuthal index $l = 3$ (helical structure) and intensity pattern when interfered with a plane wave. The phase fronts describe a triple start intertwined helix that repeats its shape every λ but only rotates fully after $l\lambda$. In **(B)**, we can see the experimental forms of the interference patterns of LG beams of index $l = 2$ and $l = 3$ with plane waves used in our experiments.

pattern (and thus the trapped particles) to rotate in a controlled fashion about the axis of the spiral pattern. The rotation of the pattern occurs because of the helical nature of the phase fronts of an LG light beam. A single-ringed LG beam is described by its azimuthal mode index l, which denotes the number of complete cycles of phase upon going around the circumference of the mode. An $l = 2$ or $l = 3$ LG mode (beam) can be thought of as consisting of phase fronts that are double or triple start helices, respectively (see Fig. 1A). Interfering this beam with a plane wave will transform the azimuthal phase variation of the pattern into an azimuthal intensity variation, resulting in a pattern with l spiral arms. As we change the path length in one arm of the interferometer, these spiral arms will rotate around the beam axis. As an analogy, this is akin to considering what occurs along a

length of thick rope that consists of l intertwined cords. Now consider cutting this rope and viewing it end-on. As you move the position of the cut along the rope, any given cord rotates around the rope axis. This is analogous to altering the optical path length in the interferometer. With this technique, we rely solely on the optical gradient force to tweeze trapped particles in the spiral arms and then use the rotation of this spiral pattern under a variation of optical path length to induce particle rotation. The technique can therefore be applied in principle to any object (or group of objects) that can be optically tweezed, in contrast to the other methods listed above. This technique can be extended to the use of LG beams of differing azimuthal index, thus offering the prospect of trapping and rotating different shaped objects and groups of objects. Here, illustrative examples

of rotation with LG beams with azimuthal indices $l = 2$ and $l = 3$ are shown.

Figure 2 shows a schematic of the trapping arrangement (13). A change in the path length in one arm of the interferometer by $l \times \lambda$ will cause a full rotation of 360° of the pattern (and thus the trapped particle array) in the optical tweezers (14). We can readily change the sense of rotation by reducing the path length of one arm of the interferometer instead of increasing it. Thus, in contrast to other rotation methods, we have a very simple way of controlling both the sense and rate of rotation of our optically trapped structure.

The use of an LG $l = 2$ beam results in two spiral arms for our interference pattern, and we used this to rotate silica spheres and glass rods in our tweezers setup (using a ×100 microscope objective). In Fig. 3A, two 1-μm silica spheres are trapped and spun at a rate of 7 Hz. The minimum optical power required to rotate the 1-μm spheres (which is the minimum power required to rotate any of the structures) is 1 mW. Silica spheres coated with streptavidin can bind to biotinylated DNA, and thus one could rotationally orient DNA strands by extending this method. In Fig. 3B, a tweezed glass rod can be seen to rotate between the frames. This constitutes an all-optical microstirrer and has potential application for optically driven micromachines and motors. We also demonstrate rotation of a Chinese hamster chromosome in our tweezers using this same interference pattern (Fig. 3C), with the axis of our pattern placed over the centromere of the chromosome. This degree of flexibility could be used for suitably orienting the chromosome before, for example, the optical excision of sections for use in polymerase chain reactions. This latter demonstration shows the potential of our method for full rotational control of biological specimens.

The rotation of trapped particles in an interference pattern between an LG ($l = 3$) beam and a plane wave can be seen in Fig. 4. The number of spiral arms in the pattern is equivalent to the azimuthal index of the LG beam used. In this instance, we used a ×40 microscope objective to increase the overall size of the beam profile and thus tweeze and rotate larger structures. In Fig. 4, we see three trapped 5-μm silica spheres rotate in this pattern. One of the spheres has a slight deformity (denoted by the arrow), and the series of pictures charts the progress of this structure as the pattern is rotated. We typically achieved rotation rates in excess of 5 Hz in the above experiments, which were limited only by the amount of optical power (~13 mW) in our interference pattern at the sample plane. The use of optimized components would readily lead to rotation rates of tens to hundreds of hertz. One can envisage other fabricated microobjects being rotated in a similar fashion.

Fig. 2. The experimental arrangement for optical tweezing and subsequent particle rotation in the interference pattern. L, lens; M, mirror; H, hologram; GP, glass plate; BS, beam splitter; Nd:YVO$_4$, neodymium yttrium vanadate laser at 1064 nm; ×100 or ×40, microscope objectives; CCD, camera; and BG, infrared filter.

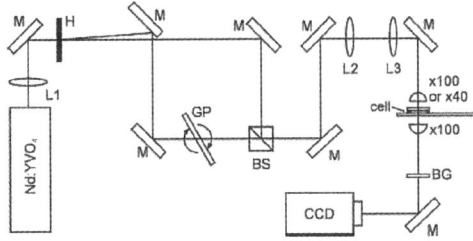

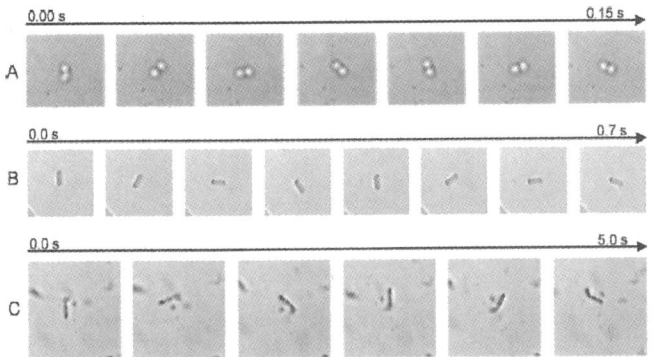

Fig. 3. Rotation of two-dimensionally trapped objects in an LG $l = 2$ interference pattern. (**A**) Rotation of two trapped 1-μm silica spheres. (**B**) Rotation of a 5-μm-long glass rod. In (**C**), we see rotation of a Chinese hamster chromosome. The elapsed time t (in seconds) is indicated by the scale at the top of each sequence of images.

Fig. 4. Rotation of three trapped silica spheres each 5 μm in diameter. The slight deformity (indicated by arrow) on one of the spheres allows us to view the degree of rotation of the structure.

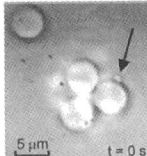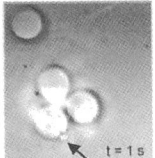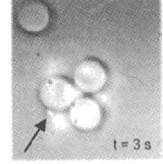

REPORTS

References and Notes

1. A. Ashkin, J. M. Dziedzic, J. E. Bjorkholm, S. Chu, *Opt. Lett.* **11**, 288 (1986).
2. A. Ashkin, J. M. Dziedzic, T. Yamane, *Nature* **330**, 769 (1987).
3. H. Liang, W. H. Wright, S. Cheng, W. He, M. W. Berns, *Exp. Cell Res.* **204**, 110 (1993).
4. P. Ormos, P. Galajda, *Appl. Phys. Lett.* **78**, 249 (2001).
5. H. He, M. E. J. Friese, N. R. Heckenberg, H. Rubinsztein-Dunlop, *Phys. Rev. Lett.* **75**, 826 (1995).
6. M. E. J. Friese, J. Enger, H. Rubinsztein-Dunlop, N. R. Heckenberg, *Phys. Rev. A* **54**, 1593 (1996).
7. N. B. Simpson, K. Dholakia, L. Allen, M. J. Padgett, *Opt. Lett.* **22**, 52 (1997).
8. L. Allen, M. W. Beijersbergen, R. J. C. Spreeuw, J. P Woerdman, *Phys. Rev. A* **45**, 8185 (1992).
9. M. E. J. Friese, T. A. Nieminen, N. R. Heckenberg, H. Rubinsztein-Dunlop, *Nature* **394**, 348 (1998).
10. M. E. J. Friese, H. Rubinsztein-Dunlop, J. Gold, P. Hagberg, D. Hanstorp, *Appl. Phys. Lett.* **78**, 547 (2001).
11. R. A. Beth, *Phys. Rev.* **50**, 115 (1936).
12. M. Padgett, J. Arlt, N. Simpson, L. Allen, *Am. J. Phys.* **64**, 77 (1996).
13. The experimental setup consisted of a Nd:YVO$_4$ laser of 300-mW power at 1064 nm. This beam is then directed through an in-house manufactured holographic element (*15*) that yielded an LG beam in its first order with an efficiency of 30%. This LG beam is then interfered with the zeroth order beam from the hologram to generate our spiral interference pattern. This pattern propagates through our optical system and is directed through either a ×40 or a ×100 microscope objective in a standard optical tweezers geometry. Typically around 1 to 13 mW of laser light was incident on the trapped structure in our optical tweezers, with losses due to optical components and the holographic element. A charge-coupled device (CCD) camera was placed above the dielectric mirror for observation purposes (Fig. 2) when the ×100 objective was used. A similar setup was used when tweezing with a ×40 objective but with the CCD camera placed below the sample slide viewing through a ×100 objective. It is important to ensure exact overlap of the light beams to guarantee that spiral arms are observed in the interference pattern—at larger angles, linear fringe patterns (with some asymmetry) can result (*15*). To set trapped structures into rotation, the relative path length between the two arms of the interferometer must be altered. We achieved this by placing a glass plate on a tilt stage in one arm. Simply by tilting this plate, we can rotate accordingly the pattern in the tweezers.
14. The tilting of the glass plate to rotate the interference pattern has a limitation when the plate reaches its maximum angle. One can, however, envisage more advanced implementations for continuous rotation using, for example, a liquid crystal phase modulator in the arm of the interferometer containing the plane wave.
15. M. A. Clifford, J. Arlt, J. Courtial, K. Dholakia, *Opt. Commun.* **156**, 300 (1998).
16. We thank the UK Engineering and Physical Sciences Research Council for supporting our work.

26 December 2000; accepted 19 March 2001

VOLUME 92, NUMBER 19

PHYSICAL REVIEW LETTERS

week ending
14 MAY 2004

Optical Torque Wrench: Angular Trapping, Rotation, and Torque Detection of Quartz Microparticles

Arthur La Porta and Michelle D. Wang

Department of Physics, Laboratory of Atomic and Solid State Physics, Cornell University, Ithaca, New York 14853, USA
(Received 14 August 2003; published 14 May 2004)

We describe an apparatus that can measure the instantaneous angular displacement and torque applied to a quartz particle which is angularly trapped. Torque is measured by detecting the change in angular momentum of the transmitted trap beam. The rotational Brownian motion of the trapped particle and its power spectral density are used to determine the angular trap stiffness. The apparatus features a feedback control that clamps torque or other rotational quantities. The torque sensitivity demonstrated is ideal for the study of known biological molecular motors.

DOI: 10.1103/PhysRevLett.92.190801

PACS numbers: 07.60.–j, 05.40.Jc, 42.79.–e, 87.80.Cc

Optical tweezers have made critical contributions to the creation of a vibrant field of biophysical research—single molecule manipulation of nucleic acids and protein complexes [1,2]. Although the possibility of manipulating biological objects was readily appreciated [3], the true potential of optical tweezers was realized only when microspheres were used as handles to displace attached biomolecules and measure the resulting force [4,5]. It would be equally useful to have a handle that could be used to *rotate* biological structures and to precisely measure the associated *torque*. Here we demonstrate angular trapping and torque detection using nominally spherical but anisotropic quartz particles. The torque acting on the particle and its deviation from the trap direction are determined by direct measurement of the change in angular momentum of the transmitted beam. The ability to measure instantaneous torque is of great importance, since it will facilitate precise measurement of the torque generated by biological structures as they rotate. The wide bandwidth and accuracy of our detection scheme allow us to measure Brownian rotational motion of the trapped particle and to use feedback to control the applied torque or particle angle.

Several other techniques have been demonstrated for rotating microscopic particles. These include use of azimuthally asymmetric beams or combinations of beams to rotate nonspherical particles [6–8], use of linearly or circularly polarized light to orient or apply torque to birefringent calcite particles [9], or use of magnetic fields to apply torque to free or optically trapped magnetic particles [10,11]. The technique we demonstrate here is similar to that employed by Friese *et al.*, but with the important advantage that angular trapping is combined with a detector allowing instantaneous measurement of the torque acting on the particle and its angular deviation from the trap direction. Using an optical power of ~10 mW, the trap is capable of rotating micron size particles with angular velocities up to 200 rad/s and generating several hundred pN · nm of torque. The reso-

lutions of torque measurement and angular confinement are limited by rotational Brownian motion of the particle.

The angular trap is based on the fact that a dielectric material subject to an external electric field **E** (constant or oscillating) generates polarization **P** given by $\mathbf{P} = \chi \mathbf{E}$, where χ is the electric susceptibility. If the material is birefringent, the susceptibility is not isotropic so that the expression for the polarization is generalized to $\mathbf{P} = \chi_x E_x \hat{x} + \chi_y E_y \hat{y} + \chi_z E_z \hat{z}$, where $\hat{x}$, $\hat{y}$, and $\hat{z}$ are unit vectors along the principal axes of the crystal and χ_x, χ_y, and χ_z are the corresponding electrical susceptibilities [12]. For typical uniaxial birefringent materials such as quartz or calcite, two of the susceptibilities are equal (χ_o ordinary) and the third is different (χ_e extraordinary).

Angular trapping occurs in particles made from materials such as quartz, in which the extraordinary axis of the crystal is more easily polarized than the ordinary axes. In this case the polarization **P** induced on a particle by an external electric field **E** will be tilted toward the extraordinary axis, as illustrated in Fig. 1(a). The misalignment between **E** and **P** results in a torque given by

$$\boldsymbol{\tau} = \int d^3 x \mathbf{P} \times \mathbf{E}$$
$$= \hat{q} \tfrac{1}{2} (\chi_o - \chi_e) \sin 2\theta \int d^3 x E_0^2(\mathbf{x}) = \hat{q} \tau_0 \sin 2\theta, \quad (1)$$

where θ is the angle between **E** and the extraordinary axis, $\hat{q}$ is a unit vector perpendicular to **E** and **P**, and τ_0 is the maximum magnitude of torque that can be exerted on the particle. (Particle shape effects are neglected in this formula.) As a result, linearly polarized light can be used to exert torque on a quartz particle. This torque tends to align the extraordinary axis of the quartz particle with the electric field direction, as shown by the quartz sphere in Fig. 1(b). Materials such as calcite, where the extraordinary axis is less polarizable than the ordinary axes, will experience torque [9] but have additional rotational degrees of freedom, as illustrated by the calcite spheres in Fig. 1(b).

0031-9007/04/92(19)/190801(4)$22.50

VOLUME 92, NUMBER 19   PHYSICAL REVIEW LETTERS   week ending
14 MAY 2004

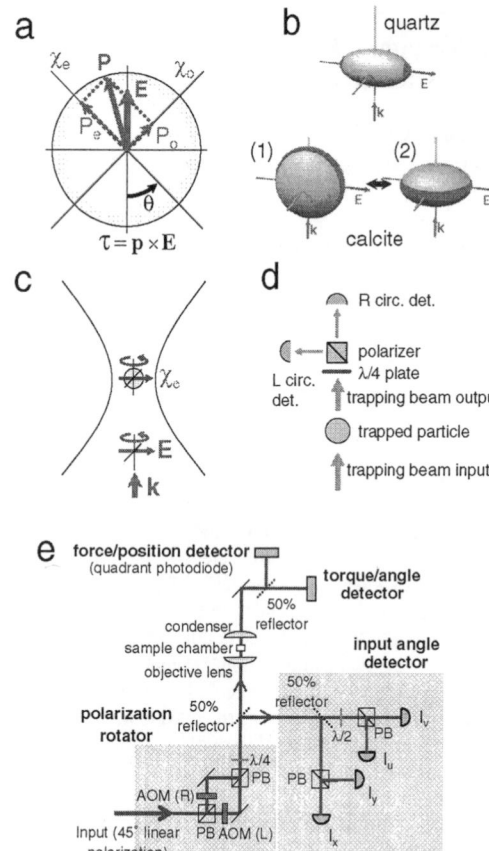

FIG. 1 (color). (a) Polarization vector **P** induced in an anisotropic particle by an external electric field **E**. Torque is generated when **P** is not aligned with the field. (b) The index ellipsoids of quartz and calcite are represented, where the region of maximum electric susceptibility is marked in red. Torque exerted on the particle tends to align the red region with the **E** vector. For quartz particles the applied torque tends to align the extraordinary axis (red areas) with the **E** vector, constraining two of three Euler angles. For calcite particles, the extraordinary axis is repelled by the **E** vector and the plane defined by the ordinary axes (red band) is attracted to the **E** field, constraining one of three Euler angles. In this case the extraordinary axis is either perpendicular [ellipsoid marked (1)] or parallel to the propagation vector **k** [ellipsoid marked (2)]. In the latter case, the birefringence of the particle is extinguished, allowing the particle to spin freely around the optical axis (along **k**). (c) The alignment of the particle with the trap polarization is illustrated. (d) The torque detector measures the difference in intensities of the right circular and left circular components of the transmitted beam. (e) Schematic representation of the apparatus. The polarization rotator is described in the text. The input detector determines the polarization angle of the input trapping field. The torque detector and quadrant detector determine the torque and force acting on the particle.

In order to detect the torque we take advantage of the conservation of angular momentum, which requires that the torque acting on the particle is equal and opposite to the rate of change of the angular momentum of the trapping beam as it passes through the particle. Since the torque is generated using polarization properties rather than the shape of the beam or particle, the angular momentum is transferred to the polarization state of the transmitted beam rather than to its spatial profile. Light with left-handed (right-handed) circular polarization contains angular momentum $+\hbar$ ($-\hbar$) and energy $\hbar\omega_o$ per photon, where $\hbar$ is the reduced Planck constant and ω_o is the optical angular frequency. The linearly polarized input trap beam contains no net angular momentum because it is composed of equal quantities of left and right circular polarization. Exertion of torque τ on a particle causes an imbalance of the power of left and right circular components (P_L and P_R) in the transmitted beam, such that $\tau = (P_R - P_L)/\omega_0$. Direct measurement of this quantity is made by the torque detector shown in Fig. 1(d). In principle, the torque is strictly determined by the angular momentum content of the transmitted beam [13]. In practice, it is impossible to collect the transmitted trap beam in its entirety, so a calibration of the detector is necessary.

To manipulate the particle angle it is necessary to have precise and rapid control of the trap polarization angle. Such control is impractical using mechanical means and is instead accomplished by the polarization rotator shown in Fig. 1(e). In this device the acousto-optic modulators (AOMs) marked L and R generate the left and right circular components of the output beam, and the relative phase of these components is determined by the relative phase of the AOM rf drives. As a result, a relative phase shift ϕ of the rf signals causes a rotation of $\phi/2$ of the output polarization. The AOM drive signals are generated by computer-controlled digital frequency synthesis, allowing the polarization angle to be changed with a response time of a few microseconds. Also shown in Fig. 1(e) are the detectors used to measure the input polarization angle, the torque, and the linear displacement of the trapped particle.

The rotation of a small quartz particle using the apparatus is shown in Fig. 2. The particle shown has a

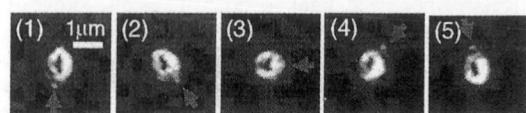

FIG. 2 (color). Video images recorded every 250 ms show a ~1 μm diameter quartz particle being rotated counterclockwise in the angular trap. Quartz particles are obtained by centrifuging and collecting fractions from quartz powder. The particle shown is close to spherical but has an irregularity on one side (indicated by arrows) that allows its orientation to be recognized in video images.

VOLUME 92, NUMBER 19   PHYSICAL REVIEW LETTERS   week ending
14 MAY 2004

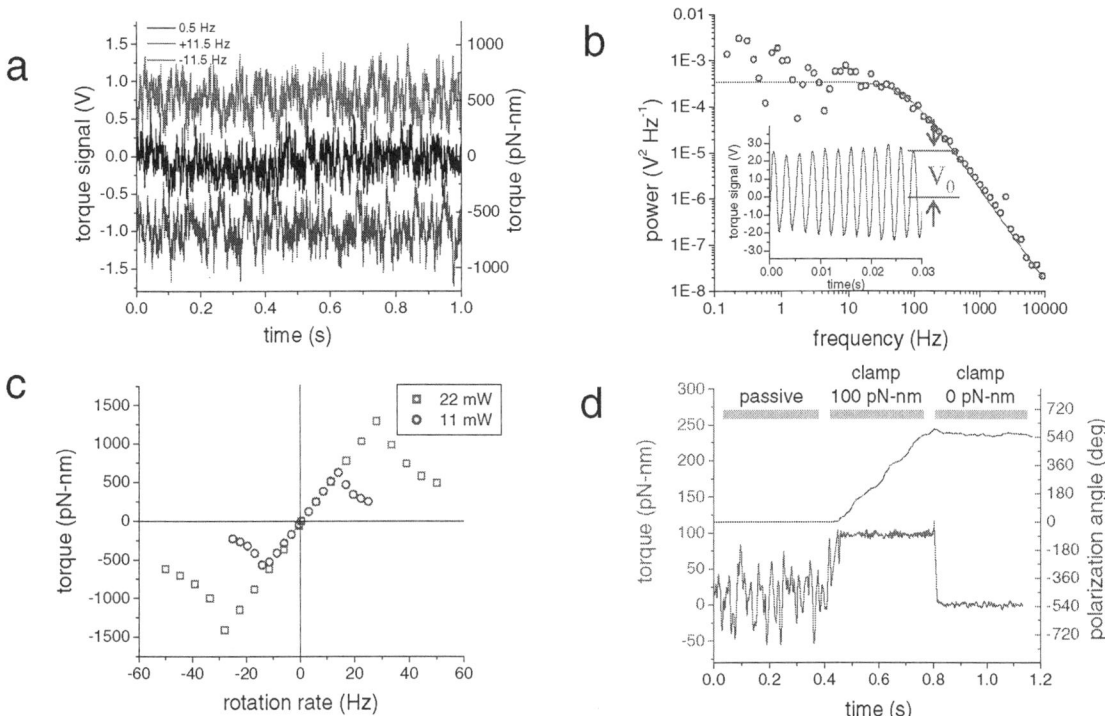

FIG. 3 (color). (a) The torque signal for a particle spinning in opposite directions at 11.5 rps, and nearly motionless at 0.5 rps. The dc offset is a measure of the drag on the spinning particle and the broadband fluctuations are primarily due to Brownian motion. (b) The power spectral density of the rotational fluctuations, and fit with Lorentzian function of the form $S(f) = A^2/(f^2 + f_0^2)$ with $A^2 = 0.059$ rad^2 (Hz) and $f_0 = 75$ Hz. The inset shows the torque signal as a quasistationary particle is scanned by a rapidly rotating polarization vector (200 Hz). The amplitude $V_0 = 2.63$ V implies a small-signal angle sensitivity of 0.19 rad/V. From these values we calculate stiffness $\kappa = 3360$ pN $\cdot$ nm/rad, damping coefficient $\xi = 7.1$ pN $\cdot$ nm $\cdot$ s, and torque sensitivity 638 pN $\cdot$ nm/V. (c) The measured torque as a function of rotation rate, converted to pN $\cdot$ nm using the calibration parameters defined in (b). From the slope of the linear regime we obtain $\xi = 7.8$ pN $\cdot$ nm $\cdot$ s. (d) Measured torque and trap polarization angle as a function of time. Feedback to stabilize the torque at 100 pN $\cdot$ nm (0 pN $\cdot$ nm) is activated at 0.4 s (0.8 s).

noticeable deviation from spherical shape, allowing the rotation of the particle to be confirmed visually. Particles which are spherical in shape can also be rotated, as can be determined using the torque detector, which also measures angular displacement.

The measurement of the torque acting on a particle spinning at uniform velocity is shown in Fig. 3(a). The three traces show broadband fluctuations with different mean values. The fluctuations arise from Brownian rotational motion of the particle, as discussed below. The mean value, which is zero in the absence of rotation and is positive or negative for counterclockwise or clockwise rotation, is a measure of the viscous drag on the spinning particle.

The first step in the calibration procedure is to relate the torque signal to the deviation of the particle from the trap polarization angle. Referring to Eq. (1) we find that the angle is given by $\theta = (1/2)\arcsin(V_\tau/V_0)$, where V_τ is the torque signal in volts and V_0 is the maximum value of

this signal, obtained at $\theta = 45°$. The value of V_0 may be determined by rotating the polarization much faster than the particle can follow, so that the polarization vector scans the quasistationary particle. The amplitude of the resulting sinusoidal modulation is V_0 [see Fig. 3(b) inset]. For small angles we can approximate $\theta \approx V_\tau/2V_0$.

Once the angular calibration is accomplished, angular deviation can be determined from the torque signal. The task remains to determine the stiffness of the angular trap and convert the torque signal to physical units of torque. Applying the standard treatment of Brownian fluctuations in a potential well to rotational motion [14], we find that the power spectral density of the angular fluctuations is of the form $S(f) = A^2/(f^2 + f_0^2)$ with corner frequency $f_0 = \kappa/2\pi\xi$ and amplitude $A^2 = k_B T/\xi\pi^2$, where k_B is the Boltzmann constant, T is the temperature in degrees kelvin, κ is the stiffness of the angular trap, and ξ is the rotational viscous damping coefficient. The damping ξ and stiffness κ are determined

VOLUME 92, NUMBER 19   PHYSICAL REVIEW LETTERS   week ending
14 MAY 2004

by fitting the predicted function to the measured power spectrum, shown in Fig. 3(b). Once the angular trap stiffness is known the torque is related to the raw torque by $\tau = V_\tau(\kappa/2V_0)$. The torque sensitivity obtained from the calibration is within experimental error of the absolute angular momentum change of the trap beam, taking into account our estimated $\sim$50% light collection efficiency.

The calibration of torque allows us to directly measure the viscous drag on a spinning particle as a function of rotation rate. Figure 3(c) shows this function measured using two different values of the trap power. As expected, the two traces show a regime of linear scaling of torque with rotation rate having identical slope. The drag coefficient obtained from this slope is within experimental uncertainty of the value obtained from Brownian motion (see the Fig. 3 caption). For a spherical particle in an infinite fluid volume, we expect $\xi = \pi\eta D^3$, where D is the particle diameter and η is the viscosity of water. Using the measured ξ we obtain $D = 1.3$ μm, which is consistent with the apparent particle size in Fig. 2. For both power levels the linear scaling fails above the rotation rate where the torque needed to spin the particle exceeds the maximum torque available.

Finally, in Fig. 3(d) we demonstrate active stabilization of the torque using feedback to the polarization angle. During the initial part of the trace the trap polarization was held constant and Brownian fluctuations of the torque were observed. At 0.4 s a computer-controlled servo loop was activated which adjusted the polarization angle to maintain a constant torque of 100 pN · nm on the particle, and at 0.8 s the setpoint was reduced to 0 pN · nm. During active feedback the polarization angle is varied rapidly to suppress torque fluctuations. Although we have demonstrated this feedback technique using free particles, it can also be used to stabilize torque (or another variable such as particle orientation) against an active load such as a molecular motor.

We anticipate that the system we have developed will be well suited for measurements of torque and rotation in single molecule biological systems. Quartz surfaces may be functionalized for biomolecule attachment using standard techniques [15], and material processing techniques may be employed to make the particles more regular in shape [16]. Molecular motors are known to generate torque ranging from tens to thousands of pN · nm

[17,18], which is well within the dynamic range of the apparatus without requiring excessive trap power. Most importantly, the instantaneous readout and feedback capabilities will allow the torque generated by a biological structure in response to the imposed rotation (or vice versa) to be continuously measured.

We thank Professor R. H. Silsbee for stimulating scientific discussions. This work was supported by grants from NIH, the Keck Foundation, and by the STC program of the NFS under Agreement No. ECS-9876771.

[1] C. Bustamante, J. J. Macosko, and G. J. Wuite, Nat. Rev. Mol. Cell. Biol. **1**, 130 (2000).
[2] M. D. Wang, Curr. Opin. Biotechnol **10**, 81 (1999).
[3] A. Ashkin and J. M. Dziedzic, Science **235**, 1517 (1987).
[4] K. Svoboda, C. F. Schmidt, B. J. Schnapp, and S. M. Block, Nature (London) **365**, 721 (1993).
[5] M. D. Wang et al., Science **282**, 902 (1998).
[6] A. T. O'Neil and M. J. Padgett, Opt. Lett. **27**, 743 (2002).
[7] L. Paterson, M. P. MacDonald, J. Arlt, W. Sibbet, P. E. Bryant, and K. Dholakia, Science **292**, 912 (2001).
[8] V. Bingelyte, J. Leach, J. Courtial, and M. J. Padgett, Appl. Phys. Lett. **82**, 829 (2003).
[9] M. E. J. Friese, T. A. Nieminen, N. R. Heckenberg, and H. Rubinsztein-Dunlop, Nature (London) **394**, 348 (1998).
[10] L. Sacconi et al., Opt. Lett. **26**, 1359 (2001).
[11] T. R. Strick, V. Croquette, and D. Bensimon, Nature (London) **404**, 901 (2000).
[12] A. Yariv, Optical Electronics (Holt, Rinehart, and Winston, New York, 1985).
[13] T. A. Nieminen, N. R. Heckenberg, and H. Rubinsztein-Dunlop, J. Mod. Opt. **48**, 405 (2001). After the submission of our manuscript a similar detector configuration was published in A. I. Bishop, T. A. Nieminen, N. R. Heckenberg, and H. Rubinsztein-Dunlop, Phys. Rev. A **68**, 033802 (2003).
[14] K. Svoboda and S. M. Block, Annu. Rev. Biophys. Biomol. Struct. **23**, 247 (1994).
[15] D. Kleinfeld, K. H. Kahler, and P. E. Hockberger, J. Neurosci. **8**, 4098 (1988).
[16] H. T. Sun, Z. T. Cheng, X. Yao, and W. Wlodarski, Sens. Actuators B **13**, 107 (1993).
[17] H. Noji, R. Yasuda, M. Yoshida, and K. Kinosita, Nature (London) **386**, 299 (1997).
[18] W. S. Ryu, R. M. Berry, and H. C. Berg, Nature (London) **403**, 444 (2000).

ELSEVIER

15 June 2002

Optics Communications 207 (2002) 169–175

OPTICS
COMMUNICATIONS

www.elsevier.com/locate/optcom

Dynamic holographic optical tweezers

Jennifer E. Curtis, Brian A. Koss, David G. Grier*

James Franck Institute and Institute for Biophysical Dynamics, University of Chicago, 5640 S. Ellis Ave., Chicago, IL 60637, USA

Received 29 March 2002; received in revised form 18 April 2002; accepted 29 April 2002

Abstract

Optical trapping is an increasingly important technique for controlling and probing matter at length scales ranging from nanometers to millimeters. This paper describes methods for creating large numbers of high-quality optical traps in arbitrary three-dimensional configurations and for dynamically reconfiguring them under computer control. In addition to forming conventional optical tweezers, these methods also can sculpt the wavefront of each trap individually, allowing for mixed arrays of traps based on different modes of light, including optical vortices, axial line traps, optical bottles and optical rotators. The ability to establish large numbers of individually structured optical traps and to move them independently in three dimensions promises exciting new opportunities for research, engineering, diagnostics, and manufacturing at mesoscopic lengthscales. © 2002 Elsevier Science B.V. All rights reserved.

PACS: 42.40.Jv; 87.80.Cc

An optical tweezer uses forces exerted by intensity gradients in a strongly focused beam of light to trap and move a microscopic volume of matter [1]. Optical tweezers' unique ability to manipulate matter at mesoscopic scales has led to widespread applications in biology [2,3], and the physical sciences [4]. This paper describes how computer-generated holograms can transform a single laser beam into hundreds of independent optical traps, each with individually specified characteristics, arranged in arbitrary three-dimensional configurations. The enhanced capabilities of such dynamic holographic optical trapping

systems offer new opportunities for research and engineering, as well as new applications in biotechnology, nanotechnology, and manufacturing.

Holographic optical tweezers (HOT) use a computer-designed diffractive optical element (DOE) to split a single collimated laser beam into several separate beams, each of which is focused into an optical tweezer by a strongly converging lens [5–7]. Originally demonstrated with microfabricated DOEs [8], holographic optical tweezers have since been implemented with computer-addressed liquid crystal spatial light modulators [9,10]. Projecting a sequence of computer-designed holograms reconfigures the resulting pattern of traps. Unfortunately, calculating the phase hologram for a desired pattern of traps is not straightforward, and the lack of appropriate algorithms has prevented dynamic holographic op-

* Corresponding author. Tel.: +1-773-702-9176; fax: +1-773-702-5863.
E-mail address: d-grier@uchicago.edu (D.G. Grier).

170   *J.E. Curtis et al. / Optics Communications 207 (2002) 169–175*

tical tweezers from achieving their potential. This paper introduces new methods for computing phase holograms for optical trapping and demonstrates their use in a practical dynamic holographic optical trapping system.

The same optical gradient forces exploited in conventional optical tweezers [1] also operate in holographic optical tweezers. A dielectric particle approaching a focused beam of light is polarized by the light's electric field and then drawn up intensity gradients toward the focal point. Radiation pressure competes with this optical gradient force and tends to displace the trapped particle along the beam's axis. For this reason, optical tweezers usually are designed around microscope objective lenses whose large numerical apertures and minimal aberrations optimize axial intensity gradients.

An optical trap can be placed anywhere within the objective lens' focal volume by appropriately selecting the input beam's propagation direction and degree of collimation. For example, a collimated beam passing straight into an infinity-corrected objective lens comes to a focus in the center of the lens' focal plane, while another beam entering at an angle comes to a focus proportionately off-center. A slightly diverging beam focuses downstream of the focal plane while a converging beam focuses upstream. By the same token, multiple beams simultaneously entering the lens' input pupil each form optical traps in the focal volume, each at a location determined by its angle of incidence and degree of collimation. This is the principle behind holographic optical tweezers.

Our implementation, shown schematically in Fig. 1, uses a Hamamatsu X7550 parallel-aligned nematic spatial light modulator (SLM) [11] to reshape the beam from a frequency-doubled Nd:YVO$_4$ laser (Coherent Verdi) into a designated pattern of beams. Each is transferred by relay optics to the entrance pupil of a 100× NA 1.4 oil immersion objective mounted in a Zeiss Axiovert S100TV inverted optical microscope and then focused into a trap. A dichroic mirror reflects the laser light into the objective while allowing images of the trapped particles to pass through to a video camera. When combined with a 0.63× widefield video eyepiece, this optical train offers a 85 × 63 μm^2 field of view.

The Hamamatsu SLM can impose selected phase shifts on the incident beam's wavefront at each 40 μm wide pixel in a 480 × 480 array. The SLM's calibrated phase transfer function offers 150 distinct phase shifts ranging from 0 to 2π at the operating wavelength of $\lambda = 532$ nm. The phase shift imposed at each pixel is specified through a computer interface with an effective refresh rate of 5 Hz for the entire array. Quite sophisticated trapping patterns are possible despite the SLM's inherently limited spatial bandwidth. The array of 400 functional optical traps shown in Fig. 1 is the largest created by any means. Improvements in the number and density of effective phase pixels, in their diffraction efficiency, in the resolution of the available phase modulation, and in the refresh rate for projecting new phase patterns will correspondingly improve the performance of dynamic holographic optical tweezer systems. Other phase modulating technologies, such as micromirror arrays could offer the additional benefit of creating optical traps in multiple wavelengths simultaneously.

Modulating only the phase and not the amplitude of the input beam is enough to establish any desired intensity pattern in the objective's focal volume and thus any pattern of traps [7]. Such intensity-shaping phase gratings are often referred to as kinoforms. Previously reported algorithms for computing optical trapping kinoforms produced only two-dimensional distributions of traps [7,9] or patterns on just two planes [10]. Moreover, the resulting traps were suitable only for dielectric particles in low-dielectric media, and could not be adapted to handle metallic particles or samples made of absorbing, reflecting, or low-dielectric-constant materials. A more general approach relaxes all of these restrictions.

We begin by modeling the incident laser beam's electric field $E_0(\vec{\rho}) = A_0(\vec{\rho}) \exp(i\psi)$, as having constant phase, $\psi = 0$ in the DOE plane, and unit intensity: $\int_\Omega |A_0(\vec{\rho})|^2 \, d^2\rho = 1$. Here $\vec{\rho}$ denotes a position in the DOEs aperture Ω, $A_0(\vec{\rho})$ is the real-valued amplitude profile of the input beam. The DOE then imposes a phase modulation $\varphi(\vec{\rho})$ onto the input beam's wavefront which, in principle, encodes the desired pattern of outgoing beams.

J.E. Curtis et al. / Optics Communications 207 (2002) 169–175   171

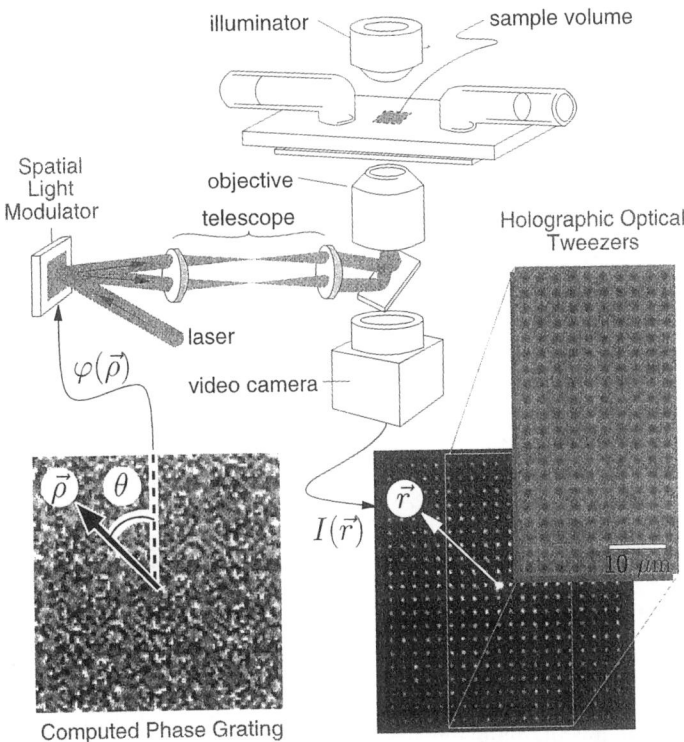

Fig. 1. Schematic implementation of dynamic holographic optical tweezers using a reflective liquid crystal spatial light modulator. The inset phase grating is 1/25 of the hologram $\varphi(\vec{\rho})$ encoding a 20 × 20 array of traps, with white regions corresponding to local phase shifts of 2π radians and black to 0. A telescope relays the diffracted beams to a high-numerical-aperture objective which focuses them into optical traps. The sample, enclosed in a glass flow cell, can be imaged through conventional video microscopy. The inset video micrograph shows the intensity $I(\vec{r}) = \sum_j |\epsilon_j|^2 \delta(\vec{r} - \vec{r}_j)$ of light from the traps reflected off a mirror temporarily placed in the object plane. The smaller inset shows these traps in action, holding two hundred colloidal polystyrene spheres, each 800 nm in diameter.

The electric field $\epsilon_j = \alpha_j \exp(\mathrm{i}\phi_j)$ at each of the discrete traps is related to the electric field $E(\vec{\rho})$ in the plane of the DOE by a generalized Fourier transform

$$
E(\vec{\rho}) = \sum_{j=1}^{N} \int \epsilon_j \delta(\vec{r} - \vec{r}_j) K_j^{-1}(\vec{r}, \vec{\rho})
$$

$$
\times \exp\left(\mathrm{i}\frac{2\pi \vec{r} \cdot \vec{\rho}}{\lambda f}\right) \mathrm{d}^3 r
$$

$$
= \sum_{j=1}^{N} \epsilon_j K_j^{-1}(\vec{r}_j, \vec{\rho}) \exp\left(\mathrm{i}\frac{2\pi \vec{r}_j \cdot \vec{\rho}}{\lambda f}\right), \tag{1}
$$

$$
\equiv A(\vec{\rho}) \exp(\mathrm{i}\varphi(\vec{\rho}), \tag{2}
$$

where f is the effective focal length of the optical train, including the relay optics and objective lens.

The kernel $K_j(\vec{r}, \vec{\rho})$ can be used to transform the jth trap from a conventional tweezer into another type of trap, and K_j^{-1} is its inverse. For conventional optical tweezers in the focal plane, $K_j = 1$.

If the calculated amplitude, $A(\vec{\rho})$, were identical to the laser beam's profile, $A_0(\vec{\rho})$, then $\varphi(\vec{\rho})$ would be the kinoform encoding the desired array of traps. Unfortunately, this is rarely the case. More generally, the spatially varying discrepancies between $A(\vec{\rho})$ and $A_0(\vec{\rho})$ direct light away from the desired traps and into ghosts and other undesirable artifacts. Despite these shortcomings, combining kinoforms with Eq. (1) is expedient and can produce useful trapping patterns [10]. Still better and more general results can be obtained by using Eqs. (1) and (2) as the basis for an iterative search for the ideal kinoform.

Following the approach pioneered by Gerchberg and Saxton (GS) [12], we treat the phase $\varphi(\vec{\rho})$ calculated with Eqs. (1) and (2) as an estimate, $\varphi^{(n)}(\vec{\rho})$, for the desired kinoform and use this to calculate the fields at the trap positions given the laser's actual profile $A_0(\vec{\rho})$

$$\epsilon_j^{(n)} = \int_\Omega A_0(\vec{\rho}) \exp(i\varphi^{(n)}(\vec{\rho})) K_j(\vec{r}_j, \vec{\rho})$$

$$\times \exp\left(-i\frac{2\pi\vec{r}_j \cdot \vec{\rho}}{\lambda f}\right) d^2\rho. \tag{3}$$

The index n refers to the nth iterative approximation to $\varphi(\vec{\rho})$.

The classic GS algorithm replaces the amplitude $\alpha_j^{(n)}$ in this estimate with the desired amplitude α_j, leaving the corresponding phase $\phi_j^{(n)}$ unchanged, and solves for the next estimate $\varphi^{(n+1)}(\vec{\rho})$ using Eqs. (1) and (2). The fraction $\sum_j |\alpha_j^{(n)}|^2$ of the incident power actually delivered to the traps by the nth approximation is useful for tracking the algorithm's convergence.

For the present application, the simple GS substitution leads to slow and non-monotonic convergence. We find that an alternate replacement scheme

$$\alpha_j^{(n+1)} = \left[(1-\xi) + \xi \frac{\alpha_j}{\alpha_j^{(n)}}\right]\alpha_j \tag{4}$$

leads to rapid monotonic convergence for $\xi \approx 0.5$. The resulting estimate for $\varphi(\vec{\rho})$ then can be discretized and transmitted to the SLM to establish a trapping pattern. In cases where the SLM offers only a few distinct phase levels, discretization can be incorporated into each iteration to minimize the associated error. In all of the examples discussed below, this algorithm yields kinoforms with theoretical efficiencies exceeding 80% in two or three iterations starting from a random choice for the traps' initial phases ϕ_j and often converges to solutions with better than 90% efficiency. Iterative optimization with Eqs. (2) and (3) is computationally efficient because discrete transforms are calculated only at the actual trap locations.

Fig. 2(a) shows 26 colloidal silica spheres 0.99 µm in diameter suspended in water and trapped in a planar fivefold pattern of optical tweezers created with Eqs. (2) and (3). Replacing this kinoform with another whose traps are slightly displaced moves the spheres into the new configuration. Projecting a sequence of trapping patterns translates the spheres deterministically into an entirely new configuration. Fig. 2(b) shows the same spheres after 16 such hops, and Fig. 2(c) after 38. Powering each trap with 1 mW of light traps the particles stably against thermal forces. Increasing the trapping power to 10 mW and updating the trapping pattern in 2 µm steps allows us to translate particles at up to 10 µm/s.

Comparable planar motions also have been implemented by rapidly scanning a single tweezer through a sequence of discrete locations, thereby creating a time-shared trapping pattern [13]. The continuous illumination of holographic optical traps offer several advantages, however. HOT patterns can be more extensive both spatially and

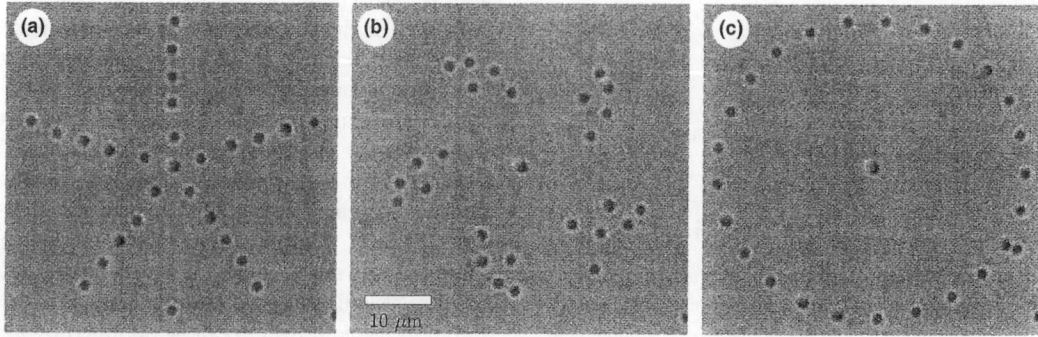

Fig. 2. A fivefold pattern of 26 colloidal silica spheres 0.99 µm in diameter is transformed into a circle using dynamic holographic optical tweezers. (a) The original configuration; (b) after 16 steps; (c) the final configuration after 38 steps.

in number of traps than time-shared arrays which must periodically release and retrieve each trapped particle. Additionally, the lower peak intensities required for continuously illuminated traps are less damaging to sensitive samples [14].

Similar rearrangements also would be possible with previous dynamic HOT implementations [10]. These studies used fast Fourier transforms to optimize the projected intensity over the entire trapping plane, and routinely achieved theoretical efficiencies exceeding 95% [7]. However, the discrete transforms adopted here allow us to encode more general patterns of traps.

Dynamic holographic optical tweezers need not be limited to planar configurations. If the laser beam illuminating the SLM were slightly diverging, then the entire pattern of traps would come to a focus downstream of the focal plane. Such divergence can be introduced with a Fresnel lens, encoded as a phase grating with

$$\varphi_z(\vec{\rho}) = \frac{2\pi\rho^2 z}{\lambda f^2} \quad \mathrm{mod}\, 2\pi, \tag{5}$$

where z is the desired displacement of the optical traps relative to the focal plane in an optical train with effective focal length f. Rather than placing a separate Fresnel lens into the input beam, the same functionality can be obtained by adding the lens' phase modulation to the existing kinoform: $[(\varphi(\vec{\rho}) + \varphi_z(\vec{\rho})] \, \mathrm{mod}\, 2\pi$. Fig. 3(a) shows a typical array of optical tweezers collectively displaced out of the plane in this manner. The accessible range of

out-of-plane motion in our system is approximately ± 10 μm.

Instead of being applied to the entire trapping pattern, separate lens functions can be applied to each trap individually with kernels

$$K_j^z(\vec{r}_j, \vec{\rho}) = \exp\left(i\frac{2\pi\rho^2 z_j}{\lambda f^2}\right) \tag{6}$$

in Eqs. (1) and (3). Fig. 3(b) shows spheres being moved independently through multiple planes in this way.

Other phase modifications implement additional functionality. For example, the phase profile

$$\varphi_\ell(\vec{\rho}) = \ell\theta \quad \mathrm{mod}\, 2\pi \tag{7}$$

converts an ordinary Gaussian laser beam into a Laguerre–Gaussian mode [15], and its corresponding optical tweezer into a so-called optical vortex [16–18]. Here θ is the polar coordinate in the DOE plane (see Fig. 1) and the integer ℓ is the beam's topological charge [15].

Because all phases are present along the circumference of a Laguerre–Gaussian beam, destructive interference cancels the beam's intensity along its axis. Optical vortices thus appear as bright rings surrounding dark centers. Such dark traps have been demonstrated to be useful for trapping reflecting, absorbing [19] or low-dielectric particles [18] not otherwise compatible with conventional optical tweezers.

Adding $\varphi_\ell(\vec{\rho})$ to a kinoform encoding an array of optical tweezers yields an array of identical

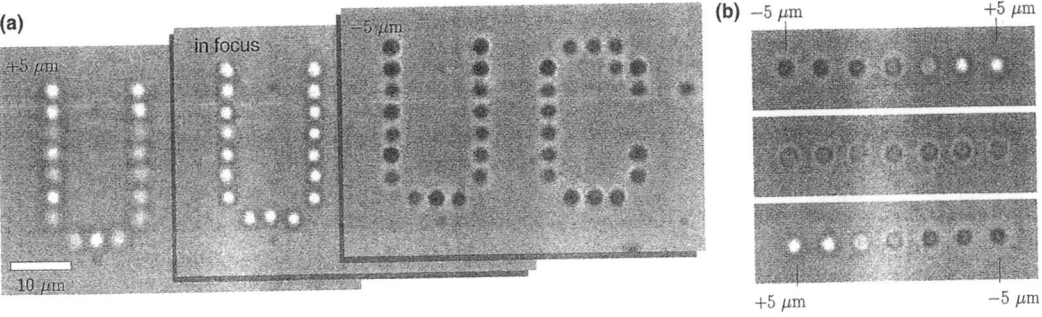

Fig. 3. Three-dimensional motion with holographic optical tweezers. The images in (a) show 34 silica spheres 0.99 μm in diameter trapped in a single plane and then displaced by ± 5 μm using Eq. (5). The spheres' images change as they move relative to the focal plane. (b) Seven spheres trapped and moved independently through seven different planes using kinoforms calculated with Eq. (6).

optical vortices, as shown in Fig. 4(a). Here, the light from the array of traps is imaged by reflection off a front-surface mirror placed in the microscope's focal plane. The vortex-forming phase function also can be applied to individual traps through

$$K_j^l(\vec{r}_j, \vec{\rho}) = \exp(i\ell_j\theta) \tag{8}$$

as demonstrated in the mixed array of optical tweezers and optical vortices shown in Fig. 4(b).

Previous reports of optical vortex trapping have considered Laguerre–Gaussian modes with relatively small topological charges, $\ell \leqslant 5$. The $\ell = 30$ examples in Fig. 4(b) are thus the most highly charged optical vortices so far reported, and traps with $\ell > 100$ are easily created with the present system.

Fig. 4(c) shows multiple colloidal particles trapped on the bright circumferences of a 3×3 array of $\ell = 15$ vortices. Because Laguerre–Gaussian modes have helical wavefronts, particles trapped on optical vortices experience tangential forces [19]. Optical vortices are useful, therefore, for driving motion at small length scales, for example in microelectromechanical systems (MEMS). Particles trapped on a vortex's bright circumference, such as the examples in Fig. 4(c) circulate rapidly around the ring, entraining cir-

culating fluid flows as they move. The resulting hydrodynamic coupling influences particles' motions on single vortices and leads to cooperative motion in particles trapped on neighboring vortices. The resulting fluid flows can be reconfigured dynamically by changing the topological charges, intensities and positions of optical vortices in an array, and may be useful for microfluidics and lab-on-a-chip applications.

The vortex-forming kernel K^ℓ can be combined with K^z to produce three-dimensional arrays of vortices. Such heterogeneous trapping patterns are useful for organizing disparate materials into hierarchical three-dimensional structures and for exerting controlled forces and torques on extended dynamical systems.

While the present study has demonstrated how a single Gaussian laser beam can be modified to create three-dimensional arrays of optical tweezers and optical vortices, other generalizations follow naturally, with virtually any mode of light having potential applications. For example, the axicon phase profile $\varphi_\gamma(\vec{\rho}) = \gamma\rho$ creates an approximation of a Bessel mode which focuses to an axial line trap whose length is controlled by γ [20]. These and other generalized trapping modes will be discussed elsewhere. Linear combinations of optical vortices and conventional tweezers have been shown to

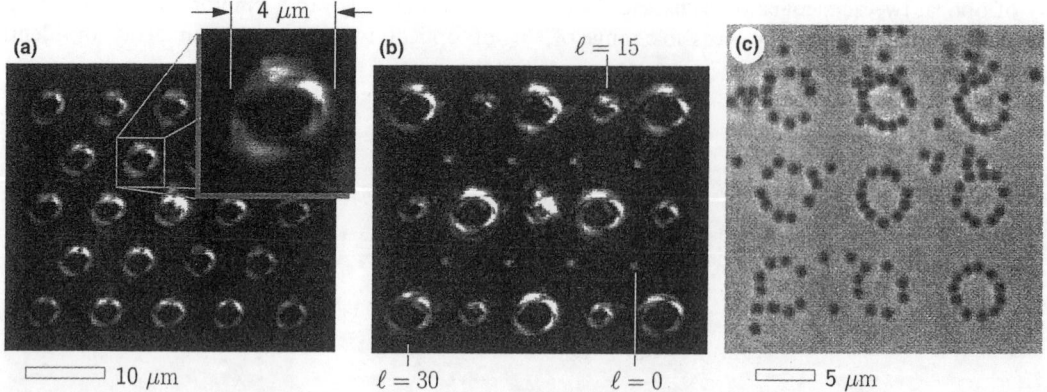

Fig. 4. (a) Triangular array of optical vortices with topological charge $\ell = 20$ created from an equivalent array of tweezers using Eq. (7). Light from the focused optical vortices is imaged by reflection from a mirror in the sample plane. The inset shows a more detailed view of one vortex's structure. (b) Mixed array of optical tweezers ($\ell = 0$) and optical vortices with $\ell = 15$ and $\ell = 30$ in the same configuration as (a), calculated with Eq. (8). The traps' amplitudes α_j were adjusted for uniform brightness. (c) Colloidal polystyrene spheres 800 nm in diameter trapped in 3×3 square array of $\ell = 15$ optical vortices.

J.E. Curtis et al. / Optics Communications 207 (2002) 169–175 175

operate as optical bottles [21] and controlled rotators [22]. All such trapping modalities can be combined dynamically using the techniques described above.

The complexity of realizable trapping patterns is limited in practice by the need to maintain three-dimensional intensity gradients for each trap, and by the maximum information content that can be encoded accurately in the SLM. For example, the former consideration precludes forming a three-dimensional cubic optical tweezer array with a lattice constant much smaller than 10 μm, while the latter limits our optical vortices to $\ell \approx 200$.

Within such practical bounds, dynamic holographic optical tweezers are highly reconfigurable, operate non-invasively in both open and sealed environments, and can be coupled with computer vision technology to create fully automated systems. A single apparatus thus can be adapted to a wide range of applications without modification. Dynamic holographic optical tweezers have a plethora of potential biotechnological applications including massively parallel high throughput screening, sub-cellular engineering, and macro-molecular sorting. In materials science, the ability to organize materials into hierarchical three-dimensional structures constitutes an entirely new category of fabrication techniques. As research tools, dynamic holographic optical tweezers combine the demonstrated utility of optical tweezers with unprecedented flexibility and adaptability.

Acknowledgements

This work was funded by a sponsored research grant from Arryx Inc., using equipment purchased under Grant Number 991705 from the W.M. Keck Foundation. The spatial light modulator used in this study was made available by Hamamatsu Corp., as a loan to The University of Chicago. Additional funding was provided by the National Science Foundation through Grant Number DMR-9730189, and by the MRSEC program of the National Science Foundation through Grant Number DMR-980595.

References

[1] A. Ashkin, J.M. Dziedzic, J.E. Bjorkholm, S. Chu, Opt. Lett. 11 (1986) 288.
[2] K. Svoboda, S.M. Block, Annu. Rev. Biophys. Biomol. Struct. 23 (1994) 247.
[3] A. Ashkin, IEEE J. Sel. Top. Quantum Electron. 6 (2000) 841.
[4] D.G. Grier, Curr. Opin. Colloid Interface Sci. 2 (1997) 264.
[5] E.R. Dufresne, D.G. Grier, Rev. Sci. Instrum. 69 (1998) 1974.
[6] D.G. Grier, E.R. Dufresne, US Patent 6,055,106, The University of Chicago (2000).
[7] E.R. Dufresne et al., Rev. Sci. Instrum. 72 (2001) 1810.
[8] D.G. Grier, Nature 393 (1998) 621.
[9] M. Reicherter, T. Haist, E.U. Wagemann, H.J. Tiziani, Opt. Lett. 24 (1999) 608.
[10] J. Liesener, M. Reicherter, T. Haist, H.J. Tiziani, Opt. Commun. 185 (2000) 77.
[11] Y. Igasaki et al., Opt. Rev. 6 (1999) 339.
[12] R.W. Gerchberg, W.O. Saxton, Optik 35 (1972) 237.
[13] K. Sasaki et al., Opt. Lett. 16 (1991) 1463.
[14] K.C. Neuman et al., Biophys. J. 77 (1999) 2856.
[15] N.R. Heckenberg et al., Opt. Quantum Electron. 24 (1992) S951.
[16] H. He, N.R. Heckenberg, H. Rubinsztein-Dunlop, J. Mod. Opt. 42 (1995) 217.
[17] N.B. Simpson, L. Allen, M.J. Padgett, J. Mod. Opt. 43 (1996) 2485.
[18] K.T. Gahagan, G.A. Swartzlander Jr., Opt. Lett. 21 (1996) 827.
[19] H. He, M.E.J. Friese, N.R. Heckenberg, H. Rubinsztein-Dunlop, Phys. Rev. Lett. 75 (1995) 826.
[20] J. Arlt, V. Garces-Chavez, W. Sibbett, K. Dholakia, Opt. Commun. 197 (2001) 239.
[21] J. Arlt, M.J. Padgett, Opt. Lett. 25 (2000) 191.
[22] L. Paterson et al., Science 292 (2001) 912.

VOLUME 67, NUMBER 2   PHYSICAL REVIEW LETTERS   8 JULY 1991

Atomic Interferometry Using Stimulated Raman Transitions

Mark Kasevich and Steven Chu

Departments of Physics and Applied Physics, Stanford University, Stanford, California 94305
(Received 23 April 1991)

The mechanical effects of stimulated Raman transitions on atoms have been used to demonstrate a matter-wave interferometer with laser-cooled sodium atoms. Interference has been observed for wave packets that have been separated by as much as 2.4 mm. Using the interferometer as an inertial sensor, the acceleration of a sodium atom due to gravity has been measured with a resolution of 3×10^{-6} after 1000 sec of integration time.

PACS numbers: 32.80.Pj, 07.60.Ly, 35.80.+s, 42.50.Vk

The potential utility of atom interferometers has been previously discussed [1,2]. As sensitive accelerometers they can be used for a variety of precision measurements such as a search for a net charge on atoms, fifth-force experiments, and tests of the equivalence principle and general relativity. Atom interferometers can also be used in Berry's phase measurements or in studies of the Aharonov-Casher effect [3]. Finally, since atom interferometers are also sensitive to the atom's internal degrees of freedom, experiments not accessible to neutron or electron interferometry are possible.

Atomic diffraction from microfabricated matter gratings [1] and from intense standing waves of light [4] have been experimentally demonstrated, and discussed as a potential means of coherently splitting an atomic wave function. It has also been suggested that the photon recoil acquired by an atom when making an optical transition be used as an atomic beam splitter [5]. More recently, an atom interferometer in a Young's double-slit diffraction geometry has been demonstrated [6].

In this Letter we describe an atom interferometer where the atomic wave function of a sodium atom is coherently split by a two-photon Raman transition between the $F=1$, $m_F=0$ and $F=2$, $m_F=0$ ground states (hyperfine splitting ≈ 1.77 GHz) of the atom. With the laser beams tuned near the $3S_{1/2}$-$3P_{3/2}$ sodium transition, we have previously demonstrated mechanical effects of this two-photon transition: If the two laser beams are counterpropagating, atoms making the Raman transition will experience a velocity recoil $v_r = 2\hbar k/M \approx 6$ cm/sec [7]. With laser-cooled atoms in an atomic fountain [8], transit times through the apparatus can approach 0.5 sec, yielding separations on the order of 1 cm.

Our interferometer separates and recombines an atom by using a $\pi/2$-π-$\pi/2$ sequence of Raman pulses. The use of NMR concepts [9] is appropriate for our situation, and if the laser detunings from either of the hyperfine ground states to the excited $3P_{3/2}$ state are large compared to the Rabi frequencies of the laser beams and the linewidth of the $3P_{3/2}$ state, the three-level system discussed here can be reduced to an equivalent two-level system [10]. First consider a wave packet with mean momentum p (p is the momentum component along the direction of the laser beams) and internal state $|1\rangle$. The first $\pi/2$ pulse puts the original state $|1,p\rangle$ into a superposition of states $|1,p\rangle$

and $|2,p+2\hbar k\rangle$. After a time Δt, the wave packets will have separated by an amount $2\hbar k \Delta t/M$. The π pulse then induces the transitions $|1,p\rangle \rightarrow |2,p+2\hbar k\rangle$ and $|2,p+2\hbar k\rangle \rightarrow |1,p\rangle$, and after another interval Δt, the two wave packets merge again. By adjusting the phase of the final $\pi/2$ pulse, the atom can be put into either of the hyperfine states. If the laser beams are aligned to be perpendicular to the motion of the atoms, we have the atomic analog to an optical Mach-Zehnder interferometer as shown in Fig. 1(a). Figure 1(b) illustrates a configuration where the atomic separation is along the direction of motion. We used this form of the interferometer to measure the acceleration of an atom due to gravity.

The final state of the atom depends on the phase difference between the two paths $1 \rightarrow 2 \rightarrow 4$ and $1 \rightarrow 3 \rightarrow 4$ of Fig. 1 during the free evolution of the atom, and the phase of the atom relative to the phases of the optical driving fields. Under conditions where the action $S = \int_{\Gamma} L \, dt$ (L is the Lagrangian of the atom) is much larger than $\hbar$, the free-evolution contribution to the phase of the atom reduces to $\phi_f = \int_{\Gamma} (\mathbf{k}_a \cdot d\mathbf{x} - \omega_a \, dt)$, where $\mathbf{p}_a = \hbar \mathbf{k}_a$ and $E_a = \hbar \omega_a$ are the classical momentum and total energy of the atom, respectively, along the classical path Γ [11]. For the interferometers described here, the phase difference between the two paths is zero for an atom in a constant gravitational field. This is to be contrasted with neutron interferometers where the interaction with the beam splitter does not change the neutron's energy so the phase difference from the $\int \omega \, dt$ term is zero, but the $\int \mathbf{k} \cdot d\mathbf{x}$ term gives a nonzero contribution [12].

The phase shifts arising from the atom's interaction with the light do not vanish. The net phase shift between the two paths can be written in the form $\Delta \phi_l = \phi_1 - \phi_2 - \phi_3 + \phi_4$ (see Fig. 1), where $\phi_i \approx \mathbf{k}_l \cdot \mathbf{x}_i - \omega_i t_i$ in the limit where effects due to the finite duration of the pulse can be ignored [13]. In this expression, $\mathbf{k}_l = \mathbf{k}_1 - \mathbf{k}_2$ is the difference of the photon wave vectors, $\mathbf{x}_i$ the position of the atom, t_i the time the light is pulsed on, and ω_i the rf frequency for that pulse. Note that the $\pi/2$-π-$\pi/2$ pulse sequence ensures that the net phase shift from the $\mathbf{k}_l \cdot \mathbf{x}_i$ term is velocity independent so that the fringe contrast can be preserved in the presence of an inhomogeneous velocity distribution.

Phase shifts due to the $\omega_i t_i$ term depend on ω_i for the

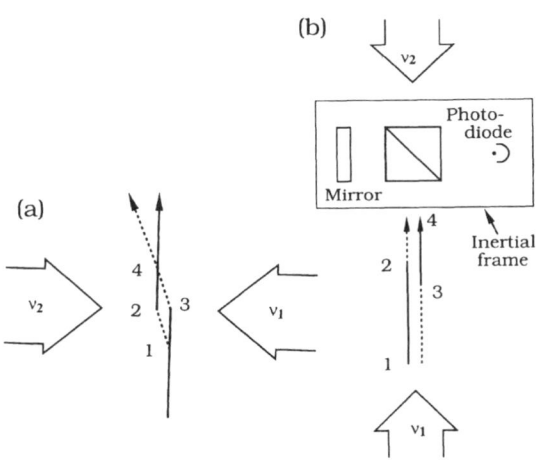

FIG. 1. Diagrams of the $\pi/2$-π-$\pi/2$ pulse interferometer described in the text. The mechanical recoil from the first $\pi/2$ pulse coherently splits (position 1) the atomic wave packet. A π pulse (positions 2 and 3) redirects each wave packet's trajectory. By adjusting the phase of the second $\pi/2$ pulse (position 4), the atom can be put into either $|1\rangle$ or $|2\rangle$. (a) The atom's mean velocity is orthogonal to the laser beams. (b) The case where the atom is traveling parallel to the beams (overlapping wave-packet trajectories are slightly displaced). The inertial frame, consisting of a beam splitter, mirror, and photodiode, is suspended just above the vacuum can. In our experiment, the atom is prepared in the $|1\rangle$ state (solid lines) and detected in the $|2\rangle$ state (dashed lines).

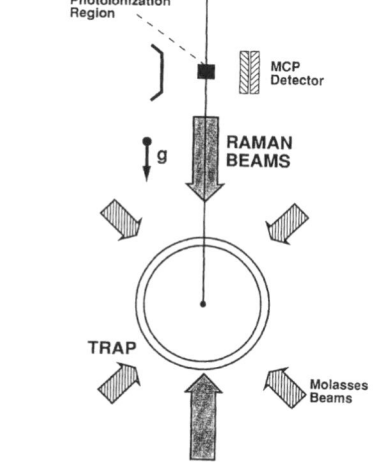

FIG. 2. A schematic of the apparatus used to demonstrate the interferometer illustrated in Fig. 1(b). Atoms were loaded into an optomagnetic trap, cooled, launched, and then optically pumped into the $F=1$ hyperfine state. Approximately 50 msec following their launch, the set of Raman beams is pulsed on three times to drive $\pi/2$-π-$\pi/2$ pulse sequence. After 135 msec, atoms in the $F=2$ state were detected by resonant photoionization. The repetition rate of the experiment was 1 Hz. Not shown are the other set of molasses beams and the atomic beam used to load the trap.

three pulses. If $\omega_i=\omega_0$ for all pulses, one can show that $\Delta\phi_l=-\mathbf{k}_l\cdot\mathbf{g}(\Delta t)^2$, where Δt is the time between pulses. This result is equivalent to an expression derived by Bordé [5]. For large accelerations and/or long measurement times Δt, it is convenient to change ω_i to keep in resonance with the atoms. (The $\pi/2$ pulse must remain a $\pi/2$ pulse as the atom accelerates.) If the frequency difference of the light is chirped to stay in resonance with the atom as it accelerates, the net phase shift is again zero. If, instead, the three frequencies of the light pulses [14] are fixed at $\omega_1=\omega_0$, $\omega_{2,3}=\omega_0+\omega_m$, $\omega_4=\omega_0+2\omega_m$, where $\omega_m\sim\mathbf{k}_l\cdot\mathbf{g}\Delta t$, then $\Delta\phi_l=-\mathbf{k}_l\cdot\mathbf{g}(\Delta t)^2+2\omega_m\Delta t$.

The experimental apparatus (see Fig. 2) has been described in prior publications [7,8]. Briefly, $\sim10^7$ atoms were loaded into an optomagnetic trap [15] from a slowed Na beam. The atomic beam was slowed by a counterpropagating, frequency-chirped laser. After 0.8 sec the magnetic field used to trap the atoms was shut off, and the atoms were further cooled in polarization-gradient optical molasses [16] to a temperature of ~30 μK. 500 μsec after the magnetic field was shut off, the $\sim$4-mm-diam ball of atoms was launched vertically in a moving molasses light field by shifting the frequencies of

the molasses beams with vertical components by ±2.9 MHz [7]. The atoms were then optically pumped into the $F=1$ hyperfine state.

The two Raman laser beams were derived from a second dye laser tuned ~2.5 GHz below the $F=2\rightarrow3$ resonance. One beam was passed twice through a ~30-MHz acousto-optic modulator while the other passed through a ~1.71-GHz electro-optic modulator. Line broadening of the transition due to mirror vibrations was eliminated with a fast servo system that monitored and corrected the beat note of the two laser beams just outside the vacuum can [see Fig. 1(b)]. A beam splitter and mirror used to overlap the two laser beams were mounted on an interferometrically stable inertial reference frame (a ~35-kg brass plate suspended from the ceiling by surgical tubing), and the measured beat note was phase locked to a stable rf reference by changing the frequency of the ~30-MHz modulator. A random shift in the position of the platform during the course of the scan by $\sim\lambda/4$ in the measurement time would destroy fringe contrast. For each Raman pulse, light at ν_L and $\nu_L-(60$ MHz) (where ν_L is the laser carrier frequency) was first turned on for 40 μsec to let the servo system phase lock to the beat note. The rf sideband from the electro-optic modulator at 1.71 GHz was then pulsed on for the time required to drive the $\pi/2$ or π pulse.

VOLUME 67, NUMBER 2　　　PHYSICAL REVIEW LETTERS　　　8 JULY 1991

Before entering the vacuum can, the Raman beams were expanded to an $\sim$2-cm $1/e^2$ diameter. Wave-front curvature was measured interferometrically to better than $\lambda/10$ in the central 0.5 cm of each beam. The vacuum-can windows were measured to be flat to at least the same degree over this area. The beams were centered on the trapping region and parallel to **g** to better than 2 mrad.

The three rf frequencies for the $\pi/2$, π, and $\pi/2$ Raman pulses were generated by mixing one of three $\sim$60-MHz synthesizers with the 1.64-GHz output of an HP8665A synthesizer. The 1.64-GHz carrier and 1.58-GHz sideband were removed with an rf filter, leaving only the 1.71-GHz sideband to drive the electro-optic modulator. All synthesizers were referenced to timing signals from a SRS FS700 LORAN-C receiver which has a short-term stability of 1 part in 10^{-11}. The relative phases of the low-frequency synthesizers were synchronized by adjusting the phase of one of the oscillators so that the initial phase difference $\Delta\phi^0 = \phi_1^0 - 2\phi_2^0 + \phi_3^0$ was constant, where $\phi_i(t) = \omega_i t + \phi_i^0$ is the phase of the ith oscillator. Phase noise of one of the low-frequency oscillators produced a $20°$ uncertainty in $\Delta\phi^0$.

The interferometer signal was derived from the measurement of the Raman transition from the $F=1$, $m_F=0$ state to the $F=2$, $m_F=0$ state. Atoms excited into the $F=2$ state were resonantly ionized and the ions were detected by a microchannel plate as shown in Fig. 2. A bias magnetic field of $\sim$85 mG was applied along the propagation axis of the light to remove the degeneracy between field-sensitive $m_f=\pm1 \to \pm1$ transitions and the field-insensitive $m_f=0 \to 0$ transition. Since the $\sim$600-kHz Doppler width of the atomic fountain (due to a spread $\Delta v_{\text{fountain}} \approx 20$ cm/sec) was much larger than the $\sim$120-kHz splitting between the field-sensitive and field-insensitive transitions, the field-sensitive transitions were also excited by the $\pi/2$ and π pulses. Atoms making the field-insensitive transition were distinguished from those in the field-sensitive state by their time of flight to the

detection region. A horizontally propagating 15-nsec pulse of 355-nm light apertured to a 4 mm (vertical)$\times$8 mm (horizontal) rectangle intersected a vertically propagating 1-μsec pulse of light resonant with the $F=2 \to 3$ transition. The resonant beam ($1/e^2$ diameter $\approx$5 mm) copropagating with one of the Raman beams created an approximately cylindrical detection region that ionized only those atoms whose trajectories were near the center of the Raman beams.

Figure 3 shows a scan of $\omega_m/2\pi$ versus ionization signal for the $\pi/2$-π-$\pi/2$ sequence. The pulses were separated by 10 msec, and the time required to drive a π pulse was 65 μsec with a 2.5-GHz detuning of the carrier from the optical resonance. The polarizations of the Raman beams were crossed linear. Each data point represents 40 fountain launches at a rate of 1 launch/sec. The 25-Hz linewidth corresponds to a Doppler resolutions $\Delta v/c = \Delta v/(v_1+v_2)$ of 7.5 μm/sec.

The fringe contrast would be 100% if the pulse length δt of the Raman $\pi/2$ and π pulses were sufficiently short to address all of the atoms in the velocity distribution of the atomic fountain, i.e., if $(1/\delta t)/(v_1+v_2) \gg \Delta v_{\text{fountain}}/c$. For the conditions in Fig. 3, the expected fringe contrast was decreased to 27% because some of the atoms in the velocity distribution were partially out of resonance with the driving laser pulses. Consequently, these atoms were excited but did not see the full $\pi/2$ and π pulses. The expected fringe contrast after background subtraction was observed for 1-ms delay times between pulses as shown in Fig. 4(a). However, the fringe contrast in Fig. 3 is 12% and a "best run" with a 40-msec delay between pulses gave a 7% contrast [Fig. 4(b)]. We suspect movement of the reference "inertial" frame and wave-front instabilities in the Raman beams to be the dominant sources of the loss of contrast at the longer delay times. With improved engineering, we believe that it should be possible to ap-

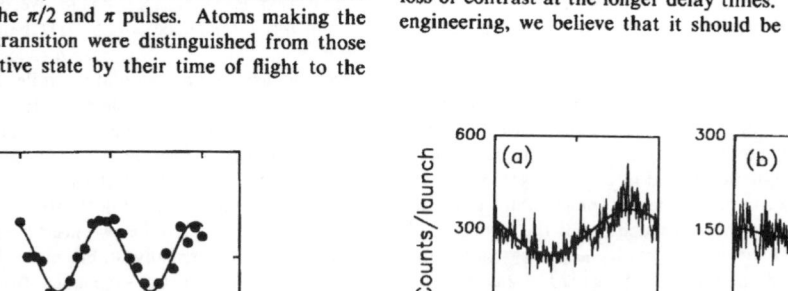

FIG. 3. Interferometer fringes from a frequency scan of the Raman laser beams when the time between pulses is 10 msec. The solid line is a nonlinear least-squares fit to the data. The linewidth of the resonance is determined by the time between the $\pi/2$ and π pulses, and is not a free parameter.

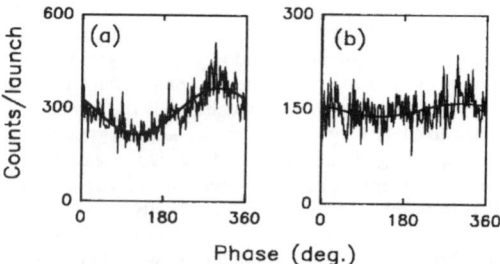

FIG. 4. Interference fringes for (a) $\Delta t = 1$ msec and (b) $\Delta t = 40$ msec pulse delay times. Phase is scanned by shifting the phase of the rf sideband used for the final $\pi/2$ pulse by the indicated amount. The heavy lines are nonlinear least-squares fits of sine functions to the data. In the case of the 40-msec delay time, the maximum wave-packet separation is 2.4 mm, and the velocity fringes have a width of 1.8 μm/sec.

proach 100% fringe contrast for measurement times on the order of $\Delta t \approx 0.1$ sec.

A least-squares fit with a sine function could determine the center of the fringe in Fig. 3 to ± 0.6 Hz. The fitted uncertainty yields a sensitivity to changes in g at the level of 3×10^{-6} after an integration time of 1000 sec for all the data shown in Fig. 3. The data in Fig. 4(b) have an inherent resolution of 6×10^{-7} uncertainty. The resolution was limited by the loss in interferometer contrast, a background counting rate, and fluctuations in the signal rate. The background counts were due to population of the $F=2$ state from spontaneous emission from the $3P_{3/2}$ state and they occurred at 20% of the peak signal rate. Sources of noise included the $\sim 10\%$ intensity fluctuations in the Nd-doped yttrium-aluminum-garnet laser, $\sim 25\%$ fluctuation in the number of trapped atoms, random phase shifts induced by motion of the inertial reference frame, and pointing instabilities in the laser used to generate the Raman beams.

An absolute measure of g can, in principle, be determined with great accuracy. Systematic phase shifts associated with magnetic-field inhomogeneities in the present apparatus were at the ~ 1-Hz level given the 8-mG field gradient in the interaction region (measured by aligning the Raman beams to be copropagating and tuning across the field-sensitive transitions). A deviation from vertical of the Raman beams by 2 mrad (the upper limit on the alignment of the beams with the local vertical) will create a systematic shift in g at the level of 2×10^{-6}. Since these systematic errors decrease quadratically with the perturbing influence, we believe that an absolute measurement of g on an atom can ultimately be done with an uncertainty better than 1 part in 10^{-10}, and relative measurements should be possible with orders of magnitude greater precision. With changes in the experimental arrangement, $\hbar/M$ can also be determined with great accuracy through a precision measurement of the recoil velocity.

We have also operated the interferometer in the Mach-Zehnder configuration. Working with 2-cm-diam laser beams, atoms with an upward velocity of 250 cm/sec were separated by 72 μm using a 1.2-msec delay between pulses. By working at the peak of a ballistic trajectory, a delay time of ~ 0.2 sec is possible, giving a wave-packet separation of 6 mm. In this geometry, it should be possible to spatially resolve the arms of the interferometer.

The precision of this type of atom interferometer follows directly from the high-frequency resolution made accessible with atomic fountains [8], and the use of optical transitions between ground states of an atom, which allows radio-frequency stability of the excitation with a Doppler sensitivity in the ultraviolet frequency range [7]. We plan to use this type of interferometer to improve the tests of the charge neutrality of atoms, tests of the equivalence principle, and for searches of gravitational anomalies.

This work was supported in part by grants from the AFOSR and the NSF. We thank D. S. Weiss for useful discussions, and D. S. Weiss and K. Gibble for a careful reading of the manuscript.

[1] D. W. Keith, M. L. Schattenburg, Henry I. Smith, and D. E. Pritchard, Phys. Rev. Lett. **61**, 1580 (1988).

[2] J. F. Clauser, Physica (Amsterdam) **151B**, 262 (1988).

[3] R. Hagen, Phys. Rev. Lett. **64**, 2347 (1990).

[4] P. L. Gould, G. A. Ruff, and D. E. Pritchard, Phys. Rev. Lett. **56**, 827 (1986).

[5] C. J. Bordé, Phys. Lett. A **140**, 10 (1989). In this paper, Bordé points out that wave-packet interference may have been inadvertently demonstrated in experiments using Ramsey fringes in the optical domain.

[6] O. Carnal and J. Mlynek, Phys. Rev. Lett. **66**, 2689 (1991).

[7] M. Kasevich, D. S. Weiss, E. Riis, K. Moler, S. Kasapi, and S. Chu, Phys. Rev. Lett. **66**, 2297 (1991).

[8] M. Kasevich, E. Riis, S. Chu, and R. G. DeVoe, Phys. Rev. Lett. **63**, 612 (1989).

[9] See, for example, L. Allen and J. H. Eberly, *Optical Resonance and Two-Level Atoms* (Wiley, New York, 1975).

[10] See, for example, R. G. Brewer and E. L. Hahn, Phys. Rev. A **11**, 1641 (1975).

[11] R. P. Feynman and A. R. Hibbs, *Quantum Mechanics and Path Integrals* (McGraw-Hill, New York, 1965).

[12] D. M. Greenberger and A. W. Overhauser, Rev. Mod. Phys. **51**, 43 (1979).

[13] The expression for $\Delta\phi$ follows from Eq. (V.26) of N. Ramsey, *Molecular Beams* (Oxford Univ. Press, London, 1956), p. 127.

[14] Most frequency synthesizers are not sufficiently phase stable during a phase continuous frequency sweep to take advantage of the inherent precision in atom interferometry.

[15] E. L. Raab, M. Prentiss, A. Cable, S. Chu, and D. E. Pritchard, Phys. Rev. Lett. **59**, 2631 (1987).

[16] J. Dalibard and C. Cohen-Tannoudji, J. Opt. Soc. Am. B **6**, 2023 (1989); P. J. Ungar, D. S. Weiss, E. Riis, and S. Chu, J. Opt. Soc. Am. B **6**, 2058 (1989).

VOLUME 74, NUMBER 19   PHYSICAL REVIEW LETTERS   8 MAY 1995

Collisions of Doubly Spin-Polarized, Ultracold ^{85}Rb Atoms

J. R. Gardner,[1] R. A. Cline,[1,*] J. D. Miller,[2,†] D. J. Heinzen,[1] H. M. J. M. Boesten,[2] and B. J. Verhaar[2]

[1]*Department of Physics, The University of Texas, Austin, Texas 78712*
[2]*Eindhoven University of Technology, Box 513, 5600MB Eindhoven, The Netherlands*
(Received 4 January 1995)

We study the collisions of doubly spin-polarized ^{85}Rb atoms at millikelvin temperatures using photoassociation spectroscopy. Because the atoms are spin polarized, only triplet collisional states are formed. This leads to photoassociation spectra of a particularly simple form, which provide a very direct probe of the ground state collision. These spectra are analyzed to yield the ground state triplet scattering length $-1000a_0 < a_T < -60a_0$ for ^{85}Rb, $+85a_0 < a_T < +140a_0$ for ^{87}Rb, and the product of the D-line dipole matrix elements $d(P_{1/2})d(P_{3/2}) = 8.75 \pm 0.25$ a.u.

PACS numbers: 32.80.–t

Rapidly developing techniques for trapping and cooling neutral atoms using laser fields are opening up a wide array of new applications. These include the construction of very precise atomic clocks [1–3], sensitive electric dipole moment searches [4], and possible studies of quantum collective phenomena such as Bose-Einstein condensation. Success in each of these applications hinges critically on understanding the long-range interactions between cold atoms. Cold collision cross sections are very sensitive to long-range atomic interactions and play a dominant role in many experiments. For example, collisional frequency shifts may limit the accuracy of cold atomic fountain clocks [2,3]. Moreover, efforts to achieve Bose-Einstein condensation in a dilute laser-cooled gas depend critically on the ground state scattering length, which must be positive and preferably large [5–8].

Despite their importance, long-range interactions between atoms have been determined by conventional molecular spectroscopy in only a limited number of cases. The lack of extensive data is due in part to the difficulty of populating long-range states starting from the molecular ground state. On the other hand, these states are readily populated in collisions between ultracold atoms, in particular, by photoassociation spectroscopy [9]. Photoassociation spectra of Na [10,11], Rb [12,13], and Li [14] have already been obtained directly yielding detailed information on their long-range excited state interactions. In this paper, we present new ^{85}Rb photoassociation data and analysis, and show for the first time that atomic ground state interaction parameters can be determined from photoassociation spectra.

A crucial aspect of our experiment is that we doubly spin polarize the colliding atoms. This is important, since the ground state collision is thereby restricted to the triplet channel, and the analysis becomes relatively straightforward. An experiment using unpolarized atoms would need to determine both singlet and triplet parameters simultaneously. As a result of the spin polarization, we are able to observe clear, quantum-statistical features of the collisions. By doubly polarizing the atoms, and choosing a suitable excited state, we obtain spectra which provide a

very simple and direct probe of the ground state collision: Each peak effectively measures the amplitude of a particular partial wave of the collision, thus yielding detailed information about the ground state interatomic potential in a narrow radial range that may be varied by the choice of excited state energy. This is, to our knowledge, the first report of doubly spin-polarized ultracold collisions of alkali atoms.

In our experiment, we load approximately 10^4 ^{85}Rb atoms into a far off-resonance optical dipole force trap (FORT) [15]. The FORT laser beam is a linearly polarized, Gaussian beam containing about 1.5 W of optical power and focused to a waist of about 10 μm. The FORT laser is tuned to 12 289 cm^{-1}, which lies between two well-resolved photoassociation peaks. To define a quantization axis, a magnetic field of 7 G is applied along the FORT laser beam propagation ($\hat{z}$) direction.

Once the atoms are loaded into the FORT, they are exposed to a combination of laser fields for 200 ms. Each 200 ms period is broken into a series of 5 μs cycles in which the atoms are irradiated by four laser fields in sequence. During the first 2.5 μs of each cycle, only the FORT laser is applied. During the next 0.6 μs, only two optical pumping (OP) beams interact with the atoms. One of these is tuned to the ^{85}Rb $5^2S_{1/2}(F = 3) \rightarrow 5^2P_{3/2}(F = 3)$ transition and is circularly polarized. It has an intensity of 100 μW/cm^2 and propagates along the z direction. The other optical pumping beam is tuned to the ^{85}Rb $5^2S_{1/2}(F = 2) \rightarrow 5^2P_{3/2}(F = 3)$ transition. During the last 1.9 μs of each cycle, only the photoassociation (PA) laser field is applied, which is linearly polarized, propagates in the z direction, and has an intensity in the range from 20 to 80 W/cm^2. The combined effect of these fields is to trap the atoms, to keep them optically pumped into the ^{85}Rb $5^2S_{1/2}(F = 3, M_F = 3)$ state, and to induce photoassociation transitions. Alternation of the fields in time prevents the light shift of the FORT beam from disrupting the optical pumping or from shifting and broadening the photoassociation resonances. At least 95% of the atoms are in the doubly spin-polarized state.

0031-9007/95/74(19)/3764(4)$06.00

VOLUME 74, NUMBER 19   PHYSICAL REVIEW LETTERS   8 MAY 1995

Photoassociation transitions induced by the PA laser promote colliding pairs of Rb atoms into specific excited bound Rb_2^* states. In order to obtain a spectrum of these states, we repeat the loading and 200 ms irradiation period for a succession of PA laser frequencies [12,13]. At the end of each cycle we detect the number of atoms remaining in the trap with laser-induced fluorescence. Because the excited Rb_2^* states decay predominantly to free pairs of atoms with kinetic energy sufficient to leave the trap, the photoassociation resonances are detectable as a reduction in the fluorescence.

A photoassociation spectrum of a single 0_g^- vibrational level at $12\,573.05$ cm^{-1} is shown in Fig. 1. In Fig. 1(a), we show the spectrum observed when the atoms are maintained in the $5^2S_{1/2}$, $F = 3$ level but are otherwise unpolarized. A pure rotational spectrum spanning the range from $J = 0$ to 4 is observed. In Fig. 1(b), we show a spectrum recorded with doubly spin-polarized atoms. The odd rotational lines disappear as a consequence of spin statistics.

In order to realize a determination of the ^{85}Rb$_2$ ground state parameters, we recorded data similar to that shown in Fig. 1 for a series of vibrational levels of the 0_g^- state that asymptotically connects to the $5^2S_{1/2} + 5^2P_{1/2}$ separated atom limit [16,17]. Five vibrational levels were used in the analysis, with $J = 0$ level energies of -3.365, -4.088, -4.901, -5.812, and -6.827 cm^{-1} with respect to the barycenter of the $5^2S_{1/2} + 5^2P_{1/2}$ dissociation limit, which corresponds to an energy of $12\,578.864$ cm^{-1} in our spectrum. The outer turning points of these states range from $41.6a_0$ to $46.7a_0$.

The photoassociation spectrum is conveniently described in the dressed-state picture. In that framework each of the Rb + Rb* rovibrational states $|\Omega JM\rangle$ is a discrete state embedded in the ground state continuum [18]. As a result of the interaction with the PA laser field, it acquires a finite partial width γ_L for decay into each of the ground state channels in addition to its spontaneous linewidth γ_0. γ_L is proportional to the PA laser intensity I_L and is given by Fermi's golden rule,

$$\gamma_L = 2\pi |\langle \Omega JM | [\mathbf{d}(1) + \mathbf{d}(2)] \cdot \mathbf{E}_L | SM_S lm_l, \epsilon\rangle|^2, \quad (1)$$

where $\mathbf{E}_L = E_L \boldsymbol{\sigma}_L$ is the PA laser field and $|SM_S lm_l, \epsilon\rangle$ is the energy-normalized continuum ground state. Accordingly, the squared S-matrix element for photoassociation followed by spontaneous emission is given by a Breit-Wigner expression. To first order in I_L

$$|S_{\Omega JM, SM_S lm_l}|^2 = \frac{\gamma_0 \gamma_L (\Omega JM \rightarrow SM_S lm_l)}{(\epsilon + E_g + \hbar\omega_L - E_e)^2 + \frac{1}{4}\gamma_0^2}. \quad (2)$$

Here E_g is the asymptotic Rb + Rb internal energy, E_e is the energy of the $|\Omega JM\rangle$ state, and ϵ is the collision energy. Equation (2) is thermally averaged and summed over J, M, l, m_l to obtain the rate coefficient. The above expressions, without the directional dependences, were first used to analyze photoassociation spectra by Napolitano et al. [19]. Expanding $|\Omega JM\rangle$ in atomic fine-structure states [16,17] coupled to total electronic angular momentum jm_j,

$$\gamma_L = \frac{2\pi I_L}{\epsilon_0 c} \left| \sum_j (J\Omega\, j\, -\Omega|l0)\,(jm_j lm_l|JM) \right.$$
$$\times \int_0^\infty dr\, c_j(r) u_{\Omega J}(r)$$
$$\left. \times \langle jm_j | [\mathbf{d}(1) + \mathbf{d}(2)] \cdot \boldsymbol{\sigma}_L | SM_S\rangle u_l(r) \right|^2. \quad (3)$$

The above-mentioned $0_g^-(S + P_{1/2})$ electronic state has a number of simplifying features that facilitate the analysis considerably. First, it is a pure triplet state [16], so that no singlet amplitude is coupled in by the excitation. Second, as with any of the $\Omega = 0$ states, it has negligible second-order hyperfine energy shifts. Third, at the relevant interatomic distances near the outer turning point, this 0_g^- state is, to very good approximation, a product of independent atomic states $S_{1/2}$ and $P_{1/2}$, coupled to form $j = 0$. As a consequence, only states for which $J = l =$ even are excited. Thus our data display directly the quantum statistics of the atoms: Because they are bosons, they may only collide in even ground state partial waves. Note that the magnetic field has negligible influence on the spectrum.

We find γ_L for the 0_g^- state to be a product of a geometrical coefficient and a squared radial integral $\int dr\, u_{\Omega J}(r) d_{eg}(r) u_l(r)$, in abbreviated notation. Notice that a $j = 0$ component of the upper state cannot be excited by σ^+ laser light starting with doubly polarized

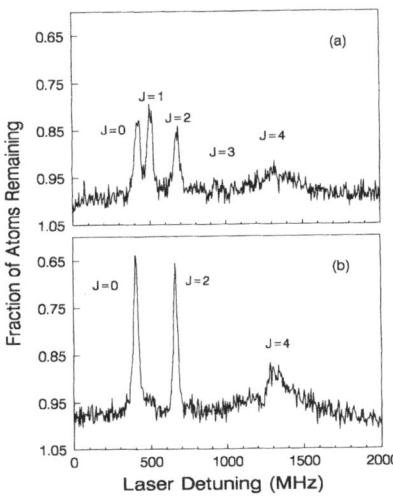

FIG. 1. Photoassociation spectrum of the 0_g^- vibrational level at $12\,573.05$ cm^{-1}. (a) Full rotational spectrum from $J = 0$ to 4 observed for atoms in $5^2S_{1/2}$, $F = 3$ level but otherwise unpolarized. (b) Spectrum recorded with doubly spin-polarized atoms. Odd rotational lines disappear as a consequence of spin statistics.

VOLUME 74, NUMBER 19   PHYSICAL REVIEW LETTERS   8 MAY 1995

ground state atoms with $S = M_S = +1$. Indeed, a full calculation of the excitation rate, taking into account all components of the 0_g^- state, shows this photoassociation rate to be smaller than that for σ^- light by 2 orders of magnitude. This prediction is confirmed by experiment.

Because of the small ranges of ϵ and l involved, cold collisions have the unique property of being insensitive to the detailed behavior of the badly known inner parts of the interatomic potential. The variation of the radial wave function with E and l is a very small first-order perturbation up to a rather large radius r_0. The only relevant information is the accumulated information contained in the phase $\phi(E, l)$ of the rapidly oscillating wave function at r_0 and its first derivatives for $E = l = 0$ [20]. Model calculations show that for $l \leq 2$ and $r_0 = 30a_0$, the calculated photoassociation rates are sufficiently insensitive to the precise values of the first derivatives that they can be taken reliably from an *ab initio* calculation [21]. This insensitivity was used and explained previously in Refs. [22,23]. A similar accumulated-phase method was adopted for the excited state. At large distances where excitation of the above-mentioned five vibrational levels occur, our calculated results are almost independent of the dispersion parameters C_{ne} ($n \geq 6$) for the excited state and C_{ng} ($n \geq 8$) for the ground state, provided that these are taken within the bounds of the present uncertainty [17,24–26].

From the energies of the five measured $J = 0$ levels we derive a value 8.75 ± 0.25 a.u. for the product $d(P_{1/2})d(P_{3/2})$ of D-line dipole matrix elements [27], which determines the strength of the resonant-dipole $1/r^3$ potential in the present r range. This value is in agreement with the most accurate previous measurements $d^2(P_{1/2}) = 8.43 \pm 0.20$ a.u. [28], $d^2(P_{3/2}) = 9.19 \pm 0.18$ a.u. [28], $d^2(P_{3/2}) = 9.08 \pm 0.28$ a.u. [29], and $d^2(P_{3/2}) = 8.68 \pm 0.16$ a.u. [30].

The analysis of the data is carried out as follows. As a first step, we use the strongest $J = 0$ and 2 rotational line shapes to determine an optimum temperature for each point of a grid of C_{6g}-ϕ_g values by a least-squares fit to the measured data points. For this purpose, C_{6g} is taken to lie between 3500 and 6000 a.u. and ϕ_g is allowed to span a full π range. Using these temperatures, we apply a least-squares fit to the ratios of the $J = 0$ peak areas over the C_{6g} and ϕ_g plane (least-squares function χ_1^2), which constrains these parameters to lie in a narrow strip and determines the temperature to be 500 ± 100 μK. We then calculate the χ_2^2 and χ_3^2 functions associated with the four $J = 2$ and one $J = 2$ to $J = 0$ ratios of peak areas, respectively, over the limited ranges of C_{6g} and ϕ_g found previously. We find that χ_2^2 sets about the same limits on C_{6g} and ϕ_g as χ_1^2. However, χ_3^2 sets a limit that corresponds to a different strip in the C_{6g}-ϕ_g plane. The intersection of these strips gives the estimated values and uncertainties of C_{6g} and ϕ_g. Figure 2 shows a contour plot of the total χ^2 surface combining the above three sets of ratios of

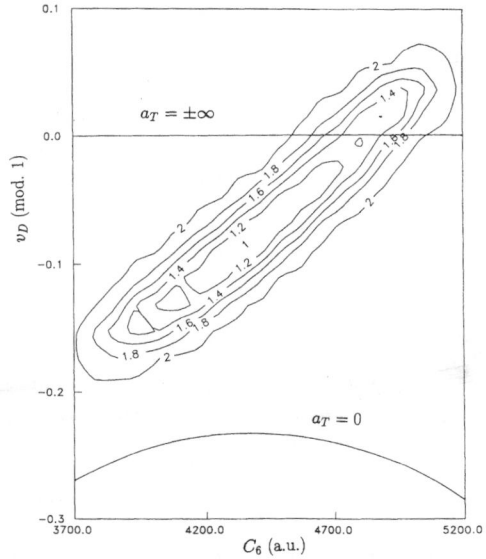

FIG. 2. Contour plot of total χ^2 surface for ratios of $J = 0$ and 2 photoassociation peak areas, as a function of C_{6g} and s-wave vibrational quantum number v_D at dissociation. Lines where a_T changes sign are indicated.

peak areas. Instead of ϕ_g, we use the more transparent (fractional) s-wave vibrational quantum number v_D at dissociation as a parameter. The lines where the triplet scattering length changes a_T sign are indicated. While C_{6g} is experimentally constrained to about the full range of theoretically predicted values [17,24,31], v_D(mod 1) is found to be in the interval between $+0.07$ and -0.19. Including the uncertainty in d^2, C_{6e}, and C_{8g}, we find a_T to be negative with at least 80% probability. For the recently predicted value $C_{6g} = 4426$ a.u. from Ref. [24], which is believed to be correct within a few percent [26], we find $v_D(^{85}\text{Rb, mod1}) = -0.09 \pm 0.07$ and $-1000a_0 < a_T(^{85}\text{Rb}) < -60a_0$. Figure 3 shows a comparison of

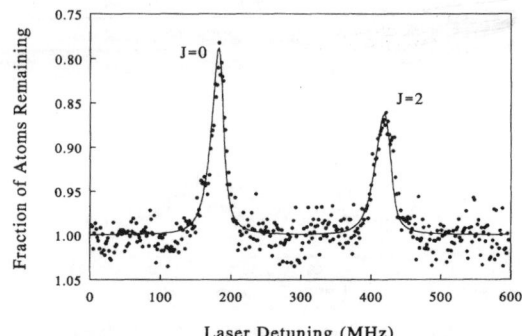

FIG. 3. Comparison of theoretical and experimental line shapes for rotational levels at $12\,573.96$ cm^{-1}. Each J peak arises only from the partial wave $l = J$.

VOLUME 74, NUMBER 19   P H Y S I C A L   R E V I E W   L E T T E R S   8 MAY 1995

theoretical and experimental line shapes of the $J = 0$ and 2 rotational levels at 12 573.96 cm^{-1} for this value of C_{6g} and $v_D(\text{mod}1) = -0.09$.

A simple $\sqrt{m}$ scaling procedure enables us to find the corresponding results for the ^{87}Rb isotope. On the basis of the triplet potential of Krauss and Stevens [21] we take the number of s-wave radial nodes within r_0 to be 34 ± 3. We then derive ϕ_g for ^{87}Rb and by radial integration to larger distances, introducing five additional nodes, find $v_D(^{87}\text{Rb}, \text{mod}1) = +0.37 \pm 0.10$ and $+85a_0 < a_T(^{87}\text{Rb}) < +140a_0$. Also for the total theoretical C_{6g} range a_T is found to be positive.

To summarize, we have used high-resolution photoassociation spectroscopy to study the collisions of ultracold ^{85}Rb atoms. By doubly spin polarizing the atoms, and choosing the particular excited state 0_g^- $(5^2S_{1/2} + 5^2P_{1/2})$, we obtain spectra which provide a very simple and direct probe of the ground state collision: Each peak effectively measures the amplitude of a particular partial wave of the ground state collision, in a narrow radial range that may be varied by the choice of excited state energy. Because of this, we obtain detailed information about the ground state interatomic potential. Analysis of these spectra reveals that the ground state triplet scattering length of ^{85}Rb is large and negative, so that no stable Bose condensate is possible for this state of this isotope. Using mass scaling arguments, we are led to the opposite conclusion for the ^{87}Rb isotope. Apparently, its triplet scattering length is positive so that it remains a viable candidate for observation of a stable Bose-Einstein condensate.

Work at the University of Texas was supported by the R. A. Welch Foundation and the National Science Foundation, work at Eindhoven by the Nederlandse Stichting voor Wetenschappelijk Onderzoek.

*Present address: Dept. of Physics, Davidson College, Davidson, NC, 28036.

†Present address: Time and Frequency Div., National Institute of Standards and Technology, Boulder, CO 80303.

[1] A. Clairon, C. Salomon, S. Guellati, and W. D. Phillips, Europhys. Lett. **12**, 683 (1991).

[2] K. Gibble and S. Chu, Phys. Rev. Lett. **70**, 1771 (1993).

[3] E. Tiesinga, B. J. Verhaar, H. T. C. Stoof, and D. van Bragt, Phys. Rev. A **45**, R2671 (1992).

[4] M. Bijlsma, B. J. Verhaar, and D. J. Heinzen, Phys. Rev. A **49**, R4285 (1994).

[5] A. L. Fetter and J. D. Walecka, *Quantum Theory of Many-Particle Systems* (McGraw-Hill, New York, 1971), p. 218.

[6] E. Tiesinga, A. J. Moerdijk, B. J. Verhaar, and H. T. C. Stoof, Phys. Rev. A **46**, R1167 (1992).

[7] H. T. C. Stoof, Phys. Rev. A **49**, 3824 (1994).

[8] C. Monroe, E. A. Cornell, C. A. Sacket, C. J. Myatt, and C. E. Wieman, Phys. Rev. Lett. **70**, 414 (1993).

[9] H. R. Thorsheim, J. Weiner, and P. S. Julienne, Phys. Rev. Lett. **58**, 2420 (1987).

[10] P. D. Lett, K. Helmerson, W. D. Phillips, L. P. Ratliff, S. L. Rolston, and M. E. Wagshul, Phys. Rev. Lett. **71**, 2200 (1993).

[11] L. P. Ratliff, M. E. Wagshul, P. D. Lett, S. L. Rolston, and W. D. Phillips, J. Chem. Phys. **101**, 2638 (1994).

[12] J. D. Miller, R. A. Cline, and D. J. Heinzen, Phys. Rev. Lett. **71**, 2204 (1993).

[13] R. A. Cline, J. D. Miller, and D. J. Heinzen, Phys. Rev. Lett. **73**, 632 (1994).

[14] W. I. McAlexander, E. R. I. Abraham, N. W. M. Ritchie, C. J. Williams, H. T. C. Stoof, and R. G. Hulet, Phys. Rev. A **51**, 871 (1995).

[15] J. D. Miller, R. A. Cline, and D. J. Heinzen, Phys. Rev. A **47**, 4567 (1993).

[16] M. Movre and G. Pichler, J. Phys. B **10**, 2631 (1977).

[17] B. Bussery and M. Aubert-Frecon, J. Chem. Phys. **82**, 3224 (1985).

[18] H. Feshbach, *Theoretical Nuclear Physics, Part 1: Nuclear Reactions* (Wiley, New York, 1992).

[19] R. Napolitano, J. Weiner, C. J. Williams, and P. S. Julienne, Phys. Rev. Lett. **73**, 1352 (1994).

[20] B. J. Verhaar, K. Gibble, and S. Chu, Phys. Rev. A **48**, R3429 (1993).

[21] M. Krauss and W. J. Stevens, J. Chem. Phys. **93**, 4236 (1993).

[22] A. J. Moerdijk, W. C. Stwalley, R. G. Hulet, and B. J. Verhaar, Phys. Rev. Lett. **72**, 40 (1994).

[23] A. J. Moerdijk and B. J. Verhaar, Phys. Rev. Lett. **73**, 518 (1994).

[24] M. Marinescu, H. R. Sadeghpour, and A. Dalgarno, Phys. Rev. A **49**, 982 (1994).

[25] M. Marinescu and A. Dalgarno (private communication).

[26] W. C. Stwalley (private commnication).

[27] H. M. J. M. Boesten *et al.* (to be published).

[28] A. Gallagher and E. L. Lewis, Phys. Rev. A **10**, 231 (1974).

[29] R. W. Schmeider *et al.*, Phys. Rev. A **2**, 1216 (1970).

[30] J. K. Link, J. Opt. Soc. Am. **56**, 1195 (1966).

[31] M. L. Manokov and V. O. Ovsiannikov, J. Phys. B **10**, 659 (1985); A. Dalgarno, Adv. Chem. Phys. **12**, 143 (1967); F. Maeder and W. Kutzelnigg, Chem. Phys. **42**, 195 (1979).

PHYSICAL REVIEW
LETTERS

VOLUME 78	10 FEBRUARY 1997	NUMBER 6

Bose-Einstein Condensation of Lithium: Observation of Limited Condensate Number

C. C. Bradley, C. A. Sackett, and R. G. Hulet

Physics Department and Rice Quantum Institute, Rice University, Houston, Texas 77005-1892
(Received 5 September 1996; revised manuscript received 23 October 1996)

Bose-Einstein condensation of ^{7}Li has been studied in a magnetically trapped gas. Because of the effectively attractive interactions between ^{7}Li atoms, many-body quantum theory predicts that the occupation number of the condensate is limited to about 1400 atoms. We observe the condensate number to be limited to a maximum value between 650 and 1300 atoms. The measurements were made using a versatile phase-contrast imaging technique. [S0031-9007(97)02369-7]

PACS numbers: 03.75.Fi, 05.30.Jp, 32.80.Pj

In a previous Letter [1] we reported evidence for Bose-Einstein condensation (BEC) in a magnetically trapped atomic gas of ^{7}Li. Unlike ^{87}Rb and ^{23}Na, the other species in which gaseous BEC has been observed [2], ^{7}Li atoms have a negative s-wave scattering length a, indicating that for a sufficiently cold and dilute gas the interatomic interactions are effectively attractive. Attractive interactions are thought to prevent BEC from occurring at all in a spatially homogeneous (i.e., untrapped) gas [3,4], and as recently as 1994, these interactions were believed to preclude BEC in a trap as well. Current theories predict that BEC can occur in a trap such as ours, but with no more than about 1400 condensate atoms [5–11]. Verification of this prediction would provide a sensitive test of many-body quantum theory. In our previous work [1], the condensate could not be directly observed, and the number of condensate atoms suggested by the measurements was overestimated. In this Letter we report quantitative measurements of the condensate number, which are consistent with the theoretical limit.

The effects of interactions on a trapped condensate are studied using mean-field theory. For densities n such that $na^3 \ll 1$, the mean-field interaction energy is given by $U = 4\pi\hbar^2 an/m$, where m is the atomic mass. For ^{7}Li, $a = (-14.5 \pm 0.4)$ Å [12]. Because $a < 0$, the interaction energy decreases with increasing n, so the condensate tends to collapse upon itself. When the confining potential is included in the theory, it is found that if U is sufficiently small compared to the trap energy-level spacing, the destabilizing influence of

the interactions is balanced by the kinetic pressure of the gas, and a metastable condensate can form. This requirement for U leads to the prediction that the number of condensate atoms N_0 is limited. As the maximum N_0 is approached, the rate for inelastic collisions increases and the gas becomes progressively less stable with respect to thermal and quantum mechanical fluctuations [7–10].

The apparatus used to produce BEC has been described in previous publications [1,13]. The magnetic trap forms a nearly symmetric harmonic oscillator potential, with oscillation frequencies $\nu_x = 150.6$ Hz, $\nu_y = 152.6$ Hz, and $\nu_z = 131.5$ Hz. The bias field at the center of the trap is 1003 G [14]. The trap is loaded using laser cooling, resulting in a number of trapped atoms $N \approx 2 \times 10^8$, and a temperature $T \approx 200$ μK. The atoms are then cooled into the quantum degenerate regime using rf-induced forced evaporative cooling [15]. Quantum degeneracy is reached with $N \approx 10^5$ atoms at $T \approx 300$ nK, after $\sim$200 sec of cooling.

After cooling, the trapped atom distribution is observed by imaging via an optical probe. This technique has previously been used to measure the spatial density distribution directly [16] and to measure the velocity distribution by first releasing the atoms from the trap and allowing them to ballistically expand [2]. In our experiment, the trap is formed by permanent magnets, so only *in situ* imaging is possible. The harmonic oscillator ground state of our trap has a Gaussian density distribution with a $1/e$ radius of 3 μm. The resolution of the imaging system must therefore be sufficient to detect

VOLUME 78, NUMBER 6   PHYSICAL REVIEW LETTERS   10 FEBRUARY 1997

such a small object. In our previous experiment [1] the imaging resolution was not sufficient, but the presence of the spatially localized condensate was deduced from distortions observed in images of quantum degenerate clouds [17].

Our improved imaging system is shown schematically in Fig. 1. With the polarizer E removed, it can be used to measure the density distribution by absorption imaging, in which the absorptive shadow of the atoms is imaged onto the camera. However, any imaging system with finite resolution will be sensitive to phase shifts caused by the index of refraction of the atom cloud, which can result in significant image distortions. In order to eliminate these distortions, it is necessary to reduce the index by using large probe detunings Δ. Since the absorption coefficient decreases as Δ^{-2}, while the index decreases as only Δ^{-1}, eliminating the distortions can leave the absorption signal too small to be detected.

Phase-contrast techniques are commonly employed to image weakly absorbing objects [18]. In passing through the cloud, the laser acquires a spatially dependent phase $\beta = \phi + i\alpha/2$, where ϕ is the dispersive phase shift and α is the optical density. The probe electric field $\mathbf{E}$ can then be written as $\mathbf{E} = \mathbf{E}_o e^{i\beta}$. In absorption imaging, the detected signal intensity I_s depends only on α: $I_s = I_o|e^{i\beta}|^2 = I_o e^{-\alpha}$, where $I_o = |\mathbf{E}_o|^2$. In the simplest phase-contrast technique, dark-field imaging, a spatially small opaque beam block is inserted at a focus of the probe laser beam (position D in Fig. 1). The resulting signal is $I_s = I_o|e^{i\beta} - 1|^2 \approx I_o\phi^2$ for $\alpha \ll |\phi| \ll 1$. Andrews et al. used this technique to image ^{23}Na Bose-Einstein condensates in situ [16]. However, minimizing refractive distortions requires $|\phi| \ll 1$, so the dark-field signal is relatively small as it is proportional to ϕ^2. Because the number of condensate atoms in ^{7}Li is limited, a more sensitive technique is required.

We use a flexible phase-contrast method which exploits the birefringence of the atoms in a strong magnetic field. The probe beam is linearly polarized perpendicular to the magnetic field axis and propagates along an axis inclined 55° with respect to the field. The electric field of the probe decomposes into two elliptical polarizations $\mathbf{E} = \mathbf{E}_c + \mathbf{E}_{nc}$, such that $\mathbf{E}_c$ couples to the σ^+ optical transition and acquires a phase shift, while $\mathbf{E}_{nc}$ does not. If the transmitted light is passed through a polarizer (E in Fig. 1), $\mathbf{E}_c$ and $\mathbf{E}_{nc}$ combine and interfere, producing a phase-contrast image. If Δ is large enough that α can be neglected, the detected intensity distribution is

$$I_s(r) = I_o\left[\cos^2\theta + \frac{\sqrt{3}}{4}\phi(r)\sin 2\theta - \frac{3}{16}\phi(r)^2\cos 2\theta\right], \quad (1)$$

where θ is the angle between the polarizer axis and the initial polarization of the probe beam. Linear phase-contrast imaging is accomplished for $\theta = 45°$ and dark-field imaging is recovered for $\theta = 90°$. By varying θ between these extremes, the relative size of the signal and background can be varied in order to maximize the signal-to-noise ratio of the image. For the data reported here, $\theta = \pm 75°$.

The cloud is probed by a pulsed laser beam with a duration of 10 μs, an intensity of 250 mW/cm^2 and with Δ in the range $20\Gamma < |\Delta| < 40\Gamma$, where $\Gamma = 5.9$ MHz is the natural linewidth of the transition. Only one image can be obtained because each atom scatters a few photons while being probed, heating the gas to several μK.

The detected signal intensity, given by Eq. (1), is related to the density distribution n of the trapped atoms by [19]

$$\beta(x', y') = \phi + i\frac{\alpha}{2}$$
$$= -\frac{\sigma_0}{2}\int dz'\, n(x', y', z')\frac{\Gamma}{2\Delta + i\Gamma}, \quad (2)$$

where $\sigma_0 = 1.43 \times 10^{-9}$ cm^2 is the resonant light scattering cross section. The z' axis is parallel to the probe propagation axis, while the x' and y' axes are perpendicular to it. Light scattering might by modified by the quantum degenerate nature of the atoms, but this effect is expected to be negligible under our conditions [20].

Because the trap is not isotropic, the density distributions are slightly ellipsoidal; the images are observed to have the expected asymmetry. The radial profiles shown in Fig. 2 are obtained by angle averaging the data around ellipses with aspect ratio 1.10, accounting for the trap asymmetry and the oblique viewing angle [21]. We assume that the gas is in thermal equilibrium, and fit T and N_0 to the data. Any two of N, T, or N_0 completely determine the density of the gas through the Bose-Einstein distribution function. The density is calculated using a semiclassical ideal-gas approximation for the noncondensed atoms [22],

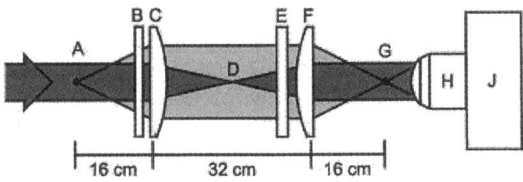

FIG. 1. A schematic of the imaging system used for in situ phase-contrast polarization imaging. A linearly polarized laser beam is directed through the cloud of trapped atoms located at A. The probe beam and scattered light field pass out of a vacuum viewport B, and are relayed to the primary image plane G by an identical pair of 3-cm diameter, 16-cm focal-length doublet lenses C and F. The light is then reimaged and magnified onto a camera J by a microscope objective H. The measured magnification is 19, and the camera pixels are 19 μm square. The linear polarizer E causes the scattered light and probe fields to interfere, producing an image sensitive to the refractive index of the cloud. The system is focused by adjusting the position of lens F, which is mounted on a translator.

VOLUME 78, NUMBER 6   PHYSICAL REVIEW LETTERS   10 FEBRUARY 1997

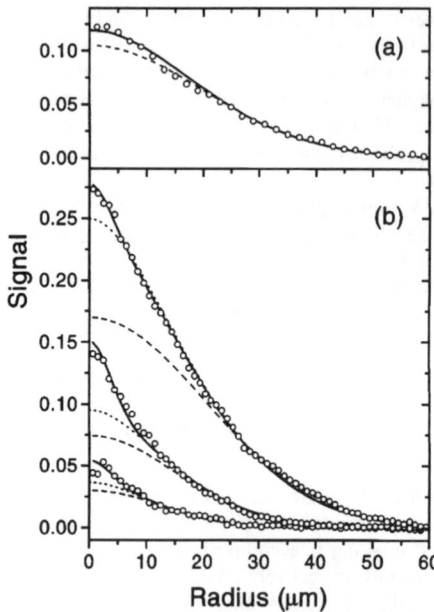

FIG. 2.   Spatial profiles of trapped ultracold ^{7}Li gas. The vertical axis gives the magnitude of the phase-contrast signal intensity relative to the probe intensity. The data points are taken from observed images. The solid curves are calculated by fitting Bose-Einstein distributions to the data, the short-dashed curves are the same distributions with the condensate atoms removed, and the long-dashed curves are classical (Gaussian) distributions fit to the tails of the data. The calculated signals are convolved with a Gaussian function to account for the limited resolution of the imaging system, assuming an effective resolution of 4 μm. For the data in (a), the probe detuning was +191 MHz, and the fitted distribution has 9.0×10^4 atoms at a temperature of 309 nK. The number of condensate atoms is ~1, indicating that the gas is just approaching degeneracy. In (b) a sequence of profiles that exhibits condensate peaks is shown. From the strongest to weakest signals, the total number of atoms and fitted temperatures are 1.01×10^5 atoms at 304 nK; 2.6×10^4 atoms at 193 nK; and 6.6×10^3 atoms at 122 nK. The corresponding numbers of condensate atoms are 500, 810, and 270, respectively. The probe detuning for these data was −130 MHz.

and a Gaussian function for the condensate. The solid lines in Fig. 2 show the results of the fits.

For temperatures sufficiently greater than the critical temperature T_c, the gas can be described by the Boltzmann distribution, which predicts a Gaussian density profile. The long-dashed lines in Fig. 2 are Gaussian functions fit to the tails of the distributions, which approximate the data only in Fig. 2(a). Figure 2(b) shows three distributions for which $T < T_c$. For these distributions, the density is distinctly non-Gaussian, due to an enhanced central peak. Comparison between Fig. 2(a) and the upper curve in Fig. 2(b) is striking, as these distributions correspond to nearly the same temperature, but differ by about 10% in number. The short-dashed curves in Fig. 2(b) show the calculated distributions with the

condensate contribution subtracted. As comparison of the short-dashed and solid curves indicates, the increase in peak signal is caused by both condensed and noncondensed atoms. The contribution of the noncondensed atoms is significant for the upper curve, but at lower temperatures, the contribution of the condensate makes up most of the enhanced peak.

Analysis of the data is complicated by the fact that the condensate size is on the order of the imaging resolution. The resolution can be included in the fit by convolving the theoretical signal with the point transfer function (PTF) of the imaging system [23]. The PTF is calculated by analyzing the propagation of light through the system, given the known lens geometry. In order to test this calculation, we used the system to image laser light emerging from an optical fiber. The intensity distribution of the light in the fiber is Gaussian with a $1/e$ radius of 1.2 μm, so that it approximates a point source. Fig. 3 shows cross sections of the images obtained with the system focused at two different points. The narrower peak shown has a $1/e$ radius of 3.0 μm, as compared with 2.5 μm expected for a diffraction-limited lens (dotted line). The curves are the results of convolving the fiber source with the calculated PTF.

Since the images of the atom cloud are produced by coherently diffracted light, the PTF convolution is performed on the electric field. The results of the coherent field convolution indicate that the primary effect of the coherence is to reduce the importance of the tail of the PTF, because the phase of the field in the tail is rapidly varying. In addition, the PTF depends on where the imaging system is focused, as Fig. 3 shows. When imaging the atom cloud, the focal position can be determined to ±200 μm by observing image distortions which occur when the system is further off focus. Given this uncertainty, and the unimportance of the tail, the PTF convolution is well approximated by an incoherent convolution of the image intensity with a Gaussian function of appropriate $1/e$ radius R. The experimental range of focal positions corresponds to values of R ranging from 2.5 to 5 μm. This uncertainty in the effective resolution is the dominant source of error in our determination of N_0.

We have observed degenerate conditions for T between 120 and 330 nK, and for N between 6800 and 135 000 atoms. In all cases, N_0 is found to be relatively small. Fitting with $R = 5$ μm, the maximum N_0 observed is about 1300 atoms. This value drops to 1000 for $R = 4$ μm, and to 650 for $R = 2.5$ μm. No systematic effects were observed as either the sign or magnitude of the detuning or the polarizer angle were varied, confirming the relations given in Eqs. (1) and (2).

In the analysis, we have assumed that the gas is ideal, but interactions are expected to alter the size and shape of the density distribution. Mean-field theory predicts that interactions will reduce the $1/e$ radius of the condensate from 3 μm for low occupation number

VOLUME 78, NUMBER 6   PHYSICAL REVIEW LETTERS   10 FEBRUARY 1997

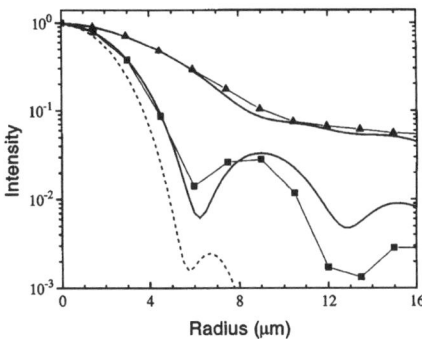

FIG. 3.   Test images of an optical fiber. The data points are obtained from cross sections through images of the light emitted by an optical fiber. The squares indicate data obtained with the imaging system at its best focus, while the triangles indicate data obtained with the system defocused by 200 μm. The solid curves are the expected intensity patterns, while the dashed curve is the expected pattern for a well-focused system in the absence of lens aberrations. The curves are obtained by calculating the PTF of the system as the Fourier transform of the system aperture function and the phase error due to the lenses [23]. The PTF is convolved with a Gaussian electric field with 1.75 μm $1/e$ radius to account for the size of the optical fiber mode, and squared to obtain the intensity. The resulting pattern is then averaged to reflect the pixel size of the camera. The phase errors are due to spherical aberration of 0.5 λ/cm^4, as calculated by ray tracing.

to ~ 2 μm as the maximum N_0 is approached [6,8–11]. If the smaller condensate radius is used in the fit, the maximum values for N_0 decrease, becoming ~ 1050 for $R = 5$ μm. The size of the condensate is not expected to change appreciably for $N_0 < 1000$, so the values obtained for $R = 2.5$ and 4 μm are not sensitive to interactions. Interactions are not expected to significantly affect the distribution of the noncondensed atoms [10,24], because at the critical density the mean interaction energy of ~ 1 nK is much smaller than T.

An estimate of our sensitivity to condensate atoms can be obtained from the fitting procedure. By fixing N_0 and fitting T to the data, χ^2 can be determined as a function of N_0. Since N_0 mostly affects the central part of the distribution, we define a restricted, unnormalized χ^2 by summing over the squares of the differences between the calculated distributions and the data for radii less than 10 μm. Calculating $\chi^2(N_0)$ for several images with large values of N_0 indicates that χ^2 is increased by a factor of 2 from its minimum value when N_0 is varied by about 150 atoms, roughly independent of R.

The sensitivity of N_0 to R could be reduced by observing a distribution consisting mainly of condensate atoms. Since the integrated intensity of an image is independent of lens aberrations, N_0 could be determined simply by measuring the number of atoms in the trap. It may be possible to produce such a distribution through a final accelerated stage of evaporative cooling once BEC has occurred, but

we have not yet successfully done so. Systematic studies of the process are technically difficult because fluctuations of the trap bias field must be less than ~ 50 μG to allow repeated production of pure condensates.

In summary, we have observed BEC in a gas with attractive interactions, by obtaining *in situ* images of degenerate clouds of atoms. The number of condensate atoms is found to be limited to a value consistent with recent theoretical predictions. The range of numbers and temperatures across which the limit is observed to hold suggests that the limit is fundamental, rather than technical. Future experiments are anticipated that will investigate the dynamics of the formation, decay, and collapse of the condensates [7–11].

We are grateful for helpful discussions with T. Bergeman, W. Ketterle, H. Stoof, E. Timmermans, N. Vansteenkiste, M. Welling, and C. Westbrook. This work is supported by the National Science Foundation and the Welch Foundation.

[1] C. C. Bradley, C. A. Sackett, J. J. Tollett, and R. G. Hulet, Phys. Rev. Lett. **75**, 1687 (1995).
[2] M. H. Anderson *et al.*, Science **269**, 198 (1995); K. B. Davis *et al.*, Phys. Rev. Lett. **75**, 3969 (1995); M.-O. Mewes *et al.*, Phys. Rev. Lett. **77**, 416 (1996).
[3] L. D. Landau and E. M. Lifshitz, *Statistical Physics* (Pergamon, London, 1958), 1st ed.
[4] H. T. C. Stoof, Phys. Rev. A **49**, 3824 (1994).
[5] P. A. Ruprecht, M. J. Holland, K. Burnett, and M. Edwards, Phys. Rev. A **51**, 4704 (1995).
[6] F. Dalfovo and S. Stringari, Phys. Rev. A **53**, 2477 (1996).
[7] Y. Kagan, G. V. Shlyapnikov, and J. T. M. Walraven, Phys. Rev. Lett. **76**, 2670 (1996).
[8] R. J. Dodd *et al.*, Phys. Rev. A **54**, 661 (1996).
[9] H . T. C. Stoof, LANL Report No. cond-mat/9601150 (to be published).
[10] M. Houbiers and H. T. C. Stoof, Phys. Rev. A **54**, 5055 (1996).
[11] T. Bergeman (to be published).
[12] E. R. I. Abraham, W. I. McAlexander, C. A. Sackett, and R. G. Hulet, Phys. Rev. Lett. **74**, 1315 (1995).
[13] J. J. Tollett, C. C. Bradley, C. A. Sackett, and R. G. Hulet, Phys. Rev. A **51**, R22 (1995).
[14] The trap frequencies and bias field differ from those reported in Ref. [1] because the magnets have been replaced.
[15] See W. Ketterle and N. J. van Druten, in *Advances in Atomic, Molecular, and Optical Physics*, edited by B. Bederson and H. Walther (Academic Press, San Diego, 1996), No. 37, p. 181, and references therein.
[16] M. R. Andrews *et al.*, Science **273**, 84 (1996).
[17] It was suggested in Ref. [1] that the distortions were caused by the scattering of probe light by the condensate, in conjunction with diffraction by the lens aperture. Subsequent analysis has shown that lens aberrations were more important than the aperture, but confirms that the distortions did indicate the presence of a condensate. This analysis will be described in a separate publication.
[18] See, for instance, E. Hecht, *Optics* (Addison-Wesley, Reading, Massachusetts, 1987), 2nd ed.

VOLUME 78, NUMBER 6   PHYSICAL REVIEW LETTERS   10 FEBRUARY 1997

[19] See, for instance, P. Meystre and M. Sargent III, *Elements of Quantum Optics* (Springer-Verlag, Berlin, 1991), 2nd ed.

[20] O. Morice, Y. Castin, and J. Dalibard, Phys. Rev. A **51**, 3896 (1995).

[21] The condensate itself is expected to have an aspect ratio of 1.05, but this difference is not discernible with our imaging resolution.

[22] V. Bagnato, D. E. Pritchard, and D. Kleppner, Phys. Rev. A **35**, 4354 (1987). The semiclassical distribution was compared to an exact calculation and found to be accurate, except for a temperature shift as noted in W. Ketterle and N. J. van Druten, Phys. Rev. A **54**, 656 (1996). All the temperatures reported here are calculated in the semiclassical approximation.

[23] See, for instance, M. Born and E. Wolf, *Principles of Optics* (Pergamon Press, New York, 1959).

[24] T. Bergeman (private communication).

PHYSICAL REVIEW
LETTERS

VOLUME 82   1 FEBRUARY 1999   NUMBER 5

Coherent Splitting of Bose-Einstein Condensed Atoms with Optically Induced Bragg Diffraction

M. Kozuma,* L. Deng,[†] E. W. Hagley, J. Wen, R. Lutwak,[‡] K. Helmerson, S. L. Rolston, and W. D. Phillips

National Institute of Standards and Technology, Gaithersburg, Maryland 20899
(Received 10 August 1998)

We have observed Bragg diffraction of a Bose-Einstein condensate of sodium atoms by a moving, periodic, optical potential. The coherent process of Bragg diffraction produced a splitting of the condensate with unidirectional momentum transfer. Using the momentum selectivity of the Bragg process, we separated a condensate component with a momentum width narrower than that of the original condensate. By repeatedly pulsing the optical potential while the atoms were trapped, we observed the trajectory of the split atomic wave packets in the confining magnetic potential. [S0031-9007(98)08316-1]

PACS numbers: 03.75.Fi

Atom optics, the manipulation of atoms in analogy to the control of light with optical elements, has seen rapid advances in recent years. Of particular interest are applications that rely on the de Broglie wave nature of atoms, such as diffraction and interferometry [1]. One such technique is Bragg diffraction by an optical standing wave [2–4], which provides coherent splitting of matter waves with unidirectional momentum transfer. With the advent of Bose-Einstein condensation of dilute atomic gases [5–8], a coherent source of matter waves analogous to an optical laser is now available. Bragg diffraction preserves the condensate's coherence properties while providing efficient, selectable momentum transfer. In this Letter, we report Bragg diffraction of a Bose-Einstein condensate (BEC) of sodium atoms by a moving, periodic optical potential.

Bragg diffraction will be a versatile technique for manipulating Bose-Einstein condensates. It will be useful as an output coupler for an atom laser [9] because the large momentum transfer produces a directed output beam and the process is coherent [10]. Bragg diffraction can manipulate a condensate in a trap, creating multiple, coherent components whose interaction and interference can be studied. The Bragg process is also sensitive to the initial momentum of an atomic wave packet, allowing one to impart a well defined momentum to the condensate while having a negligible effect on the uncondensed fraction. This may allow studies of the interactions

between the condensed and noncondensed portions of the gas, such as damping, or atom-atom scattering stimulated by bosonic enhancement.

When an atomic beam passes through a periodic optical potential formed by a standing light wave and interacts with it for a sufficiently long time, it can Bragg diffract, analogous to the Bragg diffraction of x rays from a thick crystal. In each case the incident beam must satisfy a condition on the angle of incidence. Our Bragg diffraction is instead performed on a stationary BEC. In contrast to the diffraction of an atomic beam, the interaction time is determined not by the passage of the atoms through a standing wave, but by the duration of a laser pulse. The condition on the angle of incidence becomes a condition on the frequency difference between the two beams comprising the standing wave, or equivalently, the velocity of the moving standing wave. Bragg diffraction under these conditions can also be thought of as stimulated optical Compton scattering [11], a recoil-induced resonance [12,13], or as a stimulated Raman transition between two momentum states [14–16].

nth order Bragg diffraction by a moving, optical standing wave can be viewed as a $2n$-photon stimulated Raman process in which photons are absorbed from one beam and stimulated to emit into the other [Fig. 1(a)]. The initial and final momentum states form an effective two-level system coupled by the multiphoton Raman process. Conservation of energy and momentum require

VOLUME 82, NUMBER 5   PHYSICAL REVIEW LETTERS   1 FEBRUARY 1999

$$\frac{(nP_{\text{recoil}})^2}{2M} = n\hbar\delta_n, \qquad (1)$$

where $P_{\text{recoil}} = 2\hbar k \sin(\theta/2)$ is the recoil momentum from a two-photon Raman process, $k = 2\pi/\lambda$; λ is the wavelength of the light, M is the atomic mass, and δ_n is the frequency difference between the two lasers. For our conditions first-order Bragg diffraction is resonant at $\delta_1/2\pi = 98$ kHz and higher orders at $\delta_n = n\delta_1$.

In our experiment, we produce a BEC as described in detail elsewhere [17]. Briefly, about 10^{10} Na atoms are optically cooled and trapped in a dark magneto-optical trap [18]. They are transferred into a magnetic quadrupole field where atoms in the $3S_{1/2}$ $F = 1, m_F = -1$ state are trapped, compressed, and then cooled by rf-induced evaporation. Before the atoms are lost in the zero field region in the center of the trap, a time-averaged orbiting potential (TOP) [19] trap is created by suddenly turning on a rotating bias field. The bias field rotates in the x-z plane, where x is the quadrupole axis and z is vertical along the direction of gravity. Our TOP trap differs from the design of [19] in that our bias field rotates in a plane that includes the quadrupole symmetry

axis. The ratio between spring constants along the x, y, and z directions is $K_x : K_y : K_z = 4 : 2 : 1$. The atoms are compressed in the TOP trap and cooled by evaporation to form a BEC. We obtain a condensate with about 10^6 sodium atoms having no discernible uncondensed fraction in a trap with harmonic frequencies of $\omega_x/2\pi = 360$ Hz, $\omega_y/2\pi = 250$ Hz, $\omega_z/2\pi = 180$ Hz.

Our first experiments were performed on Bose-condensed atoms released suddenly from the TOP trap. The trap is turned off in 50 μs and the BEC undergoes expansion driven by the mean-field repulsion between the atoms. After a few characteristic times $\tau = (\omega_x \omega_y \omega_z)^{-1/3}$, the mean field is negligible and the cloud expands ballistically [20]. During this ballistic expansion the condensate is exposed to a moving, periodic optical potential generated by two nearly counterpropagating ($\theta = 166°$) laser beams with parallel linear polarizations but slightly different frequencies [Fig. 1(a)]. These (phase coherent) laser beams are derived from a single laser ($\lambda = 589$ nm) using acousto-optic modulators. The intensity of each beam is 23 mW/cm^2, and the common detuning with respect to the $3S_{1/2}$, $F = 1 \rightarrow 3P_{3/2}$, $F' = 2$ transition is $\Delta/2\pi = -1.85$ GHz. To transfer all the atoms to the desired momentum state, we empirically choose laser intensities and pulse durations to give a π pulse for the effective two-level system. We use two pulses with frequencies ω and $\omega + \delta$ that overlap for 55 μs [21]. The probability of spontaneous emission is less than 0.05.

Figure 2(a) is an image taken just before the Bragg pulse is applied. The atoms are first optically pumped into the $3S_{1/2}$ $F = 2$ ground state. They are then absorption imaged [5] with probe light on the $F = 2 \rightarrow F' = 3$

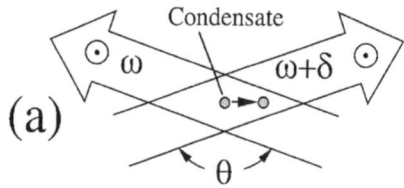

(a)

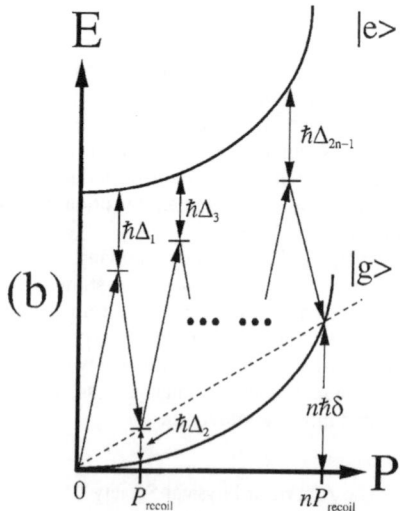

(b)

FIG. 1. Experimental arrangement of the laser beams (a) and partial transition diagram (b) for nth order Bragg diffraction. The parabolas correspond to the $P^2/2M$ kinetic energy.

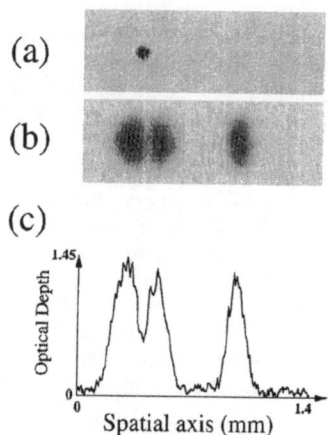

(a)

(b)

(c)

FIG. 2. Optical absorption image of a Bragg diffracted condensate. (a) An image taken just before the moving standing wave pulse is applied. (b) is taken after a time of flight of 10 ms. (c) is a line profile taken through the center of the expanding clouds.

VOLUME 82, NUMBER 5   PHYSICAL REVIEW LETTERS   1 FEBRUARY 1999

transition. The probe propagates opposite to gravity and is orthogonal to the plane that contains the Bragg beams. From different time-of-flight images taken after the mean field becomes negligible, we determine the rms momentum width of the released BEC. Along the direction of momentum transfer we measure $\Delta P_{rms} = 0.30(4)\hbar k$ (all uncertainties in this paper represent one standard deviation combined statistical and systematic uncertainties). Figure 2(b) is an image taken 10 ms after the Bragg pulse (which occurs 2 ms after the BEC is released). A slice is removed from the center of the atomic momentum distribution and is displaced by $P_{recoil} \sim 2\hbar k$. The slice appears because the resonant width of the Raman transition in momentum space is narrower than ΔP_{rms} of the released condensate. From these images, we have determined the rms momentum width of diffracted atoms along the direction of the momentum transfer to be $0.16(1)\hbar k$. The resonant width of the Bragg diffraction can be calculated from an integration of the optical Bloch equations for a two-level system with a time dependent two-photon Rabi frequency. In our case, the calculated peak two-photon Rabi frequency is $\Omega_2/2\pi = 30$ kHz, and the predicted width is $0.08\hbar k$, a factor of 2 less than measured. We do not have an explanation for this difference, although it may be due in part to residual mean-field effects [22].

This analysis implies that we can Bragg diffract the entire cloud of atoms if the pulse duration τ is short enough such that $\tau^{-1} \gg 2k\Delta P_{rms}/M$, and the intensity is sufficient for a π pulse. To achieve this we reduce the rms momentum width of the BEC by adiabatically expanding the trap. The axial gradient of the quadrupole field is linearly ramped down from 9.2 to 0.71 T/m in 2 s while the bias field is increased from 1.0 to 1.2 mT. This reduces the frequencies ω_i of the TOP trap by a factor of 14. Even for this weak trap, the size of the condensate is determined by the mean-field interaction. In this Thomas-Fermi limit, the asymptotic momentum spread of the released BEC decreases as $\omega^{3/5}$ [23]. The rms momentum width of the released condensate after adiabatic cooling should be $\frac{1}{5}$ of the width of a condensate released from the tight trap. This is in qualitative agreement with the observed expansion in time-of-flight images. (The adiabatically expanded condensate is so cold that it undergoes little expansion during the time of flight, so its momentum spread cannot be accurately measured.)

After adiabatic expansion, the BEC is released from the TOP trap and expands for 2 ms. The Bragg pulse is applied and the atoms are imaged 5.6 ms later. When the difference δ between the two laser frequencies is zero, Eq. (1) is not satisfied and diffraction does not occur [Fig. 3(a)] because the Fourier transform width of the 55 μs pulse is small compared to 98 kHz. When $\delta/2\pi = 98$ kHz, first-order Bragg diffraction [Fig. 3(b)], with up to 100% efficiency, is observed. In Fig. 3(b) we chose the diffraction efficiency to be less than 100% so that the original position of the BEC was visible.

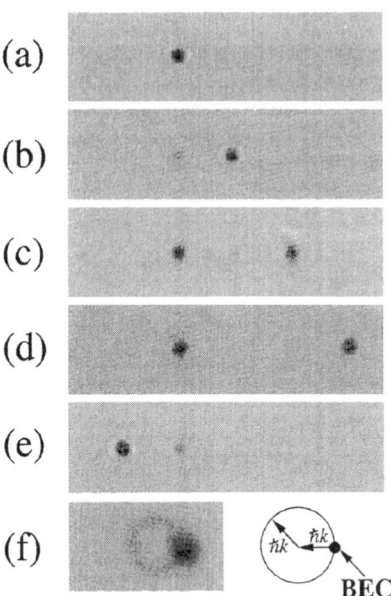

FIG. 3. Optical depth images of condensates which were first adiabatically expanded and then Bragg diffracted. (a), (b), (c), (d), and (e) are images taken 5.6 ms after Bragg pulses with frequency differences of $\delta/2\pi = 0, 98, 200, 300,$ and -98 kHz, respectively. (f) is an image where spontaneous emission occurred using a single laser beam. The width of the field of view is 2.3 mm × 0.5 mm.

Figures 3(c) and 3(d) show second- and third-order Bragg diffraction, where $\delta/2\pi$ is 200 and 300 kHz and the intensities are 230 and 340 mW/cm^2, respectively. The efficiency of third-order diffraction was as high as 45%. When the sign of δ is reversed, atoms are diffracted into the opposite direction [Fig. 3(e)]. With longer pulses we have observed up to sixth-order Bragg diffraction with a transfer of $11.9\hbar k$ (corresponding to a velocity of 0.35 m/s) and an efficiency of about 15%. For such a high-order process we cannot achieve a high efficiency with our detuning because the required pulse length and intensities make the probability of spontaneous emission near unity. Figure 3(f) provides a graphic demonstration of spontaneous emission. The ring pattern results from dipole emission of atoms illuminated by a single traveling wave. Here the direction of observation is along the laser polarization.

In Bragg diffraction the internal state remains unchanged, so the temporal variation of the Zeeman shift due to the rotating bias field of the TOP trap is unimportant. This is true provided the Zeeman shifts are small compared to Δ, as in our case. We can therefore use Bragg diffraction in the trap to study the motion of trapped atoms. In Fig. 4 we create a train of orbiting wave packets. In the adiabatically expanded TOP trap we irradiate

VOLUME 82, NUMBER 5   PHYSICAL REVIEW LETTERS   1 FEBRUARY 1999

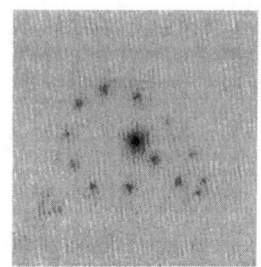

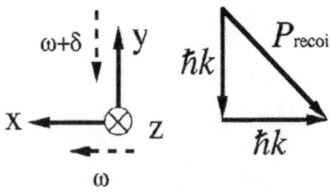

FIG. 4. Optical depth image of the condensate after 13 Bragg pulses while in the TOP trap. Here x is the quadrupole axis and z is the direction of gravity. The TOP bias field rotates in the x-z plane. The Bragg diffraction beams propagate along the x and $-y$ directions. The width of the field of view is 1.2 mm $\times$ 1.2 mm.

the BEC with a series of 13 pulses, each separated by 5 ms. The 13 wave packets and the remaining condensate are then imaged with a single probe pulse. To generate many wave packets without substantially depleting the condensate, we decrease the pulse duration to 20 μs, reducing the diffraction efficiency. Here the crossing angle of the laser beams is 90° and $\delta/2\pi = 50$ kHz to satisfy the first-order Bragg condition. Since the trap potential is noncentral, atoms launched along a direction (45° with respect to x and y in this case) other than a principal axis will not execute closed orbits, as seen in Fig. 4. The orbit of the wave packets agrees with the classical equations of motion in our harmonic trap with its incommensurate frequencies [24].

In summary, we have demonstrated coherent Bragg diffraction of Bose-Einstein condensed sodium atoms. This technique, demonstrated both with and without the trapping fields, can be used to manipulate condensates and to study fundamental aspects of their properties. For example, it is possible to measure the coherence time of a BEC. Atoms may be launched along a principal axis, so that the diffracted wave packet will return to the condensate after an integral number of half-oscillation periods. The application of a second Bragg pulse when the diffracted wave packet and the original BEC overlap would produce interference, allowing a coherence time measurement in the trap. If, instead, the trap is switched off, the phase of the condensate will start to evolve because the mean-field repulsion is no longer balanced by the trap potential. Successive

application of Bragg pulses could probe the temporal evolution of the condensate phase (here one would apply the second pulse before the first wave packet leaves the region of the condensate). Experiments on nonlinear atom optics are also possible using these techniques. For example, four-wave mixing of matter waves should occur due to the nonlinearity arising from atom-atom interactions [25,26]. Bragg diffraction can create the three different momentum states necessary to produce a fourth momentum component. High-order Bragg diffraction is ideal as an output coupler for an atom laser. Output coupling with momentum transfer should produce an extremely well collimated beam of coherent atoms.

The authors thank C.W. Clark, M. Doery, M.A. Edwards, P.S. Julienne, J. Lawall, and Y. Ovchinnikov for their valuable comments and suggestions. M.K. acknowledges the support of the Japanese Society for the Promotion of Science for Young Scientists. This work was supported in part by the Office of Naval Research and NASA.

*Current address: Department of Physics, Tokyo University, Tokyo 153-8902, Japan.
†Permanent address: Department of Physics, Georgia Southern University, Statesboro, GA 30460.
‡Current address: FTS, Inc., Beverly, MA 01915.
[1] See, for example, the special issue on atom interferometry [Appl. Phys. B **54**, 321 (1992)].
[2] P.J. Martin, B.G. Oldaker, A.H. Miklich, and D.E. Pritchard, Phys. Rev. Lett. **60**, 515 (1988).
[3] D.M. Giltner, R.W. McGowan, and S.A. Lee, Phys. Rev. A **52**, 3966 (1995); M.K. Oberthaler et al., Phys. Rev. Lett. **77**, 4980 (1996).
[4] S. Kunze, S. Dürr, and G. Rempe, Europhys. Lett. **34**, 343 (1996).
[5] M.H. Anderson et al., Science **269**, 198 (1995).
[6] K.B. Davis et al., Phys. Rev. Lett. **75**, 3969 (1995).
[7] C.C. Bradley, C.A. Sackett, and R.G. Hulet, Phys. Rev. Lett. **78**, 985 (1997); see also C.C. Bradley et al., Phys. Rev. Lett. **75**, 1687 (1995).
[8] D.G. Fried et al., Phys. Rev. Lett. **81**, 3811 (1998).
[9] M.-O. Mewes et al., Phys. Rev. Lett. **78**, 582 (1997).
[10] D.M. Giltmer, R.W. McGowan, and S.A. Lee, Phys. Rev. Lett. **75**, 2638 (1995).
[11] D.R. Meacher et al., Phys. Rev. A **50**, R1992 (1994).
[12] J.-Y. Courtois, G. Grynberg, B. Lounis, and P. Verkerk, Phys. Rev. Lett. **72**, 3017 (1994).
[13] M. Kozuma, K. Nakagawa, W. Jhe, and M. Ohtsu, Phys. Rev. Lett. **76**, 2428 (1996).
[14] J. Guo, P.R. Berman, and B. Dubetsky, Phys. Rev. A **46**, 1426 (1992).
[15] J. Guo and P.R. Berman, Phys. Rev. A **47**, 4128 (1993).
[16] P.R. Berman and B. Bian, Phys. Rev. A **55**, 4382 (1997).
[17] R. Lutwak et al. (to be published).
[18] W. Ketterle et al., Phys. Rev. Lett. **70**, 2253 (1993).
[19] W. Petrich, M.H. Anderson, J.R. Ensher, and E.A. Cornell, Phys. Rev. Lett. **74**, 3352 (1995).
[20] C.W. Clark (private communication).

VOLUME 82, NUMBER 5   PHYSICAL REVIEW LETTERS   1 FEBRUARY 1999

[21] The counterpropagating pulses are triangular with rise and fall times of 40 μs. There is a 25 μs delay between them. Although this pulse sequence was chosen for adiabatic passage experiments there is no adiabatic transfer in our case. The Raman process happens only during the 55 μs overlap.

[22] M. Doery (private communication).

[23] F. Dalfovo, S. Giorgini, L. P. Pitaevskii, and S. Stringari, Rev. Mod. Phys. (to be published).

[24] M. A. Edwards (private communication).

[25] G. Lenz, P. Meystre, and E. W. Wright, Phys. Rev. Lett. **71**, 3271 (1993).

[26] M. Trippenbach, Y. B. Band, and P. S. Julienne, Opt. Express (to be published).

VOLUME 78, NUMBER 4   PHYSICAL REVIEW LETTERS   27 JANUARY 1997

Production of Two Overlapping Bose-Einstein Condensates by Sympathetic Cooling

C. J. Myatt, E. A. Burt, R. W. Ghrist, E. A. Cornell, and C. E. Wieman

JILA and Department of Physics, University of Colorado and NIST, Boulder, Colorado 80309

(Received 20 September 1996)

A new apparatus featuring a double magneto-optic trap and an Ioffe-type magnetic trap was used to create condensates of 2×10^6 atoms in either of the $|F = 2, \; m = 2\rangle$ or $|F = 1, \; m = -1\rangle$ spin states of ^{87}Rb. Overlapping condensates of the two states were also created using nearly lossless sympathetic cooling of one state via thermal contact with the other evaporatively cooled state. We observed that (i) the scattering length of the $|1, -1\rangle$ state is positive, (ii) the rate constant for binary inelastic collisions between the two states is $2.2(9) \times 10^{-14}$ cm^3/s, and (iii) there is a repulsive interaction between the two condensates. Similarities and differences between the behaviors of the two spin states are observed. [S0031-9007(96)02208-9]

PACS numbers: 03.75.Fi, 05.30.Jp, 32.80.Pj, 51.30.+i

One of the more notable recent developments in physics has been the cooling of a trapped dilute atomic gas to below the Bose-Einstein transition temperature [1–3]. This produced a macroscopic quantum state that is both novel and readily observed. Gaseous Bose-Einstein condensation (BEC) was first reported in a cloud of atoms in a single spin state of the ground state of rubidium [1] and later in single spin states of sodium [2] and lithium [3]. To reach the necessary ultralow temperatures, these experiments used laser cooling and trapping followed by magnetic trapping and evaporative cooling. The observation of BEC led to a number of studies of the properties of these two condensates [4]. Here we report the creation of two different condensates in the same trap. The two condensates in this work correspond to two different spin states of rubidium 87, $|F = 1, \; m = -1\rangle$ and $|F = 2, \; m = 2\rangle$. We have created large samples of both condensates separately and compared their properties. We have also created mixtures of the two by using a new cooling technique. The cloud of atoms in the $|1, -1\rangle$ state was cooled by lossy evaporative cooling, as in previous work, but the $|2, 2\rangle$ state cloud was cooled only by thermal contact with $|1, -1\rangle$ atoms. Such "sympathetic" cooling of one species by another has been used at much higher temperatures to cool trapped ions [5] that have strong long-range Coulomb interactions, but this is the first time it has been applied to neutral atoms. This sympathetic evaporative cooling technique may allow the creation of degenerate Fermi gases as well as condensates in rare isotopes. We see differences in the low-temperature behaviors of the atoms in the two spin states in both condensed and uncondensed phases. When the two condensates are overlapped, additional novel features are observed in their interactions. These condensates were created using a new apparatus that incorporates a double magneto-optic trap (MOT) and a magnetic trap. It is based upon the same vapor-cell/diode-laser technology as the original JILA BEC apparatus [1]; however, it produces much larger condensates and is far more tolerant of imperfect experimental conditions [6].

The apparatus is an extension of the double MOT system that has been described previously [7]. The double MOT system has the advantage of allowing a relatively large sample of atoms to be optically trapped in a very low pressure chamber using low power lasers. It is made up of two small differentially pumped vacuum chambers connected by a 40 cm long $\times$ 1 cm diam tube that is used to transfer atoms between them. The upper vacuum chamber contains about 2×10^{-9} Torr of rubidium vapor. The transfer tube is lined with strips of permanent magnet that create a hexapole magnetic guiding field. The lower chamber, shown in Fig. 1, is made of glass and is pumped by a 60 L/s ion pump and a small titanium sublimation pump to a pressure of less than 10^{-11} Torr. Three coils outside of this chamber create a "baseball coil" magnetic trap that is similar to that used in our previous work [8] but different from that used in the recent JILA BEC

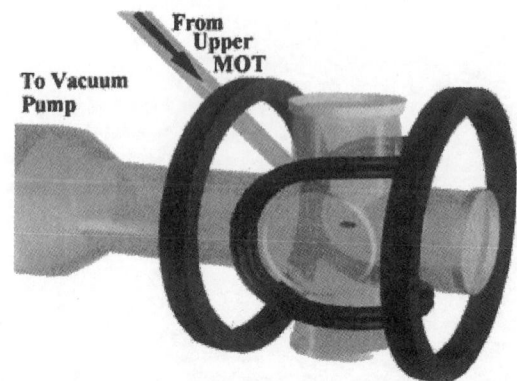

FIG. 1. The glass lower vacuum chamber is connected to the upper chamber through a narrow transfer tube and to sublimation and ion pumps as noted. It is surrounded by the three coils that comprise the magnetic trap. Small additional windows (not shown) allow the cloud to be viewed along some of the diagonals. The trapping laser beams go through the six perpendicular 2.5 cm diam windows, four of which are visible in the figure.

VOLUME 78, NUMBER 4   PHYSICAL REVIEW LETTERS   27 JANUARY 1997

studies [1,4]. The first coil is shaped like the seams on a baseball and provides field curvature in addition to a bias field, while the other two form a Helmholtz pair that can cancel all or part of the bias field. The resulting field configuration is essentially that of an Ioffe-type trap [9]. The three coils are wired in series with a variable shunt resistance across the Helmholtz pair. The trapping potential is axially symmetric with the axis nominally horizontal. There is a small additional coil for producing the adjustable-frequency rf magnetic field used in evaporative cooling. Lenses image the trapped atomic cloud onto both a CCD camera and a calibrated photodiode. The trapping and probing light is provided by low power (50 mW) diode lasers stabilized by grating feedback [10].

Single-species condensates in either the $|2, 2\rangle$ or $|1, -1\rangle$ spin states of the $5S$ ground state of rubidium are created and examined using the following procedure. First, atoms are collected in a MOT in the upper chamber for 1.0 s. This load, typically several times 10^7 atoms, is then pushed down the transfer tube using light pressure, and about 80% of them are recaptured in a second MOT in the lower chamber [7]. This procedure is repeated many times to fill the lower MOT with about 10^9 atoms. Next, this cloud of trapped atoms is compressed by increasing the MOT magnetic field gradient as in Ref. [11]. Then the MOT fields (optical and dc magnetic) are turned off and a 1 G bias field turned on. The atoms are then optically pumped into the desired spin state by applying 1 ms pulses of light from two laser beams that excite the $5S_{1/2}, F = 1$ to $5P_{3/2}, F' = 2$ and the $F = 2$ to $F' = 2$ transitions, respectively. By suitable adjustment of the polarization and relative timing of the two beams, we pump the atoms into a single state, either $|1, -1\rangle$ or $|2, 2\rangle$, with 90% efficiency. Next, the magnetic trap is turned on around the atoms with full current (200 A) flowing through the baseball coil, but with no current in the Helmholtz pair. The current in the Helmholtz pair is then ramped up in 2 s to reduce the bias field to 1 G. For the $|2, 2\rangle$ state this increases the radial frequency from 20 to 400 Hz, while leaving the 10 Hz axial frequency nearly unchanged. (The $|1, -1\rangle$ state has one-half of the $|2, 2\rangle$ magnetic moment, and so all of the $|1, -1\rangle$ frequencies are lower by $\sqrt{2}$.) This ramp compresses the cloud and raises its temperature to 250 μK. The cloud is then cooled by rf evaporation, in which the applied rf magnetic field drives the most energetic atoms to untrapped spin states. For this cooling, the frequency of the applied rf is ramped down over a period of 30 s.

The cooled cloud is probed by absorption imaging in a manner similar to that of Ref. [1]. The cloud is released from the magnetic trap, and after it expands ballistically for 20 ms it is illuminated briefly by a near resonant $5S_{1/2}F = 2$ to $5P_{3/2}F' = 3$ probe laser beam, as well as $F = 1$ to $F' = 2$ hyperfine pumping light. The probe laser is normally tuned 1.6 full linewidths off resonance because an expanded condensate cloud is tens of optical depths thick for resonant light. The resulting shadow

produced by the cloud in the illuminating beam is imaged onto the CCD camera. This absorption imaging is used to find the temperature of the cloud and the fraction of atoms in the condensate as in Ref. [1]. Also, a measurement of total fluorescence is used to accurately determine the number of atoms in the cloud. The fluorescence is measured by the calibrated photodiode after recapturing the evaporatively cooled atoms in a MOT.

To create mixtures of the two condensates by sympathetic cooling, the procedure is quite similar. If the two clouds are at the same temperature, the $|1, -1\rangle$ cloud is less tightly confined by the magnetic field than the $|2, 2\rangle$ cloud because of the difference in magnetic moments and, hence, will extend to a larger magnetic field. This causes the $|1, -1\rangle$ state to be preferentially removed by the rf field. The $|2, 2\rangle$ atoms are then cooled by elastic collisions with the evaporatively cooled $|1, -1\rangle$ atoms as long as the two clouds overlap. Because the two states have different magnetic forces but the same gravitational force, the $|2, 2\rangle$ cloud is centered slightly above the $|1, -1\rangle$ cloud. However, for the spring constants given above and a horizontal trap axis, one can easily calculate that the relative displacement of the $|1, -1\rangle$ and $|2, 2\rangle$ clouds is much less than their widths when they are not condensed. For condensates, the displacement is a significant fraction of the width, but there is still substantial overlap. By starting with the axis of the trap (the direction of the weak spring constant) slightly tilted off perpendicular to the gravitational force, it is possible to form separated condensates, if desired (Fig. 2). One can also separate the two clouds

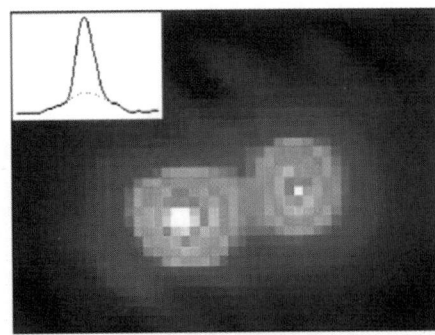

FIG. 2(color). A false-color absorption image (475 μm by 675 μm) showing condensates of both $|2, 2\rangle$ (left) and $|1, -1\rangle$ (right) states that were created simultaneously by sympathetic cooling. The condensates are separated because the trap axis was tilted 40 mrad to produce a component of the gravitational force along the weak spring constant direction. The noncondensed parts of the clouds (purple and dark blue) still overlap. The shape of both of the condensates is a function of expansion time, but the difference in their ellipticities reflects the fact that they have different initial confinements and therefore expand at different rates. The inset shows a vertical trace through the cloud on the left. The dotted line is to guide the eye in distinguishing the broad thermal background from the narrow condensate peak.

VOLUME 78, NUMBER 4   PHYSICAL REVIEW LETTERS   27 JANUARY 1997

after cooling by adiabatically lowering the trap spring constants to increase the displacement due to gravity.

For the production of two-species condensates by sympathetic cooling, the MOT portion of a cooling cycle was identical to that given above. However, in the optical pumping the polarization and timing of the two laser beams are set to produce the desired ratio of populations in the $|1, -1\rangle$ and $|2, 2\rangle$ states [12]. The evaporative cooling then proceeds exactly as it would for a pure $|1, -1\rangle$ cloud. The probing is modified slightly to obtain state selective absorption images of the mixtures. After the cloud has ballistically expanded we probe it two different ways. First, we use only the near resonant $F = 2$ to $F' = 3$ light to obtain an absorption image of only the $|2, 2\rangle$ cloud. We then take a second absorption image with the $F = 1$ to $F' = 2$ light also present [13]. This produces an absorption image of all the atoms independent of their initial state. In Figs. 2 and 3(a) we show pictures of clouds containing two simultaneous condensates.

We have compared the production of condensates for samples of either pure $|2, 2\rangle$ or pure $|1, -1\rangle$ atoms, as well as for mixtures. For the pure cases, both the total transfer efficiency from lower MOT to compressed magnetic trap ($\sim$50%) and the evaporative cooling results are similar. This is to be expected because the magnitude of the two scattering lengths are similar [14]. For each factor of 5 loss in the number of atoms during evaporation, we decrease the temperature by a factor of 10 and increase the phase space density by 200. This allows us to reach the $\sim$500 nK BEC transition temperature with 6×10^6 atoms. After further cooling down to where we can no longer see any noncondensed atoms, about 2×10^6 atoms remain. The efficiency of the evaporative cooling is relatively insensitive to initial conditions and the details of the rf ramp. This is not surprising since the initial elastic

scattering rate in the magnetic trap is about 50 times larger than the $\sim$200 collisions per trap lifetime required for runaway evaporative cooling. For the low densities obtained before evaporative cooling, this lifetime for both states is $\sim$140 s [15]. All of our cooled clouds show the three clear indications of BEC reported in Ref. [1]: a two-component velocity distribution, a nonisotropic velocity distribution, and a sudden large increase in peak density as the temperature is decreased. The first two of these are evident in Figs. 2 and 3. The observed transition temperatures for both states agree with the ideal gas value within our $\pm$20% uncertainty. Also, we find that the previously unknown sign of the scattering length for the $|1, -1\rangle$ state must be positive [16].

We also see differences between the two states. First, when the rf cooling is disabled, we observe a heating rate that is about 10 times larger for the $|2, 2\rangle$ state cloud than a comparable cloud in the $|1, -1\rangle$ state. This is true both above and below the BEC transition. Second, we see differences in how the two states are lost from the magnetic trap. At the highest densities, the lifetime of the $|1, -1\rangle$ state remains 140 s but the lifetime for the $|2, 2\rangle$ is density dependent and less than 10 s. These differences clearly involve some interesting low-temperature atomic physics that are quite relevant for the creation and study of BEC.

In Fig. 4 we illustrate the nearly lossless sympathetic cooling of the $|2, 2\rangle$ atoms in a two species cloud. Nearly one-half of the $|2, 2\rangle$ atoms remain after cooling from 250 μK to just above the BEC transition. In contrast, only about $\frac{1}{60}$ of the $|1, -1\rangle$ atoms remain. There is a small loss of $|2, 2\rangle$ atoms as the temperature approaches 1 μK (Fig. 4) and a larger loss in going from that point to a pure condensate (not shown). The loss rate depends on the densities of $|1, -1\rangle$ and $|2, 2\rangle$ atoms and is consistent with it being due to binary, inelastic collisions (presumably spin exchange) between the species. By measuring the densities and loss rates we find the total rate constant for

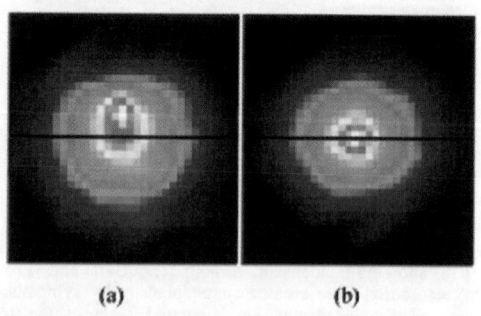

FIG. 3(color). Two 475 μm by 475 μm false-color absorption images of $|2, 2\rangle$ atoms. (a) A cloud of two overlapping condensates illuminated so that only the $|2, 2\rangle$ state atoms are visible. The condensate (white, red, and yellow) is shifted upwards relative to the center of the thermal uncondensed cloud (green, blue, and purple) due to interactions with the $|1, -1\rangle$ condensate ($|1, -1\rangle$ atoms not visible). (b) A cloud of pure $|2, 2\rangle$ atoms cooled to a comparable temperature as in (a). The black line is a guide to the eye going through the center of both thermal clouds.

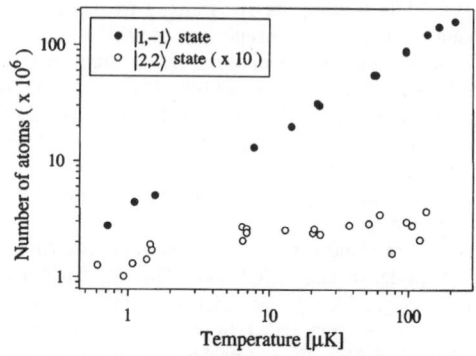

FIG. 4. Number of $|1, -1\rangle$ and $|2, 2\rangle$ atoms in a two-species cloud as a function of the temperature during the sympathetic evaporative cooling. The cloud is being cooled from the initial magnetic trap temperature to just above the condensation temperature.

VOLUME 78, NUMBER 4   P H Y S I C A L   R E V I E W   L E T T E R S   27 JANUARY 1997

inelastic processes is $2.2(9) \times 10^{-14}$ cm^3/s. When we tilt the trap, as in Fig. 2, we reduce the overlap between the clouds and the observed loss rate. We have measured the temperatures of the $|1, -1\rangle$ and $|2, 2\rangle$ clouds and find that they match closely during the evaporation process. The measured temperature of the $|2, 2\rangle$ state is consistently 5%–10% lower than that of the $|1, -1\rangle$ state; however, this is just at the limit of our resolution and may not indicate a real difference.

Finally, we have briefly examined the degree of interaction of the two overlapping condensates by comparing their ballistic expansion with that observed for single-species condensates. As can be seen in Fig. 3(a), the $|2, 2\rangle$ condensate is pushed upward from its position in a single species cloud by the interaction with the lower lying $|1, -1\rangle$ condensate. This indicates that the interaction is repulsive. We plan to make further studies about the overlap and interactions between the two condensates.

This work has demonstrated an improved apparatus and a new cooling method for producing BEC in rubidium. The large number of atoms and long magnetic trap lifetime in this setup make it relatively easy to obtain BEC [17]. It will be straightforward to use this approach to further explore the detailed interactions between the two overlapping condensates. Also, the method of sympathetic cooling will allow future experiments involving the cooling of rare and/or fermionic isotopes. It would be very difficult to cool fermionic atoms into a highly degenerate regime using normal evaporative cooling because of the requirement of a large number of elastic collisions per trap lifetime, combined with the vanishing elastic collision rate in low-temperature spin-polarized fermionic gases. However, it should be quite feasible to use bosonic atoms as a working fluid to sympathetically cool a fermionic gas into the interesting [18] degenerate regime. This technique will also allow one to cool to BEC the many species for which inelastic processes make it impossible to obtain high enough densities for conventional evaporative cooling.

We are pleased to acknowledge assistance in this work from all members of the JILA BEC group, and particularly N. Newbury and N. Claussen. This work was supported by ONR, NIST, and NSF. R. W. G. acknowledges the support of an NSF graduate fellowship.

Note added. —Recent theoretical studies of binary mixtures of Bose condensates [19] predict a rich variety of interesting behaviors for two component condensates.

[1] M. H. Anderson, J. R. Ensher, M. R. Matthews, C. E. Wieman, and E. A. Cornell, Science **269**, 198 (1995).

[2] K. B. Davis *et al.,* Phys. Rev. Lett. **75**, 3969 (1995).

[3] C. C. Bradley, C. A. Sackett, and R. G. Hulet (to be published).

[4] D. S. Jin, J. R. Ensher, M. R. Matthews, C. E. Wieman, and E. A. Cornell, Phys. Rev. Lett. **77**, 420 (1996); M.-O. Mewes *et al.,* Phys. Rev. Lett. **77**, 988 (1996); J. R. Ensher, D. S. Jin, M. R. Matthews, C. E. Wieman, and E. A. Cornell, Phys. Rev. Lett. (to be published).

[5] D. J. Larson, J. C. Bergquist, J. J. Bollinger, W. M. Itano, and D. J. Wineland, Phys. Rev. Lett. **57**, 70 (1986).

[6] Achieving BEC in the single MOT apparatus of Ref. [1] required precise optical alignment of a dark spot MOT and careful optimization of rubidium pressure, laser trapping frequencies, laser cooling and compression parameters, optical pumping, and rf evaporation rates. In the present double MOT apparatus the alignment of the MOTs is not critical and little care is required to optimize any of the other experimental conditions listed above.

[7] C. J. Myatt, N. R. Newbury, R. W. Ghrist, S. Loutzenhiser, and C. E. Wieman, Opt. Lett. **21**, 290 (1996).

[8] C. Monroe, E. Cornell, C. Sackett, C. Myatt, and C. Wieman, Phys. Rev. Lett. **70**, 414 (1993); N. Newbury, C. Myatt, E. Cornell, and C. Wieman, Phys. Rev. Lett. **74**, 2196 (1995).

[9] T. Bergeman, G. Erez, and H. J. Metcalf, Phys. Rev. A **35**, 1535 (1987). BEC has also recently been created in a trap that has a different coil configuration than ours but is the same Ioffe field geometry, by M.-O. Mewes *et al.,* Phys. Rev. Lett. **77**, 416 (1996).

[10] K. MacAdam, A. Steinbach, and C. Wieman, Am. J. Phys. **60**, 1098 (1992).

[11] W. Petrich, M. H. Anderson, J. R. Ensher, and E. A. Cornell, J. Opt. Soc. Am. B **11**, 1332 (1994).

[12] This simple pumping scheme typically results in $\frac{2}{3}$ of the atoms ending up in the other six spin states. However, all of these other states quickly leave the magnetic trap. The only state that is magnetically trapped, other than the $|1, -1\rangle$ and $|2, 2\rangle$ states, is the $|2, 1\rangle$ state. We see this state quickly depleted, presumably by spin-exchange collisions. When there is a large $|2, 2\rangle$ cloud, however, we observe a small residual $|2, 1\rangle$ population that we believe is due to the balance between $|2, 2\rangle + |2, 2\rangle$ dipole collisions populating the $|2, 1\rangle$ state and spin-exchange collisions depleting it.

[13] We take the second image on the subsequent identical cooling cycle.

[14] N. Newbury, C. Myatt, and C. Wieman, Phys. Rev. A **51**, R2680 (1995) and errata (submitted for publication) reported the absolute value of the $|1, -1\rangle$ state scattering length to be $87(10)a_0$. Note that the uncertainty given in the original publication was incorrect. H. M. J. M. Boesten *et al.,* Phys. Rev. A **55**, 636 (1997) gave the $|2, 2\rangle$ state scattering length to be $110(10)a_0$.

[15] This is only achieved after a quite careful shielding against stray laser light.

[16] If the scattering length were negative, it would not be possible theoretically to form such large condensates. For example, see P. A. Ruprecht, M. J. Holland, K. Burnett, and M. Edwards, Phys. Rev. A **51**, 4704 (1995).

[17] The critical experimental issue is the stability of the magnetic trap. The potential energy at trap center, i.e., the bias field, must not vary by much more than the trap energy level spacing during the final stages of the rf ramp.

[18] H. T. C. Stoof, M. Houbiers, C. A. Sackett, and R. G. Hulet, Phys. Rev. Lett. **76**, 10 (1996); M. Houbiers and H. T. C. Stoof, Czech. J. Phys. **47** (suppl sl) 551 (1996).

[19] T.-L. Ho and V. B. Shenoy, Phys. Rev. Lett. **77**, 3276 (1996); B. D. Esry, C. H. Greene, J. P. Burke, and J. L. Bohn (private communication).

ELSEVIER

15 February 1995

OPTICS
COMMUNICATIONS

Optics Communications 114 (1995) 421–424

Quasi-electrostatic trap for neutral atoms

T. Takekoshi, J.R. Yeh, R.J. Knize

Department of Physics and Astronomy, University of Southern California, Los Angeles, CA 90089-0484, USA

Received 26 August 1994; revised version received 10 November 1994

Abstract

We show that it is possible to trap any neutral atom using a focused high power infrared laser. The trap could hold several different species of atoms simultaneously. Small photon scattering and heating rates allow the possibility of long trap lifetimes.

Neutral atom traps are valuable tools for precision spectroscopy, the study of cold collisions, and for quantum collective effects. There have been several neutral atom traps demonstrated using either optical, magnetic, or microwave fields. Magnetostatic fields have been used to trap low field-seeking states of hydrogen [1], sodium [2], and cesium [3]. An ac magnetic trap has also been demonstrated for cesium [4]. Optical dipole force traps [5], and magneto-optical traps [6] have been demonstrated for various alkali and metastable inert gas atoms [7] using near-resonant laser fields. Recently, far off resonance optical dipole force traps have been demonstrated with sodium [8] and rubidium [9] atoms, with light detuned up to 65 nm below the first D1 resonance. A magnetic dipole force trap has been proposed for hydrogen [10], and has been recently demonstrated with cesium atoms [11] using microwave fields nearly resonant with the ground state hyperfine splitting. A dynamic electric trap using an oscillating electric field has also been proposed [12]. In this paper, we discuss a non-resonant optical dipole force trap using intense infrared light. This trap can confine all ground state atoms of several atomic species simultaneously for long periods of time with small scattering and heating rates.

An electrostatic Stark effect trap would be ideal for several reasons. First, there would be no photon scattering. Second, since the static electric polarizability is positive for any atomic ground state, multiple atomic species and ions, could be trapped at a local field maximum. Unfortunately, Maxwell's equations forbid a charge-free local electrostatic field maximum [13]. It is well known that this problem can be circumvented by creating a local rms electric field maximum using an ac field, such as in an optical dipole force trap. In this paper, we consider an optical trap where the laser frequency ω and Rabi frequency $|\Omega|$ are much smaller than the frequency of the first allowed electric dipole resonance ω_1. The trap frequency is far below all the atomic electric dipole resonances ($\omega < \omega_1/2$), that in many respects, this is essentially a quasi-electrostatic trap (QUEST). The proposed trap can be realized at the focus of a high intensity CO_2 laser beam ($\lambda = 10.6$ μm).

Previous optical dipole force traps have been analyzed and demonstrated with laser detuning δ small compared to the resonance frequency, $\delta = (\omega_1 - \omega) < \omega_1/10$. In this regime, a two-level atom in a spatially varying optical field experiences an optical dipole potential U [14]

$$U = -\frac{\hbar\delta}{2}\ln\left[1 + \frac{|\Omega|^2}{2[(\Gamma/2)^2 + \delta^2]}\right] \tag{1}$$

and the photon scattering rate S is

$$S = \frac{\Gamma |\Omega|^2 / 2}{2[(\Gamma/2)^2 + \delta^2] + |\Omega|^2}. \qquad (2)$$

Here $\Omega = d_{eg} \cdot E_\omega / \hbar$, where $|\Omega|$ is the Rabi frequency, $E = (E_\omega e^{-i\omega t} + E_\omega^* e^{i\omega t})/2$ is the electric field, d_{eg} is the dipole matrix element between the excited and ground states, and Γ is the spontaneous decay rate of the excited state.

Photon scattering causes trap losses through recoil heating, excited state collisions [15], and dipole force fluctuations. The heating rate in the trap can be estimated from the momentum diffusion constant. For a two-level atom, the excited and ground states will have equal but opposite Stark shifts for near resonance light, and the momentum diffusion constant D_p is [14]

$$D_p = \frac{\hbar^2 S}{2} [k^2 + (\nabla \ln |\Omega|)^2 + (\nabla \theta)^2], \qquad (3)$$

where $\Omega = |\Omega| e^{i\theta}$, and k is wavevector of the laser. The first optical trap [5] for neutral atoms used a focused 220 mW dye laser beam detuned about 130 GHz to the red of the sodium D1 resonance. The trap depth was about 10 mK, and the photon scattering rate was about 2600 s^{-1}. Recently, a far off resonance trap with detunings of up to $0.08\omega_1$, was demonstrated for rubidium atoms [9]. For a typical detuning of $0.023\omega_1$ below the D1 resonance and 0.8 watts of laser power, the trap depth was about 6 mK and the photon scattering rate was about 400 s^{-1}.

Since a QUEST operates with a laser frequency far below any electric dipole resonances, Eqs. (1)–(3) for an optical dipole force trap need to be modified. We estimate the saturation parameter [14] and obtain the result, that for practically attainable laser intensities, the excited state population is always extremely small. The trapping potential is therefore given by the lowest order perturbation theory expression for the Stark shift [16] of a ground state g, due to the excited states e,

$$\Delta E_g = -\frac{1}{4\hbar} \sum_e |d_{eg} \cdot E_\omega|^2 \left[\frac{1}{\omega_{eg} - \omega} + \frac{1}{\omega_{eg} + \omega} \right], \quad (4)$$

where $\omega_{eg} \equiv (E_e - E_g)/\hbar$. At QUEST frequencies this expression can be simplified, by using the ground state static scalar atomic polarizability, α_s. The overall

trapping potential including gravity can be approximated by

$$U = -\frac{1}{2} \frac{\alpha_s}{[1 - (\omega/\omega_1)^2]} |E_{rms}|^2 + mgz, \qquad (5)$$

where m is the mass of the atom, and ω_1 is the frequency of the first D1 transition. For an alkali or hydrogen atom, the discrepancy between Eq. (4) and the Stark shift term in Eq. (5) is $<0.1\%$, for $\omega < \omega_1/2$. At CO_2 laser wavelengths, the Stark shift in Eq. (5) approaches the dc Stark shift $\Delta E_g = -\alpha_s E^2/2$. In addition to the electric dipole force, there will be light shifts due to the magnetic dipole interaction [10] with the ground state hyperfine levels. This shift is a factor $2\mu_{Bohr}^2/(\hbar\alpha_s\omega)$ smaller than the Stark shift. For cesium atoms, this magnetic dipole trap potential is a factor of 10^5 smaller than the electric dipole trap potential. Since atoms are trapped without using a near-resonant laser, this trap may be able to confine several species simultaneously, provided the atoms can be cooled, and loaded into the trap. It may even be possible to confine atoms and ions simultaneously in the same QUEST.

The total scattering rate S can be calculated using the Kramers–Heisenberg formula [17]

$$S = \sum_f \frac{8\pi\alpha^2\omega^3 I}{3\hbar c^2} \left| \sum_{m,q} \frac{\langle f|r_q^{[1]}|m\rangle \langle m|r_Q^{[1]}|i\rangle}{\omega_m - \omega} \right.$$
$$\left. + \frac{\langle f|r_Q^{[1]}|m\rangle \langle m|r_q^{[1]}|i\rangle}{\omega_m + \omega} \right|^2 \qquad (6)$$

where i, m, and f are the initial, intermediate, and final states, Q indicates the incident laser polarization, q is the scattered light polarization, I is the incident light intensity, and α is the fine structure constant. It is possible to write $S = S_{Rayleigh} + S_{Raman}$. In this expression, Rayleigh scattering leaves the atom in its original state ($i = f$), whereas Raman scattering leaves the atom in a different hyperfine or Zeeman sublevel ($i \neq f$). $S_{Rayleigh}$ can be described by the usual Rayleigh formula [18]

$$S_{Rayleigh} = \frac{8\pi r_0^2 I \omega^3}{3\hbar} \left[\sum_e \frac{f_{eg}}{\omega_{eg}^2} \right]^2 \qquad (7)$$

where f_{eg} is the oscillator strength between the excited and ground states, and r_0 is the classical electron radius. At CO_2 laser frequencies, this expression is almost exact. At $\omega = \omega_1/2$, the discrepancy between Eq. (7) and the exact result from Eq. (6) is about 30%. For an alkali or hydrogen atom, where > 95% of the oscillator strength is in the first D1 and D2 transitions, the ratio between the Raman and Rayleigh scattering rates for $\omega < \omega_1/2$ can be written as

$$\frac{S_{\text{Raman}}}{S_{\text{Rayleigh}}}$$

$$= 2\left[\frac{\Delta_{\text{fs}}\omega(2\omega_1 + \Delta_{\text{fs}})}{3\omega_1^3 + 4\Delta_{\text{fs}}\omega_1^2 + (\Delta_{\text{fs}}^2 - 3\omega^2)\omega_1 - 2\Delta_{\text{fs}}\omega^2}\right]^2$$

$$\approx \frac{8}{9}\left[\frac{\Delta_{\text{fs}}\omega}{\omega_1^2}\right]^2, \qquad (8)$$

where Δ_{fs} is the fine structure frequency splitting. Eq. (8) shows that the scattering is almost entirely due to Rayleigh scattering. The factor $(\Delta_{\text{fs}}/\omega_1)^2$ appears, because the atoms do not spend enough time in the upper state to have their spin perturbed by the fine structure interaction $L \cdot S$ [19]. The factor $(\omega/\omega_1)^2$ in Eq. (8) appears, because the Raman scattering matrix elements for the anti-rotating terms $(\omega_m + \omega)^{-1}$ in Eq. (6) are equal but opposite in sign to the matrix elements for the corresponding rotating terms $(\omega_m - \omega)^{-1}$. In Eq. (8), we have neglected the excited state hyperfine splitting Δ_{hfs}, which would introduce a very small amount of additional Raman scattering on the order of $(\Delta_{\text{hfs}}\omega/\omega_1^2)^2 S_{\text{Rayleigh}}$.

Because the QUEST laser frequency is far below any electric dipole resonance frequency, the atom can no longer be treated as a simple two-level system. The Stark shift of the nearest excited state is no longer equal in magnitude and opposite in sign to the ground state's. The second term in Eq. (3) for the momentum diffusion constant may need to be modified. The order of magnitude may be calculated using the scattering rate (7) in Eq. (3). The momentum diffusion constant is expected to be very small. Velocity dependent dipole forces are negligible [14].

In order to operate a QUEST, it is desirable to have an intense infrared laser with the following characteristics: The laser frequency should be small compared to the first electric dipole transition frequency, but large compared to possible ground state hyperfine frequencies. The wavelength should be small, and the mode quality of the laser should be good enough so that the laser can be focused to a waist small enough

for a trapping potential to be obtained in the presence of gravity, Eq. (5). Commercial CO_2 lasers and associated optics satisfy these requirements. They are fairly efficient, high cw powers are available, and the strongest CO_2 transition is at 10.6 μm. There are several possible CO_2 laser configurations for producing a QUEST. A high power laser could be focused in a manner similar to previous traveling wave optical traps. As an example, it should be possible to focus a 3 kW laser down to a waist radius of 15 μm [20]. This trap would provide a potential well in excess of a mK for any neutral atom, including hydrogen and helium. It is also possible to operate a QUEST using a lower power laser with a buildup cavity, or in an intracavity configuration. In a typial 100 W CO_2 laser, there is about 2.5 kW of circulating power that could be focused intracavity to form a trapping potential. The problem of large heating rates in intense standing wave optical traps [21] is negligible because the excited state population is extremely small.

We have analyzed a CO_2 laser QUEST for alkali and hydrogen atoms. The trap is formed by focusing a horizontal TEM_{00} laser beam to a waist of radius w_0. Fig. 1 shows the trap depth U_0 for cesium atoms evaluated from Eq. (7), as a function of power and waist radius. Fig. 2 shows the total trap volume V for cesium atoms, determined numerically from Eq. (5). If the waist radius is increased, eventually a point is reached where gravity prevents the Stark shift from forming a potential well. The figures show that for laser powers from 10 to 10^4 W, the trap depth varies from 10^{-5} to 1 K, and the total trap volume varies from 10^{-5} to 1 cm^3.

As a specific example for cesium atoms, we consider a 1 kW CO_2 laser beam focused to a waist radius of 50 μm. In this case, the trap depth will be about 23 mK, and the trap volume will be about 3.5 mm^3. At the center of the trap, where maximum scattering occurs, the Rayleigh scattering rate is 0.045 s^{-1}, and the Raman scattering rate is a factor of 10^4 smaller. Using Eq. (3) we find the maximum momentum diffusion constant $D_p^{\text{max}} \sim 2 \times 10^{-50}$ erg gram/s. The lifetime of the atoms in the trap will be determined by D_p, background collisions with residual gases, and three-body recombination. Multiphoton ionization is expected to be negligible [22]. Since all ground state levels are trapped, there is no loss due to dipolar collisions such as in magnetic traps [1]. The trap life-

T. Takekoshi et al. / Optics Communications 114 (1995) 421–424

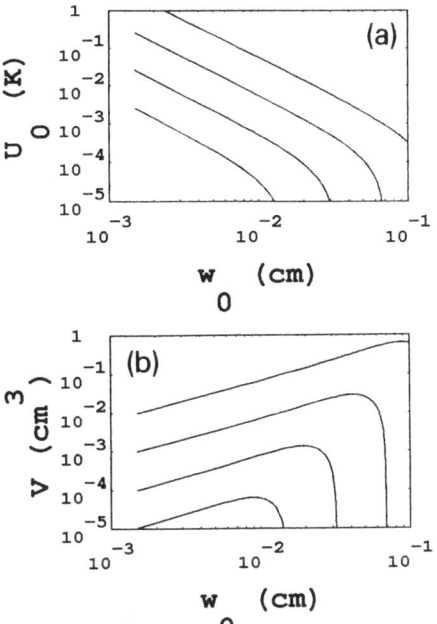

Fig. 1. (a) Trap depth U_0/k_b (including gravity) versus waist radius w_0 for cesium atoms. (b) Trap volume V versus waist radius w_0 for cesium atoms. The curves (from top to bottom) correspond to laser powers of 10 kW, 1 kW, 100 W, 10 W.

time due to momentum diffusion can be estimated from $\tau_{\mathrm{trap}} \sim mU_0/D_{\mathrm{p}}^{\mathrm{max}}$. This particular QUEST has a $\tau_{\mathrm{trap}} \sim 4 \times 10^{10}$ s which shows that the heating rate is negligible. A QUEST with the same parameters, in the standing wave configuration, will have a similar lifetime.

We have shown that it is possible to trap neutral atoms in a CO_2 laser optical trap. The laser frequency is far below any electric dipole resonances of atoms, so it is possible to simultaneously trap all the ground states of multiple atomic and ionic species, with low scattering and heating rates. Excluding background collisions, and three body recombination, the trap lifetime can be on the order of years. Two possible QUEST applications include a search for an atomic electric dipole moment (EDM), and an attempt to produce Bose–Einstein condensation of a cold atomic vapor.

This work was supported by the US Army Research Office, under grant number 30201-PH, and by the National Science Foundation.

References

[1] H. Hess, G.P. Kochanski, J.M. Doyle, N. Masuhara, D. Kleppner and T.J. Greytak, Phys. Rev. Lett. 59 (1987) 672.
[2] Alan L. Migdall, John V. Prodan, William D. Phillips, Thomas H. Bergeman and Harold J. Metcalf, Phys. Rev. Lett. 54 (1985) 2596;
K. Helmerson, A Martin and D.E. Pritchard, J. Opt. Soc. Am. B 9 (1992) 483; 9 (1992) 1988.
[3] C.R. Monroe, E.A. Cornell, C.A. Sackett, C.J Myatt and C.E. Wieman, Phys. Rev. Lett. 70 (1993) 414.
[4] Eric A. Cornell, Chris Monroe and Carl E. Wieman, Phys. Rev. Lett. 67 (1991) 2439.
[5] S. Chu, J.E. Bjorkholm, A. Ashkin and A. Cable, Phys. Rev. Lett. 57 (1986) 314.
[6] E.L. Raab, M. Prentiss, Alex Cable, Steven Chu and D.E. Pritchard, Phys. Rev. Lett. 59 (1987) 2631.
[7] Hidetori Katori and Fujio Shimizu, Phys. Rev. Lett. 70 (1993) 3545.
[8] S. Rolston, C. Gerz, K. Helmerson, P. Jessen, P. Lett, W. Phillips, R. Spreeuw and C. Westbrook, Proc. 1992 Shanghai International Symposium on Quantum Optics, eds. Yuzhu Wang, Yiqui Wang and Zugeng Wang, Proc. SPIE 1726 (1992) 205.
[9] J.D. Miller, R.A. Cline and D.J. Heinzen, Phys. Rev. A 47 (1993) R4567.
[10] Charles C. Agosta, Issac F. Silvera, H.T.C. Stoof and B.J. Verhaar, Phys. Rev. Lett. 62 (1989) 2361.
[11] R.J.C. Spreeuw, C. Gerz, Lori S. Goldner, W.D. Phillips, S.L. Rolston, C.I. Westbrook, M.W. Reynolds and Isaac F. Silvera, Phys. Rev. Lett. 72 (1994) 3162.
[12] F. Shimizu and M. Morinaga, Japn. J. Appl. Phys. 31 (1992) L1721.
[13] W. Ketterle and D.E. Pritchard, Appl. Phys. B 54 (1992) 403;
W.H. Wing, Laser Cooled and Trapped Atoms, Proc. Workshop on Spectroscopic Applications of Slow Atomic Beams, ed. W.D. Phillips, NBS Special Publication No. 653 (U.S. GPO, Washington, DC, 1983).
[14] J.P. Gordon and A. Ashkin, Phys. Rev. A 21 (1980) 1606.
[15] M. Prentiss et al., Optics Lett. 13 (1988) 452.
[16] Butcher and Cotten, Elements of Nonlinear Optics (Cambridge University Press, 1990) p. 183.
[17] Rodney Loudon, The Quantum Theory of Light, 2nd ed. (Clarendon, Oxford, 1993) p. 314.
[18] C. Cohen-Tannoudji, J. Dupont-Roc and G. Grynberg, Atom-Photon Interactions (Wiley, 1992) p. 526.
[19] R.A. Cline, J.D. Miller, M.R. Matthews and D.J. Heinzen, Optics Lett. 19 (1994) 207.
[20] Laser Power Optics Inc., San Diego CA USA, technical data 1992.
[21] J. Dalibard and C. Cohen-Tannoudji, J. Opt. Soc. Am. B 11 (1989) 1707.
[22] P. Lambropoulos, private communication.

RAPID COMMUNICATIONS

PHYSICAL REVIEW A   VOLUME 57, NUMBER 6   JUNE 1998

Cold and dense cesium clouds in far-detuned dipole traps

D. Boiron,[1] A. Michaud,[1] J. M. Fournier,[2] L. Simard,[1] M. Sprenger,[3] G. Grynberg,[1] and C. Salomon[1]

[1]*Laboratoire Kastler Brossel and Collège de France, 24 rue Lhomond, 75231 Paris, France*
[2]*Rowland Institute for Science, 100 Edwin H. Land Boulevard, Cambridge, Massachusetts 02142*
[3]*Fakultät für Physik, Universität Konstanz, D-78457 Konstanz, Germany*
(Received 4 November 1997)

We study the behavior of cesium atoms confined in far-detuned laser traps and submitted to blue Sisyphus cooling. First, in a single focused yttrium aluminum garnet (YAG) beam, the atomic cloud has a rod shape with a 6-μm transverse waist radius, a temperature of 2 μK, and a transient density of 10^{12} atoms/cm^3. For this sample, we have not detected any influence of photon multiple scattering on the atomic temperature, in contrast to previous measurements in *isotropic* samples. Second, the YAG laser is used to produce a 29-μm period hexagonal optical lattice in which we directly observe the localization of the atoms by absorption imaging. Equilibrium densities on the order of 10^{13} atoms/cm^3 are achieved in this structure. [S1050-2947(98)50706-8]

PACS number(s): 32.80.Pj

Atom manipulation using laser light has experienced a rapid growth in the recent years [1]. Laser-cooled and/or laser trapped atoms are used in a variety of applications such as cesium fountain clocks, atom interferometry, atom lithography, atom guiding in optical fibers, and the production of Bose-Einstein condensates. This manipulation can be dissipative, such as in magneto-optical traps (MOT's), optical molasses, and lattices where confinement and cooling are simultaneously present but where the density is limited to about 10^{11} atoms/cm^3. On the other hand, the use of far-detuned dipole traps leads to nearly nondissipative potentials with very low spontaneous emission rates ($\lesssim 1 s^{-1}$) in which precooled atoms can be trapped [2–4]. Raman cooling in such traps has been performed for Na atoms leading to a temperature of 1 μK at a density of 4×10^{11} atoms/cm^3 [5] and for cesium atoms to 2 μK and 1.3×10^{12} atoms/cm^3 [6]. In both experiments the size of the atomic cloud was in the $10-100$ μm range. The loading of the dipole trap may also be produced by adiabatic transfer from a superimposed nearly resonant molasses [7]. We present here another scheme for producing dense and confined samples. This scheme is very simple and it produces extremely small (micrometer range) and dense ($\sim 10^{13}$ atoms/cm^3) cesium samples. It relies on the addition of blue Sisyphus cooling (BSC) to far-red-detuned dipole traps. This BSC cooling (also called gray molasses) uses a $F_g \rightarrow F_e = F_g - 1$ transition and a positive laser detuning ($\omega_L > \omega_{at}$). It provides cesium atoms in states of the $F_g = 3$ hyperfine level that are only weakly coupled to the cooling light field, and the minimum temperature at low density is 1 μK, three times lower than the temperature in ordinary red-detuned optical molasses or MOTs [8]. We show here that this cooling is very efficient in far-detuned traps. It leads to high densities that are not limited by hyperfine changing collisions and it offers the advantage over the Raman cooling scheme that it does not require any additional laser. We have investigated first the simple case of a single focused yttrium aluminum garnet (YAG) beam. Because of the quasi-unidimensional geometry, we have not detected any influence of photon multiple scattering on the atomic temperature at the 0.5-μK level. This result constitutes an important step on the way to reaching quantum

degeneracy by purely optical methods. Then the same YAG laser is used to produce a two-dimensional hexagonal optical lattice in which 10-μK atoms are arranged in rods having a 3.5-μm waist radius, a 29-μm spacing, each containing about 10^4 atoms. We have taken absorption images that directly display the periodic ordering of the atoms in this optical structure and we show that three-dimensional (3D) trapping is obtained by means of the Talbot effect [9].

As a first test of the efficiency of the BSC in a far-detuned trap, we focus a TEM$_{00}$ YAG laser (at 1.06 μm) onto the cloud of a cesium magneto-optical trap produced in a low vapor pressure cell [Fig. 1, path (a)]. The YAG beam is horizontal and has a waist of $w_0 = 45$ μm.

In the absence of the MOT and of the BSC, the potential acting on the atoms is the sum of the dipole potential of the YAG beam and of the gravitational potential. Because the confocal parameter $b = kw_0^2 \approx 1.2$ cm$\gg w_0$, the trapping force along the YAG beam is much weaker than in the trans-

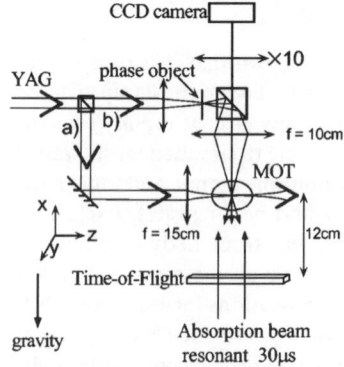

FIG. 1. Experimental setup. The YAG beam creates a horizontal dipole trap, path (a), or an optical lattice, path (b). Absorption images are taken with a vertical pulsed beam and are magnified by a $\times 10$ microscope objective. (a) The YAG power is 700 mW and the waist is 45 μm. (b) Trapping in optical lattices. The YAG beam images a periodic pattern onto the atomic sample.

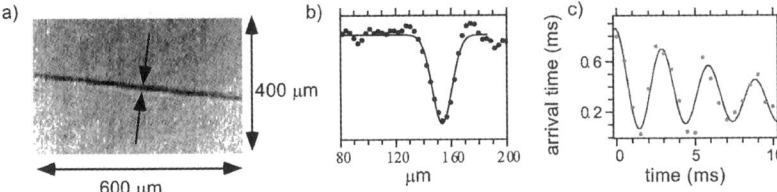

FIG. 2. (a) Absorption image of cesium atoms trapped in a single beam dipole trap. (b) Section of (a); the rms size is 6.0 μm. (c) Measurement of the vertical oscillation frequency (see text): the YAG trap is interrupted for ~ 1 ms and turned on again for an adjustable time (horizontal axis). Vertical axis: mean arrival time of the atoms in the TOF probe beam. Here $\Omega_x/2\pi = 330$ Hz.

verse direction. For a linear polarization, the depth U_0 of the dipole potential is the same for all Zeeman sublevels and is equal to $\hbar\Omega_R^2/4\delta_{eff}$, where Ω_R is the peak Rabi frequency, and δ_{eff} is an effective detuning given by $1/\delta_{eff} = 1/3\delta_1 + 2/3\delta_2$. $\delta_1 = 2\pi \ 5.3 \times 10^{13}$ rad/s (respectively, $\delta_2 = 2\pi 7.0 \times 10^{13}$ rad/s) is the detuning between the YAG laser and the first $D1$ line (respectively, $D2$) of cesium. The quantity $\eta = mgw_0/U_0$ characterizes the drop of potential due to gravity on the typical size w_0 as compared to U_0. In order to have stable trapping against gravity one should have $\eta \ll 1$. The effect of gravity is then only a mere shift of the minimum of the potential with a negligible modification of the potential depth and of the oscillation frequency at the bottom of the trap $\Omega_{osc} = (4U_0/mw_0^2)^{1/2}$. Typical values are $P = 700$ mW, $U_0/k_B = 50$ μK, $\eta = 0.13$, and $\Omega_{osc} = 2\pi 390$ rad/s. Because of the very large detuning of the YAG laser, the photon scattering rate is only 2 s^{-1}.

In order to load this trap, we first capture about 3×10^7 atoms in a MOT for 1 s while having the YAG beam continuously on. The intensity of the MOT beams is then reduced during 30 ms to contract the atomic cloud to a peak density of 5×10^{10} atoms/cm^3 and temperature of 30 μK. By a simple shift of the distributed Bragg reflector laser frequencies (via the diode current), we further apply a blue Sisyphus cooling phase with the $F_g = 3 \rightarrow F_e = 2$ component of the $D2$ line with a laser detuning of $+5\Gamma$, where $\Gamma = 2\pi \times 5.3$ MHz is the natural width of the $P_{3/2}$ state. The MOT laser now serves as a repumping laser and is tuned to the $F_g = 4 \rightarrow F_e = 4$ transition. This accumulates atoms in nearly dark states of $F_g = 3$ with a temperature typically three times lower than in conventional red-detuned optical molasses [8]. During this cooling phase, which lasts about 30 ms, part of the atoms are dragged by the YAG potential toward the bottom of the YAG trap. After switching off the BSC, only these atoms remain trapped for much longer times.

The size of the atomic cloud is measured by an absorption imaging technique [10]. A 1-mm-diam probe beam resonant on the $F_g = 4 \rightarrow F_e = 5$ transition is sent vertically through the sample and its intensity distribution is recorded on a charge-coupled-device (CCD) camera. A lens with a 100-mm focal length at a distance of 200 mm from the trap makes an image with a $\times 1$ magnification of the trap outside the vacuum chamber with a numerical aperture of 0.1. A $\times 5$ or a $\times 10$ microscope objective enlarges the image on a CCD camera with a pixel size of 15 μm. With this magnification, the rescaled pixel size is 3 μm (1.5 μm) for the $\times 5$ objective ($\times 10$) and is usually smaller than the images of the

distribution of trapped atoms. In order to have an instant picture of the atoms, the probe beam is pulsed for ~ 30 μs and its intensity is close to the saturation intensity. If the atoms are in the $F_g = 3$ hyperfine level, a repumping pulse (100 μs) of resonant light on $F_g = 3 \rightarrow F_g = 3$ is applied to transfer the atoms from $F_g = 3$ to $F_g = 4$.

The temperature and number of cooled atoms in the YAG trap (or MOT) are measured by a standard time-of-flight (TOF) technique. After switching off suddenly the YAG beam, the atoms fall through a 1-mm-high probe beam located 12 cm below the trap. The probe contains two frequencies, one resonant with the $F_g = 4 \rightarrow F_e = 5$ cycling transition, and one resonant with the $F_g = 3 \rightarrow F_e = 3$ transition to detect atoms in $F_g = 4$ or $F_g = 3$. The atomic fluorescence is collected on a low noise photodiode. The temperature of the atoms along the vertical is deduced from the width of the time-of-flight peak and the number of atoms by its area. In order to distinguish the atoms trapped in the YAG trap from those that are only sustained by the BSC, we switch off the YAG laser a time τ_{YAG} after the BSC lasers are turned off. The TOF contains then two peaks (separated by τ_{YAG}), one associated with the atoms in the BSC and one associated with the trapped atoms.

The efficiency of the loading of the dipole trap as a function of the duration τ_{BSC} of the BSC phase passes through an optimum for $\tau_{BSC} = 30$ ms. The atom number in the YAG trap is then 2×10^5 atoms (3.7 % of the atoms in the BSC molasses), a factor of 4 higher than at short times. The temperature of the atoms in the YAG trap and in the BSC are the same to within 1 μK. They range between 1 and 3 μK, depending on the BSC parameters. This demonstrates that the cooling mechanism involved in BSC is not perturbed by the YAG potential. With traditional $F_g = 4$ to $F_g = 5$ molasses, the number of trapped atoms is 5 to 10 times smaller. The CCD camera image recorded at the optimum ($\tau_{BSC} = 30$ ms, $\tau_{YAG} = 30$ ms) is presented in Fig. 2.

As expected, the atom cloud has a rod shape. Its transverse size is well fitted by a Gaussian with rms radius $\sigma_y = 6^{+1}_{-2}$ μm. The longitudinal size is $\sigma_z = 300$ μm. The corresponding peak atomic density $n = N/(2\pi)^{3/2}\sigma_x\sigma_y\sigma_z$ is $\sim 1 \times 10^{12}$ atoms/cm^3 with a global uncertainty of about a factor of 3. In these conditions the measured temperature T is $2.0(5)$ μK. Note that this density is a transient density because, in the longitudinal axis, the oscillation frequency being only 1.8 Hz, the sample has not reached its thermal equilibrium value of 950 μm. In order to compare the observed size with theory, we measure the radial oscillation frequency as follows. We switch off the YAG beam for a short time

D. BOIRON *et al.*

(~ 1 ms), during which the atoms acquire by gravity a common vertical velocity $v_x \sim 1$ cm/s. When the YAG beam is turned on again the cloud oscillates with a frequency Ω_x. Switching off again the YAG beam after a variable delay gives access to v_x through the mean arrival time of the atoms in the probe beam of the TOF detection. From Fig. 2, we obtain $\Omega_{\rm osc}^{\rm (expt)}/2\pi = 330(30)$ Hz. This frequency is in reasonable agreement with the calculated one 390(120) Hz. For a harmonic potential, one has at thermal equilibrium

$$1/2 k_B T = 1/2 m \Omega_x^2 \sigma_x^2, \quad \text{i.e.,} \quad \sigma_x = v_x^{\rm rms}/\Omega_x. \tag{1}$$

Therefore one expects $\sigma_x^{\rm (expt)} = 4.8(7)$ μm. This value is also in good agreement with the value deduced from the image $\sigma_y = 6_{-2}^{+1}$ μm, assuming cylindrical symmetry, and considering the finite resolution ($\sigma_{\rm res} \sim 2.5$ μm) of our imaging system.

These data exhibit an interesting effect: our previous measurements on isotropic clouds of atoms in BSC with a rms radius of 700 μm showed a clear dependence of the temperature on the density, varying as $T = 1 + 0.6 n/10^{10}$, where T is in μK and n in atoms/cm^3 [8]. This effect was attributed to photon multiple scattering within the cold-atom cloud. Consequently, in the present high-density YAG trap ($n = 10^{12}$ atoms/cm^3) we could expect a temperature of ~ 60 μK, 30 times higher than what we actually measure. We believe that this large difference is due to the geometry of our atomic sample. In a simple model, the temperature increase is proportional to the photon density n_ν within the cloud. n_ν results from a balance between the number of emitted photons that is proportional to the number of atoms nV (where V is the trap or molasses volume) and the photon escape proportional to the surface S of the trap. Because $V/S \sim R$ for a sphere of radius R and $V/S \sim r$ for a cylinder having a radius r much smaller than its height, we expect that at a given atomic density the excess temperature in the cylindrical geometry is reduced by a factor on the order of R/r, typically 10^2, as compared to our previous spherical geometry. In this cylindrical geometry the expected slope dT/dn is thus $10^{-9}r$ cm^2 μK [Eq. (1)], where $r \sim \sigma_y$ is in centimeters. For $r = 6$ μm and $n = 10^{12}$ atoms/cm^3, we find $\delta T = 0.6$ μK, which is on the order of our experimental detectivity.

For the trapping experiments using optical lattice potentials, we use the optics described in Fig. 1, path (b). The YAG beam images a phase grating on the atomic cloud with a $f = 100$ mm lens, also used for the absorption imaging beam. With this method, arbitrary intensity patterns can be reproduced on the atoms. The grating is hexagonal with a period $p = 29$ μm and is illuminated over a waist of 140 μm by 2.5 W of the linearly polarized YAG beam. The power of the input beam is diffracted into a zeroth-order beam (power ratio, 0.07), six first-order beams (0.41), and 12 second-order beams (0.36). Higher orders ($N > 2$) are apertured by our optical system. These 19 beams are recombined and interfere in the image region to create a hexagonal two-dimensional (2D) optical potential. The loading of atoms in this periodic structure is the same as seen previously. Figure 3 is an absorption image taken just after having switched off the BSC beams. The atoms are trapped in periodically spaced sites in a hexagonal pattern of 29-μm periods. In

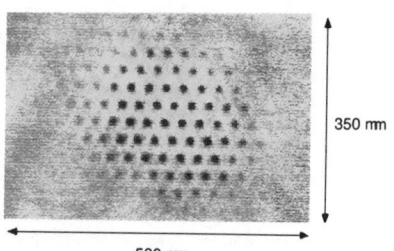

FIG. 3. Absorption image of atoms trapped in a far-detuned hexagonal lattice of period 29 μm. Each spot contains $\sim 10^4$ atoms.

each site the atoms form a vertical rod with a radial confinement of $\sigma_y = \sigma_z = 4.5(1.5)$ μm. The peak absorption coefficient is 20%. Each tube located near the center of the pattern contains about 10^4 atoms. This number remains almost constant up to a radius of 80 μm and then decreases because of the Gaussian spatial profile of the YAG beam. On the edges the potential depth is only on the order of the atomic temperature $\sim 2-5$ μK.

In this two-dimensional structure there are no vacancies, i.e., all sites are macroscopically occupied. By turning on the beam creating the lattice at variable delays, we find that it takes about 10 ms to fill this 2D structure. When the YAG beam alone is left on for a time longer than ~ 20 ms, the images vanish because the atoms fall along the tubes.

However, there is a periodic structure along x due to the Talbot effect [9], which, for high enough intensity, is able to compensate gravity. To understand the periodicity along the vertical axis, consider the interference pattern created by the zeroth-order beam propagating along x having a field amplitude A_0 and the first-order diffracted beams of amplitude A_j ($j = 1$ to 6) that propagate along a direction that makes the angle θ with Ox. The interfering structure is of the form $I = |A_0 e^{ikx} + \Sigma_j A_j e^{ik_x x + \mathbf{k}_\perp \cdot \mathbf{r}_\perp}|^2$, where $|k_x| = |k \cos \theta|$. The interference pattern is periodic along x with period $p_x = 2\lambda/\theta^2 = 1.2$ mm for small θ. More generally, the second-order beams also contribute by interfering with the zeroth-order and first-order beams, giving a periodic pattern with a more complex substructure.

We have produced such 3D traps with a YAG beam waist of $w_0 = 60$ μm on the grating. Absorption images of atoms trapped in this 3D potential are very similar to Fig. 3 with fewer occupied sites. The $1/e$ lifetime of this sample is 0.35 s. The temperature is measured by TOF along x and by TOF imaging along y and z after suddenly switching off the YAG beam. We find an isotropic temperature of $T = 10(3)$ μK. The radial oscillation frequency is 5(1) kHz more than 15 times larger than in the single beam trap and twice the recoil frequency at 852 nm (2 kHz). The measured radial rms size is 3.5 μm, a value close to our resolution limit of 2.5 μm. From the measured temperature and oscillation frequency, we make another determination of this size, 0.8 μm, well below our resolution. We have numerically calculated the 3D optical potential created by the phase grating and gravity. For 2.5 W the depth is $U_0/k_B \sim 400$ μK. Because of the limited vertical extension of the MOT cloud (300 μm), at-

RAPID COMMUNICATIONS

oms cannot be trapped in more than one or two potential wells along x, and we deduce an equilibrium size along x of $\sigma_x \sim 20$ μm. From these data, we obtain a rough estimate of the atomic peak density, which, in each tube, is larger than $\sim 10^{13}$ atoms/cm^3 and corresponds to a phase-space density larger than $\sim 10^{-3}$. Note that this estimate is in agreement with Eq. (1), which gives $n = 5 \times 10^{13}$ atoms/cm^3 for $\Delta T \sim 5$ μK. Even if the density has a large uncertainty[1] because of the uncertainty on the size of the sample, this cylindrical geometry is very appealing for future experiments with high density. These include photonic band-gap studies,

---

[1]This density could be independently measured by transferring quickly the atoms in $F = 4$ and recording the lifetime of the sample which should be in the ms range with the known value of the spin-exchange relaxation rate of $\sim 10^{-11}$ cm^3/s [10–12].

evaporative cooling in order to enhance the phase-space density, or collision studies.

As a last example of trapping in a far-detuned potential, we have recorded absorption images of atoms trapped in speckle light when the grating is replaced by a holographic speckle pattern imaged onto the MOT. The observed random distribution of potential sites and the wide distribution of sizes and shapes are a clear signature of the disordered light pattern. These devices open the way to the study of atom transport in periodic or disordered systems in the classical or quantum regime.

The authors wish to thank K. Amar, Y. Castin, C. Cohen-Tannoudji, J.-Y. Courtois, J. Dalibard, and P. Jacquot. This work has been supported by CNES, BNM, CNRS, Collège de France, NEDO (Japan), and TMR Contract No. FMRX-CT96-0002 of the E.U. Laboratoire Kastler Brossel is Unité de Recherche de l'Ecole Normale Supérieure et de l'Université Pierre et Marie Curie, associée au CNRS.

[1] See, for instance, *Laser Manipulation of Atoms and Ions*, edited by E. Arimondo, W.D. Phillips, and F. Strumia (North-Holland, Amsterdam, 1992); C. Adams and E. Riis, Prog. Quantum Electron. **21**, 1 (1997).

[2] J. D. Miller, R. A. Cline, and D. J. Heinzen, Phys. Rev. A **47**, R4567 (1994).

[3] T. Takekoshi and R. Knize, Opt. Lett. **21**, 77 (1996).

[4] S. Friebel, C. D'Andrea, J. Walz, M. Weitz, and T. Hänsch, Phys. Rev. A **57**, R20 (1998).

[5] H. Lee, C. Adams, M. Kasevich, and S. Chu, Phys. Rev. Lett. **76**, 2658 (1996).

[6] A. Kuhn, H. Perrin, W. Hänsel, and C. Salomon, in *OSA TOPS on Ultracold Atoms and BEC*, edited by K. Burnett (Optical Society of America, Washington, D.C., 1996), Vol. 7, p. 58.

[7] D. L. Haycock, S. E. Hamman, G. Klose, and P. S. Jessen, Phys. Rev. A **55**, R3991 (1997).

[8] D. Boiron, A. Michaud, P. Lemonde, Y. Castin, C. Salomon, S. Weyers, K. Szymaniec, L. Cognet, and A. Clairon, Phys. Rev. A **53**, R3734 (1996); D. Boiron, C. Triché, D. R. Meacher, P. Verkerk, and G. Grynberg, *ibid.* **52**, R3425 (1995).

[9] H. F. Talbot, Philos. Mag. **9**, 401 (1836).

[10] M. H. Anderson, J. R. Ensher, M. R. Matthews, C. E. Wieman, and E. A. Cornell, Science **269**, 198 (1995).

[11] C. R. Monroe, E. A. Cornell, C. A. Sackett, C. J. Myatt, and C. E. Wieman, Phys. Rev. Lett. **70**, 414 (1992).

[12] J. Dalibard (private communication).

VOLUME 81, NUMBER 26   PHYSICAL REVIEW LETTERS   28 DECEMBER 1998

Degenerate Raman Sideband Cooling of Trapped Cesium Atoms at Very High Atomic Densities

Vladan Vuletić, Cheng Chin, Andrew J. Kerman, and Steven Chu

Department of Physics, Stanford University, Stanford, California 94305-4060
(Received 25 August 1998)

We trap 10^7 cesium atoms in a far red detuned 1D optical lattice. With degenerate Raman sideband cooling we achieve a vibrational ground state population of 80% for the steep trapping direction. Collisional coupling enables us to cool the spin-polarized gas in 3D without loss of atoms to a peak phase space density of $1/180$ at a mean temperature of 2.8 μK and a density of 1.4×10^{13} cm^{-3}. [S0031-9007(98)08002-8]

PACS numbers: 32.80.Pj

Evaporative cooling of an optically precooled atomic gas in a magnetic trap to Bose-Einstein condensation (BEC) has become a well-established technique [1]. Although laser cooling methods contribute the bulk of the phase space compression in these experiments, severe limitations at densities approaching one atom per cubic wavelength [2] make it necessary to use evaporative cooling to reach the degenerate quantum regime. In spite of the great success of evaporative cooling in this role, however, the attainment of BEC by optical means remains an important goal. Optical cooling can be much faster than evaporation and does not remove the majority of the originally trapped atoms, which should allow more atoms to be condensed faster. In addition, evaporative cooling relies heavily on a favorable ratio of elastic to inelastic collision rates, which may preclude the production of a condensate by evaporative cooling in a magnetic trap altogether for some elements [3,4]. Finally, for the purpose of sympathetic cooling [5], a dense, continuously optically cooled atomic gas in thermal contact with the sample of interest would constitute a powerful refrigerator, with a cooling power as large as one quantum of vibration energy per oscillation period. For a trap oscillation frequency of 200 kHz and equal numbers of "coolant" and "thermal load" atoms, the cooling rate corresponds to 1 K/s.

The density limitations associated with conventional polarization gradient cooling are believed to be due to the reabsorption of spontaneously emitted photons at high optical densities and to light-induced atom-atom interaction effects [2]. While it had previously been suggested that these effects would prevent Bose-Einstein condensation by laser cooling in weakly confining traps [6], recent theoretical works have explored ways to overcome these limitations. Apart from using strongly anisotropic traps with larger surface area-to-volume ratios and therefore larger photon escape probabilities, it has been proposed that the reabsorption problem could be overcome by making the photon scattering rate smaller than the trap vibration frequency [7]. In that case most of the scattering processes are elastic and do not heat the sample.

Lamb-Dicke traps, where the ground state spread is much smaller than the wavelength of the cooling light, are particularly promising in this context since their high vibrational frequencies permit a fast optical cooling rate. These traps offer the additional advantage that the condensation temperature can easily be higher than the recoil energy. Since the critical temperature for BEC in a 3D harmonic trap is proportional to the geometric mean of the oscillation frequencies in the three spatial dimensions [8], even a trap with just one tightly bound direction, e.g., a 1D optical lattice, can have a significantly higher condensation temperature than conventional magnetic traps. However, this higher critical temperature comes only at the expense of a higher critical density. It is therefore particularly important to suppress density-dependent heating mechanisms such as inelastic two-body collisions, and to develop a laser cooling method that remains efficient at very high atomic densities.

In steep traps where the vibrational structure can be spectrally resolved it is possible to cool the center-of-mass motion using sideband cooling techniques. Single ions have been cooled to the motional ground state in traps where the vibrational frequencies exceed the natural linewidth of the atomic transition [9]. For neutral atoms and the correspondingly weaker trapping forces, the vibrational levels can be resolved with two-photon Raman transitions, and Raman sideband cooling has recently been demonstrated in 2D and 1D [10,11].

In this Letter we present a Raman sideband technique with which we have cooled trapped neutral atoms to 3D temperatures similar to those achieved in free space using polarization gradient cooling, but at densities almost 2 orders of magnitude higher. Our method uses only the two lowest-energy atomic ground states, resulting in an enormous suppression of the heating and trap loss normally caused by inelastic two-body collisions. Although this is particularly important for Cs, which has exceptionally large spin-exchange and dipolar cross sections [3], limitations on the laser cooling of other atoms due to hyperfine-changing collisions have also been observed [12]. Our cooling technique works at densities of 10^{13} cm^{-3} and we observe no cooling-induced loss of atoms. The strong collisional coupling at these high densities allows us to cool all three dimensions using

0031-9007/98/81(26)/5768(4)$15.00

VOLUME 81, NUMBER 26   PHYSICAL REVIEW LETTERS   28 DECEMBER 1998

Raman sideband cooling only along a single tightly bound Lamb-Dicke direction.

Our cooling method is similar to that used in Ref. [10]. We begin with a spin-polarized atom in the state $|F = 3, m_F = 3; \nu\rangle$, where $\nu > 0$ denotes an excited vibrational state in the Lamb-Dicke direction (Fig. 1). An external magnetic field **B** is applied to shift this level into degeneracy with $|3, 2; \nu - 1\rangle$; since the vibration frequencies of the $m = 3$ and $m = 2$ potentials are approximately the same, this condition holds for all ν in the harmonic region of the trap [13]. A cooling cycle then consists of a degenerate Raman transition from $|3, 3; \nu\rangle$ to $|3, 2; \nu - 1\rangle$, followed by optical pumping back to $|3, 3\rangle$. Since the atom is tightly bound, the recoil momentum from the scattered photon is unlikely to change its vibrational state, and the atom is preferentially returned to $|3, 3; \nu - 1\rangle$, with one quantum of vibration energy removed. After subsequent cooling cycles $|3, 3; \nu - 1\rangle$ to $|3, 2; \nu - 2\rangle$, etc., the atom reaches the vibrational ground state which is dark to both the optical pumping and degenerate Raman transitions.

Since the Raman transitions use two isoenergetic photons, they can be driven by the trapping light itself, whose large intensity and detuning allow a sizable coupling in combination with low heating by spontaneous Raman processes. The coupling strength is given by the off-diagonal matrix element of the light shift operator [13], $\langle 3, 2; \nu - 1|U(\mathbf{r})|3, 3; \nu\rangle$, which for general polarization is nonzero only when the external magnetic field **B** is oriented at a nonzero angle β relative to the wave vector **k**. Different center-of-mass wave functions are coupled by the two-photon recoil associated with the spatial dependence of U, whose parity with respect to the Lamb-Dicke potential wells is determined by the polarization configuration of the standing wave. To drive transitions with $\Delta\nu = -1$, we use two counter-propagating running waves whose linear polarizations

subtend an angle α [14]. The coupling strength for this configuration is approximately $\langle 3, 2; \nu - 1|U|3, 3; \nu\rangle = \nu^{1/2}(6^{1/2}/2)\varepsilon U_0 \eta \sin\alpha \sin\beta$, to leading order in the Lamb-Dicke parameter $\eta = kx_0$ [15]. Here U_0 is the trap depth for linearly polarized light, $\varepsilon = 2.34 \times 10^{-2}$ characterizes the relative strength of the Raman transitions for our detuning, and x_0 is the rms width of the ground state in position space. Note that both the radial and axial vibration frequencies are also functions of the angle α.

Our 1D lattice trap is produced by the TEM_{00} output of an optically injection-locked Nd:YAG laser system which we constructed. It produces a single frequency output power of 21 W at $\lambda = 1064$ nm that is used to form a vertical standing wave with a beam waist of 260 μm and a running wave power of 17 W at the position of the atomic cloud. For a linearly polarized standing wave the calculated trap depth [16] is $U_0/h = 3.2$ MHz or 160 μK, while the calculated axial and radial vibration frequencies are $f_{\mathrm{ax}} = 130$ kHz and $f_r = 120$ Hz, respectively. The estimated scattering rate induced by the trapping light is 2 s^{-1}.

We begin with a magneto-optical trap (MOT) that is loaded from a background cesium vapor with a time constant of 4.5 s. We collect 3×10^7 atoms in 500 ms in the MOT and then decrease the total repumping light intensity to 3.2 mW/cm^2 for 38 ms to compress the cloud. Subsequently, blue detuned Sisyphus cooling [16] is performed in the trap for 5 ms on the $6S_{1/2}$, $F_g = 3$ to $6P_{3/2}$, $F_e = 2$ transition with a detuning of $2\pi \times 18$ MHz. The MOT laser here serves as a repumper from the $F_g = 4$ ground state on the $6S_{1/2}$, $F_g = 4$ to $6P_{3/2}$, $F_e = 4$ transition. In the 1D lattice we trap a total of 1.0×10^7 atoms in a cigar-shaped cloud with a vertical FWHM of 2.5 mm. This length corresponds to 4700 individual pancake shaped traps each with an aspect ratio of 1000, spaced by 532 nm. The vertical density distribution is approximately Gaussian, yielding a population of 2.0×10^3 atoms per trap in the central region.

The lifetime of the trapped gas depends strongly on the hyperfine level. For atoms prepared in the lower hyperfine level $F = 3$ the decay is purely exponential with a background-pressure limited time constant of $\tau = 2.0$ s. For the upper hyperfine level we observe a much faster density-dependent loss due inelastic two-body collisions. Figure 2 shows the decay for an unpolarized sample prepared by optical pumping at $t_0 = 400$ ms in the upper hyperfine state $F = 4$ at $T = 14$ μK and $n = 1.1 \times 10^{12}$ cm^{-3}. The data are well described by a two-body loss process with a rate coefficient in agreement with a previously published value [17]. For atoms in the lower hyperfine level at zero magnetic field there are no exothermic two-body collisions.

We measure the axial and radial kinetic energy distributions of the trapped atoms using a time-of-flight method. The atoms are dropped by extinguishing the trapping light

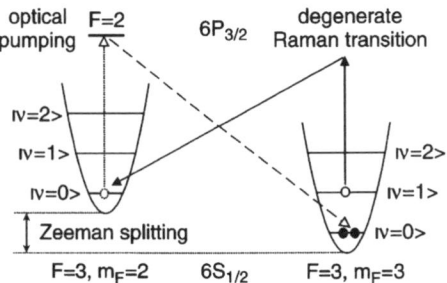

FIG. 1. Degenerate Raman sideband cooling in a Lamb-Dicke trap using the two lowest-energy magnetic levels. One cooling cycle consists of a vibration-changing Raman transition followed by optical pumping back to the $m_F = 3$ sublevel. The atoms accumulate in the vibrational ground state of the $m_F = 3$ level (black dots) which is dark to both the optical pumping light and the Raman transitions.

VOLUME 81, NUMBER 26　　　PHYSICAL REVIEW LETTERS　　　28 DECEMBER 1998

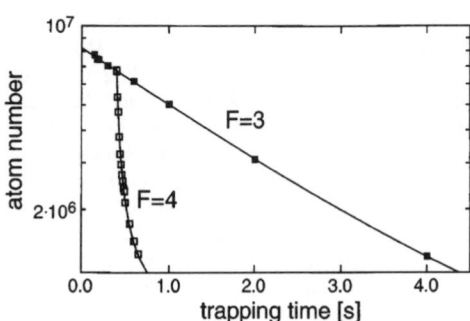

FIG. 2. Decay of the trapped gas for different hyperfine ground states. The open squares show the atom number upon preparation in the upper hyperfine state at the time $t_0 = 400$ ms. The initial density is 1.1×10^{12} cm^{-3}. The solid lines are fits including a two-body loss.

in 2 μs with an acousto-optic modulator. A light sheet is located 12 cm below the trap and the fluorescence from the falling cloud is imaged onto a photodiode and a CCD camera. From the temporal and spatial widths of these two signals, the axial and radial velocity distributions are extracted. By varying the turn-off time of the trapping light, we have verified that negligible adiabatic cooling occurs during the release. To measure the trap vibration frequencies, we use a parametric excitation technique similar to the one used in Ref. [18]. A 0.2% modulation is applied to the intensity of the trapping light, and a small heating is observed when this modulation is tuned to twice the atomic oscillation frequency. This procedure yields frequencies that agree to better than 10% with the calculated values, and gives accurate results for atoms near the bottom of the trapping potential. From the measured frequencies we calculate the potential curvatures, which in combination with measurements of temperature, atom number, and vertical cloud size are used to calculate the spatial and phase-space densities for the trapped gas.

Immediately after filling the trap, the axial temperature is 3 μK. The radial directions each have an initial mean kinetic energy of $W/h = 33$ kHz, and cannot be assigned a temperature for storage times shorter than the radial oscillation period. Within 150 ms the axial and the radial temperatures take on steady-state values of 15 μK. We believe that the observed increase is mainly due to the addition of trap potential energy during loading which is subsequently converted into kinetic energy.

For the degenerate Raman sideband cooling we apply a magnetic field of magnitude $B = 4hf_{ax}/\mu_B$, typically 230 mG, where μ_B is the Bohr magneton. The optical pumping beam is applied along the z direction with elliptical polarization such that it has both σ^+ and π components along the quantization axis defined by the external magnetic field. This mixture is necessary because the optical pumping is performed on the $6S_{1/2}$, $F_g = 3$ to $6P_{3/2}$, $F_e = 2$ transition to minimize excitation

to the $F_g = 4$ ground state. We have experimented with cooling on the $F_g = 3$ to $F_e = 3$ transition, and have observed a strongly reduced efficiency, most likely because of the longer time the atoms spend in this state. In both cases, an additional repumping beam is applied on the $F_g = 4$ to $F_e = 3$ transition. We observe cooling for large ranges of parameters, however, the lowest 3D temperatures were obtained for $\alpha = 70°$ and $\beta = 8°$, which corresponds to a Rabi frequency of $\Omega/2\pi = 1.4$ kHz for the Raman transitions. The optical pumping rate from $F = 3$, $m_F = 2$ to $m_F = 3$ was $\Gamma_{32} = 70$ s^{-1} at an intensity of 8 μW/cm^2. Figure 3 shows the cooling resonance as a function of the Zeeman splitting for 200 ms of cooling time. The asymmetric shape is due to inhomogeneous broadening of the axial vibrational levels in the Gaussian beam profile of the trap. The second peak corresponds to twice the axial vibration frequency and represents cooling on the degenerate $|3,3;\nu\rangle$ to $|3,2;\nu - 2\rangle$ transition.

The evolution of the axial and radial temperatures vs cooling time is shown in Fig. 4. The initial cooling rate in the Lamb-Dicke direction is 11 mK/s. This rate increases with optical pumping intensity, but only at the expense of a less efficient radial cooling, probably resulting from increased radial recoil heating. When the axial temperature is cooled below the radial, the latter follows with a time constant of 50 ms. This coupling has been observed directly by heating the axial direction for a short time and monitoring the subsequent rethermalization. When the density is reduced this time constant becomes longer, and we therefore conclude that the coupling is due to collisions rather than to anharmonic mixing of the axial and radial motion. It should be noted, however, that the thermalization time is much longer than the estimated collision time of 1 ms at our highest densities. One possible explanation is that at temperatures comparable to or smaller than the axial vibrational spacing it may take a larger number of collisions to repopulate higher axial vibrational states.

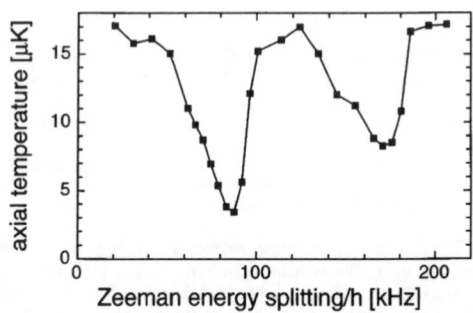

FIG. 3. Axial temperature after 200 ms of degenerate Raman sideband cooling as a function of the Zeeman energy difference between $m = 2$ and $m = 3$.

VOLUME 81, NUMBER 26   PHYSICAL REVIEW LETTERS   28 DECEMBER 1998

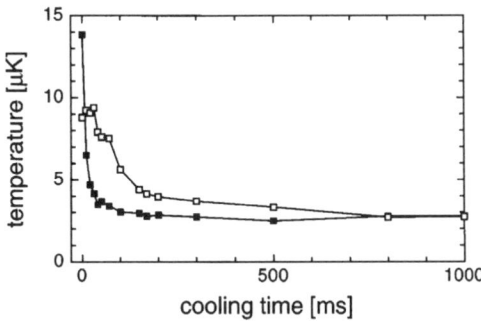

FIG. 4. Evolution of the axial (solid squares) and radial (open squares) temperatures. The degenerate Raman sideband cooling is applied only along the steep axial direction.

The lowest axial and radial temperatures obtained are $T_z = 2.5$ μK and $T_\rho = 3.0$ μK, respectively, after 500 ms of cooling. With the directly measured values for the vibration frequencies of 80 kHz and 115 Hz we calculate 80% population in the ground state of the Lamb-Dicke direction and a peak phase space density of 1/180 for the traps in the central region. To our knowledge, this is the highest phase space density yet obtained by an optical method. The final peak spatial density is 1.4×10^{13} cm^{-3}. The estimated uncertainties for the temperature, phase space density, and spatial density are 10%, 40%, and 25%, respectively. Finally, the axial motion can be further cooled to 340 nK by adiabatic expansion with a turn-off time constant of 130 μs.

Atoms were previously sideband cooled in a 2D lattice without confinement or temperature measurement in the third dimension, yielding a 2D kinetic temperature of 0.97 μK and a 2D ground state population exceeding 95% [10]. In a 1D lattice trap a 1D temperature of 3.6 μK has been achieved, at the expense of heating the two radial directions to 24 μK [11]. In contrast, relying on collisional coupling at high densities, we cool all three dimensions to 2.8 μK by 1D sideband cooling. Our cooling performance may therefore be limited by density-related processes such as reabsorption of spontaneous photons that compromise the optical pumping and thus the sideband cooling. In addition, we observe a small heating rate of 4 μK/s independent of the cooling, which in conjunction with our low final cooling rates might be determining our present temperature limit. We have investigated possible heating sources associated with trapping light fluctuations [19]. All of the heating rates as calculated from measured noise spectra are smaller than 0.1 μK/s. We are currently investigating higher-order contributions form noise at subharmonics of the trap frequencies.

In conclusion, we have cooled a tightly trapped gas of neutral atoms in 3D at a density for which the calculated elastic collision rate exceeds 1 kHz. The optical cooling of a very dense, spin-polarized gas may represent an important step towards attaining BEC by all-optical means and provides favorable starting conditions for evaporative cooling.

This work was supported in part by grants from the AFOSR and the NSF. V. V. acknowledges support from the Humboldt Foundation and C. C. from the Taiwan government.

[1] M. H. Anderson *et al.,* Science **269**, 198 (1995); C. C. Bradley *et al.,* Phys. Rev. Lett. **75**, 1687 (1995); K. B. Davis *et al.,* Phys. Rev. Lett. **75**, 3969 (1995).

[2] T. Walker, D. Sesko, and C. Wieman, J. Opt. Soc. Am. B **8**, 946 (1991).

[3] J. Söding *et al.,* Phys. Rev. Lett. **80**, 1869 (1998); D. Guéry-Odelin *et al.,* Opt. Express **2**, 323 (1998).

[4] S. J. J. M. F. Kikkelmans, B. J. Verhaar, and K. Gibble, Phys. Rev. Lett. **81**, 951 (1998).

[5] C. J. Myatt *et al.,* Phys. Rev. Lett. **78**, 586 (1997); F. A. van Abeelen, B. J. Verhaar, and A. J. Moerdijk, Phys. Rev. A **55**, 4377 (1997).

[6] M. Ol'shanii, Y. Castin, and J. Dalibard, in *Proceedings of the 12th International Conference on Laser Spectroscopy,* edited by M. Inguscio, M. Allegrini, and A. Lasso (World Scientific, Singapore, 1996).

[7] U. Janicke and M. Wilkens, Europhys. Lett. **35**, 561 (1996); J. I. Cirac, M. Lewenstein, and P. Zoller, Europhys. Lett. **35**, 647 (1996); G. Morigi *et al.,* Europhys. Lett. **39**, 13 (1997); Y. Castin, J. I. Cirac, and M. Lewenstein, Phys. Rev. Lett. **80**, 5305 (1998).

[8] V. Bagnato, D. E. Pritchard, and D. Kleppner, Phys. Rev. A **35**, 4354 (1987).

[9] C. Monroe *et al.,* Phys. Rev. Lett. **75**, 4011 (1995).

[10] S. E. Hamann *et al.,* Phys. Rev. Lett. **80**, 4149 (1998).

[11] H. Perrin *et al.,* Europhys. Lett. **42**, 395 (1998).

[12] H. J. Lee *et al.,* Phys. Rev. Lett. **76**, 2658 (1996); N. R. Newbury *et al.,* Phys. Rev. Lett. **74**, 2196 (1995).

[13] I. H. Deutsch and P. S. Jessen, Phys. Rev. A **57**, 1972 (1998).

[14] We have also observed $\nu \to \nu - 2$ cooling in a circularly polarized intensity lattice with final temperatures roughly twice as large as those obtained with $\nu \to \nu - 1$ cooling.

[15] Here we have neglected effects at the 10% level due to the tensor light shift [12], which introduces small differences between the potentials for different m levels.

[16] D. Boiron *et al.,* Phys. Rev. A **57**, R4106 (1998).

[17] P. Lemonde *et al.,* Europhys. Lett. **32**, 555 (1995).

[18] S. Friebel *et al.,* Phys. Rev. A **57**, R20 (1998).

[19] T. Savard, K. M. O'Hara, and J. E. Thomas, Phys. Rev. A **56**, R1095 (1997).

VOLUME 80, NUMBER 10   PHYSICAL REVIEW LETTERS   9 MARCH 1998

Optical Confinement of a Bose-Einstein Condensate

D. M. Stamper-Kurn, M. R. Andrews, A. P. Chikkatur, S. Inouye, H.-J. Miesner, J. Stenger, and W. Ketterle

Department of Physics and Research Laboratory of Electronics, Massachusetts Institute of Technology,
Cambridge, Massachusetts 02139
(Received 7 November 1997)

Bose-Einstein condensates of sodium atoms have been confined in an optical dipole trap using a single focused infrared laser beam. This eliminates the restrictions of magnetic traps for further studies of atom lasers and Bose-Einstein condensates. More than 5×10^6 condensed atoms were transferred into the optical trap. Densities of up to 3×10^{15} cm^{-3} of Bose condensed atoms were obtained, allowing for a measurement of the three-body loss rate constant for sodium condensates as $K_3 = 1.1(3) \times 10^{-30}$ cm^6 s^{-1}. At lower densities, the observed $1/e$ lifetime was longer than 10 s. Simultaneous confinement of Bose-Einstein condensates in several hyperfine states was demonstrated. [S0031-9007(98)05537-9]

PACS numbers: 03.75.Fi, 05.30.Jp, 32.80.Pj, 64.60.–i

The recent realization of Bose-Einstein condensation [1–3] and of an atom laser [4,5] have sparked many theoretical and experimental studies of coherent atomic matter [6]. Yet, these studies are limited by the magnetic traps used by all experiments so far. For example, in the first demonstration of an atom laser, coherent atomic pulses were coupled out into an inhomogeneous magnetic field, which served to confine the remaining condensate. Thus, during propagation, the pulses were exposed to Zeeman shifts. While these shifts were mitigated by producing $m_F = 0$ atoms, quadratic Zeeman shifts may preclude precision experiments on such pulses. Magnetic trapping also imposes limitations on the study of Bose-Einstein condensates, because only the weak-field seeking atomic states are confined. Since the atomic ground state is always strong-field seeking, weak-field seeking states can inelastically scatter into the ground state (dipolar relaxation), resulting in heating and trap loss. Furthermore, trap loss may dramatically increase through spin relaxation collisions when different hyperfine states are simultaneously trapped, restricting the study of multicomponent condensates. Although in ^{87}Rb this increase is less dramatic due to a fortuitous cancellation of transition amplitudes [7], spin relaxation is still the dominant decay mechanism for double condensates in this system.

All these problems are avoided if Bose-Einstein condensation is achieved in an optical trap based on the optical dipole force which confines atoms in all hyperfine states. This has been one motivation for the development of subrecoil cooling techniques [8,9], the development of various optical dipole traps [10–14], and for pursuing Raman cooling [15,16] and evaporative cooling [17] in such traps. The highest phase space density achieved by purely optical means was a factor of 400 below that required for Bose-Einstein condensation [15].

In this paper, we report the successful optical trapping of a Bose-Einstein condensate using a different approach: first evaporatively cooling the atoms in a magnetic trap, and then transferring them into an optical trap. This approach circumvents many difficulties usually encountered with optical dipole traps. Since the temperature of atoms is reduced through rf evaporation by a factor of 100, only milliwatts of laser power are needed as compared with several watts used to directly trap laser-cooled atoms. This ameliorates trap loss from heating processes in an optical dipole trap which are proportional to laser power, such as off-resonant Rayleigh scattering, and heating due to fluctuations in the intensity and position of the laser beam [18]. Furthermore, since the cloud shrinks while being cooled in the magnetic trap, the transfer efficiency into the small trapping volume of an optical dipole trap is increased.

The experimental setup for creating Bose-Einstein condensates was similar to our previous work [19]. Sodium atoms were optically cooled and trapped, and transferred into a magnetic trap where they were further cooled by rf-induced evaporation. The transition point was reached at densities of $\sim 1 \times 10^{14}$ cm^{-3} and temperatures of 1–2 μK. Further evaporation produced condensates containing $5 - 10 \times 10^6$ atoms in the $F = 1, m_F = -1$ electronic ground state. The atom clouds were cigar-shaped with the long axis horizontal, and had a typical aspect ratio of 15 due to the anisotropic trapping potential of the cloverleaf magnetic trap.

The optical trap was formed by focusing a near-infrared laser beam into the center of the magnetic trap along the axial direction. For this, the output of a diode laser operating at 985 nm was sent through a single-mode optical fiber and focused to a spot with a beam-waist parameter w_0 ($1/e^2$ radius for the intensity) of about 6 μm. This realized the simple single-beam arrangement for an optical dipole trap [10–13]. The infrared laser focus and the atom cloud were overlapped in three dimensions by imaging both with a CCD camera. It was necessary to compensate for focal and lateral chromatic shifts of the imaging system which were measured using an optical test pattern illuminated either at 589 or 985 nm.

VOLUME 80, NUMBER 10   PHYSICAL REVIEW LETTERS   9 MARCH 1998

The parameters of the optical trapping potential are characterized by the total laser power P and the beam-waist parameter w_0. The trap depth is proportional to P/w_0^2. For a circular Gaussian beam with $w_0 = 6\ \mu m$, the trap depth is $1\ \mu K/mW$ [20], and the aspect ratio of the atom cloud is 27. At $P = 4\ mW$, the geometric mean trapping frequency $\bar{\nu}$ is 670 Hz. The measured frequencies [see Eq. (1)] were about half the expected values, presumably due to imperfect beam quality and the coarse measurements of P and w_0 (an underestimation of w_0 by 40% would account for this discrepancy). Finally, due to the large detuning, the spontaneous scattering rate is small ($5 \times 10^{-3}\ s^{-1}$ per μK trap depth), leading to an estimated trapping time of 400 s.

Condensates were transferred into the optical trap by holding them in a steady magnetic trap while ramping up the infrared laser power, and then suddenly switching off the magnetic trap. A ramp-up time of 125 ms was chosen as slow enough to allow for adiabatic transfer, yet fast enough to minimize trap loss during the ramp-up due to high densities in the combined optical and magnetic traps. The highest transfer efficiency (85%) was observed for a laser power of about 4 mW, with a measured mean trapping frequency $\bar{\nu} = 370\ Hz$ [see Eq. (1)]. The transfer efficiency dropped for higher laser power due to trap loss during the ramp-up, and decreased rapidly for smaller laser power due to the smaller trap depth. The sudden switch-off of the magnetic fields was necessitated by imperfections in the trapping coils which displaced the center of the magnetic trap during a slow switch-off. This caused transient oscillations of the atom cloud and can be overcome in the future with auxiliary steering coils.

After 500 ms of purely optical trapping, the transferred atoms were probed by suddenly switching off the optical trap, and observing the freely expanding cloud using absorption imaging (Fig. 1). The strong anisotropic expansion after 40 ms time of flight is characteristic of Bose-Einstein condensates in strongly anisotropic trapping potentials.

We also loaded the optical trap with magnetically trapped atoms at higher temperatures and lower densities than those used in Fig. 1. In this case, depending on the temperature of the atoms loaded into the optical trap, we observed the sudden onset of a dense, low energy core of atoms amidst a broad background of noncondensed atoms (Fig. 2). Two aspects are worth noting. First, the number of thermal atoms ($\sim 10^4$) is small compared both to the number of atoms before transfer ($\sim 10^8$) and to the number of condensed atoms transferred under the optimum conditions described above ($\sim 10^6$). This is due to the small trapping volume and shallow trap depth of the optical trap which leads to a very small transfer efficiency for thermal atoms. In comparison, in magnetic traps, the number of noncondensed atoms at the transition temperature is much larger than the largest number of condensed atoms eventually produced [19,21]. The maxi-

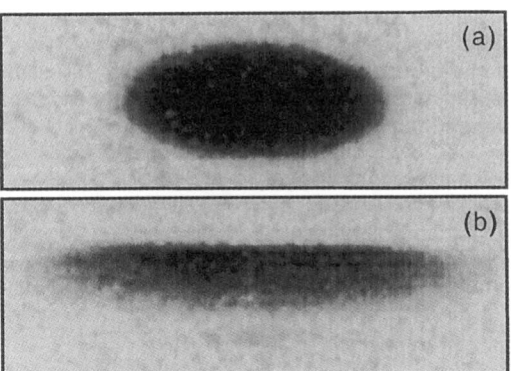

FIG. 1. Absorption images of expanding Bose-Einstein condensates, released (a) from the magnetic trap, and (b) from the optical trap ($\bar{\nu} = 370\ Hz$, $P \sim 4\ mW$). The faster expansion in (b) is indicative of the higher densities of the optical trap. The time of flight was 40 ms. The field of view for each image is 2.2 by 0.8 mm.

mum number of thermal atoms in Fig. 2 was measured to be 24 000, in agreement with a prediction based on the observed trap depth and trapping frequencies, and the assumption that the thermal atoms arrive at a temperature 1/10 of the trap depth by evaporation. Second, condensates were observed in the optical trap even when there was no condensate in the magnetic trap from which it was loaded. This was due to the partial condensation of the gas during the adiabatic deformation of the trapping potential when the infrared light was ramped up. During this deformation, the entropy remained constant through collisional equilibrium while the phase space density increased [22]. A detailed study of this effect will be reported elsewhere.

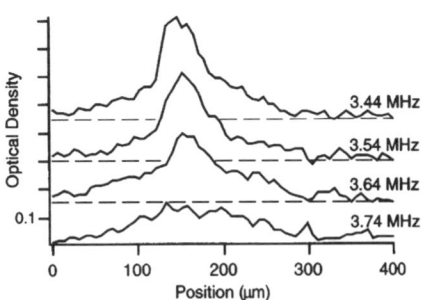

FIG. 2. Optical density profiles of optically trapped atoms at the onset of Bose-Einstein condensation. Because of the short time of flight (1 ms), the profiles show the spatial distribution along the long axis of the optical trap. Data were obtained after 125 ms of purely optical trapping. Labels give the final rf frequency used in the evaporative cooling cycle. The bimodality observed below 3.74 MHz indicates the presence of a small ($\sim 10^4$ atoms) Bose-Einstein condensate.

VOLUME 80, NUMBER 10   PHYSICAL REVIEW LETTERS   9 MARCH 1998

After the trap is switched off, the internal repulsive (mean-field) energy of the condensate is transformed into kinetic energy of the expanding cloud. This allows for the determination of peak densities n_0 and mean trapping frequencies $\bar{\nu}$ from time-of-flight data [19]. For a harmonic trapping potential in the Thomas-Fermi approximation, the average mean-field energy per atom is $2/7\, n_0 \tilde{U}$, where $\tilde{U} = 4\pi\hbar^2 a/m$ is proportional to the scattering length $a = 2.75$ nm [23]. Assuming a predominantly radial expansion, the peak density was determined from the maximum velocity observed in time-of-flight images v_{max} by $n_0 \tilde{U} = m v_{max}^2/2$. The number of condensed atoms N was measured by integrating the optical density in time-of-flight images. The mean trapping frequencies $\bar{\nu}$ are related to N and n_0 by [19]

$$\bar{\nu} = 0.945 \frac{\hbar \sqrt{a}}{m} n_0^{5/6} N^{-1/3}. \qquad (1)$$

The density of condensates in the optical trap was varied by either doubling or halving the infrared power in the all-optical trap, after having transferred the atoms at settings which maximized the initial transfer efficiency (see above). Thereafter, the infrared power was kept constant for lifetime studies. The peak densities achieved in this manner ranged from 3×10^{14} cm^{-3} in the weakest optical trap to 3×10^{15} cm^{-3} in the tightest. For the lowest infrared power used ($P \sim 2$ mW), atoms were observed spilling out of the optical trap, indicating that the depth of the trap was comparable to the 200 nK mean-field energy of the condensate which remained.

The lifetime of condensates was studied by measuring the number of condensed atoms in time-of-flight images after a variable storage time in the optical trap. Results are shown in Fig. 3 and compared to those for the magnetic trap. The lifetime in the magnetic trap was

very short unless the trap depth was lowered by "rf shielding" [19,24], allowing collisionally heated atoms to escape. Similarly, the long lifetimes observed in the optical trap were made possible by its limited trap depth. The observed loss rates per atom in the optical trap ranged from 4 s^{-1} at a peak density $n_0 = 3 \times 10^{15}$ cm^{-3} to less than $1/10$ s^{-1} at $n_0 = 3 \times 10^{14}$ cm^{-3}.

The decay curves in Fig. 3 are described by

$$\frac{dN}{dt} = -K_1 N - K_3 N \langle n^2 \rangle, \qquad (2)$$

where K_1 accounts for density independent loss processes such as residual gas scattering, Rayleigh scattering, and other external heating processes, and K_3 is the loss rate constant for three-body decay. For a harmonic trap, the mean squared density $\langle n^2 \rangle$ is related to the measured peak density by $\langle n^2 \rangle = 8/21\, n_0^2$ [25].

Three-body decay was found to be the dominant loss mechanism in both the optical and the magnetic trap. By fitting the solution of Eq. (2) to the decay curves for the various optical traps we obtained $K_1 = 0.03(2)$ s^{-1} and $K_3 = 1.1(3) \times 10^{-30}$ cm^6 s^{-1}. This three-body loss rate constant for ^{23}Na is a factor of 5 smaller than for ^{87}Rb [24], and can be ascribed completely to collisions among condensed atoms due to the small number of noncondensed atoms in the optical trap. Our result lies between two theoretical predictions for the loss rate constant of $K_3 = 3 \times 10^{-29}$ cm^6 s^{-1} [26] and $K_3 = 3.9\hbar a^4/2m = 3 \times 10^{-31}$ cm^6 s^{-1} [27,28]. The loss rate due to dipolar relaxation (two-body decay) was predicted to be negligible at the densities considered [29]. While the decay curves show three-body decay to be the dominant loss mechanism, they do not exclude two-body decay rates comparable to K_1.

One major advantage of the optical trap over magnetic traps is its ability to confine atoms in arbitrary hyperfine states. To demonstrate this, the atoms were put into a superposition of $F = 1$ hyperfine states by applying an rf field which was swept from 0 to 2 MHz in 2 ms. Parameters were chosen in such a way that the sweep was neither adiabatic nor diabatic, similar to our work on the rf output coupler [5]. The distribution over hyperfine states was analyzed through Stern-Gerlach separation by pulsing on a magnetic field gradient of a few G/cm during the 40 ms time of flight. Figure 4 demonstrates that all three states were optically trapped. By extending the time between the rf sweep and the probing, we confirmed that all $F = 1$ hyperfine states were stored stably for several seconds. The optical potential for the three sublevels is nearly identical, the relative difference being less than the ratio of the fine structure splitting to the large detuning (2.5×10^{-3}).

In conclusion, we have realized an optical trap for Bose-Einstein condensates. Because of the low energy of the condensates, just milliwatts of far-detuned laser radiation were sufficient to provide tight confinement.

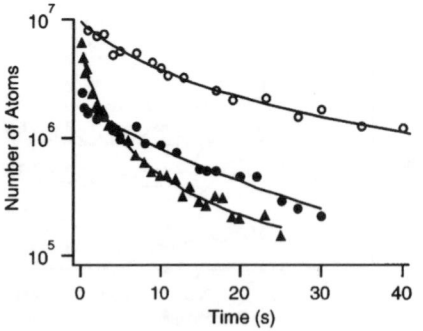

FIG. 3. Lifetime of Bose-Einstein condensates in the optical and magnetic traps. Shown is the number of condensed atoms vs trapping time. Closed triangles and circles represent data for the optical traps with the best transfer efficiency ($\bar{\nu} = 370$ Hz, $P \sim 4$ mW) and the slowest decay (weakest trap, $P \sim 2$ mW), respectively. Open circles represent data for the rf-shielded magnetic trap. Error in the number measurements is estimated as 10%. Lines are fits based on Eq. (2).

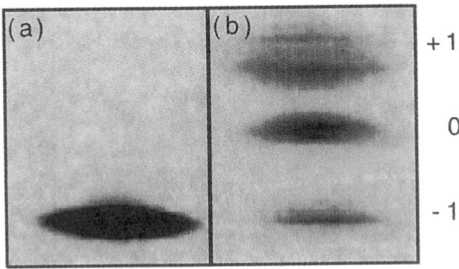

FIG. 4. Optical trapping of condensates in all $F = 1$ hyperfine states. Shown are absorbtion images after (a) 250 ms, and (b) 340 ms of optical confinement. Hyperfine states were separated by a magnetic field gradient pulse during the 40 ms time of flight. Atoms remain spin polarized in the optical trap (a). In (b) the atoms were exposed to an rf sweep which populated all hyperfine states, 90 ms before release from the trap. The absorption of the σ^- probe light is weaker for the $m_F = 0, +1$ states than for the $m_F = -1$ state. The field of view for each image is 1.6 by 1.8 mm.

More than five million condensed atoms were trapped, and lifetimes comparable to those in our dc magnetic trap were observed. Densities of 3×10^{15} cm^{-3} were achieved, unprecedented for both Bose condensates and for optically trapped atomic clouds. High densities and high condensate fractions allowed for a determination of the three-body loss rate constant in sodium as $K_3 = 1.1(3) \times 10^{-30}$ cm^6 s^{-1}. This trap offers many new opportunities to study Bose-Einstein condensates and atom lasers. Since the optical trap works at arbitrary external magnetic fields, Feshbach resonances in the scattering length [30] might now be observed for strong-field seeking states. One can study condensates in superpositions of degenerate hyperfine sublevels, and, since the spin degree of freedom is no longer constrained by magnetic trapping, it may be possible to study spin waves [31] in a Bose-condensed gas. The shallow and well controlled trap depth allows for new output-coupling schemes [32]. Finally, the optical trap may also serve as an "optical tweezers" to move condensates, and, for example, place them in optical and microwave cavities and close to surfaces.

We are grateful to Günter Steinmeyer, Erik Thoen, and Erich Ippen for their help with the infrared laser. This work was supported by the Office of Naval Research, NSF, Joint Services Electronics Program (ARO), and the David and Lucile Packard Foundation. A. P. C. and D. M. S.-K. would like to acknowledge support from the NSF, and J. S. from the Humboldt Foundation.

[1] M. H. Anderson *et al.,* Science **269**, 198 (1995).
[2] K. B. Davis *et al.,* Phys. Rev. Lett. **75**, 3969 (1995).
[3] C. C. Bradley, C. A. Sackett, and R. G. Hulet, Phys. Rev. Lett. **78**, 985 (1997); see also C. C. Bradley *et al.,* Phys. Rev. Lett. **75**, 1687 (1995).
[4] M. R. Andrews *et al.,* Science **275**, 637 (1997).
[5] M.-O. Mewes *et al.,* Phys. Rev. Lett. **78**, 582 (1997).
[6] Proceedings of the Workshop on Bose-Einstein Condensation, Castelvecchio, Italy, 1997, Book of Abstracts (unpublished).
[7] C. J. Myatt *et al.,* Phys. Rev. Lett. **78**, 586 (1997).
[8] A. Aspect *et al.,* Phys. Rev. Lett. **61**, 826 (1988).
[9] M. Kasevich and S. Chu, Phys. Rev. Lett. **69**, 1741 (1992).
[10] S. Chu, J. E. Bjorkholm, A. Ashkin, and A. Cable, Phys. Rev. Lett. **57**, 314 (1986).
[11] W. D. Phillips, in *Laser Manipulation of Atoms and Ions,* Proceedings of the International School of Physics "Enrico Fermi," Course CXVIII, edited by E. Arimondo, W. D. Phillips, and F. Strumia (North-Holland, Amsterdam, 1992), p. 289.
[12] J. D. Miller, R. A. Cline, and D. J. Heinzen, Phys. Rev. A **47**, R4567 (1993).
[13] T. Takekoshi and R. J. Knize, Opt. Lett. **21**, 77 (1996).
[14] T. Kuga *et al.,* Phys. Rev. Lett. **78**, 4713 (1997).
[15] H. J. Lee *et al.,* Phys. Rev. Lett. **76**, 2658 (1996).
[16] A. Kuhn, H. Perrin, W. Hänsel, and C. Salomon, in *Ultracold Atoms and Bose-Einstein Condensation,* edited by K. Burnett, OSA Trends in Optics and Photonics Series Vol. 7 (Optical Society of America, Washington, DC, 1996), p. 58.
[17] C. S. Adams *et al.,* Phys. Rev. Lett. **74**, 3577 (1995).
[18] T. A. Savard, K. M. O'Hara, and J. E. Thomas, Phys. Rev. A **56**, R1095 (1997).
[19] M.-O. Mewes *et al.,* Phys. Rev. Lett. **77**, 416 (1996).
[20] 25% of the trap depth comes from the "counter-rotating" term usually neglected in the rotation-wave approximation.
[21] D. S. Jin *et al.,* Phys. Rev. Lett. **78**, 764 (1997).
[22] P. W. H. Pinske *et al.,* Phys. Rev. Lett. **78**, 990 (1997).
[23] E. Tiesinga *et al.,* J. Res. Natl. Inst. Technol. **101**, 505 (1996).
[24] E. A. Burt *et al.,* Phys. Rev. Lett. **79**, 337 (1997).
[25] The assumption of a harmonic trap holds when the mean-field energy is much smaller than the trap depth. In determining K_3, we considered only data for which this was true. We estimate the effects of anhamonicity on our measured K_3 to be less than 10%.
[26] A. J. Moerdijk, H. M. J. M. Boesten, and B. J. Verhaar, Phys. Rev. A **53**, 916 (1996).
[27] P. O. Fedichev, M. W. Reynolds, and G. V. Shlyapnikov, Phys. Rev. Lett. **77**, 2921 (1996).
[28] References [26] and [27] give the three-body *event* rate constant for thermal atoms which is $2K_3$, since the event rate is 6 times larger for thermal atoms than for condensate atoms [24], while three atoms are lost per collision event.
[29] H. M. J. M. Boesten, A. J. Moerdijk, and B. J. Verhaar, Phys. Rev. A **54**, R29 (1996).
[30] E. Tiesinga, B. J. Verhaar, and H. T. C. Stoof, Phys. Rev. A **47**, 4114 (1993).
[31] B. R. Johnson *et al.,* Phys. Rev. Lett. **52**, 1508 (1984); P. J. Nacher *et al.,* J. Phys. (Paris), Lett. **45**, L441 (1984); W. J. Gully and W. J. Mullin, Phys. Rev. Lett. **52**, 1810 (1984).
[32] G. M. Moy, J. J. Hope, and C. M. Savage, Phys. Rev. A **55**, 3631 (1997).

letters to nature

Spin domains in ground-state Bose–Einstein condensates

J. Stenger, S. Inouye, D. M. Stamper-Kurn, H.-J. Miesner, A. P. Chikkatur & W. Ketterle

Department of Physics and Research Laboratory of Electronics, Massachusetts Institute of Technology, Cambridge, Massachusetts 02139, USA

Bose–Einstein condensates—a low-temperature form of matter in which a macroscopic population of bosons occupies the quantum-mechanical ground state—have been demonstrated for weakly interacting, dilute gases of alkali-metal[1–3] and hydrogen[25] atoms. Magnetic traps are usually employed to confine the condensates, but have the drawback that spin flips in the atoms lead to untrapped states. For this reason, the spin orientation of the trapped alkali atoms cannot be regarded as a degree of freedom. Such condensates are therefore described by a scalar order parameter, like the spinless superfluid ^{4}He. In contrast, a recently realized optical trap[4] for sodium condensates confines atoms independently of their spin orientations. This offers the possibility of studying 'spinor' condensates in which spin comprises a degree of freedom, so that the order parameter is a vector rather than scalar quantity. Here we report the observation of equilibrium states of sodium spinor condensates in an optical trap. The freedom of spin orientation leads to the formation of spin domains in an external magnetic field, which can be either miscible or immiscible with one another.

A variety of new phenomena are predicted[5–7] for spinor condensates, such as spin textures, propagation of spin waves and coupling between superfluid flow and atomic spin. To date, such effects could only be studied in superfluid ^{3}He, which can be

letters to nature

described by Bose–Einstein condensation of Cooper pairs of quasi-particles having both spin and orbital angular momentum[8]. Compared to the strongly interacting [3]He, the properties of weakly interacting Bose–Einstein condensates of alkali-metal gases can be calculated by mean field theories in a much more straightforward and simple way.

Other systems which go beyond the description with a single scalar order parameter are condensates of two different hyperfine states of [87]Rb confined in magnetic traps. Recent experimental studies have explored the spatial separation of the two components[9,10] and their relative phase[11]. Several theoretical papers describe their structure[12–18] and their collective excitations[19–22].

Compared to these two-component condensates, spinor condensates have several new features, including the vector character of the order parameter and the changed role of spin relaxation collisions which allow for population exchange among hyperfine states without trap loss. In contrast, for [87]Rb experiments trap loss due to spin relaxation severely limits the lifetime.

Here we consider an $F = 1$ spinor condensate subject to spin relaxation, in which two $m_F = 0$ atoms can collide and produce an $m_F = +1$ and an $m_F = -1$ atom and vice versa. (Here, F denotes the angular momentum per atom and m_F its magnetic quantum number.) We investigate the distribution of hyperfine states and the spatial distribution in equilibrium assuming conservation of the total spin.

The ground state spinor wavefunction is found by minimizing the free energy[5];

$$K = \int d^3 r\, n \left[V + \frac{c_0 n}{2} + \frac{c_2 n}{2}\langle \mathbf{F} \rangle^2 + E_{ze} - p_0 \langle F_z \rangle \right] \quad (1)$$

where kinetic energy terms are neglected in the Thomas–Fermi approximation which is valid as long as the dimension of spin domains (typically 50 μm) is larger then the penetration depth[18] (typically 1 μm). V is trapping potential, n is the density, r is the spatial coordinate and E_{ze} is the Zeeman energy in an external magnetic field. The Lagrange multiplier p_0 accounts for the total spin conservation. The mean field energy in equation (1) consists of a spin-independent part proportional to c_0 and a spin-dependent part proportional to $c_2 \langle \mathbf{F} \rangle^2$. The coefficients c_0 and c_2 are related to the scattering lengths a_0 and a_2 for two colliding atoms with total angular momentum $F_{tot} = 0$ or $F_{tot} = 2$ by $c_0 = 4\pi\hbar^2 \bar{a}/M$, $c_2 = 4\pi\hbar^2 \Delta a/M$ with $\bar{a} = (2a_2 + a_0)/3$, $\Delta a = (a_2 - a_0)/3$, and M for the atomic mass (ref. 5). The spin-dependent interaction originates from the term $c_2 \mathbf{F}_1 \cdot \mathbf{F}_2$ in the interaction of two atoms, which is ferromagnetic for $c_2 < 0$ and antiferromagnetic for $c_2 > 0$.

In the Bogoliubov approach, the many-body ground-state wavefunction is represented by the spinor wavefunction;

$$\Psi(\mathbf{r}) = \sqrt{n(\mathbf{r})}\,\zeta(\mathbf{r}) = \sqrt{n(\mathbf{r})}(\zeta_+(\mathbf{r}), \zeta_0(\mathbf{r}), \zeta_-(\mathbf{r})) \quad (2)$$

where ζ_+, ζ_0, ζ_- denote the amplitudes for the $m_F = +1, 0, -1$ states, respectively, and $|\zeta|^2 = 1$.

The Zeeman energy E_{ze} is given by;

$$E_{ze} = E_+|\zeta_+|^2 + E_0|\zeta_0|^2 + E_-|\zeta_-|^2 = E_0 - \tilde{p}\langle F_z \rangle + q\langle F_z^2 \rangle \quad (3)$$

where E_+, E_0, E_- are the Zeeman energies of the $m_F = +1, 0, -1$ states, $2q \equiv E_+ + E_- - 2E_0$ is the Zeeman energy difference in a spinflip collision, and $2\tilde{p} \equiv E_- - E_+$. The E_0 term can be included in the trapping potential V. The parameter $\tilde{p}$ can be combined with the Lagrange multiplier p_0 to give $p \equiv \tilde{p} + p_0$.

In the following we determine the spinor which minimizes the spin-dependent part K_s of the free energy:

$$K_s = c\langle \mathbf{F} \rangle^2 - p\langle F_z \rangle + q\langle F_z^2 \rangle \quad (4)$$

where $c = c_2 n/2$. The minimization of equation (4) for different values of the parameters c, p and q is straightforward, and is shown

graphically in the form of spin-domain diagrams in Fig. 1.

Experimentally, the values of c, p and q can be varied arbitrarily, representing any region of the spin-domain diagram. The magnitude (but not the sign) of the coefficient c is varied by changing the density n, either by changing the trapping potential, or by studying condensates with different numbers of atoms. In this study, the axial length of the trapped condensate is more than 60 times larger than its radial size, and thus we consider the system one-dimensional, and integrate over the radial coordinates, obtaining $n = 2n_0/3$ where n_0 is the density at the radial centre. This integration assumes a parabolic density profile within the Thomas–Fermi approximation. The value of q can be changed by applying a weak external bias field B_0: q then corresponds to the quadratic Zeeman shift which is proportional to B_0^2. The coefficient p arises both from the linear Zeeman shift and from the Lagrange multiplier p_0 which is determined by the total spin of the system. For a system with zero total spin in a homogeneous bias field B_0, p_0 cancels the linear Zeeman shift due to B_0, yielding $p = 0$. Positive (negative) values of p are achieved for condensates with a positive (negative) overall spin. Finally, the coefficients can be made to vary spatially across the condensate. In particular, applying a field gradient B' along the axis of the trapped condensate causes p to vary along the condensate length. For a condensate with zero total spin, p is proportional to $B'z$ where z is the axial coordinate with $z = 0$ at the centre of the condensate. Thus, the condensate samples a vertical line in the spin-domain diagrams of Fig. 1. The centre of this line lies at $p = 0$, and its length is given by the condensate length scaled by B'.

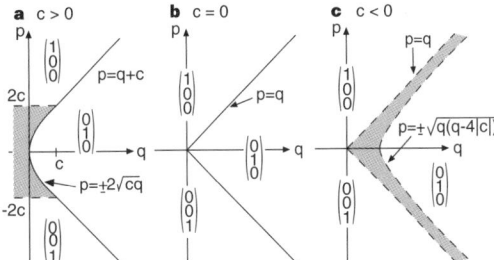

Figure 1 Spin-domain diagrams for condensates with $F = 1$. The structure of the ground-state spinor is shown as a function of the linear ($\sim p$) and quadratic ($\sim q$) Zeeman energies. Hyperfine components are mixed inside the shaded regions. Solid lines indicate a discontinuous change of state populations whereas dashed lines indicate a gradual change. The behaviour for $q < 0$ is also shown, although it is not relevant for the experiment reported here. For $c = 0$, the Zeeman energy causes the cloud to separate into three domains with $m_F = +1, 0, -1$ and with boundaries at $|p| = q$, as shown in **b**. For $c_2 \neq 0$, the mean field energy shifts the boundary region between domains and leads to regions of overlapping spin components. In the antiferromagnetic case (**a**), the $m_F = 0$ component and the $m_F = \pm 1$ components are immiscible (including the kinetic energy terms in equation (1) would lead to a thin boundary layer) and the boundary occurs at $|p| = q + c$. For small bias fields, with $q < c$ and $|p| < 2c$, the $m_F = 0$ domain is bordered by domains in which $m_F = \pm 1$ components are mixed. The ratio of the $m_F = \pm 1$ populations in these regions does not depend on q, and is given by $|\zeta_+|^2/|\zeta_-|^2 = (2c + p)/(2c - p)$. In this region of small fields, the boundary to the $m_F = 0$ component lies at $|p| = 2\sqrt{cq}$. In the ferromagnetic case (**c**) all three components are generally miscible, and have no sharp boundaries. Pure $m_F = 0$ domains occur for $|p| \leq \sqrt{q(q - 4|c|)}$ and pure $m_F = \pm 1$ domains for $|p| > q$. Here, in contrast to the antiferromagnetic case, a pure $m_F = 0$ condensate is skirted by regions where it is mixed predominantly with either the $m_F = -1$ or the $m_F = +1$ component. The contribution of the third component is very small (<2%). In all mixed regions, the $m_F = 0$ component is never the least populated of the three spin components. This qualitative feature can be used to rule out the possibility that $F = 1$ sodium atoms have ferromagnetic interactions.

Nature © Macmillan Publishers Ltd 1998

letters to nature

Our experimental study of spinor condensates required techniques to selectively prepare and probe condensates in arbitrary hyperfine states. Spinor condensates were prepared in several steps. Laser cooling and evaporative cooling were used to produce sodium condensates in the $m_F = -1$ state in a cloverleaf magnetic trap[23]. The condensates were then transferred into an optical dipole trap consisting of a single focused infrared laser beam[4]. After the spin preparation, a bias field B_0 and a field gradient B' were applied for a variable amount of time (as long as 30 s), during which the atoms relaxed towards their equilibrium distribution (Fig. 2).

The profiles in Fig. 3 were obtained from vertical cuts through absorption images. They provide clear evidence of antiferromagnetic interaction. (The opposite case is predicted for the [87]Rb $F = 1$ spin multiplet[24].) The spin structure is consistent with the corresponding spin-domain diagram in Fig. 1a. Overlapping $m_F = \pm 1$ clouds as observed are incompatible with the assumption of ferromagnetic interaction.

The strength $c = (50 \pm 20)$ Hz of the antiferromagnetic interaction was estimated by determining z_b, the location of the $m_F = 0$ to the $m_F = \pm 1$ boundary, and by plotting $p = \mu B' z_b$ versus the quadratic Zeeman shift $q = \hat{q} B_0^2$ (Fig. 4). The constants μ and $\hat{q}$ are defined in the figure legend. With $n = (2.9 \pm 0.5) \times 10^{14}$ cm^{-3}, the difference between the scattering

lengths can be determined to $a_2 - a_0 = 3\Delta a = (3.5 \pm 1.5)a_B = (0.19 \pm 0.08)$ nm where a_B is the Bohr radius. This result is in rough agreement with a theoretical calculation of $a_2 - a_0 = (5.5 \pm 0.5)a_B$ (ref. 24). The antiferromagnetic interaction energy corresponds to 2.5 nK in our condensates. Still, the magnetostatic (ferromagnetic) interaction between the atomic magnetic moments is about ten times weaker. We note that the optically trapped samples in which the domains were observed were at a temperature of the order of 100 nK, far larger than the antiferromagnetic energy. The formation of spin domains occurs only in a Bose–Einstein condensate.

Figure 3c shows a profile of the density distribution for a cloud at $B_0 = 20$ mG and almost-cancelled gradient ($B' < 2$ mG cm^{-1}). No $m_F = 0$ region can be identified. The cloud was prepared with a small total angular momentum. Due to the almost-zero gradient and the non-zero angular momentum, the cloud corresponds to a point in the shaded region in Fig. 1a, rather than a vertical line with no offset as discussed before with finite gradients and zero angular momentum. The different widths of the profiles are probably caused by residual field inhomogeneities. Figure 3c demonstrates the complete miscibility of the $m_F = \pm 1$ components.

For a homogenous two-component system the criterion for miscibility (immiscibility) is[14,15,18] $a_{ab} < (>) \sqrt{a_a a_b}$, when the mean

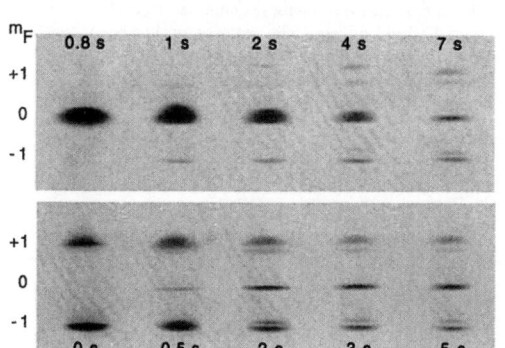

Figure 2 Formation of ground-state spin domains. Absorption images of ballistically expanding spinor condensates show both the spatial and hyperfine distributions. Arbitrary populations of the three hyperfine states were prepared using r.f. transitions (Landau-Zener sweeps)[4]. At a bias field of ~40 G, the transitions from $m_F = -1$ to $m_F = 0$ and from $m_F = 0$ to $m_F = +1$ differ in frequency by ~0.9 MHz due to the quadratic Zeeman shift, and they could be driven separately. The images of clouds with various dwell times in the trap show the evolution to the same equilibrium for condensates prepared in either a pure $m_F = 0$ state (upper row) or in equally populated $m_F = \pm 1$ states (lower row). Between 5 and 15 s dwell time, the distribution did not change significantly, although the density decreased due to three-body recombination. The bias field during the dwell time was $B_0 = 20$ mG, and the field gradient was $B' = 11$ mG cm^{-1}. These images were taken after the optical trap was suddenly switched off, and the atoms were allowed to expand. Due to the large aspect ratio (typically 60), the expansion was almost purely in the radial directions. All the mean-field energy was released after <1 ms, after which the atoms expanded as free particles. Thus, a magnetic field gradient, which was applied after 5 ms time of flight to yield a Stern-Gerlach separation of the cloud, merely translated the three spin components without affecting their shapes. In this manner, the single time-of-flight images provided both a spatial and spin-state description of the trapped cloud. Indeed, the shapes of the three clouds fit together to form a smooth total density distribution. After a total time of flight of 25 ms, the atoms were optically pumped into the $F = 2$ hyperfine state and observed using the $m_F = +2$ to $m_F = +3$ cycling transition. This technique assured the same transition strength for atoms originating from different spin states. The size of the field of view for a single spinor condensate is 1.7 × 2.7 mm.

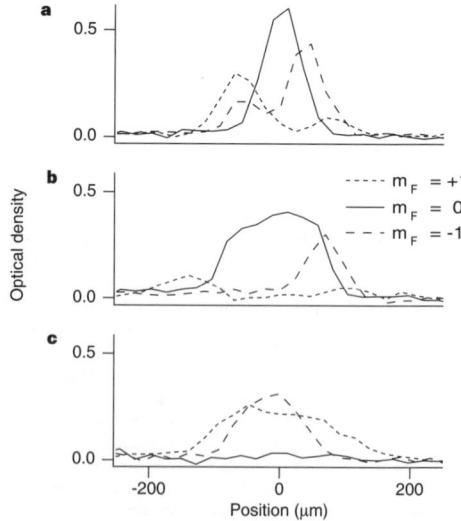

Figure 3 Miscible and immiscible spin domains. Axial column density profiles of spinor Bose-Einstein condensates are shown, obtained from time-of-flight absorption images as in Fig. 2. The profiles of the $m_F = \pm 1$ components were shifted to undo the Stern-Gerlach separation. At low bias fields (**a**), the $m_F = 0$ component was skirted on both sides by $m_F = \pm 1$ components with significant $m_F = \mp 1$ admixtures, thus demonstrating the antiferromagnetic interaction (also visible in Fig. 2). At higher fields (**b**), the $m_F = \pm 1$ components are pushed apart further by a larger $m_F = 0$ component, and the $m_F = \mp 1$ admixtures vanish. They could not be resolved for quadratic Zeeman energies $q > 20$ Hz. The antiferromagnetic interaction leads to immiscibility of the $m_F = 0$ and the $m_F = \pm 1$ components. The kinetic energy in this boundary region, which is small compared to the total mean-field energy, is released in the axial direction. Due to this axial expansion of the cloud in the time of flight, and due to imperfections in the imaging system including the limited pixel resolution, the $m_F = 0$ to $m_F = \pm 1$ boundary is not sharp. Panel **c** demonstrates the complete miscibility of the $m_F = \pm 1$ components. The magnetic field parameters were $B_0 = 20$ mG, $B' = 11$ mG cm^{-1} in **a**, $B_0 = 100$ mG, $B' = 11$ mG cm^{-1} in **b**, and $B_0 = 20$ mG, $B' < 2$ mG cm^{-1} in **c**.

letters to nature

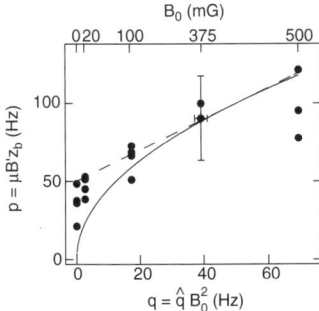

Figure 4 Estimate of the antiferromagnetic interaction energy c. Plotted is the linear Zeeman energy $p = |\mu B' z_b|$ at the boundary between the $m_F = 0$ and $m_F = \pm 1$ regions versus the quadratic Zeeman shift $q = \hat{q} B_0^2$. Here $\hat{q} = g_s^2 \mu_B^2 / 16 h^2 \nu_{hfs} = 278\,\text{Hz}\,\text{G}^{-2}$, and $\mu = g_s \mu_B / 4h = 700\,\text{kHz}\,\text{G}^{-1}$, g_s denotes the electron g-factor, ν_{hfs} the hyperfine splitting frequency, and μ_B the Bohr magneton. The solid line is a fit of the function $|p| = 2\sqrt{qc}$ for $q < c$ and $|p| = q + c$ for $q > c$. Extrapolating the linear part to zero bias field (dashed line) yields $c = (50 \pm 20)\,\text{Hz}$. The data points at a given bias field represent $p = \mu B' z_b$ for different gradient fields B' and thus $m_F = 0$ regions of different size. The scatter of these points is mainly due to residual magnetic field inhomogeneities, resulting in small deviations of the local gradient B'. The error bar represents the relative error of all data points of 30% in p and 5% in q, as estimated from the uncertainties in the magnetic-field calibration. Furthermore, the limited pixel resolution and contributions of the kinetic energy in the condensate to the axial expansion enhance the errors for the determination of z_b of small $m_F = 0$ regions.

field energy is parametrized as $(2\pi\hbar^2/M)(n_a^2 a_a + n_b^2 a_b + 2n_a n_b a_{ab})$. Here, $n_{a,b}$ and $a_{a,b}$ are densities and scattering lengths for the components a and b, and the scattering length a_{ab} characterizes the interactions between particles a and b. In our spinor condensate with mixtures of the $m_F \pm 1$ components, we have $a_{-1} = a_{+1} = \bar{a} + \Delta a$ and $a_{-1+1} = \bar{a} - \Delta a$. Thus $\Delta a > 0$, as experimentally observed, implies miscibility. For a mixture of the $m_F = 1$ and $m_F = 0$ components, we find $a_0 = \bar{a}$, $a_{+1} = \bar{a} + \Delta a$ and $a_{0+1} = \bar{a} + \Delta a$, corresponding to immiscibility. For the ^{87}Rb experiments[9,10], it is not clear whether the two components are miscible or overlap only in a surface region due to kinetic energy (E. A. Cornell, personal communication).

All regions in the spin-domain diagrams are accessible with our experimental technique, and thus any combination of the three hyperfine components can be realized by applying small external magnetic fields. Of special interest for future work is the zero-magnetic-field case, where the rotational symmetry should be spontaneously broken. We observed both miscibility and immiscibility of hyperfine components. Thus the dynamics and possible metastable configurations[7] of two interpenetrating, miscible superfluid components ($m_F = \pm 1$) with arbitrary admixtures of an immiscible component ($m_F = 0$) can now be studied. □

Received 18 June; accepted 3 September 1998.

1. Anderson, M. H., Ensher, J. R., Matthews, M. R., Wieman, C. E. & Cornell, E. A. Observation of Bose-Einstein condensation in a dilute atomic vapor. *Science* **269**, 198–201 (1995).
2. Davis, K. B. *et al.* Bose-Einstein condensation in a gas of sodium atoms. *Phys. Rev. Lett.* **75**, 3969–3973 (1995).
3. Bradley, C. C., Sackett, C. A. & Hulet, R. G. Bose-Einstein condensation of lithium: Observation of limited condensate number. *Phys. Rev. Lett.* **78**, 985–989 (1997).
4. Stamper-Kurn, D. M. *et al.* Optical confinement of a Bose-Einstein condensate. *Phys. Rev. Lett.* **80**, 2027–2030 (1998).
5. Ho, T.-L. Spinor Bose condensates in optical traps. *Phys. Rev. Lett.* **81**, 742–745 (1998).
6. Ohmi, T. & Machida, K. Bose-Einstein condensation with internal degrees of freedom in alkali atom gases. *J. Phys. Soc. Jpn* **67**, 1822–1825 (1998).
7. Law, C. K., Pu, H. & Bigelow, N. P. Quantum spins mixing in spinor Bose-Einstein condensates. *Phys. Rev. Lett.* (submitted); preprint cond-mat/9807258 at ⟨http.xxx.lanl.gov⟩ (1998).
8. Vollhardt, D. & Wölfle, P. *The Superfluid Phases of ³He* (Taylor-Francis, London, 1990).
9. Myatt, C. J., Burt, E. A., Ghrist, R. W., Cornell, E. A. & Wieman, C. E. Production of two overlapping Bose-Einstein condensates by sympathetic cooling. *Phys. Rev. Lett.* **78**, 586–589 (1997).
10. Hall, D. S., Matthews, M. R., Ensher, J. R., Wieman, C. E. & Cornell, E. A. The dynamics of component separation in a binary mixture of Bose-Einstein condensates. *Phys. Rev. Lett.* **81**, 1539–1542 (1998).
11. Hall, D. S., Matthews, M. R., Wieman, C. E. & Cornell, E. A. Measurements of relative phase in binary mixtures of Bose-Einstein condensates. *Phys. Rev. Lett.* **81**, 1543–1546 (1998).
12. Siggia, E. D. & Ruckenstein, A. E. Bose condensation in spin-polarized atomic hydrogen. *Phys. Rev. Lett.* **44**, 1423–1426 (1980).
13. Ho, T.-L. & Shenoy, V. B. Binary mixtures of Bose condensates of alkali atoms. *Phys. Rev. Lett.* **77**, 3276–3279 (1996).
14. Timmermans, E. Phase separation in Bose-Einstein condensates. *Phys. Rev. Lett.* (submitted); preprint cond-mat/9709301 at ⟨http://xxx.lanl.gov⟩ (1997).
15. Esry, B. D., Greene, C. H., Bruke, J. P. & Bohn, J. L. Hartree-Fock theory for double condensates. *Phys. Rev. Lett.* **78**, 3594–3597 (1997).
16. Öhberg, P. & Stenholm, S. Hartree–Fock treatment of the two-component Bose-Einstein condensate. *Phys. Rev. A* **57**, 1272–1279 (1998).
17. Pu, H. & Bigelow, N. P. Properties of two-species Bose condensates. *Phys. Rev. Lett.* **80**, 1130–1133 (1998).
18. Ao, P. & Chui, S. T. Binary Bose–Einstein condensate mixtures in weakly and strongly segregated phases. *Phys. Rev. A* **58(6)** (in the press).
19. Busch, T., Cirac, J. I., Pérez-Garcia, V. M. & Zoller, P. Stability and collective excitations of a two-component Bose-Einstein condensed gas: a moment approach. *Phys. Rev. A* **56**, 2978–2983 (1997).
20. Graham, R. & Walls, D. Collective excitations of trapped binary mixtures of Bose-Einstein condensed gases. *Phys. Rev. A* **57**, 484–487 (1998).
21. Pu, H. & Bigelow, N. P. Collective excitations, metastability, and nonlinear response of a trapped two-species Bose-Einstein condensate. *Phys. Rev. Lett.* **80**, 1134–1137 (1998).
22. Esry, B. D. & Greene, C. H. Low-lying excitations of double Bose-Einstein condensates. *Phys. Rev. A* **57**, 1265–1271 (1998).
23. Mewes, M.-O. *et al.* Bose-Einstein condensation in a tightly confining dc magnetic trap. *Phys. Rev. Lett.* **77**, 416–419 (1996).
24. Burke, J. P., Greene, C. H. & Bohn, J. L. Multichannel cold collisions: simple dependencies on energy and magnetic field. *Phys. Rev. Lett.* **81**, 3355–3358 (1998).
25. Fried, D. G. *et al.* Bose–Einstein condensation of atomic hydrogen. *Phys. Rev. Lett.* (in the press).

Acknowledgements. We acknowledge discussions with J. Ho and C. Greene. This work was supported by the Office of Naval Research, NSF, Joint Services Electronics Program (ARO), NASA, and the David and Lucile Packard Foundation. J.S. acknowledges support from the Alexander von Humbolt foundation, D.M.S.-K. from the JSEP Graduate Fellowship Program, and A.P.C. from the NSF.

Correspondence and requests for materials should be addressed to J.S. (e-mail: stenger@amo.mit.edu).

Observation of Feshbach resonances in a Bose–Einstein condensate

S. Inouye*, M. R. Andrews*†, J. Stenger*, H.-J. Miesner*, D. M. Stamper-Kurn* & W. Ketterle*

* *Department of Physics and Research Laboratory of Electronics, Massachusetts Institute of Technology, Cambridge, Massachusetts 02139, USA*

It has long been predicted that the scattering of ultracold atoms can be altered significantly through a so-called 'Feshbach resonance'. Two such resonances have now been observed in optically trapped Bose–Einstein condensates of sodium atoms by varying an external magnetic field. They gave rise to enhanced inelastic processes and a dispersive variation of the scattering length by a factor of over ten. These resonances open new possibilities for the study and manipulation of Bose–Einstein condensates.

Bose–Einstein condensates of atomic gases offer new opportunities for studying quantum-degenerate fluids[1-5]. All the essential properties of Bose condensed systems—the formation and shape of the condensate, the nature of its collective excitations and statistical fluctuations, and the formation and dynamics of solitons and vortices—are determined by the strength of the atomic interactions. In contrast to the situation for superfluid helium, these interactions are weak, allowing the phenomena to be theoretically described from 'first principles'. Furthermore, in atomic gases the interactions can be altered, for instance by employing different species, changing the atomic density, or, as in the present work, merely by varying a magnetic field.

At low temperatures, the interaction energy in a cloud of atoms is proportional to the density and a single atomic parameter, the scattering length a which depends on the quantum-mechanical phase shift in an elastic collision. It has been predicted that the scattering length can be modified by applying external magnetic[6-10], optical[11,12] or radio-frequency[13] (r.f.) fields. Those modifications are only pronounced in a so-called "Feshbach resonance"[14], when a quasibound molecular state has nearly zero energy and couples resonantly to the free state of the colliding atoms. In a time-dependent picture, the two atoms are transferred to the quasibound state, 'stick' together and then return to an unbound state. Such a resonance strongly affects the scattering length (elastic channel), but also affects inelastic processes such as dipolar relaxation[6,7] and three-body recombination. Feshbach resonances have so far been studied at much higher energies[15] by varying the collision energy, but here we show that they can be 'tuned' to zero energy to be resonant for ultracold atoms. The different magnetic moments of the free and quasibound states allowed us to tune these resonances with magnetic fields, and as a result, minute changes in the magnetic field strongly affected the properties of a macroscopic system.

Above and below a Feshbach resonance, the scattering length a covers the full continuum of positive and negative values. This should allow the realization of condensates over a wide range of interaction strengths. By setting $a \approx 0$, one can create a condensate with essentially non-interacting atoms, and by setting $a < 0$ one can make the system unstable and observe its collapse. Rapid tuning of an external magnetic field around a Feshbach resonance will lead to sudden changes of the scattering length. This opens the way to studies of new dynamical effects such as novel forms of collective oscillations or the sudden collapse of a large condensate when the scattering length is switched from positive to negative[16].

Theoretical predictions

Calculations for Feshbach resonances in external magnetic fields have been reported for the lower hyperfine states of the atoms Li (ref. 8), K (ref. 10), Na (ref. 8), Rb (ref. 9) and Cs (refs 6, 7). They are typically spaced by several hundred gauss, and for Li and Na occur outside the range where states in the lower hyperfine manifold are weak-field-seeking and can be magnetically trapped. Recent experimental efforts to observe Feshbach resonances have concentrated on ^{87}Rb (ref. 17) and on ^{85}Rb (ref. 18 and C. E. Wieman, personal communication) where Feshbach resonances have been predicted at relatively low magnetic fields[9]. However, our recently demonstrated all-optical confinement of a Bose condensate[19] opened the possibility of observing Feshbach resonances for strong-field-seeking states which cannot be trapped in a d.c. magnetic trap. The optical trapping potential is unaffected by magnetic fields and is independent of the hyperfine ground state. We report here the observation of two Feshbach resonances of sodium in a strong-field-seeking state.

Several Feshbach resonances in sodium are caused by quasibound hyperfine states of the second highest vibrational level, $\nu = 14$, of the triplet potential of the sodium dimer. The lowest magnetic field value B_0 for a strong Feshbach resonance in sodium was predicted to lie in the range $760 < B_0 < 925$ G (B. J. Verhaar and F. A. van Abeelen, personal communication). It occurs in collisions between atoms in the lowest hyperfine state $|m_S = -1/2, m_I = +3/2\rangle$, which correlates with the $|F = 1, m_F = +1\rangle$ state at low fields (S, I and F are the usual quantum numbers for the electronic, nuclear and total spin, respectively). This Feshbach resonance is due to a quasibound molecular state $|S = 1, m_S = +1, I = 1, m_I = +1\rangle$. A much weaker resonance due to a $|S = 1, m_S = +1, I = 3, m_I = +1\rangle$ state (which is almost degenerate with the other quasibound state) was predicted to occur 50 to 75 G below.

Near a Feshbach resonance, the scattering length a should vary dispersively as a function of magnetic field B (ref. 8):

$$a = \bar{a}\left(1 - \frac{\Delta}{B - B_0}\right) \qquad (1)$$

where Δ parametrizes the width of the resonance at $B = B_0$, and $\bar{a}$ is the scattering length outside the resonance. For sodium, $\bar{a}$ was found spectroscopically to be 2.75 nm at zero field, and increases to the triplet scattering length of 4.5 nm (ref. 20) at high magnetic fields. The widths Δ for the strong and weak resonance were predicted to be 1 G and 0.01 G, respectively (B. J. Verhaar and F. A. van Abeelen, personal communication).

---

† Present address: Bell Laboratories, Lucent Technologies, Murray Hill, New Jersey 07974, USA.

articles

Experimental set-up

Bose–Einstein condensates in the $|F = 1, m_F = -1\rangle$ state were produced as in our previous work by laser cooling, followed by evaporative cooling in a magnetic trap[21]. The condensates were transferred into an optical dipole trap formed at the focus of an infrared laser beam[19]. Atoms were then spin-flipped with nearly 100% efficiency to the $|F = 1, m_F = +1\rangle$ state with an adiabatic r.f. sweep while applying a 1 G bias field. Without large modifications of our magnetic trapping coils, we could provide bias fields of up to $\sim 1,200$ G, but only by using coils producing axial curvature[21], which for high-field-seeking states generated a repulsive axial potential. At the highest magnetic fields, this repulsion was stronger than the axial confinement provided by the optical trap. To prevent the atoms from escaping, two 'end-caps' of far-off-resonant blue-detuned laser light were placed at the ends of the condensate, creating a repulsive potential, and confining the atoms axially (Fig. 1a). For this, green light at 514 nm from an argon-ion laser was focused into two sheets about 200 μm apart. The focus of the optical trap was placed near the minimum of the bias field in order to minimize the effect of the destabilizing magnetic field curvature. The axial trapping potential at high fields was approximately "W"-shaped (Fig. 1b), and had a minimum near one of the end-caps as observed by phase-contrast imaging[22] (Fig. 1c, d).

The calibration factor between the current (up to ~ 400 A) in the coils and the magnetic bias field was determined with an accuracy of 2% by inducing r.f. transitions within the $|F = 1\rangle$ ground-state hyperfine manifold at about 40 G. Additionally, an optical resonance was found around 1,000 G, where the Zeeman shift equalled the probe light detuning of about 1.7 GHz and led to a sign-reversal of the phase-contrast signal. These two calibrations agreed within their uncertainties.

The condensate was observed in the trap directly using phase-contrast imaging[22] or by using time-of-flight absorption imaging[1,2,21]. In the latter case, the optical trap was suddenly switched off, and the magnetic bias field was shut off 1–2 ms later to ensure that the high-field value of the scattering length was responsible for the acceleration of the atoms. After ballistic expansion of the condensate (either 12 or 20 ms), the atoms were optically pumped into the $|F = 2\rangle$ ground state and probed using resonant light driving the cycling transition. The disk-like expansion of the

cloud and the radial parabolic density profile were clear evidence for the presence of a Bose condensed cloud.

Locating the resonances

When the magnetic field is swept across a Feshbach resonance one would expect to lose a condensate due to an enhanced rate of inelastic collisions (caused either by the collapse in the region of negative scattering length or by an enhanced rate coefficient for inelastic collisions). This allowed us to implement a simple procedure to locate the resonances: we first extended the field ramp until the atoms were lost and then used successively narrower field intervals to localize the loss. This procedure converged much faster than a point-by-point search. As we could take many non-destructive phase-contrast images during the magnetic field ramp, the sharp onset of trap loss at the resonance was easily monitored (Fig. 1c, d).

The most robust performance was obtained by operating the optical dipole trap at 10 mW laser power focused to a beam waist of 6 μm, resulting in tight confinement of the condensate and therefore rather short lifetimes owing to three-body recombination[19]. This required that the magnetic field be ramped up in two stages: a fast ramp at a rate of ~ 100 G ms^{-1} to a value slightly below that expected for a Feshbach resonance, followed by a slow ramp at a rate between 0.05 and 0.3 G ms^{-1} to allow for detailed observation. Near 907 G, we observed a dramatic loss of atoms, as shown in Figs 1c and 2a. This field value was reproducible to better than 0.5 G and had a calibration uncertainty of ± 20 G.

To distinguish between an actual resonance and a threshold for trap loss, we also approached the resonance from above. Fields above the Feshbach resonance were reached by ramping at a fast rate of 200 G ms^{-1}, thus minimizing the time spent near the resonance and the accompanying losses. The number of atoms above the resonance was typically three times smaller than below. Approaching the resonance from above, a similarly sharp loss phenomenon was observed about 1 G higher in field than from below (Fig. 2a), which roughly agrees with the predicted width of the resonance. A second resonance was observed 54 ± 1 G below the first one, with the observed onset of trap loss at least a factor of ten sharper than for the first. As the upper resonance was only reached by passing through the lower one, some losses of atoms were unavoidable; for example, when the lower resonance was crossed at 2 G ms^{-1},

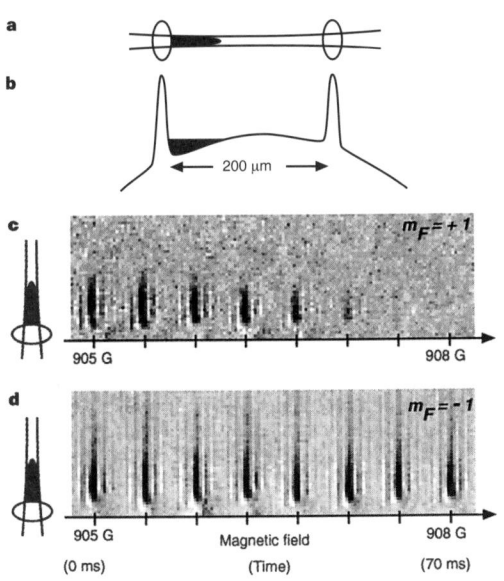

Figure 1 Observation of the Feshbach resonance at 907 G using phase-contrast imaging in an optical trap. A rapid sequence (100 Hz) of non-destructive, *in situ* phase-contrast images of a trapped cloud (which appears black) is shown. As the magnetic field was increased, the cloud suddenly disappeared for atoms in the $|m_F = +1\rangle$ state (see images in **c**), whereas nothing happened for a cloud in the $|m_F = -1\rangle$ state (images in **d**). The height of the images is 140 μm. A diagram of the optical trap is shown in **a**. It consisted of one red-detuned laser beam providing radial confinement, and two blue-detuned laser beams acting as end-caps (shown as ovals). The minimum of the magnetic field was slightly offset from the centre of the optical trap. As a result, the condensate (shaded area) was pushed by the magnetic field curvature towards one of the end-caps. The axial profile of the total potential is shown in **b**.

articles

about 80% of the atoms were lost. This, coupled with the stability and finite programming speed of the power supplies, limited the ramp rates to those given above.

The observation of twin resonances separated by 54 ± 1 G, with the weaker one at lower field, exactly matches the theoretically predicted pattern and thus strongly confirms our interpretation. No resonance phenomena were observed in the $|m_F = -1\rangle$ state at any field up to 1,000 G, in agreement with theory which predicted resonances for this state only at much higher fields.

Changing the scattering length

The trap loss measurements easily located the Feshbach resonances. To measure the variation of the scattering length around these resonances, we determined the interaction energy of a trapped condensate. This was done by suddenly switching off the trap, allowing the stored interaction energy to be converted into the kinetic energy of a freely expanding condensate and measuring it by time-of-flight absorption imaging[1,2,21]. The interaction energy is proportional to the scattering length and the average density of the condensate $\langle n \rangle$:

$$E_I/N = \frac{2\pi\hbar^2}{m} a \langle n \rangle \tag{2}$$

where N is the number of condensed atoms of mass m. For a large condensate the kinetic energy in the trap is negligible (Thomas–Fermi limit), and E_I is equal to the kinetic energy E_K of the freely expanding condensate $E_K/N = mv_{rms}^2/2$, where v_{rms} is the root-mean-square velocity of the atoms. For a three-dimensional harmonic oscillator potential one finds $\langle n \rangle \propto N(Na)^{-3/5}$ (ref. 23) (We note that, for a general power-law potential $\Sigma_i c_i x_i^{p_i}$, one obtains $\langle n \rangle \propto N(Na)^{k-1}$, where $k = 1/(1 + \Sigma_i 1/p_i)$). Thus, the value of the scattering length scales as:

$$a \propto \frac{v_{rms}^5}{N} \tag{3}$$

Both v_{rms} and N can be directly evaluated from absorption images of freely expanding condensates. For a cigar-shaped condensate the free expansion is predominantly radial, and so the contribution of the axial dimension to v_{rms} could be neglected. The quantity v_{rms}^5/N (equation (3)), normalized to unity outside the resonance, should be identical to $a/\bar{a}$ (equation (1)). This quantity was measured around the resonance at 907 G and is shown in Fig. 2b together with the theoretical prediction of a resonance with width $\Delta = 1$ G. The data clearly displays the predicted dispersive shape and shows evidence for a variation in the scattering length by more than a factor of ten.

We now discuss the assumptions for equation (3) and show that it is approximately valid for our conditions. (1) We assumed that the condensate remains in equilibrium during the magnetic field ramp. This is the case if the adiabatic condition $\dot{a}/a \ll \omega_i$ holds for the temporal change of the scattering length[16], and a similar condition for the loss of atoms (the ω_i are the trapping frequencies). For the condensate's fast radial dynamics ($\omega_r \approx 2\pi \times 1.5$ kHz) this condition is fulfilled, whereas for the slower axial motion ($\omega_z \approx 2\pi \times 0.1$ kHz) it breaks down close to or within the resonance. In this case the density would approach the two-dimensional scaling $N(Na)^{-1/2}$, but the values for $a/\bar{a}$ (Fig. 2b) would differ by at most 50%. (2) The second assumption was a three-dimensional harmonic trap. If the axial potential has linear contributions, the density scales instead like $N(Na)^{-2/3}$ resulting in at most a 50% change for $a/\bar{a}$. (3) We assumed that contributions of collective excitations to the released energy were small. Axial striations were observed in free expansion for both $|m_F = +1\rangle$ and $|m_F = -1\rangle$ atoms (probably created by the changing potential during the fast magnetic field ramp). However, the small scatter of points outside the resonance in Fig. 2b, which do not show any evidence of oscillations, suggests that the contribution of excitations to the released energy is negligible. (4) We assumed a sudden switch-off of the trap and ballistic expansion. The inhomogeneous bias field during the first 1–2 ms of free expansion accelerated the axial expansion, but had a negligible effect on the expansion of the condensate in the radial direction, which was evaluated for Fig. 2b.

None of the corrections (1)–(4) discussed above affect our conclusion that the scattering length varies dispersively near a Feshbach resonance. More accurate experiments should be done with a homogeneous bias field. In addition, an optical trap with larger volume and lower density would preclude the need to ramp the field quickly because three-body recombination would be reduced.

The trap losses observed around the Feshbach resonances merit further study as they might impose practical limits on the possibilities for varying the scattering length. An increase of the dipolar relaxation rate near Feshbach resonances has been predicted[6,7], but for atoms in the lowest hyperfine state no such inelastic binary collisions are possible. Therefore, the observed trap loss is probably due to three-body collisions. In this case the loss rate is characterized by the coefficient K_3, defined as $\dot{N}/N = -K_3\langle n^2 \rangle$. So far, there is no theoretical work on K_3 near a Feshbach resonance. An analysis based on Fig. 2 shows that K_3 increased on both sides of the resonance, because the loss rate increased while the density decreased or stayed constant. In any case, the fact that we observed Feshbach resonances at high atomic densities ($\sim 10^{15}$ cm^{-3}) strongly enhanced this loss process, which can be avoided with a condensate at lower density in a modified optical trap. Control of the bias field with a precision better than $\sim 10^{-4}$ will be necessary to achieve negative or extremely large values of the scattering length in a stable way.

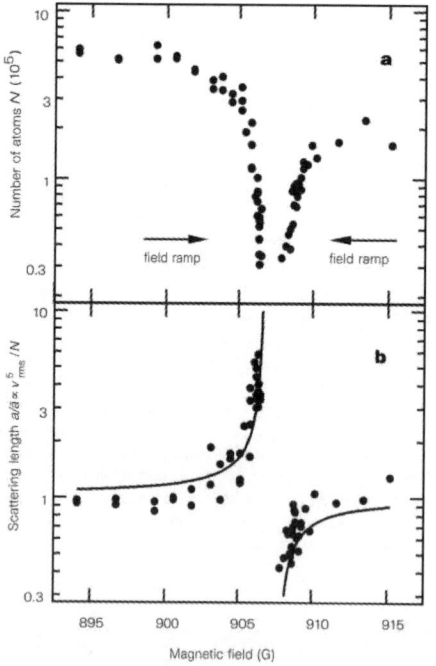

Figure 2 Observation of the Feshbach resonance at 907 G using time-of-flight absorption imaging. **a**, Number of atoms in the condensate versus magnetic field. Field values above the resonance were reached by quickly crossing the resonance from below and then slowly approaching from above. **b**, The normalized scattering length $a/\bar{a} \propto v_{rms}^5/N$ calculated from the released energy, together with the predicted shape (equation (1), solid line). The values of the magnetic field in the upper scan relative to the lower one have an uncertainty of <0.5 G.

articles

A tunable condensate

We have observed two Feshbach resonances for Bose–Einstein condensates of sodium through the abrupt loss of atoms, and obtained strong evidence for a dispersive variation of the scattering length by a factor of more than ten. 'Tuning' of the scattering length should become an important tool for 'designing' atomic quantum gases with novel properties; for example, to create ideal Bose–Einstein condensates with nearly zero scattering length, and to obtain a detailed picture of the collapse of a condensate with negative scattering length, which is so far not fully understood. Tuning the scattering length can also be used to vary interactions between different species[24] and thus control the phase diagram of multi-component condensates, possibly switching from interpenetrating superfluids to phase separation[25]. Feshbach resonances may also be important in atom optics, for modifying the atomic interactions in an atom laser, or more generally, for controlling nonlinear coefficients in atom optics with coherent beams of atoms. ☐

Received 18 February; accepted 19 February 1998.

1. Anderson, M. H., Ensher, J. R., Matthews, M. R., Wieman, C. E. & Cornell, E. A. Observation of Bose-Einstein condensation in a dilute atomic vapor. *Science* **269**, 198–201 (1995).
2. Davis, K. B. *et al.* Bose-Einstein condensation in a gas of sodium atoms. *Phys. Rev. Lett.* **75**, 3969–3973 (1995).
3. Bradley, C. C., Sackett, C. A. & Hulet, R. G. Bose-Einstein condensation of lithium: Observation of limited condensate number. *Phys. Rev. Lett.* **78**, 985–989 (1997).
4. Georgia Southern University BEC home page, http://amo.phy.gasou.edu/bec.html.
5. Bradley, C. C., Sackett, C. A., Tollett, J. J. & Hulet, R. G. Evidence of Bose-Einstein condensation in an atomic gas with attractive interactions. *Phys. Rev. Lett.* **75**, 1687–1690 (1995).
6. Tiesinga, E., Moerdijk, A. J., Verhaar, B. J. & Stoof, H. T. C. Conditions for Bose-Einstein condensation in magnetically trapped atomic cesium. *Phys. Rev. A* **46**, R1167–R1170 (1992).
7. Tiesinga, E., Verhaar, B. J. & Stoof, H. T. C. Threshold and resonance phenomena in ultracold ground-state collisions. *Phys. Rev. A* **47**, 4114–4122 (1993).
8. Moerdijk, A. J., Verhaar, B. J. & Axelsson, A. Resonances in ultracold collisions of ^{6}Li, ^{7}Li and ^{23}Na. *Phys. Rev. A* **51**, 4852–4861 (1995).
9. Vogels, J. M. *et al.* Prediction of Feshbach resonances in collisions of ultracold rubidium atoms. *Phys. Rev. A* **56**, R1067–R1070 (1997).
10. Boesten, H. M. J. M., Vogels, J. M., Tempelaars, J. G. C. & Verhaar, B. J. Properties of cold collisions of ^{39}K atoms and of ^{41}K atoms in relation to Bose-Einstein condensation. *Phys. Rev. A* **54**, R3726–R3729 (1996).
11. Fedichev, P. O., Kagan, Yu., Shlyapnikov, G. V. & Walraven, J. T. M. Influence of nearly resonant light on the scattering length in low-temperature atomic gases. *Phys. Rev. Lett.* **77**, 2913–2916 (1996).
12. Bohn, J. L. & Julienne, P. S. Prospects for influencing the scattering lengths with far-off-resonant light. *Phys. Rev. A* **56**, 1486–1491 (1997).
13. Moerdijk, A. J., Verhaar, B. J. & Nagtegaal, T. M. Collisions of dressed ground-state atoms. *Phys. Rev. A* **53**, 4343–4351 (1996).
14. Feshbach, H. A unified theory of nuclear reactions. II. *Ann. Phys.* **19**, 287–313 (1962).
15. Bryant, H. C. *et al.* Observation of resonances near 11 eV in the photodetachment cross-section of the H⁻ ion. *Phys. Rev. Lett.* **38**, 228–230 (1977).
16. Kagan, Yu., Surkov, E. L. & Shlyapnikov, G. V. Evolution and global collapse of trapped Bose condensates under variations of the scattering length. *Phys. Rev. Lett.* **79**, 2604–2607 (1997).
17. Newbury, N. R., Myatt, C. J. & Wieman, C. E. s-wave elastic collisions between cold ground-state ^{87}Rb atom. *Phys. Rev. A* **51**, R2680–R2683 (1995).
18. Courteille, P. & Heinzen, D. Paper presented at *SPIE Photonics West*, 24–30 Jan., San Jose, California (1998).
19. Stemper-Kurn, D. M. *et al.* Optical confinement of a Bose-Einstein condensate. *Phys. Rev. Lett.* (in the press).
20. Tiesinga, E. *et al.* A spectroscopic determination of scattering lengths for sodium atom collisions. *J. Res. Natl Inst. Stand. Technol.* **101**, 505–520 (1996).
21. Mewes, M.-O. *et al.* Bose-Einstein condensation in a tightly confining d.c. magnetic trap. *Phys. Rev. Lett.* **77**, 416–419 (1996).
22. Andrews, M. R. *et al.* Propagation of sound in a Bose-Einstein condensate. *Phys. Rev. Lett.* **79**, 553–556 (1997).
23. Baym, G. & Pethick, C. J. Ground-state properties of magnetically trapped Bose-condensed rubidium gas. *Phys. Rev. Lett.* **76**, 6–9 (1996).
24. van Abeelen, F. A., Verhaar, B. J. & Moerdijk, A. J. Sympathetic cooling of ^{6}Li atoms. *Phys. Rev. A* **55**, 4377–4381 (1997).
25. Ho, T.-L. & Shenoy, V. B. Binary mixtures of Bose condensates of alkali atoms. *Phys. Rev. Lett.* **77**, 3276–3279 (1996).

Acknowledgements. We thank J. M. Vogels for discussions, A. P. Chikkatur for experimental assistance, and B. J. Verhaar and F. A. van Abeelen for providing updated theoretical predictions. We also thank D. Kleppner, D. E. Pritchard and R. A. Rubenstein for a critical reading of the manuscript. This work was supported by the Office of Naval Research, NSF, Joint Services Electronics Program (ARO), and the David and Lucile Packard Foundation. J.S. acknowledges support from the Alexander von Humboldt-Foundation, and D.M.S.-K. was supported by the JSEP graduate fellowship program.

Correspondence and requests for materials should be addressed to W.K.

VOLUME 81, NUMBER 1   PHYSICAL REVIEW LETTERS   6 JULY 1998

Observation of a Feshbach Resonance in Cold Atom Scattering

Ph. Courteille, R. S. Freeland, and D. J. Heinzen

Department of Physics, The University of Texas, Austin, Texas 78712

F. A. van Abeelen and B. J. Verhaar

Eindhoven University of Technology, P.O. Box 513, 5600MB Eindhoven, The Netherlands
(Received 13 March 1998)

We probe s-wave collisions of laser-cooled ^{85}Rb($f = 2, m_f = -2$) atoms with Zeeman-resolved photoassociation spectroscopy. We observe that these collisions exhibit a magnetically tunable Feshbach resonance, and determine that this resonance tunes to zero energy at a magnetic field of 164 ± 7 G. This result indicates that the self-interaction energy of an ^{85}Rb Bose-Einstein condensate can be magnetically tuned. We also demonstrate that Zeeman-resolved photoassociation spectroscopy provides a useful new tool for the study of ultracold atomic collisions. [S0031-9007(98)06510-7]

PACS numbers: 32.80.Pj, 03.75.Fi, 34.50.−s

The observation of Bose-Einstein condensation (BEC) in dilute, magnetically trapped alkali gases has created exciting new opportunities for studies of macroscopic quantum phenomena [1–7]. An important aspect of dilute gas BEC is that two-body interactions dominate, and give rise to a condensate self-energy proportional to the two-body scattering length a. The self-energy strongly influences most of the important properties of a condensate, including its stability, formation rate, size and shape, and collective excitations. There has been considerable interest in finding ways to experimentally modify the scattering length, because that could make possible studies of a BEC with a very strong, very weak, positive, negative, or even time-dependent interaction strength, all within a single experiment. One promising proposal to do this relies on the strong variation of a that occurs if a Feshbach collision resonance is tuned through zero energy [8]. Such a tunable resonance could be induced optically, but this method introduces undesired effects of optical spontaneous emission into the condensate [9,10]. Magnetically tunable Feshbach resonances that arise from the coupling between different spin channels in an atomic collision can also result in a tunable value of a [8,11,12]. A previous search for this type of resonance [13] did not detect one. Interest in this topic increased with a prediction of a zero-energy Feshbach resonance in collisions of ^{85}Rb($f = 2, m_f = -2$) atoms [12]. In this paper, we report the observation of this resonance, which we find tunes to zero energy at a magnetic field of 164 ± 7 G. From the observed position and width of the resonance, we are able to precisely determine ^{85}Rb interaction parameters. Our work, along with a recent report of a similar resonance in an atomic ^{23}Na BEC [14], constitute the first observations of this important cold collision phenomenon.

In order to detect this resonance, we use photoassociation spectroscopy [15] to probe the collisions of laser-cooled ^{85}Rb atoms in a magnetic field. The concept of the experiment is illustrated in Fig. 1. To be concrete, we specialize to our particular case. Free, ground-state ^{85}Rb

atoms collide in the $|f = 2, m_f = -2\rangle + |f = 2, m_f = -2\rangle$ entrance channel. Here, $f = 2$ or 3 is the hyperfine state (combined electron and nuclear spin) of an atom and m_f is the spin projection quantum number of that atom. The entrance channel has a total angular momentum projection quantum number $M_F = -4$, equal to the sum of the two atomic m_f values. It is coupled to other $M_F = -4$ channels at a small internuclear distance by

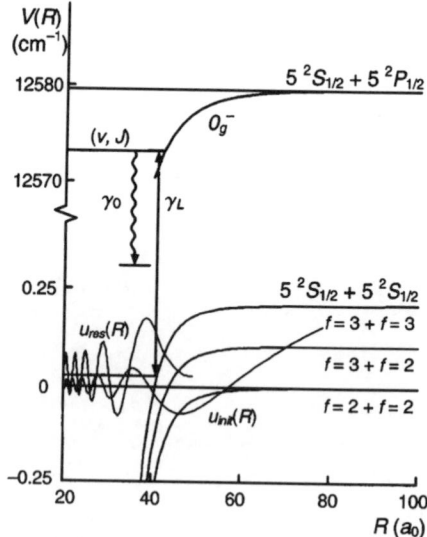

FIG. 1. Photoassociation method for detecting a Feshbach resonance in collisions of ultracold ^{85}Rb($f = 2, m_f = -2$) atoms. The entrance channel wave function $u_{\text{init}}(R)$ couples to a quasibound state with wave function $u_{\text{res}}(R)$. A laser field induces photoassociation of this state to an excited, bound $0_g^-(v, J)$ molecular state at a rate γ_L, which then decays back to free atoms at a rate γ_0. As a magnetic field is varied, the quasibound state tunes through zero energy, producing a Feshbach resonance for ultracold collisions. The resulting enhancement of $u_{\text{res}}(R)$ produces an enhancement of γ_L that we detect with a trap loss method.

VOLUME 81, NUMBER 1   PHYSICAL REVIEW LETTERS   6 JULY 1998

the electronic exchange interaction. The other $M_F = -4$ potential curves all correlate to the higher energy $f = 2 + f = 3$ or $f = 3 + f = 3$ dissociation limits. They support multichannel quasibound states at positive energies, where we take the zero of energy to be the threshold of the entrance channel. If the energy of the incoming atoms matches the energy of one of these states, a Feshbach resonance occurs in which a large wave-function amplitude builds up in the quasibound state. The resonance energy can be tuned to zero with a magnetic field because the quasibound state and threshold energy Zeeman shift at different rates. In that case the resonance strongly affects ultracold collisions. In order to detect the resonance, we drive photoassociation transitions to the excited $^{85}Rb_2$ 0_g^- bound molecular vibrational state at an energy 5.9 cm^{-1} below the $5^2S_{1/2} + 5^2P_{1/2}$ dissociation limit [16]. As discussed below, we are able to isolate a single component of the spectrum which originates from the s-wave, $M_F = -4$, collisional resonance state. Its transition rate is proportional to the square of the wave-function overlap between the collisional state and the excited state, and therefore shows an enhancement when the Feshbach resonance is tuned near zero energy.

We detect the photoassociation with a trap loss method [16–19]. About 10^4 ^{85}Rb atoms are transferred from a magneto-optical trap into a far-off resonance optical dipole force trap (FORT) [20], created by a 1.7 W, 835 nm wavelength laser beam focussed to a waist of 20 μm. The atoms are laser cooled to a temperature between 30 and 100 μK, and have a density between 10^{11} and 10^{12} cm^{-3}. We then switch on a magnetic field B and allow it to stabilize for 300 ms. After this, we continuously illuminate the atoms with a near-resonance laser beam that optically pumps them into their $f = 2$ ground hyperfine state, and with a tunable probe (PA) laser beam which induces the photoassociation transitions. In some cases we also apply an additional near-resonance σ^--polarized (OP) laser to pump the atoms into their $m_f = -2$ state. After an additional 700 to 1000 ms, we switch off these laser beams and the magnetic field, and probe the atoms remaining in the trap with laser-induced fluorescence. The photoassociation rate is detectable as reduced atomic fluorescence, because most pairs of atoms which absorb a PA laser photon return to the ground state by spontaneous emission as free atoms with a kinetic energy that is too high to remain in the trap. In the plots below, we show this measured fluorescence signal vs PA laser frequency, inverted so that photoassociation-induced trap loss produces upward going peaks.

A typical spectrum, recorded with the OP laser beam off and with no magnetic field, is shown in Fig. 2. We observe a simple spectrum that arises from the 0_g^- excited state $J = 0$, 1, and 2 rotational levels. Figure 2 also shows the spectrum with the OP laser beam off and with $B \approx 195$ G. In this case, the $f = 2 + f = 2$ dissociation limit Zeeman splits into 15 different limits for the even partial waves, and 10 different limits for the odd partial

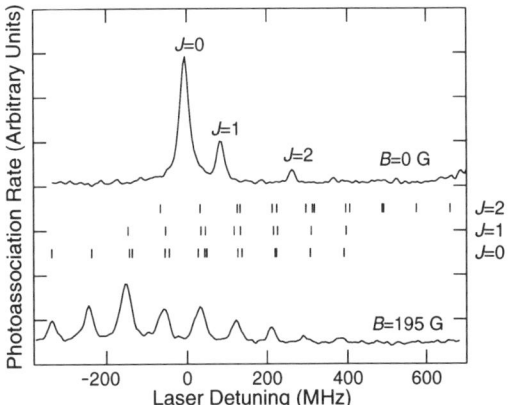

FIG. 2. $^{85}Rb_2$ photoassociation spectra for excitation from lower ($f = 2 + f = 2$) hyperfine state collisions to a single excited vibrational level, at a laser intensity of 20 W/cm^2. Upper curve: spectrum at zero magnetic field. Lower curve: spectrum at a magnetic field of 195 G. Each of the zero field components splits into 10 or 15 distinct components due to Zeeman splitting of the ground state atoms; calculated splittings are shown by the vertical dashed marks. The successive peaks in the lower spectrum correspond mainly to $J = 0$, and (from left) $M_F = -4, -3, -2, -1, 0, 1$, and 2.

waves, which correspond to the various possible combinations of the two atomic m_f quantum numbers. Without optical pumping all of these combinations are populated. The excited state does not show a significant Zeeman splitting. Because some of the splittings are not resolved, the $J = 0$ rotational peak splits into nine Zeeman components corresponding to $M_F = -4, \ldots, +4$. The leftmost peak in the spectrum arises only from $|f = 2, m_f = -2\rangle + |f = 2, m_f = -2\rangle$ ($M_F = -4$) collisions. Further, this peak arises only from s-wave collisions because the selection rule $J = l$ is obeyed for this transition, where l is the orbital angular momentum of the initial state [16]. Therefore the leftmost peak probes exclusively the desired collision channel.

In Fig. 3, we show repeated scans over the $J = 0$, $M_F = -4$ peak at many different field values. The data clearly show the effect of the Feshbach resonance. For these scans we also turn on the OP laser beam, which enhances the intensity of the $M_F = -4$ peak by a factor of 5. The PA laser intensity $I = 0.1$ W/cm^2. As the magnetic field is increased, the signal emerges from the noise, reaches a maximum strength near 167 G, and then disappears again into the noise. The field magnitude is calibrated using the Zeeman-resolved spectra. Our interpretation of this enhancement as a Feshbach resonance is supported by several factors. First, previous studies of ultracold Rb collisions have fairly strongly constrained its ground state interaction potentials [4,13,16–19,21,22], and allowed for predictions of this resonance [12,22]. We observe a resonance in the correct channel near the predicted field. Finally, we observe these photoassociation

VOLUME 81, NUMBER 1   PHYSICAL REVIEW LETTERS   6 JULY 1998

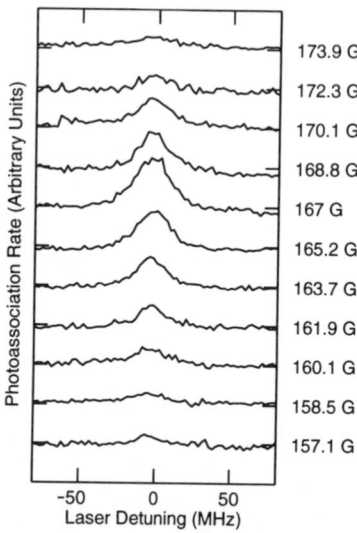

FIG. 3. Photoassociation spectra showing the $J = 0$, $M_F = -4$ peak at a succession of magnetic field values, with a laser intensity of 0.1 W/cm². The relative Zeeman shift of the successive peaks is removed so that they appear at the same laser tuning.

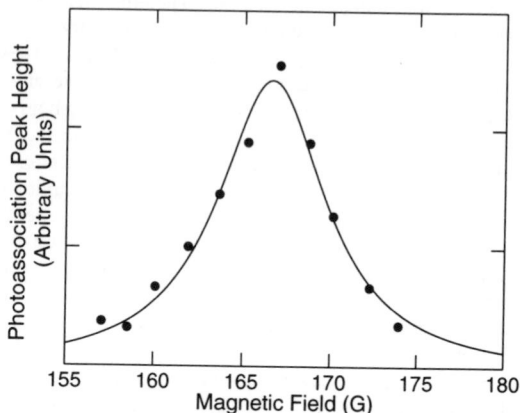

FIG. 4. Height of the photoassociation peaks shown in Fig. 3, as a function of magnetic field, showing clearly the Feshbach resonance. The solid curve shows a Lorentzian fit to the data.

peaks at an anomalously low laser intensity, at which other features in the spectrum are not visible. This can occur only with resonance enhancement of the rate γ_L due to the enhanced wave-function amplitude.

In Fig. 4, we plot the heights of the photoassociation peaks, determined from Lorentzian fits, as a function of magnetic field. We find that this curve is also well fit by a Lorentzian line shape, which yields a resonance field $B_{0,PA}(I)$ and a width (FWHM) $\Delta B_{PA}(I)$. Theoretical calculations [23] show that departures from Lorentzian line shapes should be small for our conditions. We find that optical power broadening is significant. In order to account for this, we repeat the measurements at laser intensities I ranging from 0.1 to 0.54 W/cm², and plot $B_{0,PA}(I)$ and $\Delta B_{PA}(I)$ vs I. $B_{0,PA}(I)$ varies by less than 1.5 G over this range, and $\Delta B_{PA}(I)$ varies from about 8 to about 15 G. By fitting these data, we determine zero-intensity intercepts of $B_{0,PA} = 166.6 \pm 6$ G and $\Delta B_{PA} = 5.9 \pm 2.1$ G. The error in $B_{0,PA}$ is mainly due to errors in the magnetic field calibration. We searched for and did not find any additional Feshbach resonances in the field range between 100 and 195 G.

In order to further analyze these results, we have calculated the resonance field $B_{0,PA}$ and width ΔB_{PA} using an accurate model for the atomic Rb interaction potential [23]. These quantities depend most sensitively on the Rb₂ ground state Van der Waals interaction coefficient C_6, and the two parameters $v_{DS}(\text{mod } 1)$ and $v_{DT}(\text{mod } 1)$. v_{DS} and v_{DT} correspond to the (fractional) number of bound states in the lowest singlet and triplet ^{85}Rb₂ molecular potential wells, respectively. Further, near our

parameter range the position of the resonance depends mostly on C_6 and on the sum $v_{DS}(\text{mod } 1) + v_{DT}(\text{mod } 1)$, whereas its width depends mostly on the difference $v_{DS}(\text{mod } 1) - v_{DT}(\text{mod } 1)$. Taking a fixed $C_6 = 4550$ a.u. [18] we determine from our measured value of $B_{0,PA}$ that $v_{DS}(\text{mod } 1) + v_{DT}(\text{mod } 1) = -0.082 \pm 0.011$. Allowing for a 50 a.u. uncertainty in C_6 increases the uncertainty of $v_{DS}(\text{mod } 1) + v_{DT}(\text{mod } 1)$ to ± 0.016. From the measured value of ΔB_{PA}, we determine $v_{DS}(\text{mod } 1) - v_{DT}(\text{mod } 1) = 0.058 \pm 0.016$. (We rule out an opposite sign for $v_{DS} - v_{DT}$ because it conflicts with previous measurements [19].) Combining these results, we determine $v_{DS}(\text{mod } 1) = -0.012$ and $v_{DT}(\text{mod } 1) = -0.070$, with uncertainties for both their sum and difference of ± 0.016.

The best previous determination of these quantities followed from our measurements of the highest bound levels of the ^{85}Rb₂ molecule [19]. Taking again a fixed $C_6 = 4550$ a.u., those measurements yield $v_{DS}(\text{mod } 1) = -0.006 \pm 0.008$ and $v_{DT}(\text{mod } 1) = -0.047 \pm 0.006$. Plotting the allowed regions in the $v_{DS} - v_{DT}$ plane at fixed C_6 for both the Feshbach and the bound state measurements, we find that they nearly contact each other near the point corresponding to the lower limits for both the Feshbach resonance width and resonance field. The difference between the parameters derived from the two experiments is somewhat larger than would be expected from their respective uncertainties; a possible explanation is that errors in the bound state measurements due to line-shape effects were underestimated. The uncertainty in C_6 also increases the uncertainties of the parameters derived from the bound state experiment [19], but it does not significantly change the level of agreement between the two experiments because their allowed v_{DS}-v_{DT} regions display similar shifts with C_6.

Based on our Feshbach resonance measurements, we calculate the scattering length $a_{2,-2}$ for collisions

VOLUME 81, NUMBER 1   PHYSICAL REVIEW LETTERS   6 JULY 1998

of ^{85}Rb($f = 2, m_f = -2$) atoms as a function of field strength shown in Fig. 5. The resonance in the scattering length has the dispersive form $a_{2,-2} = a_{2,-2}^0[1 - \Delta/(B - B_0)]$. For the same parameters that yield the observed values of $B_{0,PA}$ and ΔB_{PA}, we find that $a_{2,-2}^0 = -295 \pm 80\ a_0$, $\Delta = 8.2 \pm 3.8$ G, and $B_0 = 164 \pm 7$ G. B_0 is a few Gauss lower than $B_{0,PA}$ due to the fact that close to the crossing of the Rb$_2$ bound state and the $|f = 2, m_f = -2\rangle + |f = 2, m_f = -2\rangle$ threshold, the PA phenomenon is influenced to a significant extent by interference of the Feshbach resonance and the strong background (potential) scattering associated with the large background value of $a_{2,-2}$. The measured resonance field is in moderate disagreement with our previous prediction of 142 ± 10 G [12], which was based on the bound state measurements [19], for the reasons discussed above.

In summary, we have detected a zero-energy Feshbach resonance in collisions of ^{85}Rb($f = 2, m_f = -2$) atoms at a magnetic field of 164 ± 7 G. Our method, based on Zeeman-resolved photoassociation spectroscopy of ultracold atoms, allows us to search for resonances in any hyperfine, Zeeman, and partial wave channel by simply looking for an enhancement of the appropriate spectral component as the magnetic field is tuned. This method may therefore prove more generally useful as a new probe of ultracold atomic collisions. ^{85}Rb($f = 2, m_f = -2$) atoms can be magnetically trapped, and are expected to exhibit a very low two-body inelastic collision rate [22]. Evaporative cooling of this isotope is somewhat difficult due to a suppression of its elastic cross section at temperatures above 10 μK [22,24,25], but it is feasible [25]. Therefore it should be possible to study a magnetically trapped ^{85}Rb BEC with an adjustable scattering length. One attractive feature of this resonance is that its ratio Δ/B_0, which governs the degree of magnetic field control

needed to stably produce a very large scattering length, is relatively large. Two and three body collisional loss rates are also enhanced by a Feshbach resonance [14,22], and this may limit the tuning range achievable in practice. A further interesting possibility is that it should be possible to form a mixed ^{87}Rb-^{85}Rb condensate, with the ^{87}Rb and cross-species scattering length positive [17,22], and the ^{85}Rb scattering length tunable. Other Feshbach resonances in both single and multicomponent gases could play important roles in many future BEC and coherent atom optics experiments.

We gratefully acknowledge useful discussions with Carl Wieman and Chris Greene, including a preliminary indication of a resonance field value from experiments at JILA. Work at Texas was supported by the R.A. Welch Foundation, the National Science Foundation, and the NASA Microgravity Research Division.

[1] M. H. Anderson et al., Science **269**, 198 (1995).

[2] K. B. Davis et al., Phys. Rev. Lett. **75**, 3969 (1995).

[3] C. C. Bradley et al., Phys. Rev. Lett. **78**, 985 (1997).

[4] C. J. Myatt et al., Phys. Rev. Lett. **78**, 586 (1997).

[5] U. Ernst et al., Europhys. Lett. **41**, 1 (1998).

[6] L. V. Hau et al. (to be published).

[7] D. J. Han, R. H. Wynar, Ph. Courteille, and D. J. Heinzen, Phys. Rev. A **57**, R4114 (1998).

[8] E. Tiesinga, A. J. Moerdijk, B. J. Verhaar, and H. Stoof, Phys. Rev. A **46**, R1167 (1992); E. Tiesinga, B. J. Verhaar, and H. Stoof, Phys. Rev. A **47**, 4114 (1993).

[9] P. O. Fedichev et al., Phys. Rev. Lett. **77**, 2913 (1996).

[10] J. L. Bohn and P. S. Julienne, Phys. Rev. A **56**, 1486 (1997).

[11] A. J. Moerdijk et al., Phys. Rev. A **51**, 4852 (1995).

[12] J. M. Vogels et al., Phys. Rev. A **56**, R1067 (1997).

[13] N. R. Newbury et al., Phys. Rev. A **51**, R2680 (1995).

[14] S. Inouye, M. R. Andrews, J. Stenger, H.-J. Miesner, D. M. Stamper-Kurn, and W. Ketterle, Nature (London) **392**, 151 (1998).

[15] D. J. Heinzen, in *Atomic Physics 14*, edited by D. J. Wineland, C. E. Wieman, and S. J. Smith, AIP Conf. Proc. No. 323 (AIP, New York, 1995), p. 211.

[16] J. R. Gardner et al., Phys. Rev. Lett. **74**, 3764 (1995).

[17] H. M. J. M. Boesten et al., Phys. Rev. A **55**, 636 (1997).

[18] H. M. J. M. Boesten et al., Phys. Rev. Lett. **77**, 5194 (1996).

[19] C. C. Tsai et al., Phys. Rev. Lett. **79**, 1245 (1997).

[20] J. D. Miller et al., Phys. Rev. A **47**, R4567 (1993).

[21] S. J. J. M. F. Kokkelmans et al., Phys. Rev. A **55**, R1589 (1997); P. S. Julienne et al., Phys. Rev. Lett. **78**, 1880 (1997); J. P. Burke et al., Phys. Rev. A **55**, R2511 (1997).

[22] J. P. Burke et al., Phys. Rev. Lett. **80**, 2097 (1998); Chris Greene (private communication).

[23] F. A. van Abeelen, D. J. Heinzen, and B. J. Verhaar, Phys. Rev. A **57**, R4102 (1998).

[24] S. J. J. M. F. Kokkelmans, B. J. Verhaar, K. Gibble, and D. J. Heinzen, Phys. Rev. A **56**, R4389 (1997).

[25] C. Wieman (private communication).

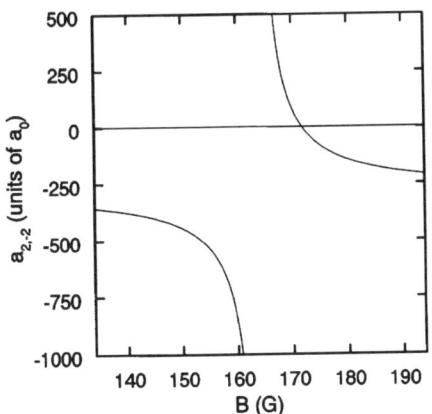

FIG. 5. Calculated field dependence of scattering length $a_{2,-2}$, corresponding with the resonance field value and width observed in this experiment.

REPORTS

Bose-Einstein Condensation of Potassium Atoms by Sympathetic Cooling

G. Modugno, G. Ferrari, G. Roati, R. J. Brecha, A. Simoni, M. Inguscio

We report on the Bose-Einstein condensation of potassium atoms, whereby quantum degeneracy is achieved by sympathetic cooling with evaporatively cooled rubidium. Because of the rapid thermalization of the two different atoms, the efficiency of the cooling process is high. The ability to achieve condensation by sympathetic cooling with a different species may provide a route to the production of degenerate systems with a larger choice of components.

Since the first realizations of Bose-Einstein condensation (BEC) in a dilute gas of alkali atoms (*1–3*), research in the interdisciplinary fields of atom optics and quantum fluids has flourished (*4*). BEC has been observed in five atomic species: H (*5*), ^{4}He (*6*), ^{7}Li (*3*), ^{23}Na (*2*), and the two isotopes ^{85}Rb (*7*) and ^{87}Rb (*1*). Direct forced evaporative cooling of the bosonic isotopes of potassium has been prevented by limitations in the temperature and density ranges achievable by laser cooling (*8*). We show that by taking advantage of thermalization between a small dilute sample of potassium (^{41}K) and evaporatively cooled rubidium (^{87}Rb), these limitations can be overcome. The technique of sympathetic cooling had been proposed for the cooling of ions (*9*) and, in the case of neutral trapped atoms, has been used to obtain quantum degeneracy, but only for two different internal states of the same atom (*10*) or for two isotopes of the same species (*11–13*). The mixing of two different atomic species turned out to be a successful strategy, although their interaction properties could be discovered only by attempting the experiment. In the case of K, sympathetic cooling of its fermionic isotope ^{40}K (*14*) with Rb represents the natural extension of this technique, and may be a new way to explore the physics below Fermi temperature (*15*), as demonstrated in the case of Li (*12, 16*).

The experimental apparatus is based on a conventional double magneto-optical trap (MOT) apparatus, although the complexity is increased because of the requirement for the simultaneous trapping and cooling of two different atomic species. K and Rb atoms are captured from a vapor background in the first MOT (MOT1) and then transferred by resonant laser beams to a second cell with a much lower background pressure, where they are recaptured in a second MOT (MOT2) and

European Laboratory for Nonlinear Spectroscopy (LENS), Università di Firenze, and Istituto Nazionale per la Fisica della Materia (INFM), Largo Enrico Fermi 2, 50125 Firenze, Italy.

loaded in a magnetic trap. Evaporative cooling of Rb is performed, and the evolution of both samples is monitored by means of absorption imaging.

The laser system to manipulate the two atomic species consists of three sources: a titanium:sapphire laser operating on the K optical transitions (767 nm) and two diode lasers operating on the Rb transitions (780 nm). The two pairs of frequencies at different wavelengths necessary for magneto-optical trapping are then injected simultaneously in a semiconductor tapered amplifier (TA), which provides the required power for the two MOTs (*17*).

The experimental sequence begins with the loading of Rb in MOT2 for 30 s. During this phase, the TA power is totally dedicated to Rb, and 10^9 Rb atoms are loaded into MOT2. Half of the TA power is then switched to K, and about 10^7 K atoms are loaded into MOT2 in 8 s. The overall efficiency of the Rb MOT in this phase is strongly reduced because of nonlinear processes in the TA, resulting in a loss of about 50% of the initial Rb sample (*18*).

The magnetic trap consists of a Ioffe-Pritchard potential created by three coils in quadrupole Ioffe configuration (QUIC) (*19*). Both species are optically pumped into the low-field seeking state $|F = 2, m_F = 2>$, before magnetic trapping. The typical axial and radial oscillation frequencies of Rb in the harmonic trap are $\nu_{ax} = 16$ Hz and $\nu_{rad} = 200$ Hz, respectively, whereas those of K are larger by a factor $(M_{Rb}/M_K)^{1/2} = 1.46$, where M_{Rb} and M_K are the masses of the two species. In the QUIC, we typically load 2×10^8 Rb atoms and 2×10^6 K atoms, with both at a temperature of about 300 μK.

Evaporative cooling of Rb is done with a microwave knife tuned to the hyperfine transition at 6.8 GHz, which induces transitions from the trapped state to the untrapped $|F = 1, m_F = 1>$ state, without affecting the K sample. Thus, the evaporation reduces the temperature of both trapped samples, in principle keeping the K population constant. Actually, we observe losses of K atoms in the first part of the evaporation, which can be minimized by forcing the speed of the evaporation ramp. We attribute these losses

REPORTS

to inelastic collisions with an initial tiny fraction of Rb atoms in the trapped $|F = 2, m_F = 1\rangle$ state.

In the evolution of the two species during the whole evaporation over 50 s (Fig. 1), the temperature of the K cloud follows that of Rb (Fig. 1A), thus indicating a very efficient interspecies thermalization. It should be noted that once the populations of Rb and K become similar (Fig. 1B), the efficiency of the evaporation decreases: The ratio η between the evaporation threshold and the temperature of the two samples is reduced from about 7 to 5 because of the relative increase of the K heat capacity, which is proportional to the number of atoms, with respect to that of Rb. Even though the evaporation removes only Rb atoms from the trap, the two heat capacities remain comparable in this final stage of evaporation because of relatively large losses of K atoms, and the cooling continues. Such losses are probably due to inelastic collisions within the K sample, which become relevant as the density increases with the lowering of temperature (Fig. 2).

In this crucial phase of the evaporation, the efficiency of the sympathetic cooling process is sustained by the large collisional interaction between the two species. This observation is confirmed by our determination of the elastic K-Rb collisional cross-section in a mixed sample containing a comparable number of atoms of the two species, at a temperature of 13 μK. Taking advantage of the different trap frequencies for the two species, we have performed a selective parametric excitation of the motion of Rb, recording the subsequent increase of the temperature of the K cloud as a consequence of the collisional energy exchange. From the measured

thermalization rate, using the model presented in (20), we obtain a large value for the zero-energy s-wave triplet scattering length, determined to within a sign as $a_{K41\text{-}Rb87} = 206^{+35}_{-38}\ a_0$ or $a_{K41\text{-}Rb87} = -266^{+55}_{-50}\ a_0$ (21). These values determine a corresponding range of values for the triplet scattering length of the pair ^{40}K-^{87}Rb, via mass scaling: We obtain $a_{K40\text{-}Rb87} = -95^{+33}_{-98}\ a_0$ or $a_{K40\text{-}Rb87} = 30^{+3}_{-2}\ a_0$, for positive or negative signs of the ^{41}K-^{87}Rb scattering length, respectively. At low temperature, the collisional cross-section is proportional to the square of the scattering length (21), and so the two different cases result in completely different scenarios for sympathetic cooling of the fermionic K with Rb.

At the end of the evaporation stage, we observe the formation of a BEC of K out of the thermal sample (22) (Fig. 3). The density of the central part of the K cloud increases as the evaporation threshold is lowered below 40 kHz, indicating the onset of the quantum degeneracy regime. Further evaporation of Rb reduces the number of K atoms in the broad thermal component, which eventually is undetectable, leaving an almost pure condensate. The typical number of atoms in the condensate is $N \approx 10^4$, and a Gaussian fit to the wings of the bimodal density distribution gives a temperature $T \approx 160$ nK (Fig. 4). This observation is consistent with

the theoretical expectation for the transition temperature, $T_C = h/k_B\ (N\,\nu_{ax}\,\nu_{rad}^2/1.202)^{1/3} = 150$ nK, where h and k_B are the Planck's and Boltzman's constants (23).

Due to the different gravitational sag of the K and Rb samples in the magnetic trap, their spatial overlap can be reduced at low temperatures. However, we expect the vertical separation of the two clouds to become larger than their thermal radius only below the K critical temperature. As discussed in (24) and confirmed by our experimental observation, this results in a good thermal contact between the two species all the way down to the BEC of K.

We have also measured the K-K triplet scattering length with a conventional cross-dimensional thermalization technique (25), on a pure K sample at 13 μK, obtaining $a_{K\text{-}K} = 78 \pm 20$ a_0. The repulsive character of the interaction between K atoms (that is, the positive sign of $a_{K\text{-}K}$) results from our observation of stable condensates containing about 10^4 atoms (23). Our direct determination of the scattering length is in agreement with the prediction $a_{K\text{-}K} = 60 \pm 2\ a_0$ from molecular photoassociation measurement on the isotope ^{39}K (26).

The K-Rb mixture is an interesting candidate for the formation of ultracold polar molecules, using recently developed schemes

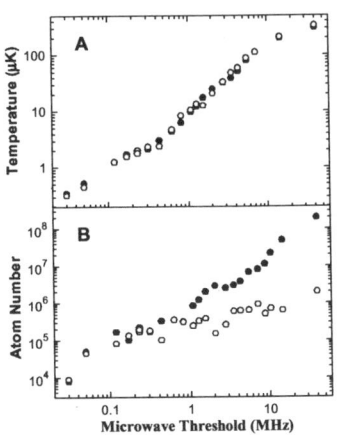

Fig. 1. Evolution of the number of atoms (**A**) and temperature (**B**) of the two atomic samples in the magnetic trap as a function of the microwave evaporation threshold of Rb. The solid circles correspond to ^{87}Rb and the open circles to ^{41}K.

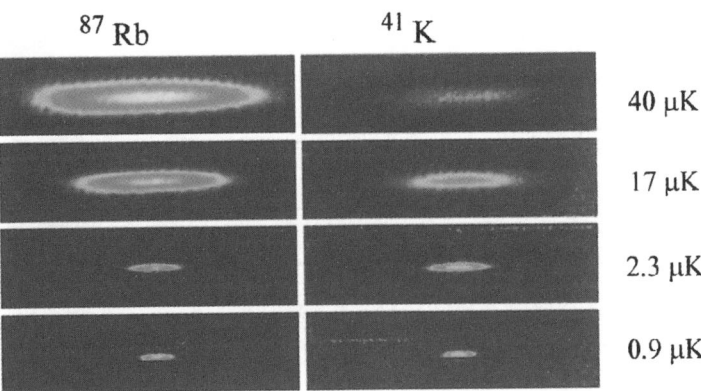

Fig. 2. False color absorption images of Rb (**left**) and K (**right**) at four different stages of the sympathetic cooling. The density of the K sample increases by more than two orders of magnitude, going from 4×10^9 cm^{-3} to 6×10^{11} cm^{-3}, when the temperature is lowered from 40 to 0.9 μK. The density of the Rb sample is instead approximately constant during the evaporation.

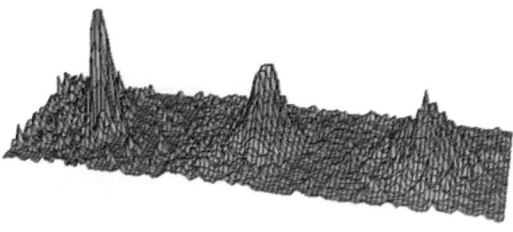

Fig. 3. False color density profiles of the ^{41}K cloud after 15 ms of ballistic expansion, across the phase transition to BEC. From right to left, profiles are as follows: thermal cloud at $T > T_C$; partially condensed sample at $T \approx T_C$; and almost pure condensate at $T < T_C$, containing about 10^4 atoms.

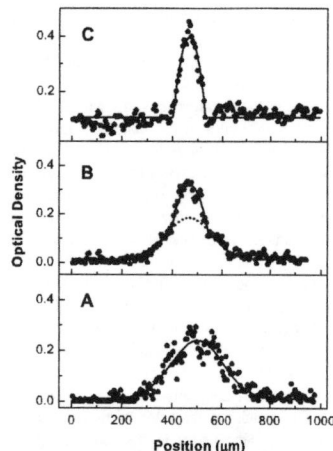

Fig. 4. Density profiles of three samples of ^{41}K after 15 ms of expansion, showing the transition to BEC. (**A**) Thermal sample at $T = 250$ nK. (**B**) Mixed sample at $T = 160$ nK. (**C**) Almost pure condensate. The lines are the best fit with a Gaussian for the thermal component and with an inverted parabola for the condensate component.

($27, 28$), which may represent a new system for quantum computing (29).

The possibility of sympathetic cooling to quantum degeneracy with a different species broadens the spectrum of coolable particles to include molecules as well (30). Because Rb is the workhorse for experiments on cold atoms, one could take advantage of recently demonstrated techniques for a simultaneous trapping of the partner species. For example, a BEC of ^{87}Rb has been produced in an optical dipole trap (31), which is an ideal tool to trap a large variety of atoms and molecules lacking a magnetic moment in their ground state. Sympathetic cooling in this kind of trap would have repercussions for high-resolution spectroscopy and metrology (32), tests of fundamental theories (33), and ultracold chemistry.

References and Notes
1. M. H. Anderson, J. R. Ensher, M. R. Mattews, C. E. Wieman, E. A. Cornell, *Science* **269**, 198 (1995).
2. K. B. Davis *et al.*, *Phys. Rev. Lett.* **75**, 3969 (1995).
3. C. C. Bradley, C. A. Sackett, J. J. Tollet, R. G. Hulet, *Phys. Rev. Lett.* **75**, 1687 (1995).
4. M. Inguscio, S. Stringari, C. E. Wieman, Eds., *Bose-Einstein Condensation in Atomic Gases*, Proceedings of the International School of Physics "Enrico Fermi," Course CXL (IOS Press, Amsterdam, 1999).
5. D. G. Fried *et al.*, *Phys. Rev. Lett.* **81**, 3811 (1998).
6. A. Robert *et al.*, *Science* **292**, 461 (2001); published online 22 March 2001 (10.1126/science.1060622).
7. S. L. Cornish *et al.*, *Phys. Rev. Lett.* **85**, 1795 (2000).
8. M. Prevedelli *et al.*, *Phys. Rev. A* **59**, 886 (1999).
9. D. J. Wineland, R. E. Drullinger, F. L. Walls, *Phys. Rev. Lett.* **40**, 1639 (1978).
10. C. J. Myatt *et al.*, *Phys. Rev. Lett.* **78**, 586 (1997).
11. F. Schreck *et al.*, *Phys. Rev. A* **87**, 011402(R)(2001).
12. G. Truscott, K. E. Strecker, W. I. McAlexander, G. B. Partridge, R. G. Hulet, *Science* **291**, 2570 (2001).
13. I. Bloch, M. Greiner, O. Mandel, T. W. Hänsch, T. Esslinger, *Phys. Rev. A* **64**, 021402(R) (2001).
14. F. S. Cataliotti *et al.*, *Phys. Rev. A* **57**, 1136 (1998).
15. B. De Marco, D. S. Jin, *Science* **285**, 1703 (1999).
16. F. Schreck *et al.*, *Phys. Rev. Lett.* **87**, 080403 (2001).
17. This is a development of the scheme introduced in G. Ferrari, M.-O. Mewes, F. Schreck, C. Salomon, *Opt. Lett.* **24**, 151 (1999).
18. A near coincidence between the K ground-state and the Rb excited-state hyperfine splittings induces losses of Rb atoms from the MOTs, when K and Rb lights are injected simultaneously in the TA.
19. T. Esslinger, I. Bloch, T. W. Hänsch, *Phys. Rev. A* **58**, R2664 (1998).
20. A. Mosk *et al.*, preprint available at http://xxx.lanl.gov/abs/physics/0107075.
21. The interaction properties of ultracold atoms are described by a single parameter, namely the scattering length a, which we give in atomic units ($a_0 = 0.0529$ nm). The zero-energy collisional cross-section between distinguishable particles can be expressed as $\sigma = 4\pi a^2$. At finite energy, σ depends also on the sign of the scattering length. The mean-field interaction energy in a Bose-Einstein condensate can be expressed as $E = h^2 na/(\pi M)$, where n is the gas density.
22. Because the evaporation ramp is optimized for K, at this stage the Rb sample typically contains less than 10^4 atoms and still follows a thermal distribution. Further reduction of the evaporation threshold results in a complete loss of all the Rb atoms.
23. S. Dalfovo, S. Giorgini, L. P. Pitaevskii, S. Stringari, *Rev. Mod. Phys.* **71**, 463 (1999).
24. G. Delannoy *et al.*, *Phys. Rev. A* **63**, 051602(R) (2001).
25. C. Monroe, E. A. Cornell, C. A. Sackett, C. J. Myatt, C. E. Wieman, *Phys. Rev. Lett.* **70**, 414 (1993).
26. H. Wang *et al.*, *Phys. Rev. A* **62**, 052704 (2000).
27. A. Fioretti *et al.*, *Phys. Rev. Lett.*, **80**, 4402 (1998).
28. S. J. J. M. F. Kokkelmans, H. M. J. Vissers, B. J. Verhaar, *Phys. Rev. A* **63**, 031601(R) (2001) and references therein.
29. D. DeMille, preprint available at http://xxx.lanl.gov/abs/quant-ph/0109083.
30. Trapped molecular samples at a few mK have recently been produced [H. L. Bethlem *et al.*, *Nature* **406**, 491 (2000)].
31. M. D. Barrett, J. A. Sauer, M. S. Chapman, *Phys. Rev. Lett.* **87**, 010404 (2001).
32. C. W. Oates, E. A. Curtis, L. Hollberg, *Opt. Lett.* **25**, 1603 (2000).
33. E. A. Hinds, K. Sangster, in *Time Reversal-The Arthur Rich Memorial Symposium*, M. Skalsey, P. H. Bucksbaum, R. S. Conti, D. W. Gidley, Eds. (American Institute of Physics, New York, 1993).
34. We benefited from stimulating discussions with all the colleagues of the laser cooling and BEC group at LENS. We thank W. Jastrzebski, N. Poli, F. Riboli, and L. Ricci for their contribution to the experiment; I. Bloch for useful hints for the construction of the magnetic trap; and R. Ballerini, M. DePas, M. Giuntini, A. Hajeb, and A. Orlando for technical assistance. Supported by the Ministero dell'Università e della Ricerca Scientifica; by the European Community under contract HPRICT1999-00111; and by the Istituto Nazionale per la Fisica della Materia, Progetto di Ricerca Avanzata "Photonmatter." G.R. is also at Dipartimento di Fisica, Università di Trento; R.B. is presently at the Department of Physics, University of Dayton, OH; A.S. is also at Dipartimento di Chimica, Università di Perugia; M.I. is also at Dipartimento di Fisica, Università di Firenze.

1 October 2001; accepted 11 October 2001
Published online 18 October 2001;
10.1126/science.1066687
Include this information when citing this paper.

R E P O R T S

Josephson Junction Arrays with Bose-Einstein Condensates

F. S. Cataliotti,[1,2,3] S. Burger,[1,2] C. Fort,[1,2] P. Maddaloni,[1,2,4]
F. Minardi,[1,2] A. Trombettoni,[2,5] A. Smerzi,[2,5] M. Inguscio[1,2,3]*

We report on the direct observation of an oscillating atomic current in a one-dimensional array of Josephson junctions realized with an atomic Bose-Einstein condensate. The array is created by a laser standing wave, with the condensates trapped in the valleys of the periodic potential and weakly coupled by the interwell barriers. The coherence of multiple tunneling between adjacent wells is continuously probed by atomic interference. The square of the small-amplitude oscillation frequency is proportional to the microscopic tunneling rate of each condensate through the barriers and provides a direct measurement of the Josephson critical current as a function of the intermediate barrier heights. Our superfluid array may allow investigation of phenomena so far inaccessible to superconducting Josephson junctions and lays a bridge between the condensate dynamics and the physics of discrete nonlinear media.

The existence of a Josephson current through a potential barrier between two superconductors or between two superfluids is a direct manifes-

[1]European Laboratory for Non-Linear Spectroscopy (LENS) and [2]Istituto Nationale per la Fisica della Materia (INFM), L.go E. Fermi 2, I-50125 Firenze, Italy. [3]Dipartimento di Fisica, Università di Firenze, L.go E. Fermi 2, I-50125 Firenze, Italy. [4]Dipartimento di Fisica, Università di Padova, via F. Marzolo 8, I-35131 Padova, Italy. [5]International School for Advanced Studies (SISSA), via Beirut 2/4, I-34014 Trieste, Italy.

*To whom correspondence should be addressed. E-mail: inguscio@lens.unifi.cit

tation of macroscopic quantum phase coherence (1, 2). The first experimental evidence of a current-phase relation was observed in superconducting systems soon after the Josephson effect was proposed in 1962 (3), whereas verification in superfluid helium has been presented only recently owing to the difficulty of creating weak links in a neutral quantum liquid (4, 5). The experimental realization of Bose-Einstein condensates (BEC) of weakly interacting alkali atoms (6, 7) has provided a route to study neutral superfluids in a controlled and tunable environment (8, 9) and to implement novel

REPORTS

geometries for the connection of several Josephson junctions so far unattainable in charged systems. The possibility of loading a BEC in a one-dimensional periodic potential has allowed the observation of quantum phase effects on a macroscopic scale such as quantum interference (10) and the study of superfluidity on a local scale (11).

A Josephson junction (JJ) is a simple device made of two coupled macroscopic quantum fluids (2). If the coupling is weak enough, an atomic mass current I flows across the two systems, driven by their relative phase $\Delta\phi$ as

$$I = I_c \sin \Delta\phi \qquad (1)$$

where I_c is the "Josephson critical current," namely the maximal current allowed to flow through the junction. The relative phase dynamics, on the other hand, is sensitive to the external and internal forces driving the system:

$$\hbar \frac{d}{dt} \Delta\phi = \Delta V \qquad (2)$$

where $\hbar$ is Planck's constant divided by 2π, t is time, and ΔV is the chemical potential difference between the two quantum fluids. Arrays of JJs are made of several simple junctions connected in various geometrical configurations. In the past decade, such systems have attracted much interest, because of their potential for studying quantum phase transitions in systems where the external parameters can be readily tuned (12). Recently, the creation of simple quantum-logic units and more complex quantum computer schemes (13) has been discussed. A great level of accuracy has been reached in the realization of two- and three-dimensional superconducting JJ arrays (12). One-dimensional (1D) geometries are more difficult to realize, because of the unavoidable presence of on-site frustration charges that substantially modify the ideal phase diagram. 1D JJ arrays with neutral superfluids (such as BEC), on the other hand, can be accurately tailored and open the possibility of directly observing several remarkable phenomena not accessible to other systems (14). First experiments with BECs held in a vertical optical lattice have shown the spatial and temporal coherence of condensate waves emitted at different heights of the gravitational field (10). More recently, the degree of phase coherence among different sites of the array (15) has been explored in the BEC ground state configuration.

We report on the realization of a 1D array of JJs by loading a BEC into an optical lattice potential generated by a standing-wave laser field. The current-phase dynamics, driven by an external harmonic oscillator potential provided by an external magnetic field, maps on a pendulumlike equation, and we performed a measurement of the critical Josephson current as a function of the interwell potentials created by the light field.

The experimental apparatus has been described in detail elsewhere (11). We produce BECs of ^{87}Rb atoms in the Zeeman state $m_F = -1$ of the hyperfine level $F = 1$ state confined by a cylindrically symmetric harmonic magnetic trap and a blue detuned laser standing wave, superimposed on the axis of the magnetic trap. In essence, the cylindrical magnetic trap is divided into an array of disk-shaped traps by the light standing wave. The axial and radial frequencies of the magnetic trap are $\omega_x = 2\pi \times 9$ Hz and $\omega_r = 2\pi \times 92$ Hz, respectively. By varying the intensity of the superimposed laser beam (detuned 150 GHz to the blue of the D1 transition at $\lambda = 795$ nm) up to 14 mW/mm^2, we can vary the interwell barrier energy V_0 from 0 to $5E_R$, where $E_R = h^2/2m\lambda^2$ is the recoil energy of an atom (of mass m) absorbing one of the lattice photons (16). The BEC is prepared by loading ~5×10^5 atoms in the magnetic trap and cooling the sample through radiofrequency-forced evaporation until a substantial fraction of condensed atoms is produced. We then switch on the laser standing wave and continue the evaporation ramp until no thermal component is experimentally visible. This ensures that the system reaches the ground state of the combined trap. The BEC splits in the wells of the optical array: The distance between the wells is $\lambda/2$ and ~200 wells are typically occupied, with ~1000 atoms in each well. The interwell barrier energy V_0, and therefore the tunneling rate, are controlled by varying the intensity of the laser, which is chosen to be much higher than the condensate chemical potential μ. μ ranges between $\mu \approx 0.10\ V_0$ for $V_0 = 2E_R$ and $\mu \approx 0.04\ V_0$ for $V_0 = 5E_R$. Each couple of condensates in neighboring wells therefore realizes a bosonic JJ, with a critical current I_c depending on the laser intensity.

In a more formal way, we can decompose the condensate order parameter that depends on position $\vec{r}$ and time t as a sum of wave functions localized in each well of the periodic potential (tight binding approximation):

$$\Psi(\vec{r},t) = \sqrt{N_T} \sum_j \psi_j(t)\Phi_j(\vec{r}) \qquad (3)$$

where N_T is the total number of atoms and $\psi_j = \sqrt{n_j(t)}\ e^{i\phi_j(t)}$ is the jth amplitude, with the fractional population $n_j = N_j/N_T$ and the number of particles N_j and the phase ϕ_j in the trap j. This assumption relies on the fact that the height of the interwell barriers is much higher than the chemical potential. We will show by a variational calculation that this assumption is verified in most of the range of our experimental parameters (17). The wave function $\Phi_j(\vec{r})$ of the condensate in the jth site of the array overlaps in the barrier region with the wave functions $\Phi_{j\pm1}$ of the condensates in the neighbor sites. Therefore, the system realizes an array of weakly coupled condensates, whose equation of motion satisfies a discrete nonlinear Schrödinger equation (18):

$$i\hbar \frac{\partial\psi_n}{\partial t} = -K(\psi_{n-1} + \psi_{n+1})$$
$$+ (\varepsilon_n + U|\psi_n|^2)\psi_n \qquad (4)$$

where $\varepsilon_n = \Omega n^2$, $\Omega = (1/2)m\omega_x^2(\lambda/2)^2 = 1.54 \times 10^{-5}E_R$ and $U = g_0 N_T \int d\vec{r},\ \Phi_j^4$. The tunneling rate is proportional to $K = -\int d\vec{r}[(\hbar^2/2m)\vec\nabla\Phi_j \cdot \vec\nabla\Phi_{j+1} + \Phi_j V_{ext}\Phi_{j+1}]$. A simple variational estimate, assuming a gaussian profile for the condensates in each trap, gives for $V_0 = 3E_R$ the values $K \sim 0.07E_R$, $U \sim 12E_R$, and a chemical potential $\mu \sim 0.06V_0$ that is much lower than the interwell potential V_0. We observe that the wave functions Φ_j, as well as K, depend on the height of the energy barrier V_0.

Equation 4 is a discrete nonlinear Schrödinger equation (DNLSE) in a parabolic external potential, conserving both the Hamiltonian $\mathscr{H} = \sum_j[-K(\psi_j\psi_{j+1}^* + \psi_j^*\psi_{j+1}) + \varepsilon_j|\psi_j|^2 + (U/2)|\psi_j|^4]$ and the norm $\sum_j n_j = 1$.

Although we can approximate the condensates in each lattice site as having their own wave functions, tunneling between adjacent wells locks all the different condensates in phase. As a result, when the condensates are released from the combined trap, they will show an interference pattern. This pattern consists of a central peak plus a symmetric comb of equally spaced peaks separated by $\pm 2\hbar k_l t_{exp}/m$, where k_l is the wave vector of the trapping laser and t_{exp} is the expansion time. In practice, one can think of the far field intensity distribution of a linear array of dipole antennas all emitting with the same phase. A complementary point of view is to regard the density distribution after expansion as the Fourier transform of the trapped one, i.e., the momentum distribution (19). It is easy to show that the sum of De Broglie waves corresponding to momentum states integer multiples of $\pm 2\hbar k_l$ is the sum of localized wave functions of Eq. 3. The expanded cloud density distribution (Fig. 1) consists of three distinct atomic clouds spaced by ~306 μm $\simeq 2\hbar k_l t_{exp}/m$, with the two external clouds corresponding to the first-order interference peaks, each containing roughly 10% of the total number of atoms. The interference pattern therefore provides us with information about the relative phase of the different condensates (15, 20); indeed, by repeating the experiment with thermal clouds, even with a temperature considerably lower than the interwell potential, we did not observe the interference pattern.

REPORTS

This situation is different from the Bragg diffraction of a condensate released from a harmonic magnetic trap (21) where the condensate is diffracted by a laser standing wave. In our case, it is the ground state of the combined magnetic harmonic trap plus optical periodic potential that by expansion produces an interference pattern. For the time scales of our experiment, the relative intensities of the three interference peaks do not depend on the time the atoms spend in the optical potential, indicating that the steady state of the system has been reached. In absence of external perturbations, the condensate remains in the state described by Eq. 3 with a lifetime of ~0.3 s at the maximum light power, limited by scattering of light from the laser standing wave.

In the ground state configuration, the BECs are distributed among the sites at the bottom of the parabolic trap. If we suddenly displace the magnetic trap along the lattice axis by a small distance of ~30 μm (the dimension of the array is ~100 μm), the cloud will be out of equilibrium and will start to move. As the potential energy that we give to the cloud is still smaller than the interwell barrier, each condensate can move along the magnetic field only by tunneling through the barriers. A collective motion can only be established at the price of an overall phase coherence among the condensates. In other words, the relative phases among all adjacent sites should remain locked together to preserve the ordering of the collective motion. The locking of the relative phases will again show up in the expanded cloud interferogram.

For displacements that are not very large, we observe a coherent collective oscillation of the condensates; i.e., we see the three peaks of the interferogram of the expanded condensates oscillating in phase, thus showing that the quantum mechanical phase is maintained over the entire condensate (Fig. 2). In Fig. 2A, we show the positions of the three peaks as a function of time spent in the combined trap after the displacement of the magnetic trap, compared with the motion of the condensate in the same displaced magnetic trap but in the absence of the optical standing wave (we refer to this as "harmonic" oscillation). The motion performed by the center of mass of the condensate is an undamped oscillation at a substantially lower frequency than in the "harmonic" case. We will comment on this frequency shift later in the text; we would like now to further stress the coherent nature of the oscillation. To do so, we repeat the same experiment with a thermal cloud. In this case, although individual atoms are allowed to tunnel through the barriers, no macroscopic phase is present in the cloud and no motion of the center of mass should be observed. The center of mass positions of the thermal clouds are reported in Fig. 2B together with the "harmonic" oscillation of the same cloud in the absence of the optical potential. As can be seen, the thermal cloud does not move from its original position in the presence of the optical lattice. Indeed, if a mixed cloud is used, only the condensate fraction starts to oscillate while the thermal component remains static, with the interaction of the two eventually leading to a damping of the condensate motion.

We now turn back to the discussion of the frequency reduction observed in the oscillation of the pure condensate in the presence of the optical lattice. The current flowing through the junction between two quantum fluids has a maximum value, the critical Josephson current I_c, which is directly proportional to the tunneling rate. The existence of such a condition essentially limits the maximum velocity at which the condensate can flow through the interwell barriers and therefore reduces the frequency of the oscillations. As a consequence, we expect a dependence of the oscillation frequency on the optical potential through the tunneling rate.

To formalize the above reasoning, we rewrite the DNLSE (Eq. 4) in terms of the canonically conjugate population/phase variables, therefore enlightening its equivalence with the Josephson equations for a 1D junction array:

$$\hbar \dot{n}_j = 2K \sqrt{n_j n_{j-1}} \sin(\phi_j - \phi_{j-1}) - 2K \sqrt{n_j n_{j+1}} \sin(\phi_{j+1} - \phi_j) \quad (5a)$$

$$\hbar \dot{\phi}_j = -U n_j - \Omega j^2 + K \sqrt{n_{j-1}/n_j} \cos(\phi_j - \phi_{j-1}) + K \sqrt{n_{j+1}/n_j} \cos(\phi_{j+1} - \phi_j) \quad (5b)$$

It is useful to introduce collective coordinates (18): The center of mass $\xi(t)$ and the dispersion $\sigma(t)$ are defined, respectively, as $\xi(t) = \Sigma_j j n_j$

Fig. 1. (A) Combined potential of the optical lattice and the magnetic trap in the axial direction. The curvature of the magnetic potential is exaggerated by a factor of 100 for clarity. (B) Absorption image of the BEC released from the combined trap. The expansion time was 26.5 ms and the optical potential height was $5E_R$.

A

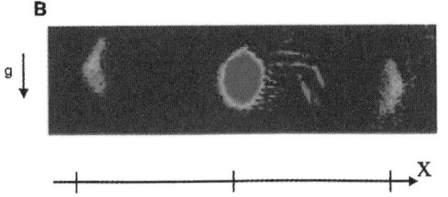

B

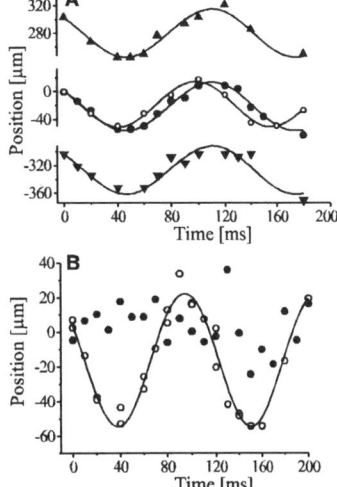

Fig. 2. (A) Center of mass positions of the three peaks in the interferogram of the expanded condensate as a function of the time spent in the combined trap after displacement of the magnetic field. Up and down triangles correspond to the first-order peaks; filled circles correspond to the central peak. Open circles show the center of mass position of the BEC in the absence of the optical lattice. The continuous lines are the fits to the data. (B) Center of mass positions of the thermal cloud as a function of time spent in the displaced magnetic trap with the standing wave turned on (filled circles) and off (open circles).

REPORTS

Fig. 3. The frequency of the atomic current in the array of Josephson junctions as a function of the interwell potential height. Experimental data (circles) are compared with the calculated values (triangles). Each experimental data point was taken after a complete oscillation in the displaced magnetic trap. The oscillation was then fitted with a sine function giving the corresponding frequency (error bars are the standard deviation of the data from the fit).

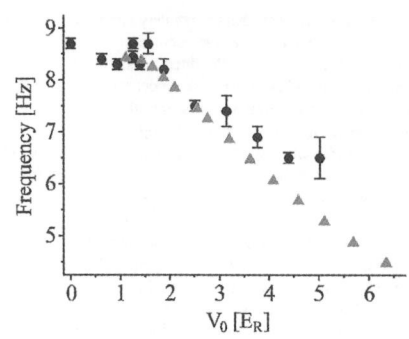

3. B. D. Josephson, *Phys. Lett.* **1**, 251 (1962).
4. O. Avenel, E. Varoquaux, *Phys. Rev. Lett.* **60**, 416 (1988).
5. S. V. Pereverzev, S. Backaus, A. Loshak, J. C. Davis, R. E. Packard, *Nature* **388**, 449 (1997).
6. M. Inguscio, C. E. Wieman, S. Stringari, Eds., *Bose-Einstein Condensation in Atomic Gases* (IOS Press, Amsterdam, 1999).
7. S. Martellucci, A. N. Chester, A. Aspect, M. Inguscio, Eds., *Bose-Einstein Condensates and Atom Lasers* (Kluwer Academic/Plenum, New York, 2000).
8. F. Dalfovo, S. Giorgini, L. P. Pitaevskii, S. Stringari, *Rev. Mod. Phys.* **71**, 463 (1999).
9. A. J. Legget, *Rev. Mod. Phys.* **73**, 307 (2001).
10. B. P. Anderson, M. A. Kasevich, *Science* **282**, 1686 (1998).
11. S. Burger et al., *Phys. Rev. Lett.* **86**, 4447 (2001).
12. R. Fazio, H. van der Zant, *Phys. Rep.*, in press (preprint available at xxx.lanl.gov/abs/cond-mat/0011152).
13. Y. Makhlin, G. Schön, A. Shnirman, *Rev. Mod. Phys.* **73**, 357 (2001).
14. A. Smerzi, S. Fantoni, S. Giovannazzi, S. R. Shenoy, *Phys. Rev. Lett.* **79**, 4950 (1997).
15. C. Orzel, A. K. Tuchman, M. L. Fenselau, M. Yasuda, M. A. Kasevich, *Science* **291**, 2386 (2001).
16. The value of the optical potential used in all the variational calculations was calibrated by performing Bragg diffraction experiments on the BEC released from the harmonic trap. The experimental result deviates from the potential calculated because of alignment imperfections.
17. The validity of the tight binding approximation is also based on the fact that the tunneling of atoms in the higher energy band is energetically forbidden: Because the gap is $\sim 3E_R$, the potential energy (1/2) $m\omega_x^2(\lambda/2)^2 j^2$ for that would require $j \sim 500$, i.e., displacements three times larger than the condensate dimensions.
18. A. Trombettoni, A. Smerzi, *Phys. Rev. Lett.* **86**, 2353 (2001).
19. The expanded density distribution reproduces the momentum distribution for expansion times much longer than the inverse of the trapping frequencies if the nonlinear terms in the Schrödinger equations (the mean field) can be neglected during the expansion. The trapping frequencies of the single traps in the array are on the order of a few kHz and the expansion time is 26.5 ms, so the first assumption is readily verified. The question of neglecting the mean field in the first part of the expansion when the density is still comparable to the original condensate is more delicate. However, this will only affect the shape of the single interference peaks and not the overall interference pattern.
20. M. Greiner, I. Bloch, O. Mandel, T. W. Haensch, T. Esslinger, preprint available at xxx.lanl.gov/abs/cond-mat/0105105.
21. M. Kozuma et al., *Phys. Rev. Lett.* **82**, 871 (1999).
22. This is the discrete analog of the "Thomas-Fermi" approximation for the continuous Gross-Pitaevski equation with an external parabolic potential. In this limit, as will be shown below, the dynamics does not depend explicitly on the nonlinear interatomic interaction, which only governs the overall shape. Our collective mode, indeed, can be seen as the discrete analog of the dipole mode in the continuous Gross-Pitaevskii equation, whose frequency depends only on the parameters of the external harmonic trap.
23. A. C. Scott, *Nonlinear Science: Emergence and Dynamics of Coherent Structures* (Oxford Univ. Press, Oxford, 1999).
24. W. Zurek, *Phys. Today* **44**, 36 (1991).
25. A. Sanchez, A. R. Bishop, *SIAM Rev.* **40**, 579 (1998).
26. F. Kh. Abdullaev, B. B. Baizakov, S. A. Darmanyan, V. V. Konotop, M. Salerno, *Phys. Rev. A*, in press (preprint available at xxx.lanl.gov/abs/cond-mat/0106042).
27. This work has been supported by the Cofinanziamento Ministero dell'Università e della Ricerca Scientifica e Tecnologica, by the European Community under contract HPRI-CT-1999-00111 and HPRN-CT-2000-00125, and by the Istituto Nazionale per la Fisica della Materia Progetto di Ricerca Avanzata "Photonmatter." We thank M. Kasevich and S. Stringari for helpful discussions. A.S. and A.T. wish to thank the LENS for the kind hospitality during the realization of this work.

17 May 2001; accepted 29 June 2001

and $\sigma^2(t) = \Sigma_j j^2 n_j - \xi^2$. From Eqs. 5a and 5b, we have $\hbar\dot\xi = 2K\Sigma_j \sqrt{n_j n_{j+1}} \sin(\phi_{j+1} - \phi_j)$. As the number of atoms is large, the "kinetic" energy term of DNLSE is small with respect to the potential and nonlinear terms, and the population density profile is simply given by an inverted discrete parabolic profile, centered around ξ (22): $n_j(t) = [\mu - \Omega(j - \xi)^2]/U$; furthermore, $(d/dt)\sigma^2 = 0$.

During the dynamical evolution, the relative phases across the junctions $\phi_{j+1} - \phi_j \equiv \Delta\phi(t)$ remain locked together to the same (oscillating) value. This has been verified by numerically studying the Fourier transform $\hat\psi_k = \Sigma_j \psi_j e^{ikj}$; from the experimental point of view, this means that the expanded condensate continues to show the three peaks of the interferogram of Fig. 1. Therefore in these collective coordinates the current-phase relation is given by

$$\hbar\frac{d}{dt}\xi(t) = 2K \sin\Delta\phi(t) \tag{6a}$$

$$\hbar\frac{d}{dt}\Delta\phi(t) = -m\omega_x^2\left(\frac{\lambda}{2}\right)^2 \xi(t) \tag{6b}$$

which, in analogy with the case of a superconducting Josephson junction [in the resistively shunted junction model (1, 2)] and with the case of ³He (5), is a pendulum equation with the relative phase $\Delta\phi$ corresponding to the angle to a vertical axis and the center of mass ξ being the corresponding angular momentum. The current-phase dynamics does not depend explicitly on the interatomic interaction. This allows us to study regimes with a number of condensate atoms spanning over different orders of magnitude, which is different from the configuration considered in (10) where nonlinear effects would dephase the collective dynamics. However, it is clear that the nonlinear interaction is crucial to determining the superfluid nature of the coupled condensates, by locking the overall phase coherence against perturbations.

From Eqs. 6a and 6b, we can see that the small-amplitude oscillation frequency ω_I of the current $I \equiv N_T(d/dt)\xi$ gives a direct measurement of the critical Josephson current $I_c \equiv 2KN_T/\hbar$ and, therefore, of the atomic tunneling

rate of each condensate through the barriers. The critical current is related to the frequency ω of the atomic oscillations in the lattice and to the frequency ω_x of the condensate oscillations in the absence of the periodic field by the relation

$$I_c = \frac{4\hbar}{m\lambda^2}\frac{N_T}{}\left(\frac{\omega}{\omega_x}\right)^2 \tag{7}$$

Figure 3 shows the experimental values of the oscillation frequencies together with the result of a variational calculation based on Eqs. 5a and 5b. It must be noted that, because of mean field interactions, in our system a bound state exists in the lattice only for potentials higher than $\sim E_R$; frequency shifts for lower potential heights are better explained in terms of the effective mass $1/m_{\text{eff}} = \partial^2\mathcal{H}/\partial k^2$ of the system (11).

Increasing the initial angular momentum ξ_0, the pendulum librations become anharmonic and can eventually reach the value $\Delta\phi_{\max} = \pi/2$. The system becomes dynamically unstable, and the phase coherence is lost after a transient time (the interference patterns wash out). In this regime, the pendulum analogy breaks down and a different dynamical picture would emerge.

With this work, we have verified that the BEC's dynamics on a lattice is governed by a discrete, nonlinear Schrödinger equation. This equation is common to a large class of discrete nonlinear systems, including polarons, optical fibers, and biological molecules (23), thus opening up interdisciplinary research. The phase rigidity among different wells can be probed against thermal fluctuations to test various theories of decoherence (24). One could study the role of collective dynamical modes in the creation of solitons and kinks of the type described in (18, 25, 26) [see also (23)] and more generally the routes to quantum phase transitions in nonhomogeneous, low-dimensional systems.

References and Notes
1. A. Barone, in *Quantum Mesoscopic Phenomena and Mesoscopic Devices in Microelectronics*, I. O. Kulik, R. Ellialtioglu, Eds. (Kluwer Academic, Dordrecht, Netherlands, 2000), pp. 301–320.
2. ———, G. Paterno, *Physics and Applications of the Josephson Effect* (Wiley, New York, 1982).

Quantum phase transition from a superfluid to a Mott insulator in a gas of ultracold atoms

Markus Greiner*, Olaf Mandel*, Tilman Esslinger†, Theodor W. Hänsch* & Immanuel Bloch*

* *Sektion Physik, Ludwig-Maximilians-Universität, Schellingstrasse 4/III, D-80799 Munich, Germany, and Max-Planck-Institut für Quantenoptik, D-85748 Garching, Germany*
† *Quantenelektronik, ETH Zürich, 8093 Zurich, Switzerland*

For a system at a temperature of absolute zero, all thermal fluctuations are frozen out, while quantum fluctuations prevail. These microscopic quantum fluctuations can induce a macroscopic phase transition in the ground state of a many-body system when the relative strength of two competing energy terms is varied across a critical value. Here we observe such a quantum phase transition in a Bose–Einstein condensate with repulsive interactions, held in a three-dimensional optical lattice potential. As the potential depth of the lattice is increased, a transition is observed from a superfluid to a Mott insulator phase. In the superfluid phase, each atom is spread out over the entire lattice, with long-range phase coherence. But in the insulating phase, exact numbers of atoms are localized at individual lattice sites, with no phase coherence across the lattice; this phase is characterized by a gap in the excitation spectrum. We can induce reversible changes between the two ground states of the system.

A physical system that crosses the boundary between two phases changes its properties in a fundamental way. It may, for example, melt or freeze. This macroscopic change is driven by microscopic fluctuations. When the temperature of the system approaches zero, all thermal fluctuations die out. This prohibits phase transitions in classical systems at zero temperature, as their opportunity to change has vanished. However, their quantum mechanical counterparts can show fundamentally different behaviour. In a quantum system, fluctuations are present even at zero temperature, due to Heisenberg's uncertainty relation. These quantum fluctuations may be strong enough to drive a transition from one phase to another, bringing about a macroscopic change.

A prominent example of such a quantum phase transition is the change from the superfluid phase to the Mott insulator phase in a system consisting of bosonic particles with repulsive interactions hopping through a lattice potential. This system was first studied theoretically in the context of superfluid-to-insulator transitions in liquid helium[1]. Recently, Jaksch *et al.*[2] have proposed that such a transition might be observable when an ultracold gas of atoms with repulsive interactions is trapped in a periodic potential. To illustrate this idea, we consider an atomic gas of bosons at low enough temperatures that a Bose–Einstein condensate is formed. The condensate is a superfluid, and is described by a wavefunction that exhibits long-range phase coherence[3]. An intriguing situation appears when the condensate is subjected to a lattice potential in which the bosons can move from one lattice site to the next only by tunnel coupling. If the lattice potential is turned on smoothly, the system remains in the superfluid phase as long as the atom–atom interactions are small compared to the tunnel coupling. In this regime a delocalized wavefunction minimizes the dominant kinetic energy, and therefore also minimizes the total energy of the many-body system. In the opposite limit, when the repulsive atom–atom interactions are large compared to the tunnel coupling, the total energy is minimized when each lattice site is filled with the same number of atoms. The reduction of fluctuations in the atom number on each site leads to increased fluctuations in the phase. Thus in the state with a fixed atom number per site phase coherence is lost. In addition, a gap in the excitation spectrum appears. The competition between two terms in the underlying hamiltonian

(here between kinetic and interaction energy) is fundamental to quantum phase transitions[4] and inherently different from normal phase transitions, which are usually driven by the competition between inner energy and entropy.

The physics of the above-described system is captured by the Bose–Hubbard model[1], which describes an interacting boson gas in a lattice potential. The hamiltonian in second quantized form reads:

$$H = -J \sum_{\langle i,j \rangle} \hat{a}_i^\dagger \hat{a}_j + \sum_i \epsilon_i \hat{n}_i + \frac{1}{2} U \sum_i \hat{n}_i (\hat{n}_i - 1) \quad (1)$$

Here $\hat{a}_i$ and $\hat{a}_i^\dagger$ correspond to the bosonic annihilation and creation operators of atoms on the ith lattice site, $\hat{n}_i = \hat{a}_i^\dagger \hat{a}_i$ is the atomic number operator counting the number of atoms on the ith lattice site, and ϵ_i denotes the energy offset of the ith lattice site due to an external harmonic confinement of the atoms[2]. The strength of the tunnelling term in the hamiltonian is characterized by the hopping matrix element between adjacent sites i,j $J = -\int d^3x \, w(x - x_i)(-\hbar^2 \nabla^2/2m + V_{lat}(x)) w(x - x_j)$, where $w(x - x_i)$ is a single particle Wannier function localized to the ith lattice site (as long as $n_i \approx O(1)$), $V_{lat}(x)$ indicates the optical lattice potential and m is the mass of a single atom. The repulsion between two atoms on a single lattice site is quantified by the on-site interaction matrix element $U = (4\pi\hbar^2 a/m)\int |w(x)|^4 d^3x$, with a being the scattering length of an atom. In our case the interaction energy is very well described by the single parameter U, due to the short range of the interactions, which is much smaller than the lattice spacing.

In the limit where the tunnelling term dominates the hamiltonian, the ground-state energy is minimized if the single-particle wavefunctions of N atoms are spread out over the entire lattice with M lattice sites. The many-body ground state for a homogeneous system ($\epsilon_i = \text{const.}$) is then given by:

$$|\Psi_{SF}\rangle_{U=0} \propto \left(\sum_{i=1}^{M} \hat{a}_i^\dagger \right)^N |0\rangle \quad (2)$$

Here all atoms occupy the identical extended Bloch state. An important feature of this state is that the probability distribution

articles

for the local occupation n_i of atoms on a single lattice site is poissonian, that is, its variance is given by $\mathrm{Var}(n_i) = \langle \hat{n}_i \rangle$. Furthermore, this state is well described by a macroscopic wavefunction with long-range phase coherence throughout the lattice.

If interactions dominate the hamiltonian, the fluctuations in atom number of a Poisson distribution become energetically very costly and the ground state of the system will instead consist of localized atomic wavefunctions with a fixed number of atoms per site that minimize the interaction energy. The many-body ground state is then a product of local Fock states for each lattice site. In this limit, the ground state of the many-body system for a commensurate filling of n atoms per lattice site in the homogeneous case is given by:

$$|\Psi_{\mathrm{MI}}\rangle_{J=0} \propto \prod_{i=1}^{M} (\hat{a}_i^\dagger)^n |0\rangle \qquad (3)$$

This Mott insulator state cannot be described by a macroscopic wavefunction like in a Bose condensed phase, and thus is not amenable to a treatment via the Gross-Pitaevskii equation or Bogoliubov's theory of weakly interacting bosons. In this state no phase coherence is prevalent in the system, but perfect correlations in the atom number exist between lattice sites.

As the strength of the interaction term relative to the tunnelling term in the Bose–Hubbard hamiltonian is changed, the system reaches a quantum critical point in the ratio of U/J, for which the system will undergo a quantum phase transition from the superfluid state to the Mott insulator state. In three dimensions, the phase transition for an average number of one atom per lattice site is expected to occur at $U/J = z \times 5.8$ (see refs 1, 5, 6, 7), with z being the number of next neighbours of a lattice site. The qualitative change in the ground-state configuration below and above the quantum critical point is also accompanied by a marked change in the excitation spectrum of the system. In the superfluid regime, the excitation spectrum is gapless whereas the Mott insulator phase exhibits a gap in the excitation spectrum[5–8]. An essential feature of a quantum phase transition is that this energy gap Δ opens up as the quantum critical point is crossed.

Studies of the Bose–Hubbard hamiltonian have so far included granular superconductors[9,10] and one- and two-dimensional Josephson junction arrays[11–16]. In the context of ultracold atoms, atom number squeezing has very recently been demonstrated with a Bose–Einstein condensate in a one-dimensional optical lattice[17]. The above experiments were mainly carried out in the limit of large boson occupancies n_i per lattice site, for which the problem can be well described by a chain of Josephson junctions.

In our present experiment we load $^{87}\mathrm{Rb}$ atoms from a Bose–Einstein condensate into a three-dimensional optical lattice potential. This system is characterized by a low atom occupancy per lattice site of the order of $\langle n_i \rangle \approx 1 - 3$, and thus provides a unique testing ground for the Bose–Hubbard model. As we increase the lattice potential depth, the hopping matrix element J decreases exponentially but the on-site interaction matrix element U increases. We are thereby able to bring the system across the critical ratio in U/J, such that the transition to the Mott insulator state is induced.

Experimental technique

The experimental set-up and procedure to create $^{87}\mathrm{Rb}$ Bose–Einstein condensates are similar to those in our previous experimental work[18,19]. In brief, spin-polarized samples of laser-cooled atoms in the $(F = 2, m_F = 2)$ state are transferred into a cigar-shaped magnetic trapping potential with trapping frequencies of $\nu_{\mathrm{radial}} = 240\,\mathrm{Hz}$ and $\nu_{\mathrm{axial}} = 24\,\mathrm{Hz}$. Here F denotes the total angular momentum and m_F the magnetic quantum number of the state. Forced radio-frequency evaporation is used to create Bose–Einstein condensates with up to 2×10^5 atoms and no discernible thermal component. The radial trapping frequencies are then relaxed over a

period of 500 ms to $\nu_{\mathrm{rad}} = 24\,\mathrm{Hz}$ such that a spherically symmetric Bose–Einstein condensate with a Thomas–Fermi diameter of $26\,\mu\mathrm{m}$ is present in the magnetic trapping potential.

In order to form the three-dimensional lattice potential, three optical standing waves are aligned orthogonal to each other, with their crossing point positioned at the centre of the Bose–Einstein condensate. Each standing wave laser field is created by focusing a laser beam to a waist of $125\,\mu\mathrm{m}$ at the position of the condensate. A second lens and a mirror are then used to reflect the laser beam back onto itself, creating the standing wave interference pattern. The lattice beams are derived from an injection seeded tapered amplifier and a laser diode operating at a wavelength of $\lambda = 852\,\mathrm{nm}$. All beams are spatially filtered and guided to the experiment using optical fibres. Acousto-optical modulators are used to control the intensity of the lattice beams and introduce a frequency difference of about 30 MHz between different standing wave laser fields. The polarization of a standing wave laser field is chosen to be linear and orthogonal polarized to all other standing waves. Due to the different frequencies in each standing wave, any residual interference between beams propagating along orthogonal directions is time-averaged to zero and therefore not seen by the atoms. The resulting three-dimensional optical potential (see ref. 20 and references therein) for the atoms is then proportional to the sum of the intensities of the three standing waves, which leads to a simple cubic type geometry of the lattice:

$$V(x, y, z) = V_0(\sin^2(kx) + \sin^2(ky) + \sin^2(kz)) \qquad (4)$$

Here $k = 2\pi/\lambda$ denotes the wavevector of the laser light and V_0 is the maximum potential depth of a single standing wave laser field. This depth V_0 is conveniently measured in units of the recoil energy $E_r = \hbar^2 k^2/2m$. The confining potential for an atom on a single lattice site due to the optical lattice can be approximated by a harmonic potential with trapping frequencies ν_r on the order of $\nu_r \approx (\hbar k^2/2\pi m)\sqrt{V_0/E_r}$. In our set-up potential depths of up to $22\,E_r$ can be reached, resulting in trapping frequencies of approximately $\nu_r \approx 30\,\mathrm{kHz}$. The gaussian intensity profile of the laser beams at the position of the condensate creates an additional weak isotropic harmonic confinement over the lattice, with trapping frequencies of 65 Hz for a potential depth of $22\,E_r$.

The magnetically trapped condensate is transferred into the optical lattice potential by slowly increasing the intensity of the lattice laser beams to their final value over a period of 80 ms using an exponential ramp with a time constant of $\tau = 20\,\mathrm{ms}$. The slow ramp speed ensures that the condensate always remains in the many-body ground state of the combined magnetic and optical trapping potential. After raising the lattice potential the condensate has been distributed over more than 150,000 lattice sites ($\sim$65 lattice sites in a single direction) with an average atom number of up to 2.5 atoms per lattice site in the centre.

In order to test whether there is still phase coherence between different lattice sites after ramping up the lattice potential, we suddenly turn off the combined trapping potential. The atomic wavefunctions are then allowed to expand freely and interfere with each other. In the superfluid regime, where all atoms are delocalized over the entire lattice with equal relative phases between different lattice sites, we obtain a high-contrast three-dimensional interference pattern as expected for a periodic array of phase coherent matter wave sources (see Fig. 1). It is important to note that the sharp interference maxima directly reflect the high degree of phase coherence in the system for these experimental values.

Entering the Mott insulator phase

As we increase the lattice potential depth, the resulting interference pattern changes markedly (see Fig. 2). Initially the strength of higher-order interference maxima increases as we raise the potential height, due to the tighter localization of the atomic wavefunctions at a single lattice site. Quite unexpectedly, however, at a potential depth of around 13 E_r the interference maxima no longer increase in strength (see Fig. 2e): instead, an incoherent background of atoms gains more and more strength until at a potential depth of 22 E_r no interference pattern is visible at all. Phase coherence has obviously been completely lost at this lattice potential depth. A remarkable feature during the evolution from the coherent to the incoherent state is that when the interference pattern is still visible no broadening of the interference peaks can be detected until they completely vanish in the incoherent background. This behaviour can be explained on the basis of the superfluid–Mott insulator phase diagram. After the system has crossed the quantum critical point $U/J = z \times 5.8$, it will evolve in the inhomogeneous case into alternating regions of incoherent Mott insulator phases and coherent superfluid phases[2], where the superfluid fraction continuously decreases for increasing ratios U/J.

Restoring coherence

A notable property of the Mott insulator state is that phase coherence can be restored very rapidly when the optical potential is lowered again to a value where the ground state of the many-body system is completely superfluid. This is shown in Fig. 3. After only 4 ms of ramp-down time, the interference pattern is fully visible again, and after 14 ms of ramp-down time the interference peaks have narrowed to their steady-state value, proving that phase coherence has been restored over the entire lattice. The timescale for the restoration of coherence is comparable to the tunnelling time $\tau_{\text{tunnel}} = \hbar/J$ between two neighbouring lattice sites in the system,

which is of the order of 2 ms for a lattice with a potential depth of 9 E_r. A significant degree of phase coherence is thus already restored on the timescale of a tunnelling time.

It is interesting to compare the rapid restoration of coherence coming from a Mott insulator state to that of a phase incoherent state, where random phases are present between neighbouring lattice sites and for which the interference pattern also vanishes. This is shown in Fig. 3b, where such a phase incoherent state is created during the ramp-up time of the lattice potential (see Fig. 3 legend) and where an otherwise identical experimental sequence is used. Such phase incoherent states can be clearly identified by adiabatically mapping the population of the energy bands onto the Brillouin zones[19,21]. When we turn off the lattice potential adiabatically, we find that a statistical mixture of states has been created, which homogeneously populates the first Brillouin zone of

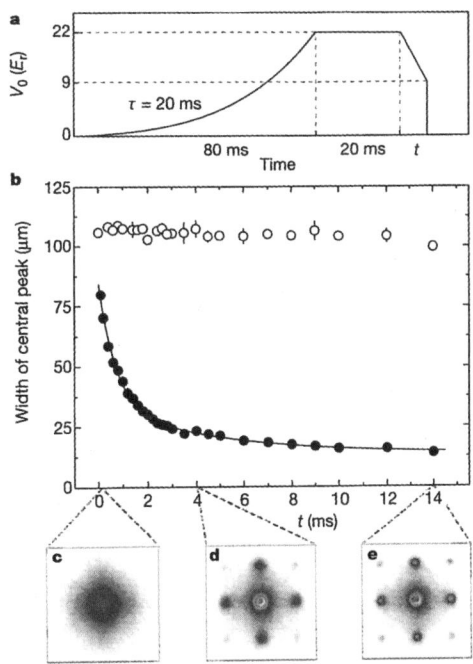

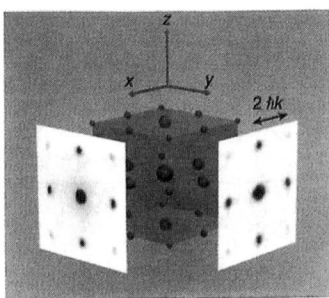

Figure 1 Schematic three-dimensional interference pattern with measured absorption images taken along two orthogonal directions. The absorption images were obtained after ballistic expansion from a lattice with a potential depth of $V_0 = 10E_r$ and a time of flight of 15 ms.

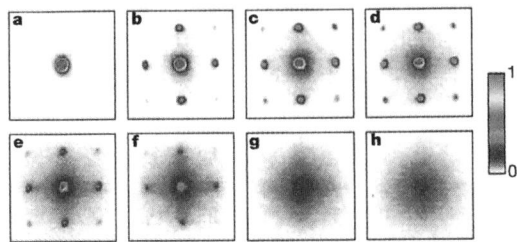

Figure 2 Absorption images of multiple matter wave interference patterns. These were obtained after suddenly releasing the atoms from an optical lattice potential with different potential depths V_0 after a time of flight of 15 ms. Values of V_0 were: **a**, 0 E_r; **b**, 3 E_r; **c**, 7 E_r; **d**, 10 E_r; **e**, 13 E_r; **f**, 14 E_r; **g**, 16 E_r; and **h**, 20 E_r.

Figure 3 Restoring coherence. **a**, Experimental sequence used to measure the restoration of coherence after bringing the system into the Mott insulator phase at $V_0 = 22E_r$ and lowering the potential afterwards to $V_0 = 9E_r$, where the system is superfluid again. The atoms are first held at the maximum potential depth V_0 for 20 ms, and then the lattice potential is decreased to a potential depth of 9 E_r in a time t after which the interference pattern of the atoms is measured by suddenly releasing them from the trapping potential. **b**, Width of the central interference peak for different ramp-down times t, based on a lorentzian fit. In case of a Mott insulator state (filled circles) coherence is rapidly restored already after 4 ms. The solid line is a fit using a double exponential decay ($\tau_1 = 0.94(7)$ ms, $\tau_2 = 10(5)$ ms). For a phase incoherent state (open circles) using the same experimental sequence, no interference pattern reappears again, even for ramp-down times t of up to 400 ms. We find that phase incoherent states are formed by applying a magnetic field gradient over a time of 10 ms during the ramp-up period, when the system is still superfluid. This leads to a dephasing of the condensate wavefunction due to the nonlinear interactions in the system. **c–e**, Absorption images of the interference patterns coming from a Mott insulator phase after ramp-down times t of 0.1 ms (**c**), 4 ms (**d**), and 14 ms (**e**).

articles

the three-dimensional lattice. This homogeneous population proves that all atoms are in the vibrational ground state of the lattice, but the relative phase between lattice sites is random. Figure 3b shows that no phase coherence is restored at all for such a system over a period of 14 ms. Even for evolution times t of up to 400 ms, no reappearance of an interference pattern could be detected. This demonstrates that the observed loss of coherence with increasing potential depth is not simply due to a dephasing of the condensate wavefunction.

Probing the excitation spectrum

In the Mott insulator state, the excitation spectrum is substantially modified compared to that of the superfluid state. The excitation spectrum has now acquired an energy gap Δ, which in the limit $J \ll U$ is equal to the on-site interaction matrix element $\Delta = U$ (see refs 5–8). This can be understood within a simplified picture in the following way. We consider a Mott insulator state with exactly $n = 1$ atom per lattice site. The lowest lying excitation for such a state is the creation of a particle–hole pair, where an atom is removed from a lattice site and added to a neighbouring lattice site (see Fig. 4a). Due to the on-site repulsion between two atoms, the energy of the state describing two atoms in a single lattice site is raised by an amount U in energy above the state with only a single atom in this lattice site. Therefore in order to create an excitation the finite amount of energy U is required. It can be shown that this is also true for number states with exactly n atoms per lattice site. Here the energy required to make a particle–hole excitation is also U. Hopping of particles throughout the lattice is therefore suppressed in the Mott insulator phase, as this energy is only available in virtual processes. If now the lattice potential is tilted by application of a potential gradient, tunnelling is allowed again if the energy difference between neighbouring lattice sites due to the potential gradient equals the on-site interaction energy U (see Fig. 4b). We thus expect a resonant excitation probability versus the applied

energy difference between neighbouring lattice sites for a Mott insulator phase.

We probe this excitation probability by using the experimental sequence shown in Fig. 5a. If excitations have been created during the application of the potential gradient at the potential depth $V_0 = V_{max}$, we will not be able to return to a perfectly coherent superfluid state by subsequently lowering the potential to a depth of $V_0 = 9E_r$. Instead, excitations in the Mott insulator phase will lead to excitations in the lowest energy band in the superfluid case. These excitations are simply phase fluctuations between lattice sites, and cause a broadening of the interference maxima in the interference pattern (see Fig. 5b). Figure 5c–f shows the width of the interference peaks versus the applied gradient for four different potential depths V_{max}. For a completely superfluid system at 10 E_r, the system is easily perturbed already for small potential gradients and for stronger gradients a complete dephasing of the wavefunctions leads to a saturation in the width of the interference peaks. At a potential depth of about 13 E_r two broad resonances start to appear in the excitation spectrum, and for a potential depth of 20 E_r a dramatic

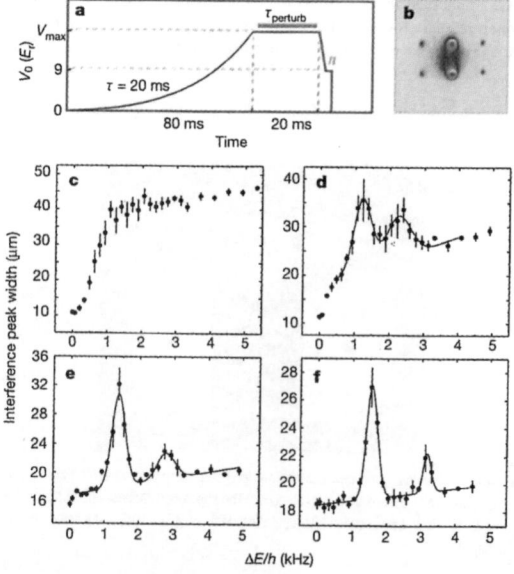

Figure 5 Probing the excitation probability versus an applied vertical potential gradient. **a**, Experimental sequence. The optical lattice potential is increased in 80 ms to a potential depth $V_0 = V_{max}$. Then the atoms are held for a time of 20 ms at this potential depth, during which a potential gradient is applied for a time $\tau_{perturb}$. The optical potential is then lowered again within 3 ms to a value of $V_0 = 9E_r$, for which the system is superfluid again. Finally, a potential gradient is applied for 300 μs with a fixed strength, such that the phases between neighbouring lattice sites in the vertical direction differ by π. The confining potential is then rapidly turned off and the resulting interference pattern is imaged after a time of flight of 15 ms (**b**). Excitations created by the potential gradient at a lattice depth of $V_0 = V_{max}$ will lead to excitations in the superfluid state at $V_0 = 9E_r$. Here excitations correspond to phase fluctuations across the lattice, which will influence the width of the observed interference peaks. **c–f**, Width of interference peaks versus the energy difference between neighbouring lattice sites ΔE, due to the potential gradient applied for a time $\tau_{perturb}$. **c**, $V_{max} = 10E_r$, $\tau_{perturb} = 2$ ms; **d**, $V_{max} = 13E_r$, $\tau_{perturb} = 6$ ms; **e**, $V_{max} = 16E_r$, $\tau_{perturb} = 10$ ms; and **f**, $V_{max} = 20E_r$, $\tau_{perturb} = 20$ ms. The perturbation times $\tau_{perturb}$ have been prolonged for deeper lattice potentials in order to account for the increasing tunnelling times. The solid lines are fits to the data based on two gaussians on top of a linear background.

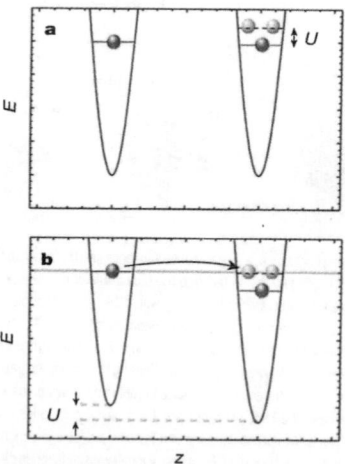

Figure 4 Excitation gap in the Mott insulator phase with exactly $n = 1$ atom per lattice site. **a**, The lowest lying excitations in the Mott insulator phase consist of removing an atom from a lattice site and adding it to neighbouring lattice sites. Owing to the on-site repulsion between the atoms, this requires a finite amount U in energy and hopping of the atoms is therefore suppressed. **b**, If a potential gradient is applied to the system along the z-direction, such that the energy difference between neighbouring lattice sites equals the on-site interaction energy U, atoms are allowed to tunnel again. Particle–hole excitations are then created in the Mott insulator phase.

articles

change in the excitation spectrum has taken place. Two narrow resonances are now clearly visible on top of an otherwise completely flat excitation probability. The slightly higher offset of the excitation probability for a deep optical lattice (Fig. 5e, f) compared to the initial width of the interference peaks in Fig. 5c, is due to the fact that after 3 ms ramp down time from a deep optical lattice, the system is still in the dynamical process of restoring coherence coming from the Mott insulator phase. For longer hold times this offset approaches almost the same initial width as in Fig. 5c, showing that we are not able to excite the system at all except for the two resonance gradients. At these large potential depths, the narrow resonances show that the energy gap Δ of the system, which is measured here as the minimum energy difference between neighbouring lattice sites for which the system can be perturbed, is almost equal to the centre position of the resonance.

We have in fact found the Mott insulator state to be extremely robust to external perturbations, such as a modulation of the trapping potential or a modulation of the gradient potential, as long as the resonance gradients are avoided. The first resonance can be directly attributed to the creation of single particle–hole excitations in the Mott insulator state, and directly proves that we have indeed entered the Mott insulator regime. The second, weaker resonance occurs at exactly twice the energy difference of the first, stronger resonance. It can most probably be attributed to at least one of the following processes: (1) simultaneous tunnelling of two particles in a Mott insulator phase with $n > 1$ atoms, (2) second-order processes, in which two particle–hole pairs are created simultaneously, with only one in the direction of the applied gradient, and (3) tunnelling processes occurring between lattice sites with $n = 1$ atom next to lattice sites with $n = 2$ atoms. In comparison, a two-dimensional lattice at a maximum potential depth of $V_{max} = 20E_r$, which we still expect to be in the superfluid regime, shows no resonances but a smooth excitation spectrum, similar to Fig. 5c.

The position of the resonances in the three-dimensional lattice can be seen to shift with increasing potential depth due to the tighter localization of the wave packets on a lattice site (see Fig. 6). We have compared the position of the first resonance versus the potential depth V_{max} to an *ab initio* calculation of U based on Wannier functions from a band structure calculation, and find good agreement within our experimental uncertainties (see Fig. 6).

Transition point

Both the vanishing of the interference pattern and the appearance of resonances in the excitation spectrum begin to occur at potential depths of $V_0 = 12(1)$–$13(1)\, E_r$, indicating the transition to the Mott insulator phase. We therefore expect the experimental transition point to lie above $V_0 = 10(1)\, E_r$, where no resonances are visible, and below $V_0 = 13(1)\, E_r$. It is important to compare this parameter range to the theoretical prediction based on the expected critical value $U/J = z \times 5.8$. In our simple cubic lattice structure, six next neighbours surround a lattice site. J and U can be calculated numerically from a band structure calculation for our experimental parameters, from which we find that $U/J \approx 36$ for a potential depth of 13 E_r. The theoretical prediction for the transition point is therefore in good agreement with the experimental parameter range for the transition point.

Outlook

We have realized experimentally the quantum phase transition from a superfluid to a Mott insulator phase in an atomic gas trapped in an optical lattice. The experiment enters a new regime in the many-body physics of an atomic gas. This regime is dominated by atom–atom interactions and it is not accessible to theoretical treatments of weakly interacting gases, which have so far proved to be very successful in describing the physics of Bose–Einstein condensates[22]. The experimental realization of the Bose–Hubbard model with an atomic gas now allows the study of strongly correlated many-body quantum mechanics with unprecedented control of parameters. For example, besides controlling mainly the tunnelling matrix element, as done in this work, it should be possible in future experiments to control the atom–atom interactions via Feshbach resonances[23,24].

The atoms in the Mott insulator phase can be considered as a new state of matter in atomic gases with unique properties. Atom number fluctuations at each lattice site are suppressed, and a well-defined phase between different lattices sites no longer exists. These number states have been proposed for the realization of a Heisenberg-limited atom interferometer[25], which should be capable of achieving improved levels of precision. The Mott insulator phase also opens a new experimental avenue for recently proposed quantum gates with neutral atoms[26]. □

Received 26 October; accepted 29 November 2001.

1. Fisher, M. P. A., Weichman, P. B., Grinstein, G. & Fisher, D. S. Boson localization and the superfluid-insulator transition. *Phys. Rev. B* **40**, 546–570 (1989).
2. Jaksch, D., Bruder, C., Cirac, J. I., Gardiner, C. W. & Zoller, P. Cold bosonic atoms in optical lattices. *Phys. Rev. Lett.* **81**, 3108–3111 (1998).
3. Stringari, S. Bose-Einstein condensation and superfluidity in trapped atomic gases. *C.R. Acad. Sci.* **4**, 381–397 (2001).
4. Sachdev, S. *Quantum Phase Transitions* (Cambridge Univ. Press, Cambridge, 2001).
5. Sheshadri, K., Krishnamurthy, H. R., Pandit, R. & Ramakrishnan, T. V. Superfluid and insulating phases in an interacting-boson model: Mean-field theory and the RPA. *Europhys. Lett.* **22**, 257–263 (1993).
6. Freericks, J. K. & Monien, H. Phase diagram of the Bose Hubbard model. *Europhys. Lett.* **26**, 545–550 (1995).
7. van Oosten, D., van der Straten, P. & Stoof, H. T. C. Quantum phases in an optical lattice. *Phys. Rev. A* **63**, 053601-1–053601-12 (2001).
8. Elstner, N. & Monien, H. Dynamics and thermodynamics of the Bose-Hubbard model. *Phys. Rev. B* **59**, 12184–12187 (1999).
9. Orr, B. G., Jaeger, H. M., Goldman, A. M. & Kuper, C. G. Global phase coherence in two-dimensional granular superconductors. *Phys. Rev. Lett.* **56**, 378–381 (1986).
10. Haviland, D. B., Liu, Y. & Goldman, A. M. Onset of superconductivity in the two-dimensional limit. *Phys. Rev. Lett.* **62**, 2180–2183 (1989).
11. Bradley, R. M. & Doniach, S. Quantum fluctuations in chains of Josephson junctions. *Phys. Rev. B* **30**, 1138–1147 (1984).
12. Geerligs, L. J., Peters, M., de Groot, L. E. M., Verbruggen, A. & Mooij, J. E. Charging effects and quantum coherence in regular Josephson junction arrays. *Phys. Rev. Lett.* **63**, 326–329 (1989).
13. Zwerger, W. Global and local phase coherence in dissipative Josephson-junction arrays. *Europhys. Lett.* **9**, 421–426 (1989).

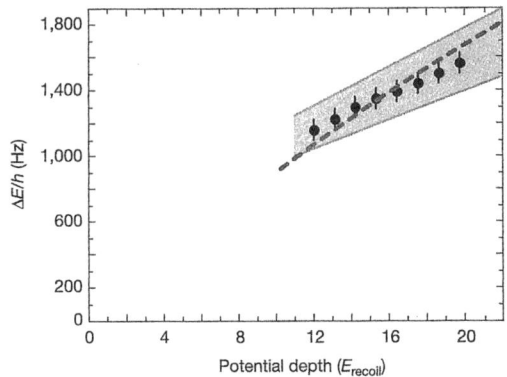

Figure 6 Energy difference between neighbouring lattice sites ΔE for which the Mott insulator phase can be resonantly perturbed versus the lattice potential depth V_{max}. Experimental data points are shown as filled circles, and the shaded grey area denotes the possible variation of experimental values due to systematic uncertainties in the calibration of the potential depth and the applied gradient. The dashed line is the theoretical prediction for the on-site interaction matrix element U, based on Wannier functions from a band structure calculation.

articles

14. van der Zant, H. S. J., Fritschy, F. C., Elion, W. J., Geerligs, L. J. & Mooij, J. E. Field-induced superconductor-to-insulator transitions in Josephson-junction arrays. *Phys. Rev. Lett.* **69**, 2971–2974 (1992).

15. van Oudenaarden, A. & Mooij, J. E. One-dimensional Mott insulator formed by quantum vortices in Josephson junction arrays. *Phys. Rev. Lett.* **76**, 4947–4950 (1996).

16. Chow, E., Delsing, P. & Haviland, D. B. Length-scale dependence of the superconductor-to-insulator quantum phase transition in one dimension. *Phys. Rev. Lett.* **81**, 204–207 (1998).

17. Orzel, C., Tuchman, A. K., Fenselau, M. L., Yasuda, M. & Kasevich, M. A. Squeezed states in a Bose-Einstein condensate. *Science* **291**, 2386–2389 (2001).

18. Greiner, M., Bloch, I., Hänsch, T. W. & Esslinger, T. Magnetic transport of trapped cold atoms over a large distance. *Phys. Rev. A* **63**, 031401-1–031401-4 (2001).

19. Greiner, M., Bloch, I., Mandel, O., Hänsch, T. W. & Esslinger, T. Exploring phase coherence in a 2D lattice of Bose-Einstein condensates. *Phys. Rev. Lett.* **87**, 160405-1–160405-4 (2001).

20. Grimm, R., Weidemüller, M. & Ovchinnikov, Yu. B. Optical dipole traps for neutral atoms. *Adv. At. Mol. Opt. Phys.* **42**, 95–170 (2000).

21. Kastberg, A., Phillips, W. D., Rolston, S. L., Spreeuw, R. J. C. & Jessen, P. S. Adiabatic cooling of cesium to 700 nK in an optical lattice. *Phys. Rev. Lett.* **74**, 1542–1545 (1995).

22. Dalfovo, F. D., Giorgini, S., Pitaevskii, L. P. & Stringari, S. Theory of Bose-Einstein condensation in trapped gases. *Rev. Mod. Phys.* **71**, 463–512 (1999).

23. Inouye, S. *et al.* Observation of Feshbach resonances in a Bose–Einstein condensate. *Nature* **392**, 151–154 (1998).

24. Donley, E. A. *et al.* Dynamics of collapsing and exploding Bose–Einstein condensates. *Nature* **412**, 295–299 (2001).

25. Bouyer, P. & Kasevich, M. Heisenberg-limited spectroscopy with degenerate Bose-Einstein gases. *Phys. Rev. A* **56**, R1083–R1086 (1997).

26. Jaksch, D., Briegel, H.-J., Cirac, J. I., Gardiner, C. W. & Zoller, P. Entanglement of atoms via cold controlled collisions. *Phys. Rev. Lett.* **82**, 1975–1978 (1999).

Acknowledgements

We thank W. Zwerger, H. Monien, I. Cirac, K. Burnett and Yu. Kagan for discussions. This work was supported by the DFG, and by the EU under the QUEST programme.

Competing interests statement

The authors declare that they have no competing financial interests.

Correspondence and requests for materials should be addressed to I.B. (e-mail: imb@mpq.mpg.de).

letters to nature

....

Bose–Einstein condensation on a microelectronic chip

W. Hänsel, P. Hommelhoff, T. W. Hänsch & J. Reichel

*Max-Planck-Institut für Quantenoptik and Sektion Physik der
Ludwig-Maximilians-Universität, Schellingstr. 4, D-80799 München, Germany*

....

Although Bose–Einstein condensates[1-3] of ultracold atoms have
been experimentally realizable for several years, their formation
and manipulation still impose considerable technical challenges.
An all-optical technique[4] that enables faster production of Bose–
Einstein condensates was recently reported. Here we demonstrate
that the formation of a condensate can be greatly simplified using
a microscopic magnetic trap on a chip[5]. We achieve Bose–Einstein
condensation inside the single vapour cell of a magneto-optical
trap in as little as 700 ms—more than a factor of ten faster than
typical experiments, and a factor of three faster than the all-
optical technique[4]. A coherent matter wave is emitted normal to
the chip surface when the trapped atoms are released into free fall;
alternatively, we couple the condensate into an 'atomic conveyor
belt'[6], which is used to transport the condensed cloud nondestructively over a macroscopic distance parallel to the chip
surface. The possibility of manipulating laser-like coherent matter
waves with such an integrated atom-optical system holds promise
for applications in interferometry, holography, microscopy, atom
lithography and quantum information processing[7].

Some of the advantages of microscopic magnetic traps on a chip
have been pointed out before. Modest electric currents can produce
large magnetic field gradients and curvatures in close proximity to a
planar arrangement of wires[8]. In an experiment which was realized
simultaneously with the results reported here, microfabricated
parallel conductors were used in a last stage of evaporative cooling
to achieve Bose–Einstein condensation (BEC)[9]. Lithographic fabrication techniques now make it possible to integrate even complex
systems of many microscopic traps, waveguides[10,11], and other
atom-optical devices[12-14] on a single 'atom chip'.

The use of such microtraps for BEC appears, in hindsight, only
natural, as quantum-mechanical phenomena tend to be more
readily observable on a smaller scale. A tight trap permits fast
adiabatic changes of the confining potential, and it becomes easy to
magnetically compress a trapped atom cloud so that elastic collision
times of the order of milliseconds are reached even with just a few
million trapped atoms. The resulting fast thermalization makes it
possible to drastically shorten the time for radio-frequency-assisted
evaporative cooling[15]. It is also advantageous that a tight magnetic
confinement positions the atom cloud near the centre of the
magnetic trap, despite the pull of gravity, so that a rather uniform
evaporation is achieved throughout the evaporation process. Collisions with background gas atoms become less important during
such a fast cooling cycle, so that the previously very stringent
requirements on the vacuum may be greatly relaxed.

Figure 1a shows the chip that is used in our BEC experiments. It
features 50-μm-wide conductors, which reproducibly support
continuous currents in excess of 3 A. The chip was fabricated

using thin-film hybrid technology, a standard microelectronics process[14]. Although a variety of fabrication techniques can be used (several groups have developed custom processes[16,25]), a standard process has the advantage of being available to non-microtechnology laboratories.

In our chip traps, the trapping potential is produced by superposing the field of the lithographic conductors with a homogeneous, external bias field B_0. A Ioffe–Pritchard potential (non-zero field in the minimum, and quadratic dependence close to this minimum) is created using the 'Z'-shaped conductor shown in yellow in Fig. 1a; the central part of this conductor has a length of 1.95 mm. Figure 1b shows the magnetic potentials created by this wire with a current of $I_0 = 2$ A for two different values of the external bias field, $B_0 = B_x e_x + B_y e_y$ (where e_x and e_y are unit vectors): $(B_x = 0, B_y = 8$ G$)$ and $(B_x = 1.9$ G$, B_y = 55$ G$)$, corresponding to the two extreme values used in the experiment. With increasing bias field, the trap centre moves closer to the surface, from $z_0 = 445$ μm to $z_0 = 70$ μm. The potentials vary quadratically with the distance from the trap centre for very small distances. In the main part of the trap, the transverse dependence can be approximately characterized by the gradient of the wire field at z_0. This gradient rises by a factor of 37—from 200 G cm^{-1} to 7,300 G cm^{-1}—for the two potentials in the figure. The central part of the longitudinal potential, on the other hand, changes very little while its steep walls grow higher with trap compression.

The chip that creates these potentials is mounted face down in a glass cell of inner dimensions $30 \times 30 \times 110$ mm; this cell is part of a simple vacuum system pumped by a 25 l s^{-1} ion pump and a small titanium sublimator (Fig. 2). The magnetic trap lifetime is up to 5 s, which indicates a background pressure in the 10^{-9} mbar

range. Thermal rubidium vapour is produced from a rubidium dispenser[17]. We operate it at low, constant, current, so that the rubidium partial pressure remains low and reduces the magnetic trap lifetime by less than 20%. During the magneto-optical trap (MOT) loading phase, a 30-W halogen reflector bulb is switched on to temporarily increase the rubidium pressure by light-induced desorption[18]. The MOT loaded in this way contains about 5×10^6 atoms of ^{87}Rb after a loading time of 3–7 s.

Trap loading is a crucial step in realizing a magnetic microtrap. We use the mirror-MOT technique developed in our group and described in detail in refs 5 and 14. This variant of the well-known MOT provides a sample of laser-cooled atoms at a distance of less than 1 mm from the substrate surface, and enables *in situ* transfer into the magnetic microtrap without requiring an intermediate, macroscopic magnetic trap. In this way we typically load 3×10^6 atoms into the magnetic trap after optical pumping into the $F = 2$, $m_F = 2$ ground state.

The initial magnetic trapping potential is shown in dashed lines in Fig. 1b. This trap has frequencies $\nu_x = 28$ Hz and $\nu_{y,z} = 220$ Hz. The temperature and peak density, measured 200 ms after transfer from the MOT, are ∼45 μK and ∼5×10^{10} cm^{-3}, respectively. Immediately after the transfer, we increase the bias field B_y in 300 ms to a final value of 55 G, leading to the potential shown in solid lines in Fig. 1b. Thus, strong compression occurs in the transverse (y, z) plane, whereas the longitudinal potential undergoes a comparatively small change. The very anisotropic final potential has a transverse curvature of 2.4×10^7 G cm^{-2} near the centre, leading to a calculated transverse oscillation frequency of $\nu_{y,z} = 6.2$ kHz, while the longitudinal frequency is only $\nu_x = 17$ Hz. Experimentally, we have used two different techniques (direct observation of the oscillation and parametric heating) to measure these frequencies, and found good agreement of calculated and measured frequencies.

Immediately after compression, we initiate forced evaporative cooling with a linear radio-frequency sweep from 30 MHz to 8 MHz. This sweep is typically performed in 900 ms, and reduces the number of atoms to 5×10^5. From here, we can proceed to BEC by two different routes, both of which involve an adiabatic change in the potential. In the first case, the trap is simply decompressed by reducing B_0—leading to a very elongated condensate centred at the position C1 in Fig. 1a. In the second approach, the wire currents are

Figure 1 The chip and the magnetic potentials that it creates. **a**, Layout of the lithographic gold wires on the substrate. The inset shows the relevant part of the conductor pattern. I_0, I_1, I_{M1} and I_{M2} create the various magnetic potentials for trapping (see **b**) and transport, as described in the main text, I_Q is used only during trap loading (intermediate MOT step[5]). **b**, Potentials created by a wire current $I_0 = 2$ A for two different values of the external bias field, $(B_x = 0, B_y = 8$ G$)$ (dashed lines) and $(B_x = 1.9$ G$, B_y = 55$ G$)$ (solid lines).

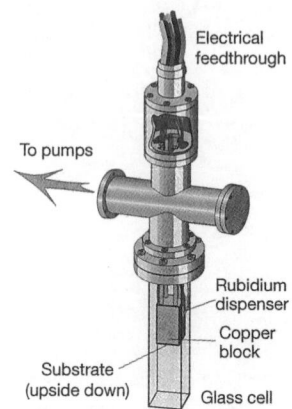

Figure 2 Vacuum system. Because of the efficient evaporation in the chip trap, this very simple set-up is sufficient to achieve Bose–Einstein condensation (BEC). The substrate is mounted facing downwards, so that the atom cloud may be released to expand in free fall (time-of-flight analysis).

letters to nature

also modified to move the trap centre to position C2, where a more spherical condensate arises.

We now describe both cases in more detail. In the first case, the initial radio-frequency ramp is immediately followed by a second, exponential ramp with 800 ms duration. At the end of this ramp, $\sim 7 \times 10^4$ atoms are left at a temperature of $\sim 6\,\mu$K, a density of 5×10^{13} cm^{-3} and a phase space density in the lower 10^{-2} range. The external bias field is now reduced to $(B_x = 1.2\,$G, $B_y = 40\,$G) within 150 ms to decompress the trap. This is done to avoid an excessively high density (which would lead to trap loss by three-body collisions), and to reduce the heating rate, which was measured to be $2.7\,\mu$K s^{-1} in the compressed trap at $B_y = 55\,$G, and reduced to $1.1\,\mu$K s^{-1} at $B_y = 40\,$G. (The origin of this heating is discussed below.) The decompressed trap has frequencies $\nu_{x,y,z} = (20, 3{,}900, 3{,}900)$ Hz. After a final radio-frequency sweep to ~ 1.6 MHz in 300 ms, a condensate appears. Figure 3 shows time-of-flight absorption images after a ballistic expansion time of 21 ms. These images were taken for descending values of the final radio frequency, and show the appearance of the sharp, non-isotropic peak in the momentum distribution which is a key signature of BEC. A bimodal distribution is first observed for a temperature of $T \approx 630$ nK, in agreement with the theoretical value of the transition temperature $T_c = 670$ nK (for 11,000 atoms). The condensate lifetime is of the order of 500 ms and can be prolonged to ~ 1.3 s when the radio-frequency power is left on at a frequency just above resonance with the condensed atoms. The number of condensate atoms in a distribution well below the transition temperature is typically around 3,000.

The high trap frequencies in the compressed traps currently prevent us from measuring directly the temperature and density in these traps. Instead, we infer these values from analysing the cloud after adiabatic decompression, using a bias field B_y of 19.4 G for the elongated trap, and of 16 G for configuration C2. Detection can then be accomplished by shutting off the wire current, leaving

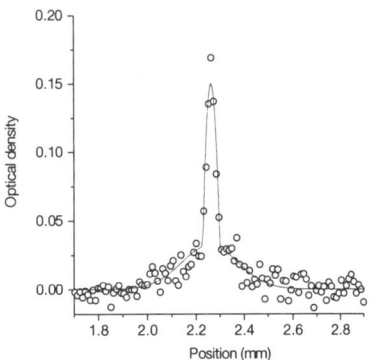

Figure 4 Axial column density profile near the transition temperature, taken in configuration C2. The curve is obtained from a two-dimensional best fit to a bimodal distribution, which corresponds to 1,600 atoms in the condensate and 10,400 atoms in the thermal component at a temperature of 1.7 µK.

the homogeneous field switched on and tuning the probe laser frequency to the Zeeman-shifted atomic resonance. If the atom cloud is cold enough to remain in the quadratic region of the potential, the temperature in the compressed trap can be calculated from the relation $T' = (\nu_1'\nu_2'\nu_3'/(\nu_1\nu_2\nu_3))^{1/3}T$, where ν_i indicates a trap frequency along the i-th axis, and ν_i' designates a value after the transformation. The temperatures we cite are inferred in this way, with the exception of the heating rate measurements below, where all values are given for the detection trap, so that they can be compared directly.

Owing to its very elongated shape, the condensate in position C1 is well inside the regime of fluctuating phase. Such 'quasicondensates' are at present receiving much attention[19,20]. We also reach the phase transition in a more spherical configuration by transferring the cloud to the position marked C2 in Fig. 1a. This is accomplished by ramping the currents and external fields to final values $I_1 = 2$ A, $I_{M1} = 1$ A, $B_x = 5$ G and $B_y = 40$ G. This transformation takes 250 ms, and results in a trap with frequencies $\nu_{x,y,z} = (300, 3{,}400, 3{,}500)$ Hz—the ratio of transverse and longitudinal frequencies is now only 12:1, in contrast to configuration C1, where it was 200:1. A single additional radio-frequency ramp with a duration of 500 ms and a final frequency of ~ 1.0 MHz is enough to achieve condensation in this trap. The number of

Figure 3 Time-of-flight absorption images of the atom cloud after 21 ms of free expansion. (Profile height and false colours are used to encode the optical density of the cloud.) The images were taken after a final radio-frequency sweep ending at frequencies $\nu_s = \nu_0 + 66$ kHz, 44 kHz and 15 kHz (from top to bottom), where ν_0 is the resonance frequency in the centre of the trap. The images show the appearance of the sharp, non-isotropic peak in the momentum distribution which is a key signature of BEC.

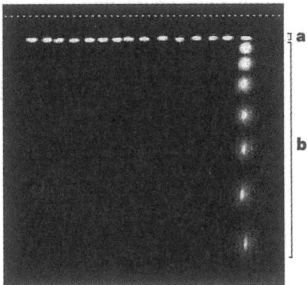

Figure 5 Transport of a BEC on the 'magnetic conveyor belt'. **a**, Superposed absorption images taken at fixed time intervals during transport. The distance between the first and last image is 1.6 mm, the transport time is 100 ms. The line of sight is parallel to the y axis (see Fig. 1); the dotted line marks the edge of the substrate. **b**, Time-of-flight images of the atom cloud after release at the final position, exhibiting the bimodal structure characteristic of a BEC. The maximum expansion time (bottom image) is 19.3 ms.

letters to nature

condensed atoms is typically slightly higher than in configuration C1. A section through a condensed cloud, taken close to the condensation threshold (Fig. 4), reveals the expected bimodal structure with a broad gaussian background of thermal atoms and a sharp peak due to the condensed fraction. When using this potential, we even observe condensation when both radio-frequency ramps are shortened and overlapped with the magnetic field ramp, so that the total evaporation time is as short as 700 ms.

The short evaporation time entails a very fast cycle time of 10 s or less, including MOT loading and detection. Moreover, it relaxes the ultrahigh-vacuum restrictions that have been one of the main restrictions of 'traditional' BEC experiments, and thus enables us to use the very simple vacuum system of Fig. 2.

With its trap–substrate distance in the 100-μm range, the microtrapped condensate approaches a room-temperature surface much more closely than do usual condensates. Little is known about condensate–surface interactions at very small distances. Surface-induced heating of trapped atom clouds has been studied theoretically[21], and is predicted to become important for distances in the 10-μm range. However, there are many other sources that could explain our observed heating rate, such as electrical current noise, mechanical vibrations, and various effects associated with collisions[22]. We have measured the heating rate after slowly reducing both I_0 and B_y by the same factor α, starting from $I_0 = 2\,A$, $B_y = 40\,G$. This operation reduces both the transverse and longitudinal gradients and curvatures by α while leaving unchanged the position of the cloud centre. We find that the heating rate decreases with α, and reduces to $0.5 \pm 0.3\,\mu K\,s^{-1}$ for $\alpha = 0.3$. Although this dependence imposes some restrictions on the nature of the heating mechanism, further measurements are required to fully understand and possibly eliminate it.

A decisive advantage of the chip trap lies in the versatility of the lithographic wire structures. With the wire layout of Fig. 1, many more complicated potentials could be realized. As a first demonstration of these capabilities, we have created a condensate in configuration C2 and transported it over a distance of 1.6 mm along the chip surface using a 'magnetic conveyor belt'[6]. This is done by applying periodically modulated currents I_{M1} and I_{M2} with a relative phase shift of $\pi/2$. Figure 5a shows the cloud during transport over two conveyor periods; images were taken separately and then superposed. Figure 5b shows a series of time-of-flight images in 2.4-ms intervals after release at the final position: the expanding cloud shows the bimodal distribution of a BEC even after transport. This is encouraging for applications such as trapped-atom interferometry[23,24] with condensates. Because of this robustness, the simplicity of the set-up and the possibilities of chip-based potentials, we expect fruitful results from microchip BEC experiments in the near future. □

Received 27 July; accepted 11 September 2001.

1. Anderson, M. H., Ensher, J. R., Matthews, M. R., Wieman, C. E. & Cornell, E. A. Observation of Bose–Einstein condensation in a dilute atomic vapor. *Science* **269**, 198–201 (1995).
2. Davis, K. B. *et al.* Bose–Einstein condensation in a gas of sodium atoms. *Phys. Rev. Lett.* **75**, 3969–1690 (1995).
3. Bradley, C. C., Sackett, C. A. & Hulet, R. G. Bose–Einstein condensation of lithium: observation of limited condensate number. *Phys. Rev. Lett.* **78**, 985–989 (1997).
4. Barrett, M. D., Sauer, J. A. & Chapman, M. S. All-optical formation of an atomic Bose–Einstein condensate. *Phys. Rev. Lett.* **87**, 010404-1–010404-4 (2001).
5. Reichel, J., Hänsel, W. & Hänsch, T. W. Atomic micromanipulation with magnetic surface traps. *Phys. Rev. Lett.* **83**, 3398–3401 (1999).
6. Hänsel, W., Reichel, J., Hommelhoff, P. & Hänsch, T. W. Magnetic conveyer belt for transporting and merging trapped atom clouds. *Phys. Rev. Lett.* **86**, 608–611 (2001).
7. Calarco, T. *et al.* Quantum gates with neutral atoms: Controlling collisional interactions in time-dependent raps. *Phys. Rev. A* **61**, 022304-1–022304-11 (2000).
8. Weinstein, J. D. & Libbrecht, K. G. Microscopic magnetic traps for neutral atoms. *Phys. Rev. A* **52**, 4004–4009 (1995).
9. Ott, H., Fortagh, J., Schlotterbeck, G., Grossmann, A. & Zimmermann, C. Bose–Einstein condensation in a surface microtrap. *Phys. Rev. Lett.* (in the press).
10. Müller, D., Anderson, D. Z., Grow, R. J., Schwindt, P. D. D. & Cornell, E. A. Guiding neutral atoms around curves with lithographically patterned current-carrying wires. *Phys. Rev. Lett.* **83**, 5194–5197 (1999).
11. Dekker, N. H. *et al.* Guiding neutral atoms on a chip. *Phys. Rev. Lett.* **84**, 1124–1127 (2000).
12. Müller, D. *et al.* Waveguide atom beamsplitter for laser-cooled neutral atoms. *Opt. Lett.* **25**, 1382–1384 (2000).
13. Cassettari, D., Hessmo, B., Folman, R., Maier, T. & Schmiedmayer, J. Beam splitter for guided atoms. *Phys. Rev. Lett.* **85**, 5483–5487 (2000).
14. Reichel, J., Hänsel, W., Hommelhoff, P. & Hänsch, T. W. Applications of integrated magnetic microtraps. *Appl. Phys. B* **72**, 81–89 (2001).
15. Ketterle, W. & van Druten, N. J. in *Advances in Atomic, Molecular and Optical Physics* Vol. 37 (eds Bederson, B. & Walther, H.) 181–236 (Academic, San Diego, 1996).
16. Drndić, M., Johnson, K. S., Thywissen, J. H., Prentiss, M. & Westervelt, R. M. Micro-electromagnets for atom manipulation. *Appl. Phys. Lett.* **72**, 2906–2908 (1998).
17. Fortagh, J., Ott, H., Grossmann, A. & Zimmermann, C. Miniaturized magnetic guide for neutral atoms. *Appl. Phys. B* **70**, 701–708 (2000).
18. Anderson, B. P. & Kasevich, M. A. Loading a vapor-cell magneto-optic trap using light-induced atom desorption. *Phys. Rev. A* **63**, 023404-1–023404-4 (2001).
19. Petrov, D. S., Shlyapnikov, G. V. & Walraven, J. T. M. Phase-fluctuating 3D Bose–Einstein condensates in elongated traps. *Phys. Rev. Lett.* **87**, 050404-1–050404-4 (2001).
20. Dettmer, S. *et al.* Observation of phase fluctuations in Bose–Einstein condensates. *Phys. Rev. Lett.* (in the press); preprint cond-mat/0105525 at ⟨http://xxx.lanl.gov⟩ (2001).
21. Henkel, C., Pötting, S. & Wilkens, M. Loss and heating of particles in small and noisy traps. *Appl. Phys. B* **69**, 379–387 (1999).
22. Cornell, E. A., Ensher, J. R. & Wieman, C. E. in *Proc. Int. School of Physics "Enrico Fermi", Course CXL* (eds Inguscio, M., Stringari, S. & Wieman, C. E.) 15–66 (IOS, Amsterdam, 1999).
23. Hinds, E. A., Vale, C. J. & Boshier, M. G. Two-wire waveguide and interferometer for cold atoms. *Phys. Rev. Lett.* **86**, 1462–1465 (2001).
24. Hänsel, W., Reichel, J., Hommelhoff, P. & Hänsch, T. W. Trapped-atom interferometer in a magnetic microtrap. *Phys. Rev. A* (in the press).
25. Folman, R. *et al.* Controlling cold atoms using nanofabricated surfaces: Atom chips. *Phys. Rev. Lett.* **85**, 5483–5487 (2001).

Acknowledgements

This work was supported in part by the European Union under the IST programme (ACQUIRE project).

Correspondence and requests for materials should be addressed to J.R. (e-mail: jakob.reichel@physik.uni-muenchen.de).

VOLUME 88, NUMBER 12   PHYSICAL REVIEW LETTERS   25 MARCH 2002

All-Optical Production of a Degenerate Fermi Gas

S. R. Granade, M. E. Gehm, K. M. O'Hara, and J. E. Thomas

Physics Department, Duke University, Durham, North Carolina 27708-0305
(Received 19 November 2001; published 8 March 2002)

We achieve degeneracy in a mixture of the two lowest hyperfine states of ^{6}Li by direct evaporation in a CO_2 laser trap, yielding the first all optically produced degenerate Fermi gas. More than 10^5 atoms are confined at temperatures below 4 μK at full trap depth, where the Fermi temperature for each state is 8 μK. This degenerate two-component mixture is ideal for exploring mechanisms of superconductivity ranging from Cooper pairing to Bose-Einstein condensation of strongly bound pairs.

DOI: 10.1103/PhysRevLett.88.120405   PACS numbers: 05.30.Fk, 32.80.Pj

Degenerate two-component Fermi gases offer tantalizing possibilities for precision studies of pairing interactions in systems for which the density, temperature, and interaction strength are widely variable. Of particular interest are certain two-component mixtures of ^{40}K and ^{6}Li which exhibit magnetically tunable Feshbach resonances, enabling variation of the s-wave scattering interaction from strongly repulsive to strongly attractive. Attractive mixtures in these systems are analogs of superconductors, since they have been predicted to undergo a superfluid transition as a result of Cooper pairing at experimentally accessible temperatures [1,2]. Recently, two groups have predicted the possibility of superfluidity arising from strong pairing in the vicinity of the Feshbach resonance [3,4]. Transition temperatures of up to half the Fermi temperature are predicted to result from the strong coupling of the two-state Fermi gas to the bosonic molecular state which causes the resonance. Since most high temperature superconductors achieve transition temperatures of only a few percent of the Fermi temperature, two-state Fermi gases may be the highest temperature Fermi superfluids ever studied [5]. Further, these systems may permit observation of the transition from weak Bardeen-Cooper-Schrieffer superfluidity to Bose condensation of strongly bound pairs [6].

In contrast to Bose-Einstein condensates, which can be prepared and studied in magnetic traps, two-component Fermi superfluids must be prepared in state-independent, optical dipole traps, since the required pairs of hyperfine states in ^{6}Li and ^{40}K are high-field seeking [1,2,7]. A degenerate Fermi gas has been produced by direct evaporation of a two-state mixture of ^{40}K in a magnetic trap, using a dual radio-frequency–knife method [8]. Sympathetic cooling of fermionic ^{6}Li to degeneracy also has been achieved by using mixtures of ^{6}Li with bosonic ^{7}Li in a magnetic trap [9,10]. However, to explore superfluidity in these systems, transfer to an optical trap and subsequent state preparation is required. The procedure for preparing an optically trapped two-state Fermi gas can be greatly simplified by direct evaporation in an optical trap.

In this Letter, we demonstrate all-optical production of a degenerate mixture of the two lowest hyperfine states of fermionic ^{6}Li in a stable, CO_2 laser trap [11]. The trap is loaded from a magneto-optical trap (MOT) at an initial

temperature of 150 μK. Degeneracy is obtained by forced evaporation, accomplished by continuously lowering the trap depth; the trap is then adiabatically recompressed to full depth. At this stage, more than 10^5 atoms remain at temperatures below 4 μK, less than half of the Fermi temperature of 8 μK. These results are consistent with scaling laws we have derived for the phase-space density as a function of trap depth [12].

Our ^{6}Li experiments employ a CO_2 laser trap with a single focused beam, rather than a crossed-beam geometry as used recently to produce a ^{87}Rb Bose-Einstein condensate (BEC) by forced evaporation [13]. Nevertheless, after free evaporation at full trap depth, we achieve a very high initial phase-space density of $\simeq 8 \times 10^{-3}$, somewhat larger than that obtained after free evaporation in the ^{87}Rb BEC experiments.

A commercial, radio-frequency–excited CO_2 laser (Coherent-DEOS LC100-NV) provides 140 W at $\lambda = 10.6$ μm for the trapping laser beam. An Agilent (6573A) power supply produces stable current for the radio-frequency source, yielding a very stable laser intensity. The laser output is deflected by an acousto-optic (A/O) modulator to control the power. A cylindrical ZnSe telescope corrects the output of the A/O for ellipticity, and the beam is expanded by a factor of 10 before passing through an aspherical 19.5 cm focal length lens. This lens focuses the beam into the vacuum system, yielding a $1/e^2$ intensity radius of 47 μm. The corresponding Rayleigh length is $z_0 = 660$ μm. With an incident power of 65 W in the trap region, the trap depth is estimated to be 690 μK. The corresponding radial (axial) oscillation frequency for ^{6}Li is predicted to be 6.6 kHz (340 Hz), with a geometric mean of $\nu = (\nu_x \nu_y \nu_z)^{1/3} = 2400$ Hz.

The radial oscillation frequency is measured by modulating the frequency of the A/O to produce a sinusoidal displacement at the trap focus with an amplitude of 0.2 μm. After the sample is initially prepared at a temperature of $\simeq 15$ μK, the modulation is applied for 1 s. The number of remaining atoms is measured by resonance fluorescence. Repeating this procedure as a function of modulation frequency reveals a resonance in the trap loss at 6.5 kHz, in close agreement with predictions. Parametric resonance methods [14] yield results consistent with the expected

VOLUME 88, NUMBER 12   PHYSICAL REVIEW LETTERS   25 MARCH 2002

radial and axial oscillation frequencies after correction for the expected resonance frequency shift [15].

Extremely low residual heating rates are attained in the experiments. At the maximum trap intensity of 1.9 MW/cm^2, the optical scattering rate is 2 photons per hour as a consequence of the 10.6 μm wavelength [16], yielding a recoil heating rate of only 16 pK/sec. At the background pressure of $<10^{-11}$ Torr, heating arising from diffractive background gas collisions [17,18] is <5 nK/sec. For the trap radial oscillation frequency of 6.6 kHz, the intensity noise heating time constant is estimated to be $>2.3 \times 10^4$ sec based on the measured laser intensity noise power spectrum [11]. A residual heating rate <5 nK/sec is measured at full trap depth over 200 sec. Trap $1/e$ lifetimes of 400 sec are observed.

The CO_2 laser trap is continuously loaded from a ^{6}Li MOT. The MOT is loaded from a Zeeman slower for 5 sec, after which the MOT laser beams are tuned ≈ 6 MHz below resonance and lowered in intensity to $0.1I_{sat} = 0.25$ mW/cm^2 to obtain a Doppler-limited temperature of ≈ 150 μK at a density of $10^{11}/\text{cm}^3$. Following this loading stage, the MOT gradient magnets are extinguished and the upper $F = 3/2$ hyperfine state is emptied to produce a 50–50 mixture of atoms in the lower $|F = 1/2, M = \pm 1/2\rangle$ states [7].

The $|F = 1/2, M = \pm 1/2\rangle$ mixture is of particular interest, as it is predicted to exhibit a Feshbach resonance near 850 G [19]. A convenient feature of this mixture is that the s-wave scattering length vanishes in the absence of a bias magnetic field [19]. However, the scattering length varies between 0 and $-300a_0$ as the bias magnetic field is tuned between 0 and 300 G [19]. Hence, rapid evaporation can be turned on and off simply by applying or not applying a bias magnetic field.

The number of trapped atoms is enhanced by increasing the intensity of the CO_2 laser during the loading stage [20]. To accomplish this, the beam which emerges from the trap is recollimated after a ZnSe exit window by a 19.5 cm focal length ZnSe lens, and then retroreflected and orthogonally polarized using a rooftop mirror oriented at 45° to the incoming polarization. The resulting backward-propagating beam is refocused into the trap region through the exit lens. After passing through the trap region, this beam is diverted by a thin film polarizer to a beam dump to avoid feedback into the laser. Typically 1.5×10^6 atoms are confined in the forward propagating trap beam alone. The backward-propagating beam increases this number to 3.5×10^6.

After the CO_2 laser trap is loaded, the atoms are precooled by free evaporation. To initiate evaporative cooling, we apply a bias magnetic field of 130 G by reversing the current in one of the MOT gradient coils, yielding a scattering length of $\approx -100a_0$. During free evaporation, a pneumatically controlled mirror slowly blocks the backward-propagating beam by diverting the power into a 100 W power meter. Since this beam is refocused, the trap region is Fourier-transform related to the plane of

the blocking mirror, and the trap smoothly evolves into a single beam configuration. After 6 sec of free evaporation, the single beam trap contains $N = 1.3 \times 10^6$ atoms at a temperature $T = 50$ μK. This precooling procedure provides excellent initial conditions for the forced evaporation experiments, since the resulting phase-space density for each state at full trap depth, $\rho_i = (N/2)(h\nu)^3/(k_B T)^3$, is 8×10^{-3}, which is extremely high.

In all of our experiments, we characterize the velocity distribution of the trapped gas by time-of-flight imaging. We use the A/O modulator to turn off the CO_2 laser trap abruptly ($\Delta t < 1$ μs), permitting the gas to expand for a time between 400 μs and 1.2 ms in zero bias magnetic field. Residual A/O leakage is reduced to less than 2×10^{-5} of the maximum intensity by extinguishing the radio-frequency synthesizer output prior to the amplifier. Then a linearly polarized probe laser pulse with a resonant intensity of $0.1I_{sat}$ and a detuning of 3 half linewidths (≈ 9 MHz) illuminates the gas for 10 μs. Simultaneously, a noncopropagating repumper, resonant with the D2 lines starting from the $F = 3/2$ state, suppresses optical pumping into the upper $F = 3/2$ hyperfine state. The probe detuning reduces sensitivity to the unresolved excited state hyperfine structure and light shifts from the resonant repumper. For the selected 9 MHz detuning, the expansion time is chosen so that the imaged cloud has a small optical absorption $<35\%$. An achromat at the vacuum system exit window produces a 1:1 image of the atomic distribution in an intermediate plane. This plane is imaged onto a CCD camera (Andor) using a microscope objective to produce a net magnification of ≈ 4. The magnification is calibrated by moving the axial position of the trap focus through ± 1.25 mm using a micrometer-controlled translation stage. Fitting the central peak of the distribution to a straight line yields a magnification of 3.9.

The images are processed to obtain the transverse spatial distribution by integrating the measured optical depth in the axial direction. In typical measurements, the cloud expands ballistically by 100–200 μm in 400 μs, much larger than its initial transverse dimension. In the classical regime, we assume ballistic expansion with a Maxwellian distribution. In this case, the temperature is readily determined from the transverse $1/e$ width of the cloud: $a(t) = v\sqrt{1/(2\pi\nu_r)^2 + t^2}$, where $v = \sqrt{2k_B T/M}$ is the thermal velocity and t is the time. Since $\nu_r = 6.6$ kHz, for $t \gg 24$ μs, $a(t) = vt$. Measurements of the cloud radius for several expansion times between 100 and 600 μs fit very well to a straight line.

The number of atoms is determined from the spatially integrated optical depth of the absorption image and the absorption cross section σ. For each of the $M = \pm 1/2$ magnetic sublevels of the populated $F = 1/2$ state, σ is taken to be $(\lambda^2/\pi)/[1 + (2\Delta/\gamma)^2]$, 2/3 of that of the cycling transition. Since the excited hyperfine states are unresolved compared to the linewidth $\gamma = 5.9$ MHz, this cross section contains contributions from both allowed transitions. Results for the number are consistent within

10% for several detunings Δ between 9 to 30 MHz and -30 to -9 MHz, and for variation of the camera focal plane over ± 1 mm from the plane which gives the sharpest image.

After precooling by free evaporation, further cooling is accomplished by lowering the trap depth, producing forced evaporation. We have developed scaling laws for the number of atoms, collision rate, and phase-space density as a function of trap depth U for an optical trap which is continuously lowered [12]. These scaling laws are valid for a fixed $\eta = U/(k_B T) \gg 1$. For $\eta = 10$, we find that the ratio of the final to initial phase-space density increases according to $\rho/\rho_i = (U_i/U)^{1.3}$. This result shows that lowering the trap depth by a factor of 100 should increase the phase-space density by a factor of 400, producing a degenerate sample for $\rho_i > 2.5 \times 10^{-3}$. To maintain a constant value of η, the trap should be lowered from its initial depth U_i according to the formula

$$U(t) = U_i/(1 + t/\tau)^\beta, \qquad (1)$$

which assures us that the lowering rate slows as the collision rate decreases [12]. Taking $\eta = 10$, we have $\beta = 1.45$ and $1/\tau = 2.0 \times 10^{-3}\gamma_i$, where γ_i is the initial elastic collision rate. For a 50–50 mixture of fermions, $\gamma_i = \pi N_i M \sigma \nu_i^3/(k_B T_i)$ with N_i the initial total number of atoms. Note that γ_i is reduced by a net factor of 4 compared to a single-component Bose gas with the same parameters. For a scattering length of $a \simeq -100a_0$, the elastic cross section is $\sigma = 8\pi a^2 = 0.7 \times 10^{-11}$ cm^2. Using $\nu_i = (\nu_x\nu_y\nu_z)^{1/3} = 2.4$ kHz, $N_i = 1.0 \times 10^6$, and $T_i = 50$ μK, we obtain $\gamma_i = 4.4 \times 10^2$ s^{-1} and $\tau = 1.1$ sec.

Unfortunately, the A/O modulator that controls the CO_2 laser intensity produces an ellipticity which varies as the radio-frequency (rf) power is varied. The ellipticity is corrected at maximum rf power by a cylindrical telescope. However, the telescope provides only fixed compensation. Hence, as the rf power is decreased to lower the trap depth, the beam becomes elliptical, reducing $\nu_x\nu_y\nu_z$ by a factor of 2 compared to that expected on the basis of the laser power alone. Further, we find that the direction of the beam changes by 3 mrad as the rf power is reduced by a factor of 100, causing vignetting. We align the trap beam to minimize this vignetting, but beam distortion still occurs. For this reason, we cannot accurately compare our evaporation results to the scaling law model. To compensate for the loss of confinement arising from the beam distortion as the trap is lowered, we increase τ to 3 sec. The trap laser intensity is lowered using an Agilent (33120A) arbitrary waveform generator, the output of which is filtered with a time constant of 0.2 sec before being applied to the multiplier input of the A/O radio-frequency generator.

We have measured atomic velocity distributions after forced evaporative cooling for a variable time t_f. To provide a calibrated reference trap, time-of-flight images are recorded after adiabatic recompression to full trap depth over 11 sec. This also increases the spatial den-

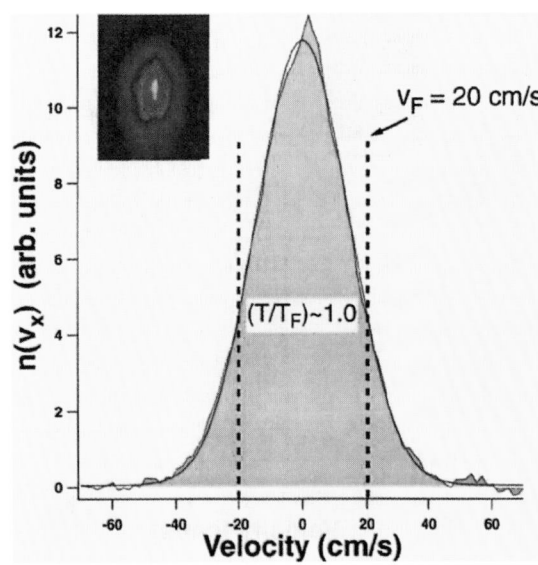

FIG. 1. Absorption image (inset) and velocity distribution after 10 sec of forced evaporative cooling followed by recompression to full trap depth. An average of five trials is shown. $T/T_F = 1$, as determined by a fit to a Maxwellian distribution. $\nu_F = 20$ cm/s is the Fermi velocity for a total $N = 8 \times 10^5$ atoms.

sity and hence the Fermi temperature, while preserving the phase-space density. Figure 1 shows the velocity distribution for $t_f = 10$ sec. The total number of atoms remaining is $N = 8 \times 10^5$, corresponding to a Fermi temperature of $T_F = h\nu(6N/2)^{1/3}/k_B = 15$ μK for each state. Assuming a Maxwellian distribution, the gas is at a temperature of 15 μK, yielding $T/T_F = 1$. At this temperature, a substantial number of atoms have velocities greater than the Fermi velocity of 20 cm/sec.

Near degeneracy, the energy of the atoms contains a contribution from the Fermi energy so that the true temperature is lower than that obtained using a Maxwell-Boltzmann (MB) distribution which assumes that all of the energy is thermal. Hence, the low temperature absorption images are fit using a Thomas-Fermi (TF) approximation to determine T/T_F [21,22], where the Fermi temperature T_F is calculated using the measured trap frequencies and integrated atom number. At the lowest temperatures achieved in the experiments, the MB temperature is $\simeq 10\%$ higher than the TF approximation.

Degeneracy is attained for $t_f = 40$ sec, where $T \simeq 5.8$ μK and $T/T_F = 0.55$ with 3×10^5 atoms remaining. At this temperature, the gas is degenerate, and $\rho \simeq (T_F/T)^3/6 \simeq 1$ [21]. We have also measured the temperature of the atoms in the lowered trap without recompression to full trap depth. We obtain temperatures a factor of $\simeq 10$ lower, i.e., $\simeq 580$ nK, as expected for a harmonic trap which is lower in depth by a factor of $\simeq 100$.

Figure 2 shows the velocity distribution for $t_f = 60$ sec. The total number of atoms is reduced to 10^5,

VOLUME 88, NUMBER 12   PHYSICAL REVIEW LETTERS   25 MARCH 2002

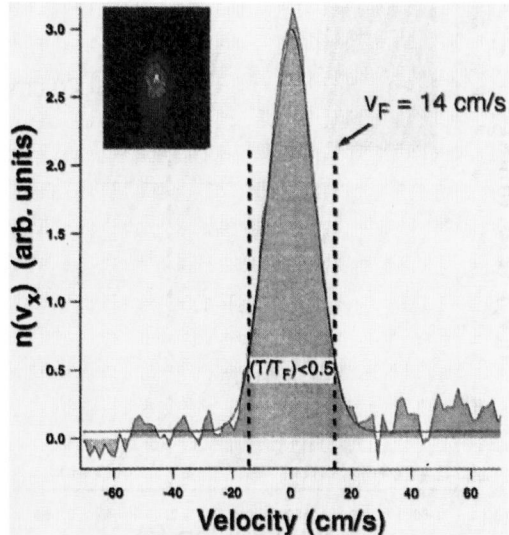

FIG. 2. Absorption image (inset) and velocity distribution after 60 sec of forced evaporative cooling followed by recompression to full trap depth. An average of five trials is shown. $T/T_F < 0.5$ as determined by a fit using a Thomas-Fermi approximation. $v_F = 14$ cm/s is the Fermi velocity for a total $N = 10^5$ atoms.

This research is supported by the physics divisions of the Army Research Office and the National Science Foundation, the Fundamental Physics in Microgravity Research program of the National Aeronautics and Space Administration, and the Chemical Sciences, Geosciences and Biosciences Division of the Office of Basic Energy Sciences, Office of Science, U.S. Department of Energy.

[1] H. T. C. Stoof, M. Houbiers, C. A. Sackett, and R. G. Hulet, Phys. Rev. Lett. **76**, 10 (1996); see also M. Houbiers *et al.*, Phys. Rev. A **56**, 4864 (1997).
[2] J. Bohn, Phys. Rev. A **61**, 053409 (2000).
[3] M. Holland, S. J. J. M. F. Kokkelmans, M. L. Chiofalo, and R. Walser, Phys. Rev. Lett. **87**, 120406 (2001).
[4] E. Timmermans, K. Furuya, P. W. Milonni, and A. K. Kerman, Phys. Lett. A **285**, 228 (2001).
[5] R. Combescot, Phys. Rev. Lett. **83**, 3766 (1999).
[6] M. Randeria, in *Bose-Einstein Condensation*, edited by A. Griffin, D. Snoke, and S. Stringari (Cambridge University Press, Cambridge, 1995), pp. 355–392.
[7] K. M. O'Hara, M. E. Gehm, S. R. Granade, S. Bali, and J. E. Thomas, Phys. Rev. Lett. **85**, 2092 (2000).
[8] B. DeMarco and D. S. Jin, Science **285**, 1703 (1999).
[9] A. G. Truscott, K. E. Strecker, W. I. McAlexander, G. B. Patridge, and R. G. Hulet, Science **291**, 2570–2572 (2001).
[10] F. Schreck, L. Khaykovich, and K. L. Corwin, G. Ferrari, T. Bourdel, J. Cubizolles, and C. Salomon, Phys. Rev. Lett. **87**, 080403 (2001).
[11] K. M. O'Hara, S. R. Granade, M. E. Gehm, T. A. Savard, S. Bali, C. Freed, and J. E. Thomas, Phys. Rev. Lett. **82**, 4204 (1999).
[12] K. M. O'Hara, M. E. Gehm, S. R. Granade, and J. E. Thomas, Phys. Rev. A **64**, 051403(R) (2001).
[13] M. D. Barrett, J. A. Sauer, and M. S. Chapman, Phys. Rev. Lett. **87**, 010404 (2001).
[14] S. Friebel, C. D'Andrea, J. Walz, M. Weitz, and T. W. Hänsch, Phys. Rev. A **57**, R20 (1998).
[15] We find by Monte Carlo modeling that the parametric resonance frequency is shifted downward in a Gaussian trap compared to that of a harmonic trap. The shift is temperature dependent and arises because the restoring force in a Gaussian trap decreases as the radius increases. For an analytic treatment, see R. Jáuregui, Phys. Rev. A **64**, 053403 (2001).
[16] T. Takekoshi and R. J. Knize, Opt. Lett. **21**, 77 (1996).
[17] S. Bali, K. M. O'Hara, M. E. Gehm, S. R. Granade, and J. E. Thomas, Phys. Rev. A **60**, R29 (1999).
[18] For a complete treatment of background gas collision-induced heating including multiple scattering, see H. C. W. Beijerinck, Phys. Rev. A **62**, 063614 (2000).
[19] M. Houbiers, H. T. C. Stoof, W. I. McAlexander, and R. G. Hulet, Phys. Rev. A **57**, R1497 (1998).
[20] K. M. O'Hara, S. R. Granade, M. E. Gehm, and J. E. Thomas, Phys. Rev. A **63**, 043403 (2001).
[21] D. A. Butts and D. S. Rokhsar, Phys. Rev. A **55**, 4346 (1997).
[22] B. DeMarco, Ph.D. thesis, University of Colorado, Boulder, 2001.
[23] W. Ketterle and N. J. Van Druten, Adv. At. Mol. Opt. Phys. **37**, 181 (1996).

corresponding to a Fermi temperature of 8 μK. The measured temperature is below 4 μK, yielding $T/T_F = 0.48$. Nearly all atoms have velocities less than the Fermi velocity of 14 cm/sec.

In the experiments, we achieve high evaporation efficiency $\chi \equiv \ln(\rho_f/\rho_i)/\ln(N_i/N_f)$ [23]. For example, after precooling, but prior to forced evaporation, $N_i = 1.3 \times 10^6$ and $\rho_i = 8 \times 10^{-3}$ per state. After 40 sec of forced evaporation, $N_f = 0.3 \times 10^6$ and $\rho_f \simeq 1$. Hence, $\chi \simeq 3.3$. The overall evaporation efficiency is similar. Starting with the loading conditions where the total number of atoms is $N_i = 3.5 \times 10^6$ at a temperature of 150 μK, we obtain $\chi = 2.9$ after 40 sec of forced evaporation. Despite the trap distortion described above, these results are comparable to the best achieved in magnetic traps [23].

In conclusion, we have produced a degenerate, two-component ^{6}Li Fermi gas in a single beam all-optical trap by direct evaporative cooling. By using a stable CO_2 laser trap at a background pressure of $<10^{-11}$ Torr, efficient evaporation over time scales of 85 sec is achieved. In future experiments, it will be possible to attain scattering lengths of $\simeq -300a_0$ by increasing the bias magnetic field to 300 G, thereby increasing the elastic cross section at low temperature by nearly a factor of 10. This should enable preparation of a degenerate sample in just a few seconds, producing substantially lower temperatures by reducing the detrimental effects of any residual heating. We are currently preparing for a systematic study of the Feshbach resonance at higher magnetic field, and hope to observe superfluid pairing in a two-state Fermi gas.

REPORTS

Bose-Einstein Condensation of Cesium

Tino Weber, Jens Herbig, Michael Mark, Hanns-Christoph Nägerl, Rudolf Grimm

Bose-Einstein condensation of cesium atoms is achieved by evaporative cooling using optical trapping techniques. The ability to tune the interactions between the ultracold atoms by an external magnetic field is crucial to obtain the condensate and offers intriguing features for potential applications. We explore various regimes of condensate self-interaction (attractive, repulsive, and null interaction strength) and demonstrate properties of imploding, exploding, and non-interacting quantum matter.

Cesium (Cs) is an atom of particular interest in physics. It serves as our primary frequency standard (1) and has various important applications in fundamental metrology, such as measurements of the fine-structure constant (2), the electric dipole moment of the electron (3), parity violation (4), and the Earth's gravitational field (5). Cs, a heavy alkali atom with small photon recoil, is well suited for laser cooling and trapping methods. However, because of quantum-mechanical scattering resonances, collisions between Cs atoms at ultralow energy exhibit unusual properties with drastic consequences, such as large frequency shifts in atomic clocks (1). A further consequence of this resonant scattering is the fact that Cs has so far resisted all attempts to produce Bose-Einstein condensation (BEC) (6–13). In contrast, all other stable alkali species—^{87}Rb (14), ^{23}Na (15), ^{7}Li (16), ^{85}Rb (17), and ^{41}K (18)—have been condensed, along with hydrogen (19) and metastable ^{4}He (20).

We report the achievement of BEC of Cs (^{133}Cs, the only stable isotope) using optical trapping methods in combination with magnetic tuning of the scattering properties. With the condensate, we explore the tunability of its self-interaction that results from low-field

Institut für Experimentalphysik, Universität Innsbruck, Technikerstraße 25, 6020 Innsbruck, Austria.

Feshbach resonances (21, 22). Across such a Feshbach resonance, the s-wave scattering length a shows a dispersive variation as a function of the applied magnetic field from very large positive to negative values. The magnetic tunability of the self-interaction has been exploited in experiments with magnetically trapped ^{85}Rb (17) and optically trapped ^{7}Li (23, 24). Cs offers further flexibility because of a unique combination of resonances at low magnetic fields: One broad Feshbach resonance allows for precise tuning, whereas several narrow resonances enable very rapid control. With the magnetic field being a free parameter in our optical trapping approach to BEC, we can take full advantage of this tunability. By switching a to zero, we realize a non-interacting condensate that undergoes minimum expansion when released from the trap. By variations of the magnetic field, we observe an imploding BEC at negative a, a gently expanding BEC at moderate positive values of a, and an exploding condensate on a narrow Feshbach resonance.

Early experiments attempting BEC of Cs (6–8) followed magnetic trapping approaches similar to those used before in the realization of BEC in Rb and Na (14, 15). Resonant elastic scattering with a large cross section was observed, but a rapid loss of atoms due to inelastic two-body collisions had a detrimental effect on evaporative cooling and prevented the experiments from reaching BEC (6, 7). More recent magnetic trapping experiments have explored a range of magnetic fields, where two-body loss is relatively small, and have approached the phase-space density required for BEC to within a factor of 10 (13).

The use of optical dipole forces allows atoms to be trapped in their lowest internal state (25), where the absence of internal energy leads to full suppression of inelastic two-body loss processes. Because atoms in this state seek out regions of high magnetic fields, they are in general not trappable by magnetic forces. Experiments in optical dipole traps (9–12) have explored cooling of Cs in the corresponding state, which has a total angular momentum $F = 3$ and a magnetic quantum number $m_F = 3$. Evaporative cooling (12) has approached BEC to within a factor of 2, before running into limitations by three-body processes (26).

Knowledge of the s-wave scattering length of the lowest internal state of Cs originates from precise measurements of low-field Feshbach resonances (21) and their theoretical analysis (22). The scattering length a shows a pronounced dependence on the magnetic field B (27, 28): It varies from a large negative value at zero field ($-3000\,a_0$ at 0 G) to large positive values at higher fields ($1000\,a_0$ at 55 G), going through zero at 17 G; $a_0 = 0.0529$ nm denotes Bohr's radius. This smooth variation over a wide magnetic field range can be interpreted as a result of a broad Feshbach resonance at about -8 G; here the negative field corresponds to a resonance position of $+8$ G (22) in the magnetically trappable state with $m_F = -3$. In addition to this broad one, several narrow Feshbach resonances occur (27, 29), with the most prominent one at 48 G.

In our experiment, the starting point for evaporative cooling is a large-volume dipole trap that we realize by horizontally crossing two 100-W CO_2 laser beams (30). The trap has an effective volume of about 1 mm³, is ~10 μK deep, and has a geometrically averaged trap frequency of $\bar{\omega}/2\pi = 14$ Hz. In contrast to the all-optical approach to BEC of

(31), which uses tightly focused CO_2 laser beams and trap frequencies of the order of 1 kHz, our trap is shallow and designed for loading a large number of atoms rather than providing tight confinement. The CO_2 laser trap is loaded by adiabatical release of Cs atoms from an optical lattice. After Raman sideband cooling in the lattice according to (32) we start with a sample of 2×10^7 atoms with predominant polarization in the state $F = 3$, $m_F = 3$, a temperature of ~1 μK, and a phase-space density of a few 10^{-3}. Transfer losses and spatial mismatches, together with an excess potential energy of the trap, reduce the initial phase-space density to ~10^{-4}.

A magnetic levitation field (30) is applied to support the atoms against gravity (12, 14). This field is essential to operate the large-volume CO_2 laser trap, which features optical forces far below the gravitational force on a Cs atom. Moreover, it is crucial for implementing efficient three-dimensional evaporation without being limited by gravitational sag. We combine the inhomogeneous levitation field with a freely adjustable bias field B_0 (30) to tune the scattering properties.

Evaporative cooling toward BEC proceeds in three stages (Fig. 1). In the first 10 s (Fig. 1A), the sample is cooled down by evaporation out of the CO_2 laser trap at constant trap depth. To obtain a sufficient collision rate at the rather low density of a few 10^{11} cm⁻³, the scattering length is tuned to a large magnitude of $a = 1200\,a_0$ by setting the bias field to $B_0 = 75$ G. After 10 s, 2×10^6 atoms remain trapped at a temperature of ~1 μK and a phase-space density of about 10^{-3}.

In the following 5 s (Fig. 1B), an additional horizontal laser beam derived from a Yb fiber laser at a wavelength of 1064 nm is focused into the sample (waist, 30 μm) to create a narrow and deep potential well in the center of the CO_2 laser trap. Within these 5 s, the power is ramped up linearly from zero to 90 mW. The adiabatically deformed potential provides a strong local increase in the number density with a minor increase in the temperature, and a substantial gain in phase-space density is achieved (33, 34). The magnetic

bias field is set to $B_0 = 23$ G ($a = 300\,a_0$), which suppresses three-body loss as observed at higher values of a. Finally, one of the CO_2 lasers is turned off and about 3×10^5 atoms remain trapped in the combined field of the 1064-nm beam intersecting the CO_2 laser beam at an angle of 30°. This results in a cigar-shaped trap with tight radial confinement provided by the 1064-nm beam (radial trap frequency, 320 Hz). The axial confinement (axial frequency, 6 Hz) is essentially provided by the remaining CO_2 laser beam. In this way, a dense sample of Cs atoms (a few 10^{12} cm⁻³) is prepared with a phase-space density on the order of 10^{-2}.

In the final stage (Fig. 1C), forced evaporative cooling is performed by ramping down the power of the 1064-nm beam to values of a few milliwatts within 17 s (30). The magnetic bias field is kept at 23 G ($a = 300\,a_0$), which is found to optimize the ratio of elastic collisions and inelastic three-body processes. We find that, for the given ramp, evaporative cooling toward BEC is possible only in a narrow magnetic field range between 21 and 25 G. Below 21 G, the cross

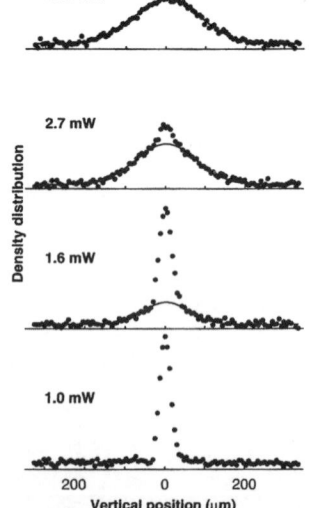

Fig. 2. Vertical density profiles of the released atom cloud observed after 50 ms of free expansion in the magnetic levitation field. The variable laser power at the end of the ramp is shown at left (in milliwatts). The profiles are obtained by horizontally integrating the measured column density in a 130-μm-wide region of interest. In these measurements, the magnetic bias field is switched to 17 G at the time of release to suppress the condensate expansion, which makes the appearance of the two-component distribution more pronounced. Gaussian fits to the thermal part of the distribution (solid lines in the upper three graphs) yield the temperatures.

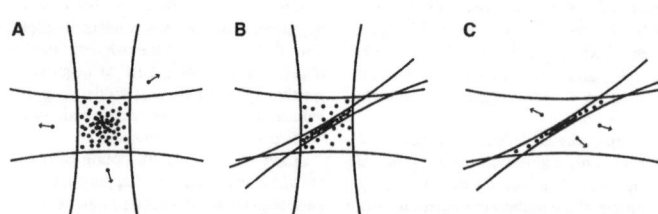

Fig. 1. Illustration of the three stages of evaporative cooling toward BEC of Cs. (**A**) Ten seconds of plain evaporation in two crossed CO_2 laser beams at a magnetic bias field of 75 G (scattering length tuned to 1200 a_0). (**B**) Five seconds of collisional loading of a tightly focused 1064-nm laser beam at 23 G (300 a_0). (**C**) Forced evaporative cooling by ramping down the power of the 1064-nm beam over 17 s, with the magnetic bias field kept at 23 G.

REPORTS

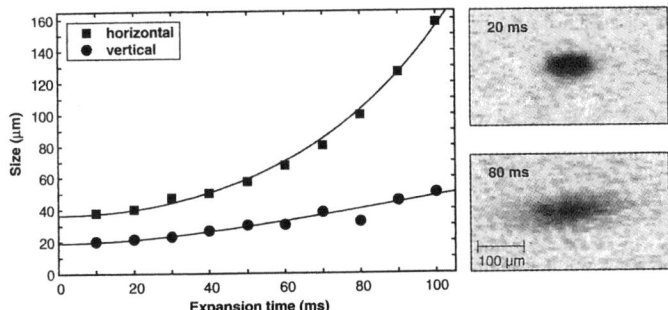

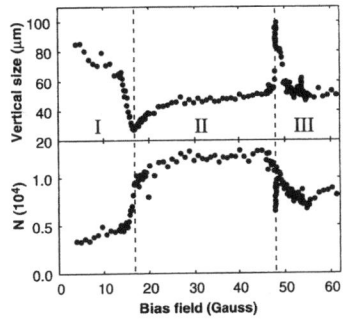

Fig. 3. Expansion of a non-interacting condensate at 17 G in the magnetic levitation field. The images at right show the condensate 20 ms (top) and 80 ms (bottom) after release. The data points at left show the 1/e half-widths of Gaussian fits to absorption images of the expanding cloud. The fit to the horizontal expansion shows the function $A\cosh(\alpha t)$, which describes the expansion with a calculated time constant of $\alpha^{-1} = 47$ ms (30) and an initial width $A = 36$ μm as the only adjustable parameter. The fit to the vertical expansion yields a mean kinetic energy of $k_B \times 600$ pK.

Fig. 4. Vertical extension (1/e half-width of a Gaussian fit) and atom number N of the expanding cloud as a function of the variable bias field applied in the first 10 ms after release. The measurements are taken after a total expansion time of 50 ms. The three different regions refer to (I) negative scattering length; (II) positive scattering length; and (III) a regime dominated by narrow Feshbach resonances, a prominent one at 48 G and a weaker one at 53 G.

section for elastic scattering is too small; and above 25 G, an increased rate of three-body processes in combination with the hydrodynamic collision regime leads to strong loss.

At the end of the ramp, the sample is released from the trap by simultaneously turning off both the CO_2 laser and the 1064-nm laser. In the magnetic levitation field, the atoms remain at a fixed vertical position. The observation time is typically limited to 100 ms by the horizontal spreading of the cloud that results from the weak transverse antitrapping force of the levitation field (30). For detection, absorption images are taken in a standard way $(14, 15)$ with a resonant probe beam that propagates horizontally at an angle of 30° with respect to the 1064-nm laser beam and 60° with respect to the CO_2 laser beam.

The BEC phase transition becomes obvious in the vertical density profiles of Fig. 2, which were taken 50 ms after release. Down to a final power of 3.45 mW (potential depth, 550 nK), a thermal cloud is observed; the corresponding profile shows a cloud of 65,000 atoms at temperature $(T) = 46$ nK. At final powers of 2.7 mW (450 nK) and 1.6 mW (260 nK), the profiles show a partially condensed cloud with $T = 36$ nK and 21 nK, respectively. At 1.0 mW (160 nK), an almost pure condensate is realized with 16,000 atoms. When the condensate with a peak density of 1.3×10^{13} cm^{-3} is kept in the trap, we measure a lifetime of ~15 s and infer an upper bound for the three-body loss coefficient of 2×10^{-27} cm^6 s^{-1}. This highlights the stability of the Cs BEC at the applied magnetic field.

We explore the magnetic tunability of the Cs condensate in the range between 5 and 50 G. The zero-crossing of the scattering length at 17 G allows us to turn off the self-interaction and thus to realize a "frozen" condensate with minimum internal energy. Figure 3 shows the measured expansion of the conden-

sate, when B_0 is switched to 17 G at the time of release. In the vertical direction, a very slow expansion is observed, with a mean kinetic energy as low as $k_B \times 600$ pK (where k_B is Boltzmann's constant), which we attribute to residual self-interaction effects within the finite switching time. In the horizontal direction, the increasing width is fully determined by the horizontal magnetic force, which magnifies the initial cloud size according to the expected cosine hyperbolicus function (30). The non-interacting condensate can also be observed when the magnetic levitation field is turned off. Then, however, the observation of the falling BEC is limited to about 35 ms. We expect that with a few straightforward technical improvements in switching, the interaction energy of the "frozen" condensate could be reduced down to values of a few pK.

In order to demonstrate the other regimes of self-interaction, we switch to a variable magnetic field synchronously with the release of the condensate. The variable field is applied for a short time interval of 10 ms, which is sufficiently long for the mean-field dynamics to take place. For subsequent expansion at fixed conditions, the magnetic bias field is switched to 17 G. After a total expansion time of 50 ms, the vertical extension of the cloud and the total number of atoms are measured. The corresponding results (Fig. 4) show a marked dependence on the magnetic field with three distinct regions: Below 17 G (region I), the scattering length is negative and the BEC implodes, leading to a large momentum spread and a substantial loss of atoms. Between 17 and 48 G (region II), the scattering length is positive and varies smoothly from zero to about 1000 a_0. The expansion shows the minimum width at 17 G and a subsequent increase in width, in agreement with the predicted behavior of the scattering

length. For higher fields (region III), the behavior is dominated by a narrow Feshbach resonance at 48 G. On this resonance, the cloud expands very rapidly as the condensate explodes in response to the strong sudden increase of its internal energy. In addition, a sharp loss feature is observed, which may indicate the formation of molecules in the BEC (35). At even higher fields, the data show an asymmetry and broadening of the resonance together with a loss of atoms, which we attribute to the finite ramp speed over the resonance (17).

The unique tunability offered by a Cs condensate is of great interest in various respects. In the field of quantum gases, it may serve as the experimental key to explore new regimes beyond standard mean-field theory, such as the strongly interacting regime with an interparticle spacing on the order of the scattering length, systems in reduced dimensionality, and the Mott insulator phase $(36, 37)$. Moreover, Cs is an interesting candidate for the creation of cold molecules and a molecular BEC $(35, 38)$. For applications in metrology, a "frozen" BEC without internal energy would represent an ideal source of cold atoms; for example, for precision measurements of the photon recoil (39). For future atomic clocks, one may envisage loading a very weak optical lattice with Cs atoms from a BEC. With one atom per site, such a system would provide long observation times with suppressed collisional frequency shifts.

Besides this particular interest in Cs, our experiments demonstrate that optical trapping methods allow BEC to be achieved with a species for which conventional magnetic trapping approaches have met severe difficulties.

References and Notes

1. C. Salomon et al., Proceedings of the 17th International Conference on Atomic Physics (ICAP 2000), E. Arimondo, P. D. Natale, M. Inguscio, Eds. (American Institute of Physics, Melville, NY, 2001), pp. 23–40.
2. J. M. Hensley, A. Wicht, B. C. Young, S. Chu, Proceedings of the 17th International Conference on Atomic Physics (ICAP 2000), E. Arimondo, P. D. Natale, M. Inguscio, Eds. (American Institute of Physics, Melville, NY, 2001), pp. 43–57.
3. C. Chin, V. Leiber, V. Vuletic, A. J. Kerman, S. Chu, Phys. Rev. A 63, 033401 (2001).
4. C. E. Wieman, Proceedings of the 16th International Conference on Atomic Physics (ICAP 1998), W. E. Baylis, G. W. Drake, Eds. (American Institute of Physics, Woodbury, NY, 1999), pp. 1–13.
5. M. J. Snadden, J. M. McGuirk, P. Bouyer, K. G. Haritos, M. A. Kasevich, Phys. Rev. Lett. 81, 971 (1998).
6. J. Söding, D. Guéry-Odelin, P. Desbiolles, G. Ferrari, J. Dalibard, Phys. Rev. Lett. 80, 1869 (1998).
7. D. Guéry-Odelin, J. Söding, P. Desbiolles, J. Dalibard, Europhys. Lett. 44, 26 (1998).
8. J. Arlt et al., J. Phys. B 31, L321 (1998).
9. H. Perrin, A. Kuhn, I. Bouchoule, C. Salomon, Europhys. Lett. 42, 395 (1998).
10. A. J. Kerman, V. Vuletić C. Chin, S. Chu, Phys. Rev. Lett. 84, 439 (2000).
11. D.-J. Han et al., Phys. Rev. Lett. 85, 724 (2000).
12. D. Han, M. T. DePue, D. Weiss, Phys. Rev. A 63, 023405 (2001).
13. S. L. Cornish, S. Hopkins, A. M. Thomas, C. J. Foot, abstract from the 7th Workshop on Atom Optics and Interferometry, 28 September to 2 October 2002, Lunteren, Netherlands (book of abstracts).
14. M. Anderson, J. Ensher, M. Matthews, C. Wieman, E. Cornell, Science 269, 198 (1995).
15. K. Davis et al., Phys. Rev. Lett. 75, 3969 (1995).
16. C. Bradley, C. Sackett, J. Tollett, R. Hulet, Phys. Rev. Lett. 75, 1687 (1995).
17. S. Cornish, N. Claussen, J. Roberts, E. Cornell, C. Wieman, Phys. Rev. Lett. 85, 1795 (2000).
18. G. Modugno et al., Science 294, 1320 (2001); published online 18 October 2001 (10.1126/science.1066687).
19. D. G. Fried et al., Phys. Rev. Lett. 81, 3811 (1998).
20. A. Robert et al., Science 292, 461 (2001); published online 22 March 2001 (10.1126/science.1060622).
21. C. Chin, V. Vuletic, A. J. Kerman, S. Chu, Phys. Rev. Lett. 85, 2717 (2000).
22. P. J. Leo, C. J. Williams, P. S. Julienne, Phys. Rev. Lett. 85, 2721 (2000).
23. L. Khaykovich et al., Science 296, 1290 (2002).
24. K. Strecker, G. Partridge, A. Truscott, R. Hulet, Nature 417, 150 (2002).
25. R. Grimm, M. Weidemüller, Yu. B. Ovchinnikov, Adv. At. Mol. Opt. Phys. 42, 95 (2000).
26. D. S. Weiss, personal communication.
27. A. J. Kerman et al., C. R. Acad. Sci. Paris IV 2, 633 (2001).
28. P. S. Julienne, E. Tiesinga, C. J. Williams, personal communication.
29. We observe the narrow Feshbach resonances at 11.0, 14.3, 15.0, 19.9, 47.7, and 53.4 G (accuracy ±0.2 G) as loss resonances in three-body decay of an uncondensed gas.
30. Materials and methods are available as supporting material on Science Online.
31. M. Barrett, J. Sauer, M. Chapman, Phys. Rev. Lett. 87, 010404 (2001).
32. P. Treutlein, K. Y. Chung, S. Chu, Phys. Rev. A 63, 051401 (2001).
33. D. Stamper-Kurn et al., Phys. Rev. Lett. 81, 2194 (1998).
34. M. Hammes, D. Rychtarik, H.-C. Nägerl, R. Grimm, Phys. Rev. A 66, 051401(R) (2002).
35. E. A. Donley, N. R. Claussen, S. T. Thompson, C. E. Wieman, Nature 417, 529 (2002).
36. D. Jaksch, C. Bruder, J. Cirac, C. Gardiner, P. Zoller, Phys. Rev. Lett. 81, 3108 (1998).
37. M. Greiner, O. Mandel, T. Esslinger, T. W. Hänsch, I. Bloch, Nature 415, 39 (2002).
38. F. Masnou-Seeuws, P. Pillet, Adv. At. Mol. Opt. Phys. 47, 53 (2001).
39. S. Gupta, K. Dieckmann, Z. Hadzibabic, D. E. Pritchard, Phys. Rev. Lett. 89, 140401 (2002).
40. We thank all members of the cold-atom group for support and a great team spirit. In particular, we thank the Cs surface trapping team, M. Hammes, D. Rychtarik, and B. Engeser, for fruitful interactions and for sharing the 1064-nm laser; A. Noga for assistance; and J. Hecker Denschlag for useful discussions. We are indebted to R. Blatt and his ion trapping group for support during our start-up phase in Innsbruck. We acknowledge financial support by the Austrian Science Fund (FWF) within project P15114 and within SFB 15 (project part 16).

Supporting Online Material
www.sciencemag.org/cgi/content/full/1079699/DC1
Materials and Methods
Figs. S1 to S3
References

23 October 2002; accepted 26 November 2002
Published online 5 December 2002;
10.1126/science.1079699
Include this information when citing this paper.

An Optical Spoon Stirs Up Vortices in a Bose–Einstein Condensate

Since the early days of Bose–Einstein condensates (BECs) in atomic gases, comparisons have been drawn to the other familiar bosonic systems: liquid helium-4 and the Cooper pairs of superconductors. Two of the early questions asked of BECs were whether they had similar coherence properties and whether they were superfluids. In ^{4}He, it is rather difficult to study properties of the condensate, such as the condensate fraction, whereas the measurement of superfluidity is fairly straightforward. The opposite has proved true for atomic condensates: The condensate fraction as a function of temperature was found early on, but only recently has their superfluidity been placed on firm footing (see PHYSICS TODAY, November 1999, page 17).

One property associated with superfluids is the ability to support quantized circulation. Normal fluids, such as a stirred cup of coffee, rotate like a rigid body. In contrast, the velocity of a one-component superfluid like ^{4}He is related to the gradient of the phase of its wavefunction. Consequently, such superfluids only support flow with zero curl, which precludes rigid-body rotation. But superfluids can support quantized circulation through the introduction of vortices. Each vortex is characterized by a node where the condensate wavefunction (and hence its density) goes to zero. Along any closed path surrounding one vortex, the wavefunction phase changes by 2π—and thus the fluid circulates around the vortex.

Last fall, researchers at JILA in Boulder, Colorado, succeeded in creating a vortex in a BEC using microwaves and a rotating laser beam to "imprint" the characteristic 2π phase winding in a two-component condensate.[1] Now Kirk Madison, Frédéric Chevy, Wendel Wohlleben, and Jean Dalibard at Ecole Normale Supérieure (ENS) in Paris have demonstrated another method for generating vortices: rotating an asymmetric trapping potential.[2,3] This new technique can readily produce multiple vortices, and it also allows quantitative studies of vortex nucleation and decay.

Optical spoon

In the classic "rotating bucket" experiments on superfluid ^{4}He, a cylindrical container is rotated as the liquid within it is cooled. (See PHYSICS

> Multiple vortices form regular arrangements in a condensate in a rotating trap.

TODAY's special issue on superfluid helium, February 1987.) Imperfections in the walls provide the transfer of angular momentum that is required to introduce vortices into the liquid. But a cylindrically symmetric magnetic trap is perfectly smooth, so another route is needed to get a BEC spinning.

The technique implemented by the ENS team in many ways resembles the rotating-bucket approach of liquid He experiments. Dalibard and company start with a Ioffe–Pritchard magnetic trap, which produces cigar-shaped condensates—in this case with a radius of 2.5 μm and a length of 100 μm. Except for a small static inhomogeneity, the trap is axially symmetric about the length of the cigar.

The researchers introduce a transverse asymmetry using a laser tuned far from the atomic resonance. The laser beam, with a diameter of 20 μm, induces a dipole moment in the atoms, which then couples to the laser's electric field. The beam is aligned parallel to the long axis of the trap but isn't centered in the trap. Instead, it is toggled rapidly back and forth between two points symmetrically located about 8 μm from the center. With a sufficiently high toggle rate, the atoms feel the time-averaged potential of two laser beams.

The additional potential from the toggled laser beam squashes the condensate's cigar shape, giving it a slightly elliptical cross section. By rotating the toggling about the long axis of the trap, the researchers rotate the asymmetric potential, which stirs the condensate.

"Toggling the laser beam is essential," notes Dalibard. Otherwise, the laser would merely shift the trap's center—the trap would remain axisymmetric. Rotating the beam would then just move the trap in a circle, which would not stir the condensate. "It's the difference between moving a coffee cup around in a circle and inserting a spoon and rotating it," explains Madison.

To introduce vortices into condensates, the researchers load their trap with about half a billion spin-polarized rubidium atoms. With the stirring laser on and rotating at a fixed frequency, they evaporatively cool the atoms through the BEC transition temperature down to a temperature of about 100 nK, at which most of the remaining 10^5 atoms are in the condensate.

Early surprises

Dalibard and coworkers have found

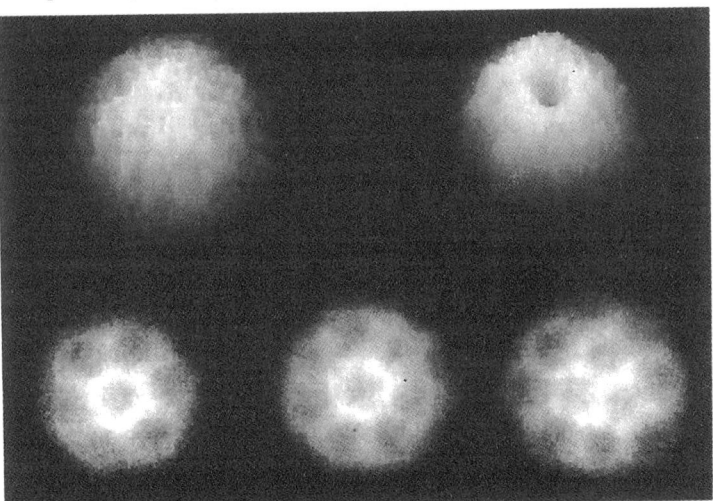

STIRRED BOSE–EINSTEIN CONDENSATES show no vortices when the stirring frequency is below the critical frequency for vortex nucleation (top left). Once the critical frequency is crossed, a vortex appears at the center of the condensate (top right). Arrays of vortices are found for higher stirring frequencies. Shown below are images of 7, 8, and 11 vortices. (Courtesy of K. Madison and F. Chevy.)

that below a critical stirring frequency, which depends on the steepness of the magnetic trapping potential, no vortices appear in the condensate. For a range of frequencies above the critical frequency, a single vortex is observed as a pronounced density dip in the center of the condensate, shown in figure on page 19. Recent measurements of the angular momentum of the stirred condensate confirm a jump from 0 to $\hbar$ per atom at the critical frequency, as expected for a single vortex in the center of the trap.[3] Surprisingly, the critical frequency determined experimentally is about 50% higher than expected from calculations of when the one-vortex state becomes the lowest-energy state of the rotating condensate.

At even higher stirring frequencies, multiple vortices appear and form regular arrangements that resemble those found in rotating superfluid ^{4}He and the triangular Abrikosov vortex lattices found in type-II superconductors in a magnetic field. The cores of the BEC vortices are relatively large compared to the size of the trapped condensate, and so only a small number of vortices can be cleanly observed. At sufficiently high stirring frequencies, a turbulent structure is seen instead in the condensate. Ultimately, at stirring frequencies approaching the restoring frequencies of the trapping potential, the condensate is lost.

In their more recent experiments, Dalibard and company have varied their condensate preparation protocol: Instead of cooling with the stirring laser on, they have begun stirring after the condensate is formed. In another surprise, they find the nucleation of vortices with essentially the same critical frequency as when they cool while stirring. In contrast, a higher rotation frequency is needed to nucleate vortices in superfluid ^{4}He if the spinning starts after the liquid is cooled than when the liquid is cooled while the bucket is rotating.

Open questions

The ENS work joins three other results that demonstrate the superfluid properties of condensates. By dragging a laser beam through a condensate, Wolfgang Ketterle's group at MIT has demonstrated frictionless flow below a critical dragging velocity.[4] Chris Foot and coworkers at the University of Oxford[5] have observed undamped irrotational oscillation of the condensate—the so-called scissors mode—when a slightly asymmetric trap is given a quick twist, as predicted by David Guéry-Odelin and Sandro Stringari of the University of Trento in Italy.[6] Meanwhile, the JILA researchers have continued to explore vortices with the phase imprinting technique, and have recently observed vortex precession in condensates.[7]

Theorists and experimenters alike have been given much to ponder and explore with the ongoing revelations into the behavior of condensates. Why the critical frequency for vortex nucleation is so much higher than predicted remains an outstanding puzzle, as is why no vortices are observed in condensates too close to the transition temperature. Vortices may help elucidate the interactions between the condensed and noncondensed atoms. The nature of the excitations of the vortices themselves is also being examined.

The behavior of superfluid ^{4}He is dominated by strong interactions between atoms and with the walls. Detailed models and simulations are consequently difficult. In contrast, BECs have weak interactions that are well described by microscopic theories, and magnetic traps are inherently clean. Furthermore, the vortex cores in BECs are about a thousand times larger than those in ^{4}He. With BECs, notes David Feder of NIST in Gaithersburg, Maryland, "many of the questions that have dogged the liquid helium community for years can be addressed directly for the first time." Eric Cornell of JILA adds, "Two of the most interesting things in a vortex's life are its birth and death. Now we can look at both."

RICHARD FITZGERALD

References

1. M. R. Matthews, B. P. Anderson, P. C. Haljan, D. S. Hall, C. E. Wieman, E. A. Cornell, Phys. Rev. Lett. **83**, 2498 (1999).
2. K. W. Madison, F. Chevy, W. Wohlleben, J. Dalibard, Phys. Rev. Lett. **84**, 806 (2000); http://xxx.lanl.gov/abs/cond-mat/0004037, to be published in J. Mod. Opt.
3. F. Chevy, K. W. Madison, J. Dalibard, http://xxx.lanl.gov/abs/cond-mat/0005221.
4. C. Raman, M. Köhl, R. Onofrio, D. S. Durfee, C. E. Kuklewicz, Z. Hadzibabic, W. Ketterle, Phys. Rev. Lett. **83**, 2502 (1999). R. Onofrio, C. Raman, J. M. Vogels, J. Abo-Shaeer, A. P. Chikkatur, W. Ketterle, http://xxx.lanl.gov/abs/cond-mat/0006111.
5. O. M. Maragò, S. A. Hopkins, J. Arlt, E. Hodby, G. Hechenblaikner, C. J. Foot, Phys. Rev. Lett. **84**, 2056 (2000).
6. D. Guéry-Odelin, S. Stringari, Phys. Rev. Lett. **83**, 4452 (1999).
7. B. P. Anderson, P. C. Haljan, C. E. Wieman, E. A. Cornell, http://xxx.lanl.gov/abs/cond-mat/0005368. ∎

R E P O R T S

Observation of Vortex Lattices in Bose-Einstein Condensates

J. R. Abo-Shaeer, C. Raman, J. M. Vogels, W. Ketterle

Quantized vortices play a key role in superfluidity and superconductivity. We have observed the formation of highly ordered vortex lattices in a rotating Bose-condensed gas. These triangular lattices contained over 100 vortices with lifetimes of several seconds. Individual vortices persisted up to 40 seconds. The lattices could be generated over a wide range of rotation frequencies and trap geometries, shedding light on the formation process. Our observation of dislocations, irregular structure, and dynamics indicates that gaseous Bose-Einstein condensates may be a model system for the study of vortex matter.

The quantization of circulation has a profound effect on the behavior of macroscopic quantum systems. Magnetic fields can penetrate type-II superconductors only as quantized flux lines. Vorticity can enter rotating superfluids only in the form of discrete line defects with quantized circulation. These phenomena are direct consequences of the existence of a macroscopic wavefunction, the phase of which must change by integer multiples of 2π around magnetic flux or vortex lines. In superconductors, magnetic flux lines arrange themselves in regular lattices that have been directly imaged (1). In superfluids, direct observation of vortices has been limited to small arrays (up to 11 vortices), both in liquid ^{4}He (2) and, more recently, in rotating gaseous Bose-Einstein condensates (BECs) (3, 4).

We report the observation of vortex lattices in a BEC. We are now able to explore the properties of bulk vortex matter, which includes local structure, defects, and long-range order. In contrast, the properties of small arrays are strongly affected by surface and finite size effects. The vortex lattices are highly excited collective states of BECs with an angular momentum of up to 60 $\hbar$ per particle. Our experiments show that such states can be prepared and are much more stable than predicted (5).

Vortices in BECs have been the subject of extensive theoretical study (6). Experimental progress began only recently with the observation of quantized circulation in a two-component condensate by a phase engineering technique (7) and of vortex arrays in a single-component BEC (3). A condensate can be subjected to a rotating perturbation by revolving laser beams around it. This technique was used to study surface waves in a trapped BEC (8), and subsequently for the creation of vortices (3). In 1997, we tried unsuccessfully to detect quantized circulation as a "centrifugal hole" in ballistic expansion of the gas (9, 10). Theoretical calculations (11–13) and ultimately the pioneering experimental work (3) showed that vortices can indeed be detected through ballistic expansion, which magnifies the spatial structure of the trapped condensate.

BECs of up to 5×10^7 Na atoms with a negligible thermal component (condensate fraction $\geq$ 90%) were produced by a combination of laser and evaporative cooling techniques (8, 10, 14). A radio-frequency "shield" limited the magnetic trap depth to 50 kHz (2.3 μK), preventing high-energy atoms from heating the condensate. Experiments were performed in cylindrical traps with widely varying aspect ratios. Most of the results and all of the images were obtained in a weak trap, with radial and axial frequencies of ν_r = 84 Hz and ν_z = 20 Hz (aspect ratio 4.2), respectively. In this weak trap inelastic losses were suppressed, resulting in larger condensates of typically 5×10^7 atoms. Such clouds had a chemical potential (μ) of 310 nK (determined from time-of-flight imaging), a peak density of 4.3×10^{14} cm^{-3}, a Thomas-Fermi radius along the radial direction (R_r) of 29 μm, and a healing length (ξ) of about 0.2 μm.

Vortex lattices were produced by rotating the condensate around its long axis with the optical dipole force exerted by blue-detuned

laser beams at a wavelength of 532 nm. A two-axis acousto-optic deflector generated a pattern of two laser beams rotating symmetrically around the condensate at variable drive frequency Ω (8). The two beams were separated by one Gaussian beam waist (w = 25 μm). The laser power of 0.7 mW in each beam corresponded to an optical dipole potential of 115 nK. This yielded a strong, anharmonic deformation of the condensate.

After the condensate was produced, the stirring beam power was ramped up over 20 ms, held constant for a variable stirring time, and then ramped down to zero over 20 ms. The condensate equilibrated in the magnetic trap for a variable hold time (typically 500 ms). The trap was then suddenly switched off, and the gas expanded for 35 ms to radial and axial sizes of $l_r \cong$ 1000 μm and $l_z \cong$ 600 μm, respectively. We probed the vortex cores using resonant absorption imaging. To avoid blurring of the images due to bending of the cores near the edges of the condensate, we pumped a thin, 50- to 100-μm slice of atoms in the center of the cloud from the F = 1 to the F = 2 hyperfine state (15). This section was then imaged along the axis of rotation with a probe pulse resonant with the cycling F = 2 $\to$ 3 transition. The duration of the pump and probe pulses was chosen to be sufficiently short (50 and 5 μs, respectively) to avoid blurring due to the recoil-induced motion and free fall of the condensate.

We observed highly ordered triangular lattices of variable vortex density containing up to 130 vortices (Fig. 1). A striking feature is the extreme regularity of these lattices, free of any major distortions, even near the boundary. Such "Abrikosov" lattices were first predicted for quantized magnetic flux lines in type-II superconductors (16). Tkachenko showed that their lowest energy structure should be triangular for an infinite system (17). A slice through images shows the high visibility of the vortex cores (Fig. 2), which was as high as 80%. For a trapped condensate with maximum vortex density, we infer that the distance between the vortices was $\cong$ 5 μm. The radial size of the condensate in the time-of-flight images was over 10% larger when it was filled with the maximum number of vortices, probably due to centrifugal forces.

When a quantum fluid is rotated at a frequency Ω, it attempts to distribute the vorticity as uniformly as possible. This is similar to a rigid body, for which the vorticity

Department of Physics, Center for Ultracold Atoms at Massachusetts Institute of Technology (MIT) and Harvard University, and Research Laboratory of Electronics, MIT, Cambridge, MA 02139, USA.

*To whom correspondence should be addressed. E-mail: jamil@mit.edu

REPORTS

Fig. 1. Observation of vortex lattices. The examples shown contain approximately (**A**) 16, (**B**) 32, (**C**) 80, and (**D**) 130 vortices. The vortices have "crystallized" in a triangular pattern. The diameter of the cloud in (D) was 1 mm after ballistic expansion, which represents a magnification of 20.

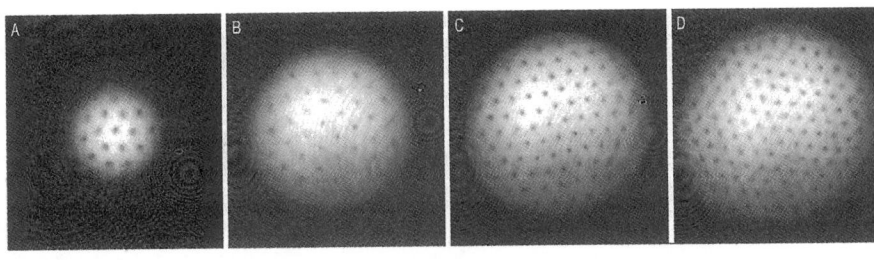

Slight asymmetries in the density distribution were due to absorption of the optical pumping light.

is constant $\nabla \times \vec{v} = 2\vec{\Omega}$. For a superfluid, the circulation of the velocity field, $\vec{v}$, is quantized in units of $\kappa = h/M$, where M is the atomic mass and h is Planck's constant. The quantized vortex lines are distributed in the fluid with a uniform area density (18)

$$n_v = 2\Omega/\kappa \qquad (1)$$

In this way the quantum fluid achieves the same average vorticity as a rigidly rotating body, when "coarse-grained" over several vortex lines. For a uniform density of vortices, the angular momentum per particle is $N_v \hbar/2$, where N_v is the number of vortices in the system.

The number of observed vortices is plotted as a function of stirring frequency Ω for two different stirring times (Fig. 3). The peak near 60 Hz corresponds to the frequency $\Omega/2\pi = v_r/\sqrt{2}$, where the asymmetry in the trapping potential induced a quadrupolar surface excitation, with angular momentum $l = 2$, about the axial direction of the condensate (the actual excitation frequency of the surface mode $v = \sqrt{2}v_r$ is two times larger due to the twofold symmetry of the quadrupole pattern). The same resonant enhancement in the vortex production was observed for a stiff trap, with $v_r = 298$ Hz and $v_z = 26$ Hz (aspect ratio 11.5), and has recently been studied in great detail for small vortex arrays (19).

Far from the resonance, the number of vortices produced increased with the stirring time. By increasing the stir time up to 1 s, vortices were observed for frequencies as low as 23 Hz ($\cong 0.27 v_r$). Similarly, in a stiff trap we observed vortices down to 85 Hz ($\cong 0.29 v_r$). From Eq. 1 one can estimate the equilibrium number of vortices at a given rotation frequency to be $N_v = 2\pi R^2 \Omega/\kappa$. The observed number was always smaller than this estimate, except near resonance. Therefore, the condensate did not receive sufficient angular momentum to reach the ground state in the rotating frame. In addition, because the drive increased the moment of inertia of the condensate (by weakening the trapping potential), we expect the lattice to rotate faster after the drive is turned off.

Looking at time evolution of a vortex lattice (Fig. 4), the condensate was driven near the

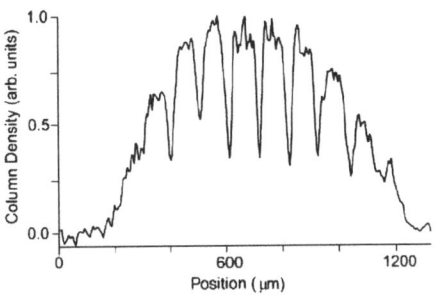

Fig. 2. Density profile through a vortex lattice. The curve represents a 5-μm-wide cut through a two-dimensional image similar to those in Fig. 1 and shows the high contrast in the observation of the vortex cores. The peak absorption in this image is 90%.

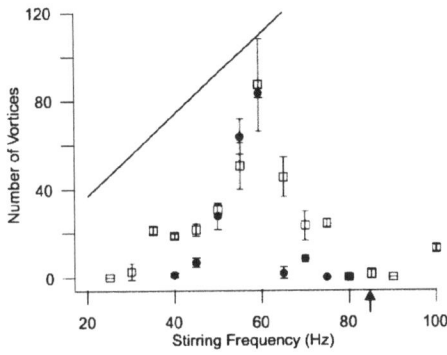

Fig. 3. Average number of vortices as a function of the stirring frequency Ω for two different stirring times, (●) 100 ms and (□) 500 ms. Each point represents the average of three measurements with the error bars given by the standard deviation. The solid line indicates the equilibrium number of vortices in a radially symmetric condensate of radius $R_r = 29$ μm, rotating at the stirring frequency. The arrow indicates the radial trapping frequency.

quadrupole resonance for 400 ms and then probed after different periods of equilibration in the magnetic trap. A blurry structure was already visible at early times. Regions of low column density are probably vortex filaments that were misaligned with the axis of rotation and showed no ordering (Fig. 4A). As the dwell time increased, the filaments began to disentangle and align with the axis of the trap (Fig. 4, B and C), and finally formed a completely ordered Abrikosov lattice after 500 ms (Fig. 4D). Lattices with fewer vortices could be generated by rotating the condensate off resonance. In these cases, it took longer for regular lattices to form. Possible explanations for this observation are the weaker interaction between vortices at lower vortex density and the larger distance

they must travel to reach their lattice sites. In principle, vortex lattices should have already formed in the rotating, anisotropic trap. We suspect that intensity fluctuations of the stirrer or improper beam alignment prevented this.

The vortex lattice had lifetimes of several seconds (Fig. 4, E to G). The observed stability of vortex arrays in such large condensates is surprising because in previous work the lifetime of vortices markedly decreased with the number of condensed atoms (3). Theoretical calculations predict a lifetime inversely proportional to the number of vortices (5). Assuming a temperature $k_B T \cong \mu$, where k_B is the Boltzmann constant, the predicted decay time of $\cong 100$ ms is much shorter than observed. After 10 s, the number of vortices

REPORTS

Fig. 4. Formation and decay of a vortex lattice. The condensate was rotated for 400 ms and then equilibrated in the stationary magnetic trap for various hold times. (**A**) 25 ms, (**B**) 100 ms, (**C**) 200 ms, (**D**) 500 ms, (**E**) 1 s, (**F**) 5 s, (**G**) 10 s, and (**H**) 40 s. The decreasing size of the cloud in (E) to (H) reflects a decrease in atom number due to inelastic collisions. The field of view is ~1 mm by 1.15 mm.

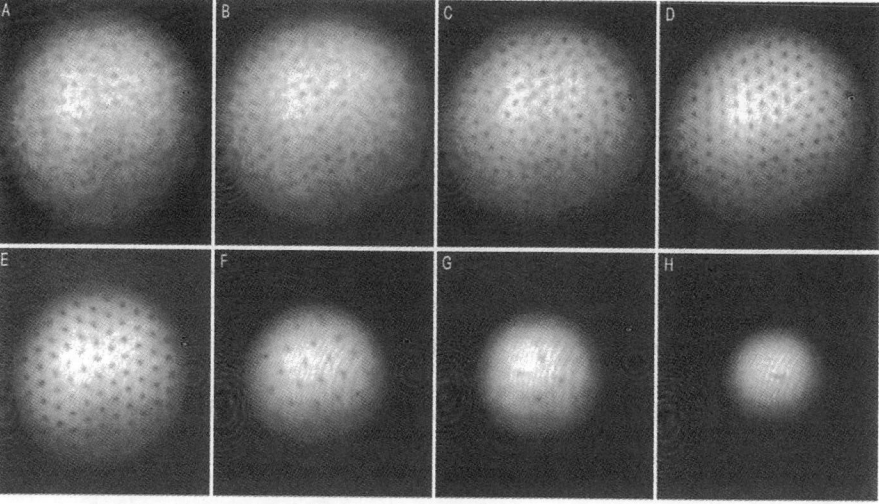

Fig. 5. Vortex lattices with defects. In (**A**), the lattice has a dislocation near the center of the condensate. In (**B**), there is a defect reminiscent of a grain boundary.

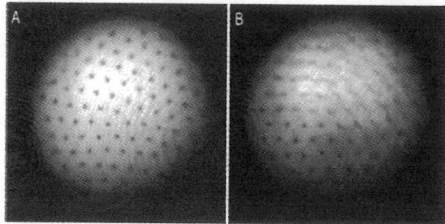

had diminished considerably. In several cases we observed a single vortex near the condensate center after 30 to 40 s (Fig. 4H). This dwell time is much longer than that observed for elongated clouds (1 to 2 s) (3) and for nearly spherical condensates (15 s) (20). We estimate that during its lifetime, the superfluid flow field near the central vortex core had completed more than 500,000 revolutions and the lattice itself had rotated ~100 times.

A feature of the lattices is their almost perfect triangular shape. Deformations are less pronounced than in theoretical calculations that predicted circular distortions (13, 21). It was shown (21) that a configuration with perfect triangular symmetry can lower its energy by rearranging the outermost ring of vortices into a circle. This distortion is caused by the cylindrical symmetry of rotation and not by boundary effects (which were neglected in the calculations). In condensates of finite size and inhomogeneous density, one may expect even larger distortions. However, in images containing large numbers of vortices (Fig. 1, C and D), the lattice is triangular throughout the condensate even up to the edge. We have observed more complex lattice configurations in a fraction of

the images. Some show patterns characteristic of partial crystallization (Fig. 4B), dislocations (Fig. 5A), and grain boundaries (Fig. 5B).

Our experiments may shed light on the ongoing discussion of vortex formation (6). A single vortex is expected to be thermodynamically stable when the rotation frequency exceeds (22)

$$\Omega_C = \frac{5}{2} \frac{\hbar}{MR_r^2} \ln \left[\frac{0.671 R_r}{\zeta} \right] \qquad (2)$$

For our trap this yields a critical angular frequency of $\cong 2\pi \times 6$ Hz, or $0.08\omega_r$. Vortices were only observed at rotation frequencies much higher than Ω_C. Some have suggested that the critical frequency of $\cong 0.7\ \omega_r$ observed in other experiments was related to the suppression of unstable excitations (so-called anomalous modes) of rectilinear vortices (6, 23, 24). These anomalous modes depend strongly on the aspect ratio of the trap. Our observation of vortices at ~0.25ω_r in traps where the aspect ratio varied by a factor of 2.7 seems to rule out a crucial role of the anomalous modes, possibly because their frequencies are lower for dense condensates.

Apart from stability conditions, vortices must be nucleated. Several theoretical pa-

pers have discussed the barrier for the entry of a vortex from the edge (6, 25, 26). The condensate becomes unstable against surface deformations when the stirring frequency exceeds

$$\Omega_s = \min_l (\omega_l / l) \qquad (3)$$

where ω_l is the frequency of a surface excitation with angular momentum l in the axisymmetric trap (27, 28). For our condensates, the instability is predicted to occur at $l \cong 18$, yielding $\Omega_s \cong 0.3\omega_r$, which is in fair agreement with our observations (29). The instabilities occur on a practical time scale only when the cloud is sufficiently deformed to excite high-l modes. In other experiments (3, 19), a sizable deformation was achieved only by resonantly driving the $l = 2$ mode near $0.7\omega_r$. In contrast, we strongly deformed the condensate over a broad range of frequencies. For such deformations, l is no longer a good quantum number, allowing for coupling into higher-order modes. Our technique is therefore well suited to studying the threshold for vortex formation.

The surface instability criterion of Eq. 3 applies to rotational excitation. In the spirit of our earlier stirring experiments on critical velocities (14, 30), we also explored linear motion of the stirrer. When we moved a small laser beam ($\cong 10\ \mu$m in diameter) at a velocity of 2 mm/s once through the condensate along the radial direction, we observed large lattices containing $\cong 50$ vortices after an equilibration time of 1 s, indistinguishable from those generated with the rotating anisotropy. We assume that imperfect alignment of the stirring beam imparted a torque on the condensate. However, the mechanism of vortex formation warrants further study.

REPORTS

Properties of vortex lattices are of broad interest in superfluids, superconductors, and even astrophysics. Fluctuations in the rotation rate of pulsars are attributed to the dynamics of the vortex lattice in a superfluid neutron liquid (5, 31). Our experiments show that vortex formation and self-assembly into a regular lattice is a robust feature of rotating BECs. Gaseous condensates may serve as a model system to study the dynamics of vortex matter, in analogy to work in type-II superconductors (32). Of particular interest are collective modes of the lattice. In liquid helium, transverse oscillations in a vortex lattice (Tkachenko oscillations) have already been investigated (33, 34). Further studies may address the nucleation, ordering, and decay of lattices, in particular to delineate the role of the thermal component (5), and possible phase transition associated with melting and crystallization.

References and Notes

1. H. Träuble, U. Essmann, *Phys. Lett.* **24A**, 526 (1967).
2. E. J. Yarmchuk, M. J. V. Gordon, R. E. Packard, *Phys. Rev. Lett.* **43**, 214 (1979).
3. K. W. Madison, F. Chevy, W. Wohlleben, J. Dalibard, *Phys. Rev. Lett.* **84**, 806 (2000).
4. ———, *J. Mod. Opt.* **47**, 2715 (2000).
5. P. O. Fedichev, A. E. Muryshev, preprint available at http://arXiv.org/abs/cond-mat/0004264.
6. Reviewed in A. L. Fetter, A. A. Svidzinsky, *J. Phys. Condens. Matter* **13**, R135 (2001).
7. M. R. Matthews *et al.*, *Phys. Rev. Lett.* **83**, 2498 (1999).
8. R. Onofrio *et al.*, *Phys. Rev. Lett.* **84**, 810 (2000).
9. M. R. Andrews, thesis, Massachusetts Institute of Technology (1998).
10. Reviewed in W. Ketterle, D. S. Durfee, D. M. Stamper-Kurn, in *Bose-Einstein Condensation in Atomic Gases, Proceedings of the International School of Physics Enrico Fermi, Course CXL*, M. Inguscio, S. Stringari, C. Wieman, Eds. (International Organisation Services B.V., Amsterdam, 1999), pp. 67–176.
11. E. Lundh, C. J. Pethick, H. Smith, *Phys. Rev. A* **58**, 4816 (1998).
12. F. Dalfovo, M. Modugno, *Phys. Rev. A* **61**, 023605 (2000).
13. Y. Castin, R. Dum, *Eur. Phys. J. D* **7**, 399 (1999).
14. R. Onofrio *et al.*, *Phys. Rev. Lett.* **85**, 2228 (2000).
15. M. R. Andrews *et al.*, *Science* **275**, 637 (1997).
16. A. A. Abrikosov, *J. Exp. Theor. Phys.* **5**, 1174 (1957) [*Zh. Eksp. Teor. Fiz.* **32**, 1442 (1957)].
17. V. K. Tkachenko, *J. Exp. Theor. Phys.* **22**, 1282 (1966) [*Zh. Eksp. Teor. Fiz.* **49**, 1875 (1965)].
18. P. Nozières, D. Pines, *The Theory of Quantum Liquids* (Addison-Wesley, Redwood City, CA, 1990).
19. K. W. Madison, F. Chevy, V. Bretin, J. Dalibard, preprint available at http://arXiv.org/abs/cond-mat/?0101051.
20. P. C. Haljan, B. P. Anderson, I. Coddington, E. A. Cornell, preprint available at http://arXiv.org/abs/cond-mat/0012320.
21. L. J. Campbell, R. M. Ziff, *Phys. Rev. B* **20**, 1886 (1979).
22. E. Lundh, C. J. Pethick, H. Smith, *Phys. Rev. A* **55**, 2126 (1997).
23. D. L. Feder, A. A. Svidzinsky, A. L. Fetter, C. W. Clark, *Phys. Rev. Lett.* **86**, 564 (2001).
24. J. J. Garcia-Ripoll, V. M. Pérez-García, preprint available at http://arXiv.org/abs/cond-mat/0012071; ———, preprint available at http://arXiv.org/abs/con-mat/0101219.
25. T. Isoshima, K. Machida, *Phys. Rev. A* **60**, 3313 (1999).
26. A. A. Svidzinsky, A. L. Fetter, *Phys. Rev. Lett.* **84**, 5919 (2000).
27. F. Dalfovo, S. Giorgini, M. Guilleumas, L. P. Pitaevskii, S. Stringari, *Phys. Rev. A* **56**, 3840 (1997).
28. F. Dalfovo, S. Stringari, *Phys. Rev. A* **63**, 011601 (2001).
29. J. R. Anglin, personal communication.
30. C. Raman *et al.*, *Phys. Rev. Lett.* **83**, 2502 (1999).
31. R. J. Donnelly, *Quantized Vortices in Helium II* (Cambridge Univ. Press, Cambridge, 1991).
32. G. Blatter, M. V. Feigel'man, V. B. Geshkenbein, A. I. Larkin, V. M. Vinokur, *Rev. Mod. Phys.* **66**, 1125 (1994).
33. V. K. Tkachenko, *J. Exp. Theor. Phys.* **23**, 1049 (1966) [*Zh. Eksp. Teor. Fiz.* **50**, 1573 (1966)].
34. S. J. Tsakadze, *Fiz. Nizk. Temp.* **4**, 148 (1978) [*Sov. J. Low Temp. Phys.* **4**, 72 (1978)].

35. We thank J. R. Anglin, A. Görlitz, R. Onofrio, and L. Levitov for useful discussions and critical readings of the manuscript and T. Rosenband for assistance with the two-axis deflector. Supported by NSF, Office of Naval Research, Army Research Office, NASA, and the David and Lucile Packard Foundation.

26 February 2001; accepted 14 March 2001
Published online 22 March 2001;
10.1126/science.1060182
Include this information when citing this paper.

VOLUME 85, NUMBER 14   PHYSICAL REVIEW LETTERS   2 OCTOBER 2000

Vortex Precession in Bose-Einstein Condensates: Observations with Filled and Empty Cores

B. P. Anderson,* P. C. Haljan, C. E. Wieman, and E. A. Cornell*

JILA, National Institute of Standards and Technology and Department of Physics, University of Colorado, Boulder, Colorado 80309-0440

(Received 19 May 2000)

We have observed and characterized the dynamics of singly quantized vortices in dilute-gas Bose-Einstein condensates. Our condensates are produced in a superposition of two internal states of ^{87}Rb, with one state supporting a vortex and the other filling the vortex core. Subsequently, the state filling the core can be partially or completely removed, reducing the radius of the core by as much as a factor of 13, all the way down to its bare value of the healing length. The corresponding superfluid rotation rates, evaluated at the core radius, vary by a factor of 150, but the precession frequency of the vortex core about the condensate axis changes by only a factor of 2.

PACS numbers: 03.75.Fi, 67.90.+z, 67.57.Fg, 32.80.Pj

The dynamics of quantized vortices in superfluid helium and superconductors have been fascinating and important research areas in low-temperature physics [1,2]. Even at zero temperature, vortex motion within a superfluid is intricately related to the quantization of current around the vortex core. Besides these superfluid systems, studies of the dynamics of optical vortices have also become an active area of research [3]. More recently, demonstrations of the creation of quantized vortices in dilute-gas Bose-Einstein condensates (BEC) [4,5] have emphasized the similarities between the condensed matter, optical, and dilute-gas quantum systems. Because of the observational capabilities and the techniques available to manipulate the quantum wave function of the condensates, dilute-gas BEC experiments provide a unique approach to studies of quantized vortices and their dynamics. This paper reports direct observations and measurements of singly quantized vortex core precession in a BEC.

Numerous theoretical papers have explored the expected stability and behavior of vortices in BEC [6–17]. One interesting expected effect is vortex core precession about the condensate axis [7,9–16]. Radial motion of the core within the condensate can also occur, and may be understood as being due to energy dissipation and damping processes. Core precession may be described in terms of a Magnus effect—a familiar concept in fluid dynamics and superfluidity [1]. An applied force on a rotating cylinder in a fluid leads to cylinder drift (due to pressure imbalances at the cylinder surface) that is orthogonal to the force. Analogously, a net force on a vortex core in a superfluid results in core motion perpendicular to both the vortex quantization axis and the force. In the condensate vortex case, these forces can be due to density gradients within the condensate, for example, or the drag due to thermal atoms. The density-gradient force may be thought of as one component of an effective buoyancy: just as a bubble in a fluid feels a force antiparallel to the local pressure gradient, a vortex core in a condensate will feel a force towards lower

condensate densities. The total effective buoyancy, however, is due less to displaced mass (the "bubble") than it is to dynamical effects of the velocity-field asymmetry, a consequence of a radially offset core. Typically, the total buoyancy force is away from the condensate center, and the net effect is an azimuthal precession of the core via the Magnus effect. Drag due to the motionless (on average) thermal atoms opposes core precession, causing the core to spiral outwards towards the condensate surface. In the absence of this drag (for temperature ~0), radial drift of the core may be negligible.

Our techniques for creating and imaging a vortex in a coupled two-component condensate are described in Refs. [4,18]. The two components are the $|F = 1, m_F = -1\rangle$ and $|F = 2, m_F = 1\rangle$ internal ground states of ^{87}Rb, henceforth labeled as states $|1\rangle$ and $|2\rangle$, respectively. We start with a condensate of 10^6 $|2\rangle$ atoms, confined in a spherical potential with oscillator frequency 7.8 Hz. A near-resonant microwave field causes some of the $|2\rangle$ atoms to convert to $|1\rangle$ atoms. The presence of a rotating, off-resonant laser beam spatially modulates the amplitude and phase of the conversion. The net result is a conversion of about half of the sample into an annular ring of $|1\rangle$ atoms with a continuous quantum phase winding from 0 to 2π about the circumference—a singly quantized vortex. The balance of the sample remains in the nonrotating $|2\rangle$ state and fills the vortex core. With resonant light pressure we can selectively remove as much of the core material as we desire. In the limit of complete removal, we are left with a single-component, bare vortex state.

In this bare-core limit, the core radius is on the order of the condensate healing length $\xi = (8\pi n_0 a)^{-1/2}$, where n_0 is the peak condensate density and a is the scattering length. For our conditions, $\xi = 0.65$ μm, well under our imaging resolution limit. The bare core can be observed after ballistic expansion [5] of the condensate, but this is a destructive measurement. On the other hand, if we leave some of the $|2\rangle$-state atoms filling the core, the pressure of the filling material opens up the radius of the $|1\rangle$ vortex core to the point where we can resolve the core in a time series of nondestructive phase-contrast images.

Filled-core dynamics.—We first discuss vortex dynamics in two-component condensates, where 10%–50% of the atoms were in the $|2\rangle$ fluid filling the $|1\rangle$ vortex core. We took successive images of the $|1\rangle$ atoms in the magnetic trap, with up to 10 images of each vortex. The vortex core is visible as a dark spot in a bright $|1\rangle$ distribution, as shown in Fig. 1(a). Instabilities in our vortex creation process usually resulted in the creation of off-center vortex cores, allowing us to observe precession of the cores. We observed precession out to ~ 2 s, after which the $|2\rangle$ fluid had decayed to the point that the vortex core was too small to be observed in the trapped condensate.

The recorded profile of each trapped condensate was fit with a smooth Thomas-Fermi distribution. Each vortex core profile was fit with a Gaussian distribution to determine its radius and position within the condensate. From the fits, we determined the overall radius R_t of the trapped condensate (typically 28 μm), the HWHM radius r of the filled vortex core, and the displacement d_t and angle θ_t of the core center with respect to the condensate center. Core angles and radii for the images in Fig. 1(a) are shown in Figs. 1(c) and 1(d). The vortex core is seen precessing in a clockwise direction, which is the same direction as the vortex fluid flow around the core. The angular precession frequency was determined from the time dependence of θ_t [Fig. 1(c)]. This and other similar data sets showed no reproducible radial motion of the core over the times and parameters examined. However, consistent decrease in the

size of the core was observed, which we interpret as being due to known decay of the $|2\rangle$ fluid through inelastic atomic collisional processes.

For each data set, we determined a mean core radius and displacement. The data cover a range of core radii ($r = 7\xi$ to 13ξ), displacements ($d_t = 0.17R_t$ to $0.48R_t$), and percentage of atoms in the core (10% to 50%). Except for a few "rogue vortices" (discussed below), the measured precession frequencies are clustered around 1.4 Hz, as shown in Fig. 2, precessing in the same direction as the fluid rotation. The data [Fig. 2(a)] suggest a slight increase in frequency for cores farther from the condensate center. We also see [Fig. 2(b)] a slight decrease in precession frequency for larger cores. These measurements are in qualitative agreement with two-dimensional numerical simulations for two-component condensates [16].

As indicated in Fig. 2, a few vortex cores exhibited precession opposite to that of the fluid flow (negative frequencies). The quality of the corresponding vortex images was routinely lower than for the positive-frequency precession points, with vortices looking more like crescents and "D" shaped objects rather than like the images of Fig. 1. We only found such occurrences with our two-component (intrap) measurements. We speculate that this "inverse precession" may be due to rare events in which total angular momentum may be distributed in a complicated way among both internal states. Such cases might arise due to position instabilities of the rotating laser beam during the vortex creation process. Recent theoretical attention has addressed the possibly related situation of nonsymmetric configurations of vortices in two-component condensates [17].

Bare-core dynamics.—To examine the dynamics of bare vortices, our procedure consisted of taking a nondestructive phase-contrast picture of the partially filled $|1\rangle$ vortex

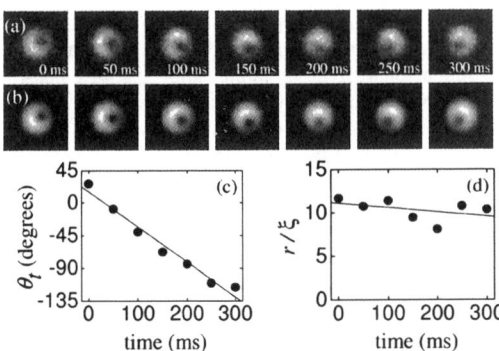

FIG. 1. (a) Seven successive images of a condensate with a vortex and (b) their corresponding fits. The 75 μm-square nondestructive images were taken at the times listed, referenced to the first image. The vortex core is visible as the dark region within the bright condensate image. (c) The azimuthal angle of the core is determined for each image, and is plotted vs time held in the trap. A linear fit to the data indicates a precession frequency of 1.3(1) Hz for this data set. (d) Core radius r in units of healing length ξ. The line shown is a linear fit to the data.

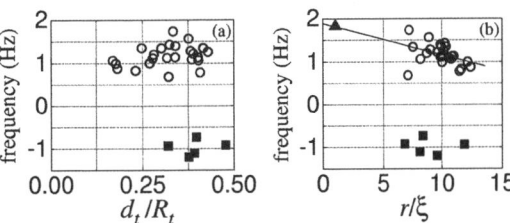

FIG. 2. Compiled data for filled vortex core precession, with each data point extracted from a series (as in Fig. 1) of nondestructive images of a single vortex. Precession frequency is plotted vs (a) core displacement d_t in units of condensate radius R_t, and (b) core radius r in units of healing length ξ. Circles correspond to positive frequencies and filled squares to negative frequencies. (Positive frequency is defined as core precession having the same handedness as the vortex angular momentum.) The triangle at $r = \xi$ shows for reference the average measured precession frequency of many bare vortices [see text and Fig. 3(b)]. A line is drawn as a guide to the trend in frequency vs core size.

VOLUME 85, NUMBER 14   PHYSICAL REVIEW LETTERS   2 OCTOBER 2000

distribution [Fig. 3(a) inset], as previously discussed, followed by complete removal of the core filling [19]. We then held the bare vortex in the trap for a variable hold time t_h, after which the condensate was released from the trap. We took a final near-resonance phase-contrast image [20] of the atomic distribution [Fig. 3(a)] after the condensate had ballistically expanded by a factor of $\sim$3.5 [21] and the empty core had expanded [22] to a fit radius of $\sim$9 μm.

Displacements d_t and angular positions θ_t of the cores for the in-trap images were extracted as described before. The images of the expanded clouds were fit with identical distributions, and the Thomas-Fermi radius R_e of the expanded cloud and the vortex core displacement d_e and angle θ_e were obtained for each image. For each pair of images, we determined the angular difference $\Delta\theta_{et} \equiv \theta_e - \theta_t$ between the cores in the expanded and in-trap images. We also determined the core displacement ratio d_e/d_t, an indicator of the radial motion of the core during the hold time t_h.

From the measurements of $\Delta\theta_{et}$ at different hold times t_h [Fig. 3(b)], we find a bare-core precession frequency of 1.8(1) Hz, slightly faster than the precession of filled cores and consistent with the trend shown in Fig. 2(b) for filled cores. To emphasize that our measurements of filled and empty cores are different limits in a continuum of filling

material, we indicate the measured bare-core precession frequency in Fig. 2(b) with a point at $r = \xi$.

From Fig. 2(b) it is apparent that the structure and content of the vortex core have a relatively modest effect on precession frequency. One can calculate, for instance, the fluid rotation rate ν_r at the inner core radius. The value of ν_r is given by the quantized azimuthal superfluid velocity evaluated at the radius of the core, divided by the circumferential length at that radius. For bare vortices, ν_r is about 260 Hz, while for the largest filled cores of Fig. 2(b) (for which nearly half of the sample mass is composed of core filling), ν_r is only about 1.7 Hz. Thus between vortices whose inner-radius fluid rotation rates vary by a factor of 150, we see only a factor of 2 difference in precession frequency.

The slower precession of filled cores can be understood in terms of our buoyancy picture. Because of its slightly smaller scattering length, $|2\rangle$ fluid has negative buoyancy with respect to $|1\rangle$ fluid, and consequently tends to sink inward towards the center of the condensate [23]. With increasing amounts of $|2\rangle$ material in the core, the inward force on the core begins to counteract the outward buoyancy of the vortex velocity field, resulting in a reduced precession velocity. It is predicted that with a filling material of sufficiently negative buoyancy in the core, the core precession may stop or even precess in a direction opposite to the direction of the fluid flow [16], but our data do not reach this regime.

Various theoretical techniques involving two- and three-dimensional numerical and analytical analyses have been explored to calculate the precession frequency of a vortex core within a condensate. We briefly compare those most readily applied to our physical parameters, assuming a spherical, single-component condensate with 3×10^5 atoms ($R_t = 22$ μm) in a nonrotating trap. Where relevant, we assume a core displacement of $d_t = 0.35 R_t$. A two-dimensional hydrodynamic image vortex analysis [24] has been analytically explored in the homogeneous gas [11] and two-dimensional harmonic confinement [14] limits. The latter of these predicts a bare-core precession frequency of $\sim$0.8 Hz. Svidzinsky and Fetter's three-dimensional [12] solution to the Gross-Pitaevskii equation predicts a precession frequency of 1.58 Hz. Jackson et al. [10] have obtained results in close agreement with this analytical solution using a numerical solution to the Gross-Pitaevskii equation. Finally, a path-integrals technique by Tempere and Devreese [13] predicts a 1.24 Hz precession, and a two-dimensional simulation by McGee and Holland [16] using a steepest-descents technique predicts a precession frequency of 1.2 Hz.

Measurements of d_e/d_t for different hold times t_h show the radial motion of the bare cores and are a probe of energy dissipation of the vortex states. The plot of Fig. 3(c) displays no trend of the core towards the condensate surface during t_h, indicating that thermal damping is negligible on the 1 s time scale [25]. However, we notice a

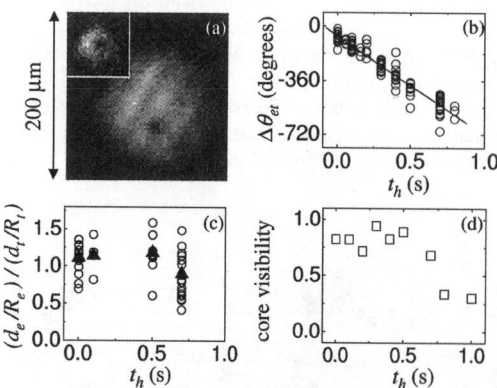

FIG. 3. (a) Ballistic expansion image of a vortex after all $|2\rangle$ atoms have been removed. The dark spot is the bare vortex core. Inset—the corresponding, preceding in-trap nondestructive image of the partially filled core. (b) Angular differences $\Delta\theta_{et}$ between vortex cores from the in-trap and expansion images, plotted against hold time t_h in the magnetic trap. The line is a fit through the data, indicating a bare-core precession frequency of 1.8(1) Hz. (c) Radial core motion is determined by $(d_e/R_e)/(d_t/R_t)$, the ratio of the fractional core displacements from the expansion and in-trap images of each data set. The data are shown as open circles, with the average of all data at each given hold time plotted as a filled triangle. (d) Core visibility of an expanded vortex, defined as the conditional probability for observing a vortex in an expanded image given the observation of a vortex in the corresponding, preexpansion in-trap image. Visibility drops dramatically for hold times $t_h > 1$ s.

VOLUME 85, NUMBER 14   PHYSICAL REVIEW LETTERS   2 OCTOBER 2000

sharply decreasing visibility of expanded bare vortices for hold times greater than $t_h \gtrsim 1$ s, as indicated in Fig. 3(d). The absence of radial core motion suggests that decreased visibility is due to imaging limitations rather than true decay. One hypothesis is that the vortex core may be tilting away from the imaging axis [26], suppressing contrast in optical depth below our signal-to-noise threshold. Such a situation may arise if the trap is not perfectly spherical, and if the vortex is not aligned along a principle axis of the trap. Evidence in support of this hypothesis [27] will be presented in a future paper [28].

Through a combination of destructive and nondestructive imaging techniques we have obtained measurements of vortex dynamics in bare- and filled-core vortices in dilute-gas BEC. Vortex precession frequencies show only modest dependence on the radius and content of the vortex core. Further measurements of vortex dynamics in condensates may reveal the rate of loss of angular momentum at finite temperatures, an indication of energy dissipation. Such measurements may suggest interpretations for "persistence of current" in condensates, further strengthening ties between BEC and superfluidity. In order to pursue these goals, we plan to extend the studies reported here to investigate higher-order dynamical behavior and to characterize the dissipative effects of finite temperatures.

We gladly acknowledge helpful discussions with Murray Holland and Sarah McGee. This work was supported by funding from NSF, ONR, and NIST.

---

*Quantum Physics Division, National Institute of Standards and Technology.

[1] D. R. Tilley and J. Tilley, *Superfluidity and Superconductivity* (IOP Publishing Ltd, Bristol, 1990), 3rd ed.; R. Donnelly, *Quantized Vortices in Helium II* (University Press, Cambridge, 1991).

[2] D. J. Thouless *et al.,* Int. J. Mod. Phys. B **13**, 675 (1999).

[3] G. A. Swartzlander, Jr. and C. T. Law, Phys. Rev. Lett. **69**, 2503 (1992); Y. S. Kivshar *et al.,* Opt. Commun. **152**, 198 (1998).

[4] M. R. Matthews *et al.,* Phys. Rev. Lett. **83**, 2498 (1999).

[5] K. W. Madison, F. Chevy, W. Wohlleben, and J. Dalibard, Phys. Rev. Lett. **84**, 806 (2000).

[6] T. Isoshima and K. Machida, J. Phys. Soc. Jpn. **66**, 3502 (1997).

[7] D. S. Rokhsar, Phys. Rev. Lett. **79**, 2164 (1997).

[8] D. A. Butts and D. S. Rokhsar, Nature (London) **397**, 327 (1999); H. Pu, C. K. Law, J. H. Eberly, and N. P. Bigelow,

Phys. Rev. A **59**, 1533 (1999); T. Isoshima and K. Machida, Phys. Rev. A **59**, 2203 (1999); J. J. García-Ripoll and V. M. Pérez-García, Phys. Rev. A **60**, 4864 (1999).

[9] E. L. Bolda and D. F. Walls, Phys. Rev. Lett. **81**, 5477 (1998).

[10] B. Jackson, J. F. McCann, and C. S. Adams, Phys. Rev. A **61**, 013604 (1999).

[11] P. O. Fedichev and G. V. Shlyapnikov, Phys. Rev. A **60**, R1779 (1999).

[12] A. A. Svidzinsky and A. L. Fetter, Phys. Rev. Lett. **84**, 5919 (2000); (private communication).

[13] J. Tempere and J. T. Devreese, Solid State Commun. **113**, 471 (2000).

[14] E. Lundh and P. Ao, Phys. Rev. A **61**, 63612 (2000).

[15] D. V. Skryabin, cond-mat/0003041.

[16] S. A. McGee and M. J. Holland, cond-mat/0007143.

[17] J. J. García-Ripoll and V. M. Pérez-García, Phys. Rev. Lett. **84**, 4264 (2000); V. M. Pérez-García and J. J. García-Ripoll, cond-mat/9912308.

[18] J. E. Williams and M. J. Holland, Nature (London) **401**, 568 (1999).

[19] The core-filling material is removed slowly enough (100 ms) that the vortex core has time to shrink adiabatically to its final bare size, but rapidly enough that the initial location of the bare core is well correlated with the location of the original filled core.

[20] We use phase-contrast imaging for both the in-trap and expansion images to simplify imaging procedures. When imaging the expanded atom cloud, we obtain good signal when the probe is detuned three linewidths from resonance.

[21] Because of our low trapping frequencies (7.8 Hz), expansions are correspondingly slow. We get larger final spatial distributions by giving the atoms a 6 ms preliminary "squeeze" at high spring constant, followed by a 50 ms expansion period.

[22] F. Dalfovo and M. Modugno, Phys. Rev. A **61**, 023605 (2000).

[23] Tin-Lun Ho and V. B. Shenoy, Phys. Rev. Lett. **77**, 3276 (1996); D. S. Hall *et al.,* Phys. Rev. Lett. **81**, 1539 (1998).

[24] G. B. Hess, Phys. Rev. **161**, 189 (1967).

[25] Our measurements were performed at temperatures of $T = 23(6)$ nK, and relative temperatures of $T/T_c = 0.8(1)$, where T_c is the BEC critical temperature. Temperatures were measured using fits to trapped bare vortices where the core was not resolvable.

[26] A. A. Svidzinsky and A. L. Fetter, cond-mat/0007139.

[27] P. C. Haljan, B. P. Anderson, C. E. Wieman, and E. A. Cornell, in *Proceedings of the Quantum Electronics and Laser Science Conference 2000, San Francisco, 2000,* Abstract No. QPD1 (Optical Society of America, Washington, DC, 2000).

[28] P. C. Haljan, B. P. Anderson, C. E. Wieman, and E. A. Cornell (to be published).

VOLUME 81, NUMBER 23   PHYSICAL REVIEW LETTERS   7 DECEMBER 1998

Observation of Optically Trapped Cold Cesium Molecules

T. Takekoshi, B. M. Patterson, and R. J. Knize

Laser and Optics Research Center, Department of Physics, United States Air Force Academy, Colorado 80840
(Received 30 June 1998)

We report the first observation of optically trapped cold neutral molecules. Cesium dimers in the electronic ground state are produced directly in a magneto-optical trap and transferred to a dipole trap formed at the focus of a CO_2 laser beam ($\lambda = 10.6\ \mu m$). These neutral molecules were detected using photoionization and time-of-flight spectroscopy. Initial experiments indicate a cold molecule trap lifetime on the order of half a second. [S0031-9007(98)07817-X]

PACS numbers: 33.80.Ps, 32.80.Pj

There have been many experiments demonstrating the production and trapping of cold atoms. Atoms with temperatures less than 1 mK can be produced using laser cooling on the alkalis, alkaline earths, metastable inert gases, and a few other elements. Once cooled through collisions with photons, these cold atoms can be confined in various optical and magnetic traps and can be used for a variety of applications [1]. In order to create large scattering forces, laser cooling requires a closed two-level transition, or a relatively simple structure that can be closed with the use of a few additional lasers to avoid optical pumping. This requirement has prevented laser cooling of most atoms and of all molecules. There is interest in producing cold molecules for Doppler-free spectroscopy, cold atom-molecule and molecule-molecule collisions, molecular quantum collective effects, alkali cluster studies, frequency standards [2], molecular optics [3], as well as many other applications already realized for atoms. An alternative cooling method is to sympathetically cool molecules by using a cold buffer gas. They can then be confined in a strong magnetic trap. Cold Eu atoms have already been trapped using this method [4], which could be extended to paramagnetic molecules.

An all-optical method for producing cold molecular vapors is difficult because of vibrational and rotational degrees of freedom, which greatly complicate direct laser cooling [5]. However, it was recently demonstrated by Fioretti and co-workers at Orsay [6] that cold electronic ground state Cs dimers can be produced from laser-cooled atoms [7]. We have independently verified this result [8] and report here the trapping of these cold Cs dimers in an optical dipole force trap formed at the focus of a CO_2 laser beam ($\lambda = 10.6\ \mu m$). This demonstration of the ability to trap cold molecules will allow ultracold molecular vapors to be observed for long times without the necessity of large magnetic fields. Many of the interesting experiments already done with laser cooled atoms may be possible with molecules as well.

We produced cold molecules using a vapor cell magneto-optical trap (MOT) [9]. Our initial detection of the cold molecules produced continuously in a MOT was straightforward. The trapping beams were suddenly switched off, and both atoms and molecules were detected using photoionization. A frequency doubled Nd:YAG laser ($\lambda = 532$ nm) producing about 13 mJ in a 3 ns pulse was utilized. This pulse creates atomic and molecular cesium ions by two-photon processes [10],

$$Cs + 2\nu \rightarrow Cs^+ + e^-, \tag{1}$$

$$Cs_2 + 2\nu \rightarrow Cs^+ + e^- + Cs, \tag{2}$$

$$Cs_2 + 2\nu \rightarrow Cs_2^+ + e^-, \tag{3}$$

$$Cs_2 + 2\nu \rightarrow Cs^+ + Cs^-. \tag{4}$$

At a wavelength of 532 nm, the cross section for the process given in Eq. (2) is about the same as that for Eq. (3). The cross section for the last process is negligible. The photoionization laser was focused so that the beam diameter was about 0.1 cm at the MOT. The ions were detected directly using a model 4721 channeltron (Galileo Corporation). The channeltron input was biased at -2700 V inside the grounded stainless steel vacuum chamber, allowing time-of-flight spectroscopy to be used. We observe an ion signal attributed to molecules because it arrives at a time $\sqrt{2}$ times that of the atom ion signal. The molecules must either be in the triplet ground state (as observed by Fioretti and co-workers [6]) or in the singlet ground state because the delay (a few ms) between turning off the trapping light and photoionizing is much longer than the excited state molecular lifetime (15 ns).

Unlike the Orsay group [6], we do not use a separate photoassociation laser to produce cold electronic ground state molecules. We enhance the small molecular signal seen from a normal MOT by increasing its density. The simplest way to demonstrate this is to suddenly ramp up the MOT's magnetic field gradient. This increased the molecular signal by an order of magnitude, which is expected in light of the fact that the rates of the two most probable mechanisms for dimer formation vary as the atomic density squared (photoassociation by the MOT lasers) and the atomic density cubed (three-body recombination). By varying the delay between switching off the MOT beams and photoionizing, the resulting decay due to falling and expansion is observed. From this, we calculate a molecular translational temperature of about $100~\mu$K. From the temperature and number of molecules detected, we infer a molecular production rate of about $600~\text{s}^{-1}$, which is the same order of magnitude as the rate calculated using the triplet three-body recombination rate of Tiesinga and co-workers [11].

We have also demonstrated that these molecules can be loaded into a dipole force trap [12]. We use a variant called a quasielectrostatic trap [13]. This type of trap is most easily implemented using a focused CO_2 laser beam ($\lambda = 10.6~\mu$m). It is suitable for trapping electronic ground state homonuclear diatomic molecules because vibrational transitions between ground states, as well as dissociative transitions to the continuum, are electric dipole forbidden. Molecules with large permanent electric dipole moments may require alternate optical trapping methods [14]. Photoionization and electronic transitions to excited Cs_2 states require many photons and are expected to be insignificant [15]. The potential energy of the trapped species can be written to good approximation as $U = -\alpha_s E^2/2$, where α_s is the electrostatic polarizability and E is the rms electric field. The relevant polarizabilities for our experiment are $\alpha_s(\text{Cs}) = 59.6 \times 10^{-24}~\text{cm}^3$ and $\alpha_s(^1\Sigma_g{}^+\text{Cs}_2) = 104 \times 10^{-24}~\text{cm}^3$ [16] [$\alpha_s(^3\Sigma_u{}^+\text{Cs}_2)$ should be comparable].

In our trapping experiment, cold molecules were loaded into the dipole trap by overlapping the waist of the CO_2 beam with a "dark MOT" (Fig. 1) [17]. The MOT beam diameters were 2 cm and the maximum intensity per beam was $3.5~\text{mW cm}^{-2}$. The frequency was detuned by $\delta = -4\Gamma$ from the $6S_{1/2}F = 4$ to $6P_{3/2}F = 5$ cycling transition (852 nm), where Γ is the natural linewidth. Repumping light was provided by a σ^+-σ^- retroreflected

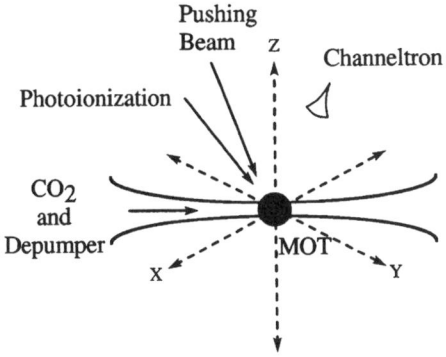

FIG. 1. Schematic of the experimental geometry. The dipole trap beam and depumping beam are combined. Their waists ($\approx 130~\mu$m $1/e^2$ diameter) intersect with the MOT, which is roughly twice as large. The MOT beams travel along the axes shown, with the repumper along the x axis. The "pushing" beam is used to remove atoms from the dipole trap. The trapped atoms and molecules are ionized by the pulsed photoionization beam ($\lambda = 532$ nm) and are then detected by the channeltron.

2 cm diameter laser beam tuned to the $6S_{1/2}F = 4$ to $6P_{1/2}F = 4$ transition (895 nm). Because Cs has such a large hyperfine splitting, an additional linearly polarized depumping beam tuned to the $6S_{1/2}F = 4$ to $6P_{3/2}F = 4$ transition (852 nm) was combined with the CO_2 laser beam to further reduce the $F = 4$ ground state population in the dipole trap region, thus increasing the atomic density. The 17 W linearly polarized CO_2 laser beam was focused to a waist radius of about $64~\mu$m and was shuttered with an acousto-optic modulator (AOM). This creates a dipole trap which, including gravity, is about $200~\mu$K deep for Cs atoms and about $350~\mu$K deep for Cs dimers. The dipole trap also has the effect of locally changing the frequencies of the MOT trapping, repumping, and depumping atomic transitions. The atomic $6P$ state is also trapped and because of the neighboring $5D$ state, actually has a larger dc Stark shift than the $6S$ ground state. At the center of the trap, the $D2$ ($6S_{1/2}$ to $6P_{3/2}$) transition was Stark shifted by about -2.9 natural linewidths [18] and the $D1$ ($6S_{1/2}$ to $6P_{1/2}$) transition was Stark shifted by -2.5 natural linewidths [19]. The CO_2 beam was roughly guided onto the MOT by using the depumper as an alignment beam. Finer alignment was achieved by chopping the CO_2 beam with its AOM and maximizing the effect seen on the MOT fluorescence signal due to the Stark shifts, when it was fed into a lock-in amplifier.

A typical loading and detection cycle works as follows: About 7×10^6 atoms are loaded into the $200-250~\mu$m diameter MOT with the trapping and repumping beams at maximum intensity. There is no CO_2 or depumping beam present. Then the trapping and repumping beam intensities are reduced at the same time the CO_2 and depumping beams are turned on. This loading period lasts for about 100 ms, after which all of the atoms are pumped into the $F = 3$ ground state by extinguishing the repumping laser for 0.5 ms. Then the MOT trapping laser, the depumping laser, and the magnetic field are all turned off. Some of the atoms and molecules remain in the dipole trap as the untrapped atoms and molecules fall away. After a variable time delay, the atoms and molecules are released. They are then detected by photoionization. Figure 2a shows the time-of-flight ion spectrum observed 110 ms after the MOT is turned off. The large signal that occurs at 44.4(2) μs arises primarily from Cs^+ created from two-photon photoionization of atoms [Eq. (1)]. The smaller signal at 62.6(2) μs arises from $Cs_2{}^+$ created from neutral cesium dimers [Eq. (3)]. The ratio of arrival times of molecular to atomic ions is 1.410(8), which agrees with the expected ratio of $\sqrt{2}$. Figure 2c shows that without the CO_2 laser, both the atomic and molecular signals are negligible.

In order to show that the dimer signal was not a result of continuous molecule formation from three-body recombination of atoms in the dipole trap, the number of atoms was reduced by using a pushing laser beam. This linearly polarized, 1 cm diameter, $1~\text{mW cm}^{-2}$ beam was the same

VOLUME 81, NUMBER 23   PHYSICAL REVIEW LETTERS   7 DECEMBER 1998

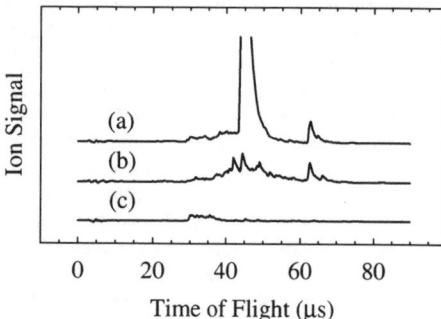

FIG. 2. Time-of-flight spectra of atoms and molecules in the dipole trap (100 shot averages). (a) Time-of-flight spectrum 110 ms after the MOT is switched off. The photoionization pulse occurs at 0 μs. A large atomic signal and a small molecular signal are observed at the appropriate times. (b) Spectrum observed at 110 ms, when a pushing beam is used to remove the atoms and leave the same number of trapped molecules. The small scattered signals near 44 μs are partly due to atomic ions created by the process in Eq. (2). (c) Spectrum observed at 110 ms when the CO_2 laser is not turned on. The residual signal is due to photoionization of Cs on the windows, and Cs from the background gas.

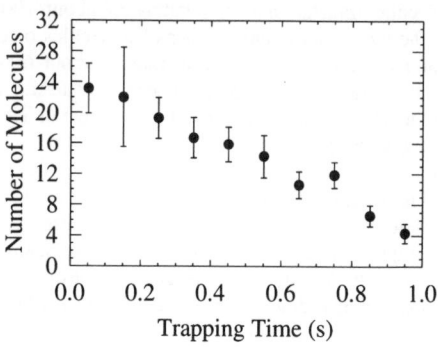

FIG. 3. Molecular trap lifetime (300 shots per point). The number of trapped molecules, as a function of time after the pushing laser is shuttered off. The pushing laser is shuttered on 50 ms after the MOT is turned off, and left on for 10 ms.

frequency as the MOT trapping light. It was shuttered on (simultaneously with the repumper beam) 50 ms after the MOT was turned off and shuttered off 10 ms later. Figure 2b shows that the number of atoms remaining was reduced by at least a factor of 50, while the number of dimers remained about the same. Because of this, we can rule out three-body recombination of trapped atoms as a source for the molecular signal. Using the molecular ion signal size, the photoionization branching ratio, the fact that the molecular photoionization was nearly saturated [10], and an estimate for the channeltron gain at high mass (266 amu), we calculate that the molecular peak in Fig. 2b corresponds to something on the order of 6 molecules. The actual number of molecules trapped is probably about a factor of 4 higher, because the photoionization beam illuminates only the central $\frac{1}{4}$ of the trapping volume.

The decay of the molecules from the trap as a function of time is illustrated in Fig. 3. The data show that the molecules remain in the trap for times on the order of 0.5 s. Since this is the same order of magnitude as the atomic trap lifetimes we have observed previously, we assume this is due mainly to collisions with thermal background gas. Further data will clarify the molecular loss mechanism.

In this experiment, the fraction of singlet and triplet state molecules is unknown but could be measured using state selective two-photon photoionization of the ground state molecules. The rotational and vibrational state distributions are also unknown. However, both possible dimerization mechanisms are expected to primarily populate the highest vibrational states [20]. In future experiments, we will measure the density dependence of

the molecular formation rate to determine whether photoassociation or three-body recombination is the dominant mechanism for molecule formation. Also, in order to do future experiments on cold molecular vapors, it is important to greatly increase the number of molecules trapped. This can be addressed by use of an additional photoassociation laser in the MOT, as well as by using denser atom traps such as those used to produce Bose-Einstein condensation. It may even be possible to use the recent proposal by Julienne and co-workers [21] to produce molecules directly from a Bose-Einstein condensate. Possible future applications using the optical trapping technique to produce cold molecular vapors are clearly numerous.

We acknowledge the support of the United States Air Force Academy, the Air Force Office of Scientific Research, and the Research Corporation. Tetsu Takekoshi gratefully acknowledges the support of the National Research Council Research Associateship Program. We also thank B. J. Verhaar for informative discussions and Cadet David A. Jones for help constructing our apparatus.

[1] For example, see H. Metcalf and P. van der Straten, Phys. Rep. **244**, 203 (1994), and references therein.
[2] E. Inbar, V. Mahal, and A. Arie, J. Opt. Soc. Am. B **13**, 1598 (1996).
[3] M. S. Chapman et al., Phys. Rev. Lett. **74**, 4783 (1995).
[4] Jinha Kim et al., Phys. Rev. Lett. **78**, 3665 (1997).
[5] J. T. Bahns, W. C. Stwalley, and P. L. Gould, J. Chem. Phys. **104**, 9689 (1996).
[6] A. Fioretti et al., Phys. Rev. Lett. **80**, 4402 (1998).
[7] C. C. Tsai et al., Phys. Rev. Lett. **79**, 1245 (1997); J. J. Blange et al., Phys. Rev. Lett. **78**, 3089 (1997); H. Wang et al., Phys. Rev. A **55**, R1569 (1997).
[8] T. Takekoshi, B. M. Patterson, and R. J. Knize, Phys. Rev. A (to be published).
[9] C. Monroe, W. Swann, H. Robinson, and C. Wieman, Phys. Rev. Lett. **65**, 1571 (1990).

[10] J. Morellec, D. Normand, G. Mainfray, and C. Manus, Phys. Rev. Lett. **44**, 1394 (1980); E. H. A. Granneman, M. Klewer, K. J. Nygaard, and M. J. Van der Wiel, J. Phys. B **9**, 865 (1976).

[11] E. Tiesinga, A. J. Moerdijk, B. J. Verhaar, and H. T. C. Stoof, Phys. Rev. A **46**, R1167 (1992).

[12] S. Chu, J. E. Bjorkholm, A. Ashkin, and A. Cable, Phys. Rev. Lett. **57**, 314 (1986); S. Rolston *et al.,* Proc. SPIE Int. Soc. Opt. Eng. **1726**, 205 (1992); J. D. Miller, R. A. Cline, and D. J. Heinzen, Phys. Rev. A **47**, R4567 (1993).

[13] T. Takekoshi, J. R. Yeh, and R. J. Knize, Opt. Commun. **114**, 421 (1995); T. Takekoshi and R. J. Knize, Opt. Lett. **21**, 77 (1996).

[14] Bretislav Friedrich and Dudley Herschbach, Phys. Rev. Lett. **74**, 4623 (1995).

[15] S. L. Chin, Y. Liang, J. E. Decker, F. A. Ilkov, and M. V. Ammosov, J. Phys. B **25**, L249 (1992).

[16] V. Tarnovsky, M. Bunimovicz, L. Vuskovic, B. Stumpf, and B. Bederson, J. Chem. Phys. **98**, 3894 (1993). This is an average of the axial and radial molecular polarizabilities measured for a thermal population.

[17] W. Ketterle, K. B. Davis, M. A. Joffe, A. Martin, and D. E. Pritchard, Phys. Rev. Lett. **70**, 2253 (1993); C. G. Townsend *et al.,* Phys. Rev. A **53**, 1702 (1996).

[18] C. Tanner and C. E. Wieman, Phys. Rev. A **38**, 162 (1988). We ignore the 2nd order Stark shift in this calculation.

[19] L. R. Hunter, D. Krause, Jr., K. E. Miller, D. J. Berkeland, and M. G. Boshier, Opt. Commun. **94**, 210 (1992).

[20] A. J. Moerdijk, H. M. J. M. Boesten, and B. J. Verhaar, Phys. Rev. A **53**, 916 (1996). Because the MOT laser frequency is so close to the atomic resonance frequency, the excited state molecules formed through photoassociation would be very weakly bound. Thus, the ground state molecules formed from their spontaneous decay would also be weakly bound.

[21] P. S. Julienne, K. Burnett, Y. B. Band, and W. C. Stwalley, Phys. Rev. A **58**, R797 (1998).

Chemical physics

Molecules are cool

John M. Doyle and Bretislav Friedrich

Under ordinary conditions, atoms and molecules of a gas zigzag in all directions and have a wide distribution of speeds related to temperature. At room temperature, for instance, gaseous atoms or molecules are most likely to move at the speed of rifle bullets. Such restlessness puts a limit on the level of detail at which atomic and molecular properties can be studied.

The emergence of methods for slowing and trapping gaseous species has led to a renaissance in atomic physics, which is now progressing into molecular/chemical physics as well. The latest developments come from Bethlem et al.[1] and Maddi et al.[2] — they have independently devised kindred techniques for slowing molecules that potentially provide new approaches to subsequent trapping and spectroscopy in particular.

Control of atomic behaviour has already become pretty sophisticated. For instance, a class of atoms, best represented by the alkali metals, can now be routinely slowed with light and loaded into traps (constructed from light or magnetic fields)[3]. This 'laser cooling' relies on the rapid absorption and emission of photons by an atom placed in light tuned near the atom's resonant frequency. With a careful arrangement of the laser beams, the atoms can preferentially absorb photons from a single direction. Due to the momentum of the photon, the atom gets small velocity 'kicks' opposite to its direction of travel. This can be used to slow a beam of atoms so they can be caught in traps and cooled to very low temperatures (for example, evaporative cooling is applied to magnetically trapped atoms to attain Bose–Einstein condensation at temperatures below 1 μK). Unfortunately, the energy-level structure of molecules prevents the maintenance of the requisite simple absorption–emission cycle and so makes laser slowing impractical. In consequence, molecules (and many complex atoms) were relegated to the sidelines.

Last year, however, two very different approaches to cooling and trapping of molecules were successfully implemented. In one, CaH molecules were cooled with a cryogenically refrigerated helium buffer gas and loaded into a magnetic trap[4]. In the other, Cs_2 molecules were formed from laser-cooled Cs atoms and then confined in a light trap[5]. Now we have the techniques devised by Bethlem et al.[1] and Maddi et al.[2]. Like laser slowing of a beam, they produce velocity 'kicks' opposite to the direction of travel. They work with bunches of atoms or molecules all travelling in one direction, and make

no recourse to lasers or cryogenics. Instead these new methods use time-varying inhomogeneous electric fields to provide the 'kicks' and the resultant slowing.

Inhomogeneous electric fields that can be turned on or off quickly are produced by pairs of electrodes. The potential energy of an atom or a molecule varies in a characteristic way with the strength of the electric field. Depending on whether the electric dipoles are on average antiparallel or parallel to the electric field, the particle's energy either increases (these are low-field seekers) or decreases (these are high-field seekers) with increasing field strength. The energy of low-field seekers is lowest at minimum field strength and, therefore, in an inhomogeneous field they seek these low-field regions. In contrast, high-field seekers are forced towards the maximum field strength.

Polar molecules possess both high- and

low-field-seeking states in an electric field. Atoms, on the other hand, become polarized only in the presence of an electric field and so their electric dipoles are always parallel to the field; they can only be high-field seekers in an electric field. The technique of Bethlem et al. relies on low-field seekers and, therefore, is well suited for slowing polar molecules; that of Maddi et al. relies on high-field-seeking states and so is potentially suitable for slowing both atoms and molecules.

Figure 1 illustrates the two approaches. In the scheme of Bethlem et al. (Fig. 1a), molecules enter an inhomogeneous electric field from a field-free region. As they approach the region of maximum field strength between the electrodes, the potential energy of low-field seekers is increasing. This occurs at the expense of their translational energy which is being correspondingly reduced (a consequence of energy conservation). To prevent the molecules from regaining the translational energy as they leave the field, the field is quickly switched off. The molecules thus end up in a homogeneous, field-free region where their potential (and thus translational) energy remains constant. To maximize the loss of translational energy, the field has to be

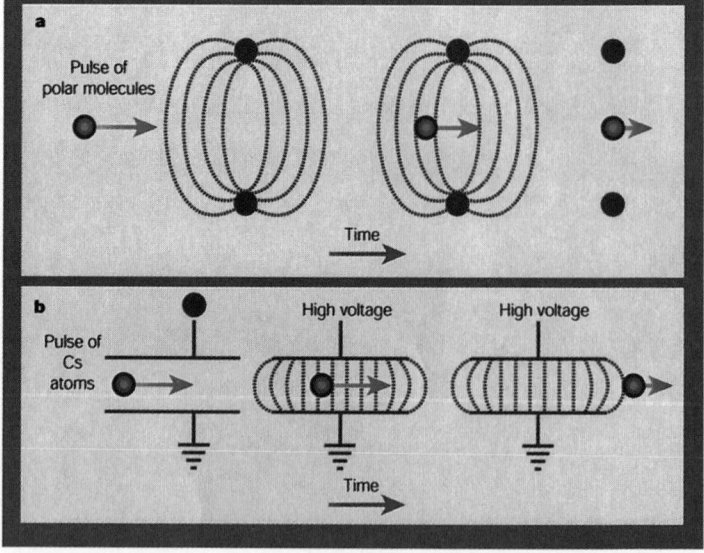

Figure 1 **Principle of molecular and atomic slowing by time-varying inhomogeneous fields. Length of orange arrows indicates speed. a,** One of the 63 stages used by Bethlem et al.[1] to decelerate a pulsed beam of $CO(a^3\Pi)$ molecules from 225 m s^{-1} to 98 m s^{-1}. The two opposing rods are simultaneously switched by two independent high-voltage switches. The decelerating stages are stacked in alternating vertical and horizontal configurations, keeping the molecules focused. They gain potential energy as they approach the field maximum, and so lose kinetic energy. When the field is switched off, the molecules are left with reduced velocity. **b,** A deceleration stage used by Maddi et al.[2] to slow a pulse of Cs atoms (released from a magneto-optic trap) from 2 m s^{-1} to 0.2 m s^{-1}. The two opposing condenser plates are simultaneously energized by two independent high-voltage power supplies. The electric field is turned on when the Cs pulse reaches the uniform field region, thereby inducing an electric dipole in the atoms. As the pulse leaves the plates it passes through an area of decreasing field strength. Its potential energy increases at the expense of kinetic energy, resulting in pulse deceleration.

news and views

switched off just at the point where the molecules reach the field's maximum.

Maddi *et al.* (Fig. 1b) used atoms in their experiments, but their scheme could, in principle, be applied to molecules. The atoms enter an inhomogeneous electric field from a region of a large homogeneous electric field. The homogeneous field is formed by two condenser electrodes that are energized after the atoms' entry. Atoms in their high-field-seeking states then gain potential energy as they pass through the inhomogeneous field at the exit of the condenser, again at the expense of their translational energy.

In both versions, the atoms or molecules arrive in pulses: a pulsed beam of metastable molecules is created from a supersonic beam expansion (Bethlem *et al.*), or a pulsed beam of ground-state atoms is launched from a magneto-optic trap (Maddi *et al.*). The pulsing is crucial, because the switching of the fields must be accurately synchronized with the arrival of the atoms or molecules. These techniques may enable atoms or molecules to be slowed enough for capture in traps.

Yet another new approach[6] to slowing beams of atoms and molecules is based on supersonic expansion from a nozzle placed at the end of a spinning armature. The nozzle is oriented so that it moves in the opposite direction to the discharging gas, and its speed roughly equals the speed of the expanding atoms or molecules. As a result, it cancels the particles' speed in the laboratory frame, leaving the atoms or molecules essentially at rest.

Cold molecules should prove useful in spectroscopy and the study of molecular structure, especially in ultra-high-resolution spectroscopy, which requires cold (slow) and trapped (long-interaction-time) samples. An especially promising area for study is collisions of ultra-cold molecules, when the molecules behave like waves, perhaps giving rise to a new chemistry. The technique may also enable the study of collective quantum effects in molecular systems, including Bose–Einstein condensation. Just as atom cooling is opening up new avenues of research, it is likely that the same will happen with molecular cooling — with repercussions for chemistry and even, perhaps, biology. ∎

John M. Doyle is in the Department of Physics, and Bretislav Friedrich is in the Department of Chemistry and Chemical Biology, and the Department of Physics, Harvard University, Cambridge, Massachusetts 02138, USA.
e-mails: jd@jsbach.harvard.edu
brich@chemistry.harvard.edu

1. Bethlem, H. L., Berden, G. & Meijer, G. *Phys. Rev. Lett.* **83**, 1558–1561 (1999).
2. Maddi, J. A., Dinneen, T. P. & Gould, H. *Phys. Rev. A* (in the press). http://xxx.lanl.gov/abs/physics/9909027
3. Wieman, C. E., Pritchard, D. E. & Wineland, D. J. *Rev. Mod. Phys.* **71**, S253–S262 (1999).
4. Weinstein, J. D. *et al. Nature* **395**, 148–150 (1998).
5. Takekoshi, T., Patterson, B. M. & Knize, R. J. *Phys. Rev. Lett.* **81**, 5105–5108 (1998).
6. Herschbach, D. *Rev. Mod. Phys.* **71**, S411–S418 (1999).

751

PERSPECTIVES: ULTRACOLD MATTER

Molecules at Rest

C. J. Williams and P. S. Julienne

In this issue (*1*), a group from the University of Texas reports producing rubidium dimers that are essentially at rest, by assembling them from ultracold Rb atoms in an atomic Bose-Einstein condensate (BEC). The report (see page 1016) contains several important accomplishments: the first observation of molecule formation in a BEC, an ultraprecise measurement of a molecular binding energy, and the first measurement of the interaction energy between a condensate and a molecule.

The development of cold or monoenergetic sources of matter has led to revolutionary breakthroughs in fundamental science and applications alike. Nowhere is this more obvious than in the use of lasers as light sources. Monoenergetic sources of atoms, neutrons, electrons, and ions have also provided new tools with wide-ranging applications in physics, chemistry, and biology. More recently, the production of cold, trapped neutral atoms, after a decade of progress in laser and evaporative cooling that reduced temperatures from 1 K to 1 nK,

The authors are at the Atomic Physics Division, Stop 8423, National Institute of Standards and Technology, Gaithersburg, MD 20899–8423, USA. E-mail: paul.julienne@nist.gov

Stimulated Raman production of molecules from an atomic Bose-Einstein condensate. The graph shows the ground and excited state potential energy curves for a pair of atoms. Two atoms in their ground state are optically coupled by lasers of frequency v_1 and v_2 to a bound dimer vibrational level with a binding energy of $h(v_1 - v_2)$. In the experiment, the frequency difference was controlled to much better than 1 kHz. Neither laser was resonant with an excited state vibrational level, and loss of coherence by excited state spontaneous radiative decay was therefore not a problem. The left and right pictures indicate a complete coherent interconversion between atomic and molecular forms of the condensate. Wynar et al. (*1*) only converted a fraction of atoms to molecules.

led to the observation of Bose-Einstein condensation of dilute atomic gases (*2*). An atom laser (*3*)—a coherent matter wave analogous to the photon laser—has now al-

so been demonstrated. Cold, neutral trapped atoms and BECs are proving to be powerful tools for extending our understanding of atomic physics and of the collective properties of quantum fluids and are finding applications in precision atomic clocks and gyroscopes and in atom lithography.

Unlike atoms, molecules have complicated internal vibrational and rotational structure and are therefore poor candidates for laser cooling. Although molecules are routinely cooled internally to a few kelvin with the use of supersonic expansions, they still have high translational velocities. Recently, several approaches have been developed for producing and detecting translationally cold molecules (*4*). At Harvard, Doyle's group (*5*) has developed an approach to trap cold molecules using a magnetic trap in a cryogenic helium refrigerator. The resulting molecules are not only translationally but also vibrationally and rotationally cold. With this method, 10^8 CaH molecules have been trapped at 400 mK. Meijer's group at Nijmegen (*6*) has used deceleration of neutral dipolar molecules using time-varying inhomogeneous fields to decelerate CO molecules to around 15 K. Further reductions in temperature are expected with this general technique for polar molecules. Much colder molecules, with translational temperatures of a few hundred microkelvin, can be made by photoassociation of a pair of colliding atoms in a magneto-optical trap (7–10). These molecules are typically rotationally cold but tend to be vibrationally excited.

Wynar et al. (1) assembled their molecules by using a coherent two-photon Raman process that photoassociates a pair of atoms in a ^{87}Rb BEC (see the figure). This approach produces molecules in a single quantum state, in contrast to molecule formation in a magneto-optical trap, where spontaneous radiative decay of an excited photoassociated molecule results in a distribution over many ground rotational and vibrational states. In (1), the molecules are produced without mechanical rotation in a single, weakly bound vibrational state, only 636 MHz below the energy of two free atoms. The Raman process is such that the molecules receive negligible photon recoil momentum.

Because the atoms in the BEC are nearly at rest, with velocities of only a few millimeters per second, the molecules are also essentially at rest. Their temperature, which was not measured, is on the order of 100 nK or less. The extremely low velocities of the atoms and molecules yield a high-precision, nearly Doppler-free, molecular spectroscopy. Such precision spectroscopy of molecular energy levels could be used to accurately determine van der Waals coefficients for long-range interatomic potentials, to refine the scattering lengths that determine condensate properties, and to look for weak relativistic effects in long-range potentials.

One limitation of the current experiment is that once produced, the molecule is likely to undergo an inelastic collision with an atom in the condensate. This will result in ejection of an atom and a molecule from the trap by converting vibrational energy of the molecule into translational energy of the atom-molecule pair. Calculations on ultracold collisions of helium with molecular hydrogen have shown that rate constants for such collisions can be exceptionally large when the vibrational quantum number is large but are relatively small when the vibrational quantum number is small (11). Thus, it will be very important to determine the extent to which vibrationally excited molecules are stable with respect to collisional loss processes.

A recent set of experiments (12, 13), although not designed to make or detect molecules, sheds light on these collisional loss processes. A magnetic field was used to tune a molecular bound state close to the energy of a pair of atoms in a ^{23}Na

BEC. These results have been interpreted in terms of collisional formation of molecules in the condensate (14–16). In this case, the "molecule" is only a transient scattering resonance at the same energy as the atoms. The exceptionally large three-body rate constants measured for condensate destruction are explained by assuming a large rate constant for vibrational relaxation of the "molecule" when it collides with a condensate atom. This is consistent with the calculation (11) of large relaxation rate constants for highly excited molecular vibrations. Under appropriate conditions, a time-dependent ramp of the magnetic field may actually make stable molecular states from colliding atom pairs (17). A similar possibility exists with the use of frequency-chirped light (18). However, the collisional relaxation of such molecules would still be problematic.

The Raman photoassociation technique may ultimately be able to produce a pure molecular condensate. In this case, scientists will have succeeded in taking a coherent set of atoms in a single macroscopic quantum wave function and, with a couple of laser pulses, converting them reversibly into a set of molecules. Exotic properties have been predicted for such condensates, such as molecular solitons (19) or liquid-like chacteristics (20), but destructive inelastic collisions may prevent experimental realization. However, more complex laser pathways or additional laser intensity might enable the production of molecules in either the ground vibrational state or a

state with small inelastic rates.

Sources of cold molecules are now at our disposal; study of their properties and development of applications are soon to come. For example, Wynar et al.'s techniques (1) could be modified to give two units of photon recoil momentum to the molecules. This might be a way to produce a coherent "molecular laser" analogous to the atom laser (3).

References

1. R. Wynar, R. S. Freeland, D. J. Han, C. Ryu, D. J. Heinzen, Science 287, 1016 (2000).
2. M. H. Anderson, J. R. Ensher, M. R. Matthews, C. E. Wieman, E. A. Cornell, Science 269, 198 (1995).
3. K. Helmerson, D. Hutchinson, K. Burnett, W. D. Phillips, Phys. World 12 (no. 8), 31(1999).
4. J. M. Doyle and B. Friedrich, Nature 401, 749 (1999).
5. J. D. Weinstein, R. deCarvalho, T. Guillet, B. Friedrich, J. M. Doyle, Nature 395, 148 (1998); B. Friedrich, J. D. Weinstein, R. deCarvalho, J. M. Doyle, J. Chem. Phys. 110, 2376 (1999).
6. H. L. Bethlem, G. Berden, G. Meijer, Phys. Rev. Lett. 83, 1558 (1999).
7. A. Fioretti et al., Phys. Rev. Lett. 80, 4402 (1998).
8. T. Takekoshi, B. M. Patterson, R. J. Knize, Phys. Rev. Lett. 81, 5105 (1998).
9. A. N. Nikolov et al., Phys. Rev. Lett. 82, 703 (1999).
10. A. N. Nikolov et al., Phys. Rev. Lett. 84, 246 (2000).
11. N. Balakrishnan, R. C. Forrey, A. Dalgarno, Chem. Phys. Lett. 280, 1 (1997).
12. S. Inouye et al., Nature 392, 151 (1998).
13. J. Stenger et al., Phys. Rev. Lett. 82, 2422 (1999).
14. E. Timmermans, P. Tommasini, M. Hussein, A. Kerman, Phys. Rep. 315, 199 (1999).
15. V. A. Yurovsky, A. Ben-Reuven, P. S. Julienne, C. J. Williams, Phys. Rev. A 60, R765 (1999).
16. F. A. van Abeelen and B. J. Verhaar, Phys. Rev. Lett. 83, 1550 (1999).
17. F. H. Mies, P. S. Julienne, E. Tiesinga, Phys. Rev. A 61, 022721-1 (2000).
18. J. Javanainen and M. Mackie, Phys. Rev. A 59, R3186 (1999).
19. P. D. Drummond, K. V. Kheruntsyan, H. He, Phys. Rev. Lett. 81, 3055 (1998).
20. E. Timmermans et al., Phys. Rev. Lett. 83, 2691 (1999).

VOLUME 85, NUMBER 10   PHYSICAL REVIEW LETTERS   4 SEPTEMBER 2000

Stable, Strongly Attractive, Two-State Mixture of Lithium Fermions in an Optical Trap

K. M. O'Hara, M. E. Gehm, S. R. Granade, S. Bali, and J. E. Thomas

Physics Department, Duke University, Durham, North Carolina 27708-0305
(Received 16 March 2000; revised manuscript received 6 June 2000)

We use an all-optical trap to confine a strongly attractive two-state mixture of lithium fermions. By measuring the rate of evaporation from the trap, we determine the effective elastic scattering cross section $4\pi a^2$ to show that the magnitude of the scattering length $|a|$ is very large, in agreement with predictions. We show that the mixture is stable against inelastic decay provided that a small bias magnetic field is applied. For this system, the s-wave interaction is widely tunable at low magnetic field, and can be turned on and off rapidly via a Raman π pulse. Hence, this mixture is well suited for fundamental studies of an interacting Fermi gas.

PACS numbers: 32.80.Pj

Trapped, ultracold atomic vapors offer exciting new opportunities for fundamental studies of an interacting Fermi gas in which the temperature, density, and interaction strength can be independently controlled. Recently, a degenerate gas of fermionic ^{40}K has been produced by using a two-state mixture to enable s-wave scattering and evaporation in a magnetic trap [1]. By removing one species, the properties of the noninteracting degenerate gas were measured, demonstrating that the momentum distribution and the total energy obey Fermi-Dirac statistics [1]. However, the properties of interacting two-state fermionic vapors have not been explored experimentally.

Theoretical treatments of an interacting Fermi gas have focused extensively on ^{6}Li [2–9]. Certain two-state ^{6}Li mixtures are predicted to be strongly attractive, i.e., they have anomalously large and negative scattering lengths [10] arising from a near-zero energy resonance in the triplet state [11]. It has been predicted that these strongly attractive mixtures can undergo a transition to a superfluid state at a relatively high transition temperature [2,4]. In addition, the two-state effective interaction potential is widely tunable in a magnetic field, permitting systematic studies of fundamental phenomena such as collective oscillations for both the normal and superfluid phases [3,5,6], as well as new tests of superconductivity theory [4].

Unfortunately, magnetically trappable mixtures in ^{6}Li with large s-wave scattering lengths are not stable, since there are correspondingly large spin-exchange and dipolar decay rates [2,7,10]. Hence, the methods employed to study degenerate ^{40}K are not applicable. For this reason, we developed an ultrastable CO_2 laser trap to confine a stable mixture of the two lowest ^{6}Li hyperfine states [12]. However, attaining a large and negative scattering length in this mixture requires high magnetic fields $B \geq 800$ G to exploit either a Feshbach resonance or the triplet scattering length [7,10].

In this Letter, we show that there exists another stable hyperfine state mixture in ^{6}Li which has the following unique properties. First, we predict that the scattering length a is large, negative, and widely tunable at *low*

magnetic field B. By monitoring the rate of evaporation from the CO_2 laser trap at a fixed well depth, we measure $|a| = 540^{+210}_{-100}a_0$ at $B = 8.3$ G. This result confirms for the first time that very large scattering lengths exist in ^{6}Li mixtures. The predicted scattering length is $-490a_0$ at $B = 8.3$ G, consistent with our observations, and is expected to increase to $-1615a_0$ as $B \rightarrow 0$. Second, we find that this system is stable against spin exchange collisions provided that $B \neq 0$. In addition, the dipolar decay rate is predicted to be very small [13], consistent with our observations. Finally, in the experiments, a Raman π pulse is employed to abruptly create an interacting mixture from a noninteracting one, a desirable feature for studies of many-body quantum dynamics [14].

Figure 1 shows the hyperfine states for ^{6}Li labeled $|1\rangle$–$|6\rangle$, in order of increasing energy in a magnetic field. At low field, the states $|1\rangle$ and $|2\rangle$ correspond to the $|F = 1/2, m\rangle$ states, while states $|3\rangle$ through $|6\rangle$ correspond to states $|F = 3/2, m\rangle$. At nonzero magnetic field,

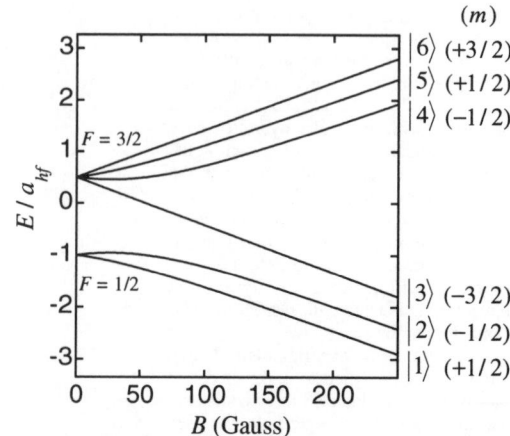

FIG. 1. ^{6}Li hyperfine states, labeled $|1\rangle$ to $|6\rangle$ in order of increasing energy in a magnetic field. The magnetic quantum number of each state is denoted by m. The hyperfine constant $a_{hf} = 152.1$ MHz.

VOLUME 85, NUMBER 10   PHYSICAL REVIEW LETTERS   4 SEPTEMBER 2000

only the magnetic quantum number m is conserved. The subject of this paper is the $|3\rangle-|1\rangle$ mixture.

Figure 2 shows the scattering length a_{31} for the $|3\rangle-|1\rangle$ mixture as a function of magnetic bias field B. We estimate $a_{31}(B)$ by using the asymptotic boundary condition (ABC) approximation [10]. This calculation incorporates the singlet and triplet scattering lengths [11], $a_S = 45.5 \pm 2.5a_0$ and $a_T = -2160 \pm 250a_0$, and a boundary radius which we take to be $R = 40a_0$ [10]. The scattering length varies from $-1620a_0$ ($\approx 3a_T/4$ as $B \to 0$) to $-480a_0$ at $B = 10$ G. The results of our approximate calculation for $B = 0$ to $B = 200$ G are confirmed within 10% by van Abeelen and Verhaar [13] using a coupled channel calculation which includes the uncertainties in the potentials. At higher fields, near 800 G, we believe the scattering length exhibits a Feshbach resonance (not shown). Above this resonance, the scattering length approaches the triplet scattering length of $-2160a_0$.

The $|3\rangle-|1\rangle$ mixture is stable against spin-exchange collisions provided that a small bias magnetic field is applied. Spin-exchange inelastic collisions conserve the two-particle total magnetic quantum number M_T, where $M_T = -1$ for the $|\{3,1\}\rangle$ state. Note that $\{,\}$ denotes the antisymmetric two-particle spin state, as required for s-wave scattering which dominates at low temperatures. There are no lower-lying antisymmetric states with $M_T = -1$. Hence, exothermic collisions are precluded. The only other states with $M_T = -1$ are $|\{4,2\}\rangle$ and $|\{5,3\}\rangle$. Without an adequate bias magnetic field, transitions to these states lead to population in level $|4\rangle$. Then, exothermic $|\{3,4\}\rangle \to |\{3,2\}\rangle$ and $|\{4,1\}\rangle \to |\{1,2\}\rangle$ collisions can occur. With an adequate bias magnetic field, the energy of states $|\{4,2\}\rangle$ and $|\{5,3\}\rangle$ can be increased relative to that of state $|\{3,1\}\rangle$ by more than the maximum relative kinetic energy, i.e., twice the well depth during evaporative cooling. By energy conservation,

spin-exchange transfer is then suppressed. In this case, the inelastic rate is limited to magnetic dipole-dipole (dipolar) interactions which contain a rank 2 relative coordinate operator of even parity [2]. Since parity is conserved and p-wave $\to$ p-wave scattering is frozen out at low temperature, the dominant dipolar process is a small $s \to d$ rate in which $|\{3,1\}\rangle \to |\{1,2\}\rangle$ [13].

In the experiments, the CO_2 laser trap is initially loaded from a magneto-optical trap (MOT) [12]. At the end of the loading period, the MOT laser beams are tuned near resonance and the intensity is lowered to decrease the temperature. Then, optical pumping is used to empty the $F = 3/2$ state to produce a 50/50 mixture of the $|1\rangle-|2\rangle$ states. These states are noninteracting at low magnetic field, i.e, the scattering amplitude vanishes as a result of an accidental cancellation [10]. With a CO_2 laser trap depth of 330 μK, up to 4×10^5 atoms are confined in the lowest-lying hyperfine states at an initial temperature between 100 and 200 μK. A bias magnetic field of 8.3 G is applied to split the two-particle energy states by ≈ 16 MHz. This is twice the maximum attainable energy at our largest well depth of 400 μK = 8 MHz. After a delay of 0.5 sec relative to the loading phase, a pair of optical fields is pulsed on to induce a Raman π pulse. This pulse transfers the population in state $|2\rangle$ to state $|3\rangle$ in two microseconds, initiating evaporative cooling in the resulting $|3\rangle-|1\rangle$ mixture. The optical fields are detuned from resonance with the $D2$ transition by ≈ 700 MHz to suppress optical pumping. If the Raman pulse is not applied, the trapped atoms remain in the noninteracting $|1\rangle-|2\rangle$ mixture and exhibit a purely exponential decay with a time constant ≈ 300 sec.

An acousto-optic modulator (A/O) in front of the CO_2 laser controls the laser intensity, which is reduced to yield a shallow trap depth of 100 μK. By using a shallow well, we avoid the problem that the elastic cross section becomes independent of the scattering length at high energy, as described below. In addition, the shallow well greatly reduces the number of loaded atoms and makes the sample optically thin, simplifying calibration of the number of trapped atoms. To determine the trap parameters, the laser power is modulated and parametric resonances [15] are observed at drive frequencies of 2ν for three different trap oscillation frequencies ν: At 100 μK well depth, $\nu_x = 2.4$ kHz, $\nu_y = 1.8$ kHz, and $\nu_z = 100$ Hz, where the trap laser beam propagates along $\mathbf{z}$. Using the measured total power as a constraint, we obtain the trap intensity $1/e^2$ radii, $w_x = 50$ μm and $w_y = 67$ μm, and the axial intensity $1/e^2$ length, $z_f \approx 1.13$ mm, where z_f is consistent with the expected Rayleigh length within 15%.

The number of atoms in the trap $N(t)$ is estimated using a calibrated photomultiplier. The detection system monitors the fluorescence induced by pulsed, retroreflected, $\sigma_\pm$ probe and repumper beams which are strongly saturating ($I/I_{sat} = 26$ for the strongest transition). To simplify calibration, only the isotropic component of the fluorescence angular distribution is measured: The collecting lens is

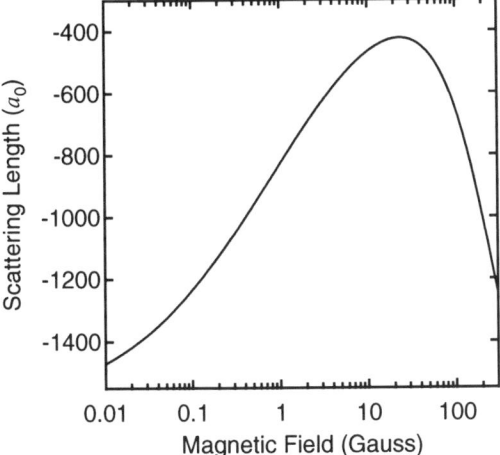

FIG. 2. Magnetic field dependence of the scattering length a_{31} for a mixture of the $|3\rangle$ and $|1\rangle$ hyperfine states of ^{6}Li.

VOLUME 85, NUMBER 10   PHYSICAL REVIEW LETTERS   4 SEPTEMBER 2000

placed at the magic angle [16] of 55° $[P_2(\cos\theta) = 0]$ with respect to the propagation direction of the probe beams. The net efficiency of the detection system is determined using laser light of known power. The primary uncertainty in the calibration arises from the excited state population fraction, which we estimate lies between $1/4$ and $1/2$.

Figure 3 shows the number of trapped atoms $N(t)$ measured for the $|3\rangle$–$|1\rangle$ mixture at a well depth $U_0 = 100\ \mu$K and a bias field of 8.3 G as a function of time between 5 ms and 20 sec after evaporation is initiated. For times beyond 50 sec (not shown), the evaporation stagnates, and we observe an exponential decay of the cooled $|3\rangle$–$|1\rangle$ mixture with a time constant of 370 sec over a period of a few hundred seconds. The error bars are the standard deviation of the mean of ten complete runs through the entire time sequence.

A model based on the s-wave Boltzmann equation [17] is used to predict $N(t)$ for comparison to the experiments. This equation is modified to include the density of states for a Gaussian potential well [12] and to include the energy dependence of the elastic cross section. Assuming a short range potential and a symmetric (s-wave) spatial state, the cross section takes the form

$$\sigma(k) = \frac{8\pi a_{31}^2}{1 + k^2 a_{31}^2}, \qquad (1)$$

where $\hbar k$ is the relative momentum. For $k|a_{31}| \ll 1$, the cross section is maximized. When $k|a_{31}| \gg 1$, the cross section approaches the unitarity limit $8\pi/k^2$ which is independent of a_{31}. Note that $k|a_{31}| = 1$ corresponds to a relative kinetic energy of $\epsilon = \hbar^2/(2\mu a_{31}^2)$, where $\mu = M/2$ is the reduced mass. For $|a_{31}| = 500a_0$, $\epsilon = 115\ \mu$K.

For a two-state mixture of fermions, the effective cross section is reduced from that of Eq. (1) by a factor of 2 since

pairs of colliding atoms are in an antisymmetric hyperfine state with a probability of $1/2$. This effective cross section is used in a Boltzmann collision integral for each state $i = 1, 3$. A decay term $-N_i(t)/\tau$ with $\tau = 370$ sec is added to account for the measured trap lifetime. A detailed description of our coupled Boltzmann equation model will be published elsewhere.

The coupled s-wave Boltzmann equations for the two states are numerically integrated to determine $N(t)$ using the well parameters as fixed inputs. From the calibrated photomultiplier signal, assuming that $1/3$ of the atoms is in the excited state, we obtain an initial total number $N_0 = 44\,000$. For this case, the initial collision rate in Hz is estimated to be $1/(2\pi\tau_c) \simeq N_0 M \sigma_0 \nu^3/(k_B T)$, where $\nu^3 = \nu_x \nu_y \nu_z$, $\sigma_0 = 8\pi a_{31}^2$, and M is the ^{6}Li mass. Assuming $|a_{31}| = 500a_0$, $\tau_c \simeq 30$ ms. Hence, for $t > 0.3$ sec, when on average ten collisions have occurred, the sample should be thermalized as assumed in the theory.

The best fit to the data starting with 22 000 atoms in each state is shown as the solid curve in Fig. 3. The χ^2 per degree of freedom for this fit is 1.4 and is found to be very sensitive to the initial temperature T_0 of the atoms in the optical trap. From the fit, we find $T_0 = 46\ \mu$K, which is less than the well depth. We believe that this low temperature is a consequence of the MOT gradient magnet, which is turned off after the MOT laser beams. The effective well depth of the optical trap is therefore reduced until the gradient is fully off, allowing hotter atoms to escape before the Raman pulse is applied to create the $|3\rangle$–$|1\rangle$ mixture. The fit is most sensitive to data for $t > 0.5$ sec, where the thermal approximation is expected to be valid. From the fit, we obtain the scattering length $|a_{31}| = 540 \pm 25a_0$, which is within 10% of the predictions of Fig. 2. The quoted error corresponds to a change of 1 in the total χ^2.

We determine the systematic errors in a_{31} due to the uncertainties in the calibration and in the population imbalance as follows. The data are fit for an initial number of atoms N_0 of 58 000 and 29 000, corresponding to an excited state fraction of $1/4$ and $1/2$. This yields $|a_{31}| = 440 \pm 20a_0$ and $|a_{31}| = 750 \pm 42a_0$, respectively. Note that for the larger scattering lengths, the cross section given by Eq. (1) approaches the unitarity limit and the error increases. We assume that the initial population imbalance for states $|3\rangle$ and $|1\rangle$ is comparable to that of states $|2\rangle$ and $|1\rangle$ in the optically pumped MOT. To estimate the latter population imbalance, we use state-selective Raman π pulses to excite $|2\rangle \rightarrow |3\rangle$ or $|1\rangle \rightarrow |6\rangle$ transitions in the MOT. Probe-induced fluorescence signals from states $|3\rangle$ or $|6\rangle$ show that the initial $|1\rangle$ and $|2\rangle$ populations are equal within 10%. Note that residual population in state $|2\rangle$ is expected to be stable and weakly interacting, since we estimate $|a_{32}| < 30a_0$ for $0 \le B \le 50$ G using the ABC method, and $a_{12} \simeq 0$ [10]. Using the parameters for the fit shown in Fig. 3, but changing the initial mixture from 50/50 to 60/40, we find a slight increase in the fitted scattering length from $540a_0$ to $563a_0$. Thus, the uncertainty in

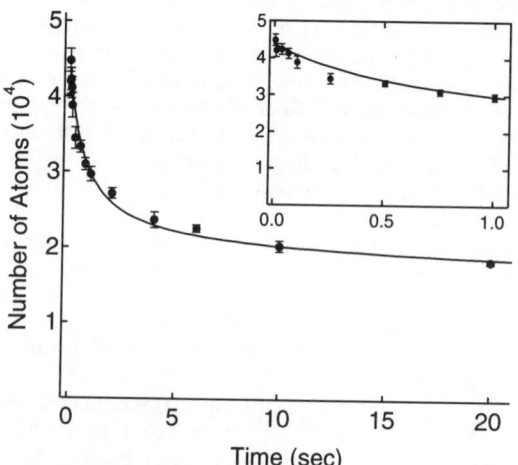

FIG. 3. Number of trapped atoms versus time for evaporation of a $|3\rangle$–$|1\rangle$ mixture of ^{6}Li at a fixed well depth of 100 μK. The solid curve shows the s-wave Boltzmann equation fit for a scattering length of $|a_{31}| = 540a_0$. Inset: 0–1 sec.

VOLUME 85, NUMBER 10   PHYSICAL REVIEW LETTERS   4 SEPTEMBER 2000

the calibration of the number of atoms produces the dominant uncertainty and $|a_{31}| = 540^{+210}_{-100}a_0$.

To demonstrate that evaporative cooling is occurring, rather than just trap loss, we have also measured the final temperature of the mixture using release and recapture [18] from the CO_2 laser trap. We obtain 9.8 ± 1 μK, which is within 10% of the final temperature of 8.7 μK predicted by the Boltzmann equation model. An excellent fit to the data is obtained for the final temperature, which describes a thermal distribution. However, the initial temperature is not so readily measured, as it is nonthermal before evaporation is initiated, and is rapidly changing during evaporation, unlike the final temperature, which stagnates.

Good fits to the evaporation data are obtained neglecting inelastic collisions, suggesting that the dipolar rate for the $|3\rangle-|1\rangle$ mixture is small, in contrast to the scattering length. A limit on the dipolar loss rate for the $|3\rangle-|1\rangle$ mixture can be estimated from the $\tau = 370$ sec lifetime of the mixture after evaporation stagnates. For equal populations in both states, dipolar decay results in an initial loss rate $\dot{n} = -Gn^2/4$, where G is the dipolar rate constant and n is the total density. To obtain a high density, the trap is loaded at a well depth of 330 μK and the temperature of the atoms is reduced by evaporation to $T \simeq 30 \pm 1$ μK. The number of atoms remaining in each state after evaporation is estimated to be $N = 6.5 \pm 2.2 \times 10^4$, where the uncertainty is in the calibration. We cannot rule out the possibility that one state is depleted on a long time scale, since we do not directly measure the individual state populations. However, we believe that, after evaporation stagnates in the deep well, a $|3\rangle-|1\rangle$ mixture remains, since subsequent reduction of the well depth yields final temperatures consistent with evaporative cooling. Note that the mixture ratio is not critical: An 80/20 mixture yields an initial loss rate $\dot{n} = -0.16Gn^2$, $\simeq 2/3$ that of a 50/50 mixture. For a fixed 330 μK trap depth, $\nu^3 = 2.6 \pm 0.3$ kHz^3, and the phase space density for one state in the harmonic approximation is then $\rho_{ph} = N/(k_B T/h\nu)^3 = 7 \times 10^{-4}$. This corresponds to a maximum total density of $n = 2\rho_{ph}/\lambda_B^3 = 6.4 \times 10^{11}/\text{cm}^3$, where $\lambda_B \equiv h/\sqrt{2\pi M k_B T}$. Since the exponential decay time of the $|3\rangle-|1\rangle$ mixture is similar to that obtained in the noninteracting $|1\rangle-|2\rangle$ mixture, we assume the loss is dominated by background gas collisions. Thus, we must have $Gn/4 \ll 1/\tau$, which yields $G \ll 2 \times 10^{-14}$ cm^3/sec. This result is consistent with the value $G \simeq 2 \times 10^{-15}$ cm^3/sec predicted for the dipolar rate constant at 30 μK by van Abeelen and Verhaar [13].

Future experiments will employ continuous evaporation by slowly reducing the well depth [19]. In this case, very large scattering lengths can be obtained at low temperatures and small well depths by using a reduced bias magnetic field B. By adiabatically recompressing the well, experiments can be carried out with the precooled atoms in a deep trap to obtain high density as well. In such experiments, the final low temperature limits the number of atoms in the high energy tail of the energy distribution, exponentially suppressing spin-exchange collisions for $B \neq 0$. For example, if a total of 3×10^5 atoms were contained in our trap at a well depth of 400 μK, the Fermi temperature $T_F = 7$ μK and the Fermi density is $4 \times 10^{13}/\text{cm}^3$. At a temperature of $T = 0.1T_F = 0.7$ μK, a bias field of $B = 0.16$ G would split the two-particle hyperfine states by $k_B T_F + 12 k_B T$, suppressing the spin-exchange rate by $\exp(-12)$, and giving $a_{31} \simeq -1200a_0$. Alternatively, as shown in Fig. 2, large a_{31} can be obtained at moderate $B \simeq 300$ G.

In conclusion, we have observed that an optically trapped $|3\rangle-|1\rangle$ mixture of ^{6}Li atoms has a very large scattering length at low magnetic field. This mixture is stable against spin-exchange collisions provided that a small bias magnetic field is applied. The evaporation curves measured for this mixture are in good agreement with a model based on an s-wave Boltzmann equation which neglects inelastic processes. We have predicted that the scattering interactions are strongly attractive and widely tunable at low magnetic field. If the parameters described above for deep wells can be attained, the system will be close to the threshold for superfluidity [2] and ideal for investigating frequency shifts and damping in collective oscillations [3,5]. Further, since s-wave interactions can be turned on and off in a few microseconds, this system is well suited for studies of many-body quantum dynamics.

This research has been supported by the Army Research Office and the National Science Foundation.

[1] B. DeMarco and D. S. Jin, Science **285**, 1703 (1999).
[2] H. T. C. Stoof et al., Phys. Rev. Lett. **76**, 10 (1996); see also, M. Houbiers et al., Phys. Rev. A **56**, 4864 (1997).
[3] L. Vichi and S. Stringari, Phys. Rev. A **60**, 4734 (1999).
[4] R. Combescot, Phys. Rev. Lett. **83**, 3766 (1999).
[5] G. M. Bruun and C. W. Clark, Phys. Rev. Lett. **83**, 5415 (1999).
[6] G. M. Bruun and C. W. Clark, cond-mat/9906392.
[7] M. Houbiers and H. T. C. Stoof, Phys. Rev. A **59**, 1556 (1999).
[8] G. Bruun et al., Eur. Phys. J. D **7**, 433 (1999).
[9] M. Houbiers and H. T. C. Stoof, cond-mat/9808171.
[10] M. Houbiers et al., Phys. Rev. A **57**, R1497 (1998).
[11] E. R. I. Abraham et al., Phys. Rev A **55**, R3299 (1997).
[12] K. M. O'Hara et al., Phys. Rev. Lett. **82**, 4204 (1999).
[13] We are indebted to F. A. van Abeelen and B. J. Verhaar who calculated the inelastic $|\{3, 1\}\rangle \rightarrow |\{1, 2\}\rangle$ dipolar rate and confirmed our calculations of the magnetic field dependence of a_{31}.
[14] P. Törmä and P. Zoller, Phys. Rev. Lett. **85**, 487 (2000).
[15] S. Friebel et al., Phys. Rev. A **57**, R20 (1998).
[16] Atomic, Molecular, and Optical Physics Handbook, edited by G. W. Drake (AIP Press, New York, 1996), p. 176.
[17] O. J. Luiten et al., Phys. Rev. A **53**, 381 (1996).
[18] S. Chu et al., Phys. Rev. Lett. **55**, 48 (1985).
[19] C. S. Adams et al., Phys. Rev. Lett. **74**, 3577 (1995).

VOLUME 92, NUMBER 4   PHYSICAL REVIEW LETTERS   week ending
30 JANUARY 2004

Observation of Resonance Condensation of Fermionic Atom Pairs

C. A. Regal, M. Greiner, and D. S. Jin*

*JILA, National Institute of Standards and Technology and University of Colorado, and Department of Physics,
University of Colorado, Boulder, Colorado 80309-0440, USA*
(Received 13 January 2004; published 28 January 2004)

We have observed condensation of fermionic atom pairs in the BCS-BEC crossover regime. A trapped gas of fermionic ^{40}K atoms is evaporatively cooled to quantum degeneracy and then a magnetic-field Feshbach resonance is used to control the atom-atom interactions. The location of this resonance is precisely determined from low-density measurements of molecule dissociation. In order to search for condensation on either side of the resonance, we introduce a technique that pairwise projects fermionic atoms onto molecules; this enables us to measure the momentum distribution of fermionic atom pairs. The transition to condensation of fermionic atom pairs is mapped out as a function of the initial atom gas temperature T compared to the Fermi temperature T_F for magnetic-field detunings on both the BCS and BEC sides of the resonance.

DOI: 10.1103/PhysRevLett.92.040403
         PACS numbers: 03.75.Ss, 05.30.Fk

Ultracold quantum gases of fermionic atoms with tunable interactions offer the unique possibility to experimentally access the predicted crossover between BCS-type superfluidity of momentum pairs and Bose-Einstein condensation (BEC) of molecules [1–7]. Magnetic-field Feshbach resonances provide the means for controlling both the strength of cold atom interactions, characterized by the s-wave scattering length a, as well as whether they are, in the mean-field approximation, effectively repulsive ($a > 0$) or attractive ($a < 0$) [8,9]. For magnetic-field detunings on the $a > 0$, or BEC, side of the resonance there exists an extremely weakly bound molecular state whose binding energy depends strongly on the detuning from the Feshbach resonance [10,11]. In Fermi gases this state can be long lived [12–15]. BEC of these molecules represents one extreme of the predicted BCS-BEC crossover and recently has been observed for both $^{40}K_2$ and $^6Li_2$ molecules [16–18].

In atomic Fermi gas systems condensates have not previously been observed beyond this molecular BEC extreme [19]. In discussing condensation of a Fermi gas throughout the BCS-BEC crossover, terms such as Cooper pairs, molecules, BEC, and fermionic condensates often have ambiguous meanings. In this Letter, we define "condensation of fermionic atom pairs," or equivalently fermionic condensates, as condensation (i.e., the macroscopic occupation of a single quantum state) in which the underlying Fermi statistics of the paired particles play an essential role [23]. In the BCS extreme this is more commonly termed condensation of Cooper pairs. Fermionic condensates are distinct from the BEC extreme where there remains no fermionic degree of freedom because all fermions are bound into bosonic molecules [24]. In this Letter, we report the observation of condensation of fermionic atom pairs near and on both sides of the Feshbach resonance, which corresponds to the BCS-BEC crossover regime. We observe condensation on the $a < 0$, or BCS, side of the Feshbach resonance. Here the

two-body physics of the resonance no longer supports the weakly bound molecular state; hence, only cooperative many-body effects can give rise to this condensation of fermion pairs [4–7,25].

Demonstrating condensation of fermionic atom pairs on the BCS side of the resonance presents significant challenges. Observation of pairing of fermions [26] is insufficient to demonstrate condensation, and rather a probe of the momentum distribution is required [7]. For example, the standard technique developed for observing BEC relies on time-of-flight expansion images [27,28]. However, this method is problematic on the BCS side of the resonance because the pairs depend on many-body effects and are not bound throughout expansion of the gas. In this work we introduce a technique that takes advantage of the Feshbach resonance to pairwise project the fermionic atoms onto molecules. We probe the system by rapidly sweeping the magnetic field to the $a > 0$, or BEC, side of the resonance, where time-of-flight imaging can be used to measure the momentum distribution of the weakly bound molecules. The projecting magnetic-field sweep is completed on a time scale that allows molecule formation but is still too brief for particles to collide or move significantly in the trap. This projection always results in 60% to 80% of the atom sample appearing as molecules. However, we find that there is a threshold curve of T/T_F versus detuning from the Feshbach resonance below which we observe a fraction of the molecules to have near zero momentum. We interpret this as reflecting a preexisting condensation of fermionic atom pairs.

Our basic experimental procedures have been discussed in prior work [20,29]. We trap and cool a dilute gas of the fermionic isotope ^{40}K, which has a total atomic spin $f = 9/2$ in its lowest hyperfine ground state and thus ten available Zeeman spin states $|f, m_f\rangle$ [29,30]. We use a far-off resonance optical dipole trap that can confine atoms in any spin state as well as the molecules we create from these atoms. The optical trap is characterized by

     0031-9007/04/92(4)/040403(4)$22.50   © 2004 The American Physical Society

VOLUME 92, NUMBER 4 PHYSICAL REVIEW LETTERS week ending
30 JANUARY 2004

radial frequencies ranging between $\nu_r = 320$ and 440 Hz, with the trap aspect ratio ν_r/ν_z fixed at 79 ± 15.

Experiments are initiated by preparing atoms in a nearly equal, incoherent mixture of the $|9/2, -7/2\rangle$ and $|9/2, -9/2\rangle$ spin states at a low T/T_F. We access an s-wave Feshbach resonance between these states located at a magnetic field near 200 G [31]. A precise determination of the magnetic-field location of the two-body resonance is an essential ingredient for exploring the BCS-BEC crossover regime. In our previous work the location of the resonance was determined from the peak in the resonantly enhanced elastic collision rate [15,32,33]. In the work reported here we have more precisely determined the location of the resonance by measuring the magnetic field B_0, above which the two-body physics no longer supports the shallow bound state.

Figure 1 shows the result of such a measurement. Molecules created by a slow magnetic-field sweep across the resonance are dissociated by raising the magnetic field to a value B_{probe} near the resonance (inset in Fig. 1). Note that, to avoid many-body effects, we dissociate the molecules after allowing the gas to expand from the trap to much lower density. After a total expansion time of 17 ms atoms not bound in molecules are selectively detected at near zero field [11]. The measured number of atoms increases sharply at $B_0 = 202.10 \pm 0.07$ G. This more precise measurement of the resonance position agrees well with previous results [15,32,33]. As an additional check, we have located the resonance by creating, rather than dissociating, molecules. The measured number of molecules decreases sharply at $B = 202.14 \pm 0.11$ G in good agreement with the molecule dissociation result [34].

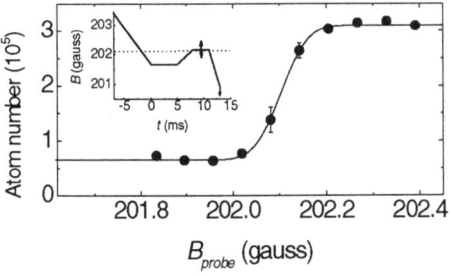

FIG. 1. Measurement of the Feshbach resonance position B_0. Shown in the inset is a schematic of the magnetic field as a function of time t measured with respect to the optical trap turn off at $t = 0$. Molecules are first created by a slow magnetic-field sweep across the resonance (dotted line) and then dissociated if B_{probe} (indicated by the arrow in the inset) is beyond the magnetic field where the two-body physics supports a new bound state. The number of atoms, measured at $t = 17$ ms, is shown as a function of B_{probe}. The two error bars indicate the spread in repeated points at these values of B. A fit of the data to an error function reveals $B_0 = 202.10 \pm 0.07$ G, where the uncertainty is given conservatively by the 10%–90% width.

In order to investigate the BCS-BEC crossover regime, we initially prepare the ultracold two-component atom gas at a magnetic field of 235.6 G, far above the Feshbach resonance. Here the gas is not strongly interacting, and we measure T/T_F through surface fits to time-of-flight images of the Fermi gas [20,29]. The field is then slowly lowered at typically 10 ms/G to a value B_{hold} near the resonance. This sweep is slow enough to allow the atoms and molecules sufficient time to move and collide in the trap. This was shown previously in Ref. [16] where, for a Fermi gas initially below $T/T_F = 0.17$, a magnetic-field sweep at 10 ms/G to a final B 0.56 G below the resonance produced a molecular condensate. In this Letter, we now explore the behavior of the sample when sweeping slowly to values of B_{hold} on either side of the Feshbach resonance.

To probe the system we pairwise project the fermionic atoms onto molecules and measure the momentum distribution of the resulting molecular gas. This projection is accomplished by rapidly lowering the magnetic field by ~ 10 G at a rate of typically $(50~\mu s/G)^{-1}$ while simultaneously releasing the gas from the trap. This puts the gas far on the BEC side of the resonance, where it is weakly interacting. The total number of molecules after the projection N corresponds to 60% to 80% of the original atom number in each spin state. After a total of typically 17 ms of expansion the molecules are selectively detected using rf photodissociation immediately followed by spin-selective absorption imaging [16]. To look for condensation, these absorption images are surface fit to a two-component function that is the sum of a Thomas-Fermi profile for a condensate and a Gaussian function for noncondensed molecules [16].

Figures 2–4 present the main result of this Letter. In Fig. 2 we plot the measured condensate fraction N_0/N as a function of the magnetic-field detuning from the resonance, $\Delta B = B_{hold} - B_0$ [39]. The data in Fig. 2 were taken for a Fermi gas initially at $T/T_F = 0.08$ and for two different wait times at B_{hold}. Condensation is observed on both the BCS ($\Delta B > 0$) and BEC ($\Delta B < 0$) sides of the resonance. We further find that the condensate on the BCS side of the Feshbach resonance has a relatively long lifetime. The lifetime was probed by increasing t_{hold} to 30 ms (triangles in Fig. 2). We find that for the BEC side of the resonance no condensate is observed for $t_{hold} = 30$ ms except very near the resonance. However, for all data on the BCS side of the resonance the observed condensate fraction is still >70% of that measured for $t_{hold} = 2$ ms. Finally, we note that the appearance of the condensate is accompanied by a significant (as large as 20% at the resonance) decrease in the measured width of the noncondensed fraction. This effect will be a subject of future investigations.

Figure 3 displays sample time-of-flight absorption images for the fermionic condensate. Figure 4 is a phase diagram created from our data; here we plot the measured condensate fraction as a function of ΔB as well as of the initial Fermi gas degeneracy T/T_F. The condensate forms

VOLUME 92, NUMBER 4   PHYSICAL REVIEW LETTERS   week ending
30 JANUARY 2004

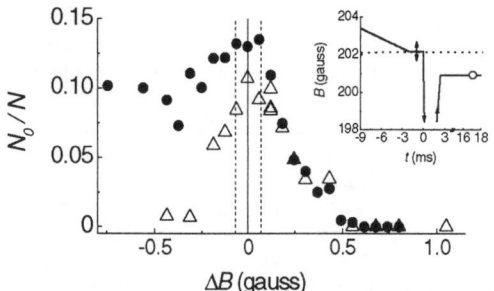

FIG. 2. Measured condensate fraction as a function of detuning from the Feshbach resonance $\Delta B = B_{hold} - B_0$. Data here were taken for $t_{hold} = 2$ ms (●) and $t_{hold} = 30$ ms (△) with an initial cloud at $T/T_F = 0.08$ and $T_F = 0.35$ μK. The area between the dashed lines around $\Delta B = 0$ reflects the uncertainty in the Feshbach resonance position based on the 10%–90% width of the feature in Fig. 1. Condensation of fermionic atom pairs is seen near and on either side of the Feshbach resonance. Comparison of the data taken with the different hold times indicates that the pair condensed state has a significantly longer lifetime near the Feshbach resonance and on the BCS ($\Delta B > 0$) side. The inset shows a schematic of a typical magnetic-field sweep used to measure the fermionic condensate fraction. The system is first prepared by a slow magnetic-field sweep towards the resonance (dotted line) to a variable position B_{hold}, indicated by the two-sided arrow. After a time t_{hold} the optical trap is turned off and the magnetic field is quickly lowered by ~10 G to project the atom gas onto a molecular gas. After free expansion, the molecules are imaged on the BEC side of the resonance (O).

at lower initial T/T_F with increasing ΔB, an effect predicted in [4–7].

An essential aspect of these measurements is the fast magnetic-field sweep that pairwise projects the fermionic atoms onto molecules. It is a potential concern that the condensation might occur during this sweep rather than at B_{hold}. However, in our previous work it was shown that a magnetic-field sweep with an inverse speed less than 800 μs/G was too fast to produce a molecular condensate

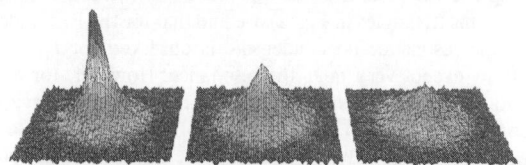

FIG. 3 (color online). Time of flight images showing the fermionic condensate. The images, taken after the projection of the fermionic system onto a molecular gas, are shown for $\Delta B = 0.12$, 0.25, and 0.55 G (left to right) on the BCS side of the resonance. The original atom cloud starts at $T/T_F = 0.07$, and the resulting fitted condensate fractions are $N_0/N = 0.10$, 0.05, and 0.01 (left to right). Each image corresponds to $N = 100\,000$ particles and is an average over 10 cycles of the experiment.

when starting with a Fermi gas 0.68 G on the BCS side of the resonance [16]. Thus, the inverse sweep speed we use in this Letter of typically 50 μs/G, while sufficiently slow to convert 60% of the sample to weakly bound molecules [11], is much too fast to produce a molecular condensate.

In addition, we have checked that the observation of a condensate on the BCS side of the resonance does not depend on this sweep speed. As seen in Fig. 5(a), much faster sweeps result in fewer molecules. This is consistent with our previous study of the molecule creation process [11]. However, we find that the measured condensate fraction is independent of the sweep rate [Fig. 5(b)]. Even with the lower number of molecules, and therefore a lower phase space density of the molecular gas, we observe an essentially unchanged condensate fraction.

Finally, we note that, as in our previous measurements performed in the BEC limit, the measured condensate fraction always remains well below one [16]. As part of our probing procedure the magnetic field is set well below the Feshbach resonance where the molecule lifetime is only on the order of milliseconds [15,40]. This results in a measured loss of 50% of the molecules and may also reduce the measured condensate fraction.

In conclusion, we have introduced a method for probing the momentum distribution of fermionic atom pairs and employed this technique to observe fermionic condensates near a Feshbach resonance. By projecting the system onto a molecular gas, we map out condensation of

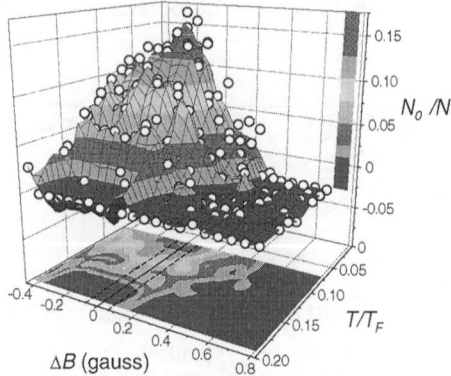

FIG. 4 (color online). Transition to condensation as a function of both ΔB and T/T_F. The data for this phase diagram were collected with the same procedure as shown in the inset in Fig. 2 with $t_{hold} \sim 2$ ms. The area between the dashed lines around $\Delta B = 0$ reflects the uncertainty in the Feshbach resonance location from the width of the feature in Fig. 1. The surface and contour plots are obtained using a Renka-Cline interpolation of approximately 200 distinct data points (O) [36]. One measure of when the gas becomes strongly interacting is the criterion $|k_F a| > 1$, where $\hbar k_F$ is the Fermi momentum [20,21,37,38]. For these data, $|\Delta B| < 0.6$ corresponds to $|k_F a| > 1$.

VOLUME 92, NUMBER 4   PHYSICAL REVIEW LETTERS   week ending
30 JANUARY 2004

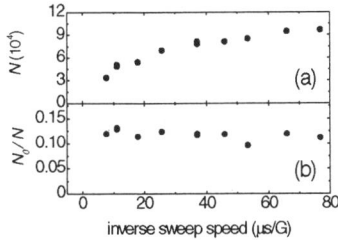

FIG. 5. Dependence of molecule number and condensate fraction on the speed of the fast magnetic-field sweep from the atomic gas onto the molecular gas. Here $\Delta B = 0.12$ and the initial T/T_F is 0.08. (a) Total number of molecules as a function of inverse sweep speed. For the fastest sweep speeds fewer molecules are created, consistent with studies in Ref. [11]. (b) Condensate fraction as a function of the inverse sweep speed. Even for the fastest sweeps and lowest molecule number, we observe an unchanged condensate fraction.

fermionic pairs as a function of both the magnetic-field detuning from the resonance and the initial Fermi degeneracy T/T_F. The fermionic condensates seen in this work occur in the BCS-BEC crossover regime, far from the perturbative BCS limit. As predicted, the system is observed to vary smoothly in the BCS-BEC crossover regime. Further, the lifetime of the condensed state is found to be significantly longer in the crossover regime than it is in the BEC limit. As in the case of BEC, one expects the resonance fermionic condensation observed here to correspond to superfluidity. The experimental realization of condensation in the BCS-BEC crossover regime demonstrated in this Letter follows more than two decades of theoretical investigation and initiates experimental study of this physics.

We thank E. A. Cornell, C. E. Wieman, M. Holland, K. Levin, E. Altman, and L. Radzihovsky for stimulating discussion and J. T. Smith for experimental assistance. This work was supported by NSF and NIST; C. A. R. acknowledges support from the Hertz Foundation.

*Permanent address: Quantum Physics Division, National Institute of Standards and Technology.

[1] A. J. Leggett, J. Phys. (Paris), Colloq. **41**, 7 (1980).
[2] P. Nozieres and S. Schmitt-Rink, J. Low Temp. Phys. **59**, 195 (1985).
[3] M. Randeria, in *Bose-Einstein Condensation*, edited by A. Griffin, D.W. Snoke, and S. Stringari (Cambridge University, Cambridge, 1995), pp. 355–392.
[4] M. Holland, S. J. J. M. F. Kokkelmans, M. L. Chiofalo, and R. Walser, Phys. Rev. Lett. **87**, 120406 (2001).
[5] E. Timmermans, K. Furuya, P.W. Milonni, and A. K. Kerman, Phys. Lett. A **285**, 228 (2001).
[6] Y. Ohashi and A. Griffin, Phys. Rev. Lett. **89**, 130402 (2002).

[7] J. Stajic *et al.*, cond-mat/0309329.
[8] W. C. Stwalley, Phys. Rev. Lett. **37**, 1628 (1976).
[9] E. Tiesinga, B. J. Verhaar, and H. T. C. Stoof, Phys. Rev. A **47**, 4114 (1993).
[10] E. A. Donley, N. R. Claussen, S. T. Thompson, and C. E. Wieman, Nature (London) **417**, 529 (2002).
[11] C. A. Regal, C. Ticknor, J. L. Bohn, and D. S. Jin, Nature (London) **424**, 47 (2003).
[12] K. E. Strecker, G. B. Partridge, and R. G. Hulet, Phys. Rev. Lett. **91**, 080406 (2003).
[13] J. Cubizolles *et al.*, Phys. Rev. Lett. **91**, 240401 (2003).
[14] S. Jochim *et al.*, Phys. Rev. Lett. **91**, 240402 (2003).
[15] C. A. Regal, M. Greiner, and D. S. Jin, Phys. Rev. Lett. (to be published).
[16] M. Greiner, C. A. Regal, and D. S. Jin, Nature (London) **426**, 537 (2003).
[17] S. Jochim *et al.*, Science **302**, 2101 (2003).
[18] M.W. Zwierlein *et al.*, Phys. Rev. Lett. **91**, 250401 (2003).
[19] Several groups have reported attaining Fermi gases on the BCS side of a Feshbach resonance at T/T_F below theoretical predictions for the critical condensation temperature; so far there has been no experimental evidence for condensate formation [12,13,20–22].
[20] C. A. Regal and D. S. Jin, Phys. Rev. Lett. **90**, 230404 (2003).
[21] K. M. O'Hara *et al.*, Science **298**, 2179 (2002).
[22] M. Bartenstein *et al.*, cond-mat/0401109.
[23] H. T. C. Stoof, M. Houbiers, C. A. Sackett, and R. G. Hulet, Phys. Rev. Lett. **76**, 10 (1996).
[24] We reserve the term molecule for a two-body bound state.
[25] See also R. A. Duine and H. T. C. Stoof, J. Opt. B Quantum Semiclassical Opt. **5**, S212 (2003).
[26] M. Greiner *et al.*, cond-mat/0308519.
[27] M. H. Anderson *et al.*, Science **269**, 198 (1995).
[28] K. B. Davis *et al.*, Phys. Rev. Lett. **75**, 3969 (1995).
[29] B. DeMarco and D. S. Jin, Science **285**, 1703 (1999).
[30] G. Roati, F. Riboli, G. Modugno, and M. Inguscio, Phys. Rev. Lett. **89**, 150403 (2002).
[31] J. L. Bohn, Phys. Rev. A **61**, 053409 (2000).
[32] T. Loftus *et al.*, Phys. Rev. Lett. **88**, 173201 (2002).
[33] C. A. Regal, C. Ticknor, J. L. Bohn, and D. S. Jin, Phys. Rev. Lett. **90**, 053201 (2003).
[34] We calibrate the magnetic field for gases in the trap and during expansion using rf transitions between Zeeman levels. The magnetic field is reproducible to <15 mG. Mean-field shifts near the resonance are avoided through transfer between the resonant states [35].
[35] M.W. Zwierlein, Z. Hadzibabic, S. Gupta, and W. Ketterle, Phys. Rev. Lett. **91**, 250404 (2003).
[36] R. L. Renka and A. K. Cline, Rocky Mt. J. Math. **14**, 223 (1984).
[37] T. Bourdel *et al.*, Phys. Rev. Lett. **91**, 020402 (2003).
[38] S. Gupta *et al.*, Science **300**, 1723 (2003).
[39] The scattering length corresponding to ΔB can be calculated from $a = a_{bg}(1 - \frac{w}{\Delta B})$, where $a_{bg} = 174a_0$ and $w = 7.8 \pm 0.6$ G [33].
[40] D. S. Petrov, C. Salomon, and G. V. Shlyapnikov, cond-mat/0309010.

VOLUME 92, NUMBER 15   PHYSICAL REVIEW LETTERS   week ending
16 APRIL 2004

Evidence for Superfluidity in a Resonantly Interacting Fermi Gas

J. Kinast, S. L. Hemmer, M. E. Gehm, A. Turlapov, and J. E. Thomas

Physics Department, Duke University, Durham, North Carolina 27708-0305
(Received 21 March 2004; published 13 April 2004)

We observe collective oscillations of a trapped, degenerate Fermi gas of ^{6}Li atoms at a magnetic field just above a Feshbach resonance, where the two-body physics does not support a bound state. The gas exhibits a radial breathing mode at a frequency of 2837(05) Hz, in excellent agreement with the frequency of $\nu_H \equiv \sqrt{10\nu_x\nu_y/3} = 2830(20)$ Hz predicted for a *hydrodynamic* Fermi gas with unitarity-limited interactions. The measured damping times and frequencies are inconsistent with predictions for both the collisionless mean field regime and for collisional hydrodynamics. These observations provide the first evidence for superfluid hydrodynamics in a resonantly interacting Fermi gas.

DOI: 10.1103/PhysRevLett.92.150402

PACS numbers: 03.75.Ss, 32.80.Pj

Strongly-interacting two-component Fermi gases provide a unique testing ground for the theories of exotic systems in nature, ranging from super-high temperature superconductors to neutron stars and nuclear matter. The feature that all of these systems have in common is a strong interaction between pairs of spin-up and spin-down particles. In atomic Fermi gases, tunable, strong interactions are produced using a Feshbach resonance [1–3]. Near the resonance, the zero-energy s-wave scattering length a exceeds the interparticle spacing, and the interparticle interactions are unitarity limited and universal [4–6]. In this region, high temperature Cooper pairing has been predicted [7–10].

In this Letter, we present measurements of the frequencies and damping times for the radial hydrodynamic breathing mode of a trapped, highly degenerate gas of ^{6}Li atoms just above a Feshbach resonance at 822(3) G [11]. A cross-check of the measurement method is provided by observing the breathing mode of a noninteracing gas.

Hydrodynamic behavior in a collisionless quantum gas at very low temperature is known to be a hallmark of superfluidity. Previously, we observed hydrodynamic, anisotropic expansion of a strongly interacting, ultracold, two-component Fermi gas [12]. However, an initially collisionless gas could have become collisionally hydrodynamic as the Fermi surface significantly deformed during the expansion [13]. Thus, the observations suggested superfluid hydrodynamics, but were not conclusive.

Superfluidity has been observed in Bose-Einstein condensates (BECs) of molecular dimers, which have been produced from a two-component strongly interacting Fermi gas. The first experiments produced the dimer BECs at a magnetic field below the Feshbach resonance, where the atoms have a nonzero binding energy [14–17]. Although the two-body physics does not support a bound state at magnetic fields above the Feshbach resonance, for fermionic atoms, the many-body physics does. Observations of BECs originating from such dimers are

consistent with the existence of preformed pairs [11,18]. These experiments indirectly explore the microscopic structure, while our experiments are complementary in that they directly measure the macroscopic dynamics.

One method for distinguishing between a BEC and a superfluid Fermi gas is to examine their collective hydrodynamic modes at a low temperature where the trapped gas is collisionless [19]. For a weakly repulsive BEC contained in a nearly cylindrically symmetric trap, the radial breathing mode occurs at a frequency of $\nu_B = 2\sqrt{\nu_x\nu_y}$, where ν_i is the harmonic oscillation frequency in Hz of a noninteracting gas in the ith direction of the trap. In contrast to a weakly repulsive BEC, a superfluid Fermi gas in a cigar-shaped trap is predicted to have a radial breathing mode at the hydrodynamic frequency

$$\nu_H = \sqrt{\frac{10}{3}\nu_x\nu_y}. \tag{1}$$

This result is obtained in the unitarity limit, where the shift from the interparticle interactions vanishes for a hydrodynamic gas [20,21]. In general, hydrodynamics with the frequency ν_H can arise from superfluidity or from collisions in a normal fluid. However, at low temperature, Pauli-blocking is expected to suppress the collision rate in a Fermi gas, *increasing* the damping rate of the collisionally hydrodynamic modes as the temperature is lowered.

In our experiments with a trapped Fermi gas, Pauli-blocking of collisions is expected to be effective at the lowest temperatures achieved. Nevertheless, a weakly damped radial mode at precisely the hydrodynamic frequency of ν_H is observed, and the damping rate *decreases* strongly as the temperature is lowered below $\simeq$30% of the Fermi temperature for a noninteracting gas. These observations provide the first evidence for superfluid hydrodynamics in a resonantly interacting Fermi gas.

We prepare a degenerate 50-50 mixture of the two lowest spin states of ^{6}Li atoms by forced evaporation in an ultrastable CO_2 laser trap [12], at a chosen magnetic

VOLUME 92, NUMBER 15   PHYSICAL REVIEW LETTERS   week ending
16 APRIL 2004

field in the range 770–910 G. The trap depth is lowered by a factor of $\simeq 580$ over 4 s, then recompressed to 4.6% of the full trap depth in 1 s and held for 1 s to assure equilibrium. For our parameters, the typical Fermi temperature T_F for a noninteracting gas is $\simeq 2.5$ μK, small compared to the final trap depth of 35 μK.

Absorption images of the cloud use a probe pulse of 5 μs duration and a two-level optical transition [12]. The entire imaging system has a measured resolution of 5.5 μm. For low temperatures T where $T/T_F \leq 0.4$, the ratio T/T_F is determined by fitting a Thomas-Fermi profile for a noninteracting Fermi gas to the transverse (x) distribution obtained by integrating the column density in the axial (z) direction. At 910 G, this procedure yields temperatures as low as $T/T_F = 0.06$ and excellent fits. However, at fields closer to resonance, slightly higher temperatures are obtained, and the shape may not be precisely Thomas-Fermi due to many-body effects. For measurements at the highest temperatures $T/T_F = 0.5–1.2$, the expansion dynamics of the gas may not be perfectly hydrodynamic, and hence temperature estimates are less precise. Here, we assume hydrodynamic expansion of a Maxwell-Boltzmann spatial distribution for a noninteracting gas. The measured temperatures therefore indicate the trend, but not necessarily the absolute temperature. We find consistency between the measured number of atoms, temperature, and the initial cloud size obtained by hydrodynamic scaling [12]. For the strongly interacting gas, we include a reduction of the cloud radius arising from the mean field [5]. By correcting selected images for our estimated saturation $I/I_{\text{sat}} = 0.2$, we estimate that the true temperatures are lower by 0.03 T_F and the true atom numbers are increased by a factor $\simeq 1.15$ compared to the values given in Table I.

Trap oscillation frequencies at 4.6% of the full well depth are measured by parametric resonance in a weakly-interacting sample. The gas is cooled by forced evaporation over 25 s to temperatures of 0.3 T_F at a field of 300 G, and the trap depth is then modulated by 0.1% for 4 s. During this period, the low collision rate produces little damping, but permits the gas to thermalize. After modulation, imaging at 526 G is used to measure the release energy versus drive frequency. Well-resolved resonances are obtained at $2\omega_x = 2\pi \times 3200(20)$ Hz and $2\omega_y = 2\pi \times 3000(20)$ Hz. Because of the low frequency, the axial resonance is measured at full trap depth. We obtain $2\omega_z = 2\pi \times 600(20)$ Hz, yielding $\omega_z = 2\pi \times 70(3)$ Hz at 4.6% of full trap depth, including a quadratically combined magnetic field curvature contribution of 21 Hz at 870 G. From these measurements, $\nu_\perp \equiv \sqrt{\nu_x \nu_y} = 1550(20)$ Hz.

To excite the transverse breathing mode, the trap is turned off abruptly (≤ 1 μs) and turned back on after a delay of $t_0 = 50$ μs. Then the sample is held for a variable time t_{hold}. Finally, the trap is extinguished suddenly, releasing the gas which is imaged after 1 ms. To show that our excitation is a weak perturbation, we estimate the

TABLE I. Breathing mode frequencies ν and damping times τ_{damp}. B is the applied magnetic field, T/T_F is the initial temperature, and N is the total number of atoms, uncorrected for saturation. x_{rms} is the time-averaged root-mean-square size of the oscillating cloud. Error estimates are from the fit only.

B(G)	T/T_F	$N(10^3)$	$x_{\text{rms}}(\mu\text{m})$	ν(Hz)	τ_{damp}(ms)
526[b]	0.30(0.02)	288(18)	35.2	3212(30)	2.04(0.4)
770	0.13(0.03)[a]	138(20)	29.3	3000(150)	2.00(1.1)
815	0.14(0.04)[a]	198(24)	24.0	2931(19)	3.60(1.5)
860	0.14(0.04)	294(26)	28.6	2857(16)	3.67(1.1)
870[b]	0.17(0.06)	288(30)	32.0	2837(05)	3.85(0.4)
870[c]	0.15(0.03)	225(36)	33.5	2838(06)	6.01(1.4)
870[d]	0.18(0.04)	207(28)	41.8	5938(18)	1.44(0.2)
870[b]	0.33(0.02)	379(24)	46.1	2754(14)	2.01(0.3)
870[b]	0.50(0.06)	290(32)	45.1	2775(08)	1.39(0.1)
870	1.15(0.10)	244(10)	41.7	2779(50)	1.08(0.4)
880	0.12(0.04)	258(30)	30.0	2836(16)	3.95(1.5)
910	0.11(0.06)	268(17)	27.8	2798(15)	3.30(1.1)

[a]From the tails of a bimodal distribution.
[b]Shown in the figures.
[c]For $t_0 = 25$ μs.
[d]At 18.8(0.9)% trap depth, $t_0 = 25$ μs and 0.8 ms expansion time.

energy increase, which arises principally from the change in potential energy in the transverse directions. Assuming approximately ballistic expansion, $\Delta E_\perp = E_\perp (\omega_\perp t_0)^2/2 = 0.1$ $E_\perp$. For initial temperatures of 0.1–0.15 T_F, the corresponding temperature change is $\Delta T/T_F \simeq 0.05$ when the gas thermalizes, consistent with our measurements.

To study a noninteracting sample, breathing modes are excited at 526 G, where the scattering length is nearly zero [22,23], after cooling at 300 G as described above. Figure 1 shows $\sqrt{\langle x^2 \rangle}$ for the expanded gas ($\langle x \rangle \equiv 0$), plotted versus t_{hold}. Fitting with a damped sinusoid $x_{\text{rms}} + A \exp(-t/\tau_{\text{damp}}) \sin(2\pi\nu t + \varphi)$, we obtain $\nu = 3212(30)$ Hz, in excellent agreement with the frequency $3200(20)$ Hz measured by the parametric resonance method for the x direction which is imaged in the experiments. The damping time $\tau_{\text{damp}} = 2.04(0.4)$ ms is consistent with a small anharmonicity from the Gaussian profile of the trap potential.

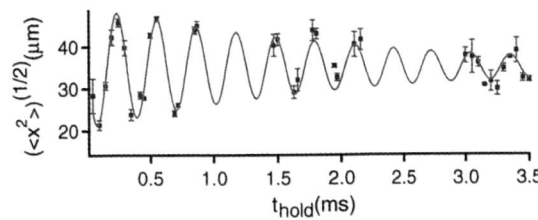

FIG. 1 (color online). Excitation of the breathing mode in a noninteracting Fermi gas of ^{6}Li at 526 G. Error bars = 68% confidence interval.

VOLUME 92, NUMBER 15   PHYSICAL REVIEW LETTERS   week ending
16 APRIL 2004

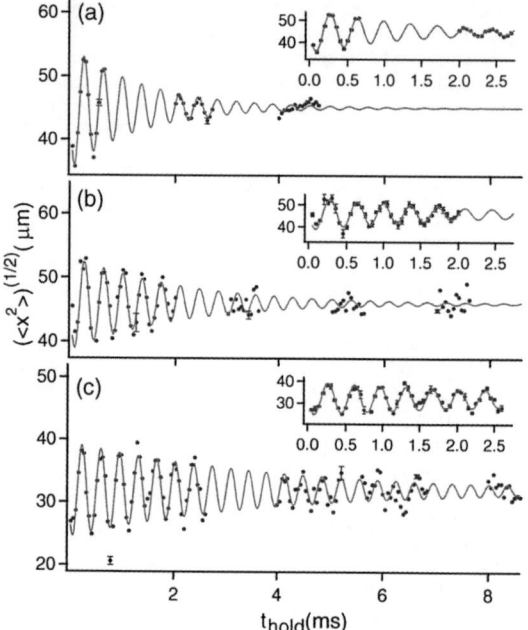

FIG. 2 (color online). Excitation of the breathing mode in a strongly interacting Fermi gas of ^{6}Li at 870 G. (a) $T/T_F = 0.50$. (b) $T/T_F = 0.33$. (c) $T/T_F = 0.17$. Error bars = 68% confidence interval.

Results for the strongly interacting gas above resonance at 870 G are shown in Figs. 2(a)–2(c) and summarized for different temperatures and magnetic fields in Table I. For the strongly interacting gas at 870 G and $T/T_F = 0.17$, the measured radial breathing mode frequency 2837(05) Hz is in excellent agreement with the prediction of Eq. (1), $\nu_H = 2830(20)$ for a hydrodynamic Fermi gas, and it differs significantly from that of the noninteracting gas and the weakly-interacting Bose gas, $\nu_B = 3100$ Hz. For the axial direction, the measured amplitude of the oscillation is consistent with zero.

On the molecular side just below resonance, at 815 G at $T/T_F = 0.14$, we obtain $\nu = 2931(19)$, which is near ν_H, consistent with predictions that the response near resonance is fermionic [20,21]. At a much lower field of 770 G, we find $\nu = 3000(150)$ closer to the predicted Bose frequency of 3100(20) Hz. However, the data is not of as high quality as that shown in Fig. 2.

Over the range of magnetic fields studied, 770–910 G, our measured oscillation frequencies $\nu(B)$ at the lowest temperatures show the same magnetic field dependence as those of M. Bartenstein et al. [24]. However, our data show a much smaller shift with respect to ν_H of Eq. (1).

We estimate the frequency shifts $\Delta\nu \equiv \nu(\text{meas}) - \nu(\text{actual})$ arising from anharmonicity in the trapping potential [25]. For the hydrodynamic frequency, $\Delta\nu_H = -(32/25)\sqrt{10/3}\,\nu_\perp M\omega_\perp^2 x_{\text{rms}}^2/(b_H^2 U)$, where U is the trap

depth, M is the ^{6}Li mass, and $\omega_\perp = 2\pi\nu_\perp$. Here, b_H is the hydrodynamic expansion factor, 11.3 after 1 ms [26]. The shift in the geometric mean of the transverse frequencies $\Delta\nu_\perp = -(6/5)\nu_\perp M\omega_\perp^2 x_{\text{rms}}^2/(b_B^2 U)$, where $b_B = 10.3$ is the ballistic expansion factor at 1 ms. Using Table I, these results yield a net $\Delta\nu \equiv \Delta\nu_H - \sqrt{10/3}\Delta\nu_\perp = +24$ Hz at 870 G and $T/T_F = 0.17$. For the three higher temperatures at 870 G, we find $\Delta\nu \simeq -35$ Hz at $T/T_F = 0.33$ and 0.5, and -16 Hz at $T/T_F = 1.15$, consistent with the measured $\simeq -60$ Hz shift below the lowest temperature data.

We have also investigated the effect of decreasing the oscillation amplitude to less than 10% by reducing t_0 to 25 μs at 870 G and $T/T_F = 0.15$. For small amplitudes, the aspect ratio of the cloud changes very little. Hence, we expect that the deformation of the Fermi surface is very small, so that collisional behavior is not induced. We obtain $\nu = 2838(06)$ Hz, in precise agreement with the results for $t_0 = 50$ μs. The damping time is somewhat increased to 6.00(1.4) ms, presumably due to the smaller energy input rather than reduced anharmonicity (since the frequency is unchanged).

The measured damping time shows a rapid increase with decreasing temperature, Fig. 3. The open circle shows the result at a trap depth increased by a factor of 4.1. The measured hydrodynamic frequency scales as $\sqrt{4.1}$ within 3% and the product $\nu_{\text{meas}}\tau_{\text{damp}}$ is consistent with those at lower trap depth. This is consistent with a damping rate which scales linearly with trap frequency, as expected for unitarity-limited interactions, where the rate scales with the Fermi energy. Note that anharmonicity cannot make a major contribution to the temperature dependence of our damping rates: The anharmonic contribution to $1/\tau_{\text{damp}}$ would be proportional to the frequency shift and therefore independent of trap depth for fixed T/T_F and N. Then, $\nu_{\text{meas}}\tau_{\text{damp}} \propto \nu_{\text{meas}}$, and the vertical position of the open circle would increase by a factor of 2. Further, the anharmonic contribution is identical for the three highest temperature points where the mean cloud sizes are nearly the same.

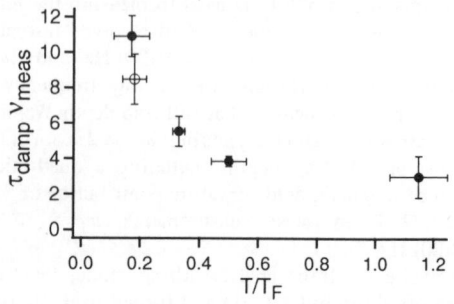

FIG. 3. Product of the damping time and the measured breathing mode frequency versus temperature. The open circle shows the result when the trap depth is increased by a factor of 4.1.

VOLUME 92, NUMBER 15   PHYSICAL REVIEW LETTERS   week ending
16 APRIL 2004

We have attempted to model the data at 870 G without invoking superfluidity. A first scenario is that the gas is nearly collisionless at the lowest temperature, and the long damping time and measured frequency are the result of collisionless mean field evolution. A second scenario is collisional hydrodynamics.

The collisionless mean field scenario requires a large negative mean field shift to explain the difference between the frequencies of 3212(30) and 2837(20) Hz measured for the noninteracting and strongly interacting samples, respectively. However, for a unitarity-limited interparticle interaction with a negative $\beta = -0.55$ [27], a Vlasov equation model [28] yields a +90 Hz shift relative to 3200 Hz, while we observe a −400 Hz shift. Also, for our trap, the same model shows that the coupling of the collisionless transverse modes by the interaction would produce a noticeable beat at 370 Hz with an amplitude minimum at 1 ms, which is not observed. Hence, the data are inconsistent with the collisionless scenario.

To investigate the second scenario, we considered a collisional hydrodynamic model, including two-body Pauli blocking [19,29]. A small negative shift at the higher temperatures might arise from the mean field in the hydrodynamic limit [20,21] or from the anharmonic shift described above. Neglecting these shifts, a relaxation approximation model [30] can be used to determine both the breathing mode frequency and the damping time in terms of the measured trap oscillation frequencies, for an arbitrary momentum relaxation rate. We find that a very large momentum relaxation rate is needed to fit the 4 ms damping time of the $T/T_F = 0.17$ data in a collisionally hydrodynamic regime. Then, the predicted damping time is large over a broad temperature range, inconsistent with the observed rapid decrease in damping time with temperature. Lowering the maximum relaxation rate, we can fit the damping times at the two highest temperatures. In this case, however, obtaining a 4 ms damping time requires a temperature below $T/T_F = 0.1$, i.e., a nearly collisionless regime, inconsistent with observations as described above.

In conclusion, at our lowest temperatures, we observe a breathing mode at precisely the hydrodynamic frequency as well as highly anisotropic hydrodynamic expansion, as in our previous experiments [12,28]. The damping time increases rapidly as the temperature is lowered [31], consistent with a transition from collisional to superfluid hydrodynamics at a temperature between 0.3 and 0.2 T_F. On the basis of the above arguments, we believe the data are not consistent with either collisionless mean field evolution or collisional hydrodynamics. It is therefore difficult to see how the observations can be explained without invoking superfluidity.

Recent theory describes the BCS-BEC crossover regime in terms of very large fermionic pairs, comparable in size to the interparticle spacing [32–34]. Falco and Stoof [33] predict BEC-like or BCS-like behavior *above* the Feshbach resonance, depending on whether $\epsilon_b \equiv \hbar^2/(a^2 m_{atom})$ is $\leq 2k_B T_F$ or $\geq 2k_B T_F$ [35]. Near resonance, where $\epsilon_b \ll 2k_B T_F$, the majority of fermionic pairs (either Bose molecules or Cooper pairs) are very large. Hence, one expects that the response of the system to compression is fermionic, and scales with density as $n^{2/3}$ [36], consistent with our measurements and with predictions [20,21].

This research is supported by DOE, ARO, NSF, and NASA.

[1] M. Houbiers et al., Phys. Rev. A **57**, R1497 (1998).
[2] W. C. Stwalley, Phys. Rev. Lett. **37**, 1628 (1976).
[3] E. Tiesinga et al., Phys. Rev. A **47**, 4114 (1993).
[4] H. Heiselberg, Phys. Rev. A **63**, 043606 (2001).
[5] M. E. Gehm et al., Phys. Rev. A **68**, 011401 (2003).
[6] T.-L. Ho, Phys. Rev. Lett. **92**, 090402 (2004).
[7] BCS pairing of ^{6}Li atoms was predicted by H. T. C. Stoof et al., Phys. Rev. Lett. **76**, 10 (1996); M. Houbiers et al., Phys. Rev. A **56**, 4864 (1997).
[8] M. Holland et al., Phys. Rev. Lett. **87**, 120406 (2001).
[9] E. Timmermans et al., Phys. Lett. A **285**, 228 (2001).
[10] Y. Ohashi and A. Griffin, Phys. Rev. Lett. **89**, 130402 (2002).
[11] M. Zwierlein et al., Phys. Rev. Lett. **92**, 040403 (2004).
[12] K. M. O'Hara et al., Science **298**, 2179 (2002).
[13] S. Gupta et al., Phys. Rev. Lett. **92**, 100401 (2004).
[14] M. Greiner et al., Nature (London) **426**, 537 (2003).
[15] S. Jochim et al., Science **302**, 2101 (2003).
[16] M. Zwierlein et al., Phys. Rev. Lett. **91**, 250401 (2003).
[17] T. Bourdel et al., cond-mat/0403091.
[18] C. A. Regal et al., Phys. Rev. Lett. **92**, 040403 (2004).
[19] L. Vichi, J. Low Temp. Phys. **121**, 177 (2000).
[20] H. Heiselberg, cond-mat/0403041.
[21] S. Stringari, Europhys. Lett. **65**, 749 (2004).
[22] K. M. O'Hara et al., Phys. Rev. A **66**, 041401 (2002).
[23] S. Jochim et al., Phys. Rev. Lett. **89**, 273202 (2002).
[24] M. Bartenstein et al., cond-mat/0403716.
[25] S. Stringari (private communication).
[26] From the frequency measured at 18.8% trap depth, $b_H[0.8\,\text{ms}] = 19.3$.
[27] J. Carlson et al., Phys. Rev. Lett. **91**, 050401 (2003).
[28] C. Menotti et al., Phys. Rev. Lett. **89**, 250402 (2002).
[29] M. E. Gehm et al., Phys. Rev. A **68**, 011603(R) (2003).
[30] D. Guery-Odelin et al., Phys. Rev. A **60**, 4851 (1999).
[31] The measured damping times at 870 G are well fit by $\tau[\text{ms}] = 0.88 \exp(+0.25\, T_F/T)$.
[32] G. M. Bruun, cond-mat/0401497.
[33] G. M. Falco and H. T. C. Stoof, Phys. Rev. Lett. **92**, 130401 (2004).
[34] H. Xiong and S. Liu, cond-mat/0403336.
[35] ϵ_b may play additional roles, e.g., relative to the thermal energy, the trap energy level spacing, trap laser recoil energy, etc.
[36] W. Ketterle (private communication).

nature   Vol 438|24 November 2005|doi:10.1038/nature04268

ARTICLES

Direct observation of base-pair stepping by RNA polymerase

Elio A. Abbondanzieri[1]*, William J. Greenleaf[1]*, Joshua W. Shaevitz[2]†, Robert Landick[4] & Steven M. Block[1,3]

During transcription, RNA polymerase (RNAP) moves processively along a DNA template, creating a complementary RNA. Here we present the development of an ultra-stable optical trapping system with ångström-level resolution, which we used to monitor transcriptional elongation by single molecules of *Escherichia coli* RNAP. Records showed discrete steps averaging 3.7 ± 0.6 Å, a distance equivalent to the mean rise per base found in B-DNA. By combining our results with quantitative gel analysis, we conclude that RNAP advances along DNA by a single base pair per nucleotide addition to the nascent RNA. We also determined the force–velocity relationship for transcription at both saturating and sub-saturating nucleotide concentrations; fits to these data returned a characteristic distance parameter equivalent to one base pair. Global fits were inconsistent with a model for movement incorporating a power stroke tightly coupled to pyrophosphate release, but consistent with a brownian ratchet model incorporating a secondary NTP binding site.

Processive molecular motors tend to move in discrete steps[1]. Recent advances in single-molecule techniques have made it possible to observe such steps directly at length scales of a few nanometres or greater. The ability to detect individual catalytic turnovers, as monitored through motor displacement, while simultaneously controlling the force, substrate concentration, temperature or other parameters, provides a means to probe the mechanisms responsible for motility. Single-molecule measurements of stepping have supplied fresh insight into the mechanisms responsible for motion in motor proteins such as myosin, kinesin, dynein and the F_1-ATPase[2–9]. A number of processive nucleic acid-based enzymes, such as lambda exonuclease[10,11], RecBCD helicase[12–14] and RNAP[15–19], have also been studied successfully by single-molecule methods, but the comparatively small size of their steps has been experimentally inaccessible up to this point. Movements through a single base pair along double-stranded DNA correspond to a displacement of just ~3.4 Å (ref. 20), which is more than 20-fold smaller than the 8-nm kinesin step[4] and sevenfold smaller than the 2–3-nm resolution limit attained in most previous work[2,3,14,19,21].

During transcription, *E. coli* RNAP translocates along DNA while following its helical pitch[22], adding ribonucleoside triphosphates (NTPs) successively to the growing RNA. The basic reaction cycle consists of binding the appropriate NTP, incorporation of the associated nucleoside monophosphate into the RNA, and release of pyrophosphate. In addition to following the main reaction pathway, RNAP can reversibly enter any of several off-pathway paused states. For example, RNAP may backtrack by several bases along DNA, displacing the RNA 3′ end from the catalytic centre and temporarily inactivating the enzyme[23–25]. Single-molecule studies have shown that backtracking pauses are enhanced under hindering loads and can be triggered by misincorporation of non-complementary NTPs[19]. In a separate class of pauses, hairpin structures formed in the nascent RNA can induce pausing via a distinct mechanism[26]. A third class of pauses occurs commonly and is independent of backtracking[18] or hairpin formation. Regardless of the cause, paused states complicate the interpretation of biochemical studies of RNAP

elongation because these kinetic measurements convolve on- and off-pathway events into overall rates. Single-molecule techniques avoid ensemble averaging and permit, in principle, observations that may distinguish between elongation and off-pathway states. The resolution of individual enzymatic turnovers would be especially helpful in unravelling the behaviour of complex reaction cycles for enzymes such as RNAP.

The mechanism that leads to translocation during transcriptional elongation continues to be debated[27–32], and at least two classes of models have been proposed. The first class postulates that a power stroke tightly coupled to pyrophosphate release drives motion[29]. In the second class, reversible diffusion of the enzyme along the DNA template between its pre- and post-translocated states is directionally rectified through the binding of the incoming NTP, in a brownian ratchet mechanism[30–32]. By detecting individual translocation events during transcription and characterizing the force and nucleotide sensitivity of the corresponding motions, one can distinguish between the different classes of mechanism.

Construction of an ultra-stable optical trap

Because the distance spanned by a base pair is so small, it was necessary to construct a stable optical trapping system capable of ångström-level resolution. Sources of noise that hamper optical measurements include drift of the microscope stage (or other nominally stationary components), pointing fluctuations leading to relative motions of the laser beams used for trapping and position detection, and brownian motions of the trapped bead itself. To minimize the noise associated with stage motions, we used a dual-trap 'dumbbell' arrangement, described previously[19]. In this experimental geometry (Fig. 1a), all components of the assay are optically levitated above the coverglass surface and thereby decoupled from stage drift. A stalled transcription complex[18] containing a biotin tag on the carboxy terminus of the β′-subunit was specifically attached via an avidin linkage to the surface of a 600-nm diameter polystyrene bead. Depending on the desired direction of applied load, either the transcriptionally upstream or downstream end of the DNA was then

[1]Department of Applied Physics, [2]Department of Physics, [3]Department of Biological Sciences, Stanford University, Stanford, California 94305, USA. [4]Department of Bacteriology, University of Wisconsin, Madison, Wisconsin 53706, USA. †Present address: Department of Integrative Biology, University of California, Berkeley, California 94720, USA.
*These authors contributed equally to this work.

NATURE|Vol 438|24 November 2005

bound via a digoxigenin antibody linkage to a 700-nm diameter bead, forming a bead–DNA–RNAP–bead dumbbell. Dumbbells were suspended ~1 μm above the microscope coverglass by two independently steered traps, T_{weak} and T_{strong}.

To isolate the detection and trapping beams from the effects of random air currents, which introduce density fluctuations that perturb the positional stability of laser beams, we enclosed all optical elements external to the microscope in a sealed box filled with helium gas at atmospheric pressure. Because the refractive index of helium is closer to unity than that of air ($n_{He} = 1.000036$ compared with $n_{air} = 1.000293$), density fluctuations introduce smaller deflections.

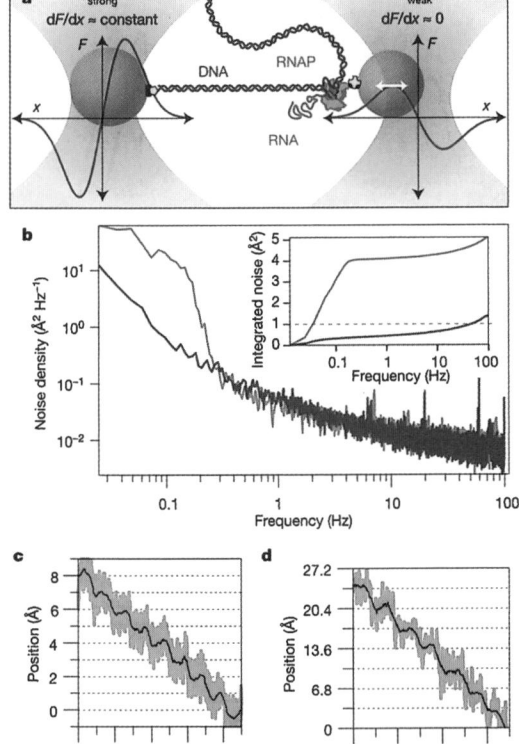

Figure 1 | Experimental set-up, passive force clamp and sensitivity of the RNAP dumbbell assay. a, Cartoon of the dumbbell geometry with schematic force versus position curves (dark red) shown for both trap beams (not drawn to scale). A single, transcriptionally active molecule of RNAP (green) is attached to a bead (blue) held in trap T_{weak} (pink, right) and tethered via the upstream DNA (dark blue) to a larger bead held in trap T_{strong} (pink, left). The right bead is maintained at a position near the peak of the force-extension curve of T_{weak}, where trap stiffness vanishes (white arrow), creating a force clamp (trap stiffness $k = dF/dx$). During elongation, the DNA tether lengthens and the beads move apart. Owing to the force clamp arrangement, only the right bead moves: displacement is measured for this bead. **b,** Power spectrum acquired for a stiffly trapped bead with external optics under air (red) or helium (blue). Inset: integrated noise spectra for air (red) and helium (blue) showing a tenfold reduction in power. **c,** Steps resolved for a stiffly trapped bead moved in 1-Å increments at 1 Hz. Data were median filtered with a 5-ms (pink) and 500-ms (black) window. **d,** Steps resolved for a bead–DNA–bead dumbbell held at 27 pN of tension, produced by moving T_{strong} in 3.4-Å increments at 1 Hz and measuring the corresponding displacements in T_{weak}.

Using helium, we realized a tenfold reduction in the noise spectral density at 0.1 Hz for a stiffly trapped, 700-nm diameter bead ($k = 1.9\,pN\,nm^{-1}$), and the integrated system noise power remained below ~1 Å over the bandwidth of interest (Fig. 1b). To illustrate the resolution achieved, we moved a trapped bead in increments of 1 Å at 1-s intervals by displacing T_{strong} with an acousto-optic deflector (AOD)[33]. Steps were clearly resolved, with signal-to-noise ratio of ~1 over a 100-Hz bandwidth (Fig. 1c).

Finally, we implemented a recently developed method to maintain constant force on trapped beads using an all-optical arrangement, without the need for computer feedback[34]. Such a passive force clamp eliminates artefacts associated with feedback loops and provides exceedingly high bandwidth. To create this clamp, the bead in T_{weak} was maintained in a ~50-nm region near the maximum of the force-extension curve, where the force is independent of bead position. Force clamps eliminate the need for elastic corrections due to either the compliance of T_{strong} or the stretching of the DNA along with its associated linkages, so that all molecular displacements are registered in the motion of the bead in T_{weak}. To demonstrate the resolution achieved in our set-up, a dumbbell consisting of beads connected by a DNA tether (but no RNAP enzyme) was stepped in increments of 3.4 Å at 1 Hz (Fig. 1d).

RNA polymerase takes single-base-pair steps

To resolve individual translocation events, we required that RNAP transcribe slowly enough to time-average to the ångström level over positional uncertainties caused by brownian motions, but quickly enough so that long-term drift did not obscure motion. Because RNAP has different average rates of addition for the four species of nucleotide[35], we determined by gel analysis the concentration ratios at which each species becomes equally rate limiting for elongation on our template. Unless otherwise noted, our experiments were conducted at $[NTP]_{eq} = 10\,μM$ GTP, $10\,μM$ UTP, $5\,μM$ ATP and $2.5\,μM$ CTP, concentrations that produce a mean elongation rate of ~1 base pair (bp) s^{-1} under our conditions (see Supplementary Fig. S1).

In single-molecule records of transcription by RNAP selected for their low noise and drift, we observed clear, stepwise advancements. Figure 2a shows six representative traces obtained under 18 pN of assisting load. Although dwells at some expected positions (Fig. 2a, dotted lines) were missed or skipped, steps were uniform in size, corresponding to nearly integral multiples of a common spacing. To estimate this fundamental spacing, we performed a periodogram analysis[36]. The position histograms for 37 segments derived from transcription records for 28 individual RNAP molecules were computed and the autocorrelation function calculated for each of these[37]. These autocorrelations were combined into a global average that displays a series of peaks near multiples of the mean spacing (Fig. 2b), with the first and strongest peak at 3.4 ± 0.8 Å. The power spectral density of this function measures the corresponding spatial frequencies and displays a prominent peak at the inverse of 3.7 ± 0.6 Å (Fig. 2c). This distance is consistent with the crystallographic spacing between neighbouring base pairs in B-DNA[38] (3.4 ± 0.5 Å). Although the foregoing analysis was performed on selected traces, a fully automated procedure was also conducted on a continuous, ~300-bp record of elongation, and returned a similar spacing of 3.7 ± 1.5 Å (Supplementary Fig. S2).

In our records, RNAP does not dwell at every base-pair position along the template. Were certain bases skipped altogether, or merely missed due to finite time resolution? Because RNAP is linked to the bead via the C terminus of its $β'$-subunit, skipped steps might be explained, in principle, by relative motions of this point of attachment with respect to the catalytic core of the enzyme, allowing the active site to undergo one or more rounds of nucleotide addition before translocation of the attachment point in a single jump, constituting a form of 'inchworming' movement. Alternatively, if the upstream DNA exists in a 'scrunched' state within the enzyme after templating the production of RNA, as proposed to occur during

ARTICLES

NATURE|Vol 438|24 November 2005

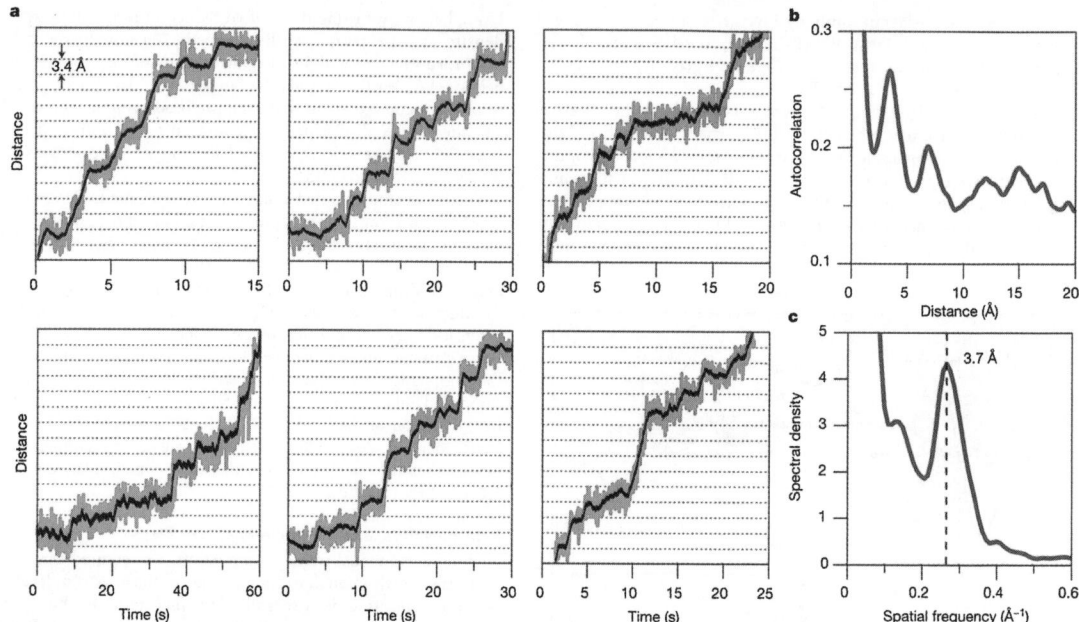

Figure 2 | RNAP moves in discrete steps. a, Representative records for single molecules of RNAP transcribing at [NTP]$_{eq}$ under 18 pN of assisting load, median-filtered at 50 ms (pink) and 750 ms (black). Horizontal lines (dotted) are spaced at 3.4-Å intervals. **b,** The average autocorrelation function derived from position histograms ($N = 37$) exhibits periodicity at multiples of the step size. **c,** The power spectrum of **b** shows a peak at the dominant spatial frequency, corresponding to the inverse of the fundamental step size, 3.7 ± 0.6 Å.

initiation and during certain regulatory pauses[39–42], then the periodic release of variable amounts of scrunched DNA might also lead to discontinuous enzyme advancement. However, a more parsimonious explanation of our stepping data is that RNAP exhibits heterogeneity in individual nucleotide addition rates during transcription, causing some dwells to occur on timescales too fast to be resolved in our recordings, but maintaining tight coupling between the transcript length and position along the template.

We tested the latter explanation by performing quantitative gel assays at [NTP]$_{eq}$, to measure transcript lengths produced by RNAP over a portion of the same DNA template used in our single-molecule studies. Analysis of these gels indicated a variability of more than an order of magnitude in local rates of NTP addition. Expected dwell-time distributions derived from gel analysis are well fitted to the actual distribution of dwell-times measured independently in single molecule records (Supplementary Fig. S3). The close correspondence between variability in next-nucleotide addition rates (measured biochemically) and variability in next-base translocation rates

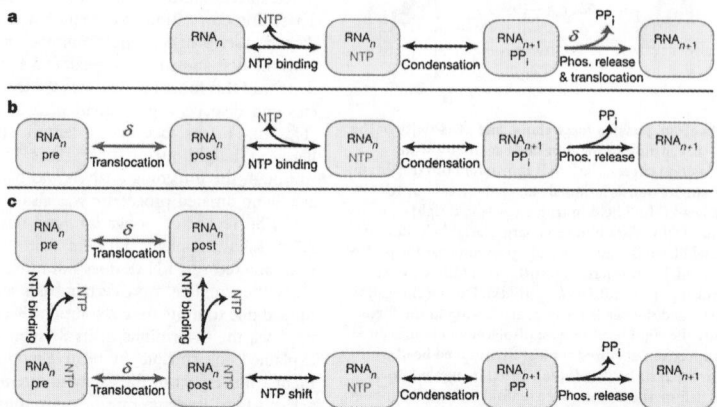

Figure 3 | Alternative kinetic models for RNAP translocation. a, A power stroke model where translocation (δ, red) is driven by irreversible PP$_i$ release. **b,** A brownian ratchet model where reversible oscillation between pre- and post-translocated enzyme states can occur before NTP binding (blue). **c,** A brownian ratchet model where translocation and NTP binding can occur in either order. This model postulates the existence of a secondary NTP site to accommodate the possibility of nucleotide binding when the enzyme is in its pre-translocated state.

(observed in single molecules) is therefore fully consistent with tight coupling between our attachment point to RNAP on the template and the length of the resulting RNA transcript. Moreover, the observation that RNAP can frequently be seen to step by single base increments (multiple examples of which are found in Fig. 2a) is incompatible with an obligate scrunching mechanism, or with any alternative mechanism where the catalytic core moves with respect to our point of enzyme attachment.

Force dependence of the nucleotide addition cycle

The application of load selectively modulates rates within the biochemical cycle involving motion. A simple Boltzmann relation describes the resulting force–velocity relationship for many mechanoenzymes[3,43,44]:

$$v(F) = \frac{v_{max}}{1 + \exp\left[-\frac{(F - F_{1/2})\delta}{k_B T}\right]} \quad (1)$$

where v_{max} is the velocity at large assisting load, $F_{1/2}$ is the force at which velocity reaches half its maximal value, k_B is Boltzmann's constant, T is the temperature, and δ is a parameter that represents the effective distance over which force acts. The physical interpretation of δ depends on the underlying model[43]. In power stroke models, δ typically represents the distance from the pre-translocated position to the transition state. In such a model for RNAP inspired by ref. 29, this transition state is located between the PP_i-bound and PP_i-released states, where PP_i is pyrophosphate (Fig. 3a), and δ corresponds to some fraction of a single-base-pair separation. In brownian ratchet models, δ typically represents the characteristic

distance associated with fluctuations between pre- and post-translocated states. For RNAP, this distance generally corresponds to one base pair. We considered two instances of a brownian ratchet: a model inspired by ref. 32, where translocation precedes NTP binding (Fig. 3b), and a model where translocation can either precede or follow NTP binding (Fig. 3c). Because the enzyme active site is occluded by the 3′ end of RNA in the pre-translocated state, the latter model requires an incoming NTP to occupy a secondary binding site before being loaded into the active site. Such a secondary binding site might represent, for example, the 'E site' observed in crystal structures of polymerase II[45,46], or the templated binding sites proposed in biochemical studies[27,28,47]. These binding sites are structurally distinct, but the binding of an NTP to either type of site may be modelled by the simplified kinetic pathway shown in Fig. 3c (see also Supplementary Information).

To distinguish between the models of Fig. 3 on the basis of force–velocity behaviour, it was critical to remove load-dependent, off-pathway events from records. In particular, we observed occasional pauses associated with both backtracking[19] and backstepping (Fig. 4). These off-pathway events decrease overall elongation rates by differing amounts depending on load, obscuring the on-pathway load dependence. With the improved resolution obtained, it was possible to identify and remove all rearward motions greater than 1 nm from traces. The run velocity for each molecule was computed by dividing

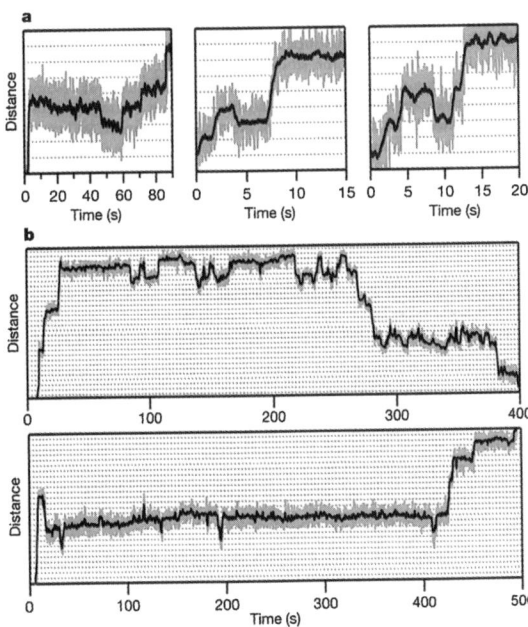

Figure 4 | RNAP backstepping and backtracking resolved at high resolution. a, Backstepping observed under assisting loads of 18 pN at $[NTP]_{eq}$. Molecules were occasionally found to move backward by one base pair (left and middle panels) or by two base pairs (right panel) before resuming elongation. **b**, Backtracking (>3 bp) under a hindering load of 9 pN. Molecules dwelled at specific preferred locations on the template before irreversibly backtracking (top) or recovering (bottom). Horizontal gridlines (dotted) are spaced at 3.4-Å intervals. (See Supplementary Information for a discussion.)

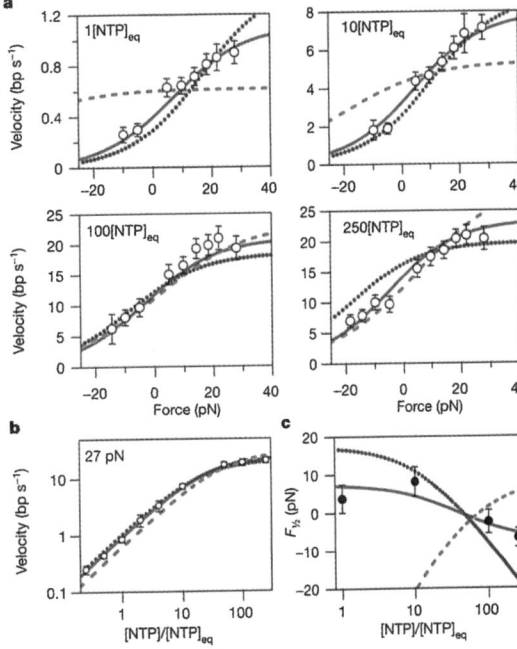

Figure 5 | Force–velocity and Michaelis–Menten relationships along with model fits. a, b, Measured force–velocity relationships for RNAP at $[NTP]_{eq}$, $10[NTP]_{eq}$, $100[NTP]_{eq}$ and $250[NTP]_{eq}$, with off-pathway events removed (see text) (**a**), and single-molecule velocity as a function of [NTP] measured at 27 pN assisting load (open circles) with associated errors (see Methods) (**b**). Negative forces correspond to hindering loads; positive forces to assisting loads. Global fits to the power stroke model of Fig. 3a (green dashed line), the brownian ratchet model of Fig. 3b (blue dotted line) and the brownian ratchet model of Fig. 3c (solid red line) are shown. **c**, The $F_{1/2}$ value as a function of [NTP] derived from free fits of the data in **a** to equation (1) (Supplementary Fig. S4), shown together with predictions from the three models (coloured lines) (Supplementary Fig. S5).

the total distance of advance by the total elapsed time minus any time spent in backtracked states. We note that off-pathway pauses that are insensitive to load (and therefore have no associated motion) do not affect the force dependence of nucleotide addition, because they occur with a fixed probability per unit time.

Elongation velocities were measured over a wide range of assisting and hindering loads (-18 to 28 pN) at four NTP concentrations (1, 10, 100 and 250 × [NTP]$_{eq}$), with an average of 32 molecules per data point (Fig. 5a). With backtracking included, the velocity interpolated to zero load at [NTP]$_{eq}$ was in excellent agreement with an independent estimate obtained through gel-based measurements. Once backtracking pauses were removed, each force–velocity curve was fit individually to equation (1). These unconstrained fits returned a characteristic distance parameter $\delta = 3.4 \pm 0.5$ Å (Supplementary Fig. S4).

We generated global fits of all three models to our entire data set of velocities ($N = 40$; Fig. 5a, b). Note that more force is required to hinder elongation with increasing [NTP] (that is, $F_{1/2}$ decreases; Fig. 5c), a trend opposite to that predicted for a power stroke model coupled to PP$_i$ release. The global fit to the power stroke model generated a poor fit ($\chi_\nu^2 = 6.03$; $\nu = 35$; $p(\chi_\nu^2) = 5.3 \times 10^{-27}$; five parameters; ν is the number of degrees of freedom). These findings, together with a computed distance parameter corresponding to a full base-pair displacement (rather than some fraction of a base pair expected for a power stroke model) and previous results showing that elongation velocity is insensitive to PP$_i$ concentration[43], all argue against the mechanism of Fig. 3a.

A global fit of our data to the simple brownian ratchet model of Fig. 3b returned better results ($\chi_\nu^2 = 2.67$; $\nu = 37$; $p(\chi_\nu^2) = 1.6 \times 10^{-7}$; three parameters), and qualitatively predicted the behaviour of $F_{1/2}$ as a function of [NTP]. An even better fit was obtained for the ratchet model of Fig. 3c, which invokes a secondary NTP binding site. This model predicted all unconstrained $F_{1/2}$ values to within error, and was statistically consistent with the complete data set ($\chi_\nu^2 = 0.64$; $\nu = 36$; $p(\chi_\nu^2) = 0.956$; four parameters). The fit parameters suggest the presence of a small energetic penalty (~ 1 kT) associated with nucleotide binding to the secondary site when the molecule is pre-translocated, compared with the post-translocated binding energy (under standard conditions). Saturating NTP concentrations tend to ensure that the secondary binding site remains occupied and thereby bias the enzyme towards the post-translocated state. However, in contrast to the ratchet mechanism of ref. 32, hindering loads can overcome this bias by forcing the NTP-bound form into a pre-translocated state, increasing the force sensitivity at higher NTP levels. This model seems attractive for its simplicity (four free parameters), its ability to fit all available force–velocity data, and the close correspondence to recent structural and biochemical studies supplying evidence for a secondary site[27,28,45,46]. Clearly, alternative kinetic schemes may be formulated to fit the data presented here and elsewhere.

The marked improvement in resolution obtained in this optical trapping study has led to direct measurements of base-pair stepping by an individual enzyme and supplied insights into the molecular mechanism of transcription by RNAP. Our data argue directly against any power stroke mechanism that is tightly coupled to PP$_i$ release. Furthermore, although other recent publications have supplied independent biochemical and biophysical evidence in support of various forms of a brownian ratchet mechanism[30,31,48], we propose a specific model incorporating a secondary NTP binding site that is consistent with our data and others[27,45,46]. The techniques presented here are broadly applicable. In particular, it may be possible to use our approach to relate the behaviour of a nucleic acid-based enzyme directly to the underlying DNA sequence to which it is bound, facilitating studies of sequence-dependent effects in replication, transcription and translation, and possible use in single-molecule DNA sequencing. The ability to detect motions at the ångström scale in single enzymes opens new avenues for the study of biomolecules.

METHODS

Optical trapping. The overall design of the apparatus has been described[19,34]. Salient modifications to our apparatus include the construction of a sealed optics enclosure where helium gas at atmospheric pressure replaces ambient air and the implementation of a passive, all-optical force clamp[34]. The force clamp was calibrated by measuring the relaxation rate of a 700-nm polystyrene bead (Bangs Labs) after release from a point $\sim$300 nm from the trap centre[49]. In this low-Reynolds-number regime, the velocity is proportional to the force acting on the bead: using this relationship, we found that force remained constant within 5% over a 50-nm-wide clamp region located $\sim$220 nm from the trap centre. During single-molecule transcription experiments, the bead held in T$_{weak}$ was maintained in this zero-stiffness zone by occasionally moving T$_{strong}$ by 20 nm whenever the 600-nm bead in T$_{weak}$ approached the outer limit of the clamp region.

For collection of force–velocity data (only), we used an active, AOD-based force clamp[33], which allowed us to alter rapidly the load on an individual enzyme during a single run. To cover the entire range of forces, each RNAP molecule included for analysis was subjected to a cycle of hindering (-5, -10, -14, -18 pN) or assisting (5, 10, 14, 18, 22, 29 pN) loads until it either terminated or stalled. By generating data for each molecule over a range of forces, we minimized variations due to intrinsic velocity heterogeneity[18].

Data analysis. Unless noted, bead displacement data were filtered with a 1-kHz, 4-pole Bessel filter, digitally acquired at 2 kHz, and median-filtered at 50 ms or 750 ms. To determine RNAP step size, we selected 37 segments of transcription through 51-Å-wide windows from 28 molecules at 18 pN or 27 pN of assisting load, then generated position histograms for each of these traces using a bin size of 0.1 Å. Histograms were autocorrelated, normalized by the number of data points, and averaged to form a global autocorrelation function. The power spectrum derived from this autocorrelation function was smoothed with a 5-point binomial filter.

To generate the graphs of Fig. 5, raw data were smoothed with an 800-ms boxcar filter and decimated to 2.5 Hz. Transcriptional pauses associated with $\geq$1 nm backward motion were removed by an automated algorithm implemented in Igor (Wavemetrics). Errors in velocity were computed as follows. First, the contribution to velocity of the positional uncertainty was estimated from the levels of statistical fluctuation occurring in regions where the enzyme had stalled. Next, the stochastic variation of RNAP motion was estimated using a randomness parameter[50] of 10 (randomness was estimated by single-molecule analysis; data not shown). Finally, a heterogeneity in the population velocity equivalent to 50% of the mean was assumed[18]. These three sources of error were assumed to be independent, combined in quadrature, and used to compile a weighted average of the velocity data and associated uncertainty in the mean, displayed as error bars in Fig. 5a, b.

Received 14 June; accepted 26 September 2005.
Published online 13 November 2005.

1. Vale, R. D. & Milligan, R. A. The way things move: looking under the hood of molecular motor proteins. *Science* **288**, 88–95 (2000).
2. Purcell, T. J., Morris, C., Spudich, J. A. & Sweeney, H. L. Role of the lever arm in the processive stepping of myosin V. *Proc. Natl Acad. Sci. USA* **99**, 14159–14164 (2002).
3. Altman, D., Sweeney, H. L. & Spudich, J. A. The mechanism of myosin VI translocation and its load-induced anchoring. *Cell* **116**, 737–749 (2004).
4. Svoboda, K., Schmidt, C. F., Schnapp, B. J. & Block, S. M. Direct observation of kinesin stepping by optical trapping interferometry. *Nature* **365**, 721–727 (1993).
5. Schnitzer, M. J. & Block, S. M. Kinesin hydrolyses one ATP per 8-nm step. *Nature* **388**, 386–390 (1997).
6. Asbury, C. L., Fehr, A. N. & Block, S. M. Kinesin moves by an asymmetric hand-over-hand mechanism. *Science* **302**, 2130–2134 (2003).
7. Yildiz, A., Tomishige, M., Vale, R. D. & Selvin, P. R. Kinesin walks hand-over-hand. *Science* **303**, 676–678 (2004).
8. Mallik, R., Carter, B. C., Lex, S. A., King, S. J. & Gross, S. P. Cytoplasmic dynein functions as a gear in response to load. *Nature* **427**, 649–652 (2004).
9. Yasuda, R., Noji, H., Yoshida, M., Kinosita, K. Jr & Itoh, H. Resolution of distinct rotational substeps by submillisecond kinetic analysis of F$_1$-ATPase. *Nature* **410**, 898–904 (2001).
10. Perkins, T. T., Dalal, R. V., Mitsis, P. G. & Block, S. M. Sequence-dependent pausing of single lambda exonuclease molecules. *Science* **301**, 1914–1918 (2003).
11. van Oijen, A. M. et al. Single-molecule kinetics of lambda exonuclease reveal base dependence and dynamic disorder. *Science* **301**, 1235–1238 (2003).
12. Dohoney, K. M. & Gelles, J. Chi-sequence recognition and DNA translocation by single RecBCD helicase/nuclease molecules. *Nature* **409**, 370–374 (2001).
13. Handa, N., Bianco, P. R., Baskin, R. J. & Kowalczykowski, S. C. Direct visualization of RecBCD movement reveals cotranslocation of the RecD motor after chi recognition. *Mol. Cell* **17**, 745–750 (2005).

NATURE|Vol 438|24 November 2005

14. Perkins, T. T., Li, H. W., Dalal, R. V., Gelles, J. & Block, S. M. Forward and reverse motion of single RecBCD molecules on DNA. *Biophys. J.* **86,** 1640–1648 (2004).

15. Yin, H. *et al.* Transcription against an applied force. *Science* **270,** 1653–1657 (1995).

16. Forde, N. R., Izhaky, D., Woodcock, G. R., Wuite, G. J. & Bustamante, C. Using mechanical force to probe the mechanism of pausing and arrest during continuous elongation by *Escherichia coli* RNA polymerase. *Proc. Natl Acad. Sci. USA* **99,** 11682–11687 (2002).

17. Adelman, K. *et al.* Single molecule analysis of RNA polymerase elongation reveals uniform kinetic behaviour. *Proc. Natl Acad. Sci. USA* **99,** 13538–13543 (2002).

18. Neuman, K. C., Abbondanzieri, E. A., Landick, R., Gelles, J. & Block, S. M. Ubiquitous transcriptional pausing is independent of RNA polymerase backtracking. *Cell* **115,** 437–447 (2003).

19. Shaevitz, J. W., Abbondanzieri, E. A., Landick, R. & Block, S. M. Backtracking by single RNA polymerase molecules observed at near-base-pair resolution. *Nature* **426,** 684–687 (2003).

20. Watson, J. D. & Crick, F. H. Molecular structure of nucleic acids; a structure for deoxyribose nucleic acid. *Nature* **171,** 737–738 (1953).

21. Li, L., Huang, H. H., Badilla, C. L. & Fernandez, J. M. Mechanical unfolding intermediates observed by single-molecule force spectroscopy in a fibronectin type III module. *J. Mol. Biol.* **345,** 817–826 (2005).

22. Harada, Y. *et al.* Direct observation of DNA rotation during transcription by *Escherichia coli* RNA polymerase. *Nature* **409,** 113–115 (2001).

23. Nudler, E., Mustaev, A., Lukhtanov, E. & Goldfarb, A. The RNA-DNA hybrid maintains the register of transcription by preventing backtracking of RNA polymerase. *Cell* **89,** 33–41 (1997).

24. Reeder, T. C. & Hawley, D. K. Promoter proximal sequences modulate RNA polymerase II elongation by a novel mechanism. *Cell* **87,** 767–777 (1996).

25. Komissarova, N. & Kashlev, M. RNA polymerase switches between inactivated and activated states by translocating back and forth along the DNA and the RNA. *J. Biol. Chem.* **272,** 15329–15338 (1997).

26. Artsimovitch, I. & Landick, R. Pausing by bacterial RNA polymerase is mediated by mechanistically distinct classes of signals. *Proc. Natl Acad. Sci. USA* **97,** 7090–7095 (2000).

27. Zhang, C. & Burton, Z. F. Transcription factors IIF and IIS and nucleoside triphosphate substrates as dynamic probes of the human RNA polymerase II mechanism. *J. Mol. Biol.* **342,** 1085–1099 (2004).

28. Gong, X. Q., Zhang, C., Feig, M. & Burton, Z. F. Dynamic error correction and regulation of downstream bubble opening by human RNA polymerase II. *Mol. Cell* **18,** 461–470 (2005).

29. Yin, Y. W. & Steitz, T. A. The structural mechanism of translocation and helicase activity in T7 RNA polymerase. *Cell* **116,** 393–404 (2004).

30. Bai, L., Shundrovsky, A. & Wang, M. D. Sequence-dependent kinetic model for transcription elongation by RNA polymerase. *J. Mol. Biol.* **344,** 335–349 (2004).

31. Bar-Nahum, G. *et al.* A ratchet mechanism of transcription elongation and its control. *Cell* **120,** 183–193 (2005).

32. Guajardo, R. & Sousa, R. A model for the mechanism of polymerase translocation. *J. Mol. Biol.* **265,** 8–19 (1997).

33. Lang, M. J., Asbury, C. L., Shaevitz, J. W. & Block, S. M. An automated two-dimensional optical force clamp for single molecule studies. *Biophys. J.* **83,** 491–501 (2002).

34. Greenleaf, W. J., Woodside, M. T., Abbondanzieri, E. A. & Block, S. M. A passive all-optical force clamp for high-resolution laser trapping. *Phys. Rev. Lett.* (in the press).

35. Rhodes, G. & Chamberlin, M. J. Ribonucleic acid chain elongation by *Escherichia coli* ribonucleic acid polymerase. I. Isolation of ternary complexes and the kinetics of elongation. *J. Biol. Chem.* **249,** 6675–6683 (1974).

36. Gelles, J., Schnapp, B. J. & Sheetz, M. P. Tracking kinesin-driven movements with nanometre-scale precision. *Nature* **331,** 450–453 (1988).

37. Block, S. M. & Svoboda, K. Analysis of high resolution recordings of motor movement. *Biophys. J.* **68,** 2305S–2395S 2395S–2415S (1995).

38. Yanagi, K., Prive, G. G. & Dickerson, R. E. Analysis of local helix geometry in three B-DNA decamers and eight dodecamers. *J. Mol. Biol.* **217,** 201–214 (1991).

39. Cheetham, G. M. & Steitz, T. A. Structure of a transcribing T7 RNA polymerase initiation complex. *Science* **286,** 2305–2309 (1999).

40. Marr, M. T., Datwyler, S. A., Meares, C. F. & Roberts, J. W. Restructuring of an RNA polymerase holoenzyme elongation complex by lambdoid phage Q proteins. *Proc. Natl Acad. Sci. USA* **98,** 8972–8978 (2001).

41. Mukherjee, S., Brieba, L. G. & Sousa, R. Discontinuous movement and conformational change during pausing and termination by T7 RNA polymerase. *EMBO J.* **22,** 6483–6493 (2003).

42. Pal, M., Ponticelli, A. S. & Luse, D. S. The role of the transcription bubble and TFIIB in promoter clearance by RNA polymerase II. *Mol. Cell* **19,** 101–110 (2005).

43. Wang, M. D. *et al.* Force and velocity measured for single molecules of RNA polymerase. *Science* **282,** 902–907 (1998).

44. Block, S. M., Asbury, C. L., Shaevitz, J. W. & Lang, M. J. Probing the kinesin reaction cycle with a 2D optical force clamp. *Proc. Natl Acad. Sci. USA* **100,** 2351–2356 (2003).

45. Westover, K. D., Bushnell, D. A. & Kornberg, R. D. Structural basis of transcription: nucleotide selection by rotation in the RNA polymerase II active center. *Cell* **119,** 481–489 (2004).

46. Temiakov, D. *et al.* Structural basis of transcription inhibition by antibiotic streptolydigin. *Mol. Cell* **19,** 655–666 (2005).

47. Holmes, S. F. & Erie, D. A. Downstream DNA sequence effects on transcription elongation. Allosteric binding of nucleoside triphosphates facilitates translocation via a ratchet motion. *J. Biol. Chem.* **278,** 35597–35608 (2003).

48. Thomen, P., Lopez, P. J. & Heslot, F. Unravelling the mechanism of RNA-polymerase forward motion by using mechanical force. *Phys. Rev. Lett.* **94,** 128102 (2005).

49. Simmons, R. M., Finer, J. T., Chu, S. & Spudich, J. A. Quantitative measurements of force and displacement using an optical trap. *Biophys. J.* **70,** 1813–1822 (1996).

50. Schnitzer, M. J. & Block, S. M. Statistical kinetics of processive enzymes. *Cold Spring Harb. Symp. Quant. Biol.* **60,** 793–802 (1995).

Supplementary Information is linked to the online version of the paper at www.nature.com/nature.

Acknowledgements We thank J. Gelles for general discussions and continued inspiration, D. Bushnell and C. Kaplan for discussions relating to RNAP secondary binding sites, P. Fordyce, N Guydosh, A. Meyer, A. La Porta and M. Woodside for comments on the manuscript, and R. Byer for discussions about the use of helium. W.J.G. acknowledges the support of a Predoctoral Fellowship from the NSF. This work was supported by grants to S.M.B. from the NIH-NIGMS.

Author Information Reprints and permissions information is available at npg.nature.com/reprintsandpermissions. The authors declare no competing financial interests. Correspondence and requests for materials should be addressed to S.M.B. (sblock@stanford.edu).

Vol 435|23 June 2005

nature

NEWS & VIEWS

LOW-TEMPERATURE PHYSICS

A quantum revolution

Rudolf Grimm

Tiny quantum tornadoes observed in ultracold gases of fermionic atoms provide definitive evidence of superfluidity, and open up new vistas in the modelling of quantum many-body systems.

Almost exactly ten years after the first observation of a Bose–Einstein condensate (BEC) in ultracold atomic gases consisting of so-called bosons[1,2], a similar revolution is now unfolding. Evidence has piled up that atoms of the class of particles known as fermions can also be cooled down to a superfluid state. On page 1047 of this issue, Zwierlein *et al.*[3] present a final, spectacular proof for superfluidity — frictionless flow — in an ultracold gas of fermionic atoms.

Fundamental particles are divided into bosons and fermions depending on their internal angular momentum, or 'spin'. If the total spin is an integer multiple of Planck's constant, *h*, divided by 2π, the particle is a boson. An ultracold ensemble of these particles can condense into the lowest possible quantum energy state, where it forms a BEC. The building blocks of matter such as electrons, protons and neutrons are, however, particles with half-integer spin — fermions. Fermions obey Pauli's exclusion principle, which forbids two or more particles to occupy the same quantum state. The formation of an ultracold condensate similar to a BEC is thus not allowed for a system of single fermions.

Fermions can, however, condense into a macroscopic quantum state and form a superfluid if they pair up, forming compound objects with whole-integer spin and bosonic character. Bardeen–Cooper–Schrieffer (BCS) theory[4], for example, describes the frictionless

transport of electrons in superconductors in terms of composites known as Cooper pairs. The great interest in ultracold Fermi gases[5,6] is due to their unique properties for modelling the physics of quantum matter in general — table-top experiments promise insights not only into the mechanisms of high-temperature superconductivity, but also into the physics underlying neutron stars and the quark–gluon plasma, the state of matter thought to have dominated at a critical stage in the early development of the Universe.

A crucial parameter in these situations is the interaction strength between two fermions, as this determines the binding energy and size of the pairs, and thus the macroscopic properties of the quantum system. In an ultracold gas, the pair interaction can be varied conveniently with a magnetic field if a so-called Feshbach resonance is present. Below this resonance, a regime of strong pairing can be realized, in which fermionic atoms bind together to form bosonic molecules that eventually condense into a molecular BEC. Far above the Feshbach resonance, weak Cooper pairing leads to a BCS-type system. The ability continuously to vary the system properties in between these two extremes has opened up the possibility of studying the long-elusive crossover from a BEC to a BCS state. The fundamental properties of this crossover have been studied with lithium (⁶Li) and potassium (⁴⁰K) gases. But although the observation of pair condensa-

tion[7,8], and measurements of collective oscillation modes[9,10], pairing energy[11] and heat capacity[12], together with supporting theory, provided compelling evidence for superfluidity, a final proof — a 'smoking gun' — was still missing.

So what is an unambiguous signature for frictionless flow in a macroscopic quantum state? A striking possibility results from the discrete nature of angular momentum in quantum mechanics. A superfluid cannot rotate like a classical fluid, but arranges itself in a system of vortices, with each of these tiny quantum tornadoes carrying a separate chunk of the total angular momentum of the system. The vortices expel particles from their centres, forming filament-like empty cores that penetrate the superfluid. The vortices also repel each other, leading, in thermal equilibrium, to their crystallization into a regular lattice, known as an Abrikosov lattice. Such structures are a well established signature of superfluid flow in some kinds of superconductors[4], and have also been observed for rotating BECs.

Zwierlein and colleagues' experiment[3] with a rotating Fermi gas posed great challenges. After first using a BEC consisting of sodium atoms to optimize their setup for vortex creation (Fig. 1a), the authors primed a laser trap, carefully optimized to be as round as possible, with an ultracold Fermi gas of ⁶Li atoms. Two additional laser beams, swirling like spoons in the quantum fluid, stirred the lithium gas vigorously to introduce angular momentum

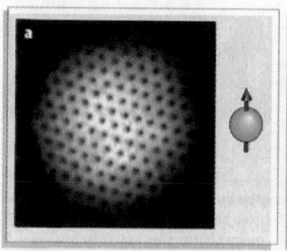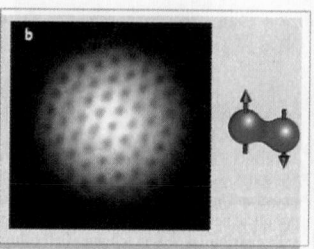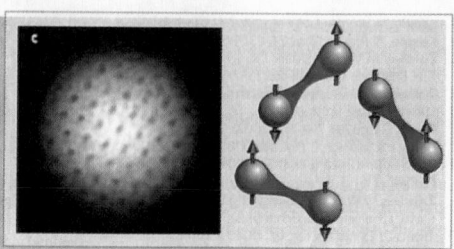

ANDRE SCHIROTZEK

Figure 1 | Cool rotations. Vortex structures observed in rotating superfluids by Zwierlein *et al.*[3]. **a,** A Bose–Einstein condensate (BEC) of bosonic (integer spin) sodium atoms. **b,** A BEC formed of two fermionic (half-integer spin) ⁶Li atoms bound together tightly to form a gas of bosonic molecules. **c,** A Fermi gas of loosely bound pairs of ⁶Li atoms in the strongly interacting regime, the first unambiguous sign of superfluidity seen in a fermionic gas.

NEWS & VIEWS

NATURE|Vol 435|23 June 2005

to the system, which was then given a variable time for formation and crystallization of the vortices.

The vortex cores are far too small to be resolved by optical imaging. So Zwierlein and colleagues magnified the vortex cores and the whole vortex lattice by turning off the laser trap and releasing the system into free space, where it expanded. They also increased the size of the vortex cores, and thus their visibility, by changing the interaction strength during the expansion.

The authors first demonstrated the formation of vortex lattices in the lithium gas in the molecular BEC regime. Here the size of the fermion pairs is small compared with the typical interparticle distances, and a closely bound, bosonic molecule is formed (Fig. 1b). In the strongly interacting regime close to the Feshbach resonance on the BCS side, the pair size is comparable to typical interparticle distances. Here, the fermion pairs cannot bind together to form isolated molecules — yet similar vortex patterns were observed (Fig. 1c). The time required for the formation of the vortex lattice was about a hundred times longer than the expansion timescale — ruling out the possibility that vortices are formed during expansion.

The spectacular observation of vortices in a Fermi gas heralds the advent of a new era of research reaching far beyond Bose–Einstein condensation. As an immediate experimental step, interfering light fields can be used to simulate a crystal lattice[13], providing a unique tool for solving problems in condensed-matter physics[14]. And the amazing level of control demonstrated in the work of Zwierlein *et al.*[3] can be extended to more sophisticated systems — mixed Fermi systems could be used to simulate a nucleus of protons and neutrons, or exotic superconductors. This final proof of superfluidity in a Fermi system opens fantastic new prospects for many different fields of many-body quantum physics. ∎

Rudolf Grimm is at the Institute of Experimental Physics, University of Innsbruck, and the Institute of Quantum Optics and Quantum Information, Austrian Academy of Sciences, 6020 Innsbruck, Austria.
e-mail: rudolf.grimm@ultracold.at

1. Cornell, E. & Wieman, C. *Rev. Mod. Phys.* **74**, 875–893 (2002).
2. Ketterle, W. *Rev. Mod. Phys.* **74**, 1131–1151 (2002).
3. Zwierlein, M. W., Abo-Shaeer, J. R., Schirotzek, A., Schunck, C. H. & Ketterle, W. *Nature* **435**, 1047–1051 (2005).
4. Tinkham, M. *Introduction to Superconductivity* 2nd edn (Dover, Mineola, NY, 2004).
5. Cho, A. *Science* **301**, 750–752 (2003).
6. Chevy, F. & Salomon, C. *Phys. World* **18** (3), 43–47 (2005).
7. Regal, C., Greiner, M. & Jin, D. S. *Phys. Rev. Lett.* **92**, 040403 (2004).
8. Zwierlein, M. W. *et al. Phys. Rev. Lett.* **92**, 120403 (2004).
9. Kinast, J., Hemmer, S. L., Gehm, M. E., Turlapov, A. & Thomas, J. E. *Phys. Rev. Lett.* **92**, 150402 (2004).
10. Bartenstein, M. *et al. Phys. Rev. Lett.* **92**, 203201 (2004).
11. Chin, C. *et al. Science* **305**, 1128–1130 (2004).
12. Kinast, J. *et al. Science* **307**, 1296–1299 (2005).
13. Köhl, M., Moritz, H., Stöferle, T., Günter, K. & Esslinger, T. *Phys. Rev. Lett.* **94**, 080403 (2005).
14. Jaksch, D. & Zoller, P. *Ann. Phys.* **315**, 52–79 (2005).